BIBLIOGRAPHY OF SEEDS

COMPILED AND EDITED BY

LELA V. BARTON

1967

COLUMBIA UNIVERSITY PRESS

NEW YORK AND LONDON

The preparation of this bibliography was carried out with the assistance of a grant from the American Seed Research Foundation, and the publication was aided by Grant GN138 of the National Science Foundation.

Dr. Barton is Plant Physiologist at the Boyce Thompson Institute for Plant Research in Yonkers, New York

Library of Congress Catalog Card Number 66-20492
Printed in the United States of America

FOREWORD

This bibliography was undertaken to make available a comprehensive reference to seed literature for worldwide use in seed research, seed technology, and agricultural practices. It has been prepared from an existing file at Boyce Thompson Institute for Plant Research, Inc., of some 30,000 citations. This file comprises references to seed literature from all parts of the world from early reports down to the present.

The nucleus of the bibliography was brought to the Institute in 1924 by the late director, William Crocker, seed physiologist. Dr. Crocker began immediately to catalog information on every aspect of seed research and technology to supplement a card file he had developed previously at the University of Chicago and the U.S. Department of Agriculture. Since then the file has been maintained and expanded to facilitate our own research program and to serve as source material for many general articles and for three books.

Inquiries received and personal references made to the file by interested persons from around the world encouraged us to undertake publication. It was discovered early that it would not be profitable or practical to publish all citations, since many were merely notes and since there were often replicate citations describing the same material in the same or different publications. This necessitated sorting and editing, with the inevitable result of some unfortunate omissions or additions. New citations were taken from current literature up to June 1, 1964. Editing resulted in the selection of 20,163 references to published literature with 6,320 coauthor entries.

Subject and plant indexes are included. It is hoped that they will serve as guides to the information desired.

Such a stupendous undertaking could not have been consummated except for the advice, patience, and encouragement of many people. Chief among these are George L. McNew, Managing Director of Boyce Thompson Institute for Plant Research, Inc., and Clyde Chandler and Bettie M. Brooks of the Institute. Florence Flemion has contributed substantially to the literature review over the years. The technical assistance of Dorothy Pardini, Elaine Cole, Mildred Gill, Dorothy Kirkpatrick, and Annette Hassler is gratefully acknowledged.

Yonkers, New York
November, 1966

LELA V. BARTON

A

1. Aalto-Setälä, J. E. 1937: Über die Dauer der Keimversuche mit verschiedenen Samen. Proc. Intern. Seed Testing Assoc. 9:123–30.

2. —— 1939: Über die Forderungen an Keimfähigkeit der Malzgerste und Feststellung derselben. Mallasjuomat, Helsinki 1939:393–403. Proc. Intern. Seed Testing Assoc. 12:192–96. 1940.

3. —— 1939: Über die Keimzeit bei der Prüfung der Malzgerste. Mallasjuomat, Helsinki 1939: 445–50. Proc. Intern. Seed Testing Assoc. 12:196–200. 1940.

4. Aamisepp, A. 1959: Effect of chlorinated phenoxyacetic acids on vegetative development, seed formation and germination biology of wild oats. Växtodling (Plant Husbandry) 10:58–67.

5. —— 1961: The occurrence of 2,4-D in seeds from cultivated plants sprayed with chlorinated phenoxyacetic acids. A preliminary report. Kungl. Lant. Ann. 27:445–49.

6. Aamodt, O. S. 1935: Germination of Russian pigweed seeds in ice and on frozen soil. Sci. Agr. 15:507–8. Expt. Sta. Rec. 74:336. 1936.

7. —— 1942: The seed situation and the war. Seed World 52:10–11, 18–19. J. Am. Soc. Agron. 35:85–91. 1943.

8. —— and A. W. Platt. 1934: Kälteresistenz von Wildhafer und verschiedenen allgemeinen Getreidesorten. Sci. Agr. 14:12. Proc. Intern. Seed Testing Assoc. 7:58–59. 1935.

9. —— and J. H. Torrie. 1934: A simple method for determining the relative weight per bushel of the grain from individual wheat plants. Can. J. Res. 11:589–93. Expt. Sta. Rec. 72:615. 1935.

10. —— and —— 1935: Kernel texture as an indicator of quality in hard red spring wheats. Can. J. Res. Sect.C 13:79–88. Expt. Sta. Rec. 74:195–96. 1936.

11. —— and —— 1935: Studies on the inheritance of and the relation between kernel texture and protein content in several spring wheat crosses. Can. J. Res. Sect.C 13:202–19. Expt. Sta. Rec. 74:473. 1936.

12. ——, ——, and A. Wilson. 1935: Studies on the inheritance of and the relationships between kernel texture, grain yield, and tiller-survival in crosses between Reward and Militurum spring wheats. J. Am. Soc. Agron. 27:456–66. Biol. Abstr. 11:17845. 1937.

Aamodt, O. S. 18588.

13. Aarkrog, A. 1960: Strontium 90 in Danish grain in 1959. Tidsskr. Planteavl 64:914–20. Biol. Abstr. 36:65757. 1961.

14. Aase, H. C. 1946: Embryo sac, embryo, and endosperm. Botan. Rev. 12:301–4.

15. Aastveit, K., and E. Strand. 1954/55: Investigations on the respiration rate of small grain seeds of different moisture content and harvested at different stages of maturity. Acta Agr. Scand. 5:76–84.

16. Abad-Manrique, J. B. 1949: Antiseptics as stimulants of germination. II. Hydrogen peroxide. Ion 9:590–97. Chem. Abstr. 44:4969. 1950.

17. —— 1949: Calcium hypochlorite-chloride as germination stimulant. Farmacognosia (Madrid) 9(14):13–38. Chem. Abstr. 44:3653. 1950.

18. Abakumova, M. 1938: On the problem of photoperiodical and temperature aftereffects. Compt. Rend. Acad. Sci. URSS 18:105–9.

19. Abbe, C. 1894/95: The influence of cold on plants—a resumé. Expt. Sta. Rec. 6:777–81.

20. Abbey, F. M. 1930: Habits of the fringed gentian. Horticulture 8(3):49.

21. Abbiss, H. W., and S. P. Craze. 1952: Influence of storage and production temperatures on some horticultural crops. Gt. Brit. J. Min. Agr. 58:471–77.

22. Abbott, D. L. 1955: Temperature and the dormancy of apple seeds. 14th Intern. Hort. Congr. The Hague-Scheveningen Rept. 1:746–53.

23. Abbott, E. V. 1950: Some observations on sugarcane flowering and seed production in Louisiana and an experiment on production and storage of true sugarcane seed. Sugar Bull. 28:329–32. Biol. Abstr. 25:25006. 1951.

24. Abdelghani, A. 1938: Separation of eelworm infected grains in wheat. Proc. Intern. Seed Testing Assoc. 10:240–43.

25. Abdel-Wahab, M. F., and S. A. El-Kinawi. 1960: The acid-soluble nucleotides of *Vicia faba*. II. Acta Chem. Scand. 14:1667–68.

Abdel-Wahab, M. F. 1881.

26. Abdoh, Y., and M. M. Nafissi. 1963: Free amino acid content of tobacco-seeds. Nature 199:178–79.

Abdul-Wahab, A. S. 10544.

27. Abdurashitov, S. A. 1957: Raising the resistance to cold in corn by means of pre-sowing treatment of seeds with solutions of tracer elements. J. Botan. (URSS) 42:1099–106. Biol. Abstr. 35:45561. 1960.

28. Abdymanapov, S., and V. Chebolda. 1961: Growing of sugar beets for seeds without transplanting. Sel'skoe Khoz. Kirgizii 1961(8):20–23.
Abe, M. 18253.
29. Abe, S., H. Katsumata, and H. Kanki. 1951: On the poor strains of "Senshu-ki" onion due to improper seed growing systems. Japan. J. Breed. 1:103–9. Biol. Abstr. 27:28371. 1953.
30. —— and H. Nagayoshi. 1956: Studies on the strawberry breeding. II. On germination of seeds. Kyushu Agr. Res. 17:88.
Abe, S. 5824.
31. Abe, W. 1960: Studies on germination and growth of seedlings of soybean plants. I. Water content, dry matter content, and respiration of whole and dehulled seeds and of excised embryonic tissues. Yamagata Daigaku Kiyo, Shizen Kagaku 5(1):209–39. Chem. Abstr. 60:7149. 1964.
32. Abel, G. H., and A. J. MacKenzie. 1964: Salt tolerance of soybean varieties (*Glycine max* L.) during germination and later growth. Crop Sci. 4:157–61.
33. Abele, C. 1927: Keimungsstimulationsversuche an den Samen von *Sinapis alba* L. Zell-Stim.-Forsch. 2:277–84.
34. Aberconway, L. 1946: The viability of seed. J. Roy. Hort. Soc. 71:370.
35. Aberg, E. 1940: Vernalization of winter rye. Lantbrukshögsk. Ann. Sweden 8: 71–86.
36. —— 1940: Viviparous forms of *Agrostis*. Lantbrukshögsk. Ann. Sweden 8:461–63.
37. —— 1944: Hormones and their use for practical purposes. Lantbrukshögsk. Inst. Växt. Sweden 1:62–63.
38. —— 1945: Effect of vernalization on the development of stripe in barley. Phytopathology 35:367–68.
39. —— 1946: Practical value of miscolored clover seeds. Lantbrukshögsk. Ann. Sweden 13:223–38. Biol. Abstr. 21:15259. 1947.
40. —— 1950: Barley and wheat from the Saqqara pyramid in Egypt. Lantbrukshögsk. Ann. Sweden 17:59–63. Biol. Abstr. 25:26373. 1951.
41. —— and G. Bjälfve. 1940: Scarification of hard-boiling peas. Lantbrukshögsk. Ann. Sweden 8:457–59. Biol. Abstr. 16:5292. 1942.
42. —— and S. Fall. 1950: Mechanical counting of kernels and other large-sized seeds. Statens Centrala Frökontrollanstalt Medd. 25:50–58.
43. —— and C. Ljungdahl. 1937: Hydrogen peroxide for seed treatment. Lantbrukshögsk. Ann. Sweden 4:113–30. Chem. Abstr. 32:5144. 1938.
44. ——, R. H. Porter, and W. A. Robbins. 1945: Further experiments with the Iowa air blast seed separator for the analysis of small-seeded grasses. Iowa Agr. Expt. Sta. Res. Bull. 340:765–804.
45. ——, G. A. Wiebe, and A. D. Dickson. 1945: Ash content of barley awns and kernels as influenced by location, season, and variety. J. Am. Soc. Agron. 37:583–86.
Aberg, E. 14886.
Abeyratna, E. F. L. 16159.
46. Abichandani, C. T., and K. Ramiah. 1951: Treatment of rice seed in nutrient solutions as a means of increasing yield. Current Sci. (India) 20:270–71.
47. Abidin, Z. 1956: Über die Beeinflussung der Samenkeimung durch den Temperaturwechsel. Z. Botan. 44:207–19.
Able, C. F. 18572.
48. Abolina, G. 1938: Significance of mineral elements in the yarovization process. Compt. Rend. Acad. Sci. URSS 18:199–202. Chem. Abstr. 32:8471–72. 1938.
49. Abolina, G. I. 1949: Importance of mineral nutrition at the phase of vernalization. Selek. i Semen. 16(9):59–61.
50. —— 1951: Importance of increased phosphorus nutrition during the yarovization phase in wheat. Dokl. Akad. Nauk SSSR 79:161–64.
51. —— 1951: On the yarovization phase of branching wheats. Selek. i Semen. 18(11): 11–19.
52. —— 1951: Viability of spring wheat seeds at various phases of ripeness. Agrobiologiya 1951(1):150–53.
53. —— 1956: The effect of Dorogov's biological preparation on seed germination and plant growth and development. Dokl. Akad. Nauk Uz. SSR 1956(11):49–53. Biol. Abstr. 35:27549.
54. Aboul-Ela, M. M. 1952: Physiological changes and freezing injury in maturing maize. Plant Physiol. 27:778–86.
Aboul-Ela, M. M. 9344.
55. Abraham, A., and K. Ramachandran. 1960: Growing *Colocasia* embryos in culture. Current Sci. (India) 29:342–43.
56. —— and K. J. Thomas. 1962: A note on the in vitro culture of excised coconut embryos. Indian Coconut J. 15:84–88.
Abraham, K. J. 2124.
57. Abraham, M. 1884: Bau und Entwickelungsgeschlecht der Wandverdickungen

in den Samenoberhautzellen einiger Cruciferen. Pringsh. Jahrb. 16:599–637.

58. Abrahao, J., and A. C. Andrade. 1961: Experiments on seed-treatments of cotton during 1955–59. Arquiv. Inst. Biol. 28:133–48.

59. Abramov, I. N. 1931: Disinfection of seeds with sawdust impregnated with formalin. Plant Protect. 8:155–59.

60. Abramov, N. 1955: The effect of certain biological factors on the duration of seed dormancy of apple and pear trees. Izv. Akad. Nauk SSSR Ser. Biol. 1955(1): 53–66. Biol. Abstr. 32:9905. 1958.

61. Abramov, N. A. 1952: Effect of pollinator [varieties] on time of apple seed germination. Agrobiologiya 1952(5):111–15.

62. —— 1956: The causes of the quiescent period in the seeds of fruit plants. Tr. Plodoovoshchn. Inst. Imeni I.V. Michurina 1956(9):111–27. Biol. Abstr. 35:24901. 1960.

63. —— 1957: Some factors determining the length of the "dormant period" in apple and pear tree seeds. Agrobiologiya 1957(3):59–66. Biol. Abstr. 35:35990. 1960.

64. Abramova, E. I. 1955: Good seeds—a necessary condition for obtaining high yields of flax. Vozdelyvanie l'na-dolguntsa Moskva, Gosudarstvennoe Izdatel'stvo Sel'skok-hoziaistvennoi Literatury, pp. 52–59.

65. Abramova, Z. V. 1962: The process of the formation of the embryo of winter wheat depending on temperature and variety. Zap. Leningr. Sel'skokhoz. Inst. 84:138–42. Biol. Abstr. 45:12359. 1964.

66. —— 1963: Effect of planting date on development of wheat caryopsis. Selek. i Semen. 2:41–44.

Abrams, G. J. von; *see* Von Abrams, G. J.

67. Abrams, R. 1957: Effects of X-ray radiation on oat seed. J. Agr. Univ. Puerto Rico 41:242–58.

68. —— 1960: Effects of gibberellic acid on dormant seeds and subsequent crops of Pigeonpeas (*Cajanus cajan*). J. Agr. Univ. Puerto Rico 44:21–27.

69. —— and K. J. Frey. 1957: The relationship between moisture content and X-ray sensitivity of oat seeds, *Avena sativa*. Proc. Iowa Acad. Sci. 64:155–59.

70. —— and —— 1958: Effect of recurrent X-radiation on germination and seedling vigor in oats. Proc. Iowa Acad. Sci. 65:174–83.

71. Abu-Nasr, A. M., and W. H. Potts. 1953: The analysis and characterization of the oil from the seed of *Citrullus colocynthis*. J. Am. Oil Chemists Soc. 30:118–20. Ber. Wiss. Biol. 85:138. 1953.

72. Abutalybov, M. G. 1955: Effect of minor elements on development and yields of cotton and alfalfa seeds. Akad. Nauk Azerb. SSR Inst. Pachvoved. Agrokhim. Tr. 7:53–70.

73. Acatay, A. 1938: Investigations on the quality and quantity of seed in the different portions of the crowns of native forest trees. Tharandter Forstl. Jahrb. 89:265–364. U.S. Dept. Agr. Forest Serv. 5:182. 1938.

74. Accorsi, W. R. 1953: Seeds with opercula. Escola Super. Agr."Luiz de Queiroz" An. 10:95–96.

Acharya, B. N. 5290.

75. Acharya, C. N. 1934: Studies on the hydrolysis of starch by the enzymes in Cholam malt, *Sorghum vulgare*. Indian J. Agr. 4:476–536. Biol. Abstr. 10:15493. 1936.

76. Achaya, K. T., and S. A. Saletore. 1958: Proximate composition of some Hyderabad and Bombay cottonseed. Papers Symposium Cottonseed and By-products, Hyderabad, India, Sec. 1:39–41. Chem. Abstr. 53:19042–43. 1959.

Achmed, H. 12507.

77. Achterberg, H. H. 1950: Die Prüfung des Forstlichen Saatgutes. Forstwirtsch. Holzwirtsch. 4:216–18.

78. Achterberg, H. H. 1958: Anwendung von Feuchtsubstraten bei Fruhtesten. (Samenprüfung der Holzarten Eiche und Linde.) Kurzbericht über einen Forschungsauftrag. Forst. Jagd. 8:15–16.

Ackerman, A. 12527.

79. Ackerman, R. F. 1957: The effect of various seed-bed treatments on the germination and survival of white spruce and lodgepole pine seedlings. Can. Dept. Northern Affairs Natl. Resources, Forest Res. Div. Tech. Note 63. 23 pp.

80. Ackerman, W. L. 1957: After-ripening requirements for germination of seeds of *Acer truncatum* Bunge. Proc. Am. Soc. Hort. Sci. 69:570–73.

81. —— 1961: Flowering, pollination, self-sterility, and seed development of Chinese jujube. Proc. Am. Soc. Hort. Sci. 77: 265–69.

82. Ackermann, D., and W. Appel. 1939:

The nitrogenous constituents of jack bean. Z. Physiol. Chem. 262:103–10. Chem. Abstr. 34:2335–36. 1940.

Ackerson, C. W. 2769–2772.

83. Acosta-Solis, M. 1948: Tagua or vegetable ivory—a forest product of Ecuador. Econ. Botany 2:46–57.

Acree, F., Jr. 9646.

84. Acton, E. H. 1893: Changes in the reserve materials of wheat on keeping. Ann. Botany 7:383–87.

Acuna, J. B. 4518, 4519.

85. Acuna, R. A. 1938: The effect of storage on the germination of West Indian cedar. Philippine J. Forestry 1:293–99. Biol. Abstr. 13:8518. 1939.

Adachi, A. 19927.

86. Adachi, T. 1955: Studies on the relation between hulls and shape of rice caryopsis. Bull. Div. Plant Breed. Cult., Tokai-Kinki Natl. Agr. Expt. Sta. 2:207–14.

87. Adair, C. R. 1940: Effect of time of seeding on yield, milling quality, and other characters in rice. J. Am. Soc. Agron. 32:697–706.

88. ——, E. M. Cralley, and T. H. Johnston. 1958: Long-time trends in grain yields of five rice varieties. Agron. J. 50:233–35.

Adair, C. R. 9958.

89. Adam, W. B. 1938: The composition and texture of dried peas. II. Univ. Bristol, Fruit Veg. Preserv. Res. Sta. Ann. Rept. 1936/37:41–50. Chem. Abstr. 33:2641–42. 1939.

90. —— and G. S. Siddappa. 1937: The composition and texture of dried peas. Univ. Bristol, Fruit Veg. Preserv. Res. Sta. Ann. Rept. 1935/36:34–50. Chem. Abstr. 32:669. 1938.

Adam, W. B. 17398.

91. Adams, A. B. 1931: The germination of crimson clover seed (*Trifolium incarnatum*). J. Agr. W. Australia 8:227. Biol. Abstr. 6:19775. 1932.

92. —— 1943: A note on the longevity of "hard" sub-clover seed. J. Agr. W. Australia 20:155.

Adams, A. F. R. 19315.

93. Adams, C. E. 1937: Growing tree peonies from seed. Flower Grower 24(11):527.

94. —— 1957: The value of seed tests vs. the cost of seed testing. Assoc. Offic. Seed Analysts News Letter 31(2):5–10.

95. —— 1957: What's wrong with the seed testing? Southern Seedsman 20(6):19, 38–42.

Adams, C. L. 17470.

96. Adams, E. N. 1938: Poinsettia from seed. Horticulture 16(24):479.

97. Adams, J. 1913: On the germination of the seeds of some dicotyledons. Sci. Proc. Roy. Dublin Soc. N.S. 13:467–99. Expt. Sta. Rec. 29:828. 1913.

98. —— 1927: The germination of the seeds of some plants with fleshy fruits. Am. J. Botany 14:415–28.

99. Adams, J. D. 1957: Modification of X-ray effects in dormant barley seeds by past-treatments of oxygen and nitrogen. Dissertation Abstr. 17:7.

100. ——, R. A. Nilan, and H. M. Gunthardt. 1955: After-effects of ionizing radiation in barley. I. Modification by storage of X-rayed seeds in oxygen and nitrogen. Northwest Sci. 29:101–8. Nucl. Sci. Abstr. 9:7632. 1955.

101. Adams, J. E., H. M. Boggs, and E. M. Roller. 1937: Effect of fertilizers on composition of soybean hay and seed and of crop management on carbon, nitrogen and reaction of Norfolk sand. U.S. Dept. Agr. Tech. Bull. 586. 34 pp.

102. Adams, L. 1962: Planting depths for seeds of three species of *Ceanothus*. U.S. Forest Serv. Pacific Southwest Forest & Range Expt. Sta. Res. Note 194. 3 pp.

103. ——, E. Stefanescu, and D. J. Dunaway. 1961: Gibberellin and thiourea break seed dormancy in California *Ceanothus*. U.S. Forest Serv. Pacific Southwest Forest & Range Expt. Sta. Res. Note 178. 4 pp.

104. Adams, M. 1931: Conditions influencing the use of the citric acid enzyme in cucumber seeds for the quantitative determination of citric acid. Proc. Staff Meetings Mayo Clinic 6:252. Chem. Abstr. 25:4296–97. 1931.

Adams, M. 17881.

Adams, M. E. 10268, 10269.

105. Adams, M. W. 1953: Cross- and self-incompatibility in relation to seed-setting in *Bromus inermis*. Botan. Gaz. 115:95–105.

Adams, M. W. 15645.

106. Adams, R. E., J. E. Wolfe, M. Milner, and J. A. Shellenberger. 1954: Detection of internal insect infestation in grain by sound amplification. Cereal Chem. 31:271–76.

107. Adams, S. L., W. H. Stark, and P. Kolachov. 1943: Reduction of the fermentable carbohydrate content of corn by kiln drying. Cereal Chem. 20:260–66. Expt. Sta. Rec. 92:164. 1945.

108. Adams, W. R. 1934: Studies in tolerance of New England forest trees. XI. The influence of soil temperature on the germination and development of white pine seedlings. Vermont Agr. Expt. Sta. Bull. 379. 18 pp.

109. Adamson, R. M. 1952: Effects of various growing methods in leek seed production. Sci. Agr. (Ottawa) 32:634–37. Biol. Abstr. 27:20496. 1953.

110. —— 1961: Acceleration of seed maturity in slow-bolting lettuce by vernalization. Can. J. Plant Sci. 41:394–400.

111. Adamson, W. F. 1941: Growing orchids artificially and references to nature's lessons. Australian Orchid Rev. 6:70–72.

112. Adbel-Kader, M. M. 1944: Vitamin C in germinated *Vicia faba* and *Trigonella foenum-graecum*. J. Roy. Egypt. Med. Assoc. 27:197–98. Chem. Abstr. 39:3089. 1945.

113. Ader, F. 1962: The determination of tetrazolium viability of grains. Saatgut-Wirtsch. 14:338–40.

114. Adhikari, A. K. 1930: Notes on germination and sowing of bonsum (*Phoebe hainesiana*) seeds under shelter-wood and taungya. Indian Forester 56:455–59.

115. Adiga, P. R., G. Padmanaban, S. L. N. Rao, and P. S. Sarma. 1962: The isolation of a toxic principle from *Lathyrus sativus* seeds. J. Sci. Ind. Res. (India) 21C:284–86. Chem. Abstr. 58:3687. 1963.

Adiga, P. R. 15967, 15968.

116. Adler, E., and M. Sreenivasaya. 1937: Über die Komponenten der Dehydrasesysteme. XVI. Zur Kenntnis der Formicodehydrase und der Alkoholdehydrase in Pflanzensamen. Z. Physiol. Chem. 249(1): 24–39.

Adler, Y. 19285.

117. Admiralskaya, S. A. 1960: Fertilization and development of the embryo and endosperm of the allopolyploid variety of peppermint (*Mentha piperita* L.). Dokl. Botan. Sci. Sect. (English Transl.) 130: 12–14.

118. Adolph, W. H., and H. C. Chiang. 1935: The proteins of the cowpea (*Vigna sinensis*). Chinese J. Physiol. 9:347–53. Expt. Sta. Rec. 75:739. 1936.

119. Adriaens, E. L., and E. J. Bigwood. 1954: Amino acid composition of the protein material of cotton seed. Bull. Soc. Chim. Biol. 36:579:83. Chem. Abstr. 49:1114. 1955.

120. Adriaens, L. 1932: Etude chimique des graines de *Sterculia bequaerti* de Wild. (*Sterculia tragacantha* Lindl. var. *cruciata* Verm.). Mat. Grasses 24:9386–89. Chem. Abstr. 26:2488–89. 1932.

121. —— 1933: Chemical study of the seed of *Hydnocarpus anthelmintica* Pierre, from the Belgian Congo. Mat. Grasses 25:9798–99. Chem. Abstr. 28:4176–77. 1934.

122. —— 1936: Chemical study of the seeds of two palm trees of the Belgian Congo. Bull. Agr. Congo Belge 27:456–71. Chem. Abstr. 31:4365. 1937.

Adrian, J. 3499.

123. Adriance, G. W., and F. R. Brison. 1931: Plant propagation and orcharding. Edwards Bros., Ann Arbor, Mich. 69 pp.

124. Adrianowsky, A. 1883: Wirkung des Lichtes auf das erste Keimungsstadium der Samen. Nachr. Petrowskischen Agr. Forstakad. Moskow 6:171–92. Botan. Centralbl. 5:73–75. 1884.

125. —— 1884: Influence de la lumière sur la première period de germination des graines. Bull. Soc. Botan. France

126. Afanasiev, M. 1937: A physiological study of dormancy in seed of *Magnolia acuminata*. Cornell Univ. Agr. Expt. Sta. Mem. 208. 37. pp.

127. —— 1938: Seed propagation problems. Methods of meeting difficulties encountered in propagating plants by means of seeds. Am. Nurseryman 68(8):9–10.

128. —— 1939: Acid treatment of seeds. Florists Exchange Hort. Trade World 92(22):11.

129. —— 1940: New seed-handling methods facilitate growing native trees. Oklahoma Agr. Expt. Sta. Biennial Rept. 1938/40: 124–26.

130. —— 1942: Propagation of trees and shrubs by seed. Oklahoma Agr. Expt. Sta. Circ. C-106. 43 pp. Expt. Sta. Rec. 88:194. 1943.

131. —— 1943: Germinating *Nandina domestica* seeds. Am Nurseryman 78(9):5–6.

132. —— 1944: A study of dormancy and germination of seeds of *Cercis canadensis*. J. Agr. Res. 69:405–20.

133. —— 1955: Storage of afterripened seed of eastern red cedar (*Juniperus virginiana*). Tree Planters Notes 21:28–30.

134. —— and M. Cress. 1942: Changes within the seeds of *Juniperus scopulorum* during the processes of after-ripening and germination. J. Forestry 40:798–801. Expt. Sta. Rec. 88:342. 1943.

135. —— and —— 1942: Producing seedlings of eastern red cedar (*Juniperus virginiana*

L.). Oklahoma Agr. Expt. Sta. Bull. 256. 21 pp.

136. —— and H. E. Morris. 1947: Effect of different soil and seed treatments on the control of seedling diseases of sugar beets under controlled conditions. Am. Soc. Sugar Beet Technol. Proc. 4(1946):331–40.

Afanasiev, M. 17777.

137. Afanasijev, D. 1958: Seed stimulation. Translation from Sumarski List 82:163–76.

138. Afanassieva, A. S. 1938: The effect of convallarine upon the seeds of summer wheat. Compt. Rend. Acad. Sci. URSS 21:144–46.

139. —— 1939: The action of X-rays on air-dry, moist, and germinated wheat grains. Byul. Mosk. Obshchestva Ispytatelei Prirody Otd. Biol. 48:19–28.

Afanassieva, A. S. 2996, 2998.

140. Afanseve, L. A., T. Kolgushkina, and S. Vobshchina. 1956: Effect of dates of planting spring wheat on the quality of seeds. Agrobiologiya 1956(3):143–44.

141. Affourtit, M. F. A., and H. C. C. La Rivière. 1915: On the ribbing of the seeds of Ginkgo. Ann. Botany 29:591–95.

142. Afrikian, S. V. 1955: Effect of seed stratification on the development of spring forms of wheat. Akad. Nauk Armianskoi SSR Izv. Biol. Sel'skokhoz. Nauk 8:127–29.

143. Afrikyan, E. K., V. G. Tumanyan, and R. A. Bobikyan. 1961: The distribution and retention of some antibiotics in seeds and plants. Primenenie Antibiotikov v Rastenievodstve Akad. Nauk Armyansk. SSR: Erevan. 57–66. Biol. Abstr. 44: 24921. 1964.

144. Afzelius, K. 1932: Zur Kenntnis der Fortpflanzungsverhältnisse und Chromosomenzahlen bei *Nigritella nigra*. Svensk Botan. Tidsskr. 26:365–69. Biol. Abstr. 10:19448. 1936.

145. —— 1954: Embryo-sac development in *Epipogium aphyllum*. Svensk Botan. Tidsskr. 48:513–20.

146. Agababian, V. G. 1956: Concerning the development of salt-resistant corn varieties. Akad. Nauk Armianskoi SSR Izv. Biol. Sel'skokhoz. Nauk 9:85–92.

147. Agadzhanian, A. M. 1954: Influence of the transplant of the embryo upon the setting of seeds in inbreeding of rye. Akad. Nauk Armianskoi SSR Izv. Biol. Sel'skokhoz. Nauk 7:77–80.

148. Agagino, G., and S. Feinhandler. 1957: Amaranth seeds from San Jose site in New Mexico. Texas J. Sci. 9:155.

149. —— 1962: Accelerating effect of low temperatures on germination of cotton seeds on salinized soils. Plant Physiol. (USSR) (English Transl.) 8:505–6.

150. Agakishiev, D. 1962: Effect of low temperatures on accelerated growth, development, and germination of seeds of the cotton plant. Plant Physiol. (USSR) (English Transl.) 8:633–35.

151. Agarkov, V. A. 1948: Thermic drying of [wheat] seed treated with hot water. Soviet Agron. 6:62–65.

152. —— 1957: Dry method of treatment of seed sugar beet. Agrobiologiya 1957(3): 146–48.

153. —— 1958: Seed treatment of barley with granosan against helminthosporiosis. Agrobiologiya 1958(1):125–28.

154. Agarwal, R. R. 1936: Chemical examination of *Cuscuta reflexa*, Roxb. IV. Isolation of a new yellow flavonic coloring matter from the seeds. J. Indian Chem. Soc. 13:531–36. Chem. Abstr. 31:1807–8. 1937.

155. —— 1937: On the amylase from the Indian water chestnut (*Trapa bispinosa*, Roxb.). Proc. Natl. Acad. Sci. India 7:29–34.

156. —— and S. Dutt. 1936: Chemical examination of *Cuscuta reflexa*, Roxb. III. Constituents of the oil from the seeds. J. Indian Chem. Soc. 13:264–67. Chem. Abstr. 30:6327. 1936.

Agarwal, R. R. 9665.

157. Agerst, G., and D. Budd. 1960: Effect of moisture content on the storage of Brazil nuts. J. Sci. Food Agr. 11:390–96.

158. Aggarwal, J. S. 1955: Kamala seed oil. J. Sci. Food Agr. 6:364–68.

Aggarwal, J. S. 2415, 7860, 17746.

159. Agharkar, S. P., and I. Banerji. 1930: Studies in the pollination and seed formation of water-hyacinth (*Eichhornia speciosa* Kunth). Agr. J. India 25:286–96. Indian Sci. Congr. 17th, Allahabad, Proc., pp. 298–99. 1930.

160. Aginian, A. A. 1950: The dynamics of the yarovization processes in winter rye seeds collected at various phases of embryonic development. Akad. Nauk Armianskoi SSR Izv. Biol. Sel'skokhoz. Nauk 3:295–307.

161. —— 1950: Vernalization of [wheat] seeds in relation to their embryonic development. Agrobiologiya 1950(3):57–66.

162. Aglibut, A. P., and J. M. Lawas. 1955: Effects of harvesting and shelling methods and length of storage on the germination of college yellow flint seed corn. Philippine Agriculturist 38 :521–29.

Agnayeva, E. J. 7753.

163. Agnihotri, C. B. 1952: Effect of different dates of germination on growth and quality of sugar cane. Sugar Technol. Assoc. India Proc. 21 :95–98.

164. Agnistikova, V. N. 1962: Effect of moisture of acorns on their germination capacity. Akad. Nauk SSSR Laboratoriya Lesovedeniya Fiziol. drevesnyka rastenii, pp. 104–11.

165. Agnoli, R. T. 1932: Studies on hypophysis hormones. V. Influence of pituitary hormones on the germination of seeds of *Lupinus albus* L. J. Pharmacol. 44 :55–62. Ber. Wiss. Biol. 22 :795. 1932.

166. Agrawal, J. S. 1952: The embryology of *Lilaea subulata* H. B. K. with a discussion on its systematic position. Phytomorphology 2 :15–29.

167. —— 1954: Female gametophyte, endosperm, and embryo of *Scurrula* Linn. Indian Congr. Assoc. Proc. 41 :137. [Abstract.]

168. Agren, G. 1952: Microbiological determinations of amino acids in foodstuffs. III. Acta Chem. Scand. 6 :608–9.

169. Agronomov, E. A., A. A. Bundel, A. N. Goryachikh, and N. A. Korenev. 1936: Data on the spontaneous heating of grains. J. Botan. URSS 21 :196–205.

170. ——, M. S. Dunin, A. A. Bundel, A. N. Goryachikh, and N. A. Korenev. 1934: Biochemistry and microbiology of stored wheat grain infected by *Fusarium*. Food Suppl. Tech. Publ. Office, Leningrad. Separate, 96 pp. Chem. Abstr. 29:6960. 1935.

Aguilar, C. 18632.

Aguirre, A. 9770.

Aguirre, F. 9770.

171. Agulhon, H., and T. Robert. 1915: The action of radium and ratio-activity on germination in the higher plants. Ann. Inst. Pasteur 29 :261–73. Expt. Sta. Rec. 34 :730. 1916.

172. Ahlberg, E. 1957: The determination of the viability of some slowly germinating seeds through tetrazolium treatment. Maataloustieteellinen Aikakauskirja 29 :143–49. Biol. Abstr. 35 :64324. 1960.

173. Ahlberg, G. 1935: The vitamin C content of seeds. Nord. Hyg. Tidskr. 16 : 98–102. Ber. Ges. Physiol. Exptl. Pharmakol. 88 :413. Chem. Abstr. 31 :7950–51. 1937.

174. Ahlfvengren, F. E. 1898: Effect of inductive electricity on the germination of seed. Ofvers. K. Svenska Vetensk. Akad. Forhandl. 1898(8). 22 pp. Botan. Centralbl. 79 :53–54. 1899. Expt. Sta. Rec. 11 :462. 1899–1900.

175. Ahlgren, G. H. 1941: The effect of adding vitamin B_1 (thiamin) to several grass species. J. Am. Soc. Agron. 33 :572–76. Expt. Sta. Rec. 86 :319. 1942.

176. —— 1941: The influence of thiamin additions on the germination and growth of certain grasses and of white clover. Doctoral Dissertations Accepted by American Universities 1940/41(8) :48. New Jersey Agr. Expt. Sta. Bull. 692. 19 pp. 1942.

177. ——, J. G. Fiske, and A. Dotzenko. 1950: Viability of Bromegrass seed as affected by dehulling and by storage in fertilizer. Agron. J. 42 :336–37.

Ahlgren, H. L. 18501.

178. Ahlmark, A., and B. Tunblad. 1949: The toxicity of organic mercurial compounds used as seed dressings. Internat. Congr. Crop Protect. Abstr. Comm. 2 :78–79.

179. Ahluwalia, K. S. 1956: Germination of pretreated seeds of medicinal plants. Indian Forester 82 :573–76.

180. —— 1958: Test on the viability of seeds of medicinal plants. Indian Forester 84 :740–44.

181. Ahmad, A. 1962: A preliminary report on germination of Tendu (*Diospyros melanoxylon*) seed. Pakistan J. Forestry 12 :138–44. Biol. Abstr. 44 :20947. 1963.

182. Ahmad, B. 1936: Vitamins, hormones, and plant stimulants. Soc. Biol. Chemists, India, Ann. Rev. Biochem. Allied Res. India 7 :15–28.

183. —— and A. N. Bahl. 1946: Relative digestibility of common edible fats. II. Hydrolysis by pancreatic lipase. J. Sci. Ind. Res. (India) 5 :1–3. Biol. Abstr. 22 :229. 1948.

184. ——, A. A. Qurashi, I. Babbar, and P.C. Sawhney. 1946: Observations on ascorbic acid. IV. The effect of certain factors on ascorbic acid production during germination of seeds. Ann. Biochem. Exptl. Med. (Calcutta) 6 :29–34. Biol. Abstr. 22 :12372. 1948.

185. —— and R. N. Sareen. 1946: Relative digestibility of common edible fats. I. By *Ricinus* lipase. J. Sci. Ind. Res. (India) 4 :710–12. Biol. Abstr. 22 :228. 1948.

Ahmad, B. 14009.

Ahmad, K. 10132.

Ahmad, M. S. 6227, 6228, 6230.

186. Ahmad, S. A., and D. R. Dhingra. 1945: Chemical study of the seeds of sarda. J. Indian Chem. Soc. 22:337–38. Chem. Abstr. 40:4115. 1946.

187. Ahmad, S. H., and M. M. Qurashi. 1959: The development of rapid methods for the estimation of the oil-content of single cotton seeds. Part II. A preliminary examination of some non-destructive methods. Pakistan J. Sci. Ind. Res. 2:223–28.

188. —— and —— 1960: Investigations on the statistical frequency distribution of some physical characteristics of cottonseed kernels. Pakistan J. Sci. Ind. Res. 3:183–86.

Ahmad, S. H. 2429.

189. Ahmad, S. N., S. A. K. Lodhi, and M. M. Qurashi. 1960: Development of rapid methods for the estimation of the oil content of single cottonseeds. Part III. Critical examination of the oil-expression method. Pakistan J. Sci. Ind. Res. 3:87–92.

190. —— and M. M. Qurashi. 1959: Development of rapid methods for the estimation of the oil content of single cottonseeds. Part I. An oil-expression technique. Pakistan J. Sci. Ind. Res. 2:13–18.

Ahmed, A. 348.

Ahmed, M. 10503.

Ahmed, Z. [F.]. 8086, 17404.

191. Ahmed-Basha, A. 1953: Karanja (*Pongamia glabra*) seed and oil. Oils Oilseed J. (Bombay) 5(10/12):31–32. Soils Fertilizers 17:2584. 1954.

Ahnstroem, G. 14231.

192. Ahola, V. K. 1951: The duration of germinability in pine and spruce seed. Metsätaloudellinen Aikakauslehti 2/3:47–48.

193. Ahring, R. M. 1959: A suggested method for determining purity of certain chaffy-seeded grasses. Agron. Abstr., 51st Ann. Meeting, p. 76.

194. —— 1962: Storageability under laboratory conditions of seed of blue grama, side-oats grama and smooth bromegrass. Oklahoma Agr. Expt. Sta. Tech. Bull. T-97. 11 pp.

195. ——, N. L. Dunn, Jr., and J. R. Harlan. 1963: Effect of various treatments in breaking seed dormancy in Sand Lovegrass, *Eragrostis trichodes* (Nutt.) Wood. Crop Sci. 3:131–33.

196. —— and J. R. Harlan. 1961: Germination characteristics of some accessions of *Bothriochloa ischaemum* (L.) Keng. Oklahoma Agr. Expt. Sta. Tech. Bull. T-89. 19 pp.

197. —— and —— 1961: Germination studies on the *Dicanthium annulatum* complex. Oklahoma Agr. Expt. Sta. Tech. Bull. T-90. 15 pp.

198. ——, R. D. Morrison, and M. L. Wilhite. 1959: Uniformity trials on germination of Switchgrass seed. Agron. J. 51:734–37.

Ahring, R. M. 8180.

199. Ahsan, A. M., and G. Hahn. 1958: Chemical investigation of neem seeds. Pakistan J. Sci. Ind. Res. 1:146–50.

200. Aichele, F. 1931: Keimung von Gramineensamen in Medien verschiedener Wasserstoffionenkonzentration und die damit verbundenen Reaktionsveränderungen. Botan. Arch. 33:406–500.

Aichele, F. 12046.

Aichele, M. D. 6454.

201. Aicher, L. C. 1944: A buffalo grass seed drier. Agr. Eng. 25:394, 396, 402. Biol. Abstr. 19:14037. 1945.

202. Aiken, G. D. 1931: Success with fringed gentians. Home Acres & Countryside Mag. 19(10):6.

203. Aikman, J. M. 1933: Germination of corn in the field as affected by soil and air temperature. Proc. Iowa Acad. Sci. 40:84.

204. —— 1935: The effect of low temperature on the germination and survival of native oaks. Proc. Iowa Acad. Sci. 41:89–93. Biol. Abstr. 10:21879. 1936.

205. —— and E. L. Bargar. 1944: Hillculture studies in Iowa. Iowa Agr. Expt. Sta. Rept. Agr. Res. 1943/44(pt. 1):158–64.

206. Aimi, R., and K. Fujimaki. 1960: Physiological investigation on the ripening mechanism in crop plants. X. The behavior of inorganic phosphorus in rice kernel during ripening. Proc. Crop Sci. Soc. Japan 29:11–14.

207. —— and S. Konno. 1958: Behaviour of phosphorus compounds during the ripening in rice kernel. Proc. Crop Sci. Soc. Japan 26:228–30.

208. Airan, J. W. 1948: Oil from the seeds of *Zizyphus xylopyra* (Willd.). Current Sci. (India) 17:150.

209. —— and S. V. Shah. 1942: The fat from the seeds of *Vangueria spinosa* (N. O. Rubiaceae). Current Sci. (India) 11:400.

Airan, J. W. 6249, 9655, 14051.

210. Aitken, Y. 1939: The problem of hard seeds in subterranean clover. Proc. Roy. Soc. Victoria 51:187–210.

211. Aiyadurai, S. G. 1956: Observation on

germination of dry seed coconuts. Madras Agr. J. 43:464–66. Biol. Abstr. 31:36004. 1957.

212. —— 1959: A note on the germination of neem seeds (*Azadirachta indica* A. Juss.). Madras Agr. J. 46:127–29. Biol. Abstr. 34:2549. 1959.

213. ——— 1962: When and where to collect seed cocounts in Madras State. Coconut Bull. 16:311–14. Biol. Abstr. 42:19456. 1963.

214. ——, P. A. M. Ibrahim, and T. Ramanthan. 1959: Multiple seedlings from a germinating coconut (*Cocos nucifera* Linn.). Sci. Cult. (Calcutta) 24:339–40.

Aiyadurai, S. G. 16823.

Aizenshtat, I. S. 3027.

215. Ajrekar, S. L., and J. D. Oza. 1935: The nature and germination of seeds of *Tinospora cordifolia.* Current Sci. (India) 3:379.

216. Ajroldi, R. 1938: Influenza di micromiceti sulla germinazione e sullo sviluppo di piante ortensi. Ital. Agr. 75:519–87. Rev. Appl. Mycol. 18:49–50. 1939.

Akahoshi, R. 16646.

217. Akamine, E. K. 1940: Germination of grass seeds. Hawaii Agr. Expt. Sta. Rept. 1940:79–81.

218. —— 1942: Methods of increasing the germination of koa haole seed. Hawaii Agr. Expt. Sta. Circ. 21:1–14. Biol. Abstr. 20:1709. 1946.

219. —— 1943: The effect of temperature and humidity on viability of stored seeds in Hawaii. Hawaii Agr. Expt. Sta. Bull. No. 90. 23 pp.

220. —— 1943: Seed studies. Germination of koa haole. Hawaii Agr. Expt. Sta. Rept. 1941/42:66–67.

221. —— 1944: Germination of Hawaiian range grass seeds. Hawaii Agr. Expt. Sta. Tech. Bull. No. 2. 60 pp.

222. —— 1947: Germination of *Asystasia Gangetica* L. seed with special reference to the effect of age on the temperature requirement for germination. Plant Physiol. 22:603–7.

223. —— 1948: Seed germination and viability. Hawaii Agr. Expt. Sta. Rept. 1946/48:132.

224. ——1950: Factors affecting the germination and viability of seeds in Hawaii. Hawaii Agr. Expt. Sta. Biennial Rept. 1948/50:133.

225. —— 1951: Viability of Hawaiian forest tree seeds in storage at various temperatures and relative humidities. Pacific Sci. 5:36–46.

226. —— 1952: Germination of seed of koa haole (*Leucaena glauca* [L.] Benth.). Pacific Sci. 6:51–52. Biol. Abstr. 26:13214. 1952.

227. —— and J. C. Ripperton. 1938: Germination and vitality of crop seeds in Hawaii. Hawaii Agr. Expt. Sta. Ann. Rept. 1937:10–12.

228. —— and —— 1939: Germination of treated koa haole, *Leucaena glauca*, seed after storage. Hawaii Agr. Expt. Sta. Ann. Rept. 1938:59–60.

229. Akazawa, T., and H. Beevers. 1957: Some chemical components of mitochondria and microsomes from germinating castor beans. Biochem. J. 67:110–14.

230. —— and —— 1957: Mitochondria in the endosperm of the germinating Castor bean: a developmental study. Biochem. J. 67:115–18.

231. Akbar, K., and S. M. Khan. 1962: Control of tea seed bugs (*Poeciliocoris* latus). Pakistan J. Biol. Agr. Sci. 5:30–36. Biol. Abstr. 43:20885. 1963.

Akemine, H. 454, 455.

232. Akemine, M. 1913: Beitrag zur Kenntnis der Keimung von *Oryza sativa.* Oesterr. Botan. Z. 63:194–200. Botan. Centralbl. 126:212. 1914.

233. —— 1914: On the physiology of germination of *Oryza sativa.* Fühling's Landwirtsch. Ztg. 63:78–93. Expt. Sta. Rec. 31:529. 1914.

234. —— 1928: Variation in spinach seeds. Leptokurtic curve indicates that two-spined character is little affected by environment. J. Heredity 19:17–21.

235. Akerberg, E. 1936: Studien über die Samenbildung bei *Poa pratensis* L. Botan. Notiser 1936:213–80.

236. —— 1938: Seed production of the *Poa* species. Herbage Rev. 6:228–33.

237. —— 1939: Apomictic and sexual seed formation in *Poa pratensis.* Hereditas 25:359–70. Biol. Abstr. 15:29. 1941.

238. —— 1943: Further studies of the embryo- and endosperm-development in *Poa pratensis.* Hereditas 29:199–201. Royal Swedish Acad. Agr., Agr. Chem. 1941/42:56. 1944.

239. —— 1952: Better seed—give more hay! Försök och Forsk. 9:17–19.

240. —— 1955: Investigation of the seed production of red clover. Nord. Bitidskr. 7:71–74.

241. —— 1955: Seed production in Sweden.

Organisation for European Economic Cooperation. European Productivity Agency. Development of Seed Production and the Seed Trade in Europe, pp. 17–24. Paris.

242. —— 1961: The activities and future aims of the Swedish Seed Association. Sveriges Utsädesfören. Tidskr. 71 :646–55.

Akerberg, E. 10053, 12117.

243. Akerman, A. 1912: Einfluss der Meerwassers auf die Keimfähigkeit der Samen einiger skandinavischen Pflanzen. Botan. Notiser 1912 :191–203. Botan. Centralbl. 122 :167–68. 1913.

244. —— 1936: Über die Keimungsverhältnisse und Auswuchsneigung Rot- und Weisskörniger Wiezensorten. Züchter 8 : 25–29. Biol. Abstr. 11 :4225. 1937.

245. —— 1943: Annual report of the work of the Swedish Seed Association for 1942. Sveriges Utsädesfören. Tidskr. 53 : 117–67.

246. —— 1951: The Swedish Seed Association. Organization, methods, and results in plant breeding. Natl. Inst. Agr. Botan. J. 6 :118–30.

247. —— 1951: Working plan of the Swedish Seed Association for the year 1952. Sveriges Utsädesfören. Tidskr. 61 :273–341.

248. —— 1951: The new laboratory building of the Swedish Seed Association. Sveriges Utsädesfören. Tidskr. 61 :356–73.

249. —— 1953: Annual report of the Swedish Seed Association in 1952. Sveriges Utsädesfören. Tidskr. 63 :83–148. Biol. Abstr. 28 :21833. 1954.

250. —— and J. Jakobsson. 1943: Storage experiments with bread-corn. Sveriges Utsädesfören. Tidskr. 53 :82–89. Kgl. Lantbruksakad. Stockholm. Author's summ. Swedish Papers Agr. Publ. 1943 : 34–35. 1945.

251. Akh, Z. 1938: Parthenocarpic forms of Caucasian persimmon. Sov. Subtrop. (Moskva) 1938(6) :87–88.

252. Akhinyan, G. M. 1963: The effect of gibberellin on the germination of seeds and growth of trees and shrubs grown on the dry slopes of Armenia. Izv. Akad. Nauk Arm. SSR Biol. Nauki 16(8) :91–98. Chem Abstr. 60 :6149. 1964.

253. Akhmetov, K. 1962: The field germination of seeds and the thickness of the growth of the plants in plantings of corn. Kukuruza 1962(5) :28–29. Biol. Abstr. 40 :12281. 1962.

254. Aki, S. 1959: Studies on the storage of the seeds of Kintoki-carrot. III. On the artificial drying of the seeds. Tech. Bull. Fac. Agr. Kagawa Univ. 11 :163–67.

255. —— 1960: Effect of gibberellin on breaking the dormancy of Kintoki-carrot seeds. Tech. Bull. Fac. Agr. Kagawa Univ. 12 :73–77.

256. —— 1961: Germination-inhibiting substances in carrot seeds. Nogyo Oyobi Engei 36 :559–60. Chem. Abstr. 56 :14661. 1962.

257. —— and S. Watanabe. 1955: Studies on the storage of the seeds of Kintoki carrot. I. Influence of storage temperature and seed moisture on the germination of seeds. Tech. Bull. Kagawa Agr. Coll. 7 :31–35.

258. —— and —— 1960: Studies on the storage of the seeds of Kintoki carrot. II. Influence of seed moisture on the germination of stored seeds. Tech. Bull. Fac. Agr. Kagawa Univ. 10 :27–28.

259. —— and —— 1961: On germination of, Kintoki-carrot seeds. VI. Isolation of self-inhibitor. Japan Soc. Hort. Sci. 30: 311–17.

260. —— and —— 1962: On germination of Kintoki carrot seeds. VII. The relations between the carrotol contents and delayed germination phenomena. Tech. Bull. Fac. Agr. Kagawa Univ. 14 :14–19.

Aki, S. 19398, 19399.

Akimochkina, T. A. 11370.

261. Akins, I. M., and M. J. Norris. 1955: The influence of awns on yield and certain morphological characters of wheat. Agron J. 47 :218–20.

Akita, S. 10300, 11112–11114.

262. Akivis, S. I. 1958: Effect of lowered temperature on the sowing qualities of seeds. Kukuruza 1958(2) :24–25.

263. Akkerman, A. M., and H. Veldstra. 1947: The chemical nature of Kockemann's blastocholine from *Lycopersicon esculentum* Mill. Rec. Trav. Chim. 66 :411–12. Biol. Abstr. 23 :12727. 1949.

Akour, A. A. 7687.

264. Akselrod, D. M. 1939: Die Wirkung der Jarowisation auf das Wachstum und die Entwicklung der blauen Lupine. Ber. (Dokl.) Allruss. Akad. Wiss. 8 :11–12. Botan. Z. 175 :125. 1939.

265. Akune, S., and S. Takagi. 1962: Studies on the metabolism of azuki seed protein. Part V. Purification of azuki protease. Agr. Biol. Chem. (Tokyo) 26 :63–71.

266. Al-Aish, M., and W. V. Brown. 1958:

Grass germination responses to isopropyl-phenyl carbamate and classifications. Am. J. Botany 45:16–23.

Alam, M. 17280.

Alam, N. 17510.

Alam, Z. 13656.

Alascio, F. D. 14363 .

267. Alban, E. K., H. W. Ford, and F. S. Howlett. 1948: A preliminary report on the effect of various cultural practices with greenhouse tomatoes on the respiration rate of the harvested fruit. Proc. Am. Soc. Hort. Sci. 52:385–90.

Alban, L. A. 9225.

268. Albaum, H. G. 1940: The growth of oat coleoptiles after seed exposure to different oxygen concentrations. Biol. Bull. 79:360–61.

269. —— and P. P. Cohen. 1943: Transamination and protein synthesis in germinating oat seedlings. J. Biol. Chem. 149:19–27.

270. ——, J. Donnelly, and S. Korkes. 1942: The growth and metabolism of oat seedlings after seed exposure to oxygen. Am. J. Botany 29:388–95.

271. ——, S. Kaiser, and B. Eichel. 1940: Coleoptile growth after preliminary soaking of oat grains in various solutions. Am. J. Botany 27:619–23. Chem. Abstr. 35: 1453–54.

272. ——, ——, C. Guttentag, and B. Eichel. 1941: The effects of various respiratory stimulants and poisons on the growth of oat coleoptiles. Am. J. Botany Suppl. 28:10s.

273. ——, M. Ogur, and A. Hirshfeld. 1949: Isolation of adenosine triphosphate from plant tissue. Federation Proc. 8:179.

274. —— and W. W. Umbreit. 1943: Phosphorus transformations during the development of the oat embryo. Am. J. Botany 30:553–58.

Albaum, H. G. 15667.

275. Alben, A. O. 1946: Nitrogen deficiency of pecans in the spring of 1946. Proc. Texas Pecan Growers Assoc. 25:24–27.

276. Alberizzi, E., and U. Hoepli. 1943: Preparation for promoting seed germination and plant growth. Ger. Patent 730,233. Chem. Abstr. 38:402. 1944.

Albers, C. C. 9974.

Albert, W. B. 1363.

277. Alberts, A. A. 1953: Use of fish emulsion for the germination of orchid seed. Orchid J. 2:464–66.

278. Alberts, H. W. 1926: Moisture content of corn in relation to relative humidity and temperature of the atmosphere. J. Am. Soc. Agron. 18:1029–34.

279. —— 1927: A method for storing small quantities of seed corn. J. Am. Soc. Agron. 19:567–68.

280. —— 1927: Relation of endosperm character in corn to absorption of hygroscopic moisture. J. Am. Soc. Agron. 19:590–95.

281. Albizzati, C. M. 1937: Presence of glutathione in wheat germ and its influence on flour. Anales Soc. Cienti. Arg. 124: 194–202. Chem. Abstr. 32:2227. 1938.

282. Albo, G. 1908: Les enzymes et la faculté germinative des graines. Bibliotheca Univ. Arch. Sci. Phys. Nat. Ser. 4, 25: 45–52.

283. Albrecht, H. R. 1942: Earliness of maturity as a factor influencing seed production in vetch. J. Am. Soc. Agron. 34:662–67.

Albrecht, H. R. 4387.

284. Albrecht, J. S. D. 1740: Germinatibus in fruchtu *Melopeponis seminibus*. Acta Phys. Med. Acad. Caes. Nat. Cir. 5:94–97.

285. Albrecht, R. E., L. G. Gavrilova, and V. N. Lubimenko. 1931: Influence de la lumière sur le développement des fruits et des graines du *Nicotiana tabacum* L. News of the Nikitski Botanical Garden 3/4:305–25.

286. Albrecht, W. A. 1941: Calcium as a factor in seed germination. J. Am. Soc. Agron. 33:153–55.

287. —— and W. H. Allison. 1931: Changes in composition of soy beans toward maturity as related to their use as a green manure. Soil Sci. 32:271–82. Chem. Abstr. 26:551. 1932.

Albrecht, W. A. 6589.

Albright, L. E. 11898.

Alcala, J. M. R. de Barbolla-y-; *see* De Barbolla-y-Alcala, J. M. R.

Alcock, A. W. 545, 8903, 8904.

288. Alcock, N. L. 1928: Seed of *Trifolium repens* L. carrying a fungus resembling *Sclerotinia trifoliorum*, clover stem rot. Proc. Intern. Seed Testing Assoc. 6:31–36.

289. —— 1931: Notes on common diseases sometimes seed-borne. Trans. Proc. Roy. Soc. Edinburgh 30:332–37.

290. Alcock, Mrs. N. L., and M. D. Martin. 1928: A seed-borne disease of clover, *Trifolium repens* L. Trans. Proc. Botan. Soc. Edinburgh 30:13–18. Rev. Appl. Mycol. 8:110. 1929.

Alcontres, G. S. d'; *see* D'Alcontres, G.S.

291. Alcorn, G. B., P.C. Berryman, and R. R. Parks. 1948: Portable cleaners for seed grain. Calif. Agr. Expt. Sta. Ext. Serv. Circ. 141:1–20. Biol. Abstr. 22:19074. 1948.

292. Alcorn, S. M., and E. B. Kurtz, Jr. 1959: Some factors affecting the germination of seed of the Saguaro cactus (*Carnegiea gigantea*). Am. J. Botany 46:526–29.

Alcorn, S. M. 11579.

Aldred, W. H. 17750.

Aldrich, R. J. 12195.

293. Aldrich, S. R. 1943: Maturity measurements in corn and an indication that grain development continues after premature cutting. J. Am. Soc. Agron. 35:667–80.

294. Aleeva, M. M. 1961: The treatment of corn seed. Tr. Chuvashsk. Sel'skokhoz. Inst. 5:61–63. Biol. Abstr. 43:3242. 1963.

295. Alekperov, S. A. 1961: The effect of concentrated solutions of salts on the physiology of the swelling and on the germinating capacity of seeds of some trees and shrubs. Izv. Akad. Nauk Azerb. SSR Ser. Biol. Med. Nauk 1961(1):15–23.

296. —— and T. E. Khrzhanovskaya. 1961: The effect of salts on certain physiological characteristics of seeds and sprouts of *Sophora japonica*. Izv. Akad. Nauk Azerb. SSR Ser. Biol. i Med. Nauk 1961(12):3–15. Biol. Abstr. 44:7906. 1963.

Aleksandrov, S. V. 3028.

Aleksandrovskaia, V. 13808.

Aleksandrovskaya, Z. V. Ivanov-; *see* Ivanov-Aleksandrovskaya, Z. V.

Aleksanyan, A. K. 10096.

297. Aleksanyan, S. V. 1956: The chemical composition of corn at different stages of its growth. Sotsial. Tvarinnitsvo 1956(7): 28–30. Chem. Abstr. 51:8909. 1957.

298. Alekseenko, I. K. 1950: Viability of hybrid tomato seeds. Selek. i Semen. 17:72–73.

299. Alekseenko, M. P. 1952: Effect of conditions for growing buckwheat seeds on its yields. Selek. i. Semen. 19:64–67.

300. Alekseev, V. A. 1962: Stimulating effect of hydroxytetracycline on sprouting of corn seeds. Byul. Nauchn.-Tekhn. Inform. po Sel'skokhoz. Mikrobiol. 1962 (2):6–7. Chem. Abstr. 60:1041. 1964.

301. Alekseeva, M. V. 1958: The seed proteins of *Cucurbita*. Tr. po Khim. Prirodn. Soedin. Kishinevsk. Gos. Univ. 1958(1):103–8. Chem. Abstr. 54:1667. 1960.

302. —— 1961: A comparative study of nitrogen-containing substances in the seeds of some representatives of the gourd family. Tr. po Khim. Prirodn. Soedin. Kishinevsk. Gos. Univ. 1961(4):37–43. Chem. Abstr. 58:10514. 1963.

303. —— 1962: Paper electrophoresis of salt-soluble proteins of seeds of certain representatives of Cucurbitaceae. Tr. po Khim. Prirodn. Soedin. Kishinevsk. Gos. Univ. 1962(5):69–74. Chem. Abstr. 59: 14294. 1963.

Alekseeva, M. V. 10806.

304. Aleshin, E. P. 1958: Effect of germination conditions on oxidases and catalases of rice. Plant Physiol. (USSR) (English Transl.) 5:359–60. Chem. Abstr. 52: 20444. 1958.

305. —— 1959: Water conditions in a rice field during germination of the rice seeds. Plant Physiol. (USSR) (English Transl.) 6:744–46.

306. —— 1961: Physiological characteristics of germination of rice seeds. Plant Physiol. (USSR) (English Transl.) 7:562–66.

307. —— and A. I. Aprod. 1960: The effect of temperature on germination of seed of Kuban commercial varieties of rice. Plant Physiol. (USSR) (English Transl.) 7:297–300.

308. —— and B. V. Filin-Koldakov. 1961: Terminal oxidases of germinating rice grains. Dokl. Akad. Nauk SSSR 134:208–9.

309. —— and M. I. Ryazantseva. 1960: Effect of storage conditions on the quality of high-fat-containing sunflower seeds. Soobshch. Ref. Vses Nauchn. Issled. Inst. Zerna Produkt. Ego Pererabotiki 1960(3): 14–15. Chem. Abstr. 56:9134. 1962.

310. Aleskerov, K. 1961: Effect of the developmental conditions on the dynamics of the formation and shedding of the seed pods in cotton. Izv. Akad. Nauk Azerb. SSR Ser. Biol. Med. Nauk 10:55–66. Biol. Abstr. 43:11654. 1963.

Alessi, J. 3611.

311. Alexander, D. E. 1963: The "Lysenko Method" of increasing oil content of the sunflower. Crop Sci. 3:279–80.

312. Alexander, J. C., and D. C. Hill. 1952: The effect of heat on the lysine and methionine in sunflower seed oil meal. J. Nutr. 48:149–59.

313. Alexander, M. W., and C. F. Genter. 1962: Production of corn and soybeans in alternate pairs of rows. Agron. J. 54: 233–34.

Alexander, T. L. 13774, 13775.

314. Alexandrescu, V. 1960: Biochemistry

of the seeds of wheat, corn, and cotton stimulated with potassium bromide, hydroquinone and glutamic acid. Acad. Rep. Populare Romine Studii Cercetari Biochim. 3:171–79. Chem. Abstr. 55:4667. 1961.

315. Alexandrov, V. G. 1937: On the morphology of the grain of cereals. Compt. Rend. Acad. Sci. URSS 17:389–91. Bull. Soc. Botan. France 87:171. 1940.

316. —— 1943: History of development and morphology of caryopsis of some grasses of the type of Hordeae. Sov. Botan. 1943(4):24–35.

317. —— 1943: On the structure of pericarp in wheat kernel. Compt. Rend. Acad. Sci. URSS 40:81–84.

318. —— and O. G. Alexandrova. 1934: The distribution of pigments in the testa of some varieties of soybeans, *Glycine hispida* Maxim. Bull. Appl. Botany, Genet., Plant Breed. (USSR) Ser. 3(4):3–47. Chem. Abstr. 29:5149. 1935.

319. —— and —— 1936: Anatomy of endosperm of wheat. Flora 30:21–38. J. Roy. Microscop. Soc. Ser. III 56:182. 1936.

320. —— and —— 1935: Anatomy of flower, fruit, and seed of peas. Bull. Appl. Botany, Genet., Plant Breed. (USSR) Ser. 3(9):1–151. Biol. Abstr. 10:22030. 1936.

321. —— and —— 1936: On the method of anatomic study of the endosperm structure in the wheat grain. Bull. Appl. Botany, Genet., Plant Breed. (USSR) Ser. 5A(2): 22–23.

322. —— and —— 1936: On the anatomical characters determining the flintiness of the wheat grain. Bull. Appl. Botany, Genet., Plant Breed. (USSR) Ser. 5A(2):41–42.

323. —— and —— 1936: Anatomical analysis of some moments in the filling and ripening of the wheat grain. Bull. Appl. Botany, Genet., Plant Breed. (USSR) Ser. 5A(2):56–57.

324 —— and —— 1937: On the mosaic of wheat endosperm. Compt. Rend. Acad. Sci. URSS 17:495–98. Bull. Soc. Botan. France 87:171. 1940.

325 —— and —— 1938: On the flintiness and flouriness of wheat endosperm. Compt. Rend. Acad. Sci. URSS 18:111–14. Biol. Abstr. 14:11038. 1940.

326. —— and —— 1938: On the structure of the grain in *Aegilops*. J. Botan. URSS 23:396.

327. —— and —— 1938: Structure of shrivelled ("meagre") grains of hard wheat. Compt. Rend. Acad. Sci. URSS 18:307–10.

328. —— and —— 1938: Full and meagre grain in soft wheat. Compt. Rend. Acad. Sci. URSS 18:613–16. Bull. Soc. Botan. France 87:172. 1940.

329. —— and —— 1938: On the endosperm nucleus and its role in the filling and ripening of the grains of cereals. Compt. Rend. Acad. Sci. URSS 20:613–16.

330. —— and —— 1938: On the structure of furrow in wheat kernel. Compt. Rend. Acad. Sci. URSS 21:81–84.

331. —— and —— 1939: On the state of the cell nuclei in the pericarp and the endosperm of wheat during the period of maturation of the grain. Compt. Rend. Acad. Sci. URSS 22:194–97.

332. —— and —— 1939: Anatomical characters of the achenes of the early maturing and late maturing wheat. Compt. Rend. Acad. Sci. URSS 23:388–91.

333. —— and —— 1939: Development of endosperm in cereal kernel and its morphology. Compt. Rend. Acad. Sci. URSS 24:796–99.

334. —— and —— 1939: Formation of endosperm in hard and soft wheats. Compt. Rend. Acad. Sci. URSS 25:320–23.

335. —— and —— 1940: Peculiarities of the endosperm development in wheats with short and long periods of ripening. Compt. Rend. Acad. Sci. URSS 26:283–86.

336. —— and —— 1944: On the structure of spermoderm of the wheat kernel and on protandry. Compt. Rend. Acad. Sci. URSS 42:91–95.

337. —— and —— 1954: On the dying off and disintegration of nuclei in cereal endospermic cells as one of the most important factors which condition the ripening of the kernel. Izv. Akad. Nauk SSSR Ser. Biol. 5:88–103.

338. —— and M. S. Iakovlev. 1935: Die Morphologie des Korns und der Bau des Endosperms bei verschiedenen Formen von *Zea mais* L. J. Botan. URSS 20:245–81.

339. —— and —— 1937: On the starch granules in the oat grain. J. Botan. URSS 22:544–45.

340. —— and N. V. Pervukhina. 1950: Ovules, ovaries, and nature of seed of Umbelliferae. Dokl. Akad. Nauk SSSR 70:113–16.

341. Alexandrova, O. G. 1936: Pigmentation of the grain of wild einkorns. Bull. Appl. Botany, Genet., Plant Breed. (USSR) Ser. 5A(2):62.

342. —— 1937: The anatomy of various types of wheat grain. Compt. Rend. Acad.

Sci. URSS 17:385–88. Bull. Soc. Botan. France 87:171. 1940.
Alexandrova, O. G. 318–337.
343. Alexandrow, W., and M. Jakowlew. 1938: Über die morphologische Bedeutung der Sackförmigen Zellen der Karyopse von Gramineen. Acad. Sci. RSS Ukraine Inst. Botan. Kiev. Symposium dedicated to the memory of A. V. Fomin 1938:239–40.
Alexeievsky, A. 14601.
344. Alferov, V. A. 1952: Experience in propagation of lilies from seeds. Byul. Mosk. Gl. Botan. Sad. 12:70–73.
Alfin-Slater, R. B. 10222.
345. Alfnes, A. 1961: Pelleted carrot seed for one seed equipment. Gartneryrket 51: 1287.
346. Alghisi, P. 1959: Principal diseases of vegetable plants transmitted by seeds and vegetative organs. Agr. delle Venezie 13:345–53.
347. Ali, G., and Karimullah. 1958: Germination of the seeds of guara (*Cyamopsis tetragonoloba*); chemistry and analysis. Pakistan J. Sci. Res. 10:93–97. Chem. Abstr. 53:6361. 1959.
Ali, M. E. 15796.
348. Ali, M. I., A. Ahmed, and M. H. Khundkar. 1958: Evaluation of fungicidal properties of lignin acetic acid. Part II. Studies on treated jute seeds in infested soil. Pakistan J. Sci. Ind. Res. 1:113–15.
349. Ali, M. K., and S. D. Chaudhuri. 1962: New technique for growing tops of jute plants for seed. Botan. Gaz. 123:259–61.
350. Ali, N. 1959: Studies on seed germination in rootstock species of citrus. Punjab Fruit J. 22(79):22–32.
351. Ali, S. B., and M. Nayeemullah. 1959: Seed treatment of groundnut. Agr. Pakistan 10:185–94.
352. Alikonis, J. J. 1955: Quick way to seed coatings. Food Eng. 27(6):110, 115.
353. Alim, A. 1959: The programme of breeding of improved varieties of field crops and their seed multiplication, certification and distribution in East Pakistan. Agr. Pakistan 10:374–82.
354. —— and S. M. Hasanuzzaman. 1952: Effect of soaking in water and delayed sowing on the germination percentage of "shail" paddy seeds. [Abstract.] Pakistan Sci. Conf. Proc. 4(3):8–9.
355. —— and J. L. Sim. 1956: Vernalisation in spring paddy. Agr. Pakistan 7:322–26.
Alimchandani, R. L. 18831.
Aliod, M. Claver-. 13937.
356. Aljawdina, A. A. 1931: Key to the determination of the fruits and seeds of the weed species and cultivated species of Cruciferae, occurring in the European part of USSR. Bull. Appl. Botany, Genet., Plant Breed. (USSR) 25(2):213.
357. —— 1931: Bedeutung der Anatomie der Frucht und des Samens für die Systematik der Familie Cruciferae. J. Soc. Botan. Russ. 16:85–100.
358. —— 1935: Zur Kenntnis der Funktion der Samenschale bei der Samenkeimung. Botan. Centralbl. Beih. Abt. A 53:200–16.
359. —— 1937: Mikroskopische Untersuchung der Keimenden Samen und des Isolierten Endosperms von *Linum usitatissimum* L. J. Botan. URSS 22:51.
360. —— 1939: Outer integuments of the wheat kernel as a conducting system. Compt. Rend. Acad. Sci. URSS 25:524–26.
361. —— 1941: Zur Frage über die Grenze des Perikarps und der Samenschale in den Karyopsen der Getreide. Compt. Rend. Acad. Sci. URSS 31:953–55.
362. Alkon, P. U., and D. L. Schierbaum. 1962: Germination of staghorn sumac (*Rhus typhina*) seeds. N.Y. Fish & Game J. 9:69–71.
363. Allan, J. M. 1934: Destruction of wild turnip seed in second-hand bags by chemical means. J. Agr. W. Australia 11:608–10. Chem. Abstr. 30:5354. 1936.
364. Allan, R. E., O. A. Vogel, and J. C. Craddock, Jr. 1961: Effect of gibberellic acid upon seedling emergence of slow and fast emerging wheat varieties. Agron. J. 53:30–32.
365. Allan, W. 1931: Diplodia infection and the treatment of maize seed. Northern Rhodesia Ann. Bull. Dept. Agr. 1:20–30. Biol. Abstr. 9:289. 1935.
366. Allard, H. A. 1943: Length-of-day behavior of *Nicotiana gossei*. J. Agr. Res. 67:459–64.
367. —— 1948: Length of day in the climates of past geological eras and its possible effects upon changes in plant life. *In* Murneek, A. W., and R. O. Whyte, Vernalization and photoperiodism: A Symposium. Chronica Botan., Waltham, Mass., pp. 101–19.
368. —— and M. W. Evans. 1941: Growth and flowering of some tame and wild grasses in response to different photoperiods. J. Agr. Res. 62:193–228.
Allard, H. A. 6988, 6989.
369. Allard, R. W. 1953: Inheritance of some seed-coat colors and patterns in lima beans. Hilgardia 22:167–77.

370. ——, H. R. DeRose, and C. P. Swanson. 1946: Some effects of plant growth-regulators on seed germination and seedling development. Botan. Gaz. 107:575–83.

371. —— and F. L. Smith. 1959: Dry edible bean production in California. Calif. Agr. Expt. Sta. Ext. Serv. Circ. 436. 28 pp.

Allayyanavaramath, S. B. 10450.

372. Alldridge, N. A., and W. G. Whaley. 1956: The pattern of respiration and carbohydrate level in germinating maize. Plant Physiol. (Suppl.) 31:33.

373. Allegri, A. 1935: Sunflower-seed diet in experimental polyneuritis. Boll. Soc. Ital. Biol. Sper. 10:839–43. Chem. Abstr. 30:513. 1936.

374. Allen, C. E. 1949: Pelleted seed panel. A comparison of two types of Crested wheatgrass pellets. Proc. Assoc. Offic. Seed Analysts 39:73–74.

Allen, F. W. 11755.

375. Allen, G. S. 1941: A basis for forecasting seed crops of some coniferous trees. J. Forestry 39:1014–16.

376. —— 1941: A standard germination test for Douglas fir seed. Forestry Chron. 17:75–78. Biol. Abstr. 15:19629. 1941.

377. —— 1941: Light and temperature as factors in the germination of the seed of Douglas fir [*Pseudotsuga taxifolia* (Lamb.) Britt.]. Forestry Chron. 17:99–109. Biol. Abstr. 16:5324. 1942.

378. —— 1942: Parthenocarpy, parthenogenesis, and self-sterility of Douglas fir. J. Forestry 40:642–44.

379. —— 1942: Douglas fir seed from young trees. J. Forestry 40:722–23.

380. —— 1943: The embryogeny of *Pseudotsuga taxifolia* (Lamb.) Britt. Am. J. Botany 30:655–61.

381. —— 1947: Embryogeny and the development of the apical meristems of *Pseudotsuga*. II. Late embryogeny. Am. J. Botany 34:73–80.

382. —— 1947: Mold-free germination of coniferous seeds. J. Forestry 45:51.

383. —— 1956: The effect of date of cone collection upon the viability, germination behavior, and storage characteristics of Western hemlock (*Tsuga heterophylla*) seed. Forestry Chron. 32:262–63.

384. —— 1957: Storage behavior of conifer seeds in sealed containers held at 0° F., 32° F., and room temperature. J. Forestry 55:278–81.

385. —— 1958: Factors affecting the viability and germination behavior of coniferous seed. I–III. Forestry Chron. 34:266–98.

386. —— 1958: Germination conditions for western hemlock. J. Forestry 56:770–71.

387. —— 1960: Factors affecting the viability and germination behavior of coniferous seed. IV. Stratification period and incubation temperature, *Pseudotsuga menziesii* (Mirb.) Franco. Forestry Chron. 36:18–29.

388. —— 1961: Testing Douglas fir seed for provenance. Proc. Intern. Seed Testing Assoc. 26:388–403.

389. —— 1962: Factors affecting the viability and germination behavior of coniferous seed. VI. Stratification and subsequent treatment, *Pseudotsuga menziesii* (Mirb.) Franco. Forestry Chron. 38:485–96.

390. —— and W. Bientjes. 1954: Studies on coniferous tree seed at the University of British Columbia. Forestry Chron. 39: 183–96. Biol. Abstr. 29:6785. 1955.

391. Allen, H. T. 1945: Germination of seed of trees and shrubs. W. Can. Soc. Hort. Rept. Proc. 2:56. Processed.

392. Allen, J. D. 1961: Hollow heart of pea seed. New Zealand J. Agr. Res. 4:286–88.

393. Allen, Mrs. J. D. 1940: Growing primulas from seed. Horticulture 18(14):314.

394. —— and H. C. Smith. 1961: Dry-rot (*Leptosphaeria maculans*) of Brassicas: seed transmission and treatment. New Zealand J. Agr. Res. 4:676–85.

Allen, N. K. 3052.

Allen, O. N. 1464.

395. Allen, P. H., and K. B. Trousdell. 1961: Loblolly pine seed production in the Virginia–North Carolina coastal plain. J. Forestry 59:187–90.

396. Allen, R. L. 1860: The American farm book or compend of American agriculture. C. M. Saxton, Barker & Co., New York. 325 pp.

397. Allen, R. M. 1951: Cotton seeds are capable of carrying Verticillium. Plant Disease Reptr. 35:11–12.

398. Allen, S. E., and F. Skoog. 1952: Stimulation of seedling growth by seed treatments with N-phenyl succinimide derivatives. Plant Physiol. 27:179–83.

399. Allen, T. C., and L. K. Brunn. 1945: Increased toxicity of lime-treated Sabadilla seed in dust suspensions. J. Econ. Entomol. 38:291–93. Rev. Appl. Entomol. Ser. A 34:301. 1946.

400. ——, F. J. Dexheimer, and E. Cole. 1945: Reduction of certain insects infesting alfalfa by use of Sabadilla. J. Econ. Entomol. 38:389–90. Rev. Appl. Entomol. Ser A 34:312. 1946.

401. —— and E. Fisher. 1943: Increase yield of wax beans with "hormone" insecticide dusts. Canner 96(22):12–13. Expt. Sta. Rec. 89:310. 1943.

402. ——, K. P. Link, M. Ikawa, and L. K. Brunn. 1945: Relative effectiveness of the principal alkaloids of Sabadilla seed. J. Econ. Entomol. 38:293–96.

Allen, T. C. 9359.

403. Allen, W. S. 1959: Grain storage. Clemson Agr. Coll. Circ. 462. 11 pp.

404. Allerup, S. 1958: Effect of temperature on uptake of water in seeds. Physiol. Plantarum 11:99–105.

Alley, H. P. 19492.

405. Allington, H. V. 1952: Exposure of *Camellia* seed to X-rays. Am. Camellia Yearbook 1952:120–22.

Allison, D. C. 2601.

406. Allison, H. W. S., and R. H. Kenten. 1963: Seasonal variation in the fermentation of West African *Amelonado cocoa.* Trop. Agr. (London) 40:217–21.

407. Allison, J. L., and J. H. Torrie. 1944: Effect of several seed protectants on germination and stands of various forage legumes. Phytopathology 34:799–804.

Allison, J. L. 19512.

Allison, W. H. 287.

408. Allman. 1878: Dissemination of *Pinguicula grandiflora* through the agency of birds. J. Linn. Soc. London 17:157–58.

409. Allman, D. 1940: Snapdragon seed sowing. Florists Exchange Hort. Trade World 95(6):11.

410. Allman, S. L. 1952: Mango seed weevil. Agr. Gaz. N.S. Wales 63:206.

Allman, S. L. 5715.

411. Allnutt, R. B. 1936: The establishment and maintenance of a pure supply of white sesame seed. E. African Agr. J. 1:369–71, 420.

412. Allred, K. R., and D. C. Tingey. 1964: Germination and spring emergence of dodder as influenced by temperature. Weeds 12:45–48.

Allred, K. R. 18586.

413. Allred, S. E. 1963: The effect of ammonia, mono- and diammonium phosphate on the germination and emergence of corn. Dissertation Abstr. 23:2648.

414. —— and A. J. Ohlrogge. 1964: Principles of nutrient uptake from fertilizer bands. VI. Germination and emergence of corn as affected by ammonia and ammonium phosphate. Agron. J. 56:309–13.

415. Allsop, F. 1951: First production of fertile seed by the principal exotic conifers in New Zealand. Forest Res. Inst. Forest Res. Notes 1(4):13–14.

416. Almazova, D. I., and T. A. Rabotnov. 1953: In regard to the biology of sorrel (*Rumex confertus* Willd.). I. Reproduction of horse sorrel by seeds. Byul. Mosk. Obshch. Ispytatelei Prirody Otd. Biol. 58(6):47–54.

Almeida, A. d'. 7120.

Almeida, A. O. de; *see* De Almeida, A. O.

Almeida-Costa, O. de; *see* De Almeida Costa, O.

Almendinger, V. 11204–11207.

417. Almeyda, N., and E. P. Hume. 1949: Seed storage. Puerto Rico Fed. Expt. Sta. Rept. 1948:14.

Almgard, G. 14641.

Almquist, H. J. 7594, 12436.

418. Aloi, A. 1881: Una piccola prova sulla germinazione dei vinacciuoli americani. Riv. Viticult. Enol. Ital. 14:438–41. Botan. Jahresber. 9:33. 1881.

419. —— 1891: Influence of electricity on germination of seeds. Malpighia 5:116–25.

420. Aloni, R., and A. Poljakoff-Mayber. 1962: Adenosine triphosphatase activity in germinating lettuce seeds. Plant Cell Physiol. 3:105–10.

Alonso, R. E. 4519.

421. Alov, A. S. 1943: The after-effect of fertilizers through the seeds. Vsesoiuzn. Akad. Sel'sk. Khoz. Nauk im V. I. Lenina Dok. 8(4):19–21.

422. —— 1946: On the question of interaction between germinating seeds and external medium. Lenin Acad. Agr. Sci. USSR 1946(7/8):14–21.

423. Alpatev, A. V. 1953: Productivity of tomatoes in relation to methods of growing seeds. Agrobiologiya 1953(2): 48–54.

424. Alpatov, V. V. 1949: Increasing the germination of seeds under the influence of high temperature and ultra-violet rays. Priroda 38(12):41–43.

Alsberg, C. L. 3023, 7668.

425. Alston, C. 1943: Treated cotton seed yields 800 pounds per acre—untreated 600 pounds. Agr. News Letter 11:50.

426. Alten, F., and E. Schulte. 1941: The effect of fertilizers on the speed of germination of cereal grains. Ernähr. Pflanze 37:13–21, 25–28. Chem. Abstr. 36:6732–33. 1942.

Altena, A. C. J. van; *see* Van Altena, A. C. J.

427. Althaus, H. C. 1949: Seeds with

overcoats. Pelletizing allows precise planting with seeds. Southern Seedsman 12(2):13, 40–41.

428. Altman, R. F. A. 1956: Chemical studies of Amazonian plants. VI. The seed of *Euterpe oleracea*. Bol. Tec. Inst. Agron. Norte 31:109–11. Chem. Abstr. 52:506. 1958.

429. Altschul, A. M. 1949: Chemical treatments of seeds to prevent heating and deterioration during storage. Cotton Gin Oil Mill Press 50(1):A2–A10.

430. —— 1949: Sweet and unspoiled. Chemical treatment prevents deterioration of cottonseed under storage. Chem. Ind. 64:211.

431. —— 1951: Research on cottonseed storage and processing at the Southern Regional Research Laboratory. Cotton Gin Oil Mill Press 52(16):24, 26, 28–29, 30, 32. Cotton Dig. 23(46):6–7, 42–43, 46. 1951. U.S. Dept. Agr. List Publ. & Patents, p. 8. 1951.

432. —— 1962: Seed proteins and world food problems. Econ. Botany 16:2–3.

433. ——, M. Z. Condon, and M. G. Lambou. 1951: Method of prevention of deterioration in seeds. U.S. Patent 2,571,095.

434. —— and M. L. Karon. 1947: A manometric method for the determination of peroxidase activity. Application to an investigation of the peroxidase of germinating cottonseed. Arch. Biochem. 13: 161–73.

435. ——, ——, and P. J. Fynn. 1946: An apparatus and method for the measurement of the respiration of cottonseed. Plant Physiol. 21:410–15.

436. ——, ——, and L. Kyame. 1948: Catalase and peroxidase in germinating cottonseed. Arch. Biochem. 18:161–70.

437. ——, ——, ——, and M. Caravella. 1943: The storage of cottonseed. II. The effect of ammonia treatment on the free fatty acids and color of the seed oil, and on the rate and degree of heating of the seed. Oil Soap 20:258–62. Expt. Sta. Rec. 92:756. 1945.

438. ——, ——, ——, and C. M. Hall. 1946: Effect of inhibitors on the respiration and storage of cottonseed. Plant Physiol. 21:573–87.

Altschul, A. M. 1663, 4329–4331, 4333, 5352, 6102, 9784, 10270, 10271, 11618, 11715, 11716, 14852, 19970.

Altstatt, G. E. 12896.

439. Alvahydo, M., and E. Vellasco. 1931: A germinação das sementes de algodão e o indice de acidez dos oleos. Brazil Min. Agr. Serv. Algodão (Seccao Tec.) 5 pp.

Alvarez, C. Carbona. 14837.

Alvarez, L. A. 1460.

440. Alves, I. S. A., and A. A. D. Dos Santos. 1953: Efeito da germinação no valor nutritivo e teor vitamínico de algumas sementes de leguminosas. Melhoramento 6:75–113.

441. Alvey, D. D., and J. W. Pendleton. 1960: Pre-harvest drying of grain sorghum. Agron. J. 52:669.

442. Alvim-Carneiro, P. de T. 1948: Vernalization of garden plants. A review of the literature. Rev. Ceres. 7:297–320.

443. —— 1958: A simple method of conserving the germinating power of cacao seeds. Agronomía (Peru) 25(92):43–48.

444. Alvin, K. L. 1960: On the seed of *Vesquia tournaisii* C. E. Bertrand, from the Belgian Wealden. Ann. Botany 24:508–15.

445. Alvistur, C. E., P. White, and C. Collazos. 1953: Biological value of quinoa. Bol. Soc. Quim. Peru 19:197–209. Chem. Abstr. 48:10243. 1954.

Alyakrinskaya, E. A. 11375.

446. Alyanskaya, N. S. 1962: Some data on the germination of seeds of Bergenia. Byul. Mosk. Gl. Botan. Sad. 46:105–8.

Amadori, M. E. 6848.

447. Aman, J., L. Farkas, M. H. Ben-Shamai, and M. Plaut. 1946: Experiments on the use of ethylene dibromide as a fumigant for grain and seed. Ann. Appl. Biol. 33:389–95.

448. Amar, S. S. 1962: The use of Cellager for the study of germination. Progr. Agr. (Bologna) 8:244–51.

Amato, F. d'; *see* D'Amato, F.

449. Ambe, K. S., and K. Sohonie. 1956: Trypsin inhibitor in Indian foods. III. Trypsin inhibitor in germinating double bean and field bean and their growing plants. J. Sci. Ind. Res. 15C:136–40. Chem. Abstr. 51:1391. 1957.

Ambe, K. S. 17709.

450. Ambe, L., and K. Sohonie. 1963: Purification and properties of glutamate decarboxylase from field bean (*Dolichos lablab*). Enzymologia 26:98–107.

Ambegaokar, K. B. 15311–15313.

451. Ambrose, C. 1951: Vertical or horizontal? Which is the correct method of planting out seed-nuts? Ceylon Coconut Quart. 2:169–71.

452. Ameen, O. A., A. D. Day, and K. C. Hamilton. 1960: Effect of 1,2,4,5-tetrachlorobenzene on the germination and seedling vigor of barley, oats, and wheat. Agron. J. 52:87–89.

453. Amelinckx, F. 1952: Harvesting, treatment, storage, and planting of tree seeds. Cult. Handel 18:526–28.

454. Amemiya, A., H. Akemine, and K. Toriyama. 1956: Cultural conditions and growth of immature embryo in rice plant. (Studies on the embryo culture in rice plant. 1.) Natl. Inst. Agr. Sci. Ser. D 6:1–40.

455. ——, ——, and —— 1956: The first germinative stage and varietal differences in growth response of cultured embryos in rice plant. (Studies on the embryo culture in rice plant. 2.) Natl. Inst. Agr. Sci. Ser. D 6:41–60.

456. Amen, R. D. 1963: Germination and dormancy in two species of Alpine *Carex* from the Colorado front range. Dissertation Abstr. 23:3603.

457. Ames, O. 1963: The significance of the angiosperm seed. Econ. Botany 17:3–9.

458. Amick, C. H., W. F. Crosier, W. D. Hay, D. Isley, and M. A. Willard. 1948: Seed treatments with DDT and DDT-fungicide mixtures. Assoc. Offic. Seed Analysts News Letter 22(2):46–48.

459. Amin, J. V., and H. E. Joham. 1960: The molybdenum of cotton seed as an indicator of molybdenum availability in soils. Proc. Ann. Conv. Assoc. Southern Agr. Workers 57:245–46.

460. Amirkhanov, N. A. 1961: Germination of the seed of *Crambe kotshiana*. Tr. Samarkandsk. Univ. 115:255–61. Biol. Abstr. 43:20639. 1963.

461. Amirshahi, M. C., and F. L. Patterson. 1956: Development of a standard artificial freezing technique for evaluating cold resistance of oats. Agron. J. 48:181–84.

462. Amlong, H. U. 1941: The influence of treating sugar-beet seed with plant hormones. Angew. Botan. 23:289–303. Chem. Abstr. 37:3795. 1943.

463. —— and G. Naundorf. 1937: A practical application of plant growth substances. Forschungsdienst 4:417–27. Chem. Abstr. 32:5030. 1938.

464. —— and —— 1938: Neue Wege der Pflanzenstimulation. Forschungsdienst 5:292–303.

465. —— and —— 1942: Treating beet seed. Ger. Patent 722,198. Chem. Abstr. 37:5187–88. 1943.

466. Amma, B. S. 1964: Preliminary studies on the effect of micronutrients on the germination of coconut seednuts. Current Sci. (India) 33:49–50.

467. Ammons, N. P. 1926: Date-seed germination. Proc. West V. Acad. Sci. 1:23.

468. Amsler, M. 1945: The effect of artificial frost on seed germination. J. Roy. Hort. Soc. 70:316–17.

469. —— and W. Lawrence. 1933: Synonymity and culture of the commoner European gentians. Gardeners Chron. 93:58.

470. Anagnostopoulos, P. T. 1936: Preparation to germination of seeds of some trees. Hort. Res. (Athens) A:62–75.

471. —— 1956: La printanisation de l'olivier et la germination de ses noyaux. Fruits Primeurs de l'Agr. Nord 26:205–17.

Anaguchi, I. 18140.

472. Anand, S. C. 1962: Inheritance of yield components and seed coat mottling and smudginess in soybeans. Dissertation Abstr. 23:1148.

473. —— and J. H. Torrie. 1964: Heritability of frequency and intensity of seed coat mottling and smudginess and interrelationships with other traits in soybeans. Crop Sci. 4:185–86.

Anandan, A. P. 5027.

474. Ananichev, A. V. 1963: Separation of furocoumarins from the seeds of *Ammi majus* with the aid of paper chromatography. Med. Prom. SSSR 17(8):44–46. Chem. Abstr. 60:850. 1964.

475. Anantaswamy-Rau, M. 1954: The development of the embryo of *Cyamopsis*, *Desmodium*, and *Lespedeza* with a discussion on the position of the Papilionaceae in the system of embryogenic classification. Phytomorphology 4:418–30. Biol. Abstr. 31:39201. 1957.

Anavaradham, L. 17838.

476. Ancel, S. 1924: Action de faibles doses de rayons X sur des graines sèches. Compt. Rend. Soc. Biol. 91:1435–36.

477. —— 1925: Les rayons X appliqués sur des graines sèches n'ont aucune influence sur l'époque d'apparition du germe. Bull. Soc. Botan. France 72:195–99.

478. —— 1926: De la conservation de la lésion produite par les rayons X chez les graines sèches. Bull. Soc. Botan. France 72:991–95. Biol. Abstr. 1:6534. 1926.

479. —— 1926: Recherche du meilleur test de la radio-réaction des graines de Légumineuses. Bull. Soc. Botan. France 73:71–73. Biol. Abstr. 1:8845. 1926.

480. —— 1927: Sur les variations de sensibilité aux rayons X de graines germées suivant le stade du développement. Compt. Rend. Soc. Biol. 96:986–89.

481. —— 1927: Influence de la temperature sur la radiosensibilité de graines de Lentilles au cours de la germination. Compt. Rend. Soc. Biol. 97:800–2.

482. —— 1927: Le retard de développement determiné par l'application des rayons X sur les graines est-il dû à une action sur les réserves? Bull. Soc. Botan. France 74:12–17.

483. —— 1927: Sur la cumulation des effets lésants determinés par l'action successive de deux facteurs sur des grains de Légumineuses. Bull. Soc. Botan. France 74:346–49.

484. —— 1928: De l'effet du fractionnement des doses de rayons X sur des graines germées. Compt. Rend. Soc. Biol. 98:223–24. Biol. Abstr. 3:17829. 1929.

Ancelin, R. 15255.

Anchori, H. 6114.

Andel, O. M. van. 5135.

485. Andersen, A. L., and D. J. De Zeeuw. 1952: Seed treatment studies for damping-off control in garden and canning beans. (Report of progress.) Mich. Agr. Expt. Sta. Quart. Bull. 34:357–64.

486. —— and —— 1952: Pea seed treatment investigations. Mich. Agr. Expt. Sta. Quart. Bull. 35:25–33.

Andersen, A. L. 5315, 5322–5324, 5326.

487. Andersen, A. M. 1927: Development of the female gametophyte and caryopsis of *Poa pratensis* and *Poa compressa*. J. Agr. Res. 34:1001–18.

488. —— 1932: The effect of removing the glumes on the germination of seeds of *Poa compressa*. [Abstract.] Am. J. Botany 19:835–36.

489. —— 1933: The effect of carbon dioxide and some other gases on the germination of seeds of *Poa compressa*. Abstr. Paper, Botan. Soc. Am. Physiol. Sect., Boston, Dec. 28–30. Am. J. Botany 20:678–79. 1933.

490. —— 1935: The effect of soil in relation to nitrate solution on the germination of seed of *Poa compressa*. Am. J. Botany 22:906.

491. —— 1938: Comparison of methods used in germinating seeds of *Poa compressa*. Proc. Intern. Seed Testing Assoc. 10:307–15.

492. —— 1939: Germination of seed of *Poa compressa* L. and *Poa pratensis* L. at different alternating temperatures. Am. J. Botany 26:18s.

493. —— 1939: Moistening and drying as a pretreatment in germinating seeds of *Poa compressa*. Proc. Assoc. Offic. Seed Analysts 30:246.

494. —— 1941: Germination of freshly harvested seed of Kentucky bluegrass. Proc. Assoc. Offic. Seed Analysts 33:96–98.

495. —— 1941: The effect of different temperatures on the germination of freshly harvested and mature seed of *Axonopus compressus*. Proc. Assoc. Offic. Seed Analysts 33:99–102.

496. —— 1944: Germination of freshly harvested seed of Western grown Astoria bentgrass. Proc. Assoc. Offic. Seed Analysts 1943/44:138–46.

497. —— 1946: The effect of light, temperature, and potassium nitrate on the germination of *Agrostis tenuis* Sibth. and *A. tenuis* var. Highland seed. Proc. Assoc. Offic. Seed Analysts 1945/46:112–25.

498. —— 1947: Some factors influencing the germination of seed of *Poa compressa* L. Proc. Assoc. Offic. Seed Analysts 37:134–43.

499. —— 1947: The effect of alternating temperatures, light intensities, and moistening agents of the substratum on the germination of freshly harvested seed of Oregon grown ryegrass (*Lolium* spp.). Proc. Assoc. Offic. Seed Analysts 37:152–61.

500. —— 1948: The effect of pretreatment and substrata on the germination of sugar beet (*Beta vulgaris* L.) seedballs. Proc. Am. Soc. Sugar Beet Tech. 5:95–99.

501. —— 1948: Some factors associated with the browning of sugar beet (*Beta vulgaris* L.) seedlings in laboratory germination tests. Proc. Assoc. Offic. Seed Analysts 38:52–58

502. —— 1949: Some factors affecting the germination of 1- and 2-year-old domestic ryegrass (*Lolium* spp.) seed. Am. J. Botany 36:820–21.

503. —— 1950: The swollen-seed problem in suckling clover, *Trifolium dubium* Sibth. Proc. Assoc. Offic. Seed Analysts 40:77–82.

504. —— 1953: Effect of the glumes of *Paspalum notatum* (Flugge) on germination. Assoc. Offic. Seed Analysts News Letter 27(3):35–36.

505. —— 1953: Germination of buffel grass, *Pennisetum ciliare* (L.) (Link) seed. Assoc. Offic. Seed Analysts News Letter 27(3):36–37.

506. —— 1954: Some factors affecting the germination of 1- and 2-year-old ryegrass (*Lolium*) seed. Proc. Intern. Seed Testing Assoc. 19:5–13.

507. —— 1955: Fungi associated with germination tests of Merion bluegrass. [Abstract.] Plant Physiol. Suppl. 30:10.

508. —— 1957: Evaluation of normal and questionable seedlings of species of *Melilotus, Lotus, Trifolium* and *Medicago* by greenhouse tests. Proc. Intern. Seed Testing Assoc. 22:237–58.
509. —— 1960: Comparative value of normal and questionable seedlings of peanuts and soybeans by greenhouse tests. Proc. Intern. Seed Testing Assoc. 25:365–78.
510. —— 1962: Effect of gibberellic acid, kinetin-like substance, ceresan, and phenacridane chloride on the germination of *Panicum ramosum* seeds. Proc. Intern. Seed Testing Assoc. 27:730–41.
511. —— 1963: Germination of seed of Texas needlegrass (*Stipa leucotricha*). Assoc. Offic. Seed Analysts News Letter 37(2):18.
512. —— and V. C. Drake. 1944: Preliminary study of seed of crested wheatgrass exhibiting delayed germination. Proc. Assoc. Offic. Seed Analysts 1943/44:146–52.
513. —— and O. L. Justice. 1948: Germination of seeds of five kinds of cucurbits at two temperatures. Proc. Assoc. Offic. Seed Analysts 38:45–47.
514. —— and C. M. Leach. 1961: Testing seeds for seed borne organisms. Yearbook Agr. U.S. Dept. Agr. 1961:453–57.
Andersen, E. M. 10896.
515. Andersen, J. C. 1957: Seed crop culture, 1958. Tidsskr. Froavl 46(545): 305–12.
516. Andersen, R. N. 1958: Control of wild oats by prevention of normal seed development with sodium 2,2-dichloropropionate. Weeds 6:263–70.
Anderson, A. J. 12405.
517. Anderson, C. H., and A. M. F. Hennig. 1964: The effect of date of seeding and fertility level on the yield of wheat, oats, and barley in Northwestern Alberta. Can. J. Plant Sci. 44:15–20.
518. Anderson, D., L. P. Hamilton, H. G. Reynolds, and R. R. Humphrey. 1953: Reseeding desert grassland ranges in southern Arizona. Ariz. Agr. Expt. Sta. Bull. 249. 32 pp.
519. Anderson, D. A., and R. H. Walker. 1931: Residual effects of some germicides used in sterilizing legume seeds. Proc. Iowa Acad. Sci. 38:321–25. Chem. Abstr. 27:1978. 1933.
Anderson, D. B. 10480.
520. Anderson, D. G., and A. G. McCalla. 1960: Sedimentation of crystalline globulin from cucurbit seeds. Can. J. Biochem. Physiol. 38:275–86.
521. Anderson, D. M. W., and C. T. Greenwood. 1955: An investigation of the polysaccharide content of oats, *Avena sativa* L. J. Sci. Food Agr. 6:587–92.
522. ——, ——, and J. S. M. Robertson. 1957: Carbohydrates of the seeds of the rubber tree, *Hevea brasiliensis*. J. Chem. Soc. 1957:401–5.
Anderson, D. T. 5592.
523. Anderson, E. 1943: The seeds of *Tradescantia micrantha*. Ann. Missouri Botan. Garden 30:69.
524. —— and M. Fireman. 1935: The mucilage from Psyllium seed, *Plantago psyllium*, L. J. Biol. Chem. 109:437–42. Chem. Abstr. 29:4409. 1935.
525. ——, L. A. Gillette, and M. G. Seeley. 1941: The mucilage from Indian wheat, *Plantago fastigiata*. J. Biol. Chem. 140: 569–74. Chem. Abstr. 35:6622–23. 1941.
526. ——, J. Hechtman, and M. Seeley. 1938: Hemicelluloses from cottonseed hulls. J. Biol. Chem. 126:175–79.
527. —— and S. Kinsman. 1931: Studies on the hemicelluloses. II. The composition of the hemicellulose from cottonseed hulls. J. Biol. Chem. 94:39–47. Chem. Abstr. 26:754. 1932.
528. —— and H. J. Lowe. 1947: The composition of flaxseed mucilage. J. Biol. Chem. 168:289–97. Biol. Abstr. 21:20713. 1947.
529. Anderson, E. G. 1948: The life history of *Cuscuta campestris* Yuncker. Quebec Soc. Protect. Plants, 29th Rept., pp. 1–4.
530. —— 1949: Field dodder—a menace to important crops. Quebec Soc. Protect. Plants, 30th Rept., pp. 74–75.
531. —— 1952: A chromosomal technique used in biochemical studies on the developing endosperm of maize. Agron. J. 44:560–61.
Anderson, E. G. 11322, 18428.
532. Anderson, E. J., and M. Wood. 1944: Honeybees and red clover pollination. Am. Bee J. 84:156–57. Biol. Abstr. 18:12725. 1944.
533. Anderson, F. W. 1888: Dispersion of seeds of *Euphorbia serpyllifolia*. Botan. Gaz. 13:271.
534. Anderson, H. B. 1945: The activation of jack-bean arginase by cobalt, manganese, and iron. Biochem. J. 39:139–42. Biol. Abstr. 20:5721. 1946.
535. Anderson, H. W. 1946: The relation of antibiotics research to horticulture. Trans. Illinois State Hort. Soc. 79:264–70.
Anderson, H. W. 10108.
536. Anderson, J. A. 1932: The yellow

coloring matter of Khapli wheat, *Triticum dicoccum*. II. The constitution of tricetin. Can. J. Res. 7:285–92. Biol. Abstr. 7:20702. 1933.

537. —— 1934: Studies on the nature of rust resistance in wheat. VI. Effect of hydrogen-ion concentration, phenolic compounds, and host extracts on the germination of urediniospores of *Puccinia graminis tritici*, form 21. Can. J. Res. 11:667–86. Chem. Abstr. 29:1463. 1935.

538. —— 1941: Effect of insecticides on wheat quality. Dept. Trade & Commerce, Can., Board Grain Commissioners Grain Res. Lab., Winnipeg, Ann. Rept. 15:29–30. Chem. Abstr. 36:4206. 1942.

539. —— 1941: Bleached and sprouted wheats. Dept. Trade & Commerce, Can., Board Grain Commissioners Grain Res. Lab., Winnipeg, Ann. Rept. 15:34. Chem. Abstr. 36:4205. 1942.

540. —— 1941: Peeled wheat. Dept. Trade & Commerce, Can., Board Grain Commissioners Grain Res. Lab., Winnipeg, Ann. Rept. 15:34–35. Chem. Abstr. 36: 4205. 1942.

541. —— 1944: Canadian research on malting barley. I. Malting tests and qualities. Empire J. Exptl. Agr. 12:219–32. Biol. Abstr. 19:17187. 1945.

542. —— 1944: Factors related to the protein content of malting barley. Wallerstein Lab. Commun. 7(22):179–91. 1944. Biol. Abstr. 19:9163. 1945.

543. —— 1945: Canadian research on malting barley. II. The relations between certain barley and malt properties. Empire J. Exptl. Agr. 13:1–10. Biol. Abstr. 19:17188. 1945.

544. —— 1945: Canadian research on malting barley. Wallerstein Lab. Commun. 8(23):5–22. Biol. Abstr. 19:17189. 1945.

545. —— and A. W. Alcock, editors. 1954: Storage of cereal grains and their products. Am. Assoc. Cereal Chemists, Monograph Ser., St. Paul, Minn., 2. 515 pp.

546. ——, J. D. Babbitt, and W. O. S. Meredith. 1943: The effect of temperature differential on the moisture content of stored wheat. Can. J. Res. Sec. C 21: 297–306.

547. —— and W. J. Eva. 1943: The protein content of corresponding grades of wheat drawn from the northern and southern portions of western Canada. Can. J. Res. Sec. C 21:276–88.

548. —— and —— 1943: Starch content of western Canadian wheat. II. Its estimation from protein content, and some estimated data. Can. J. Res. Sec. C 21:323–31. Biol. Abstr. 18:5178. 1944.

549. —— and —— 1943: Variations in the protein content of western Canadian wheat, 1927–1938. A study based on twelve annual protein surveys. Bull. Board Grain Comm. Can. 4:1–39. Biol. Abstr. 20:20860. 1946.

550. —— and K. Hlynka. 1948: Protein survey of western Canadian wheat 1948 crop. Can. Grain Res. Lab. Crop Bull. 30. 30 pp.

551. ——, W. O. S. Meredith, W. J. Eva, and A. C. Heise. 1944: Oil seeds in western Canadian grain screenings. Can. J. Res. Sec. F 22:19–27. 1944.

552. ——, K. W. Neatby, and I. Levi. 1944: A preharvest survey of the protein content of western Canadian wheat of the 1943 crop. Cereal Chem. 21:283–95. Expt. Sta. Rec. 92:51. 1945.

Anderson, J. A. 9443–9445, 12919, 16666, 16667.

Anderson, J. C. 6845, 17511.

553. Anderson, J. E. 1935: The determination of moisture in wheat and wheat products. Northwest. Miller 184(2, Sec. 2):25, 28. Expt. Sta. Rec. 77:441. 1937.

554. —— 1936: Some facts concerning vacuum-oven moisture determinations. Cereal Chem. 13:437–52.

555. Anderson, J. F. 1933: To germinate alpine seeds. Gardeners Chron. 37:29.

Anderson, J. F. 17769.

Anderson, L. J. 13212, 13213.

556. Anderson, M. L. 1953: Prerequisites for natural seeding. Quart. J. Forestry 47:159–69. Biol. Abstr. 28:4323. 1954.

557. Anderson, R. A. 1962: Protein content of North Dakota wheat. N. Dakota Farm Res. 22(4):24–27.

558. ——, D. E. Uhl, W. L. Deatherage, and E. L. Griffin, Jr. 1962: Composition of the component parts of two hybrid high-amylose corns. Cereal Chem. 39: 282–86.

559. ——, C. Vojnovich, and E. L. Griffin, Jr. 1961: Wet-milling high-amylose corn containing 66- to 68-percent-amylose starch. Cereal Chem. 38:84–93.

560. ——, ——, and —— 1961: A note on the effect of steeping time on wet-mulling high-amylose corn containing 57 percent-amylose starch. Cereal Chem. 38:94–97.

561. Anderson, R. E., and R. A. Brink. 1952: Kernel pattern in variegated pericarp maize and the frequency of self-colored offspring. Am. J. Botany 39:637–44.

Anderson, R. J. 2276, 2277.

562. Anderson, S. R. 1955: Development of pods and seeds of Birdsfoot Trefoil, *Lotus corniculatus* L., as related to maturity and to seed yields. Agron. J. 47:483–87.

563. —— and D. S. Metcalfe. 1957: Seed yields of Birdsfoot Trefoil (*Lotus corniculatus* L.) as affected by pre-harvest clipping and by growing in association with three adapted grasses. Agron. J. 49:52–55.

Anderson, S. R. 1648.

564. Anderson, T. 1929: Committee on hard seeds and broken seedlings. Actes Congr. Intern. Essais Semences 5:239–58.

565. Anderson, W. A. 1954: Seed production on the farm. Crops Soils 7(3):18–19.

566. Anderson, W. H., R. L. Carolus, and D. P. Watson. 1953: The germination of okra seed as influenced by treatment with acetone and alcohol. Proc. Am. Soc. Hort. Sci. 62:427–32.

567. Andersson, A. E. 1960: Seed orchards in Swedish forestry. Tidsskr. Kungl. Skogs-Och Lantbruksakad. 99:65–87.

568. Andersson, B. 1932: Über hexosediphosphatdehydrogenase und carboxylase in pflanzensamen. Z. Physiol. Chem. 210: 15–29.

569. Andersson, G. 1961: Variety control of present day seed. Svensk Frötidn. 30:122–24, 126.

Andersson, G. 5831.

570. Andersson, J. A. 1902: A contribution to the question of the germination capacity of hulled oats. Landtmannen 13:676–78.

571. Andeweg, J. M., and A. Van Kooten. 1957: The practical importance of an identification of garden pea varieties in the seedling stage. Euphytica 6:237–41.

572. Ando, H. Absorption of water by rice seed. Imp. Univ. Coll. Agr. (Tokyo) Bull. 3:474–78. Expt. Sta. Rec. 10:1049. 1898–99.

573. —— On the specific gravity of rice seeds at different stages of ripening. Imp. Univ. Coll. Agr. (Tokyo) Bull. 3:479–81. Expt. Sta. Rec. 10:967. 1898–99.

Andrade, A. C. 58.

Andrade, E. N. de; *see* De Andrade, E. N.

574. Andrasovszky, J. 1915: Die Bedeutung der Traubensamen für die Unterscheidung der Sorten. Intern. Agr.-Techn. Rundschau. 7:138–41. 1916. Botan. Centralbl. 135:30. 1917.

575. André, E. 1928: L'aleurone des graines oléagineuses, substance alimentaire. Compt. Rend. Acad. Agr. France 14:1292–303.

576. —— and C. Besse. 1931: Castor oil. II. Variations in oil from different varieties of *Ricinus communis*. Ann. Offic. Nat. Combust. Liquid 6:243–59. Soc. Chem. Ind. J. 52:75B. 1933.

577. —— and H. Canal. 1931: Castor oil. III. Rotatory power. Ann. Offic. Nat. Combust. Liquid 6:459–70. Soc. Chem. Ind. J. 52:75B–76B. 1933.

578. —— and M. Carbouères. 1952: Contribution à l'étude des substances sulfurées volatiles de quelque graines de Crucifères (choux fourragers). Compt. Rend. 235:816–18.

579. —— and —— 1955: Sur la présence dans les graines de certaines crucifères d'un enzyme qui provoque la disparition des sénevols. Compt. Rend. 240:1468–70.

580. ——, ——, and M. Maille. 1953: Contribution à l'étude des substances insaponifiables des huiles de graines de Crucifères. II. Compt. Rend. 236:1695–97.

581. —— and M. Durou-Carbouères. 1957: Presence in grains of black mustard and rapeseed of an enzyme which has a destructive action on isothiocyanates. Compt. Rend. 244:2546–48. Chem. Abstr. 51: 13082. 1957.

582. —— and —— 1957: Sur la présence dans les graines de moutarde noire, de colza et de navette d'un enzyme qui possède une action destructrice sur les sénevals. Compt. Rend. 244:3080–82.

583. —— and P. Delaveau. 1954: Recherches chimiques sur la composition des essences sulfurées des graines de colza. Oléagineux 9:591–600.

584. —— and K. Hou. 1932: Sur la présence d'une oxydase des lipides ou lipoxydase dans la graine de soja, *Glycine soja* Sieb. & Zucc. Compt. Rend. 194:645–47.

585. —— and —— 1932: Sur les lipoxydases des graines de *Glycine soja* (Sieb. & Zucc.) et de *Phaseolus vulgaris* L. Compt. Rend. 195:172–74.

586. —— and M. Kogane-Charles. 1944: The existence, in the oil of the seeds of colza and other Cruciferae, of volatile organic sulfur compounds directly soluble in ether. Compt. Rend. 218:850–52. Chem. Abstr. 40:3800. 1946.

587. —— and —— 1946: The determination and identification of the essence of black mustard seed, *Brassica nigra* Koch. Compt. Rend. 222:103–5. Chem. Abstr. 40:3225–26. 1946.

588. —— and —— 1946: Sur la teneur en

huile et en linamaroside des graines d'une série de variétés de lin. Compt. Rend. 223:457–59.

589. —— and —— 1948: Contribution à l'étude des composés sulfurés volatils des graines de crucifères, 3ième Memoire—Sur les essences sulfurées d'origine glucosidique: Allyl et crotonyl-sénevols. Ann. Agron. 18:547–58. Biol. Abstr. 23:19706. 1949.

590. —— and M. Maille. 1951:Contribution à l'étude des essences sulfurées des graines de Crucifères. IV. Dosage de l'essence fournie par les graines de *Brassica juncea*, Czerny et Cosson. Ann. Inst. Natl. Rech. Agron. Ser. A, Ann. Agron. 2:442–55.

591. —— and —— 1957: Sur la détermination quantitative de l'acide cyanhydrique existant en puissance dans les graines et les tourteaux de lin. Compt. Rend. 244: 2091–92.

592. André, G. 1903: Sur la maturation des plantes privé'es de leurs cotyledons. Compt. Rend. 136:1401–4.

593. —— 1904: Sur les variations que presente la composition des graines pendant leur maturation. Compt. Rend. 138:1510–12.

594. —— 1904: Étude de la variation des matières minerales pendant la maturation des graines. Compt. Rend. 138:1712–14.

595. —— 1905: The transformation of nitrogenous material during the ripening of seeds. Compt. Rend. 140:1417–19. Expt. Sta. Rec. 17:541. 1906.

596. —— 1912: The displacement by water of nutrient substances contained in seeds. Compt. Rend. 154:1103–5.

597. Andree, S. M. 1892: Comparative investigations of the weight and specific gravity of the different peas in a pod. Tidskr. Allm. Svenska Utsadesfor. 2: 132–44.

598. Andreeva, R. A., and M. Shchetinina. 1956: The influence of seed treatment with heteroauxin on yield of sugar beets. Udobr. i Urozhai 1(8):27–30. Chem. Abstr. 51:1518–19. 1957.

599. Andren, F. 1943: Some seed-treatment results. Statens Växtskyddanstalt, Växtskyddsnotiser 1943(4):5–10. Chem. Abstr. 39:1496. 1945.

600. —— 1944: Some results of the 1943 cereal seed grain disinfection experiments. Statens Växtskyddanstalt, Växtskyddsnotiser 1944(8):19–23. Chem. Abstr. 39:1497. 1945.

601. —— 1945: Storage experiments with disinfected seed. Statens Växtskyddanstalt, Växtskyddsnotiser 1945(1):1–6. Rev. Appl. Mycol. 25:207. 1946.

602. —— 1945: Inhibitory effect of growth substances on the germination of brown beans. Statens Växtskyddanstalt, Växtskyddsnotiser 1945(2):22–25.

603. —— 1945: Does oil treatment prevent germination of seed? Statens Växtskyddanstalt, Växtskyddsnotiser 1945(4):61–62.

604. —— 1946: A case of germination injury in autumn seed. Statens Växtskyddanstalt, Växtskyddsnotiser 1946(2):25–27. Rev. Appl. Mycol. 25:495–96. 1946.

605. —— 1947: Flax and hemp seed treatment. Statens Växtskyddanstalt, Växtskyddsnotiser 1947:85–87.

606. —— 1952: Treat fall crop seeds! Lantmannen 36:625–26.

607. —— 1953: Preliminary disinfection experiment against the brush mould on seed. Statens Växtskyddanstalt, Växtskyddsnotiser 1953(3):38–39. Rev. Appl. Mycol. 23:291. 1954.

608. Andres, L. A., V. E. Burton, R. F. Smith, and J. E. Swift. 1955: DDT tolerance by lygus bugs on seed alfalfa. J. Econ. Ent. 48:509–13.

Andresen, K. Gad-; *see* Gad-Andresen, K.

609. Andrew, R. H. 1952: Volume changes and germination of sweet corn kernels at different temperature sequences. Agron. J. 44:473–76.

610. —— 1953: The influence of depth of planting and temperature upon stand and seedling vigor of sweet corn strains. Agron. J. 45:32–35.

611. ——, R. A. Brink, and N. P. Neal. 1944: Some effects of the waxy and sugary genes on endosperm development in maize. J. Agr. Res. 69:335–71.

612. ——, F. P. Ferwerda, and A. M. Strommen. 1956: Maturation and yield of corn as influenced by climate and production technique. Agron. J. 48:231–36.

613. Andrew, W. D. 1957: Interaction of the moisture and nutrition regimes on hard seed production in barrel medic, *Medicago tribuloides* Desr. Australian Plant Nutr. Conf. Proc. 1:79–91. 1956.

614. Andrew, W. T. 1954: Seedstalk development on lettuce as affected by growth regulators, vernalization, temperature, and photoperiod. Dissertation Abstr. 14: 432–33.

615. Andrews, C. H., and H. D. Bunch. 1959: The effect of kernel dimensional characteristics and planter plate speed on the planting accuracy of hybrid seed corn. Agron. Abstr. 1959:78.

616. Andrews, F. M., and C. C. Beals. 1919: The effect of soaking in water and of aeration on the growth of *Zea mays*. Bull. Torrey Botan. Club 46:91–100.

Andrews, F. R. 4329–4331, 9784, 11716.

617. Andrews, F. S. 1936: Physiological factors associated with the fruiting habits of the bush lima bean. Proc. Am. Soc. Hort. Sci. 33:473–76.

618. Andrews, F. W. 1940: A study of nut grass (*Cyperus rotundus* L.) in the cotton soil of the Gezira. I. The maintenance of life in the tuber. Ann. Botany (London) 4:177–93.

619. —— 1946: Parasitism of *Striga* sp. on *Dolichos lablab* Linn. Nature 157:515.

620. —— 1946: A study of nut grass (*Cyperus rotundus* L.) in the cotton soil of the Gezira. II. The perpetuation of the plant by means of seed. Ann. Botany (London) 10:15–30.

621. —— 1947: The parasitism of *Striga hermonthica* Benth. on leguminous plants. Ann. Appl. Biol. 34:267–75.

622. Andrews, H. N. 1941: *Dichophyllum moorei* and certain associated seeds. Ann. Missouri Botan. Garden 28:385–84.

623. —— 1963: Early seed plants. Science 142:925–31.

624. Andrews, J. E. 1958: Controlled low temperature tests of sprouted seeds as a measure of cold hardiness of winter wheat varieties. Can. J. Plant Sci. 38:1–7.

Andrews, J. E. 4889, 15435.

625. Andrews, J. S., and C. H. Bailey. 1934: The amylases of normal and germinated wheat flours. Cereal Chem. 11:551–56.

626. ——, H. M. Boyd, and D. E. Terry. 1942: Riboflavin analysis of cereals: Application of the microbiological method. Ind. Eng. Chem. Anal. Ed. 14:271–74.

627. ——, ——, and —— 1942: The riboflavin content of cereal grains and bread and its distribution in products of wheat milling. Cereal Chem. 19:55–64.

628. —— and C. Felt. 1941: The iron content of cereals. Cereal Chem. 18: 819–26.

Andrews, J. S. 3420, 14569, 17317.

629. Andrews, P., L. Hough, and J. K. N. Jones. 1952: Mannose-containing polysaccharides. II. The galactomannen of Fenugreek seed (*Trigonella foenum-graecum*). J. Chem. Soc. 1952:2744–50.

630. ——, ——, and —— 1953: Mannose-containing polysaccharides. III. The polysaccharides in the seeds of *Iris ochroleuca* and *I. sibirica*. J. Chem. Soc. 1953: 1186–92.

631. Andronescu, D. I. 1919: Germination and further development of the embryo of *Zea mays* separated from the endosperm. Am. J. Botany 6:443–52.

632. Androsova, M. 1940: The effect on yield of subjecting germinating seeds and young shoots of cucumbers to low temperatures. Vestn. Ovoshchevodstvo i Kartofel 1940 (3):98–100. Biol. Abstr. 18:14423. 1944.

633. Andrus, C. F. 1938: Seed transmission of *Macrophomina phaseoli*. Phytopathology 28:620–34.

Andrzejewski, M. 9727.

634. Aneli, N. A. 1940: Inhibiting action of blue light upon germination of tomato seeds. Compt. Rend. Acad. Sci. URSS 28:267–69.

635. Anema, P. 1959: World seed year 1961. Maandblad Landbouwvoorlichtingsdienst 16:655–63.

Angelis, E. de. 11633.

636. Angell, H. R. 1950: Seedling blight. I. Seed colour and soil in relation to preemergence blight of William Massey peas. Australian J. Agr. Res. 1:33–42.

637. Angelo, S. 1937: Photosynthesis of pigments of the coffee seed. Chemistry of the seed anthocyanins. Rev. Brasil. Chim. 4:368–71. Chem. Abstr. 32:1752. 1938.

Anghel, A. 19142.

638. Anghel, G., and M. Raianu. 1960: Germination of wild-oat grains under laboratory and field conditions. Ann. Inst. Cerc. Agron. Bucuresti, Ser. C 27:83–95. Weed Abstr. 12:40. 1963.

639. ——, ——, and H. Slusanschi. 1957: La faculté germinative des glomerules de betterave à sucre en raison de la phase de maturation. Ann. Inst. Cerc. Agron. Bucuresti, N.S. 24:549–58. 1956.

Anghel, G. 2775.

Angladette, A. 4491.

640. Anhaeusser, H. 1954: Zur Echtheitbestimmung von Raps- und Rübensaatgut. Angew. Botan. 28:13–40.

641. Anisimov, A. A., and O. N. Popova. 1954: The effect of the phosphorus nutrition of tomatoes on the quality of the seeds. Agrobiologiya 1954(5):110–11. Soils Fertilizers 18:501. 1955.

642. Aniskin, V. I., and E. G. Saprykina. 1962: Water permeability of seeds of forage legumes and lupine. Selek. i Semen. 4:18–22.

643. Anker, C. A., and W. F. Geddes. 1942: A study of the net weight changes and moisture content of wheat flour at various relative humidities. Cereal Chem. 19: 128–50.

644. Annett, H. E. 1917: Occurrence of raffinose in the seed of the jute plant (*Corchorus capsularis*). Biochem. J. 11: 1–6. Expt. Sta. Rec. 37:710. 1917.

645. Anonymous. 1843: Mummy wheat. Gardeners Chron. 1843:787–88.

646. —— 1882: Latent vitality of seeds. Botan. Gaz. 7:121.

647. —— 1885: Curious observations upon germination. N.Y. Agr. Expt. Sta. Ann. Rept. 4:71–73.

648. —— 1915: The selection of coconuts for germination. Kew Roy. Botan. Gardens Bull. Misc. Inf. 2:72–76.

649. —— 1915: A study of the influence of pod position upon viability and vigor of seedlings. New Jersey Agr. Expt. Sta. Rept. 35(1913/14):317–21.

650. —— 1919: Procedimiento para hacer germinar las semillas de dificil germinación. Inform. Agr. (Madrid) 9:11. Botan. Abstr. 3:192. 1920.

651. —— 1919: The storage of onion seed in the tropics. Agr. News (Barbados) 18:375. Botan. Abstr. 7:46. 1921.

652. —— 1927: Seed Trade Buyer's Guide. [Annual.]

653. —— 1928: The content of hard seeds in unthreshed samples; influence of conditions of growth on hard seeds grown in the open; germination of hard seed sown in the field. Intern. Seed Testing Conf. Rome 5: 253–56.

654. —— 1928: Effect of sulphuric acid on cotton seeds. Nature 122:329.

655. —— 1928: Facilitating germination of seeds. Brit. Patent 355,102. Chem. Abstr. 25:1325. 1931.

656. —— 1928: Noble fir seed keep well in cold storage. Forest Worker 4(4):14–15.

657. —— 1929: Algunos datos interesantes acerca de la estimulación quimica de las simillas. Rev. Agr. Puerto Rico 22(7):34–35, 39.

658. —— 1929: Hollyhocks grown from seed. Horticulture 7(16):409.

659. —— 1929: Seed-bearing *Viburnum carlesi*. Am. Nurseryman 50(8):159.

660. —— 1930: Chinese tung oil seeds. Gt. Brit. Imp. Inst. Bull. 28:267–72. Soc. Chem. Ind. J. 50:30-B. 1931.

661. —— 1930: Dhupa kernels (*Vateria indica*) from India. Gt. Brit. Imp. Inst. Bull. 28:279–81. Soc. Chem. Ind. J. 50:31-B. 1931.

662. —— 1930: Disinfection of Swede seed. Ugeskrift Landmaend. 75:200. Rev. Appl. Mycol. 9:755. 1930.

663. —— 1930: Distribution of *Arum* seeds. Am. Botan. 36:42.

664. —— 1930: Dusting and steeping of cereal seeds. Intern. Bull. Plant Protec. 4:37–41, 53–58.

665. —— 1930: The fat and oil of *Sapium sebiferum* seeds. Gt. Brit. Imp. Inst. Bull. 28:429–31. Chem. Abstr. 25:1693–94. 1931.

666. —— 1930: The germinating power of flower seeds. Horticulture 8(2):30.

667. —— 1930: Hastening the germination of conifer seeds. Gardeners Chron. 87:379.

668. —— 1930: Keimkraft und Keimdauer der Gemüse-samen. Möller's Deut. Gärtner-Ztg. 45:146–47.

669. —— 1930: *Moringa aptera* seed from Egypt. Gt. Brit. Imp. Inst. Bull. 28 :276–79. Soc. Chem. Ind. J. 50:31-B. 1931.

670. —— 1930: Northern white pine seed germinates better after exposure to low temperatures. Forest Worker 6(5): 11–12.

671. —— 1931: Accelerating germination. Am. Botan. 37:129.

672. —— 1931: Brown Co. offers hardwood seed of certified origin. Forest Worker 7(5):19.

673. —— 1931: Collection, care, and planting of tree seeds. Oklahoma Forest Serv. Publ. 10. 16 pp.

674. —— 1931: Dormancy in basswood seed. N.Y. Agr. Expt. Sta. (Ithaca) Ann. Rept. 44(2):56.

675. —— 1931: Germination of damaged seeds. Nature 128:307.

676. —— 1931: Germination of pine seeds. Forest Worker 7(1):18.

677. —— 1931: Growing succulents from seed. Horticulture 9(9):203.

678. —— 1931: Longleaf pine seed stored in a paper bag at a temperature of 32° F. at the Southern Forest Experiment Station. Forest Worker 7(6):15.

679. —— 1931: Regulations for warehousemen storing seeds under the United States Warehouse Act. U.S. Dept. Agr., Bur. Agr. Econ. S.R.A. 122. 21 pp.

680. —— 1931: Seed propagation, 130. J. Dept. Agr. Ireland 30:282–300.

681. —— 1931: Seed characteristics of Western white pine. Forest Worker 7(5):11.

682. —— 1931: Sprouting peony seeds. Am. Botan. 37:163.
683. —— 1931: The storage of seed. Skögen 18:14.
684. —— 1931: The suitability of rain-damaged wheat for seed. Agr. Gaz. N.S. Wales 42:387.
685. —— 1931: Sur la germination des graines de palmier à huile. Agron. Coloniale 20(164):54.
686. —— 1929: Warren-Teed Seed Co.—Treatment of seeds to secure uniform germination. Brit. Patent 337,601. Soc. Chem. Ind. J. 50:132-B. 1931.
687. —— 1932: Die Beizung des Saatgutes. Wien. Landwirtsch. Z. 82:259. Rev. Appl. Mycol. 12:16.
688. —— 1932: British Columbia study of influences affecting seed production. Forest Worker 8(1):21.
689. —— 1932: Cold and germination. Am. Botan. 38:32–33.
690. —— 1932: Delay in germination due to mercuric bichloride soak. Nursery Disease Notes 4(10):3–4.
691. —— 1932: The effect of time and method of planting cotton on germination. Georgia Agr. Expt. Sta. Ann. Rept. 45: 16–18.
692. —— 1932: Fungicide for oat seed. Nature 129:464.
693. —— 1932: Germinating seeds. Florists Rev. 70(1808):27.
694. —— 1932: The germination of cotton seed at low temperature. Georgia Agr. Expt. Sta. Ann. Rept. 45:18–19.
695. —— 1932: Growing dahlias from seed. Horticulture 10(5):99.
696. —— 1932: Growing gloxinias from seed. Horticulture 10(2):32.
697. —— 1932: Helping nature to seed the forest. J. Forestry 30:508–9.
698. —— 1932: Influence of age of pumpkin seed on yield. Queensland Agr. J. 37:74.
699. —— 1932: Instructions for carrying out "Yarovization" of wheat (*Triticum durum*). (Ukrain. Inst. Genet.) Bull. Yarovization 1:62–71.
700. —— 1932: Is light important immediately after germination? J. Forestry 30: 509–10.
701. —— 1932: Longevity of seeds. Flower Grower 19(10):439.
702. —— 1932: Oilseeds and vegetable oils. Gt. Brit. Empire Market. Bd. Statis. Intell. Branch, H.M. Stationery Office, London. 66 pp. Malayan Agr. J. 21:73–74. 1933.
703. —— 1932: Polyembryony in the seeds of conifers. Gardeners Chron. 91:231.
704. —— 1932: Premature seeding of celery. N.Y. Arg. Expt. Sta. (Ithaca) Ann. Rept. 45:160.
705. —— 1932: Report on *Psyllium* seed from Cyprus. Anon. Cyprus Agr. J. 27:117–18. Chem. Abstr. 29: 292. 1935.
706. —— 1932: Seeds buried in soil 30 years show amazing vitality when dug. Seed World 32(1):17.
707. —— 1932: Shea nuts from the Gold Coast. Gt. Brit. Imp. Inst. Bull. 30:282–93. Soc. Chem. Ind. J. 52:75-B. 1933.
708. —— 1932: Some interesting statistics on the longevity of flower seeds. Seed World 31(11):17.
709. —— 1932: Sowing seeds of shrubs. Horticulture (10)1:12.
710. —— 1932: Tung-seed and oil from Empire sources. Gt. Brit. Imp. Inst. Bull. 30:271–82. Soc. Chem. Ind. J. 52:75-B. 1933.
711. —— 1932: Urease content of leguminous seeds. Current Sci. (India) 1:47.
712. —— 1933: Beschleunigung der Keimung von Baumsamen. Deut. Dendrol. Ges. Mitt. 45:405.
713. —— 1933: Composition of Nipa palm seeds. Gt. Brit. Imp. Inst. Bull. 31:3–5. Chem. Abstr. 27:3352. 1933.
714. —— 1933: Erfahrungen über Luzernesamenbau. Mitt. Deut. Landwirtsch. Ges. 1933:181. Fortschr. Landwirtsch. 8:256. 1933.
715. —— 1933: Experiments revive hope for broadcast seeding. Forest Worker 9(1): 10–11.
716. —— 1933: Fruits and seeds of "Caper Spurge" are poisonous. Agr. Gaz. N.S. Wales 44:438.
717. —— 1933: Gloxinias from seeds. Florists Rev. 71(1841):23.
718. —— 1933: Lilies from seeds. Florists Rev. 71(1838):13.
719. —— 1933: Lilies from seeds. Florists Rev. 72(1847):23.
720. —— 1933: Samen aus mumien. Deut. Dendrol. Ges. Mitt. 45:345.
721. —— 1933: Seeds undrowned after 23 years under water. Sci. News Letter 24:131.
722. ——1933: Snapdragons from seed. Horticulture 11(7):124–25.
723. —— 1933: Starting hollyhocks from seed. Horticulture 11(14):240.
724. —— 1933: Viability of seeds. Am. Botan. 39:179–81.

725. —— 1934: Age does not determine the vitality of forage and grain seeds. Seed World 36(11):9.
726. —— 1934: Mummy wheat. Nature 134:730.
727. —— 1934: Mutations stimulated by aging seeds. Carnegie Inst. Wash. Yearbook 33:37–38.
728. —— 1935: Germination tests for fruit tree seeds. N.Y. Agr. Expt. Sta. Ann. Rept. 1934/35:73–74.
729. —— 1935: Hastening seed germination. Florists Rev. 75(1944):52.
730. —— 1935: Hydrogen peroxide rids seeds of disease spores. Sci. News Letter 27:232.
731. —— 1935: Inspection of agricultural seeds. Indiana Agr. Expt. Sta. Circ. 213. 93 pp.
732. —— 1935: Propagation of game food and cover plant of the Lake States. U.S. Dept. Agr. Forest Serv. Lake States Forest Expt. Sta. 81 pp. Processed.
733. —— 1935: The question of seed vitality. Horticulture 13(14):341.
734. —— 1935: Seed storage and the effects of origin of the seed on germination and internal infection by organisms. S. Carolina Agr. Expt. Sta. Ann. Rept. 1934/35:41.
735. —— 1935: Seed treatments for shelterbelt species. Forest Res. Dig. 1935(8):6–7.
736. —— 1935: The seeds of *Strophanthus emini.* Quart. J. Pharm. Pharmacol. 8:61–70.
737. —— 1935: Some African oil seeds. Gt. Brit. Imp. Inst. Bull. 33:271–93. Analyst 61:48–49. 1936.
738. —— 1935: Trees that sprang up from seeds scattered by airplanes. Science 82:9S.
739. —— 1935: Valuable information for seedsmen (Bodgers Seeds, Ltd., El Monte, Calif.). Florists Exchange Hort. Trade World 85(1):5.
740. —— 1936: *Granadilla* (Passion fruit) seed from Kenya. Gt. Brit. Imp. Inst. Bull. 35:22–23. Analyst 62:471. 1937.
741. —— 1936: Hastening germination of *Acacia* seeds by soaking in boiling water. Current Sci. (India) 4:757.
742. —— 1936: Influence of storage conditions on White pine seed. Forest Res. Dig. 1936(5/6):9–10.
743. —— 1936: Keimung von *Gleditschia*-Samen. Deut. Dendrol. Ges. Mitt. 48:280.
744. —— 1936: Light works on seeds through chlorophyll. Sci. News Letter 30:67.
745. —— 1936: More about Jack pine seeds. Forest Res. Dig. 1936(9/10):5.
746. —— 1936: Oil is extracted from tobacco seeds. Ind. Eng. Chem. News Ed. 14:467.
747. —— 1936: The propagation of forest trees in South Africa. S. Africa Dept. Agr. Forestry Bull. 165. 26 pp.
748. —— 1936: Propagation of mountain ash. Forest Res. Dig. 1936(3/4):13–14.
749. —— 1936: Seed source influences drought resistance of green ash. Forest Res. Dig. 1936(9/10):8–10.
750. —— 1936: Seed statistics. U.S. Dept. Agr. Bur. Agr. Econ. 63 pp. Expt. Sta. Rec. 75:197. 1936.
751. —— 1936: Seed sterilization. Exptl. Res. Sta. Tutner's Hill, Cheshunt, Herts, 22nd Ann. Rept. 1936:57–58. Chem. Abstr. 31:7584. 1937.
752. —— 1937: A biological test for the boron content of seeds. New Jersey Agr. Expt. Sta. Ann. Rept. 1936/37:85–86.
753. —— 1937: Determination of moisture content of grain. Mühle 74:1198. Soc. Chem. Ind. J. 57:574-B. 1938.
754. —— 1937: Determination of winter hardiness by germinating seeds in sugar and salt solutions. Herbage Rev. 5:161.
755. —— 1937: Division of seed investigations. N.Y. Agr. Expt. Sta. Ann. Rept. 1936/37:88–95.
756. —— 1937: Endosperm and cmbryo in reciprocal cereal crosses. Nature 140:511.
757. —— 1937: The English daisy from seed. Horticulture 15(14):311.
758. —— 1937: Germination inhibitors hinted in Middle Ages. Sci. News Letter 32:170.
759. —— 1937: Growing the amaryllis from seed. Horticulture 15(20):429.
760. —— 1937: Is there a foolproof method of germinating fine seeds? Florists Exchange Hort. Trade World 88(18):25.
761. —— 1937: Longevity of flower seeds. Seed Trade Buyers Guide 1937:150–51.
762. —— 1937: *Martynia proboscidea.* Seed World 41(1):32.
763. —— 1937: Production of grass seed. Herbage Publ. Ser. Aberystwyth: Imp. Bur. Plant Genet. Bull. 19. 46 pp. Ann. Appl. Biol. 25:440. 1938.
764. —— 1937: Production of legume seed. Herbage Publ. Ser. Aberystwyth: Imp. Bur. Plant Genet. Bull. 23. 48 pp. Ann. Appl. Biol. 25:440. 1938.
765. —— 1937: Rapid testing with ultraviolet light. Sci. Dig. 2(1):19–22.
766. —— 1937: Report of the seed propagation division, 1936. J. Dept. Agr. Ireland 34:248–61.

767. —— 1937: Sandalwood regeneration and propagation. Perfumery Essent. Oil Record 28:338. Pharm. Abstr. 4:6. 1938.

768. —— 1938: Active principle of Kurjan seed (*Myrsine africana*). Gt. Brit. Imp. Inst. Bull. 36:319–22. Analyst 63:900. 1938.

769. —— 1938: Chewings fescue seed reported low in germination. Seed World 44(10):15.

770. —— 1938: Croton seed. Pharm. J. 141:24. Pharm. Abstr. 5:71. 1939.

771. —— 1938: The germination of clover seed. Tasmanian J. Agr. 9:109.

772. —— 1938: Horticultural seed growing. Intern. Hort. Congr. Berlin 12(4). 9 pp.

773. —— 1938: Juniper seeds. Am. Nurseryman 68(11):18.

774. —— 1938: Kapok seed from British Honduras. Gt. Brit. Imp. Inst. Bull. 36:468–70.

775. —— 1938: Mercurial dusts and seed germination. Fertiliser Feeding Stuffs J. 23:627, 629. Rev. Appl. Mycol. 18:239–40. 1939.

776. —— 1938: New Croton seed from Nyasaland. Gt. Brit. Imp. Inst. Bull. 36:151–53. Analyst 63:619. 1938.

777. —— 1938: A note on embryo-rejuvenation. J. Heredity 29:419–22.

778. —— 1938: Pea seed treatment results reported from Oregon. Agr. News Letter 6:82.

779. —— 1938: *Perilla* seed from Burma. Gt. Brit. Imp. Inst. Bull. 36:467–68.

780. —— 1938: Raising cyclamens from seeds. Gardeners Chron. 104:28.

781. —— 1938: Rules and recommendations for testing seeds. U.S. Dept. Agr. Circ. 480. 24 pp.

782. —— 1938: Seed germination and growth. Florists Exchange Hort. Trade World 90(17):13.

783. —— 1938: Seed germination studies. Cawthron Inst. New Zealand Ann. Rept. 1938:24–25.

784. —— 1938: Seed origin affects survival of green ash in the nursery. J. Forestry 36:434.

785. —— 1938: Seeds and germination. Gardners Chron. 103:261.

786. —— 1938: Seeds which germinate slowly. Horticulture 16(10):222.

787. —— 1938: Tea-seed oil-Japanese. Pharm. Abstr. 4:362.

788. —— 1938: Tonka bean. Pharm. Abstr. 4:230.

789. —— 1938: Water-lily seed from the Sudan. Gt. Brit. Imp. Inst. Bull. 36:470–72. Chem. Abstr. 33:3015. 1939.

790. —— 1938: Wormseed oil. Perfumery Essent. Oil Record 29:309. Pharm. Abstr. 5:489. 1939.

791. —— 1939: Association of Official Seed Analysts of North America. Proc. 1938. 279 pp.

792. —— 1939: Colchicine treatment of pine seeds. J. Heredity 30:389–90.

793. —— 1939: Double snapdragons from seeds. Florists Rev. 84(2163):33.

794. —— 1939: The germination of ripe and unripe weed seeds. Gardners Chron. 105:17.

795. —— 1939: Law prohibits the export of *Oiticica* seeds. Ind. Eng. Chem. News Ed. 17:319.

796. —— 1939: Less seed disinfectant needed with storage time lengthened. Illinois Agr. Expt. Sta. Ann. Rept. 1936/37:71–73.

797. —— 1939: Lettuce seed storage. Ariz. Agr. Expt. Sta. 49th Ann. Rept. 1937/38: 51.

798. —— 1939: Low temperatures for snapdragon seed. Florists Exchange Hort. Trade World 93(11):15.

799. —— 1939: *Macadamia*. Hawaii Agr. Expt. Sta. Rept. 1938:15–20.

800. —— 1939: *Mesua ferrea* seed oil found good for soap making. Ind. Eng. Chem. News Ed. 17:292.

801. —— 1939: Mustard and rape seed industry in India. Current Sci. (India) 8:239.

802. —— 1939: Oil palm seed germination. Gt. Brit. Imp. Inst. Bull. 37:211–12.

803. —— 1939: Papaya. Hawaii Agr. Expt. Sta. Rept. 1938:11–15.

804. —— 1939: The propagation of lilies. (Roy. Hort. Soc.) Lily Yearbook 8:66–76.

805. —— 1939: Quinine seed was germinated readily at the experiment station. Puerto Rico Agr. Expt. Sta. Rept. 1937/38:27–29.

806. —— 1939: Seed-germination studies. New Zealand Dept. Sci. Ind. Res. Ann. Rept. 1939:79.

807. —— 1939: Seed treatment: its origin and application. Bayer Semesan, Wilmington, Del. 23 pp.

808. —— 1939: Seedless watermelon developed in Michigan. Sci. News Letter 35:136.

809. —— 1939: Sowing lily seeds. Florists Rev. 83(2146):29–30.

810. —— 1939: Sterile culture of peach seed. New Jersey Agr. Expt. Sta. Ann. Rept. 1937/38 59:47–49.

811. —— 1939: Summer storage of vetch seed may affect its germination. Seed World 46(1):34.
812. —— 1939: Value of seed from roasted pimientos for planting. Georgia Agr. Expt. Sta. Ann. Rept. 1938/39:75.
813. —— 1939: Warm water improved germination of ambrette seeds. Puerto Rico Agr. Expt. Sta. Rept. 1937/38:43–44.
814. —— 1940: Air-condition seeds. Am. Nurseryman 72(5):31–32.
815. —— 1940: Booklet about weeds and their seeds, published by Seed World. Horticulture 18(23):467.
816. —— 1940: The Brazil nut. Trop. Agr. (London) 17:16–18.
817. —— 1940: Chemical composition of corn. Iowa State Dept. Agr. Yearbook Agr. 41:443.
818. —— 1940: Cinchona seeds lose viability rapidly. Puerto Rico (Mayaguez) Agr. Expt. Sta. Rept. 1940:18–19.
819. —— 1940: Cultivation of castor seed in Brazil. Gt. Brit. Imp. Inst. Bull. 38:321–24. Gt. Brit. Imp. Bur. Soil Sci. 4:42. 1941.
820. —— 1940: Dormancy in vegetable seed. Natl. Canners Assoc. Bur. Raw Prod. Res. 5:53.
821. —— 1940: Effect of roasting and scalding perfection pimiento fruits on the planting value of the seed. Georgia Agr. Expt. Sta. Ann. Rept. 1939/40:67–68.
822. —— 1940: Effect of season on germination and early survival. Lake States For. Expt. Sta. Tech. Notes, No. 168. 1 p.
823. —— 1940: Effects of storage on proteins of grains and seeds. U.S. Dept. Agr. Rept. Chief Bur. Agr. Chem.Eng.1939/40:36–37.
824. —— 1940: Fall sowing of conifer seed best. Florists Exchange Hort. Trade World 95(13):17.
825. —— 1940: French grape-seed oil. Chem. Age 43:207. Soap 17:62. 1941.
826. —— 1940: Germination of wattle seeds. Queensland Agr. J. 53:241.
827. —— 1940: Glass wool used in seed testing. Horticulture 18(20):404.
828. ——1940: Grapefruit seed oil is a new byproduct. U.S. Dept. Agr. Clip Sheet 1152:3.
829. —— 1940: Inspection of agricultural seeds. Indiana Agr. Expt. Sta. Circ. 259. 125 pp.
830. —— 1940: Lettuce seed storage. Ariz. Agr. Expt. Sta. Ann. Rept. 1938/39:68.
831. —— 1940: Longevity of flower seeds. Seed Trade Buyers Guide 1940:52–53.
832. —— 1940: Mangosteen seeds differ from ordinary dicotyledonous seeds. Puerto Rico Agr. Expt. Sta. Rept. 1939/40: 55–57.
833. —— 1940: Peculiarities of flower seed germination. Florists Exchange Hort. Trade World 95(23):9.
834. —— 1940: Preharvest germination in relation to date of harvest. Ariz. Agr. Expt. Sta. Ann. Rept. 1938/39:74–75.
835. —— 1940: Raising begonias from seed. Flower Grower 27(2):65.
836. —— 1940: Relationship of temperature to dispersal of Jack pine seeds. Lake States For. Expt. Sta. Tech. Notes. No. 164. 1 p.
837. —— 1940: Rules and regulations under the federal seed act. U.S. Dept. Agr. AMS, Serv. & Regulatory Announcements 156. (Reprinted with Amendments 1960.)
838. —— 1940: A satisfactory method of packing palm seed for distant transportation was developed. Puerto Rico Agr. Expt. Sta. Rept. 1939/40:62–63.
839. —— 1940: Seed-germinating tests. Florists Rev. 87(2246):82.
840. —— 1940: Seed storage studies. Puerto Rico Trop. Forest Expt. Sta. Ann. Rept. 1940:3.
841. —— 1940: Seeds of *Araujia sericifera* poisonous to fowls. Gardeners Chron. 108:151.
842. —— 1940: Sowing the seeds of moonflowers. Horticulture 18(10):238.
843. —— 1940: Tear gas kills weed seeds. Ohio Agr. Expt. Sta. Bull. 617:14–15.
844. —— 1940: Use of salt bags as seed containers. Queensland Agr. J. 53:106.
845. —— 1940: Winter seed production for northern crops. Puerto Rico Agr. Expt. Sta. Rept. 1938/39:46–51.
846. —— 1940: Zamia-seeds poisonous to stock. Queensland Agr. J. 54:158.
847. —— 1941: Are there any particular difficulties in the germination of *Trollius* seeds? I never have any luck. Horticulture 19(14):332.
848. —— 1941: Can *Achimenes* be grown from seed? If so, how? Horticulture 19(7):179.
849. —— 1941: Cold for delphinium seeds. Florists Rev. 87(2256):17.
850. —— 1941: The Commission on Forest Seeds and Tree Race Problems of the International Union of Forest Research Organizations. Chronica Botan. 6:207.
851. —— 1941: Daylilies from seed. Florists Exchange Hort. Trade World 97(8):13.

852. —— 1941: Does Pfitzer seed? Florists Exchange Hort. Trade World 97(11):23.
853. —— 1941: Effect of excessive applications of manure and fertilizers on germination of tobacco seed and growth of the seedlings. Kentucky Agr. Expt. Sta. Ann. Rept. 1940(1):32.
854. —— 1941: Effect of temperature on germination. U.S. Dept. Agr. Lake States Forest Expt. Sta., Univ. Farm, St. Paul, Minn. Tech. Notes. 1 p.
855. —— 1941: Factors influencing germination of Buffalo grass seed. Texas Agr. Expt. Sta. Ann. Rept. 1940:199.
856. —— 1941: Fumigation of coffee seed. E. African Agr. J. 6:126.
857. —— 1941: Germinating celery seeds. Florists Rev. 87(2259):19.
858. —— 1941: Germination of orchid seeds under sterile conditions. Nature 147:578.
859. —— 1941: The germination of seeds. Am. Botanist 47:11–13.
860. —— 1941: Growing plants from seed. Bull. Missouri Botan. Garden 29:81–94.
861. —— 1941: How can I make lupin and heavenly blue morning glory seeds germinate faster and better? Horticulture 19(11):258.
862. —— 1941: *Hyptis* seed oil from Nigeria and the Sudan. Gt. Brit. Imp. Inst. Bull. 39:97–98. Analyst 66:381–82. 1941.
863. —— 1941: Inspection of agricultural seeds. Indiana Agr. Expt. Sta. Circ. 267. 168 pp.
864. —— 1941: Lettuce seed storage. Ariz. Agr. Expt. Sta. 41st Ann. Rept. 1939/40:64.
865. —— 1941: Lilies from seeds. Florists Rev. 89(2297):29.
866. —— 1941: Lily propagation by seed. Florists Exchange Hort. Trade World 96(17):14.
867. —— 1941: Lily seeds do not carry mosaic. Florists Exchange Hort. Trade World 96(12):18.
868. —— 1941: Live seed 500 years old. Florists Exchange Hort. Trade World 97(17):44.
869. —— 1941: Midwest now grows coriander seed. Sci. News Letter 40:284.
870. —— 1941: A note on the distribution of vitamin B_1 (thiamin) in the rice grain. Malayan Agr. J. 29:127–28.
871. —— 1941: Oil from grape seeds. Chemist Druggist 135:76. Pharm. Abstr. 9:14. 1943.
872. —— 1941: Oyster nut oil. Perfumery Essent. Oil Record 32:93. Pharm. Abstr. 9:14. 1943.
873. —— 1941: Papaw seed. Queensland Agr. J. 55:157.
874. —— 1941: Planes taking rubber seed from Canal Zone to Brazil. Chronica Botan. 6:325.
875. —— 1941: Poppy seed as a commercial crop. Florists Exchange Hort. Trade World 96(11):8.
876. —— 1941: Red pine seed can be kept in tight containers 10 years. Florists Exchange Hort. Trade World 96(8):16.
877. —— 1941: Seed research in cooperation with the seed laboratory of the production service. Can. Minister Agr. Rept. 1940/41:22–23.
878. —— 1941: Seeds of the Zamia palm. Analyst 66:64.
879. —— 1941: Sowing perennial seeds. Florists Exchange Hort. Trade World 97(15):16.
880. —— 1941: Special physiological seed studies. N.Y. Agr. Expt. Sta. 60th Ann. Rept. 1940/41:54–55.
881. —— 1941: Storing coffee seed. Trop. Agr. 18:25.
882. —— 1941: Testing cone ripeness. U.S. Dept. Agr. Lake States Forest Expt. Sta. Univ. Farm, St. Paul, Minn. Tech. Notes. 1 p.
883. —— 1941: Vanilla vines from seed. Perfumery Essent. Oil Record 32:347. Pharm. Abstr. 8:226. 1942.
884. —— 1941: Viability studies on tree and shrub seed. N.Y. Agr. Expt. Sta. Ann. Rept. 1939/40:41.
885. —— 1941: Vitamin B_1 and phytohormones. N.Y. Agr. Expt. Sta. Ann. Rept. 1939/40:41.
886. —— 1941: What happened to the cyclamen seeds. Florists Exchange Hort. Trade World 97(2):17.
887. —— 1942: Castor seeds. Chemist Druggist 137:120. Pharm. Abstr. 9:32. 1943.
888. —— 1942: Germination of peas improved by dusting seed with copper carbonate. J. Agr. W. Australia 19:239–40.
889. —— 1942: Lilies from seeds. Florists Rev. 89(2302):23.
890. —— 1942: National Peanut Council Yearbook, including Proceedings of Second Annual Meeting, Pensacola, Fla. 125 pp. Biol. Abstr. 19:11383. 1945.
891. —— 1942: Oil extracted from tomato seeds. Agr. Americas 2:57.

892. —— 1942: Recent work on germination. Nature 149:658–59.

893. —— 1942: Seed storage problems. Puerto Rico Agr. Expt. Sta. Rept. 1941/42:39.

894. —— 1942: Seeds germinate after 39 years. Florists Exchange Hort. Trade World 99(4):15.

895. —— 1942: Some seeds long lived. Chemurgic Dig. 1(15):116.

896. —— 1942: Sowing seeds in sphagnum. Flower Grower 29(4):198.

897. —— 1942: Special seed studies. N.Y. Agr. Expt. Sta. Ann. Rept. 1941/42:78–79.

898. —— 1943: Botanists in London are watching areas of bomb-scarred and fire-scorched soil in the city .... Science Suppl. 98(2535):9S.

899. —— 1943: Chemical composition of some South African cereals and their milling products. Analyst 68:152–54.

900. —— 1943: Germination of treated and untreated cotton seeds under severe conditions. Ariz. Agr. Expt. Sta. Ann. Rept. 1941/42:76.

901. —— 1943: Mexican poppy seed (*Argemone Mexicana*) from Nigeria. Gt. Brit. Imp. Inst. Bull. 41:227–28.

902. —— 1943: New method of seed extraction. Queensland Agr. J. 57:213.

903. —— 1943: Peanut seed treatment. Georgia Agr. Expt. Sta. Bull. 36:35.

904. —— 1943: Pecans as a source of vitamin B_1 and B_2. Ariz. Agr. Expt. Sta. Ann. Rept. 1941/42:67–69.

905. —— 1943: Plants and light: germination of seeds. Gardeners Chron. 114:65.

906. —— 1943: The production of seed of root crops and vegetables. Imp. Agr. Bur. Joint Publ. No. 5. *In* Sci. Agr. 24:52. 1943.

907. —— 1943: Scientific utilisation of Indian oil-seeds. Current Sci. (India) 12:97–98.

908. —— 1943: Variations in the vitamin B_1 (thiamin) content of pinto beans. Ariz. Agr. Expt. Sta. Ann. Rept. 1941/42:69.

909. —— 1944: Cleaning tomato seeds. Horticulture 22(15):338.

910. —— 1944: Dehulling seeds. Food Ind. 17:161–62. 1945.

911. —— 1944: Disinfected garden seed. Horticulture 22(6):127.

912. —— 1944: Extracting oil from tomato seed. Seed World 56(9):44.

913. —— 1944: Extraction of tomato seeds with hydrochloric acid. Gardeners Chron. 116:29. Biol. Abstr. 19:5253. 1945.

914. —— 1944: Germination of Belladonna seed. Florists Exchange Hort. Trade World 102(2):17.

915. —— 1944: Greenhouse method for testing dust seed treatments to control certain cereal smuts. Phytopathology 34:401–4.

916. —— 1944: Growth of cereal embryos. Nature 154:681.

917. —— 1944: Life history of seed. Can. Minister Agr. Rept. 1943/44:29.

918. —— 1944: Lupines from seed. Am. Botanist 50:18–19.

919. —— 1944: "Magni-focuser" eye-loop for seed analysts. Southern Seedsman. 7(11):34.

920. —— 1944: Nutritional value of wheat germ and corn germ. J. Am. Med. Assoc. 125:848–49. Chem. Abstr. 38:5546–47. 1944.

921. —— 1944: Peach kernel oil. Food Ind. 16:143.

922. —— 1944: Pheasants scatter seeds of own food plants. Sci. News Letter 46:361.

923. —— 1944: Rules for testing seeds. Proc. Assoc. Offic. Seed Analysts 1943/44: 17–42.

924. —— 1944: Seed germination studies. Can. Minister Agr. Rept. 1943/44:29.

925. —— 1944: Seed storage. Puerto Rico Agr. Expt. Sta. Ann. Rept. 1942/43:34.

926. —— 1944: Seeds treated against birds. Chem. Industries 54:582.

927. —— 1944: Studies on vegetable seed treatments in 1943. Plant Disease Reptr. 145. 97 pp. Expt. Sta. Rec. 91:168–69. 1944.

928. —— 1944: Test seed germination by removing embryos. Am. Nurseryman 80(10):28–29.

929. —— 1944: Trials on the storage of undried and kiln-dried spring wheat seed at the Department's Cereal Station, Ballinacurra, Co. Cork. J. Dept. Agr. Ireland 41:181–87.

930. —— 1944: Unusual plant seed oils. Australasian J. Pharm. 25:209. Pharm. Abstr. 10:338. 1944.

931. —— 1944: What percentage of germination can be expected from salvia seeds? Horticulture 22(8):185.

932. —— 1945: Chemicals injure bean seeds. Wyoming Agr. Expt. Sta. 55th Ann. Rept. 1944/45:21.

933. —— 1945: A conference on plant embryo culture U.S.A. Chronica Botan. 9:193–94.

934. —— 1945: Geraniums from seeds. Flower Grower 32(7):376.

935. —— 1945: Good bread made from twenty-year-old wheat. Sci. News Letter 47:184.

936. —— 1945: Inspection of agricultural seeds. Indiana Agr. Expt. Sta. Circ. 308. 147 pp.

937. —— 1945: Plastic coatings for seeds. Sci. Dig. 18(2):76.

938. —— 1945: Production of high quality tomato seed. For farm and commercial use. Agr. Gaz. N.S. Wales 56:545–48.

939. —— 1945: Repellents slow germination. Horticulture 23(10):266.

940. —— 1945: Results of cooperative corn, flax, and soybean seed treatment tests in 1944. Plant Disease Reptr. Suppl. 159: 201–24. Expt. Sta. Rec. 94:772. 1946.

941. —— 1945: Seed germination of milkweed. Can. Minister Agr. Rept. 1944/45:27.

942. —— 1945: Some seed disinfectants kill disease organisms by contact—others release toxic vapors or fumes. Agr. News Letter 13:39.

943. —— 1945: Sowing pansy seeds. Horticulture 23(14):359.

944. —— 1945/46: Peanut seed treatment. Georgia Agr. Expt. Sta. Ann. Rept. 58:70–71.

945. —— 1946: Carryover of vegetable seeds much smaller. Seed World 60(5):40.

946. —— 1946: Early planted treated seed produces higher sorghum yields. S. Dakota Agr. Expt. Sta. Bull. 381.

947. —— 1946: Fuchsias from seed. Horticulture 24(19):538.

948. —— 1946: Germination of tobacco seed. Kentucky Agr. Expt. Sta. Ann. Rept. 1945:45.

949. —— 1946: Grains supply protein. Sci. News Letter 50:239.

950. —— 1946: "Paraseeds" fly far. Sci. News Letter 50:229.

951. —— 1946: Red cedar germination. Am. Nurseryman 84(12):18, 19.

952. —— 1946: Saving vegetable seed. Horticulture 24(14):399.

953. —— 1946: Seed treatments for fall- and winter-sown onions. Georgia Coastal Plain Expt. Sta. Ann. Rept. Bull. 43:50–53.

954. —— 1946: Studies on vegetable seed treatments in 1944. Plant Disease Reptr. Suppl. 161. 66 pp. Rev. Appl. Mycol. 25:590–91. (1946) 1947.

955. —— 1946: Test seed germination. Florists Exchange Hort. Trade World 106(17):15.

956. —— 1946/47: Cotton seed treatment. Georgia Agr. Expt. Sta. Ann. Rept. 59:11–12.

957. —— 1947: Agricultural seed production as a new industry in Welsh farming. Gt. Brit. Imp. Bur. Pastures Field Crops Bull. 38:47–51.

958. —— 1947: Arkansas experiments show seed treatment increases corn yields. Agr. News Letter 15:93.

959. —— 1947: Chile's sunflower-seed acreage shows expansion. Agr. Americas 7:77.

960. —— 1947: Embryoless soft chess. Assoc. Offic. Seed Analysts News Letter 21(4):12.

961. —— 1947: New antibiotic attacks seed-dwelling bacteria. Sci. News Letter 51:307.

962. —— 1947: New anti-germ compound found in radish seed. Sci. News Letter 52:213.

963. —— 1947: Process prevents spoilage of seeds stored in bulk. Sci. News Letter 51:360.

964. —— 1947: Seed treatment gives better stands and bigger yields of rice. Agr. News Letter 15:92–93.

965. —— 1947: Studies in seed germination behaviour. Can. Minister Agr. Rept. 1946/47:36–37.

966. —— 1947: Test seeds by electricity. Am. Nurseryman 85(12):6.

967. —— 1947: Uranium solution boosts yield from treated seed. Sci. News Letter 52:185.

968. —— 1947: Western Ways. Planting by air. Ariz Highways 23(5):28–31.

969. — 1947/48: Cotton seed treatment. Georgia Agr. Expt. Sta. Ann. Rept. 60:18–19.

970. —— 1947/48: DDT protects seeds from stored-grain pests. Illinois Agr. Expt. Sta. Rept. 1947/48:115–16.

971. —— 1947/48: Seed treatment for the control of cereal smuts. Can. Minister Agr. Rept. 1948:49–50.

972. —— 1948: After-ripening of seeds in ice cube trays. Flower Grower 35(9):722.

973. —— 1948: The amino acids and the world food supply. Down to Earth 3(4):6–7.

974. —— 1948: Corn seed-coat injuries increase seedling blight. Illinois Agr. Expt. Sta. Rept. 1938/47:28–29.

975. —— 1948: Corn-seed disinfectants raise yields. Illinois Agr. Expt. Sta. Rept. 1938/47:29–30.

976. —— 1948: Corn seeds exposed to atomic rays. Sci. Dig. 24(4):77.

977. —— 1948: Don't spit on your seed. Assoc. Offic. Seed Analysts News Letter 22(4):17.
978. —— 1948: Fertilizer coated seed. Agr. Chem. 3(2):49.
979. —— 1948: New circular issued by Oklahoma Station advises growers to treat oat seed against Victoria blight. Agr. News Letter 16:56–58.
980. —— 1948: Report of the seed propagation division, 1947. J. Dept. Agr. Ireland 45:88–95.
981. —— 1948: Results with seed treatments. Plant Disease Reptr. Suppl. 176:139.
982. —— 1948: Seed of metasequoias will be planted in U.S. Sci News Letter 53:87.
983. —— 1948: Seed problems. Gardeners Chron. 123:189.
984. —— 1948: Seed sown successfully from the air. Indian Farming 9:261.
985. —— 1948: Sphagnum moss useful in preventing seedling disease. Minn. Horticulturist 76:21.
986. —— 1948: Spoilage of moist cottonseed in storage reduced by chemical treatments. U.S. Dept. Agr. Bur. Agr. Ind. Chem. Rept. 1948:11.
987. —— 1948: The *Stillingia* tree. . . . Sci. News Letter 54:140.
988. —— 1948: Tests for seeds. Chem. Industries 62:604, 606.
989. —— 1948: Vitality of buried weed seeds. Rothamsted Agr. Expt. Sta. Rept. 1947:41.
990. —— 1948: Woody-plant seed manual. U.S. Dept. Agr. Misc. Publ. 654. 416 pp.
991. —— 1948: *Zoysia matrella*, one of the wonder-grasses.". . . Assoc. Offic. Seed Analysts News Letter 22(2):7.
992. —— 1948/49: Diseases of field peas, lupines, and soybeans, and lupine seed storage problems. Georgia Agr. Expt. Sta. Ann. Rept. 61:39–40.
993. —— 1949: Airplane pellet reseeding. Natl. Wool Grower 39(10):21.
994. —— 1949: Chemical seed treatment and inoculation with nitrogen-fixing bacteria can both be applied prior to planting without affecting nodulation. Seed World 65(6):22.
995. —— 1949: Dry cold storage of seed corn. U.S. Dept. Agr. Rept. Agr. Expt. Sta. 1948:43–44.
996. —— 1949: Effect of dry storage on peach pit viability. Am. Nurseryman 90(10):58–60.
997. —— 1949: Finland State Seed Testing Station, fiscal year 1947–48. Helsinki. 31 pp.
998. —— 1949: Germination of government-approved seeds. New Zealand J. Agr. 79:272.
999. —— 1949: Germination of Yezo spruce seed. Nature 164:648.
1000. —— 1949: Penick warns of research needed on *Strophanthus* seed. Oil Paint Drug Reptr. 156(8):7.
1001. —— 1949: Plant seed found "elixir of life." Chemurgic Dig. 8(8):15.
1002. —— 1949: Rats plant seed of range land weed. Sci. News Letter 56:359.
1003. —— 1949: Rose reports human amino acid requirements. Chem. Eng. News 27:1364–65.
1004. —— 1949: Rules for testing seeds—Association of Official Seed Analysts. Proc. Assoc. Offic. Seed Analysts 39:23–59.
1005. —— 1949: Seed pellets and methods of preparing the same. Filtrol Corp. (Brit. Patent 690,813, Addn. to 685,217.) Brit. Abstr. B 3:343. 1953.
1006. —— 1949: A study of mechanical injury to seed beans. Asgrow Monograph No. 1. 46 pp.
1007. —— 1949: Vegetable seed treatment. Agr. Gaz. N.S. Wales 60:367–69.
1008. —— 1950: Anent lily seeds. N. & S. Dakota Hort. 23:50.
1009. —— 1950: Asgrow, Germain's will distribute coated seed. Southern Seedsman 13(1):51.
1010. —— 1950: The astounding progress made in treating seed. Seed World 66(4):8–9, 44–45.
1011. —— 1950: Beet seed germination. Horticulture 28(12):451.
1012. —— 1950: Chemical slows ripening but lowers quality of snap beans. Crops & Soils 3(3):23.
1013. —— 1950: Control of certain seed-borne diseases. Idaho Agr. Expt. Sta. Ann. Rept. 280:30–32.
1014. —— 1950: Cottonseed treatment: Its effect on seedling emergence, seedling survival, plant stands, and yields. U.S. Dept. Agr. Tech. Bull. 1025. 134 pp.
1015. —— 1950: Crop inspection and seed examination for the control of seed-borne diseases. Can. Minister Agr. Rept. 1949/50:77–78.
1016. —— 1950: Effect of 2,4-D on germination of cottonseed. Texas Agr. Expt. Sta. Agr. Res. Texas 1950:52.
1017. —— 1950: Filtrol's color coating of seeds. Seed World 66(12):10.
1018. —— 1950: Find fumigant harmless to seeds. Seed World 66(3):50.

1019. —— 1950: Flax seed. Texas Agr. Expt. Sta. Agr. Res. Texas 1950:68.
1020. —— 1950: Germination tests on newly harvested rice may not be reliable. Rice J. 53(10):28, 32. Biol. Abstr. 25: 31826.
1021. —— 1950: Grass seeds with "overcoats" make thicker lawns. Sci. Dig. 27(6):17.
1022. —— 1950: Hammer mill cuts germination. Crops & Soils 3(2):26.
1023. —— 1950: Handling nut seeds for germination. Am. Nurseryman 91(8):53.
1024. —— 1950: The influence of rate of seeding on the yield of wheat and oats. J. Dept. Agr. Ireland 47:97–103.
1025. —— 1950: Pellet range seeding fails to live up to advance publicity. Crops & Soils 2(9):25.
1026. —— 1950: Rice to be sown from the air. Indian Farming 11:389.
1027. —— 1950: Seed corn keeps best when stored near zero in tight, dry container. Crops & Soils 2(9):29.
1028. —— 1950: Seeds of *Agropyron*, eleven species of economic interest compared, by A. F. Musil. U.S. Bur. Plant Ind. Soils Agr. Eng., Div. Forage Crops Diseases, Beltsville. 6 pp.
1029. —— 1950: Sound waves help seed-yields, kill insects and bacteria. Chemurgic Dig. 9(7):16.
1030. —— 1950: Storage and drying of cereals for use as seed. Natl. Inst. Agr. Botany Seed Notes 37. 7 pp.
1031. —— 1950: Use of the airplane in rice production. Rice J. 53(11):14–15. Biol. Abstr. 25:31827.
1032. —— 1950: Variation in protein content of barley varieties studied. S. Dakota Agr. Expt. Sta. Ann. Rept. 1949/50:104–6.
1033. —— 1950/51: Crops in peace and war. Yearbook Agr. U.S. Dept. Agr. 942 pp.
1034. —— 1950/51: Inspection, seed sale. N.Y. Dept. Agr. Ann. Rept. 1950/51:147–48.
1035. —— 1951: Ancient seeds not viable. Sci. News Letter 59:180.
1036. — 1951: Approved vegetable seed, Sept. 1951. Agr. Gaz. N.S. Wales 47:452.
1037. —— 1951: Bushel weight of seed. Tasmanian J. Agr. 22:239.
1038. —— 1951: Certain native grass seeds need dormancy before they germinate. Crops & Soils 3(5):24.
1039. —— 1951: Chemical kills weed seeds in crop fields. Sci. News Letter 60:376.
1040. —— 1951: Chlorine treatment improves germination of hybrid strawberry, raspberry, and blackberry seeds. U.S. Dept. Agr. Rept. Administrator Agr. Res. 1951:402.
1041. —— 1951: Deep freezing prolongs viability of sugarcane seed. U.S. Dept. Agr. Rept. Administrator Agr. Res. 1951:379.
1042. —— 1951: Doubt case on revival of dead seeds by X-rays. Sci. Dig. 29(1):44.
1043. —— 1951: The effect of 2,4-D on yields of inbreds, single crosses, and double crosses of Kentucky 203 corn in succeeding generation after treatment. Kentucky Agr. Expt. Sta. 64th Ann. Rept., p. 14.
1044. —— 1951: Farming from the sky. Sci. News Letter 60:170–71.
1045. —— 1951: Germination of *Dichondra repens*. Seed World 68(8):34.
1046. —— 1951: Hardy weed seeds live many years. Sci. News Letter 59:34.
1047. —— 1951: Incidence of twinning in tomato seed. W. Can. Soc. Hort. Rept. Proc. 7:62–64.
1048. —— 1951: Longevity of *Nelumbo* fruits. Nature 168:905.
1049. —— 1951: Maize bibliography for the years 1937 to 1945, inclusive, compiled by J. C. Cunningham. Iowa Agr. Expt. Sta. Corn Res. Inst. Contrib. 3:194–298.
1050. —— 1951: Mango seed unlocks adventure. Florida Grower 59(3):12,17,28.
1051. —— 1951: New germinator. Seed World 69(5):38.
1052. —— 1951: New grain drying method dries kernels from inside out, says USDA engineer. Crops & Soils 4(1):27.
1053. —— 1951: Paint oil from tobacco seed. Chemurgic Dig. 10(10):19.
1054. —— 1951: Pelleted seeds germination tests. Horticulture 29(8):300.
1055. —— 1951: Recent literature 1940–48. Proc. Intern. Seed Testing Assoc. 17:52–259.
1056. —— 1951: Rubber-seed oil. Paint, Oil Colour J. 120:676. Brit. Abstr. B 2:273.
1057. —— 1951: Sampling ryegrass crops for blind seed examination. New Zealand J. Agr. 83:338.
1058. —— 1951: Seed germination tests. Florists Rev. 108(2792):80.
1059. —— 1951: Seed treatment will control anthracnose in *Digitalis*. U.S. Dept. Agr., Agr. Res. Admin. Rept. 1951:386.
1060. —— 1951: Seeds. Diagnosis and control of seed-borne bacterial diseases. Can. Minister Agr. Rept. 1951:34.

1061. —— 1951: Tung production. U.S. Dept. Agr. Farmers Bull. 2031. 41 pp.
1062. —— 1951: Weed handbook. Kansas State Bd. Agr. Rept. 70(306). 38 pp.
1063. —— 1951/52: Plant physiology. (c) Germination. 4th Ann. Rept., Commonwealth Sci. Ind. Res. Org. 1951/52:23.
1064. —— 1952: Better seed—better crops; New York certified seed. N.Y. Certified Seed Growers Coop., Ithaca. 56 pp.
1065. —— 1952: Collecting seed of tropical water-lilies. Bull. Missouri Botan. Garden 40:89–90.
1066. —— 1952: Deseeder for fiber-flax. Chemurgic Dig. 11(5):23.
1067. —— 1952: Direct seeding of cabbage tested. Seed World 70(8):41.
1068. —— 1952: Farm production, farm disposition, and value of principal field crops. Revised estimates 1944–49. By states. U.S. Dept. Agr. Bur. Agr. Econ. Statist. Bull. 115. 91 pp.
1069. —— 1952: First flower seed to sow is petunia in a seed box. Minn. Horticulturist 80:45.
1070. —— 1952: For kids of 8 to 80. Seed travellers. Green Thumb 9(10):21.
1071. —— 1952: Germinating lettuce seed. Seed World 70(9):42–43.
1072. —— 1952: Handbook of seed certification. Ohio Seed Improvement Assoc., Columbus. 55 pp.
1073. —— 1952: Inspection of agricultural seeds. Purdue Univ. Agr. Expt. Sta. Circ. 381. 64 pp.
1074. —— 1952: Locust seed treatment. Am. Nurseryman 96(4):31.
1075. —— 1952: Low temperature storage keeps sugarcane seed viable many months. Crops & Soils 4(8):27.
1076. —— 1952: Manual for testing agricultural and vegetable seeds. U.S. Dept. Agr., Agr. Handbook 30. 440 pp.
1077. —— 1952: National Institute of Agricultural Botany. Seed Production Committee. 8th Rept., June 1950 to May 1951. Cambridge.
1078. —— 1952: Nursery research at Ohio State. Am. Nurseryman 95(6):15, 94–96.
1079. —— 1952: Oil from groundnut kernels. Chem. Age 64:409–10.
1080. —— 1952: Quelques aspects du problème de l'application des oudes ultracourtes dans le sèchage et la désinfection des graines par le chauffage par pertes diélectriques. Rev. Hort. 124:725–26.
1081. —— 1952: Revised estimates 1944–1949. Acreage, yield, and production of principal field crops by states. U.S. Dept. Agr. Crop Production, Crop Reptg. Bd. Bur. Agr. Econ. Statist. Bull. 108. 43 pp.
1082. —— 1952: Seed analyses report. Penn. Dept. Agr. Gen. Bull. 641. [35(2)]. 14 pp.
1083. —— Seed crops. Farm production, farm disposition, and value of field seed crops. Revised estimates 1939–1950. By states. U.S. Dept. Agr. Bur. Agr. Econ. Statist. Bull. 119. 65 pp.
1084. —— 1952: Seed treatment with peroxygen chemicals. Can. Chem. Process. 36:84, 86, 87, 98.
1085. —— 1952: Soy bean seed harmed by 2,4-D. Crops & Soils 4(8):27.
1086. —— 1952: Soak blackberry seeds. Am. Nurseryman 95(9):56–57.
1087. —— 1952: Station féderale d'essais et de controle de semences, Mont Calme, Lausanne. Rapport d'activité 1932–1950. Landwirtsch. Jahrb. Schweiz N.S. 1:17–97.
1088. —— 1952: Steinlite tests moisture faster. Southern Seedsman 15(1):72.
1089. —— 1952: Traces. Ind. Lab. 3(12):5.
1090. —— 1952: A wild flower seed collection. Wild Flower 28:68.
1091. —— 1952/53: Cold germination tests and dates of planting sorghum. S. Dakota Agr. Expt. Sta. 66th Ann. Rept. 107–8.
1092. —— 1952/53: Seed certification. Iowa State Dept. Agr., Iowa Book of Agr. 1952/53:212–17.
1093. —— 1953: Abnormality in subterranean clover seedlings. Agr. Gaz. N.S. Wales 64:70.
1094. —— 1953: Acid treatment for hulling *Zoysia* seed. U.S. Dept. Agr., Agr. Res. Admin. Rept. 1953:362–63.
1095. —— 1953: Better seeds: Better crops. Natl. Inst. Agr. Botany 34th Rept. & Accounts.
1096. —— 1953: Biennials and perennials—they come easily from seed. Horticulture 31(7):298–308.
1097. —— 1953: Cotton seed viability and seedling vigor. U.S. Dept. Agr., Agr. Res. Admin. Rept. 1953:369.
1098. —— 1953: Electronic destroyer of stored-grain insects. Agr. Res. 2(4):10–11.
1099. —— 1953: Forage crop seeds enough. Agr. Res. 2(5):8–10.
1100. —— 1953: Germination in *Rosa.* John Innes Hort. Inst. 44th Ann. Rept. 1953:27–28.
1101. —— 1953: Holly from seed. Flower Grower 40(8):53.
1102. —— 1953: International rules for seed

testing. Proc. Intern. Seed Testing Assoc. 18:1–69.
1103. —— 1953: Longevity of *Halogeton* seeds. Calif. Dept. Agr. Monthly Bull. 42:290–91.
1104. —— 1953: The longevity of seeds. Gardeners Chron. 133:127.
1105. —— 1953: Nebraska Crop Improvement Association. Nebraska Cert. Seed Handbook, 44th Ann. Rept. Lincoln. 72 pp.
1106. — 1953: New South Wales trials with pelleted seed. Agr. Gaz. N.S. Wales 64:595–96.
1107. —— 1953: Presence of insects in grain shown by X-ray examinations. Agr. News Letter 21:82–83.
1108. —— 1953: Protein survey of wheat 1953. Can. Bd. Grain Commiss. Crop Bull. 50:1–35. Biol. Abstr. 28:14435. 1954.
1109. —— 1953: Publications on the identification of seeds. U.S. Dept. Agr. Office of Information. 4 pp.
1110. —— 1953: Recent research on drying and storage of rough rice. Southern Coop. Ser. Bull. 29. 29 pp.
1111. —— 1953: Seed analyses report. Penn. Dept. Agr. Bull. 32(2). 13 pp. [General Bull. 647.]
1112. —— 1953: Seed-borne diseases and their control. Agr. Gaz. N.S. Wales 64:30–33.
1113. —— 1953: Seed regulations, 1953. Act 21/52. Suppl. S. Rhodesia Govt. Gaz., April 2, 1953, printed by the Government printer.
1114. —— 1953: Storage of valuable cotton seed stocks. Report on the Agricultural Experiment Stations. U.S. Dept. Agr. Office Expt. Sta., p. 77.
1115. —— 1953: The test weight per bushel of grain: methods of use and calibration of apparatus. U.S. Dept. Agr. Circ. 921. 11 pp.
1116. —— 1953: Tetrazolium test. Seed World 73(10):14.
1117. —— 1953: Twelve amino acids found in barley seed. U.S. Dept. Agr., Agr. Res. Admin. Rept. 1953:354.
1118. —— 1953: The use of aircraft in hill farming. Dept. Agr. Scotland Hill Farm Res. Rept. 2:73–81. Soils Fertilizers 17:356.
1119. —— 1954: Aleurone layer in barley. Nature 173:344.
1120. —— 1954: Better seeds, better crops. Natl. Inst. Agr. Botany 35th Ann. Rept. 50 pp.
1121. —— 1954: Bibliography of weed investigations. Weeds 3:1–15.
1122. —— 1954: Chemical treatment of summer grains. Netherlands Plantenziektenk Dienst. Ber. 1068. 2 pp.
1123. —— 1954: Cyclamen seeds. Gardeners Chron. 135:70.
1124. —— 1954: Désinfection des semences des végétaux. Compt. Rend. Congr. Protect. Veg. Leurs Prod. Climats Chaud. 1954:312–13.
1125. —— 1954: Dormancy and some other problems in seed testing. New Zealand J. Agr. 88:207, 209.
1126. —— 1954: Fertilizer seed treatment. J. Agr. Food Chem. 2:1055.
1127. —— 1954: Freesias from seed. J. Roy. Hort. Soc. 79:217–18.
1128. —— 1954: Good seed for better field crops. Quebec Dept. Agr. Bull. 174-A. 100 pp.
1129. —— 1954: Lindane seed treatments harm seed at high rates. Wisconsin Agr. Expt. Sta. Bull. 511:28.
1130. —— 1954: Losses in agriculture. U.S. Dept. Agr., Agr. Res. Ser. 20:95–101.
1131. —— 1954: A new seed treatment. World Crops 6:134.
1132. —— 1954: Pasture seed certification—purity and germination. Standards of certified pasture seeds. Agronomy division Tasmanian J. Agr. 25:334–35.
1133. —— 1954: The preservation of viability and vigor in vegetable seed. Asgrow Monograph 2. 32 pp.
1134. —— 1954: Seed Bahia grass at 10 to 15 pounds per acre, Georgia Station says. Crops & Soils 6(6):25.
1135. —— 1954: Seedcoat injury in soybeans is index of poor germination. Crops & Soils 6(5):20.
1136. —— 1954: Seed production, testing and distribution in European countries. Nature 173:1030.
1137. —— 1954: Seed treatment fungicides. Farm Chem. 117(2):11–13.
1138. —— 1954: Trials with pelleted seed in New South Wales. World Crops 6:162.
1139. —— 1954: You can store grain safely on the farm. U.S. Dept. Agr. Farmers Bull. 2071. 15 pp.
1140. —— 1955: Ancient seed plants: the Cycads. Bull. Missouri Botan. Garden 43:65–80.
1141. —— 1955: Antibiotic seed disinfection. World Crops 7:339–40.
1142. —— 1955: Collection and storage of ash (*Fraxinus excelsior*), sycamore, and

maple (*Acer*) seed. Gt. Brit. Forestry Comm. Leaflet 33. 9 pp.

1143. —— 1955: France. Direction général des eaux et forêts. Reboisement par semis directs. Nancy École Natl. Eaux Forêts, 9. 16 pp.

1144. —— 1955: Growing alfalfa for seed. Utah Agr. Expt. Sta. Circ. 135. 60 pp.

1145. —— 1955: *Hemerocallis* germination. Am. Nurseryman 102(7):113–14.

1146. —— 1955: Hot water seed treatment helps barley growers. Forward Together. N. Carolina State Coll. School Agr. Ann. Rept. 1955:22.

1147. —— 1955: Loblolly pine seed tree requirements. U.S. Dept. Agr. Forest Serv. Southeast. Forest Expt. Sta. Ann. Rept. 1955:10–13.

1148. —— 1955: New machine for seed lab. New Jersey Agr. 37(2):9.

1149. —— 1955: Noxious weed seeds in feed wheat. Tasmanian J. Agr. 26:38, 46.

1150. —— 1955: Plant virus stopped by chemical in seed. Science 68(2):23.

1151. —— 1955: Recent advances in science: Plant Physiology by Walter Stiles. Seed Germination. Sci. Progr. (London) 43: 696–705.

1152. —— 1955: Tamarind seed has many uses. Indian Farming 5(8):21–22, 40.

1153. —— 1955: Temperature dips, rises stimulate germination in Ladino clover seed. Crops & Soils 8(4):23.

1154. —— 1955: This is the inside story of a wheat berry. Progr. thru Res. 9(3):4–5.

1155. —— 1955: The tree peonies. Natl. Hort. Mag. 34:40–44.

1156. —— 1955/56: Annual Report on the research activities of the Plant Physiological Research Institute, University of Pretoria.

1157. —— 1956: Alpine Garden Society. Seed distribution. 18 pp.

1158. —— 1956: Atoms for seed. World Crops 8:1–2.

1159. —— 1956: Bean storage and yielding ability; bean storage and eating quality. Agr. Res. 5(6):12.

1160. —— 1956: Chemical seed treatment. J. Agr. Food Chem. 4:586–87.

1161. —— 1956: Chemicals cut losses in harvesting seeds. Crops & Soils 8(9):26.

1162. —— 1956: Comparative germination tests on certain *Agropyron* spp. Assoc. Offic. Seed Analysts News Letter 30(4):39–46.

1163. —— 1956: Danish grass and legume seeds uncertain. (Foreign Agr. Serv.) Seed World 79(4):18.

1164. —— 1956: Forest Seed Directory. Food Agr. Organ. U.N., Rome. 178 pp.

1165. —— 1956: Hot water bath makes cotton seeds sprout. Sci. News Letter 69:279.

1166. —— 1956: International rules for seed testing. Proc. Intern. Seed Testing Assoc. 21:1–80.

1167. —— 1956: Metabolism of fresh green peas in relation to their storage. Gt. Brit. Dept. Sci. Ind. Res. Food Invest. 60(1955).

1168. —— 1956: On the threshold of our second century. Asgrow Bull. 48 pp.

1169. —— 1956: Seed analysis report—1955. Penn. Dept. Agr. Bull. 39(2). 18 pp. [General Bull. 665.]

1170. —— 1956: Seed inspection in Kentucky, 1955–56. Kentucky Agr. Expt. Sta. Regulatory Bull. 129. 59 pp.

1171. —— 1956: Suggested guide for chemical control of weeds. U.S. Dept. Agr. Spec. Rept. ARS 22–23. 38 pp.

1172. —— 1956: Tree planting practices in temperate Asia: Japan. Food Agr. Organ. U.N., Rome. 156 pp.

1173. —— 1956/57: Germination and storage of seed. W. African Inst. Oil Palm Res. Ann. Rept. 5:112–13.

1174. —— 1956/57: Report of the seed propagation division. J. Dept. Agr. Ireland 43:180–88.

1175. —— 1957: *Acer* seed treatment. Am. Nurseryman 106(9):111–12.

1176. —— 1957: Agronomy handbook for South Carolina. Clemson Agr. Coll. Bull. 104.

1177. —— 1957: Device shocks seeds into germination. Sci. News Letter 71:39.

1178. —— 1957: Disinfection of cereal seed. N. Ireland Minister Agr. Leaflet 96. 3 pp.

1179. —— 1957: Dwarf growth of seedlings from unstratified peach seeds. Plant Physiol. Res. Inst. Univ. Pretoria Ann. Rept. 1957:9–11.

1180. —— 1957: Fruit sap and seed germination. Nature 179:1056.

1181. —— 1957: Germination of weed seeds. Plant Physiol. Res. Inst. Univ. Pretoria Ann. Rept. 1957:11–19.

1182. —— 1957: The influence of storage conditions on the viability of *Hibiscus cannabinus* ("stokroos") seed. Plant Physiol. Res. Inst. Univ. Pretoria Ann. Rept. 1957:19–20.

1183. —— 1957: Insect control by seed treatment. Agr. Res. 5(8):10.

1184. —— 1957: Internationale Vorschriften für die Prüfung von Saatgut. Proc.

Intern. Seed Testing Assoc. 22:581–671.

1185. —— 1957: News briefs. Science 126:693.

1186. —— 1957: Radio wave treatment softens legume seed. Crops & Soils 9(9):25.

1187. —— 1957: Rainfall affects viability of cotton seed. S. Carolina Agr. Expt. Sta. 69th Ann. Rept., pp. 34–35.

1188. —— 1957: Report on the work of the work groups of the Oil Seed Foundation with linseed, rapeseed and poppy seed in 1955 and 1956. Sticht. Oliehoudende Zaden. Meded. 5. 62 pp.

1189. —— 1957: Seed germination: biochemical studies. Plant Physiol. Res. Inst. Univ. Pretoria Ann. Rept. 1957:54–55.

1190. —— 1957: Slingshot seed sorter. Crops & Soils 9(9):25.

1191. —— 1957/58: Direct seeding of pine. Agr. News Letter 26:16.

1192. —— 1957/58: Germination and storage of seeds. W. African Inst. Oil Palm Res. Ann. Rept. 6:117–19.

1193. —— 1957/58: Samenprüfstelle der Landwirtschaftskammer Westfalen-Lippe. Jahresber. 18 pp.

1194. —— 1957/58: Seed germination. Commonwealth Sci. Ind. Res. Org. 1957/58: 25–26.

1195. —— 1958: Annual report of the activities of the Swedish Seed Association, 1957. Sveriges Utsädesfören. Tidskr. 98:99–195.

1196. —— 1958: 10th Annual Report from the Government Plant Protection Service of Denmark on seed pathological testing. Kobenhavn. 1957/58. 15 pp.

1197. —— 1958: High amylose corn. Agr. Food Chem. 6:641–42.

1198. —— 1958: New national seed storage laboratory nears completion. Seed World 83(2):18–19.

1199. —— 1958: New seed test separates alfalfa, sweetclover. Crops & Soils 11(1): 20.

1200. —— 1958: Poplars in forestry and land use. Food Agr. Organ. U.N., Rome. 511 pp.

1201. —— 1958: Seed treatment is low-cost insurance. Crops & Soils 10(5):20–22.

1202. —— 1958: Stored-grain pests. U.S. Dept. Agr. Farmers Bull. 1260. 46 pp.

1203. —— 1958: The viability of seeds grown in Denmark, July 1 to November 30, 1957. Dansk Froavl 41:86–88.

1204. —— 1958/59: Gibberellic acid and breaking of seed dormancy. Bose Inst. Rept. 1958/59:50–51.

1205. —— 1958/60: Samenprüfstelle der Landwirtschaftskammer Westfalen-Lippe. Jahresber. 22 pp.

1206. —— 1959: Dwarf mistletoe seeds are faster than rocket. Sci. News Letter 76:168.

1207. —— 1959: Embryological studies in angiospermae. Tr. Mosk. Gl. Botan. Sad. 6. 225 pp. Bibliog. Agr. 25:17923. 1961.

1208. —— 1959: How to store seed safely. Crops & Soils 12(1):9–11, 18.

1209. —— 1959: International rules for seed testing. Proc. Intern. Seed Testing Assoc. 24:475–584.

1210. —— 1959: Magnetic seed cleaners. Crops & Soils 11(10):20–21.

1211. —— 1959: Methods for determining moisture content as specified in the official grain standards of the United States and in the United States standards for beans, peas, lentils, and rice. U.S. Dept. Agr., Service and Regulatory Announcements No. 147 (revised). 3 pp.

1212. —— 1959: Physicochemical examination of *Phaseolus glabra* seeds. Indian J. Appl. Chem. 22:37. Chem. Abstr. 54: 6887. 1960.

1213. —— 1959: Precautions to be observed in the hot water treatment of seeds. Tasmanian J. Agr. 30:106.

1214. —— 1959: Swedish seed production. Kgl. Lantbrukshögsk. Stat. Lantbruksförsök Stat. Jordbruksförsök Särtryck och smaskrifter 109. 66 pp.

1215. —— 1959/60: Annual Report and Minutes of the 1960 Annual Meeting held at the University of New Brunswick, Fredirikton, N.B., June 15–17, 1960. Can. Seed Growers Assoc. Ann. Rept. 1959/60. 74 pp.

1216. —— 1959/60: Endosperm nuclei. Bose Inst. Rept. 1959/60:57–58.

1217. —— 1960: Certified subterranean clover seed. Agr. Gaz. N.S. Wales 71:206, 219–20.

1218. —— 1960: Collection and storage of ash, sycamore and maple seed. Gt. Brit. Forestry Comm. Leaflet (revised) 33. 11 pp.

1219. —— 1960: Control of seed-borne diseases by fungicidal treatment. Farm J. Brit. Guiana 21:13–14.

1220. —— 1960: Dormancy in common ryegrass. Assoc. Offic. Seed Analysts News Letter 34(4):31–33.

1221. —— 1960: Electrically treated cotton. Current Sci. (India) 29:117.

1222. —— 1960: For chaffy grass seed: mechanical analysis. Agr. Res. 9(3):7.

1223. —— 1960: Infra-red drying for farm crops and seeds. Crops & Soils 13(2): 14–15.

1224. —— 1960: International rules for seed testing. Proc. Intern. Seed Testing Assoc. 25:873–993.

1225. —— 1960: International rules for seed testing. Proc. Intern. Seed Testing Assoc. 25:997–1121.

1226. —— 1960: Principal seed and plant characters for distinguishing varieties of twelve crops. Assoc. Offic. Seed Analysts News Letter 34(3):10–18.

1227. —— 1960: Seed testing station has international reputation. Farming S. Africa 35(10):10. Biol. Abstr. 35:35798. 1960.

1228. — 1960: Seed treatment and sowing methods for cover crops. Rubber Res. Inst. Malaya Planters Bull. 50:94–98. Biol. Abstr. 36:50137. 1961.

1229. —— 1960: Viability of weed seed after storage in silage. Tidsskr. Planteavl. 64:715–16.

1230. —— 1960: World seed campaign. Agr. 58:661.

1231. —— 1960: World seed campaign. New Zealand J. Agr. 100:418.

1232. —— 1960: X-rays reveal seed properties of sugarbeets. Agr. Res. 9(3):5.

1233. —— 1961: Bibliography of weed investigations for October, November, December 1960. Weeds 9:499–514.

1234. —— 1961: Germination. W. African Inst. Oil Palm Res. Ann. Rept. 9:89–93.

1235. —— 1961: Radio impulses help germination of seeds. Sci. News Letter 80:104.

1236. —— 1961: Seedling vigor and rate of emergence in two sources of Newport Kentucky bluegrass. Crop Sci. 1:305–6.

1237. —— 1961: Seeds. Yearbook Agr. U.S. Dept. Agr. 591 pp.

1238. —— 1961: Society of American Foresters report on a study of seed certification conducted by the committee on forest tree improvement. J. Forestry 59:656–61.

1239. —— 1961: *Trichodesma zelandicum* as an oilseed. Trop. Sci. 3:132.

1240. —— 1961: Who wants tree seed certification and why? J. Forestry 59:831–32.

1241. —— 1962: Applicación de la energia nuclear a la agricultura. Informe anual de la comisión de Energía Atomica de los Estados Unidos bajo el Contrato AT (30-1)-2043. Turrialba, Costa Rica, Inst. Interamer. Ciencias Agr. OEA, 1962:13–14.

1242. —— 1962: Bacteria inhibitor found. Sci. News Letter 82:172.

1243. —— 1962: Field and seed crops. Production, farm use, sales, value. United States, 1949–59; by states, 1954–59. Revised estimates. U.S. Dept. Agr. Statist. Reptg. Serv. Crop Reptg. Bd. Statist. Bull. 311. 104 pp.

1244. —— 1962: Germination and root establishment in seedlings of curly mesquitegrass. Texas Agr. Expt. Sta. Misc. Publ. MP-615.

1245. —— 1962: Germination inhibitions in native grasses (*Lasirus hirsutus*, *Cenchrus ciliaris*, *Cenchrus setigereus*). Agr. Res. 12(1):58–59. Biol. Abstr. 42:23673. 1963.

1246. —— 1962: Germination inhibitor. Agr. Res. 11(3):6.

1247. —— 1962: H_2O_2 speeds pine seed germination. Agr. Res. 11(4):15.

1248. —— 1962: Jonathan spot: Incidence in relation to seed numbers. J. Australian Inst. Agr. Sci. 28:289.

1249. —— 1962: Panwar seeds. Hamdard Med. Dig. 6(7/8):39. Biol. Abstr. 43:3451. 1963.

1250. —— 1962: Pelleted lettuce seeds look good in greenhouse tests. What's New in Farm Science. Wisconsin Agr. Expt. Sta. Bull. 558:32.

1251. —— 1962: Peppers "adopt" eggplant seeds. What's New in Farm Science. Wisconsin Agr. Expt. Sta. Bull. 558:83.

1252. —— 1962: Seed setting of forest trees in 1961. Tidsskr. Skogbruk 70:240–46.

1253. —— 1962: Standardized names of weeds—common and scientific names (also terminology and herbicide nomenclature). Weeds 10:255–71.

1254. —— 1962: Symposium on Plant Embryology, University of Delhi, 1960. New Delhi Council Sci. Ind. Res. 1962. 274 pp.

1255. —— 1962: Wheel track planting OK with fall plowing. What's New in Farm Science. Wisconsin Agr. Expt. Sta. Bull. 558:44.

1256. —— 1963: Amendements apportés en 1962 aux règles internationales pour les essais de semences. Proc. Intern. Seed Testing Assoc. 28:173–93.

1257. —— 1963: Effect of seed size on vigor and yield of runner peanuts. Auburn Univ. Agr. Expt. Sta. Bull. 18 pp.

1258. —— 1963: Ergänzung zu den Internationalen Vorschriften für die Prüfung von Saatgut 1962. Proc. Intern. Seed Testing Assoc. 28:496–516.

1259. —— 1963: Foundation seed works for you! Crops & Soils 15(9):15.

1260. —— 1963: Lime pelleting legume seed to improve nodulation in acid soil. Agr. Gaz. N.S. Wales 74:137–40.

1261. —— 1963: Savory herbs—seeds. Can. Dept. Agr. Publ. 1158:6.

1262. —— 1963: The seed we use: Part I. What we need to know about it. SAF Seed Certification Subcommittee. J. Forestry 61:181–84.

1263. —— 1963: The seed we use: Part II. How to assure reliable information about it. SAF Seed Certification Subcommittee. J. Forestry 61:265–69.

1264. —— 1963: Seeding of flax. Growing Flax, Can. Dept. Agr. Publ. 545:6–7.

1265. —— 1963: Soybeans. Agr. Gaz. N.S. Wales 74:209–12.

1266. —— 1963: A summary of range grass seeding trials in Colorado. Colorado Agr. Expt. Sta. Tech. Bull. 73:1–81.

1267. —— 1963: Technique of tobacco seed production. Southern Rhodesia Tobacco Seed Assoc. 5 pp. Rev. Appl. Mycol. 42:572. 1963.

1268 .—— 1964: New Jersey state seed law and rules and regulations. New Jersey Agr. Expt. Sta. Bull. 809. 23 pp.

1269. —— 1964: Radio frequency doubles seed viability. Agr. News Letter 31:7.

1270. Anpilogov, M. Z. 1951: Treatment of seeds against smut. Selek. i Semen. 18(12): 47–50.

1271. Anselme, C. 1958: Étude comparative de differents produits utilisés pour la désinfection de semences de lin. Phytiat.-Phytopharm. 7:3–13.

1272. —— and R. Champion. 1962: L'analyse sanitaire des semences de pois. Proc. Intern. Seed Testing Assoc. 27:829–42.

Anselmi, S. 12877.

1273. Ansiaux, J. R. 1962: The cation composition of fruits and seeds. Bull. Soc. Franc. Physiol. Vegetale 8:81–84. Chem. Abstr. 59:5492. 1963.

1274. Ansin, B. N., and E. P. Kushelewa. 1934: The temperature of drying and germination of apple seeds. Trans. Res. Inst. North Fruitsfarming I. W. Mitchurin 1:43.

1275. Anslow, R. C. 1958: Note on an improved technique in establishing *Setaria sphacelata* from seed. Empire J. Exptl. Agr. 26:55–56.

1276. —— 1962: A quantitative analysis of germination and early seedling growth in perennial ryegrass. J. Brit. Grassland Soc. 17:260–63. Biol. Abstr. 42:11820. 1963.

Anstrand, G. 14574.

1278. Anthistle, M. J., D. F. Ashdown, and D. Dickinson. 1959: Ripening of broad beans. J. Sci. Food Agr. 10:412–14.

1279. Anthony, D. S., F. W. Quackenbush, A. J. Ihde, and H. Steenbock. 1943: Antiacrodynic potency of seed oils. J. Nutr. 26:303–7.

1280. Anthony, K. R. M., and S. A. J. Tarr. 1952: The causes of deterioration of cotton seed in the equatoria province of the Anglo-Egyptian Sudan. Empire J. Exptl. Agr. 20:56–65.

Antic, M. 13433, 13434.

1281. Antipov, V. G. 1957: The effect of gases discharged by industrial plants on the seeds of trees and shrubs. J. Botan. URSS 42:1230–32.

1282. Antoniani, C. 1935: Enzymic transformation of glycerophosphoric acid into pyruvic acid by means of the germinated seeds of *Sorghum saccharatum*. Atti Accad. Lincei 21:192–95. Chem. Abstr. 29:6255. 1935.

1283. —— 1961: Embryo transplanting as a method of obtaining "vegetative hybrids" or of favoring interspecific and intergeneric crosses in cereals. Ann. Sper. Agrar. (Rome) 15:207–23.

1284. —— and R. M. Cioffi. 1947: Variation in vitamin B_1 content of germinated barley. Acta Chim. Appl. 37:492–94. Chem. Abstr. 42:8884. 1948.

1285. ——, G. Serini, and L. Federico. 1954: Acetoinic condensation process in the Graminaceae and Leguminosae plants. II. Rend. Ist. Lombardo Sci. Lettere, Pt. I: Classe Sci. Mat. Nat. 87:530–36. Chem. Abstr. 49:14910. 1955.

1286. —— and F. Zanelli. 1932: The sitosterol group. The sterols from grape seed oil. Atti Accad. Lincei 15:284–86. Chem. Abstr. 26:5960. 1932.

Antoniani, C. 6447.

Antonini, E. 16430.

1287. Anttinen, O. 1962: Results of timothy seed production trials. Maatalouden Tutkimuskeskus, Maatalous ja Koeb. 16:7–12.

1288. Antunes, H. 1953: Seeds of the "moca" and "concha" type in new world coffee.

Sao Paulo Supt. Serv. Café Bull. 28(317): 8–16.

1289. Anzai, I. 1945: Protease activity of the resting and germinating soybean. Tohoku J. Exptl. Med. 48:31–39. Chem. Abstr. 42:6895. 1948.

1290. Aoyama, S. 1931: Saponin of seeds of *Thea sinensis* L. and a comparison of *Camellia*, *Sasanqua*, and *Thea* saponins. Yakugaku Zasshi 5:430–34. Japan. J. Med. Sci. Biochem. 2:(68). 1934.

Apelard, J. 15221.

Aponte, E. 17791.

1291. Apostol, T., C. Balan, A. Balasoiu, I. Berenghi, L. Buda, and T. Popa. 1957: Ann. Inst. Cerc. Agron. Bucuresti, N.S. 24:179–94. 1956.

1292. Appala-Naidu, B., and B. Venkateswarlu. 1962: Studies on the effects of magnesium chloride and sodium carbonate on the germination and growth of rice seedlings, *Oryza sativa* L. Andhra Agr. J. 9:82–89.

Appel, W. 82.

1293. Appelqvist, L. A. 1961: Fat quality of Swedish oil crops. I. Variation in content of free fatty acids with different environmental factors. Sveriges Utsädesfören. Tidskr. 71:74–97.

1294. —— 1962: Hereditary variation in content of isothiocyanates and thiooxazolidones in seeds of rape and turnip rape. Acta Chem. Scand. 16:1284–85.

1295. Appleman, C. O. 1919: Carbohydrate metabolism on sweet corn during storage at different temperatures. J. Agr. Res. 17:137–52.

1296. —— and S. V. Eaton. 1921: Evaluation of climatic temperature efficiency for the ripening processes in sweet corn. J. Agr. Res. 20:795–805.

1297. Appleman, D. 1952: Catalase-chlorophyll relationship in barley seedlings. Plant Physiol. 27:613–21.

1298. Appleman, M. D. 1941: Effect of seed treatment on nodulation of soybeans and peas. Soil Sci. Soc. Am. Proc. 6:200–3. Expt. Sta. Rec. 87:666. 1942.

Appleman, W. 13083.

1299. Appleton, J. M. 1937: Influence of external factors on seed-setting in Lucerne. Herbage Rev. 5:87–90.

Applewhite, T. H. 2470.

1300. Aprod, A. I. 1960: The effect of harvesting times on the industrial quality of rice. Zemledeli 1960(8):89–91.

1301. —— 1961: Effect of block disintegration on the germination of rice seed. Kubansk. Ris. Opyt. Stantsiya: Krasnodar 112–15. Biol. Abstr. 43:3243. 1963.

Aprod, A. I. 307.

1302. Apsit, J., and E. Gain. 1909: Les graines tuees par anesthesie conservent leur proprietes diastasique. Compt. Rend. 149:58–60.

Aquin, E. L. d'; *see* D'Aquin, E. L.

Arai, K. 19950.

1303. Arai, M., and H. Chisaka. 1961: Ecological studies on *Alopercurus aequalis* Sobol., a noxious weed in winter cropping. 9. On the secondary dormancy of the seed. Proc. Crop Sci. Soc. Japan 30:43–46.

1304. —— and T. Kataoka. 1956: Ecological studies on *Alopecurus aequalis* Sobol. (1) On the influence of soil moisture and of the depth of soil over the seeds upon germination and emergence of them. (2) On the temperature for germination and emergence, and seasonal variation in emerging capacity. Proc. Crop Sci. Soc. Japan 24:275–78.

1305. —— and —— 1956: Ecological studies on *Alopecurus aequalis* Sobol. (3) Influence of soil moisture on the dormancy and longevity of seeds. (4) Seasonal variation in the viable seed population and its vertical distribution in the soil. Proc. Crop Sci. Soc. Japan 24:319–23.

1306. —— and M. Miyahara. 1962: Physiological and ecological studies on barnyard grass (*Echinocloa crus-galli* Beauv. var. *oryzicola* Ohwi). II. On the primary dormancy of the seed. (2) On the dormancy broken of the seed in the soil. Proc. Crop Sci. Soc. Japan 31:73–79.

1307. —— and —— 1962: Physiological and ecological studies on barnyard grass (*Echinocloa crus-galli* Beauv. var. *oryzicola* Ohwi). III. On the secondary dormancy of the seed. Proc. Crop Sci. Soc. Japan 31:186–89.

1308. —— and —— 1962: Physiological and ecological studies on barnyard grass (*Echinocloa crus-galli* Beauv. var. *oryzicola* Ohwi). IV. On the death of seeds in the process of dormancy awakening. Proc. Crop Sci. Soc. Japan 31:190–94.

1309. —— and —— 1963: Physiological and ecological studies on barnyard grass (*Echinocloa crus-galli* Beauv. var. *oryzicola* Ohwi). V. On the germination of the seed. Proc. Crop Sci. Soc. Japan 31:362–66.

Arai, M. 6830.

1310. Arakeri, H. R. 1956: A note on storage and germination requirements of seeds of surai (*Amorphophallus campanulatus*). Indian J. Genet. Plant Breeding 16:60–61.

1311. —— and R. S. Dunham. 1950: Environmental factors relating to the pre-emergence treatment of corn with 2,4-D and soybeans with TCA. Minn. Agr. Expt. Sta. Tech. Bull. 190. 28 pp.

1312. —— and S. S. Patil. 1956: Effect of bulb size, spacing, and time of planting on the yield of onion seed. Indian J. Agron. 1:75–79.

1313. —— and A. R. Schmid. 1949: Cold resistance of various legumes and grasses in early stages of growth. J. Am. Soc. Agron. 41:182–85.

Araki, S. 9490.

1314. Arakwa, S. 1955: Some observations on *Fraxinus* seeds. J. Japan Forestry Soc. 37:1–5.

1315. Arango, H. 1945: Germination of orchid seed in Colombia. Am. Orchid Soc. Bull. 14:212. Biol. Abstr. 20:16138.

Araujo, A. A. de; *see* De Araujo, A. A.

1316. Araratian, A. G. 1956: On the biological role of different nutrients received by plant embryo. Izv. Akad. Nauk Arm. SSR Biol. i Sel'skokhoz. Nauki 9(5):3–13.

1317. Arasimovich, V. V. 1933: Biochemical study of *Cucurbita*. Bull. Appl. Botany, Genet., Plant Breed. (USSR) Ser. 3(1):73–101. Biol. Abstr. 9:13348. 1935.

Arata, H. 14515.

1318. Arawinko, Z. M., and E. L. Nielsen. 1956: A rubbing-type thresher for small-seeded grasses and legumes. Agron. J. 48:532–33.

1319. Arceneaux, G., and L. P. Hebert. 1950: Pre-emergence control of Johnson grass seedlings with 2,4-D. Sugar Bull. 29:5, 11–12.

1320. Arceneaux, T. J. 1945: Viability of Louisiana produced Johnson grass seed; a factor in eradication. Sugar Bull. 23:123–24.

1321. Archangeli, G. 1888: Germination of the seeds of *Euryale ferox*. Nuovo Giorn. Botan. Ital. 20:467–73. J. Roy. Microscop. Soc. 1889:250–51.

1322. —— 1889: Seeds of Nymphaeceae. Nuovo Giorn. Botan. Ital. 21:125–26. J. Roy. Microscop. Soc. 1889:407.

1323. —— 1889: Seed of Victoria. Nuovo Giorn. Botan. Ital. 21:286–89.

1324. Archbold, H. K. 1945: Some factors concerned in the process of starch storage in the barley grain. Nature 156:70–73. Expt. Sta. Rec. 94:179. 1946.

1325. Archer, S. G., and C. E. Bunch. 1953: The American grass book. A manual of pasture and range practices. Univ. of Oklahoma Press, Norman, Okla. 330 pp.

1326. Archer, W. A. 1938: Seeds sleep 60 years, sprout when houses fall. Sci. News Letter 33:173.

1327. Archibald, E. E. A. 1939: The development of the ovule and seed of jointed cactus (*Opuntia aurantiaca*, Lindley). S. African J. Sci. 36:195–211.

1328. Archibald, R. G. 1927: Sulfuric acid treatment of cotton seed. Soil Sci. 23:1–3.

1329. Archibald, R. M., and P. B. Hamilton. 1943: Removal of canavanine from preparations of jack bean urease. J. Biol. Chem. 150:155–58.

1330. Archie, D. W. 1932: Alpine plants from seeds. Flower Grower 19(3):95.

1331. Arcichovslij, V. 1913: Die Wirkung der Giftstoffe verschiedener Konzentrationen auf die Samen. Ein Beitrag zum Studien der biochemischen Wirkung der höchst Konzentrierten Lösungen. Biochem. Z. 50:233–44.

Ard, J. S. 16243.

1332. Ardashev, B. I. 1933: Constants of the materials obtained in delinting with gaseous hydrogen chloride at ordinary temperatures. Trest Khlopkochistitelnoi Prom. (Cotton Ind. Trust, Collection of Papers) 1:47–83. Chem. Abstr. 28:5228. 1934.

1333. —— and B. I. Leonov. 1933: Delinting cottonseed with hydrogen chloride gas. Trest Khlopkochistitelnoi Prom. (Cotton Ind. Trust, Collection of Papers) 1:55–73. Chem. Abstr. 28:5227–28. 1934.

1334. Arekal, G. D. 1961: Contribution to the embryology of *Hoppea dichotoma* Willd. (Gentianaceae). Can. J. Botany 39:1001–6.

1335. —— 1961: Embryology of *Klugia notoniana*. Botan. Gaz. 123:144–50.

1336. Arena, M. 1960: À propos du pouvoir germinatif des graines et de la vitalité des semis d'*Abies nebrodensis* (Lojac) Mattei. Ital. Forest. e Mont. 15:247–50.

1337. Arends, G. 1932: Stauden-Anzucht aus Samen. Von welchen Stauden kann Man mit Erfolg Nachzucht betreiben? Möllers Deut. Gärtner-Ztg. 47:75–76.

1338. Arenkova, D. N. 1940: Acenaphthene-induced tetraploidy in muskmelons. Compt. Rend. Acad. Sci. URSS 27:1028–29. Chem. Abstr. 35:1452–53. 1941.

1339. Arens, K. 1951: Anatomical study of the seeds of Ucuuba (*Virola surinamensis* Warb.). Rodriguesia 13:251–56. 1950.

1340. Arens, P. 1912: The life-history of the flowers of *Manihot glaziovii* and the securing of pure seed of this tree. Mededel. Proefsta. Malang. 1912:4–16.

1341. Arevalo, J. J. 1957: Pumpkin seeds. Studia, Rev. Univ. Atlantico, Barranquilla 2:138–209. Chem. Abstr. 52:15648. 1958.

1342. Arghirescu, V., and I. Vladescu. 1936: Über die industrielle Verwertung der Tabaksamen. Bul. Cult. Ferment. Tutunului 25:34–40.

1343. Argle, R. H. 1945: Treatment of primrose seed. Horticulture 23(6):163.

Argüello, G. 5137.

1344. Arima, J., and M. Ninomiya. 1938: Poppy seeds—use of, in Manchoukuo. I. Rept. Inst. Sci. Res. Manchoukuo 2:45–46. Soc. Chem. Ind. J. 58:776. 1939. Pharm. Abstr. 6:23. 1940.

Arima, K. 10921.

1345. Arisimovich, V. V. 1951: Biochemical indexes in taxonomy of cultivated plants. Biokhim. Plodov i Ovoshchei Sbornik 2:53–66. Chem. Abstr. 46:8200. 1952.

1346. Arisumi, T. 1951: The germination of early peach seeds. Abstr. Theses, Univ. Illinois, Urbana. 4 pp.

1347. —— 1962: Effects of forcing temperatures on the growth, flowering, and seed set of *Hemerocallis* cv. Purity. Proc. Am. Soc. Hort. Sci. 81:518–20.

1348. —— and J. M. Harrison. 1961: The germination of Rock elm seeds. Am. Nurseryman 114(7):10.

1349. Ariyanayagam, D. V. 1952: Preliminary studies on the pre-sowing treatment of paddy. Trop. Agriculturist 108:11–20.

1350. —— 1953: Growth and developmental changes caused by pre-sowing treatment of seed paddy. Trop. Agriculturist 109:4–15.

1351. —— 1953: Saline paddy trials in the eastern division. Trop. Agriculturist 109: 16–18.

Ariyoshi, U. 4467.

1352. Ark, P. A., and L. D. Leach. 1946: Seed transmission of bacterial blight of sugar beet. Phytopathology 36:549–52.

1353. —— and J. P. Thompson. 1958: Antibiotic properties of the seeds of wheat and barley. Plant Disease Reptr. 42:959–62.

1354. Arkhangelsky, M. P. 1929/30: On the percentage of chaff in oat grains in dependence on their position in the spikelet. Bull. Appl. Botany, Genet., Plant Breed. (USSR) 24(2):43–55.

1355. —— 1936: On the chemical composition of the oat grain and the energetic principle in the evaluation of the variety. Bull. Appl. Botany, Genet., Plant Breed. (USSR) Ser. 3, 15:111–22.

1356. —— and V. N. Suchkina. 1931: The process of the accumulation of oil and other principal nutritious substances in the grain of flax for fiber and for seed. Bull. Appl. Botany, Genet., Plant Breed. (USSR) 25:199–219. Chem. Abstr. 26:756–57. 1932.

1357. Arkhimovich, A. 1940: Causes of biological disparity in seeds from the same beetroot glomerule. Acad. Sci. RSS Ukraine Inst. Botan. J. 31:165–66.

1358. Arkhipov, V. D. 1951: On the treatment of acorns (with Granosan). Lesn. Kohz. 4:80.

Armbruster, G. A. 13972.

1359. Armenise, V. 1957: Contributo alla conoscenza dello sviluppo embriologico di *Phillyrea latifolia* L. Nuovo Giorn. Botan. Ital. 64:297–318.

1360. Armer, A. 1948: A mechanical spray treater for sugar beet seed. Proc. Am. Soc. Sugar Beet Technol. 5:108–13.

1361. Armolik, N., and J. G. Dickson. 1956: Minimum humidity requirement for germination of conidia of fungi associated with storage of grain. Phytopathology 46:462–65.

1362. ——, ——, and A. D. Dickson. 1956: Deterioration of barley in storage by microorganisms. Phytopathology 46:457–61.

Armstrong, G. A. 5575.

1363. Armstrong, G. M., and W. B. Albert. 1930: A study of the factors which influence the growth and development of cotton buds and bolls. S. Carolina Agr. Expt. Sta. Ann. Rept. 43:40–42.

1364. Armstrong, J. M., and W. J. White. 1935: Factors influencing seed-setting in alfalfa. J. Agr. Sci. 25:161–79.

1365. Armstrong, K. S. 1940: Starting plants from seed simplified. Horticulture 18(5):91.

1366. Armstrong, W. D. 1936: New varieties and pollination of Muscadine grapes. Proc. Am. Soc. Hort. Sci. 33:450–52.

1367. Army, T. J., and E. B. Hudspeth, Jr. 1960: Alteration of the microclimate of the seed zone. Agron. J. 52:17–22.

Arnaldo, S. R. 3592.

Arnarez, M. 6405.

1368. Arnason, T. J. 1943: Female sterility in potatoes. Can. J. Res. Sect.. C 21:41–56.

1369. ——, E. Cumming, and J. W. T. Spinks. 1948: Chromosome breakage induced by absorbed radioactive phosphorus, P^{32}. Can. J. Res. Sect. C 26:109–14. Biol. Abstr. 23:2524. 1949.

1370. Arnaud, G. 1929: Essais de traitement de la carie du blé. Compt. Rend. Acad. Agr. France 15:844–51.

1371. —— and M. Gaudineau. 1930: Le traitement de la carie du blé (1929–30). Compt. Rend. Acad. Agr. France 16:1029–35. Chem. Abstr. 25:2234. 1931.

1372. —— and —— 1931: Le traitement de la carie du blé. II. Ann. Agron. 1:61–81. Chem. Abstr. 25:4349. 1931.

1373. —— and —— 1932: Le traitement de la carie du blé. III. Action comparée des produits cupriques et du formol. Ann. Agron. 2:229–46. Chem. Abstr. 27:1707. 1933.

1374. Arnaudov, N. 1929: Beitrag zur Keimungsphysiologie des Tabaksamens. Jahrb. Univ. Sofia 25:3. Ber. Wiss. Biol. 14:576. 1930.

1375. —— and A. Ghegov. 1933/34: Autour du problème de la germination au clair. Ann. Univ. Sofia, Faculte Phys. Math. Tome 30. K.N.Z. Hist. Nat. Proc. Intern. Seed Testing Assoc. 7:107–8. 1935.

1376. Arnborg, T. 1947: Gathering and sowing of birch seed, with some points on birch cultivation. Skogen 1947(8):7–8. Biol. Abstr. 22:17137. 1948.

1377. —— 1958: Forest tree seeds. Svensk Papperstid. 61:611–17.

Arnd, T. 3194.

1378. Arndt, C. H. 1931: The internal infection of cotton seed. S. Carolina Agr. Sta. 44th Ann. Rept. 1930/31:51–52.

1379. —— 1932: A study of some of the factors which may influence cotton seed germination and seedling growth. S. Carolina Agr. Expt. Sta. 45th Ann. Rept., pp. 46–49.

1380. —— 1933/34: Germination of cotton seed and growth of the seedlings at low temperatures. S. Carolina Agr. Expt. Sta. 47th Ann. Rept., pp. 33–38.

1381. —— 1935: Notes on polyembryony and multiple shoots from the seed in *Mangifera indica*. Am. J. Botany 22: 26–30.

1382. —— 1945: Viability and infection of light and heavy cotton seeds. Phytopathology 35:747–53.

1383. —— 1946: Effect of storage conditions on survival of *Colletotrichum gossypii*. Phytopathology 36:24–29.

1384. —— 1946: The internal infection of cotton seed and the loss of viability in storage. Phytopathology 36:30–37.

1385. —— 1948: An evaluation of certain substituted phenol esters for the treatment of cotton seed. Phytopathology 38:978–87.

1386. —— 1953: Survival of *Colletotrichum gossypii* on cotton seeds in storage. Phytopathology 43:220.

1387. —— 1956: Cotton seed produced in South Carolina in 1954 and 1955, its viability and infestation by fungi. Plant Disease Reptr. 40:1001–4.

Arndt, M. 2798, 2799.

Arney, S. E. 9696.

1388. Arnold, C. A. 1935: On seedlike structures associated with *Archaeopteris*, from the Upper Devonian of northern Pennsylvania. Contrib. Museum Paleontol. Univ. Mich. 4:283–86. Biol. Abstr. 10:22080. 1936.

1389. —— 1938: Paleozoic seeds. Botan. Rev. 4:204–34.

1390. —— 1939: *Lagenospermum impariameum* sp. nov., a seedbearing fructification from the Mississippian of Pennsylvania and Virginia. Bull. Torrey Botan. Club 66: 297–303.

1391. —— 1948: Paleozoic seeds. II. Botan. Rev. 14:450–72.

1392. —— 1948: Some cutinized seed membranes from the coalbearing rocks of Michigan. Bull. Torrey Botan. Club 75:131–46.

1393. Arnold, C. G. 1958: Antibiotische Stoffe aus Pflanzen-Samen. Z. Botan. 46:516–49.

1394. Arnold, F. R. 1944: Plant your Ginkgo seeds before snow comes. Gardeners Chron. 48:284–85.

1395. Arnold, H. A. 1945: Seed scarifiers. Tenn. Agr. Expt. Sta. Bull. 194.

1396. Arnold, L. E. 1950: Seed improvement—looking ahead. Intern. Crop Improvement Assoc. Ann. Rept. 32:72–79.

1397. Arnold, L. K., and R. B. R. Choudhury. 1962: Ethanol extraction of soybean oil. J. Am. Oil Chemists Soc. 39:379–80.

1398. —— and W. G. Juhl. 1954: Solvent extraction of cottonseed meats. J. Am. Oil Chemists Soc. 31:613–18.

1399. —— and B. J. Schriver. 1958: Comparative solubility of protein in cottonseed flakes extracted by hexane and by ethanol. Proc. Iowa Acad. Sci. 65:230–33.

1400. Arnold, R. E. 1959: The effect of harvest damage on the germination of barley. J. Agr. Eng. Res. 4:24–29.

1401. —— 1963: Effects of harvest damage on the rate of fall in viability of wheat stored at a range of moisture levels. J. Agr. Eng. Res. 8:7–16.

1402. Arnold, W. 1912: The fatty oil of the seeds of *Balanites aegyptica.* Z. Untersuch. Nahrungs.- u. Genussmittel 23:391.

1403. Arnold, Z. 1927: Die Entwicklung und die Ausgabe des Aleurons bei einigen Getreidearten. Acta Botan. Inst. Botan. Univ. Zagreb. 2:57–77. Botan. Centralbl. 10:258. 1927.

Arnoldi, V. A. Poddubnaja; *see* Poddubnaja-Arnoldi, V. A.

Arnon, D. I. 7985.

1404. Arnott, D. A. 1956: Some factors reducing carrot seed yields in British Columbia. Proc. Entomol. Soc. Brit. Columbia 52:27–30. Biol. Abstr. 30: 32517. 1956.

1405. Arnott, H. J. 1962: The seed, germination, and seedling of Yucca. Univ. Calif. (Berkeley) Publ. Botany 31(1):1–164.

Arnott, H. J. 3720.

1406. Arnoux, M. 1962: Traitements des semences de plantes maraichères. Phytiat.-Phytopharm. 11:41–53.

Arnulfo, M. 4125.

1407. Arny, A. C., and R. J. Garber. 1918: Variation and correlation in wheat, with special reference to weight of seed planted. J. Agr. Res. 14:359–92.

Arny, A. C. 13969.

1408. Arny, D. C., and C. Leben. 1954: Vapor action of certain mercury seed treatment materials. Phytopathology 44: 380–83.

1409. —— and —— 1955: The effect of the water-soak treatment on the germination of certain barley varieties grown at different locations. Phytopathology 45:518–19.

1410. —— and —— 1956: Control of several small grain diseases by the water-soak seed treatment. Phytopathology 46:344–45.

Arny, D. C. 11948–11952.

1411. —— and E. K. Wade. 1952: Seed treatment for oats, barley, wheat, rye, and corn. Wisconsin Agr. Coll. Ext. Circ. 416. 10 pp.

1412. Arora, B. K. 1957: Effects of α-naphthaleneacetic acid on partially vernalized seeds of *Linum usitatissimum.* Vikram, J. Vikram Univ. 1 :93–94. Chem. Abstr. 54:12281. 1960.

Arora, U. 16599.

1413. Arroyave, G., J. Mendez, and R. Bressani. 1955: Contenido de diversos nutrientes en alimentos procedentes de centro-America. 2. Cereals, leguminosas, y otras semillas. Arch. Venezolanos Nutr. 6:101–9. Nutr. Abstr. Rev. 27:3111. 1957.

Arroyave, G. 9770.

Arruda, H. V. de; *see* De Arruda, H. V.

Arruda, L. F. de. 13736.

1414. Arscott, G. H., and J. A. Harper. 1963: Relationship of 2,5-Diamino-4,6-diketopyrimidine, 2,4-Diaminobutyric acid and a crude preparation of β-cyano-L-alanine to the toxicity of common and hairy vetch seed fed to chicks. J. Nutr. 80:251–54.

1415. Arthur, B. W. 1956: Insects in stored peanuts and their seasonal abundance. J. Econ. Entomol. 49:119–20.

1416. Arthur, J. C. 1882: Prolonged vitality of seeds. Botan. Gaz. 7:88.

1417. —— 1887: Germination of weed seeds. N.Y. Agr. Expt. Sta. Rept. 1887: 362–63.

1418. —— 1893: Deviation in development due to the use of unripe seeds. Botan. Gaz. 18:341–42.

1419. —— 1895: Delayed germination of cocklebur and other paired seeds. Proc. Soc. Promotion Agr. Sci. 16:70–79.

1420. Arthur, J. M., J. D. Guthrie, and J. M. Newell. 1930: Some effects of artificial climates on the growth and chemical composition of plants. Am. J. Botany 17:416–82. Contrib. Boyce Thompson Inst. 2:445–511. 1930.

1421. Arthur, P. 1957: Growing gesneriads from seed. Horticulture 35(2):83.

1422. Artschwager, E. 1927: Development of flowers and seed in the sugar beet. J. Agr. Res. 34:1–25.

1423. ——, E. W. Brandes, and R. C. Starrett. 1929: Development of flower and seed of some varieties of sugar cane. J. Agr. Res. 39:1–30.

1424. —— and R. C. Starrett. 1933: The time factor in fertilization and embryo development in the sugar beet. J. Agr. Res. 47:823–43.

Artsikhovskaya, E. V. 16498.

1425. Artsixovsky, V., and I. Stom. 1917: Disinfection of seeds with bromin. Estac. Ensayo Semillas (Madrid) 1917:51–64. Expt. Sta. Rec. 40:443. 1919.

1426. Arvy, L., and J. Lhoste. 1946: Sur l'action des solutions aqueuses de 2–4 dichlorophenoxyacetate de sodium sur la germination du blé. Compt. Rend. Soc. Biol. 140:74.

1427. Arya, H. C., and R. Prasada. 1953: *Alternaria* blight of linseed. Indian Phytopathologist 5:33–39. 1952.

1428. Arzeni, C. B. 1963: *Octoblepharum* as a seeding medium. Econ. Botany 17:11–15.

1429. Arzt, T. 1933: Über die Embryobildung von Pseudomonokotylen (*Podophyllum emodi* und *Eranthis hiemalis*). Botan. Centralbl. Beih. Abt. I 50:671–96.

1430. Asada, K., and Z. Kasai. 1962: Formation of *myo*-inositol and phytin in ripening rice grains. Plant Cell Physiol. (Tokyo) 3:397–406.

Asahira, T. 18800.

Asakawa, M. 14142.

1431. Asakawa, S. 1954: Preliminary studies on the growth inhibitors in *Fraxinus* fruits. J. Japan. Forestry Soc. 36:153–59.

1432. —— 1955: Some observations on *Fraxinus* seeds. J. Japan. Forestry Soc. 37:1–5.

1433. —— 1955: Hastening the germination of *Pinus koraiensis* seeds. Preliminary study on chemical changes during pretreatments. J. Japan. Forestry Soc. 37: 127–32.

1434. —— 1955: Preliminary paper on hastening the germination of *Pinus pentaphylla* seeds. J. Japan. Forestry Soc. 37:508–9.

1435. —— 1956: Further investigation on hastening the germination of *Pinus koraiensis* seeds. J. Japan. Forestry Soc. 38:1–4.

1436. —— 1956: On the water absorption by *Pinus koraiensis* seeds. J. Japan. Forestry Soc. 38:125–29.

1437. —— 1956: Studies on the delayed germination of *Fraxinus mandshurica* var. *japonica* seeds. (1) Historical and preliminary observations. Bull. Govt. Forestry Expt. Sta. (Japan) (English Transl.) 83:1–18.

1438. —— 1956: Studies on the delayed germination of *Fraxinus mandshurica* var. *japonica* seeds. (2) Pre-germination of *F. mandshurica* var. *japonica* seeds. Physiological properties of embryos of *Fraxinus* seeds. Bull. Govt. Forestry Expt. Sta. (Japan) (English Transl.) 83:19–28.

1439. —— 1956: Studies on the delayed germination of *Fraxinus mandshurica* var. *japonica* seeds. (3) On the growth inhibitors in *Fraxinus* fruits. Bull. Govt. Forestry Expt. Sta. (Japan) (English Transl.) 83:29–38.

1440. —— 1956: Studies on the delayed germination of *Fraxinus mandshurica* var. *japonica* seeds. (4) On water absorption. Bull. Govt. Forestry Expt. Sta. (Japan) (English Transl.) 87:75–84.

1441. —— 1956: Thermoperiodic control of germination of *Fraxinus mandshurica* var. *japonica* seeds. J. Japan. Forestry Soc. 38:269–72.

1442. —— 1957: Studies on the delayed germination of *Fraxinus mandshurica* var. *japanica* seeds. (6) The experiments with special reference to the effect of burying-storage on germination. Bull. Govt. Forestry Expt. Sta. (Japan) (English Transl.) 100:23–29.

1443. —— 1957: Studies on hastening the germination of the seeds of five-leaved pines. Bull. Govt. Forestry Expt. Sta. (Japan) (English Transl.) 100:41–54.

1444. —— 1959: Physiology of embryo dormancy. J. Japan. Forestry Soc. 41: 389–95.

1445. —— 1959: Germination behavior of several coniferous seeds. J. Japan. Forestry Soc. 41:430–35.

1446. —— 1960: Outline of the international rules related to testing of tree seeds. J. Japan. Forestry Soc. 42:417–22.

1447. —— 1961: A preliminary study concerning the relationship between an embryo and endosperm during stratification. J. Japan. Forestry Soc. 43:195–97.

1448. —— 1961: Light-sensitivity in the germination of *Fraxinus mandshurica* var. *japonica* seeds. J. Japan. Forestry Soc. 43:310–11.

1449. —— 1963: Germination behavior of *Alnus inokumai* seeds. J. Japan. Forestry Soc. 45:331–34.

1450. —— and T. Inokuma. 1961: Light-sensitivity in the germination of *Pinus thunbergii* and *Picea glehnii* seeds. J. Japan. Forestry Soc. 43:331–35.

Asakawa, S. 9414, 14070.

1451. Asami, Y. 1954: The influence of ammonium sulphate and sodium nitrate on the germination of flax seed in the case of pot culture. Sapporo Res. Inst. Flax & Flax Fiber Bull. 5:12–16.

1452. Asana, R. D., V. S. Mani, and Vedprakash. 1955: The effect of auxins on the growth and yield of wheat. Physiol. Plantarum. 8:279–87.

1453. —— and A. D. Saini. 1962: Studies in physiological analysis of yield. V. Grain development in wheat in relation to temperature, soil moisture, and changes with age in the sugar content of the stem and in the photosynthetic surface. Indian J. Plant Physiol. 5:128–72.

Asano, H. 19400.

1454. Asarov, K. K. 1961: Effect of scarification on germinating power of seeds of sundial lupine. Izv. Timiryazev. Sel'skokhoz. Akad. 39:224–27.

1455. Ascherson, P. 1875: Keimpflanze von *Pirus malus* im Inneren eines kleinen Borsdorfer Apfels. Verhandl. Brandenburg Botan. Vereins 17:79–80.

1456. —— 1889: Dissemination of the seeds of *Harpagophyton*. Verhandl. Brandenburg Botan. Vereins 30:2–5. J. Roy. Microscop. Soc. 1891:69.

1457. Aseeva, I. V., and A. N. Belozerskii. 1957: The effect of light and starvation on the biosynthesis of nucleic acids in wheat seedlings. Vest. Mosk. Univ. Ser. Biol. Pochvoved. Geol. i Geogr. 4:17–23. Biol. Abstr. 45:12361. 1964.

Aseeva, I. V. 2258.

1458. Asen, S. 1948: Embryo culture of rose seeds. Am. Rose Ann. 151–52.

1459. —— and R. E. Larson. 1951: Artificial culturing of rose embryos. Penn. Agr. Expt. Sta. Progr. Rept. 40. 4 pp.

1460. Asenjo, C. F., D. H. Cook, M. De Fernandez, and L. A. Alvarez. 1943: Chemical changes of the papaya plant during development, with special reference to its proteolytic activity. J. Agr. Univ. Puerto Rico 27:1–15. Expt. Sta. Rec. 91:373–74. 1944.

1461. —— and M. de Fernandez. 1942: A new protease from *Bromella* [sic.] *pinguin* L. Science 95:48–49.

1462. —— and J. A. Goyco. 1943: Characteristics and composition of Guanabana seed oil. J. Am. Chem. Soc. 65:208–9.

1463. —— and —— 1952: Puerto Rican fatty oils. VIII. Characteristics and composition of the oils from the fatty covering and the kernel of the anacaguitas seed (*Sterculia apetala*, Jacq. Karst). J. Agr. Univ. Puerto Rico 36:1–4. Brit. Abstr. B 3:165. 1953.

1464. Ash, C. G., and O. N. Allen. 1949: A comparison of methods recommended for the surface sterilization of leguminous seed. Soil Sci. Soc. Am. Proc. 13:279–83.

1465. Ashby, E. 1937: Studies in the inheritance of physiological characters. III. Hybrid vigor in the tomato. Part I. Manifestations of hybrid vigor from germination to the onset of flowering. Ann. Botany (London) 1:11–41.

1466. —— 1939: Correlation between seed weight and "adult" weight in tomatoes· Nature 144:712.

1467. —— 1946: Plant physiology in the U.S.S.R. Nature 157:558–61.

1468. Ashby, W. C. 1962: Germination capacity in American basswood. Trans. Illinois State Acad. Sci. 55:120–23.

1469. —— and H. Hellmers. 1955: Temperature requirements for germination in relation to wild-land seeding. J. Range Management 8:80–83.

Ashdown, D. F. 1278.

1470. Ashman, R.B. 1960: Stippled aleurone in maize. Genetics 45:19–34.

1471. Ashton, F. M., and H. Beevers. 1963: Proteolytic enzymes from germinating seeds of *Cucurbita maxima*. Proc. Plant Physiol. Meetings 38:xxxi.

Ashton, G. C. 4515.

Ashton, W. M. 6178.

1472. Ashworth, L. J., B. C. Langley, and W. H. Thames. 1962: A seed-borne disease of Spanish peanuts. Texas Agr. Progr. 8(5):19–20.

1473. Asikari, H. 1940: Estrone-like substance present in rice embryo. Arb. Med. Fak. Okayama 6:448–56. Chem. Abstr. 36:5496. 1942.

1474. Aslam, M. 1958: Examination of safflower oil from indigenous sources with references to its use in the manufacture of paints and varnishes. Pakistan J. Sci. Ind. Res. 1:312–15.

1475. Aso, K., and T. Watanabe. 1962: Nioigome. I. Composition of nioigome (perfumed rice). Tohoku J. Agr. Res. 13:381–87. Chem. Abstr. 59:3086. 1963.

1476. Aso, T. 1960: Studies on the germination of agricultural seeds. I. Effects of sulphuric acid upon the germination of hard seeds of Renge (*Astragalus sinicus* L.). Seiken Ziho 11:55–62.

1477. —— 1962: Studies on the germination of agricultural seeds. II. Seed extraction of tomato and cucumber with sulphuric acid treatment. Seiken Ziho 13:100–4.

Aspinall, D. 14984.

Assche, C. van. 18707.

Assemien, P. 13426.

Associated Seed Growers, Inc. 1006, 1168.

1478. Astbury, W. T., S. Dickinson, and K. Bailey. 1935: The X-ray interpretation of denaturation and the structure of the seed globulins. Biochem. J. 29:2351–60. Physiol. Abstr. 21:297. 1936.

Atabekian, E. A. 1643.

1479. Atabekova, A. I. 1936: Effect of X-rays

on pea seeds and seedlings. Biol. Zh. 5:99–116.

1480. Atabekowa, A. J. 1936: Die Wirkung der Röntgenbestrahlung ruhender und keimender Samen. Protoplasma 25:234–60. Biol. Abstr. 11:18239. 1937.

1481. Atal, C. K. 1960: Microscopic differentiation of cruciferous and non-cruciferous seeds containing mucilage. Current Sci. (India) 29:56.

Atal, C. K. 13289, 17241.

Atanas, L. G. 13529.

1482. Atanasoff, D., and A. G. Johnson. 1920: Treatment of cereal seeds by dry heat. J. Agr. Res. 18:379–90.

1483. Atanasova, E. 1962: Changes in the free and combined amino acids during germination of wheat. Selskostopanska Nauka 1:43–48.

1484. Atay, I. 1959: Research on the seeds of black pine (*Pinus nigra* var. *pallasiana*). Istanbul Univ. Orman Fak. Dergisi Ser. A 9:48–96.

1485. Atherton, D., and M. L. Meara. 1939: Component fatty acids and glycerides of some *Myristica* fats. Soc. Chem. Ind. J. 58:353–57. Analyst 65:517. 1940.

1486. —— and —— 1940: The fatty acids and glycerides of solid seed fats. IX. *Mimusops heckelii* (Baku) kernel fat. Soc. Chem. Ind. J. 59:95–96.

1487. Athow, K. L., and R. M. Caldwell. 1948/49: Influence of seed treatment on stand and yield of soybeans. Indiana Agr. Expt. Sta. Ann. Rept. 62:46–47.

1488. —— and —— 1956: The influence of seed treatment and planting rate on the emergence and yield of soybeans. Phytopathology 46:91–95.

1489. —— and F. A. Laviolette. 1961: The relation of seed-transmitted tobacco ringspot virus to soybean yield. Phytopathology 51:341–42.

1490. Atkeson, F. W., H. W. Hulbert, and T. R. Warren. 1934: Effect of bovine digestion and manure storage on the viability of weed seeds. J. Am. Soc. Agron. 26:390–97. Expt. Sta. Rec. 71: 772–73. 1934.

1491. Atkin, J. D. 1957: Bean seed injury and germination. Farm Res. (N.Y.) 23(2): 10–11.

1492. —— 1958: Relative susceptibility of snap bean varieties to mechanical injury of seed. Proc. Am. Soc. Hort. Sci. 72:370–73.

Atkin, J. D. 19402.

Atkins, G. A. 4997.

1493. Atkins, I. M., and M. J. Norris. 1955: The influence of awns on yield and certain morphological characters of wheat. Agron. J. 47:218–20.

Atkins, M. D. 4390, 12416.

1494. Atkins, R. E., G. R. Pesho, and F. F. Dicke. 1963: Yield and seed size reduction in grain sorghum infested by the European corn borer (*Ostrinia nubilalis* hbn.). Iowa State J. Sci. 37:447–52.

Atkins, R. E. 3613, 6382, 7698, 7699, 17867.

1495. Atkins, W. R. G. 1909: The absorption of water by seeds. Sci. Proc. Roy. Dublin Soc. N.S. 12:35–46. Expt. Sta. Rec. 21:725–26. 1909.

1496. —— 1921: Note on the deterioration of indigo seed during storing. Agr. Res. Inst. Pusa Indigo Publ. No. 10.

Atkinson, R. W. 6845.

1497. Atri, S. P. 1951: Treatment of jowar seed (*Andropogon sorghum*) in Delhi State during 1949. Plant Protect. Bull. (India) 3:20. Chem. Abstr. 48:316. 1954.

1498. Atsatt, P. R., and L. C. Bliss. 1963: Some effects of emulsified hexaoctadecanol on germination, establishment, and growth of Kentucky bluegrass. Agron. J. 55: 533–37.

1499. Atsmon, D. 1958: Germination inhibition in castor beans. Bull. Res. Council Israel Sect. D 6:260.

1500. Atsumi, K. 1936: Sur la nenorn ala germado de malnovaj de *Brassica chinensis* L. Hort. Assoc. Japan J. 7:72–85.

Atta, G. R. van; *see* Van Atta, G. R.

1501. Atterberg, A. 1899: The actual germinating capacity of poorly mature grain and its usefulness as to planting. Tidsskr. Landtman 20:477–82.

1502. —— 1899: Concerning after-ripening of grain. Kgl. Landtbruks-Akad. Handl. Tidskr. 38:227–50.

1503. —— 1906: The influence of light and alternating temperatures on the germination of seeds of cultivated plants, especially pine. Kgl. Landtbruks-Akad. Handl. Tidskr. 45:299–315.

1504. —— 1907: The after-ripening of grains. Landwirtsch. Versuchssta. 67:129–43.

1505. Attia, M. S., and M. H. Bahr. 1958: Study on the effect of bulb size and some storage treatments on the seed yield of onions. Agr. Res. Rev. (Cairo) 36:506–14. Biol. Abstr. 35:64551. 1960.

1506. Atwater, B. R. 1939: Notes on flower seed germination. Proc. Assoc. Offic. Seed Analysts 31:111–13.

1507. —— 1940: Seed-germinating tests. Florists Rev. 87(2246):82–83.
1508. —— 1947: The importance of standardized methods in the marketing of sugar beet seed. Proc. Am. Soc. Sugar Beet Technol. 1946:281–83.
Atwater, B. R. 5566.
1509. Atwood, W. M. 1914: A physiological study of the germination of *Avena fatua*. Botan. Gaz. 57:386–414.
1510. —— 1922: Physiological studies of effects of formaldehyde on wheat. Botan. Gaz. 74:233–63.
1511. Aubertin, M. P., and D. F. Dayton. 1961: Some observations on early embryogeny in two "giant" sports of yellow transparent apple. Proc. Am. Soc. Hort. Sci. 77:54–59.
Auchter, E. C. 18753.
1512. Auckland, A. K. 1961: The influence of seed quality on the early growth of cashew. Trop. Agr. (London) 38:57–67.
1513. Audus, L. J. 1953: Plant growth substances. Leonard Hill, Ltd., London. 465 pp.
1514. —— 1956: An efficient aerating washer for seeds or tissue discs. Ann. Botany (London) 20:191–92.
1515. —— and J. H. Quastel. 1947: Toxic effects of amino acids and amines on seedling growth. Nature 160:222–23.
1516. Auer, S. 1962: Investigations on the epidermal development of the cotyledon in *Pulsatilla vulgaris*. Z. Botan. 50:128–53.
Auerman, L. Y. 11375, 11389.
1517. Aufhammer, G., and U. Simon. 1957: Die Samen landwirtschaftlicher Kulturpflanzen im Grundstein des ehemaligen Nürnberger Stadttheaters und ihre Keimfähigkeit. Z. Acker- Pflanzenbau 103: 454–72.
1518. —— and H. Weinmann. 1932: Catalase contents of different varieties of barley. Wochschr. Brau 49:57–59, 68–70. Chem. Abstr. 26:5322. 1932.
Auge, G. 15527.
1519. Augsten, H. 1956: Wachstumsversuche mit isolierten Weizen-Embryonen. Planta 48:24–46.
1520. —— 1958: Die Physiologie der Vernalisation. Urania 21:11–15.
1521. Augustin, B. 1926: Das Volumgewicht der Samen und Früchte ungarischer Heilpflanzen. Heil- Gewürz Pflanzen 8:174–76.
1522. —— 1935: Beitrage zur Keimung der Diachenien von "*Coriandrum sativum*" und "*Pimpinella anisum*." Z. Botan. 32:208. Bull. Soc. Botan. France 83:95. 1936.
1523. Augustin-Bachmanowa, S. 1961: Preliminary investigations into the chemical content, yield, and growth of two varieties of chicory obtained from seeds acted upon with ultrasonic field. Plant Breed., Acclimatization, Seed Prod. 5:275–83.
1524. Augustowska, D. 1956: Fat content in *Ricinus* seeds, degree of its acidity, and sprouting force of seeds. Acta Polon. Pharm. 13:155–58. Chem. Abstr. 50: 14054. 1956.
Augustyniak, J. 19665.
1525. Aurand, L. W., R. C. Miller, and L. L. Huber. 1950: Influence of heredity on carotene and protein contents of corn. Penn. Agr. Expt. Sta. Bull. 526. 17 pp.
Aurand, L. W. 13500.
Ausherman, L. E. 3064.
1526. Austin, A., D. Singh, and K. V. Jhamb. 1962: Protein and gluten contents of some improved Indian wheats as influenced by varietal and seasonal differences. Current Sci. (India) 31:391–92.
1527. Austin, R. B. 1963: Yield of onions from seed as affected by place and method of seed production. J. Hort. Sci. 38:277–85.
1528. Austin, T. 1959: Seed setting of forest trees 1958. Tidsskr. Skogbruk 67:48–58.
1529. —— 1960: Seed supply in Norwegian forestry. Proc. Intern. Seed Testing Assoc. 25:627–35.
1530. Austwick, P. K. C. 1963: Groundnut microflora and toxicity. Chem. Ind. (London) 1963:55–61.
1531. Auterhoff, H. 1955: Sabadilla-Nebenalkaloide. 8. Mitt. Veratrin-Veratrumalkaloide. Arch. Pharm. 288:549–60. Biol. Abstr. 31:29465. 1957.
Auwermellen, E. van der. 12961.
Avadhani, P. N. 15938.
1532. Avakian, A. A. 1936: Iarovization of rice. Iaroviz. Zh. Biol. Razv. Rastenie 1936(1):47–52.
1533. —— and M. G. Iastreb. 1949: On the irreversibility of phasic processes. Agrobiologiya 1949(6):41–50.
1534. Avasthi, B. K., and J. D. Tewari. 1955: Chemical investigation of seeds from *Desmodium gangeticum*. Arch. Pharm. 288:272–75. Chem. Abstr. 50:439. 1956.
1535. Avery, A. G., and A. F. Blakeslee. 1936: Visible mutations from aged seeds. [Abstract of papers.] Am. Naturalist 70(726):36–37.
1536. —— and —— 1943: Mutation rate in

Datura seed which had been buried 39 years. [Abstract.] Genetics 28:69–70.

Avery, A. G. 16792.

1537. Avery, G. A. 1963: Lucerne seed production in Marlborough. New Zealand J. Agr. 106:381–83.

1538. Avery, G. S., Jr. 1933: Structure and germination of tobacco seed and the developmental anatomy of the seedling plant. Am. J. Botany 20:309–27.

1539. —— 1937: The growth hormones found in plants. Ohio J. Sci. 37:317–32.

1540. ——, J. Berger, and B. Shalucha. 1942: Total auxin extraction from wheat. Am. J. Botany 29:612–16.

1541. ——, ——, and —— 1942: Auxin content of maize kernels during ontogeny, from plants of varying heterotic vigor. Am. J. Botany 29:765–72.

1542. ——, ——, and —— 1942: Auxin storage as related to endosperm type in maize. Botan. Gaz. 103:806–8.

1543. ——, H. B. Creighton, and B. Shalucha 1940: Extraction methods in relation to hormone content of maize endosperm. Am. J. Botany 27:289–300.

Avery, G. S., Jr. 2310, 2311, 11831.

1544. Avery, H. C. 1938: Seed-germinating and plant-growing medium. Brit. Patent 490,298. Chem. Abstr. 33:1084–85. 1939.

1545. Avetisian, A. A., and V. M. Gulanian. 1954: The influence of α-naphthylacetic acid treatment of seed on the growth and yield of *Vicia ervilia*. Izv. Akad. Nauk Arm. SSR Biol. i Sel'skokhoz. Nauki 7(11):69–74. Chem. Abstr. 49:11785–86. 1955.

1546. —— and —— 1957: Study of the effect of certain preparations in preplanting seed treatment on the growth and development of bitter vetch (*Vicia ervillia*). Izv. Akad. Nauk Arm. SSR Biol. i Sel'skokhoz. Nauki 10(6):97–100.

Avetisian, A. A. 7008.

1547. Avézard, J. 1939: Structure et développement du tegument seminal de quelques solanées. Rev. Gen. Botan. 51(602): 65–88.

Avramescu, A. 1748.

1548. Avramov, L., and G. Jelenkovic. 1961: The effect of parents on the germination of hybrid seeds from grape vines. Arhiv Poljoprivredne Nauke 14(44):60–73.

1549. Avtonomov, A. 1959: On the problem of transmission of Fusarium wilt by seeds. Khlopkovodstvo 1959(12):32–34.

1550. Awakjan, A. A. 1936: Bestimmung von Zwiebelarten nach morphologisch-anatomischen Merkmalen der Samen. Bull. Appl. Botany, Genet., Plant Breed. (USSR) Ser 4(1):85.

1551. Awano, J., and S. Iizumi. 1956: On the analysis of the seed-population present in the arable soil. Japan. J. Ecol. 5:140–44. Biol. Abstr. 31:33680. 1957.

1552. Awe, W. 1937: Seed oil of common field poppy. Naturwissenschaften 25:366. Soc. Chem. Ind. J. 56:806 B. 1937. Pharm. Abstr. 4:77. 1938.

Axelrod, B. 1819, 6104, 16180.

1553. Axentiev, B. N. 1927: Influence of light on the seed-germination of weeds. Ukraine Agr. Sci. Comm. J. Agr. Botany 1:104–25.

1554. —— 1928: Über die Wirkung von Samenauszügen auf das Keimen von Samen. J. Soc. Botan. Russ. Acad. Sci. U.S.S.R. 12:291–306.

1555. —— 1929: Über den Einfluss einiger Salze auf die Keimung der Samen von *Amaranthus retroflexus* L. Biochem. Z. 211:454–67.

1556. —— 1929: Über die Rolle der Schalen von Samen und Fruechten, die bei der Keimung auf Licht reagieren. Botan. Centralbl. Beih. 46:119–202.

1557. —— 1930: Über die Entwicklung der Keimlinge aus mit Nitratloesungen behandelten Samen. Biochem. Z. 223:387–93.

1558. —— 1930: Über die Wirkung von Samenextrakt auf das Keimen der Samen. [Abstract.] Proc. Intern. Seed Testing Assoc. 2:149.

1559. —— 1931: Über die Entwickelung der Keimlinge aus vorher in Nitratloesungen eingeweichten Samen. J. Soc. Botan. Russ. 16:257–66. Ber. Wiss. Biol. 21:648–49. 1932.

1560. —— 1931: Zur Frage der Bestimmung der Keimungsgeschwindigkeit von Samen und Früchten. J. Soc. Botan. Russ. 16: 267–78.

1561. —— 1932: Über die Entwicklung der Rapskeimlinge, die aus mit Kalisalpeter behandelten Samen wachsen. Botan. Zh. 17:125–30.

1562. —— 1933: Über die stimulierende Wirkung von Salpeter auf Weizensamen und über die Ursachen derselben. J. Botan. URSS 18:107–25.

1563. —— 1934: Über die Veränderung der Eigenschaften der Samen unter dem Einfluss chemischer Behandlung. J. Botan. URSS 19:419–33.

1564. —— 1936: Über die Wachstumseigentümlichkeiten der Keime von regeinigten

Weizensorten. J. Botan. URSS 21:175.
Axtman, G. 2746.
Axtmayer, J. 4354.

1565. Ay, H. 1930: Presence and effects of phosphatese and phosphatase in germinating oats. Z. Ges. Brauw. 53:125–31. Chem. Zentr. 2:2141–42. 1930. Chem. Abstr. 25:5688. 1931.

1566. Aycock, R. 1951: Watermelon seed treatment. S. Carolina Agr. Expt. Sta. 63rd Ann. Rept. 1951:147–48.

1567. —— 1958: Effect of seed treatments on development of anthracnose in watermelon seedlings. Plant Disease Reptr. 42:134–41.

1568. Ayers, A. D. 1952: Seed germination as affected by soil moisture and salinity. Agron. J. 44:82–84.

1569. ——1953: Germination and emergence of several varieties of barley in salinized soil cultures. Agron. J. 45:68–71.

1570. —— and H. E. Hayward. 1948: Study of salt effect on plant growth. Chemurgic Dig. 7(11):12.

1571. —— and —— 1949: A method for measuring the effects of soil salinity on seed germination with observations on several crop plants. Soil Sci. Soc. Am. Proc. 13:224–26.

1572. Ayers, C., and A. W. Young. 1953: The effects of certain plant growth stimulants used as seed treatments on the yield of cotton and a grain sorghum. Texas J. Sci. 5:424–31.

1573. Ayers, J. 1960: Embryo culture of *Phalaenopsis* seedlings. Am. Orchid Soc. Bull. 29:518–19.

1574. Ayerst, G., and D. Budd. 1960: Effect of moisture content on the storage of Brazil nuts. J. Sci. Food Agr. 11:390–96.
Ayerst, G. 15437.

1575. Ayfer, M., and E. F. Serr. 1961: Effects of gibberellin and other factors on seed germination and early growth in *Pistacia* species. Proc. Am. Soc. Hort. Sci. 77:308–15.

1576. Aylward, F., and A. J. Showler. 1962: Plant lipids. IV. The glycerides and phosphatides in cereal grains. J. Sci. Food Agr. 13:492–96.
Ayres, A. S. 6369.

1577. Ayres, J. C. 1929: Relation of plumpness and viability to color of Grimm alfalfa seed. Seed World 25(2):17.

1578. Ayyangar, G. N. R., and M. A. S. Ayyar. 1930: Double seeds in sorghum. Madras Dept. Agr. Yearbook 1929:1–6.

1579. —— and —— 1935: Germination test on sorghum seeds preserved in earhead. Madras Agr. J. 23:179–80. Plant Breed. Abstr. 6:14. 1935.

1580. —— and N. Krishnaswami. 1930: Polyembryony in *Eleusine coracana*, Gaertn. (Ragi). Madras Agr. J. 18:593–95.

1581. —— and V. Narayanan. 1938: The occurrence of hollow grains and hollow peduncles in sorghum. Current Sci. (India) 7:185.

1582. —— and V. P. Rao. 1935: Vivipary in sorghum. Current Sci. (India) 3:617–19. Plant Breed. Abstr. 6:13. 1935.

1583. —— and —— 1936: Multiple seededness in sorghum and consequent repercussions. Madras Agr. J. 24:15–18. Expt. Sta. Rec. 76:183. 1937.

1584. ——, ——, and T. V. Reddy. 1936: Cracked grains in sorghum. Current Sci. (India) 4:874. Expt. Sta. Rec. 76:183. 1937.

1585. ——, U. A. Wariar, and D. S. Rajabhooshanam. 1938: The response of Ragi-*Eleusine coracana* (Gaertn.), the finger millet, to sowings in the different seasons of the year. Madras Agr. J. 26:279–84. Biol. Abstr. 16:5184. 1942.
Ayyar, M. A. S. 1578, 1579.

1586. Ayyar, P. R. 1928: Lignoceric acid from the seeds of *Adenanthera pavonina* L. Proc. Indian Sci. Congr. 15th:161. Chem. Abstr. 25:2971. 1931.

1587. —— 1928: Thevetin, a crystalline glucoside from the seeds of *Thevetia neriifolia* L. Proc. Indian Sci. Congr. 15th: 161. Chem. Abstr. 25:3007. 1931.

1588. —— and S. M. Patel. 1930: Oil from the seeds of *Schleichera trijuga*, Willd. Proc. Indian Sci. Congr. 17th:186.

1589. —— and V. A. Patwardhan. 1930: Oil from the seeds of *Ocimum basilicum* (Linn.). Proc. Indian Sci. Congr. 17th: 185–86.

1590. —— and —— 1935: The fixed oil from the seeds of *Amoora rohituka* (W. & A.). J. Indian Inst. Sci. 18(5):19–24.
Ayyar, P. R. 15048.
Azevedo, L. H. P. 13697.

1591. Azin, L. A., A. M. Basov, F. I. Izakov, and V. N. Shmigel. 1961: Presowing treatment of seeds in an electrical field. Vestn. Sel'skokhoz. Nauki 1961(4):109–15. Biol. Abstr. 40:3423. 1962.
Aziz, M. A. 6228.

B

1592. B., A. F. 1945: A conference on plant embryo culture. Chronica Botan. 9:193–94.

1593. B., G. T. 1944: Isano oil. Gt. Brit. Imp. Inst. Bull. 42:250–53.

1594. Ba, L. T. 1960: Embryogenie des Rubiacées. Développement du proembryon chez le *Galium aparine* L. Compt. Rend. 251:1812–14.

1595. —— 1961: Embryogenie des Rubiacées Développement de l'embryon chez le *Spermacoce tenuior* Gaertn. Compt. Rend. 252:927–29.

1596. —— 1961: Embryogenie des Renonculacées. Développment de l'embryon chez l'*Isopyrum fumarioides* L. Compt. Rend. 253:1603–5.

1597. —— 1961: Embryogenie des Renonculacées. Développement du proembryon chez le *Trollius europaeus* L. Compt. Rend. 253:2753–55.

1598. —— 1962: Embryogenie de Calycanthacées. Développement de l'embryon chez le *Chimonanthus fragrans* Lindl. Compt. Rend. 254:1323–25.

1599. —— 1962: Embryogenie des Renonculacées. Développement de l'embryon chez le *Nigella damascena* L. Compt. Rend. 254:3119–20.

1600. Baar, H. 1912: Über den Einfluss des Lichtes auf die Samenkeimung und seine Abhängigkeit von anderen Faktoren. Sitzber. Kais. Akad. Wiss. Wien 1, 121:667–705.

1601. —— 1913: Zur Anatomie und Keimungsphysiologie heteromorpher Samen von *Chenopodium album* und *Atriplex nitens*. Sitzber. Kais. Akad. Wiss. Wien 1, 122:21–40.

1602. Babadzhanidi, D. I. 1963: Physiological differences in *Dianthus* seeds. Agrobiologiya 1963(1):118–19. Biol. Abstr. 44: 25159. 1964.

1603. Babaian, V. O. 1954: On the viability of unripe wheat seeds. Izv. Akad. Nauk Arm. SSR Biol. i Sel'skokhoz. Nauki 7(4):23–29.

1604. —— 1954: On transforming winter wheat into spring wheat by sowing unripe seeds. Izv. Akad. Nauk Arm. SSR Biol. i Sel'skokhoz. Nauki 7(10):11–19.

1605. —— 1960: The productivity of plants grown from seeds of waxy maturity. Selek. i Semen. 1960(3):23–25.

1606. —— and L. A. Muradyan. 1961: Some data on infructescence of seeds in the potato under various conditions. Izv. Akad. Nauk Arm. SSR Biol. i Sel'skokhoz. Nauki 14(9):97–100.

1607. Babamov, L., and P. Popovska. 1962: Calibration of wheat seed as a condition of potential variety value. Arhiv Poljoprivredne Nauke 15(50):211–19.

1608. Babayan, A. A. 1956: Centralized dipping treatment of cottonseed. Izv. Akad. Nauk Arm. SSR Biol. i Sel'skokhoz. Nauki 9(8):67–77. Chem. Abstr. 51: 10813. 1957.

1609. —— and K. A. Karapetyan. 1954: Centralized disinfection of cotton seeds in Armenia and tests of some new treating agents. Izv. Akad. Nauk Arm. SSR Biol. i Sel'skokhoz. Nauki 7(5):15–25. Chem. Abstr. 48:11705. 1954.

1610. Babayan, G. B. 1958: Increasing the protein content in wheat seeds and hereditary reproduction of this ability. Izv. Akad. Nauk Arm. SSR Biol. i Sel'skokhoz. Nauki 11(3):77–80.

1611. Babayan, M. N. 1960: Carotenoid pigment content of the endosperm of wheat depending on the species and conditions of growth. Izv. Akad. Nauk Azerb. SSR Ser. Biol. i Med. Nauk 1960(3):69–75. Chem. Abstr. 55:3736. 1961.

1612. Babb, M. F. 1959: Propagation of woody plants by seed with notes on other methods of reproduction. Alaska Agr. Expt. Sta. Bull. 26. 12 pp.

Babbar, I. 184.

1613. Babbitt, J. D. 1940: Observations on the permeability of hygroscopic materials to water vapour. I. Observations at relative humidities less than 75%. Can. J. Res. Sec. A 18:105–21. Biol. Abstr. 14:12961. 1940.

1614. —— 1945: Hysteresis in the adsorption of water vapour by wheat. Nature 156: 265–66.

1615. —— 1945: The thermal properties of wheat in bulk. Can. J. Res. Sect. F 23: 388–401. Biol. Abstr. 20:13526. 1946.

1616. —— 1949: Observations on the adsorption of water vapor by wheat. Can. J. Res. Sec. F 27:55–72.

Babbitt, J. D. 546.

1617. Babcock, E. B., and G. L. Stebbins, Jr. 1938: The American species of *Crepis*. Their interrelationships and distribution as affected by polyploidy and apomixis. Carnegie Inst. Wash. Publ. 504:1–199. Biol. Abstr. 13:4777. 1939.

Babcock, E. B. 17888.

1618. Babcock, S. M. 1912: Metabolic water: its production and role in vital phenomena. Wisconsin Agr. Expt. Sta. Res. Bull. 22:87–181.

1619. Babel, A. 1935: Neuere Versuche zur Lein-Beizung. Nachr. Schädlingsbekämpf. 10(2):70–73.

Babichev, I. A. 15723.

Babintseva, M. B. 15726, 15727.

1620. Babushkin, L. N. 1959: Effect of temperature variation on tomato seeds. Plant Physiol. (USSR) (English Transl.) 6:218–20.

1621. Bacchi, O. 1955: Seca da semente de cafe ao sol. Bragantia 14:225–36.

1622. —— 1956: Novos ensaios sobre a seca da semente de café ao sol. Bragantia 15:83–90.

1623. —— 1958: Estudos sobre a conservação de sementes. II. Citros. Bragantia 17:157–66.

1624. —— 1958: Estudos sobre a conservação des sementes. III. Trigo. Bragantia 17:205–12.

1625. —— 1958: Estudos sobre a conservação de sementes. IV. Café. Bragantia 17:261–70.

1626. —— 1960: Seed storage studies. V. Lettuce. Bragantia 19:41–45.

1627. —— 1960: Seed storage studies. VI. Onion. Bragantia 19:69–72.

Bacchi, O. 6641.

1628. Bacchi, S. P., N. C. Ganguli, and S. C. Roy. 1955: Amino acid composition of some Indian pulses: Quantitative determination by paper chromatography. Ann. Biochem. Exptl. Med. (Calcutta) 15:149–54. Biol. Abstr. 31:2318. 1957.

1629. Bach, A., and A. Oparin. 1923: Über die Fermentbildung in keimenden Pflanzensamen. Biochem. Z. 134:183–89.

1630. ——, ——, and R. Wähner. 1927: Untersuchungen über den Fermentgehalt von reifenden, ruhenden und keimenden Weizensamen. Biochem. Z. 180:363–70.

1631. Bach, D. 1938: Respiratory complex of dormant seeds. Dehydrogenases and coenzyme of *Pisum sativum*. Compt. Rend. Soc. Biol. 127:175–77. Chem. Abstr. 32: 3453. 1938.

1632. —— 1938: The respiratory complex of dormant seeds. Separation of coenzyme by ultrafiltration of extracts of dried garden peas. Compt. Rend. Soc. Biol. 127:1063–65. Chem. Abstr. 32:5442–43. 1938.

1633. —— 1938: Respiratory complexes of dormant seeds. Distribution of hexosediphosphate dehydrogenase in the Leguminosae. Compt. Rend. Soc. Biol. 128:36–39. Chem. Abstr. 32:6291. 1938.

1634. Bachmann, E. 1885: Biologische Bedeutung des Arillus einiger Leguminosen. Ber. Deut. Botan. Ges. 3:25–29.

1635. Bachmann, F. 1962: Der Wassergehalt von Saatgetreide. Schweiz. Landwirtsch. Mitt. 10:72–79.

1636. —— 1947: Notas sobre drogas, plantas y alimentos mexicanos. VII. Almidon do Chamal (Dioon edule Lindl.). Ciencia (Mex.) 8:109–10. Biol. Abstr. 23:12634. 1949.

Bachmanowa, S. Augustin- ; *see* Augustin-Bachmanowa, S.

1637. Bachthaler, G. 1957: Untersuchung zur Keimungsphysiologie des Flughafers (*Avena fatua* L.) im besonderen Hinblick auf sein zunehmendes auftreten in Bayern und seine derzeitigen wirtschaftlichen Bekämpfungsmöglichkeiten. Z. Acker-Pflanzenbau 103:128–56.

Back, E. 14800.

1638. Back, E. A., R. T. Cotton, H. D. Young, and J. H. Cox. 1930: The use of ethylene oxide-carbon dioxide mixture for treating stored grain. U.S. Dept. Agr. Leaflet. 10 pp. Rev. Appl. Entomol. Ser. A 18:647–48. 1930.

1639. Backes, C. L. 1960/61: Production and marketing of certified seed. Intern. Crop Improvement Assoc. Ann. Rept. 42/43:101–8.

1640. Backgaard, H. C. 1955: Examinations of the content of fluorescent seeds in Danish strains of perennial ryegrass. Proc. Intern. Seed Testing Assoc. 20:89–93.

Bacon, J. A. 14873, 14874.

1641. Bacon, J. S. D. 1959: The trisaccharide fraction of some monocotyledons. Biochem. J. 73:507–14.

1642. Bacon, O. G., W. D. Riley, J. R. Russell, and W. C. Batiste. 1964: Experiments on control of the alfalfa seed chalcid, *Bruchophagus roddi*, in seed alfalfa. J. Econ. Entomol. 57:105–10.

Bacq, Z. M. 8696, 13372, 13834.

1643. Badalian, V. S., and E. A. Atabekian. 1958: Pre-planting treatment of tomato seeds with the purpose of increasing their drought resistance and yields. Izv. Akad. Nauk Arm. SSR Biol. i Sel'skokhoz. Nauki 11(5):97–100.

1644. Badami, R. C., and F. D. Gunstone. 1962: Vegetable oils. X. Examination of component acids of *Argemone mexicana* seed oil by reversed-phase chromatography. J. Sci. Food Agr. 13:255–57.

1645. —— and —— 1963: Vegetable oils. XI. The component acids of *Torresea cearensis* seed oil. J. Sci. Food Agr. 14: 479–81.

1646. —— and —— 1963: Vegetable oils. XII. *Vernonia* seed oils. J. Sci. Food Agr. 14:481–84. 1963.

1647. Badea, M. 1961: The direct sowing of fir by means of planes as an economical method to introduce resinous species in beech woods. Rev. Padurilor (Bucharest) 76:402–5.

1648. Bader, K. L., and S. R. Anderson. 1962: Effect of pollen and nectar collecting honeybees on the seed yield of Birdsfoot trefoil, *Lotus corniculatus* L. Crop Sci. 2:148–49.

1649. Badhwar, R. L., and B. K. Sharma. 1963: A note on the germination of *Podophyllum* seeds. Indian Forester 89: 445–47.

1650. Badolato, F., and I. Bucci. 1960: Ascorbic acid (A.A.) in a free and bound state in the seeds of *Phaseolus vulgaris*. Ric. Sci. Suppl. 30:97–102.

1651. —— and —— 1960: Free and bound ascorbic acid (A.A.) in the seeds of *Phaseolus vulgaris*. II. Ric. Sci. Suppl. 30:207–10.

Badolato, F. 5147, 5151, 14844.

1652. Badruddoja, K. M., and M. Ishaque. 1952: Control of stem-rot of jute by seed treatments and fungicidal sprays. [Abstract.] Pakistan Sci. Conf. Proc. 4:43–44.

1653. Baekgaard, H. C. 1952: Comparison between laboratory and field germination of cereal seed. Tidsskr. Planteavl 55:476–510.

1654. —— 1955: Examination of the content of fluorescent seeds in Danish strains of perennial ryegrass. Proc. Intern. Seed Testing Assoc. 20:89–93.

1655. —— 1960: Seedling colour as a means of determining the genuineness of varieties and strains with beets. Proc. Intern. Seed Testing Assoc. 25:308–20.

1656. —— 1962: Variety committee. Proc. Intern. Seed Testing Assoc. 27:92–103.

1657. —— 1962: Continued examinations of the content of fluorescent seeds in Danish varieties of perennial ryegrass (*Lolium perenne*). Proc. Intern. Seed Testing Assoc. 27:562–72.

1658. Baenitz, C. 1911: Die Keimpflanzen der Holzgewächse. Deut. Botan. Monatsschr. 22:145–49, 161–66, 177–81.

1659. Baens, L. 1941: Tannin from kernels of green betel nuts. Philippine J. Sci. 75:363–67.

Baenziger, H. 7465.

1660. Bagaev, V. B. 1953: The effect of the conditions of phosphate nutrition on the composition of soybean seeds. Dokl. Akad. Nauk SSSR 91:961–64. Chem. Abstr. 48:819. 1954.

Bagchi, P. 13881.

Bagchi, R. 4046–4048.

1661. Bagdasarashvili, Z. G. 1951: Influence of certain salts upon the germination of seeds of Gruziyan grape varieties. Akad. Nauk Gruzinsk. SSR Inst. Vinograd i Vinodel Tr. 7:99–124.

1662. Baghdadi, H. A., F. A. Minessy, and M. K. Higazy. 1956: Effect of agrosan and 2,4-D on germination albinism and seedling growth in sweet lime and balady lime. Alexandria J. Agr. Res. 4:49–68.

1663. Bagley, B. W., J. H. Cherry, M. L. Rollins, and A. M. Altschul. 1963: A study of protein bodies during germination of peanut (*Arachis hypogaea*) seed. Am. J. Botany 50:523–32.

1664. Baglioni, T. 1956: Action of some substances on the growth of roots of cucumber seeds (*Cucumis sativus*). Soc. Ital. Sci. Vet. Atti 9:491–93. 1955.

1665. Bagni, I. 1951: The germinating quality of seed grain. Agr. Ferrarese 55: 169–70.

Bahl, A. N. 183.

1666. Bahme, R. B. 1949: Nicotinic acid as a growth factor for certain orchid embryos. Science 109:522–23.

Bahr, M. H. 1505.

1667. Bahrt, G. M., and G. F. Potter. 1947: Effects of nitrogen, phosphorus, and potassium on growth and yield of tung trees and composition of fruits. Proc. Am. Soc. Hort. Sci. 50:137–41.

1668. —— and —— 1948: Effects of fertilizer on oil content of nuts. Tung World 3:16–17, 19.

Bailes, J. P. 17669.

1669. Bailey, A. E. 1945: The new chemistry and technology of oils and fats. Chronica Botan. 9:128–35. Biol. Abstr. 21:2814. 1947.

1670. —— 1948: Cottonseed and cottonseed products. Their chemistry and chemical technology. Interscience, New York. 936 pp.

1671. Bailey, C. H. 1917: The catalase activity of American wheat flours. J. Biol. Chem. 32:539–45. Expt. Sta. Rec. 38:712. 1918.

1672. —— 1917: The moisture content of heating wheat. J. Am. Soc. Agron. 9:248–51.
1673. —— 1917: The handling and storage of spring wheat. J. Am. Soc. Agron. 9:275–81.
1674. —— 1921: Respiration of shelled corn. Minn. Agr. Expt. Sta. Tech. Bull. 3. 44 pp.
1675. —— 1938: Germ content of American wheats. Cereal Chem. 15:102–6. Expt. Sta. Rec. 80:191. 1939.
1676. —— 1940: Respiration of cereal grains and flaxseed. Plant Physiol. 15: 257–74.
1677. —— 1941: Protein surveys of American hard spring and soft winter wheats. Minn. Agr. Expt. Sta. Tech. Bull. 147. 46 pp.
1678. —— 1944: Constituents of wheat and wheat products. Reinhold, New York. 332 pp. Biol. Abstr. 18:17258. 1944.
1679. —— 1960: A note on the vitamin B content of Korean rice. Cereal Chem. 37:103–4.
1680. —— and A. M. Gurjar. 1918: Respiration of stored wheat. J. Agr. Res. 12:685–713.
Bailey, C. H. 625, 7716, 8812, 8813, 12866, 12867, 17297, 17317, 18157.
1681. Bailey, C. V. 1961: Early collection and immediate sowing increase germination of basswood (*Tilia americana*) seed. Tree Planters Notes 46:27–28.
1682. Bailey, D. M., and R. M. Bailey. 1939: The relation of the pericarp to tenderness in sweet corn. Proc. Am. Soc. Hort. Sci. 36:555–59.
1683. Bailey, J. E., and J. G. Woodroof. 1932: Propagation of pecans. Georgia Agr. Expt. Sta. Bull. 172. 22 pp. Expt. Sta. Rec. 68:765. 1933.
Bailey, J. E. 19844.
1684. Bailey, K. 1935: Cress-seed mucilage. Biochem. J. 29:2477–85. Chem. Abstr. 30:3025. 1936.
1685. —— and F. W. Norris. 1932: Nature and composition of the mucilage of the seed of white mustard [*Brassica alba*]. Biochem. J. 26:1609–23. Chem. Abstr. 27:808. 1933.
Bailey, K. 1478.
Bailey, L. F. 9970.
1686. Bailey, L. H. 1889: On the influence of certain conditions upon the sprouting of seeds. Cornell Univ. Agr. Expt. Sta. Bull. 7. 43 pp.
1687. Bailey, R. M. 1941: The effect of plant spacing on yield, ear size, and other characters in sweet corn. Proc. Am. Soc. Hort. Sci. 38:546–53.
Bailey, R. M. 1682.
1688. Bailey, S. D. 1960: Studies on the volatile components of different varieties of cocoa beans. J. Food Sci. 27:165–70. Anal. Abstr. 10:2051. 1963.
1689. Bailey, S. W. 1955: Air-tight storage of grain; its effects on insect pests. I: *Calandra granaria* L. (Coleoptera, Curculionidae). Australian J. Agr. Res. 6:33–51.
1690. —— 1956: Air-tight storage of grain; its effects on insect pests. II: *Calandra oryzae* (small strain). Australian J. Agr. Res. 7:7–19.
1691. —— 1957: Air-tight storage of grain; its effects on insect pests. III: *Calandra oryzae* (large strain). Australian J. Agr. Res. 8:595–603.
1692. —— 1962: The effects of percussion on insect pests of grain. J. Econ. Entomol. 55:301–5.
1693. Bailey, W. K., E. H. Toole, V. K. Toole, and M. F. Drowne. 1958: Influence of temperature on the after-ripening of freshly harvested Virginia bunch peanut seeds. Proc. Am. Soc. Hort. Sci. 71:422–24.
Bailey, W. K. 8733.
1694. Bailey, W. M. 1933: Structural and metabolic after-effects of soaking seeds of *Phaseolus*. Botan. Gaz. 94:688–713. Expt. Sta. Rec. 69:653. 1933.
Bailey, W. W. 14343.
1695. Baillon, H. 1882: Germination of walnut and almond. Botan. Gaz. 7:91–92.
1696. Bain, D. C. 1939: Effect of sulphuric-acid treatment on fungi and bacteria present on cotton seed from diseased bolls. Phytopathology 29:879–84.
1697. —— 1950: Fungi recovered from seed of *Sorghum vulgare* Pers. Phytopathology 40:521–22.
Bain, D. C. 3939.
Bain, J. A. 17051.
1698. Bainer, R. 1948: Present status of sugar beet seed processing. Proc. Am. Soc. Sugar Beet Technol. 5:120–24.
1699. —— and H. A. Borthwick. 1934: Thresher and other mechanical injury to seed beans of the lima type. Calif. Agr. Expt. Sta. Bull. 580. 30 pp.
1700. —— and L. D. Leach. 1947: Processing sugar beet seed by decortication, burr reduction, and segmentation. Proc. Am. Soc. Sugar Beet Technol. 1946:625–39. Biol. Abstr. 22:14748. 1948.
1701. —— and J. S. Winters. 1937: New

principle in threshing lima bean seed. Agr. Eng. 18:205–6.

1702. Baines, A. E. 1921: Germination in its electrical aspect. Dutton, New York. Routledge, London. 185 pp.

Baines, G. S. 14211.

1703. Bair, R. A. 1947/48: Seed storage investigations. Florida Agr. Expt. Sta. Ann. Rept. 1948:204.

1704. Baird, E. W. 1950: The commercial production of green manure seed in the Cairns hinterland. Queensland Agr. J. 71:143–53.

1705. Baird, P. D., M. M. MacMasters, and C. E. Rist. 1950: Studies on a rapid test for the viability of corn for industrial use. Cereal Chem. 27:508–13.

Baird, P. D. 7051.

1706. Baisted, D. J., and W. R. Nes. 1963: The relative efficacy of mevalonic and dimethylacrylic acids as isopentenoid precursors in germinating seeds of *Pisum sativum*. J. Biol. Chem. 238:1947–52.

1707. Bajai, J. 1959: A contribution to the seed testing methods and germination tests of sorghum seeds. Acta Agron. (Budapest) 9:185–208.

1708. Bajer, A. 1954: Cine-micrographic studies on mitosis in endosperm. I. Acta Soc. Botan. Polon. 23:383–412.

Bajer, J. Mole-; *see* Mole-Bajer, J.

1709. —— and J. Mole-Bajer. 1954: Endosperm, material for study on the physiology of cell division. Acta Soc. Botan. Polon. 23:69–98.

1710. —— and —— 1956: Cine-micrographic studies on mitosis in endosperm. II. Chromosoma 7:558–607.

Bajesco, N. 2775.

1711. Bajpai, P. N., and R. K. Trivedi. 1961: Storage of mango seedstone. Hort. Advan. 5:228–33.

Bajwa, B. S. 15920.

1712. Baker, A. Z., and M. D. Wright. 1937: Note on the vitamin B activity of raw wheat germ. J. Hyg. 37:303–6. Chem. Abstr. 31:4703–4. 1937.

Baker, B. E. 10509.

1713. Baker, B. Y., and E. V. Lynn. 1953: Examination of the seed of *Pachyrrhizus erosus*. J. Am. Pharm. Assoc. 42:117–18.

1714. —— and M. W. Quimby. 1953: A morphological study of the seeds of *Pachyrrhizus erosus* (L.) Urban, the yam bean. J. Am. Pharm. Assoc. 42:471–78.

1715. Baker, C. G. 1936: Tigridias from seed. Flower Grower 23(12):633.

1716. Baker, D. 1962: Fatty acid composition of oil from damaged corn and wheat. Cereal Chem. 39:393–97.

1717. ——, M. H. Neustadt, and L. Zeleny. 1957: Application of the fat acidity test as an index of grain deterioration. Cereal Chem. 34:226–33.

1718. ——, ——, and —— 1959: Relationships between fat acidity values and types of damage in grain. Cereal Chem. 36: 308–11.

Baker, F. E. 6456.

1719. Baker, F. J. 1903/4: On the productivity of seeds. J. Roy. Hort. Soc. 28:385–88.

1720. Baker, F. S. 1950: Principles of silviculture. McGraw-Hill, New York. 414 pp.

1721. Baker, G. A. 1942: Mathematical model of embryo abortion in Phillips cling peaches. J. Agr. Res. 64:173–78.

1722. —— 1945: Test of the significance of the differences of percents of emergence of seedlings in multiple field trials. J. Am. Statist. Assoc. 40:93–97.

1723. —— and R. M. Brooks. 1947: Effects of fruit thinning on almond fruits and seeds. Botan. Gaz. 108:550–56. Biol. Abstr. 22:2178. 1948.

1724. Baker, G. N., T. Göksu, and M. L. Baker. 1960: Undecorticated safflower meal as a protein supplement for wintering calves. Nebraska Agr. Expt. Sta. SB 458. 7 pp.

1725. Baker, H. G. 1960: Apomixis and polyembryony in *Pachira oleaginea* (Bombacaceae). Am. J. Botany 47:296–302.

1726. Baker, J. L. 1902: Action of ungerminated barley diastase on starch. J. Chem. Soc. 81:1177–85.

1727. —— and H. F. E. Hulton. 1921: Amylases of the cereal grains—rye. J. Chem. Soc. 119:805–9.

1728. —— and —— 1929: Amylases of the cereal grains—oats. J. Chem. Soc. 218: 1655–60.

1729. Baker, J. N. 1958: Effect of gibberellic acid, 2,4-D and indole-acetic acid on seed germination and epicotyl and radicle growth of intermediate and pubescent wheatgrasses. J. Range Management 11: 227–30. Biol. Abstr. 33:23450. 1959.

1730. Baker, K. F. 1947: Seed transmission of *Rhizoctonia solani* in relation to control of seedling damping-off. Phytopathology 37:912–24.

1731. —— 1948: The coat-bound condition of germinating pepper seeds. Am. J. Botany 35:192–93.

1732. —— 1952: Seed-borne parasites. Seed World 70(11):38, 40, 44, 46–47.
1733. —— 1956: Development and production of pathogen-free seed of three ornamental plants. Plant Disease Reptr. Suppl. 238:68–71.
1734. —— and L. H. Davis. 1951: An unusual occurrence of sclerotia of *Sclerotinia* spp. with seed of *Centaurea cyamus*. Plant Disease Reptr. 35:39–41.
1735. —— and W. F. Locke. 1946: Perithecia of powdery mildew on zinnia seed. Phytopathology 36:379–80.
1736. Baker, L. O., and D. H. Leighty. 1958: Germination studies with wild oat seeds. Proc. Western Weed Control Conf. 16: 69–74.
Baker, M. L. 1724.
1737. Baker, R. E. 1943: Induced polyploid, periclinal chimeras in *Solanum tuberosum*. Am. J. Botany 30:187–95.
Baker, R. S. 6925.
1738. Bakhalbashian, D. A. 1955: Seed formation in wheat during pollination of spikes on different tiers of the bush. Izv. Akad. Nauk Arm. SSR Biol. i Sel'şkokhoz. Nauki 8(2):39–47.
1739. Bakhteyev, F., P. Kalinin, and D. Goryunov. 1936: Efficiency of the vernalization of barley grown beyond the arctic circle, in Detskoye Selo and in the Crimea. Bull. Appl. Botany Genet. & Pl. Breed. Ser. A(19):37–40.
1740. Bakke, A. L., and N. L. Noecker. 1933: The relation of moisture to respiration and heating in stored oats. Iowa Agr. Expt. Sta. Res. Bull. 165:317–36.
1741. ——, H. W. Richey, and K. Reeves. 1926: Germination and storage of apple seeds. Iowa Agr. Expt. Sta. Res. Bull. 97: 243–55.
1742. —— and H. Stiles. 1935: Thermal conductivity of stored oats with different moisture content. Plant Physiol. 10: 521–27.
Bakke, A. L. 3394.
1743. Bakker, M. 1951: Control of the Phomopsis-disease in seed umbels of carrot. Tijdschr. Plantenziekten 57:157–66. Biol. Abstr. 26:16573. 1952.
1744. Balabaew, G. A. 1927: Unkrautelemente und deren Verbreitung in Getreidesamen in Gebirgsteile von Zerawschan. Ann. Essais Semences 5(4):9–32.
Balabanova, L. A. 18805.
1745. Balaeva, A. P. 1957: Plantings (of buckwheat) for attraction (of bees) and seed productivity of (red) clover. Nauka i Peredovoi Opyt v Sel'sk. Khoz. 7(2): 25–27.
1746. Balakrishna, K. J., and T. R. Seshadri. 1947: Colouring matter of tambul seeds. I. Tambuletin. Proc. Indian Acad. Sci. Sect. A 25:449–52. Biol. Abstr. 22:12373. 1948.
1747. —— and —— 1947: Colouring matter of tambul seeds. II. Constitution of tambuletin: Synthesis of o-tetraethyl tambuletin. Proc. Indian Acad. Sci. Sect. A 26:72–76. Biol. Abstr. 22:12374. 1948.
Balan, C. 1291.
1748. Balan, D., and A. Avramescu. 1958: Vernalisation des semences de piments et de tomates. Ann. Inst. Cerc. Agron. Bucuresti 25:307–16. 1957.
1749. Balandin, D. A. 1940: Fatty oil from kernels of *Schizandra chinensis*. Compt. Rend. Acad. Sci. URSS 26:584–86.
1750. Balansard, J., and P. Bernard. 1952: Action of magnesium marrubinate and magnesium marrubylmarrubinate on germination. Bull. Soc. Pharm. Marseille Bull. Trimest., pp. 31–33. Chem. Abstr. 49:7069. 1955.
1751. —— and F. Pelissier. 1943: Action of Quillaya saponin on germination of wheat. Compt. Rend. Soc. Biol. 137:246–47. Chem. Abstr. 38:1766. 1944.
1752. —— and —— 1943: Action de diverses saponines sur le bouturage épiphylle de quelques Begoniacées. Compt. Rend. Soc. Biol. 137:454.
1753. —— and —— 1943: Action of Quillaja saponin on absorbent capacity of pea seeds. Compt. Rend. Soc. Biol. 137:455. Chem. Abstr. 39:2779. 1945.
1754. —— and —— 1943: Action of Quillaja saponin on isolated wheat germs. Compt. Rend. Soc. Biol. 137:461–62. Chem. Abstr. 39:2779–80. 1945.
1775. —— and —— 1944: Action of lauryl alcohol sulfate on germination and growth of wheat. Compt. Rend. Soc. Biol. 138: 596–97. Chem. Abstr. 39:4916. 1945.
1756. —— and —— 1944: Action of saponins on growth. Influence of treatment of seeds on later phases of vegetation. Compt. Rend. Soc. Biol. 138:622–23. Chem. Abstr. 40:624. 1946.
1757. —— and —— 1945: Saponins and the growth of the tomato plant. Compt. Rend. Soc. Biol. 139:1008–10. Chem. Abstr. 40:5805. 1946.
1758. ——, ——, and S. Conil. 1946: Action of Quillaja and Sapindus saponins on water absorption by the seeds and on

germination and growth of *Zea mays*. Compt. Rend. Soc. Biol. 140:140–42.

1759. Balashov, I. S. 1963: Maximum permissible moisture content of wheat seeds infected with fusariosis. Selek. i Semen. 28(1):79–80. Biol. Abstr. 43:16256. 1963.

Balasoiu, A. 1291.

1760. Balasubramaniam, G., and D. D. Sundararaj. 1955: Causes of poor viability in a few grass seeds. Madras Agr. J. 42: 271–74. Biol. Abstr. 30:32448. 1956.

1761. Balasubramanian, R. 1931: Parthenogenesis in cotton. Madras Agr. J. 19: 509–16.

1762. —— and V. R. Mudaliar. 1933: A physiological study of delayed germination in cotton. Madras Agr. J. 21:147–62. Chem. Abstr. 27:3970. 1933.

1763. Balasubramanian, S. C., M. Ramachandran, T. Viswanatha, and S. S. De. 1952: Amino acid composition of Indian foodstuffs. II. Indian J. Med. Res. 40: 219–34. Biol. Abstr. 28:3103. 1954.

1764. Balavoine, P. 1946: The mucilage content of some muce-pectic plants. Pharm. Acta Helv. 21:19–22. Chem. Abstr. 40:4768. 1946.

Balbach, P. 15790, 17058, 17059.

Balbi, G. 2937.

Balcazar, M. R. 3690.

Balce, L. V. 19016.

1765. Baldacci, E. 1941: Sterility of rice panicles. Intern. Bull. Plant Protect. (Rome) 15:114M–16M. Expt. Sta. Rec. 86:795. 1942.

1766. ——, T. Scotti, and L. Thalmann. 1950: The treatment of rice seeds (experimental research). Risicoltura 38:214–21.

Baldauski, F. A. 9783.

Baldev, B. 12696.

1767. Baldoni, R., and A. Canova. 1952: Rapid detection of infection of sugar-beet seeds by *Phoma betae*. Ann. Sper. Agrar. (Rome) 6:435:42. Brit. Abstr. B 111:459. 1952.

1768. Baldovinos, G., and A. Echegaray. 1954: Possible determining factors in the pigmentation of the pericarp of *Phaseolus vulgaris*. Bol. Tec. Soc. Agron. Mex. 1:39–61. Chem. Abstr. 49:4093. 1955.

Baldridge, D. E. 13239.

1769. Baldwin. 1944: Report on apricot-kernel oils. Gt. Brit. Imp. Inst. Bull. 42:153–160.

1770. Baldwin, H. I. 1930: The effect of after-ripening treatment on the germination of Eastern hemlock seed. J. Forestry 28: 853–57.

1771. —— 1930: An individual Jacobsen germinator. Science 72:345–46.

1772. ——1932: Alcohol separation of empty seed, and its effect on the germination of red spruce. Am. J. Botany 19:1–11.

1773. —— 1932: Comment on cutting tests for seeds. J. Forestry 30:746-47.

1774. —— 1933: The importance of the origin of forest seeds. Empire Forestry J. 12:198–210. Plant Breed. Abstr. 4:185. 1934.

1775. —— 1934: Effect of after-ripening treatment on germination of white pine seeds of different ages. Botan. Gaz. 96:372–76.

1776. —— 1934: Further notes on the germination of hemlock seed. J. Forestry 32:99–100. Expt. Sta. Rec. 70:788. 1934.

1777. —— 1934: Germination of the red spruce. Plant Physiol. 9:491–532.

1778. —— 1935: Seasonal variations in the germination of red spruce. Am. J. Botany 22:392–94.

1779. —— 1935: Catalase activity as a measure of viability of tree seeds. Am. J. Botany 22:635–44.

1780. —— 1936: X-ray treatment of tree seeds. J. Forestry 34:1069–70.

1781. —— 1939: Some new aspects of seed certification. J. Forestry 37:28–34.

1782. —— 1942: Forest tree seed of the north temperate regions with special reference to North America. Chronica Botanica Company, Waltham, Mass. 240 pp.

1783. —— 1949: New seed-extraction apparatus. J. Forestry 47:199.

1784. —— 1954: Handling tree seeds. Unasylva 8:66–70.

1785. —— 1954: Seed certification and forest genetics. J. Forestry 52:654–55.

1786. —— and F. Flemion. 1941: Rapid method for testing white pine germination. New Hampshire Forestry & Recreation Dept. Fox Forest Notes 30.

1787. —— and M. Leloup. 1955: Handling forest tree seed. Food Agr. Organ. U.N., Rome. 110 pp.

1788. —— and W. E. Percival. 1931: Polyembryony in red spruce. Science 74: 203.

1789. —— and H. L. Shirley. 1936: Forest seed control. J. Forestry 34:653–63.

1790. —— and —— 1936: A forest seed program for the United States. J. Forestry 34:766–70.

1791. Baldwin, J. T., Jr. 1949: Loss of oil from *Hevea* seed, a variation having phyletic and economic implications. J. Heredity 40:47–49.

1792. —— and R. Culp. 1941: Polyploidy in *Diospyros virginiana* L. Am. J. Botany 28:942–44.

1793. Balian, V. P. 1957: Methods of increasing the germination of seeds of the wild growing clovers of Armenia. Erevansk. Arm. Nauchn.-Issled. Inst. Zhivotn. i Vet. Tr. Zhivotn. i Kormodobyvanie 6:257–74.

Balieva, M. A. 20022.

Baliga, B. R. 11414.

Ball, C. D. 3478, 10655, 14892, 14893.

Ball, C. D., Jr. 10654.

1794. Ball, C. P. 1936: Raising freesias from seed. Horticulture 14(4):69.

1795. Ball, E. 1956: Growth of the embryo of *Ginkgo biloba* under experimental conditions. I. Origin of the first root of the seedling *in vitro*. Am. J. Botany 43:488–95.

1796. —— 1957: Growth of the embryo of *Ginkgo biloba* under experimental conditions. II. Effects of a longitudinal split in the tip of the hypocotyl. Am. J. Botany 43:802–10.

1797. —— 1959: Growth of the embryo of *Ginkgo biloba* under experimental conditions. III. Growth rates of root and shoot upon media absorbed through the cotyledons. Am. J. Botany 46:130–39.

1798. Ball, G. J. 1928: Orchids from seed at Los Angeles range. Florists Rev. 62(1595): 27–30.

1799. —— 1932: Germinating flower seed. Florists Rev. 71(1823):23–24, 59–60.

1800. ——1935: Serious losses in germination of seed can generally be avoided. Florists Rev. 75(1937):19–20, 49–51.

1801. —— 1936: Snapdragon seed—its cost, quality, and results. Florists Exchange Hort. Trade World 87(20):15.

1802. —— 1942: White queen lily from seed. Florists Rev. 90(2324):16.

1803. Ball, W. S. 1931: Germination tests on puncture vine. Calif. Dept. Agr. Bull. 20:316–18.

1804. —— 1944: Germination of buried weed seed. Calif. Dept. Agr. Bull. 33: 105–7.

1805. —— and M. K. Bellue. 1935: A new means of dissemination of nut grass (*Cyperus esculentus* L.). Calif. Dept. Agr. Bull. 24:235–37. Expt. Sta. Rec. 74:197. 1936.

1806. —— and W. W. Robbins. 1932: Puncture vine (*Tribulus terrestris* L). Calif. Dept. Agr. Bull. 21:211–13.

Ball, Y. S. 11371.

Ballantyne, D. J. 18761.

1807. Ballard, F. 1942: Cleistogamy in grasses. J. Ecol. 30:232–33.

1808. Ballard, J. C., S. H. Wittwer, M. J. Bukovac, and F. J. McArdle. 1959: Effect of gibberellin on the yield and maturity of processing peas. Mich. Agr. Expt. Sta. Quart. Bull. 41:624–29.

1809. Ballard, L. A. T. 1958: Studies of dormancy in the seeds of subterranean clover (*Trifolium subterraneum* L.). I. Breaking of dormancy by carbon dioxide and by activated carbon. Australian J. Biol. Sci. 11:246–60.

1810. —— 1961: Studies of dormancy in the seeds of subterranean clover (*Trifolium subterraneum* L.). II. The interaction of time, temperature, and carbon dioxide during passage out of dormancy. Australian J. Biol. Sci. 14:173–86.

1811. —— and A. E. G. Lipp. 1959: Differential specificity exhibited by two germination inhibitors present in *Echium plantagineum* L. Australian J. Biol. Sci. 12:343–47.

Ballard, L. A. T. 12304, 12305.

1812. Ballard, W. W. 1937: Pure seed requirements in the production of Sea Island cotton. Georgia Agr. Expt. Sta. Circ. 113. 10 pp. Expt. Sta. Rec. 77: 777–78. 1937.

1813. Ballatore, G. P. 1951: First results of some vernalization tests of beans "fava da granella." Palermo Univ. Fac. Sci. Agr. Ann. 1:181–98. 1950.

1814. —— and G. Mirto. 1956: Experimental research on the vernalization of the horse bean (*Vicia faba major*). Ann. Sper. Agrar. (Rome) 10:1457–86.

Ballatore, G. P. 20058.

1815. Ballinger, R. J. 1944: Extraction of tomato, cucumber, and marrow seed with hydrochloric acid. Gt. Brit. Min. Agr. J. 51:178–84.

Ballinger, R. J. 1956.

1816. Ballion, G. 1928: The difficulty of germinating orchid seed. Orchid Rev. 36(417):84–88.

1817. —— and M. Ballion. 1924: The nonsymbiotic germination of orchid seed in Belgium. Orchid Rev. 32:305–8.

1818. —— and —— 1928: Asymbiotic germination of orchid seed. Orchid Rev. 36(418):103–11.

Ballion, M. 1817, 1818.

1819. Balls, A. K., B. Axelrod, and M. W. Kies. 1943: Soybean lipoxidase. J. Biol. Chem. 149:491–504.

1820. —— and W. S. Hale. 1932: The estimation of catalase in agricultural products. J. Assoc. Offic. Agr. Chemists 15:483–90.

1821. —— and —— 1938: The preparation and properties of wheat proteinase. Cereal Chem. 15:622–28.

1822. —— and —— 1940: The effect of ethylene on freshly harvested wheat. Cereal Chem. 17:490–94.

1823. ——, ——, and T. H. Harris. 1942: A crystalline protein obtained from a lipoprotein of wheat flour. Cereal Chem. 19:279–88.

1824. —— and T. H. Harris. 1944: The inhibitory effect of a protamine from wheat flour on the fermentation of wheat mashes. Cereal Chem. 21:74–79.

1825. —— and S. Schwimmer. 1944: Digestion of raw starch. J. Biol. Chem. 156:203–10.

Balls, W. L. 7551.

Balode, A. [A.] 1836, 1838.

1826. Balovoine, P., and J. Terrier. 1944: Seeds of *Carpinus betulus*. Pharm. Acta Helv. 19:230. Pharm. Abstr. 11:11. 1945.

Balsac, F. H. de; *see* De Balsac, F. H.

Balsac, H. H. de; *see* De Balsac, H. H.

1827. Baltensperger, A. A., and R. R. Kalton. 1959: Variability in reed canarygrass, *Phalaris arundinaceae* L. II. Seed shattering. Agron. J. 51:37–38.

Baltensperger, A. A. 19888.

Baltes, J. 10339.

Balu, V. 15407.

1828. Balutsa, G., and I. V. Brad. 1960: The biochemical processes which take place during the vernalization of winter wheat under normal circumstances and after treatment with solutions of various substances. Biokhim. Zerna, Sb. 5:87–99. Biol. Abstr. 40:20524. 1962.

1829. Bamann, E., R. Boshart, and E. Ullmann. 1959: Die Konfigurationsspezifität der Lipase der Samens von *Chelidonium majus*. 2. Arch. Pharm. 292:602–10.

1830. —— and O. Schimke. 1941: α-Peptid-Spaltung durch Enzympräparate aus wachsenden Keimpflanzen; Beeinflussung durch natürliche und zusätzliche Aktivatoren. Zur Kenntnis der Peptidasen. II. Biochem. Z. 310:119–30.

1831. —— and N. Tietz. 1953: Über das Verhalten gereinigter Samenlipasepräparate bei der Elektrophorese; 5. Mitteilung zur Kenntnis der Lipase höherer Pflanzen. Biochem. Z. 324:502–3.

1832. —— and E. Ullman. 1942: Untersuchungen über die Lipase höherer Pflanzen. Biochem. Z. 312:9–40.

1833. Bambacioni-Mezzetti, V. 1931: Nuove ricerche sull'embriologia delle gigliacée. Ann. Botan. (Turin) 19:365–82.

1834. —— 1933: Morphology and microchemistry of developing seed of *Tulipa gesneriana* L. Ann. Botan. (Turin) 20: 1–11. Biol. Abstr. 9:12325. 1935.

1835. —— 1934: Sulla presenza di albume e perisperma nel seme in via di sviluppo di *Laurus nobilis* L., della Dott. Nuovo Giorn. Botan. Ital. 41:768.

1836. Bamberg, K. K., and A. A. Balode. 1961: Effect of seed treatment and concurrent powdering of seeds with trace elements on germinating capacity in field conditions and yields. Akad. Nauk Latviisk. SSR Inst. Biol. Tr. 21:319–35.

1837. Bambergs, K. 1949: Data on chemical composition of grains of Latvia. Latvijas PSR Zinatnu Akad. Vestis 1949(6):45–54. Chem. Abstr. 48:231. 1954.

1838. ——, A. Balode, and A. Zarina. 1957: Effect of microelement dust spraying of seeds on the crop. Latvijas PSR Zinatnu Akad. Vestis 1957(6):33–44. Chem. Abstr. 52:6508a. 1958.

1839. —— and M. Knavina. 1952: Content of mineral substances and microelements in legumes. Latvijas PSR Zinatnu Akad. Vestis 1952(7):55–65. Chem. Abstr. 48: 13824. 1954.

Bami, H. L. 4937.

1840. Banasik, O. J. 1960: Quick wheat protein test. N. Dakota Farm Res. 21(8):13–16.

1841. Bancher, E., and J. Holzl. 1959: Beobachtungen zur Färbung und Fluorochromierung von Stärkekörnern der Kartofel (*Solanum tuberosum*). Österr. Botan. Z. 106:571–76.

Bancroft, T. A. 9686.

Banerjee, B. N. 15899.

1842. Banerjee, H. K., and B. K. Kar. 1939/41: Studies in the physiology of some Indian fruits. II. Catalase and oxidase activity in *Mangifera indica*. Trans. Bose Res. Inst. (Calcutta) 14:171–82. Biol. Abstr. 16:23095. 1942.

Banerjee, H. N. 14058–14061.

Banerjee, M. 1849, 16353.

1843. Banerjee, R. M., and N. B. Das. 1954: The diastatic activity of certain Indian wheats. Indian J. Agr. Sci. 24:45–49. Chem. Abstr. 49:1228. 1955.

Banerjee, R. [M.] 1844, 4923.

1844. Banerjee, S., and R. Banerjee. 1950: Studies on the biosynthesis of nicotinic acid. I. Biosynthesis of nicotinic acid by germinating pulses. Indian J. Med. Res. 38:153–60. Biol. Abstr. 25:32043. 1951.

1845. ——, N. C. Ghosh, and N. Nandi. 1951: Biosynthesis of nicotinic acid. III. Effect of germination on the nicotinic acid, nicotinuric acid, N′-methylnicotinamide, and trigonelline values of pulses and cereals. Indian J. Med. Res. 39:447–51. Chem. Abstr. 47:629. 1953.

1846. —— and S. Lahiri. 1955: Studies on the biosynthesis of nicotinic acid. V. Further observations on the biosynthesis of nicotinic acid in germinating pulses. Indian J. Med. Res. 43:331–35.

1847. ——, P. Mitra, and H. Chattopadhyay. 1958: Bound ascorbic acid contents of germinated leguminous seeds. Indian J. Med. Res. 46:778–81. Biol. Abstr. 33: 46960. 1959.

1848. —— and N. Nandi. 1949: Effect of germination on phytin content and phytase activity of some common Indian pulses. Proc. Soc. Exptl. Biol. Med. 71:588–89.

1849. ——, K. Rohatgi, M. Banerjee, D. Chattopadhyay, and H. Chattopadhyay. 1955: Pyridoxine, inositol, and vitamin K contents of germinated pulses. Food Res. 20:545–47.

1850. ——, ——, and S. Lahiri. 1954: Pantothenic acid, folic acid, biotin and niacin content of germinated pulses. Food Res. 19:134–37.

1851. ——, A. R. G. Roy, and P. K. Ghosh. 1959: Folic acid metabolism in germinating seeds. Food Res. 24:332–34.

Banerjee, S. 3853–3857, 13609, 16353, 17508.

1852. Banerjee, S. N., and R. M. Datta. 1960: Genesis of embryosac in *Vigna catjang* Engl. Indian Agriculturist 4: 90–94.

1853. Banerji, E. A. R. 1953: A polyembryonic grain of paddy (*Oryza sativa*). Sci. Cult. (Calcutta) 1:296–97.

1854. Banerji, I. 1932: The development of the embryo-sac and fertilisation in jute. J. Indian Botan. Soc. 11:228–40.

1855. —— 1938: A note on the embryology of the ground nut (*Arachis hypogaea* L.). J. Bombay Nat. Hist. Soc. 40:539–43.

1856. —— 1961: The endosperm in Scrophulariaceae. J. Indian Botan. Soc. 40:1–11.

1857. —— and P. N. Bhaduri. 1933: Polyembryony in Solanaceae. Current Sci. (India) 1:310–11.

1858. —— and M. C. Das. 1937: A note on the development of thc cmbryo-sac in *Trichosanthes dioica* Roxb. Current Sci. (India) 5:427–28.

Banerji, I. 159.

1859. Bang, Y. H. 1962: Effectiveness of malathion in protecting stored polished rice from damage by several species of stored-grain insects. J. Econ. Entomol. 55:188–90.

1860. Banga, O. 1948: Vernalisation and devernalisation of beets. Mededel. Directeur Tuinbouw 11:324–46. Biol. Abstr. 24:34265. 1950.

1861. —— 1948: Differences in bolting tendency of different varieties and strains of red table beets. Mededel. Directeur Tuinbouw 11:724–28.

1862. Bangs, C. C. 1936: Yucca from seed. Flower Grower 23(12):642.

1863. —— 1950: Raising lilies from seed. North Am. Lily Soc. Yearbook 3:98–99.

1864. Banham, F. L. 1951: The effect of certain insecticides on the germination and growth of onions. Proc. Entomol. Soc. Brit. Columbia 48:67–69. Biol. Abstr. 27:17950. 1953.

1865. Banlaki, S., and I. Dobrovszky. 1952: Use of Mitchurinian methods in the cultivation of beet shoots and seeds. Növenytermeles 1:209–28. Brit. Abstr. B 3:369. 1953.

1866. Banna, A. 1954: Oilseeds, fats, and oils, and their products. U.S. Dept. Agr. Statist. Bull. 147. 234 pp.

1867. Bannan, M. W. 1947: Tetraploid *Taraxacum kok-saghyz*. III. Achene weight, flowering, and plant development. Can. J. Res. Sect. C 25:59–72.

1868. Bannerjee, K. K. 1942: An attempt on quick germination of *Tectona grandis* seeds. Indian Forester 68:240–44.

1869. Bannister, M. H., and A. J. McQuire. 1956: Studies of tree seed (extraction, germination, and storage). Forest Res. Inst. Ann. Rept. 1956:42–43.

1870. Bant, J. H., A. Beaumont, and I. F. Storey. 1950: Hot-water treatment of broccoli seed. Gt. Brit. Natl. Agr. Advan. Serv. Quart. Rev. 9:43–46.

1871. —— and I. F. Storey. 1952: Hot-water treatment of celery seed in Lancashire. Plant Pathology 1:81–83.

1872. Banting, J. D. 1962: The dormancy behavior of *Avena fatua* L. in cultivated soil. Can. J. Plant Sci. 42:22–39.

1873. ——, Y. S. Wu, and L. H. Shebeski. 1961: The seed value of frost-damaged wheat as determined by its commercial grade, bushel weight, and seedling development. Can. J. Plant Sci. 41:137–52.

1874. Banu, C., and C. Constantinescu. 1935: Les parasites cryptogamiques dans les semis du tabac de l'institut au courant de l'année 1935. Bull. Inst. Cultiv. Ferment. Tabac 24:171–80.

1875. ——, E. Tomita, and C. Constantinescu. 1935: Contributions à l'étude de la faculté germinative des semences du tabac. Bull. Inst. Cultiv. Ferment. Tabac 24:347–54.

1876. Baptist, N. G. 1951: Amino acids in finger seed (Ragi kurakkan) *Eleusine coracana*. Current Sci. (India) 20:268.

1877. —— 1952: Amino acids in legumes. Nature 170:76–77.

1878. —— 1963: Free amino-acids in the endosperm of the developing coconut (*Cocos nucifera*). J. Exptl. Botany 14: 29–41.

1879. —— and B. P. M. Perera. 1956: Essential amino-acids of some tropical cereal millets. Brit. J. Nutr. 10:334–37. Nutr. Abstr. Rev. 27:1611. 1957.

1880. Baquiran, J. P. 1953: Effect of rate of seeding on yield and quality of fiber of *Corchorus capsularis* and *C. olitorius*. Philippine Agriculturist 37:159–74.

Barai, S. C. 5061.

1881. Barakat, M. Z., M. F. Abdel-Wahab, and S. A. El-Kinawi. 1962: Changes in the ascorbic acid content of seeds of *Vicia faba* induced by *beta* radiation. Monatsh. Chem. 93:754–60. Nucl. Sci. Abstr. 17: 2041. 1963.

Barakat, S. D. Y. 5922.

Bar-Akiba, A. 15446.

1882. Baram, B. M., and A. L. Markman. 1963: *Poterium polygamum* seed oil. Uzbeksk. Khim. Zh. 7(4):47–52. Chem. Abstr. 60:7038. 1964.

Baranova, T. A. 10811.

1883. Baranovskii, P. M. 1947: Biochemical processes in sprouting of soybean seeds. Izv. Akad. Nauk Kaz. SSR Ser. Fiziol. i Biokhim. Rast. 39:44–52. Chem. Abstr. 46:6706. 1952.

Baranowska, H. Dalkiewicz-; *see* Dalkiewicz-Baranowska, H.

1884. Baranowska, I., and H. Kozaczenko. 1960/61: Effect of herbicides on vegetable seed germination. Biul. Warzywniczy 5: 145–52. Chem. Abstr. 58:3834. 1963.

Baranska, K. Olechowska-; *see* Olechowska-Baranska, K.

1885. Barbat, I., and I. Puia. 1958: Stimulating and inhibiting action of some growth substances, antibiotics, and vegetable extracts on the germination of corn, wheat, and barley seeds. Acad. Rep. Populare Romine Filiala Cluj Studii Cercetari Agron. 9:51–56. Chem. Abstr. 53:15216. 1959.

1886. Barbee, O. E., and E. G. Schafer. 1939: Seed flax in eastern Washington. Wash. Agr. Expt. Sta. Bull. 370. 16 pp. Expt. Sta. Rec. 81:367. 1939.

1887. Barber, A. 1956: My selection of perennial flowers from seeds. Gardeners Chron. 139:710–11.

1888. Barber, J. C. 1957: A vacuum sowing device for experimental nurseries. J. Forestry 55:907.

Barber, J. C. 3723.

1889. Barber, J. T., and D. Boulter. 1963: Argininosuccinic acid in germinating seeds of *Vicia faba* L. Nature 197:1112.

Barber, J. T. 2856.

Barber, S. A. 11887.

1890. Barbera, C. E., D. Barbera, and F. Messina. 1956: Identification of cysteine, methionine, and proline in raw and roasted coffee. Coffee Tea Ind. 79:12–14.

Barbera, D. 1890.

Barbier, E. 2631.

1891. Barbieri, N. A. 1930: The physiological cultivation of orchids. Orchid Rev. 38(440):35–48.

1892. Barbour, E. H. 1932: Fossil seeds significant. Sci. News Letter 22:118.

1893. Barcinskij, R. 1934: On the germination of the seeds of *Orobanche cumana*. Compt. Rend. Acad. Sci. URSS 1:343–46.

Barclay, A. S. 5745.

Bardier, M. 4429.

Bardin, R. 7707, 20116.

1894. Bardinskaya, M. S., and T. A. Shubert. 1962: Phenolic compounds of cereals. Biochemistry (USSR) (English Transl.) 27:46–51.

1895. Bardner, R. 1960: Effect of formulation on toxicity to plants and insects of some systemic insecticidal seed dressings. J. Sci. Food Agr. 11:736–44.

1896. —— 1963: Some systemic insecticides compared as seed treatments against aphids. Plant Pathology 12:72–79.

1897. Bar-Droma, M. 1954: Harvesting

sunflower seeds for eating. Hassadeh 34: 605–6.

1898. Barenbrug, T. H. 1935: Veld improvement; seed production; seed testing. S. African J. Sci. 32:257–67.

1899. Barendse, G. W. M. 1963: Devernalization in *Cheiranthus allionii*. Proc. Biol. Med. Sci. Ser. C 66:183–88.

Baret, R. 13826–13828.

1900. Barford, R. A., S. F. Herb, F. E. Luddy, P. Magidman, and R. W. Riemenschneider. 1963: Alcoholysis of *Veronia anthelmintica* seed oil and isolation of methyl epoxyoleate. J. Am. Oil Chemists Soc. 40:136–38. Chem. Abstr. 59:844. 1963.

Bargar, E. L. 205.

1901. Barger, G., and J. J. Blackie. 1936: Alkaloids of *Senecio*. II. Senecionine and squalidine. J. Chem. Soc. 1936:743–45.

Barham, W. S. 11178.

Barinova, G. A. 10260.

1902. Bariola, R. 1913: Sull'anatomia del seme dell' *Abrus precatorius* L. (Jequirity) e dei semi usati per sofisticarlo. Atti Ist. Botan. Pavia 2, XVI. 16 pp. Botan. Centralbl. 125:418. 1914.

1903. Barkenbus, C., and C. F. Krewson. 1932: The oil of the bittersweet seed. J. Am. Chem. Soc. 54:3993–97.

1904. —— and S. T. Thorn. 1935: The oil of the rose mallow (*Hibiscus moscheutos* × *H. coccineus*) seed. J. Am. Chem. Soc. 57: 728–29.

1905. Barker, G. R., and T. Douglas. 1960: Function of ribonuclease in germinating peas. Nature 188:943–44.

Barker, M. B. 18874.

1906. Barker, M. F. 1932: Development of oil in the seed of the flax fiber crop and the variations in its character with maturity. J. Soc. Chem. Ind. 51:218T–222T. Chem. Abstr. 26:4971–72. 1932.

1907. Barkley, W. D. 1950: Do you know your seeds? Flower Grower 37(7):44–45, 65.

1908. —— 1951: Hitch-hiking seeds. Nat. Mag. 44:420–24.

Barkley, W. D. 12895.

1909. Barlow, A. L. 1954: Thomas walnut seed treatment with radioactive cobalt. Northern Nut Growers Assoc. Ann. Rept. 45:56–59.

1910. Barlow, K. E. 1955: Seeds from organic soils. Mother Earth 9:80–82.

1911. Barmenkov, I. P. 1961: The role of flaxseed chlorophyll in the process of oil synthesis. Tr. Orenburgsk. Sel'skokhoz. Inst. 12:369–78. Chem. Abstr. 58:4818. 1963.

1912. Barmington, R. D. 1962: Planter tests and seed preparation using monogerm sugar beet seed. J. Am. Soc. Sugar Beet Technologists 11:469–81.

1913. Barmore, M. A. 1947: Some characteristics of gliadin and glutenin indicated by dispersion and viscosity. Cereal Chem. 24:49–58. Biol. Abstr. 21:18252. 1947.

1914. Barna, A. K., K. Sachindra, Chakrabarte, and A. K. Ray. 1958: Chemical examination of the seeds of *Luffa acutangula* Roxb. J. Indian Chem. Soc. 35:480–82.

1915. Barnard, D. M. 1951: An evaluation of pellet seeding as it applies to range land. [Abstracts of theses and graduate degrees awarded January, 1949—June, 1951.] Wyoming Univ. Publ. 16:81–82.

1916. Barnard, P. D. W. 1959: On *Eosperma oxroadense* gen. et sp. nov.: a new lower carboniferous seed from East Lothian. Ann. Botany (London) 23:285–96.

1917. Barnard, R. C. 1950: Fruit collection and germination of certain Dipterocarpaceae. Malayan Forester 13:164–65. Biol. Abstr. 26:32202. 1952.

1918. Barner, H. 1958: Seed supply and forest tree breeding. Dansk Skovfor. Tidsskr. 43:1–84.

1919. —— and F. Dalskov. 1955: Notes on storage of Douglas fir seed. Proc. Intern. Seed Testing Assoc. 20:57–61.

1920. Barnes, A. S. L. 1933: The relation between collecting date and maturity of red pine seed. Forestry Chron. 9:48–59.

1921. Barnes, H. 1943: Growing vegetables for seed. Queensland Agr. J. 57:84–89.

1922. Barnes, J. E. 1940: Some experimental results with "hard seeds." Seed World 48(10):6–7.

1923. —— 1953: Soybean germination and the evaluation of soybean seedlings. Seed World 72(10):34–36.

1924. Barnes, J. H., and A. J. Grove. 1916: Insects attacking stored wheat in the Punjab, and the methods of combating them, including a chapter on the chemistry of respiration. India Dept. Agr. Chem. Ser. Mem. 4:165–280.

Barnes, O. L. 16559.

1925. Barnes, R. L. 1961: Adenine catabolism in pine embryos and tissue cultures. Botan. Gaz. 123:141–43.

1926. —— and A. W. Naylor. 1961: Utilization of radioactive ornithine by germinating embryos of longleaf and slash pines. Forest Sci. 7:130–35.

Baron, J. 7779.

Barone, G. 5148–5150.

1927. Baron-Hay, G. K. 1939: The sowing of pasture seeds. J. Agr. W. Australia 16:274–78.

1928. Baroni, E. 1891: Structure of seed of *Euonymus.* Nuovo Giorn. Botan. Ital. 23:513–21. J. Roy. Microscop. Soc. 1891:764.

1929. —— 1892: Seeds of *Hemerocallis.* Bull. Soc. Botan. Ital. 1:61–65.

1930. —— 1892: Fruit and seed of *Eugenia.* Bull. Soc. Botan. Ital. 1:275–83.

Barr, H. T. 4370.

1931. Barre, H. J. 1963: Important equipment for drying seeds in North America. Proc. Intern. Seed Testing Assoc. 28:815–26.

Barre, H. J. 8959.

1932. Barre, R., and G. Wormser. 1955: Proteins of the seeds of the almond tree. Ann. Pharm. Franc. 13:736–44. Chem. Abstr. 50:13187. 1956.

1933. Barreiro, E., and A. Charro. 1953: A nutritional study of the chestnut. Anales Bromatol. (Madrid) 5:453–90. Chem. Abstr. 48:7814. 1954.

1934. Barreiros, H. A. 1951: Production of rice seed. Gaz. Agr. 3:290–91.

1935. Barrer, P. R. 1952: Seed production in New Zealand. Seed cleaning: grasses and clovers. New Zealand J. Agr. 84:13–25.

1936. Barreto, M. B. 1952: Preliminary tests on seed germination. Gaz. Agr. 4:125–26.

1937. Barrett, L. I. 1931: Influence of forest litter on the germination and early survival of chestnut oak, *Quercus montana*, Willd. Ecology 12:476–84.

Barrett, M. S. 18826.

1938. Barrett, P. 1943: Gladiolus from seed. Gladiolus 18:76–78.

1939. Barrett, R. E. 1960: Germinating seeds of the American holly. Bull. Holly Soc. Am. 28:7–8.

1940. —— 1962: Embryogeny of *Ilex opaca* Ait. Dissertation Abstr. 23:1888.

1941. —— and J. H. McLaughlin. 1954: Disease-resistance factors in wheat. Electrophoretic and chromatographic analysis of protein extracts of wheat seedlings. J. Agr. Food Chem. 2:1026–29.

1942. Barrette, R. J. W. 1936: Germinating *Camellia* seeds. Horticulture 14(1):13.

1943. Barriga, C. 1961: Effects of mechanical abuse of Navy bean seed at various moisture levels. Agron. J. 53:250–51.

1944. Barring, V. 1956: The germination percentage of the seed of the coastal pine compared to that of the inland pine in Holland. Svenska Skogsvardsforen. Tidskr. 54:423–32.

1945. Barritt, N. W. 1929: The structure of the seed coat in *Gossypium* and its relation to the growth and nutrition of the lint hairs. Ann. Botany (London) 43:483–89.

Barron, D. H. 7235.

1946. Barron, E. S. G., editor. 1952: Modern trends in physiology and biochemistry. Academic Press, New York. 538 pp.

Barron, G. L. 12215.

Barron, R. K. Le.; *see* Le Barron, R. K.

1947. Barrons, K. C., and D. M. McLean. 1945: A study of the causes of low germination of radish seed crops. Mich. Agr. Sta. Quart. Bull. 27:398–408.

Barrons, K. C. 2894.

1948. Barros, F. 1949: A study on the effects of removing half of the seed upon the germination, growth, and yield of peanut. Philippine J. Agr. 14:13–17.

1949. Barrow, E. H. R. 1915: The heating of cottonseed—its causes and prevention. Ind. Eng. Chem. 7:709–12.

1950. Barrows, F. L. 1936: Propagation of *Epigaea repens* L. I. Cuttings and seeds. Contrib. Boyce Thompson Inst. 8:81–97.

1951. Barrus, M. F., and H. W. Rankin. 1932: Seed treatment of *Alpha* barley to control loose smut. Cornell Univ. Agr. Expt. Sta. Ann. Rept. 45:143–44.

1952. Barskaya, E. I., and E. Z. Oknina. 1959: Nucleic acids during the formation and growth of seeds in fruit crops. Plant Physiol. (USSR) (English Transl.) 6:98–100.

Barskaya, E. I. 14740.

1953. Barskaya, E. N., and E. Z. Oknina. 1959: The role of nucleic acids in the processes of growth and bud dormancy of fruit crops. Plant Physiol. (USSR) (English Transl.) 6:470–76.

1954. Barskii, N. 1932: Oil from the seeds of tomatoes. Maslob.-Zhir Delo 6:56. Chem. Abstr. 27:3352. 1933.

1955. —— 1939: Eine wirksame Methode zur bekämpfung der Milben auf den Sonnenblumensamen. Öl- Fett-Ind. 1939 (1):24–26. Centralbl. Bakteriol. Parasitol. Infektions. Abt. II 102:414. 1940.

1956. Barson, D. M., and R. J. Ballinger. 1944: Extraction of tomato, cucumber, and marrow seed with HCl. Agriculture (London) 51:178–84. Chem. Abstr. 38: 6479. 1944.

1957. Barsukov, L. N., and K. M. Zabavskaia. 1953: Effect of high-frequency vibrations on germination of seeds and development of plants. Agrobiologiya 1953(5):80–85.

1958. Barszczak, T., and F. Fafara. 1963: Influence of soaking seeds in borax solutions on their energy and germinating power and on their boron content. Roczniki Nauk Rolniczych: Ser. A 87:409–15. Chem. Abstr. 59:8060. 1963.

1959. —— and F. Stec. 1963: Influence of soaking seeds in manganese sulfate solutions on their energy and germinating power and on their manganese content. Roczniki Nauk Rolniczych: Ser. A 87:417–21. Chem. Abstr. 59:8060. 1963.

1960. Bartel, A. T. 1941: Green seeds in immature small grains and their relation to germination. J. Am. Soc. Agron. 33:732–38.

1961. Bartels, F. 1960: On the development of the embryo of *Epilobium hirsutum*. II. The cell divisions of the spical meristem during a plastochron. Flora (Jena) 149: 206–24. Biol. Abstr. 39:11735. 1962.

1962. —— 1960: On the development of the embryo of *Epilobium hirsutum*. III. Growth-tendencies which lead to the formation of a leaf primordium and of a cotyledon. Flora (Jena) 149:225–42. Biol. Abstr. 39:11736. 1962.

1963. Bartels, H. 1956: Über die Abhängigkeit der Keimlingsgrösse vom Gewicht und vom Entwicklungszustand des Samens bei Fichte und Kiefer. Allgem. Forst-Jagdzeitung 127:164–68.

1964. —— 1956: Über den Wassergehalt der Nadelholzsamen. Forstwiss. Centralbl. 75:65–128.

1965. —— 1957: Kultur isoliertir Koniferenembryonen. II. Naturwissenschaften 44: 595–96.

1966. —— 1958: Enzyme im keimenden Samen von *Pinus nigra* Arn. Naturwissenschaften 45:495.

1967. —— 1958: Untersuchungen über die Hitzetoleranz der Koniferensamen. Forstwiss. Centralbl. 77:287–94.

1968. —— 1960: Enzymaktivitäten in ruhenden und keimenden Samen von *Pinus nigra* Arn. Planta 55:573–97.

1969. —— 1960: Über die Physiologie der Samenkeimung. Forst- Holzwirt 15:363–66.

1970. Barth, A. 1935: The influence of the time of collection on the quality of spruce seeds and the development of spruce seedlings. Forstlige Forsok Grondahl & Sons Boktrykkeri, Oslo. 55 pp.

1971. Barthasarathy, S. V. 1940: Physiological studies during vernalization of rice. Madras Agr. J. 28:133–37. Chem. Abstr. 34:6976. 1940.

1972. Barthodeiszky, A. 1962: Grundlagen für eine Diskussion über die Reinheit, den Reinigungsgrad, die Beurteiling der äusseren Erscheinung und die Bestimmung der Herkunft des Saatgutes. Proc. Intern. Seed Testing Assoc. 27:498–511.

1973. Bartholomew, R. P. 1944: Seed treatment with plant hormones in crop production. Arkansas Agr. Expt. Sta. Bull. 444.

1974. ——, L. C. Carter, and J. E. Casey. 1942: Deterioration of bur clover seed under irrigated conditions. J. Am. Soc. Agron. 34:104–6.

1975. Bartl, K., P. Curth, H. E. Fisher, and H. Schneider. 1957: Untersuchungen über die Keimfähigkeit von polyploiden Zuckerrübensaatgut. Zucker 10:142–47.

1976. ——, ——, ——, and —— 1957: Untersuchungen über die Keimfähigkeit von polyploiden Zuckerrübensaatgut (Conclusion). Zucker 10:163–68.

1977. Barton, A. W. 1929: Studies in autolysis. I. Autolysis in seeds. Trans. Kansas Acad. Sci. 32:34–36. Biol. Abstr. 6:6563. 1932.

1978. —— 1929: The effect of long-continued soaking on seed germination. Trans. Kansas Acad. Sci. 32:37.

1979. Barton, D. W. 1964: Recent research on seed testing. Seed World 94(8):5–8.

1980. Barton, L. V. 1928: Hastening the germination of southern pine seeds. J. Forestry 26:774–85. Boyce Thompson Inst. Prof. Pap. 1:58–69. 1928.

1981. —— 1930: Hastening the germination of some coniferous seeds. Am. J. Botany 17:88–115. Contrib. Boyce Thompson Inst. 2:315–42. 1930.

1982. —— 1932: Germination of bayberry seeds. Contrib. Boyce Thompson Inst. 4:19–25.

1983. —— 1932: Effect of storage on the vitality of delphinium seeds. Contrib. Boyce Thompson Inst. 4:141–53.

1984. —— 1933: Germination and storage of delphinium seed. Bull. Am. Delphinium

Soc. 2:12–14. Boyce Thompson Inst. Prof. Pap. 1:248–50. 1933.
1985. —— 1933: Seedling production of tree peony. Contrib. Boyce Thompson Inst. 5:451–60.
1986. —— 1934: Dormancy in *Tilia* seeds. Contrib. Boyce Thompson Inst. 6:69–89.
1987. —— 1935: Storage of vegetable seeds. Contrib. Boyce Thompson Inst. 7:323–32.
1988. —— 1935: Storage of some coniferous seeds. Contrib. Boyce Thompson Inst. 7:379–404.
1989. —— 1935: Germination of delphinium seeds. Contrib. Boyce Thompson Inst. 7:405–9.
1990. —— 1936: Seedling production in *Carya ovata* (Mill.) K. Koch, *Juglans cinerea* L., and *Juglans nigra* L. Contrib. Boyce Thompson Inst. 8:1–5.
1991. —— 1936: Germination of some desert seeds. Contrib. Boyce Thompson Inst. 8:7–11.
1992. —— 1936: Germination and seedling production in *Lilium* sp. Contrib. Boyce Thompson Inst. 8:297–309.
1993. —— 1939: Experiments at Boyce Thompson Institute on germination and dormancy in seeds. Sci. Hort. 7:186–93.
1994. —— 1939: A further report on the storage of vegetable seeds. Contrib. Boyce Thompson Inst. 10:205–20.
1995. —— 1939: Germination and storage of lily seeds. Natl. Hort. Mag. 18:193–94.
1996. —— 1939: Storage of elm seeds. Contrib. Boyce Thompson Inst. 10:221–33.
1997. —— 1939: Storage of some flower seeds. Contrib. Boyce Thompson Inst. 10:399–427.
1998. —— 1940: Germination of seeds. Sci. Monthly 51:542–44.
1999. —— 1940: Some effects of treatment of non-dormant seeds with certain growth substances. Contrib. Boyce Thompson Inst. 11:181–205.
2000. —— 1940: Some effects of treatment of seeds with growth substances on dormancy. Contrib. Boyce Thompson Inst. 11:229–40.
2001. —— 1940: Toxicity of ammonia, chlorine, hydrogen cyanide, hydrogen sulphide, and sulphur dioxide gases. IV. Seeds. Contrib. Boyce Thompson Inst. 11:357–63.
2002. —— 1941: Do your garden seeds fail to grow? Sci. Dig. 9(2):65.
2003. —— 1941: Relation of certain air temperatures and humidities to viability of seeds. Contrib. Boyce Thompson Inst. 12:85–102.
2004. —— 1942: Some special problems in seed dormancy. Torreya 42:63–64.
2005. —— 1943: Effect of moisture fluctuations on the viability of seeds in storage. Contrib. Boyce Thompson Inst. 13:35–45.
2006. —— 1943: The storage of citrus seeds. Contrib. Boyce Thompson Inst. 13:47–55.
2007. —— 1944: Some seeds showing special dormancy. Contrib. Boyce Thompson Inst. 13:259–71.
2008. —— 1945: A note on the viability of seeds of maga, *Montezuma speciosissima*. Contrib. Boyce Thompson Inst. 13:423–26.
2009. —— 1945: Problems in seed germination. Plants & Gardens 1:174–81.
2110. —— 1945: Respiration and germination studies of seeds in moist storage. Ann. N.Y. Acad. Sci. 46:185–208.
2011. —— 1945: Viability of seeds of *Fraxinus* after storage. Contrib. Boyce Thompson Inst. 13:427–32.
2012. —— 1946: Storage and germination of delphinium seeds. Am. Delphinium Soc. 1946:27–30.
2013. —— 1947: Special studies on seed coat impermeability. Contrib. Boyce Thompson Inst. 14:355–62.
2014. —— 1947: Effect of different storage conditions on the germination of seeds of *Cinchona ledgeriana* Moens. Contrib. Boyce Thompson Inst. 15:1–10.
2015. —— 1948: Storage of seeds of the regal lily. North Am. Lily Soc. Yearbook 1947/48:51–53. Boyce Thompson Inst. Prof. Pap. 2:45–51. 1948.
2016. ——1949: Seed packets and onion seed viability. Contrib. Boyce Thompson Inst. 15:341–52.
2017. —— 1950: Relation of different gases to the soaking injury of seeds. Contrib. Boyce Thompson Inst. 16:55–71.
2018. —— 1951: Germination of seeds of *Juniperus virginiana* L. Contrib. Boyce Thompson Inst. 16:387–93.
2019. —— 1951: A note on the germination of *Viburnum* seeds. Arboretum Bull. (Wash. Univ.) 14(3):13–14, 27.
2020. —— 1952: Relation of different gases to the soaking injury of seeds. II. Contrib. Boyce Thompson Inst. 17:7–34.
2021. —— 1953: Dormancy in seeds. 13th Intern. Hort. Congr., 1952 Rept. 2:1001–12.
2022. —— 1953: Seed storage and viability. Contrib. Boyce Thompson Inst. 17:87–103.
2023. —— 1954: Effect of presoaking on dormancy in seeds. Contrib. Boyce Thompson Inst. 17:435–38.

2024. —— 1954: Effect of subfreezing temperatures on viability of conifer seeds in storage. Contrib. Boyce Thompson Inst. 18:21–24.

2025. —— 1954: Storage and packeting of seeds of Douglas fir and Western hemlock. Contrib. Boyce Thompson Inst. 18:25–37.

2026. —— 1955: Seeds. Dispersal, germination, dormancy, storage, testing, seed industry. Encyclopedia Americana 24: 527b–530b.

2027. —— 1956: Growth response of physiologic dwarfs of *Malus Arnoldiana* Sarg. to gibberellic acid. Contrib. Boyce Thompson Inst. 18:311–17.

2028. —— 1956: 6 Tables. *In* Handbook of biological data, W. S. Spector, editor.

2029. —— 1956: Seed sowing and germination. Roy. Hort. Soc. Suppl. Dictionary Gardening. Clarendon Press, Oxford, England, pp. 313–16.

2030. —— 1958: The gibberellins: powerful plant growth regulators. Trans. N.Y. Acad. Sci. Ser. II 20:717–32.

2031. —— 1960: Storage of seeds of *Lobelia cardinalis* L. Contrib. Boyce Thompson Inst. 20:395–401.

2032. —— 1960: Life-span of frost-damaged corn seeds. Contrib. Boyce Thompson Inst. 20:403–8.

2033. —— 1961: Anatomy of seeds. *In* Effects of ionizing radiations on seeds (Proc. Symp. Karlsruhe, 8–12 August 1960 jointly sponsored by the IAEA and FAO) Intern. Atomic Energy Agency, Vienna, pp. 24–42.

2034. —— 1961: Biochemical studies of dormancy and after-ripening of seeds. II. Changes in oligobasic organic acids and carbohydrates. Contrib. Boyce Thompson Inst. 21:147–62.

2035. —— 1961: Experimental seed physiology at Boyce Thompson Institute for Plant Research, Inc., Yonkers, N.Y., U.S.A., 1924–1961. Proc. Intern. Seed Testing Assoc. 26:561–96.

2036. —— 1961: Seed preservation and longevity. Leonard Hill, London. 216 pp.

2037. —— 1962: The germination of weed seeds. Weeds 10:174–82.

2038. —— and J. L. Bray. 1962: Biochemical studies of dormancy and after-ripening of seeds. III. Nitrogen metabolism. Contrib. Boyce Thompson Inst. 21:465–72.

2039. —— and C. Chandler. 1957: Physiological and morphological effects of gibberellic acid on epicotyl dormancy of tree peony. Contrib. Boyce Thompson Inst. 19:201–14.

2040. —— and W. Crocker. 1948: Twenty years of seed research. Faber & Faber, London. 148 pp.

2041. —— and J. M. Fine. 1958: Non-interference between effects of gibberellic acid and fungicides. Contrib. Boyce Thompson Inst. 19:291–94.

2042. —— and H. R. Garman. 1946: Effect of age and storage condition of seeds on the yields of certain plants. Contrib. Boyce Thompson Inst. 14:243–55.

2043. —— and A. Goenaga. 1952: An electrical moisture meter for the determination of moisture in coffee beans. Contrib. Boyce Thompson Inst. 16:461–68.

2044. —— and J. E. Hotchkiss. 1951: Germination of seeds of *Eichhornia crassipes* Solms. Contrib. Boyce Thompson Inst. 16:215–20.

2045. —— and J. MacNab. 1954: Effect of antibiotics on plant growth. Contrib. Boyce Thompson Inst. 17:419–34.

2046. —— and —— 1956: Relation of different gases to the soaking injury of seeds. III. Some chemical aspects. Contrib. Boyce Thompson Inst. 18:339–56.

2047. —— and E. M. Schroeder. 1942: Dormancy in seeds of *Convallaria majalis* L. and *Smilacina racemosa* (L.) Desf. Contrib. Boyce Thompson Inst. 12:277–300.

2048. —— and M. L. Solt. 1948: Growth inhibitors in seeds. Contrib. Boyce Thompson Inst. 15:259–78.

2049. —— and N. C. Thornton. 1947: Germination and sex population studies of *Ilex opaca* Ait. Contrib. Boyce Thompson Inst. 14:405–10.

2050. —— and S. F. Trelease. 1927: Stimulation, toxicity, and antagonism of calcium nitrate and manganese chloride as indicated by growth of wheat roots. Bull. Torrey Botan. Club 54:559–77.

Barton, L. V. 3656, 3802, 4600, 4601, 6377, 6983, 6984, 7221, 17029.

2051. Barton-Wright, E. C. 1938: Observations of the nature of the lipides of wheat flour, germ, and bran. Cereal Chem. 15:723–38.

2052. ——, R. G. Booth, and W. J. S. Pringle. 1944: Analysis of barley from King Tuthankhamen's tomb. Nature 153:288.

2053. —— and T. Moran. 1946: The microbiological assay of amino acids. II.

The distribution of amino acids in the wheat grain. Analyst 71:278–80.

Barton-Wright, E. [C.] 8050, 12973.

Bartoszewska, M. Veillet-; *see* Veillet-Bartoszewska, M.

2054. Bartsinskii, R. M. 1935: The agent stimulating sprouting of *Orobanche cumana* Wallr. seeds. Compt. Rend. Acad. Sci. URSS 2:306–13. Biol. Abstr. 11:13228. 1937.

2055. Barua, A. K. 1955: Triterpenoids. II. Entagenic acid. A new triterpene acid from the seeds of *Entada phaseoloides*. Sci. Cult. (Calcutta) 21:44–45. Chem. Abstr. 50:11284. 1956.

2056. ——, S. K. Chakraborti, and A. Kumar. 1958: Chemical examination of the seed of *Luffa acutangula*. J. Indian Chem. Soc. 35:480–82. Chem. Abstr. 53:8329. 1959.

2057. Barua, D. N. 1961: The significance of seed-size in cultivated tea (*Camellia sinensis* L.). Empire J. Exptl. Agr. 29:143–52.

Baruah, H. K. 4920.

Baruah, J. N. 3861.

2058. Baruch and Shimeon. 1954: Avocado germination experiments. Hassadeh 34:489–90.

Barwood, H. G. 8000.

Basak, M. 7203, 7204.

2059. Basak, M. N. 1951: Behaviour of jute seeds towards saline solutions during the process of germination. Sci. Cult. (Calcutta) 17:38–40. Biol. Abstr. 26:7190. 1952.

Basha, M. K. A. 8342.

2060. Bashaw, E. C. 1962: Apomixis and sexuality in Buffelgrass. Crop Sci. 2:412–15.

2061. —— and E. C. Holt. 1958: Megasporogenesis, embryo sac development, and embryogenesis in Dallisgrass, *Paspalum dilatatum*, Poir. Agron. J. 50:753–56.

Bashaw, E. C. 8982.

Basile, J. V. 6300.

2062. Baslas, K. K. 1955: Essential oil from seeds of *Aglaia odorotissima* [sic] Blume, Bijdr. J. Indian Chem. Soc. 32:445–49. 1955. Biol. Abstr. 30:32718. 1956.

2063. Basov, A. M., N. D. Potanina, and G. A. Iasnov. 1960: The electric field as a stimulating factor in improving the sowing qualities of grain seeds. Vestn. Sel'skokhoz. Nauki 1960(2):113–15.

Basov, A. M. 1591.

2064. Bass, A. A., and F. Saunders. 1933: A comparative study of the proteins of citrus seeds. Proc. Soc. Exptl. Biol. Med. 30:445.

2065. Bass, C. M. 1939: Sub-committee on viability of injured weed seeds. Proc. Assoc. Offic. Seed Analysts 31:68–72.

2066. Bass, L., and D. Isely. 1949: A proposed change in the rules for germination of Kentucky bluegrass. Assoc. Offic. Seed Analysts News Letter 23(1):19–22.

2067. Bass, L. N. 1948: Germination of freshly harvested oats. Proc. Assoc. Offic. Seed Analysts 38:47–52.

2068. —— 1950: Dormancy in redtop. Assoc. Offic. Seed Analysts News Letter 24(3):42.

2069. —— 1950: Effect of wave length bands of filtered light on germination of seeds of Kentucky bluegrass (*Poa pratensis*). Proc. Iowa Acad. Sci. 57:61–71.

2070. —— 1953: Notes on some proposed rule changes. Assoc. Offic. Seed Analysts News Letter 27(2):8–10.

2071. —— 1954: Factors affecting germination of Kentucky bluegrass seed. Iowa State Coll. J. Sci. 28:503–19.

2072. —— 1957: The effect of various concentrations of gibberellic acid upon germination of common Kentucky bluegrass and Merion Kentucky bluegrass seed. Agron. Abstr. 49:76.

2073. —— 1960: Packaging and storage of Kentucky bluegrass and creeping red fescue seed. Proc. Assoc. Offic. Seed Analysts 49:173–81.

2074. ——, and D. C. Clark, and E. James. 1963: Vacuum and inert-gas storage of safflower and sesame seeds. Crop Sci. 3:237–40.

2075. ——, ——, and —— 1963: Vacuum and inert-gas storage of Crimson clover and sorghum seeds. Crop Sci. 3:425–28.

2076. —— and E. James. 1961: Device for vacuum and inert-gas sealing of tin cans. Agron. J. 53:278.

2077. Bassarskaya, M. A. 1932: Prevention of germination of seeds during "yarovization." (Ukr. Inst. Genet.) Bull. Yarovization 2/3:87–104.

2078. —— 1951: Yarovization of rice. Selek. i Semen. 18(12):50–56.

2079. Bassiotis, K. 1960: Die Samen und Pflanzen Beschaffung in Griechenland. Dassika Chron. 2:958–69.

2080. Basson, J. K. 1959: A simple apparatus for the production of radiation-induced mutations in seeds. Intern. J. Appl. Radiation Isotopes 5:69–71.

2081. Bastin, R. 1952: Les substances

freinantes de la germination. Agricultura (Louvain) 50:66–81.

2082. Basu, A. K. 1962: Effects of chronic γ radiation in rice (*Oryza sativa* L.). Trans. Bose Res. Inst. (Calcutta) 25:49–52.

2083. Basu, K. P. 1941: Quality in pulses and cereals. Calcutta Med. J. 38:106–7. Chem. Abstr. 36:4155–56. 1942.

2084. —— and J. N. Karkun. 1943: Glucose dehydrogenase from germinated seeds of green and black grams (*Phaseolus radiatus* and *Phaseolus mungo*, Linn.). J. Indian Chem. Soc. 20:229–38. Biol. Abstr. 18:14634.

2085. —— and —— 1943: Succinic acid dehydrogenase from cucumber seeds. J. Indian Chem. Soc. 20:277–81. Biol. Abstr. 18:14635.

2086. ——, M. C. Nath, M. O. Ghani, and R. Mukherjee. 1937: Extraction and chemical analysis of the proteins of green gram (*Phaseolus mungo*), lentil (*Lens esculenta*), and *Lathyrus sativa* (Khesari). Indian J. Med. Res. 24:1027–42. Chem. Abstr. 31:8562. 1937.

2087. ——, ——, and R. Mukherjee. 1937: Biological value of the proteins of soybean, field pea, and *Lathyrus sativa*. By the balance-sheet and growth methods. Indian J. Med. Res. 24:1001–26. Chem. Abstr. 31:8562. 1937.

2088. Basu, R. K. 1954: Preliminary investigations with radioactive isotope (phosphorus^{32}) on jute seeds. Sci. Cult. (Calcutta) 19:515–16. Chem. Abstr. 49: 437. 1955.

Basu, S. K. 13889.

Basu, T. 4938.

Basu, U. P. 6058.

2089. Bates, C. G. 1930: Why nurserymen prefer southern seeds. J. Forestry 28:232–33.

2090. —— 1930: One-year storage of white pine seed. J. Forestry 28:571–72.

2091. —— 1930: Production, extraction, and germination of lodgepole pine seed. U.S. Dept. Agr. Tech. Bull. 191. 92 pp.

2092. —— 1931: A new principle in seed collecting for Norway pine. J. Forestry 29:661–78.

2093. Bates, E. N., and G. P. Bodnar. 1929: Smut on Pacific Coast wheat can profitably be removed by washing. Yearbook Agr. U.S. Dept. Agr. 1928:542–43.

2094. Bates, G. R. 1940: Growth behaviour of plants following seed treatment of organic mercury compounds. Nature 145:262–63.

2095. —— 1963: Certified tobacco seed. Commonwealth Phytopathol. News 9:1–3. Tobacco Abstr. 7:393. 1963.

2096. Bates, J. C. 1939: Seed germination and development of the seedling in *Commelina virginica*. Trans. Kansas Acad. Sci. 42:109–12.

2097. —— 1943: Varietal differences in anatomy of cross-section of wheat grain. Botan. Gaz. 104:490–93.

Bates, J. D. 8693.

2098. Bates, R. P. 1955: Effects of photoperiods on plant growth, flowering, seed production, and tannin content of *Lespedeza cuneata* Don. Agron. J. 47:564–67.

2099. Bate-Smith, E. C., and P. Ribereau-Gayon. 1959: Leucoanthocyanins in seed. Qualitas Plant. Mater. Vegetabiles 5:189–98. Chem. Abstr. 53:16293–94. 1959.

Bati, M. Ercoli-; *see* Ercoli-Bati, M.

Batisto, W. C. 1642.

2100. Batra, H. N., B. R. S. Rao, and S. K. Bhatia. 1959: A new record of phytophagous chalcidoid in fennel seed in India. Current Sci. (India) 28:451–52.

Batra, P. C. 15756.

2101. Battaglia, E. 1940: Embriologia di *Ligularia kaempferi* sieb. et Zucc. Nuovo Giorn. Botan. Ital. 47:271–86. Biol. Abstr. 15:11166. 1941.

2102. —— 1941: Contributo all'embriologia della Tamaricaceae. Nuovo Giorn. Botan. Ital. 48:575–612.

2103. —— 1947: Ricerche cariologiche e embriologiche sul genere *Rudbeckia* (Asteraceae). XII. Il gametofito femminile e maschile di *Rudbeckia flava* Greene, con particolare riguardo al suo comportamento di ibrido strutturale. Nuovo Giorn. Botan. Ital. 54:560–67.

Battaglini, A. 13091, 13092.

2104. Battandier, J. A. 1911: Experiences sur la germination d'une plante aquatique, le *Damasonium Bourgaei* Cosson. Compt. Rend. 152:1495–97.

Batten, E. T. 17914, 17915.

2105. Batterson, V. C., and W. M. Potts. 1951: The analysis and characterization of the oil from the seed of *Stillingia sylvatica*. J. Am. Oil Chemists Soc. 28:87–88. Biol. Abstr. 26:26051. 1952.

2106. Battiato, C. 1942: Sulla struttura del pistillo di *Citrus limonum* L. e sui processi istolitici che si compiono nelle ghiandole oleifere dell'epicarpo e nelle emergenze dell'endocarpo. Nuovo Giorn. Botan. Ital. 49:499–502.

2107. Battle, W. R. 1948: Effect of scarification on longevity of alfalfa seed. J. Am. Soc. Agron. 40:758–59.
2108. —— 1949: Seed production on excised red clover stems. J. Am. Soc. Agron. 41:141–43.
2109. Batzer, H. O. 1961: Jack pine from Lake States seed sources differ in susceptibility to attack by the White-pine weevil. U.S. Dept. Agr. Forest Serv. Lake States Forest Expt. Sta. Tech. Notes 595. 2 pp.
Baude, E. Hartl-. 7954.
2110. Baudet, J., and J. Mosse. 1962: Étude de l'extraction des protéines des graines. I. Extraction par l'eau des proteines de la farine de blé. Ann. Physiol. Végétale 4:315–31.
2111. —— and —— 1963: Extraction des protéines des graines. III. Extraction par l'eau et par les solutions salines des protéines de la farine de blé. Ann. Physiol. Végétale 5:303–19.
2112. Bauer, G. 1928: Die Keimungsbedingungen der Orchidaensamen. Gartenwelt 32:421–22.
2113. Bauer, K. H. 1934: Change in the composition of sunflower oil during ripening of the seed. Fettchem. Umschau 41:1–2. Chem. Abstr. 28:2208. 1934.
2114. —— and R. Neu. 1942: Coffee oil. II. Coffeol. Fette Seifen Anstrichmittel 49:419–28. Biol. Abstr. 18:9299. 1944.
Bauer, W. 3287.
Baughman, W. F. 9708.
Baum, F. 2800.
2115. Baum, H. 1948: Die Stellung der Samenanlagen am Karpell bei *Asclepias syriaca*, *Cynanchum vincetoxicum* und *Erythraea centaurium*. Österr. Botan. Z. 95:251–56.
2116. Bauman, L. F., T. F. Conway, and S. A. Watson. 1963: Heritability of variations in oil content of individual corn kernels. Science 139:498–99.
2117. Bauman, M. 1932: Hempseed oil. Vses. Akad. Sel'skokhoz. Nauk Lenina Inst. Novogo Lubyan. Suir'ya 2:113–24. Chem. Abstr. 27:3839. 1933.
2118. Baumann, L. A. 1959: Evaluating refined cottonseed oils in storage. J. Am. Oil Chemists Soc. 36:28–34.
Baumann, L. A. 19632.
2119. Baumeister, G. 1959: Eine Methode zur Keimungsbeschleunigung bei *Rubus*-Samen. Züchter 29:185–87.
2120. Baumeister, W. 1939: The influence of mineral fertilization upon yield and composition of grain of spring wheat. Bodenkunde Pflanzenernähr. 12:175–222. Chem. Abstr. 33:3953. 1939.
2121. Baur, F. 1880: Über Gewicht und Körnerzahl einiger Waldsamen pro Liter. Forstwiss. Centralbl. 2:341.
2122. Baur, F. J., Jr., and J. B. Brown. 1945: The fatty acids of corn oil. J. Am. Chem. Soc. 67:1899–900.
2123. Bautista, G. M., and P. Linko. 1962: Glutamic acid decarboxylase activity as a measure of damage in artificially dried and stored corn. Cereal Chem. 39:455–59.
Bautista, G. M. 10055.
2124. Bavappa, K. V. A., and K. J. Abraham. 1961: Influence of seed weight on the quality of seedlings in arecanut. Arecanut J. 12:129–35.
Bavappa, K. V. A. 18911.
2125. Bawdin, R. F. 1946: Plant hormone-like substances as related to barley germination. Proc. Am. Soc. Brewing Chemists 11:10–13. Biol. Abstr. 22:9974. 1948.
2126. Bawolski, S. 1958: Biology of germination of *Onobrychis viciaefolia* Scop. seeds. Plant Breed. Acclimitiz. Inst. Sci. Publ. 2:633–45.
2127. —— 1961: The influence of sun ray heat on the germination of sainfoin seeds. Inst. Uprawy Nawozenia Gleboznawstwa. Pamiet. Pulawski 1:193–206.
2128. Bayfield, E. G. 1946: The effect of wheat variety upon maltose values. J. Am. Soc. Agron. 38:624–29.
2129. —— and W. W. O'Donnell. 1945: Thiamine content of stored wheat. Food Res. 10:485–88.
Bayfield, E. G. 4443, 11694, 14670.
Bayle, L. 8480.
2130. Bayles, B. B., and F. A. Coffman. 1929: Effects of dehulling seed and of date of seeding on germination and smut infection in oats. J. Am. Soc. Agron. 21:41–51.
2131. —— and J. F. Martin. 1931: Growth habit and yield in wheat as influenced by time of seeding. J. Agr. Res. 42:483–500.
Bayles, B. B. 13412, 18409.
2132. Bayley, G. 1933: Direct seeding in Ontario. Forestry Chron. 9:66–78.
2133. Baylis, G. T. S. 1941: Fungi which cause pre-emergence injury to garden peas. Ann. Appl. Biol. 28:210–18.
2134. —— 1941: Viability of dusted wheat after storage. New Zealand J. Sci. Tech. Sect. A 23:126–30.
2135. ——, R. S. Deshpande, and I. F. Storey. 1943: Effect of seed treatment on

emergence of peas. Ann. Appl. Biol. 30:19–26.

2136. Bayliss, R. A. 1938: On the cytology and embryology of *Solanum citrullifolium* A. Br. and *Solanum Balbisii* Hort. Acad. Sci. RSS Ukraine Inst. Botan. Kiev 25:116–17.

2137. —— 1940: On the embryology and cytology of rye. I. Ontogenic development of winter rye (*Secale cereale* L.). Acad. Sci. RSS Ukraine Inst. Botan. Kiev 1:126–28.

Bayly, L. 4138.

Bazarnova, V. M. 11375.

2138. Bazhanova, A. P. 1957: Seed selection from the cotton plant, *Gossypium barbadense* L., as a method of increasing its productivity. Akad. Nauk Turkm. SSR Inst. Zemled. Tr. 1:19–34.

2139. Beach, G. R., Jr. 1950: The seed dormancy problem. Am. Iris Soc. Bull. 118:39–43.

Beachell, H. M. 9820, 9958.

2140. Beadle, N. C. W. 1940: Soil temperature during forest fires and their effect on the survival of vegetation. J. Ecol. 28:180–92.

2141. —— 1952: Studies in halophytes. I. The germination of the seed and establishment of the seedlings in five species of *Atriplex* in Australia. Ecology 33:49–62.

Beadles, J. R. 13588–13590.

Beal, J. L. 16841.

2142. Beal, J. M. 1943: Histological studies on parthenocarpic fruits of *Lilium regale* induced by growth substances. Botan. Gaz. 105:25–34.

2143. —— 1951: Negative results following exposure of several kinds of seeds to cosmic rays and other radiations at high altitudes. Botan. Gaz. 112:533–34.

2144. Beal, W. J. 1885: The vitality of seeds buried in soil. Mich Agr. Coll. Bull. 5:1–3.

2145. —— 1905: The vitality of seed. Botan. Gaz. 40:140–43.

Beals, C. C. 616.

Bean, R. C. 6598.

Bear, R. P. 5072.

2146. Beard, B. H., F. A. Haskins, and C. O. Gardner. 1958: Comparison of effects of X-rays and thermal neutrons on dormant seeds of barley, maize, mustard, and safflower. Genetics 43:728–36.

Beard, B. H. 3513, 4323.

2147. Beard, D. F., and E. A. Hollowell. 1952: The effect of performance when seed of forage-crop varieties is grown under different environmental conditions. 6th Intern. Grassland Congr. 1952 Proc., pp. 860–66.

2148. —— and W. Marshall. 1951: Do certified seeds or crop varieties run out? Seed World 68(4):14, 16.

2149. Beard, J. S. 1955: Seed research on *Liquidambar styraciflua* [the American sweetgum]. J. S. African Forestry Assoc. 26:77–78. Biol. Abstr. 32:35087. 1958.

Beardow, J. 6473.

Beasley, J. O. 16079, 16080.

Beaton, J. D. 16038.

Beattie, H. G. 7349.

2150. Beattie, J. H. 1931: Peanut seed may be kept for several years under proper conditions. Yearbook Agr. U.S. Dept. Agr. 1931:426–27.

2151. —— and V. R. Boswell. 1939: Longevity of onion seed in relation to storage conditions. U.S. Dept. Agr. Circ. 512. 23 pp.

2152. ——, A. M. Jackson, and R. E. Currin. 1932: Effect of cold storage and age of seed on germination and yield of peanuts. U.S. Dept. Agr. Circ. 233. 12 pp.

2153. —— and L. J. Kushman. 1947: Effect of moisture contents of peanuts on shelling damage. Peanut J. Nut World 27:35. Biol. Abstr. 23:16431. 1949.

2154. ——, F. W. Poos, and B. B. Higgins. 1954. Growing peanuts. U.S. Dept. Agr. Farmers Bull. 2063. 53 pp.

2155. —— and E. C. Tatman. 1950: Long-term storage of parsnip seed. Proc. Am. Soc. Hort. Sci. 55:435–37.

Beattie, J. H. 11586, 17914.

Beaucamp, G. 19540.

2156. Beauchesne, G. 1961: Research on growth substances of immature maize. Rev. Gen. Botan. 68:493–580.

2157. —— and J. P. Jouanneau. 1960: Progrès dans la technique de séparation et l'étude de la fraction newtre des substances de croissance des graines de Maio immature. Compt. Rend. 251:2396–99.

2158. —— and G. Netien. 1952: Séparation des acides amines d'un extrait de graines de mais immature a l'aide d'une résin synthetique et identification par chromatographie de partage sur papier. Compt. Rend. Soc. Biol. 146:1342–44.

2159. Beaudreau, G. S., and L. F. Remmert. 1955: Krebs cycle activity of particles from bean seedlings. Arch. Biochem. Biophys. 55:469–85.

2160. Beaudry, J. R. 1951: Seed development following the mating *Elymus virginicus*

L. × *Agropyron repens* (L.) Beauv. Genetics 36:109–33.
2161. Beaufait, W. R. 1960: Some effects of high temperatures on the cones and seeds of Jack pine. Forest Sci. 6:194–99.
2162. Beaufils, E. R. 1956: Physiological considerations on the composition of the fruit of *Hevea*. Oleagineaux 11:379–84. Chem. Abstr. 51:6788. 1957.
2163. Beaumont, A. 1954: Hot water treatment of seeds. Gardeners Chron. 136:216.
Beaumont, A. 1870.
2164. Beaumont, A. B., and G. J. Larsinos. 1928: Method for growing small-seeded plants under sterile conditions. Science 67:350–51.
2165. Beaumont, J. H. 1937: The evaluation of certain nut characters used in selecting varieties of *Macadamia*. Proc. Am. Soc. Hort. Sci. 34:235–37.
2166. Beauverie, J. 1919: Les méthodes de selection appliquées aux céréales de semences. État actuel de la question. Rev. Gen. Sci. Pures Appl. 30:79–87, 108–14. Botan. Abstr. 3:87. 1920.
Bechtel, W. G. 3527.
2167. Beck, F. V. 1944: The field seed industry in the United States. Univ. of Wisconsin Press, Madison. 230 pp. J. Am. Soc. Agron. 36:785–86. 1944.
2168. —— 1944: History and future of the seed business. Univ. of Wisconsin Press, Madison. 230 pp. N.Y. Botan. Garden J. 46:250. 1945.
2169. Beck, P., and J. S. Horton. 1932: Microsporogenesis and embryogeny in certain species of *Bromus*. Botan. Gaz. 93:42–54.
2170. Beck, W. A., and M. P. Schroeder. 1938: The relation of pigment production to the size of seeds. Studies Inst. Divi Thomae (Cincinnati) 2:101–5.
2171. Beckel, A. C., and J. L. Cartter. 1943: The effect of variety and environment on the equilibrium moisture content of soybean seed. Cereal Chem. 20:362–68.
2172. —— and F. R. Earle. 1941: A study of the moisture in soybeans. Ind. Eng. Chem. Anal. Ed. 13:40–43. Expt. Sta. Rec. 86:726. 1942.
2173. Becker, A. 1926: Über den Einfluss der Samenbehandlung mit Reizchemikalien auf die Keimung und das Wachstum. Landwirtsch. Jahrb. 63:501–61. Biol. Abstr. 1:11450. 1926.
2174. —— 1958: Zur Frage der Lagerung von gebeiztem Leguminosensaatgut. Saatgut-Wirtsch. 10:132–33.
2175. Becker, E., and O. Stelling. 1952: The protein of cacao bean. Zucker- Süsswarenwirtsch. 5:936–38. Chem. Abstr. 49:1988. 1955.
2176. Becker, H. 1912: Über die Keimung verschiedenartiger Früchte und Samen bei desselben Species. Inaug. Dissertation, Münster, pp. 1–129.
2177. Becker, H. A., and H. R. Sallans. 1955: A study of internal moisture movement in the drying of the wheat kernel. Cereal Chem. 32:212–26.
2178. Becker, K. E. 1928: Beobachtungen bei der Bestimmung der Triebkraft in Zink kästen. Prakt. Bl. Pflanzenbau Pflanzenschutz 28:181–88.
2179. Becker, Y., and L. Guyot. 1951: Sur les toxines racinaires des sols incultes. Compt. Rend. 232:105–7. Biol. Abstr. 26:19384. 1952.
2180. Becker-Dillingen, H. 1950: Pillensaatgut. Saatgut-Wirtsch. 2:276–77. Biol. Abstr. 25:24942. 1951.
2181. Beckett, E. 1937: Orchid propagation by seeds. Gardeners Chron. 41:309–10.
2182. Beckmann, S. 1963: Keimungs- und Entwicklungshemmende Wirkung von Bicyclo-[1.2.2]-heptan-lactonen. Naturwissenschaften 50:47.
Beckord, O. C. 16718.
2183. Beckwith, M. H. 1895: Effect of carbon bisulphide on the vitality of seed and plants. Delaware Agr. Expt. Sta. Rept. 1895:152–53. Expt. Sta. Rec. 8:497. 1896–97.
2184. Becquerel, P. 1904: Résistance de certaines graines a l'action se l'alcool absolu. Compt. Rend. 138:1179–81.
2185. —— 1904: Sur la permeabilité aux gaz de l'atmosphere, du tegument de certaines graines dessèchées. Compt. Rend. 138: 1347–49.
2186. —— 1904: De l'extraction complète de l'eau et de gaz de la graine a l'état de vie ralentie. Compt. Rend. 138:1721–23.
2187. —— 1905: Action de l'ether et du chloroforme sur des graines sèches. Compt. Rend. 140:1049–52.
2188. —— 1905: Action de l'air liquide sur la vie de la graine. Compt. Rend. 140:1652–54.
2189. —— 1906: Action de l'acide carbonique sur la vie latente de quelques graines dessèchées. Compt. Rend. 142:843–45.
2190. —— 1906: Sur la longevité des graines. Compt. Rend. 142:1549–51.

2191. —— 1906: Sur la respiration des graines à l'état de vie latente. Compt. Rend. 143:974–77.
2192. —— 1906: Sur la nature de la vie latente des graines et sur les véritables caractères de la vie. Compt. Rend. 143:1177–79.
2193. —— 1907: Recherches sur la vie latente des graines. Ann. Sci. Nat. Botan. 5/6:193–311.
2194. —— 1909: Sur la suspension momentanée de la vie chez certaines graines. Compt. Rend. 148 :1052–54.
2195. —— 1925: La suspension de la vie des graines dans le vide à la temperature de l'helium liquide. Compt. Rend. 181:805–7.
2196. —— 1932: La réviviscence des plantules dessèchées soumises aux actions du vide et des très basses temperatures. Compt. Rend. 194:2158–59.
2197. —— 1934: La longevité des graines macrobiotiques. Compt. Rend. 199 :1662–64.
2198. —— 1947: Uranium solution boosts yield from treated seed. Sci. News Letter 52:185.
2199. —— and J. Rousseau. 1947: Action de quelques substances de croissance sur la germination et la productivité des graines du *Pisum sativum*. Compt. Rend. 224: 700–2.
2200. —— and —— 1947: Action de quelques substances de croissance additionées soit de nitrate d'urane, soit de sulfate de manganese, sur la germination et la productivité du *Pisum sativum*. Compt. Rend. 224:773–75.
2201. —— and —— 1947: Action de quelques substances de croissance sur la germination et la productivité des graines du *Triticum vulgare*. Compt. Rend. 225:911–13.
2202. —— and —— 1947: Action de quelques substances de croissance additionnées soit de nitrate d'urane, soit de sulfate de manganese sur la germination et la productivité du blé. Compt. Rend. 225:1106–9.
2203. Becvarova, H., S. Leblova-Svobodova, and V. Jiracek. 1961: The metabolism of germinating rape. II. Sugar and phosphorus metabolism. Sb. Česk. Akad. Zeměděl. Ved Rostlinna Vyroba 7:1591–94. Chem. Abstr. 57:5041. 1962.
2204. Beddows, A. R. 1935: Seed setting and flowering in *Brachypodium pinnatum* Beauv., *B. sylvaticum* R. and S., *Desmazeria sicula* Dum., and *Lamarckia aurea* Moench. New Phytologist 34:326–38.
Bednar, K. 3354.
Bedri, A. 11762.
Beecher, F. S. 5467.
Beer, J. F. de; *see* De Beer, J. F.
2205. Beeson, C. F. C. 1946: The moon and plant growth. Nature 158:572–73.
Beeson, K. C. 9248.
Beeson, K. E. 15088.
2206. Beetle, D. E. 1951: The effect of DDT, triton, and xylene upon the germination of some crop plants. Wyoming Univ. Publ. 16:50–54.
2207. Beevers, H., and D. A. Walker. 1956: The oxidative activity of particulate fractions from germinating castor beans. Biochem. J. 62:114–20.
Beevers, H. 229, 230, 1471, 2271, 2272, 11223, 14267, 14654, 14655.
2208. Begemann, H. 1930: New methods for disinfecting coffee seed. II. Disinfection by fumigation with turpentine. Mededel. Proefsta. Malang. 76:11. Rev. Appl. Entomol. Ser A 19:193. 1931.
2209. Behaeghe, T. 1960: Conservation et utilisation des semences de graminées; premier essais. Bull. Agr. Congo Belge 51:1223–40.
2210. Behrendt, S. 1962: Possibilities of acceleration of the germination of perennial grass species by gibberellic acid. Eigenschaften Wirkungen Gibberelline, Symp. Giessen, Ger. 1960:195–99. Chem. Abstr. 60:6148. 1964.
2211. —— 1963: Effects of gibberellic acid on germination, growth, and development of perennial grass species. Z. Acker-Pflanzenbau 117:289–312. Chem. Abstr. 60:1044. 1964.
2212. Behrens, H. 1934: Beiträge zur Kenntnis der Hartschaligkeit von Leguminosensamen. Dissertation Hamburgh Inst. Angew. Botan. Proc. Intern. Seed Testing Assoc. 7:70–71. 1935.
2213. Behrens, J. 1906: Über die Beeinflussung der Keimfähigkeit gewisser Samen durch Narkose und Verwindung. Ber. Grossherzogisch Badischen Landwirtsch. Versuchsanstalt Augustenberg 1906:60–64. Botan. Centralbl. 107:538. 1908.
2214. Behrent, D. F. 1960: Seed aging. Seed World 86(10):8, 10, 12, 14, 17.
2215. Beijerinck, W., and J. Scurti. 1954: Two keys for the identification of "seeds." Res. Council Ontario, Toronto, Can. 22 mimeo. pp.
Beiler, A. 6801.

2216. Beilis, R. A. 1938: On the cytology and embryology of *Solanum citrullifolium* A. Br. and *Solanum Balbisii* Hort. Acad. Sci. RSS Ukraine Inst. Botan. Kiev 25:113–19.

Beilis, R. A. 13644.

Beinhart, E. G. 16201.

Bejnar, W. 6364, 6843.

2217. Bejowski, A. 1962: Physiology and biology of ripening wheat grain. II. Saccharides in the developing wheat grain. Roczniki Nauk Rolniczych Ser. A 85:447–60. Biol. Abstr. 39:16260. 1962.

2218. Beketovsky, A. 1932: Zur Frage über die Parthenokarpie der *Salix alba* L., *S. caprea* L., *Populus alba* L., and *Ulmus campestris* L. Botan. Zh. 17:385–400.

2219. Bekker, Z. 1933: The influence of carbonic acid on the microflora of the wheat grain. Microbiology (USSR) 2:360–68.

2220. —— 1934: Influence of preservation of wheat upon its microflora. Botan. Zh. 19:126–40. Biol. Abstr. 11:6735. 1937.

2221. Bela, A. 1929: Das absolute Gewicht der Heilpflanzensamen. Kiserl. Közlemen. 32:664–67. Ber. Ges. Physiol. 54:302. 1930.

Belajeva, G. M. 14254.

2222. Belderok, B. 1961: Studies on dormancy in wheat. Proc. Intern. Seed Testing Assoc. 26:697–760.

2223. —— 1962: Histochemical determination of bound disulfide groups in barley and wheat with regard to germ dormancy. Ber. Getreidechemiker-Tagung, Detmold 1962:21–25. Chem. Abstr. 58:9410. 1963.

2224. Belenovskii, V. N. 1959: Seed growing on collective and state farms of Kuibyshev Region in the seven-year-plan. Selek. i Semen. 1959(6):18–21.

2225. Belevtsev, D. N. 1962: Effect of agricultural engineering procedures on the accumulation of oil in sunflower seeds. Maslob.-Zhir. Prom. 28:8–10. Chem. Abstr. 57:11570. 1962.

Belgovskaya, A. A. Prokovea-. 10550.

Belia, B. P. 3563.

2226. Beliaiev, N. 1929: The seeds and oil of *Sinapis dissecta* Lag. Maslob.-Zhir. Delo 6:25–26. Chem. Abstr. 25:1631. 1931.

2227. —— 1929: Pumpkin-seed oil. Maslob.-Zhir. Delo 7:38–40. Soc. Chem. Ind. J. 50:210B. 1931.

2228. Belik, V. F. 1963: The effect of cucumber seed hardening under fluctuating temperatures on physiological features and cold tolerance. Plant Physiol. (USSR) (English Transl.) 10:351–57. Biol. Abstr. 45:21586. 1964.

2229. Belikov, I. F., R. Z. Im, I. V. Grusvitzky, and E. Khon. 1960: Preliminary results of testing seed stratification and light regime of ginseng. Akad. Nauk SSST Sibirskoe Otd. Dal'nevostoch. Filial im V. L. Komarova Marwe. k Izuch Zhen'shenya i Limonnika 4:97–104.

2230. —— and E. I. Nedel'ko. 1956: Changes in the chemical composition of soya seed during maturation under the Maritime Province environment. Biokhim. Zerna, Sb. 3:162–70. Chem. Abstr. 51:8914. 1957.

2231. Belikova, N. L. 1952: Polyembryony in beans. Byul. Mosk. Obshch. Ispytatelei Prirody Otd. Biol. 57(5):65–66.

2232. Belitser, N. V. 1963: On the embryology of *Zizania aquatica* L. Acad. Sci. RSS Ukraine Kiev J. Botan. 20:7–15.

2233. Beliz, J. V. C. M. 1948: Laboratory tests of rice seed germination compared with field results. Agros (Lisbon) 31:17–26.

2234. Belizin, A. 1936: Influence of vernalization on the comparative infestation of cereals with insect pests. Plant Protect. 1936(9):54–67. Rev. Appl. Entomol. Ser. A 25:393–94. 1937.

2235. Beljdenkova, A. F., V. F. Koriakina, and A. I. Smetannikova. 1945: How seed from some biennial vegetable crops can be produced in one year. Sov. Botan. (Leningrad) 13(5):29–35. Biol. Abstr. 21:23325. 1947.

Belkhode, M. L. 14235.

Bell, C. R. 17936.

2236. Bell, E. A. 1961: Isolation of a new amino acid from *Lathyrus tingitanus.* Biochim. Biophys. Acta 47:602–3.

2237. —— 1962: Associations of ninhydrin-reacting compounds in the seeds of 49 species of *Lathyrus*. Biochem. J. 83:225–29.

2238. —— 1962: The isolation of L-homorginine from seeds of *Lathyrus cicera.* Biochem. J. 85:91–93.

Bell, E. W. 18434.

Bell, F. J. 13413.

Bell, F. K. 11328.

2239. Bell, G. D. H. 1935: Preliminary experiments on vernalisation. J. Agr. Sci. 25:245–57.

2240. —— 1936: Experiments on vernalisation. J. Agr. Sci. 26:155–71.

2241. —— 1937: The effect of low-temperature grain pre-treatment on the development, yield, and grain of some varieties of wheat and barley. J. Agr. Sci. 27:377–93.

2242. —— 1946: Induced bolting and anthesis in sugar beet and the effect of selection of physiological types. J. Agr. Sci. 36:167–83.

Bell, H. K. 3307.

2243. Bell, H. P. 1957: The development of the blueberry seed. Can. J. Botany 35:139–54.

2244. Bell, J. M. 1955: The nutritional value of rapeseed oilmeal: A review. Can. J. Agr. Sci. 35:242–51.

2245. Bell. M. E. 1954: The development of the embryo of *Zea* in relation to position on the ear. Iowa State Coll. J. Sci. 29:133–40.

2246. Bell, P. M., and D. H. Simmonds. 1963: The protein composition of different flours and its relationship to nitrogen content and baking performance. Cereal Chem. 40:121–28.

2247. Bell, R. S., and E. Larssen. 1960: Experiments with the germination of northern nutgrass (*Cyperus esculentus*) seed. Proc. Northeast. Weed Control Conf. 14:45–48.

2248. Bell, R. W., L. S. Robertson, and R. L. Cook. 1945: The effect of shearing sugar-beet seed on stand of beets, on labor requirements at the time of blocking and thinning, and on yield of roots. Mich. Agr. Expt. Sta. Quart. Bull. 28:157–64.

Bell, R. W. 16288.

Bell, W. 18191, 19334.

2249. Bellando, M. 1963: Nonphosphorylated precursors of nucleic acids from cotyledons of *Phaseolus vulgaris* in the dormant period. Atti Accad. Nazl. Lincei Rend. Classe Sci. Fis. Mat. Nat. 34:560–62. Chem. Abstr. 60:7141. 1964.

Bellando, M. 3739.

Bellini, J. 13827.

2250. Bellini, P. 1958: Vernalization of perennial fodder grasses: preliminary research on Italian ryegrass (*Lolium italicum* Al. Br.). I. Ann. Sper. Agrar. (Rome) 12:1103–13.

2251. —— 1959: Vernalization of poly-annual fodder grasses. II. Ann. Sper. Agrar. (Rome) 13:143–56.

2252. Bellinzaghi, F. 1947: Comparative investigations on the biological action of electromagnetic radiation. II. Boll. Soc. Ital. Biol. Sper. 23:103–7. Biol. Abstr. 23:24810. 1949.

2253. Bellmann, K., and G. Meinl. 1961: Study on the relationship between seed size and yield of diploid and tetraploid red clover (*Trifolium pratense* L.). Züchter 31:358–62.

Bellrose, F. C., Jr. 12455.

Bellue, M. K. 1805.

2254. Belokhonov, J. V. 1935: Drying fruit trees seeds and preparing them for planting. Res. Inst. Fruit Growing I. W. Mitchurin, Mitschurinsk, Sci. Fruit Growing 4:18.

2255. Belonozhko, I., and T. Chebotareva. 1951: Yarovization of perennial grass seeds in order to obtain a full-value grass-legume mixture. Sov. Agron. 9:44–58.

Beloserova, N. 11260.

2256. Belostokov, G. P. 1962: Some data on the seed structure of certain tree species. J. Botan. URSS 47:1611–29.

2257. Belozerov, P. I. 1956: The germination of weed seeds contained in the arable stratum of the soil. Zemledelie 1956(12): 110–12. Biol. Abstr. 35:11412. 1960.

2258. Belozerskii, A. N., and I. V. Aseeva. 1958: Biosynthesis of nucleic acids during germination of wheat and pea seeds. Biokhim. Zerna, Sb. 4:22–23. Chem. Abstr. 53:509–10. 1959.

2259. —— and L. A. Chernomordikova. 1940: The nucleoproteids and nucleic acid of wheat germs. Biochemistry (USSR) (English Transl.) 5:133–39. Chem. Abstr. 35:1457. 1941.

2260. —— and I. I. Dubrovskaya. 1936: Proteins and thymonucleic acid in horse chestnut seeds. Biochemistry (USSR) (English Transl.) 1:665–75. Chem. Abstr. 31:3100. 1937.

2261. —— and I. S. Kornev. 1937: Comparative study of phosphatides of the embryos and cotyledons of soybeans. Biochemistry (USSR) (English Transl.) 2:894–901. Chem. Abstr. 32:7957. 1938.

Belozerskii, A. N. 1457.

Belozersky, A. 10575.

Belpaire, G. 12956.

Belschner, H. G. 17163.

2262. Belskaja, T. N. 1933: Vererbung und ontogenetische Verteilung von Samen veränderter Färbung bei *Phaseolus*. Sammelband "Phänogenetische Variabilität" Moskau 2:11–49. Botan. Centralbl. 168: 254. 1935.

Belsunce, M. de. 7824.

2263. Belton, W. E., and C. A. Hoover. 1948: Investigations on the mung bean (*Phaseolus aureus* Roxburgh). I. The determination of eighteen amino acids in the mung bean hydrolysate by chemical and microbiological methods. J. Biol. Chem. 175:377–83.

2264. Belyaev, N. V. 1962: The influence of cooling of tomato seeds on the biochemical processes in plants. Vopr. Fiziol. i Biokhim, Kul't. Rast., Akad. Nauk Moldavsk. SSR, Inst. Fiziol. i Biokhim. Rast. 1962(1):12–27. Chem. Abstr. 60:7147. 1964.

Bement, R. E. 9318.

2265. Bemis, W. P. 1957: Inheritance of a base seed-coat color factor in lima beans. J. Heredity 48:124–27.

2266. —— and W. A. Huelsen. 1955: Dehydration and rehydration in relation to endosperm fracturing of popcorn. Proc. Am. Soc. Hort. Sci. 65:371–80.

Bemis, W. P. 9183.

2267. Benary, E. 1923: Die Anzucht der Pflanzen aus Samen im Gartenbau; ein Handbuch für Gartenfreunde, Gärter und Samenhändler. P. Parey, Berlin. 355 pp.

2268. Bendana, F. E., and R. D. Powell. 1960: Germination of *Coffea arabica*. Assoc. South. Agr. Workers Proc. Ann. Conv. 57:244–45.

2269. Bendicenti, A., M. Bogliolo, P. Montenero, and M. A. Spadoni. 1957: Amino acid content of some varieties of Italian wheat with different nitrogen content. Quaderni Nutriz. 17:149–58. Chem. Abstr. 53:3394. 1959.

Benea, V. 9202.

2270. Benedetti, E. 1926: Action of a high frequency oscillating electromagnetic field on seeds. Atti Accad. Lincei Rend. 4:324–32. Biol. Abstr. 3:831. 1929.

2271. Benedict, C. R., and H. Beevers. 1961: Formation of sucrose from malate in germinating castor beans. I. Conversion of malate to phosphoenol-pyruvate. Plant Physiol. 36:540–44.

2272. —— and —— 1962: Formation of sucrose from malate in germinating castor beans. II. Reaction sequence from phosphoenol-pyruvate to sucrose. Plant Physiol. 37:176–78.

2273. Benedict, H. M., and H. Kersten. 1934: Effect of soft X-rays on germination of wheat seeds. Plant Physiol. 9:173–78.

2274. —— and J. Robinson. 1946: Studies on the germination of Guayule seed. U.S. Dept. Agr. Tech. Bull. 921. 48 pp.

Benedict, H. M. 6014, 13594, 13598, 13599, 19773.

Benedict, R. G. 8621.

Benedictis, G. de. 3717.

2275. Benediktov, I. A. 1939: Organization of the fight for high quality elite seeds and for the production of new varieties. Selek. i Semen. 9(5):4–7. Biol. Abstr. 15:5506. 1941.

Benero, J. R. 16699.

2276. Bengis, R. O., and R. J. Anderson. 1932: The chemistry of the coffee-bean. I. Concerning the unsaponifiable matter of the coffee-bean oil. Preparation and properties of Kahweol. J. Biol. Chem. 97:99–113.

2277. —— and —— 1934: The chemistry of the coffee-bean. II. The composition of the glycerides of the coffee-bean oil. J. Biol. Chem. 105:139–51.

Bengtsen, K. Korsan-. 14501.

Beni, G. 18640.

2278. Benlloch, M. 1931: Experiencias sobre la disinfección de semillas de remolacha. Bol. Patol. Veg. Entomol. Agr. (Madrid) 5:81–85. Biol. Abstr. 8:324. 1934.

2279. —— 1946: Contribución al establecimiento de un metodo para ensayer los desinfectantes contra el "Tizon" del trigo (*Tilletia tritici* (Bjerk) Wint. y *T. laevis* Kuhn). Bol. Patol. Veg. Entomol. Agr. (Madrid) 14:157–68.

Benne, E. J. 5315, 6732, 11842, 14567.

2280. Bennett, C. A., and G. N. Franks. 1953: Cottonseed handling with small air pipes. U.S. Dept. Agr. Circ. 768. 8 pp.

2281. Bennett, C. W., and A. S. Costa. 1961: Sowbane mosaic caused by a seed-transmitted virus. Phytopathology 51:546–50.

2282. Bennett, E. 1947: The hemicelluloses of maize cobs and rye straw. J. Agr. Res. 75:43–47.

Bennett, E. [H.] 9889, 11481.

2283. Bennett, H., E. Brown, and H. T. Islip. 1950: *Licania mollis* and *L. venosa* fruits from British Guiana. Colonial Plant Animal Prod. 1:232–36. Biol. Abstr. 25:37492. 1951.

Bennett, H. O. 14659.

2284. Bennett, H. W. 1959: The effectiveness of selection for the hard-seeded character in Crimson clover. Agron. J. 51:15–16.

2285. Bennett, M. K. 1954: The world's food. Harper Brothers, New York. 282 pp.

2286. Bennett, N., and W. E. Loomis. 1949: Tetrazolium chloride as a test reagent for freezing injury of seed corn. Plant Physiol. 24:162–74.

2287. Bennion, E. B. 1924: Effect of germination on the aleurone layer. Cereal Chem. 1:179–83.

Ben-Shamai, M. H. 447.

2288. Benson, B. 1949: "Shot in the arm"

for alfalfa. Makers guarantee hormone spray increases seed yields by holding bloom longer. Southern Seedsman 12(8): 23, 43.

Benson, L. C. 19695.

2289. Benson, S. R., and H. N. Calderwood. 1936: The chemical examination of the seed of *Abies balsamea* (L.) Miller. J. Am. Chem. Soc. 58:523–25.

2290. Bentancur, M. O. 1939: Diferentes metodos de cura contra el carbon (*Ustilago bromivora* [Tul.]) de la cebadilla australiana (*Bromus unioloides* [Willd.] H.B.K.). Arch. Fitotec. (Uruguay) 3:170–73. Biol. Abstr. 18:1685. 1944.

Bentley, C. F. 16131.

2291. Bentley, H. R., J. A. Henry, D. A. Irvine, and F. S. Spring. 1953: Triterpene resinols and related acids. XXVIII. The non-saponifiable fraction from *Strychnos nux-vomica* seed fat: The structure of *cyclo* Artenol. J. Chem. Soc. 1953:3673–78.

2292. Bentley, J. A., S. Housley, and G. Britton. 1956: Hormones and hormone precursors in leaves, roots, and seeds. Chem. Mode Action Plant Growth Substances Proc. Symp. Wye 1955:40–51. Chem. Abstr. 53:18198. 1959.

Bentley, J. A. 6240.

Benze, R. 6752.

2293. Benzie, J. W. 1960: Viability of Balsam fir seed depends on age of tree. U.S. Dept. Agr. Forest Serv. Lake States Forest Expt. Sta. Tech. Notes 591. 1 p.

2294. Ber, A. 1949: New ideas on the mode of action of auxins. Acta Biol. Exptl. Suppl. (Warsaw) 15:24–27. Biol. Abstr. 24:13457. 1950.

2295. Beran, V., and Z. Krofta. 1962: Experiments with the desiccation of seed alfalfa in middle Bohemia in 1961. Socialist. Zeměděl. 12:704–10. Chem. Abstr. 58:1718. 1963.

Berardi, L. C. 12922.

Berdichevskii, E. I. 9558.

2296. Berducou, J., R. Mache, and M. Kiredjian. 1962: Influence du molybdene sur la germination du mais et pénétration de cet élément dans la plantule. Compt. Rend. 254:3890–92.

2297. Berducou, J., and J. Phipps. 1961: Action of $CuSO_4$ on the germination of lettuce. Bull. Soc. Hist. Nat. Toulouse 96:75–79. Biol. Abstr. 45:3146. 1964.

Berenbom, M. 16668.

Berenghi, I. 1291.

2298. Beresford, S. W., and A. A. Jackson. 1961: The nutrition of garden beet in relation to seed production. J. Hort. Sci. 36:153–59.

Berezina, N. M. 2997.

2299. Bereznegovskaya, L. N. 1962: Nitrogen metabolism of belladonna seed during the process of stratification. Sb. Tr. Posvyashch. 20-Letiyu (Dvadtsatiletiyu) Farmatsevt. Fak. 1962:97–99. Chem. Abstr. 59:10479. 1963.

2300. Bereznitskaya, N. I. 1957: The effect of microelements in seed soaking on the growth of corn. Zap. Khar'kovsk. Sel'skoe Khoz. Inst. 13(50):83–91. Biol. Abstr. 35:48827. 1960.

2301. Berezovikov, A. D., and V. G. Klimenko. 1962: On the proteins of broad bean seeds (*Vicia faba* L.). Dokl. Akad. Nauk SSSR 144:475–77.

2302. —— and —— 1962: Proteins of seeds of feed beans. Dokl. Akad. Nauk SSSR 144:659–61. Chem. Abstr. 57:11553. 1962.

Berezovikov, A. D. 10807, 10808.

Berg, A. 11035.

Berg, M. A. 13238, 13239, 13241.

2303. Berg, R. Y. 1954: Development and dispersal of the seed of *Pedicularis silvatica*. Nytt Mag. Botan. 2:1–60. Biol. Abstr. 29:26861. 1955.

2304. —— 1962: Contribution to the comparative embryology of the Liliaceae: *Scoliopus*, *Trillium*, *Paris*, and *Medeola*. Oslo Univ. Press, Oslo. 64 pp.

Bergal, G. 3705, 3706, 17868.

Bergamaschi, M. 15506.

2305. Bergan, J. 1958: Sowing experiments with seeds of Scotch pine at different times during the growing season in Tromms and Finnmark. Norsk Skogbruk 4:69–71.

Bergdoll, M. S. 5503, 5504.

Bergen, H. J. 17606.

2306. Berger, C. A., E. J. Feeley, and E. R. Withus. 1956: The cytology of *Xanthisma texanum* D.C. IV. Megasporogenesis and embryo sac formation, pollen mitosis, and embryo formation. Bull. Torrey Botan. Club 83:428–34.

2307. Berger, F. 1938: Contribution to the knowledge of some tropical seeds and their composition. Sci. Pharm. 9:122. Pharm. Abstr. 6:304. 1940.

2308. —— 1938: Tropical seeds and their contents—contribution to the knowledge of some. Sci. Pharm. 11:122–24. Pharm. Abstr. 5:366. 1939.

2309. —— 1958: The anatomy of the seeds of the genera *Elettaria*, *Amomum*, and

Aframomum. Sci. Pharm. 26:224–58. Biol. Abstr. 34:14198. 1959.

2310. Berger, J., and G. S. Avery, Jr. 1944: Isolation of an auxin precursor and an auxin (indoleacetic acid) from maize. Am. J. Botany 31:199–203.

2311. —— and —— 1944: Chemical and physiological properties of maize auxin precursor. Am. J. Botany 31:203–8.

Berger, J. 1540–1542.

Berger, K. C. 6425, 10752.

2312. Berger, P. 1959: Hot air treatment of tobacco seed. Ber. Inst. Tabakforsch. Dresden 6:47–60. Tobacco Abstr. 3:467. 1959.

2313. —— 1962: How germination of tobacco seeds is influenced by physical and chemical pretreatments. Ber. Inst. Tabakforsch. Dresden 9:75–103.

2314. Berger, X. 1953: Untersuchungen über die Embryologie Parteill apomiktischer Rubusbastarde. Ber. Schweiz. Botan. Ges. 63:224–66.

Bergeret, B. 3446.

2315. Bergersen, F. J., J. Brockwell, and J. A. Thompson. 1958: Clover seed pelleted with Bentonite and organic material as an aid to inoculation with nodule bacteria. J. Australian Inst. Agr. Sci. 24:158–60.

2316. Berggren, G. 1960: Seed characters of certain cultivated and wild growing *Brassica* and *Sinapis* species and a key for the determination of these species on seed samples. Sweden Stat. Cent. Frökontanst. Medd. 35:28–64.

2317. —— 1962: Reviews on the taxonomy of some species of the genus *Brassica*, based on their seeds. Svensk Botan. Tidskr. 56:65–135.

2318. Bergh, T. K. 1949: A method of seed cleaning. Am. Nurseryman 89(8):7–8.

2319. —— 1960: Will seed from northern plants produce plants hardier than those from southern regions? Proc. Plant Propagators Soc. 10:141–54.

2320. Bergman, B. 1941: Studies on the embryo sac mother cell and its development in *Hieracium* Subg. *Archieracium*. Svensk Botan. Tidskr. 35:1–42.

2321. —— 1944: A contribution to the knowledge of the embryo sac mother cell and its development in two apomicts. Svensk Botan. Tidskr. 38:249–59.

2322. Bergman, F. 1955: Experiments with forced seed production of pine, spruce, and birch. Svenska Skogsvardsforen. Tidskr. 53:275–304.

2323. —— 1960: Attempts at increasing the growth energy and viability of forest seeds. Svenska Skogsvardsforen. Tidskr. 58:15–36.

2324. Bergman, H. F. 1932: Intracarpellary fruits and other central proliferations of the floral axis in *Hibiscus*. Am. J. Botany 19:600–3.

Bergman, R. J. 15298.

Bernal, D. Mesa-; *see* Mesa-Bernal, D.

2325. Bergquist, A. M. S., and N. Nielsen. 1959: The influence of gibberellic acid on the transaminase content of germinating barley seeds. Physiol. Plantarum 12:386–89.

Bergquist, S. 11308.

Bergqvist, G. 14398.

Bergsteinsson, H. N. 11802.

2326. Bergstrom, F. T. 1950: Growing camellias from seed. Am. Camellia Yearbook 1950:236–37.

2327. Bergtheil, C., and D. L. Day. 1907: On the cause of the "hardness" in the seeds of *Indigofera arrecta*. Ann. Botany (London) 21:57–60.

Beri, R. M. 15757.

Beridze, R. K. 10219.

2328. Berkeley, C. J. 1927: "Hard seeds" in Leguminosae. Nature 119:198.

2329. Berkeley, G. H., and G. O. Madden. 1932: Transmission of streak and mosaic diseases of tomato through seed. Sci. Agr. 13:194–97.

2330. —— and —— 1933: Transmission of streak and mosaic diseases of tomato through seed. II. Sci. Agr. 13:455–57.

2331. Berkeley, M. J. 1855: Action of sea water on the vitality of seeds. Gardeners Chron. 1855:580.

2332. Berkner, F. 1935: Der Einfluss der Herkunft des Samens auf den Ertrag der Zuckerrübe. Zuckerrübenbau 17:33–40.

2333. —— and W. Schlimm. 1929: Der Einfluss der Herkunft von Weizenkörnern auf ihre Keimenergie. Wiss. Arch. Landwirtsch. 2:545–62.

2334. —— and —— 1930: Physikalische und chemische Untersuchungen über den Einfluss der Ernte in den verschiedenen Reifestadien auf die Wertgebenden. Bestandteile des Getreidekorns. Landwirtsch. Jahrb. 72:269–91. Biol. Abstr. 6:9312. 1932.

2335. —— and —— 1932: Kritische Beitrage zur Frage der Saugkraftmessungen an unseren Getreidearten. Landwirtsch. Jahrb. 75:499–530.

2336. Berlyn, P. G. 1962: Developmental

patterns in pine polyembryony. Am. J. Botany 49:327–33.
Berman, A. L. 19677.

2337. Bernal, D. M. 1953/54: The use of 2,3,5-triphenyltetrazolium chloride for the determination of the viability of seeds. Colombia Min. Agr. Div. Invest. Inform. Tec. 1:79–129. Chem. Abstr. 49:11782. 1955.

2338. Bernard, C. 1917: The germination and selection of tea seed. Indian Tea Assoc. Sci. Dept. Quart. J. 1:1–12.

2339. Bernard, M. 1899: On the germination of *Neottia nidus avis*. Compt. Rend. 128:1253–55.

2340. Bernard, N. 1903: La germination des Orchidées. Compt. Rend. 137:483–85.

2341. —— 1909: L'évolution dans la symbiose. Les orchidées et leur champignons commensaux. Ann. Sci. Nat. Botan. 9:1–191.
Bernard, P. 1750.
Bernardini, J. V. 11061.

2342. Bernardini, L. B., and F. Gallucis. 1912: The pentosans in the germination of seeds. Staz. Sper. Agr. Ital. 45:874–84. Expt. Sta. Rec. 29:525. 1913.
Bernardo, F. A. 18886.
Bernhard, K. 3888–3890.

2343. Bernhard, R. A. 1959: Some studies of coumarin and coumarin analogues as germination inhibitors of radish seeds. Botan. Gaz. 121:17–21.

2344. Berninger, E. 1962: Germ contents of maize seeds, adaptation factor. Ann. Amelioration Plantes 12:117–22.

2345. Bernstein, L. 1943: Amylases and carbohydrates in developing maize endosperm. Am. J. Botany 30:517–26.

2346. —— 1943: Hybrid vigor in corn and the mobilization of endosperm reserves. Am. J. Botany 30:801–9.
Bernstein, L. 5837, 19981.

2347. Bernstrom, P. 1950: Cleisto- and chasmogamic seed setting in di- and tetraploid *Lamium amplexicaule*. Hereditas 36:492–506.
Bernton, H. S. 17801.

2348. Beroza, M., and M. L. Kinman. 1955: Sesamin, sesamolin, and sesamol content of the oil of sesame seed as affected by strain, location grown, ageing, and frost damage. J. Am. Oil Chemists Soc. 32:348–50.

2349. Berrie, A. M. M. 1960: Abnormal growth and development in pea resulting from exposure to adverse conditions during germination. Nature 185:626–27.
Berrie, A. M. M. 5575.

2350. Berry, E. W. 1929: Seeds of a new species of Vitaceae from the Wilcox Eocene of Texas. J. Wash. Acad. Sci. 19:39–41.

2351. —— 1929: A palm nut of *Attalea* from the upper Eocene of Florida. J. Wash. Acad. Sci. 19:252–55.

2352. Berry, F. H. 1960: Germination of Chinese chestnut (*Castanea mollissima*) seed. Northern Nut Growers Assoc. Ann. Rept. 51:40–42.

2353. Berry, G. H. 1953: Germination of alpine seeds. Alpine Gardeners Soc. Quart. Bull. 21:259–63.

2354. —— 1954: *Campanula morettiana* seed. Alpine Gardeners Soc. Quart. Bull. 22:260–61.

2355. Berry, W. E. 1948: Fossil seeds from South Carolina. Bull. Geol. Soc. Am. 59:1311–12. Biol. Abstr. 25:34338. 1951.
Berry, W. H. 9196.
Berryman, G. H. 6700, 12231.
Berryman, P. C. 291.

2356. Bertagni, P. 1939: Study on vitamins in sweet sorghum. I. Determination of aneurine in seeds of *Sorghum saccaratum*. Riv. Biol. 28:282–86. Chem. Abstr. 37:6003. 1943.

2357. Berthoud, E. L. 1892: Dissemination of plants by buffaloes. Botan. Gaz. 17:321–26.

2358. Bertrand, D. 1959: Lithium content of seed. Compt. Rend. 249:333–35. Chem. Abstr. 54:3600. 1960.
Bertrand, D. 2360, 2361.

2359. Bertrand, G. 1942: Recherches sur la teneur en bore des graines. Ann. Agron. 12:189–92. Biol. Abstr. 20:1667. 1946.

2360. —— and D. Bertrand. 1946: Sur la présence générale du rubidium chez les plantes. Ann. Agron. 16:1–6. Biol. Abstr. 21:4387. 1947.

2361. —— and —— 1951: Additional note on the rubidium content of seeds. Compt. Rend. 233:1539. Chem. Abstr. 46:5672d. 1952.

2362. —— and L. Silberstein. 1933: Le soufre et le phosphore dans les diverses parties de grain de blé. Compt. Rend. 197:285–88.

2363. —— and —— 1941: Boron content of seeds. Compt. Rend. 213:221–24. Chem. Abstr. 37:2772. 1943.

2364. —— and —— 1942: Distribution of boron in various parts of seeds. Compt. Rend. 214:41–45. Chem. Abstr. 37:5108. 1943.

2365. —— and —— 1944: The distribution

of boron in different parts of seed. Ann. Agron. 14:257–60. Chem. Abstr. 40:2501. 1946.

2366. —— and —— 1945: Le bore dans le grain de blé, la farine et le pain. Ann. Inst. Pasteur 71:445–48. Biol. Abstr. 21:25991. 1947.

2367. —— and —— 1953: Sur la teneur des graines en manganese. Compt. Rend. 237:481–82.

2368. Bertrand, H. W. R. 1941: The storage of rice in steel drums. Trop. Agriculturist 97:188–89.

2369. Besemer, A. F. H. 1950: Control of diseases in seed field cultures. II. Culture of onion and shallot seeds. Zaadbelangen 4:370–72.

Besse, C. 576.

2370. Bessey, C. E. 1893: Seed of *Salsola kali tragus*. Nebraska Agr. Expt. Sta. Bull. 6:67–71.

2371. —— 1904: Botanical notes. The number and weight of cottonwood seeds, and weight of dandelion down. Science 20:118–19.

2372. —— and E. Hyatt. 1894: Structure of the wheat grain. Nebraska Agr. Expt. Sta. Bull. 6:100–10.

2373. Bessonov, I. 1953: Our experience in processing cotton seeds for planting. Khlopkovodstvo 1953(2):56–62.

Best, K. F. 6551.

2374. Beth, K. 1938: Untersuchungen über die Auslösung von Adventivembryonie durch Wundreiz. Planta 28:296–343.

2375. Bethke, R. M. 1947: Niacin and pantothenic acid content of corn. Farm science and practice. Ohio Agr. Expt. Sta. Bull. 673, p. 43.

Bethke, R. M. 5399, 9215–9218.

2376. Bethmann, W., and S. Langer. 1962: Keimprüfung von Saatgut auf Aktivkohle-Filterpapier-Prüfung von β-Rübensamen. Proc. Intern. Seed Testing Assoc. 27:672–74.

Bethmann, W. 5807.

2377. Bethune, J. E., and E. R. Roth. 1960: Source of seed affects growth of longleaf pine (*Pinus palustris*), fifth year results. U.S. Dept. Agr. Forest Serv. Forest Prod. Lab. Rept. 146. 2 pp.

2378. Betzema, J. 1960: Seeding of cauliflowers. Tuinbouwberichten 24:210–12. Bibliog. Agr. 25:19197. 1961.

2379. Beumer, H. 1933: Über die Sterinbildung bei der Keimung. Biochem. Z. 259:469–70

2380. Beuster, K. H. 1958: Höhere Samenleistung beim spätschossenden Deutschen Weidelgras. Praxis Forsch. Fortschrittl. Landwirt 10:45–47.

2381. Bevacqua, R. 1949: The phenomenon of interaction in germinating seeds. Natura (Milan) 40:7–10. Biol. Abstr. 24:5011. 1950.

2382. Bevensee, E. 1914: Über die Fehler der Keimprüfungen. Kiel. 53 pp.

Bevenue, A. 19703.

Bever, D. N. 3943.

Bever, W. M. 10469, 10972, 18372.

2383. Beveridge, J. L., and C. P. Wilsie. 1959: Influence of depth of planting, seed size, and variety on emergence and seeding vigor in alfalfa. Agron. J. 51:731–34.

2384. Bevilacqua, L. R. 1955: Latest contribution to the enzymatic physiology of germination. Atti Accad. Naz. Lincei Rend. Classe Sci. Fis. Mat. Nat. 18:214–22.

Bevilacqua, L. Riggio-; *see* Riggio-Bevilacqua, L.

Bevilacqua, L. Rodriquez-; *see* Rodriquez-Bevilacqua, L.

2385. —— 1957: Physiological action of hydrazine on higher plants. II. Effect on seed germination and seedling growth. Atti Accad. Ligure Sci. Lettere (Genoa) 13:241–49. Chem. Abstr. 52:20451. 1958.

2386. —— and L. C. Pignato. 1961: Dehydrogenase activity in quiescent seeds of Graminaceae and Leguminosae. Atti Accad. Ligure Sci. Lettere (Genoa) 17:227–36. Chem. Abstr. 57:11553–54. 1962.

2387. Bevilotti, V. 1944: Biological action of radioactive substances. II. Action of colloidal thorium oxide on germination of seeds. Boll. Soc. Ital. Biol. Sper. 19:283–85. Chem. Abstr. 41:1733–34. 1947.

2388. —— 1945: Biological action of radioactive substances. VI. Action of uranium on the germination of vegetable seeds. Boll. Soc. Ital. Biol. Sper. 20:558–59. Gt. Brit. Imp. Bur. Soil. Sci. Soils Fertilizers 10:160. 1947.

Bevoise, F. de; *see* De Bevoise, F.

Bexon, D. 8944.

2389. Beyer, A. 1927: Zur Keimungsphysiologie von *Avena sativa*. Ber. Deut. Botan. Ges. 45:179–87.

2390. Beyers, E. 1939: Relationship between "drop" and seedlessness in Waltham Cross grapes. S. African Dept. Agr. Forestry Div. Plant Ind. Low Temp. Res. Lab. Capetown Rept. 1937/38:87–90.

2391. Beyle, M. 1928: Seltene Früchte und Samen der norddeutschen Flora. Verhandl.

Ver. Naturwiss. Unterhalt. Hamburg 20: 78–92. Biol. Abstr. 6:8141. 1932.

2392. Beyliss-Wyrowaja, R. A. 1941/45: On the embryology and cytology of rye. II. Observations on the ripening and germination of immature caryopses of spring and winter rye (*Secale cereale* L.). Acad. Sci. RSS Ukraine Inst. Botan. Kiev 2:55–57.

2393. Bhaduri, P. N. 1936: Studies on the embryogeny of the Solanaceae. I. Botan. Gaz. 98:283–95.

Bhaduri, P. N. 1857.

2394. Bhagvat, K., and K. K. P. N. Rao. 1942: Vitamin C in germinating grains. Indian J. Med. Res. 30:493–504. Chem. Abstr. 38:2074. 1944.

2395. Bhagwat, V. Y. 1959: Disinfection of cotton seed and its effect on seed viability. Poona Agr. Coll. Mag. 50:191–94.

2396. Bhakuni, D. S. 1959: Chemical examination of the seed of *Oldenlandia biflora* Linn. J. Sci. Ind. Res. (India) 18B:445–46.

2397. —— 1959: Chemical examination of the seed of *Phyllanthus maderaspatensis* Linn. J. Sci. Ind. Res. (India) 18B:446–47.

2398. ——, V. N. Sharma, S. N. Srivastava, and K. N. Kaul. 1961: Chemical constituents of the seeds of *Luffa echinata* variety *longistyla*. J. Sci. Ind. Res. (India) 20B:556–57.

2399. ——, S. N. Srivastava, V. N. Sharma, and K. N. Kaul. 1962: Chemical examination of the fruits of *Coccinia indica* W. & A. J. Sci. Ind. Res. (India) 21B:237–38.

Bhakuni, D. S. 17174.

2400. Bhalerao, V. R., and N. N. Dastur. 1946: Composition of "raintree" fruits. Current Sci. (India) 15:250–51. Biol. Abstr. 21:12604. 1947.

2401. Bhandari, P. R., J. L. Bose, and S. Siddiqui. 1948: Constituents of Indian soya bean germs. J. Sci. Ind. Res. (India) 7B:105. Biol. Abstr. 23:19707. 1949.

Bhandarkar, A. P. 17710.

2402. Bhapkar, D. G., and J. A. Patil. 1963: Inheritance of foilage, flower, and seed-coat colors in gram. Crop Sci. 3:361.

2403. Bhar, C. 1948: Crystalline components of the seeds of *Heracleum nepalense* (N. O. Umbelliferae). I. J. Indian Chem. Soc. 25:139–44. Biol. Abstr. 23:12640. 1949.

2404. Bharani, S. P., Y. S. Shah, and M. Sreenivasan. 1953: Biosynthesis of vitamin C during germination. III–IV. Proc. Indian Acad. Sci. Sect. B 37:33–71.

2405. Bhardwaj, S. N. 1962: Studies on early seedling metabolism in wheat. J. Indian Botan. Soc. 41:24–32.

2406. —— 1962: Influence of pretreating the seeds with synthetic phytohormones on yield and growth of wheat. J. Indian Botan. Soc. 41:326–35.

2407. —— and I. M. Rao. 1954: Relative toxicity of sulpha-drugs to wheat. Current Sci. (India) 23:296.

2408. —— and —— 1955: Effect of sodium sulfate and sodium chloride on germination, growth, and maturity of gram. Agra Univ. J. Res. Sci. Suppl. 4:767–75. Chem. Abstr. 51:15710. 1957.

2409. —— and —— 1955: Effects of pre-sowing seed treatment with a sulfonamide on germination, growth, and maturity of wheat. Sci. Cult. (Calcutta) 20:600–1. Chem. Abstr. 49:16088. 1955.

2410. ——, V. Santhanam, and R. Krishnamourthy. 1963: Influence of pretreating the seeds with NAA on yield and growth of cotton. Indian Cotton Growing Rev. 17:5–11.

2411. Bhargava, P. N. 1946: Chemical examination of the unsaponifiable matter of the fat from the fleshy arils of *Celastrus paniculata*. Proc. Indian Acad. Sci. Sect. A 24:506–9. Biol. Abstr. 21:18253. 1947.

2412. Bhargava, P. P., and C. N. Haksar. 1959: Examination of essential oil from Ajowan seeds. Perfumery Essent. Oil Record 50:204.

2413. Bharucha, F. R., and V. S. Shirke. 1947: A study in the important katabolic changes in the seedlings of *Avicennia officinalis* Linn. J. Univ. Bombay 40(21): 1–14. Biol. Abstr. 22:7473. 1948.

2414. Bharucha, K. E., and F. D. Gunstone. 1956: Vegetable oils. V. The component acids of *Cephalocroton cordofanus* (Muell.-Arg.) seed oil. J. Sci. Food Agr. 7:606–9.

2415. Bhasin, M. M., A. S. Gupta, and J. S. Aggarwal. 1950: Chemical examination of the seeds of *Cucumis utilissimus* Roxb. J. Sci. Ind. Res. (India) 9B:230–34.

2416. Bhaskaran, S., A. T. Natarajan, and M. S. Swaminathan. 1957: Effect of mutagens on the content of nucleic acids in wheat. Experientia 13:282–84. Chem. Abstr. 52:3034L. 1958.

Bhaskaran, S. 2820.

2417. Bhat, N. R., and M. H. Patel. 1957: Effects of some pre-sowing treatments of seed on growth and yield of cotton. Indian Cotton Growing Rev. 11:111–16.

Bhat, N. R. 14974.
Bhatia, D. S. 13726, 14211, 18125.
Bhatia, S. K. 2100.

2418. Bhatnagar, G. S. 1960: Reproduction by seeds (fruits) in *Jasminum* spp. Sci. Cult. (Calcutta) 26:192.

Bhatnagar, M. P. 6943.
Bhatnagar, S. P. 9914.
Bhatnagar, V. B. 4043.

2419. Bhatt, J. G. 1963: Ascorbic acid and fusarium-wilt in linseed. Indian J. Plant Physiol. 6:60–64.

Bhattacharyya, S. 3758, 15108.
Bhattacharyya, U. C. 17267.

2420. Bhatty, M. K., and Karimullah. 1954: Chemical investigations on *Sesbania aegyptica*. Pakistan J. Sci. Res. 6:70–73. Chem. Abstr. 49:4096. 1955.

Bhatty, M. K. 9356, 10514, 10515.

2421. Bhatty, R. S. 1964: Influence of nitrogen fertilization on the yield, protein, and oil content on two varieties of rape. Can. J. Plant Sci. 44:215–17.

2422. Bheemeswar, B., and M. Sreenivasaya. 1953: Ribonuclease. I. Preparation and properties of ribonuclease from castor bean (*Ricinus communis*). J. Sci. Ind. Res. (India) 12B:529–33. Chem. Abstr. 48: 6478. 1954.

Bhide, B. V. 5331, 9203, 17257.

2423. Bhide, N. V., and D. L. Sahasrabuddhe. 1943: The enzymes present in germinating seeds. J. Univ. Bombay 12A:81–84. Chem. Abstr. 38:1765. 1944.

2424. —— and —— 1943: Enzymes present in germinating seeds. II. Optimum pH for enzymes from germinating bajri and gram. Rasayanam 1:258–60. Chem. Abstr. 38:4280–81. 1944.

2425. Bhide, V. P., M. K. Desai, and M. S. Rane. 1957: Control of seedborne infection of anthracnose of cotton in Bombay State. Indian Cotton Growing Rev. 11:496–98.

2426. Bhojraj-Naidu, N., and S. A. Saletore. 1954: Studies on *Annona squamosa* (sitaphal) seed. III. Production and collection in Hyderabad. Indian J. Agr. Sci. 24:253–59. Biol. Abstr. 31:35852. 1957.

2427. Bhola, M. P. 1918: Germination of *Cupressus torulosa* seed. Indian Forester 44:175–76. Botan. Abstr. 1:48–49. 1918.

2428. Bhowmick, T. P., and S. B. Chattopadhyay. 1960: Germination of seeds of larger cardamon. Sci. Cult. (Calcutta) 26:185–86. Biol. Abstr. 36:50164. 1961.

2429. Bhuiya, A. K., S. H. Ahmad, and M. M. Qurashi. 1960: Development of rapid methods for the estimation of the oil content of single cottonseeds. IV. Investigation of an electrical circuit for measuring the electrical response of one cottonseed at 1,000 cycles/second. Pakistan J. Sci. Ind. Res. 3:143–48.

Bhutiani, R. C. 15408.
Bialek, H. F. 5737.

2430. Bialosouknia, R. 1919: Sur la germination des semences d'*Orobanche*. Ann. Essais Semences 4(3):3–8.

2431. —— 1919: Sur les ferments des semences grasses pendant la germination et la maturation. Ann. Essais Semences 4(3):9–19.

2432. Bianchi, A. 1960: Genetic damage in seeds of maize treated with X-rays and with neutrons under different physiological conditions. Atti Assoc. Genet. Ital. 5:137–48. Nucl. Sci. Abstr. 17:1342. 1963.

Bianculli, J. A. 7393.

2433. Bibbey, R. O. 1935: The influence of environment upon the germination of weed seeds. Sci. Agr. (Ottawa) 16:141–50.

2434. —— 1948: Physiological studies of weed seed germination. Plant Physiol. 23:467–84.

2435. —— 1960: Inherent and environmental dormancies of weed seeds and their relationship to effectiveness of tillage and chemical weed control practices. Agr. Pesticide Technol. Soc. Proc. 7:58–62. Bibliog. Agr. 25:19250. 1961.

2436. Bibby, K. M. 1953: Laboratory germination tests of *Nothofagus* seed. New Zealand Forestry Res. Notes 1(7):1–8. Biol. Abstr. 31:8889. 1957.

2437. —— 1953: Experiments in pre-germination treatment of tree seed. New Zealand Forestry Res. Notes 1(7):9–16.

2438. —— and R. W. M. Williams. 1953: The effect of various over-winter storage conditions on *Nothofagus menziesii* seed. New Zealand Forestry Res. Notes 1(7):29–33. Biol. Abstr. 31:8891. 1957.

2439. Bice, C. W., M. M. MacMasters, and G. E. Hilbert. 1945: Wheat starch properties in relation to grain maturity. Cereal Chem. 22:463–76.

2440. Bickel, A. F. 1947: On the structure of leucaenine (leucaenol) from *Leucaena glauca* Bentham. I–II. J. Am. Chem. Soc. 69:1801–6.

2441. —— 1948: On the structure of leucaenine (leucaenol) from *Leucaena glauca* Benthem. J. Am. Chem. Soc. 70:326–27.

2442. Bickenbach, K. 1932: Zur Anatomie und Physiologie einiger Strand- und

Duenenpflanzen. Beiträge zum Halophytenproblem. Beitr. Biol. Pflanzen 19:334–70.
2443. Bidwell, G. L., and A. T. Semple. 1936: Analyses of feeds for farm animals. U.S. Dept. Agr. Bur. Animal Ind. (Animal Husbandry Div.) 20. 6 pp. Expt. Sta. Rec. 75:821. 1936.
2444. Biebl, R. 1958: The radiosensitive phase in plant germination. 2nd U.N. Intern. Congr. Peaceful Uses of Atomic Energy. Isotopes in Agriculture 27:299–304 (Columbia Univ. Press, New York).
Biehler, W. E. 11856.
2445. Bieliukiene, S. 1961: Effect of preplanting treatment of barley seeds with vitamins on catalase activity. Akad. Nauk Lit. SSR. Inst. Botan. Sb. Statei 1:147–60.
Bientjes, W. 390.
2446. Bier, A. 1925: Über Keimverzug und seine Bedeutung nach Versuchen an Samen des gelben Lupine. Angew. Botan. 7:335–56.
2447. —— 1938: Neue Gesichtspunkte in der Vererbung. Springer, Berlin. 70 pp. J. Heredity 30:80. 1939.
Bierling, E. 17079.
2448. Bieter, R. N. 1954: Trends in producing grass and legume seeds. Crops & Soils 6(4):12–14.
2449. —— 1955: The hard seed problem in alfalfa. Assoc. Offic. Seed Analysts News Letter 29(4):33–35.
Bievre, M. de. 12959.
Bige. 12783.
2450. Bigelow, P. W. 1935: Lilies from seed for beginners. Horticulture 13(4):76.
2451. Bigger, J. H. 1935: Insects in relation to production of red clover seed. Trans. Illinois State Acad. Sci. 28:60–63.
2452. Biggs, R. H., and M. C. Langan. 1963: Effect of temperature on germination of Okinawa peach seeds. Proc. Florida State Hort. Soc. 75(1962):379–81.
Biggs, R. H. 6995.
2453. Bighi, C., and G. Mantovani. 1955: The determination of the Cu, Mn, B, Zn, and Mo contents in the seeds and the corresponding roots of some sugar-beet varieties. Ind. Saccar. Ital. 48:335–39. Chem. Abstr. 54:13275. 1960.
Bigwood, E. J. 119, 5689.
2454. Bihlmeier, M. 1927: Der Einfluss der Vorquellung und der Samenschale auf die Keimung lichtgeförderter Samen. Jahrb. Wiss. Botan. 67:702–36.
2455. —— 1928: Zur Kenntnis der Keimungsphysiologie einiger Labiaten und Cruciferen-Samen. Botan. Centralbl. Beih. 45:83–96.
Bilfinger, F. 12104.
2456. Bilimovich, O. F. 1935: Über den Bau der Perikarps bei Labiaten von systematischen Gesichtspunkte. Tr. Voronezhsk. Gos. Univ. 7:68–84.
2457. Bille, O. 1958: Perennials from seeds. Norsk Hagetid. 74:52–56.
2458. Billi, L. 1954: Transaminase, amylase, and phosphatase activities and content of reducing sugars in the course of the ripening of the caryopses of wheat var. Roma. Ann. Sper. Agrar. (Rome) 8:1097–102. Chem. Abstr. 49:436. 1955.
2459. Billings, F. H. 1901: Beiträge zur Kenntnis der Samenentwicklung. Flora 88:253–318.
2460. —— 1932: Microsporogenesis in *Phoradendron.* Ann. Botany (London) 46:979–92.
2461. —— 1933: Development of the embryo-sac in *Phoradendron.* Ann. Botany (London) 47:261–78.
2462. —— 1937: Some new features in the reproductive cytology of angiosperms, illustrated by *Isomeris arborea* Nutt. New Phytologist 36:301–26.
2463. Bilokin, P. 1955: The effect of volatile organic substances of the soil on the formation of chlorophyll in the cotyledons of the squash. Nauk. Zap. Kiivs'k. Derzh. Univ. 13:187–90. Chem. Abstr. 51:11487. 1957.
2464. Biloni, J. S. 1953: Sobre la diseminación y el valor de las semillas del "falso café" (*Manihot flabellifolia*) en Buenos Aires. Darwiniana 10:258–59.
2465. Bilous, V. I. 1961: Effect of depth and rate of seeding on seed germination and on good rooting of beech seedlings. Ukr. Akad. Sil's'kohospodar. Nauk. Uchbova Chastyna. Nauk Pratsi Aspirantiv 6:237–42.
2466. Bilquez, A. F., and J. P. Martin. 1961: Varietal differences of peanuts in sensitivity to X-rays. J. Agr. Trop. Botan. Appl. 8:30–43.
2467. Bils, R. F. 1961: Biochemical and cytological changes in the developing soybean cotyledon. Dissertation. Abstr. 21:2444–45.
2468. —— and R. W. Howell. 1963: Biochemical and cytological changes in developing soybean cotyledons. Crop Sci. 3:304–8.
2469. Binard, L. 1958: Résultats de quelques essais sur la germination de *Panicum*

maximum. Agricultura (Louvain) 6:305–10.

2470. Binder, R. G., T. H. Applewhite, M. J. Diamond, and L. A. Goldblatt. 1964: Chromatographic analysis of seed oils. II. Fatty acid composition of *Dimorphotheca* oil. J. Am. Oil Chemists Soc. 41:108–11.

2471. Bindloss, E. A. 1938: Nuclear size in plumular meristems of inbred and hybrid maize. Am. J. Botany 25:738–43.

2472. Binet, L., and J. Magrou. 1931: Soufre et croissance. Compt. Rend. 193:115–17. Biol. Abstr. 7:10302. 1933.

2473. Binet, P. 1956: Influence de la lumière sur la germination de *Zygophyllum album* L. Rev. Gen. Botan. 63:537–44.

2474. —— 1957: Les obstacles à la germination chez *Zilla macroptera* Coss. Compt. Rend. 245:345–48.

2475. —— 1958: Étude de quelques aspects morphologiques, anatomiques et physiologiques de la germination des semences de *Zilla macroptera* Coss. Rev. Gen. Botan. 65:129–85.

2476. —— 1958: La germination des semences de *Zilla macroptera* Coss. dans des sables d'humidité variable. Compt. Rend. 246:3093–95.

2477. —— 1959: Dormance et vieillisement des graines d'*Hyocyamus muticus* L. Compt. Rend. 248:2628–30.

2478. —— 1959: Étude physiologique et écologique de la germination des semences de *Zilla macroptera* dans des sables d'humidité variable. Rev. Gen. Botan. 66:232–54.

2479. —— 1960: La dormance des semences de *Suaeda vulgaris* Mog. et de *Suaeda macrocarpa* Mog. Bull. Soc. Botan. France 107:159–61.

2480. —— 1960: L'eau de mer et le froid, facteurs de levée de la dormance pour les semences de *Triglochin maritimum* L. Compt. Rend. 251:422–24.

2481. —— 1961: Action de la temperature et de l'eau de mer sur la germination des graines de *Cochlearia anglica* L. Compt. Rend. 253:895–97.

2482. —— 1962: Induction d'une dormance chez les graines de *Juneus maritimus* Lamk. Rôle protecteur de l'eau de mer. Rev. Gen. Botan. 69:5–13.

Bingefors, S. 14500, 15212.

Binkley, A. M. 6552.

Binnington, J. P. 17785.

Biol, M. I. 12644.

Biosio, M. L. 4475.

2483. Birch, E. O. 1944: Raising *Azalea schlippenbachii* from seed. N.Y. Botan. Garden J. 45:88–91.

2484. Birch, W. R. 1957: The seeding and germination of some Kenya weeds. J. Ecol. 45:85–91.

2485. Birckner, V. 1913: Beiträge zur Kenntnis der Gerstenkeimung. Biol. Centralbl. 33:180–89.

2486. Bird, J. N. 1944: Seed setting in red clover. J. Am. Soc. Agron. 36:346–57.

Bird, L. S. 6007.

2487. Birecka, H., and M. Wlodkowski. 1961: Influence of phosphorus content of seeds on nitrogen accumulation and the growth of beans and yellow lupine. Roczniki Nauk Rolniczych Ser. A 84:347–67. Biol. Abstr. 39:16523. 1962.

2488. ——, ——, and J. Skupinska. 1962: Phosphorus metabolism in pea plants grown from seeds of different phosphorus content. Roczniki Nauk Rolniczych Ser. A 85:29–46. Biol. Abstr. 41:11913.

2489. Birger, S. 1907: Über den Einfluss des Meerwassers auf die Keimfähigkeit der Samen. Botan. Centralbl. Beih. 21:268–80. Botan. Centralbl. 105:532. 1907.

2490. Birk, Y., A. Gertler, and S. Khaleb. 1963: A pure trypsin inhibitor from soya beans. Biochem. J. 87:281–84.

Birks, F. T. 15232.

2491. Bisalputra, T., and K. Esau. 1964: Polarized light study of phloem differentiation in embryo of *Chenopodium album*. Botan. Gaz. 125:1–7.

2492. Biscaro, G., and L. DeCaro. 1935: Chemical composition and nutritive value of wheat and maize-germ flour. Quaderni Nutriz. 2:189–96. Chem. Abstr. 30:6460–61. 1936.

Biserte, G. 2855.

2493. Bishop, C. T. 1955: Carbohydrates of sunflower heads. Can. J. Chem. 33:1521–29.

2494. Bishop, F. A. 1950: Delphiniums from seed. Brit. Delphinium Soc. Yearbook 1950:42–44.

2495. Bishop, G. 1953: Seeds in sawdust for goodness' sake! Home Garden 21:35.

2496. Bishop, G. W. 1954: Life history and habits of a new seed midge, *Dasymura gentneri* Pritchard. J. Econ. Entomol. 47:141–46.

Bishop, J. [C.] 16194, 19279.

2497. Bishop, L. R. 1944: Barley germination. J. Inst. Brewing 50:166–85. Brit. Chem. Physiol. Abstr. B 3:205. 1944.

2498. —— 1945: Second memorandum on

barley germination. J. Inst. Brewing 51:215–24. Chem. Abstr. 40:427. 1946.

2499. —— 1946: Third memorandum on barley germination. J. Inst. Brewing 52:273. Biol. Abstr. 21:23326. 1947.

2500. —— 1948: Notes on the dormancy of varieties of Scotch common barley. J. Incorp. Brewers Guild 34:427–28. Biol. Abstr. 23:16434. 1949.

2501. —— 1957: Ultra-rapid method for measurement of the germinative capacity of barley grain. J. Inst. Brewing 63:516–20.

2502. —— 1958: Barley dormancy as a varietal characteristic. J. Inst. Brewing 64:484–88.

2503. —— 1963: Objectives and definitions for germination and other tests for malting barley. J. Inst. Brewing 69:237–38.

Bisivas, T. S. 4924.

2504. Bisset, N. G. 1961: Cardiac glycosides. IV. Apocynaceae. A preliminary paper chromatographic study of the glycosides from *Thevetia peruviana.* Ann. Bogor. 4:145–52. Chem. Abstr. 58:14438. 1963.

2505. Bisset, S. K. 1954: The non-protein nitrogen of extracts of *Pisum sativum.* Biochem. J. 58:225–27.

Bissi, L. 18865.

2506. Bissinger, G. H. 1931: Double-seed vs. single-seed planting. Sugar News 12:289–93. Biol. Abstr. 8:9280. 1934.

2507. Bisson, C. S., and H. A. Jones. 1932: Changes accompanying fruit development in the garden pea. Plant Physiol. 7:91–105.

Bisson, C. S. 9950.

Biswas, A. K. 5333.

2508. Biswas, H. G., and K. L. Das. 1937: A comparative study of vitamin C in a few germinated oilseeds. Sci. Cult. (Calcutta) 3:176–77. Chem. Abstr. 32:218. 1938.

2509. —— and —— 1938: Comparative study of vitamin C in a few germinated cereals. Sci. Cult. (Calcutta) 4:360. Chem. Abstr. 33:2182–83. 1939.

2510. —— and A. R. Ghosh. 1936: Comparative study of vitamin C in the germinated peas commonly found in Bengal. Sci. Cult. (Calcutta) 1:778–79. Chem. Abstr. 30:6041. 1936.

2511. Biswas, I. 1957: Embryological studies in *Daemia extensa* Br. J. Indian Botan. Soc. 36:207–26.

Biswas, M. 17534.

Biswell, H. H. 17060.

2512. Bitjukow, K. 1937: Die Jarowisation des Hanfes. Lein Hanf. 2:49–52. Botan. Centralbl. 174:111. 1939.

2513. Bitter, C. R. 1936: The effect of high frequency irradiation upon barley seeds. Univ. Colo. Studies 23:209–15.

Bittera, N. von; *see* Von Bittera, N.

Bjälfve, G. 41.

Björling, C. O. 17945.

2514. Bjorling, K. 1961: Influence of beet seed culture on beet diseases. Sveriges Betodlares Centför. Tidskr. 24:59–62.

2515. Bjornseth, I. P. 1960: Germinative energy of spruce seeds and its use as an expression of seed quality. Arsskr. Norske Skogplantesk. 1959:97–99, 101–7.

2516. ——, J. Goksoyr, and K. Mikaelsen. 1957: Experiments on the respiration of neutron-irradiated barley seeds. II. Respiration in relation to growth and nitrogen metabolism. Physiol. Plantarum 10:328–39.

Bjornseth, I. P. 13439.

2517. Black, C. A. 1945: Effect of commercial fertilizers on the sex expression of hemp. Botan. Gaz. 107:114–20.

Black, C. A. 7894.

2518. Black, J. N. 1955: The influence of depth of sowing and temperature on pre-emergence weight changes in subterranean clover (*Trifolium subterraneum* L.). Australian J. Agr. Res. 6:203–11.

2519. —— 1956: The influence of seed size and depth of sowing on pre-emergence and early vegetative growth of subterranean clover (*Trifolium subterraneum* L.). Australian J. Agr. Res. 7:98–109.

2520. —— 1957: The early vegetative growth of three strains of subterranean clover (*Trifolium subterraneum* L.) in relation to size of seed. Australian J. Agr. Res. 8:1–14.

2521. —— 1957: Seed size as a factor in the growth of subterranean clover (*Trifolium subterraneum* L.) under spaced and sward conditions. Australian J. Agr. Res. 8:335–51.

2522. —— 1958: Competition between plants of different initial seed sizes in swards of subterranean clover (*Trifolium subterraneum* L.) with particular reference to leaf area and the light microclimate. Australian J. Agr. Res. 9:299–318.

2523. —— 1959: Seed size in herbage legumes. Herbage Abstr. 29:235–41.

2524. Black, M. 1956: Interrelationships of germination inhibitors and oxygen in the dormancy of seed of *Betula.* Nature 178:924–25.

2525. —— 1959: Dormancy studies in seed of *Avena fatua.* I. The possible role of germination inhibitors. Can. J. Botany 37:393–402.

2526. —— and J. M. Naylor. 1959: Prevention of the onset of seed dormancy by gibberellic acid. Nature 184:468–69.

2527. —— and P. F. Wareing. 1954: Photo-periodic control of germination of seed of Birch. (*Betula pubescens*). Nature 174:705.

2528. —— and —— 1955: Growth studies in woody species. VII. Photo-periodic control of germination in *Betula pubescens* Ehrh. Physiol. Plantarum 8:300–16.

2529. —— and —— 1957: Sensitivity of light-inhibited seeds to certain spectral regions. Nature 180:395.

2530. —— and —— 1959: The role of germination inhibitors and oxygen in the dormancy of the light-sensitive seed of *Betula* spp. J. Exptl. Botany 10:134–45.

2531. —— and —— 1960: Photoperiodism in the light-inhibited seed of *Nemophila insignis*. J. Exptl. Botany 11:28–39.

2532. Black, M. A. 1946: Effect of ceresan on the germination of stored linen-flax seed. New Zealand J. Sci. Technol. Sect. A 28:217–18.

2533. —— 1951: The effect of hot-water treatment on the germination of dreadnought wheat. New Zealand J. Sci. Technol. Sect. A 33:44–48.

2534. Black, R. H. 1949: Cause and prevention of "sick wheat." Milling Production 14:11–12.

Black, R. H. 10439.

2535. Black, W. H. 1951: Culture flask design for seed germination. Australian Orchid Rev. 16:109–12.

2536. Blackhurst, H. T., C. C. Singletary, and A. S. Nemec. 1962: Seeding mechanically through plastic mulch. Proc. Am. Soc. Hort. Sci. 80:556–64.

Blackhurst, H. T. 11837, 15110.

Blackie, J. J. 1901.

2537. Blackith, R. E. 1962: The protection of seed in storage. Indian Farming 11(12):6–7.

2538. —— and O. F. Lubatti. 1960: The influence of oil content on the susceptibility of seeds to fumigation with methyl bromide. J. Sci. Food Agr. 11:253–58.

Blackith, R. E. 12474.

2539. Blackman, G. E. 1952: Oil seed crops in temperate climates. Chem. Ind. (London) 1952:169–70.

2540. —— and E. S. Bunting. 1951: Studies in oil-seed crops. I. Factors controlling seed production in linseed (*Linum usitatissimum*). J. Agr. Sci. 41:256–70.

2541. —— and —— 1954: Studies in oil-seed crops. II. An assessment of the interrelationships between plant development and seed production in linseed (*Linum usitatissimum*). J. Agr. Res. 45:3–9.

Blackman, G. E. 3339, 3340, 12327.

2542. Blackman, V. H. 1904: On the relation of fertilization, "apogamy" and patheno-genesis. New Phytologist 3:149–58.

2543. —— 1928: Report on the sulphuric acid treatment of cotton seed. Empire Cotton Growing Rev. 5:240–41.

2544. Blackmore, L. W. 1952: Drilling seed into tussock grass land. New Zealand J. Agr. 84:221–22.

2545. Blackstone, J. H., H. S. Ward, Jr., J. L. Butt, I. F. Reed, and W. F. McCreery. 1954: Factors affecting germination of runner peanuts. Alabama Agr. Expt. Sta. Bull. 289. 31 pp.

2546. Blackwell, G. 1960: Some seed testing techniques. Gardeners Chron. 147:397.

2547. —— 1960: Seed analysis. Some distinguishing features. Gardeners Chron. 148:45.

2548. Blagoveshchenskii, A. V. 1925: Untersuchungen über die Samenreifung. Biochem. Z. 157:201–19.

2549. —— 1934: On the influence of cotyledons damages on the sterility of cotton-plant. Acta Acad. Sci. USSR Inst. Botan. Leningr. Ser. 4 1:273–76.

2550. —— 1936: Über die Lokalisation der organbildenden Stoffe in den Kotyledonen der Pflanzen. Bull. Biol. Med. Exptn. URSS 2:367–69.

2551. —— 1946: A study of the temperature effect on the quality of enzymes in cucumber seeds and seedlings. Tr. Inst. Fiziol. Rast. Akad. Nauk SSSR 4:193–99.

2552. —— 1951: Substances that delay the sprouting of seeds. Byul. Mosk. Gl. Botan. Sad. 9:54–58.

2553. —— 1953: Biochemistry in difficult germination of seed. Tr. Mosk. Gl. Botan. Sad. 3:3–57.

2554. —— 1955: Proteolytic ferment of delphinium seeds. Byul. Mosk. Gl. Botan. Sad. 20:95–98.

2555. —— 1958: Crystalline proteinase of seeds of *Phaseolus aureus* Roxb. Dokl. Akad. Nauk SSSR 120:356–58.

2556. —— 1962: Evolution of protein complexes in seeds and in phanerogamous plants. Izv. Akad. Nauk SSSR Ser. Biol.

27:845–56. Chem. Abstr. 58:9409. 1963.

2557. —— and I. P. Ivanova. 1955: Nitrogen metabolism in stratification of seeds. Dokl. Akad. Nauk SSSR 100:511–13. Chem. Abstr. 49:8399. 1955.

2558. —— and G. A. Kirillova. 1955: The effect of temporary cooling on the nitrogen-containing substances of the embryo and endosperm of winter wheat. Dokl. Akad. Nauk SSSR 100:171–73. Chem. Abstr. 49:8394. 1955.

2559. ——, E. V. Koblobkova, and N. A. Kudriasheva. 1951: Alternation of ferment action and quality in the germination of seeds. Tr. Mosk. Gl. Botan. Sad. 2:73–94.

2560. —— and A. I. Kologrivova. 1945: Activation of catalase by biogenic stimulants. Compt. Rend. Acad. Sci. URSS 48: 571–74. Chem. Abstr. 40:5080–81. 1946.

2561. —— and N. A. Kudriasheva. 1952: Germination inhibitors in ripening seeds. Byul. Mosk. Gl. Botan. Sad. 13:37–39.

2562. —— and R. M. Melamed. 1934: Die proteolytischen Fermente der Samen einiger Pflanzen. Biochem. Z. 273:435–45.

2563. —— and U. A. Petrochenko. 1959: Effect of succinic and fumaric acid treatment of seeds on certain physiological processes in plants. Plant Physiol. (USSR) (English Transl.) 6:49–56. Biol. Abstr. 34:14438. 1959.

2564. —— and T. A. Schubert. 1934: Amino acids in the sunflower-seed globulin. Biochem. Z. 269:375–78. Chem. Abstr. 28:4085. 1934.

2565. —— and N. I. Sossiedov. 1933: The gluten-dissolving ferment of wheat and barley seeds. Biochem. J. 27:1575–77.

2566. —— and —— 1934: Regarding the possibility of characterizing hard and soft wheats according to the ratio between total and amino nitrogen of their seeds and energy of autolysis of the latter. Cereal Chem. 11:117–20.

2567. —— and M. P. Yurgenson. 1937: On the changes of leucosin, gliadin, and glutenin under the action of flour and yeast proteinases. Cereal Chem. 14:103–7. Biol. Abstr. 12:399. 1938.

Blaignan, M. S. 4838.

2568. Blaim, K. 1954: Oxidation-reduction potentials of germinating winter and spring wheats. Roczniki Nauk Rolniczych Ser. A 68:539–53. Chem. Abstr. 49:12613. 1955.

2569. —— 1955: Biochemical factors controlling the germination of seeds. I. Factors inhibiting the germination of seeds. Postepy Nauk Rolniczych 2: 99–104. Chem. Abstr. 53:15226. 1959.

2570. —— 1956: Studies on the biochemistry of germinating grain of winter and spring wheat. Roczniki Nauk Rolniczych Ser. A 72:9–37. Biol. Abstr. 31:36007. 1957.

2571. —— 1956: Tocopherol (vitamin E) content in oleaginous seeds. Roczniki Nauk Rolniczych Ser. A 73:407–10. Biol. Abstr. 33:3400. 1959.

2572. —— 1958: A contribution to studies of biochemistry of *Melilotus* and *Trigonella* seeds. Plant Breed. Acclimitiz. Inst. Sci. Publ. 2:201–12.

2573. —— 1960: The effect of coumarin on uptake of water by seeds. J. Exptl. Botany 11:377–80.

2574. —— 1962: About the occurrence of betane and choline in seeds. Roczniki Nauk Rolniczych Ser. A 86:527–31. Biol. Abstr. 41:3441. 1963.

2575. Blair, B. O. 1951: Mesquite seed and seedling response to 2,4-D and 2,4,5-T. Botan. Gaz. 112:518–21.

2576. Blair, F. M. 1959: Raising large-flowered clematis from seed. Garden J. 9:11, 14–15, 29.

2577. Blair, I. D. 1936: Sprouted grain for seed-wheat purposes. New Zealand J. Agr. 53:37–39.

2578. —— 1938: Mangel-seed treatment. New Zealand J. Sci. Technol. Sect. A 20: 78–81.

2579. —— 1938: Seed-disinfection trials with cereals. New Zealand J. Sci. Technol. Sect. A 20:128–32.

2580. —— and L. G. Copp. 1953: Pea seed disinfection: a statistical study. New Zealand J. Sci. Technol. Sect. A 34:397–404.

2581. Blake, A. K. 1935: Viability and germination of seeds and early life history of prairie plants. Ecol. Monographs 5:405–60. Biol. Abstr. 11:18009. 1937.

2582. Blake, M. A., and L. J. Edgerton. 1946: The value of stone markings in peach varietal identification. Proc. Am. Soc. Hort. Sci. 48:100–4.

2583. Blakeslee, A. F. 1944: Determination of the factors involved in chemical regulation of embryo development in plants with the possibility of their ultimate control. Am. Phil. Soc. Yearbook 1943:158–59.

2584. —— 1944: Micrografting, a method of securing seedlings from excised embryos which fail to develop roots. Am. J. Botany 31:1s.

2585. —— 1944: Smith College conference on plant embryo culture. Science 100:497–98.

2586. —— 1953: New hybrids in *Datura* from culture of excised embryos. Proc. Intern. Congr. Botany 7:324–25. 1950.

2587. —— 1954: The aging of seeds and mutation rates. Ann. N.Y. Acad. Sci. 57:488–99.

2588. —— and S. Satina. 1944: New hybrids from incompatible crosses in *Datura* through culture of excised embryos on malt media. Science 99:331–34.

2589. —— and —— 1947: Ovular tumors associated with hybrid embryos in *Datura*. Science 105:633.

2590. —— and H. E. Warmke. 1938: Increase seed size. Sci. News Letter 34:292–93.

2591. —— and —— 1938: Size of seed and other criteria of polyploids. Science 88:440.

Blakeslee, A. F. 1535, 1536, 3657–3660, 15062, 15986, 16205–16208, 16792, 16793, 17789, 17790, 17792, 18987, 18988.

2592. Blanc, A. 1938: Essais de conservation de blé en atmosphere confinée. Compt. Rend. Acad. Agr. France 24 (Suppl. of 18). 71 pp.

2593. —— 1939: Les premiers résultats d'une nouvelle serie d'essais de conservation de blé en atmosphere confinée. Compt. Rend. Acad. Agr. France 25:1005–15.

Blanchard, K. L. 8595.

2594. Blanchard, M. 1941: Conservation des semences de soja. Compt. Rend. Acad. Agr. France 27:500–3.

2595. —— 1953: Contribution à la determination des espèces et variétés de blés cultivées en Algerie par la réaction des graines, des epis et des pailles à l'acide phénique. Proc. Intern. Seed Testing Assoc. 18:161–66.

2596. —— 1953: Action des produits à base de derives organo-mercuriques sur la germination des graines d'*Orobanche*. Méthode de lutte contre l'*Orobanche*. Proc. Intern. Seed Testing Assoc. 18:249–52.

2597. —— 1957: Contribution à l'étude du phénomène de dormance chez *Hordeum vulgare* L. Proc. Intern. Seed Testing Assoc. 22:192–95.

Blanco, J. Garcia-; *see* Garcia-Blanco, J.

2598. Blanco, L. C. 1957: Comparative effects of arasan, granosan, and semesan dust treatments on vegetable seeds as a control for damping-off. Araneta J. Agr. 4:57–64.

2599. Blanford, H. R. 1931: Experiments in connection with sowing and planting teak in Taungya plantations. Burma Forestry Bull. 24, Silvicult. Ser. 14. Nature 129:134. 1932.

2600. Blank, F. 1940: The hypocotyl in seeds. Chronica Botan. 6:101–2.

2601. Blank, L. M., and D. C. Allison. 1963: Frequency of polyembryony in certain strains of *Gossypium hirsutum* L. Crop Sci. 3:97–98.

2602. Blaringhem, L. 1925: Sur la germination des hybrides sur pied. Rev. Pathol. Végétale Entomol. Agr. France 12:306–16.

2603. —— and K. C. Chin. 1938: Nouveaux cas de xénie chez des hybrides de blés (*Triticum monodurum*, *T. aegilopoides*, *T. vulgare*). Compt. Rend. 207:1141–44.

2604. Blas, L. 1936: Chemical study of pimento seed (*Capsicum annum*). Mon. Farm. 42:97–98. Chem. Abstr. 31:3583. 1937.

2605. Blasdale, W. C. 1898: Seed production. Asa Gray Bull. 6:656.

2606. Blaser, R. E., and G. W. Killinger. 1950: Life history studies of Louisiana white clover (*Trifolium repens* L.). I. Seed germination as related to temperature, pasture management, and adaptation. Agron. J. 42:215–20.

Blaser, R. E. 3103, 19350.

2607. Blatchford, S. M., and D. W. Hall. 1963: Methods of drying groundnuts. I. Natural methods (literature survey). Trop. Sci. 5:6–33.

2608. Blattny, C., and V. Osvald. 1954: Transmission of hop viruses by seed. Preslia 26:1–26. Biol. Abstr. 29:9641. 1955.

Blauth, J. Opienska-; *see* Opienska-Blauth, J.

2609. Bleak, A. T. 1959: Germinative characteristics of grass seed under snow. J. Range Management 12:298–302. Biol. Abstr. 35:27731. 1960.

2610. —— and A. C. Hull. 1958: Seeding pelleted and unpelleted seed on four range types. J. Range Management 11:28–33.

2611. —— and T. A. Phillips. 1950: Seedling stands from airplane broadcasting of pelleted and unpelleted seed in southeastern Utah. U.S. Dept. Agr. Forest Serv. Intermountain Forest Range Expt. Sta. Res. Paper 22. 14 pp.

2612. Bledsoe, R. P. 1929: Multiple kernels in wheat-rye hybrids. J. Heredity 20:137–42. Biol. Abstr. 6:646. 1932.
2613. —— and J. G. Futral. 1946: An efficient scarifier for small seed samples. J. Am. Soc. Agron. 38:1111–14.
2614. Bledsoe, R. W., C. L. Comar, and H. C. Harris. 1949: Absorption of radioactive calcium by the peanut fruit. Science 109:329–30.
2615. Bless, A. A. 1937: Effects of the length of X-ray waves on seeds. Proc. Natl. Acad. Sci. U.S. 23:194–96.
2616. —— 1938: Effects of X-rays on seeds. Plant Physiol. 13:209–11.
Bless, A. A. 9975.
2617. Blessin, C. W. 1962: Carotenoids of corn and sorghum. I. Analytical procedure. Cereal Chem. 39:236–42.
2618. ——, R. J. Dimler, and O. J. Webster. 1962: Carotenoids of corn and sorghum. II. Carotenoid loss in yellow-endosperm sorghum grain during weathering. Cereal Chem. 39:389–92.
2619. ——, C. H. Van Etten, and R. Wiebe. 1958: Carotenoid content of the grain from yellow endosperm-type sorghums. Cereal Chem. 35:359–65.
2620. Blish, M. J., and C. E. Bode. 1935: Catalase activity in wheat flour. Cereal Chem. 12:133–42.
2621. ——, R. M. Sandstedt, and E. Kneen. 1938: The cereal amylases with reference to flour and malt behavior. Cereal Chem. 15:629–57.
2622. Bliss, C. I. 1947: Biological measurement of the depth dose of X-rays of lettuce seedlings. [Abstract.] Biometrics 3:60.
2623. Bliss, L., and N. M. Naylor. 1946: The phosphorylase of waxy maize. Cereal Chem. 23:177–86.
2624. Bliss, L. C. 1959: Seed germination in arctic and alpine species. Arctic 11:180–88. Biol. Abstr. 33:35050. 1959.
Bliss, L. C. 1498.
2625. Bloch, I. 1946: A theory of membrane permeability. II. Diffusion in the presence of waterflow. Bull. Math. Biophys. 8:21–28. Biol. Abstr. 20:11985. 1946.
2626. Blociszewski, T. 1876: Physiologische Untersuchungen über die Keimung und weitere Entwickelung einiger Samenteile bedecktsamiger Pflanzen. Landwirtsch. Jahrb. 5:145–61. Just's Jahresber. 1876: 878–79.
Block, H. 5654.
2627. Block, R. J., and D. Bolling. 1945: Amino acids yielded by yeast, sunflower seed meal, and sesame seed after hydrolysis of the fat-free tissue. Arch. Biochem. 6:277–79.
2628. —— and —— 1951: The amino acid composition of proteins and foods. Analytical methods and results. 2nd ed. Charles C Thomas, Springfield, Ill. 576 pp.
2629. Blodgett, E. C., and G. KenKnight. 1945: A practical hot water seed treater. Seed World 58(8):8–10.
Blodgett, E. C. 6454.
2630. Blommaert, K. L. J. 1960: New plant hormone replaces stratification of deciduous fruit seed. Farming S. Africa 35(11):13–14.
Blondeau, R. 4520.
Blondel, B. 16205.
2631. Blondel, L., and E. Barbier. 1963: Seeds in the fruits of the citrus variety Clementine. Fruits (Paris) 18:153–56. Chem. Abstr. 59:4485. 1963.
2632. Blood, H. L. 1933: Fermentation frees seed from bacterial diseases. Sci. News Letter 24:14. 1933.
2633. —— 1937: A possible acid seed soak for the control of bacterial canker of the tomato. Science 86:199–200.
2634. Blood, P. T., and G. M. Dunn. 1953: An efficient scarifier for small lots of legume seed. Agron. J. 45:576.
2635. Bloom, J. R. 1963: Effect of temperature extremes on the wheat seed gall nematode, *Anguina tritici*. Plant Disease Reptr. 47:938–40.
2636. Bloomquist, H. L. 1923: Dormancy in the seeds of the persimmon. J. Elisha Mitchell Sci. Soc. 29:83–85.
2637. Blossfeld, R. 1933: Kultur-erfahrungen mit *Kalanchoe globulifera* var. *coccinea* aus Samen. Möllers Deut. Gärtner-Ztg. 48:183–84.
2638. Blowers, J. W. 1960: More about raising orchids from seeds. Gardeners Chron. 148:586.
2639. —— 1961: Orchids from seed. Gardeners Chron. 150:128.
Blum, A. E. 9060.
2640. Blum, P. H. 1954: Grain spoilage: A review. Wallerstein Lab. Commun. 17(56): 35–45. Biol. Abstr. 28:28300. 1954.
2641. Blumenthal-Goldschmidt, S. 1958: Some further investigations on the growth active substances extractable from dry lettuce seeds. Bull. Res. Council Israel Sect. D 6:91–98.
2642. —— 1960: Changes in the content of germination regulating substances in lettuce seeds during imbibition. Bull. Res. Council Israel Sect. D 9:93–99.

2643. —— 1961: The inhibition of IAA oxidase by phenolic substances present in lettuce seeds. Bull. Res. Council Israel Sect. D 9:187–89.

2644. —— and A. Lang. 1960: Presence of gibberellin-like substances in lettuce seed. Nature 186:815–16.

2645. —— and G. Stein. 1954: Carotene in lettuce seeds. Bull. Res. Council Israel 4:84–85. Biol. Abstr. 30:6064. 1956.

Blumenthal-Goldschmidt, S. 6122.

2646. Blumer, J. C. 1910: The vitality of pine seeds in serotinous cones. Torreya 10:108–11.

2647. Blumer, S., and A. Harder. 1953: Über die Beizung von Gemüsesamen. Landwirtsch. Jahrb. Schweiz 2:315–35.

2648. —— and J. Kundert. 1960: Über den Einfluss der Temperatur auf die Wirkung von Fungiziden. Landwirtsch. Jahrb. Schweiz 9:465–95.

Blundell, J. B. 9630.

2649. Boa, W. 1953: Pelleted seed trials, a report on field trials made in 1949 and 1950 with pelleted sugar beet, onion, carrot, and lettuce seeds. Natl. Inst. Agr. Eng. Tech. Mem. 91. 12 pp. J. Sci. Food Agr. 5:76. 1954.

2650. Boarder, A. 1946: Germination of cacti seeds. Cactus Succulent J. Gt. Brit. 8:80–81.

2651. —— 1954: Seed raising. Cactus Succulent J. Gt. Brit. 16:66–68.

2652. Boatner, C. H. 1944: The pigments of cottonseed. Oil Soap 21:10–15. Chem. Abstr. 38:1266. 1944.

2653. ——, M. Caravella, and C. S. Samuels. 1944: Orange-colored pigment of cottonseed. J. Am. Chem. Soc. 66:838–39. Chem. Abstr. 38:3692. 1944.

2654. ——, L. E. Castillon, C. M. Hall, and J. W. Neely. 1949: Gossypol and gossypurpurin in cottonseed of different varieties of *Gossypium barbadense* and *G. hirsutum* and variation of the pigments during storage of the seed. J. Am. Oil Chemists Soc. 26:19–25. Chem. Abstr. 43:1998–99. 1949.

2655. —— and C. M. Hall. 1946: Pigment glands of cottonseed. I. Behavior of the glands toward organic solvents. Oil Soap 23:123–28. Chem. Abstr. 40:3624. 1946.

2656. ——, ——, R. T. O'Connor, and L. E. Castillon. 1947: Pigment glands of cottonseed. III. Distribution and some properties of cottonseed pigments. Botan. Gaz. 109:108–20.

2657. ——, ——, M. L. Rollins, and L. E. Castillon. 1947: Pigment glands of cottonseed. II. Nature and properties of gland walls. Botan. Gaz. 108:484–94.

2658. ——, R. T. O'Connor, M. C. Curet, and C. S. Samuels. 1947: The pigments of cottonseed. III. Gossyfulvin, a native cottonseed pigment related to gossypol. J. Am. Chem. Soc. 69:1268–73. Biol. Abstr. 22:19262. 1948.

Bobikyan, R. A. 143.

2659. Bobrisheva, A. M., and E. Z. Oknina. 1961: Transformation of reserve substance in flower buds of black current during the annual cycle. Plant Physiol. (USSR) (English Transl.) 7:484–88.

2660. Bocchi, A. 1953: Action of eosin and β-indoleacetic acid on the germinating power and the lengthening of the primary roots of some Leguminosae. Nuovo Giorn. Botan. Ital. 60:336–41.

2661. Bochantseva, Z. P. 1944: Contributions to the biology of florescence and embryology of *Haloxylon ammodendron* Bge. J. Botan. URSS 29:48.

2662. —— 1955: Germination of tulip seeds subjected to effect of ultra-sound. Dokl. Akad. Nauk Uz. SSR 1955(4):43–45. Biol. Abstr. 33:7480. 1959.

2663. Bockstahler, H. W., and R. F. Seamans. 1940: Threshing and cleaning equipment for sugar beet seed. J. Am. Soc. Agron. 32:794–802.

Bocsa, E. 10686.

2664. Bodalski, T., and M. Woroszczuk. 1962: Phytochemical analysis of the seeds of the Hungarian plum (*Prunus domestica* L.). I. Dissertationes Pharmaceut. 14:339–45. Biol. Abstr. 41:20131. 1963.

2665. Bodard, M. 1954: Polyembryonie chez *Cola nitida*. Congr. Intern. Botan. Paris Rappt. Commun. 8:259–61.

Bode, C. E. 2620, 13412.

2666. Bode, G. 1929: Die Bedeutung des Rohrzuchers in der Gerste. Fortschr. Landwirtsch. 4:545–47.

2667. Bode, H. R. 1940: Über die Blattausscheidungen des Wermuts und ihre Wirkung auf andere Pflanzen. Planta 30:567–89.

2668. Bode, W. 1915: Schutzeinrichtungen von Pflanzensamen gegen das Meerwasser. Naturwissenschaften 6:413–16.

2669. Bodekar, F. W. T. 1930: A few observations on the flowering of kyathaungwa (*Bambusa polymorpha* Munro). Indian Forester 56:404–5.

2670. Boden, R. W. 1957: Some aspects of seed dormancy in *Eucalyptus*. Australian Forestry 21:81–85.

Bodger Seeds, Ltd. 739, 831.

2671. Bodnar, J., and I. Kiss. 1948: Influence of tobacco on the enzyme of carbohydrate metabolism. Rev. Intern. Tabacs 23:132. Chem. Abstr. 43:8407. 1949.

2672. —— and L. E. Roth. 1932: Die Wirkung der Quecksilbersalze auf die Samenkeimung. Biochem. Z. 248:375–82.

Bodo, G. 17181.

2673. Bodson, M. T. 1936: Influence du dichlorethylene sur la germination et la croissance de quelques végétaux. Bull. Ste. Royale Botan. Belg. 19:40–48. Ann. Agron. 7:305. 1937.

2674. Boe, K. N. 1954: Periodicity of cone crops for five Montana conifers. Proc. Montana Acad. Sci. 14:5–9.

2675. Boehm, J. 1873: Über das Keimen von Samen in reinen Sauerstoffgase. Sitzber. Akad. Wiss. Abt. I 48:1–10.

2676. Boehm, K. 1931: Embryologische Untersuchungen an Zingiberaceen. Planta 14:411–40.

2677. Böhmer, K. 1928: Die Bedeutung der Samenteile für die Lichtwirkung und die Wechselbezichung von Licht und Sauerstoff bei der Keimung lichtempfindlicher Samen. Jahrb. Wiss. Botan. 68:549–601.

2678. Böhmig, F. 1930: Zwei Versuche an Cyklamensamen. Gartenwelt 34:202.

2679. Böhnert, E. 1930: Hat das Samengewicht Einfluss auf das Grössen-Wachstum der Pflanze? Blumen Pflanzenb. 45:37–39.

2680. —— 1931: Hat die Kleinkörnigkeit des Cyclamensamens Einfluss auf die gute Entwicklung der Pflanzen? Gartenflora 80:47–48.

Böjtös, Z. 12117.

2681. Boekenoogen, H. A. 1937: Oil from seeds of *Onguekoa gore*, Engler. Fette Seifen 44:344. Soc. Chem. Ind. J. 56:1233. 1937. Pharm. Abstr. 4:203. 1938.

Boekholt, K. 8698.

2682. Boelcke, O. 1951: The seed production of *Trifolium repens* and *Bromus cartharticus* in the Pergamino region of Argentina. Rev. Arg. Agron. 18:193–99. Biol. Abstr. 26:16155. 1952.

2683. Boening, K. 1932: Die Bekämpfung der Brennfleckenkrankheit des Tabaks (*Colletotrichum tabacum*) durch Beizung des Samens und vorbeugende Behandlung der Pflanzen mit chemischen Mitteln. Prakt. Bl. Pflanzenbau Pflanzenschutz 10:89–106.

2684. Boenjamin, H. I. 1960: The influence of gibberellic acid upon germination in cabbage (*Brassica oleracea* var. *capitata*), lettuce (*Lactuca sativa*) and rice (*Oryza sativa*). Bogor. Balai Besar Penjelidikan Pertanian Pemberitaan 156. 19 pp.

2685. Böös, G. 1917: Über Parthenogenesis in der Gruppe Aphanes der Gattung *Alchemilla* nebst einigen im Zusammenhang damit stehenden Fragen. Lunds Univ. Arsskr. Avd. 2. 13(4):3–37.

Boer, H. W. de; *see* De Boer, H. W.

Boeri, E. 7381.

2686. Boerker, R. H. I. 1916: Ecological investigations upon the germination and early growth of forest trees. Thesis, Nebraska Univ. 89 pp.

2687. Boerlage, M. J. G. 1898: Germination of *Heritiera littoralis*. Ann. Jard. Botan. Buitenzorg 1898:136–42. J. Roy. Microscop. Soc. 1899(5):507.

2688. Börner, H. 1955: Organic compounds from seeds of rye, wheat, and barley during swelling. Naturwissenschaften 42:48. Chem. Abstr. 49:16077. 1955.

2689. ——, W. Morgenstern, and B. Rademacher. 1960: Markierung von Unkrautsamen (*Sinapis arvensis* L.) mit S-35. Naturwissenschaften 47:116.

2690. Boesman, G. 1962: Resultaten van Kiemproeven met Orchideezaad. Gent Med. Landhogs. Opzoekings Sta. 27: 619–42.

2691. Boev, N. 1958: Les oiseaux et le sol. Izv. Bulgar. Akad. Nauk Poch. Inst. 5:223–50.

2692. Bogdan, A. V. 1961: The depth of planting Rhodes grass seed. E. African Agr. Forestry J. 26:187.

2693. Bogdan, I., Z. Papp, and M. Szabo. 1963: Two hundred years old wheat seeds in the National Archives. Agrartorteneti Szemle 5:50–66.

2694. Bogdanov, B. P. 1954: Effect of acorn size on growth of young oaks. Botan. Zh. 39:584–88.

2695. —— 1962: Effect of high temperature on pine seeds, with reference to drying of the cones under reduced pressure. Lesn. Zh. 6:20–23.

2696. Bogdanovich, M. V. 1961: Preplanting treatment of sugar beet seeds. Veselopodolyanskaya Opytn.-Selek. Sta. Nauch. Tr. 1961:151–57.

2697. Bogdashevskaia, O. V. 1952: Formation of ricinine in castor beans. Dokl. Akad. Nauk SSSR 82:1001–3.

Boggs, H. M. 101.

2698. Bogh, H. 1936: The seed-ripening of the most important herbage legumes and the influence of harvest-time on quantity and quality of the seed yield. Kgl. Vet. Landbohojskole Arsskr. 1936:117–21.

2699. —— and I. Jensen. 1941/43: Some morphological characters of red clover (*Trifolium pratense*) and their use in the control of genuineness. Proc. Intern. Seed Testing Assoc. 13:92–112.

Bogliolo, M. 2269.

2700. Bogojavlenskii, A. A. 1930: Effect of the duration of formalin disinfection on the germination of grains of wheat, oats, and millet. Plant Protect. 1930(7):179–81.

Bogoljubowa, A. M. 10190.

2701. Bogolyubova, V. G. 1954: A case of seed germination within the growing tomato fruit and greening of seed areas. Botan. Zh. 39:908–10. Chem. Abstr. 49:10447. 1955.

2702. Bogomolov, A. M. 1951: Use of large grains in selection and seed production of rye. Selek. i Semen. 18(6):52–53.

Bogomolova, R. I. 7825.

2703. Bogomolova, T. D. 1959: Fertilization and development of the chicory seed. Moskov. Ordena Lenina Sel'skokhoz. Akad. Im. K. A. Timiriazeva Dokl. 46: 227–32.

2704. Bogorad, L., and W. J. McIlrath. 1959: A correlation of photoperiodic response of *Xanthium* and germination of implanted lettuce seed. Publ. Am. Assoc. Advan. Sci. 55:301–9. Chem. Abstr. 55:2819. 1961.

Bogorad, L. 13191.

2705. Bogoslovskaya, E. V. 1950: Some peculiarities of metabolism in cotton plant that was adapted, before planting, to salinity. Tr. Inst. Fiziol. Rast. Akad. Nauk SSSR 7:174–86. Chem. Abstr. 46:8723. 1952.

2706. Boguslawski, E. V., A. Vömel, and G. Reichelt. 1954: Nutrient ratios in fertilization and yield. Z. Acker- Pflanzenbau 97:267–98. Chem. Abstr. 49:551. 1955.

Bohart, G. E. 8419, 15171.

2707. Bohling, H. 1959: Papierchromatographische Untersuchungen über die Wuchs- und Hemmestoffe des Haferkeimlings und des Maisscutellums. Planta 53:69–108.

Bohmont, D. W. 19492.

Boichenko, E. A. 7293.

2708. Boichuk, O. B. 1960: Dynamics of auxins in germinating tomato seeds. Ukr. Botan. Zh. 17(5):23–31. Biol. Abstr. 38:3205. 1962.

Boiko, A. P. 11216.

2709. Boischot, P., and J. Hebert. 1947: Influence d'extraits végétaux sur la germination de l'orge. Ann. Agron. 17:867.

2710. —— and J. Lajon. 1956: Diffusion des P_2O_5 des graines en germination dans la solution du sol. Compt. Rend. 242: 550–52.

2711. Boiteau, P., and A. R. Ratsimamanga. 1958: Effects of asiaticoside on germination and growth of plants. Compt. Rend. Soc. Biol. 152:1106–7.

2712. —— and —— 1961: Effects of some triterpenic and sterolic substances on seed germination and plant growth. Intern. Hort. Congr. Proc. 15:66–73.

2713. Bok, R. 1941: The influence of oxygen-nitrogen mixtures upon the dwarfing of *Ardisia crispa* (Thunb.) A.DC. Konink. Ned. Akad. Wetenschap. Proc. Sect. Sci. 44:1128–30. Biol. Abstr. 22: 5542. 1948.

2714. Bokareva, L. I., and A. N. Fogel. 1957: Peculiar features of blossoming and seed formation in *Orthosiphon stamineus.* Benth. Botan. Zh. 42:1205–20.

2715. Boke, N. H. 1963: Anatomy and development of the flower and fruit of *Pereskia pititache*. Am. J. Botany 50: 843–58.

2716. Bokorny, T. 1913: Über den Einfluss verschiedener Substanzen auf die Keimung der Pflanzensamen. Biochem. Z. 50:1–118. Expt. Sta. Rec. 29:421. 1913.

2717. —— 1914: On various treatments for the disinfection of seeds. Biochem. Z. 62:58–88. Expt. Sta. Rec. 31:738. 1914.

Bolas, B. D. 7443, 7444.

2718. Bolhuis, G. G. 1958: Observations of the flowering and fructification of the groundnut, *Arachis hypogaea*. Neth. J. Agr. Sci. 6:18–23. Biol. Abstr. 33:3031. 1959.

2719. —— and W. De Groot. 1959: Observations on the effect of varying temperatures on the flowering and fruit set in three varieties of groundnut. Neth. J. Agr. Sci. 7:317–26.

2720. Bolin, D. W., and W. Schroeder. 1939: The calcium-phosphorus ratio of the skins of canning peas and its relation to maturity. J. Agr. Res. 58:631–36.

2721. Boll, W. G. 1953: Stimulation of the growth of the radicle after treatment of tomato seeds with bromine water. Nature 172:963.

Bollen, W. B. 7062.

2722. Bollenbacher, K., N. D. Fulton, and B. E. McCutchen. 1963: Behavior of artificially deteriorated cottonseed after various storage periods. Agron. J. 55: 521–23.

Bollens, W. F. 9784.

2723. Bolley, H. L. 1912: Hard seeds of clover and alfalfa. N. Dakota Agr. Expt. Sta. Rept. 22:81–96.

2724. —— and T. F. Manns. 1932: Fungi of flaxseed and of flax-sick soils. N. Dakota Agr. Expt. Sta. Tech. Bull. 259: 2–57. Chem. Abstr. 27:802. 1933.

2725. Bolli, M. 1949: Action of talc on the germination of seeds. Ann. Fac. Agrar. Univ. Studi Perugia 6:229–30. Chem. Abstr. 46:4615i, 4616a. 1952.

Bolling, D. 2627, 2628.

2726. Bolotov, K. D. 1953: Effect of seeding norms on yields and quality of yellow lupine seeds. Zemledelie 1953 (7):100–1.

2727. Bolssunov, I. I. 1927: Zur Frage über die Möglichkeit Kulturrübenvarietäten nach deren Samen zu unterscheiden. Ann. Essais Semences 5(1):1–18.

2728. Bolsunov, I. 1927: Selection of wheat seed by testing in salt solutions. Mem. Agr. Inst. Kiev (English Transl.) 2:33–41. Expt. Sta. Rec. 63:124. 1930.

2729. Bolsunova, O. 1936: Effect of varying degrees of infection of flax seed with anthracnose on the incidence and development of the disease in the crop. Plant Protect. 1936(11):58–67. Rev. Appl. Mycol. 16:386–87. 1937.

2730. Bolton, A. 1955: Experiments on tobacco-seed production. Empire J. Exptl. Agr. 23:240–43.

2731. —— 1956: Experiments on tobacco-seed production. II. The relationship between rate of applied nutrients, chemical composition of leaf, and seed yield. Empire J. Exptl. Agr. 24:161–66.

2732. Bolton, J. L., and J. R. Fryer. 1937: Inter-plant variations in certain seed-setting processes in alfalfa. Sci. Agr. (Ottawa) 18:148–60.

Bolton, J. L. 6740, 6741, 15170.

Bolyakina, Y. P. 18198.

2733. Bompard, A. 1958: Les semences de riz; leurs qualités et leurs origines. Bull. Inform. Rizicult. France 55:24–27.

2734. Bonaventura, G. 1932: La machine à decuscuter "Bedell" et la faculté germinative de quelques semences de légumineuses fourragères decuscutées avec cette même machine. Ital. Agr. 69. Proc. Intern. Seed Testing Assoc. 7:110–11. 1935.

Bonaventura, G. 19085.

2735. Bonazzi, A. 1936: Estudios sobre la fisiolgia del tabacco. II. La bioquimica de la germinación de la semilla. Bol. (Tec.) Est. Exptl. Agron. Santiago, de las Vegas (Habana) No. 59. Boll. Tec. Scafati Ist. Sper. Colt. Tabac. 34:280–82. 1937.

2736. Bond, A. B., and R. L. Glass. 1963: The sugars of germinating corn (*Zea mays*). Cereal Chem. 40:459–66.

Bonde, E. K. 13724.

2737. Bond, L. 1942: Colchicine stimulation of seed germination in *Petunia axillaris*. J. Heredity 33:200–1.

2738. Bondar, G. 1942: Penao, *Cnidoscolus marcgravii* Polh.: Novo recurso oleifero da Bahia. Bol. Inst. Cent. Fomento Econ. Bahia 12:16. Expt. Sta. Rec. 88:740. 1943.

2739. Bondev, I., and K. Kochev. 1961: The biology of the germination of the seeds of some high-mountain grasses. Izv. Bulgar. Akad. Nauk. Botan. Inst. 8:117–28.

Bondi, A. 14766, 15446.

2740. Bondoc, L. O., and R. P. Creencia. 1962: Studies on cacao seed germination. Philippine Agriculturist 46:38–44.

2741. Bongini, V. 1935: Ricerche sulla germinabilità dei semi delle cuscute inquinanti le semenzine foraggere piemontesi. Difesa Piante Torino 12:86–106, 117–39.

2742. Bonisteel, W. J. 1940: Polyploidy in relation to chemical analysis. J. Am. Pharm. Assoc. 29:404–8.

Bonjour, S. 5253.

Bonnafous, J. Uieu-. 13334.

2743. Bonnemaison L. 1960: Traitement des semences de betterave avec des produits endothérapiques aphicides. Phytiat.-Phytopharm. 9:241–58.

2744. —— and P. Jourdheuil. 1951: Influence of various insecticides on the germination and growth of several cultivated plants. Parasitica (Gembloux) 7:81–108. Chem. Abstr. 46:6790–91. 1952.

Bonner, D. 2747.

2745. Bonner, J. 1938: Nicotinic acid and the growth of isolated pea embryos. Plant Physiol. 13:865–68.

2746. —— and G. Axtman. 1937: The growth of plant embryos in vitro. Preliminary experiments on the role of accessory substances. Proc. Natl. Acad. Sci. U.S. 23:453–57.

2747. —— and D. Bonner. 1938: Ascorbic acid and the growth of plant embryos. Proc. Natl. Acad. Sci. U.S. 24:70–75.
Bonner, J. 6925, 12121, 12354.
Bonnet, F. R. 18450, 18451.
Bonnet, J. 9287.

2748. Bonnet, J. A., and M. A. Lugo-Lopez. 1953: Effect of different quantities of velvetbean green manure on corn yields in Puerto Rico. J. Agr. Univ. Puerto Rico 37:96–101.
Bonnet, N. 10572.

2749. Bonnet, R. 1929: L'évolution de l'azote au cours de la germination. Bull. Soc. Chim. Biol. 11:1025–61.

2750. —— and R. Weiss. 1952: Évolution des differentes formes de phosphore au cours de la germination des graines. Compt. Rend. 234:2004–6.

2751. Bonnett, O. T. 1938: Seed setting and average seed weight as affected by two methods of opening barley flowers for emasculation. J. Am. Soc. Agron. 30: 501–6.

2752. —— 1940: Development of the staminate and pistillate inflorescences of sweet corn. J. Agr. Res. 60:25–37.
Bonnett, O. T. 19856.

2753. Bonney, V. B., and H. Fischbach. 1945: Comparative chemical studies on pea seed and canned soaked dry peas. J. Assoc. Offic. Agr. Chemists 28:409–17.
Bonnichsen, R. K. 7381.

2754. Bonnier, C. 1952: Sur les propriétés antibactériennes de la graine d'*Arachis hypogaea* L. Compt. Rend. Soc. Biol. 146:309–10.

2755. Bonnier, G. 1892: Regermination of seeds. Rev. Gen. Botan. 4:193–201.

2756. —— and L. Mangin. 1884: Sur les variations de la respiration des graines germant avec le développement. Bull. Soc. Botan. France 31:306–9.
Bonnier, G. 19003.
Bonomo, R. 5145.

2757. Bonvicini, M. 1931: Sur la durée de la faculté germinative dans le blé. Action de la naphtaline. Ital. Agr. 68. Proc. Intern. Seed Testing Assoc. 7:109. 1935.

2758. Boodle, L. A., and A. W. Hill. 1929: *Typhonodorum lindleyanum:* The development of the embryo and germination of the seed. Ann. Botany (London) 43:437–50.
Booer, J. R. 3012.

2759. Boonstra, A. E. H. R. 1936: Der Einfluss der verschiedenen assimilierenden Teile auf den Samenertrag von Weizen. Z. Zücht. Reihe A. Pflanzenzücht. 21:115–47. Biol. Abstr. 11:18226. 1937.

2760. Boosalis, M. G. 1950: Studies on the parasitism of *Rhizoctonia solani* Kuehn on soybeans. Phytopathology 40:820–31.

2761. Booster, D. E. 1956: Caged lightning sorts seeds. Western Feed Seed 11:51–52. Biol. Abstr. 31:35776. 1957.

2762. Booth, C. E. 1951: Try sphagnum moss for seed germination in flats. Org. Gardener 18:32–33.

2763. Booth, E. G. 1929: Daily growth of the oat kernel and effect on germination of immaturity and controlled low temperatures. Minn. Agr. Expt. Sta. Tech. Bull. 62. 42 pp.

2764. Booth, R. G. 1944: Cereal phosphatases. I. The assay of free wheat phosphomonoesterase and characterization of the free phosphatases of wheat. Biochem. J. 38:355–62.

2765. —— 1948: Some characteristics of the nuclear enzymes of wheat. Biochem. J. 42:liv–lv.
Booth, R. G. 2052, 9272.

2766. Boothroyd, C. W. 1958: New approach to corn seed treatment. Farm Res. N.Y. 24(2):15.

2767. Borasio, L. 1935: The specific gravity of cereals; method of determination. Giorn. Risicoltura 25:149–55. Chem. Abstr. 29:7506–7. 1935.
Borbolla-y-Alcala, J. M. R. de; *see* De Borbolla-y-Alcala, J. M. R.

2768. Borchers, R. 1961: Counteraction of the growth depression of raw soybean oil meal by amino acid supplements in weanling rats. J. Nutr. 75:330–34.

2769. —— and C. W. Ackerson. 1950: The nutritive value of legume seeds. X. Effect of autoclaving and the trypsin inhibitor test for 17 species. J. Nutr. 41:339–45.

2770. —— and —— 1951: The nutritive value of legume seeds. XI. Counteracting the growth inhibitor of raw soybeans. Proc. Soc. Exptl. Biol. Med. 78:81–83.

2771. ——, ——, and L. Kimmett. 1947: Trypsin inhibitor. IV. Occurrence in seeds of the Leguminosae and other seeds. Arch. Biochem. 13:291–93.

2772. ——, ——, and F. E. Mussehl. 1948: Trypsin inhibitor. VIII. Growth inhibiting properties of a soybean trypsin inhibitor. Arch. Biochem. 19:317–22.

2773. Bordas, J. 1937: Le soja et son role alimentaire. III. Actualités Sci. Ind.

(Hermann, Paris) 557:1–36. Biol. Abstr. 12:10457. 1938.

2774. Bordas, M. 1932: Estudio de la semilla del garbanzo (*Cicer arietinum* L.). Bol. Real. Soc. Espan. Hist. Nat. 32: 393–416.

2775. Bordeiano, T., G. Constantinesco, G. Anghel, D. Teodoresco, N. Bajesco, I. Poenaru, and M. Mihalco. 1958: Contributions to the problem of radiostimulation of fruit-tree seeds. Proc. Intern. Conf. Peaceful Uses Atomic Energy 2:282–85.

2776. Borgeson, C., and H. K. Hayes. 1941: The Minnesota method of seed increase and seed registration for hybrid corn. J. Am. Soc. Agron. 33:70–74.

2777. Borghi, D. C. 1959: The effects on seeds of induced neutron microchemistry. Ann. Fitopatol. 4:71–78.

Boril, J. 3735.

2778. Boringer, D. 1963: New possibilities of seed production in foreign countries. Saatgut-Wirtsch. 15:70–71.

2779. Borisenko, S. I. 1950: Heat treatment of seeds. Selek. i Semen. 17:68–73.

2780. —— 1950: Treatment of pea seed in the control of Ascochyta. Sov. Agron. 8:81–82.

2781. —— 1953: Treatment of wheat and barley seeds with Granosan during vernalization. Zemledelie 1953(6):118–19.

2782. Borovskaya, V. A., A. V. Gurzhieva, A. K. Stankevich, and N. V. Sharapova. 1960: Influence of pre-sowing seed vernalization and growing conditions locality on the phenotypical composition of forage grass populations. Tr. Prikl. Botan. Genet. i Selek. 33:127–48.

2783. Born, W. H. V., and W. G. Corns. 1959: Studies on seed dormancy, plant development, and chemical control of Tartary buckwheat (*Fagopyrum tataricum* [L.] Gaertn.). III. Chemical control and effects of competition. Can. J. Plant Sci. 39:63–70.

2784. Bornebusch, C. H. 1939: Germination of forest tree seed. Dansk Skovfor. Tidsskr. 1939(1):1–26. Biol. Abstr. 14: 7490. 1940.

2785. Bornemann, G. 1886: Versuche über Erhaltung der Keimfähigkeit bei importirten Samen von Wasserpflanzen während des Transportes. Gartenflora 35:532–34.

Bornscheuer, E. 14274.

2786. Boroaik, S. A., and T. F. Boroaik-Romanova. 1944: Determination of microelements in different parts of fruits and seeds. Trav. Lab. Biogeochim. Acad. Sci. URSS 7:111–13. Chem. Abstr. 40:5473. 1946.

Boroaik-Romanova, T. F. 2786.

2787. Borodin, D. N. 1933: Yarovization formulas for winter oats and barleys. [Abstract of Paper.] Botan. Soc. Am. Physiol. Sect. Boston. Am. J. Botany 20:677–78. 1933.

2788. —— 1934: Yarovization of winter barleys. Am. J. Botany 21:708.

2789. Boroughs, H., and J. R. Hunter. 1963: The effect of temperature on the germination of *Cacao* seeds. Proc. Am. Soc. Hort. Sci. 82:222–24.

2790. —— and Y. C. Labarca. 1962: Factores que afectan la germinación del cacao. Turrialba 12:210–12.

Borovikova, N. V. 3333.

2791. Borowski, B., and J. Lutomski. 1960: Chromatographische Untersuchungen der Alkaloidfraktion aus dem Kraute und den Samen des *Tribulus terrestris* L. Poznan. Inst. Przem. Zielarskiego Bull. 6:220–26.

2792. Borriss, H. 1936: Zusammenwirken von Substrat, Temperatur und Licht bei der Keimung des Samens von *Vaccaria pyramidata*. Ber. Deut. Botan. Ges. 54:12–13.

2793. —— 1936: Über das Wesen der keimungsfördernden Wirkung der Erde. Ber. Deut. Botan. Ges. 54:472–86.

2794. —— 1941: Über die inneren Vorgange bei der Samenkeimung und ihre Beeinflussing durch Aussefaktoren. Jahrb. Wiss. Botan. 89:254–339.

2795. —— 1953: Keimungsphysiologische Untersuchungen an Caryophyllaceensamen. Ber. Deut. Botan. Ges. 65:(6)–(7). 1952. Biol. Abstr. 29:1800. 1955.

2796. —— 1955: Relation between metabolism and the physiological behavior of the embryo of *Agrostemma* seeds. Abhandl. Deut. Akad. Wiss. Berlin Kl. Chem. Geol. Biol. 1955(5):90–100. Chem. Abstr. 52: 2175. 1958.

2797. —— 1956/57: Über einige Ergebnisse und Probleme der Keimungsphysiologie. Wiss. Z. Ernst Moritz Arndt-Univ. Griefswald, Math.-Naturw. Reihe 6:251–65.

2798. —— and M. Arndt. 1956: Die Entwicklung isolierter *Agrostemma*-Embryonen in Abhängigkeit vom Reife- und Nachreifezustand der Samen. Flora 143: 492–98.

2799. —— and —— 1956: Die Temperaturabhängigkeit der Stärkebildung in *Agrostemma*-Embryonen und ihre Beziehung zur Keimungsbereitschaft der Samen. Naturwissenschaften 43:255–56.

2800. —— and F. Baum. 1957: Der Aminosäuregehalt von *Agrostemma*-Samen verschiedener Keimungsbereitschaft. Naturwissenschaften 44:404–5.

2801. —— and G. Schneider. 1955: Die Beziehungen zwischen N-Haushalb und keimungsphysiologischem Verhalten von *Agrostemma*-Samen. Naturwissenschaften 42:102–3.

2802. —— and —— 1955: Über den Nachweis eines Peptids in den Samen von *Agrostemma githago* L. und seine keimungsphysiologische Bedeutung. Naturwissenschaften 42:103.

2803. Borthwick, H. A. 1925: Factors influencing the rate of germination of the seed of *Asparagus officinalis*. Calif. Agr. Expt. Sta. Tech. Paper 18.

2804. —— 1931: Carrot seed germination. Proc. Am. Soc. Hort. Sci. 28:310–14.

2805. —— 1931: Development of the macrogametophyte and embryo of *Daucus carota*. Botan. Gaz. 92:23–44.

2806. —— 1936: Retarded germination in the seed of *Hypericum perforatum* caused by calcium. Botan. Gaz. 98:270–82.

2807. —— 1957: Light effects on tree growth and seed germination. Ohio J. Sci. 57: 357–64.

2808. —— and S. B. Hendricks. 1960: Photoperiodism in plants. Science 132: 1223–28.

2809. ——, ——, M. W. Parker, E. H. Toole, and V. K. Toole. 1952: A reversible photoreaction controlling seed germination. Proc. Natl. Acad. Sci. U.S. 38:662–66.

2810. ——, ——, E. H. Toole, and V. K. Toole. 1954: Action of light on lettuce seed germination. Botan. Gaz. 115:205–25.

2811. —— and W. W. Robbins. 1928: Lettuce seed and its germination. Hilgardia 3:275–89.

Borthwick, H. A. 1699, 5533, 5945, 7986, 8628, 15417, 16259, 18660, 18666, 18668, 18679–18681, 18683, 18695–18697.

Borys, M. 19659.

2812. Borzini, G. 1935: Influence of thallium ions on the germination of various seeds and on the initial development of the seedling. Boll. Staz. Patol. Veg. 15:200–31. Biol. Abstr. 11:13248. 1937.

2813. —— 1935: The action of growth substances on the germination of the seeds and on the formation of cork in potato tubers. Boll. Staz. Patol. Veg. 15:323–37. Chem. Abstr. 29:8051. 1935.

2814. —— 1936: Ricerche sull'effetto di soluzioni di zuccheri e di glicerina sulla germinazione di semi e sull'ulteriore sviluppo di alcune piante. Boll. Staz. Patol. Veg. 16:155–65.

2815. —— 1949: Undamaged vs. mechanically damaged "cryptolesioned" grain. Notiz. Malattie Piante 6:15–18. Biol. Abstr. 26:7318. 1952.

2816. —— 1950: Treatment of seeds for caries. Humus (Milan) 6:20–22.

2817. —— and R. Tomaselli. 1949: Seed treatment and mechanical lesions of wheat. Intern. Congr. Crop Protect. Abstr. Commun. 2:76.

2818. Bos, H. 1929: Die Kontrolle der Samen auf Sortenechtheit. Fortschr. Landwirtsch. 4:713–18.

2819. Bos, L., and J. P. H. Van der Want. 1962: Early browning of pea, a disease caused by a soil and seed-borne virus. T. Pl.-Ziekten 68:368–90.

Bose, A. 7788.

Bose, A. B. 2826.

Bose, A. C. 16752.

Bose, A. K. 3766.

2820. Bose, B. C., H. N. De, S. S. Gupta, and S. Bhaskaran. 1957: Mechanism of induction of polyploidy by colchicine. II. Cytochemical studies on dehydrogenase activity in gram (*Cicer arietinum*) after induction of polyploidization by colchicine. Arch. Intern. Pharmacodyn. 109:462–67. Chem. Abstr. 51:16731. 1957.

Bose, J. L. 2401, 11413.

2821. Bose, P. K., and S. N. Dutt. 1940: Oil from the fruit of *Ferula alliacea*. J. Indian Chem. Soc. 17:49–52. Analyst 65:421. 1940.

2822. Bose, S. M., M. Swaminathan, and V. Subrahmanyan. 1953: Nutritive value of tamarind seed (*Tamarindus indicus*). Bull. Central Food Technol. Res. Inst. Mysore (India) 3:67–68. J. Sci. Food Agr. 5:ii, 32. 1954.

2823. Bose, S. R. 1947: Hereditary (seed-borne) symbiosis in *Casuarina equisetifolia* Forst. Nature 159:512–14.

2824. —— 1955: Hereditary (seed-borne) symbiosis in *Ardisia humilis* Vahl. Nature 175:395.

2825. —— 1956: *Alternaria* within the pericarp of the wheat seed. Nature 178:640–41.

2826. ——, A. B. Bose, and K. L. Dey. 1948: Effect of crude polyporin on seed germination and root growth: a preliminary study. Science 107:63.

Boshart, R. 1829.

Bosian, G. 19213.

2827. Bosticco, A. 1952: Carotenoid content of corn grains. Changes of the carotene and cryptoxanthine as influenced by the variety and the duration of the vegetative cycle of the plant. Atti Soc. Ital. Sci. Vet. 6:174–76. Chem. Abstr. 48:8886. 1954.

2828. —— and A. Ubertalle. 1954: Germinative energy in seeds of "*Avena sativa*, L." and "*Lupinus albus*, L." treated with numerous solutions of meat extract from the trade. Torino Univ. Fac. Med. Vet. Ann. 4:327–34.

2829. Bostick, V. B. 1951: Seeding burned over Sacramento Mountain forest lands by airplane proves successful. New Mex. Stockman 16(12):40, 89.

2830. Boswell, F. C., and W. L. Parks. 1957: The effect of soil potassium levels on yield, lodging, and mineral composition of corn. Soil Sci. Soc. Am. Proc. 21:301–5.

2831. Boswell, J. G. 1936: Constitution of certain nutshells. I. The seed-coat of *Bertolletia excelsa* (Brazil-nut). Biochem. J. 30:971–76.

2832. Boswell, V. R. 1925: A study of some environmental factors influencing the shooting to seed of wintered over cabbage. Proc. Am. Soc. Hort. Sci. 1925:380–93.

2833. —— 1955: Effects of insecticides on flavor and quality of food products. J. Econ. Entomol. 48:495–99.

2834. ——, E. H. Toole, and D. F. Fisher. 1939: A study of rapid deterioration of vegetable seeds and methods for its prevention. Proc. Am. Soc. Hort. Sci. 36:655–59.

2835. ——, ——, V. K. Toole, and D. F. Fisher. 1940: A study of rapid deterioration of vegetable seeds and methods for its prevention. U.S. Dept. Agr. Tech. Bull. 708. 48 pp.

Boswell, V. R. 2151.

2836. Botha, P. J. 1948: The parasitism of *Alectra vogelii* Benth. with special reference to the germination of its seeds. S. African J. Sci. 44:119.

2837. —— 1950: The germination of the seeds of angiospermous root parasites. I–II. J. S. African Botany 16:23–38. Biol. Abstr. 25:29206. 1951.

2838. ——, A. P. Goossens, H. H. Cornell, and N. J. Viljoen. 1939: A study of the organic nitrogen content of three varieties of the cowpea at different stages of growth. S. Africa Dept. Agr. Forestry Sci. Bull. 177. 43 pp.

2839. Botkin, C. W. 1929: Composition of the cotton seed produced in New Mexico. New Mexico Agr. Expt. Sta. Bull. 175. 15 pp.

2840. —— 1935: The protein and moisture content of wheat grown in New Mexico. New Mexico Agr. Expt. Sta. Bull. 230. 16 pp.

2841. Bottazzi, G. B. 1947: Esperienze di jarovizzazione del frumento. Genet. Agr. 1:147–66.

2842. Bottelier, H. P., H. Holter, and K. Linderstrom-Lang. 1943: Studies on enzymatic histo-chemistry. XXXVI. Determination of peptidase activity, nitrogen content, and reduced weight in roots of the barley, *Hordeum vulgare*. Compt. Rend. Trav. Lab. Carlsberg Ser. Chim. 24: 289–313.

2842-A. Bottini, E., and C. Tarantola. 1952/54: Protein metabolism during the budding of some seeds studied by means of chromatography. Ist. Sper. Chim. Agrar. Ann. 18:285–311.

2843. —— and —— 1954: The metabolism of proteins during the germination of caryopses and seeds studied by means of chromatography. Ann. Sper. Agrar. (Rome) 8:809–35. Chem. Abstr. 48:12240. 1954.

2844. Bottini, O. 1937/38: Sulla utilizzazione deisemi di *Plantago psyllium* L. Portici Ist. Super. Agrar. Ann. 9:75–82.

2845. Bottomly, R. A., C. M. Christensen, and W. F. Geddes. 1950: Grain storage studies. IX. The influence of various temperatures, humidities, and oxygen concentrations on mold growth and biochemical changes in stored yellow corn. Cereal Chem. 27:271–96.

2846. ——, ——, and —— 1952: Grain storage studies. X. The influence of aeration, time, and moisture content on fat acidity, nonreducing sugars, and mold flora of stored yellow corn. Cereal Chem. 29:53–64.

Botts, R. R. 6839.

2847. Boudon, S. 1953: Étude de la lipolyse au cours de la germination des graines de pin maritime. Rev. Gen. Botan. 60:284–313.

2848. Bouges, Mlle. 1934: Sur quelques resultats de la suralimentation et de la sous-alimentation embryonnaires chez l'avoine. Compt. Rend. 198:606–8. Biol. Abstr. 10:360. 1936.

2849. Bouharmont, J. 1961: Embryo culture of rice on sterile medium. Euphytica 10:283–93.

2850. —— 1962: Fécondation de l'ovule et developpement de la graine après croisement et autopollinisation chez *Hevea brasiliensis* Muell. Agr. Cellule 62:119–30.

2851. Bouharmont, P. 1959: La culture d'embryons d'*Elaeis guineensis* Jacq. Agricultura (Louvain) 7:297–323.

2852. Bouillenne, R., and M. Bouillenne-Walrand. 1947: Contribution à la théorie de la Rhizocaline. Lejeunia 11:17–36. Biol. Abstr. 23:5123. 1949.

2853. Bouillenne-Walrand, M. 1943: Action de l'hétéroauxine appliquée aux graines de *Zea mays* sur le rendement en épis. Bull. Soc. Roy. Sci. Liege 3:161–75. Biol. Abstr. 23:12561. 1949.

2854. —— 1943: Wirkung von Heteroauxin auf die Atmung der Keimpflanzen von *Zea mais*. Bull. Soc. Roy. Sci. Liege 12:175–79. Chem. Abstr. 38:5257. 1944.

Bouillenne-Walrand, M. 2852.

Boulakia, F. Cohen-; *see* Cohen-Boulakia, F.

2855. Boulanger, P., J. Claveau, and G. Biserte. 1955: Évolution des acides amides libres des graines de haricot (*Phaseolus vulgaris*) au cours de la germination. Compt. Rend. 241:577–79.

2856. Boulter, D., and J. T. Barber. 1963: Amino-acid metabolism in germinating seeds of *Vicia faba* L. in relation to their biology. New Phytologist 62:301–16.

Boulter, D. 1889, 6584, 9697.

2857. Bouquet, J. 1935: Criminal poisoning with seeds of *Datura* and *Mandragora*. Bull. Sci. Pharmacol. 42:456–59. Chem. Abstr. 30:769. 1936.

2858. Bourcet, P. 1933: Analysis of areca nuts. Bull. Sci. Pharmacol. 40:98–100. Chem. Abstr. 27:2250–51. 1933.

2859. Bourdillon, J. 1951: A crystalline bean seed protein in combination with phytic acid J. Biol. Chem. 189:65–72.

2860. —— 1953: Bean protein and phytic acid. N.Y. State Dept. Health Ann. Rept. Div. Lab. Res. 1953:23. Chem. Abstr. 48:12823. 1954.

2861. Bourdon, D., and M. Quillet. 1958: Sur l'évolution des galactosides pendant les periodes de formation et de germination des graines de *Soja hispida* (Moench). Compt. Rend. 247:504–6.

2862. Bourdouil, C. 1932: Relation, chez les hybrides de *Pisum*, entre la synthese de l'amidon et le poids des graines. Compt. Rend. 195:1317–19.

2863. —— 1933: Comparative germination of varieties of pea. Bull. Soc. Chim. Biol. 15:1121–23. Chem. Abstr. 28:4098. 1934.

Bourdouil, C. 3034.

Bourdouil, C. Sosa-; *see* Sosa-Bourdouil, C.

2864. Bourdu, R. 1963: Inulin and oligosaccharides in *Dahlia achenes* and young seedlings. Physiol. Vegetale 1:77–94. Chem. Abstr. 59:14298. 1963.

2865. —— and R. Gorenflot. 1961: Sur les galactosides des graines dans le genre *Plantago* L. Compt. Rend. 253:698–700.

2866. Boureau, E. 1949: Le depôts lacustres pliocenes-pleistocenes de l'Ankaratra (Madagascar). La flore fossile du bassin lacustre de l'Ankaratra (Madagascar). Ann. Geol. Serv. Mines (Madagascar) 18:117–37. Biol. Abstr. 25:28636. 1951.

Bourgeat, J. Créveaux-; *see* Créveaux-Bourgeat, J.

Bourget, S. J. 14394.

2867. Bouriquet, G. 1947: Sur la germination des graines de vanillier (*Vanilla planifolia* And.). Agron. Trop. (Nogent-sur-Marne) 2:150–64.

2868. —— 1948: La germination des graines de vanillier. Agron. Trop. (Nogent-sur-Marne) 3:498–99.

2869. —— 1949: La transplantation des jeunes vanilliers issus de semis asymbiotiques. Agron. Trop. (Nogent-sur-Marne) 4:614–17.

2870. Bourlakow, G. G. 1897: Sur la question de la respiration du germe de froment. Kharkof. Univ. Obshch. Ispytatelei Prirody 31:i–xv. Botan. Centralbl. 74:323–24. 1898.

Bourne, E. J. 15164.

2871. Bournot, K. 1914: Über das Enzym der *Chelidoniumsamen*. II. Biochem. Z. 65:140–57.

2872. Boursnell, J. G. 1950: The symbiotic seed-borne fungus in the Cistaceae. I. Distribution and function of the fungus in the seeding and in the tissues of the mature plant. Ann. Botany (London) 14:217–43.

2873. Bouvarel, P. 1960: Note sur la résistance au froid de quelques provenances de pin maritime. Rev. Forestry (France) 7:495–508.

2874. —— 1961: Le problème de l'identification de l'origine des graines. Suggestions pour quelques méthodes de recherches. Proc. Intern. Seed Testing Assoc. 26: 404–10.

2875. —— 1962: The influence of the provenance of spruce seeds on growth in the nursery, the precocity and frequency of Lammas shoots. Ann. École Natl. Eaux Forêts Nancy 19:415–39.

2876. Bouvarel, P., and M. Lemoine. 1959: L'examen des graines forestières aux rayons X. Rev. Forestry (France) 8/9: 618–24.

2877. Bouygues, H. 1926: Les blessures des téguments et le gonflément des graines. Actes Soc. Linn. Bordeaux 78(Proc. Verb.): 67–71.

2878. —— 1926: Sur le gonflément des graines de *Phaseolus vulgaris* (var. *soissons*). Actes Soc. Linn. Bordeaux 78(Proc. Verb.): 72–77.

2879. Bouyoucos, G. J., and M. M. McCool. 1921: Measurement of the amount of water that seeds cause to become unfree and their water-soluble material. J. Agr. Res. 20:587–93.

2880. Bowen, H. D. 1961: Oxygen and physical impedance affecting germination. Cotton Gin Oil Mill Press 62(2):37–38.

2881. Bowen, H. J. M., and P. A. Cawse. 1962: The effects of deficiencies in essential elements on the radiosensitivity of tomato seeds. Radiation Botany 1:215–22.

2882. —— and J. Thick. 1960: Factors from seed extracts that modify radiosensitivity. Radiation Res. 13:234–41. Chem. Abstr. 55:1823. 1961.

2883. —— and —— 1961: Effects of seed extracts on radiosensitivity. Symp. Effects Ionizing Radiations Seeds Significance Crop Improv. Proc. 1960:75–82.

2884. Bowers, C. G. 1954: Rhododendron seed germination in agar nutrient solution. Natl. Hort. Mag. 33:206–8.

Bowler, E. 17136.

2885. Bowles, W. G. 1943: Hybrid amaryllis from seed to flower. Flower Grower 30(5):225–26.

2886. Bowling, C. C. 1957: Seed treatment for control of the rice water weevil. J. Econ. Entomol. 50:450–52.

2887. —— 1960: A comparison of the effectiveness of aldrin seed treatment for rice water weevil control on water- and drill-seeded rice. J. Econ. Entomol. 53:1135–37.

2888. —— 1964: Studies of the effects of aldrin on seed rice germination and emergence. J. Econ. Entomol. 57:83–85.

2889. Bowman, D. E. 1943: The ether soluble fraction of navy beans and the digestion of starch. Science 98:308–9. Biol. Abstr. 18:3501. 1944.

2890. —— 1944: Fractions derived from soy beans and navy beans which retard tryptic digestion of casein. Proc. Soc. Exptl. Biol. Med. 57:139–40.

2891. —— 1945: Amylase inhibitor of navy beans. Science 102:358–59.

2892. —— 1949: Antithromboplastic activity of navy bean trypsin inhibitor in vivo. Federation Proc. 8:186–87.

2893. Bowman, D. H. 1943: Chemical seed treatments for corn. Plant Disease Reptr. 27:141–43. Biol. Abstr. 18:1686. 1944.

Bowman, R. I. 16189.

2894. Bowser, P., F. N. Hewetson, and K. C. Barrons. 1944: Production of onion seed from wintered-over "scallions." Seed World 56(2):12–13. Biol. Abstr. 19: 7247. 1945.

2895. Box, M. M. 1935: El analisis de simillas mediante los rayos ultravioletas. Econ. Tec. Agr. 4:377–83.

2896. Boy, G. 1935: Energy of growth. XIV. Action of toxic concentrations of zinc and manganese salts on energy yields of germinating seeds. Bull. Soc. Chim. Biol. 17:1414–26. Chem. Abstr. 30:739. 1936.

Boyadjiev, V. 9377.

2897. Boyce, D. S. 1960: The determination of moisture in parchment coffee and green coffee beans with a dielectric-type moisture-meter. J. Agr. Univ. Puerto Rico 44: 176–93.

2898. Boyce, J. S. 1915: Some methods in the germination tests of coniferous tree seeds. Nebraska Univ. Forest Club Ann. 6:71–88. Expt. Sta. Rec. 33:645. 1915.

2899. Boyce, S. G., and J. F. Hosner. 1963: Alternating storage temperatures increase the germination of yellow-poplar seed. J. Forestry 61:731–33.

2900. —— and M. Kaeiser. 1961: Why yellow-poplar (*Liriodendron tulipifera*) seeds have low viability. U.S. Dept. Agr. Forest Serv. Central States Forest Expt. Sta. Tech. Paper 186. 16 pp.

Boyce, S. G. 10116.

2901. Boyd, D. A. 1952: The effect of seed-rate on yield of cereals. Empire J. Exptl. Agr. 20:115–22.

2902. Boyd, F. T. 1940: Seed storage investigations. Florida Agr. Expt. Sta. Ann. Rept. 1939/40:183.

2903. Boyd, H. 1961: Australian orchids. Their pollination and seed collection. Australian Plants 1(9):29. Biol. Abstr. 43:11880. 1963.

Boyd, H. M. 626, 627.

2904. Boyd, I. L. 1943: Germination tests on four species of sumac. Trans. Kansas Acad. Sci. 46:85–86.

Boyd, J. S. 11841.

2905. Boyd, L. 1932: Monocotylous seedlings: morphological studies in the post-seminal development of the embryo. Trans. Proc. Botan. Soc. Edinburgh 31:5–224. Biol. Abstr. 8:4429. 1934.

2906. Boyd, R. J. 1954: Effects of seed treatments upon the germination of certain browse species of Colorado. [Abstract.] J. Colo.-Wyo. Acad. Sci. 4:43–44.

Boyd, W. C. 12898, 16905.

2907. Boyenval, J. 1947: Recherches sur les teguments des grains de haricots. Rev. Hort. (Paris) 30:248–50.

2908. Boyer, W. D. 1958: Longleaf pine seed dispersal in South Alabama. J. Forestry 56:265–68.

2909. Boyes, J. W. 1939: Embryo-sac development in *Plumbagella*. Proc. Natl. Acad. Sci. U.S. 25:141–45. Biol. Abstr. 15:19302. 1941.

2910. —— and W. P. Thompson. 1937: Développement de l'endosperme et de l'embryon dans des croisements interspécifiques réciproques chez les céréales. J. Genet. 34:203–27. Ann. Epiphyties Phytogenetique 3:651. 1937.

Boyes, J. W. 15639, 15640.

2911. Boyle, L. M. 1946: Orchid town is America's foremost Cymbidium orchid grower. El Rancho Rinconada, Ojai, Calif. 105 pp.

2912. Boyles, B. B., and J. F. Martin. 1931: Growth habit and yield in wheat as influenced by time of seeding. J. Agr. Res. 42:483–500.

2913. Boynton, M. F. 1895: Observations upon the dissemination of seed. Botan. Gaz. 20:502–3.

2914. Boysen-Jensen, P. 1915: Über synthetische Vorgänge im pflanzlichen Organismus. II. Vorkommen, Bedeutung und Bildung des Rohrzuckers bei der Keimungs von *Pisum sativum*. Jahrb. Wiss. Botan. 56:431–46.

2915. —— 1958: Untersuchungen über determination und Differenzierung. V. Über die Wirkungsweise des Wuchsstoffes in dem Epikotyl von *Phaseolus*. Kgl. Danske Videnskab. Selskab Biol. Medd. 23(8):1–51.

Bozhenko, A. 7396–7398.

2916. Bozzini, A., R. S. Caldecott, and D. T. North. 1962: The relation of seedling height to genetic injury in X-irradiated barley seeds. Nucl. Sci. Abstr. 16:2277.

2917. Braak, J. P. 1955: Effects of some internal and external factors on the embryo and seedling development of the cherry. 14th Intern. Hort. Congr. 1955 Rept. 1:845–52.

2918. —— 1962: Influence of growth conditions on fruit and embryo development in apple and cherry. Wageningen Inst. Veredeling Tuinbouwgewassen Mededel. 182:57–65.

Braak, J. P. 10531.

2919. Braathe, P. 1960: Effect of the moss and humus cover on the germination and development of spruce plants. Norsk Skogbruk 6:463–65.

2920. Bracey, B. O. 1932: Raising seedling orchids at Sawtelle. Am. Orchid Soc. Bull. 1:60–64.

2921. Bracken, A. F. 1934: Effect of various smut treatments on yield of winter wheat. J. Am. Soc. Agron. 26:748–51. Biol. Abstr. 9:19410. 1935.

2922. —— 1940: Variation in protein percentage of wheat the result of environment: moisture, temperature, and soil are important factors in determining quality in wheat. Farm Home Sci. (Utah Agr. Expt. Sta.) 1(4):9. Expt. Sta. Rec. 84:616. 1941.

Bracken, A. F. 7612.

2923. Brad, I., I. Laszlo, and G. Valuta. 1959: Physiological and biochemical study of the behavior of some varieties of winter barley during dormancy. Ann. Inst. Cerc. Agron. Bucuresti Ser. C, Fiziol. Genet. Amelior. Protect. Plantelor Technol. Agr. 27:35–45.

2924. ——, ——, ——, and C. Miclea. 1962: The influence of the transplantation of the embryo of seeds of rye on some biochemical processes of the hybrids, in correlation with the resistance to frost. Acad. Rep. Populare Romine, Studii Cercetari Biochim. 5:551–59. Chem. Abstr. 58:12855. 1963.

Brad, I. V. 1828

2925. Bradbeer, J. W., and V. M. Floyd. 1964: Nucleotide synthesis in hazel seeds during after-ripening. Nature 201:99–100.

2926. Bradbury, D., I. M. Cull, and M. M. MacMasters. 1956: Structure of the mature wheat kernel. I. Gross anatomy and relationships of parts. Cereal Chem. 33:329–42.

2927. ——, M. M. MacMasters, and I. M. Cull. 1956: Structure of the mature wheat kernel. II. Microscopic structure of pericarp, seed coat, and other coverings of the endosperm and germ of hard red winter wheat. Cereal Chem. 33:342–60.

2928. ——, ——, and —— 1956: Structure of the mature wheat kernel. III. Microscopic structure of the endosperm of hard red winter wheat. Cereal Chem. 33:361–73.

2929. ——, ——, and —— 1956: Structure of the mature wheat kernel. IV. Microscopic structure of the germ of hard red winter wheat. Cereal Chem. 33:373–91.

2930. ——, M. J. Wolf, and R. J. Dimler. 1962: The hilar layer of white corn. Cereal Chem. 39:72–78.

2931. Bradbury, F. R. 1963: The systemic action of benzene hexachloride seed dressings. Ann. Appl. Biol. 52:361–70.

Bradley, A. 19635–19638.

2932. Bradley, M. V., and J. C. Crane. 1957: Effects of auxins on development of apricot seeds and seedlings. Am. J. Botany 44:164–75.

Bradley, W. B. 3527.

2933. Brady, N. C. 1947: The effect of period of calcium supply and mobility of calcium in plant on peanut fruit filling. Soil Sci. Soc. Am. Proc. 12:336–41.

2934. ——, J. F. Reed, and W. E. Colwell. 1948: The effect of certain mineral elements on peanut fruit filling. J. Am. Soc. Agron. 40:155–67.

Brady, N. C. 4311, 4312.

2935. Brady, S. E. 1938: Phytochemical study. Seed of the *Magnolia grandiflora*. J. Am. Pharm. Assoc. 27:407–17. Biol. Abstr. 13:1315. 1939.

2936. Brage, B. L., W. R. Zich, and L. O. Fine. 1959: The germination of small grain and corn as influenced by urea and other nitrogenous fertilizers. Agron. Abstr. 1959:23.

Bragonier, W. H. 9467.

2937. Brambilla, M., and G. Balbi. 1936: Tomato-seed oil. Chim. Ind. (Milan) 18:399. Analyst 61:855. 1936.

2938. Bramble, W. C. 1946: Natural reproduction of jack pine in Pennsylvania. J. Forestry 44:204. Expt. Sta. Rec. 95:205. 1946.

2939. Bramesfeld, B. 1955: Untersuchungen zur Frage der β-Indolylessigsäure-Bindung im Maisscutellum. Z. Botan. 43:103–24.

2940. Brand, C. J., and W. A. Sherman. 1913: Behavior of seed cotton in farm storage. U.S. Dept. Agr. Bur. Plant Ind. Circ. 123:11–20.

2941. Brand, K., and G. Westerburg. 1931: Über das ätherische öl der Früchte von *Phellodendron japonicum* Maxim. I. Arch. Pharm. 269:369–83. Ber. Ges. Physiol. 64:237–38. 1932. Ber. Wiss. Biol. 19: 261–62. 1932.

2942. Brandenburg, N. R. 1961: Separating seeds with electricity. Crops & Soils 14(1):12–13.

Brandenburg, N. R. 8193.

Brandes, E. W. 1423.

Brandrup, W. 13693.

Brandt, L. E. 5979.

2943. Brandwein, P. F. 1938: The emergence of smut-inoculated oat seedlings through sand and loam soil. Bull. Torrey Botan. Club 65:477–83.

2944. —— 1944: Seedling invasion of the covered smut of oats. Phytopathology 34:481–84.

2945. Brandza, M. M. 1891: Development of the integument of the seed. Rev. Gen. Botan. III, vol. 2:1–32, 71–84, 105–26, 150–65, 229–40.

2946. Branscheidt, P. 1940: Befruchtungsphysiologische Untersuchungen an *Taxus baccata* L. Ber. Deut. Botan. Ges. 57:495–505.

Branton, C. I. 10766.

2947. Brase, K. D. 1948: Field nursery tests with newly harvested and one- or two-year stored Lovell peach seed. Proc. Am. Soc. Hort. Sci. 51:258–62.

Brase, K. D. 7258.

Brassington, N. 7928.

2948. Braun, A. J. 1940: The development of the seed of *Lycopersicon esculentum* Mill. with special reference to the differentiation of the seed coat. Am. J. Botany 27:706.

2949. Braun, E., R. M. MacVicar, D. R. Gibson, P. Pankiw, and J. Guppy. 1953: Studies in red clover seed production. II. Can. J. Agr. Sci. 33:437–47.

Braun, E. 12662.

2950. Braun, H. 1919: Presoaking as a means of preventing seed injury due to disinfectants and of increasing germicidal efficiency. Science 49:544–45.

2951. —— 1920: Presoak method of seed treatment: a means of preventing seed injury due to chemical disinfectants and of increasing germicidal efficiency. J. Agr. Res. 19:363–92.

2952. —— 1924: A gradient of permeability to iodin in wheat seed coats. J. Agr. Res. 28:225–26.

2953. —— 1931: Untersuchungen über den Einfluss von Kohlensäure und Sauerstoff auf Keimung und Pflanzgutwert der Kartoffelknolle. Arb. Biol. Reichsanst. Land- Forstwirtsch. Berlin-Dahlem 19: 17–93. Biol. Abstr. 7:20638. 1933.

2954. Brauner, L. 1930: Untersuchungen über die Electrolyt-Permeabilitat und Quellung einer leblosen natürlichen Membran. Jahrb. f. Wiss. Bot. 73:513–632.

Bravo, G. 16550.

2955. Bray, G. T., and F. Major. 1945: Sunflower seed from Nigeria. Gt. Brit. Imp. Inst. Bull. 43:83–86. Brit. Abstr. B 3:241. 1945.

2956. Brayshaw, T. C. 1953: An approach to the determination of moisture requirements for the germination of seeds of Douglas fir (*Pseudotsuga taxifolia*), Ponerosa pine (*Pinus ponderosa*), and Engleman spruce (*Picea engelmanni*). Brit. Columbia Univ. Fac. Forestry Res. Notes 8. 6 pp.

2957. Breakwell, E. 1921: Some germination tests of prickly pear seeds. Agr. Gaz. N.S. Wales 32:579–80.

Breal, E. 5123.

2958. Brebion, G., and R. Scuflaire. 1957: Excitation et inhibition de la germination et des premières phases de la croissance par diverses substances chimiques. Bull. Soc. Franc. Physiol. Vegetale 3:125.

2959. Brecher, J. 1949: Seeds and fruits deprived of their outer cases in view of those who are concerned in seed-culture. Sb. Cesk. Akad. Zemedel. Ved 22:574–85.

2960. ——, A. Szabo, and M. Vinsova. 1960: Einiger Fragen zur laboratoriumsnässigen Prüfung des Saatgutes der Saaterbse *Pisum sativum* L. spp. *hortense* (Neillr.) A. Gr. in der Tschechoslowakischen Republik. Proc. Intern. Seed Testing Assoc. 25:379–90.

2961. Brecht, F. 1936: Der Einfluss von Wuchsstoff- und Säurepasten auf das Wachstum von *Avena*- und *Helianthus*-keimlingen und seine Abhängigkeit von Sauerstoffgehalt der Luft. Jahrb. Wiss. Botan. 82:581–612.

Breda, N. G. van; *see* Van Breda, N. G.

2962. Breddy, N. C. 1952: The asymbiotic raising of orchid seedlings. Australian Orchid Rev. 17:99–100.

2963. —— 1952: The asymbiotic raising of orchid seedlings. (Cont'd). Australian Orchid Rev. 17:119–20.

2964. —— 1953: Observations on the raising of orchids by asymbiotic cultures. Am. Orchid Soc. Bull. 22:12–17.

2965. Bredemann, G. 1924: Versuche über Saatgutbehandlung mit Reizmitteln (Saatgut-Stimulierung). Pflanzenbau. 1: 161–69.

2966. —— 1926: Weitere Versuche über Saatgut-Stimulierung. Landwirtsch. Jahrb. 63:369–86.

2967. —— 1931: Neuere Untersuchungen über die Hartschaligkeit bei Leguminosen. Proc. Intern. Seed Testing Assoc. 3:148–59.

2968. —— 1932: Zur Schwefelsäurestimulation hartschaliger Samen. Deut. Landwirtsch. Presse 59:526.

2969. —— 1938: Zur Frage der Bewertung hartschaliger Luzernesamen. Proc. Intern. Seed Testing Assoc. 10:123–45.

2970. —— and W. Kötter. 1931: Zur Entwicklungsgeschichte der Wacholderbeeren (*Fructus juniperi*). Arch. Pharm. 269:167–75. Botan. Centralbl. 164:390. 1933.

2971. Bredereck, H., G. Caro, and F. Richter. 1938: Über die fermentative Aufspaltung der Hefe- und Thymo-nucleinsäure (Nucleinsäuren, X.). Ber. Deut. Chem. Ges. 71B:2389–91. Biol. Abstr. 13:12103. 1939.

Bredow, K. von. 3292.

2972. Breese, M. H. 1955: Hysteresis in the hygroscopic equilibria of rough rice at 25°C. Cereal Chem. 32:481–87.

2973. Brega, C. 1927–28: The influence of seeding time on the development of cereal rusts. Riv. Patol. Vegetale 17:153–56; 18:153–60. Expt. Sta. Rec. 63:245–46.

2974. Breggar, T. 1931: Waxy endosperm in Argentina maize. J. Heredity 19:111. Biol. Abstr. 6:6120. 1932.

2975. Breiholz, H. 1878: Untersuchungen über die Ölgehalt einige landwirtschaftlich wichtigen Grasfrüchte und dessen Beziehung, oder Verhältniss zur Keimungsenergie und zur mittlaren Keimdauer. Dissertation, Jena. 42 pp.

2976. Bremer, A. H. 1926/27: Germination of garden seeds at various temperatures. Nord. Jordbrugsforskn. 8/9:377–90.

2977. —— 1928: Soaked or unsoaked seed. Norg. Lantbrukshögskol. 8:108–22.

2978. Bremer, H. 1929: Die Abhängigkeit der Zuckerrübenkeimung von der Temperatur. Angew. Botan. 11:112–15.

2979. —— 1932: Zur Frage der Beizung von Tomatensaatgut. Nachr. Deut. Pflanzenschutzdienst (Berlin) 12:2.

2980. —— 1957: Zur Behandlung von Bohnensaatgut mit kombinierten Beizmitteln. Anz. Schädlingskunde 30:84–85.

2981. —— 1961: Pathology of germination of vegetables. Z. Pflanzenkrankh. Pflanzenschutz 68:587–91.

Bremer, J. W. 5737.

2982. Bremner, P. M., R. N. Eckersall, and R. K. Scott. 1963: The relative importance of embryo size and endosperm size in causing the effects associated with seed size in wheat. J. Agr. Sci. 61:139–45.

2983. Brenchley, W. E. 1909: On the strength and development of the grain of wheat (*Triticum vulgare*). Ann. Botany (London) 23:117–39.

2984. —— 1918: Buried weed seeds. J. Agr. Sci. 9:1–31.

2985. —— 1944: Weed control and the dormancy of weed seeds. Gt. Brit. Min. Agr. J. 50:452–55.

2986. —— and K. Warington. 1930: The weed seed population of arable soil. I. Numerical estimation of viable seeds and observations on their natural dormancy. J. Ecol. 18:235–72. Biol. Abstr. 9:1371. 1935.

2987. —— and —— 1933: The weed seed population of arable soil. II. Influence of crop, soil and methods of cultivation upon the relative abundance of viable seeds. J. Ecol. 29:103–27.

2988. —— and —— 1936: The weed seed population of arable soil. III. J. Ecol. 24:479–501.

2989. —— and —— 1945: Influence of periodic fallowing on prevalence of viable weed seeds in arable soil. Ann. Appl. Biol. 32:285–96. Brit. Abstr. B 3:197. 1947.

2990. Brender, E. V. 1958: A 10-year record of pine seed production on The Hitchiti Experimental Forest. J. Forestry 56:408–10.

2991. Brennan, M., and J. Doyle. 1956: The gametophytes and embryogeny of *Athrotaxis*. Sci. Proc. Roy. Dublin Soc. N.S. 27:193–252.

2992. Brenner, E. 1943: Berried shrubs from seed. Horticulture 21(13):271–72.

2993. Brentzel, W. E. 1956: Seed treatments. N. Dakota Agr. Expt. Sta. Bull. 402. 31 pp.

Brentzel, W. E. 18026.

2994. Bresaola, M. 1919: Devitalization of the seeds of *Cuscuta*. Staz. Sper. Agr. Ital. 52:193–207.

2995. Breslavetz, L. P. 1960: Effect of chronic irradiation on dried rye seeds. Biophysics (USSR) (English Transl.) 5: 48–53.

2996. —— and A. S. Afanassieva. 1937: The action of X-rays on the rye. II. X-radiation of seeds. Cytologia (Tokyo) 8:110–27.

2997. ——, N. M. Berezina, G. I. Shchibria, M. L. Romanchikova, V. A. Iazykova, and Z. F. Mileshko. 1960: Increase in the yield of radishes and carrots by X- or γ-irradiation of the seeds before sowing. Biophysics (USSR) (English Transl.) 5: 87–91. Nucl. Sci. Abstr. 14:2751. 1960.

2998. ——, G. B. Medwedewa, and A. S. Afanassieva. 1935: Die Wirkung der Röntgenstrahlen auf Roggen. Protoplasma 23:520–33. Biol. Abstr. 10:7571. 1936.

2999. —— and Z. F. Mileshko. 1958: An investigation of the effect of neutrons on dry seeds of diploid and tetraploid winter rye. Dokl. Akad. Nauk SSSR 120:146–47.

3000. Bressani, R. 1960: I. Corn protein studies. II. A factor in corn which improves the growth of rats on amino acid diets. III. Studies on amino acid requirements. Dissertation Abstr. 21:290–91.

3001. ——, A. A. Campos, R. L. Squibb, and N. S. Scrimshaw. 1953: Nutritive value of Central American corns. IV. The carotene content of thirty-two selections of Guatemalan corn. Food Res. 18:618–24.

3002. —— and R. Condé. 1961: Changes in the chemical composition and in the distribution of nitrogen of maize at different stages of development. Cereal Chem. 38:76–84.

3003. ——, L. G. Elias, N. S. Scrimshaw, and M. A. Guzman. 1962: Nutritive value of Central American corns. VI. Varietal and environmental influence on the nitrogen, essential amino acid, and fat content of ten varieties. Cereal Chem. 39:59–67.

3004. ——, E. Marcucci, C. E. Robles, and N. S. Scrimshaw. 1954: Nutritive value of Central American beans. I. Variation in the nitrogen, tryptophane, and niacin content of ten Guatemalan black beans (*Phaseolus vulgaris* L.), and the retention of the niacin after cooking. Food Res. 19:263–68.

3005. —— and E. T. Mertz. 1958: Studies on corn proteins. IV. Protein and amino acid content of different corn varieties. Cereal Chem. 35:227–35.

3006. —— and B. J. Rios. 1962: The chemical and essential amino acid composition of twenty-five selections of grain sorghum. Cereal Chem. 39:50–58.

Bressani, R. 1413, 5893, 17831.

Bressler, R. 5885.

3007. Bressman, E. N. 1929: The effect of land plaster applied as a dust to seed corn. Phytopathology 19:1131–33.

Bressman, E. N. 17635.

Bretschneider, L. H. 5960.

3008. Brett, C. C. 1939: The production, handling, testing, and diseases of seeds. Ann. Appl. Biol. 26:616–27.

3009. Brett, C. C. 1951: The official seed testing station for England and Wales thirteenth conference of seed analysts, 18th May, 1951. Natl. Inst. Agr. Botany J. 6:59–66.

3010. —— 1953: Factors affecting the viability of grass and legume seed in storage and during shipment. Proc. 6th Intern. Grassland Congr., Univ. Park, Pa. 1952:878–84.

3011. —— and W. A. R. D. Weston. 1941: Seed disinfection. IV. Loss of vitality during storage of grain treated with organo-mercury seed disinfectants. J. Agr. Sci. 31:500–17.

3012. ——, ——, and J. R. Booer. 1937: Seed disinfection. III. Experiments on the germination of peas—seed protection by the use of disinfectant dusts containing mercury. J. Agr. Sci. 27:53–66. Expt. Sta. Rec. 77:57. 1937.

Brett, C. C. 9273.

3013. Brett, P. G. C. 1946: Seed setting of sugar cane in South Africa. Nature 157:657–58.

3014. Bretz, T. W. 1950: Seed transmission of the elm mosaic virus. [Abstract.] Phytopathology 40:3–4.

Brevis, J. G. 19005.

3015. Brewbaker, H. E. 1942: Mosaic and seed production. Proc. Am. Soc. Sugar Beet Technol. 1942:381–85. Biol. Abstr. 18:14713. 1944.

3016. —— and H. L. Bush. 1947: Graded whole vs. segmented, pelletted, and single- and double-germ seed in field germination tests. Proc. Am. Soc. Sugar Beet Technol. 1947:188–93.

3017. —— and R. R. Wood. 1947: Single- and double-germ seed. Proc. Am. Soc. Sugar Beet Technol. 1947:172–75.

3018. ——, ——, and H. L. Bush. 1946: Single-germ seed. Proc. Am. Soc. Sugar Beet Technol. 1946:259–62. Biol. Abstr. 21:23982. 1947.

3019. Brewer, H. E., and J. L. Butt. 1950: Hygroscopic equilibrium and viability of naturally and artificially dried blue lupine seeds. Plant Physiol. 25:245–68.

3020. Brewer, L. 1961: Germinating seeds in vermiculite. Plant Life 17:143–45.

3021. Brewster, A. A. 1920: Germination of choko seed. Australian Nat. 4:121. Botan. Abstr. 5:1881. 1920.

3022. Brewster, D. R., and J. A. Larsen. 1925: Studies in Western yellow pine nursery practice. J. Agr. Res. 31:1101–20.

3023. Brewster, J. F., and C. L. Alsberg. 1919: Determination of the distribution of nitrogen in certain seeds. J. Biol. Chem. 37:367–71. Botan. Abstr. 2:1318. 1919.

3024. Breymann, C. 1914: Beiträge zur Anatomie der Samenschale einiger Cuscuta-Arten. Mitt. Kaiser Wilhelms Inst. Landwirtsch. Bromberg 6:95–114. Botan. Centralbl. 126:514. 1914.

3025. Brezhnev, D. D. 1950: Biological methods for improving the varietal characteristics of seeds. Tr. Prikl. Botan. Genet. i Selek. Vses. Inst. Rasteni. 28: 197–98.

3026. —— 1951: Effect of conditions for tomato growing on natural characteristics of seeds. Agrobiologiya 1951(1):154–56.

3027. —— and I. S. Aizenshtat. 1951: Some problems in increasing the viability of hybrid seeds. Izv. Akad. Nauk SSSR Ser. Biol. 3:40–51.

3028. —— and S. V. Aleksandrov. 1951: Seed production of greenhouse tomato and cucumber varieties. Selek. i Semen. 18(12):59–62.

3029. Brian, P. W. 1952: Antibiotics as systemic fungicides and bactericides. Ann. Appl. Biol. 39:434–38.

3030. ——, M. Jamieson, and J. C. McGowan. 1948: Toxicity of sulfhydryl compounds to seeds. Nature 162:780.

3031. Briccoli, M. 1939: Trials of yarovization upon certain varieties of domestic wheat. Ital. Agr. 76:119–23. Biol. Abstr. 14:11194.

3032. Brice, P. B. 1957: Reproduction and growth in various lily species following X- and γ-irradiation of pollen and X-irradiation of seeds. Dissertation Abstr. 17:1650.

3033. Brickell, C. 1958: The use of polythene covers for propagating cuttings and seeds. J. Roy. Hort. Soc. 83:515–16.

3034. Bridel, M., and C. Bourdouil. 1931: Sur l'évolution des glucides au cours de la formation de la graine de deux variétés des pois. Compt. Rend. 193:949–51. Chem. Abstr. 26:1640. 1932.

3035. Bridge, R. E., M. M. Chakrabarty, and T. P. Hilditch. 1951: Composition of poppyseed oils. J. Oil Colour Chemists Assoc. 34:354–60. Chem. Abstr. 46:8394. 1952.

3036. ——, A. Crossley, and T. P. Hilditch. 1951: The influence of environment upon the composition of sunflower seed oils. III. Oils from sunflower seeds grown in different regions of Australia. J. Sci. Food Agr. 2:472–76.

3037. Briggs, D. E. 1962: Development of enzymes by barley embryos in vitro. J. Inst. Brewing 68:470–75.

3038. —— 1963: Effects of gibberellic acid on barley germination and its use in malting: a review. J. Inst. Brewing 69: 244–48.

Briggs, D. R. 6096.

3039. Briggs, F. N. 1927: Dehulling barley seed with sulphuric acid to induce infection with covered smut. J. Agr. Res. 35: 907–14.

Briggs, G. M. 18715.

3040. Briggs, G. W. G. 1955: Production of high-quality seeds of field crops in relation to climatic and other conditions. Organ. European Econ. Cooperation European Productivity Agency. High Quality Seed, pp. 73–84.

3041. Briggs, H. M., W. D. Gallup, and A. E. Darlow. 1946: The nutritive value of cottonseed meal, soybean meal, and peanut meal when used separately and together to supplement the protein of prairie hay in experiments with steers. J. Agr. Res .73:167–76.

3042. ——, V. G. Heller, A. E. Darlow, W. D. Gallup, and J. A. Hoefer. 1946: The nutritive value of cottonseed meal, soybean meal, and peanut meal when used separately and together to supplement the protein of prairie hay in experiments with lambs. J. Agr. Res. 73:359–66.

3043. Briggs, L. H., and J. L. Mangan. 1948: *Sophora* alkaloids. V. The alkaloids of the seeds of a possibly new species from Anawhata, New Zealand. J. Chem. Soc. 1948:1889–91. Biol. Abstr. 23:19708. 1949.

3044. —— and J. Ricketts. 1937: *Sophora* alkaloids. I. The alkaloids of *S. microphylla*, Ait. J. Chem. Soc. 1937:1795–98.

3045. —— and W. E. Russell. 1942: *Sophora* alkaloids. The alkaloids of the seeds of *Sophora chrysophylla*. J. Chem. Soc. 1942:507–9.

3046. —— and —— 1942: *Sophora* alkaloids. IV. The alkaloids from the seeds of the Chatham Islands species. J. Chem. Soc. 1942:555–56.

3047. —— and W. S. Taylor. 1938 : *Sophora* alkaloids. II. The alkaloids of the seeds of *S. tetraptera*. J. Chem. Soc. 1938:1206–7.

3048. Briggs, L. J. 1908: An electrical resistance method for the rapid determination of the moisture content of grain. Science 28:810–13.

3049. Brigham, R. D., and M. M. Hoover, Jr. 1956: A scarifying cup for small lots of legume seed. Agron. J. 48:531–32.

3050. —— and C. P. Wilsie. 1955: Seed setting and vegetative vigor of Ladino clover (*Trifolium repens* Leyss) clones and their diallel crosses. Agron. J. 47:125–27.

3051. Brightwell, W. T. 1941: Germination of Muscadine grape seeds and growth of seedlings. S. Carolina Agr. Expt. Sta. 54th Ann. Rept. 1940/41:49–52.

3052. Brilliantine, L., and N. K. Allen. 1961: Studies on hemaglutinins in California wild flower seeds. J. Immunol. 86:575–77.

Brim, C. A. 9134.

Brimhall, B. 13495, 13496, 17812, 17815.

3053. Brindley, T. A., and F. G. Hinman. 1937: Effect of growth of pea weevil on weight and germination of seed peas. J. Econ. Entomol. 30:644–70. Expt. Sta. Rec. 77:823. 1937.

3054. Bringhurst, R. S., and V. Voth. 1957: Effect of stratification on strawberry seed germination. Proc. Am. Soc. Hort. Sci. 70:144–49.

3055. Brink, R. A. 1934: Self-incompatibility in yellow sweet clover, *Melilotus officinalis*. J. Am. Soc. Agron. 26:307–12.

3056. —— 1948: The angiosperm seed as a genetic mosaic. Science 107:458–59.

3057. —— 1958: A stable somatic mutation to colorless from variegated pericarp in maize. Genetics 43:435–47.

3058. —— and D. C. Cooper. 1939: Learn basic factors affecting seed formation in alfalfa. Wisconsin Agr. Expt. Sta. Bull. 443:65–66.

3059. —— and —— 1940: Double fertilization and development of the seed in angiosperms. Botan. Gaz. 102:1–25.

3060. —— and —— 1944: The antipodals in relation to abnormal endosperm behavior in *Hordeum jubatum* × *Secale* *cereale* hybrid seeds. Genetics 29:391–406.

3061. —— and —— 1947: Effect of the DE17 allele on development of the maize caryopsis. Genetics 32:350–68.

3062. —— and —— 1947: The endosperm in seed development. Botan. Rev. 13: 423–77.

3063. —— and —— 1947: The endosperm in seed development. Botan. Rev. 13: 479–541.

3064. ——, ——, and L. E. Ausherman. 1944: A hybrid between *Hordeum jubatum* and *Secale cereale*, reared from an artificially cultivated embryo. J. Heredity 35:67–75. Biol. Abstr. 18:12620. 1944.

3065. —— and R. A. Nilan. 1952: The relation between light variegated and medium variegated pericarp in maize. Genetics 37:519–44.

Brink, R. A. 561, 611, 4381–4387, 7626, 10466, 20101, 20102.

3066. Brink, V. C. 1958: Note on a fluorescence test to distinguish seeds of alfalfa and sweet clover in mixtures. Can. J. Plant Sci. 38:120–21.

3067. Brinkerhoff, L. A., and R. E. Hunter. 1963: Internally infected seed as a source of inoculum for the primary cycle of bacterial blight of cotton. Phytopathology 53:1397–1401.

Brinkerhoff, L. A. 9236, 11966.

3068. Briscoe, C. B. 1961: Germination of cherrybark and Nuttall oak acorns following flooding. Ecology 42:430–31.

3069. —— and A. P. Du Barry. 1959: A simplified germination tests for American sycamore. Tree Planters Notes 35:21

3070. Brison, F. R. 1942: Influence of storage conditions upon the germination of onion seed. Proc. Trans. Texas Acad. Sci. 25:69–70. Biol. Abstr. 17:12232. 1943.

3071. —— 1945: The storage of shelled pecans. Texas Agr. Expt. Sta. Bull. 667. 16 pp.

3072. —— 1952: Influence of storage conditions on the germination of onion seed. Texas Agr. Expt. Sta. Progr. Rept. 1492:1–4. Biol. Abstr. 27:17660. 1953.

Brison, F. R. 123.

3073. Britikov, E. A., and R. N. Petropavlovskaya. 1954: Effect of enzymes, vitamins, and auxin-type stimulants on binding of seeds in castrated rye. Dokl. Akad. Nauk SSSR 97:341–44. Chem. Abstr. 48:13832. 1954.

3074. Brittain, W. H., and C. C. Eidt. 1933: Seed content, seedling production, and fruitfulness in apples. Can. J. Res. 9:307–33.

3075. Britten, E. J. 1947: The effect of naphthaleneacetic acid on the developing maize caryopsis. Am. J. Botany 34: 211–18.

3076. —— 1950: Natural and induced parthenocarpy in maize and its relation to hormone production by the developing seed. Am. J. Botany 37:345–52.

3077. Brittingham, W. H. 1942: The nature and extent of variation in Kentucky bluegrass as criteria for apomictic seed formation. [Abstract.] Genetics 27:134–35.

3078. —— 1943: Type of seed formation as indicated by the nature and extent of variation in Kentucky bluegrass, and its practical implications. J. Agr. Res. 67:225–64.

3079. —— 1950: The inheritance of date of pod maturity, pod length, seed shape, and seed size in the southern pea, *Vigna sinensis*. [Abstract.] Assoc. South. Agr. Workers Proc. 47:104.

Brittingham, W. H. 13794, 16181.

3080. Britton, G., and S. Housley. 1961: Chromatography of water-soluble auxins in immature corn kernels. Phyton 16:1–19.

Britton, G. 2292, 6240.

3081. Brixhe, A., and A. Gillet. 1961: Effects of the bites of *Hemiptera on the* quality of cotton seeds. Ghent Landbhogesch. Mededel. 26:1713–22.

3082. Brizicky, G. K. 1964: Polyembryony in *Euonymus* (Celastraceae). J. Arnold Arboretum 45:251–59.

3083. Broadbent, J. H., and G. Shone. 1963: The composition of *Pachyrrhizus erosus* (yam bean) seed oil. J. Sci. Food Agr. 14:524–27.

Broadfoot, W. C. 12602.

3084. Brock, R. D. 1955: Chromosome balance and endosperm failure in hyacinths. Heredity 9:199–222.

3085. Brockelsby, C. F. 1945: An electrical moisture meter. J. Sci. Instr. 22:243–44. Biol. Abstr. 21:25763. 1947.

Brockelt, G. 18931.

3086. Brockington, S. F., H. C. Dorin, and H. K. Howerton. 1949: Hygroscopic equilibria of whole kernel corn. Cereal Chem. 26:166–73.

3087. Brockwell, J. 1963: Seed pelleting as an aid to legume seed inoculation. World Crops 15:334–38.

3088. —— and F. W. Hely. 1962: Relationship between viability and availability of

Rhizobium trifolii introduced into seeds of *Trifolium subterraneum* L. by imbibition and pressure. Australian J. Agr. Res. 13:1041–53.

Brockwell, J. 2315.

3089. Brocq-Rousseau, and E. Gain. 1909: Oxydases et peroxydiastases des graines. Rev. Gen. Botan. 21:55–62.

3090. Brodie, N. 1937: Indian vegetable oils. Bull. Indian Ind. Res. 10. 16 pp. Current Sci. (India) 6:361. 1938.

3091. Brodskis, B. 1949: Catalase, oxydation biologique et vitalité du grain. Bull. Soc. Sci. Nat. Phys. Maroc 35/37:9–12. Biol. Abstr. 23:30230. 1949.

Brodskis, G. 13424.

3092. Brodzikowski, W. 1957: Effect of injuring acorns in germination stage upon their subsequent growth and seedling development. Sylwan 101:46–60. Biol. Abstr. 32:31718. 1958.

3093. Broekhuizen, J. T. M. 1960: Provenance and nomenclature of some poplar cultivars. Ned. Bosbouw Tijdschr. 32: 153–59.

3094. Broekmans, A. F. M. 1957: The production of improved oil palm seed in Nigeria. W. African Inst. Oil Palm Res. J. 2:116–32.

3095. Broertjes, C. 1957: Application of atomic energy in aboriculture. Mededel. Directeur Tuinbouw 20:127–30, 190.

3096. Brohult, S., and E. Sandegren. 1954: Seed proteins. *In* The Proteins, H. Neurath and K. Bailey, editors. 2A: 487–512.

Brokaw, W. C. 5641.

3097. Broman, T. 1932: The use of an extract of *Echinocystis lobata* seeds as a test for citric acid. Skand. Arch. Physiol. 64:171–76. Chem. Abstr. 26:3207. 1932.

Brombacher, W. G. 19572.

3098. Bronckers, F. 1958: Recherches sur la conservation des semences de coton. Bull. Agr. Congo Belge 49:923–33.

Bronckers, F. 14818.

3099. Broniewski, S. 1962: Unterscheide im Verlaufe des Keimprozesses der Samen auf Grund eines Beispieles mit Samen des Cyklamen (*Cyclamen persicum* Mill.). Proc. Intern. Seed Testing Assoc. 27: 627–48.

Bronovitskaya, Z. S. 16903, 17595.

Bronskaya, L. M. 5920.

Bronstein, K. 3391.

3100. Bronwell, A. F. 1947: Effect of irradiation on delphinium seeds. Am. Delphinium Soc. Yearbook 12:34–36.

3101. Brooke, J. P. 1961: Protection of grain in storage. World Crops 13:27–30.

3102. Brooker, S. G. 1932: Chemical changes during the germination of seeds of white clover (*Trifolium repens*). Thesis, Univ. New Zealand, Wellington. 29 pp. Biol. Abstr. 10:361. 1936.

3103. Brooks, C., R. E. Blaser, and W. W. Moschler. 1959: The effect of fertilizer solutions on seed germination. Agron. Abstr. 1959:88.

3104. Brooks, G. 1934: Sur les vitamines de la graine de coton. Coton Cult. Coton 9:99–107.

3105. —— 1954: The biochemical research of urease in Brazilian beans. Bol. Inst. Quim. Agr. (Rio de Janeiro) 30:7–39. Chem. Abstr. 49:14906. 1955.

3106. Brooks, H. J., and L. F. Hough. 1958: Vernalization studies with peach embryos. Proc. Am. Soc. Hort. Sci. 71:95–102.

Brooks, J. S. 3907.

Brooks, R. M. 1723.

3107. Brother, G. H., A. K. Smith, and S. J. Circle. 1940: Soybean protein: resume and bibliography. U.S. Dept. Agr. ACE 62. 64 pp. Expt. Sta. Rec. 86:150. 1942.

Broughton, L. B. 19467.

Brouwer, G. 6680.

3108. Brouwer, W. 1927: Landwirtschaftliche Samenkunde. J. Newmann, Newdamm. 130 pp.

3109. —— and A. Stählin. 1955: Handbuch der Samenkunde für Landwirtschaft, Gartenbau und Gorstwirtschaft. DLG-Verlags-GMBH, Frankfurt. 656 pp.

3110. Brown, A. H. 1933: Effects of sulphuric-acid delinting of cotton seeds. Botan. Gaz. 94:755–70.

3111. —— and D. R. Goddard. 1941: Cytochrome oxidase in wheat embryos. Am. J. Botany 28:319–24.

3112. Brown, A. J. 1907: On the existence of a semipermeable membrane enclosing the seeds of some of the Gramineae. Ann. Botany 21:79–87.

3113. —— 1909. The selective semi-permeability of the covering of the seeds of *Hordeum vulgare*. Proc. Roy. Soc. London Ser. B 81:82–93.

3114. —— and F. Tinker. 1916: Selective permeability: the absorption of phenol and other solutions by the seeds of *Hordeum vulgare*. Proc. Roy. Soc. London Ser. B 89:373–79.

3115. —— and F. P. Worley. 1912: The influence of temperature on the absorption of water by seeds of *Hordeum vulgare* in relation to the temperature coefficient of chemical change. Proc. Roy. Soc. London Ser. B 85:546–53.

Brown, A. M. 12600, 12601.

3116. Brown, B. E., E. M. Meade, and J. R. Butterfield. 1962: The effect of germination upon the fat of the soybean. J. Am. Oil Chemists Soc. 39:327–30.

Brown, B. I. 13870.

Brown, C. A. 19624.

3117. Brown, C. L., and E. M. Gifford, Jr. 1958: The relation of the cotyledons to root development of pine embryos grown in vitro. Plant Physiol. 33:57–64.

3118. Brown, C. M. 1962: Effect of age of seed on performance of several oat varieties. Agron. J. 54:519–20.

3119. —— and H. L. Shands. 1956: Factors influencing seed set of oat crosses. Agron. J. 48:173–77.

3120. Brown, C. S. 1955: Hard seed in Birdsfoot trefoil. Cornell Univ. Publ. 14:344. Dissertation Abstr. 15:1961. 1955.

3121. Brown, C. W., and C. A. Moore. 1963: On-farm storage and disposal of sorghum grain. Texas Agr. Expt. Sta. Bull. 997. 8 pp.

3122. —— and A. Myers. 1960: Long term cereal storage. Agr. Gaz. N.S. Wales 71:348–55.

Brown, C. W. 13700.

3123. Brown, E. 1902: Germination of Kentucky bluegrass. U.S. Dept. Agr. Office Expt. Sta. Bull. 115:105–10.

3124. —— 1928: The interpretation of germination tests. Proc. Intern. Seed Testing Assoc. 3:67–70.

3125. —— 1932: Longevity of wheat seeds. Gardeners Chron. 92:288.

3126. —— 1935: Age and vitality of seeds. J. Franklin Inst. 219:132–33.

3127. —— 1939: Preserving the viability of Bermuda onion seed. Science 89:292–93.

3128. —— and W. L. Goss. 1908: The germination of vegetable seeds. U.S. Dept. Agr. Bur. Plant Ind. Bull. 131:5.

3129. —— and —— 1912: The germination of packeted vegetable seeds. U.S. Dept. Agr. Bur. Plant Ind. Circ. 101. 9 pp.

3130. —— and A. L. Robert. 1943: *Alternaria* sp. on grain kernels killed by high-temperature storage. Phytopathology 33: 333–35.

3131. ——, T. R. Stanton, G. A. Wiebe, and J. H. Martin. 1948: Dormancy and the effect of storage on oats, barley, and sorghum. U.S. Dept. Agr. Tech. Bull. 953:1–30.

3132. —— and E. H. Toole. 1931: The evaluation of seed tests. Proc. Intern. Seed Testing Assoc. 3:203–7.

3133. ——, ——, and W. L. Goss. 1928: A seed counter. U.S. Dept. Agr. Circ. 53. 3 pp.

3134. ——, ——, and —— 1934: Moisture content important factor in international trade in seeds. Proc. Intern. Seed Testing Assoc. 6:405–9.

Brown, E. 2283, 7502, 7503, 15375, 15545, 18661.

3135. Brown, E. E., and J. F. Fudge. 1960: Detection of *Crotalaria* seed in mixed feeds. Texas Agr. Expt. Sta. Misc. Publ. 474. 4 mimeo. pp.

3136. Brown, E. G. 1962: The acid-soluble nucleotides of mature pea seeds. Biochem. J. 85:633–40.

3137. —— 1963: Purine and pyrimidine derivatives in mature pea seeds. Biochem. J. 88:498–504.

3138. Brown, E. O. 1943: A laboratory method for testing seeds of Buffalo grass (*Buchloe dactyloides* [Nutt.] Engelm.). Assoc. Offic. Seed Analysts News Letter 17(4):3.

3139. —— 1952: Note on germination of buffelgrass. Assoc. Offic. Seed Analysts News Letter 26(4):17.

3140. —— and R. H. Porter. 1936: An improved method of testing seeds of Kentucky bluegrass (*Poa pratensis* L.). Proc. Assoc. Offic. Seed Analysts 1935: 44–49.

3141. —— and —— 1942: The viability and germination of seeds of *Convolvulus arvensis* L. and other perennial weeds. Iowa Agr. Expt. Sta. Res. Bull. 294: 473–504.

3142. Brown, F. R. 1942: Influence of storage conditions on germination of onion seed. Proc. Am. Soc. Hort. Sci. 40:501–3.

3143. Brown, G. B. 1950: Carrot and beet seed production as affected by steckling planting date and delayed planting after removal from storage. Proc. Am. Soc. Hort. Sci. 56:271–74.

3144. Brown, H. D. 1924: Size of seed in tomatoes in relation to plant growth and yields. Proc. Am. Soc. Hort. Sci. 21:57–60.

3145. —— 1945: Value of field tests as a supplement to laboratory tests in evaluation of vegetable seeds. Proc. Assoc. Offic. Seed Analysts 35:79–80.

Brown, H. E. 7705.

Brown, H. M. 7100, 10667, 10789.

3146. Brown, H. T., and F. Escombe. 1897/98: Note on the influence of very low temperatures on the germinative power of seeds. Proc. Roy. Soc. London 62:160–65.

3147. —— and —— 1898: On the depletion of the endosperm of *Hordeum vulgare* during germination. Proc. Roy. Soc. London 63:3–25.

3148. —— and G. H. Morris. 1890: The germination of some of the Gramineae. J. Chem. Soc. 57:458–528.

Brown, J. B. 2122.

3149. Brown, J. C. 1954: Some observations on the reduction of 2,3,5-triphenyltetrazolium chloride in plant tissue as influenced by mineral nutrition. Plant Physiol. 29:104–7.

3150. —— and P. M. Harmer. 1951: The influence of copper compounds on the yield, growth pattern, and composition of spring wheat and corn grown on organic soil. Soil Sci. Soc. Am. Proc. 15:284–91.

3151. Brown, J. G. 1936: A comparison of linted and acid-delinted cotton seed. Am. Phytopath. Soc. Abstr. Papers 28th Meeting, Atlantic City, N.J., p. 2.

3152. —— 1939: Permeability of the testa of normal and treated cotton seeds. [Abstract.] Phytopathology 29:753.

3153. —— and F. Gibson. 1925: A machine for treating cotton seed with sulphuric acid Ariz. Agr. Expt. Sta. Bull. 105. 15 pp.

3154. —— and E. F. Serr. 1953: Copper deficiency, one cause of shriveled kernels. Diamond Walnut News 35:20–21.

Brown, J. H., Jr. 6012.

3155. Brown, J. W. 1939: Respiration of acorns as related to temperature and after-ripening. Plant Physiol. 14:621–45.

Brown, J. W. 13595.

3156. Brown, L. C., and A. H. Hyer. 1954: Chemical defoliation of cotton. III. A study of seed and fiber from cotton plants treated with amino triazole. Agron. J. 46:580.

3157. —— and E. B. Kurtz, Jr. 1959: The *in vitro* synthesis of fats in cotton-seed. Agron. J. 51:49–50.

Brown, L. E. 15526.

3158. Brown, M. S. 1950: Cotton from Bikini; chromosome irregularities found in plants grown from seed exposed to γ-radiation. J. Heredity 41:115–21.

3159. Brown, R. 1931: The absorption of water by seeds of *Lolium perenne* (L.) and certain other Gramineae. Ann. Appl. Biol. 18:559–73.

3160. —— 1932: The absorption of the solute from aqueous solutions by the grain of wheat. Ann. Botany (London) 46: 571–82.

3161. —— 1940: An experimental study of the permeability to gases of the seed-coat membranes of *Cucurbita pepo*. Ann. Botany (London) 4:379–95.

3162. —— 1941: The growth of isolated cotyledons of *Cucurbita pepo*. Ann. Botany (London) 5:175–92.

3163. —— 1942: The gaseous exchange of seeds and isolated cotyledons of *Cucurbita pepo*. Ann. Botany (London) 6:293–321.

3164. —— 1943: Studies on germination and seedling growth. I. The water content, gaseous exchange, and dry weight of attached and isolated embryos of barley. Ann. Botany (London) 7:93–113.

3165. —— 1943: Studies on germination and seedling growth. II. The effect of the environment during germination on the subsequent growth of the seedling of barley. Ann. Botany (London) 7:275–96.

3166. —— 1946: Biological stimulation in germination. Nature 157:64–69.

3167. —— 1946: Studies on germination and seedling growth. III. Early growth in relation to certain aspects of nitrogen metabolism in the seedling of barley. Ann. Botany (London) 10:73–96.

3168. —— and M. Edwards. 1944: The germination of the seed of *Striga lutea*. I. Host influence and the progress of germination. Ann. Botany (London) 8: 131–48.

3169. —— and —— 1945: Effects of thiourea and allylthiourea on the germination of the seed of *Striga lutea*. Nature 155:455–56.

3170. —— and —— 1946: The germination of the seed of *Striga lutea*. II. The effect of time of treatment and of concentration of the host stimulant. Ann. Botany (London) 10:133–42.

3171. ——, A. D. Greenwood, A. W. Johnson, A. R. Lansdown, A. G. Long, and N. Sunderland. 1952: The *Orobanche* germination factor. III. Concentration of the factor by counter-current distribution. Biochem. J. 52:571–74.

3172. ——, ——, ——, and A. G. Long. 1951: The stimulant involved in the germination of *Orobanche minor* Sm. I. Assay technique and bulk preparation of the stimulant. Biochem. J. 48:559–64.

3173. ——, ——, ——, ——, and G. J. Tyler. 1951: The stimulant involved in the germination of *Orobanche minor* Sm. II.

Chromatographic purification of crude concentrates. Biochem. J. 48:564–68.

3174. ——, A. W. Johnson, E. Robinson, and A. R. Todd. 1949: Stimulant involved in the germination of *Striga hermonthica*. Proc. Roy. Soc. London Ser. B 136:1–12. Chem. Abstr. 43:6285. 1949.

3175. ——, ——, ——, and G. J. Tyler. 1952: The *Striga* germination factor. II. Chromatographic purification of crude concentrates. Biochem. J. 50:596–600.

3176. —— and E. Robinson. 1949: Effect of the *Striga* germination stimulant on the respiration of *Striga* seeds. Nature 164:1057.

3177. ——, ——, and A. W. Johnson. 1949: Effect of *Striga* germination stimulant on extension growth on the roots of peas. Nature 163:842–43.

3178. Brown, R. C. 1948: Experience with decorticated seed on 30,000 acres. Proc. Am. Soc. Sugar Beet Technol. 5:114–16.

Brown, R. L. 12069.

Brown, R. M. 16070.

Brown, S. A. 3478.

3179. Brown, W. B., and E. H. Farmer. 1935: Unsaturated acids of natural oils. III. The highly unsaturated acid of the kernels of *Parinarium macrophyllum*. J. Chem. Soc. 1935:761–63.

3180. Brown, W. G., and C. P. Lentz. 1956: Moisture absorption by some packaging materials at subfreezing temperatures. Can. J. Technol. 34:366–72.

3181. Brown, W. L. 1942: The practical aspects of sexual and asexual reproduction in turf grasses. Turf Cult. 2:246–52. Biol. Abstr. 17:6729. 1943.

3182. —— 1942: The threonine, serine, cystine, and methionine content of peanut proteins. J. Biol. Chem. 142:299–301.

3183. —— 1944: The tryptophane and tyrosine content of peanut proteins. J. Biol. Chem. 154:57–61.

3184. Brown, W. N. 1953: Developmental morphology of pericarp blistering in sweet corn (*Zea mays* L.). Dissertation Abstr. 13:3.

Brown, W. N. 9184, 9185.

3185. Brown, W. V. 1955: A species of grass with liquid endosperm. Bull. Torrey Botan. Club 82:284–85.

3186. —— 1960: The morphology of the grass embryo. Phytomorphology 10:215–23.

Brown, W. V. 266, 5934.

Browning, G. M. 11900.

Bruce, A. C. 15126.

Bruce, C. F. 4657.

3187. Bruch, E. C. 1949: Seed control and seed certification in California. Intern. Crop Improvement Assoc. Proc. 31:81–82.

3188. —— 1961: California Department of Agriculture, buried seed project 1932–1960. Calif. Dept. Agr. Bull. 50:29–32.

3189. Brücher, H. 1948: Eine Schnellmethode zur Bestimmung der Keimfähigkeit von Samen. Physiol. Plantarum 1:343–58.

3190. —— 1950: A new biochemical method of testing the germinating power of grain. Sveriges Utsadesforen. Tidskr. 60:210–14.

3191. Brückner, G. 1937: Specific gravity of grain, hectoliter basis. Z. Ges. Getreide-, Mühlen- Bäckereiw. 24:262–67. Chem. Abstr. 32:2228. 1938.

3192. —— 1941: Changes in superficial grain characteristics on drying. Z. Ges. Getreide-, Mühlen- Bäckereiw. 28:163–68. Chem. Abstr. 37:5790. 1943.

Bruehl, G. W. 8293.

3193. Brühl, P., and J. S. Gupta. 1927: Commentationes phytomorphologicae et phytophysiologicae. IV. *Eichhornia* studie. J. Dept. Sci. Univ. Calcutta 8(5). 8 pp.

3194. Bruene, F., T. Arnd, E. Gunther, and E. Poock. 1932: Die Umgestaltung der Keimpflanzenmethode zum Zwecke der Untersuchung von Moorboeden. Z. Pflanz. A. 26:271–83. Ann. Agron. 3:118. 1933.

Bruere, P. 16470.

Brugovitzky, E. 15228, 15229.

3195. Bruhin, A. 1951: Auslösung von Mutationen in ruhenden Samen durch hohe Temperaturen. Naturwissenschaften 38:565–66.

3196. —— 1955: Über die polyploidisierende Wirkung eines Samenbeizmittels. Phytopathol. Z. 23:381–94.

3197. Bruijning, F. F., and J. Van Haarst. 1899: On the presence of hydrocyanic acid in the seeds of *Vicia*. Rec. Trav. Chim. 18:468–71. Expt. Sta. Rec. 11:748. 1899–1900.

Bruijning, W. H. 6622–6631.

Bruin, H. J. de; *see* De Bruin, H. J.

3198. Bruinsma, J., and S. S. Patil. 1963: The effects of 3-indoleacetic acid, gibberellic acid, and vitamin E on flower initiation in unvernalized Petkus winter rye plants. Naturwissenschaften 50:505.

3199. Brukhlei, M. A. 1957: Effect of the prolonged vernalization of seeds and the lighting regime on frost resistance of seedlings. Akad. Nauk Ukr. RSR Dopvidi 1957(4):402–5.

3200. —— 1959: Effect of pre-planting vernalization of seeds at low temperatures and various light regime on the morphogenetic processes in some fruit crops. Tr. Soveshch. po Morfogenezu Rast. 1: 272–74.

3201. Bruman, A. J. 1937: The place of iarovization in plant breeding. J. Heredity 28:31–33.

3202. Brun, J., and L. Tronstad. 1935: Some germination experiments with peas in heavy water. Kgl. Norske Videnskab. Selskabs Forh. 7:171–73. Chem. Abstr. 29:5137. 1935.

3203. Brunel, A., and R. Echevin. 1937: Evolution of nitrogen, appearance of allantoinase and urease in the germination of Nielle (*Agrostemma githago* L.). Compt. Rend. 205:81–83. Chem. Abstr. 31:7472. 1937.

3204. —— and —— 1938: La présence, l'origine et le rôle physiologique des ureides glyoxyliques dans les germinations de *Soja hispida* Mnch. Rev. Gen. Botan. 50:72–93.

Brunel, A. 5766, 17760.

3205. Brunk, E. L., and T. Hansbrough. 1960: Effect of temperature and light intensity on the germination of sycamore, sweetgum, and American elm seeds. Louisiana State Univ. Forestry Notes 39. 2 pp.

Brunn, L. K. 399, 402.

3206. Brunner, G. 1932: Beiträge zur Entwicklungsphysiologie der Kiefernkeimlinge. Jahrb. Wiss. Botan. 76:407–40.

3207. Bruno, F. 1933: Spectroscopical study of the green pigment of the funiculus and of the immature seeds of *Yacaranda mimosaefolia*. Lavori Ist. Botan. Palermo 4:155–67. Chem. Abstr. 28:17107. 1934.

3208. Bruno, P. 1953: On the presence and distribution at different depths of viable seeds in laboured and unlaboured soil. Allionia (Turin) 1:257–65. Biol. Abstr. 29:1699. 1955.

Brunori, M. 16430.

3209. Bruns, E. 1892: Embryo of grasses. Flora (Jena) 76:1–33. J. Roy. Microscop. Soc. 1893:350.

3210. Bruns, V. F., and L. W. Rasmussen. 1953: The effects of fresh water storage on the germination of certain weed seeds. I. White top, Russian knapweed, Canada thistle, morning glory, and poverty weed. Weeds 2:138–47.

3211. —— and —— 1957: The effects of fresh water storage on the germination of certain weed seeds. II. White top, Russian knapweed, Canada thistle, morning glory, and poverty weed. Weeds 5:20–24.

3212. —— and —— 1958: The effects of fresh water storage on the germination of certain weed seeds. III. Quackgrass, green bristlegrass, yellow bristlegrass, watergrass, pigweed, and *Halogeton*. Weeds 6:42–48.

3213. Brunson, A. M. 1946: The effect of storage conditions on the longevity of seed corn. Indiana Agr. Expt. Sta. Rept. Director 59:8–9.

3214. —— 1949: The effect of storage conditions on the longevity of seed corn. Purdue Univ. Agr. Expt. Sta. Ann. Rept. 62:42.

3215. ——, F. R. Early, and J. J. Curtis. 1948: Interrelations among factors influencing the oil content of corn. J. Am. Soc. Agron. 40:180–85.

3216. —— and J. G. Willier. 1929: Correlations between seed ear and kernel characters and yield of corn. J. Am. Soc. Agron. 21:912–22.

Brunson, A. M. 4769, 5504.

Brusca, J. N. 7920.

3217. Bruschi, D. 1908: Researches on the vitality and self-digestion of the endosperm of some Graminaceae. Ann. Botany (London) 22:449–63.

3218. Bruttini, A. 1894: The influence of certain salts on germination. Staz. Sper. Agr. Ital. 27:30. Expt. Sta. Rec. 6:904. 1894/95.

3219. Bruyning, F. F., Jr. 1893: Beiträge zur Kenntnis unseren Landbausamereien (Die Hartschaligkeit der Samen des Stechginsters [*Ulex europalus* L.]). J. Landwirtsch. 41:85–94.

3220. —— 1930: Verband zwischen Keimkraft und fermentatives Vermögen ungekeimte Samen. Landbouwk. Tijdschr. 1930:97–108.

Bruzzesi, M. R. 16430.

Bryan, A. A. 9187.

3221. Bryan, W. E. 1918: Hastening the germination of Bermuda grass seed by the sulfuric acid treatment. J. Am. Soc. Agron. 10:279–81.

3222. Bryant, L. H., and F. R. Morrison. 1942: Composition of soya beans grown in New South Wales. J. Australian Inst. Agr. Sci. 8:117–18. Biol. Abstr. 17:24788. 1943.

3223. Bryant, L. R. 1935: A study of the factors affecting the development of the embryo-sac and the embryo in the

McIntosh apple. New Hampshire Agr. Expt. Sta. Tech. Bull. 61. 38 pp.

3224. Bryder, O. 1961: A polyethylene bag for sowing seed. Am. Orchid Soc. Bull. 30:812.

Bryushkova, K. 11577.

3225. Brzeski, W. 1959: Time of ripening of the castor bean and its influence upon the yield, physical characteristics, and fat content. Polskie Towar. Botan. Acta 28:665–75.

3226. Brzostowski, H. W. 1962: Influence of pH and superphosphate on establishment of *Cenchrus ciliaris* from seed. Trop. Agr. (London) 39:289–96.

Bubar, J. S. 3329.

3227. Bublitz, W. 1953: Keimhemmung, Nachreife, endogene Jahresrhythmik und ihre Bedeutung für die Pflanzenzucht. Allgem. Forstz. 8:446–48.

3228. —— 1953: Über den Einfluss von Aussenfaktoren auf die Keimung und ihre Bedeutung für die Naturverjüngung. Allgem. Forstz. 8:594–96.

3229. —— 1953: Über die keimhemmende Wirkung der Fichtenstreu. Naturwissenschaften 40:275–76.

Bucci, I. 1650, 1651.

3230. Buch, T. G. 1955: The physiology of seed germination in ginseng. Byul. Mosk. Gl. Botan. Sad. 20:109–14.

3231. —— 1955: In regard to the physiology of seed germination and growth of certain plants. Byul. Mosk. Gl. Botan. Sad. 22:74–76.

3232. —— 1957: The storage of seeds which rapidly lose viability. Lesn. Kh-vo. 1957(8):81–82. Biol. Abstr. 35:25118. 1960.

3233. —— 1960: Problems in the storage of seeds of willows and poplars. Tr. Mosk. Gl. Botan. Sad. 7:219–39.

3234. —— 1961: Comparative study of biochemical characteristics in seeds of coltsfoot, poplar, and willow. Byul. Mosk. Gl. Botan. Sad. No. 41:66–73. Chem. Abstr. 57:1276. 1962.

Buch, T. G. 15579–15583.

3235. Bucha, H. C. 1942: Seed treatment as an aid in the wartime production of peanuts. Agr. News Letter 10:114–22. Chem. Abstr. 36:6298. 1942.

3236. Buchannon, K. W., and H. A. H. Wallace. 1962: Note on the effect of leaf disease on yield, bushel weight, and thousand-kernel weight of Parkland barley. Can. J. Plant Sci. 42:534–36.

3237. Buchert, J. G. 1959: The application of a nuclear-cytoplasmic genetic interaction in a new method of sweet corn seed production. Proc. Penn. Acad. Sci. 33:46–49. Biol. Abstr. 35:11313. 1960.

3238. Buchheim, A. N. 1935: Effect of freezing on the germinability of wheat seeds infected with loose smut, and on the development of the plants raised from them. Plant Protect. 6:134–37. Rev. Appl. Mycol. 15:637. 1936.

Buchholtz, K. P. 8132.

3239. Buchholtz, W. F. 1943: A comparison of dosages of copper carbonate and ethyl mercuric phosphate with chloranil and sulfur as sorghum seed treatments. Proc. S. Dakota Acad. Sci. 23:56–64. Expt. Sta. Rec. 93:729. 1945.

3240. Buchholz, J. T. 1931: The pine embryo and the embryos of related genera. Trans. Illinois State Acad. Sci. 23:117–125.

3241. —— 1931: The suspensor of *Sciadopitys.* Botan. Gaz. 92:243–62.

3242. —— 1932: The embryogeny of *Chamaecyparis obtusa.* Am. J. Botany 19:230–38.

3243. —— 1933: Determinate cleavage polyembryony, with special reference to *Dacrydium.* Botan. Gaz. 94:579–88. Biol. Abstr. 7:22112. 1933.

3244. —— 1936: Embryogeny of species of *Podocarpus* of the subgenus *Stachycarpus.* Botan. Gaz. 98:135–46.

3245. —— 1938: Cone formation in *Sequoia gigantea.* I. The relation of stem size and tissue development to cone formation. II. The history of the seed cone. Am. J. Botany 25:296–305.

3246. —— 1939: The embryogeny of *Sequoia sempervirens* with a comparison of the sequoias. Am. J. Botany 26:248–57.

3247. —— 1940: The embryogeny of *Cunninghamia.* Am. J. Botany 27:877–83.

3248. —— 1940: The embryogeny of *Torreya*, with a note on *Austrotaxus.* Bull. Torrey Botan. Club 67:731–54.

3249. —— 1941: Embryogeny of the Podocarpaceae. Botan. Gaz. 103:1–37.

3250. —— 1941: The embryogeny of *Pseudolarix* and *Keteleeria.* Am. J. Botany Suppl. 28:1s–2s.

3251. —— 1941: Multi-seeded acorns. Trans. Illinois State Acad. Sci. 34:99–101.

3252. —— 1942: A comparison of the embryogeny of *Picea* and *Abies.* Madrono 6:156–67.

3253. —— 1946: Volumetric studies of seeds, endosperms, and embryos in *Pinus ponderosa* during embryonic differentiation. Botan. Gaz. 108:232–44.

3254. —— and M. Kaeiser. 1940: A statistical study of two variables in the sequoias, pollen grain size and cotyledon number. Am. Naturalist 74(752):279–83.

3255. —— and E. M. Old. 1933: The anatomy of the embryo of *Cedrus* in the dormant stage. Am. J. Botany 20:35–44.

3256. —— and M. L. Stiemert. 1946: Development of seeds and embryos in *Pinus ponderosa*, with special reference to seed size. Trans. Illinois State Acad. Sci. 38:27–50.

Buchholz, J. T. 3474.

3257. Buchinger, A. 1927: Saugkraftmessungen ("Osmotisches verhalten") verschiedener Gerstensorten. Fortschr. Landwirtsch. 2:344–50.

3258. —— 1929: Der Einfluss hoher Anfangstemperaturen auf die Keimung, dargestellt an *Trifolium pratense*. Jahrb. Wiss. Botan. 71:149–53.

3259. —— 1929: Die Keimung von *Oryza sativa* zwischen Glasstäben. Fortschr. Landwirtsch. 4:46–47.

3260. —— 1931: Beiträge zur Kenntnis der Physiologie von *Convolvulus arvensis* L. J. Landwirtsch. 79:3–14.

3261. —— 1931: Können die Saugkraftmessungen zur Sortenechtheitsbestimmung herangezogen werden? Proc. Intern. Seed Testing Assoc. 3:167–74.

3262. —— 1932: Welchen Anteil haben Embryo und Endosperm an der Saugkraft der Getreidefrüchte? Dargestellt an *Triticum spelta muticum*. Proc. Intern. Seed Testing Assoc. 4:46–64.

3263. —— 1937: Keimungsversuche mit eigenartig verletzten Erbsen-samen! Proc. Intern. Seed Testing Assoc. 9:153–56.

3264. —— 1937: Keimungsversuche mit eigenartig verletzten Fisolen-samen! Proc. Intern. Seed Testing Assoc. 9:254–56.

Buchman, R. G. 3267.

Buchner, A. 9238.

3265. Buchstez, M., and A. M. C. Gaja. 1948: Notas sabre drogas, plantas y alimentos mexacanas. VIII. Accite de las semillas de mamey (*Lucuma mammonos* Gr.). Ciencia (Mex.) 8:264–65.

Buckingham, W. H. 11698.

3266. Buckley, J. P., S. T. Coker, and R. Goldfield. 1955: A pharmacological investigation of the seeds of *Pyrularia pubera* Michx. J. Am. Pharm. Assoc. 44:249–51.

3267. Buckman, R. E., and R. G. Buchman. 1962: Red pine plantation with 48 sources of seed shows little variation in total height at 27 years of age. U.S. Dept. Agr. Forest Serv. Lake States Forest Expt. Sta. Tech. Notes 616. 2 pp.

3268. Buckner, G. D. 1921: Comparative utilization of the mineral constituents in the cotyledons of bean seedlings grown in soil and in distilled water. J. Agr. Res. 20:875–80.

3269. —— and J. H. Kastle. 1917: The growth of isolated plant embryos. J. Biol. Chem. 29:209–13.

3270. Bucksteeg, W. 1938: Keimungsförderung durch *Soja*extrakt. Germany Biol. Reichsanstalt Land . . . Nachrbl. 18:87–90.

Buda, L. 1291.

3271. Budagovskii, V. I. 1949: To improve the breeding qualities of wild apple seeds. Sad i Ogorod 1949(7):17–21.

3272. Budd, A. C., W. S. Chepel, and J. L. Doughty. 1954: Germination of weed seeds. The influence of crops and fallow on the weed seed population of the soil. Can. J. Agr. Sci. 34:18–27.

Budd, D. 157, 1574.

3273. Buduryan, N. N. 1962: The effect of presowing cooling of the seeds on the physiological and biochemical processes in melons of the species Melo zard. Vopr. Fiziol. i Biokhim. Kul't. Rast. Shtiintsa: Kishinev 1:52–70. Biol. Abstr. 45:26051. 1964.

3274. —— and G. V. Kandina. 1962: Biochemical changes in corn seeds exposed to the cold. Vopr. Fiziol. i Biokhim. Kul't. Rast. Akad. Nauk Moldavsk. SSR, Inst. Fiziol. i Biokhim. Rast. 1962(1):33–44. Chem. Abstr. 59:11899. 1963.

Bührer, H. Daimser-: *see* Daimser-Bührer, H.

3275. Buell, K. M. 1952: Developmental morphology in *Dianthus*. I. Structure of the pistil and seed development. Am. J. Botany 39:194–210.

3276. —— 1952: Developmental morphology in *Dianthus*. II. Starch accumulation in ovule and seed. Am. J. Botany 39:458–67.

3277. Buell, M. F. 1935: Seed and seedling of *Acorus calamus*. Botan. Gaz. 96:758–65.

3278. —— 1937: Endosperm development in the Araceae. Elisha Mitchell Sci. Soc. J. 53:224–25.

3279. —— 1938: Embryogeny of *Acorus calamus*. Botan. Gaz. 99:556–68.

3280. Bünning, E. 1931: Die Abhängigkeit der Entwicklung etiolierter Bohnen (*Phaseolus multiflorus*) vom Zeitpunkt der Samenreife. Ber. Deut. Botan. Ges. 49:494–96.

3281. —— 1935: Über die Wirkung der Aussenbedingungen auf reifende Samen von *Phaseolus multiflorus*. Flora (Jena) 129:120–39.

3282. —— 1947: Die endogene Ruheperiode der Samen. Planta 35:352–59.

3283. —— 1949: Zur Physiologie der endogenen Jahresrhythmik in Pflanzen, speziell in Samen. Z. Naturforsch. 4:167–76.

3284. —— 1954: Parental age and germinative characters of the seeds. Ann. N.Y. Acad. Sci. 57:484–87.

3285. —— 1956: Endogenous rhythms in plants. Ann. Rev. Plant Physiol. 7:71–90.

3286. —— 1957: Endogenous diurnal cycles of activity in plants. *In* Rudnick, D. Rhythmic and synthetic processes in growth. Princeton Univ. Press, Princeton, N.J. 217 pp.

3287. —— and E. W. Bauer. 1952: Über die Ursachen endogener Keimfähigkeitsschwankungen in Samen. Z. Botan. 40:67–76.

3288. —— and L. Müssle. 1951: Der Verlauf der endogenen Jahresrhythmik in Samen unter dem Einfluss verschiedenartiger Aussenfactoren. Z. Naturforsch. 6b:108–12.

3289. —— and M. Tazawa. 1957: Über den Temperatureinfluss auf die endogene Tagesrhythmik bei *Phaseolus*. Planta 50:107–21.

Bueno, P. J. 7412.

3290. Bünsow, R., J. Penner, and R. Harder. 1958: Blütenbildung bei *Bryophyllum* durch Extrakt aus Bohnensamen. Naturwissenschaften 45:46–47.

3291. —— and C. Seiferth. 1959: Förderung der Samenkeimung durch Gibberellin in Abhängigkeit vom Reifezustand. Naturwissenschaften 46:153–54.

3292. —— and K. Von Bredow. 1958: Wirkung von licht und Gibberellin auf die Samenkeimung der Kurztagpflanze *Kalanchoë blossfeldiana*. Biol. Centralbl. 77: 132–41.

3293. Bünsow, R. C. 1960. The circadian rhythm of photoperiodic responsiveness in Kalanchoë. Cold Spring Harbor Symp. Quant. Biol. 25:257–60.

3294. Buer, A. 1925: Über Keimzug und seine Bedeutung nach versuchen an Samen der gelben Lupine. Angew. Botan. 7:335–56.

Bürke, R. P. 17100.

3295. Buerkle, B. 1929: Physiologische Untersuchungen über Umwandlungen des Öles im reifenden Sonnenblumensamen. Botan. Arch. 26:385–436.

Büter, H. 10339.

Buettner, G. 10068.

3296. Bugaeva, M. G. 1957: Practices in growing cabbage seeds in Kazakhstan. Sad i Ogorod 95(7):20–22.

3297. Bugajewsky, M. F. 1931: Alkoholverbrauch bei der Atmung des Erbsensamens. Biochem. Z. 238:60–66.

3298. Bugini, F. 1938: Partenocarpia e apogamia nelle piante arboree da frutto. Riv. Frutticult. (Ravenna) 2:183–200.

3299. Buglass, E. 1964: Seed production of crested wheatgrass as influenced by various management practices. Can. J. Plant Sci. 44:66–74.

3300. Bugnon, P. 1937: Une question de terminologie: la coléoptile et la piléole chez les Graminées. Bull. Soc. Botan. France 84:563–65.

3301. —— and A. Parrot. 1932: Sur la valeur morphologique du cotyledon chez les Ombellifères monocotyledones. Compt. Rend. 195:332–34.

Bugueno, V. Valdivia-; *see* Valdivia-Bugueno, V.

3302. Buianov, I. A. 1952: Growing high yields of seeds of winter crops (rye and wheat) on an occupied fallow. Selek. i Semen. 19(2):53–56.

3303. Buie, T. S. 1951: Plentiful seed for soil conserving. Better Crops Plant Food 35(7):22–26.

Buila, L. 7177.

3304. Buis, R. 1956: Action of low concentrations of coumarin and of maleic hydrazide on the development of maize. Rev. Gen. Botan. 63:29–37. Chem. Abstr. 51:3759. 1957.

3305. Bukeikhanov, G. K. 1962: Effect of treatment of seeds of *Larix sibirica* prior to planting on their ground germination in the Transilian Alatau. Vestn. Sel'skokhoz. Nauki (Alma-Ata) 3:45–53.

3306. Buker, R. J., and A. S. Carter. 1959: Shifts in the fall dormancy and bacterial wilt reaction of certified and non-certified Ranger alfalfa. Agron. Abstr. 1959:86.

3307. Bukovac, M. J., R. P. Larsen, and H. K. Bell. 1960: Effect of gibberellin on berry set and development of Concord grapes. Mich. Agr. Expt. Sta. Quart. Bull. 42:503–10.

3308. —— and S. H. Wittwer. 1958: Reproductive responses of lettuce (*Lactuca* *sativa*, variety Great Lakes) to gibberellin as influenced by seed vernalization, photoperiod, and temperature. Proc. Am. Soc. Hort. Sci. 71:407–11.

Bukovac, M. J. 1808, 16215, 19790.

3309. Bulajewski, M. 1932: Biological and chemical investigation of pumpkin seeds. Wiadomosci Farm. 59:315–16. Chem. Abstr. 26:5610. 1932.

3310. Bulard, C. 1948: Inversion of geotropism by heteroauxin of *Thuja orientalis*. Compt. Rend. 227:443–44. Chem. Abstr. 43:280. 1949.

3311. —— 1951: Accélération et amélioration de la germination par l'action de l'hétéro-auxine chez *Picea excelsa*. Toulouse Soc. Hist. Nat. Bull. 85(1950):185–91.

3312. —— 1952: Culture aseptique d'embryons de *Ginkgo biloba:* Róle des cotyledons dans l'absorption du sucre et la croissance de la tige. Compt. Rend. 235:739–41.

3313. —— 1956: Physiologie et evolution à propos de la germination de quelques gymnospermes. Compt. Rend. 243:599–600.

3314. —— 1960: Cultures adeptiques de plantules de blé separées de l'albumen: action de l'acide gibberellique. Compt. Rend. 250:3716–18.

3315. —— and J. Monin. 1960: Action de l'acide gibberellique sur des embryons dormants d'*Evonymus europaeus* cultivés *in vitro*. Compt. Rend. 250:2922–24.

3316. —— and —— 1960: Graines et embryons dormants d'*Evonymus europaeus*: Differentes modalités dans l'eveil de leur dormance par l'acide gibberellique. Compt. Rend. 250:4197–201.

3317. —— and —— 1963: Ètude du comportement d'embryons de *Fraxinus excelsior* L. prelevés dans des graines dormantes et cultivés in vitro. Phyton 20:115–25.

3318. —— and —— 1964: Influence d'une lyophilisation sur la germination ulterieure de caryopses de blé et d'embryons d'*Evonymus vulgaris*. Compt. Rend. 258:1613–15.

3319. Bulat, H. 1960: Experimentelle Beiträge zur Feststellung der Keimfähigkeit bei hitzebehandeltem Mais. Saatgut-Wirtsch. 12:134–35.

3320. —— 1961: Reduktionsvorgänge in lebendem Gewebe, Formazan, Tetrazoliumsalze und ihre Bedeutung als Redoxindicatoren im ruhenden Samen. Proc. Intern. Seed Testing Assoc. 26:686–96.

3321. —— 1962: Keimversuche, Triebkraftversuche und Tetrazoliumprüfungen mit Samen von *Gossypium* spp. Proc. Intern. Seed Testing Assoc. 27:779–94.

3322. —— 1963: Das allmähliche, durch ungünstige Lagerungsbedingungen beschleunigte, Absterben der Samen bzw. Rückgang der Keimfähigkeit im Bilde des topographischen Tetrazoliumverfahrens. Proc. Intern. Seed Testing Assoc. 28:713–51.

3323. —— and W. Lindenbein. 1961: Zwanzig Jahre Erfahrung mit der Tetrazolium-Methode an Gehölzsamen. Proc. Intern. Seed Testing Assoc. 26:453–71.

Bulat, H. 11679, 12258.

Bulatowa, M. N. 5814.

3324. Bulgakova, Z. P. 1937: Die Lagerung der Getreide. Izv. Nauchn. Inst. P. F. Lesgafta 20:69–104.

3325. —— 1951: Biology of the dormancy period in seeds of some tree species. Byul. Mosk. Obshch. Ispytatelei Prirody Otd. Biol. 57(6):77–81.

Bulger, C. S. 19631.

Bulger, J. W. 9845.

Bulinaru, V. 11907.

3326. Bull, W. C., and T. H. Hopper. 1941: The composition and yield of crude lipids obtained from soybeans by successive solvent extractions. Oil Soap 18:219–22. Biol. Abstr. 16:5187. 1942.

3327. Bullard, E. T., and E. C. Stevenson. 1953: Production of hybrid tomato seed. Proc. Am. Soc. Hort. Sci. 61:451–58.

3328. Buller, R. E. 1957: La semilla de alfalfa. Bull. Inform. Torrion. Cam. Agr. Ganad. 4(46):10–15.

3329. ——, J. S. Bubar, H. R. Fortmann, and H. L. Carnahan. 1955: Effects of nitrogen fertilization and rate and method of seeding on grass seed yields in Pennsylvania. Agron. J. 47:559–63.

Bullis, D. E. 6215.

Bullock, J. A. 7566.

3330. Bu'Lock, J. D., and G. N. Smith. 1963: Acetylonic fatty acids in seeds and seedlings of sweet quandong. Phytochemistry 2:289–96.

3331. Bunakov, D. I. 1949: Accelerated method for determining the viability of alfalfa and sainfoin after wintering. Selek. i Semen. 16(3):55–57.

Bunch, C. E. 1325.

3332. Bunch, H. D. 1960: Relationship between moisture content of seed and mechanical damage in seed conveying. Seed World 86(5):14, 16–17.

Bunch, H. D. 615.

Bunday, G. 12475.

3333. Bundel, A. A., and N. V. Borovikova. 1963: Incorporation of N^{15}-hydroxylamine into the proteins of wheat seedlings. Life Sci. 2:184–89. Chem. Abstr. 59: 13118. 1963.

Bundel, A. A. 169, 170, 11372–11375, 11383.

3334. Bunge, R. 1935: The toxicity of ripe poppy seeds. Arch. Exptl. Pathol. Pharmakol. 179:465–74. Chem. Abstr. 30:3080. 1936.

3335. Bunker, H. J. 1946: Living seeds from Pharaoh's tomb? Sci. Dig. 20(1):55–56.

3336. Bunn, J. M. 1957: Effect of temperature, moisture, and light on the germination probability of Virginia type tobacco. Thesis, North Carolina State College, Raleigh. Tobacco Abstr. 3:231–32. 1959.

Bunnelle, P. R. 9966.

3337. Bunting, A. H., and W. O. James. 1941: Carboxylase and cocarboxylase in barley. New Phytologist 40:262–67.

3338. Bunting, E. S. 1956: Winter rape as an oil seed crop. Agriculture (London) 63:17–20.

3339. —— and G. E. Blackman. 1951: An assessment of the factors controlling the productivity of maize in England. J. Agr. Sci. 41:271–81.

3340. —— and —— 1951: Varietal differences in linseed. Natl. Inst. Agr. Botany J. 6:45–48.

Bunting, E. S. 2540, 2541, 12501.

3341. Bunting, R. H. 1932: Actinomyces in cacao-beans. Ann. Appl. Biol. 19:515–17.

3342. Bunzell, H. H. 1943: A new constituent in wheat germ oil. Bull. Torrey Botan. Club 70:599–604.

3343. Burbidge, N. T. 1945: Germination studies of Australian Chenopodiaceae with special reference to the conditions necessary for regeneration. I. *Atriplex vesicarium* Heward. Trans. Roy. Soc. S. Australia 69:73–85. Biol. Abstr. 19: 21041. 1945.

3344. Burchard, O. 1891: Über sogenannte chemische Keimkraftprüfungen. Sachs Landwirtsch. Z. 39:181–83.

3345. —— 1891: Untersuchungen von Kleesaaten aus verschiedenen Staaten Nordamerikas. Deut. Landwirtsch. Presse 18:367–68.

3346. —— 1892: Keimversuche mit entspelzten Grassaaten. Deut. Landwirtsch. Presse 19:755.

3347. —— 1892: Über die Temperatur bei Keimversuchen. Österr.-Landwirtsch. Wochbl. 18:274–75.

3348. —— 1894: Über den Bau der Samenschale einiger *Brassica* und *Sinapis*-arten. J. Landwirtsch. 42:125–36.

3349. Burchfield, H. P., and G. L. McNew. 1948: Quantitative determination of tetrachloro-p-benzoquinone on treated seed. Phytopathology 38:299–306.

3350. —— and —— 1948: The colorimetric determination of 2,3-dichloro-1,4-naphthoquinone on seed. Phytopathology 38:665–69.

3351. Burczyk, H. 1958: The influence of methods of storage and treatment of seed of lupine on its value and yields. Roczniki Nauk Rolniczych Ser. A 79:391–99.

Burda, E. J. 12558.

Burden, E. H. W. J. 7687.

3352. Burdick, A. B. 1951: Evidence concerning the structure of an ear of corn (*Zea mays*. L.). Am. Naturalist 85(825): 353–56.

3353. —— 1963: The relationship between somatic and genetic response of tomato seed and pollen to X-irradiation using "water protection" and mutations at specific loci. Contract A.T. (11-1)-856. 10 pp. Nucl. Sci. Abstr. 17:2315. 1963.

3354. Bures, E., and K. Bednar. 1938: Ash seed oil. Casopis Lekaru Ceskych 18 :107–13. Pharm. Abstr. 6:344. 1940.

3355. Burgeff, H. 1934: Pflanzliche Avitaminose und ihre Behebung durch Vitaminzufuhr. Ber. Deut. Botan. Ges. 52:384–90.

3356. —— 1954: Samenkeimung und Kultur europäischer Erdorchideen, nebst versuchen zu ihrer Verbreitung. Fischer, Stuttgart. 48 pp.

3357. Burger, A. W., and W. F. Campbell. 1961: Effect of rates and methods of seeding on the original stand, tillering, stem diameter, leaf-stem ratio, and yield of Sudan grass. Agron. J. 53:289–91.

3358. Burger, W. C. 1963: Glutamic-aspartic transaminase in the germinating barley embryo. Cereal Chem. 40:220–30.

3359. Burgerstein, A. 1895: On the duration of the vitality of the seeds of various grains. Verhandl. Zool.-Botan. Ges. Wien 45:414–21. Expt. Sta. Rec. 7 :777. 1896.

3360. —— 1906: Über die Wirkung anaestisierenden Substanzen auf Lebenserscheinungen der Pflanzen. Verhandl. Zool.-Botan. Ges. Wien 56:243–62.

3361. —— 1913: Keimversuche mit Getreidefrüchten im Lichte und bei Lichtabschluss. Z. Landwirtsch. Versuchs. Deut. Österr. 16:849–61. Chem. Abstr. 8:152. 1914.

3362. —— 1915: Triebkraftversuche bei Gramineen und Leguminosen. Z. Landwirtsch. Versuchs. Deut. Österr. 18:559–70.

3363. Burges, H. D. 1960: A spear for sampling bulk grain by suction. Bull. Entomol. Res. 51:1–5.

3364. —— and N. J. Burrell. 1964: Cooling bulk grain in the British climate to control storage insects and to improve keeping quality. J. Sci. Food Agr. 15:32–50.

3365. ——, D. M. Edwards, N. J. Burrell, and M. E. Cammell. 1963: Effect of storage temperature and moisture content on the germinative energy of malting barley, with particular reference to high temperature. J. Sci. Food Agr. 14:580–83.

Burges, H. D. 3408.

3366. Burgess, A. H., and G. Smith. 1941: Mould fungi as a possible cause of deterioration of stored hops. J. Inst. Brewing 47:53–55.

3367. Burgess, C. H. 1936: A note on the germination of padi. Malayan Agr. J. 24:541–42.

3368. Burgess, J. L. 1919: Relation of varying degrees of heat to the viability of seeds. J. Am. Soc. Agron. 11:118–20.

3369. —— 1938: Report on project to determine the percentage and duration of viability of different varieties of soybeans grown in North Carolina. Proc. Assoc. Offic. Seed Analysts 23:69.

3370. Burgesser, F. W. 1950: Coated seeds in modern agriculture. Seed World 66(5):23–24.

3371. Burgis, D. S., and A. J. Overman. 1955: Factors affecting germination of seed and growth of vegetable plants in seedbeds in sandy soil. Florida Agr. Expt. Sta. Ann. Rept. 1955:279–81.

3372. Burgos, J. S. 1949: Sulphuric acid treatment of seeds of *Gymnocladus dioica.* Rev. Fac. Cienc. Agrar. 1:61–66. Brit. Abstr. B III:134. 1951.

3373. Burian, A. 1952: The vernalization and storing of sugar beet seed. Ann. Czech. Acad. Agr. (English Transl.) 25:185–98.

3374. Burk, L. G. 1957: Overcoming seed dormancy in *Nicotiana.* Agron. J. 49:461.

3375. —— and S. O. Nelson. 1964: Effects of radiofrequency electric fields on seeds of *Nicotiana tabacum* L. Crop Sci. 4:100–3.

Burk, P. C. 14364.

3376. Burka, A. 1957: An experiment in treating flax seeds with boron and copper micronutrients before sowing. Latvijas PSR Zinatnu Akad. Vestis 1957(4):69–71. Biol. Abstr. 35:27803.

3377. Burkart, A., and M. N. Carera. 1953: Estudios sobre malezas. Morfologia vegetativa y germinación del "sunchillo" (*Wedelia glauca*). Darwiniana 10:113–44. Biol. Abstr. 29:19583. 1955.

3378. Burke, D., and G. H. Starr. 1948: Direct measures used on control tests of bacterial blight of beans. J. Colo. -Wyo. Acad. Sci. 3:43. Rev. Appl. Mycol. 29:345. 1950.

Burke, D. W. 17574.

3379. Burke, T. W. L. 1930: Studies of water absorption and germination with varieties of *Triticum vulgare* and *T. durum.* Sci. Agr. (Ottawa) 10:369–88.

3380. Burkert, G. M., and A. D. Dickson. 1941: The action of the amylases of two barley varieties on the barley starches. Cereal Chem. 18:605–15.

3381. Burkhart, B. A. 1939: The electrometric determination of diastatic power of malts. Cereal Chem. 16:652–57.

Burkhart, B. A. 5346, 5348.
Burkhart, L. 14322.
Burkhead, C. E. 1083.

3382. Burkholder, P. R. 1943: Vitamins in dehydrated seeds and sprouts. Science 97:562–64.

3383. —— 1943: Vitamins in edible soybeans. Science 98:188–90.

3384. —— and I. McVeigh. 1942: The increase of B vitamins in germinating seeds. Proc. Natl. Acad. Sci. U.S. 28:440–46.

3385. —— and —— 1945: Vitamin content of some mature and germinated legume seeds. Plant Physiol. 20:310–6.

3386. ——, ——, and D. Moyer. 1944: Niacin in maize. Yale J. Biol. Med. 16:659–63. Expt. Sta. Rec. 92:44. 1945.

Burkholder, P. R. 16710.

3387. Burkill, I. H. 1920: Twin nutmeg seeds. Garden Bull. Straits Settlements 2:158. Botan. Abstr. 4:1809. 1920.

3388. —— 1939: The trigger-mechanism in the germination of the seed of "*Tamus communis.*" Bull. Soc. Botan. France 88:772. 1941.

3389. —— 1949: An interruption in the after-ripening of *Tamus* which alters the range of temperature controlling germination. Proc. Linn. Soc. London 161:151–55. Biol. Abstr. 24:21584. 1950.

3390. —— 1958: Inhibition of germination of the white mustard (*Brassica hirta*) by bryony (*Tamus communis*) juice. Proc.

Linn. Soc. London 169:62–63. Biol. Abstr. 32:42008. 1958.

3391. Burkser, E., W. Kondoguri, W. Milgewska, and K. Bronstein. 1931: Versuche einer Bestimmung von Radiumelementen in Pflanzen. III. Biochem. Z. 233:58–61. Biol. Abstr. 6:24558. 1932.

Burlage, H. M. 17652.

3392. Burleson, C. A., W. R. Cowley, and G. Otey. 1956: Effect of nitrogen fertilization on yield and protein content of grain sorghum in the Lower Rio Grande valley of Texas. Agron. J. 48:524–25.

3393. Burlison, W. L., C. A. Van Doren, and J. C. Hackleman. 1940: Eleven years of soybean investigations. Illinois Agr. Expt. Sta. Bull. 462. 167 pp.

Burlison, W. L. 6802, 10974, 15357, 16524.

3394. Burnett, L. C., and A. L. Bakke. 1930: The effect of delayed harvest upon yield of grain. Iowa Agr. Expt. Sta. Res. Bull. 130:301–40.

3395. —— and C. S. Reddy. 1931: Seed-treatment and date-of-sowing experiments with six varieties of flax. Phytopathology 21:985–89.

3396. Burnett, R. S. 1945: Peanut protein hydrates: preparation and properties. Ind. Eng. Chem. 37:861–64. Expt. Sta. Rec. 95:163–64. 1946.

3397. —— and T. D. Fontaine. 1944: Commerical peanut meal: peptization and extraction of nitrogenous constituents and the color comparison of protein solutions. Ind. Eng. Chem. 36:284–88. Expt. Sta. Rec. 94:714. 1946.

3398. ——, E. D. Parker, and E. J. Roberts 1945: Peanut protein hydrates: utilization as tacky and remoistening adhesives. Ind. Eng. Chem. 37:980–82. Expt. Sta. Rec. 95:164. 1946.

3399. ——, E. J. Roberts, and E. D. Parker. 1945: Viscosity patterns of peanut protein solutions: comparison with other vegetable proteins and with casein. Ind. Eng. Chem. 37:276–81. Expt. Sta. Rec. 95:2. 1946.

Burnett, R. S. 6532.

Burnham, C. R. 3444.

3400. Burns, E. E., and J. W. Winzer. 1962: Pigment characteristics of the southern pea, *Vigna sinensis*. Proc. Am. Soc. Hort. Sci. 80:449–56.

3401. Burns, P. Y., D. M. Christisen, and J. M. Nichols. 1954: Acorn production in Missouri Ozarks. Missouri Agr. Expt. Sta. Bull. 611. 8 pp.

3402. Burns, R. E. 1959: Effect of acid scarification on lupine seed impermeability. Plant Physiol. 34:107–8.

3403. ——, J. L. Weimer, and P. R. Henson. 1958: Factors affecting the longevity of blue lupine seed. U.S. Dept. Agr. ARS 34–5. 5 pp.

3404. Burns, R. M. 1960: Response of selected coniferous seeds to gibberellic acid. U.S. Dept. Agr. Forest Serv. Southeast. Forest Expt. Sta. Southeast. Forest Res. 1:13–16.

3405. Burnside, O. C., and W. L. Colville. 1964: Yield components and composition of soybeans as affected by mechanical, cultural, and chemical weed control practices. Agron. J. 56:348–51.

Burr, H. S. 14316, 14317.

3406. Burrage, A. C., Jr. 1952: Grow your asparagus from seed. Horticulture 30(2): 47, 62, 69.

3407. Burrage, R. H. 1963: Emergence of Chinook wheat plants grown from seed treated with Aldrin powder, or Aldrin or Heptachlor liquid dressings, with or without a mercury fungicide. J. Econ. Entomol. 56:32–34.

3408. Burrell, N. J., and H. D. Burges. 1964: Cooling bulk grain by aeration. World Crops 16:33–38.

Burrell, N. J. 3364, 3365.

3409. Burrell, R. C., and E. D. Walter. 1935: A saponin from the soybean. J. Biol. Chem. 108:55–60. Chem. Abstr. 29:1458. 1935.

3410. —— and A. C. Wolfe. 1940: A comparative study of the chemical composition of five varieties of soybeans. Food Res. 5:109–13.

Burrell, R. C. 5891, 19799.

3411. Burri, L., M. Fireman, and O. A. Harvey. 1958: Soil salinity hazard to seeds. Calif. Agr. 12(9):7.

Burris, R. H. 12089.

Burstroem, D. 19200.

Burström, H. 12528, 12529.

3412. Burton, C. L. 1932: Variation in characteristics of black locust seeds from two regions. J. Forestry 30:29–33.

3413. Burton, G. W. 1938: A useful seed blower for the grass breeder. J. Am. Soc. Agron. 30:446–48.

3414. —— 1939: Scarification studies on southern grass seeds. J. Am. Soc. Agron. 31:179–87.

3415. —— 1940: Factors influencing the germination of seeds of *Trifolium repens*. J. Am. Soc. Agron. 32:731–38.

3416. —— 1943: Factors influencing seed setting in several southern grasses. J. Am. Soc. Agron. 35:465–74. Chem. Abstr. 37:5181–82. 1943.

3417. —— 1944: Seed production of several southern grasses as influenced by burning and fertilization. J. Am. Soc. Agron. 36:523–29.

3418. —— 1948: The method of reproduction in common Bahia grass, *Paspalum notatum*. J. Am. Soc. Agron. 40:443–52.

3419. —— 1952: The immediate effect of gametic relationship upon seed production in Pearl millet, *Pennisetum glaucum*. Agron. J. 44:424–27.

3420. —— and J. S. Andrews. 1948: Recovery and viability of seeds of certain southern grasses and lespedeza passed through the bovine digestive tract. J. Agr. Res. 76:95–103.

Burton, V. E. 608.

3421. Buruiana, L. M., V. Pop, E. Pop, and L. Popa. 1959: Variation of catalase and of citric acid during germination of some seeds. Acad. Rep. Populare Romine Studii Cercetari Biochim. 2:395–402. Chem. Abstr. 54:22863. 1960.

3422. ——, ——, L. Popa, and P. Niculescu. 1957: A biochemical method for evaluating the effect of a substance on germination. Acad. Rep. Populare Romine Baza Cercetari Stiint. Timisoara Ser. Stiinte Agr. 4:119–24. Chem. Abstr. 53:11524. 1959.

3423. ——, ——, ——, and E. Pop. 1957: The variation of catalase and the concentration of citric acid in seeds during germination. Acad. Rep. Populare Romine Baza Cercetari Stiint. Timisoara Ser. Stiinte Agr. 4:127–34. Chem. Abstr. 53:11524. 1959.

3424. Burunjik, M. 1933: The conditions necessary for the germination of the seeds of some plants native to the Chicago region. Dissertation, Univ. Chicago. 31 pp. Biol. Abstr. 9:2382. 1935.

Burvil, G. H. 13506.

Buscalioni, L. 13025, 13026.

Buschlen, M. J. 16073.

3425. Bush, C. D. 1953: Nut grower's handbook; a practical guide to the successful propagation, planting, cultivation, harvesting, and marketing of nuts. Rev. ed. Orange Judd, New York. 199 pp.

3426. Bush, H. L. 1948: Field compared with blotter germinations for processed, graded, single- and double-germ seed. Proc. Am. Soc. Sugar Beet Technol. 5:70–77.

Bush, H. L. 3016, 3018.

3427. Bushell, W. J., and T. P. Hilditch. 1938: The fatty acids and glycerides of solid seed fats. IV. Seed fat of *Madhuca*, Butyracea (Pulwara butter). Soc. Chem. Ind. J. 57:48–49. Pharm. Abstr. 4:362. 1938.

3428. —— and —— 1938: The fatty acids and glycerides of solid seed fats. VI. Borneo tallow. Soc. Chem. Ind. J. 57:447–49.

3429. —— and —— 1939: The fatty acids and glycerides of solid seed fats. VII. Dika fat. Soc. Chem. Ind. J. 58:24–26. Chem. Abstr. 33:2745–46. 1939.

3430. Bushinskaya, I. S. 1961: Effect of treating corn seed with succinic acid on growth and development. Dokl. Velikoluksk. Sel'skokhoz. Inst. 43–45. Biol. Abstr. 44:24985. 1964.

3431. Bushnell, E. P. 1936: Development of the Macrogametophyte in certain Labiatae. Botan. Gaz. 98:190–97.

Bushuk, W. 9444, 9445.

3432. Buslova, E. 1938: On the physiology of embryo development during seed germination. Acad. Sci. RSS Ukraine Inst. Botan. J. 25:81–82.

3433. Buslova, E. D. 1937: Contribution à la méthode de la culture des embryons des plantes superieures, dépourvus de réserves nutritives de la graine. Compt. Rend. Acad. Sci. URSS 14:389–92.

3434. Busnel, R. G., and G. Obolensky. 1954: Action des ultrasons sur la vitesse de germination et la croissance de l'orge. Compt. Rend. 239:777–78.

3435. —— and —— 1955: Étude microcalorimetrique de l'accélération de la germination des graines traitées aux ultrasons. Compt. Rend. 240:1358–60.

3436. Bussard, L. 1935: Contribution à l'étude des variations de la faculté germinative des semences au cours de leur conservation. Ann. Agron. 5:249–77.

3437. Busse, I. 1931: Einfluss des Alters der Mutterkiefer auf ihre Nachkommenschaft. Mitt. Deut. Dendrol. Ges. 43:61–74. Botan. Centralbl. 163:345–46. 1932.

3438. —— 1935: Samenaufbewahrung im Vakuum. Z. Forst.- Jagdw. 67:321–26.

3439. Busse, J. 1913: Ein Weg zur verbesserung unseres Kiefernsaatgutes. Z. Forst-Jagdw. 45:300–13. Botan. Centralbl. 125:76–77. 1913:

3440. —— 1937: Welchen Einfluss übt das Alter der Mutterkiefer auf die Nachkommenschaft? Z. Forst- Jagdw. 69:448–52. Biol. Abstr. 12:1696. 1938.

3441. Busse, W. 1904: Über den Einfluss des Naphthalins auf die Keimkraft der Getreidesamen. Tropenpflanzen Z. Trop. Landwirtsch. 8:61–68.

3442. —— 1926: Die Keimung des Tabaksamens in ihren Beziehungen zum Licht. Z. Botan. 18:65–97.

3443. Busse, W. F. 1930: Effect of low temperatures on germination of impermeable seeds. Botan. Gaz. 89:169–79.

3444. —— and C. R. Burnham. 1930: Some effects of low temperatures on seeds. Botan. Gaz. 90:399–411.

3445. —— and F. Daniels. 1929: Some effects of cathode rays on seeds. Am. J. Botany 16:139–53.

3446. Busson, F., R. Carbiener, and B. Bergeret. 1959: Étude de la fraction protidique des graines de *Vigna unguiculata* Walp. Qualitas Plant. Mater. Vegetabiles 6:11–15. Biol. Abstr. 35: 14195. 1960.

3447. ——, ——, A. Georgin, J. Lanza, and H. Dubois. 1960: Amino acids from seeds of legumes as influenced by genotype and by environment. Ann. Nutr. Aliment. 14:171–76. Chem. Abstr. 54:25069. 1960.

3448. ——, J. Perisse, and P. Jaeger. 1958: The chromatographic determination of amino acids in the seeds of *Parkia biglobosa* on resin columns. Z. Physiol. Chem. 310:1–3. Chem. Abstr. 52:12107. 1958.

3449. ——, ——, M. Rolland, and H. Meal. 1957: Composition of the proteins of baobab seeds. Med. Trop. 17:437–40. Chem. Abstr. 52:607. 1958.

Busson, F. 4266, 5369, 11791, 16087, 18867.

3450. Bustinza, F. 1955: Contribución al estudio de la actividad antibacteriana del jugo de la porción carmosa de las semillas de *Punica granatum* L. Anales Jardin Botan. Madrid 14:5–18.

3451. —— and A. Del Amo. 1945: Contribución al estudio de la influencía de las sulfonamides sobre la germinación. Anales Inst. Espan. Edafol. Ecol. Fisiol. Vegetale 41:185–97. Biol. Abstr. 23:30202. 1949.

3452. —— and M. L. C. Santamaria. 1953: Preliminary note on the phytotoxic activity of isonicotinic acid hydrazide. Antibiot. Chemotherapia 3:793–97. Biol. Abstr. 29:1846. 1955.

3453. Bustinza-Lachiondo, F. 1929: Catalasa y poder germinativo de las semillas. Bol. Real Soc. Espan. Hist. Nat. 29:227–30.

3453-A. —— and A. C. Lopez. 1947: Contribución al estudio de la influencía de los antibioticos en la germinación de las semillas. Anales Jardin Botan. Madrid 7:177–96.

3454. Buston, H. W., and S. K. Roy. 1949: The physiological activity of some simple unsaturated lactones. II. Effect on certain tissues of higher plants. Arch. Biochem. 22:269–74.

3455. Buszczak, T. 1961: Caraway (*Carum carvi* L.). I. Investigations into the biology of germination. Hodowla Roslin Aklim. Nasiennictwo 5:503–20.

3456. Buszewicz, G. 1961: The longevity of beechnuts in relation to storage conditions. Proc. Intern. Seed Testing Assoc. 26:504–15.

3457. —— 1962: A comparison of methods of moisture determination for forest tree seeds. Proc. Intern. Seed Testing Assoc. 27:952–61.

3458. —— and G. D. Holmes. 1956: Preliminary trials of a quick method for estimating the seed content of cones. Gt. Brit. Forestry Comm. Rept. Forest Res. 1956:98–106.

3459. —— and —— 1956: Summary of five years' experience of seed testing for larches. Gt. Brit. Forestry Comm. Rept. Forest Res. 1956:106–13.

3460. —— and —— 1956/57: Seven years' seed testing experience with the tetrazolium viability test for conifer species. Gt. Brit. Forestry Comm. Rept. Forest Res. 1956/57:142–51.

3461. —— and —— 1961: A summary of ten years' seed testing experience with western hemlock, *Tsuga heterophylla*. Gt. Brit. Forestry Comm. Rept. Forest Res. 1960: 110–19. Biol. Abstr. 37:11306. 1962.

Buszewicz, G. 8970–8973.

3462. Butenandt, A., and H. Jakoby. 1934: A substance identical with a female sex hormone, theelin, obtained from palm nuts. Science Suppl. 79(2038):8.

Butkova, G. B. 10197.

3463. Butler, F. C. 1949: Diseases of peas carried in the seed. Agr. Gaz. N.S. Wales 60:294–95.

3464. Butler, J. B., and J. M. Sheridan. 1915: A preliminary account of a new ocdanometer for measuring the expansive force of single seeds, or similar small bodies when wetted. Sci. Proc. Roy. Dublin Soc. 14:462–80.

Butler, J. R. 10943.

Butler, R. D. 11756, 11757.

3465. Butler, W. L. 1960: Chlorogenic acid of lettuce seeds. Nature 185:856–57.

Butman, L. A. 11310.

Butov, K. N. 18835.

Butt, J. L. 2545, 3109, 19356, 19357.

Butterfield, J. R. 3116.

3466. Button, E. F. 1959: Effect of gibberellic acids on laboratory germination of creeping red fescue (*Festuca rubra*). Agron. J. 51:60–61.

3467. Buttress, F. A., and R. W. G. Dennis. 1947: The early history of cereal seed treatment in England. Agr. Hist. 21:93–103. Rev. Appl. Mycol. 26:384. 1947.

3468. Buttrose, M. S. 1960: Submicroscopic development and structure of starch granules in cereal endosperms. J. Ultrastruct. Res. 4:231–57.

3469. —— 1962: Formation of rice starch granules. Naturwissenschaften 49:307–8.

3470. —— 1963: Ultrastructure of the developing wheat endosperm. Australian J. Biol. Sci. 16:305–17.

3471. —— 1963: Ultrastructure of the developing aleurone cells of wheat grain. Australian J. Biol. Sci. 16:768–74.

3472. ——, A. Frey-Wyssling, and K. Mühlethaler. 1960: Intracisternal granules in the endosperm cells of the barley grain. J. Ultrastruct. Res. 4:258–63.

3473. —— and L. H. May. 1959: Physiology of cereal grain. I. The source of carbon for the developing barley kernel. Australian J. Biol. Sci. 12:40–52. Chem. Abstr. 53:8314. 1959.

Buttrose, M. S. 13062, 14985.

3474. Butts, D., and J. T. Buchholz. 1940: Cotyledon numbers in conifers. Trans. Illinois State Acad. Sci. 33:58–62.

Butts, J. S. 6215.

3475. Butusov, V. A., and I. A. Lysenko. 1962: Fluorescent method of determining the vital capacity of seeds of cereal legumes. Selek. i Semen. 5:59–60.

Buuter, W. A., Jr. 14834.

3476. Buxton, L. O. 1939: Carotinoids in yellow corn. Ind. Eng. Chem. Anal. Ed. 11:128–29. Biol. Abstr. 15:19838. 1941.

3477. Buzinova, A. I. 1956: The relation between the productivity of the European spindle tree (*Euonymus europaeus*) and the weight of the seed. Sb. Rabot Lesn. Kh-vu. 1956(32):157–60. Biol. Abstr. 35:25120. 1960.

3478. Byerrum, R. U., S. A. Brown, and C. D. Ball. 1950: The action of electrolytes on oxalacetic descarboxylase from *Cucurbita* seeds. Arch. Biochem. 26:442–56.

3479. Byford, W. J. 1963: Experiments on the use of ethyl mercury phosphate and other materials for treating sugar-beet seed. Ann. Appl. Biol. 51:41–49.

3480. Bylda, A. Z. 1962: On the stratification of cherry seeds. Sadovodstvo 9:23–24.

3481. Bykon, M. M., and S. I. Dapina. 1956: The influence of micro γ macro elements in solution on the yield of quality seeds, growth, and development of spring wheat. Dokl. Vses. Akad. Sel'skokhoz. Nauk 21(11):12–16. Chem. Abstr. 51:9805. 1957.

3482. Bykovskii, V. A. 1952: Preplanting treatment of acorns with zinc phosphide in order to protect them from siberian marmots and other rodents. Dokl. Vses. Akad. Sel'skokhoz. Nauk 17(12):21–23.

Byran, H. F. 12615.

Byrom, M. H. 17649.

Bystrom, B. G. 17136.

Bystrov, A. I. 5581.

3483. Bystrov, B. A., and A. P. Pavlova. 1955: The intensity of respiration of squash seeds in connection with the method of pollination. Plant Physiol. (USSR) (English Transl.) 2:444–46. Biol. Abstr. 32:10133. 1958.

3484. Bytschikhina, E. A. 1928: Die Gegend der Kultur und die Geschwindigkeit der Nachreifung der Winterweizen. Ann. Essais Semences 5(5):33–43.

3485. —— 1929/30: Winter wheats of Ukraina and their after-ripening. Bull. Appl. Botany, Genet., Plant Breed. (USSR) 23:299–347.

3486. Byvshikh, N. A. 1956: Changes and conversion of nitrogenous substances and of proteins in seeds of ripening watermelon. Tr. Plodoovoshch. Inst. I. V. Michurina 9:225–44. Chem. Abstr. 52: 10296. 1958.

3487. —— 1960: Amino acid composition of various maturity watermelon seed globulins. Plant Physiol. (USSR) (English Transl.) 7:274–77.

C

3488. C., H. C. 1945: Propagation of nut trees. Home Garden 6:62.

Cabanillas, E. 16905, 17791.

3489. Cable, C. C., Jr. 1955: Retailing and wholesaling cotton planting seed in Arkansas. Arkansas Agr. Expt. Sta. Bull. 554. 32 pp.

3490. Cabrajic, T. 1960: Einige Eigenschaften des Schwarzkiefersamens. Nar. Sumar 14:141–44.

Cabral, A. 15817.

3491. Cabral, J. F. 1954: Scarification, an efficient treatment to counteract coat hardness in seeds of "*Melilotus sejetalis*" Ser., "*M. indica*" (L.) All., "*Scorpiurus sulcata*" L. and of other leguminous forage species. Proc. Intern. Seed Testing Assoc. 19:32–43.

3492. —— 1962: Essai et certification de semences au Portugal. Proc. Intern. Seed Testing Assoc. 27:27–35.

3493. —— and M. R. Cruz. 1960: The topographical tetrazolium method for determining germinating capacity of *Lupinus luteus* L. Proc. Intern. Seed Testing Assoc. 25:474–83.

Cabulea, I. 11902.

3494. Cacciari, P. 1914: Influence of naphthalin on germination, growth, and nitrification in plants. Staz. Sper. Agr. Ital. 47:347–67. Expt. Sta. Rec. 33:523. 1915.

3495. Cadwallder, S. E. 1933: Germinating seeds of cardinal climber. Flower Grower 20(3):134.

Cady, F. B. 6140.

3496. Caesar, J. 1927: Untersuchungen über den Einfluss der Triebkraft des Saatgutes und der Verteilung der Pflanzen auf Versuchsflächen auf den Ertrag und Versuchsfehler. Inaug. Dissertation Halle, Roemer: Halle Saale. 67 pp. Biol. Abstr. 5:21026. 1931.

Caestecker, K. de; *see* De Caestecker, D.

3497. Caffrey, M., and P. T. Carroll. 1936: Vernalization, its principles and practice. J. Dept. Agr. Ireland 34:53–62.

Cagle, A. W. 19693.

Cagniant, P. 7001.

Cagninacci, C. 10573.

3498. Caguioa, V. 1938: Germination and early development of white lauan under various media. Philippine J. Forestry 1:65–79. Biol. Abstr. 12:16689. 1938.

3499. Cailleau, R., J. Levy, T. Terroine, J. Adrian, and R. Jacquot. 1949: Les variations de la riboflavine, des acides nicotinique, pantothénique et ascorbique dans les débuts de la germination du blé et du pois. Compt. Rend. 228:1044–45.

3500. Cain, R. F., and J. W. Groves. 1948: Notes on seed-borne fungi. VI. Sordaria. Can. J. Res. Sect. C 26:486–95.

3501. Cairns, T. C. 1937: The oil palm (*Elaeis guineensis* Jacq.) in the Kigoma district. E. African Agr. J. 2:451–54.

Cairns, H. 14001.

3502. Cajlachjan, M. C. 1934: Investigations of the physiological nature of the differences between spring and winter plants. I. The influence of environmental factors on the development of spring and winter plants. Thesis (Ph.D.), Acad. Sci., Leningrad. 75 pp.

3503. —— and L. P. Zdanova. 1938: The role of growth hormones in form-building processes. III. Effect of heteroauxin treatment of seeds upon growth and development of plants. Compt. Rend. Acad. Sci. URSS 19:303–6.

3504. Cajola, R. 1930: The properties and composition of mustard seed. Riv. Ital. Essenze-Profumi 12:22–23. Chem. Abstr. 25:1637. 1931.

3505. Cajori, F. A. 1942: The effect of various substances on the activity of purified yeast carboxylase. J. Biol. Chem. 143:357–61. Biol. Abstr. 16:17562. 1942.

3506. Calabrese, A. M. 1953: On the activity of lipases of "*Ricinus communis*" in particular experimental conditions. I. Method of determination of lipase activity. Ric. Sci. Suppl. 23:1963–74. Biol. Abstr. 28:22059. 1954.

3507. —— 1954: The lipase activity of *Ricinus communis* under particular experimental conditions. II. Variations of the activity during the cycle of the maturation of the seed. Ric. Sci. Suppl. 24:1016–20. Chem. Abstr. 49:2570. 1955.

Calavan, E. C. 19427.

Calcandi, J. 18382.

3508. Caldecott, R. S. 1948: Effects of combination heat and X-ray treatments of seeds of barley and Einkorn. Thesis, Washington State Coll., Pullman. 42 pp.

3509. —— 1953: The influence of oxygen tension on dormant seeds of barley irradiated with X-rays, fast neutrons and slow neutrons. [Abstract.] Genetics 38:658.

3510. —— 1954: Inverse relationship between the water content of seeds and their sensitivity to X-rays. Science 120:809–10.

3511. —— 1955: Effects of ionizing radiations on seeds of barley. Radiation Res. 2:339–50. Biol. Abstr. 30:8642. 1956.

3512. —— 1958: Post-irradiation modification of injury in barley—its basic and applied significance. 2nd U.N. Intern. Conf. Peaceful Uses Atomic Energy, "Isotopes in Agriculture" 27:260–69.

3513. ——, B. H. Beard, and C. O. Gardner. 1954: Cytogenetic effects of X-ray and thermal neutron irradiation on seeds of barley. Genetics 39:240–59.

3514. ——, E. F. Frolik, and R. Morris. 1952: A comparison of the effects of X-rays and thermal neutrons on dormant seeds of barley. Proc. Natl. Acad. Sci. U.S. 38: 804–9.

3515. ——, C. F. Konzak, and E. B. Johnson. 1957: Post treatment modification of X-ray induced injury in barley by aerobic and anaerobic hydration. Radiation Res. 7:308. Nucl. Sci. Abstr. 12:8265. 1958.

3516. —— and D. T. North. 1961: Factors modifying the radio-sensitivity of seeds and the theoretical significance of the acute irradiation of successive generations. Proc. Symp. Mutation Plant Breed. 1960:365–404.

3517. —— and L. Smith. 1948: Resuscitation of heat-inactivated seeds with X-radiation. J. Heredity 39:195–98.

3518. —— and —— 1952: The influence of heat treatments on the injury and cytogenetic effects of X-rays in barley. Genetics 37:136–57.

Caldecott, R. S. 2916, 4763, 17658.

Calder, R. A. 7963–7965.

Calderon, M. 15446.

Calderwood, H. N. 2289.

3519. Caldwell, E. L., and L. A. Sciuchetti. 1963: Effect of gibberellin and other treatments on the germination and subsequent biogenesis of alkaloids in *Datura stramonium*. J. Pharm. Sci. 52:1062–65. Chem. Abstr. 60:3425. 1964.

3520. Caldwell, F. 1961: Entry of fungi and chemical substances in solution into mature cereal grains. J. Sci. Food Agr. 12:169–74.

3521. —— 1963: Germination of dried barley after prolonged storage when treated with sulphuric acid. J. Sci. Food. Agr. 14:765–69.

3522. Caldwell, F. Y. K., and A. C. W. Davies. 1955: The effect of heating and chilling as a means of overcoming dormancy in barley. Natl. Inst. Agr. Eng. Tech. Memo 123. 5 pp.

3523. Caldwell, J. 1962: Seed-transmission of viruses. Nature 193:457–59.

Caldwell, R. M. 1487, 1488, 15127.

3524. Caldwell, S. Y. 1942: The quickest lily from seed. Flower Grower 29(2):74, 93.

3525. —— 1953: Seed-bearing *Lycoris radiata* (Extr.). Plant Life 9:90–91.

3526. Cale, M. T. 1953: Phosphatase activity of *Ricinus communis* seeds. Quaderni Nutriz. 13:324–29. Chem. Abstr. 50:7968. 1956.

3527. Calhoun, W. K., W. G. Bechtel, and W. B. Bradley. 1958: The vitamin content of wheat, flour and bread. Cereal Chem. 35:350–59.

3528. Calistri, I., and S. Murazio. 1961: Effect of gibberellin on the germination of the seeds of the flowering ash (*Fraxinus ornus* L.). Ital. Forest. Mont. 16:98–101.

3529. Callaham, R. Z. 1960: Selecting the proper seed source of Ponderosa pine. Proc. Soc. Am. Foresters 1959:26–27. Biol. Abstr. 35:39355. 1960.

3530. Callan, I. W., and D. G. Sharp. 1960: The pattern of seed dressing development. World Crops 12:19–21.

Callan, I. W. M. 9703.

3531. Calma, V. C., and C. L. Castelo. 1951: The effect of distance of planting upon the yield of headgreen soybean. Phillippine Agriculturist 35:25–31.

3532. —— and R. P. Martinez. 1951: The influence upon the yield of sweet corn of the number of plants in a hill. Philippine Agriculturist 34:169–73.

3533. ——, L. G. Paderna, and M. A. Palo. 1929: A study of certain chemical treatments in relation to seed-borne diseases of Calauan yellow flint maize. Philippine Agriculturist 17:499–506.

3534. Calo, A. 1935: Phosphorus compounds in germinated oats. Ann. Chim. Appl. 25:49–53. Chem. Abstr. 29:5147. 1935.

Calo, A. 12878.

3535. Calvert, E. L., and A. E. Muskett. 1944: Blind seed disease of rye-grass. Nature 153:287–88.

3536. Calvert, J. 1934: The effect of arsenical sheep dip on the germination of Noogoora burr seeds. J. Council Sci. Ind. Res. Australia 7:50–52. Biol. Abstr. 9:17910. 1935.

3537. Calvert, O. H., and C. A. Thomas. 1954: Some factors affecting seed transmission of safflower rust. Phytopathology 44:609.

Calvet, E. 15661–15663.

3538. Calvino, E. M. 1931: Seed germination in roses. Intern. Hort. Congr. London Rept. Proc. 1930:150–53.

Cama, H. R. 15957, 15958.

3539. Camara, A. 1943: Transplantation of embryos. Agron. Lusitana 5:375–86. Biol. Abstr. 20:11735. 1946.

3540. —— 1950: Germination of a grain of wheat. Agros (Lisbon) 33:43–50.

3541. Camefort, H. 1957: Développement de l'embryon du *Pinus laricio* (var. *austriaca*): résultats généraux de l'analyse cytologique. Compt. Rend. 245:2088–91.

3542. Cameron, D. R. 1949: Inheritance in *Nicotiana tabacum*. XXII. Investigations on multiple seedlings. Am. J. Botany 36:526–29.

Cameron, D. R. 13639.

3543. Cameron, J. W., and D. A. Cole, Jr. 1959: Effects of the genes su_1, su_2, and du on carbohydrates in developing maize kernels. Agron. J. 51:424–27.

3544. ——, ——, and E. M. Nauer. 1960: Fruit size in relation to seed number in the valencia orange and some other Citrus varieties. Proc. Am. Soc. Hort. Sci. 76:170–80.

3545. ——, R. K. Soost, and H. B. Frost. 1958: Behavior of nucellar seedling lines. Calif. Citrograph 43:290–93.

3546. ——, ——, and —— 1959: The horticultural significance of nucellar embryony in citrus. Univ. Calif. Citrus Expt. Sta. Paper 1111:191–96.

3547. —— and H. J. Teas. 1948: The relation between nicotinic acid and carbohydrates in a series of maize endosperm genotypes. Proc. Natl. Acad. Sci. U.S. 34:390–98.

3548. ——, A. VanMaren, and D. A. Cole, Jr. 1962: Seed size in relation to plant growth and time of ear maturity of hybrid sweet corn in a winter planting area. Proc. Am. Soc. Hort. Sci. 80:481–84.

Cameron, J. W. 14932, 15085, 18428, 18429, 18431.

Cameron, S. H. 8891, 18267.

Camien, M. N. 5654.

Cammell, M. E. 3365.

3549. Camp, A. F., H. Mowry, and K. W. Loucks. 1933: The effect of soil temperature on the germination of citrus seeds. Am. J. Botany 20:348–57.

3550. Camp, W. H., and M. M. Hubbard. 1963: Vascular supply and structure of the ovule and aril in peony and of the aril in nutmeg. Am. J. Botany 50:174–78.

3551. Campacci, C. A. 1951: The problem of onion seed. Biologico 17:13–15. Biol. Abstr. 26:2000. 1952.

3552. Campbell, A. G. 1959: A germination inhibitor and root-growth retarder in Chou Moellier (*Brassica oleracea* var.). Nature 183:1263–64.

3553. Campbell, C. M. 1964: Influence of seed formation of corn on accumulation of vegetative dry matter and stalk strength. Crop Sci. 4:31–34.

3554. Campbell, C. W. 1961: Comparison of yields of polyembryonic and monoembryonic mangos. Proc. Florida State Hort. Soc. 74:363–65. Biol. Abstr. 39:3520. 1962.

3555. Campbell, E. O. 1936: The embryo and stelar development of *Histiopteris incisa*. Trans. Proc. Roy. Soc. New Zealand 66:1–11.

Campbell, G. F. 14863–14869.

Campbell, J. A. 8641.

3556. Campbell, J. D. 1958: The effect of mechanical damage to wheat grains during scouring on their subsequent absorption of water during washing. Cereal Chem. 35:47–56.

3557. —— and C. R. Jones. 1955: On the physical basis of the response of endosperm density to change in moisture content. I. Densities of dissected and of milled endosperm particles. Cereal Chem. 32:325–33.

3558. —— and —— 1955: On the physical basis of the response of endosperm density to change in moisture content. II. Reversibility of endosperm density on drying in relation to behavior of whole-wheat grains. Cereal Chem. 32:333–43.

3559. —— and —— 1957: The rates of penetration of moisture to different points in the central cross-section of the endosperm in damped Manitoba wheat grains. Cereal Chem. 34:110–16.

Campbell, J. D. 9930.

Campbell, L. E. 6376.

3560. Campbell, R. S., and J. G. Keller. 1932: Growth and reproduction of *Yucca elata*. Ecology 13:364–74.

Campbell, W. F. 3357.

3561. Campbell, W. P. 1958: A cause of pink seeds in wheat. Plant Disease Reptr. 42:1272.

Campos, A. A. 3001.

Campos, F. A. de Moura-; *see* De Moura-Campos, F. A.

3562. Campos, F. F., and L. G. Espiritu. 1959: Correlation of the endosperm and embryo size of the Philippine recommended rice varieties. Philippine Agriculturist 43:478–79.

Canak, M. 4421, 4422.

Canal, H. 577.

Canale, A. 12946.

Cancel, M. M. 16699.

Candolle, A. de; *see* De Candolle, A.

Canfield, A. P. 14630.

3563. Canic, V. D., D. I. Hadzijev, B. P. Belia, and S. A. Grujic. 1962: Ribonucleic acid in the kernel of yellow maize. Savremena Poljoprivreda 10:931–45.

Canillot, R. 14345.

3564. Cannon, C., and F. A. Kummerow. 1957: A comparison of plant and grain wax from two varieties of sorghum. J. Am. Oil Chemists Soc. 34:519–20.

Cannon, H. B. 16574.

3565. Cannon, J. A., M. M. MacMasters, M. J. Wolf, and C. E. Rist. 1952: Chemical composition of the mature corn kernel. Trans. Am. Assoc. Cereal Chemists 10:74–97. Chem. Abstr. 46:5741. 1952.

3566. Cannon, J. R., N. H. Corbett, K. P. Haydock, J. G. Tracey, and L. J. Webb. 1962: An investigation of the effect of the dehydroangustione present in the leaf litter of *Backhousia angustifolia* on the germination of *Araucaria cunninghamii*—an experimental approach to a problem in rain-forest ecology. Australian J. Botany 10:119–28.

3567. Cannon, W. A. 1904: Observations on the germination of *Phoradendron villosum* and *P. californicum*. Bull. Torrey Botan. Club 31:435–43.

3568. Canode, C. L., and E. V. Horning. 1961: A grass seed nursery harvester. Agron. J. 53:407–8.

3569. ——, ——, and J. D. Maguire. 1963: Seed dormancy in *Dactylis glomerata* L. Crop Sci. 3:17–19.

Canode, C. L. 5307, 9068.

3570. Canova, A. 1961: The adversities of plants seen with the lens: seeds. Progr. Agr. (Bologna) 7:632–33.

Canova, A. 1767.

Cantlon, J. E. 4760, 4761.

3571. Canvin, D. T. 1963: Formation of oil in the seed of *Ricinus communis* L. Can. J. Biochem. Physiol. 41:1879–85.

3572. —— and L. E. Evans. 1963: A method of inducing seed production in dwarf wheat plants. Can. J. Plant Sci. 43:419–21.

3573. —— and G. Friesen. 1959: Cytological effects of CDAA and IPC on germinating barley and peas. Weeds 7:153–56.

Canvin, D. T. 11642.

Capdevila, C. 6303.

3574. Capen, R. G., and J. A. LeClerc. 1948: Wild rice and its chemical composition. J. Agr. Res. 77:65–79. Biol. Abstr. 23:1763. 1949.

Caperci, J. F. 5979.

3575. Capinpin, J. M., and V. F. Guevara. 1952: The floral biology and fructification of peanut, *Arachis hypogaea*. Philippine Agriculturist 35:137–42.

3576. —— and T. A. Irabagon. 1950: A genetic study of pod and seed characters in *Vigna*. Philippine Agriculturist 33:263–77.

3577. —— and V. B. Silva. 1951: The influence of hybridization on the protein content of maize. Philippine Agriculturist 35:86–89.

3578. Capitaine, L. 1912: Contributions à l'étude morphologique des graines de Légumineuses. Thesis (Ph.D.), Sci. Natur., Paris. 436 pp. Botan. Centralbl. 125:578. 1914.

3579. Capizzano, N. 1926: On the biological action of X-rays. Effect of X-rays produced by high tension current and high milliamperage on the germination of seed. Bol. Inst. Med. Exptl. Estud. Tratamiento Cancer (Buenos Aires) 2:922–28.

3580. Cappelletti, C. 1931: Sulla presenza di miceli nei tegumenti seminali di alcune liliaceae e particolarmente nel genere tulipa. Nuovo Giorn. Botan. Ital. 38:479–508.

3581. —— 1935: Observations on the asymbiotic and symbiotic germination of some Orchidaceae. Nuovo Giorn. Botan. Ital. 42:436–57. Rev. Appl. Mycol. 15: 243. 1936.

3582. —— 1935: Ulteriori osservazioni sulla germinazione asimbiotica di *Cymbidium*. Nuovo Giorn. Botan. Ital. 42:685–90.

3583. —— 1936: Osservazioni sulla germinazione dei semi di *Orobanche gracilis* Sm. e *Orobanche crenata* Forsk. Nuovo Giorn. Botan. Ital. 43:263–67.

3584. —— 1957: Physiological researches on symbiosis in orchids. Lavori Botan. Torino 8:56–76. Biol. Abstr. 22:12331. 1948.

3585. —— 1959: Germination of seeds of *Xanthium italicum* Moretti. Nuovo Giorn. Botan. Ital. 66:345–47.

3586. —— 1960: Research on the gradual germination of seeds of *Xanthium italicum*

Moretti. Atti Accad. Nazl. Lincei, Rend. Classe Sci. Fiz. Mat. Nat. 28:572–81.

3587. Capuler, B. 1949: The germination problem of durra. Hassadeh 29:239–40.

3588. Caputa, J. 1952: Essais de semences de trèfle violet de diverses provenances en 1950–1952. Rev. Romande Agr. Viticult. Arboricult. 8:85–87. Biol. Abstr. 27: 28296. 1953.

3589. —— 1953: Influence de la profondeur de semis sur la levée des graines de plantes fourragères. Rev. Romande Agr. Viticult. Arboricult. 9:53–55. Biol. Abstr. 29: 24564. 1955.

3590. —— 1957: Étude de la conservation des épis de mais hybride pour semences en séchoir à claire-voie "crib." Landwirtsch. Jahrb. Schweiz 6:775–83. Biol. Abstr. 33:30944. 1959.

3591. —— and J. Rapin. 1951: Influence de la provenance des semences sur le développement et le rendement de la luzerne et de quelques graminées fourragères. Rev. Romande Agr. Viticult. Arboricult. 7:4–6.

Caputa, J. 15982.
Caputo, A. 16430.
Caraman, M. de; *see* De Caraman, M.

3592. Carangal, A. R., Jr., S. R. Arnaldo, and E. A. Tucay. 1956: Chemical studies on coffee grown in the Philippines. I. The nonnitrogenous organic acids of the leaves, shells, and beans. Philippine Agriculturist 40:99–103. Chem. Abstr. 51:9804. 1957.

Caravella, M. 437, 2653.
Carbiener, R. 3446, 3447.
Carbona-Alvarez, C. 14837.

3593. Carbonnier, H. 1937: Problem of seed source with spruce. Skogen 24(11):229–33. Biol. Abstr. 13:8525. 1939.

Carbouères, M. 578–580.
Carbouères, M. Durou-. 581, 582.
Cardwell, A. B. 10332, 10333.

3594. Care, A. D. 1954: Goitrogenic activity in linseed. New Zealand J. Sci. Technol. Sect. A 36:321–27.

Care, M. 5119.
Carera, M. N. 3377.

3595. Carey, G., and L. Fraser. 1932: The embryology and seedling development of *Aegiceras majus* Gaertn. Proc. Linnean Soc. N.S. Wales 57:341–60. Biol. Abstr. 8:6929. 1934.

3596. Carew, D. P., and A. E. Schwarting. 1958: Production of rye embryo callus. Botan. Gaz. 119:237–39.

3597. Carew, H. J. 1949: Pelleted seeds. Horticulture 27(7):267.

Carillo, J. 16381.
Cariss, H. G. 18493.

3598. Carl, W. 1914: Über den Einfluss des Quecksilberdampflichtes auf die Keimung und das erste Wachstum von Pflanzen. Beitr. Biol. Pflanz. 12:435–37.

3599. Carles, J., and F. Lattes. 1959: L'acide malonique et l'acide quinique au cours de la germination du grain de blé et du grain de lupin. Compt. Rend. 249:447–49.

Carleton, M. A. 8863.

3600. Carleton, R. M. 1933: Growing iris from seed. Flower Grower 20(9):407.

3601. —— 1936: Seeds of difficult germination. Horticulture 14(3):40.

3602. —— 1936: Germination of ancient seeds. Horticulture 14(6):136.

3603. —— 1936: Strawberries from seed. Horticulture 14(10):229.

3604. —— 1937: The refrigerator as a garden tool. Horticulutre 15(19):412.

3605. —— 1946: Vermiculite, the garden stepchild. Flower Grower 33(2):87–88, 138–39.

3606. —— 1951: What makes seeds grow? The germination requirements for various seeds differ. Flower Grower 38(3):56, 62.

3607. —— 1953: Perennials from seed. Flower Grower 40(1):60–62.

3608. —— 1955/56: Seed germination. Plants Gardens 11:261–64.

Carleton, W. M. 8113.

3609. Carlow, C. A., and J. A. Irvine. 1961: An electronic seed counter and group number discriminator. J. Agr. Eng. Res. 6:130–36.

3610. Carlquist, S. 1962: *Tremtolobelia*: seed dispersal; anatomy of fruit and seeds. Pacific Sci. 16:126–34. Biol. Abstr. 39:3254. 1962.

3611. Carlson, C. W., J. Alessi, and R. H. Mickelson. 1959: Evapotranspiration and yield of corn as influenced by moisture level, nitrogen fertilization, and plant density. Soil Sci. Soc. Am. Proc. 23:242–45.

Carlson, E. C. 11769, 11770.

3612. Carlson, G. E. 1960: Counter-planter for grain sorghum in greenhouse flats. Agron. J. 52:367.

3613. —— and R. E. Atkins. 1960: Effect of freezing temperatures on seed viability and seedling vigor of grain sorghum. Agron. J. 52:329–33.

3614. Carlson, J. W. 1930: Artificial tripping of flowers in alfalfa in relation to seed production. J. Am. Soc. Agron. 22:780–86.

3615. —— 1940: Lygus bug damage to alfalfa in relation to seed production. J. Agr. Res. 61:791–815.

3616. —— 1946: Pollination, Lygus infestation, genotype, and size of plants as affecting seed setting and seed production in alfalfa. J. Am. Soc. Agron. 38:502–14.

3617. —— and G. Stewart. 1931: Alfalfa-seed production. Utah Agr. Expt. Sta. Tech. Bull. 226. 54 pp.

Carlson, J. W. 17995.

3618. Carlson, L. E., and T. H. King. 1963: The effect of environmental factors on the quality of stored pea seed. Proc. Minn. Acad. Sci. 30:132–35.

3619. Carlson, M. C. 1935: The germination of the seed and development of the seedling of *Calopogon pulchellus* (SW.) R.BR. Trans. Illinois State Acad. Sci. 28:85–86.

3620. —— 1940: Formation of the seed of *Cypripedium parviflorum*. Botan. Gaz. 102:295–301.

3621. —— 1945: Megasporogenesis and development of the embryo sac of *Cypripedium parviflorum*. Botan. Gaz. 107:107–14.

3622. Carlson, R. F. 1946: Seasonal variation of peach seeds. Am. Nurseryman 83(9):40.

3623. —— 1946: Treatment of peach seed with fungicides for increased germination and improved stand of peach seedlings in the nursery. Proc. Am. Soc. Hort. Sci. 48:105–13.

3624. —— and H. B. Tukey. 1945: Differences in after-ripening requirements of several sources and varieties of peach seed. Proc. Am. Soc. Hort. Sci. 46:199–202.

Carlson, R. F. 18827.

3625. Carlsson, G. 1959: Correlation between seed size, cotyledon size and earliness, and between seed size and *Alternaria*-attacks in radish varieties for forcing. Their use in plant breeding and their practical significance. Medd. Hammenhog. Gullakers Växtforadlingsanst. 15:144–59.

3626. Carmer, S. G., and J. A. Jackobs. 1963: Establishment and yield of late-summer alfalfa seedlings as influenced by placement of seed and phosphate fertilizer, seeding rate, and row spacing. Agron. J. 55:28–30.

3627. Carmichael, R. L. 1957: Relation of seeding date to germination of Douglas fir seed. Northwest Sci. 31:177–82.

3628. —— and J. Dick. 1956: A seeding of Sitka spruce and western hemlock in southwestern Washington. Northwest Sci. 30:56–60.

3629. Carnahan, G., and A. C. Hull, Jr. 1962: The inhibition of seeded plants by tarweed. Weeds 10:87–90.

Carnahan, H. L. 3329.

Carneiro, P. de T. Alvim-; *see* Alvim-Carneiro, P. de T.

Carney, W. B. 6102.

Carns, H. R. 11719.

Caro, G. 2971.

Caro, L. de. 2492.

3630. Carolus, R. L. 1931: Some chemical and physical changes observed in green lima beans subjected to various storage conditions. Proc. Am. Soc. Hort. Sci. 28:367–74.

3631. —— 1949: Possibilities with the use of pelleted seed. Ohio Veg. Potato Growers Assoc. Ann. Proc. 34:56–62. Biol. Abstr. 24:21379. 1950.

3632. —— 1954: Do you plant by the moon? Am. Veg. Grower 2(5):13–15.

Carolus, R. L. 566.

3633. Carpenter, D. C., and F. E. Lovelace. 1933: Isoelectric point of orange-seed globulin. J. Am. Chem. Soc. 55:3738–40. Chem. Abstr. 27:4987. 1933.

3634. Carpenter, I. W., and A. T. Guard. Some effects of cross-pollination on seed production and hybrid vigor of tuliptree. J. Forestry 48:852–55.

3635. Carpenter, J. A. 1958: Production and use of seed in sea-shore *Paspalum*. J. Australian Inst. Agr. Sci. 24:252–56.

3636. Carquennes, H. 1953: Considérations sur la germination des graines de coton dans l'Ubangi. Bull. Inform. INEAC (Inst. Natl. Étude Agron. Congo Belge) 2:51–62. Biol. Abstr. 28:11804. 1954.

3637. Carr, A. R. 1958: Certified tomato seed. Queensland Agr. J. 84:633–35.

Carr, C. J. 11328.

3638. Carr, D. J. 1961: Chemical influences of the environment. Encyclopedia Plant Physiol. 16:737–94.

3639. —— and K. G. M. Skene. 1961: Diauxic growth curves of seeds, with special reference to French beans (*Phaseolus vulgaris* L.). Australian J. Biol. Sci. 14:1–12.

Carr, D. J. 6370, 17562.

3640. Carr, R. H. 1938: Removal of the bran from cereals. Cereal Chem. 15:658–62.

3641. Carra, P., and P. Desvernes. 1956: Longévité des semences. Rev. Hort. Algerie 60:17–23.

3642. Carrasco, M. O. 1954: Evaluation of riboflavine in seeds of germinating legumes. Anales Fac. Farm. Bio-Quim. 5:136–40. Chem. Abstr. 51:3756. 1957.

3643. Carrière, M. 1946: Tomato-seed oil. Ind. Corps Gras. 2:152–54. Chem. Abstr. 40:5582. 1946.

3644. —— and A. Coulier. 1942: Watermelon seed oil. Compt. Rend. Fac. Sci. Marseille 1:83–85. Soap 23:94. 1947.

3645. Carroll, J. C., and C. L. Peng. 1951: A comparison of the niacin and pantothenic acid content of certain cereal grains grown under the same environmental conditions. Science 113:211–12.

Carroll, P. T. 3497.

Carroll, W. E. 6192.

3646. Carruthers, W. 1911: On the vitality of farm seeds. J. Roy. Agr. Soc. Engl. 72:168–83.

Carsner, E. 15685.

3647. Carson, G. P. 1957: Variety and its identification. Natl. Inst. Agr. Botany J. 8:194–97.

Carson, R. B. 12634, 16615.

3648. Carsten, H. A., T. P. Hilditch, and M. L. Meara. 1945: The component acids of the testa and kernel fats of the oil palm. Soc. Chem. Ind. J. 64:207–9.

3649. Carter, A. S. 1949: Seed control and seed certification. Intern. Crop Improvement Assoc. Proc. 31:77–81.

3650. —— 1962: Report of the sampling and bulking committee. Proc. Intern. Seed Testing Assoc. 27:37–47.

Carter, A. S. 3306. 13470, 15790

3651. Carter, D. G., and L. E. Holman. 1952: Storing soybeans on the farm. Univ. Illinois Coll. Agr. Circ. 692. 15 pp.

Carter, D. G. 8960.

3652. Carter, E. P., and G. Y. Young. 1945: Effect of moisture content, temperature, and length of storage on the development of "sick" wheat in sealed containers. Cereal Chem. 22:418–28.

Carter, G. F. 19590, 19591.

Carter, H. 6912.

3653. Carter, J. E., and J. Pace. 1964: Distribution of dehydroascorbic acid reductase in the wheat grain. Nature 201: 503.

Carter, J. L. 13219.

Carter, J. N. 5751.

Carter, L. C. 1974.

3654. Carter, M. C., and L. Jones. 1962: The effect of hydrogen peroxide on the germination of loblolly and slash pine seed. U.S. Dept. Agr. Forest Serv. Southeast. Forest Expt. Sta. Paper 141. 12 pp.

3655. Carter, W. B. 1964: How to grow cactus from seed. Horticulture 42(4):44, 57.

3656. Cartledge, J. L., L. V. Barton, and A. F. Blakeslee. 1936: Heat and moisture as factors in the increased mutation rate from *Datura* seeds. Proc. Am. Phil. Soc. 76:663–85.

3657. —— and A. F. Blakeslee. 1933: Evolutionary changes speeded by aging seeds. Sci. News Letter 24:345.

3658. —— and —— 1934: Mutation rate increased by ageing seeds as shown by pollen abortion. Proc. Natl. Acad. Sci. U.S. 20:103–10.

3659. —— and —— 1935: Mutation rate from old *Datura* seeds. Science 81:492–93.

3660. ——, M. J. Murray, and A. F. Blakeslee. 1935: Increased mutation rate from aged *Datura* pollen. Proc. Natl. Acad. Sci. U.S. 21:597–600.

3661. Cartter, J. L. 1941: Effect of environment on composition of soybean seed. Soil Sci. Soc. Am. Proc. 5:125–30. 1940.

3662. —— 1942: Equipment for maintaining controlled temperature and low humidity in a seed storage room. J. Am. Soc. Agron. 34:1017–27.

3663. —— and T. H. Hopper. 1942: Influence of variety, environment and fertility level on the chemical composition of soybean seed. U.S. Dept. Agr. Tech. Bull. 787. 66 pp. Chem. Abstr. 36:6199. 1942.

Cartter, J. L. 2171, 4302, 9130, 9131, 14877.

3664. Cartwright, P. M., J. T. Sykes, and R. L. Wain. 1956: Distribution of natural hormones in germinating seeds and seedling plants. Proc. Symp. Chem. Mode Action Growth Substances, London 1955: 32–39. Chem. Abstr. 53:18198. 1959.

3665. Carvalho, A., and C. A. Krug. 1944: A quineira (*Cinchona* sp.) origen, classificação, exploração econômica no mundo e tentativas de sua aclimatação no Brasil. 141 pp. Biol. Abstr. 19:5295. 1945.

3666. —— and —— 1949: Coffee genetics. XII. Inheritance of yellow seed color. Bragantia 9:193–202.

3667. Carvell, K. L., and C. F. Korstian. 1955: Production and dissemination of yellow-poplar seed. J. Forestry 53:169–70.

3668. Carver, W. A. 1936: The effect of certain environmental factors on the development of cotton seed, germinating ability, and resultant yield of cotton. [Abstract.] Proc. Florida Acad. Sci. 1:150.
3669. Casaburi, V. 1930: Mercury salts as disinfectants of seeds. Ind. Chim. 5:1362–68. Chem. Abstr. 25:1942. 1931.
3670. —— 1934: Employment of synthetic tannins and their metallic salts for disinfecting and stimulating seeds. Intern. Rev. Agr. N.S. 25:T342–T344. Rev. Appl. Mycol. 14:114. 1935.
Casady, A. J. 16422.
3671. Casares, R., and C. Lopez-Herrera. 1959: Bromatological study of Spanish sunflower seeds. Anales Bromatol. (Madrid) 11:477–81. Biol. Abstr. 37:7117. 1962.
3672. ——, R. G. Olmedo, and T. Peralta. 1954: Composition of the seeds of *Trigonella foenum-graecum*. 10th Intern. Congr. Ind. Agr. Aliment., Madrid, pp. 358–63. Chem. Abstr. 50:14882. 1956.
3673. Casey, J. E. 1939: Seed price versus seed quality. Proc. Assoc. Offic. Seed Analysts 30:99–103.
3674. —— 1939: Boiling-water treatment an aid in germinating bur clover (*Medicago arabica* or *M. maculata*) seed. Proc. Assoc. Offic. Seed Analysts 31:114–15.
3675. —— 1947: Apparent dormancy in sorghum seed. Assoc. Offic. Seed Analysts News Letter 21(2):34–36.
3676. —— 1948: Analyses of *Lespedeza* seed samples drawn with 9-inch and 30-inch triers. Proc. Assoc. Offic. Seed Analysts 38:40–45. Biol. Abstr. 28:6743. 1954.
Casey, J. E. 1974.
3677. Cashmore, A. B. 1939: A note on the germination of St. John's wort seed. J. Council Sci. Ind. Res. Australia 12:181–82.
3678. Cashmore, W. H. 1934: A rapid method for measuring the moisture content of wheat. Proc. Intern. Seed Testing Assoc. 6:36–37.
3679. Casier, J., and W. Pulzeyt. 1961: Vergelijkende studie over de gluciden bij harde en zachte tarwesoorten. Agricultura (Louvain) 9:537–59. Landb. Doc. 18:153. 1962.
Casier, J. 18707.
3680. Casimir, J., and L. Detroux. 1955: The effect of humidity, temperature, and disinfectants on the free amino acids of wheat. Bull. Inst. Agron. Sta. Rech. Gembloux 23:368–77. Chem. Abstr. 50:8812. 1956.
3681. Caskey, C., Jr., and W. D. Gallup. 1931: Changes in the sugar, oil and gossypol content of the developing cotton boll. J. Agr. Res. 42:671–73. Biol. Abstr. 7:15444. 1933.
3682. Caso, O. H., and A. Marzocca. 1959: Preliminary tests on the action of gibberellic acid on *Taraxacum kok-saghyz*. Bol. Soc. Botan. Agr. 8:19–23. Biol. Abstr. 42:7606. 1963.
Caso, O. H. 12937.
3683. Caspary, R. 1854: Über Samen, Keimung, Specien und Nährpflanzen der Orobanchen. I. Flora (Jena) 37:577–92.
3684. —— 1854: Über Samen, Keimung, Specien und Nährpflanzen der Orobanchen. II. Flora (Jena) 38:593–603.
3685. —— 1867: Untersuchungen über den Samen und die Keimung von *Pinguicula vulgaris* var. Schriften Physik.-Ökonom. Ges. Königsberg 8:16–17.
3686. Cassera, J. D. 1935: Origin and development of the female gametophyte, endosperm and embryo in *Orobanche uniflora*. Bull. Torrey Botan. Club 62:455–66.
3687. Cassidy, N. G. 1961: Can flotation of coconuts be used to indicate quick germination? Agr. J. Fiji Dept. Agr. 31:10–11.
Cast, E. M. 9014.
3688. Castagne, E. 1933: Chemical composition of the fruits of two species of *Dialium*. Bull. Agence Gen. Colonies 26: 1315–23. Chem. Abstr. 28:2032–33. 1934.
3689. —— 1935: Contribution à l'étude chimique des akenes de *Treculia africana* Decne. Bull. Inst. Roy. Colonies Belge 6:694–710.
3690. Castaneda, M., F. F. Gavarron, and M. R. Balcazar. 1942: On a new protease from *Pileus mexicanus*. Science 96:365–66.
3691. Castellani, E. 1935: The influence of the excretory products of certain fungi isolated from the root system of beet on the germination of maize. Nuovo Giorn. Botan. Ital. 42:614–22. Rev. Appl. Mycol. 15:673–74. 1936.
3692. Castellanos, A. 1928: Germinación del *Trichocereus pasacana* (Web.) Brit. et Rose. Physis. 9:141–45. Botan. Centralbl. 158: 471. 1930.
3693. Castelli, L. 1927: Résultats des essais de germination des graines d'*Elaeis* effectués en 1922 et au début de 1923. Agron. Coloniale 16(109):13–18.
Castelo, C. L. 3531.

3694. Castiblanco G., L. E. 1960: Metodos de siembra y cantidad de semilla en alfalfa. Acta Agron. (Palmira, Colombia) 10:17–39. Biol. Abstr. 36:28449. 1961.

3695. Castille, A. 1939: Zur Kenntnis des fetten Öls der Samen von *Ongokea klaineana* Pierre. Ann. Chem. 543:104–10. Chem. Abstr. 34:2329. 1940.

Castillo, M. T. 7278.

3696. Castillo Z., J. 1959: Observations upon size and abnormalities of the beans of some commercial varieties of coffee. Cenicafe 10:397–418. Biol. Abstr. 36: 28534. 1961.

Castillon, L. E. 2654, 2656, 2657, 15526.

3697. Castle, H., and L. G. Nickell. 1942: The aerosol-hypochlorite technique for the sterilization of orchid seeds. Am. Orchid Soc. Bull. 11:200–1. Rev. Appl. Mycol. 22:171. 1943.

Castro, F. S. de; *see* De Castro, F. S.

3698. Caswell, G. H., and H. T. Clifford. 1958: The effect of ethylene dichloride and carbon tetrachloride on the germination and early growth of maize. Empire J. Exptl. Agr. 26:365–72.

3699. —— and —— 1960: The effect of moisture content on the germination and growth of fumigated maize grain. Empire J. Exptl. Agr. 28:139–44.

Cathcart, W. H. 5532, 7349.

3700. Cation, D. 1949: Transmission of cherry yellows virus complex through seeds. Phytopathology 39:37–40.

3701. —— 1960: Negative transmission of an apple virus through seeds. Plant Disease Reptr. 44:695.

Catron, D. 6192.

3702. Cattaneo, P. 1945: Chemical composition of oil from the seeds of *Erythrina cristogalli*. Anales Asoc. Quim. Arg. 33:5–18. Chem. Abstr. 40:1051. 1946.

3703. —— 1946: Composición quimica del aceite de semilla ce *Chorisia insignis* "Palo borracho." Anales Asoc. Quim. Arg. 34:5–14. Biol. Abstr. 22:12382. 1948.

3704. ——, G. Karman, and L. M. Darnet. 1945: The chemical composition of the oil of the seeds of *Maytenus disticha*. Rev. Med. Cienc. Afines (Buenos Aires) 7:865–69. Chem. Abstr. 40:3280. 1946.

Cattaneo, P. 16216.

3705. Caujolle, F., and G. Bergal. 1949: Influence de l'acide phénylborique sur la germination de quelques apétales. Compt. Rend. 228:1249–51.

3706. —— and —— 1950: Influence des acides borique et phénylborique sur la germination et le développment des plantules de quelques dicotylédones. Compt. Rend. 231:1550–52. Biol. Abstr. 26: 16403. 1952.

3707. Cavazza, L. 1950: Differenze biologiche tra semi duri e semi permeabili in *Hedysarum coronarium* (Sulla). Nuovo Giorn. Botan. Ital. 57:238–46.

3708. —— 1950: Hot water treatment of hard seeds of certain Leguminosae. Ann. Bari Univ. Fac. Agr. 7:187–202.

3709. —— 1950: Observations on the determination of germinative energy. Ann. Bari Univ. Fac. Agr. 7:211–21.

3710. —— 1950: Ricerche sull'influenza del guscio nella germinazione del seme di Lupinella (*Onobrychis viciaefolia*). Nuovo Giorn. Botan. Ital. 57:257–62.

3711. —— 1951: Sul differente comportamento dei semi duri di tre diverse leguminose. Nuovo Giorn. Botan. Ital. 58:13–26.

3712. —— 1951: Osservazioni sull'azione stimolante del borace. Nuovo Giorn. Botan. Ital. 58:392–93.

3713. —— 1951: L'azione dell'alcool sin seme duri. Nuovo Giorn. Botan. Ital. 58:393–97.

3714. —— 1951: Influenza reciproca di cariossidi di frumento in germinazione. Nuovo Giorn. Botan. Ital. 58:397–404.

3715. —— 1953: Considerazioni sull'impiego dell'acido solforico nel trattamento ai semi duri. Nuovo Giorn. Botan. Ital. 60:750–59.

3716. —— 1953: L'influenza die basse tensioni dell'acqua sulla germinazione die alcuni semi. Nuovo Giorn. Botan. Ital. 60:759–62.

3717. —— and G. DeBenedictis. 1959: Increasing hard seed percentage with hot water treatment. Nuovo Giorn. Botan. Ital. 66:151–60.

3718. Cave, M. S. 1941: Megasporogenesis and embryo sac development in *Calochortus*. Am. J. Botany 28:390–94.

3719. —— 1948: Sporogenesis and embryo sac development of *Hesperocallis* and *Leucocrinum* in relation to their systematic position. Am. J. Botany 35:343–49.

3720. ——, H. J. Arnott, and S. A. Cook. 1961: Embryogeny in the California peonies with reference to their taxonomic position. Am. J. Botany 48:397–404.

Cawse, P. A. 2881.

3721. Cayford, J. H. 1959: Germination and survival of jack pine and red pine after scarification in southeastern Manitoba.

Can. Dept. Northern Affairs Natl. Resources Forest Res. Div. Tech. Note 78. 14 pp.

3722. Cecchetti, A. 1939: Degree of ripening and viability of seeds of French honeysuckle. Photoprint: Ann. Tec. Agrar. 12:153–58.

3723. Cech, F. C., J. C. Barber, and B. J. Zobel. Comments on "Who wants tree seed certification and why?" J. Forestry 60:208–10.

Cecil, S. R. 19843.

Cederquist, D. C. 6133.

3724. Cederstam, B. 1963: Supplies of pine seed in the higher parts of Norrland. Skogen 50:88–90. Forestry Abstr. 24:411. 1963.

3725. Celli, O. 1943: Studio embriologico sulla vite di Almeria. Ann. Botan. 22:70–90.

3726. Carbulis, J. 1955: Carbohydrates in cacao beans. II. Sugars in Caracas cacao beans. Arch. Biochem. Biophys. 58:406–13.

Cerca-Miguel, M. C. de Sousa-Monteiro; *see* De Sousa.

3727. Cerceau-Larrival, M. T. 1957: Morphologie des cotyledons d'Ombellifères. Compt. Rend. 244:659–60.

3728. Cercos, A. P. 1951: Efecto de la bacilomicina y fungocina sobre la germinación del trigo. Rev. Arg. Agron. 18:33–35. Biol. Abstr. 26:4750. 1952.

3729. Cerighelli, R. 1926: Influence de la lumière et de la temperature sur la germination des graines en absence de calcium. Compt. Rend. 182:483–85.

3730. —— 1926: Recherches sur l'action combinée de la temperature et du calcium sur la croissance des racines pendant la periode germinative. Bull. Soc. Botan. France 73:729–32.

3731. —— and R. David. 1941: Germination in pure water. I. Preliminary experiments. Compt. Rend. Fac. Sci. Marseille 1:55–58. Chem. Abstr. 40:5473. 1946.

3732. —— and —— 1932: Influence du calcium sur la germination. Comportement des graines à germination epigée. Rev. Sci. 80:134–36. Biol. Abstr. 23:8829. 1949.

3733. —— and V. Durand. 1954: Influence du chlorure de sodium sur la germination et le développement du riz. Journées Riz 4:34–45.

Cerighelli, R. 12832.

3734. Cernohorsky, Z. 1950: Anatomie des graines des espèces tchèques du genre *Thlaspi* L. Bull. Intern. Cl. Sci. Math. Nat. Med. Acad. Tchèque Sci. 47(1946):5–10. Biol. Abstr. 26:6838. 1952.

3735. Černy, J., and J. Boril. 1956: Yields of Savoy cabbage seeds—19.50 centners per hectare. Socialist. Zeměděl. 6:732–34.

3736. Ceruti, A. 1947: The metabolism of the chlorophyll and carotenoid pigments during the ripening of seeds of *Pisum sativum*. Lavori Botan. Torino 8:97–106. Chem. Abstr. 44:5433–34. 1950.

3737. —— 1947: The metabolism of nucleic acid in embryonic tissues. Lavori Botan. Torino 8:107–25. Biol. Abstr. 22:12314. 1948.

3738. —— 1947: Starch in bean seeds during germination. Nuovo Giorn. Botan. Ital. 54:370–71. Biol. Abstr. 23:24841. 1949.

3739. —— and M. Bellando. 1961: The histochemical characters of the protoplasm of the promeristematic and proparenchymatic cells of bean embryos during dormancy. Allionia (Turin) 7:195–208.

Cesaris-Coromaldi, L. de; *see* De Cesaris-Coromaldi, L.

3740. Cescon, I., and G. L. Giovetti. 1962: Fatty acids and physical-chemical characteristics of the oil extracted from different kinds of tomato seeds. Riv. Ital. Sostanze Grasse 39:349–52. Chem. Abstr. 60:3206. 1964.

3741. Cetl, I. 1963: The effect of sterile conditions on the survival of maize kernels in cold soil. Biol. Plant. 5:211–15.

3742. Chabelska-Frydman, C. 1959: *Cephaliophora tropica* Thaxter, a rare hypomycete isolated from seeds in Israel. Bull. Res. Council Israel Sect. D7:81–83.

Chabrol, J. Desveaux-. 18453.

3743. Chabrol, P. 1936: Notes sur le développement des racines laterales du chêne en pépinière. Bull. Trimest. Soc. Forest. Franche-Comte Prov. Est. 21: 419–25. Biol. Abstr. 11:9402. 1937.

3744. Chabrolin, C. 1934: La germination des graines d'*Orobanche*. Compt. Rend. 198:2275–77.

3745. —— 1934: La germination des graines de *Thesium humile* exige l'intervention de champignons saprophytes. Compt. Rend. 199:225–26.

3746. ——1934: Les graines d'*Oxalis cornua* Thunb. en Tunisie. Bull. Soc. Hist. Nat. Afr. Nord 25:396–98.

3747. —— 1937/38: Germination of the seeds of broom rape with beans. Ann. Serv. Botan. Agron. Tunisie 14/15:91–144. Chem. Abstr. 35:491–92. 1941.

3748. Chadha, J. S., and T. R. Seshadri. 1962: Leucocyanidins from the seeds of *Litchi chinensis*. Current Sci. (India) 31:56–57.

3749. Chadwick, L. C. 1943: Handling *Nandina* seeds. Am. Nurseryman 78(12):8.

3750. —— 1946: Seeds of red cedar. Am. Nurseryman 83(9):10.

3751. —— and J. C. Swartley. 1941: Further studies on the effects of synthetic growth substances on cuttings and seeds. Proc. Am. Soc. Hort. Sci. 38:690–94.

Chadwick, L. C. 11872, 17618, 18224.

3752. Chai, T. S., and H. L. Hansen. 1952: Characteristics of black spruce seeds from cones of different ages. Minn. Forestry Notes 2:1–2. Biol. Abstr. 28:4329. 1954.

Chaikina, K. V. 10549.

3753. Chailakhian, M. C., and L. P. Zdanova. 1938: Vernalization of plants and the alteration of growth hormones. Bull. Acad. Sci. URSS Cl. Sci. Math. Nat. Ser. Biol. (English Transl.) 1938(2):523–38.

3754. Chailakhian, M. K., and L. P. Zdanova. 1938: Influence of heteroauxine on the growth and development of plants by treatment of seeds. Bull. Acad. Sci. URSS Cl. Sci. Math. Nat. Ser. Biol. (English Transl.) 1938(5/6):1281–96.

Chailakhian, M. K. 11332.

Chais, L. 6656.

3755. Chak, I. M., and K. V. Giri. 1947: Blood clotting factor in castor seed (Ricinus communis). Sci. Cult. (Calcutta) 12:598–99.

Chakrabarte. 1914.

Chakrabarti, J. K. 17223.

3756. Chakrabartty, T., and A. Chatterjee. 1957: Constitution of swietenin, the non-bitter principle of the seeds of *Swietenia macrophylla*. II. J. Indian Chem. Soc. 34:117–20. Chem. Abstr. 51:14903. 1957.

Chakrabartty, T. 7201.

3757. Chakrabarty, M. K., and M. M. Chakrabarty. 1961: The nutritional aspects of hydrogenated tobacco seed and safflower seed oils. J. Sci. Ind. Res. (India) 20C:150–52. Chem. Abstr. 60:7217. 1964.

3758. Chakrabarty, M. M., S. Bhattacharyya, M. J. Desai, and S. A. Patel. 1956: Studies on the seed fats of "Cucurbitaceae" family rich in conjugated acids (the component fatty acids of *Momordica dioica* and *Trichosanthes cucumerina*). Naturwissenschaften 43:523–24.

3759. —— and D. K. Chowdhuri. 1957: The fatty acid composition of the seed fat from *Swietenia macrophylla*. J. Am. Oil Chemists Soc. 34:489–90.

3760. ——, ——, and B. K. Mukerji. 1955: Studies on some seed fats of Cucurbitaceae family. Naturwissenschaften 42:344–45.

Chakrabarty, M. M. 3035, 3757, 4049, 7846, 7847, 7848, 15108, 16750.

3761. Chakraborti, S. K. 1962: Triterpenes from the seeds of *Albizzia lucida*. Sci. Cult. (Calcutta) 28:385–86. Chem. Abstr. 58: 1722. 1963.

3762. —— and A. K. Roy. 1962: Triterpenoid sapogenins from the seeds of *Albizzia lucida* Benth. Trans. Bose Res. Inst. (Calcutta) 25:69–71.

Chakraborti, S. K. 2056.

3763. Chakravarthy, R. S. 1935: Polyembryony in *Murraya koenigi* (Spreng.). Current Sci. (India) 3:361–62.

3764. —— 1936: Nucellar polyembryony in the Rutaceae. Current Sci. (India) 5: 202–3.

Chakravarti, D. 3767.

3765. Chakravarti, J. K., and N. K. Sen. 1951: Jute seeds—*Corchorus olitorius*. The composition of oil olitorius. I. J. Indian Chem. Soc. 28:390–94.

3766. Chakravarti, K. K., A. K. Bose, and S. Siddiqui. 1948: Chemical examination of the seeds of *Psoralea corylifolia* Linn. (Babehi). J. Sci. Ind. Res. (India) 7B:24–26.

3767. Chakravarti, R. N., D. Chakravarti, M. N. Mitra, B. D. Gupta, and P. C. Maiti. 1955: Sterols from seeds of *Cassia* plants. Bull. Calcutta School Trop. Med. 3:163–64. Biol. Abstr. 31:2473. 1957.

3768. ——, P. C. Maiti, and N. K. Chakravarty. 1954: Variation in the composition of total alkaloids of the seeds of *Argemone mexicana* with storage. Bull. Calcutta School Trop. Med. 1:11–12. Chem. Abstr. 49:11789. 1955.

3769. Chakravarti, S. C. 1948: A simple method of isolating mustard embryos and their cultivation. Current Sci. (India) 17:299.

3770. —— 1950: Anatomical changes in the embryo of mustard during vernalisation. Current Sci. (India) 19:319–21.

3771. —— 1954: Inhibition of vernalization in *Linum usitissimum* Linn. by certain synthetic hormones. Nature 174:461–62.

3772. —— 1946: Effects of certain synthetic growth regulators on the process of vernalization of *Sesamum indicum*. Agra Univ. J. Res. 5:343–45. Chem. Abstr. 52:5562. 1958.

3773. —— 1958: After effect of seed treatment with low temperature and certain growth regulators on the length of the hypocotyl and area of the cotyledonary leaves of *Phaseolus radiatus* L. Indian J. Plant Physiol. 1:39–43.

3774. —— 1958: Some effects of gibberellic acid on *Sesamum indicum* L. Phyton 11:75–83.

3775. —— 1959: Effect of high temperature and dessication on the vernalization of certain Indian crop plants. Proc. Natl. Acad. Sci. India 29:151–56. Biol. Abstr. 36:21608. 1961.

3776. —— and G. D. Pershad. 1953: Studies on the viability of immature weed seeds. Indian J. Agr. Sci. 23:153–55.

3777. —— and V. N. K. Pillai. 1955: Studies in auxin-vernalization relationships. I. The effects of certain synthetic auxins and their antagonists on the vernalization of *Brassica campestris* L. Phyton 5:1–17.

Chakravarti, S. C. 17214–17217.

Chakravarty, K. 5060.

3778. Chakravarty, M. N., A. N. Ghosh, and S. K. Roy. 1955: Research communications. The effect of seed treatment of paddy with infra-red and ultra-violet rays on the germination, growth, and yield. Bihar Acad. Agr. Sci. Proc. 4:97–100. Biol. Abstr. 31:29234. 1957.

Chakravarty, N. K. 3768.

Chakravarty, T. 17541.

3779. Chakraverti, I. B., S. Nag, and W. G. MacMillan. 1961: Isolation, purification and fractionation of tamarind kernel polysaccharide. J. Sci. Ind. Res. (India) 20D:380–84. Biol. Abstr. 38:15651. 1962.

3780. Chakravorti, A. K. 1947: The development of female gametophyte and seed of *Coccinia indica* W. & A. J. Indian Botan. Soc. 26:95–104. Biol. Abstr. 22:12005. 1948.

3781. Chalabuda, T. W. 1935: Die Saugkraft als Indikator von physiologischen Eigenschaften der Machorka-Tabaksorten. Acad. Sci. RSS Ukraine Inst. Botan. J. 4:67.

3782. Chalam, G. V. 1949: Germination of double grained paddy in relation to the anatomy of lemma and palea. Current Sci. (India) 18:401–2.

Chalmot, G. de; *see* De Chalmot, G.

3783. Chaloner, W. G., and J. M. Pettitt. 1963: A Devonian seed megaspore. Nature 198:808–9.

Chaloupka, J. 4115.

Chalupova-Janovicova, J. 18567.

3784. Chamberlain, C. J. 1907: Preliminary note on *Ceratozamia*. Botan. Gaz. 43:137.

3785. —— 1920: The living cycads and the phylogeny of seed plants. Am. J. Botany 7:146–53. Botan. Abstr. 5:1885. 1920.

Chamberlain, C. J. 4454.

3786. Chamberlain, E. E., and P. R. Fry. 1950: Influence of method on tomato seed extraction on seed transmission of tobacco mosaic and tomato-streak. New Zealand J. Sci. Technol. A32:19–23.

3787. —— and C. H. Proctor. 1947: Investigations on growing *Pyrethrum* in New Zealand. I. Methods of propagation, cultivation, harvesting and drying. New Zealand J. Sci. Technol. A28:353–61.

Chambers, D. C. 17801.

3788. Chambers, S. C. 1963: Bare patch and poor emergence of cereals. IV. Quality of the seed. J. Agr. W. Australia 4:230–36.

3789. Chamblee, D. S., and R. L. Lovvorn. 1953: The effect of rate and method of seeding on the yield and botanical composition of alfalfa-orchardgrass and alfalfa-tall fescue. Agron. J. 45:192–96.

3790. Chamney, N. P. 1928: The distribution of white and pink seeds in the cola plant (*Cola acuminata*). Gold Coast Dept. Agr. (Yearbook) Bull. 13:121–22.

3791. Champion, H. G. 1930: Storage of seed of Chir pine (*Pinus longifolia*). Indian Forester 56:481–83.

3792. —— 1933: The importance of the origin of seed used in forestry. Indian Forest Records 17(5):1–76. Biol. Abstr. 8:16234. 1934.

3793. —— 1943: Seed selection in forestry. Nature 152:354.

3794. —— 1950: Tree breeding and seed supplies in northwest America with special reference to Douglas fir (*Pseudotsuga taxifolia* Brit.). Forestry 23:69–77.

3795. —— and B. D. Pant. 1931: Investigations on the seed and seedlings of *Shorea robusta* Gaertn.-f. Indian Forest Records 16(5):1–33. Biol. Abstr. 7:14142. 1933.

3796. —— and —— 1932: Notes on *Pinus longifolia* Roxb. The plantations in Dehra Dun and the Central Provinces and miscellaneous seed studies. Indian Forest Records 16(7):1–25. Biol. Abstr. 7:14143. 1933.

Champion, R. 1272.

3797. Champness, S. S. 1949: Notes on the buried seed populations beneath different types of ley in their seeding year. J. Ecol. 37:51–56.

3798. —— and K. Morris. 1948: The population of buried viable seeds in relation to contrasting pasture and soil types. J. Ecol. 36:149–73.

Champy, C. 5090.

3799. Chan, C. T. 1937: Study of the relation of different pH value of nutrient solution and tree seed germination. J. Agr. Assoc. China (Taipei) 158:21–47. Biol. Abstr. 12:10473. 1938.

Chanda, R. 10413.

3800. Chandler, C. 1951: Flowering habits and fertility of some *Cinchona* species in Guatemala. Contrib. Boyce Thompson Inst. 16:249–59.

3801. —— 1953: Seed germination for some species of *Plantago*. Contrib. Boyce Thompson Inst. 17:265–71.

3802. —— and L. V. Barton. 1955: Morphological and physiological studies of diploid and tetraploid *Plantago ovata* Forsk. Contrib. Boyce Thompson Inst. 18:193–214.

Chandler, C. 2039.

Chandler, P. 7671.

3803. Chandler, R. F., Jr. 1938: The influence of nitrogenoûs fertilizer applications upon seed production of certain deciduous forest trees. J. Forestry 36:761–66.

3804. Chandler, S. E. 1956: Botanical aspects of *Pyrethrum*. III. The natural history of the secretory organs; the pyrethrins content of the fertile achenes ("seed"). Pyrethrum Post 4:10–15.

3805. Chandraratna, M. F., L. H. Fernando, and C. Wattegedera. 1952: Seed dormancy in rice. Trop. Agriculturist 108:261–64.

Chandraratna, M. F. 16159.

3806. Chandrasekhara, M. R., and M. Swaminathan. 1953: Enzymes of ragi (*Eleusine coracana*) and ragi malt. I. Amylases. J. Sci. Ind. Res. (India) 12B:51–56. Chem. Abstr. 48:5887. 1954.

3807. —— and —— 1954: Enzymes of ragi (*Eleusine coracana*) and ragi malt. III. Pyro- and glycerophosphates. J. Sci. Ind. Res. (India) 13B:492–96. Chem. Abstr. 49–399. 1955.

3808. —— and —— 1957: The enzymes of pearl millet (*Pennisetum typhoideum*) malt. I. Amylases. J. Sci. Ind. Res. (India) 16C:35–38. Chem. Abstr. 51:10679. 1957.

3809. Chaney, R. W. 1951: How old are the Manchurian lotus seeds? Garden J. N.Y. Botan. Garden 1:137–39.

3810. Chang, C. W. 1963: Comparative growth of barley embryos *in vitro* and *in vivo*. Bull. Torrey Botan. Club 90:385–91.

3811. Chang, C. Y., E. Witschi, and I. V. Ponseti. 1955: Teratogenic effects of *Lathyrus odoratus* seeds on development and regeneration of vertebrate limbs. Proc. Soc. Exptl. Biol. Med. 90:45–50.

Chang, C. Y. 17045.

Chang, H. C. 18367.

3812. Chang, H. S. 1962: The seed certification program in Taiwan. Proc. Intern. Seed Testing Assoc. 27:581–91.

3813. —— 1962: General introduction of the Taiwan provincial seed testing laboratory, Taichung, Taiwan. Proc. Intern. Seed Testing Assoc. 27:983–88.

3814. Chang, H. T. 1962: Problems concerning multiplication and distribution of pure seeds in countries with small farms, with special reference to problems encountered in Taiwan. Proc. Intern. Seed Testing Assoc. 27:515–29.

3815. Chang, K. C., and Y. S. Chao. 1936: Studies on cottonseeds. I. Nutritive value of cottonseed meal. J. Chem. Eng. (China) 3:30–38. Chem. Abstr. 30:4902. 1936.

Chang, K. J. 3912.

Chang, P. L. 12191.

3816. Chang, S. C. 1943: Length of dormancy in cereal crops and its relationship to after-harvest sprouting. J. Am. Soc. Agron. 35:482–90.

3817. Chang, S. T. 1960: The effect of freezing temperatures of varying intensities and durations on the germination of maize harvested at different stages of maturity. Dissertation Abstr. 21:406.

Chang. T. M. 9147.

Chang, W. Y. 16808.

3818. Chang, Y. F., T. C. Cheng, and H. Ho. 1963: Biochemical studies on fat and carbohydrate metabolism during peanut seed germination. Taiwan Univ. Mem. Coll. Agr. 7:12–25.

3819. Chang, Y. O., and C. W. McAnelly. 1961: Essential amino acids in dry beans. Wyoming Univ. Publ. 25:1–12. Biol. Abstr. 39:7828. 1962.

3820. Chang, Z. Y. 1959: The effect of moisture content on the germination percentage of tobacco seeds during storage. J. Agr. Assoc. China (Taipei) 26:49–51.

Channappagowder, B. B. 18831.

3821. Channon, H. J., and C. A. M. Foster. 1934: The phosphatides of the wheat germ. Biochem. J. 28:853–64.

Chantegrel, P. 16087.

3822. Chanturiia, N. N., and M. A. Kakuliia. 1956: Testing certain preparations for treating mulberry seeds. Akad. Nauk Gruzinsk. SSR Inst. Zashch. Rast. Tr. 11:9–19.

Chao, S. H. 7095.

3823. Chao, T. F., F. T. Wang, S. Wang, and T. P. Tsai. 1959: A study of the physiology of after-ripening of wheat grains. Acta Botan. Sinica 8:230–38. Biol. Abstr. 35:64325. 1960.

3824. Chao, T. Y., Y. L. Chou, P. T. Young, and T. Q. Chou. 1962: Studies on the alkaloids of embryo loti, *Nelumbo nucifera* Gaertn. I. Sci. Sinica (Peking) 11:215–19.

Chao, Y. S. 3815.

3825. Chapas, L. C., P. B. H. Tinker, and C. O. Ziboh. 1957: The determination of the oil content of the oil plam fruit. W. African Inst. Oil Palm Res. 2:230–36.

3826. Chapin, J. S., and F. W. Smith. 1960: Germination of wheat at various levels of soil moisture as affected by applications of ammonium nitrate and muriate of potash. Soil Sci. 89:322–27.

3827. Chapman, A. G. 1936: Scarification of black locust seed to increase and hasten germination. J. Forestry 34:66–74.

3828. —— 1941: Tolerance of shortleaf pine seedlings for some variations in soluble calcium and H-ion concentration. Plant Physiol. 16:313–26.

3829. Chapman, A. L., and M. L. Peterson. 1962: The seedling establishment of rice under water in relation to temperature and dissolved oxygen. Crop Sci. 2:391–95.

3830. Chapman, H. C. 1960: Stored-grain insects and their control in New Jersey. J. Econ. Entomol. 53:536–39.

3831. Chapman, H. H. 1936: Effect of fire in preparation of seedbed for longleaf pine seedlings. J. Forestry 34:852–54.

3832. —— 1946: Origin and results of the seed-tree experiment with Norway pine on the Chippewa National Forest. J. Forestry 44:178–83.

Chapman, J. L. 10424.

3833. Chapman, M. 1933: The ovule and embryo sac of *Saxifraga virginiensis*. Am. J. Botany 20:151–58.

Chapman, P. R. 12915.

3834. Chapman, W. H. 1961: The effects of spacing, hybrids and varieties, fumigation and mulches on yields of highly fertilized corn. Soil Crop Sci. Soc. Florida Proc. 21:206–13.

Chapon, L. 18896, 18897.

3835. Chapot, H. 1955: Remarques sur la germination des pépins de *Poncirus trifoliata* (Rafin). Fruits (Paris) 10:465–68.

3836. —— and J. C. Praloran. 1955: Les graines de citrus (première contribution à l'étude de leur morphologie avec référence particulière a leur degré de polyembryonie). 14th Intern. Hort. Congr. Rept. 1955(2): 1294–323.

3837. Chappuzeau, B. 1930: Untersuchungen über die Bedeutung von Licht, Feuchtigkeit und Korngrösse bei der Kleekeimung. Angew. Botan. 12:99–162.

3838. Charavanapavan, C. 1960: A note on storage tests on paddy. Trop. Agriculturist 116:27–29.

3839. Charconnet-Harding, F., and R. Jacquot 1949: La synthèse du troptophane par les plantules isolées. Comparaison avec le pouvoir de synthése de *Sterigmatocystis nigra*. Compt. Rend. 228:498–500.

Charetschko, E. I. 17847.

Charles, M. Kogane-. 586–589, 5261–5265.

Charley, H. 4220.

Charlton, F. B. 15568.

Charmillon, R. 12784.

3840. Charpentier, A. 1887: Wirkung von Cocain auf die alkoholische Gährung und die Keimung. Just's Botan. Jahresber. 15:128.

Charro, A. 1933.

3841. Chase, A. 1935: Some seeds caught in the upper air. J. Washington Acad. Sci. 25:93.

3842. Chase, S. B. 1942: The influence of delayed hulling on the color and quality of eastern black walnut kernels. Proc. Am. Soc. Hort. Sci. 41:131–35. Expt. Sta. Rec. 88:637. 1943.

3843. —— 1947: Eastern black walnut germination and seedbed studies. J. Forestry 45:661–68.

3844. Chase, S. S. 1947: Polyploidy in an immersed aquatic angiosperm. Am. J. Botany 34:581–82. Biol. Abstr. 23:42. 1949:.

3845. Chatak, N. 1934: Chemical examination of the seeds of *Abrus precatorius* Linn. II. The colouring matter of the seed-coat. Allahabad Univ. Studies 10:289–95.

3846. Chatfield, C. 1934: Sesame seeds have high nutritive value; very rich in calcium. Yearbook Agr. U.S. Dept. Agr. 1934:316–17.

3847. Chatt, E. M. 1953: Cocoa. Interscience, New York. 302 pp.

Chatterjee, A. 3756, 7201.

3848. Chatterjee, N. R., S. Ghosh, and R. N. Chopra. 1938: Studies on the enzymes of the seeds of *Butea frondosa*. I. J. Indian Chem. Soc. 15:101–6, 107–9.

3849. Chatterjee, S. K. 1960: Effect of pre-sowing treatments of tung (*Aleurites fordii*, Hemsl.). Sci. Cult. (Calcutta) 26:130–31.

3850. Chatterji, B., and S. K. Mukerji. 1953/54: Studies in germination performance of some crops seeds under laboratory and field tests. Bihar Acad. Agr. Sci. Proc. 4:68–72. Biol. Abstr. 31:29223. 1957.

3851. Chatterji, N. G., and A. C. Gupta. 1937: Oil of Nageshwar (Ironwood) seed (*Mesua ferrea*). J. Oil Colour Trades 91:1656. Soc. Chem. Ind. J. 56:696B. 1937. Pharm. Abstr. 4:361. 1938.

3852. Chatters, R. M. 1938: The effect of fast neutrons on dry seeds. Science 87:262–63.

Chatters, R. M. 16926.

Chattopadhyay, D. 1849.

3853. Chattopadhyay, H., and S. Banerjee. 1951: Effect of germination on the carotene content of pulses and cereals. Science 113:600–1.

3854. —— and —— 1951: Studies on the choline content of some common Indian pulses and cereals both before and during the course of germination. Food Res. 16:230–32.

3855. —— and —— 1952: The ascorbic acid oxidase activity of some common Indian pulses during germination. Indian J. Med. Res. 40:439–42. Chem. Abstr. 47:8193. 1953.

3856. —— and —— 1952: Effect of germination on the total tocopherol content of pulses and cereals. Food Res. 17:402–3.

3857. —— and —— 1953: Effect of germination on biological values of proteins and trypsin-inhibitor activity of common Indian pulses. Indian J. Med. Res. 41:185–89. Chem. Abstr. 49:4794. 1955.

Chattopadhyay, H. 1847, 1849.

Chattopadhyay, S. B. 2428.

Chattopadhyay, S. K. 7853.

3858. Chaudhari, B. B. 1956: Tiny seed and tiny pod segregates in Bengal gram (*Cicer arietinum*). Current Sci. (India) 25:330–32.

3859. —— 1958: Occurrence of tri-cotyledonary seedlings in some dicots. Sci. Cult. (Calcutta) 23:654–55. Biol. Abstr. 33: 23051. 1959.

3860. Chaudhri, I. I. 1956: Beziehung zwischen der photoperiodischen Beeinflussung der Samenkeimung und anderer Entwicklungsvorgänge. Z. Botan. 44:319–36.

Chaudhri, I. I. 10504.

3861. Chaudhuri, S. B., J. N. Baruah, and P. R. Rao. 1964: Amino-acids of Nahor seed oilcake. Current Sci. (India) 33:16.

Chaudhuri, S. D. 349.

3862. Chauveaud, G. 1894: Moyen d'assurer et de rendre très hative la germination des vignes. Compt. Rend. 118:211–12.

3863. —— 1894: Sur les caractères internes de la graine des vignes et leur emploi dans la determination des espèces et la distinction des hybrides. Compt. Rend. 118:485–87.

Chavalier, J. M. 5464.

3864. Chavan, A. R., and P. N. Trivedi. 1962: Seed germination in some weeds. Indian Forester 88:436–39. Weed Abstr. 11:1921. 1962.

3865. Chavan, V. M. 1960: Better crops through better seeds. Farmer (Bombay) 11:36–40, 48.

3866. Chavarriaga, M. E. 1955: Producción y certificación de semillas. Agr. Trop. (Bogotá) 11:598.

Chaves, J. M. 15168.

3867. Chaze, J. 1937: Production of choline in the seed and seedling of rye grass, in relation to parasitism. Compt. Rend. 204: 1443–45. Chem. Abstr. 31:6287. 1937.

Cheadle, J. N. 11702.

Chebolda, V. 28.

Chebotareva, T. 2255.

3868. Cheema, G. S. 1956: Seed variations in frangipani (*Plumeria acutifolia*). Indian J. Hort. 13:24–25. Biol. Abstr. 31:35902. 1957.

3869. —— and P. G. Dani. 1929: Seedlessness in papayas. Agr. J. India 24:206–7.

3870. Cheldelin, V. H., and R. L. Lane. 1943: B vitamins in germinating seeds. Proc. Soc. Exptl. Biol. Med. 54:53–55.

3871. Cheliadinova, A. I. 1953: Germination of seeds of tree species. Lesn. Khoz. 6:50–54.

3872. —— 1957: Adaptive significance of the dormancy period in seeds of woody plants. Agrobiologiya 1957(2):94–100.

3873. Chellappa, T. 1959: The physiology of strawberry seed germination. Dissertation Abstr. 20:859–60.

3874. Chemin, E. 1931: Les graines de *Lathraea clandestina* peuvent germer sans l'assistance d'aucune autre plante. Réponse à E. Heinricher. Bull. Soc. Botan. France 78:708–21.

Chen, A. L. 3878.

3875. Chen, C. C. 1920: Internal fungous parasites of agriculture seeds. Maryland Agr. Expt. Sta. Bull. 240:81–110. Botan. Abstr. 11:1720. 1922.

3876. Chen, C. H., G. P. Hanson, and J. G. Ross. 1962: Endosperm color inheritance in tetraploid corn. Proc. S. Dakota Acad. Sci. 41:128–30.

Chen, C. J. 16372.

3877. Chen, F. H. 1940: A study of *Primula* seeds with reference to the criteria of sections. Bull. Fan Mem. Inst. Biol. 10:69–81.

3878. Chen, K. K., and A. L. Chen. 1934: The constituents of be-still nuts, *Thevetia neriifolia*. J. Biol. Chem. 105:231–40.

Chen, H. K. 15615.

Chen, H. M. 11976.

Chen, S. M. 12372–12374, 18793.

Chen, T. M. 17300.

3879. Cheng, C. F. 1946: Self-fertility studies in three species of commerical grasses. J. Am. Soc. Agron. 38:873–81.

Cheng, C. P. 9290.

3880. Cheng, T. 1955: The effect of seed treatment with microelements upon the germination and early growth of wheat. Sci. Sinica (Peking) 4:129–35.

Cheng, T. C. 3818.

3881. Cheng, Y. W. 1962: The effects of X-ray and thermal neutron irradiation on soybeans. J. Agr. Assoc. China (Taipei) 37:23–33.

3882. Cheng, Y. Y., P. Linko, and M. Milner. 1960: On the nature of glutamic acid decarboxylase in wheat embryos. Plant Physiol. 35:68–71.

Cheng, Y. Y. 12282.

3883. Cheo, P. C. 1955: Effect of seed maturation on inhibition of southern bean mosaic virus in bean. Phytopathology 45:17–21.

3884. Cheo, T. Y. 1947: Observations on seed germination of *Ormosia taiana*. Acad. Sinica Inst. Botan., Botan. Bull. 1:131–32.

3885. Cheoa-Sakul, P., and A. C. Santos. 1939: Alkaloids of, and the fixed oil of the seeds of *Datura alba* Ness, Phillippine. Natl. Res. Council Philippines Bull., Proc. 5th Sci. Conv. 23:163. Pharm. Abstr. 7:164. 1941.

3886. Chepil, W. S. 1946: Germination of weed seeds. I. Longevity, periodicity of germination, and vitality of seeds in cultivated soil. Sci. Agr. (Ottawa) 26:307–46.

3887. —— 1946: Germination of weed seeds. II. The influence of tillage treatments on germination. Sci. Agr. (Ottawa) 26:347–57.

3888. Cherbuliez, E., and K. Bernhard. 1932: Croton seed. I. Crotonoside (2-hydroxy-6-aminopurine-δ-riboside). Helv. Chim. Acta 15:464–71.

3889. ——, ——, and E. Ehninger. 1932: Recherches sur la graine de croton. III. Sur l'improbabilité de l'existence de l'alcoloide de tuson, et sur la presence de saccharose, et de glucosides du glucose, dans la graine de croton. Helv. Chim. Acta 15:855–56.

3890. ——, E. Ehninger, and K. Bernhard. 1932: Croton seed. II. The vesicant principle. Helv. Chim. Acta 15:658–70.

Cherednichenko, S. V. 19165.

3891. Cheremisinov, N. A. 1957: Fungus infection of corn seeds and measures for preventing it. Vses. Akad. Sel'skokh. Nauk V. I. Lenina Dokl. 22(8):40–43.

3892. Cherewick, W. J. 1947: Determination of smut spore load on cereal seed grain. [Abstract.] Proc. Can. Phytopathol. Soc. 15:14. Rev. Appl. Mycol. 27:315–16. 1948.

3893. —— 1948: Relation between spore load on the seed and amount of smut infection in cereal grains. [Abstract.] Phytopathology 38:4–5.

3894. —— 1954: Studies on seed-borne microflora and the effect of seed treatment of rice. Malayan Agr. J. 37:169–72.

Cherewick, W. J. 12602.

3895. Chernavina, I. A. 1952: The effect of molybdenum on the crop and chemical composition of the bean family plants. Mikroelementy v Zhizni Rast. i Zhivotn. Akad. Nauk SSSR Tr. Konf. po Mikroelementam 1950:546–60. Chem. Abstr. 49:441. 1955.

3896. Chernetskii, A. M. 1936: Experiments with vernalization of sugar beet seed. Selek. i Semen. 1936(8):51–56.

3897. Chernomaz, P. A. 1950: Effect of size of seeds and supplementary artificial pollination on yields of sunflowers. Selek. i Semen. 17(12):72–73.

3898. —— 1956: Presowing treatment of winter wheat grains. Zemledelie 4(8):40–48. Chem. Abstr. 51:11474. 1957.

Chernomordikova, L. A. 2259.

3899. Chernov, G. N. 1955: The seed-properties of the hyaline and floury grains of soft wheat. Zemledelie 1955(3):83–85. Biol. Abstr. 32:9769. 1958.

3900. Chernov, P. I. 1953: Effect of warming on field germination of seeds and productivity of plants. Agrobiologiya 1953(4): 117–24.

3901. Chernyshev, I. A. 1956: Experiment in aerial seeding of pine in the Ural area. Lesn. Khoz. 9:80–81.

3902. Cherry, J. H. 1961: Biochemical and physiological aspects of growth in X-irradiated corn seed. Dissertation Abstr. 22:1383–84.

3903. —— 1963: Nucleic acid changes in the storage time of seeds during germination. Biochim. Biophys. Acta 68:193–98.

3904. ——, R. H. Hageman, F. I. Collins, and D. Flesher. 1961: Effects of X-irradiation on corn seed. Plant Physiol. 36:566–72.

Cherry, J. H. 1663.

3905. Chesnais, F. 1943: Sur la formation de la cicatrice de la graine chez les Sapotacées. Bull. Soc. Botan. France 90:177–81.

3906. Chessin, M. 1960: Germination studies of seeds of Northern Rocky Mountain plants. Proc. Montana Acad. Sci. 19:112–17.

3907. Chessmore, R., and J. S. Brooks. 1948: Should hybrid corn be planted at a higher rate than open-pollinated varieties? Oklahoma Agr. Expt. Sta. Bull. 319:32–41.

3908. Chester, K. S. 1938: Gravity-grading, a method for reducing seed-borne disease in cotton. Phytopathology 28:745–49.

3909. —— 1941: Custom operation of the Oklahoma hot-water grain seed treater. Phytopathology 31:5.

3910. —— 1948: Vapor-heat treats seed inside. New method when followed by dust or slurry treatment, assures complete seed protection. Southern Seedsman 11(4): 28, 32, 36.

3911. Chevalier, A. 1949: La cortisone et sa production par les graines de *Strophanthus sarmentosus*. Rev. Intern. Botan. Appl. Agr. Trop. 29:605–7.

Chevelii, M. 6528.

3912. Chi, Y. F., Y. S. Kao, and K. J. Chang. 1936: The alkaloids of *Fritillaria roylei*. I. Isolation of peimine. J. Am. Chem. Soc. 58:1306–7.

Chiang, H. C. 118.

Chiang, S. M. 18360.

3913. Chiappelli, R. 1932: Wood's rays and the physiological selection of rice seed. Giorn. Risicoltura 22:171–73. Chem. Abstr. 27:1657. 1933.

3914. —— 1935: Selection of rice and seed control. Giorn. Risicoltura 25:145–47. Biol. Abstr. 11:6740. 1937.

3915. —— 1951: Germinability of rice. Risicoltura 39:9–13. Brit. Abstr. B III: 324. 1951.

3916. —— 1951: Germinating energy of rice and the new rice season. Risicoltura 39:115–16.

3917. Chiappini, M. 1954: Research on the embryonic development of some species of the genus *Centaurea* L. (Asteraceae). Sassari Univ. Ist. Botan. Boll. 4:274–89. 1951.

3918. —— 1962: Caryoembryological studies in *Ornithogalum* (Liliaceae) of Sardinia. I. Sporogenesis and development of the gametophytes in *Ornithogalum caudatum* (Ait.). Nuovo Giorn. Botan. Ital. 69:91–102. Biol. Abstr. 44:7853. 1963.

3919. Chiarugi, A. 1928/30: Osservazioni embriologiche sulla partenocarpia in *Zizyphus sativa* Gaertn. Atti Soc. Ital. Prog. Sci., 18th Riunione, Florence, September, 1928; Partenocarpia in *Zizyphus sativa* Gaertn. Nuovo Giorn. Botan.Ital. 37:287–312. 1930. Biol. Abstr. 6:5177. 1932.

3920. Chiasson, T. C. 1961: Awn brittleness in barley. Can. J. Plant Sci. 41:759–62.

3921. Chick, H., and E. M. Delf. 1919: The anti-scorbutic value of dry and germinated seeds. Biochem. J. 13:199–217.

3922. Chidambaram, N., and R. R. Iyer. 1945: Chemical examination of the seeds of *Amaranthus gangeticus*. I. Fatty oil from the seeds. J. Indian Chem. Soc. 22:117–18.

3923. Chien, H. 1963: "Lard fruit" domesticated in China. Euphytica 12:261–62.

3924. Chien, N. F., and C. N. Kang. 1960: Some observations on the endosperm development in angiosperms. Acta Botan. Sinica 9:107–11. Biol. Abstr. 35:64124. 1960.

3925. Chigirev, S. D. 1955: Influence of deep freezing during irradiation on the synthesis of glutathione and ascorbic acid in wheat seeds. Dokl. Akad. Nauk SSSR 101:1055–57. Chem. Abstr. 50:2763. 1956.

Chigirev, S. D. 17595.

Chih, F. C. 18358.

3926. Chikalov, P. M. 1950: Improvement of seeding characteristics of coriander seeds. Agrobiologiya 1950(4):119–25.

3927. Chikalova, E. A. 1959: Effect of soaking seeds in vitamin solutions on ascorbic acid accumulation. Vitaminy Akad. Nauk Ukr. SSR Inst. Biokhim. 4:206–8. Chem. Abstr. 54:10075. 1960.

3928. Chikasuye, M. 1954: Biochemical studies on cereal seeds. V. Changes of catalase activity during the germination and maturation of cereal seeds. J. Agr.

Chem. Soc. Japan (English Transl.) 28: 105–11. Chem. Abstr. 50:7958. 1956.

3929. Chikubu, S. 1963: Studies on thiamine of cereals and beans by the photographic method. VIII. The distribution of thiamine in barley and naked barley seedlings at different stages of germination in the dark. Tokyo Food Res. Inst. Rept. 17:75–79.

3930. —— and H. Horiuchi. 1955: Thiamine of cereals and legumes studied by the photographic method. IV. Distribution of thiamine in the seed of kidney beans. J. Agr. Chem. Soc. Japan (English Transl.) 29:225–29. Chem. Abstr. 51:8903. 1957.

3931. ——, ——, M. Furuya, and S. Niizeki. 1956: Thiamine of cereals and legumes studied by the photographic method. V. Distribution of thiamine in kidney bean seedlings at different stages of germination in the dark. J. Agr. Chem. Soc. Japan (English Transl.) 30:7–11. Chem. Abstr. 51:8904. 1957.

3932. ——, ——, and S. Yanai. 1958: Thiamine of cereals and legumes studied by the photographic method. VI. Distribution of thiamine in rice seedlings at different stages of germination in the dark. J. Agr. Chem. Soc. Japan (English Transl.) 32:272–75. Chem. Abstr. 54: 21343. 1960.

3933. Child, R. 1935: The seed oil of *Aegle marmelos*, Corr. J. Am. Chem. Soc. 57:356–57.

3934. —— and W. R. N. Nathanael. 1942: The seed fat of *Litsea longifolia* Bth. & Hk. J. Am. Chem. Soc. 64:1079–81.

3935. —— and —— 1944: *Hydnocarpus* oils in Ceylon. II. Trop. Agriculturist 99:140–42. Brit. Abstr. B III:259. 1945.

3936. Childs, J. F. L., and G. Hrnciar. 1949: A method of maintaining viability of citrus seed in storage. Proc. Florida State Hort. Soc. 61:64–69.

3937. Chilton, S. J. P. 1940: The occurrence of *Helminthosporium turcicum* in the seed and glumes of Sudan grass. Phytopathology 30:533–36.

3938. —— 1942: Some pathogenic fungi occurring in the seed of red and subterranean clover. Phytopathology 32:738–39.

3939. ——, D. C. Bain, and L. H. Person. 1942: Effect of seed treatment on stands of ornamental plants. Proc. Louisiana Acad. Sci. 7:36. Biol. Abstr. 18:11998. 1944.

3940. —— and R. J. Garber. 1941: Effect of seed treatment on stands of some forage legumes. J. Am. Soc. Agron. 33:75–83.

3941. —— and T. C. Ryker. 1947: Seed treatment of rice. Louisiana Agr. Expt. Sta. Bull. 412. 8 pp.

Chima, I. S. 12549.

3942. Chin, C. L., and S. Y. Shen. 1937: A study of rice borer injury and its effect on the quality of rice and grain germination. J. Agr. Assoc. China (Taipei) 162:15–26. Lingnan Sci. J. 16:647. 1937.

Chin, K. C. 2603.

3943. Ching, K. K., and D. N. Bever. 1960: Provenance study of Douglas fir in the Pacific Northwest region. I. Nursery performance. Proc. Soc. Am. Foresters 1959:39. Biol. Abstr. 35:39360. 1960.

Ching, K. K. 3954.

3944. Ching, T. M. 1956: Seed quality as affected by moisture and temperature in storage. Oregon Seed Growers League Proc. 16:35–36.

3945. Ching, T. M. 1958: Safer seed storage. Crops & Soils 11(3):16–17.

3946. —— 1958: Some experiments on the optimum germination conditions for Western hemlock (*Tsuga heterophylla* Sarg.). J. Forestry 56:277–79.

3947. —— 1959: Activation of germination in Douglas fir seed by hydrogen peroxide. Plant Physiol. 34:557–63.

3948. —— 1959: Keeping field seed quality in storage and in transit. Western Feed Seed 14:21–22.

3949. —— 1960: Seed production from individual cones of Grand fir (*Abies grandis* Lindl.). J. Forestry 58:959–61.

3950. —— 1961: Respiration of forage seed in hermetically sealed cans. Agron. J. 53:6–8.

3951. —— 1961: Science in the garden: Hydrogen peroxide helps seed germination. Calif. Hort. Soc. J. 2:156–57.

3952. —— 1963: Changes of chemical reserves in germinating Douglas fir seed. Forest Sci. 9:226–31. Chem. Abstr. 59: 6722. 1963.

3953. —— 1963: Fat utilization in germinating Douglas fir seed. Plant Physiol. 38:722–28.

3954. —— and K. K. Ching. 1962: Physical and physiological changes in maturing Douglas fir cones and seed. Forest Sci. 8:21–31.

3955. —— and W. H. Foote. 1961: Post-harvest dormancy in wheat varieties. Agron. J. 53:183–86.

3956. —— and M. C. Parker. 1958: Hydrogen peroxide for rapid viability tests of some coniferous tree seeds. Forest Sci. 4:128–34.

3957. ——, ——, and D. D. Hill. 1959: Interaction of moisture and temperature on viability of forage seeds stored in hermetically sealed cans. Agron. J. 51:680–84.

3958. ——, I. Schoolcraft, P. Rowell, H. L. Taylor, and B. Davidson. 1963: Change of forage seed quality in commercial warehouses in Western Oregon. Agron. J. 55:379–82.

3959. ——, H. L. Taylor, and P. T. Rowell. 1960: Change of forage seed quality under different simulated shipping conditions. Agron. J. 52:37–40.

Ching, T. M. 4612.

3960. Chinn, S. H. F., and R. C. Russell. 1958: The control of soaking injury of barley seed. Phytopathology 48:553–56.

Chinn, S. H. F. 16572.

3961. Chinoy, D., and S. N. Sharma. 1957: The development of the seed in cereals. Variety differences in the growth of wheat seeds in three directions and the effect of environment. Agrobiologiya 1957(3):32–41. Biol. Abstr. 35:16993. 1960.

3962. Chinoy, J. J. 1942: Pre-sowing treatment and phasic development. Current Sci. (India) 11:400–2.

3963. —— 1950: Effect of vernalization and photoperiodic treatments on growth and development of wheat. Nature 165:882–83.

3964. —— 1956: Determination of photothermic and vernalization quanta for the vegetative period of wheat. Physiol. Plantarum 9:1–18.

3965. —— 1962: Formation and utilization of ascorbic acid in the shoot apex of wheat as factors of growth and development. Indian J. Plant Physiol. 5:172–201.

Chinoy, J. J. 14168–14173.

3966. Chinte, F. O. 1951: Pre-treatment of leguminous seeds of hasten germination. Filipino Forester 3:29–44.

3967. Chippindale, H. G. 1931: "Suction-force" measurements on the seeds of some strains of grasses. Welsh J. Agr. 7:168–82. Chem. Abstr. 25:5238–39. 1931.

3968. —— 1933: The effect of soaking in water on the "seeds" of *Dactylis glomerata* L. Ann. Botany (London) 47:841–49.

3969. —— 1933: The effect of some chemicals on germination in cocksfoot (*Dactylis glomerata* L.). Ann. Appl. Biol. 20:369–76.

3970. —— 1934: The effect of Ceresan on the germination of grass seeds. Welsh J. Agr. 10:284–88. Chem. Abstr. 29:1204. 1935.

3971. —— 1934: The effect of soaking in water on the "seeds" of some Gramineae. Ann. Appl. Biol. 21:225–32.

3972. —— 1934: Note on a device for supporting seeds in contact with liquids. Proc. Intern. Seed Testing Assoc. 6:1–3.

3973. —— 1949: Environment and germination in grass seeds. Brit. Golf Unions Board Greenkeeping Res. J. 7:283–84.

3974. —— and W. E. J. Milton. 1934: On the viable seeds present in the soil beneath pastures. J. Ecol. 22:508–31.

3975. Chirilei, H. 1954: Accélération par traitement thermique de la germination des semences de quelques essences forestières. Acad. Rep. Populare Romine Bul. Stiint. 6:1037–45.

3976. Chirkov, V. I. 1950: Seed germination of some species of apples. Botan. Zh. 35:387–94.

3977. Chirkovskii, V. I. 1939: Effect of temperature and atmospheric humidity on the seed of tobacco and of *Nicotiana rustica* during storage. Vses. Nauchn.-Issled. Inst. Tabach. Makhoroch. Promy. 138:78–92. Chem. Abstr. 34:6329. 1940.

3978. —— 1952: How to safeguard germination of tobacco and makhorka seed in storage. Tabak (USSR) 13(5):43–46.

3979. —— 1953: Influence of the age of seed on development of the plant in tobacco. Dokl. Akad. Nauk SSSR 92: 439–42. Biol. Abstr. 29:1673. 1955.

3980. —— 1956: On the role of the seed in studying the evolution of tobacco. Sb. Nauchn.-Issled. Rabot Krasnodar Vses. Nauchn.-Issled. Inst. Tabaka Makhorki A. I. Mikoiana 149:131–46.

3981. —— 1956: Effect of DDT on germination of tobacco seed. Tabak (USSR) 16(2):22–23. Tobacco Abstr. 1:425. 1957.

3982. —— 1958: On the frost resistance of tobacco and makhorka seeds. Tabak (USSR) 19(4):56–58.

3983. —— 1959: Improving the germination of newly harvested tobacco seed. V. Tabak (USSR) 20(4):41–48.

3984. —— 1960: Obtaining tobacco and Aztec tobacco seeds in zones of uncertain success in raising these crops for seeds. Agrobiologiya 1960(4):539–41.

3985. —— 1960: The role of ecological factors in the light-sensitivity of tobacco seeds during germination. Vestn. Sel'-skokhoz. Nauki 1960(5):124–25.

3986. —— 1961: Overcoming non-germination of the seeds of tobacco and of other *Nicotiana* species by means of gibberellin.

Tabak (USSR) 22(4):42–45. Tobacco Abstr. 6:347. 1962.

Chisaka, H. 1303.

3987. Chisholm, M. J., and C. Y. Hopkins. 1953: Fatty acids of filbert oil and nasturtium seed oil. Can. J. Chem. 31:1131–37.

3988. —— and —— 1956: Composition of delphinium seed oil. Can. J. Chem. 34:459–64.

3989. —— and —— 1957: An oxygenated fatty acid from the seed oil of *Hibiscus esculentus*. Can. J. Chem. 35:358–64.

3990. —— and —— 1958: Fatty acids of the seed oil of *Cardiospermum halicacabum*. Can. J. Chem. 36:1537–40.

3991. —— and —— 1960: 11-Octadecenoic acid and other fatty acids of *Asclepias syriaca* seed oil. Can. J. Chem. 38:805–12.

Chisholm, M. J. 9017–9023.

3992. Chistoni, A. 1931: The oil and alkaloid content of the seeds of *Papaver somniferum*. Arch. Farmacol. Sper. 52: 29–32. Chem. Abstr. 25:4084. 1931.

3993. Chistov, S. A. 1939: The physicochemical properties of wheat which is affected by frost. Izv. Tomsk. Mukomol'no-Elevator-nogo Inst. 4:97–135. Chem. Abstr. 37:1469. 1943.

Chistyakova, L. D. 10393.

3994. Chitaly, S. D. 1958: Seeds of *Viracarpon hexaspermum* Sahni from the intertrappean beds of Mohgaon Kalan, India. J. Indian Botan. Soc. 37:408–11.

3995. Chitre, R. G., D. B. Desai, and V. S. Raut. 1956: Niacin biosynthesis in germinating cereals and pulses. I. Changes in niacin metabolites during germination. Indian J. Med. Res. 44:217–26. Chem. Abstr. 50:13179–80. 1956.

3996. ——, S. Ganapathy, D. B. Desai, J. S. Kumana, and S. M. Vallury. 1956: Nutritive value of purebred strains of cereals and pulses. III. Essential amino-acid contents of purebred strains of cereals and pulses. Indian J. Med. Res. 44:573–76. Chem. Abstr. 51:3753. 1957.

Chitre, R. G. 5225, 16014.

3997. Chitrowo, W. 1914: Atlas von Samen und Früchten der Feldunkräuter aus Mittelrussland. Bull. Angew. Botan., St. Petersburg 7:181–202. Botan. Centralbl. 126:487. 1914.

3998. Chmelar, F. 1921: L'influence de la sécheresseanormale de l'année 1921 sur la faculté germinative de céréales. Publ. Inst. Rech. Agron. Brno. C 75, Publ. d. 1. Sect. Essais Semences C. 1:25–27.

3999. —— 1925: Organisation des essais collectifs de longue durée et leur importance pour l'agriculture. Tirage Apart du Compt. Rend. 12th Intern. Congr. Agr. Varsovie, pp. 93–99.

4000. —— 1928: Keimungsverlauf von quellungsunfähigen (harten) Samen bei Rotklee und Luzerne, ausgesät im Keimapparat und auf dem Felde. Tschech. Land. Akad. 4(2):1–10.

4001. —— 1934: The possibilities of accelerating seed analysis and the determination of variety by employing luminescence tests in ultra-violet light. Proc. Intern. Seed Testing Assoc. 5:435–45.

4002. —— 1947: Germination of imbibition resistant (hard) seeds of red clover (*Trifolium pratense*) placed into the germinator before 26 years. Sb. Česk. Akad. Zeměděl. 19:231–32.

4003. —— and F. Mikolasek. 1931: The growing ability and yields of plants from some leguminous grains resisting to swelling imbibition. Mitt. Tschech. Akad. Landwirtsch. 7(6/7). 10 pp.

4004. —— and K. Mostovoj. 1933: A fast laboratory method for distinguishing of American alpha sweet clover from the unimproved white sweet clover after their growth under permanent light. Věstn. Česk. Akad. Zeměděl. 9:510–15.

4005. —— and —— 1934: A quick method for distinguishing of soybean varieties and clover species after luminescence of germinated grains. Věstn. Česk. Akad. Zeměděl. 10:289–95.

4006. —— and —— 1938: On the application of some old and on the introduction of new methods for testing genuineness of variety in the laboratory. Proc. Intern. Seed Testing Assoc. 10:68–74.

4007. —— and J. Simon. 1934: Ungewöhnliches Auftreten von Ausschossern bei Hackfrüchten im Jahre 1933 und dessen Ursachen. Mitt. Tschech. Akad. Landwirtsch. 10:302–6c.

4008. Chmyr, A. D. 1959: The effect of Co-60 λ-ray radiation on some biochemical and physiological properties of corn kernels. Ukr. Biokhim. Zh. 31:671–79. Chem. Abstr. 54:13284. 1960.

4009. Cho, J. 1938: The anatomical observation of the embryo in the rice. Botan. Mag. (Tokyo) 52:520–31. Botan. Centralbl. 176:101. 1940.

4010. —— 1956: Double fertilization in *Oryza sativa* L. and development of the endosperm with special reference to the aleuron layer. Bull. Natl. Inst. Agr. Sci. Ser. D 1956(6):61–101.

4011. Choate, H. A. 1921: Chemical changes in wheat during germination. Botan. Gaz. 71:409–25.

4012. —— 1940: Dormancy and germination in seeds of *Echinocystis lobata.* Am. J. Botany 27:156–60.

4013. Chodat, F., and C. Weber. 1942: Influence de la fluorescence sur la germination. Compt. Rend. Soc. Phys. Hist. Nat. Genève 59:28–32.

Chodat, F. 6898.

4014. Choe, C. E., C. H. Kim, P. S. Song, and T. Y. Lee. 1959: Variation of vitamin in soybean during germination. Bull. Sci. Res. Inst. (Seoul) 4:181. Chem. Abstr. 54:7835. 1960.

Choe, K. W. 10619.

Chohan, J. G. 15826.

4015. Cholnoky, B. J. 1948: Summary of germination methods for garden and flowerseeds. N.V. Sluis en Groot's Konin. Zaadteelt en Zaadhandel: Enkhuizen, the Netherlands. 98 pp.

4016. —— 1949: Supplement of methods of laboratory germination. N.V. Sluis en Groot's Konin. Zaadteelt en Zaadhandel: Enkhuizen, the Netherlands. 23 pp.

4017. Cholodny, N. 1935: Über das Keimungshormon von Gramineen. Planta 23:289–312.

4018. Cholodny, N. G. 1935: Sur la question du rôle des hormones dans la germination des semences. Sov. Botan. (Leningrad) 2(1):19–38.

4019. —— 1936: Hormonization of grains. Compt. Rend. Acad. Sci. URSS 3:439–42.

4020. —— and E. C. Sankewitsch. 1937: Influence of weak electric currents upon the growth of the coleoptile. Plant Physiol. 12:385–408.

4021. Chopinet, R. 1953: La conservation des semences dans les jardins botaniques. Ann. Biol. 57:283–90.

4022. —— 1953: La conservation des semences en atmosphère sèche; son intérêt en sélection. Intern. Hort. Congr. London 1952 Rept. 13:1033–43.

4023. —— and G. Trebuchet. 1949: Essai de classification et d'identification des principales variétés de haricots cultivées en France. Rev. Hort. 121:102–3.

Chopra, I. C. 10243, 14239.

4024. Chopra, R. N. 1953: The endosperm in some Cucurbitaceae. Current Sci. (India) 22:383–84.

4025. —— 1962: Effect of some growth substances and calyx on fruit and seed development of *Althaea rosea* Cav. *In* Plant Embryology. Symposium held at the Univ. of Delhi. Council Sci. Ind. Res., New Delhi 1962:170–81. Biol. Abstr. 43: 7507. 1963.

Chopra, R. N. 3848, 12704, 16600.

4026. Chou, S. L. 1956: Experiment on the germination of seeds of beef-wood (*Casuarina equisetifolia*, Forst.). J. Agr. Forestry 5:56–67.

Chou, T. Q. 3824.

Chou, Y. L. 3824.

4027. Chouard, P. 1950/51: Débat sur la vernalisation. Bull. Soc. Botan. France Mem. 1950/51:67–81.

4028. —— 1954: Dormances et inhibitions des graines et des bourgeons; preparation au forçage; thermoperiodisme. Les Cours de Sorbonne, pp. 1–157.

4029. —— 1956: Colloqué consacré à la vernalisation. Bull. Soc. Franc. Physiol. Végétale 2:125–73.

4030. —— 1957: Réversibilité de l'état reproductif a l'état végétatif par le photoperiodisme, et production expérimentale de fleurs prolifères chez *Anagallis arvensis.* Compt. Rend. 245:2351–53.

4031. —— 1957: La journée courte ou l'acide gibberellique comma succédanés du froid pour la vernalisation d'une plante vivace en rosette, le *Scabiosa succisa* L. Compt. Rend. 245:2520–22.

4032. —— 1958: Diversité des mécanismes des dormances, de la vernalisation et du photopériodisme, revelée notamment par l'action de l'acide gibberellique. Bull. Soc. Botan. France Mem. 1956/57:51–64.

4033. —— 1960: Vernalization and its relations to dormancy. Ann. Rev. Plant Physiol. 11:191–238.

4034. —— and P. Poignant. 1951: Recherches preliminaries sur la vernalisation en presence d'inhibiteurs de germination et de respiration. Phyton 4:11–13.

4035. —— and G. Teissier. 1932: Relations entre la croissance des diverses parties de la plantule du melon et la quantité de réserves laissées a la disposition de l'embryon. Compt. Rend. 194:1513–15.

4036. —— and —— 1933: Variations de l'intensité de croissance chez les plantules de melon au cours du développement et en fonction de la quantité de réserves disponsibles. Compt. Rend. 194:1976–78. Biol. Abstr. 7:7822. 1933.

4037. —— and M. R. Weber. 1956: Un nouveau type de vernalisation chez une plante vivace, la benoîte, *Geum urbanum* L. Compt. Rend. 243:1659–61.

4038. Choudbury, B. 1955: Embryo culture technique. I. Growth of immature tomato embryo in vitro. Indian J. Hort. 12:143–51. Chem. Abstr. 51:3750. 1957.

4039. —— 1955: Embryo culture technique. II. "Embryo factors" and immature tomato embryo. Indian J. Hort. 12:152–54. Chem. Abstr. 51:3750. 1957.

4040. —— 1955: Embryo culture technique. III. Growth of hybrid embryos (*Lycopersicon esculentum* × *L. peruvianum*) in culture medium. Indian J. Hort. 12:155–56. Chem. Abstr. 51:3750. 1957.

4041. —— 1958: Development of seed and its parts. I. Of *Lycopersicon esculentum* Linn. Indian J. Hort. 15:72–75.

4042. Choudhari, B. L. 1953: Seedless papaya. Indian Farming 3(9):14–15.

4043. Choudhri, R. S., and V. B. Bhatnagar. 1953: Effect of salts of certain minor elements on seed germination. J. Sci. Res. Banaras Hindu Univ. 3:76–85.

4044. —— and B. K. Gaur. 1954/55: Effect of certain phytohormones on the germination of tomato seeds. J. Sci. Res. Banaras Hindu Univ. 5:60–68. Chem. Abstr. 49:8396. 1955.

Choudhri, R. S. 17492–17494.
Choudhury, D. 13881.
Choudhury, J. M. 4927, 4928.

4045. Choudhury, M. R. 1961: On the structure of embryo-sac in 4N *Corchorus capsularis* L. (strain: D154). Indian Agriculturist 5:29–31.

Choudhury, R. B. R. 1397.
Choumakov, B. A. 12319.
Chow, A. Y. 17473.
Chowdhuri, D. K. 3759, 3760.

4046. Chowdhury, D. K., and R. Bagchi. 1956: Composition of the seed fat of *Cucumis melo*. Naturwissenschaften 43: 350–51.

4047. —— and —— 1956: Composition of the seed fat of *Amaranthus gangeticus*. Naturwissenschaften 43:538.

4048. —— and —— 1957: Composition of the seed fats of *Vigna catjang*. Naturwissenschaften 44:263.

4049. ——, M. M. Chakrabarty, and B. K. Mukherji. 1955: Studies on some seed fats of Cucurbitaceae family. J. Am. Oil Chemists Soc. 32:384–86.

4050. Chowdhury, N. P. 1952: A note on the multiple seeded rice of India. Indian Forester 75:494–99. 1949.

4051. Christensen, B. W., and A. Kjaer. 1963: A mustard oil of *Hesperis matronalis* seed, 6-methylsulphinylhexyl isothiocyanate. Acta Chem. Scand. 17:846–47.

Christensen, B. W. 10745.

4052. Christensen, C. M. 1947: The mold population of corn and wheat and its relation to heating. Am. J. Botany (Suppl.) 34(10):15a.

4053. —— 1951: Fungi on and in wheat seed. Cereal Chem. 28:408–15.

4054. —— 1955: Grain storage studies. XVIII. Mold invasion of wheat stored for 16 months at moisture contents below 15 per cent. Cereal Chem. 32:107–16.

4055. —— 1955: Grain storage studies. XXI. Viability and moldiness of commercial wheat in relation to the incidence of germ damage. Cereal Chem. 32:507–18.

4056. —— 1957: Deterioration of stored grains by fungi. Botan. Rev. 23:108–34.

4057. —— 1962: Invasion of stored wheat by *Aspergillus ochraceus*. Cereal Chem. 39:100–6.

4058. —— and R. F. Drescher. 1954: Grain storage studies. XIV. Changes in moisture content, germination percentage, and moldiness of wheat samples stored in different portions of bulk wheat in commercial bins. Cereal Chem. 31:206–16.

4059. —— and D. R. Gordon. 1948: The mold flora of stored wheat and corn and its relation to heating of moist grain. Cereal Chem. 25:40–51.

4060. —— and P. Linko. 1963: Moisture contents of hard red winter wheat as determined by meters and by oven drying, and influence of small differences in moisture content upon subsequent deterioration of the grain in storage. Cereal Chem. 40:129–37.

4061. —— and L. C. Lopez F. 1963: Pathology of stored seeds. Proc. Intern. Seed Testing Assoc. 28:701–11.

4062. ——, J. H. Olafson, and W. F. Geddes. 1949: Grain storage studies. VIII. Relation of molds in moist stored cottonseed to increased production of carbon dioxide, fatty acids, and heat. Cereal Chem. 26:109–28.

4063. —— and S. A. Qasem. 1959: Note on a rapid method of detecting germ damage in wheat and corn. Cereal Chem. 36: 461–63.

Christensen, C. M. 2845, 2846, 5168, 5678, 7068, 7300, 7415, 7447, 7677,

9208, 11060, 12546, 13513–13515, 14742, 15039–15041, 15244, 15788, 15789, 17752, 18812–18814.

4064. Christensen, J. J. 1936: Associations of microorganisms in relation to seedling injury arising from infected seed. Phytopathology 26:1091–105.

4065. —— 1949: The relation of seed-borne microorganisms to certification. Intern. Crop Improvement Assoc. Ann. Rept. 30:38–40.

4066. —— 1963: Variability of the microflora in barley kernels. Plant Disease Reptr. 47:635–38.

4067. —— 1963: Longevity of fungi in barley kernels. Plant Disease Reptr. 47:639–42.

4068. —— and M. B. Moore. 1942: Variability in response of flax to seed treatment. Phytopathology 32:2.

4069. —— and E. C. Stakman. 1935: Relation of *Fusarium* and *Helminthosporium* in barley seed to seedling blight and yield. Phytopathology 25:309–27.

Christensen, J. J. 11057–11059.

4070. Christensen, N. A. 1954: Seed crop trials. Tidsskr. Froavl 42:282–90.

4071. —— 1956: Seed crop field trials. Tidsskr. Froavl 44:330–35.

Christensen, R. 13099.

4072. Christenson, K. 1957: The future of the seed business in North America. Seed World 81(5):8, 10–11, 30, 32.

4073. Christian, B. C., and T. P. Hilditch. 1929: Seed fats of the Umbelliferae. II. The seed fats of some cultivated species. Biochem. J. 23:327–38.

4074. Christian, W. E., Jr., and B. K. Doyle. 1950: Mississippi seed industry. Mississippi Agr. Expt. Sta. Circ. 153. 14 pp.

4075. Christiansen, M. N., and N. Justus. 1963: Prevention of field deterioration of cottonseed by an impermeable seed coat. Crop Sci. 3:439–40.

4076. —— and R. P. Moore. 1958: Seed coat structural differences that influence water uptake and seed quality in hard seed cotton. Agron. J. 51:582–84.

4077. ——, ——, and C. L. Rhyne. 1960: Cotton seed quality preservation by a hard seed coat characteristic which restricts internal water uptake. Agron. J. 52:81–84.

4078. ——, ——, and —— 1960: Potential of hard-seeded cotton as a field and storage protectant against loss of seed quality. Cotton Improvement Conf. Proc. 12:44–51.

4079. Christidis, B. G. 1936: Cottonseed treatment with sulfuric acid. J. Agr. Sci. 26:648–63. Chem. Abstr. 31:2340–41. 1937.

4080. —— 1940: The viability of cotton seed as affected by its moisture-content. Empire J. Exptl. Agr. 8:148–58.

4081. —— 1947: Cotton seed treatment for controlling seedling diseases. Cotton Res. Inst. (Sindos, Greece) Sci. Bull. 1:1–32. Biol. Abstr. 23:24985. 1949.

4082. —— 1945: Seed vitality and other cotton characters as affected by the age of seed. Plant Physiol. 29:124–31.

4083. —— 1955: Dormancy in cotton seed. Agron. J. 47:400–3.

4084. Christie, B. R., and R. R. Kalton. 1959: Sampling for seed weight determinations in bromegrass, *Bromus inermis* Leyss. Can. J. Plant Sci. 39:158–63.

4085. —— and —— 1960: Inheritance of seed weight and associated traits in bromegrass, *Bromus inermis* Leyss. Can. J. Plant Sci. 40:353–65.

4086. —— and —— 1960: Recurrent selection of seed weight in bromegrass, *Bromus inermis* Leyss. Agron. J. 52:575–78.

Christie, W. 9991.

4087. Christisen, D. M. 1955: Yield of seed by oaks in the Missouri Ozarks. J. Forestry 53:539–441.

Christisen, D. M. 3401.

4088. Christoph, H. 1921: Untersuchungen über die Mykotrophen Verhältnisse der "Ericales" und die Keimung von Pirolaceen. Botan. Centralbl. Beih. 38:115–57.

4089. Chroboczek, E. 1934: A study of some ecological factors influencing seed-stalk development of beets (*Beta vulgaris* L.). Cornell Univ. Agr. Expt. Sta. Mem. 154. 84 pp.

4090. —— 1955: Influence of the age of plants at the beginning of cold treatment and of the length of cold treatment on seedstalk development and yield of seed of two white cabbage varieties. 14th Intern. Hort. Congr. 1955, Rept. 1:429–39.

4091. —— and J. Kostecki. 1937: Influence of a relatively low temperature in seedling stages in the greenhouse in comparison with seed vernalization on "bolting" in beets (*Beta vulgaris* L.). Roczniki Nauk Ogrodn. Warsaw 4:153–75.

4092. Chroust, M. 1961: The effect of covering depth on soil germinating power and necessary (normalized) seed volume of larch and pine seeds. Sb. Vysoke Skoly Zeměděl. Lesnicke Fak. Prague 4:121–60.

4093. Chrzaszcz, T., and J. Janicki. 1932: Das Stärkeverflüssigungsvermögen der Amylase and seine Bestimmungsmethoden. Biochem. Z. 256:252–91.

4094. —— and —— 1933: Die wirkliche und scheinbare Amylasemenge in verschiedenen Getreidearten vor und während der Keimung. Biochem. Z. 265:260–81.

4095. —— and —— 1934: Simultaneous action of proteolytic enzymes in increasing the amounts of amylase of various kinds of grain. Polish Agr. Forestal Ann. 33:191–200. Chem. Abstr. 29:1110. 1935.

4096. —— and —— 1936: Amylase during the growth and ripening of grains. Biochem. J. 30:1298–302.

4097. —— and —— 1936: Increase in the amount of active amylase in ungerminated cereals by means of hydrogen sulfide and papain. Biochem. Z. 285:47–54. Chem. Abstr. 30:4881. 1936.

4098. —— and —— 1936: Amylase of rye during the development and ripening of the seeds. Biochem. Z. 286:13–19.

4099. Chu, A. C., G. S. King, and G. D. Sherman. 1953: *Macadamia* storage studies. Hawaii Agr. Expt. Sta. Progr. Notes 90:1–9. Biol. Abstr. 29:2963. 1955.

4100. Chu, C., and P. S. Tang. 1960: Studies on plant respiration. IV. Some morphological, physiological and biochemical changes during germination of rice at different partial pressures of oxygen. Acta Botan. Sinica 8:201–14. Biol. Abstr. 35:64326. 1960.

4101. Chu, K. H., and J. O. Culbertson. 1952: Studies of inheritance of seed size and other characters in a cross between an Indian and a North American variety of flax. Agron. J. 44:26–30.

Chu, P. S. 6213.

4102. Chu, T. H. 1963: The proteolytic enzyme of mung bean seeds, *Phaseolus aureus*. Acta Biochim. Biophys. Sinica 3:139–46. Chem. Abstr. 59:10397. 1963.

4103. Chu, Y., S. M. To, and L. Wang. 1957: The proper use of arsenic for treating wheat seed. Acta Phytopathol. Sinica 3:177–82. Rev. Appl. Mycol. 43:405. 1958.

Chudinui, E. 4104.

4104. Chudinui, O., and E. Chudinui. 1909: The influence of various salts on the germination of corn. Izv. Moskov Sel'skokhoz. Inst. 15:230–34. Expt. Sta. Rec. 22:228. 1910.

4105. Chudjakoff, J. P., and E. A. Raznicina. 1939: Die Anwendung mykolytischer Bakterien zur Bakterisation des Saatgutes bei Jarowisation. Ber. Akad. Wiss. SSSR Biol. Ser. 1939(1):117–19. Centralbl. Bakteriol. Parasitenk. Abt. II. 102:281–82.

4106. Chudoba, Z. 1963: The influence of temperature of the medium on seed vitality of some cereal plants and grasses. Howdowla Roslin Aklim. Nasiennictwo 7:105–15.

Chudoba, Z. 12339.

4107. Chujo, H. 1960: Studies in short day vernalization of barleys. I. Differences in short day vernalization effects due to the degree of plant development. Crop Sci. Soc. Japan Proc. 29:114–16.

4108. —— 1961: The effect of temperature on the low temperature vernalization of barley. Bull. Univ. Osaka Ser. B12:55–64.

Chujo, M. 14854.

Chukanov, V. I. 11331.

4109. Chukhnin, I. A. 1955: Early winter sowing as a method for improvement of sunflower seed quality. Agrobiologiya 1955(2):114–15.

4110. Chumakov, A. E., and V. A. Smirnov. 1961: Rust in cereal grains. Distribution of pests and diseases of agricultural crops in USSR in 1960 and a prediction of their appearance in 1961. Leningrad, pp. 108–12. Biol. Abstr. 43:25020. 1963.

Chumbalov, T. K. 18378.

4111. Chung, D. S., L. T. Fan, and J. Shellenberger. 1961: Volume increase of wheat kernels accompanying absorption of liquid water. J. Biochem. Microbiol. Technol. Eng. 3:377–93.

Chung, D. S. 6214.

4112. Church, T. W., Jr., and E. I. Sucoff. 1960: Virginia pine seed viable two months before natural cone opening. U.S. Dept. Agr. Forest Serv. Northeast. Forest Expt. Sta. Forest Res. Notes 102. 4 pp.

Church, T. W., Jr. 18136.

4113. Churchill, B. R. 1944: Smooth bromegrass seed production in Michigan. Mich. Agr. Expt. Sta. Circ. 192:1–22.

4114. Chute, H. M. 1930: The morphology and anatomy of the achene. Am. J. Botany 17:703–23.

4115. Chvojka, L., L. Travnicek, J. Chaloupka, and I. Rihova. 1962: Study of the metabolism of exogenous tryptophane in the apple tree and the estimation of bound tryptophane in apple and pear seeds. Biol. Plant. 4:315–18.

4116. Chytrinski, V. F. 1943: Phenomenon of one side yellowishness of wheat seed and means of its removal. Proc. Lenin. Acad. Agr. Sci. 1943(2):36–42.

4117. Chzhou, C. TS. 1956: Effect of seed position on plants on cabbage productivity. Sb. Stud. Nauchn.-Issled. Rabot Moskov. Ordena Lenina Sel'skokh. Akad. im. K. A. Timiriazeva 6:126–33.

4118. Ciaccio, C. 1954: Histochemical investigations on aleuron. Caryologia 6: 124–27. Biol. Abstr. 30:32721. 1956.

4119. Ciantelli, E. M. 1942: Ricerche istologische ed embriologiche durante l'evoluzione dell'ovuto di *Campsis radicans* (L.) Seem. (Bignonaceae). Nuovo Giorn. Botan. Ital. 49:269–81.

4120. Cielens, E. 1949: Total choline content in seeds of local (Latvian) leguminous plants. Latvijas PSR Zinatnu Akad. Vestis 1949(7):71–76. Chem. Abstr. 48: 231. 1954.

4121. Cieleszky, V. 1947: Determination of copper, arsenic, and mercury in disinfected seeds. Kiserl. Kozl. 47:70–83. Chem. Abstr. 42:3124–25. 1948.

4122. Cieslar, A. 1883: Untersuchungen über den Einfluss des Lichtes auf die Keimung der Samen. Forsch. Gebiete Agrikulturphysik 6:270–95.

4123. —— 1885: Versuche mit Nädelholzsamen. Centralbl. Ges. Forstwesen 11: 510–18.

4124. —— 1897: Versuche über Aufbewahrung von Nädelholzsamen unter Luftdichtem Verschlusse. Centralbl. Ges. Forstwesen 23:162–74.

4125. Ciferri, F., and M. Arnulfo. 1947: The effect of photosensitivity of coumarin in light and darkness on the germination of some seeds. Ist. Botan. Univ. Lab. Crittogam. Pavia Atti 3:315–21. Chem. Abstr. 43:8010. 1949.

Ciferri, F. 4128.

4126. Ciferri, R. 1947/48: The biochemistry of tobacco. V. Effect of treatment of seeds with auxin upon the growth of tobacco. Ist. Botan. Univ. Lab. Crittogam. Pavia, Il Tabacc, Miscellanea 1947/48:1–6. Chem. Abstr. 44:3093. 1950.

4127. —— 1952: Growth of rice and wheat seedlings under microaerobic culture. Notiz. Malattie Piante 19:79–84. Biol. Abstr. 27:26122. 1953.

4128. —— and F. Ciferri. 1950: Activities of 2,4-D and isopropyl-N-phenyl carbamate on the germination of the grain and development of the seedlings of *Panicum crus-galli* and *Oryza sativa*. Risicoltura 38:156–60.

4129. —— and E. Marcelli. 1951: Toxic effect of a mercurial zeolite on tobacco seedlings. Notiz. Malattie Piante 14:11–16. Biol. Abstr. 26:7296. 1952.

4130. ——, G. Scaramuzzi, and A. Vidali. 1950: Preliminary investigation on the germination of sugar beet seeds treated with fungicides and insecticides. Ist. Mal. Piante 10:45–52.

4131. Cincerova-Hejna, A. 1957: Dynamics of free glucides in embryo and endosperm of wheat grains during germination at low temperature. Biologia 12:8–17. Chem. Abstr. 51:8223. 1957.

Cintra, G. 13689.

Ciobanu, I. 16238.

4132. Cioffi, R. M. 1949: Prevention of the decrease of amino acids during alcoholic fermentation. Riv. Viticolt. Enol. (Conegliano) 2:121–23. Chem. Abstr. 43:4807. 1949.

Cioffi, R. M. 1284.

4133. Cionga, E., and C. Iliescu. 1941: The alkaloids from the seeds of *Delphinium consolida* L. Ber. 74B:1031–34. Chem. Abstr. 35:7408–09. 1941.

Circle, S. J. 3107.

Cirino, V. O. 13115.

4134. Cirkovskij, V. I. 1953: Der Einfluss des Alterns der Samen auf die Entwicklung der Pflanzen beim Tabak. Compt. Rend. Acad. Sci. URSS 92:439–42. Ber. Wiss. Biol. 90:78. 1954.

Cist, F. M. 19256.

4135. Cizkova-Pisarovicova, J. 1958: Die Beeinflussung der Keimung bei Erbsen und Maiskörnern durch Thyroxin. Naturwissenschaften 20:496.

4136. Claassen, C. E., W. G. Ekdahl, and G. M. Severson. 1950: The estimation of oil percentage in safflower seed and the association of oil percentage with hull and nitrogen percentages, seed size, and degree of spininess of the plant. Agron. J. 42: 478–82.

4137. Claeys, R. 1950: Changes in the food reserves of germinating peas as a result of treatment with ammonium 2,4-dichlorophenoxyacetate. Natuurw. Tijdschr. (Ghent) 32:115–24. Chem. Abstr. 46: 8195. 1952.

Clagett, C. O. 10848, 17043.

Claire, F. P. Le. 8479.

4138. Clandinin, D. R., and L. Bayly. 1963: Effect of variety and growing conditions on the protein and amino acid composition of rapeseed. Can. J. Animal Sci. 43:65–67. Chem. Abstr. 60:851. 1964.

Clandinin, D. R. 13169, 16132, 16204.

4139. Clare, L. E. A., and E. J. Underwood. 1948: The thiamin and riboflavin contents of wheat and its mill products. Australian J. Exptl. Biol. Med. Sci. 26:419–23.

Clare, L. E. A. 11985.

4140. Clare, T. S., and G. R. Johnstone. 1931: Polyembryony and germination of polyembryonic coniferous seeds. Am. J. Botany 18:674–83.

Clare, T. S. 9909.

Clark, A. 12092.

4141. Clark, B. E. 1943: Comparative laboratory and field germination of onion seed. Proc. Assoc. Offic. Seed Analysts 1942:90–99.

4142. —— 1948: Nature and causes of abnormalities in onion seed germination. Cornell Univ. Agr. Expt. Sta. Mem. 282. 27 pp.

4143. —— 1950: How many plants should a trueness-to-type trial contain? Proc. Assoc. Offic. Seed Analysts 40:89–91.

4144. —— 1954: Factors affecting the germination of sweet corn in low-temperature laboratory tests. N.Y. Agr. Expt. Sta. Bull. 769. 23 pp.

4145. —— 1960: Tentative list of preferred temperatures for germination tests. Assoc. Offic. Seed Analysts News Letter 34(2): 8–11.

4146. ——, W. F. Crosier, B. Cullinan, C. E. Heit, M. T. Munn, and S. R. Patrick. 1949: Germination testing of seeds. N.Y. Agr. Expt. Sta. Ann. Rept. 68:41–42.

4147. —— and C. E. Heit. 1958: Flower seed research bears "fruit." Farm Res. (N.Y.) 24(1):12.

4148. —— and H. L. Page. 1963: The quality and labeling of seeds in New York as revealed by sampling and testing in 1962. N.Y. Agr. Expt. Sta. Bull. 802. 44 pp.

4149. —— and S. H. Wittwer. 1949: Effect of certain growth regulators on seed stalk development in lettuce and celery. Plant Physiol. 24:555–76.

4150. Clark, C. F. 1940: Longevity of potato seed. Am. Potato J. 17:147–52. Expt. Sta. Rec. 83:767. 1940.

Clark, D. C. 2074, 2075.

4151. Clark, D. J., H. H. Hatt, and W. J. Troyahn. 1951: The seed fat of the okari nut (*Terminalia kaernbachii*). Australian J. Appl. Sci. 2:378–84. Biol. Abstr. 27:20460. 1953.

4152. Clark, E. P. 1943: Rotenone and related substances in Berebera seeds. Chromatographic separation of Deguelin and Tephrosin. J. Am. Chem. Soc. 65: 27–29. Analyst 68:158. 1943.

4153. Clark, E. R. 1941: Viability of injured seeds of *Avena fatua* L. Proc. Assoc. Offic. Seed Analysts 33:46–47.

4154. Clark, E. S. 1926: Effects of mutilating soybean seeds on the subsequent growth of the plants. New Jersey Agr. Expt. Sta. Ann. Rept. 46:322–32.

4155. Clark, E. W. 1957: Effect of hot-water treatment for hard cottonseed on a pink bollworm infestation. J. Econ. Entomol. 50:795–96.

4156. Clark, F. B. 1962: White ash, hackberry, and yellow-poplar seed remain viable when stored in the forest litter. Proc. Indiana Acad. Sci. 72:112–14.

4157. —— and F. G. Liming. 1953: Sprouting of blackjack oak in the Missouri Ozarks. U.S. Dept. Agr. Forest Serv. Central States Forest Expt. Sta. Tech. Paper 137. 22 pp.

4158. Clark, F. J., and F. C. Copeland. 1940: Chromosome aberrations in the endosperm of maize. Am. J. Botany 27:247–51.

4159. Clark, F. R. 1931: Sprouting peony seeds. Am. Botanist 37:163.

4160. Clark, F. S. 1952: The development of an isolated area for the production of smut-free barley seed. Agr. Inst. Rev. 7:37–38.

4161. Clark, G. L., and J. L. Gring. 1937: Carotenoids in yellow corn. Quantitative estimation from modified absorption spectra measurements of carotenoids in yellow corn of different varieties and qualities. Ind. Eng. Chem. (Anal. Ed.) 9:271–74.

4162. Clark, H. E., and K. R. Kerns. 1943: Effects of growth regulating substances on a parthenocarpic fruit. Botan. Gaz. 104:639–44.

4163. Clark, J. A. 1956: Collection, preservation and utilization of indigenous strains of maize. Econ. Botany 10: 194–200.

Clark, J. C. 12268.

4164. Clark, J. D., L. W. W. Lehrle, and J. H. G. Smith. 1954: Scarification in Engelmann spruce–alpine fir forests. Brit. Columbia Forest Serv. Res. Notes 24. 7 pp.

4165. Clark, R. B. 1959: Hollies from seed. Bull. Hort. Soc. N.Y. 9(11):5–6.

4166. Clark, R. H. 1958: Direct seeding of

cherrybark red oak (*Quercus pagadaefolia*). Louisiana Univ. School Forestry Proc. Forestry Symp. 7:5–9.

4167. Clark, R. V., and F. J. Zillinsky. 1962: The influence of several fungicidal treatments on yields of oats infected by *Septoria*. Can. J. Plant Sci. 42:620–27.

4168. Clark, T. F., and E. C. Lathrop. 1953: Nutshells and fruit pits: their composition, availability, agricultural and industrial uses. U.S. Dept. Agr. Bur. Agr. Ind. Chem. Northern Regional Res. Lab., Peoria, Ill., AIC Rept. 352. 39 pp.

4169. Clarke, A. E., W. C. Edmundson, and P. M. Lombard. 1941: Seed-setting in potatoes as affected by spraying with α-naphthaleneacetamide and by light. Am. Potato J. 18:273–79.

4170. —— and J. R. Fryer. 1930: Seed setting in alfalfa. Sci. Agr. (Ottawa) 11: 38–44.

4171. —— and P. M. Lombard. 1939: Relation of length of day to flower and seed production in potato varieties. Am. Potato J. 16:236–44. Biol. Abstr. 14:1516. 1940.

4172. —— and L. H. Pollard. 1949: The amount of self-pollination in male-sterile onion lines. Proc. Am. Soc. Hort. Sci. 53:299–301.

4173. ——, ——, and L. R. Hawthorn. 1952: Effect of time of seeding on the winter survival and subsequent seed stalk development of onions. Proc. Am. Soc. Hort. Sci. 59:439–44.

4174. —— and F. J. Stevenson. 1943: Factors influencing the germination of seeds of the potato. Am. Potato J. 20:247–58.

Clarke, A. E. 9951.

Clarke, B. 18268.

4175. Claudel, L. 1890: Sur la localisation des matières colorantes dans les téguments. Compt. Rend. 110:298–300.

4176. Clausen, J. 1930: Male sterility in *Viola orphanidis*. Hereditas 14:53–72.

4177. ——, D. D. Keck, and W. M. Hiesey. 1947: Partial apomixis: An evolutionary labyrinth. Carnegie Inst. Washington Yearbook 46:101–3.

4178. Clausen, K. E., and P. O. Rudolf. 1958: Germination of 29-year-old red pine (*Pinus resinosa*) seed. Minn. Forestry Notes 72. 2 pp.

Clausen, K. E. 16512.

Clausen, R. T. 8124.

Claveau, J. 2855.

Claver-Aliod, M. 13937.

4179. Clay, D. W. T. 1964: Germination of the kola nut [*Cola nitida* (Vent.) Schott & Endl.]. Trop. Agr. (London) 41:55–60.

Claybrook, J. R. 17563.

4180. Claypool, L. L. 1932: Influence of fertilizer treatment on lettuce head and seed production. Proc. Am. Soc. Hort. Sci. 29:438–41.

4181. —— and P. Esau. 1956: Frost damage to walnut kernels; low temperatures during harvest season may cause injury to kernels resulting in chemical changes that produce rancidity. Calif. Agr. 10(6):7, 16.

Claypool, L. L. 7679.

4182. Clayton, C. N. 1948: Effect of several seed protectants on emergence and stand of okra. Phytopathology 38:102–5.

4183. —— and C. J. Nusbaum. 1945: Vegetable-seed protectants. S. Carolina Agr. Expt. Sta. Bull. 361. 24 pp. Chem. Abstr. 41:1793. 1947.

4184. Clayton, E. E. 1931: Effect of seed treatments on seed longevity. Phytopathology 21:105–6.

4185. —— 1931: Vegetable seed treatment with special reference to the use of hot water and organic mercurials. N.Y. Agr. Expt. Sta. Tech. Bull. 183. 44 pp.

Clayton, J. S. 11803.

4186. Clayton, J. T. 1956: Cooling stored grain to prevent spoilage in the top layers. Illinois Agr. Expt. Sta. Ext. Serv. Circ. 764. 8 pp.

4187. Clayton, L. C. W., and D. B. MacKay. 1962: An automatic germination chamber. Proc. Intern. Seed Testing Assoc. 27: 623–26.

Cleland, R. E. 14367.

4188. Clemens. 1911: Beiträge zur forstlichen Samenkunde. III. Naturw. Z. Forst- Landwirtsch. 9:402–9.

Clemens, H. 5347.

4189. Clement, E. 1924: Germination of *Odontoglossum* and other seed without fungal aid. Orchid Rev. 32(374):233–38.

4190. —— 1929: Non-symbiotic and symbiotic germination of orchid seeds. Orchid Rev. 37:68–75.

4191. Clement, E. O. 1938: Germination of lily seed. Lily Yearbook R. Hort. Soc. 1938:156–57.

4192. Clement, R. H. 1964: Does cottonseed high in free fatty acid deteriorate in storage? Assoc. Offic. Seed Analysts News Letter 38(1):17.

Clemente, L. J. 12594.

4193. Clementi, A. 1944: Ricerche di embriochimica vegetale. Nota II. La formazione di ammoniaca e di azoto amidico non proteico durante la germogliazione dei semi di *Lupinus albus* in soluzioni saline prive di sali azotati. Boll. Soc. Ital. Biol. Sper. 19:129–31.

4194. —— and D. Torrisi. 1934: The action of lupanine alkaloid and of the infusion of the seeds of *Lupinus albus* on normal glucemia and on diabetic glucosuria and hyperglucemia. Boll. Soc. Ital. Biol. Sper. 9:1004–8. Chem. Abstr. 29:1882. 1935.

4195. ——, ——, and C. Ruta. 1935: Synthesis of amino-acids in corn germinating in solutions containing no nitrogen. Arch. Sci. Biol. Napoli 21:483–94. Physiol. Abstr. 21:280. 1936.

4196. Clemo, G. R., W. M. Morgan, and R. Raper. 1936: The lupine alkaloids. X. The synthesis of dl-oxysparteine. J. Chem. Soc. 1936:1025–28.

4197. Clepper, H. 1957: The marvelous seeds of trees. Nat. Mag. 50:36–38.

Clerc, J. A. Le. 3574, 7659.

Clerg, E. L. Le; *see* Le Clerg, E. L.

4198. Clevenger, J. F. 1934: Volatile oil in cardamom seed. J. Assoc. Offic. Agr. Chemists 17:283–85.

4199. —— 1935: Volatile oils in mace and nutmegs. J. Assoc. Offic. Agr. Chemists 18:611–16.

4200. —— 1942: Volatile oils in fennel seed. J. Assoc. Offic. Agr. Chemists 25:962–63. Analyst 68:119. 1943.

4201. Clifford, E. D. 1963: The effects of soil sterilant chemicals on the germination and development of conifer seedlings and weed control. Tree Planters Notes 58: 5–11.

4202. Clifford, H. T. 1953: A note on the germination of Eucalyptus seed. Australian Forestry 17:17–20.

4203. —— 1956: Seed dispersal on footwear. Botan. Soc. Brit. Isles Proc. 2:129–31.

4204. —— 1959: Seed dispersal by motor vehicle. J. Ecol. 47:311–15.

Clifford, H. T. 3698, 3699.

4205. Climenco, V. N. 1936: Studies on the use of biochemical method of selection in the breeding of citrus fruits and Tung-oil tree. Batum Subtrop. Botan. Garden Bull. 1:126.

4206. Clinch, F. A. 1951: African violets have wide popularity; steps in growing them from seeds are traced in detail. Farm Garden (Watertown) 1951:11.

4207. Clinch, P. E. M., and J. B. Loughnane. 1948: Seed transmission of virus yellows of sugar beet (*Beta vulgaris* L.) and the existence of strains of the virus in Eire. Sci. Proc. Roy. Dublin Soc. N.S. 24:307–18. Biol. Abstr. 23:16938. 1949.

4208. Clint, K. L. 1958: Germination of *Parmongaia* seeds. Plant Life 14:94–95.

Clinton, P. K. S. 7212.

4209. Clode, J. J. 1959: Note on gibberellin-like excretions from germinating beans. Port. Acta Biol. Ser. A 6:75–76.

4210. Clopton, J. R., and A. Roberts. 1949: Chemical studies on oil-bearing seeds. II. Hubam clover seed. J. Am. Oil Chemists Soc. 26:11–13. Chem. Abstr. 43:1998. 1949.

4211. —— and H. O. Triebold. 1944: Fanweed, a potential major source of vegetable oil. Chemurgic Dig. 3(12): 160–61.

4212. —— and R. W. VonKorff. 1945: *Typha* (cattail) seed oil. Oil Soap 22:330–31. Chem. Abstr. 40:1051. 1946.

4213. Clos, D. 1857: Des graines de l'*Atriplex hortensis* et de leur germination. Bull. Soc. Botan. France 4:441.

4214. Clos, M. D. 1890: Singulier cas de germination des graines d'une Cactée dans leur pericarpe. Compt. Rend. 111:954–56.

4215. —— 1891: Interprétation des parties germinatives du *Trapa natans*, de quelques Guttifères et des *Nelumbium*. Bull. Soc. Botan. France 38:271–76.

4216. Close, A. W. 1938: Use of live sphagnum in seed germination. Proc. Am. Soc. Hort. Sci. 35:858–59.

4217. —— 1941: Easy germination of seeds of Amaryllidaceae in sphagnum moss. Herbertia 8:122–24.

Close, A. W. 9012, 18075, 18076.

4218. Clowes, G. J. 1945: The tyrosine and tryptophane contents of some legume seed proteins. Australian J. Exptl. Biol. Med. Sci. 23:227–29.

Clowes, G. J. 12514.

4219. Clum, H. H., and A. Nason. 1958: Triphosphopyridine nucleotide diaphorase from wheat germ. Plant Physiol. 33:354–60.

4220. Cluskey, J. E., N. W. Taylor, H. Charley, and F. R. Senti. 1961: Electrophoretic composition and intrinsic viscosity of glutens from different varieties of wheat. Cereal Chem. 38:325–35.

4221. Clute, W. N. 1932: Seed distribution in *Fraxinella*. Am. Botanist 38:167–69.

4222. —— 1934: Seed dispersal in *Oenothera*. Am. Botanist 40:173–74.

4223. Cluzet, J., and T. Kofman. 1929: Action des rayons ultraviolets, seuls ou associés aux rayons X, sur la germination. Compt. Rend. Soc. Biol. 101:820–21.

4224. Cmelik, S. H. W. 1963: The component fatty acids of the glyceride and phospholipid fractions of the baobab seed (*Adansonia digitata*). J. Sci. Food Agr. 14:287–91.

4225. Coaker, T. H. 1958: Palatability of maize under storage in Uganda. E. African Agr. J. 24:57–60.

4226. —— 1959: "Insack" treatment of maize with insecticide for protection against storage pests in Uganda. E. African Agr. J. 24:244–50.

4227. Coates, C. E., and M. M. Vick. 1934: Magnolia seed oil. Oil Soap 11:174–75. Chem. Abstr. 28:6007. 1934.

4228. Cobb, F. E. 1932: Growing trees from seeds. N. & S. Dakota Hort. 4:127–28.

4229. Cobb, N. A. 1903: Effect of engine-boiler steam on the vitality of seed and spores. Effect of steam on vitality of wheat grains. Agr. Gaz. N.S. Wales 14:26–29.

4230. Cobb, R. D. 1954: The identification of seed of the dodder, *Cuscuta australis* R. Brown. Calif. Dept. Agr. Bull. 43:23–24.

4231. —— 1958: Seed germination after fumigation with methyl bromide for Khapra beetle control. Calif. Dept. Agr. Bull. 47:1–16.

4232. —— 1959: Recent advances in seed analysis. Seed World 84(10):18–20.

4233. —— and T. C. Fuller. 1961: Identification of the seed of *Ipomoea eriocarpa* R. Br. Assoc. Offic. Seed Analysts News Letter 35(3):14–15.

4234. —— and L. G. Jones. 1962: Germinating dormant seeds of *Avena fatua* (wild oat). [Abstract.] Am. J. Botany 49:658–59.

4235. ——, D. C. Sumner, and L. G. Jones. 1961: Some effects of temperature and hydrogen peroxide treatment on the germination of Switch grass (*Panicum virgatum*). Assoc. Offic. Seed Analysts News Letter 35(1):26–29.

Cobb, R. D. 9965, 18165.

4236. Cobin, M. 1945: Weight of mangosteen seeds correlated with seedling vigor. Puerto Rico Agr. Expt. Sta. Rept. 1943/44: 23–24.

Cobin, M. 9206.

4237. Cochran, H. L. 1936: Some factors which influence the germination of pepper seeds. Proc. Am. Soc. Hort. Sci. 33:477–80. Biol. Abstr. 11:5554. 1937.

4238. —— 1938: A morphological study of flower and seed development in pepper. J. Agr. Res. 56:395–419. Proc. Am. Soc. Hort. Sci. 35:696. 1938.

4239. —— 1940: Effect of roasting and scalding pimiento pepper fruits to remove the skins prior to canning on the subsequent germination of the seed. [Abstract.] Proc. Am. Soc. Hort. Sci. 37:679.

4240. —— 1941: Better methods of pimiento production. Georgia Agr. Expt. Sta. Bull. 218. 41 pp.

4241. —— 1943: Effect of stage of fruit maturity at time of harvest and method of drying on the germination of pimiento seed. Proc. Am. Soc. Hort. Sci. 43:229–34.

Cocke, R. P. 9267.

4242. Cocker, H. 1955: A remarkable seed pod. Gardeners Chron. 137:134.

4243. Cockerline, D. W. 1961: Studies on the development of barley embryos in culture. Dissertation Abst. 22:1804.

4244. Cocucci, A. E. 1961: Embryology of *Trianthema argentina* (Aizoaceae). Kurtziana 1:105–22.

4245. Coe, F. M., and R. K. Gerber. 1934: Preliminary studies of the after-ripening and germination of cherry seeds. Proc. Utah Acad. Sci. Arts Letters 11:185–87. Biol. Abstr. 10:15441. 1936.

4246. Coe, G. E. 1961: Failure of fertilization in apomictic angiosperms. Intern. Botan. Congr. Recent Advan. Botany 9:702–6. 1959.

Coenradie, J. 14812.

4247. Coetzee, D. J., and J. E. Pons. 1961: Grass seed germinates better in consolidated soil. Farming S. Africa 36(12): 22–23.

Coffey, L. C. 10649, 13369.

4248. Coffman, F. A. 1923: The minimum temperature of germination of seeds. J. Am. Soc. Agron. 15:257–70.

4249. —— 1937: Factors influencing seed set in oat crossing. J. Heredity 28:297–303.

4250. —— and T. R. Stanton. 1938: Variability in germination of freshly harvested *Avena*. J. Agr. Res. 57:57–72.

4251. —— and —— 1940: Dormancy in fatuoid and normal oat kernels. J. Am. Soc. Agron. 32:459–66.

4252. —— and H. Stevens. 1951: Relation of temperature and time of day of pollination to seed set in oat crossing. Agron. J. 43:498–99.

4253. —— and J. W. Taylor. 1936: Widespread occurrence and origin of fatuoids in Fulghum oats. J. Agr. Res. 52:123–31.

Coffman, F. A. 2130, 13413, 18410, 18662.

4254. Coggeshall, R. G. 1954: Propagating some rarer plants from seed. Arnoldia 14:25–28.
4255. Cognee, M. 1960: Premières observations sur les fontes de sernis de cotonniers a la station de Bambari. Phytiat.-Phytopharm. 9:207–22.
4256. Cohen, A. 1956: Studies on the viability of citrus seeds and certain properties of their coats. Bull. Res. Council Israel Sect. D 5:200–9.
Cohen, A. 15445.
4257. Cohen, C. 1937: Some aspects of the reproduction of *Stoebe vulgaris*, and certain local grass and weed seeds. S. African J. Sci. 33:351–54.
4258. Cohen, D. 1958: The mechanism of germination stimulation by alternating temperatures. Bull. Res. Council Israel Sect. D 6:111–17.
Cohen, D. 11030.
4259. Cohen, L. I. 1963: Studies on the ontogeny of the dwarf mistletoes, *Arceuthobium*. I. Embryogeny and histogenesis. Am. J. Botany 50:400–7.
Cohen, O. Heller-. 15445.
4260. Cohen, P. P. 1946: The carboxylase activity of Jack beans (*Canavalia ensiformis*) and soybeans (*Glycine hispida*). J. Biol. Chem. 164:685–89.
Cohen, P. P. 269.
4261. Cohen, R. Z., and T. W. Goodwin. 1961: 3-Nucleotidase activity of rye grass seeds. Phytochemistry 1:47–51. Chem. Abstr. 56:3819. 1962.
4262. Cohen-Boulakia, F., and J. Polonovski. 1960: The different lipide fractions in the course of germination of soybeans. Biochem. Lipids, Proc. 5th Intern. Conf. Vienna 1958:39–46. Chem. Abstr. 55: 1824. 1961.
4263. Cohn, A. E., and D. J. DeZeeuw. 1950: Response of certain varieties of snap bean (*Phaseolus vulgaris*) to seed treatments. Mich. Agr. Expt. Sta. Quart. Bull. 32:386–401.
4264. Cohn, F. 1849: Beiträge zur Physiologie der Samen. Flora (Jena) 32:481–512.
4265. Coïc, Y., and J. Duranton. 1959: Influence comparée du traitement vernalisant (temperature basse) d'un blé d'hiver et d'un blé de printemps sur les échanges gazeux respiratoires après traitement. Ann. Physiol. Veg. 1 (Année 2): 113–20.
4266. ——, G. Fauconneau, R. Pion, F. Busson, C. Lesaint, and F. Labonne. 1963: Influence de l'alimentation minérale sur la composition des protides des graines de céréales (blé et orge). Ann. Physiol. Veg. 5:281–92.
4267. ——, C. Lesaint and M. Grandjean. 1963: Influence de la deficience en phosphore sur la composition minérale de l'orge à la récolte. Ann. Physiol. Veg. 5:277–80.
4268. Coile, T. S. 1934: Influence of the moisture content of slash pine seeds on germination. J. Forestry 32:468–69.
4269. Cojeneanu, N. 1959: Researches on the connection between the period of dormancy and the need of cold temperatures for apricots. Acad. Rep. Populare Romine Filiala Iasi Studii Cercetari Stiint. Biol. Stiinte Agr. 10:275–89.
Coker, S. T. 3266.
4270. Coker, W. C. 1909: Vitality of pine seeds and the delayed opening of cones. Am. Naturalist 43:677–81.
4271. Colbeth, I. M. 1951: Castor beans—paying crop for peace or war. Chemurgic Dig. 10(12):4–8.
4272. Colbry, V. L. 1953: Germination of imported orchard grass. Assoc. Offic. Seed Analysts News Letter 27(3):37–40.
4273. —— 1963: Testing annual (Itálian) and perennial ryegrass seed. Assoc. Offic. Seed Analysts News Letter 37(3): 20–27.
Cole, D. A., Jr. 3543, 3544, 3548.
4274. Cole, D. E. 1962: Harvesting the first certified pine cones. Forest Farmer 21(5):6–8.
Cole, E. 400.
4275. Cole, E. W., and M. Milner. 1953: Colorimetric and fluorometric properties of wheat in relation to germ damage. Cereal Chem. 30:378–91.
4276. Cole, G. M. 1926: A discussion of an abnormal sprout occurring in onion seed germination. Proc. Assoc. Offic. Seed Analysts 18:56–62.
4277. —— 1929: Problems in germination testing of flower and vegetable seeds. Seed World 26(2):16–17.
4278. Cole, J. S., and O. R. Mathews. 1940: Relation of the depth to which the soil is wet at seeding time to the yield of spring wheat on the Great Plains. U.S. Dept. Agr. Circ. 563:1–19. Biol. Abstr. 14:12492. 1940.
4279. Cole, L. N., and B. M. Craig. 1953: The fatty acid composition of *Caragana* seed oil. Can. J. Technol. 31:196–201.
4280. Coleman, C. H. 1953: Potato seed retains germination 13 years. Crops & Soils 5(5):21.

4281. Coleman, D. A., and H. C. Fellows. 1925: Hygroscopic moisture of cereal grains and flaxseed exposed to atmospheres of different relative humidity. Cereal Chem. 2:275–87.
4282. —— and —— 1928: A simple method for determining the oil content of seeds and other oil-bearing materials. U.S. Dept. Agr. Tech. Bull. 71. 13 pp.
4283. ——, B. E. Rothgeb, and H. C. Fellows. 1928: Respiration of sorghum grains. U.S. Dept. Agr. Tech. Bull. 100. 16 pp.
Coleman, D. A. 20084.
4284. Coleman, F. B. 1955: Identifying pigeon pea seed. Queensland Agr. J. 80:315.
4285. —— and A. C. Peel. 1952: Storage of seeds. Queensland Agr. J. 74:265–76.
4286. —— and —— 1952: Seed testing explained. Queensland Agr. J. 75:153–62.
4287. Coleman, R. G., G. M. Windrum, and E. M. Hutton. 1960: Studies on the toxicity of *Indigofera endecaphylla*. J. Nutrition 70:267–70.
4288. Coleno, P. 1950: The disinfection of peanut seeds. Oleagineux 5:149–55. Chem. Abstr. 44:9104. 1950.
4289. Colhoun, J. 1945: The prevention of seed-borne diseases of flax. III. The dusting, short wet, and fixation methods of seed disinfection in relation to the storage of the seed. Ann. Appl. Biol. 32:34–37. Rev. Appl. Mycol. 24:370–71. 1945.
4290. —— 1946: Observations on the effects of browning (*Polyspora lini* Laff.) of flax on seed production. Ann. Appl. Biol. 33: 255–59.
4291. —— 1948: Varietal resistance of flax to disease. Ann. Appl. Biol. 35:582–97.
4292. —— and A. E. Muskett. 1948: A study of the longevity of the seed-borne parasites of flax in relation to the storage of the seed. Ann. Appl. Biol. 35:429–34.
Colhoun, J. 14002.
Colin, M. 5787.
Colindres, R. 5893.
4293. Colla, S. 1931: Sulla nutrizione di embrioni di Graminacee mediante albumi di specie diversa. Boll. Soc. Ital. Biol. Sper. 6:74–77.
4294. —— 1936: Endocarpic germination. Arch. Botan. N.S. 2:1–16. J. Roy. Microscop. Soc. 56:181–82.
Collazos, C. 445.
Colle. 5189.
4295. Colli, L. U. F., G. Guidotti, R. D. Lonati, M. Orsenigo, and D. Sommariva. 1955: Further measurements on the bioluminescense of the seedlings. Experientia 11:479–81. Biol. Abstr. 30:32693. 1956.
4296. Collier, J. W. 1963: Caryopsis development in several grain sorghum varieties and hybrids. Crop Sci. 3:419–22.
4297. Collin, G. 1933: The fatty acid and glyceride structure of the seed fat of *Myristica malabarica*. Soc. Chem. Ind. J. 52:100T.
4298. —— 1933: The kernel-fats of some members of the Palmae: *Acrocomia sclerocarpa* Mart. (gru-gru palm), *Manicaria saccifera* Gaertn., *Astrocaryum tucuma* Mar., *Maximiliana caribaea* Grisel, *Attalea excelsa* Mart. (Pallia palm) and *Cocos nucifera* Linn. (coconut). Biochem. J. 27:1366–72.
4299. —— and T. P. Hilditch. 1929: Regularities in the glyceride structure of vegetable seed-fats. Biochem. J. 23:1273–89. Biol. Abstr. 5:26975. 1931.
4300. Collings, H. 1960: Hermetic sealing of a stack of maize with bituminous roofing felt. Trop. Agr. (London) 37:53–60.
4301. Collins, E. J. 1918: The structure of the integumentary system of the barley grain in relation to localized water absorption and semi-permeability. Ann. Botany (London) 32:381–414.
4302. Collins, F. I., and J. L. Cartter. 1956: Variability in chemical composition of seed from different portions of the soybean plant. Agron. J. 48:216–19.
4303. —— and R. W. Howell. 1957: Variability of linolenic and linoleic acids in soybean oil. J. Am. Oil Chemists Soc. 34:491–93.
4304. —— and O. A. Krober. 1945: Some factors which affect the determination of oil in soybeans. Oil Soap 22:307–10. Expt. Sta. Rec. 95:621. 1946.
Collins, F. I. 3904, 9132, 11426, 11427.
Collins, N. D. 10332, 10333.
4305. Collins, W. B., and C. F. Everett. 1955: Summary report of the 1955 weed control investigations at the Experimental Farm, Fredericton, N.B. Proc. 9th Meeting Eastern Sect. Natl. Weed Comm. 1955: 67–77.
4306. Collis-George, N., and J. E. Sands. 1959: The control of seed germination by moisture as a soil and physical property. Australian J. Agr. Res. 10:628–36.
4307. —— and —— 1962: Comparison of the effects of the physical and chemical components of soil water energy on seed

germination. Australian J. Agr. Res. 13:575–84.

4308. Colquhoun, T. T. 1939: Seed pickling. J. Dept. Agr. S. Australia 42:704–8.
Colon, F. Rodriquez-. 19763.
Colon, J. E. N. 9059.
Columbus, A. 12805.

4309. Colville, W. L., and D. P. McGill. 1962: Effect of rate and method of planting on several plant characters and yield of irrigated corn. Agron. J. 54:235–38.
Colville, W. L. 3405.

4310. Colvin, R., B. M. Craig, and H. R. Sallans. 1947: Hygroscopic equilibria for hulls and kernels of sunflower-seed and oats. Can. J. Res. Sect. F 25:111–18.

4311. Colwell, W. E., and N. C. Brady. 1945: The effect of calcium on certain characteristics of peanut fruit. J. Am. Soc. Agron. 37:696–708.

4312. ——, ——, and J. R. Piland. 1945: Composition of peanut shells of filled and unfilled fruits as affected by fertilizer treatments. J. Am. Soc. Agron. 37:792–805.
Colwell, W. E. 2934, 14322.
Comar, C. L. 2614.

4313. Comber, J. 1932: Magnolias from seeds. Gardeners Chron. 92:228.

4314. —— 1951: Raising rhododendrons from seeds. Gardeners Chron. 130:151.
Comber, R. 4609.

4315. Combs, W. B. 1944: Measuring storage qualities by grain standards. Trans. Am. Assoc. Cereal Chemists 2: 3–13. Biol. Abstr. 18:22518. 1944.

4316. Comé, D. 1961: Étude expérimentale de la permeabilité a differents gaz du tégument de quelques graines de taille réduite. Compt. Rend. Acad. Agr. France 47:55–62.

4317. —— 1962: Comment l'oxygene necessaire a la germination des graines parvient —il a l'embryon? Rev. Gen. Botan. 69:563–73.

4318. —— 1962: Diffusion of oxygen and carbon dioxide through seed coats. J. Rech. Centre Natl. Rech. Sci. Lab. Bellevue (Paris) No. 59:187–94. Chem. Abstr. 58:14445. 1963.
Comé, M. Daniel-; *see* Daniel-Comé, M.

4319. Comeaux, D. J. 1947: The effect of cold storage on the viability of rice seed. Proc. Louisiana Acad. Sci. 10:184–87.
Comhaire. 6390.

4320. Comini, A. 1927: Germination of orchid seed. Nature (Milano) 18:131–36. Biol. Abstr. 6:15593. 1932.

4321. Common, R. H. 1945: The composition and digestibility of northern Irish ryegrass seed and ryegrass seed cleanings. J. Agr. Sci. 35:56–63.
Common, R. H. 12711, 16580.
Compton, L. E. 15127.
Comstock, R. E. 6967, 8136.

4322. Comstock, V. E., and J. O. Culbertson. 1958: A rapid method of determining the oil content of the seed and iodine values of the oil from small samples of flaxseed. Agron. J. 50:112–14.

4323. ——, J. H. Ford, and B. H. Beard. 1963: Association among seed and agronomic characteristics in isogenic lines of flax. Crop Sci. 3:171–72.
Comstock, V. E. 4726.

4324. Conagini, C. H. T. M. 1957: Cytology of seed development in the peanut plant. Bragantia 16:15–33.

4325. Conard, E. C. 1946: The effect of harvesting method on germination of the seed of Russian wild rye, *Elymus junceus*, Fisch. J. Am. Soc. Agron. 38:842–44.

4326. Conard, H. S. 1911: Seed distribution by surface tension. Science 34:845.
Condé, R. 3002.

4327. Condit, I. J. 1939: Parthenocarpy in the fig. Proc. Am. Soc. Hort. Sci. 36: 401–4.

4328. —— 1948: Parthenocarpy in caprifigs. Proc. Am. Soc. Hort. Sci. 52:233–36.
Condit, I. J. 17999.

4329. Condon, M. Z., F. R. Andrews, and A. M. Altschul. 1952: The measurement of oxidation-reduction potentials of cottonseed slurries. Plant Physiol. 27:500–6.

4330. ——, ——, M. G. Lambou, and A. M. Altschul. 1947: Inhibition of heating and lipolysis in seeds. Science 105:525–27.

4331. ——, ——, ——, and —— 1947: Process prevents spoilage of seeds stored in bulk. Sci. News Letter 51:360.

4332. ——, E. A. Jensen, A. B. Watts, and C. W. Pope. 1954: Nutritional variation in cottonseed. Effect of autoclaving in presence and absence of gossypol on solvent extracted cottonseed meal. J. Agr. Food Chem. 2:822–26.

4333. ——, M. G. Lambou, J. L. Vignes, J. B. Loe, and A. M. Altschul. 1949: Inhibitors of heating and deterioration in seeds. I. Ethylene chlorhydrin and related compounds. Plant Physiol. 24:241–54.
Condon, M. Z. 433, 11715–11717.
Condorelli, C. 7274, 7275.

4334. Conger, A. D. 1961: Biological after-effect and long-lived free radicals in irradiated seeds. J. Cellular Comp. Physiol. 58:27–32.

4335. —— and M. L. Randolph. 1959: Magnetic centers (free radicals) produced in cereal embryos by ionizing radiation. Radiation Res. 11:54–66. Biol. Abstr. 34:14456. 1959.

Conil, S. 1758.

4336. Conkerton, E. J., W. H. Martinez, G. E. Mann, and V. L. Frampton. 1957: Cottonseed processing, changes induced by auto-claving a solvent-extracted cottonseed meal. J. Agr. Food Chem. 5:460–63.

Conklin, M. E. 18987, 18988.

Conn, E. E. 17871.

4337. Conner, J. F. 1931: Western yellow pine seed extraction in the Black Hills. J. Forestry 29:1165–67.

4338. Conner, R. T., and G. J. Straub. 1941: The thiamin and riboflavin contents of wheat and corn. Cereal Chem. 18:671–77.

4339. Conners, C. H. 1928: Germination of cyclamen seed. New Jersey Agr. Expt. Sta. Rept. 49:230–32.

4340. Connor, E. C. 1947: The storage and germination of apple seed. Agr. Gaz. N.S. Wales 58:414–16.

4341. Connors, C. H. 1945: Growing holly. New Jersey Agr. Expt. Sta. Circ. 493:1–8.

4342. Conrad, E. C. 1946: The effect of harvesting method on germination of the seed of Russian wild rye, *Elymus junceus* Fisch. J. Am. Soc. Agron. 38:842–44.

Conrat, H. Fraenkel-; *see* Fraenkel-Conrat, H.

Constantin, M. J. 14875.

4343. Constantinesco, D. 1943: Sur l'évolution du chonfriôme du sac embryonnaire de *Digitalis purpurea* L. Compt. Rend 216:206–7.

Constantinesco, D. 17719.

Constantinesco, G. 2775.

Constantinesco, M. 17719.

4344. Constantinescu, C. 1935: Disinfection of tobacco seed against cryptogams and its relation to the preservation of germinative power. Bul. Cult. Ferment. Tutunului 24:314–20. Chem. Abstr. 30:2308. 1936.

Constantinescu, C. 1874, 1875.

4345. Constantinescu, D. G., and M. Retezeanu. 1958: The action of cumarinylacetic acids on germination of seeds of *Nicotiana rustica*. Lucrarile Prezentate Conf. Natl. Farm. Bucharest 1958:488–92. Chem. Abstr. 53:9372. 1959.

Constantinescu, D. G. 18382.

Constantinescu, O. 9202.

Constantinou, P. T. 5176.

4346. Conti, G. 1929: Sulla funzione delle reste nei grani duri. Ann. Tec. Agrar. 1:419–22.

Conti, J. 9787, 10746–10748.

4347. Conti, M. 1953: Preserving wheat and rye embryos. Ital. Patent 478,786. Chem. Abstr. 48:13125. 1954.

4348. Cook, A. A., R. H. Larson, and J. C. Walker. 1952: Relation of the black rot pathogen to cabbage seed. Phytopathology 42:316–20.

4349. Cook, A. H., and J. R. A. Pollock. 1952: Chemical aspects of malting. II. Method for assaying germination-inhibitory activity of barley steeping liquors. J. Inst. Brewing 58:407–13. Brit. Abstr. B III:191. 1953.

4350. —— and —— 1953: Chemical aspects of malting. III. Presence of acetic acid in steeping liquors, and germination-inhibitory activities of simple carboxylic acids. J. Inst. Brewing 59:313–16. Brit. Abstr. B III:451. 1953.

4351. Cook, C. W. 1959: The effect of site on the palatability and nutritive content of seeded wheatgrasses. J. Range Management 12:289–92. Biol. Abstr. 35:27744. 1960.

4352. —— and L. A. Stoddart. 1953: The halogeton problem in Utah. Utah Agr. Expt. Sta. Bull. 364. 44 pp.

4353. Cook, D. B. 1940: Cone production by young trees in New York. J. Forestry 38:369.

4354. Cook, D. H., and J. Axtmayer. 1932: A new plant source of vitamin A activity. Science 75:85–86.

Cook, D. H. 1460.

4355. Cook, H. T. 1937: Spinach and cabbage seed treatment. Virginia Truck Expt. Sta. Bull. 96:1491–510. Rev. Appl. Mycol. 17:219–20. 1938.

4356. —— 1937: Vegetable seed treatment experiments and practices in Virginia. Proc. Assoc. Offic. Seed Analysts 29: 105–11.

Cook, J. A. 6849.

4357. Cook, M. T. 1903: Polyembryony in Ginkgo. Botan. Gaz. 36:142.

4358. —— 1924: Development of seed of *Linaria vulgaris*. Botan. Gaz. 77:225–27.

4359. —— 1941: Germination failures of the magnolia in Puerto Rico. J. Agr. Univ. Puerto Rico 25:51–53.

4360. Cook, O. F., and C. B. Doyle. 1916: Germinating coconuts. J. Heredity 7: 148–57. Expt. Sta. Rec. 35:344. 1916.

4361. Cook, P. L. 1939: A new type of embryogeny in the conifers. Am. J. Botany 26:138–43.

Cook, R. L. 2248, 16288.

Cook, S. A. 3720.

4362. Cook, W. H., J. W. Hopkins, and W. F. Geddes. 1934: Rapid determination of moisture in grain. I. Comparison of 130° C. air oven and Brown-Duvel methods with vacuum oven method. Can. J. Res. 11:264–89.

4363. ——, ——, and —— 1934: Rapid determination of moisture in grain. II. Calibration and comparison of electrical moisture meters with vacuum oven for hard red spring wheat. Can. J. Res. 11:409–47.

Cook, W. H. 14378.

Cooke, D. A. 10900.

4364. Cooke, G. B., and C. F. Schmidt, Jr. 1945: Million cork acorns distributed. Chemurgic Dig. 4(11):195–96.

4365. Cooke, H. L., C. D. Peedin, and R. P. Moore. 1952: Measured crop performance. N. Carolina Agr. Expt. Sta. Bull. 379. 44 pp.

4366. Cooler, E. W. 1962: Seed storage temperature signal system. Tree Planters Notes 52:31–32.

4367. Cooley, G. E. 1895: On the reserve cellulose of the seeds of Liliaceae and of some related orders. Mem. Boston Soc. Nat. Hist. 5:1–29.

Coolidge, M. H. 17740.

Coombe, B. G. 14985.

4368. Coombs, S. V. 1935: Sun-ripened South African seed. Horticulture 13(7): 160.

4369. Coomes, T. J., H. T. Islip, and W. S. A. Matthews. 1955: *Aframomum augustifolium* seed from Zanzibar. Colonial Plant Animal Prod. (Gt. Brit.) 5:68–77. Chem. Abstr. 50:436–37. 1956.

4370. Coonrod, L. G., H. T. Barr, and R. K. Walker. 1949/50: Rice drying and storage. Louisiana Agr. Expt. Sta. Ann. Rept., p. 31.

4371. Cooper, C. C. 1959: Polyethylene protective seed packages. Proc. Am. Soc. Hort. Sci. 75:569–79.

Cooper, C. C. 10029.

4372. Cooper, D. C. 1933: Macrosporogenesis and embryology of *Melilotus*. Botan. Gaz. 95:143–55.

4373. —— 1934: Development of the embryo sac of *Lilium henryi*. Proc. Natl. Acad. Sci. U.S. 20:163–66. Expt. Sta. Rec. 76:18. 1937.

4374. —— 1935: Macrosporogenesis and embryology of *Medicago*. J. Agr. Res. 51:471–77.

4375. —— 1937: Macrosporogenesis and embryo-sac development in *Euchlaena mexicana* and *Zea mays*. J. Agr. Res. 55:539–51. Expt. Sta. Rec. 78:326. 1938.

4376. —— 1938: Embryology of *Pisum sativum*. Botan. Gaz. 100:123–32. Expt. Sta. Rec. 80:170. 1939.

4377. —— 1940: Macrosporogenesis and embryology of *Portulaca oleracea*. Am. J. Botany 27:326–30.

4378. —— 1946: Double fertilization in petunia. Am. J. Botany 33:54–57.

4379. —— 1949: Flower and seed development in *Oxybaphus nyctagineus*. Am. J. Botany 36:348–55.

4380. —— 1961: The embryo and endosperm in hybrids. Intern. Botan. Congr. Recent Advan. Botany 9:672–75.

4381. —— and R. A. Brink. 1940: Partial self-incompatibility and the collapse of fertile ovules as factors affecting seed formation in alfalfa. J. Agr. Res. 60: 453–72.

4382. —— and —— 1940: Somatoplastic sterility as a cause of seed failure after interspecific hybridization. Genetics 25: 593–617.

4383. —— and —— 1942: The endosperm as a barrier to interspecific hybridization in flowering plants. Science 95:75–76.

4384. —— and —— 1944: Collapse of the seed following the mating of *Hordeum jubatum* × *Secale cereale*. Genetics 29:370–90.

4385. —— and —— 1945: Seed collapse following matings between diploid and tetraploid races of *Lycopersicon pimpinellifolium*. Genetics 30:376–401.

4386. —— and —— 1949: The endosperm-embryo relationship in an autonomous apomict, *Taraxacum officinale*. Botan. Gaz. 111:139–53.

4387. ——, ——, and H. R. Albrecht. 1937: Embryo mortality in relation to seed formation in alfalfa (*Medicago sativa*). Am. J. Botany 24:203–13.

Cooper, D. C. 3058–3064, 5416, 20094.

4388. Cooper, G. O. 1942: Development of the ovule and the formation of the seed in *Plantago lanceolata*. Am. J. Botany 29:577–81.

4389. —— 1942: Microsporogenesis and development of seed in *Lobelia cardinalis*. Botan. Gaz. 104:72–81.

4390. Cooper, H. W., J. E. Smith, Jr., and M. D. Atkins. 1957: Producing and harvesting grass seed in the Great Plains. U.S. Dept. Agr. Farmers Bull. 2112. 30 pp.

4391. Cooper, J. M. 1941: Isolation of a toxic principle from the seeds of *Macrozamia spiralis*. J. Proc. Roy. Soc. N.S. Wales 74:450–54.

4392. Cooper, S. C. 1940: A note on the maturation period of cacao pods in Grenada. Trop. Agr. (Trinidad) 17:165. Expt. Sta. Rec. 84:335. 1941.

4393. Cooper, T. P. 1943: Hemp seed beater developed in Kentucky. Agr. Leaders Dig. 24(5):25–26. Biol. Abstr. 17:22342. 1943.

4394. Cooper, W. C., E. O. Olson, N. Maxwell, and A. V. Shull. 1958: Nursery and orchard performance of nucellar seedling clones of citrus in the Rio Grande Valley of Texas. J. Rio Grande Valley Hort. Soc. 12:44–52.

Coover, C. A. 15213.

Copeland, F. C. 4158.

4395. Copeland, H. F. 1933: The development of seeds in certain Ericales. Am. J. Botany 20:513–17.

4396. Copertini, S. 1939: Chemical and technological examination of the Lybian barley. Agr. Coloniale 33:626–34. Chem. Abstr. 34:2939. 1940.

4397. Copin, H. 1942: Action des ondes électriques très courtes sur la germination. Bull. Soc. Linn. Lyon. 1942(4):56. Bull. Soc. Botan. France 91:128. 1944.

4398. Copp, L. G. 1943: The influence of "weathering" on the yield of a ripe crop of wheat. New Zealand J. Sci. Technol. Sect. A 25:174–76. Biol. Abstr. 20:3563. 1946.

Copp, L. G. 2580.

4399. Coppens, N. 1956: Biosynthesis of fatty acids by seeds of *Ricinus communis*. Nature 177:279.

Coppock, G. E. 16057.

4400. Coppola, S. 1960: The importance of seeds in agricultural technique and economy. Tec. Agr. 12:673–77.

4401. Corbett, L. C. 1896: Soaked seeds of *Lycopersicum esculentum*. S. Dakota Agr. Expt. Sta. Bull. 47:24–25.

4402. —— 1896: Ripe vs. sun ripened and green fruits for seed of *Lycopersicum esculentum*. S. Dakota Agr. Expt. Sta. Bull. 47:26–30.

Corbett, N. H. 3566.

4403. Corbetta, G. 1956: A rot of rice seeds germinated in the field. Riso 5(5):15–17.

4404. Corcilius, F. 1955: Paper chromatographic investigations of hydroxy-coumarins and flavones from horse chestnut (*Aesculus hippocastanum*). Planta Med. 3:55–61. Chem. Abstr. 50:7396. 1956.

4405. Corcoran, M. R., and B. O. Phinney. 1962: Changes in amounts of gibberellin-like substances in developing seed of *Echinocystis*, *Lupinus* and *Phaseolus*. Physiol. Plantarum 15:252–62.

Cordner, H. B. 18010.

4406. Cordy, C. B., and I. C. MacSwan. 1961: Some evidence that pear bark measles is seed-borne. Plant Disease Reptr. 45:891.

Corkill, L. 9651.

4407. Cormack, M. W. 1961: Longevity of the bacterial wilt organism in alfalfa hay, pod debris, and seed. Phytopathology 51:260–61.

4408. Cormack, R. G. H. 1952: A note on the dormancy of tartary buckwheat seeds. Sci. Agr. (Ottawa) 32:170–72.

4409. Cormany, C. E., and D. F. Peterson. 1948: Stand count comparisons of segmented and decorticated sugar beet seed. Proc. Am. Soc. Sugar Beet Technol. 5:117–19.

4410. Cornelius, D. R. 1947: The effect of source of little bluestem grass seed on growth, adaptation, and use in re-vegetation seedings. J. Agr. Res. 74:133–43.

4411. —— 1950: Seed production of native grasses under cultivation in eastern Kansas. Ecol. Monographs 20:1–29.

4412. —— and N. C. Melcher. 1942: Estimating the yield of blue grama grass seed. Trans. Kansas Acad. Sci. 45:71–74.

4413. Cornelius, J. A., W. D. Raymond, and J. B. Ward. 1960: Wild sesame seed from Kenya. Trop. Sci. 1:212–14.

Cornell, H. H. 2838.

4414. Corner, E. J. H. 1949: The annonaceous seed and its four integuments. New Phytologist 48:332–64.

4415. —— 1951: The leguminous seed. Phytomorphology 1:1–34.

Cornini, M. G. 11792, 11793.

4416. Corns, W. G. 1959: Effects of seed treatments with gibberellin and dates of seeding on winter survival and vegetative yield of Kharkov wheat. Can. J. Plant Sci. 39:293–97.

4417. —— 1960: Effects of gibberellin treatments on germination of various species of weed seeds. Can. J. Plant Sci. 40:47–51.

4418. —— and R. J. Schraa. 1962: Dormancy and germination of seeds of silverberry (*Elaeagnus commutata* Bernh.). Can. J. Botany 40:1051–55.

Corns, W. G. 2783, 13119.

4419. Cornu, A., and G. Trabut. 1962: Study of the modalities of seeds of three varieties of hard wheat. Al Awamia Rev. Rech. Agron. Marocaine 2:11–33. Biol. Abstr. 39:16216. 1962.

4420. Cornwell, P. B. 1959: Effects of γ-radiation on the the taste and manufacturing properties of soft wheat. J. Sci. Food Agr. 10:411–12.

Coromaldi, L. de Cesaris-; *see* De Cesaris-Coromaldi, L.

4421. Corovic, M., and M. Canak. 1957: Germination of *Hyssopus officinalis* L. seeds and changes in young plants when radiating seeds by X-rays. Arhiv Poljoprivredne Nauke 10(28):3–13.

4422. —— and —— 1958: The effect of different doses of X-ray irradiation on the seed germination of *Ricinus communis* L. Arhiv Poljoprivredne Nauke 11(33):24–33

4423. Corrales-Macedo, A. 1955: Ensayo de aclimatación de semilla importada de tomate. Estac. Expt. Agr. Molina Informe Mens. 29:15–16.

4424. Correale, P., and E. Cortese. 1954: Über das Vorkommen von Hydroxyphenylalkylanisen in den Samen und Keimen zweier Ginsterarten. Naturwissenschaften 41:139–40.

4425. Correns, C. 1906: Das Keimen der beiderlei Früchte der *Dimorphotheca pluvialis*. Ber. Deut. Botan. Ges. 24:173–76.

4426. —— 1927: Der Unterschied in der Keimungsgeschwindigkeit der Männchensamen bei *Melandrium*. Hereditas 9:33–44.

4427. Correvon, H. 1900: Raising alpine plants from seed. Gardeners Chron. 27:84.

Corrin, W. R. 11769.

4428. Corsini, E., and R. Indovina. 1937: Fruit of *Sterculia lurida* F. Muell. Atti Accad. Nazl. Lincei Rend. Classe Sci. Fis. Mat. Nat. 25:263–68. Chem. Abstr. 31:7933–34. 1937.

4429. Cortaborria, E. J., M. Bardier, and G. J. Fischer. 1949: Differential susceptibility of wheat samples to anticryptogamic treatment. Arch. Fitotec Uruguay 4:55–60.

Cortese, E. 4424.

4430. Cortesi, R. 1955: Toxicity of antibiotics and their influence on the germination of some seeds. Boll. Chim. Farm. 94:45–54. Chem. Abstr. 49:8400. 1955.

4431. —— and R. Girard. 1954: Action of several antibiotics on bean plants. IV. Bull. Soc. Pharm. Bordeaux 92:22. Chem. Abstr. 49:12608–9. 1955.

4432. Corti, R. 1946: Embriologia ed embriogenesi in "Genista Monosperma." Nuovo Giorn. Botan. Ital. 53:284–320.

4433. —— 1953: Significance of the nucellar tissue and its function during the development of the seed of *Quercus*. Atti Accad. Nazl. Lincei Rend. Classe Sci. Fis. Mat. Nat. 14:530–39.

Cory, V. L. 16155.

4434. Cosgrove, D. J., D. G. H. Daniels, E. N Greer, J. B. Hutchinson, T. Moran, and J. K. Whitehead. 1952: Isolation of methoxy- and 2:6-dimethoxy-*p*-benzoquinone from fermented wheat germ. Nature 169: 966–67.

4435. Cossins, E. A., and E. R. Turner. 1963: The metabolism of ethanol in germinating pea seedlings. J. Exptl. Botany 14:290–98.

4436. Costa, A. S. 1939: Experiences de germinacoes com *Gossypium thurberi* Tod. Rev. Agr. (Piracicaba) 14:315–20.

Costa, A. S. 2281.

Costa, da; *see* Da Costa.

4437. Costa, G. 1931: La reaction du terrain et le développement de l'*Orobanche* de la fève. Ann. Tec. Agrar. 5/6: Proc. Intern. Seed Testing Assoc. 7:109. 1935.

Costa, O. De Almeida-; *see* De Almeida-Costa, O.

4438. Costantin, J. 1935: Conséquences pratiques de la germination des graines de pomme de terre en montagne. Compt. Rend. 200:177–81.

Costea, G. 5956.

4439. Cotrufo, C. 1963: Stimulation by citric acid of germination of Eastern red cedar (*Juniperus virginiana* L.). Nature 199:92–93.

Cotton, R. H. 19279.

4440. Cotton, R. T. 1960: Pests of stored grain and grain products. Burgess Publishing, Minneapolis. 306 pp.

4441. —— and J. C. Frankenfeld. 1945: Protecting stored seed from insect attack. U.S. Dept. Agr., Agr. Res. Admin. Bur. Entomol. Plant Quarantine E-677. 15 pp.

4442. —— and —— 1948: Only dormant seeds store safely. Southern Seedsman 11(11):8.

4443. ——, ——, E. G. Bayfield, and J. A. Johnson. 1946: Wheat viability loss due to fumigation and its relation to baking quality. Northwest. Miller 227(2, Sect. 2): 4a–5a. Expt. Sta. Rec. 95:813. 1946.

4444. —— and H. E. Gray. 1948: Preservation of grains and cereal products in storage from insect attack. Food Agr. Organ. U.N., FAO Agr. Studies No. 2, pp. 35–71.

4445. ——, H. H. Walkden, G. D. White, and D. A. Wilbur. 1953: Causes of outbreaks of stored-grain insects. Kansas Agr. Expt. Sta. Bull. 359. 36 pp.

4446. —— and H. D. Young. 1929: The use of carbon dioxide to increase the insecticidal efficacy of fumigants. Proc. Entomol. Soc. Wash. 31:97–102.

Cotton, R. T. 1638, 8959.

4447. Cottrell, H. J. 1948: Tetrazolium salt as a seed germination indicator. Ann. Appl. Biol. 35:123–31.

Cottrin, L. 19412.

Couch, D. 14778.

4448. Couch, H. B. 1955: Studies on seed transmission of lettuce mosaic virus. Phytopathology 45:63–70.

4449. Couch, J. F. 1936: Lupine studies. IX. Monolupine, a new alkaloid from *Lupinus caudatus* Kellog. J. Am. Chem. Soc. 58:686–87.

4450. —— 1936: Lupine studies. XI. The alkaloids of *Lupinus barbiger* S. Wats. J. Am. Chem. Soc. 58:1296–99.

4451. —— 1940: Lupine studies. XV. The alkaloids of *Lupinus sericeus* Pursh. J. Am. Chem. Soc. 62:554–56.

4452. Coukos, C. J. 1944: Seed dormancy and germination in some native grasses. J. Am. Soc. Agron. 36:337–45.

Coulier, A. 3644.

Coulier, V. 19073.

Coulon, J. de; *see* De Coulon, J.

Coulson, C. B. 11050.

Coulson, E. J. 17800–17802.

Coulson, J. G. 12383.

4453. Coulter, F. C. 1942: How the seed trade began in America. Seed World 51(11):50–54.

4454. Coulter, J. M., and C. J. Chamberlain. 1903: Morphology of angiosperms. (Morphology of spermatophytes. Part II.) Appleton, New York, pp. 205–10.

4455. Coupin, H. 1897: Mucilage exomosed from seeds. Rev. Gen. Botan. 9:241–44.

4456. —— 1897: Sur la structure du micropyle des graines des Légumineuses. Rev. Gen. Botan. 9:175–80.

4457. —— 1898: Sur la résistance des graines a l'immersion dans l'eau. Compt. Rend. 126:1365–68.

4458. —— 1899: Action des vapeurs anesthésique sur la vitalité des graines sèches et des grains humides. Compt. Rend. 129:561–62.

4459. —— 1911: Upon the localization of pigments in the seed-coats of the bean. Compt. Rend. 153:1489–92.

4460. Court, D. 1914: Enzymatic peptolysis in germinating seeds. Proc. Roy. Soc. Edinburgh 34:113–27.

4461. Courtaulds, Ltd., and R. L. Wormell. 1944: Extraction of casein from seeds. Brit. Patent 575,402. Brit. Abstr. B 3:143. 1946.

4462. Courtois, J. 1934: Effect of the reaction of the medium on the hydrolysis of α- and β-glycerophosphoric acids by various seed phosphatases. Compt. Rend. 199:1252–54. Chem. Abstr. 29:1444. 1935.

4463. —— 1938: Comparative hydrolysis of α- and β-glycerophosphoric acids by different plant phosphatases. VI. Phosphatase of white mustard seed. VII. Phosphatase of sweet almonds. Bull. Soc. Chim. Biol. 20:1359–75, 1376–91. Pharm. Abstr. 7:166. 1941.

4464. —— 1947: Recherches sur la phytase. V. Étude préliminaire de l'hydrolyse des inositohexaphosphates et glycerophosphates par les graines de Moutarde blanche. Bull. Soc. Chim. Biol. 29:944–1950. Biol. Abstr. 23:8853. 1949.

4465. —— 1948: Recherches sur la phytase. VII. Action de preparations phosphatasiques purifiées de Moutarde-blanche sur divers esters phosphoriques. Bull. Soc. Chim. Biol. 30:36–43. Biol. Abstr. 23:8855. 1949.

4466. Courtois, J. E. 1957: Studies on natural oligosaccharides of vegetable origin containing galactose associated with sucrose. Biochemistry (USSR) (English Transl.) 22:233–42.

4467. —— and U. Ariyoshi. 1962: Galactosides of the sucrose of Caryophyllaceae seeds. Bull. Soc. Chim. Biol. 44:31–37.

4468. —— and P. LeDizet. 1963: Study of the galactomannans of the seeds of gleditschia and clover. Bull. Soc. Chim. Biol. 45:447–48. Biol. Abstr. 45:21382. 1964.

4469. —— and —— 1963: Galactomannans. III. Some properties of galactomannans of seeds of *Gleditsia ferox*. Bull. Soc.

Chim. Biol. 45:731–41. Chem. Abstr. 59:14295. 1963.

4470. ——, F. Petek, and T. Dong. 1961: Synthesis of planteose by the transferring action of the α-galactosidase of *Plantago* seeds. Bull. Soc. Chim. Biol. 43:1189–96. Chem. Abstr. 56:7718. 1962.

4471. ——, ——, and —— 1963: Separation of two α-galactosidases with distinct properties from the seeds of *Plantago ovata*. Bull. Soc. Chim. Biol. 45:95–100. Chem. Abstr. 58:11620. 1963.

4472. Cousin, E., and P. Creach. 1937: Lipase from the seed of *Ricinus communis*. Bull. Soc. Pharm. Bordeaux 75:12–16. Chem. Abstr. 31:6681. 1937.

Coustou, M. 19159.

4473. Couto, F. A. A. 1956: Production of carrot seeds in the United States. Rev. Ceres 9:452–57.

Coutts, J. 19835.

4474. Couturier, J. 1962: The sunflower. Technology and characteristics of the oil and derivatives. Bull. Soc. Sci. Hyg. Aliment. 50:279–95. Chem. Abstr. 59: 6721. 1963.

4475. Covello, M., and M. L. Biosio. 1953: Variation in the amylolytic power of barley during germination as related to the presence of biocatalysts in the water of maceration. Boll. Soc. Ital. Biol. Sper. 29:466–69. Chem. Abstr. 49:4805. 1955.

4476. Coville, P. 1929: Improved forest tree seed. A suggested study. J. Heredity 20:459–67.

4477. —— 1932: Growing Christmas holly on the farm. U.S. Dept. Agr. Farmers Bull. 1693. 21 pp.

Cowan, C. B., Jr. 15050.

Cowan, J. C. 18434.

Cowart, F. F. 5826.

4478. Cowen, M. E. H. 1942: Weed seeds in flax. J. Dept. Agr. Victoria 40:35–42.

4479. Cowgill, W. H. 1944: Studies on the germination, early seedling growth and nutrition of *Cinchona*. Thesis (Ph.D.), Maryland Univ. 142 pp.

4480. Cowles, R. B. 1936: The relation of birds to seed dispersal of the desert mistletoe. Madrono 3:352–56.

Cowley, W. R. 3392, 9887.

4481. Cowly, B., and B. Towers. 1941: Report of work on Johnson grass, crabgrass, wild morning glory, yellow star thistle. Proc. Assoc. Offic. Seed Analysts 33:35–39.

4482. Cox, C. H. 1933: Report of seed analysis committee. Oil Soap 10:9–11. Chem. Abstr. 27:1222. 1933.

4483. Cox, E. A., G. Stotzky, and R. D. Goos. 1960: *In vitro* culture of *Musa balbisiana* Colla embryos. Nature 185:403–4.

Cox, E. A. 18064, 18065.

4484. Cox, E. H. M. 1933: Growing *Nomocharis* from seed. Horticulture 11(16): 275.

4485. Cox, F. R., and P. H. Reid. 1964: Calcium-boron nutrition as related to concealed damage in peanuts. Agron. J. 56:173–76.

4486. Cox, H. C., and J. H. Lilly. 1952: Effects of aldrin and dieldrin on germination and early growth of field crop seeds. J. Econ. Entomol. 45:421–28.

4487. Cox, J. F., and G. E. Starr. 1927: Seed production and marketing. Wiley, New York. 450 pp.

Cox, J. H. 1638.

4488. Cox, L. G. 1942: A physiological study of embryo dormancy in the seed of native hardwoods and iris. Thesis (Ph.D.), Cornell Univ. 287 pp. Doctoral Dissertation Accepted by Am. Univ. 1941/ 42(9):51. 1942.

4489. ——, H. M. Munger, and E. A. Smith. 1945: A germination inhibitor in the seed coats of certain varieties of cabbage. Plant Physiol. 20:289–94.

Cox, L. G. 15922.

4490. Coyama, N. 1954: The growth-inhibiting substances contained in the leaf-litter of the trees. I. The inhibiting effect on germination of the coniferous seeds. J. Japan. Forestry Soc. 36:38–41.

4491. Coyaud, Y., A. Angladette, V. D. Hien, G. Huet, R. Philippe, L. Roger, and D. K. Vuong. 1951: Les semences de riz. French Indochina Off. Indochin. Riz. Arch. 31. 94 pp.

4492. Coyola, G., and J. Etchebarne. 1952: La production des semences de mais en France. Potasse 193:60–64.

4493. Cozzo, D. 1962: Repetition of an experiment to find the relationship between size and weight of seed of *Araucaria angustifolia* and its germinative capacity and the height of seedlings. Rev. Forestry (Argentina) 6:99–106. Forestry Abstr. 24:412. 1963.

4494. Crabb, A. R. 1947: The hybrid-corn makers. Rutgers Univ. Press, New Brunswick, N.J. 331 pp.

Craddock, J. C., Jr. 364.

4495. Crafa, J. A., J. J. Stansbrey, and B. L. Scallet. 1949: Statistical evaluation of growth and staining methods for determination of germinative capacity of barley; a proposed whole-grain staining method. Am. Soc. Brewing Chemists Proc. 1949:39–52.

4496. Craft, J. H. 1943: Preliminary studies of the physiology and morphology of the germinating foliar embryos of *Bryophyllum calycinum*. Proc. Iowa Acad. Sci. 50: 171–79.

Cragg, J. M. 9698.

4497. Craig, B. M. 1953: The fatty acid composition of safflower seed oil. Can. J. Technol. 31:202–7.

4498. —— 1956: Comparison of the fatty acid composition of rapeseed and mustard-seed oils. Can. J. Technol. 34:335–39.

4499. —— 1961: Varietal and environmental effects on rapeseed. III. Fatty acid composition of 1958 varietal tests. Can. J. Plant Sci. 41:204–10.

Craig, B. M. 4279, 4310, 11805, 11806, 19570, 20013.

4500. Craig, F. N. 1937: The respiratory quotient of seedlings of *Lupinus albus* during the early stages of germination. J. Gen. Physiol. 20:449–53.

Craig, L. C. 9643.

4501. Craig, W. N. 1946: Growing lilies from seed. Horticulture 24(7):195.

Craig, W. T. 12451.

4502. Craigie, J. H. 1937: Seed-borne diseases and their control. Can. Seed Growers Assoc. Ann. Rept. 1936/37:45–51.

4503. Craigmiles, J. P., and H. B. Harris. 1963: Dosage evaluations for γ irradiation of Rhodesian sudangrass, *Sorghum arundinaceum*, seed. Crop. Sci. 3:452.

4504. Craine, E. M., and K. E. Fahrenholtz. 1958: The proteins in water extracts of corn. Cereal Chem. 35:245–59.

4505. Crall, J. M. 1959: Effect of seed source on watermelon maturity. Proc. Am. Soc. Hort. Sci. 74:555–57.

4506. Cralley, E. M. 1947: Maize seed treatment. Arkansas Agr. Expt. Sta. Bull. 466. 10 pp.

4507. —— and E. C. Tullis. 1937: Effect of seed treatment on seedling emergence, severity of seedling blight, and yield of rice. Arkansas Agr. Expt. Sta. Bull. 345. 24 pp. Rev. Appl. Mycol. 16:771. 1937.

Cralley, E. M. 88.

4508. Cram, R. H. 1930: The fringed gentian in Maine. Horticulture 8(6):131.

4509. Cram, W. H. 1951: Spruce seed viability: dormancy of seed from four species of spruce. Forestry Chron. 27:1–9.

4510. —— 1956: Maturity of Colorado spruce cones. Forest Sci. 2:26–30.

4511. —— and C. H. Lindquist. 1963: Germination and growth of three tree species under four photoperiods. Forest Sci. 9:279–82.

4512. —— and O. Vaartaja. 1955: Toxicity of eight pesticides to spruce and *Caragana* seed. Forestry Chron. 31:247–49.

4513. —— and H. A. Worden. 1957: Maturity of white spruce cones and seed. Forest Sci. 3:263–69.

4514. Cramer, P. J. S. 1958: Germination and growth of the coconut. SPC Quart. Bull. 8:39, 58.

4515. Crampton, E. W., and G. C. Ashton. 1943: The role of the germ in the nutritive properties of the cereal grains. Sci. Agr. 23:445–50.

4516. Crandall, H., and H. D. Tate. 1947: The bee, *Calliopsis andreniformis*, as a factor in alfalfa seed setting. J. Am. Soc. Agron. 39:161–63.

4517. Crane, H. L., and M. B. Hardy. 1934: Interrelation between cultural treatment of pecan trees, the size and degree of filling of the nuts, and the composition of kernels. J. Agr. Res. 49:643–61.

Crane, H. L. 7504, 17289.

4518. Crane, J. C., and J. B. Acuna. 1945: Effect of plant spacing and time of planting on seed yield of kenaf, *Hibiscus cannabinus* L. J. Am. Soc. Agron. 37:969–77.

4519. ——, ——, and R. E. Alonso. 1946: Effect of plant spacing and time of planting on fiber yield of kenaf, *Hibiscus cannabinus* L. J. Am. Soc. Agron. 38:46–59.

4520. —— and R. Blondeau. 1948: The use of growth-regulating chemicals to induce parthenocarpic fruit in the Calimyrna fig. Proc. Am. Soc. Hort. Sci. 52:236.

4521. —— and P. Punsri. 1956: Comparative growth of the endosperm and the embryo in unsprayed and 2,4,5-trichlorophenoxyacetic acid sprayed Royal and Tilton apricots. Proc. Am. Soc. Hort. Sci. 68:96–104.

Crane, J. C. 2932.

Crane, L. E. 17748.

4522. Crane, P. L., S. R. Miles, and J. E. Newman. 1959: Factors associated with

varietal differences in rate of field drying in corn. Agron. J. 51:318–20.
Cranston, D. M. 7507.
Craven, P. M. 6098.
Cravens, W. W. 16204.
Cravioto, R. O. 6737.

4523. Crawford, C. G. 1951: Germinating hardy cyclamen. Am. Rock Gard. Soc. B 9:16–17.

4524. —— 1955: Your own primroses from seed. Popular Gardening 6(2):19–20, 75.
Crawford, J. H. 12900.
Crawford, W. N. 8262.
Craze, S. P. 21.
Creach, P. 4472.

4525. Creasy, W. D. 1961: Development of the megaspore, megagametophyte, and embryo of *Arenaria fendleri* var. *fendleri*. Castanea 26:140–46. Biol. Abstr. 38: 15556. 1962.

4526. Creech, J. L., and R. F. Dowdle. 1951: Propagation of *Strophanthus*. Econ. Botany 6:48–54.
Creech, J. L. 8412.
Creencia, R. P. 2740.
Creighton, H. B. 1543.
Creighton, J. W. 5316.

4527. Creighton, M., and N. M. Naylor. 1933: A study of the starch-digesting and the sugar-forming enzymes of wheat. Iowa State Coll. J. Sci. 7:253–60.
Cremer, C. L. 10874.

4528. Crescini, F. 1930: A cruciferous weed. Ital. Agr. 3:4–7. Biol. Abstr. 6:15594. 1932.

4529. —— 1935: Mutations of seeds and induced modifications. Z. Zücht. Reihe A. Pflanzenzücht. 20:492–502. Biol. Abstr. 11:2995. 1937.

4530. —— 1936: Sur les limites de la germination et du développement de certaines espèces. Arch. Botan. 12:91–93. Bull. Soc. Botan. France 85:620. 1938.

4531. —— 1938: Significato pratico di alcuni aspetti della germinazione del riso (*Oryza sativa* L.). Ital. Agr. 75:123–28.

4532. —— 1943: Forme di Canapa (*Cannabis sativa* L.). III. La germinabilità. Nuovo Giorn. Botan. Ital. 50:209–31. Biol. Abstr. 20:16214. 1946.

4533. —— and L. Spreafico. 1953: Cultivated plants versus weeds. I. Researches on the biology of the germination of the achenes of *Artemisia vulgaris* L. Ann. Sper. Agrar. (Rome) 7:1597–610.
Cress, M. 134, 135.

4534. Cresta, M., M. A. Spadoni, and G. Tecce. 1953: Action of 2,4-dinitrophenol (DNP) on the respiration of endosperms of *Ricinus communis* seeds. Quaderni Nutriz. 13:143–51. Chem. Abstr. 50:7968. 1956.
Cretcher, L. H. 16125.

4535. Crête, P. 1936: Développement et structure du tégument seminal chez le *Reseda luteola* L. Bull. Soc. Botan. France 83:43–46.

4536. —— 1936: Transformation de l'ovule en graine chez l'*Androsaemum officinale* All. Bull. Soc. Botan. France 83:654–57.

4537. —— 1937: Développement et structure du tégument seminal chez le *Radiola linoides* Roth. Bull. Soc. Botan. France 84:655–59.

4538. —— 1938: Embryogénie des Lobeliacées. Développement de l'embryon chez le *Lobelia syphilitica* L. Compt. Rend. 207:177–79.

4539. —— 1939: La polyembryonie chez le *Lobelia syphilitica* L. Bull. Soc. Botan. France 85:580–83.

4540. —— 1943: Recherches histologiques et physiologiques sur l'embryologie des labiatiflores. Contribution à l'étude des formations haustoriales. Ann. Sci. Nat. Botan. Biol. Végétale 4:122–23.

4541. —— 1943: Recherches histologiques et physiologiques sur l'embryologie des labiatiflores. Embryologie du *Globularia vulgaris* L. I–II. Bull. Soc. Botan. France 90:29–32, 36–39.

4542. —— 1944: Polyembryonie chez l'*Actinidia chinensis* Planch. Bull. Soc. Botan. France 91:89–92.

4543. —— 1944: Recherches anatomiques sur la séminogenèse de l'*Actinidia chinensis* Planch. Afinités des Actinidiacées. Bull. Soc. Botan. France 91:153–60.

4544. —— 1946: Embryogénie des Loasacées Développement de l'embryon chez le *Loasa lateritia* Gill. Compt. Rend. 222: 920–21.

4545. —— 1946: Embryogénie des Crassulacées. Développement de l'embryon chez le *Cotyledon umbilicus* L. Compt. Rend. 222:1311–13.

4546. —— 1946: Embryogénie des Hydrophyllacées. Développement de l'embryon chez le *Phacelia tanacetifolia* Benth. Compt. Rend. 223:459–60.

4547. —— 1947: Un cas de polyembryonie chez le *Phytolacca decandra* L. Bull. Soc. Botan. France 94:304–5.

4548. —— 1948: Développement de l'embryon chez le *Calandrinia procumbens* Moris. Remarques embryographiques sur quelques Portulacacées. Compt. Rend. 227:81–83

4549. —— 1951: Embryogénie des Capparidacées. Développement de l'embryon chez le *Cleome graveolens* Rafin. Compt. Rend. 233:562–64.

4550. —— 1954: Développement du gametophyte femelle, de l'albumen et de l'embryon chez l'*Erinus alpinus* L. (Scrofulariacées). Congr. Intern. Botan. Paris Rappt. Commun. 8:243–44.

4551. —— 1954: Embryogenie des Solanacées. Les premiers stades du développement de l'embryon chez le *Physalis peruviana* L. Compt. Rend. 239:552–54.

4552. —— 1955: Embryogénie des Gentianacées. Développement de l'embryon chez le *Chlova parfoliata* L. Compt. Rend. 241: 1825–27.

4553. —— 1955: L'origine du sac embryonnaire et le développement de l'albumen chez l'*Alloplectus sanguineus* Mart. (Gesneriacées). Bull. Soc. Botan. France 102:205–8.

4554. —— 1956: Contribution a l'étude de l'albumen et de l'embryon chez les Campanulacées et les Lobeliacées. Bull. Soc. Botan. France 103:446–54.

4555. —— 1956: Lentibulariacées. Développement de l'embryon chez le *Pinguicula leptoceras* Rchb. Compt. Rend. 242:1063–66.

4556. —— 1956: Lentibulariacées. Développement de l'albumen chez les *Pinguicula*. Compt. Rend. 242:1346–52.

4557. —— 1956: Menyanthacées. Développement de l'embryon chez le *Limnanthemum nymphoides* Hoffg. et Link. Compt. Rend. 242:3110–13.

4558. —— 1956: À propos de l'embryogénie du *Sonerila wallichi* Benn. Bull. Soc. Botan. France 103:599–603.

4559. —— 1956: Rubiacées. Développement de l'embryon chez le *Diodia dasycephala* Chem. et Schlecht. Compt. Rend. 243:861–63.

4560. —— 1957: Embryogénie des Melastomacées. Développement de l'embryon chez le *Clidemia hirta* D. Don. Compt. Rend. 245:374–76.

4561. —— 1957: Papaveracées. Le développement de l'embryon chez le *Dicranostigma franchetianum* (Prain Fedde). Compt. Rend. 245:720.

4562. —— 1957: Nolanacées. Développement de l'embryon chez le *Nolana prostrata* L. Compt. Rend. 245:853–55.

4563. —— 1957: Papilionacées. Développement de l'embryon chez le *Bonaveria securidaca* Desv. (*Securigera coronilla* DC.). Compt. Rend. 245:1741–43.

4564. —— 1958: Cucurbitacées. Développement de l'albumen et de l'embryon chez le *Sicyos angulata* L. Compt. Rend. 246:456–59.

4565. —— 1958: Papilionacées. Développement de l'albumen et de l'embryon chez le *Cantharospermum barbatum* Koorders (*Atylosia barbata* Baker). Compt. Rend. 246:821–23.

4566. —— 1959: Embryogénie des Gentianacées. Développement de l'embryon chez le *Gentiana punctata* L. Compt. Rend. 248:3594–96.

4567. —— 1959: Tovariacées. Embryogénie du *Tovaria pendula* Ruiz et Pav. Compt. Rend. 249:748–50.

4568. —— 1960: Embryogénie du *Calvoa orientalis* Taub. Compt. Rend. 250:3710–12.

4569. —— 1960: Embryogénie des Solanacées. Développement de l'embryon chez le *Saracha jaltomata* Schlecht. Compt. Rend. 250:4194–96.

4570. —— 1960: Embryogénie des Cucurbitacées. Développement de l'embryon chez l'*Ecballium elaterium* A. Rich. Compt. Rend. 251:968–70.

4571. —— 1960: Embryogénie des Melastomacées. Développement de l'embryon chez le *Bertolonia maculata* D.C. Compt. Rend. 251:1907–9.

4572. —— 1961: Embryogénine des Solanacées. Développement de l'embryon chez la *Datura tatula* L. Compt. Rend. 252:2128–30.

4573. —— 1961: Embryogénie des Solanacées. Développement de l'embryon chez le *Browallia demissa* L. Compt. Rend. 252:2921–23.

4574. —— 1961: Embryogénie des Solanacées. Développement du proembryon chez le *Capsicum annuum* L. Compt. Rend. 252:3104–6.

4575. —— 1961: Embryologie des Solanacées. Développement de l'embryon chez le *Salpiglossis sinuata* Ruiz et Pav. Compt. Rend. 253:3032–34.

4576. —— 1962: Embryogénie des Cucurbitacées. Développement de l'embryon chez le *Cyclanthera explodens* Naud. Compt. Rend. 254:3411–12.

4577. —— and M. M. Janot. 1949: Action de deux substances de croissance sur l'embryogénie du *Lychnis githago* Scop. Compt. Rend. 228:336–38.

4578. Créveaux-Bourgeat, J., T. Terroine, and R. Jacout. 1948: Le facteur responsable de l'inaptitude du maïs à la croissance. Compt. Rend. 227:305–7.

Crews, L. T. 5217.

4579. Cribb, A. B. 1958: Seed dispersal in the blue waterlily. Queensland Nat. 16:12–13.

4580. Crider, F. J., and M. M. Hoover, 1937: Collection of native grass seed in the Great Plains, U.S.A. Imp. Bur. Plant Genet. Herb. Plants (Aberystwyth) Bull. 24. 15 pp. Expt. Sta. Rec. 78:481. 1938.

Cristau, B. 19111.

4581. Cristobal, U. L. 1940: The preservation of stored grain. The "new" method of siliceous dusts. Bol. Lab. Zool. Agr. Fac. Agron. Univ. Nacl. La Plata 5. 8 pp. Chem. Abstr. 36:4611. 1942.

Cristy, F. Llavat-. 10462.

4582. Crnjakovic, D. 1944: Beitrag zur Untersuchung der qualitativen Eigenschaften unserer Bohne (*Phaseolus vulgaris* L.). Rev. Sci. Agr. Inst. Hort. Fac. Agron.-Forest. Univ. Zagrebiensis No. 2, Zagreb. Proc. Intern. Seed Testing Assoc. 14:74. 1948.

4583. Crochetelle, J. 1913: Influence of manganese sulphate on germination. J. Agr. Prat. 26:398–99. Expt. Sta. Rec. 30:332. 1914.

4584. Crocioni, A. 1934: Influence de l'âge de la graine sur le développement de la plante. Nuovi Ann. Agr. (Rome) 14(2). Proc. Intern. Seed Testing Assoc. 8:93. 1936.

4585. —— 1934: Stimulating action of some substances used for disinfecting seed wheat. Nuovi Ann. Agr. (Rome) 14(2): 279–90. Biol. Abstr. 9:11031. 1935.

4586. —— 1947: Seed germination tests by the staining method. Humus (Milan) 3(8):15–16.

4587. —— 1950: Duration of viability of hemp seed in relation to storage conditions. Bologna Univ. Cent. Studi Ric. Lavoraz. Coltivaz. Econ. Canapa Quaderni 9. 30 pp.

4588. Crocker, W. 1906: Role of seed coats in delayed germination. Botan. Gaz. 42:265–91.

4589. —— 1909: Longevity of seeds. Botan. Gaz. 47:69–72.

4590. —— 1916: Mechanics of dormancy in seeds. Am. J. Botany 3:99–120.

4591. —— 1925: Seeds: Their tricks and traits. N.Y. Botan. Garden J. 26:178–87.

4592. —— 1926: After-ripening and germination of rose seeds. Am. Rose Ann. 11:34–37.

4593. —— 1927: Dormancy in hybrid seeds. Mem. Hort. Soc. N.Y. 3:33–38. Boyce Thompson Inst. Prof. Pap. 1:36–41. 1927.

4594. —— 1928: Storage, after-ripening, and germination of apple seeds. Am. J. Botany 15:625–26.

4595. —— 1930: Harvesting, storage and stratification of seeds in relation to nursery practice. Boyce Thompson Inst. Prof. Pap. 1:114–20.

4596. —— 1936: Effect of the visible spectrum upon the germination of seeds and fruits. *In* B. M. Duggar, editor, Biological effects of radiation. McGraw-Hill, New York, pp. 791–827.

4597. —— 1938: Life-span of seeds. Botan. Rev. 4:235–74.

4598. —— 1945: Longevity of seeds. N.Y. Botan. Garden J. 46:26–35, 48.

4599. —— 1948: Growth of plants: Twenty years' research at Boyce Thompson Institute. Reinhold, New York. 459 pp.

4600. —— and L. V. Barton. 1931: After-ripening, germination, and storage of certain roseaceous seeds. Contrib. Boyce Thompson Inst. 3:385–404.

4601. —— and —— 1953: Physiology of seeds: An introduction to the experimental study of seed and germination problems. Chronica Botanica Co., Waltham, Mass. 267 pp.

4602. —— and W. E. Davis. 1914: Delayed germination in seed of *Alisma plantago*. Botan. Gaz. 58:285–321.

4603. —— and J. F. Groves. 1915: A method of prophesying the life duration of seeds. Proc. Natl. Acad. Sci. U.S. 1:152–55.

4604. —— and G. T. Harrington. 1918: Catalase and oxidase content of seeds in relation to their dormancy, age, vitality and respiration. J. Agr. Res. 15:137–74.

4605. ——, N. C. Thornton, and E. M. Schroeder. 1946: Internal pressure necessary to break shells of nuts and the role of the shells in delayed germination. Contrib. Boyce Thompson Inst. 14:173–201.

Crocker, W. 2040, 7222, 8209–8211.

Croix, L. J. La; *see* La Croix, L. J.

4606. Crombie, B. 1940: Wheat ripening as influenced by time of sowing and other factors. J. Dept. Agr. Ireland 37:391–53.

4607. Crombie, W. M. 1956: Fat metabolism in the West African oil palm (*Elaeis guineensis*). Part I. Fatty acid formation in the maturing kennel. J. Exptl. Botany 7:181–93.

4608. —— 1959: Fat metabolism in the West African oil palm (*Elaeis guineensis*). Part IV. Respiration in the developing seed. W. African Inst. Oil Palm Res. J. 3:96–101.

4609. —— and R. Comber. 1956: Fat metabolism in germinating *Citrullus vulgaris*. J. Exptl. Botany 7:166–80.

4610. —— and E. E. Hardman. 1958: Fat metabolism in the West Africa oil palm (*Elaeis guineensis*). Part III. Fatty acid formation in the maturing exocarp. J. Exptl. Botany 9:247–53.

Crombie, W. M. 8165.

4611. Cropp, C. 1940: Pansies and violas from seed. Flower Grower 27(8):359–60, 385.

4612. Cropsey, M. G., C. M. Leach, T. M. Ching, L. A. Sather, and D. E. Wiant. 1962: The effects of cathode ray irradiation of wheat on surface moulds, respiration, flour flavour and odour, and germination. Ann. Appl. Biol. 50:487–505.

4613. —— and D. E. Wiant. 1959: The effect of cathode ray irradiation on the respiration rate, milling tests, and baking properties of wheat. Mich. Agr. Expt. Sta. Quart. Bull. 42:164–75.

Cropsey, M. G. 12410.

Crosbie, C. J. 16706.

4614. Crosier, W. F. 1934: Black spot of germinating pea seed. Phytopathology 24:827–29.

4615. —— 1936: *Ascochyta pisi viciae* in seeds of vetch. Proc. Assoc. Offic. Seed Analysts 1935:82–86.

4616. —— 1936: Detection and identification of seed-borne parasites. Proc. Assoc. Offic. Seed Analysts 27:87–92.

4617. —— 1936: Procedure used in an analytical and mycological study of seed wheat. Proc. Assoc. Offic. Seed Analysts 1936:89–91.

4618. —— 1936: Prevalence and significance of fungous associates of pea seeds. Proc. Assoc. Offic. Seed Analysts 28:101–7. Expt. Sta. Rec. 77:499. 1937.

4619. —— 1937: Influence of chemical and soil factors upon the germination of seeds of small grains. Proc. Assoc. Offic. Seed Analysts 29:126–29.

4620. —— 1938: The planting value of oats and barley collected from farmers' drills and granaries. N.Y. Agr. Expt. Sta. Bull. 681. 46 pp.

4621. —— 1939: Occurrence and longevity of *Ascochyta pisi* in seeds of hairy vetch. J. Agr. Res. 59:683–97.

4622. —— 1940: Treated grain displays little injury during storage. Farm Res. (N.Y.) 6(2):6, 14.

4623. —— 1941: Oats and barley seed germinate well after eight years. Farm Res. (N.Y.) 7(4):8–9.

4624. —— 1941: Split seeds in peas and beans. Proc. Assoc. Offic. Seed Analysts 33:66–69. Biol. Abstr. 19:3584. 1945.

4625. —— 1942: Effect of seed-borne microorganisms on germination. Intern. Crop Improvement Assoc. Ann. Rept. 23:22–25.

4626. —— 1943: Materials and methods in controlling seed-contaminating microorganisms. Proc. Assoc. Offic. Seed Analysts 34:104–8.

4627. —— 1943: Chemical treatment of cabbage seed. Assoc. Offic. Seed Analysts News Letter 17(3):10–11.

4628. —— 1943: Baldheads in beans, occurrence and influence on yields. Proc. Assoc. Offic. Seed Analysts 34:118–23.

4629. —— 1944: Control of seed-borne organisms during germination of cucurbit seeds: value of recently introduced chemicals. Assoc. Offic. Seed Analysts News Letter 18(2):3–6.

4630. —— 1944: *Diplodia gossypina* and other fungi in cotton seed. Assoc. Offic. Seed Analysts News Letter 18(2):13–15.

4631. —— 1944: Some fungi found in seedstocks during recent years. Proc. Assoc. Offic. Seed Analysts 1943/44:99–102.

4632. —— 1946: Beet seedballs vary widely in size and weight: smallest seed has greatest bushel weight but lowest viability-germination increased by resizing seedstocks. Farm Res. (N.Y.) 12(1):15, 17.

4633. —— 1946: Chemical control of seed-borne fungi during germination testing of peas and sweet corn. Phytopathology 36:92–99.

4634. —— 1946: Effect of storage on treated seed wheat. Farm Res. (N.Y.) 12(2):4, 10.

4635. —— 1946: Germinating freshly harvested cereals. Farm Res. (N.Y.) 12(4): 4–5.

4636. —— 1946: Germination and fungus content of Vicland oats. Proc. Assoc. Offic. Seed Analysts 1945/46:152–58.

4637. —— 1946: Some observations on spergon-treated seeds. Proc. Assoc. Offic. Seed Analysts 1945/46:159–64.

4638. —— 1948: Germination of pea seeds treated with bird repellents and/or fungicides. Assoc. Offic. Seed Analysts News Letter 22(4):7–10.

4639. —— 1949: The germination of high-moisture content seeds fumigated with carbacryl. Proc. Assoc. Offic. Seed Analysts 39:117–21.

4640. —— 1949: Influence of temperature upon germination of corn seed. Assoc. Offic. Seed Analysts News Letter 23(1):9–12.

4641. —— 1949: Planting value of dehulled seed oats. Seed World 64(4):20, 22.

4642. —— 1951: Influence of storage conditions upon germination of Vicland oats. Assoc. Offic. Seed Analysts News Letter 25(3):5–7.

4643. —— 1952: Some observations on the germination of flax seed. Assoc. Offic. Seed Analysts News Letter 26(1):9–12.

4644. —— 1953: Report of the committee on seed-borne diseases and pests. Proc. Intern. Seed Testing Assoc. 18:189–95.

4645. —— 1953: Black kernel smudge of wheat seed. Proc. Intern. Seed Testing Assoc. 18:205–17.

4646. —— 1954: Seed testing and control in Sweden. Seed World 75(4):12–14.

4647. —— 1954: Seed testing and control in Sweden (Conclusion). Seed World 75(5): 40–41.

4648. —— 1954: Seed testing in Europe stresses quality control. Farm Res. (N.Y.) 20(3):12–13.

4649. —— 1956: Soil sterilization as an aid in seed germination testing. Assoc. Offic. Seed Analysts News Letter 30(4):13–17.

4650. —— 1957: Germination of *Botrytis* infected seed of China aster. Assoc. Offic. Seed Analysts News Letter 31(1):3–4.

4651. —— 1957: The plant pathologist views seed quality. Assoc. Offic. Seed Analysts News Letter 31(2):40–44.

4652. —— 1957: Research and the state seed law protect users of treated seed. Farm Res. (N.Y.) 23(3):12–13.

4653. —— 1958: Corn seed and seed coat damage. Farm Res. (N.Y.) 24(2):14.

4654. —— 1960: Longevity of seeds. Assoc. Offic. Seed Analysts News Letter 34(4):10–11.

4655. —— 1961: The paradoxical value of seed-borne fungi. Assoc. Offic. Seed Analysts News Letter 35(4):14–17.

4656. —— 1962: *Glomerella cingulata* and *Myrothecium* spp. in seeds of small grains. Plant Disease Reptr. 46:111–13.

4657. —— and C. F. Bruce. 1960: Detection of fungi and fungicides on seeds of winter grains. Proc. Assoc. Offic. Seed Analysts 50:144–51. Chem. Abstr. 55:11742. 1961.

4658. —— and B. Cullinan. Some observations in the germination of grass seeds. Proc. Assoc. Offic. Seed Analysts 33:69–74.

4659. —— and C. E. Heit. 1941: Response of germinants from beans, corn and peas to seed treatments with hormones and mercurials. Proc. Assoc. Offic. Seed Analysts 32:88–92.

4660. —— and —— 1948: Some seed-borne fungi of flowers. Proc. Assoc. Offic. Seed Analysts 38:73–76. Biol. Abstr. 28:4447. 1954.

4661. —— and —— 1954: *Pleospora herbarum* Wallr. and other fungi in seeds of sweet alyssum. Plant Disease Reptr. 38:286–89.

4662. —— and —— 1955: Some interesting fungi infecting flower seeds. Assoc. Offic. Seed Analysts News Letter 29(4):25–29.

4663. ——, ——, M. T. Munn, and S. R. Patrick. 1947: Germination testing of seeds. N.Y. Agr. Expt. Sta. Ann. Rept. 66:48–49.

4664. —— and F. Krapf. 1942: Reduced germination in beans and peas. Farm Res. (N.Y.) 8(2):5, 11.

4665. —— and C. Nelson. 1939: Chemical control of molds when germinating lima beans. Proc. Assoc. Offic. Seed Analysts 30:189–94.

4666. —— and S. R. Patrick. 1937: The value of chemical seed treatments in germination studies. Proc. Assoc. Offic. Seed Analysts 29:117–21.

4667. —— and —— 1939: Chemical elimination of saprophytes during laboratory germination of seed peas. J. Agr. Res. 58:397–422.

4668. —— and —— 1939: Injuries to seedlings induced by chemicals dissolving from germinator trays. Proc. Assoc. Offic. Seed Analysts 30:253–54.

4669. —— and —— 1943: Persistent hardseededness in Early Perfectah peas. Proc. Assoc. Offic. Seed Analysts 34:114–18.

4670. —— and —— 1946: Arasan for control of fungi in germinating corn seed. Phytopathology 36:162–64.

4671. —— and —— 1952: Longevity of hard seeds in winter vetch. Proc. Assoc. Offic. Seed Analysts 42:75–80.

4672. —— and —— 1952: Mushrooms on seeds of side-oats grama (*Bouteloua* *curtipendula*). Assoc. Offic. Seed Analysts News Letter 26(4):7–8.

4673. ——, ——, and M. T. Munn. 1937: Abnormal germination resulting from improperly galvanized trays. Phytopathology 27:867–68.

4674. ——, ——, and D. Weimer. 1941: Influence of phytohormones on germination and growth of cereals. Proc. Assoc. Offic. Seed Analysts 32:83–86.

4675. ——, ——, and C. O. Willits. 1937: Zinc injury in germination tests. Proc. Intern. Seed Testing Assoc. 9:257–62.

4676. —— and E. C. Waters. 1959: *Fusarium graminearum* and other fungi in seedstocks of small grains. Plant Disease Reptr. 43:1013–15. Biol. Abstr. 35:30434. 1960.

4677. —— and D. Weimer. 1940: Some fungi associated with grass seed. Proc. Assoc. Offic. Seed Analysts 31:120–24. Biol. Abstr. 15:1430. 1941.

Crosier, W. F. 458, 4146, 5279, 7297, 13492, 13934, 15119, 15120.

Cross, A. T. 9082.

4678. Cross, D. O. 1936: A rare occurrence. Germination of Kikuyu grass seed. Agr. Gaz. N.S. Wales 47:485.

Cross, F. 8836.

4679. Cross, H. 1938: Laboratory germination of weed seeds. Proc. Assoc. Offic. Seed Analysts 24:125–28.

4680. Cross, J. E., and T. R. Richmond. 1959: The use of glandless seed to determine the amount of natural crossing in *Gossypium hirsutum*. Agron. J. 51:511–12.

4681. Cross, R. J. 1940: Raising primulas and meconopses from seeds. Gardeners Chron. 107:273.

4682. —— 1950: Propagating species *Rhododendron* from seed. Northern Gardener 4:14–15, 18.

4683. Crossley, A., and A. Thomas. 1964: Keeping properties of edible oils. III. Identity of trace materials absorbed from peanut oil by chromatography on alumina. J. Am. Oil Chemists Soc. 41:95–100.

Crossley, A. 3036.

4684. Crossley, D. I. 1953: Seed maturity in white spruce (*Picea glauca*). Can. Forest Res. Div. Silvicult. Res. Note 104. 16 pp.

4685. —— 1955: The production and dispersal of Lodgepole pine seed. Can. Dept. Northern Affairs Natl. Resources Forest Res. Div. Tech. Note 25. 12 pp.

4686. —— 1955: Viability of the seed of Lodgepole pine after 20 years in artificial storage. Forestry Chron. 31:250–53. Biol. Abstr. 31:8899. 1957.

4687. —— 1956: Effect of crown cover and slash density on the release of seed from slash-borne Lodgepole pine (*Pinus contorta*) cones. Can. Dept. Northern Affairs Natl. Resources Forest Res. Div. Tech. Note 41. 51 pp.

Croston, C. B. 6077.

4688. Crovetto, A. 1957: The action of individual nutritive substances on the germination of orchid seeds. Riv. Ortoflorofrutticol. Ital. 41:637–38.

4689. Crow, J. F. 1948: Alternative hypothesis of hybrid vigor. Genetics 33:477–87. Biol. Abstr. 23:5923. 1949.

4690. Crow, W. D., and N. V. Riggs. 1955: The alkaloids of *Lupinus varius* L. I. Isolation of the alkaloids. Australian J. Chem. 8:136–39.

4691. Crowder, J. P., Jr., and B. VonSchmeling. 1958: A device for seeding greenhouse flats. Weeds 6:59–60.

Crowder, L. V. 12551.

4692. Crowfoot, D., and I. Fankuchen. 1938: Molecular weight of a tobacco-seed globulin. Nature 141:522–23.

4693. Crowl, E. D. 1935: Growing delphiniums from seed in California. Flower Grower 22(10):443–44.

Crowley, M. A. 17046.

4694. Crowley, N. C. 1955: The effect of seed extracts on the infectivity of plant viruses and its bearing on seed transmission. Australian J. Biol. Sci. 8:56–67.

4695. —— 1957: The effect of developing embryos on plant viruses. Australian J. Biol. Sci. 10:443–48.

4696. —— 1957: Studies on the seed transmission of plant virus diseases. Australian J. Biol. Sci. 10:449–64.

4697. Crowther, E. M., and H. L. Richardson. 1932: Studies on calcium cyanamide. I. The decomposition of calcium cyanamide in the soil and its effects on germination, nitrification and soil reaction. J. Agr. Sci. 22:300–34. Ber. Ges. Physiol. 68:96. 1932.

4698. Croxall, H. E., and L. Ogilvie. 1940: The incorporation of growth hormones in seed dressings. J. Pomol. Hort. Sci. 17:362–84.

4699. ——, R. W. Sidwell, and J. E. E. Jenkins. 1953: White rot (*Sclerotium cepivorum*) of onions in Worcestershire with special reference to control by seed treatment with calomel. Ann. Appl. Biol. 40:166–75.

4700. Crüger, O. 1932: Ein Anbauversuch mit Rotkleeherkünften. "Georgine," Königsberg (Ostpreussen) 109(68). Proc. Intern. Seed Testing Assoc. 4:208–9. 1932.

4701. Cruess, W. V. 1942: Debittering of apricot kernels. Fruit Products J. Am. Vinegar Ind. 21:365. Expt. Sta. Rec. 88:441. 1943.

Cruess, W. V. 5924, 12933.

4702. Cruickshank, D. H., and F. A. Isherwood. 1958: Glutamic-alanine and glutamic-aspartic transaminases of wheat germ. Biochem. J. 69:189–95.

4703. Cruickshank, I. A. M. 1954: Thermochemical seed treatment. Nature 173:217–18.

4704. —— and H. Jacks. 1953: Seed disinfection. VII. Effect of seed treatments on emergence of linseed and linen flax. New Zealand J. Sci. Technol. Sect. A 35:33–38.

Cruickshank, I. A. M. 9625.

4705. Crum, C. R. 1941: Bees on clover, the value of honeybees as pollinators. Am. Bee J. 81:270–72. Biol. Abstr. 16:13190. 1942.

Crum, R. A. 5325.

4706. Cruz, A. O., and A. P. West. 1931: Composition of Philippine kapok-seed oil. Philippine J. Sci. 46:131–39.

4707. —— and —— 1932: Composition of Philippine rice oil (Ramai variety). Philippine J. Sci. 48:5–12.

4708. —— and —— 1932: Composition of Philippine talisay oil from the seeds of *Terminalia catappa* Linnaeus. Philippine J. Sci. 48:13–19.

4709. —— and —— 1932: Composition of Philippine soybeans and soybean oil. Philippine J. Sci. 48:77–88.

4710. —— and —— 1936: Philippine tobacco seed oil. Philippine J. Sci. 61:161–68. Analyst 62:619. 1937.

4711. —— and —— 1936: Composition of Philippine physic-nut oil. Philippine J. Sci. 61:437–45.

4712. ——, ——, and N. B. Mendiola. 1932: Composition of Philippine rice oil (Hambas variety). Philippine J. Sci. 47:487–95. Ber. Ges. Physiol. 67:633. 1932.

Cruz, A. O. 18806.

4713. Cruz, E. E. 1952: Growing kenaf for fiber and seeds. Philippine Agr. Eng. J. 3:17–18.

4714. —— 1953: A brief review of abaca breeding work in the Bureau of Plant Industry with special reference to directions for handling abaca seeds. Philippines Bur. Plant Ind., Plant Ind. Dig. 16(5):12–15.

Cruz, M. R. 3493.

4715. Cruz, P. I. 1926: A viability test for some tropical seeds. Philippine Agriculturist 14:636–41.

Cruz, P. I. 18714.

Cruz, R. d'; *see* D'Cruz, R.

Cruzado, H. J. 12448.

4716. Csepregi, P. 1959: Germination results of some table grape varieties. Kertesz. Szölesz. Foiskola Evkönyve 7:83–103.

4717. Csonka, F. A. 1939: Amino acids in the corn kernel. J. Agr. Res. 59:765–68.

4718. —— and C. A. Denton. 1946: Methionine determination in proteins and foods. J. Biol. Chem. 163:329–38. Expt. Sta. Rec. 95:773–74. 1946.

4719. —— and D. B. Jones. 1933: Differences in the amino acid content of the chief protein (glycinin) from seeds of several varieties of soybean. J. Agr. Res. 46:51–55.

4720. Cuchtmanowna, S. 1930: Sur la structure du sac embryonnaire chez *Pentunia violacea* Lindl. Acta Soc. Botan. Polon. 7:197–204. Botan. Centralbl. 19: 451. 1931.

4721. Cuculescu, I. 1927: Analytical constants of the seed oil of pumpkin from the Bukowina. Bul. Fac. Stiinte Cernauti 1:49–52. Chem. Abstr. 26:3947. 1932.

Cucullu, A. F. 13115, 17872.

4722. Cuddy, T. F. 1960: Studies on the germination of sugar beet seed. Proc. Assoc. Offic. Seed Analysts 49:98–102. 1959.

Cuddy, T. F. 14393.

4723. Cueller, R., J. Lopez, and R. Toler. 1957: Tratamientos de semillas que se recomiendan para Panama. Divisa Inst. Nacl. Agr. Bol. Inform. 8. 6 pp.

4724. —— and R. W. Toller. 1956: Tratamiento de la semilla de frijol para el control del mal de almacigo. Divisa Inst. Nacl. Agr. Fol. 28. 7 pp.

Cuendet, L. S. 7068, 7414, 7415, 9208, 15244, 17752, 17753.

Cugnac, A. de; *see* De Cugnac, A.

4725. Culbertson, J. O. 1943: Segmented sugar-beet seed. Minn. Univ. Agr. Ext. Serv. Ext. Bull. 240. 8 pp.

4726. ——, V. E. Comstock, and R. A. Frederiksen. 1960: Further studies on the effect of seed coat color on agronomic and chemical characters and seed injury in flax. Agron. J. 52:210–12.

4727. —— and T. Kommedahl. 1956: The effect of seed coat color upon agronomic and chemical characters and seed injury in flax. Agron. J. 48:25–28.

Culbertson, J. O. 4101, 4322, 10586, 11058.

Cull, I. M. 2926–2929.

4728. Cullinan, B. 1947: Germinating certain southern grass seeds. Proc. Assoc. Offic. Seed Analysts 37:162–63.

Cullinan, B. 4146, 4658.

Cullinan, F. P. 17133.

Culp, R. 1792.

4729. Culp, T. W. 1959: Inheritance and association of oil and protein content and seed coat type in sesame, *Sesamum indicum* L. Genetics 44:897–909.

4730. —— 1960: Inheritance of papershell capsules, capsule number and plant color in sesame. J. Heredity 51:146–48.

4731. —— 1960: Inheritance of plant height and capsule length in sesame, *Sesamum indicum* L. Agron. J. 52:101–3.

4732. Culpepper, C. W. 1940: Attempts to grow *Lycoris* from seeds. Herbertia 7:171–72.

4733. —— and C. A. Magoon. 1928: Relation of kernel density to table and canning quality in different varieties of maize. U.S. Dept. Agr. Tech. Bull. 97.

4734. —— and —— 1930: Effects of defoliation and root pruning on the chemical composition of sweet-corn kernels. J. Agr. Res. 40:575–83.

4735. —— and H. H. Moon. 1941: Effect of stage of maturity at time of harvest on germination of sweet corn. J. Agr. Res. 63:335–43.

4736. Culpin, C. 1959: Developments in farm grain drying and storage. Agriculture (London) 66:226–30.

4737. Culter, S. H. 1930: Phytochemical notes. CV. A preliminary study of *Digitalis* seed. Am. J. Pharmacy 102:545–47. Ber. Ges. Physiol. 58:450. 1931.

4738. Culvenor, E. C. J. 1954: The alkaloids of *Heliotropium europaeum* L. II. Isolation and structures of the third major alkaloid and two minor alkaloids and isolation of the principal N-oxides. Australian J. Chem. 7:287–97.

4739. Culver, C. 1951: Johnny ladyslipper seed. Am. Orchid Soc. Bull. 20:609–11.

Cumings, G. A. 15075.

4740. Cumming, B. G. 1963: The dependence of germination on photoperiod, light quality, and temperature, in *Chenopodium* spp. Can. J. Botany 41:1211–33.

4741. —— and J. R. Hay. 1958: Light and dormancy in wild oats (*Avena fatua* L.). Nature 182:609–10.

Cumming, B. G. 8428.

Cumming, E. 1369.

4742. Cumming, W. A. 1960: Germination studies with *Cotoneaster lucida* being conducted at the Canada Experimental Farm, Morden, Man. Western Can. Soc. Hort. Rept. Proc. 16:43–44.

4743. Cummings, M. B. 1914: Large seed a factor in plant production. Vermont Agr. Expt. Sta. Bull. 177:89–123.

4744. ——, E. W. Jenkins, and R. G. Dunning. 1933: Rootstock effects with cherries: seed and phyton propagation. Vermont Agr. Expt. Sta. Bull. 352. 36 pp. Expt. Sta. Rec. 69:805. 1933.

4745. Cummings, W. H. 1952: Loblolly pine shows early differences with source of seed and locality of planting. J. Forestry 50:626–27.

4746. —— and T. G. Zarger. 1945: Nut propagation. Am. Nurseryman 81(8):12.

4747. Cummins, D. G., and W. L. Parks. 1961: The germination of corn and wheat as affected by various fertilizer salts at different soil temperatures. Soil Sci. Soc. Am. Proc. 25:47–49.

4748. Cummins, M. P. 1929: Development of the integument and germination of the seed of *Eleusine indica*. Bull. Torrey Botan. Club 56:155–62.

Cueno, C. 5008.

Cunha, A. G. da; *see* Da Cunha, A. G.

4749. Cunningham, H. S. 1944: Lima bean seed treatment on Long Island. Phytopathology 34:790–98.

4750. Cunningham, J. C., compiler. 1941: Maize bibliography for the years 1917 to 1936, inclusive. Contrib. Iowa Corn Res. Inst. 2(1). 364 pp.

Cunningham, J. C. 1049.

4751. Cunningham, R. H. 1950: A mechanical shaking device. Sci. Agr. (Ottawa) 30:396–97.

4752. —— and A. E. Hannah. 1956: A roller thresher for experimental plot material. Can. J. Agr. Sci. 36:208–9.

4753. Cunningham, T. M. 1957: Seed production and seed fall of *Eucalyptus regnans* (F. Muell.). Australian Forestry 21:30–39.

4754. —— 1960: Seed and seedling survival of *Eucalyptus regnans* and the natural regeneration of second-growth stands. APPITA 13:124–31.

4755. Cupcea, E., and L. Stoicovici. 1961: Influence of 2,4-dichlorophenoxyacetic and of iodoacetic acid on seed repiration in the course of germination. Acad. Rep. Populare Romine Studii Cercetari Biol. Ser. "Biol. Veg." 13:357–68. Chem. Abstr. 57:5053. 1962.

4756. Curcio, M. 1949: Observations on the germination of tobacco seeds. Tobacco 53:49–53.

Curet, M. C. 2658.

Curet, M. E. 9784.

4757. Curin, B. 1938: Food value of Czechoslovakian maize. Sb. Česk. Akad. Zeměděl. Věd 13:154–60. Chem. Abstr 33:3468. 1939.

4758. Curl, A. L., and E. K. Nelson. 1944: A water-soluble Mannan from the seeds of *Daubentonia drummondii*. J. Am. Chem. Soc. 66:1227.

Curnow, D. H. 16434.

4759. Currence, T. M. 1944: A combination of semi-sterility with two simply inherited characters that can be used to reduce the cost of hybrid tomato seed. Proc. Am. Soc. Hort. Sci. 44:403–6.

Currin, R. E. 2152.

Curth, P. 1975, 1976.

Curtis, B. C. 16926.

4760. Curtis, E. J. C., and J. E. Cantlon. 1963: Germination of *Melampyrum lineare:* effects of gibberellic acid and duration of storage at 20°C. Proc. Plant Physiol. Meetings 38:liv.

4761. —— and —— 1963: Germination of *Melampyrum lineare:* interrelated effects of afterripening and gibberellic acid. Science 140:406–8.

4762. Curtis, H. 1961: Radiation effects in dry seeds. Nucl. Sci. Abstr. 15:2317.

4763. Curtis, H. J., N. Delihas, R. S. Caldecott, and C. F. Konzak. 1958: Modification of radiation damage in dormant seeds by storage. Radiation Res. 8:526–34. Biol. Abstr. 33:3314. 1959.

4764. —— and H. H. Smith. 1963: Corn seeds affected by heavy cosmic ray particles. Science 141:158–60.

Curtis, H. J. 11188.

4765. Curtis, J. D. 1952: Effect of pregermination treatments on the viability of *Ceanothus seed.* Ecology 33:577–78.

4766. —— 1955: Effects of origin and storage method on the germinative capacity of Ponderosa pine seed. U.S. Dept. Agr. Forest Serv. Intermountain Forest Range Expt. Sta. Res. Note 26. 5 Mimeo. pp.

4767. —— 1958: Germinative capacity of Englemann spruce seed. U.S. Dept. Agr. Forest Serv. Intermountain Forest Range Expt. Sta. Res. Note 58. 3 Mimeo. pp.

4768. —— and G. P. Steinbauer. 1947: Effects of variations in moisture and depth of planting on germination of seeds of Northern white cedar. J. Forestry 45: 592–93.

Curtis, J. D. 14763.

4769. Curtis, J. J., A. M. Brunson, J. E. Hubbard, and F. R. Earle, 1956: Effect of the pollen parent on oil content of the corn kernel. Agron. J. 48:551–55.

Curtis, J. J. 3215, 5743, 12076, 12077, 18362.

4770. Curtis, J. T. 1936/37: Some phases of symbiotic and nonsymbiotic orchid seed germination. Thesis (Ph.D.) Univ. of, Wisconsin.

4771. —— 1939: The relation of specificity of orchid mycorrhizal fungi to the problem of symbiosis. Am. J. Botany 26:390–99.

4772. —— 1943: Germination and seedling development of five species of *Cypripedium* L. Am. J. Botany 30:199–206.

4773. —— 1947: Studies on the nitrogen nutrition of orchid embryos. I. Complex nitrogen sources. Am. Orchid Soc. Bull. 16:654–60. Biol. Abstr. 23:5070. 1949.

4774. —— 1947: Undifferentiated growth of orchid embryos on media containing barbiturates. Science 105:128.

4775. —— and M. A. Nichol. 1948: Culture of proliferating orchid embryos in vitro. Bull. Torrey Botan. Club 75:358–73.

4776. Curtis, L. C. 1946: The possibilities of using species of perennial cucurbits as source of vegetable fats and protein. Chemurgic Dig. 5(13):221, 223–24.

4777. —— 1948: The use of naked seed in *Cucurbita pepo* as a source of high quality liquid vegetable fat, as a high analysis protein, as a new confection, and as a sandwich spread. Proc. Am. Soc. Hort. Sci. 52:403–6.

4778. —— and J. Scarchuk. 1948: Seedless peppers. A single Mendelian recessive character. J. Heredity 39:159–60.

4779. Curtis, W. C. 1942: Propagation of wild flowers. II. Wild Flower 19:34–40.

4780. Curtler, E. A. 1926: Experiments on the germination of African oil palm seeds. Malayan Agr. J. 14:84–87.

Cushman, H. E. 9225.

4781. Cutler, A. M. 1945: Butterfly weeds from seed. Horticulture 33(15):390.

4782. —— 1946: Wild Flowers from seed. Horticulture 24(14):390.

4783. Cutler, G. H. 1932: Comparison of chamber and field germination tests of soybeans. J. Am. Soc. Agron. 24:544–50.
4784. —— 1940: Effect of "clipping" or rubbing the oat grain on the weight and viability of the seed. J. Am. Soc. Agron. 32:167–75.
Cutler, H. G. 19166.
Cutler, J. R. 5610.
4785. Cutler, L. H. 1934: Starting lilies from seed indoors. Horticulture 12(11):194.
4786. Cutter, V. M., Jr., and B. Freeman. 1954: Development of the syncytial endosperm of *Cocos nucifera*. Nature 173:827–28.
4787. ——, K. S. Wilson, and B. Freeman. 1955: Nuclear behavior and cell formation in the developing endosperm of *Cocos nucifera*. Am. J. Botany 42:109–15.
Cutler, V. M., Jr. 19736.
4788. Cuzin, J., A. Fardy, and D. Schwartz. 1951: Microchirurgie de la graine de tabac technique et premiers resultats. Ann. Inst. Exptl. Tabac Bergerac 1(2):36–54. Biol. Abstr. 26:9590. 1952.
Cuzin, J. 6217.
Cuzzocrea, G. 4827.
4789. Cynberg, D. I. 1928: Recherches sur la catalase. II. Recherches sur la transformation de la matière chez le *Raphanus*. Thesis No. 669, Univ. of Geneva. 35 pp. Soc. Anonyme, Geneva. Biol. Abstr. 6:15557. 1932.
4790. Czabator, F. J. 1962: Germination value: An index combining speed and completeness of pine seed germination. Forest Sci. 8:386–96.
Czabator, F. J. 13229.
4791. Czaja. 1957: Pectins from tamarind seed (*Tamarindus indicus*). Materiae Veg. 2:258–59. Chem. Abstr. 52:12094. 1958.
4792. Czaja, A. T. 1947: Untersuchungen über die Samenproduktion der gebauten Yuccapflanzen. Planta 35:117–31.
4793. —— 1963: Neue Untersuchungen an der Testa der Tomatensamen. Planta 59:262–79.
4794. Czapik, R. 1961: Embryological studies in the genus *Potentilla* L. I. *P. crantzii*. Acta Biol. Cracov. Ser. Botan. 4:97–119.
Czapski, J. 20038.
4795. Czepa, A. 1924: The effect of X-rays on the germination process of plant seeds. Fortschr. Gebiete Röntgenstrahlen Nuklearmed. 32:65–75.
Czerny, H. 11040.
4796. Czosnowski, J. 1962: Metabolism of excised embryos of *Lupinus luteus* L. I. Effect of metabolic inhibitors and growth substances on the water uptake. Acta Soc. Botan. Polon. 31:135–52.
4797. —— 1962: Metabolism of excised embryos of *Lupinus luteus* L. II. The water uptake as influenced by external concentration. Acta Soc. Botan. Polon. 31:683–91.
4798. —— 1962: Metabolism of excised embryos of *Lupinus luteus* L. III. Comparative study of cultured embryos and normal seedling axes. Acta Soc. Botan. Polon. 31:693–702.
4799. Czyzewska, S. 1958: Mycological and phytopathological studies on rape seeds (*Brassica napus* L. var. *oleifera* D.C.). Roczniki Nauk Rolniczych Ser. A 78: 283–307.
4800. Czyzewski, T. 1958: Causes of the loss of germination capacity in *Daucus carota* seeds. Zeszyty Nauk. Wyzszej Szkoly Rolniczej Wroclawiu Rolnictwo 17:69–94. Chem. Abstr. 53:8314. 1959.

D

Dabney, J. C. 8724.
4801. Dacherman, M. 1931: The tomato-seed oil. Maslob.-Zhir. Delo 12:30–36. Chem. Abstr. 26:4727. 1932.
4802. Da Costa-Primo, S. 1936: The behavior of seed germination in presence of some alcohols. Bull. Soc. Port. Sci. Nat. 12:167–70. Chim. Ind. (Paris) 39:559. Chem. Abstr. 32:4631. 1938.
4803. Da Costa-Sacco, J. 1952: Composition of *Canavalia bonariensis* seeds. Agros (Pelotas) 6:118–30. Chem. Abstr. 49:5594. 955.
4804. Da Cunha, A. G. 1928: La méthode des impregnations argentiques dans l'étude des grains pendant la germination. Compt. Rend. Hebd. Séances Mem. Soc. Biol. (Paris) 98:1017–19.
4805. —— 1929/30: Sur l'origine mitochondriale de la diastase pendant la germination des graines. Bol. Soc. Broteriana Ser. II 6:116–23.
4806. —— 1932: Remarques sur la cytologie du développement du la graine de blé. Compt. Rend. Soc. Biol. 110:1043–44.

4807. —— 1932: Sur le dépôt de l'amidon dans les cellules de l'albumen pendant le développement des graines. Compt. Rend. Soc. Biol. 110:1045–46.

4808. Dadd, A. H., and S. E. Jacobs. 1958: The uptake of mercury by seeds treated with mercuric chloride solution, and its subsequent release. J. Appl. Bacteriol. 21:97–99.

Dadd, A. H. 9642.

4809. Dadykin, V. P. 1951: On swelling of seeds in frozen soil. Dokl. Akad. Nauk SSSR 77:333–36.

4810. Daessler, H. G., and W. Zentsch. 1959: The germination-inhibiting action of constituents of fir cone oil (templin oil). Pharmazie 14:111–12. Biol. Abstr. 34: 5489. 1959.

4811. Da Fano, A. 1916: On the germinability of rice (*Oryza sativa*) and corn (*Zea mays*) in relation to temperature and humidity. Atti Ist. Botan. R. Univ. Pavia 2 Ser. 16:17–39. Expt. Sta. Rec. 37:736. 1917.

Dag, J. 12834.

Dagand, G. S. 5075.

4812. Daggs, R. G., and H. S. Halcro-Wardlaw. 1933: The conversion of fat to carbohydrate in the germinating castor bean. II. The combustion respiratory quotient as determined by a modified oxycalorimeter. J. Gen. Physiol. 17:303–9.

4813. Daglish, C., and F. Wokes. 1948: Hydrojuglone and apparent vitamin C in walnuts. Nature 162:179–80.

4814. Dagys, J. 1935: Wuchstoffe der Mikroorganismen in Embryonalen Geweben und im Blutungssaft. Protoplasma 24:14–91.

4815. Dahl, C. G. 1930: Root stocks from seeds of known parents. 9th Intern. Hort. Congr., London Rept. Proc. 1930: 141–49. Expt. Sta. Rec. 66:334. 1932.

4816. Dahlberg, R. C. 1917: The germination of seeds of *Agropyron repens*. Proc. Assoc. Offic. Seed Analysts 9/10:21–24.

4817. Dahlgren, B. E. 1938: Plants from seeds centuries old. Sci. Dig. 3(6):82.

4818. Dahlgren, K. V. O. 1931: Mehrere Pflanzen aus einem Samen. Naturens Liv i Ord Och Bild. Stockholm 54:180–88. Botan. Centralbl. 164:260. 1933.

4819. —— 1934: Die Embryologie von *Impatiens roylei*. Svensk Botan. Tidskr. 28:103–25. Biol. Abstr. 9:10139. 1935.

4820. —— 1934: Die Embryosackentwicklung von *Echinodorus macrophyllus* und *Sagittaria sagittifolia*. Z. Wiss. Biol. Abt. E 21:602–12. Biol. Abstr. 9:5841. 1935.

4821. —— 1937: Die Entwicklung des Embryosackes bei *Plumbago zeylanica*. Botan. Notiser 1937:487–98.

4822. —— 1939: Sur la présence d'amidon dans le sac embryonnaire chez les angiospermes. Botan. Notiser 1939:221–31. Botan. Centralbl. 175:216. 1940.

4823. —— 1939: Endosperm- und Embryobildung bei *Zostera marina*. Botan. Notiser 1939:607–15.

Dahlmann, M. Weber-; *see* Weber-Dahlmann, M.

4824. Dahm, P. 1924: Untersuchungen über die Abhängigkeit der Endospermentleerung bei *Zea mais* von verschiedenen Salzen. Jahrb. Wiss. Botan. 63:273–320.

Dahmen, M. Moutschen-; *see* Moutschen-Dahmen, M.

4825. Daigo, Y. 1942: Hormone treatment of lowland rice. Crop Sci. Soc. Japan Proc. 13:275–78.

Daimer, J. 13342.

Dalberg, L. Y. 6926.

4826. Dalby, D. H. 1963: Seed dispersal in *Salicornia pusilla*. Nature 199:197–98.

4827. D'Alcontres, G. S., and G. Cuzzocrea. 1955/56: The chemical and physicochemical characteristics of proteins extractable from seeds of *Jatropha curcas*. Att. Soc. Peloritana Sci. Fis. Mat. Nat. 2:71–84. Chem. Abstr. 52:8230c. 1958.

4828. Dale, W. T., and L. I. Scott. 1943: Structural characteristics of the testa in *Capsella* and *Sinapis*. Proc. Leeds Phil. Lit. Soc. Sci. Sect. 4:111–22.

4829. Dales, H. 1953: Technique of the Gentner fluorescence test. A suggested modification. Proc. Intern. Seed Testing Assoc. 18:263–66.

4830. Daletzkaya, T. V. 1963: On the possible causes of the physiological dwarfing of plant embryos. J. Botan. URSS 48:1361–68.

4831. Dalkiewicz-Baranowska, H. 1962: Hard seeds. Szkola Glowna Gosp. Wiejsk. Roln. Warsaw 5:3–29.

Dall'Olio, G. 12881.

4832. Dalmatov, V. 1933: The "proto acid" of the field nut. Tr. Lab. Izucheniyu Belka Belkovogo Obmena Organizme 1933:50–53. Chem. Abstr. 29:6613. 1935.

Dalle-Rose, A. Dona-; *see* Dona-Dalle-Rose, A.

D'Almeida, A. 7120.

Dalskov, F. 1919.

4833. Damanski, A. F., and N. V. Radej. 1954: L-Threonine in wheat, white and yellow maize, beans and peas. Acta Pharm. Jugoslav. 4:109–14. Chem. Abstr. 49:4805. 1955.

Damanski, A. F. 13557.

4834. D'Amato, F. 1939: Ricerche embryiologiche e cariologiche sul genere "*Euphorbia*." Nuovo Giorn. Botan. Ital. 46:470–509.

4835. —— 1953: Presence of antimitotic substances in the seeds of *Pisum sativum*. Caryologia 5:217–22. 1952. Chem. Abstr. 48:13837. 1954.

4836. —— 1954: On some physiological and genetical aspects of seed aging. A contribution to the problem of aging and of spontaneous mutability in plants. Caryologia 6:217–40. Biol. Abstr. 29:29695. 1955.

4837. —— 1955: Protective action of cysteine against the inhibition of germination and growth of wheat by triethylenemelamine. Atti Accad. Nazl. Lincei Rend. Cl. Sci. Fis. Mat. Nat. 18:222–26. Chem. Abstr. 49:16075. 1955.

D'Amato, F. 13306.

Damaty, A. H. El-. 10173.

Damblon, J. 13372.

Dameron, W. H. 16155.

4838. Damiens, A., and M. S. Blaignan. 1931: The normal bromine (vegetable kingdom): edible seeds, wheat, bread. Compt. Rend. 193:1460–62. Chem. Abstr. 26:2488. 1932.

4839. Damisch, W. 1963: Über den Einfluss biologischer Faktoren auf den Samenansatz bei Rotklee. Biol. Centralbl. 82:303–41.

4840. Dammer, U. 1897: Verpackung und Versand von Samen, welche ihre Keimkraft schnell verlieren. Z. Trop. Landwirtsch. 1:40–41.

4841. —— 1900: Über die Keimung der *Lodoicea sechellarum* Labil. Notiz. Kgl. Botan. Gartens Museums Berlin 22.

4842. Damodaran, M., and K. R. Nair. 1936: Amino acid dehydrogenases in germinating seedlings. Current Sci. (India) 5:134. Chem. Abstr. 31:735. 1937.

4843. —— and —— 1938: Glutamic acid dehydrogenase from germinating seeds. Biochem. J. 32:1064–75.

4844. —— and R. Ramaswamy. 1937: Isolation of l-3:4-dihydroxyphenylalanine from the seeds of *Mucuna pruriens*. Biochem. J. 31:2149–52.

4845. —— and —— 1940: Succinic dehydrogenase in germinating seeds. Current Sci. (India) 9:319–20. Chem. Abstr. 34:7992–93. 1940.

4846. ——, ——, T. R. Venkatesan, S. Mahadevan, and K. Ramdas. 1946: Amide synthesis in plants. II. Amino-acid changes in germinating seedlings. Proc. Indian Acad. Sci. Sect. B 23:86–89. Expt. Sta. Rec. 95:491–92. 1946.

4847. —— and P. N. Rangachari. 1946: Tamarind seed "pectin." Current Sci. (India) 15:20–21.

4848. —— and S. S. Subramanian. 1948: Amide synthesis in plants. IV. Aspartase in germinating seedlings. Proc. Indian Acad. Sci. Sect. B 27:47–53. Biol. Abstr. 22:25388. 1948.

4849. —— and T. R. Venkatesan. 1948: Amide synthesis in plants. III. Urea formation in seedlings. Proc. Indian Acad. Sci. Sect. B 27:26–32. Biol. Abstr. 22:25365. 1948.

4850. Danard, M. C. 1953: Germination tests of red top and bentgrass. Assoc. Offic. Seed Analysts News Letter 27(1): 18–20.

4851. Dankwortt, P. W. 1935: Toxicity and fluorescence of vetch seeds. Z. Lebensm. Untersuch.-Forsch. 69:458–63. Chem. Abstr. 29:7521–22. 1935.

Dandliker, W. B. 7916.

4852. Dangeard, P. 1921: Sur la formation des grains d'aleurone dans l'albumen du ricin. Compt. Rend. 173:857–59.

4853. —— 1922: Sur l'origine des vacuoles aux depens de l'aleurone pendant germination des graminées. Compt. Rend. 174: 319–21.

4854. —— 1946: Sur le moment de la fecondation chez le *Ginkgo*. Bull. Soc. Botan. France 93:19–20.

4855. Dangoumau, A. 1935: Recherches sur l'insaponifiable de l'huile de froment. Libr. Delmas, Bordeaux. 147 pp. Expt. Sta. Rec. 80:439. 1939.

4856. Dang-Van-Liem, M. 1959: Embryogenic des Buxacées. Développement de l'embryon chez le *Buxus sempervirens* L. Compt. Rend. 248:1844–47.

4857. —— 1959: Embryogénie des Euphorbiacées. Développement de l'embryon chez l'*Euphorbia palustris* L. Compt. Rend. 249:2378–80.

4858. —— 1961: Embryogénie des Euphorbiacées. Développement de l'embryon chez le *Phyllanthus grandifolius* L. Compt. Rend. 252:443–45.

Dani, P. G. 3869.

Daniel, A. Faludi-; *see* Faludi-Daniel, A.

4859. Daniel, L. 1929: Production de variétés nouvelles par le semis des graines du topinambour greffe sur soleil annuel. Rev. Bretonne Botan. 1929:119–30.

4860. —— 1938: Sur deux nouveaux cas de Xenie chez le pommier. Compt. Rend. 207:1360–62.

4861. —— 1939: Sur les variations de la couleur des gousées et des graines des haricots. Compt. Rend. 209:499–501.

4862. Daniel, W. H., and N. R. Goetze. 1957: Germination of turf cover species as influenced by periods of soaking in various fertilizer solutions. Agron. Abstr. 4:74.

4863. Daniel-Comé, M. 1961: An experimental study of the permeability to different gases of the integument of some seeds. Compt. Rend. Hebd. Séances Acad. Agr. France 47:55–61. Biol. Abstr. 45:25891. 1964.

4864. Daniels, D. G. H. 1959: The estimation and location of methoxyhydroquinone glycosides in the wheat grain. Cereal Chem. 36:32–41.

4865. ——, H. G. C. King, and H. F. Martin. 1963: Antioxidants in oats: esters of phenolic acids. J. Sci. Food Agr. 14:385–90.

Daniels, D. G. H. 4434, 18755.

Daniels, F. 3445.

4866. Danielsson, B. 1950: Embryo cultures of stone fruits. Sveriges Pomol. För. Äraskr 1950:200–6.

4867. Danielsson, C. E. 1949: Seed globulins of the Gramineae and Leguminosae. Biochem. J. 44:387–400. Chem. Abstr. 43:8456. 1949.

4868. —— 1950: An electrophoretic investigation of vicilin and legumin from seeds of peas. Acta Chem. Scand. 4:762–71.

4869. —— 1951: The breakdown of the high-molecular reserve proteins of peas during germination. Acta Chem. Scand. 5:541–54.

4870. —— 1951: Studies on a proteolytic enzyme from seeds of peas. Acta Chem. Scand. 5:791–804.

4871. —— 1952: A contribution to the study of the synthesis of the reserve protein in ripening pea seeds. Acta Chem. Scand. 6:149–59.

4872. —— 1952: The seed proteins of the Gramineae and Leguminosae. Svensk Kem. Tidskr. 64:43–63. Chem. Abstr. 46:6204–5. 1952.

4873. —— 1956: Starch formation in ripening pea seeds. Physiol. Plantarum 9:212–19.

4874. —— 1959: Sugar and starch in ripening peas. Acta Agr. Scand 9:181–88.

4875. —— 1960: Sugar and water in ripening peas (*Pisum sativum* L.). Physiol. Plantarum 13:342–47.

4876. —— and H. Halina. 1952: Differences in the chemical composition of some pea proteins. Acta Chem. Scand. 6:139–48. Biol. Abstr. 27:14193. 1953.

4877. —— and B. Igelman. 1947: A comparative study of globulins from seeds of various types of beet. Svensk Kem. Tidskr. 59:162–65. Biol. Abstr. 23:12650. 1949.

Danielsson, C. E. 17681.

4878. Danilina, A. N. 1959: The embryology of *Lathyrus annuus* L. Nauchn. Dokl. Vysshei Shkoly Biol. Nauki 1959(4):108–13. Biol. Abstr. 42:3020. 1963.

4879. Danilova, T. A. 1950: The influence of the processing of seeds with heteroauxin on the yield and the course of biochemical processes in the sugar beet. Dokl. Akad. Nauk SSSR 72:801–4. Chem. Abstr. 48:5939. 1954.

4880. Danjo, T. 1952: On the consumption of the stored foods in the seeds in germination of cereals. III. Relation to the growth of corn. Proc. Crop Sci. Soc. Japan 20: 268–70.

4881. —— 1953: On the consumption of the stored foods in the seeds during germination of cereals. IV. Relation to sowing depth of wheat, barley and oats. Proc. Crop Sci. Soc. Japan 21:203–4.

4882. —— 1953: On the consumption of the stored foods in the seeds in germination of cereals. V. Effect of hot water treatment for seed of wheat and barley. Proc. Crop Sci. Soc. Japan 22:67–68.

4883. —— 1955: On the consumption of the stored foods in the seeds during germination of cereals. VIII. Expansion of epithelium tissue of scutellum in oat seed. Proc. Crop Sci. Soc. Japan 23:242.

4884. —— and M. Inosaka. 1960: On the tissue of scutellum concerning with absorption of nutrients of endosperm in rice and oat seeds. Proc. Crop Sci. Soc. Japan 29:100–2.

4885. ——, H. Shin, and Y. Shikamoto. 1960: On the germination of rice seeds in ears in early season cultivation. Bull. Fac. Agr. Univ. Miyazaki 6:137–43. Biol. Abstr. 36:46059. 1961.

Dankert, M. 16041.

4886. Dann, B., and L. R. Waldron. 1933: A statistical study of certain characters of the alfalfa plant concerned with seed production. J. Am. Soc. Agron. 25:184–90.

4887. Dann, W. J. 1931: The temperature coefficients and energy exchanges of the citric acid dehydrogenase of cucumber seeds. Biochem. J. 25:177–89.

Danon, J. 20119.

4888. Danon, M., and V. Lusin. 1961: The influence of seed disinfectants on the germination of stored corn. Nachrbl. Deut. Pflanzenschutzdienst (Berlin) 13:23–25. Chem. Abstr. 59:6716. 1963.

4889. Dantuma, G., and J. E. Andrews. 1960: Differential response of certain barley and wheat varieties to hardening and freezing during sprouting. Can. J. Botany 38:133–51.

4890. Danuser-Bührer, H. 1960: Vergleiche von Samenquantum und Keimprozent bei Lärchensamen verschiedener Provenienz der Ernte 1958/1959 sowie Erfahrungen bei der Aussaat. Schweiz. Z. Forstw. 111:169–72.

4891. Daoud, K. M. 1932: The reserve polysaccharide of the seeds of fenugreek; its digestibility and its fate during germination. Biochem. J. 26:255–63.

Dapina, S. I. 3481.

4892. Da Ponte, J. J. 1962: On the germinative power of seeds of moco cotton, *Gossypium hirsutum* Marie-Galante Hutch., treated with sulfuric acid. Turrialba 12:47–49. Biol. Abstr. 39:24762. 1962.

4893. D'Aquin, E. L., J. Pominski, H. L. E. Vix, and E. A. Gastrock. 1961: Filtration-extraction of safflower seed on a bench scale. J. Am. Oil Chemists Soc. 38:44–47.

Darab, C. 15264.

4894. Daragan-Sushchova, A. I. 1952: The effect of pre-sowing treatment of seeds with biogenous stimulators on the biochemical indexes of plants. Dokl. Akad. Nauk SSSR 82:469–72. Chem. Abstr. 49:5586–87. 1955.

4895. Darbre, A., and F. W. Norris. 1956: Vitamins in germination. Determination of free and combined inositol in germinating oats. Biochem. J. 64:441–46.

4896. —— and —— 1957: Determination of free and combined inositol in the ungerminated bean seed and the young plant. Biochem. J. 66:404–7.

4897. Darby, S. P. 1957: The Georgia seed certification program. Southern Conf. Forest Tree Impr. Proc. 4:102–4.

4898. D'Arcy, R. A. 1954: A routine method for estimating the starch content of wheat by-products. Cereal Chem. 31:37–42.

4899. Dariyev, A. S. 1963: Morphologo-anatomical structure of cotyledons of some cotton (*Gossypium* L.) species. J. Botan. URSS 48:1355–61.

4900. Darlington, H. T. 1922: Dr. W. J. Beal's seed-viability experiment. Am. J. Botany 9:266–69.

4901. —— 1931: The 50 year period for Dr. Beal's seed viability experiment. Am. J. Botany 18:262–65.

4902. —— 1941: The sixty-year period for Dr. Beal's seed viability experiment. Am. J. Botany 28:271–73.

4903. —— 1951: The seventy-year period for Dr. Beal's seed viability experiment. Am. J. Botany 38:379–81.

4904. —— and G. P. Steinbauer. 1961: The eighty-year period for Dr. Beal's seed viability experiment. Am. J. Botany 48:321–25.

Darlow, A. E. 3041, 3042.

Darnet, L. M. 3704.

4905. Darpaux, H., J. Louvet, and J. Ponchet. 1957: Essais de traitement des semences de crucifères contre le *Phoma lingam* (Tode) Desm. et l'*Alternaria brassicae* (Berk.) Sacc. Ann. Epiphyties 4:545–57.

4906. —— and J. Ponchet. 1957: Controle de l'état sanitaire et disinfection des semences études realisées en France. Proc. Intern. Seed Testing Assoc. 22:375–92.

4907. Darquennes, H. 1953: The germination of cotton seeds in Ubangi. Bull. Inform. INEAC (Inst. Natl. Étude Agron. Congo Belge) 2:51–62. Chem. Abstr. 48:934. 1954.

4908. Darragh, W. H. 1931: Fumigation of seed maize. Agr. Gaz. N.S. Wales 42:377–78. Biol. Abstr. 6:10953. 1932.

4909. —— 1931: The viability of sugar cane seed. Agr. Gaz. N.S. Wales 42:852–54.

4910. Darrah, W. C. 1941: Fossil embryos in Iowa coal balls. Chronica Botan. 6:388–89. Biol. Abstr. 16:10364. 1942.

4911. Darrow, B. S. 1936: A fossil Araucarian embryo from the Cerro Cuadrado of Patagonia. Botan. Gaz. 98:328–38.

4912. Darrow, G. M. 1940: Seed size in strawberries. Proc. Am. Soc. Hort. Sci. 37:564–66.

4913. —— 1941: Seed size in blueberry and related species. Proc. Am. Soc. Hort. Sci. 38:438–40.

4914. —— 1958: Seed number in blueberry fruits. Proc. Am. Soc. Hort. Sci. 72:212–15.

4915. —— and D. H. Scott. 1954: Longevity of blueberry seed in cool storage. Proc. Am. Soc. Hort. Sci. 63:271.

4916. —— and H. Sherwood. 1931: Seed and berry size of cane fruits. Proc. Am. Soc. Hort. Sci. 28:194–99.

Darrow, G. M. 13790.

4917. Darsie, M. L., C. Elliott, and G. J. Peirce. 1914: A study of the germinating power of seeds. Botan. Gaz. 58:101–36.

4918. Darwin, C. 1857: On the action of sea-water on the germination of seeds. J. Linn. Soc. London 1:130–40.

4919. Das, B. C. 1963: Effects of β-naphthoxyacetic acid on tillering and grain production in rice. Botan. Gaz. 124:270–74.

4920. Das, C. R., and H. K. Baruah. 1944/46: V. Experimental studies on the parasitism of rice by *Helminthosporium oryzae* Breda de Haan and its control in field and storage. Trans. Bose Res. Inst. (Calcutta) 16:31–46.

Das, D. C. 15797–15799.

4921. Das, H. H., and S. C. Roy. 1959: Oxidation of some intermediates of the tricarboxylic acid cycle by a mitochondrial preparation from the germinating *Vigna sinensis*. Sci. Cult. (Calcutta) 25:317–18.

Das, K. L. 2508, 2509.

Das, M. C. 1858.

4922. Das, M. L., and B. C. Guha. 1960: Isolation and chemical characterization of bound niacin (niacinogen) in cereal grains. J. Biol. Chem. 235:2971–76. Chem. Abstr. 55:4668. 1961.

4923. Das, B. N., and R. M. Banerjee. 1952: Utilization of mango-seed kernel as a source of starch. Sci. Cult. (Calcutta) 17:339–40. Biol. Abstr. 26:25919. 1952.

4924. —— and T. S. Bisivas. 1953: Identification of amino acids in wheat, gluten and mango-seed kernel by paper chromatoggraphy. Sci. Cult. (Calcutta) 19:158–59. Biol. Abstr. 29:4386. 1955.

Das, N. B. 1843, 7861, 7862.

4925. Das, S. 1945: Apricot seed cake as a nitrogenous manure. Indian J. Agr. Sci. 15:30–36. Biol. Abstr. 20:5555. 1946.

4926. Das, S. B. C. 1961: Polyembryony in mulberry. Current Sci. (India) 30:203.

4927. Das, T. M., and J. M. Choudhury. 1960: The effect of light on translocation of hydrolysed substances from endosperm to embryo and increase of dry weight in rice seedlings. Current Sci. (India) 29:64–65.

4928. —— and —— 1963: Effect of periodic distribution of low irradiance upon accumulation of dry matter in different parts of the rice embryo and changes in the dry weight of the endosperm during post germination stages. Österr. Botan. Z. 110:33–43.

Das, T. M. 17535–17537.

4929. Das, V., S. Rama, and M. P. Rao. 1962: Respiratory and acid metabolism in the developing seeds of *Hibiscus esculentus* L. Indian J. Plant Physiol. 5:230–36.

4930. Dashkevich, M. D. 1961: Effect of size of Scots pine cones on the yield of seeds, their quality, and the growth of seedlings in the nursery. Lesn. Zh. 1961: 159–61.

DaSilva, A. R. 6406.

4931. Da Silva, F. I. R. 1960: A short contribution to the biometric study of the seeds of certain strains of rice cultivated in Portugal. Proc. Intern. Seed Testing Assoc. 25:355–63.

4932. —— 1962: Les graines de mauvaises herbes dans les lots certifies pour l'exportation. Proc. Intern. Seed Testing Assoc. 27:489–97.

4933. Daskalov, K. 1956: Seed stimulation as a method of raising crop yields. Selskostpanska Misul 1:321–36.

4934. Dass, C. M. 1943: The production and storage of vegetable seed. Agr. J. Fiji Dept. Agr. 14:38–40.

Dastidar, A. G. 17538.

Dastugue, G. 7354.

Dastur, N. N. 2400.

4935. Dastur, R. H., and R. M. Desai. 1935: The carbon dioxide carbohydrate ratio in the aerobic and the anaerobic respiration of rice. Ann. Botany (London) 49:53–69.

4936. —— and L. T. Mone. 1958: Scientific aspects of the method of pre-soaking seeds in solutions of salts for getting increased yields in crop plants. Indian J. Agr. Sci. 28:1–12. Biol. Abstr. 35:17068. 1960.

Datta, C. 13607.

4937. Datta, D. D., and H. L. Bami. 1962: Oil from seeds of *Lepidium sativum* Linn. Sci. Cult. (Calcutta) 28:283. Biol. Abstr. 44:12546. 1963.

Datta, D. D. 4944.

Datta, P. M. 1852.

4938. Datta, R. L., T. Basu, and P. K. Ghose. 1960: Kusum tree, seed and oil (conclusion). Oils Oilseeds J. 12(7):14–16.

4939. Datta, R. M. 1934: On the development of the embryo sac and the pollen grain in *Cassia tora*, Linn. J. Indian Botan. Soc. 13:277–99.
4940. —— 1941: Tricotyledony in *Morus multicaulis* Perr. Current Sci. (India) 10:440.
4941. —— 1960: Polyembryony in *Corchorus olitorius* Linn. (Wild.). Sci. Cult. (Calcutta) 25:642.
4942. Datta, S. 1934: Embryological and cytological studies in *Nolana atriplicifolia* and *Nolana prostrata*. J. Indian Botan. Soc. 12:131–53.
4943. Datta, S. C. 1961: Germination studies on the seeds of two desert plants. Indian Agriculturist 5:111–12.
4944. —— and D. D. Datta. 1961: Pharmacognostical study of the seeds of *Lepidium sativum* Linn. J. Sci. Ind. Res. (India) 20:60–62. Biol. Abstr. 38:3519. 1962.
Datta, S. C. 5062, 5063.
Datta, Y. P. 13292.
4945. Dattilo, M. 1955: Research on the germinative power of the seeds of *Scorpiurus muricata* L. var. *subvillosa* (L.). Ann. Ist. Sper. Zootec. Roma 5:335–47. Biol. Abstr. 32:42009. 1958.
4946. Datuashvili, G. A. 1960: Effect of seed sorting on yields of corn. Kukuruza 1960(2):20–21.
Daubney, C. G. 9314.
4947. Daugherty, P. M., H. H. Sineath, and T. A. Wastler. 1958: Industrial raw materials of plant origin. IV. A survey of *Simmondsia chinensis* (Jojoba). Econ. Botany 12:296–306.
4948. Dauphine, A., and S. Rivière. 1946: Les tubes cribles de l'embryon et leur mise en evidence. Rev. Gen. Botan. 53:19–24.
Davenport, M. G. 17749, 17750.
4949. Davern, C. I., J. W. Peak, and F. H. W. Morley. 1957: The inheritance of flowering time in *Trifolium subterraneum* L. Australian J. Agr. Res. 8:121–34.
4950. Davey, A. J. 1935: Note on the structure of the epicotyl in *Juglans nigra*. New Phytologist 34:201–10.
Davich, T. B. 12268.
4951. David, E. 1938: Embryologische Untersuchungen an Myoporaceen, Salvadoraceen, Sapindaceen und Hippocrateaceen. Planta 28:680–703.
4952. David, R. 1937: Étude experimentale des effets toxiques de l'urine humaine sur la germination. Rev. Gen. Botan. 49:705–29.
4953. —— 1938: Compléments a l'étude de l'influence des temperatures elevées sur la vitalité des graines oléagineuses. Bull. Mat. Grasses Inst. Colon. Marseille 22:183–93.
4954. —— 1938: Influence of the gonadotropic hormone of the anterior hypophysis on the germination of seeds. Compt. Rend. Soc. Biol. 128:569–72.
4955. —— 1938: L'influence des enzymes sur la germination de *Pisum sativum*. Compt. Rend. Soc. Biol. 129:274–76.
4956. —— 1939: Existe-t-il un "gradient axial" chez les graines? Compt. Rend. Soc. Biol. 131:421–24.
4957. —— 1940: Sur la résistance des semences aux hautes temperatures. Rev. Sci. 78:168–72.
4958. —— 1944: Contribution a l'étude biologique du froid essais de printanisation du mais. Rev. Sci. 82:106–8. Biol. Abstr. 25:29111. 1951.
4959. —— 1944: Sur les phénomènes biologiques provoqués chez les céréales par la printanisation des semences. Compt. Rend. Soc. Biol. 138:849.
4960. —— 1945: L'aleurone et l'amidon transitoire de l'embryon de blé printanisé. Compt. Rend. Soc. Biol. 139:560–62.
4961. —— 1947: L'évolution des inclusions lipidiques du germe de blé pendant le traitement de printanisation. Compt. Rend. 224:146–47.
4962. —— 1948: Essais de stimulation de la germination du pin maritime. Compt. Rend. Soc. Biol. 142:1545–46.
4963. —— 1948: La printanisation des végétaux superieurs. Assoc. Franc. Avan. Sci. Paris I(1945) 64:118–32.
4964. —— 1949: L'influence de la printanisation sur les glucides de l'albumen du grain de blé. Compt. Rend. 228:1242–43.
4965. —— 1950: La printanisation des semences; sou importance scientifique, ses possibilités d'application. Ann. Biol. 54:413–46.
4966. —— 1951: L'influence de la printanisation sur les lipides de la plantule et de l'albumen de la semence de blé. Compt. Rend. 233:428–33.
4967. —— 1952: L'influence de la dessiccation des semences de blé printanisé sur le développement des plantes qui en proviennent. Compt. Rend. 234:982–84.
4968. —— 1954: Printanisation des semences préalablement soumises à l'action des hautes temperatures. Congr. Intern. Botan. Paris Rappt. Commun. 8:293–95.
4969. —— and A. Guerindon. 1951: Influence stimulante exercée par le froid sur

la germination des graines de pin maritime. Compt. Rend. Soc. Biol. 145:715–17.

4970. —— and —— 1952: Influence of desiccation on "activated" seeds of the maritime pine on their germination. Compt. Rend. Acad. Agr. France 39:278–80. Brit. Abstr. B III:612. 1952.

4971. —— and J. Sechet. 1947: L'influence de la printanisation sur les composés azotés de l'orge d'hiver. Compt. Rend. Soc. Biol. 141:459–60. Biol. Abstr. 22:19257. 1948.

4972. —— and —— 1947: Printanisation [cold treatment of seeds] of cultivated and indigenous vegetables. Compt. Rend. 225:763–65. Brit. Abstr. B III:202. 1948.

4973. —— and —— 1948: Le rôle respectif de l'albumen et de l'embryon dans le développement du blé d'hiver printanisé. Compt. Rend. Soc. Biol. 142:72–74. Biol. Abstr. 23:19693. 1949.

4974. —— and —— 1948: L'influence du traitement de printanisation sur les glucides de l'embryon du grain de blé. Compt. Rend. 227:537–39.

4975. —— and —— 1951: Nouvelles observations sur la vernalisation du blé et de l'orge. Compt. Rend. Acad. Agr. France 37:45–48.

David, R. 3731, 3732.

4976. David, W. A. L., and B. O. C. Gardiner. 1955: The aphicidal action of some systemic insecticides applied to seeds. Ann. Appl. Biol. 43:594–614.

4977. Davide, R. G. 1957: Effects of several fungicides for seed treatment of corn. Philippine Agriculturist 41:295–305.

Davidson, B. 3958.

Davidson, F. E. 16115.

4978. Davidson, J. 1926: Changes in nitrogen, potassium and phosphorus content of wheat seedlings during germination and early stages. Botan. Gaz. 81:87–94.

4979. —— 1941: Wheat storage problems in South Australia. J. Dept. Agr. S. Australia 44:391–95.

4980. —— 1945: Total and free amylase content of dormant cereals and related seeds. J. Agr. Res. 70:175–200.

4981. Davidson, O. W. 1935: Growing trees from "non-viable" peach seeds. Proc. Am. Soc. Hort. Sci. 32:308–12.

4982. Davidson, Mrs. T. 1932: Dahlias from seed. Flower Grower 19(3):107.

4983. Davidson, W. A. 1939: Fundamentals of seed control. Proc. Assoc. Offic. Seed Analysts 30:87–90.

4984. —— 1942: Weights for working samples. Assoc. Offic. Seed Analysts News Letter 16(2):4–6.

4985. —— 1953: Blending seed to uniformity. Proc. Intern. Seed Testing Assoc. 18:106–11.

4986. —— 1955: Seed laws protect the farmer. Crops & Soils 8(2):12–13.

4987. —— and O. L. Justice. 1959: Comments on tetrazolium testing. Seed World 84(2):16.

Davies, A. C. W. 3522.

Davies, D. L. 5737, 5738.

4988. Davies, D. R. 1960: The embryo culture of inter-specific hybrids of *Hordeum.* New Phytologist 59:9–14.

4989. —— and E. T. Wall. 1960: The influence of oxygen and temperature on the radiosensitivity of soaked barley seeds. Intern. J. Radiation Biol. 2:261–67. Biol. Abstr. 36:11846. 1961.

Davies, D. W. 16691.

4990. Davies, J. C. 1960: Storage of maize in a prefabricated aluminum silo in tropical conditions. E. African Agr. J. 25:225–28.

4991. Davies, M. E. 1964: Acetolactate and acetoin synthesis in ripening peas. Plant Physiol. 39:53–59.

4992. Davies, P. A. 1928: High pressure and seed germination. Am. J. Botany 15:149–56.

4993. —— 1928: The effect of high pressure on the percentages of soft and hard seeds of *Medicago sativa* and *Melilotus alba.* Am. J. Botany 15:433–36.

4994. Davies, P. W. A., and J. Winstanley. 1957: The use of 2,3,5-triphenyltetrazolium chloride as a measure of the seed viability of Mexican grass, *Ixophorus unisetus* (Presl.) Schlecht. Trop. Agr. (London) 34:144–48.

4995. Davies, R. M. 1932: The rapid determination of moisture in seeds and other granular substances. Proc. Phys. Soc. (London) 44:231–45. Chem. Abstr. 26:3752. 1932.

4996. Davies, R. P. 1939: The chemical composition of the seed and straw of various grasses. Welsh J. Agr. 15:250–60.

4997. Davies, W., G. A. Atkins, and P. C. B. Hudson. 1937: The effect of ascorbic acid and certain indole derivatives on the regeneration and germination of plants. Ann. Botany (London) 1:329–51.

4998. Davies, W. E. 1956: Storage of clover seed. I. First interim report, 1947–1954. J. Brit. Grassland Soc. 11:224–29.

Davis, A. R. 19161, 19162.

4999. Davis, B. A. 1956: Growing day lilies and irises from seed. Flower Grower 43(8):18.

5000. Davis, B. H., and C. M. Haenseler. 1943: Tests with some vegetable-seed protectants. Plant Disease Reptr. 27:170–73. Chem. Abstr. 37:4523. 1943.

5001. Davis, B. M. 1915: A method of obtaining complete germination of seeds in *Oenothera* and of recording the residue of sterile seed-like structures. Proc. Natl. Acad. Sci. U.S. 1:360–63.

5002. Davis, C. A. 1943: Methods of germinating rose seeds. Am. Rose Mag. 5(4):68–69.

Davis, C. F. 11863.

Davis, D. G. 8596.

Davis, D. J. 13240.

5003. Davis, E. F. 1928: The toxic principle of *Juglans nigra* as identified with synthetic juglone and its toxic effect on tomato and alfalfa plants. Am. J. Botany 15:620.

5004. Davis, G. L. 1962: Embryological studies in the Compositae. I. Sporogenesis, gametogenesis, and embryony in *Cotula australis* (Less.) Hook. F. Australian J. Botany 10:1–12.

5005. Davis, G. N. 1936: Laboratory detection of smut in oats. Proc. Assoc. Offic. Seed Analysts 1935:93.

5006. —— 1937: Germination of treated and untreated pea seeds in autoclaved and unautoclaved soil. Proc. Assoc. Offic. Seed Analysts 29:121–25.

5007. —— 1937: Germination of cucurbit seed in sand. Proc. Assoc. Offic. Seed Analysts 29:133–35.

5008. —— and C. Cuneo. 1949: A thresher for small lots of seed. Proc. Am. Soc. Hort. Sci. 53:291–93.

5009. —— and R. H. Porter. 1936: Comparative absorption of water by endosperm and embryo of corn kernels. Proc. Assoc. Offic. Seed Analysts 28:62–67.

Davis, G. N. 20122.

5010. Davis, J. F. 1945: The effect of some environmental factors on the set of pods and yield of white pea beans. J. Agr. Res. 70:237–49.

5011. —— 1951: The effect of seed source and spacing of plants in the row on the yield of sugar beets grown on muck soils. Proc. Am. Soc. Sugar Beet Technol. 6:117–23.

5012. —— 1952: Some factors and their interactions affecting the yield of sweet corn grown on organic soil. Mich. Agr. Expt. Sta. Quart. Bull. 34:253–61.

5013. Davis, J. H., H. H. Kramer, and R. L. Whistler. 1955: Expression of the gene *du* in the endosperm of maize. Agron. J. 47:232–35.

Davis, J. R. 14269.

Davis, J. W. 15050.

5014. Davis, L. D. 1933: Size and growth relations of fruit in splitting of peach pits. Proc. Am. Soc. Hort. Sci. 30:195–200. Expt. Sta. Rec. 72:190. 1935.

5015. —— 1935: Some ash constituents of alternate-bearing sugar prune trees. Proc. Am. Soc. Hort. Sci. 32:125–30.

5016. —— 1940: Size of aborted embryos in the Phillips cling peach. Proc. Am. Soc. Hort. Sci. 37:198–202.

5017. —— 1941: Split-pit of peaches, estimation of time when splitting occurs. Proc. Am. Soc. Hort. Sci. 39:183–89.

Davis, L. H. 1734.

5018. Davis, L. L. 1940: Germination and growth of crab-apple seedlings (*Pyrus baccata*) as influenced by fungicidal treatment. Proc. Am. Soc. Hort. Sci. 37:359–60.

Davis, L. L. 9958.

Davis, M. E. 12616.

5019. Davis, N. D. 1961: Peanut storage studies: effect of storage moisture on three varieties of runner peanuts. J. Am. Oil Chemists Soc. 38:516–17.

5020. —— 1963: Growth and kojic acid production by *Aspergillus flavus* growing on peanut oil. Econ. Botany 17:263–69.

5021. Davis, N. S., G. J. Silverman, and W. H. Keller. 1963: Combined effects of ultrahigh vacuum and temperature on the viability of some spores and soil organisms. Appl. Microbiol. 11:202–10.

5022. Davis, O. H. 1926: Germination of seeds of certain horticultural plants. Florists Exchange Hort. Trade World 63(1):917, 922.

5023. —— 1927: Germination and early growth of *Cornus florida*, *Sambucus canadensis*, and *Berberis thunbergii*. Botan. Gaz. 84:225–63.

Davis, R. A. 5325, 5327.

5024. Davis, R. L. 1929: Sugar cane fuzz viability and December rainfall. Plant. Sugar Mfg. 82:343–44.

5025. —— and R. O. Hammons. 1956: Reaction of alfalfa seedlings from dormant seeds subjected to various dosages of X-rays. Agron. J. 48:528–29.

Davis, R. L. 16317, 18593.

5026. Davis, S. H., Jr., and J. B. Harry.

1942: A Xylaria pathogenic to *Ginkgo biloba* L. seeds. Phytopathology 32:91–93.
5027. Davis, T. A., A. P. Anandan, and K. P. V. Menon. 1953: A case of polyembryony in coconut. Indian Coconut J. 6:67–71.
5028. Davis, W. B. 1939: A proteolytic enzyme of lima bean (*Phaseolus lunatus* L.). Food Res. 4:613–19. Biol. Abstr. 14: 9391. 1940.
Davis, W. B. 17384.
5029. Davis, W. C. 1931: Phenolase activity in relation to seed viability. Plant Physiol. 6:127–38.
Davis, W. C. 9959.
5030. Davis, W. E. 1917: Resistance of seed coats of *Abutilon theophrasti* to intake of water. Botan. Gaz. 64:166–67.
5031. —— 1925: The use of catalase as a means of determining the viability of seeds. Proc. Assoc. Offic. Seed Analysts 18th Ann. Meeting, pp. 33–39. Boyce Thompson Inst. Prof. Pap. 1:6–12. 1926.
5032. —— 1930: Primary dormancy, after-ripening, and the development of secondary dormancy in embryos of *Ambrosia trifida*. Am. J. Botany 17:58–76.
5033. —— 1930: The development of dormancy in seeds of cocklebur (*Xanthium*). Am. J. Botany 17:77–87.
5034. —— 1933: Seeds undrowned after 23 years under water. Sci. News Letter 24:131.
5035. —— 1939: An explanation of the advantage of alternating temperatures over constant temperatures in the germination of certain seeds. Am. J. Botany 26:17s–18s.
5036. —— and R. C. Rose. 1912: The effects of external conditions upon the after-ripening of the seeds of *Crataegus mollis*. Botan. Gaz. 54:49–62.
Davis, W. E. 4602.
5037. Davis, W. H., G. K. Middleton, and T. T. Herbert. 1961: Inheritance of protein, texture, and yield in wheat. Crop Sci. 1:235–38.
Davis, W. H. 13416.
5038. Davison, D. C. 1949: Distribution of formic and alcoholic dehydrogenases in the higher plants—their variation in the pea plant during its life cycle. Proc. Linnean Soc. N.S. Wales 74:26–36. Chem. Abstr. 43:8462. 1949.
5039. —— 1949: Formic dehydrogenase in the oxidation mechanisms of *Pisum sativum*. Proc. Linnean Soc. N.S. Wales 74:37–56. Chem. Abstr. 43:8462. 1949.
5040. Davshan, A., and P. Sukhankin. 1953: In regard to methods of laboratory evaluation of cotton seed germination. Khlopkovodstov 3(6):15–23.
Davtyan, L. V. 10176.
5041. Davydoff, A. W. 1930: Über das Verhältnis zwischen der Keimfähigkeit der Sämereien im Keimapparat und der Keimzahl im Saatgarten. Russ. Staatsinst. Wiss. Forsch. Gebiet Forstwirts. Mitt. Forstl. Versuchswesen 6:23–58.
5042. Dawis, V. M. 1938: Variability among *Anthurium* plants produced from seeds. Philippine Agriculturist 27:495–501.
Dawson, E. C. 9931.
Dawson, E. H. 12329.
5043. Dawson, J. H. 1963: Effects of EPTC on barnyardgrass seeds. Weeds 11:184–86.
5044. Dawson, M. D., and D. H. Heinrichs. 1952: The effects of various germination techniques to overcome dormancy in green Stipagrass seed. Sci. Agr. (Ottawa) 32:266–71.
5045. Dawson, O. 1956: Primroses and Polyanthuses. Gardeners Chron. 140:488.
5046. Dawson, R. B., and N. L. Ferguson. 1938: The effect of lead arsenate on the germination of seed and subsequent development in certain grasses. Brit. Golf Unions Board Greenkeeping Res. J. 5(19):274–81. Biol. Abstr. 15:19359. 1941.
5047. —— and —— 1939: Seed production of a turf grass: red fescue. Brit. Golf Unions Board Greenkeeping Res. J. 6(20): 20–34. Biol. Abstr. 15:19358. 1941.
Dawson, R. F. 16184, 16185.
5048. Dawson, T. R., and T. H. Messenger. 1932: Products from rubber seed. I. Preparation and properties. J. Res. Assoc. Brit. Rubber Mfrs. 1:33–44. Soc. Chem. Ind. J. 51:697B. 1932.
Dawson, V. L. 14262.
5049. Dawydow, G. K. 1940: Wirkung des Ultraschalles auf die Samen der Zuckerrübe. Compt. Rend. Acad. Sci. URSS 29:491–93.
5050. Dax, P. J., and F. R. Graesser. 1946: Comparison of "unit of weight" and "unit of number" method for tracing the history of β-amylase in barley and malt. Wallerstein Lab. Commun. 9:49–56. Brit. Abstr. BIII:171. 1946.
5051. Daxenbichler, M. E., C. H. Van Etten, and I. A. Wolff. 1961: Identification of a new, naturally occurring, steam-volatile isothiocyanate from *Lesquerella lasiocarpa* seed. J. Org. Chem. 26:4168–69.

5052. ——, ——, and —— 1964: (R)-2-Hydroxy-3-butenylglucosinolate. A new thioglucoside from *Crambe* seed. Abstr. Papers Presented at the 147th Meeting, Am. Chem. Soc. 1A: No. 2.

5053. Day, A. D. 1957: Effect of lodging on yield, test weight and other seed characteristics of spring barley grown under flood irrigation as a winter annual. Agron. J. 49:536–39.

5054. —— 1963: Inheritance of seed dehiscence in a cross between *Avena byzantina* C. Koch and *Avena fatua* L. Crop Sci. 3:165–66.

5055. ——, E. E. Down, and K. J. Frey. 1955: Association between diastatic power and certain visible characteristics and heritability of diastatic power in barley. Agron. J. 47:163–65.

Day, A. D. 452, 15896.

5056. Day, C. 1891: The influence of temperature on germinating barley. J. Chem. Soc. 59:664–77.

Day, D. L. 2327.

Day, K. M. 11893, 11894.

5057. Day, M. W. 1964: Results of a study on direct seeding. Mich. Agr. Expt. Sta. Quart. Bull. 46:412–15.

5058. Day, W. R. 1958: Variations in susceptibility of European larch of differing seed origin in Scotland to injury by experimental freezing. Scot. Forestry 12:143–46.

Dayton, D. F. 1511.

5059. Dayton, E. W., Jr. 1960: Biological and chemical evaluation of corn bred for higher-oil and higher-protein content. Univ. Microfilms, Ann Arbor, Mich., L.C. Card No. Mic 60–3900. 96 pp. Dissertation Abstr. 21:1311. Chem. Abstr. 55:5664. 1961.

5060. D'Cruz, R., and K. Chakravarty. 1961: Seed development in *Gossypium allopolyploid*. Poona Agr. Coll. Mag. 51:41–42.

5061. De, H. N., and S. C. Barai. 1949: Mechanism of biosynthesis of ascorbic acid during germination. Indian J. Med. Res. 37:101–11. Chem. Abstr. 44:706. 1950.

5062. —— and S. C. Datta. 1951: Biosynthesis of nicotinic acid during germination of cereals and pulses. Indian J. Agr. Sci. 21:375–94. Chem. Abstr. 47:8840. 1953.

5063. —— and —— 1951: Effect of colchicine on the synthesis of nicotinic acid in germinating seeds. Indian J. Agr. Sci. 21:39–4050. Chem. Abstr. 47:8840. 1953.

De, H. N. 2820.

De, N. N. 16636.

5064. De, R. N. 1955: Peculiar germination of *Cucurbita pepo* D.C. Sci. Cult. (Calcutta) 20:342.

De, S. S. 1763, 5245, 5246, 15871.

De, T. S. 6058.

5065. Deacon, B. D., and A. C. Wamble. 1954: Vitamin E analytical procedure for cottonseed and its products. J. Am. Oil Chemists Soc. 3:284–87.

5066. De Almeida, A. O., M. D. Goulart, M. Ielpo, and A. V. Pinto. 1941: Estudo da ação inhibidora do suco de "*Solanum lycopersicum*" sobre a germinação de sementes e crescimento de plantas. Rev. Brasil. Biol. 1:345–54. Biol. Abstr. 16:8009. 1942.

5067. De Almeida-Costa, O. 1931: Analysis of the wheat germ. Rev. Brasil. Chim. 2:457. Chem. Abstr. 26:1986–87. 1932.

Dean, J. G. 5508.

Dean, L. L. 19743.

5068. Dean, W. S. 1960: Planting and growing lily seeds of quick germination. North Am. Lily Soc. Yearbook 13:79–83.

5069. De Andrade, E. N. 1915: Experiments in the germination of coffee carried out in Brazil. Fazenderio (São Paulo) 8:3–8. Expt. Sta. Rec. 33:841. 1915.

De Angelis, E. 11633.

Dear, E. 15851.

5070. De Araujo, A. A. 1950: Sowing of seeds mixed with rice hulls. Chacaras Quintais 81:75–76.

5071. De Arruda, H. V. 1957: Correlation between the weight of plants and the weight of seeds in bean varieties. Bragantia 16:385–88.

De Arruda, L. F. 13736.

5072. Deatherage, W. L., M. M. MacMasters, M. L. Vineyard, and R. P. Bear. 1954: A note on starch of high amylose content from corn with high starch content. Cereal Chem. 31:50–52.

Deatherage, W. L. 558, 20130–20132.

5073. Deavenport, B. M. 1945: Two-level seed planter insures germination. Sci. News Letter 47:24.

5074. Deb, N. C. 1939: Chemical examination of the stearin from the seeds of *Mesua ferrea* Linn. (Nag-Keshor). Indian Soap J. 5:211–13. Chem. Abstr. 33:4445. 1939.

5075. De Balsac, F. H., G. S. Dagand, H. H. De Balsac, L. Lefevre, and J. Maheu. 1929: La graine oléifère et l'huile de thé (*Thea sinensis* L.). Bull. Agence Gen.

Colonies 22:853–72. Biol. Abstr. 5:2089. 1931.

5076. ——, E. Delhotel, H. H. De Balsac, L. Lefevre, and A. Parveaud. 1930: Fruit, graine et huile de "cam" d'Annam (*Parinarium annamense* Hance—Rosacées). Bull. Agence Gen. Colonies 23:310–21.

De Balsac, F. H. 11629.

5077. De Balsac, H. H., and N. Mayaud. 1930: Études d'écologie ornithologique. Compléments a l'étude de la propagation du gui (*Viscum album* L.) par les Oiseaux. Alauda (1)2:474–93. Ann. Biol. Partie 2nd Fascicule IV. Morphol. Biol. Gen. 35:561. 1931.

De Balsac, H. H. 5075, 5076.

5078. De Beer, J. F. 1963: Influences of temperature on *Arachis hypogaea* L. with special reference to its pollen viability: germination and development of the seedling. Verslag. Landbouwk. Onderzoek. 69.2:7–10.

De Belsunce, M. 7824.

De Benedictis, G. 3717.

5079. De Bevoise, F. 1932: Alpines from seeds. Horticulture 10(2):29.

De Bievre, M. 12959.

5080. Debney, E. W. 1960: Dynamic seed sampling spears in the United Kingdom. Proc. Intern. Seed Testing Assoc. 25:174–81.

5081. De Boer, H. W. 1959: Structure of the seed capsule of *Jensenobotrya lossowiana* Herre. Succulenta 6:93–95.

5082. De Borbolla-y-Alcala, J. M. R., M. J. Fernandez-Diez, and F. G. Pelisso. 1955: Grasas Aceites (Seville) 6:5–22. Chem. Abstr. 50:1986. 1956.

5083. De Bruin, H. J. 1959: Seed treatment and control of weeds of grain. Sticht. Ned. Graan-Cent. Mededel. 6:62–75.

5084. Debuch, H. 1957: Enzymic splitting of soybean lecithins. Z. Physiol. Chem. 306:279–86. Chem. Abstr. 52:11992. 1958.

5085. De Caestecker, K., and L. De Leenheer. 1954: Stubble density as criterion of germination and tillering in cereals. Mededel. Landbouwhogeschool Gent 19: 88–97. Soils Fertilizers 18:348. 1955.

5086. De Candolle, A. 1835: Pflanzen-Physiologie (Ger. ed.). Vol. 2, Ch. 6, "Von der Keimung," pp. 268–328.

5087. —— 1946: Sur la durée relative de la faculté de germination, germer dans des graines appartenant à divers familles. Ann. Sci. Nat. 3 Ser. 6:373–82.

5088. —— 1879: De l'effet des temperatures tres-basses sur la faculté germinative des graines. Verhandl. Schweiz. Naturforsch. Ges. Jahresber. 1877–78. (Bern, 1879.) Botan. Ztg. 38:64. 1880.

5089. —— 1897: The latent vitality of seeds. Popular Sci. Monthly 51:106–11.

5090. De Caraman, M., and C. Champy. 1936: Action of radium emanation on the germination of seeds. Compt. Rend. Soc. Biol. 121:750–52. Chem. Abstr. 30:3856. 1936.

De Caro, L. 2492.

5091. De Castro, F. S. 1962: Selection and treatment of coffee seed. Rev. Cafétal. (ser. 4) 1(5):43–47.

De Cesaris-Coromaldi, L. 18894.

5092. De Chalmot, G. 1894: The influence of nitrates on germinating seeds. Agr. Sci. 12:463–64. Expt. Sta. Rec. 7:36. 1895/96.

Dechary, J. M. 6102.

5093. Decker, A. E., and L. P. Reitz. 1948: Germination tests with flax stored at different moisture and temperature levels. Proc. Intern. Seed Testing Assoc. 14:27–34.

Decker, A. E. 16115.

5094. Decker, A. M., H. J. Retzer, and F. G. Swain. 1964: Improved soil openers for the establishment of small-seeded legumes in sod. Agron. J. 56:211–14.

Decker, M. 6293.

5095. Decker, P. 1946: The effect of depth of planting on the emergence and survival of blue lupine. Phytopathology 36:479–80.

Decker, R. W. 5199.

5096. Decossas, K. M., S. P. Koltun, J. J. Spadaro, R. O. Feuge, E. F. Pollard, and E. L. Patton. 1962: Cocoa butter-like fat from cottonseed oil: Preliminary cost study. J. Am. Oil Chemists Soc. 39:146–50.

5097. De Coulon, J. 1923: *Nardus stricta*, étude physiologique, anatomique et embryologique. Mem. Soc. Sci. Nat. 1923:247–52.

5098. Decoux, L., G. Roland, and L. Ernould. 1942: pH et germination de la betterave. Publ. Inst. Belge Amélioraț. Betterave (Tirlemont) 10:535–40. Ann. Agron. 13:200. 1943.

5099. —— and J. Vanderwaeren. 1944: Hormones and the sugar beet. Publ. Inst. Belge Amélioraț. Betterave (Tirlemont) 12:695–709. Chem. Abstr. 40:974. 1946.

5100. De Cugnac, A. 1932: Quelques données sur le contingent glucidique soluble des grains de blé. Bull. Soc. Botan. France 79:231–36.

5101. Dedolph, R. R. 1962: Effect of benzothiazole-2-oxyacetate on flowering and fruiting of papaya. Botan. Gaz. 124:75–78.

Dedolph, R. R. 8730, 14836.

5102. Deen, J. L. 1933: Effect of weight class on germination in longleaf pine. J. Forestry 31:434–35.

5103. Deeter, W. T. 1938: Old delphinium seeds. Horticulture 16(9):204.

5104. Deeva, V. E. 1956: Effect of sulfurous anhydride on enzyme respiration of barley grains. Tr. Leningr. Tekhnol. Inst. Pishchevoi Prom. 13:48–58. Chem. Abstr. 52:20439. 1958.

5105. Deevers, C. L. 1937: Structure of palaeozoic seeds of the Trigonocarpales. Botan. Gaz. 98:572–85. Botan. Centralbl. 173:282. 1938.

5106. De Fazi, R. 1930: Action of ultraviolet rays on the germination of barley in malting. Atti Accad. Lincei Ser. 6:1022–24. Soc. Chem. Ind. J. 50:86B. 1931.

De Felice, D. 11981.

De Fernandez, M. d'C. 1460, 1461.

5107. De Fina, A. L. 1939: The germination of flax. Rev. Fac. Agron. Univ. Nacl. La Plata 23(1938):27–38. Biol. Abstr. 14:16652. 1940.

5108. De Fluiter, H. J. 1939: Observations concerning the storage of seed coffee. Bergcultures 13:1506–12.

De Fontaine, D. 6590.

5109. De France, J. A. 1941: Seed treatments with phytohormones and talc. Proc. Am. Soc. Hort. Sci. 38:679–82.

5110. —— 1943: Effect of certain chemicals on the germination of crabgrass seed when plants are treated during the period of seed formation. Proc. Am. Soc. Hort. Sci. 43:331–35.

5111. —— 1943: The killing of weed seed in compost by the use of certain fertilizers and chemicals. Proc. Am. Soc. Hort. Sci. 43:336–42.

5112. —— and J. A. Simmons. 1952: Death knell for weeds. Flower Grower 39(4):56, 97–98.

De G. Paula, R. Descartes; *see* Descartes, R. de G. Paula.

Degen, A. von; *see* Von Degen, A.

5113. Degman, E. S. 1953: Effect of boron sprays on fruit set and yield of Anjou pears. Proc. Am. Soc. Hort. Sci. 62:167–72.

5114. De Graaf, J. 1960: Lilies from seed. Gardeners Chron. 147:366–67.

5115. De Graeve, P. 1937: Évolution de l'azote purique au cours de la germination. Compt. Rend. 204:445–47.

5116. —— 1937: Évolution de l'azote purique au cours de la germination. Compt. Rend. 204:798–800.

De Graeve, P. 6571.

De Groot, W. 2719.

5117. De Grotte, R. 1951: La vermiculite dans la culture des plantes horticoles. Bull. Hort. (N.S.) 6:103–6.

5118. De Haas, P. G., and H. Schander. 1952: Keimungsphysiologische Studien an Kernobst. I. Samen und Keimung. Z. Pflanzenzücht. 31:457–512.

5119. Dehay, C., and M. Care. 1958: Étude de la composition des excretions radicellaires chez quelques légumineuses d'origine africaine. Compt. Rend. 247:336–38.

De Heen, P. 13400.

5120. De Heer, A. N. 1930: Determination of moisture-content in cacao beans. Gold Coast Dept. Agr. Bull. 23:53–59.

5121. De Hemptinne, Y., and W. Vegult 1954: Chromatographic determination of glutamic acid (in maize gluten). Fermentatio 4:190–93. Anal. Abstr. 2:1664. 1955.

5122. Deherain, P. P. 1895: Experiments with wheat and oats at Grignon in 1895. Ann. Agron. 21:545–65. Expt. Sta. Rec. 7:952. 1895–96.

5123. —— and E. Breal. 1883: Untersuchungen über den Einfluss der Mineralstoffe auf die Keimung. Ann. Agron. 9:58–77. Botan. Jahresb. Jahr. 11:40. 1883.

5124. —— and E. Demoussy. 1900: Culture of white lupine. Ann. Agron. 26:57–77. Expt. Sta. Rec. 12:143. 1900–1.

5125. Deighton, B. 1957: Judas trees from seed. J. Roy. Hort. Soc. 82:304.

5126. De Jamblinne, A. T., and A. A. Elens. 1953: Adaptation à la température et germination des graines chez *Pinus sylvestris* L. et chez *Pinus nigra* Arnold en Campine. Agricultura (Louvain) 1:101–17.

5127. De Janczewski, M. E. 1892: Germination of anemone. Rev. Gen. Botan. 4:243–44, 289–301. J. Roy. Microscop. Soc. 1892:211.

5128. De Jong, J. J. 1960: F.A.O. world seed campaign. Proc. Intern. Seed Testing Assoc. 25:71–76.

5129. —— 1960: Good seed doesn't cost—it pays! Euphytica 9:191–95.

5130. —— 1962: Accomplishments of the F.A.O. world seed campaign. Proc. Intern. Seed Testing Assoc. 27:315.

De Jong, J. J. 19639.

5131. De Jong, P. 1952: The effect of seed dressings on the rate of germination of tea seed. Plant. Chron. 47:177–81. Rev. Appl. Mycol. 31:578. 1952.

De Ketelaere, F. 12957–12960.

5132. Dekker, J. 1955: Internal seed disinfection by an antibiotic from *Streptomyces rimosus*. Nature 175:689–90.

5133. —— 1957: Beet seed disinfection with antibiotics. [Abstract.] Tijdschr. Plantenziekten 63:25.

5134. —— 1957: Internal seed disinfection of peas infected by *Ascochyta pisi* by means of the antibiotics rimocidin and pimaricin, and some aspects of the parasitism of this fungus. Tijdschr. Plantenziekten 63:65–144.

5135. ——, O. M. Van Andel, and A. K. Sijpesteijn. 1958: Internal seed disinfection with pyridine-2-thiol-N-oxide and a derivative. Nature 181:1017.

Dekker, J. 14786.

Dekker, L. P. van der Mijll; *see* Van der Mijll Dekker, L. P.

5136. De Kock, P. C., and R. F. Hunter. 1950: A germination inhibitor from sugar beet. Nature 166:440–41.

5137. De la Loma, J. L., G. Argüello, and F. Zertuche. 1955: Influencia de la escarificación de las semillas de trebol y alfalfa en la velocidad de germinación. Chapingo 8:102–6.

5138. De Lamarlière, M. L. G. 1893: Germination of Umbelliferae. Rev. Gen. Botan. 5:159–71, 224–29, 258–64. J. Roy. Microscop. Soc. 1893:758.

Del Amo, A. 3451.

5139. De Lanye, O., and A. L. Guyot. 1928: Sur la desinfection des graines de betterave. Rev. Pathol. Vegetale Entomol. Agr. France 15:160–72.

5140. Delassus, M. 1911: Influence de la suppression partielle des reserves de la graine sur le développement de la plante. Compt. Rend. 153:1494–97.

5141. Delavan, C. C. 1916: The relation of the storage of the seeds of some of the oaks and hickories to their germination. Mich. Acad. Sci. Rept. 17(1915):161–63.

Delaveau, P. 583.

5142. Delay, C., and G. Mangenot. 1960: Le développement de la graine chez *Allanblackia floribunda* Oliv. Ann. Sci. Nat. Botan. Biol. Végétale 12:387–439.

Del Bano, F. 16673.

Delcourt, L. 11626.

5143. Deleanu, N. T., N. Popovici, and I. Ionescu. 1938: Über den Einfluss einiger Alkaloide auf die Katalase des Weizens. Beitr. Biol. Pflanz. 25:261–72.

De Leenheer, L. 5085.

5144. De Leo, A. 1962: Action of indolebutyric acid, gibberellic acid, and eosin on *Arachis hypogaea*. Lavori Ist. Botan. Palermo 18:97–111. Chem. Abstr. 59: 14507. 1963.

5145. —— and R. Bonomo. 1962: On behavior of soybean plants and of principal components of their seeds in relation to fertilization, nature of the soil and planting date. Palermo Studi Inform. Giard. Colon. Bull. 23:56–74.

5146. De Leo, T. 1954: The amylolytic activity of the seeds of *Canavalia ensiformis* under particular experimental conditions. I. Optimum conditions for the extraction and activity of α-amylase. Ric. Sci. 24:1445–55. Chem. Abstr. 49:2570. 1955.

5147. ——, F. Badolato, and L. Foti. 1956: The amylolytic activity of the seeds of *Canavalia ensiformis* under particular experimental conditions. VI. An inhibitor present in the germs. Ric. Sci. 26:2726–34. Chem. Abstr. 51:12235. 1957.

5148. —— and G. Barone. 1955: Amylolytic activity of seeds of *Cvnavalia ensiformis*. III. α-Amylolytic activity of seeds of *Canavalia ensiformis* with time. Ric. Sci. 25:3060–68. Chem. Abstr. 50:9530. 1956.

5149. —— and —— 1955: Amylase activity of seeds of *Canavalia ensiformis* under particular experimental conditions. IV. Activity of the α-amylase of *C. ensiformis* against time in relation to retrogradation of starch solutions. Ric. Sci. 25:3350–55. Chem. Abstr. 50:8757. 1956.

5150. —— and —— 1956: Amylolytic activity of *Canavalia ensiformis* seeds. V. Amylase activity of *Canavalia ensiformis* and that of malt as a function of time. Ric. Sci. 26:131–37. Chem. Abstr. 50: 13178. 1956.

5151. ——, M. R. De Marco, and F. Badolato. 1957: Components of *Triticum vulgare* and *Triticum durum*. II. Iodine-binding capacity of starch of soft and hard wheat. Ric. Sci. 27:2120–27. Chem. Abstr. 52:1367f.

5152. —— and U. Manfredi. 1957: On the amylolytic activity of the seeds of *Canavalia ensiformis* under particular experimental conditions. Ric. Sci. 27:2426–42.

De Leo, T. 14844.

5153. De Leon, N. J. 1961: Viability of palm seeds. Hort. Mag. 40:131–32.

Delevoryas, T. 18416.

5154. Delevoy, G. 1936: À propos de semis de graines charnues. Bull. Soc. Centr. Forest Belg. 43:285–308. Biol. Abstr. 11:14586. 1937.

Delf, E. M. 3921.

Delgado, M. L. Ortega-; *see* Ortega-Delgado, M. L.

5155. Del Giudice, E. 1935: The bean orobanche. I. Germination of the seed. Staz. Sper. Granicolt. "B. Mussolini," Sicilia-Catania Publ. No. 7. 26 pp. Chem. Abstr. 29:5153–54. 1935.

Delhotel, E. 5076.

Delihas, N. 4763, 11188.

5156. De Lima, A. P. 1930: Action des rayons ultra-violets, de quelques colorants et de quelques alcaloides sur la germination du seigle. Compt. Rend. Soc. Biol. 105: 493–95.

5157. Delizo, T. 1938: Methods of hastening the germination of the seeds of akle. Philippine J. Forestry 1:99–102. Biol. Abstr. 12:16691. 1938.

5158. Dellepiane, G., and F. Gioelli. 1930: An asserted hormonal influence exercised upon growth of plants by extracts of endocine glands. Boll. Soc. Ital. Biol. Sper. 5:243–45. Biol. Abstr. 6:9321. 1932.

5159. Deloffre, G. 1936: Effects of glycerol and mannose on the metabolism of the cell nuclei of the lupine-seed embryo. Compt. Rend. Soc. Biol. 121:1100–3. Chem. Abstr. 30:4899. 1936.

Delone, N. L. 10549.

5160. De Long, H. H. 1958: Aeration of wheat in prolonged storage for quality control. S. Dakota Agr. Expt. Sta. Tech. Bull. 19. 36 pp.

5161. —— 1962: Corn drying. S. Dakota Agr. Expt. Sta. Agr. Eng. Dept. Circ. 152. 24 pp.

5162. —— and A. J. Schwantes. 1942: Mechanical injury in threshing barley. Agr. Eng. 23:99–101.

Delong, R. A. 10172.

5163. De Loose, R. 1961: Sensitivity to X-rays of seeds of pasture type English ryegrass of different hygrometric degrees in relation to heat treatment. Rev. Agr. 14:1196–200.

5164. Delouche, J. C. 1954: Influence of moisture and temperature levels on the germination of corn soybeans and watermelons. Iowa State Coll. J. Sci. 28:427–33.

5165. —— 1956: Dormancy in seeds of *Agropyron smithii*, *Digitaria sanguinalis* and *Poa pratensis*. [Abstract.] Iowa State Coll. J. Sci. 30:348–49.

5166. —— 1963: Seed deterioration. Seed World 92(4):14–15.

5167. ——, T. W. Still, M. Raspet, and M. Lienhard. 1962: The tetrazolium test for seed viability. Mississippi Agr. Expt. Sta. Tech. Bull. 51. 63 pp.

Delouche, J. C. 19034.

Delphaut, J. 13828.

5168. Del Prado, F. A., and C. M. Christensen. 1952: Grain storage studies. XII. The fungus flora of stored rice seed. Cereal Chem. 29:456–62.

5169. De Luca, G. 1951: Phosphorus metabolism in plants. Studies with carob seeds. Palermo Univ. Fac. Sci. Agr. Ann. 1:199–204. 1950.

De Marco, M. R. 5151.

5170. Demaree, D. 1932: Plant responses to sawdust. Proc. Indiana Acad. Sci. 4:125–26. Biol. Abstr. 7:18002. 1933.

5171. —— 1932: Submerging experiments with *Taxodium*. Ecology 13:258–62.

5172. Demaree, G. E., and V. E. Tyler, Jr. 1956: The accumulation of hordenine in the seedlings of *Panicum miliaceum*. J. Am. Pharm. Assoc. 45:421–22.

5173. —— and M. S. Wilcox. 1938: The black-seed disease of strawberry. [Abstract.] Phytopathology 28:6.

5174. De Matos, A. G., Jr., and F. Riveca. 1958: Amino acids of Bahia coconut (*Cocos nucifera*). Anais Fac. Med. Univ. Recife 18:251–56. Chem. Abstr. 54:4788. 1960.

5175. De Menezes, O. B. 1955: Vernalization of corn. Pernambuco Sec. Agr. Ind. Com. Bol. 21:6–10. 1954.

5176. Demetriades, S. T., and P. T. Constantinou. 1956: Sur les variations de la concentration en aspargine chez l'*Hibiscus esculentus* L. au cours du développement de plantules normal et privées de fer. Compt. Rend. 242:2384–86.

5177. Demic, V. B., Jr. 1939: The effect of depth of coverings on the germination of Douglas fir seed. Publs. Theses Ser. 4:157–59. Expt. Sta. Rec. 82:778. 1940.

Demick, B. van, Jr., *see* Van Demick, B., Jr.

5178. Demina, O. K., and E. K. Pogosov. 1955: Preplanting treatment of tomato seed as a method of increasing their drought resistance and productivity. Plant Physiol. (USSR) (English Transl.) 2:459–64.

5179. Deming, G. W., and D. W. Robertson.

1933: Dormancy in small-grain seeds. Colo. Agr. Expt. Sta. Tech. Bull. 5. 12 pp.
5180. De Miranda, H. 1962: Germination and purity tolerances. Discussion of general principles. Proc. Intern. Seed Testing Assoc. 27:373–85.
De Miranda, H. 8847, 18960.
5181. Demirlicakmak, A., M. L. Kaufmann, and L. P. V. Johnson. 1963: The influence of seed size and seeding rate on yield and yield components of barley. Can. J. Plant Sci. 43:330–37.
5182. Demiröz, S. 1962: Seed production and certification in Turkey. Proc. Intern. Seed Testing Assoc. 27:1003–4.
5183. Demkin, A. P. 1959: Results of scientific research work on the study of the most important problems of growing hemp seeds. Tr. Vses. Nauchn.-Issled. Inst. Lub. Kul't. 24:219–36.
5184. Demkovski, P. E. 1932: Biochemical processes connected with "Yarovization." (Ukr. Inst. Genet.) Bull. Yarovization 1:42–46.
5185. —— 1932: Some data for the study of bio-chemical reactions connected with "Yarovization." (Ukr. Inst. Genet.) Bull. Yarovization 2/3:105–8.
5186. De Moura-Campos, F. A. 1935: Pharmacological action of Tonka bean on the muscular system. Rev. Flora Med. 1:565–84. Chem. Abstr. 29:7502. 1935.
Demoussy, E. 5124.
Demoussy, M. E. 12833.
5187. Dempsey, W. H., and J. F. Harrington. 1951: Red cotyledon of lettuce. Calif. Agr. 5(7):4. Biol. Abstr. 26:7297. 1952.
5188. De Muelenaere, H., and R. Feldman. 1960: Availability of amino acids in maize. J. Nutr. 72:447–50.
De Mytt, L. E. 10666.
5189. Denaiffe, Colle, and Flandrin. 1936: Anticryptogamic treatments and germination. J. Agr. Prat. 100:381–82, 401–3. Chem. Abstr. 30:5712. 1936.
5190. Dendrinos, A. 1955: A brief report on the accomplishments of the Seed Testing and Control Station at Chalandri, for the year 1954. Georg. Delt. (Ser. 3) 10:9–16.
5191. Dendy, A., and H. D. Elkington. 1918: Report on the effect of air-tight storage upon grain insects. Roy. Soc. (London) Grain Pests (War) Com. Rept. Part II 3:1–18.
5192. —— and —— 1919: On the prevention of heating in wheat by means of air-tight storage. Roy. Soc. (London) Grain Pests (War) Com. Rept. 5:3–10.
5193. Denes, G. 1953: The enzymic synthesis of glutamine with extracts from seeds of *Lupinus albus*. Experimentia 9:24–25. Chem. Abstr. 47:3891. 1953.
Denffer, D. von; *see* Von Denffer, D.
5194. Dengler, A. 1925: Über die Wirkung der Bedeckungstiefe auf das Auflaufen und die erste Entwicklung des Kiefersamens. Z. Forst-Jagdw. 57:385–408.
Den Hartog, G. T. 17872, 17875.
5195. Denisova, G. A. 1960: Some questions of the growth biology of the seeds of *Caropodium platycarpum* (Boiss. et Hausstn.) Schischk. and *Echinophora trichophylla* Smith. Botan. Zh. 45:249–52. Biol. Abstr. 38:7395. 1962.
5196. —— 1962: Specific characteristics of embryo development in seeds of certain Umbelliferae. Dokl. Akad. Nauk SSSR 139:127–28.
Denisova, G. A. 18328.
5197. Denisova, L. V. 1962: The effect of daylength on the formation of the period of dormancy in the seeds of *Chenopodium album* L. Botan. J. (Botan. Soc. USSR) 47:1349–54.
Denna, D. 8996.
5198. Dennis, F. G., Jr., and J. P. Nitsch. 1963: Extraction and purification of growth substances from apple seeds. Proc. Plant Physiol. Meeting 38:xxxviii.
5199. Dennis, N. M., and R. W. Decker. 1962: A method and machine for detecting living internal insect infestation in wheat. J. Econ. Entomol. 55:199–203.
5200. Dennis, R. W. G. 1939: Notes on seed transmission of *Phoma lingam* in relation to dry rot of swedes in Scotland. Ann. Appl. Biol. 26:627–30.
Dennis, R. W. G. 3467.
5201. Dennison, P. R. 1958: Experimentation with camellia seeds. Am. Camellia Yearbook 1958:178–89.
Dennison, R. A. 9010.
5202. Denny, F. E. 1917: Permeability of certain plant membranes to water. Botan. Gaz. 63:373–97.
5203. —— 1917: Permeability of membranes as related to their composition. Botan. Gaz. 63:468–85.
5204. —— 1948: The role of the surface micro-flora in measurements of the respiration rate of germinating seeds. Contrib. Boyce Thompson Inst. 15:211–27.
5205. Denoga, N. 1939: Germination of teak. Philippine J. Forestry 2:173-83. Biol. Abstr. 13:17647. 1939.
Dent, K. W. 11924.

5206. Dent, T. V. 1947: Seed storage with particular reference to the storage of seed of Indian forest plants. Indian Forest Records Silvicult. 7(1):1–134. J. Forestry 47:656–57. 1949.

Denton, C. A. 4718.

5207. Deodhar, G. W. 1954: Bleaching tamarind seed or seed powder. Indian 47:457. Chem. Abstr. 48:7816. 1954.

5208. Deokule, A. G. 1961: Flowering and seed setting in sweet potato. Poona Agr. Coll. Mag. 51:13–17.

5209. De Oliveira, M. A. 1954: Tratamento de sementes de especies olericolas com fungicidas para cultivos sob condições de campo. Inst. Agron. Sul (Pelotas, Brazil) Bol. Tec. 11. 23 pp.

5210. De Ong, E. R. 1919: Effect of excessive sterilization measures on the germination of seeds. J. Econ. Entomol. 12:343–45.

5211. De Paolis, C. 1931: Experiments on the action of the excretory and metabolic products of "*Pythium*" sp. upon the germination of wheat. Boll. Staz. Patol. Veg. 11:138–43. Rev. Appl. Mycol. 11:164–65. 1932.

5212. —— 1931: Esperienze sul trattmento del grano con anticrittogamici a base di sali di mercurio. Boll. Staz. Patol. Veg. 11:158–64. Biol. Abstr. 7:7795. 1933.

De Raggio, N. M. 15865, 15866.

5213. Derbyshire, D. M. 1961: A study of seed-borne infection of tomato by *Didymella lycopersici* Kleb. Proc. Intern. Seed Testing Assoc. 26:61–67.

5214. Derenne, P. 1954: Application du test du chlorure de 2,3,5-triphenyltetrazolium aux plantules de céréales endommagées par le froid. Bull. Inst. Agron. Sta. Rech. Gembloux 22:155–56.

5215. —— 1960: La transplantation embryonnaire des céréales: problème génétique ou non? Bull. Inst. Agron. Sta. Rech. Gembloux 28:33–38.

5216. De Rey-Pailhade, J. 1895: Rôles respectifs du philothion et de la laccase dans les graines en germination. Compt. Rend. 121:1162–64.

5217. Dermer, O. C., and L. T. Crews. 1939: The composition of the seeds of *Sapindus drummondii*. J. Am. Chem. Soc. 61:2697–98.

5218. De Ropp, R. S. 1939: Studies in the vernalisation of cereals. IV. The effect of preliminary soaking of the grain on the growth and tropic responses of the excised embryo of winter rye. Ann. Botany (London) 3:243–52.

De Ropp, R. S. 7641.

De Rose, H. R. 370.

5219. De Roubaix, J., and O. Lazar. 1947: Vernalization of sugar beet. Nature 159:442.

5220. —— and —— 1957: Respiratory metabolism of the sugat beet. VII. Inhibiting substances contained in the glomerules of sugar beets. Sucr. Belge 76:185–97. Chem. Abstr. 51:5222. 1957.

5221. ——, ——, and L. Massart. 1956: Respiratory metabolism of the sugar beet. VI. Liberation and utilization of the cellular substrates of the seeds and envelopes (perianth and pericarp) upon germination. Sucr. Belge 75:261–80. Chem. Abstr. 50:9532. 1956.

De Roubaix, J. 12956.

5222. Derr, H. J. 1958: Direct seeding: a fast, reliable method of regenerating longleaf pine. Tree Planters Notes 32:15–20.

5223. Derscheid, L. A., and R. E. Schultze. 1960: Achene development of Canada thistle and perennial sow thistle. Weeds 8:55–62.

5224. Desai, C. M., and M. T. Vyas. 1949: Chemical investigation of the seeds of *Jatropha curcas* Linn. Current Sci. (India) 18:49.

5225. Desai, D. B., and R. G. Chitre. 1957: Niacin biosynthesis in germinating cereals and pulses. II. Further observations on the effect of tryptophan on alkali-labile precursor of niacin during germination in cereals. Indian J. Med. Res. 45:461–64. Chem. Abstr. 52:1367d. 1958.

Desai, D. B. 3995, 3996.

Desai, M. J. 3758.

Desai, M. K. 2425.

5226. Desai, M. V., and C. S. Reddy. 1958: Toward a uniform seed corn cold test. Phytopathology 48:386–87.

Desai, R. M. 4935.

Desai, R. S. 13290.

5227. De Santos, I., and A. P. West. 1929: Resins in the seed coats of Philippine chaulmoogra seeds (*Hydnocarpus alcalae*). Philippine J. Sci. 40:485–89.

5228. ——, ——, and P. D. Esguerra. 1931: Philippine pine-needle oil from *Pinus insularis* (Endlicher). Philippine J. Sci. 46:1–7. Ber. Ges. Physiol. 64:656. 1932.

5229. De Saussure, T. 1827: De l'influence du dessechement sur la germination du plusieurs graines alimentaire. Ann. Sci. Nat. 10:68–93.

5230. Descartes de G. Paula, R. 1936: Botanical and chemical study of the seeds of cipoabacate (*Salacia brachypoda*). Rev. Flora Med. 2:397–413. Chem. Abstr. 30:6040–41. 1936.

5231. —— 1938: Carpotroside, a new glucoside or heteroside isolated from the seed of *Carpotroche brasiliensis* Endl. Rev. Soc. Brasil. Quim. 7:129–40. Pharm. Abstr. 7:14. 1941.

5232. —— 1941: Chemistry of Brazilian kola nut. Rev. Quim. Ind. (Rio de Janeiro) 10(110):13–14. Chem. Abstr. 36:4214. 1942.

5233. Deshmukh, A. D., and K. Sohonie. 1961: Isolation and purification of pipecolic acid from the double bean, *Vicia faba* (Moench.). J. Sci. Ind. Res. (India) 20C:328–29. Biol. Abstr. 38:23596. 1962.

5234. —— and —— 1961: Distribution of pipecolic acid in germinating legumes and in their growing plants. J. Sci. Ind. Res. (India) 20C:329–30. Biol. Abstr. 38: 23597. 1962.

5235. Deshmukh, N. Y. 1960: A case of seeded variety of banana (*Musa* sp.). Nagpur Agr. Coll. Mag. 34(1/2):28–29.

5236. Deshmukh, V. K., and U. K. Kanitkar. 1954: Hormones and development of vitamin C in pea. Univ. Poona Sci. Technol. 6:1–4. Chem. Abstr. 49:14916. 1955.

5237. Deshpande, B. R., and V. G. Kulkarni. 1962: Studies in germination of coconut. I. Coconut Bull. 16:336–38, 343. Biol. Abstr. 44:3808. 1963.

5238. Deshpande, P. D., and M. V. R. Rao. 1954: Nitrogen complex and amino-acid composition of (I) amaranth (*Amaranthus gangeticus*) and (II) aconite bean (*Phaseolus aconitifolius*). Indian J. Med. Res. 42:77–84. Biol. Abstr. 30:8685. 1956.

5239. Deshpande, P. K. 1959: Contribution to the embryology of *Kirganelia reticulata* Baill. J. Biol. Sci. (Bombay) 2:76–83.

5240. —— 1960: Morphology of the endosperm in *Caesulia axillaris* Roxb. Current Sci. (India) 29:56–57.

5241. —— 1962: Contribution to the embryology of *Caesulia axillaris* Roxb. J. Indian Botan. Soc. 41:540–50.

5242. —— 1963: Fertilization and development of endosperm, embryo and fruit in *Flaveria repanda* Lag. J. Indian Botan. Soc. 41:505–9. 1962.

5243. Deshpande, R. B., and A. K. Mallik. 1938: Studies in Indian oil-seeds. VI. Some correlations between oil-content and other characters in the Pusa linseed hybrids. Indian J. Agr. Sci. 7:841. Agr. Live-stock India 8:75. 1938.

Deshpande, R. S. 2135.

5244. Desikachar, H. S. R. 1956: Changes leading to improved culinary properties of rice on storage. Cereal Chem. 33:324–28.

5245. —— and S. S. De. 1947: Role of inhibitors in soybean. Science 106:421–22.

5246. —— and —— 1950: The tryptic inhibitor and the availability of cystine and methionine in raw and germinating soya beans. Biochim. Biophys. Acta 5:285–89.

5247. ——, S. K. Majumber, S. V. Pingale, and V. Subrahmanyan. 1959: Discoloration in rice: Some studies on its nature and effect on nutritive value. Cereal Chem. 36:78–83.

5248. —— and V. Subrahmanyan. 1960: The relative effects of enzymatic and physical changes during storage on the culinary properties of rice. Cereal Chem. 37:1–8.

Desikachar, H. S. R. 10154.

De Silva, D. S. 6474.

5249. Desjardins, P. R., R. L. Latterell, and J. E. Mitchell. 1954: Seed transmission of tobacco-ringspot virus in Lincoln variety of soybean. Phytopathology 44:86.

Desmarteau, R. 11875.

5250. De Smedt, A. 1950: La récolte et la conservation de semences d'essences forestières. Ann. Agr. Hort. [Gand, Belgium] (1950/51) 3:100–3.

5251. Desneux, R. 1957: La germination des graines d'Elaeis en chambre chaude à la station de Kiyaka. Bull. Inform. INEAC (Inst. Natl. Etude Agron. Congo Belge) 6:11–19. Bull. Agr. Congo Belge 48:1957.

5252. —— and C. Mahy. 1960: À propos d'une méthode simplifiée de germination des graines du palmier à huile. Bull. Inform. INEAC (Inst. Natl. Etude Agron. Congo Belge) 9:43–48. Bull. Agr. Congo Belge 51:1960.

5253. Desnuelle, P., J. Molines, and S. Bonjour. 1951: Thermal treatment of oil seeds and lecithin content of the oil. Oléagineux 6:264–67. Chem. Abstr. 46: 8394. 1952.

5254. De Sousa - Monteiro - Cerca - Miguel, M.C. 1953: Identification of lettuce (*Lactuca sativa* L.) varieties on the basis of seed and young plant characteristics. Proc. Intern. Seed Testing Assoc. 18:180–87.

5255. Despois, R. 1958: Coumarins and phenolic acids in relation to inhibition and activation of germination. Proc. Intern. Congr. Biochem. Vienna 4(5):33–36.

Des-Rochettes, A. Mercier-; *see* Mercier-Des-Rochettes, A.

5256. Dessureaux, L. 1950: Variation in the seed setting ability of Ladino white clover. Sci. Agr. (Ottawa) 30:509–17.

5257. —— 1951: Ovule formation as a factor influencing seed-setting. Sci. Agr. (Ottawa) 31:373–82.

5258. —— 1962: Study of the heredity of the absence of chlorophyll in lucerne cotyledons. Natl. Can. 89:341–55.

5259. De Stigter, H. C. M. 1957: The germination of rose seeds for stocks. Mededel. Directeur Tuinbouw 20:356–62.

5260. De Suqueira, R., E. Pechnik, and O. Guernelli. 1953: Researches in applied biochemistry. I. Study of the rubber tree seed. Rev. Quim. Farm. Rio de Janeiro 18:41–53. Chem. Abstr. 48:5313. 1954.

5261. Desveaux, R., and M. Kogane-Charles. 1952: Étude sur la germination de quelques graines oléagineuses. Ann. Inst. Rech. Agron. Ser. A, Ann. Agron. 3:385–416.

5262. —— and —— 1963: Étude de la germination des graines oléagineuses. Ann. Agron. 4:861–62.

5263. —— and —— 1954: The acids formed during the germination of castor bean. Congr. Intern. Botan. Paris Rappt. Commun. 8:37. Chem. Abstr. 48:11562. 1954.

5264. —— and —— 1955: Germination of oil-containing seeds. II. Water-soluble acids formed during germination. Ann. Agron. 6:337–61. Chem. Abstr. 54: 12279. 1960.

5265. —— and —— 1956: Sur la présence de glycerol dans les graines de *Ricinus communis* et de *Linum usitatissimum* en germination. Compt. Rend. 243:1929–30.

Desveaux-Chabrol, J. 18453.

Desvernes, P. 3641.

5266. De Tempe, J. 1953: Proposal for the preparation of a new handbook of seed-borne diseases and pests. Proc. Intern. Seed Testing Assoc. 18:197–204.

5267. —— 1957: Routine investigation of seeds for their health condition in the Dutch testing station at Wageningen. Proc. Intern. Seed Testing Assoc. 22:354–70.

5268. —— 1958: Three years of field experiments on seed-borne diseases and seed treatment of cereals. Proc. Intern. Seed Testing Assoc. 23:38–67.

5269. —— 1958: Germination and health of radish seed. Proc. Intern. Seed Testing Assoc. 23:332–39.

5270. —— 1961: Seed weakness (general information). Proc. Intern. Seed Testing Assoc. 26:3–11.

5271. —— 1961: Routine methods for determining the health condition of seeds in the seed testing station (general information). Proc. Intern. Seed Testing Assoc. 26:27–60.

5272. —— 1962: Hot-water treatment as a means for determining seed weakness. Proc. Intern. Seed Testing Assoc. 27:773–78.

5273. —— 1962: Comparison of methods for seed health testing. Proc. Intern. Seed Testing Assoc. 27:819–28.

5274. —— 1963: On methods of seed health testing principles and practice. Proc. Intern. Seed Testing Assoc. 28:97–105.

5275. —— 1963: Health testing of flax seed. Proc. Intern. Seed Testing Assoc. 28:107–31.

5276. —— 1963: The blotter method for seed health testing. Proc. Intern. Seed Testing Assoc. 28:133–51.

5277. —— 1963: Inspection of seeds for adhering pathogenic elements. Proc. Intern. Seed Testing Assoc. 28:153–65.

5278. —— 1963: The use of correlation coefficients in comparing methods for seed vigor testing. Proc. Intern. Seed Testing Assoc. 28:167–72.

5279. —— and W. F. Crosier. 1961: Pesticide treatment (general information). Proc. Intern. Seed Testing Assoc. 26:12–26.

De Tempe, J. 11198, 14541.

5280. Detjen, L. R. 1943: The influence of the rootstock on seeds and on seedling progenies of tomato grafts. Proc. Am. Soc. Hort. Sci. 43:147–48.

5281. —— and E. M. Rahn. 1948: Effect on the progeny resulting from grafting tomato scions onto Jimson weed (*Datura stramonium* L.) stocks. Delaware Agr. Expt. Sta. Bull. 271:29.

5282. Detmer, W. 1880: Vergleichende Physiologie des Keimungsprocesses der Samen. Gustav Fischer, Jena. 565 pp.

5283. De Toledo, F. F. 1960: Germination of cotton seeds treated with fungicides and systemic insecticides. Rev. Agr. (Piracicaba) 35:173–78.

5284. De Toledo, T. A. N., and R. Wasicky. 1954: Formação dos alcaloides no *Pilocarpus pennatifolius* Lemaire, durante a germinação da semente e primeiros estagios vegetativos. Anales Fac. Farm. Odontol. Univ. Sao Paulo 12:83–92. Biol. Abstr. 31:29476. 1957.

Detroux, L. 3680.

De Turk, E. E. 5752, 16988, 18114.

5285. Detwiler, S. B. 1931: The effect of ultra-violet light on germination of seeds and growth of seedlings of *Ribes rotundifolia* Michx. J. Forestry 29:131–33.

5286. —— 1943: Better acorns from a heavily fertilized white oak tree. J. Forestry 41:915–16. Biol. Abstr. 18:9284. 1944.

Detwiler, S. B., Jr. 6533.

5287. Deuber, C. G. 1931: Chemical treatments to shorten the rest period of tree seeds. Science 73:320–21.

5288. —— 1932: Chemical treatments to shorten the rest period of red and black oak acorns. J. Forestry 30:674–79.

Deuel, H. 9089.

Deuel, H. J., Jr. 12229.

5289. Deuse, P. 1952/53: Ecologie de *Molinia coerulea*, Moench. germinations. Liege Univ. Inst. Botan. Arch. 22:53–59.

5290. Devadatta, S. C., B. N. Acharya, and S. B. Nadkarni. 1951: Effect of germination on the nutritional qualities of some of the vegetable proteins. Proc. Indian Acad. Sci. Sect. B 33:150–58.

5291. Devaux, H., and J. Sechet. 1949: Experiences sur la latence à la germination observée chez des grains de blé semés en pleine terre immediatement après la récolte et ensuite à intervalles de temps reguliers, pendant trois mois. Compt. Rend. 228:1542–45.

5292. De Vay, J. E. 1952: A note on the effect of mold growth and increased moisture content on the free amino acids in hard red spring wheat. Cereal Chem. 29:309–11.

De Vay, J. E. 11060, 15242.

De Veer, E. J. van A.; *see* Van A. de Veer, E. J.

Devi, Y. V. S. 17399, 17400.

5293. De Villiers, F. J. 1931: Citrus byproducts: Orange and lemon seeds. Farming S. Africa 6:132. Chem. Abstr. 26:865. 1932.

5294. De Vilmorin, A. 1960: Les problèmes de semences dans le Marche Commun. Agriculture (Paris) 23:179–81.

5295. De Vilmorin, R. 1944: Note sur la conservation de la faculté germinative des graines d'*Abies pectinata*. Compt. Rend. Acad. Agr. France 30:509–12.

5296. Devine, V. 1947: Embryology of *Lychnis alba*. Am. J. Botany (Suppl.) 34:4a.

5297. —— 1948: Note on the culture of *Lychnis* embryos. Proc. Iowa Acad. Sci. 55:95–97.

5298. De Vos, M. P. 1961: On the embryology and relationships of the South African genera of the Haemodoraceae. Intern. Botan. Congr. Recent Advan. Botany 9:694–98. 1959.

5299. —— 1963: Studies on the embryology and relationships of South African genera of the Haemodoraceae: *Lanaria* Ait. J. S. African Botany 29:79–90.

5300. De Vries, H. 1879: Keimunggeschichte der Zuckerruben. Neue Z. Rubenzucker-Ind. 2:65–85.

5301. —— 1894: Variation in cotyledons. Ber. Deut. Botan. Ges. 12:25–39.

5302. —— 1915: Über Künstliche Beschleunigung der Wasseraufnahme in Samen durch Druck. Biol. Centralbl. 35:161–76.

De Vries, J. E. 6827.

5303. De Waal, D. 1940: Differentiation of the different varieties of Dutch wheat by the color reaction of the seed when treated with phenol. Landbouw. Tijdschr. 52: 753–60. Chem. Abstr. 37:5148. 1943.

5304. Dewey, D. R. 1962: Germination of Crested wheatgrass in salinized soil. Agron. J. 54:353–55.

5305. Dewez, J. 1964: Water uptake and heat evolution by germinating cotton seed. Plant Physiol. 39:240–44.

5306. De Wildeman, E. 1947: "Vitalité" des graines. Bull. Classe Sci. Acad. Roy. Belg. 33:77–88. Biol. Abstr. 22:19261A. 1948.

De Winter, F. 15358.

5307. De Witt, J. L., C. L. Canode, and J. K. Patterson. 1962: Effects of heating and storage on the viability of grass seed harvested with high moisture content. Agron. J. 54:126–29.

De Wolfe, T. A. 10851.

Dexheimer, F. J. 400.

5308. Dexter, S. T. 1943: Commercial hormone dusts for seed treatment: a second report. Mich. Agr. Expt. Sta. Quart. Bull. 25:279–82.

5309. —— 1951: The germination of sugarbeet seedballs. Amer. Soc. Sugar Beet Technol. East. U.S. & Canada. Proc. Region Mtg. 6:45–46.

5310. —— 1953: Environmental and cultural variables that influence the establishment and yield of meadow seedings. Mich. Agr. Expt. Sta. Quart. Bull. 36:138–47.

5311. —— 1955: The vapor-pressure or relative humidity approach to moisture-testing for safe farm-storage of harvested crops. Agron. J. 47:267–70.

5312. —— 1955: Alfalfa seedling emergence from seed lots varying in origin and hard seed content. Agron. J. 47:357–60.

5313. —— 1957: Moisture equilibrium values in relation to mold formation of several grasses and small-seeded legumes. Agron. J. 49:485–88.

5314. —— 1957: Storage of alfalfa seed in the presence of herbicides. Mich. Agr. Expt. Sta. Quart. Bull. 39:647–49.

5315. ——, A. L. Anderson, P. L. Pfahler, and E. J. Benne. 1955: Responses of white pea beans to various humidities and temperatures of storage. Agron. J. 47:246–50.

5316. —— and J. W. Creighton. 1948: A method of curing farm products by the use of drying agents. J. Am. Soc. Agron. 40:70–79.

5317. —— and E. G. McKibben. 1945: Vacuum-type harvester for white clover seed. Mich. Agr. Expt. Sta. Quart. Bull. 27:294–97.

5318. —— and T. Miyamoto. 1959: Acceleration of water uptake and germination of sugarbeet seedballs by surface coatings of hydrophilic colloids. Agron. J. 51:388–89.

Dexter, S. T. 9234, 13623.

Dey, B. 17539.

Dey, K. L. 2826.

Deyoe, C. W. 13486.

5319. Deysson, G., M. Deysson, and N. Mazel. 1952: Effect of meso-inositol on the germination and growth of various angiosperms. Bull. Soc. Botan. France Mem. 1952:143–55. Chem. Abstr. 47: 10068. 1953.

5320. Deysson, M. 1953: Action de l'uracile sur la germination et la croissance de diverses angiospermes. Bull. Soc. Botan. France 100:14–17.

Deysson, M. 5319.

5321. De Zeeuw, D. J. 1954: Fungicide treatment of table beet and spinach seeds for the prevention of damping-off. Mich. Agr. Expt. Sta. Quart. Bull. 37:105–18.

5322. —— and A. L. Andersen. 1952: Response of pea varieties to dry and slurry methods of seed treatment. Phytopathology 42:52–56.

5323. —— and —— 1953: Lima bean seed treatment trials in Michigan, 1951–1952. Plant Disease Reptr. 37:69–70.

5324. ——, ——, L. G. Merrill, and W. Zurakowski. 1954: Fungicide-insecticide combination seed treatment of beans for the control of damping-off and seed corn maggot. Mich. Agr. Expt. Sta. Quart. Bull. 37:204–18.

5325. ——, R. A. Davis, and R. A. Crum. 1961: Storage of fungicide-treated pea and cucumber seeds. Plant Disease Reptr. 45:544–47.

5326. ——, G. E. Guyer, and A. L. Andersen. 1956: Fungicide and insecticide seed treatments of peas and beans, 1953–1955. Plant Disease Reptr. 40:727–33.

5327. ——, ——, A. L. Wells, and R. A. Davis. 1959: The effects of storage of vegetable seeds treated with fungicides and insecticides on germination and field stand. Plant Disease Reptr. 43:213–20.

De Zeeuw, D. J. 485, 486, 4263, 10789.

5328. Dhaliwah, A. S., D. K. Salunkhe, and A. P. Lorz. 1960: Starch grain characteristics and amylose content in relation to pollination time and seed size, and germination behaviour in *Phaseolus* species. Indian J. Genet. Plant Breeding 20:118–25. Biol. Abstr. 36:21662. 1961.

5329. Dhar, D. C., M. L. Dhar, and D. L. Shrivastava. 1951: Seeds of *Emblica officinalis* Gaertn. I. The fatty oil and its component fatty acids. J. Sci. Ind. Res. (India) 10B:88–91. Brit. Abstr. BII:273. 1952.

5330. Dhar, M. L., and R. Gopalachari. 1952: Chemical examination of seeds of *Achyranthes aspera* Linn. J. Sci. Ind. Res. (India) 11B:209. Biol. Abstr. 27:4700. 1953.

Dhar, M. L. 5329.

Dharmani, P. L. C. 11748.

5331. Dhekne, V. V., and B. V. Bhide. 1951: Chemical investigation of the seeds of *Balanites roxburgii*, Planch (hingora seeds). I. Saponin 28:588–90. Biol. Abstr. 27:4701. 1953.

5332. Dhillon, P. S., and P. S. Johnson. 1962: Germination and subsequent hypocotyl growth of Western larch seed following pretreatment with hydrogen peroxide. Proc. Montana Acad. Sci. 21:18–23.

5333. Dhingra, D. R., and A. K. Biswas. 1945: Component fatty acids of oil of *Citrullus vulgaris*, Schrad (water-melon) seeds. J. Indian Chem. Soc. 22:119–22. Analyst 71:291. 1946.

5334. —— and T. P. Hilditch. 1931: The fatty acids of some Indian seed oils: the seed fats of *Mesua ferrea*, *Calophyllum inophyllum*, and *Pastacia vera*. Soc. Chem. Ind. J. 50:9T–12T.

5335. —— and P. Narain. 1945: Component fatty acids of melon (*Cucumis melo*, Linn.) seed oil. J. Indian Chem. Soc. 22:123–26. Analyst 71:291. 1946.

5336. —— and K. Pershad. 1945: Component fatty acids of Kahu (*Lactuca scariola*, Linn.) seed oil. J. Indian Chem. Soc. 22:127–29. Analyst 71:330. 1946.

5337. ——, G. L. Seth, and P. C. Speers. 1933: Indian seed fats—mowha (*Bassia latifolia*) and tamal (*Garcinia morella*) fats. Soc. Chem. Ind. J. 52:116T–118T. Chem. Abstr. 27:3097. 1933.

Dhingra, D. R. 186.

Dhodapkar, D. R. 16186.

5338. Diaconita, I. 1930: Étude morphologique et biometrique des graines de *Nicotiana tabacum*, variété Ghimpatzi. Bul. Cult. Ferment. Tutunului 19:462–98.

Diamond, M. J. 2470.

Dianova, V. 15468.

Diaz, M. A. Jimenez-; *see* Jimenez-Diaz, M. A.

5339. Dick, J., J. M. Finnis, L. O. Hunt, and N. B. Kverno. 1958: Treatment of Douglas-fir seed to reduce loss to rodents. J. Forestry 56:660–61.

Dick, J. 3628.

Dicke, F. F. 1494.

Dicke, R. J. 9359.

5340. Dickerson, L. M. 1931: The inheritance of ruby seed coat color in peas. J. Heredity 22:319–21.

5341. Dickey, R. D. 1942: The importance of tung seed selection. Proc. Am. Soc. Hort. Sci. 41:127–30.

5342. —— 1953: Germination of pigmy date palm (*Pheonix roebeleni*) seed as affected by treatment with sulphuric acid. [Abstract.] Proc. Assoc. Southern Agr. Workers 50:139.

5343. —— and W. Reuther. 1940: Flowering, fruiting, yield, and growth habits of tung trees. Florida Agr. Expt. Sta. Bull. 343. 28 pp. Expt. Sta. Rec. 83:341. 1940.

Dickey, R. D. 8072.

5344. Dickhart, W. H. 1932: Quince-seed oil. Am. J. Pharm. 104:335–36. Chem. Abstr. 26:3948.

Dickinson, D. 1278.

Dickinson, S. 1478.

5345. Dickson, A. 1883: On the germination of *Streptocarpus caulescens*. Trans. Proc. Botan. Soc. Edinburgh 14:362–64.

5346. Dickson, A. D., and B. A. Burkhart. 1942: Changes in the barley kernel during malting: chemical comparisons of germ and distal portions. Cereal Chem. 19:251–62.

5347. ——, H. Clemens, and N. Standridge. 1960: The influence of the wild oat herbicide, barban, on barley composition, germination, and malt quality. U.S. Dept. Agr. ARS 34-13. 7 pp.

5348. ——, H. L. Shands, and B. A. Burkhart. 1949: Influence of growing-regulating chemicals on the malting of barley and the composition of malt. Cereal Chem. 26:13–23.

Dickson, A. D. 45, 1362, 3380.

Dickson, J. G. 1361, 1362, 8999, 12137, 14392, 15742.

5349. Dickson, M. H., and C. E. Peterson. 1958: Hastening greenhouse seed production for carrot breeding. Proc. Am. Soc. Hort. Sci. 71:412–15.

5350. —— and —— 1960: The influence of gibberellin on the flowering of carrots. Can. J. Plant Sci. 40:468–73.

5351. Dickson, R. E., J. F. Hosner, and L. Kahler. 1961: Polyethylene bag storage versus stratification in sand as a pregermination treatment for white and loblolly pine. Tree Planters Notes 48:15–16.

Dickson, R. E. 9087.

5352. Dieckert, J. W., J. E. Snowden, Jr., A. T. Moore, D. C. Heinzelman, and A. M. Altschul. 1962: Composition of some subcellular fractions from seeds of *Arachis hypogaea*. J. Food Sci. 27:321–25.

5353. Diemair, W., H. Janecke, and D. Heusser. 1952: Untersuchungen über die Farbstoffe aus *Bixa orellana* L. Naturwissenschaften 39:211.

5354. Diener, U. L. 1960: The mycoflora of peanuts in storage. Phytopathology 50: 220–23.

Diener, U. L. 19358.

5355. Dieterich, K. 1924: Über Kultur von Embryonen ausserhalb des Samens. Flora (Jena) 117:379–417.

Dietrich, W. C. 18951.

5356. Dietz, A. A. 1960: Enhancement of the nutritional value of cereals through germination. Trans. Illinois State Acad. Sci. 53:106–14.

5357. Dietz, R. 1931: Versuche zur Abkürzung des Keimpflanzenverfahrens von Neubauer. Ernähr. Pflanze 27:429–32.

5358. Dieudonné, P., and R. Vanderwalle. 1935: Recherches sur la desinfection des semences d'orge dans la lutte contre *Ustilago nuda* (Jensen) Rostrup. Bull. Inst. Agron. Sta. Rech. Gembloux 4:366–76. Biol. Abstr. 11:18181. 1937.

Diez, M. J. Fernandez. 5082.

5359. Di Fazio, I. L. 1954: Lipase of *Ricinus communis*, in the quiescent and in the germinating seed. Tumori 40:638–42. Chem. Abstr. 49:14923. 1955.

5360. Di Fonzo, M. A. 1942: El tratamiento de la semilla del Algodonero con productos anticriptogamicos. Bol. Junta Nacl. Algodon. (Buenos Aires) 1942(81–82):43–51. Rev. Appl. Mycol. 22:64. 1943.

5361. —— 1942: La humedad de la semilla del Algodonero relacionada con su poder germinativo y con la presencia del *Aspergillus wentii*. Bol. Junta Nacl. Algodon. (Buenos Aires) 1942(85–86):156–58. Rev. Appl. Mycol. 22:64. 1943.

5362. Dijkstra, A. 1958: Influence of growth factors on the hardness of covering of clover seed. Neth. Rlandbvoorlichtingsdienst Landbvoorlichting 15:538–42.

5363. —— 1959: The influence of germinating power of summer wheat on the development and yield of the crop. Landbouwk. Tijdschr. 71:176–80.

5364. —— 1959: The influence of germinating power of summer wheat on the development and yields of the crop. Landbouwk. Tijdschr. 71:197–201.

5365. —— 1961: Influence of some growth factors on the hard seeds of clover. Neth. Rlandbvoorlichtingsdienst Landbvoorlichting 18:95–99.

Dijkstra, D. L. J. 9201.

5366. Dikshit, N. N. 1952: Vernalization of peas. Current Sci. (India) 21:48–49.

5367. —— and U. P. Singh. 1952: Vernalisation of cabbage seeds (*Brassica oleracea* var. *capitata*). Current Sci. (India) 21:249–50.

5368. Dikusar, I. G., and A. F. Kalinkevich. 1940: Mineral nutrition of kok-saghyz during various phases of its growth, as a factor influencing the metabolism and increasing the yield of seeds and rubber. Vestn. Sel'skokhoz. Nauki Tekhn. Kul'tury 1940(1):9–19. Chem. Abstr. 37:3126. 1943.

5369. Dilhac, P., F. Busson, and A. Georgin. 1959: Chemical study of the seed of *Voandzeia subterranea*. Riz. Rizicult. 5:120. Chem. Abstr. 53:19042. 1959.

5370. Dillge, A. 1931: Die einjahrige Kultur der Edel-Monatserdbeere "Ruegen" aus Samen. Möllers Deut. Gärtner-Ztg. 46:379.

5371. Dilling, W. J. 1926: Influence of lead and the metallic ions of copper, zinc, thorium, beryllium and thallium on the germination of seeds. Ann. Appl. Biol. 13:160–67.

Dillingen, H. Becker-; *see* Becker-Dillingen, H.

5372. Dillman, A. C. 1928: Daily growth and oil content of flaxseeds. J. Agr. Res. 37:357–77.

5373. —— 1930: Hygroscopic moisture of flax seed and wheat and its relation to combine harvesting. J. Am. Soc. Agron. 22:51–74.

5374. —— and T. H. Hopper. 1943: Effect of climate on the yield and oil content of flaxseed and on the iodine number of linseed oil. U.S. Dept. Agr. Tech. Bull. 844:1–69.

5375. —— and E. H. Toole. 1937: Effect of age, condition, and temperature on the germination of flaxseed. J. Am. Soc. Agron. 29:23–29.

Dima, M. 14891.

5376. Di Maggio, G. 1951: Pharmacological study of thiazoles. VIII. Action of 2-amino-5-methylthiazole-4-carboxylic acid (TC) on germination of wheat seeds and development of seedlings. Acta Vitaminol. 5:199–201. Chem. Abstr. 48:5934. 1954.

5377. —— 1952: Vitamin-like activity of quercetin on development and growth of plants. Boll. Soc. Ital. Biol. Sper. 28:36–39. Chem. Abstr. 47:10632. 1953.

Dimakov, S. 15274.

5378. Di Mattei, P. 1928: Enzymes of the seed of *Strophanthus kombe* and their influence on strophanthin. Arch. Intern. Pharmacodyn. 35:113–35. Biol. Abstr. 4:23891. 1930.

5379. Dimentshtein, F. I. 1958: Composition of the grain of *Elymus*. Selek. i Semen. 1958(5):65–67. Chem. Abstr. 53:19052. 1959.

5380. Dimitriewicz, N. 1875/77: Quellungsversuche mit einigen Samenarten. Wiss.-Prakt. Gebiete Pflanzenbaues 1:75–79.

5381. —— 1875/77: Wie lange bewahren die Samen unserer Kulturpflanzen ihre Keimfähigkeit. Wiss.-Prakt. Gebiete Pflanzenbaues 1:98–104.

5382. —— 1876: Über die Methoden der Samenprüfung landwirtschaflicher Kulturpflanzen. Inaug. Dissertation Univ. Leipzig. 34 pp.

Dimitrov, K. 9377.

5383. Dimitrov, M. 1956: The radiobiological reaction at vita minima. Izv. Bulgar. Akad. Nauk. Inst. Biol. "Metodii Popov" 7:169–92.

5384. Dimitrov, P., and S. Topchiiski. 1957: Die Bedeutung der Stratifikation des Erdbeersamens für die Selektion der Erdbeere. Nauchni Tr. Vissh Selskostopanski Inst. "Georgi Dimitrov," Agron. Fak. (Sofia) 4:101–10.

5385. Dimitrov, T. 1929: The seed of some imported forest trees. Ann. Fac. Agrar. Univ. Sofia 7:354–76. Biol. Abstr. 5: 11671. 1931.

5386. Dimitrowa, A. 1935: Über die Beschleunigung der Zellteilungsrate durch Einwirkung von Koleoptilenspitzenextrakten. Arch. Protistenk. 84:415–18. Botan. Centralbl. 168:346. 1935.

Dimler, R. J. 2618, 2930, 17160.

5387. Dimmock, F. 1947: The effects of immaturity and artificial drying upon the quality of seed corn. Can. Dept. Agr. Tech. Bull. 58. 66 pp.

5388. —— and F. S. Warren. 1953: The influence of time of planting on the yield and composition of soybean seed. Can. J. Agr. Sci. 33:550–58.

Dimmock, F. 19377.

5389. Dimock, A. W. 1950: Aster seed treatment. Florists Rev. 105(2729):105–6.

5390. Dingler, H. 1887: Über die Bewegung rotirender Flügelfrüchte und Flügelsamen. Ber. Deut. Ges. 5:430–34.

5391. —— 1912: The dissemination and germination of rose fruits. Botan. Jahrb. Beibl. 46:41–45.

5392. Dinnis, E. R. 1938: The treatment of "hard" seeds of sea pea to induce germination. South-Eastern Agr. Coll. J. (Wye, Kent) 42:126–29.

5393. —— and S. Jordan. 1939: The germination of freshly-harvested and of stored seeds of sea pea. South-Eastern Agr. Coll. J. (Wye, Kent) 44:140–42.

5394. Dionne, L. A. 1958: 2,4-Dichlorophenoxyacetic acid as an aid to seed production when widely-separated *Solanum* species are crossed. Nature 181:360–61.

Dionne, L. A. 17798.

5395. Di Pascal, E. C. 1935: Acción del extracto alcalino de lobulo anterior de hipofisis sobre la germinación. Rev. Soc. Agr. Biol. 11:68–74. Biol. Abstr. 10: 2897. 1936.

5396. D'Ippolito, G. 1912: Determination of germinating energy of seeds according to the average time required for germination. Staz. Sper. Agr. Ital. 45:302–20.

Dirks, V. A. 7565.

Diskin, T. 11035.

5397. Disney, R. W. 1954: The specific heat of some cereal grains. Cereal Chem. 31:229–39.

5398. Ditmer, W. P. 1963: Seed analysis report—1963. Penn. Dept. Agr. Bull. 46(4). 21 pp.

5399. Ditzler, L., C. H. Hunt, and R. M. Bethke. 1948: Effect of heredity on the niacin and pantothenic acid content of corn. Cereal Chem. 25:273–79.

Ditzler, L. 9215.

5400. Divekar, C. B. 1960: Effect of treatment of wheat seeds with 2,4-dichlorophenoxyacetic acid. I. Mysore Agr. J. 35:146–57. Chem. Abstr. 57:17120. 1962.

Divine, J. P. 9934.

5401. Dix, W. 1935: Versuch zur Beeinflussung der Pflanzenbestandes einer Wiese durch verschiedenzeitiges aussäen der Samen der verschiedenen Wiesenpflanzen. Landwirtsch. Jahrb. 81:45–69.

Dixit, S. N. 12700.

5402. Dixon, H. B., and S. R. Snider. 1941: The reliability of an automatic kernel counter for 1,000 kernel-weight determination. Cereal Chem. 18:209–20.

5403. Dixon, H. H. 1901: Vitality of seeds. Nature 64:256–57.

5404. —— 1902: On the germination of seeds after exposure to high temperatures. Trinity Coll. Dublin Botan. School Notes 5:176–86.

5405. —— 1902: Resistance of seeds to poisons. Trinity Coll. Dublin Botan. School Notes 5:187–88.

Dixon, K. 12545.

5406. Dixon, S. 1932: Relation of food to disease. Weed seeds. Soc. Chem. Ind. J. 51:789.

5407. Djakonow, W. 1880: Zur Frage über die Mittel, die Keimung der Samen, insbesondere der Coniferer Samen, zu beschleunigen. Mitt. Landwirtsch. Forstw. Akad. Petrowskoë Rasu Nowskoë, pp. 1–21. Just's Jahresber. 1882:31–32.

Djatschkow, N. 14801.

Dmitriev, N. 11937.

5408. Dmitrieva, A. N. 1958: Biochemistry of grains of hybrid corn. Byul. Ukr. Nauchn.-Issled. Inst. Rastenievod. Selek. Genet. 1958(2):51–53. Chem. Abstr. 53:8321. 1959.

5409. Dmochowski, J. 1934: Quantity of oil in linseed and hemp seed during the different periods of maturation and germination as well as in the birch and linden trees during the different phases of winter. Polish Agr. Forestal Ann. 32:35–77. Chem. Abstr. 28:6769. 1934.

5410. Dnyansagar, V. R. 1951: Embryological studies in the Leguminosae. III. A contribution to the embryology of *Pithecolobium saman* Benth. syn. *Enterolobium saman* Prain. Proc. Indian Acad. Sci. Sect. B 34:188–98. Biol. Abstr. 26: 19088. 1952.

5411. —— 1954: Embryological studies in the Leguminosae. VII. Endosperm and embryo development in *Neptunia triquetra* Benth. and *Prosopis spicigera* Linn. J. Indian Botan. Soc. 33:247–53.

5412. —— 1954: Embryological studies in the Leguminosae. IX. Development of the endosperm and embryo in *Dichrostachys cinerea* W. &. A. and *Parkia biglandulosa* W. & A. J. Indian Botan. Soc. 33:423–32.

5413. —— 1954: Embryological studies in the Leguminosae. X. Supplementary observations on the development of the endosperm and embryo in *Leucaena glauca* Benth. and *Mimosa hamata* Willd. J. Indian Botan. Soc. 33:433–42.

5414. —— 1955: Embryological studies in the Leguminosae. XI. Embryological features and formula and taxonomy of the Mimosaceae. J. Indian Botan. Soc. 34:362–74.

5415. —— 1957: Embryological studies in the Leguminosae. V. *Prosopis spicigera* and *Desmanthus virgatus*. Botan. Gaz. 118:180–86.

5416. —— and D. C. Cooper. 1960: Development of the seed of *Solanum phureja*. Am. J. Botany 47:176–86.

5417. —— and S. R. Malkhede. 1963: Development of the seed of *Trianthema portulacastrum* Linn. Proc. Indian Acad. Sci. Sect. B 57:343–55.

Doak, B. W. 9175.

5418. Doak, C. C. 1937: Embryogeny of *Cupressus*. Botan. Gaz. 98:808–15.

5419. —— 1937: The pistil anatomy of cotton as related to experimental control of fertilization under varied conditions of pollination. Am. J. Botany 24:187–94.

Doane, W. M. 19585.

Dobashi, K. 18307.

5420. Dobrohleb, I. F. 1941: The overwintering weeds and the means of controlling them. Dokl. Vses. Akad. Sel'skokhoz. Nauk 1941(1):12–14. Biol. Abstr. 16:18849. 1942.

5421. Dobrolyubskiy, O. K. 1958: Presowing treatment of seeds with trace elements. Priroda 1958:95–97. Biol. Abstr. 33:46852. 1959.

5422. Dobrovol'skii, N. A. 1938: Naphthalene treatment of *Nicotiana rustica* seeds. Tabak (USSR) 8(5):49–50. Chem. Abstr. 34:3418. 1940.

Dobrovszky, I. 1865.

5423. Doby, G. 1909: The role of oxalate in the germination of beet seed. Landwirtsch. Versuchssta. 70:155–58. J. Chem. Soc. 1909:ii, 256. 1909.

Dockins, J. O. 9959.

5424. Dodds, M. E. 1957: The effect of swathing at different stages of maturity on the bushel weight and yield of wheat. Can. J. Plant Sci. 37:149–56.

5425. —— 1961: The effect of swathing at different stages of maturity on the bushel weight and yield of oats. Can. J. Plant Sci. 41:401–6.

5426. —— and J. W. White. 1960: Seed cleaners and separators. Can. Dept. Agr. Res. Branch Publ. 1061.

5427. Dode, L. A. 1907: Observations sur la germination de certaines graines d'arbres. Bull. Soc. Dendrol. France Fasc. 3:18–22. Botan. Centralbl. 113:226. 1910.

5428. Dodge, F. N. 1944: A method of measuring the degree of kernel development of samples of pecans. Proc. Am. Soc. Hort. Sci. 45:151–57.

5429. Dodonova, E. V. 1941: Variations in the dehydrogenase content of pea seeds. Enzymologia 9:373–79. Chem. Abstr. 36:798. 1942.

5430. —— and N. N. Ivanov. 1939: The biochemical properties of spring wheat and winter wheat. Selek. i Semen. 1939(2/3): 16–21. Chem. Abstr. 34:5114–15. 1940.

5431. —— and —— 1939: Rapid colorimetric method for the determination of protein in seeds. J. Appl. Chem. USSR (English Transl.) 12:1575–81. Chem. Abstr. 34:7022–23. 1940.

Dodt, E. 8139.

Dölberg, U. 10142.

5432. Doerfel, F. 1930: Über den Einfluss des Frostes und intermittierender Temperaturen auf die Keimung verschiedener Samen. Botan. Arch. 30:1–50.

5433. Doerpinghaus, S. L. 1947: Differences between species of *Datura* in utilization of five carbohydrates. Am. J. Botany (Suppl.) 34:5a.

5434. Doessler, H. G., and W. Zentsch. 1959: Germination inhibiting action of constituents of pine turpentine oil. Pharmazie 14:46–48. Biol. Abstr. 33:42986. 1959.

5435. Dogger, J. R., and J. H. Lilly. 1949: Seed treatment may control wireworm in corn. Wisconsin Agr. Expt. Sta. Bull. 489:8–11.

5436. Dogonadze, D. S. 1956: Growing dahlias from seeds and its importance in ornamental gardening. Vestn. Tbilisskii Botan. Sad. 1956:197–202.

5437. Doguet G. 1953: Polyembryonie chez le *Cytisus laburnum* L. Bull. Soc. Botan. France 100:141–42.

5438. Doherty, N. W. 1963: Making a bed for a seed to lie in. Queensland Agr. J. 89:368–70.

5439. Doi, K., and S. Kakimoto. 1959: Studies on cracked grains produced by the early-sowing cultivation of paddy rice. II. Kyushu Agr. Res. 21:91–94.

5440. Dolcher, T. 1959: Observations on the embryology of *Zygopetalum* (Orchidaceae). Nuovo Giorn. Botan. Ital. 66:702–3. Biol. Abstr. 38:11203. 1962.

5441. —— 1961: Hormonal relations in the development of orchid ovary. I–II. Nuovo Giorn. Botan. Ital. 68:213–19.

Dolgiy, S. N. 19165.

5442. Dolgova, A. E. 1960: Causes of low germination in grape seeds. Agrobiologiya 1960(2):302–3.

Dollahite, J. W. 17797.

5443. Dollear, F. G., P. Krauczunas, and K. S. Markley. 1940: The chemical composition of some high iodine number soybean oils. Oil Soap 17:120–21. Expt. Sta. Rec. 85:293–94. 1941.

5444. Dolman, J., Jr. 1935: Some notes on germination. Am. Iris Soc. Bull. 59:16–18.

5445. Dolya, N. I. 1957: Determining the degree of readiness for germination of seeds of *Fraxinus excelsior* L. Zap. Khar'kovsk. Sel'skoe Khoz. Inst. 16:233–42. Biol. Abstr. 35:24903. 1960.

Domanska, H. 18087.

Dombrovskii, G. D. 13529.

Domenigg, M. Sorger-. 7415, 17752, 17753.

Domingo, C. E. 19109.

5446. Domingo, N. T., Jr. 1954: Germination of vanilla seeds. Philippine Agriculturist 37:543–49.

5447. Domingo, W. E. 1945: Inheritance of number of seeds per pod and leaflet shape in the soybean. J. Agr. Res. 70:251–68.

5448. —— 1945: Oil percentage through test weight. Seed World 57(4):8–9. Biol. Abstr. 19:17198. 1945.

Domingo, W. E. 18980.

Dominguez, F. Sanz-. 14980.

5449. Dominguez, X. A., and C. Villarreal. 1963: Phytochemical investigations. II. Study of the oil of pods and seeds of *Nerium oleander* pink flower variety. Ciencia (Mexico) 22:163–66. 1963. Chem. Abstr. 40:5887. 1964.

5450. Dommen, M. E. 1958: Availability of seeds of Nigerian forest trees. Nigeria Forestry Dept. Inform. Bull. 47. 2 pp.

Domsch, K. 15042.

5451. Domsch, K. H. 1956: Beitrag zur Beizmittelprüfung an Samen von Zierpflanzen. Nachrbl. Deut. Pflanzenschutzdienst (Berlin) 8:69–72.

5452. Dona-Dalle-Rose, A. 1948: Disinfection and viability of wheat seed. Ann. Sper. Agrar. (Rome) 2:693–700. Biol. Abstr. 24:5105. 1950.

5453. —— 1957: Germination of sugar beets in the field compared with germination values determined in the laboratory. Agr. Venezie 11:479–505.

5454. —— 1959: Germination of sugar beet glomerules exposed to ionizing radiations. Sementi Eletti 5(6):26–35.

5455. Donald, C. M. 1954: Competition among pasture plants. II. The influence of density on flowering and seed production in annual pasture plants. Australian J. Agr. Res. 5:585–97.

5456. —— 1959: The production and life span of seed of subterranean clover (*Trifolium subterraneum* L.). Australian J. Agr. Res. 10:771–87.

5457. —— and D. Spencer. 1951: The control of molybdenum deficiency in subterranean clover by pre-soaking the seed in sodium molybdate solution. Australian J. Agr. Res. 2:295–301.

5458. Donald, D. G. M. 1963: Results of tetrazolium bromide viability tests on seed of three *Pinus* species grown commercially in southern Africa. Forestry S. Africa 2:99–107, 109–10.

5459. Donald, H. P. 1935: Sulfuric acid treatment of mangel seed. New Zealand J. Agr. 50:336–40. Chem. Abstr. 29:6351–52. 1935.

5460. Donard, E., and H. Labbe. 1936: Germinated barley yields diabetes remedy like insulin. Sci. News Letter 29:117.

5461. Donchenko, V. S. 1957: Peculiarities of swelling (imbibition) and germination of seeds of the major weeds of millet. Uch. Zap. Omsk. Gos. Ped. Inst. 1957(6):43–61. Biol. Abstr. 35:48766. 1960.

5462. Doneen, L. D., and J. H. MacGillivray. 1943: Germination (emergence) of vegetable seed as affected by different soil moisture conditions. Plant Physiol. 18: 524–29.

Dong, T. 4470, 4471.

5463. Donnelly, E. D. 1955: The effects of outcrossing on forage and seed yields in Sericea Lespedeza, *L. cuneata*. Agron. J. 47:466–67.

Donnelly, J. 270.

5464. Donnet, V., J. M. Chavalier, and A. Pruneyre. 1960: Action de la cortisone sur la métamorphose et la croissance des têtards et la germination de certains végétaux. Compt. Rend. Soc. Biol. 154: 1451–53.

5465. Donoho, C. W., Jr., A. E. Mitchell, and H. M. Sell. 1962: Enzymatic destruction of C^{14} labeled indoleacetic acid and naphthaleneacetic acid by developing apple and peach seeds. Proc. Am. Soc. Hort. Sci. 80:43–49.

5466. Doolette, J. B., and H. C. Nourse. 1956: Seed wheat from research centres; production, distribution, use. J. Dept. Agr. S. Australia 59:346–49.

5467. Doolittle, S. P., and F. S. Beecher. 1937: Seed transmission of tomato mosaic following the planting of freshly extracted seed. Phytopathology 27:800–1.

Doolittle, S. P. 8335.

5468. Doppler, C. L. 1936: Investigations on some constituents of cocoa, especially on a few of the cellwall materials contained therein. Dissertation, Delft. Haarlem. (Enschede & Zonen) 149S. Botan. Centralbl. 171:131–32. 1937.

5469. Doran, W. L. 1938: Germination of seeds and damping-off and growth of seedlings of ornamental plants as affected by soil treatments. Mass. Agr. Expt. Sta. Bull. 351. 42 pp.

5470. Dore, J. 1955: Dormancy and viability of padi seed. Malayan Agr. J. 38: 163–73.

5471. —— and P. V. Thevan. 1959: Studies on seed germination and seedling growth in rice (*Oryza sativa* L.). Trop. Agr. (London) 36:15–34.

5472. Dore, W. G. 1956: Some grass genera with liquid endosperm. Bull. Torrey Botan. Club 83:335–37.

5473. —— and L. C. Raymond. 1942: Pasture studies. XXIV. Viable seeds in pasture soil and manure. Sci. Agr. (Ottawa) 23:69–79.

Doren, C. A. van. 3393.

5474. Dorety, H. A. 1908: The embryo of ceratozamia: a physiological study. Botan. Gaz. 45:412–16.

5475. —— 1919: Embryo and seedling of *Dioon spinulosum*. Botan. Gaz. 67:251–57.

Dorin, H. C. 3086.

5476. Dorland, R. E., and F. W. Went. 1947: Plant growth under controlled conditions. VIII. Growth and fruiting of the chili pepper (*Capsicum annuum*). Am. J. Botany 34:393–401.

Dorman, K. 8170.

5477. Dorne, A. S. 1910: The absorption of water by certain leguminous seeds. Brit. Assoc. Advan. Sci. Rept. 1910:786.

5478. Dorofeev, P. I. 1957: Seeds of *Ampelopsis* from the tertiary deposits of the territory of USSR. Botan. Zh. 42:643–48.

5479. Doroganevskaia, E. A. 1950: In regard to geographic variation in chemical composition of plants. Dokl. Akad. Nauk SSSR 71:387–90.

5480. Dorogin, G. 1927: Notwednigkeit die Samen auf die Pilzparasiten bei der Samenkontrolle zu Untersuchen. Lenin. Jardin Botan. Inst. Essais Semences Ann. Semences 5(2):19–27.

5481. Dorokhov, B. L. 1953: Forms of nitrogen in the seeds of reciprocal corn hybrids. Uch. Zap. Kishinevsk. Univ. 8:159–64. Chem. Abstr. 50:5104. 1956.

5482. Doroshenko, A. V. 1933: The method of staining the seeds of Umbelliferae to determine their germination capacity. Bull. Appl. Botany, Genet., Plant Breed. (USSR) Ser. A No. 7:185–93. Chem. Abstr. 28:3760. 1934.

5483. —— 1937: Plasmolytic method of determining the germinating capacity of seeds. Bull. Appl. Botany, Genet., Plant Breed. Ser. 4. Seed Sci. & Seed Testing 2:119.

5484. Dorph-Petersen, K. 1910: Kurze Mitteilungen über Keimuntersuchungen mit Samen verschiedener wildwachsenden Pflanzen. Jahresber. Ver. Angew. Botan. 8:239–47.

5485. —— 1910: Some researches on weed seed appearance and vitality. Tidsskr. Planteavl 17:584–626.

5486. —— 1924: How long do the various seed species retain their germination power? Intern. Inst. Agr., Intern. Rev. Sci., Pract. Agr. (N.S.) 2:283–301.

5487. —— 1925: Germination tests in the laboratory and in soil of cereal seed which is not "germinating-ripe." 4th Intern. Seed Testing Congr., Cambridge, July 7–12, 1924, pp. 76–82.

5488. —— 1925: Examinations of the occurrence and vitality of various weed seed species under different conditions, made at the Danish State Seed Testing Station during the years 1896–1923. 4th Intern. Seed Testing Congr., Cambridge, July 7–12, 1924, pp. 124–38.

5489. —— 1928: Comment les plus importantes des espèces de semences cultivées gardent-elles leur faculté germinative dans les magasins de semences ordinaires? Proc. Intern. Seed Testing Assoc. 1:57–71.

5490. —— 1928: Combien de temps les semences de *Tussilago farfara* gardent-elles leur faculté germinative sous de differentes conditions de temperature? Proc. Intern. Seed Testing Assoc. 1:72–76.

5491. —— 1931: Die Bestimmung der anormalen Keimlinge. Proc. Intern. Seed Testing Assoc. 3:125–26.

5492. —— 1936: The occurrence of abnormal growths during the years 1931–1935. Proc. Intern. Seed Testing Assoc. 8:18–32.

5493. —— and D. Lauesen. 1931: Untersuchungen von Weisskleeproben dänischer und ausländischer (besonders polnischer) Herkunft. Proc. Intern. Seed Testing Assoc. 3:42. Fortschr. Landwirtsch. 7: 284–85. 1932.

5494. Dorsey, M. J. 1929: The relation between embryo-sac development and the set of fruit in the apple. Proc. Am. Soc. Hort. Sci. 26:56–61.

5495. —— 1937: A record of peach seed germination tests. Proc. Am. Soc. Hort. Sci. 34:257–63.

5496. —— and R. L. McMunn. 1933: Seed size in relation to fruit size in the peach: the fourth report on the Illinois thinning investigations. Proc. Am. Soc. Hort. Sci. 29:13–22.

5497. Dorst, H. E. 1964: Pelleted seed containing insecticides controls insects in sugar beets. Utah Farm Home Sci. 25:20–21.

Dort, H. W. van; *see* Van Dort, H. W.

5498. Dorywalski, J. 1935: The cooperation of external agents in germination. Roczniki Nauk Rolniczych Lesnych 35:79–140. Biol. Abstr. 10:15443. 1936.

5499. —— 1960: Influence of the depth of seeding maize on seedling development and yields. Roczniki Nauk Rolniczych Ser. A 80:513–23.

5500. Doshchinskaya, N. V. 1962: Certain characteristics of dehydrogenase activity and respiration in henbane and belladonna seeds under conditions of stratification. Sb. Tr. Posvyashch. XX-letiyu (Dvadtsatiletiyu) Farmatsevt. Fak. 1962:85–90. Chem. Abstr. 60:7150. 1964.

Dosman, E. J. 18999.

Dos Santos, A. A. D. 440.

5501. Dostal, L. 1956: Germination of acorns and cultivation of red oak (*Quercus borealis*). Les (Bratislava) 12:384–85.

Dostert, R. 11627.

5502. Doty, D. M. 1941: Methods for estimation of some amino acids in corn grain. Ind. Eng. Chem. Anal. Ed. 13: 169–72.

5503. ——, M. S. Bergdoll, and S. R. Miles. 1943: The chemical composition of commercial hybrid and open-pollinated varieties of dent corn and its relation to soil, season, and degree of maturity (a preliminary report). Cereal Chem. 20: 113–20.

5504. ——, ——, H. A. Nash, and A. M. Brunson. 1946: Amino acids in corn grain from several single cross hybrids. Cereal Chem. 23:199–209.

5505. ——, L. C. Schenberger, and S. Hicks. 1948/49: Enzymes in relation to seed germination. Indiana Agr. Expt. Sta. Ann. Rept. 62:98–99.

5506. ——, G. M. Smith, J. R. Roach, and J. Sullivan. 1945: The effect of storage on the chemical composition of some inbred and hybrid strains of sweet-corn. Indiana Agr. Expt. Sta. Bull. 503. 31 pp.

Doty, D. M. 8727.

5507. Doty, R. E. 1928: The preservation of cane seed. Hawaiian Planters Record 32:307–20.

5508. Dotzenko, A. D., and J. G. Dean. 1959: Germination of six alfalfa varieties at three levels of osmotic pressure. Agron. J. 51:308–9.

5509. —— and T. E. Haus. 1960: Selection of alfalfa lines for their ability to germinate under high osmotic pressure. Agron. J. 52:200–1.

Dotzenko, A. [D.] 177.

Dotzler, F. 12664.

Doughty, H. N. 12807.

Doughty, J. L. 3272.

5510. Douglas, A. G., and J. B. Weaver, Jr. 1963: Comparison of open-pollinated seed with registered seed of three cotton varieties. Crop Sci. 3:515–16.

5511. Douglas, C. F. 1962: Some environmental factors influencing plant development and ear placement of sweet and dent corn. Dissertation Abstr. 22:2125.

5512. Douglas, M. 1947: Jojoba—an oil-producing plant of the southwestern states. N.Y. Botan. Garden J. 48:29–32.

Douglas, T. 1905.

5513. Douglas, T. H. 1897: The germination of conifer seeds. Gardens & Forests 10:407.

5514. Douglass, C. D., W. L. Howard, and S. H. Wender. 1949: The isolation of isoquercitrin from the seed pods of *Cercis canadensis*. J. Am. Chem. Soc. 71:2658–59.

5515. Douglass, J. 1929: Production of cauliflower seed. Agr. Gaz. N.S. Wales 40:207–10.

5516. —— 1932: Tomato seed production. A commercial proposition for New South Wales growers. Agr. Gaz. N.S. Wales 43:245–49.

5517. —— 1939: Lettuce seed production. Agr. Gaz. N.S. Wales 50:136–38.

5518. —— 1943: The harvesting and threshing of cucurbit seed. Agr. Gaz. N.S. Wales 54:499–500, 516.

5519. Douine, M. S. 1929: À propos de la question des erreurs dans l'analyse des semences. 5th Intern. Seed Testing Conf., Rome. Actes 1928:418–31. 1929.

5520. —— and M. N. Mjasdrikowa. 1931: Zur Frage de Bestimmung des Wasserminimums für die Samenkeimung. Angew. Botan. 13:49–81. Biol. Abstr. 6:18453. 1932.

5521. Dove, W. E. 1954: How to protect stored wheat from insect damage. Pest Control 22(6):9–10, 14, 60.

Dow, R. B. 16249.

Dowdle, R. F. 4526.

5522. Dowling, R. E. 1933: The reproduction of *Plantago coronopus:* an example of morphological and biological seed dimorphism. Ann. Botany (London) 47:861–72.

Dowling, T. I. 12307.

5523. Down, E. E. 1931: Alfalfa seed production studies in Michigan. J. Am. Soc. Agron. 23:983–99.

Down, E. E. 5055, 10995.

5524. Downey, R. K., J. E. R. Greenshields, and W. J. White. 1954: The nature and inheritance of seed-coat spotting in permeable-seeded strains of sweet clover (*Melilotus alba* L.). Can. J. Agr. Sci. 34:514–27.

Downey, R. K. 8291.

Downie, A. R. 5538, 5539, 9858.

5525. Downie, D. G. 1940: On the germination and growth of *Goodyera repens*. Trans. Proc. Botan. Soc. Edinburgh 33: 36–51. Biol. Abstr. 15:7581. 1941.

5526. —— 1941: Notes on the germination of some British orchids. Trans. Proc. Botan. Soc. Edinburgh 33:94–103. Biol. Abstr. 16:17493. 1942.

5527. —— 1943: Notes on the germination of *Corallorhiza innata*. Trans. Proc. Botan. Soc. Edinburgh 33:380–82. Rev. Appl. Mycol. 24:28. 1945.

5528. —— 1950: The germination of *Goodyera repens* (L.) R. Br. in fungal extract. Trans. Proc. Botan. Soc. Edinburgh 35: 120–25. 1948/49.

5529. —— 1950: The germination of *Listera ovata* (L.) R. Br. Trans. Proc. Botan. Soc. Edinburgh 35:126–30. 1948/49.

5530. —— 1959: *Rhizoctonia solani* and orchid seed. Trans. Proc. Botan. Soc. Edinburgh 37:279–85.

5531. Downs, A. A., and W. E. McQuilkin. 1944: Seed production of southern Appalachian oaks. J. Forestry 42:913–20.

5532. Downs, D. E., and W. H. Cathcart. 1941: Thiamin content of commercial wheats of the 1940 crop. Cereal Chem. 18:796–801.

5533. Downs, R. J., H. A. Borthwick, and A. A. Piringer. 1961: Light and plants. A series of experiments demonstrating light effects on seed germination, plant growth, and plant development. U.S. Dept. Agr. Misc. Publ. 879:1–26.

5534. —— and A. A. Piringer. 1958: Seed germination in the Bromeliaceae. Bromeliad Soc. Bull. 8(3):36–38.

5535. ——, ——, and G. A. Wiebe. 1959: Effects of photoperiod and kind of supplemental light on growth and reproduction of several varieties of wheat and barley. Botan. Gaz. 120:170–77.

Downs, R. J. 15417.

Downs, S. C. 8339.

Dox, A. W. 14997.

5536. Doxtator, C. W. 1947: Experiments on beet seed sizing and segmenting. Proc. Am. Soc. Sugar Beet Technol. 1946:594–99. Biol. Abstr. 22:12089. 1948.

5537. —— 1958: Gibberellic acid effects on seed and seedlings of sugar beets. J. Am. Soc. Sugar Beet Technol. 10:117–23.

5538. —— and A. R. Downie. 1947: Comparison of yields from two beet hills originating from one and two seeds. Proc. Am. Soc. Sugar Beet Technol. 1946:154–56. Biol. Abstr. 22:12090. 1948.

5539. —— and —— 1948: Beet seed germination methods. Proc. Am. Soc. Sugar Beet Technol. 5:81–82.

5540. —— and R. E. Finkner. 1958: Sugar beet germination selection results in osmotic pressure solutions. I. Germination methodology and osmotic selection effects on germination of four varieties. J. Am. Soc. Sugar Beet Technol. 10:237–46.

5541. ——, P. C. Hanzas, and R. F. Olson. 1958: Sugar beet germination selection results in osmotic pressure solutions. II. Yield and chemical constituents of progenies, of four varieties from osmotic selection. J. Am. Soc. Sugar Beet Technol. 10:247–53.

5542. —— and R. H. Helmerick. 1962: Selection for seed size in monogerm varieties. J. Am. Soc. Sugar Beet Technol. 12:268–72. Biol. Abstr. 45:7715. 1964.

Doxtator, C. W. 16042.

5543. Doyer, L. C. 1930: Untersuchungen über den Gesundheitszustand des Saatgutes. Proc. Intern. Seed Testing Assoc. 13/14:1–41.

5544. —— 1932: Möhrensamen. Nachr. Schädlingsbekämpf. 7:107–9.

5545. —— 1938: Manual for the determination of seed-borne diseases. H. Veenman, Wageningen. 59 pp. Biol. Abstr. 13:12163. 1939.

Doyle, B. K. 4074.

Doyle, C. B. 4360.

5546. Doyle, J. 1957: Aspects and problems of conifer embryology. Advan. Sci. 14:120–30.

5547. —— and W. J. Looby. 1939: Embryogeny in *Saxegothaea* and its relation to other podocarps. Sci. Proc. Roy. Dublin Soc. N.S. 22:127–47.

Doyle, J. 2991, 12423, 12424.

5548. Drabkin, B. 1936: Sur la position de l'hormone de l'endosperme dans une semence "yarovisée." Bull. Sci. Rec. Biol. Univ. Kiev 2:145–49. Biol. Abstr. 14:9360. 1940.

5549. Drachal, T. 1958: A new machine for larch seed extraction. Sylwan 102:83–86.

Drachovska, M. 16703.

5550. Draghetti, A. 1936: Physiology of the ripening of wheat in different environments. Ann. Staz. Sper. Agrar. Modena (N.S.) 5:245–69. Chem. Abstr. 32:3454. 1938.

5551. Dragone-Testi, G. 1931: Action of alkaloids on seed germination. Ann. Botan. (Turin) 19:345–52.

5552. —— 1933: Osservazioni sul significato fisiologico dei lipoidi nel seme di *Viburnum tinus* L. Ann. Botan. (Turin) 20:255–66.

5553. —— 1934: Experiments on the germination of detached wheat embryos. Ann. Botan. (Turin) 20:534–67. Biol. Abstr. 10:18001. 1936.

5554. —— 1937: Germination and development of embryos of *Arum italicum* depleted of the residual seed. Boll. Soc. Ital. Biol. Sper. 12:736–38. Chem. Abstr. 32:7957–58. 1938.

Dragoo, R. A. 11783.

5555. Drahorad, T. 1947: Wie lange sind Samen keimfähig? Landwirtsch. 1947: 234–35. Biol. Abstr. 23:12595. 1949.

Drake, A. D. 8350.

Drake, R. J. 19324.

5556. Drake, R. M. 1945: The development of viable seed in yellow star thistle plants under various conditions. Calif. Dept. Agr. Bull. 34:89–91. Expt. Sta. Rec. 93:713. 1945.

5557. Drake, V. C. 1941: The effect of different temperature conditions on the germination of freshly harvested seed of crested wheatgrass, *Agropyron cristatum.* Assoc. Offic. Seed Analysts News Letter 15(7):3–4.

5558. —— 1942: Report on germination requirements of *Alysicarpus vaginalis* (L.) DC. Assoc. Offic. Seed Analysts News Letter 16(4):16–18.

5559. —— 1943: Some data on the field value of baldhead bean plants. Assoc. Offic. Seed Analysts News Letter 17(2): 15–20.

5560. —— 1946: Yield in snap beans of normal plants. Seed World 60(7):8–12, 53.

5561. —— 1947: Effect of temperature and aging on the germination of Alyce clover (*Alysicarpus vaginalis* [L.] DC.). Proc. Assoc. Offic. Seed Analysts 37:143–52.

5562. —— 1947: Germination response at two sets of alternating temperatures of New Zealand-grown Colonial bentgrass. Assoc. Offic. Seed Analysts News Letter 21(2):32–33.

5563. —— 1948: Decline in viability of lettuce seed during laboratory storage. Assoc. Offic. Seed Analysts News Letter 22(2):31–33.

5564. —— 1948: Germination of certain dormant oat and barley samples. Assoc.

Offic. Seed Analysts News Letter 22(4): 13–16.

5565. —— 1949: Germination of rescue grass seed as affected by temperature, substrata, light, and removal of glumes. Assoc. Offic. Seed Analysts News Letter 23(2): 42–46.

5566. —— and B. R. Atwater. 1939: Observations on lima bean germination. Proc. Assoc. Offic. Seed Analysts 30: 218–20.

Drake, V. C. 512.

5567. Dranitsyna, I. A. 1961: Fat oil of the seeds of *Elsholtzia patrini*. Tr. Botan. Inst. Akad. Nauk SSSR Ser. 5 Rast. Syr'e 1961(8):32–34. Chem. Abstr. 56:10586. 1962.

5568. —— 1961: Furocoumarins from seeds of *Archangelica decurrens*. Tr. Botan. Inst. Akad. Nauk SSSR Ser. 5 Rast. Syr'e 1961(8):43–48. Chem. Abstr. 56:9134. 1962.

Drapala, W. J. 5779–5781.

Draper, R. P. 8423.

5569. Drawert, H. 1948: Beiträge zur Stimulation des Pflanzenwachstums. I. Der Einfluss von Extrakten aus *Folia digitalis* auf die Keimung der Früchte von *Cichorium endivia* L. Planta 35:555–78.

5570. —— 1949: Stimulation of plant growth. II. Effect of heteroauxin on the germination of seeds. Planta 37:1–5.

Drechsler, C. 19555.

5571. Drees, E. M. 1941: Germination tests and seedlings of *Acacia*. Tectona 34:1–45. Biol. Abstr. 15:15482. 1941.

5572. —— 1959: The use of quantitative characteristics in the identification of *Rumex* seeds. Mededel. Wageningen Proefsta. Zaadcontr. 10:85–96.

5573. —— 1962: Minimum size of working sample for provenance estimation in clover seed. Proc. Intern. Seed Testing Assoc. 27:431–44.

Dremova, O. B. 10809.

5574. Drennan, D. S. H. 1962: Physiological studies of germination in the genus *Avena*. II. Changes in some metabolites during the germination of grains of *A. sativa*. New Phytologist 61:261–65.

5575. ——, A. M. M. Berrie, and G. A. Armstrong. 1961: Prevention of manganese deficiency in oat by soaking the grain in solutions of manganous chloride. Nature 190:824.

5576. Drenth, J., and E. H. Wiebenga. 1955: Excelsin, edestin, and tobacco-seed globulin crystals; preparation, density and X-ray measurements. Rec. Trav. Chim. 74:813–31. Chem. Abstr. 49:15999. 1955.

Drescher, R. F. 4058.

Drews, M. 7358.

5577. Dreyer, H. V. 1928: Is the sickliness of red alder due to the source of the seed? Dansk Skovfor Tidsskr. 13:329–76. Biol. Abstr. 5:8776. 1931.

5578. Driessche, T. V. 1961: Content and quality of oil from the seed of cotton grown in controlled mineral cultures. Ann. Physiol. Végétale Univ. Bruxelles 6(2): 27–98. Biol. Abstr. 41:20249. 1963.

Driessche, T. vanden; *see* Vanden Driessche, T.

5579. Droit, S. 1932: Examination of curcas seed and oil (from *Jatropha curcas*, L.). Bull. Mat. Grasses 16:270–73. Soc. Chem. Ind. J. 52:27B. 1933.

5580. Drokova, I. G., E. V. Lakhno, and R. S. Fidman. 1953: Vitamins of some varieties of wheat and rye. Vitaminy Akad. Nauk Ukr. SSR Inst. Biokhim. 1:125–29. Chem. Abstr. 49:1228. 1955.

Drowne, M. F. 1693.

5581. Drozdov, N. A., and A. I. Bystrov. 1960: Wetting fiber flax seeds with succinic acid raises the yield and quality of the product. Lën i Konoplia 1960(4):40–41.

5582. —— and I. N. Shcherbakov. 1955: Treatment of flax seed with copper sulfate. Zemledelie 1955(4):46–47. Biol. Abstr. 31:26018. 1957.

5583. Drozdova, O. I. 1957: On certain peculiarities of corn development from seeds grown in different areas. Agrobiologiya 1957(1):61–66.

Drozdova, T. V. 11372, 11389.

5584. Drummitt, M. 1946: The germination of dodder seed occurring in *Lespedeza*. Proc. Assoc. Offic. Seed Analysts 1945/46: 125–31.

5585. Drummond, J. C., E. Singer, and R. J. MacWalter. 1935: A study of the unsaponifiable fraction of wheat germ oil with special reference to vitamin E. Biochem. J. 29:456–71.

5586. ——, ——, and —— 1935: Further observations on the constituents of the unsaponifiable fraction of wheat germ oil with particular reference to vitamin E. Biochem. J. 29:2510–21.

Drummond, P. L. 18663.

5587. Drummond, W. C. 1955: Begonia seed growing. Begonian 22:14.

5588. Druzhinin, D. 1958: Conditions which influence the nitrogen content in the kernels of spring cereals. Sel'skoe Khoz. Sibiri 1958(1):27–29.
Duarte, C. 7120.
5589. Du Barry, A. P. 1963: Germination of bottomland tree seeds while immersed in water. J. Forestry 61:225–26.
Du Barry, A. P. 3069.
Dubbs, A. L. 13239.
5590. Dubeck, M., and S. Kirkwood. 1952: The origin of the O- and N-methyl groups of the alkaloid ricinine. J. Biol. Chem. 199:307–12.
5591. Dubetz, S. 1958: The effect of fertilizers and osmotic pressure on germination. J. Am. Soc. Sugar Beet Technol. 10:212–19.
5592. ——, G. C. Russell, and D. T. Anderson. Effect of soil temperature on seedling emergence. Can. J. Plant Sci. 42:481–87.
5593. ——, R. L. Smith, and G. C. Russell. 1959: The effect of fertilizers and osmotic pressure on germination. Can. J. Soil Sci. 39:157–64.
Dubey, A. S. 16681.
5594. Dubinskaia, M. M. 1949: On dormancy of *Atropa belladonna* seeds. Selek. i Semen. 16(8):74–75.
5595. Dublyanskaya, N. [F.] 1931: The quality and composition of peanut oil from nuts of different degrees of maturity. Maslob.-Zhir. Delo. 1931(11):55–57. Chem Abstr. 26:4492. 1932.
5596. —— 1956: The botanical oil content of husks of the sunflower seed in relation to the oil content of the kernel. Kratkii Otchet o Nauchn.-Issled. Rabote Vses. Nauchn.-Issled. Inst. Maslichn. i Efiromasl. Kul'tur 1955:89–94. Chem. Abstr. 52:3044. 1958.
5597. —— 1957: Biochemical changes during the ripening of Chinese mustard seeds. Kratkii Otchet o Nauchn.-Issled. Rabote Vses. Nauchn.-Issled. Inst. Maslichn. i Efiromasl. Kul'tur 1957:102–7. Chem. Abstr. 55:3739. 1961.
5598. —— 1959: Chemical composition of seeds of present day varieties of Sareptan mustard. Maslob.-Zhir. Prom. 25(9):8–11.
Dubois, H. 3447.
5599. Dubois, M., W. F. Geddes, and F. Smith. 1960: The carbohydrates of the Gramineae. Cereal Chem. 37:557–68.
Dubrovskaya, I. I. 2260.
5600. Du Buy, H. G. 1938: Factors causing early development and vernalization. [Abstract.] Am. J. Botany 25:11s–12s.
Du Buy, H. G. 18011.
Ducay, E. D. 6598.
Duchartre, M. Favre-; *see* Favre-Duchartre, M.
5601. Duchartre, M. P. 1872: Quelques mots sur la germination du *Delphinium nudicaule*. Bull. Soc. Botan. France 19:183–85.
5602. —— 1891: Germination of *Freesia refracta*. J. Soc. Hort. France 1891:152–60, 215–30.
5603. Duchartre, P. 1874: Observations sur la germination et sur la formation première de l'ognon chez diverses espèces de Lis. Paris, 8vo. 30 pp. J. Soc. Cent. Hort. France 1874:554–81.
5604. Du Châtelier, G. G. 1949: Contribution à la connaissance morphologique du grain d'amidon par l'acide lactique. Rev. Gen. Botan. 56:189–204.
Duchon, F. 14330.
Duczmal, K. 11798.
5605. Dudar, A. K. 1956: The collection and sowing of *Agnuscastus* seeds. Ovtsevodstvo 1956(10):28–30.
5606. Dude, M. 1903: Über den Einfluss des Sauerstoffentzuges auf pflanzliche Organismen. Flora (Jena) 92:205–52. Botan. Centralbl. 93:18. 1903.
Dudeck, A. E. 5624.
5607. Dudik, F. S., and E. D. Kharchenko. 1955: Growing cannas from seeds. Akad. Nauk Ukr. SSR Botan. Tr. 3:151–53.
5608. Dudkin, M. S. 1961: Comparative study of the xylan structure of oat and millet seed hulls. Tr. Vses. Nauchn.-Issled. Inst. Zerna i Produktov ego Pererabotki 1961(41):115–20. Chem. Abstr. 58:9414–15. 1963.
Dudkin, M. S. 19240.
5609. Duerden, J. C. 1958: The storage of groundnuts under tropical conditions. II. The effects of storage in paper sacks. J. Sci. Food Agr. 9:483–87.
5610. —— and J. R. Cutler. 1957: The storage of groundnuts under tropical conditions. I. The effects of prolonged storage on undecorticated and decorticated groundnuts. J. Sci. Food Agr. 8:600–4.
5611. Duffee, F. W. 1937: Drying seed corn with electricity. Agr. Eng. 18:149–51.
5612. Duffield, J. W. 1960: A simple device for weight-sorting seeds. Forest Sci. 6:362.
Duffield, J. W. 17023, 18056.
5613. Duffield, P. C. 1952: Combination

insecticide-fungicide seed treatments for corn. J. Econ. Entomol. 45:672–74.

5614. Dufour, L. 1887: Experiments on the germination of *Faba vulgaris*. Bull. Soc. Botan. France 34:463.

5615. Dufrenoy, J. 1928: Sur des phenomenes preparant la germination. Compt. Rend. Hebd. Séances Mem. Soc. Biol. (Paris) 98:1497–98.

5616. —— and R. Pratt. 1948: Responses of cuttings, seeds, and flowers to dithiobiuret. Science 108:715–16.. Chem. Abstr. 43:2678–79. 1949.

5617. —— and A. Radoeff. 1932: Effets du nitrate d'argent et de l'hexylresorcine sur la germination du tabac. Compt. Rend. Soc. Biol. 110:195–97.

5618. —— and —— 1932: Effets du formol sur la germination et la croissance des plantules de tabac. Compt. Rend. Soc. Biol. 110:386–88.

Dufrenoy, J. 7817, 15668.

5619. Duggar, B. M. 1920: The nutritive value of the food reserves in cotyledons. Ann. Missouri Botan. Garden 7:291–98.

5620. —— 1930: The problem of seed transmission of the typical mosaic of tobacco. J. Bacteriol. 19:20.

5621. Duggar, J. F. 1935: Depth of planting *Crotalaria* seed. Alabama Agr. Expt. Sta. Ann. Rept. 1934/35:27.

5622. —— 1935: The distance to which *Crotalaria* seeds are thrown by pods. Alabama Agr. Expt. Sta. Ann. Rept. 1934/35:27.

5623. —— 1938: Stands from delayed planting of scalded *Crotalaria* seed. Alabama Agr. Expt. Sta. Ann. Rept. 49:26.

5624. Duich, J. M., B. R. Fleming, A. E. Dudeck, and G. J. Shoop. 1962: The effect of certain preemergence chemicals on grass germination and seedling grasses. Proc. Northeast. Weed Control Conf. 16: 479–83.

5625. Duka, J. 1930: Zur Frage von der Bedeutung der Sammengrösse in den ersten Entwicklungsperioden der Feldpflanzen. Angew. Botan. 2:63–70.

5626. Duke, J. 1960: The first flower of our dawn redwood. Bull. Missouri Botan. Garden 48:134–36.

Duley, F. L. 13123, 13124.

5627. Dumanovic, J., and M. Pencic. 1962: The effect of planting rate and nitrogen levels on the grain yield and protein content in the maize kernel. Savremena Poljoprivreda 10:240–50.

5628. Dumas, L. 1895: Germinatif pouvoir du *Isatis tinctoria* (Pastel). J. Agr. Prat. 1895:623–34.

5629. Dumazert, C., and L. Vo-Van. 1963: Acid-soluble compounds of germinating *Phaseolus aureus*. IV. Action of phytotoxic cations on the nucleotides liberated during germination. Compt. Rend. Soc. Biol. 157:329–31. Chem. Abstr. 59:14298. 1963.

5630. ——, ——, and J. Fayolle. 1963: Sur l'évolution des proteines au cours de la germination. Compt. Rend. Soc. Biol. 157:1463–65.

5631. ——, ——, and R. Marchetti. 1963: Changes in the carbohydrates during germination [of *Phaseolus aureus*]. Compt. Rend. Soc. Biol. 157:576–77. Chem. Abstr. 59:13111. 1963.

5632. Dumbleton, L. J. 1952: New genus of seed infesting micropteryged moths. Pacific Sci. 6:17–29. Brit. Abstr. B III:265. 1952.

Dumlao, R. C. 18874.

5633. Dumont, J. 1906: Les radiations lumineuses et la richesse azotée du blé. Compt. Rend. 143:1179–81.

Dunaway, D. J. 103.

5634. Dunbar, R. H. 1952: Taping gladiolus seed. Gladiolus Mag. 16:21–23. Biol. Abstr. 27:17686. 1953.

5635. Duncan, J. R., and A. R. Marston. 1925: Curing and storing of seed corn. The effect of different temperatures in drying seed corn of different stages of maturity on germination. Mich. Agr. Expt. Sta. Quart. Bull. 8:66–68.

5636. Duncan, R. E., and M. D. Persidsky. 1958: The achromatic figure during mitosis in maize endosperm. Am. J. Botany 45:719–29.

5637. Duncomb, Mrs. R. J. 1938: Geraniums are easy from seed. Flower Grower 25(2):97.

5638. Dungan, G. H. 1924: Some factors affecting the water absorption and germination of seed corn. J. Am. Soc. Agron. 16:473–81.

5639. —— 1940: Influence of age on the value of seed corn. Trans. Illinois State Acad. Sci. 33:28–29.

5640. —— 1944: Yield and bushel weight of corn grain as influenced by time of planting. J. Am. Soc. Agron. 36:166–70. Biol. Abstr. 18:9143. 1944.

5641. —— and W. C. Brokaw. 1941: Relation between moisture content of the soil and the optimum depth of planting corn. Trans. Illinois State Acad. Sci. 34:46–47.

5642. —— and B. Koehler. 1944: Age of seed corn in relation to seed infection and yielding capacity. J. Am. Soc. Agron. 36:436–43.

Dungan, G. H. 7051, 10973, 10974, 11763, 13588, 15192.

Dunham, R. S. 1311, 6139.

5643. Dunham, W. E. 1943: Honeybees increase clover seed production 15 times. Am. Bee J. 83:310–11. Biol. Abstr. 18:3246. 1944.

5644. Dunin, I. S., and M. N. Mjazdrikowa. 1930: Method of determining the minimum water quantity for seed germination. Jard. Botan. Princ. Ann. Essais Semences (Leningrad) 7(2):71–103. Biol. Abstr. 6:12860. 1932.

5645. Dunin, M. S. 1928: Über Einwirkung einiger Faktoren auf die Aufkeimung der Samen von *Hibiscus cannabinus* L. Jard. Botan. Ann. Essais Semences (Leningrad) 5(5):20–49.

5646. —— 1929: Das Öl des Kenaphsamens (*Hibiscus cannabinus* L.). Angew. Botan. 11:569–78.

5647. —— and L. S. Gitman. 1931: Experiments on drying hemp seed. Tr. Inst. Nov. Lub. Syr. (Trans. Lenin Acad. Agr. Sci. USSR New Bast-Fibers Res. Inst.) 1: 243–53. Expt. Sta. Rec. 72:177. 1935.

5648. —— and M. N. Mjasdrikowa. 1931: Zur Frage der Bestimmung des Wasserminimums für die Samenkeimung. Angew. Botan. 13:49–81.

5649. —— and F. M. Schemjakin. 1928: Über den Einfluss der Salze auf die Viskosität des Leinsamenschleims. Kolloid-Z. 45:146–52.

Dunin, M. S. 170.

Dunker, M. F. W. 8739.

5650. Dunkle, E. C., and F. G. Merkle. 1943: The conductivity of soil extracts in relation to germination and growth of certain plants. Soil Sci. Soc. Am. Proc. 8:185–88. Chem. Abstr. 39:375. 1945.

5651. Dunlap, A. A., and A. D. McDonnell. 1939: Testing germination in sand. J. Forestry 37:330–32.

Dunlap, A. A. 13192.

Dunn, E. 14856.

Dunn, G. M. 2634.

5652. Dunn, H. C., T. P. Hilditch, and J. P. Riley. 1948: The composition of seed fats of West Indian citrus fruits. Soc. Chem. Ind. J. 67:199–203.

Dunn, J. S. 13034.

5653. Dunn, L. E. 1939: Influence of low temperature treatments on the germination of seeds of sweet clover and smooth vetch. J. Am. Soc. Agron. 31:687–94.

5654. Dunn, M. S., M. N. Camien, S. Shankman, and H. Block. 1948: Amino acids in lupine and soybean seeds and sprouts. Arch. Biochem. 18:195–200.

Dunn, N. L., Jr. 195.

5655. Dunn, S. 1944: Treatment of seeds with hormones and hormone-like preparations. New Hampshire Agr. Expt. Sta. Bull. 354:54–55.

5656. Dunnewald, T. J. 1934: Soil factors in germination of pine seeds. Wyoming Univ. Publ. Sci. 1:267–74.

5657. Dunnill, P. M., and L. Fowden. 1963: γ-L-Glutamyl-β-pyrazol-1-yl-L-alanine, a peptide from cucumber seeds. Biochem. J. 86:388–91.

5658. Dunning, J. W. 1954: Processing of castor beans. J. Am. Oil Chemists Soc. 31:290–91.

Dunning, R. G. 4744.

5659. Dunster, B. P. 1939: Raising rhododendrons from seeds. Gardeners Chron. 105:198.

5660. Duperon, P. 1963: Les composés des cotyledons et des "parties axiales" de la graine de haricot (*Phaseolus vulgaris*). Compt. Rend. Soc. Biol. 157:2268–72.

5661. Duperon, R. 1949: Influence de la printanisation sur l'évolution des lipides et des glucides chez le *Sinapis alba*. Compt. Rend. 228:192–94.

5662. —— 1950: Influence de la vernalisation sur l'évolution des glucides au cours de la germination de *Raphanus sativus*. Compt. Rend. 230:225–27.

5663. —— 1951: À propos de la réversibilité des processus de vernalisation. Compt. Rend. 233:1053–55.

5664. —— 1952: The glucides of the seeds of crucifers. Compt. Rend. 235:1331–33. Chem. Abstr. 47:4436. 1953.

5665. —— 1953: Influence de la vernalisaton sur le métabolisme (Suite). Rev. Gen. Botan. 60:33–78.

5666. —— 1953: Influence de la vernalisation sur le métabolisme. Rev. Gen. Botan. 60:90–122.

5667. —— 1954: Les glucides des graines oléagineuses. Leur rôle au cours de la germination. Rev. Gen. Botan. 61:261–84.

5668. —— 1955: Répartition, formation et utilization du raffinose et du stachyose dans les graines. Compt. Rend. 241:1817–18.

5669. —— 1956: Influence du froid sur l'évolution des acides organiques hydrosolubles, non volatils, au cours de la germination des graines de *Brassica oleracea* var. *acephala*. Compt. Rend. 243:1139–41.

5670. —— 1958: Évolution des acides organiques hydrosolubles, non volatils, au cours de la croissance et de la maturation de semences. Compt. Rend. 247:342–44.

5671. —— 1960: Présence de l'acide malonique dans les graines de *Phaseolus vulgaris*. Évolution de cet acide au cours de la germination. Compt. Rend. 251:260–62.

5672. —— 1961: Fixation du CO_2 a l'obscurité, par les graines de *Phaseolus vulgaris*, lors de l'imbibition. Compt. Rend. 253:1488–90.

5673. —— 1963: Recherches sur l'origine des acides malique et malonique synthetisés dans les graines de *Phaseolus vulgaris* en germination. Compt. Rend. 256:2906–9.

Duperon, R. 17756.

Durand, E. Michel-; *see* Michel-Durand, E.

5674. Durand, V. 1953: Germination comparée du riz dans l'eau et dans l'air humide a differentes temperature. Compt. Rend. Soc. Biol. 147:457–60.

5675. —— 1954: Action du chlorure de sodium sur la faculté et l'energie germinations des grains de riz immergés. Compt. Rend. Soc. Biol. 148:123–25.

Durand, V. 3733.

Durand, W. D. 18725.

Duranton, J. 4265.

5676. Durbin, R. D. 1956: A method of separating cactus seed. Cactus Succulent J. Gt. Brit. 28:50. Biol. Abstr. 31:25731. 1957.

5677. —— 1957: *Strelitzia* seed treatment. Florists Rev. 119(3086):115–16.

5678. —— and C. M. Christensen. 1963: A technique for measuring $C^{14}O_2$ evolution from seeds of low moisture content. Intern. J. Appl. Radiation Isotopes 14:528. Nucl. Sci. Abstr. 18:923. 1964.

5679. ——, E. F. Frolich, and G. A. Zentmyer. 1957: Hot water treatment of avocado seed to eradicate root rot. Calif. Citrograph 43:39–40.

5680. Dure, L. S. 1960: Aspects of germination in maize. Dissertation Abstr. 21: 1344.

5681. —— 1960: Gross nutritional contributions of maize endosperm and scutellum to germination growth of maize axis. Plant Physiol. 35:919–25.

5682. —— 1960: Site of origin and extent of activity of amylases in maize germination. Plant Physiol. 35:925–34.

5683. —— and W. A. Jensen. 1957: The influence of gibberellic acid and indoleacetic acid on cotton embryos cultured *in vitro*. Botan. Gaz. 118:254–61.

5684. Durfee, B. 1948: Carrot seed germination observations. Assoc. Offic. Seed Analysts News Letter 22(3):17–18.

Durivaux, C. Exbrayat-; *see* Exbrayat-Durivaux, C.

5685. Durkee, A. B., and P. A. Poapst. 1962: Possible presence of indolyl-3-acetic acid in *Malus*. Science 193:273–74.

Durou-Carbouères, M. 581, 582.

Durrell, M. 15610.

5686. Du Sablon, L. 1893: Germination of *Ricinus*. Compt. Rend. 117:524–27.

5687. —— 1895: Recherches sur la germination des graines oléagieuses. Rev. Gen. Botan. 7:145–65, 205–15, 258–69.

5688. —— 1897: Sur la germination des amandes. Rev. Gen. Botan. 9:5–16.

5689. Dustin, J. P., E. Schram, S. Moore, and E. J. Bigwood. 1953: Dosage chromatographique des acides amines d'une graine de céréale (orge), d'un foin, et d'un tourteau de lin. Bull. Soc. Chim. Biol. 35:1137–47.

Dutcher, R. A. 13027.

5690. Duthie, D. W. 1938: The determination of tannins in cacao kernel. Analyst 63:27–31.

5691. Du Toit, J. P. 1961: Identifying seed samples. Farming S. Africa 37(2):23–26. Biol. Abstr. 36:82138. 1961.

5692. Dutrieu, J. 1957: Modifications chimiques accompagnant le développement embryonnaire d'*Artemia salina* L. Compt. Rend. 244:2650–53.

5693. Dutt, B. K., and A. Guha-Thakurta. 1935: Automatic record of embryonic growth on germination of the seed. Trans. Bose Res. Inst. (Calcutta) 9:58–76. Quart. Rev. Biol. 11:224–25. 1936.

5694. —— and —— 1937/38: Catalase activity of *Cajanus* seed at different stages—pre-resting, resting and post-resting. Trans. Bose Res. Inst. (Calcutta) 13:93–103. Chem. Abstr. 34:7995. 1940.

5695. —— and —— 1953/55: IV. Viability of vegetable seeds in storage. Trans. Bose Res. Inst. (Calcutta) 19:27–36.

Dutt, B. K. 7791–7793.

5696. Dutt, B. S. M. 1959: Ovule and embryo-sac of *Crinum latifolium* L., a reinvestigation. Current Sci. (India) 28: 293–94.

5697. —— 1964: Ovule and embryo sac of *Pancratium maritumum* L., a reinvestigation. Current Sci. (India) 33:150–51.

Dutt, K. M. 19031.

5698. Dutt, N. L., M. K. Krishnaswami, and K. S. S. Rao. 1938: Seed setting and seed germination in certain sugar canes. Indian J. Agr. Sci. 8:429–39. Chem. Abstr. 33:687. 1939.

Dutt, S. 156, 7855, 7856, 13581, 14192, 15055, 15195.

Dutt, S. N. 2821.

5699. Dutta, A., J. C. Gupta, S. Ghosh, and B. H. Kahali. 1947: *Rauwolfia serpentina* Benth. Indian J. Pharm. 9:54–57. Biol. Abstr. 24:13360. 1950.

Dutta, A. T. 7852.

5700. Dutta, P. K. 1962: Germination and chemical composition of *Rauwolfia serpentina* seeds. Indian J. Pharm. 24:61–63. Chem. Abstr. 57:1277. 1962.

Dutta, R. M. 13608.

Duuren, A. J. van; *see* Van Duuren, A. J.

Duval, L. 5705.

5701. Duvel, J. W. T. 1904: The vitality and germination of seeds. U.S. Dept. Agr. Bur. Plant Ind. Bull. 58. 96 pp.

5702. —— 1905: Vitality of buried seeds. U.S. Dept. Agr. Bur. Plant Ind. Bull. 83. 22 pp.

5703. —— 1905: The storage and germination of wild rice seed. U.S. Dept. Agr. Bur. Plant Ind. Bull. 90, Pt. 1. 6 pp.

5704. —— 1909: The deterioration of corn in storage. U.S. Dept. Agr. Bur. Plant Ind. Circ. 43. 12 pp.

5705. —— and L. Duval. 1913: Shrinkage of shelled corn while in cars in transit. U.S. Dept. Agr. Bull. 48. 21 pp.

5706. Duvick, D. N. 1952: Free amino acids in the developing endosperm of maize. Am. J. Botany 39:656–61.

5707. —— 1953: Phosphorylase in the maize endosperm. Botan. Gaz. 115:82–85.

5708. —— 1955: Cytoplasmic inclusions of the developing and mature maize endosperm. Am. J. Botany 42:717–25.

5709. —— 1961: Protein granules of maize endosperm cells. Cereal Chem.38:374–85.

5710. Duym, C. P. A., J. G. Komen, A. J. Ultee, and B. M. van der Weide. 1947: The inhibition of germination, caused by extracts of seedballs of the sugarbeet (*Beta vulgaris*). Koninkl. Ned. Akad. Wetensch. Proc. 50:527–35.

5711. Dvorakovskii, M. S. 1961: Growth of the *Acer tataricum* L. from the seeds of various origin. Vestn. Mosk. Univ. Ser. 6, 1961(3):35–40.

5712. Dworak, L. 1935: The composition and ripening of corn. Mezögazdasagi Kutatasok 8:147–56. Chem. Abstr. 29: 6273. 1935.

5713. Dwornikow, W. 1928: Über Nachreife der Samen des Flachses. J. Landwirtsch. Wiss. Mosk. 1928:106–11. Botan. Centralbl. 156:207. 1929.

5714. Dwyer, R. E. [P.] 1931: Seed setting in lucerne. Agr. Gaz. N.S. Wales 42: 703–8.

5715. —— and S. L. Allman. 1932: Further observations on pollination and seed setting in lucerne. Agr. Gaz. N.S. Wales 43:141–46.

5716. Dyakova, E. V. 1952: Effect of boron fertilizers on the seed organs of clover and alfalfa in different soils. Mikroelementy v Zhizni Rast. i Zhivotn. Akad. Nauk SSSR, Tr. Konf. po Mikroelementam 1950: 189–92. Chem. Abstr. 49:440. 1955.

5717. Dybing, C. D. 1963: Determination of oil and fatty acid contents of small samples of immature flax seed. Crop Sci. 3:280–82.

5718. Dyer, F. C. 1938: Analysis of soil for seeds. A method of concentrating the seeds from soil for the purpose of counting and naming the seeds in a soil survey. Sci. Agr. (Ottawa) 18:641–46.

5719. —— and H. L. McClelland. 1937: The development of a flotation process for the cleaning of clover seeds. Sci. Agr. (Ottawa) 18:60–101.

5720. Dyer, H. J. 1947: Influence of potassium and sodium on metabolism of peanut cotyledons during germination. Botan. Gaz. 108:570–81.

5721. Dyer, R. A. 1935: The seed germination of certain species of *Euphorbia*. S. African J. Sci. 32:313–19.

5722. —— 1937: A note on the occurrence of vivipary in *Haworthia truncata* Schonl. S. African J. Sci. 33:335–36.

Dyer, W. Thiselton-; *see* Thiselton-Dyer, W.

5723. Dykes, W. R. 1933: Raising irises from seed. Flower Grower 20(10):426.

5724. Dyksterhuis, E. J. 1945: Axillary cleistogenes in *Stipa leucotricha* and their role in nature. Ecology 26:195–99.

5725. Dykstra, T. P. 1961: Production of disease-free seed. Botan. Rev. 27:445–50.

Dykyj, J. 5728–5732.

5726. Dykyj-Sajfertova, D. 1952: Influence of growth stimulants upon enzymes of germinating seeds. I. Česk. Botan. Ges. Ber. 62:123–63.

5727. —— 1956: The influence of growth substances on the enzymes of the germinating seed. II. Catalase activity in germinating, stimulated wheat. Sb. Česk. Akad. Zeměděl. Věd Rostlinna Vyroba 29:181–92. Chem. Abstr. 53:22288. 1959.

5728. —— and J. Dykyj. 1942: The influence of some hormone derivatives on sugar beets. Listy Cukrovar. 60:221–28. Chem. Abstr. 42:3121–22. 1948.

5729. —— and —— 1943: The effect of water on the effectiveness of growth solutions in seed hormonization. Sb. Česk. Akad. Zeměděl. 18:15–23. Chem. Abstr. 43:1837. 1949.

5730. —— and —— 1943: Influence of illuminating gas on the germination and yield of sugar beets. Z. Zuckerind. Böhmen Mähren 66:193–97. Chem. Abstr. 38:5880. 1944.

5731. —— and —— 1943: Untersuchungen über Samenkeimung und synthetische Wuchsstoffe. I. Einfluss von Quellungstemperatur und Wuchsstoffkoncentration auf die Keimung des Weizens. Angew. Botan. 25:274–300.

5732. —— and —— 1956: Effect of plant stimulators in various ionization form on the germination of wheat. Preslia 28:1–22. Chem. Abstr. 51:8219. 1957.

Dylla, U. 19219.

Dymchishina, T. D. 10810.

5733. Dymes, T. A. 1931/32: Some seeds taken from the nest of the harvesting ant, "Messor barbarus." Proc. Linn. Soc. London 1931/32:62–65. Bull. Soc. Botan. France 80:278. 1933.

5734. Dymond, J. R. 1917: Paper packet seeds. Proc. Assoc. Offic. Seed Analysts 9:30–35.

5735. —— 1920: Color characteristics of red clover seed. Seed World 7:23. Botan. Abstr. 7:607. 1921.

Dzhorogyan, G. A. 11375.

Dziedzianowicz, W. 11352.

E

5736. Eagle, E. 1960: A review of some physiological effects of gossypol and cottonseed pigment glands. J. Am. Oil Chemists Soc. 37:40–43.

5737. ——, H. F. Bialek, D. L. Davies, and J. W. Bremer. 1954: Detoxification of cottonseed by salts and alkalies. J. Am. Oil Chemists Soc. 31:121–24.

5738. —— and D. L. Davies. 1958: Effect of long-term storage on acute oral toxicity and gossypol content of cottonseed pigment glands. J. Am. Oil Chemists Soc. 35:36–37.

5739. Eagles, C. F., and P. F. Wareing. 1963: Experimental induction of dormancy in *Betula pubescens*. Nature 199:874–75.

5740. Eames, A. J. 1955: The seed and ginkgo. J. Arnold Arboretum 36:165–70.

Eames, A. J. 19685, 19686.

5741. Earhart, R. W. 1958: Pathological aspects of a seed survey. Intern. Crop Improvement Assoc. Ann. Rept. 40: 126–33.

5742. Earle, F. R. 1954: Protein level of corn falls in cool seasons; rises in warmer years. Crops & Soils 7(1):23.

5743. ——, J. J. Curtis, and J. E. Hubbard. 1946: Composition of the component parts of the corn kernel. Cereal Chem. 23:504–11.

5744. —— and Q. Jones. 1962: Analyses of seed samples from 113 plant families. Econ. Botany 16:221–50.

5745. ——, K. L. Mikolajczak, I. A. Wolff, and A. S. Barclay. 1964: Search for new industrial oils. X. Seed oils of the Calenduleae. J. Am. Oil Chemists Soc. 41:345–47.

5746. —— and R. T. Milner. 1946: Improvements in the determination of starch in corn and wheat. Cereal Chem. 21: 567–75. Expt. Sta. Rec. 95:472–73. 1946.

Earle, F. R. 2172, 3215, 4769, 7051, 9161, 12077, 13185, 13453, 13454, 13613, 18972.

5747. Earle, T. 1944: Improved wheat-peeling obtained by new method. Sci. News Letter 45:137.

5748. Earle, T. T. 1938: Origin of the seed coats in magnolia. Am. J. Botany 25: 221–22.

5749. —— 1941: Embryo and endosperm formation in *Illicium floridanum* Ellis. Am. J. Botany (Suppl.) 28:2s.

Earle, T. T. 8861, 8862, 15198.

5750. Earley, E. B. 1952: Percentage of carbohydrates in kernels of Station Reid yellow dent corn at several stages of development. Plant Physiol. 27:184–90.

5751. ——, J. N. Carter, and B. C. Johnson. 1952: Nicotinic acid, thiamine, and carotene content of kernels of medium and high protein corn. Agron. J. 44:326–28.

5752. —— and E. E. De Turk. 1944: Time and rate of synthesis of phytin in corn grain during the reproductive period. J. Am. Soc. Agron. 36:803–14.

Earley, E. B. 12569, 16988.

5753. Easley, L. T. 1963: Growth of Loblolly pine seed produced in a seed production area vs. nursery run stock. J. Forestry 61:388–89.

5754. Eastham, A. 1914: The longevity of some common seeds. Agr. Gaz. Can. 1:544–46.

5755. Eaton, E. L. 1957: The spread of blueberry seed through manure and by migrating robins. Proc. Am. Soc. Hort. Sci. 69:293.

Eaton, H. C. van; *see* Van Eaton, H. C.

5756. Eaton, R. J. 1929: Propagation by seed in hydrangea. Rhodora 31:254–55.

5757. Eaton, S. V. 1942: Sulphur content of seeds and seed weight in relation to effects of sulphur deficiency on growth of sunflower plants. Plant Physiol. 17:422–34.

Eaton, S. V. 1296.

Eaves, P. H. 17771.

5758. Ebata, M. 1961: Studies on white-core rice kernel. IV. Effects of night temperature in ripening time upon the occurrence and types of white-core kernels. Proc. Crop Sci. Soc. Japan 29:409–11.

Ebata, M. 14076–14078.

5759. Ebata, Y. 1934: Chemical constituents of the seeds of *Eriobotria japonicum* L. J. Agr. Chem. Soc. Japan (English Transl.) 10:1224–27. Chem. Abstr. 29:1133. 1935.

Ebbe, T. Schjelderup-; *see* Schjelderup-Ebbe, T.

5760. Ebeling, H. 1895: The influence of the weight of seed upon the grain production of yellow and blue lupines, common vetch, and buckwheat. Inaug. Dissertation, Leipzig. 65 pp. Botan. Centralbl. Beih. 5:537–39.

5761. Eberhart, C. 1906: Untersuchungen über das Vorquellen der Samen. Inaug. Dissertation, Univ. Jena. Tuhl. Land. Ztg. 56:159. 1907.

5762. Ebes, K. 1936: Die Methode Gurewitschs zur Bestimmung der Keimfähigkeit ohne Keimprüfung mittels Dinitrobenzen. Mededeel. Landbouwhogeschool Wageningen 40:19–21.

5763. Ebling, M. 1885: Absorbing organs of albuminous seeds. Flora (Jena) 68: 179–202.

5764. Ebner, H. 1924: Keimungs physiologie von *Draba verna, Thlaspi perfoliatum, Holosteum umbellatum* und *Veronica hederifolia*. Österr. Botan. Z. 73:23–41.

5765. Ebner, L. 1959: Untersuchungen über den Einfluss der Herkunft auf Keimfähigkeit und Beizempfindlichkeit von Zucherrübensaatgut. Saatgut-Wirtsch. 11: 324–26.

Echegaray, A. 1768.

5766. Echevin, R., and A. Brunel. 1937: Nitrogen metabolism during the germination of lupine (*Lupinus albus* L.). Compt. Rend. 204:881–83.

5767. —— and M. C. El Chichini. 1955: Sur la germination des graines d'Euphorbe, en particulier *Euphorbia exigua* L. et *E. lathyris* L. Compt. Rend. 241:92–94.

Echevin, R. 3203, 3204.

Eckersall, R. N. 2982.

5768. Eckerson, S. H. 1913: A physiological and chemical study of after-ripening. Botan. Gaz. 55:286–99.

5769. —— 1917: Microchemical studies in the progressive development of the wheat plant. Wash. Agr. Expt. Sta. Bull. 139. 20 pp.

5770. Eckles, M. L. 1941: Megasporogenesis and development of the megagametophyte in *Nothoscordum bivalve*. Botan. Gaz. 102:806–9.

5771. Edelman, J., V. Ginsburg, and W. Z. Hassid. 1955: Conversion of monosaccharides to sucrose and cellulose in wheat seedlings. J. Biol. Chem. 213:843–54.

5772. ——, S. I. Shibko, and A. J. Keys. 1959: The role of the scutellum of cereal seedlings in the synthesis and transport of sucrose. J. Exptl. Botany 10:178–89.

5773. Edelman, N. M. 1935: Effect of hydrogen sulfide on seeds. Tr. Zashchite Rast. Ser. III, pp. 65–67. Chem. Abstr. 31:1147. 1937.

5774. Edelmann, O., C. Tamm, and T. Reichstein. 1956: Die Glykoside der Samen von *Strophanthus schuchardtii* Pax. Glykoside und Aglykone, 155. Mitteilung. Helv. Chim. Acta 39:16–33.

5775. Eden, D. R. A. 1963: Coconut improvement by seed selection. World Crops 15:429–32.

Edfeldt, O. 14711.

Edgar, J. 19627.

Edgerton, L. J. 2582.

Edler, G. C. 1083.

5776. Edmond, J. B. 1933: The influence of planting depth on the shape of the scarlet globe radish. Proc. Am. Soc. Hort. Sci. 30:554–57.

5777. —— 1959: The influence of certain storage conditions and calcium oxide on the storage life of Golden Cross Bantam sweet corn seed. Proc. Am. Soc. Hort. Sci. 73:415–21.

5778. —— 1962: Studies of storage of vegetable seed under warm-humid conditions. Mississippi Agr. Expt. Sta. Tech. Bull. 50. 24 pp.

5779. —— and W. J. Drapala. 1958: The effects of temperature, sand and soil, and acetone on germination of okra seed. Proc. Am. Soc. Hort. Sci. 71:428–34.

5780. —— and —— 1959: The effect of temperature, immersion in acetone, and sulfuric acid on germination of five varieties of okra seed. Proc. Am. Soc. Hort. Sci. 74:601–6.

5781. —— and —— 1960: Studies of the germination of okra seed. Mississippi Agr. Expt. Sta. Tech. Bull. 47. 15 pp.

5782. Edmonds, A. C., and S. S. Patrudu. 1929: Sugarcane seed in Vizagapatam district: Its improvement. Agr. J. India 24:365–72.

5783. Edmondson, C. H. 1941: Viability of coconut seeds after floating in sea. Bernice P. Bishop Mus. Occas. Pap. 16:293–304. Expt. Sta. Rec. 87:353. 1942.

Edmundson, W. C. 4169, 17987.

5784. Edwards, D. C. 1933: "Hard" seeds in *Panicum coloratum*, Stapf. Nature 132:209.

Edwards, D. M. 3365.

5785. Edwards, E. J. 1936: Maize seed selection and disease control. The problem of internal seed-born infection. Agr. Gaz. N.S. Wales 47:303–6. Rev. Appl. Mycol. 15:793–94. 1936.

5786. Edwards, E. T. 1941: Internal grain infection in maize due to *Gibberella fujikuroi* and *Gibberella fujikuroi* var. *sublutinans*. Australian Inst. Agr. Sci. 7:74–82.

Edwards, F. W. 10563.

Edwards, G. 8974.

5787. Edwards, M., and M. Colin. 1834: De l'influence de la temperature sur la germination. Ann. Sci. Nat. Ser. II Botan. 6:257–70.

Edwards, M. 3168–3170.

5788. Edwards, M. M. 1945: Germination of seeds of *Striga lutea*. New Phytologist 44:104.

5789. —— 1959: The germination of charlock (*Sinapis arvensis*) seeds. Nottingham Univ. School Agr. Rept. 1958:36–38.

5790. Edwards, M. V. 1962: Choosing seed origins of European larch. Scot. Forestry 16:167–69.

Edwards, R. A. 8471.

5791. Edwards, R. S. 1957: Studies of the growth of cereals on seedbeds prepared by dry-sieving of the soil. II. Post-emergence survival and development. Empire J. Exptl. Agr. 25:304–20.

5792. —— 1958: Studies on the growth of cereals on seedbeds prepared by dry-sieving of the soil. III. The yields of grain and of straw. Empire J. Exptl. Agr. 26:323–43.

5793. Edwards, T. I. 1932: Temperature relations of seed germination. Quart. Rev. Biol. 7:428–43.

5794. —— 1933: The germination and growth of *Peltandra virginica* in the absence of oxygen. Bull. Torrey Botan. Club. 60:573–81.

5795. —— 1934: Relations of germinating soy beans to temperature and length of incubation time. Plant Physiol. 9:1–30.

5796. —— 1934: Seed frequencies in *Cytisus* and *Peltandra*. Am. Naturalist 68:283–86.

Edwards, T. I. 7515, 7516.

5797. Edwards, W. R., Jr., and J. C. Miller. 1947: Okra seed oil. Chemurgic Dig. 6(2):29, 31–33. Biol. Abstr. 21:20715. 1947.

5798. Efejkin, A. K. 1939: Wirkung hoher Temperaturen auf Vernalisierten Winterweizen. Ber. Akad. Wiss. USSR N.F. 25:308–10. Botan. Centralbl. 176:201–2. 1940.

5799. —— 1941: Devernalization of vernalized winter wheat. Compt. Rend. Acad. Sci. URSS 30:661–63.

5800. Effmann, H. 1963: Beurteilung anomaler Keime bei *Lupinus luteus*. Proc. Intern. Seed Testing Assoc. 28:61–69.

Efimov, V. N. 12734.

5801. Efimova, N. I. 1934: Experiments on the vernalization of makhorka (*Nicotiana rustica* L.) and tobacco. Physiol. Res. Makhorka Plants Kiev 1:47–81.

5802. Efros, A. M. 1961: The effect of benzimidazole derivatives on the germination of seeds and development of plants. Primenenie Gerbitsidov i Stimulyatorov Rosta Rast. (Akad. Nauk Belorussk. SSR) Sb. 1961:251–54. Chem. Abstr. 58:6134. 1963.

5803. Egami, F., K. Ohmachi, K. Lida, and S. Taniguchi. 1957: Nitrate reducing systems in cotyledons and seedlings of bean seed embryos, *Vigna sesquipedalis*, during the germinating stage. Biochemistry (USSR) (English Transl.) 22:115–27.

5804. Egerer, A. 1962: On the question of transmission of *Peronospora tabacina* Adam by seeds. Dresden Inst. Tabakforsch. Ber. 9:50–57.

5805. Eggebrecht, H. 1941: Methodik der Samenuntersuchung. Forschungsdienst Sonderheft 16:240–45.

5806. —— 1953/54: Zur Arten- und Sorten-Diagnostik in der Samenprüfung. Wiss. Z. Friedrich-Schiller-Univ. Jena 3:431–37.

5807. —— and W. Bethmann. 1939: Das Selenfärbeverfahren im Vergleich zu der üblichen Keimprüfung insbesondere bei Wintergerste mit Keimruhe. Angew. Botan. 21:448–55.

5808. Eggers, E. R. 1937: Germinating power of avocado seed from frost-injured fruit. Calif. Avocado Assoc. Yearbook 1937:120. Expt. Sta. Rec. 79:198. 1938.

5809. —— 1942: Effect of the removal of the seed coats on avocado seed germination. Calif. Avocado Assoc. Yearbook 1942: 41–43. Expt. Sta. Rec. 89:217. 1943.

Eggers, E. R. 8892, 8893.

5810. Eggleston, H. 1949: Safe keeping for seeds. Here's how to build seed storage to preserve seed viability even in moist southern climate. Southern Seedsman 12(1):14, 54.

Egle, K. 17248.

5811. Eglitis, M., and F. Johnson. 1957: Control of hard seed of alfalfa with high-frequency energy. [Abstract.] Phytopathology 47:9.

5812. Eglitis, V. 1954: Seed treatment with benzene hexachloride—method of increasing yields of grain crops. Latvijas PSR Zinatnu Akad. Vestis 2:53–59.

5813. Eglits, H. 1932: Flax diseases and flax seed disinfection experiments during the period 1929 to 1931. Acta Inst. Defens. Plantarum Latviensis, Riga 2:5–32. Chem. Abstr. 27:1978. 1933.

5814. Egorova, A. A., M. N. Bulatowa, and Z. M. Schulz. 1930: Materialien für Kenntnis der Keimung und Reife von Samen. Zap. Semenoved. (Ann. Essais Semences Leningr.) 7:58–63.

Egorova, A. A. 7682.

5815. Eguchi, H., M. Yamamoto, and H. Toda. 1962: Studies on the formation of seed in Satsuma orange fruit. Saga Agr. Expt. Sta. Bull. 3:161–66.

5816. Eguchi, T. 1960: Influence of nitrogenous fertilizer applied at different stages of growth on seed production in cabbage and Chinese cabbage. Proc. Am. Soc. Hort. Sci. 76:425–35.

5817. —— and A. Kagawa. 1955: Studies on the seed production in cabbage (*Brassica oleracea* L.). I. On the differentiation and development of terminal flower buds in cabbage. Bull. Natl. Inst. Agr. Sci. Ser. E 1955(4):217–24.

5818. ——, T. Kato, and S. Watanabe. 1954: Studies on the physiological sterility of Japanese radishes. I. Effects of vernalization on the bolting, flowering and fruiting. Bull. Natl. Inst. Agr. Sci. Ser. E 1954(3): 113–26.

5819. ——, T. Matsumura, and T. Koyama. 1963: The effect of low temperature on flower and seed formation in Japanese radish and Chinese cabbage. Proc. Am. Soc. Hort. Sci. 82:322–31.

5820. ——, Y. Oshika, T. Matsumura, H. Yamada, and T. Koyama. 1958: Studies on the seed production in Welsh onion (*Allium fistulosum* L.). III. Influence of nitrogenous fertilizers applied at different stages on the yield and quality of seeds in Welsh onion. Bull. Natl. Inst. Agr. Sci. Ser. E 1958(7):133–44.

5821. —— and H. Yamada. 1958: Studies on the effect of maturity on longevity in vegetable seeds. Bull. Natl. Inst. Agr. Sci. Ser. E 1958(7):145–65.

5822. Eguchi, Y., Y. Oshika, E. Yamada, and J. Watahara. 1962: Seed production of eggplant under different climatic conditions. J. Japan Soc. Hort. Sci. 31:33–38.

5823. Ehara, K. 1959: Effect of atmospheric humidity during ripening on the pod setting and seed germination in excised alsike clover (*Trifolium hybridum* L.). Kyushu Univ. Fac. Agr. J. 11:99–107.

5824. —— and S. Abe. 1952: Studies on the wild Japanese barnyard millet as a weed in the lowland rice-field. VI. Physiological investigations on the germination. Proc. Crop Sci. Soc. Japan 21:61–62.

5825. Eheart, J. F., C. F. Genter, and W. N. Linkous. 1949/50: Factors affecting the protein and other extract (oil) content of inbred, hybrid, and open-pollinated corn grain. Virginia Agr. Expt. Sta. Rept. 1949/50:23.

5826. ——, J. T. Wakeley, M. Speirs, F. F. Cowart, J. Miller, P. H. Heinze, M. S. Kanapaux, A. Sheets, L. McWhirter, M. Gieger, and R. C. Moore. 1951: Effects of different planting dates, bean maturity, and location on the vitamin contents of lima beans. Southern Coop. Ser. Bull. 12. 15 pp.

Eheart, J. F. 7101.

Ehle, H. Nilsson-; *see* Nilsson-Ehle, H.

5827. Ehlers, M. 1932: Eine Rose mit Seltsamen Früchten. Gartenwelt 36:639.

Ehninger, E. 3889, 3890.

5828. Ehrenberg, A., and L. Ehrenberg. 1958: The decay of X-ray induced free radicals in plant seeds and starch. Arkiv Fysik 14:133–41.

5829. Ehrenberg, C., A. Gustafsson, C. P. Forshell, and M. Simak. 1955: Seed quality and the principles of forest genetics. Hereditas 41:291–366.

5830. Ehrenberg, L. 1955: The mechanism of action of ionizing radiations in plant seeds. Svensk Kem. Tidskr. 67:207–24. Chem. Abstr. 50:8806–7. 1956.

5831. —— and G. Andersson. 1954: Probable side effect of nuclear reactions in the biological action of fast protons. Nature 173:1086. Nucl. Sci. Abstr. 8:5098. 1954.

5832. ——, A. Gustafsson, U. Lundqvist, and N. Nybom. 1953: Irradiation effects, seed soaking and oxygen pressure in barley. Hereditas 39:493–504.

5833. ——, ——, and N. Nybom. 1952: Biochemical aspects of the plant injury caused by ionizing radiations. Acta Chem. Scand. 6:1554–55. Nucl. Sci. Abstr. 7:332. 1953.

5834. —— and M. Jaarma. 1955: Effects of X-rays and water content on sugars in barley seeds. Acta Chem. Scand. 9:539–40.

5835. —— and E. Saeland. 1954: Effects of pile radiation on barley seeds. J. Nucl. Energy 1:150–69. Chem. Abstr. 49: 6340. 1955.

Ehrenberg, L. 5828, 8454, 12005, 13835.

5836. Ehrendorfer, K. 1955: Vergleichende Vorkeimversuche mit Licht- und Dunkelkeimen. Bodenkultur 8:168–81.

5837. Ehrler, W., and L. Bernstein. 1958: Effects of root temperature, mineral nutrition and salinity on the growth and composition of rice. Botan. Gaz. 120: 67–74.

5838. Eichberg, J. 1947: Soybean lecithin comes of age. Chemurgic Dig. 6:109–11. Biol. Abstr. 22:17038. 1948.

Eichel, B. 271, 272.

5839. Eichinger, A. 1910: Polyembryony in plants. A case of polyembryony in wheat. Naturw. Wochschr. 9:769–73.

5840. —— 1928: Die Beeinflussung der Länge der Winterroggenähren und der Zahl der Ährchen durch Düngung und Aussaatzeit. Eine Methode zur Unterscheidung von Winter- und Sommersaatgut. Angew. Botan. 10:66–79.

5841. Eidam, E. 1877: Über die schädlichkeit allzugrosser Wasserzufuhr für keimende Samen. Jahresber. Schlesische Ges. Vaterländ. Kultur, Breslau, pp. 13–16.

5842. —— 1883: Über den Einfluss wechselnder feuchtigkeit und Temperature auf die Keimung der Grassamen und der Runkelknauel. Ber. Botan. Sec. Schlesische Ges., pp. 232–38.

5843. Eide, E. 1930: Influence of summer temperature on the seed germination of Norway spruce. Medd. Norsk. Skogforsöks. 3:475–508.

5844. Eidemanis, R. 1954: Methods for increasing yields of clover seeds. Latvijas PSR Zinatnu Akad. Vestis 1954(4):41–48.

5845. Eidmann, F. E. 1934: The germination of teak (*Tectona grandis* L. f.). Tectona 27:233–87. Biol. Abstr. 9:3537. 1935.

5846. —— 1936: Saatgutprüfung auf biochemischem Wege. Z. Forst-Jagdw. 68: 422–43.

5847. —— 1938: Eine neue biochemische Methode zur Erkennung des Aussaatwertes von Samen. Proc. Intern. Seed Testing Assoc. 10:203–11.

Eidt, C. C. 3074.

5848. Eifert, V. S. 1959: The seed of a (water) lotus. Wild Flower 35:21–22.

5849. Eifrig, H. 1940: Untersuchungen über Keimverzögerungen bei *Brassica*. Agnew. Botan. 22:418–19.

5850. —— 1950: Der Einfluss verschiedener Temperaturen auf die Keimfähigkeit von *Brassica*-Samen in Beziehung zum Feuchtigkeitsgehalt derselben. Saatgut-Wirtsch. 2:280–82. Biol. Abstr. 25:25030. 1951.

5851. —— 1951: Untersuchungen zur kurzfristigen Unterscheidung von Kopfsalat und Pflück-Schnittsalat durch Laboratoriumversuche. Saatgut-Wirtsch. Fachzeitschr. Samen Saaten 12(21/12):280–82. Proc. Intern. Seed Testing Assoc. 17:144. 1951/52.

5852. —— 1951: Beitrag zur Unterscheidung von *Brassica*-Samen. Landwirtsch.

Forsch. 11(3). Proc. Intern. Seed Testing Assoc. 17:143–44. 1951/52.

5853. —— 1952: Echtheitsbestimmung der Varietäten von *Brassica oleracea* im Laboratorium. Saatgut-Wirtsch. Fachzeitschr. Samen Saaten 26(6):142–43.

5854. —— 1953: Echtheitsuntersuchungen bei *Brassica* und *Lactuca* im Laboratorium. Proc. Intern. Seed Testing Assoc. 18:167–79.

5855. —— 1953: Die Verwendung von Leuchtstoffröhrem im Artenechtheits laboratorium. Saatgut-Wirtsch. Fachzeitschr. Samen Saaten 28(2):41–44.

5856. —— 1954: Beobachtung von Fluoreszenzerscheinungen bei der Keimung von Brassiceae. Ber. Deut. Botan. Ges. 67:300–10.

5857. —— 1956: Ergebnisse der Freilandprüfung von Importsaatgut 1954–1955 auf Grund von Beobachtungen der Samenprüfstelle Münster. Saatgut-Wirtsch. Fachzeitschr. Samen Saaten 31(1):13–17.

5858. —— 1957: Untersuchungen über Erhaltung der Keimfähigkeit von *Beta*-Samen. Proc. Intern. Seed Testing Assoc. 22:1–7.

5859. —— 1960: Zur Artendiagnose bei *Trifolium*, *Medicago*, *Melilotus* und *Lotus*. Proc. Intern. Seed Testing Assoc. 25:321–27.

5860. —— 1960: Zur Methodik der Unterscheidung von *Trifolium*- und *Medicago*-Arten auf Grund von Fluoreszenzerscheinungen. Saatgut-Wirtsch. Fachzeitschr. Samen Saaten 1960(7):194–95.

5861. —— 1961: Keimungsphysiologische Probleme bei der Einkeimung frisch geernteter Getreidefrüchte. Proc. Intern. Seed Testing Assoc. 26:761–71.

5862. —— 1961: Results of trials on storing normal and monogerm beet seed with and without seed treatment performed through several years. Zucker 14:633–37.

5863. —— 1962: Die Festellung grünhülsiger und gelbhülsiger Busch- und Stangenbohnen im Laboratorium versuch. Proc. Intern. Seed Testing Assoc. 27:573–75.

5864. —— 1962: Zur Methodik der Keimprüfung einiger Gemüsesämereien. Proc. Intern. Seed Testing Assoc. 27:649–56.

5865. —— and S. K. Kamra. 1960: On the distinction of sugar beets, sugar-fodder beets (*Beta vulgaris* L.) and mangold in the laboratory. Proc. Intern. Seed Testing Assoc. 25:298–307.

5866. Eikema, J. S. 1941: Kiem- en bewaarproeven van *Hevea* zaden. Bergcultures 15:1049–60.

Einset, J. 15665.

5867. Einset, O. 1939: Correlation of number of seed to weight of fruit in the apple. Gartenbauwissenschaft 13:351–53.

5868. Eisler, M., and L. Portheim. 1926: Über die Bedeutung die Samenschale und ihre Beeinflussung durch Salze bei Vergiftung von Tabaksamen durch Nikotin. Planta 2:542–54.

5869. —— and —— II. Weitere Untersuchungen über die Nikotinvergiftung von Früchten und Samen. Planta 8:1–26.

5870. Eitingen, G. R. 1946: Growth of oak in relation to the weight of acorns. Sov. Agron. 4:67–75.

5871. Ekbote, R. B. 1947: Vernalization, an aid in crop breeding. Indian J. Agr. Sci. 17:301. Biol. Abstr. 23:28103. 1949.

Ekbote, R. B. 17507.

5872. Ekdahl, I. 1941: Die Entwicklung von Embryosack und Embryo bei *Ulmus glabra* Huds. Svensk Botan. Tidskr. 35:143–56.

Ekdahl, W. G. 4136.

5873. Eklund, B. 1956: An experiment in sowing and planting pine with different spacings. An account of the experience gained in the course of a 50-year regeneration experiment in the crown forest of Granvik. Statens Skogsforsk. Med. 46: 96–98.

5874. Eklund, O. 1925/26: Versuche über das Keimungs- und Schwimmvermögen einiger Samen und Früchte in Ostseewasser. Soc. Fauna Flora Fennica Mem. 2:13–29.

5875. —— 1926/27: Weitere Versuche über Keimung in Meerwasser. Soc. Fauna Flora Fennica Mem. 3:45–88.

5876. —— 1929: On the resistability of some seeds against sea salt. Soc. Fauna Flora Fennica Mem. 5:6–11. Botan. Centralbl. 157:414. 1929.

5877. Ekster, I. E. 1957: The connection between fat and tannin substances in grape seeds. Tr. Odessk. Sel'skokhoz. Inst. 8:89–92. Chem. Abstr. 53:507. 1959.

5878. Ekstrand, H. 1955: Germination trials with seeds of *Berberis vulgaris* L., with special regard to the effect of sodium chlorate on the germination. Statens Vaxtskyddsanstalt Medd. 69. 29 pp.

Ekstrand, H. 12529.

5879. Elagin, I. N. 1953: Fertility of seeds and characteristics of alfalfa development. Selek. i Semen. 20(4):57–64.
5880. —— 1953: Über die Gewinnung hybrider Buchweizensamen. Dokl. Vses. Akad. Sel'skokhoz. Nauk 18(4):11–15. Ber. Wiss. Biol. 87:264. 1954.
5881. —— 1958: Harvesting and storage of sorghum seeds. Sovkhoz. Proizvodstvo 1958(7):62–63.
5882. —— 1958: Storage of maize seeds in winter. Mashino-traktornaia Sta. 18(2): 29–30.
Elbe, J. von. 19449.
El Chichini, M. C. 5767.
5883. Elcock, H. A., and J. C. Overpeck. 1933: Methods of producing sugar-beet seed in southern New Mexico. New Mex. Agr. Expt. Sta. Bull. 207. 29 pp.
El-Damaty A. H. 10173.
Elder, A. H. 15187.
5884. Elder, A. L., and D. M. Rathmann. 1962: Seed oils in human nutrition. Econ. Botany 16:196–205.
5885. Elder, S., and R. Bressler. 1875/77: Auslaugungsversuche mit verschiedenen Samen. Wiss.-Prakt. Gebiete Pflanzenbaues 1:41–46.
5886. Eldredge, J. C., and I. J. Johnson. 1953: Cold storage of popcorn. Ice & Refrig. 124(3):3–16. J. Sci. Food Agr. 5:260. 1954.
5887. Eldridge, A. C., W. J. Wolf, A. M. Nash, and A. K. Smith. 1963: Protein purification, alcohol washing of soybean protein. J. Agr. Food Chem. 11:323–28.
5888. Elehwany, N. E. 1955: Chemical and physical changes as related to maturity of raw and processed sweet peas. Univ. Maryland Publ. 14:187. Dissertation Abstr. 15:2156. 1955.
5889. Eleizalde, L., M. Mezquita, and A. Vich. 1932: Über die phytotoxische Wirkung pathologischer menschlicher Seren. I. Vorbemerkungen. Über Keimung und Wasseraufnahme der Samen von *Lupinus albus*. Arch. Cardiol. 13:18–28. Ber. Wiss. Biol. 21:809. 1932.
5890. Elford, F. G. 1930: The palisade cells of the seed coat of *Albizzia lophantha*. Proc. Roy. Soc. Victoria 42:92–98. Biol. Abstr. 6:5181. 1932.
5891. El Gindy, M. M., C. A. Lamb, and R. C. Burrell. 1957: Influence of variety, fertilizer treatment, and soil on the protein content and mineral composition of wheat, flour, and flour fractions. Cereal Chem. 34:185–95.
5892. El-Helaly, A. F. 1940: The effect of *Rhizoctonia solani* Kühn on the germination of lettuce seed at somewhat high temperatures. Proc. Linn. Soc. London 1939/40, Session 152, Pt. 1. Rev. Appl. Mycol. 20:510–11. 1941.
5893. Elias, L. G., R. Colindres, and R. Bressani. 1964: The nutritive value of eight varieties of cowpea (*Vigna sinensis*). J. Food Sci. 29:118–22.
Elias, L. G. 3003.
5894. Eliason, E. J., and C. E. Heit. 1940: The size of Scotch pine cones as related to seed size and yield. J. Forestry 38:65–66.
5895. —— and —— 1940: The results of laboratory tests as applied to large scale extraction of red pine seed. J. Forestry 38:426–29.
5896. —— and —— 1941: The effect of light and temperature on the dormancy of Scotch pine seed. Proc. Assoc. Offic. Seed Analysts 32:92–102.
Eliason, E. J. 8585.
5897. Elion, E. 1943: The action of glutathione and wheat germ on dough in relation to proteolytic enzymes in wheat flour. Cereal Chem. 20:234–50.
5898. —— 1945: The cereal amylases. Bakers Dig. 19:95–96, 112. Biol. Abstr. 19:22816. 1945.
5899. Elizarova, S. S. 1937: Inheritance of enzyme characters; catalase of barley. Bull. Acad. Sci. URSS Cl. Sci. Math. Nat. Ser. Biol. 1937:1781–87.
5900. —— 1937: An investigation of the quantitative catalase index in barley. Biochemistry (USSR) (English Transl.) 2:442–53.
5901. —— 1940: The β-amylase of wheat and barley. Compt. Rend. Acad. Sci. URSS 26:698–701.
5902. El-Kilany, M. A. 1939: Flax in Egypt. II. Factors influencing seed germination. III. Flower and anther colour inheritance. Egypt Min. Agr. Tech. Sci. Serv. Bull. 205. 7 pp.
El-Kinawi, S. A. 25, 1881.
Elkington, H. D. 5191, 5192.
5903. Ellerström, S., and A. Hagberg. 1954: Competition between diploids and tetraploids in mixed rye populations. Hereditas 40:535–37.
5904. —— and —— 1962: The cyto-genetic department of the Swedish Seed Association 1931–1961. A brief survey of work carried out during the past 30 years. Sveriges Utsadesforen. Tidskr. 72:192–209.

5905. Elliot, C. S., H. N. Greenwood, and B. C. R. Reiss. 1953: Spring sowing of winter cereals. Gt. Brit. Min. Agr. J. 60:378–83.

5906. Elliott, B. B., and A. C. Leopold. 1953: An inhibitor of germination and of amylase activity in oat seeds. Physiol. Plantarum 6:65–77.

Elliott, C. 4917, 14778.

Elliott, E. S. 11914.

Elliott, F. C. 8161.

5907. Elliott, G. A. 1953: The forest tree seed committee. Proc. Intern. Seed Testing Assoc. 18:253–55.

5908. Elliott, H. G. 1929: Subterranean clover seed and its impurities, with a comparison between machine cleaned seed and seed in the burr. J. Agr. W. Australia 6:274–79.

5909. —— 1930: Agricultural seeds and their weed seed impurities. J. Agr. W. Australia Ser. 2, 7:574–77.

5910. Elliott, J. A. 1923: Cotton-wilt, a seed-borne disease. J. Agr. Res. 23:387–93.

Elliott, J. R. 19715.

5911. Elliott, R. F. 1957/58: Studies on the interaction of temperature and light on *Mimulus* germination. Carnegie Inst. Wash. Yearbook 57:272.

5912. —— and C. S. French. 1959: Germination of light sensitive seed in crossed gradients of temperature and light. Plant Physiol. 34:454–56.

5913. Ellis, G. H., L. F. Randolph, and G. Matrone. 1945: A comparison of the chemical composition of diploid and tetraploid corn. J. Agr. Res. 72:123–30.

Ellis, I. D. 19751.

5914. Ellis, J. H. 1943: Report on effect of certain cultural and fertility practices on the protein content of wheat. Sci. Agr. (Ottawa) 23:474–95.

Ellis, J. W. 20055.

5915. Ellis, M. M. 1913: Seed production in *Yucca glauca*. Botan. Gaz. 56:72–78.

Ellison, J. H. 16891.

5916. Ells, J. E. 1963: The influence of treating tomato seed with nutrient solutions on emergence rate and seedling growth. Proc. Am. Soc. Hort. Sci. 83:684–87.

5917. Elmore, J. C. 1953: Treatment of bean seed for the control of the seed-corn maggot. J. Econ. Entomol. 46:1054–59.

5918. El Murabaa, A. I. M. 1957: Factors affecting seed set in Brussell sprouts, radish, and cyclamen. Mededel. Landbouwhogeschool Wageningen 57(4):1–33.

5919. Elomaa, A. 1952: Results of cereal seed treatment experiments with disinfectants during the years 1928–1950. Valtion Maatalouskoetoiminnan Tiedonantoja 226:1–43. Biol. Abstr. 28:7016. 1954.

5920. Elpiner, I. E., and L. M. Bronskaya. 1959: Some physiochemical bases for the stimulating action of ultrasound on the sprouting of corn seeds. Dokl. Akad. Nauk SSSR 128:1073–75. Chem. Abstr. 54: 13281. 1960.

El-Rafie, M. S. 8343.

Elsherif, S. M. 7170.

5921. El-Shishiny, E. D. H. 1953: Effect of temperature and desiccation during storage on germination and keeping quality of *Kochia indica* seeds. J. Exptl. Botany 4:403–6.

5922. ——, M. A. Noseir, and S. D. Y. Barakat. 1961: Viability of *Kochia indica* seeds under different storage conditions. Bull. Inst. Desert Egypt. 8:23–32. 1958.

5923. —— and D. Thoday. 1953: Inhibition of germination in *Kochia indica*. J. Exptl. Botany 4:10–22.

Elstad, V. B. 10780, 17364.

5924. El Tabey, A. M., and M. V. Cruess. 1949: The oxidase of the apricot. Plant Physiol. 24:307–16.

5925. Elvehjem, C. A. 1961: The significance of basic research. Seed World 88(4):8–12.

Elvehjem, C. A. 8194, 8654, 9096, 16204.

Elvin-Lewis, M. 12188.

5926. Elze, D. L. 1931: Die Übertragbarkeit mit dem Samen von Aukuba-Mosaik sowie Blattroll (Phloemnekrose) der Kartoffel. Phytopathol. Z. 3:449–60. Ber. Wiss. Biol. 19:491. 1931.

5927. —— 1949: Germination of citrus seeds in relation to certain nursery practices. Palestine J. Botany 7:69–80. Biol. Abstr. 25:25460. 1951.

5928. Elzenga, G. 1952: Investigation of viability and germination rate of *Atropa belladonna*, *Digitalis purpurea*, *Digitalis lanata*, *Valeriana officinalis* and *Viola tricolor* in connection with different seed treatment. Herba 11:73–78.

5929. Emack, E. P. 1938: Changes in weight and percentage composition of seed samples. Proc. Assoc. Offic. Seed Analysts 23:56–60.

5930. Emden, L. 1947: Über Kürbis-Katalase. Helv. Chim. Acta 30:15–19.

Emerson, G. A. 5931, 5932, 6093, 6094.

Emerson, J. N. 15436.

5931. Emerson, O. H., G. A. Emerson, and H. M. Evans. 1936: The isolation from cottonseed oil of an alcohol resembling α-tocopherol from wheat-germ oil. Science 83:421.

5932. ——, ——, and —— 1939: The occurrence of γ-tocopherol in corn embryo oil. Science 89:183.

Emerson, O. H. 6094.

5933. Emery, F. E. 1897: Test of effect of chemical manure and cottonseed meal on the germination of wheat. North Carolina Agr. Expt. Sta. Bull. 146:252–54.

5934. Emery, W. H. P., and W. V. Brown. 1957: Extra-ovular development of embryos in two grass species. Bull. Torrey Botan. Club 84:361.

5935. Emmanuel, F. 1935: Semis de pommes de terre. Rev. Botan. Appl. 15:551–53.

5936. Emmel, M. W. 1935: The toxicity of *Glottidium vesicarium* (Jacq.) Harper seeds for the fowl. J. Am. Vet. Med. Assoc. 40:13–21. Biol. Abstr. 10:4122. 1936.

5937. —— 1945: The toxic principle of the tung tree. Ann. Proc. Am. Tung Oil Assoc. 1945:38–42. Chem. Abstr. 39:4165. 1945.

5938. Emmerich, H. 1942: Der Einfluss zusätzlicher Wirkstoffbehandlung auf Wachstum und Entwicklung der Pflanzen. Jahrb. Wiss. Botan. 90:99–140. Biol. Abstr. 23:30181. 1949.

5939. Emmons, A. H. 1934: Growing dahlias from seed. Flower Grower 21(2): 55.

5940. Emonson, K. H. 1957: Certified French bean seed, 1956–1957. J. Agr. (Victoria) 55:647–48.

Emoto, Y. 18617.

5941. Emparan, P. R., and H. M. Tysdal. 1957: The effect of light and other factors on breaking the dormancy of guayule seed. Agron. J. 49:15–19.

5942. Emsweller, E. 1951: The effect of age and temperature on germination of seed of *Lilium speciosum rubrum*. North Am. Lily Soc. Yearbook 4:49–52.

5943. Emsweller, S. L. 1950: How to propagate lilies. Horticulture 28(1):17, 28–29.

5944. —— 1952: The lily research programme at the Beltsville Station. Lily Yearbook 1953(16):11–21.

5945. —— and H. A. Borthwick. 1938: Effects of short periods of low temperature on flower production in stocks (*Matthiola*). Proc. Am. Soc. Hort. Sci. 35:755–57.

5946. Enachescu, G., O. Iordachescu, and R. Marinescu. 1959: Dynamics of some chemical components during the growth and development of seeds in the garden pea. Comun. Acad. Rep. Populare Romine 9:473–78. Chem. Abstr. 53:22283. 1959.

5947. Enari, T. M., T. Linnahalme, and M. Linko. 1961: Effects of air and carbon dioxide in the steeping of barley. J. Inst. Brewing 67:358–61.

5948. ——, M. Nummi, J. Mikola, and V. Makinen. 1962: Extraction of albumins and globulins of barley grain. Finska Kemistsamfundets Medd. 71:44–52. Chem. Abstr. 58:747. 1962.

5949. Endemann, W. 1953: Wird der Nikotingehalt du Samennachkommenschaften durch Pfropfungen beeinflusst? Züchter 23:206–7. Ber. Wiss. Biol. 87:59. 1954.

5950. Enders, C., G. Nowak, F. Schneebauer, and A. Pfahler. 1940: Hydrogen peroxide steeps improve germinating capacity in malting barley. [Abstract.] Food Ind. 12:87.

5951. ——, A. Pfahler, and F. Schneebauer. 1937: Rapid method for estimating the germinative capacity of barley and malt. Wochschr. Brau. 54:297–300. Chem. Abstr. 32:2571. 1938.

5952. Endresz-Hajas, M., and J. Hamori-Szabo. 1961: Nitrogen, phosphorus and potassium mobilization in corn seedlings during germination. Gödöllo. Agrartud. Egyet. Mezogazd. Kar. Közlemen. 1961: 287-93.

5953. Eneroth, O. 1926/27: A study on the risks of using in a particular district pine-seed from other sources. Sweden Skogsförsföksanst. Medd. 23:1–62.

5954. —— 1928: Contribution to the question of the origin of seeds of *Pinus sylvestris*. Intern. Rev. Agr. 19:739–40.

5955. —— 1931: The influence of ash of burned slash on the germination of conifer seed and the early development of seedlings. Suomen Metsätieteellinen Seura Commentat. Forest. 5:1–67. Biol. Abstr. 7:4124. 1933.

5956. Enescu, V., and G. Costea. 1962: The possibility of raising the value of the qualitative indices of *Gleditschia triacanthos* seeds and the problem of regional indices of fir. Rev. Padurilor (Bucharest) 77:718–21.

5957. —— and V. Enescu. 1957: Investigations on the germination of elm seeds. Rev. Padurilor (Bucharest) 71:230–35.

5958. Engel, C. 1947: The distribution of the enzymes in resting cereals. I. The distribution of the saccharogenic amylase in wheat, rye, and barley. Biochim. Biophys. Acta 1:42–49.

5959. —— 1947: The distribution of the enzymes in resting cereals. III. The distribution of esterase in wheat, rye, and barley. Biochim. Biophys. Acta 1:278–79.

5960. —— and L. H. Bretschneider. 1947: The distribution of the enzymes in resting cereals. IV. A comparative investigation of the distribution of enzymes and mitochondria in wheat grains. Biochim. Biophys. Acta 1:357–63.

5961. —— and J. Heins. 1947: The distribution of the enzymes in resting cereals. II. The distribution of the proteolytic enzymes in wheat, rye, and barley. Biochim. Biophys. Acta 1:190–96.

Engel, C. 12261.

5962. Engel, O. S. 1950: Effect of winter wheat seed vernalization on sensitivity of sprouts to X-ray action. Dokl. Akad. Nauk SSSR 71:1151–53.

5963. —— 1951: Effect of Roentgen rays on wheat seeds in relation to degree of their ripeness. Dokl. Akad. Nauk SSSR 78:811–14.

5964. —— 1952: Change in sensitivity of wheat seeds to radiation in relation to duration of swelling. Dokl. Akad. Nauk SSSR 85:229–31.

5965. —— 1957: Mobilization of reserve substances in germinating sunflower seeds. Plant Physiol. (USSR) (English Transl.) 4:514–19. Chem. Abstr. 52:6509h. 1957.

5966. —— and A. A. Prokofev. 1960: Effect of water content in seeds on the mobilization of reserve substances during germination. Plant Physiol. (USSR) (English Transl.) 7:44–48. Chem. Abstr. 55:1822. 1961.

Engel, R. E. 6844.

5967. Engelbert, V. 1932: A study of various factors influencing seed production in alfalfa (*Medicago sativa*). Dissertation, Toronto. Sci. Agr. (Ottawa) 12:593–603. Proc. Intern. Seed Testing Assoc. 4: 93–103. 1932.

5968. —— 1940: Reproduction in some *Poa* species. Can. J. Res. Sec. C 18:518–21. Expt. Sta. Rec. 84:467. 1941.

5969. —— 1941: The development of twin embryo sacs, embryos, and endosperm in *Poa arctica* R. BR. Can. J. Res. Sec. C 19:135–44.

5970. Engelbrecht, L. 1954: Über Allantoinsäure und Allantoin. III. Ihre Rolle bei der Keimung des Ahorns. Flora (Jena) 141:501–22. Biol. Abstr. 29:29696. 1955.

5971. —— 1954: Über Allantoinsäure und Allantoin. IV. Ihre Beziehungen zu den Säureamiden bei der Keimung von *Phaseolus vulgaris* L. Flora (Jena) 142:25–44.

5972. Engelhardt, M., M. Vicente, and K. Silberschmidt. 1961: Notes on the germination of the seeds of various Brazilian members of the Solanaceae. Acad. Brasil. Ciene. 33:55–68.

5973. ——, ——, and —— 1962: The stimulating effect of light and potassium nitrate on the germination of seeds of "*Amaranthus hybridus*" L. Rev. Brasil. Biol. 22:1–7.

Engelhardt, M. 19101, 19102.

5974. Engelmann, C. 1962: On the germinating power of moist seed. Z. Landwirtsch. Versuchs- Untersuchungsw. 8:218–33.

5975. Engelmann, G. 1880: Acorns and their germination. Tràns. Acad. Sci. St. Louis 4:190–92.

5976. Enggas, C. E. 1946: Delphinium germination and growth in Missouri, in temperatures over 100°. Am. Delphinium Soc. Yearbook 1946:48–50.

5977. Engle, L. G., and R. D. Williams. 1957: Scarifying seedbed boosts yellow poplar (*Liriodendron tulipifera*) germination. U.S. Dept. Agr. Forest Serv. Central States Forest Expt. Sta., Sta. Note 110. 2 pp.

Englehorn, A. J. 7894.

5978. Engleman, E. M. 1960: Ovule and seed development in certain cacti. Am. J. Botany 47:460–67.

5979. English, C. S., Jr., E. H. English, J. F. Caperci, W. R. Hill, and L. E. Brandt. 1952: The raising of rhododendrons from seed. Arboretum Bull. 15(3):12–14, 25.

English, E. H. 5979.

5980. English, J. G. 1948: Decorticated seed—one year's experience. Proc. Am. Soc. Sugar Beet Technol. 5:102–5.

5981. Engstrom, A. 1948: Growing cottonwood from seed. J. Forestry 46:130–32.

5982. —— 1950: Mulching seedbeds with cellophane. For quicker and more uniform germination of tree seeds. Plants Gardens 6:204.

5983. Engstrom, W. H. 1955: Direct seeding of western red cedar (*Thuja plicata*). Oregon State Bd. Forestry Res. Note 20. 11 pp.

5984. Enikeev, M. G. 1960: The effect of thermal and chemical treatment of seeds on their germinating ability. Selek. i Semen. 1960(1):60–61.

5985. —— 1960: Sanitation of wheat seeds which have reduced their germinating ability. Zemledelie 1960(3):86–87.

5986. Enikeev, S. G. 1952: Effect of boron on seed production of leguminous feed grasses under conditions of Kirgiz S.S.R. Mikroelementy Zhizni Rast. i Zhivotn. Akad. Nauk SSSR Tr. Konf. po Mikroelementam 1950:193–96. Chem. Abstr. 49:440. 1955.

5987. —— 1952: On pre-planting treatment of sugar beet seeds with α-naphthaleneacetic acid. Sov. Agron. 10:94–96.

5988. Enileev, K. K. 1954: Effect of preplanting soaking of cotton seeds in a 25% solution of ammonium nitrate on growth, plant development and yields of raw cotton. Sotsial. Sel'skoe Khoz. Uzbek. 15:22–27.

5989. —— and V. P. Solovev. 1960: A study of the causes of diversity in germination behavior of cotton seeds. Plant Physiol. (USSR) (English Transl.) 7: 20–24.

Enloe, J. A. 18555.

5990. Enomoto, N. 1929: Mutation of the endosperm character in rice plant. Japan. J. Genetics 5:49–72. Japan. J. Botany 5:(1)–(2). 1930.

5991. —— and T. Nagai. 1942: Studies on salt-resistance of cotton species at the germination stage and influences of inorganic salts to the cell content. I. Proc. Crop Sci. Soc. Japan 14:5–14. Biol. Abstr. 23:30205. 1949.

5992. Enser, K. 1935: Untersuchungen über die Umwandlung höherer Fettsäuren in Kohlehydrate bei der Keimung von Kürbissamen. II. Kohlehydratbestimmungen. Jahrb. Wiss. Botan. 82:158–69. Biol. Abstr. 10:10354. 1936.

5993. Enyi, B. A. C. 1963: The effect of seedling age, depth of planting and fertilizer placement on the growth and yield of rice (*Oryza sativa* L.). J. Agr. Sci. 61:291–97.

5994. —— 1964: Note on the effect of fertilizers and soil cover on submerged pregerminated rice seeds. Empire J. Exptl. Agr. 32:92–96.

5995. Enzenberg, U. 1961: Beiträge zur Karyologie des Endosperms. Österr. Botan. Z. 108:245–85.

5996. Enzie, J. V. 1943: Experiments in the production of carrot seed. New Mexico Agr. Expt. Sta. Bull. 308. 11 pp. Expt. Sta. Rec. 89:666. 1943.

5997. Enzie, W. D. 1942: The relation of spacing to yield and to plant and ear development of some yellow sweet corn hybrids in New York. N.Y. Agr. Expt. Sta. Bull. 700. 19 pp. Expt. Sta. Rec. 87:58. 1942.

5998. Eperjessy, G. 1928: Keimung einiger Weizensorten in sauren und alkalischen Losüngen. Mezogazdasagi Kutatasok 1: 95. Int. Seed Test. Assoc. Proc. 11–12: 145–46. 1930.

5999. —— 1930: Unterschiede bei der Keimung von auf alkalischem und saurem Boden gezüchteten Weizensorten. Fortschr. Landwirtsch. 5:345–50.

6000. —— 1932: Comparative experiments on the phosphoric acid and potash absorption capacities of germs of a Petkus rye and of a Hungarian type rye. Mezogazdasagi Kutatasok 5:1–8. Chem. Abstr. 26:2813. 1932.

Epshtein, I. A. 7160.

6001. Erb, J. A. 1938: Zur Gewinnung von Lärchensamen. Forstwiss. Centralbl. 60: 162–63.

6002. Ercoli-Bati, M. 1941: The chemical composition of the soil and the gluten content of wheat. Ital. Agr. 78:546–48. Biol. Abstr. 22:14698. 1948.

6003. Erdelsky, K. 1960: Einfluss einer γ-Bestrahlung auf die Keimfähigkeit und die ersten Entwicklungsphasen des Maises. Biologia 15:890–94.

Erdman, L. W. 19314.

6004. Erdmann, K. 1936: Investigations of the dependency of the X-ray effect upon the water content of protoplasm. Protoplasma 26:557–76.

6005. Eremenko, L. L. 1950: On difference in quality of blossoms and seeds in the carrot umbel. Agrobiologiya 1950(6): 123–27.

6006. Eremenko, V. T. 1935: Effect of chemical stimulants on duration of iarovization and on length of the vegetative period. Sov. Botan. (Leningrad) 1935(6): 36–45. Biol. Abstr. 11:6744. 1937.

6007. Ergle, D. R., and L. S. Bird. 1958: Preliminary experiments employing gibberellic acid as a cottonseed treatment. Plant Disease Reptr. 42:320–23.

6008. —— and G. Guinn. 1959: Phosphorus compounds of cotton embryos and their changes during germination. Plant Physiol. 34:476–81.

Ergle, D. R. 13192.

6009. Erhart, H. 1934: Sur l'influence de la provenance pédologique des semences dans la culture des céréales. Compt. Rend. 199:1349–51.

Erickson, A. E. 9235.

6010. Erickson, Mrs. E. H. 1943: Growing dahlias from seed. Horticulture 21(4): 86, 87.

6011. Erickson, E. L., and R. H. Porter. 1938: Laboratory tests of field crop seeds as indicators of seeding value. Iowa Agr. Expt. Sta. Res. Bull. 239. 34 pp.

6012. Erickson, J. L. E., and J. H. Brown, Jr. 1942: Study of the toxic properties of tung nuts. J. Pharmacol. 74:114–17. Pharm. Abstr. 8:206. 1942.

6013. Erickson, L. C. 1946: Effect of lucerne seed size and depth of seeding on the subsequent procurement of stand. J. Am. Soc. Agron. 38:964–73.

6014. —— and H. M. Benedict. 1947: Origin of the seed coats in guayule. J. Agr. Res. 74:329–35.

6015. ——, C. I. Seely, and K. H. Klages. 1948: Effect of 2,4-D upon the protein content of wheats. J. Am. Soc. Agron. 40:659–60. Biol. Abstr. 22:25117. 1948.

Erickson, L. C. 19862.

6016. Erikkson, C., and E. VonSydow. 1964: Postharvest metabolism of green peas (*Pisum sativum*) with special reference to glutamic acid and related compounds. J. Food Sci. 29:58–64.

Erikkson, S. 16939.

6017. Erikson, J. 1892: Influence of water content on the germination and keeping qualities of seeds. Kgl. Lantbruks-Akad. Handl. Tidskr. 31:180–91.

6018. Erith, A. G. 1930: The inheritance of colour, size and form of seeds, and of flower colour in *Vicia faba* L. Genetica 12:477–510. Ber. Wiss. Biol. 19:475. 1931.

6019. Erkama, J., and N. Pettersson. 1950: Vitamin K in germinating peas. Acta Chem. Scand. 4:922–25.

6020. Erlander, S. R. 1960: The production of amylose and amylopectin in corn endosperms and in potato tubers. Cereal Chem. 37:81–93.

6021. Ermakov, A. I. 1931: The first stages of seed germination in oleiferous plants. Bull. Appl. Botany, Genet., Plant Breed. (USSR) 25:135–62. Chem. Abstr. 26: 756. 1932.

6022. —— 1932: A method of determining the oil in a portion of the seed. Bull. Appl. Botany, Genet., Plant Breed. (USSR) Ser. A No. 3:88–96. Chem. Abstr. 27:2323. 1933.

6023. —— 1933: Influence of the date of sowing on amount and quality of the oil from the seeds of oleiferous plants. Bull. Appl. Botany, Genet., Plant Breed. (USSR) Ser. 3(1):31–69. Chem. Abstr. 27:5370. 1933.

6024. —— 1934: Elaboration of methods for determining quality of oil seeds. Bull. Appl. Botany, Genet., Plant Breed. (USSR) Ser. 3(5):33–71. Biol. Abstr. 10:12624. 1936.

6025. —— 1937: Individual and intraspecies variability of the oil content of seeds. Bull. Acad. Sci. URSS Cl. Sci. Math. Nat. Ser. Biol. 1937:1853–70.

6026. —— 1958: Chemical compositions of flax seeds as affected by growing conditions in different soil and climatic zones. Tr. Prikl. Botan. Genet. i Selek. 31:36–60. Biol. Abstr. 43:11706. 1963.

6027. —— and N. N. Ivanov. 1931: Über die Atmung der Samen von Ölpflanzen. Biochem. Z. 231:79–91.

6028. ——, N. P. Yarosh, and A. I. Fragina. 1959: Effect of nutritional conditions on the synthesis of oils and other organic substances in the flax plant. Tr. Prikl. Botan. Genet. i Selek. 31:277–87. Chem. Abstr. 54:16556. 1960.

Ermakov, A. I. 9551, 13975.

Ermakova, E. A. 9428.

6029. Ermilov, G. B. 1952: On dissemination and concentration of seeds. Dokl. Akad. Nauk SSSR 84:599–601.

6030. —— 1960: Some peculiarities of water absorption by maize seeds. Plant Physiol. (USSR) (English Transl.) 7:36–41.

6031. Ermolaev, E. J. 1941: The effect of drying seeds on the growth and development of cucumbers. Sov. Botan. (Leningrad) 1941(1/2):86–89. Biol. Abstr. 21: 23331. 1947.

6032. Ernould, L. 1949: Treatment of beet seed with gammexane against wire worms. Publ. Inst. Belge Ameliorat. Betterave (Tirlemont) 17:45–61. Chem. Abstr. 43: 8088. 1949.

Ernould, L. 5098.

6033. Ernst, A. 1905: Das Ergrünen der Samen von *Eriobotrya japonica*. Botan. Centralbl. Beih. 19:118–30.

6034. —— 1906: Das Keimen der dimorphen Fruchtchen von *Synedrella nodiflora*. Ber. Deut. Botan. Ges. 24:450–58.

6035. —— 1932: Stauden-Anzucht aus Samen. Möllers Deut. Gärtner-Ztg. 47:67.

Erofeev, N. G. 6149.

6036. Erskine, A. J., and J. K. N. Jones. 1957: The structure of linseed mucilage. I. Can. J. Chem. 35:1174–82.

6037. Erwin, A. T. 1935: Alegria-a popping seed used in Mexico as a substitute for popcorn (*Amaranthus caudatus* L. var. *leucospermus* Th.). Iowa State Coll. J. Sci. 9:661–65.

Erxleben, H. 10958.

6038. Erygin, P. S. 1936: Change in activity of enzymes, soluble carbohydrates, and intensity of respiration of rice seeds germinating under water. Plant Physiol. 11:821–32.

Esashi, Y. 14071.

6039. Esau, A., and E. Klapp. 1932: Über Wachstums- und Ertragsbeeinflussende Wirkungen von Ultrakurzwellen auf Kulturpflanzen. Pflanzenbau 9:129–32.

6040. Esau, K. 1960: Anatomy of seed plants. Wiley, New York. 376 pp.

Esau, K. 2491.

Esau, P. 4181.

6041. Esbo, H. 1954: Vitality of unhulled and hulled timothy seeds under ordinary storing conditions in whole bags. Kgl. Lantbruksakad. Tidskr. 93:123–48.

6042. —— 1959: Vitality of timothy seed during long storage. Växtodling 12. 101 pp.

6043. —— 1962: Purity committee. Proc. Intern. Seed Testing Assoc. 27:48–69.

6044. —— and H. Kolk. 1953: Quality and seed value of spring wheat damaged by black stem rust. Med. Stat. Cent. Frökontr. 28:29–43.

6045. Eschenbecher, F. 1960: Zur Kenntnis der Samenhaare von *Epilobium hirsutum*. Planta 54:314–25.

6046. Eschenwald, A., and C. W. Hall. 1961: Coefficient of friction, angle of repose, specific gravity, and bulk density of coffee fruits and coffee beans. J. Agr. Univ. Puerto Rico 45:19–25.

6047. Escombe, F. 1897: Germination of seeds. I. The vitality of dormant and germinating seeds. Sci. Progr. N.S. 1:585–608.

Escombe, F. 3146, 3147.

6048. Escribens, M. R., and M. Maurtua C. 1952: The oil of *Jatropha curcas*. Agronomia (Peru) 17(70):117–19. Chem. Abstr. 47:10174. 1953.

6049. Escudero, P., and P. B. Landabure. 1943: Food value of pinons of the araucaria (*Pinus araucanus* Mol, *Araucana imbricata* Pav., Patagonian pine). Rev. Asoc. Agr. Dietol. 1:25–28. Chem. Abstr. 39:4997. 1945.

6050. Esdorn, I. 1928: Die Feststellung der Wirkung von Trockenbeizmitteln im im Laboratoriumsversuch. Angew. Botan. 10:178–86.

6051. —— 1928: Beiträge zur Keimungsphysiologie hartschaliger Samen. Angew. Botan. 10:469.

6052. —— 1928: Der Einfluss der Lagerung auf die Keimfähigkeit der gelben Lupine. Fortschr. Landwirtsch. 3:346–53.

6053. —— 1930: Untersuchungen über die Hartschaligkeit der gelben Lupine. Wiss. Arch. Landwirtsch. Abt. A Pflanzen. 4:497–549. Proc. Intern. Seed Testing Assoc. 15/16/17:181–91. 1931.

6054. —— 1932: Zur Schwefelsäurestimulation hartschaliger Samen. Deut. Landwirtsch. Presse 59:526.

6055. —— 1940: Untersuchungen an *Arnica montana* L., die Lebensdauer ihres Saatgutes und ihre Anbaubedingungen. Deut. Heilpflanze 6(4):23–24.

6056. —— and H. Stütz. 1932: Die Bewertung harter Leguminosensamen. Landwirtsch. Versuchssta. 114:137–47. Fortschr. Landwirtsch. 8:134. 1933.

6057. Esdorn, J. 1931: Untersuchungen über anomale Kleekeimungen. Landwirtsch. Versuchssta. 112:103–14.

Esguerra, P. D. 5228.

6058. Esh, G. C., T. S. De, and U. P. Basu. 1959: Influence of genetic strain and environment on the protein content of pulses. Science 129:148–49.

6059. Esitashvili, G. L. 1956: Age and preplanting drying of water- and muskmelon seeds. Akad. Nauk Gruzinsk. SSR Inst. Polevod. Tr. 1956(9):235–42.

Eskew, E. B. 15695.

Eslick, R. F. 7361.

6060. Espino, R. B., and F. T. Pantaleon. 1935: Effects of heat upon the viability of bunga (*Aeginetia indica* Linn.) seeds. Philippine Agriculturist 24:439–50.

Espino, R. B. 6315.

Espiritu, L. G. 3562.

Essam, E. D. M. 16655.

6061. Essenburg, F. W., and A. F. Schoorel. 1961: Het verband tussen de kiemkrachtsbepaling van zaaizaden in het laboratorium en de opkomst te velde. Zaaizaadonderzoek II, Centr. Orgaan, Wageningen, pp. 42–66.

6062. Esser, K. 1963: Bildung und Abbau von Callose in den Samenanlagen der *Petunia hybrida*. Z. Botan. 51:32–51.
6063. Esser, M. M. 1933: The percentage and persistence of polyembryony in certain citrus fruits. Trans. Illinois State Acad. Sci. 25:120.
6064. Essery, R. E., and J. R. A. Pollock. 1957: Studies in barley and malt. X. Note on the effect of desiccation and of heating on the germinative behavior of dormant barley. J. Inst. Brewing 63:221–22.
Essery, R. E. 15517.
6065. Esson, J. G. 1933: Hardy plants from seed. Horticulture 11(3):43–44.
6066. Esteban-Velasco, E. 1960: Experiencias de vernalización de trigo. Anales Edafol. Agrobiol. (Madrid) 19:155–62.
6067. Estreicher, E. 1914: Über Kalteresistenz und den Kaltetod der Samen. Bull. Intern. Acad. Sci. Cracovie Cl. Math. Nat. Ser. Biol., pp. 844–79. Botan. Centralbl. 140:386. 1919.
Etchebarne, J. 4492.
Eto, T. 10998.
Eto, W. H. 8159.
Etten, C. H. van; *see* Van Etten, C. H.
6068. Etter, A. G. 1948: Seeds that ride livestock. Bull. Missouri Botan. Garden 36:170–72.
6069. Ettlinger, M. G., and J. E. Hodgkins. 1956: The mustard oil of papaya seed. J. Org. Chem. 21:204–5.
6070. —— and A. J. Lundeen. 1956: The mustard oil of *Limnanthes douglasii* seed, *m*-methoxybenzyl isothiocyanate. J. Am. Chem. Soc. 78:1952–54.
6071. Eubel, J. 1956: Allgemeine Zusammensetzung und Aminosäuren des Eiweisses von Hasel- und Walnüssen. Z. Lebensm.-Untersuch.-Forsch. 104:429–34. Nutr. Abstr. Rev. 27:1627. 1957.
6072. Eugster, J., and W. Hauptmann. 1934: Durchdringende Umgebungsstrahlung und Zellwachstum. Mit besonderer Berücksichtigung der hesschen Ultrastrahlung und der Votgänge im strahlenarmen Milieu. Strahlentherapie 49:223–37. Biol. Abstr. 11:2945. 1937.
Eulenberg, B. E. G. von; *see* Von Eulenberg, B. E. G.
6073. Eulenburg, B. E. 1932: Keimversuche mit Getreidearten und ihre Beeinflussung durch die Art des Keimbettes. Dissertation, Albertus Univ., Königsburg. 74 pp.
Euler, H. von; *see* Von Euler, H.
6074. Eunus, A. M. 1955: The effects of X-rays on the embryonal growth and development of *Hordeum vulgare* L. J. Exptl. Botany 6:409–21.
Eunus, A. M. 13337.
Euw, J. von; *see* Von Euw, J.
6075. Eva, W. J., and I. Levi. 1943: A note on the chemical composition of Manitoba shelled corn, 1941 crop. Sci. Agr. (Ottawa) 23:362–64.
6076. —— and E. E. Rankin. 1945: A comparison of five methods for determining starch content of wheat. Can. J. Res. Sect. B 23:260–68.
Eva, W. J. 547–549, 551.
6077. Evans, C. D., R. J. Foster, and C. B. Croston. 1945: Preparation of zein by precipitation method. Ind. Eng. Chem. 37:175–77. Expt. Sta. Rec. 94:707. 1946.
6078. —— and R. H. Manley. 1944: Ternary solvents for zein. Ind. Eng. Chem. 36:408–10. Expt. Sta. Rec. 94:707. 1946.
6079. Evans, C. R. 1922: Effect of temperature on germination of *Amaranthus retroflexus*. Botan. Gaz. 73:213–25.
6080. —— 1933: Germination behavior of *Magnolia grandiflora*. Botan. Gaz. 94: 729–54.
6081. Evans, G. 1933: Fertilizing red clover for seed. Welsh J. Agr. 9:135–41. Chem. Abstr. 27:5136. 1933.
6082. —— 1937: Technique of grass seed production at the Welsh Plant Breeding Station. Imp. Bur. Plant Genet. Herb. Plants (Aberystwyth) Bull. 22. 36 pp. Expt. Sta. Rec. 78:481. 1938.
6083. —— 1939: Perennial ryegrass: seed production of bred strains. Gt. Brit. Min. Agr. J. 46:355–62.
6084. —— 1949: Conditions for certification of herbage seed crops. Welsh Plant Breed. Sta. Seed Growers L. 14. 4 pp.
6085. —— 1953: Advances in herbage seed production in the United States of America. J. Brit. Grassland Soc. 8:27–29. Biol. Abstr. 28:26963. 1954.
6086. —— 1957: Red clover seed storage for 23 years. J. Brit. Grassland Soc. 12:171–77.
6087. —— 1957: The viability over a period of fifteen years of severely dried ryegrass seed. J. Brit. Grassland Soc. 12:286–89.
6088. —— 1959: Seed rates of grasses for seed production. I. Pasture varieties of ryegrass, cocksfoot, Timothy. Empire J. Exptl. Agr. 27:263–75.
6089. —— 1962: Seed rates of grasses for seed production. II. Bred varieties of tall fescue, meadow fescue, cockboot (Haytype)

and phalaris hybrid. Empire J. Exptl. Agr. 30:181–91.

6090. —— 1963: Seed rates of grasses for seed production. III. Early perennial ryegrass S. 24. Empire J. Exptl. Agr. 31:34–40.

6091. ——, H. M. Roberts, and J. Lewis. 1958: Maintenance of life in seeds. Welsh Plant Breed. Sta. Aberystwyth Rept. 1950/56:152–55.

Evans, G. 19707.

6092. Evans, H. 1950: Results of some experiments on the preservation of cacao seed in viable condition. Trop. Agr. (Trinidad) 27:48–55.

6093. Evans, H. M., and G. A. Emerson. 1939: Growth-stimulating action of wheat-germ oil treated with ferric chloride. Proc. Soc. Exptl. Biol. Med. 41:170–72. Chem. Abstr. 33:8247. 1939.

6094. ——, O. H. Emerson, and G. A. Emerson. 1936: The isolation from wheat germ oil of an alcohol, α-tocopherol, having the properties of vitamin E. J. Biol. Chem. 113:319–32.

Evans, H. M. 5931, 5932.

6095. Evans, J. W. 1941: Changes in the biochemical composition of the corn kernel during development. Cereal Chem. 18:468–73.

6096. —— and D. R. Briggs. 1941: The fatty acid composition of the lipids of corn starch at various stages during the development of the corn kernel. Cereal Chem. 18:465–68.

Evans, L. E. 3572.

Evans, L. T. 13761.

Evans, M. W. 368.

6097. Evans, R. A., B. L. Kay, and C. M. McKell. 1963: Herbicides to prevent seed set or germination of Medusahead. Weeds 11:273–76.

Evans, R. E. 19841.

6098. Evans, R. J., J. L. St. John, P. M. Craven, J. L. Haddock, D. G. Wells, and S. P. Swenson. 1947: Some factors influencing the protein, cystine, and methionine content of dry peas. Cereal Chem. 24:150–56. Biol. Abstr. 21:25828. 1947.

Evans, R. J. 18037.

6099. Evans, S. A. 1956: Chemical desiccation of red clover for seed. Agriculture (London) 63:280–83.

6100. Evans, T. A. 1954: The effect of nitrogen applications at different dates on the seed yield of pedigree grasses. J. Brit. Grassland Soc. 9:53–60. Biol. Abstr. 30:32476. 1956.

6101. Evans, W. F., and F. C. Stickler. 1961: Grain sorghum seed germination under moisture and temperature stresses. Agron. J. 53:369–72.

6102. Evans, W. J., A. A. Woodham, W. B. Carney, J. M. Dechary, and A. M. Altschul. 1963: Fractionation of particulate proteins of the peanut cotyledon. Proc. Seed Protein Conf. New Orleans 1963:218–28. Chem. Abstr. 60:4460. 1964.

6103. Evans, W. R. 1961: The partial purification and properties of a germination inhibitor(s) isolated from apple seed. Dissertation Abstr. 22:49. 1961.

6104. —— and B. Axelrod. 1961: Pyrimidine metabolism in germinating seedlings. Plant Physiol. 36:9–13.

Evenari, L. 6113.

6105. Evenari, M. 1940: On germination inhibitors. I. Introduction. Palestine J. Botany 2:1–5.

6106. —— 1949: Germination inhibitors. Botan. Rev. 15:153–94.

6107. —— 1952: The germination of lettuce seeds. I. Light, temperature and coumarin as germination factors. Palestine J. Botany 5:138–59.

6108. —— 1957: The physiological action and biological importance of germination inhibitors. Symp. Soc. Exptl. Biol. 11:21–43.

6109. —— 1957: Les problèmes physiologiques de la germination. Bull. Soc. Franc. Physiol. Végétale 3:105–24.

6110. —— 1961: Chemical influences of other plants (allelopathy). Encyclopedia Plant Physiol. 16:691–736.

6111. —— 1961: A survey of the work done in seed physiology by the Department of Botany, Hebrew University, Jerusalem (Israel). Proc. Intern. Seed Testing Assoc. 26:597–658.

6112. —— 1962: Plant physiology and arid zone research. Arid zone research XVIII. The problems of the arid zone. Proc. Paris Symp. (1960) UNESCO, Paris, pp. 175–95.

6113. —— and L. Evenari. 1951: Laboratory experiments in the germination of sugar beet seeds. Bull. Res. Council Israel 1:132–36. Biol. Abstr. 26:13215. 1952.

6114. ——, S. Klein, H. Anchori, and N. Feinbrun. 1957: The beginning of cell division and cell elongation in germinating lettuce seed. Bull. Res. Council Israel Sect. D 6:33–37.

6115. ——, E. Konis, and D. Zirkin. 1947: On the germination of some rosaceous seeds. I. The germination of apple seeds. Palestine J. Botany 4:86–89. Biol. Abstr. 27:7722. 1953.

6116. ——, ——, and —— 1948: On the germination of some rosaceous seeds. II. The germination of Kerassi seeds. Palestine J. Botany 4:166–70.

6117. —— and A. M. Mayer. 1954: The effect of auxin on the germination of lettuce seeds. Bull. Res. Council Israel Sect. B 4:81–82.

6118. ——, ——, and A. Poljakoff-Mayber. 1954: Action of coumarin as a germination inhibitor. Congr. Intern. Botan. Paris Rappt. Commun. 8:173–74. Chem. Abstr. 48:10852. 1954.

6119. —— and G. Neumann. 1952: The germination of lettuce seed. II. Bull. Res. Council Israel 2:75–78.

6120. —— and —— 1953: The germination of lettuce seeds. III. The effect of light on germination. Bull. Res. Council Israel 3:136–44. Biol. Abstr. 28:29641. 1954.

6121. —— and —— 1953: The germination of lettuce seeds. IV. The influence of relative humidity of the air on light effect and germination. Palestine J. Botany 6:96–100.

6122. ——, ——, S. Blumenthal-Goldschmidt, A. M. Mayer, and A. Poljakoff-Mayber. 1958: The influence of gibberellic acid and kinetin on germination and seedling growth of lettuce. Bull. Res. Council Israel Sect. D 6:65–72.

6123. ——, ——, and S. Klein. 1955: The influence of red and infrared light on the respiration of photoblastic seeds. Physiol. Plantarum 8:33–47.

6124. ——, ——, and A. Poljakoff-Mayber. 1958: The respiration of lettuce seeds germinated at a high temperature. Bull. Res. Council Israel Sect. D 6:108–10.

6125. ——, ——, and —— 1958: Respiration studies with lettuce seeds imbibed for very long periods. Bull. Res. Council Israel Sect. D 6:125–26.

6126. ——, ——, and G. Stein. 1953: Factors modifying the influence of light on germination. Nature 172:452–53.

6127. ——, ——, and —— 1957: Action of blue light on the germination of seeds. Nature 180:609–10.

Evenari, M. 12107, 12108, 12153, 13074–13076, 15498, 15499, 19961.

Everett, C. F. 4305.

Everett, H. L. 16490.

6128. Everett, T. H. 1932: *Shortia galacifolia* from seed. Florists Exchange Hort. Trade World 80(18):14.

6129. Everly, R. T., and K. T. Payne. 1959: Effect of pre-seeding insecticide treatments of the seed bed on red clover stands and forage yields. J. Econ. Entomol. 52: 1145–54.

6130. —— and R. C. Pickett. 1960: The effect of phorate applied to seed on the growth, development and insects attacking grain sorghum. J. Econ. Entomol. 53:154–60.

6131. Evers, A., and A. Sunneveld. 1953: Early cutting in seed production of selected Italian ryegrass. (Project 246.) Verslag. Inst. Landbouwk. Onderzoek. Wageningen 1953:38–45. Soils Fertilizers 18:414. 1955.

Evers, H. W. G. J. 6632–6636.

6132. Everson, E. H., and E. F. Seeborg. 1958: The heritability of milling quality in wheat as measured by the separation of the bran and endosperm. Agron. J. 50:511–13.

Everson, E. H. 13624, 13625.

6133. Everson, G. J., H. Steenbock, D. C. Cederquist, and H. T. Parsons. 1944: The effect of germination, the stage of maturity and the variety upon the nutritive value of soybean protein. J. Nutr. 27:225–29. Chem. Abstr. 38:3378. 1944.

6134. Everson, L. E. 1949: Preliminary studies to establish laboratory methods for the germination of weed seeds. Proc. Assoc. Offic. Seed Analysts 39:84–89.

6135. —— 1950: Further studies on the effect of 2,4-D on seeds. Proc. Assoc. Offic. Seed Analysts 40:84–87.

6136. —— 1953: A comparison of two methods for the classification of weed seeds. Proc. Assoc. Offic. Seed Analyst 43:109–14.

6137. —— 1956: Noxious weed seeds in Iowa oats and red clover samples. Proc. Iowa Acad. Sci. 63:252–58.

6138. —— 1958: A comparison of methods and blowers for the purity analysis of Kentucky bluegrass seed. Iowa State Coll. Agr. Home Econ. Expt. Sta. Res. Bull. 464:398–407.

6139. —— and R. S. Dunham. 1951: The effect of 2,4-dichlorophenoxyacetic acid on certain weed and crop seeds. Minn. Agr. Expt. Sta. Tech. Bull. 197. 24 pp.

6140. ——, C. S. Shih, and F. B. Cady. 1962: A comparison of the "hand" and "uniform" methods for the purity analysis of Kentucky bluegrass (*Poa pratensis*) seed. Proc. Intern. Seed Testing Assoc. 27:476–88.

6141. Evrard, R. 1955: Contribution à l'étude des propriétés germinatives de faines par la méthode au chlorure de 2-3-5 triphenyltetrazolium. Bull. Inst. Agron. Sta. Rech. Gembloux 23:49–54.

6142. —— 1957: L'analyse de la qualité des semences par la méthode aux rayons X. Ann. Gembloux 63:81–87.

6143. —— 1957: Aperçu des méthodes indirectes d'appreciation des propriétés germinatives des semences. Ann. Gembloux 63:253–67.

6144. Evreinov, M. G., I. S. Smirnova, N. F. Kozhevnikova, and M. V. Kotlyarov. 1961: On the problem of pre-planting treatment of seeds with ultraviolet rays and electric current. Tr. Vses. Nauchn.-Issled. Inst. Elektrif. Sel'skoe Khoz. Nauchn. 10:147–67.

6145. ——, V. G. Tikhomirov, I. S. Smirnova, V. A. Tyutyunnikova, and N. F. Kozhevnikova. 1962: Pre-sowing seed treatment of spring cereals with electric high-tension alternating current. Vestn. Sel'skokhoz Nauki 1962(8):95–98.

6146. Evreinov, V. A. 1945: Le jujubier. Rev. Hort. 29:231–33.

6147. —— 1948: Le pistachier. Fruits d'Outre-Mer 3:45–50. Biol. Abstr. 23: 8716. 1949.

6148. Evreinova, T. N. 1956: Proteolytic enzymes of acorns. Vestn. Mosk. Univ. 11, No. 3, Ser. Fiz.-Mat. i Estestven. Nauk No. 2, pp. 95–100. Chem. Abstr. 50:13184. 1956.

6149. —— and N. G. Erofeev. 1956: Amylase of acorns stored under different conditions. Vestn. Mosk. Univ. Ser. Biol. Pochvoved. Geol. i Geogr. 11:39–43. Chem. Abstr. 51:16745. 1957.

Evstigneeva, Z. G. 11376, 11377.

6150. Evtushenko, G. A. 1939: On the physiology of tobacco seed germination. Krasnodar Inst. Tobacco Ind. Bull. 138:3–30.

6151. Ewart, A. J. 1894: Observations on the vitality and germination of seeds. Trans. Liverpool Biol. Soc. 8:207–47.

6152. —— 1896: Additional observations on the vitality and germination of seeds. Trans. Liverpool Biol. Soc. 10: 185–93.

6153. —— 1897: On the power of withstanding desiccation in plants. Trans. Liverpool Biol. Soc. 11:151–59.

6154. —— 1908: On the longevity of seeds. Proc. Roy. Soc Victoria 21:1–210.

6155. —— 1909: Longevity of seeds. Botan. Gaz. 48:463.

6156. —— 1912: The influence of super phosphates on the germination of wheat. J. Dept. Agr. Victoria 10:256–58.

6157. Ewart, A. J. 1912: The influence of radio-active mineral on the germination and on the growth of wheat. J. Dept. Agr. Victoria 10:417–21.

Ewing, A. L. Orr-; *see* Orr-Ewing, A. L.

6158. Exbrayat-Durivaux, C. 1930: Notes sur la germination des *Moringa malgaches.* Ann. Mus. Colonial Marseille 8(2):5–21. Biol. Abstr. 6:11200. 1932.

6159. Exell, A. W. 1931: The longevity of seeds. Gardeners Chron. III 89:283.

6160. Eynard, I. 1957: Effect of liquid nitrogen and oxygen on the germination of the hard seeds of alsike clover, lucerne and red clover. Allionia (Turin) 3:23–32.

6161. Eyre, F. H. 1936: Dispersal of jack-pine seed from seed trees and slash. Papers Mich. Acad. Sci. 21:279–84. Biol. Abstr. 11:14558. 1937.

6162. —— 1938: Can jack-pine be regenerated without fire? J. Forestry 36: 1067–72.

Eyre, F. H. 11928.

6163. Eyre, J. V. 1931: Notes on oil development in the seed of a growing plant. Biochem. J. 25:1902–8.

6164. Eyster, H. C. 1938: Conditioning seeds to tolerate submergence in water. Am. J. Botany 25:33–36.

6165. —— 1940: The cause of decreased germination of bean seeds soaked in water. Am. J. Botany 27:652–59.

6166. —— 1942: Factors which affect absorption of water by seeds. Proc. S. Dakota Acad. Sci. 22:17–27. Biol. Abstr. 18:14612. 1944.

Eyster, H. C. 17996.

6167. Eyster, W. H. 1931: Vivipary in maize. Genetics 16:574–90.

Ezell, B. D. 13596.

Ezhov, I. S. 14611–14613.

F

6168. Fa, W. C., and S. Roy. 1944: The effect of light on the synthesis of ascorbic acid by germinating seeds. Sci. Cult. (Calcutta) 9:564. Biol. Abstr. 19:1613. 1945.

6169. Fabbri, A. 1932: Action du chlorure de sodium sur la germination des graines. Ann. Tech. Agrar. 3:203. Proc. Intern. Seed Testing Assoc. 7:114. 1935.

6170. Faberge, A. C. 1958: Relation between chromatid-type and chromosome-type breakage-fusion-bridge cycles in maize endosperm. Genetics 43:737–49.

Fabianova, Nemtsova-; *see* Nemtsova-Fabianova.

Fabini, R. Pohloudek. 18261, 18262.

6171. Fabricius, L. 1926: Ein Versuch mit Samenstimulation. Forstwiss, Centralbl. 48:113–19.

6172. —— 1928: Forstliche Versuche. I. Der forstliche Wert verschiedener Mittel gegen Wildverbiss. Forstwiss. Centralbl. 50:610–19.

6173. —— 1928: Forstliche Versuche. II Keimfähigkeit des Samens alter Tannen. III. Wirkung ultravioletter Strahlen auf die Keimung. Forstwiss. Centralbl. 50: 694–701. Biol. Abstr. 4:23906. 1930.

6174. —— 1929: Forstliche Versuche. V. Die Einwirkung von Waldbrandasche auf Samenkeimung und erste Pflanzenentwicklung. Forstwiss. Centralbl. 51:269–76. Botan. Centralbl. 157:401. 1929.

6175. —— 1931: Forstliche Versuche. X. Die Samenkeimung von *Sorbus aucuparia* L. Forstwiss. Centralbl. 53:413–18. Biol. Abstr. 6:1941. 1932.

6176. —— 1954: Versuch zur Klärung der Frage: Hat der Mond Einfluss auf die Keimung und erstjährige Entwicklung von Fichten und Kiefersamen? Forstwiss. Centralbl. 73:142–51.

6177. Fabry, A., M. Stefl, and J. Tulach. 1957: Home-growing of *Perilla* and biochemical evaluation of the seed quality. Sb. Česk. Akad. Zeměděl. Věd. Rostlinna Vyroba 3:979–92. Chem. Abstr. 52:12096. 1958.

Fafara, F. 1958.

6178. Fagan, T. W., E. T. Jones, and W. M. Ashton. 1943: The yield and chemical composition of cereals under different systems of management. Welsh J. Agr. 17:103–7. Biol. Abstr. 20:20864. 1946.

6179. Fagerlind, F. 1936: Embryologische Beobactungen über die Gattung *Phyllis*. Botan. Notiser 1936:577–84.

6180. —— 1939: Perisperm oder Endopserm bei *Coffea*? Svensk Botan. Tidskr. 33: 303–9.

6181. —— 1940: Die Entwicklung des Embryosackes bei *Peperomia pellucida* Kunth. Ark. Botan. 29A. 15 pp.

6182. —— 1940: Stempelbau und Embryosackentwicklung bei einigen Pandanazeen. Ann. Jardin Botan. Buitenzorg. 49:55–78.

6183. —— 1940: Die Terminologie der Apomixis. Prozesse. Hereditas 26:1–22. Biol. Abstr. 15:11769. 1941.

6184. —— 1941: Die Embryosackentwicklung bei *Tanacetum vulgare* und einigen Chrysanthemum-Arten. Svensk Botan. Tidskr. 35:157–76.

6185. —— 1941: Die Embryosackbildung bei *Camassia*. Svensk Botan. Tidskr. 35:258–60.

6186. —— 1945: Bildung und Entwicklung des Embryosacks bei sexuellen und agamospermischen *Balanophora*-Arten. Svensk Botan. Tidskr. 39:65–82.

6187. —— 1945: Bau des Gynöceums, der Samenanlage und des Embryosackes bei einigen Repräsentanten der Familie Icacinaceae. Svensk Botan. Tidskr. 39:346–64.

6188. —— 1946: Sporogenesis, Embryosackentwicklung und Pseudogame Samenbildung bei *Rudbeckia laciniata* L. Acta Horti Bergiani 14(3):39–90.

6189. —— 1951: Influence of pollen parent on the production of hips, achenes and seeds in the canina roses. Acta Horti Bergiani 16(4):121–68. Biol. Abstr. 27: 2270. 1953.

Fahrenholtz, K. E. 4504.

Failes, G. D. 12125.

6190. Fainbron, D. B. 1961: Variability of wheat following the sowing of embryonically young seed. Dokl. Mosk. Sel'skokhoz. Akad. 62:119–24. Biol. Abstr. 43:24912. 1963.

6191. Fairbairn, J. W. 1945: Structural and numerical features of the testa-Cardamom. Quart. J. Pharm. Pharmacol. 18:214–27. Pharm. Abstr. 12:144. 1946.

6192. Fairbanks, B. W., J. L. Krider, D. Catron, and W. E. Carroll. 1947: Value of soybean meal prepared from frosted-field-damaged soybeans for growing-fattening swine. J. Agr. Res. 75:155–59.

6193. Fairbrother, T. H. 1929: The influence of environment on the moisture content of flour and wheat. Cereal Chem. 6:379–95.

6194. [Fairburn], D. C. 1941: Raising magnolias from seed. Bull. Missouri Botan. Garden 29:43–47.

6195. —— 1943: Raising *Nepenthes* from seed. Bull. Missouri Botan. Garden 31:174–80.

6196. —— 1946: Germination period for garden seeds. Bull. Missouri Botan. Garden 34:66–69.

Fairchild, H. W. 16288.

6197. Fairchild, P. L. K., R. M. Johnson, and W. D. Raymond. 1955: Melon seed (*Citrullus vulgaris*) oil from the Sudan. Colonial Plant Animal Prod. (Gt. Brit.) 5:62–63.

6198. ——, W. D. Raymond, and R. G. W. Spickett. 1955: The extraction of oil from conophor nuts. Colonial Plant Animal Prod. (Gt. Brit.) 5:63–68.

6199. Fairchild, R. S. 1953: Comparative development of the embryos of inbred and hybrid maize. Iowa State Coll. J. Sci. 27:381–405.

6200. Faithful, W. 1948: Seed production in New Zealand. Chewings fescue. New Zealand J. Agr. 76:331–36.

6201. Faivre, M. E. 1878/79: Studies on the latex tubes and latex during the normal germination of the embryo of *Tragopogon porrifolius* L. Mem. Acad. Sci. Belles (Lyon) 23:361–419.

6202. Faleev, A. V., and F. T. Sukhenko. 1941: Activity of vegetable catalase as influenced by structure of water. Compt. Rend. Acad. Sci. URSS 33:301–3. Chem. Abstr. 38:5857. 1944.

6203. Falk, K. G. 1915: Studies on enzyme action. XIII. The lipase of soybeans. J. Am. Chem. Soc. 37:649–53.

6204. —— 1917: Studies on enzyme action. XIV. Further experiments on lipolytic actions. J. Biol. Chem. 31:97–123.

6205. —— and M. L. Hamlin. 1931: Studies on enzyme action. III. The action of manganese sulphate on castor bean lipase. J. Am. Chem. Soc. 35:210–19.

6206. —— and K. Sugiura, K. 1915: Studies on enzyme action. XII. The esterase and lipase of castor beans. J. Am. Chem. Soc. 37:217–30.

Fall, S. 42.

6207. Falqui, M. C. 1962: Some effects of coumarin on the germination of "*Triticum durum*" Desf. Ric. Sci. Rend. Sect. B 2:106–8.

6208. Faludi-Daniel, A. 1957: Changes in the organic acid and amino nitrogen contents of peas and maize during germination. Acta Botan. Acad. Sci. Hung. 3:243–51. Chem. Abstr. 52:10300. 1958.

6209. Famiani, V. 1932: The nutritive value of wheat germ. Atti Accad. Lincei 16:275–78. Chem. Abstr. 27:2178. 1933.

6210. —— 1932: The food value of the embryo of various cereal and leguminous kernels. Atti Accad. Lincei 16:360–65. Chem. Abstr. 27:2179. 1933.

Famiani, V. 20041.

6211. Famin, A. 1931: Action de la temperature sur le noyau et la karyokinese chez *Vicia faba*. Compt. Rend. 193:745–48. Biol. Abstr. 7:1784. 1933.

6212. Famintzin, A. 1893: Concerning the fate of chlorophyll grains in seeds and sprouts. Arb. Botan. Lab. Akad. St. Petersburg No. 5. 16 pp. Botan. Centralbl. 58:378–79. 1894.

6213. Fan, L. T., P. S. Chu, and J. A. Shellenberger. 1963: Diffusion of water in kernels of corn and sorghum. Cereal Chem. 40:303–13.

6214. ——, D. S. Chung, and J. A. Shellenberger. 1961: Diffusion coefficients of water in wheat kernels. Cereal Chem. 38:540–47.

Fan, L. T. 4111.

Fan, S. S. 7096.

Fanelli, A. Rossi-; *see* Rossi-Fanelli, A.

6215. Fang, S. C., D. E. Bullis, and J. S. Butts. 1953: Investigation of Barcelona and Duchilly filbert nuts. III. Amino acid and B vitamin contents of oil-free filbert nut meals. Food Res. 18:555–59.

6216. Fankhauser, F. 1931: Concerning the significance of the seed source of our forest trees. J. Forestry 29:652–60.

Fankuchen, I. 4692.

Fano, A. da; *see* Da Fano, A.

Fant, G. W. 12034.

6217. Fardy, A., J. Cuzin, and D. Schwartz. 1959: Greffes embryonnaires de *Nicotiana tabacum* sur quelque solanées; conséquences sur le développement et la nicotinogenese dans l'epibiote et sur la descendance issue de celui-ci. Bull. Soc. Sci. Bretagne 34:156–60.

6218. Faris, D. G., and F. L. Smith. 1964: Effect of maturity at time of cutting on quality of dark red kidney beans. Crop Sci. 4:66–69.

Faris, J. A. 12138.

Farkas, G. L. 10687.

Farkass, E. 10770.

6219. Farley, H. M., and A. H. Hutchinson. 1941: Seed development in *Medicago* (alfalfa) hybrids. I. The normal ovule. Can. J. Res. Sect. C 19:421–37.

Farley, T. M. 10849.

6220. Farmer, E. H., and E. Sunderland. 1935: Unsaturated acids of natural oils. II. The high unsaturated acid of the kernels of *Parinarium laurinum*. J. Chem. Soc. 1935:759–61.

6221. —— and F. A. Van den Heuvel. 1936: Highly unsaturated compounds. VI. The triene acid from seeds of pomegranates. J. Chem. Soc. 1936:1809–11. Analyst 62:221. 1937.

Farmer, E. H. 3179.

6222. Farnham, R. B. 1931: Germination of seeds of annuals. New Jersey Agr. Expt. Sta. Ann. Rept. 52:296–301.

6223. —— 1940: Coal ashes for seed germination. Florists Exchange Hort. Trade World 94(1):12.

6224. Farooq, M. 1958: The development of the embryo in *Utricularia stellaris* Linn. f. var. *inflexa*. Sci. Cult. (Calcutta) 23:479–80. Biol. Abstr. 33:2847. 1959.

6225. —— 1960: The embryology of *Galium asperifolium* Wall. J. Indian Botan. Soc. 39:171–75.

6226. —— 1960: The embryology of some species of *Utricularia*. Sci. Cult. (Calcutta) 25:428–30.

6227. Farooq, M. O., M. S. Ahmad, and M. A. Malik. 1954: Chemical investigation of seed oil of *Sesbania aegyptica*. J. Sci. Food Agr. 5:498–500.

6228. ——, M. A. Aziz, and M. S. Ahmad. 1956: Seed oils from *Cassia fistula*, *C. occidentalis* and *C. Tora* (Indian varieties). J. Am. Oil Chemists Soc. 33:21–23.

6229. ——, S. R. Gupta, M. Kiamuddin, W. Rahman, and T. R. Seshadri. 1953: Chemical examination of celery seeds. J. Sci. Ind. Res. (India) 12B:400–7. Chem. Abstr. 48:3483. 1954.

6230. ——, S. M. Osman, and M. S. Ahmad. 1953: The fixed oil from the seeds of *Carum copticum* Benth. (*Ptychotis ajowan*). J. Sci. Food Agr. 4:132–34.

6231. —— and M. S. Siddiqui. 1954: Chemical investigation of the seed-oil of *Leucaena glauca*, Benth. J. Am. Oil Chemists Soc. 31:8–9.

6232. ——, I. P. Varshney, and Z. Naim. 1962: Saponins and sapogenins. X. The occurrence of acacic acid in the seeds of *Acacia concinna* DC. and *Acacia intsia* Willd. Arch. Pharm. 295:12–14.

6233. ——, ——, and W. Rahman. 1958: On the presence of apiin in Indian celery seeds (*Apium graveolens*). Naturwissenschaften 45:265.

6234. ——, ——, M. S. Siddiqui, and H. Hasan. 1959: Étude chimique de l'huile de la graine d'*Albizzia procera* Benth. Bull. Soc. Chim. Biol. 41:901–6.

6235. Farooqi, M. I. H., and K. N. Kaul. 1964: Qualitative studies of sugars and amino-acids from the aril of *Litchi* fruit. Current Sci. (India) 33:183–84.

6236. Farooqui, H. M., and J. P. McCollum. 1954: Relation of morphological structure and development to seed coat rupture in beans (*Phaseolus vulgaris* L.). Proc. Am. Soc. Hort. Sci. 63:333–41.

6237. Farquharson, L. I. 1955: Apomixis and polyembryony in *Tripsacum dactyloides*. Am. J. Botany 42:737–43.

6238. Farr, B. K. 1945: *Achimenes* grown from seed. Horticulture 23(8):204.

Farrar, J. L. 6675.

6239. Farrar, K. R. 1962: The zeanins: a group of growth-promoting compounds from immature maize kernels. Biochem. J. 83:220–24.

6240. ——, J. A. Bentley, G. Britton, and S. Housley. 1958: Auxins in immature maize kernels. Nature 181:553–54.

6241. Farrar, L. L. 1958: Control of covered kernel smut (*Sphacelotheca sorghi*) of grain and black loos smut (*Ustilago avenae*) of oats with gibberellic acid in greenhouse studies. Plant Disease Reptr. 42:1254–61.

6242. —— and S. V. Stacy. 1958: Diseases of small grains observed in Georgia during the 1957–1958 season. Plant Disease Reptr. 42:1262–67.

6243. Farrar, M. D., and W. P. Flint. 1939: How to recognize some common insect enemies of stored grain. Illinois Agr. Expt. Sta. Circ. 497. 8 pp.

6244. —— and J. M. Wright. 1946: Insect damage and germination of seed treated with DDT. J. Econ. Entomol. 39:520–22.

6245. Farrar, R. F. 1956: An investigation into the seed content of triploid hops. Wye Coll. Dept. Hop Res. Ann. Rept. 1955:68–77.

Farrell, E. P. 17298.

6246. Farringer, D. E. 1946: Sesame. Agriculture Americas 6:160–63.

6247. Fassbender, P. 1925: Lichtkeimung und Sauresubstrat. Botan. Centralbl. Beih. 41:239–86.

6248. Fatidar, A. N. 1939: Naked ovule of

Galphimia. J. Indian Botan. Soc. 18:59–63. J. Roy. Microscop. Soc. 60:72.

6249. Fatterpekar, S. S., and J. W. Airan. 1960/61: Variations in the amino acid compositions during the germination of seeds of *Lens esculenta*. Bombay Univ. J. (N.S.) 29(3/5, Sect. B):6–10.

Fauconneau, G. 4266.

6250. Faulkner, R., and J. D. Matthews. 1961: The management of seed stands and seed orchards. Proc. Intern. Seed Testing Assoc. 26:366–87.

6251. Faure, R. 1954: Action sur la germination du grain de mais de certaines substances repulsives et adhesives. Phytiat.-Phytopharm. 3:153–58.

6252. Faust, M. E. 1936: Germination of *Populus grandidentata* and *P. tremuloides*, with particular reference to oxygen consumption. Botan. Gaz. 97:808–21.

6253. Fauth, A. 1903: Beiträge zur Anatomie und Biologi der Früchte und Samen einiger einheimischer Wasser- und Sumpfpflanzen. Dissertation, Jena. 47 pp.

6254. Favarger, C. 1953: Sur la germination des gentianes. Phyton 4:275–89.

6255. —— 1954: Sur une fonction curieuse de l'albumen pendant la germination. Ber. Schweiz. Botan. Ges. 64:84–93. Biol. Abstr. 30:23752. 1956.

6256. Favilli, R. 1939: Su alcune caratteristiche dei semi di *Trifoglio pratense* in relazione al colore del loro tegumento. Ann. Fac. Agrar. Univ. Pisa 2:598–607.

6257. —— 1950: The use of tetrazolium salts for the rapid determination of the germinative capacity of seeds. Ann. Fac. Agrar. Univ. Pisa 11:57–84. Chem. Abstr. 46:8196. 1952.

Favorskaya, T. N. 10393.

6258. Favre-Duchartre, M. 1943: Sur le comportement des ovules de *Ginkgo biloba*. Bull. Soc. Botan. France 90:111–16.

6259. Favretti, R. J. 1964: Germinating very fine seeds. Horticulture 42(3):52–55, 63.

6260. Fawcett, H. S. 1908: Viability of weed seeds under different conditions of treatment, and a study of their dormant periods. Proc. Iowa Acad. Sci. 15:25–45.

6261. —— 1929: Temperature experiments in germinating orange seeds. Calif. Citrograph 14:515.

Fawcett, R. G. 12405.

6262. Fay, C. L., and J. H. Miller. 1963: Influence of dalapon on maturity, yield, and seed and fiber properties of cotton. Weeds 11:31–36.

6263. Fayemi, A. A. 1957: Effect of temperature on the rate of seed swelling and germination of legume seeds. Agron. J. 49:75–76.

Fayolle, J. 5630.

6264. Fazecas, I. 1956: Recherches concernant l'établissement du régime de vernalization du ricin. Probl. Agr. (Bucharest) 8:13–25.

Fazi, R. de; *see* De Fazi, R.

Fazio, I. L. di; *see* Di Fazio, I. L.

Feaster, C. V. 18833.

6265. Feather, M. S., and R. L. Whistler. 1962: Isolation and characterization of the principal hemicullulose from corn germ. Arch. Biochem. Biophys. 98:111–15.

6266. Fedeli, A. 1936: Oligodynamic action of manganous and cuprous chlorides on the germination of the seed of *Medicago sativa*. Boll. Staz. Sper. Gelsicolt. Bachicolt. 15:17–25. Chim. Ind. 38:372. 1936. Chem. Abstr. 32:218. 1938.

Federer, W. T. 9106, 17813.

Federico, L. 1285.

6267. Federov, A. K. 1955: Terminating the vernalization phase by winter plants under natural conditions. Akad. Nauk SSSR Inst. Genet. Tr. 22:104–15.

6268. Federov, P. S. 1955: Relation of the productivity of seeds of vernalized wheat to their protein content. Zemledelie 1955(9): 103–4. Biol. Abstr. 32:10087. 1958.

6269. —— 1957: Protein content in wheat grains of plants having different biological and physiological properties. Tr. Kirgizsk. Nauchn.-Issled. Inst. Zemled. 1957(1):114–32. Chem. Abstr. 53:10397. 1959.

6270. —— 1958: Biochemical indexes of seeds of corn hybrids. Byul. Kirgizsk. Nauchn.-Issled. Inst. Zemledeliya 1958(2): 25–30. Chem. Abstr. 53:14244. 1959.

6271. Fedorowa-Ssarkissowa, O. W. 1931: Über Apogamie bei der Weide. Staatsinst. Wiss. Forsch. Mitt. 1931(10):57–64.

6272. Fedortschuk, W. 1931: Embryologische Untersuchung von *Cuscuta monogyna* Vahl und *Cuscuta epithymum* L. Planta 14:94–111. J. Roy. Microscop. Soc. 52:72. 1932.

6273. Fedoseev, I. V. 1958: Effect of sun treatment on germination of cherry seeds. Sad i Ogorod 96(11):48–51.

6274. Fedoseeva, A. I. 1936: The anatomy of the seed coats of some *Brassica* hybrids. Tr. Voronezhsk. Gos. Univ. 9:65–82.

6275. Fedotov, V. S. 1932: Preliminary results of the searches for alkaloidless lupins. *In* N. N. Ivanov, editor. Problem of the alkaloidless lupin. Bull. Appl. Botany, Genet., Plant Breed. Suppl. No. 54:29–42.

6276. Fedotova, T. I. 1948: Significance of individual seed proteins in manifesting resistance of plants to diseases. Tr. Leningr. Inst. Zashch. Rast. Sb. 1:62–71.

Feeley, E. J. 2306.

Feeley, J. 12848, 12849.

6277. Feenstra, W. J. 1959: Chemical aspects of three seedcoat colour genes of *Phaseolus vulgaris* L. Nederland. Akad. van Wetensch. Proc. Ser. C. Biol. and Med. Sci. 62(2):119–30.

6278. —— 1960: Biochemical aspects of seedcoat colour inheritance in *Phaseolus vulgaris* L. H. Veenman & Zonen N. V., Wageningen. 53 pp.

6279. Feher, D., and S. Vagi. 1925: Untersuchungen über die Einwirkung von Na_2CO_3 auf Keimung und Wachstum der Pflanzen. I. Biochem. Z. 158:357–65.

6280. Feinbrun, N., and S. Klein. 1962: ^{3}H-Thymidine incorporation into cell nuclei in germinating lettuce seeds. Plant Cell Physiol. (Tokyo) 3:407–13.

Feinbrun, N. 6114.

Feinhandler, S. 148.

6281. Feinstein, L., and J. R. Hart. 1959: A simple method for determining the protein content of wheat and flour samples. Cereal Chem. 36:191–93.

Feinstein, L. 8274.

6282. Feitsarenko, A. I. 1937: Effect of NaOH on the germination of sugar beet and the speeding up of development in sugar beet plants. Nauchn. Zap. Sakharn. Prom. Kiev 14:54–58.

6283. Fejer, S. O. 1960: Response of some New Zealand pasture species vernalisation. New Zealand J. Agr. Res. 3:656–62.

6284. Fekete, L., and I. Korpaczy. 1955: Variations in the nutritive value of sunflower-seed protein. Elelmezesi Ipar 9:89–91. Chem. Abstr. 51:15025. 1957.

6285. Feldman, J. M. 1963: The inhibitory activity of extracts of pepper seed (*Capsicum annuum* L.) on the infectivity of some plant viruses. Ann. Epiphyties 14:377–85.

Feldman, R. 5188.

6286. Feldmeier, I., and H. VonGuttenberg. 1953: Über das Vorkommen von Wuchs und Hemmstoffen in Samen und Keimpflanzen von *Phaseolus vulgaris*. Planta 42:1–18.

Felice, D. de. 11981.

6287. Felja, M. 1951: Ratio of the weight of the product to the number of seeds. Biljna Proizvodnja 4:263–66.

6288. Felke, J. 1913: Über die Giftstoffe der Samen von *Jatropha curcas*. Landwirtsch-Versuchssta. 82:427–64.

6289. Fellers, C. R. 1919: Longevity of *P. radicola* on legume seeds. Soil Sci. 7:217–32. Botan. Abstr. 3:2934. 1920.

Fellows, H. 19448.

Fellows, H. C. 4281–4283.

Felt, C. 628.

6290. Fenner, L. M. 1931: Bacterial canker of tomato and its distribution with the seed from infected fruit. J. Econ. Entomol. 24:544–47.

6291. Fenton, F. A., and W. W. Waite. 1932: Detecting pink bollworms in cottonseeds by the X-ray. J. Agr. Res. 45:347–48.

6292. Fenton, F. C. 1941: Storage of grain sorghums. Agr. Eng. 22:185–88.

6293. ——, M. Decker, and G. H. Larson. 1961: Aeration of wheat in storage. Kansas Agr. Expt. Sta. Tech. Bull. 24 pp.

6294. —— and C. O. Swanson. 1930: Studies on the qualities of combined wheats as affected by type of bin, moisture and temperature conditions. I. Cereal Chem. 7:428–48.

Fenton, S. W. 12860.

6295. Feofanova, N. D. 1954: The dependence of intensity of respiration and viability of seeds on the conditions of their reproduction. Dokl. Akad. Nauk SSSR 96:853–56. Chem. Abstr. 48:11555. 1954.

6296. —— 1961: Effect of supersonic vibrations and radioactive phosphorus on the growth and development of some vegetables. Tr. Prikl. Botan. Genet. i Selek. 34:149–53.

Ferando, M. 9433.

6297. Ferenczy, L. 1956: Antibacterial substances in seeds. Nature 178:639–40.

6298. —— 1957: Examination of ether-extractable germination inhibitors of fleshy fruits separated by paper chromatography. Acta Biol. Acad. Sci. Hung. 8:31–37. Biol. Abstr. 33:42988. 1959.

6299. Fergason, V. L., and M. S. Zuber. 1962: Influence of environment on amylose content of maize endosperm. Crop Sci. 2:209–11.

Fergason, V. L. 20131, 20132.

Fergus, E. N. 17793, 18101, 18412.

Ferguson, M. H. 6540.

Ferguson, N. L. 5046, 5047.

6300. Ferguson, R. B., M. M. Furniss, and J. V. Basile. 1963: Insects destructive to bitterbrush flowers and seeds in southwestern Idaho. J. Econ. Entomol. 56:459–62.

6301. Ferguson, W. 1950: A comparison between pelleted and unpelleted vegetable seed. Agr. Inst. Rev. 5(3):17.

6302. Fernandez, M., and S. Grisolia. 1960: Phosphoglyceric acid mutase activity without added 2,3-diphosphoglycerate in preparations purified from rice germ. J. Biol. Chem. 235:2188–90.

Fernandez, M. d'C. de; *see* De Fernandez, M. d'C.

6303. Fernandez, O., C. Capdevila, A. Pizarroso, and M. Ortega. 1956: The supposed leucoanthocyanidin of certain seed coats. Farm. Nueva (Madrid) 21:61–66. Chem. Abstr. 50:10195. 1956.

6304. —— and M. D. Lara. 1954: Chemical phylogeny of lignin. VI. Elastic coverings of seeds. Rev. Real Acad. Cienc. Exact. Fis. Nat. Madrid 48:147–50. Chem. Abstr. 49:8394. 1954.

6305. —— and M. Santaolalla. 1953: Copper in elastic seed coats. Rev. Real Acad. Cienc. Exact. Fis. Nat. Madrid 47:553–55. Chem. Abstr. 48:8341. 1954.

6306. —— and —— 1960: Phylogenetic chemistry of lignin. X. Analysis of the seed covering of *Phoenix dactylifera*. Rev. Real Acad. Cienc. Exact. Fix. Nat. Madrid 54:201–7. Chem. Abstr. 54:25082. 1960.

Fernandez-Diez, M. J. 5082.

Fernando, L. H. 3805.

6307. Fernaroli, L. 1950: Production of corn seed in Italy. Congr. Intern. Ind. Agr. (Rappt.) 8:121–24.

Fernholz, D. L. 11799.

6308. Fernholz, E., and H. E. Stavely. 1940: Brassicasterol, the characteristic sterol of rapeseed oil. J. Am. Chem. Soc. 62:1875–77. Analyst 65:572. 1940.

Fernqvist, I. 14496.

6309. Ferrand, M. 1954: Effects de la désinfection des graines d'arachides sur la conservation de leur pouvoir germinatif et sur les rendements de la culture. Congr. Protect. Vég. Leurs Prod. Climats Chaud. (Compt. Rend.) 1954:314–16.

6310. Ferrao, J. E. M. 1960: The culture of the African oilpalm (*Elaeis guineensis* Jacq.). I. Seed germination. Gaz. Agr. Angola 4:408–11.

6311. Ferrara, A. 1943: Chemical-technological analyses of oil seeds of Italian East Africa. Agron. Coloniale 27:85–98. Chem Abstr. 40:1330–31. 1946.

Ferraz, E. C. 13601, 13602.

6312. Ferraz, F. G. P. 1959/60: Quantitative relation of amino acids to the evolutive state of alfalfa. Arquiv. Port. Bioquim. 4:43–47. Chem. Abstr. 56:721. 1962.

6313. Ferraz, M. d'B., and A. d'A. Vega. 1960: A better beverage and higher germinating capacity in coffee. Bol. Super. Serv. Café São Paulo 35(398):5–11; 35(399):6–18. Hort. Abstr. 31:414. 1961.

6314. Ferreira, C. 1949: Toxic effects of 2,4-D on germination and initial growth of beans, *Phaseolus vulgaris*. Anais Fac. Farm. Odontol. Univ. São Paulo 7:387–96. Chem. Abstr. 44:10990. 1950.

Ferreira, C. 19384.

Ferrel, R. E. 9109, 10494, 16113.

6315. Ferrer, L. G., and R. B. Espino. 1923: A study of the germination of *Abaca* seeds. Philippine Agriculturist 12:101–10.

Ferrer-Monge, J. A. 19374.

6316. Ferri, F. 1963: Seed-transmitted hemp diseases. Progr. Agr. (Bologna) 9:346–51.

6317. Ferri, S. 1961: The embryological development of *Erigeron speciosus* DC. Nuovo Giorn. Botan. Ital 68:269–84.

6318. Fertig, S. N., J. H. Huddleston, and A. H. Furrer. 1963: The effect of stage of maturity and ensiling on the viability of yellow rocket (*Barbarea vulgaris*) seed. Proc. Northeast. Weed Control Conf. 17:352–56.

6319. Ferweda, J. D. 1956: Germination of oil palm seeds. Trop. Agr. (London) 33:51–66.

Ferwerda, F. P. 612.

Festa, M. 7407.

Festa, N. I. 7408.

6320. Fetre, J., and A. Lebegue. 1962: Embryogénie des Crassulacées. Développement de l'embryon chez le *Kalanchoe verticillata* Jacq. Compt. Rend. 255:1462–64.

6321. Fetter, E. 1938: Quelques notes sur la jarovisation des graines de betterave. Ann. Gembloux 44:319–26.

Fetzer, G. 16890.

Feuge, R. O. 5096.

Fevold, H. L. 10844–10846, 12229–12231.

6322. Feytsarenko, A. I. 1937: Effect of NaOH on the germination of sugar beet and the accelerated development of beet plants. Nauchn. Zap 14:54–58.

6323. Fezer, K. D. 1959: Survival of loose smut in barley seed. Plant Disease Reptr. 43:1287.

6324. —— 1962: Differential incidence of loose smut among seed-size classes within barley seed lots. Soil Sci. 2:162–64.

6325. Fick, G. L., and R. P. Hibbard. 1925: A method for determining seed viability by electrical conductivity measurements. Mich. Acad. Sci. Arts Letters 5:95–103.

6326. Fidler, J. H., and D. P. Webley. 1956: Seed dressings for the control of fruit fly in oats. Nature 178:1480.

Fidman, R. S. 5580.

6327. Fiedler, J. 1954: Results of pre-planting preparation of sugar beet seeds. Socialist. Zeměděl. 4:228–32.

6328. Fiedler, U. 1955: New compounds isolated from *Crataegus oxyacantha*. Flavone derivatives. Arzneimittel-Forsch. 5: 609–10. Chem. Abstr. 50:2761. 1956.

6329. Fiedziusko, M. 1927: Action of disinfectants on seed germination. Roczniki Nauk Rolniczych Lesnych 18:279–99. Biol. Abstr. 6:21619. 1932.

6330. Fieger, E. A., and V. R. Williams. 1949/50: Storage experiments with brown rice. Louisiana Agr. Expt. Sta. Ann. Rept. 1949/50:11.

6331. Fielding, J. M. 1945: Cone holes in Monterey pine. Australia Commonwealth Forestry Bur. Leaflet 55. 7 pp.

Fielding, W. L. 12587.

6332. Fields, R. W., and T. H. King. 1962: Influence of storage fungi on deterioration of stored pea seed. Phytopathology 52:336–39.

6333. Fifield, C. C., and D. W. Robertson. 1952: Milling, baking, and chemical properties of Marquis and Kanred wheat grown in Colorado and stored 19 to 27 years. Agron. J. 44:555–59.

6334. ——, G. S. Smith, and J. F. Hayes. 1937: Quality in durum wheats and a method for testing small samples. Cereal Chem. 14:661–73.

Fifield, C. C. 9060, 18409.

6335. Figdor, W. 1907: Über den Einfluss des Lichtes auf die Keimung der Samen einiger Gesneriaceen. Ber. Deut. Botan. Ges. 25:582–85.

6336. —— 1912: Die Beeinflussung der Keimung von Gesneriaceensamen durch das Licht. Ber. Deut. Botan. Ges. 30:648–53.

6337. Fikry, M. A. 1936: The influence of size and weight of seed upon the course of subsequent growth and upon yield of wheat. Bull. Roy. Agr. Soc. Egypt 23:1–54. Biol. Abstr. 11:1806. 1937.

6338. Filan, F. 1961: Production of perennial ryegrass seed. Agr. Gaz. N.S. Wales 72:534, 535–36, 550.

6339. Filatov, F. I. 1949: Increasing the yields of forage plant and grass seeds. Sov. Agron. 7:19–28.

6340. Filatova, T. A. 1939: The diagnosis of vernalized seeds of sugar beet. Osnov. Vyvod. Rabot V. N. I. S. 1937:342–43. Chem. Abstr. 36:6584. 1942.

Filho, O. A. Gurgel-; *see* Gurgel-Filho, O. A.

6341. Filimonov, M. A. 1951: A simple and rapid method for determining the germination of legume seeds. Kormovaia Baza 2:34–38.

6342. —— 1952: Peculiarities of swelling of viable and nonviable seeds of leguminous plants. Dokl. Akad. Nauk SSSR 84:377–79.

6343. —— 1954: Characteristics of the germination of seeds of grasses and legumes in hydroponic culture. Byul. Mosk. Obshch. Ispytatelei Prirody Otd. Biol. 59(2):33–38.

6344. —— 1955: Effect of soil moisture on seed germination of forage crops. Byul. Mosk. Obshch. Ispytatelei Prirody Otd. Biol. 60(2):79–89.

6345. —— 1958: Swelling of the seeds and their vitality. Byul. Mosk. Obshch. Ispytatelei Prirody Otd. Biol. 63(5):41–51.

6346. —— 1959: The requirements to sowing qualities of seeds of wild perennial forage grasses. Tr. Botan. Inst. Akad. Nauk SSSR Ser. 6, Introduktsiya Rast. i Zelenoe Stroit 7:248–53.

6347. —— 1961: Determination of the preservation of germinating ability of stored red clover seeds. Selek. i Semen. 1961(1):61–63.

6348. —— 1961: Estimation of the ability of seed to germinate under field conditions. Kukuruza 3:33–34. Biol. Abstr. 38:7302. 1962.

6349. —— 1962: The germination of (corn) seeds treated with a synthetic drying oil. Kukuruza 5:33–34. Biol. Abstr. 40:12287. 1962.

6350. Filimonova, T. G. 1962: Use of gibberellin for determining the vital capacity of seeds. Selek. i Semen 1962(5): 60–61.

Filimonova, V. D. 15669.

Filin-Koldakov, B. V. 308.

6351. Filippenko, A. 1937: Formation of bios in the yarovized embryos of winter wheat. I. Compt. Rend. Acad. Sci. URSS 17:329–32. Chem. Abstr. 32:2983. 1938.

6352. Filippenko, I. A. 1939: Yarovization of seed of agricultural plants by immersion in water. Compt. Rend. Acad. Sci. URSS 23:580–82.

6353. —— 1940: Inhibition of developmental processes in vernalized plants that have suffered partial anaerobiosis. Compt. Rend. Acad. Sci. URSS 28:167–69.

6354. —— 1957: The effect of 2,4-D on the quality of wheat seeds. Plant Physiol. (USSR) (English Transl.) 4:444–49.

Filippovych, T. M. 9141.

Fillippovich, T. N. 7702, 7703.

6355. Fillmore, R. H. 1949: Growing rhododendrons from seeds. Arnoldia 9:45–51.

6356. Filmer, R. S., P. Marucci, and H. Moulter. 1958: Seed counts and size of cranberries. Am. Cranberry Growers Assoc. Proc. Ann. Meeting 88:22, 26–30.

6357. Filov, A. 1962: On the quality of seeds from unripe and overripe melons. Kartofel i Ovoshchi 7:40–41.

6358. Filter, P. 1911: Über einen Fall von Lichtkeimung bei Italienischem Raygras. Illustr. Landwirtsch. Ztg. 31:205.

6359. —— 1914: Über die Wasseraufnahme und Keimung der Samen unter verschiedenen namentlich erschwerenden Bedingungen der Wasserzufuhr. Disseration, Berlin. Phil. Fak. Friedrich Wilhelm Univ. 60 pp.

6360. —— 1932: Untersuchungen über die Lebensdauer von Handels- und anderen Saaten, mit besonderer Berücksichtung der harten Samen und der Altersverfärbung bei den Leguminosen. Landwirtsch. Versuchssta. 114:149–70. Fortschr. Landwirtsch. 8:134–35. 1933.

6361. —— 1933: Untersuchung von *Lolium perenne* L. deutscher Herkunft. Proc. Intern. Seed Testing Assoc. 5:163-65.

6362. —— and W. Laschke. 1909: Vergleichende Untersuchungen über den Einfluss von Temperaturen und Aufbewahrungsart auf die Keimfähigkeit lagernder Sämereien. Landwirtsch. Jahrb. 38:759–66.

6363. Filutowicz, A. 1961: The role of the size and shape of the seeds of monogerm diploid and polyploid sugar beet varieties on their seed value. Bull. Warsaw Inst. Hodowli Aklim. Roslin. 3/4:61–65.

6364. —— and W. Bejnar. 1954: The influence of the age of sugar beet seeds on their seeding value. Roczniki Nauk Rolniczych 69:323–40. Biol. Abstr. 30:14577. 1956.

6365. Finch, A. H. 1937: The use of ethylene to improve pecan harvesting. Proc. Am. Soc. Hort. Sci. 34:74–77.

6366. —— and C. W. Van Horn. 1936: The physiology and control of pecan nut filling and maturity. Ariz. Agr. Expt. Sta. Tech. Bull. 62:421–72.

6367. —— and —— 1940: Studies on the evaluation of factors influencing oil content and filling of pecan nuts. Proc. Am. Soc. Hort. Sci. 37:481–83.

6368. —— and —— 1940: Notes on the relation of warm winter temperatures to blossoming and nut setting of the pecan. Proc. Am. Soc. Hort. Sci. 37:493–97.

6369. ——, ——, and A. S. Ayres. 1937: The quality and maturity of pecan nuts as affected by date of picking and treatment with ethylene gas. Proc. Am. Soc. Hort. Sci. 35:238.

Fina, A. L. da; *see* Da Fina, A. L.

6370. Finch, L. R., and D. J. Carr. 1956: Nucleic acid content of Petkus rye embryos in relation to vernalization and devernalization. Australian J. Biol. Sci. 9:355–63.

6371. Fincke, A. 1952: The vitamin B_1 content of sweet and bitter almonds, hazelnuts, peanuts, and apricot pits. Zucker-Süsswarenwirtsch. 5:513–14. Chem Abstr. 49:1237. 1955.

6372. —— 1952: The vitamin B_1 content of finished and unfinished products from sweet and bitter almonds, apricot pits, hazelnuts and peanuts. Zucker-Süsswarenwirtsch. 5:553–55. Chem. Abstr. 49:1238. 1955.

6373. —— 1963: The occurrence of caffeine and theobromine in cocoa fats. Fette, Seifen, Anstrichmittel 65:647–50. Chem. Abstr. 60:3205. 1964.

6374. Findeis, M. 1917: Über das Wachstum des Embryos im ausgesäten Samen vor der Keimung. Sitzber. Math.-Naturw. Kgl. Akad. Wiss. Wien Abt. 1, 126:77–102. Botan. Centralbl. 42:242. 1918.

Findlay, W. I. 6841.

6375. Findley, W. M. 1919: The size of seed. N. Scotland Coll. Agr. Bull. 23:1–16. Expt. Sta. Rec. 41:536–37. 1919.

6376. Findley, W. R., Jr., and L. E. Campbell. 1953: Ultrasonic treatments of dormant hybrid corn seed. Agron. J. 45:357–58.

6377. Fine, J. M., and L. V. Barton. 1958: Biochemical studies of dormancy and after-ripening in seeds. I. Changes in free

amino acid content. Contrib. Boyce Thompson Inst. 19:483–500.
Fine, J. M. 2041.
Fine, L. O. 2936.
Fink, E. 10221.
6378. Fink, H., H. Haehn, and E. Zenger. 1936: Formation of respiration enzymes during germination of barley. Wochschr. Brau. 53:65–69, 73–77, 83–87, 93–95, 101–3. Chem. Abstr. 30:7587–88. 1936.
6379. —— and E. Schwieger. 1950: Über die Schnellmethoden zur Bestimmung der Keimfähigkeit, besonders nach dem Tetrazoliumverfahren. Brauwissenschaft 3:68–73.
6380. ——, K. Wöger, and H. Heider. 1952: Determination of germinating power by tetrazolium. IV. Brauwissenschaft 1952: 165–69. Brit. Abstr. BIII:141. 1953.
Fink, H. 12510.
Fink, H. von; *see* Von Fink, H.
6381. Fink, J. 1960: Lability of phosphoric compounds of corn seeds in acidy reagents. II. Hodowla Roslin Aklim. Nasiennictwo 4:33–68.
6382. Finkner, R. E., H. C. Murphy, R. E. Atkins, and D. W. West. 1954: Varietal reaction and inheritance of fluoroscence in oats. Agron. J. 46:270–74.
Finkner, R. E. 5540.
6383. Finlay, K. W. 1960: Genetic study of barley germination behaviour. I. Genetic and environmental variation. J. Inst. Brewing 66:51–57.
6384. —— 1960: Genetic study of barley germination behaviour. II. General and specific combining ability. J. Inst. Brewing 66:58–64.
6385. Finlayson, D. G. 1952: The effects of certain insecticides on the germination and growth of onions. II. Insecticides applied to the soil. Proc. Entomol. Soc. Brit. Columbia 48:70–76. Biol. Abstr. 27: 17954. 1953.
6386. Finlayson, R. A. 1928: The identification of the seeds of some species of *Brassica*. Proc. Intern. Seed Testing Assoc. 6:37–40.
Finn, R. F. 10628, 12244.
6387. Finnerty, P. 1956: Note on the raising of birch (*Betula*) from seed in the nursery. Irish Forestry 13:17–18.
Finney, K. F. 11664.
6388. Finnis, J. M. 1950: Seed maturity in Douglas fir. Brit. Columbia Forest Serv. Res. Notes 18. 4 pp.
6389. —— 1959: The pattern of Douglas fir (*Pseudotsuga menziesii*) seed-dispersal from a helicopter. Brit. Columbia Forest Serv. Res. Rev. 1959:17–18.
Finnis, J. M. 5339.
Finn-Kelcey, P. 9627.
Finrock, D. C. 10494.
Finzenhagen, H. 17103.
Fior, G. 7179, 7180.
Firch, J. G. 15791.
Firea, S. Mihailescu-. 18382.
Fireman, M. 524, 3411.
Firestone, D. 9572.
6390. Firket, J., and Comhaire. 1929: Recherches experimentales sur la teneur en glutathion des pois au début de la germination. Bull. Acad. Roy. Med. Belg. 9:93–122. Biol. Abstr. 6:6515. 1932.
Firket, J. 8696.
6391. Firsova, M. K. 1936: Germination of different varieties of plants used in USSR. Bull. Appl. Botany, Genet., Plant Breed. Ser. A(17):67–88.
6392. —— 1937: Special characteristics of the germination of potato seed. Bull. Appl. Botany, Genet., Plant Breed. Ser. 4(2):56.
6393. —— 1937: A natural method of accelerating the germination of mature wheat seeds based on a study of its various phases. Bull. Appl. Botany, Genet., Plant Breed. Ser. 4(2):65–66.
6394. —— 1937: Effect of low temperature and freezing on the germinating capacity of seeds of the opium poppy (*Papaver somniferum* L.). Bull. Appl. Botany, Genet., Plant Breed. Ser. 4(2):124.
6395. —— 1937: Results of an investigation of the quality of seeds of forest trees and shrubs. Bull. Appl. Botany, Genet., Plant Breed. Ser. 4(2):189.
6396. —— 1949: In regard to germination in the field of immature barley seeds in Chita region. Selek. i Semen. 16(8):45–47.
6397. —— 1954: The effect of fertilizers on the growth and development of wheat seedlings. Agrobiologiya 1954(5):137–39. Soils Fertilizers 18:357. 1955.
6398. Fischbach, H. 1937: Bayern und die Getreidebeizung. Nachr. Schädlingsbekämpf. Leverkusen 12:160–69. Rev. Appl. Mycol. 17:20–21. 1938.
Fischbach, H. 2753.
6399. Fischbein, G., and B. Klugman. 1956: Mechanization of seed plant packaging. Southern Seedsman 19(5):17, 67.
6400. Fischer, A. 1907: Wasserstoff- und Hydroxylionen als Keimungsreize. Ber. Deut. Botan. Ges. 25:108–22.

6401. —— 1949: The application of soybean inhibitor in tissue cultivation. Science 109:611–12.

6402. Fischer, A. C. 1943: Growth of seeds promoted by adhesive envelopes. Sci. News Letter 43:234.

6403. Fischer, E. 1944: Über die Polyembryonie beim Hanf. Beitr. Pflanzen 27:372–91.

Fischer, E. B. 14362.

Fischer, E. E. 7314, 7315.

6404. Fischer, F. 1952: Saatversuche mit Eschenhainbuchen- und Feldahorn-samen. Allgem. Forstz. 7:89–90.

6405. Fischer, G. J., R. Santoro, and M. Arnarez. 1938: Germination experiments with wheat subjected to fungicidal treatments. Arch. Fitotec. Uruguay 3:69–85. Chem. Abstr. 34:8167. 1940.

6406. ——, G. V. Ubisch, and A. R. Da Silva. 1949: Germinación de semillas tratadas con desinfectantes y estimulantes. Arch. Fitotec. Uruguay 4:20–22.

Fischer, G. J. 4429.

6407. Fischer, G. W. 1944: The blind-seed disease of ryegrass (*Lolium* spp.) in Oregon. Phytopathology 34:934–35.

6408. —— and J. P. Meiners. 1952: Comparative value of several fungicides in the control of head and stripe smuts in certain forage grasses. Wash. Agr. Expt. Sta. Bull. 532. 18 pp.

6409. Fischer, H. 1926: Nachwirkung ungemügender Samenreife. Angew. Botan. 8:99–101.

6410. —— 1930: Eine einfache Vorrichtung für Samenkeimung. Angew. Botan. 12: 171.

Fischer, H. 17118.

Fischer, H. E. 15923.

6411. Fischer, W. 1930: Wie weit kommt heute Grassamenbau auf leichten Sandböden in Frage? (Knaulgras- Rotschwingel, Schaffschwingel- Samenbau.) Deut. Landwirtsch. Presse 57:247.

6412. —— 1932: Zur Vereinfachung der Berechnung der Saatgemische bei Grünlandanlagen. Pflanzenbau 9:136–47.

6413. Fischnich, O., A. Grahl, and M. Thielebein. 1962: Ausprägung der primdren Keimruhe innerhalf einer Wintergerstensorte. Naturwissenschaften 49:41.

6414. —— and H. Grimm. 1958: Aufhebung der Keimruhe von Kartoffelsamen durch Gibberellin. Vorläufige Mitteilung. Landbauforsch. Völkenrode Mittbl. Forsch. Landwirtsch. Braunschweig-Völkenrode 8:95–96.

6415. —— and F. Heilinger. 1958: Applications of ultrasonics in biology and agriculture. Ber. Oberhess. Ges. Natur-Heilk. Giessen Naturw. Abt. 29:94–99. Chem. Abstr. 54:695. 1960.

6416. —— and G. Lubbert. 1955: Fruiting of potatoes and promotion of the rapidity of germination of their seeds. Beitr. Biol. Pflanz. 31:179–206. Chem. Abstr. 49: 11784. 1955.

6417. ——, M. Thielebein, and A. Grahl. 1957: Brechung der Keimruhe bei Gerste durch Gibberellinsäure und Rindite. Naturwissenschaften 44:642.

6418. ——, ——, and —— 1961: Sekundäre Keimruhe bei Getreide. Proc. Intern. Seed Testing Assoc. 26:89–114.

6419. ——, ——, and —— 1962: Influencing grain seed quality by outer factors with special consideration on dormancy. Saatgut-Wirtsch. 14:12–14.

6420. ——, ——, and —— 1962: Influencing the value of grain seed by external factors with particular consideration of dormancy (Conclusion). Saatgut-Wirtsch. 14:39–42.

Fischnich, O. 7575.

6421. Fischer, A. 1907: Wasserstoff- und Hydroxylionen als Keimungsreize. Ber. Deut. Botan. Ges. 25:108–22. Botan. Gaz. 43:387. 1909.

6422. Fisher, C. 1929: How nature plants her flowers. The many ways that flowers and trees scatter their seed. How the wind aids some. How birds and animals carry others. The extraordinary diversity of these natural methods. Nat. History 29:65–74.

6423. Fisher, D. E. 1954: Internal infection of tomato seed by *Didymella lycopersici* Kleb. Nature 174:656–57.

Fisher, D. F. 2834, 2835.

Fisher, E. 401.

6424. Fisher, E. A., and C. R. Jones. 1939: A note on moisture interchange in mixed wheats, with observations on the rate of absorption of moisture by wheat. Cereal Chem. 16:573–83.

6425. Fisher, E. H., and K. C. Berger. 1951: Alfalfa seed production as influenced by insecticide and fertilizer application. J. Econ. Entomol. 44:113–14.

6426. Fisher, G. C. 1914: Seed development in the genus *Peperomia*. Bull. Torrey Botan. Club 41:137–56, 221–41.

6427. Fisher, G. M. 1935: Comparative germination of tree species on various kinds of surface-soil material in the western white pine type. Ecology 16:606–11.

6428. —— 1954: Buffalograss from seed. Green Thumb 11(8):34–35.
6429. Fisher, G. S. 1945: Determination of γ-tocopherol in vegetable oils. Ind. Eng. Chem. Anal. Ed. 17:224–27. Expt. Sta. Rec. 95:160–61. 1946.
6430. Fisher, H., P. Griminger, G. A. Leveille, and R. Shapiro. 1960: Quantitative aspects of lysine deficiency and amino acid imbalance. J. Nutr. 71:213–20.
Fisher, H. E. 1975, 1976.
6431. Fisher, J. E. 1957: Effect of gravity on flowering of soybeans. Science 125:396.
6432. Fisher, J. R. 1945: Analysis of F.A.Q. wheat for the season 1944–1945. Agr. Gaz. N.S. Wales 56:239–42. Biol. Abstr. 20:5561. 1946.
Fisher, M. E. 8138.
6433. Fisher, P. L. 1941: Germination reduction and radicle decay of conifers caused by certain fungi. J. Agr. Res. 62:87–95.
6434. Fisher, W. B., Jr., and W. L. Parks. 1958: Influence of soil temperature on urea hydrolysis and subsequent nitrification. Soil Sci. Soc. Am. Proc. 22:247–48.
Fisk, E. L. 7805.
6435. Fiske, J. G. 1935: Fruits and seeds of some common New Jersey trees. New Jersey Agr. Expt. Sta. Circ. 355. 52 pp. Expt. Sta. Rec. 74:38. 1936.
6436. —— 1939: Abnormalities in the germination of lima beans. Proc. Assoc. Offic. Seed Analysts 30:214–17.
6437. —— 1943: How long will they live? New Jersey Agr. 25(4):1–2.
Fiske, J. G. 177.
Fitch, J. B. 20043.
6438. Fitzgerald, P. H. 1959: Germination induced by excision of the endosperm of immature wheat grains. New Zealand J. Agr. Res. 2:735–40.
6439. Fitzgibbon, M. 1943: Seed disinfection. The determination of the adhesiveness of seed dressings to cereal seeds. Soc. Chem. Ind. J. 62:8–11.
6440. Fitzpatrick, Mrs. F. W. 1940: Raising hardy asters from seed. Flower Grower 27(10):462.
Fitzpatrick, H. F. 15841.
6441. Fivaz, A. E. 1931: Longevity and germination of seeds of *Ribes*, particularly *R. rotundifolium*, under laboratory and natural conditions. U.S. Dept. Agr. Tech. Bull. 261. 40 pp.
6442. Fjellstect, G. L. 1955: Shelling of timothy seed. Lantmannen 39:908.
Flach, W. R. 13500.
6443. Flaig, W., and E. Saalbach. 1955: Zur Kenntnis der Huminsäuren. XI. Die Beeinflussung der Keimung von Getreide durch Thymohydrochinons als Modellsubstanz von Vorstufen bzw. Abbauprodukten von Huminsäuren. Z. Pflanzenernähr. Düng. Bodenk. 71:215–24.
6444. —— and —— 1956: Humic acids. XII. The influence of thymohydroquinone on the respiration of cereals. Z. Pflanzenernähr. Düng. Bodenk. 72:1–7. Chem. Abstr. 50:7232. 1956.
6445. —— and —— 1956: Humic acids. XIII. The effect on initial development of cereal in Neubauer vessels of thymohydroquinone. Z. Pflanzenernähr. Düng. Bodenk. 72:7–15. Chem. Abstr. 50:7232. 1956.
6446. Flaksberger, C. A. 1932: Protein in the grain of the wheats of the world. Bull. Appl. Botany, Genet., Plant Breed. Ser. A(2):15–31. Fortschr. Landwirtsch. 8: 230. 1933.
6447. Flamini, B., and C. Antoniani. 1959: Research on the treatment of sugar beet seeds. Progr. Agr. (Bologna) 5:58–67.
Flandrin. 5189.
6448. Flanzy, M., and J. Reingpach. 1943: Grape seeds as a source of oil. Ann. Agron. 13:60–71. Chem. Abstr. 39:5101–2. 1945.
Fleckinger, J. 15827.
Fleet, D. S. van; *see* Van Fleet, D. S.
Fleet, W. van; *see* Van Fleet, W.
6449. Fleischer, E. 1929: Zur Biologie feilspanförmiger Samen. Botan. Arch. 26:86–132. Botan. Centralbl. 158:91–92. 1930.
6450. Fleischer, F. 1851: Beiträge zur Lehre von dem Keimen der Samen der Gewachse, insbesondere der Samen ökonomischer Pflanzen. Gebrüdern Mäntier, Stuttgart. 159 pp.
6451. Fleischmann, L. 1959: The action of refrigeration of some biochemical and enzymatic aspects of seeds in the course of germination. Ric. Sci. 29:287–94. Biol. Abstr. 33:38933. 1959.
6452. —— 1959: Action of low temperature on some biochemical and enzymic aspects of seeds during germination. II. Ric. Sci. 29:2151–57. Chem. Abstr. 54:13272. 1960.
6453. Fleischmann, R. 1930: Keimversuche mit Getreidekörnern aus Entwicklungsstadien vor der Ernte. Pflanzenbau, Pflanzenschutz, Pflanzenzucht 7:18–20.

6454. Fleisher, Z., E. C. Blodgett, and M. D. Aichele. 1964: Presence of virus (necrotic ring spot group) in Mazzard and Mahaleb cherry seedlings grown in the Pacific Northwest from various seed sources. Plant Disease Reptr. 48:280–81.

Fleming, B. R. 5624.

6455. Fleming, J. R., and J. A. Johnson. 1964: Wheat gibberellins. Science 144: 1021–22.

6456. Fleming, W. E., F. E. Baker, and L. Koblitsky. 1937: Effect of applying acid lead arsenate for control of Japanese beetle larvae on the germination and development of evergreen seedlings. J. Forestry 35:679–88.

6457. Flemion, F. 1931: After-ripening, germination, and vitality of seeds of *Sorbus aucuparia* L. Contrib. Boyce Thompson Inst. 3:413–39.

6458. —— 1933: Physiological and chemical studies of after-ripening of *Rhodotypos kerrioides* seeds. Contrib. Boyce Thompson Inst. 5:143–59.

6459. —— 1933: Dwarf seedlings from non-after-ripened embryos of *Rhodotypos kerrioides*. Contrib. Boyce Thompson Inst. 5:161–65.

6460. —— 1934: Physiological and chemical changes preceding and during the after-ripening of *Symphoricarpos racemosus* seeds. Contrib. Boyce Thompson Inst. 6:91–102.

6461. —— 1934: Dwarf seedlings from non-after-ripened embryos of peach, apple, and hawthorn. Contrib. Boyce Thompson Inst. 6:205–9.

6462. —— 1936: A rapid method for determining the germinative power of peach seeds. Contrib. Boyce Thompson Inst. 8:289–93.

6463. —— 1937: After-ripening at 5° C. favors germination of grape seeds. Contrib. Boyce Thompson Inst. 9:7–15.

6464. —— 1938: A rapid method for determining the viability of dormant seeds. Contrib. Boyce Thompson Inst. 9:339–51.

6465. —— 1938: Breaking the dormancy of seeds of *Crataegus* species. Contrib. Boyce Thompson Inst. 9:409–23.

6466. —— 1941: Further studies on the rapid determination of the germinative capacity of seeds. Contrib. Boyce Thompson Inst. 11:455–64.

6467. —— 1941: Rapid method for testing white pine germination. Fox Forest Notes No. 30. 1 mimeo p.

6468. —— 1942: Effect of the addition of nitrogen upon germination of seeds of *Symphoricarpos racemosus*. Contrib. Boyce Thompson Inst. 12:485–89.

6469. —— 1948: Reliability of the excised embryo method as a rapid test for determining the germinative capacity of dormant seeds. Contrib. Boyce Thompson Inst. 15:229–41.

6470. —— 1957: Physiological and chemical changes occurring prior to, during and subsequent to germination of some rogaceous seeds. 14th Intern. Hort. Congr. The Hague-Scheveningen 1955 Rept., pp. 1197–200.

6471. —— 1959: Effect of temperature, light and gibberellic acid on stem elongation and leaf development in physiologically dwarfed seedlings of peach and *Rhodotypos*. Contrib. Boyce Thompson Inst. 20:57–70.

6472. —— 1962: Insect damage as a factor affecting fruit set. Proc. Campbell Soup Co. Plant Sci. Symp., pp. 163–71.

6473. —— and J. Beardow. 1963: Histological studies of physiologically dwarfed peach seedlings. I. Structure of anomalous leaves. Contrib. Boyce Thompson Inst. 22:117–31.

6474. —— and D. S. De Silva. 1960: Bioassay and biochemical studies of extracts of peach seeds in various stages of dormancy. Contrib. Boyce Thompson Inst. 20:365–79.

6475. —— and E. T. Henrickson. 1949: Further studies on the occurrence of embryoless seeds and immature embryos in the Umbelliferae. Contrib. Boyce Thompson Inst. 15:291–97.

6476. —— and B. T. MacNear. 1951: Reduction of vegetative growth and seed yield in umbelliferous plants by *Lygus oblineatus*. Contrib. Boyce Thompson Inst. 16:279–83.

6477. —— and J. Olson. 1950: *Lygus* bugs in relation to seed production and occurrence of embryoless seeds in various umbelliferous species. Contrib. Boyce Thompson Inst. 16:39–46.

6478. —— and E. Parker. 1942: Germination studies of seeds of *Symphoricarpos orbiculatus*. Contrib. Boyce Thompson Inst. 12:301–7.

6479. —— and H. Poole. 1948: Seed viability tests with 2,3,5-triphenyltetrazolium chloride. Contrib. Boyce Thompson Inst. 15:243–58.

6480. ——, ——, and J. Olson. 1949: Relation of *Lygus* bugs to embryoless seeds in dill. Contrib. Boyce Thompson Inst. 15:299–310.

6481. —— and P. L. Prober. 1960: Production of peach seedlings from unchilled seeds. I. Effect of nutrients in the absence of cotyledonary tissue. Contrib. Boyce Thompson Inst. 20:409–19.

6482. —— and C. Topping. 1961: Cytochemical studies of the shoot apices of normal and physiologically dwarfed peach seedlings. I. Cell wall components. Contrib. Boyce Thompson Inst. 21:233–45.

6483. —— and —— 1963: Cytochemical studies of the shoot apices of normal and physiologically dwarfed peach seedlings. II. Starch distribution. Contrib. Boyce Thompson Inst. 22:17–22.

6484. —— and G. Uhlmann. 1946: Further studies of embryoless seeds in the Umbelliferae. Contrib. Boyce Thompson Inst. 14:283–93.

6485. —— and E. Waterbury. 1941: Embryoless dill seeds. Contrib. Boyce Thompson Inst. 12:157–61.

6486. —— and —— 1945: Further studies with dwarf seedlings of non-after-ripened peach seeds. Contrib. Boyce Thompson Inst. 13:415–22.

Flemion, F. 1786.

6487. Flensberg, R. 1949: Untersuchungen über die Warmwasserbeize unter besonderer Berücksichtigung des Warmwasserdauerbades. Phytopathol. Z. 16:1–40.

6488. Flerov, A. F., and E. I. Kovalenko. 1947: Effect of growth substances and alkaloids on the development of cuttings and on the sprouting of seeds of the grape. Dokl. Akad. Nauk SSSR 58:677–79. Chem. Abstr. 46:4723. 1952.

Flesher, D. 3904.

Fletchall, O. H. 13764.

6489. Fletcher, H. H. 1960: Growing rose seeds. Am. Rose Mag. 15(21):10.

6490. Fletcher, J. 1899: The worst weeds of the northwest. Can. Expt. Farms Rept. 1899:184–94. Expt. Sta. Rec. 12:565. 1900–1.

6491. Fletcher, W. F. 1915: The native persimmon. U.S. Dept. Agr. Farmers Bull. 685:1–28.

6492. Fletcher, W. W., and D. J. Martin. 1962: The effect of gibberellic acid on the germination, growth and development of *Trifolium repens* L. J. Agr. Sci. 58:235–41.

6493. Fleurent, M. E. 1898: Recherches sur la composition immediate et élémentaire des matières albuminoides extraites du grain des céréales et des graines de légumineuses. Ann. Sci. Agron. 1:371–471.

6494. Fleuret, H. 1930: De l'influence de la germination sur la valeur nutritive de la graine. Compt. Rend. Soc. Biol. 104:304–6.

6495. Flick, O. R. 1943: Amaryllis from seed. Horticulture 21(12):253.

6496. Flint, L. H. 1935: Sensitivity of dormant lettuce seed to light and temperature. J. Wash. Acad. Sci. 25:95–96.

6497. —— 1942: Note on the germination of the seeds of spider lily, *Hymenocallis occidentalis*. Proc. Louisiana Acad. Sci. 7:20–23. Biol. Abstr. 18:11765. 1944.

6498. —— and E. D. McAlister. 1935: Wave lengths of radiation in the visible spectrum inhibiting the germination of light-sensitive lettuce seed. Smithsonian Inst. Publ. 3334. 11 pp.

6499. —— and C. F. Moreland. 1939: Response of lettuce seedlings to 7600A radiation. Am. J. Botany 26:231–33.

6500. —— and —— 1943: Note on photosynthetic activity in seeds of the spider lily. Am. J. Botany 30:315–17.

Flint, R. N. 10901, 15322.

Flint, W. P. 6243.

6501. Flintoff, R. J. 1940: The germination of the seeds of some abnormal forms of plants. I. *Plantago lanceolata* L., *Epilobium montanum* L. Northwest Nat. 15:135–38.

6502. Flor, H. H. 1936: Flax seed-treatment tests. Phytopathology 26:429–38.

6503. —— 1944: Relation of rust damage in seed flax to seed size, oil content, and iodine value of oil. Phytopathology 34:348–49.

6504. Florell, V. H. 1929: Effect of date of seeding on yield, lodging, maturity, and nitrogen content in cereal varietal experiments. J. Am. Soc. Agron. 21:725–31.

6505. Florence, A. 1951: Graines maraichères, leur longevité et leur germination. Rev. Hort. Algerie 55:117–19.

6506. Florencio, M. d'L. F. d'S. 1962: On the occurrence of *Piricularia oryzae* Cav. in certified rice seeds in Portugal. Proc. Intern. Seed Testing Assoc. 27:862–69.

6507. Flores, F. B. 1938: Viability of seeds of cotton as affected by moisture and age under different methods of storing. Philippine J. Agr. 9:347–56.

6508. Florescu, I. I. 1961: Some considerations on the water soaking of larch seed (*Larix decidua* Mill.) prior to sowing. Rev. Padurilor (Bucharest) 76:77–80.

6509. Flory, W. S., Jr. 1939: Parthenogenesis in Zephyrantheae. Herbertia 6:196–202. Texas Agr. Expt. Sta. Circ. 93:18. 1941.
6510. Floyd, E. H. 1961: Effectiveness of malathion dust as a protectant for farm-stored corn in Louisiana. J. Econ. Entomol. 54:900–4.
6511. —— and L. D. Newsom. 1956: Protection of stored corn with lindane-impregnated sawdust. J. Econ. Entomol. 49:753–57.
Floyd, V. M. 2925.
Fluiter, H. J. de; *see* De Fluiter, H. J.
6512. Flynn, L. M., M. S. Zuber, D. H. Leweke, R. B. Grainger, and A. G. Hogan. 1954: Relation between protein content of corn and concentration of amino acids and nicotinic acid. Cereal Chem. 31:217–28.
Flynn, L. M. 8717.
6513. Focke, W. O. 1889: Distribution of seeds by birds. Abhandl. Naturw. Ver. Bremen 10:140.
Foda, H. A. 19369, 19370.
6514. Fodor, A., and L. Frankenthal. 1932: Fortgesetzte Studien über die Atmungsvorgänge in Erbsen- und Getreidesamen. IX. Biochem. Z. 246:414–30.
6515. —— and N. Lichtenstein. 1935: The activity of the dehydrases in pea seeds. Fermentforsch. 14:413–21. Chem. Abstr. 29:8009. 1935.
6516. —— and A. Reifenberg. 1925: Studies on the nature of the process of germination. A method for the determination of proteins by means of adsorption applied to the decomposition of proteins in germinating pea seeds. Biochem. J. 19:188–93.
6517. Förster, R. 1957: Über den Einfluss von Gerbstoffen auf Keimung und Wachstum von höheren Pflanzen. Beitr. Biol. Pflanz. 33:279–311.
6518. Fogel, A. A. 1955: Improving the seeding characteristics of wheat seeds by high-frequency heating. Akad. Nauk Inst. Biol. Fiz. Tr. Nauchn. Sessii, pp. 180–200.
Fogel, A. N. 2714.
6519. Fogle, H. W. 1958: Effects of duration of after-ripening, gibberellin and other pretreatments on sweet cherry germination and seedling growth. Proc. Am. Soc. Hort. Sci. 72:129–33.
6520. —— and C. S. McCrory. 1960: Effect of cracking, after-ripening and gibberellin on germination of Lambert cherry seeds. Proc. Am. Soc. Hort. Sci. 76:134–38.
6521. Foley, T. A. 1950: Rate of fall of kauri (*Agathis australis* Salisb.) seeds in quiet air. Forest Res. Inst. Forest Res. Notes 1:32–40.
6522. Folkers, K., and F. Koniuszy. 1939: *Erythrina* alkaloids. III. Isolation and characterization of a new alkaloid, erythramine. J. Am. Chem. Soc. 61:1232–35. Biol. Abstr. 13:12020. 1939.
6523. —— and —— 1940: *Erythrina* alkaloids. VII. Isolation and characterization of the new alkaloids, erythraline and erythratine. J. Am. Chem. Soc. 62:436–41.
6524. Folkes, B. F., and E. W. Yemm. 1956: The amino acid content of the proteins of barley grains. Biochem. J. 62:4–11.
6525. —— and —— 1958: The respiration of barley plants. X. Respiration and the metabolism of amino-acids and proteins in germinating grains. New Phytologist 57:106–31.
6526. Follman, G., and M. Nakagava. 1963: Keimhemmung von Angiospermensamen durch Flechtenstoffe. Naturwissenschaften. 50:696–97.
6527. Folsom, S. 1952: Pertinent facts about seed certification. Seed World 71(7):12, 14.
6528. Fomin, E., and M. Chevelii. 1935: Methods of germination test of the seed to be treated with formalin. Ukr. Sci. Res. Inst. Grain Cult. Bull. 1935(1):46.
6529. —— and K. Nozdrachev. 1935: Dependence of wheat infestation with bunt on different factors, and differentiation of fungicides. Ukr. Sci. Res. Inst. Grain Cult. Bull. 1935(1):65.
6530. —— and —— 1935: The treatment of vernalized spring wheat seeds with mordants. Ukr. Sci. Res. Inst. Grain Cult. Bull. 1935(1):66–76. Chem. Abstr. 30: 2687. 1936.
6531. Fonshtein, L. M. 1961: Effect of irradiation of the endosperm of the seed on the growth and development of the plant in wheat. Tr. Inst. Genet. Akad. Nauk SSSR 28:153–57. Biol. Abstr. 42:23633. 1963.
Fontaine, D. de. 6590.
6532. Fontaine, T. D., and R. S. Burnett. 1944: Peptization of nitrogenous constituents of solvent-extracted peanut meal. Ind. Eng. Chem. 36:164–67. Expt. Sta. Rec. 94:583. 1946.
6533. ——, S. B. Detwiler, Jr., and G. W. Irving, Jr. 1945: Improvement in the color of peanut and cottonseed proteins. Ind.

Eng. Chem. 37:1232–36. Expt. Sta. Rec. 95:764. 1946.
6534. —— and G. W. Irving, Jr. 1945: Electrophoretic investigation of peanut proteins. II. Composition of several peanut protein fractions. Arch. Biochem. 8:239–49. Biol. Abstr. 21:25992. 1947.
6535. ——, W. A. Pons, Jr., and G. W. Irving, Jr. 1946: Protein-phytic acid relationship in peanuts and cottonseed. J. Biol. Chem. 164:487–507.
6536. ——, C. Samuels, and G. W. Irving, Jr. 1944: Properties of peanut meal: influence of processing factors. Ind. Eng. Chem. 36:625–27.
Fontaine, T. D. 3397, 9446, 14745–14748.
Fonzo, M. A. di; *see* Di Fonzo, M. A.
6537. Food and Agriculture Organization of the United Nations. 1961: Agricultural and horticultural seeds. Their production, control and distribution. Food Agr. Studies No. 55, Rome. 531 pp.
Foote, H. E. 9906.
Foote, N. 16195.
6538. Foote, P. A., and L. G. Gramling. 1940: A chemical investigation of the seeds of *Glottidium vesicarium* (Jacq.) Harper. J. Am. Pharm. Assoc. 29:311–12.
Foote, W. H. 3955.
Forbes, E. B. 19282.
6539. Forbes, I., Jr. 1960: A rapid enzyme smear technique for the detection and study of plural embryo sacs in mature ovaries in several *Paspalum* species. Agron. J. 52:300–1.
6540. —— and M. H. Ferguson. 1948: Effect of strain differences, seed treatment, and planting depth on seed germination of *Zoysia* spp. J. Am. Soc. Agron. 40: 725–32.
6541. Forbes, R. D., editor. 1956: Forestry handbook. Ronald Press, New York. 1,095 pp.
6542. Ford, E. 1955: Tree planting in arid regions. World Crops 7:353–56.
Ford, H. W. 267.
6543. Ford, J. H. 1964: Influence of time of flowering on seed development of flax. Crop Sci. 4:52–54.
Ford, J. H. 4323.
Ford, O. W. 17302.
6544. Fordham, A. [J.] 1960: Germination of double-dormant seeds. Plant Propagators Soc. Proc. 10:206–10.
6545. —— 1960: Propagation of woody plants by seed. Arnoldia 20:33–40.
6546. Forester, J. 1938: *Iris tectorum* from seed. Horticulture 16(9):204.
6547. Forlani, R. 1950: Occurrence of parthenocarpy in *Hordeum*. Genet. Agr. 2:193–98.
6548. Formanowicz, H. 1958: Biologie des Keimens der Samen des Adonisröchens (*Adonis vernalis* L.). Poznan. Panst. Inst. Nauk Leczn. Surowcow Roslin. Byul. 4:237–46.
6549. Formanowiczowa, H. 1962: Influence of some medicinal plant substances on plant seed germination. Wiadomosci Botan. 6:333–38.
6550. Fornerod, W. P. 1963: Testing of packages and packaging materials for sowing seeds on their behaviour to water vapour permeability. Proc. Intern. Seed Testing Assoc. 28:935–42.
6551. Forsberg, D. E., and K. F. Best. 1964. The emergence and plant development of wild buckwheat (*Polygonum convolvulus*). Can. J. Plant Sci. 44:100–3.
6552. Forsberg, J. L., and A. M. Binkley. 1947: The effect of seed treatments, commercial fertilizers, and minor elements on root rot, stand, and yield of pod peas. Phytopathology 37:650–56.
6553. Forshell, C. P. 1953: The development of cones and seeds in the case of self- and cross-pollination in *Pinus silvestris* L. Medd. Statens Skogsforskningsinst. 43(10): 1–42.
Forshell, C. P. 5829.
6554. Forshey, C. G. 1953: Blossom structure and its relation to fruit set in delicious and several other commercially important apple varieties. Proc. Am. Soc. Hort. Sci. 62:154–58.
6555. Forster, H. C. 1935: Response of English and Australian wheats to length of day and temperature. J. Australian Inst. Agr. Sci. 1:27–28. Plant Breed. Abstr. 5:280–81.
6556. Forster, I. W., G. L. Jeffery, and T. Palmer-Jones. 1962: Factors causing losses of clover seed. New Zealand J. Agr. Res. 5:326–30.
6557. Forster, R. 1957: Über den Einfluss von Gerbstoffen auf Keimung und Wachstum von höheren Pflanzen. Beitr. Biol. Pflanz. 33:279–311.
6558. Forsyth, D. D. 1952: Specification of quality seed. Assoc. Offic. Seed Analysts News Letter 26(1):30–32.
6559. —— 1961: A five-year survey of oat seed quality in Wisconsin. Trans. Wisconsin Acad. Sci. 50:163–74.

6560. —— 1962: Seed dividers can cause errors. Assoc. Offic. Seed Analysts News Letter 36(1):22.

6561. —— 1963: Observations on the cause of stubby roots in germination tests of sweetclover seed. Assoc. Offic. Seed Analysts News Letter 37(2):15.

6562. —— and O. A. Vogel. 1945: Effect of seedcoat injuries during threshing on emergence of flax seedlings. J. Am. Soc. Agron. 37:387–93.

6563. Forsyth, W. G. C. 1952: Cacao polyphenolic substances. I. Fractionation of the fresh bean. Biochem. J. 51:511–16.

6564. —— 1955: Cacao polyphenolic substances. III. Separation and estimation on paper chromatograms. Biochem. J. 60: 108–11.

6565. Fortak, G. 1931: Die Cytologie der Keimung einer Zingiberacee und einer Piperacee. Botan. Arch. 33:97–135.

6566. Fortini, S. 1950. Catalases in the seedcoat of wheat during maturation subject to different water regimes. Ric. Sci. 20:1278–87.

Fortini, S. 18630.

Fortmann, H. R. 3329.

6567. Forward, B. F. 1947: Chemical tests for seed viability. Assoc. Offic. Seed Analysts News Letter 21(2):25–26.

6568. —— 1958: Studies of germination in oats. Proc. Intern. Seed Testing Assoc. 23:5–37.

6569. Foschum, O. 1928: Über die Saugkraft verschiedener Hafer- und Maissorten. Fortschr. Landwirtsch. 3:1014–17.

6570. Fosler, G. M. 1958: An easy method for germinating flower seeds—plus seed-storage hints. Illinois Agr. Expt. Sta. Circ. 796. 11 pp.

6571. Fosse, R., A. Brunel, P. De Graeve, P. E. Thomas, and J. Sarazin. 1930: Destruction, dans la graine de *Soja hispida*, de l'un des ferments sans suppression de l'activité de deux autres. Compt. Rend. 191:1025–28.

6572. Foster, A. A. 1947: Acceleration and retardation of germination of some vegetable seeds resulting from treatment with copper fungicides. Phytopathology 37:390–98.

Foster, C. A. M. 3821.

Foster, E. O. 7631.

6573. Foster, R. E., and C. W. Van Horn. 1957: Lettuce seed production in Arizona. Ariz. Agr. Expt. Sta. Bull. 282. 22 pp.

Foster, R. E. 18810.

Foster, R. J. 6077.

Foster, W. I. 17292.

Foth, H. D. 10667.

Foti, L. 5147.

6574. Fotidar, A. N. 1939: An example of a naked ovule in *Galphimia gracilis*. J. Indian Botan. Soc. 18:59–63.

Fouarge, J. 15626.

Foucher, M. 9734.

6575. Foulds, F. E. 1929: A study of the comparative morphology of the seeds of *Agropyron*. Sci. Agr. (Ottawa) 10:200–19.

Fourbert, C. L. 6989.

6576. Fourbet, F. 1962: Production of forage grass seeds in Vendée. Potasse 36:85–88.

6577. Fourmont, R. 1938: Étude sur la désinfection des semences de betteraves par le formol. Ann. Epiphyties Phytogenetique 4:1–19. Biol. Abstr. 12:12726. 1938.

6578. Fowden, L., and M. Wolfe. 1957: The protein composition of some East African seeds. E. African Agr. J. 22:207–12. Biol. Abstr. 33:7512. 1959.

Fowden, L. 5657, 7598, 19800.

6579. Fowells, H. A. 1953: The effect of seed and stock sizes on survival and early growth of Ponderosa and Jeffrey pine. J. Forestry 51:504–7.

6580. —— and G. H. Schubert. 1956: Seed crops of forest trees in the pine region of California. U.S. Dept. Agr. Tech. Bull. 1150. 48 pp.

6581. Fowlds, M. 1939: Seed color studies in biennial white sweet clover, *Melilotus alba*. J. Am. Soc. Agron. 31:678–86.

6582. Fowler, D. 1939: Grow lady-slippers from seed. Flower Grower 26(7):327.

6583. Fowler, H. D. 1956: The characterization of phytin in peas. J. Sci. Food Agr. 7:381–85.

6584. Fox, D. J., D. A. Thurman, and D. Boulter. 1964: Studies on the proteins of seeds of the Leguminosae. I. Phytochemistry 3:417–19.

6585. Fox, H. M. 1931: Lilies from seed in the East. Horticulture 9(16):344.

6586. ——, M. J. Geoghegan, J. A. Silk, and L. A. Summers. 1963: Azobenzene derivatives as seed treatments. Ann. Appl. Botany 52:33–44.

Fox, H. M. 12288, 12289.

6587. Fox, L. E. 1934: The effects of high temperatures on seeds. Am. J. Botany 21:711.

6588. —— 1936: The effects on bean seeds of heating to high temperatures. [Abstract.] Am. J. Botany (Suppl.) 23:692.

6589. Fox, R. L., and W. A. Albrecht. 1957: Soil fertility and the quality of seeds. Missouri Agr. Expt. Sta. Res. Bull. 619. 23 pp.

6590. Fox, S. W., and D. De Fontaine. 1956: Sequential assays of four amino acids in grains of wheat, rye, and in their hybrid. Proc. Soc. Exptl. Biol. Med. 92:503–6. Chem. Abstr. 51:1382. 1957.

6591. ——, C. Warner, and T. L. Hurst. 1955: Quantitative determination of terminal methionine, leucine and lysine in raw and toasted soybean oil meal. J. Agr. Food Chem. 3:704–6.

6592. Foy, N. R. 1926: Agricultural seeds. Notes and figures on various aspects. New Zealand J. Agr. 33:393–97.

6593. —— 1927: The official seed testing station. New Zealand J. Agr. 34:186–94.

6594. —— 1927: Germination of hulled oats. New Zealand J. Agr. 35:412.

6595. —— 1931: Use of filtered ultra-violet light in diagnosis of the various types of rye-grass in New Zealand. New Zealand J. Agr. 43:389–400.

6596. —— 1932: Delayed germination in cereals. New Zealand J. Agr. 44:129.

6597. —— 1934: Deterioration problems in New Zealand chewings fescue. New Zealand J. Agr. 49:10–24.

6598. Fraenkel-Conrat, H., R. C. Bean, E. D. Ducay, and H. S. Olcott. 1952: Isolation and characterization of a trypsin inhibitor from lima beans. Arch. Biochem. Biophys. 37:393–407.

Fragina, A. I. 6028.

Frakes, M. G. 13284.

6599. Frampton, V. L., W. A. Pons, Jr., and T. Kerr. 1960: A comparison of chemical properties of seeds of *Gossypium* species. Econ. Botany 14:197–99.

Frampton, V. L. 4336, 10659, 11555, 12923, 15531.

France, J. A. de; *see* De France, J. A.

France, W. G. 17319.

6600. Franchet, M. A. 1889: *Primula* with anatropous seeds. J. Botan. (Morot) 3:49–52. J. Roy. Microscop. Soc. 1889: 663.

6601. Francis, C. M., and J. S. Gladstones. 1963: Studies on the use of mutagenic agents in plant breeding. I. The effect of seed moisture content on sensitivity to X-rays in *Lupinus angustifolius*. Australian J. Agr. Res. 14:12–19.

6602. Francis, T. 1938: Hollyhocks from seeds. Gardeners Chron. 103:428.

6603. —— 1939: Dahlias from seeds. Gardeners Chron. 105:179.

6604. Francis, W. D. 1927: The anatomy of the Australian bush nut (*Macadamia ternifolia*). Proc. Roy. Soc. Queensland 39:43–53. Biol. Abstr. 6:11203. 1932.

6605. Franck, O. 1931: Fertilization, ripening time and grain quality. Nord. Jordbrugsforskn. No. 6/7:282. Chem. Abstr. 27:2241–42. 1933.

6606. Franck, W. J. 1921: Examination of beet seed in the government agricultural experiment stations for seed-control. Cultura 33:155–68. Botan. Abstr. 10:2.

6607. —— 1925: Germination tests at low temperature, with particular reference to seeds which are not fully after-ripened. 4th Intern. Seed Testing Congr., Cambridge 1924:59–76.

6608. —— 1928: Bibliography of "Germination of Seed." Wageningen. 242 mimeo. pp.

6609. —— 1929: The germination test regarded from a biological point of view. Proc. Intern. Seed Testing Assoc. 9/10:1–33.

6610. —— 1929: Germination tests considered from a biological viewpoint. III. Landbouwk. Tijdschr. 41:512–26.

6611. —— 1929: Germination tests as viewed from a biological standpoint. VI. (Conclusion) Landbouwk. Tijdschr. 41: 613–25.

6612. —— 1929: Observations with regard to the bibliography of germination of seed. Actes Congr. Intern. Essais Semences 5:334–37. 1928.

6613. —— 1929: Recent literature 1927. Proc. Intern. Seed Testing Assoc. 2:71–116.

6614. —— 1930: Tests and cleaning of contract-seed in the laboratory. Ned. Dept. Binnenlandsche Versl. Landbouwk. Onderzoek. 35:99–142.

6615. —— 1930: The behavior of carrot-seed by germination in laboratory and in the field. Ned. Dept. Binnenlandsche Versl. Landbouwk. Onderzoek. 35:143–47.

6616. —— 1930: Recent literature. Proc. Intern. Seed Testing Assoc. 2:157–78.

6617. —— 1930: Which direction must be followed in judging the germination capacity of seeds. Proc. Intern. Seed Testing Assoc. 2:42–48.

6618. —— 1931: Considerations concerning the composition of a germination and a general bibliography, presented by the publications committee to the members of the International Seed Testing Congress. Proc. Intern. Seed Testing Assoc. 3:175–79.

6619. —— 1932: Modern way of judgment of the germinating power of seeds and its influence on the value of seed testing for practice. Proc. Intern. Seed Testing Assoc. 4:207–8.

6620. —— 1934: Moisture content of seed. Nieuwe Veldbode, pp. 1–24. Proc. Intern. Seed Testing Assoc. 7:64. 1935.

6621. —— 1949: Recent literature 1939–1948. Proc. Intern. Seed Testing Assoc. 15:32–168.

6622. —— and W. H. Bruijning. 1929: Recent literature. Proc. Intern. Seed Testing Assoc. 2:67–95.

6623. —— and —— 1930: Recent literature. Proc. Intern. Seed Testing Assoc. 2:120–41.

6624. —— and —— 1931: Recent literature. Proc. Intern. Seed Testing Assoc. 2:204–35.

6625. —— and —— 1932: Recent literature 1930–1931. Proc. Intern. Seed Testing Assoc. 4:104–42.

6626. —— and —— 1933: Recent literature 1931–1932. Proc. Intern. Seed Testing Assoc. 5:87–112.

6627. —— and —— 1933: Recent literature 1931–1932–1933. Proc. Intern. Seed Testing Assoc. 5:205–47.

6628. —— and —— 1934: Recent literature 1932–1933. Proc. Intern. Seed Testing Assoc. 6:102–34.

6629. —— and —— 1937: Recent literature 1935–1936. Proc. Intern. Seed Testing Assoc. 9:316–51.

6630. —— and —— 1938: Recent literature 1936–1938. Proc. Intern. Seed Testing Assoc. 10:674–722.

6631. —— and —— 1940: Recent literature [on seed testing], 1939. Proc. Intern. Seed Testing Assoc. 12:120–36. Chem. Abstr. 35:2930. 1941.

6632. —— and H. W. G. J. Evers. 1955: Recent literature 1950–1952. Proc. Intern. Seed Testing Assoc. 20:103–204.

6633. —— and —— 1956: Recent literature. Proc. Intern. Seed Testing Assoc. 21:142–261.

6634. —— and —— 1958: Recent literature. Proc. Intern. Seed Testing Assoc. 23:105–262.

6635. —— and —— 1961: Recent literature 1954–1959. Proc. Intern. Seed Testing Assoc. 26:185–352.

6636. —— and —— 1963: Recent literature 1957–1962. Proc. Intern. Seed Testing Assoc. 28:218–495.

6637. —— and G. Wieringa. 1926: Effect of artificial drying and low temperatures on germination. Seed World 20(6):8–9.

6638. —— and —— 1928: Artificial drying and low temperature as means employed in obtaining an increase in germination of some vegetable seeds. Proc. Assoc. Offic. Seed Analysts 19/20:24–27.

6639. Franco, C. M. 1943: The preservation of seeds. I. Respiration of cotton seeds in various relative humidities. Bragantia 3:137–49. Chem. Abstr. 38:4011. 1944.

6640. —— 1954: Multiplication of orchids by germination of seeds. (Conclusion) Flores Brasil. 1:5–8.

6641. —— and O. Bacchi. 1960: Influence of the relative air humidity on the viability of tobacco seed. Bragantia 19:105–7. Tobacco Abstr. 5:141–42. 1961.

Franco, C. M. 13322.

6642. François, E. 1938: Germination asymbiotique des graines de vanille. Rev. Hort. 110:130.

6643. François, L. 1927: La détermination de la provenance des semences. II. Nouvelles recherches de geographie botanique 1926. Ann. Sci. Agron. 44:45–70.

6644. —— 1928: La provenance des semences, dernières recherches de géographie botanique. Ann. Sci. Agron. 45:48–55.

6645. —— 1928: Les semences des plantes adventices dans les céréales. Ann. Sci. Agron. 45:543–67.

6646. —— 1929: Les semences des plantes adventices dans les céréales. Ann. Sci. Agron. 46:176–93.

6647. —— 1929: Provenance des semences. Études de géographie botanique. Ann. Sci. Agron. 46:719–32.

6648. —— 1937: Quelques procédés speciaux relatifs aux analyses de semences. Ann. Agron. 7:257–69.

6649. —— 1937: Semences et premières phases du développement des plantes se rencontrant a peu près partout dans les céréales. Ann. Épiphyties Phytogénétique 3:1–21.

6650. —— 1938: Semences et premières du développement des plantes se rencontrant a peu près partout dans les céréales. Ann. Épiphyties Phytogénétique 4:535–63.

6651. —— 1939: Semences et premières phases du développement des plantes se rencontrant a peu près partout dans les céréales. Ann. Épiphyties Phytogénétique 5:537–54.

6652. —— 1941: Semences et premières phases du développement des plantes caractéristiques de la provenance des céréales. Ann. Épiphyties Phytogénétique 7(Sp. No.):1–32.

6653. —— 1941: Les semences des plantes commensales des légumineuses fourragères et des graminées de prairies. Ann. Épiphyties Phytogénétique 7(Sp. No.):33–58.

François, L. E. 19980, 19981.

6654. François M. J. 1957: Les relations thermiques de la germination chez diverses espèces du genre *Gossypium* L. Acad. Roy. Sci. Coloniales Bull. Seances 3:1145–47.

6655. François, M. T. 1936: Sur l'analyse de graines de tung (*Aleurites fordii*) cultivées au Maroc (Recolte 1933). Agron. Coloniale 25:89–91.

6656. —— and L. Chais. 1960: Rutin in the seeds of *Brassica campestris*. Bull. Soc. Pharm. Nancy 1960(46):21–23. Chem. Abstr. 55:3744. 1961.

François, M. T. 15093.

Franke, G. 10764.

Franke, 7026, 7027, 13288.

6657. Frankel, E. M. 1955: Clear solutions of fermentable carbohydrates from grains. U.S. Patent 2,698,825. Chem. Abstr. 49:4229. 1955.

Frankenfeld, J. C. 4441–4443.

Frankenthal, L. 6514, 16096.

6658. Frankforter, G. B., and A. W. Martin. 1904: A chemical study of the seed of *Rhus glabra*. Am. J. Pharm. 76:151–58.

6659. Frankland, B. 1961: Effect of gibberellic acid, kinetin and other substances on seed dormancy. Nature 192:678–79.

6660. —— and P. F. Wareing. 1960: Effect of gibberellic acid on hypocotyl growth of lettuce seedlings. Nature 185:255–56.

6661. —— and —— 1962: Changes in endogenous gibberellins in relation to chilling of dormant seeds. Nature 194: 313–14.

6662. Franklin, D. F. 1948: Using hormone sprays to facilitate bolting and seed production of hard-headed lettuce varieties. Proc. Am. Soc. Hort. Sci. 51:453–56.

6663. —— 1953: Growing carrot seed in Idaho. Idaho Agr. Expt. Sta. Bull. 294. 35 pp.

6664. Franklin, M. C., and R. F. Powning. 1942: Studies on the chemical composition of pods and seeds of certain species of Medicago, with a note on the apparent digestibility of cluster clover (*Trifolium glomeratum* L.) seed. Australian Council Sci. Ind. Res. J. 15:191–200.

6665. Franklin, W. W. 1951: Insects affecting alfalfa seed production in Kansas. Kansas Agr. Expt. Sta. Tech. Bull. 70. 64 pp.

6666. Franks, G. M. 1951: Not dry enough! Seed moisture guardian stops ravisher of germination, profits. Southern Seedsman 14(5):25, 37.

Franks, G. N. 2280, 17279.

6667. Franzen, G. 1930: Pharmakologische untersuchungen über Semen Vincetoxici. Arch. Exptl. Pathol. Pharmakol. 148:211. Physiol. Abstr. 15:124. 1930.

6668. Franzke, C. 1956: Experiments on ripening sunflower seeds. Fette, Seifen, Anstrichmittel 58:340–42. Chem. Abstr. 51:10665. 1957.

Franzke, C. J. 9205.

Fraps, G. S. 9935, 12803.

6669. Fraser, C. R. 1944: The air jet method of dehulling seeds. Can. J. Res. Sect. F 22:157–62.

Fraser, C. R. 910.

6670. Fraser, J. G. C. 1941: The relative values of seed injured by rust, frost or drought. Sci. Agr. (Ottawa) 21:307–14.

6671. —— 1950: Survey of progress in the improvement of elite and foundation seed stocks. Can. Seed Growers Assoc. Ann. Rept. 1949/50:20–24.

6672. —— and F. Gfeller. 1936: Inheritance and use of phenol color reaction in hard red spring wheats. Sci. Agr. (Ottawa) 17:243–49. Chem. Abstr. 31:3100. 1937.

6673. —— and L. M. Pidgeon. 1933: Electrolysis of seeds of cereals. Sci. Agr. (Ottawa) 14:141–48.

6674. Fraser, J. W. 1959: The effect of sunlight on the germination and early growth of Jack pine and red pine. Can. Dept. Northern Affairs Natl. Resources Forest Res. Div. Tech. Note 71. 7 pp.

6675. —— and J. L. Farrar. 1955: Effect of watering, shading, seed-bed medium, and depth of sowing on red pine (*Pinus resinosa*) germination. Can. Dept. Northern Affairs Natl. Resources Forest Res. Div. Tech. Note 15. 3 pp.

Fraser, L. 3595.

6676. Fraser, M. T. 1916: Parallel tests of seeds by germination and by electrical response. Preliminary experiments. Ann. Botany (London) 30:181–89.

6677. Fraser, W. D. 1941: Little seeds build great empires. Seed World 49(6):7–8.
Fratoni, A. 14845.
Frazier, J. C. 17176, 17177.
6678. Frazier, W. A. 1939: Some factors influencing field germination and seedling vigor of Imperial 152 lettuce seeds. Proc. Am. Soc. Hort. Sci. 37:701.
Frazier, W. A. 19574.
6679. Frear, D. W. 1961: A half century of seed testing, 1910–1960. Minn. Agr. Expt. Sta. 16 pp.
Frecerro, J. R. Muxi-; *see* Muxi-Frecerro, J. R.
6680. Frechmann, W., and G. Brouwer. 1927: Atlas der Samenkunde. J. Neumann, Neudamm.
6681. Fred, E. B. 1916: Relation of green manures to the failure of certain seedlings. J. Agr. Res. 5:1161–76.
6682. —— 1918: The effect of certain organic substances on seed germination Soil Sci. 6:333–49.
Fred, E. B. 7921.
6683. Frederiksen, P. S. 1947: Effect of soil temperature on germination of flax. Lin, Dansk Svensk Hortidsskr. 1:105–8.
6684. —— 1949: Germination of fibre flax seed injured by threshing. Proc. Intern. Seed Testing Assoc. 14:144–69.
6685. —— 1950: Use of chemical germination indicators on flax seed. Lin, Dansk Svensk Hortidsskr. 4:27–30.
6686. —— 1955: Germination rest of flax seed. Lin, Dansk Svensk Hortidsskr. 9:85–87.
6687. —— 1957: Investigations concerning storage of fiber flax seeds. Lin, Dansk Svensk Hortidsskr. 11:1–4.
Frederiksen, R. A. 4726.
6688. Frédou, C. 1945: La préparation des graines de résineux. Un nouveau modèle de sécherie. Rev. Eaux Forêts 83:641–55. Biol. Abstr. 20:8971. 1946.
6689. Free, M. 1949: Nature's wondrous way with seeds. Home Garden 14:46–54.
6690. —— 1955: African violets grown from seed. Flower Grower 42(10):40–43.
6691. Free, R. A. 1951: Germination and emergence of eucalypt seed when sown at depth. Australian Forestry 15:115–20.
6692. Freedman, R. 1945: Heavier stand, higher yield attributed to freezing seed. Southern Seedsman 8(5):12.
Freeman, B. 4786, 4787.
6693. Freeman, E. M. 1902: The seed fungus of *Lolium temulentum* L., the Darnel. Proc. Roy. Soc. London 71:27–30.
6694. Freise, F. W. 1931: Little-known Brazilian oil seeds. Seifensieder-Ztg. 58: 131–32. Chem. Abstr. 25:2582. 1931.
6695. —— 1932: Über Beziehungen zwischen verschiedenen Eigenschaften von Obstbaumsamen und dem alter der Stammpflanzen. Gartenbauwissenschaft 7:196–201.
6696. —— 1938: Composition of the seeds of *Cajanus* sp. and *Canavalia* sp., Brazilian native foods. Z. Untersuch. Lebensm. 75:566–68. Analyst 63:605–6.
6697. —— 1938: Pumpkin kernels as anthelmintics. Pharm. Zentralhalle 79:97–99. Soc. Chem. Ind. 57:588B. 1938.
6698. Freitas, A. P. C. e. 1945: Cariopse do milho (*Zea mays* L.). Variação de peso e volume consoante a posição na espiga. Bol. Soc. Broteriana Ser. II 19:655–88. Biol. Abstr. 21:25764. 1947.
6699. —— 1946: E esterilidade do centeio e as suas relaçoes com o comprimento da espiga e com o volume e peso dos cariopses. Agron. Lusitana 8:165–203.
6699A. —— 1953: A alteração da textura do albumen no cariopse do trigo. Seu significado em melhoramento. Agron. Lusitana 15:67–93. Biol. Abstr. 28:24330. 1954.
6700. French, C. E., G. H. Berryman, J. T. Goorley, H. A. Harper, D. M. Harkness, and E. J. Thacker. 1944: The production of vitamins in germinated peas, soybeans, and other beans. J. Nutr. 28:63–70. Biol. Abstr. 18:18944. 1944.
French, C. S. 5912.
6701. French, D., R. W. Youngquist, and A. Lee. 1959: Isolation and crystallization of planteose from mint seeds. Arch. Biochem. Biophys. 85:471–73. Chem. Abstr. 54:11166. 1960.
6702. French, H. M. 1933: The Regal lily from seed. Flower Grower 20(3):109.
6703. French, M. H. 1938: The nutritive value of cowpea seeds. Ann. Rept. Dept. Vet. Sci. Animal Husbandry, Tanganyika Territ. 1937:83–85. Chem. Abstr. 33: 1828. 1939.
6704. —— 1944: Composition and nutritive value of pulp and seeds in the fruit of the Baobab. E. African Agr. J. 9:144–45.
6705. French, R. B. 1962: Analysis of pecan, peanut, and other oils by gas-liquid chromatography and ultra-violet spectophotometry. J. Am. Oil Chemists Soc. 39:176–78.

6706. French, R. C. 1959: Formation of embryo starch during germination as an indicator of viability and vigor in heat-damaged barley. Plant Physiol. 34:500–5.
6707. —— and C. H. Kingsolver. 1964: The effect of excessive heat during artificial drying of corn on reducing sugar content and diastatic activity. Cereal Chem. 41: 47–58.
6708. ——, J. A. Thompson, and C. H. Kingsolver. 1962: Indoxyl acetate as an indicator of cracked seed coats of white beans and other light colored legume seeds. Proc. Am. Soc. Hort. Sci. 80:377–86.
French, S. A. W. 19407.
6709. Frenyo, V. 1949: Agents which stimulate and inhibit seed germination. Agrartudomany 1:164–70. Chem. Abstr. 43:9169. 1949.
Frenyo, V. 7263, 7264.
6710. Fressanges, E. 1954: Germination et premier développement des plantules de cafeirs. Dir. Agr. Elevage Forets Sect. Tech. Agr. Trop. Bull. Sci. (France) 5:221–22.
6711. Freston, W. H. 1941: Germinating delphinum seeds. Horticulture 19(22):484.
6712. Frets, G. P. 1936: The heredity of the size and the form of the seeds of *Phaseolus vulgaris*. The segregation of the F_2-generation. Proc. Akad. Wetenschap. Amsterdam 39:897–901.
6713. —— 1938: Dimension and form with the growth of the seeds of *Phaseolus vulgaris*. Proc. Akad. Weterschap. Amsterdam 41:324–32.
6714. —— 1938: Dimension and form with the growth of the seeds of *Phaseolus vulgaris*. II. Proc. Akad. Wetenschap. Amsterdam 41:431–39.
6715. —— 1939: The growth curves of the dimensions and of the weight of the seeds of *Phaseolus vulgaris*. Proc. Akad. Wetenschap. Amsterdam 42:215–23.
6716. —— 1939: A second group of observations regarding dimension and form with the growth of the seeds of *Phaseolus*. Proc. Akad. Wetenschap. Amsterdam 42:224–37.
6717. —— 1947: Inheritance of dimensions, shape and weight of the seeds of *Phaseolus vulgaris*. Genetica 24:18–56. Biol. Abstr. 22:10590. 1948.
6718. —— 1954: The heredity of the dimensions, the weight and the indices (size and form) of the seeds of *Phaseolus vulgaris*. Nijhoff, The Hague. 80 pp.
6719. —— and G. Wanrooy. 1937: Die Erblichkeit der Bohnenform und des Bohnengewichtes bei *Phaseolus vulgaris*. IV. Die F_3 Samengeneration (die F_2 Pflanzen). Genetica 19:156–87. Biol. Abstr. 11:17854. 1937.
Freudl, A. 19137.
Freundlich, M. 12929.
6720. Frey, E. 1953: Reinheits- und Keimfähigkeitswerte einiger wichtiger Klee- und Grassämereien. Landwirtsch. Jahrb. Schweiz 2:589–96.
6721. —— 1960: Der Wassergehalt und die Keimfähigkeit von Weizen unter verschiedenen Lagerungsbedingungen und an verschiedenen Stellen innerhalb der Säcke. Proc. Intern. Seed Testing Assoc. 25:555–66.
6722. —— 1962: On some old problems of seed research. Schweiz. Landwirtsch. Mitt. 10:225–33.
Frey, E. 19907.
6723. Frey, G. 1954: Activity and localization of acid phosphatase in the vegetative parts of some angiosperms and in some seeds. Ber. Schweiz. Botan. Ges. 64:390–452. Chem. Abstr. 49:16092. 1955.
6724. Frey, K. G., R. L. Grindeland, and H. C. Murphy. 1961: A flotation method for separating dehulled oats from grain samples. Crop Sci. 1:464–65.
6725. Frey, K. J. 1948: Inheritance of protein, zein, tryptophane, valine, leucine, and isoleucine in two maize hybrids. Iowa State Coll. J. Sci. 23:33–36.
6726. —— 1949: The inheritance of protein and certain of its components in maize. Agron. J. 41:113–17.
6727. —— 1953: Effects of variety and location on thiamine, pantothenic acid, riboflavin, and niacin contents in oats. Mich. Agr. Expt. Sta. Quart. Bull. 36:13–17.
6728. —— 1959: Yield components in oats. IV. Effect of delayed application of nitrogen. Proc. Iowa Acad. Sci. 66:137–42.
6729. —— 1962: Influence of leaf-blade removal on seed weight of oats. Iowa State Coll. J. Sci. 37:17–22.
6730. —— 1963: Inheritance of seed weight and its relation to grain yield of oats. 74th Ann. Meeting Iowa Acad. Sci. 1962. Proc. Iowa Acad. Sci. 69:165–69. Biol. Abstr. 44:16585. 1963.
6731. ——, E. Ruan, and S. C. Wiggans. 1958: Dry weights and germination of developing oat seeds. Agron. J. 50:248–51.

6732. ——, M. C. Shekleton, H. H. Hall, and E. J. Benne. 1954: Inheritance of niacin, riboflavin and protein in two oat crosses. Agron. J. 46:137–39.

6733. —— and S. C. Wiggans. 1956: Growth rates of oats from different test weight seed lots. Agron. J. 48:521–23.

Frey, K. J. 69, 70, 5055, 7426, 13965, 16496, 19671.

Freyberg, W. von; *see* Von Freyberg, W.

6734. Freyer, E. 1934: Additional data on the relation of moisture content to the increase of free fatty acid content of cottonseed in storage. Oil Soap 11:162–64. Chem. Abstr. 28:6006–7. 1934.

6735. —— 1935: A modified procedure for determining amount of lint on cottonseed. Oil Soap 12:208–10. Chem. Abstr. 29: 7105. 1935.

6736. —— 1938: Moisture in oilseeds—rapid determination of, and their products. Oil Soap 15:236–39. Pharm. Abstr. 5:480. 1939.

6737. Freytag, G. F., R. O. Cravioto, J. Guzman, G., and G. Massieu H. 1956: Estudio sobre las propiedades nutritivas del frijol. I. El contenido total de proteina de los tipos de frijol. Mex. Sec. Agr. Ganader. Estud. Esp. Foll. Tec. 19:1–31. Biol. Abstr. 33:23518. 1959.

6738. Freytag, K. 1958: Quellende Schleimzellen der Samenepidermis von *Sinapis alba* im polarisierten Licht. Protoplasma 50: 166–72.

Frey-Wyssling, A. 3472.

6739. Frezal, P., and B. Gerbinot. 1951: Action de la monochlorhydrine du glycol et du bromure de méthyle utilisés sous pression atmospherique et sous vide partiel sur la periode de vie latente et sur la germination des semences de pommes de terre. Compt. Rend. Acad. Agr. France 37:556–61. Biol. Abstr. 27:2335. 1953.

6740. Fridriksson, S., and J. L. Bolton. 1963: Development of the embryo of *Medicago sativa* L. after normal fertilization and after pollination by other species of *Medicago*. Can. J. Botany 41:23–33.

6741. —— and —— 1963: Preliminary report on the culture of alfalfa embryos. Can. J. Botany 41:439–40.

6742. Friedberg, L. 1957: Détermination et conservation de la faculté germinative des graines. J. Agr. Trop. Botan. Appl. 4:303–14. Biol. Abstr. 33:7483. 1959.

6743. Friederich, J. C. 1955: Results of research in disinfection of flax seeds for planting. Ghent Landbhogesch. Mededel. 20:399–410.

6744. Friedrich, H. 1942: The significance of the dissimilar results of growth-substance treatment of seeds. Nachrbl. Deut. Pflanzenschutzdienst (Berlin) 22:69–70. Chem. Abstr. 38:3686. 1944.

6745. —— 1958: Untersuchungen über die phenolischen Inhaltsstoffe von *Pyrus communis* L. IV. Verhalten des Arbutins beim Keimen von Birnensamen und Verteilung in jungen Pflanzen. Pharmazie 13:153–55. Biol. Abstr. 33:3408. 1959.

6746. Friedrichs, G. 1933: The determination of the adhesiveness of dusts to treated cereal seed grain in the supervision of cooperative disinfection plants. Nachrbl. Deut. Pflanzenschutzdienst (Berlin) 13:25–27. Chem. Abstr. 28:248. 1934.

6747. Friend, D. J. C., and F. G. Gregory. 1953: Acceleration of flowering in partially vernalized grain of Petkus winter rye by subsequent treatment at high temperature. Nature 172:667–68.

6748. —— and O. N. Purvis. 1963: Studies in vernalisation of cereals. XIV. The thermal reactions in vernalisation. Ann. Botany (London) 27:553–79.

6749. Fries, N. 1954: Chemical factors controlling the growth of the decotylized pea seedling. Symbolae Botan. Upsalienses 13:1–83. Biol. Abstr. 29:29674. 1955.

6750. —— 1955: Further experiments on the response of decotylised pea seedlings to arginine. Physiol. Plantarum 8:164–73.

6751. —— 1957: Some experiments aiming at an isolation of biochemical mutations in *Pisum sativum* L. J. Madras Univ. 27B:9–16.

6752. Friese, H., R. Benze, H. Pommer, and R. Wiebeck. 1942: The components of rye germ. Ber. 75B:1996–2003. Chem. Abstr. 38:1544. 1944.

6753. Friese, W. 1951: The yew tree (*Taxus baccata*). Pharm. Zentralhalle 90:259–62, 289–91. Chem. Abstr. 46:2634. 1952.

6754. Friesen, G. 1929: Neue Untersuchungen über Samenvorbehandlung und ihre folgen für die Keimpflanzen. Ber. Deut. Botan. Ges. 47:(69)–(77).

6755. —— 1929: Untersuchungen über die Wirkung der Phenylurethans auf Samenkeimung und Entwicklung. Planta 8:666–79.

6756. —— and L. H. Shebeski. 1961: The

influence of temperature on the germination of wild oat seeds. Weeds 9:634–38.
Friesen, G. 3573.
Friesen, H. A. 7259, 10900.
6757. Freitinger, G. 1927: Untersuchungen über die Kohlensäureabgabe und Sauerstoffaufnahme bei keimenden Samen. Flora (Jena) 122:167–201.
Frimmel, F. von; *see* Von Frimmel, F.
6758. Frimmel, G. 1961: Importance of germination inhibition of substances in soils. Bodenkultur 12:116–22.
6759. Frisak, A. 1923: Correlation between the appearance of seeds and its germinating and yielding capacity. Norges Landbr. Hoiskole Meldinger 4:189–236. Expt. Sta. Rec. 50:636. 1924.
6760. —— 1929: Comparative analyses of sweet clover by American and European methods. Proc. Assoc. Offic. Seed Analysts 1929:43–47.
Frisch, A. 10307.
6761. Frischenschlager, B. 1937: Versuche über die Keimstimmung an einigen Gemüsearten. Gartenbauwissenschaften 11:159–66.
6762. Frischknecht, N. C. 1951: Seedling emergence and survival of range grasses in central Utah. Agron. J. 43:177–82.
6763. —— 1959: Effects of presowing vernalization on survival and development of several grasses. J. Range Management 12:280–86. Biol. Abstr. 35:27751. 1960.
6764. Frith, J. L. 1957: The germination of *Sporobolus virginicus*. J. Australian Inst. Agr. Sci. 23:69–75.
6765. Fritz, J. A. 1940: Growing freesias from seed. Horticulture 18(16):346.
6766. Fritz, J. C. 1954: Thiamine content of soybeans. Trans. Am. Assoc. Cereal Chem. 12:60–62. Biol. Abstr. 28:28295. 1954.
6767. Fritzsche, R., and H. Wolffgang. 1959: Beeinflussung des Saatgutwerkes und der Backqualität des Weizens durch Weizengallmückenbefall. Z. Pflanzenkrankh. Pflanzenschutz 66:645–53.
6768. Fröier, K. 1941: Keimung und Triebkraft bei Hafer und Weizen nach verschiedenen Röntgendosen. Hereditas 27:360–70. Biol. Abstr. 16:3563. 1942.
6769. —— and A. Gustafsson. 1941: The influence of X-ray dosage on germination and sprouting ability in barley and wheat. Svensk Botan. Tidskr. 35:43–56.
6770. —— and —— 1944: The influence of seed size and hulls on X-ray sensitivity in cereals. Hereditas 30:583–89.
6771. Frölich, A. 1953: Experiments with different treatments of rape seed meal. I. Review of literature and preliminary findings. Lant.-Högs. Ann. Uppsala 20:105–16.
6772. Fröschel, P. 1939: Onderzoekingen over de physiologie van de kieming. I. Remstoffen. Natuurw. Tijdschr. (Ghent) 21:93–116.
6773. —— 1941/43: Neuere Untersuchungen zur Physiologie der Keimung und ihre Bedeutung für die Samenkontrolle. Proc. Intern. Seed Testing Assoc. 13:113–27.
6774. —— 1955: Over de Remstoffen in de zaden van *Beta saccharifera*. Natuurw. Tijdschr. (Ghent) 37:97–100.
6775. —— 1956: Hemmstoffe und Wachstum. Acta Botan. Neerl. 5:264–67.
6776. —— and G. L. Funke. 1939: Een poging tot experimenteele plantensociologie. Natuurw. Tijdschr. (Ghent) 21:348–55.
6777. —— and E. Tilemans. 1942: Investigations on plant inhibitors. Bières Boissons 3:171–73. Chem. Abstr. 38:2786. 1944.
6778. Fröst, S. 1963: Number of accessory chromosomes and protein content in rye seeds. Hereditas 50:150–60.
6779. Frohberger, P. E. 1958: Studies on the question of the influence of seed treatment and water content of grain on latter's storage property. Höfchen Briefe (English Ed.) 11:103–14.
Frolich, E. 8044.
Frolich, E. F. 5679, 9900.
6780. Frolick, E. F. 1960/61: The experiment station and its relation to the foundation seed organization. Intern. Crop Improvement Assoc. Ann. Rept. 42/43:60–64.
6781. —— and R. D. Lewis. 1944: Seed certification in the United States and Canada. J. Am. Soc. Agron. 36:183–93. Biol. Abstr. 18:11546. 1944.
Frolik, E. F. 3514, 16966.
Frolova, S. L. 16604.
6782. Fromantin, J. 1957: Influence de l'acide 2,4,5-trichlorophenoxyacétic et de l'acide β-indoleacétique sur la germination et la croissance des plantules de quelques conifères. J. Exptl. Botany 9:1–14.
6783. —— 1958: Réactions comparées des cotyledones de quelques Cupressacées et Abiesacées (du genre *Pinus*) à l'action de l'acide 2,4,5-trichlorophenoxy acétique et de l'acide indol-β-acétique. Bull. Soc. Hist. Nat. Toulouse 93:61–69. Biol. Abstr. 35:11204. 1960.

6784. —— 1960: Combined action of cold and β-indoleacetic acid on the germination of *Larix europaea.* Bull. Soc. Hist. Nat. Toulouse 95:370–76. Chem. Abstr. 56: 6405.

6785. Fromme, F. D. 1928: The control of cereal smuts by seed treatment. Virginia Agr. Expt. Sta. Bull. 262. 16 pp.

6786. Fron, G. 1935: The disinfection of beet seeds. Compt. Rend. Acad. Agr. France 21:427–33. Chem. Abstr. 29:4885. 1935.

6787. —— 1936: Action de certains derivés de la quinoleine sur la germination du blé. Ann. Épiphyties Phytogénétique 2:333–40.

Fronsacq-Collin, M. 18898.

Froschel, P. 18981.

6788. Frost, H. B. 1938: Nucellar embryony and juvenile characters in clonal varieties of citrus. J. Heredity 29:423–32.

6789. —— 1943: Seed reproduction: development of gametes and embryos (in citrus). Genetics and breeding. Citrus Industry, Chapters 8 & 9:767–913. (Edited by H. J. Webber and L. D. Batchelor.) Biol. Abstr. 18:9825. 1944.

Frost, H. B. 3545, 3546, 13130.

6790. Frota, A. M. 1957: Production of tobacco seeds. Gaz. Agr. 9(92):9–11.

Fry, P. C. 12288, 12289.

Fry, P. R. 3786.

6791. Frydenberg, O. 1963: Some theoretical aspects of the scoring of mutation frequencies after mutagenic treatment of barley seeds. Radiation Botany 3:135–43.

Frydman, C. Chabelska-; *see* Chabelska-Frydman, C.

6792. Frydman, R. B., and E. F. Neufeld. 1963: Synthesis of galactosylinositol by extracts from peas. Biochem. Biophys. Res. Commun. 12:121–25.

6793. Fryer, J. R. 1922: The influence of light and of fluctuating temperatures on the germination of *Poa compressa* (L.). Sci. Agr. (Ottawa) 2:225–30.

6794. —— 1939: The maternal-line selection method of breeding for increased seed-setting in alfalfa. Sci. Agr. (Ottawa) 20:131–39.

6795. —— 1945: The alfalfa seed setting problem. Pr. Bull. Alberta 30:10–12. Plant Breeding Abstr. 16:125. 1946.

Fryer, J. R. 2732, 4170.

6796. Fryxell, P. A. 1963: Morphology of the base of seed hairs of *Gossypium.* I. Gross morphology. Botan. Gaz. 124:196–99.

6797. Fu, C. S. 1958: Effects of light, temperature, oxygen and storage upon the seed germination of *Pistia stratiotes* L. K'o Hsueh T'ung Pao 12:377–78.

Fu, S. Y. 12444.

6798. Fuchs, A. 1938: Beiträge zur Embryologie der Thymelaeaceae. Österr. Botan. Z. 87:1–41. Biol. Abstr. 12:12323. 1938.

Fuchs, F. J., Sr. 12563.

6799. Fuchs, K. 1934: Zytologische Beiträge zur Reifung der Samen, dargestellt an *Helianthus annuus, Triticum vulgare* und *Lilium martagon.* Botan. Arch. 36:444–96.

6800. Fuchs, L. 1932: Examination of *Fructus papaveris* at various stages of growth. Pharm. Monatsh. 13:223–25. Chem. Abstr. 27:561–62. 1933.

6801. Fuchs, W. H., and A. Beiler. 1943: Über die heisswasser-Empfindlichkeit der Karyopsen des Weizens. I. Ber. Deut. Botan. Ges. 61:164–74. Biol. Abstr. 24:21587. 1950.

Fudge, J. F. 3135.

6802. Fuelleman, R. F., and W. L. Burlison. 1942: Germination studies of bromegrass seed, *Bromus inermis* Leyss. Trans. Illinois State Acad. Sci. 35:32–34. Expt. Sta. Rec. 89:307. 1943.

6803. Fürste, K. 1958: Möglichkeiten zur Abkürzung der Keimruhe bei Keimlingen von Getreidearten durch Entfernen der Kornhülle. (Conclusion). Saatgut-Wirtsch. 10:276–79.

6804. Fürtauer, R. 1939: Keimungsförderung von Beerenobstsamen. Gartenbauwissenschaften 14:141–50.

Fürtauer, R. 10715.

6805. Fuhr, C. 1938: A comparison of the blotter test with the daylight test for germinating orchard grass seed. Proc. Assoc. Offic. Seed Analysts 23:75.

6806. Fuja, M. C. 1929: On the formation and development of roots and shoots on the isolated cotyledons of *Cucurbita, Cucumis* and *Lupinus.* Bull. Acad. Polon. Sci. Lettres 1929:209–18. Botan. Centralbl. 20:6. 1932.

6807. Fujii, M. 1962: Studies on the characters of rice seeds from the plants grown on reclaimed tidal paddy fields. II. Influence upon the germination force under saline water of the seeds exerted by saline treatments of their parent plants. Proc. Crop Sci. Soc. Japan 30:201–5.

Fujii, M. 8871.

6808. Fujii, R. 1956: Paper-chromatographic survey of soluble sugars in germinating bean seeds. Botan. Mag. (Tokyo) 69:34–38.

Fujii, R. 14790, 14791, 14797.

6809. Fujii, S. 1956: Studies on oil formation in rape-seed. III. On the spring-seeded rape at high lands. Proc. Crop. Sci. Soc. Japan 25:8–10.

6810. —— and Y. Haniu. 1959: Influence of the treatment with maleic hydrazide on the seed stalk development in rape. Bull. Univ. Osaka Prefect. Ser. B 9:67–75. Biol. Abstr. 36:25169. 1961.

6811. Fujii, T. 1962: Chlorophyll mutations induced by radiation in *einkorn* wheat and their occurrence in clusters. Seiken Ziho 14:1–11.

6812. —— 1962: Studies on photoperiodic responses involved in the germination of *Eragrostis* seeds. Botan. Mag. (Tokyo) 75:56–62.

6813. —— 1963: Inhibitory effect of 5-bromouracil and 5-fluorouracil on photoperiodically induced germination of *Eragrostis* seed. Plant Cell Physiol. 4:277–83.

6814. —— 1963: On the anaerobic process involved in the photoperiodically induced germination of *Eragrostis* seed. Plant Cell Physiol. 4:357–59.

6815. —— and S. Isikawa. 1961: The effects of temperature after the light-exposure on the germination of *Oenothera* seeds. Botan. Mag. (Tokyo) 74:414–18.

6816. —— and —— 1961: Successive processes involved in the germination response of *Nasturtium* seeds. Plant Cell Physiol. 2:77–86.

6817. —— and —— 1962: Effects of after-ripening on photoperiodic control of seed germination in *Eragrostis ferruginea* Beauv. Botan. Mag. (Tokyo) 75:296–301.

6818. ——, ——, and A. Nakagawa. 1960: The effects of gibberellin on the germination of the seeds of *Sedum kamtschaticum* Fisch. Botan. Mag. (Tokyo) 73:404–11.

Fujii, T. 9491, 9492.

Fujikawa, M. 10120.

Fujimaki, K. 206.

Fujita, K. 8849.

6819. Fujiwara, A., and M. Kadowaki. 1957: Phosphorus metabolism in higher plants. I. Phosphorus distribution in rice and barley seedlings. Nippon Dojo-Hiryogaku Zasshi 28:157–60. Tohoku J. Agr. Res. 8:55–63. 1957: Chem. Abstr. 52:20427. 1958.

6820. Fujiwara, M. 1949: The influence of storage temperature and humidity on the thiamine contents of various polished rices. Eiyo To Shokuryo 2:12–17. Chem. Abstr. 46:9737. 1952.

Fujiwara, T. 19388.

6821. Fukai, T., and M. Kushijaki. 1952: The role of nitrogen, phosphorus and potassium in the yield of rice-plant in relation to the meteorological conditions. Hokkaido Natl. Agr. Expt. Sta. Res. Bull. 63:37–42.

6822. Fukaki, S. 1930: On the retarding effects of fruit juice upon the germination of seeds. Proc. Crop Sci. Soc. Japan 2:134–38.

6823. Fukuba, H. 1953: Higher fatty acid dehydrogenase in plant seeds. Symp. Enzyme Chem. (Tokyo) 8:85–86. Chem. Abstr. 47:11359. 1953.

6824. —— 1957: Higher-fatty acid dehydrogenase in plant seeds. II. Higher-fatty acid dehydrogenase in germinating soybeans. J. Agr. Chem. Soc. Japan (English Transl.) 31:67–69. Chem. Abstr. 51:15709. 1957.

6825. —— 1957: Higher-fatty acid dehydrogenase in plant seeds. III. The cofactor. J. Agr. Chem. Soc. Japan (English Transl.) 31:69–72. Chem. Abstr. 51:15709. 1957.

6826. —— and T. Mitsuzawa. 1957: Higher fatty acid dehydrogenase in plant seeds. VI. Lecithin dehydrogenase. J. Agr. Chem. Soc. Japan (English Transl.) 31:151–53. Chem. Abstr. 52:12001. 1958.

6827. Fukui, H. N., J. E. De Vries, S. H. Wittwer, and H. M. Sell. 1957: Ethyl-3-indoleacetate: an artefact in extracts of immature corn kernels. Nature 180:1205.

6828. ——, L. E. Weller, S. H. Wittwer, and H. M. Sell. 1958: Natural growth substances in vernalized and non-vernalized lettuce seedlings. Am. J. Botany 45: 73–74.

6829. Fukui, J. 1960: On the changes of seed-weight and seed-shape of soybean under different seed-ripening conditions. Kanto-Tosan Agr. Expt. Sta. J. 15:47–56.

6830. —— and M. Arai. 1951: Ecological studies on Japanese soybean varieties. I. Classification of soybean varieties on the basis of the days from germination to blooming and from blooming to ripening with special reference to their geographical differentiation. Japan. J. Breeding 1:27–39. Biol. Abstr. 27:20372. 1953.

6831. ——, M. Ojima, and H. Yarimizu. 1956: After-effect of day-length and temperature treatment given during the seed ripening period of soybeans on the plant growth and yield in the following generation. Japan. J. Breeding 6:5–10.

6832. —— and H. Ono. 1955: Studies on the adaptability of green manure and forage crops to paddy conditions. II. Germination of the seeds in relation to the length of the periods of their soaking. Proc. Crop Sci. Soc. Japan 24:108–10.

6833. Fukui, T., and Z. Nikuni. 1959: Changes in sugar content during the germination of rice seeds. Application of ion-exchange resin chromatography. J. Agr. Chem. Soc. Japan (English Transl.) 33:72–78. Chem. Abstr. 54:3507. 1960.

Fukumoto, S. 14071.

Fukunaga, K. 19968.

6834. Fulkerson, R. S. 1959: The effects of seeding rate and row width in relation to seed production in orchard grass, *Dactylis glomerata* L. Can. J. Plant Sci. 39:355–63.

6835. —— and W. E. Tossell. 1955: Seed treatment of forage legumes and grasses with three antibiotics. Can. J. Agr. Sci. 35:259–63.

6836. —— and —— 1961: Row width and seeding rate in relation to seed production in timothy. Can. J. Plant Sci. 41:549–58.

6837. Fuller, G. D. 1912: Germination and growth of the cottonwood upon the sand dunes of Lake Michigan near Chicago. Trans. Illinois State Acad. Sci. 5:137–43.

Fuller, H. J. 7600.

Fuller, T. C. 4233.

6838. Fuller, W. H., N. C. Gomness, and L. V. Sherwood. 1953: The influence of soil aggregate stabilizers on stand, composition, and yield of crops on calcareous soils of southern Arizona. Ariz. Agr. Expt. Sta. Tech. Bull. 129:257–80.

6839. Fulmer, J. L., and R. R. Botts. 1951: Analysis of factors influencing cotton yields and their variability. U.S. Dept. Agr. Tech. Bull. 1042. 183 pp.

6840. Fulton, H. J. 1941: Some factors that influence the immediate effects of pollen on boll characters in cotton. J. Agr. Res. 63:469–80.

6841. Fulton, J. M., and W. I. Findlay. 1960: The nutrient composition of hybrid corn as influenced by fertilization. I. The nitrogen percentage in leaf and grain. Can. J. Soil Sci. 40:42–48.

6842. Fulton, N. D. 1962: Viability of deteriorated cottonseed. Arkansas Farm Res. 11:4.

Fulton, N. D. 2722.

Fulton, R. W. 17309.

6843. Fulutowicz, A., and W. Bejnar. 1954: The influence of the age of sugar beet seeds on their seeding value. Roczniki Nauk Rolniczych Ser. A 69:323–40.

6844. Fulwider, J. R., and R. E. Engel. 1959: The effect of temperature and light on germination of seed of goosegrass, *Eleusine indica*. Weeds 7:359–61.

Funahashi, S. 14065, 14134, 14135.

6845. Funk, C. R., J. C. Anderson, M. W. Johnson, and R. W. Atkinson. 1962: Effect of seed source and seed age on field and laboratory performance of field corn. Crop Sci. 2:318–30.

Funke, G. L. 6776.

6846. Funke, H. 1939: Beiträge zur Kenntnis von Keimung und Bau der Mistel. Botan. Centralbl. Beih. 59:235–74.

6847. —— 1943: Über Hemm- und Wuchsstoffe des keimenden Maiskorns. Jahrb. Wiss. Botan. 91:54–82.

Funke, H. 17705.

6848. Furman, L., J., E. Amadori M., and O. Guzman F. 1959: Direct extraction of sunflower seed by filtration-extraction. J. Am. Oil Chemists Soc. 36:454–57.

Furniss, M. M. 6300.

6849. Furr, J. R., and J. A. Cook. 1952: Nitrogen content of pinnae, fruit and seed of Deglet Noor and Khadrawy date palms as related to nitrogen fertilization. Date Growers Inst., Coachella Valley, Calif., Rept. 29:13–14. Chem. Abstr. 47:3410. 1953.

Furrer, A. H. 6318.

6850. Fursov, V. I. 1963: Localization of organic substances in wheat grain. Vestn. Sel'skokhoz. Nauki, Min. Sel'sk. Khoz. Kaz. SSR 6(4):3–10. Chem. Abstr. 59:11891. 1963.

6851. —— and V. M. Inyushin. 1962: Deoxyribonucleic acid content of the wheat embryo nucleus. Tsitologiya 4:574–77. Chem. Abstr. 58:7136. 1963.

6852. Furuichi, M. 1928: On the physical and chemical properties of the various proteins in the nuts of *Ginkgo biloba* L. Tottori Soc. Agr. Sci. Japan Trans. 1(2):105–25.

6853. Furukawa, S. 1934: Constituents of *Ginkgo biloba* L. fruits. I. Sci. Papers Inst. Phys. Chem. Res. (Tokyo) 24:304–13. Chem. Abstr. 29:464. 1935.

6854. Furukawa, T. 1956: The light germination of forest tree seeds. III. The effects of light illumination to stored seeds of Japanese black and red pine. J. Japan. Forestry Soc. 38:103–4.

Furukawa, T. 8300, 8301.

6855. Furusato, K., and Y. Ohta. 1957: Notes on seedless *Citrus* sp. Seiken Ziho Rept. Kihara Inst. Biol. Res. No. 8:100–2.
6856. ——, Y. Ohta, and K. Ishibashi. 1957: Studies on polyembryony in *Citrus*. Seiken Ziho Rept. Kihara Inst. Biol. Res. No. 8:40–48.
Furusato, K. 14717.
6857. Furuta, T. 1961: Influence of gibberellin on germination of seeds. Am. Camellia Yearbook 1961:141–45.
6858. Furutani, Y., and M. Kukii. 1951: On the fruiting habits of soybeans. II. The effects of defoliation at the different stages at the pod and seed development. Kyushu Noji Shikinjo 8:51–52.
6859. Furuya, M. 1956: Effects of the vapour of methyl 2,4-dichlorophenoxyacetate on growth and differentiation in *Phaseolus vulgaris* L. II. Behavior of decotylated embryos in germination *in vitro* and the role of cotyledons in formative response. Japan. J. Botany 15:270–84.
6860. —— and S. Osaka. 1955: Effects of the vapor of methyl ester of 2,4-dichlorophenoxyacetate on growth and differentiation in *Phaseolus vulgaris*. I. Formative effects induced in the seedling after various grades of application on dry seeds. Japan. J. Botany 15:117–39. Chem. Abstr. 49:11104. 1955.
6861. —— and K. Soma. 1957: The effects of auxins on the development of bean embryos cultivated *in vitro*. J. Fac. Sci. Univ. Tokyo Sect. III 7:163–98. Chem. Abstr. 54:19868. 1960.
Furuya, M. 3931.
6862. Fuseya, H., and Y. Nagata. 1960: Physiological chemistry of decrepitude of seeds. I. Gifu Univ. Nat. Sci., Sci. Rept. Fac. Liberal Arts Educ. 2:406–9.
6863. Fushtey, S. G. 1961: Control of seed-borne diseases. Ontario Soil Crop Improvement Assoc. Addresses Proc. 1961: 65–67.
6864. Fusillo, M. H., J. F. Metzger, and D. M. Kuhns. 1952: Effect of chloromycetin and streptomycin on embryonic tissue growth in *in vitro* tissue culture. Proc. Soc. Exptl. Biol. Med. 79:376–77.
Futral, J. G. 2613, 6996.
Fynn, P. J. 435.

G

6865. Gaal, G. 1930: The fatty oil of seeds of *Colchicum autumnale* L. Magy. Gyogyszeresztud Tarsasag Ertesitöje 6:149–67. Chem. Abstr. 25:2582. 1931.
6866. Gabrichidze, N. M. 1938: Catalase as an indicator of the quality of dried wheat. Biochemistry (USSR) (English Transl.) 3:813–20. Chem. Abstr. 33:3899. 1939.
6867. Gabrielsen, E. K. 1942: The CO_2 assimilation and the dry-substance production of heads of wheat. Kgl. Veterin. Landsbohojskole Aarskr. 1942:28–54. Chem. Abstr. 38:2997. 1944.
6868. Gad-Andresen, K., and E. Jarlov. 1934: Can sexual hormones influence the germination and growth of plants? Acta Med. Scand. 84:241–52. Biol. Abstr. 10:354. 1936.
6869. Gadd, I. 1926: Some observations on broken seedlings in red clover. Medd. Stat. Cent. Frökontr. 1:39–44.
6870. —— 1926: A research concerning immature grain. Medd. Stat. Cent. Frökontr. 1:51–59.
6871. —— 1927: Treatment of barley and winter-wheat of reduced vitality with fungicides. Medd. Stat. Cent. Frökontr. 2:87–100.
6872. —— 1928: Comparing germination tests of dill and other garden plants. Medd. Stat. Cent. Frökontr. 3:70–82.
6873. —— 1928: Two years' storage of wheat. Medd. Stat. Cent. Frökontr. 3:83–93.
6874. —— 1930: The importance of the method as regarding to germination of grain. Medd. Stat. Cent. Frökontr. 5:69–121.
6875. —— 1931: Schwedische Erfahrungen über die beste Keimmethodik für Getreide und grosssamige Leguminosen. Proc. Intern. Seed Testing Assoc. 3:270–86.
6876. —— 1931: Über Wassergehaltsbestimmung bei Getreide. Medd. Stat. Cent. Frökontr. 6:113–19.
6877. —— 1932: Untersuchungen über die Korrelationsverhältnisse zwischen Keimfähigkeit auf dem Laboratorium und Auflauf auf dem Felde. Versuche mit Getreide. Medd. Stat. Cent. Frökontr. 7:87–133.
6878. —— 1933: Über anormale Keimlinge und ihren Wert. Proc. Intern. Seed Testing Assoc. 5:137–61.
6879. —— 1934: The significance of water content in storing grain in sacks. Medd. Stat. Cent. Frökontr. 9:58–69.

6880. —— 1936: Studien über Keimungsmethodik bei Erbsen, spez. Gartenerbsen, und den Zusammenhang zwischen ihrer Keimfähigkeit und dem Aufgang auf dem Felde. Proc. Intern. Seed Testing Assoc. 8:159–210.

6881. —— 1938: Über die Natur der Hartschaligkeit der kleinsamigen Leguminosen und den Einfluss der Lagerung auf dieselbe. Proc. Intern. Seed Testing Assoc. 10:146–74.

6882. —— 1939: On methods for the elimination of seed dormancy in seed control work. Proc. Intern. Seed Testing Assoc. 11:108–18.

6883. —— 1944: Vital colouring of pea seeds by means of malachite green. Proc. Intern. Seed Testing Assoc. 13:5–76.

6884. —— 1950: Hot-water treatment, a new method of controlling *Ascochyta* spp. on pea seeds. Proc. Intern. Seed Testing Assoc 16:1–94.

6885. —— 1953: Report of the Biochemical and Seedling Vigour Test Committee. Proc. Intern. Seed Testing Assoc. 18:289–304.

6886. —— 1954: Storage of dressed wheat seed. Medd. Stat. Cent. Frökontr. 29:62–78.

6887. —— 1955: Continued laboratory experiments with dressing of wheat seed and its storing. Medd. Stat. Cent. Frökontr. 30:45–55.

6888. —— 1955: Germination of seed of New Zealand browntop, *Agrostis tenuis* Sibth. Proc. Intern. Seed Testing Assoc. 20:29–45.

6889. —— 1958: The influence of mercury compound dressing on the storage capacity of spring wheat. Medd. Stat. Cent. Frökontr. 33:31–59.

6890. —— 1959: The sowing value of seed infected by moulds elucidated by means of laboratory and field experiments. Medd. Stat. Cent. Frökontr. 34:26–82.

6891. —— and A. Kjaer. 1940: Über die Verwendbarkeit der Selen- und Indigokarminmethoden bei der Prüfung von Frost- und Fusariumgeschädigtem Getreide. Proc. Intern. Seed Testing Assoc. 12:140–49.

6892. Gaertner, E. E. 1950: Studies of seed germination, seed identification, and host relationships in dodders, *Cuscuta* spp. Cornell Univ. Agr. Expt. Sta. Mem. 294. 56 pp.

6893. —— 1954: Key to "seeds" of fleshy fruits. Trans. Roy. Can. Inst. 30:33–46. 1953. Biol. Abstr. 29:17210. 1955.

6894. —— 1956: Dormancy in the seed of *Cuscuta europea*. Ecology 37:389.

6895. Gaessler, W. G., R. M. Hixon, and E. S. Haber. 1940: The quantity of pericarp in several hybrid and inbred strains of sweet corn. Iowa State Coll. J. Sci. 14:379–83.

Gaessler, W. G. 7939.

6896. Gäumann, E. 1932: The influence of the temperature of germination on the chemical composition of grain seedlings. I. Z. Botan. 25:385–461. Chem. Abstr. 26:3819. 1932.

6897. Gaffman, I. I., and I. A. Vaintraub. 1961: The use of paper electrophoresis in checking homogeneity of seed proteins. Biochemistry (USSR) (English Transl.) 25:820–25.

6898. Gagnebin, F., L. Wavrin, and F. Chodat. 1951: Effets sur le pouvoir germinatif de traitments antérieurs à la germination. Arch. Sci. (Geneva) 4:65–71. Biol. Abstr. 26:3176. 1952.

6899. Gaidukevich, L. I. 1957: Seeding of dry-valley meadows. Zemledelie 5(4):84–86.

6900. Gain, E. 1895: The physiological role of water in plants. Ann. Sci. Nat. Ser. 7 Botan. 20:63–215. Expt. Sta. Rec. 7:367–70. 1895–96.

6901. —— 1900. Tests of the value of seeds of first and second flowering. Sta. Agron. Nancy Bull. 2:42–46. Expt. Sta. Rec. 11:143. 1899–1900.

6902. —— 1900. On the embryo of Mummy wheat and barley. Compt. Rend. 130: 1643–46. Expt. Sta. Rec. 12:825–26. 1900–1.

6903. —— 1922: Temperature ultra-maxima supportée par les embryons d'*Helianthus annuus* L. Compt. Rend. 174:1031–33.

6904. —— 1922: Sur la résistance comparative à la chaleur des points végétatifs de l'embryon du grand-soleil. Compt. Rend. 174:1557–59.

6905. —— 1924: Anomalies des *Helianthus* issus de graines chauffées de 120° a 150° C. Compt. Rend. 178:865–67.

6906. —— 1927: Action des temperatures elevées sur les graines, et morphologie des plantes issues d'embryons chauffés de 115° a 155° C. Rev. Gen. Botan. 39:234–53.

6907. —— 1927: Action des temperatures elevées sur les graines, et morphologie des plantes issues d'embryons chauffés de 115°

à 155° C. (Conclusion). Rev. Gen. Botan. 39:306–29.

6908. —— and A. Jungelson. 1915: Sur les graines de maies issus de la végétation d'embryons libres. Compt. Rend. 160: 142–44.

Gain, E. 1302, 3089.

6909. Gaitonde, M. K., and K. Sohonie. 1951: Trypsin inhibitor in field bean (*Dolichos lablab*). Current Sci. (India) 20:217–18.

Gaja, A. M. C. 3265.

Gajic, D. 8197.

6910. Gal, E. 1938: Effect of organic acids on germination, growth and ascorbic acid content of wheat seedlings. Nature 142: 1119.

6911. Galachalova, Z. 1957: Use of high temperatures in the drying of wheat seeds. Sel'skoe Khoz. Sibiri 1957(8):40–42.

Galachalova, Z. N. 11375.

6912. Galanos, D. S., T. Nakayama, L. Law, E. Weber, and H. Carter. 1957: Studies on the complex lipides of plant seeds. Chim. Chronika (Athens) 22:237–43. Chem. Abstr. 52:3928e. 1958.

6913. Galimberti, Z. 1963: Embryology of *Codiaeum variegatum* (L.) Blume var. *pectum* (Lodd.) Müll. Nuovo Giorn. Botan. Ital. 70:21–32.

6914. Galinovsky, F., O. Vogt, and H. Nesvadba. 1953: A new alkaloid from *Cytisus laburnum*. Sci. Pharm. 21:256–58. Biol. Abstr. 28:29552. 1954.

6915. Galitz, D. S. 1962: A physiological study of the developing soybean seed radicle. Dissertation Abstr. 22:3358.

Galitz, D. S. 10498.

Gallagher, M. C. 8247.

Gallatin, M. H. 12535.

6916. Gallego, A., R. Segura, and T. Menzel. 1955: Biochemical study of seed of the genus *Lupinus*. Anales Inst. Farmacol. Espan. 4:129–36. Chem. Abstr. 52:1379d. 1958.

Galleymore, H. B. 9315.

6917. Galligar, G. C. 1938: Correlation between growth of excised root tips and types of food stored in the seed. Plant Physiol. 13:599–609.

6918. Gallo, P., and H. Valeri. 1954: The antibiotic activity of some extracts of *Solanum mammosum*, *Cassia occidentalis*, and *Cecropia* species. Rev. Med. Vet. Parasitol. (Maracay) 12:119–24. 1953. Chem. Abstr. 48:12872. 1954.

6919. —— and —— 1954: The antibiotic activity of the seed of avocado pear (*Persea americana*). Rev. Med. Vet. Parasitol. (Maracay) 12:125–29. Chem. Abstr. 48:12873. 1954.

6920. Gallot, R. 1930: Experience internationale sur l'origine des graines du pin sylvestre. Bull. Soc. Centr. Forest Belg. 37:287–97. Biol. Abstr. 8:16254. 1934.

6921. Galloway, B. T. 1894: The growth of radishes as affected by the size and weight of the seed. Agr. Sci. 8:557–67.

Gallucis, F. 2342.

6922. Gallup, W. D. 1927: The gossypol content and chemical composition of cottonseeds during certain periods of development. J. Agr. Res. 34:987–92.

6923. —— 1928: A chemical study of the development of cotton bolls and the rate of formation of gossypol in the cotton seed. J. Agr. Res. 36:471–80.

6924. —— 1936: The oil:gossypol ratio of cottonseeds and its possible effect on the refining loss of crude cottonseed oil. Proc. Oklahoma Acad. Sci. 16:90–93. Expt. Sta. Rec. 78:156. 1938.

Gallup, W. D. 3041, 3042, 3681.

6925. Galston, A. W., J. Bonner, and R. S. Baker. 1953: Flavoprotein and peroxidase as components of the indoleacetic acid oxidase system of peas. Arch. Biochem. Biophys. 42:456–70.

6926. —— and L. Y. Dalberg. 1954: The adaptive formation and physiological significance of indoleacetic acid oxidase. Am. J. Botany 41:373–80.

Galston, A. W. 7387.

6927. Galt, R. 1953: Methods of germinating oil palm seeds. W. African Inst. Oil Palm Res. J. 1:76–87.

6928. Galun, E. 1955: The effect of mechanical injury, seed treatment, soil types and soil micro-organisms on the germination of flax seed in Israel. Proc. Intern. Seed Testing Assoc. 20:46–56.

6929. —— 1956: Effect of seed treatment on sex expression in the cucumber. Experientia 12:218–19. Chem. Abstr. 51:556. 1957.

Galyas, E. 11378.

6930. Gamalii, G. M. 1951: Effect of seed vernalization and supplementary feeding of seedlings on yield and earliness of early cabbage. Sb. Stud. Nauchn.-Issled. Rabot Umanskii Sel'skokhoz. Inst. 1:68–72.

6931. Gambhur, I. R., and S. S. Joshi. 1952: Chemical examination of the seeds of *Abutilon indicum*. J. Indian Chem. Soc. 29:451–54. Chem. Abstr. 48:12252. 1954.

6932. Gamborg, O. L., and L. R. Wetter. 1963: An aromatic amino acid transaminase from mung bean. Can. J. Biochem. Physiol. 41:1733–40. Chem. Abstr. 59:13052. 1963.

6933. —— and S. Zalik. 1958: Studies on a lipoxidase system from sunflower seeds and seedlings. Can. J. Biochem. Physiol. 36:1149–57.

6934. Gammage, J. S., and L. C. Maisenhelder. 1960: Refrigeration prolongs viability of cottonwood seed. Tree Planters Notes 43:5–6.

6935. Gammon, E. T. 1952: Atmospheric fumigation of various seeds with methyl bromide. Calif. Dept. Agr. Bull. 41:27–30.

Gamperl, G. 18259.

6936. Ganapathy, P. M. 1956: Floral morphology and embryology of *Hydrophylax maritima* L. f. J. Madras Univ. 26B:263–75. Biol. Abstr. 31:39212. 1957.

6937. —— 1956: Female gametophyte and endosperm of *Ophiorrhiza mungos* L. J. Madras Univ. 26B:589–92. Biol. Abstr. 32:28236. 1958.

Ganapathy, P. M. 18205.

Ganapathy, S. 3996.

6938. Ganapathy, S. N., K. Krishnamurthy, M. Swaminathan, and V. Subrahmanyan. 1958: Studies on the non-protein nitrogenous constituents of resting and germinated *Lathyrus sativus* seeds. Ann. Biochem. Exptl. Med. (Calcutta) 18:157–60. Biol. Abstr. 34:8654. 1959.

6939. Gancedo, A. P. G. 1950: Efectos comparados del 2,4-D y la naftalenacetamida sobre la germinación del *Triticum vulgare* Vill var. *mentana*. Ann. Jardin Botan. Madrid 1:181–91.

6940. Ganchev, A. 1957: Herbstsaat und Stratifikation der Samen von gemeiner Esche, Bergahorn und Spitzahorn in verschiedenem Reifezustant. Izv. Bulgar. Akad. Nauk Inst. Gorata 2:437–50.

6941. —— and L. Ploshtakova. 1959: Untersuchungen über die Stratifikation von Lindensamen in verschiedenem Reifestadium. Izv. Bulgar. Akad. Nauk Inst. Gorata 5:297–342.

6942. Gandar, J. C. 1960: Evolution of growth substances of the seed and pods of green peas in the course of their development. Ann. Physiol. Veg. 2:255–60.

6943. Gandhi, S. M., and M. P. Bhatnagar. 1961: Effect of certain hormones on germination, flowering, fruiting, branching and yield of cumin (*Cuminum cyminum*). J. Indian Botan. Soc. 40:628–34.

6944. Gane, R. 1935: Respiration and water content of seeds. Gt. Brit. Dept. Sci. Ind. Res. Food Invest. Bd. Rept. 1935:135–37. Chem. Abstr. 31:6692–93. 1937.

6945. —— 1941: The water content of wheats as a function of temperature and humidity. Soc. Chem. Ind. J. 60:44–46.

6946. —— 1948: The water content of the seeds of peas, soybeans, linseed, grass, onion and carrot as a function of temperature and humidity of the atmosphere. J. Agr. Sci. 38:81–83.

6947. —— 1948: Effect of temperature, water content, and composition of the atmosphere on viability of carrot, onion, and parsnip seeds in storage. J. Agr. Sci. 38:84–89.

6948. —— 1948: The effect of temperature, humidity and atmosphere on the viability of Chewing's Fescue grass seed in storage. J. Agr. Sci. 38:90–92.

6949. —— and A. J. M. Smith. 1938: Water relations of seeds. Gt. Brit. Dept. Sci. Ind. Res. Food Invest. Bd. Rept. 1937:214–22.

Gane, R. 17609–17611.

6950. Ganesan, D. 1942: Hybrid vigor in cotton. I. The manifestation of hybrid vigor in the seed. Indian J. Genet. Plant Breeding 2:134–50. Biol. Abstr. 18:5835. 1944.

6950A. —— 1950: Sizes and densities of seeds sown and their effect on vigour and yield in cotton. Indian Cotton Growing Rev. 4:37–45.

6950B. —— 1950: "Agronomic effect" on the quality of cotton seed. Indian Cotton Growing Rev. 4:167–72.

Gangstad, E. O. 17140.

Gangulee, H. 12697.

6950C. Gangulee, H. C. 1938: On the development of the embryo-sac and pollen grains in *Canavalia ensiformis* D.C. J. Dept. Sci. Univ. Calcutta 1:73–94.

6950D. —— 1955: Studies on the date of ear emergence in rice. I. Relation between sowing time and date of ear emergence. Botan. Gaz. 117:1–10.

Ganguli, B. N. 10346.

6950E. Ganguli, N. C. 1954: Studies on changes in free amino acid composition of germinating *Phaseolus radiatus* by paper chromatography. Naturwissenschaften 41: 140–41.

6950F. —— 1955: Studies on the free amino

acid composition of germinating *Phaseolus radiatus*. Naturwissenschaften 42:18.
Ganguli, N. C. 1628.
Ganguly, A. K. 14178.
Ganguly, D. 15022.
6950G. Ganhao, J. F. P. 1957: Notes on seed disinfection. Agros (Lisbon) 40:269–83.
6950H. Gannon, R. 1960: Grow houseplants from seed. . . . It's a thrill. Horticulture 38(10):527.
6950I. Ganong, W. F. 1898: Embryo and seedling of the Cactaceae. Ann. Botany (London) 12:423–74.
6950J. Ganong, W. F. 1898: Polyembryony and its morphology in *Opuntia vulgaris*. Botan. Gaz. 25:221–28.
Ganter, W. 15230.
Ganyushkina, L. G. 11313.
6950K. Gapochko, M. P. 1930/31: Chemical characteristic of the fruits of Crimean almonds. Bull. Appl. Botany, Genet., Plant Breeding 25:196–98.
Gapochko, M. P. 9552.
6950L. Gaponenkov, T. K. 1934: Formation of phenols in plants. J. Gen. Chem. USSR (English Transl.) 4:466–71. Chem. Abstr. 29:2197. 1935.
Garay, A. S. 16625.
6951. Garbari, C. 1930: Wie ich zur Cymbidiumzucht kam. Orchideen-Anzucht aus Samen mit Hilfe von Wurzelpilz. Möllers Deut. Gärtner-Ztg. 45:229–30.
6952. —— 1930: Wie ich zur Cymbidiumzucht kam. Orchideen-Anzucht aus Samen mit Hilfe von Wurzelpilz. Möllers Deut. Gärtner-Ztg. 45:242–43.
6953. —— 1930: Über Cymbidiumzucht. Möllers Deut. Gärtner-Ztg. 45:275–76.
6954. —— 1930: Zur Cymbidiumzucht. Über Orchideen-Keimung mit hilfe von Wurzelpilz. Möllers Deut. Gärtner-Ztg. 45:325–26.
6955. —— 1930: Zur Cymbidiumzucht. Möllers Deut. Gärtner-Ztg. 45:392–94.
6956. —— 1930: Zur Cymbidiumzucht. Orchideen-Keimung mit hilfe von Wurzelpilz. Möllers Deut. Gärtner-Ztg. 45:402–3.
6957. Garber, L. F. 1922: Scarification as it effects longevity of alfalfa seed. J. Am. Soc. Agron. 14:298–302.
6958. Garber, R. H., and B. R. Houston. 1959: An inhibitor of *Verticillium alboatrum* in cotton seed. Phytopathology 49:449–51.
Garber, R. J. 1407, 3940, 11357.
6959. Garbowski, L., and P. Leszczenko. 1929: Essais concernant le traitement anticryptogamique des semences. Prace Wydz. Chorob. Roslin Bydgoszczy Panst. Inst. Nauk. Gos. 8:1–43.
6960. Garbusova, A. P. 1941: Regarding methods for determining ripeness of Lupine seeds. Proc. Lenin Acad. Agr. Sci. USSR 1941(10):30–33.
Garbuzova, G. I. 11208.
6961. Garcia, G. M. 1956: A preliminary study of the production of cauliflower seed. Philippine J. Agr. 19:143–52.
6962. Garcia, M. B. 1931: Weeds in rice paddies: germination of seeds and resistance of the young plants to submergence in water. Philippine Agriculturist 20:217–32.
Garcia, P. R. 6999.
6963. Garcia-Blanco, J., and S. Grisolia. 1943/44: Action of some indole derivatives on the initial growth of certain plants on an aqueous medium. Trabajos Inst. Nacl. Cienc. Med. (Madrid) 2:549–58. Chem. Abstr. 42:8887. 1948.
6964. —— and —— 1943/44: Action of sulfonamides on germination and growth of *Raphanus sativus*. Trabajos Inst. Nacl. Cienc. Med. (Madrid) 3:347–52. Chem. Abstr. 42:8885. 1948.
Garciduenas, M. Rojas-; *see* Rojas-Garciduenas, M.
6965. Garde, A. A. 1961: Effects of ionising radiations on seeds and their application in plant improvement. Agricultura (Lisbon) 9:39–40.
Garde, N. Malheiros-; *see* Malheiros-Garde, N.
Gardiner, B. O. C. 4976.
6966. Gardner, C. O. 1961: An evaluation of effects of mass selection and seed irradiation with thermal neutrons on yield of corn. Crop Sci. 1:241–45.
6967. ——, P. H. Harvey, R. E. Comstock, and H. F. Robinson. 1953: Dominance of genes controlling quantitative characters in maize. Agron. J. 45:186–91.
Gardner, C. O. 2146, 3513, 16016.
6968. Gardner, F. E., and E. J. Kraus. 1937: Histological comparison of fruits developing parthenocarpically and following pollination. Botan. Gaz. 99:355–76.
6969. —— and P. C. Marth. 1937: Parthenocarpic fruits induced by spraying with growth promoting compounds. Botan. Gaz. 99:184–95.
6970. —— and —— 1938: Germination and seedling vigor of peach varieties for understocks. Proc. Am. Soc. Hort. Sci. 35:409–14.

6971. —— and —— 1939: Effectiveness of several growth substances on parthenocarpy in holly. Botan. Gaz. 101:226–29.
Gardner, F. E. 8400, 9053, 15416.
6972. Gardner, F. P., and M. J. Kasperbauer. 1961: Seedling emergence and growth responses of dwarf grain sorghum as affected by gibberellic acid. Iowa State J. Sci. 35:311–18.
Gardner, F. P. 10308, 10309, 19672.
Gardner, K. E. 14364.
Gardner, R. 16286.
6973. Gardner, R. C. B. 1937: Storage of acorns. Quart. J. Forestry 31:32–33. Biol. Abstr. 12:5383. 1938.
6974. Gardner, W. A. 1921: Effect of light on germination of light-sensitive seeds. Botan. Gaz. 71:249–88.
6975. Garese, P. 1962: Germination of weeds in relation to time of planting the crops. Idia (Buenos Aires) 179:36–40.
6976. Garg, N. K., and C. R. K. Murti. 1962: Tocopherol and ubiquinone changes during germination. J. Sci. Ind. Res. (India) 21C:103–5.
6977. Garilli, D. 1932: Peroxidase activity as an index of seed vitality. Giorn. Biol. Appl. Ind. Chim. 2:206–16. Biol. Abstr. 10:17976. 1936.
6978. Garin, V. 1916: Germination of *Yerba*. Bol. Min. Agr. (Argentina) 20:568–77. Expt. Sta. Rec. 36:445. 1917.
6979. Garina, K. P. 1956: The effect of different methods of winter rye pollination on the percent of seed germination and viability of the hybrid progeny. Izv. Akad. Nauk SSSR Ser. Biol. 4:58–66.
6980. Garman, H. 1890: *Phelipaea ramosa* L. (broom rape). Kentucky Agr. Expt. Sta. Bull. 24:8–11.
6981. —— 1903: The broom rapes. *Orobanche* (germination and vitality of the seed). Kentucky Agr. Expt. Sta. Bull. 105:13.
6982. —— and E. C. Vaughn. 1916: The curing of blue grass seeds as affecting their viability. Kentucky Agr. Expt. Sta. Bull. 198. 3 pp.
6983. Garman, H. R., and L. V. Barton. 1946: Germination of seeds of *Panicum anceps* Michx. Contrib. Boyce Thompson Inst. 14:117–22.
6984. —— and —— 1946: Response of lettuce seeds to thiourea treatments as affected by variety and age. Contrib. Boyce Thompson Inst. 14:229–41.
Garman, H. R. 2042.
Garner, C. F. 10649.
6985. Garner, F. H., and H. G. Sanders. 1932: Investigations in crop husbandry. I. The effects of seed treatments on the germination and yield of sugar beet. J. Agr. Sci. 22:551–59.
6986. —— and —— 1933: Sulphuric-acid treatment of sugar-beet seed. Gt. Brit. Min. Agr. J. 39:986–87.
6987. —— and —— 1935: Investigations in crop husbandry. II. On the age of seed beans. J. Agr. Sci. 25:361–68.
6988. Garner, W. W., and H. A. Allard. 1920: Effect of the relative length of day and night and other factors of the environment on growth and reproduction in plants. J. Agr. Res. 18:553–606.
6989. ——, ——, and C. L. Fourbert. 1914: Oil content of seeds as affected by the nutrition of the plant. J. Agr. Res. 3: 227–49.
6990. Garnet, J. R. 1947: On the germination of peas and other seeds at low temperature. Victorian Nat. 63:263–65.
Garnier, J. 10573.
6991. Garofalo, F. 1955: Observations on seeds of eggplant infected by Fusarium. Nuovo Giorn. Botan. Ital. 62:545–46.
6992. Garrard, A. 1954: The effects of β-indoleacetic acid on the germination and root growth of certain members of the Cruciferae. New Phytologist 53:165–76.
6993. —— 1955: The germination and longevity of seeds in an equatorial climate. Gardens Bull. (Singapore) 14:534–45.
6994. —— 1956: The effect of fruit sap on the germination of four species of tropical plants. Gardens Bull. (Singapore) 15: 276–84. Biol. Abstr. 32:6415. 1958.
6995. Garrard, L. A., and R. H. Biggs. 1962: Influence of thiourea on the germination of Lovell peach seed. Proc. Assoc. Southern Agr. Workers Ann. Conv. 59: 225.
6996. Garren, K. H., B. B. Higgins, and J. G. Futral. 1947: Blue-black discoloration of Spanish peanuts. Phytopathology 37:669–79.
Garren, R., Jr. 11346.
6997. Garrett, H. E. 1957: Small seeds in New Zealand farm management. Canterbury Agr. Coll., Lincoln, N.Z. 78 pp.
6998. Garrido, T. G. 1954: On the cotton seed project. Philippines Bur. Plant Ind., Plant Ind. Dig. 17(9/10):19–22.
6999. —— and P. R. Garcia. 1953: Observations on growing kenaf for fiber and seed. Philippines Bur. Plant Ind., Plant Ind. Dig. 16:11–14.

7000. Garrigues, R. 1960: Action inhibiteur sur la croissance radiculaire d'une substance extraite de la graine de Caroubier. Compt. Rend. 251:269–71.

7001. ——, B. Hoï, and P. Cagniant. 1952: Action de quelques carbures polycycliques sur la germination. Compt. Rend. 234: 553–55.

7002. Garrison, C. S. (coordinator). 1960: Technical advances in grass and legume seed production and testing. Advan. Agron. 12:41–125.

7003. —— 1963: Certification now goes international. Crops & Soils 15(9):13–14.

Garrison, C. S. 15088.

7004. Garrison, R. H. 1960: Sow seeds with certainty. Crops & Soils 12(5):10–11.

7005. Garside, S., and S. Lockyer. 1930: Seed dispersal from the hygroscopic fruits of *Carpanthea* (*Mesembryanthemum*) *pomeridiana* N.E. Br. Ann. Botany (London) 44:639–55.

Garz, J. 12050.

7006. Gaskill, J. I. 1957: A preliminary report on the use of gibberellic acid to hasten reproductive development in sugar beet seedlings. J. Am. Soc. Sugar Beet Technol. 9:521–28.

7007. Gaskins, M. H. 1958: Gibberellins and their effects on yield and quality of beans (*Phaseolus vulgaris*) and Southern peas (*Vigna sinensis*). Dissertation Abstr. 19:927–28.

7008. Gasparian, M. G., and A. A. Avetisian. 1958: Influence of certain physiologically active substances on enzyme activities in seeds of Kyurushna grass. Izv. Akad. Nauk Arm. SSR, Biol. i Sel'skokhoz. Nauki 11(5):67–72. Chem. Abstr. 52:18698. 1958.

7009. Gasperini, M. 1941: The influence of early frosts on the germination and on the further development of wheat. Ital. Agr. 78:171–75. Biol. Abstr. 22:9751. 1948.

7010. Gassian, S., M. Marchand, and J. Sechet. 1960: Effets de divers traitements appliqués aux semences de betterave sur la composition biochimique des tubercules. Compt. Rend. 251:433–35.

7011. Gassner, G. 1910: Über Keimungsbedingungen einiger suedamerikanischer Gramineensamen. I. Ber. Deut. Botan. Ges. 28:350–64.

7012. —— 1910: Über Keimungsbedingungen einiger suedamerikanischer Gramineensamen. II. Ber. Deut. Botan. Ges. 28: 504–12.

7013. —— 1911: Untersuchungen über die Wirkung des Lichtes und des Temperaturwechsels auf die Keimung von *Chloris ciliata*. Jahrb. Hamburg. Wiss. Anstalten 29(3):1–121.

7014. —— 1911: Vorläufige Mitteilung neuerer Ergebnisse meiner Keimungsuntersuchungen mit *Chloris ciliata*. Ber. Deut. Botan. Ges. 29:708–22.

7015. —— 1915: Altes und neues zur Frage des Zusammenwirkes von Licht und Temperatur bei der Keimung lichtempfindlicher Samen. Ber. Deut. Botan. Ges. 33:203–17.

7016. —— 1915: Beiträge zur Frage der Lichtkeimung. Z. Botan. 7:609–61.

7017. —— 1915: Einige neue Fälle von Keimungauslösender Wirkung der Stickstoffverbindungen auf lichtempfindliche Samen. Ber. Deut. Botan. Ges. 33:217–32. Botan. Centralbl. 131:72. 1916. Expt. Sta. Rec. 35:222. 1916.

7018. —— 1915: Über die Keimungsauslösende Wirkung der Stickstoffsalze auf lichtempfindliche Samen. Jahrb. Wiss. Botan. 55:259–342.

7019. —— 1926: Der gegenwärtige Stand der Stimulationsfrage. Ber. Deut. Botan. Ges. 44:341–67.

7020. —— 1930: Untersuchungen über die Wirkung von Temperatur und Temperaturkombinationen auf die Keimung von *Poa pratensis* und anderen *Poa*-arten. Z. Botan. 23:767–838.

7021. —— 1938: Über die Hartschaligkeit von Robiniensamen und eine Methode zu ihrer Beseitigung. Angew. Botan. 20:293–303.

7022. —— 1943: Methods for the laboratory testing of seed disinfectants. Phytopathol. Z. 14:303–9.

7023. —— 1950: Die chemotherapeutische Bewertung von Quecksilber verbindungen in den verschieden Beizverfahren. Phytopathol. Z. 17:1–35.

7024. —— 1952: Keimungstemperatur und flugbrandbefall. Tijdschr. Plantenziekten 58:219–23.

7025. —— 1952: Zur Frage der Übertragung von *Cercospora beticola* durch das Rübensaatgut. Angew. Botan. 26:55–59. Biol. Abstr. 27:4743. 1953.

7026. —— and W. Franke. 1935: Einige Versuche über den Stickstoffhaushalt lichtkeimender Samen im dunklen Keimbett. Z. Botan. 28:446–63.

7027. —— and —— 1938: Die Wirkung

tiefer Keimtemperaturen auf den Stickstoffhaushalt junger Weizenkeimblätter. Phytopathol. Z. 11:297–318.

7028. —— and K. Hassebrauk. 1927: Blausäurebegasungen als Mittel zur schnellen erzielung voller Keimreife. Pflanzenbau 4:1–4.

Gast, L. E. 18434.

7029. Gastler, G. F., and A. W. Halverson. 1952: The riboflavin and protein contents of some South Dakota barleys. Proc. S. Dakota Acad. Sci. 31:182–83. Biol. Abstr. 28:4419. 1954.

Gastler, G. F. 19622.

Gastrock, E. A. 4893, 17771.

7030. Gates, B. N. 1940: Dissemination by ants of the seeds of *Trillium grandiflorum*. Rhodora 42:194–96.

7031. —— 1941: Observations in 1940 on the dissemination by ants of the seeds of *Trillium grandiflorum*. Rhodora 43:206–7.

7032. —— 1942: The dissemination by ants of the seeds of bloodroot, *Sanguinaria canadensis*. Rhodora 44:13–15.

7033. Gates, L. F. 1959: Seed and soil treatment with systemic insecticides to prevent aphid colonization of emerging beet seedlings. Ann. Appl. Biol. 47:492–501.

7034. Gates, R. R. 1932: Heterospory and the angiosperms. Nature 129:793.

7035. —— 1951: Epigeal germination in the Leguminosae. Botan. Gaz. 113:151–57.

7036. Gathercoal, E. N. 1942: Drug plant cultivation. Drug Cosmetic Ind. 51:264–72, 319. Pharm. Abstr. 9:93. 1943.

7037. Gatherum, G. E. 1961: Pellet seeding on sagebrush (*Artemisia*) range. J. Range Management 14:160.

Gatherum, G. E. 9788, 10470.

7038. Gathy, P. 1960: L'experience internationale sur l'origine des graines d'epica (*Picea abies* Karst.). Resultats en Belgique. Groenendaal Sta. Rech. Trav. Ser. B 24. 30 pp.

7039. Gatin, C. L. 1904: Observations sur la germination et la formation de la première racine de quelques palmiers. Rev. Gen. Botan. 16:177–88.

7040. —— 1905: Contribution a l'étude chimique de la germination de *Borassus flabelliformis* L. Bull. Soc. Botan. France 52:558–61.

7041. —— 1908: Transformation diastasique du mannose en glucose au cours du la germination du *Borassus flabelliformis*. Bull. Soc. Botan. France 8:383–86.

7042. —— 1912: Upon the structure of the embryo of the "Zingiberacee" and the "Marantacee." Compt. Rend. 154:35–37.

7043. —— 1919: La maturation artificielle des fruits. J. Agr. Trop. 19:256–60.

7044. Gauba, E., and L. D. Pryor. 1958: Seed coat anatomy and taxonomy in *Eucalyptus*. I. Proc. Linnean Soc. N.S. Wales 83:20–32. Biol. Abstr. 33:30898. 1959.

7045. —— and —— 1959: Seed coat anatomy and taxonomy in *Eucalyptus*. II. Proc. Linnean Soc. N.S. Wales 84:278–91.

7046. —— and —— 1961: Seed coat anatomy and taxonomy in *Eucalyptus*. III. Proc. Linnean Soc. N.S. Wales 86:96–111.

Gauch, H. G. 13168.

Gaudineau, M. 1371–1373.

7047. Gaul, H. 1957: Effect of X-rays in association with CO_2, colchicine, and heat on barley. Z. Pflanzenzücht. 38:397–429. Chem. Abstr. 54:25071. 1960.

7048. —— 1959: Über die Chimärenbildung in Gerstenpflanzen nach Röntgenbestrahlung von Samen. Flora (Jena) 147:207–41.

7049. Gaume, R. 1912: Germination, développement, et structure anatomique de quelques Cistenées. Rev. Gen. Botan. 24:273–95.

Gaur, B. K. 4044.

7050. Gaus, G. E., C. S. Shaw, and W. H. Kliever. 1941: A practical seed cotton moisture tester for use at gins. U.S.D.A. Circ. No. 621. 25 pp.

7051. Gausman, H. W., J. H. Ramser, G. H. Dungan, F. R. Earle, M. M. MacMasters, H. H. Hall, and P. D. Baird. 1952: Some effects of artificial drying of corn grain. Plant Physiol. 27:794–802.

Gausseran, M. T. 15094.

Gautheret, R. 12414.

Gavarron, F. F. 3690.

7052. Gavrielit-Gelmond, C. 1951/52: Experiments on dormancy of *Lathyrus ochrus*. Proc. Intern. Seed Testing Assoc. 17:139–42.

Gavrielit-Gelmond, C. 15443.

7053. Gavrielit-Gelmond, H. 1956: Effect of sulphuric acid upon germination of cotton seeds. Hassadeh 36:620–22.

7054. —— 1957: The effect of environment and agro-technical methods on dormancy in *Lathyrus ochrus* seeds. Proc. Intern. Seed Testing Assoc. 22:276–81.

7055. —— 1957: Optimal sowing time for cotton seeds. Hassadeh 38:230–33.

7056. —— 1962: A review of problems associated with testing of peanut seed (*Arachis hypogaea*). Proc. Intern. Seed Testing Assoc. 27:357–72.

7057. —— and S. Hurwitz. 1950: The viability period of field crop seeds (Cont'd). Hassadeh 30:593–95.

Gavrielit-Gelmond, H. 9249.

7058. Gavrilenko, B. D. 1958: On the dispersion of seeds of *Parrotia persica* C.A.M. Botan. Zh. 43:867–69.

Gavrilova, L. G. 285

7059. Gavrilova, L. V., and E. S. Morgunova. 1958: Some physiological and biochemical data on rye Vyatka. II. Uch. Zap. Kirovsk. Gos. Ped. Inst. 1958(14):68–74. Chem. Abstr. 54:1668. 1960.

7060. Gavrilova, M. 1935: On the reversibility of the vernalization process. Compt. Rend. Acad. Sci. URSS 6:562–63.

7061. Gawda, J. 1937: Influence of storage conditions on certain seeds. Roczniki Nauk Rolniczych Lesnych 43:12–17.

7062. Gawell, L. J., and W. B. Bollen. 1960: Sterilization of grass seeds with peracetic acid. Agron. J. 52:718–19.

7063. Gay, F. J. 1941: Some moisture relations of Australian wheats. Australian Council Sci. Ind. Res. J. 14:117–20.

7064. —— 1946: The effect of temperature on the moisture content-relative humidity equilibria of wheat. Australian Council Sci. Ind. Res. J. 19:187–89. Biol. Abstr. 21:18024. 1947.

Gayon, P. Ribereau-. 2099.

Gayte-Sorbier, A. 11791.

Gaywala, P. M. 7380.

7065. Geddes, W. F. 1935: Some physical and chemical characteristics of a sample of "mummy" wheat. Can. Dept. Trade Commerce, Dominion Grain Res. Lab. 9th Ann. Rept. 1935:63–64. Chem. Abstr. 30:7703. 1936.

7066. —— 1936: Studies of oleaginous seeds. Can. Dept. Trade Commerce, Dominion Grain Res. Lab. 10th Ann. Rept. 1936: 75–82. Chem. Abstr. 31:7175. 1937.

7067. —— 1943: Factors affecting the keeping quality of grain in storage. Am. Miller 71(8):40, 93. Biol. Abstr. 17:24804. 1943.

7068. ——, L. S. Cuendet, and C. M. Christensen. 1955: Recent investigations on grain storage. 3rd Intern. Bread Congr. Hamburg 1955:136–39. Chem. Abstr. 51:15025. 1957.

7069. —— and F. H. Lehberg. 1936: Flax-studies. I. The relation between weight per measured bushel, weight per thousand kernels, and oil content of flaxseed. Can. J. Res. Sect. C 14:45–47.

7070. —— and —— 1936: Flax-studies. II. An improved refractometric method for estimating the oil content of flaxseed. Can. J. Res. Sect. C 14:48–61.

Geddes, W. F. 643, 2845, 2846, 4062, 4362, 4363, 5599, 7298–7300, 7414, 7415, 8903, 8904, 9208, 11804, 12030, 12159, 12241, 12562, 12769, 13513–13519, 13523, 14310, 14311, 14742, 15244, 15915, 15916, 17752, 17753.

7071. Gedeon, J. 1955: Triterpenoid sapogenins: sapogenins of the seed of *Achyranthes bidentata*. Arch. Pharm. 288:417–18. Chem. Abstr. 51:1393. 1957.

Gehenio, P. M. 12552, 12553.

7072. Gehle, H. 1933: The physical behavior of wheat grains upon drying and conditioning. Mühle 70:1155–58. Chem. Zentr. 1:627. 1934. Chem. Abstr. 29:5537. 1935.

7073. Gehlen, W. 1934: Zur Chemie und Pharmakologie der Kolanuss. Arch. Exptl. Pathol. Pharmakol. 174:695–714. Physiol. Abstr. 19:682. 1935.

7074. Gehlsen, C. A. 1931: Über Mittel zur Behebung der Hartschaligkeit bei *Lupinus luteus* und einigen anderen Leguminosen (mit besonderer Berücksichtigung ihrer Verwendbarkeit in der Landwirtschaft). Dissertation, Hamburg. 38 pp. Proc. Intern. Seed Testing Assoc. 4:83–84. 1932.

7075. —— 1937: A germinação demorada das sementes das leguminosas e seus remedios. Bol. Sec. Agr. Ind. Com. Pernambuco 2:327–31.

7076. —— 1937: Observaçoes sobre o babassu (*Orbignya speciosa*) e sua germinação—a germinação da oiticica. Bol. Sec. Agr. Ind. Com. Pernambuco 2:428–33.

Gehrke, C. W. 20133.

7077. Geiger, E., and E. B. Hagerty. 1949: Physiological availability of amino acids present in zein. Am. Chem. Soc. Div. Agr. Food Chem. Abstr. Papers, San Francisco, pp. 4A–5A.

7078. Geiger, M. 1928: Beitrag zur Kenntnis der Physiologie keimender Samen. I. Einfluss der Quellungsbedingungen auf den Gasaustausch. Jahrb. Wiss. Botan. 69:331–56.

Geiger, M. 14737, 14738.

7079. Geise, C. E., P. G. Homeyer, and R. G. Tischer. 1951: A comparison of three methods for determination of moisture in sweet corn. Food Technol. 5:250–53.

7080. Geissler, E. 1960: Über die Bedeutung des Bestrahlungsmilieus beim photodynamic Effekt. Intern. J. Radiation Biol. 2:91–96. Biol. Abstr. 35:45624. 1960.

7081. Geitler, L. 1955: Reisenkerne im Endosperm von *Allium ursinum*. Österr. Botan. Z. 102:460–75.

7082. —— 1962: Embryological studies on *Coleochaeta* species. Österr. Botan. Z. 109:495–509.

7083. —— 1962: Embryology of *Coleochaeta nitellarum* and the right-left problem. Österr. Botan. Z. 109:529–39.

Gelemy, S. A. 20079.

7084. Gelin, O. E. V. 1948: The influence of some hormone derivatives upon the yield and quality of seed of meadow fescue and red fescue. Agr. Hort. Gen. 6:163–66.

7085. Geller, I. A. 1953: Influence of oxidation-reduction properties of soil on seed germination. Dokl. Akad. Nauk SSSR 89:355–58. Chem. Abstr. 47:10793. 1953.

7086. Gelman, A. A. 1961: Effect of phytoncides from certain wild pasture plants on seed germination and initial growth in plant species sowed in pasture grass mixtures. Tr. Vologodsk. Molochn. Inst. 44:78–91. Biol. Abstr. 43:11615. 1963.

Gelman, N. 14802.

Gelmond, C. Gavrielet-; *see* Gavrielet-Gelmond, C.

Gelmond, H. Gavrielet-; *see* Gavrielet-Gelmond, H.

7087. Geltser, F. I. 1951: Methods for treatment of acorns with mycorhizal fungus and sources for obtaining it. Sov. Agron. 9:54–62.

7088. —— and E. A. Voronkova. 1953: In regard to the necessity of treating acorns with mycorhizal fungi. Tr. Vses. Nauchn. Issled. Inst. Sel'skokhoz. Mikrobiol. 13: 156–68.

Gemignani, G. 7272.

7089. Gemma, T. 1957: On the cause of non-germination phenomena in *Beta* spp. Yamagata Agr. Forestry Soc. J. 1957(11): 3–5.

7090. Gen, S. 1937: A comparative study on the influence of various light and soil moisture conditions upon the germination of seeds and the growth of seedlings of *Pinus densiflora* and *Chamaecyparis obtusa*. Kjusu Imp. Univ. Fak. Terk. Bul. Sci. 7:404–5.

7091. Genazzani, E., and F. Miele. 1952: Action of synthetic antihistamines on germination of grain caryopses. Boll. Soc. Ital. Biol. Sper. 28:439–41. Brit. Abstr. B III:218. 1953.

7092. Genders, R. 1956: Annuals from seed. Foyles Handbooks, London. 112 pp.

7093. Generalov, G. F. 1946: Protein and starch content of peas in relation to geographical and meteorological conditions. Selek. i Semen. 13(4/5):56–59.

7094. Geneves, L. 1956: Synthèse de substance mitochondriale pendant la germination des embryons d'*Allium cepa* L. variété (Lil.). Compt. Rend. 242:2030–32.

7095. Genkel, P. A., and S. H. Chao. 1959: The role of plasmodesmata in the loss of germination ability of coffee (*Coffea robusta* Linn.) seeds. Plant Physiol. (USSR) (English Transl.) 5:303–7. 1958. Biol. Abstr. 34:2554. 1959.

7096. —— and S. S. Fan. 1958: On the physiological significance of vivipary in mangrove plants. Acta Botan. Sinica 7:51–70.

7097. ——, A. P. Sarycheva, and O. A. Sitnikova. 1955: The effect of treating seeds with variable temperatures on the growth and ripening of maize. Plant Physiol. (USSR) (English Transl.) 2: 447–53. Biol. Abstr. 32:9772. 1958.

7098. —— and I. V. Tsvetkova. 1955: Increasing the heat resistance of plants. Dokl. Akad. Nauk SSSR 102:383–86.

7099. Gentcheff, G., and A. Gustafsson. 1940: Parthenogenesis and pseudogamy in Potentilla. Botan. Notiser 1940(2):109–32.

7100. Genter, C. F., and H. M. Brown. 1941: X-ray studies on the field bean. J. Heredity 32:39–44.

7101. ——, J. F. Eheart, and W. N. Linkous. 1956: Effects of location, hybrid, fertilizer and rate of planting on the oil and protein contents of corn grain. Agron. J. 48: 63–67.

Genter, C. F. 313, 5825.

Gentile, A. C. 11309.

7102. Gentner, G. 1928: Beiträge zu einer Monographie der Provenienzen der Klee- und Grassaaten. 5th Intern. Seed Testing Conf., Rome, pp. 103–25.

7103. —— 1928: Über die Verwendbarkeit von ultravioletten Strahlen bei der Samenprüfung. Prakt. Bl. Pflanzenbau Pflanzenschutz 28:166–72.

7104. —— 1929: Eine Methode zum Nachweis der Sporen des Steinbrandes und anderen Pilzarten am Saatgut. Fortschr. Landwirtsch. 4:353–56.

7105. —— 1929: Zur Frage der Verwendung von Zinkkästen bei der Triebkraftbestimmung. Prakt. Bl. Pflanzenbau Pflanzenschutz 7:25–31.

7106. —— 1930: Die Untersuchung des Getreides auf Korndicke (Sortierung). Prakt. Bl. Pflanzenbau Pflanzenschutz 7:300–11.

7107. —— 1930: Der Einfluss der Witterung auf die Beschaffenheit des Weizen- und Roggensaatgutes. Prakt. Bl. Pflanzenbau Pflanzenschutz 8:139–42.

7108. —— 1934: Beiträge zue einer Monographie der Klee- und Grassaaten. Proc. Intern. Seed Testing Assoc. 6:11–20.

7109. —— 1937: Die Herkunftsbestimmung der Kleesaaten. Proc. Intern. Seed Testing Assoc. 9:1–81.

7110. Gentry, H. S. 1955: Apomixis in black pepper and jojoba? J. Heredity 46:8.

7111. —— 1958: The natural history of jojoba (*Simmondsia chinensis*) and its cultural aspects. Econ. Botany 12:261–95.

Geoghegan, M. J. 6586.

7112. George, E. J. 1937: Storage and dewinging of American elm seed. J. Forestry 35:769–72.

7113. George, H. 1929/30: Studies in the germination of the seeds of native herbaceous plants. Abstr. Theses, Univ. Chicago Sci. Ser. 7:295–98.

George, N. Collis-; *see* Collis-George, N.

7114. Georgi, C. D. V. 1932: Malayan palm kernels. Malayan Agr. J. 20:53–60.

7115. —— 1932: Centrifugal extraction of palm oil at Serdang. Malayan Agr. J. 20:446–59. Soc. Chem. Ind. J. 52:27B. 1933.

7116. ——, V. R. Greenstreet, and G. L. Teik. 1932: Storage of rubble seeds. Malayan Agr. J. 20:164–76.

7117. —— and G. L. Teik. 1931: Tonkabean oil. Chem. Ind. 50:318T. Biol. Abstr. 8:9419. 1934.

7118. Georgiev, G. N., and K. Nikolov. 1959: Embryologische Untersuchungen an nichtviviparen Ahrchen von *Poa bulbosa* L. var. *vivipara* Koeler. Nauchn. Tr. Vissh Selskostopanski Inst. "Georgi Dimitrov," Agron. Fak. (Sofia) 7:195–207.

Georgin, A. 3447, 5369.

7119. Georlette, R. 1948: La production de tomates sans graines. Ann. Gembloux 54:219–23.

7120. Geraldes, C. d'M., A. D'Almeida, and C. Duarte. 1930: Oleaginous seeds from the forest of Angola. Bull. Mat. Grasses Inst. Colon. Marseille 14:332–42. Chem. Abstr. 25:4140–41. 1931.

Gerard, C. J. 9225.

7121. Gerashchenko, S. G. 1950: On increasing the seed germination of sandy sainfoin. Selek. i Semen. 17(2):62–63.

Gerasimov, B. A. 12721.

7122. Gerasimova, H. 1935: The nature and causes of mutations. II. Transmission of mutations arising in aged seeds: occurrence of "homozygous dislocants" among progeny of plants raised from aged seeds. Cytologia (Tokyo) 6:431–37. Biol. Abstr. 10:7638. 1936.

Gerasimova, H. 14253, 14254.

7123. Gerasimova-Navashina, E. N. 1954: Embryosac development, double fertilization and the origin of angiosperms. Botan. Zh. 39:655–79. Biol. Abstr. 32:9735. 1958.

7124. —— 1960: The effect of temperature conditions on the course of embryological processes in plants. Dokl. Botan. Sci. Sect. (English Transl.) 131:81–84.

7125. —— 1962: Cytological data on stimulation to development of embryo sac cells. Tr. Botan. Inst. Akad. Nauk SSSR Ser. 7, 1962(5):238–49. Biol. Abstr. 44:3564. 1963.

7126. —— and S. N. Korobova. 1959: On the role of synergids in fertilization. Byul. Mosk. Obshchestva Ispytatelei Prirody Otd. Biol. 64:69–76.

7127. Gerber, C. 1900: Étude comparée de la respiration des graines oléagineuses perdant leur développement et pendant leur germination. Actes Congr. Intern. Botan., Paris 1900:59–101.

Gerber, R. K. 4245.

7128. Gerberg, E. J., and S. L. Goldheim. 1957: Weight loss in stored corn and beans caused by insect feeding. J. Econ. Entomol. 50:391–93.

Gerbinot, B. 6739.

7129. Gerbracht, J. H. 1937: Evergreens from seed. N. & S. Dakota Hort. 10:46, 48.

Gerdes, F. L. 16552.

7130. Gerdes, G. 1961: Comparative germination ability and germination potency research with artificially dried and overstored wheat (*Triticum aestivum* L.) and rye (*Secale cereale* L.) seeds. Z. Landwirtsch. Versuchs- Untersuchungsw. 7:231–40.

7131. Gerhardt, F. 1931: Distribution and isolation of hemicellulose cleaving enzymes in seeds. Iowa Agr. Expt. Sta. Rept. Agr. Res., p. 53.

7132. Gerhold, H. D. 1960: Coloration and seed source of pines. Am. Christmas Tree Growers J. 4(4):21–22, 24.

7133. Gericke, W. F. 1918: Effects of rest and no-rest periods upon growth of *Solanum*. Botan. Gaz. 65:344–53.

7134. —— 1929: Some relations of maintained temperatures to germination and the early growth of wheat in nutrient solutions. Philippine J. Sci. 38:215–39.

7135. —— 1933: Bread quality of wheat produced in aqueous culture media. Science 77:229–32. Biol. Abstr. 9:6650. 1935.

7136. Gering, K. 1959: Changes of fermentative systems of respiration during the germination of corn seeds. Tr. Soveshch. po Morfogenezu Rast. 1:520–23.

Gerken, H. 16893.

7137. Gerlai, A. 1937: Keimungsphysiologische Untersuchungen mit Verwendung künstlicher Lichtquellen. Erdeszeti Kiserl. 39:163–264. Proc. Intern. Seed Testing Assoc. 10:656–57. 1938.

7138. Gerloff, U. 1936: Beiträge zur Kenntnis des Zusammensetzung Fetter Öle in verschiedenen Teilen einer Pflanze. Planta 25:667–88.

7139. —— and H. Oestermann. 1937: Hydrocyanic acid in flaxseed. Pharm. Zentralhalle 78:437–38. Chem. Abstr. 31:7088. 1937.

7140. Germ, H. 1935: Keimversuche mit Gerste. Erfahrungen aus den Bundesgerstenschauen der Jahre 1934/1935 und aus eigenen Versuchen. Landeskultur Vienna 2:168–72. Proc. Intern. Seed Testing Assoc. 8:86.

7141. —— 1936: Zur Unterscheidung der Samen von *Vicia pannonica* Cr. (*Vicia pannonica* Cr. var. *typica* Beck) und *Vicia striata* M.B. (*Vicia pannonica* Cr. var. *purpurascens* Ser.). Proc. Intern. Seed Testing Assoc. 8:133–35.

7142. —— 1937: *Vicia lutea* L. als Verunreinigung in Wickensaatgut. Ein weiterer Beitrag zur Unterscheidung von Wickensamen durch Phenolfärbung. Proc. Intern. Seed Testing Assoc. 9:272–74.

7143. —— 1943: Die Keimfähigkeit der Sojabohne. Forschungsdienst 15:68–78. Biol. Abstr. 23:12599. 1949.

7144. —— 1952: Über die Triebkraft von Samen landwirtschaflicher Kulturpflanzen und die Keimrollenmethode. Bodenkultur Sonderheft 3:28–62. Proc. Intern. Seed Testing Assoc. 17:143. 1951/52.

7145. —— 1953: Ein Weg zur physiologischen Prüfung der Samen. Proc. Intern. Seed Testing Assoc. 18:277–88.

7146. —— 1955: The distinction between seeds of *Trifolium incarnatum*, *T. alexandrinum*, and *T. squarrosum*. Saatgut-Wirtsch. 7:349–50. Chem. Abstr. 50: 5098. 1956.

7147. —— 1957: Report of the biochemical viability test committee. Proc. Intern. Seed Testing Assoc. 22:302–21.

7148. —— 1960: Report of activities of the committee for the tetrazolium test (1956–1959). Proc. Intern. Seed Testing Assoc. 25:492–97.

7149. —— 1960: Methodology of the vigour test for wheat, rye and barley in rolled filter paper. Proc. Intern. Seed Testing Assoc. 25:515–18.

7150. —— 1962: Ausschuss für Herkunft. Proc. Intern. Seed Testing Assoc. 27: 253–67.

7151. —— and M. Kietreiber. 1952: Vitalitat des Endosperms und Feldaufgang von *Zea mays*. Bodenkultur Österr. Zentralorgan Landwirtsch. Ernährungsforsch. 4 Suppl.:14–16. Biol. Abstr. 28: 17007. 1954.

7152. —— and —— 1954: Die Prüfung der Vitalität des Maiskornes. Jahrb. Bundesanstalt Pflanzenbau Samenprüfung Vienna 1953:29–47.

7153. Gernert, W. B. 1936: Possibility of parthenogenesis in grass. Science 84: 441–42.

7154. Gerny, B. 1935: A new method of sowing seed in forest nurseries. Lesnická Práce 14:155–67. Biol. Abstr. 9:16623. 1935.

7155. Gerola, F. M. 1940: Gametogenesis among alpine plants during winter. Nuovo Giorn. Botan. Ital. 47:323–48. Biol. Abstr. 15:11171. 1941.

7156. Gerre, H. 1954: *Welwitschia mirabilis* Hook. f. from seed to seed in the botanic garden of the Universich of Stellenbosch, C.P. J. S. African Botany 20:23–24.

7157. Gerritsen, T. 1956: Lupin seed proteins. IV. Amino acid composition of the globulins from *Lupinus angustifolius* and *Lupinus luteus*. Biochim. Biophys. Acta 22:269–73. Nutr. Abstr. Rev. 27:1620. 1957.

7158. Gersdorff, C. E. F. 1920: Germination of gladiolus seed. Flower Grower 7(5):73.

7159. —— 1936: Japanese and Siberian iris from seed. Flower Grower 23(11):597.
Gersdorff, C. E. F. 9934, 9936–9938.

7160. Gershanovich, V. N., G. L. Mednik, and I. A. Epshtein. 1956: Antagonism between sulfanilamide and folic acid in germination and chlorophyll synthesis of wheat. Biochemistry (USSR) (English Transl.) 21:633–36. Biol. Abstr. 33:31294. 1959.

7161. Gershtein, L. A. 1950: Transformation of substances in the seeds of tung (*Aleurites fordii*) during sprouting. Dokl. Akad. Nauk SSSR 72:1091–94. Chem. Abstr. 44:10057–58. 1950.
Gertler, A. 2490.

7162. Gertler, A. M. 1949: The comparatively new vermiculite growing medium produces more and sturdier seedlings. Seed World 64(10):16.
Gertler, S. I. 9709.
Gerzhoi, A. P. 11375.

7163. Geslin, H., and P. Jonard. 1946: Courbes caracteristiques du développement du grain de blé au point de vue physique. Compt. Rend. 222:410–12.

7164. —— and —— 1946: Courbes caracteristiques du développement du grain de blé au point de vue physique. Compt. Rend. 222:513–15.

7165. Gesslein, S. 1959: Ripening process of cereals and harvesting methods. Växtodling 13. 164 pp.

7166. Gessner, F. 1956: Die Wasseraufnahme durch Blätter und Samen. Handbuch Pflanzenphysiol. 3: Water Relations of Plants, pp. 215–44.

7167. Gesundheit, Z., and A. Poljakoff-Mayber. 1962: Some qualitative changes in the distribution of phosphorus compounds during germination of lettuce seeds. Bull. Res. Council Israel Sect. D 11:25–30.

7168. Geza, L. 1927: Bericht über die Tätigkeit der Kgl. ung. Samenkontroll-station in Budapest in den Jahren 1922–1927. Hungary, Samenkontroll-station. Bericht über die Tätigkeit. 68 pp.

7169. Gfeller, F., and F. Svejda. 1960: Inheritance of post-harvest seed dormancy and kernel colour in spring wheat lines. Can. J. Plant Sci. 40:1–6.
Gfeller, F. 6672.

7170. Ghamrawy, A. K., M. N. Hashad, and S. M. Elsherif. 1956: Effect of delayed planting on yield of flax and formation of oil in the linseed. Ann. Agr. Sci. (Cairo) 1:131–42.
Ghani, M. O. 2086.

7171. Ghatak, N. 1932: Composition of seeds of *Thevetia neriifolia* Juss. I. Bull. Acad. Sci. United Prov. Agra Oudh, India 2:79–84. Chem. Abstr. 27:5471–72. 1933.

7172. —— 1934: Chemical examination of the kernels of the seeds of *Caesalpinia bonducella*. Proc. Acad. Sci. United Prov. Agra Oudh, India 4:141–46. Chem. Abstr. 29:7578–79. 1935.

7173. —— 1934: Coloring matter of the seed-coat of *Abrus precatorius* Linn. (scarlet variety). Current Sci. (India) 2:380. Chem. Abstr. 28:4426. 1934.
Ghegov, A. 1375

7174. Gheorghe, V. 1959: The exosmosis of electrolytes in wheat seeds during vernalization. Acad. Rep. Populare Romine Studii Cercetari Biochim. 2:163–68. Chem Abstr. 53:22281. 1959.
Gheorghe, V. 18193.
Gheorghiu, P. 14774.

7175. Ghetie, V. 1959: The proteins of beans (*Phaseolus vulgaris*). Characteristics of the seed proteins. Acad. Rep. Populare Romine Studii Cercetari Biochim. 2: 169–79. Chem. Abstr. 53:22280. 1959.

7176. —— 1959: Proteins of the bean (*Phaseolus vulgaris*). II. Modification of the proteins of the seed during germination. Acad. Rep. Populare Romine Studii Cercetari Biochim. 2:293–98. Biol. Abstr. 36:69651. 1961.

7177. —— and L. Buila. 1961: The reserve and protoplasmatic proteins from seeds of green beans. Acad. Rep. Populare Romine Studii Cercetari Biochim. 4:241–55. Chem. Abstr. 56:6370. 1962.

7178. Ghillini, C. A. 1959: Principal diseases of vegetables transmissable by seeds and vegetative organs. Agr. Venezie 13:269–77.

7179. —— and G. Fior. 1956: Effect of antiparasite substances on the germination and productivity of wounded corn seeds. Notiz. Malotti Piante 35/36:129–50. Chem. Abstr. 51:5219. 1957.

7180. —— and —— 1957: Practical laboratory method for research in fungicides in treatment of corn seeds. Notiz. Malotti Piante 39:73–78.

7181. Ghimpu, V. 1942: La desinfection thermique des semences de tabac. Bull.

Inst. Expt. Cult. Ferment. Tabac (Bucarest) 31:138–47. Ann. Agron. 13:333. 1943.

7182. Ghisleni, P. L. 1948: L'azione dei raggi U.V. sul grano. I. L'azione sulla germinabilità di cariossidi nate da piante trattate con raggi U.V. Nuovo Giorn. Botan. Ital. 55:358–74.

7183. —— 1948: On the organic components of *Oryza sativa* L. Risicoltura 36:96–103.

7184. —— 1949: Effect of ultrasonics on grain germination. Ital. Agr. 86:528–32.

7185. —— 1955: Continual and modulated action in the ultrasonic treatment of *Cynara scolymus* seed. Nuovo Giorn. Botan. Ital. 62:549–51. Biol. Abstr. 32:21105. 1958.

7186. —— 1956: Further research concerning the action of ultrasounds, both of continuous and impulse type: Effect on asparagus seeds (*Asparagus officinalis* L.). Allionia (Turin) 3:69–74. Biol. Abstr. 40:16452. 1962.

7187. —— 1957: L'azione dell'ossigeno liquido sulla germinabilità dei semi duri di ginestrino (*Lotus corniculatus* L.) e di ladino (*Trifolium repens* L.). Nuovo Giorn. Botan. Ital. 64:145–52.

7188. Ghose, B. N. 1939: Oil from *Meconopsis* seeds. Indian Forester 65:742–46. Biol. Abstr. 15:6302. 1941.

7189. —— 1947: Germination of *Primula petiolaris* seeds. Gardeners Chron. 122: 110.

Ghose, P. K. 4938.

7190. Ghose, T. P., and S. Krishna. 1942: Tamarind seed. A valuable source of commercial pectin. J. Indian Chem. Soc. Ind. News Ed. 5:114–20. Chem. Abstr. 37:1796–97. 1943.

7191. —— and —— 1945: Tamarind-seed pectin. Current Sci. (India) 14:299–300. Chem. Abstr. 40:2480. 1946.

Ghosh, A. N. 3778.

Ghosh, A. R. 2510, 7789.

7192. Ghosh, B. N. 1949: A study on vernalization of rice by temperature treatments. J. Indian Botan. Soc. 28: 124–33.

7193. —— 1950: Physiological studies on the effect of colchicine on rice. II. Proc. Natl. Inst. Sci. India 16:135–45.

7194. —— 1954: Studies on the physiology of rice. VII. Effect of varying water levels on growth of rice in relation to nitrogen absorption. Proc. Natl. Inst. Sci. India 20:371–87.

7195. —— 1960: Studies on germination of spring (Boro) and summer (Aus) paddy to determine dormancy and viability. Indian J. Agr. Sci. 30:129–34.

7196. —— 1962: Agro-meteorological studies on rice. I. Influence of climatic factors on dormancy and viability of paddy seeds. Indian J. Agr. Sci. 32:235–41.

Ghosh, B. N. 8474, 17540.

7197. Ghosh, J. J. 1952: Changes in the phytin-phosphorus content of rice grains during mold attack. Indian J. Physiol. Allied Sci. 6:102–6. Chem. Abstr. 47: 4971. 1953.

7198. Ghosh, M. K., and K. C. Mazumder. 1950: Spectrochemical analysis of molybdenum in green peas. Indian J. Phys. 24:67–72. Chem. Abstr. 46:5741i, 5742a. 1952.

7199. Ghosh, M. N. 1924: Oil content of castor seeds as affected by climate and other conditions. Agr. J. India 19:81–84.

Ghosh, N. C. 1845.

Ghosh, P. K. 1851.

7200. Ghosh, R. B. 1962: A note on the nucellar polyembryony in *Aphanamixis polystachya* (Wall) Parker. Current Sci. (India) 31:165.

7201. Ghosh, S., T. Chakrabartty, and A. Chatterjee. 1960: Swietenine, the non-bitter principle of the seeds of *Swietenia macrophylla* King. J. Indian Chem. Soc. 37:440–41.

Ghosh, S. 3848, 5699, 7852, 11662.

7202. Ghosh, S. S., and R. Shahi. 1957: A case of polyembryony in *Shorea robusta* Gaertn. [Dipterocarpaceae]. Sci. Cult. (Calcutta) 23:254–56. Biol. Abstr. 32: 24460. 1958.

7203. Ghosh, T., and M. Basak. 1958: Method of storing jute seed and effect of age of seed on yield of fibre. Indian J. Agr. Sci. 28:235–42. Biol. Abstr. 35:17069. 1960.

7204. ——, ——, and B. C. Kundu. 1951: Effect of insolation and chemical treatment in relation to storing of jute seeds. Indian Phytopathol. 4:38–44.

Ghosh, T. 13880.

7205. Ghouseuddin, M., M. Sayeeduddin. 1935: Abnormal fruits- viviparous germination in *Coccinia indica* W. & A. (family Cucurbitacea). Current Sci. (India) 3: 434–35.

7206. Giacanelli, E. 1951: Researches on the content in catalase of various races of wheat. II. Ann. Sper. Agrar. (Rome) 5:499–516.

Giamalva, M. 9061.

7207. Gianfagna, A. J., and A. M. S. Pridham. 1951: Some aspects of dormancy and germination of crabgrass seed, *Digitaria sanguinalis* Scop. Proc. Am. Soc. Hort. Sci. 58:291–97.

7208. Gibault, G. 1929: Note sur les graines que conservent leur faculté germinative pendant plusieurs siècles. Bull. Mens. Soc. Nat. Hort. France 2:226.

7209. Gibbins, L. N., and F. W. Norris. 1963: Vitamins in germination. Distribution of inositol during the germination of the dwarf bean, *Phaseolus vulgaris*. Biochem. J. 86:64–67.

7210. Gibler, J. W. 1951: Pelleting soybean seed with fungicides for control of damping-off and root rot. [Abstract.] Phytopathology 41:13–14.

Gibson, D. R. 2949, 12662, 12663.

Gibson, F. 3153.

7211. Gibson, I. A. S. 1953: Crown rot, a seedling disease of groundnuts caused by *Aspergillus niger*. II. An anomalous effect of organo-mercurial seed dressings. Brit. Mycol. Soc. Trans. 36:324–34.

7212. —— and P. K. S. Clinton. 1953: Pre-emergence seed-bed losses in groundnuts at Urambo, Tanganyika Territory. Empire J. Exptl. Agr. 21:226–35.

7213. Gichner, T. 1961: The activity of trypsin inhibitor and its correlation with the oil content in *Glycine*. Biol. Plant. 3:305–11.

7214. Gideon, C. A., and V. L. N. Rao. 1957: Action of auxins on germination of groundnut. Sci. Cult. (Calcutta) 22: 632–33. Biol. Abstr. 32:24546. 1958.

7215. Gieger, M. 1941: The effect of fertilization and cultural practices on the oil and ammonia content of cottonseed grown on Yazoo-Mississippi delta soils. J. Agr. Res. 63:49–54.

Gieger, M. 5826.

7216. Gielsdorf, K. 1953: Anzucht von Kakteen aus Samen. Gartenwelt 53:355–56.

Gierczak, M. 18244.

7217. Giersbach, J. 1934: After-ripening and germination of *Cotoneaster* seeds. Contrib. Boyce Thompson Inst. 6:323–38.

7218. —— 1937: Germination and seedling production of *Arctostaphylos uva-ursi*. Contrib. Boyce Thompson Inst. 9:71–78.

7219. —— 1937: Germination and seedling production of species of *Viburnum*. Contrib. Boyce Thompson Inst. 9:79–90.

7220. —— 1937: Some factors affecting germination and growth of gentian. Contrib. Boyce Thompson Inst. 9:91–103.

7221. —— and L. V. Barton. 1932: Germination of seeds of the silver bell, *Halesia carolina*. Contrib. Boyce Thompson Inst. 4:27–38.

7222. —— and W. Crocker. 1932: Germination and storage of wild plum seeds. Contrib. Boyce Thompson Inst. 4:39–52.

7223. Giesbrecht, A. M. 1960: The occurrence of bufotenine in seeds of *Piptadenia falcata*, Benth. Anales Assoc. Bras. Quim. 19:117–19.

Gifford, E. M., Jr. 3117.

Gifford, R. H. 17395.

7224. Giglioli, I. 1895: Latent vitality in seed. Nature 52:544, 545.

7225. Gik, Y. 1944: Production of vitamin C in the USSR. Science 100:10s.

7226. Gilbert, S. G., and F. S. Lagasse. 1948: Accelerating germination of seeds in winter. Tung World 3(9):11, 17–18.

7227. —— and H. M. Sell. 1958: Biochemical changes during germination of the tung seed. Plant Physiol. 32:668–74.

Gilbert, S. G. 8073.

7228. Gilgut, C. J. 1939: Seed and seedling troubles. Horticulture 17(6):144–45.

7229. —— 1951: Propagation of woody plants. Plants Gardens 7:98–106.

Gilkeson, A. L. 8406.

7230. Gill, A. H. 1933: Pine-nut oil. Oil Soap 10:7–8. Chem. Abstr. 27:1223. 1933.

7231. Gill, N. T. 1938: The viability of weed seeds at various stages of maturity. Ann. Appl. Biol. 25:447–56.

Gill, W. D. 14856.

7232. Gillemet, R., and A. Guilbot. 1945: Separation of the protein-cell layer of the wheat grain by a chemical method. Composition of the cell walls. Bull. Soc. Chim. Biol. 27:49–55. Chem. Abstr. 40:3197. 1946.

7233. Gilles, E. 1932: Effets de faibles irradiations au moyen de la lampe à vapeur de mercure sur la germination. Compt. Rend. Soc. Biol. 109:739–41. Physiol. Abstr. 17:690. 1933.

7234. —— 1932: Influence des rayons ultra-violets sur la germination des graines irradiées à l'état sec. Compt. Rend. Soc. Biol. 110:841–43.

7235. —— 1932: Effets d'irradiations de diverses durées sur des semences a differents états de gonflement. Compt. Rend. Soc. Biol. 110:843–44.

Gillespie, G. T. 8938.

Gillet, A. 3081.

Gillette, L. A. 525.

7236. Gillier, P. 1963: Germination and disinfection of peanut seeds. Oléagineux 18:243–45.

7237. Gillies, E. 1946: Inhibitory action of the vital stain, neutral red, even in very small concentrations, on the growth of seedlings. Bull. Mens. Soc. Linnéenne Lyon 15:97–100. Chem. Abstr. 43:1084. 1949.

Gilliland, C. B. 9880.

7238. Gillis, J. 1940: La composition d'extraits aqueux de la semence de *Beta saccharifera*. Rec. Trav. Chim. 59:526–33. Biol. Abstr. 15:13535. 1941.

7239. —— 1940: Water extracts of the seeds of *Beta saccharifera*. Natuurw. Tijdschr. (Ghent) 21:263–70. Chem. Abstr. 34: 4114–15. 1940.

7240. Gillot, J., and J. Philouze. 1962: Gibberellic acid action on seed yield of lettuce. Ann. Amelioration Plantes 12: 215–17. Chem. Abstr. 58:7303. 1963.

7241. Gillot, P. 1925: Observations sur la germination des graines der *Mercurialis annua* L. Bull. Soc. Botan. France 72: 139–53.

7242. —— 1926: Sur la composition chimique des graines des mercuriales. Ann. Sci. Agron. 43:389–93.

7243. —— 1926: De l'innocurte des graines de *Mercuriale annuelle*. Ann. Sci. Agron. 43:394–96.

7244. —— 1927: Recherches sur les graines d'*Euphorbia amygdaloides* L. Bull. Sci. Pharmacol. 34:139–42.

7245. —— 1927: Recherches sur les graines de l'*Euphorbia cyparissias* L. Bull. Sci. Pharmacol. 34:429–33.

7246. —— 1927: Sur la composition chimique des graines des mercuriales. Bull. Mat. Grasses Inst. Colon. Marseille 1927: 59–62.

7247. —— 1928: Recherches sur les graines de l'*Euphorbia platyphylla* L. Bull. Sci. Pharmacol. 35:107–10.

7248. —— 1928: Recherches sur les graines de l'*Euphorbia paralias* L. Matières Grasses 20:8307–8.

7249. —— 1928: Recherches sur les graines de l'*Euphorbia verrucosa* Jacq. Bull. Sci. Pharmacol. 35:288–91.

7250. —— 1929: Recherches sur les graines de l'*Euphorbia esula* L. Matières Grasses 21:8390–91.

7251. —— 1932: The seeds of *Euphorbia stricta* L. Bull. Sci. Pharmacol. 39:529–32. Chem. Abstr. 27:324. 1933.

7252. —— 1933. The seeds of *Euphorbia exigua* L. Bull. Sci. Pharmacol. 40:449–53. Chem. Abstr. 28:667. 1934.

7253. Gilman, J. C., and D. H. Barron. 1930: Effect of molds on temperature of stored grain. Plant Physiol. 5:565–73.

Gilman, J. C. 17196.

7254. Gilman, W. 1950: A new way to start your garden; sow your vegetable and flower seeds in sphagnum moss; you'll have more luck than ever before. Farm J. 74(2):33.

7255. Gilmartin, A. J. 1958: Post-fertilization seed and ovary development in *Passiflora edulis* Sims. Trop. Agr. (London) 35:41–58.

7256. Gilmer, R. M. 1955: Imported mahaleb seeds as carriers of necrotic ring spot virus. Plant Disease Reptr. 39:727–28.

7257. —— 1964: Longevity of sour cherry yellows virus in infected cherry seeds. Plant Disease Reptr. 48:338–39.

7258. —— and K. D. Brase. 1962: Evidence against seed transmission of green ring mottle virus in mazzard cherry seedlings. Plant Disease Reptr. 46:356.

7259. Gilmer, W. E., H. A. Friesen, and J. B. Harrington. 1946: The resistance of wheat varieties to seed bleaching. Sci. Agr. (Ottawa) 26:437–47.

7260. Gilmore, A. E. 1950: A technique for embryo culture of peaches. Hilgardia 20:147–70.

Gilmore, A. R. 12557.

7261. Gilshtein, N. N., and V. A. Moshkin. 1957: On raising the castorbean seed production. Maslob.-Zhir. Prom. 23(5): 6–9.

7262. Gilyarov, M. S. 1942: Naphthalene as a means to control seed pests of biennial kok-saghyz. Compt. Rend. Acad. Sci. URSS 37:109–12. Chem. Abstr. 37:4850. 1943.

Gimeno N., F. 18934, 18935.

7263. Gimesi, N., and V. Frenyo. 1949: Method for the numerical evaluation of the effect of agents inhibiting germination. Agrartudomany 1:170–72. Chem. Abstr. 43:8009. 1949.

7264. ——, ——, M. Maroti, and B. Pozsar· 1952: Physiology of the germination of the tobacco seed. Acta Biol. Acad. Sci. Hung. 3:419–41. Chem. Abstr. 47:4437. 1953.

7265. —— and B. Pozsar. 1957: Physiological effect of heat stimulus on the germination of light- and dark-sensitive seeds.

Acta Biol. Acad. Sci. Hung. 7:135–51. Tobacco Abstr. 3:468. 1959.

7266. Gindele, F. J. 1929: Untersuchungen über die Wirkung chemischer Stoffe auf die Atmung keimender Samen. Botan. Arch. 23:532–78.

7267. Ginieis, C. 1956: Considerations generales sur la morphologie et l'anatomie des embryons de palmiers. Bull. Museum Natl. Hist. Nat. Paris 28:550–54.

7268. —— 1957: Étude morphologique et anatomique des embryons de quelques espèces du genre *Phoenix*. Bull. Museum Natl. Hist. Nat. Paris 29:439–45. Biol. Abstr. 33:2849. 1959.

7269. Ginns, R. 1954: Seeds and seed raising. Alpine Gardeners Soc. Quart. Bull. 22:245–48.

Ginsburg, V. 5771.

7270. Gioelli, F. 1931: Il gametofito femmineo e l'evoluzione dell'ovulo in seme in *Chamaerops humilis* L. Lavori R. Ist. Botan. Palermo 2:3–21. Botan. Centralbl. 163:264. 1932.

7271. —— 1935: Ricerche embriologiche sulla *Phyllostachys nigra* Munro. Ann. di Botan. 21:51–60.

Gioelli, F. 5158.

7272. Giordano, E., and G. Gemignani. 1962: Germination tests with some species of Eucalyptus and with a number of species obtained from nurseries. Cellulosa Carta (Rome) 13:17–20.

Gioshvili, V. D. 18398.

7273. Giovannini, E. 1958: Phosphorus metabolism in plants. Research on the seeds of *Pistacia vera.* Atti Simp. Intern. Agro-Chim. 2 Procchio 1958:322–28. Chem. Abstr. 54:7830. 1960.

7274. —— and C. Condorelli. 1956: Metabolism of phosphorus in the plants. Researches on the seeds of pistachio tree. Chimica (Milan) 12:194–99. Chem. Abstr. 50:16985. 1956.

7275. —— and —— 1956: Vitamin content of seeds of *Pistacia vera.* Giorn. Biochim. 5:542–60. Chem. Abstr. 51:15715. 1957.

7276. Giovannozzi-Sermanni, G., and M. E. Scarascia-Venezian. 1956: Analytical researches in *Nicotiana rustica* with paper-chromatography. Modifications of some reserve substances in seeds of different age. Tabacco 60:71–77. Biol. Abstr. 31:36032. 1957.

Giovannozzi-Sermanni, G. 16854, 16855.

Giovetti, G. L. 3740.

7277. Giral, F. 1947: Observaciónes sobre el aislamiento de hipaforina del colorin (*Erythrina americana* Mill.). Ciencia (Mexico) 8:19–20. Biol. Abstr. 23:12656. 1949.

7278. Giral, J., and M. T. Castillo. Determinations of copper in Mexican foods. Mem. Congr. Cient. Mex. IV Centenario Univ. Mex. 2:232–34. Chem. Abstr. 49:1981. 1955.

7279. —— and M. G. Mejia. 1953: Mineral elements in Mexican corn. Mem. Congr. Cient. Mex. IV Centenario Univ. Mex. 2: 167–74. Chem. Abstr. 49:2574. 1955.

7280. —— and L. Ortega. 1953: Lipophanerosis in cereal flours and brewer's yeast. Mem. Congr. Cient. Mex. IV Centenario Univ. Mex. 2:177–81. Chem. Abstr. 49:1982. 1955.

7281. Girard, R. 1939: Microchemical assay of *Strophanthus* seeds. Bull. Trav. Soc. Pharm. Bordeaux 77:23–25. Pharm. Abstr. 5:488–89. 1939.

Girard, R. 4431.

7282. Girgidov, D. A. 1961: The meteorological method of forecasting the yield of pine seeds. Akad. Nauk Latv. SSR Inst. Les'kokhoz. Prob. Tr. 22:182–87.

7283. Giri, K. V., and A. Sreenivasan. 1937: Das Amylasesystem des Reiskornes während des Reifens und Keimens. Biochem. Z. 289:155–66.

7284. —— and —— 1938: Die Änderung der amyloytischen Wirksamkeit beim Reiskorn während der Reifung und Keimung. Biochem. Z. 296:428–33.

7285. —— and V. Subrahmanyan. Studies in electro-culture. II. Influence of electrical treatment on the germination of barley and the diastatic activity of malt. J. Indian Inst. Sci. 14A:78–97. Chem. Abstr. 26:1383. 1932.

Giri, K. V. 3755, 14236–14238, 14454, 15845, 16044, 16754, 16756.

7286. Girton, R. E. 1936: Sterilization of corn grains with sodium hypochlorite. Plant Physiol. 11:635–39.

7287. —— and E. R. Park. 1942: Respiration studies on germinating white-oak acorns. Proc. Indiana Acad. Sci. 57:83–86.

7288. Gisevius. 1914: Untersuchungen über Keimfähigkeit und Triebfähigkeit. Fühl. Land. Ztg. 63:297–318.

7289. Gisquet, P., H. Hitier, C. Izard, and A. Mounat. 1951: Mutations naturelles observées chez *Nicotiana tabacum* L. et mutations experimentales provoquées par l'extrait à froid de graines vielles prematurement. Ann. Inst. Exptl. Tabac Bergerac 1(2):5–36. Biol. Abstr. 26:9591. 1952.

7290. Gisvold, O. 1934: Sterols from *Stramonium* seed. J. Am. Pharm. Assoc. 23:106–8. Chem. Abstr. 28:6522. 1934.

7291. —— 1934: Digitonin and phytosterol from the seed of *Digitalis purpurea*. J. Am. Pharm. Assoc. 13:664–66. Chem. Abstr. 28:5927. 1934.

7292. —— 1940: Crystalline xanthophyll from wheat germ. J. Am. Pharm. Assoc. 29:312–13.

7293. Gitman, L. S., and E. A. Boichenko. 1935: Injuries due to bacteriosis in the seeds of *Hibiscus cannabinus*. Botan. Zh. 20:487–93. Expt. Sta. Rec. 76:62. 1937.

Gitman, L. S. 5647.

Giumarro, C. 17414.

7294. Giza, C. A., and S. M. Siegel. 1963: Growth promoting derivatives of 5-hydroxybenzofuran. The effects of selected derivatives on germination and seedling growth. Physiol. Plantarum 16:52–56.

7295. Gladstones, J. S. 1958: The influence of temperature and humidity in storage on seed viability and hard-seededness in the West Australian blue lupin, *Lupinus digitatus* Forsk. Australian J. Agr. Res. 9:171–81.

Gladstones, J. S. 6601.

7296. Glamazdin, A. A. 1952: Germination of black locust (*Robinia pseudoacacia*) seeds. Les i Step 4(10):80–81.

7297. Glass, E. H., and W. F. Crosier. 1949: The viability of seeds fumigated with an acrylonitrile-carbon tetrachloride mixture. J. Econ. Entomol. 42:646–49.

7298. Glass, R. L., and W. F. Geddes. 1959: Grain storage studies. XXVII. The inorganic phosphorus content of deteriorating wheat. Cereal Chem. 36:186–90.

7299. —— and —— 1960: Grain storage studies. Cereal Chem. 37:568–72.

7300. ——, J. G. Ponte, Jr., C. M. Christensen, and W. F. Geddes. 1959: Grain storage studies. XXVIII. The influence of temperature and moisture level on the behavior of wheat stored in air or nitrogen. Cereal Chem. 36:341–56.

Glass, R. L. 2736, 12562, 14310, 14311.

7301. Glasscock, H. H. 1940: Blind seed disease of rye-grass. Nature 146:368–69.

7302. —— and R. L. Wain. 1940: Distribution of manganese in the pea seed in relation to marsh spot. J. Agr. Sci. 30:132–40.

7303. Glasziou, K. T., J. A. Sacher, and D. R. McCalla. 1960: On the effects of auxins on membrane permeability and pectic substances in bean endocarp. Am. J. Botany 47:743–52.

Glasziou, K. T. 16608.

7304. Gleisberg, W. 1924: Saatgutstimulierung, ein neuer Weg zu höheren Erträgen. Deut. Obst. Gemüsebau Ztg. 70:28–29.

7305. —— 1928: Der Einfluss der Samengrösse bei Radüschen auf Keim und Lebenleistung. (Eine Beitrag zur Gemüsesamenkontrolle.) Gartenbauwissenschaft 1:81–92.

7306. —— 1954: Saatgut- und Samenkeimungsuntersuchungen bei Heil- und Gewürzpflanzen. I. Gartenbauwissenschaft 19:181–217.

7307. Glendening, G. E. 1942: Germination and emergence of some native grasses in relation to litter cover and soil moisture. J. Am. Soc. Agron. 34:797–804. Expt. Sta. Rec. 89:52. 1943.

7308. —— and H. A. Paulsen. 1950: Recovery and viability of mesquite seeds fed to sheep receiving 2,4-D in drinking water. Botan. Gaz. 111:486–91.

7309. Glendinning, D. R. 1963: The inheritance of bean size, pod size, and number of beans per pod in cocoa (*Theobroma cacao* L.), with a note on bean shape. Euphytica 12:311–22.

7310. Glenny, W. W. 1897: The vitality and the dissemination of seeds. J. Roy. Agr. Soc. England 3 Ser. 8:324–34. Expt. Sta. Rec. 9:758. 1897–98.

7311. Glick, D. 1937: Quantitative distribution of ascorbic acid in the developing barley embryo. Compt. Rend. Trav. Lab. Carlsberg Ser. Chim. 21:203–9. Chem. Abstr. 31:4699. 1937.

7312. —— 1941: The enzymes of barley. Some recent contributions to our knowledge of their nature and properties. Wallerstein Lab. Commun. 4(11):20–24. Biol. Abstr. 15:19820. 1941.

7313. —— 1944: Simplified procedures for the determination of thiamine in wheat flours and bread by the thiochrome method. Cereal Chem. 21:119–26.

7314. —— and E. E. Fischer. 1946: Studies in histochemistry. XVII. Localization of phosphatases in the wheat grain and in the epicotyl and roots of the germinated grain. Arch. Biochem. 11:65–79.

7315. —— and —— 1946: Studies in histochemistry. XVIII. Localization of arginine in the wheat grain and in the epicotyl and roots of the germinated grain. Arch. Biochem. 11:81–87. Biol. Abstr. 22:7485. 1948.

7316. Glimcher, J. 1938: The germination of *Viscum cruciatum* Sieb. Palestine J. Botany Ser. J 1:103–5.

7317. —— 1951: On the biology of germination in the *Melanocrommyum* section of the genus *Allium*. Palestine J. Botany Ser. J 5:115–18.

7318. Glimm, E., and S. Halasa. 1931: Über die Verteilung des Jods im Getreidekorn. Biochem. Z. 243:88–96.

7319. Glinka, Z., and L. Reinhold. 1962: Rapid changes in permeability in cell membranes to water brought about by carbon dioxide and oxygen. Plant Physiol. 37:481–86.

7320. Glinyany, N. P. 1938: Vernalization of grain at the time of the formation of the embryo. Compt. Rend. Acad. Sci. URSS 20:189–94.

7321. —— 1940: Vernalization of seeds in the period of embryo formation. Compt. Rend. Acad. Sci. URSS 27:714–17.

7322. Glisic, L. M. 1928: Development of the female gametophyte and endosperm in *Haberlea rhodopensis* Friv. Bull. Inst. Jard. Botan. Univ. Belgrade 1:1–13.

7323. —— 1928: Zur Entwicklungsgeschichte der Solanaceen. Die Endospermbildung von *Datura metel* L. Bull. Inst. Jard. Botan. Univ. Belgrade 1:75–85.

7324. —— 1929: Über die Endosperm- und Haustorienbildung bei *Orobanche hederae* Duby und *Orobanche gracilis* Sm. Bull. Inst. Jard. Botan. Univ. Belgrade 1:106–41.

7325. —— 1934: Zur Kenntnis der Samentwicklung der Gesneriaceen. Über die Endosperm- und Haustorienbildung von Roettlera. Bull. Inst. Jard. Botan. Univ. Belgrade 3:94–111.

7326. —— 1937: Embryogeny of *Datura tatula* L. Bull. Inst. Jard. Botan. Univ. Belgrade 4:1–23. J. Roy. Microscop. Soc. Ser. III 58:220. 1938.

7327. —— 1937: Ein Versuch der Verwertung der Endospermmerkmale für typologische und phylogenetische Zwecke innerhalf der Scrophulariaceen. Bull. Inst. Jard. Botan. Univ. Belgrade 4:42–73.

7328. Glowczynski, Z. 1938: Vitamin A content of the seeds of canary grass (*Phalaris canariensis*). Bull. Intern. Acad. Polon. Sci. Classe Sci. Math. Nat. 2B: 115–28. Chem. Abstr. 33:2943. 1939.

7329. Gloyer, S. W., and H. A. Schuette. 1939: The sterols of rye germ oil. J. Am. Chem. Soc. 61:1901–3.

7330. Gloyer, W. O. 1921: Sclerema and hardshell, two types of hardness of the bean. Proc. Assoc. Offic. Seed Analysts 12/13:60.

7331. —— 1928: Hardshell of beans: its production and prevention under storage conditions. Proc. Assoc. Offic. Seed Analysts 20:52–55.

7332. —— 1928: Two new varieties of red kidney bean: Geneva and York. N.Y. Agr. Expt. Sta. Tech. Bull. 145. 51 pp.

7333. —— 1931: China aster seed treatment and storage. N.Y. Agr. Expt. Sta. Tech. Bull. 177. 41 pp.

7334. —— 1932: Percentage of hardshell in pea and bean varieties. N.Y. Agr. Expt. Sta. Tech. Bull. 195. 20 pp.

7335. Glückman, S. A. 1929/30: Materials to the question of the seed-stimulation. II. The present state of the problem of stimulation. Bull. Appl. Botany, Genet., Plant Breeding 23(2):231–56.

7336. —— 1929/30: Materials to the question of the seed-stimulation. III. The changes of P_H and P_{Cl} of the solutions during the soaking of seeds. Bull. Appl. Botany, Genet., Plant Breeding 23(2):257–98.

Glückman, S. 11558, 11559.

7337. Glushchenko, G. I. 1957: Contribution to the cyto-embryology of *Allium cepa* L. Izv. Akad. Nauk SSSR Ser. Biol. 1957(2):220–33.

Gluskina, B. G. 13669.

Glynn, L. M. 8938.

7338. Gmelin, R. 1959: Free amino acids in seeds of *Acacia willardiana*. Isolation of willardine, a new plant amino acid, probably L-uracil-1-(α-aminopropionic acid). Z. Physiol. Chem. 316:164–69. Chem. Abstr. 54:12285. 1960.

7339. —— 1962: Mimosacean amino acids. VI. Dichrostachnic acid, a new sulfur containing amino acid from the seeds of *Dichrostachys glomerata* (Forsk.) Hutch and Dalz (Mimosaceae). Z. Physiol. Chem. 327:186–94. Biol. Abstr. 43:7562. 1963.

7340. ——, G. Strauss, and G. Hasenmaier. 1958: Isolierung von 2 neuen pflanzlichen Aminosäuren: S(β-Carboxyäthyl)-L-Cystein und Albizzin aus den Samen von *Albizzia julibrissin* Durazz. (Mimosaceae). Z. Naturforsch. 13b:252–56.

Gmelin, R. 10749.

7341. Gnevyshev, N. 1956: Pre-sowing processing of seeds with microelements. Sad i Ogorod 1956(3):31–32.

7342. Goa, J., and L. Strid. 1959: Amino-acid content of leguminous proteins as affected by genetic and nutritional factors. III. Arch. Mikrobiol. 33:253–59. Chem. Abstr. 54:13278. 1960.

Goad, L. J. 16486, 16487.

7343. Goas, G. 1959: Sur le métabolisme des germinations de pois privées de leurs cotyledons. Croissance en présence de différentes formes d'azote. Compt. Rend. 249:2383–85.

Gobjila, M. 11907.

Gocho, H. 14116.

Godaishi, K. 13628–13630.

7344. Godbole, N. N., B. G. Gunde, and P. D. Srivastava. 1941: Seed fat of *Buchanania latifolia*. J. Indian Chem. Soc. 18:557–62. Analyst 67:199. 1942.

7345. Goddard, D. R. 1944: Cytochrome C and cytochrome oxidase from wheat germ. Am. J. Botany 31:270–76.

Goddard, D. R. 3111, 13351.

Goddard, R. E. 17435.

7346. Godel, G. L. 1935: Relation between rate of seeding and yield of cereal crops in competition with weeds. Sci. Agr. (Ottawa) 16:165–68.

7347. —— 1938: Cereal growing on weedy land in northeastern Saskatchewan. Effect of heavy seeding with the use of fertilizer on the development of weeds and crops. Sci. Agr. (Ottawa) 19:21–32.

Godfrey, G. 19962.

Godin, C. 19472.

7348. Godineau, J. C. 1962: Étude cytologique du caryopse d'orge au cours de la germination. Rev. Gen. Botan. 69: 577–622.

7349. Godkin, W. J., H. G. Beattie, and W. H. Cathcart. 1951: Retardation of rancidity in pecans. Food Technol. 5:442–47.

7350. Godman, R. M. 1953: Seed dispersal in southeast Alaska. U.S. Forest Serv. Alaska Forest Res. Center Tech. Notes 16. 2 pp.

7351. Godnev, T. N., and A. V. Mironenko. 1953: The amount of alkaloids in different varieties of lupine grown in Byelorussian S.S.R. Izv. Akad. Nauk Byelorussk. SSR 1953(4):63–71. Chem. Abstr. 49:10587. 1955.

7352. —— and E. F. Shabelskaya. 1963: Effects of low temperature treatment of seeds on the formation of photosynthesizing apparatus in tomatoes. Vestsi Akad. Navuk Belarusk. SSR, Ser. Biyal. Navuk 1963(3):49–55. Chem. Abstr. 60:2050. 1964.

7353. Godneva, M. T. 1958: Effect of gases present in the air on the accumulation of oil in seeds *Papaver somniferum*. Akad. Nauk Belorussk. SSR, Inst. Biol., Biol. Fiz. 1958:152–59.

7354. Godonneche, J., and G. Dastugue. 1934: Étude de l'action des eaux minérales arsenicales de la Bourboule sur la germination des graines et sur la croissance et les métamorphoses de larves de batraciens. Bull. Soc. Chim. Biol. 16:248–56.

7355. Godoy, C. 1954: Sand germinators for coffee. Rev. Agr. (Piracicaba) 29: 291–96.

7356. Goebelez, M. 1961: Un procédé peu utilisé pour trier les semences suivant les differences de densité. Proc. Intern. Seed Testing Assoc. 26:115–21.

7357. Goedert, R. D. 1960: Planting instructions for *Amaryllis seeds*. Plant Life 16:129–30.

Goedkoop, W. 14812.

7358. Goehler, F., and M. Drews. 1962: Direct seed germination as a rapid method for determining substances injurious to plants in horticultural soils. Arch. Gartenbau 10:333–46. Chem. Abstr. 59: 15597. 1963.

7359. Göksin, A. 1942: Altersermittlung beim Saatgut der Fichte und Kiefer. Forstwiss. Zentr. 64:111–17. Biol. Abstr. 17:22456. 1943.

Göksu, T. 1724.

Goenaga, A. 2043.

7360. Goerdeler, J. 1948: Extraction of lichenase from barley. Z. Naturforsch. 3b:403–6. Chem. Abstr. 43:8407. 1949.

7361. Goering, K. J., R. F. Eslick, and C. A. Ryan, Jr. 1957: Some effects of environment and variety on the amylose content of barley. Cereal Chem. 34: 437–43.

7362. Goettsch, H. B. 1955: Viability and germination in the field. Mededeel. Algem. Keuringsdienst Landbzaden Aardappelpootgoed 12:64–67.

Görlich, B. 15701, 15702.

Goetze, N. R. 4862.

7363. Goff, E. S. 1887: Vitality of seeds as influenced by age. N.Y. Agr. Expt. Sta. Rept. 5:50–53.

7364. —— 1890: The comparative vitality of hulled and unhulled timothy seeds. Wisconsin Agr. Expt. Sta. Ann. Rept. 7:202–4.

7365. —— 1892: A breeding experiment with tomatoes. Wisconsin Agr. Expt. Sta. Rept. 8:152–59.

7366. —— 1897: Influence of varying amounts of water on the germination of beet seed. Wisconsin Agr. Expt. Sta. Rept. 1897:299. Expt. Sta. Rec. 10:759. 1899.

7367. —— 1900: The effects of continued use of immature seed. Wisconsin Agr. Expt. Sta. Rept. 17:295–99.

7368. —— 1901: The relation of seediness to quality in melons. Science 12:310.

Gofman, Y. Y. 10811.

7369. Gogin, A. F. 1949: Duration period of cherry seed stratification. Sad i Ogorod 1949(6):91–92.

7370. Gogina, E. E. 1960: Seed productivity of some dominant species of the high-mountain meadows of South Ossetia. Botan. Zh. 45:1330–36.

7371. Gogiya, V. T. 1950: Formation of tannins and dynamics of oxidizing enzymes during sprouting of tea seeds. Biokhim. Chain. Proizv. Akad. Nauk SSSR Sb. 6:70–81. Chem. Abstr. 46:2631. 1952.

7372. Gogolev, F. T. 1957: Effect of variable heat regime of drying on seed quality. Selek. i Semen. 22(4):52–57.

7373. Goh, C. 1950: Etiolated seedlings. III. Determination of ascorbic acid content in dry seeds (*Phaseolus mungo*). J. Ferment. Technol. 28:79–86. Chem. Abstr. 47: 12531. 1953.

7374. —— 1950: Etiolated seedlings. IV. Determination of ascorbic acid and glutathione content at the time of germination and growth of seedlings. J. Ferment. Technol. 28:110–16. Chem. Abstr. 48: 1486. 1954.

7375. Gohrn, V. 1962: List of some areas suitable for import of forest seed to be used in Denmark. Dansk Skovfor. Tidsskr. 47:401–27.

7376. Goiko, V. A. 1951: On seed germination during fermentation of tomatoes. Sad i Ogorod 1951(7):70–71.

7377. —— 1953: On seedlessness of beet seed balls. Sad i Ogorod 1953(7):54–55.

7378. —— 1955: On measures for controlling seedlessness of vegetable seed plants. Sad i Ogorod 1955(7):18–19.

7379. Goiko, W. 1930: Der Einfluss der Erwärmung der Samen auf die Entwicklung der Pflanzen. Ukr. Inst. Angew. Botan. 1:344.

Gojkovic, N. Plavsic-; *see* Plavsic-Gojkovic, N.

7380. Gokhale, V. G., and P. M. Gaywala. 1933: The effect of the contact of chemical fertilizers with seeds on their germination. Agr. Livestock India 3:256–63. Chem. Abstr. 27:5134. 1933.

7381. Goksoyr, J., E. Boeri, and R. K. Bonnichsen. 1953: The variation of ADH (alcohol dehydrogenase) and catalase activity during the germination of the green pea (*Pisum sativum*). Acta Chem. Scand. 7:657–62.

Goksoyr, J. 2516.

7382. Gola, G. 1905: Ricerche sulla biologia e sulla fisiologia dei semi a tegumento impermeabile. Mem. Accad. Reale Sci. Torino 55:237–70. Botan. Centralbl. 101:438, 439. 1906.

7383. —— 1906: Über die Respirationstätigkeit einiger Samen wahrend der Ruheperiode. Jahresber. Forschr. Tierchemi. 36:763–64.

7384. Golaszewska, Z. 1934: Die Entwicklung des Embryosackes bei *Aspidistra elatior*. Acta Soc. Botan. Polon. 11:399–407.

7385. Golcz, L., and Z. Kowalewski. 1962: The glycoside content in *Erysimum*-seeds (*Erysimum perofskianum* Fisch. et May.) in different stages of their maturity. Poznan. Inst. Przem. Zielarskiego Bull. 8:35–38.

7386. Gold, A. H., C. A. Suneson, B. R. Houston, and J. W. Oswald. 1954: Electron microscopy and seed and pollen transmission of rod-shaped particles associated with the false stripe virus disease of barley. Phytopathology 44:115–17.

7387. Goldacre, P. L., A. W. Galston, and R. L. Weintraub. 1953: The effect of substituted phenols on the activity of the indoleacetic acid oxidase of peas. Arch. Biochem. Biophys. 43:358–73.

7388. Goldberg, G. A. 1961: Significance of the temperature factors in pre-sowing soaking of cotton plant seeds. Uzbeksk. Biol. Zh. 4:38–42. Biol. Abstr. 42:11856. 1963.

7389. Goldberg, H. S., editor. 1959: Antibiotics: Their chemistry and non-medical uses. Van Nostrand, Princeton, N.J. 608 pp.

7390. Goldberg, M. J. 1899: Sur la formation des matières proteiques, pendant la germination du blé à l'obscurité. Rev. Gen. Botan. 11:337–40.

7391. Goldberg, R. 1963: Évolution des glucides et plus specialement des composés à mannose au cours de la maturation des graines d'*Endymion non-scriptus* L. Compt. Rend. 256:3495–98.

Goldblatt, L. A. 2470.

Goldblith, S. A. 11718.

7392. Golden, A. M., and T. Shafer. 1959: Influence of germinating seeds of sugar beet (*Beta vulgaris*) on emergence of larvae from cysts of the sugar-beet nematode (*Heterodera schachtii*). Plant Disease Reptr. 43:1103–4.

7393. Goldfield, R., J. A. Bianculli, and R. W. Sager. 1954: Analytical constants of the fixed oil from the seed of *Pyrularia pubera*. Am. J. Pharm. 126:62–64. Chem. Abstr. 48:8481. 1954.

Goldfield, R. 3266.

Goldheim, S. L. 7128.

7394. Goldin, M. I. 1939: Tobacco mosaic virus propagation by tomato seeds. Microbiology (USSR) (English Transl.) 8:613–18. Chem. Abstr. 35:2565. 1941.

7395. Goldner, K. J., and C. H. Rogers. 1939: The preparation of the mannonic lactones from the seeds of the date palm (*Phoenix dactylifera*) and a study of their action and that of new derivatives upon gastric mucin smears. J. Am. Pharm. Assoc. 28:364–69.

7396. Goldovski, A., and A. Bozhenko. 1932: The carbohydrates of sunflower seeds. Maslob.-Zhir. Delo. 1932(7):30–34. Chem. Abstr. 27:3498. 1933.

7397. —— and —— 1932: The phosphorus-containing compounds in the sunflower seed. Maslob.-Zhir. Delo. 1932(10):21–27. Chem. Abstr. 27:3498. 1933.

7398. —— and —— 1933: The water-soluble organic acids of sunflower seeds. Maslob.-Zhir. Delo. 1933(1):13–20. Chem. Abstr. 27:5997. 1933.

7399. —— and M. I. Lishkevich. 1937: Seeds—composition of the oil phase of. Maslob.-Zhir. Delo. 1937(6):7–8. J. Soc. Chem. Ind. 57:546. 1938. Pharm. Abstr. 5:44. 1939.

7400. —— and —— 1938: Localization of some chemical constituents in the vegetable cell. Biochemistry (USSR) (English Transl.) 3:9–15. Chem. Abstr. 32:3785–86. 1938.

7401. —— and M. Podolskaya. 1932: The tannins of sunflower seeds. Maslob.-Zhir. Delo. 1932(12):20–27. Chem. Abstr. 27:3499. 1933.

Goldschmidt, S. Blumenthal-; *see* Blumenthal-Goldschmidt, S.

Goldschmidt-Blumenthal, S. 15499.

Goldsmith, R. E. 14954.

7402. Golenkov, V. F. 1961: Gluten in the seeds of *Ceratonia siliqua*. Tr. Vses. Nauchn.-Issled. Inst. Zerna i Produktov ego Pererabotki 41:101–6. Biol. Abstr. 42:11750. 1963.

7403. Golik, M. G. 1954: Accumulation of fat and its distribution in the grain of maize. Biokhim. Zerna Sb. 2 1954(2):226–35.

7404. —— 1954: Changes in the enzymic activity of ripening corn seeds. Tr. Vses. Nauchn.-Issled. Inst. Zerna 27:66–72. Chem. Abstr. 50:7238–39. 1956.

7405. —— 1956: Dehydrogenases of developing grain of maize. Biokhim. Zerna, Sb. 1956(3):156–61. Chem. Abstr. 51:8914. 1957.

7406. —— 1962: Effect of the degree of maturity of corn seed on its germination. Selek. i Semen. 1962(1):72–73. Biol. Abstr. 44:3657. 1963.

7407. —— and M. Festa. 1962: A comparative evaluation of methods of determining germination in seeds. Kukuruza 1962(3):25–28. Biol. Abstr. 39:24766. 1962.

7408. —— and N. I. Festa. 1962: Germination of corn seeds on various soils. Kukuruza 1962(2):38–39.

7409. Golinska, J. 1928: Germination des semences de *Solanum lycopersicum* dans le fruit. Acta Soc. Botan. Polon. 5:1–5.

7410. —— 1929: Effect of size of seed on development of *Brassica oleracea* var. *caulorapa* and *Raphanus sativus*. Roczniki Nauk Rolniczych Lesnych 21:23–46. Biol. Abstr. 6:21485. 1932.

7411. —— 1930: Étude de la valeur des semences de *Solanum lycopersicum* de differants degrés de maturation. Roczniki Nauk Rolniczych Lesnych 24:151–200.

7412. Gollan, J., P. J. Bueno, and M. Silverstrini. 1932: Determination of oil and moisture in oil seeds. Rev. Fac. Quim. Ind. Agr. Univ. Nacl. Litoral 2:152–55. Chem. Abstr. 27:5997. 1933.

Goller, A. Kipp-; *see* Kipp-Goller, A.

7413. Golubchuk, M. 1957: Studies on the relation of molds and mold products to "sick wheat" and other types of damage in stored wheat. Dissertation Abstr. 17:1889.

7414. ——, L. S. Cuendet, and W. F. Geddes. 1960: Grain storage studies. XXX. Chitin content of wheat as an index of mold contamination and wheat deterioration. Cereal Chem. 37:405–11.

7415. ——, H. Sorger-Domenigg, L. S. Cuendet, C. M. Christensen, and W. F.

Geddes. 1956: Grain storage studies. XIX. Influence of mold infestation and temperature on the deterioration of wheat during storage at approximately 12 per cent moisture. Cereal Chem. 33:45–52.

7416. Golubinski, I. N. 1941: Influence of temperature alternation on germinating power of hop seeds, *Humulus lupulus* L. Compt. Rend. Acad. Sci. URSS 32:85–86.

7417. Golubinskii, I. N. 1949: On hereditary differences in quality of seeds which sprouted at different times. Selek. i Semen. 16(10):24–26.

7418. —— 1950: Influence of seeds on each other during germination. Priroda 39: 58–59.

7419. —— and M. S. Schloss. 1939: A contribution to the problem of parthenogenesis in *Humulus lupulus* L. Compt. Rend. Acad. Sci. URSS 23:486–87.

Golumbic, C. 8275, 8276, 8278.

Gomes, F. P. 13689.

Gomez, A. Munoz-; *see* Munoz-Gomez, A.

Gomness, N. C. 6838.

7420. Gomolyako, S. E. 1960: The structure of seeds in monospermous sugar beets. Vestn. Sel'skokhoz. Nauki 11:30–35. Biol. Abstr. 37:10944. 1962.

7421. Gonchar, A. L. 1956: Increasing the germination of freshly harvested tobacco seeds. Dokl. Akad. Nauk Azerb. SSR 12:427–32.

7422. Gondar, J. 1963: High frequency treatment of oil seeds. Fette, Seifen, Anstrichmittel 65:196–202. Chem. Abstr. 59:845. 1963.

7423. Gonderman, R. L., and F. L. S. O'Rourke. 1961: Factors affecting the germination of *Koelreuteria* seed. Plant Propagators Soc. Proc. 11:98–100.

7424. Gonian, G. G. 1957: Effect of scarification on the germination of seeds of the wild growing perennial vetch in Armenia. Erivan. Armianskii Nauchn.-Issled. Inst. Zhivotn. i Vet. Tr. Zhivotn. i Kormodobyvanie 6:275–80.

7425. Gontaev, A. F. 1940: Viability of seeds of clover, *Cuscuta* isolated from manure. Proc. Lenin. Acad. Agr. Sci. 1940(10):29–32.

7426. Gonzalez, C. L., and K. J. Frey. 1959: Effect of seed size and hulls upon X-ray sensitivity of oat seeds. Proc. Iowa Acad. Sci. 66:123–28.

7427. Gonzalez-Molina, C. L. 1959: Factors that govern the sensitivity of oat seed to X-rays. J. Agr. Univ. Puerto Rico 43:194–211. Biol. Abstr. 35:14235. 1960.

7428. Gonzalez-Rios, P. 1929: La germinación de la semilla de café. Rev. Agr. Puerto Rico 22(10):147–48.

7429. Goo, M. 1948: Effects of individual seed weight and seed coat on the germination of the seeds collected from the young and old mother trees in *Cryptomeria japonica* D. Don. Bull. Tokyo Univ. Forests 36:1–10.

7430. —— 1951: Water absorption by tree seeds. Bull. Tokyo Univ. Forests 39: 55–60.

7431. —— 1951: The location and the velocity of entrance of water to the seeds of *Pinus thunbergii* Parl. Bull. Tokyo Univ. Forests 41:51–56.

7432. —— 1952: When cell division begins in the germinating seeds of *Pinus thunbergii* Parl. J. Japan. Forestry Soc. 34:3–4.

7433. —— 1954: Is germination per cent of sugi (*Cryptomeria japonica*) seeds affected by the position in the crown of their cones? J. Japan. Forestry Soc. 36:24–26.

7434. —— 1956: Effect of weight of seed on germination of *Cryptomeria japonica* and *Pinus thunbergii*. J. Japan. Forestry Soc. 38:114.

7435. —— 1956: Effect of seed coat on the germination of *Pinus thunbergii* and *P. densiflora*. J. Japan. Forestry Soc. 38:276.

7436. —— 1956: A physiological study of germination of coniferous seed by the application of water absorption curve. Bull. Tokyo Univ. Forests 51:159–236.

7437. —— 1962: Germination of *Alnus maximowiczii* Call. seed. J. Japan. Forestry Soc. 44:217–18.

7438. —— and H. Hiramatu. 1951: Seed maturity and germination power. Bull. Tokyo Univ. Forests 48:103–14. Biol. Abstr. 30:14811. 1956.

7439. —— and K. Negisi. 1952: Changes of reserve foods in seeds of kuromatu (*Pinus thunbergii* Parl.). Bull. Tokyo Univ. Forests 43:83–89.

7440. —— and H. Tusihasi. 1958: Hastening the germination of *Pinus densiflora* seeds by gibberellin. J. Japan. Forestry Soc. 40:509–11.

Goo, M. 14143, 16804, 16805.

7441. Good, R. D'O. 1924: The germination of *Hippuris vulgaris* L. J. Linn. Soc. London 46:443–48.

7442. Goodacre, A. G. 1932: Dahlia seed growing. Flower Grower 19(10):439.

7443. Goodall, D. W., and B. D. Bolas. 1940: Vernalisation of tomatos. Gardeners Chron. III 108:261–62.

7444. —— and —— 1942: The vernalization of tomato seed. Ann. Appl. Biol. 29:1–10.

7445. Goode, G. P. 1932: The formation of vitamin A in corn sprouts by light and the transfer of the vitamin from the sprout to the grain. Bull. Basic Sci. Res. 4:55–58. Chem. Abstr. 27:2174. 1933.

7446. Goodey, J. B. 1949: Control of *Anguilluina dipsaci* on the seed of teazle and red clover by fumigation with methyl bromide. J. Helminthol. 23:171–74. Chem Abstr. 44:5511. 1950.

Gooding, J. A. 13521.

7447. Goodman, J. J., and C. M. Christensen. 1952: Grain storage studies. XI. Lipolytic activity of fungi isolated from stored corn. Cereal Chem. 29:299–308.

7448. —— and A. W. Henry. 1947: Action of subtilin in reducing infection by a seed-borne pathogen. Science 105:320–21.

Goodrum, P. D. 16095.

7449. Goodsell, S. F. 1948: Triphenyltetrazolium chloride for viability determination of frozen seed corn. J. Am. Soc. Agron. 40:432–42.

7450. —— 1957: Germination of dormant sorghum seed. Agron. J. 49:387–89.

7451. ——, G. Huey, and R. Royce. 1955: The effect of moisture and temperature during storage on cold test reaction of *Zea mays* seed stored in air, carbon dioxide, or nitrogen. Agron. J. 47:61–64.

7452. Goodson, J. A. 1943: Alkaloids of the seeds of *Delphinium elatum* L. J. Chem. Soc. 1943:139–41. Chem. Abstr. 37: 4401–2. 1943.

7453. —— 1945: Delphinium alkaloids. IV. The alkaloids of the seeds of *Delphinium ajacis*. J. Chem. Soc. 1945:245–46.

7454. Goodspeed, T. H. 1915: Notes on the germination of tobacco seed. II. Univ. Calif. (Berkeley) Publ. Botany 5:233–48.

7455. —— 1915: Parthenogenesis, parthenocarpy and phenospermy in *Nicotiana*. Univ. Calif. (Berkeley) Publ. Botany 5:249–72.

7456. —— 1919: Notes on the germination of tobacco seed. III. Note on the relation of light and darkness to germination. Univ. Calif. (Berkeley) Publ. Botany 5:451–55.

7457. —— 1929: The effects of X-rays and radium on species of the genus *Nicotiana*. J. Heredity 20:243–59.

7458. Goodwin, R. H. 1948: How to grow dogwood from seed. Plants Gardens 4: 236–38.

Goodwin, T. W. 4261.

7459. Goor, A. Y. 1955: Tree planting practices for arid areas. Food Agr. Org. U.N. FAO Forestry Develop. Paper 14. 126 pp.

Goorley, J. T. 6700.

7460. Goos, A., and L. G. Ri. 1959: The influence of dressing with DDT and BHC preparations on the germination power of some seeds. Wroclaw. Wyzsza Szkola Roln. Zeszyty Nauk Roln. 8:64–76.

Goos, R. D. 4483, 18065.

Goossens, A. P. 2838.

7461. Gopalachari, N. C. 1963: Changes in the activities of certain oxidizing enzymes during germination and seedling development of *Phaseolus mungo* and *Sorghum vulgare*. Indian J. Exptl. Biol. 1:98–100. Chem. Abstr. 59:7862. 1963.

Gopalachari, N. C. 14972–14974.

Gopalachari, R. 5330.

7462. Gopalakrishma, K. S., P. A. Kurup, and P. L. N. Rao. 1954: Antibiotic principles from *Moringa pterygosperma*. III. Action of pterygospermin on germination of seed and filamentous fungi. Indian J. Med. Res. 42:97–99. Chem. Abstr. 48:10083. 1954.

Gopalakrishnan, P. 17839.

7463. Gopalakrishnan, S., and L. Veerannah. 1962: Studies on the germinating groundnut seed (TMV. 2) in red and black soils treated with micro-nutrients. Madras Agr. J. 49:405–11.

7464. Gopalkrishnan, K. S., and J. A. Jump. 1952: The activity of thiolutin against certain fungi and seed-borne diseases. Proc. Indian Acad. Sci. 61:97–102.

Gopalkrishnan, S. 14073.

7465. Goplen, B. P., J. E. R. Greenshields, and H. Baenziger. 1957: The inheritance of coumarin in sweet clover. Can. J. Botany 35:583–93.

7466. Gopp, K., and A. Graf. 1960: Über die Dauer der Keimruhe bei Gersten-Zuchtstämmen. Braugersten Jahrb. 1958/59:73–82.

7467. Gorbunov, M. G. 1954: Grape seeds form pliocene deposits of Chulym River lowlands in western Siberia. Dokl. Akad. Nauk SSSR 97:527–30.

Gordin, A. 15445.

Gordon, D. R. 4059.

7468. Gordon, E. M. 1951: Light- and temperature-sensitiveness in germinating

seed of timothy (*Phleum pratense* L.). Sci. Agr. (Ottawa) 31:71–84.
Gordon, J. J. 14559.
7469. Gordon, M. 1922: The development of endosperm in cereals. Proc. Roy. Soc. Victoria 34:105–16.
7470. Gordon, S. A. 1946: Auxin-protein complexes of the wheat grain. Am. J. Botany 33:160–69.
Gordon, S. A. 18182.
7471. Gordon, S. M. 1928: The seeds of *Nepeta cataria*. Am. J. Pharm. 100:155–56.
7472. Gordon, W. L. 1952: The occurrence of *Fusarium* species in Canada. II. Prevalence and taxonomy of *Fusarium* species in cereal seed. Can. J. Botany 30:209–51.
7473. Gordon, W. T. 1941: On *Salpingostoma dasu*, a new carboniferous seed from East Lothian. Trans. Roy. Soc. Edinburgh 60:427–64. Botan. Centralbl. 178:153–54. 1943.
Gore, H. C. 10039.
7474. Gore, U. R. 1932: Development of the female gametophyte and embryo in cotton. Am. J. Botany 19:795–807.
7475. —— 1943: Delinting and treating cottonseed in Georgia. Georgia Agr. Expt. Sta. Circ. 141. 18 pp. Chem. Abstr. 37:4192. 1943.
Gorelik, M. 16108.
Gorelkina, T. M. 10886.
7476. Gorenflot, R. 1954: Germinations et plantules des formes tétraploides de *Plantago coronopus* L. Compt. Rend. 238:505–7.
Gorenflot, R. 2865.
7477. Gorfinkel, L. 1948: The influence of crossing on the nicotinic acid content of maize. J. Agr. Sci. 38:339–42.
7478. Gori, C. 1956: Aberrations and retarded development in plantules and plants from aged seed of *Vicia faba* L. Caryologia 9:76–81.
7479. —— 1957: On the embryology and cytology of some species of the genus *Reseda*. Caryologia 10:391–401.
7480. Gorin, D. P. 1955: On some problems of growing elite seeds. Zemledelie 3(5): 81–83.
7481. Gorlach, A. A. 1960: Biotypes of winter wheat based on length of the period of post-harvest ripening of the seeds. Agrobiologia 1960(3):447–52.
7482. Gorlenko, M. V., and I. I. Shneider. 1956: Disinfection of bean seeds from bacteriosis by the method of summer plantings. Dokl. Vses. Akad. Sel'skokhoz. Nauk 21(8):38–40.
Gorman, E. A. 18682.
7483. Gorman, L. W. 1939: Low germination susceptibility of perennial rye-grass strains. New Zealand J. Sci. Technol. Sect. A 20:392–401.
7484. —— 1940: Blind seed disease investigations. New Zealand J. Sci. Technol. Sect. A 22:79–83.
7485. —— and R. M. Greenwood. 1951: Effect of moisture content on the viability of Italian ryegrass seed in storage. New Zealand J. Sci. Technol. Sect. A 33:58–61.
7486. Gorostiaga, A., and R. F. Guerreros. 1943: The carob tree. Rev. Mens. Asoc. Rural Uruguay 70(3):9–11. Chem. Abstr. 37:6758. 1943.
7487. Gorski, M., and A. Ostrowska. 1959: Influence of seed-soaking in radioactive calcium (^{45}Ca) and strontium (^{90}Sr) salts on plant crops. Roczniki Gleboznawcze 7:3–25.
7488. Gorter, G. J. M. A. 1961: A guide to South African literature on plant diseases. Pretoria Div. Plant Pathol. Dept. Agr. Tech. Serv. 106 pp.
7489. Gorter, J. 1955: In vitro culture of *Cyclamen* embryos. Koninkl. Ned. Akad. Wetenschap. Proc. Ser. C 58:377–85.
7490. Gortner, R. A. 1929: Outlines of biochemistry. Wiley, New York. 793 pp.
7491. —— and R. T. MacDonald. 1944: Studies on the fractionation of zein. Cereal Chem. 21:324–33.
Gortner, R. A. 12166.
Goryachikh, A. N. 169, 170.
7492. Goryaev, M. I., and L. V. Katsova. 1959: Isolation of glycosides from squash seeds. Tr. Alma-Atinsk. Zoovet. Inst. 11:217–19. Chem. Abstr. 56:717. 1962.
Goryunov, D. 1739.
Gorz, H. J. 8340, 11764, 16861.
Goss, J. A. 10131.
Goss, J. R. 18166.
Goss, M. J. 15321.
Goss, O. M. 17673.
7493. Goss, W. H. 1948: Oil from grain. U.S. Dept. Agr., Agr. Res. Admin. AIC Rept. 208. 15 pp.
7494. —— and J. E. Ruckman. 1944: Oil from tumbling mustard seed. Oil Soap 21:234–36. Expt. Sta. Rec. 95:459. 1946.
7495. Goss, W. L. 1923: The germination of Kentucky blue grass seed in blotters, bell jars and Petri dishes. Proc. Assoc. Offic. Seed Analysts 14/15:119–20.

7496. —— 1924: The vitality of buried seeds. J. Agr. Res. 29:349–62.
7497. —— 1931: Viability of seed as shown by laboratory tests. Calif. Dept. Agr. Bull. 20:319–21.
7498. —— 1933: Buried seed experiment. Calif. Dept. Agr. Bull. 22:302–4.
7499. —— 1937: Germination of flower seeds stored for ten years in the California State Seed Laboratory. Calif. Dept. Agr. Bull. 26:326–33.
7500. —— 1939: Germination of buried weed seeds. Calif. Dept. Agr. Bull. 28: 132–35.
7501. —— 1939: Vegetable seed growing, particularly onion, and laboratory testing for germination. Calif. Dept. Agr. Bull. 28:253–65.
7502. —— and E. Brown. 1939: Buried red rice seed. J. Am. Soc. Agron. 31:633–37.
7503. —— and —— 1940: Buried red rice seed. J. Am. Soc. Agron. 32:974.
Goss, W. L. 3128, 3129, 3133, 3134, 18664, 18665.
7504. Gossard, A. C., and H. L. Crane. 1955: Germination of pecan nuts. Northern Nut Growers Assoc. Ann. Rept. 46:84–89.
Gossard, A. C. 13203.
7505. Gostick, K. G. 1963: Control of seed-borne lucerne stem eelworm with phorate. Ann. Appl. Biol. 51:503–7.
7506. —— 1963: Effect of temperature on methyl bromide fumigation of lucerne seed for control of stem eelworm. Plant Pathol. 12:62–64.
7507. —— and D. M. Cranston. 1964: The sorption of methyl bromide by lucerne seed. Plant Pathol. 13:1–6.
Gothan, W. 20111.
7508. Goto, Y., M. Yamamoto, M. Yatazawa, and K. Okami. 1954: Boron deficiency in crops. I. Obstacle in the seed maturation in rape. Shiga Kenritsu Nogyo Tankidaigaku, Gakujutsu Hokoku Ser. I No. 6:25–29. Chem. Abstr. 51:9992. 1957.
Goto, Y. 11000, 11303.
7509. Gott, M. B., F. G. Gregory, and O. N. Purvis. 1955: Studies in vernalisation of cereals. XIII. Photoperiodic control of stages in flowering between initiation and ear formation in vernalised and unvernalised Petkus winter rye. Ann. Botany (London) 19:87–126.
7510. Gottl, R. 1963: Hybrid tomato seed production. Agr. Gaz. N.S. Wales 74: 153–56.
Goulart, M. D. 5066.
7511. Gould, C. J. 1949: Vegetable seed-treatment in western Washington. Wash. Agr. Expt. Sta. Bull. 510. 45 pp.
7512. —— and C. E. Reynolds. 1949: Treating Douglas fir seed with "Spergon" before stratification. J. Forestry 47:984.
Gould, C. J. 9169.
7513. Gould, G. E. 1955: Studies on germination of seed corn treated to control soil insects. Purdue Univ. Agr. Expt. Sta. Bull. 624. 10 pp.
7514. Gould, M. F., and R. N. Greenshields. 1964: Distribution and changes in the galactose-containing oligosaccharides in ripening and germinating bean seeds. Nature 202:108–9.
7515. Gould, S. A., R. Pearl, and T. I. Edwards. 1934: On the effects of the partial removal of the cotyledons upon the growth and duration of life of canteloupe seedlings without exogenous food. Ann. Botany (London) 48:575–99.
7516. ——, ——, ——, and J. R. Miner. 1933: Available food, relative growth and duration of life in seedlings of *Cucumis melo*. Proc. Natl. Acad. Sci. U.S. 19: 228–33.
7517. Goulden, C. H. 1945: A seed dispenser-device for measuring seed by volume for rod row plots. Sci. Agr. (Ottawa) 25:707–10.
7518. —— 1951: The Canadian system of collecting and maintaining genetic stocks of cereals. Proc. U.N. Sci. Conf. Conserv. Util. Resources 6:286–89. 1949.
7519. —— and W. J. Mason. 1958: An electronic seed counter. Can. J. Plant Sci. 38:84–87.
Goulden, C. H. 19627.
7520. Goursat, M. J. 1961: Développement de l'embryon chez le *Lathyrus latifolius* L. Compt. Rend. 252:3858–60.
7521. —— 1961: Embryogénie des Legumineuses. Développement de l'embryon chez le *Scorpiurus vermiculata* L. Compt. Rend. 253:307–9.
7522. —— 1961: Embryogénie des Légumineuses. Développement de l'embryon chez le *Pacockia*. Compt. Rend. 253:960–62.
7523. —— 1962: Embryogénie des Légumineuses. Développement de l'embryon chez le *Bisserula pelecinus* L. Compt. Rend. 254:153–55.
7524. —— 1962: Embryogénie des Légumineuses. Développement de l'embryon chez le *Spartium junceum* L. Compt. Rend. 254:2050–52.

7525. —— 1963: Développement de l'embryon chez le *Cercis siliquastrum* L. Compt Rend. 257:3452–54.

7526. Gover, R., C. H. Lindquist, and E. W. Martin. 1962: Seed and propagation studies: elm seed viability. Extr. Summ. Rept. Forest Nursery Sta. Sask. 1962:21–22. Forestry Abstr. 24:412. 1963.

7527. Govindappa, D. A. 1956: The development of endosperm in *Scilla indica* Baker. Current Sci. (India) 25:198–99.

7528. —— and T. R. B. Naidu. 1956: The embryo sac and endosperm of *Blyxa oryzetorum* Hook. f. J. Indian Botan. Soc. 35:417–22.

7529. —— and K. Shamakumari. 1957: Development of the embryo in *Hypoxis aurea* Lour. J. Indian Botan. Soc. 36: 324–27.

7530. Govindarajan, V. S., and A. G. Mathew. 1963: Polyphenolic substances of arecanut. I. Chromatographic analysis of fresh mature nut. Phytochemistry 2: 321–26.

7531. Govindaswamy, C. V. 1955: Studies on seed-borne fungi of rice. Madras Agr. J. 42:56–65.

7532. —— 1957: Seed treatment and seedling vigour. Madras Agr. J. 44:63–64. Biol. Abstr. 32:2858. 1958.

7533. —— 1957: Seed-borne fungi and seed treatment in sorghum. Madras Agr. J. 44:247–48.

7534. Govindu, H. C. 1950: Studies in the embryology of some members of the Bignoniaceae. Proc. Indian Acad. Sci. Sect. B 32:164–78. Biol. Abstr. 25:28836. 1951.

Govindu, H. C. 16622.

Goyco, J. A. 1462, 1463.

7535. Gozdetzkii, N. 1935: Spontaneous ignition of seeds. Maslob.-Zhir. Delo. 11: 612–13. Chem. Abstr. 30:4346. 1936.

Graaf, J. de; *see* De Graaf, J.

7536. Grabe, D. F. 1956: Maturity in smooth bromegrass. Agron. J. 48:253–58.

7537. —— 1957: Artificial drying and storage of smooth bromegrass seed. Agron. J. 49:161–65.

7538. —— 1957: Laboratory methods for identification of varieties of soybeans and oats. Iowa State Coll. J. Sci. 32:179.

7539. —— 1963: Seed corn-storage and vigor. Seed World 92(10):12, 14.

7540. Graber, L. F. 1922: The effect of scarifying on the keeping power of alfalfa seed. J. Am. Soc. Agron. 14:298–302.

7541. Grable, A. R. 1964: The effects of carbon dioxide on germination and growth of corn and soybeans in soil containing different levels of oxygen and moisture. Dissertation Abstr. 24:2644.

7542. Gracanin, M. 1926: Ein Beitrag zur Kenntnis der Katalasewirkung bei autotrophen Pflanzen. Biochem. Z. 168:429–42.

7543. —— 1927: Über das Verhältnis zwischen der Katalase-aktivität und der Samenvitalität. Biochem. Z. 180:205–10.

7544. —— 1928: Orthophosphorsäure als Stimulator der Keimungsenergie und Aktivator der Keimfähigkeit der Samen. Biochem. Z. 195:457–68.

7545. Grace, N. H. 1937: Physiologic curve of response to phytohormones by seeds, growing plants, cuttings, and lower plant forms. Can. J. Res. Sect. C 15:538–46.

7546. —— 1938: Effect of phytohormones on seeds damaged by formaldehyde and other disinfectants. Can. J. Res. Sect. C 16:313–29.

7547. —— 1939: Effects of plant and animal hormones on seeds damaged by formaldehyde. Can. J. Res. Sect. C 17:445–51.

7548. —— 1940: The effect of nutrient salts in organic mercurial seed disinfectants on the germination and early growth of wheat. Can. J. Res. Sect. C 18:151–57.

7549. —— 1940: Effects of two preparations of naphthylacetic acid on the germination and early growth of wheat seed damaged by formaldehyde. Can. J. Res. Sect. C 18:215–18.

7550. —— 1941: Effects of phytohormone, potassium nitrate, and ethyl mercuric bromide on the germination and early growth of wheat. Can. J. Res. Sect. C 19:211–15.

Grace, N. H. 8913, 13257.

7551. Gracie, D. S., and W. L. Balls. 1939: Dibble-sowing of cotton. Method, effects and profits. Egypt, Min. Agr. Tech. Sci. Serv. Bull. 229:1–49. Biol. Abstr. 14: 3366. 1940.

7552. Gracio, J. T. M., and N. O. Lima. 1961: Effect of seed phosphorus on "L values." Agron. Lusitana 23:207–27. Biol. Abstr. 45:25816. 1964.

7553. Gracza, L. 1958: Paper-chromatographic examination of germination- and growth-inhibiting substances found in the dry fruits of *Gleditsia.* Acta Univ. Szeged. Acta Biol. (N.S.) 4:51–57. Chem. Abstr. 53:14252. 1959.

7554. —— 1958: Relation between the germination vigor of cereals and the auxin-sensitivity of their coleoptiles. Acta Agron. Acad. Sci. Hung. 8:161–65. Chem. Abstr. 53:10388. 1959.

7555. Gradi, A. 1963: Study of the water content of strobiles and seeds, determining factor for proper preparation of Coniferae seeds and their preservation. Monti e Boschi 14:195–208.

7556. Gradinarov, L. 1957: Studies on the absorption of water in relation to the quality of the seeds of field beans. Izv. Inst. Biol. "Metodii Popov" Bulgar. Akad. Nauk. 8:185–206.

7557. —— 1958: Non-simultaneous swelling as symptom of qualitative difference in bean seeds. Compt. Rend. Acad. Bulgare Sci. 11:217–20.

7558. Graef, P. E. 1955: Ovule and embryo sac development in *Typha latifolia* and *Typha angustifolia*. Am. J. Botany 42: 806–9.

7559. —— 1958: The ovule and embryo sac of *Galax aphylla*. Virginia J. Sci. (N.S.) 9:319–22.

Graesser, F. R. 5050.

7560. Graetz, K. E. 1952: A small thresher-dehuller-seed counter. Agron. J. 44:386.

7561. —— 1953: A continuous seed huller. Agron. J. 45:577.

Graeve, P. de; *see* De Graeve, P.

Graf, A. 7466.

Graf, L. H. 20102.

Graf, V. 16143.

7562. Grafe, V. 1924: Methodik der Beeinflussung der Samenkeimung und des Wachstums von Keimpflanzen. *In* E. Abderhalden, Handbuch der biologischen Arbeitsmethoden, pp. 445–548.

7563. —— 1936: The relation of the bitter principle of *Chichorium intybus* to the other contents of the plant and its migration in the germinating plant. Beitr. Biol. Pflanz. 24:138–49. Chem. Abstr. 31:2638–39. 1937.

7564. Grafius, J. E., and R. L. Kiesling. 1960: The production of the relative yields of different oat varieties based on known environmental variables. Agron. J. 52: 396–99.

7565. ——, W. L. Nelson, and V. A. Dirks. 1952: The heritability of yield in barley as measured by early generation bulked progenies. Agron. J. 44:253–57.

Graham, G. J. 9626.

7566. Graham, J. F., J. A. Bullock, and A. W. Peers. 1959: Seed-dressing against the black wheat beetle (*Heteronychus* spp.) in Kenya. E. African Agr. J. 25:23–24.

7567. Graham, J. S. D., A. C. Jennings, R. K. Morton, B. A. Palk, and J. K. Raison 1962: Protein bodies and protein synthesis in developing wheat endosperm. Nature 196:967–69.

7568. —— and R. K. Morton. 1963: Proteins of developing wheat endosperm; separation by starch-gel electrophoresis and incorporation of sulfate-S^{35}. Australian J. Biol. Sci. 16:357–65. Chem. Abstr. 59:5490. 1963.

7569. ——, ——, and J. K. Raison. 1963: Isolation and characterization of protein bodies from developing wheat endosperm. Australian J. Biol. Sci. 16:375–83. Biol. Abstr. 45:7601. 1964.

7570. ——, ——, and D. H. Simmons. 1963: Studies of proteins of developing wheat endosperm fractionation by ion-exchange chromatography. Australian J. Biol. Sci. 16:350–56. Biol. Abstr. 45:7602. 1964.

Graham, R. P. 9024.

Graham, R. W. 18498.

7571. Graham, W. M. 1959: Fumigation of maize in a large shell-concrete building. J. Sci. Food Agr. 10:478–83.

7572. Grahl, A. 1960: Die Reduktion der Kälteresistenz im Koleoptilenstadium des Weizens durch Vernalisation. Beitr. Biol. Pflanz. 35:447–71.

7573. —— and H. Grimm. 1959: Beeinflussung der Keimruhe von Gerstenkörnern durch Chemikalien. Naturwissenschaften 46:336. 1959.

7574. —— and M. Thielebein. 1959: Einfluss von Licht auf die Keimung der Gerste. Naturwissenschaften 46:336–37.

7575. ——, ——, and O. Fischnich. 1962: Keimung von Gersten-Karyopsen während der primären Keimruhe in Abhängigkeit von ihrem Sitz in der Ähre. Naturwissenschaften 49:41–42.

Grahl, A. 6413, 6417–6420, 18479.

7576. Graillot, G. 1951: Action sur la germination des graines de chicorée de quelques corp à fonction lactone derivés de la coumarine. Étude spectrale de ces derivés. Bull. Soc. Chim. Biol. 33:1584–90.

7577. ——, J. Guerillot, and A. Guerillot-Vinet. 1948: Action inhibitrice de la coumarine sur la germination, relations avec la vitamine C des graines. Grignon. Ann. Ecole Natl. Agr. Ser. 3, 6:5–15.

7578. Grainger, J., and D. E. Simpson. 1950: Electronic heating and control of seed-borne diseases. Nature 165:532–33.
Grainger, R. B. 6512.
Gralen, N. 13120.
7579. Gralley, E. M., and E. C. Tullis. 1937: Effect of seed treatments on seedling emergence, severity of seedling blight and yield of rice. Arkansas Agr. Expt. Sta. Bull. 345. 24 pp.
7580. Gram, E. 1929: Seed disinfection. III. Cereals and graminaceous seeds. Tidsskr. Planteavl 35:141–268. Expt. Sta. Rec. 66:643. 1932.
Gramling, L. G. 6538.
7581. Grande, F. 1934: Occurrence of palmito- and stearo-dehydrogenases in certain plant seeds. Skand. Arch. Physiol. 69:189–96. Chem. Abstr. 28:5499–500. 1934.
7582. Grandel, F. 1940: The grain germ—a valuable source of biologically active substances. Umschau 44:100. Chem. Abstr. 36:3532. 1942.
7583. —— and H. Neumann. 1940: Fat-chemistry studies concerning rancidity processes in cereals, germ, flour and bread. I. Mühlenlab. 10:1–6. Chem. Abstr. 36:175. 1942.
7584. Grandfield, C. O. 1945: Alfalfa seed production as affected by organic reserves, air temperature, humidity, and soil moisture. J. Agr. Res. 70:123–32.
7585. —— 1950: How to double yields of alfalfa seed. Intern. Crop Improvement Assoc. Ann. Rept. 32:69–71.
Grandjean, M. 4267.
7586. Graner, E. A. 1945: The yellow-orange endosperm of maize. Am. Naturalist 79(781):187–92.
7587. Granick, S. 1937: Urease distribution in *Canavalia ensiformis*. Plant Physiol. 12:601–23.
7588. —— 1938: Urease distribution in *Soja max*. Plant Physiol. 13:29–54.
7589. Granitova, O. N. 1955: Effect of temperature and moisture on seed germination of certain central Asiatic (desert) plants. Tr. Akad. Nauk Uzbeksk. SSR Inst. Botan. 3:63–101.
7590. Grano, C. X. 1958: Tetrazolium chloride to test loblolly pine seed viability. Forest Sci. 4:50–53.
7591. —— 1961: Germination of stratified latex-coated loblolly pine seed after submergence. J. Forestry 59:452.
Granström, G. 17449.
Grant, R. 13782.
7592. Grasso, V. 1949: Bunt control with copper and mercury compounds in relation to their influence on the germination capacity of wheat kernels. Ann. Sper. Agrar. (Rome) 3:451–88. Chem. Abstr. 44:2689. 1950.
7593. Gratkowski, H. J. 1962: Heat as a factor in germination of seeds of *Ceanothus velutinus* var. *laevigatus* T. & G. Dissertation Abstr. 23:1890–91.
7594. Grau, C. R., and H. J. Almquist. 1945: Value of sunflower-seed protein. Proc. Soc. Exptl. Biol. Med. 60:373–74. Chem. Abstr. 40:1564. 1946.
Gravatt, A. Rathbun-; *see* Rathbun-Gravatt, A.
7595. Graves, H. C. H. 1946: The nutritive value of wheat flour proteins. Chem. Ind. (London) 1946:261–63.
7596. Graves, H. S. 1915: The place of forestry among natural sciences. Science 41:117–27.
7597. Gray, A. 1893: Germination of *Aralia quinquefolia*. Kew Bull. Misc. Inform., pp. 71–75.
7598. Gray, D. O., and L. Fowden. 1962: α-(Methylenecyclopropyl) glycine from *Litchi* seeds. Biochem. J. 82:385–89.
Gray, E. 18411.
7599. Gray, E. K. 1934: How to grow amaryllis from seed. Flower Grower 21(7):310.
Gray, H. E. 4444.
7600. Gray, N. E., and H. J. Fuller. 1942: Effects of mercury vapor upon seed germination. Am. J. Botany 29:456–59.
7601. Gray, S. G. 1942: Increased earliness of flowering in lettuce through vernalization. Australian Council Sci. Ind. Res. J. 15:211–12.
7602. —— 1962: Hot water seed treatment for *Leucaena glauca* (L.) Benth. Australian J. Exptl. Agr. Animal Husbandry 2:178–80.
7603. —— and L. Sharp. 1943: The production of Swede turnip seed at Canberra, A. C. T. Australian Council Sci. Ind. Res. J. 16:104–6.
7604. Grayson, J. M. 1948: Germination of fumigated peanuts. J. Econ. Entomol. 41:816–17.
7605. Greaney, F. J. 1946: Cooperative flax seed treatment tests in 1945. Plant Disease Reptr. 30:113–19.
7606. —— and J. E. Machacek. 1939: The epidemiology of seed-borne microorganisms in cereals. Phytopathology 29:8.
7607. —— and —— 1942: Prevalence of seed-borne fungi on cereals in certain seed

inspection districts of Canada. Sci. Agr. (Ottawa) 22:419–37.

7608. —— and —— 1946: The prevalence and control of seed-born diseases of cereals in Manitoba. Sci. Agr. (Ottawa) 26:59–78.

7609. —— and H. A. H. Wallace. 1943: New and standard seed treatments in the control of certain seed-borne diseases of wheat, oats, and barley. Phytopathology 33:1064–70.

Greaney, F. J. 12603.

7610. Greaney, J. B. 1946: Cooperative flax seed treatment tests in 1945. Plant Disease Reptr. 30:113–19. Rev. Appl. Mycol. 25:560. 1947.

7611. Greathead, A. 1962: New mosaic-free lettuce seed program promising in Salinas Valley. Calif. Agr. 16(12):15. Biol. Abstr. 43:16353. 1963.

Greaves, J. D. 10844.

7612. Greaves, J. E., and A. F. Bracken. 1937: The sulfur content of wheat. Cereal Chem. 14:578–81.

7613. —— and C. T. Hirst. 1929: The mineral content of grains. Utah Agr. Expt. Sta. Bull. 210. 38 pp.

Greaves, J. E. 7763, 9885, 20004.

7614. Green, D. H. 1962: Further simple experiments on the greenhouse germination of some East African weed species. Misc. Rept. Trop. Pesticides Res. Inst. 320. 6 pp. Weed Abstr. 11:1922. 1962.

7615. —— 1963: Further simple experiments on the greenhouse germination of some East African weed species. Pesticides Abstr. News Summ. Sect. C 9:43–46.

Green, E. L. 10488.

7616. Green, H. E. 1961: Some notes on seed dispersal. Botan. Soc. Brit. Isles Proc. 4:153–54.

7617. Green, J. G. 1962: High temperature storage of dry maize. Dissertation Abstr. 22:3358–59.

7618. Green, J. R. 1887. On the changes in the proteids in the seed which accompany germination. Phil. Trans. Roy. Soc. London Ser. B 178:39–59.

7619. —— 1891: On the germination of the seed of the castor-oil plant (*Ricinus communis*). Proc. Roy. Soc. (London) Ser. B 48:370–92.

7620. —— and H. Jackson. 1905: Further observations on the germination of the seeds of the castor oil plant (Ricinus communis). Proc. Roy. Soc. (London) Ser. B 77:69–85.

7621. Green, T. G., and T. P. Hilditch. 1938: The fatty acids and glycerides of solid seed fats. V. Shea butter. Soc. Chem. Ind. J. 57:49–53.

7622. ——, ——, and W. J. Stainsby. 1936: Seed wax of *Simmondsia californica*. J. Chem. Soc. 1936:1750–55.

Green, V. E., Jr. 8263.

7623. Green, W. J. 1890: Asparagus. I. Seed bearing and non-seed bearing asparagus. Ohio Agr. Expt. Sta. Bull. (Ser. 2) 3:241–44.

7624. Greenall, A. F. 1943: Low germination of perennial ryegrass seed in South Otago. New Zealand J. Agr. 67:79, 81.

7625. Greenberg, D. 1948: A quantitative and qualitative evaluation of official and unofficial species of *Psyllium* seeds and their mucilages. J. Am. Pharm. Assoc. Sci. Ed. 37:139–45.

7626. Greenblatt, I. M., and R. A. Brink. 1962: Twin mutations in medium variegated pericarp maize. Genetics 47:489–501.

7627. Greene, H. C. 1948: Differences in achene characters and germination in some species of *Cacalia* L. Am. Midland Naturalist 39:758–60.

7628. Greene, J. T. 1962: A seed source study of slash pine (Pinus elliottii) within the state of Georgia. Tree Planters Notes 51:11–14.

7629. Greene, R. A. 1932: Composition of the pulp and seeds of *Adansonia digitata*. Botan. Gaz. 94:215–20.

7630. —— 1932: Composition of the beans of *Parkinsonia aculeata*. Botan. Gaz. 94:411–15. Chem. Abstr. 27:1027–28. 1933.

7631. —— and E. O. Foster. 1933: The liquid wax of seeds of *Simmondsia californica*. Botan. Gaz. 94:826–38.

7632. Greenham, C. 1934: A simple soybean counter. J. Am. Soc. Agron. 26:75–76.

7633. Greenshields, J. E. R. 1951: Polyembryony in alfalfa. Sci. Agr. (Ottawa) 31:212–22.

Greenshields, J. E. R. 5524, 7465.

Greenshields, R. N. 7514.

Greenstreet, V. R. 7116.

Greenwood, A. D. 3171–73.

7634. Greenwood, C. T. 1963: Starches of cereals. Ind. Aliment. Agr. (Paris) 80: 11–15. Chem. Abstr. 59:6722. 1963.

7635. —— and J. S. M. Robertson. 1954: Physicochemical studies on starches. I. The characterization of the starch present in the seeds of the rubber tree, *Hevea*

brasiliensis. J. Chem. Soc. 1954:3769–78.
Greenwood, C. T. 521, 522.
7636. Greenwood, H. N., and D. H. Peto. 1961: Production of high quality seed. World Crops 13:13–16.
Greenwood, H. N. 5905.
7637. Greenwood, R. M. 1950: Physiological studies on the germination and early seedling growth of clovers. New Zealand Grassland Assoc. Proc. Conf. 12:99–106.
Greenwood, R. M. 7485, 9988.
Greer, E. N. 4434, 9273, 9274.
7638. Greer, M. A. 1956: Isolation from rutabaga seed of progoitrin, the precursor of the naturally occurring antithyroid compound, goitrin (L-5-vinyl-2-oxazolidinethione). J. Am. Chem. Soc. 78:1260–61. Chem. Abstr. 50:7402. 1956.
7639. Gregg, B. R. 1964: Rapid and uniform measurement of seeds and other plant materials. Crop Sci. 4:113.
Gregoire, J. 19247, 19248.
7640. Gregor, H. P., and K. Sollner. 1946: Improved methods of preparation of "permselective" collodion membranes combining extreme ionic selectivity with high permeability. J. Phys. Chem. 50: 53–70.
7641. Gregory, F. G., and R. S. DeRopp. 1938: Vernalization of excised embryos. Nature 142:481–82.
7642. —— and O. N. Purvis. 1936: Vernalization. Nature 138:249.
7643. —— and —— 1936: Vernalization of winter rye during ripening. Nature 138: 973.
7644. —— and —— 1936: Devernalization of winter rye by high temperature. Nature 138:1013–14.
7645. —— and —— 1937: Devernalization of spring rye by anaerobic conditions and revernalization by low temperature. Nature 140:547.
7646. —— and —— 1938: Studies in vernalisation of cereals. II. The vernalisation of excised mature embryos, and of developing ears. Ann. Botany (London) 2:237–51.
7647. —— and —— 1938: Studies in vernalisation of cereals. III. The use of anaerobic conditions in the analysis of the vernalising effect of low temperature during germination. Ann. Botany (London) 2:753–64.
7648. —— and —— 1948: Reversal of vernalization by high temperature. Nature 161:859–60.
Gregory, F. G. 6747, 7509, 8378, 15776–15778.
7649. Grekov, P. I. 1939: Double supply of endosperm to spring wheat seed embryo. Compt. Rend. Acad. Sci. URSS 24:496–98.
7650. —— 1940: Verstärkte (verdoppelte) Ernährung des Embryo der Sommerweizen mit Endosperm und ihr Einfluss auf die erste Generation. Compt. Rend Acad. Sci. URSS 27:834–37.
7651. Grenot, C. 1960: Sur quelques substances de croissance (derivés indoliques) contenues dans le lait de mais immature. Compt. Rend. 251:771–73.
7652. Greshnova, V. N. 1959: Raising yields and resistance to dust smut of wheat by means of pre-sowing treatment of seeds with trace elements. Sel'skoe Khoz. Povolzhia 1959(11):65–67.
7653. —— 1961: Effect of solar radiation of seeds on wheat resistance to smut. Agrobiologiya 1961(1):129–31.
7654. Gressel, J. B. 1963: Chemical inhibition of germination by weed seeds and the nature of inhibition by *Abutilon theophrasti*. Dissertation Abstr. 24:2238–39.
7655. —— and L. G. Holm. 1964: Chemical inhibition of crop germination by weed seeds and the nature of inhibition by *Abutilon theophrasti*. Weed Res. 4:44–53.
7656. Grether, E. 1947: How to grow delphiniums from seed. Flower Grower 34(8):630, 647.
7657. Greulach, V. A., and S. Singh. 1949: Some effects of nonherbicidal concentrations of 2,4-D on the development of the bean plant. Science 109:336–37.
7658. Grewal, K. S., and B. D. Kochbar. 1943: *Luffa acutangula*: the chemical and pharmacological investigation of luffa seeds. Indian J. Med. Res. 31:63–66. Chem. Abstr. 38:5003. 1944.
7659. Grewe, E., and J. A. LeClerc. 1943: Commercial wheat germ, its composition. Cereal Chem. 20:423–34.
7660. Gribanov, L. N. 1956: Energy of seed germination of Scotch pine (*Pinus sylvestris* and its importance in evaluating the quality of seeds. Tr. Akad. Nauk Kazakhskol SSR Botan. Sad. 3:49–53.
7661. Gribkov, V. V. 1960: Pre-planting treatment of conifer seeds by microelements. Vestn. Sel'skokhoz. Nauki 4:129–31. Biol. Abstr. 36:46275. 1961.
7662. —— 1962: Pre-sowing treatment of conifer seed with ultraviolet and solar radiation. Lesn. Khoz. 15:36–38. Biol. Abstr. 43:16218. 1963.

7663. Griebel, C. 1940: Beiträge zur Kenntnis der Ginkgosamen. Deut. Apotheker-Z. 54:603–8. Pharm. Abstr. 6:500. 1940.
7664. —— 1942: Mountain almonds. Z. Untersuch. Lebensm. 84:134–37. Chem. Abstr. 37:5508. 1943.
7665. Gries, G. A. 1943: Juglone—the active agent in walnut toxicity. Northern Nut Growers Assoc. Proc. 34:52–55. Expt. Sta. Rec. 92:659. 1945.
Gries, G. A. 16191.
7666. Griesbach, R. A., and P. D. Voth. 1957: On dormancy and seed germination in *Hemerocallis*. Botan. Gaz. 118:223–37.
7667. Griffeth, W. L., and C. M. Harrison. 1954: Maturity and curing temperatures and their influence on germination of Reed canary grass seed. Agron. J. 46:163–67.
Griffin, E. L., Jr. 558–560.
7668. Griffing, E. P., and C. L. Alsberg. 1931: Composition of kapok seed. Ind. Eng. Chem. 23:908–9.
7669. Griffith, A. L. 1945: Storage of seed of *Prosopis* species. Indian Forester 71:251–52.
7670. Griffith, E., and C. B. Link. 1957: Germination and growth response of *Cattleya* seeds as influenced by the use of certain organic materials in the nutrient media. Am. Orchid Soc. Bull. 26:184–92.
7671. Griffith, G., and P. Chandler. 1939: The conservation of the nitrogen of cotton seed. E. African Agr. J. 5:143–45.
7672. Griffith, J. P. 1959: You can grow your own amaryllis. Horticulture 37(6): 350.
Griffith, R. B. 9967.
7673. Griffiths, A., Jr. 1955: Culture and classification of *Camellia* and related genera. Florida Agr. Expt. Sta. Ann. Rept. 1954/55:111.
7674. Griffiths, A. E. 1938: Observations on the germination of lettuce seed. Contrib. Boyce Thompson Inst. 9:329–37.
7675. —— 1942: The viability of lettuce seed: a physiological and microchemical study. Cornell Univ. Agr. Expt. Sta. Mem. 245. 39 pp.
7676. Griffiths, D. 1934: Bulbs from seed. U.S. Dept. Agr. Circ. 311. 32 pp. Expt. Sta. Rec. 72:55. 1935.
7677. Griffiths, D. A., A. C. Hodson, and C. M. Christensen. 1959: Grain storage fungi associated with mites. J. Econ. Entomol. 52:514–18.
7678. Griffiths, L. A. 1960: Comparative study of the seed polyphenols of *Theobroma*. Biochem. J. 74:362–65.
Griffiths, N. M. 9108.
7679. Griggs, W. H., B. T. Iwakiri, and L. L. Claypool. 1957: A comparison of growth, maturity, and quality of seedless and seeded Bartlett pears. Proc. Am. Soc. Hort. Sci. 70:74–84.
Grigorieva, W. F. 9550, 9551.
Grigsby, B. 17925–17928.
7680. Grillo, M. A. 1955: Adenosinetriphosphatase in the cotyledon of *Amygdalus communis*. Boll. Soc. Ital. Biol. Sper. 31:998–1000. Chem. Abstr. 50:7241. 1956.
Grimball, P. C. 15539.
Griminger, P. 6430.
Grimm, H. 6414, 7573.
7681. Grimm, K. 1928: Über die Keimung des Klees und aussere Einfluss auf Diese. Botan. Arch. 21:344–445.
7682. Grimm, M. G., A. A. Predtetshenskaja, A. M. Tschizova, and A. A. Egorova. 1928: Über die Anwendung der "Vitalfärbung" zur Bestimmung der Keimfähigkeit der Samen. Actes 5th Intern. Seed Testing Conf., Rome, pp. 400–3.
Grimm, M. H. 9511.
7683. Grimsley, G. 1930: Some observations on *Tilia* (basswood) seeds. J. Elisha Mitchell Sci. Soc. 46:72–73.
Grinberg, I. P. 10812.
Grindeland, R. [L.] 6724, 13966.
7684. Grindley, D. N. 1946: Investigation of the seed oils of some Sudan Caesalpinioideae. Soc. Chem. Ind. J. 65:118–19. Chem. Abstr. 40:4899. 1946.
7685. —— 1948: Investigations of some new Sudan seed oils. Soc. Chem. Ind. J. 67:230–31.
7686. —— 1952: Sunflower seed oil: the influence of temperature on the composition of the fatty acids. J. Sci. Food Agr. 3:82–86.
7687. ——, E. H. W. J. Burden, and A. A. Akour. 1954: The seed oils of *Clitoria ternatea* and of *Entada phaseoloides*. J. Sci. Food Agr. 5:278–80.
Grindley, D. N. 8638, 8639.
7688. Grinfeld, E. G. 1958: Catalase activity in centrifuged seeds. Dokl. Akad. Nauk SSSR 120:158–61. Biol. Abstr. 33:46930.
Gring, J. L. 4161.
7689. Gris, A. 1864: Recherches anatomiques et physiologiques sur la germination. Ann. Sci. Nat. Ser. 5 Botan. 2:5–123.
7690. Grisch, A. 1935: *Plantago rugelii* Dcne. *Plantago media* L. und *Plantago major* L. Proc. Intern. Seed Testing Assoc. 7:49–53.

7691. —— 1941/43: Die Herkunftsbestimmung der Klee- und Grassamen. Proc. Intern. Seed Testing Assoc. 13:147–416.

7692. —— and R. Koblet. 1931: Zur Frage der Beurteilung der Keimkraft von Coniferensamen auf Grund der Katalasebestimmung. Proc. Intern. Seed Testing Assoc. 3:60–74.

7693. —— and —— 1938: Vergleichende Untersuchungen über die Keimung von Grassämereien im Laboratorium, in Erde im Glashaus und im Freiland. Proc. Intern. Seed Testing Assoc. 10:176–98.

7694. Grischenko, A. P. 1935: The percentage composition of the wheat kernel and the ash content of its constituents. Mühlenlab. 5:182–88. Chem. Abstr. 30: 3894. 1936.

Grischenko, A. P. 10165.

7695. Grishchenko, G. L., and A. M. Slovtsov. 1952: To store forest seeds only of high quality. Lesn. Khoz. 5(8):46–50.

Grisolia, S. 6302, 6963, 6964.

7696. Grison, P., and D. Martouret. 1951: Destruction of insects in seeds by high-frequency currents. Compt. Rend. Acad. Agr. France 37:422–24. Brit. Abstr. BIII:260. 1952.

7697. Griswold, S. M. 1936: Effect of alternate moistening and drying on germination of seeds of western range plants. Botan. Gaz. 98:243–69.

7698. Gritton, E. T., and R. E. Atkins. 1963: Germination of grain sorghum as affected by freezing temperatures. Agron. J. 55:139–42.

7699. —— and —— 1963: Germination of sorghum seed as affected by dormancy. Agron. J. 55:169–74.

Griuntukh, R. N. 16033.

7700. Grizzard, A. L., and E. M. Matthews. 1942: The effect of boron on seed production of alfalfa. J. Am. Soc. Agron. 34:365–68.

Grobler, J. H. 19005.

7701. Grobnic, G. 1963: The results of some experiments concerning the preservation of spruce seeds. Rev. Padurilor (Bucharest) 78:7–11.

7702. Grodzinsky, A. M., N. I. Mokhova, L. D. Pilipenko-Yurchak, and T. N. Fillippovich. 1962: On inhibiting substances in field crop residues and weeds. I. Effect of water-soluble inhibitors on the germination of seeds and plant growth. Acad. Sci. RSS Ukraine Kiev J. Botan. 19:30–38.

7703. ——, ——, ——, and —— 1963: On inhibiting substances in afterharvest residues of field crops and in weeds. II. Effect of volatile excretions on seed germination and photosynthesis of plants. Acad. Sci. RSS Ukraine Kiev J. Botan. 20:66–72.

7704. Grogan, C. O. 1958: Interaction of insecticides and fungicides on the germination of stored seed corn. Am. Seed Trade Assoc. Hybrid Corn Div. Rept. Hybrid Corn Ind. Res. Conf. 13:7–12.

7705. ——, M. S. Zuber, H. E. Brown, M. D. Whitehead, and V. M. Stanway. 1959: Effect of fungicides and insecticides on the germination of corn after storage. Plant Disease Reptr. 43:1132–37.

Grogan, C. O. 20130, 20132.

7706. Grogan, F. O. 1949: Wheat yields and variability in western Australia, 1929/1930 to 1946/1947. Australian Inst. Agr. Sci. J. 15:135–41.

Grogan, R. 13493.

7707. Grogan, R. G., J. E. Welch, and R. Bardin. 1952: Common lettuce mosaic control; use of mosaic-free seed effectively reduced the seed-borne aphid-transmitted disease in large-scale field plantings. Calif. Agr. 6(8):5, 14.

Grogan, R. G. 18414, 20116.

7708. Groh, H. 1940: Turkestan alfalfa as a medium of weed introduction. Sci. Agr. (Ottawa) 21:36–43. Biol. Abstr. 15:5590. 1941.

7709. —— 1943: Notes on common milkweed, *Asclepias syriaca*. Sci. Agr. (Ottawa) 23:625–32.

7710. Grohne, U. 1952: Untersuchungen zur Frage der Lichtkeimung von *Digitalis purpurea* L. Biol. Centralbl. 71:10–42.

Groot, W. de. 2719.

7711. Grose, R. J., and W. J. Zimmer. 1958: A description of the seeds of 70 Victorian eucalypts (illustrated). Victoria Forests Commun. Bull. 8. 24 pp.

7712. —— and —— 1958: Influence of seed size on germination and early growth of seedlings of *Eucalyptus maculata* Hook. f. and *Eucalyptus sieberiana* F. v. M. Victoria Forests Commun. Bull. 9. 10 pp.

7713. —— and —— 1958: The collection and testing of seed from some Victorian eucalypts with results of viability tests. Victoria Forests Commun. Bull. 10. 14 pp.

7714. —— and —— 1958: Some laboratory germination responses of the seeds of river red gum, *Eucalyptus camaldulensis* Dehn. syn. *Eucalyptus rostrata* Schlecht. Australian J. Botany 6:129–53.

7715. Gross, E. 1917: Veränderungen der Getreidesamen bei 10 jähriger Lagerung. Z. Landwirtsch. Versuchs. Deut. Österr. 20:471–87.

Gross, H. 17119.

7716. Gross, R. A., and C. H. Bailey. 1937: Chemical constitution of the oils from superior and inferior flaxseeds. Oil Soap 14:260–63. Expt. Sta. Rec. 81:470–71.

Gross, W. G. 8600.

7717. Grossfeld, J., and A. Splittgerber, editors. 1923: Anhang zu Ölsamen. Zusammensetzung der Pflanzenlichen Nahrungs und Genussmittel, pp. 117–80.

Grossman, E. F. 9200.

7718. Groszmann, A. 1949: Inheritance of the weight and component parts of the seed in a dent x pop corn cross. Lilloa 19:75–88. Biol. Abstr. 26:20303. 1952.

7719. —— and G. F. Sprague. 1948: Comparative growth rates in a reciprocal maize cross. I. The kernel and its component parts. J. Am. Soc. Agron. 40:88–98.

7720. Groszmann, H. M. 1952: Pulse crops (beans and peas). Queensland Agr. J. 74:249–63.

Grotte, R. de; *see* De Grotte, R.

7721. Grove, A. 1930: Germination of seeds of *Lilium giganteum*. Gardeners Chron. 88:251.

7722. —— 1933: Fruits of lilies. I. Gardeners Chron. 93:204–5.

7723. —— 1933: Fruits of lilies. II. Gardeners Chron. 93:294.

7724. —— 1933: Fruits of lilies. III. Gardeners Chron. 93:419–20.

7725. —— 1933: Fruits of lilies. IV. Gardeners Chron. 94:44.

7726. —— 1938: The seminal propagation of lilies. Gardeners Chron. 104:389–90.

Grove, A. J. 1924.

7727. Grover, R. 1962: Effect of gibberellic acid on seed germination of elm, Scotch pine, Colorado and white spruce. Forest Sci. 8:187–90.

7728. —— 1962: Germination of three coniferous species after treatment with simazine. Weeds 10:246.

7729. ——, E. W. Martin, and C. H. Lindquist. 1963: Maturity and storage of Siberian elm seed. Forest Sci. 9:493–96.

Grover, R. 14173.

7730. Groves, J. F. 1917: Temperature and life duration of seeds. Botan. Gaz. 63:169–89.

Groves, J. F. 4603.

7731. Groves, J. W., and A. J. Skolko. 1944: Notes on seed-borne fungi. Can. J. Res. Sect. C 22:190–99.

7732. —— and —— 1944: Notes on seed-borne fungi. II. *Alternaria*. Can. J. Res. Sect. C 22:217–34.

7733. —— and —— 1945: Notes on seed-borne fungi. III. *Curvularia*. Can. J. Res. Sect. C 23:94–104.

7734. —— and —— 1946: Notes on seed-borne fungi. IV. *Acremoniella*, *Chlamydomyces* and *Trichocladium*. Can. J. Res. Sect. C 24:74–80.

Groves, J. W. 3500, 17571, 17572.

7735. Groves, K. 1940: Poisonous constituents of *Amsinckia intermedia*. Wash. Agr. Expt. Sta. Bull. 394:30–31.

Gruber, M. 17704.

Gruber, F. von. 19186.

7736. Grümmer, G. 1953: Die Wirkung von *Alternaria*-Infektionen an Mohnkapeln auf die Keimungsbereitschaft ihrer Samen. Flora (Jena) 140:298–306. Ber. Wiss. Biol. 86:391. 1954.

7737. Grünfeld, O. 1926: Über die Entleerung und Wiederauffüllung isolierter Getreideendosperme, insbesondere von Mais, unter aseptischen Bedingungen. Botan. Centralbl. Beih. Abt. I 42:355–74.

7738. —— 1926: Über die Entleerung und Wiederauffüllung isolierter Getreideendosperme, insbesondere von Mais, unter aseptischen Bedingungen. Botan. Centralbl. Beih. Abt. I 43:167–203.

7739. Gruenhagen, R. H. 1940: Growth substances of doubtful benefit for treatment of pine seeds. J. Forestry 38:739–40.

7740. Grüss, J. 1896: Beiträge zur Physiologie der Keimung. Landwirtsch. Jahrb. 25:385–452.

7741. —— 1929: Bau und Semipermeabilität der Gerstenfrucht- und Samenschale. I. Wochschr. Brau. 46:61–66, 74–80. Biol. Abstr. 6:15599. 1932.

7742. —— 1930: Semipermeabilität der Gerstenfrucht und Samenschale. III. Wochschr. Brau. 47:249–52. Biol. Abstr. 6:6435. 1932.

7743. —— 1931: Über die Beschaffenheit der durch Schwefelsäure entspelzten Gerste. Wochschr. Brau. 48:342–45. Biol. Abstr. 9:9974. 1935.

Grujic, S. A. 3563.

7744. Grundbacher, F. J. 1963: The physiological function of the cereal awn. Botan. Rev. 29:366–81.

7745. Grunder, M. S., and C. R. Megee. 1929: Maturity influences quality of alfalfa seed. Mich. Agr. Expt. Sta. Quart. Bull. 12:61–62.

7746. Grunert, C. 1932: Stauden aus Semen—Staudensamenzucht. Gartenwelt 36:733.

7747. Grusheva, Z. G. 1963: Significance of the supply of seeds in the soil for the regeneration of forests in Transbaikalia. Lesn. Khoz. 4:67–68.

7748. Grushevoi, S. E. 1939: The effect of temperature conditions upon the germination of *Orobanche* seeds. Krasnodar Inst. Tobacco Ind. Bull. 137:45–46.

7749. —— 1939: The effect of medium reactions on the germination of *Orobanche* seeds. Krasnodar Inst. Tobacco Ind. Bull. 137:50.

7750. —— and I. P. Khudyna. 1938: Methods for obtaining tobacco seeds free of infection. Krasnodar Inst. Tobacco Ind. Bull. 135:47–48.

7751. ——, ——, and A. A. Popova. 1940: Thermal method for disinfecting tobacco and *Nicotiana rustica* seeds. Krasnodar Inst. Tobacco Ind. Bull. 141:3–29. Biol. Abstr. 15:15686. 1941.

7752. Grusvitzky, I. V. 1963: Subterranean germination and the functions of cotyledons. Botan. Zh. 48:906–15.

7753. ——, E. J. Agnayeva, and E. F. Kuzina. 1963: On the diversity of ripe carrot seeds with respect to the size of the embryo. Botan. Zh. 48:1484–89.

7754. —— and R. S. Limar. 1961: Hastening germination in genseng by means of ultrasound. Botan. Zh. 46:1282–85. Biol. Abstr. 42:3235. 1963.

Grusvitzky, I. V. 2229.

7755. Grzesiuk, S. 1961: Research on the physiology of the ripening grain of cereals. Olsztyn. Wyzsza Szkola Roln. Zeszyty Nauk. 11(104):1–82.

7756. —— and K. Kulka. 1962: Mono- and oligosaccharides in vernalization process of winter rye (*Secale cereale* L.) grains. Acta Soc. Botan. Polon. 31:83–93.

7757. —— and —— 1963: Free amino acids in vernalized winter rye. Acta Soc. Botan. Polon. 32:313–25. Chem. Abstr. 60:3268. 1964.

7758. ——, T. Mierzwinska, and E. Sojka. 1962: Physiology and biochemistry of the development of the seeds of *Vicia faba* ssp. *minor*. Plant Physiol. (USSR) (English Transl.) 9:682–92. Chem. Abstr. 58:7129. 1963.

7759. —— and E. Sojka. 1961: Studies on the physiology of maturing small field bean seeds (*Vicia faba* L. ssp. *minor*). Roczniki Nauk Rolniczych Ser. A 83:735–70.

7760. Gstirner, F. 1932: Estimation of alkaloids in sabadilla seed. Pharm. Ztg. 77:509. Chem. Abstr. 26:3871. 1932.

7761. Guard, A. T. 1943: Development of the seed of the tulip tree. [Abstract.] Proc. Indiana Acad. Sci. 58:34.

7762. —— and R. E. Wean. 1941: Seed production in the tulip poplar. J. Forestry 39:1032–33.

Guard, A. T. 3634, 19424.

Gube, M. 7911, 10142.

7763. Gubler, C. J., and J. E. Greaves. 1942: Cystine content of wheat and its determination. Food Res. 7:405–13. Chem. Abstr. 37:2084–85. 1943.

7764. Gude, H. 1955: Characteristics of the seed oils of two plants growing in waste places, *Sisymbrium sinapistrum* and *Verbascum thapsus*. Pharm. Zentralhalle 94:46–47. Chem. Abstr. 52:13892. 1958.

7765. Gudkov, A. N. 1936: A comparison of the chemical and morphological methods of identifying *Pisum arvense* L. and *Pisum sativum* L. Bull. Appl. Botany, Genet., Plant Breed. Ser. 4(1):87–92. Chem. Abstr. 31:5842–43. 1937.

7766. —— 1937: The phenomenon of embryolessness in wheat seeds and its causes. Bull. Appl. Botany, Genet., Plant Breed. Ser. 4(2):166–67.

7767. —— 1940: Formation of pigment in glumes and kernels of wheat. Compt. Rend. Acad. Sci. URSS 27:841–45.

7768. Gudtsev, I. I. 1951: Pre-planting treatment of Coniferae seeds which were stored in hermetically sealed containers. Lesn. Khoz. 4(2):37–39.

7769. Gümbel, H. 1912: Untersuchungen über die Keimungsverhältnisse verschiedener Unkräuter. Landwirtsch. Jahrb. 43:251–321.

7770. Günther, E. 1937: Oil of coriander seed. Am. Perfumer 35:45–47, 91–92. Pharm. Abstr. 4:358. 1938.

7771. —— 1943: Oil of parsley. Chemurgic Dig. 2:57, 59–60.

7772. —— 1943: Oil of celery. Chemurgic Dig. 2(11):88–91.

7773. —— 1943: Oil of *Calamus*. Chemurgic Dig. 2(16):138–40.

7774. —— 1943: U.S. Production of oil of dill. II. Food Ind. 15:72–73.

7775. Günther, G. 1958: Über Zusammenhänge zwischen Atmungsintensität und keimungsphysiologischem Verhalten von

Caryophyllaceen-Samen. Flora (Jena) 145:421–60.

7776. —— 1961: Der Einfluss von Inhibitoren auf die Wirksamkeit einer Jarowisationbehandlung von Winterweizen. Naturwissenschaften 48:385–86.

Günther, G. 19197.

7777. Guenther, O. 1927: Beiträge zur Kenntnis der Entwicklung des Getreideendosperma und seines Verhaltens bei der Keimung, Botan. Arch. 18:299–319.

7778. Guenzi, W. D., and T. M. McCalla. 1962: Inhibition of germination and seedling development by crop residues. Soil Sci. Soc. Am. Proc. 26:456–58.

7779. Gúerillot, J., A. Guérillot-Vinet, and J. Baron. 1953: Action de certains oligo element sur la faculté germinative de graines de semence anormales. Ann. Agron. 4:863–64.

7780. ——, ——, and R. Morel. 1949: Comparative germination in the laboratory and in the open field of coated and uncoated segmented beet glomerules. Ann. Agron. 19:574–77. Chem. Abstr. 44:194. 1950.

Gúerillot, J. 7577.

Guérillot-Vinet, A. 7577, 7779, 7780.

7781. Guérin, P. 1930: Le développement de l'oeuf et la polyembryonie chez l'*Erythronium dens-canis* L. (Liliacées). Compt. Rend. 191:1369–72.

7782. —— 1937: L'ovulet et la graine des *Erythronium* et des *Calochortus*. Ann. Sci. Nat. 19:255–65.

Guérindon, A. 4969, 4970.

Guernelli, O. 5260.

7783. Guerrant, N. B. 1927: Some relations of the phospholipins in seeds to other constituents. J. Agr. Res. 35:1001–19.

Guerreros, R. F. 7486.

7784. Guest, P. 1942: The relationship between chlorosis of macadamia seedlings and certain chemical consitituents of macadamia seeds. Proc. Am. Soc. Hort. Sci. 41:61–64.

7785. Gütlin-Schmitz, P. H. 1957: Über die Anderung des Aneuringehaltes während der Keimung von Samen mit verschiedenartigen Reservestoffen. Ber. Schweiz. Botan. Ges. 67:5–25. Biol. Abstr. 33: 11479. 1959.

7786. Guevara, R. 1940: Pharmacodynamic study of Lansones resin, Tangan-Tangan oil and Palo Santo seeds. Rev. Filipina Med. Farm. 31:143. Pharm. Abstr. 7:335. 1941.

7787. Guevara, V. F. 1961: The occurrence of twin embryos in *Cocos nucifera* L. Philippine Agriculturist 44:520–22.

Guevara, V. F. 3575.

Guggenheim, K. 15448.

7788. Guha, B. C., and A. Bose. 1944: Synthesis of thiamin by germinating seeds. Sci. Cult. (Calcutta) 10:260. Biol. Abstr. 19:14453.

7789. —— and A. R. Ghosh. 1934: Effect of germination on the reducing power of *Phaseolus mung*. Current Sci. (India) 2:390. Chem. Abstr. 28:4449. 1934.

Guha, B. C. 4922, 18178.

7790. Guha, S. 1962: In vitro production of onion seeds. *In* Plant Embryology, a symposium held at the Univ. of Delhi, 1960: Council Sci. Ind. Res., New Delhi, pp. 182–87. Biol. Abstr. 43:7486. 1963.

Guha, S. 16601.

7791. Guha-Thakurta, A., and B. K. Dutt. 1938: Investigation on the oxygen consumption of the seed of *Cajanus*. Trans. Bose Res. Inst. (Calcutta) 11:39–53.

7792. —— and —— 1937/39: Investigation on the catalase activity of jute seeds (*Olitorius* and *Capsularis*). Trans. Bose Res. Inst. (Calcutta) 13:83–92.

7793. —— and —— 1944/46: IV. Vegetative propagation of *Cinchona ledgeriana* from gootes (Marcotte) and cuttings by treatment with auxins. Trans. Bose Res. Inst. (Calcutta) 16:23–29.

Guha-Thakurta, A. 5693–5695.

7794. Guichard, F. 1945: Préparation directe, en partant des graines, des esters d'*Hydnocarpus anthelmintica* Pierre, en vue du traitement de la lèpre. Bull. Soc. Pathol. Exotique 38:9–12. Biol. Abstr. 26:9818. 1952.

Guidotti, G. 4295.

7795. Guignard, J. L. 1961: Cyperacées. Développement de l'embryon chez le *Cyperus vegetus* Willd. Compt. Rend. 252:2125–27.

7796. —— 1961: Embryogenie des Graminées. Développement de l'embryon chez le *Setaria verticillata* Beauv. Bull. Soc. Botan France 108:212–17.

7797. —— 1961: Embryogenie des Palmiers. Développement de l'embryon chez le *Chamaerops humilis* L. Compt. Rend. 253:1834–36.

7798. —— 1961: Research on embryogeny of the Gramineae; relationship with other monocotyledons. Ann. Sci. Nat. Botan. Biol. Vegetale (Ser. 12) 2:491–610.

7799. —— 1962: Embryogenie des Iridacées. Développement de l'embryon chez l'*Iris pseudacorus* L. Compt. Rend. 255:2161–63.

7800. —— 1963: Développement de l'embryon chez le *Mibora minima* Desv. (Graminées). Bull. Soc. Botan. France 110:193–95.

7801. —— 1963: Embryogenie des Caryophyllacées. Développement de l'embryon chez le *Dianthus armeria* L. Compt. Rend. 256:756–58.

7802. —— 1963: Embryogenie des Dioscoreacées. Développement de l'embryon chez le *Tamus communis* L. Compt. Rend. 256:3172–75.

7803. —— 1963: Embryogenie des Aracées. Développement de l'embryon chez le *Pistia stratiotes* L. Compt. Rend. 257: 1139–42.

7804. —— 1963: Développement de l'embryon chez le *Moehringia pentandra* J. Gay (Carophyllacées). Compt. Rend. 257: 2694–97.

Guilbot, A. 7232, 7813.

7805. Guilford, V. B., and E. L. Fisk. 1952: Megasporogenesis and seed development in *Mimulus tigrinus* and *Torenia fowinieri*. Bull. Torrey Botan. Club 71:6–24.

7806. Guillaume, A. 1930: Migration des alcaloides au cours de la germination des graines et de la formation des plantules: Recherches sur *Lupinus mutabilis* var. Cruiskanks. Compt. Rend. 190:1068–70.

7807. —— 1942: A lupine useful as a substitute for wheat. Compt. Rend. 215: 33–35. Chem. Abstr. 38:5984. 1944.

7808. Guillaumin, A. 1928: Le maintien des graines dans un milieu privé d'oxygene moyen de prolonger leur faculté germinative. Compt. Rend. 187:571–72.

7809. Guillemat, J. 1951: Les maladies vehiculées par les semences. Leur importance: Les moyens pour obtenir des semences saines. Bull. Tech. Inform. Ingénieurs Serv. Agr. 60:383–91. Biol. Abstr. 26:13338. 1952.

7810. —— and J. Lambert. 1960: Action du traitement aux organo-mercuriques de la teneur en eau et de la durée du stockage sur la faculté germinative des semences de blé. Phytiat.-Phytopharm. 9:63–74.

7811. Guillemet, R. 1931: Sur le pouvoir lipolytique de differentes varietés de ricin; facteurs susceptibles de le modifier. Compt. Rend. Soc. Biol. 108:779–81.

7812. —— 1931: Sur le pH optimum d'hydrolyse des huiles végétales par la lipase de la graine de ricin. Compt. Rend. Soc. Biol. 108:781–83.

7813. —— and A. Guilbot. 1945: Separation of the protein-cell layer of the wheat grain by a chemical method. Composition of the cell walls. Bull. Soc. Chim. Biol. 27:49–55. Chem. Abstr. 40:3197. 1946.

7814. —— and R. Jacquot. 1944: Sur la valeur nutritive des differentes régions du grain de blé. Bull. Soc. Biol. 26:324–34.

7815. Guilleminot, H. 1907: De l'action des rayons du radium et des rayons X sur la germination. Compt. Rend. Assoc. Franc. Avance Sci. 37:1344–53.

7816. Guilliermond, A. 1914: The formation of starch in the embryo before the maturity of the seed. Compt. Rend. Soc. Biol. 76:567–71.

7817. ——, J. Dufrenoy, and F. Labrousse. 1930: Germination des graines de tabac dans des milieux additionnés de rouge neutre et coloration du vacuome pendant le développement des plantules. Compt. Rend. 190:1439–42.

7818. Guillinta, M. J. G. 1953: Chemical study of *Chenopodium pallidicaule*. Bol. Soc. Quim. Peru 19:163–80. Chem. Abstr. 48:10243. 1954.

7819. Guillochon, L. 1926: Observations sur la mode de germination de l'*Araucaria imbricata* Pav. Bull. Soc. Botan. France 73:444–45.

Guinn, G. 6008.

7820. Guinaudeau, J. 1954: Germination du pin maritime (*Pinus pinaster*) resultats des experiences realisées par l'annexe de la Station de Recherches a Bordeaux. Rev. Forestry (France) 5:320–30.

7821. Guinet, P. 1959: Essai d'identification des graines de chenopodes commensaux des cultures ou cultivés en France. J. Agr. Trop. Botan. Appl. 6:241–66.

7822. Guitard, A. A., J. A. Newman, and P. B. Hoyt. 1961: The influence of seeding rate on the yield and the yield components of wheat, oats and barley. Can. J. Plant Sci. 41:751–58.

7823. Guitton, Y. 1957: Nitrogen metabolism of gymnosperms. Presence of arginase in the seeds. Compt. Rend. 245:1157–60. Chem. Abstr. 52:4749i. 1958.

7824. —— and M. De Belsunce. 1963: On nitrogen metabolism of gymnosperms. Evidence of an aminopeptidase in seeds and young plants of *Pinus pinea*. Compt. Rend. 256:2689–91. Chem. Abstr. 58: 14447. 1963.

7825. Gukova, M. M., and R. I. Bogomolova. 1963: The effect of nutrition on the growth and storage of nitrogen by broad beans in a mixed crop with corn. Dokl. Akad. Nauk SSSR 149:516–18.

Gulanian, V. M. 1545, 1546.

7826. Gulati, A. N. 1934: A note on the differentiation of hairs from the epidermis of cotton seeds. Indian J. Agr. Sci. 4:471–75. Expt. Sta. Rec. 72:177. 1935.

Gulati, K. C. 10294.

7827. Guliaev, V. V. 1952: Treatment of pine seeds during storage. Agrobiologiya 1952(2):119–22.

7828. Gulkanyan, V. O. 1962: On the significance of phytological technique in the growing of wheat for seeds. Izv. Akad. Nauk Arm. SSR, Biol. Nauki 15(5):3–10.

Gulnina, K. I. 11007.

7829. Gulyaev, V. A. 1963: Comparative embryology of the Cucurbitaceae and its importance in the taxonomy of this family. Botan. Zh. 48:80–85. Biol. Abstr. 44:12247. 1963.

7830. Guminska, Z. 1958: Combinaisons humiques oxydées comme stimulateurs de la germination des graines. Acta Soc. Botan. Polon. 27:501–22.

7831. Gunckel, J. E., and A. H. Sparrow. 1961: Ionizing radiations: biochemical, physiological and morphological aspects of their effects on plants. *In* Encyclopedia Plant Physiol. 16. Springer, Berlin, pp. 555–611.

7832. Gunde, B. G., and T. P. Hilditch. 1938: Seed and fruit-coat fats on *Neolitsea involucrata*. J. Chem. Soc. 1938:1610–14. Chem. Abstr. 33:690. 1939.

7833. —— and —— 1938: Seed and fruit-coat fats of *Celastrus paniculatus*. J. Chem. Soc. 1938:1980–85. Analyst 64:355–56. 1939.

7834. —— and —— 1939: The seed fats of *Salvadora oleoides* and *Salvadora persica*. J. Chem. Soc. 1939:1015–16.

7835. —— and —— 1940: The mixed unsaturated glycerides of liquid seed fats. I. Some "non-drying" oils. Soc. Chem. Ind. J. 59:47–53.

Gunde, B. G. 7344.

7836. Gundy, L. J. 1959: Variations in grain size in corn due to variety and environment. Agron. Abstr. 1959:81.

7837. Gunstone, F. D. 1952: Vegetable oils. I. The component acids of *Strophanthus sarmentosus* seed oil. J. Sci. Food Agr. 3:185–89.

7838. —— 1953: Vegetable oils. II. Further studies of seed oils of various Strophanthus species. J. Sci. Food Agr. 4:129–32.

7839. —— and P. J. Sykes. 1961: Vegetable oils. IX. Application of reversed-phase chromatography to the analysis of seed oils. J. Sci. Food Agr. 12:115–23.

Gunstone, F. D. 1644–1646, 2414, 8765.

7840. Gunthardt, H., and J. McGinnis. 1957: Effects of nitrogen fertilization on amino acids in whole wheat. J. Nutr. 61:167–76.

7841. ——, L. Smith, M. E. Haferkamp, and R. A. Nilan. 1953: Studies on aged seeds. II. Relation of age of seeds to cytogenetic effects. Agron. J. 45:438–41.

Gunthardt, H. M. 100, 14485.

Gunther, E. 3194.

7842 Guppy, H. B. 1882: Dispersal of seeds ("The gizzard content of some oceanic birds"). Nature 26:12.

7843. Guppy, H. B. 1897. On the postponement of germination of seeds of aquatic plants. Proc. Roy. Phys. Soc. Edinburgh 13:344–59.

7844. —— 1912: Studies in seeds and fruits. Williams & Norgate, London. 528 pp.

7845. —— 1917: Plants, seeds, and currents in the West Indies and Azores. Williams & Norgate, London, pp. 12–531. Expt. Sta. Rec. 38:125. 1918.

Guppy, J. 2949.

Gupta, A. C. 3851.

7846. Gupta, A. S., and M. M. Chakrabarty. 1957: Studies on the seed fat composition of desert plants. I. The component fatty acids of "*Gynandropsis pentaphylla*" seed fat(N.O.-*Cappridacea*). Sci.Cult.(Calcutta) 23:306–7. Biol. Abstr. 33:3434. 1959.

7847. —— and —— 1964: Composition of the seed fats of the Capparidaceae family. J. Sci. Food Agr. 15:69–73.

7848. —— and —— 1964: The component fatty acids of *Citrullus colocynthis* seed fat. J. Sci. Food Agr. 15:74–77.

7849. ——, S. C. Mutha, and A. P. Waghrey. 1963: The component fatty acids of chaulmoögra oil (*Taraktogenos kurzii*, King). J. Sci. Food Agr. 14:457–63.

Gupta, A. S. 2415.

Gupta, B. D. 3767.

7850. Gupta, D. K., and B. T. R. Iyengar. 1956: Seed fat from *Caesalpinia digyna*. Sci. Cult. (Calcutta) 21:682–83. Biol. Abstr. 32:35324. 1958.

7850A. Gupta, D. R., and N. C. Mahajan. 1962: Amino acid content of the seeds of

Litsea consimilis. Proc. Natl. Acad. Sci. India Sect. A 32:48–49. Chem. Abstr. 57:8909. 1962.

7851. Gupta, G. P. 1962: Synthesis of oil at various stages of seed development in linseed. Indian Oilseeds J. 6:207–10. Chem. Abstr. 58:7134. 1963.

7852. Gupta, J. C., S. Ghosh, A. T. Dutta, and B. S. Hall. 1947: A note on the hypnotic principle of *Rauwolfia serpentina*. J. Am. Pharm. Assoc. Sci. Ed. 36:416. Biol. Abstr. 24:7294. 1950.

Gupta, J. C. 5699.

7853. Gupta, J. C. S., and S. K. Chattopadhyay. 1955: Effect of presowing treatments with hormones on germination of some common Indian fibre crops. Bull. Botan. Soc. Bengal 9:10–13. Chem. Abstr. 51:4500. 1957.

7854. Gupta, J. N. S. 1937: Seed weights, plant per cents, etc., for forest plants in India. Indian Forest Records 2(5):175–221.

Gupta, J. S. 3193.

7854A. Gupta, M. L. 1953: Comparative study of improved varieties of wheat versus "local" at different seed rates and sowing dates in Agra. Allahabad Farmer 27: 102–5.

7855. Gupta, M. P., and S. Dutt. 1936: Oil from the seeds of *Solanum xanthocarpum*. J. Indian Chem. Soc. 13:613–18. Analyst 62:397. 1937.

7856. —— and —— 1938: Constituents of the seeds of *Cleome viscosa*. J. Indian Chem. Soc. 15:532–36. Analyst 64:208–9. 1939.

7857. —— and J. B. Lal. 1937: Chemical examination of the seeds of *Physalis peruviana* or Cape gooseberry. II. Proc. Natl. Acad. Sci. India 7:131–36.

Gupta, P. C. 18471.

7858. Gupta, P. K. 1962: The occurrence of indigenous auxins and inhibitors and their effect on the germination of strawberry seeds. Dissertation Abstr. 22:4148.

Gupta, P. S. 10522.

Gupta, R. K. 18600, 18601.

7859. Gupta, S. C. 1962: Development of exembryonate seeds in *Foeniculum vulgare* Mill. *In* Plant Embryology, a symposium held at Univ. of Delhi, 1960. Council Sci. Ind. Res., New Delhi, pp. 188–91. Biol. Abstr. 43:16291. 1963.

7860. ——, S. S. Gupta, and J. S. Aggarwal. 1954: The component acids of Kamala oil (*Mallotus philippinensis*, Muell. Arg.). J. Am. Oil Chemists Soc. 31:287–89.

Gupta, S. C. 17746.

Gupta, S. K. S. 10522.

Gupta, S. M. 14169.

Gupta, S. R. 6229.

Gupta, S. S. 2820, 7860.

7861. Gupta, Y. P., and N. B. Das. 1955: Amino-acid content of pure strains of Indian pulses. I. Methionine, cystine and tryplophane. Ann. Biochem. Exptl. Med. 15:75–78. Nutr. Abstr. Rev. 27:230. 1957.

7862. —— and —— 1956: The quality of wheat as affected by manures and fertilisers. II. Thiamine, riboflavin and niacin contents. J. Indian Soc. Soil. Sci. 4:125–28.

7863. Gurchot, C. 1935: Apricot seeds as a source of dehydrogenases. Proc. Soc. Exptl. Biol. Med. 33:285–87. Biol. Abstr. 10:15498. 1936.

7864. Gurevich, M. 1935: Anatomy and microchemistry of the cotton seed. Maslob.-Zhir. Delo. 1935(11):301–2. Chem. Abstr. 29:8051. 1935.

7865. Gurewitsch, A. 1929: Untersuchungen über die Permeabilität der Hülle des Weizenkorns. Jahrb. Wiss. Botan. 70:657–706.

7866. —— 1935: Erwiderung auf W. Freybergs Kritik meiner Untersuchungen über die Permeabilität der Hülle des Weizenkorns. Botan. Arch. 37:65–74.

7867. —— 1935: Über eine Methode zur Bestimmung der Keimfähigkeit ohne Keimprüfung. Ber. Deut. Botan. Ges. 53:303–18.

7868. —— 1937: Über die Dinitrobenzol-Methode zur Bestimmung der Keimfähigkeit ohne Keimprüfung. II. Mitteilung. Ber. Deut. Botan. Ges. 55:54–58.

7869. Gurgel, J. T. A., and J. Mitidieri. 1955: Pretreatment of seeds of *Hibiscus esculentus* to accelerate germination and make it uniform. Rev. Agr. (Piracicaba) 30:173–84.

7870. —— and J. Soubihe. 1951: Analysis of polyembryony in citrus, especially pummelos. Escola Super. Agr. "Luis de Queiroz" Anales 8:727–46.

Gurgel, J. T. A. 13736.

7871. Gurgel-Filho, O. A. 1954: Processos para auxiliar a germinação das sementes florestais. Dusenia 5:95–103. Biol. Abstr. 29:19842. 1955.

7872. Gurgenidze, M. G. 1956: How to promote the germination of stony seeds.

Sad i Ogorod 1956(6):45–46. Biol. Abstr. 33:30900. 1959.
Gurjar, A. M. 1680.
Gurzhieva, A. V. 2782.
7873. Guseinov, S. F. 1958: Effect of sodium bicarbonate on germination and yield of winter wheat. Akad. Nauk Azerb. SSR Inst. Zemled. Tr. 4:7–10.
7874. Guseinov, S. G. 1960: Contribution to the problem of the effect of minor elements on the germination of seeds, growth, and vital capacity of *Pinus eldarica*. Izv. Akad. Nauk Azerb. SSR Ser. Biol. i Med. Nauk 1960(1):111–18.
7875. Gustafson, F. G. 1937: Parthenocarpy induced by pollen extracts. Am. J. Botany 24:102–7. Biol. Abstr. 11:18204. 1937.
7876. —— 1938: Further studies on artificial parthenocarpy. Am. J. Botany 25:237–44. Chem. Abstr. 32:5870. 1938.
7877. —— 1939: The cause of natural parthenocarpy. Am. J. Botany 26:135–38. Biol. Abstr. 13:10300. 1939.
7878. —— 1940: Parthenocarpic and normal fruits compared as to percentage of setting and size. Botan. Gaz. 102:280–86.
7879. —— 1941: Probable causes for the difference in facility of producing parthenocarpic fruits in different plants. Proc. Am. Soc. Hort. Sci. 38:479–81.
7880. —— 1942: Parthenocarpy: natural and artificial. Botan. Rev. 8:599–654.
7881. —— 1949: Tryptophane as an intermediate in the synthesis of nicotinic acid by green plants. Science 110:279–80.
7882. Gustafson, P. R. 1951: A rapid method for determining the viability of barley. Cereal Chem. 28:343–46.
7883. Gustafsson, A. 1934: Entwicklungswege der parthenogenetischen Embryosackmutterzelle. Botan. Notiser 1934(4):333–38. Biol. Abstr. 10:19462. 1936.
7884. —— 1937: Der Tod als ein nuklearer Prozess. Hereditas 23:1–37.
7885. —— 1944: The X-ray resistance of dormant seeds in some agricultural plants. Hereditas 30:165–78.
7886. —— and J. Schröderheim. 1944: Ascorbic acid and hip fertility in *Rosa* species. Nature 153:196–97.
7887. —— and M. Simak. 1956: X-ray diagnostics and seed quality in forestry. Intern. Union Forest Res. Organ. Congr. Proc. 12(1, Sect. 22):398–413.
7888. —— and —— 1958: Effect of X- and γ-rays on conifer seed. Medd. Statens Skogsforskningsinst. 48:1–20.
7889. —— and —— 1963: X-ray photography and seed sterility in *Phragmites communis* Trin. Hereditas 49:442–50.
Gustafsson, A. 5829, 5833, 6769, 6770, 7099, 13858, 17446–17449.
Gustafsson, H. 5832.
Guterman, C. E. F. 9074.
7890. Guthrie, F. E., and R. R. Kincaid. 1957: The effect of methyl bromide fumigation on the germination of tobacco seed. Tobacco 144(5):18–19. Biol. Abstr. 31:39558. 1957.
Guthrie, J. D. 1420, 8928, 15532, 16082, 17863, 17874.
7891. Gutmanis, K. 1959: The influence of medium constituents on biosynthesis of riboflavine in seeds. Latvijas PSR Zinatnu Akad. Vestis 1959(1):55–58. Chem. Abstr. 54:6887. 1960.
7892. —— 1959: Biosynthesis of riboflavine and enzymic processes in seeds. Latvijas PSR Zinatnu Akad. Vestis 1959(3):73–75. Chem. Abstr. 54:11165. 1960.
7893. Gutsalyuk, T. 1963: Treatment of seeds with minor elements. Kartofel i Ovoshchi 3:53–54.
7894. Guttay, J. R., J. A. Stritzel, A. J. Englehorn, and C. A. Black. 1957: Treatment of corn seed with phosphate. Agron. J. 49:98–101.
Guttenberg, H. von; *see* Von Guttenberg, H.
Guttentag, C. 272.
Guyer, G. E. 5326. 5327.
7895. Guyon, G. 1939: Variations in the nitrogen content of wheat grain during growth. Rech. Fertilisation Sta. Agron. Ministere Agr. (France) 12:55–56. Chem. Abstr. 33:7842. 1939.
Guyot, A. L. 5139.
Guyot, J. L. 11659.
7896. Guyot, L. 1960: Sur la présence dans les terres cultivées et incultes des semences dormantes des espèces adventices. Bull. Serv. Carte Phytogeog. (B) 5:197–254. Weed Abstr. 11:222–23. 1962.
Guyot, L. 2179.
7897. Guzhev, Y. L. 1962: Effect of thermal and light factors on passing of period of dormancy by woody species. Akad. Nauk SSSR Inst. Genet. Tr. 29:149–55.
7898. Guzhevskii, A. 1960: Hydrothermal warming of seeds. Sel'skoe Khoz. Sibiri 1960(5):22–23.
Guzman, M. A. 3003.
Guzman, F. O. 6848.
Guzman, G. J. 6737.

7899. Guzowska, I. 1964: Reinvestigation of embryo sac development, fertilization and early embryogeny in *Cytinus hypocistis* L. Acta Soc. Botan. Polon. 33:157–66.

7900. Gvaladze, G. E. 1960: Embryogenesis of *Allium atroviolaceum* Boiss. (Geo.). Akad. Nauk Gruzinsk. SSR Inst. Botan. Sb. Rabot 1960:25–32.

7901. —— 1961: The embryology of the genus *Allium* L. Soobshch. Akad. Nauk Gruzinsk. 26:193–200. Biol. Abstr. 42: 7696. 1963.

7902. Gvozdyak, P. I. 1957: Influence of aerobic space bacteria isolated from the rhizospheres of wheat and barley on germination of seeds and growth of these plants. Mikrobiol. Zh. 19(4):40–44. Biol. Abstr. 35:35703. 1960.

7903. Gwynne, M. D. 1959: A quick method for extracting plant embryos from certain types of seed. Nature 184:1588–89.

7904. Gysel, L. W. 1957: Acorn production on good, medium, and poor oak sites in southern Michigan. J. Forestry 55:570–74.

7905. Gzhesyuk, S. 1962: Pre-sowing vernalization of cereal grain seeds at various stages of ripeness. Agrobiologiya 1962(4): 547–53. Biol. Abstr. 42:3156. 1963.

7906. —— and T. Mezhvinskaya. 1961: Physiologic and biochemical features of ripening corn grain. Agrobiologiya 1961(1):37–49. Biol. Abstr. 38:15607. 1962.

H

7907. H., A. G. 1940: Can seeds of European vegetables be produced successfully in the tropics? E. African Agr. J. 5:476–78.

7908. H., F. N. 1930: Seedless pawpaws. Kew Roy. Botan. Gardens Bull. Misc. Inform. 1930(5):224.

7909. —— 1931: *Psyllium* seeds. Kew Roy. Botan. Gardens Bull. Misc. Inform. 1931(1):62–63.

7910. H., R. L. 1930: Germinating hybrid seed. Florists Rev. 65(1679):45–46.

7911. Haack, E., F. Kaiser, M. Gube, and H. Spingler. 1956: Die genuinen Glykoside der Blätter und Samen von *Digitalis purpurea*. Naturwissenschaften 43:301–2.

Haack, E. 10142.

7912. Haack, O. 1905: Untersuchung über den Einfluss verschieden höher Darchitze auf das Keimprozent des Kiefernsamens. Z. Forst-Jagdw. 37:296–312.

7913. —— 1906: Über die Keimung und Bewertung des Kiefernsamens nach Keimproben. Z. Forst-Jagdw. 38:440–75.

7914. —— 1909: Der Kiefernsamen. Z. Forst-Jagdw. 41:353–81.

7915. —— 1912: Die Prüfung des Keifernsamens. Z. Forst-Jagdw. 44:193–222, 273–307.

7916. Haagen-Smit, A. J., W. B. Dandliker, S. H. Wittwer, and A. E. Murneek. 1946: Isolation of 3-indoleacetic acid from immature corn kernels. Am. J. Botany 33:118–20.

7917. ——, R. Siu, and G. Wilson. 1945: A method for the culturing of excised, immature corn embryos in vitro. Science 101:234.

Haagen-Smit, A. J. 10957, 10958, 18989.

Haar, A. W. van den; *see* Van den Haar, A. W.

7918. Haarer, A. E. 1964: Selecting Arabian coffee seed. World Crops 16:49–51.

7919. Haarring, R. J., A. T. Wallace, A. J. Norden, and S. C. Schank. 1964: The sensitivity of castorbean (*Ricinus communis* L.) seeds to treatments with ethyl methane sulfonate and γ rays as measured by M_1-seedling response. Radiation Botany 4:43–51.

Haarst, J. van. 3197.

7920. Haas, A. R. C., and J. N. Brusca. 1954: 2,4-D treatment of citrus seed. Calif. Agr. 8(2):8, 12. Biol. Abstr. 28: 21933. 1954.

7921. —— and E. B. Fred. 1919: Effect of soybean germination upon the growth of its nodule-forming bacteria. Soil Sci. 7:237–45.

7922. Haas, P., and T. G. Hill. 1929: An introduction to the chemistry of plant products. Vol. II. Metabolic processes. 2d ed., Longmans, Green, London. 220 pp.

Haas, P. G. de; *see* De Haas, P. G.

7923. Haase, H. 1963: For the protection of seed. Tech. Landwirtsch. 15:76–78.

7924. Haasis, F. W. 1928: Germinative energy of lots of coniferous-tree seed as related to incubation temperature and to duration of incubation. Plant Physiol. 3:365–412.

7925. —— and A. C. Thrupp. 1931: Temperature relations of lodgepole-pine seed germination. Ecology 12:728–44.

Haasis, F. W. 12357.

Haba, M. 19990.

7926. Habel, W. 1957: Über die Wirkungsweise der Eggen gegen Samenunkräuter sowie deren Empfindlichkeit gegen den Eggvorgang. Z. Acker- Pflanzenbau 104: 39–70.

7927. Haber, A. H. 1959: Rendering the germination of light-insensitive lettuce seeds sensitive to light. Physiol. Plantarum 12:456–64.

7928. —— and N. Brassington. 1959: Non-respiratory gas released from seeds during moistening. Nature 183:619–20.

7929. —— and H. J. Luippold. 1959: Dormancy resulting from γ-irradiation of lettuce seed. Intern. J. Radiation Biol. 1:317–27. Biol. Abstr. 35:45639. 1960.

7930. —— and —— 1960: Action of 6-(substituted) purines on mitotic activity in "dormant" lettuce seeds. Physiol. Plantarum 13:298–307.

7931. —— and —— 1960: Effects of gibberellin on γ-irradiated wheat. Am. J. Botany 47:140–44.

7932. —— and —— 1960: Separation of mechanisms initiating cell division and cell expansion in lettuce seed germination. Plant Physiol 35:168–73.

7933. —— and —— 1960: Effects of gibberellin, kinetin, thiourea, and photomorphogenic radiation on mitotic activity in dormant lettuce seed. Plant Physiol. 35:486–94.

7934. —— and N. E. Tolbert. 1959: Effects of gibberellic acid, kinetin, and light on the germination of lettuce seed. Publ. Am. Assoc. Advan. Sci. 55:197–206. Chem. Abstr. 55:705. 1961.

7935. —— and —— 1959: Metabolism of C^{14}-bicarbonate, P^{32}-phosphate, or S^{35}-sulfate by lettuce seed during germination. Plant Physiol. 34:376–80.

7936. —— and J. D. White. 1960: Action of maleic hydrazide on dormancy, cell division, and cell expansion. Plant Physiol. 35:495–99.

Haber, A. H. 12411.

7937. Haber, E. S. 1929: The effect of delayed pollination on size of cob and kernel in Country Gentleman sweet corn. Proc. Am. Soc. Hort. Sci. 26:119–22.

7938. —— 1950: Longevity of the seed of sweet corn inbreds and hybrids. Proc. Am. Soc. Hort. Sci. 55:410–12.

7939. —— and W. G. Gaessler. 1942: Sugar content of sweet-corn pollen and kernels of inbred and hybrid strains susceptible to tassel infestation by aphis. Proc. Am. Soc. Hort. Sci. 40:429–31. Chem. Abstr. 36:6580. 1942.

7940. —— and E. Swift. 1941: Vitamin B_1 treatment of cuttings and seeds. Ann. Rept. (Trans.) Iowa State Hort. Soc. 75:153–56.

Haber, E. S. 6895, 12708.

7941. Haberg, A., and G. Julen. 1962: Analyses of the frequency of diploids in different seed size fractions of lots of tetraploid clover. Sveriges Utsadesforen. Tidskr. 72:50–57.

7942. Haberlandt, F. 1875: Die Aufnahme von garformigem Wasser durch Samen. Wiss.-Prakt. Gebiete Pflanzenbaues 1:63–75.

7943. —— 1875: Über den Einfluss der Theilweisen Entfernung der Samenschale oder des Endosperms auf die Wasseraufnahme und das Keimen der Samen. Fühling's Landwirtsch. Ztg. 24:14. Beidermann's Centralbl. Agr.-chem. 8:24–27. 1875. Just's Botan. Jahresber. 3:857. 1975.

7944. —— 1875/77: Wie Verhalten sich luftleer gemachte Samen beim Keimen? Wiss.-Prakt. Gebiete Pflanzenbauses 1: 104–9.

7945. —— 1875/77: Über die untere Grenze der Keimungstemperatur der Samen unserer Culturpflanzen. Wiss.-Prakt. Gebiete Pflanzenbaues 1:109–16.

7946. —— 1939: Über experimentelle adventivembryonie. Naturwissenschaften 27: 818.

Habermann, H. M. 19327.

Habib, A. 8343.

Habib, D. 11785.

7947. Haborg, W. A. F., G. M. Warner, and N. A. Phillips. 1950: Use of 2,4-D as an inhibitor of germination in routine examinations of beans for seed-borne infection. Science 111:91.

7948. Haccius, B. 1952: Die Embryoentwicklung bei *Ottelia alismoides* und das Problem des terminalen Monokotylen-Keimblatts. Planta 40:443–60.

7949. —— 1957: Regenerationserscheinungen an pflanzlichen Embryonen nach Behandlung mit antimitotisch wirksamen Substanzen. Beitr. Biol. Pflanzen 34:3–18.

7950. —— 1957: Über die Regenerationsfähigkeit junger Embryonen von *Eranthis hiemalis* nach Colchicin-Behandlung. Naturwissenschaften 44:18–19.

7951. —— 1959: Morphoregulatorische Beeinflussung pflanzlicher Embryonen durch Phenylborsäure. Naturwissenschaften 46: 153.

7952. —— 1959: Über die unterschiedliche Antimitotica-Empfindlichkeit der Zellen nach undifferenzieter Embryonen von *Eranthis hiemalis*. Z. Naturforsch. 14b: 206–9. Biol. Abstr. 35:32715. 1960.
7953. —— 1962: Anatomie der Gramineen. Z. Botan. 50:201–2.
7954. —— and E. Hartl-Baude. 1957: Embryologische und histogenetische Studien an "Monokotylen Dikotylen." II. *Pinguicula vulgaris* L. und *Pinguicula alpina* L. Botan. Z. 103:567–87.
7955. Hachisuka, S. 1959: Fundamental studies on testing seed and on shortening dormancy of vegetable crop, ornamental flower and pasture grass seeds. J. Agr. Sci. (Setagaya) 5:40–52.
7956. Hack, H. R. B. 1963: Tomato seed germination and emergence at controlled soil-moisture tensions from 0.5 to 20 cm. mercury. Ann. Appl. Biol. 51:473–84.
7957. Hacke-Oudemans, J. J. 1959: Seed kiln in the forest of La Joux in the Jura. Ned. Bosbouw Tijdschr. 31:318–22.
Hackett, T. E. 14631–14633.
7958. Hackleman, J. C. 1960/61: Distribution of foundation seed in other states. Intern. Crop Improvement Assoc. Ann. Rept. 42/43:53–58.
Hackleman, J. C. 3393.
7959. Hacskaylo, J., and C. D. Ranney. 1961: Emergence of phorate-treated cotton seed as affected by substrate moisture and temperature. J. Econ. Entomol. 54:296–98.
Hacskaylo, J. 12268.
7960. Haddad, E. S. 1931: Morphological development of sweet corn pericarp in two inbred lines and their F_1 hybrid. Indiana Agr. Expt. Sta. Bull. 347. 24 pp.
Haddock, J. L. 6098.
7961. Haddock, P. G. 1942: A study of the rest period in seeds and buds of *Pinus lambertiana* Dougl. and *Pinus jeffreyi* Murr. Thesis (Ph.D.), Univ. of California. 229 pp.
7962. —— 1954: Sapling sugar pines grown from excised mature embryos. J. Forestry 52:434–37.
7963. Hadfield, J. W., and R. A. Calder. 1933: Smooth seeds arising in dimpled varieties of marrowfat peas. New Zealand J. Sci. Technol. 14:210–19. Biol. Abstr. 8:5208. 1934.
7964. —— and —— 1935: A genetic analysis of the seed characters wrinkled, dimpled and smooth in *Pisum*. J. Agr. Sci. 25:264–77.
7965. —— and —— 1936: Lucerne. Pollination and seed production. New Zealand J. Agr. 53:28–33.
7966. Hadfield, W. 1958: The raising of bamboo from seed. Tocklai Expt. Sta. News Letter 5(2):9–10.
7967. Hadley, E. B. 1961: Influence of temperature and other factors on *Ceanothus megacarpus* seed germination. Madrono 16:132–38. Weed Abstr. 12:39. 1963.
Hadley, H. H. 15540.
7968. Hadley, P. 1949: "Flying seeds"; many wild honey plants depend upon the wind to scatter their seeds. Mod. Beekeeping 33:407–8.
7969. Hadorn, H. 1945: Sugar determination in wheat germs and in wheat-germ flours. Mitt. Lebensm. Hyg. 36:386–92. Chem. Abstr. 40:4450. 1946.
Hads, H. L. 10878.
Hadzijev, D. [I.] 3563, 13433, 13434.
Haehn, H. 6378.
7970. Hähne. 1939: Untersuchungen über die Gewinnung von Samen aus fleischigen Früchten. Landwirtsch. Jahrb. 87:669–70.
Haekgarrd, H. C. 8671.
7971. Hänlein, H. 1880: Über die Keimkraft von Unkrautsamen. Landwirtsch.-Versuchssta. 25:465–70.
Hänlein, H. 14538.
7972. Hänninen, P. 1956: Mixed cropping of peas and vetches with oats in Finland legume-seed yields and some seed characteristics. Publ. Finnish State Agr. Res. Bd. 153. 131 pp.
7973. Hänsel, H. 1953: Vernalisation of winter rye by negative temperatures and the influence of vernalisation upon the lamina length of the first and second leaf in winter rye, spring barley, and winter barley. Ann. Botany (London) 17:417–32.
7974. Hänsel, H. D., J. G. Ross, and C. C. Huang. 1963: Irradiation-induced mutations in a colchicine-reactive genotype in sorghum. Crop Sci. 3:242–45.
Haenseler, C. M. 5000.
Haertl, E. J. 9765.
Häussermann, C. Riek-; *see* Riek-Häussermann, C.
7975. Hafekost, G. 1930: Saugkraftmessungen an Zucker- und Futterrüben. Fortschr. Landwirtsch. 5:175–77.
7976. —— 1930: Zur Theorie der Saugkraftmessungen an Kulurpflanzen im Keimlingsstadium. Biol. Generalis (Wien) 6: 633–50. Ber. Ges. Physiol. Exptl. Pharmakol. 61:67. 1931.

7977. Hafen, L., and E. C. Stevenson. The relationship of temperature, method of pollination, and male sterility to the time required for producing hybrid tomato seed. Proc. Am. Soc. Hort. Sci. 67:355–60.

Hafenrichter, A. L. 13893, 17122.

7978. Haferkamp, M. 1949: Germination and other tests on superannuated cereal seeds. Proc. Assoc. Offic. Seed Analysts 39:111–14.

7979. ——, L. Smith, and R. A. Nilan. 1953: Studies on aged seeds. I. Relation of age of seed to germination and longevity. Agron. J. 45:434–37.

Haferkamp, M. E. 7841.

7980. Hafiz, A. 1953: Seed borne diseases of plants and their control. Agr. Pakistan 2:137–42. 1953.

7981. —— and A. Sattar. 1952: Control of flag smut of wheat by seed disinfectants. Pakistan J. Sci. Res. 4:40–44. Chem. Abstr. 47:8956.

7982. Haftorn, S. 1956: Synzoic seed dispersal by birds in Norway. Blyttia 14:103–21.

Haga, P. R. van; *see* Van Haga, P. R.

7983. Hagberg, A., and G. Julen. 1962: Analyses of the frequency of diploids in different seed size fractions of lots of tetraploid clover. Sveriges Utsadesforen. Tidskr. 72:50–57.

Hagberg, A. 5903, 5904.

7984. Hagedorn, D. J. 1950: The effect of drought on the seeds of Delwiche Early Scotch field pea. Agron. J. 42:412–13.

Hagedorn, D. J. 19449.

7985. Hageman, R. H., and D. I. Arnon. 1955: The isolation of triosephosphate dehydrogenase from pea seeds. Arch. Biochem. Biophys. 55:162–68.

Hageman, R. H. 3904, 13141.

7986. Hagen, C. E., H. A. Borthwick, and S. B. Hendricks. 1954: Oxygen consumption of lettuce seed in relation to photocontrol of germination. Botan. Gaz. 115:360–64.

7987. Hagen, G. L., Jr. 1957: Physiological and morphogenetic responses of *Nicotiana glauca*, *N. langsdorffii* and their interspecific hybrid. Dissertation Abstr. 17:2297–98.

Hagen, I. K. 13797.

7988. Hager, G., and K. Ihle. 1931: Die Nährstoffverluste des Getreides durch Auswuchs. Fortschr. Landwirtsch. 6:259–63.

7989. Hager, H. 1953: Growing cattleyas from seed to flower in 2½ years. Orchid Dig. 17:12–16.

7990. Hager, O. 1931: Das Keimfloss. Fortschr. Landwirtsch. 6:644–45.

Hagerty, E. B. 7077.

7991. Hagerup, O. 1942: The morphology and biology of the *Corylus*-fruit. Kgl. Danske Videnskab. Selskab, Biol. Medd. 17(6):1–32.

7992. —— 1945: Facultative parthenogenesis and haploidy in *Epipactis latifolia*. Kgl. Danske Videnskab. Selskab, Biol. Medd. 19(11):1–13. Biol. Abstr. 21:2524. 1947.

7993. Hagihara, B., K. Tagawa, I. Morikawa, M. Shin, and K. Okunuki. 1958: Two crystalline haemoproteins of wheat. Nature 181:1656–58.

7994. Hagiwara, J., M. Kageyama, and T. Hiraoka. 1949: Difference in germination between oil-coated and untreated radish seeds in soils of various moisture content. Hort. Assoc. Japan J. 18:187–88.

7995. Hagiwara, T. 1931: Genetico-physiological studies in the seed-colour of Japanese morning glory. Japan. J. Genetics 7:1–16. Japan. J. Botany 6: Abstr. p. (1). 1932.

7996. —— 1937: On a mutable gene of the seed coat colour in *Pharbitis nil*. Japan. J. Genetics 13:185–92.

7997. Hagiya, K. 1949: Studies on the delayed germination in spinach seed. Hort. Assoc. Japan J. 18:198–201.

7998. —— 1952: Studies on the so-called "home" and "foreign" crops of seeds of Tokinashi radish, a Japanese spring variety. III. On the physiological mechanism of the fasciation, and its selective action upon the degeneration of the variety. Japan. J. Breeding 1:215–18. Biol. Abstr. 27:20506. 1953.

7999. —— and Y. Tazima. 1951: Studies on the bolting behavior of radish varieties sowed in autumn after vernalization, and its application of breeding. Hort. Assoc. Japan J. 20:51–52.

8000. Hagler, T. B., W. A. Johnson, and H. G. Barwood. 1953: Factors affecting pecan yields. Alabama Agr. Expt. Sta. Circ. 115. 19 pp.

8001. Hagner, S. 1958: On the production of cones and seed in Swedish conifer forests. Medd. Statens Skogsforskningsinst. 47(8): 1–120. Biol. Abstr. 33:19273. 1959.

8002. —— and M. Simak. 1958: Ein Stratifizierungsversuch mit Samen von *Pinus cembra*. Särtryck Norrlands Skogsvardsförbunds Tidsskr. 2:227–75.

8003. Hagsand, E. 1957: Timothy seed culture. Svensk Frötidn. 26(4):42–45.

8004. Hague, N. G. M. 1963: Fumigation of agricultural products. XVIII. Effect of methyl bromide on the Bent grass nematode, *Anguina agrostis* (Steinbuch 1799) Filipjev 1936, and on the germination of Bent grass, *Agrostis tenuis*. J. Sci. Food Agr. 14:577–79.

Hague, N. G. M. 14954.

8005. Hahn, E. 1955: Sowing bromeliad seeds upon peat slabs. Bromeliad Soc. Bull. 5(1):10–11.

8006. Hahn, F. L. 1943: The analysis of plant products. I. Algarrobilla (Huarango) Ciencia (Mex.) 4:242–43. Chem. Abstr. 39:2538. 1945.

Hahn, F. V. von; *see* Von Hahn, F. V.

Hahn, G. 199.

8007. Haig, I. T. 1932: Premature germination of forest tree seed during natural storage in duff. Ecology 13:311–12.

8008. Haigh, J. C. 1934: Vernalization. Trop. Agriculturist 83:329–38.

8009. —— 1940: The propagation of water hyacinth (*Eichhornia crassipes* Solms.) by seed. Trop. Agriculturist 94:296–97.

8010. —— 1952: A note on the viability of potato seeds. Ann. Botany (London) 16:317–19.

8011. Hair, J. B. 1952: Blind seed of ryegrass. III. Effect of fungicides on the disease. New Zealand J. Sci. Technol. Sect. A 34:117–21.

8012. Hajas, J. 1962: Effects of mechanical influences on the seed corn. *In* I. Iso, editor, Kukorica-Termesztesi Kiserletek, 1958–1960, pp. 191–94.

Hajas, M. Endresz-; *see* Hajas-Endresz, M.

8013. Hakansson, A. 1943: Die Entwicklung des Embryosacks und die Befruchtung bei *Poa alpina*. Hereditas 29:25–61. Expt. Sta. Rec. 91:397–98.

8014. —— 1947: Some observations on the seed development in Eucadorian cacao. Hereditas 33:526–38. Ann. Biol. 51:320. 1947.

8015. —— 1952: Seed development in *Bowlesia tenera*. Botan. Notiser 1952(1): 33–45. Biol. Abstr. 27:14536. 1953.

8016. —— 1953: Die Samenbildung bei *Nothoscordum fragrans*. Botan. Notiser 1953(1):129–39.

8017. —— 1954: Endosperm formation in *Salix*. Botan. Notiser 1954(3):326–32. Biol. Abstr. 30:23753. 1956.

8018. —— 1955: Endosperm formation in *Myrica gale*. Botan. Notiser 1955(1):6–16. Biol. Abstr. 31:8749. 1957.

8019. —— 1956: Seed development of *Picea abies* and *Pinus silvestris*. Medd. Statens Skogsforskningsinst. 46(2). 23 pp.

8020. —— 1957: Notes on endosperm formation in *Betula*. Botan. Notiser 110: 201–4. Biol. Abstr. 33:2850. 1959.

8021. —— 1959: Seed development of pine grafts. Botan. Notiser 112:65–72.

8022. —— 1960: Seed development in *Larix*. Botan. Notiser 113:29–40.

8023. Haksar, C. N., and S. G. Kendurkar. 1960: *Strychnos potatorum* seeds. I. Vikram, J. Vikram Univ. 4:141–45. Chem. Abstr. 57:2596. 1962.

Haksar, C. N. 2412.

Halasa, S. 7318.

Halazon, G. 19681.

Halbrook, N. J. 13220.

Halcro-Wardlaw, H. S. 4812.

8024. Halden, W., G. Knauer, and H. Lieb. 1951: Veränderungen des Aneurin-, Phytinsäureund Phosphatitgehaltes bei der Keimung von Gramineensamen. Bodenkultur Österr. Zentralorgan Landwirtsch. Ernährungsforsch. 5:437–48. Biol. Abstr. 26: 32342. 1952.

Hale, C. R. 19434.

Hale, F. 12559, 12560.

8025. Hale, G. A. 1936: The effect of variety, planting date, spacing, and seed treatment on cotton yields and stands. J. Am. Soc. Agron. 28:364–69.

8026. Hale, W. S. 1939: The proteinase in wheat flour. Cereal Chem. 16:695–702. Expt. Sta. Rec. 83:441. 1940.

Hale, W. S. 1820–1823.

8027. Haley, D. E., and J. F. Lyman. 1921: Castor bean lipase, its preparation and some of its properties. J. Am. Chem. Soc. 43:2664–70.

Haley, D. E. 12415.

Halfon, A. 15444, 15445.

Halina, H. 4876.

8028. Hall, A. P., P. Wheeler, A. Thielen, and A. F. Morgan. 1953: A new B vitamin study of walnuts. Food Res. 18:574–77.

8029. Hall, B. A. 1947: The nature of the ovary in the genus *Rhus*. Am. J. Botany (Suppl.) 34:6a.

Hall, B. S. 7852.

8030. Hall, C. B. 1948: Culture of *Solanum nigrum* embryos. Proc. Am. Soc. Hort. Sci. 52:343–46.

Hall, C. M. 438, 2654–2657.

8031. Hall, C. W. 1956: Drying temperatures and storage problems of sugar beet seeds. J. Am. Soc. Sugar Beet Technol. 9:160–66.

8032. —— 1958: Calcium chloride for moisture absorption in corn storage. Mich. Agr. Expt. Sta. Quart. Bull. 40:561–67.

8033. —— 1960: Equilibrium moisture content of cacao beans. Acta Agron. (Budapest) 10:53–56.

Hall, C. W. 6046, 15215, 15216.

8034. Hall, D. W., and M. B. Hyde. 1954: The modern method of hermetic storage. Trop. Agr. (Trinidad) 31:149–60.

Hall, D. W. 2607.

Hall, E. E. 17915.

8035. Hall, G. 1937: Propagation of primulas. Gardeners Chron. 102:397.

8036. Hall, G. E., and H. K. Heizer. 1964: Sedimentation values of artificially dried and heated wheat. Ohio Agr. Expt. Sta. Res. Circ. 127. 14 pp.

Hall, H. H. 6732, 7051.

8037. Hall, O. L. 1954: Hybridization of wheat and rye after embryo transplantation. Hereditas 40:453–58.

8038. Hall, P. J. 1963: Alternate methods for germinating range grass seed. Assoc. Offic. Seed Analysts News Letter 37(2):9.

Hall, R. J. 15231, 19360.

8039. Hall, W. C. 1949: Effects of photoperiod and nitrogen supply on growth and reproduction in the gherkin. Plant Physiol. 24:753–69.

Hall, W. E. 9225.

8040. Hallauer, A. R., and W. A. Russell. 1961: Effect of selected weather factors on grain moisture reduction from silking to physiologic maturity in corn. Agron. J. 53:225–29.

8041. Halle, T. G. 1929: Some seed-bearing Pteridosperms from the Permian of China. Kgl. Svenska Vetenskapsakad. Handl. Ser. 6, 3. 24 pp. Botan. Centralbl. 159:56–57. 1930.

Haller, H. L. 9646.

8042. Hallett, G. F. 1936: Greenhouse plants from seeds. Gardeners Chron. 99:106.

8043. Hallock, D. L. 1962: Effect of time and rate of fertilizer application on yield and seed size of Jumbo runner peanuts. Agron. J. 54:428–30.

8044. Halma, F. F., and E. Frolich. 1949: Storing avocado seeds and hastening germination. Calif. Avocado Assoc. Yearbook 34:136–38.

8045. Halperin, W., and D. F. Wetherell. 1964: Adventive embryony in tissue cultures of the wild carrot, *Daucus carota*. Am. J. Botany 51:274–83.

Halstead, R. L. 14394.

8046. Halsted, B. D. 1917: Weight of seeds as related to their number and position in the pod. Torreya 17:102–3.

8047. —— 1918: Possible correlations concerning position of seeds in the pod. Botan. Gaz. 67:243–50.

8048. Haltiwanger, W. L., C. C. Wernham, and L. L. Huber. 1953: The effect of temperature, incubation period, disease organisms, and cultural media on the germination of six corn hybrids. Northeast. Corn Improvement Conf. Rept. 8:8–11.

8049. Haltmeier, O. 1937: Water absorption by grain during washing. Mühle 74:1163–66. Soc. Chem. Ind. J. 57:574B. 1938.

8050. Halton, P., and E. Barton-Wright. 1943: The endosperm content of wheat. Soc. Chem. Ind. J. 62:123–24.

Halton, P. 13477, 13823.

8051. Halverson, A. W., and O. E. Olson. 1953: The protein composition of barley grown in South Dakota. S. Dakota Agr. Expt. Sta. Tech. Bull. 13. 18 pp.

8052. ——, D. F. Petersen, and H. L. Klug. 1951: Fractionation of the selenium protective factor in flaxseed. Proc. S. Dakota Acad. Sci. 30:97–102. Chem. Abstr. 48:12824. 1954.

Halverson, A. W. 7029, 17043.

8053. Halverson, J. O., and F. H. Smith. 1941: Estimation of gossypol in crude cottonseed oil. Ind. Eng. Chem. Anal. Ed. 13:46–48. Expt. Sta. Rec. 86:734. 1942.

8054. Halvorsen, H. 1955: The gas exchange of flax seeds in relation to temperature. I. Experiments with immature seeds and capsules. Physiol. Plantarum 8:501–11.

8055. —— 1956: The gas exchange of flax seeds in relation to temperature. II. Experiments with germinating seeds. Physiol. Plantarum 9:412–20.

Halvorsen, H. 13439.

8056. Ham, W. E., R. M. Sandstedt, and F. E. Mussehl. 1945: The proteolytic inhibiting substance in the extract from unleated soybean meal and its effect upon growth in chicks. J. Biol. Chem. 161:635–42.

8057. Hamada, A. 1958: Effect of seed treatment with phosphorus-32 for *Atriplex hortensis* L. Japan. J. Breeding 8:202–3.

8058. Hamada, H. 1933: Wachstumsverhältnisse der Keimorgane von verschiedenen Gramineen im Dunkel und bei Belichtung mit besonderer Berücksichtigung ihrer systematischen Stellung. Mem.

Coll. Sci. Univ. Kyoto Ser. B 9:71–128.

8059. —— 1937: Physiologisch-systematische Untersuchungen über das Wachstum der Keimorgane von *Oryza sativa* L. Kyoto Univ. Botan. Inst. Publ. 45. 51 pp.

8060. Hamann, U. 1962: On structure and development of the endosperm of Philydraceae and on the concepts "mealy nutrient tissue" and "Farinosae." Botan. Jahrb. 81:397–407.

8061. —— 1963: Die Embryologie von *Philydrum lanuginosum* (monocotyledonae —Philydraceae). Ber. Deut. Botan. Ges. 76:203–8.

8062. Hamant, C. 1935: Hydrocyanic acid and nitrates (formed) during the course of germination of the Sorgho. Compt. Rend. 201:1503–5.

8063. Hamashima, S. 1957: On the dispersion of seed of *Tetragonia expansa* Murr. Hokuriku J. Botan. 6(3):62–74.

8064. Hambleton, J. I. 1944: The role of bees in the production of fruit and seed. J. Econ. Entomol. 37:522–25.

8065. Hamblin, S. F. 1937: Hardy aquatics from seed. Horticulture 15(11):271.

8066. Hambrouck, K. 1953: Identification of horticultural seeds. Cult. en Handel 19:106.

8067. Hameed, A., and Karimullah. 1955: The chemistry of germination of seeds. I. Cottonseed. Pakistan J. Sci. Res. 7:170–71.

8068. Hamet, R. 1911: Observations sur la germination des Crassulacies. Bull. Soc. Bot. France 58:1–13.

Hamet, Raymond-; *see* Raymond-Hamet.

8069. Hamid, A. 1938: Longevity of cotton seed delinted with sulfuric acid. Empire Cotton Growing Rev. 15:312–14. Chem. Abstr. 33:5026–27. 1939.

8070. —— and N. Mohammad. 1936: Preliminary studies on the effect of delinting cotton seed with sulfuric acid on germination and yield. Agr. Livestock India 6:653–61. Chem. Abstr. 31:3101. 1937.

Hamilton, H. C. 8900, 8901.

8071. Hamilton, H. S., and A. W. Hofer. 1940: The selective disinfection of seeds. J. Bacteriol. 40:165–66.

8072. Hamilton, J., and R. D. Dickey. 1943: Growing and transplanting nursery trees of *Aleurites montana* in Florida. Proc. Am. Soc. Hort. Sci. 42:375–79.

8073. —— and S. G. Gilbert. 1947: Rapid method for determining oil content of tung kernels. Anal. Chem. 19:453–56.

Hamilton, J., Jr. 16397.

Hamilton, K. C. 452, 19694.

8074. Hamilton, L. P., and W. M. Wooton. 1950: Grass seed production. Ariz. Agr. Expt. Sta. Bull. 228. 40 pp.

Hamilton, L. P. 518.

Hamilton, P. B. 1329.

8075. Hamilton, R. A. 1957: A study of germination and storage life of *Macadamia* seed. Proc. Am. Soc. Hort. Sci. 70:209–12.

Hamilton, R. A. 18061.

8076. Hamilton, S. W. 1943: Ratios of clean and rough seed. J. Forestry 41:63–64.

8077. Hamilton, T. S., and F. I. Nakamura. 1940: The cystine content of eleven varieties of soybeans. J. Agr. Res. 61:207–14.

Hamilton, T. S. 13589, 13590.

Hamlin, M. L. 6205.

8078. Hamly, D. H. 1932: Softening of the seeds of *Melilotus alba*. Botan. Gaz. 93:345–75.

8079. —— 1935: The light line in *Melilotus alba*. Botan. Gaz. 96:755–57.

8080. —— 1945: A small separator for the recovery of milkweed floss and seed. I. Construction and operation. Can. J. Res. Sect. F 23:313–25. Biol. Abstr. 20:1475. 1946.

Hamm, P. C. 12161.

8081. Hammar, H. E., and J. H. Hunter. 1946: Some physical and chemical changes in the composition of pecan nuts during kernel filling. Plant Physiol. 21:476–91.

Hammar, H. E. 9231.

8082. Hammar, S. 1901: Observations on the germination capacity of Red clover. Kgl. Landtbr.-Akad. Handl. Tidskr. 40: 269–72.

8083. Hammerton, J. L. 1961: Studies on the effects of soil aggregate size on the emergence and growth of beet (*Beta vulgaris* L.). I. Seedling emergence. J. Agr. Sci. 56:213–28.

8084. —— 1963: The effects of soil aggregate size on the emergence and growth of beet (*Beta vulgaris* L.). III. Seedling growth. Empire J. Exptl. Agr. 31:83–96. Biol. Abstr. 43:3400. 1963.

8085. Hammond, B. L. 1959: Effect of gibberellin, sodium hypochlorite, light, and planting depth on germination of Guayule seed. Agron. J. 51:621–23.

Hammons, R. O. 5025.

8086. Hammouda, F. M., and Z. F. Ahmed. 1962: A phytochemical study of the seeds of *Pisum sativum*. I. Congr. Sci. Farm. Conf. Comun. 21, Pisa, 1961, pp. 551–64. Chem. Abstr. 59:903. 1963.

8087. Hammer, C. L., J. E. Moulton, and H. B. Tukey. 1946: Effect of soil and seed treatment with 2,4-dichlorophenoxyacetic acid on germination of seeds and development of seedlings. Botan. Gaz. 107:352–61.

Hamner, K. C. 17740.

Hamori-Szabo, J. 5952.

8088. Hampe, O. 1932: Poinsettien aus Samen? Gartenwelt 36:59.

8089. Hampson, C. P. 1960: Uniform treatment of seed with liquid dressings. World Crops 12:74.

8090. Hampton, R. E., W. H. Sill, Jr., and C. D. Hansing. 1957: Barley stripe mosaic virus in Kansas and its control by a greenhouse seed-lot testing technic. Plant Disease Reptr. 41:735.

Hamsago, R. S. 10340.

8091. Hanada, K. 1955: On the change of the resistance to the toxic action of potassium chlorate in wheat seeds during their germination period. Proc. Crop Sci. Soc. Japan 23:202–3.

Hanada, K. 1943.

8092. Hanaoka, T. 1953: The role of the cotyledons of the kidney-bean seeds in relation to the growth-phases of seedlings during germination. Hokkaido Natl. Agr. Expt. Sta. Res. Bull. 64:33–41.

8093. —— 1956: Studies on the production of seed in spinach. II. The effects of vernalization on growth and flowering of spinach. Hokkaido Natl. Agr. Expt. Sta. Bull. 71:128–34.

8094. Hanausek, I. F. 1895: Über symmetrische und polyembryonische Samen von *Coffea arabica*. Ber. Deut. Botan. Ges. 13:73–78.

8095. Hanawa, J. 1953: The embryo sac formation and the embryogeny of *Sesamum indicum* L. in the diploid and tetraploid strains. Botan. Mag. (Tokyo) 66:777–78.

8096. Hanchey, R. H., and W. D. Kimbrough. 1954: Mimosa seed germination tests. Proc. Assoc. Southern Agr. Workers 1954:139.

8097. —— and —— 1954: *Magnolia grandiflora* seed germination tests. Proc. Assoc. Southern Agr. Workers 1954:140.

8098. ——, J. R. King, and W. D. Kimbrough. 1956: Polyembryony in *Camellia japonica*. Am. Camellia Yearbook 1956: 140–42.

8099. Hancock, N. I. 1938: A new method of delinting cottonseed with sulfuric acid. Tenn. Agr. Expt. Sta. Circ. 61. 8 pp.

8100. —— 1940: Poor germination of mechanically dehulled oats. Tenn. Agr. Expt. Sta. Circ. 67. 4 pp.

8101. Hancox, N. C., and H. H. Hatt. 1955: The seed fat of *Mallotus philippinensis* (kamala oil). An examination of some Australian material. Queensland J. Agr. Sci. 12:57–60. Chem. Abstr. 7234. 1956.

Hand, D. B. 15924.

Hand, M. E. 19185.

8102. Hanes, C. S. 1940: Enzymic synthesis of starch from glucose-l-phosphate. Nature 145:348–49.

8103. Haney, G. P. 1962: Seedbed scarification aids regeneration of Shortleaf pine. J. Forestry 60:400–2.

8104. Haney, T. G. 1954: Tobacco seedling production. J. Agr. W. Australia 3:371–74.

8105. Hanf, M. 1941: Keimung und Entwicklung des Klettenlabkrautes (*Galium aparine* L.) in verschiedener Aussaattiefe. Angew. Botan. 23:152–63.

8106. —— 1943: Keimung von Unkraütern unter verschiedenen Bedingungen im Boden. Landwirtsch. Jarhb. 93:169–254.

8107. —— 1957: Wuchsstoffebedingte anomalien bei Getreideinflorescenzen. Beitr. Biol. Pflanz. 34:19–33.

Haniu, Y. 6810.

8108. Hanks, R. J., and F. C. Thorp. 1956: Seedling emergence of wheat as related to soil moisture content, bulk density, oxygen diffusion rate, and crust strength. Soil Sci. Soc. Am. Proc. 20:307–10.

8109. —— and —— 1957: Seedling emergence of wheat, grain sorghum, and soybeans as influenced by soil crust strength and moisture content. Soil Sci. Soc. Am. Proc. 21:357–59.

8110. Hanley, F. 1934: Experiments on the value of decorticated sugar-beet seed. Gt. Brit. Min. Agr. J. 41:21–28. Expt. Sta. Rec. 71:317. 1934.

8111. —— 1944: Sowing seed and fertilizer together. Gt. Brit. Min. Agr. J. 50:493–96.

8112. —— and R. M. Woodman. 1930: Effect of sulphuric acid treatment on the germination of sugar beet seed. Soc. Chem. Ind. J. 49:215t–220t.

8113. Hanna, G. B., W. H. Sheldon, and W. M. Carleton. 1952: Methods and equipment for harvesting Ladino clover seed. Mich. Agr. Expt. Sta. Quart. Bull. 34:320–26.

8114. Hanna, W. F. 1936: Effect of vernalization on the incidence of loose smut in wheat. Sci. Agr. (Ottawa) 16:404–7.
8115. —— and W. Pope. 1935: Experiments on the control of cereal smuts by seed treatments. Sci. Agr. (Ottawa) 15:745–53.
Hannah, A. E. 4752.
8116. Hannemon, I. 1931: Physico-chemical investigation on proteins of the millet, the Japanese barnyard millet, and the Italian millet grains. Res. Bull. Gifu Imp. Coll. Agr. 17:1–168. Biol. Abstr. 7:10369. 1933.
8117. Hannibal, L. S. 1943: *Alstroemeria* and *Bomarea* from seeds. Herbertia 9:183–87.
8118. —— 1955: Seed color inheritance and breeding patterns in *Brunsvigia* x *multiflora* hybrids. Plant Life 11:67–75.
8119. —— 1955: Unusual seeding habits of *Amaryllis belladonna.* J. Roy. Hort. Soc. 80:130–31.
8120. —— 1958: Germinating seeds in vermiculite. Plant Life 14:95–96.
8121. Hannig, E. 1904: Zur Physiologie pflanzlicher Embryonen. I. Über die Cultur von Cruciferen-Embryonen ausserhalb. des Embryosacks. Botan. Ztg. 62:45–80.
8122. —— 1906: Zur Physiologie pflanzlicher Embryonen. II. Über das Zustandekommen der Lagerung der Keimlinge bei den Cruciferen-Embryonen. Botan. Ztg. 64:1–14.
8123. —— 1907: Zur Physiologie pflanzlicher Embryonen. III. Assimilieren Cruciferen-Embryonen in Künstlicher Kultur die Nitrate der Nährlösung? Botan. Ztg. 65:39–44.
Hans, M. 19419.
8124. Hansberry, R., R. T. Clausen, and L. B. Norton. 1947: Variations in the chemical composition and insecticidal properties of the yam bean (*Pachyrrhizus*). J. Agr. Res. 74:55–64.
Hansberry, R. 14593.
Hansbrough, T. 3205.
8125. Hansche, R., and E. U. Still. 1933: A study of the properties of psyllium seed. Am. J. Pharm. 105:433–35. Biol. Abstr. 9:16665. 1935.
8126. Hansel, H. 1962: Research on the depletion of the endosperm in winter wheat (*Triticum aestivum* L.) seeds. Bodenkultur 13:294–309.
8127. Hansel, V. 1876: Über die Keimung der *Preissia commutata* N. ab E. Sitzber. Wien. Wiss. Akad. Math.-Naturw. Kl. 73:89–97.
8128. Hansen, C. R. 1947: Germinating *Cobaea.* Flower Grower 34(8):640.
8129. Hansen, E. 1943: Ascorbic acid content of walnut hulls. Proc. Am. Soc. Hort. Sci. 42:265–66.
8130. —— 1963: Control of CO_2 concentration in sealed polyethylene pear box liners by use of packaged hydrated lime inserts. Proc. Am. Soc. Hort. Sci. 83:210–16.
Hansen, H. L. 3752, 16999.
8131. Hansen, J. H., and D. W. Muelder. 1963: Testing of redwood seed for silvicultural research by X-ray photography. Forest Sci. 9:470–76.
8132. Hansen, J. R., and K. P. Buchholtz. 1950: Germination and seedling responses of inbred lines of corn to 2,4-dichlorophenoxyacetic acid. Agron. J. 42:452–55.
Hansen, J. Roll-; *see* Roll-Hansen, J.
8133. Hansen, K. 1954: Harvesting and storage of grain. Jydsk Landbr. 36:183–91.
Hansen, T. Schantz-; *see* Schantz-Hansen, T.
Hansing, E. D. 8090, 16115, 16862, 18225.
8134. Hanson, C. H. 1943: Cleistogamy and the development of the embryo sac in *Lespedeza stipulacea.* J. Agr. Res. 67:265–72.
8135. —— and R. P. Moore. 1959: Viability of seeds of eight forage crop plants stored under subfreezing conditions. Agron. J. 51:627.
8136. ——, H. F. Robinson, and R. E. Comstock. 1956: Biometrical studies of yield in segregating populations of Korean *Lespedeza.* Agron. J. 48:268–72.
Hanson, G. P. 3876.
8137. Hanson, H. S. 1952: *Megastigmus* seed flies. Quart. J. Forestry 46:261–64. Biol. Abstr. 27:14029.
8138. Hanson, J. B., A. E. Vatter, M. E. Fisher, and R. F. Bils. 1959: The development of mitochondria in the scutellum of germinating corn. Agron. J. 51:295–301.
Hanson, J. B. 10134.
8139. Hanssen, E., and E. Dodt. 1955: Untersuchungen über die Eigenfluoreszenz bei Gramineen und Leguminosen. Naturwissenschaften 42:76.
Hanssen, E. 8684.
8140. Hansson, N. 1924: The feeding value and utilization of rape seed and hemp seed

cake. Medd. Centralanst. Försöks. Jordbrucks. (Sweden) 274. 18 pp. Expt. Sta. Rec. 53:579. 1925.

Hanway, D. G. 16016, 19701.

Hanzas, P. C. 5541.

8141. Hao, K. S. 1939: Über Saatgutprüfung auf biochemischem Wege. Z. Forst-Jagdw. 71:141–56.

8142. —— 1939: Über Saatgutprüfung auf biochemischem Wege. Z. Forst-Jagdw. 71:187–204.

8143. —— 1939: Über Saatgutprüfung auf biochemischem Wege. Z. Forst-Jagdw. 71:249–69.

Haque, A. 12698.

8144. Haque, M. N. 1949: The pharmacological properties of chaksine, an alkaloid of the seeds of *Cassia absus*, Linn. [Excerpts.] Pakistan J. Sci. Res. 1:91–95.

Hara, F. 13903.

8145. Hara, H., and H. Kawahara. 1951: Swelling of seeds and effects of supersonics. Bull. Exptl. Biol. 1:36–39. Biol. Abstr. 27:20640. 1953.

Hara, H. 10363.

8146. Hara, M., and K. Tanaka. 1957: On the reforestation by seeding of Japanese black pine (*Pinus thunbergii* Parl.) by the use of capsules on the sand dunes. I. Trans. Tottori Soc. Agr. Sci. 11:125–30.

Hara, M. 8743.

Harada, A. 13903.

8147. Harada, H. 1954: Influence of moisture condition in seed bed upon the water absorption and the germination of the seed of *Cryptomeria japonica* D. Don. Proc. Tottori Soc. Agr. Sci. 10:87–94.

8148. —— 1962: Étude de substances naturelles de croissance en relation avec la floraison. Isolément d'une substance de moutaison. Rev. Gen. Botan. 69:201–97.

8149. Harada, K. 1934: Diagnosis of minute seeds by means of the impressions upon celluloid-membrane ("sump"-figures). II. J. Japan. Botany 10:792–94.

8150. —— and K. Nakamura. 1953: The effect of the soil layer between seed and basic fertilizer on growth and root development of young wheat in light diluvial soils. J. Kanto-Tosan Agr. Expt. Sta. 4:27–31. Soils Fertilizers 17:369. 1954.

Harada, K. 10152.

8151. Harada, N. 1961: Studies on chestnuts in storage. III. On the polyphenolic substances and the oxidizing enzymes of chestnuts in storage. J. Japan. Soc. Hort. Sci. 30:126–29.

8152. Harada, Y. 1938: Einfluss der täglichen Belichtungdauer für Keimung des Samens und für das Leben der einjährigen Jungwüchse von Sachalintannen. Botany Zoology 6:1511–22. Biol. Abstr. 14:9377. 1940.

8153. —— and T. Yanagisawa. 1952: Turnout of seed storing experiments for some trees. Hokkaido Forestry Expt. Sta. Sapporo Br. Rept. 71:26–46.

8154. —— and —— 1952: A long-time storage of tree-seed in northern Japan. Hokkaido Forestry Expt. Sta. Sapporo Br. Rept. 71:47–68.

8155. Haralamb, A., and D. Varga. 1963: The depositing of spruce seeds in refrigerating chambers and air tight closed jars. Rev. Padurilor (Bucharest) 78:289–93.

8156. Harashima, S. 1937: Comparative studies of upland and lowland rice varieties in regard to the germination of seeds at low temperatures. Proc. Crop Sci. Soc. Japan 9:407–17.

8157. Harbord, W. L. 1949: Seed production in New Zealand. New Zealand J. Agr. 78:277–80.

8158. Hardenburg, E. V. 1944: Effect of hormone dust on pod-set and yield in beans. Proc. Am. Soc. Hort. Sci. 45:367–70.

8159. —— and W. H. Eto. 1948: The role of snakehead plants in beans. Proc. Am. Soc. Hort. Sci. 51:486–92.

Harder, A. 2647.

8160. Harder, R., and H. Von Witsch. 1941: Wirkung von Photoperiodismus und Yarowisation auf die Blutenbildung von *Kalanchoe blossfeldiana*. Gartenbauwissenschaft 15:226–46. Biol. Abstr. 23:8824. 1949.

Harder, R. 3290.

8161. Hardesty, B., and F. C. Elliott. 1956: Differential post-ripening effects among seeds from the same parental wheat spike. Agron. J. 48:406–9.

8162. Hardin, E. 1962: A laboratory model velvet-roll machine is a useful piece of equipment in a seed laboratory. Assoc. Offic. Seed Analysts News Letter 36(1):21–22.

Harding, F. 9652.

Harding, F. Charconnet-; *see* Charconnet-Harding, F.

8163. Harding, J. H. 1940: *Acacia* germination trials. Australian Forestry 5:53–56. Biol. Abstr. 15:6155. 1941.

8164. Hardison, J. R. 1948: Field control of blind disease of perennial ryegrass in Oregon. Phytopathology 38:404–19.

8165. Hardman, E. E., and W. M. Crombie. 1958: Fat metabolism in germinating *Citrullus vulgaris*. II. J. Exptl. Botany 9:239–46.

Hardman, E. E. 4610.

8166. Hardon, H. J., and J. A. Nijholt. 1941: Composition of indigenous castor seeds. Landbouw. 17:476–85. Chem. Abstr. 36:1795. 1942.

Hardy, M. B. 4517.

8167. Hare, W. W. 1955: Seed treatment of cowpeas. Plant Disease Reptr. 39:580–81.

8168. Hareb, E., and A. M. Mayer. 1963: Interference of phenolase in the determination of cytochrome oxidase activity in lettuce seeds. Phytochemistry 2:417–19.

8169. Hargrave, P. D. 1937: Seed germination of the Saskatoon and pincherry. Sci. Agr. (Ottawa) 17:736–39.

8170. Hargreaves, L. A., and K. Dorman. 1956: Administrative and technical aspects of establishing pine seed orchards. Proc. Soc. Am. Foresters 1956:8–16.

8171. Haring, P., and D. Varga. 1962: Extraction of seed from Larch cones. Rev. Padurilor (Bucharest) 77:650–52. Forestry Abstr. 24:411. 1963.

Harjanne, K. 16127.

8172. Harker, K. W. 1957: A note on *Digitaria scalarum* seed. E. African Agr. J. 23:109.

8173. Harkness, B. 1941: Sowing seeds in snow. Flower Grower 28(3):133–34.

Harkness, D. M. 6700.

8174. Harlan, H. V. 1925: The germination of barley under late spring malting conditions in India. Science 61:566–67.

8175. —— and M. N. Pope. 1925: Some cases of apparent single fertilization in barley. Am. J. Botany 12:50–53.

8176. —— and —— 1926: Development of immature barley kernels removed from the plant. J. Agr. Res. 32:669–78.

Harlan, H. V. 12813, 12924.

8177. Harlan, J. R. 1946: The development of Buffalo grass seed. J. Am. Soc. Agron. 38:135–41.

8178. —— 1949: Apomixis in side-oats grama. Am. J. Botany 36:495–99.

8179. —— 1956: Theory and dynamics of grassland agriculture. Van Nostrand, Princeton, N.J. 281 pp.

8180. —— and R. M. Ahring. 1960: A suggested method for determining purity of certain chaffy-seeded grasses. Agron. J. 52:223–26.

8181. —— and W. R. Kneebone. 1953: Effect of various methods and rates of nitrogen application on seed yield of switchgrass (*Panicum virgatum* L.). Agron. J. 45:385–86.

Harlan, J. R. 195–197.

8182. Harland, S. S. 1936: Haploids in polyembryonic seeds of Sea Island cotton. J. Heredity 27:229–31.

8183. Harler, C. R. 1950: Tea seed gardens. Nyasaland Agr. Quart. J. 8:97–103. 1949: Biol. Abstr. 25:21439. 1951.

8184. Harley, A. 1928: Meconopses and gentians at Devonhall, Perthshire. J. Roy. Hort. Soc. 53:220–34.

8185. Harling, G. 1946: Studien über den Blütenbau und die Embryologie der Familie Cyclanthaceae. Svensk Botan. Tidskr. 40:257–72.

8186. —— 1951: Embryological studies in the Compositae. II–III. Acta Horti Bergiani 16:1–56, 74–120.

8187. —— 1954: The embryo-sac development of *Vittadinia triloba* (Gaud.) DC. Svensk Botan. Tidskr. 48:489–96.

8188. Harmath, J. 1960: Determining diseases and pests on dry seeds. Magy. Mezogazdasag 15(2):16–17.

8189. Harmeling, S. J. 1935: Viable seed of Madonna lilies. Horticulture 13(21):486.

Harmer, P. M. 3150.

8190. Harmon, F. N., and E. Snyder. 1936: A seeded mutation of the Panariti grape. J. Heredity 27:77–78.

8191. —— and J. H. Weinberger. 1959: Effects of storage and stratification on germination of *Vinifera* grape seeds. Proc. Am. Soc. Hort. Sci. 73:147–50.

8192. Harmon, G. W., and F. D. Keim. 1934: The percentage and viability of weed seeds recovered in the feces of farm animals and their longevity when buried in manure. J. Am. Soc. Agron. 26:762–67.

8193. Harmond, J. E., L. M. Klein, and N. R. Brandenburg. 1961: Seed cleaning and handling. U.S. Dept. Agr., ARS, Agr. Handbook 179. 38 pp.

8194. Harper, A. E., B. D. Punekar, and C. A. Elvehjem. 1958: Effect of alkali treatment on the availability of niacin and amino acids in maize. J. Nutr. 66:163–72.

Harper, F. A. 15157.

Harper, H. A. 6700.

Harper, H. J. 9166.

Harper, J. A. 1414.

8195. Harper, J. L. 1956: Studies in seed and seedling mortality. V. Direct and indirect influences of low temperatures on the mortality of maize. New Phytologist 55:35–44.

8196. —— 1959: The ecological significance of dormancy and its importance in weed control. Proc. 4th Intern. Congr. Crop Protect. 1:415–20.

8197. —— and D. Gajic. 1961: Experimental studies of the mortality and plasticity of a weed. Weed Res. 1:91–104.

8198. —— and P. A. Landragin. 1955: The influence of the environment on seed and seedling mortality. IV. Soil temperature and maize grain mortality with especial reference to cold test procedure. Plant Soil 6:360–72.

8199. ——, ——, and J. W. Ludwig. 1955: The influence of environment on seed and seedling mortality. I. The influence of time of planting on the germination of maize. New Phytologist 54:107–18.

8200. ——, ——, and —— 1955: The influence of environment on seed and seedling mortality. II. The pathogenic potential of the soil. New Phytologist 54:119–31.

Harper, J. L. 12501–12503.

8201. Harpstead, D. D. 1962: The effects of γ ray and thermal neutron irradiation on dormant flax seeds as measured by oil quality and seed yield. Dissertation Abstr. 22:2551. 1962.

Harrel, C. G. 9113.

Harrell, D. C. 19367.

8202. Harrington, G. T. 1916: Agricultural value of impermeable seeds. J. Agr. Res. 6:761–96.

8203. —— 1917: Further studies on the germination of Johnson grass seed. Proc. Assoc. Offic. Seed Analysts 10:71–76.

8204. —— 1919: Comparative chemical analyses of Johnson grass seeds and Sudan grass seeds. Proc. Assoc. Offic. Seed Analysts 11:58–63.

8205. —— 1921: Optimum temperatures for flower seed germination. Botan. Gaz. 72:337–58.

8206. —— 1923: Forcing the germination of freshly harvested wheat and other cereals. J. Agr. Res. 23:79–100.

8207. —— 1923: Respiration of apple seeds. J. Agr. Res. 23:117–30.

8208. —— 1923: Use of alternating temperatures in the germination of seeds. J. Agr. Res. 23:295–332.

8209. —— and W. Crocker. 1918: Resistance of seeds to desiccation. J. Agr. Res. 14:525–32.

8210. —— and —— 1923: A new and efficient respirometer for seeds and other small objects: directions for its use. J. Agr. Res. 23:101–15.

8211. —— and —— 1923: Structure, physical characteristics, and composition of the pericarp and integument of Johnson grass seed in relation to its physiology. J. Agr. Res. 23:193–222.

8212. —— and B. C. Hite. 1923: After-ripening and germination of apple seeds. J. Agr. Res. 23:153–61.

Harrington, G. T. 4604.

8213. Harrington, J. B. 1935: Character combinations in relation to endosperm develpment in inter-specific *Triticum* crosses. Can. J. Res. Sect. C 13:388–99.

8214. —— 1935: Decreasing the doubts [in the production of registered seed]. Ann. Rept. Can. Seed Growers Assoc. 1934/35:49–55.

8215. —— 1949: Testing cereal varieties for dormancy. Sci. Agr. (Ottawa) 29:538–50.

8216. —— and P. F. Knowles. 1940: Dormancy in wheat and barley varieties in relation to breeding. Sci. Agr. (Ottawa) 20:355–64.

8217. —— and —— 1940: The breeding significance of after-harvest sprouting in wheat. Sci. Agr. (Ottawa) 20:402–13.

8218. —— and J. B. Marshall. 1934: F_2 endosperm development in relation to breeding technique with inter-specific wheat crosses. Can. J. Res. 11:333–46.

Harrington, J. B. 7259.

8219. Harrington, J. F. 1949: Hard seed in beans and other legumes. Seed World 64(1):42, 44.

8220. —— 1951: Effect of spacing and size of root on carrot seed yield and germination. Proc. Am. Soc. Hort. Sci. 58:165–67.

8221. —— 1958: Moisture-proof packaging of seeds. Seed World 83(4):8, 10–11.

8222. —— 1959: Effect of fruit maturity and harvesting methods on germination of muskmelon seed. Proc. Am. Soc. Hort. Sci. 73:422–30.

8223. —— 1960: Drying, storing and packaging seed to maintain germination, vigor. [Part I.] Seedsmens Dig. 11(1):16, [56–57], 64, [66], 68.

8224. —— 1960: Germination of seeds from carrot, lettuce and pepper plants grown under severe nutrient deficiencies. Hilgardia 30:219–35.

8225. —— 1960: Methods of prolonging viability of seeds. Calif. Hort. Soc. J. 21:127–29.

8226. —— 1960: Seed storage and seed packages. Seed World 87 (July 22):4–6.

8227. —— 1960: Thumb rules of drying seed. Crops & Soils 13(1):16–17.

8228. —— 1960: The use of gibberellic acid to induce bolting and increase seed yield of tight-heading lettuce Proc. Am. Soc. Hort. Sci. 75:476–79.

8229. —— 1963: Practical instructions and advice on seed storage. Proc. Intern. Seed Testing Assoc. 28:989–94.

8230. —— 1963: The value of moisture-resistant containers in vegetable seed packaging. Calif. Agr. Expt. Sta. Bull. 792. 23 pp.

8231. —— and G. M. Kihara. 1960: Chilling injury of germinating muskmelon and pepper seed. Proc. Am. Soc. Hort. Sci. 75:485–89.

8232. —— and R. C. Thompson. 1952: Effect of variety and area of production on subsequent germination of lettuce seed at high temperatures. Proc. Am. Soc. Hort. Sci. 59:445–50.

Harrington, J. F. 5187.

Harris, A. W. 10666.

8233. Harris, C. J. 1935: The development of the flower and seed in *Galinsoga ciliata* (Raf.) Blake. Univ. Pittsburgh Bull. 32:131–37. Biol. Abstr. 10:22043. 1936.

8234. Harris, F. S., and G. Thomas. 1914: The change of weight of grain in arid regions during storage. Utah Agr. Expt. Sta. Bull. 130:305–13.

8235. Harris, G., and J. C. MacWilliam. 1958: A note on the development of the starch of the ripening barley ear. Cereal Chem. 35:82–83.

8236. Harris, G. P. 1956: Amino acids as sources of nitrogen for the growth of isolated oat embryos. New Phytologist 55:253–68.

8237. Harris, G. S. 1961: The periodicity of germination in some grass species. New Zealand J. Agr. Res. 4:253–60.

Harris, G. S. 9313.

Harris, H. B. 4503.

Harris, H. C. 2614, 19549.

8238. Harris, J. A. 1912: The influence of the seed upon the size of the fruit in *Staphylea*. Botan. Gaz. 53:204–18, 396–414.

8239. —— 1912: On differential mortality with respect to seed weight occurring in field cultures of *Phaseolus vulgaris*. Am. Naturalist 46:512–25.

8240. —— 1913: A first study of the relationship between the weight of the bean seed, *Phaseolus vulgaris*, and the time required for its germination. Plant World 16:267–74.

8241. —— 1913: Supplementary studies on the differential mortality with respect to seed weight in the germination of garden beans. II. Am. Naturalist 47:739–59.

8242. —— 1914: Further observations on the relationship between the number of ovules formed and the number of seeds developing in *Cercis*. Bull. Torrey Botan. Club 41:533–49.

8243. —— 1914: On differential mortality with respect to seed weight occurring in field cultures of *Pisum sativum*. Am. Naturalist 48:83–86.

8244. —— 1915: The influence of position in the pod upon weight of the bean seed. Am. Naturalist 49:44–47.

8245. —— 1916: A quantitative study of the factors influencing the weight of the bean seed. II. Correlation between number of pods per plant and seed weight. Bull Torrey Botan. Club 43:485–94.

8246. —— 1916: A tetracotyledonous race of *Phaseolus vulgaris*. N.Y. Botan. Garden Mem. 6:229–44.

8247. Harris, L. E., and M. C. Gallagher. 1931: The seed of *Euphorbia marginata* Pursh. J. Am. Pharm. Assoc. 20:1281–86.

8248. Harris, R. H., E. A. Helgeson, and L. D. Sibbitt. 1943: The effect of maturity upon the quality of hard red spring and durum wheats. Cereal Chem. 20:447–63.

8249. ——, M. Johnson, and E. Jesperson. 1946: Some properties of a flour proteinase. Arch. Biochem. 10:267–76.

8250. —— and L. D. Sibbitt. 1941: Micromethods in comparison with the standard method of determining the test weight of hard red spring wheat. N. Dakota Agr. Expt. Sta. Bimonthly Bull. 3:23–27. Chem. Abstr. 36:4204. 1942.

8251. ——, G. S. Smith, and L. D. Sibbitt. 1943: The effect of sprout damage on the quality of durum wheat, semolina, and macaroni. Cereal Chem. 20:333–45.

8252. —— and L. R. Waldron. 1946: The relation of test weight and protein content to the milling and baking quality of hard red spring wheat hybrids. J. Agr. Res. 72:131–36.

8253. —— and H. L. Walster. 1953: Observations on a sixty-four-year-old wheat sample. Cereal Chem. 30:58–62.

8254. Harris, S. W., and W. H. Marshall. 1960: Experimental germination of seed

and establishment of seedlings of *Phragmites communis*. Ecology 41:395.
Harris, T. H. 1823, 1824.
8255. Harris, T. M. 1954: Mesozoic seed cuticles. Svensk Botan. Tidskr. 48:281–91.
8256. —— 1958: The seed of *Caytonia*. Palaeobotanist 7:93–106. Biol. Abstr. 36:2456. 1961.
Harris, W. 16753.
8257. Harris, W. W., and J. R. Lawless. 1963: Interrelations between post-harvest dormancy and soil moisture in germination of wheat. Agron. J. 55:340–43.
8258. Harrison, A. L. 1935: Mosaic of the Refugee bean. N.Y. Agr. Expt. Sta. Bull. 656. 19 pp.
8259. Harrison, B. F., M. D. Thomas, and G. R. Hill. 1944: Radioautographs showing the distribution of sulphur in wheat. Plant Physiol. 19:245–57.
8260. Harrison, B. J. 1956: Seed storage. John Innes Hort. Inst. Rept. 46:15–16.
8261. —— and J. McLeish. 1954: Abnormalities of stored seed. Nature 173:593–94.
8262. Harrison, C. M., and W. N. Crawford. 1941: Seed production of smooth brome grass as influenced by applications of nitrogen. J. Am. Soc. Agron. 33:643–51.
Harrison, C. M. 7667.
8263. Harrison, D. S., and V. E. Green, Jr. 1960: A three-year study of drying high moisture snapped corn with heated air. Soil Crop Sci. Soc. Florida Proc. 20:19–22.
8264. Harrison, G. J. 1939: Treatment of cotton seed with organic mercury dust and sulphuric acid. [Abstract.] Phytopathology 29:755.
Harrison, G. J. 16518, 16519.
8265. Harrison, J. E. 1931: "Hard seeds" in subterranean clover. J. Dept. Agr. Victoria 29:287–88, 299.
Harrison, J. Heslop-; *see* Heslop-Harrison, J.
Harrison, J. M. 1348.
8266. Harrison, S. C. 1954: The technique of ultra-violet testing in New Zealand. Proc. Intern. Seed Testing Assoc. 19:44–49.
8267. —— 1957: The effect of weedkillers on seed germination. Proc. New Zealand Weed Control Conf. 10:55–58.
8268. Harrold, M. E. 1935: Domestic pansy seeds. Florists Rev. 75(1948):58–59, 66–67.
8269. Harrold, T. J. 1935: Comparative study of the developing and aborting fruits of *Prunus persica*. Botan. Gaz. 96:505–20.
8270. Harron, J. W. 1953: Study of seed production, seed identification, and seed germination of *Chenopodium* spp. Cornell Univ. Agr. Expt. Sta. Mem. 320. 24 pp.
Harry, J. B. 5026.
Harshbarger, K. E. 14364.
8271. Harshberger, J. W. 1916: A new method of germinating acorns for forest planting. Am. Forestry 22:687–88.
8272. Hart, H. T. 1928: Delayed germination in seeds of *Peltandra virginica* and *Celastrus scandens*. Puget Sound Biol. Sta. Publ. 6:255–61.
8273. Hart, J. H., editor. 1904: Seeds, loss of vitality. Botan. Dept. Trinidad Bull. Misc. Inform. 44:142–43.
8274. Hart, J. R., and L. Feinstein. 1958: A method for the determination of the surface area of wheat kernels. Cereal Chem. 35:395–99.
8275. —— and C. Golumbic. 1962: A comparison of basic methods for moisture determination in seeds. Proc. Intern. Seed Testing Assoc. 27:907–19.
8276. —— and —— 1963: Methods of moisture determination in seeds. Proc. Intern. Seed Testing Assoc. 28:911–33.
8277. —— and M. H. Neustadt. 1957: Application of the Karl Fischer method to grain moisture determination. Cereal Chem. 34:26–37.
8278. ——, K. H. Norris, and C. Golumbic. 1962: Determination of the moisture content of seeds by near-infrared spectrophotometry of their methanol extracts. Cereal Chem. 39:94–99.
Hart, J. R. 6281, 9224.
8279. Hart, L. 1940: Poisonous to fowls. Seeds of the plant *Araujia sericifera*. Agr. Gaz. N.S. Wales 51:472–74.
8280. Hart, L. P. 1933: Notes on rubber-seed oil. Am. Paint Varnish Mfrs. Assoc. Circ. 438:232–33. Chem. Abstr. 27:4433. 1933.
Hart, R. B. 19898.
8281. Hartig. 1890: Über frühzeitiges Keimen der Buchen- und Ahornsamen. Botan. Centralbl. 41:285–86.
8282. Hartig, H. E., and B. Sullivan. 1933: Value of electrical methods for estimating the moisture content of wheat. Ind. Eng. Chem. Anal. Ed. 5:107–9.
8283. Hartisch, J. 1939: Über die Wirkung der Keimstimmung auf landwirtschaftliche Nutzpflanzen. Pflanzenbau 15:265–88.
Hartisch, J. 16521.
8284. Hartl, D. 1959: Das alveolierte Endosperm bei Scrophulariaceen, seine

Entstehung, Anatomie und taxonomische Bedeutung. Beitr. Biol. Pflanz. 35:95–110.

Hartl-Baude, E. 7954.

8285. Hartleb, R., and A. Stutzer. 1897: Untersuchungen über die Methode der Samenprüfung, insbesondere diejenige der Grassämereien. J. Landwirtsch. 45:43–60.

8286. Hartley, A. 1928: Meconopses and gentians at Devonhall, Perthshire. J. Roy. Hort. Soc. 53:220–34.

8287. —— 1949: Rate of occurrence of mottled seed in Madrid sweetclover. Proc. Assoc. Offic. Seed Analysts 39:77–78.

Hartman, G. P. 13239.

8288. Hartman, T. A. 1957: Investigations on jarowisation 1956: Mededel. Wageningen Inst. Biol. Scheik. Onderz. Landbgewassen 1957(20):117–21.

Hartmann, W. 17106.

Hartog, G. T. den. 17872, 17875.

8289. Hartsema, A. M., and I. Luyten. 1955: Flower formation and seed production of silverskin onions (var. Pompei). Mededel. Directeur Tuinbouw 18:49–56.

Hartwig, E. E. 9134.

8290. Hartung, M. E., and W. B. Storey. 1939: The development of the fruit of *Macadamia ternifolia*. J. Agr. Res. 59:397–406.

8291. Harvey, B. L., and R. J. Downey. 1964: The inheritance of erucic acid content in rapeseed (*Brassica napus*). Can. J. Plant Sci. 44:104–11.

8292. Harvey, B. T. 1915: Notes on the dissemination of Virginia creeper seeds by English sparrows. Plant World 18:217–19.

Harvey, O. A. 3411.

Harvey, P. H. 6967, 13414.

Harvey, R. B. 11821, 18916.

Harwood, L. H. 19716.

8293. Harwood, R. F., and G. W. Bruehl. 1961: Seed systemics for control of aphids on oats and barley. J. Econ. Entomol. 54:883–85.

8294. ——, W. L. Nelson, and H. S. Telford. 1957: Seed treatments of wheat for control of the great basin wireworm. J. Econ. Entomol. 50:702–3.

8295. Harz, C. O. 1885: Landwirtschaftliche Samenkunde. Handbuch für Botaniker, Landwirthe, Gärtner, Droguisten, Hygieniker. Bd. I, II. IX. P. Parey, Berlin. 1,362 pp. Botan. Centralbl. 26:262–69. 1886.

Hasan, H. 6234.

8296. Hasan-Iljas, B. 1960: The influence of gibberellic acid upon germination in cabbage (*Brassica oleracea* var. *capitata*), lettuce (*Lactuca sativa*) and rice (*Oryza sativa*). Pemberitaan Balai Besar Penjelidikan Pertanian 156:1–19. Biol. Abstr. 36:86314. 1961.

Hasanuzzaman, S. M. 354.

8297. Hasegawa, K. 1925: The determination of viability of seed by reagents. Keiyjo (Japan) Forest Expt. Sta. Bull. 1:1–16.

8298. —— 1935: On the determination of vitality in seed by reagents. Proc. Intern. Seed Testing Assoc. 7:148–53.

8299. —— 1936: On a method of determining seed vitality by a certain reagent. J. Japan. Botany 8:1–4.

8300. Hasegawa, M., and T. Furukawa. 1953: The light germination of forest tree seeds. I. The differences of germination behaviors between Kuromatsu (*Pinus thunbergii*) seed and Akamatsu (*Pinus densiflora*) seed. J. Japan. Forestry Soc. 35:382–84.

8301. —— and —— 1955: The light germination of forest tree seeds. II. The effects of light illumination and low temperature on germination of Japanese black and red pine seeds. J. Japan. Forestry Soc. 37:6–7.

8302. —— and H. Hirokawa. 1955: Organic acids contained in the seeds of *Pinus densiflora*. I. J. Japan. Forestry Soc. 37:530–32.

8303. ——, Y. Iwakura, and A. Nabeya. 1956: Growth inhibitor in the pericarp of spinach fruit. J. Agr. Chem. Soc. Japan (English Transl.) 77:1320–24. Chem. Abstr. 53:3393. 1959.

8304. ——, M. Takayama, and S. Yoshida. 1953: Carbohydrases of coniferous seeds. J. Japan. Forestry Soc. 35:156–57. Chem. Abstr. 48:9482. 1954.

8305. ——, T. Takayama, and T. Shiroya. 1951: The carbohydrates contained in seeds. Kagaku (Tokyo) 21:593–94. Chem. Abstr. 48:5297. 1954.

Hasegawa, S. 14143, 16805.

8306. Hasenfratz, V. 1918: The determination of cellulose in wheat. Compt. Rend. Soc. Biol. 81:457–58.

8307. —— 1941: χ-Tanghinin, a new crystalline substance extracted from the kernels of *Tanghinia venenifera*. Compt. Rend. 213:404–6. Chem. Abstr. 37:2008. 1943.

Hasenmaier, G. 7340.

Hashad, M. N. 7170.

8308. Hashikawa, K. 1962: Studies on the ripening of wheat grain. IV. Influence of temperature upon the development of the endosperm. Proc. Crop Sci. Soc. Japan 30:228–31.

8309. Hashimoto, N., I. Shihira, and S. Ishikawa. 1954: Effect of colored light on the germination of forest tree seeds. J. Japan. Forestry Soc. 36:63–65.

8310. Hashimoto, R., and T. Hirano. 1961: The period required for the completion of vernalization under the outdoor condition in wheat sown late autumn. Proc. Crop Sci. Soc. Japan 29:249–50.

8311. Hashimoto, T. 1958: Increase in percentage of gibberellin-induced dark germination of tobacco seeds by N-compounds. Botan. Mag. (Tokyo) 71: 430–31.

8312. —— 1961: Influence of inorganic nitrogenous compounds on tobacco seed germination induced by gibberellin or A_3, kinetin and ammonium salts of organic acids. Plant Cell Physiol. (Tokyo) 2:463–69.

8313. —— and M. Ikeda. 1962: Effects of applications of basic nutrients on swelling velocity of leguminous seeds under water. I. Effects of applications of potassium fertilizers. Japan. Soc. Grassland Sci. J. 8(2):1–5.

8314. —— and M. Okamoto. 1954: The magnesium nutrition of crops. III. Amounts of magnesium and calcium in pods and seeds of soybeans. J. Sci. Soil Manure Japan 24:281–82. Chem. Abstr. 48:7710. 1954.

8315. —— and T. Yamaki. 1962: Interactions of gibberellin or A_3 and inorganic phosphate in tobacco seed germination. Plant Cell Physiol. (Tokyo) 3:175–87.

Hashimoto, T. 18289, 18293, 18313.

8316. Hashimoto, Y. 1962: Influence of chlorine content in fertilizer on the germination of barley. Yamaguti Univ. Fac. Agr. Bull. 13:349–60.

8317. Hashioka, Y., and H. Ikegami. 1956: Further experiments on the seed dressing of Chinese milk vetch against the sclerotia of *Sclerotinia trifoliorum*. Res. Bull. Fac. Agr. Gifu Univ. 6:50–55. Biol. Abstr. 31:29655. 1957.

Hashizume, A. 8440.

8318. Hashizume, H. 1960: The effect of gibberellin upon sex differentiation in *Cryptomeria japonica* strobiles. J. Japan. Forestry Soc. 42:176–80.

8319. —— 1960: The germination of *Chaemacyparis obtusa* seeds collected from cones borne by spraying with gibberellin. J. Japan. Forestry Soc. 42:190–91.

8320. —— 1960: The effect of gibberellin upon flower-bud formation in *Cryptomeria japonica*. II. The germination of seeds collected from cones born by spraying with gibberellin. J. Japan. Forestry Soc. 42:226–28.

Hashmi, M. A. 10505.

8321. Hasitschka-Jenschke, G. 1957: Die Entwicklung der Samenanlage von *Allium ursinum* mit besonderer Berücksichtigung der endopolyploiden Kerne in Synergiden und Antipodin. Österr. Botan. Z. 104:1–24.

8322. —— 1958: Zur Karyologie der Samenanlage dreier *Allium*-Arten. Österr. Botan. Z. 105:71–82.

8323. Haskell, D. A. 1959: The use of chlorazol black E in staining angiosperm embryos. [Abstract.] Proc. Indiana Acad. Sci. 68:92. 1958.

8324. Haskell, G. 1948: Effect of low temperatures on the germination of inbred lines of sweet corn. Science 107:150.

8325. —— 1949: Cold hardiness and seed shape in maize. Nature 163:907–8.

8326. —— 1949: Presoaking and cold hardiness in maize. Plant Soil 1:342–45.

8327. —— 1949: Studies with sweet corn. Plant Soil 2:49–57.

8328. —— and G. G. Selman. 1950: Studies with sweet corn. III. The primary effects of treating seeds with ultrasonics. Plant Soil 2:359–73.

8329. —— and W. R. Singleton. 1949: Use of controlled low temperature in evaluating the cold hardiness of inbred and hybrid maize. Agron. J. 41:34–40.

8330. Haskell, R. J. 1936: The present status of seed treatment discussed at American seed trade association convention. Agr. News Letter 4:135–39.

8331. —— 1936: The treatment of vegetable seed by seedsmen is becoming a practice in certain sections. Agr. News Letter 4:177–80.

8332. —— 1939: Recent developments and future trends in cereal seed treatment. U.S. Dept. Agr. Ext. Pathol. 37:41–45.

8333. —— 1949: The role of alkly mercury as a fungicide. Agr. Chem. 4(8):30–32.

8334. —— 1950: More attention to seeds called urgent in limiting spread of plant diseases. Florists Exchange Hort. Trade World 114(25):11, 45, 49.

8335. —— and S. P. Doolittle. 1940: Vegetable seed treatments. U.S. Dept. Agr. Farmers Bull. 1862.

8336. Haskins, C. P., and C. N. Moore. 1935: Growth modifications in citrus seedlings grown from X-rayed seed. Plant Physiol. 10:179–85.

8337. Haskins, F. A. 1955: Changes in the activities of several enzymes during germination and seedling development in corn (*Zea mays* L.). Plant Physiol. 30:74–77.

8338. —— 1961: Chromatographic study of ninhydrin-positive substances in etiolated shoots from control and irradiated maize seeds. Crop Sci. 1:219–21.

8339. —— and S. C. Downs. 1961: Activities of acid phosphatase, peroxidase, and polyphenolase in etiolated shoots from control and irradiated maize seeds. Agron. J. 53:90–93.

8340. —— and H. J. Gorz. 1963: Glucosides of coumarinic and *o*-coumaric acids in the Tonka bean. Science 139:496–97.

Haskins, F. A. 2146, 16861.

8341. Hasler, A., and A. Maurizio. 1949: Die Wirkung von Bor auf Samenansatz und Nektarsekretion bei Raps (*Brassica napus* L.). Phytopathol. Z. 15:193–207.

8342. Hassan, A., and M. K. A. Basha. 1932: A preliminary note on a prolamin isolated from a legume seed. Biochem. J. 26:1843–45.

8343. Hassan, A. A. G., A. Habib, M. Zaki, and M. S. el-Rafie. 1959: The effect of carbon disulphide on the viability of seed wheat. III. The effect of period of fumigation and dise of CS_2. Ann. Agr. Sci. 4:35–50.

Hassan, M. V. 15974, 15975.

8344. Hasserbrauk, K. 1928: Über den Einfluss der Balusäure auf die Keimreife von Samen. Angew. Botan. 10:407–68.

8345. —— 1937: Zur Frage der Verbreitung des Löwenmaulrostes durch das Saatgut. Blumen- Pflanzenbau Gardenwelt 41:261–62. Rev. Appl. Mycol. 16:752. 1937.

8346. —— 1959: Die Getreiderostsituation in Europa, kein kontinentales, sondern ein interkontinentales Problem. Acad. Rep. Populare Romine, Omagiu lui Traian Savulescu cu prilejul implinirii a 70 de ani, pp. 275–81.

Hassebrauk, K. 7028.

Hasselquist, H. 19201.

8347. Hassid, W. Z. 1936: Comparison of the total nitrogen in wheat seeds by the Gunning (modified Kjeldahl) and the Dumas combustion methods. J. Am. Chem. Soc. 58:2075.

8348. —— and R. M. McCready. 1941: The molecular constitution of glycogen and starch from the seed of sweet corn (*Zea mays*). J. Am. Chem. Soc. 63:1632–35.

Hassid, W. Z. 5771.

8349. Hastie, E. L. 1958: Store bean seed the right way. Queensland Agr. J. 84: 501–4.

8350. Hastings, A., and A. D. Drake. 1962: Pelleting inoculated clover seed. New Zealand J. Agr. 104:330–31, 333.

Hastings, G. T. 16482.

8351. Hastings, L. 1950: Estudios sobre la viabilidad de la semilla en Costa Rica. Turrialba 1:86–88.

8352. —— 1951: Seed treatment studies for control of rice *Helminthosporium*. Mex. Ofic. Estud. Espec. Fol. Misc. 4:90–99.

8353. Haswell, G. A. 1954: A note on the specific heat of rice, oats, and their products. Cereal Chem. 31:341–42.

Hata, C. 10117–10121.

Hata, K. 19389.

Hata, T. 10322.

8354. Hatamura, M., K. Saio, M. Horie, A. Ita, R. Kawashima, and K. Mikoshiba. 1962: Experimental studies on the breeding technique for high-protein-content soybeans. I. Relations between grain weight and protein-content of soybeans under various cultivating conditions. 22nd Meeting Japan. Soc. Breeding 1962: 90–99. Japan. J. Breeding 13:54. 1963. Biol. Abstr. 44:20868. 1963.

8355. Hatanaka, S. 1959: Oligosaccharides in the seeds of *Sesamum indicum* L. Arch. Biochem. Biophys. 82:188–94.

8356. Hatano, K. 1950: On the germination of hazenoki (*Rhus succedanea* Linn.) seeds treated with sulphuric acid. I. J. Japan. Forestry Soc. 32:185–88.

8357. —— 1951: On the germination of hazenoki (*Rhus succedanea* Linn.) seeds treated with sulphuric acid. II. J. Japan. Forestry Soc. 33:218–21.

8358. —— 1951: On the absorption of water by seeds of *Larix kaempferi* and *Pinus thunbergii*. J. Japan. Forestry Soc. 33:426–30.

8359. —— 1952: Tetrazolium salts as seed germination testers. J. Japan. Forestry Soc. 34:37–41.

8360. —— 1953: On the amino acids in the seeds and seedlings of *Ginkgo biloba*. Bull. Tokyo Univ. Forests 44:221–27.

8361. —— 1953: On the amino acids of coniferous seeds at the germination stage. Bull. Tokyo Univ. Forests 45:145–54.

8362. —— 1954: On the amino acids in the seeds of *Carya pecan*. Trans. 63rd Meeting Japan. Forestry Soc., pp. 84–86. Tokyo Univ. Fac. Agr. Univ. Tokyo Rec. Res. 4(1953–54):68. 1955.
8363. —— 1955: Freie Aminosäuren in den reifenden *Ginkgo*-Samen. J. Japan. Forestry Soc. 37:527–29.
8364. —— 1957: α-Keto acids in pine seeds. Bull. Tokyo Univ. Forests 53:135–38. Biol. Abstr. 32:13930. 1958.
8365. —— 1957: The chemistry of tree seeds. J. Japan. Forestry Soc. 39:185–93.
8366. —— 1957: Chemistry of tree seeds. J. Japan. Forestry Soc. 39:238–44.
8367. —— 1957: Über die Abhängigkeit der Atmungsintensität von Wassergehalt bei keimenden Kiefernsamen verschiedener Keimfähigkeit. Forstwiss. Centralbl. 76: 376–81.
8368. —— 1961: Über einige Enzyme des oxydativen Abbaues in verschiedenen Teilen der Samen von Pisum densiflora und *Pinus thunbergii*. Flora (Jena) 151: 231–42.
8369. —— 1962: The pigments in the germinating seeds of conifer. J. Japan. Forestry Soc. 44:213–16.
8370. —— 1963: Respiration of germinating pine seeds. Plant Cell Physiol. (Tokyo) 4:129–34.
8371. —— and T. Kano. 1952: A brief report on the after-ripening of the seeds of *Ginkgo biloba*. J. Japan. Forestry Soc. 34:369–70.
Hatano, K. 10230.
Hatano, T. 11070.
8372. Hatch, M. D., and P. K. Stumpf. 1961: Fat metabolism in higher plants. XVI. Acetyl coenzyme A carboxylase and acyl coenzyme A-malonyl coenzyme A transcarboxylase from wheat germ. J. Biol. Chem. 236:2879–85.
8373. —— and J. F. Turner. 1958: Glycolysis by an extract from pea seeds. Biochem. J. 69:495–501.
8374. —— and —— 1960: A protein disulphide reductase from pea seeds. Biochem. J. 76:556–62.
8375. Hatcher, E. S. J. 1940: Studies in the inheritance of physiological characters. V. Hybrid vigour in the tomato. Part III. A critical examination of the relation of embryo development to the manifestation of hybrid vigour. Ann. Botany (London) 4:735–64. Plant Breeding Abstr. 11:88. 1941.
8376. —— 1943: Auxin production during development of the grain in cereals. Nature 151:278–79. Chem. Abstr. 37: 3795. 1943.
8377. —— 1945: Studies in the vernalisation of cereals. IX. Auxin production during development and ripening of the anther and carpel of spring and winter rye. Ann. Botany (London) 9:235–66.
8378. —— and F. G. Gregory. 1941: Auxin production during the development of the grain of cereals. Nature 148:626.
8379. —— and O. N. Purvis. 1945: On the behaviour in the field of small grain obtained by premature harvesting. J. Agr. Sci. 35:177–83.
Hatcher, E. S. J. 15779.
Hatem, J. Nieto-; *see* Nieto-Hatem, J.
8380. Hathaway, I. L., F. D. Yung, and T. A. Kiesselbach. 1952: The effect of drying temperature upon the nutritive value and commercial grade of corn. J. Animal Sci. 11:430–40. Biol. Abstr. 26:2295. 1952.
Hathazi, C. 11903.
Hatheway, W. H. 16550.
8381. Hatt, H. H., and A. Z. Szumer. 1954: The presence of an acetylenic acid in the seed fat of plants of the Santalaceae family. Chem. Ind. (London) 1954:962–63.
8382. —— and —— 1954: The seed fat of *Omphalea queenslandiae*. 1954: J. Sci. Food Agr. 5:534–36.
8383. —— and W. J. Troyahn. 1946: The seed oil of the saffron thistle (*Carthamus lanatus* L.). Australian Council Sci. Ind. Res. J. 19:86–95.
Hatt, H. H. 4151, 8101.
8384. Hattori, S., and T. Shiroya. 1951: The sugars in the seeds and seedlings of *Pinus thunbergii*. Arch. Biochem. Biophys. 34:121–34.
Hattori, Y. 14138, 14139.
8385. Hauch, L. A. 1923: The cultivation of the oak (*Quercus pedunculata*) in Denmark. Quart. J. Forestry 17:35–50, 88–97.
8386. Haugen, G., and M. Pijoan. 1946: Certain factors affecting vitamin C content of bean sprouts. Arch. Biochem. 10:227–34.
8387. Haun, F., and W. Leistner. 1940: Hydrocyanic acid content of flax seed. Z. Tierernähr. Futtermittelk. 4:201–6. Chem. Abstr. 36:585. 1942.
8388. Haunold, A., V. A. Johnson, and J. W. Schmidt. 1962: Variation in protein content of the grain in four varieties of *Triticum aestivum* L. Agron. J. 54:121–25.

8389. ——, ——, and —— 1962: Genetic measurements of protein in the grain of *Triticum aestivum* L. Agron. J. 54:203–6.

8390. —— and M. F. Lindsey. 1964: Amylose analysis of single kernels and its implication for breeding of high-amylose corn (*Zea mays* L.). Crop Sci. 4:58–60.

8391. Haupt, A. W. 1934: Ovule and embryo sac of *Plumbago capensis*. Botan. Gaz. 95:649–59.

Hauptmann, W. 6072.

Haus, T. E. 5509, 16283.

8392. Hauschka, T., G. Toennies, and A. P. Swain. 1945: The mechanism of growth inhibition by hexenolactone. Science 101:383–85.

Hausen, S. S. von. 19147, 19148.

8393. Hauser, J. 1934: Effect of seed disinfectants on germination. Mezogazdasagi Kutatasok 7:69–76. Chem. Abstr. 29:1927. 1935.

8394. Hausmann, O. K., and H. P. Iwanissowa. 1909: Zur Frage der Beziehung zwischen Keimfähigkeit und Atmungsintensität der Samen. Bull. Jardin Imp. Botan. St. Petersbourg 9:97–106. Botan. Centralbl. 113:274. 1910.

8395. Hauss, H. 1927: Beiträge zur Kenntnis der Entwicklungsgeschichte von Flugeinrichtungen bei höheren Samen. Botan. Arch. 20:74–108.

8396. Haut, I. C. 1933: The influence of drying on the after-ripening and germination of fruit tree seeds. Proc. Am. Soc. Hort. Sci. 29:371–74.

8397. —— 1933: Catalase activity in relation to the after-ripening of fruit tree seeds. Proc. Am. Soc. Hort. Sci. 29:375–79.

8398. —— 1934: The effect of various low temperatures upon the after-ripening of fruit tree seeds. Proc. Am. Soc. Hort. Sci. 30:365–67.

8399. —— 1938: Physiological studies on after-ripening and germination of fruit-tree seeds. Maryland Agr. Expt. Sta. Bull. 420. 54 pp.

8400. —— and F. E. Gardner. 1935: The influence of pulp disintegration upon the viability of peach seeds. Proc. Am. Soc. Hort. Sci. 32:323–27.

8401. Havas, L. 1935: Ascorbic acid (vitamin C) and the germination and growth of seedlings. Nature 136:435.

8402. —— 1935: Some effects of radioactive mud upon germination of seeds and growth of seedlings. J. Agr. Sci. 25:198–216.

8403. —— 1937: Effects of colchicin and of *Viscum album* preparations upon germination of seeds and growth of seedlings. Nature 139:371–72.

8404. Havas, L. J. 1946: Parthenocarpy and accompanying hormonal syndromes induced by unrelated chemicals. Nature 157:629–30.

8405. Havis, L. 1938: Seedless peaches as a result of freezing injury. Ohio Agr. Expt. Sta. Bimonth. Bull. 23:214–19.

8406. —— and A. L. Gilkeson. 1949: Starting seedlings of Montmorency cherry. Proc. Am. Soc. Hort. Sci. 53:216–18.

8407. Hawk, V. B. 1957: Native grass seed. Crops & Soils 9(9):18–19.

Hawke, F. 16998.

8408. Hawkins, R. P. 1956: A preliminary survey of red clover seed production. Ann. Appl. Biol. 44:657–64.

8409. —— 1959: Problems of red clover seed production. Natl. Inst. Agr. Botany J. 8:525–29. Biol. Abstr. 36:25253. 1961.

8410. Hawksworth, F. G. 1959: Ballistics of dwarf mistletoe seeds. Science 130:504.

8411. —— 1961: Abnormal fruits and seeds in *Arceuthobium*. Madrono 16:96–101. Biol. Abstr. 36:81966. 1961.

Hawksworth, F. G. 8832.

8412. Hawley, W. O., and J. L. Creech. 1953: Some seed germination studies with Kenaf. Turrialba 3:156–59. Biol. Abstr. 29:26918. 1955.

Hawley, R. 19449.

8413. Haworth, R. D. 1932: New alkaloid from *Holarrhena antidysenterica* seeds. J. Chem. Soc. 1932:631–34.

8414. Haws, A. 1956: Alfalfa seed production in central Iowa as influenced by honey bees and certain other factors. [Abstract.] Iowa State Coll. J. Sci. 30:378–79.

8415. Hawthorn, L. R. 1937: Seedlessness in tomatoes. Science 85:199.

8416. —— 1937: Texas summer heat makes seedless tomatoes. Sci. News Letter 31:152.

8417. —— 1951: Cultural studies on carrot stecklings in relation to seed production. U.S. Dept. Agr. Circ. 877. 19 pp.

8418. —— 1951: Studies of soil moisture and spacing for seed crops of carrots and onions. U.S. Dept. Agr. Circ. 892. 26 pp.

8419. ——, G. E. Bohart, and E. H. Toole. 1956: Carrot seed yield and germination as affected by different levels of insect pollination. Proc. Am. Soc. Hort. Sci. 67:384–89.

8420. —— and L. H. Pollard. 1951: Selection of Great Lakes lettuce strains for higher seed yields. Proc. Am. Soc. Hort. Sci. 57:323–28.

8421. —— and —— 1954: Vegetable and flower seed production. Blakiston, New York. 626 pp.

8422. —— and —— 1956: Production of lettuce seed as affected by soil moisture and fertility conditions. Utah Agr. Expt. Sta. Bull. 386. 23 pp.

8423. ——, ——, and R. P. Draper. 1949: A device for handling small experimental seed lots on a gravity separator. Proc. Am. Soc. Hort. Sci. 54:339–41.

8424. ——, E. H. Toole, and V. K. Toole. 1962: Yield and viability of carrot seeds as affected by position of umbel and time of harvest. Proc. Am. Soc. Hort. Sci. 80: 401–7.

8425. —— and R. Wellington. 1930: "Geneva," a greenhouse cucumber that develops fruit without pollination. N.Y. Agr. Expt. Sta. Bull. 580. 11 pp.

Hawthorn, L. R. 4173, 12682.

Hay, G. K. Baron-; *see* Baron-Hay, G. K

8426. Hay, J. G. 1942: The distribution of phytic acid in wheat and a preliminary study of some of the calcium salts of this acid. Cereal Chem. 19:326–33.

8427. Hay, J. R. 1962: Experiments on the mechanism of induced dormancy in wild oats, *Avena fatua* L. Can. J. Botany 40:191–202.

8428. —— and B. G. Cumming. 1959: A method for inducing dormancy in wild oats (*Avena fatua* L.). Weeds 7:34–40.

Hay, J. R. 4741.

Hay, R. J. 19589.

8429. Hay, T. 1930: *Meconopsis* in winter. New Flora Silva. 2:167–70.

8430. Hay, W. D. 1930: Impurities commonly found in Montana grown alfalfa seed. Proc. Assoc. Offic. Seed Analysts 22: 39–42.

8431. —— 1936: Germination of crested wheat grass (*Agropyron cristatum*): preliminary studies. Proc. Assoc. Offic. Seed Analysts 27:66–70.

8432. —— 1936: Further studies with the germination of crested wheat grass. Proc. Assoc. Offic. Seed Analysts 28:86–88.

8433. —— 1938: Laboratory and field germination of infected beans: effectiveness of seed treatments. Proc. Assoc. Offic. Seed Analysts 23/26:80–82.

8434. —— 1938: Canada thistle seed production and viability in Montana and its occurrence in crop seeds. Proc. Assoc. Offic. Seed Analysts 23/26:152–55.

8435. —— 1939: Laboratory germination studies with *Agropyron smithii*—preliminary results. Proc. Assoc. Offic. Seed Analysts 30:244–45.

8436. ——, moderator. 1949: Problems in testing coated and pelleted seeds. Panel discussion. Proc. Assoc. Offic. Seed Analysts 39:72–73.

Hay, W. D. 458.

Hayakawa, I. 11067.

Hayashi, J. 9579, 14544–14546.

8437. Hayashi, K., and T. Mizuno. 1953: The carbohydrate in the seed of Japanese persimmons. J. Agr. Chem. Soc. Japan (English Transl.) 26:569–72. 1952. Chem. Abstr. 49:4803. 1955.

8438. ——, Y. Nagata, and T. Inoue. 1954: Preliminary report on the nucleic acid in the embryo of *Zea mays*. Res. Bull. Fac. Agr. Gifu Univ. 3:95–97.

8439. ——, ——, ——, and I. Yanagisawa. 1958: Studies on the nucleoprotein in the embryo of *Zea mays*. J. Agr. Chem. Soc. Japan (English Transl.) 32:610–14. Bull. Agr. Chem. Soc. Japan 22(5):9–10. 1958.

Hayashi, K. 9519, 9520, 11071.

8440. Hayashi, M., A. Hashizume, S. Shinozuka, and G. Igarashi. 1960: Studies on the influence of water percolation in rice plants during ripening period in ill-drained paddy-field. Proc. Crop Sci. Soc. Japan 29:43–46.

Hayashi, S. 9525, 9526.

Hayashi, T. 13945.

8441. Hayashi, Y. 1935: An experiment in reforestation by the direct seeding of the principal forest trees indigenous to Chosen (Korea). Keijo(Chosen) Forest Expt. Sta. Bull. 19:103–11.

8442. —— 1963: The embryology of the family Magnoliaceae Sens. Lat. II. Megasporogenesis, female gametophyte and embryogeny of *Schisandra repanda* Radlkoper and *Kadsura japonica* Dunal. Sci. Rept. Tohoku Univ. Fourth Ser. 29: 403–11.

8443. Hayat, M. A. 1963: Morphology of seed germination and seedling in *Annona squamosa*. Botan. Gaz. 124:360–62.

8444. Hayden, A. 1934: Distribution and reproduction of Canada thistle in Iowa. Am. J. Botany 21:355–73.

8445. —— 1937: Notes on seed production in smooth perennial sow thistle (*Sonchus arvensis* var. *glabrescens* Wimm. and

Graeb.). [Abstract.] Proc. Iowa Acad. Sci. 44:105.

8446. Hayden, B., and L. Smith. 1949: The relation of atmosphere to biological effects of X-rays. Genetics 34:26–43.

Haydock, K. P. 3566.

Hayes, H. K. 2776, 9864.

Hayes, J. F. 6334.

Hayes, P. 13504.

8447. Hayes, T. R. 1937: Some aspects of groundnut production. E. African Agr. J. 2:455–58.

8448. Haynes, B. C., Jr. 1961: Vapor pressure determination of seed hygroscopicity. U.S. Dept. Agr. Tech. Bull. 1229. 20 pp.

8449. Haynes, F. L. 1954: Potato embryo culture. Am. Potato J. 31:282–88.

8450. Haynes, J. L., and L. E. Thatcher. 1950: Band seeding gives better legume stands. Crops & Soils 2(7):15–17.

8451. —— and —— 1953: Band seeding. Farm Home Res. 38(281):27–28.

Haynes, J. L. 19177.

Hays, W. D. 13183.

8452. Hayward, H. E. 1938: The structure of economic plants. Macmillan, New York. 674 pp.

Hayward, H.E. 1570, 1571.

8453. Hayward, W. 1958: Cold storing amaryllis seeds. Plant Life 14:80–81.

Hazama, K. 8454.

8454. Hazama, Y., K. Hazama, and L. Ehrenberg. 1963: The influence of paramagnetic ions on the radiation sensitivity of plant seeds. Radiation Botany 3:7–18.

8455. Hazard, R. 1944: p-Aminobenzoic acid as an antagonist of the inhibiting action of sulfanilamide on germination and growth of flax seed. Compt. Rend. Soc. Biol. 138:972–73.

8456. Headden, W. P. 1919: The vitality of alfalfa seed as affected by age. Proc. Colo. Sci. Soc. 9:239–49. Botan. Abstr. 9:1867. 1920.

8457. Heady, H. F. 1954: Viable seed recovered from fecal pellets of sheep and deer. J. Range Management 7:259–61. Biol. Abstr. 30:304. 1956.

Heald, F. D. 16996.

8458. Heady, A. J. 1943: Seed dispersal by human activity. Nature 151:140.

Heard, C. R. C. 9699.

8459. Heaslip, M. B. 1959: Effects of seed irradiation on germination and seedling growth of certain deciduous trees. Ecology 40:383–88.

8460. Heathcote, J. G. 1950: Protein quality of oats. Brit. J. Nutr. 4:145–54. Brit. Abstr. B III:308. 1951.

Hebert, J. 2709.

8461. Hebert, L. P. 1953: A method of measuring the quantity of viable Johnson grass seed in the soil. Proc. Am. Soc. Sugar Beet Technol. 1946/50:71–74.

Hebert, L. P. 1319.

Hebert, T. T. 5037, 12591, 13415, 13416.

8462. Hebrard, J. 1938: Contribution à l'étude de la iarovisation de quelques céréales sous le climat Mediterranéen. Ann. École Natl. Agr. Montpellier 25:43–55.

8463. Hecht, W. 1929: Beobachtungen bei der Keimung von *Origanum majorana* L. und *Ocimum basilicum* L. im mistbeete. Heil-Gewürzpflanz. 11:178–84. Ber. Wiss. Biol. 12:340. 1929.

Hechtman, J. 526.

8464. Heckel, E. 1878: De l'influence des acides salicylique, theymique, et de quelques essences sur la germination. Compt. Rend. 87:613–14.

8465. —— 1880: De l'action des températures elevées et humides et de quelques substances chimiques (benzoate de soude, acid benzoique, acid sulfureux) sur la germination. Compt. Rend. 91:129–31.

8466. —— 1890: Transformation of the alkaloids during germination. Compt. Rend. 110:88–90.

8467. —— 1892: Germination of *Araucaria bidwilli*. Rev. Gen. Botan. 4:243–44, 289–301.

8468. —— 1898: Sur quelques phenomes morphologiques de la germination dans le *Ximenia americana*. Bull. Soc. Botan. France 45:438.

8469. Heckendorn, W. 1956: Seed storage facility. Chemurgic Dig. 15(7/8):24–27.

8470. —— 1961: Responsibility vs. regulation. Seed World 89(9):18–21.

8471. —— and R. A. Edwards. 1961: The four types of seed trade associations. Yearbook Agr. U.S. Dept. Agr. 1961: 517–21.

8472. Hector, J. M. 1936: Introduction to the botany of field crops. Vol. I. Cereals. Central News Agency, Johannesburg, pp. 1–478.

8473. —— 1936: Introduction to the botany of field crops. Vol. II. Non-cereals. Central News Agency, Johannesburg, pp. 479–1127.

8474. Hedayetullah, S., and B. N. Ghosh. 1946: Vernalization of rice. Nature 157:194.
8475. Hedden, E. N. 1929: Bittersweet for winter beauty. Horticulture 7(2):31.
8476. —— 1931: Pot-grown gentians from seed. Horticulture 9(4):80.
8477. Heddle, R. G., and J. B. D. Herriott. 1955: The establishment, growth and yield of ultra-simple grass seeds. II. The effects of varying grass and clover seed rates. J. Brit. Grassland Soc. 10:157–67.
8478. Hedemann, E. 1931: Über experimentelle Erzeugung von Adventivembryonen bei *Mirabilis uniflora* und *Mirabilis froebelii*. Biol. Centralbl. 51:647–52.
Hedgcoth, C. 17564.
8479. Hedlin, R. A., R. E. Smith, and F. P. Le Claire. 1957: Effect of crop residues and fertilizer treatments on the yield and protein content of wheat. Can. J. Soil Sci. 37:34–40.
8480. Hee, A., and L. Bayle. 1932: Chemical research on germination. I. Changes in fat and phospholipide content of *Lupinus albus* during the first stages of development. Bull. Soc. Chim. Biol. 14:758–82. Chem. Abstr. 26:5122. 1932.
Heen, P. de. 13400.
Heer, A. N. de; *see* De Heer, A. N.
8481. Heftmann, E. 1947: Antioxidant properties of carrot oil. J. Am. Oil Chemists Soc. 24:404–9. Chem. Abstr. 42:1069–70. 1948.
8482. Hegarty, C. P. 1955: Observations on the germination of orchid seed. Am. Orchid Soc. Bull. 24:457–64.
8483. —— and O. Rahn. 1934: Growth retardation by freshly distilled water. J. Bacteriol. 28:20–30. Biol. Abstr. 9:4676. 1935.
Hegedüs, H. 19202.
8484. Hegelmaier, F. 1874: Zur Entwicklungsgeschichte monokotyledoner Keime nebst Bemerkungen über die Bildung der Samendeckel. III. *Triticum vulgare*. Botan. Ztg. 32:657–68.
8485. —— 1895: Über Orientirung des Keimes im Angiospermemsamen. Botan. Ztg. 53:143–73.
8486. —— 1897: Zur Kenntniss der Polyembryonie von *Allium odorum* L. Botan. Ztg. 55:133–43.
8487. —— 1903: Zur Kenntnis der Polyembryonie von *Euphorbia dulcis* Jacqn. E. (*purpurata* Thuill). Ber. Deut. Botan. Ges. 21:6–19.
Heger, B. 8930.
Heggeness, H. G. 12125, 12126.
8488. Hegnaur, R. 1957: The endosperm of Leguminosae. Pharm. Zentralhalle 96: 141–48. Biol. Abstr. 32:38556. 1958.
8489. Heiba, A. S. 1948: Multicellular hairs in *Gossypium*. Science 107:650–51.
8490. Heiberg, S. O. 1934: Briquette planting. J. Forestry 32:333–36.
8491. Heiden, E. 1859: Über das Keimen der Gerste. Dissertation, Griefswald. 108 pp.
Heider, A. 16933.
8492. Heiduschka, A., and C. Shu-Sheng. 1931: Über Teesamenöl. Arch. Pharm. 269: 456–68. Ber. Ges. Physiol. 64:656. 1932.
8493. —— and H. Tettenborn. 1927: Über Galaktoaraban der Lupinensamen. Biochem. Z. 189:203–7.
8494. —— and R. Wallenreuter. 1914: Zur Kenntnis des Öles der Samen von *Strychnos nux vomica* L. Arch. Pharm. 252:202.
8495. Heiken, A., and B. Soegaard. 1962: A study of the variation of seed weight in *Larix decidua* on the basis of radiographs. Uppsala Vetensksamhallet Arsb. 6:55–58.
Heikkenen, H. J. 9883.
8496. Heil, H. 1927: Vergleichend-anatomische Studien an Samen von *Chamaegigas* und verwandten Gattungen. Ber. Deut. Botan. Ges. 45:555–61.
8497. Heilborn, O. 1931: Studies on the taxonomy, geographical distribution and embryology of the genus *Siparuna* Aubl. Svensk Botan. Tidskr. 25:202–28. Bull. Soc. Botan. France 78:752. 1931.
Heilinger, F. 6415.
8498. Heilpern, E. 1914: Keimungsphysiologische Untersuchungen. Österr. Botan. Z. 64:286–93. Botan. Centralbl. 128: 265–66. 1915.
8499. Heinen, E. 1954: Report on sodium in plants. J. Assoc. Offic. Agr. Chemists 37:714–16.
8500. Heinis, F. 1943: Zur Kenntnis der Lebensdauer resp. Keimfähigkeit von Samen. Ber. Geobotan. Forschungsinst. Rübel Zürich 1942:75. Biol. Abstr. 20: 9046. 1946
8501. Heinisch, O. 1931: Der Einfluss der Kornlage auf die Resultate des Keimversuchs. Fortschr. Landwirtsch. 6:44–47.
8502. —— 1931: Die Keimreifung der Gerste. Wochschr. Brau. 48:293–97, 307–11. Biol. Abstr. 8:19545. 1934.
8503. —— 1932: Der Bogenamarant (*Amaranthus retroflexus* L.), ein wenig Beachtetes Unkraut. Fortschr. Landwirtsch. 7:344–47.

8504. —— 1933: Die Bestimmung der Keimfähigkeit physiologisch unreifer Gerste. Wochschr. Brau. 50:305–8. Proc. Intern. Seed Testing Assoc. 7:67–68. 1935.
8505. —— 1933: Der Einfluss der Düngung auf die Keimreifung von Braugerste. Z. Pflanzenernähr. Düng. Bodenk. Teil B 12:503–24. Biol. Abstr. 10:21817. 1936.
8506. —— 1935: Beitrag zur Kenntnis der Ursachen der Keimreifung von Braugerste. J. Landwirtsch. 83:1–42. Proc. Intern. Seed Testing Assoc. 8:101–2. 1936.
8507. —— 1937: Die Dauer der Keimreifung der Getreidearten als erbliche Sorteneigenschaft. Z. Zücht. Reihe A, Pflanzenzücht. 21:294–305. Proc. Intern. Seed Testing Assoc. 9:293–94. 1937.
8508. —— 1937: Der Einfluss des Klimas auf die Dauer der Keimreifung von zweizeiliger Sommergerste. Z. Zücht. Reihe A, Pflanzenzücht. 21:451–65. Proc. Intern. Seed Testing Assoc. 10:647–48. 1938.
8509. —— 1938: Die Bedeutung der Keimwurzelanzahl der Getreidearten für den Saatgutwert. Z. Zücht. Reihe A, Pflanzenzücht. 22:209–32. Proc. Intern. Seed Testing Assoc. 10:648. 1938.
8510. —— 1955: Samenatlas der wichtigsten Futterpflanzen und ihrer Unkräuter. Deut. Akad. Landwirtsch.-Wiss., Berlin. 122 tables.
8511. —— 1958: Die Keimfähigkeit des Zuckerrübensaatgutes, ihre Bedeutung und die sie beeinflussenden Faktoren. Z. Zuckerind. 8:63–67.
8512. —— 1958: Die Untersuchung der Keimfähigkeit von Rübensamen mit Hilfe des topographischen Tetrazolium-Verfahrens. Z. Zuckerind. 8:487–89. Chem. Abstr. 53:3398. 1959.
8513. Heinrich, M. 1912: Einige Erfahrungen bei Keimprüfungen im Jahre 1910/1911. Landwirtsch. Versuchssta. 78: 165–78.
8514. —— 1913: Die Einfluss der Luftfeuchtigkeit der Wärme und des Sauerstoffs der Luft auf lagerndes Saatgut. Landwirtsch. Versuchssta. 81:289–376.
8515. —— 1914: Vierzigjährige Ergebnisse der Samenkontrolle. Landwirtsch. Versuchssta. 85:269–357.
8516. —— 1915: Grössenverhältnis zwischen Klee- und Seidesamen in trockenem und gequollenem Zustande. Landwirtsch. Versuchssta. 87:395–408. Botan. Centralbl. 135:176. 1917.
8517. —— 1918: Aufgaben und Ziele der Samenkontrolle. Jahresber. Angew. Botan. 16:116–23. Botan. Abstr. 7:1549. 1921.
8518. —— 1919: Beiträge über die Keimung Bespelzter und Nackter Timothyfrüchte. Landwirtsch. Versuchssta. 93:259–76.
8519. —— 1921: Die Abhängigkeit der Keimtriebkraft vom Keimmedium und ihre Beeinflussung durch verschiedene Beizmittel. Landwirtsch. Versuchssta. 98: 65–115. Botan. Abstr. 11:2120. 1922.
8520. Heinrich, R. 1894: On the germination of various colored clover seed. Zweiter Ber. Landwirtsch. Versuchssta. Rostock, pp. 96, 97. Expt. Sta. Rec. 7:509. 1895/96.
8521. —— 1894: On the presence of hard seed among clover seed. Zweiter Ber. Landwirtsch. Versuchssta. Rostock, pp. 97–102. Expt. Sta. Rec. 7:509. 1895/96.
8522. —— 1894: Effect of size of seed upon their productivity. Zweiter Ber. Landwirtsch. Versuchssta. Rostock, pp. 141–54. Expt. Sta. Rec. 7:510. 1895/96.
8523. —— 1894: Chemical composition of light and heavy oats. Zweiter Ber. Landwirtsch. Versuchssta. Rostock, pp. 206–13.
8524. Heinricher, E. 1898: Notez über die Keimung von *Lathraea squamaria* L. Ber. Deut. Botan. Ges. 16:1–5.
8525. —— 1899: Ein Fallbeschleunigender Wirkung des Lichtes auf die Samenkeimung. Ber. Deut. Botan. Ges. 17:308–11.
8526. —— 1901: Die grünen Halbschmarotzer. III. Jahrb. Wiss. Botan. 36:665–749.
8527. —— 1903: Notwendigkeit des Lichtes und befördernde Wirkung desselben bei der Samenkeimung. Botan. Centralbl. Beih. 13:164–72.
8528. —— 1908: Die Samenkeimung und das Licht. Ber. Deut. Botan. Ges. 26A: 298–301.
8529. —— 1909: Die Keimung von *Phacelia tanacetifolia* Benth. und das Licht. Botan. Ztg. 67:45–66. Botan. Centralbl. 113:360–61. 1910.
8530. —— 1912: Über Versuche, die Mistel (*Viscum album* L.) auf monocotylen und auf sukkulenten Gewächshauspflanzen zu ziehen. Sitzber. Wien. Wiss. Akad. Math.-Naturw. Kl. 121:541–72.
8531. —— 1912: Samenreife und Samenruhe der Mistel (*Viscum album* L.) und die Umstände, welche die Keimung beeinflussen. Sitzber. Wien. Wiss. Akad. Math.-Naturw. Kl. 121:573–613.
8532. —— 1915: Keimung und Entwicklungsgeschichte der Wacholdermistel, *Arceuthobium oxycedri*, auf Grund durch

geführter Kulturen geschildert. Sitzber. Wien. Wiss. Akad. Math.-Naturw. Kl. 124:319–25.

8533. —— 1915: Über besondere Keimungsbedingungen, welche die Samen der Zwerg-Mistel *Arceuthobium oxycedri* (DC.) M. Bieb. beanspruchen. Centralbl. Bakteriol. Parasitenk. Abt. II 42:705–11.

8534. —— 1917: Über tötende Wirkung des Mistelschleims auf das Zellgewebe von Blättern und Sprossen. Sitzber. Wien. Wiss. Akad. Math.-Naturw. Kl. 126: 815–36.

8535. —— 1930: Wie steht es mit den Beweisen für die Behauptung E. Chemin's, dass eine chemische Reizung durch ein lebendes Nährobjekt für die Samen von *Lathraea clandestina* unnötig sei? Beitr. Biol. Pflanz. 18:1–16.

8536. Heinrichs, D. H. 1959: Germination of alfalfa varieties in solutions of varying osmotic pressure and relationship to winter hardiness. Can. J. Plant Sci. 39:384–94.

Heinrichs, D. H. 5044, 13212, 13213.

Heins, J. 5961.

8537. Heinstein, P. F., F. H. Smith, and S. B. Tove. 1962: Biosynthesis of C^{14}-labeled gossypol. J. Biol. Chem. 237: 2643–46.

8538. Heintze, A. 1914: Beobachtungen über Chionochore Samenverbreitung. Botan. Notiser 1914:193–215. Botan. Centralbl. 128:259–60. 1915.

8539. —— 1917: In welchem Masse werden Samen durch die kleineren Sumpfvogel verzehrt und verbreitet. Fauna Flora 1917:116–28. Botan. Centralbl. 137:113. 1918.

8540. —— 1917/18: Endo- and syn-zoic dispersal of seeds by European Corvidae. Botan. Notiser 1917:209–40, 297–300; 1918:1–47. Botan. Abstr. 4:230. 1920.

8541. Heinze, K. 1942: Die Feldbereinigung der Sojakulturen als Schutzmassnahme gegen die Ausbreitung des virösen Sojamosaiks. Vorläufige Mitteilung. Züchter 14:254–58. Biol. Abstr. 22:25477. 1948.

Heinze, P. H. 5826.

8542. Heinzelman, D. C., and R. T. O'Connor. 1951: Trace metals in cottonseed kernels. J. Am. Oil Chemists Soc. 28: 373–74. U.S. Dept. Agr. List Publ. & Patents, p. 11. 1951.

Heinzelman, D. C. 5352, 13115.

8543. Heise, A. C. 1941: Germination of green foxtail seeds. Proc. Assoc. Offic. Seed Analysts 33:43–44.

8544. —— 1943: Unusual types of secondary dormancy in seeds of certain species of Gramineae. Assoc. Offic. Seed Analysts News Letter 17(4):6–7.

8545. —— 1943: Germination of New Zealand spinach seed. Assoc. Offic. Seed Analysts News Letter 17(4):7.

Heise, A. C. 551.

Heiser, F. 16995.

8546. Heiss, R. 1934: Plant growth hastened by means of electricity. Ind. Eng. Chem. News Ed. 12:27.

8547. Heit, C. E. 1940: Testing seeds of trees and shrubs. Seed World 47(3):44.

8548. —— 1941: Germination of seeds. Am. Nurseryman 74(11):26.

8549. —— 1941: Laboratory germination of belladonna (*Atropa belladonna*) seed. Proc. Assoc. Offic. Seed Analysts 33:84–87.

8550. —— 1941: Storage of conifer seed. N.Y. Nursery Notes 10:2–3.

8551. —— 1942: Pretreating seed of black locust. Farm Res. (N.Y.) 8(2):11, 14, 17.

8552. —— 1943: Germination of seeds. Am. Nurseryman 77(6):32.

8553. —— 1943: Germination of seeds. Am. Nurseryman 77(9):28.

8554. —— 1943: Removing embryos of many seeds hastens germination tests. Farm Res. (N.Y.) 9(3):11, 20.

8555. —— 1943: Snow - on - the - mountain (*Euphorbia marginata*) seed data—a rapid method of detecting viability. Proc. Assoc. Offic. Seed Analysts 34:78–82.

8556. —— 1944: Germination studies with coniferous seed. N.Y. Agr. Expt. Sta. Ann. Rept. 1942/43:55.

8557. —— 1944: Recent findings on laboratory germination and longevity of New Zealand spinach seed. Proc. Assoc. Offic. Seed Analysts 1943/44:115–19.

8558. —— 1944: Some seed production problems as seen in the laboratory. Farm Res. (N.Y.) 10(3):3, 13.

8559. —— 1945: Fall planting of hardwood tree seed. Farm Res. (N.Y.) 11(3):14–15.

8560. —— 1945: Recent findings on laboratory germination and longevity of New Zealand spinach seed. Proc. Assoc. Offic. Seed Analysts 35:115–20.

8561. —— 1945: Physiology of germination. N.Y. Agr. Expt. Sta. Ann. Rept. 1943/44: 45.

8562. —— 1945: Viability testing of extremely dormant seeds by embryo excision. N.Y. Agr. Expt. Sta. Ann. Rept. 1943/44:46.

8563. —— 1946: Fruit tree seed testing aids the nurseryman. Farm Res. (N.Y.) 12(4):11–12.
8564. —— 1964: Germination and vitality in sweet pea seedstocks. Proc. Assoc. Offic. Seed Analysts 36:137–40.
8565. —— 1946: Laboratory germination results with certain flower seeds. Proc. Assoc. Offic. Seed Analysts 36:141–49.
8566. —— 1946: Summarized laboratory germination data on over 100 flower seed species. Assoc. Offic. Seed Analysts News Letter 20(4):13–16.
8567. —— 1947: Germinating *Dracaena indivisa* seed. Proc. Assoc. Offic. Seed Analysts 37:163–66.
8568. —— 1947: Physiology of germination. N.Y. Agr. Expt. Sta. Ann. Rept. 1946: 58–59.
8569. —— 1947: Physiology of germination. N.Y. Agr. Expt. Sta. Ann. Rept. 1947: 49–50.
8570. —— 1948: Physiology of germination. N.Y. Agr. Expt. Sta. Ann. Rept. 67:43–45.
8571. —— 1948: Report of subcommittee on dormancy in seeds. Proc. Assoc. Offic. Seed Analysts 38:25–26.
8572. —— 1948: Laboratory germination results with herb and drug seed. Proc. Assoc. Offic. Seed Analysts 38:58–62.
8573. —— 1949: Germinating castor-bean seed in the laboratory. Proc. Assoc. Offic. Seed Analysts 39:114–17.
8574. —— 1949: Physiology of germination. N.Y. Agr. Expt. Sta. Ann. Rept. 68:42–43.
8575. —— 1950: Nasturtium seed germination. Proc. Assoc. Offic. Seed Analysts 40:100–4.
8576. —— 1950: Germination of sensitive flower seed kinds and varieties with suggested methods for testing in the laboratory. Proc. Assoc. Offic. Seed Analysts 40:107–17.
8577. —— 1955: The excised embryo method for testing germination quality of dormant seed. Proc. Assoc. Offic. Seed Analysts (Ann. Meeting) 45:108–17.
8578. —— 1957: Laboratory germination and vigor as compared to soil tests and field performance in flower seed. Proc. Assoc. Offic. Seed Analysts 47:126–35.
8579. —— 1958: The effect of light and temperature on germination of certain hard pines and suggested methods for laboratory testing. Proc. Assoc. Offic. Seed Analysts 48:111–17.
8580. —— 1958: Laboratory germination results with *Calendula* seed. Assoc. Offic. Seed Analysts News Letter 32(1):56–60.
8581. —— 1959: The testing of tree seed in the laboratory and in the nursery. Morris Arboretum Bull. 10:61–66.
8582. —— 1961: Abnormal germination during laboratory testing in coniferous tree seed. Proc. Intern. Seed Testing Assoc. 26:419–27.
8583. —— 1961: Shorter germination test durations for tree seeds. Proc. Intern. Seed Testing Assoc. 26:428–36.
8584. —— 1963: *Verbena* and *Coreopsis* seed sensitiveness to moisture during laboratory germination. Assoc. Offic. Seed Analysts News Letter 37(2):17.
8585. —— and E. J. Eliason. 1940: Coniferous tree seed testing and factors affecting germination and seed quality. N.Y. Agr. Expt. Sta. Tech. Bull. 255. 45 pp.
8586. —— and C. N. Legnini. 1945: Germination problems. N.Y. Agr. Expt. Sta. 44th Ann. Rept. 1944/45:60–61.
8587. —— and M. T. Munn. 1941: New Zealand spinach germination studies. Proc. Assoc. Offic. Seed Analysts 33:90–95.
8588. —— and —— 1947: Effect of low seed vitality on yield. Farm Res. (N.Y.) 13(2):14.
8589. —— and C. Nelson. 1941: Approximate germination tests of dormant seeds by excising embryos. Proc. Assoc. Offic. Seed Analysts 33:87–89.
8590. —— and G. L. Slate. 1950: Treatment of blackberry seed to secure first year germination. Proc. Am. Soc. Hort. Sci. 55:297–301.
Heit, C. E. 4146, 4147, 4659–4663, 5894–5896, 13935, 15120.
Heitz, G. A. O. 19203.
8591. Heitz, S. 1959: Les saponosides des graines de Fenugrec, *Trigonella foenum graecum* L. Compt. Rend. 248:283–86.
Heizer, H. K. 8036.
Hejna, A. Cincerova-; *see* Cincerova-Hejna, A.
8592. Helgason, S. B., and M. Rommel. 1963: Seed development in relation to aneuploidy in autotetraploid *Hordeum vulgare* L. Can. J. Genet. Cytol. 5:189–96. Biol. Abstr. 44:13285. 1963.
8593. Helgeson, E. A. 1932: Impermeability in mature and immature sweet clover seeds as affected by conditions of storage. Trans. Wisconsin Acad. Sci. 27:193–206.
8594. —— 1956: Quackgrass root extract has puzzling effect on seeds of crop plants. N.

Dakota Agr. Expt. Sta. Bimonthly Bull. 18:191–92.

8595. —— and K. L. Blanchard. 1942: The relation of moisture content to freezing injury in Rival and Mindum wheats. N. Dakota Agr. Expt. Sta. Bimonthly Bull. 4:15–16.

8596. —— and D. G. Davis. 1963: Studies of the phytotoxic properties of some extracts from wild oat seeds. Proc. N. Dakota Acad. Sci. 17:66. Chem. Abstr. 60:7376.

Helgeson, E. A. 8248.

8597. Hellbo, E. 1926: Determination of the contents of introrse-grain of oats and of unhulled seeds of timothy. Medd. Statens Centr. Frökontr. 1:45–50.

8598. —— 1933: Über die Fluoreszenz des Hafers bei Beleuchtung mit ultravioletten Strahlen. Medd. Statens Centr. Frökontr. 8:52–56.

8599. Heller, C. 1961: Monogerm seed from prepared sugar beet seed. Landtech. Forsch. 11:111–14.

8600. Heller, V. G., and W. G. Gross. 1946: Chemical and nutritive properties of *Centaurea americana* Nutt. (star thistle). Proc. Oklahoma Acad. Sci. 26:39–40. Biol. Abstr. 21:4451. 1947.

Heller, V. G. 3042.

Heller-Cohen, O. 15445.

8601. Hellings, M. L. 1932: Propagation of unusual trees. Gardeners Chron. 36:266.

8602. —— 1932: Treating seeds difficult to germination. Horticulture 10(5):100.

Hellmers, H. 1469.

8603. Hellot, R., and M. Mâcheboeuf. 1947: Les protéides de la graine d'arachide. Identification des impurétés phosphorées. Bull. Soc. Chim. Biol. 29:817–22. Biol. Abstr. 23:12658. 1949.

8604. Hellqvist, H. 1958: Control of fungus diseases of barley transmitted through seeds and of wireworms. Lantbrtidskr. Jämtland Härjedalen 11(2):5–9.

8605. Helmerick, R. H., and R. P. Pfeifer. 1954: Differential varietal responses of winter wheat germination and early growth to controlled limited moisture conditions. Agron. J. 46:560–62.

Helmerick, R. H. 5542.

8606. Helms, A., and C. A. Jörgensen. 1925: The birches on "Maglemose." Botan. Tidsskr. 39:57–134. Botan. Abstr. 15: 701–2.

Hely, F. W. 3088.

8607. Hemberg, T. 1955: Studies on the balance between free and bound auxin in germinating maize. Physiol. Plantarum 8:418–32.

8608. —— 1958: Auxins and growth-inhibiting substances in maize kernels. Physiol. Plantarum 11:284–311.

Hemerick, G. A. 10475.

Hemmi, S. 10387, 14115.

8609. Hemming, E. S. 1943: Oak trees for food crop. Am. Nurseryman 77(11):16.

8610. —— 1945: Germinating nuts. Am. Nurseryman 82(2):30–31.

8611. —— 1948: Sawdust mulch in seed germination. Am. Nurseryman 87(11):38.

8612. —— 1949: Some seed mysteries. Am. Nurseryman 91(5):54.

Hemptinne, Y. de; *see* De Hemptine, Y.

8613. Hemsley, W. B. 1891: Vitality of seeds. Ann. Botany (London) 5:406–7.

8614. —— 1895: The vitality of seeds. Nature 52:5–7.

8615. —— 1903: On the germination of the seeds of *Davidia involucrata* Baill. J. Linn. Soc. London 35(247). Botan. Centralbl. 95:113. 1904.

8616. Hemstad, C. A. 1960: A study of the inheritance of several characteristics of wheat. Proc. N. Dakota Acad. Sci. 14:7–17. Biol. Abstr. 36:38654. 1961.

8617. Henckel, P. A. 1959: Seeds resist drought. Science News Letter 76:147.

8618. —— and S. S. Kolotova. 1938: Hardening before sowing and frost-hardiness of the plants. Acad. Sci. Ukr. Inst. Botan. Symp. dedicated to the memory of V. N. Lubimenko, p. 206.

8619. —— and —— 1940: Increasing the salt resistance of plants by pre-sowing treatment. Bull. Acad. Sci. URSS Cl. Sci. Math. Nat. Ser. Biol. 1940(4):473–83. Biol. Abstr. 17:6753. 1943.

8620. —— and P. S. Nikitin. 1935: On the pre-sowing hardening of oats for drought under conditions of vegetation experiment. Bull. Perm. Inst. Rech. Biol. 9:334–35.

8621. Hendershot, W. F., C. W. Hesseltine, T. G. Pridham, R. G. Benedict, and R. W. Jackson. 1962: Ramulosin: Inhibitory effect against plant seeds and various fungi. Arch. Biochem. Biophys. 96:166–70.

Henderson, C. F. 18500.

8622. Henderson, F. Y., and T. A. Oxley. 1944: The properties of grain in bulk. II. The coefficient of diffusion of carbon dioxide through wheat. Soc. Chem. Ind. J. 63:52–53. Biol. Abstr. 19:11343. 1945.

8623. Henderson, K. V. 1963: Stored feed stabilizes income from livestock in central

Queensland. Queensland Agr. J. 89: 259–65.

8624. Henderson, M. T., and E. L. Le Clerg. 1943: Studies of some factors affecting fruit setting in *Solanum tuberosum* in the field in Louisiana. J. Agr. Res. 66:67–76.

8625. Henderson, R. G., E. M. Matthews, and W. A. Jenkins. 1945: Tobacco plant-bed management. Virginia Agr. Expt. Sta. Bull. 384. 12 pp. Expt. Sta. Rec. 94: 758–59.

8626. Henderson, W. J. 1941: Demonstrations show value of sorghum seed treatment in Colorado. Extension Pathol. 46:30–32.

8627. Hendricks, S. B. 1958: Photoperiodism. Agron. J. 50:724–29.

8628. ——, E. H. Toole, V. K. Toole, and H. A. Borthwick. 1959: Photocontrol of plant development by the simultaneous excitations of two interconvertible pigments. III. Control of seed germination and axis elongation. Botan. Gaz. 121:1–8.

Hendricks, S. B. 2808–2810, 7986, 15417, 18660, 18666, 18680, 18681, 18683, 18695, 18697.

8629. Hendrickson, H. S. 1963: Chemistry of flaxseed lipids. Structural studies of dehydrophytosphingosine. Dissertation Abstr. 24:2251.

Hennig, A. M. F. 517.

8630. Hennig, L. 1930: Beiträge zur Kenntnis der Resedaceenblüte und -Frucht. Planta 9:507–63.

8631. Henning, E. 1916: The possibilities of combating plant diseases by means of sharp seed selection. Kungl. Landtbruks-Akad. Handl. Tidskr. 55:282–300.

8632. Henning, P. D. 1933: The chief weeds of grain lands. III. Wild vetch (*Vicia* spp.). Farming S. Africa 8(87):220.

8633. Henrici, M. 1935: Germination of Karroo bush seeds. S. African J. Sci. 32: 223–34.

8634. —— 1939: Germination of Karroo bush seeds. II. S. African J. Sci. 36:212–19.

8635. —— 1943: On the possibility of using sodium selenite for seed testing of Karroo bushes. S. African J. Sci. 39:152–54.

Henrickson, E. T. 6475.

8636. Henriet, J. 1956: Contribution a l'analyse des produits phytopharmaceutiques. VIII. Determination de la disinfection des graines de lin au disulfure de tetramethylthiurame (TMTD). Parasitica (Gembloux) 12:16–23.

8637. Henriksson, L. E. 1951: Asymbiotic germination of orchids and some effects of vitamins on *Thunia marshalliana*. Svensk Botan. Tidskr. 45:447–59.

8638. Henry, A. J., and D. N. Grindley. 1943: Investigation of the oil of the seeds of *Cephalocroton cordofanus*. Soc. Chem. Ind. J. 62:60. Pharm. Abstr. 10:143. 1944.

8639. —— and —— 1944: The oils of the seeds of *Ocimum kilimandscharicum*, *Euphorbia calycina*, *E. erythraeae*, *Sterculia tomantosa*, and *Trichilia emetica*. Soc. Chem. Ind. J. 63:188–90.

8640. Henry, A. W. 1943: On the value of Spergon for seed treatment in small-grain crops. Phytopathology 33:332–33.

8641. —— and J. A. Campbell. 1938: Inactivation of seed-borne plant pathogens in the soil. Can. J. Res. Sect. C 16:331–38.

8642. ——, R. L. Millar, and E. A. Peterson. 1952: Control of covered smut of wheat by rapid seed treatment with an antibiotic. Science 115:90–91.

8643. ——, E. A. Peterson, and R. L. Millar. 1953: Compatibility of Acti-dione and Orthocide in the treatment of seeds. Phytopathology 43:461.

8644. ——, ——, ——, and J. S. Horricks. 1951: Control of covered smut of oats by seed treatment with an antibiotic. Science 113:390.

Henry, A. W. 7448.

8645. Henry, C. H., E. B. Wilcox, L. H. Pollard, D. K. Salunkhe, and F. E. Lindquist. 1956: Evaluation of certain methods to determine maturity in relation to yield and quality of yellow sweet corn for processing. Food Technol. 10:374–80.

8646. Henry, E. M. 1935: The germination of strawberry seeds and the technic of handling the seedlings. Proc. Am. Soc. Hort. Sci. 32:431–33.

Henry, H. H. 8809, 8810.

Henry, J. A. 2291.

8647. Henry, Mrs. J. N. 1931: Growing lilacs from seed. Horticulture 9(10): 224–25.

8648. Henry, M. V. 1945: The identification of oat varieties by seed characters. Quart. J. Florida Acad. Sci. 8:171–74.

8649. Henry, P. 1951: La germination des graines d'*Elaeis*. Rev. Intern. Botan. Appl. Agr. Trop. 31:565–91.

8650. —— 1956: Étude cytologique du lait de coco au cours du développement de la noix. Compt. Rend. 243:401–4.

8651. Henson, L. 1950: Determining means whereby the seed yields of red clover can be increased. Intern. Crop Improvement Assoc. Ann. Rept. 32:99–102.

8652. Henson, P. R., and L. A. Tayman. 1961: Seed weights of varieties of birdsfoot trefoil as affecting seedling growth. Crop Sci. 1:306.

Henson, P. R. 3403, 10449.

8653. Hentrich, W. 1961: The action of the radioactive phosphorus isotope P^{32} on barley under various conditions of germination. Z. Pflanzenzücht. 46:417–32. Nucl. Sci. Abstr. 17:3286. 1963.

8654. Hepburn, F. N., E. W. Lewis, Jr., and C. A. Elvehjem. 1957: The amino acid content of wheat, flour and bread. Cereal Chem. 34:312–22.

8655. Hepburn, J. S., and T. L. Laughlin, Jr. 1930: A chemical study of psyllium seed. Am. J. Pharm. 102:565–68. Chem. Abstr. 25:171. 1931.

8656. Hepperly, I. W. 1949: A corn with odd-rowed ears. J. Heredity 40:63–64.

Herb, S. F. 1900.

Herbel, C. H. 10892.

Herber, E. 14607.

8657. Herbert, A. 1911: Upon the composition of various oleaginous seeds of French West Africa. Bull. Soc. Chim. France 9/10:662–72.

8658. Herele, R. 1934: Einfluss des Lichtes auf Keimung und Blüten. Blumen- Pflanzenbau 38:255. Proc. Intern. Seed Testing Assoc. 7:69. 1935.

8659. Herich, R. 1961: Influence of gibberellic acid on metabolism of amino acids. Physiol. Plantarum 14:111–14.

8660. Herissey, H., and G. Poirot. 1942: Oil of the seed of the Portuguese laurel, *Cerasus lusitanica* Lois. J. Pharm. Chim. [9](2):97. Pharm. Abstr. 10:265. 1944.

8661. Herman, H. A., and A. C. Ragsdale. 1942: Korean *Lespedeza* seed as a protein supplement for milk production. Missouri Agr. Expt. Sta. Bull. 451. 7 pp.

8662. Hermann, E. M., and W. Hermann. 1939: The effect of maturity at time of harvest on certain responses of seed of crested wheat-grass, *Agropyron cristatum* (L.) Gaertn. J. Am. Soc. Agron. 31: 876–85.

8663. Hermann, R. 1960: The influence of seedbed microenvironments upon the establishment of Douglas fir (*Pseudotsuga menziesii* [Mirb.] Franco) seedlings. Dissertation Abstr. 21:403–4.

8664. Hermann, S., and R. Neiger. 1934: Beeinflussung der Keimfähigkeit von Weizen durch Beizung mit Salizylsäure und Salizylsäureverbindungen. Botan. Centralbl. Beih. Abt. A 52:577–83.

8665. Hermann, W. 1928: Die Unterscheidung von Weizensorten durch Phenolfärbung der Samen. Kühn-Arch. 19: 11–65.

Hermann, W. 8662, 13200.

8666. Hermansen, J. E. 1953: Effects on the seed quality of spraying of red clover, white clover, and alfalfa with 4K-2M at different times. Dansk Frøavl 36:98.

8667. Hermelin, J. 1958: Germination of north-Swedish pine seed of different embryo types after treatment with varying temperatures and air humidities. Svenska Skogsvardsforen. Tidskr. 56:447–54.

8668. Hernandez, B. 1926: Germination of rice seeds: the effect of soaking in water and delayed sowing on the rate, percentage and uniformity of germination. Philippine Agriculturist 14:553–56.

8669. Herner, M. C. 1937: Relative vitamin A contents of three varieties of northwestern dent corn—Manitoba-grown Red, Yellow, and American Yellow. Sci. Agr. (Ottawa) 18:32–36.

8670. Herno, A. 1944: Examination of germinating capacity and moisture content, etc. in seeds of a number of important crop plants, stored for several years in an ordinary warehouse. Tidsskr. Planteavl 48:598–601.

8671. —— and H. C. Haekgarrd. 1938: Varietal characters of some of the two-row barley varieties cultivated in Denmark. Statsfrokontrollen (Suppl.) Beretning 1937/ 38:22–24.

8672. Herold, G. 1941: Sortenprüfung bei Erdbeeren unter besonderer Berücksichtigung der Befruchtungsverhältnisse. Gartenbauwissenschaften 16:216–62. Biol. Abstr. 18:17442. 1944.

8673. Herold, H. 1929: Keimungfördernde und Keimunghemmende Einflüsse. Gartenwelt 33:101–2.

8674. Herr, J. M., Jr. 1959: The development of the ovule and megagametophyte in the genus *Ilex* L. Elisha Mitchell Sci. Soc. J. 75:107–28. Biol. Abstr. 35:11150. 1960.

8675. —— 1961: Endosperm development and associated ovule modifications in the genus *Ilex* L. Elisha Mitchell Sci. Soc. J. 77:26–32.

8676. Herre, H. 1954: *Welwitschia mirabilis* Hook. f. from seed to seed in the Botanic Garden of the University of Stellenbosch,

C. P. J. S. African Botany 20:23–24. Biol. Abstr. 29:7601. 1955.

Herrera, C. Lopez-. 3671.

Herriott, J. B. D. 8477.

8677. Herron, J. W. 1953: Study of seed production, seed identification, and seed germination of *Chenopodium* spp. Cornell Univ. Agr. Expt. Sta. Mem. 320. 24 pp.

8678. Hervert, V. 1946/47: An experiment on seed treatments to control wheat bunt. Ochrona Rostlin 19/20:78–81. Biol. Abstr. 23:27582. 1949.

Hervey, G. E. R. 13170, 13171, 14246.

8679. Herzfeld, S. 1928: Über die Kernteilungen im Proembryo von *Ginkgo biloba*. Jahrb. Wiss. Botan. 69:264–94.

8680. Herzner, R. 1928: Über die Natur des wasserlöslichen Proteins im Weizensamen. Biochem. Z. 202:320–28.

8681. Heslop-Harrison, J. 1961: Apomixis, environment, and adaptation. Intern. Botan. Congr. Recent Advan. Botany 9:891–95. 1959.

8682. Hesmer, H. 1935: Samen- und Knospenschuppenanalysen in Mooren. Ein weiterer Beitrag zur Feststellung der natürlichen Beziehungen zwischen Standort und Bestand im nordostdeutschen Kieferngebiet mittels neuer Methoden. Z. Forst-Jagdw. 67:600–21. Biol. Abstr. 10: 12765. 1936.

Hess, H. 19203.

8683. Hess, K. 1960: Wheat proteins and lipoproteins. Qualitas Plant. Mater. Vegetabiles 6:312–26. Chem. Abstr. 55:5674. 1961.

8684. ——, H. Kiessig, and E. Hanssen. 1952: Die Samenproteine der Gramineen und ihre Wandlungen beim Teigen und Backen. Naturwissenschaften 6:135–36.

Hess, K. 14156.

8685. Hess, O. 1957: Die Entwicklung von Halbkeimen bei *Limnaea stagnalis*. Naturwissenschaften 44:383.

8686. Hesse, C. O., and D. E. Kester. 1955: Germination of embryos of *Prunus* related to degree of embryo development and method of handling. Proc. Am. Soc. Hort. Sci. 65:251–64.

Hesse, C. O. 10493.

8687. Hesse, O. 1924: Untersuchungen über die Einwirkung chemischer Stoffe auf die Keimung lichtempfindlicher Samen. Botan. Arch. 5:133–71.

8688. Hesse, R. 1952: Zur Wirkung der Ultraschall-Behandlung von Samen auf die Keimung und das nachfolgende Wachstum von Pflanzen. Flora (Jena) 139:565–85.

8689. Hesselman, H. 1932/34: Einige Beobachtungen über die Beziehung zwischen der Samenproduktion von Fichte und Kiefer und der Besamung der Kahlhiebe. Statens Skogsforsk. Med. 27:174–82.

8690. —— 1938/39: Further studies of pine and spruce seed dispersal, and direct seeding of clear-cut forest areas. Statens Skogsforsk. Med. 31:1–58. Biol. Abstr. 16:21004. 1942.

Hesseltine, C. W. 8621.

Hester, J. B. 20014.

8691. Hester, J. P. 1941: Cacti: By their seeds ye shall know them. Desert (Pasadena) 13:189–92.

8692. Hester, W. 1933: Yucca from seed. Flower Grower 20(9):405.

8693. Heuberger, J. W., J. D. Bates, and R. K. Jones. 1963: Apple scab. IV. Effect of temperature and relative humidity on the viability of conidia of *Ventura inaequalis*. Plant Disease Reptr. 47: 826–30.

8694. Heumann, W. 1944: Über den Fettstoffwechsel keimender Kürbissamen. Planta 34:1–16. Biol. Abstr. 21:20701. 1947.

8695. Heursel, J. 1961: Health conditions of tobacco seeds. Ghent Landbhogesch. Mededel. 26:1222–37.

8696. Heuschem, C., J. Firket, and Z. M. Bacq. 1947: Action de la chloropicrine sur la germination des pois. Bull. Soc. Chim. Biol. 29:453–59.

8697. Heuser, W. 1933: Untersuchungen über den Einfluss verschieden Später Saatzeiten auf die Erträge und den Entwicklungsrhythmus von Lupinen, Erbsen und Gerste im lichteder Lehre des Photoperiodismus. Pflanzenbau 9:241–49. Fortschr. Landwirtsch. 8:301. 1933.

8698. ——, K. Boekholt, and G. Ulich. 1934: Der Gehalt der Samen von *Lupinus albus* an Eiweiss, Fett und Alkaloiden im Vergleich zu anderen Lupinenarten und unter dem Einfluss äusserer Bedingungen. Pflanzenbau 11:129–38. Proc. Intern. Seed Testing Assoc. 7:77. 1935.

8699. —— and W. Zeiner. 1936: Der Einfluss der Keimstimmung durch Temperatur und Tageslänge auf den Entwicklungsrhythmus und die Ertragsstruktur verschiedener Sommerweizensorten und die Bedeutung für deren Saatzeitempfindlichkeit. Pflanzenbau 13:106–19.

Heusser, D. 5353.

Heuvel, F. A. van den. 6221.

Hewetson, F. N. 2894.

8700. Hewett, P. D. 1958: Effects of heat and loss of moisture on the dormancy of barley. Nature 181:424–25.

8701. —— 1962: Methods for detecting viable seed-borne infection with celery leaf spot. Natl. Inst. Agr. Botany J. 9:174–78.

8702. —— 1963: Prevalence and importance of viable spores of *Septoria* spp. in samples of commercial celery seed. Nature 197: 918–19.

8703. Hewitt, E. J., and P. Miles. 1952: The effects of seed reserves on the production of deficiencies of copper, zinc and molybdenum in some large-seeded leguminous crops. Ann. Rept. Long Ashton Agr. Hort. Res. Sta. 1952:199–203.

Hewitt, W. B. 17645.

8704. Hewitt, W. C. 1939: Seed development of *Lobelia amoena*. Elisha Mitchell Sci. Soc. J. 55:63–82.

8705. Hewlett, C. H., and J. H. Hewlett. 1933: Hot-water treatment of seed barley. Crop results in Canterbury, seasons 1930–1933. New Zealand J. Agr. 47:33–37.

Hewlett, J. H. 8705.

8706. Heydecker, W. 1956: Establishment of seedlings in the field. I. Influence of sowing depth on seedling emergence. J. Hort. Sci. 31:76–88.

8707. —— 1958: Effect of anaerobic conditions on imbibed lettuce seeds. Nature 181:1140–41.

8708. —— 1958: The germination of *Crocus* seeds. Gardeners Chron. 144:299.

8709. —— 1960: Can we measure seedling vigour? Proc. Intern. Seed Testing Assoc. 25:498–512.

8710. —— 1962: From seed to seedling: factors affecting the establishment of vegetable crops. Ann. Appl. Biol. 50: 662–27.

8711. —— 1962: Seedling Vigour Test Committee. Proc. Intern. Seed Testing Assoc. 27:211–22.

8712. —— 1964: Physiology, ecology and biochemistry of germination. Nature 201:979–81.

Heydecker, W. 12818.

8713. Heydeman, M. T. 1958: Isocitritase in germinating marrow seeds. Nature 181:627–28.

8714. Heydt, A. 1932: *Acacia verticillata*, eine lohnende Kulturpflanze für die Frühjahrblüte. Vorbehandeln ("Kochen") des Samens. Möllers Deut. Gärtner-Ztg. 47:280.

8715. —— 1933: Eine alte lohnende Kultur. Die Anzucht von *Ficus elastica* aus Samen. Möllers Deut. Gärtner-Ztg. 48:61–62.

8716. —— 1933: Palmen-Samen billig! Mehr deutsche Palmen-Kultur eine Notwendigkeit. Möllers Deut. Gärtner-Ztg. 48:81.

Heyne, E. G. 16422.

8717. Hibbard, A. D., and L. M. Flynn. 1945: Effect of maturity on the vitamin content of green snap beans. Proc. Am. Soc. Hort. Sci. 46:350–54.

8718. Hibbard, R. P., and E. V. Miller. 1928: Biochemical studies on seed viability. I. Measurements of conductance and reduction. Plant Physiol. 3:335–52.

8719. —— and O. E. Street. 1929: Biochemical studies on seed viability. II. Chemical constituents operating in reduction. Mich. Acad. Sci. Papers 9:139–62.

Hibbard, R. P. 6325, 18572.

8720. Hibberd, N. V. 1954: Abnormal developments in germinating French bean seed. Queensland Agr. J. 78:33–36.

8721. Hickel, R. 1926: À propos de la germination des *Araucaria*. Bull. Soc. Botan. France 73:968–70.

8722. Hicks, C. S. 1957: The basis of mankind's food; fascinating history of the seven grass-seeds which feed the overcrowded earth. Power Farming Australia, New Zealand & Better Farming Dig. 66(7):3–4.

8723. Hicks, G. H. 1900: The germination of seeds as affected by certain chemical fertilizers. U.S. Dept. Agr. Div. Botany Bull. 24. 15 pp.

8724. —— and J. C. Dabney. 1896: The superior value of large heavy seed. Yearbook Agr. U.S. Dept. Agr. 1896:305–22.

8725. —— and S. Key. 1897: Additional notes on seed testing. Yearbook Agr. U.S. Dept. Agr. 1897:441–52.

Hicks, M. 8929.

Hicks, S. 5505.

8726. Hicks, S. S. 1951: Initial activity of certain enzyme systems as related to seed germination. Am. Chem. Soc. Abstr. Papers, 119th Meeting, Boston, Cleveland, pp. 31A–32A.

8727. ——, D. M. Doty, and F. W. Quackenbush. 1951: A germination inhibitor from apple and celery seeds. [Abstract.] Federation Proc. 10:196–97.

8728. Hieatt, F. L. 1947: Germinating rose seeds. Am. Rose Ann. 32:125–27.

8729. Hieke, K. 1956: Bemerkungen zum Keimen von Reis. Biologia 11:257–71.
Hiemeleers, J. 18707.
Hien, V. D. 4491.
Hiesey, W. M. 4177, 10049.
8730. Higaki, T., and R. R. Dedolph. 1963: The storage of macadamia nuts (*Macadamia integrifolia*, Maiden and Betche). Proc. Am. Soc. Hort. Sci. 83:359–69.
Higazy, M. K. 1662.
8731. Higgins, B. B. 1940: Inheritance of seed-coat color in peanuts. J. Agr. Res. 61:745–52.
8732. —— 1949: Some factors affecting viability of peanut seed. Phytopathology 39:496.
8733. —— and W. K. Bailey. 1959: Peanut seed and seed treatments. Georgia Agr. Expt. Sta. Mimeo. Ser. N.S. 70. 93 mimeo. pp.
8734. ——, K. T. Holley, T. A. Pickett, and C. D. Wheeler. 1941: II. Thiamin chloride and nicotinic acid content of peanuts and peanut products. Georgia Agr. Expt. Sta. Bull. 213:11–14.
Higgins, B. B. 2154, 6996.
8735. Higgins, E. C. 1948: Frequency of multiple sprouts in some seeds. Proc. Assoc. Offic. Seed Analysts 1948:68–69.
Higgins, E. C. 18026.
8736. Higgins, G. M., and C. Sheard. 1927: Germination and growth of seeds as dependent upon selective irradiation. Plant Physiol. 2:325–35.
Higgins, G. M. 17291, 17292.
8737. Higgins, J. E. 1916: Germination of pineapple seeds. Hawaii Agr. Expt. Sta. Rept. 1916:15–17.
8738. Higgins, V. 1944: On the germination of difficult seeds by the amateur. J. Roy. Hort. Soc. 69:340–44.
8739. Higgins, W. M., and M. F. W. Dunker. 1945: Isolation of mannitol from the seeds of *Citrullus vulgaris*. J. Am. Chem. Soc. 67:153–54.
8740. Highkin, H. R. 1955: Flower-promoting activity of pea seed diffusates. Plant Physiol. 30:390.
8741. —— 1956: Vernalization in peas. Plant Physiol. 31:399–403.
8742. —— 1959: Effect of vernalization on heat resistance in two varieties of peas. Plant Physiol. 34:643–44.
Highkin, H. R. 16290.
Higuchi, Y. 14701.
8743. Higuti, T., M. Kondo, M. Hara, S. Narita, and K. Matuzawa. 1928: The relation between the grade of polishing of rice and its chemical constituents. Rept. Inst. Nutr. 2:129–36. Japan. J. Med. Sci. II. Biochem. 3:(100).
8744. Hijimoto, S. 1954: A study on the seeds of Kentucky 31 fescue. I. About the yielding efficiency and utility value. Kagawa Agr. Coll. Tech. Bull. 6:52–56.
8745. Hilbe, J. J. 1941: The effects of ultracentrifuging germinating seeds of onion and rye. Proc. Iowa Acad. Sci. 48:457–66.
8746. Hilbert, G. E. 1948: Soybean studies at the Northern Regional Research Laboratory. Chemurgic Dig. 7(2):11–14.
8747. —— and M. M. MacMasters. 1946: Pea starch, a starch of high amylose content. J. Biol. Chem. 162:229–38.
Hilbert, G. E. 2439, 12650–12652.
8748. Hildebrand, A. A., and L. W. Koch. 1950: Observations on six years' seed treatment of soybeans in Ontario. Sci. Agr. (Ottawa) 30:112–18.
Hildebrand, E. M. 15841.
8749. Hildebrand, F. 1881: Die Samenverbreitung von *Aponogeton distachyum*. Flora (Jena) 39:497–501.
8750. —— 1884: Über die Schutzeinichtungen bei den *Oxalis*-zwiebeln. Ber. Deut. Botan. Ges. 2:108–11.
8751. —— 1892: Einige Beobachtungen an Keimlingen und Stecklingen. Botan. Z. 50:1–11, 33–42.
8752. —— 1899: Germination of *Anemone apennia*. Ber. Deut. Botan. Ges. 17:161–66.
Hildebrandt, A. C. 20094.
8753. Hildebrandt, W. 1959: Keimungsphysiologische Studien an Steinobst. I. Die Ursachen der Keimruhe von Steinobstsaatgut. Gartenbauwissenschaft 24:161–76.
8754. —— 1960: Keimungsphysiologische Studien an Steinobst. II. Über den Einfluss der Stratifikationstemperatur auf Nachreife und Keimung verschiedener Steinobstarten. Gartenbauwissenschaft 24:411–29.
8755. —— 1960: Keimungsphysiologische Studien an Steinobst. III. Die Feuchtigkeitsaufnahme des Steinobstsaatgutes bei der Quellung und der Einfluss der Feuchtigkeit auf die Keimung. Gartenbauwissenschaft 25:15–32.
8756. Hilditch, T. P. 1928: The chemical composition of vegetable seed fats in relation to the natural orders of plants. Proc. Roy. Soc. (London) Ser. B 103:111–17.

8757. —— 1933: A note on certain general features common to most fruit-coat fats. Soc. Chem. Ind. J. 52:169T–171T.
8758. —— 1947: The chemical constitution of natural fats. 2d ed., Wiley, New York. 554 pp.
8759. —— 1949: Current work at the University of Liverpool on potential sources of drying oils. J. Oil Colour Chemists Assoc. 32:5–23. Chem. Abstr. 43:4025–26. 1949.
8760. —— 1951: Biosynthesis of unsaturated fatty acids in ripening seeds. Nature 167:298–301.
8761. —— 1951: Rational grading of seed oils. Chem. Ind. (London) 1951(41): 846–49.
8762. —— 1952: The drying oil in the seeds of *Mercurialis perennis* (dog's mercury) Chem. Ind. (London) 1952(40):981.
8763. —— 1952: The seed and fruit fats of plants. Endeavour 11:173–82.
8764. —— and collaborators. 1951: Variations in the composition of some linolenic-rich seed oils. J. Sci. Food Agr. 2: 543–47.
8765. ——, F. D. Gunstone, and J. P. Riley. 1948: African vine studied for high yield of oil. Chemurgic Dig. 7(10):18.
8766. —— and M. B. Ichaporia. 1936: Composition of some Solanaceous seed-fats. Soc. Chem. Ind. J. 55:189T–190T.
8767. —— and —— 1938: Solid, fatty acids and glycerides of seed fats. III. Seed fat of *Madhuca* (*Bassia*) *latifolia* (Mowrah fat). Soc. Chem. Ind. J. 57:44–48.
8768. —— and H. Jasperson. 1939: The constitution of the linoleic acid of seed fats. Soc. Chem. Ind. J. 58:233–41.
8769. —— and E. C. Jones. 1932: The component glycerides of partially hydrogenated fats. I. The alterations in glyceride structure produced during progressive hydrogenation of live and cottonseed oils. J. Chem. Soc. 1932:805–20.
8770. —— and M. L. Meara. 1938: The seed fat of the annual *Nasturtium* (*tropaeolum* var.). J. Chem. Soc. 1938:1608–10.
8771. —— and —— 1944: The fatty acids and glycerides of solid seed fats. XII. *Lophira alata* kernel fat (Niam fat). Soc. Chem. Ind. J. 63:114–15.
8772. ——, ——, and W. H. Pedelty. 1939: Seed fat of *Hodgsonia capniocarpa*. Soc. Chem. Ind. J. 58:26–29.
8773. ——, ——, and O. A. Roels. 1947: The composition of commercial palm oils. VI. Component acids and glycerides of a Belgian Congo palm oil (studied with the aid of low-temperature crystallization). Soc. Chem. Ind. J. 66:284–88.
8774. ——, ——, and Y. A. H. Zaky. 1941: The component acids of *Sterculia foetida* seed fat (*Sterculia* oil): A correction of work previously reported. Soc. Chem. Ind. J. 60:198–203.
8775. —— and K. S. Murti. 1941: Fatty acids and glycerides of solid seed fats. X. Seed fats of *Garcinia morella* and *G. indica*. Soc. Chem. Ind. J. 60:16–18.
8776. —— and W. H. Pedelty. 1937: The component fatty acids of the phosphatides of soya bean and rape seeds. Biochem. J. 31:1964–72.
8777. —— and H. M. Thompson. 1937: Some further observations on the component glycerides of olive and tea-seed oils. Soc. Chem. Ind. J. 56:434T–438T.
8778. —— and Y. A. H. Zaky. 1942: Component fatty acids of some vegetable seed phosphatides. Biochem. J. 36:815–21.
8779. —— and —— 1942: The fatty acids and glycerides of solid seed fats. XI. *Shorea robusta* kernel fat. Soc. Chem. Ind. J. 61:34–36.
Hilditch, T. P. 3035, 3036, 3427–3429, 3648, 4073, 4299, 5334, 5652, 7621, 7622, 7832–7835.
8780. Hildreth, R. J., and C. A. Moore. 1958: Profits and losses from on-farm drying and storage of grain sorghum in central Texas and the coastal bend. Texas Agr. Expt. Sta. Bull. 887. 12 pp.
8781. —— and J. W. Sorenson, Jr. 1957: Profits and losses from on-farm drying and storage of rice in Texas. Texas Agr. Expt. Sta. Bull. 865. 15 pp.
8782. Hiley, W. E. 1921: Recent investigations on the germination and culture of forest seeds. Quart. J. Forestry 15:160–68.
8783. Hilf, R., and E. Rohmeder. 1955: Untersuchungen über das Verhalten von Tetrazoliumchlorid bei der Keimfähigkeitprüfung forstlicher Sämereien. Forstwiss. Centralbl. 74:279–92.
Hilfman, L. J. 14578.
8784. Hilkenbäumer, F. 1936: Versuche zur Behebung des Keimverzugs bei Steinobstsamen und zur klärung seiner Ursache. Landwirtsch. Jahrb. 82:883–924.
8785. Hill, A. W. 1916: Studies in seed germination. The genus *Marah* (*Megarrhiza*), Cucurbitaceae. Ann. Botany (London) 30:215–22.
8786. —— 1920: Studies in seed germination. Ann. Botany (London) 34:417–29.

8787. —— 1929: The original home and mode of dispersal of the coconut. Nature 124:133–34, 151–53.

8788. —— 1933: Germinating coconuts on a new volcanic island, Krakatoa. Nature 132:674.

8789. —— 1933: The method of germination of seeds enclosed in a stony endocarp. Ann. Botany (London) 47:873–87.

8790. —— 1934: Germination of seeds. Nature 133:858–59.

8791. —— 1934: Germination of seeds. Nature 133:896–98.

8792. —— 1937: The method of germination of seeds enclosed in a stony endocarp. II. Ann. Botany (London) N.S. 1:239–56.

8793. —— 1941: The genus *Calvaria*, with an account of the stony endocarp and germination of the seed and description of a new species. Ann. Botany (London) N.S. 5:587–606.

Hill, A. W. 2758.
Hill, B. 10845.
Hill, D. C. 312.

8794. Hill, D. D. 1933: The chemical composition and grades of barley and oat varieties. J. Am. Soc. Agron. 25:301–11.

Hill, D. D. 3957, 13680, 15071, 19582.

8795. Hill, E. J. 1891: "Sling-fruit" of *Cryptotaenia*. Botan. Gaz. 16:299–302.

8796. Hill, E. R., W. H. Lachman, and D. N. Maynard. 1963: Reproductive potential of yellow nutsedge by seed. Weeds 11:160–61.

8797. ——, ——, and —— 1963: Translocation of amitrole in yellow nutsedge and its effect on seed germination. Weeds 11:165–66.

8798. ——, ——, ——, and W. C. Lincoln. 1962: The effect of foliar applications of amino triazole on the germination of northern nutgrass (*Cyperus esculentus*) seed. Proc. Northeast. Weed Control Conf. 16:64–68.

Hill, G. R. 8259.

8799. Hill, K. 1934: Alfalfa seed emulsin. Ber. Verhandl. Sächs. Akad. Wiss. Leipzig Math.-Physik. Klasse 86:115–28. Chem. Abstr. 28:5843. 1934.

8800. Hill, K. W. 1947: Standard methods of laboratory germination of sugar beet seed in Canada. Proc. Am. Soc. Sugar Beet Technol. 1946:283–84. Biol. Abstr. 22:12094. 1948.

8801. —— 1964: Protein content of wheat as affected by agronomic practices. Can. J. Plant Sci. 44:115–22.

8802. Hill, R. 1936: The growing of canary seed (*Phalaris canariensis*) in South Australia. J. Dept. Agr. S. Australia 39:1138.

Hill, T. G. 7922.
Hill, W. R. 5979.
Hillel, D. 11039.
Hillery, B. E. 10272.

8803. Hilli, A. 1958: Concerning factors which decrease the germinating power of seed. J. Sci. Agr. Soc. Finland 30:85–96. Biol. Abstr. 35:60844. 1960.

8804. —— 1960: Germinating ability of summer cereals after a dry growth period. Maataloustieteellinen Aikakauskirja 32: 131–43. Biol. Abstr. 39:24772. 1962.

8805. Hillman, F. H. 1928: The botany of seed testing. Proc. Assoc. Offic. Seed Analysts 19:15–24.

8806. —— 1930: Identifying turf-grass seed. Bull. U.S. Golf Assoc. Green Sect. 10(3): 39–43.

8807. —— 1931: *Agrostis*. Proc. Intern. Seed Testing Assoc. 3:235–36.

8808. —— 1931: Weed seeds in crop seeds aid in origin determination. Seed World 30(12):16–17.

8809. —— and H. H. Henry. 1928: The incidental seeds found in commercial seed of alfalfa and red clover. Proc. Intern. Seed Testing Assoc. 6:1–22.

8810. —— and —— 1935: Photographs of drawings of seeds: the more important forage-plant seeds and incidental seeds commonly found with them. U.S. Dept. Agr. Bur. Plant Ind. 19 pp. Expt. Sta. Rec. 74:335. 1936.

Hillman, W. S. 15627.

8811. Hills, B. 1947: *Cymbidium* seed-pods. Gardeners Chron. 122:209.

8812. Hills, C. H., and C. H. Bailey. 1938: The nature of the increase in amylase activity of germinating barley. Cereal Chem. 15:273–81.

8813. —— and —— 1938: The amylases of different barley varieties. Cereal Chem. 15:351–62.

Hills, F. J. 11918, 19716.

8814. Hills, K. L. 1942: Dormancy and hardseededness in *T. subterraneum*. I. The effect of time of harvest and of certain seed storage conditions. Australian Council Sci. Ind. Res. J. 15:275–84.

8815. —— 1944: Dormancy and hardseededness in *T. subterraneum*. II. The progress of after-harvest ripening. Australian Council Sci. Ind. Res. J. 17:186–90.

8816. —— 1944: Dormancy and hardseededness in *T. subterraneum*. III. The effect upon dormancy of germination at three different constant temperatures. Australian Council Sci. Ind. Res. J. 17:191–96.

8817. —— 1944: Dormancy and hardseededness in *T. subterraneum.* IV. Variation between varieties. Australian Council Sci. Ind. Res. J. 17:242–50.

8818. —— 1944: Dormancy and hardseededness in *T. subterraneum.* V. The effect of the condition of the seed coat upon embryo dormancy. Australian Council Sci. Ind. Res. J. 17:251–54.

8819. —— 1944: Dormancy and hardseededness in *T. subterraneum.* VI. Application of the results to the problem of reestablishment in the field and to seed testing and marketing. Australian Council Sci. Ind. Res. J. 17:255–57.

8820. Hilpert, F. 1937: Über Entstehung und Wirtschaftlichkeit der verschiedenen Typen unreifer *Ricinus*samen. Botan. Centralbl. Beih. Abt. A 57:277–96.

8821. Hilpert, R. S., and W. Krüger. 1939: Untersuchungen über pflanzliche Samenschalen. Ber. Deut. Chem. Ges. 72:400–4. Chem. Abstr. 33:3841–42. 1939.

8822. Hiltner, E. 1927: Die Beizung des Saatgutes. Landwirtsch. Jahrb. Bayern 7/9:328–51.

8823. Hiltner, L. 1898/99: Über die Keimung gärtnerischer Samereien. Flora Sitzber. Abhandl. 3:85–103.

8824. —— 1902: Die Keimungsverhältnisse der Leguminosensamen und ihre Beeinflussung durch Organismenwirkung. Germany K. Gesundheitsamt.-Biol. Abt. Arb. 3:1–102.

8825. —— 1910: Die Prüfung des Saatgutes auf Frische und Gesundheit. Jahresber. Vereinigung Angew. Botan. 8:219–38.

8826. —— and W. Kinzel. 1906: Über die Ursachen und die Beseitigung der Keimungshemmungen bei verschiedenen praktisch wichtigeren Samenarten. Naturwiss. Z. Land- Forstwirtsch. 4:193–204.

8827. Hilton, J. W. 1941: Effects of certain micro-ecological factors on the germinability and early development of *Eurotia lanata.* Northwest Sci. 15:86–92. Biol. Abstr. 16:8931. 1941.

8828. Hilton, R. J. 1944: Parthenocarpic fruit production in horticultural plants. Sci. Agr. (Ottawa) 24:451–55.

8829. —— and B. J. E. Teskey. 1963: Quick viability tests for rootstock seeds. Fruit Var. Hort. Dig. 17:66. Biol. Abstr. 45:26114. 1964.

8830. Hin, G. S., and M. J. Lovkova. 1959: Nicotine and amino acid transformation in ripening tobacco seeds. Biochemistry (USSR) (English Transl.) 24:838–41.

8831. Hindmarsh, W. L. 1937: Poisonous to stock. The unripe seeds of the Moreton Bay chestnut (*Castanospermum australe*). Agr. Gaz. N.S. Wales 48:608.

8832. Hinds, T. E., F. G. Hawksworth, and W. J. McGinnies. 1963: Seed discharge in *Arceuthobium*: a photographic study. Science 140:1236–38.

8833. Hinkle, D. A. 1952: Pre-harvest drying of rice with chemicals. Arkansas Farmer 54(2):8–9.

Hinman, F. G. 3053.

8834. Hino, I., and K. Katumoto. 1953: The virus in the seeds of tobacco and radish plants. Yamaguti Univ. Fac. Agr. Bull. 4:199–202.

8835. Hinrichs, C. T. 1933: Keimversuche mit Ölpalmensaaten (*Elaeis guineensis*). Tropenpflanzer 36:513–21. Biol. Abstr. 9:12206. 1935.

8836. Hinrichs, H., and F. Cross. 1943: Pecan cracking test. Proc. Am. Soc. Hort. Sci. 42:259–64.

8837. Hinson, E. 1935: Collection of yellow poplar seed. J. Forestry 33:1007–8.

8838. Hinsvark, O. N., W. H. Houff, S. H. Wittwer, and H. M. Sell. 1954: The extraction and colorimetric estimation of indole-3-acetic acid and its esters in developing corn kernels. Plant Physiol. 29:107–8.

8839. Hinton, J. J. C. 1942: The vitamin-B_1 and riboflavin contents of wheat germ. Soc Chem. Ind. J. 61:143–44.

8840. —— 1944: The chemistry of wheat germ with particular reference to the scutellum. Biochem. J. 38:214–17.

8841. —— 1951: Distribution of vitamin B_1 in the rice grain. Ital. Cereali 6:82–90. Chem. Abstr. 49:2575. 1955.

8842. —— 1951: Distribution of vitamin B_1 and of nitrogen in the wheat grain. Ital. Cereali 6:261–67. Chem. Abstr. 49:2575. 1955.

8843. —— 1955: Resistance of the testa to entry of water into the wheat kernel. Cereal Chem. 32:296–306.

8844. —— 1959: The distribution of ash in the wheat kernel. Cereal Chem. 36:19–31.

8845. —— and T. Moran. 1957: Methionine sulphoximine and the growth of the wheat embryo. Brit. J. Nutr. 11:323–28. Biol. Abstr. 32:28570. 1958.

8846. Hintringer, A. 1927: Über die Ablösung der Samen von der Plazenta, Beziehungsweise vom Perikarp. Sitzber. Wien. Wiss. Akad. Math.-Naturw. Kl. 136: 257–79.

8847. Hintzer, H. M. R., and H. de Miranda. 1954: Investigations on the quality of diploid and tetraploid rye for breadmaking. Cereal Chem. 31:405–16.

Hiramatu, H. 7438.

Hirano, J. 18147.

8848. Hirano, T., H. Yarimizu, and S. Onodera. 1960: Studies on the relationship between ripening of rice and the accumulation of carbohydrates and nitrogen compounds. Proc. Crop. Sci. Soc. Japan 29:7–10.

Hirano, T. 8310.

Hiraoka, T. 7994.

Hirata, K. 14524.

Hirata, M. 18443.

8849. Hirata, T., and K. Fujita. 1961: On amino acid constitution of *Poncirus* seed protein. Botan. Mag. (Tokyo) 74:313–14.

8850. Hirayoshi, I., S. Kuroda, and K. Nishikawa. 1955: Growth inhibiting substance excreted by plants. Preliminary report. Inhibition of seed germination and seedling growth of Petsai (*Brassica pekinensis*) by the extract of the residue solutions of water culture. Agr. Hort. (Tokyo) 30:453–54. Biol. Abstr. 31:39559. 1957.

8851. ——, ——, and —— 1957: Growth-inhibiting substance produced by plants. I. Biological tests on the inhibitory substance extracted from water culture medium. Kagaku (Tokyo) 27:92–93. Chem. Abstr. 51:14905. 1957.

Hirayoshi, I. 13029.

8852. Hirohata, R. 1932: Globulins of the seeds of some Cucurbitaceae. Z. Physiol. Chem. 212:1–6. Chem. Abstr. 27:119. 1933.

Hirokawa, H. 8302.

8853. Hirose, T., and T. Nakagawa. 1955: Effect of stage of fruit maturity and after-ripening on the germination of pepper seed. Saikyo Univ. Fac. Agr. Sci. Rept. 7:117–20.

Hirota, K. 10152.

8854. Hirsch, W. 1890: Transport of reserve-materials from the endosperm to the embryo. Ber. Deut. Botan. Ges. 8:1–8.

8855. Hirschowitz, S. 1924: Der Nachweis abgetöteter Knäule im Rübensamen. Z. Zuckerind. 74(N.S. 61):115–34. Chem. Abstr. 18:1847–48. 1924.

Hirshfeld, A. 273.

8856. Hirsjärvi, V. P. 1953: Über die Wasseraufnahmefähigkeit der Samen von Winterrübsen, *Brassica rapa* f. *oleifera* subf. *biennis*. J. Sci. Agr. Soc. Finland 24:55–67. Ber. Wiss. Biol. 86:238. 1954.

Hirst, C. T. 7613.

8857. Hirst, E. L., and J. K. N. Jones. 1948: The galactomannan of carob-seed gum (gum gatto). J. Chem. Soc. 1948:1278–82.

8858. ——, ——, and W. O. Walder. 1947: Pectic substances. VII. The constitution of the galactan from *Lupinus albus*. J. Chem. Soc. 1947:1225–29.

8859. ——, E. G. V. Percival, and C. B. Wylam. 1954: Studies on seed mucilages. VI. The seed mucilage of *Plantago arenaria*. J. Chem. Soc. 1954:189–98.

8860. Hisar, R. S. 1952: Mahalab seed. Turk. Bull. Hyg. Exptl. Biol. 12:227–40. Chem. Abstr. 47:7598. 1953.

8861. Hitchcock, A. E., P. W. Zimmerman, H. Kirkpatrick, Jr., and T. T. Earle. 1949: Water hyacinth: Its growth, reproduction, and practical control by 2,4-D. Contrib. Boyce Thompson Inst. 15:363–401.

8862. ——, ——, ——, and —— 1950: Growth and reproduction of water hyacinth and alligator weed and their control by means of 2,4-D. Contrib. Boyce Thompson Inst. 16:91–130.

8863. Hitchcock, A. S., and M. A. Carleton. 1893: The effect of fungicides upon the germination of corn. Kansas Agr. Expt. Sta. Bull. 41:63–79.

8864. Hite, B. C. 1919: Forcing the germination of blue-grass. Proc. Assoc. Offic. Seed Analysts 11:53–58.

8865. —— 1923: Effect of storage on the germination of bluegrass seed. Proc. Assoc. Offic. Seed Analysts 14/15:97.

Hite, B. C. 8212.

8866. Hitier, H., and C. Izard. 1951: Application des substances de croissance a la culture du tabac. Ann. Inst. Exptl. Tabac Bergerac 1(2):97–125. Biol. Abstr. 26:9592. 1952.

8867. —— and —— 1955: Effets de la streptomycine, de la framycétine, des hydrazides maléique et isonicotinique sur la germination des graines de tabac. Ann. Inst. Exptl. Tabac Bergerac 2:55–60.

Hitier, H. 7289, 9599, 9600.

8868. Hitt, R. G. 1962: Forest tree seed verification and certification. Central States Forest Tree Impr. Conf. Proc. 2: 29–37. 1960.

Hittle, C. N. 9612.

8869. Hixon, R. M. 1920: Effect of the reaction of a nutritive solution on germination and the first stages of plant growth. Medd. Kgl. Vetenskaps-Akad. Nobelinst. 4(9):1–28.

8870. —— and G. F. Sprague. 1942: Waxy starch of maize and other cereals. A possible competitor for tapioca. Ind. Eng. Chem. Ind. Ed. 34:959–62. Biol. Abstr. 17:6754. 1943.

Hixon, R. M. 6895.

8871. Hizukuri, S., M. Fujii, and Z. Nikuni. 1961: Effect of temperature during germination on the crystalline type of starch in soybean seedlings. Nature 192:239–40.

8872. Hjelmqvist, H. 1953: The embryo sac development of *Quercus robur* L. Phytomorphology 3:377–83. Biol. Abstr. 31: 22145. 1957.

8873. —— 1956: The embryology of some African *Alchemilla* species. Botan. Notiser 109:21–32. Biol. Abstr. 31:29199. 1957.

8874. —— 1957: Some notes on the endosperm and embryo development in Fagales and related orders. Botan. Notiser 110: 173–95.

8875. —— 1959: Studien über Embryologie und Variabilität bei einigen *Aphanes*-Arten. Botan. Notiser 112:17–64. Biol. Abstr. 35:11151. 1960.

8876. —— 1959: On the embryology of two *Malus* hybrids. Botan. Notiser 112:453–64.

8877. —— 1962: The embryo sac development of some *Cotoneaster* species. Botan. Notiser 115:208–36.

8878. —— and U. Holmberg. 1961: The development of the embryo sac in *Sanvitalia procumbens*. Botan. Notiser 114: 353–60. Biol. Abstr. 38:19879. 1962.

8879. Hladik, J. 1937: Determination of crude fibre in oleaginous seeds and pressed cakes made from them. Z. Untersuch. Lebensm. 73:140–42. Analyst 62:404–5. 1937.

8880. Hlavaty, J. 1929: Sur la disparition de la vitamine B des graines germant dans un germoir ou dans le sol. Compt. Rend. Soc. Biol. 100:587–89.

Hlynka, K. 550.

Ho, C. H. 12346.

Ho, H. 3818.

8881. Ho, K., S. Wan, and S. H. Wen. 1935: Iodine value of tung oil. Ind. Eng. Chem. Anal. Ed. 7:96. Analyst 60:569–70. 1935.

Hoadley, A. D. 20014.

8882. Hobbs, M. H. 1955: Growing bromeliads from seed. Bromeliad Soc. Bull. 5:76–77.

8883. Hochaptel, H. 1953: Sources of error in the testing of fruit seed with tetrazolium. Deut. Baumsch. 5:224–28. J. Sci. Food Agr. 5:75. 1954.

8884. Hocking, B. 1949: Effect of crude benzene hexachloride on wheat seedlings. Nature 163:731.

8885. Hocombe, S. D. 1961: Simple experiments on the greenhouse germination of some East African weed species. Pesticides Abstr. News Summ. Sect. C 7:149–54.

8886. Hocquette, M., C. Lespagnol, and P. Navrez. 1956: Benzoxazolone et dormance des semences. Compt. Rend. 242:541–43.

8887. ——, J. Montreuil, and S. Paul. 1952: Variations quantitatives du phosphore total, du phosphore des acides ribo- et desoxyribo-nycleiques dans l'axe hypocotyle de *Phaseolus vulgaris* au courade la germination normale, du jeune glucidique apres ablation des cotyledons et de la regeneration cellulaire. Compt. Rend. 253:1525.

8888. Hodge, W. H. 1949: The Goliath of seeds. Nat. History 58:34–35.

Hodgdon, A. R. 15696.

Hodgkins, J. E. 6069.

8889. Hodgson, H. J. 1949: Effect of heat and acid scarification on germination of seed of Bahia grass, *Paspalum notatum*, Flugge. Agron. J. 41:531–33.

8890. —— 1956: Effect of seeding rates and time of harvest on yield and quality of oat-pea forage. Agron. J. 48:87–90.

8891. Hodgson, R. W., and S. H. Cameron. 1938: Effects of reproduction by nucellar embryony on clonal characteristics in citrus. J. Heredity 29:417–19.

8892. —— and E. R. Eggers. 1938: Correlation between seed, seedling and budling in the avocado. Proc. Am. Soc. Hort. Sci. 35:372–74.

8893. —— and —— 1938: Correlations between size of seed, seedling, and nursery tree in the avocado. Calif. Avocado Assoc. Yearbook 1938:92–96. Biol. Abstr. 13:15598. 1939.

8894. Hodnett, G. E. 1958: The effect of temperature and moisture during storage on the viability of grass seed. Trop. Agr. (London) 35:208–12.

Hodson, A. C. 7677.

Hoefer, J. A. 3042.

8895. Hoefle, O. M. 1929: Results of a study of *Daucus carota* seeds. Proc. Assoc. Offic. Seed Analysts 21:35–36.

8896. —— 1932: Weed seeds found in vegetable seeds. N.Y. Agr. Expt. Sta. Bull. 616. 15 pp.

8897. Höhn, K. 1952: Nachreifstudien an *Datura*-Samen. Planta 40:407–18.

8898. Hoehnel, F. 1875/77: Über die Ursache der Quellungsunfahigkeit von Leguminosensamen und der Einfluss der chemisch-physicalischen Beschaffenheit der Pallisadenschicht auf die Keimfähigkeit derselben. Wiss.-Prakt. Gebiete Pflanzenbaues 1:80–88.

8899. —— 1877: Welche wärmegrade Trockene Samen ertragen ohne ihre Keimfähigkeit einzubssen. Wiss.-Prakt. Gebiete Pflanzenbaues 2:77–78.

Höhnel, F. von; *see* Von Höhnel, F.

Hoener, I. R. 17680.

Hoepli, U. 276.

Hoesbergen, M. 17125.

8900. Hofer, A. W., and H. C. Hamilton. 1938: Further studies on seed "sterilization." Soil Sci. Soc. Am. Proc. 3:167, 168.

8901. —— and —— 1940: Bacterial contamination of seed. Soil Sci. Soc. Am. Proc. 5:264–65.

Hofer, A. W. 8071.

8902. Hoff, D. J., and H. J. Mederski. 1960: Effect of equidistant corn plant spacing on yield. Agron. J. 52:295–97.

Hoff, J. C. 16283.

Hoff, J. K. 14371.

8903. Hoffer, A., A. W. Alcock, and W. F. Geddes. 1944: The effect of variations in Canadian spring wheat on the thiamine and ash of long extraction flours. Cereal Chem. 21:210–22.

8904. ——, ——, and —— 1946: The distribution of thiamine in wheat seedlings at different stages of germination in the dark. Cereal Chem. 23:76–83. Biol. Abstr. 20:11161. 1946.

8905. Hoffman, E. J., F. G. Lum, and A. L. Pitman. 1945: Retention of carotene in alfalfa stored in atmospheres of low oxygen content. J. Agr. Res. 71:361–73.

Hoffman, G. O. 17797.

8906. Hoffman, H. 1880: Influence of thick seeding on the sex of the plant. Ber. Oberhess. Ges. Natur- Heilk. Sitzber. 19:165. Abstr. Botan. 3/4:1294. 1880.

8907. Hoffman, I. 1962: Spectrophotometric determination of nickel in wheat fractions. Analyst 87:650–52.

8908. Hoffman, I. C. 1925: The relation of size of kernels in sweet corn to evenness of maturity. J. Agr. Res. 31:1043–53.

8909. Hoffman, M. 1897: Preparation of seed to prevent its being eaten by the birds. Deut. Landwirtsch. Presse 24(82):750.

8910. Hoffman, O. L., and E. P. Sylwester. 1952: Top killing alfalfa for seed production. Down to Earth 7(4):14–15.

8911. Hoffman, R. 1936: Versuche zur Klärung des Keimverlaufs bei der Pappel. Forstwiss. Centralbl. 58:573–81. Biol. Abstr. 11:4303. 1937.

8912. Hoffman, W. D. 1948: Observations in testing Bahia grass (*Paspalum notatum*, Flügge) for germination in the Alabama laboratory. Assoc. Offic. Seed Analysts News Letter 22(4):28.

Hoffman, W. D. 13705.

8913. Hoffman, W. H., A. Zuckerman, and N. H. Grace. 1952: Canadian erucic acid oil. VIII. Component fatty acids of the oil from weed seed screenings, largely charlock. J. Am. Oil Chemists Soc. 28:522–24. Biol. Abstr. 26:26055. 1952.

8914. Hoffmann, A., and J. Kummerow. 1962: Aspectos anatomicas, morfologicas y de la fisiologia de germinación de semillas de *Maytenus boaria* Mol. Phyton 18:51–56.

Hoffmann, A. 11533.

8915. Hoffmann, E. 1933: Untersuchungen über die Wirksamen physikalischen Eigenschaften der zur Bekämpfung der Kiefernschutte gebräuchlichen Kupferkalk- und Kupfersodabrühen. Deut. Forst-Z. 48: 521–24.

8916. Hoffmann, G. M., and F. Schickedanz. 1962: Seed treatment of the China aster (*Callistephus chinensis*). Saatgut-Wirtsch. 14:218–21.

8917. Hoffmann, J. 1962: Hitherto existing provenance trails with beech. Allgem. Forstz. 17:121–23.

8918. Hoffmann, J. F. 1907: Einige Betrachtungen über die Keimfähigkeit von Getreide und anderen Samen. Wochschr. Brau. 24:255–58.

8919. Hoffmann, O. L. 1961: Breaking wild oat dormancy with gases at high pressure. Weeds 9:493.

8920. —— 1962: Chemical seed treatments as herbicidal antidotes. Weeds 10:322–23.

8921. Hoffmann, R. 1865: Beiträge zum Keimungsprocess. Landwirtsch.-Versuchssta. 7:47–54.

8922. —— 1942: Untersuchungen über die Keimung und das Jugendwachstum der Schwarz- und Walnuss. Forstwiss. Centralbl. 64:59–90. Biol. Abstr. 17:19113. 1943.

8923. Hoffmann, W. 1934: Über das Auswachsen des Getreides, speziell der Gerste. Angew. Botan. 16:396–424. Proc. Intern. Seed Testing Assoc. 7:90–91.

8924. —— 1937: Physiologische Untersuchungen an Gersten und Betrachtungen über Winterfestigkeit im Hinblick auf die Züchtung winterfester Gerstenformen. Z. Zücht. Reihe A, Pflanzenzücht. 21:277–93. Biol. Abstr. 11:18218. 1937.

8925. —— 1944: Die Vererbung der Winter-Sommer-Form und der Winterfestigkeit der Gerste. Z. Pflanzenzücht. 26:56–91. Biol. Abstr. 21:16312. 1947.

8926. Hoffmannowa, A. 1964: Some aspects of sugar nutrition of excised embryos of *Lupinus luteus* L. and *Brassica oleracea* L. Acta Soc. Botan. Polon. 33:193–210.

Hoffmaster, T. E. 12617.

8927. Hoffpauir, C. L. 1945: Determination of moisture in peanut kernels. Oil Soap 22:283–86.

8928. ——, D. H. Petty, and J. D. Guthrie. 1947: Germination and free fatty acid in individual cotton seeds. Science 106: 344–45.

8929. ——, S. E. Poe, L. U. Wiles, and M. Hicks. 1950: Germination and free fatty acids in seed stock lots of cottonseed. J. Am. Oil Chemists Soc. 27:347–48.

Hoffpauir, C. L. 13114, 13115, 15530, 17876.

8930. Hofman, J., and B. Hegar. 1962: The sowing time and pretreatment of Douglas fir seeds. Prace Vyzkum. Ustavu Lesn. CSSSR 24:37–69.

8931. Hofmann, J. V. 1918: The importance of seed characteristics in the natural reproduction of coniferous forests. Univ. Minn. Stud. Biol. Sci. 2. 25 pp.

8932. —— 1925: Laboratory tests on effect of heat on seeds of noble and silver fir, Western white pine and Douglas fir. J. Agr. Res. 31:197–99.

8933. Hofmaster, R. N., and T. J. Nugent. 1952: Sweet corn, snap bean, and lima bean seed treatments with insecticide-fungicide combinations. Peninsula Hort. Soc. Trans. 42(5):21–24.

8934. —— and —— 1956: Insecticide-fungicide seed treatments for control of seed-corn maggot and seed decay organisms. Virginia Truck Expt. Sta. Bull. 116. 31 pp.

8935. Hofmeyer, J. 1921: A note on the germination of the seed of *Elephantorrhiza burchellii*. S. African J. Nat. History 3:215–16. Botan. Abstr. 10:1825. 1922.

8936. Hofmeyr, J. D. J. 1945: Further studies of tetraploidy in *Carica papaya* L. S. African J. Sci. 41:225–30. Biol. Abstr. 20:9456. 1946.

Hofstee, B. H. J. 17481.

Hofsten, C. G. von; *see* Von Hofsten, C. G.

8937. Hogaboam, G. J. 1961: Radiographing as a method for observing some seed characters in monogerm sugar beet fruits. J. Am. Soc. Sugar Beet Technol. 11:605–9.

8938. Hogan, A. G., G. T. Gillespie, O. Koctürk, B. L. O'Dell, and L. M. Flynn. 1955: The percentage of protein in corn and its nutritional properties. J. Nutr. 57: 225–39.

Hogan, A. G. 6512.

Hogan, J. T. 13177.

8939. Hohendorf, E. 1953: The influence of electric short waves on the germination of seeds and the development of certain plants after application of rays. Roczniki Nauk Rolniczych Ser. A 66:147–48.

Hoi, B. 7001.

8940. Hojesky, J. 1921: Über hartschaliges Kleesaatgut. Z. Landwirtsch. Versuchsw. Deut. Österr. 24:101–19.

Hoji, K. 13627.

Hojito, S. 17179.

8941. Holbert, J. R., and B. Koehler. 1931: Results of seed-treatment experiments with yellow dent corn. U.S. Dept. Agr. Tech. Bull. 260. 64 pp.

Holcmanaite, M. 17884.

8942. Holden, D. J. 1952: Pod dehiscence in *Astragalus canadensis* L. [Abstract.] Proc. S. Dakota Acad. Sci. 31:143–46.

8943. Holden, H. S. 1954: The morphology of a new species of pteridosperm seed from the Yorkshire coal measures. Ann. Botany (London) 18:407–15.

8944. —— and D. Bexon. 1918: Observations on the anatomy of teratological seedlings. I. On the anatomy of some polycotylous seedlings of *Cheiranthus cheiri*. Ann. Botany (London) 32:513–30.

8945. Holdridge, L. R. 1940: A rapid method of extracting balsa seed. Caribbean Forester 1(2):25.

8946. —— and J. Marrero. 1940: Preliminary notes on the silviculture of the big-leaf mahogany. Caribbean Forester 2(1): 20–23.

8947. Holdsworth, M. 1945: A comparative study of onion varieties in relation to

bolting and yield when grown from sets. Ann. Appl. Biol. 32:22–34.
Holdt, M. M. Von. 12236.
Holland, B. R. 12559.
8948. Holland, T. H. 1927: Germination tests with Robusta and Liberian coffees. Ceylon Dept. Agr. Yearbook 1927:60.
8949. Holle, M., and L. C. Peirce. 1960: Plot technique for field evaluation of earliness, pod number, and total yield in the lima bean. Proc. Am. Soc. Hort. Sci. 76:403–8.
Holley, K. T. 8734.
8950. Hollis, L. 1951: Hybrid musk roses from seed. Natl. Rose Soc. Rose Ann. 1951:37–39.
8951. Holloway, J. T. 1937: Ovule anatomy and development and embryogeny in *Phyllocladus alpinus* (Hook.) and in *P. glaucus* (Carr.). Trans. Proc. Roy. Soc. New Zealand 67:149–65.
8952. Hollowell, E. A. 1929: Influence of atmospheric and soil moisture upon seed setting in red clover. J. Agr. Res. 39: 229–47.
8953. —— 1938: The establishment of low hop clover, *Trifolium procumbens*, as affected by time of seeding and growth of associated grass. J. Am. Soc. Agron. 30:589–98.
8954. —— 1940: The pubescent characteristic of red clover, *Trifolium pratense*, as related to the determination of origin of the seed. J. Am. Soc. Agron. 32:1–11. Expt. Sta. Rec. 83:50. 1940.
Hollowell, E. A. 2147, 18667.
8955. Hollrung, M. 1919: Die Krankhaften Zustände des Saatgutes. Kühn-Arch. 8. 352 pp.
8956. —— 1933: Die Werschätzung der Getreidesaat durch die Bewurzelungsstärke im Keimversuch. Pflanzenbau 10:17–21.
Holm, F. 13847.
Holm, L. G. 7655.
8957. Holm, T. 1897: Germination of Cyperaceae. Am. J. Sci. 1897:14–181.
8958. Holman, L. E. 1951: Artificial drying of grain crops. Crops & Soils 4(1):9–13.
8959. ——, H. J. Barre, R. T. Cotton, and H. H. Walkden. 1949: Storage of dry shelled corn in farm-type bins. U.S. Dept. Agr. Circ. 826. 36 pp.
8960. —— and D. G. Carter. 1952: Soybean storage in farm-type bins; a research report. Illinois Agr. Expt. Sta. Bull. 553:449–96.
Holman, L. E. 3651.
8961. Holman, R. T. 1948: Lipoxidase activity and fat composition of germinating soybeans. Arch. Biochem. 17:459–66.
8962. Holmberg, J., and G. Sellman. 1952: The component acids and glycerides of *Camelina sativa* seed oil and its relation to other cruciferous oils. Svensk Kem. Tidskr. 64:270–79. Biol. Abstr. 27:14199. 1953.
Holmberg, U. 8878.
Holmen, H. 12437.
8963. Holmes, A. D. 1953: Germination of seeds removed from mature and immature butternut squashes after seven months of storage. Proc. Am. Soc. Hort. Sci. 62:433–36.
8964. ——, C. T. Smith, J. W. Kuzmeski, and W. H. Lachman. 1948: Hormone treatment of field-grown tomatoes. Food Technol. 2:252–55. Brit. Abstr. B III:437. 1949.
8965. Holmes, E. 1961: Seed dressing policy. World Crops 13:459–60.
8966. Holmes, F. S. 1934: Weight of 1,000 seeds of some agricultural seeds. Maryland Agr. Expt. Sta. Bull. 362:378–79. Expt. Sta. Rec. 72:45. 1935.
8967. Holmes, G. D. 1957: Forest tree seed investigations. Gt. Brit. Forestry Comm. Rept. Forest Res. 1956:15–19.
8968. —— 1960: Report of the forest seeds committee on revision of the international rules relating to testing of tree seeds. Proc. Intern. Seed Testing Assoc. 25:673–84.
8969. —— 1960: Supplementary report of the forest seeds committee. Referee testing of *Picea sitchensis* and *Pseudotsuga taxifolia.* Proc. Intern. Seed Testing Assoc. 25: 685–705.
8970. —— and G. Buszewicz. 1955: Experiments with cold-wet pretreatment as a method of increasing the germination rate of seed of Douglas fir, Sitka spruce, and lodgepole pine. Gt. Brit. Forestry Comm. Rept. Forest Res. 1954:84–91.
8971. —— and —— 1958: The storage of seed of temperate forest tree species. Forestry Abstr. 19:313–22.
8972. —— and —— 1958: The storage of seed of temperate forest tree species. II. Forestry Abstr. 19:455–576.
8973. —— and —— 1959: The assessment of seed sowing value. Quart. J. Forestry 53:220–27. Biol. Abstr. 35:5774. 1960.
8974. ——, G. Edwards, and E. Magini. 1962: Forest Seeds Committee. Proc. Intern. Seed Testing Assoc. 27:119–210.
Holmes, G. D. 3458–3461.

8975. Holmes, L. E. 1933: Studies in the morphology and biochemistry of the pineapple. II. Reserves in the seeds of two genera of the Bromeliaceae and of various pineapple hybrids. New Phytologist 32: 382–92.

8976. Holmes, P. 1951: The arginine content of certain seed proteins. Australian J. Exptl. Biol. Med. Sci. 29:63–66. Chem. Abstr. 45:9621. 1951.

8977. —— 1953: The amino-acid composition of certain seed proteins. Australian J. Exptl. Biol. Med. Sci. 36:595–602.

8978. Holmes, R. L., J. C. Minor, and R. S. McKinney. 1953: The rate of development of acidity in stored tung seeds and kernels. J. Am. Oil Chemists Soc. 30:137–39. Biol. Abstr. 27:20647. 1953.

8979. —— and E. T. Rayner. 1958: Isolation of two nitrogen-free toxins from tung kernels. J. Am. Oil Chemists Soc. 35: 586–89.

Holmes, R. M. 14394.

Holmgren, R. C. 9196.

8980. Holmsgaard, E., and A. Kjaer. 1951: Investigation on the germination in laboratory and tree nursery of four *Abies* and two *Picea* species. Dansk Skovfor. Tidsskr. 36:203–26.

8981. Holst, M. J., and C. W. Yeatman. 1961: A provenance study in *Pinus banksiana* Lamb. Intern. Botan. Congr. Recent Advan. Botany 9:1612–16. 1959.

8982. Holt, E. C., and E. C. Bashaw. 1963: Factors affecting seed production of Dallisgrass. Texas Agr. Expt. Sta. MP662. 8 pp.

Holt, E. C. 2061, 9849.

Holt, J. 18057.

8983. Holten, A. 1920: Storing acorns two years. Dansk Skovfor. Tidsskr. 5:191–94.

Holter, H. 2842.

Holtz, H. W. 9878.

8984. Holynska, M., and E. Pawelczyk. 1960: Die Allelopathie im Verhältnis zur Keimung der Samen von Arzneipflanzen. Poznan. Panst. Inst. Nauk Leczn. Surowcow. Roslin. Byul. 6:185–91.

Holzl, J. 1841.

8985. Holzner. 1902: Die äussere Samenhaut der deutschen Drosera-Arten. Flora (Jena) 90:342–43.

8986. —— 1902: Die Caruncula der Samen von *Polygala*. Flora (Jena) 90:343–44.

Homeyer, P. G. 7079.

8987. Honda, K., and T. Kitamura. 1958: Studies on mitochondriogenesis during plant germinating period. Bull. Agr. Chem. Soc. Japan 22:316–19.

8988. Honecker, L. 1931: Untersuchungen über den Verlauf der Wasseraufnahme bei Quellung und Keimung der Getreide. Auswertung der Befunde für die Pflanzenzüchtung nach ökologischen Gesichtspunkten. Buchdruckerei Thomas & Hans Sedlmair, Freising. 88 pp.

8989. Honeywell, E. M., and H. Steenbock. 1924: The synthesis of vitamin C by germination. Am. J. Physiol. 70:322–32.

8990. Honing, J. A. 1916: The influence of light on the germination of the seed of varieties of *Nicotiana tabacum*. Bul. Deli Proefsta. Medan. 7:1–14. Expt. Sta. Rec. 38:127. 1918.

8991. —— 1916: Der Einfluss einer Warmwasserbehandlung auf das Keimprozent der Samen von *Albizzia moluccana*, *Pithecolobium* Samen, *Mimosa invisa* und *Crotalaria striata*. Bul. Deli Proefsta. Medan. 7:14–24. Botan. Centralbl. 135: 153.

8992. —— 1928: Dominanzwechsel bei der Lichtkeimung. Z. Induktive Abstammungs- Vererbungslehre Suppl. b 2:861–65. Biol. Abstr. 5:16425. 1931.

8993. —— 1930: Nucleus and plasma in the heredity of the need of light for germination in *Nicotiana* seeds. Genetica 12:441–68. Biol. Abstr. 8:17168. 1934.

8994. —— 1946: Seed germination and light. Nature 158:454.

8995. Honma, S. 1955: A technique for artificial culturing of bean embryos. Proc. Am. Soc. Hort. Sci. 65:405–8.

8996. —— and D. Denna. 1962: Moisture movement in impermeable snap bean seed. Mich. Agr. Expt. Sta. Quart. Bull. 44: 726–30.

8997. Hooker, A. L. 1951: The corn embryo as a factor in resistance in *Pythium* during germination. [Abstract.] Phytopathology 41:17.

8998. —— 1955: Additional seed factors affecting stands of corn in cold soils. Agron. J. 47:582–85.

8999. —— and J. G. Dickson. 1952: Resistance to *Pythium* manifest by excised corn embryos at low temperatures. Agron. J. 44:443–47.

9000. Hooks, J. A., and M. S. Zuber. 1963: Effect of soil and soil moisture levels on cold-test germination of corn. Agron. J. 55:453–55.

9001. Hoon, R. C., H. N. K. Nadig, and J. T. Karira. 1955: Soil conditioner evolved from tamarind (*Tamarindus indica*) seed powder. Irrig. Power 12:79–89.

Hoop, D. J. N. van der; *see* Van der Hoop, D. J. N.

9002. Hootman, J. G. 1939: Thinks old seeds are best. Horticulture 17(13):305.

9003. —— 1940: Poinsettias grown from seed. Horticulture 18(21):426.

9004. —— 1941: Germinating lupine seeds. Horticulture 19(16):370.

9005. Hooven, E. F. 1956: Field test of tetramine treated Douglas fir (*Pseudotsuga menziesii*) seed. Oregon State Bd. Forestry Res. Note 29. 11 pp.

9006. —— 1957: Field test of endrin-treated Douglas fir seed. Oregon State Bd. Forestry Res. Note 28. 8 pp.

Hoover, C. A. 2263.

Hoover, M. M. 4580.

Hoover, M. M., Jr. 3049.

9007. Hoover, M. W. 1955: Influence of maturity and storage treatments upon the ascorbic acid content of the seeds of southern peas. Food Res. 20:469–73.

9008. —— 1955: Some effects of temperature upon the growth of southern peas. Proc. Am. Soc. Hort. Sci. 66:308–14.

9009. —— 1957: Influence of maturity and storage on the seed shell-out of southern peas. Proc. Am. Soc. Hort. Sci. 70:291–96.

9010. —— and R. A. Dennison. 1953: The correlation of stages of maturity with certain physical measurements in the southern pea, *Vigna sinensis*. Proc. Am. Soc. Hort. Sci. 62:391–96.

9011. Hope, A. 1927: The dissemination of weed seeds by irrigation water in Alberta. Sci. Agr. (Ottawa) 7:268–76.

Hope, A. 15415.

9012. Hope, C., V. T. Stoutemeyer, and A. W. Close. 1941: The control of damping-off by the use of sphagnum for seed germination. Proc. Am. Soc. Hort. Sci. 39:397–406.

Hope, C. 15173, 18076.

9013. Hopf, L. 1932: The improvement of grain by proper drying. Mühlenlab. 58–59, 64–65. Chem. Abstr. 27:141. 1933.

9014. Hopkins, C. G., L. H. Smith, and E. M. Cast. 1903: The structure of the corn kernel and the composition of its different parts. Illinois Agr. Expt. Sta. Bull. 87: 77–112.

9015. Hopkins, C. Y. 1936: Thermal death point of certain weed seeds. Can. J. Res. Sect. C 14:178–83.

9016. —— 1938: A sulphur-containing substance from the seed of *Conringia orientalis*. Can. J. Res. Sect. B 16:341–44.

9017. —— and M. J. Chisholm. 1957: Fatty acids of asparagus seed oil. J. Am. Oil Chemists Soc. 34:477.

9018. —— and —— 1959: Fatty acids of Kenaf seed oil. J. Am. Oil Chemists Soc. 36:95–96.

9019. —— and —— 1960: Epoxy acid in seed oils of Malvaceae and preparation of (+) *threo*-12,13-dihydroxyoleic acid. J. Am. Oil Chemists Soc. 37:682–84.

9020. —— and —— 1961: Development of oil in the seed of *Asclepias syriaca* L. Can. J. Biochem. Physiol. 39:829–35.

9021. —— and —— 1961: Development of oil in the seed of *Helianthus annuus* L. Can. J. Biochem. Physiol. 39:1481–87.

9022. —— and —— 1962: Occurrence of octadeca-trans-10,trans-12-dienoic acid in a seed oil. Chem. Ind. (London) 49:2064.

9023. —— and —— 1964: Isolation of a natural isomer of linoleic acid from a seed oil. J. Am. Oil Chemists Soc. 41:42–44.

9024. —— and R. P. Graham. 1935: Starch content of some samples of Canadian wheat. Can. J. Res. 12:820–24.

Hopkins, C. Y. 3987–3991.

9025. Hopkins, E. F. 1924: Studies on lettuce seed germination. Proc. Assoc. Offic. Seed Analysts 16:68–69.

9026. —— 1928: Further studies of celery seed germination. Proc. Assoc. Offic. Seed Analysts 20:69–70.

9027. ——, F. J. Ramirez-Silva, V. Pagan, and A. G. Villafane. 1947: Investigations on the storage and preservation of seed in Puerto Rico. Puerto Rico Agr. Expt. Sta. Bull. 72. 47 pp.

9028. Hopkins, E. R. 1960: Germination stimulation in *Pinus pinaster* Ait. Western Australia Forests Dept. Bull. 66. 10 pp.

9029. Hopkins, F. G., and E. J. Morgan. 1943: Appearance of glutathione during the early stages of the germination of seeds. Nature 152:288–90.

9030. Hopkins, H. 1928: Seed gathering in Minnesota's pines. Am. Forests Forest Life 34:743–44.

Hopkins, H. H. 12306.

9031. Hopkins, H. T., and E. H. Toole. 1950: Effect of DDT on germination of certain seeds. Botan. Gaz. 112:130–32.

9032. Hopkins, J. C. F. 1932: Diplodia in maize. The benefits of seed treatment. Rhodesia Agr. J. 29:349–51.

9033. Hopkins, J. W. 1932: Effect of rate of seeding upon comparison of varieties of oats. Can. J. Res. 7:1–50.
9034. —— 1935: Influence of weather conditions on the nitrogen content of wheat. II. Can. J. Res. Sect. C 13:127–33.
9035. —— 1936: Effect of fertilizers on yield and malting quality of Manitoba barley. Sci. Agr. (Ottawa) 17:250–59.
9036. —— 1938: Influence of air temperature and soil moisture subsequent to flowering on the nitrogen content of wheat. Can. J. Res. Sect. C 16:135–42.
9037. —— 1940: Effect of phytohormone dust seed treatment on growth and yield of barley under greenhouse conditions. Can. J. Res. Sect. C 18:507–12.
Hopkins, J. W. 4362, 4363, 13257.
9038. Hopkins, R. H., R. H. Murray, and A. R. Lockwood. 1946: The β-amylase of barley. Biochem. J. 40:507–12.
9039. Hopp, H. 1957: Untersuchungen über die Lebensdauer von Unkrautsamen mit neueren Methoden. Arbeit Inst. Pflanzenschutz Landwirtsch. Hochschule Hohenheim. 72 pp.
9040. Hopp, R. 1947: Internal temperatures of plants. Proc. Am. Soc. Hort. Sci. 50:103–8.
Hopp, V. 16376.
9041. Hoppe, P. E. 1943: Comparison of certain mercury and non-metallic dusts for corn seed treatment. Phytopathology 33:602–6.
9042. —— 1944: Gladiolus seed treatment. Gladiolus 19:126–27.
9043. —— 1948: Seed treatment with mercury dusts injurious to corn with mechanical injuries near embryo. Phytopathology 38:82.
9044. —— 1949: Differences in *Pythium* injury to corn seedlings at high and low soil temperatures. Phytopathology 39:77–84. Commonwealth Bur. Soil Sci. (Gt. Brit.) Soil Fertilizers 12:125.
9045. —— 1951: A new technique for incubating seed corn in cold soil for disease tests. Phytopathology 41:747–51.
9046. —— 1951: A glass-tumbler-paper-doll technique for seed corn incubation and germination tests. Phytopathology 41: 856–58.
9047. —— 1955: Cold testing seed corn. Wisconsin Agr. Expt. Sta. Bull. 507. 5 pp.
9048. —— 1956: Cold testing. Am. Seed Trade Assoc. Hybrid Corn Div. Rept. Hybrid Corn Ind. Res. Conf. 11:68–74.
9049. —— 1958: Correlation between laboratory cold tests and field stands of corn. Plant Disease Reptr. 42:367–72.
9050. —— and A. H. Wright. 1955: How well is corn seed treated in commercial practice?
9051. Hopper, T. H. 1935: Report on refractometric determination of oil in seeds. J. Assoc. Offic. Agr. Chemists 18:433–34.
Hopper, T. H. 3326, 3663, 5374, 17874.
9052. Hopper, W. E. R. 1941: Seed formation, germination, and post-germination development in certain Cichorieae. Trans. Illinois State Acad. Sci. 34:70–72.
Hora, F. B. 9700.
9053. Horanic, G. E., and F. E. Gardner. 1958: Effects of sub-freezing temperatures on the viability of citrus seeds. Proc. Florida State Hort. Soc. 71:79–81.
Horavka, F. 20085.
Hori, H. 18154.
9054. Hori, T., and I. Nakamura. 1953: Respiratory enzymes and growth hormone in bean sprouts. I. Catalase and peroxidase in sprouts of *Vicia faba*. Botan. Mag. (Tokyo) 66:173–76. Chem. Abstr. 48: 8883. 1954.
9055. Hori, Y., and T. Sugiyama. 1953: Germination of vegetable seeds under various concentrations of oxygen and carbon dioxide. Hort. Assoc. Japan J. 22:72–80.
9056. —— and —— 1954: Dormancy of the seeds of leaf mustards. II. Hort. Assoc. Japan J. 22:223–29.
Horie, M. 8354.
Horike, K. 10370.
Horiuchi, H. 3930–3932.
9057. Horn, C. L. 1940: Existence of only one variety of cultivated mangosteen explained by asexually formed "seed." Science 92:237–38.
9058. —— 1943: The frequency of polyembryony in twenty varieties of mango. Proc. Am. Soc. Hort. Sci. 42:318–20.
9059. —— and J. E. N. Colon. 1942: Acid scarification of the seed of two Cuban fiber plants. J. Am. Soc. Agron. 34:1137–38.
Horn, C. W. van; *see* Van Horn, C. W.
Horn, D. H. S. 12236.
Horn, D. L. van; *see* Van Horn, D. L.
9060. Horn, M. J., C. C. Fifield, A. E. Blum, and H. W. Warren. 1958: The distribution of amino acids in wheat and certain wheat products. Cereal Chem. 35:411–21.
9061. Horn, N. L., W. F. Wilson, and M. Giamalva. 1957: Seed and insect

transmission of cucumber anthracnose. Plant Disease Reptr. 41:69–71.
Horn, N. L. 18515.
9062. Horn, P. J., and H. M. Schwartz. 1961: Kaffircorn malting and brewing studies. IX. Amino-acid composition of Kaffircorn grain and malt. J. Sci. Food Agr. 12:457–59.
9063. Horne, F. R. 1962: The O.E.C.D. herbage certification scheme and the I.S.T.A. Proc. Intern. Seed Testing Assoc. 27:324–42.
Horne, F. R. 14506.
9064. Horne, M. 1952: Seed panels. Green Thumb 9(11):22.
9065. Horner, H. T., Jr. 1963: Anatomy of reserve food tissue in some agavaceous seeds. [Abstract.] Am. J. Botany 50:622.
9066. Horner, W. H. 1937: The effect of date of sowing upon the comparative performance of two wheat varieties. Sci. Agr. (Ottawa) 17:502–14. Expt. Sta. Rec. 78:480. 1938.
9067. Horning, E. S., and A. H. K. Petrie. 1928: Über die enzymatische Funktion der Mitochondrien bei der Keimung der Getreidearten. Wochschr. Brau. 45:265–70, 285–87, 302–3. Botan. Centralbl. 158:8. 1930.
9068. Horning, E. V., and C. L. Canode. 1963: Effect of harvest method and moisture content on seed quality of smooth bromegrass. Agron. J. 55:337–40.
Horning, E. V. 3568, 3569.
9069. Horning, T. R., and M. M. Oveson. 1962: Cloddy and conventional seedbeds for wheat production in the Northwest. Agron. J. 54:229–32.
9070. Horovitz, H. 1958: Data on sugar and amino acid content of germinating barley. Acad. Rep. Populare Romine Studii Cercetari Biol. Ser. Biol. Veg. 10:187–203. Chem. Abstr. 53:15227. 1959.
9071. Horovitz-Vlasova, L. M., and N. V. Novotelnov. 1932: Obtaining fats and oils from oil seeds. Schriften Zentral. Biochem. Forschungsinst. Nahr. Genussmittelind. (Moscow) 2:294–302. Chem. Abstr. 27: 5562. 1933.
Horowitz, B. 13532.
9072. Horr, W. H. 1927: Some studies on the structure and behavior during germination of *Gymnocladus canadensis* Lam. Univ. Kansas Sci. Bull. 17:331–65.
Horricks, J. S. 8644.
9073. Horsfall, J. G., and Z. I. Kertesz. 1933: Abnormal enlargement of peas from plants affected with root-rot. N.Y. Agr. Expt. Sta. Bull. 621. 20 pp.
9074. ——, A. G. Newhall, and C. E. F. Guterman. 1934: Dusting miscellaneous seeds with red copper oxide to combat damping-off. N.Y. Agr. Expt. Sta. Bull. 643. 39 pp.
Horsfall, J. G. 10489.
Horski, J. 16534.
9075. Horstmann, H. J. 1960: Untersuchungen zum Stoffwechsel der Lungenschnecken. I. Glykolyse bei den Embryonen von *Lynnaea stagnalis* L. Z. Physiol. Chem. 319:110–19.
9076. Horton, J. S., F. C. Mounts, and J. M. Kraft. 1960: Seed germination and seedling establishment of phreatophyte species. U.S. Dept. Agr. Forest Serv. Rocky Mt. Forest Range Expt. Sta., Sta. Paper 48. 26 pp.
Horton, J. S. 2169.
9077. Horvath, M. H. 1937: A veteran's experience with rose seeds. Am. Rose Ann. 1937:37–42.
Horvath, S. 9565.
Horwitz, W. 9572.
9078. Hoshikawa, K. 1960: Observations on the embryo sac containing double egg apparatus in *Triticum aestivum* L. Botan. Mag. (Tokyo) 73:107–12.
9079. —— 1961: Studies on the ripening of wheat grain. II. Development of the endosperm tissue. III. Development of the starch grain and reserve protein particle in the endosperm. Proc. Crop. Sci. Soc. Japan 29:415–20.
9080. Hoshino, M., J. Ikeda, and F. Matsumoto. 1959: Studies on the environmental factors affecting the seed germination and seedling growth of several pasture species. II. Effect of soil moisture on the seed germination of some grasses and legumes. Proc. Crop Sci. Soc. Japan 28:92–93.
9081. ——, F. Matsumoto, and J. Ikeda. 1960: Some experiments on the storage of seeds of leguminous plants. Proc. Crop Sci. Soc. Japan 29:177–78.
9082. Hoskins, J. H., and A. T. Cross. 1946: Some basic problems in the classification of paleozoic seeds. Am. J. Botany 33:226.
9083. Hoskins, T. H. 1895: Germination of nuts. Gardens & Forests 8:448. Expt. Sta. Rec. 7:407. 1895.
9084. Hoskinson, P. E., and T. S. Osborne. 1961: Radiosensitivity of seeds. II. Effects of soaking, storage, and gamma rays. Nucl. Sci. Abstr. 15:2321.

9085. Hosner, J. F. 1957: Effects of water upon the seed germination of bottomland trees. Forest Sci. 3:67–70.
9086. —— 1959: Multi-seeded acorns. J. Forestry 57:127.
9087. ——, R. E. Dickson, and L. Kahler. 1959: Storing loblolly pine seed in polyethylene bags as a substitute for stratification. J. Forestry 57:495–96.
Hosner, J. F. 2899, 5351.
9088. Hosoda, T. 1951: Maternal inclination observed in the bolting in inter-varietal F_1 hybrids of Japanese radish. Bull. Coll. Agr. Utxonomiya Univ. 1:179–84.
Hosoda, Y. 14114.
Hostetter, A. E. 10332.
9089. Hostettler, F., and H. Deuel. 1951: Über die Aldobiosäure aus dem Samenschleim von *Plantago arenaria* Waldst. et Kit. Helv. Chim. Acta 34:2440–45. Biol. Abstr. 26:22639. 1952.
Hotchkiss, J. E. 2044.
9090. Hotta, Y. 1957: Morphogenetical studies in *Vigna sesquipedalis*. II. Roles of the cotyledon in the morphological differentiation of bean seedlings. Botan. Mag. (Tokyo) 70:383–90.
9091. Hottenroth, B., and W. Mohr. 1956: Water-soluble bitter compounds in oats. Deut. Lebensm. Rundschau 52:225–28. Chem. Abstr. 51:8214. 1957.
9092. Hotter, E. 1892: Über die Vorgänge bei der Nachreife von Weizen. Landwirtsch. Versuchssta. 40:356–64.
9093. —— 1935: Physiology of seed germination. [Abstract.] Trans. Illinois State Acad. Sci. 28:49.
9094. —— and W. A. Huelsen. 1927: The determination of quality in sweet corn seeds by means of the optical measurements of leached materials. J. Agr. Res. 35: 147–66.
9095. —— and H. K. Wilson. 1930: Resistance of wheat to high temperatures under different methods of heating. J. Am. Soc. Agron. 22:108–12.
Hottes, C. F. 19731.
9096. Hou, H. C., W. H. Riesen, and C. A. Elvehjem. 1949: Influence of heating on the liberation of certain amino acids from whole soybeans. Proc. Soc. Exptl. Biol. Med. 70:416–19.
Hou, K. 584, 585.
Houff, W. H. 8838.
9097. Houget, J. 1942: La transformation des lipides en glucides lors de la germination du *Ricin*. I. La formation des glucides. II. Mécanisme de l'attaque des acides gras. Bull. Soc. Chim. Biol. 24:346–52, 352–57.
9098. —— 1943: The mechanism of the transformation of lipides into glucides during germination of *Ricinus* seeds. Compt. Rend. 216:821–22. Chem. Abstr. 38:4647. 1944.
9099. Hough, A. F. 1952: Preliminary results of red pine (*Pinus resinosa*) seed-source tests in northwestern Pennsylvania. U.S. Dept. Agr. Forest Serv. Northeast. Forest Expt. Sta., Sta. Paper 49. 28 pp.
9100. —— 1952: Preliminary results of red pine (*Pinus resinosa*) seed source, seed weight, seedling weight, and height growth, in Kane Test Plantation. U.S. Dept. Agr. Forest Serv. Northeast. Forest Expt. Sta., Sta. Paper 50. 14 pp.
Hough, L. 629, 630.
Hough, L. F. 3106, 18477, 19411, 19430.
9101. Houghton, K. W. 1939: Growing gloxinias from seed. Horticulture 17(8): 205.
9102. —— 1940: Growing primroses from seed. Horticulture 18(8):204.
9103. —— 1953: Japanese iris from seed. Horticulture 31(3):222–23.
9104. Houk, W. G. 1936: The ovule and seed of *Coffea arabica* L. Science 83: 464–65.
9105. —— 1938: Endosperm and perisperm of coffee with notes on the morphology of the ovule and seed development. Am. J. Botany 25:56–61.
9106. Houseman, E. E., C. R. Weber, and W. T. Federer. 1946: Pre-harvest sampling of soybeans for yield and quality. Iowa Agr. Expt. Sta. Res. Bull. 341:805–26.
9107. Houser, H. A. 1940: Methods of raising garden lilies from seed. Am. Lily Yearbook 1940:20–26.
9108. Housley, S., and N. M. Griffiths. 1962: The occurrence of 3-indolylacetic acid in seeds and shoots of pea. Phyton 19:85–93.
Housley, S. 2292, 3080, 6240.
Houston, B. R. 6958, 7386, 15242, 18166.
9109. Houston, D. F., R. E. Ferrel, I. R. Hunter, and E. B. Kester. 1959: Preservation of rough rice by cold storage. Cereal Chem. 36:103–7.
9110. ——, I. R. Hunter, and E. B. Kester. 1951: Effect of steaming fresh paddy rice on the development of free fatty acids during storage of brown rice. Cereal Chem. 28:394–99.

9111. ——, R. P. Straka, I. R. Hunter, R. L. Roberts, and E. B. Kester. 1957: Changes in rough rice of different moisture content during storage at controlled temperatures. Cereal Chem. 34:444–56.

Houston, D. F. 9230.

9112. Hove, E. L. 1944: Gossypol as a carotene-protecting antioxidant, in vivo and in vitro. J. Biol. Chem. 156:633–42.

9113. —— and C. G. Harrel. 1943: The nutritive value of wheat-germ protein. Cereal Chem. 20:141–48. Chem. Abstr. 37:3130–31. 1943.

9114. Hoveland, C. S. 1964: Germination and seedling vigor of clovers as affected by grass root extracts. Crop Sci. 4:211–13.

9115. Howard, H. W. 1939: The size of seeds in diploid and autotetraploid *Brassica oleracea* L. J. Genet. 38:325–40.

9116. —— 1942: The effect of polyploidy and hybridity on seed size in crosses between *Brassica chinensis*, *B. carinata*, amphidiploid *B. chinensis-carinata*, and autotetraploid *B. chinensis*. J. Genet. 43:105–19.

9117. —— 1947: Seed size in crosses between diploid and autotetraploid *Nasturtium officinale* and allotetraploid *N. uniseriatum*. J. Genet. 48:111–18.

9118. —— and A. G. Lyon. 1951: Effect of light on the germination of watercress seeds. Nature 168:253–54.

9119. Howard, W. L. 1915: An experimental study of the rest period in plants. IV. Seeds. Missouri Agr. Expt. Sta. Res. Bull. 17. 58 pp.

Howard, W. L. 5514.

9120. Howarth, G. 1954: Comparison between the germination of peas in the fields and in the laboratory. Tidsskr. Planteavl 57:251–70.

9121. Howe, M. 1947: Proteolytic enzymes of milled fractions of wheat. Bakers Dig. 21:57–58, 61. Biol. Abstr. 21:25770. 1947.

Howe, M. A. 18158, 18159.

Howe, O. W. 9777.

9122. Howe, R. W. 1962: A study of the heating of stored grain caused by insects. Ann. Appl. Biol. 50:137–58.

9123. —— and T. A. Oxley. 1944: The use of carbon dioxide production as a measure of infestation of grain by insects. Bull. Entomol. Res. 35:11–22.

Howe, R. W. 14919.

9124. Howe, T. D. 1963: Development of the embryo sac in *Gutierrezia sarothrae*. [Abstract.] Am. J. Botany 50:615.

9125. Howe, V. K. 1962: Culture of excised embryos of *Pinus ponderosa*. Proc. Iowa Acad. Sci. 69:137–42.

9126. Howe, W. L., W. T. Schroeder, and K. G. Swenson. 1952: Seed treatment for control of seed-corn maggot and seed decay organisms. N.Y. Agr. Expt. Sta. Bull. 752. 34 pp.

9127. Howell, D. W. 1958: Gibberellin raises bean pods for harvest. Crops & Soils 11(3):18.

9128. Howell, R. W. 1960: Physiology of the soybean. Advan. Agron. 12:265–310.

9129. —— 1961: Changes in metabolic characteristics of mitochondria from soybean cotyledons during germination. Physiol. Plantarum 14:89–97.

9130. —— and J. L. Cartter. 1953: Physiological factors affecting composition of soybeans. I. Correlation of temperatures during certain portions of the pod filling stage with oil percentage in mature beans. Agron. J. 45:526–28.

9131. —— and —— 1958: Physiological factors affecting composition of soybeans. II. Responses of oil and other constituents of soybeans to temperature under controlled conditions. Agron. J. 50:664.

9132. ——, F. I. Collins, and V. E. Sedgwick. 1959: Respiration of soybean seeds as related to weathering losses during ripening. Agron. J. 51:677–79.

9133. —— and F. Skoog. 1955: Effect of adenine and other substances on growth of excised *Pisum* epicotyls cultured in vitro. Am. J. Botany 42:356–60.

9134. ——, C. J. Wargel, C. A. Brim, E. E. Hartwig, J. W. Lambert, J. R. Thompson, B. F. Stefansson, J. K. Park, W. E. Seigler, and B. K. Webb. 1960: Response of soybeans to seed-treatment with gibberellin under simulated commercial conditions. Agron. J. 52:144–46.

Howell, R. W. 2468, 4303, 10134, 14714, 14715.

Howerton, H. K. 3086.

9135. Howland, J. E. 1944: Preliminary studies on low-temperature vernalization of column stocks, *Mathiola incana*. Proc. Am. Soc. Hort. Sci. 44:518–20.

Howlett, F. S. 267.

9136. Hoyko, V. 1938: The influence of the wheat grain integuments on its germination and its ripening after harvesting. Acad. Sci. Ukr. Inst. Botan. Symp. Dedicated to the Memory of V. N. Lubimenko, p. 408.

9137. Hoyt, G. M. 1960: Seed germination made simple. Am. Orchid Soc. Bull. 29:597–99.

9138. —— 1960: Seed germination made simple. (Cont'd.) Am. Orchid Soc. Bull. 29:678–80.

Hoyt, P. B. 7822.

9139. Hoyt, W. D. 1928: The periodic fruiting of *Dictyota*—an acquired character. Am. Naturalist 62:546–53.

Hozyo, Y. 13635, 13636.

9140. Hren, J. A. 1939: Raising of lupines from seed. Horticulture 17(7):174.

Hrnciar, G. 3936.

9141. Hrodzinskyi, A. M., N. I. Mokhova, L. D. Pylypenkoyurchak, and T. M. Filippovych. 1962: Inhibiting substances in field crop residues and weeds. I. Effect of water-soluble inhibitors on the germination of seeds and plant growth. Ukr. Botan. Zh. 19(6):30–38. Biol. Abstr. 43:20605. 1963.

9142. Hrsel, I. 1958: The influence of different physiological conditions on the cytoplasm of germinating rice. Comparison of the cytoplasmatic structures of plant and animal cells. Rozpravy Česk. Akad. Věd 68(10):1–67. Biol. Abstr. 34:8647. 1959.

9143. Hruby, K. 1938: Embryo sac development in *Erythronium dens canis*. Chronica Botan. 4:20–21.

9144. —— 1954: Growing of interspecific *Prunus* hybrids from excised embryos. Congr. Intern. Botan. Paris Rappt. Commun. 8:210–11.

9145. Hsi, C. H. 1956: Toxicity of mercurial seed disinfectants to wheat. Plant Disease Reptr. 40:1065–70.

9146. Hsi, D. C. H., and H. M. Reeder. 1953: Dormancy of uplant and American-Egyptian Cottonseed. Agron. J. 45:454.

Hsieh, K. M. 17300.

9147. Hsieh, S. C., and T. M. Chang. 1962: Genic analysis for purple pericarps and other characters in rice. Agr. Assoc. China N.S. 40:14–39.

9148. Hsu, C. F. 1932: A chemical study of the seed of *Leonurus sibiricus*, L. (I-maotsao) and the composition of its oil. Acad. Sinica Mem. Natl. Res. Inst. Chem. 8:1–32. Chem. Abstr. 27:3287. 1933.

9149. —— 1934: The alkaloid of the seed of *Leonurus sibiricus* L. (I-mao-tsao). J. Chinese Chem. Soc. 2:337–39. Chem. Abstr. 29:2659. 1935.

9150. Hsueh, Y. L., H. H. Liu, and T. E. Huang. 1960: The growth response of rice seedlings to seed-treatment with $NaHCO_3$ as influenced by concentration temperature and the physiological state of the seeds. Acta Agr. Sinica 11:48–56. Biol. Abstr. 35:54061. 1960.

9151. —— and C. H. Lou. 1947: Effects of 2,4-D on seed germination and respiration. Science 105:283–85.

Hsueh, Y. L. 12443.

9152. Hu, C. C. 1953: A review of literature on *Citrus* nucellar embryony, and some new data on old and young lines of Valencia orange. Taiwan Univ. Mem. Coll. Agr. 3:20–32.

Hu, J. H. 18272.

9153. Hu, S. Y. 1960: Studies in the polyembryony of *Hosta caerulea* Tratt. I. Observations on the embryogeny and the origin of polyembryony. Acta Botan. Sinica 9:147–52.

Huang, C. C. 7974.

9154. Huang, H. S. 1946: Variations in yield of wheat as affected by depth of soil. Soil Sci. 62:227–31.

Huang, R. C. 19016, 20002.

Huang, S. M. 11989.

Huang, T. E. 9150.

9155. Huart, A. 1957: Absorption of oxygen by seeds of Scotch pine (*Pinus sylvestris*) from various sources. Agricultura (Louvain) 5:3–13. Chem. Abstr. 51:9804. 1957.

Huart, R. 11970.

9156. Hubac, C. 1958: Variation de l'aptitude à la survie après déssèchement, au cours de la germination des graines de rutabaga. Compt. Rend. 246:2649–51.

9157. —— 1961: Adaptability of embryos and seedlings to anhydrobiosis and its natural and experimental variants. U.N. Educ. Sci. Cult. Organ. Arid Zone Res. 16: 271–74.

9158. —— 1962: Recherche de substances inhibitrices dans les caryopses immature de blé "Fylgia." Compt. Rend. 255:758–60.

9159. —— 1963: Recherche de substances hydrosolubles inhibitrices la germination dans les caryopses immatures de blé "Fylgia." Compt. Rend. 257:3204–7.

9160. Hubbard, F. C. 1936: Tigridias and lilies from seed. Flower Grower 23(11): 586.

9161. Hubbard, J. E., F. R. Earle, and F. R. Senti. 1957: Moisture relations in wheat and corn. Cereal Chem. 34:422–33.

9162. —— and E. R. Leng. 1955: Leaf number in mature embryos of inbred lines of dent maize. Agron. J. 47:40–42.

Hubbard, J. E. 4769, 5743, 13525, 19798.
Hubbard, M. M. 3550.
9163. Hubbard, R. L. 1958: Hot water bath and thiourea break dormancy of wedgeleaf *Ceanothus* seed. U.S. Dept. Agr. Forest Serv. Calif. Forest Range Expt. Sta. Forest Res. Notes 143. 4 pp.
9164. —— and B. O. Pearson. 1958: Germination of thiourea-treated butterbrush (*Purshia tridentata*) seed in the field. U.S. Dept. Agr. Forest Serv. Calif. Forest Range Expt. Sta. Forest Res. Notes 138. 6 pp.
9165. Hubbard, V. C. 1941: Irregular germination of wheat in a dry soil. J. Am. Soc. Agron. 33:577–79.
9166. —— and H. J. Harper. 1949: Effect of clipping small grains on composition and yield of forage and grain. J. Agron. 41:85–92.
9167. Hubbell, R. B., H. B. Vickery, and L. S. Nolan. 1943: Cucurbit-seed globulins. II. Use as substitutes for edestin in experimental diets. J. Nutr. 25:99–101. Chem. Abstr. 40:2507–8. 1946.
Hubbell, R. B. 19106.
Huber, A. 9250.
9168. Huber, B., and G. Merkenschlager. 1951: Über die Wirkung kleinster Saugkräfte auf die Samenkeimung. Planta 40:112–20. Biol. Abstr. 27:17890. 1953.
9169. Huber, G. A., and C. J. Gould. 1949: Cabbage seed treatment. Phytopathology 39:869–75.
9170. Huber, H. 1951: Über den Einfluss der Belichtung auf die Wuchsstoffempfindlichkeit der Keimstengel von *Cucumis sativus* L. Inaug. Dissertation Bern. Buchdruckerei Büchler & Co., pp. 499–539.
9171. Huber, J. U. 1928: Schlüssel zur Bestimmung der Früchte und Samen der wichtigsten Wiesenpflanzen. Datterer, Freising-Munich. 45 pp. Biol. Abstr. 5:28988. 1931.
Huber, L. L. 1525, 8048.
9172. Huber, R. E., and S. Zalik. 1963: Lipid and protein in germinating and developing flaxseed, *Linum usitatissimum* L. Can. J. Biochem. Physiol. 41:745–54.
9173. Huber, W. 1954: Praktische Fragen des Kleesamenbaues. Mitt. Arbgemeinsch. Ford. des Futterbaues 44:11–17.
9174. Hubert, B., and J. Maton. 1939: Parthenocarpie en Groeistof. Natuurw. Tijdschr. (Ghent) 21:339–48.
Huddleston, J. H. 6318.
9175. Hudson, A. W., J. W. Woodcock, and B. W. Doak. 1937: The effect of some phosphatic fertilizers and superphosphate-lime mixtures on turnip-seed germination. New Zealand J. Sci. Technol. 18:739–49.
9176. Hudson, H. G. 1941: Population studies with wheat. III. Seed rates in nursery trials and field plots. J. Agr. Sci. 31:138–44.
9177. Hudson, J. P. 1947: The growing of raspberries from suckers and seeds. New Zealand J. Agr. 75:449–50.
Hudson, P. C. B. 4997.
9178. Hudson, P. S. 1936: Vernalization in agricultural practice. Gt. Brit. Min. Agr. J. 43:536–43. Expt. Sta. Rec. 76:301. 1937.
Hudson, P. S. 19640.
Hudspeth, E. B., Jr. 1367.
9179. Hueber, F. 1929: Untersuchungen über die Saugkraft verschiedener Roggen- und Weizensorten. Fortschr. Landwirtsch. 4:97–100.
9180. Hübner, R. 1938: Untersuchungen über die Hartschaligkeit der Zottelwicke und ihre Behebung auf züchterischem Wege. Landwirtsch. Jahrb. 85:751–89.
9181. Hueck, H. J., and G. Vermeulen. 1949: The sprouting damage of brewing barley. Landbouwk. Tijdschr. 61:227–32. Chem. Abstr. 43:5091. 1949.
9182. Huelsen, W. A. 1954: Sweet corn. Interscience, New York. 409 pp.
9183. —— and W. P. Bemis. 1958: Changes in popcorn kernels and cobs while maturing. Illinois Agr. Expt. Sta. Bull. 625. 59 pp.
9184. —— and W. N. Brown. 1952: Physical damage to sweet corn seed caused by mechanical harvesting and subsequent processing. Illinois Agr. Expt. Sta. Bull. 561. 45 pp.
9185. —— and —— 1960: Experiments in artificial drying of sweet corn seed. Illinois Agr. Expt. Sta. Bull. 656. 48 pp.
Huelsen, W. A. 2266, 9094.
Huet, G. 4491.
Huey, E. 11893.
Huey, G. 7451.
9186. Hugard, J., and R. Saunier. 1961: The effect of 2,4,5-TP on early ripening, coloration, size, and seed germination in peaches. Ann. Amelioration Plantes 11: 5–24. Chem. Abstr. 56:3841. 1962.
9187. Hughes, H. D., J. L. Robinson, and A. A. Bryan. 1931: Ear and kernel characteristics of seed corn in relation to

yield. Iowa Agr. Expt. Sta. Rept. Agr. Res. (Ann. Rept.) 1930/31:65–66.

Hughes, H. D. 9468.

Hughes, J. H. 18221.

9188. Hughes, R. 1958: The agronomic characteristics of lines of maize from different sources. W. African Sci. Assoc. J. 4:117–27. Biol. Abstr. 35:5563. 1960.

9189. —— 1958: The agronomic characteristics of lines of millet (*Pennisetum* spp.) from different sources. W. African Sci. Assoc. J. 4:128–40. Biol. Abstr. 35:5564. 1960.

Hujishita, M. 9390.

9190. Hukill, W. V. 1963: Storage of seeds. Proc. Intern. Seed Testing Assoc. 28:871–83.

Hukill, W. V. 16965.

Hukuhara, N. 10324.

9191. Hulbert, H. W., and G. M. Whitney. 1934: Effect of seed injury upon the germination of *Pisum sativum*. J. Am. Soc. Agron. 26:876–84.

Hulbert, H. W. 1490.

9192. Hulett, G. K., and G. W. Tomanek. 1961: Effects of some environmental factors on germination of salt cedar (*Tamarix pentandra* Pall.). Trans. Kansas Acad. Sci. 64:96–101. Biol. Abstr. 36: 69841. 1961.

9193. Hulewiczowa, T. 1954: Research on the influence of external agents on the development of lupine. II. The influence of vernalization applied at various dates of seeding on the development and yields of lupine. Roczniki Nauk Rolniczych Ser. A 68:555–57.

Hull, A. C. 2610.

9194. Hull, A. C., Jr. 1960: Seeding coated seed pellets on rangelands of southern Idaho. Dissertation Abstr. 20:2999. Biol. Abstr. 35:39271. 1960.

9195. —— 1960: Winter germination of intermediate wheatgrass on mountain lands. J. Range Management 13:257–60. Biol. Abstr. 36:11933. 1961.

9196. ——, R. C. Holmgren, W. H. Berry, and J. A. Wagner. 1963: Pellet seeding on western rangelands. U.S. Dept. Agr. Res. Serv. & Forest Serv. in coop. with Bur. Indian Affairs & Bur. Land Management U.S. Dept. Interior Misc. Publ. 922. 34 pp.

9197. —— and W. M. Johnson. 1955: Range seeding in the Ponderosa pine zone in Colorado. U.S. Dept. Agr. Circ. 953. 40 pp.

Hull. A. C., Jr. 3629.

9198. Hull, E. D. 1935: The orchid embryo. Am. Botanist 41:112–13.

9199. Hull, F. H. 1937: Inheritance of rest period of seeds and certain other characters in the peanut. Florida Agr. Expt. Sta. Bull. 314. 46 pp.

9200. —— and E. F. Grossman. 1932: Heat-induced chlorophyll mutations in maize. Preliminary report on chlorophyll deficiencies induced by heating dormant seeds. J. Heredity 23:123–27.

Hull, W. W. 14738.

9201. Hulshoff, A. J. A., and D. L. J. Dijkstra. 1961: Warmwaterbehandelingsmethoden ter bestrijding van stuifbrand bij tarwe en gerst. Zaaizaadonderzoek 2:136–51.

Hulton, H. F. E. 1727, 1728.

9202. Huluta, C., C. Iordache, H. D. Schnell, O. Constantinescu, and V. Benea. 1961: The sprouting of seeds and the metabolism of young plants of *Fraxinus excelsior*, whose seeds were treated with solutions of radioactive P^{32}. Acad. Rep. Populare Romine Studii Cercetari Biochim. 4:527–36. Chem. Abstr. 57:3815. 1962.

9203. Hulyalkar, R. K., T. R. Ingle, and B. V. Bhide. 1956: Studies in carbohydrates. VI. Composition of the mucilage from seeds of *Mimosa pudica*. J. Indian Chem. Soc. 33:864–66. Biol. Abstr. 32:28553. 1958.

9204. Humbert, J. 1938: Développement et structure du tégument de la graine de la linaire cymbalaire (*Linaria cymbalaria* Mill.). Rev. Gen. Botan. 50:309–33.

9205. Hume, A. N., and C. J. Franzke. 1933: The germination of seed corn and its relation to the occurrence of molds during germination. S. Dakota Agr. Expt. Sta. Bull. 275. 19 pp.

9206. Hume, E. P., and M. Cobin. 1946: Relation of seed size to germination and early growth of mangosteen. Proc. Am. Soc. Hort. Sci. 48:298–302.

Hume, E. P. 417.

9207. Hume, P. F. 1955: Storage of tea seed. Tea Quart. 26:93–95. Biol. Abstr. 32: 2697. 1958.

Hummel, B. 15244.

9208. Hummel, B. C. W., L. S. Cuendet, C. M. Christensen, and W. F. Geddes. 1954: Grain storage studies. XIII. Comparative changes in respiration, viability, and chemical composition of mold-free and mold-contaminated wheat upon storage. Cereal Chem. 31:143–50.

9209. Hummel, D. 1930: Aus der Biologie des Samentragens der Waldbäume. Z. Forst.-Jagdw. 62:365–71.
Hummer, R. W. 13899.

9210. Humphrey, J. E. 1896: On the development of the seed in the Scitamineae. Ann. Botany (London) 10(37):1–40.
Humphrey, R. R. 518.
Humphreys, M. E. 14787.

9211. Humphries, E. C. 1960: Effects of indoleacetic acid, naphthaleneacetic acid, and kinetin on phosphorus fractions in hypocotyls of dwarf bean (*Phaseolus vulgaris*). Ann. Botany (London) 24:311–16. Biol. Abstr. 36:69628. 1961.
Humphries, K. P. 9946.

9212. Humphries, W. R., and W. M. Hurst. 1935: Moisture changes in some agricultural products due to atmospheric conditions. Agr. Eng. 16:8–12.
Humphries, W. R. 9246.

9213. Hung, L. B. 1958: Preliminary study on the seed of teak. Taiwan Forestry Res. Inst. Bull. 59. 10 pp.
Hunger, A. 13878.

9214. Hunsperger, O., and K. Leupin. 1944: Contribution to the identification of *Strophanthus* seeds in filtered ultraviolet light. Pharm. Acta Helv. 19:355–58. Pharm. Abstr. 12:223. 1946.

9215. Hunt, C. H., L. Ditzler, and R. M. Bethke. 1947: Niacin and pantothenic acid content of corn hybrids. Cereal Chem. 24:355–63.

9216. ——, L. D. Rodriguez, and R. M. Bethke. 1950: The environmental and agronomical factors influencing the thiamine, riboflavin, niacin, and pantothenic acid content of wheat, corn, and oats. Cereal Chem. 27:79–96.

9217. ——, ——, and —— 1950: The niacin and pantothenic acid content of certain Ohio corn hybrids. Cereal Chem. 27:149–56.

9218. ——, ——, and —— 1950: The effect of maturity on the niacin and pantothenic acid content of the stalks and leaves, tassels, and grain of four sweet corn varieties. Cereal Chem. 27:157–61.
Hunt, C. H. 5399.

9219. Hunt, C. M. 1962: A White pine seed orchard layout. Northeast. Forest Tree Improvement Conf. 9:58–63.

9220. Hunt, H. E. 1958/59: Merchandising certified seed. Intern. Crop Improvement Assoc. Ann. Rept. 40/41:293–97.
Hunt, L. E. 5339.

9221. Hunt, O. J. 1961: Low-temperature germination, a possible strain response of Tall wheatgrass, *Agropyron elongatum* (Host) Beauv. Agron. J. 53:277.

9222. Hunt, T. F. 1890: The vitality of Kentucky blue grass seed, *Poa pratensis*. Agr. Sci. 4:4–7.

9223. Hunt, W. H., and M. H. Neustadt. 1953: An infrared, forced-draft, drying unit. Chem. Anal. 42:72.

9224. ——, ——, J. R. Hart, and L. Zeleny. 1952: A rapid dielectric method for determining the oil content of soybeans. J. Am. Oil Chemists Soc. 29:258–61. Chem. Abstr. 46:8394. 1952.
Hunt, W. H. 9886.

9225. Hunter, A. S., C. J. Gerard, H. M. Waddoups, W. E. Hall, H. E. Cushman, and L. A. Alban. 1958: The effect of nitrogen fertilizers on the relationship between increases in yields and protein content of pastry-type wheats. Agron. J. 50:311–14.

9226. Hunter, A. W. S. 1941: The experimental induction of parthenocarpic strawberries. Can. J. Res. Sect. C 19:413–19.
Hunter, G. 18808.

9227. Hunter, H. 1946: The drying of grain for seed purposes. J. Roy. Agr. Soc. Engl. 107:134–41. Biol. Abstr. 21:25756. 1947.

9228. —— 1952: The barley crop. Crosby Lockwood & Son, Ltd., London. 187 pp.

9229. —— 1953: Identification of barley varieties grown in the British Isles by examination of threshed grain. Wallerstein Lab. Commun. 16(55):327–37. Biol. Abstr. 28:21847. 1953.

9230. Hunter, I. R., D. F. Houston, and E. B. Kester. 1951: Development of free fatty acids during storage of brown (husked) rice. Cereal Chem. 28:232–39.
Hunter, I. R. 9109–9111.
Hunter, J. J. 16243.

9231. Hunter, J. H., and H. E. Hammar. 1952: Influence of cultivation, mulching, and fertilizers on chemical composition of pecan leaves and their relation to yield and quality of nuts. Soil Sci. Soc. Am. Proc. 16:346–49.
Hunter, J. H. 8081, 12186.

9232. Hunter, J. R. 1959: Germination in *Theobroma cacao*. Cacao Columbia 4(4):1–8.

9233. —— 1961: A comparison of four methods for propagating *Theobroma cacao*. Proc. Am. Soc. Hort. Sci. Caribbean Region 9:94–99.

9234. —— and S. T. Dexter. 1951: Some experiments with seed processing for faster germination of sugar beet seeds (*Beta vulgaris* L.). Proc. Am. Soc. Sugar Beet Technol. East U.S. & Can. Region. Mtg. 6:39–44.

9235. —— and A. E. Erickson. 1952: Relation of seed germination to soil moisture tension. Agron. J. 44:107–9.

Hunter, J. R. 2789.

Hunter, R. 18214.

9236. Hunter, R. E., and L. A. Brinkerhoff. 1964: Longevity of *Xanthomonas malvacearum* on and in cotton seed. Phytopathology 54:617.

Hunter, R. E. 3067.

Hunter, R. F. 5136.

9237. Hunziker, A. T. 1943: Las especies alimenticias de *Amaranthus y Chenopodium* cultivadas por los indios de America. Rev. Arg. Agron. 10:297–354. Biol. Abstr. 19:17206. 1945.

9238. Huppert, V., and A. Buchner. 1951: Stickstoffdüngung und Eiweissgehalt beim Getreide. Z. Pflanzenbau Pflanzenschutz 2:115–22.

9239. Huratado, C. A. 1960: Metodo de conservación de semilla de cacao. Vita Agr. (Lima) 37:653, 655, 657–58.

9240. Hurd, A. M. 1920: Injury to seed wheat resulting from drying after disinfection with formaldehyde. J. Agr. Res. 20:209–44.

9241. —— 1921: Seed coat injury and viability of seeds of wheat and barley as factors in susceptibility to molds and fungicides. J. Agr. Res. 21:99–122.

9242. Hurd, E. A. 1959: Inheritance of blue kernel colour in wheat. Can. J. Plant Sci. 39:1–8.

9243. —— 1961: Induction of seed set in dwarf-spring and winter wheat. Can. J. Plant Sci. 41:623–30.

9244. Hurd, R. G. 1964: The effect of restricted germination during vernalization on first leaf length of winter rye. J. Exptl. Botany 15:381–90.

Hurich, J. 14616, 15736, 15738.

Hurlburt, W. C. 19272.

9245. Hurst, C. C. 1931: Embryo-sac formation in diploid and polyploid species of Roseae. Proc. Roy. Soc. (London) Ser. B 109:126–48.

Hurst, T. L. 6591, 13496.

9246. Hurst, W. M., and W. R. Humphries. 1936: An absorptive agent for drying grain. Agr. Eng. 17:62. Expt. Sta. Rec. 74:864 1936.

Hurst, W. M. 9212.

9247. Hurtado, C. A. 1960: A note on the storage of cacao seeds. Inter-Am. Cacao Conf. Proc. 8:116–20.

9248. Hurwitz, C., and K. C. Beeson. 1944: Cobalt content of some food plants. Food Res. 9:348–57. Biol. Abstr. 19:5235. 1945.

9249. Hurwitz, S., and H. Gavrielit-Gelmond. 1952: Hardening of sweet lupine seeds. Israel Agr. Res. Sta. Ktavim 2/3:57–64.

9250. ——, D. Lachover, and A. Huber. 1951/52: Farm manures, soils, and seeds of cultivated crops as sources of weed seed infestation under subtropical conditions. Proc. Intern. Seed Testing Assoc. 17:39–43.

Hurwitz, S. 7057.

9251. Husain, A. 1960: Effect of certain fungicides on germination and growth of rape and mustard. Indian Oilseeds J. 4:174–75.

9252. —— 1960: Effectiveness of certain fungicides and gibberellic acid as seed treatment for pearl millet. Plant Disease Reptr. 44:444–45.

Husain, A. 16713.

9253. Huss, E. 1951: On dewinging injuries to forest seeds and its effect on the resulting plant. Skogen (Stockholm) 38:68–71.

9254. —— 1951: On changes in viability because of storage of cones and seeds. Skogen (Stockholm) 38:151–54.

9255. —— 1951: Procedure in seed testing at the Forest Research Institute. Medd. Statens Skogsforskningsinst. 40(6):1–82.

9256. —— 1953: On the handling of cones and seeds of conifers. Svenska Skogsvardsforen. Tidskr. 51:230–41.

9257. —— 1954: Studies of the importance of water content for the quality of conifer seed during storage. Medd. Statens Skogsforskningsinst. 44(7):1–60.

9258. —— 1956: On the quality of forest tree seed and other factors affecting the sowing result. Medd. Statens Skogsforskningsinst. 46(9):55–59.

9259. —— 1957: Dewinging of forest seed in modern seed extraction equipment. Svenska Skogsvardsforen. Tidskr. 55:359–74.

9260. —— 1963. Viability of Scots pine and Norway spruce seed, 1962. Skogen (Stockholm) 50:18–19.

9261. Huss, M. 1890: Über Quellungsfahigkeit von Leguminosensamen und

Mittel zu deren Abhilfe. Inaug. Dissertation, Halle. Jahresber. Agr.-Chem. 14:245–65.

9262. Hussain, A., and A. Saleem. 1957: Cotton-seed pigments and the toxicity of gossypol. A review. Agr. Pakistan 8:35–43. Chem. Abstr. 52:9317. 1958.

9263. ——, A. Wahhab, and A. Saleem. 1955: Some studies in the protein quality of cottonseed and cottonseed cake. Texas J. Sci. 7:227–32.

Hussain, A. 19274.

9264. Hussey, G. 1958: An analysis of the factors controlling the germination of the seed of the oil palm, *Elaeis guineensis* (Jacq.). Ann. Botany (London) 22:259–84.

9265. —— 1959: The germination of oil palm seed: experiments with *Tenera* nuts and kernels. W. African Inst. Oil Palm Res. J. 2:331–54.

9266. Hutchence, P. 1957: Raising biennials and perennials from seed. Gardeners Chron. 141:424–25.

9267. Hutcheson, T. B., and R. P. Cocke. 1941: Effects of boron on yield and duration of alfalfa. Virginia Agr. Expt. Sta. Bull. 336. 9 pp. Expt. Sta. Rec. 86:616–17.

9268. Hutchings, S. S. 1932: Light in relation to the seed germination of *Mimulus ringens*, L. Am. J. Botany 19:632–43.

9269. Hutchins, L. M. 1926: Studies on the oxygen-supplying power of the soil together with quantitative observations on the oxygen-supplying power requisite for seed germination. Plant Physiol. 1:95–150.

9270. Hutchins, R. E. 1953: Flight secrets from a jungle seed (*Zanonia*). Nat. History 62:416–19.

Hutchinson, A. H. 6219.

9271. Hutchinson, J. B. 1944: The drying of wheat. III. The effect of temperature on germination capacity. Soc. Chem. Ind. J. 63:104–7.

9272. —— and R. G. Booth. 1946: The drying of wheat. IV. Phosphatase activity as an index of heat damage in cereals. Soc. Chem. Ind. J. 65:235–37.

9273. ——, E. N. Greer, and C. C. Brett. 1948: Resistance of wheat to sprouting in the ear: preliminary investigations. Empire J. Exptl. Agr. 16:23–32.

9274. ——, ——, and P. T. Thomas. 1946: Heat damage in cereal seeds. Nature 158:120–21.

9275. ——, N. L. Kent, and H. F. Martin. 1953: The kernel content of oats. Comparisons of kernel content and thousand kernel weight for winter and spring varieties. Natl. Inst. Agr. Botany J. 6:443–53.

9276. ——, ——, and —— 1954: Oat kernels with decayed embryos: A factor in the germination capacity of oats. Natl. Inst. Agr. Botany J. 7:24–32.

9277. —— and H. F. Martin. 1955: The chemical composition of oats. I. The oil and free fatty acid content of oats and groats. J. Agr. Sci. 45:411–18.

9278. —— and —— 1955: The chemical composition of oats. II. The nitrogen content of oats and groats. J. Agr. Sci. 45:419–27.

Hutchinson, J. B. 4434, 18755.

9279. Huth, E. 1890: Schleuderfruchte. Samml. Naturw. Vortrage 3. 23 pp. J. Roy. Microscop. Soc. 1891:763.

9280. Hutin, C. 1962: Analyses de routine de la polyploidie des semences de betterave. Proc. Intern. Seed Testing Assoc. 27:542–56.

Hutinel, P. 14346.

9281. Hutnik, R. J. 1954: Effect of seedbed condition on paper birch reproduction. J. Forestry 52:493–95.

9282. Hutt, P. A. 1927: Conifer seed collection and extraction. Quart. J. Forestry 51:116–20. Biol. Abstr. 32:6226. 1958.

9283. Hutton, E. M. 1943: A new method for tomato and cucumber seed extraction. Australian Council Sci. Ind. Res. J. 16:97–103.

9284. —— 1944: The field emergence and yield of garden peas as affected by treatment of the seed with fungicidal dusts. Australian Council Sci. Ind. Res. J. 17:71–74.

9285. —— 1954: Ploidy and virus reaction in *Physalis floridana* Rydb. Australian J. Agr. Res. 5:350–55. Biol. Abstr. 29:6969. 1955.

Hutton, E. M. 4287, 16545.

9286. Hutton, M. E. J., and R. H. Porter. 1937: Seed impermeability and viability of native and introduced species of Leguminosae. Iowa State Coll. J. Sci. 12:5–24.

9287. Huyge, C., and J. Bonnet. 1939: Essais de germination de l'orge sous rayons ultra-violets. Bull. Inst. Agron. Sta. Rech. Gembloux 8:33–42.

9288. Huyskes, J. A. 1959: The influence of seed diameter on the development of asparagus seedlings. Mededel. Directeur Tuinbouw 22:636–39.

9289. Hvidsten, L. 1953: The action of ensilage upon the germinating ability of weed seeds. Tidsskr. Norske Landbruk 60:242–60. Chem. Abstr. 48:5398. 1954.

9290. Hwang, C. C., S. A. Yu, and C. P. Cheng. 1955: Effect of chemicals on the germination of vegetable seeds. Agr. Res. 5(3/4):23–34.

9291. Hwang, Y., and H. L. Pearse. 1940: The response of seeds and seedlings to treatment with indolylacetic acid. Ann. Botany (London) 4:31–37.

Hyatt, E. 2372.

9292. Hyatt, J. 1875: Periodicity in vegetation. Proc. Am. Soc. Hort. Sci. 24:257–64.

Hwiswe, H. 6380.

9293. Hyde, A. E. 1935: Potentillas raised from seed. Horticulture 13(4):81.

9294. Hyde, E. O. C. 1932: Germinating capacity of perennial rye-grass seed. New Zealand J. Agr. 44:316–19.

9295. —— 1935: Chewings fescue seed. The influence of temperature and moisture content upon the rate of loss of its germinating capacity. New Zealand J. Agr. 51:40–42.

9296. —— 1935: The germination of perennial rye-grass in the soil and in the laboratory. New Zealand J. Agr. 51:224–28.

9297. —— 1935: Observations on the germination of newly harvested Algerian oats. New Zealand J. Agr. 51:361–67.

9298. —— 1937: Germination of "New Season's" Algerian oats. New Zealand J. Agr. 54:30–31.

9299. —— 1938: Germinating capacity of Italian rye-grass seed. New Zealand J. Agr. 56:357.

9300. —— 1939: Germination of seed wheat. New Zealand J. Agr. 58:434.

9301. —— 1939: Observations on the germination and seedling establishment of peas. New Zealand J. Sci. Technol. Sect. A 21:61–70.

9302. —— 1942: Red clover seed damaged by faulty threshing. New Zealand J. Agr. 64:351.

9303. —— 1945: Weed seeds in agricultural seed. New Zealand J. Agr. 70:29–32.

9304. —— 1945: Ryegrass seed. New Zealand J. Agr. 70:271–76.

9305. —— 1947: Weed seeds in agricultural seed. New Zealand J. Agr. 74:64–65.

9306. —— 1947: Weed seeds in agricutltural seed. New Zealand J. Agr. 74:150–51.

9307. —— 1947: Weed seeds in agricultural seed. New Zealand J. Agr. 74:262–63.

9308. —— 1949: Methods for determining the viability of various seeds by tetrazolium staining. I. Chewing's fescue. New Zealand J. Sci. Technol. Sect. A 31:13–20.

9309. —— 1950: Studies on the development of white clover seed. New Zealand Grassland Conf. Papers 12:80–86.

9310. —— 1952: Methods for determining the viability of various seeds by tetrazolium staining. II. Perennial ryegrass. New Zealand J. Sci. Technol. Sect. A 34:195–201.

9311. —— 1954: The function of the hilum in some Papilionaceae in relation to the ripening of the seed and the permeability of the testa. Ann. Botany (London) 18:241–56.

9312. —— 1955: Methods for determining the viability of various seeds by tetrazolium staining. III. Italian ryegrass and short-rotation ryegrass. New Zealand J. Sci. Technol. 37:36–39.

9313. ——, M. A. McLeavey, and G. S. Harris. 1959: Seed development in ryegrass, and in red and white clover. New Zealand J. Agr. Res. 2:947–52.

Hyde, E. O. C. 14297, 18527.

9314. Hyde, M. B., and C. G. Daubney. 1960: A study of grain storage in Fossae in Malta. Trop. Sci. 2:115–29.

9315. —— and H. B. Galleymore. 1951: The subepidermal fungi of cereal grains. II. The nature, identity and origin of the mycelium in wheat. Ann. Appl. Biol. 38:348–56.

9316. —— and T. A. Oxley. 1960: Experiments on the airtight storage of damp grain. I. Introduction, effect on the grain and the intergranular atmosphere. Ann. Appl. Biol. 48:687–710.

Hyde, M. B. 8034, 15437.

Hyde, M. E. 14917.

9317. Hyde, T. G. 1953: Nitrogen metabolism in *Pisum sativum.* Biochem. J. 55:21–22.

Hyer, A. H. 3156.

9318. Hylton, L. O., Jr., and R. E. Bement. 1961: Effects of environment on germination and occurrence of six weeks fescue. J. Range Management 14:257–60. Biol. Abstr. 38:15708. 1962.

9319. Hymowitz, T., and R. S. Matlock. 1961: Variations in seedcoat color of Groehler guar (*Cyamopsis tetragonoloba* [L.] Taub.)—genetic or environmental? Crop Sci. 1:465.

9320. —— and —— 1964: Guar: Seed, plant and population studies. Oklahoma Agr. Expt. Sta. Tech. Bull. B-108. 35 pp.

9321. Hynes, H. J., and R. D. Wilson. 1939: Fungicidal treatment of pea seed. Agr. Gaz. N.S. Wales 50:657–59. Chem. Abstr. 34:3864. 1940.

9322. Hyslop, G. R. 1933: Seed production in relation to mosaic diseases. Seed World 33(13):22–24.

9323. Hyun, S. K. 1961: Problems in seed orchard establishment. Intern. Botan. Congr. Recent Advan. Botany 9:1675–81. 1959.

9324. —— 1962: Mass production of control-pollinated seed of conifers. Proc. World Forestry Congr. 5:787–92. 1960.

I

9325. Iachan, A., and R. Roitman. 1959: Free amino acids and peptides in varieties of peanut. Anais Assoc. Brasil. Quim. 18:189–90. Chem. Abstr. 54:11163. 1960.

9326. Iagovoi, P. N. 1958: Effect of minor elements on the germination of clover seeds. Zemledelie 6(2):68–69.

9327. Iakobson, I. I. 1934: Action of an electric current on cottonseed. Bor'ba Khlopok 1/2:156–57. Expt. Sta. Rec. 72:321. 1935.

9328. Iakovlev, M. S. 1950: Structure of endosperm and embryo in cereals as a systematic characteristic. Tr. Botan. Inst. Akad. Nauk SSSR Ser. 7, Morf. i Anat. Rast. 1950:121–218.

9329. —— 1957: Principal types of polyembryony in higher plants. Tr. Botan. Inst. Akad. Nauk SSSR Ser. 7, Morf. i Anat. Rast. 4:201–10.

9330. —— and D. P. Snegirev. 1954: Effect of growth substances on the formation of polyembryonic kernels in wheat. Botan. Zh. 39:187–94.

Iakovlev, M. S. 338, 339.

9331. Iakovleva, N. 1956: Disinfecting spring wheat by a complex system of seed treatment. Ardena Lenina Sel'skokhoz. Akad. im. K. A. Timiriazeva Sb. Studencheskikh Nauch.-Issled. Rabot 5:74–84.

Iakovleva, V. I. 11379.

9332. Iakovuk, A. S., and A. S. Komel. 1958: Effect of age of tobacco seeds planted in late autumn on yield of seedlings suitable for planting. Tabak (USSR) 19(3):55–57. Tobacco Abstr. 3:232–33. 1959.

9333. Iancu, S. S. 1959: Some data regarding the repeated stimulation (with hydrogen sulfide) of the wheat seeds of spring Lutescens 62, and Hordeiforme 432. Analele Univ. "C. I. Parhon" Bucuresti, Ser. Stiint. Nat. 21:117–23. Chem. Abstr. 54:13288. 1960.

9334. Ianovskii, D. M. 1956: Sun heat-accumulator for increasing the germinative capacity of seeds. Sel'skoe Khoz. Povolzhia 1956(3):41–43.

9335. Iarkin, S. A. 1936: Studies on different ripening stages of tobacco seeds. [Biull.] Vses. Inst. Tabachn. Makhor. Promyshl. Krasnodar 132:217–51.

9336. Iarosh, N. P. 1958: Intraspecific and interspecific variability in cotton seed chemical composition. Tr. Prikl. Botan. Genet. i Selek. 31:170–84. Biol. Abstr. 43:11622. 1963.

Iasnov, G. A. 2063.

Iastreb, M. G. 1533.

Iavorskii, A. G. 15622.

Iazykova, V. A. 2997.

9337. Ibanez, M. L. 1963: A reversal of cacao seed sensitivity to cold. Turrialba 13:31–32. Biol. Abstr. 44:16667. 1963.

9338. —— 1964: Role of the cotyledon in sensitivity to cold of cacao seed. Nature 201:414–15.

Ibert, E. R. 10663.

9339. Ibragimov, B. B. 1931: Causes of death of plants from frost, and the nature of cold hardiness in wheat seed. Wiss. Ber. Biol. Fak. Tomsker Staats-Univ. 1:71–96. Biol. Abstr. 7:5302. 1933.

9340. Ibragimov, M. K. 1961: Changes of pH and E_h taking place in cotton seeds soaked in water. Uzbeksk. Biol. Zh. 1961(6):26–31. Chem. Abstr. 56:14643. 1962.

9341. —— 1962: The exosmosis of various compounds from cotton seed in water medium. Uzbeksk. Biol. Zh. 6(1):21–26. Chem. Abstr. 57:6326. 1962.

9342. —— 1962: Conversion of phosphorus compounds in germinating and ripening cotton seeds. Nauchn. Tr. Tashkentsk. Univ. 1962(210):75–81. Chem. Abstr. 59:14305. 1963.

9343. Ibragimov, S. I., L. A. Nezgovorov, and A. K. Solovev. 1963: The application of tetramethylthiuram sulfide for seed rot prevention. Khlopkovodstvo 13(4):34–36. Chem. Abstr. 59:10480. 1963.

Ibragimov, S. I. 14381.

9344. Ibrahim, I. A., and M. M. Aboul-Ela. 1961: The effect of seed treatment with some fungicides on the stand and yield of

certain cotton varieties in Egypt. Alexandria J. Agr. Res. 9:49–62.
Ibrahim, P. A. M. 214.
Ichaporia, M. B. 8766, 8767.

9345. Ichiba, A. 1937: Phytosterol in wheat germ oil. II. Preliminary report on a new sterol. Sci. Papers Inst. Phys. Chem. Res. 34:116–20. Pharm. Abstr. 6:68. 1940.

9346. —— 1937: Unsaponifiable matter of wheat germ oil. Sci. Papers Inst. Phys. Chem. Res. 34:121–31. Pharm. Abstr. 6:72. 1940.

9347. Ichihara, J. 1954: Dormancy of the seeds of herbaceous flowers. Hort. Assoc. Japan J. 23:115–22.
Ichihara, J. 14153.
Ichihara, K. 19944.

9348. Ichii, T. 1960: Growth studies of chestnut fruits. I. Developmental changes in chemical compositions of chestnut fruits. Hort. Assoc. Japan J. 29:281–87.
Ichikawa, I. 14699–14701.

9349. Ichimura, K. 1961: Sedimentation analyses of ribonucleic acids from germinating bean seeds. Plant Cell Physiol. (Tokyo) 2:339–42.

9350. ——, M. Izawa, and Y. Oota. 1960: Ribonucleic acids in soluble cytoplasm of germinating bean. Plant Cell Physiol. (Tokyo) 1:317–25.
Ideda, T. 10121.
Ielpo, M. 5066.

9351. Ienger, N. G. C., V. Jayaraman, and Y. V. S. Rau. 1955: Thiamine contents of Indian foodstuffs. II. Thiamine content of a few leguminous seeds during germination. Ann. Biochem. Exptl. Med. (Calcutta) 15:49–54. Chem. Abstr. 50:6596–97. 1956.
Igarashi, G. 8440.

9352. Igaue, I. 1960: Q enzyme in rice seed at the milky stage. IV. Chromatography of the Q enzyme fraction. J. Agr. Chem. Soc. Japan (English Transl.) 34:1028–31. Chem. Abstr. 59:10396–97. 1963.
Igelman, B. 4877.

9353. Igolchenko, M. I., and V. M. Kopeikovskii. 1962: Dependence of size and heat treatment on the equilibrium humidity of sunflower seeds. Izv. Vysshikh Uchebn. Zavedenii Pishchevaya Tekhnol. 1962(6): 12–15. Chem. Abstr. 58:12856. 1963.

9354. Ihde, A. J., and H. A. Schuette. 1941: The chemistry of the rye germ. VI. Allantoin and other acetone-extractives. J. Am. Chem. Soc. 63:2486–87.
Ihde, A. J. 1279.
Ihle, K. 7988.

9355. Ihlenfeldt, H. D. 1959: Über die Samentaschen in den Früchten der Mesembryanthemen. Ber. Deut. Botan. Ges. 72:333–42.

9356. Ihsan-Ul-Haque, M. I. Qureshi, M. K. Bhatty, and Karimullah. 1963: Composition of mazri (*Nannorhops ritchieana*) seed oil. Pakistan J. Sci. Ind. Res. 6:77–78.
Iinuma, H. 19989.

9357. Iizuka, H. 1957: Studies on the microorganisms found in Thai rice and Burma rice. I. On the microflora of Thai rice. J. Gen. Appl. Microbiol. 3:146–61.

9358. —— 1958: Studies on the microorganisms found in Thai rice and Burma rice. II. On the microflora of Burma rice. J. Gen. Appl. Microbiol. 4:108–19.
Iizumi, S. 1551, 11575.

9359. Ikawa, M., R. J. Dicke, T. C. Allen, and K. P. Link. 1945: The principal alkaloids of sabadilla seed and their toxicity to *Musca domestica* L. J. Biol. Chem. 159:517–24. Chem. Abstr. 39: 4705. 1945.
Ikawa, M. 402.
Ikeda, C. 13473, 13474.

9360. Ikeda, H. 1961: The effect of temperature on the variation of amylase activity in the germinating seeds of wheat. Kyushu Univ. Fac. Agr. Sci. Bull. 18:371–79.

9361. —— and T. Nagamatsu. 1958: Studies on the devernalization in wheat and barley. Proc. Crop Sci. Soc. Japan 27:283–84.
Ikeda, H. 14066.
Ikeda, J. 9080, 9081.
Ikeda, M. 8313.

9362. Ikeda, N., and M. Nakamura. 1956: Effects of various treatments on the germination of mint seeds. Okayama Univ. Fac. Agr. Sci. Rept. 9:39–45.

9363. ——, I. Saisho, and S. Minakata. 1960: Studies on the storage of mint seed. Okayama Univ. Fac. Agr. Sci. Rept. 16:1–5.

9364. Ikeda, T. 1956: On the crystalloidal property of endosperm in hard wheat. Proc. Crop Sci. Soc. Japan 25:88–89.
Ikegami, H. 8317.

9365. Ikenberry, R. W. 1962: The isolation of *Fusarium moniliforme* Sheld. from corn kernels. Proc. Iowa Acad. Sci. 60:100–2.

9366. Ikeshima, S., and K. Uchida. 1936: The diuretic action of catalpa seed. Japan. J. Med. Sci. 4. Pharmacol. Trans. 9:37. Chem. Abstr. 31:8021. 1937.
Ikonen, E. V. 15273.

Ikonnikova, M. I. Smirnova-; *see* Smirnova-Ikonnikova, M. I.

9367. Ikramov, K. U. 1963: Determination of limits of low temperatures for germination of sorghum seeds. Zemledelie 1963(3): 77–79.

9368. Ikramova, M. M. 1958: Dependence of quality and quantity of oil in flax seeds on temperature and moisture of the soil. Dokl. Akad. Nauk Tadzh. SSR 1:43–46.

9369. —— 1963: The content of oil in some nuts grown in Tadzhikistan. Dokl. Akad. Nauk Tadzh. SSR 6:44–47. Chem. Abstr. 59:7851. 1963.

9370. Ikuma, H., and K. V. Thimann. 1959: Photosensitive site in lettuce seeds. Science 130:568–69.

9371. —— and —— 1960: Action of gibberellic acid on lettuce seed germination. Plant Physiol. 35:557–66.

9372. —— and —— 1963: Action of kinetin on photosensitive germination of lettuce seed as compared with that of gibberellic acid. Plant Cell Physiol. (Tokyo) 4:113–28.

9373. —— and —— 1963: Activity of gibberellin "D" on the germination of photosensitive lettuce seeds. Nature 197: 1313–14.

9374. —— and —— 1963: The role of the seed-coats in germination of photosensitive lettuce seeds. Plant Cell Physiol. (Tokyo) 4:169–85.

Ikuzawa, M. 18618.

Ilchev, D. 9377.

9375. Ilieff, P. 1942: Genetische Untersuchungen an der Erdnuss (*Arachis hypogaea* L.). Bohnenfarbe- und Fruchtgrössenverteilung. Züchter 14:141–45. Biol. Abstr. 22:10595. 1948.

9376. Iliesco, E. 1947: La stimulation des semences de chanvre par la vanilline. Acad. Roumaine Sect. Sci. Bull. 29:405–9.

Iliescu, C. 4133.

9377. Iliev, P., M. D. Popov, P. Mazdrakov, E. Popov, L. Velikov, G. Kalupchiev, D. Ilchev, K. Dimitrov, and V. Boyadjiev. 1956: Application of the methods of stimulation in agriculture and results of the stimulation experiments during 1954. Izv. Inst. po Biol. "Metodii Popov" Bulgar. Akad. Nauk 7:3–40.

9378. Ilin, A. I. 1952: Effect of size of pine seeds on their quality. Lesn. Khoz. 5(7):54–55.

9379. Ilin, G. S. 1934: Proteins of the tobacco seed. Biochem. Z. 273:350–53. Chem. Abstr. 29:190. 1935.

9380. —— and M. I. Lovkova. 1959: Transformations of nicotine and amino acids in the ripening tobacco seed. Biochemistry (USSR) (English Transl.) 24: 769–72.

9381. Ilina, A. I. 1954: On the reciprocal relations between embryo and endosperm in the early phases of seed-formation in olive flax. Dokl. Akad. Nauk SSSR 95:167–69. Biol. Abstr. 32:13460. 1958.

9382. —— 1954: Causes for dihiscence of seeds in *Lallemantia iberica* v. *typica* (Ster.) F. et M. Dokl. Akad. Nauk SSSR 96:1253–56.

9383. —— 1958: On the interrelationship of embryo and endosperm during the early formative stages in sesame seeds (*Sesamum indicum* L.). Dokl. Akad. Nauk SSSR 120:123–27.

9384. Ilina, G. M. 1961: An embryological study of *Papaver somniferum* L. Byul. Mosk. Obshchestva Ispytatelei Prirody Otd. Biol. 66:13–25. Biol. Abstr. 41: 20012. 1963.

Ilina, V. N. 11310.

Iljas, B. Hasan-; *see* Hasan-Iljas, B.

9385. Iljin, G. 1934: Die Umwandlung des Nicotins beim Reifen der Tabaksamen Biochem. Z. 268:253. Physiol. Abstr. 20:352. 1935.

9386. Illies, R. 1952: The fat and ash content of almonds and apricot pits. Zucker- Süsswarenwirtsch. 5:293–94. Chem. Abstr. 49:1237. 1955.

Illner, E. 18591.

9387. Ilnicki, R. D., and M. W. Johnson. 1959: Temperature, light, and seed size and their effects on germination of dog fennel (*Anthemis cotula*). Proc. Northeast. Weed Control Conf. 13:440.

Ilnicki, R. D. 18594.

Im, R. Z. 2229.

Imai, T. 14516.

9388. Imamura, H. 1956: Effect of MH on ripening tobacco seed and its after-effects. Kyushu Agr. Res. 1956(18):70–71.

Imamura, S. 11494, 14689.

Imamura, S. C. 18388.

9389. Imazu, T. 1961: Morphological, ecological, and cytological studies on cultivated and wild butterburs, *Petasites japonicus* Maxim. I. On the morphology of leaf, sprouting- and flowering-habits, and germination of seed. J. Japan. Soc. Hort. Sci. 30:233–40.

9390. —— and M. Hujishita. 1962: Morphological, ecological, and cytological studies on cultivated and wild butterburs, *Petasites* *japonicus* Maxim. III. On the pollen and seed-fertilities. J. Japan. Soc. Hort. Sci. 31:17–22.

9391. —— and T. Osawa. 1958: Studies on udo, *Aralia cordata* Thunb. II. On the dormancy of domestic and wild udo seeds. Hort. Assoc. Japan J. 27:149–53.

9392. Immer, F. R. 1941: Relation between yielding ability and homozygosis in barley crosses. J. Am. Soc. Agron. 33:200–6.

9393. Inaba, T., and K. Kitagawa. 1935: China jute seed and oil. J. Soc. Chem. Ind. Japan 38:77B. Analyst 60:335. 1935.

9394. Inada, Y., and K. Shibata. 1960: Vinyl pheoporphyrin A_5 in the inner seed coats of pumpkins. Plant Cell Physiol. (Tokyo) 1:311–16.

Inamori, Y. 14519.

Inamura, H. 19968.

9395. Incze, F. 1961: Relation between the number of fruits and the weight of seeds of Cyclamen persicum. Kertesz. Szolesz. Foiskola Evkonyve 9:189–94.

9396. Indin, V. G. 1959: Physiological studies on the processes of seed maturation in certain species of maple. Botan. Zh. 44:559–62.

Indovina, R. 4428.

9397. Ingalls, R. A. 1946: A study of the occurrence of baldheads in beans and a comparison of their field performance with that of normal plants. Proc. Assoc. Offic. Seed Analysts 36:177–83.

Ingalls, R. A. 12029.

9398. Ingber, E. 1931: Ricerche eseguite sulla *Vicia faba* var. *minor*. Morfologia esterna. Intorno alla radioeccitazione e alla radiolesione. Azione biologica dei raggi Röntgen duri (memoria I)-semi-molli (Mem. II)-molli (Mem. III). Atti Ist. Botan. "Giovanni Briosi" e lab. crittogam. Ital. R. Univ. Pavia 2:173–268. 1930:3:95–145, 223–307. 1932. Biol. Abstr. 8:10539. 1934.

9399. Ingle, J. 1963: The extraction and estimation of nucleotides and nucleic acids from plant material. Phytochemistry 2:353–70.

9400. Ingle, L. 1933: Effects of environmental conditions on longevity. Science 78:511–13.

Ingle, T. R. 9203.

9401. Ingold, C. T. 1960: Dispersal by air and water—the take-off. Plant Pathol. 4:137–68.

9402. Ingold, J. 1938: Growing columbines from seed. Horticulture 16(6):132–33.

9403. Ingold, M. 1955: Dix ans de production de semences de céréales en Suisse romande (1945/1954). Rev. Romande Agr. Viticult. Arboricult. 11:41–43. Biol. Abstr. 31:25592. 1957.

9404. —— 1957: Étude préliminaire de l'utilisation d'une soufflerie pour l'analyse des graminées. Proc. Intern. Seed Testing Assoc. 22:106–22.

9405. —— 1958: Le "cold test" appliqué à la germination des semences de mais. Landwirtsch. Jahrb. Schweiz 7:445–56.

9406. —— 1960: Contribution a l'étude de la germination des semences d'*Allium cepa* L. et *Cucurbita pepo* L. Proc. Intern. Seed Testing Assoc. 25:785–836.

9407. —— and J. Rapin. 1958: Resultats de germination de semences de betteraves sucrières. Rev. Romande Agr. Viticult. Arboricult. 14:94–96.

Ingram, D. J. E. 10831.

9408. Ingram, J. 1935: Development of lily seedlings. Lily Yearbook 1935:83–86.

9409. Ingwersen, W. 1932: Raising alpine plants from seed. Gardeners Chron. 36:271, 282.

9410. —— 1956: Seeds from early flowering alpines. Gardeners Chron. 139:712.

Ink, D. P. 17128–17131.

Inman, R. 19464.

9411. Innamorati, M. 1957: Vernalisation and "tenderization" (yellow berry) in *Triticum durum*, Desf. Ric. Sci. 27:2711–26.

9412. —— 1961: Starch and endosperm of vitreous and starchy (yellow berries) kernels of *Triticum durum* Desf. Nuovo Giorn. Botan. Ital. 68:314–34.

9413. —— 1963: Rate of plant development and endosperm texture in *Triticum durum* Desf. Nuovo Giorn. Botan. Ital. 70:48–61.

9414. Inokuma, T., and S. Asakawa. 1961: Germination behavior of *Picea jezoensis* and *P. glehnii* seeds. J. Japan. Forestry Soc. 43:166–68.

Inokuma, T. 1450.

Inokuti, A. 18870.

Inosaka, M. 4884.

9415. Inoue, S., and M. Yahiro. 1956: Physiological changes in the germinating seeds during low temperature treatment. II. On the activity of catalase and peroxidase. Botan. Mag. (Tokyo) 69:215–18.

Inoue, S. 10999.

9416. Inoue, T., and K. Ogata. 1958: Studies on the physiological mechanism of maleic hydrazide to the subterranean crops and seeds. III. The effect of MH on the sprout inhibition, growth substances, and respiration in onion bulbs and green gram seeds. Hort. Assoc. Japan J. 27:49–52.

Inoue, T. 8438, 8439, 19911.

9417. Inoue, Y. 1937: On some seed characters and their relations to the classification of varieties in *Cucumis melo*. Hort. Assoc. Japan J. 8:72–86.

9418. ——, M. Shibuya, and Y. Suzuki. 1955: The effect of moisture on the vernalization of radish seed. Japan. J. Breeding 5:54–58.

9419. —— and G. Shintani. 1942: Seed disinfectants. I. Effect of C_1-C_5 alkylmercuric chlorides on the germination of rice. J. Agr. Chem. Soc. Japan (English Transl.) 18:1141–45. Chem. Abstr. 43:1138. 1949.

9420. —— and Y. Suzuki. 1962: Studies on the effect of maturity and after-ripening of seeds upon the seed germination in snap bean, *Phaseolus vulgaris* L. J. Japan. Soc. Hort. Sci. 31:146–50.

9421. Inouye, C. 1927: Die Keimungsunterschiede zwischen Bergreis und Wasserreis. Ber. Deut. Botan. Ges. 45:187–97.

9422. —— 1935: Studies of germination temperature of seed in certain crop plants. I. Rice. Proc. Crop Sci. Soc. Japan 7:200–17.

9423. —— 1952: Influence of temperature on the germination of seeds. VI. Maize (*Zea mays* L.). Proc. Crop Sci. Soc. Japan 21:79–80.

9424. —— 1953: Influence of temperature on the germination of seeds. IX. Soybean. Proc. Crop Sci. Soc. Japan 21:276–77.

9425. Inskip, W. L. 1951: Easy does it: vermiculite for lily seed germination. North Am. Lily Soc. Yearbook 4:29–31.

9426. International Atomic Energy Agency. 1961: Proceedings symposium on the effects of ionizing radiations on seeds and their significance for crop improvement, Karlsruhe, Germany, Aug. 8–12, 1960. Intern. Atomic Energy Agency, Vienna. 655 pp.

9427. International Seed Testing Association. 1959: Handbook on seed health testing. 159 pp.

Inyushin, V. M. 6851.

9428. Ioffe, K. G., and E. A. Ermakova. 1954: Protein similarity in mungbean (*Phaseolus aureus*) and soybean. Biochemistry (USSR) (English Transl.) 19:561–64. Chem. Abstr. 49:1883. 1955.

9429. Ioffe, M. D. 1957: Development of embryo and endosperm in wheat, horse beans and radishes. Tr. Botan. Inst. Akad. Nauk SSSR Ser. 7, Morf. i Anat. Rast. 4:211–69.

9430. —— 1962: Embryology of *Trochodendron aratioides* Sieb. et Zucc. (pollen and embryo sac development). Tr. Botan. Inst. Akad. Nauk SSSR Ser. 5, Rast. Syr'e 9:250–59. Biol. Abstr. 44:7857. 1963.

Ioffe, M. D. 19914.

9431. Ioffe, R. K. 1950: Peculiarities of post-harvest ripening and germination of sainfoin seeds. Selek. i Semen. 17(12):47–49

9432. Ionesco, St. 1932: Effect of zinc on the respiration of germinating seeds of *Lupinus albus*. Compt. Rend. Soc. Biol. 110:655–57. Chem. Abstr. 26:4627. 1932.

Ionescu, C. N. 19142.

Ionescu, I. 5143.

9433. Ionescu, M., M. Ivanescu, and M. Ferando. 1959: Biochemical characteristics of the double hybrids of corn cultivated at Fundulea (Romania) in 1957. Probl. Agr. (Bucharest) 11:13–22. Chem. Abstr. 53:22283. 1959.

9434. Ionut, V. 1962: Contributions concerning the harvesting, handling and storing of the Douglas seeds. Rev. Padurilor (Bucharest) 77:581–83.

Iordache, C. 9202, 18193.

Iordachescu, O. 5946.

Ippolito, G. d'; *see* D'Ippolito, G.

Irabagon, T. A. 3576.

9435. Irani, R. J. 1946: Chemistry of Kurchi seeds. I. Isolation of a crystalline glyco-alkaloid. Current Sci. (India) 15:106–7.

9436. —— 1946: Chemistry of Kurchi seeds. II. Isolation of the bromide of a linoleo-dilinolenin from the fatty oil. Current Sci. (India) 15:161.

9437. —— 1946: Chemistry of Kurchi seeds. III. A new and simple method of analysis of bromoglycerides. Current Sci. (India) 15:191–92.

9438. —— 1946: Chemistry of Kurchi seeds. IV. Isolation of galactose from the picric acid hydrolysis of the glyco-alkaloid. Current Sci. (India) 15:229–30.

9439. Ireland, J. C. 1951: Give your seeds electronic boost. High frequencies stimulate plant growth—produce some amazing yields. Southern Seedsman 14(10):29, 40, 48.

9440. Irisawa, T. 1930: Influence of the ultra-violet ray irradiation upon the germinating seeds and the effect of photosensitizers on the action of light. Hokuetsu

Ig. Kw. Z. 45:428–47. Japan. J. Med. Sci Biochem.2:(75). 1934.

9441. Iritani, W. M., and G. W. Woodbury. 1954: Use of radio-frequency heat in seed treatment. Idaho Agr. Expt. Sta. Res. Bull. 25. 19 pp.

9442. Irmak, A. 1961: The seed-fall of firs and their germination in the snow. Istanbul Univ. Orman Fak. Dergisi A 11:1–6.

Irvine, D. S. 2291.

9443. Irvine, G. N., and J. A. Anderson. 1953: Variation in principal quality factors of Durum wheats with a quality prediction test for wheat or Semolina. Cereal Chem. 30:334–42.

9444. Irvine, G. N., W. Bushuk, and J. A. Anderson. 1954: Kinetic studies of the catalase system of wheat. Cereal Chem. 31:256–66.

9445. ——, ——, and —— 1954: Note on the catalase activity of North American wheats. Cereal Chem. 31:267–70.

Irvine, J. A. 3609.

9446. Irving, G. W., Jr., T. D. Fontaine, and R. C. Warner. 1945: Electrophoretic investigation of peanut proteins. I. II. Arch Biochem. 7:475–89; 8:239–49.

Irving, G. W., Jr. 6533–6536.

9447. Irwin, J. O. 1931: On the influence of soil temperature on the germination interval of crops. J. Agr. Sci. 21:241–50. Expt. Sta. Rec. 66:325. 1932.

9448. Irwin, M. R., and H. L. Kempster. 1943: Korean *Lespedeza* seed as a protein supplement in chick rations. Missouri Agr. Expt. Sta. Bull. 462. 11 pp.

9449. Isaac, L. A. 1930: Seed flight in the Douglas fir region. J. Forestry 28:492–99.

9450. —— 1930: Seedling survival on burned and unburned surfaces. J. Forestry 28:569–71.

9451. —— 1934: Cold storage prolongs the life of noble fir seed and apparently increases germinative power. Ecology 15:216–17.

9452. —— 1935: Life of Douglas fir seed in the forest floor. J. Forestry 33:61–66.

9453. —— 1938: Factors affecting establishment of Douglas fir seedlings. U.S. Dept. Agr. Circ. 486. 46 pp. Expt. Sta. Rec. 80:348–49. 1939.

9454. —— 1943: Reproductive habits of Douglas fir. Charles Lathrop Pack Forestry Foundation, Washington, D.C. 107 pp. Biol. Abstr. 18:9287. 1944.

9455. —— 1962: Problems and proposals for international certification of tree seed origin and stand quality with particular reference to western North American species. Proc. World Forestry Congr. 5:690–96. 1960.

9456. Isaac, W. E. 1938: Protein breakdown during germination of *Lathyrus odoratus*. Ann. Botany (London) 2:23–31.

9457. —— and N. H. Winch. 1947: The guaiacol-hydrogen peroxide and benzidine-hydrogen peroxide colour reactions of the bean (*Phaseolus vulgaris* L.) pod. J. Pomol. Hort. Sci. 23:23–37.

9458. Isaev, S. Z. 1953: Effect of summer planting dates on fruit-formation and yields of alfalfa seeds. Vestn. Akad. Nauk SSR 10:69–77.

9459. Isajev, F., J. Kolousek, and Z. Müller. 1954: The feeding value of *Gelega officinalis*. Sb. Česk. Akad. Zeměděl. Ved Ser. A 27:363–84. Chem. Abstr. 49:4203. 1955.

9460. Isakova, A. A. 1937: On the true influence of bacteriorhizal microorganisms on the germination of seeds. II. Compt. Rend. Acad. Sci. URSS 14:463–65.

9461. —— 1940: Bacterial treatment of seeds as a means of controlling the synthesis of vitamins C and B_1 in the seedlings. Compt. Rend. Acad. Sci. URSS 28:170–73.

9462. Isely, D. 1944: A study of conditions that affect the germination of *Scirpus* seeds. Cornell Univ. Agr. Expt. Sta. Mem. 257. 28 pp.

9463. —— 1951: Seeds of *Bromus secalinus* and *commutatus*. Proc. Iowa Acad. Sci. 58:155–63.

9464. —— 1955: Observations on seeds of the Leguminosae: Mimosoideae and Caesalpinioideae. Proc. Iowa Acad. Sci. 62:142–45.

9465. —— 1955: Key to seeds of Caesalpinoideae and Mimosoideae of north central states. Proc. Iowa Acad. Sci. 62:146–49.

9466. —— 1961: Seed technology. Econ. Botany 15:332–46.

9467. —— and W. H. Bragonier. 1962: Seeds of Iowa noxious and common weeds. Iowa Agr. & Home Econ. Expt. Sta. Bull. P-131. 23 pp.

9468. ——, W. E. Loomis, and H. D. Hughes. 1948: Application of chemical methods to determine seed viability. Iowa Agr. Expt. Sta. Rept. 1947/48:195.

9469. ——, D. West, and R. W. Pohl. 1951: Seeds of agricultural and weedy *Bromus*. Iowa State Coll. J. Sci. 25:531–48.

9470. —— and W. H. Wright. 1954:

Noxious weed seeds. Iowa State Coll. J. Sci. 28:521–86.
Isely, D. 13871.
Isensee, H. 12581.
Ishaque, M. 1652.
Isherwood, F. A. 4702.
Ishibashi, H. 6856.

9471. Ishibashi, I. 1952: The influence of electric current on the yield of crop seeds. I. Soybean. Proc. Crop Sci. Soc. Japan 20:305–6.
Ishida, R. 10380–10384.

9472. Ishidoya, K., K. Nishimura, and Y. Muraoka. 1953: Studies on the change of chemical components in growing tobacco plant. I. On the change of free amino acids and sugars at the stages of young seedling and of flower formation indicated by the paper partition chromatography method. Proc. Crop Sci. Soc. Japan 21:306–8.

9473. Ishihara, A. 1957: Physiological studies on the vernalization of wheat plants. I. The retention of vernalized condition in excised stem tips. Proc. Crop Sci. Soc. Japan 26:20–23.

9474. —— 1958: The effect of gibberellin on the spike initiation in wheat plants. Proc. Crop Sci. Soc. Japan 27:285–88.

9475. Ishii, J.: Mannane in the seeds of *Diospyrous kaki* L. Bull. Coll. Agr. Tokyo 2:101–2. Expt. Sta. Rec. 6:386–87. 1896.
Ishii, J. 12389.

9476. Ishikawa, E., S. Takeuchi, and K. Shiraishi. 1955: On the germination and growth of weeds sown to a field monthly throughout the year. Shikoku Agr. Expt. Sta. Bull. 2:103–12.

9477. Ishikawa, S. 1955: A note on the effects of coumarin on the germination of seeds. Kumamoto J. Sci. Ser. B 2:97–103. Biol. Abstr. 33:3482. 1959.

9478. —— 1960: Physiological studies on seed germination. III. Germination inhibition by 8-hydroxyquinoline and its prevention with metal ions (preliminary report). Kumamoto J. Sci. Ser. B Sect. 2, 5:108–12. Biol. Abstr. 36:65696. 1961.
Ishikawa, S. 8309.

9479. Ishikawa, T. 1963: On the regermination of rice seeds which were dried after beginning germination. Bull. Fac. Agr. Miyazaki Univ. 8:358–62.
Ishikawa, T. 9493.
Ishiwa, H. 11162, 11163.
Ishiyama, T. 9527.

9480. Ishizaki, H. 1950: Effect of old cultures of *Ophiobolus heterostrophus* on the germination of seeds. Igaku To Seibutsugaku 16:215–18. Chem. Abstr. 45:1208. 1951.
Isikawa, H. 18609.

9481. Isikawa, I. 1959: Hard seed of *Acacia mollissima* Willd. in relation to the position of seeds in legume. Bull. Tokyo Univ. Forests 55:171–79.

9482. —— 1960: Germination and hard seed of *Acacia mollissima* Willd. in relation to the date of collecting the seeds. Misc. Inform. Tokyo Univ. Forests 13:46–56.

9483. Isikawa, S. 1952: On the light-sensitivity of tobacco seeds. I. Change of the light-sensitivity with the time of imbibition. Botan. Mag. (Tokyo) 65:257–62.

9484. —— 1953: On the light-sensitivity of tobacco seeds. II. A very small dosage of light and the lapse of long time of imbibition. Botan. Mag. (Tokyo) 66:37–42.

9485. —— 1954: Light-sensitivity against the germination. I. "Photoperiodism" of seeds. Botan. Mag. (Tokyo) 67:51–56.

9486. —— 1955: Light-sensitivity against the germination. II. Interaction of light and darkness on the germination of *Epilobium cephalostigma* seeds. Botan. Mag. (Tokyo) 68:173–79.

9487. —— 1957: Interaction of temperature and light in the germination of *Nigella* seeds. I. Botan. Mag. (Tokyo) 70:264–70.

9488. —— 1957: Interaction of temperature and light in the germination of *Nigella* seeds. II. Botan Mag. (Tokyo) 70:271–75.

9489. —— 1962: Light sensitivity against the germination. III. Studies on various partial processes in light sensitive seeds. Japan. J. Botany 18:105–32.

9490. —— and S. Araki. 1955: Effect of light upon the germination of forest tree seeds. II. On the photoperiodic tendency. J. Japan. Forestry Soc. 37:485–87.

9491. —— and T. Fujii. 1961: Photocontrol and temperature dependence of germination of *Rumex* seeds. Plant Cell. Physiol. (Tokyo) 2:51–62.

9492. ——, ——, and Y. Yokohama. 1961: Photoperiodic control of the germination of *Eragrostis* seeds. Botan. Mag. (Tokyo) 74:14–18.

9493. —— and T. Ishikawa. 1960: Requirement of low temperature treatment following illumination in the germination of seed of *Elsholtzia*. Plant Cell Physiol. (Tokyo) 1:143–49.

9494. —— and G. Shimogawara. 1954: Effects of light upon the germination of forest tree seeds. I. Light-sensitivity and its degree. J. Japan. Forestry Soc. 36: 318–23.

9495. —— and Y. Yokohama. 1962: Effect of "intermittent irradiations" on the germination of *Epilobium* and *Hypericum* seeds. Botan. Mag. (Tokyo) 75:127–32.

Isikawa, S. 6815–6818, 11538.

Isimaru, K. 14112–14114.

9496. Isip, I. N. 1939: Effect of protective substances presented in vegetable organisms on the growth of seed. Sov. Botan. (Leningrad) 3:8. Chem. Abstr. 34:4418. 1940.

9497. —— 1950: Autumn sowing of sunflower with freshly-harvested seeds. Sov. Agron. 8:90–93.

9498. Iskhakov, N. I., A. I. Ismailov, A. S. Sadykov, and A. M. Yakubov. 1963: The effect of some factors on the oil content and the fatty acids composition in cotton plant seeds. Uzbeksk. Khim. Zh. 1(3):52–56. Chem. Abstr. 59:13120. 1963.

9499. —— and A. S. Sadykov. 1961: Fatty acid composition of cottonseed oil. Dokl. Akad. Nauk Uz. SSR 1961(11):21–23. Chem. Abstr. 57:6318. 1962.

9500. Islam, A. S. 1964: A rare hybrid combination through application of hormone and embryo culture. Nature 201: 320.

9501. —— and M. I. Khan. 1957: Studies on the seed germination of *Corchorus* spp. Biologia 3:165–67. Biol. Abstr. 33:15363. 1959.

Isley, D. 458, 2066.

Islip, H. T. 2283, 4369.

Ismailov, A. I. 9498.

9502. Isogai, Y. 1953: Inhibitory effect on seed germination of an aliphatic β-keto-aldehyde isolated from the rhizome of *Houttuynia cordata*. Sci. Papers Coll. Gen. Educ. Univ. Tokyo 3:39–40.

9503. —— 1956: Studies on the growth promoting agents separated from etiolated bean seedlings. III. Separation of another active fraction. Sci. Papers Coll. Gen. Educ. Univ. Tokyo 6:167–73.

9504. —— 1957: Studies on the growth promoting agents separated from etiolated bean seedlings. IV. Purification of three active fractions. Sci. Papers Coll. Gen. Educ. Univ. Tokyo 7:89–95.

Israelstam, G. F. 11447.

9505. Issatchenko, B. 1918: Sur la germination des graines de certaines plantes medicinalis. Jard. Botan. Inst. Essais Semences Ann. Essais Semences (Leningrad) 4:3–10.

9506. —— 1918: Étude comparative sur l'influence de la litière sur la germination des graines des céréales. Jard. Botan. Inst. Essais Semences Ann. Essais Semences (Leningrad) 4:15–22.

9507. —— 1929: Berichte über die Arbeiten der Abteilung für Samenprüfung (gegr. 1877) des Botanischen Hauptgartens in Leningrad. Actes Congr. Intern. Essais Semences 5:384–404.

9508. —— 1931: Resultate vergleichender Untersuchungen von Samenproben im 1929: von Kontrollstationen der RSFSR. Jard. Botan. Inst. Essais Semences Ann. Essais Semences (Leningrad) 8:53–67.

9509. —— 1931: Über die Verwendung von Farblösungen zur Untersuchung der Keimfähigkeit der Samen. Fortschr. Landwirtsch. 6:257–58.

9510. Issatchenko, B. L. 1945: Sur la germination des graines des plantes sauvages. Sov. Botan. (Leningrad) 13(3):53–56.

9511. —— and M. H. Grimm. 1931: Die Dauer der Keimversuche. Jard. Botan. Inst. Essais Semences Ann. Essais Semences 8:47–52.

9512. —— and T. W. Lipskaja. 1934: Zur Frage der Samenkeimung arktischer Pflanzen. Arctica 2:101–11. Botan. Centralbl. 168:369–70.

9513. —— and A. A. Predtetschenskaja. 1936: Über die Anwendung von Vitalfärbung als methode zur Bestimmung der Lebensfähigkeit von Samen. Botan. Experimentalis Ser. IV 2:347–79.

Isshiki, S. 11094–11097, 11158.

9514. Issipe, I. N. 1939: L'influence des substances protestrices de l'organisme végétal sur la germination des semences. Sov. Botan. (Leningrad) 1939(3):3–8.

9515. Istatkov, S. 1963: Chemical composition of the most common Bulgarian beans. Izv. Nauchnoizsled. Inst. Rastenievudstvo Akad. Selskostopansk. Nauki Bulgar. 14:189–201. Chem. Abstr. 59: 5495. 1963.

Istvany, A. Szent-. 18636.

Ita, A. 8354.

9516. Itaru, T. 1954: Germination of orchid seeds from premature pods. Pua Okika Hawaii Nei 4:130–34.

9517. Itijimoto, S. 1954: A study on the seeds of Kentucky 31 fescue. I. About the yielding efficiency and utility value. Viagawa Agr. Coll. Tech. Bull. 6:52–56.

9518. Ito, H. 1931: Physicochemical investigation of proteins of the millet, the Japanese barnyard millet and the Italian millet grains. Gifu Coll. Agr. Res. Bull. 17:1–68. Chem. Abstr. 26:484. 1932.

9519. —— and K. Hayashi. 1960: Studies on the storage of rice seeds. I. Influences of temperature and moisture content on the longevity of rice seeds and methods of drying seeds. Proc. Crop Sci. Soc. Japan 28:363–64.

9520. —— and —— 1960: Studies on the storage of rice seeds. II. The moisture content, viability, and longevity of rice seeds in different methods of processing. Proc. Crop Sci. Soc. Japan 29:97–99.

9521. ——, M. Kitahara, and M. Kawamura. 1931: Differences in the amylolytic activity of the germinating grains of 15 strains of *Setaria italica*, Kunth. Kagami Bull. Imp. Coll. Agr. Gihu 16:1–33. Japan. J. Med. Sci. Biochem. 2:(15).

9522. Ito, I. 1952: Über die Förderung der Samenkeimung einiger *Cuscuta-* und verwandten Arten. Saikyo Univ. Fac. Agr. Sci. Rept. 2:10–16.

9523. —— 1955: Germination of seeds from immature pod and subsequent growth of seedlings in *Dendrobium nobile* Lindl. Saikyo Univ. Fac. Agr. Sci. Rept. 7:35–41.

Ito, K. 18338.

9524. Ito, R. 1936: The existence of a substance which controls the hydrolytic and synthetic actions of lipase. I. J. Biochem. (Tokyo) 23:299–304. Chem. Abstr. 30:4882. 1936.

9525. Ito, S., and S. Hayashi. 1958: Studies on the organic acid metabolism in the germinating stage of rice seeds. Proc. Crop Sci. Soc. Japan 26:242.

9526. —— and —— 1961: Effect of limited oxygen supply on activities of catalase, peroxidase and cytochrome oxidase in germinating rice seeds. Proc. Crop Sci. Soc. Japan 30:97–100.

9527. —— and T. Ishiyama. 1929: On the internal fungus parasites of rice seeds (preliminary report). Sapporo Soc. Agr. Forestry J. 21:218–35.

9528. ——, Y. Koyama, and Y. Toyama. 1963: The separation and structure determination of an eicosatrienoic and an eicosadienoic acid in Nagi seed oil. Bull. Chem. Soc. Japan 36:1439–44. Chem. Abstr. 60:7037. 1964.

9529. —— and T. Nakamura. 1953: Studies in the critical reaction of the protoplasm of crop plants (1). I. Changes of critical reaction in the germinating stages of the seed of the wheat and barley. II. The occurrence of "viviparous head" in the wheat plant. Proc. Crop Sci. Soc. Japan 22:79–80.

9530. —— and H. Takenaga. 1958: Studies on the sugar metabolism in the germinating stage of the lowland rice seeds. Proc. Crop Sci. Soc. Japan 26:241.

9531. Ito, T. 1960: Preparation of a hexokinase from soybean. J. Agr. Chem. Soc. Japan (English Transl.) 34:275–78. Chem. Abstr. 54:18694. 1960.

9532. Itoh, R. 1936: On the existence of a substance which controls the splitting and synthetic action of lipase. I. Report. J. Biochem. (Tokyo) 23:299–304.

9533. —— 1937: Studies on lipase. II. Report. J. Biochem. (Tokyo) 25:167–76.

9534. Iudin, V. G. 1956: Biology of maple seed germination. Tr. Botan. Inst. Akad. Nauk SSSR Ser. 4, Eksperim. Botan. 11:351–69.

9535. —— 1958: Effects of boron and hydrogen peroxide on seed germination of some trees. Tr. Botan. Inst. Akad. Nauk SSSR Ser. 4, Eksperim. Botan. 12:290–98.

9536. —— 1959: Physiological studies on the processes of seed maturation in certain species of maple. Botan. Zh. 44:559–62. Biol. Abstr. 36:42307. 1961.

9537. —— 1959: The dependency of depth of dormancy of seeds of some maple (*Acer* L.) species on the degree of their maturity. Botan. Zh. 44:1748–53.

Ivanescu, M. 9433.

Ivanitskaia, E. F. 19179.

Ivannikova, N. A. 11375.

9538. Ivanoff, S. S. 1947: Effect of presoaking unshelled peanuts on fungus control, germination, and emergence. [Abstract.] Phytopathology 37:433–34.

9539. —— 1950: A laboratory technique for testing cold resistance in oats. [Abstract.] Phytopathology 40:13–14.

9540. —— 1951: The use of activated oat seeds in the study of winter-hardiness. Botan. Gaz. 113:90–94.

9541. —— 1952: A rag-doll method for mass testing of oat seeds for resistance to *Helminthosporium victoriae*. [Abstract.] Phytopathology 42:12.

9542. —— 1956: Radiosensitivity factors in oat seeds: dormancy, water, and development. Science 123:1125–26.

9543. —— 1958: The water-soak method of plant disease control in relation to microbial activities, oxygen supply, and food availability. Phytopathology 48:502–8.

9544. Ivanoff, Z. W., and J. A. Parfentieff. 1927: Über die Einwirkung der Arsenverbindungen auf die Aufkeimung der Samen. Jard. Botan. Inst. Essais Semences Ann. Essais Semences (Leningrad) 5:1–6.

9545. Ivanov, I. K. 1961: Specific features of seeds and seedlings of grapevine (*Vitis* spp.) obtained from grafted plants and those on their own roots. Agrobiologiya 1961(6):814–20. Biol. Abstr. 43: 11838. 1963.

Ivanov, J. 12778.

9546. Ivanov, N. N. 1932: Biochemical search of the alkaloidless lupin. *In* N. N. Ivanov, editor, Problem of the alkaloidless lupin. Bull. Appl. Botany, Genet., Plant Breed. Suppl. 54:5–28.

9547. —— 1932: Über die Veränderlichkeit des Fermentgehaltes in Samen und Früchten. Biochem. Z. 254:71–87.

9548. —— 1933: Cause of the chemical variation of chickpea seeds. Bull. Appl. Botany, Genet., Plant Breed. Ser. 3(1): 3–11. Chem. Abstr. 27:5370. 1933.

9549. —— 1936: Microchemical analysis of seeds without loss of their germinable ability. Mikrochem. Festschr. Hans Molisch. 1936:243–58. Chem. Abstr. 31:4694 1937.

9550. —— and W. F. Grigorieva. 1928: Über die Unveränderlichkeit der ätherischen öles beim Keimen der Anisfrüchte. Biochem. Z. 202:284–93.

9551. ——, ——, and A. I. Ermakov. 1929: Content of essential oil during the process of maturation and germination. Bull. Appl. Botany, Genet., Plant Breed. 21(4):321–41. Chem. Abstr. 26:5605. 1932.

9552. ——, M. N. Lavrova, and M. P. Gapochko. 1931: Die chemische Zusammensetzung der Samen der ölpflanzen in den geographischen Versuchen. Tr. Prikl. Botan. 25:1–86. Chem. Abstr. 26:756. 1932.

9553. —— and M. I. Lishkevich. 1928/29: Catalase in barley seed. Bull. Appl. Botany, Genet., Plant Breed. 21(4):3–46. Biol. Abstr. 6:24631. 1932.

Ivanov, N. N. 5430, 5431, 6027.

9554. Ivanov, S. 1912: Über die Verwandlung des Öls in der Pflanze. Jahrb. Wiss. Botan. 50:375–86.

9555. —— 1932: Climatic influence on the quality of oil from ripening linseed. Allgem. Öl. Fett Ztg. 29:149–50. Chem. Abstr. 27:2053–54. 1933.

Ivanov, S. 15554.

9556. Ivanov, S. L. 1913: Reversibility of the physiological process in the ripening of seeds. Zh. Opytn. Agron. 14:64–74. Expt. Sta. Rec. 29:526–27. 1913.

9557. —— 1913: Respiration coefficient for oil-containing seeds. Zh. Opytn. Agron. 14:75–87.

9558. —— and E. I. Berdichevskii. 1933: Influence of various degrees of humidity on the preservation of oily seeds and composition of oil. Botan. Zh. 18:321–30. Biol. Abstr. 11:6866. 1937.

9559. ——, E. S. Popova, and L. S. Komarova. 1931: Cereal germ oils and their commercial utilization. Tr. Biokhim. Inst. Pichtch. Prom. 1:26–39. Chem. Abstr. 26:4492. 1932.

9560. Ivanov, V. I. 1950: Determining seed germination with the air of fuchsin. Selek. i Semen. 17(2):60–61.

9561. Ivanov, V. K. 1960: How to obtain large sunflower seeds from seed crops. Selek. i Semen. 1960(2):46–48.

9562. Ivanov, Y. A., and B. N. Kulikov. 1962: Response of certain varieties of wheat and barley to Co^{60} irradiation. Nucl. Sci. Abstr. 16:1269.

Ivanova, A. N. 10813.

Ivanova, G. A. 16553.

Ivanova, G. F. 17318.

Ivanova, I. P. 2557.

9563. Ivanova-Aleksandrovskaya, Z. V. 1941: Effect of chloropicrin on the germination capacity of cotton seeds. Proc. Lenin. Acad. Agr. Sci. USSR 6(11):24–26. Chem. Abstr. 37:6809. 1943.

9564. Ivanovic, M. 1954: A study of hard seeds of domestic red clover (*Trifolium pratense* L.) and lucerne (*Medicago sativa* L.). Zbornik Radova Belgrade Univ. Poljoprivredni Fak. 2:105–11.

9565. Ivanovics, G., and S. Horvath. 1947: Isolation and properties of raphanin, an antibacterial substance from radish seed. Proc. Soc. Exptl. Biol. Med. 66:625–30.

9566. Ivanovskaia, E. V. 1946: Hybrid embryos of cereals grown on artificial nutrient medium. Compt. Rend. Acad. Sci. URSS 54:445–48.

9567. —— 1960: Structure of the ovule and the nature of polyembryony of the Papirovka variety of apple. Nauchn. Dokl.

Vysshei Shkoly Biol. Nauki 1960(1):99–102. Biol. Abstr. 41:11899. 1963.

9568. Ivanovskaia, T. 1939: Effect of soaking cottonseed in a nutritive solution on the rate of germination. Sov. Khlopok. 1939(2): 56–57. Chem. Abstr. 34:3000. 1940.

9569. Ivanovskaia, T. L. 1943: The influence of various external conditions on the germination of seeds of wild oats. Proc. Lenin. Acad. Agr. Sci. 1943(3):28–33.

9570. —— 1944: The effect of temperature and moisture on buckwheat seed germination. Proc. Lenin. Acad. Sci. 1944(8/9): 19–28.

9571. —— 1951: Effect of depth of planting on germination of *Caragana* arborescens seeds. Agrobiologiya 1951(2):88–94.

9572. Iverson, J. L., D. Firestone, and W. Horwitz. 1963: Fatty acid composition of oil from roasted and unroasted peanuts by gas-liquid chromatography. J. Assoc. Offic. Agr. Chemists 46:718–25.

9573. Ives, S. A. 1923: Maturation and germination of seeds of *Ilex opaca*. Botan. Gaz. 76:60–77.

9574. Iveshin, D. S. 1960: Reproduction from seed and vegetative reproduction of *Arnica montana* L. and *Gentiana lutea* L. in the Ukrainian carpathians. Botan. J. (USSR) 45:1039–44.

9575. Ivliev, L. A., and D. G. Kononov. 1962: Some mass pests of conifer seeds in Kamchatka. Soobshch. Dal'nevost. Filiala Sibirsk. Otd. Akad. Nauk SSSR 15:83–88. Biol. Abstr. 43:12030. 1963.

9576. Iwai, K., S. Nakagawa, and O. Okinaka. 1962: The growth inhibition of the germinating seeds by sulfonamides and its reversal by folic acid analogues. J. Vitaminol. (Kyoto) 8:20–29.

9577. Iwakawa, M., and S. Kotani. 1954: Effect of colored light on the germination of akamatsu (*Pinus densiflora*) and kuromatsu (*Pinus thunbergii*) seeds. J. Japan. Forestry Soc. 36:249–52.

9578. Iwaki, S., and K. Ota. 1952: Studies on the salt injury in rice plant (preliminary report). I. Absorption and germination of rice seeds in varying concentrations of NaCl. Proc. Crop Sci. Soc. Japan 21:87–89.

Iwakiri, B. T. 7679

Iwakura, Y. 8303.

9579. Iwamoto, K., J. Hayashi, and T. Tanaka. 1952: Thiamine of rice. Sci. Rept. Shiga Agr. Col. Ser. I(2):40–43. Chem. Abstr. 50:6598. 1956.

9580. Iwamoto, Y. 1930: The oil and the lipase-like enzyme in Para rubber seed. J. Soc. Chem. Ind. Japan (Suppl.) 33:409–11. Chem. Abstr. 25:2019–20. 1931.

Iwanissowa, H. P. 8394.

9581. Iwanoff, L. 1911: Über die Wirkung des Sauerstoffs auf die alkoholische Gärung der Erbsensamen. Ber. Deut. Botan. Ges. 29:622–29.

9582. Iwanow, L. 1902: Über Umwandlung des Phosphors beim Keimen der Wickensamen. J. Exp. Landwirtsch. St. Petersburg 1:44. Botan. Centralbl. 95:295–96. 1904.

Iwao, H. 10261.

9583. Iwasaki, F. 1958: Influence of the short-day treatment on germination and early growth in rice seed. [Abstract.] Japan. J. Breeding 8:195–96.

9584. Iwasaki, T. 1952: Investigation on seeds and seedlings of citrus. Bull. Hort. Div. Tokai Kinki Natl. Agr. Expt. Sta. 1:34–48.

Iwata, H. 14161

9585. Iwata, I., and Y. Sakurai. 1956: Action of fumigants on the components of cereals. Eiyo To Shokuryo 9:140–42. Chem. Abstr. 51:11482. 1957.

9586. Iwata, T., and K. Ogata. 1959: Studies in the storage of chestnuts treated with γ radiation. Bull. Univ. Osaka Prefect. Ser. B 9:59–65. Biol. Abstr. 36:38526. 1961.

Iweama, V. O. 18223.

9587. Iyama, S., H. Morishima, and H. I. Oka. 1958: Inheritance of grain size in rice analyzed by statistical genetic methods. [Abstract.] Japan. J. Breeding 8:203–4.

Iyemperumal, S. 14277.

Iyengar, B. A. S. 16782.

Iyengar, B. T. R. 7850.

9588. Iyengar, C. V. K. 1937: Development of embryo-sac and endosperm-Haustoria in some members of the Scrophularineae. I. An account of *Sopubia delphinifolia* G. Don. and Alonsoa sp. J. Indian Botan. Soc. 16:99–109.

9588A. —— 1939: Development of embryo-sac and endosperm-Haustoria in some members of Scrophularineae. II. *Iso plexis canariensis* Lindl. and *Celsia coromandeliana* Vahl. J. Indian Botan. Soc. 18:13–20.

9588B. —— 1939: Development of embryo-sac and endosperm-Haustoria in some members of Scrophularineae. III. *Limnophila heterophylla* Benth., and *Stemodia* *viscosa*, Roxb. J. Indian Botan. Soc. 18:35–42.

9589. —— 1940: Development of embryo-sac and endosperm-Haustoria in some members of Scrophularineae. IV. *Vandellia hirsuta* Ham and *Vandellia scabra* Benth. J. Indian Botan. Soc. 18:179–90. Biol. Abstr. 15:19314. 1941.

9590. —— 1940: Development of embryo-sac and endosperm-Haustoria in some members of Scrophularineae. V. *Ilysanthes hyssopioides* Benth., and *Bonnaya tenuifolia* Spreng. J. Indian Botan. Soc. 19:5–18. Biol. Abstr. 15:19315. 1941.

9591. —— 1940: Structure and development of seed in *Sopubia trifida* Ham. J. Indian Botan. Soc. 19:251–61.

9592. —— 1944: Variation in the rate of respiration of a germinating seed. J. Indian Botan. Soc. 23:9–20.

Iyengar, J. R. 18125.

Iyengar, M. A. S. 18332.

9593. Iyengar, N. K. 1943: Standards for papain and its preparations. Indian J. Med. Res. 31:211–14. Chem. Abstr. 38:6499. 1944.

Iyengar, R. L. N. 15651.

Iyer, B. H. 10314, 13978.

9594. Iyer, R. D., K. Sulbha, and S. Ramanujam. 1960: Embryo culture studies in jute and tomato. Mem. Indian Botan. Soc. 2:30–35. 1959.

Iyer, R. R. 3922.

9595. Iyer, Y. V. S. 1935: Characteristics of sandalwood seeds and seed oil (Mysore). Analyst 60:319–20.

Izakov, F. Y. 1591.

9596. Izard, C. 1956: Recherches sur la physiologie de la germination des graines immatures de *N. tabacum* L. en présence de sels d'urane. Ann. Inst. Exptl. Tabac Bergerac 2(3):61–95. Tobacco Abstr. 2:153–54. 1958.

9597. —— 1956: Sur la germination des graines immatures de certaines variétés de tabac, en présence du lumière et de sels d'urane. Compt. Rend. 242:2027–30.

9598. —— 1958: Biochemical investigations of the seeds of *Phelypaea ramosa* (*Orobanche ramosa*), a parasite of tobacco. I. Changes in amino acids and sugars. II. Isolation of a flavonoid pigment. Ann. Inst. Exptl. Tabac Bergerac 3(1):77–83. Chem. Abstr. 53:14239. 1959.

9599. —— and H. Hitier. 1954: Nouvelles recherches sur la germination des graines de l'Orobanche parasite du tabac. Ann. Inst. Exptl. Tabac Bergerac 2:9–13.

9600. —— and —— 1958: Action du 1.3-dichloropropane 1.2-dichloropropene, de la rindité et de la gibberelline, sur la germination des graines de l'Orobanche, parasite du tabac. Compt. Rend. 246: 2659–61.

9601. —— and J. Masquelier. 1958: Flavonoid pigment extracted from the seed of *Phoelipaea ramosa* L. Compt. Rend. 246:1454–56. Chem. Abstr. 52:12101. 1958.

9602. —— and P. Schiltz. 1961: Observations sur la germination des graines de quelques *Nicotianae* résistantes au mildew du tabac (*Peronospora tabacina* Adam.). Compt. Rend. 252:3113–15.

Izard, C. 7289, 8866, 8867.

9603. Izawa, G., and S. Okamoto. 1962: Transport of chemical constituents in germinating plant seeds. I. Transport of nitrogenous constituents in germinating soybean seeds. Hyogo Noka Daigaku Kenkyu Hokoku, Nogei-kagaku Hen 5:101–8. Chem. Abstr. 60:2052. 1964.

Izawa, M. 9350.

9604. Izvoshchikov, V. P. 1935: The influence of hydrogen-ion concentration on the germination of tobacco seeds. Vses. Nauchn.-Issled. Inst. Tabach. Makhoroch. Promy. 118:143–48. Chem. Abstr. 32: 4195. 1938.

J

Jaarma, M. 5834.

Jabbar, J. 19275.

9605. Jablonski, M. 1962: Viability and real value of *Crambe abyssinica* Hochst. fruits. Pamietnik Pulawski 8:145–240. Biol. Abstr. 45:7729. 1964.

Jabolonski, J. R. 13485.

9606. Jacalne, D. V. 1957: Summary of some germination studies in the Division of Forest Investigations, Bureau of Forestry College, Laguna. Forestry Leaves 9(3/4): 11–17.

9607. —— 1962: Further study on hastening the germination of akle (*Albizzia acle*) seeds. Philippine J. Forestry 15:63–72. 1959. Biol. Abstr. 44:3885. 1963.

9608. Jack, H. W., and R. B. Jagoe. 1932: Variation in fruiting ability of oil palms. Malayan Agr. J. 20:14–19.

9609. Jack, J. G. 1893: Fructification of *Juniperus*. Botan. Gaz. 18:369–75.

9610. —— 1914: Plural seeds in acorns. Rhodora 16:141–144.

9611. Jackobs, J. A. 1947: Factors affecting seed rotting caused by *Pythium* spp. in sweet-clover with preliminary tests in alfalfa and red clover. J. Am. Soc. Agron. 39:702–18. Rev. Appl. Mycol. 26:549. 1947.

9612. —— and C. N. Hittle. 1958: Variations among seedlots of certified Ladino clover and other white clovers. Agron. J. 50:327–30.

Jackobs, J. A. 3626, 15621.

9613. Jacks, H. 1951: The efficiency of chemical treatments of vegetable seeds against seed-borne and soil-borne organisms. Ann. Appl. Biol. 38:135–68.

9614. —— 1951: Seed disinfection. I. Preliminary selection of vegetable-seed protectants. New Zealand J. Sci. Technol. Sect. A 33:27–38.

9615. —— 1952: Seed disinfection. II. Effect of various chemicals on the emergence of lettuce and pea seedlings. New Zealand J. Sci. Technol. Sect. A 34:122–25.

9616. —— 1952: Seed disinfection. III. Effect of seed dressing applied at different concentrations on emergence of vegetable seedlings. New Zealand J. Sci. Technol. Sect. A34:202–5.

9617. —— 1952: Seed disinfection. IV. Effect of seed dressings, applied separately or as mixtures, on emergence of certain vegetable seeds. New Zealand J. Sci. Technol. Sect. A 34:206–8.

9618. —— 1953: Disinfection of vegetable seeds. New Zealand J. Agr. 87:217–24.

9619. —— 1953: Seed disinfection. V. A seed-dressing apparatus for the smaller seedsman. New Zealand J. Sci. Technol. Sect. A 34:249–51.

9620. —— 1953: Seed disinfection. VIII. Control of weevil damaging French beans and maize seed in storage. New Zealand J. Sci. Technol. Sect. A 35:159–63.

9621. —— 1954: Seed disinfection. X. Effect of seed protectants on emergence of spinach. New Zealand J. Sci. Technol. Sect. A 36:129–33.

9622. —— 1956: Seed disinfection. XII. Glasshouse tests for control of damping-off of *Pinus radiata* seed. New Zealand J. Sci. Technol. Sect. A 37:427–31.

9623. —— 1959: Seed disinfection. XV. Glasshouse tests for control of damping-off of pea seed. New Zealand J. Agr. Res. 2:306–11.

9624. —— 1960: Seed disinfection. XVI. A note on the effects of fungicides on viability of onion seed. New Zealand J. Agr. Res. 3:194–95.

9625. —— and I. A. M. Cruickshank. 1956: Seed disinfection. XIII. The effect of fungicidal seed treatments on emergence and covered smut (*Ustilago kolleri* Wille) of oats. New Zealand J. Sci. Technol. Sect. A 38:27–29.

9626. —— and G. J. Graham. 1955: Seed disinfection. XI. Control of head smut (*Sorosporium reilianum* [Kuhn] McAlpine) of maize. New Zealand J. Sci. Technol. Sect. A 37:141–45.

Jacks, H. 4704.

9627. Jackson, A. A., and P. Finn-Kelcey. 1956: Drying hybrid sweet-corn seed. Agriculture (London) 63:222–24.

Jackson, A. A. 2298.

Jackson, A. M. 2152.

9628. Jackson, C. R. 1964: Field comparisons of seed-treatment fungicides for control of *Aspergillus* crown rot and other seedling diseases of peanut. Plant Disease Reptr. 48:264–67.

9629. Jackson, C. V. 1928: Seed germination in certain New Mexico range plants. Botan. Gaz. 86:270–94.

9630. Jackson, G. A. D., and J. B. Blundell. 1963: Germination in *Rosa*. J. Hort. Sci. 38:310–20.

Jackson, H. 7620.

9631. Jackson, J. F., and K. Lindahl-Kiessling. 1963: Mitogenic activity of bean endosperm, embryo and sprout extracts in human leucocyte cultures. Hereditas 49:427–32.

Jackson, J. R. 13960.

9632. Jackson, L. W. R. 1952: Influence of littleleaf on quality of shortleaf pine seed. Phytopathology 42:57–58.

Jackson, R. W. 8621.

9633. Jackson, S. H., and A. G. O. Whiteside. 1942: Thiamine content of Canadian wheat and flour. Sci. Agr. (Ottawa) 22:366–77. Chem. Abstr. 36:4918. 1942.

Jackson, T. W. 10666.

9634. Jackson, W. E. 1955: Seeding and growing black (yellow) locust trees. Kentucky Coll. Agr. & Home Ex. Agr. Ext. Serv. Circ. 531. 14 pp.

9635. Jacob, K. T. 1949/50: X-ray studies in jute. II. A comparative study of the

germination percentage, size and external morphology, with different doses of X-rays. Trans. Bose Res. Inst. (Calcutta) 18:23–29.
Jacob, K. T. 13880.

9636. Jacobi, G. 1928: Untersuchungen über die Wirkung des ultravioletten Lichtes auf Keimung und Wachstum. Beitr. Biol. Pflanz. 16:405–64.

9637. Jacobi, H. 1910: Über den Einfluss der Verletzung von Kotyledonen auf das Wachstum von Keimlingen. Flora (Jena) 101:279–89.

9638. Jacobs, A. W. 1924: Polyembryonism in sugar pine. J. Forestry 22:573–74.

9639. —— 1925: Hastening the germination of sugar pine seeds. J. Forestry 23:919–31.

9640. Jacobs, M. R. 1937: Tropical water lilies from seed. Horticulture 15(5):118.

9641. —— 1940: Spider lilies grown from seed. Horticulture 18(20):416.

9642. Jacobs, S. E., and A. H. Dadd. 1959: Antibacterial substances in seed coats and their role in the infection of sweet peas by *Corynebacterium fascians* (Tilford) Dowson. Ann. Appl. Biol. 47:666–72.
Jacobs, S. E. 4808.

9643. Jacobs, W. A., and L. C. Craig. 1943: The aconite alkaloids. J. Biol. Chem. 147:567–69, 571–72.

9644. Jacobsen, I. 1929: Untersuchungen von Gehölzsamen an der dänischen Staatssamenkontrolle. Z. Forst.-Jagdw. 61:141–56.
Jacobson, A. 16347.

9645. Jacobson, H. G. M. 1948: A technique for introducing radioactive phosphorus into the grain of growing corn. Plant Physiol. 23:636–37.

9646. Jacobson, M., F. Acree, Jr., and H. L. Haller. 1944: Determination of sesamin. Ind. Eng. Chem. Anal. Ed. 16:166–67. Expt. Sta. Rec. 95:155. 1946.

9647. Jacobsson, K. 1955: The effect of urea on the inhibition of trypsin by soybean trypsin inhibitor. Biochim. Biophys. Acta 16:264–67.

9648. Jacobus, M. R. 1944: Rose seed germination and growing the young seedlings. Am. Rose Ann. 29:86–88.

9649. Jacoby, B., and H. R. Oppenheimer. 1962: Pre-sowing treatment of sorghum grains and its influence on drought resistance of the resulting plants. Phyton 19:109–13. XII.
Jacoby, H. 3462.

9650. Jacques, R. 1957: Quelques données sur la photopériodisme de *Chenopodium polyspermum* L.; influence sur la germination des graines. Colloq. Intern. Photothermopériodism Ser. B (Colloq.) No. 34.

9651. Jacques, W. A., and L. Corkill. 1932: Colour of seed in crested dogstail (*Cynosurus cristatus*). Its relation to weight, vitality and germination-capacity. New Zealand J. Sci. Technol. 13:221–25.

9652. Jacquot, R., and F. Harding. 1947: Les variations du tryptophane au cours de la germination de *Phaseolus multiflorus.* Compt. Rend. 224:1576–78.

9653. —— and A. Mayer. 1926: Hydratation et respiration des graines. Ann. Physiol. Physicochim. Biol. 2:408–25.
Jacquot, R. 3499, 3839, 4578, 7814.

9654. Jadhav, A. S. 1960: Canna seeds can be germinated. Poona Agr. Coll. Mag. 51:28–29.

9655. Jadhav, K. A., and J. W. Airan. 1960/61: Amino acids during germination of *Cicer arietinum.* Bombay Univ. J. (N.S.) 29(3/5, Sect. B):11–15. Chem. Abstr. 57:17076. 1962.
Jaeger, C. M. 12653.
Jäger, H. 11294.

9656. Jaeger, P. 1942: À propos de la germination des graines de *Gentiana lutea.* Bull. Soc. Bot. France 89(7/9):145–49; (10/11):210–14.

9657. —— 1948: Sur la germination de la graine de l'*Androcymbium gramineum* Macbride en présence de colchicine. Compt. Rend. 227:730–31.
Jaeger, P. 3448.

9658. Jaffe, W. G. 1949: Limiting essential amino acids of some legume seeds. Proc. Soc. Exptl. Biol. Med. 71:398–99.

9659. —— 1962: Hemagglutinating and toxic protein fractions from beans. Experientia 18:76–77.
Jagendorf, T. G. 12837.

9660. Jagoe, R. B. 1954: An experiment in the storage of under-milled rice. Malayan Agr. J. 37:154–68.
Jagoe, R. B. 9608.

9661. Jahnel, H. 1955: Beiträge zum Stratifizieren von Forstsaatgut. I. Angew. Botan. 29:139–51.

9662. —— 1956: Beiträge zum Stratifizieren von Forstsaatgut. II. Angew. Botan. 30:185–201. Biol. Abstr. 32:6227. 1958.

9663. —— 1961: Zur Frage der Bestimmung der Keimfähigkeit in der Saatgutprüfung überliegenden Forstsaatgutes (Schnittprobe, Keimprüfung, Tetrazolium). Angew. Botan. 35:107–16.

9664. —— and W. Zentsch. 1961: Beiträge zum Stratifizieren von Forstsaatgut. IV. *Fraxinus excelsior* L. Angew. Botan. 34:221–45.

9665. Jain, D. N., and R. R. Agarwal. 1938: On the amylase from the Indian water chestnut (*Trapa bispinosa* Roxb.). II. Studies with the purified powdered form of the enzyme. Proc. Indian Acad. Sci. Sect. A 7:384–89. Biol. Abstr. 14:3610. 1940.

Jain, K. 14204.

9666. Jain, S. S. 1952: Some important seed-borne diseases of cereal crops in the Uttar Pradesh and their control. Agr. Animal Husbandry (India) 3:34–36.

Jakabsons, V. 14954.

Jakobsson, J. 250.

9667. Jakovleo, V. G., and M. D. Yoffe. 1957: On the embryogenesis in *Paeonia* L. J. Botan. URSS 42:1491–502.

9668. Jakovlev, M. S. 1938: The endosperm structure of the principal selection varieties of wheat of the URSS. Compt. Rend. Acad. Sci. URSS 18:203–6. Bull. Soc. Botan. France 87:172. 1940.

9669. —— 1939: Functional significance of epiblast in wheat embryo. Compt. Rend. Acad. Sci. URSS 22:637–39.

9670. Jakovuk, A. S. 1961: Methods of germination and the cold resistance of tobacco seeds. Krasnodar Vses. Nauchn.-Issled. Inst. Tad. Makhorki Sb. Nauchn.-Issled. Rabot (152):71–73. Tobacco Abstr. 8:52–53. 1964.

Jakowlew, M. 343.

9671. Jakushkin, I. V., and N. S. Torman. 1940: On the influence of regions of cultivation on seed qualities of wheat. Proc. Lenin. Acad. Agr. Sci. USSR 6:3–11.

Jalan, S. 10239.

9672. Jalander, Y. W. 1911: Zur Kenntnis der *Ricinus*lipase. Biochem. Z. 36:435–76.

9673. Jamalainen, E. A. 1936: Internal necrosis of pea seeds. Valtion Maatalouskoetoiminnan Julkaisuja 76. 6 pp. Rev. Appl. Mycol. 16:83. 1937.

9674. —— 1953: The significance of cereal seed dressing in Finland. Höfchen Briefe 6:80–83.

Jamblinne, A. T. de; *see* De Jamblinne, A. T.

9675. Jambor, R. 1935: Beiträge zur Kenntnis der Hartschaligkeit der Luzernesamen. Mat. Termeszett. Ertesitö Magyar Tud. Akad. 52:474–88.

9676. —— 1943: The germination of spinach seed. Kiserl. Kozl. 46:301. Biol. Abstr. 21:10020. 1947.

9677. —— 1960: Über die Keimfähigkeit von gelagerten Arzneifpflanzensamen. Planta Med. 8:152, 153–59.

9678. —— 1962: Tasks of evaluation of the apparently well-preserved seeds of some drug and spice plants. Z. Landwirtsch. Versuchs- Untersuchungsw. 8:459–63.

9679. James, A. L. 1940: The carbohydrate metabolism of germinating barley. New Phytologist 39:133–44.

James, A. L. 9701.

9680. James, E. 1945: A clover huller for head samples. J. Am. Soc. Agron. 37:158–60.

9681. —— 1949: The effect of inbreeding on crimson clover seed-coat permeability. Agron. J. 41:261–66.

9682. —— 1961: An annotated bibliography on seed storage and deterioration. A review of 20th Century literature reported in the English language. U.S. Dept. Agr. Agr. Res. Serv. ARS 34-15-1. 81 pp.

9683. —— 1961: Perpetuation and protection of germ plasm as seed. *In* R. E. Hodgson, Germ plasm resources. AAAS Publ. No. 66, pp. 317–26.

9684. —— 1961: Policy on seed deposits at National Seed Storage Laboratory. Seed World 89(8):28–29.

9685. —— 1963: An annotated bibliography on seed storage and deterioration. A review of 20th Century literature reported in foreign languages. U.S. Dept. Agr. Agr. Res. Serv. ARS 34-15-2. 30 pp.

9686. —— and T. A. Bancroft. 1951: The use of half-plants in a balanced incomplete block in investigating the effect of calcium, phosphorus, and potassium, at two levels each, on the production of hard seed in crimson clover, *Trifolium incarnatum*. Agron, J. 43:96–98.

James, E. 2074–2076.

9687. James, E. W. 1958: The storage of acorns in water. Forestry 31:163–66.

9688. James, G. M., and W. O. James. 1940: The formation of pyruvic acid in barley respiration. New Phytologist 39:266–70.

James, G. M. 9699.

9689. James, L. H. 1942: Poisoning by mowrah seed (*Bassia latifolia*). New Zealand J. Sci. Technol. Sect. B 23:200.

9690. James, N., and A. R. Lejeune. 1952: Microflora and the heating of damp stored wheat. Can. J. Botany 30:1–8.

9691. James, W. M. 1938: Vernalization and phasic development with special reference to amaryllids. Herbertia 5:155–56.
9692. —— 1939: Treatment of amaryllid and other seeds with hormone powder. Herbertia 6:206–39.
9693. —— 1940: Growing *Alstroemerias* from seeds. Herbertia 7:169–70.
9694. James, W. O. 1947: Biosynthesis of the belladonna alkaloids. Nature 159:196–97. Biol. Abstr. 21:25996. 1947.
9695. —— 1949: The amino-acid precursors of the belladonna alkaloids. New Phytologist 48:176–85.
9696. —— and S. E. Arney. 1939: Phosphorylation and respiration in barley. New Phytologist 38:340–51.
9697. —— and D. Boulter. 1955: Further studies on the terminal oxidases in the embryos and young roots of barley. New Phytologist 54:1–12.
9698. —— and J. M. Cragg. 1941: Ascorbic acid system in barley. Nature 148:726–27.
9699. ——, C. R. C. Heard, and G. M. James. 1944: On the oxidative decomposition of hexosediphosphate by barley: the role of ascorbic acid. New Phytologist 43:62–74.
9700. —— and F. B. Hora. 1940: The effect of cyanide on the respiration of barley. Ann. Botany (London) 4:107–18.
9701. —— and A. L. James. 1940: The respiration of barley germinating in the dark. New Phytologist 39:145–76.
James, W. O. 3337, 9688.
9702. Jameson, H. R. 1950: Control of turnip flea beetle by benzene hexachloride drilled with the seed. Nature 165:980.
9703. —— and I. W. M. Callan. 1951: Effect of γ-BHC seed dressings on the germination of cereals. Nature 167:490.
9704. ——, F. J. D. Thomas, and C. C. Tanner. 1951: The control of wireworm by γ benzene hexachloride; the development of a seed dressing for cereals. Ann. Appl. Biol. 38:121–34.
Jamieson, C. A. 12662.
9705. Jamieson, G. S. 1939: Ben (Moringa) seed oil. Oil Soap 16:173, 174. Expt. Sta. Rec. 83:150. 1940.
9706. —— 1943: Vegetable fats and oils. Reinhold, New York. 2nd ed. 508 pp.
9707. —— 1946: The oiticica tree of Brazil and the oil from its seeds. N.Y. Botan. Garden J. 47:243–46.
9708. ——, W. F. Baughman, and R. S. McKinney. 1933: Oil content of nine varieties of soybeans and the characteristics of the extracted oils. J. Agr. Res. 46:57–58.
9709. —— and S. I. Gertler. 1930: American cherry-kernel oil. Oil Fat Ind. 7:371–72, 387. Soc. Chem. Ind. J. 50:31B. 1931.
9710. —— and R. S. McKinney. 1932/33: Determination of free fatty acids in cottonseed. Oil Soap 9:228–31; 10:9–11. Soc. Chem. Ind. J. 52:876B. 1933.
9711. —— and —— 1933: California apricot [-kernel] oil. Oil Soap 10:147–49. Soc. Chem. Ind. J. 52:877B. 1933.
9712. —— and W. G. Rose. 1940: *Cocos pulposa* palm kernel oil. Oil Soap 17:144. Expt. Sta. Rec. 85:294. 1941.
9713. —— and —— 1943: Mexican prickly poppy seed oil. Oil Soap 20:33–35.
9714. —— and —— 1943: The seed oil of *Garcia nutans*. Oil Soap 20:202–3. Biol. Abstr. 18:3451. 1944.
Jamieson, G. S. 13219, 13221.
Jamieson, M. 3030.
9715. Jaminet, F. 1954: Biochemical study of *Sarothamnus scoparius*. II. Degradation of sparteine during maturation of seeds. J. Pharm. Belg. 8:9–19. Chem. Abstr. 48:11553. 1954.
9716. Janchen, E. 1937: Der morphologische Wert der Gramineen-vorspelze. Österr. Botan. Z. 87:51–61.
Janczewski, M. E. de; *see* De Janczewski, M. E.
9717. Jane, F. W., and B. Russell-Wells. 1935/36: Observations on the seeds and seedlings of *Utricularia vulgaris*, L. I. Development and morphology of the seedling by F. W. Jane. II. Growth of seedlings under different conditions by F. W. Jane and B. Russell-Wells. Trans. Norfolk Norwich Nat. Soc. 14:31–54. Biol. Abstr. 11:2905. 1937:
Janecke, H. 5353.
9718. Janes, B. E. 1941: Some chemical differences between artificially produced parthenocarpic fruits and normal seeded fruits of tomato. Am. J. Botany 28:639–46.
9719. —— 1942: A comparison of the chemical composition of artificially produced parthenocarpic fruits and normal seeded fruits of peppers. Proc. Am. Soc. Hort. Sci. 40:432–36.
Jang, R. 9736.
9720. Janick, J., and E. C. Stevenson. 1955: Environmental influences on sex expression in monoecious lines of spinach. Proc. Am. Soc. Hort. Sci. 65:416–22.

9721. Janicki, J. 1937: Das Oxydationsreduktionspotential der wässerigen Auszüge der keimenden Gerste. Enzymologia 4:107–10.

9722. —— and E. Kaminski. 1962: Analysis of vitamins in cereals and cereal products. Results obtained by the vitamin committee of the International Association for Cereal Chemistry. Roczniki Technol. i Chem. Zywnosci 9:5–16. Anal. Abstr. 10:2475. 1963.

Janicki, J. 4093–4098.

9723. Janisevskii, D. E., and N. V. Pervuhina. 1941: Prolonging the life of seeds which lose their viability quickly. Sov. Botan. (Leningrad) 1941(3):80–86. Biol. Abstr. 21:25989. 1947.

9724. Janjatovic, V., and J. Stankovic. 1956: Effect of herbicides on the germination per cent and growth of plants. (*Zea mays*, *Convolvulus arvensis* and *Agrostema githago*.) Matica Srpska Zb. 1956(11):150–56.

9725. Jankovic, M. M. 1955: (*Trapa* L.). I. Die Frage der Absezung der individuellen Entwicklung der Wassernus. Arch. Biol. Nauka (Beograd) 7(3/4):15–23. Biol. Abstr. 33:23070. 1959.

9726. Jankowski, F. 1960: Effect of storage conditions on the viability of tobacco seed. Roczniki Nauk Rolniczych Ser. A 81:739–55. Tobacco Abstr. 5:209–10. 1961.

9727. Jankowski, S., and M. Andrzejewski. 1961: Chemical and technological properties of grain of several tetraploid and nexaploid wheat hybrids. Plant Breeding, Acclimatiz., Seed Production 5:74–75.

9728. Jannaccone, A. 1935: La printanisation des plantes cultivées, en theorie et en pratique. Ann. Tec. Agrar. 8:216–25, 270–87. Ann. Agron. 6:336. 1936.

9729. —— 1938: Sulla conservazione del potere germinativo del seme di canapa. Ann. Tec. Agrar. 11:57–63.

9730. —— 1956: Hard wheats and the evaluation of "shriveling." Ital. Agr. 93:509–17.

9731. Jannicke, W. 1889: Gekeimte Samen in Früchten von *Impatiens longicomis*. Ber. Deut. Botan. Ges. 7:318.

9732. Janos, W., F. Nagymihaly, and A. M. Plosz. 1950: Researches on the biology of germination. II. Agrartudomany 2: 206–9.

9733. Janot, M. M., and R. Mouton. 1930: The toxicity of the seeds from *Artemisia maritima* L. compared with that of santonin. Bull. Sci. Pharmacol. 37:593–99. Chem. Abstr. 25:743. 1931.

9734. ——, E. Saïas, and M. Foucher. 1953: Chromatographie sur papier de quelques glucosides végétaux. Bull. Soc. Chim. Biol. 35(10):1101–10.

Janot, M. M. 4577.

Janovicova, J. Chalupova-; *see* Chalupova-Janovicova, J.

9735. Jansen, E. C. 1952: Quality improvement of trees for seed production. Ned. Heidemaatsch. Tijdschr. 63:310–12.

9736. Jansen, E. F., R. Jang, and L. R. MacDonnell. 1947: Citrus acetylesterase. Arch. Biochem. 15:415–31.

9737. Janson, L. 1960: The effect of depth on sowing pine seed. Sylwan 104:51–60. Biol. Abstr. 45:12655. 1964.

9738. Jansson, G. 1956: An investigation of moisture content and enzyme activities in barley seedlings in relation to their growth rate. Arkiv Kemi 9:139–45. Biol. Abstr. 30:29751. 1956.

9739. —— 1959: Germination experiments with water-sensitive barley. Arkiv Kemi 14:161–69. Biol. Abstr. 35:11318. 1960.

9740. —— 1959: Chemically induced water-sensitivity in barley seeds. Arkiv Kemi 14:279–89. Biol. Abstr. 35:11225. 1960.

9741. —— 1960: Breaking of water sensitivity in barley seeds by treatment with chemicals. Arkiv Kemi 15:439–50. Chem. Abstr. 55:2818. 1961.

9742. —— 1961: Induction of water-sensitivity in barley by treatment with water. Arkiv Kemi 17:281–89. Biol. Abstr. 36:82055. 1961.

9743. —— 1962: Respiration and phosphorylation in water-sensitive barley. Arkiv Kemi 19:141–48. Chem. Abstr. 57: 5035. 1962.

9744. —— 1962: A study on 3-indoleacetic acid in barley seeds in relation to water-sensitivity. Arkiv Kemi 19:149–59. Biol. Abstr. 42:11676. 1963.

9745. —— 1962: Some properties of a substance in barley seeds which induce a lag period in the enzymic oxidation of 3-indoleacetic acid. Arkiv Kemi 19:161–71.

9746. —— 1962: Water-sensitivity in barley seeds. Svensk Kem. Tidskr. 74:181–94. Chem. Abstr. 57:6315. 1962.

Jantz, O. K. 19631.

9747. Japing, H. W. 1935: Keimingsonderzoek van zaden. Bergcultures 9:8–21.

9748. Jaranowski, J. 1961: Haploid-diploid twin embryos in *Melilotus*. Genet. Polon. 2:129–37.

9749. —— 1962: Development of embryos and seeds of certain species and species crosses in *Melilotus*. Genet. Polon. 3:45–59. Biol. Abstr. 43:11626. 1963.
9750. —— 1962: Fertilization and embryo development in the genus *Lupinus*. I. Seed development in cases of autogamy. Genet. Polon. 3:209–46. Biol. Abstr. 44:3518.
9751. Jaretzky, R., and H. Ulbrich. 1934: Intraplasmatic processes of mucilage formation in the seed of *Linum usitatissimum* L. and in the roots of *Althaea officinalis* L. Arch. Pharm. 272:796–811. Chem. Abstr. 29:830. 1935.
Jarlov, E. 6868.
9752. Jaroschewskij, P. E. 1938: Über die Jarowisation und andere Methoden zur Vorbereitung der Zuckerrübensamen zur Saat. Zuckerrübenbau 1938(3):36–43. Botan. Centralbl. 174:287. 1939.
9753. Jarosz, W. 1933: The hydrolysis of phytic compounds derived from seeds of hemp, horse-bean, flax, horse-chestnut and wheat and from embryos of rye. Bull. Intern. Acad. Polon. Sci. Classe Sci. Math. Nat. B 1:123–33. Chem. Abstr. 28:5501. 1934.
9754. Jart, A. 1959: The fatty acid composition of shea butter and olive kernel oil. Acta Chem. Scand. 13:1723–24.
9755. Jaruis, P. M. 1946: Industry makes markets for cottonseed. Proc. Cotton Res. Congr. 7:76–82. Biol. Abstr. 22:17044. 1948.
9756. Jarvis, P. G. 1963: The effects of acorn size and provenance on the growth of seedlings of sessile oak. Quart. J. Forestry 52:11–19. Biol. Abstr. 44:3886. 1963.
Jasinska, Z. 19793.
9757. Jasnowski, S. 1934: On the inheritance of weight of grains in *Triticum vulgare*. Roczniki Nauk Rolniczych Lesnych 33: 59–68.
Jasperson, H. 8768.
9758. Jaubert, P. 1952: Poursuite des études sur l'emploi depondres fongicides pour le traitement des semences d'Arachides. Bull. Agron. Fr. d'Out. Mer 7:167–76. Rev. Appl. Mycol. 33:334. 1954.
9759. Jauerka, O. 1921: Die ersten Stadien der Kohlensaureausscheidung bei quellenden Samen. Beitr. Biol. Pflanz. 2:193–248. Botan. Centralbl. 123:19–20. 1913.
9760. Jauhare, O. S., and A. Prasad. 1963: Effect of 2,4,5-trichlorophenoxyacetic acid on fruit set, size, ripening, and quality of *Phalsa*. Sci. Cult. (Calcutta) 29:42–43. Chem. Abstr. 59:5705. 1963.
Jauregg, T. Wagner-; *see* Wagner-Jauregg, T.
Jayaraman, V. 9351.
9761. Jayasekera, E. W. H. 1956: Techniques of seed and seedling production with sugar cane. Trop. Agriculturist 112:262–66.
Jedlitschka, F. 15701, 15702.
9762. Jedrychowska, A., and A. Sroczynska. 1934: On the cytology and embryology of *Ceratophyllum submersum*. Acta Soc. Botan. Polon. 11:423–41.
Jeffery, G. L. 6556.
9763. Jeffrey, E. C. 1895: Polyembryony in *Erythronium americanum*. Ann. Botany 9(26):537–41.
9764. —— 1948: Hormones in relation to parthenogenesis. Genetics 33:615. Biol. Abstr. 23:13708. 1949.
9765. —— and E. J. Haertl. 1939: Apomixis in *Trillium*. Cellule 48:77–88. Biol. Abstr. 14:1887. 1940.
9766. Jeffreys, M. D. W. 1959: Length of viability of luffa seeds. S. African J. Sci. 55:193.
Jelenkovic, G. 1548.
9767. Jelitto, K. A. 1956: Die Reaktion von Samen verschiedener Kulturarten bei tiefen Lagertemperaturen. Saatgut-Wirtsch. 8: 112–15.
9768. Jelitto, L. 1933: Aussaat der Kopfprimeln. Gartenwelt 37:234.
9769. Jelley, M. E. 1937: Eastern red cedar. J. Forestry 35:865–67.
9770. Jelliffe, D. B., G. Arroyave, F. Aguirre, A. Aguirre, and N. S. Scrimshaw. 1959: Amino-acid composition of certain tropical cereal grains and legumes. Bol. Ofic. Sanit. Panam. Suppl. No. 3:197–99. Chem. Abstr. 53:19044. 1959.
9771. Jemison, G. M., and C. F. Korstian. 1944: Loblolly pine seed production and dispersal. J. Forestry 42:734–41.
9772. Jenkin, T. J. 1936: Uncertified New Zealand rye-grass seed. Gt. Brit. Min. Agr. J. 43:109–16.
Jenkins, E. W. 4744.
Jenkins, J. A. 17889.
Jenkins, J. E. E. 4699.
9773. Jenkins, J. M., Jr. 1949: Pelleted seed improves lettuce stand, requires less thinning. Research & Farming 1949: N. Carolina Agr. Expt. Sta. Ann. Rept. 72:20. 1950.
9774. Jenkins, M. E. 1961: The effect of gibberellin on the germination rate of Eastern white pine. J. Forestry 59:210–11.
Jenkins, M. T. 17814.

9775. Jenkins, R. R., and C. B. Sayre. 1936: Chemical studies on the open-pollinated, top-cross, and hybrid inbred strains of yellow sweet corn in relation to the quality of the canned product. Food Res. 1:199–216. Expt. Sta. Rec. 77:46. 1937.

Jenkins, R. R. 11981.

Jenkins, W. A. 8625.

Jenkins, W. H. 19367.

9776. Jenkins, W. R. 1951: The effect of 2,4-dichlorophenoxyacetic acid on the germination and seedling development of *Allium fistulosum.* Virginia J. Sci. 2:318–19.

9777. Jenne, E. A., H. F. Rhoades, C. H. Yien, and O. W. Howe. 1958: Change in nutrient element accumulation by corn with depletion of soil moisture. Agron. J. 50:71–74.

9778. Jennings, A. C. 1961: Determination of the nitrogen content of cereal grain by colorimetric methods. Cereal Chem. 38: 467–78.

9779. —— and R. K. Morton. 1963: Changes in carbohydrate, protein, and non-protein nitrogenous compounds of developing wheat grain. Australian J. Biol. Sci. 16:318–31.

9780. —— and —— 1963: Amino acids and protein synthesis in developing wheat endosperm. Australian J. Biol. Sci. 16: 384–94. Biol. Abstr. 45:7605. 1964.

Jennings, A. C. 7567.

9781. Jennings, P. R., and R. K. Zuck. 1955: The cotyledon in relation to photoperiodism in cocklebur. Botan. Gaz. 116:199–200.

Jenschke, G. Masitschka-; *see* Masitschka-Jenschke, G.

9782. Jensen, C. 1941: Is it possible that seeds through treatment with light may keep their germinating power through a longer span of years than normal? J. D. Quist & Co., Bogtrykkeri Aktieselskab, Copenhagen. 16 pp. Expt. Sta. Rec. 85:343. 1941.

9783. Jensen, C. O., W. Sacks, and F. A. Baldauski. 1951: The reduction of triphenyltetrazolium chloride by dehydrogenases of corn embryos. Science 113:65–66.

Jensen, C. O. 13027, 16613.

Jensen, E. 11716.

9784. Jensen, E. A., M. G. Lambou, F. R. Andrews, R. Y. Mayne, M. L. Karon, M. E. Curet, F. B. Wilcox, A. M. Altschul, W. Newby, and W. F. Bollens. 1951: The storage of cottonseed. IX. Behavior of cottonseed during storage under mill conditions. J. Am. Oil Chemists Soc. 28:241–45.

Jensen, E. A. 4332.

9785. Jensen, H. 1954: The establishment of forest tree seed orchards at Ramlosa, 1941–1954. Acta Horti Gotobirgensis 19:157–92. Biol. Abstr. 29:9491. 1955.

Jensen, I. 2699.

9786. Jensen, J. L. 1904: Preliminary treatment of seed grain as a source of vegetative energy. Tidsskr. Landbr. Planteavl 10:46–63. Expt. Sta. Rec. 15:1086. 1904.

9787. Jensen, K. A., J. Conti, and A. Kjaer. 1953: Isothiocyanates. II. Volatile isothiocyanates in seeds and roots of various Brassicae. Acta Chem. Scand. 7:1267–70.

Jensen, K. A. 10747.

9788. Jensen, K. F., and G. E. Gatherum. 1964: Scotch pine provenance trial in northeast Iowa. Iowa State J. Sci. 38:377–84.

Jensen, L. 15368–15370.

9789. Jensen, L. A. 1953: Some problems of delayed germination in Oregon-grown Bentgrass. Assoc. Offic. Seed Analysts News Letter 27(2):11–20.

9790. —— 1962: Experiences using the climax blowing point technique. Proc. Assoc. Offic. Seed Analysts 51:186–93.

9791. —— 1964: Germination of *Abies concolor* and *Pinus contorta.* Assoc. Offic. Seed Analysts News Letter 38(1):12–13.

9792. Jensen, L. P. 1937: Growing woody plants from seeds. Bull. Missouri Botan. Garden. 25:53–58.

9793. Jensen, N. F. 1962: A world germ plasm bank for cereals. Crop Sci. 2:361–63.

Jensen, N. F. 1189.

Jensen, P. Boysen-; *see* Boysen-Jensen, P.

9794. Jensen, R. 1952: Isolation and amino acid composition of β-globulin extracted from the seeds of barley (*Hordeum vulgare*). Acta Chem. Scand. 6:771–81. Chem. Abstr. 47:2234. 1953.

9795. —— 1953: Results of comparative ultracentrifugal experiments made on protein fractions isolated from barley samples received from the EBC trials in 1950 and 1951. European Brewery Conv., Proc. Congr. 1953:74–84. Chem. Abstr. 48:11578. 1954.

Jensen, R. A. 16879.

Jensen, W. A. 5683, 15513.

9796. Jentsch, M. S., and A. F. Morgan. 1949: Thiamin, riboflavin, and niacin content of walnuts. Food Res. 14:40–53.

Jerchel, D. 11487, 11488.

9797. Jermstad, A. 1931: The fatty oil from the seeds of currant (*Ribes rubrum* L.). Norg. Apotekerforen. Tidsskr. 39:92–93. Chem. Abstr. 26:3947. 1932.

9798. Jesenko, F. 1911: Einige neue Verfahren, die Ruhe-periode der Holzgewächse abzukürzen. I. Ber. Deut. Botan. Ges. 29:273–84.

9799. —— 1912: Einige neue Verfahren, die Ruhe-periode der Holzgewächse abzukürzen. II. Ber. Deut. Botan. Ges. 30:81–93.

Jesperson, E. 8249.

9800. Jessel, W. H. 1962: Techniques in harvesting and sowing orchid seed for amateurs. Am. Orchid Soc. Bull. 31:559–61.

9801. Jessen, W. 1949: Die Wirkung der Phosphorsäuredüngung bei verschiedenen Aussaatzeiten von Sommerraps. Z. Pflanzenernähr. Düng. Bodenk. Abt. A und B 47:161–64. Biol. Abstr. 25:21400. 1951.

9802. —— 1949: Über die Wirkung steigender Stickstoff und Phosphorsäuregaben auf das Wachstum sowie die Eiweiss und Fetterträge der Öelrauke. Z. Pflanzenernähr. Düng. Bodenk. Abt. A und B 47: 164–67. Biol. Abstr. 25:21401. 1951.

9803. Jevtic, J. 1960: Quelques observations sur le rendement de semences de *Picea omorica* Pancic dans le Massif Tara. Sumarstvo 13:79–87.

Jewell, H. 11889.

Jha, J. D. 11686.

Jhamb, K. V. 1526.

9804. Jimenez-Diaz, M. A. 1947: A study of the oxidizing enzymes of the guava. Rev. Soc. Cubana Ingr. 44:861–74. Chem. Abstr. 43:4729. 1949.

9805. Jinarajadasa, C. 1934: Dissemination of the brazil nut. Nature 134:376.

9806. Jiracek, V., J. Kutova, and S. Leblova. 1962: Metabolism of sprouting peas. II. Metabolism of phosphorus and sugar in germinating pea seeds. Czech. Min. Zemedel. Lesn. Vodniho Hospodar. Ust. Vedtech. Inform. Rostlinna Vyroba 8:903–18.

9807. ——, ——, and S. Leblova-Svobodova. 1962: Metabolism of germinating pea plants. I. Nitrogen metabolism of germinating seeds. Sb. Česk. Akad. Zeměděl. Věd Rostlinna Vyroba 8:345–64. Chem. Abstr. 57:7637.

Jiracek, V. 2204.

9808. Joarma, M. 1955: Effects of chronic γ irradiation on some sugars in *Vicia faba*. Suomen Kemistilehti 28B:40–45. Biol. Abstr. 30:11438. 1956.

9809. Jochems, S. C. J. 1927: On viability of tobacco seed buried in the soil on Sumatra's east coast. Deli Proefsta. Mededeel. (Sumatra) 2nd Ser. No. 50. 27 pp.

9810. Jodidi, S. L. 1927: The nitrogen compounds of the rice kernel as compared with those of other cereals. J. Agr. Res. 34:309–25.

9811. —— 1934: Distribution of the non-protein nitrogen in the Alaska pea. J. Franklin Inst. 218:581–91.

9812. —— 1937: A new maturity test of peas. J. Franklin Inst. 223:593–607. Expt. Sta. Rec. 78:44. 1938.

9813. —— 1938: Preliminary biochemical studies on effects of certain environmental factors on development and composition of the peanut. J. Agr. Res. 57:301–11.

9814. —— and J. Peklo. 1929: Symbiotic fungi of cereal seeds and their relation to cereal proteins. J. Agr. Res. 38:69–91.

9815. Jodin, V. 1896: Vie latente des graines. Compt. Rend. 122:1349–51.

9816. —— 1897: Recherches sur la germination. I. Ann. Agron. 23:433–71.

9817. —— 1899: Sur la résistance des graines aux temperatures elevées. Compt. Rend. 129:893–94.

9818. —— 1902: Sur la durée germinative des graines exposées à la lumière solaire. Compt. Rend. 135:443–44.

9819. Jodon, N. E. 1953: Growing period of leading rice varieties when sown on different dates. Louisiana Agr. Expt. Sta. Bull. 476. 8 pp.

9820. —— and H. M. Beachell. 1938: Row spacing and rate of seeding for rice nursery plats. J. Am. Soc. Agron. 30:212–19.

Jodon, N. E. 9958.

9821. Joelsson, G. 1961: Determination of microscopical injuries on the pericarps of cereals by means of indigo carmine. Medd. Statens Frökontr. Inst. 36:43–53.

9822. —— 1961: The tetrazolium method of estimating the viability of cereals. Nord. Jordbrugforskn. 43:89–107.

Joerges, E. Lehle-. 19211.

Joessel, P. H. 18450, 18451.

Joham, H. E. 459.

9823. Johann, H. 1935: Histology of the caryopsis of yellow dent corn, with reference to resistance and susceptibility to kernel rots. J. Agr. Res. 51:855–83.

9824. —— 1942: Origin of the suberized semipermeable membrane in the caryopsis of maize. J. Agr. Res. 64:275–82.

Johann, H. 15742.

9825. Johannsen, W. 1897: Studies on periodicity in plants. I. Antagonistic activities in the change of matter, especially during the ripening and rest. Kgl. Danske Vidensk. Selskab Skrifter. Naturvid. Math. Afd. 6. Räkke, 8. Bind, pp. 273–394.

9826. —— 1899: The relation between grain weight and percentage nitrogen content of barley. Chem. Ztg. 23:275. Expt. Sta. Rec. 11:633. 1899.

9827. Johannson, H., and C. E. Rich. 1942: The variability in the thiamin content of western Canadian hard red spring wheat of the 1940 crop. Cereal Chem. 19:308–13.

9828. Johansen, D. A. 1928: The hypostase and seed sterility in the Onagraceae. Madrono 1:165–67.

9829. —— 1936: Morphology and embryology of *Fouquieria*. Am. J. Botany 23:95–99.

9830. —— 1941: Use of the term "embryo sac." Chronica Botan. 6:200–1.

9831. —— 1945: A critical survey of the present status of plant embryology. Botan. Rev. 11:87–107.

9832. —— 1951: Plant embryology. Embryogeny of the Spermatophyta. Chronica Botanica Co., Waltham, Mass., and Stechert-Hafner, New York. xvi + 305 pp.

9833. Johansen, D. C. 1945: Classification of the types of angiospermic embryo development. Chronica Botan. 9:139–40.

9834. Johansen, E. L., and B. W. Smith. 1956: *Arachis hypogaea* x *A. diogoi*. Embryo and seed failure. Am. J. Botany 43:250–58.

9835. Johansen, Y. T., and H. Mygind. 1956: Experiments in seed-borne disease, *Ascochyta imperfecta* on black medic. Tidsskr. Planteavl 60:20–42.

9836. Johanson, R. 1948: The cystine plus cysteine and methionine contents of the seed proteins, and sulphur distributions of seeds of *Vicia* spp. (vetches), with notes on their cyanogenesis. Australian J. Exptl. Biol. Med. Sci. 26:259–70.

9837. —— and J. W. H. Lugg. 1946: Sulphur distributions, with special reference to the cystine plus cysteine and methionine contents, of the seeds and seed proteins of two varieties of soya beans. Australian J. Exptl. Biol. Med. Sci. 24:321–25.

9838. John, C. A., and C. Sova. 1955: Incidence of tobacco mosaic virus on tomato seed. Phytopathology 45:636.

9839. John, C. M. 1941: Industrial uses of cashew and its products. Madras Agr. J. 29(5). E. African Agr. J. 8:59–60. 1942.

9840. John, C. M., C. R. Seshadri, and M. B. S. Rao. 1948: Dormancy of the seed in the groundnut. Madras Agr. J. 35:159–67. Biol. Abstr. 24:5012. 1950.

9841. Johns, I. B., and W. Oldham. 1938: The isolation of the compound giving yellow corn its characteristic odor. Cereal Chem. 15:377–79.

9842. Johnsen, T. N. 1959: Longevity of stored juniper seeds. Ecology 40:487–88.

Johnsen, V. L. 17612.

9843. Johnson, A. A. 1952: Methods of teaching the value of good seed. Intern. Crop Improvement Assoc. Ann. Rept. 32/33:44–54. 1950/51.

9844. —— 1961: We need a seed industry now. Ind. Sugar (Calcutta) 11:513.

9845. Johnson, A. C., J. S. Pinckney, J. W. Bulger, and A. M. Phillips. 1940: Fumigation of vetch seed for the betch bruchid. U.S. Dept. Agr. Circ. 555. 11 pp.

9846. Johnson, A. G. 1947: Sexual maturity in two white pine species. J. Forestry 45:826.

9847. —— 1948: The role of fungi in the deterioration of grains in storage. Food Agr. Organ. U.N., FAO Agr. Studies No. 2, pp. 88–91.

Johnson, A. G. 1482, 12137.

9848. Johnson, A. M. 1936: Polyembryony in *Eugenia hookeri*. Am. J. Botany 23: 83–88.

Johnson, A. W. 3171–3175, 3177.

Johnson, B. C. 5751, 18787.

9849. Johnson, B. J., and E. C. Holt. 1962: A study of techniques for evaluating seed set in dallisgrass. Crop Sci 2:447–48.

9850. Johnson, D. R. 1949: Chemical treatment of seed corn for control of *Agonoderus* comma. J. Econ. Entomol. 42:801–5.

9851. Johnson, D. S. 1900: Endosperm and embryo of *Peperomia*. Botan. Gaz. 30:1–11.

9852. —— 1918: The fruit of *Opuntia fulgida*. Carnegie Inst. Wash. Publ. 269. 62 pp.

9853. Johnson, E. 1937: Seeds of small spring bulbs. Horticulture 15(9):224.

Johnson, E. B. 3515.

9854. Johnson, E. L. 1928: Growth and germination of sunflowers as influenced by X-rays. Am. J. Botany 15:65–76.

9855. —— 1939: Growth of wheat plants from dry and soaked irradiated grains. Plant Physiol. 14:493–504.

9856. Johnson, E. M., and W. D. Valleau. 1946: Germination of tobacco seed prevented by cyanamid. Plant Disease Reptr. 30:214–15.

Johnson, F. 5811.

9857. Johnson, F. M. 1942: Mechanical injury to onion seeds. Assoc. Offic. Seed Analysts News Letter 16:8.

9858. Johnson, H. P. H., and A. R. Downie. 1954: Sugar beet seed, a new industry. Reclamation Era 40:63–64.

9859. Johnson, H. W. 1951: Soybean seed treatment. Soybean Dig. 11(7):17–20. Rev. Appl. Mycol. 31:220. 1952.

9860. —— 1959: Treatment of forage legume seeds with fungicides in the south. Plant Disease Reptr. 43:1016–18.

Johnson, H. W. 10617, 12002.

9861. Johnson, I. J. 1932: Correlation studies with strains of flax with particular reference to the quantity and quality of the oil. J. Am. Soc. Agron. 24:537–44.

9862. —— 1932: Relation of agronomic practice to the quantity and quality of oil in flax seed. J. Agr. Res. 45:239–55.

9863. —— 1960/61: Certified seed and privately developed grass and legume varieties. Intern. Crop Improvement Assoc. Ann. Rept. 42/43:218–23.

9864. —— and H. K. Hayes. 1938: The inheritance of pericarp tenderness in sweet corn. J. Am. Soc. Agron. 30:220–31.

9865. —— and E. S. Miller. 1939: Immediate effect of cross pollination on the carotenoid pigments in the endosperm of maize. Cereal Chem. 16:88–92.

Johnson, I. J. 5886.

9866. Johnson, I. M. 1952: Growing amaryllis from seed. Horticulture 30(1): 12, 22.

9867. Johnson, J. 1919: The influence of heated soils on seed germination and plant growth. Soil Sci. 7:1–87.

9868. ——, H. F. Murwin, and W. B. Ogden. 1930: The germination of tobacco seed. Wisconsin Agr. Expt. Sta. Res. Bull. 104. 15 pp.

9869. —— and W. B. Ogden. 1945: Tobacco seedbeds. Wisconsin Agr. Expt. Sta. Bull. 467. 28 pp. Expt. Sta. Rec. 94:758. 1946.

Johnson, J. A. 4443, 6455.

9870. Johnson, J. C. 1913: Influence of "pickling" on germination of cereals. Gt. Brit. Board Agr. J. 20:120–24. Expt. Sta. Rec. 29:346. 1915.

9871. Johnson, K. W., and L. A. Tatum. 1954: The relationship between corn seed factors affecting germination under adverse conditions and seed permeability. Agron. Abstr. 46:79.

9872. Johnson, L. P. V. 1935: General preliminary studies on the physiology of delayed germination in *Avena fatua*. Can. J. Res. Sect. C 13:283–300.

9873. —— 1935: The inheritance of delayed germination in hybrids of *Avena fatua* and *A. sativa*. Can. J. Res. Sect. C 13:367–87.

9874. —— 1946: Effect of chemical treatments on the germination of forest tree seeds. Forestry Chron. 22:17–24.

9875. —— 1946: Effect of humidity on the longevity of *Populus* and *Ulmus* seeds in storage. Can. J. Res. Sect. C 24:298–302.

9876. —— 1946: A practical method of overcoming seed dormancy in *Tilia americana* L. Forestry Chron. 22:182–90.

9877. —— 1947: Embryonic reaction to sodium biselenite as a test of seed vitality. J. Am. Soc. Agron. 39:943–47.

9878. —— and H. W. Holtz. 1946: Colchicine treatment techniques for sprouted seeds and seedlings. Can. J. Res. Sect. C 24:303–4.

9879. ——, G. A. Young, and J. B. Marshall. 1945: A note on the production of vitamin C by sprouting seeds. Sci. Agr. 25:499–503.

Johnson, L. P. V. 5181.

Johnson, L. R. 18541.

Johnson, M. 8249.

9880. Johnson, M. B., and C. B. Gilliland. 1960: Changes in farmers' stock peanuts in storage. Marketing significance. U.S. Dept. Agr. Marketing Res. Rept. 381. 34 pp.

9881. Johnson, M. P. 1952: Real gardens from seed packets. Home Garden 19: 28–31.

9882. —— 1957: Angels-trumpet: Spectacular yet easy from seed. Flower Grower 44(4):60.

Johnson, M. W. 6845, 9387, 16891.

9883. Johnson, N. E., and H. J. Heikkenen. 1958: Damage to the seed of Douglas fir by the Douglas fir cone midge. Forestry Sci. 4:274–82.

Johnson, N. E. 19752.

9884. Johnson, P. 1946: The proteins of the ground-nut (*Arachis hypogaea*). I. The isolation and properties of the proteins. II. The relationships between the different sedimenting species isolated from extracts of the ground-nut. Trans. Faraday Soc. 42:28–45.

Johnson, P. S. 5332.

9885. Johnson, R. M., and J. E. Greaves. 1940: The quantity and nature of the oil in *Conringia orientalis*. Proc. Utah Acad. Sci. 17:85–87.

9886. ——, W. H. Hunt, M. H. Neustadt, and L. Zeleny. 1956: A rapid dielectric method for determining the oil content of safflower and sunflower seed. J. Am. Oil Chemists Soc. 33:314–16.

Johnson, R. M. 6197.

Johnson, R. T. 19577.

9887. Johnson, S. P., and W. R. Cowley. 1956: A preliminary study of the effect of low concentrations of phenoxytype growth regulators on early physiological shedding in Deltapine D10, TPSA, cotton. I. Seed treatment with dusts. Agron. J. 48: 412–15.

9888. Johnson, T. 1906/7: Principles of seed-testing. Sci. Progr. 1:483–96. Botan. Centralbl. 104:461–62.

Johnson, V. A. 8388, 8389, 18115.

Johnson, W. A. 8000.

9889. Johnson, W. B., and E. Bennett. 1952: A turbidity test for measuring the maturity of sweet corn. Proc. Am. Soc. Hort. Sci. 59:400–4.

9890. Johnson, W. H. 1959: Harvesting and drying of high moisture wheat. Ohio Agr. Expt. Sta. Res. Bull. 841. 46 pp.

Johnson, W. M. 9197.

9891. Johnsson, H. 1958: Germination maturity and storage maturity of spruce seed harvested early. Svenska Skogsvardsforen. Tidskr. 56:213–24.

9892. Johnston, A. 1949: The germination of malvaceous seeds. Trop. Agr. (Trinidad) 26:63.

9893. —— 1958: A note on fungicidal seed dressing of soya bean, groundnut and long bean. Malayan Agr. J. 41:152–55.

9894. —— 1958: Fungicidal treatment of padi seed. Malayan Agr. J. 41:282–89.

9895. —— 1961: Some factors affecting germination, emergence, and early growth of three range grasses. Can. J. Plant Sci. 41:59–70.

9896. Johnston, C. O., and L. E. Melchers. 1928: The control of sorghum kernel smut and the effect of seed treatments on vitality of sorghum seed. Kansas Agr. Expt. Sta. Tech. Bull. 22. 37 pp.

Johnston, F. A. 16391.

9897. Johnston, F. A., Jr., and H. M. Sell. 1944: Changes in chemical composition of tung kernels during germination. Plant Physiol. 19:694–98.

Johnston, F. A., Jr. 17184.

9898. Johnston, F. B., M. Nasatir, and H. Stern. 1957: The isolation of nuclei and oxidative particles from viable wheat embryos. Plant Physiol. 32:124–29.

9899. ——, G. Setterfield, and H. Stern. 1959: The isolation of nucleoli from ungerminated pea embryos. J. Biophys. Biochem. Cytol. 6:53–55.

Johnston, F. B. 17242.

9900. Johnston, J. C., and E. F. Frolich. 1957: Avocado propagation. Calif. Agr. Expt. Sta. Ext. Serv. Circ. 463. 19 pp.

9901. Johnston, M. E. H. 1960: Investigations into seed setting in cocksfoot seed crops in New Zealand. New Zealand J. Agr. Res. 3:345–57.

9902. —— and J. G. Miller. 1960: Optimum conditions for germination tests of *Trifolium repens* L., white clover. Proc. Intern. Seed Testing Assoc. 25:416–21.

9903. —— and —— 1962: Investigation into germination techniques for rye-grass, *Lolium* spp. Proc. Intern. Seed Testing Assoc. 27:345–56.

9904. —— and —— 1963: Optimum germination conditions for some species of the genus *Brassica*. Proc. Intern. Seed Testing Assoc. 28:39–44.

Johnston, T. H. 88.

9905. Johnston, W. H. 1934: Studies on the dehulling of barley kernels with sulphuric acid and on the inheritance of reaction to covered smut *Ustilago hordei* (Pers.) K & S. infection in crosses between Glabron and Trebi barleys. Can. J. Res. 11:458–73.

9906. Johnston, W. R., and H. E. Foote. 1951: Development of a new process for curing coffee. Food Technol. 5:464–67.

9907. Johnstone, A. 1889: The colouring matter of the testa of the seeds of *Brassica napus*. Nature 1889:15.

9908. Johnstone, G. R. 1940: Further studies on polyembryony and germination of polyembryonic pine seeds. Am. J. Botany 27:808–11.

9909. Johnstone, G. R., and T. S. Clare. 1931: Hastening the germination of Western pine seeds. J. Forestry 29:895–906.

Johnstone, G. R. 4140.

9910. Johnstone, J. H. 1956: Nitrogen metabolism in Jack bean (*Canavalia ensiformis*). Biochem. J. 64:21P.

9911. Johri, B. M. 1938: The embryo sac of *Hydrocleis nymphoides* Buchen. Botan. Centralbl. Beih. Abt. A 58:165–72.

9912. —— 1938: The embryo sac of *Limnocharis emarginata* L. New Phytologist 37:279–85.

9913. —— 1962: Nutrition of the embryo sac. *In* Maheshwar, Johri, and Vasil, editors, Proc. Summer School Botany, June 2–15, 1960, Darjeeling. Min. Sci. Res. Cult. Affairs, Govt. India, New Delhi, pp. 106–18. Biol. Abstr. 41:20087. 1963.

9914. —— and S. P. Bhatnagar. 1955: Does polyembryony occur in *Limnocharis emarginata* L. (syn. *L. flava* Buchen.)? Sci. Cult. (Calcutta) 21:276.

9915. —— and D. Kak. 1954: The embryology of *Tamarix* Linn. Phytomorphology 41:230–47.

9916. —— and R. N. Konar. 1955: A contribution to the morphology and embryology of *Ficus religiosa* Linn. Current Sci. (India) 24:382–85.

9917. —— and C. B. Sehgal. 1963: Chemical induction of polyembryony in *Anthemum graveolens* L. Naturwissenschaften 50: 47–48.

Johri, B. M. 12699, 12700.

9918. Jois, H. S., and B. L. Manjunath. 1930: Chemical examination of the seeds of *Cassia tora* (Linn.). Proc. Indian Sci. Congr. Allahabad 17:183–84.

9919. —— and —— 1936: Identity of isopsoralene, a constituent of the seeds of *Psoralea corylifolia* L., with agelicin from the roots of *Angelica archangelica* L. Ber. Deut. Chem. Ges. 69B:964–67. Chem. Abstr. 30:4855. 1936.

9920. ——, ——, and S. V. Rao. 1933: Chemical examination of the seeds of *Psoralea corylifolia* (Linn.). I. J. Indian Chem. Soc. 10:41–46. Chem. Abstr. 27:3030. 1933.

9921. Jokela, M. 1952: On the weed seed content of timothy and red clover seed in Finland and the other northern countries. Nord. Jordbrugsforskn. 34:49–59.

9922. Jolliffe, W. H. 1955: Trees on the farm. Collection, extraction, and storage of seed. New Zealand J. Agr. 91:557.

9923. Jolly, R. G. 1957: Seed production in New Zealand. Italian ryegrass and short-rotation ryegrass. New Zealand J. Agr. 94:373, 375, 377, 379–81.

9924. —— 1958: Seed production in New Zealand: white clover. New Zealand J. Agr. 96:7–14.

9925. Jonap, L. 1962: Effect of seed corn injuries on germination. *In* I. Iso, editor, Kukoricatermesztesi kiserletek 1958/60: 195–99.

Jonard, P. 7163, 7164.

9926. Jonas, H. 1952: Some effects of radio frequency irradiations on small oilbearing seeds. Physiol. Plantarum 5:41–51.

9927. —— 1953: R-F irradiation of seeds. Electronics 26:161–63.

9928. Jonas, K. 1961: Keimstimulation des Samens von *Poa pratensis*. Bayer. Landwirtsch. Jahrb. 38:209–37.

9929. Joneidi, M. J. 1953: An examination of mucilage-containing seeds of *Alyssum campestre* L. J. Pharm. Pharmacol. 5:625–31. Ber. Wiss. Biol. 86:384. 1954.

9930. Jones, C. R., and J. D. Campbell. 1953: Micro-determination of endosperm density as a means of mapping moisture distribution in wheat grains. Cereal Chem. 30:177–89.

9931. —— and E. C. Dawson. 1945: The arsenic content of grain dried directly with flue gas. Analyst 70:256–57.

Jones, C. R. 3557–3559, 6424.

9932. Jones, D. A. 1961: Lectins in old seeds. Vox Sanguinis 6:502–5. Chem. Abstr. 56:718. 1962.

9933. Jones, D. B. 1947: Proteins of the castor bean—their preparation, properties, and utilization. J. Am. Oil Chemists Soc. 24:247–51.

9934. ——, J. P. Divine, and C. E. F. Gersdorff. 1942: The effect of storage of corn on the chemical properties of its proteins and on its growth-promoting value. Cereal Chem. 19:819–30.

9935. ——, G. S. Fraps, B. H. Thomas, and L. Zeleny. 1943: The effect of storage of grains on their nutritive value. Natl. Res. Council Repr. & Circ. Ser. 116. 14 pp.

9936. —— and C. E. F. Gersdorff. 1939: The effect of storage on the proteins of seeds and their flours. Soybeans and wheat. J. Biol. Chem. 128:49–50.

9937. —— and —— 1941: The effect of storage on the protein of wheat, white flour, and whole wheat flour. Cereal Chem. 18:417–34.

9938. ——, ——, and S. Phillips. 1938: Proteins of the black bean of the mayas, *Phaseolus vulgaris*. J. Biol. Chem. 122: 745–55.

Jones, D. B. 4719.

9939. Jones, D. F. 1944: Growth changes in maize endosperm associated with the relocation of chromosome parts. Genetics 29:420–27.

9940. ——, J. Macmillan, and M. Radley. 1963: Plant hormones. III. Identification of gibberellic acid in immature barley and immature grass. Phytochemistry 2:307–14. Chem. Abstr. 60:4705. 1964.

9941. —— and P. C. Mangelsdorf. 1951: The production of hybrid corn seed without detasseling. Conn. Agr. Expt. Sta. Bull. 550. 21 pp.

Jones, D. L. 10276, 19360.

9942. Jones, D. R. H., and J. S. Waid. 1963: Influence of fungal isolates on germination and growth of perennial ryegrass. Plant Soil 19:139–50.

Jones, E. C. 8769.

Jones, E. T. 6178.

9943. Jones, E. W. 1943: The cone production of Scots pine seed bearers. Forestry 17:52–54.

9944. —— 1945: The regeneration of Douglas fir, *Pseudotsuga taxifolia* Britt., in the new forest. J. Ecology 33:44–56.

9945. —— 1958: The storage of acorns in water. Forestry 31:163–66.

9946. Jones, F. G. W., and K. P. Humphries. 1954: The use of seed dressings containing γ-BAC in the establishment of sugar-beet seedlings. Ann. Appl. Biol. 41:562–77.

9947. Jones, F. R. 1938: A seed-borne disease of sweetclover. Phytopathology 28:661–62.

9948. Jones, G. H. 1939: Systematic and automatic warm water steeping to control loose smut of wheat. Egypt Min. Agr. Bull. 220. 12 pp.

9949. Jones, H. A. 1920: Physiological study of maple seeds. Botan. Gaz. 69:127–52.

9950. —— and C. S. Bisson. 1932: Changes in the composition of the garden pea after harvest. Plant Physiol. 7:273–83.

9951. —— and A. E. Clarke. 1943: Inheritance of male sterility in the onion and production of hybrid seed. Proc. Am. Soc. Hort. Sci. 43:189–94.

Jones, H. A. 2507.

9952. Jones, J. A. 1928: Overcoming delayed germination of *Nelumbo lutea.* Botan. Gaz. 85:341–43.

Jones, J. D. 14918.

Jones, J. K. N. 629, 630, 6036, 8857, 8858.

9953. Jones, J. P. 1928: Physiological study of dormancy in vetch seeds. Cornell Univ. Agr. Expt. Sta. Mem. 120. 50 pp.

9954. —— 1959: Purple stain of soybean seeds incited by several *Cercospora* species. Phytopathology 49:430–32.

Jones, J. R. D. 10410

9955. Jones, J. S. 1939: A comparison of the regular method and a fractional method of analyzing orchard grass seed for purity. Proc. Assoc. Offic. Seed Analysts 30:118–19.

9956. Jones, J. W. 1926: Germination of rice seed as affected by temperature, fungicides, and age. J. Am. Soc. Agron. 18:576–92.

9957. —— 1933: Effect of reduced oxygen pressure on rice germination. J. Am. Soc. Agron. 25:69–81.

9958. ——, C. R. Adair, N. E. Jodon, H. M. Beachell, and L. L. Davis. 1947: Effect of environment and source of seed on yield and other characters in rice. J. Am. Soc. Agron. 39:874–86.

9959. ——, J. O. Dockins, R. K. Walker, and W. C. Davis. 1952: Rice production in the southern states. U.S. Dept. Agr. Farmers Bull. 2043. 36 pp.

9960. Jones, L. 1960: Rapid moisture determination of tree seed with an electronic meter. Tree Planters Notes 43:7.

9961. —— 1961: Effect of light on germination of forest tree seed. Proc. Intern. Seed Testing Assoc. 26:437–52.

9962. —— 1962: Ninth Annual Report, Eastern Tree Seed Laboratory. U.S. Dept. Agr. Forest Serv. Southern Region. 18 mimeo. pp.

9963. —— 1962: Recommendations for successful storage of tree seed. Tree Planters Notes 55:9–20.

Jones, L. 3654, 18231.

9964. Jones, L. G. 1959: Seed crop production. Calif. Univ. Ext. Div. Farm Manag. Seminar Bank of America Off., pp. 268–74.

9965. —— and R. D. Cobb. 1963: A technique for increasing the speed of laboratory germination testing. Assoc. Offic. Seed Analysts News Letter 37(2):19.

9966. ——, V. P. Osterli, P. R. Bunnelle, and A. D. Reed. 1953: Red clover seed production. Calif. Agr. Expt. Sta. Ext. Serv. Circ. 432. 11 pp.

9967. ——, F. P. Zscheile, and R. B. Griffith. 1953: Carotene and protein contents of alfalfa as influenced by variety and certain environmental factors. Hilgardia 22:179–202.

Jones, L. G. 4234, 4235.

9968. Jones, L. M., and P. J. Olson. 1943: Seed setting in alfalfa. III. Sci. Agr. 23:315–21.

9969. Jones, M. A., and G. C. Vincente. 1949: Inactivation and vacuum infiltration of vanilla enzyme systems. J. Agr. Res. Res. 78:435–43.

9970. Jones, M. B., and L. F. Bailey. 1956: Light effects on the germination of seeds of henbit (*Lamium amplexicaule* L.). Plant Physiol. 31:347–49.

9971. Jones, M. D., C. Puentes, and R. Suarez. 1955: Isolation of kenaf for seed increase. Agron. J. 47:256–57.

9972. —— and J. B. Sieglinger. 1952: Effect of the waxy gene on grain yields of sorghum. Oklahoma Agr. Expt. Sta. Tech. Bull. T-44. 15 pp.

9973. Jones, M. G., and M. A. H. Tincker. 1926: Yield studies in oats: the effect of pre-treatment of the parent crop upon the seed produced, its germination and subsequent growth. Ann. Appl. Biol. 13:535–59.

9974. Jones, M. J., and C. C. Albers. 1955: Further studies on Texas *Plantago* seeds. J. Am. Pharm. Assoc. 44:100–5.

9975. Jones, M. W., B. Kivel, and A. A. Bless. 1951: The bioelectric potential of seeds as a function of growth and of X-ray dosage. Plant Physiol. 26:19–29.

Jones, Q. 5744, 17628, 18972, 18973.

Jones, R. K. 8693.

9976. Jones, R. L., T. P. Metcalfe, and W. A. Sexton. 1949: The relationship between the constitution and the effect of chemical compounds on plant growth. I. 2-Phenoxyethylamine derivatives. Biochem. J. 45:143–49.

9977. Jones, R. M. 1963: IV. Preliminary studies of the germination of seed of *Acacia cyclops* and *Acacia cyanophylla*. S. African J. Sci. 59:296–98.

Jones, T. Palmer-. 6556.

9978. Jones, W. W. 1938: The physiology of oil production in the macadamia (*Macadamia integrifolia*, Maiden et Betche). Proc. Am. Soc. Hort. Sci. 35:239–45.

9979. —— 1939: A study of developmental changes in composition of the macadamia. Plant Physiol. 14:755–68.

9980. —— and L. Shaw. 1943: The process of oil formation and accumulation in the macadamia. Plant Physiol. 18:1–7.

9981. Jonesco, S. 1932: Influence du zinc sur la respiration des graines germees de *Lupinus albus*. Compt. Rend. Soc. Biol. 110:655–57. Botan. Centralbl. 164:151. 1933.

Jong, J. J. de; *see* De Jong, J. J.

Jong, P. de; *see* De Jong, P.

9982. Jono, Y. 1931: The enzyme content of dormant and germinating seeds. Acta Schol. Med. Univ. Kioto 13:211–38. Chem. Abstr. 25:3030. 1931.

9983. Jönsson, B. 1893: Observations of the importance of light on the germination of seeds. Kgl. Fysiograf. Sallskap. i Lund Handl. (Särtryck). 47 pp.

9984. —— 1896: On the influence of arsenic on the germination of seeds. Kgl. Landtbruks-Akad. Handl. Tidskr. 35:95–112. Expt. Sta. Rec. 8:232–33. 1896.

Jordan, S. 5393.

9985. Jorgensen, C. A. 1934: Some investigations on plant diseases from seed infection. Tidsskr, Planteavl 40:119–47. Rev. Appl. Mycol. 13:740–41. 1934.

Jorgensen, C. A. 8606.

Jorgensen, H. 19555.

9986. Jorgensen, J. H. 1964: Seed grading in polyploid varieties of beet. Kgl. Veterin. Landsbohojskole Yearbook, pp. 121–32.

9987. Jorissen, M. A. 1886: Les phénomènes chimiques de la germination. Bruxelles. 140 pp. J. Roy. Microscop. Soc. 1887: 615–16.

9988. Jorman, L. W., and R. M. Greenwood. 1951: Effect of moisture content on the viability of Italian ryegrass seed in storage. New Zealand J. Sci. Technol. 33:58–61.

9989. Jornlin, F. M. 1948: The atomic bomb's effects on plants and soils. Fertilizer Rev. 23(4):7–11, 14; (5):7–9, 11–13. Chem. Abstr. 42:8607. 1948.

9990. Jorritsma, J. 1953: Seed bed and germination of sugar beets. Landbouwk. Tijdschr. 65:684–88.

9991. Jorstad, I., and W. Christie. 1926: Treatment with hot water of spring wheat in the control of bunt of wheat in the years 1924 and 1925. Norg. Landbrugs-Höisk. Meld. 6:405–36.

9992. —— and A. E. Traaen. 1930: Seed grain treatment experiments with disinfectants in the years 1925–1929. Norg. Landbrugs-Höisk. Meld. 10:1–50.

9993. —— and —— 1930: The hot water treatment against covered smut of barley. Norg. Landbrugs-Höisk. Meld. 10:51–56.

9994. Josan, F. 1957: Nouvelle méthode pour la détermination rapide de la germination du mais. Probl. Agr. (Bucharest) 9:99–102.

9995. Josefsky, K. 1927: Anzucht der Rhododendron aus Samen. Gartenwelt 31:285–86.
9996. Joseph, G. H. 1947: Essential oils of citrus. Econ. Botany 1:417–19.
9997. Joseph, H. C. 1928: Growing bittersweet (*Celastrus scandens*) from seed. Florists Exchange Hort. Trade World 68:499.
9998. —— 1929: Germination and keeping quality of parsnip seeds under various conditions. Botan. Gaz. 87:195–210. Contrib. Boyce Thompson Inst. 2:115–30. 1929.
Joseph, T. C. 14100, 14101.
Josepovits, G. 18138.
Joshi, A. B. 14968.
9999. Joshi, A. C. 1932: Vivipary in *Atriplex crassifolia* and *Sueda fruticosa*. J. Indian Botan. Soc. 11:71–72.
10000. —— 1933: Megaspore formation and embryo-sac of *Argemone mexicana*, Linn. J. Indian Botan. Soc. 12:83–91.
10001. —— 1933: A suggested explanation of the prevalence of vivipary on the seashore. J. Ecology 21:209–12.
10002. —— 1936: A contribution to the embryology and cytology of *Rivina humilis* Linn. J. Indian Botan. Soc. 15:91–103.
10003. —— 1937: Contributions to the embryology of the Menispermaceae. I. *Cocculus villosus* DC. Proc. Indian Acad. Sci. 5B:57–63.
10004. —— 1938: Parthenocarpy in *Dodonaea* (parthenocarpy in *Dodonaea viscosa*). J. Indian Botan. Soc. 17:97–99. J. Roy. Microscop. Soc. 59:53. 1939.
10005. —— 1938: Morphology of gynoecium of *Psoralea*. (A note on the morphology of the gynoecium, ovule, and embryo-sac of *Psoralea corylifolia* L.) J. Indian Botan. Soc. 17:169–72. J. Roy. Microscop. Soc. 59:53. 1939.
10006. —— 1938: A note on the morphology of the ovule of Rubiaceae with special reference to *Cinchona* and coffee. Current Sci. (India) 7:236–37.
10007. —— 1939: Morphology of *Tinospora cordifolia*, with some observations on the origin of the single integument, nature of Synergidae, and affinities of the Menispermaceae. Am. J. Botany 26:433–39. J. Roy. Microscop. Soc. 59:301. 1939.
10008. —— 1940: Development of the embryo-sac of *Gagea fascicularis* Salisb. Bull. Torrey Botan. Club 67:155–58.
10009. —— and L. B. Kajale. 1936: A note on the structure and development of the embryo-sac, ovule, and fruit of *Tamarix dioica* Roxb. Ann. Botany (London) 50:421–25.
10010. —— and —— 1937: Fertilisation and seed development in Amarantaceae. Proc. Indian Acad. Sci. 5B:91–100.
10011. —— and J. V. Pantulu. 1939: Origin of the inferior ovary in the Amaryllidaceae. Current Sci. (India) 8:212–13.
10012. —— and V. R. Rao. 1934: Fruit and seed development in *Tinospora cordifolia* Miers without fertilisation and embryo formation inside. Current Sci. (India) 3:62–63.
10013. —— and —— 1936: The embryology of *Gisekia pharnaceoides* Linn. Proc. Indian Acad. Sci. 3B:71–92.
10014. —— and J. Venkateswarlu. 1935: A case of reversed polarity in the embryo-sac. Ann. Botany (London) 49:841–43.
10015. —— and —— 1936: Embryological studies in the Lythraceae. III. Proc. Indian Acad. Sci. 3B:377–400.
10016. Joshi, B. C., and J. D. Tewari. 1957: Chemical investigation of the seeds of *Lepidium iberis* (var. *alb.*). II. Analysis of mucilages. Arch. Pharm. 290:257–60. Chem. Abstr. 51:15707. 1957.
10017. Joshi, B. N., and V. P. Varma. 1955: Castor seed proteins and their viscosities. J. Am. Oil Chemists Soc. 32:553–55.
Joshi, H. C. 17218.
10018. Joshi, K. V. 1940: Cashew nut cultivation in South Kanara. J. Mysore Agr. Exptl. Union 19:14–19.
10019. Joshi, M. C., and T. M. Varghese. 1962: Presence of twin-embryos and twin-ovules in *Anticharis linearis* Hochst., a member of Scrophulariaceae. Sci. Cult. (Calcutta) 28:489. Biol. Abstr. 43:7441. 1963.
10020. Joshi, P. C. 1962: In vitro growth of cotton ovules. *In* Plant Embryology. Symposium held at Univ. Delhi, 1960. Council Sci. Ind. Res.: New Delhi, India, pp. 199–204. Biol. Abstr. 43:7487. 1963
Joshi, S. S. 6931.
Jouanneau, J. P. 2157.
10021. Joubert, F. J. 1955: Lupin seed proteins. I. A physicochemical study of the proteins from blue lupin seed (*Lupinus angustifolius*). Biochim. Biophys. Acta 16:370–76.
10022. —— 1955: Sunflower seed proteins. Biochim. Biophys. Acta 16:520–23.
10023. —— 1955: A physicochemical study of the proteins from the oyster nut. J.

S. African Chem. Inst. 8:43–47. Chem. Abstr. 50:14885. 1956.

10024. —— 1955: Chemistry of narras seed (*Acanthosicyos horrida*). II. A physicochemical study of narras seed proteins in sodium dodecyl sulfate solutions. J. S. African Chem. Inst. 8:48–58. Chem. Abstr. 50:14885. 1956.

10025. —— 1955: Ultracentrifuge studies on seed proteins of Leguminosae. I. Yellow lupine seed (*Lupinus luteus*) proteins. J. S. African Chem. Inst. 8:68–74. Chem. Abstr. 50:14885–86. 1956.

10026. —— 1956: Lupine seed proteins. III. Physicochemical study of the proteins from white lupine seed (*Lupinus albus*). Biochim. Biophys. Acta 19:172–73. Chem. Abstr. 50:7964. 1956.

10027. —— 1957: Ultracentrifuge studies on seed proteins of legumes. III. Proteins from various species. J. S. African Chem. Inst. 10:16–20. Chem. Abstr. 51:16730. 1957.

10028. —— 1957: Ultracentrifuge studies on seed proteins of legumes. IV. Isolation and physicochemical examination of the globulins from white lupine seed. J. S. African Chem. Inst. 10:21–28. Chem. Abstr. 51: 16730. 1957.

10029. —— and C. C. Cooper. 1954: Narras seed (*Acanthosicyos horrida*). I. A physicochemical study of the proteins from narras seed. J. S. African Chem. Inst. 7:99–106. Chem. Abstr. 49:8394. 1954.

10030. Joukov, N. I. 1934: Influence of X-rays on the rate and percentage of germination of tobacco seed. Krasnodar Inst. Tobacco Ind. Bull. 110:171–77.

Jourdheuil, P. 2744.

Jovanovic, B. 10033.

10031. Jovanovic, M. 1960: L'influence de la grosseur des graines du pin noir (*Pinus nigra* Arn.) sur la germination et le développement des semis pendant la première année de végétation. Rev. Forestry (France) 5:301–8.

10032. —— 1961: Influence of the size of seeds of *Pinus nigra* Arn. of different provenances, on the germination and development in the seedbed during the first year of growth. Arhiv Biol. Nauka 13:67–73.

10033. Jovanovic, Z., and B. Jovanovic. 1961: Results of observations on the dormancy period of seeds of domestic and foreign varieties of wheat. Zbornik Radova Belgrade Univ. Poljoprivredni Fak. 9. 15 pp.

10034. Joxe, A. 1912: Sur l'ouverture des fruits indehiscents a la germination. Ann. Sci. Nat. Ser. 9 Botan. 15:257–375.

10035. Joyner, F. A. 1933: A seed aeroplane. *Macrozonania macrocarpa.* Gardeners Chron. 93:417.

Joyner, J. F. 17140.

10036. Jozefowicz, M. 1930: The effect of certain treatments on the germination of tomato seeds. Ann. Appl. Biol. 17:504–13.

10037. —— 1930: Some observations on tomato plants from seed submitted to high temperatures. Ann. Appl. Biol. 17: 514–21.

10038. —— 1938: Keimstimmungsversuche mit tomaten. Gartenbauwissenschaft 12: 70–76.

10039. Jozsa, S., and H. C. Gore. 1932: Development of diastatic enzymes of malt and flour. Effect of certain salts, carbohydrates, and papain. Ind. Eng. Chem. 24:95–98.

10040. Juby, D. V., and J. H. Pheasant. 1933: On intermittent germination as illustrated by *Helianthemum guttatum* Miller. J. Ecology 21:442–51.

10041. Judd, W. H. 1930: The fruiting of Franklinia. Horticulture 8(5):103.

10042. —— 1932: Raising the dove tree from seed. Horticulture 10(20):386.

10043. Judkins, W. P. 1946: The influence of kernel size, age, location in panicle, and variety of oat, on the variability of the *Avena* test. Am. J. Botany 33:181–84.

10044. Juel, H. O. 1903: Ein Beitrag zur Entwickelungsgeschichte der Samenanlage von *Casuarina*. Flora (Jena) 92:284–93.

10045. —— 1903: Zur Entwickelungsgeschichte des Samens von Cynomarium. Botan. Centralbl. Beih. 13:194–202.

10046. Juel, I. 1941: Der Auxingehalt in Samen verschiedenen Alters, sowie einige Untersuchungen betreffend der Haltbarkeit der Auxine. Planta 32:227–33.

10047. Jugenheimer, R. W. 1961: Breeding for oil and protein content in maize. Euphytica 10:152–56. Chem. Abstr. 56: 13256. 1962.

Jugenheimer, R. W. 19857.

10048. Juhans, J. 1934: Meie seemnehaigustest. Mitt. Phytopath. Versuchssta. Univ. Tartu No. 19. 13 pp. Rev. Appl. Mycol. 13:389–90. 1934.

Juhl, W. G. 1398.

Juhren, G. 18058, 19532.

10049. Juhren, M., W. M. Hiesey, and F. W. Went. 1953: Germination and early

growth of grasses in controlled conditions. Ecology 34:288–300.
Juhren, M. C. 19532.
10050. Juillet, A. 1952: Étude de la germination d'*Urena lobata.* Agron. Trop. 7:487–507.
10051. Julen, G. 1956: Alfalfa breeding and seed growing. Jord-Groda-Djur 12:97–103.
10052. —— 1956: Recent investigations concerning seeds from tetraploidy red clover. Svensk Frötidn. 25:118–21.
10053. —— and E. Akerberg. 1948: Insects and seed formation of field crops. Svensk Jordbr. Forskning 1948:86–97.
Julen, G. 7941, 7983.
10054. Juliani, M. 1931: Germinative capacity in embryos removed from seeds. Ann. Botan. (Rome) 19:353–64. Expt. Sta. Rec. 67:660. 1932.
10055. Juliano, B. O., G. M. Bautista, J. C. Lugay, and A. C. Reyes. 1964: Studies on the physicochemical properties of rice. Agr. Food Chem. 12:131–38.
10056. Juliano, J. B. 1932: The cause of sterility in *Spondias purpurea* Linn. Philippine Agriculturist 21:15–24. Biol. Abstr. 7:14334. 1933.
10057. —— 1934: Origin of embryos in the strawberry mango. Philippine J. Sci. 54:553–59. 1934.
10058. —— 1935: Preliminary experiments on the inoculation of potted plants with Bunga seeds. Philippine Agriculturist 24:262–71.
10059. —— 1936/37: Embryos of Carabao mango (*Mangifera indica* Linn.). Philippine Agriculturist 25:749–58.
10060. —— 1940: Viability of some Philippine weed seeds. Philippine Agriculturist 29:313–26.
10061. Julien, C. 1937: The relation existent between the number of productive plants yielding a harvest and the quantity of seed sown. Intern. Rev. Agr. 28:149T–154T.
10062. Jump, J. A. 1952: Grow desert plants from seed. . . . Flower Grower 39(11): 32–33, 46.
Jump, J. A. 7464.
10063. Junell, S. 1937: Die Samenentwicklung bei einigen Labiaten. Svensk Botan. Tidskr. 31:67–110.
10064. —— 1962: Embryology of *Hebenstreitia*, *Dischisma*, *Sutera*, and *Zaluzianskya* (Scrophulariaceae). Acta Horti Gotoburgensis 25:91–101.
10065. Jungelson, A. 1917: Abnormalities of the ear in corn produced by treatment of the seed with copper. Rev. Gen. Botan. 29:244–48, 261–85. Expt. Sta. Rec. 39:526. 1918.
Jungelson, A. 6908.
10066. Junges, W. 1954: Erhöhung der Vitalität von Gurkensaatgut durch Vorbehandlung des Saatgutes mit Wärme und Feuchtigkeit. Arch. Gartenbau 2:173–77.
10067. —— 1956: Jarowisation von Spinat mit verschiedenen Temperaturen und der Einfluss von Dauerlicht während der Jarowisation. Gartenbauwissenschaft 21: 429–38.
10068. —— and G. Buettner. 1955: Seed treatment and cultivation techniques for castor bean plant. Pharmazie 10:615–20. Biol. Abstr. 31:5642. 1957.
10069. —— and H. Ludwig. 1963: Einfluss von Wasserversorgung, Düngung und Licht während der Samenbildung auf die Samenruhe bei *Impatiens balsamina* L. Proc. Intern. Seed Testing Assoc. 28:71–96.
Jungkunz, R. 15709–15712.
10070. Junkersdorf, P., and K. Meyer. 1932: The nutrient content of a Central American bean. Biochem. Z. 256:100–4. Chem. Abstr. 27:1058. 1933.
Jurgelsky, W., Jr. 16905.
Jurgens, J. F. 13115.
10071. Juscafresa, B. 1952: Preparación de semillas para siembra. Cult. Mod. 35: 98–99.
10072. —— 1955: La influencía de la profundidad en la germinación de las semillas. Ceres (Valladolid) 20(225):41, 43.
10073. Juska, F. V. 1959: Effects of several pre-emergence crabgrass chemicals on germination and growth of some turfgrasses. Agron. Abstr. 1959:89–90.
10074. Just, L. 1875: Wirkungen höherer Temperaturen auf die Keimfähigkeit der Samen von *Trifolium pratense.* Botan. Ztg. 33:52–63.
10075. Justham, M. C. D., and L. Ogilvie. 1952: Tomato mosaic in relation to source of seed. Plant Pathol. 1:64–65.
10076. Justice, O. L. 1940: Methods of breaking dormancy in isolated embryos of *Polygonum scandens.* Am. J. Botany 27:16s.
10077. —— 1941: A study of dormancy in seeds of *Polygonum.* Cornell Univ. Agr. Expt. Sta. Mem. 235. 43 pp.
10078. —— 1944: Germination of seed of Florida beggarweed. Proc. Assoc. Offic. Seed Analysts 1943/44:103–4.
10079. —— 1944: Viability and dormancy in seeds of *Polygonum amphibium* L., *P.*

coccineum Muhl., and *P. hydropiperoides* Michx. Am. J. Botany 31:369–77.

10080. —— 1946: An air-conditioned seed germinator and fluorescent light grids to replace daylight. Proc. Assoc. Offic. Seed Analysts 1945/46:107–11.

10081. —— 1947: Further notes on the use of fluorescent lights in germination chambers. Proc. Assoc. Offic. Seed Analysts 37:101–3.

10082. —— 1947: Observations on the germination of New Zealand spinach seed. Proc. Assoc. Offic. Seed Analysts 37: 120–27.

10083. —— 1948: New method of testing bluegrass seed for purity. Assoc. Offic. Seed Analysts News Letter 22(2):34–35.

10084. —— 1949: A comparison of some ultraviolet lamps in testing ryegrass for fluorescence. Proc. Assoc. Offic. Seed Analysts 39:89–92.

10085. —— 1953: Report of the committee on methods and rules for countries with temperate climates. Proc. Intern. Seed Testing Assoc. 18:317–24.

10086. —— 1960: Report of referee testing on a regional basis. Proc. Intern. Seed Testing Assoc. 25:636–48.

10087. —— 1960: Report of the rules committee, International Seed Testing Association. Revision of the international rules for seed testing. Proc. Intern. Seed Testing Assoc. 25:716–54.

10088. —— 1961: The science of seed testing. Yearbook Agr. U.S. Dept. Agr. 1961:407–13.

10089. —— 1962: Referee testing on regional basis. Proc. Intern. Seed Testing Assoc. 27:274–83.

10090. —— and R. W. Marks. 1945: Germination of unscarified and scarified seed of *Lathyrus hirsutus* L., under laboratory, greenhouse, and field conditions. Proc. Assoc. Offic. Seed Analysts 35: 104–15.

10091. —— and M. D. Whitehead. 1942: Viability of velvetbean, *Stizolobium* spp., seed as affected by date of harvest, weathering, storage, and lodging. J. Am. Soc. Agron. 34:1000–9.

10092. —— and —— 1946: Seed production, viability, and dormancy in the nutgrasses, *Cyperus rotundus* and *C. esculentus*. J. Agr. Res. 73:303–18.

Justice, O. L. 513, 4987, 13872, 19546.

10093. Justus, N. 1964: Mechanical separation of hybrid and self-pollinated seed as a means of increasing percentage hybrids in upland cotton. Crop Sci. 4:161–63.

Justus, N. 4075.

10094. Jutchinson, J. B. 1944: The drying of wheat. III. The effect of temperature on germination capacity. Soc. Chem. Ind. J. 63:104–7.

Jutros, M. W. 19988.

K

10095. K., P. A. 1941: Growing plants from seed. Bull. Missouri Botan. Garden 29: 81–94.

10096. Kabulov, D. T., M. M. Muminov, and A. K. Aleksanyan. 1962: Irradiation of seeds with radioactive cobalt increases the cotton yield. Khlopkovodstvo 1962(3): 62.

10097. Kachioni-Walther, L. S. 1927: Germination and seedling growth of tobacco in relation to pH of the medium. Bull. Sta. Acclimatiz. Leningrad Agr. Inst. Detskoe Selo 4:11–30. Biol. Abstr. 6:18363. 1932.

10098. Kaden, N. N. 1951: Fruits and seeds of middle-Russian Nymphaeaceae and Berberidaceae. Mosk. Obshch. Isp. Prirody Byul. Otd. Biol. 56:81–90.

10099. Kadkol, S. B., V. Srinivasamurthy, and M. Swaminathan. 1954: Nutritive value of the seeds of *Paspalum scrobiculatum*. J. Sci. Ind. Res. (India) 13B:744–45. Chem. Abstr. 49:6403. 1955.

10100. Kadman, A. 1957: Germination of avocado seeds. Hassadeh 38:167–68.

10101. Kadman-Zahavi, A. 1955: The effect of light and temperature on the germination of *Amaranthus blitoides* seeds. Palestine J. Botany 6:140–44.

10102. —— 1955: Notes on the germination of *Atriplex rosea*. Palestine J. Botany 6:145–48.

10103. —— 1960: Effects of short and continuous illuminations on the germination of *Amaranthus retroflexus*. Bull. Res. Council Israel Sect. D 9:1–20.

10104. Kadota, T. 1937: Vigor of F_1 plants as related to age of seeds in eggplants. Hort. Assoc. Japan J. 8:351–55.

10105. —— 1941: Germination test of the seeds preserved by canning. Hort. Assoc. Japan J. 12:321.

10106. —— 1942: Relations of size and maturity to germination of spinach seeds. Hort. Assoc. Japan J. 13:52.

10107. Kadow, K. J. 1934: Seed transmission of *Verticillium* wilt of eggplants and tomatoes. Phytopathology 24:1265–68.

10108. —— and H. W. Anderson. 1937: Damping-off control; an evaluation of seed and soil treatments. Illinois Agr. Expt. Sta. Bull. 439:291–348.

Kadowski, M. 6819.

10109. Kadry, A. E. R. 1946: Embryology of *Cardiospermum halicacabum* L. Svensk Botan. Tidskr. 40:111–26. Biol. Abstr. 21:9720. 1947.

10110. —— 1955: The development of endosperm and embryo in *Cistanche tinctoria* (Forssk.) G. Beck. Botan. Notiser 108:231–43.

10111. —— 1960: The seed of *Cardiospermum halicacabum* L. A criticism. Acta Botan. Neerl. 1:330–32.

10112. —— and H. Tewfic. 1956: A contribution to the morphology and anatomy of seed germination in *Orobanche crenata*. Botan. Notiser 109:385–99. Biol. Abstr. 32:6026. 1958.

10113. —— and —— 1956: Seed germination in *Orobanche crenata* Forssk. Svensk Botan. Tidskr. 50:270–86. Biol. Abstr. 32:38316. 1958.

10114. Kaeiser, M. 1940: Note on embryo development in *Hippuris*. Trans. Illinois State Acad. Sci. 33:52–54.

10115. —— 1949: Embryo development of the pond cypress (*Taxodium asendens* Brongn.). Trans. Illinois State Acad. Sci. 42:63–67.

10116. —— and S. G. Boyce. 1962: X-ray negatives for sorting yellow-poplar samaras with filled and empty seeds. J. Forestry 60:410–11.

Kaeiser, M. 2900, 3254.

10117. Kafuku, K., and C. Hata. 1933: Formosan plant-seed oils. VII. Oil in *Mukuna capitata* and *Melia azedarach*. J. Chem. Soc. Japan 54:169–73. Chem. Abstr. 27:3098. 1933.

10118. —— and —— 1933: Formosan plant-seed oils. VIII. Oil and the lipase-like enzyme in *Bombax malabaricum* seed. J. Chem. Soc. Japan 54:174–77. Chem. Abstr. 27:3098. 1933.

10119. —— and —— 1935: Seed oils of Formosan plants. XII. Sapotaceae oil. J. Chem. Soc. Japan 56:1081–83. Chem. Abstr. 30:314. 1936.

10120. ——, ——, and M. Fujikawa. 1932: Formosan plant-seed oils. V. J. Chem. Soc. Japan 53:1115–19. Chem. Abstr. 27:1532–33. 1933.

10121. ——, T. Ikeda, and C. Hata. 1932: I. Formosan plant-seed oils. II. Akagi-seed oil. J. Chem. Soc. Japan 53:388–94; 395–98. Chem. Abstr. 27:201–2. 1933.

10122. Kagawa, A. 1955: Studies on the effect of low temperature induction in radish plant. I. Studies on the transmission of thermo-inducing stimulus and the effect of removing cotyledon upon the flowering induction of Japanese radishes. Bull. Natl. Inst. Agr. Sci. Ser. E 1955(4):265–74.

10123. —— 1956: Studies on the floral induction of slow bolting varieties in spinach. I. Effect of chemical vernalization and low temperature treatments of seeds. Hort. Assoc. Japan J. 25:173–80.

10124. —— 1957: Studies on the floral induction of slow bolting spinach. II. Effects of the temperature for seed vernalization and of drying of chilled seeds on flowering of spinach. Hort. Assoc. Japan J. 26:230–35.

10125. —— 1958: Studies on the floral induction of slow bolting spinach. III. The influence of the time of planting and interaction between vernalization and photoperiod upon flowering of spinach. Hort. Assoc. Japan J. 27:53–59.

10126. —— and M. Nakamura. 1958: Studies on the effect of low temperature induction in radish plant. III. On the effect of high temperature to reproductive development of vernalized and unvernalized radish plant. Gifu Univ. Fac. Agr. Res. Bull. 9:32–40.

Kagawa, A. 5817, 6000.

10127. Kagawa, F. 1939: Studies on the inheritance of a type of large-grained, partially sterile rice plant. Japan. J. Botany 10:1–33. Biol. Abstr. 14:7980. 1940.

Kageyama, M. 7994.

Kahali, B. H. 5699.

Kahler, L. 5351.

10128. Kahler, L. H. 1955: Pelleting conifer seeds for control of damping-off. Tree Planters Notes 21:3–6.

10129. Kahn, A. 1960: An analysis of "dark-osmotic inhibition" of germination of lettuce seeds. Plant Physiol. 35:1–7.

10130. —— 1960: Promotion of lettuce seed germination by gibberellin. Plant Physiol. 35:333–39.

10131. ——, J. A. Goss, and D. E. Smith. 1957: Effect of gibberellin on germination of lettuce seed. Science 125:645–46.

10132. Kahn, N. H., and K. Ahmad. 1957: Chemical investigation of the fatty oil from the seeds of *Sapium indicum* (Willd.). Pakistan J. Biol. Agr. Sci. 1:59–62. Biol. Abstr. 37:3387. 1962.

10133. Kahn, R. P. 1956: Seed transmission of the tomato-ringspot virus in the Lincoln variety of soybeans. Phytopathology 46:295.

10134. Kahn, V., R. W. Howell, and J. B. Hanson. 1960: Fat metabolism in germinating soybeans. I. Physiology of native fat. Plant Physiol. 35:854–60.

10135. Kahre, L. 1955: The influence of harvesting methods on seed quality of six-rowed barley. Preliminary report. Medd. Statens Cent. Frökontr. 30:33–44.

10136. —— 1956: Mechanical damages on kernels at direct combining of six-rowed barley and spring wheat. Medd. Statens Cent. Frökontr. 31:27–37.

10137. —— 1962: Seed storage in jute and paper bags. Preliminary report. Medd. Statens Cent. Frökontr. 37:30–36.

10138. ——, H. Kolk, and H. Wiberg. 1962: Note on dormancy-breaking in seeds (cereals and timothy). Proc. Intern. Seed Testing Assoc. 27:679–83.

10139. Kaihara, H. 1951: Comparative storage-tolerance of some cereal grains in Japan. Ber. Ohara Inst. Landwirtsch. Forsch. Ber. 9:435–40.

10140. —— 1951: Quality of wheat. III. Change of the protein content with nitrogenous fertilizers. Rept. Ohara Inst. Agr. Res. 39:139–42. Chem. Abstr. 46:2734. 1952.

Kaihara, H. 11159.

10141. Kain, R. M. 1964: Testing shows true value of seed. J. Agr. S. Australia 67:232–37.

10142. Kaiser, F., E. Haack, M. Gube, U. Dölberg, and H. Spingler. 1959: Neue Glykoside aus den Samen von *Strophanthus kombe*. Naturwissenschaften 46:670.

Kaiser, F. 7911.

10143. Kaiser, S. 1935: The factors governing shape and size in *Capsicum* fruits; a genetic and developmental analysis. Bull. Torrey Botan. Club 62:433–54.

Kaiser, S. 271, 272.

10144. Kajale, L. B. 1937: A case of polyembryony in the Nyctaginaceae. Current Sci. (India) 5:429.

10145. —— 1938: Embryo and seed development in the Nyctaginaceae. I. Studies in the genus *Boerhaavia*. J. Indian Botan. Soc. 17:243–54.

10146. —— 1944: Development of the embryo-sac of *Zizyphus jujuba* Lamk. Current Sci. (India) 13:49.

10147. —— 1954: A contribution to the embryology of the Phytolaccaceae. II. Fertilization and the development of embryo, seed and fruit in *Rivina humilis* Linn. and *Phytolacca diocca* Linn. J. Indian Botan. Soc. 33:206–25.

10148. —— 1954: Development of the embryo, endosperm, and seed coat in two species of the Phytolaccaceae. Proc. Indian Sci. Congr. 41:140.

10149. —— 1954: Fertilization and the development of embryo and seed in *Euphorbia hirta* Linn. Proc. Natl. Acad. Sci. India 20:353–60.

10150. —— and K. S. N. Murthy. 1954: The embryo sac of *Acalypha ciliata* Forsk. J. Indian Botan. Soc. 33:417–22.

Kajale, L. B. 10009, 10010.

10151. Kajanus, B. 1913: On the continuous violet colored seed of *Pisum arvense*. Fühling's Landwirtsch. Ztg. 62:153–60; 849–952. Botan. Centralbl. 123:162. 1913.

Kaji, J. 18330.

Kajima, M. 11074, 11075.

10152. Kajimoto, Y., K. Harada, and K. Hirota. 1952: Some active principles of the seed of *Illicium religiosum*. Folia Pharmacol. Japan. 48:4–5. Chem. Abstr. 47:5072. 1953.

10153. Kajizuka, S. 1960: The effect of micro elements on the oil content of plant seeds. Yamagata Agr. Forestry Soc. J. 16:52–54.

Kak, D. 9915.

Kakimoto, S. 5439.

10154. Kakol, S. B., H. S. R. Desikachar, and M. Srinivasan. 1961: The mucilaginous principles in black gram. J. Sci. Ind. Res. (India) 200:252–53. Chem. Abstr. 56: 10583. 1962.

Kakuliia, M. A. 3822.

10155. Kalab, J. 1961: A new method of extracting larch seed without crushing the cones and its use in extraction plants in Czechoslovakia. Sb. Česk. Akad. Zeměděl. Věd Lesnictvi 7:475–94. Forestry Abstr. 24:412. 1963.

10156. Kalantyr, M. S. 1951: Peculiarities of blooming and fruit-bearing of *Eucommia* and methods for increasing the yields of seeds. Agrobiologiya 1951(3):34–46.

10157. —— and D. S. Zabotkin. 1956: Large-seed form of *Eucommia* [*ulmoides*]. Pushkino Vses. Nauchn.-Issled. Inst. Lesov.

i Mekh. Lesn. Khoz. Sb. Rabot po Lesn. Khoz. 1956(33):52–62.
Kalashnikova, M. I. 17296.
10158. Kalbe, L., and H. Vogel. 1960: Untersuchungen über die Wirkungssteigerung von Fungiziden durch oberflächen aktiven Substanzen. Zbh. Bakt., 2, 113:22–25; 766–72. Rev. Appl. Mycol. 40:336. 1961.
Kalbfleisch, W. 10447, 10448, 12671.
10159. Kaleysaraj, R., and P. A. Kurup. 1963: Anthelmintic principle of the seeds of *Butea frondosa*—isolation of an active fraction. Current Sci. (India) 32:456–57.
10160. Kalinin, F. L. 1949: Cultivation of immature radish (*Raphanus sativus* L.) embryos under artificial conditions. Dokl. Akad. Nauk SSSR 66:1191–94.
10161. —— 1950: Some regularities in the development of radish, *Raphanus sativus*, embryo on the mother plant. Dokl. Akad. Nauk SSSR 71:945–48.
10162. —— 1959: Dehydrogenase activity in developing wheat grains. Nauchn. Tr. Ukr. Inst. Fiziol. Rast. 1959(20):32–36. Chem. Abstr. 55:5672. 1961.
10163. —— 1959: Periodicity of the oxidation-reduction regimen of developing plant embryos. Doklady (Transl.) 125: 104–7.
Kalinin, P. 1739.
10164. Kalinina, L. V. 1959: Development of the embryo in corn grains. Kukuruza 6:23–24. Biol. Abstr. 38:3179. 1962.
Kalinina, L. V. 10813.
Kalinkevich, A. F. 5368.
10165. Kalitajew, V. V., and A. P. Grischenko. 1937: Über die Brauchbarkeit der zerschlagenen Weizenkörner als Saatgut. Proc. Intern. Seed Testing Assoc. 9:185–206.
10166. Kallio, T. K. 1963: Stratification of apple seeds. Ann. Agr. Fenn. 2:7–11.
10167. Kaloshina, Z. M. 1958: Comparative evaluation of the methods for the determination of viability of seeds of winter cereal crops. Mosk. Ordena Lenina Sel'skokhoz. Akad. im K. A. Timiriazeva Dokl. TSKHA 39:77–82.
10168. —— 1959: Effect of low temperature on rate of germination to freshly-harvested winter wheat seeds. Mosk. Ordena Lenina Sel'skokhoz. Akad. im K. A. Timiriazeva Dokl. 46:91–95.
10169. Kaloyereas, S. A. 1955: A gas storage for preserving rough rice. Rice J. 58(2):40–42.
10170. —— 1958: Rancidity as a factor in the loss of viability of pine and other seeds. J. Am. Oil Chemists Soc. 35:176–79.
10171. Kalton, R. R. 1959: Seed production problems in isolated recombination blocks. Agron. J. 51:629.
10172. ——, R. A. Delong, and D. S. McLeod. 1959: Cultural factors in seedling vigor of smooth bromegrass and other forage species. Iowa State J. Sci. 34:47–80.
Kalton, R. R. 1827, 4084–4086, 14400.
Kalupchiev, G. 9377.
10173. Kamal, M. A. M., and A. H. El-Damaty. 1957: Effect of fertilizer treatments and water levels on yield and composition of sunflower seed varieties. Ann. Agr. Sci. (Cairo) 2:137–49. Biol. Abstr. 35:5614. 1960.
10174. ——, M. I. Naguib, and F. Y. Refai. 1956: Carbohydrate fractions in different varieties of Egyptian wheat kernels with reference to the influence of some fertilizer treatments and locations. Ann. Agr. Sci. (Cairo) 1:95–110. Biol. Abstr. 35:5493. 1960.
10175. ——, F. Y. Refai, and M. I. Naguib. 1956: Study on the effect of certain fertilizer treatments on the biochemical constituents of Egyptian wheat kernel varieties. Effect on the nitrogen content and gluten quality with reference to the weight of kernels. Ann. Agr. Sci. (Cairo) 1:127–44. Biol. Abstr. 35:5338. 1960.
10176. Kamalyan, G. V., and L. V. Davtyan. 1963: Effect of some amino alcohols on the phosphorus metabolism is sprouting seeds. Dokl. Akad. Nauk Arm. SSR 37:83–87. Chem. Abstr. 60:3428. 1964.
10177. Kamel, A. E. H. 1959: The effect of carbon bisulphide and ethylene dichloride-carbon tetrachloride mixture on the germination of some seeds of certain crops. Agr. Res. Rev. (Cairo) 37:330–50.
Kamel, M. 18258.
Kamemoto, H. 18061.
10178. Kamenskii, K. V., and T. A. Orekhova. 1937: Stimulating the germination of maturing seeds by means of ultraviolet. Bull. Appl. Botany, Genet., Plant Breed. USSR Ser. 4(2):67–71. Chem. Abstr. 31:5843. 1937.
10179. Kamensky, K. 1928: The separation of seeds of *Convolvulus* from clover by action of electromagnet. Congr. Russ. Botanists Abstr. Papers, State Russ. Botan. Soc. Leningrad., p. 282.
10180. Kamensky, K. W. 1920: Eine Erforschung über die Befeuchtung bei

Keimung der Weizensamen. Jard. Bot. Inst. Essais Semences, Leningrad. Ann. Essais Semences 1920(4):14–20.

10181. —— 1923: Die morphologischen Unterschiede der Samen einiger arten Caryophyllaceae. Jard. Bot. Inst. Essais Semences, Leningrad. Ann. Essais Semences 1923(4):1–13.

10182. —— 1927: Unterscheidung der "Samen" von *Poa pratensis* L. und *P. trivialus* L. sowie von anderen verwandten *Poa*-arten. Jard. Bot. Inst. Essais Semences, Leningrad. Ann. Essais Semences 1927(5):17–38.

10183. —— 1928: Das stimulieren der Samenaufkeimung durch Einwirkung von kochendem Wasser und "falsche Keimung." Jard. Bot. Inst. Essais Semences, Leningrad. Ann. Essais Semences 1928(5): 1–19.

10184. ——1928: Anatomische Unterschiede der Samen der *Cuscuta*-Arten. Jard. Bot. Inst. Essais Semences, Leningrad. Ann. Essais Semences 1928(6):117–21.

10185. —— 1929: Methoden zur Feststellung der Panzerschicht bei Achänen verschiedener Arten von *Helianthus annuus*. Jard. Bot. Inst. Essais Semences, Leningrad. Ann. Essais Semences 1929(1):13–18.

10186. —— 1929: Neues "Unkraut" des Sibirischen Weizens. Jard. Bot. Inst. Essais Semences, Leningrad. Ann. Essais Semences 1929(1):19–22.

10187. —— 1930/31: Morphologisch-anatomische Unterscheidungsmerkmale der Unkrautsamen aus den familien Liliaceae und Iridaceae. Bull. Appl. Botany 25(4): 103–8.

10188. —— 1932: Über den Bau keimloser Getreidesamen. Proc. Intern. Seed Testing Assoc. 4:30–40.

10189. —— 1937: Results of five years' work (1931–36) of the Department of Seed Science and Testing of the Institute of Plant Industry. Bull. Appl. Botany, Genet., Plant Breed. (USSR) Ser. 4(2):5–17.

10190. —— and A. M. Bogoljubowa. 1931: Über den Einfluss von Frost auf die Keimfähigkeit von Rotkleesamen. Proc. Intern. Seed Testing Assoc. 3:246–60.

10191. —— and T. A. Orekhova. 1932: Embryoless seeds of cereals. Bull. Appl. Botany, Genet., Plant Breed. (USSR) Ser. 5(1):199–206.

10192. —— and —— 1937: Stimulation of maturing seeds to germination by ultraviolet irradiation. Bull. Appl. Botany, Genet., Plant Breed. (USSR) Ser. 4(2): 71–72.

10193. ——, ——, and Z. M. Schulz. 1931: Einige Beobachtungen und Schlussfolgerungen zur Frage von den zerbrochenen Rotkleesamen und Keimen. Proc. Intern. Seed Testing Assoc. 3:261–69.

10194. Kaminski, C. 1959: Phenoloxidases of the hypocotyl of *Impatiens balsamina*. Acad. Roy. Belg. Bull. Classe Sci. 45:154–68. Chem. Abstr. 54:13289. 1960.

10195. Kaminski, E. 1958: Biochemical characteristics of rye varieties grown in Poland. Biokhim. Zerna, Sb. 1958(4): 108–27. Chem. Abstr. 53:510. 1959.

Kaminski, E. 9722.

Kamiyama, K. 16800.

10196. Kammacher, P., and J. B. Roux. 1960: A new case of polyembryony in *G. hirsutum*. Coton Fibres Trop. 15:23–26. Biol. Abstr. 36:25263. 1961.

10197. Kamnev, I. Y., M. K. Zakiyan, G. B. Butkova, and N. P. Novik. 1956: Viability changes in the seeds of sunflowers and castor oil plants during storage. Uch. Zap. Rostovsk. na Donu Gos. Univ. 1956(26):85–92. Biol. Abstr. 36:38542. 1961.

10198. Kamra, O. P. 1959: Radiation response of barley seeds in relation to hydration and to chemical substances lost by leaching. Dissertation Abstr. 20: 839–40.

10199. —— and S. K. Kamra. 1962: Factors influencing X-ray tolerance of *Haplopappus gracilis* seeds. Can. J. Genet. Cytol. 4:255–62. Nucl. Sci. Abstr. 17: 1015. 1963.

10200. ——, ——, R. A. Nilan, and C. F. Konzak. 1960: Radiation response of soaked barley seeds. I. Substances lost by leaching. Hereditas 46:152–70.

10201. ——, ——, ——, and —— 1960: Radiation response of soaked barley seeds. II. Relation of radiobiological damage to substances lost by leaching. Hereditas 46:261–73.

10202. Kamra, S. K. 1959: The effect of treatment with absolute alcohol on seeds of different species. Indian J. Plant Physiol. 2:132–40.

10203. —— 1961: The improved seed in India—its problems and prospects. Proc. Intern. Seed Testing Assoc. 26:170–75.

10204. —— 1961: Modern trends in seed testing. Indian J. Agr. Sci. 31:183–91.

10205. —— 1961: Seed testing and certification in Germany. Sci. Cult. (Calcutta) 27:220–24.

10206. —— 1963: Determination of mechanical damage on Scots pine seed with X-ray contrast method. Studia Forestalia Suecica, Skogshögskolan, Stockholm, No. 8. 20 pp.

10207. —— 1963: The place of seed testing in agriculture. Indian Farming 13(6):9, 29.

10208. —— 1963: Studies on a suitable contrast agent for the X-ray radiography of Norway spruce seed (*Picea abies*). Proc. Intern. Seed Testing Assoc. 28:197–201.

10209. —— and K. L. Mehra. 1961: The problems of seed certification in tropical agriculture. Proc. Intern. Seed Testing Assoc. 26:176–82.

Kamra, S. K. 5865, 10199–10201, 14968, 17450, 18203.

10210. Kamson-Rappaport, M. 1958: Changes in the ascorbic acid content of germinating lettuce seeds. Bull. Res. Council Israel Sect. D 6:126–27.

10211. Kan, R. 1948: A comparative study between seed-coat tissues of Asiatic cotton and those of Upland cotton in the process of development. Bull. Physiograph. Sci. Res. Inst. Tokyo Univ. 1:1–16. Biol. Abstr. 27:10617. 1953.

10212. —— 1951: Studies on the origin of fiber cell on the seed-coat of Asiatic and Uplant cotton. Bull. Physiograph. Sci. Res. Inst. Tokyo Univ. 7:45–51. Biol. Abstr. 27:23232. 1953.

Kanabu, F. 13755.

Kanapaux, M. S. 5826, 19257.

10213. Kanash, S. S. 1957: Achievements in the selection and study of cotton seeds. Izv. Akad. Nauk Uz. SSR Ser. Biol. 1957(4):13–20.

10214. Kanda, M., and N. Takahashi. 1952: Studies on the germination of rice seed. I. The speed of germination and the appearance of plumule and radicle. Proc. Crop Sci. Soc. Japan 21:176–77.

10215. Kanda, T. 1931: The inheritance of seed-coat colour in the watermelon. Japan. J. Genetics 7:30–48. Japan. J. Botany 6:(5). 1932.

10216. Kandaswamy, K., and C. L. Subramanian. 1954: Seed treatment against black arm disease (*Xanthomonas malvacearum*) of cotton. Madras Agr. J. 41:134–40.

10217. Kandatu, M. 1940: Chemical studies of the seeds of *Aesculus turbinata* Blume. III. Chemical properties of crystalline saponin. J. Agr. Chem. Soc. Japan 16:741–50. Chem. Abstr. 35:1794–95. 1941.

Kandatu, M. 16766.

10218. Kandelaki, G. V. 1962: Cytoembryological study of induced parthenogenesis. Akad. Nauk Gruzinsk. SSR Soobshcheniya 29:571–78.

10219. —— and R. K. Beridze. 1956: Transplantation [of embryos on endosperm] in cereals. Tr. Tbilissk. Botan. Inst. 18: 181–205.

Kandina, G. V. 3274.

10220. Kandler, O. 1953: Über den "Synthetischen Wirkungsgrad" in vitro Kultivierter Embryonen, Wurzeln und Sprosse. Z. Naturforsch. 8b:109–17.

10221. —— and E. Fink. 1955: Über die Abhängigkeit des synthetischen Wirkungsgrade in vitro kultivierter Embryonen von verschiedenen Aussenfaktoren. Planta 45: 289–306.

Kando, Y. 19990.

10222. Kaneda, T., and R. B. Alfin-Slater. 1963: A comparison of the effects of the polyunsaturated fatty acids of cuttlefish liver oil and cottonseed oil on cholesterol metabolism. J. Am. Oil Chemists Soc. 40:336–38.

Kaneki, Y. 13610.

10223. Kaneko, K. 1957: Studies of the embryo culture on the interspecific hybridization of chrysanthemum. Japan. J. Genetics 32:300–5.

10224. Kanevskaia, Z. E., and N. I. Kzelkova. 1955: The determination of the quality of seeds on the basis of their viability. Zemledelie 1955(5):99–101. Biol. Abstr. 31:29225. 1957.

Kang, C. N. 3924, 18581.

Kanga, D. D. 15308.

10225. Kangas, E. 1942: The influence of kiln temperature on shedding and germination of pine seed. Acta Forest. Fennica 50(14):1–42. Biol. Abstr. 21:25946. 1947.

10226. Kanipe, L. A. 1939: The germination of the seed of *Alsine media*. Proc. Assoc. Offic. Seed Analysts 30:249–52.

10227. —— 1941: The plant producing value of injured seeds of *Rumex* spp. Proc. Assoc. Offic. Seed Analysts 32:32–39.

10228. —— 1941: Viability of injured weed seed, *Bromus tectorum*. Proc. Assoc. Offic. Seed Analysts 33:48–50.

10229. —— 1947: Open 2,4-D packages endanger nearby legume seed viability. Southern Seedsman 10(3):46.

Kanitkar, U. K. 5236.

Kanki, H. 29.

Kano, K. 18801.

10230. Kano, T., and K. Hatano. 1953: On the amino acids in the seeds of *Pinus thunbergii* at the germination stage. J. Japan. Forestry Soc. 35:292–98.

Kano, T. 8371.

10231. Kantack, E. J. 1955: Carbowax as a carrier of systemic insecticides for treating wheat seed. J. Econ. Entomol. 48:486.

10232. Kantor, J. 1952: Effect of soaking seeds of *Pinus sylvestris* and *Picea excelsa* in growth substances on the sprouting and development of seedlings. Brunn. Vysoka Sk. Zemedel. Lesn. Sb. 3/4:291–97.

10233. —— 1959: Der Einfluss der Aufkeimung, der verschiedenen Keimbeschädigung und des Samengewichtes auf die Entwicklung der Sämlinge der Edelkastenie (*Castanea vesca*). Sb. Česk. Akad. Zeměděl. Věd Lesnictvi 32:347–52.

10234. —— 1961: Germinative capacity of European larch (*Larix decidua*) seed from artificial pollination. Sb. Česk. Akad. Zeměděl. Věd Lesnictvi 34:21–34.

10235. Kantor, T. S. 1959: Comparative embryogenesis of some flax varieties in respect to their phylogenic interrelations. Tr. Soveshch. po Morfogenezu Rast. 1:441–44.

10236. Kanyak, K. 1962: Pine selection and seed production in Czechoslovakia. Lesn. Khoz. 15(10):81–85. Biol. Abstr. 43: 16224. 1963.

10237. Kanzler, L. 1925: Beiträge zur Physiologie der Keimung und der Keimlunge. Botan. Centralbl. Beih. 41:185–238.

10238. Kao, H. C. 1936: Lipoid metabolism in the mung bean during germination. Biochem. J. 30:202–7.

Kao, Y. S. 3912.

10239. Kapil, R. N., and S. Jalan. 1962: Studies in the family Ranunculaceae. I. The embryology of *Caltha palustris* L. Symp. Plant Embryol., Univ. Delhi, 1960. (Papers), pp. 205–14.

10240. —— and M. R. Vyayaraghavan. 1962: Embryology and systematic position of *Pentaphragma horsfieldii* (Miq.) Airy Shaw. Current Sci. (India) 31:270–72.

Kapil, R. N. 12707.

10241. Kapinos, G. E. 1960: Flowering, pollination, and embryonic development of *Sternbergia lutea* (L.) Ker.-Sawl. and *S. fischeriana* (Herb.) Roem. Botan. J. (USSR) 45:1044–55.

Kaplan, J. M. 19677.

10242. Kaplan, R. W. 1951: Chromosomen- und Faktormutationsraten in Gerstenkörnern bei verschiedenartigen Quellungsbehandlungen oder Kälte während oder nach der Röntgensowie bei Dosisfrakionierung. Z. Induktive Abstammungs- Vererbungslehre 83:347–82. Biol. Abstr. 25: 29935. 1951.

Kapoor, B. M. 18349, 18350.

10243. Kapoor, L. D., I. C. Chopra, and S. Nath. 1952: A note on the germination of seeds of Indian belladonna (*Atropa belladonna*). Indian Forester 78:34–35.

Kapoor, L. D. 10345.

10244. Kapoor, M., and E. R. Waygood. 1962: Initial steps of purine biosynthesis in wheat germ. Biochem. Biophys. Res. Commun. 9:7–10.

Kapoor, M. 16602, 16603.

Kapoor, S. L. 17493–17495.

10245. Kaposts, V. 1961: Effect of minor elements on germination of pine and fir seeds and on the subsequent growth of seedlings. Latvijas PSR Zinatnu Akad. Vestis 1961(3):111–16.

10246. Kapp, L. C. 1947: The effect of common salt on rice production. Arkansas Agr. Expt. Sta. Bull. 456, p. 1–7. Biol. Abstr. 22:17045. 1948.

10247. Kapper, O. G. 1931: Samenertrag der Kiefer in Suedrussland. Z. Forst-Jagdw. 63:734–40.

10248. Kappert, H. 1933: Erbliche Polyembryonie bei *Linum usitatissimum*. Biol. Centralbl. 53:276–307.

10249. Kapralova, N. P. 1958: On jute sowing for seed. Len i Konoplia 1958(3): 28–29. Biol. Abstr. 44:20857. 1963.

10250. Kar, B. K. 1940: Vernalisation of Indian crops. Current Sci. (India) 9: 233–35.

10251. —— 1942/43: Vernalization of Indian crops. I. Some observations on wheat (*Triticum vulgare*). Trans. Bose Res. Inst. (Calcutta) 15:105–26. Biol. Abstr. 19: 5323. 1945.

10252. —— 1944: Vernalization of Indian crops. II. Photostage in wheat (*Triticum vulgare*) and oat (*Avena* sp.). Proc. Indian Sci. Congr., Delhi 31:87–88. Plant Breeding Abstr. 15:102. 1945.

10253. —— 1944: Vernalization of jute. Current Sci. (India) 13:130–31.

10254. —— 1946: Vernalization of crops cultivated in India. Nature 157:811–12.

Kar, B. K. 1842.

10255. Karabaki, A. K. 1963: An apparatus for cleaning and sorting conifer seeds. Lesn. Khoz. 6:68–70.

Karabashev, N. 15558.

Karabasheva, T. 15558.

10256. Karaerov, P. G. 1957: Seed production of vetch. Vestn. Sel'skokhoz. Nauki 2(5):53–55.

Karapetyan, K. A. 1609.

10257. Karapetyan, V. K. 1962: Mutation of plants grown from embryonically young seeds. Akad. Nauk SSSR Inst. Genet. Tr. 29:435–41.

10258. Karashchuk, I. M. 1950: Summer and autumn seeding of sainfoin with freshly-harvested seeds. Sov. Agron. 8: 48–57.

Karashchuk, I. M. 13567.

10259. Karchevski. 1901: Influence of the fluctuation of temperature on the respiration of seeds and embryos of wheat. Mem. Warsaw Univ. 9. 114 pp. Expt. Sta. Rec. 14:839. 1903.

Karel, V. 12967.

Kargl, T. E. 15791.

10260. Kariakin, A. V., D. N. Lazarev, and G. A. Barinova. 1956: Luminescence method for determining viability in seeds of agricultural plants. Dokl. Akad. Nauk SSSR 106:739–42.

Karimullah. 347, 2420, 8067, 9356, 10514, 10515.

Karira, J. T. 9001.

10261. Kariyone, T., and H. Iwao. 1939: Fat of the seed of *Cinnamomum loureirii*. J. Pharm. Soc. Japan 59:113–15. Chem. Abstr. 33:4445. 1939.

Karkun, J. N. 2084, 2085.

10262. Karlberg, S. 1951: Quick method for determining viability of certain kinds of forest seeds. Skogen (Stockholm) 38:47.

10263. Karlberg, S. 1953: Treatment of pine and spruce seed to stimulate germination. Bull. Roy. School Forestry Sweden No. 11:1–42.

10264. Karmakar, D. V. 1930: Chemical study of the germination of groundnut. Proc. Indian Sci. Congr., Allahabad 17: 191.

10265. —— and V. N. Patwardhan. 1930: Amylase from wheat. J. Indian Inst. Sci. 13A:159–64. Chem. Abstr. 25:2450. 1931.

Karman, G. 3704, 16216.

10266. Karnatz, A. 1962: Fruit without seeds. Umschau 62:405–8.

10267. Karon, M. L. 1947: Hygroscopic equilibrium of cottonseed. J. Am. Oil Chemists Soc. 24:56–58.

10268. —— and M. E. Adams. 1948: Note on the hygroscopic equilibrium of cottonseed and cottonseed products. J. Am. Oil Chemists Soc. 25:21–22.

10269. —— and —— 1949: Hygroscopic equilibrium of rice and rice fractions. Cereal Chem. 26:1–12.

10270. —— and A. M. Altschul. 1944: Effect of moisture and of treatments with acid and alkali on rate of formation of free fatty acids in stored cottonseed. Plant Physiol. 19:310–25.

10271. —— and —— 1946: Respiration of cottonseed. Plant Physiol. 21:506–21.

10272. —— and B. E. Hillery. 1949: Hygroscopic equilibrium of peanuts. J. Am. Oil Chemists Soc. 26:16–19.

Karon, M. L. 434–438, 9784.

10273. Karper, R. E. 1928: Longevity and viability of kafir seed. J. Am. Soc. Agron. 20:527.

10274. —— 1931: Multiple seeded spikelets in sorghum. Am. J. Botany 18:189–94.

10275. —— 1933: Inheritance of waxy endosperm in sorghum. J. Heredity 24: 257–62.

10276. —— and D. L. Jones. 1936: Longevity and viability of sorghum seed. J. Am. Soc. Agron. 28:330–31.

10277. —— and J. R. Quinby. 1947: The inheritance of callus formation and seed shedding in sorghum. J. Heredity 38:211–14. Biol. Abstr. 22:5426. 1948.

10278. —— and —— 1963: Sugary endosperm in sorghum. J. Heredity 54:121–26.

10279. Karpovich, I. 1961: Dependence of seed germination on seed moisture at the time of threshing. Sel'skoe Khoz. Sibiri 1961(4):24–28.

10280. Karrer, P., and H. Keller. 1939: Potentiometrische Titration von Weizenkeimlingsölen. Helv. Chim. Acta 22: 617–18. Biol. Abstr. 13:12029. 1939.

10281. —— and H. Salomon. 1937: Constituents of plant germs. I. New compounds from the unsaponifiable portion of wheat-germ oil. Helv. Chim. Acta 20:424–36. Chem. Abstr. 31:6250–51. 1937.

10282. —— and —— 1938: Isolation of tocopherols from wheat-germ oil. Helv. Chim. Acta 21:514–19. Chem. Abstr. 32:7035. 1938.

Karrer, P. 16942, 16943.

10283. Karschon, R. 1960: Lime tolerance and seed provenance in eucalypts. Food Agr. Organ., Jt. Subcomm. Mediterranean Forestry Probl., Session Working Party Eucalyptus, 4th, Lisbon FAO/SCM/EU/60-9b. 7 pp. Chem. Abstr. 59:6938. 1963.

10284. Karstad, O. 1951: Self-seeded spruce in western Norway. Tidsskr. Skogbruk. 59:410–11.

10285. Kartaschoff, R. 1958: Die Keimverzögerung der Samen von *Iris pseudacorus* L. Ber. Schweiz. Botan. Ges. 68:145–82. Biol. Abstr. 35:5437. 1960.

10286. —— 1958: Die Keimverzögerung der Samen von *Iris pseudacorus* L. Inaug.-Dissertation zur Erlangung Philosophischen Doktorwürde vorgelogt Philosophisch-Naturwiss. Fak. Univ. Basel. 38 pp.

10287. Kartha, A., R. Sukumaran, and K. N. Menon. 1944: The oil of *Mimusops elangi* (Linn.). Proc. Indian Acad. Sci. Sect. A 19:1–4. Biol. Abstr. 19:1580. 1945.

10288. Kartha, A. R. S. 1963: Variations in the proportions and iodine values of fats at different locations in the endosperm or embryo. J. Sci. Food Agr. 14:515–19.

10289. —— and R. Narayanan. 1962: The component acid composition and glyceride structure of *Myristica attenuata* seed fat. J. Sci. Ind. Res. (India) 21:494–98. Biol. Abstr. 45:21673. 1964.

10290. —— and A. S. Sethi. 1958: Mobilization of fat in germinating seed. J. Sci. Ind. Res. (India) 17C:34–35. Chem. Abstr. 52:20421. 1958.

10291. —— and —— 1958: Occurrence of free higher fatty acids in germinating seeds. J. Sci. Ind. Res. (India) 17C:182–85. Chem. Abstr. 53:10389. 1959.

10292. —— and —— 1960: Effect of polyploidy on biosynthesis of fat in seeds. J. Sci. Ind. Res. (India) 19C:126. Chem. Abstr. 54:25081. 1960.

10293. —— and —— 1963: Biosynthesis of erucic acid in Cruciferae and Tropaeolaceae seed fats. Indian J. Agr. Sci. 33: 38–43. Biol. Abstr. 45:25880. 1964.

10294. ——, ——, and K. C. Gulati. 1955: Rapid estimation of yield and iodine value of oils in small samples of oilseeds. Indian J. Agr. Sci. 25:79–84. Biol. Abstr. 31:2494. 1957.

Kartha, A. R. S. 14212, 14213.

10295. Kartuzova, M. A. 1953: Strength of initial growth as a method of biological evaluation of (wheat) seed quality. Selek. i Semen. 20(2):30–36.

10296. Karuna, V. A. 1959: A tricotyledonous sugar beet. Botan. Zh. 44:1340–41.

10297. Karunakar, P. D., and T. Rajagopalan. 1936: Azotobacter inoculation of seeds of cereals. Experiments with sorghum. Proc. Assoc. Econ. Biologists (Coimbatore) 4:64–73. Chem. Abstr. 32:5983. 1938.

10298. Karve, A. D. 1962: A note on the germination of seeds of *Cassia tora*. Res. Bull. Panjab. Univ. 13:185–86. Biol. Abstr. 43:11453. 1963.

10299. Kasahara, Y. 1944: Experiments on substances promoting and inhibiting germination. I. Rice. Rept. Ohara Agr. Inst. 36:1–101. Biol. Abstr. 23:30206. 1949.

10300. —— and S. Akita. 1951: The abnormality in the germination of squash, *Cucurbita moschata* Duch., caused by light. Ohara Inst. Landwirtsch. Forsch. Ber. 9:451–53.

Kasahara, Y. 11098–11117.

Kasai, Z. 1430.

10301. Kasaka, H. 1925: Über den Einfluss der Frucht auf die Samenreife bei einigen Kulturpflanzen. Kyushu Univ. Dept. Agr. J. 1:197–206.

10302. Kasasky, C. 1930: Trockenstimulierungsversuche bei Saatwicke. Deut. Landwirtsch. Presse 57:130.

10303. —— 1934: Ein Versuch mit einer neuen Samenstimulierungs-methode. Bulgar. Zemledelie (Ackerbau) 8. Proc. Intern. Seed Testing Assoc. 7:108. 1935.

10304. —— 1951: Samenstimulationsversuche an Winterweisen. Izv. Bulgar. Akad. Nauk. Biol. Inst. 2:265–82.

10305. Kaserer, H. 1945: Acceleration of germination of sugar and forage beets and other plants by powdered bypsum. Bodenkunde Pflanzenernähr 36:74–92. Chem. Abstr. 41:4542. 1947.

10306. —— [no year]: Der Einfluss des Fruchtbereiches auf die Entwicklung der Keim-Pflanzen. Zugleich ein Beitrag zur Frage der Synaptospermie. Biol. Generalis (Wien) 18:243–53. Proc. Intern. Seed Testing Assoc. 14:74–75. 1948.

10307. —— and A. Frisch. 1940: Über den Einfluss der Spelzen auf die Keimung bei bespelztem Getreide. Anz. Akad. Wiss. Wien Math.-Naturw. Kl. No. 9. Proc. Intern. Seed Testing Assoc. 14:75–76. 1948.

Kashida, T. 14160.

10308. Kasperbauer, M. J., and F. P. Gardner. 1959: The effect of gibberellic acid and depth of seeding on emergence and growth of grain sorghum. Agron. Abstr. 1959:71.

10309. ——, F. P. Gardner, and W. E. Loomis. 1962: Interaction of photoperiod and vernalization in flowering of sweet clover (*Melilotus*). Plant Physiol. 37: 165–70.

Kasperbauer, M. J. 6972.

Kassab, R. 13826, 13828.

10310. Kassanis, B., and A. Kleczkowski. 1948: The isolation and some properties of a virus-inhibiting protein from *Phytolacca esculenta*. J. Gen. Microbiol. 2:143–53.

10311. Kastberg, W. 1939: Chrysanthemums from seed. Gardeners Chron. 43:18.

10312. Kastle, J. H., and A. S. Loevenhart. 1900: Concerning lipase, the fat-splitting enzyme, and the reversibility of its action. Am. Chem. J. 24:491–525.

Kastle, J. H. 3269.

10313. Kastoff, D. 1935: Mutations and the ageing of seeds. Nature 135:107.

Kastrubin, M. 10576.

10314. Kasturi, T. R., and B. H. Iyer. 1955: Fixed oil from *Elettaria cardoamomum* seeds. J. Indian Inst. Sci. 37:106–12. Biol. Abstr. 30:8695. 1956.

10315. Katagiri, H., and N. Mugibayasi. 1938: Relation between nitrogen compounds of crop seeds and diastatic power of their malts. III. Glutinous and non-glutinous rice. J. Agr. Chem. Soc. Japan (English Transl.) 14:1332–34. Chem. Abstr. 33:7843. 1939.

10316. —— and —— 1939: Relation between the nitrogenous compounds of seeds of various crops and the diastatic power of their malts. IV. Barley and naked barley. J. Agr. Chem. Soc. Japan (English Transl.) 15:309–16. Chem. Abstr. 34:137. 1940.

Katagiri, K. 11585.

10317. Kataoka, T., and F. Nonaka. 1951: How to germinate orchid seeds. Hawaii Farm & Home 14(6):25.

Kataoka, T. 1304, 1305.

Katayama, M. 10385, 10386.

Katayama, T. 14067, 14068.

10318. Katayama, Y. 1931: Über den Ansatz und die Keimungsfähigkeit der Körner in reziproken intra- und inter generischen Bestäubungsversuchen im Weizen- Aegilops-Kreise und die Sterilität der F_1-Pflanzen. Agr. Hort. 6:1909–24. Japan. J. Botany 6:(5)–(6). 1932.

10319. Kater, J. M. 1927: A cytological study of dormancy in the seed of *Phaseolus vulgaris*. Ann. Botany (London) 41: 629–41.

Katerov, K. 12778.

Katkeirch, Y Y. 14476.

10320. Kato, I., and S. Sakaguchi. 1953: Morphological and physiological studies on occurrence of aborted seed in soybean. I. Development of flower and seed. II. Growth of seed and pod. Proc. Crop Sci. Soc. Japan 21:271–72.

10321. ——, ——, and Y. Naito. 1954: Development of flower parts and seed in the soybean plant, *Glycine max* M. Tokai Kinki Natl. Agr. Expt. Sta. Div. Plant Breed. Cult. Bull. 1:96–114.

10322. Kato, K., and T. Hata. 1932: Seed-oil of Formosan plants. VI. Seed-oil of *Jatropha curcas* Linn. Nippon Kgk. Kw. Si. 53:1120–23. Japan. J. Med. Sci. II. Biochem. 4:(59). 1938.

Kato, T. 5818.

10323. Kato, Y. 1960: Automutagenic action of the extract from immature *Allium fistulosum* seeds on mitosis. Japan. J. Genetics 35:251–56.

10324. Kato, Y., N. Hukuhara, and R. Kobayashi. 1960: Stimulation of differentiation of flower buds in Sugi by gibberellin. IV. Germination of the seeds. J. Japan. Forestry Soc. 42:323–25.

10325. Katsarov, A. 1954: Machine for cleaning and sorting tobacco seed. Visshch. Selsk. Inst. "Vasil Kolarov" Nauchn. 3:315–24. Tobacco Abstr. 1:1622. 1957.

Katsova, L. V. 7492.

Katsumata, H. 29.

10326. Katsunuma, S., and H. Nakamura. 1932: Iron distribution in the earlier stages of growth of soybean and chick embryo. Nagoya J. Med. Sci. 6:107–12. Chem. Abstr. 27:3490. 1933.

10327. Katsuta, M. 1959: Ripening and germinating processes of pine seeds. I. Changes of seed proteins. Tokyo Daigaku Nogakubu Enshurin Hokoku 55:125–59. Chem. Abstr. 58:1722. 1963.

10328. —— 1961: The breakdown of reserve protein of pine seeds during germination. J. Japan. Forestry Soc. 43:241–44.

10329. —— 1963: Nitrogen distribution in *Pinus thunbergii* seedlings arising from seeds of different weight. J. Japan. Forestry Soc. 45:104–6.

Katsuura, T. 18801.

Katti, M. C. Tummin-; *see* Tummin-Katti, M. C.

Katulska, C. 19658.

Katumoto, K. 8834.

10330. Katura, K., and T. Narasaki. 1954: Fundamental studies on the utilization of olive fruits. I. Change of the crude components of flesh during the growth of olive fruits. Kagawa Agr. Coll. Tech. Bull. 6:1–7.

10331. Katz, J. R. 1917: Die Gesetze der Quellung. I. Kolloidchem. Beih. 9:1–182. Chem. Abstr. 11:3141–42. 1917.

Katz, M. W. 18712.

10332. Katz, R., A. B. Cardwell, N. D. Collins, and A. E. Hostetter. 1959: A new grain hardness tester. Central Chem. 36:393–401.

10333. ——, N. D. Collins, and A. B. Cardwell. 1961: Hardness and moisture content of wheat kernels. Cereal Chem. 38:364–68.

Katz, R. 13520.

Katzenellenbogen, W. Mejbaum-; *see* Mejbaum-Katzenellenbogen, W.

10334. Katznelson, H., E. A. Peterson, and J. W. Rouatt. 1962: Phosphate-dissolving microorganisms on seed and in the root zone of plants. Can. J. Botany 40:1181–86.

10335. Kaufman, C. M., and H. G. Posey. 1953: Production and quality of pond pine seed in a pocosin area of North Carolina. J. Forestry 51:280–82.

10336. Kaufman, I. M. 1951: Most effective method for storing acorns. Les. i Step 3(11):35–37.

10337. Kaufman, W. H. 1954: Alfalfa seed dermatitis. J. Am. Med. Assoc. 155:1058–59.

10338. Kaufmann, E. 1943: The physiology of germination of *Cucumis sativus* as related to the problem of growth-promoting substances. Planta 33:516–45. Chem. Abstr. 38:5881–82. 1944.

10339. Kaufmann, H. P., J. Baltes, and H. Büter. 1937: Composition of the seed fat of *Trichosanthes cucumeroides*. Ber. Deut. Chem. Ges. 70:2535–37. Analyst 63:212. 1938.

10340. —— and R. S. Hamsago. 1962: Lipids of coffee seeds. I. Fatty acid esters of cafestol. Fette Seifen Anstrichmittel 64:206–13. J. Am. Oil Chemists Soc. 39:26. 1962.

10341. —— and M. C. Keller. 1942: Analysis of oil seeds. III. The simultaneous estimation of fat and water. Fette Seifen Anstrichmittel 49:272–75. Chem. Abstr. 37:6917. 1943.

10342. —— and F. Y. Liu. 1941: Pine seed oil. Soap 17:61.

10343. Kaufmann, M. L. 1958: Seed size as a problem in genetic studies of barley. [Abstract.] Proc. Genet. Soc. Can. 3: 30–32.

Kaufmann, M. L. 5181.

10344. Kaukis, K., and O. J. Webster. 1956: Effects of thermal neutrons on dormant seeds of *Sorghum vulgare* Per. Agron. J. 48:401–6.

10345. Kaul, B. K., and L. D. Kapoor. 1962: Preliminary studies on the effect of gibberellic acid on dill seed. Planta Med. 10:91–97. Chem. Abstr. 57:15551. 1962.

Kaul, K. N. 2398, 2399, 6235, 13575, 17174.

10346. Kaul, R. N., and B. N. Ganguli. 1962: Seed studies on eucalypts. Indian Forester 88:101–8.

10347. Kausar, A. G. 1955: Seed treatment for the control of seed-borne diseases of wheat and barley in Pakistan. Agr. Pakistan 6:25–29.

10348. Kausch, W. 1952: Physiological effects of very small suction forces. Planta 41:59–63.

10349. —— 1955: Die Samenkeimung im Tensiostaten. Planta 45:82–93.

Kausch, W. 11477.

10350. Kausche, G. A. 1940: Über eine das Virusprotein inaktivierende Substanz im Samen von *Nicotiana tabacum* Samsun. Biol. Centralbl. 60:423–38.

10351. Kausik, S. B. 1938: Embryogeny of *Lobelia*. (Gametogenesis and embryogeny in *Lobelia nicotianaefolia* Heyne.) J. Indian Botan. Soc. 17:161–68. J. Roy. Microscop. Soc. 59:53. 1939.

10352. —— 1938: The endosperm in *Grevillea robusta* Cunn. Current Sci. (India) 6:332–33.

10353. —— 1939: Dissection and preparation of whole mounts of endosperm from the seeds of *Grevillea* (Proteaceae). Stain Technol. 14:43–46.

10354. —— 1940: Studies in the Proteaceae. IV. Structure and development of the ovule of *Hakea saligna* Knight. Ann. Botany (London) 4:73–80.

10355. —— and M. V. S. Raju. 1956: Variations in the development of proembryo in *Urtricularia coerulea* L. Current Sci. (India) 25:296–97.

10356. —— and K. Subramanyam. 1945: A contribution to the embryology of *Lobelia trialata* Buch.-Ham. J. Indian Botan. Soc. 24:175–81.

10357. —— and —— 1946: Development of

endosperm in *Lobelia nicotianaefolia* Heyne. Current Sci. (India) 15:78–79.

10358. —— and —— 1947: Embryogeny of *Isotoma longiflora* Presl. Proc. Indian Acad. Sci. Sect. B 26:164–67.

10359. Kavina, K. 1932: Die Inversion des Embryos bei unseren Coniferen. Vestn. Česk. Akad. Zeměděl. 8:373–74. Ber. Wiss. Biol. 22:760. 1932.

Kawabata, S. 18233.

10360. Kawaguti, R., and K. W. Kim. 1940: Constituents of the seeds of *Zizyphus vulgaris* Lamark var. *spinosus* Bunge. J. Pharm. Soc. Japan 60:343–52. Chem. Abstr. 35:1396–97. 1941.

10361. Kawahara, A., and H. Takada. 1961: The germination of *Trapella* seed. I. Some factors influencing stimulation of germination. Indian J. Plant Physiol. 4:156–68.

10362. Kawahara, E., and T. Motogi. 1952: Studies on the quality of upland rice. II. On the germination of glutinous hulled upland rice grains of different transparency. Proc. Crop Sci. Soc. Japan 20:311–12.

10363. Kawahara, H., and H. Hara. 1951: On the effects of supersonics and heat upon the germination of seeds and growth of sprouts. Bull. Exptl. Biol. 1:32–35. Biol. Abstr. 27:20641. 1953.

Kawahara, H. 8145, 18304.

Kawai, T. 12939.

10364. Kawamata, M., and Y. Tsukijima. 1956: Studies on flowering, fruiting, and germination of lupine plants. I. On the flowering and fruiting of yellow lupine. [Abstract.] Proc. Crop Sci. Soc. Japan 25:51–52.

10365. —— and —— 1956: Studies on flowering, fruiting, and germination of lupine plants. II. Germination of yellow lupine seeds produced under cover against rain-fall in the ripening stage. [Abstract.] Proc. Crop Sci. Soc. Japan 25:52.

10366. —— and —— 1957: Studies on flowering, fruiting and germination of lupine plants. IV. Effect of drier "Desiccan" sprayed at ripening stage on time of maturity and germination of seed. Proc. Crop Sci. Soc. Japan 26:137.

Kawamura, M. 9521.

10367. Kawamura, S. 1953: Studies on soybean carbohydrates. II. The carbohydrate content of defatted soybeans. Kagawa Agr. Coll. Tech. Bull. 5:1–14.

10368. —— 1954: The components of the seed of *Vicia faba*. Kagawa Agr. Coll. Tech. Bull. 6:8–19.

10369. ——, T. Kobayashi, M. Osima, and M. Mino. 1955: Studies on the carbohydrates of soybeans. V. A hemicellulose fraction soluble in hot water. Isolation and component sugar determinations. Bull. Agr. Chem. Soc. Japan 19:69. Records Res. Fac. Agr. Univ. Tokyo 5:9. 1954/55.

10370. ——, H. Nakamura, and K. Horike 1954: The carbohydrates of the seed of *Vicia faba*. I. The determination of carbohydrates. Kagawa Agr. Coll. Tech. Bull. 6:20–26.

10371. —— and T. Narasaki. 1959: Carbohydrates of the seed of *Vicia faba*. V. Fucose as a constituent of hemicellulose B_1 from broad bean seeds. J. Agr. Chem. Soc. Japan (English Transl.) 33:817–21. Chem. Abstr. 54:16553. 1960.

10372. ——, ——, R. Kusuhara, and T. Sato. 1957: Studies on the carbohydrates from the seeds of *Vicia faba*. IV. Ribose from hemicellulose hydrolysis. Tech. Bull. Fac. Agr. Kagawa Univ. 8:201–11. Biol. Abstr. 34:8656. 1959.

10373. ——, H. Suzuki, and T. Matsumoto. 1955: The relation of maturity to the composition of broad bean seeds. II. On the carbohydrates. Kagawa Agr. Coll. Tech. Bull. 7:81–86.

10374. ——, Y. Tuboi, and H. K. Nakamura. 1954: The carbohydrates of the seed of *Vicia faba*. II. Sugars of broad beans. Kagawa Agr. Coll. Tech. Bull. 6:27–31.

Kawamura, S. 18195, 18251.

Kawana, A. 10634, 13430–13432.

10375. Kawano, K. 1955: Influence of pH of mediums on the germination of mulberry seeds. Igaku To Seibutsugaku 35:156–58. Chem. Abstr. 52:2185b. 1958.

10376. Kawasaki, C. 1962: Biosynthesis of thiamine in radish during germination of its seeds. Vitamins (Nagoya) 26:293–97.

10377. Kawase, K., and H. Nakano. 1943: Germicidal power of calcium cyanamide. I. Effect on the germination of wheat and the lethal effect on *Tilletia tritici*. J. Agr. Chem. Soc. Japan (English Transl.) 19:909–13. Chem. Abstr. 43:1137–38. 1949.

10378. Kawase, M. 1958: The after-ripening of dormant apple seeds in relation to auxin content. Hort. Assoc. Japan J. 27: 256–64.

Kawashima, R. 8354.

Kawashima, S. 11164.

Kawata, J. 14508.

10379. Kawata, S. 1942: The activities of synthetic growth substances. Treatment of

Leguminosae seeds. Science (Japan) 12: 210–14. Chem. Abstr. 45:10309. 1951.
10380. Kawatake, M., and R. Ishida. 1955: studies on the developmental processes of hard seed of some legumes. II. On the seed weight upon germination. Tokai Kinki Natl. Agr. Expt. Sta. Bull. Div. Plant Breed. Cult. No. 2:183–88.
10381. —— and —— 1955: Studies on the developmental processes of hard seed of some legumes. III. On the flowering date and the number of days after anthesis in relation to germination. Tokai Kinki Natl. Agr. Expt. Sta. Bull. Div. Plant Breed. Cult. No. 2:189–92.
10382. ——, K. Shimura, and R. Ishida. 1956: On the analogy of the relation of the Rachilla-length of rice plant caryopsis and that of the funiculus-length of the seed of wild legume to their germination. Proc. Crop Sci. Soc. Japan 24:297–98.
10383. ——, ——, and —— 1956: Studies on the developmental processes of hard seed of some legumes. V. The length of funiculus and germination of *Vicia sativa* L. var. *normalis* Makino. Tokai Kinki Natl. Agr. Expt. Sta. Bull. Div. Plant Breed. Cult. No. 3:123.
10384. ——, K. Takizawa, and R. Ishida. 1955: Studies on the developmental processes of hard seed of some legumes. IV. Anatomical observations on the growth of the ovule. Tokai Kinki Natl. Agr. Expt. Sta. Bull. Div. Plant Breed. Cult. No. 2:193–202.
10385. ——, ——, and M. Katayama. 1955: Effects of temperature upon germination of some forage and wild plants. Tokai Kinki Natl. Agr. Expt. Sta. Bull. Div. Plant Breed. Cult. No. 2:169–76.
10386. ——, ——, and —— 1955: Studies on the developmental processes of hard seed of some legumes. I. Colour of seed coat in relation to germination and hard seed. Tokai Kinki Natl. Agr. Expt. Sta. Bull. Div. Plant Breed. Cult. No. 2:177–82.
10387. Kawatani, T., and S. Hemmi. 1962: Effects of γ-ray radiation on the germination and growth of *Rauwolfia serpentina* Benth. Bull. Natl. Inst. Hyg. Sci. 80:166–67. Biol. Abstr. 44:25119. 1963.
10388. —— and T. Ohno. 1962: Effects of X-ray radiation on the germination and growth in earlier stages of *Atropa belladonna* L. Bull. Natl. Inst. Hyg. Sci. 80: 165–67. Biol. Abstr. 45:7741. 1964.
Kawin, B. 18456.
Kay, B. L. 6097.
10389. Kayser, G. 1891: Seeds of Umbelliferae. Ber. Pharm. Ges. 1:152–62.
10390. —— 1892: Integument of the seed of Euphorbiaceae. Ber. Pharm. Ges. 2:5–19.
10391. —— 1893: Beitrag zur Kenntnis der Entwickelungsgeschichte der Samen, mit besonderen Berücksichtigung des histogenetischen Aufbaues der Samenschalen. Pringsh. Jahrb. 25:79–146.
10392. —— 1893: Über das Verhalten des Nucellus in den Samenlagen von *Croton flavens*. Ber. Deut. Botan. Ges. 11:61.
10393. Kazakevich, L. I., B. M. Smirnov, L. D. Chistyakova, and T. N. Favorskaya. 1961: Development of measures for the control of immature weeds. Nauchn. Tr. Nauchn.-Issled. Inst. Sel'skokhoz. Yugo-Vostoka 21:24–88. Biol. Abstr. 43:24947. 1963.
10394. Kazakov, E. D., and I. A. Sakharova. 1960: Change in ash content of endosperm after hydrothermic treatment of wheat. Dokl. Akad. Nauk SSSR 132:1438–40. Chem. Abstr. 54:18690–91. 1960.
10395. Kazakova, A. A. 1950: Improving the termination of onion seeds. Selek. i Semen. 17(12):31–35.
10396. Kaziev, M. Z. 1963: Pre-planting soaking of alfalfa seed in molybdenum solution. Khlopkovodstvo 1963(3):22–23.
10397. Kazimierz, B. 1962: The occurrence of betaine and choline in seeds. Roczniki Nauk Rolniczych Ser. A 86:527–31. Chem. Abstr. 58:4811. 1963.
10398. Kearney, T. H. 1928: Variation in seed fuzziness on individual plants of Pima cotton. J. Agr. Res. 37:465–72.
10399. Kearns, V., and E. H. Toole. 1938: Temperature and other factors affecting the germination of the seed of fescues. Proc. Intern. Seed Testing Assoc. 10:337–41.
10400. —— and —— 1938: The relation of temperature and moisture content to the longevity of chewings fescue seed. Proc. Intern. Seed Testing Assoc. 10:342–44.
10401. —— and —— 1939: Temperature and other factors affecting the germination of fescue seed. U.S. Dept. Agr. Tech. Bull. 638. 36 pp.
10402. —— and —— 1939: Relation of temperature and moisture content to longevity of chewings fescue seed. U.S. Dept. Agr. Tech. Bull. 670. 27 pp.
Keck, D. D. 4177.
10403. Kee, W. Y. 1938: Metabolic studies on germinating soybeans. I. General metabolic changes in the germinating

soybean. J. Chinese Chem. Soc. 6:15–22. Chem. Abstr. 35:5151. 1941.

10404. Keeble, F. W. 1896: Observations on the Loranthaceae of Ceylon. Trans. Linn. Soc. London Ser. 2, Botany 5(pt. 3): Botan. Gaz. 22:184–87. 1896.

10405. Keech, B. 1933: Cardinal climber from seed. Flower Grower 20(6):276.

10406. Keenan, G. L. 1946: Microscopic differentiation of some common mustard seeds with polarized light. J. Assoc. Offic. Agr. Chemists 29:173–75. Pharm. Abstr. 13:98. 1947.

10407. Keese, H. 1945: Über den Einfluss verschiedener Wasserund Kaliversorgung auf die Eiweissbildung im Hafer. Bodenkunde Pflanzenernähr 36:1–9. Biol. Abstr. 26:12996. 1952.

Keever, C. 15464.

10408. Keffer, C. A. 1889: Germination of frosted grain. S. Dakota Agr. Expt. Sta. Bull. 10:3–8.

10409. Kefford, N. P. 1962: Auxin-gibberellin interaction in rice coleoptile elongation. Plant Physiol. 37:380–86.

10410. —— and J. R. D. Jones. 1956: Onion seed production. J. Dept. Agr. Victoria 54:297–303. Rev. Appl. Mycol. 35:862. 1956.

10411. Kegl, L. 1961: Studies on nutrient storage in wheat grain during germination. Buzatermesztesi Kierletek 1952/59:115–21.

10412. —— 1961: Relation between the germination changes of wheat grain and the degree of lodging. Buzatermesztesi Kiserletek 1952/59:122–28.

10413. Kehar, N. D., and R. Chanda. 1946: Mango-seed kernel—a new source of food. Current Sci. (India) 15:48.

10414. Keil, J. 1939: Die Bedeutung der Blausäure bei der Samenkeimung. Jahrb. Wiss. Botan. 88:345–72.

10415. —— 1940: A new nutrient medium for germination tests with cereal fungus spores (*Ustilago avenae, U. tritici, U. hordei, Tilletia tritici,* and *Helminthosporium gramineum*). Arch. Mikrobiol. 11:85–88. Abstr. Rev. Appl. Mycol. 19:522. 1940.

10416. —— 1940: Neuere Ergebnisse der keimungsphysiologischen Forschung. Umschau 44:667–70.

Keil, J. 11669.

10417. Keilholz, F. J. 1948: Hybrid corn goes high-protein. Country Gentleman 118(5):15, 106–9.

Keim, F. D. 8192.

10418. Keim, W. F. 1953: An embryo culture technique for forage legumes. Agron. J. 45:509–10.

10419. —— 1957: Seed set failures in zigzag clover, *Trifolium medium.* Agron. Abstr. 49:74. 1957.

Keitt, G. W. 11950, 11951.

Kelcey, P. Finn-. 9627.

10420. Kelkar, S. S. 1958: A contribution to the embryology of *Lannea coromandelica* (Houtt.) Merr. J. Univ. Bombay Sect. B 26:152–59. Biol. Abstr. 35:71532. 1960.

10421. —— 1960/61: The development of endosperm and embryo in *Lannea coromadelica* (Houtt.) Merr. J. Univ. Bombay Sect. B 29:1–5. Biol. Abstr. 42:7426. 1963.

10422. Kellackey, M. C. 1941: The auxin content of germinating seeds as determined by different methods of extraction. Univ. Notre Dame Press, Notre Dame, Ind. 14 pp.

10423. Kelleher, J. S. 1954: Damage to sweet clover seed by the sweet clover weevil, *Sitonia cylindricollis* Fahr. (Coleoptera: Curculionidae). Can. Entomol. 86:179–80.

10424. Kellenbarger, S., V. Silveira, R. M. McCready, H. S. Owens, and J. L. Chapman. 1951: Inheritance of starch content and amylose content of the starch in peas (*Pisum sativum*). Agron. J. 43:337–40.

Kellenbarger, S. 13972.

10425. Keller, D. 1937: Magnolias raised from seed. Flower Grower 24(10):465–66.

Keller, H. 10280.

10426. Keller, I. A. 1895: The jelly-like secretion of the fruit of *Peltandra undulata* Rafn. Proc. Phil. Acad. Nat. Sci. 1895: 287–90.

Keller, J. G. 3560.

10427. Keller, K. R. 1953: Seed germination in hops, *Humulus lupulus* L. Agron. J. 45:146–50.

10428. —— and R. A. Magee. 1954: The effect of application rates of nitrogen, phosphorus, and potash on some chemical constituents in two varieties of hops. Agron. J. 46:388–91.

10429. Keller, L., and C. Tamm. 1959: Die Glykoside der Samen von *Strophanthus hispidus* P.DC. Helv. Chim. Acta 42:2467–68.

Keller, M. C. 10341.

10430. Keller, W. 1943: Seed production on grass culms detached prior to pollination. J. Am. Soc. Agron. 35:617–24.

Keller, W. H. 5021.

10431. Kellerman, W. A. 1893: Experiments in germination of treated seed. Ohio Agr. Expt. Sta. Tech. Ser. Bull. 1:201–5.

10432. Kellermayer, I., and L. Zechner. 1953: Das Samen Öl von *Cynanchicum vincetoxicum*. Sci. Pharm. 21:161-68 Biol. Abstr. 29:1825. 1955.

Kelley, J. B. 13508.

10433. Kelley, W. R. 1953: Study of seed identification and seed germination of *Potentilla* spp. and *Veronica* spp. Cornell Univ. Agr. Expt. Sta. Mem. 317. 29 pp.

10434. Kellogg, L. F. 1936: Collection, extraction, and yield of black locust seed. U.S. Dept. Agr. Forest Serv. Central States Forest Expt. Sta., Sta. Note 28. 5 pp.

10435. Kelly, A. F. 1955: Cereal seed multiplication at the National Institute of Agricultural Botany. Natl. Inst. Agr. Botany J. 7:237–39.

10436. —— 1963: Seed—the key to better food supplies. World Crops 15:301–5, 311.

10437. Kelly, C. F. 1940: Methods of ventilating wheat in farm storages. U.S. Dept. Agr. Circ. 544. 74 pp.

10438. —— 1941: Temperatures of wheat in experimental farm-type storages. U.S. Dept. Agr. Circ. 587. 61 pp.

10439. ——, B. M. Stahl, S. C. Salmon, and R. H. Black. 1942: Wheat storage in experimental farm-type bins. U.S. Dept. Agr. Circ. 637. 245 pp.

Kelly, K. O. 14110.

10440. Kelly, R. B. 1953: From seed to bulb. Gladiolus 28:129–32.

10441. Kelly, S. 1947: The relationship between respiration and water uptake in the oat coleoptile. Am. J. Botany 34: 521–26.

10442. Kelly, T. K. 1963: Ten tons of *Glycine* (*javanica*) seed from Atherton. Queensland Agr. J. 89:227–28.

10443. Kelso, W. T. 1960: Measuring grain moisture before storage. Queensland Agr. J. 86:551–54.

10444. Keltner, C. H. 1947/48: Laboratory ranks high. Modern seed testing equipment attracts national interest—purity and germination determined by expert analysts. Illinois Dept. Agr. Ann. Rept. 31:182–204.

10445. Kemmer, E., and I. Thiele. 1955: Frostresistenzprüfungen an keimenden Kernobstsamen. Züchter 25:57–60.

10446. Kemp, H. K. 1946: Artificial parthenocarpy in the tomato. J. Australian Inst. Agr. Sci. 12:147. Biol. Abstr. 21:20690. 1947.

10447. Kemp, J. G., and W. Kalbfleisch. 1962: Note on the evaluation of the New Zealand space-plant seeder. Can. J. Plant Sci. 42:554–57.

10448. ——, R. Loiselle, and W. Kalbfleisch. 1962: Note on a hot water treater for small samples of loose smut infected seed. Can. J. Plant Sci. 42:396–97.

10449. Kemp, W. B., and P. R. Henson. 1934: The effect of type of endosperm upon carbohydrate distribution in the mature corn plant. J. Am. Soc. Agron. 26:519–24.

10450. Kempanna, C., and S. B. Allayyanavaramath. 1958: Presowing treatment of seeds of *crotalaria striata* D.C. Mysore Agr. J. 33:63–65. Biol. Abstr. 33:31038. 1959.

10451. Kemper, A. 1962: Experience in the application of chemical soil treatments on the seed beds of forest tree nurseries. Z. Pflanzenkrankh. Pflanzenschutz. 69:394–401. Chem. Abstr. 57:14224. 1962.

Kemper, W. C. 18062.

10452. Kempner, A. H. 1956: Grow primulas from seeds. Flower Grower 43(10):65.

Kempster, H. L. 9448.

10453. Kempton, J. H., and L. R. Maxwell. 1941: Effect of temperature during irradiation on the X-ray sensitivity of maize seed. J. Agr. Res. 62:603–18.

10454. —— and J. W. McLane. 1942: Hybrid vigor and weight of germs in the seeds of maize. J. Agr. Res. 64:65–80.

Kempton, J. H. 13058, 13059.

10455. Kendall, C. 1933: Germinating delicate flower seeds. Flower Grower 20(3):119.

Kendall, K. A. 14364.

10456. Kendall, W. A. 1963: Effect of amino acids on germination of red clover seeds. Agron. J. 55:51–52.

Kendall, W. A. 18411, 18412.

10457. Kendrick, E. L., and L. H. Purdy. 1959: Influence of environmental factors on the development of wheat bunt in the Pacific Northwest. II. Effect of soil temperature and soil moisture on infection by seed-borne spores. Phytopathology 49:433–34.

10458. Kendrick, J. B. 1934: Cucurbit mosaic transmitted by muskmelon seed. Phytopathology 24:820–23.

10459. —— 1938: A seed-borne bacterial disease of garden stocks, *Matthiola incana*. [Abstract.] Phytopathology 28:12.

10460. —— 1944: Fruit invasion and seed carriage of tomato *Fusarium* wilt. [Abstract.] Phytopathology 34:1005–6.
Kendrick, J. B., Jr. 11918.
Kendurkar, S. G. 8023.
KenKnight, G. 2629.
10461. Kennard, W. C. 1955: Development of the fruit, seed, and embryo of the Paheri mango. Botan. Gaz. 117:28–32.
10462. —— and F. Llavat-Cristy. 1953: Macadamia seed. Puerto Rico Agr. Expt. Sta. Rept. 1953:6.
Kennedy, C. 15287.
10463. Kennedy, D. 1932: The fumigation of Douglas fir seed. New Zealand J. Sci. Technol. 14:58–61.
10464. Kennedy, K. 1962: The "crab's eye" seed (*Abrus precatorius*). Australian Nat. Hist. 14:68. Biol. Abstr. 41:20157. 1963.
10465. Kennedy, W. K., and J. Unrau. 1949: A rapid method for determining the oil content of safflower and sunflower seeds. J. Am. Soc. Agron. 41:93–95.
Kennedy, W. K. 10791.
10466. Kent, N., and R. A. Brink. 1947: Growth *in vitro* of immature *Hordeum* embryos. Science 106:547–48.
Kent, N. L. 9275, 9276.
10467. Kenten, R. H., and B. D. Powell. 1960: Production of heat during fermentation of cacao beans. J. Sci. Food Agr. 11:396–400.
Kenten, R. H. 406.
10468. Kentzer, T. 1960: Dynamics of growth regulators during vernalization of wheat. I. Contents of growth substances in different stages of vernalization of seeds. II. Changes in the free and bound auxin content in kernels during vernalization. III. Chromatographic analysis of growth regulators during vernalization of seeds and growth of winter wheat. Zeszyty Nauk. Uniw. Toruniu Nauki Mat.-Przyrod. Biol. 6:27–29, 31–46, 47–63. Chem. Abstr. 57:5058. 1962.
Kentzer, T. 12208.
10469. Keohler, B., and W. M. Bever. 1950: Oat and wheat seed treatments in 1950 in relation to dosage, storage before planting, and thoroughness of mixing. Plant Disease Reptr. 34:259–61.
10470. Kepler, J. E., and G. E. Gatherum. 1964: Japanese larch provenance trial in northeast Iowa. Iowa State J. Sci. 38:393–403.
10471. Kerbosch, M. G. J. 1920: Het kiemvermogen van kinazaad. Dept. Landb. Nijr. en Handel, Mededeel. v. h. Gouv. Kina. Proefsta. 8. 50 pp. Bandoeng.
10472. Kerenyi, E., and G. Laszay. 1959: Comparative germination studies on grass seeds of varied age. (Preliminary report.) Kertesz. Szolesz. Foiskola Evkonyve 1959 (7):157–60.
Kern, D. 16907.
10473. Kernick, M. D. 1961: Seed production of specific crops. Food Agr. Org. U.N. FAO Agr. Studies 55:181–461.
10474. Kernkamp, M. F. 1953: Seed treatment of small seeded legumes. Minn. Agr. Expt. Sta. Tech. Bull. 209. 11 pp.
10475. Kernkamp, M. F., and G. A. Hemerick. 1953: The relation of *Ascochyta imperfecta* to alfalfa seed production in Minnesota. Phytopathology 43:378–83.
10476. Kerns, K. 1927: Pineapple seed germination. Pineapple News 1:38–39.
Kerns, K. R. 4162.
10477. Kerr, A. 1930: Fruit and seeds in the drift on Kaw Tao. J. Siam Soc. Nat. Hist. Suppl. 8:103–17. Biol. Abstr. 6:21344. 1932.
10478. —— 1931: Dispersal of fruit by wind. J. Siam Soc. Nat. Hist. Suppl. 8:216–17. Biol. Abstr. 7:12732. 1933.
10479. Kerr, E. A. 1954: Seed development in blackberries. Can. J. Botany 32:654–72.
10480. Kerr, T., and D. B. Anderson. 1944: Osmotic quantities in growing cotton bolls. Plant Physiol. 19:338–49.
Kerr, T. 6599.
10481. Kerr, W. E. 1952: Tobacco seed germination. Rhodesia Agr. J. 49:266–69. Biol. Abstr. 29:24588. 1955.
10482. —— 1955: Handling and testing crop and vegetable seed in S. Rhodesia. Rhodesia Agr. J. 52:293–301.
Kershaw, B. B. 18396.
Kersten, H. 2273, 12412.
10483. Kersting, J. F., A. W. Pauli, and F. C. Stickler. 1961: Grain sorghum caryopsis development. II. Changes in chemical composition. Agron. J. 53:74–77.
10484. ——, F. C. Stickler, and A. W. Pauli. 1961: Grain sorghum caryopsis development. I. Changes in dry weight, moisture percentage, and viability. Agron. J. 53:36–38.
10485. Kertesz, Z. I. 1928: Über die Entwicklung der Mutasewirkung in keimender Gerste. Z. Physiol. Chem. 176: 144–50.
10486. —— 1933: Some carbohydrate

changes in shelled green peas. N.Y. Agr. Expt. Sta. Bull. 622. 14 pp.

10487. —— 1934: Carbon dioxide content of the gas from pea pods. Plant Physiol. 9:339–50.

10488. —— and E. L. Green. 1932: Deterioration in shelled green peas held a few days in storage prior to canning. J. Agr. Res. 45:361–70.

10489. ——, J. G. Horsfall, and A. H. Rouse. 1934: Aberrations in the chemical composition of peas from plants affected with root rot. J. Agr. Res. 49:799–814.

Kertesz, Z. I. 9073, 14272.

10490. Kery, G. 1963: The propagation of cactus from seeds. Buvar 8:179–80.

Kessinger, A. M. 16753.

10491. Kester, D. E. 1953: Factors affecting the aseptic culture of Lovell peach seedlings. Hilgardia 22:335–65.

10492. —— 1960: Seed dormancy and its relationship to nursery practices. Plant Propagators Soc. Proc. 10:256–67.

10493. —— and C. O. Hesse. 1955: Embryo culture of peach varieties in relation to season of ripening. Proc. Am. Soc. Hort. Sci. 65:265–73.

Kester, D. E. 8686.

10494. Kester, E. B., H. C. Lukens, R. E. Ferrel, A. Mohammad, and D. C. Finrock. 1963: Influence of maturity on properties of Western rices. Cereal Chem. 40:323–36.

Kester, E. B. 9109–9111, 9230.

10495. Keszycki, Z. 1935: Saugkraftmessungen verschiedener Weizensorten. Roczniki Nauk Rolniczych Lesnych 35: 55–78.

Ketelaere, F. de. 12957–12960.

10496. Kevanishvili, V. N. 1958: The chemical composition of grains of different wheat varieties. Tr. Nauchn.-Issled. Inst. Zemledel. Akad. Sel'skokhoz. Nauk Gruzin. SSR 11:151–60. Chem. Abstr. 54:13284. 1960.

10497. Kevorkian, A. G. 1941: *Cinchona* production. Puerto Rico Agr. Expt. Sta. Rept. 1941:9–10.

10498. Key, J. L., and D. S. Galitz. 1959: Growth inhibitor in immature soybean seeds and 2,4-D-sprayed soybean seedlings. Science 130:1340–41.

10499. —— and J. C. Shannon. 1964: Enhancement by auxin of ribonucleic acid synthesis in excised soybean hypocotyl tissue. Plant Physiol. 39:360–64.

Key, S. 8725.

10500. Keyes, C. G. 1952: Survival of fall and spring sown sweet pea seed when germinated before planting. Proc. Am. Soc. Hort. Sci. 59:523–26.

Keys, A. J. 5772.

10501. Keys, C. E. 1949: Observations on the seed and germination of *Searita italica* (L.) Beauv. Trans. Kansas Acad. Sci. 52:474–77.

10502. Khachatryan, M. S. 1961: Transmission of chickpea *Aschochyta* by seeds and effectiveness of treatment. Sb. Nauchn. Tr. Nauchn.-Issled. Inst. Zemledel. Arm. SSR 1961(2):147–55. Chem. Abstr. 58: 8244. 1963.

Khaleb, S. 2490.

Khalique, A. 15797–15799.

10503. Khalique, M. A., and M. D. Ahmed. 1954: Corchsularin, a new bitter from jute seeds. I. Its isolation and constitution of corchsularose. J. Org. Chem. 19:1523–28.

10504. Khan, A. A., and I. I. Chaudhri. 1959: Investigations on seed dormancy of *Asphodellus tenuifolius* Cavan. Pakistan J. Sci. Ind. Res. 2:267–71.

10505. Khan, A. H., and M. A. Hashmi. 1961: Studies on *Abrus precatorius* Linn. I. Isolation and toxic properties of "Abrulin" a protein fraction from the seeds. Pakistan J. Sci. Ind. Res. 4:53–56.

10506. Khan, A. R. 1929: Studies in Indian oil seeds. III. *Carthamus tinctorius* Linn. The types of safflower. Mem. Dept. Agr. India Botan. Ser. 18:81–87.

10507. —— and C. A. Waheed. 1959: Effect of different pollination treatments on seed formation and size of head in the next generation of cauliflower (*Brassica oleracea* var. *botrytis* L.). Pakistan J. Sci. Res. 11:12–19.

Khan, A. R. 13657, 17280.

Khan, I. U. 10852.

Khan, M. A. 13904.

Khan, M. I. 9501.

10508. Khan, M. Q. 1950: Deterioration of different grades of polished rice in storage. Indian J. Entomol. 10:279–86. 1948. Biol. Abstr. 25:29296. 1951.

Khan, M. S. 15870.

Khan, M. S. Y. 19018.

10509. Khan, N. A., and B. E. Baker. 1957: The amino-acid composition of some Pakistani pulses. J. Sci. Food Agr. 8:301–5.

10510. —— and B. D. Mukherjee. 1959: The polysaccharide in tamarind-seed kernel. Chem. Ind. (London) 1959:1413–14. Chem. Abstr. 54:6886. 1960.

10511. Khan, R. 1942: A contribution to the embryology of *Jussieua repens* Linn. J. Indian Botan. Soc. 21:267–82.

10512. —— 1954: A contribution to the embryology of *Utricularia flexuosa* Vahl. Phytomorphology 4:80–117. Biol. Abstr. 31:22203. 1957.

10513. —— and L. F. Randolph. 1960: Growth of excised embryos of wheat (*Triticum vulgare* var. Pb. 591) in different media with varying agar concentration. Proc. Natl. Acad. Sci. India Sect. B 30: 391–96. Biol. Abstr. 36:38499. 1961.

Khan, R. 12701.

10514. Khan, S. A., M. I. Qureshi, M. K. Bhatty, and Karimullah. 1961: Composition of the oil of *Asphodelus fistulosis* (Piazi) seeds. J. Am. Oil Chemists Soc. 38:452–53.

10515. ——, ——, ——, and —— 1961: A critical study of composition of *Momordica charantia* L. (Karela) seed oil. Pakistan J. Sci. Res. 13:111–12. Biol. Abstr. 43:7765. 1963.

Khan, S. M. 231.

10516. Khandaswamy, M., and A. V. Row. 1956: A note on the extraction of seeds from the pods of *Prosopis juliflora*. Madras Agr. J. 43:520–21. Biol. Abstr. 32:10024. 1958.

10517. Khandige, S. B. 1956: The influence of drying on the germination of seed arecanuts (*Areca catechu* L.). Madras Agr. J. 43:3–6.

Khanna, K. L. 13656.

10518. Khana, P. 1964: Embryology of *Trichodesma amplexicaule* Roth. Bull. Torrey Botan. Club 19:105–14.

10519. Khanna, R. N., and T. R. Seshadri. 1963: Pongaglabrone, a new component of the seeds of *Pongamia glabra:* its constitution and synthesis. Tetrahedron 19:219–25. Biol. Abstr. 43:7566. 1963.

Kharabanda, B. C. 11660, 11661.

10520. Kharach, A. M. 1963: Increasing the sugar content of radishes by presowing treatment of the seeds. Byul. Gl. Botan. Sada Akad. Nauk SSSR 48:93–94. Biol. Abstr. 45:26060. 1964.

Kharchenko, E. D. 5607.

10521. Kharchenko, S. M. 1957: The stimulating influence of culture liquid of *Penicillum* sp. on the germination of seeds of cotton and wheat. Nauk. Zap. Kiivs'k. Derzh. Univ. 16(20) Zbirnik Biol. Tak. 14:147–52. Chem. Abstr. 52:15664. 1958.

10522. Khastigir, H. N., S. K. S. Gupta, and P. S. Gupta. 1958: The sapogenin from seeds of *Achyranthes aspera* Linn. J. Indian Chem. Soc. 35:693–94. Biol. Abstr. 35:60711. 1960.

Khatipov, A. E. 10523.

10523. Khatipov, S. E., and A. E. Khatipov. 1953: Method of fruit seed stratification. Sad i Ogorod 1953(11):65–68.

10524. Khattak, G. M. 1962: Seed germination of dwarf palm (*Nannorhops ritchiana*). Pakistan J. Forestry 12:202–4. Biol. Abstr. 45:7848. 1964.

10525. —— 1962: Effect of locality of collection and seed presowing treatments on the germination of *Olea cuspidata*. Pakistan J. Forestry 12:233–35. Biol. Abstr. 45:3527. 1964.

10526. Khavzhu, I. M. 1957: Two-phase harvesting of sunflowers and its effect on the seed moisture. Maslob.-Zhir. Prom. 23(7):5–7.

10527. Kheladze, V. S., and A. B. Matinian. 1960: An experiment in the pre-planting treatment of seeds with minor elements. Byul. Gl. Botan. Sada 36:103–4.

10528. Khlopin, V., and Pasvik-Khlopin. 1932: Atomic weight of potassium extracted from pea seeds. Bull. Soc. Chim. 51:1227–31. Chem. Abstr. 27:1247. 1933.

10529. Khmelev, B. I. 1950: On transplanting cereal embryos. Dokl. Akad. Nauk SSSR 70:909–12.

10530. —— and I. Malishchuk. 1960: Presowing irradiation of seeds by hard and soft X-rays. Nauchn. Tr. Kamenets-Podolsk. Sel'skokhoz. Inst. 3:106–18. Biol. Abstr. 45:12539. 1964.

10531. Kho, Y. O., and J. P. Braak. 1956: Reduction in the yield and viability of carrot seed in relation to the occurrence of the plant bug *Lygus campestris* L. Euphytica 5:146–56.

10532. Khokhlenko, A. F. 1959: Nitrogenous substances of sunflower seeds. Izv. Vysshikh Uchebn. Zavedenii Pishchevaya Tekhnol. 1959(2):151–54. Chem. Abstr. 53:19052–53. 1959.

10533. —— 1960: Amino acids of the protein fraction of kernels of sunflower seeds. Izv. Vysshikh Uchebn. Zavedenii Pishchevaya Tekhnol. 1960(3):31–36. Chem. Abstr. 55:1821. 1961.

Khoklova, I. M. 11332.

10534. Kholodenko, B. G. 1952: Seed formation in pigweed (*Axyris amaranthoides*) and garden orach (*Atriplex hortense*). Agrobiologiya 1952(3):135–38.

10535. Kholodnyi, N. G. 1952: Artificial parthenocarpy and accelerated ripening of figs. Priroda 41:100–2.

10536. Kholodnyuk, I. I. 1935: The problem of the disinfection of wheat seeds infected by *Fusarium*. Plant Protect. (USSR) 1935(7):119–29. Chem. Abstr. 30:8501. 1936.

Kholodova, V. P.

10537. Khomentovsky, I. I. 1932: Contributions to the germinating ability of seeds as resulting from hybridization work on apple trees in 1925–1929. Nauchn.-Issled. Inst. Iuzhn. Plod. i Iagodn. Khoz. Kiev No. 16. 66 pp.

Khon, E. 2229.

10538. Khoroshkin, M. N. 1957: Effect of trace elements on the enzymic activity of seeds. Sb. Nauchn.-Issled. Rabot Azovo-Chernomorsk. Sel'skokhoz. Inst. 15:115–22. Chem. Abstr. 53:9381. 1959.

10539. Khotin, A. A. 1948: Seed formation and accumulation of essential oil in caraway. Vses. Akad. Sel'skokhoz. Nauk V. I. Lenina Dokl. 13(12):17–21.

Khripak, S. M. 12072.

10540. Khripunova, L. G. 1957: Reliable method of increasing the size of the wheat kernel. Sel'skoe Khoz. Povolzhia 2(7):36–38.

Khristov, V. 15559.

Khrzhanovskaya, T. E. 296.

10541. Khuchua, K. N. 1957: Effect of dates of planting seed beets on yields of [sugar beet] seeds. Sakharn. Svekla 2(6):41–43.

10542. Khudairi, A. K. 1956: Breaking the dormancy of *Prosopis* seeds. Physiol. Plantarum 9:452–61.

10543. —— 1958: Studies on the germination of date-palm seeds. The effect of sodium chloride. Physiol. Plantarum 11:16–22.

10544. —— and A. S. Abdul-Wahab. 1956: Studies on the germination of *Prosopis* seeds. Bull. Coll. Arts Sci. Baghdad 1:66–76.

10545. Khudyak, M. I. 1962: Morphologo-physiological characteristics of the endosperm of angiosperms in the process of its development. Acad. Sci. RSS Ukraine Kiev J. Botan. 19(5):31–41.

10546. Khudyakov, Y. P., and E. A. Raznitzyna. 1939: The use of mycolytic bacteria for the inoculation of seed during vernalization. Bull. Acad. Sci. URSS Ser. Biol. 1:117–20. Rev. Appl. Mycol. 18:659–60. 1939.

10547. —— and L. S. Zinovev. 1962: Effect of phenol compounds on dormancy and sprouting of rye and potato seeds. Byul. Nauchn.-Tekhn. Inform. po Sel'skokhoz. Mikrobiol. 1962(2):7–11. Chem. Abstr. 60:1042. 1964.

Khudyna, I. P. 7750, 7751.

Khundkar, M. H. 348.

Khvan, A. V. 17559.

10548. Khvatova, K. A. 1955: Growing spring vetch and forage peas for seed. Moskva Gosudarst. Izd. Sel'skokhoz. Literatury. 44 pp.

10549. Khvostova, V. V., N. L. Delone, O. N Sorokina, V. L. Turkov, S. P. Tselishchev, and K. V. Chaikina. 1958: Development of soft wheat seedlings from thermal-neutron irradiated seeds. Biophysics 3:440–45.

10550. ——, A. A. Prokoveva-Belgovskaya, B. N. Sidorov, and N. N. Sokolov. 1962: The effect of space flight conditions on the seeds of higher plants and actinomycetes. Problems of Space Biology 2:161–72. Nucl. Sci. Abstr. 17:2317. 1963.

Kiamuddin, M. 6229.

Kibe, M. M. 16634.

Kichunova, K. V. 15584, 15585.

10551. Kida, T. 1939: A study of bursting of corn seeds. J. Taihoku Soc. Agr. Forestry 3:239–46.

10552. Kidd, F. 1914: The controlling influence of carbon dioxide in the maturation, dormancy, and germination of seeds. I. Proc. Roy. Soc. (London) Ser. B 87:408–21.

10553. —— 1914: The controlling influence of carbon dioxide in the maturation, dormancy, and germination of seeds. II. Proc. Roy. Soc. (London) Ser. B 87:609–25.

10554. —— 1955: The controlling influence of carbon dioxide. III. The retarding effect of carbon dioxide on respiration. Proc. Roy. Soc. (London) Ser. B 89:136–56.

10555. —— and C. West. 1917: The controlling influence of carbon dioxide. IV. On the production of secondary dormancy in seeds of *Brassica alba* following treatment with carbon dioxide, and the relation of this phenomenon to the question of stimuli in growth processes. Ann. Botany (London) 31:457–87.

10556. —— and —— 1918: Physiological pre-determination: the influence of the physiological conditions of the seed upon the course of subsequent growth and upon the yield. I. The effects of soaking seeds in water. Ann. Appl. Biol. 5:1–10.

10557. —— and —— 1918: Physiological pre-determination: the influence of the physiological condition of the seed upon the course of subsequent growth and upon the yield. II. Review of literature, Chapter I. Ann. Appl. Biol. 5:112–42.

10558. —— and —— 1919: The influence of temperature on the soaking of seeds. New Phytologist 18:35–39.

10559. —— and —— 1919: Physiological pre-determination: the influence of the physiological condition of the seed upon the course of subsequent growth and upon the yield. III. Review of literature, Chapter II. Ann. Appl. Biol. 5:157–70.

10560. —— and —— 1919: Physiological pre-determination: the influence of the physiological condition of the seed upon the course of subsequent growth and upon the yield. IV. Review of literature, Chapter III. Ann. Appl. Biol. 5:220–51.

10561. —— and —— 1919: Physiological pre-determination: the influence of the physiological condition of the seed upon the course of subsequent growth and upon the yield. V. Review of literature, Chapter IV. Ann. Appl. Biol. 6:1–26.

10562. —— and —— 1920: The role of the seed-coat in relation to the germination of immature seed. Ann. Botany (London) 34:439–46.

10563. Kidd, J. D., H. R. Nanii, and F. W. Edwards. 1939: The determination of caffeine in extracts of coffee and tea. Analyst 66:240. Biol. Abstr. 15:22630. 1941.

10564. Kiece, V. 1956: The growth and development of corn in relation to seed vernalization and stimulation. Dokl. Akad. Nauk SSSR 108:349–51

10565. —— 1959: Effect of irradiation by radioactive cobalt on the germination of seeds and growth of corn. Latvijas PSR Zinatnu Akad. Vestis 1959(5):131–34.

10566. Kiellander, C. L. 1935: Apomixis bei *Poa serotina*. (Vorläuf Mitteil.) Botan. Notiser 1935(1/2):87–95. Biol. Abstr. 10:15249. 1936.

10567. Kielland-Lund, J. 1963: Some factors affecting germination of spruce. Tidskr. Skogbruk. 71:58–70.

10568. Kielpinski, J., and J. A. Soltys. 1950: Effect of osmotic pressure of the medium on germination of clover seeds. Polish Acad. Sci. Agr.-For. Pap. 52. 38 pp. Brit. Abstr. BIII:128. 1951.

10569. Kienholz, R. 1924: The effect of high temperatures on the germination and subsequent growth of corn. Philippine J Sci. 25:311–47.

10570. Kienitz, M. 1880: Über Ausführung von Keimproben. Forstliche Blätter 16: 1–6.

10571. Kienitz, P. H., and T. Tharandt. 1952: Verminderung der Keimdauer hartschaliger Samen durch beizen mit Kalipermanganat. Deut. Landwirtsch. 3: 249–50.

10572. Kientzler, L., N. Bonnet, and D. Watrelos. 1961: Action of grain straw extracts on the germination of seeds of wheat and rape and on the development of their seedlings. Nancy École Natl. Super. Agron. Bull. 3:12–22.

10573. ——, J. Garnier, and C. Cagninacci. 1962: A case of tricotyledony in *Phaseolus vulgaris* L. Nancy École Natl. Super. Agron. Bull. 4:109–32.

Kiermayer, O. 12293.

10574. Kies, M. W. 1947: Activation of soybean lipoxidase. J. Biol. Chem. 170: 121–32.

Kies, M. W. 1819.

10575. Kiesel, A., and A. Belozersky. 1934: Protoplasm. V. The nucleic acid and nucleoproteins of pea germs. Z. Physiol. Chem. 229:160–66. Chem. Abstr. 29:482–83. 1935.

10576. —— and M. Kastrubin. 1934. Variations in the composition of the proteins of ripening wheat grains. Z. Physiol. Chem. 230:216–24. Chem. Abstr: 29:1441. 1935.

Kiesling, R. L. 7564, 15200.

10577. Kiesselbach, T. A. 1924: Relation of seed size to yield of small grain crops. J. Am. Soc. Agron. 16:670–82.

10578. —— 1927: Field experiments with seed corn treatments and crop stimulants. Nebraska Agr. Expt. Sta. Bull. 218. 15 pp.

10579. —— 1930: Field tests with treated seed corn. J. Agr. Res. 40:169–89.

10580. —— 1937: Effects of age, size, and source of seed on the corn crop. Nebraska Agr. Expt. Sta. Bull. 305. 16 pp.

10581. —— 1938: Production of foundation seed for hybrid corn. Ann. Rept. Nebraska State Bd. Agr. 1938:232–42.

10582. —— 1939: Effect of artificial drying upon the germination of seed corn. J. Am. Soc. Agron. 31:489–96.

10583. —— 1943: Crop response to hormone seed treatments. J. Am. Soc. Agron. 35:321–31.

10584. —— 1948: Endosperm type as a physiologic factor in corn yields. J. Am. Soc. Agron. 40:216–36.

10585. —— 1960: The significance of xenia effects on the kernel weight of corn. Nebraska Agr. Expt. Sta. Res. Bull. 191. 30 pp.

10586. —— and J. O. Culbertson. 1931: An analysis of the effects of *Diplodia* infection and treatment of seed corn. J. Agr. Res. 42:723–49.

10587. —— and W. H. Leonard. 1932: The effect of pollen source upon the grain yield of corn. J. Am. Soc. Agron. 24:517–23.

10588. —— and W. E. Lyness. 1939: The effects of stinking smut (bunt) and seed treatment upon the yield of winter wheat. Nebraska Agr. Expt. Sta. Res. Bull. 110. 22 pp.

10589. —— and —— 1945: Is vernalized winter wheat practical in Nebraska? Nebraska Agr. Expt. Sta. Ann. Rept. 58(1944):21–22.

10590. —— and —— 1945: Simulated hail injury of corn. Nebraska Agr. Expt. Sta. Bull. 377. 22 pp.

10591. —— and J. A. Ratcliff. 1920: Freezing injury of seed corn. Nebraska Agr. Expt. Sta. Res. Bull. 16. 96 pp.

10592. —— and E. R. Walker. 1952: Structure of certain specialized tissues in the kernel of corn. Am. J. Botany 39:561–69.

Kiesselbach, T. A. 8380, 15184.

Kiessig, H. 8684.

Kiessling, K. Lindahl-. 9631.

10593. Kiessling, L. 1911: Untersuchungen über die Keimreifung der Getreide. Landwirtsch. Jahrb. Bayern. 1:449–514. Expt. Sta. Rec. 26:130–31. 1912.

10594. —— 1918: Über schädliche Nebenwirkungen der Formalinbeizung des Saatgutes auf die Keimung. J. Landwirtsch. 66:7–51. Botan. Abstr. 2:773. 1919.

10595. Kietreiber, M. 1954: Die Vitalitätsprüfung von Erbsen und Bohnen nach der Keimrollenmethode. Jahrb. Bundesanstalt Pflanzenbau Samenprüfung Vienna 1953: 48–58.

10596. —— 1958: Hitzeschädigungen bei Mais und der Tetrazoliumtest. Jahrb. Bundesanstalt Pflanzenbau Samenprüfung Vienna 1957:33–38.

10597. —— 1960: Über differenzen zwischen TTC- und Keimfähigkeitswert bei *Zea mays*. Proc. Intern. Seed Testing Assoc. 25:467–73.

10598. —— 1962: Diseases in seed tests in the year 1961. Jahrb. Bundesanstalt Pflanzenbau Samenprüfung Vienna 1961: 21–27.

10599. —— 1962: Der *Septoria*-Befall von Weizenkörnern (zur Methodik der Erkennung). Proc. Intern. Seed Testing Assoc. 27:843–55.

Kietreiber, M. 7151, 7152.

Kietse, V. T. 11304.

10600. Kiewnick, L. 1961: Untersuchungen über den Einfluss der Samen- und Bodenbürtigen Mikroflora auf die Lebensdauer der Spelzfrüchte des Flughafers (*Avena fatua* L.). Stuttgart. 77 pp.

10601. —— 1963: Untersuchungen über den Einfluss der Samen- und Bodenmikroflora auf die Lebensdauer der Spelzfrüchte des Flughafers (*Avena fatua* L.). I. Vorkommen, Artzusammensetzung und Eigenschaften der Mikroorganismen an Flughaferfrüchten. Weed Res. 3:322–32.

10602. Kiffmann, R. 1954: Samen- und Hülsenmorphologie der zugelassenen westdeutschen Erbenzuchtsorten und die Abartensystematik der gemeinen Kulturerbse (*Pisum sativum* L.). Z. Pflanzenbau Pflanzenschutz 5:49–76.

Kihara, G. M. 8231.

10603. Kihara, H. 1948: Seedless watermelons. Sci. News Letter 54:163.

10604. —— and M. Kimura. 1951: Studies on the right- and left-handedness of spikelets in Einkorn wheats. Proc. Imp. Acad. Japan 27:678–83.

10605. Kihara, Y., and S. Tada. 1956: Chemical composition of seeds of *Citrus natsudaidai*. J. Agr. Chem. Soc. Japan (English Transl.) 30:675–77. Chem. Abstr. 52:12098. 1958.

10606. Kikuchi, M. 1952: Studies on the transplantation of the immature embryo in cereals. I. On the transplanting method and some results. Utsonomiya Univ. Coll. Agr. Bull. 2:9–21.

10607. —— 1954: Studies on the decrepitude of seeds by use of embryo transplanting method. I. On the old seeds of wheat. Utsonomiya Univ. Coll. Agr. Bull. 2:277–92.

10608. —— 1956: Studies on decrepitude of seeds by the method of embryo transplanting. II. Effects of high temperature treatment of dry wheat seeds upon their endosperms. Proc. Crop Sci. Soc. Japan 24:176.

10609. —— 1957: Studies on the transplantation of the immature embryos in cereals. III. Japan. J. Breeding 7:66–75.

10610. —— 1958: Studies on the transplantation of the immature embryos in cereals. IV. Influence of endosperms upon the leaf rust susceptibility in wheat seedlings. [Abstract.] Japan. J. Breeding 8:50.

10611. —— 1959: Studies on the transplantation of the immature embryos in cereals. VI. Differentiation and growth of organs of transplanted embryos. Proc. Crop Sci. Soc. Japan 28:147–49.

10612. Kikuchi, S. 1954: Studies on the results of the natural seeding of the beech under the isolated single seed-tree. II. On the factors which affect the densities of seedlings germinating on the bare soil beds. J. Japan. Forestry Soc. 36:76–80.

Kikuo, M. 14074.

10613. Kikuta, K., L. D. Whitney, and G. K. Parris. 1938: Seeds and seedlings of the taro, *Colocasia esculenta.* Am. J. Botany 25:186–88.

10614. Kilcher, M. R., and T. Lawrence. 1960: Quality of intermediate wheatgrass seed, with and without hulls. Can. J. Plant Sci. 40:482–86.

Kilcher, M. R. 11896.

10615. Killian, C. 1951: Germination et developpement post-embryonnaire de *Genlisea africana.* Inst. Franc. Afrique Noire Bull. 13:1029–36.

Killinger, G. W. 2606.

10616. Kilpatrick, R. A. 1957: Fungi associated with the flowers, pods, and seeds of soybeans. Phytopathology 47:131–35.

10617. —— and H. W. Johnson. 1956: Purple stain of legume seeds caused by *Cercospora* species. Phytopathology 46: 201–4.

Kim, C. H. 4014.

10618. Kim, C. M. 1954: Effect of soil moisture on the germination and early development of Bengal grass (*Setaria italica*) and barn grass (*Panicum crusgalli frumentaceum*). Ecol. Rev. 13:271–76.

10619. Kim, D. W., Y. C. Kim, C. H. Lee, and K. W. Choe. 1962: Breeding experiment on mutation induced by irradiation. I. Effects of X- and γ-ray irradiation on cabbage seeds. J. Nucl. Sci. 2:91–97. Nucl. Sci. Abstr. 17:35527. 1963.

10620. Kim, E. A. 1937: Comparative tests on the applicability of laboratory control methods in identifying cruciferous seeds. Bull. Appl. Botany, Genet., Plant Breed. (USSR) Ser. 4(2):23–39. Chem. Abstr. 31:5843. 1937.

Kim, Y. C. 10619.

Kimball, K. E. 16213.

10621. Kimber, D. S. 1957: "Atomic" seeds. Agr. Merchant 37(3):64–66.

Kimble, K. A. 18414.

Kimbrough, W. D. 8096–8098.

10622. Kimm, R. H. 1935: The constituents of rice embryo. II. Vitamin E in rice embryo. J. Agr. Chem. Soc. Japan (English Transl.) 11:514–23. Chem. Abstr. 29:6928–29. 1935.

10623. —— 1935: The constituents of rice embryo. III. The nutritive values of polished and unpolished rice and rice embryo. Bull. Inst. Phys.-Chem. Res. (Tokyo) 14:424–38. Chem. Abstr. 30: 2241. 1936.

10624. —— 1938: The chemical constituents of rice embryo. Sci. Papers Inst. Phys.-Chem. Res. (Tokyo) 34:637–77. Chem. Abstr. 32:7582–83. 1938.

10625. —— and T. Noguchi. 1933: The chemical constituents of rice embryo. Sci. Papers Inst. Phys.-Chem. Res. (Tokyo) 21:1–14.

10626. —— and —— 1933: Rice embryo. I. Fats, waxes, and oils. Bull. Inst. Phys.-Chem. Res. (Tokyo) 12:271–85. Chem. Abstr. 27:2690. 1933.

10627. Kimmel, J. R., and E. L. Smith. 1959: The amino acid composition of crystalline pumpkin seed globulin, edestin, C-Phycocyanin, and R-Phycoerythrin. Soc Chim. Biol. 40:2049–65.

Kimmett, L. 2771.

10628. Kimstrom, G. A., and R. F. Finn. 1956: Seed source and nursery effects on yellow-poplar plantations. J. Forestry 54:828–31.

10629. Kimura, C. 1955: The embryo sac of *Populus sieboldii* Miquel. Sci. Rept. Tohoku Univ. Fourth Ser. 21:122–25. Biol. Abstr. 33:2854. 1959.

10630. —— 1955: On the embryo sac of three Japanese species of *Salix.* Sci. Rept. Tohoku Univ. Fourth Ser. 21:126–30. Biol. Abstr. 33:2855. 1959.

10631. Kimura, K., and N. Ota. 1950: A study on the cultivation of medicinal plants. I. Influence of vegetable hormones upon the germination of some medicinal plant seeds. Bull. Pharm. Res. Inst. (Osaka) 1:11–19. Biol. Abstr. 27:30642. 1953.

10632. Kimura, M. 1934: The nucleic acid of rice embryo (*Oryza*-nucleic acid). I. The separation of pyrimidine base (cytosine). J. Soc. Chem. Ind. Japan 37:199–200. Chem. Abstr. 28:5497. 1934.

Kimura, M. 10604.

10633. Kimura, Y. 1940: Chemical study of the seeds of *Alpinia japonica* Miq. VI. Constituents of the seeds of *Alpinia chinensis* Roscoe. I. J. Pharm. Soc. Japan 60:145–51. Chem. Abstr. 34:4063. 1940.

10634. ——, A. Kawana, and K. Migita. 1954: On the germination and the yields of seedlings of *Alpinia japonica* Miq. in a nursery. J. Japan. Forestry Soc. 36:143–45.

Kin, P. S. 19280.

10635. Kinbacher, E. J. 1962: Effect of seed source on the cold resistance of pre-emerged Dubois winter oat seedlings. Crop Sci. 2:91–93.

Kinbara, E. 19990.

10636. Kincaid, R. R. 1935: Effects of certain invironmental factors on germination of Florida cigar-wrapper tobacco seeds. Florida Agr. Expt. Sta. Bull. 277. 43 pp.

10637. —— 1936: Toxicity of mercury vapor to germinating tobacco seed. Plant Physiol. 11:654–56.

10638. —— 1943: Effect of storage conditions on the viability of tobacco seed. J. Agr. Res. 67:407–10.

10639. —— 1958: Tobacco seed storage for 25 years. Florida Agr. Expt. Sta. Tech. Bull. 593. 10 pp.

Kincaid, R. R. 7890.

10640. Kinch, D. M. 1952/53: Physical factors affecting the germination vitality of weed seeds. (Michigan State) Doctoral Dissertations 1952/53(20):91.

10641. ——, J. K. Wang, and R. E. Strohman. 1961: Equipment for husking macadamia nuts. Hawaii Agr. Expt. Sta. Bull. 126. 19 pp.

10642. Kinch, R. C., and L. E. Wiesner. 1963: Seed quality in green needlegrass. J. Range Management 16:187–90. Biol. Abstr. 45:12496. 1964.

Kinch, R. C. 11702.

10643. Kincl, F. A., J. Romo, G. Rosenkranz, and F. Sondheimer. 1956: The constituents of *Casimiroa edulis* Llave et Lex. I. The seed. J. Chem. Soc. 1956: 4163–69.

10644. King, C. C. 1947: Changes of carbohydrates of germinating wheat seeds in manganese sulphate, indoleacetic acid, and colchicine media. Acad. Sinica Botan. Bull. 1:9–24.

10645. —— 1948: Further studies on the change of carbohydrates of germinating wheat seeds in the manganese sulphate and indoleacetic acid solutions. Acad. Sinica Botan. Bull. 2:80–86.

10646. King, C. M. 1924: Viability tests of seeds of different ages. Proc. Iowa Acad. Sci. 31:257–64.

10647. —— 1932: Germination studies of woody plants, with notes on some buried seeds. Proc. Iowa Acad. Sci. 39:65–76.

King, C. M. 14998–15012.

10648. King, D. L., and A. W. Riddolls. 1960: Damage to wheat seed and pea seed in threshing. J. Agr. Eng. Res. 5:387–98.

10649. King, D. R., C. F. Garner, R. B. Metzer, and L. C. Coffey. 1960: Effect of fumigation for insect control on seed germination. Texas Agr. Expt. Sta. Misc. Publ. MP-448. 12 pp.

10650. King, G. N. 1947: Artificial parthenocarpy in *Lycopersicum esculentum;* tissue development. Plant Physiol. 22: 572–81.

King, G. S. 4099, 11717.

10651. King, H. G. C. 1962: Phenolic compounds of commercial wheat germ. J. Food Sci. 27:446–54.

King, H. G. C. 4865.

10652. King, J. 1958: Repellent treatment of pine seed for bird protection. Tree Planters Notes 1958(32):11–12.

King, J. E. 18844.

10653. King, J. R. 1947: Development of ovule and megagametophyte in pomegranate. Botan. Gaz. 108:394–98.

King, J. R. 8098.

10654. King, L. C., and C. D. Ball, Jr. 1939: The sterols of alfalfa seed oil. I. J. Am. Chem. Soc. 61:2910–12.

10655. ——, C. D. Ball, B. Riegel, C. E. Schweitzer, P. G. Smith, and E. W. Meyer. 1943: The isolation of β-amyrin from the leaves and seeds of alfalfa. J. Am. Chem. Soc. 65:1168–70. Expt. Sta. Rec. 91:116. 1944.

10656. King, L. J. 1952: Germination and chemical control of the giant foxtail grass. Contrib. Boyce Thompson Inst. 16:469–87.

10657. —— and J. A. Kramer. 1955: A review of the crabgrass (*Digitaria ischaemum* and *D. sanguinalis*) with notes on their pre-emergence control in turf. Proc. Northeast. Weed Control Conf. 9:359–63.

King, T. H. 3618, 6332.

10658. King, T. R., and H. W. Titus. 1943: Acorns of the willow oak, *Quercus phellos*, a source of vitamin A activity. Poultry Sci. 22:56–60. Chem. Abstr. 37:2784. 1943.

10659. King, W. H., and V. L. Frampton. 1961: Properties of oil extracted from cottonseed with acetone-hexane-water solvent mixture. J. Am. Oil Chemists Soc. 38:497–99.
10660. King-Li-Pin, and Woo-Ping-Soung. 1934: Contribution à l'étude de l'action diurétique des graines de *Plantago major* L. var. *asiatica.* J. Physiol. Pathol. Gen. 32:1144–47. Biol. Abstr. 10:4126. 1936.
10661. Kinges, H. 1961: Characteristics of Gramineae embryos; a contribution to the systematic of grasses. Botan. Jahrb. 81:50–93.
10661A. Kingsbury, J. M. 1964: Poisonous plants of the U.S. and Canada. Prentice Hall, Inc., Englewood Cliffs, N.J. 626 pp.
Kingsley, A. F. 18593.
Kingsolver, C. H. 6707, 6708.
10662. Kingston, A. D. 1900: Germination of *Auricula* seeds. Gardening Illus. 22(1095):8.
Kinkor, C. P. 19269.
10663. Kinman, M. L., and E. R. Ibert. 1956: Seed germination as an index of potential free fatty acid content of sesame oil. J. Am. Oil Chemists Soc. 33:637–39. Biol. Abstr. 31:36049. 1957.
10664. —— and S. M. Stark. 1954: Yield and chemical composition of sesame, *Sesamum indicum* L., as affected by variety and location grown. J. Am. Oil Chemists Soc. 31:104–8.
Kinman, M. L. 2348.
10665. Kinnard, M. A. 1951: L'influence des désinfectants sur la faculté germinative des semences. Agricultura (Louvain) 49:83–101.
10666. Kinney, C. R., T. W. Jackson, L. E. DeMytt, and A. W. Harris. 1941: Oil of *Artemisia tridentata* (American sage brush). J. Org. Chem. 6:612–25. Expt. Sta. Rec. 91:243. 1944.
Kinoshita, S. 17330.
Kinpyo, T. 13634.
10667. Kinra, K. L., H. D. Foth, L. S. Robertson, and H. M. Brown. 1963: Effect of seeding rate, row spacing, and rate and placement of fertilizer on winter wheat performance in Michigan. Agron. J. 55:24–27.
Kinsman, S. 527.
10668. Kintigh, R. G. 1949: Some effects of temperature on germination and development of Pinyon pine. J. Forestry 47:622–26.
10669. Kinzel, W. 1900: Über die Keimung Halbreifer und reifer Samen der Gattung *Cuscuta.* Landwirtsch. Versuchssta. 54:125–33.
10670. —— 1900: Über die Wirkung wechselnder Warmheit auf die Keimung einzelner Samen. Landwirtsch. Versuchssta. 54:134–39.
10671. —— 1907: Über den Einfluss des Lichtes auf die Keimung, "lichtharte" Samen. Ber. Deut. Botan. Ges. 25:269–76.
10672. —— 1908: Die Wirkung des Lichtes auf die Keimung. Ber. Deut. Botan. Ges. 26A:105–15.
10673. —— 1908: Lichtkeimung. Einige bestätigende und ergänzende Bemerkungen zu den vorläufigen Mitteilungen von 1907 und 1908. Ber. Deut. Botan. Ges. 26A:631–45.
10674. —— 1908: Lichtkeimung. Weitere bestätigende und ergänzende Bemerkungen zu den vorläufigen Mitteilungen von 1907 und 1908. Ber. Deut. Botan. Ges. 26A:654–65.
10675. —— 1909: Lichtkeimung. Erlauterungen und Ergängungen. Ber. Deut. Botan. Ges. 27:536–45.
10676. —— 1911: Über die Wirkung des durchfrierens der Samen auf die Keimung und die Beziehungen zwischen Frost und Lichtwirkung. Prakt. Bl. Pflanzenbau Pflanzenschutz 9:105–14.
10677. —— 1913: Frost und Licht als beeinflussende Kräfte bei der Samenkeimung. Stuttgart, Ulmer. 170 pp. with 2 suppl. and 1 vol. tables.
10678. —— 1915: Notes on the germination of some tree and shrub seeds. Naturw. Z. Forst- Landwirtsch. 13:129–59. Expt. Sta. Rec. 33:343. 1915.
10679. —— 1919: Über eine neue Methode des Durchfrierens und die damit erzielten Erfolge bei Zahlreichen bisher nicht oder kaum zur Keimung gebrachten Samen. Naturw. Z. Forst- Landwirtsch. 17:139–42.
10680. —— 1929: Die Samen der Erdballs, ihr Werden und ihr Vergehen. Arch. Pharm. 267:455–65.
10681. —— 1929: Über den Einfluss von Licht und Frost sowie wechselnder Temperaturen auf die Keimung der Samen. Prakt. Bl. Pflanzenbau Pflanzenschutz 28:274–80.
10682. —— 1930: Grenzen der förderlichen Einwirkung von Frost und Licht bei der Samen-keimung. Angew. Botan. 12:16–22. Biol. Abstr. 5:27193. 1931.
10683. —— 1932: Höhenkeimer. Angew.

Botan. 14:182–93. J. Forestry 30:1024–25. 1932.

Kinzel, W. 8826.

10684. Kipp. M. 1929: Die Abgabe von Kohlensäure und die Aufnahme von Sauerstoff bei der Keimung lichtgeförderter Samen von *Nicotiana tabacum*. Jahrb. Wiss. Botan. 71:533–95.

10685. Kipp-Goller, A. 1939: Über Bau und Entwicklung der viviparen Mangrovekeimlinge. Z. Botan. 35:1–40.

10686. Kiraly, Z., E. Bocsa, and J. Vörös. 1962: Effect of treatment of wheat seed with gibberellic acid on bunt infection. Phytopathology 52:171–72.

10687. —— and G. L. Farkas. 1957: On the role of ascorbic oxidase in the parasitically increased respiration of wheat. Arch. Biochem. Biophys. 66:474–85.

10688. Kirch, M. H. 1936: Some abnormalities in the development of the embryo sac of *Lilium longiflorum*. Bull. Torrey Botan. Club 63:383–94.

Kircher, V. Schaffer-; *see* Schaffer-Kircher, V.

10689. Kirchheimer, F. 1936: Über die botanische Zugehörigkeit weiterer Früchte und Samen aus dem deutschen Tertiär. Planta 25:481–90.

10690. —— 1938: Beiträge zur näheren Kenntnis von Vitaceen-samenformen Tertiären alters. Planta 28:582–98.

10691. —— 1950: Pflanzenreste aus einer Braunkohle des Nördlinger Rieses. Neues Jahrb. Mineral. Monatsh. Abt. B 1949 (11/12):354–62. 1949. Biol. Abstr. 25: 28640. 1951.

10692. Kirchhoff, H. 1932: Über den Einfluss der Keimungstemperature und anderer Keimbettfaktoren auf das Verhalten gebeizten Getreides. Angew. Botan. 14:349–89. Rev. Appl. Mycol. 11:773. 1932.

Kirchner, H. A. 16110.

Kiredjian, M. 2296.

10693. Kirillova, G. A. 1958: Free amino acids in temporarily chilled wheat. Plant Physiol. (USSR) (English Transl.) 5:182–85. Chem. Abstr. 52:14768. 1958.

Kirillova, G. A. 2558.

10694. Kirinde, S. T. W. 1949: Some effects of IPC and CIPC on groundnuts. Trop. Agriculturist 115:7–13.

10695. Kiriyama, T., and T. Konishi. 1959: Varietal adaptability to early sowing in early naked barley. Kyushu Agr. Res. 21:82–85.

10696. Kirk, J. S., and J. B. Sumner. 1932: Crystalline urease. Preparation of meal from Jack beans. Ind. Eng. Chem. 24:454–55. Biol. Abstr. 7:2859. 1933.

10697. Kirk, L. E., and T. Stevenson. 1931: Seed colour markings in white flowered sweet clover, *Melilotus alba*. Desr. Sci. Agr. (Ottawa) 11:607–11. Expt. Sta. Rec. 65:531. 1931.

10698. —— and W. J. White. 1933: Autogamous alfalfa. Sci. Agr. (Ottawa) 13:591–93.

Kirk, L. E. 14375.

10699. Kirkland, R. J. A., and J. F. Turner. 1959: Nucleoside diphosphokinase of pea seeds. Biochem. J. 72:716–20.

Kirkpatrick, H., Jr. 8861, 8862.

10700. Kirkwood, F. R., and S. P. Mercer. 1934: Note on the technique of the Buchinger method of determining "suction-force." Proc. Intern. Seed Testing Assoc. 6:483–86.

Kirkwood, S. 5590.

10701. Kirnossowa, L. 1936: Der Einfluss hoher Temperatur auf das Mutieren der ruhenden Embryonen von *Crepis tectorum* L. bei konstanter Feuchtigkeit. Planta 25:491–501.

Kirsop, B. H. 15517.

10702. Kirste, A. 1931: Das Keimvermögen des Saatgutes. Pflanzenbau 8:30–31.

10703. Kiryalov, N. P. 1936: The chemical composition of the essential oil of the new species of anise "Anisette," *Pimpinella anisetum* Boiss. Bull. Appl. Botany, Genet., Plant Breed. (USSR) Ser. 3(15): 255.

10704. Kiselev, A. K., and T. G. Sinitsyna. 1960: Treatment of forest seeds with granosan and merkuran before storing. Lesn. Khoz. 1960(2):42.

10705. Kiseleva, E. N. 1951: Effect of dusting with benzene hexachloride on seed germination of grain crops which were previously treated. Selek. i Semen. 18(8):80.

10706. Kiseleva, N. A. 1956: The secondary dormancy of seeds of the wild oat and of *Bromus secalinus*. Agrobiologiya 1956(2): 130–34. Biol. Abstr. 32:35267. 1958.

10707. Kishikawa, H. 1955: The effects of ultrasonic wave on rice, *Oryza sativa* L. (Preliminary report.) Saga Univ. Agr. Bull. 3:1–13.

10708. —— 1960: X-ray experiments on dormant seeds of rye, *Secale cereale* L. Saga Univ. Agr. Bull. 11:57–64.

10709. Kising, W. 1949: Untersuchungen über den Rotkleesamenbau unter besonderer Berücksichtigung des Ansatzes. Z. Acker- Pflanzenbau 91:65–119.
10710. Kisljakov, P. 1930: Die Kultur des Blumenkohls zur Samengewinnung. Tr. Prikl. Botan. i Pr. 24:205–34. Gartenbauwissenschaft 6:4–5. 1931.
Kiss, I. 2671.
10711. Kisser, J. 1929: Zur Analyse der chemischen Reizerfolge bei der Samenkeimung. Anz. Akad. Wiss. Wien. Math.-Naturw. Kl. 66:223.
10712. —— 1932: Kritische Betrachtungen über das Wesen und den Begriff der Samenkeimung. Biol. Centralbl. 52:534–48.
10713. —— 1932: Zur Analyse chemischer Reizerfolge auf die Samenkeimung. Beitr. Biol. Pflanz. 20:59–76.
10714. —— 1933/34: Zur Frage nach Beziehungen zwischen Keimschnelligkeit und Geschwindigkeit des Keimlingswachstums. Gartenbauwissenschaft 8:336–45.
10715. —— and R. Fürtauer. 1932: Untersuchungen über den Einfluss gewisser chemischer Agentien auf die CO_2-Abgabe keimender Samen von *Pisum sativum* und *Triticum vulgare* unter optimalen Keimungsbedingungen. Anz. Akad. Wiss. Wien Math.-Naturw. Kl. 69:202–3. Botan. Centralbl. 166:148. 1934.
10716. —— and K. Lettmayr. 1933: Untersuchungen über die Absorption von Salzen durch Samen. Z. Pflanzenernähr. Düng. Bodenk. 29A:195–210. Chem. Abstr. 27:5464. 1933.
10717. —— and —— 1934: Experiments on the facility with which absorbed salts can be washed from seeds and their significance for seed stimulation. Z. Pflanzenernähr. Düng. Bodenk. 34A:172–81. Chem. Abstr. 29:879. 1935.
10718. —— and M. Lorenz. 1933: Die Wirkung von Reizchemikalien auf die Keimung von *Pisum* und *Triticum* unter optimalen Keimungsbedingungen. Jahrb. Wiss. Botan. 78:665–750.
10719. —— and L. Portheim. 1934: Versuch über die Verwendbarkeit von Wasserstoffsuperoxyd als Saatgutbeizmittel. Phytopathol. Z. 7:409–26. Rev. Appl. Mycol. 14:48. 1935.
10720. —— and J. Possnig. 1932: Untersuchungen über den Einfluss gehemmter und geförderter Sauerstoffatmung auf Samenkeimung und Keimlingswachstum. Beitr. Biol. Pflanz. 20:77–104.
10721. —— and H. Schmid. 1932: Untersuchungen über die Permeabilität der Samenhüllen von *Pisum* und *Triticum* für Wasser sowie die Saugkräfte der Samen. Anz. Akad. Wiss. Wien Math.-Naturw. Kl. 69:197–200. Botan. Centralbl. 166: 145. 1934.
10722. —— and R. Strasser. 1930: Untersuchungen über die bei der Keimung geschälter Leguminosensamen auftretenden Wurzel- und Hypokotylkrümmungen. Beitr. Biol. Pflanz. 18:161–84.
10723. —— and R. Windischbauer. 1929: Untersuchungen über die Permeabilität der Samenschale von *Pisum sativum* für Wasser und Gase. Anz. Akad. Wiss. Wien Math.-Naturw. Kl. 66:226–28.
Kisser, J. 10772.
Kitagawa, K. 9393.
10724. Kitagawa, M., and T. Tomiyama. 1929: A new amino-compound in the Jack bean and a corresponding new ferment. I. J. Biochem. (Tokyo) 11:265–71. Ber. Ges. Physiol. 54:676–77.
10725. —— and S. Yosue. 1939: On the nature of the different forms of urease found in the seeds of various *Canavalia*. II. Kyusyu Imp. Univ. Dept. Agr. Bul. Sci. 8:313–14.
10726. Kitagiri, H., and N. Mugibayasi. 1939: Relation between the nitrogenous compounds of seeds of various crops and the diastatic power of their malts. IV. Barley and naked barley. J. Agr. Chem. Soc. Japan (English Transl.) 15:309–16. Chem. Abstr. 34:137. 1940.
Kitahara, M. 9521.
Kitamura, T. 8987.
10727. Kitchin, J. T. 1954: Pelleted seed. New Jersey State Hort. News 35:2765.
10728. —— 1955: Coated seed, soil conditioners, and high nitrogen have little effect on carrot yields. New Jersey Agr. 37(1):8–9, 11.
10729. Kittock, D. L. 1960: Relation of respiration and 2,3,5-triphenyl-2H-tetrazolium chloride reduction in germinating seeds to seedling vigor. Univ. Microfilms (Ann Arbor, Mich.), L.C. Card No. Mic. 60-2708. 58 pp. Dissertation Abstr. 21: 413. Chem. Abstr. 55:1808. 1961.
10730. —— and J. K. Patterson. 1962: Seed size effects on performance of dryland grasses. Agron. J. 54:277–78.
10731. Kitunen, E. 1929: Das Vorkommen und der Saatwert der entspelzten Timotheesamen. Acta Agral. Fenn. 20. 79 pp. Botan. Centralbl. 160:175–76. 1931.

10732. Kitzke, E. D. 1958: A method for germination of *Copernica* palm seeds. Principes J. Palm Soc. 2:5–8. Biol. Abstr. 36:86290. 1961.

10733. —— and E. A. Wilder. 1961: The cuticle wax of the Cuban palm, *Copernica hospita*. Amer. Oil Chem. Soc. J. 38:699–700.

10734. Kiuchi, T. 1951: The effect of potassium on the rice plants. I. Relation between yield, panicle formation, and periods of potassium deficiency. J. Sci. Soil Manure, Japan 22:132–36. Chem. Abstr. 46:3696fg. 1952.

Kivel, B. 9975.

10735. Kivi, E. I. 1953: Über das Vorkommen von Zwillingskeimen bei den Getriedearten und besonders über die Einwirkung einiger äusserer Faktoren auf ihre Menge. Arch. Soc. Zool. Botan. Fennicae "Vanamo" 8:45–53. Biol. Abstr. 29:29663. 1955.

10736. Kizel, A., and S. Mikhlin. 1937: The distribution of enzymes in wheat grain. Biochemistry (USSR) (English Transl.) 2:734–40. Chem. Abstr. 32:2171. 1938.

10737. Kizilova, E. G. 1961: Certain physiological peculiarities in germination of freshly gathered corn seeds. Plant Physiol. (USSR) (English Transl.) 7:450–54.

10738. Kizima, P. N. 1951: Investigation of the amino-acid composition of the protein in some varieties of winter wheat. Selek. i Semen. 18(5):8–11. Chem. Abstr. 49:11196. 1955.

10739. Kjaer, A. 1940: Germination of buried and dry stored seeds. I. 1934–1939. Proc. Intern. Seed Testing Assoc. 12:167–90.

10740. —— 1945: Laboratory methods of determining the germinating capacity of lupine seed compared with the field germination. Tidsskr. Planteavl 49:429–44.

10741. —— 1948: Germination of buried and dry stored seed. II. 1934–1944. Proc. Intern. Seed Testing Assoc. 14:19–26.

10742. —— 1949: On germination conditions in tree species with seed dormancy. Dansk Skovfor. Tidsskr. 34:409–24.

10743. —— 1950: Storage of *Larix* seeds. Proc. Intern. Seed Testing Assoc. 16:95–97.

10744. —— 1953: On the extended use of certificates of varietal identity and purity with special reference to international seed trade. Proc. Intern. Seed Testing Assoc. 18:146–51.

10745. —— and B. W. Christensen. 1959: Isothiocyanates. XXXVI. (+)-4-Methyl-2-oxazolidimethione, the enzymic hydrolysis product of a glucoside (Glucosisymbrin) in seeds of *Sisymbrium austriacum* Jacq. Acta Chem. Scand. 13:1575–84.

10746. —— and J. Conti. 1953: Isothiocyanates. V. The occurrence of isopropyl isothiocyanate in seeds and fresh plants of various Cruciferae. Acta Chem. Scand. 7:1011–12.

10747. ——, ——, and K. A. Jensen. 1953: Isothiocyanates. III. The volatile isothiocyanates in seeds of rape (*Brassica napus* L.). Acta Chem. Scand. 7:1271–75. Ber. Wiss. Biol. 89:46. 1954.

10748. ——, ——, and I. Larsen. 1953: Isothiocyanates. IV. A systematic investigation of the occurrence and chemical nature of volatile isothiocyanates in seeds of various plants. Acta Chem. Scand. 7:1276–83. Ber. Wis. Biol. 89:47. 1954.

10749. —— and R. Gmelin. 1956: Isothiocyanates. XIX. L(−)-5-Methylsulphinylpentylisothiocyanate, the aglucone of a new naturally occurring glucoside (glucoalyssin). Acta Chem. Scand. 10: 1100–10.

Kjaer, A. 4051, 6891, 8980, 9787.

Klaas, H. 12655.

10750. Klaber, D. 1957: I grow my alpines from seed. Horticulture 35(10):508, 528.

10751. Klacan, G. R. 1962: Pod and seed development in canning peas as influenced by mineral nutrition and root temperature. Dissertation Abstr. 23:782.

10752. —— and K. C. Berger. 1963: Effect of nitrogen and magnesium nutrition on pod and seed development in canning peas. Agron. J. 55:228–31.

10753. Klaehn, F. U. 1960: Seed orchard classification. J. Forestry 58:355–60.

10754. Klaembt, H. D. 1958: The auxin metabolism of developing seeds of red currant and its reaction to α-naphthaleneacetic acid. Intern. Hort. Congr. Abstr. Papers 15:57–58.

Klages, K. H. 6015.

10755. Klapov, A. A. 1952: Accelerated stratification of tree and shrub seeds. Lesn. Khoz. 5(2):85.

Klapp, E. 6039.

10756. Klas, Z. 1930: Wachstumsstudien an Bohnenfrüchten. Acta Botan. Inst. Univ. Zagreb. 5:119–43. Botan. Centralbl. 161: 77–78. 1931.

10757. —— 1953: Studien über den Fervoreffekt bei pflanzlichen Organismen. II. Komparative Untersuchungen der Einwirkung des thermalen und des fervorisierten Wassers auf das Auskeimen von Senfsamen. Trav. Inst. Biol. Expt. Acad. Yugoslave 1953:81–99. Ber. Wiss. Biol. 90:312. 1954.

10758. —— 1953: Studien über den Fervoreffekt bei pflanzlichen Organismen. III. Komparative Untersuchungen über die Einwirkung des fervorisierten und glacialisierten Wassers auf das Auskeimen von Senfsamen. Trav. Inst. Biol. Expt. Acad. Yugoslave 1953:101–12. Ber. Wiss. Biol. 90:312. 1954.

10759. Klatzkin, C., F. W. Norris, and F. Wokes. 1948: Nicotinic acid in cereals. I. The effect of germination. Biochem. J. 42:414–20.

10760. ——, ——, and —— 1949: The biochemistry of germination. Intern. Congr. Biochem. Abstr. Commun. 1:26–28.

10761. Klaus, E. I. 1958: Improved counter-distributer of seeds. Selek. i Semen. 23(6):52–53.

10762. Klawitter, G., and R. Von Sengbusch. 1947: Die Züchtung von vollkommen alkaloidfreien Süsslupinen, die sich zur Herstellung von menschlichen Nahrungsmitteln eignen. Züchter 15:10–12. Biol. Abstr. 22:10596. 1948.

10763. Klebahn, H. 1904: Verwendung der Spezialisierung des Schmarotzers zur Unterscheidung der Arten der Wirte. Klebahn. Die Wirtwechselnden Rostpilze Jour. 1904:140–41.

10764. Kleber, W., and G. Franke. 1961: Die Keimungsenergie von Gersten verschiedener Jahrgänge. Brauwissenschaft 14:120–21.

10765. ——, M. Lindemann, and P. Schmid. 1959: Verkürzung der Keimruhe von Gerste durch Gibberellinsäure. Brauwelt 91:1745–46.

10766. Klebesadel, L. J., C. I. Branton, and J. J. Koranda. 1962: Seed characteristics of bluejoint (*Calamagrostis canadensis*) and techniques for threshing. J. Range Management 15:227–29.

10767. Klečka, A., and O. Malis. 1936: Samen im Heuausfall. Sb. Česk. Akad. Zeměděl. 11:426–36.

Kleczkowski, A. 10310.

10768. Kleczkowski, K., and I. Reifer. 1959: Metabolism of ornithine in green pea, bean, and wheat seedlings. Bull. Acad. Polon. Sci. Ser. Sci. Biol. 7:53. Chem. Abstr. 54:25065. 1960.

10769. Klee, F. 1931: Fragen zur Cyklamenaussaat. Gartenwelt 35:5–6.

10770. Klein, C., and E. Farkass. 1940: Microchemical detection of alkaloids in plants. XIV. Cytisine. Österr. Botan. Z. 79:107–24.

10771. Klein, G. 1931: Contribution a l'étude histologique et chimique du pistil, du fruit et des organes vegetatifs des Dipsacees. These Doctorat, Univ. Pharm. 111 pp. Bull. Soc. Botan. France 78:752–53. 1931.

10772. —— and J. Kisser. 1924: Die sterile Kultur der höheren Pflanzen. Botan. Abhandl. 2. 64 pp.

10773. —— and K. Tauboeck. 1932: Argininstoffwechsel und Harnstoffgenese bei höheren Pflanzen. Biochem. Z. 251: 10–50.

Klein, L. M. 8193.

10774. Klein, N. I. 1956: Duration of germinative power of seeds of some shrubby plants. Anales Soc. Botan. Brasil 5:335. 1954.

10775. Klein, R. 1914: Über den mikrochemischen Nachweis von Strychnin und Brucin im Samen von *Strychnos nux vomica*. Anz. Akad. Wiss. Wien Math.-Naturw. Kl. 3:39–40. Botan. Abstr. 126:349. 1914.

Klein, R. M. 17065.

10776. Klein, S. 1956: A simple potometer for measuring rate of water uptake by small seeds. Nature 177:707–8.

10777. —— 1962: Dormancy and dormancy breaking in seeds. *In* N. Grossowicz, editor. Cryptobiotic stages in biological systems. Elsevier, Amsterdam. 232 pp.

10778. —— and J. W. Preiss. 1958: Depth controlled deuteron irradiation of *Lactuca sativa* seeds. I. Effect on germination and growth. Plant Physiol. 33:321–25.

10779. —— and —— 1958: Reversibility of the red-far-red reaction by irradiation at different sites. Nature 181:200–1.

Klein, S. 6114, 6123, 6280, 11031, 15678.

10780. Klein, W. H., R. B. Withrow, and V. B. Elstad. 1956: Response of the hypocotyl hook of bean seedlings to radiant energy and other factors. Plant Physiol. 31:289–94.

Klein, W. H. 17364.

10781. Klein, W. M. 1958: Cotyledon variations in *Juniperus occidentalis* Hook. Aliso 4:129.

10782. Kleinau, E. 1931: Untersuchungen über die Veränderungen des Korns verschiedener Roggensorten unter dem Einfluss von Boden-temperature, Niederschlägen und Sorte. Wiss. Arch. Landwirtsch. Abt. A Pflanzenbau 5:374–410.

10783. —— 1931: Untersuchungen über die Veränderungen des Korns verschiedener Roggensorten unter dem Einfluss von Boden-temperature, Niederschlägen und Sorte. II. Wiss. Arch. Landwirtsch. Abt. A Pflanzenbau 5:411–50. Fortschr. Landwirtsch. 6:505. 1931.

10784. Kleine, R. 1924: Beobachtungen bei Keimversuchen mit Rispengräsern. Grünland 42:282–90.

10785. —— 1929: Über die Keimung mit *Poa fertilis*. Pflanzenbau 5:211–13.

10786. —— 1930: Beeinflusst das Sonenlicht die Keimung von *Poa pratensis*? Pflanzenbau 7:43–44.

10787. Kleinert, J. 1963: Chemical composition of cocoa radicles. Rev. Intern. Chocolat. 18:2–4, 6–7. Chem. Abstr. 59:3086. 1963.

10788. Kleinsmith, A. W., C. L. Shrewsbury, and H. R. Kraybill. 1938/39: Separation of glycerides of soybean oil. Indiana Agr. Expt. Sta. Ann. Rept. Director 52:29.

10789. Kleis, R. W., E. H. Lucas, H. M. Brown, and D. J. De Zeeuw. 1951: Dry treatment of wheat for loose smut (a preliminary report). Mich. Agr. Expt. Sta. Quart. Bull. 34:162–65.

Klement, Z. A. 12454.

10790. Klenha, J. 1960: Ein Beitrag zur Prüfung der Keimungsfähigkeit einiger Pflanzensamen. Naturwissenschaften 47: 382.

Kliever, W. H. 7050.

10791. Kliewer, W. M., and W. K. Kennedy. 1958: Fungicides used for the storage of moist grain. Agron. J. 50:493.

10792. Klimenko, R. T. 1936: Stimulating the growth of citrus seeds. Sov. Subtropiki. 1936(7):64–67. Chem. Abstr. 32:621. 1938.

10793. Klimenko, V. G. 1950: Amino acid composition of proteins from seeds of various beans. Ukr. Biokhim. Zh. 22:20–26. Chem. Abstr. 48:12241. 1954.

10794. —— 1950: Specific differences of lupine seeds according to content of nitrogenous substances. Dokl. Akad. Nauk SSSR 71:101–3.

10795. —— 1953: Forms of nitrogen and protein in the seeds of tree and bush forms of the bean family. Ukr. Biokhim. Zh. 25:54–61. Chem. Abstr. 47:12532. 1953.

10796. —— 1953: Forms of nitrogen and proteins of the seeds of *Dolichos lablab*, *Phaseolus aureus*, and *Phaseolus vulgaris*. Ukr. Biokhim. Zh. 25:342–50. Chem. Abstr. 48:7709. 1954.

10797. —— 1953: Forms of nitrogen in the seed and proteins of some varieties of peas. Biochemistry (USSR) (English Transl.) 18:141–50. Chem. Abstr. 47:8842. 1953.

10798. —— 1953: The forms of nitrogen of seeds and proteins of lupines. Uch. Zap. Kishinevsk. Gos. Univ. 8:165–73. Chem. Abstr. 50:445. 1956.

10799. —— 1954: Nitrogen forms of the seeds and proteins of some types of soybeans. Biochemistry (USSR) (English Transl.) 19:3–10. Chem. Abstr. 48:7815. 1954.

10800. —— 1956: The proteins of the seed of white vetch (*Lathyrus sativus* L.). Biochemistry (USSR) (English Transl.) 21:191–95. Chem. Abstr. 50:15753. 1956.

10801. —— 1958: The seed proteins of ligneous and bushy varieties of legumes. Tr. po Khim. Prirodn. Soedin. Kishinevsk. Gos. Univ. 1958(1):41–52. Chem. Abstr. 54:1667. 1960.

10802. —— 1958: Nitrogen-containing substances of seeds of some Rosaceae. Tr. po Khim. Prirodn. Soedin. Kishinevsk. Gos. Univ. 1958(1):61–69. Chem. Abstr. 54: 693. 1960.

10803. —— 1958: Identical nitrogen forms of seeds of Cucurbitaceae. Tr. po Khim. Prirodn. Soedin. Kishinevsk. Gos. Univ. 1958(1):79–87. Chem. Abstr. 53:18194. 1959.

10804. —— 1959: Proteins of the corn kernel. Tr. po Khim. Prirodn. Soedin. Kishinevsk. Gos. Univ. 1959(2):3–18. Chem. Abstr. 55:5664. 1961.

10805. —— 1959: Proteins of the seeds of some representatives of the Cruciferae. Tr. po Khim. Prirodn. Soedin. Kishinevsk. Gos. Univ. 1959(2):25–30. Chem. Abstr. 56:717. 1962.

10806. —— and M. V. Alekseeva. 1957: Effect of solvents on the content of different nitrogen-containing substances and proteins of seeds of some leguminous plants. Uch. Zap. Kishinevsk. Univ. 21:11–18. Chem. Abstr. 52:12100. 1958.

10807. —— and A. D. Berezovikov. 1960: N-containing compounds of the vegetative pods and seeds of *Vicia faba* in the course of development. Tr. po Khim. Prirodn.

Soedin. Kishinevsk. Gos. Univ. 1960(3): 145–57. Chem. Abstr. 56:5126. 1962.

10808. —— and —— 1963: Protein transformation and non-protein nitrogenous compounds in the ripening of fodder bean (*Vicia faba* L.) seeds. Biochemistry (USSR) (English Transl.) 28:238–45. Biol Abstr. 45:3170. 1964.

10809. —— and O. B. Dremova. 1960: The changeability in the content of protein and non-protein nitrogen in the seeds and vegetative mass of certain varieties of cowpea. Tr. po Khim.Prirodn. Soedin. Kishinevsk. Gos. Univ. 1960(3): 173–83. Chem. Abstr. 57:8910. 1962.

10810. —— and T. D. Dymchishina. 1957: Proteins of *Phaseolus*. Uch. Zap. Kishinevsk. Univ. 28:59–70. Chem. Abstr. 52:17412. 1958.

10811. ——, Y. Y. Gofman, and T. A. Baranova. 1960: Protein and non-protein substances in seeds and vegetative tissues of certain species of *Lathyrus*. Tr. po Khim. Prirodn. Soedin. Kishinevsk. Gos. Univ. 1960(3):27–39. Chem. Abstr. 56: 2716. 1962.

10812. —— and I. P. Grinberg. 1956: The proteins of the seeds of hybrid corn. Uch. Zap. Kishinevsk. Univ. 23:31–45. Chem. Abstr. 51:11490. 1957.

10813. ——, L. V. Kalinina, and A. N. Ivanova. 1957: Effect of sowing time on the protein content and the amino acid composition in seeds of some soya varieties. Uch. Zap. Kishinevsk. Univ. 28:29–48. Chem. Abstr. 52:16499. 1958.

10814. —— and V. A. Kostyuk. 1953: Nitrogen fractions of seeds of some leguminous plants. Kishinevsk. Univ. 8:153–58. Chem. Abstr. 50:444. 1956.

10815. —— and V. E. Kozubenko. 1957: Proteins of seeds of different corn hybrids. Uch. Zap. Kishinevsk. Univ. 28:3–28. Chem. Abstr. 52:16498. 1958.

10816. —— and G. S. Mikenas. 1954: The proteins of seeds of high-fat-producing plants. Uch. Zap. Kajan. Univ. 114:179–98. Chem. Abstr. 51:11490. 1957.

10817. —— and N. N. Telegina. 1961: Variability in protein and non-protein nitrogen in seed and vegetative tissues of some lentil varieties. Tr. po Khim. Prirodn. Soedin. Kishinevsk. Gos. Univ. 1961(4):123–36. Chem. Abstr. 57:14187. 1962.

Klimenko, V. G. 2301, 2302, 15406, 18923, 19010.

10818. Klimova, A. A. 1961: Preplanting treatment of seeds with radiant energy. Sel'skoe Khoz. Povolzhia 1961(2):48–51.

Klimovskaya, Z. A. 18183.

10819. Kline, E. L. 1949: Some notes on the commercial production of lily seed. North Am. Lily Soc. Lily Yearbook 2:39–41.

10820. —— 1956: Growing *Cyclamen* from seed. Horticulture 34(7):373.

Kline, G. L. 17750.

10821. Kling, F. 1915: Beitrag zur Prüfung der Gräserkeimung. J. Landwirtsch. 63: 285–343.

10822. Kling, M. 1931: Ein Beitrag zur Keimfähigkeit der Unkrautsamen. Fortschr. Landwirtsch. 6:577–79.

10823. —— 1942/43: Ein Beitrag zur Keimfähigkeit der Unkratusamen. Prakt. Bl. Pflanzenbau Pflanzenschutz 20:43–47.

Klinga-Mayer, M. 13087.

Klingman, G. C. 19865.

10824. Klingmüller, W. [No date]: Modification of radiation damage and electron spin resonance measurements in *Vicia faba* seeds. Proc. 3d Intern. Congr. Photobiol., pp. 261–65.

10825. —— 1959: Zur Möglichkeit eines nachträglichen Strahlenschutzes bei Samen von *Vicia faba*. Z. Naturforsch. 14(B): 268–72.

10826. —— 1960: Zur Frage einer jahresperiodischen Änderung der Strahlenempfindlichkeit bei Gerste. Naturwissenschaften 47:261–62.

10827. —— 1961: Modification of radiation damage in seeds by post treatment. Nature 189:413–14.

10828. —— 1961: Die Schädigung von *Vicia faba*-Samen durch Trocknung—ein umkehrbarer Effekt. Planta 55:290–301.

10829. —— 1962: Strahlenbiologische Untersuchungen an *Vicia faba* Samen. Der Einfluss des Wassergehaltes, der Sauerstoffzufuhr und der Lagerungszeit auf primäre und sekundäre Strahleneffekte. Intern. J. Radiation Biol. 4:255–76.

10830. —— and G. R. Lane. 1960: Damaging effect of drying on *Vicia faba* seeds. Science 185:699–700.

10831. ——, ——, M. C. Saxena, and D. J. E. Ingram. 1959: Free radicals in X-rayed seeds of high and low water content, as measured by electron spin resonance. Nature 184:464–65.

10832. Klinkowski, M., H. Kohler, and H. Schrodter. 1955: Möglishkeiten der

Desinfektion fettfleckenkranken Bohnensaatgutes durch Antibiotika unter Berücksichtingung des Einflusses meterologischer Faktoren. Phytopathol. Z. 23:345–80. Rev. Appl. Mycol. 34:696. 1955.

10833. Klisiewicz, J. M., and G. S. Pound. 1961: Studies of control of black rot of crucifers by treating seeds with antibiotics. Phytopathology 51:495–500.

10834. Kliucnikova, M. I. 1937: On the physiological characteristics of yarovized and non-yarovized *Perilla*. Compt. Rend. Acad. Sci. URSS 14:219–22.

10835. Klockkova, K. P. 1950: For an extensive use of the new agrotechnical method for increasing germination and sprouting energy of (field crop) seeds. Sov. Agron. 8:27–43.

Kloeck, M. 19174.

10836. Kloen, D. 1950: Jarovisation of turnip seed. Maandblad Landbouwvoorlichtingsdienst 7:193–201.

10837. —— 1951: Yarovization of rape seed. Plattelands-Post. 7(3):7.

10838. —— 1952: Vernalization of beet seeds. Zaadbelangen 6:177–78.

10839. —— 1954: La jarovisation de la betterave, du navet fourrager, du colza et des choux. Congr. Intern. Botan. Paris Rappt. Commun. 8:291–92.

Kloke, A. 16892, 16893.

10840. Kloor, I. A. 1952: Technique of coloring sugar beet seed for treatment identification. Proc. Am. Soc. Sugar Beet Technol. 7:609–10.

10841. Klosa, J. 1948: Some plant substances that inhibit germination of seeds and growth of bacteria. Pharmazie 3: 410–13. Chem. Abstr. 43:2679. 1949.

10842. —— 1953: The influence of some tuberculostatics on the germination of seeds. Pharmazie 8:837. Biol. Abstr. 31:5831. 1957.

10843. —— 1954: Über einige die Keimung von Samen und das Wachstum von Bakterien hemmende Substanzen aus Vegetabilien. IV. Arch. Pharm. 287:313–20. Biol. Abstr. 30:6040. 1956.

10844. Klose, A. A., J. D. Greaves, and H. L. Fevold. 1948: Inadequacy of proteolytic enzyme inhibition as explanation for growth depression by lima bean protein fractions. Science 108:88–89.

10845. ——, B. Hill, and H. L. Fevold. 1946: Presence of a growth inhibiting substance in raw soybeans. Proc. Soc. Exptl. Biol. Med. 62:10–12.

10846. ——, J. Peat, and H. L. Fevold. 1948: Vitamin C content of walnuts (Persian) during growth and development. Plant Physiol. 23:133–41.

10847. Kloss. 1931: Schlangenfichten aus Samen. Gartenflora 80:401.

10848. Klosterman, H. J., and C. O. Clagett. 1954: A new carbohydrate from flaxseed. Proc. N. Dakota Acad. Sci. 8:20. Biol. Abstr. 29:22429. 1955.

10849. ——, T. M. Farley, J. L. Parsons, and G. L. Lamoureux. 1963: The antipyridoxine factor in flaxseed. Am. Chem. Soc. Abstr. Papers, 145th Meeting, New York, Sept. 8–13, 1963:1A.

10850. Klostermeyer, E. C. 1958: Alfalfa seed insects. Wash. Agr. Expt. Sta. Bull. 587. 14 pp.

10851. Klotz, L. J., T. A. De Wolfe, and P. Wong. 1957: Heat-treat citrus seed to kill *Phytophthora* brown-rot fungi. Citrus Leaves 37(5):14–15. Biol. Abstr. 32:3077. 1958.

10852. —— and I. U. Khan. 1948: A technique for growing citrus seedlings under aseptic conditions of culture. Phytopathology 38:756–57.

10853. Kloz, J., and V. Turkova. 1963: Legumin, vicilin, and proteins similar to them in the seeds of some species of the Viciaceae family (a comparative serological study). Biol. Plant. 5:29–40.

10854. Klozova, E. 1962: The effect of acute irradiation of balsam seeds (*Impatiens balsamina* L.) on the formation of anthocyanins in blossoms. Biol. Plant. 4:246–54. Biol. Abstr. 45:12400. 1964.

Klug, H. L. 8052.

10855. Kluge, K., and S. Kramer. 1956: The stimulating effect of colchicine on the germination of *Fragaria* and *Vaccinium* seeds. Arch. Gartenbau 3:355–61. Chem. Abstr. 50:16994. 1956.

10856. Klugh, G. F. 1908: Some cases of delayed germination in seeds. Science 28:816.

10857. Kluijver, H., and K. W. Smilde. 1960: Influence of the photoperiod on the oil percentage of sesame seeds (*Sesamum indicum* L.). Neth. J. Agr. Sci. 8:11–14. Biol. Abstr. 35:42489.

Klugman, B. 6399.

10858. Klykken, O. 1937: Anatomie und Entwicklungsgeschichte des Samens von *Alliaria officinalis* Andrz. Nytt Mag. Naturvidensk. 77:201–15.

10859. Klykov, A. P. 1945: The viability of the causal agent of black bacteriosis in wheat seed. Microbiology (USSR) (English Transl.) 14:413–14. Rev. Appl. Mycol. 26:240. 1947.

10860. —— 1951: Cotyledonic bacteriolysis in soybean and its control. Microbiology (USSR) (English Transl.) 20:33–40. Biol. Abstr. 26:4626. 1952.

10861. —— 1956: New forms of soybeans from slowly swelling seeds. Vses. Akad. Sel'skokhoz. Nauk V. I. Lenina Dokl. 21:13–15.

10862. Klyshev, L. K., and R. K. Moiseev. 1962: The influence of storage conditions on certain biochemical indexes and germination of *Anabasis aphylla* seeds. Vestn. Akad. Nauk Kaz. SSR 18:49–54. Chem. Abstr. 58:2654. 1963.

10863. Kmetec, E., and E. H. Newcomb. 1956: Properties of particulate fractions isolated from homogenates of peanut cotyledons. Am. J. Botany 43:333–41.

Knaap, W. P. van der; *see* Van der Knaap, W. P.

10864. Knaggs, N. S. 1944: Carnauba—Brazil's tree of life. Chem. Eng. News 22:1564–69. Biol. Abstr. 19:1581. 1945.

10865. Knapp, R. 1954: Experimentelle Soziologie der hoheren Pflanzen. Einwirkung der Pflanzen aufeinander. Soziologie der Keimung und des aufwachsenden Bestandes. Eugen Ulmer, Stuttgart. Vol. I. 202 pp. Biol. Abstr. 29:23137. 1955.

10866. —— 1955: Über den Verlauf der Keimung von einigen Gräsern. Leguminosae und *Iberis amara* bei unterschiedlichen Temperaturen. Angew. Botan. 29:125–38.

10867. —— 1956: Effekts wechselnder Lober und liefer Temperaturen bei der Keimung von *Sonchus arvensis* L. Naturwissenschaften 43:41–42.

10868. —— 1957: Über den Einfluss der Temperatur während der Keimung auf die spätere Entwicklung einiger annuellen Pflanzenarten. Z. Naturforsch. 12b:564–68.

10869. —— and H. F. Linskens. 1961: Über den Aminosäure- und Aneurin-Gehalt vor und während der Keimung bei unterschiedlichen Temperaturen. Züchter 31:172–79.

Knauer, G. 8024.

10870. Knauer, W. 1940: Anatomy and history of development of the rough-coated castor bean and the detection of *Ricinus* in feeding stuffs. Z. Tierernähr. Futtermittelk. 4:125–82. Chem. Abstr. 36:582. 1942.

Knavina, M. 1839.

10871. Knebel, C. 1932: Keimkraft von Phyllokakteen-Samen. Möllers Deut. Gärtner-Ztg. 47:77–78.

10872. Knecht, H. 1931: Über die Beziehungen zwischen Katalaseaktivität und Vitalität im ruhenden Samen. Botan. Centralbl. Beih. Abt. I 48:229–313.

10873. Kneebone, W. R. 1960: Size of caryopses in Buffalo grass (*Buchloe dactyloides* [Nutt.] Engelm.) as related to their germination and longevity. Agron. J. 52:553.

10874. —— and C. L. Cremer. 1955: The relationship of seed size to seedling vigor in some native grass species. Agron. J. 47:472–77.

Kneebone, W. R. 8181.

10875. Kneen, E. 1943: The cereal amylases. Wallerstein Lab. Commun. 6(18):101–10. Biol. Abstr. 18:5414. 1944.

10876. —— 1944: A comparative study of the development of amylases in germinating cereals. Cereal Chem. 21:304–14.

10877. —— 1945: Sorghum amylase. Cereal Chem. 22:112–34.

10878. —— and H. L. Hads. 1945: Effects of variety and environment on the amylase of germinated wheat and barley. Cereal Chem. 22:407–18. Biol. Abstr. 20:1737. 1946.

10879. ——, B. S. Miller, and R. M. Sandstedt. 1942: The influence of temperature on the development of amylase in germinating wheat. Cereal Chem. 19:11–27.

10880. —— and R. M. Sandstedt. 1941: β-Amylase activity and its determination in germinated and ungerminated cereals. Cereal Chem. 18:237–52.

10881. —— and —— 1946: Distribution and general properties of an amylase inhibitor in cereals. Arch. Biochem. 9:235–49.

Kneen, E. 2621, 12362, 13473, 13474, 13478.

10882. Kniaginichev, M. I. 1935: Variability of protein content of individual seeds in different kinds of peas. Bull. Appl. Botany, Genet., Plant Breed. (USSR) Ser. A, Plant Ind. 16:31–36.

10883. —— 1936: Intergeneric and interspecific regularities in the variation of the protein content in the ears of cereals. Bull. Appl. Botany, Genet., Plant Breed. (USSR) Ser. III, Phys., Biochem., Anat. Plants 15:55–65.

10884. —— 1936: Content and distribution of the ash in the mealy nucleus of the endosperm in Vulgare and Durum wheats. Bull. Appl. Botany, Genet., Plant Breed. (USSR)

Ser. III, Phys., Biochem., Anat. Plants 15:66–72.

10885. —— 1936: Differences in the variation of the protein content in wheat and barley grains within one ear. Compt. Rend. Acad. Sci. URSS 10:139–43. Physiol. Abstr. 21:703. 1936.

10886. —— and T. M. Gorelkina. 1938: Structural peculiarities of starch in cereal and leguminous plant seeds. Compt. Rend. Acad. Sci. URSS 19:117–21.

10887. ——, I. F. Mutul, and J. K. Palilova. 1940: Activity of carbohydrase in the ripening grains of various wheats. Biochemistry (USSR) (English Transl.) 5:288–300. Chem. Abstr. 35:4788. 1941.

10888. ——, ——, and —— 1940: Wheat species characterized according to activity and quality of amylase in their grain. Compt. Rend. Acad. Sci. URSS 27:1020–23.

10889. —— and J. K. Palilova. 1940: The activity and properties of catalase in the wheat, barley, and genera related to wheat. Biochemistry (USSR) (English Transl.) 5:55–64. Biol. Abstr. 16:5494. 1942.

10890. Knight, W. E. 1953: Interrelationships of some morphological and physiological characteristics of Ladino clover. Agron. J. 45:197–99.

10891. —— 1955: The influence of photoperiod and temperature on growth, flowering, and seed production of Dallisgrass, *Paspalum dilatatum* Poir. Agron. J. 47:555–59.

10892. Knipe, D., and C. H. Herbel. 1960: The effects of limited moisture on germination and initial growth of six grass species. J. Range Management 13:297–302. Biol. Abstr. 36:21450. 1961.

10893. Knoepfler, N. B., E. J. McCourtney, L. M. Molaison, and J. J. Spadaro. 1959: A comparison of six solvents for the extraction of jojoba seed. J. Am. Oil Chemists Soc. 36:644–48.

10894. Knott, D. R. 1958: The inheritance in wheat of a blue endosperm color derived from *Agropyron elongatum*. Can. J. Botany 36:571–74.

10895. Knott, J. E. 1925: Effects of soaking seed of some vegetables before sowing. J. Am. Soc. Agron. 17:49–54.

10896. ——, O. W. Terry, and E. M. Andersen. 1938: Vernalization of lettuce. Proc. Am. Soc. Hort. Sci. 35:644–48.

Knowles, P. F. 8216, 8217, 19982.

10897. Knowles, R. H., and S. Zalik. 1958: Effects of temperature treatment and of a native inhibitor on seed dormancy and of cotyledon removal on epicotyl growth in *Viburnum trilobum* Marsh. Can. J. Botany 36:561–66.

10898. Knowles, R. P. 1943: The role of insects, weather conditions, and plant character in seed setting of alfalfa. Sci. Agr. (Ottawa) 24:29–50.

10899. —— 1961: Seed production of perennial grasses in the greenhouse. Can. J. Plant Sci. 41:1–9.

10900. ——, H. A. Friesen, and D. A. Cooke. 1951: Brome grass seed production in western Canada. Can. Dept. Agr. Publ. 866. 13 pp.

10901. Knox, H., and R. N. Flint. 1949: The germination of centipede grass. Proc. Assoc. Offic. Seed Analysts 39:78–79.

10902. Knudson, L. 1922: Non-symbiotic germination of orchid seeds. Botan. Gaz. 73:1–25.

10903. —— 1924: Further observations on non-symbiotic germination of orchid seeds. Botan. Gaz. 77:212–19.

10904. —— 1925: Physiological study of the symbiotic germination of orchid seeds. Botan. Gaz. 79:345–79.

10905. —— 1927: Symbiosis and asymbiosis relative to orchids. New Phytologist 26:328–36.

10906. —— 1928: Is the fungus necessary for development of seedlings of *Calluna vulgaris*? Am. J. Botany 15:624–25.

10907. —— 1929: Seed germination and growth of *Calluna vulgaris*. New Phytologist 28:369–76.

10908. —— 1930: Flower production by orchid grown non-symbiotically. Botan. Gaz. 89:192–99.

10909. —— 1934: Non-symbiotic germination of orchid seed. Am. J. Botany 21:712.

10910. —— 1941: Germination of seed of *Goodyera pubescens*. Am. Orchid Soc. Bull. 9:199–201. Biol. Abstr. 16:13378. 1942.

10911. —— 1941: Investigations on delayed photosynthesis in chlorophyll-bearing embryos of orchids, and other studies on orchid seed. Cornell Univ. Agr. Expt. Sta. Ann. Rept. 54:117–18.

10912. —— 1950: Germination of seeds of vanilla. Am. J. Botany 37:241–47.

10913. —— 1951: Nutrient solutions for orchids. Botan. Gaz. 112:528–32.

10914. —— 1953: Viability of orchid seed. Am. Orchid Soc. Bull. 22:260–61.

10915. Knutsson, G. 1962: Seed culture in

Uppsala province. Uppsala Lans Hushs-allsk. Forsok Undervis. Vaxtodling 1962: 6–10.

10916. Knypl, J. S. 1960: Inhibitory effect of coumarin vapours on the growth of barley seedlings. Naturwissenschaften 47: 524.

10917. Koaze, Y. 1958: Germination promotant for plants seed, produced by micro-organisms. II. Germination promotant for rice plant seeds, produced by *Streptomyces* sp. S-580. Culture condition and extraction of the active crude syrup. Bull Agr. Chem. Soc. Japan 22:91–97.

10918. —— 1958: Germination promotant for plants seed, produced by micro-organisms. III. Germination promotant for rice plant seeds, produced by *Streptomyces* sp. S-580. Isolation and structure of an active neutral crystal. (Factor-E). Bull. Agr. Chem. Soc. Japan 22:98–103.

10919. —— 1958: Germination promotant for plants seed, produced by micro-organisms. IV. The germination promotant for rice plant seeds, produced by *Streptomyces* sp. S-580. Isolation and structure of another active crystal (Factor-D). Bull. Agr. Chem. Soc. Japan 22:238–42.

10920. —— 1960: Germination promotant for plant seeds produced by microorganisms. V. The germination promotant for rice plant seeds, produced by *Streptomyces*, S-580. (An addendum.) J. Agr. Chem. Soc. Japan (English Transl.) 34:371–74. Chem. Abstr. 58:9568–69. 1963.

10921. ——, H. Sakai, and K. Arima. 1957: Studies on the plants seed germination promoting substances which are produced by the microorganisms. I. Selection of strains. J. Agr. Chem. Soc. Japan (English Transl.) 31:338–49. Tobacco Abstr. 1: 379. 1957.

10922. Kobayashi, M. 1952: Studies on the development of seeds in the pod of the peanut. Proc. Crop Sci. Soc. Japan 21:43–44.

Kobayashi, R. 10324.

10923. Kobayashi, S., and K. Shiseki. 1953: Fate of phosphorus compounds and dehydrogenase activities in the germination process. Igaku To Seibutsugaku 27:229–32. Chem. Abstr. 48:1489. 1954.

10924. —— and —— 1954: Histochemical studies on nucleic acids and polysaccharides in germinating soybeans. I. Igaku To Seibutsugaku 30:172–75. Chem. Abstr. 50:8813. 1956.

10925. —— and —— 1954: Histochemical demonstration of glycerophosphatase and ribonucleic acid- and deoxyribonucleic acid-phosphatases in germinating soybeans. Igaku To Seibutsugaku 31:215–18. Chem. Abstr. 50:9527. 1956.

Kobayashi, T. 10369.

Kobayashi, Y. 14079, 14084.

10926. Kobel, F. 1926: Versuche zur Stimulation von Samen und Stecklingen mit besonderer Berücksichtigung der Rebe. Landwirtsch. Jahrb. Schweiz. 40:751–71.

Kobel, M. 14351, 14352.

10927. Kobelt, R. 1941: Über die Eiweissbildung im Getreide. Schweiz. Landwirtsch. Monatsh. 19:122–34. Biol. Abstr. 23:12523. 1949.

10928. Koblet, R. 1932: Über die Keimung von *Pinus strobus* unter besonderer Berücksichtigung der Herkunft des Samens. Ber. Schweiz. Botan. Ges. 41:199–283. Botan. Centralbl. 165:76–77. 1933.

10929. —— 1936: Variationsstatistisches zur Samenkeimung. Ber. Schweiz. Botan. Ges. 45:166–80.

10930. —— 1937: Untersuchungen über die Keimung von Kernobstsamen. Proc. Intern. Seed Testing Assoc. 9:82–122.

10931. —— 1937: Über die proteolytische Aktivität von Weymouthskiefern- und Weizensamen unter besonderer Berücksichtigung des Einflusses der Vorkühlung. Proc. Intern. Seed Testing Assoc. 9:228–53.

10932. —— 1940: Untersuchungen über die stofflichen Veränderungen im wachsenden und reifenden Weizenkorn. Ber. Schweiz. Botan. Ges. 50:99–232. Biol. Abstr. 15:19851. 1941.

10933. —— 1942: The influencing of plant growth by seed treatment. Landwirtsch. Jahrb. Schweiz. 56:278–88. Chem. Abstr. 36:6586. 1942.

10934. —— 1947: Über die Reservestoffbildung in landwirtschaftlichen Kulturpflanzen. Experientia 3:95–127.

10935. —— 1950: Culture du trèfle violet et du trèfle hybride en régions elvées. Importance de la variété et de la provenance des graines. Rev. Romande Agr. Viticult. Arboricult. 6:73–75. Biol. Abstr. 26: 9576. 1952.

Koblet, R. 7692, 7693.

Koblitsky, L. 6456, 6486.

Koblobkova, E. V. 2559.

10936. Koblova, M. N. 1962: The effect of ecological and geographic factors on seed germination in *Amaranthus retroflexus.*

Uch. Zap. Kabardino-Balkarsk. Univ. 12:91–102. Biol. Abstr. 44:16642. 1963.

10937. Kobuladze, O. T. 1963: Variation in the quantity of proteins and starch in foxtail millet seeds as affected by agricultural engineering and by (soil-climate) zoning. Soobshch. Akad. Nauk Gruz. 30:573–78. Chem. Abstr. 60:2046. 1964.

10938. Koch, A. 1957: The fatty oil of the seeds of *Cornus mas*. Pharm. Zentralhalle 96:448–51. Biol. Abstr. 32:28425. 1958.

10939. Koch, B. A. 1955/56: Effects of storage upon the B-complex vitamins and proteins of barley grain as measured by biological methods. Doctoral Dissertations 16(13):54.

10940. Koch, F. 1953: Beitrag zur Frage der Beizung des Rübensaatgutes gegen die Blattfleckenkrankheit (*Cercospora beticola* Sacc.). Z. Pflanzenkrankh. Pflanzenschutz 60:337–48.

Koch, G. H. 18712.

10941. Koch, G. P. 1919: *Atropa belladonna*. J. Am. Pharm. Assoc. 8:390–405. Botan. Abstr. 3:837. 1920.

10942. —— 1919: *Hyoscyamus niger*. Am. J. Pharm. 91:68–83. Botan. Abstr. 3:1687. 1920.

10943. —— and J. R. Butler. 1919: *Digitalis purpurea*. Am. J. Pharm. 91:211–21. Botan. Abstr. 3:1686. 1920.

10944. Koch, L. 1875: Germination, growth, and development of embryo in *Cuscuta* species. Landwirtsch. Versuchssta. 18:53–55.

Koch, L. W. 8748, 15121.

10945. Koch, W., and A. J. Linck. 1960: The *in vitro* culture of embryos, ovules, and ovaries of *Pisum sativum*. Proc. Minn. Acad. Sci. 28:42–44.

Kochev, K. 2739.

Kochhar, B. D. 7658.

10946. Kochi, I., and K. Matsue. 1959: Studies on the vernalization on the garlics. I. Kyushu Agr. Res. 21:134–35.

10947. Kochkar, N. T. 1950: Relationship of *Pinus* seed yields and their germination to size of cones. Lesn. Khoz. 3(8):82–83.

10948. —— 1961: The effect of ultrasonics on germination of pine and larch seeds. Lesn. Khoz. 6:38–39. Biol. Abstr. 42:3323. 1963.

10949. Kociecki, S. 1957: Pregermination and prechilling of pine seed. Sylwan 101:112–16. Biol. Abstr. 32:31732. 1958.

10950. —— 1958: Forest seed problems of the German Federal Republic in 1957. Sylwan 102:55–68.

Kock, P. C. de; *see* De Kock, P. C.

Koctürk, O. 8938.

10951. Kodama, T., and S. Suzuki. 1955: Effects of the soil structure on the growth of crop plant. I. Seed germination of wheat as affected by soil structure. II. Relation between the growth of wheat plant at early growing stage and soil structure. Proc. Crop Sci. Soc. Japan 24:59–62.

10952. Kodymsky, I. A. 1954: The effect of low temperatures on the increase in germinative ability of thermophilic agricultural crops. Akad. Nauk Ukr. RSR Inst. Ahrobiol. Pratsi 4:104–16.

10953. —— 1955: The effect of low temperature on the slightly germinated seeds of corn. Agrobiologiya 1955(5):133–35. Biol. Abstr. 32:9777. 1958.

10954. Köck, G. 1932: Vinegar as a seed disinfectant. Fortschr. Landwirtsch. 7: 226–27. Chem. Abstr. 26:4674. 1932.

10955. Köckemann, A. 1934: Über eine keimungshemmende Substanze in fleischigen Früchten. Ber. Deut. Botan. Ges. 52:523–26.

10956. —— 1936: Zur Frage der keimungshemmenden Substanzen in fleischigen Früchten. Botan. Centralbl. Beih. 55:191–96.

10957. Kögl, F., and A. J. Haagen-Smit. 1936: Biotin und Aneurin als Phytohormone. Ein Beitrag zur Physiologie der Keimung. XXIII. Mitteilung über pflanzliche Wachstumsstoffe. Z. Physiol. Chem. 243:209–26.

10958. ——, ——, and H. Erxleben. 1934: Über den Einfluss der Auxine auf das Wurzelwachstum und über die chemische Natur des Auxins der Graskoleoptilen. Z. Physiol. Chem. 228:104–12.

10959. Koehler, B. 1935: Effect of seed coat injury on germination, vigor, and yield of corn. Trans. Illinois State Acad. Sci. 28:52–54.

10960. —— 1935: Pathologic significance of seed-coat injury in dent corn. Phytopathology 25:24.

10961. —— 1935: Seed treatments for the control of certain diseases of wheat, oats, and barley. Illinois Agr. Expt. Sta. Bull. 420:499–575. Chem. Abstr. 30:2308. 1936.

10962. —— 1938: Effect of prolonged storage of treated seed corn. [Abstract.] Phytopathology 28:13.

10963. —— 1938: Fungus growth in shelled corn as affected by moisture. J. Agr. Res.

56:291–307. Expt. Sta. Rec. 79:346–47. 1938.

10964. —— 1938: Seed treatment tests with crown-injured corn. [Abstract.] Phytopathology 28:13.

10965. —— 1940: Rapid seed-corn drying checks seed infection. [Abstract.] Phytopathology 30:14.

10966. —— 1940: Treat seed oats against new disease. Seed World 47(5):36.

10967. —— 1941: Effect of storage on yields of farm seed treated for disease control. Wheat-oats-barley-corn. Illinois Agr. Expt. Sta. Bull. 476:259–76.

10968. —— 1946: Corn pericarp injuries and seedling diseases. [Abstract.] Phytopathology 36:403.

10969. —— 1946: Hemp seed treatments in relation to different dosages and conditions of storage. Phytopathology 36:937–42.

10970. —— 1954: Some conditions influencing the results from corn seed treatment tests. Phytopathology 44:575–83.

10971. —— 1957: Pericarp injuries in seed corn. Prevalence in dent corn and relation to seedling blights. Illinois Agr. Expt. Sta. Bull. 617. 72 pp.

10972. —— and W. M. Bever. 1954: Oat and wheat seed treatments: methods, injuries, and gains. Plant Disease Reptr. 38:762–68.

10973. —— and G. H. Dungan. 1940: Disease infection and field performance of bin- and hanger-dried seed corn. J. Am. Soc. Agron. 32:768–81.

10974. ——, ——, and W. L. Burlison. 1934: Maturity of seed corn in relation to yielding ability and disease infection. J. Am. Soc. Agron. 26:262–74.

Koehler, B. 5642, 8941, 13588.

10975. Koehler, Z. 1927: Sur les composés phosphorés des plantes. I. La solubilité et la répartition des composés phosphorés contenus dans les semences. Bull. Acad. Polon. Sci. Lettr. Cl. Math. Nat. Ser. B 1926:709–848. Botan. Centralbl. 155:80. 1928/29.

10976. —— 1929: Sur les composés phosphorés des plantes. V. La solubilité des composés phosphorés des embryons de seigle. Bull. Intern. Acad. Polon. Sci. Lettr. Cl. Sci. Math. Nat. Ser. B 1929:237–54. Biol. Abstr. 6:682. 1932.

10977. —— 1930: Solubility of the P compounds of rye embryos. Acta Biol. Exptl. 4:33–49. Biol. Abstr. 8:13077. 1934.

Koekum, S. 12102.

10978. Koenig, P. 1936: Olio da semi d tabacco. Chim. Ind. (Milan) 18:263–64. Ist. Sper. Coltivaz. Tabac. Scafati Boll. Tec. 33:144. 1936.

10979. —— and W. Müller. 1939: Über Tabaksamen und die Keimprüfungs-sowie Beiz-methoden. Tabac [Fed. Intern. Tech. Agron. (Rome)] 2:7–19.

10980. Koeppen, O. W. 1887: Über das Verhalten des Zellkerns im ruhenden Samen. Inaug. Dissertation Univ. Leipzig. 52 pp.

10981. Koerber, T. W. 1963: *Leptoglossus occidentalis* (Hemiptera, Coreidae), a newly discovered pest of coniferous seed. Ann. Entomol. Soc. Am. 56:229–34.

10982. Koernicke, M. 1915: Über die Wirkung verschieden starker Röntgenstrahlen auf Keimung und Wachstum bei der höheren Pflanzen. Jahrb. Wiss. Botan. 56:416–30. Expt. Sta. Rec. 35:436. 1916.

10983. Koert, J. L. 1955: Control of downy mildew on seed onions and on onion seed. Zaadbelangen 9:137–38.

10984. Körting, A. 1960: Schädigung lagernden Saatgutes durch holzschutzmittel. Nachrbl. Deut. Pflanzenschutzdienst (Berlin) 12:70–74.

Kötter, W. 2970.

10985. Köves, E. 1957: Paper chromatographic investigations of the ether-soluble germination and growth inhiibitors in oat husks. Acta Biol. (Szeged) 3:179–87. Biol. Abstr. 35:54062. 1960.

Köves, E. 19011.

10986. Kofman, T. 1933: Action, sur la germination, des tubes lumineux en quartz a haute tension. Compt. Rend. Soc. Biol. 112:895–97.

Kofman, T. 4223.

10987. Kofoed, S. S. 1959: Investigations on drying grain with slightly pre-heated air in thick layers. Roy. Vet. Agr. Coll., Copenhagen, Den. Yearbook 1959:17–49.

Kogane-Charles, M. 586–589, 5261–5265.

10988. Kohl, H. C. 1962: Notes on the development of *Cymbidium* from seed to plantlet. Am. Orchid Soc. Bull. 31:117–20.

10989. Kohl, P. A. 1941: Growing plants from seed. Bull. Missouri Botan. Garden 29:81–94.

10990. Kohlbecker, R. 1957: Die Abhängigkeit des Längenwachstums und der phototropischen Krümmungen von der Lichtqualität bei Keimwurzeln von *Sinapis alba*. Z. Botan. 45:507–24.

10991. Kohler, H. 1961: Seed incrustation of bushbeans with antibiotics. Biol. Bundesanst Land- Forstwirtsch. Mitt. (Berlin) 104:135–38.

Kohler, H. 10832.

10992. Kohler, M. 1956: Pflanzenbauliche Versuch mit Monogermsaat im Zuckerrübenbau. Mitt. Schweiz. Landwirtsch. 4:33–39.

10993. Kohls, H. L. Seed production of space-isolated vs. bagged mother beets and a discussion of some factors influencing the latter. J. Am. Soc. Agron. 27:699–706.

10994. —— 1937: The influence of some climatological factors on seed-stalk development and seed yield of space-isolated mother beets. J. Am. Soc. Agron. 29:280–85.

10995. —— and E. E. Down. 1934: Influence of inbreeding and selection on seed production of space-isolated mother beets. J. Am. Soc. Agron. 26:327–32.

10996. —— and H. C. Rather. 1941: Growing chicory seed. Mich. Agr. Expt. Sta. Quart. Bull. 23:243–50.

10997. Kohnke, H., and C. M. Vestal. 1949: The effect of nitrogen fertilization on the feeding value of corn. Soil Sci. Soc. Am. Proc. 13:299–302.

10998. Kojima, H., M. Yahiro, and T. Eto. 1957: On the influence of the auxins and the anti-auxin upon vernalization. Kyushu Univ. Fac. Agr. J. 11:25–35.

10999. ——, ——, and S. Inoue. 1954: The vernalization and the cotyledon of the seedling. Botan. Mag. (Tokyo) 67:112–21.

11000. ——, M. Yatazawa, and Y. Goto. 1953: Interconversion of amino acids at an early stage of germination of plant seeds. Sci. Rept. Shiga Agr. Coll. Ser. I 4:51–56. Chem. Abstr. 50:9527. 1956.

11001. Kokawa, S. 1958: On the discreet distribution of morphometric values of *Menyanthes* seed remains in Japan. J. Inst. Polytech. Osaka City Univ. Ser. D 9:119–23.

11002. —— 1959: Morphometry of *Menyanthes* seed remains in Japan. J. Inst. Polytech. Osaka City Univ. Ser. D 10:45–63.

11003. —— 1960: Morphometric reconstruction of the compressed seed remains of *Menyanthes* in Japan. J. Inst. Polytech. Osaka City Univ. Ser. D 11:79–89.

11004. —— 1962: Relations between morphometric values and geologic horizons of the fossil *Menyanthes* seeds in Japan represented by Szaferow's graphic method. Quaternary Res. 2:180–87. Biol. Abstr. 43:11318. 1963.

11005. Koketsu, R., and H. Kosaka. 1928: Bestimmung des "spezifischen Pulvergewichtes" verschiedener gehüllten Reiskörner mittels der "Pulvermethode" und ihre Bedeutung. Proc. Crop Sci. Soc. Japan 2:7–12. Biol. Abstr. 6:9266. 1932.

11006. Kokin, A., and V. Voskoboynikova. 1937: The effect of treating spring wheat seeds with naphthaline. Plant Protect. 1937(12):179–81.

11007. Kokin, A. Y., D. A. Rachkova, and K. I. Gulnina. 1950: Effect of trace elements on increasing the crops of grains. Mikroelementy Zhizni Rast i Zhivotn. Akad. Nauk SSSR Tr. Konf. po Mikroelementam 1950:150–66. Chem. Abstr. 49:439. 1955.

11008. Kokina, S. I. 1938: The effect of various salts and their concentrations on the germination of the seeds and the development of the sprouts of saksaul (*Haloxylon*). J. Botan. URSS 23:302–3.

11009. Koklyaeva, P. V. 1961: Effect of a pre-sowing treatment of seeds with trace elements on their sowing properties. Tr. Vses. Sel'skokhoz. Inst. Zaochn. Obrazov. Chem. Abstr. 58:3847. 1963.

Kolachov, P. 107.

Kolankiewicz, J. 18245.

11010. Kolbach, P., and E. Schild. 1939: Influence of temperature and duration of germination on protein modification during malting. Wochschr. Brau. 56:349–52. Soc. Chem. Ind. J. 59:481B. 1940.

Koldakov, B. V. Filin-. 308.

11011. Koleff, N. 1943: Untersuchungen über die Keimruhe bei Salatsamen (*Lactuca sativa* var. *capitata* L.). I. Gartenbauwissenschaft 17:263–72.

11012. Kolehmainen, V. A. 1957: Broadcast sowing on snow crust. Metsätaloudellinen Aikakauslehti 1957(4):116–19.

11013. —— 1958: Planting forest seeds. Skogen (Stockholm) 45:537–38, 541.

11014. Kolesnik, A. A., O. K. Ogneva, and G. B. Malyshkina. 1961: The volatile substances of seeds. Sb. Nauchn. Rabot Mosk. Inst. Nar. Khos. 1961(17):24–34. Chem. Abstr. 56:6383. 1962.

11015. Kolesnik, I. L. 1953: Sun, heat, and thermal drying of winter wheat seed. Agrobiologiya 1953(6):110–16. Chem. Abstr. 48:12918. 1954.

11016. Kolesnikov, P. A., and A. A. Mutuskin. 1963: Distribution of iron during seed germination and seedling growth in wheat. Biochemistry (USSR) (English Transl.) 28:216–23. Chem. Abstr. 59: 1965. 1963.

Kolesnikov, P. A. 14010, 15268, 15739.

11017. Kolesov, V. M., and M. S. Reznichenko. 1956: The amino acid content of hordein from barley and of avenin from oats. Biochemistry (USSR) (English Transl.) 21:665–68.

Kolev, V. M. 11620.

Kolgushkina, T. 140.

11018. Kolk, H. 1947: Studies in germination biology in weeds. Växtodling 2:108–67.

11019. —— 1951: Diseases of flax seed of the harvest of 1949 caused by fungi and the influence of chemical treatment. Statens Cent. Frökontanst. Medd. 26:52–63.

11020. —— 1958: Laboratory determination of fungus diseases in seeds. Some experiments with different seed kinds. Kgl. Skogs- Lantbruksakad. Tidskr. 97: 298–312.

11021. —— 1962: Viability and dormancy of dry stored weed seeds. Studies in several species occurring frequently in Swedish grassland seed lots. Plant Husbandry. Publ. from Inst. Plant Husbandry Roy. Agr. Coll. Sweden 18. 192 pp. Almqvist & Wiksells, Uppsala.

Kolk, H. 6044, 10138, 19646.

11022. Kolkwitz, R. 1901: Über die Athmung ruhenden Samen. Ber. Deut. Botan. Ges. 19:285–87.

11023. Kolle, F. 1926: Beiträge zur Kenntnis der sogenannten Kleber- oder Aleuronschicht bei den Samen der Gramineen und anderen Pflanzenfamilien. Nytt. Mag. Naturvidensk. 64:116–27.

11024. Koller, D. 1955: Germination regulating mechanism in some desert seeds. Bull. Res. Council Israel Sect. D 4:379–87. Biol. Abstr. 30:20515. 1956.

11025. —— 1955: The regulation of germination in seeds. (Review.) Bull. Res. Council Israel Sect. D 5:85–108.

11026. —— 1956: Germination regulating mechanisms in some desert seeds. III. *Calligonum comosum* L'Her. Ecology 37:430–33.

11027. —— 1957: Germination regulating mechanisms in some desert seeds. IV. *Atriplex dimorphostegia* Kar. et Kir. Ecology 38:1–13.

11028. —— 1962: Ecological problems of seed dormancy. *In* N. Grossowicz, editor. Cryptobiotic stages in biological systems. Elsevier, Amsterdam. 232 pp.

11029. —— 1962: Preconditioning of germination in lettuce at time of fruit ripening. Am. J. Botany 49:841–44.

11030. —— and D. Cohen. 1959: Germination regulating mechanisms in some desert seeds. VI. *Convolvulus lanatus* Vahl., *Convolvulus negevensis* Zoh. and *Convolvulus secundus* Desr. Bull. Res. Council Israel Sect. D 7:175–80.

11031. ——, A. M. Mayer, A. Poljakoff-Mayber, and S. Klein. 1962: Seed germination. Ann. Rev. Plant Physiol. 13:437–64.

11032. —— and M. Negbi. 1955: Germination regulating mechanisms in some desert seeds. V. *Colutea istria* Mill. Bull. Res. Council Israel Sect. D 5:73–84.

11033. —— and —— 1957: Hastening the germination of *Panicum antidotale* Retz. Bull. Res. Council Israel Sect. D 5:225–38.

11034. —— and —— 1959: The regulation of germination in *Oryzopsis miliacea.* Ecology 40:20–36.

11035. ——, A. Poljakoff-Mayber, A. Berg, and T. Diskin. 1963: Germination regulating mechanisms in *Citrullus colocynthis.* Am. J. Botany 50:597–603.

11036. —— and N. Roth. 1963: Germination regulating mechanisms in some desert seeds. VII. *Panicum turgidum* (Gramineae). Israel J. Botany 12:64–73.

11037. ——, M. Sachs, and M. Negbi. 1964: Spectral sensitivity of seed germination in *Artemisia monosperma.* Plant Cell Physiol. 5:79–84.

11038. ——, ——, and —— 1964: Germination regulating mechanisms in some desert seeds. VIII. *Artemisia monosperma.* Plant Cell Physiol. 5:85–100.

11039. ——, N. H. Tadmor, and D. Hillel. 1958: Experiments in the propagation of *Atriplex halimus* L. for desert pasture and soil conservation. Ktavim 9:83–106.

Koller, D. 14289, 18252.

11040. Koller, G., and H. Czerny. 1936: Limonin, the bitter principle of orange seeds. Monatsh. Chem. 67:248–68. Chem. Abstr. 30:3831. 1936.

11041. Kolobkova, E. V. 1954: Proteolytic enzymes of the hard germinating seeds of almonds (*Prunus amygalus*) and hawthorn (*Crataegus* sp.). Byul. Gl. Botan. Sada 19:78–85.

11042. —— 1958: The nitrogen metabolism of maturing maize seeds. Dokl. Akad.

Nauk SSSR 120:154–57. Biol. Abstr. 33:46943. 1959.

11043. —— 1960: Transformation of nitrogenous substances during the maturation of seeds of a wheat-wheatgrass hybrid and of corn. Tr. Gl. Botan. Sada Akad. Nauk SSSR 7:67–92.

11044. —— 1961: Dynamics of nitrogenous substances during ripening of leguminous seeds. Tr. Gl. Botan. Sada Akad. Nauk SSSR 8:75–96. Chem. Abstr. 56:14642. 1962.

11045. —— and N. A. Kudryashova. 1956: An inhibitor of sprouting from the seeds of the yellow acacia. Plant Physiol. (USSR) (English Transl.) 3:115–20. Biol. Abstr. 32:35377. 1958.

11046. —— and —— 1960: Seed-germination inhibitors. Tr. Gl. Botan. Sada Akad. Nauk SSSR 7:8–31.

Kolobkova, E. V. 11469.

11047. Kolobov, A. N., and N. A. Krashennikov. 1954: Postharvest ripening of hemp seeds. Zemledelie 1954(8):97–102.

Kologrivova, A. Y. 2560.

11048. Koloiarova, L. 1955: Again on laboratory methods of determining the germination of cotton seeds. Khlopkovodstvo 5:32–35.

Kolomeitseva, M. F. 11340.

Kolomiets, O. K. 14833.

11049. Kolosova, A. Y., and Z. M. Potyagilo. 1957: Chemical composition of sunflower-seed hulls of different varieties. Gidrolizn. i Lesokhim. Prom. 10(2):12–13. Chem. Abstr. 51:10674. 1957.

Kolotova, S. S. 8618, 8619.

11050. Kolousek, J., and C. B. Coulson. 1955: Plant proteins. II. Amino acid content of the seed protein of *Trigonella foenum graecum* L. and of the seed protein and thallus protein of *Galega officianalis* L. J. Sci. Food Agr. 6:203–6.

Kolousek, J. 9459.

Koloushek. 14331.

11051. Kolpikov, O. M. 1960: Regarding the yields and the quality of seeds of the larch (*Larix*) sukaczevii in the Lindulovo Grove (Karelian Isthmus). Botan. Zh. 45:730–32.

11052. Kolte, S. G. 1954: Harvesting and preparation of ground nuts marketing in Madhya Pradesh. Indian Farming 3(12): 31–32.

Koltun, S. P. 5096.

11053. Kolz, J., and V. Turkova. 1963: Legumin, vicilin, and proteins similar to them in the seeds of some species of the Vicieae tribe (a comparative serological study). Biol. Plant. Acad. Sci. Bohemoslov. 5:29–40. Biol. Abstr. 43:15840. 1963.

11054. Kolzlowski, A. 1937: A respiratory chromogen in the seeds of pea (*Pisum sativum* L.). Acta Soc. Botan. Polon. 14:49–55.

11055. Komarek, V. 1930: Zur experimentellen Beeinflussung der Korrelationstätigkeit von epigäischen Keimblättern. Flora (Jena) 24:301–14. Biol. Abstr. 6:6598 1932.

Komarov, I. A. 16497.

Komarova, L. S. 9559.

11056. Komatsu, S., and T. Okada. 1933: Chemical studies on the vitality of soy beans. J. Chem. Soc. Japan 54:570–81. Chem. Abstr. 27:5778. 1933.

Komatsu, S. 18366.

Komel, A. S. 9332, 16653.

Komen, J. G. 5710.

11057. Kommedahl, T., and J. J. Christensen. 1950: Relation of mechanical injury and seed treatment to germination of flax seed of the 1948 crop. [Abstract.] Phytopathology 40:15.

11058. ——, ——, J. O. Culbertson, and M. B. Moore. 1955: The prevalence and importance of damaged seed in flax. Minn. Agr. Expt. Sta. Tech. Bull. 215. 40 pp.

11059. ——, ——, and E. Schwinghamer. 1950: Prevalence of *Colletotrichum linicolum* of flax seed and seedling infection of flax varieties. [Abstract.] Phytopathology 40:15–16.

11060. ——, J. E. De Vay, and C. M. Christensen. 1958: Factors affecting dormancy and seedling development in wild oats. Weeds 6:12–18.

11061. ——, J. B. Kotheimer, and J. V. Bernardini. 1959: The effects of quackgrass on germination and seedling development of certain crop plants. Weeds 7:1–12.

Kommedahl, T. 4727, 11317, 16380.

11062. Kommerell, E. 1927: Quantitative Versuche über den Einfluss des Lichtes verschiedener Wellenlängen auf die Keimung von Samen. Jahrb. Wiss. Botan. 66:461–512.

11063. Komori, S., and S. Ueno. 1938: Composition of "Akebi" seed oil. Bull. Chem. Soc. Japan 13:505–7. Pharm. Abstr. 5:252. 1939.

11064. Komoti, S. 1956: Studies on temperature treatments of seeds. I. Effects of temperature treatments on germination of garden asparagus seeds. Hokkaido Natl. Agr. Expt. Sta. Res. Bull. 70:42–49.

11065. —— 1957: Studies on temperature treatments of seeds. II. Dormancy and germinating temperature in garden asparagus seeds. Hokkaido Natl. Agr. Expt. Sta. Res. Bull. 73:9–19.

11066. Komuro, H. 1924: Studies in the effect of Roentgen rays upon germination of *Oryza sativa.* Botan. Mag. (Tokyo) 38:1–20.

11067. Komuro, T., and I. Hayakawa. 1944: Osmotic pressure of the soybean-protein solution. J. Chem. Soc. Japan 65:66–68. Biol. Abstr. 22:2188. 1948.

Konar, R. N. 9916.

11068. Konarev, V. G. 1954: Der Einfluss der Vernalisation auf das Verhalten der Nucleoproteide und Nucleinsäuren in den Embryonen von Gräsern. Biochemistry (USSR) (English Transl.) 19:131–36. Ber. Wiss. Biol. 93:82. 1955.

11069. Kondak, M., and N. Vecherskii. 1962: Experience in pelleting seeds. Kartofel i Ovoshchi 3:29–31. Biol. Abstr. 42:15446. 1963.

11070. Kondo, F., and T. Hatano. 1955: Studies on the protein hydrolysis for the lycine estimation in feed. Tohoku J. Agr. Res. 5:243–49.

11071. Kondo, H., and K. Hayashi. 1956: Studien über die Aktivitäten der Enzyme in der ruhenden und keimenden Samen. Gifu Univ. Fac. Agr. Res. Bull. 1956(7): 109–14.

11072. —— and M. Sumi. 1956: Changes in the distribution of flavine in seeds during maturation. Kagaku (Tokyo) 26:254–55. Chem. Abstr. 51:2128. 1957.

11073. Kondo, K., and S. Mori. 1949: Enzymes in organisms. II. Shift of iron, copper, zinc, and manganese in soybeans during germination. Rept. Inst. Chem. Res. Kyoto Univ. 17:80–81. Chem. Abstr. 45:9628. 1951.

11074. ——, ——, and M. Kajima. 1953: Proteins. LVI. The components of soybean protein. Bull. Res. Inst. Food Sci. Kyoto Univ. No. 11:1–23. Chem. Abstr. 48:12829. 1954.

11075. ——, ——, and —— 1954: Proteins. LXI. The components of soybean protein. 2. Formation of the component protein of soybean during the ripening period. Bull. Res. Inst. Food Sci. Kyoto Univ. No. 15: 37–48. Chem. Abstr. 48:13751. 1954.

11076. —— and Y. Morita. 1954: Proteins. LVIII. Nucleic acid in rice glutelin. Bull. Res. Inst. Food Sci. Kyoto Univ. No. 15: 1–17. Chem. Abstr. 48:13750. 1954.

11077. —— and K. Sasaoka. 1954: Proteins. LIX. The molecular sizes and shapes of rice glutelins. Bull. Res. Inst. Food Sci. Kyoto Univ. No. 15:18–24. Chem. Abstr. 48:13750. 1954.

11078. —— and —— 1954: Proteins. LXII. The components of soybean protein. 3. Germination of soybean and variation of component protein. Bull. Res. Inst. Food Sci. Kyoto Univ. No. 15:49–55. Chem. Abstr. 48:13751–52. 1954.

11079. Kondo, M. 1917: Untersuchung der Samen der in Japan vertretenen Brassica-Arten. Ein Beitrag zur genauen Feststellung der Sortenunterschiede. Ber. Ohara Inst. Landwirtsch. Forsch. 1:123–50.

11080. —— 1918: Über die in der Landwirtschaft Japans gebrauchten Samen. Ber. Ohara Inst. Landwirtsch. Forsch. 1:261–324. Botan. Abstr. 5:37. 1920.

11081. —— 1919: Über Nachreife und Keimung verschieden reifer Reiskörner. Ber. Ohara Inst. Landwirtsch. Forsch. 1:361–87.

11082. —— 1926: Über die Dauer der Erhaltung der Keimkraft vei verschiedenen Samenarten in Japan. Ber. Ohara Inst. Landwirtsch. Forsch. 3:127–33.

11083. —— 1926: Über die Einwirkung des Kalks auf die Erhaltung der Keimkraft von Sämereien. Ber. Ohara Inst. Landwirtsch. Forsch. 3:135–46.

11084. —— 1926: Über die Erhaltung der Keimkraft von Sämereien und über Trocknungsmittel. Ber. Ohara Inst. Landwirtsch. Forsch. 3:147–51.

11085. —— 1926: The storage of rice and change of its physical properties during this period. Ber. Ohara Inst. Landwirtsch. Forsch. 3:153–75.

11086. —— 1928: Ein Beitrag zur genauen Feststellung der Sortenunterschiede der in der Landwirtschaft Japans gebrauchten Samen. Proc. Intern. Seed Testing Assoc. 6:41–50.

11087. —— 1929: Über die harten Samen von *Astragalus sinicus* L. und *Robinia pseudacacia* L. Ber. Ohara Inst. Landwirtsch. Forsch. 4:289–93.

11088. —— 1936: Berichte über die Tätigkeit des Ausschusses für die Samen der warmen Klimate. I. Vergleichende Untersuchung der Samen von *Nicotiana tabacum*, 1934. Ber. Ohara Inst. Landwirtsch. Forsch. 7:129–35.

11089. —— 1936: Berichte über die Tätigkeit des Ausschusses für die Samen der warmen

Klimate. II. Vergleichende Untersuchungen der Samen von *Gossypium peruvianum*, 1934. Ber. Ohara Inst. Landwirtsch. Forsch. 7:137–45.

11090. —— 1936: Berichte über die Tätigkeit des Ausschusses für die Samen der warmen Klimate. III. Vergleichende Untersuchungen der Samen von *Oryza sativa*, 1935. Ber. Ohara Inst. Landwirtsch. Forsch. 7:147–52.

11091. —— 1936: Untersuchungen über harte Samen von *Astragalus sinicus*, die in 17 bzw. 18 Jahre lang unter Wasser nicht gequollen hatten, besonders über die Vererbungsverhältnisse dieser Hartschaligkeit. Ber. Ohara Inst. Landwirtsch. Forsch. 7:321–28.

11092. —— 1936/39: Storage of inadequately dried rice at low temperatures for five and a half years. Rept. Ohara Agr. Inst. 28:1–47. 1936; 31:232–37. 1939.

11093. —— 1941: Untersuchung und Bewertung von Gerste, Weizen, Sojabohnen und Rapssamen. II. Zusatz: Reis und Mais. Ber. Ohara Inst. Landwirtsch. Forsch. 8:349–70. 1940.

11094. —— and S. Isshiki. 1935: The presence of abnormal rice kernels which are either germless or which possess two embryos. Ber. Ohara Inst. Landwirtsch. Forsch. 6:515–24.

11095. —— and —— 1936: Storage of rice. XIV. Removal of moisture from the air in a granary and the hulled rice stored therein by a desiccating material. Ber. Ohara Inst. Landwirtsch. Forsch. 7:227–37.

11096. ——, ——, and Y. Terasaka. 1935: Untersuchungen über die Ährenkeimung des Weizens, speziell in den klimatischen Verhältnissen Japans. I. Ber. Ohara Inst. Landwirtsch. Forsch. 7:449–55.

11097. ——, ——, and —— 1941: Storage of hulled rice with table-salt. Rept. Ohara Agr. Inst. 32:78–94.

11098. —— and Y. Kasahara. 1940: Some examples of the years' storage of seeds. Proc. Crop Sci. Soc. Japan 12:21–24.

11099. —— and —— 1940: on the discrimination of rice varieties by means of coloring reaction of hulled kernels to phenol-fuchsin. Proc. Crop Sci. Soc. Japan 12:122–28.

11100. —— and —— 1941: Feststellung der Sortenechtheit des Saatgutes des Weizens durch Phenolfärbung. II. Ber. Ohara Inst. Landwirtsch. Forsch. 8:305–15.

11101. —— and —— 1941: Feststellung der Sortenechtheit des Saatgutes der Gerste durch Phenolfärbung. Ber. Ohara Inst. Landwirtsch. Forsch. 8:317–24.

11102. —— and —— 1941: Ursache der Phenolfärbung von Weizen- und Gerstenkörnern Anhang: Behandlung mit Paracresol, Benzidin usw. Ber. Ohara Inst. Landwirtsch. Forsch. 8:325–47.

11103. —— and —— 1941: Untersuchungen über Unkrautsamen Japans. V. Gramineae (3). Ber. Ohara Inst. Landwirtsch. Forsch. 8:371–87.

11104. —— and —— 1941: Untersuchungen über Unkrautsamen Japans. VI. Compositae (1). Ber. Ohara Inst. Landwirtsch. Forsch. 8:390–408.

11105. —— and —— 1941: Untersuchungen über Unkrautsamen Japans. VII. Polygonaceae (1). Ber. Ohara Inst. Landwirtsch. Forsch. 8:409–20.

11106. —— and —— 1941: A report on storage of seeds for ten years with calcium chloride. Rept. Ohara Agr. Inst. 32:304–10. Biol. Abstr. 24:10581. 1950.

11107. —— and —— 1941: Studies on germination of weed seeds. I. Influence of light and change of temperature on germination. II. Relation between germination, after-ripening, and fluctuating temperature. Rept. Ohara Agr. Inst. 32:357–408. Biol. Abstr. 24:21591. 1950.

11108. —— and —— 1942: Untersuchungen über Unkrautsamen Japans. VIII. Leguminosae. Ber. Ohara Inst. Landwirtsch. Forsch. 9:1–13.

11109. —— and —— 1942: Untersuchungen über die Keimung der Unkrautsamens Japans. I. Ber. Ohara Inst. Landwirtsch. Forsch. 9:14–21.

11110. —— and —— 1943: Studies on spike-germination (viviparous germination) of wheat. IV. Relation between spike germination and its inhibiting substance. Rept. Ohara Agr. Inst. 35:131–46. Biol. Abstr. 23:30207. 1949.

11111. —— and —— 1944: Storage of seeds of forest trees. II. Seeds of the Japanese cedar, *Cryptomeria japonica*, and the Japanese cypress, *Chamaecyparis obtusa*. III. Further studies on seed of Cupuliferae. Rept. Ohara Agr. Inst. 36:497–536. Biol. Abstr. 24:4974. 1950.

11112. ——, ——, and S. Akita. 1947: Qualities of unhulled rice stored for fifteen years under hermetically sealed condition. Rept. Ohara Agr. Inst. 37:1–2. Chem. Abstr. 44:5486. 1950.

11113. ——, ——, and —— 1947: Low-temperature storage of unhulled rice for eleven years. Rept. Ohara Agr. Inst. 37:38–40. Chem. Abstr. 44:5486. 1950.

11114. ——, ——, and —— 1950: Germination of hemp seeds stored for nineteen years and their growth. Rept. Ohara Agr. Inst 39:37–39.

11115. ——, ——, and Y. Terasaka. 1935: Untersuchungen über Unkrautsamen Japans. I. Gramineen (1). Ber. Ohara Inst. Landwirtsch. Forsch. 6:525–38.

11116. ——, ——, and —— 1938: Untersuchungen über Unkrautsamen Japans. III. Cyperaceae (1). Ber. Ohara Inst. Landwirtsch. Forsch. 8:163–80.

11117. ——, ——, and —— 1938: Untersuchungen über Unkrautsamen Japans. IV. Miscellaneen (1). Ber. Ohara Inst. Landwirtsch. Forsch. 8:231–48.

11118. ——, S. Matsushima, and T. Okamura. 1929: Keimkraft, Nährstoffe und B-Vitamin der im kohlensäuren- und luftdichten Verschlusse vier Jahre lang aufbewahrten Reiskörner. Proc. Imp. Acad. Japan 5:159–60.

11119. —— and T. Mizuta. 1944: An experiment on storage of hulled rice in paper bags. Rept. Ohara Agr. Inst. 36:537–46.

11120. —— and T. Okamura. 1926: Die Beziehung zwischen dem Geschlecht der Pflanzen und der Grösse, der Form, dem Gewicht und dem spezifischen Gewicht der Samen von *Cannabis sativa* L. J. Sci. Agr. Soc. Japan 283:233–57.

11121. —— and —— 1927: Einwirkung des Kohlensäure- und luftdichten Verschlusses auf die Reiskörner während der Aufbewahrungszeit. Proc. Imp. Acad. Japan 3:706–9.

11122. —— and —— 1927: Untersuchungen der verschiedenen Reiskörner geringerer Qualität. I. Die braun gefärbten enthülsten Reiskörner "Tschamai." Ber. Ohara Inst. Landwirtsch. Forsch. 3:405–19.

11123. —— and —— 1929: On the effect of air-tight and carbon dioxide [storage] upon the storage of rice. Ber. Ohara Inst. Landwirtsch. Forsch. 4:1–18.

11124. —— and —— 1929: Untersuchungen der verschiedenen Reiskörner geringerer Qualität. II. Entstehung der braun gefärbten enthülsten Reiskörner "Tschamai." Ber. Ohara Inst. Landwirtsch. Forsch. 4:173–80.

11125. —— and —— 1930: On the influences of various temperatures in duration of storage and various moistures of rice upon the preservation of germination power of hulled rice. Ber. Ohara Inst. Landwirtsch. Forsch. 4:315–41.

11126. —— and —— 1930: Germination power, analyses and vitamin B of hulled rice stored during four years either air-tight or in carbon dioxide. Ber. Ohara Inst. Landwirtsch. Forsch. 4:343–48.

11127. —— and —— 1930: Untersuchungen der verschiedenen Reiskörner geringerer Qualität. III. Über die grün gefärbten enthülsten Reiskörner "Aomai." Ber. Ohara Inst. Landwirtsch. Forsch. 4:413–27.

11128. —— and —— 1930: Untersuchungen der verschiedenen Reiskörner geringerer Qualität. IV. Der durch die Feuchtigkeitszunahme verursachte Querriss (Doware) des Reiskörnes. Ber. Ohara Inst. Landwirtsch. Forsch. 4:429–46.

11129. —— and —— 1930: Vergleichende Untersuchungen der physikalischen Eigenschaften des enthülsten (Genmai) und bespelzten Reiskörnes (Momimai). IV. Vergleich der quantitativen Eigenschaften des bespelzten, des enthülsten Reiskörnes und der Spelze. Ber. Ohara Inst. Landwirtsch. Forsch. 4:511–27.

11130. —— and —— 1931: Untersuchungen der verschiedenen Reiskörner geringerer Qualität. V. Über die Entfärbung der grün gefärbten enthülsten Reiskörner "Aomai." Ber. Ohara Inst. Landwirtsch. Forsch. 5:57–66.

11131. —— and —— 1931: Beziehung zwischen Wassertemperatur und Wachstum der Reispflanzen. Ber. Ohara Inst. Landwirtsch. Forsch. 5:67–85.

11132. —— and —— 1931: Storage of rice. V. On the influence of a desiccating material upon the preservation of the germinating power of hulled rice having different moisture contents and stored at different temperatures. Ber. Ohara Inst. Landwirtsch. Forsch. 5:221–42.

11133. —— and —— 1931: Versuche über die Aufbewahrung von enthülsten Reiskörner. Einfluss von Temperatur, Wassergehalt der Reiskörner und Anwendung eines Trochnungsmittels auf die Erhaltung der Keimkraft von enthülsten Reiskörnern. Proc. Imp. Acad. Japan 7:327–30. Biol. Abstr. 6:24706. 1932.

11134. —— and —— 1932: Beziehung zwischen der Wassertemperatur und dem Wachstum der Reispflanzen. III. Schädliche Einfluss der Wassertemperatur auf das Wachstum der Unter Wasser stehenden Reispflanzen. Ber. Ohara Inst. Landwirtsch. Forsch. 5:347–74.

11135. —— and —— 1932: The relation between the germinating power of hulled rice and vitamin B. Rept. Ohara Agr. Inst. 18:64–77. Japan. J. Med. Sci. 3:(97). 1937.

11136. —— and —— 1932/33: Storage of rice. VI. Physical and biochemical studies of hulled rice stored in straw bags. Ber. Ohara Inst. Landwirtsch. Forsch. 5:395–406. Chem. Abstr. 27:4284. 1933.

11137. —— and —— 1932/33: Storage of rice. VII. On the influence of varying moisture content and germinating power upon the preservation of vitamin B in hulled rice. Ber. Ohara Inst. Landwirtsch. Forsch. 5:407–12.

11138. —— and —— 1932/33: Storage of rice. VIII. Studies on hulled rice stored air-tight 26 and 28 years. Ber. Ohara Inst. Landwirtsch. Forsch. 5:413–20. Chem. Abstr. 27:4284. 1933.

11139. —— and —— 1934: Storage of hulled rice grains in simple air-tight containers and in carbon dioxide, with reference to the dependence of change in quality on the water content. Proc. Imp. Acad. Japan 10:528–31. Chem. Abstr. 29:1521. 1935.

11140. —— and —— 1934: Physical and biochemical studies on the qualities of hulled rice grains stored for some years in straw sacks. Proc. Imp. Acad. Japan 10:532–35. Chem. Abstr. 29:1521. 1935.

11141. —— and —— 1934: Storage of rice. IX. Relation between varying moisture content and change in quality of hulled rice stored in containers air-tight as well as with carbon-dioxide. Ber. Ohara Inst. Landwirtsch. Forsch. 6:149–74.

11142. —— and —— 1934: Storage of rice. X. Studies on four lots of unhulled rice stored 46 to 84 years in granaries. Ber. Ohara Inst. Landwirtsch. Forsch. 6:175–85.

11143. —— and —— 1934: Storage of rice. XI. Studies on Hoshii stored hermetically sealed for 23 years on a mountain. Ber. Ohara Inst. Landwirtsch. Forsch. 6: 331–34.

11144. —— and —— 1934: Storage of rice. XII. Storage of rice in tin containers with calcium chloride, with special reference to the underdried product . 1. Ber Ohara Inst. Landwirtsch. Forsch. 6:335–39.

11145. —— and —— 1935: Storage of rice. XIII. Storage of rice in tin containers with calcium chloride, with special reference to the underdried product. 2. Ber. Ohara Inst. Landwirtsch. Forsch. 7:99–102.

11146. —— and —— 1937: Storage of rice. XVI. Storage of rice in concrete silos for five years. Ber. Ohara Inst. Landwirtsch. Forsch. 7:471–81.

11147. —— and —— 1937: Storage of rice. XVII. Comparative study of unhulled rice and hulled rice in regard to the changes of its qualities during long storage in straw bags. Ber. Ohara Inst. Landwirtsch. Forsch. 7:483–90.

11148. —— and —— 1938: Storage of rice. XVIII. Relation between varying moisture content and change in quality of hulled rice, with special reference to the underdried product. Ber. Ohara Inst. Landwirtsch. Forsch. 8:11–34.

11149. —— and —— 1938: Storage of rice. XX. Studies on unhulled rice stored about 100 years in a granary. Ber. Ohara Inst. Landwirtsch. Forsch. 8:47–52.

11150. —— and R. Takahashi. 1938: Feststellung der Sortenechtheit des Saatgutes des Weizens durch Phenolfärbung. Ber. Ohara Inst. Landwirtsch. Forsch. 8:211–21.

11151. —— and —— 1938: Untersuchungen über die Ährenkeimung des Weizens, speziell in den klimatischen Verhältnissen Japans. III. Ber. Ohara Inst. Landwirtsch. Forsch. 8:223–30.

11152. ——, ——, and Y. Terasaka. 1936: Untersuchung und Bewertung von Gersten, Weizen Soiabohnen und Rapssamen. Ber. Ohara Inst. Landwirtsch. Forsch. 7: 335–58.

11153. ——, ——, and —— 1936: Berichte über die Tätigkeit des Ausschusses für die Samen der warmen Klimate. IV. Vergleichende Untersuchungen der Samen von *Nicotiana tabacum*, 1936. Ber. Ohara Inst. Landwirtsch. Forsch. 7:429–47.

11154. ——, ——, and —— 1937: Untersuchungen über die Ährenkeimung des Weizens, speziell in den klimatischen Verhältnissen Japans. II. Ber. Ohara Inst. Landwirtsch. Forsch. 7:457–69.

11155. ——, ——, and —— 1938: Berichte über die Tatigkeit des Ausschusses für die Samen der warmen Klimate. V. Vergleichende Untersuchungen der Samen von *Chrysanthemum cinerariaefolium* Bocc. Ber. Ohara Inst. Landwirtsch. Forsch. 8:1–9.

11156. ——, ——, and —— 1938: Die Jarowisation des Weizens, speziell in Bezug auf den praktischen Wert des Verfahrens in den klimatischen Verhältnissen Südjapans. Ber. Ohara Inst. Landwirtsch. Forsch. 8:199–209.

11157. ——, ——, and —— 1941: On the storage of seeds of forest trees. I. Ann. Sci. 32:283–303. Japan. J. Botan. Abstr. 11:(141)–(142). 1941.

11158. ——, ——, ——, and S. Isshiki. 1938: Storage of rice. XIX. Removal of moisture from hulled rice by heated air. Ber. Ohara Inst. Landwirtsch. Forsch. 8:35–46.

11159. ——, ——, ——, and H. Kaihara. 1941/44: Storage of rice in concrete silos. III. Storage of insufficiently dried rice for over four years after desiccation by heating. IV. Storage for seven years. Rept. Ohara Agr. Inst. 32:95–113. 1941; 36:198–208. 1944.

11160. ——, ——, and Y. Terasawa. 1938: Experiments on the jarowization of wheat. Ann. Sci. 30:1–38. Japan. J. Botany 10:(11). 1939.

11161. —— and Y. Terasaka. 1936: Storage of rice. XV. Comparison of calcium oxide and calcium chloride as a desiccating material for rice stored in tin containers. Ber. Ohara Inst. Landwirtsch. Forsch. 7: 329–34.

Kondo, M. 8743.

11162. Kondo, S., and H. Ishiwa. 1962: Effect of surface dose change on small seeds exposed to ^{60}CO γ-ray. Seiken Ziho 14:61–62.

11163. Kondo, S., and H. Ishiwa. 1963: Calculation of absorbed dose delivered to wheat seeds soaked in ^{32}P and ^{131}I aqueous solutions. Seiken Ziho 14:78–80.

11164. ——, N. Yokota, and S. Kawashima. 1958: Diffusion of radioisotopes in soaked wheat seeds. Mishima Natl. Inst. Genet. Ann. Rept. 8:98–100. 1957.

Kondoguri, W. 3391.

Kondrashova, A. A. 15642.

11165. Kondratenko, F. 1940: Analysis of populations of winter-rye according to length of vernalization stage. Vestn. Sotsial. Rast. 1:27–34. Biol. Abstr. 16: 8019. 1942.

11166. Konigs, G. 1961: Planting *Parodia* seed. Succulenta 2:20–21.

11167. Konis, E. 1940: On germination inhibitors. II. On the action of germination inhibiting substances in the tomato fruit. Palestine J. Botany 2:6–27.

11168. —— 1947: On germination inhibitors. VI. The inhibiting action of leaf-saps on germination and growth. Palestine J. Botany 4:77–85.

Konis, E. 6115, 6116.

Konishi, T. 10695.

Koniuszy, F. 6522, 6523.

Konno, S. 207.

Kono, T. 14134, 14135.

Kono, Y. 14080.

11169. Kononkov, P. P. 1956: Characteristics of the light stage in sorrel as affected by the place of reproduction of the seed. Agrobiologiya 1956(3):128–29. Biol. Abstr. 35:11355. 1960.

Kononov, D. G. 9575.

11170. Konovalov, I. N. 1936: The effect of the yarovisation of plants upon the accumulation of organic substance. Compt. Rend. Acad. Sci. URSS 11:41–45.

11171. —— 1937: Essai de vernalisation d'embryons de blé, de graines dépouillées d'albumen. Compt. Rend. Acad. Sci. URSS 16:381–83. Bull. Soc. Botan. France 88:569. 1941.

11172. —— and I. E. Rogalev. 1937: The behavior of nitrogenous substances during yarovization of plants. Compt. Rend. Acad. Sci. URSS 16:65–68. Chem. Abstr. 31:8601. 1937.

11173. Konovalov, N. A. 1950: Experiment in utilizing vernalization methods on seeds of trees. Dokl. Akad. Nauk SSSR 71:561–63.

11174. —— 1961: Pre-planting treatment of seeds of *Larix sukaczewii.* Lesn. Khoz. 1961(4):162–64.

11175. Konovalov, Y. B. 1959: Effect of soil moisture deficiency on grain ripening in spring wheat. Plant Physiol. (USSR) (English Transl.) 6:189–95.

11176. —— 1963: Causes for different grain sizes in the spikes of barley and wheat. Dokl. Akad. Nauk SSSR 149:519–21.

Konowalowa, R. 14822.

11177. Konshakov, P. N. 1953: Effect of cobalt on the feeding properties of growing grain. Sov. Zootekh. 1953(4):91. Chem. Abstr. 48:11555. 1954.

11178. Konsler, T. R., and W. S. Barham. 1958: The inheritance of seed size in watermelon. Proc. Am. Soc. Hort. Sci. 71:480–84.

11179. Konstantinov, N. 1936: La yarovization du Cotonnier. Compt. Rend. Acad. Sci. URSS 11:299–301. Bull. Soc. Botan. France 85:471. 1938.

11180. Konstantinov, N. N., and T. I. Volkova. 1960: Difficult germination of seeds in the genus *Gossypium* and the biological significance of the hairs and the hulls. Tr. Gl. Botan. Sada Akad. Nauk SSSR 7:127–51.

11181. Konstantinova, T. 1960: Stratification of seeds of stone fruit crops and pip fruit crops in ice. Sel'skoe Khoz. Sibiri 1960(11):61–63.

Konstantinova, T. N. 11332.

11182. Konstantinovi, P. N. 1939: The influence of the place of reproduction of seeds on the yield and the principles of supplying seed-testing stations with seeds. Selek. i Semen. 9(5):18–19. Biol. Abstr. 15:5624. 1941.

11183. Konsuloff, S. 1927: Biochemische Untersuchungen an Reissamen. Annuaire Univ. Sofia 23:95–132. Botan. Centralbl. 156:143–44. 1929.

11184. —— 1929: Samenstimulation durch Druck und Vakuum nebst Bemerkungen über die Stimulationserklarungsversuche. Biol. Generalis (Wien) 5:605–26.

11185. Kontic, V. 1962: Some observations on the germination date of *Pinus heldreichi leucodermis* seeds. Nar. Sumar 16:335–46.

11186. Konzak, C. F. 1955: Radiation sensitivity of dormant and germinating barley seeds. Science 122:197.

11187. —— 1957: The influence of oxygen on the mutagenic effects of X-rays on maize endosperm loci. Radiation Res. 6:1–10.

11188. ——, H. J. Curtis, N. Delihas, and R. A. Nilan. 1960: Modification of radiation induced damage in barley seeds by thermal energy. Can. J. Genet. Cytol. 2:129–41. Nucl. Sci. Abstr. 15:320. 1961.

11189. ——, L. F. Randolph, and N. F. Jensen. 1951: Embryo culture of barley species hybrids. J. Heredity 42:125–34.

Konzak, C. F. 3515, 4763, 10200, 10201, 11431, 16764.

11190. Kooiman, P. 1957: Amyloids of plant seeds. Nature 179:107–9.

11191. —— 1960: On the occurrence of amyloids in plant seeds. Acta Botan. Neerl. 9:208–19.

11192. Koopman, M. J. F. 1955: Control of agricultural seeds in the Netherlands. Euphytica 4:25–30.

11193. —— 1960: Report of the seed moisture and storage committee. Proc. Intern. Seed Testing Assoc. 25:618–21.

11194. —— 1962: Moisture and storage committee. Proc. Intern. Seed Testing Assoc. 27:87–91.

11195. —— 1962: Referee testing on association-wide basis. Proc. Intern. Seed Testing Assoc. 27:284–90.

11196. —— 1963: Result of a number of storage experiments conducted under controlled conditions; b. other than agricultural seeds. Proc. Intern. Seed Testing Assoc. 28:853–60.

11197. —— 1963: Results of an enquiry into harvest and storage conditions in different countries. Proc. Intern. Seed Testing Assoc. 28:995–98.

11198. —— and J. De Tempe. 1954: Comparison of laboratory evaluation and field performance of flax seed. Proc. Intern. Seed Testing Assoc. 19:24–31.

11199. Koos, K. 1948: Summary of a three-year study of abnormal bean seedlings. Assoc. Offic. Seed Analysts News Letter 22(4):26.

Koos, K. 15611.

Kooten, A. van. 571.

11200. Kootz, F. W. 1955: Zur Stratifizierung von Kiefernsaatgut. Z. Forst.-Jagdw. 5:389–95.

11201. Kooyman, W. H. 1961: Storing vegetable seed. J. Agr. W. Australia 2:622–23.

11202. Kopczynska, K. 1964: Embryo sac development in *Pinguicula vulgaris* L. Acta Soc. Botan. Polon. 33:141–56.

11203. Kopečky, O. 1928: Gravimetrische Volumbestimmung einiger Getreidesamen und ihre physikalischen Veränderungen während des ununterbrochenen Weichens in destilliertem Wasser. Vestn. Česk. Akad. Zeměděl. 4:799–801. Ber. Wiss. Biol. 10:827. 1929.

11204. —— and V. Almendinger. 1929: Der Stickstoff in den Samen des Maises. Vestn. Česk. Akad. Zeměděl. 5:549–51. Botan. Centralbl. 157:463. 1929/30.

11205. —— and —— 1929: Der Stickstoff in Erbsensamen. Vestn. Česk. Akad. Zeměděl. 5:551–56. Botan. Centralbl. 157:462. 1929/30.

11206. —— and —— 1930: Über den individuellen Stickstoffgehalt der Erbsensamen. Sb. Česk. Akad. Zeměděl. 5:67–87. Ber. Ges. Physiol. 55:744. 1930.

11207. —— and —— 1930: Über den individuellen Stickstoffgehalt der Maiskörner. Sb. Česk. Akad. Zeměděl. 5: 89–100. Ber. Ges. Physiol. 55:744–45. 1930.

11208. Kopeikovskii, V. M., G. I. Garbuzova and M. I. Ryazantseva. 1963: Effect of temperature on preservation of quality of dried stored seeds. Maslob.-Zhir. Prom. 29(1): 12–16.

11209. —— and V. K. Kostenko. 1963: Changes in the acid number of oil and losses of dry matter in high-oil sunflower seed during drying. Maslob.-Zhir. Prom. 29(9): 7–13. Chem. Abstr. 60: 5757. 1964.

11210. ——, Z. K. Zabelina, and V. G. Shcherbakov. 1956: Influence of moisture and some of the biochemical factors on the conditions of storage for flax seeds of high oil content. Maslob.-Zhir. Prom. 22(6): 9–12. Chem. Abstr. 51: 2128. 1957.

Kopeikovskii, V. M. 9353.

11211. Kopelkievskii, G. V. 1952: Timely harvesting and storing of honey plant seeds. Pchelovodstvo 29: 43–45.

11212. Koperzhinskii, V. V. 1939: The effect of ethylene on the crop of alfalfa seeds. Dokl. Vses. Akad. Sel'skokhoz. Nauk 1939(2/3): 57–59. Chem. Abstr. 34: 4215. 1940.

11213. —— 1953: Use of 2,4-D for raising the yield of seeds of alfalfa and esparette. Dokl. Akad. Nauk SSSR 88: 353–56. Chem. Abstr. 47: 6593. 1953.

Koposhian, H. 15614.

Korac, M. 11908.

Koranda, J. J. 10766.

11214. Kordyum, E. J. 1959: Comparative embryological study of Ranunculaceae. Tr. Soveshch. po Morfogenezu Rast. 1: 473–77.

11215. —— 1963: Peculiarities of the embryology of *Poa bulbosa* L. var. *vivipara* Koel. Acad. Sci. RSS Ukr. Kiev J. Botan. 20: 43–53.

11216. —— and A. P. Boiko. 1962: On the embryology of *Gerbera anandria* Schultz. (Dopovidi) Akad. Nauk Ukr. RSR 8: 1109–12.

11217. —— and V. A. Zayets. 1962: On the embryology of *Euphorbia peplis* L. Akad. Sci. RSS Ukr. Kiev J. Botan. 19: 42–48.

11218. Koreisha, Z. I., and T. G. Muminov. 1960: Changes in growth and development of peach induced by subjecting seeds to low temperatures. Plant Physiol. (USSR) (English Transl.) 7: 67–69.

11219. Korenev, G. V. 1953: Pre-planting preparation of freshly-harvested seeds of perennial grasses. Zemledelie 1953(2): 66–69.

Korenev, N. A. 169, 170.

Korff, R. W. von. 4212.

11220. Koriakina, V. F. 1928: Einwirkung der Schwefelsäure auf die Samenkeimfähigkeit der wildwachsenden Leguminosenarten. Ann. Essais Semences Leningrad 1928(2): 146–51. Jard. Botan. Inst. Essais Semences, Leningrad, Ann. Essais Semences 1928(2): 146–51.

11221. —— 1937: Über den Keimungsrhythmus der Wiesengräsersamen. Botan. Zh. 22: 527–28.

Koriakina, V. F. 2235.

Korkes, S. 270.

11222. Korkina, V. N. 1962: Stratification of seeds of Manchurian ash in trenches. Lesn. Khoz. 1962(2): 70–71.

Korlakowska, K. 12873.

Kormann, W. 16612.

Korn, P. P. 19860.

11223. Kornberg, H. L., and H. Beevers. 1957: The glyoxylate cycle as a stage in the conversion of fat to carbohydrate in castor beans. Biochim. Biophys. Acta 26: 531–37.

11224. Korneev, P. K. 1962: Effect of presowing stimulation on some physiological processes and yield of corn. Izv. Inst. po Biol. "Metodii Popov," Bulgar. Akad. Nauk 12: 245–52. Chem. Abstr. 57: 15550. 1962.

Kornev, I. S. 2261.

11225. Kornfeld, A. 1930: Untersuchung der Keimfähigkeit mit Hilfe von Farblösungen. Fortschr. Landwirtsch. 5: 682–83.

11226. Kornjakow, D. G. 1937: Über die Jarowisation der Zuckerrüben. Zuckerrübenbau 2: 21–32. Botan. Centralbl. 174: 63. 1938.

Korobova, S. N. 7126.

11227. Korotkevich, A. V., and L. M. Korotkevich. 1960: Tanning components of grape seeds. Sadovodstvo Vinogradarstvo i Vinodelie Moldavii 15: 53. Chem. Abstr. 54: 25083. 1960.

Korotkevich, L. M. 11227.

Korovin, F. N. 11375.

Korpaczy, I. 6284.

Korsan-Bengtsen, K. 14501.

11228. Korsmo, E. 1911: The vitality of weed seeds often passing through digestive tract of farm animals. Tidsskr. Norske Landbruk 18: 223–30. Expt. Sta. Rec. 26: 839. 1912.

11229. —— 1912: Über die Keimfähigkeit des Queckensamens und über die Quecke (*Triticum repens*). Nytt. Mag. Naturvidensk. 50: 238–48.

11230. —— 1912: Über die Fähigkeit der Samen, den Verdauungskanal der Haustiere zu passieren, ohne ihre Keimkraft zu verlieren. Nytt. Mag. Naturvidensk. 50: 251–58.

11231. —— 1927: Size of seed experiment with cereals. Meldinger Norg. Landbrukshögskole 7:299–374. Expt. Sta. Rec. 57:826. 1927.

11232. —— 1930: Unkräuter im Ackerbau der Neuzeit. Biologische und praktische Untersuchungen. Springer, Berlin. 580 pp.

11233. —— 1933: Experiments on sowing-depths and quantity of seed-grain in spring cereals. The influence of the sowing-depths and the quantity of seed-grain on number of plants and culms, grain- and straw-yield. Johansen & Nielsens, Oslo. 157 pp. Meldinger Norg. Landbrukshögskole 13:321–477. 1933. Biol. Abstr. 8:11679. 1934.

11234. —— 1935: Weed seeds. Gyldendal Norsk Forlag., Oslo. 175 pp. Expt. Sta. Rec. 74:485. 1936.

11235. —— 1954: Anatomy of weeds. Anatomical description of 95 weed species with 2050 original drawings. Grondahl & Sons, Oslo. 413 pp.

11236. —— 1954: Ugras. I. natidens Jordbruk. A-S. Norsk Landsbruk, Oslo. 635 pp.

11237. Korstian, C. F. 1927: Factors controlling germination and early survival in oaks. Yale Univ. School Forestry Bull. 19. 115 pp.

11238. —— 1930: Acorn storage in the southern states. J. Forestry 28:858–63.

11239. —— and A. L. MacKinney. 1931: Planting versus direct seeding of yellow poplar in the southern Appalachian region. J. Forestry 29:1178–81.

Korstian, C. F. 3667, 9771, 12630, 18726.

Kortenhorst, A. T. 14812.

Korth, J. A. 17047.

11240. Kortleve, I. A. 1930: Kieming van oliepalmzaden. Bergcultures 4:729–30. Biol. Abstr. 6:4876. 1932.

11241. Koruzhev, A. 1933: Media for seed beds and their relations to tobacco seedlings in Crimea. State Inst. Tobacco Invest. (USSR) 97:35–57. Chem. Abstr. 28:3825. 1934.

11242. Korven, N. A. 1961: Note on wild oat control by delayed seeding of wheat. Can. J. Plant Sci. 41:685–86.

11243. Korytnyk, W., and E. Metzler. 1962: Formation of raffinose and stachyose in lima beans (*Phaseolus lunatus*). Nature 195:616.

11244. Korzchinsky, S. 1886: Seeds of *Aldrovanda*. Botan. Centralbl. 27:302–4, 334, 335.

11245. Kosaka, H. 1925: Über den Einfluss der Frucht auf die Samenreife bei einigen Kulturpflanzen. Kyushu Imp. Univ. Dept. Agr. J. 1:197–216.

Kosaka, H. 11005.

11246. Kosar, W. F., and R. C. Thompson. 1957: Influence of storage humidity on dormancy and longevity of lettuce seeds. Proc. Am. Soc. Hort. Sci. 70:273–76.

Kosar, W. F. 18516.

11247. Koshelenko, V. M. 1953: Stratification of seeds of some fruit species. Sad i Ogorod 1953(9):12–14.

11248. Koshimizu, T. 1929: Germination power and buoyancy of the seeds in fresh and sea water. B. Z. 7:547–51. Japan. J. Botany 10:(49). 1939.

11249. —— 1930: Carpobiological studies of *Crinum asiaticum* L. var. *japonicum* Bak. Mem. Coll. Sci. Univ. Kyoto Ser. B 5:183–227.

11250. —— 1936: On the relation between the ripening stages of the Maize-seed and its germination. Botan. Mag. (Tokyo) 50:504–13.

11251. —— 1940: Various forms of the seed of *Ginkgo biloba* L. J. Japan. Botany 16:551–52.

11252. —— 1940: Viviparous fruit of *Ardisia japonica* Blume. J. Japan. Botany 16:553.

11253. —— 1951: Studies on the germination of seeds and the growth of seedlings of *Callistemon rigidus* R. Br. J. Japan. Botany 26:211–14. Biol. Abstr. 26:13218. 1952.

11254. —— 1951: Studies on the role of growth substances in the vivparity of the seed of *Podocarpus machrophyllus* D. Don. Nara Women's Univ. J. 1:1–15.

11255. Kosikova, P. G. 1960: The germination of seeds of several species of weeds and ruderal plants after treatment with solutions of gibberellic acid of various concentrations. Dokl.-Botan. Sci. Sect. (English Transl.) 130:45–47.

11256. Kosker, O. 1955: The protein and amino-acid content of the main types of Turkish wheat and their nutritional value. Ankara Univ. Fac. Agr. Publ. 72. 74 pp.

11257. Kosovac, Z. 1960: Die Wirkung einiger Herbiciden auf die Ruheperiode und Keimfähigkeit des Weizensamens. Savremena Poljoprivreda 8:605–11.

Kostanian, B. A. 13549.

11258. Kostecka-Madalska, O. 1958: Vom Einfluss des Merkurochroms auf das Keimen und die Entwicklung der Sämlinge der *Lactuca sativa* L. Soc. Botan. Polon. 27:359–72.

Kostecki, J. 4091.

Kostenko, V. K. 11209.

11259. Kostjucenko, I. 1938: Winterhardiness of plants as influenced by yarovization of ripening seed-grains. Compt. Rend. Acad. Sci. URSS 18:589–92. Biol. Abstr. 14:9223. 1940.

11260. —— and N. Beloserova. 1944: The germinating power of seeds and its relation to the productivity of plants. Proc. Lenin. Acad. Agr. Sci. USSR 1944(10):11–23.

11261. —— and T. J. Zarubailo. 1938: Variability of biological properties of plants depending on the growth conditions of the seeds. Acad. Sci. RSS Ukraine Inst. Botan. J. 26/27:81–96.

11262. Kostjucenko, I. A., and T. J. Zarubailo. 1937: Vernalization of seed during ripening and its significance in practice. Herbage Rev. 5:146–57.

11263. Kostoff, D. 1935: Mutations and the ageing of seeds. Nature (Suppl.) 135:107.

11264. ——. 1937: The size of *Nicotiana rustica* L. × *Nicotiana tabacum* L. hybrid embryos and hybrids in respect to their parents. Compt. Rend. Acad. Sci. URSS 17:427–29.

11265. —— 1938: The effect of centrifuging upon the germinated seeds from various plants. Cytologia 8:420–42.

11266. —— 1939: The frequency of polyembryony and chlorophyll deficiency in rye. Current Sci. (India) 8:356–58.

11267. —— 1940: A case of vivipary in rye. Current Sci. (India) 9:279–80. Plant Breeding Abstr. 11:5. 1941.

11268. Kostrikova, L. N. 1955: Embryological investigation of *Vicia faba* (L.). Byul. Mosk. Obshchestva Ispytatelei Prirody Otd. Biol. 60:99–106.

11269. Kostromina, A. S. 1951: Germination of acorns in relation to dates of collection. Les. i Step 3(5):37–42.

Kostromina, A. S. 17296.

11270. Kostytschew, S., and A. Scheloumoff. 1913: Über Alkoholbildung durch Weizenkeime. Ber. Deut. Botan. Ges. 31:422–31. Botan. Centralbl. 126:125. 1914.

Kostyuk, V. A. 10814.

Kosugi, K. 18164.

Kotani, S. 9577.

11271. Kotasthane, W. V., and N. Narayana. 1937: The proteins of groundnut (peanut), *Arachis hypogaea*, Linn. Proc. Indian Acad. Sci. Sect. B 6:376–86.

11272. —— and —— 1938: Seed oils of Indian and Australian sandalwood trees. Poona Agr. Coll. Mag. 29(4):Analyst. 63:619–20. 1938.

11273. Kothary, K., and K. Sohonie. 1960: Proteins of double bean, *Faba vulgaris* (*Vicia faba*). I. Isolation, fractionation, and amino acid composition of the proteins. J. Sci. Ind. Res. (India) 19C: 14–16. Chem. Abstr. 54:18692. 1960.

Kotheimer, J. B. 11061.

11274. Kotlyar, G. I. 1953: Polarographic studies of the proteolytic activity of grain. Biochemistry (USSR) (English Transl.) 18:423–29. Chem. Abstr. 48:5298. 1954.

11275. —— 1962: Polarographic investigation of modifications of albumins of seeds on drying. Teoriya i Prakt. Polyarogr. Analiza, Akad. Nauk Moldavsk. SSR, Materialy Pervogo Vses. Soveshch. 1962:403–5. Chem. Abstr. 59:5397. 1963.

Kotlyarov, M. V. 6144.

11276. Kotowski, F. 1926: Effect of size of seed on plant production. Acta Soc. Botan. Polon. 3:253–76. Expt. Sta. Rec. 55:835. 1926.

11277. —— 1926: Chemical stimulants and germination of seeds. Proc. Am. Soc. Hort. Sci. 23:173–76.

11278. —— 1926: Temperature relations to germination of vegetable seed. Proc. Am. Soc. Hort. Sci. 23:176–84.

11279. —— 1927: Semipermeability of seed coverings and stimulation of seeds. Plant Physiol. 2:177–86.

11280. Kott, S. 1944: Certain peculiarities of weed development in Siberia. Sov. Botan. (Leningrad) 1944(4/5):81–84. Biol. Abstr. 19:18065. 1945.

11281. —— 1945: Peculiarities of weed seed germination and some questions of agricultural practices. Proc. Lenin. Acad. Agr. Sci. USSR 1945(6):40–45.

11282. Koulechoff, N. 1928: Quelques considerations sur la question de la determination de la provenance du Tourkestan des semences de luzerne (*Medicago sativa* L.). Proc. Intern. Seed Testing Assoc. 3:5–17.

Koulechoff, N. N. 17853.

11283. Kovacevic, J. 1958: Eine neue Art der Herkunftsbestimmung der Kleesaaten. Vorschlag zur Methodik. Proc. Intern. Seed Testing Assoc. 23:73–77.

11284. —— 1958: Yugoslav bibliography of weeds. Hemizacija Polj. 23:41–60. Proc. Intern. Seed Testing Assoc. 25:868. 1960.

11285. Kovachevsky, I. C. 1946: Vegetable-seed treatment with organic mercurials and copper carbonate. Bull. Chambre Culture Nat. (Sofia) Ser. Biol. Agr. Silvicult. 1: 87–102. Chem. Abstr. 41:7039. 1947.

Kovalenko, E. I. 6488.

11286. Kovalenko, Y. I. 1957: The effect of several alkaloids on the germination of grape seeds and the subsequent seedling growth. Tr. Stavropol'sk. Sel'skokhoz. Inst. 1956(7):173–80. Biol. Abstr. 35: 24905. 1960.

11287. Kovalev, N. V. 1957: Propagation of peach with seeds. Agrobiologiya 1957(2): 127–29. Biol. Abstr. 35:35994. 1960.

11288. Kovalevskii, D. V. 1947: Germination of apple seeds without stratification. Sad i Ogorod 1947(10):56–59.

11289. Koverga, A. 1944: Iarovisation of turnip for seed production. Proc. Lenin. Acad. Agr. Sci. USSR 1944(11/12):42–47.

11290. Koves, E., and M. Varga. 1961: Effect of phenol-carboxylic acid-like inhibitors of grain stubble residues on the germination of stubble crop seeds. Agrokem. Talajtan 10:135–44.

11291. Kowal, T. 1953: A key for the determination of the seeds of the genera *Chenopodium* L. and *Atriplex* L. Monogr. Botan. 1:87–163. Biol. Abstr. 29:1625. 1955.

11292. —— 1954: The morphological and anatomical features of the seeds of genus *Amaranthus* L. and keys for their determination. Monogr. Botan. 2:162–93.

11293. —— 1961: Morphology and anatomy of the seeds in Portulacaceae Rchb. Monogr. Botan. 12:3–47.

11294. Kowalewski, Z., H. Jäger, O. Schindler, and T. Reichstein. 1960: Die Cardenolide der Samen von *Erysimum perofskianum* Fisch. et Mey. Glykoside und Aglykone. Helv. Chim. Acta 43:957–72.

Kowalewski, Z. 7385.

Kowalska, M. Rychter-; *see* Rychter-Kowalska, M.

11295. Koyama, M. 1919: Forcing the germination of seeds of *Keyaki* (*Zelkowa serrata*), *Ho* (*Magnolia hypoleuca*), and *Urushi* (*Rhus vernicifera*). Bull. For. Expt. Sta. 18:1–82.

11296. —— 1920: The degree of stored seeds in relation to their vitality. Bull. For. Expt. Sta. 21:1–56.

11297. —— 1922: After-ripening and the germination of *Keyaki* (*Zelkowa serrata*) seeds. Bull. For. Expt. Sta. 22:1–30.

11298. —— 1952: Studies on empty seeds in triploid watermelon. I. Seiken Ziho 5:105–9.

Koyama, M. 17336, 17337.

Koyama, T. 5819, 5820.

Koyama, Y. 9528.

Koyanagi, H. 18799.

Kozaczenko, H. 1884.

11299. Kozak, M. P. 1933: Seed productivity and its characteristic in different forms of the guayule (*Parthenium argentatum* Gray). Bull. Appl. Botany, Genet., Plant Breed. (USSR) Ser. A No. 5/6:125–35.

11300. —— 1958: On the acceleration of spring vetch seed ripening. Zemledelie 1958(2):73–75. Biol. Abstr. 35:48881. 1960.

Kozelkova, N. I. 12376.

11301. Kozhevnikova, L. M. 1953: Exposure to sunlight of spring wheat seeds as a means of controlling loose smut of wheat (*Ustilago tritici*). Agrobiologiya 1953(3): 41–46.

Kozhevnikova, N. F. 6144, 6145.

11302. Kozhina, I. S. 1961: Study of the carbohydrate composition of the seeds of Umbelliferae by paper chromatography. Tr. Botan. Inst. Akad. Nauk SSSR Ser. 5 Rast. Syre. 1961(8):40–42. Chem. Abstr. 56:7724. 1962.

11303. Kozima, H., M. Yatazawa, and Y. Goto. 1953: Amino acids interconversion at an early stage of germination of plant seeds. Shiga Agr. Coll. Sci. Rept. 4:51–56.

Kozlova, L. M. 14477, 14478.

11304. Kozlova, N. A., E. A. Taranova, and V. T. Kietse. 1955: Critical phases in the seed development of the Low Latvian cherry. Latvijas PSR Zinatnu Akad. Vestis 1955(2):63–76.

11305. Kozlovskii, A. I. 1949: Problem in improving the ripening of forage beet and carrot seeds under conditions of Siberia. Selek. i Semen. 16(10):49–53.

11306. Kozlowski, A. 1937: A respiratory chromogen in the seeds of pea, *Pisum sativum* L. Acta Soc. Botan. Polon. 14:49–55.

11307. —— 1937: Hirsutorubrin, a rare type of anthocyanin, found in the seeds of *Lupinus hirsutus*, and its formation by oxidation of chromogen. Acta Soc. Botan. Polon. 14:167–73.

11308. Kozlowksi, T. T., and S. Bergquist. 1955/56: Metabolism of coniferous tree seed during germination. Mass. Agr. Expt. Sta. Ann. Rept. Bull. 494:20.

11309. —— and A. C. Gentile. 1959: Influence of the seed coat on germination, water absorption and oxygen uptake of eastern white pine seed. Forest Sci. 5:389–96. Biol. Abstr. 35:16878. 1960.

11310. Kozmina, N. P., L. A. Butman, V. N. Ilina, and A. T. Naumova. 1959: Data on intermediate and fixed proteins of wheat endosperm. Tr. Vses. Nauchn.-Issled. Inst. Zerna i Produktov ego Pererabotki 36:30–41. Chem. Abstr. 56:1775. 1962.

11311. —— and A. T. Naumova. 1957: Differences in hyaline and meally endosperm of wheat. Soobshch. i Referat. Vses. Nauchn.-Issled. Inst. Zerna i Produktov ego Pererabotki 1957(3):9–11. Chem. Abstr. 53:5423. 1959.

11312. —— and M. S. Romanova. 1938: Alterations of wheat gluten in the process of sprouting. Biochemistry (USSR) (English Transl.) 3:378–86. Chem. Abstr. 33:198. 1939.

Kozubenko, V. E. 10815.

11313. Kozubov, G. M., and L. G. Ganyushkina. 1963: The effect produced by ultrasound on pine and spruce seeds. Dokl. Akad. Nauk SSSR 150:421–23.

11314. Kozyr, A. T. 1952: Germination of newly harvested seeds of perennial forage plants. Selek. i Semen. 19(8):47–51.

Kraebel, C. J. 13565.

11315. Kraevoj, S. J. 1934: Einfluss der Röntgenstrahlen auf das Geschlechtssystem und die Samenproduktivität von *Pisum* und *Lathyrus*. Sov. Botan. (Leningrad) 1:74–84. Botan. Centralbl. 169:129. 1935.

11316. Kraft, D. 1932: The oil-bearing seeds of some cultivated and some wild plants. Maslob.-Zhir. Delo 1932(6):55–56. Chem. Abstr. 27:3352. 1933.

11317. Kraft, J. M., T. Kommedahl, and A. J. Linck. 1963: Histological study of malformation in flax seed after exposure to 31° C. Botan. Gaz. 124:367–71.

Kraft, J. M. 9076.

11318. Kralin, P. 1944: Testing for viability and increasing the germination of cereal seeds. Proc. Lenin. Acad. Agr. Sci. USSR 1944(5/6):3–12.

Krall, J. L. 13239.

11319. Kramer, A. 1952: A tri-metric test for seed corn quality. Proc. Am. Soc. Hort. Sci. 59:405–13.

11320. Kramer, C. 1954: Total amount of heat during the germination period of barley. Landbouwk. Tijdschr. 66:519–21.

11321. Kramer, H. A. 1962: U.S.D.A. researchers develop electronic seed counter. Seed World 90(10):9.

11322. Kramer, H. H., R. L. Whistler, and E. G. Anderson. 1956: A new gene interaction in the endosperm of maize. Agron. J. 48:170–72.

Kramer, H. H. 5013, 19586.

Kramer, J. A. 10657.

11323. Kramer, M. 1957: Inhibition of germination of black rice and red rice with 2,4-D. Biologico 23:51–55.

11324. —— and K. Silberschmidt. 1948: Human saliva as a germination inhibitor. Science 108:410.

11325. Kramer, N. W. 1944: Relationship of mottled, sweet clover seed to the percentage of yellow-flowered plants. Assoc. Offic. Seed Analysts News Letter 18(4):6–9.

11326. Kramer, P. J. 1955: Water relations of plant cells and tissues. Ann. Rev. Plant Physiol. 6:253–72.

Kramer, S. 10855.

11327. Kramer, W. 1960: Pflanzenkunde für Landwirte, Gärtner und Forstwirte. Fischer, Jena. 676 pp.

11328. Krantz, J. C., Jr., C. J. Carr, and F. K. Bell. 1943: A contribution to the pharmacology of Xanthium. J. Am. Pharm. Assoc. Sci. Ed. 32:244–47.

11329. Kranz, W. 1932: The swelling of the wheat kernel. Mühlenlab. 1932:45–46. Chem. Abstr. 26:4888. 1932.

Krapf, F. 4664.

11330. Krapivine, V. 1928: A study on certain germinating conditions of *Nicotiana* seeds. Detskoje Selo. Sta. Acclim. Lenin. Agr. Inst. Bull. 7:181–87.

Krashennikov, N. A. 11047.

11331. Krasikov, Z. D., and V. I. Chukanov. 1956: Effect of vernalization of spring wheat on yields and seeding quality of seeds. Agrobiologiya 1956(4):70–77.

11332. Krasilnikov, N. A., M. K. Chailakhian, G. K. Skriabin, I. M. Khoklova, I. V. Ulezlo, and T. N. Konstantinova. 1958: The stimulatory effect of gibberellins from different sources. Dokl. Akad. Nauk SSSR 121:214–18.

11333. Krasin, A. I. 1951: On the yarovization of (*Papaver somniferum*) and (*Salvia sclarea*) seeds. Byul. Gl. Botan. Sada 8:67–69.

11334. Krasinska, Z. 1929: Contribution a l'etude du metabolisme energetique de la germination (*Helianthus annus*). Acta Biol. Exptl. (Warsaw) 3:101–41.

11335. Krasnosselsky-Maximow, T. A. 1926: Rapidity of swelling and the size of the wheat seed. Bull. Appl. Botany, Genet., Plant Breed. (USSR) 16:241–50.

11336. Krasochkin, V. T. 1950: Vernalization of (sugar) beet seeds. Dokl. Akad. Nauk SSSR 70:895–98.

11337. —— 1958: Influence of place of reproduction of seeds and geographical zone in beet seed growing. Tr. Prikl. Botan. Genet. i Selek. 33:242–52.

11338. —— 1958: Methods of hastening beet selection and seed-growing. Tr. Prikl. Botan. Genet. i Selek. 33:253–56.

11339. Krasovsky, I. 1940: Effect of conditions at time of setting and ripening of seeds on the drought resistance of wheat. Bull. Acad. Sci. URSS Cl. Sci. Math. Nat. Ser. Biol. 1940(4):495–503. Biol. Abstr. 17:6760. 1943.

11340. Krasulin, N. P., and M. F. Kolomeitseva. 1962: Effect of microelements on tree seeds which germinate with difficulty. Pushkino Vses. Nauchn.-Issled. Inst. Lesov. i Mekh. Lesn. Khoz. Sb. Rabot po Lesn. Khoz. 1962(42):216–23.

11341. Krasulin, W. P. 1938: Die Wirkung hoher Temperaturen auf die Keimfähigkeit der Samen von *Cuscuta epilinum*. Lein Hanf. 2:16–19. Centralbl. Bakteriol. Parasitenk. Abt. II 100:477. 1939.

11342. Kratzer, F. H. 1944: The tryptophane content of feedstuff proteins. J. Biol. Chem. 156:507–9. Expt. Sta. Rec. 93:386. 1945.

11343. Kratzmann, E. 1839: Die Lehre vom Samen der Pflanzen. Borrosch und Andre, Prague. 98 pp.

Krauczunas, P. 5443.

11344. Krauia, A. I. 1958: The influence of trace elements on seed germination and some biochemical processes in plants. Akad. Nauk Latv. SSR Inst. Biol. Tr. 9:195–216.

11345. Kraus, C. 1882: Versammlung von Vorständen und Interessenten der Samenkontrole zu Salzburg. Landwirtsch. Versuchssta. 27:345–46.

11346. Kraus, E. J., and R. Garren, Jr. 1953: Propagation by seeds. Chrysanthemums for the home garden. Oregon Agr. Expt. Sta. Bull. 534:12–19.

Kraus, E. J. 6968.

11347. Kraus, J. F. 1959: Methoxone treatment of longleaf pine seed. J. Forestry 57:650.

11348. Krause, E. W. 1945: Über Keimung und das Jugendwachstum im Hinblick auf die Entwicklung der Pflanzendecke. Planta 34:138–53. Biol. Abstr. 21:18980. 1947.

11349. Krause, W. G. C. 1954: Producing cyclamen seed. Nurseryman Seedsman 119:1304–6.

11350. Krauss, L. 1933: Entwicklungsgeschichte der Früchte von *Hordeum*, *Triticum*, *Bromus*, and *Poa*, mit besonderer Berücksichtigung ihrer Samenschalen. Jahrb. Wiss. Botan. 77:733–808.

11351. Krauze, S. 1933: Investigations on stored grain. Wiadomosci Farm. 60:15–17, 43–45, 69–71, 84–87. Chem. Abstr. 27:1958. 1933.

11352. —— and W. Dziedzianowicz. 1959: Toxicity of beech seeds (*Fagus silvatica*). Nahrung 3:213–17. Chem. Abstr. 54:5842. 1960.

11353. Kravchenko, V. I. 1961: The effect of the duration of pistachio seed storage on their germinating power in the ground. Izv. Akad. Nauk Turkm. SSR Ser. Biol. Nauk 1961(5):83–87. Biol. Abstr. 42:7667. 1963.

Kraybill, H. R. 10788.

11354. Kreeb, K. 1959: Die Bodenversalzung als störender Faktor bei Feldversuchen und ihre Bedeutung für Keimung, Wachstum und Ertrag. Ber. Deut. Botan. Ges. 72:123–37.

11355. Krefting, L. W., and E. I. Roe. 1949: The role of some birds and mammals in seed germination. Ecol. Monographs 19:271–85.

11356. Kreis, W., C. Tamm, and T. Reichstein. 1957: Die Glykoside der Samen von *Corchorus capsularis* L. Helv. Chim. Acta 40:593–604.

Kreis, W. 17223.

11357. Kreitlow, K. W., and R. J. Garber. 1946: Viability of stored seeds of forage crops treated with different fungicides. [Abstract.] Phytopathology 36:403.

Kreitlow, K. W. 17314.

Kreke, C. W. 14579.

11358. Krekule, J. 1961: The length of the first leaf-blade of winter-wheat is not an indicator of vernalization. Biol. Plant. 3:85–88.

11359. —— 1961: Application of some inhibitors in studying the physiology of vernalization. Biol. Plant. Acad. Sci. Bohemoslov. 3:107–14.

11360. —— 1961: The effect of photoperiodic regime on vernalization of winter wheat. Biol. Plant. Acad. Sci. Bohemoslov. 3:180–91.

11361. —— and L. Teltscherova. 1963: Content of auxin and gibberellin-like material in vernalized and nonvernalized embryos of spring and winter wheats. Biol. Plant. Acad. Sci. Bohemoslov. 5:252–57.

Kremers, R. E. 17185.

11362. Krenke, A. N. 1961: Changes in the respiration rate and content of phosphorus compounds during the ripening of plant seeds. Tr. Gl. Botan. Sada Akad. Nauk SSSR 8:97–112. Biol. Abstr. 42:7472. 1963.

11363. Krenner, J. A. 1958: The natural history of the sunflower broomrape (*Orobanche cumana* Wallr.). I. The morphological anatomy of the sunflower broomrape seed, its germination and the infection mechanism of its seed. Acta Botan. Acad. Sci. Hung. 4:113–44. Biol. Abstr. 33:23075. 1959.

11364. Kreps, E. H. 1934: Amaryllis from seed. Flower Grower 21(3):133.

11365. —— 1942: Grow these bulbs from seed. Flower Grower 29(11):484–85, 505.

11366. —— 1943: Growing perennial phlox from seed. Horticulture 21(3):54–55.

11367. Kretovich, V. L. 1934: The distribution of sugar and nitrogen compounds in wheat grain. Sci. Inst. Cereal Res. (Moscow) 13:70–73. Chem. Abstr. 29: 490. 1935.

11368. —— 1941: Respiration in seeds of flax and hulled cereals. Compt. Rend. Acad. Sci. URSS 33:355–57. Chem. Abstr. 38:5534. 1944.

11369. —— 1945: Physiologico-biochemical bases for the storage of grain. Akad. Nauk SSSR Biokin, Moscow-Leningrad. 136 pp. Chem. Abstr. 40:1237–38. 1946.

11370. —— and T. A. Akimochkina. 1941: Biochemical changes in wheat during its maturing after the harvest. Biochemistry (USSR) (English Transl.) 6:393–400. Chem. Abstr. 36:6577. 1942.

11371. —— and Y. S. Ball. 1956: Antioxidants in seeds of oil-bearing cultures. Biokhim. Zerna, Sb. 1956(3):188–94. Chem. Abstr. 51:8914. 1957.

11372. ——, A. A. Bundel, and T. V. Drozdova. 1948: Sulfhydryl compounds and ascorbic acid in sprouting and ripening seeds. Biochemistry (USSR) (English Transl.) 13:332–36. Chem. Abstr. 42: 8888–89. 1948.

11373. ——, ——, N. A. Lepina, and S. S. Melik-Sarkisyan. 1956: Proteins of pumpkin seeds. Biokhim. Zerna, Sb. 1956(3): 5–34. Chem. Abstr. 51:13970. 1957.

11374. ——, ——, S. S. Melik-Sarkisyan, and K. M. Stepanovich. 1954: On the so-called reserve proteins of seeds. Biochemistry (USSR) (English Transl.) 19: 208–15.

11375. ——, ——, T. I. Smirnova, Z. N. Galachalova, Z. A. Plyusnina, S. D. Ptitsyn, G. A. Nikitash, A. P. Gerzhoi, G. A. Dzhorogyan, L. Y. Auerman, R. R. Tokareva, F. N. Korovin, V. M. Bazarnova, E. A. Alyakrinskaya, A. G. Kulman, and N. A. Ivannikova. 1954: The effect of drying on the grain and the technological properties of freshly harvested wheat grain. Biokhim. Zerna, Sb. 1954(2):45–111. Chem. Abstr. 49:7655. 1955.

11376. —— and Z. G. Evstigneeva. 1953: Synthesis of protein from asparagine and glutamine in wheat grains. Dokl. Akad. Nauk SSSR 93:879–81. Chem. Abstr. 48:5298. 1954.

11377. ——, ——, and M. M. Makarenko. 1954: Cleavage and synthesis of amides in growing grains of corn, alfalfa and pumpkin. Biokhim. Zerna, Sb. 1954(2):161–78. Chem. Abstr. 49:7655. 1955.

11378. —— and E. Galyas. 1961: The biosynthesis of amino acids from oxalacetic and pyruvic acids in living tissues of germinating and developing grain. Biochemistry (USSR) (English Transl.) 26:85–89.

11379. —— and V. I. Iakovleva. 1959: Biosynthesis of glutamic acid and glutamine in pea and wheat seedlings. Plant Physiol. (USSR) (English Transl.) 6:165–70.

11380. ——, E. A. Morgunova, and A. I. Starodubtseva. 1960: Effect of heating on physiological and biochemical properties of sunflower seeds. Maslob.-Zhir. Prom. 26(2):8–11. Chem. Abstr. 54: 25080. 1960.

11381. —— and I. S. Petrova. 1948: Transformation of slime (soluble pentosans) during germination and ripening of rye seed. Dokl. Akad. Nauk SSSR 59: 281–83. Chem. Abstr. 42:6418–19. 1948.

11382. —— and E. N. Riazanceva. 1935: Biochemical changes in wheat grain under the action of high temperatures. Compt. Rend. Acad. Sci. URSS 8:409–12.

11383. ——, T. I. Smirnova, A. A. Bundel, and S. S. Pechnikova. 1950: Investigation of *Euonymus verrucosus* seed germination. Dokl. Akad. Nauk SSSR 73:527–30.

11384. —— and A. I. Sokolova. 1942: The

dehydrogenases of wheat embryos. Biochemistry (USSR) (English Transl.) 7:232–37. Chem. Abstr. 38:1536. 1944.

11385. —— and A. I. Starodubtseva. 1956: Respiration and lipolytic activity of stored oil-bearing seeds. Biokhim. Zerna, Sb. 1956(3):179–87. Chem. Abstr. 51:8914. 1957.

11386. —— and P. [P.] Tokareva. 1939: Biochemical features of frost-damaged grain. Biochemistry (USSR) (English Transl.) 4:68–79. Chem. Abstr. 33:6903. 1939.

11387. —— and —— 1939: Biochemical changes in the grains of wheat damaged by the wheat bug. II. Biochemistry (USSR) (English Transl.) 4:636–47. Chem. Abstr. 34:5550. 1940.

11388. —— and R. Tokareva. 1938: Biochemical changes in the grain of wheat damaged by the wheat-bug. Biochemistry (USSR) (English Transl.) 3:387–96. Chem. Abstr. 33:683. 1939.

11389. ——, ——, I. S. Petrova, T. V. Drozdova, L. Y. Auerman, and N. Smolina. 1947: Biochemistry of the ripening of the rye kernel. Biochemistry (USSR) (English Transl.) 12:545–55.

11390. —— and E. N. Ushakova. 1940: Über kritische Feuchtigkeit und Gaswechsel des Nornem bei Lagerung. Dokl. Akad. Nauk SSSR 29:118–21.

11391. —— and Z. V. Uspenskaya. 1954: A method for isolation of chlorogenic acid from sunflower seeds. Biokhim. Zerna, Sb. 1954(2):202–7. Chem. Abstr. 49:8562. 1955.

11392. —— and —— 1958: The synthesis of phenylalanine from phenylpyruvic acid in homogenates from pea seedlings. Biochemistry (USSR) (English Transl.) 23: 232–37. Biol. Abstr. 35:16959. 1960.

11393. —— and —— 1959: The synthesis of phenylalanine and transformations of phenylpyruvic acid in the ripening wheat ear. Biochemistry (USSR) (English Transl.) 24:104–9.

11394. —— and V. I. Yakovleva. 1958: Biosynthesis of glutamic acid and glutamine in pea and wheat seedlings. Plant Physiol. (USSR) (English Transl.) 6:174–78. Biol. Abstr. 35:16889. 1960.

Kretovich, V. L. 17596.

Krewinkle, J. 15177.

11395. Krewson, C. F., and F. E. Luddy. 1964: *Veronia anthelmintica* (L.) Willd. Highly purified epoxy components from the seed oil. J. Am. Oil Chemists Soc. 41:134–36.

Krewson, C. F. 1903.

Krexner, R. 19538.

11396. Kreyger, J. 1948: The drying and storage of grain and seed. Cent. Organ. T.N.O. Alg. Tech. Afd. Rap. T.A. 239. 94 pp.

11397. —— 1954: The determination of vapor-pressure isotherms in the seeds of nine crops. Cent. Inst. Landbouwk. Onderzoek., Droogtechn. Lab. (Wageningen), Publ. 39. 7 pp.

11398. —— 1954: Drying and storage of seed. Zaadbelangen 8:6–7, 21–23.

11399. —— 1956: Results from airtight storage tests with the seeds of nine crops. Cent. Inst. Landbouwk. Onderzoek., Droogtechn. Lab. (Wageningen), Publ. 67. 13 pp.

11400. —— 1960: Drying of seeds. Proc. Intern. Seed Testing Assoc. 25:590–601.

11401. —— 1963: General considerations concerning the drying of seeds. Proc. Intern. Seed Testing Assoc. 28:753–84.

11402. —— 1963: Equipment of importance for drying seeds in Europe. Proc. Intern. Seed Testing Assoc. 28:793–814.

11403. —— 1963: General considerations concerning the storage of seeds. Proc. Intern. Seed Testing Assoc. 28:827–36.

11404. —— 1963: Slow drying during temporary or prolonged ventilated storage of cereal sowing seeds. Proc. Intern. Seed Testing Assoc. 28:861–70.

11405. —— 1963: Some general comments on the drying of seeds. Proc. Intern. Seed Testing Assoc. 28:981–87.

11406. Kreysing, M. 1924: Beiträge zur Keimungsphysiologie des Wiesenschwingels. J. Landwirtsch. 72:237–78.

11407. Kribben, F. J. 1957: Die Abkürzung der Samenruhe bei *Arabidopsis* durch Gibberellinsäure. Naturwissenschaften 44: 313.

11408. —— and H. J. Reisener. 1957: Über das Wachstum des Gurkenkotyledons unter dem Einfluss von Gibberellinsäure. Beitr. Biol. Pflanz. 34:379–93.

11409. Krickl, M. 1958: Züchtungsversuche über die Möglichkeit einer unbedingten Keimfähigkeit der Samen von *Rosa canina* imersten Frühjahr. Klosterneuburg Höhere Bundeslehr- Versanst. Wein- Obstbau. Mitt. Ser. B, Obst Gart. 8:189–207.

Krider, J. L. 6192.

11410. Kriechbaum, W. 1953: Nahrboden

für Orchideenanzucht aus Samen. Garten-Z., Flora 76(2):17–18. Biol. Abstr. 27:23346. 1953.

11411. Krier, J. P. 1949: Effects of treatments to induce germination of seeds of several species valuable for soil conservation plantings. [Abstr. Thesis.] J. Forestry 47:312.

11412. Krishna, S., and S. V. Puntambekar. 1931: The oil from the seeds of *Putranjiva roxburghii*, Wall. J. Indian Chem. Soc. 8:301–6. Chem. Abstr. 26:612. 1932.

Krishna, S. 7190, 7191, 15758–15762, 15964.

Krishnamourthy, R. 2410.

11413. Krishnamurthi, A., J. L. Bose, and S. Siddiqui. 1948: Growth promoting factors from the germs of Indian pulses. I. Influence of biochanin C on the growth of yeast and *Penicillium notatum*. J. Sci. Ind. Res. (India) 7B:39–41. Biol. Abstr. 23:5139. 1949.

11414. Krishnamurthy, K., B. R. Baliga, R. Rajagopalan, M. Swaminathan, and V. Subramanyan. 1959: Amino-acid composition of sunflower seeds (*Helianthus annuus*) and Cuddapah almonds (*Buchanania latifolia*). Food Sci. (Mysore) 8:119–20. Biol. Abstr. 37:7123. 1962.

11415. ——, P. K. Tasker, T. N. Ramakrishnan, R. Rajagopalan, and N. Swaminathan. 1960: Studies on the nutritive value of sesame seeds. I. The amino-acid composition of the proteins and the chemical composition of white and black varieties of sesame seed and meal. Ann. Biochem. Exptl. Med. (Calcutta) 20:73–76. Biol. Abstr. 35:71681. 1960.

Krishnamurthy, K. 6938.

11416. Krishnamurti, S. T. S. 1951: Improved varieties of oilseeds—*Sesamum*. Indian Farming 1(9):22–23.

11417. Krishnamurty, I. V. G. 1958: Production and processing of tobacco seed. Indian Tobacco 8:37–41.

11418. —— and V. V. R. Rao. 1959: Studies on "light" and "heavy" seed in tobacco. Indian Tobacco 9:217–20.

11419. Krishnan, P. S. and T. K. Krishnaswamy. 1939: Proteins and other nitrogenous constituents of watermelon seeds (*Citrullus vulgaris*). Biochem. J. 33:1284–90.

Krishnaswami M. K. 5698.

Krishnaswami, N. 1580.

11420. Krishnaswamy, P. 1952: Storage and germination of millet seeds. Madras Agr. J. 39:485–90. Biol. Abstr. 27:13854. 1953. Krishnaswamy, T. K. 11419.

11421. Kristensen, H. R. 1953: Lettuce mosaic, a virus disease transmitted through seeds. Gart. Tidsskr. 69:522–23.

11422. Kristic, M. 1957: Sur le changement des propriétés d'embryons de quelques pins, provoqué par la haute température et décelé par les procédés rapides. Proc. Intern. Seed Testing Assoc. 22:447–56.

11423. Krivkina, N. P. 1956: Biochemical processes in cotton seeds as affected by the pruning periods. Izv. Akad. Nauk Tadzh. SSR Otd. Estest. Nauk 1956(17):23–24.

11424. —— 1961: Carbohydrate metabolism and fat accumulation in the seeds of various strains of cotton differing in rate of maturation. Byul. Nauchn.-Tekhnol. Inform. Tadzh. Nauchn.-Issled. Inst. Sel'skokhoz 1961(1):47–54. Biol. Abstr. 44: 8025. 1963.

11425. Krkaska, S. 1928: Samen von Arzneipflanzen, ihre Keimfähigkeit und andere Eigenschaften. Intern. Congr. European Medicinal Herbs Interest—Papers Presented 2:27–28.

11426. Krober, O. A., and F. I. Collins. 1944: Effect of relative humidity on the determination of oil in soybeans. Oil Soap 21:1–5. Biol. Abstr. 22:12118. 1948.

11427. —— and —— 1948: Effect of weather damage on the chemical composition of soybeans. J. Am. Oil Chemists Soc. 25:296–98. Biol. Abstr. 23:24617. 1949.

Krober, O. A. 4304.

11428. Krochmal, A. 1952: Seeds of weedy *Euphorbia* species and their identification. Weeds 1:243–55.

11429. Krodel, F. 1959: Der Einfluss der Inkrustierung von Gemüsesaatgut auf Keimenergie und Keimfähigkeit. Saatgut-Wirtsch. 11:205–6.

11430. Kroeger, G. S. 1941: Dormancy in seeds of *Impatiens balsamina* L. Contrib. Boyce Thompson Inst. 12:203–12.

Kroeger, H. 16287.

Krofta, Z. 2295.

Krohne, H. E. 12262, 12264.

11431. Kronstad, W. E., R. A. Nilan, and C. F. Konzak. 1959: Mutagenic effect of oxygen on barley seeds. Science 129:1618.

11432. Krosby, P. 1960: Seed testing and sowing seed control in Norway. Proc. Intern. Seed Testing Assoc. 25:48–59.

11433. Krotov, A. S. 1960: Buckwheat seeds viability under different conditions of storage. Tr. Prikl. Botan. Genet. i Selek. 32:307–14.

11434. Krotova, O. A. 1957: Seed treatment with ultrasonic vibrations. Sad i Ogorod 95(9):28–29.

11435. —— 1961: Pelleting of vegetable seeds. Moskov. Ordena Lenina Sel'skokhoz. Akad. Im. K. A. Timiryazeva Dokl. 72:129–34.

11436. Krstic, M. 1950: Contribution à l'étude de la conservation des graines de l'Orme champêtre (*Ulmus campestris*). Inst. Nauchn. Istrazh. Sumarst. Srbije, Rad. 1:53–61.

11437. Krstitch, M. 1951/52: Pour une définition plus précise de l'énergie germinative des semences forestières. Proc. Intern. Seed Testing Assoc. 17:16–19.

11438. —— 1951/52: La détermination de la vitalité des embryons de *Pinus nigra* Arn. sur le milieu à la gélatine tannisée. Proc. Intern. Seed Testing Assoc. 17:20–22.

Krueger, G. J. 17050.

Krueger, K. W. 17434.

11439. Krüger, L. 1931: Ein Beitrag zur Biologie von *Chenopodium album*. Angew. Botan. 13:1–49, 97–121.

Krüger, W. 8821.

Krueger, W. C. 17520.

11440. Krümmel, K. 1932: Weitere Untersuchungen über die Befruchtungsverhältnisse bei Kirschen. Gartenbauwissenschaft 6:262–302. Botan. Centralbl. 165: 95. 1933.

11441. Krüpe, M. 1953: Hämagglutinine mit spezifischen Affinitäten in Samen von Papilionaceen. Biol. Centralbl. 72:424–31.

Krug, C. A. 3665, 3666.

11442. Krug, H. 1929: Beiträge zur Keimungsphysiologie und Bekämpfung von Samenunkräutern. Botan. Arch. 27:420–518.

11443. Kruger, W. 1960: The influence of antibiotics on the germination of tomato seed and development of the seedlings. S. African J. Agr. Sci. 3:561–71. Biol. Abstr. 36:46253. 1961.

11444. Krumbholz, G. 1930: Über Verschiedenheiten in der Embryonengrösse einiger *Oenotheren* und ihrer reziproken Bastarde. Z. Induktive Abstammungs-Vererbungslehre 56:383–92.

11445. Krupko, S. 1962: Embryological and cytological investigations in *Hypodiscus aristatus* Nees (Restionaceae). J. S. African Botany 28:21–44. Biol. Abstr. 39:3260. 1962.

11446. —— 1963: Macrosporogenesis and embryo sac development in *Chondropetalum hookerianum* (Mast.) Pillans (Restionaceae). Acta Soc. Botan. Polon. 32:171–90.

11447. ——, G. F. Israelstam, and B. Martinovic. 1954: Embryosac development and chromosome number in vanilla roscheri from Inhaca island. S. African J. Sci. 51:115–17.

11448. Kruse, A. 1960: Polyembryony in *Brassica napus* var. *rapifera* L. and *Beta vulgaris* L. Kgl. Veterin. Landbohojskole Arsskr. 1960:37–46.

11449. Kruse, A. 1961: Haploids in polyembryos of beet, *Beta vulgaris* L. Roy. Vet. Agr. Coll., Copenhagen, Den., Yearbook 1961:87–98.

Krussen, F. 18260.

11450. Krusser, O. V. 1940: Biochemistry of *Corylus*. Biokhim. Kul'tur Rast. 7:440–46. Chem. Abstr. 35:3287. 1941.

11451. Kruyt, W. 1952: Effects of some plant growth substances on early growth of pea embryos in sterile culture; a study in connection with the problem of hormonisation of seeds. Koninkl. Ned. Akad. Wetenschap. Proc. Ser. C 55:503–13.

11452. —— 1954: The problem of hormonization of seeds. Acta Botan. Neerl. 3:1–82. Chem. Abstr. 50:2748. 1956.

11453. Kruzhilin, A. S., and Z. M. Shvedskaya. 1957: Change in the sugar content of biennials during vernalization. Dokl.-Botan. Sci. Sect. (English Transl.) 116:209–16.

11454. —— and —— 1960: The role of leaves in the vernalization of winter cereals and biennials. Plant Physiol. (USSR) (English Transl.) 7:237–43.

Kruzhilin, A. S. 17396.

Krynkina, E. I. 18046.

Kryukova, L. M. 12733, 12734.

11455. Krzeszkiericz, W. 1939: Determination of germinative capacity of pine seeds by the coloring method. Warsaw Inst. Badawczy Lasow Panstwowych. Rozprawy i Sprawozdania Ser. A 44:1–45.

11456. Krznarich, P. W. 1940: The hemicelluloses from oat hulls. II. Cereal Chem. 17:457–59.

11457. Kuata, H. 1954: Seed size of Formosan linalool tree. J. Japan. Forestry Soc. 36:236–37.

11458. Kubicz, A. 1962: Comparison of the electrophoretic protein fractions of the seeds of *Allium cepa* and *Allium porrum*.

Howdowla Roslin Aklim. Nasiennictwo 6:91–103.

11459. Kucera, C. 1928: Vitamin B und C im keimenden Getreidekorn and in der jungen Pflanze. Biol. Listy 14:130–38. Ber. Ges. Physiol. Exptl. Pharm. 48:374. 1929.

11460. —— 1934: The influence of artificial radio-active water on the germination of the seeds of various plants. Biol. Spisy Vysoke Skoly Zverolekarske Brno 13:121–28. Sign. B 183. Proc. Intern. Seed Testing Assoc. 9:164–65. 1937.

11461. Kucera, M., and A. Svoboda. 1962: Seeding of the shredleaf sumac (*Rhus typhina* f. *dissecta* Rehder). Dendrol. Sb. 3:143–46. 1961.

11462. Kucheneva, G. G. 1961: Propagation of garden forms of trees and shrubs by seeds. Byul. Gl. Botan. Sad. 41:101–6.

11463. Kucherenko, N. E. 1958: Some questions dealing with storage of forage plants seeds. Moskov. Ordena Lenina Sel'skokhoz. Akad. Im. K. A. Timiryazeva Dokl. TSKHA 32:223–26.

Kucherov, E. V. 17235.

Kuck, J. C. 15531.

11464. Kudeiarov, N. I., and V. D. Smertkin. 1957: Practices of [vegetable] seed growers in Krasnodar Territory. Sad i Ogorod 95(9):15–20.

11465. Kudelka, F. 1875: Über die Entwickelung und den Bau der Frucht- und Samenschale unserer Cerealien. Landwirtsch. Jahrb. 4:461–78.

Kudriasheva, N. A. 2559, 2561.

11466. Kudrina, V. N. 1961: Chemical composition and moisture balance in vegetable seeds. Dokl. Mosk. Sel'skokhoz. Akad. 1961(62):505–8. Chem. Abstr. 58: 750. 1963.

11467. Kudryashova, N. A. 1960: On proteolytic enzymes of plants. The proteolytic enzyme of *Arachis* seeds. Tr. Gl. Botan. Sad. Akad. Nauk SSSR 7:93–126.

11468. —— 1961: Proteolytic enzymes in seeds of leguminous plants. Tr. Gl. Botan. Sad. Akad. Nauk SSSR 8:60–74. Chem. Abstr. 56:14631. 1962.

11469. —— and E. V. Kolobkova. 1953: On the content of free amino acids in dormant seeds. Dokl. Akad. Nauk SSSR 91:1365–68.

Kudryashova, N. A. 11045, 11046.

11470. Kudzin, I. K., and V. F. Melnichenko. 1958: Effect of different varieties of fertilizers on the albumin contained in wheat in the steppe zone of the Ukrainian SSR. Dokl. Vses. Akad. Sel'skokhoz. Nauk 7:31–34.

11471. Kühl, H. 1933: Die Haltbarmachung des Brotgetreides. Landwirtsch. Versuchssta. 117:217–28.

11472. —— 1937: Composition and nutritive value of grain embryos. Production of nutrient flours. Pharm. Ztg. 82:367–69. Chem. Abstr. 31:4731–32. 1937.

11473. —— 1938: Physiological and therapeutic nutritional value of grain embryos. Pharm. Zentralhalle 79:569–73. Chem. Abstr. 33:245. 1939.

11474. —— 1940: The enzymes of cereal grains and their activity at lower temperatures. Mühlenlab. 10:17–20. Chem. Abstr. 36:175. 1942.

11475. Kühle, L. 1911: Der Einfluss des Schalens von Rubesamen auf die Keimungmaschinelle Entfernung der Perigonhulle. Jahresber. Vereinigung Angew. Botan. 4:190–200.

11476. Kuehn, O. 1925: Die Hartschaligkeit bei *Lupinus angustifolius*. Kuehn-Arch. 9:332–404.

11477. Kühne, L., and W. Kausch. 1961: Die Kinetik der Samenquellung von *Pisum sativum*. Naturwissenschaften 48: 460–61.

11478. Kugler, I. 1951: Untersuchungen über das Keimverhalten einiger Rassen von *Arabidopsis thaliana* (L.) Heynh. Ein Beitrag zum Problem der Lichtkeimung. Beitr. Biol. Pflanz. 28:211–43.

11479. —— 1952: Keimfähigkeit und Abgabe floureszierender Stoffe bei Samen. Naturwissenschaften 39:213. Biol. Abstr. 27:4678. 1953.

11480. —— 1955: Zur Frage der Abgabe keimungshemmender Stoffe durch Samen. Beitr. Biol. Pflanz. 31:313–32.

11481. Kugler, O. E., and E. H. Bennett. 1947: Histo-chemical localization of acid phosphatase in germinating maize kernels. Stain Technol. 22:9–15.

11482. Kuhles, R. 1929: Untersuchung über die Atmung der Gerste während der Vermälzung. Wochschr. Brau. 46:501–3. Biol. Abstr. 5:399. 1931.

11483. Kuhlmann, J. G. 1928: Monographia das especies brasilciras dos generos da tribu Oncobeae: *Carpotroche*, *Mayna* e *Lindackeria* (Flacourtiaceae) cujas sementes contem um óleo analogo ão obtido das sementes da Chaulmoogra. Mem. Inst. Oswaldo Cruz 21:389–416.

11484. —— 1938: Alguns casos interessantes de poliembrionia. Reunion Sul-Americana de Botan. 1st, Rio de Janeiro. Anais 3:203–6. 1941.

11485. Kuhn, E. 1915: Neue Beiträge zur Kenntnis der Keimung von *Phacelia tanacetifolia* Benth. Ber. Deut. Botan. Ges. 33:367–73.

11486. —— 1916: Dunkelkeimer und Substrat. Ber. Deut. Botan. Ges. 34:369–86.

11487. Kuhn, R., and D. Jerchel. 1941: Über invertseifen. VIII. Reduktion von tetrazoliumsalzen durch bakterien, gärende Hefe und keimende Samen. Ber. Deut. Chem. Ges. 74:949–52. Botan. Centralbl. 178:40. 1942.

11488. ——, D. Jerchel, F. Moewus, and E. F. Moller. 1943: Über die chemische Natur der Blastokoline und ihre Einwirkung auf keimende Samen, Pollenkörner, Hefen, Bakterien, Epithelgewebe und Fibroblasten. Naturwissenschaften 31: 468.

11489. ——, G. Pfluderer, and W. Schulz. 1954: Über die Reduktion von TTC durch Maissamen. Ann. Chem. 578:159–70. Ber. Wiss. Biol. 86:167. 1954.

11490. Kuhn, V. 1936: Morphology, biology, and biochemistry of poppy. Sb. Česk. Akad. Zeměděl. 11:243–50. Chem. Abstr. 30:7147. 1936.

Kuhns, D. M. 6864.

11491. Kuiken, K. A., and C. M. Lyman. 1948: Availability of amino acids in some foods. J. Nutr. 36:359–68.

11492. —— and —— 1949: Essential amino acid composition of soy bean meals prepared from twenty strains of soy beans. J. Biol. Chem. 177:29–36.

Kuiken, K. A. 12560.

11493. Kujala, V. 1927: Untersuchungen über den Bau und die Keimfahigkeit von Kiefern- und Fichtensamen in Finland. Finland Metsatieteelisen Koel Julkaisuja Med. Forst. Fors. Commonic. Inst. Quaest. Forst. 12:1–68.

11494. Kujirai, C., and S. Imamura. 1958: Über die photoperiodische Empfindlichkeit der Kotyledon von *Pharbitis nil* Chois. Botan. Mag. (Tokyo) 71:408–16.

11495. Kuke, A. 1932: Physiological activity of rye seed under the influence of time and its significance in the seedling method [for determining nutrient values of soils]. Z. Pflanzenernähr. Düng. Bodenk. 25A: 24–48. Soc. Chem. Ind. J. 51:697B. 1932.

Kukii, M. 6858.

11496. Kulagina, N. N. 1943: Influence of hydrothermic treatment on the resistance of wheat proteins. Biochemistry (USSR) (English Transl.) 8:293–96. Chem. Abstr. 39:357. 1945.

11497. Kulakova, A. N. 1953: In regard to the embryology of *Lathyrus vernus*. Mosk. Obshch. Isp. Prirody Byul. Otd. Biol. 58:63–70.

11498. Kulash, W. M. 1954: Save stored grain from insect pests. N. Carolina Agr. Expt. Sta. Bull. 389. 23 pp.

11499. Kuleshov, N. N. 1927: Über die Unterscheidung von Winter- von Sommerweizen. Pflanzenbau 4:70–72.

11500. —— 1930: Germination and purity of cucurbit seeds. Bull. Appl. Botany, Genet., Plant Breed. (USSR) 23:277–96.

11501. —— 1946: On the methods for studying the period from sowing to the seedling stage under field and laboratory conditions. Compt. Rend. Acad. Sci. URSS 53:75–78.

11502. —— 1947: An accelerated method for determining the germination of seeds. Selek. i Semen. 14(3):65–68.

11503. —— 1955: The effect of germinating ability and conditions of growth of seed material on the density of stands, survival, and productivity of the plants of the next generation. Tr. Inst. Genet. i Selek. 1955(4):5–20. Biol. Abstr. 32:28262. 1958.

Kuleshov, N. N. 11608.

11504. Kulgatschewa, R. A. 1930: Zur Frage über die Wirkung von Schwefelkohlenstoff und Chlorpikrin auf Samen verschiedener Ölpflanzen. Nordkaukasische Landwirtsch. Versuchssta. Bull. 313. Z. Angew. Entomol. 19:636. 1932.

11505. Kulibekov, M. 1957: The effect of seed selection from within the cob on the corn yield. Sotsial. Sel'skoe Khoz. Azerb. 1957(4):21–23. Biol. Abstr. 35:19549. 1960.

Kulikov, B. N. 9562.

11506. Kulikov, N. V. 1957: Effect of soaking seeds in a mixture of β-radiation agents on the biomass and structure of experimental phytocoenosis. Tr. Inst. Biol. Akad. Nauk SSSR, Ural'sk. Filial Sb. Rabot 1957(9):252–91. Chem. Abstr. 53:8321. 1959.

Kulka, K. 7756, 7757.

11507. Kulkarni, L. B. 1925: Germination test of the grass seeds. Poona Agr. Coll. Mag. 17:75–80.

11508. Kulkarni, R. A., and A. Sreenivasan. 1954: Biosynthesis of vitamin C during germination. A study of certain related enzyme systems. J. Sci. Ind. Res. 13B:704–7. Biol. Abstr. 31:5845. 1957.

Kulkarni, V. G. 5237.

11509. Kullen, B. 1941: Behavior of certain enzymes during the storage of wheat and its mill products. Vorratspflege Lebensmittelforsch. 4:421–27. Chem. Abstr. 37:2084. 1943.

Kul'man, A. G. 11375.

11510. Kulpa, W. 1960: Germination biology of spring adonis (*Adonis vernalis* L.). Roszniki Nauk Rolniczych Ser. A 81:337–82.

11511. —— 1961: Teratogenic influence of 2,4-dichlorophenoxyacetic acid (2,4-D) on the fruits of *Polygonum nodosum* Pers., *Anthemis arvensis* L. and *Tripleurospermum inodorum* (L.) Schutz. Bip. Roczniki Nauk Rolniczych Ser. A 84:495–521.

11512. Kumada, H. 1953: The nitrate utilization in seed embryos of *Vigna sesqui*-pedalis. J. Biochem. (Tokyo) 40: 439–50.

11513. Kumagai, T. 1953: Studies on hygrotype and xerotype oats. I. Change in amylase activity in oat seeds germinating under different soil moisture conditions. Hokkaido Natl. Agr. Expt. Sta. Res. Bull. 64:1–6

Kumagai, T. 14071.

11514. Kumaki, Y. 1956: Vernalization experiment of carrots. Hort. Assoc. Japan J. 25:163–66.

Kumana, J. S. 3996.

Kumar, A. 2056.

11515. Kumar, K., and A. K. Rajan. 1954: Use of growth substances in the induction of parthenocarpy in *Lycopersicon esculentum* and *Capsicum annuum.* Current Sci. (India) 23:94.

11516. —— and O. S. Singh. 1961: Influence of sulpha-drugs on germination, growth, yield, and quality of early paddy (*Oryza sativa* var. *Kuari*) and tomato (*Lycopersicon esculentum* var. *Ponderosa*). Natl. Acad. Sci., India. Allahabad 2. Annual Number 1960:63. Biol. Abstr. 37:7151. 1962.

11517. —— and —— 1961: Influence of vitamins on germination, growth, yield, and quality of early paddy (*Oryza sativa* var. *Kuari*) and tomato (*Lycopersicon esculentum* var. *Ponderosa*). Natl. Acad. Sci., India. Allahabad 2. Annual Number 1960:72. Biol. Abstr. 37:7152. 1962.

11518. —— and J. P. Srivastava. 1951: Tolerance of different crop and vegetable seeds to various reactions of H ion concentrations. Indian J. Agr. Sci. 21:39–44. Biol. Abstr. 27:17891. 1953.

11519. —— and K. C. Srivastava. 1954: Effect of sulpha-drugs on germination and early root growth of *Hibiscus esculentus.* [Abstract.] Proc. Indian Sci. Congr. 40: 116–17. 1953.

Kumar, K. 17496.

11520. Kumar, L. S. S. 1942: Non-heritable polyembryony in *Andropogon sorghum.* Current Sci. (India) 11:241.

11521. —— and D. S. R. Rao. 1943: Studies in the development of seed characters in linseed (*Linum usitatissimum*). J. Univ. Bombay 11:113–19. Biol. Abstr. 18:3264. 1944.

11522. —— and S. Solomon. 1940: The influence of light on the germination of species of *Striga.* Current Sci. (India) 9:541.

Kumar, S. A. 15957, 15958.

11523. Kumaraswamy, O. N., and B. L. Manjunath. 1936: Chemical examination of the fixed oil from the seeds of *Celastrus paniculatus* Willd. J. Indian Chem. Soc. 13:353–57. Analyst 61:704. 1936.

11524. Kumazawa, M. 1931: Supplementary notes on the cotyledons of *Glaucidium palmatum.* Botan. Mag. (Tokyo) 45:537.

11525. —— 1938: On the ovular structure in the Ranunculaceae and Berberidaceae. J. Japan. Botany 14:10–25. Botan. Centralbl. 174:7. 1938.

11526. Kummer, A. P. 1951: Weed seedlings. Univ. Chicago Press. 435 pp.

11527. Kummer, H. 1932: Fett und Fettsäuregehalt bei Gramineensamen in Beziehung zur Lichtdürftigkeit bei der Keimung. Ber. Deut. Botan. Ges. 50:300–3.

11528. —— 1935: Untersuchungen über die Lokalisation der Phosphatide in den Samen der Sojabohne (*Glycine hispida* Maxim.) unter besonderer Berücksichtigung der Extraktionsleichtigkeit der Phosphatide und des Oles. Hamburg. 146 pp.

11529. —— 1953: Schädigung der Keimfähigkeit von Gemüsesamen beim Aufbewahren in Buntdrucktüten. Angew. Botan. 27:115–42.

Kummerow, F. A. 3564, 15633.

11530. Kummerow, J. 1958: Beiträge zur Kenntnis der Ruheperiode von Winterknospen und Samen. Beitr. Biol. Pflanz. 34:293–314.

11531. —— 1960: Endogene Schwankungen der Keimfähigkeit bei *Dactylis glomerata*. Naturwissenschaften 47:262.

11532. —— 1963: Endogenous fluctuations of germination capacity in *Dactylis glomerata*. Am. J. Botany 50:915–20.

11533. —— and A. Hoffmann. 1959: Influencia de la cubierta de las clusas sobre la germinacion de semillas de lavendula. Chile Estac. Exptl. Agron. Bull. Tec. 4:3–13.

11534. —— and C. Labarca. 1961: Studies on the fruit and seed of *Nothofagus alpina* (Poepp. and Endl.) Krasser. Phyton Rev. Intern. Botan. Exptl. Argentina 17:205–10. Biol. Abstr. 43:7734. 1963.

Kummerow, J. 8914.

Kundert, J. 2648.

11535. Kundu, B. C., and M. S. Sarma. 1955: Jute seed. I. Nature and characteristics. Jute Bull. 17:323–27.

11536. —— and —— 1955: Jute seed. II. Jute Bull. 17:361–63.

Kundu, B. C. 7204.

Kundu, M. 17542.

11537. Kunert, G. 1960: On lipolysis in germinating seeds of *Ricinus*. Flora (Jena) 149:320–22. Biol. Abstr. 39:11811. 1962.

Kung, C. C. 18367.

11538. Kunimi, S., Y. Yokohama, and S. Isikawa. 1960: A method of promoting germination of weed-seeds and its application to afforestation pre-chilling and subsequent drying. J. Japan. Forestry Soc. 42:415–16.

11539. Kunisch, G. 1937: Die Zusammensetzung des Gesamtstickstoffs verschiedener Braugerstensorten, ihre Veränderung- beim Reifen und Keimen und ihre Bedeutung für den Eiweisslösungsgrad. G. Kunisch, Berlin-Neukölln. 111 pp.

11540. Kunitz, M. 1946: Crystalline soybean trypsin inhibitor. J. Gen. Physiol. 29:149–54. Biol. Abstr. 20:7962. 1946.

11541. Kunkel, M. P. 1958: How to grow pansies from seed. Popular Gardening 9(7):22, 41.

Kunstmann, F. W. 15787.

11542. Kunz, E. 1934: Réactions spécifiques et localisation d'alcaloide lupanine dans la graine du *Lupinus angustifolius* v. L. Zemedel. Arch. 24:354–59. Proc. Intern. Seed Testing Assoc. 7:94. 1935.

11543. —— 1934: Les variations de l'alcaloide au cours de la germination et de la maturité de la graine du *Lupinus angustifolius* v. L. Zemedel. Arch. 24:447–51. Proc. Intern. Seed Testing Assoc. 7:94–95. 1935.

11544. Kunze, R. E. 1882: The germination and vitality of seeds. Torrey Botan. Club Separate. 14 pp.

11545. Kuo, C. Y. 1954: Inheritance and interrelationships of tryptophan, niacin, protein, and kernel weight in corn. Dissertation Abstr. 14:433–34.

11546. Kupche, A. K. 1938: Biochemistry of vetch. Biokhim. Kul't. Rast. 2:341–57. Chem. Abstr. 35:5537. 1941.

11547. Kuperman, F. M. 1948: Effect of removal of various parts of the grain kernel on the growth of sprouts. Dokl. Akad. Nauk SSSR 60:1425–28.

11548. —— 1949: Effect of mechanical injuries on the field germination of [cereal] seeds. Selek. i Semen. 26(1):62–66.

11549. —— 1950: Irregularity in the development of embryonic spike in cereals. Dokl. Vses. Akad. Sel'skokhoz. Nauk 15(3):33–38.

11550. Kuppers, J. R. 1953: Some biometric observations on cacao fruit. Science 117:354–55.

11551. Kuppuswamy, S., J. M. Rao, M. Srinivasan, and V. Subrahmanyan. 1958: Ascorbic acid in germinating seeds of *Sesbania grandiflora* Pers. Current Sci. 27:343–45.

11552. Kuptsov, A. I. 1943: Effect of large seeds on early ripening of field crops. Priroda 32(6):46–50.

11553. Kurahtanova, T. I., and A. P. Scerbakov. 1960: Vorsaatbehandlung der Samen von *Pinus silvestris* und *Caragana microphylla* mit Spurenelementen. Sb. Mosk. Akad. Nauk SSSR 1960:66–74.

11554. Kurasawa, H., and Y. Yamamoto. 1957: Biochemical studies of rice starch. II. Physico-chemical properties of paddy rice-grain starch during maturity. J. Agr. Chem. Soc. Japan (English Transl.) 31: 516–19. Chem. Abstr. 52:20294. 1958.

11555. Kurata, H., K. Ogasawara, and V. L. Frampton. 1957: Microflora of milled rice. Cereal Chem. 34:47–55.

Kurata, H. 14116.

11556. Kurata, T. 1954: Seed size of Formosan linalool tree. J. Japan. Forestry Soc. 36:236–38. Biol. Abstr. 29:17218. 1955.

11557. Kurbanov, R. I. 1962: Obtaining sprouts of scarified dry seeds of gum-bearing *Astragalus* plants in the soils of the mountain zones of Azerbaijan. Dokl. Akad. Nauk Azerb. SSR 18(7):47–51.

11558. Kurbatov, V., and S. Glückman.

1929/30: Materials to the question of the seed-stimulation. 3 parts. Bull. Appl. Botany, Genet., Plant Breed. (USSR) 23:155–298.

11559. —— and —— 1930: Influence of inorganic ions on the properties of seeds. Protoplasma 9:34–96.

11560. Kurdiani, S. Z. 1932: Biology of forest trees; experiments and observations. Tiflis Inst. Forestry. 134 pp.

11561. Kurdina, V. N. 1961: Chemical composition and equilibrium moisture content of vegetable seeds. Dokl. Mosk. Sel'skokhoz. Akad. 1961(62):505–8. Chem. Abstr. 58:9411–12.

11562. —— 1962: Changes in weight and quality of stored vegetable seeds. Mosk. Ordena Lenina Sel'skokhoz. Akad. Im. K. A. Timiryazeva Dokl. 77:443–50.

11563. Kurdzhieva, N. 1956: Analysis of the toxic substances and oil of cockle (*Agrostemma githago*) seeds. Selskostopanska Akad. "Georgi Dimitrov" Zootekh. Fak. Nauchn. Trudove 6:285–93.

11564. Kurgatnikov, M. M. 1940: Structural differences and the changes in the starch grains during the ripening of pea seeds. Biochemistry (USSR) (English Transl.) 5:417–31. Chem. Abstr. 35:4804. 1941.

11565. Kurginyan, R. G. 1962: Results of the treatment of corn seed with trace elements. Izv. Min. Sel'skokhoz. Arm. SSR Sel'skokhoz. Nauki 1962(2):69–72. Biol. Abstr. 43:20620. 1963.

11566. Kurguzov, P. I. 1955: Effect of depth of seed planting on yields of spring wheat. Zemledelie 3(3):116–18.

11567. Kurihara, H. 1936: The endosperm formation in *Cryptomeria japonica*. Sci. Rept. Tohoku Univ. Fourth Ser. 11:185–89. Japan. J. Botany 9:10. 1937.

11568. Kurimoto, T., H. Takada, and S. Nagai. 1954: Physiology of *Metasequoia glyptostroboides* and related species of conifers. I. J. Inst. Polytech. Osaka City Univ. Ser. D 5:55–65.

11569. Kurnik, E., and B. I. Pozsar. 1961: Stability in the ratio of unsaturated and saturated fatty acids during germination in soybeans. Iregszemcse. Delkeletdunantuli Mezogazdasagi Kiserlet. Intezet. Kozlem. 1961(1):11–20.

11570. ——, ——, and J. Varsanyi. 1961: The effect of gibberellic acid on the oil content of soybeans during various photoperiods. Kiserl. Kozl., A. Novenytermesz. 53(3):3–23. 1960. Chem. Abstr. 60:2263. 1964.

11571. Kuroda, C., and M. Wada. 1934: The coloring matter of the adzuki bean. Proc. Imp. Acad. Japan 10:472–74. Chem. Abstr. 29:1134. 1935.

Kuroda, S. 8850, 8851.

11572. Kurokami, T., M. Mori, T. Takematsu, and J. Takahashi. 1950: Studies on the development of keeping quality of chestnut fruits by delaying their germination with phytohormone treatments. Kagawa Agr. Coll. Tech. Bull. 2(1):39–49.

11573. —— and T. Takematsu. 1949: Experimental studies on phytohormones. II. Comparison of phytohormone treatments between spraying on foliage of radishes and seed immersion before setting in field of same. Kagawa Agr. Coll. Tech. Bull. 1(2):17–20.

11574. —— and —— 1950: Studies on the development of keeping quality of chestnut fruits by delaying their germination with phytohormone treatments. Kagawa Agr. Coll. Tech. Bull. 1(3):82–89.

11575. Kurosaki, Z., and S. Iizumi. 1960: On the relationship between grazing habits and vegetation of grassland. VI. On the dissemination of seeds by grazing cattle. Tohoku Univ.Rept.Inst.Agr.Res.11:59–67.

11576. Kurssanow, A. L. 1932: Biochemische Untersuchung über die Fruchtreife der japanischen Mispel (*Eriobotrya japonica*). Planta 15:752–66.

11577. —— and K. Bryushkova. 1939: Reversible enzyme action in germinating seeds. Biochemistry (USSR) (English Transl.) 4:566–74. Chem. Abstr. 34:5882. 1940.

11578. Kurth, H. 1956: The vernalization of some leguminous plants, especially of hairy vetches. Euphytica 5:63–70.

11579. Kurtz, E. B., Jr., and S. M. Alcorn. 1960: Some germination requirements of Saguaro cactus seeds. Cactus Succulent J. Gt. Brit. 32:72–74. Biol. Abstr. 36:21596. 1961.

Kurtz, E. B., Jr. 292, 3157.

Kurup, P. A. 7462, 10159.

11580. Kurupp, N. K. B. 1941: Viability test with paddy variety, Cherunel. Proc. Indian Sci. Congr. 28. Paper No. 35. Biol. Abstr. 16:15356. 1942.

11581. Kurzwelly, W. 1903: Über die Widerstandsfahigkeit trockener pflanzlicher Organismen genen giftige Stoffe. Jahrb. Wiss. Botan. 38:291–341.

11582. Kusano, S. 1909: Further studies on *Aeginetia indica*. Botan. Centralbl. Beih. Abt. I 24:286–300.

11583. —— 1915: Experimental studies on the embryonal development in an angiosperm. Tokyo Imp. Univ. Coll. Agr. J. 6:7–120.

11584. Kusatz, H. 1931: Die Bestimmung der Sortenreinheit und Sortenechtheit bei Futterrübensamen. Wien. Landwirtsch. Ztg. 81:98–99. Botan. Centralbl. 21:118. 1932.

Kushelewa, E. P. 1274.

11585. Kushima, K., K. Katagiri, and Y. Takahashi. 1962: Studies on vernalization of peas. Proc. Crop Sci. Soc. Japan 30:318–20.

Kushizaki, M. 6821.

11586. Kushman, L. J., and J. H. Beattie. 1948: Production of "pops" in Spanish and Virginia type peanuts as affected by nutritional level and balance under greenhouse conditions. Proc. Am. Soc. Hort. Sci. 51:547–51. Biol. Abstr. 23:16470. 1949.

Kushman, L. J. 2153.

11587. Kust, C. A. 1963: Dormancy and viability of witchweed seeds as affected by temperature and relative humidity during storage. Weeds 11:247–50.

Kust, C. A. 16295.

Kustova, A. K. 14594.

Kusuhara, R. 10372.

11588. Kutaček, M. 1964: Glucobrassicin, a potential inhibitor of unusual type affecting the germination and growth of plants; mechanism of its action. Biol. Plant. Acad. Sci. Bohemoslov. 6:88–98.

11589. ——, K. Rokosova, and R. Retovsky. 1959: Metabolism of exogenous tryptophan and β-indoleacetic acid in extirpated wheat embryos. Biol. Plant., Praha 1:54–62. Chem. Abstr. 53:17251. 1959.

11590. Kuthy, S. 1937: The changes in the content of vitamins B and C during the germination of grains. Congr. Intern. Tech. Chim. Ind. Agr. 5th, Compt. Rend. 1:119–26. Chem. Abstr. 32:3786–87. 1938.

Kutova, J. 9806, 9807.

11591. Kutscher, F., E. Müller, and W. Spahr. 1932: Extractives of *Acanthias vulgaris* embryos. Z. Biol. 93:239–40. Chem. Abstr. 27:1657. 1933.

11592. Kutty, O. V. U. 1955: External characters of seed coconuts and the quality of seedlings. Indian Coconut J. 8:74–78.

11593. Kuwabara, Y. 1957: On the seasonal fluctuation of the growth of the weeds in the germinating test practised in every other month. Japan. J. Ecol. 6:141–44. Biol. Abstr. 31:39560. 1957.

11594. Kuwada, H. 1949: Studies on the difference of physiological and ecological characteristics of the artificially raised amphidiploids in comparison with those of the parents. II. The effect of chemicals on the germination of seeds. Kagawa Agr. Coll. Tech. Bull. 1(2):1–11.

11595. —— 1952: Studies on the differences of physiological and ecological characteristics of the artificially raised amphidiploid in comparison with those of the parents. VIII. On the germination of seeds stored in dry state. Proc. Crop Sci. Soc. Japan 21:180–84.

11596. —— 1952: Studies on the differences of physiological and ecological characteristics of the artificially raised amphidiploid in comparison with those of the parents. IX. On the relation between the germination of seeds and the tissue of seed coat. Japan. J. Breeding 2:42–46.

11597. —— 1953: Studies on the differences of physiological and ecological characteristics of the artificially raised amphidiploid in comparison with those of the parents. I. On the drought resistance. Proc. Crop Sci. Soc. Japan 22:1–2.

11598. —— 1953: Studies on the differences of physiological and ecological characteristics of the artificially raised amphidiploid in comparison with those of the parents. X. On the temperature of seed germination. Proc. Crop Sci. Soc. Japan 22:3–4.

11599. —— 1956: Studies on the differences of physiological and ecological characteristics of the artificially raised amphidiploid in comparison with those of the parents. XVII. The effect of the alternative temperature on the seed germination. Kagawa Agr. Coll. Tech. Bull. 7(2):190–92.

11600. —— 1962: Studies on the pod setting and shedding of *Abelmoschus manihot* with special reference to the growth stage and the environment. Proc. Crop Sci. Soc. Japan 30:211–14.

11601. —— and K. Yamamoto. 1952: The germinating percentage of Fl seeds, fertility and other characters of the specific hybrids between *Abel* (*Abelomoschatus*) *esculentus* and *Abel manihot*. Japan. J. Breeding 2:106–8.

11602. Kuwada, S., and S. Yosiku. 1940: Sterol of the seeds of *Momordica cochinchinensis*. J. Pharm. Soc. Japan 60:581–85. Pharm. Abstr. 8:11. 1942.

11603. Kuwada, T. 1957: On the equilibrium relationship between moisture-content of brown rice and humidity in a warehouse. Proc. Crop Sci. Soc. Japan 25:143–44.

11604. Kuykendall, R. 1948: Commercial orchid culture in the United States. The development of an industry. Bull. Missouri Botan. Garden 36:127–37.

Kuzelka, T. J. 1083.

Kuzina, E. F. 7753.

Kuzmanov, B. 15549.

11605. Kuzmenko, A. 1937: The effect of irradiating germinating tobacco seeds with light of various spectral composition on the growth and development of plants. Acad. Sci. RSS Ukr. Inst. Botan. J. 21/22:194–96.

11606. Kuzmenko, A. A. 1937: Experiments on the illumination of seed by light of different wavelengths. Compt. Rend. Acad. Sci. URSS, N.S. 14:227–30. Expt. Sta. Rec. 78:313. 1938.

11607. —— 1939: Vernalization and hardening of makhorka seeds to increase drought resistance. Tabak (USSR) 9(6):25–30. Chem. Abstr. 36:5501. 1942.

Kuzmeski, J. W. 8964.

11608. Kuzmichev, V. P., and N. N. Kulechov. 1955: The use of corn seed which has not reached full maturity for sowing. Zemledelie 1955(7):36–38. Biol. Abstr. 30:35442. 1956.

11609. Kuzmin, I. A. 1961: Storage of pine cones in felling areas and using them for planting. Akad. Nauk SSSR Karel'sk. Filial. Tr. 25:34–42.

11610. Kuzmin, V. P. 1960: On the production of elite seeds and variety-renewal in Kazakhstan. Selek. i Semen. 1960(4): 20–23.

11611. Kuznetsov, P. G. 1953: Storing acorns. Lesn. Khoz. 6(2):57–58.

11612. Kuznetsov, V. P. 1952: Experiments on the effect of boron, molybdenum, and other trace elements on seed sprouting. Mikroelementy v Zhizni Rast. i Zhivotn. Akad. Nauk SSSR, Tr. Konf. po Mikroelementam 1950:561–67. Chem. Abstr. 49:441. 1955.

11613. Kuznetsova, M. S. 1939: A rapid method of determining the viability of seeds. Lesn. Khoz 6(1):19–23.

Kuznetsova, T. N. Zarudnaija-; *see* Zarudnaija-Kuznetsova, T. N.

Kuznetsova, V. S. 17233.

Kuzovkov, A. D. 15441.

11614. Kvaratskheliia, M. S. 1959: Propagation of feijoa from seed. Subtrop. Kult. 1:62–65. Biol. Abstr. 36:50231. 1961.

Kverno, N. B. 5339.

11615. Kvicala, A. 1957: The problem of seed transmission of the lettuce mosaic virus. Sb. Česk. Akad. Zeměděl. Věd Rostlinna Výrobá 3:157–62.

11616. Kvicala, B. 1957: The effect of virus mosaic on sugar-beet grown for seed in the yield and quality of seed. Sb. Česk. Akad. Zeměděl. Věd Rostlinna Výrobá 3:75–86.

11617. Kwang, Y., and H. L. Pearse. 1940: The response of seeds and seedlings to treatment with indolylacetic acid. Ann. Botany (London) 4:31–38. Botan. Centralbl. 176:137–38. 1940.

Kwon, T. W. 11987, 11988.

11618. Kyame, L., and A. M. Altschul. 1946: Comparison of respiration, free fatty acid formation, and changes in the spectrum of the seed oil during the storage of cottonseed. Plant Physiol. 21:550–61.

Kyame, L. 436–438.

11619. Kydrev, T. G. 1959: The effect of presowing drought hardening on wheat caryopses on induced dormancy and germination. Plant Physiol. (USSR) (English Transl.) 6:179–83.

11620. —— and V. M. Kolev. 1962: Thermal treatment of dry wheat seeds and its effect on the plants. Soviet Plant Physiol. 8:460–63.

11621. Kyle, J. H., and T. E. Randall. 1963: A new concept of the hard seed character in *Phaseolus vulgaris* L. and its use in breeding and inheritance studies. Proc. Am. Soc. Hort. Sci. 83:461–75.

11622. Kyurdzhiyeva, V. 1956: Several problems in the biology of rice seed germination. Sb. Nauchn.-Issled. Rabot Stud. Stavropol'sk. Selskoe-khoz. Inst. 1956(4):7–10. Biol. Abstr. 35:17009. 1960.

Kzelkova, N. I. 10224.

L

11623. Laas, E. 1962: On the quality of seeds of *Pseudotsuga*. Estonskaya Sel'skokhoz. Akad. Sb. Nauchn. Trudov 23:139–47.

11624. —— 1963: On the extraction of seeds and the seed quality of the common spruce Estonskaya Sel'skokhoz. Akad. Sb. Nauchn. Trudov 29:3–9.

11625. Labadie, E. A. 1949: A fascinating business—cacti from seed. Horticulture 27(2):49.
Labarca, C. 11534.
Labarca, Y. C. 2790.
11626. Labarre, J., and L. Delcourt. 1942: The proteins of the seed of the horse bean or broad bean (*Vicia faba*). Rev. Can. Biol. 1:72–87. Chem. Abstr. 36:2277. 1942.
11627. —— and R. Dostert. 1942: Progressive denaturation of the proteins of the seed of the horse bean or broad bean (*Vicia faba*). Rev. Can. Biol. 1:504–22. Chem. Abstr. 36:5489–90. 1942.
11628. —— and S. Pfeffer. 1946: Studies on the enzymes of the kidney bean (*Vicia faba* L.), during germination. Rev. Can. Biol. 5:233–46. Chem. Abstr. 40:4412. 1946.
11629. Labbé, H., and F. H. De Balsac. 1936: Présence de la vitamine E dans l'embryon de cacao. Compt. Rend. 203:587–89.
Labbé, H. 5460.
11630. Labo, P. 1949: Moderate and complete scarification of kudzu and perennial *Lespedeza* seed. Agron. J. 41: 491–92.
Labonne, F. 4266.
Laborey, F. 11885, 11886.
11631. Labounek, V., and N. Labounkova. 1961: Effect of provenance on the dormancy of some summer barley varieties of Czechoslovakia. Sb. Česk. Akad. Zeměděl. Věd Rostlinna Výrobá 7:1109–26.
Labounkova, N. 11631.
Labrousse, F. 7817.
11632. Labrum, M. C. 1962: Gladiolus seed culture. Can. Gladiolus Soc. Ann. 1962: 77–78.
11633. Labruto, G., and E. De Angelis. 1939: Composition of seeds and seed oil of *Brachychiton diversifolium*. Ann. Chim. Appl. 29:68–73. Soc. Chem. Ind. J. 58:742. 1939.
11634. —— and G. Reitano. 1954: Chemical composition of the seeds of *Aleurites moluccana*. Ann. Chim. (Rome) 44:397–406. Chem. Abstr. 49:435. 1955.
11635. Lacave-Beauchamp, C. 1960: Acide gibberellique et propriétés enzymatiques des germinations de *Lycopersicum esculentum* Mill. Phénoloxydases. Compt. Rend. 250:2430–32.
11636. Lach, editor. 1936: Jahrbuch des Hauptausschusses für forstliche Saatgutanerkennung e. V. für 1935. Potsdam. 209 pp. Biol. Abstr. 11:6839. 1937.
Lacharme, J. 14347.
Lachiondo, F. Bustinza-; *see* Bustinza-Lachiondo, F.
Lachman, W. H. 8796–8798, 8964.
11637. Lachman, W. L. 1938: Vernalization of vegetable crops. Mass. Agr. Expt. Sta. Bull. 347:78.
Lachover, D. 9250, 15446.
11638. Lacic, B., and B. Mysakowska-Palleolog. 1961: Germination power of *Crambe abyssinica* Hochst. in different conditions of storage. Lublin. Univ. Marii Curé-Sklodowskiej. Ann. Sect. E 14:211–35. 1959.
11639. Lackey, C. F. 1948: Chemical loosening of seed caps in relation to germination of sugar beet seed. Proc. Am. Soc. Sugar Beet Technol. 5:66–69.
11640. Lacour, L. F., and A. Rutishauser. 1956: Spontaneous chromosome breakage in endosperm. John Innes Hort. Inst. Rept. 46:18. Biol. Abstr. 31:19541. 1957.
11641. La Croix, L. J. 1961: Seed dormancy in *Abutilon theophrasti* and *Polygonum pennsylvanicum*. Dissertation Abstr. 22: 717.
11642. —— and D. T. Canoin. 1963: The role of light and other factors in the growth and differentiation of barley embryos. Proc. Plant Physiol. Meetings 38:xxiii.
11643. ——, J. Naylor, and E. N. Larter. 1962: Factors controlling embryo growth and development in barley (*Hordeum vulgare* L.). Can. J. Botany 40:1515–23.
11644. Ladhams, E. 1940: Newer perennials from seeds. Am. Nurseryman 72(11):5–6.
Laer, M. H. van; *see* Van Laer, M. H.
Laffers, A. 17453.
11645. Lafferty, H. A. 1931: The loss of vitality in stored farm seeds. Eire Dept. Agr. J. 30:237–45. Biol. Abstr. 8:11681. 1934.
11646. —— 1931: Purity tolerance and international seed analysis certificates. Proc. Intern. Seed Testing Assoc. 3:6–16.
11647. —— 1932: Purity determinations of cocksfoot by the continental and Irish methods, with special reference to the effect of "light" seeds on germination results. Proc. Intern. Seed Testing Assoc. 4:14–21.
11648. —— 1934: On the evaluation of broken seedlings which produce adventitious roots during a germination test. Proc. Intern. Seed Testing Assoc. 6:341–49.
11649. —— 1936: The effects of tar on the germination of wheat seed. Eire Dept.

Agr. J. 34:63–72. Chem. Abstr. 30:7271. 1936.

11650. —— 1938: The duration of laboratory tests with notes on purity and germination. Proc. Intern. Seed Testing Assoc. 10:212–21.

11651. —— 1940: The effect of certain hormones on barley. Proc. Intern. Seed Testing Assoc. 12:19–31.

11652. —— 1942: The moisture content of wheat in relation to its bushel weight and keeping quality. Eire Dept. Agr. J. 39:230–44. Chem. Abstr. 37:6749–50. 1943.

11653. —— 1953: The farmer's debt to the seed-analyst. Proc. Intern. Seed Testing Assoc. 18:93–104.

11654. —— 1953: Abnormal germination in cereals. Proc. Intern. Seed Testing Assoc. 18:239–47.

Lafon, R. 13360.

11655. La Garde, R. V. 1929: Non-symbiotic germination of orchids. Ann. Missouri Botan. Garden 16:499–514.

11656. Lagasse, F. S. 1944: The present status of the domestic tung industry. Proc. Florida Acad. Sci. 7:133–37. Biol. Abstr. 20:7288. 1946.

Lagasse, F. S. 7226, 17184.

11657. Lagière, R. 1952: Possibilités de transmission de la fusariose du cotonnier. Coton Fibres Trop. 7:146–48. Rev. Appl. Mycol. 32:186. 1953.

11658. —— 1954: Conservation et traitement des semences du cotonnier. II. Exposé des résultats obtenus à la Station I.R.C.T. Bambari. Coton Fibres Trop. 9:97–127. Rev. Appl. Mycol. 33:673. 1954.

11659. —— and J. L. Guyot. 1954: Classement des graines de cotonnier selon leur densité dans l'eau—qualités et infections des deux fractions. Coton Fibres Trop. 9:128–38. Rev. Appl. Mycol. 33:674. 1954.

11660. Lahiri, A. N., and B. C. Kharabanda. 1961: Dimorphic seeds in some arid zone grasses and the significance of growth differences in their seedlings. Sci. Cult. (Calcutta) 27:448–50. Biol. Abstr. 38: 11222. 1962.

11661. —— and —— 1963: Germination studies on arid zone plants. I. Germination in relation to moisture uptake by grass seeds and the probable role of coumarine in this mechanism. Proc. Natl. Inst. Sci. India Pt. B 29:287–96.

Lahiri, A. N. 17537, 17543.

11662. Lahiri, J. K., and S. Ghosh. 1942: Chemical examination of the seeds of *Barringtonia acutangula* Gaertn. J. Am. Pharm. Assoc. Sci. Ed. 31:193–94.

Lahiri, S. 1846, 1850.

11663. Lai, C. Y. 1958: Enzyme activities of *Canavalia ensiformis* and *Mucuna capitata* during germination. J. Formosan Med. Assoc. (English Transl.) 57:85–93. Chem. Abstr. 52:13010. 1958.

11664. Lai, S. P., K. F. Finney, and M. Milner. 1959: Treatment of wheat with ionizing radiations. IV. Oxidative, physical and biochemical changes. Cereal Chem. 36:401–11.

11665. Laibach, F. 1925: Das Taubwerden von Bastardsamen und die künstliche Aufzucht frü absterbender Bastardembryonen. Z. Botan. 17:416–59.

11666. —— 1929: Ectogenesis in plants. Methods and genetic possibilities of propagating embryos otherwise dying in the seed. J. Heredity 20:200–8. Biol. Abstr. 6:15577. 1932.

11667. —— 1941: Wuchsstoffbildung mit Kohlehydratstoffwechsel (nach Versuchen mit Gurkenkeimlingen). Ber. Deut. Botan. Ges. 59:257–71. Biol. Abstr. 23:5143. 1949.

11668. —— 1956: Über die Brechung der Samenruhe bei *Arabidopsis thaliana* (L.) Heynh. Naturwissenschaften 43:164.

11669. —— and J. Keil. 1937: Über die keimungshemmende Wirkung der natürlichen freien Blausäure. Ber. Deut. Botan. Ges. 55:579–83.

11670. Laidlaw, W. M. R. 1961: Extracting barley embryos for loose smut examination. Plant Pathol. 10:63–66.

11671. Laine, T. A. 1944: Investigations of the structure and enzymic splitting of the seed protein zein. Ann. Acad. Sci. Fennicae Ser. A II 11:7–97. Chem. Abstr. 40:3781. 1945.

11672. Laird, W. E. 1936: Tomato seeds and low temperature. Gardeners Chron. 99: 361.

Laj, F. Pannocchia-; *see* Laj-Pannocchia, F.

Lajon, J. 2710.

Lakhno, E. V. 5580.

11673. Lakon, G. 1940: Die topographische Selenmethode, ein neuer Verfahren zur Feststellung der Keimfähigkeit der Getreidefrüchte ohne Keimversuch. Proc. Intern. Seed Testing Assoc. 12:1–18.

11674. —— 1942: Topographischer Nachweis der Keimfähigkeit der Getreidefrüchte durch Tetrazoliumsalze. Ber. Deut. Botan. Ges. 60:299–305.

11675. —— 1942: Topographischer Nachweis der Keimfähigkeit von Mais durch Tetrazoliumsalze. Ber. Deut. Botan. Ges. 60:434–44.

11676. —— 1949: The topographical tetrazolium method for determining the germinating capacity of seeds. Plant Physiol. 24:389–94.

11677. —— 1950: Die "Triebkraft" der Samen und ihre Festellung nach dem topographischen Tetrazolium-Verfahren. Saatgut-Wirtsch. 2:37–39. Biol. Abstr. 25:21359. 1951.

11678. —— 1954: The application of a combined tetrazolium solution in topographical germination methods. Saatgut-Wirtsch. 6:207–8. Chem. Abstr. 49:9749. 1955.

11679. —— and H. Bulat. 1955: Die Festellung der Keimfähigkeit der Kompositenfrüchte nach dem Topographischen Tetrazolium-Verfahren. Saatgut-Wirtsch. 7:201–4.

11680. Lakshmanan, K. K. 1961: Embryological studies in the Hydrocharitaceae. I. *Blyxa oxtandra* Planch. J. Madras Univ. 31:133–42. Biol. Abstr. 41:7832. 1963.

11681. —— 1963: Embryological studies in the Hydrocharitaceae. II. *Halophila ovata* Gaudich. J. Indian Botan. Soc. 42:15–18.

Lakshmanan, K. K. 18206.

11682. Lakshminarasimhaiah, A., B. L. Manjunath, and B. S. Nagaraj. 1942: Chemical investigation of the seeds of *Alangium lamarckii*. I. Mysore Univ. J., Sect. B 3:113–16. Chem. Abstr. 37:1439, 1943.

11683. Lal, B. M., V. Prakash, and S. C. Verma. 1963: The distribution of nutrients in the seed parts of Bengal gram. Experientia 19:154–55. Chem. Abstr. 58:14441. 1963.

Lal, B. M. 19081.

Lal, B. N. 17497.

11684. Lal, J. B. 1936: Constituents of the seeds of *Blepharis edulis*, Pers. I. J. Indian Chem. Soc. 13:109–14.

Lal, J. B. 7857, 15196.

11685. Lal, K. N., and G. Prashad. 1947: Studies in crop physiology; growth characters and seed quality in wheat as influenced by nitrogen, phosphoric acid, and potash. Proc. Natl. Acad. Sci. India Sect. B 17:83. Biol. Abstr. 26:4316. 1952.

11686. ——, J. D. Sah, and J. D. Jha. 1946: Studies in crop physiology. Fertiliser effects upon seed quality, photosynthesis, respiration, and chlorophyll content of wheat leaves during two successive generations. J. Indian Botan. Soc. 25:77–88.

11687. Laliberte, N. 1963: Verticality in germination. Assoc. Offic. Seed Analysts News Letter 37(3):34–36.

11688. Lallemand, S. 1930: Étude de l'action des rayons X sur le développement des plantes. Arch. Electr. Med. 38:28–33, 126–40, 179–90, 232–39, 256–66, 349–61, 468–75, 516–23. Biol. Abstr. 7:7847. 1933.

11689. —— 1932: Variations de radiosensibilité des graines de *Lens esculenta* et de *Pisum sativum* au cours de leur hydration. Compt. Rend. Soc. Biol. 109:1284–87.

11690. —— 1934: Sur la faculté et l'energie germinative de graines sèches irradiées. Compt. Rend. 199:1064–66.

11691. —— and S. Lallemand. 1935: Toxicity of ethyl alcohol for dormant and germinating seeds. Bull. Soc. Chim. Biol. 17:1509–21. Chem. Abstr. 30:1088. 1936.

La Loma, J. L. de; *see* De la Loma, J. L.

11692. Laloraya, M. M., and V. K. Rai. 1962: Growth of the seedling of *Phaseolus mungo* L. as affected by auxin and gibberellic acid. Indian. J. Plant Physiol. 5:218–29.

11693. —— and —— 1962: Promotion of hypocotyl growth with gibberellic acid in epigealy growing seedlings. Physiol. Plantarum 15:649–55.

Lamarlière, M. L. G. de; *see* De Lamarlière, M. L. G.

11694. Lamb, C. A., and E. G. Bayfield. 1941: The influence of season and location on the grain of several wheat varieties. J. Am. Soc. Agron. 33:294–303. Biol. Abstr. 15:15391. 1941.

Lamb, C. A. 5891, 17509.

11695. Lamba, P. S. 1949: The relation of glume measurements to kernel shape and size in wheat. Agron. J. 41:167–68.

11696. Lambdin, R. S. 1947: Comparison of methods employed in the germination of beet seed. Proc. Am. Soc. Sugar Beet Technol. 1946:285–86. Biol. Abstr. 22: 14763. 1948.

11697. —— 1947: Processing sheared seed. Proc. Am. Soc. Sugar Beet Technol. 1946:600–2. Biol. Abstr. 22:12096. 1948.

11698. —— and W. H. Buckingham. 1948: Beet seed decortication with a commercial huller. Proc. Am. Soc. Sugar Beet Technol. 5:100–1.

11699. Lambers, H. R. 1930: Polyembryony and polyspermy of coffee. Arch. Koffiecult. Indonesie 4:32–43.

11700. Lambert, A. 1941: La conservation du blé a l'air libre par le procède des cellules métalliques. Génie Rural 44:378–79.

11701. Lambert, D. A. 1963: The influence of density and nitrogen in seed production stands of S37 cocksfoot (*Dactylis glomerata* L.). J. Agr. Sci. 61:361–73.

11702. Lambert, D. W., W. W. Worzella, R. C. Kinch, and J. N. Cheadle. 1950: Devitalization of cereal and weed seeds by high frequency. Agron. J. 42:304–6.

11703. Lambert, J. 1931: Contribution à l'étude de la radio-sensibilité des graines au début de la germination. Compt. Rend. Soc. Biol. 108:240–44.

11704. —— 1932: Nouvelles recherches sur la radio-sensibilité des graines au début de la germination. Compt. Rend. Soc. Biol. 110:580–83.

11705. —— 1932: Les facteurs de la radiosensibilité des graines avant l'apparition des phénomènes morphologiques. Compt. Rend. Soc. Biol. 110:583–86. Ber. Wiss. Biol. 23:321. 1932.

11706. —— 1933: Les variations du métabolisme de l'embryon de pois au début de la germination. Compt. Rend. Soc. Biol. 112:1564–69.

11707. —— 1933: Sur la place du métabolisme cellulaire dans les facteurs de la radiosensibilité des graines avant l'apparition des phénomènes morphollogiques. Compt. Rend. Soc. Biol. 112:1570–73.

Lambert, J. 7810.

11708. Lambert, J. P. 1956: Seed production studies. I. The effect of nitrogenous fertilizer, applied at different rates and dates, on seed production in cocksfoot (*Dactylis glomerata* L.). New Zealand J. Sci. Technol. Sect. A 37:432–42.

11709. —— 1956: Seed production studies. II. The effect of grazing on seed production in cocksfoot (*Dactylis glomerata* L.). New Zealand J. Sci. Technol. Sect. A 37:443–50.

11710. —— 1956: Seed production studies. III. The effect of nitrogenous fertilizer applied at different rates and dates on seed production in timothy (*Phleum pratense* L.). New Zealand J. Sci. Technol. Sect. A 37: 467–77.

Lambert, J. W. 9134, 12035.

11711. Lambert, R. G., and A. J. Linck. 1958: Effects of high temperature on yield of peas. Plant Physiol. 33:347–50.

11712. Lambeth, V. N. 1950: Some factors influencing pod set and yield of the lima bean. Missouri Agr. Expt. Sta. Res. Bull. 466. 60 pp.

11713. Lambin, A. Z. 1952: Effect of presowing treatment of seeds with solutions of trace elements on the crop of agricultural plants. Mikroelementy v Zhinzi Rast. i Zhivotn. Akad. Nauk SSSR, Tr. Konf. po Mikroelementam 1950:381–99. Chem. Abstr. 49:552. 1955.

11714. Lambou, M. G. 1953: 2,3,5-Triphenyltetrazolium chloride as a rapid indicator of viability in cottonseed. Science 117:690–93.

11715. ——, M. Z. Condon, and A. M. Altschul. 1951: New ideas on problems of storage. Yearbook Agr. U.S. Dept. Agr. "Crops in War and Peace" 1950/51:551–57.

11716. ——, ——, E. Jensen, F. R. Andrews, and A. M. Altschul. 1958: A rapid method for the measurement of the inhibition of deterioration in intact seeds. Plant Physiol. 23:84–97.

11717. ——, G. S. King, and M. Z. Condon. 1948: Effect of chemical treatment prior to storage on viability and growth of cottonseed. Science 108:563–64.

11718. ——, R. Y. Mayne, B. E. Proctor, and S. A. Goldblith. 1952: Effects of high-voltage cathode-ray irradiation on cottonseed. Science 115:269–70.

11719. ——, N. S. Parker, and H. R. Carns. 1956: Effect of maleic hydrazide applied to the cotton plant on the development of free fatty acids in the seed. J. Am. Oil Chemists Soc. 33:199–202. Chem. Abstr. 50:9528. 1956.

Lambou, M. G. 433, 4330, 4331, 4333, 9784.

11720. Lami, R. 1927: Influence d'une peptone sur la germination de quelques vandées. Compt. Rend. 184:1579–81.

Lamm, O. 12272.

11721. Lamm, R. 1937: A contribution to the embryology of the potato. Svensk. Botan. Tidskr. 31:217–20.

11722. Lammerts, W. E. 1942: Embryo culture an effective technique for shortening the breeding cycle of deciduous trees and increasing germination of hybrid seed. Am. J. Botany 29:166–71.

11723. —— 1943: Effect of photoperiod and

temperature on growth of embryo-cultured peach seedlings. Am. J. Botany 30:707–11.

11724. —— 1946: Use of embryo culture in rose breeding. Plants & Gardens 2:111.

11725. —— 1950: The use of embryo culture in germination of camellia seeds. Camellia Res. 7–8. (Horticultural Research Committee, Southern California Camellia Society.)

11726. —— 1958: Embryo culture in *Camellia* seed germination. Tourje, E. C., editor. Camellia Culture. Macmillan, New York, pp. 171–74.

Lamoureaux, G. L. 10849.

11727. Lampe, L. 1931: A microchemical and morphological study of the developing endosperm of maize. Botan. Gaz. 91:337–76.

11728. Lamprecht, H. 1932: Inheritance of seed coat color in *Phaseolus*. Hereditas 16:169–211. Expt. Sta. Rec. 66:727. 1932.

11729. —— 1944: On the relation between seed production and pod colour in common garden beans, with special reference to the effects of chlorophyll and xanthophyll. Agri Hort. Gen. 2(3/4):1–19.

11730. —— 1947: The inheritance of seed-coat colour of *Pisum*. Agri Hort. Gen. 5(3/4):85–105.

11731. —— 1951: The inheritance of the seed-coat colour of *Phaseolus vulgaris* L. Agri Hort. Gen. 9(1–2):18–83.

11732. —— 1951: The inheritance of the red pod colour of *Phaseolus vulgaris*. Agri Hort. Gen. 9(1–2):84–87.

11733. —— 1952: Ein Gen für Truncata-Samen bei *Phaseolus*. Agri Hort. Gen. 10(3–4):105–12.

11734. —— 1956: Estudios sobre la capacidad germinativa de las semillas de *Cedrela odorata* L. en relación con el metodo de almacenamiento. Merida. Univ. Los Andes Fac. Cienc. Forestry Bull. 3(12):40–61.

11735. —— 1961: Eine neue Testafarbe von *Pisum*-Samen: (Salmoneus). Agri Hort. Gen. 19(2):213–22.

11736. —— 1961: The inheritance of a *Phaseolus*-type with three cotyledons and on the effect of three new genes. Agri Hort. Gen. 19(3–4):333–43.

11737. —— 1961: Die Vererbung der rezessive roten Testafarbe von *Phaseolus* sowie Bermerkungen zur Manifestation und Symbolik von Testafarbgenen. Agri Hort. Gen. 19(3–4):344–59.

11738. —— 1962: A new gene for the shape of seeds of *Pisum* and a new gene for partly coloring of the testa. Agri Hort. Gen. 20(3–4):137–55.

11739. —— 1963: Die Vererbung des Samentyps einer Erbse aus dem Jordanthal in Israel. Allgemeines über Grösse und Form des Hilums sowie Koppelungs- und Strukturstudien. Agri Hort. Gen. 31(3–4): 111–36.

11740. —— 1963: Ein neuer typ von Teilfarbigkeit der Samenachale von *Pisum*. Agri Hort. Gen. 31(3–4):174–77.

11741. —— and C. Liscano. 1957: Studies on the germination and early stages of growth of *Podocarpus rospigliosii* Pilger. Bol. Inst. Forest. Latino Am. Invest. Capacitacion 2:41–72. Biol. Abstr. 36: 86364. 1961.

11742. Lampton, R. K. 1953: Developmental morphology of ovule and seed of *Asimina triloba* Dunal. [Abstract.] J. Tenn. Acad. Sci. 28:182–83.

11743. Lancaster, D. 1958/59: Cold and warm germination tests of hybrid sorghum seed. Intern. Crop Improvement Assoc. Ann. Rept. 40/41:266–68.

Land, C. E. 14578.

Landabure, P. B. 6049.

11744. Landagora, F. T. Effect of feeding Achiote seed on egg-yolk pigmentation. J. Agr. Univ. Puerto Rico 46:91–96.

11745. Landau, N. 1940: The effect of hetero-auxin on the germination of some seeds. Palestine J. Botany 1:409–13.

11746. —— 1949: Effect of spacing on the bolting of onions. Nature 164:496–97.

11747. —— 1958: Vernalization effects in mangold and sugar-beet under thermal conditions unfavorable to flowering. Bull. Res. Council Israel Sect. D 6:127–28.

11748. Lander, P. E., and P. L. C. Dharmani. 1935: Digestibility trials on Indian feeding stuffs. IX. American cotton seeds. 2. Indian J. Vet. Sci. 5:343–49. Chem. Abstr. 30:3032. 1936.

11749. Landes, M. 1946: Seed development in *Acalypha rhomboidea* and some other Euphorbiaceae. Am. J. Botany 33:562–68.

11750. Landgraf, H. 1960: Guava seed oil. Rev. Agr. Piracicaba 35:179–81.

11751. Landgraf, J. E. 1961: Über den Einfluss des Lichtes auf den Proteinstoffwechsel bei Keimlingen von *Sinapis alba* L. Planta 57:543–56.

11752. Landi, E. 1936: Studi sulla fisiologia del gelso (*Morus alba* L.). Variazione della germinabilita di semi-gelso nei varia mesi dell'anno. Padua Staz. Bacol. Sper. Ann. 48:639–66.

11753. Landi, R. 1951: Study of lesions in wheat seeds. Genet. Agr. 3:121–28.

Landragin, P. A. 8198–8200.

11754. Landrau, P., Jr., G. Samuels, and P. Rodriguez. 1953: Influence of fertilizers, minor elements, and soil pH on the growth and protein content of tropical kudyu. J. Agr. Univ. Puerto Rico 37:81–95.

Landrau, P., Jr. 16696.

11755. Lane, B. G., and F. W. Allen. 1961: The terminal residues of wheat germ ribonucleates. Biochim. Biophys. Acta 47:36–46.

11756. Lane, G. R., and R. D. Butler. 1958: Effect of a single light exposure on germinating broad bean seeds. Nature 181: 1081.

11757. —— and —— 1959: Effect of light on maturing broad bean seeds. Nature 183:1063–64.

Lane, G. R. 10830, 10831.

Lane, R. H. 18482.

Lane, R. L. 3870.

11758. Lang, A. 1949: Experiments on the mechanism of cold action in vernalisation of rye. Am. J. Botany 36:826.

11759. —— 1952: Vernalization and its interrelation with photoperiodism in "Physiology of Flowering." Ann. Rev. Plant Physiol. 3:290–96.

11760. —— 1956: Induction of flower formation in biennial *Hyoscyamus* by treatment with gibberellin. Naturwissenschaften 43:284–85.

11761. —— and G. Melchers. 1947: Vernalisation und Devernalisation bei einer zweijährigen Pflanze. Z. Naturforsch. 2b:444–49.

11762. ——, J. A. Sandoval, and A. Bedri. 1957: Induction of bolting and flowering in *Hyoscyamus* and *Samolus* by a gibberellin-like material from a seed plant. Proc. Natl. Acad. Sci. U.S. 43:960–63.

Lang, A. 2644, 16503.

11763. Lang, A. L., J. W. Pendleton, and G. H. Dungan. 1956: Influence of population and nitrogen levels on yield and protein and oil content of nine corn hybrids. Agron. J. 48:284–89.

Lang, K. Linderstrom-; *see* Linderstrom-Lang, K.

11764. Lang, R. C., and H. J. Gorz. 1960: Factors affecting embryo development in crosses of *Melilotus officinalis* × *M. alba.* Agron. J. 52:71–74.

Langan, M. C. 2452.

11765. Langdon, L. M. 1934: Embryogeny of *Carya* and *Juglans*, a comparative study. Botan. Gaz. 96:93–117.

11766. Langdon, O. G. 1958: Cone and seed size of South Florida slash pine and their effects on seedling size and survival. J. Forestry 56:122–27.

11767. Lange, A. H. 1957: Gibberellins—new plant growth regulators. Hawaii Farm Sci. 6:10. Chem. Abstr. 52:14764. 1958.

11768. —— 1961: Effect of the sarcotesta on germination of *Carica papaya.* Botan. Gaz. 122:305–11.

Lange, W. H. 11918.

11769. Lange, W. H., Jr., E. C. Carlson, and W. R. Corrin. 1951: Seed treatments for control of the seed-corn maggot in Northern California. J. Econ. Entomol. 44:202–8.

11770. ——, ——, and L. D. Leach. 1949: Seed treatments for wireworm control with particular reference to the use of lindane. J. Econ. Entomol. 42:942–55.

11771. Langeld, F. 1930: Oil content in the seeds of truck and garden crops and of the soya bean. J. Agr. Sci. S.E. USSR 8:355–65. Biol. Abstr. 6:25432. 1932.

11772. Langer, R. H. M. 1955: Ear formation in timothy grass (*Phleum pratense*) following vernalization and short day treatments. Nature 176:263.

Langer, S. 2376.

11773. Langfield, E. C. B. 1957: Time of harvest in relation to grain breakage on milling in rice. J. Australian Inst. Agr. Sci. 23:340–41.

11774. Langham, D. G. 1942: Fertile tetraploids of sesame, *Sesamum indicum* Loew, induced by colchicine. Science 95:204.

11775. —— 1945: Plantas de ajonjolí (*Sesamum indicum* L.) que se tornan amarillas antes de la madurez. Agr. Venezolano 10(111):16. Biol. Abstr. 20: 13559. 1946.

11776. Langlet, O. 1934: New light on the problem of tree seed origin. Skogen (Stockholm) 21:245–52. Biol. Abstr. 9:5684. 1935.

11777. —— 1943: Photoperiodicity and seed origin in Scotch pine. Swedish Inst. Exptl. Forestry Rept. 32:295–330.

11778. —— 1954: Importance d'une spécification rigoureuse de l'origine des semences. Congr. Intern. Botan. Paris Rappt. Commun. 8:230–31.

Langley, B. C. 1472.

11779. Langner, W. 1960: Keimungsverlauf bei den Herkünften eines Provenienzversuches mit *Laris leptolepis* (Sieb. et Zucc.) Gord. Silvae Genet. 9:165–67.

11780. Lanham, F. B. 1939: The germination of grain sorghums. Georgia Univ. Res. Invest. Activities 39(8):57–59.

11781. Lanquist, K. B. 1946: Tests of seven principal forest tree seeds in northern California. J. Forestry 44:1063–66.

11782. —— 1961: Cold treatment of seed and moisture absorption. Tree Planters Notes 48:21–22.

11783. —— and R. A. Dragoo. 1958: The use of tetrazolium for quick germination tests. J. Forestry 56:762.

11784. Lansade, M. 1948: La désinfection des semences de céréales. L'Association Générale des Productions de Blé et Autres Céréales, Paris. 11 pp. Rev. Appl. Mycol. 28:386–87. 1949.

Lansdown, A. R. 3171.

11785. Lanson, H. J., D. Habib, and P. E. Spoerri. 1945: Milkweed seed oil. Potential value in protective coating. Ind. Eng. Chem. 37:179–81.

11786. Lantelme, W. 1933: Künstliche Herbeiführung von Fruchtbildung an Waldbäumen. Z. Forst- Jagdw. 65:378–85. Biol. Abstr. 8:14207. 1934.

11787. Lantz, C. W. 1927: Respiration in corn with special reference to catalase. Am. J. Botany 14:85–105.

11788. —— 1928: Temperature and catalase activity in germinating corn. Proc. Iowa Acad. Sci. 35:147–49.

Lanuza, E. A. 18714.

Lanye, O. de; *see* De Lanye, O.

11789. Lanza, F. 1952: Vernalization of some species of *Triticum*. I. Agrobiological studies. Ann. Staz. Chim.-Agrar. Sper. Roma, Ser. III, Pubbl. 85. 19 pp. Chem. Abstr. 48:3473. 1954.

11790. —— 1957: Methodology and factors affecting the germination of maize (*Zea mays* L.) in the cold test. Maydica 2:115–34.

11791. Lanza, J., A. Gayte-Sorbier, F. Busson, and P. Lunven. Contribution à l'étude chimique des graines de Kapok (*Ceiba guineensis* Chev. var. Clausa). Oléagineux 15:719–20. Hort. Abstr. 31:132. 1961.

Lanza, J. 3447.

11792. Lanzani, G. A., M. G. Cornini, and P. Sequi. 1962: Action of low temperature on biochemical and enzymic aspects of seed germination. V. Biochim. Appl. 9:41–45. Chem. Abstr. 58:5994. 1963.

11793. ——, ——, and L. Tagliabue. 1961: Effect of low temperatures on enzymic activity. IV. Peroxidase activity in wheat germs. Biochim. Appl. 8:85–88. Chem. Abstr. 59:6719. 1963.

11794. Lapina, A. 1926: Economical optimum of temperature of seed germination. Bull. Univ. Asie Centr. Taschkent. 13:99–113.

11795. —— 1938: Contribution to the problem of an acceleration of seed germination with special reference to temperature and to the conditions of material the seeds are placed on. Bull. Univ. Asie Centr. Taschkent. 22:369–73. 1937.

11796. La Pine, L. J., and E. Milberg. 1948: A study of the effect of moisture on the longevity of imported chewings fescue seed under laboratory and warehouse conditions Proc. Assoc. Offic. Seed Analysts 38:62–65.

11797. Lapinskaya, S. M., and P. N. Petrova. 1937: How to identify wheat and rye seeds injured by excess disinfection. Bull. Appl. Botany, Genet., Plant Breed. (USSR) Ser. 4(2):129.

Lapinskaya, S. M. 14825.

11798. Lapinski, M., and K. Duczmal. 1959: Preliminary observations on respiration, β-amylases, and proteases activity in developing maize corn. Howdowla Roslin Aklim. Nasiennictwo 3:737–50.

Lara, M. D. 6304.

11799. Large, J. R., D. L. Fernholz, S. Merrill, Jr., and G. F. Potter. 1947: Longevity of tung seed as affected by storage temperatures. Proc. Am. Soc. Hort. Sci. 49:147–50.

11800. Larionoff, D. K. 1926: Über achener einiger Arten der *Centaurea* L. Jard. Botan. Inst. Essais Semences, Leningrad, Ann. Essais Semences 1926(4):1–26.

11801. —— 1926: Über die Samen der *Lycopsis orientalis* L. und *Nonnea pulla* D.C. Jard. Botan. Inst. Essais Semences, Leningrad, Ann. Essais Semences 1926(8):26.

La Rivière, H. C. C. 141.

11802. Larmour, R. K., and H. N. Bergsteinsson. 1938: The effect on wheat quality of long exposure to carbon tetrachloride. Can. J. Res. Sect. C 16:241–47.

11803. ——, J. S. Clayton, and C. L. Wrenshall. 1935: A study of the respiration and heating of damp wheat. Can. J. Res. 12:627–45.

11804. —— and W. F. Geddes. 1935: Comparison of various harvesting methods in respect to moisture content and grade of the grain (final report). Can. J. Res. 12:614–26.

11805. ——, H. R. Sallans, and B. M. Craig. 1944: Hygroscopic equilibrium of sunflowerseed, flaxseed, and soybeans. Can. J. Res. Sect. F 22:1–8.

11806. ——, ——, and —— 1944: Respiration of whole and dehulled sunflowerseed and of flaxseed. Can. J. Res. Sect. F 22:9–18.

Larmour, R. K. 12769, 16668, 16671.

11807. La Rotonda, C. 1928: Faculté germinative et concentration hydrogenionique. Ann. Tec. Agrar. 1:63–87. Proc. Intern. Seed Testing Assoc. 7:116. 1935.

11808. —— 1930: Germinability and hydrogen-ion concentration in relation to size of seed. Ann. Tec. Agrar. 3:409–21. Expt. Sta. Rec. 67:372. 1932.

11809. —— 1932: Faculté germinative et concentration hydrogenionique. Le pouvoir-tampon des substances de réserve. Ann. Tec. Agrar. 5:125–39. Proc. Intern. Seed Testing Assoc. 7:117–18.

Larrival, M. T. Cerceau-; *see* Cerceau-Larrival, M. T.

11810. Larsen, C. S. 1942: Planting of aspen and poplar. Dansk Skovfor. Tidsskr. 27:145–72. Biol. Abstr. 23:12460. 1949.

11811. Larsen, E. C. 1947: Photoperiodic responses of geographical strains of *Andropogon scoparius*. Botan. Gaz. 109: 132–49.

Larsen, I. 10748.

Larsen, I. T. 13284.

11812. Larsen, J. A. 1922: Some characteristics of seeds of coniferous trees from the Pacific Northwest. Natl. Nurseryman 30:146–49.

11813. —— 1925: Methods of stimulating germination of Western white-pine seeds. J. Agr. Res. 31:889–99.

Larsen, J. A. 3022.

11814. Larsen, P. 1936: Über einen Wuchstoffinaktivierenden Stoff aus *Phaseolus*-keimpflanzen. Planta 25:311–14.

11815. —— 1951: Enzymatic conversion of indole acetaldehyde and naphthalene acetaldehyde to auxins. Plant Physiol. 26:697–707.

11816. Larsen, P. O. 1962: Amino acids and γ-glutamel derivatives in seeds of *Lunaria annua* L. Acta Chem. Scand. 16:1511–18.

Larsen, R. P. 3307.

11817. Larsen, R. T. F. 1960: World seed year and Scotland. Scot. Forestry 14: 205–6.

11818. —— 1961: The certification of forest tree seed in Britain. Proc. Intern. Seed Testing Assoc. 26:411–18.

11819. —— 1962: The certification of forest tree seed in Britain. Proc. World Forestry Congr. 5:708–10. 1960.

11820. Larsen, W. E. 1962: Precision planting of coated seeds. Crops & Soils 14:16–17.

Larsinos, G. J. 2164.

11821. Larson, A. H., R. B. Harvey, and J. Larson. 1936: Length of the dormant period in cereal seeds. J. Agr. Res. 52:811–36.

Larson, G. H. 6293.

Larson, J. 11821.

11822. Larson, K. L., and D. Smith. 1963: Association of various morphological characters and seed germination with the winterhardiness of alfalfa. Crop Sci. 3:234–37.

11823. Larson, M. M. 1961: Seed size, germination dates, and survival relationships of Ponderosa pine in the southwest. U.S. Dept. Agr. Forest Serv. Rocky Mt. Forest Range Expt. Sta. Res. Note 66. 4 pp.

11824. —— 1963: Initial root development of Ponderosa pine seedlings as related to germination date and size of seed. Forest Sci. 9:456–60.

11825. Larson, R. E., and L. Peng-fi. 1949: Embryo size and productivity in segregating generations of tomatoes. Science 109: 567–68.

Larson, R. E. 1459, 15508, 17692.

Larson, R. H. 4348.

Larssen, E. 2247.

11826. Larter, E. N., and J. Whitehouse. 1958: A study of yield and protein response of malting barley varieties to different fertilizers. Can. J. Plant Sci. 38:430–39.

Larter, E. N. 11643.

11827. Larter, L. N. H. 1947: Maize seed storage. Trop. Agr. (Trinidad) 1947(4/6): 40–45.

11828. La Rue, C. D. 1936: The growth of plant embryos in culture. Bull. Torrey Botan. Club 63:365–82.

11829. —— 1947: Growth and regeneration of the endosperm of maize in culture. Am. J. Botany (Suppl.) 34:7a–8a.

11830. —— 1952: Growth of the scutellum of maize in culture. Science 115:315–16.

11831. —— and G. S. Avery, Jr. 1938: The development of the embryo of *Zizania aquatica* in the seed and in artificial culture. Bull. Torrey Botan. Club 65:11–21.

11832. —— and J. Merry. 1939: The development of excised maize embryos in an atmosphere of nitrogen. Am. J. Botany 26:18s.

11833. —— and T. J. Muzik. 1951: Does the mangrove really plant its seedlings? Science 114:661–62.

La Rue, C. D. 18086.

11834. La Rue, R. C. 1947: Nucellar seedlings and clones. Citrus Leaves 27(12):23–24.

Larzelère, H. E. 14836.

11835. Laschke, W. 1907: Einige vergleichende Untersuchungen über den Einfluss des Keimbettes, sowie des Lichtes auf die Keimung verschiedener Sämereien. Landwirtsch. Versuchssta. 65:295–300.

Laschke, W. 6362.

11836. Lascombes, S. 1957: Germination et développement des plantules d'*Aconitum napellus*. Bull. Soc. Hist. Nat. Toulouse 92:117–22. Biol. Abstr. 33:23441. 1959.

11837. Lashee, A. M., and H. T. Blackhurst. 1956: Biochemical changes associated with dormancy and after-ripening of blackberry seed. Proc. Am. Soc. Hort. Sci. 67:331–40.

11838. Lashuk, G. I., and D. P. Snegirev. 1952: Selection and seed growing of lavender. Selek. i Semen. 19(6):37–46.

11839. Laskowsky, N. 1891: Chemische Analyses der Samen von Runkelrüben. Landwirtsch. Versuchssta. 38:315–19.

11840. Lasser, E. 1938: Der Einfluss von Licht und Jarowisation auf den Befall von Weizen, Hafer und Gerste durch *Tilletia*, *Ustilago* und *Helminthosprorium*. Kühn-Arch. 44:161–210. Rev. Appl. Mycol. 17:512. 1938.

11841. Lassiter, C. A., J. S. Boyd, and E. J. Benne. 1960: Storage of high moisture content corn in upright silos and its feeding value for dairy cows. Mich. Agr. Expt. Sta. Quart. Bull. 43:58–66.

Laszay, G. 10472.

Laszlo, I. 2923, 2924.

11842. Latawiec, K. 1958: Investigations on electrophotometric and fluorescence methods for determination of alkaloids in the seeds of white lupine. Roczniki Nauk Rolniczych Ser. A 79:43–101. Biol. Abstr. 34:14564. 1959.

Lathrop, E. C. 4168.

11843. Latif, S. M. 1934: Viviparie en vegetatieve vermeerdering van orchideën. Orchidee (Bandoeng) 3:185–87.

11844. Latimer, L. P., and C. O. Rawlings. 1937: The occurrence of seedless apples as a result of frost. Proc. Am. Soc. Hort. Sci. 34:111–15. Expt. Sta. Rec. 79:481. 1938.

11845. Latour, H. J. 1942: Notes on the storage of red pine seed for nine years. N.Y. State Conserv. Dept. Notes Forestry Invest. 38:1.

11846. Latshaw, W. L. 1921: Report on sulphur and phosphorus in the seeds of plants. J. Assoc. Offic. Agr. Chemists 5:136–38. Botan. Abstr. 11:1905. 1922.

11847. Latter, O. H. 1941: A factor in the dispersal of Burdock (*Arctium lappa*, Linn.). Nature 148:726.

11848. Latterell, R. L., and D. M. Steffensen. 1962: Changes in sensitivity of maize chromosomes to X-rays during seed germination. Am. J. Botany 49:472–78.

Latterell, R. L. 5249.

Lattes, F. 3599.

11849. Lau, D. 1962: The relationship between quality and germination maturity of malting barley. Z. Pflanzenzücht. 48:215–29.

Laude, H. H. 13521, 15135, 15810, 18004.

11850. Laude, H. M. 1949: Delayed germination of California oatgrass, *Danthonia californica*. Agron. J. 41:404–8.

11851. —— 1956: Germination of freshly harvested seed of some western range species. J. Range Management 9:126–29. Biol. Abstr. 31:2345. 1957.

11852. —— 1956: The seedling emergence of grasses as affected by low temperature. Agron. J. 48:558–60.

11853. —— 1957: Comparative pre-emergence heat tolerance of some seeded grasses and of weeds. Botan. Gaz. 119:44–46.

11854. —— 1962: Fresh-seed dormancy in annual grasses. Calif. Agr. 16(4):3. Biol. Abstr. 41:24476. 1963.

11855. ——, E. H. Stanford, and J. A. Enloe. 1958: Photoperiod, temperature, and competitive ability as factors affecting the seed production of selected clones of Ladino clover. Agron. J. 50:223–25.

11856. ——, J. E. Shrum, Jr., and W. E. Biehler. 1952: The effect of high soil temperature on the seedling emergence of perennial grasses. Agron. J. 44:110–12.

11857. Lauder. 1920: The electrical treatment of seeds. Scot. J. Agr. 3:340–44. Botan. Abstr. 8:31. 1921.

11858. Laudureau, M. A. 1879: Du rôle des corps gras dans la germination des graines végétales. Ann. Agron. 5:554–66.

11859. —— 1880: Étude sur la composition chimique de la graine de lin. Ann. Agron. 6:215–23.

11860. Lauenstein, A. 1953: The effect of glycosides on plant growth. Arch. Gartenbau 1:416–20. Chem. Abstr. 48:8877. 1954.

11861. Lauer, E. 1953: Über die Keimtemperatur von Ackerunkräutern und deren Einfluss auf die Zusammensetzung von Unkrautgesellschaften. Flora (Jena) 120: 557–95.

Lauesen, D. 5493.

11862. Laufer, G. 1934: Nutrient assimilation in the seedling method at different seasons and its dependence upon the quality of the rye seed. Z. Pflanzenernähr. Düng. Bodenk. 34A:229–42. Chem. Abstr. 29:877. 1935.

11863. Laufer, S., H. Tauber, and C. F. Davis. 1944: The amylolytic and proteolytic activity of soybean seed. Cereal Chem. 21:267–74.

11864. Laufke, R. 1960: Paper chromatographic investigation of the seeds of *Strophanthus gratus*. Pharmazie 15:81–82. Biol. Abstr. 36:86057. 1961.

Laughland, D. H. 11865.

11865. Laughland, J., and D. H. Laughland. 1939: The effect of age on the vitality of soybean seed. Sci. Agr. (Ottawa) 20:236–37.

Laughlin, T. L., Jr. 8655.

11866. Laughnan, J. R. 1953: The effect of the sh_2 factor on carbohydrate reserves in the mature endosperm of maize. Genetics 38:485–99.

11867. Laulan, R. 1948: L'activation des semences de pommes de terre aux basses températures. Nature (Paris) 3157:134–35.

11868. Laumont, P. 1953: Organisation de la production et du commerce des semences de plantes de grand culture en Algerie. Proc. Intern. Seed Testing Assoc. 18:339–45.

11869. —— and M. Murat. 1934: Observations sur la moucheture et la mauvaise germination de quelques blés en 1933. Bull. Soc. Hist. Nat. Afr. Nord 25:253–65. Rev. Appl. Mycol. 14:91–92. 1935.

11870. Laurent, E. 1902: Sur le pouvoir germinatif des graines exposées à la lumière solaire. Compt. Rend. 135:1295–98.

11871. Laurie, A. 1943: Germination of belladonna seed and effect of winter mulches on plant mortality. Ohio Agr. Expt. Sta. Bimonthly Bull. 28:227–29.

11872. ——, V. H. Ries, L. C. Chadwick, and G. H. Poesch. 1933: Seed propagation. Ohio Agr. Expt. Sta. Bull. 525:12–13.

Laurie, A. 18018.

11873. Laurie, M. V. 1938: Branching and seed origin in Coorg teak plantations. Indian Forester 64:596–600. Biol. Abstr. 13:10256. 1939.

11874. —— 1939: Germination of nim seed (*Azadirachta indica*). Indian Forester 65:104–6. Biol. Abstr. 14:5569. 1940.

Lauter, N. M. 14624.

11875. Lavallee, E., and R. Desmarteau. 1940: Essais de traitements de semences. Rev. Inst. Agr. d'Oka 14:50–54.

11876. Lavate, W. V., and A. Sreenivasan. 1956: Absence of vitamin B_{12} in sprouted mung (*Phaseolus mungo* var. *radiatus*). J. Sci. Ind. Res. (India) 15C:213–14.

11877. Lavee, S. 1960: Effect of gibberellic acid on seeded grapes. Nature 185:395.

11878. Lavender, D. P. 1955: Ponderosa pine seed germination after exposure to varying temperatures and humidities. Oregon State Bd. Forestry Res. Note 24. 4 pp.

11879. —— 1958: Effect of seed size on Douglas fir (*Pseudotsuga taxifolia*) seedlings. Oregon Forest Lands Res. Cent. Res. Note 32. 8 pp.

11880. —— 1958: Effect of field stratification on Douglas fir (*Pseudotsuga taxifolia*) seed germination. Oregon Forest Lands Res. Cent. Res. Note 33. 8 pp.

11881. —— 1958: Seeding dates and Douglas fir germination. Oregon Forest Lands Res. Cent. Res. Note 34. 15 pp.

11882. —— 1958: Effect of ground cover on seedling germination and survival. Oregon Forest Lands Res. Cent. Res. Note 38. 32 pp.

Laver, M. L. 19805.

11883. Lavialle, P. 1929: Observations et considerations sur la germination accidentelle des graines au sein des fruits charnus. Bull. Soc. Botan. France 76:276–79.

Laviolette, F. A. 1489.

11884. Lavollay, J. 1950/51: Une méthode sensible et précise pour l'étude des inhibiteurs ou des activateurs de la germination. Mem. Soc. Botan. France 1950/51:163–73.

11885. —— and F. Laborey. 1948: Thiourée et germination. Antagonisme entre thio-urée et coumarine. Compt. Rend. 226:2015–16.

11886. —— and —— 1951: Stimulation de la croissance initale de l'orge par traitment des graines à la coumarine. Compt. Rend. 232:2348–50.

Lavrova, M. N. 9552.

11887. Lavy, T. L., and S. A. Barber. 1963: A relationship between the yield response of soybeans to molybdenum applications and the molybdenum content of the seed produced. Agron. J. 55:154–55.

11888. Law, J. H., Jr. 1957: The isolation and chemistry of phytoglycolipide, a new complex lipide from plant seeds. Dissertation Abstr. 17:2407.

Law, L. 6912.

Lawas, J. J. 162.

Lawless, J. R. 8357.

11889. Lawlor, N., Jr., J. F. Swink, H. Jewell, and D. B. Ogden. 1957: Results of seed treatment tests. Proc. Am. Soc. Sugar Beet Technol. 9:529–35.

11890. Lawrence D. B., E. G. Lawrence, and A. L. Seim. 1947: Data essential to completeness of reports on seed germination of native plants. Ecology 28:76–78.

Lawrence, E. G. 11890.

11891. Lawrence, H. 1913: The influence of the radiations of radium upon the germination of seeds (oats). Rept. Australian Assoc. Advan. Sci. 14:325, 326. Expt. Sta. Rec. 34:626. 1916.

11892. Lawrence, J. M. 1950: Formation of reducing substances in pea seeds. Arch. Biochem. 27:1–5.

11893. ——, K. M. Day, E. Huey, and B. Lee. 1958: Lysine content of wheat varieties, species, and related genera. Cereal Chem. 35:169–78.

11894. ——, ——, and J. E. Stephenson. 1959: Nitrogen mobilization in pea seedlings. Plant Physiol. 34:668–74.

11895. Lawrence, T. 1960: Quality of Russian wild ryegrass seed as influenced by time and method of harvesting. Can. J. Plant Sci. 40:474–81.

11896. —— and M. R. Kilcher. 1962: The effect of fourteen root extracts upon germination and seedling length of fifteen plant species. Can. J. Plant Sci. 42:308–13.

Lawrence, T. 10614.

Lawrence, W. 469.

11897. Lawrence, W. H., and J. H. Rediske. 1962: Fate of sown Douglas-fir seed. Forest Sci. 8:210–18.

11898. Lawshe, C. H., and L. E. Albright. 1956: A manual for the selection of competent seed analysts. Purdue Univ. Agr. Expt. Sta. 14 pp.

11899. Lawson, E. H., and R. C. Rossiter. 1958: The influence of seed size and seeding rate on the growth of two strains of subterranean clover (*Trifolium subterraneum* L.). Australian J. Agr. Res. 9:286–98.

11900. Lawton, K., and G. M. Browning. 1949: The effect of tillage practices on the nutrient content and yield of corn. Soil. Sci. Soc. Am. Proc. 13:311–17.

Lawton, K. 18456.

11901. Laycock, D. H. 1951: An experiment with sizes and weights of tea seed. Nyasaland Agr. Quart. J. 10:134–38. Biol. Abstr. 27:23300. 1953.

11902. Lazanyi, A., and I. Cabulea. 1957: Ethylene, propylene, and butylene treatment of seed in regard to the growth of oats. Studii Cercetari Agron. 8:117–28. Biol. Abstr. 35:22093. 1960.

11903. ——, A. Marki, C. Hathazi, and A. Timariu. 1961: Research on the morphogenesis of *Helianthus annuus* L. after treating the seeds with sulfamides (2,4-D and colchicine). Studii Cercetari Biol. (Cluj) 12:175–88. Biol. Abstr. 40:3438. 1962.

11904. Lazar, O. 1961: Germination of sugar beet glomerules as influenced by washing, procession, and enzymatic activities. Sucr. Belge 81:175–96.

11905. —— 1962: Germination of sugar beet seeds as a function of washing, polishing, and enzymic activity. II. Oxygen absorption activities of polyphenoloxidase and decarboxylases in non-polished and polished seeds. Sucr. Belge 81:221–33. Chem. Abstr. 58:14438. 1963.

Lazar, O. 5219–5221, 12956.

11906. Lazarescu, C. 1957: Biometrical measurements in connection with forest seeds standardization. Rev. Padurilor 71:157–59.

11907. Lazarescu, E., V. Bulinaru, and M. Gobjila. 1957: Influence du traitement des semences de mais par ultrasons sur la germination et les processus biochimiques. Probl. Agr. (Bucharest) 9:65–68.

Lazarev, D. N. 10260.

11908. Lazarevic, Z., and M. Korac. 1961: Some remarks on the effect of ionizing rays on *Picea omorika* seeds. Glasnik Sumarskog Fak. Univ. Beogradu 25:257–66.

11909. Lazenby, A. 1955: Germination and establishment of *Juncus effusus* L. The effect of different companion species and of variation in soil and fertility conditions. J. Ecol. 43:103–19.

11910. —— 1955: Germination and establishment of *Juncus effusus* L. II. The interaction effects of moisture and competition. J. Ecol. 43:595–605.

Lazurevskiy, G. V. 14829.

11911. Leach, C. M. 1958: Additional evidence for seed-borne mycelium of *Sclerotinia sclerotiorum* associated with clover seed. Phytopathology 48:388–89.

11912. —— 1962: *Kabatiella caulivora*, a seed-borne pathogen of *Trifolium incarnatum* in Oregon. Phytopathology 52:1184–90.

11913. —— and M. Purpoint. 1956: Seed transmission of *Rhizoctonia solani* in *Phaseolus vulgaris* and *P. lunatus*. Plant Disease Reptr. 40:907.

Leach, C. M. 514, 4612.

11914. Leach, J. G., and E. S. Elliott. 1951: Blackpatch of red clover is seed-transmitted and causes seedling blight. [Abstract.] Phytopathology 41:23.

11915. Leach, L. D. 1944: Incidence of Phoma infection on sugar-beet seeds and the efficiency of seed treatments. Phytopathology 34:935.

11916. —— 1948: Effects of combining fungicidal treatments and coating of seeds. [Abstract.] Phytopathology 38:916.

11917. —— 1949: Use of soluble and wettable fungicides for seed treatment. [Abstract.] Phytopathology 39:12–13.

11918. ——, W. H. Lange, F. J. Hills, and J. B. Kendrick, Jr. 1954: Lima bean seed treatment trials in California 1950–1952. Plant Disease Reptr. 38:193–99.

11919. —— and P. G. Smith. 1945: Effect of seed treatment on protection, rate of emergence, and growth of garden peas. Phytopathology 35:191–206.

Leach, L. D. 1352, 1700, 11770.

11920. Leach, W. 1936: Researches on plant respiration. IV. The relation between the respiration in air and in nitrogen of certain seeds during germination. (b) Seeds in which carbohydrates constitute the chief food reserve. Proc. Roy. Soc. (London) Ser. B 119:507–21.

11921. —— 1942: Studies on the metabolism of cereal grains. I. The output of carbon dioxide by wheat grains during absorption of water and germination. Can. J. Res. Sect. C 20:160–68.

11922. —— 1943: Studies on the metabolism of cereal grains. II. Effect of age and kernel size on the course of respiration of wheat during early germination stages. Can. J. Res. Sect. C 21:289–96.

11923. —— 1944: Studies on the metabolism of cereal grains. III. The influence of atmospheric humidity and mould infection on the carbon dioxide output of wheat. Can. J. Res. Sect. C 22:150–61.

11924. —— and K. W. Dent. 1934: Researches on plant respiration. III. The relationship between the respiration in air and in nitrogen of certain seeds during germination. (a) Seeds in which fats constitute the chief food reserve. Proc. Roy. Soc. (London) Ser. B 116:150–69.

Leach, W. 18016, 18017.

11925. Leake, H. M. 1904/5: On the occurrence of "hard seed" in *Indigofera arrecta* Hochst. J. Roy. Hort. Soc. 29:65–67.

Leavenworth, C. S. 19107.

11926. Lebard, P. 1913: Remarques sur l'ouverture à la germination de quelques akenes de Liguliflores. Rev. Gen. Botan. 25:396–413, 432–42.

11927. Le Barron, R. K. 1944: Influence of controllable environmental conditions on regeneration of Jack pine and Black spruce. J. Agr. Res. 68:97–119.

11928. —— and F. H. Eyre. 1939: The release of seeds from Jack pine cones. J. Forestry 37:305–9.

11929. —— and E. I. Roe. 1945: Hastening the extractions of Jack pine seeds. J. Forestry 43:820–21.

11930. Lebedeff, G. A. 1943: Heredity and environment in the production of hard seeds in common beans (*Phaseolus vulgaris*). Puerto Rico Agr. Expt. Sta. Res. Bull. 4. 27 pp.

11931. —— 1947: Studies on the inheritance of hard seeds in beans. J. Agr. Res. 74:205–15.

11932. —— 1958: Inheritance of loose pericarp in corn. J. Heredity 49:129–32.

11933. —— 1960: Further studies on the inheritance of loose pericarp in corn. J. Heredity 51:15–18.

11934. Lebedev, A. N., and T.V. Pereverzeva. 1931: Methods for the determination of moisture in soybeans. Schriften Zentral. Biochem. Forschungsinst. Nahr. Genussmittelind. (Moscow) 1:200–10. Chem. Zentr. 2:1986. 1932. Chem. Abstr. 27:5836. 1933.

11935. Lebedev, G. V. 1960: Rate of water interchange in swollen plant seeds. Dokl. Botan. Sci. Sect. (English Transl.) 128:246–49.

11936. Lebedeva, O. P., and N. K. Melnikova. 1929: Tobacco-makhorka plants grown for seed. Sb. Tr. po Makh. Prom. 2:5–36. Biol. Abstr. 9:9979. 1935.

11937. Lebedyantzev, A., and N. Dmitriev. 1929: The determination of the oil content of seeds in series. Maslob.-Zhir. Delo 10:3–10; 11:3–9; 12:9–17. Chem. Abstr. 25:1694. 1931.

11938. Lebegue, A. 1948: Embryologie des Crucifères: Polyembryonie chez l'*Arabis lyallii* S. Wats. Bull. Soc. Botan. France 95:250–52.

11939. —— 1948: Recherches embryogéniques sur les Saxifragacées. L'embryon du *Saxifraga caespitosa* L. Bull. Soc. Botan. France 95:327–29.

11940. —— 1950: Polyembryonie chez le *Potentilla aurea* L. Bull. Soc. Botan. France 97:199–201. Biol. Abstr. 26:32019. 1952.

11941. —— 1952: Trois exemples de polyembryonie chez le *Peltiphyllum peltatum* Engl. Bull. Soc. Botan. France 99:254–56.

11942. —— 1954: Embryogénie végétale. Embryogénie des Oenotheracées. Développement de l'embryon chez le *Gaura biennis* L. Compt. Rend. 239:616–18.

11943. —— 1956: Embryogénie des Urticacées. Développement de l'embryon chez le *Parietaria officinalis* DC. Compt. Rend. 243:817–20.

11944. —— 1956: Embryogénie des Urticacées. Développement de l'embryon chez l'*Urtica dioica* L. Compt. Rend. 242:923–26.

11945. —— 1926: Embryogénie des Boragacées. Développement de l'embryon chez le *Borago officinalis* L. Compt. Rend. 254:3413–15.

11946. —— and R. Poix. 1956: Le développement embryonnaire chez l'*Urtica urens* L. Bull. Soc. Botan. France 103:587–89.

11947. —— and P. Sarrazin. 1957: Les modalités du développment embryonnaire chez quelques Crucifères. Bull. Soc. Botan. France 104:466–72.

Lebegue, A. 6320, 18161.

Lebek, J. 12875.

11948. Leben, C., and D. C. Arny. 1952: Seed treatment experiments with helixin B. [Abstract.] Phytopathology 42:469.

11949. —— and —— 1954: Soaking treatments for the control of loose smut of barley. Phytopathology 44:329–30.

11950. ——, ——, and G. W. Keitt. 1953: Small grain seed treatment with the antibiotic, helixin B. Phytopathology 43: 391–94.

11951. ——, ——, and —— 1954: Effectiveness of certain antibiotics for the control of seed-borne diseases of small grain. Phytopathology 44:704–7.

11952. ——, R. W. Scott, and D. C. Arny. 1956: On the nature of the mechanism of the water-soak method for controlling diseases incited by certain seed-borne pathogens. Phytopathology 46:273–77.

Leben, C. 1408–1410.

11953. Leblond, D. 1948[?]: Relation entre le nombre de spores de maladies portées par les grains de semence et la manifestation de ces maladies dans le champ. Rept. Quebec Soc. Protect. Plants 1945/47:141–45. Rev. Appl. Mycol. 28:520–21. 1949.

Leblova, S. 9806.

11954. Leblova-Svobodova, S. 1962: Action of isothiocyanates on germinating plants. Experientia 18:554–55. Chem. Abstr. 58: 4979. 1963.

Leblova-Svobodova, S. 2204, 9807.

11955. Lebrun, J. 1956: Sur l'écologie de la germination chez *Arachis hypogaea* L. Bull. Classe Sci. Acad. Roy. Belg. 52:1146–58. Biol. Abstr. 32:7502. 1958.

11956. —— 1958: Sur l'écologie de la germination chez un hybride trispécifique du genre *Gossypium* L. Bull. Classe Sci. Acad. Roy. Belg. 44:423–43.

11957. Lecat, P. 1959: Effect of maleic hydrazide and potassium sulfate sprays on the morphine content of *Papaver somniferum nigrum* pods. Compt. Rend. Hebd. Séances Acad. Agr. France 12:592–94. Biol. Abstr. 41:12106. 1963.

11958. Lechevallier, D. 1962: Évolution de l'activité de l'α-galactosidase au cours de la germination et de la maturation de diverses semences. Compt. Rend. 255:3211–13.

Le Claire, F. P. 8479.

Le Clerc, J. A. 3574, 7659.

11959. Le Clerg, E. L. 1953: Seed-borne plant pathogens. Plant Disease Reptr. 37:485–92.

Le Clerg, E. L. 8624.

11960. Le Compte, S. B., Jr., J. Szasz, and A. M. Stern. 1956: Soil conditioner aids germination. New Jersey Agr. 38(3):11.

11961. Lecomte, H. 1914: Sur la constitution des graines de *Musa*. Compt. Rend. 159:94–96.

11962. Lecoq, R. 1943: L'huile de germe de blé est-elle toxique? Compt. Rend. Soc. Biol. 137:76–77.

11963. —— 1944: La teneur en vitamine E de l'huile de germe de blé. Compt. Rend. Soc. Biol. 1944:836. Bull. Soc. Botan. France 92:113. 1945.

11964. —— and A. Raffy. 1944: Chemical composition and vitamin B_2 activity of the seeds of *Parkia africana* R. Br. Compt. Rend. 218:169–71. Chem. Abstr. 40:1612. 1946.

11965. Lecrenier, A. 1950: Contribution à l'étude des raisins de serre. Bull. Inst. Agron. Sta. Rech. Gembloux 18:75–88.

Ledin, R. B. 16523.

11966. Leding, A. R., and L. A. Brinkerhoff. 1948: The effect of soaking cotton seed on the incidence of angular leaf spot in New Mexico and Arizona. Phytopathology 32: 974–77.

11967. Ledingham, R. J. 1946: The effect of seed treatment and dates of seeding on the emergence and yield of peas. Sci. Agr. (Ottawa) 26:248–57.

11968. —— 1953: Growth cracks in wheat seeds. Can. J. Agr. Sci. 33:198–99.

11969. ——, B. J. Sallans, and P. M. Simmonds. 1949: The significance of the bacterial flora on wheat seed in inoculation studies with *Helminthosporium sativum*. Sci. Agr. (Ottawa) 29:253–62.

Ledingham, R. J. 16573.

Le Dizet, P. 4468, 4469.

11970. Ledoux, L., and R. Huart. 1963: Absorption of ribonucleic acid (RNA) containing thiouracil, and of deoxyribonucleic acid (DNA) by germinating barley. Arch. Intern. Physiol. Biochem. 71:822–23. Chem. Abstr. 60:7152. 1964.

11971. Leduc, A. 1958: Sur les micromycètes alternarioides de la flore fungique des graines des lins. Rev. Gen. Botan. 65:541–80.

Lee, A. 6701.

11972. Lee, A. E. 1950: The growth in culture of intact seedlings and isolated seedling organs. Am. J. Botany 37:312–18.

11973. —— 1950: The influence of various sugars on the growth in culture of intact seedlings and isolated seedling organs. Am. J. Botany 37:528–33.

11974. —— 1952: The growth of excised immature sedge embryos in culture. Bull. Torrey Botan. Club 79:59–62.

11975. —— 1955: Growth in culture of excised portions of lupine embryos. Botan. Gaz. 116:359–64.

Lee, B. 11893.

11976. Lee, C. H., and H. M. Chen. 1957: The location of water entrances to the seeds of Armand pine (*Pinus armandi*). Taipei Natl. Taiwan Univ. Forest Expt. Sta. Tech. Bull. 17. 8 pp.

11977. —— and S. S. Wang. 1962: Effect of cone size on the germination of Luchu-pine (*Pinus luchuensis*) seeds. Formosan Sci. 16:50–52.

Lee, C. H. 10619.

11978. Lee, C. L. 1956: Fertilization in *Ginkgo biloba*. Botan. Gaz. 117:79–100.

Lee, C. Y. 11987, 11988.

11979. Lee, F. A. 1942: Determination of the maturity of frozen peas. Ind. Eng. Chem. Anal. Ed. 14:241.

11980. —— 1948: Determination of maturity of frozen lima beans. N.Y. Agr. Expt. Sta. Bull. 729. 12 pp.

11981. ——, D. De Felice, and R. R. Jenkins. 1942: Determining the maturity of frozen vegetables. Ind. Eng. Chem. Anal. Ed. 14:240–41.

11982. —— and H. B. Tukey. 1942: Chemical changes accompanying growth and development of seed and fruit of the Elberta peach. Botan. Gaz. 104:348–55.

Lee, F. A. 18828.

Lee, H. D. 19450.

Lee, J. B. 13205, 18845.

11983. Lee, J. G. 1950: Experimental lathyrism produced by feeding singletary pea (*Lathyrus pusillus*) seed. J. Nutr. 40:587–94.

11984. Lee, J. W. 1964: Gluten protein composition in individual wheat grains. Nature 201:79–80.

11985. ——, L. E. A. Clare, and E. J. Underwood. 1949: The influence of crop rotation in the thiamin, riboflavin, and nitrogen contents of wheat. Australian J. Exptl. Biol. Med. Sci. 27:533–42.

11986. —— and E. J. Underwood. 1950: The influence of variety on the thiamin and nitrogen contents of wheat. Australian J. Exptl. Biol. Med. Sci. 28:543–48.

11987. Lee, K. Y., C. Y. Lee, T. Y. Lee, and T. W. Kwon. 1959: Chemical changes during germination of soybean. Carbohydrate metabolism. Seoul Univ. J. 8:35. Chem. Abstr. 54:697. 1960.

11988. ——, ——, ——, and —— 1959: Chemical changes during germination of soybean: Carbohydrate metabolism. Seoul Univ. J. 9:12. Chem. Abstr. 54:13295. 1960.

Lee, M. R. 13520.

11989. Lee, P. N., C. W. Shen, and S. M. Huang. 1961: A study on the relation between size and production of vegetable seeds of same variety. I. Agr. Res. 10(4): 14–19.

11990. Lee, S., and Y. S. Loo. 1958: Report of some experiments on the germination of true sugarcane seeds. Rept. Taiwan Sugar Expt. Sta. 18:1–13. Biol. Abstr. 36:2530. 1961.

Lee, T. Y. 4014, 11987, 11988.

11991. Lee, W. Y. 1936: The formation and distribution of vitamin C in the germinating pea, *Pisum sativum* L. (blue bantam variety). J. Chinese Chem. Soc. 4:219–23. Chem. Abstr. 31:430–31. 1937.

11992. —— 1938: Metabolic studies on germinating soybeans. I. General metabolic changes in the germinating soybean. J. Chinese Chem. Soc. 6:15–22. Chem. Abstr. 35:5151. 1941.

11993. —— and S. L. Li. 1938: Distribution of phosphorus in the germinating soybean. Chinese J. Physiol. 13:257–64. Chem. Abstr. 33:2942. 1939.

11994. —— and B. E. Read. 1936: The effect of light on the production and distribution of ascorbic acid in germinated soybeans. J. Chinese Chem. Soc. 4:208–18. Chem. Abstr. 31:430. 1937.

11995. Leendertz, K. 1928: Means of identifying red fescue and sheep's fescue. Proc. Assoc. Offic. Seed Analysts 20:56–57.

11996. —— 1928: Short impressions about some American seed testing stations and unification of international rules for seed testing. Proc. Intern. Seed Testing Assoc. 1:15–22.

11997. —— 1948: The determination of moisture in seeds. Proc. Intern. Seed Testing Assoc. 14:38–46.

Leenheer, L. de. 5085.

11998. Leese, B. M., Jr. 1958: Identification of Asiatic species of *Phaseolus* by seed characters. Am. Midland Naturalist 60: 132–44. Biol. Abstr. 33:22971. 1959.

11999. Lefebvre, C. L. 1942: *Claviceps yanagawaensis* in imported seed of Japanese lawn grass. Phytopathology 32:809–12.

12000. Lefevre, J. 1937: Quelques effets observés comme suite au traitement de graines et de boutures par diverses substances et specialement par l'heteroauzine. Compt. Rend. 205:1437–39.

Lefevre, L. 5075, 5076.

12001. Leff, J. 1964: Interaction between kinetin and light on germination of Grand Rapids lettuce seeds. Plant Physiol. 39: 299–303.

12002. Leffel, R. C., M. G. Weiss, and H. W. Johnson. 1955: Effectiveness of recurrent and pedigree selection for increased oil content of seed in a soybean cross, Adams × Hawkeye. Agron. Abstr. 47:52.

12003. Leffert, E. M. 1960: Lady-slippers can be grown from seed. Horticulture 38(7):378.

12004. Lefort, M. 1954: Action des doses croissantes de rayons X sur les graines de *Lycopersicum esculentum*. Compt. Rend. 239:1401–3.

12005. —— and L. Ehrenberg. 1955: L'influence de la teneur en eau des graines sur leur sensibilité aux rayons X. Arkiv Botan. 3:121–24.

12006. Leggatt, C. W. 1929: The comparative value of scarified and unscarified sweet clover seed. Sci. Agr. (Ottawa) 9:611–14.

12007. —— 1929/30: Catalase activity as a measure of seed viability. Sci. Agr. (Ottawa) 10:73–110.

12008. —— 1930: Altitude as a factor in germination testing. Proc. Intern. Seed Testing Assoc. 11/12:109–16.

12009. —— 1931: Further studies on the hard seed problem: Alfalfa and sweet clover. Sci. Agr. (Ottawa) 11:418–27.

12010. —— 1933: A further note on catalase activity as a measure of seed viability. Can. J. Res. 9:571–73.

12011. —— 1933: The incidence of weed seeds in duplicate analyses. Proc. Intern. Seed Testing Assoc. 5:34–41.

12012. —— 1935: Contributions to the study of the statistics of seed testing. Proc. Intern. Seed Testing Assoc. 7:27–37.

12013. —— 1935: Contributions to the study of the statistics of seed testing. II. Proc. Intern. Seed Testing Assoc. 7:38–48.

12014. —— 1935: Contributions to the study of the statistics of seed testing. III. A theoretical study of purity tolerance, with special reference to the tolerance for pure seed. Proc. Intern. Seed Testing Assoc. 7:166–73.

12015. —— 1936: Contributions to the study of the statistics of seed testing. IV. The binomial distribution. Proc. Intern. Seed Testing Assoc. 8:5–17.

12016. —— 1936: Purity and germination tolerances fundamentally the same problem. Proc. Assoc. Offic. Seed Analysts 1935: 101–7.

12017. —— 1937: Contributions to the study of the statistics of seed testing. V

Isoprobes for the Poisson distribution. Proc. Intern. Seed Testing Assoc. 9:207–17.

12018. —— 1937: Contributions to the study of the statistics of seed testing. VI. Distribution of particles differing in specific gravity or size. Proc. Intern. Seed Testing Assoc. 9:218–27.

12019. —— 1939: Contributions to the study of the statistics of seed testing. VII. Further studies on the distribution of particles differing in specific gravity or size. Proc. Intern. Seed Testing Assoc. 11:25–39. Biol. Abstr. 16:2469. 1942.

12020. —— 1939: A new seed blower. Proc. Assoc. Offic. Seed Analysts 30:120–32.

12021. —— 1939: Statistical aspects of seed analysis. Botan. Rev. 5:505–29.

12022. —— 1941: A new seed mixer and sampler. Sci. Agr. (Ottawa) 21:233–36.

12023. —— 1941: A study of the relative efficiency of seed sampling methods. Can. J. Res. Sect. C 19:156–62.

12024. —— 1946: Germination of seeds of three species of *Agrostis*. Can. J. Res. Sect. C 24:7–21.

12025. —— 1948: A contribution to the study of dormancy in seeds. *Lactuca sativa* L. Can. J. Res. Sect. C 26:194–217.

12026. —— 1948: Germination of boron-deficient peas. Sci. Agr. (Ottawa) 28:131–39.

12027. —— 1952: Progress in developing international rules for testing forage seeds. Proc. Intern. Grassland Congr. 6:885–90.

12028. —— 1953: International trade in seeds. Proc. Intern. Seed Testing Assoc. 18:112.

12029. —— and R. A. Ingalls. 1949: Size of seed in relation to size and shape of root in Swede turnips. Sci. Agr. (Ottawa) 29:357–69.

Leggatt, C. W. 15612.

Legnini, C. N. 8586.

Le Grande, W. P. 12442.

12030. Lehberg, F. H., W. G. McGregor, and W. F. Geddes. 1939: Flax studies. IV. The physical and chemical characteristics of flaxseed at progressive stages of maturity. Can. J. Res. Sect. C 17:181–94.

Lehberg, F. H. 7069, 7070.

Lehle-Joerges, E. 19211.

12031. Lehman, S. G. 1925: Studies on treatment of cotton seed. N. Carolina Agr. Expt. Sta. Tech. Bull. 26. 71 pp.

12032. —— 1943: Vapor action of certain fungicidal materials prepared for dusting cotton seed. Phytopathology 33:431–48.

12033. —— 1947/48: Delinting seed boosts yields. N. Carolina Agr. Expt. Sta. Ann. Rept. 71:27.

12034. —— and G. W. Fant. 1929: Control of oat smut by seed treatment. N. Carolina Agr. Expt. Sta. Bull. 268. 16 pp.

12035. Lehman, W. F., and J. W. Lambert. 1960: Effects of spacing of soybean plants between and within rows on yield and its components. Agron. J. 52:84–86.

12036. Lehmann, E. 1909: Zur Keimungsphysiologie und -biologie von *Ranunculus sceleratus* L. und einigen anderen Samen. Ber. Deut. Botan. Ges. 27:476–94.

12037. —— 1910: Neuere Untersuchungen über Lichtkeimung. Jahresber. Vereinigung Angew. Botan. 8:248–57.

12038. —— 1911: Temperature und Temperaturwechsel in ihrer Wirkung auf die Keimung lichtempfindlicher Samen. Ber. Deut. Botan. Ges. 29:577–89.

12039. —— 1912: Über die Beeinflussung der Keimung lichtempfindlicher Samen durch die Temperatur. Z. Botan. 4:465–529.

12040. —— 1913: Einige neuer Keimungsarbeiten. Sammelreferat. Z. Botan. 5:365–77.

12041. —— 1913: Über katalytische Lichtwirkung bei der Samenkeimung. Biochem. Z. 50:388–92.

12042. —— 1915: Lichtkeimungsfragen. Z. Botan. 7:560–80.

12043. —— 1918: Minimum illumination interval for the germination of seed of *Lythrum*. Ber. Deut. Botan. Ges. 36:157–63.

12044. —— 1919: Über die keimfördunde Wirkung von Nitrat auf lichtgehemmte Samen von *Veronica tournefortii*. Z. Botan. 11:161–79.

12045. —— 1924: Keimungsversuche mit Samen von *Lythrum salicaria*. Ber. Deut. Botan. Ges. 42:55–60.

12046. —— and F. Aichele. 1931: Keimungsphysiologie der Gräser (Gramineen). Eine Lebensgeschichte des reifenden, ruhenden und Keimenden Grassamens. Ferdinand Enke, Stuttgart. 678 pp.

12047. —— and A. Ottenwaelder. 1913: Über katalytische Wirkung des Lichtes bei der keinunglichtempfindlicher Samen. Z. Botan. 5:337–64.

12048. —— and L. Rao. 1924: Über die Gultigkeit des Produktgesetzes bei der Lichtkeimung von *Lythrum salicaria*. Ber. Deut. Botan. Ges. 42:65–69.

12049. Lehmann, K. 1960: Changes in nitrogen, phosphorous, and potassium contents in various kinds of barley grain as dependent upon the conditions of growing. Poznan. Wysz. Szkol. Roln. Rocz. 9:255–65.

12050. —— and J. Garz. 1962: Reversal of substances containing phosphate and of protein in the cotyledons of germinating pea seeds. Flora (Jena) 152:516–22. Chem. Abstr. 59:6732. 1963.

12051. Lehmann, W. 1937: Zur Keimungsphysiologie des Buchweizens. Landwirtsch. Jahrb. 84:741–78.

Lehrle, L. W. W. 4164.

12052. Lehrman, L. 1945: The extraction of fat and phosphorus from wheat starch. J. Am. Chem. Soc. 67:1541–42.

12053. Lehto, T. V. 1960: Stratifying pine seeds in plastic bags. Am. Nurseryman 111(7):15.

12054. Leibundgut, H. 1960: Etude sur l'influence du charbon de bois sur la germination et le developpement des semis d'epiceas, de pins et de melezes. Schweiz. Z. Forstw. 111:172–78.

12055. Leighty, C. E., and J. W. Taylor. 1924: Electrochemical treatment of seed wheat. U.S. Dept. Agr. Circ. 305. 7 pp.

Leighty, C. E. 12438, 12452.

Leighty, D. H. 1736.

12056. Leijerstam, B. 1962: Studies in powdery mildew on wheat in Sweden. I. Physiological races in Scandinavia in 1960 and 1961. Stat. Växt. Med. 12:94:283–93.

12057. Leininger, Mrs. F. 1933: Growing Regal lilies from seeds. Flower Grower 20(6):250.

12058. —— 1935: Raising aconites from seed. Horticulture 13(1):16.

12059. —— 1938: Seeds—their likes and dislikes. Horticulture 16(9):212.

12060. —— 1939: Growing *Lilium cernuum* from seed. Horticulture 17(21):456.

12061. Leininger, L. N., and A. L. Urie. 1964: Development of safflower seed from flowering to maturity. Crop Sci. 4:83–87.

Leissner, G. Nilsson-; *see* Nilsson-Leissner, G.

Leistner, W. 8387.

12062. Leithe, W. 1936: Determination of the refractive index of fats in oilseeds by means of bromonaphthalene. Z. Untersuch. Lebensm. 71:33–44. Analyst 61: 274–75. 1936.

Leitsis, P. R. 14959.

12063. Leizorowitz, R., and A. Poljakoff-Mayber. 1960: The effect of gibberellin and kinetin on the germination of some photosensitive desert seeds as compared with that of light. Phyton 15:103–7.

12064. Lejeune, A. J., and P. J. Olson. 1940: Seed setting in alfalfa. Sci. Agr. (Ottawa) 20:570–73.

Lejeune, A. R. 9690.

Lejeune, G. 18898.

12065. Le Lacheur, A. G. 1962: The use of verification trials as a method of rating seed classes of climax timothy. Can. J. Plant Sci. 42:116–21.

Lelieweld, G. van; *see* Van Lelieweld, G.

12066. Leliveld, J. A. 1935: Cytological studies in the genus *Ulmus*. II. The embryo sac and seed development in the common Dutch elm. Rec. Trav. Botan. Neerl. 32:543–74.

12067. Lelley, J. 1950: Results of vernalization experiments with Hungarian winter wheat varieties. Agrartudomany 2:202–5. Biol. Abstr. 24:33863. 1950.

Leloir, L. F. 16041.

Leloup, M. 1787.

12068. Lemmon, P. E., R. L. Brown, and W. E. Chapin. 1943: Sulfuric acid seed treatment of Beach pea, *Lathyrus maritimus*, and Silvery pea, *L. littoralis*, to increase germination, seedling establishment, and field stands. J. Am. Soc. Agron. 35:177–91.

Lemmon, P. E. 13488.

12069. Lemmon, R. S. 1944: Harvest of the fields; some pertinent notes on gathering wildflower seeds. Home Gard. 3(6):38–39.

12070. Lemoigne, M., and P. Monguillon. 1930: Présence de l'acetylméthylcarbinol et du 2.3-butyleneglycol chez les plantes supérieures. Formation au cours de la germination. Compt. Rend. 190:1457–59.

12071. Lemoine, M. 1962: Essais de germination et de pretraitement sur differents milieux de graines de conifères. Proc. Intern. Seed Testing Assoc. 27:684–93.

Lemoine, M. 2876.

Lemon, H. B. 17385.

12072. Lenarskii, I. I., and S. M. Khripak. 1958: The group and amino acid composition of corn proteins. Dokl. Soobshch. Uzhgorod. Univ. 1958(2):74–76. Chem. Abstr. 54:693. 1960.

12073. —— and I. I. Lukyanova. 1961: Fractional and amino-acid composition of the seed proteins of *Vicia faba*. Tr. Vses. Nauchn.-Issled. Inst. Zerna i Produktov ego Pererabotki 41:95–100. Biol. Abstr. 42:23665. 1963.

12074. Lendle, A. 1938: Constituents of pumpkin seeds. Arch. Pharm. 276:45–53. Chem. Abstr. 32:4276. 1938.

12075. Leng, E. R. 1963: Component analysis in inheritance studies of grain yield in maize. Crop Sci. 3:187–90.

12076. ——, J. J. Curtis, and M. C. Shekleton. 1950: Niacin content of waxy, sugary, and dent F_2 segregating kernels in corn. Science 111:665–66.

12077. ——, F. R. Earle, and J. J. Curtis. 1951: Direct effect of pollen parent on protein content of the corn kernel. Cereal Chem. 28:479–82.

12078. —— and C. M. Woodworth. 1953: Effects of long-continued selection for oil and protein content in maize. Genetics 38:675.

Leng, E. R. 9162, 19857.

12079. Lengel, P. A. 1960: Development of the pollen and the embryo sac in *Capsicum frutescens* L. var. Japanese variegated ornamental. Ohio J. Sci. 60:8–12.

12080. Lengyel, G. 1938: Bericht über die Tätigkeit des Seideausschusses. Proc. Intern. Seed Testing Assoc. 10:222–29.

12081. Lenkov, L. D. 1957: Die Nachreife des Gerstensamens. Vissh/Selsk. Inst. "Georgi Dimitrov" Agron. Fak. Nauchn. Tr. 4:111–19.

12082. Lennartz, H. 1957: Über die Beeinflussung der Keimfähigkeit der Samen von Grünlandpflanzen beim Durchgang durch den Verdauungstraktus des Rindes. Z. Adker- Pflanzenbau 103:427–53.

12083. Lennox, C. G. 1928: The effect of covering on the germination of sugar cane fuzz. Hawaiian Planters Record 32:14–17.

12084. Lentz, C. P. 1954: Humidification of cold storages. Can. J. Technol. 32:156–63.

Lentz, C. P. 3180.

Lentz, J. von; *see* Von Lentz, J.

12085. Lenz, L. W. 1954: The endosperm as a barrier to intersectional hybridization in iris. Aliso 3:57–58.

12086. —— 1955: Studies in iris embryo culture. I. Germination of embryos of the subsection *Hexapogon* Benth. (Sect. *Regelia sensu* Dykes.) Aliso 3:173–82. Biol. Abstr. 30:20364. 1956.

12087. —— 1956: Development of the embryo sac, endosperm, and embryo in *Iris munzii* and the hybrid *I. munzii* × *I. sibirica* "Caesar's Brother." Aliso 3:329–43. Biol. Abstr. 31:29204. 1957.

Leo, A. de; *see* De Leo, A.

Leo, T. de; *see* De Leo, T.

Leon, N. J. de; *see* De Leon, N. J.

Leonard, E. R. 15784, 19362.

12088. Leonard, J. M. 1941: A study of soybean esterase. Contrib. Biol. Lab. Catholic Univ. Am. 39:1–81. Biol. Abstr. 17:12254. 1943.

12089. Leonard, M. J. K., and R. H. Burris. 1947: A survey of transaminases in plants. J. Biol. Chem. 170:701–9.

Leonard, O. A. 15680, 19717.

12090. Leonard, W. F. 1955: Onion seed production. New Zealand J. Agr. 90:289–93.

12091. Leonard, W. H. 1935: The relation between bushel weight and maturity in corn. J. Am. Soc. Agron. 27:928–33.

12092. —— and A. Clark. 1936: Protein content of corn as influenced by laboratory analyses and field replication. Colorado Agr. Expt. Sta. Tech. Bull. 19. 9 pp. Expt. Sta. Rec. 76:783–84. 1937.

Leonard, W. H. 10587.

12093. Leonardi, P. 1932: Contribution a la connaissance de l'action des sels sur la germination des graines et sur le developpement des petites plantes. Riv. Biol. 14:217. Proc. Intern. Seed Testing Assoc. 7:114–15. 1935.

12094. Leoncini, G. 1927: Utilisation by part of *V. faba* plants in the post-germination period of the minerals in the cotyledons. Boll. R. Ist. Sup. Agrar. Pisa 3:587–96. Biol. Abstr. 6:21561. 1932.

12095. —— 1930: Frutti con semi perfetti e frutti consemi completamente abortiti nel *Diospyros kaki*. Boll. R. Ist. Sup. Agrar. Pisa 6:541–50. Biol. Abstr. 7:14109. 1933.

12096. —— 1930: Can the cotyledonous tissues, in the post-germinating period, have a selective power to the advantage of the young plants. Boll. R. Ist. Sup. Agrar. Pisa 6:657–62. Chem. Abstr. 25:5191. 1931.

12097. —— 1931: Enzymes hydrolyzing glucosides in some seeds. Boll. R. Ist. Sup. Agrar. Pisa 7:603–17. Chem. Abstr. 26:4626. 1932.

12098. —— 1931: The action of sodium, potassium and calcium on germinating seeds. Boll. R. Ist. Sup. Agrar. Pisa 7:619–31. Chem. Abstr. 26:4626. 1932.

Leonov, B. I. 1333.

12099. Leontev, I. 1933: The physicochemical and immunological properties of the protein from the seeds of *Phaseolus aureus* R. I. Tr. Lab. Izucheniyu Belka

Belkovogo Obmena Organizme 1933(5): 62–67. Chem. Abstr. 29:6614. 1935.

12100. Leontjew, H. 1934: Über die Identität der Globuline aus den Samen einiger Cucurbitaceen. Biochem. Z. 274:163–66.

12101. Leopold, A. C. 1955: Auxins and plant growth. Univ. California Press, Berkeley. 354 pp.

Leopold, A. C. 5906.

12102. Le Pelley, R., and S. Koekum. 1954: Experiments in the use of insecticides for the protection of grain in storage. Bull. Entomol. Res. 45:295–311.

12103. Lepigre, A. 1936: The disinfection of grains with mixtures of ethylene oxide and carbon dioxide. Bull. Soc. Encour. Ind. Natl. 135:385–462. Chem. Abstr. 30: 8500–1. 1936.

Lepina, N. A. 11373.

12104. Lepine, P., and F. Bilfinger. 1934: Action comparative sur la croissance des cobayes, du ble germe, du ble boulilli et du son. Compt. Rend. Soc. Biol. 116:290–93. Physiol. Abstr. 19:462. 1934.

12105. Lepper, W. 1933: A test for mercury in treated seeds. Landwirtsch. Versuchssta. 117:1–2, 109–12. Chem. Abstr. 28:3515. 1934.

12106. Leprevost, A. 1952: Chemical study of *Dioclea violacea*. Arquiv. Biol. Tecnol. Inst. Biol. Pesquisas Tecnol. 7:137–41. Chem. Abstr. 48:6512.

12106-A. Lercari, D. 1958: Gibberellic acid used in germination tests. Am. Iris Soc. Bull. 148:57–58.

12107. Lerner, H. R., and M. Evenari. 1961: The nature of the germination inhibitor present in leaves of *Eucalyptus rostrata*. Physiol. Plantarum 14:221–29.

12108. ——, A. M. Mayer, and M. Evenari. 1959: The nature of the germination inhibitors present in dispersal units ot *Zygophyllum dumosum* and *Trigonella arabica*. Physiol. Plantarum 12:245–50.

12109. Leroux, D. 1941: Trace elements and nitrogen content of pea seeds. Compt. Rend. Acad. Agr. France 27:807–10. Chem. Abstr. 40:5184. 1946.

Leroux, D. 16930.

12110. Lesage, P. 1906: Actions indirects de l'électricité sur la germination. Compt. Rend. 143:695–97.

12111. —— 1911: Sur l'emploi des solutions de potasse à la reconnaissance de la faculté germinative de certaines graines. Compt. Rend. 152:615–17.

12112. —— 1916: Essais des graines de *Lepidium sativum* dans des conditions très diverses. Compt. Rend. 163:486–89.

12113. —— 1917: Au voisinage des limites de la germination dans les graines de *Lepidium sativum*. Rev. Gen. Botan. 29:97–112, 137–58, 181–92.

12114. —— 1917: Germination des graines dans les solutions salines. Compt. Rend. 164:639–41.

12115. —— 1922: Sur la détermination de la faculté germinative autrement que par la germination des graines. Compt. Rend. 174:766–67.

Lesaint, C. 4266, 4267.

Lesins, I. 12118.

12116. Lesins, K. 1950: Investigations into seed setting of lucerne at Ultuna, Sweden, 1945–1949. Lantbrukshögskol. Ann. 17: 441–83. Biol. Abstr. 25:31847. 1951.

12117. ——, E. Akerberg, and Z. Böjtös 1954: Tripping in alfalfa flowers. Acta Agr. Scand. 4:239–56.

12118. —— and I. Lesins. 1962: Trueness-to-species in seed samples of *Medicago*, with a note on 2n = 14 species. Can. J. Genet. Cytol. 4:337–39.

12119. Leskova, E. S. 1959: On the biology of seed germination in *Dichroa febrifuga* Lour. Botan. Zh. 44:1344–47.

12120. —— 1961: Germination of the seeds of arctic spoonwort. Agrobiologiya 1961 (1):135–36. Biol. Abstr. 38:15610. 1962.

12121. Lesley, J. W., and J. Bonner. 1952: Development of normal peach seedlings from seeds of early-maturing varieties. Proc. Am. Soc. Hort. Sci. 60:238–42.

12122. —— and M. W. Lesley. 1956: Effect of seed treatments with X-rays and phosphorus 32 on tomato plants of first, second, and third generations. Genetics 41:575–88.

Lesley, J. W. 12123, 12124, 13120.

12123. Lesley, M. M., and J. W. Lesley. 1941: Parthenocarpy in a tomato deficient for a part of a chromosome and in its aneuploid progeny. Genetics 26:374–86.

12124. —— and —— 1958: Male-sterile mutants in the tomato from seed treated with radioactive isotype P^{32}. Proc. Am. Soc. Hort. Sci. 71:339–43.

Lesley, M. M. 12122.

Lespagnol, C. 8886.

Leszczenko, P. 6959.

12125. Le Tourneau, D., G. D. Failes, and H. G. Heggeness. 1956: The effects of aqueous extracts of plant tissue on germination of seeds and growth of seedlings. Weeds 4:363–68.

12126. —— and H. G. Heggeness. 1957: Germination and growth inhibitors in leafy spurge foliage and quackgrass rhizomes. Weeds 5:12–19.

Lettmayr, K. 10716, 10717.

12127. Leukel, R. W. 1926: The present status of seed treatment, with special reference to cereals. Botan. Rev. 2:498–527.

12128. —— 1930: Seed treatment for controlling covered smut of barley. U.S. Dept. Agr. Tech. Bull. 207. 23 pp.

12129. —— 1937: Seed treatment experiments with oats naturally and artificially inoculated with smuts. U.S. Dept. Agr. Tech. Bull. 568:16.

12130. —— 1943: Chemical seed treatments for the control of certain diseases of sorghum. U.S. Dept. Agr. Tech. Bull. 849. 24 pp.

12131. —— 1948: Recent developments in seed treatment. Botan. Rev. 14:235–69.

12132. —— 1950: Use of chemicals in cereal seed treatment. Agr. Chem. 5(8):27–30, 75.

12133. —— 1950: Chemical seed treatment. II. Agr. Chem. 5(9):47–49.

12134. —— 1951: Cooperative seed-treatment tests on small grains in 1951. Plant Disease Reptr. 35:445–51.

12135. —— 1951: Cereal seed-treatment results (additions and corrections). Plant Disease Reptr. 35:485–86.

12136. —— 1953: Cooperative tests of seed-treatment chemicals on spring wheat and oats in 1953. Plant Disease Reptr. 37:587–94.

12137. ——, J. G. Dickson, and A. G. Johnson. 1926: Seed treatment experiments for controlling stripe disease of barley. Phytopathology 16:565–76.

12138. ——, J. A. Faris, and V. F. Tapke. 1938: Wheat smuts and their control. U.S. Dept. Agr. Farm Bull. 1711. 18 pp.

12139. —— and J. E. Livingston. 1945: Smut control in sorghum and effect of dust fungicides and storage on emergence. Phytopathology 35:645–53.

12140. —— and J. H. Martin. 1943: Seed rot and seedling blight of sorghum. U.S. Dept. Agr. Tech. Bull. 839. 26 pp.

12141. —— and J. W. Mitchell. 1956: Smut control in sorghum with the antibiotic complex F-17. Plant Disease Reptr. 40: 1073.

12142. —— and O. A. Nelson. 1939: Chlorine gas as a seed disinfectant. Phytopathology 29:913–14.

12143. —— and T. R. Stanton. 1934: Effect of seed treatments on yield of oats. J. Am. Soc. Agron. 26:851–57.

12144. ——, ——, and H. Stevens. 1938: Comparison of different methods of inoculating oat seed with smut. J. Am. Soc. Agron. 30:878–82.

Leukel, R. W. 12902.

12145. Leukhina, T. A. 1956: Some data on physiological characteristics of pine seedlings. Sb. Tr. Povolzhsk. Lesotekhn. Inst. 1956(51):109–14. Biol. Abstr. 35: 5372. 1960.

Leupi, H. 17935.

Leupin, K. 9214.

12146. Leus, F. P. 1930: Relation of age of farm crop seeds to production. Thesis, Coll. Agr. No. 321. Expt. Sta. Contrib. 682. Philippine Agriculturist 19:411–12. 1930.

12147. Leuschner, J. O. 1953: Culturing of the orchid seed. Orchid Dig. 17:214–17.

12148. Leuthold, P. 1934: Beiträge zur Morphologie und Keimungsphysiologie von *Telfairia pedata* Hook. Botan. Centralbl. Beih. Abt. A 52:148–204.

12149. Leutskii, K. M., and R. I. Stulnikova. 1951: Nicotinic acid and tryptophan content in bean seeds during germination and ripening. Dokl. Akad. Nauk SSSR 80:919–20.

12150. Levadoux, L. 1950: La sélection et l'hybridation chez la vigne. Ann. École Natl. Agr. Montpellier 28. 195 pp.

12151. Levan, A. 1937: Eine erbliche Anomalie der Samenanlage bei Petunia. Botan. Notiser 1937:35–55. Z. Botan. 33:426. 1938.

12152. Levari, R. 1953: The effect of 2,4-D and coumarin on respiration of germinating wheat and lettuce seeds. Palestine J. Botany J. Ser. 6:46–59.

12153. ——, A. M. Mayer, and M. Evenari. 1952: The effect of various metabolites on the action of some germination inhibitors. Israel Res. Council Bull. 2:27–36.

Leveille, G. A. 6430.

12154. Levensson, E. 1928: Observations on the germination of grain seeds. Bull. Soc. Botan. Bulgarie 2:29–34. Biol. Abstr. 5:24579. 1931.

12155. Lever, R. A. 1964: Ivory nut palms. World Crops 16:67.

Levi, I. 552, 6075.

12156. Levin, E. 1943: Corn and wheat embryo. Each year hundreds of millions of pounds of human food are lost. Natl.

Farm Chemurg. Council Chemurg. Paper 252. 10 mimeo. pp.

Levin, M. D. 15171.

12157. Levin, M. H. 1960: Studies on the ecological life history of *Saxifraga virginiensis*. Bull. Torrey Botan. Club 87:348–54.

12158. Levine, H., and R. E. Remington. 1937: Vitamin G content of some foods. J. Nutr. 13:525–42. Pharm. Abstr. 4:85.

Levine, J. 12972.

12159. Levine, M. N., and W. F. Geddes. 1957: Effect of leaf and stem rust on productivity, desiccation rate, and kernel weight of spring wheat at successive stages of development. Cereal Chem. 34:410–21.

12160. Levine, V. E. 1925: The effect of Selenium compounds upon growth and germination in plants. Am. J. Botany 12:82–90.

12161. Levitt, J., and P. C. Hamm. 1943: A method of increasing the rate of seed germination of *Taraxacum kok-saghyz*. Plant Physiol. 18:288–93.

12162. Levshin, A., and J. Sikorsky. 1939: Contribution to the problem of the dielectric coefficient, electric conductivity, and thermogenesis in seeds under the influence of an electric ultrahigh frequency field. Acad. Sci. RSS Ukraine J. 28:108–9.

12163. Levy, F. 1953: Growing primroses from seed. Am. Primrose Soc. Quart. 11:53–55.

Levy, J. 3499.

12164. Levyns, M. R. 1929: The problem of the rhenoster bush. S. African J. Sci. 26:166–69.

12165. —— 1935: Germination in some South African seeds. J. S. African Botany 1:161–70. Biol. Abstr. 10:20568. 1936.

12166. Lew, B. W., and R. A. Gortner. 1943: The gums from the carob bean, *Ceratonia siliqua* L. Arch. Biochem. 1:325–28. Chem. Abstr. 37:2776. 1943.

Leweke, D. H. 6512.

12167. Leweling, H. F. 1947: Hasten seedling growth of *Viburnum carlesi*. Am. Nurseryman 86:58–59.

12168. Lewes, D. 1951: Observations on the internal pressure of the ripening fruit of *Ecballium elaterium*. Kew Roy. Botan. Gardens Bull. 1951(3):443–44.

12169. Lewicki, S. 1929: The problem of pigmentation in the ears of wheat and Aegilops and its physiological significance. Mem. Inst. Natl. Polon. Econ. Rur. Pulawy 10:293–341.

12170. —— 1939: Studies on the vernalization of plants. Mem. Inst. Natl. Polon. Econ. Rur. Pulawy 17:113–32.

12171. Lewin, M. 1905: Über die Atmung keimender Samen unter Druck. Ber. Deut. Botan. Ges. 23:101–4.

12172. Lewis, C. H. 1958: How to grow roses from seed. Am. Rose Mag. 14(22): 16–17, 29.

12173. Lewis, D. 1942: Parthenocarpy induced by frost in pears. J. Pomol. Hort. Sci. 20:40–41. Biol. Abstr. 17:2709. 1943.

12174. —— 1946: Chemical control of fruit formation. J. Pomol. Hort. Sci. 22:175–83. Biol. Abstr. 22:7464. 1948.

12175. —— 1953: Seed storage. John Innes Hort. Inst. Ann. Rept. 43:14–15.

12176. —— 1953: Chromosomal abnormalities and reduced yield from stored seed. John Innes Hort. Inst. Ann. Rept. 44:14–15.

Lewis, E. W., Jr. 8654.

12177. Lewis, G. N. 1933: Heavy water prevents seeds from sprouting. Science News Letter 24:116.

Lewis, G. N. 12303.

12178. Lewis, I. G. 1934: Chewings fescue. The problems of germination and purity and their influence on greenkeeping and commerce in the United Kingdom. New Zealand J. Agr. 49:6–9.

12179. Lewis, J. 1958: Longevity of crop and weed seeds. Proc. Intern. Seed Testing Assoc. 23:340–54.

12180. —— 1961: The influence of water level, soil depth, and type on the survival of crop and weed seeds. Proc. Intern. Seed Testing Assoc. 26:68–85.

Lewis, J. 6091.

Lewis, M. Elvin-. 12188.

12181. Lewis, N. G. 1945: Seed science and weed control. Proc. Assoc. Offic. Seed Analysts 35:84–86.

12182. —— 1949: Measuring dormancy in oats grown in Northern Alberta. Assoc. Offic. Seed Analysts News Letter 23(2):35–36.

12183. —— 1949: Problem of irregular germination of weed seeds. Proc. Abstr. Western Can. Weed Control Conf. 3:163–66.

12184. —— and A. G. McCalla. 1937: The seed value of frosted wheat. Sci. Agr. (Ottawa) 17:431–44.

12185. Lewis, R. D. 1932: Determination of oil content of pecans. Ind. Eng. Chem. Anal. Ed. 4:296–97.

12186. —— and J. H. Hunter. 1944: Changes in some mineral constituents of pecan nuts and their supporting shoots during development. J. Agr. Res. 68:299–306.

Lewis, R. D. 6781.

12187. Lewis, R. M. 1963: Control of loose smut of barley. (Aqueous soak and slurry methods of applying chemicals to seeds.) Dissertation Abstr. 23:2287. Biol. Abstr. 43:3511. 1963.

Lewis, R. W. 12483.

12188. Lewis, W. H., and M. Elvin-Lewis. 1961: Medium for growing small rubiaceous seeds from herbarium material. Castanea 26:146–55.

12189. Lewy-van Severen, M. 1953: Aceituno seed fat. J. Am. Oil Chemists Soc. 30:124–26.

12190. Lexander, K. 1963: Relations between earliness of spring wheat varieties and some properties in connection with their germination and seedling growth. Physiol. Plantarum Suppl. 2. 66 pp.

Lhoste, J. 1426.

Li, C. H. 15925.

12191. Li, C. S., P. L. Chang, and S. L. Yang. 1950: The effect of stratification and other treatments on germination and establishment of Chinese olive seeds. Fukien Agr. J. 11:147a–56.

12192. Li, H. L. 1957: The discovery and cultivation of *Metasequoia*. Morris Arboretum Bull. 8:49–53.

12193. —— 1957: The evolutionary significance of the endosperm and its bearing on the origin of angiosperms. J. Wash. Acad. Sci. 47:33–38.

12194. Li, L. Y. 1943: The influence of stratification of tung seeds upon emergence and establishment of seedlings in the nursery. New Zealand J. Sci. Technol. Sect. A 25:43–48.

12195. Li, M., W. F. Meggitt, and R. J. Aldrich. 1959: Weed competition and its effect on the growth and yield of oats and corn. Agron. Abstr. 1959:85.

Li, S. L. 11993.

Li, T. K. 12278.

12196. Li, T. T. 1934: The development of embryo of *Ginkgo biloba*. Sci. Rept. Natl. Tsing Hua Univ. Ser. B 2:29–35. Lingnan Sci. J. 14:738–39. 1935.

12197. —— 1934: The development of *Ginkgo* embryo in vitro. Sci. Rept. Natl. Tsing Hua Univ. Ser. B 2:41–52. Expt. Sta. Rec. 74:172. 1936.

12198. —— and T. Shen. 1934: The effect of "pantothenic acid" on the growth of the yeast and on the growth of the radical of *Ginkgo* embryo in artificial media. Sci. Rept. Natl. Tsing Hua Univ. Ser. B 2:53–60. Expt. Sta. Rec. 74:172–73.

12199. Li, Y. 1940: Watering germinating seedbeds. J. Forestry 38:897–98.

12200. Liacos, L. G., and E. C. Nord. 1961: Curlleaf *Cercocarpus* seed dormancy yields to acid and thiourea. J. Range Management 14:317–20. Biol. Abstr. 38:23541. 1962.

12201. Liang, C. C. 1937: The influence of the reaction of hydrochloric acid upon the catalase activity in germinated millet seeds. Bull. Chinese Botan. Soc. 3:45–52. Chem. Abstr. 34:467. 1940.

12202. Liang, H., and T. P. Chang. 1935: The oil content of seed of some tung-oil trees (*Aleurites*). Nung Hsueh Ts'ung K'an (Nanking) 3:140–49.

12203. Lib, J. A. 1948: Wovon hängt die Blutenbildung der Samenpflanzen ab? Garten-Z., Flora 71(1):2–3. Biol. Abstr. 23:12552. 1949.

12204. Libbert, E. 1957: Interactions between auxins and inhibitors in germination. Phyton 9:81–105. Chem. Abstr. 52:9325. 1958.

12205. —— 1959: Wechselwirkungen zwischen verschiedenen synthetischen Auxinen bei Keimungsversuchen. Biol. Centralbl. 78:502–17.

12206. —— 1961: Änderungen im keimungsphysiologischen Verhalten, speziell der Hemmungsempfindlichkeit, im Verlaufe von Samen-Nachreifung und -Alterung. Flora (Jena) 151:509–17.

12207. —— 1961: Some special aspects of the function of endogenous growth inhibitors in dormancy. Intern. Botan. Congr. Recent Advan. Botany 9:1213–16.

12208. ——, T. Kentzer, and B. Steyer. 1961: The effects of some adenine compounds including ATP on germination, seedling development, and auxin transport. Flora (Jena) 15:663–69.

12209. —— and H. Luebke. 1957: Physiological effects of Scopoletin. I. Influence of Scopoletin on seed germination. Flora (Jena) 145:256–63. Tobacco Abstr. 3:468. 1959.

12210. —— and H. Zatschler. 1963: Inhibition of seed germination and bud growth with natural and synthetic enzyme inhibitors and its reversal by their antagonists. A contribution to the study of the primary action of the correlative inhibitor. Flora (Jena) 153:233–41. Chem. Abstr. 60:4706. 1964.

12211. Libby, W. F. 1951: Radiocarbon dates. II. Science 114:291–96.

12212. —— 1954: Chicago radiocarbon dates. IV. Science 119:135–40.

Lichtenstein, N. 6515.

Lichtentels, I. 16377.

12213. Lichti, H., C. Tamm, and T. Reichstein. 1956: Die Glykoside der Samen von *Strophanthus ledienii* Stein. Helv. Chim. Acta 39:1914–32.

12214. ——, ——, and —— 1956: Die Glykoside der Samen von *Strophanthus vanderijstii* Staner. Helv. Chim. Acta 39:1933–73.

12215. Lichtwardt, R. W., G. L. Barron, and L. H. Tiffany. 1958: Mold flora associated with shelled corn in Iowa. Iowa State Coll. J. Sci. 33:1–11.

12216. Lid, J. 1937: Leaves and seeds blown from the lowlands up into the mountains of Hardangervidda, Norway. Nytt Mag. Naturvidensk. 77:99–102.

Lida, K. 5803.

12217. Liddell, R. W. 1946: A simplified technique for germinating orchid seed. Am. Orchid Soc. Bull. 15:9–15, 69–73. Biol. Abstr. 21:7123. 1947.

12218. —— 1948: Further experiments with Clorox in the germination of *Cattleya* seed. Am. Orchid Soc. Bull. 17:354–57.

12219. —— 1953: Notes on germinating *Cypripedium* seed. Am. Orchid Soc. Bull. 22:195–97.

12220. —— 1953: Notes on germinating *Cypripedium* seed. II. Am. Orchid Soc. Bull. 22:580–82.

12221. —— 1955: Notes on germinating *Cypripedium* seed. III. Procedure with infected seed. Am. Orchid Soc. Bull. 24:661–63.

Lieb, H. 8024.

Liebenberg, A. R. von; *see* Von Liebenberg, A. R.

12222. Lieberman, F. V. 1945: Experiments with DDT, Sabadilla, and pyrethrum dusts for control of *Lygus* spp. on seed alfalfa. U.S. Dept. Agr. Bur. Entomol. Plant Quarantine Publ. 658.

12223. Liebsch, G. 1926: Wissenwerts von Samen und Früchten. I. Allgemeines. Gartenflora 75:10–12.

12224. —— 1926: Wissenwerts von Samen und Früchten. II. Von der Keimfähigkeit und Keimkraft der Samen. Gartenflora 75:63–67.

12225. —— 1926: Wissenwerts von Samen und Früchten. IV. Verbreitungsmittel der Fruchtorgane. Gartenflora 75:249–52.

12226. —— 1926: Wissenwerts von Samen und Früchten. V. Gartenflora 75:283–85.

Liels, D. 14036.

12227. Liener, I. E. 1963: Seed hemagglutinins. Proc. Seed Protein Conf., New Orleans 1963:69–97. Chem. Abstr. 60: 4456. 1964.

12228. —— 1964: Seed hemagglutinins. Econ. Botany 18:27–33.

12229. ——, H. J. Deuel, Jr., and H. L. Fevold. 1949: The effect of supplemental methionine on the nutritive value of diets containing concentrates of the soybean trypsin inhibitor. J. Nutr. 39:325–39.

12230. —— and H. L. Fevold. 1949: The effect of the soybean trypsin inhibitor on the enzymatic release of amino acids from autoclaved soybean meal. Arch. Biochem. 21:395–407.

12231. ——, H. Spector, H. L. Fevold, and G. H. Berryman. 1949: The effect of soybean growth inhibitors on the availability of methionine for growth and lipotropism. Arch. Biochem. 24:299–304.

Lienhard, M. 5167.

12232. Lier, O., and J. Rasten. 1954: Certified seed culture in Norway. Tidsskr. Norske Landbruk 1954(1/2):3–32.

12233. Liesegang, F. 1935: Untersuchungen über die Notwendigkeit des Kalkes für die Keimung von Samen und das weitere Wachstum der Keimpflanzen. Deut. Landwirtsch. Presse 62:370.

Lietke, D. 16531.

Lievesley, P. M. 17807.

Lifshitz, D. B. 14599.

12234. Ligachev, I. N. 1963: Form and dimensions of seed wings and their role in the dissemination of pine seeds. Lesn. Zh. 2:163–64.

12235. Ligon, L. L. 1945: Mung beans: a legume for seed and forage production. Oklahoma Agr. Expt. Sta. Bull. 284:1–12. Biol. Abstr. 20:13561. 1946.

12236. Ligthelm, S. P., D. H. S. Horn, H. M. Schwartz, and M. M. Von Holdt. 1954: A chemical study of the fruits of three South African *Ximenia* species, with special reference to the kernel oils. J. Sci. Food Agr. 5:281–88.

12237. Lihnell, D. 1942: Keimungsversuche mit Pyrolasamen. Symbolae Botan. Upsalienses 6(3):1–37.

12238. Likhonos, F. D. 1959: Some specific biological characteristic of the seeds and seedlings of fruit trees. Botan. Zh. 44: 1341–44. Biol. Abstr. 36:42476. 1961.

12239. Lilienstern, M. F. 1937: The influence of physicochemical conditions of the substrate on the germination and rapidity of germination of flax and flax dodder. Bull. Appl. Botany, Genet., Plant Breed. (USSR) Ser. 4(2):131–36. Chem. Abstr. 31:5843. 1937.

12240. Liljefors, A. 1934: Über normale und apospore Embryosackentwicklung in der gattung Sorbus, nebst einigen Bemerkungen über die Chromosomenzahlen. Svensk Botan. Tidskr. 28:290–99.

12241. Lillevik, H. A., and W. F. Geddes. 1943: Investigation of a death by asphyxiation in a grain elevator bin containing flaxseed. Cereal Chem. 20:318–28.

Lilly, J. H. 4486, 5435, 12413.

Lima, A. P. de; *see* De Lima, A. P.

Lima, I. H. 19069–19071.

Lima, N. O. 7552.

12242. Limar, R. S. 1961: Effect of ultrasound on the seed of certain plants. Data from a Symposium on the Use of Biophysics in the Field of Plant Protection, Leningrad, pp. 24–25. Biol. Abstr. 42: 23614. 1963.

Limar, R. S. 7754.

12243. Limarj, R. S. 1961: The effect of ultra-sound on the germination of the lentil (*Lens esculenta* Moench.). Botan. J. (Botan. Soc. USSR) 46:1171–77.

Liming, F. G. 4157, 15309.

Limozin, N. 19248.

12244. Limstrom, G. A., and R. F. Finn. 1956: Seed source and nursery effects on yellow-poplar plantations. J. Forestry 54:828–31.

Lin, C. C. 18122.

12245. Lin, C. I. 1955: Studies on ear-germination of rice plant. I. Varietal differences in ear-germination of rice plants. Taiwan Univ. Mem. Coll. Agr. 4:98–104.

12246. —— 1957: Investigation on the factors which affect the on-ear-germination of rice plants. Taipei Natl. Taiwan Univ. Mem. Coll. Agr. 4(4):13–26.

Linn, C. I. 18785.

12247. Lin, J. Y., and K. H. Ling. 1961: Studies on free amino acids in the seeds of *Leucaena glauca* Benth. I. Isolation and identification of mimosine. J. Formosan Med. Assoc. (English Transl.) 60:651–56. Biol. Abstr. 39:3591. 1962.

12248. —— and —— 1961: Studies on free amino acids in the seeds of *Leucaena glauca* Benth. III. Biological study of mimosine. J. Formosan Med. Assoc. (English Transl.) 60:657–65. Biol. Abstr. 39:3592. 1962.

12249. Linn, R. C. Y. 1956: The origin and distribution of acetylcholine in the Malayan jack-fruit plant, *Artocarpus integra.* Abstr. Commun. 20th Intern. Physiol. Congr., Brussels.

12250. Lin, T. 1959: The influence of insecticides on the seed germination percentage of rice, wheat, and soybean during storage. J. Agr. Assoc. China (Taipei) 27:62–73.

Lin, Y. 19894.

Linck, A. J. 10945, 11317, 11711.

Lincoln, C. 16475.

12251. Lincoln, R. 1941: Composition des riz. Rev. Agr. Ile Maurice 20:284–85. Biol. Abstr. 19:5164. 1945.

Lincoln, W. C. 8798.

Lindahl-Kiessling, K. 9631.

12252. Lindberg, J. E., and I. Sörensson. 1959: Heat treatment and hot-air drying of grain, notably wheat. Kgl. Skogs-Lantbruksakad. Tidskr. Suppl. 1. 73 pp.

12253. Lindemann, E. 1953: The change in fat content of sorghum and maize germ during storage. Stärke 5:139–43. Chem. Abstr. 48:1483. 1954.

Lindemann, M. 10765.

12254. Lindenbein, W. 1956: Beiträge zur Kenntnis der Samen und Früchte. I. *Sanguisorba minor* Scop. und *S. muricata* (Spach) Gremli. Angew. Botan. 30:107–19.

12255. —— 1958: Beiträge zur Kenntnis der Samen und Früchte. II. *Trifolium dubium* Sibth. und *T. campestre* Schreb. Flora (Jena) 146:376–85.

12256. —— 1960: Keimungsphysiologische Probleme unter besonderer Berüchsichtigung der biochemischen Keimprüfung. Saatgut-Wirtsch. 12:308–312.

12257. —— 1962: Tetrazolium test committee. Proc. Intern. Seed Testing Assoc. 27:224–34.

12258. —— and H. Bulat. 1960: Beiträge zur topographischen Tetrazolium-methode. II. Einige Versuchsergebnisse mit kunstlich getrockneten Spinatsamen. Proc. Intern. Seed Testing Assoc. 25:460–66.

12259. —— and H. Tabatabai-Yazdi. 1962: Decline of germination and germination capacity of sugar beets seeds. Saatgut-Wirtsch. 14:193–95.

Lindenbein, W. 3323.

12260. Linder, E. A. 1930: The growing of orchids from seed. Bull. Missouri Botan. Garden 18:29–36.

12261. Linderstrom-Lang, K., and C. Engel. 1937: Distribution of amylase in the outer layers of barley grains. Enzymologia 3:138–46. Chem. Abstr. 31:8565. 1937.

Linderstrom-Lang, K. 2842.

12262. Lindgren, D. L., H. E. Krohne, and L. E. Vincent. 1954: Malathion and chlorthion for control of insects infesting stored grain. J. Econ. Entomol. 47:705–6.

12263. —— and L. E. Vincent. 1960: The relation of moisture content and temperature of stored grain to the effectiveness of grain fumigants under forced circulation. J. Econ. Entomol. 53:1071–77.

12264. ——, ——, and H. E. Krohne. 1955: Fumigation of seed as affecting germination. Calif. Citrus Expt. Sta. News Letter 72. 24 pp.

Lindgren, D. L. 18091–18096.

12265. Lindner, H. 1961: Problems of determining the germination ability and vigor of seeds. Z. Landwirtsch. Versuchs-Untersuchungsw. 7:294–306.

12266. —— and H. Lindner. 1962: Origin and measurement of differences between germination power and field growth of *Pisum sativum* L. Z. Landwirtsch. Versuchs- Untersuchungsw. 8:121–26.

12267. Lindquist, C. H. 1960: Note on the moisture requirements of the stratifying media for seed of *Caragana arborescens*, Lam. Can. J. Plant Sci. 40:576–77.

Lindquist, C. H. 4511, 7526, 7729.

12268. Lindquist, D. A., J. Hacskaylo, J. C. Clark, and T. B. Davich. 1961: Systemic activity of dimethoate applied to cotton seeds. J. Econ. Entomol. 54:1132–35.

Lindquist, F. E. 8645.

12269. Lindsey, K. E., and M. L. Peterson. 1962: High temperature suppression of flowering in *Poa pratensis* L. Crop Sci. 2:71–74.

Lindsey, K. E. 18167.

Lindsey, M. F. 8390.

12270. Lindström, O. 1958: Mechanism of liquid seed treatment. J. Agr. Food Chem. 6:283–98.

12271. —— 1960: Seed disinfection. Fungicide and dye distribution in liquid seed treatment. J. Agr. Food Chem. 8:217–24.

12272. —— and O. Lamm. 1951: The chemical effects produced by ultrasonic waves. J. Phys. Colloid Chem. 55:1139–46.

12273. Linehan, P. A., and D. Mathews. 1962: Measurement of uniformity in seed bulks. II. Proc. Intern. Seed Testing Assoc. 27:423–30.

12274. —— and S. P. Mercer. 1931: A method of distinguishing certain strains of New Zealand perennial ryegrass by examination of seedlings under screened ultra-violet light. Sci. Proc. Roy. Dublin Soc. 20:75–83.

12275. —— and —— 1933: Ultra-violet rays on seed. Science Suppl. 87(2030):10.

12276. —— and —— 1936: Does presoaking accelerate laboratory germination in cocksfoot (*Dactylis glomerata* L.)? Proc. Intern. Seed Testing Assoc. 8:42–45.

12277. Lines, R. 1962: Choosing seed origins of Japanese larch. Scot. Forestry 16:164–66.

Ling, K. H. 12247, 12248.

12278. Ling, L., and T. K. Li. 1943: The fungicidal efficiency of ferrous sulphate and its application as a seed disinfectant of barley. Chinese J. Sci. Agr. 1:126–32. Rev. Appl. Mycol. 24:361. 1945.

Link, C. B. 7670.

12279. Link, F. W. 1955: Studies with pelleted lettuce seed. Proc. Am. Soc. Hort. Sci. 65:335–41.

Link, K. P. 402, 9359, 14401.

Linko, M. 5947.

12280. Linko, P. 1960: Current research on the biochemistry of grain storage. Cereal Sci. Today 5:302–6.

12281. —— 1961: Zone electrophoresis pattern of free amino acids as an index of storage condition of wheat. Cereal Chem. 38:187–94.

12282. ——, Y. Y. Cheng, and M. Milner. 1960: Changes in the soluble carbohydrates during browning of wheat embryos. Cereal Chem. 37:548–56.

12283. —— and M. Milner. 1959: Enzyme activation in wheat grains in relation to water content, glutamic acid-alanine transaminase, and glutamic acid decarboxylase. Plant Physiol. 34:392–96.

12284. —— and —— 1959: Gas exchange induced in dry wheat embryos by wetting. Cereal Chem. 36:274–79.

12285. —— and —— 1959: Free amino and keto acids of wheat grains and embryos in relation to water content and germination. Cereal Chem. 36:280–94.

12286. —— and —— 1960: Treatment of wheat with ionizing radiations. V. Effect of γ-radiation on some enzyme systems. Cereal Chem. 37:223–27.

12287. —— and L. Sogn. 1960: Relation of viability and storage deterioration to glutamic acid decarboxylase in wheat. Cereal Chem. 37:489–99.

Linko, P. 2123, 3882, 4060.

Linkous, W. N. 5825, 7101.

12288. Linkswiler, H., H. M. Fox, and P. C. Fry. 1960: Availability to man of amino acids from foods. III. Threonine from corn. J. Nutr. 72:389–96.

12289. ——, ——, and —— 1960: Availability to man of amino acids from foods. IV. Isoleucine from corn. J. Nutr. 72:397–403.

12290. Linn, M. B., and A. G. Newhall. 1948: Comparison of two methods of pelleting onion seed in the control of smut. Phytopathology 38:218–21.

Linn, M. B. 13140, 13497.

Linnahalme, T. 5947.

12291. Linnett, R. J. B. 1956: The principles of seed processing. Queensland Agr. J. 82:14–17.

12292. Linsbauer, L. 1930: Elektrische Stimulation von Gemüsesamen. Klosterneuburg 1930:108–11. Botan. Centralbl. 161:77. 1931.

12293. Linser, H., O. Kiermayer, and E. Youssef. 1958: The growth-substance content of a number of Brassicaceae at various stages of development. Planta 52:173–86. Chem. Abstr. 53:2371. 1959.

12294. ——, H. Mayr, and F. Maschek. 1954: Papierchromatographie von Zellstreckend Wirksamen Indolkörpern aus *Brassica*-arten. Planta 44:103–20.

12295. —— and E. Primost. 1952: Keimversuche mit Adsorptionsschutz der Samen in 2,4-D behandeltem Boden. Pflanzenschutz Ber. 8(5/6):65–74. Biol. Abstr. 26:32125. 1952.

12296. Linskens, H. F. 1952: Über den Trocknungsprozess an progressiven Keimstadien. Angew. Botan. 26:255–64. Biol. Abstr. 27:20642. 1953.

12297. —— 1953: Die unterschiedliche Empfindlichkeit von Samen gegenüber Heisswasserbädern. (Untersuchungen an Weizen.) Phytopathol. Z. 21:45–52.

Linskens, H. F. 10869.

12298. Linteau, A. 1948: Factors affecting germination and early survival of yellow birch (*Betula lutea* Michx.) in Quebec. Forestry Chron. 24:27–86. Biol. Abstr. 22:25282. 1948.

12299. Linz, F. 1896: Beiträge zur Physiologie der Keimung von *Zea mais* L. Jahrb. Wiss. Botan. 29:267–319.

12300. Lipkin, B. J. 1925: Der Einfluss niedriger Temperaturen auf die Keimfähigkeit von Waldsamen. Goretskii Sel'sk. Inst. Mitt. 2:182–97.

12301. Lipkin, B. J. 1927: Duration of viability of the seeds of different conifers. Ann. Weissruthenischen Staat. Akad. Landwirtsch. Gorky. 5:25–33. Biol. Abstr. 5:8792. 1931.

12302. Lipman, C. B. 1936: Normal viability of seeds and bacterial spores after exposure to temperatures near the absolute zero. Plant Physiol. 11:201–5.

12303. —— and G. N. Lewis. 1934: Tolerance of liquid-air temperatures by seeds of higher plants for 60 days. Plant Physiol. 9:392–94.

12304. Lipp, A. E. G., and L. A. T. Ballard. 1959: The breaking of seed dormancy of some legumes by carbon dioxide. Australian J. Agr. Res. 10:495–99.

12305. —— and —— 1963: Germination patterns shown by the light-sensitive seed of *Anagallis arvensis*. Australian J. Biol. Sci. 16:572–84.

Lipp, A. E. G. 1811.

12306. Lippert, R. D., and H. H. Hopkins. 1950: Study of viable seeds in various habitats in mixed prairie. Trans. Kansas Acad. Sci. 53:355–64.

12307. Lipscomb, G. F., and T. I. Dowling. 1926: Factors that influence life and germination of cotton seeds. Science 64:186–87.

Lipskaja, T. W. 9512.

Liscano, C. 11741.

12308. Lischkewitsch, M. J. 1930: Das Abschätzen der Samenqualität nach dem Fermentgehalt. Fermentforsch. 12:244–61. Biol. Abstr. 6:12839. 1932.

12309. —— and S. P. Prizemina. 1929: Über den Fermentgehalt in Samen verschiedenen Ursprungs. Biochem. Z. 212: 280–90. Biol. Abstr. 6:24632.

12310. Lishkevich, M. I. 1931: Estimation of the quality of seeds from their enzyme content. Bull. Appl. Botany, Genet., Plant Breed. (USSR) 25:103–33. Chem. Abstr. 26:756. 1932.

12311. —— 1939: Storing of phosphorus-containing substances in cotton seeds during their ripening. Tr. VNIIZH (Vses. Tsentral. Nauchn.-Issled. Inst. Zhirov "VNIIZH") 1939:106–10. Chem. Abstr. 36:4546, 1942.

12312. —— 1953: The effect of moisture content of sunflower seeds on their keeping.

Maslob.-Zhir. Prom. 18(9):1–3. Chem. Abstr. 48:820. 1954.

Lishkevich, M. I. 7399, 7400, 9553.

12313. Lisin, S. S. 1950: Timely preparation and sowing of tree and shrub seeds with difficult germination. Les i Step 2(7):29–32.

Lisina, A. I. 19153.

12314. Lisitsyn, D. I. 1959: The dynamics of the carbohydrate and fat content in ripening and germination of ash tree seeds. Biochemistry (USSR) (English Transl.) 24:779–82.

12315. Liso, A. 1935: Aumento di germinazione e densità de semina. Radiobiol. Generalis 3:841–44. Biol. Abstr. 11:8014. 1937.

12316. Lisosvskii, G. M. 1962: Hard seeds in edible beans. Agrobiologiya 1: 147–48. Biol. Abstr. 39:16541. 1962.

12317. Lissenko, T. D. 1932: "Yarovization" of agricultural plants. [Ukr. Inst. Genet.] Bull. Yarovization 1:14–29.

12318. —— 1932: "Yarovization" of maize, millet, soy. [Ukr. Inst. Genet.] Bull. Yarovization 2/3:46–64.

12319. Lissovskii, V. I., and B. A. Choumakov. 1916: Antagonistic action of salts of sodium and calcium, and their influence on germination of lucerne and of maize. Russ. J. Agr. Exptl. 17:446–71. Phys. Abstr. 2:703. 1918.

12320. List, A., Jr. 1963: Some features of embryo and endosperm development in *Gleditsia triacanthos*. [Abstract.] Am. J. Botany 50:622.

12321. Lister, R. M. 1960: Transmission of soil-borne viruses through seed. Virology 10:547–49.

12322. Lithgow, A. V. 1956: Testing of seeds. New Zealand J. Agr. 92:414–21.

12323. —— 1957: Seed: In search of quality. New Zealand J. Agr. 94:422–25, 427.

12324. —— 1957: Some aspects of germination analysis. New Zealand Grassland Assoc. Proc. Conf. 19:125–30.

12325. —— 1958: Seed production in New Zealand. Seed testing. New Zealand J. Agr. 97:473, 475–77, 479, 481.

12326. Little, A. D., Inc. 1951: Marketing potential for oilseed protein materials in industrial uses. U.S. Dept. Agr. Tech. Bull. 1043. A research and marketing act contract report. 120 pp.

12327. Little, E. C. S., and G. E. Blackman. 1963: The movement of growth regulators in plants. III. Comparative studies of transport in *Phaseolus vulgaris*. New Phytologist 62:173–97.

12328. Little, E. L., Jr. 1940: Development of cones and seeds in pinon (*Pinus edulis* Engelm.). Am. J. Botany (Suppl.) 27:6s.

12329. Little, R. R., and E. H. Dawson. 1960: Histology and histochemistry of raw and cooked rice kernels. Food Res. 25: 611–22.

12330. Little, W. R. 1958: Relationship of seedsmen and certified seed growers. Intern. Crop Improvement Assoc. Ann. Rept. 40:120–24.

12331. Littlefield, E. W. 1939: Some new aspects of seed certification. Comments. J. Forestry 37:35.

12332. Litvinov, L. 1938: Inhibitory effect of tomato juice on the germination of tomato seed. Bull. Inst. Rech. Biol. Perm. 11:163–70. Chem. Abstr. 33:5030. 1939.

12333. —— and N. Lukianov. 1938: Vernalization of tomatoes and cucumbers. Acad. Sci. Ukr. Inst. Botan. 1938:227–28.

12334. Litynski, A., S. Trzecki, and K. Wiazecka. 1961: The influence of temperatures and time of drying on seed vitality of some winter rape varieties with different starting moisture; laboratory investigations. Howdowla Roslin Aklim. Nasiennictwo 5:765–80.

12335. Litynski, M. 1955: The effect of environmental temperature on seed vitality in some vegetable species. Acta Agrobotan. (Warsaw) 3:75–128.

12336. —— 1957: The influence of moisture in the environment on vitality of certain species of vegetable seeds. Roczniki Nauk Rolniczych Ser. A 76:217–94. Biol. Abstr. 33:23305. 1959.

12337. —— 1960: Einige Versuche einer vorschriftsmassigen Beurteilung des Wassergehaltes bei einer zur Aufbewahrung bestimmten Partie von Maiskolben. Proc. Intern. Seed Testing Assoc. 25:567–72.

12338. —— 1960: Die Anwendung infraroter Strahlen zur Samentrocknung. Proc. Intern. Seed Testing Assoc. 25:602–7.

12339. —— and Z. Chudoba. 1963: Observations of red clover seeds (*Trifolium pratense* L.) during long storage. Howdowla Roslin Aklim. Nasiennictwo 7:89–102.

12340. —— and Z. Urbaniak. 1958: Observations on the influence of light on seeds of certain plant species during storage. Plant Breed. Acclimitiz. Inst. Sci. Publ. 2:21–66.

12341. —— and H. Zagorska. 1959: Observations on seeds stored in hermetically sealed containers. Bull. Warzywniczy 4: 391–96.

12342. Litynski, T. 1936: Determination of nuclear phosphorus in the seeds of the bean, *Vicia faba minor*. Bull. Intern. Acad. Polon. Sci. Classe Sci. Math. Nat. B 1:103–29. Chem. Abstr. 31:1322. 1937.

12343. Litzenberger, S. C., and C. R. Pineda. 1956: Pruebabde germinacion y tratamiento de la semilla. Nicaragua Serv. Tec. Agr. Circ. 31. 10 pp.

12344. Liu, C. H., and C. H. Lou. 1945: Flourescein-induced parthenocarpy. Nature 155:23.

Liu, F. Y. 10342.

Liu, H. H. 9150.

12345. Liu, H. L. 1949: Inheritance of defective seed coat in soybeans. J. Heredity 40:317–22.

12346. Liu, S. L., and C. H. Ho. 1960: Nature of silicon in rice hull. I. Solubility of the silicon part. J. Chinese Chem. Soc. (Taiwan) 6:141–53. Chem. Abstr. 55: 4673. 1961.

12347. Liu, S. T. 1948: Seed-borne diseases of soybean. Acad. Sinica [Inst. Botan.] Botan. Bull. 2:69–80. Biol. Abstr. 23: 16896. 1949.

12348. Liu, T. C. 1948: Cultivation of excised plumules of *Nelumbo speciosum* in vitro. Acad. Sinica [Inst. Botan.] Botan. Bull. 2:207–10.

12349. —— 1955: Experiment on intergeneric embryo grafting. Acta Botan. Sinica 4:179–81.

12350. Liubavskaia, A. I. 1952: Influence of selfpollination and cross pollination upon the planting qualities of birch seeds. Les i Step 4(7):27–34.

12351. Liubchenko, V. M. 1959: Catalase activity in seeds of *Tilia cordata* and *Euonymus europaeus* in process of stratification. Botan. Zh. 44:522–24. Chem. Abstr. 53:22285. 1959.

12352. —— 1960: Germination of the embryos of *Tila* and *Euonymus* spp. and its bearing on the stratification of seeds. Botan. Zh. 45:144–48.

12353. Liventseva, E. 1952: On determining the seed germination of *Phacelia*. Pchelovodstvo 29(3):42–44.

12354. Liverman, J. L., and J. Bonner. 1953: The interaction of auxin and light in the growth responses of plants. Proc. Natl. Acad. Sci. U.S. 39:905–16.

12355. Livingston, A. 1945: Lily notes. Lilies from seed. Natl. Hort. Mag. 24:58–60.

12356. —— 1950: Notes on the behavior of lily seeds. Natl. Hort. Mag. 29:95–96.

12357. Livingston, B. E., and F. W. Haasis. 1933: Relations of time and maintained temperature to germination percentage for a lot of rice seed. Am. J. Botany 20:596–614.

12358. Livingston, H. 1944: Quality in seed of tomato and pepper is a three-year program. Seed World 55(11):24.

12359. Livingston, J. E. 1947: Barley fertilizer and seed treatment test. Phytopathology 37:426–28.

12360. —— 1951: Effect of low temperature on the germination of artificially dried seed corn. Nebraska Agr. Expt. Sta. Res. Bull. 169. 14 pp.

12361. —— 1952: Injury and drying of seed corn in relation to emergence. Phytopathology 42:221–22.

12362. —— and E. Kneen. 1944: A rag-doll technique for the inoculation of wheat with bunt (*Tilletia levis*). Phytopathology 34: 124–28.

Livingston, J. E. 12139

12363. Liyanage, D. V. 1950: The relative merits of first and second bunch coconuts for seed purposes. Trop. Agriculturist 106:151–55. Biol. Abstr. 25:34723. 1951.

12364. —— 1953: Selection of coconut seednuts and seedlings. Ceylon Coconut Quart. 4:127–29.

12365. —— 1961: The use of isolated seed gardens for coconut seed production. Ceylon Coconut Quart. 12:121–24.

Ljungdahl, C. 43.

12366. Llamas, R. 1954: Modificaciones enzimaticas durante la germinacion. Congr. Intern. Botan. Paris Rappt. Commun. 8:107–9.

Llavat-Cristy, F. 10462.

12367. Llaveria, M. 1953: Estudio sobre semilla de papa. Agronomia (Peru) 18(73):21–26.

12368. Lloyd, A. B. 1959: The transmission of *Phoma lingam* (Tode) Desm. in the seeds of swede, turnip, chou moellier, rape, and kale. New Zealand J. Agr. Res. 2:649–58.

12369. Lloyd, F. E. 1910: Development and nutrition of the embryo, seed, and carpel in the date, *Phoenix dactylifera*, L. Missouri Botan. Garden Rept. 21:103–64.

12370. Lloyd, J. U. 1894: Weight of a grain of *Triticum vulgare*. Am. J. Pharm. 66:473–79.

12371. Lo, T. C. 1964: Control of seed borne diseases by radioactive irradiation. Acad. Sinica [Inst. Botan.] Botan. Bull. 5:1–16.

12372. Lo, T. Y., and S. M. Chen. 1944: The formation of vitamin P in some germinating legumes. J. Chinese Chem. Soc. 11:52–56. Chem. Abstr. 39:1197. 1945.

12373. —— and —— 1947: The effect of various elements on vitamin C formation in germinated mung beans. Sci. Record (China) 2:81–83. Chem. Abstr. 42:2649. 1948.

12374. —— and —— 1947: The effects of thiamine, riboflavin, pantothenic, and ascorbic acids on the diastatic and proteolytic activities of the enzymes in rice meals. J. Chinese Chem. Soc. 15:98–101. Chem. Abstr. 42:938. 1948.

12375. Lobanov, V. Y. 1961: Germination of corn seeds under cold temperatures. Kukuruza 1961(11):29–30.

12376. —— and N. I. Kozelkova. 1958: New methods for the determination of the viability of seeds. Selek. i Semen. 1958(1): 56–58. Biol. Abstr. 41:20164. 1963.

Lobay, W. 13169.

12377. Lobb, W. R. 1958: Seed pelleting. New Zealand J. Agr. 96:556.

Lochhead, A. G. 19328.

Locke, W. F. 1735.

12378. Lockett, M. C., and B. J. Luyet. 1951: Survival of frozen seeds of various water contents. Biodynamica 7(134):67–76.

Lockwood, A. R. 9038.

12379. Lockyer, S. 1932: Seed dispersal from hygroscopic *Mesembryanthemum* fruits; *Bergeranthus scapigerus*, Schw., and *Dorotheanthus bellidiformis*, N.E.Br., with a note on *Carpanthea pomeridiana*, N.E.Br. Ann. Botany (London) 46:323–42.

Lockyer, S. 7005.

12380. Loden, H. D. 1959: Progress report on development of gossypol free cottonseed. Cottonseed Process. Clinic Southern Reg. Res. Lab. Proc. 8:34–40.

12381. —— and C. C. Wilson. 1951: The effect of temperature upon the germination and subsequent growth of a number of varieties of *Gossypium hirsutum* (L.). Proc. Ann. Conv. Assoc. Southern Agr. Workers 48:162.

12382. Lodhi, S. A., and A. Naeem. 1955: Some seed-borne fungi from Pakistan. Brit. Mycol. Soc. Trans. 38:240–42.

Lodhi, S. A. K. 189.

12383. Lods, E. A., and J. G. Coulson. 1929: A preliminary report on size of seed in relation to smut infection in oats. Quebec Soc. Protect. Plants Ann. Rept. 21:80–85. Rev. Appl. Mycol. 10:179–80.

Loe, J. B. 4333.

12384. Loebel, M. 1940: Ein kleiner Aufbewahrungsversuch mit Nadelholzsamen in Humus. Forstwiss. Centralbl. 62:237–42.

Loebel, M. 16373.

Loesecke, H. von; *see* Von Loesecke, H.

Loesecke, H. W. von. 14560, 14561.

Loevenhart, A. S. 10312.

12385. Lövinson, O. 1900: Influence of salts of the fatty acids on germination and growth. Botan. Centralbl. 83:1–12, 33–43, 65–75, 97–106, 129–38, 185–95, 209–24.

12386. Loew, M. 1937: Canterbury-bells from seed. Flower Grower 24(6):306.

12387. —— 1937: Raising *Trollius* from seed. Flower Grower 24(6):317.

12388. —— 1946: Androsaces from seed. Flower Grower 33(7):414.

12389. Loew, O., and J. Ishii. 1895: Über das Vorkommen von Mannan in den Samen der Kakifrüchte. Landwirtsch.-Versuchssta. 45:435–36.

12390. Loewenberg, J. R. 1955: The development of bean seeds (*Phaseolus vulgaris* L.). Plant Physiol. 30:244–50.

Loewus, F. A. 18425.

12391. Lofland, H. B., Jr. 1950: In vitro culture of the cotton embryo. Botan. Gaz. 111:307–11.

12392. Loh, T. C. 1934: An improved method for the control of seed-borne diseases of rice. Lingnan Sci. J. 13:603–5. Chem. Abstr. 29:1571. 1935.

Lohman, R. L. 18646.

Loiselle, R. 10448.

12393. Lojkin, M. 1936: Moisture and temperature requirements for yarovization of winter wheat. Contrib. Boyce Thompson Inst. 8:237–61.

12394. Loken, A. 1958: Germination of Sitka spruce and larch. Arsskr. Norske Skogplantesk. 1957:28–35.

12395. —— 1960: Seed and seed treatment. Arsskr. Norske Skogplantesk. 1959:31–41.

12396. Loktev, N. N. 1958: The dispersion of seeds of certain weeds by the seeds of *Bromus secalinus* L. Botan. Zh. 43:1314–16.

12397. —— 1958: Viability of the seeds of *Bromus secalinus* on humus and other agricultural waste. Zemledelie 1958(10): 85–86.

12398. Lomagin, A. G. 1957: Effect of characteristics of seed material on the speed of vernalization. Vestn. Leningr. Univ. 1957(3):38–43. Biol. Abstr. 35: 27662. 1960.

12399. Lombard, P. B., and A. E. Mitchell. 1962: Anatomical and hormonal development in Redhaven peach seeds as related to the timing of naphthaleneacetic acid for fruit thinning. Proc. Am. Soc. Hort. Sci. 80:163–71.

Lombard, P. M. 4169, 4171.

12400. Lombersi, L., and M. Tarantola. 1950: Chemical composition of seeds of some plants in relation to water availabilities in the soil. II. Ann. Sper. Agrar. (Rome) 4:715–24.

12401. Lona, F. 1947: L'influenza delle condizioni ambientali, durante l'embriogenesi, sulle caratteristiche del seme e della pianta che ne deriva. Lavori Botan. 1947:313–51.

12402. —— 1956: Gibberellic acid induces germination of seeds of *Lactuca scariola* in the dark-inhibition phase. Ateneo Parmense 27:641–44. Chem. Abstr. 51: 1382. 1957.

Lonati, R. D. 4295.

12403. Londan, N. 1949: Effect of spacing on the bolting of onions. Nature 164:496–97.

12404. London, V. C. 1943: Rock plants from seed. Alpine Gard. Soc. Quart. Bull. 11:218–20.

12405. Loneragan, J. F., D. Meyer, R. G. Fawcett, and A. J. Anderson. 1955: Lime pelleted clover seeds for nodulation on acid soils. J. Australian Inst. Agr. Sci. 21:264–65.

Long, A. G. 3171–3173.

12406. Long, B. 1930: Spikelets of Johnson grass and Sudan grass. Botan. Gaz. 89:154–68.

Long, H. H. de; *see* De Long, H. H.

Long, H. O. 17982.

12407. Long, O. H. 1953: Nitrogen and spacing experiments with corn. Tenn. Agr. Expt. Sta. Bull. 232. 34 pp.

12408. —— and L. F. Seatz. 1953: Correlation of soil tests for available phosphorus and potassium with crop yield responses to fertilization. Soil Sci. 17:258–62.

12409. —— and C. D. Sherbakoff. 1951: Effect of nitrogen on yield and quality of wheat. Agron. J. 43:320–21.

12410. Long, T. E., and M. G. Cropsey. 1941: Grain storage on the farm. N. Dakota Agr. Expt. Sta. Bull. 302:1–68. Biol. Abstr. 17:9706. 1943.

12411. Long, T. J., and A. H. Haber. 1963: Comparisons between the growth of embryos on carbohydrates after excision from irradiated (500 kr) and unirradiated wheat grains. Proc. Plant Physiol. Meetings 38:xxii–xxiii.

12412. Long, T. P., and H. Kersten. 1936: Stimulation of growth of soybean seeds by soft X-rays. Plant Physiol. 11:615–21.

12413. Long, W. H., and J. H. Lilly. 1959: Effects of chemical seed treatments on wireworm activities. J. Econ. Entomol. 52:509–11.

12414. Longchamp, R., M. Roy, and R. Gautheret. 1952: Sur la stérilisation du blé par les hétéro-auxines désherbantes. Compt. Rend. 234:558–60.

12415. Longenecker, H. E., and D. E. Haley. 1935: *Ricinus* lipase, its nature and specificity. J. Am. Chem. Soc. 57:2019–21.

Longley, A. E. 15925.

12416. Longley, A. J., and M. D. Atkins. 1950: Airplane seeding of sand lovegrass (*Eragrostis trichodes*). Soil Conserv. 15: 258–60.

12417. Longo, B. 1917: Studies on polyembryony. Ann. Botan. (Rome) 14:151–62. Expt. Sta. Rec. 39:527. 1918.

12418. Longwell, J. H. 1945: Effect of storage on the feed value of hard red spring wheat. N. Dakota Agr. Expt. Sta. Bimonthly Bull. 7:3–5. Biol. Abstr. 20:1483. 1946.

12419. Lonnquist, J. H. 1953: Heterosis and yield of grain in maize. Agron. J. 45:539–42.

12420. Lonzinger, E., and R. Raskina. 1931: A study of cotton seed and of cottonseed oil at various stages of maturity. Maslob.-Zhir. Delo 1931(2/3):57–60. Chem. Abstr. 26:3127. 1932.

12421. Loo, S. W., and F. H. Wang. 1943: Culture of young conifer embryos in vitro. Science 98:544.

12422. Loo, T. L., and Y. W. Tang. 1945: Growth stimulation by manganese sulphate indole-3-acetic acid, and colchicine in the seed germination and early growth of several cultivated plants. Am. J. Botany 32:106–14.

Loo, T. L. 18361.

Loo, Y. S. 11990.

12423. Looby, W. J., and J. Doyle. 1937: Fertilization and proembryo formation in Sequoia. Sci. Proc. Roy. Dublin Soc. N.S. 21:457–76. Expt. Sta. Rec. 78:55. 1938.

12424. —— and —— 1939: The ovule, gametophytes, and pro-embryo in Saxegothoeae. Sci. Proc. Roy. Dublin Soc. N.S. 22:95–117. J. Roy. Microscop. Soc. 59:302–3. 1939.

Looby, W. J. 5547.

12425. Loomis, H. F. 1961: Culture of the palms. Preparation and germination of palm seeds. Hort. Mag. 40:128–30.

Loomis, W. E. 2286, 9468, 10309, 15863.

Loon, J. van; *see* Van Loon, J.

12426. Loos, W. 1963: Wirkungen des gibberellins in Kombination mit anderen Stoffen. Phyton 20:65–72.

Loose, R. de; *see* De Loose, R.

Lopez, A. C. 3453A.

12427. Lopez, A. P. 1961: Effects of scarification and gibberellic acid on germination of Congo watermelon seed. J. Agr. Univ. Puerto Rico 45:182–87.

Lopez, J. 4723.

12428. Lopez, M. 1938: Storage and germination of large-leaf mahogany seeds. Philippine J. Forestry 1:397–410.

Lopez, M. A. Lugo-. 2748.

Lopez, M. Rodriguez-; *see* Rodriguez-Lopez, M.

Lopez, S. L. 14363.

Lopez F., L. C. 4061.

Lopez-Herrera, C. 3671.

12429. Lopriore, G. 1913: Prove di germinazione con i semi della *Chloris gayana* Kunth. Staz. Sper. Agrar. Ital. 46:618–22.

12430. —— 1921: The germination of tricotyledonous seed. Staz. Sper. Agrar. Ital. 54:227–47. Botan. Abstr. 11:4213. 1922.

12431. Lopushanskii, P. I., and G. K. Molotkovskii. 1956: Phenomenon of polarity and dynamics of fat accumulation and vitamin C in the upper part of walnut. Byul. Mosk. Obshchestva Ispytatelei Prirody Otd. Biol. 61:61–67. Chem. Abstr. 50:8811. 1956.

12432. Lord, C. M. 1946: Telling how to grow primroses from seed. Horticulture 24(20):558.

12433. Lord, L. 1929: The germination of rice seeds in Ceylon. Ann. Roy. Botan. Garden Peradeniya, Ceylon 11:113–23.

12434. Lord, W. J., and D. G. White. 1962: The induction of parthenocarpy and an attempt to promote apomixis in the strawberry treatments with growth regulators. Proc. Am. Soc. Hort. Sci. 80:350–62.

12435. Lorensson, A. 1958: Photoperiodism—a factor to be considered in seed culture. Svensk Frötidn. 27:74–79.

12436. Lorenz, F. W., and H. J. Almquist. 1934: Effect of malvaceous seeds on stored-egg quality. Ind. Eng. Chem. 26:1311–13. Chem. Abstr. 29:848–49. 1935.

Lorenz, M. 10718.

12437. Lorenz, R. J., G. A. Rogler, and H. Holmen. 1959: Yields of bromegrass and a bromegrass-alfalfa mixture as influenced by moisture levels, fertilizer rates, and harvest frequency. Agron. Abstr. 1959:82.

Lorenz, R. J. 16859.

Lorz, A. P. 5328.

12438. Lose, H. H., and C. E. Leighty. 1912: Germination of seed as affected by sulphuric acid treatment. Cornell Univ. Agr. Expt. Sta. Bull. 312:293–336.

12439. Lotfy, A. A. M. 1958: Use of ultraviolet rays in differentiating between the seeds of *Trifolium alexandrinum* varieties; Miskawi & Fahl. Proc. Intern. Seed Testing Assoc. 23:68.

12440. Lototskii, I. S. 1953: On storing acorns in ice storehouses. Lesn. Khoz. 6(7):61–62.

12441. Lotti, T. 1955: Yellow-poplar height growth affected by seed source. Tree Planters Notes 22:3.

12442. —— and W. P. Le Grande. 1959: Loblolly pine seed production and seedling crops in the lower coastal plain of South Carolina. J. Forestry 57:580–81.

12443. Lou, C. H., L. F. Yen, and Y. L. Hsueh. 1950: The mechanism of 2,4-D action. I. 2,4-D seed treatment and its effects of subsequent growth of germinating seeds. Chinese J. Exptl. Biol. 3:41–55.

12444. ——, ——, K. H. Lou, C. Y. Yang, and S. Y. Fu. 1950: Separation of oil from sesame seeds by displacement with water. Sci. Record (China) 3:203–7.

Lou, C. H. 9151, 12344.

Lou, K. H. 12444.

Loucks, K. W. 3549.

12445. Loughman, B. C. 1961: Effect of boric acid on the phosphoglucomutase of pea seeds. Nature 191:1399–400.

Loughnane, J. B. 4207.

Louis, L. 13479.

12446. Louis-Marie, O. C. S. O. 1954: La graine du lotier (*Lotus corniculatus*). Rev. Oka Agron. Med. Vet. 28:77–86.

12447. Loustalot, A. J. 1940: Catalase activity in relation to after-ripening of apple seeds. Proc. Am. Soc. Hort. Sci. 37:361–62.

12448. ——, T. J. Muzik, and H. J. Cruzado. 1954: Studies on nutgrass (*Cyperus rotundus* L.) and its control. Puerto Rico (Mayaguez) Fed. Expt. Sta. Bull. 52. 30 pp.
Loustalot, A. J. 17622.
Louvet, J. 4905.
12449. Lovato, A. 1961: The evaluation of normal and abnormal seedlings in germination tests of "seeds" of *Cichorium intybus* L. and *C. endivia* L. Ann. Sper. Agrar. (Rome) 15:179–89.
12450. —— 1962: Evidences on laboratory methods for the recognition of seeds of the Miskawi and Fahl varieties of *Trifolium alexandrinum* L. Proc. Intern. Seed Testing Assoc. 27:530–41.
12451. Love, H. H., and W. T. Craig. 1944: A machine for dehulling oats. J. Am. Soc. Agron. 36:264–66.
12452. —— and C. E. Leighty. 1912: Germination of seed as affected by sulfuric acid treatment. Cornell Univ. Agr. Expt. Sta. Bull. 312:293–336.
Lovelace, F. E. 3633.
12453. Lovelock, B. 1952: Alpines from seed. Horticulture 30(7):286.
Lovestead, H. S. 13145.
Lovkova, M. J. 8830.
Lovkova, M. Y. 9380.
12454. Lovrekovich, L., and Z. A. Klement. 1963: A practical method to demonstrate the bacterial infection of bean seeds. Acta Agron. (Budapest) 12:83–87.
Lovvorn, R. L. 3789.
12455. Low, J. B., and F. C. Bellrose, Jr. 1944: The seed and vegetative yield of waterfowl food plants in the Illinois River Valley. J. Wildlife Management 8:7–22. Expt. Sta. Rec. 90:794. 1944.
12456. Lowe, A. E. 1940: Viability of Buffalo grass seeds found in the walls of a sod house. J. Am. Soc. Agron. 32:891–93.
12457. Lowe, C. C., and R. P. Murphy. 1955: Open-pollinated seed setting among self-sterile clones of smooth bromegrass. Agron. J. 47:221–24.
12458. Lowe, G., Jr. 1944: *Rudbeckia triloba* from seed. Horticulture 22(21):488.
Lowe, H. J. 528.
12459. Lowe, J., and O. E. Nelson, Jr. 1946: Miniature seed—a study in the development of a defective caryopsis in maize. Genetics 31:525–33.
12460. Lowe, W. H. 1960: Harvesting orchid seed. Am. Orchid Soc. Bull. 29:203–4.
12461. Lowery, J. C. 1952/53: Seed certification and a sound agronomy-extension program. Intern. Crop Improvement Assoc. Ann. Rept. 34/35:182–91.
12462. Lowewy, W. W. 1944: Removing the epidermis coverings from grain kernels such as those of wheat. U.S. Patent 2,355,810. Chem. Abstr. 39:134. 1945.
12463. Lowig, E. 1953: Beiträge zur Frage der Saatgutlagerung. Saatgut-Wirtsch. 5: 148–50.
12464. —— 1954: Beiträge zur Frage der Saatgutlagerung. II. Saatgut-Wirtsch. 6: 124–27.
12465. —— 1955: Beiträge zur Frage der Saatgutlagerung. III. Saatgut-Wirtsch. 7:14–16.
12466. —— 1958: Saatgut-Lagerschäden an Salat-Keimpflanzen. Saatgut-Wirtsch. 10: 160–62.
12467. —— 1960: Keimschutz für Saatgut durch geeingente Verpackung. Proc. Intern. Seed Testing Assoc. 25:607–17.
12468. —— 1961: Knowledge and problems, measures and means in the storage protection of seeds. II. Saatgut-Wirtsch. 13:284–85.
12469. —— 1962: Experience and problems, measures and means to protect seeds in storage. VII. Saatgut-Wirtsch. 14:336–38.
12470. —— 1962: Vorratsschutz für Saatgut Erkenntnisse aus Forschung und Entwicklung. Proc. Intern. Seed Testing Assoc. 27:929–43.
12471. —— 1963: Moderne Saatgutverpackung. Proc. Intern. Seed Testing Assoc. 28:943–79.
12472. Lowry, W. S. B. 1959: Influence of temperature and water content on the X-ray sensitivity of barley seeds. Phys. Med. Biol. 3:291–93. Chem. Abstr. 53:9390. 1959.
12473. Loza, J. 1952: Action des ultrasons sur la germination des grains, de riz, de soja, de pois et de radis. Congr. Intern. Biochim. Paris 2:453–54.
Lozova, G. I. 18137.
12474. Lubatti, O. F., and R. E. Blackith. 1957: Fumigation of agricultural products. XV. Germination and growth of cereals fumigated with methyl bromide. J. Sci. Food Agr. 8:691–97.
12475. —— and G. Bunday. 1960: The water content of seeds. I. The moisture relations of seed peas etc. J. Sci. Food Agr. 11:685–90.
12476. ——, J. Russell, and J. S. Porter. 1948: Fumigation of agricultural products.

I. A fumigation chamber for onion seeds. Soc. Chem. Ind. J. 67:309–13.

12477. —— and B. Smith. 1948: Determination of fumigants. XIX. Sorption of methyl bromide by onion seed. Soc. Chem. Ind. J. 67:297–309.

Lubatti, O. F. 2538.

Lubbert, G. 6416.

12478. Lubbock, J. 1891: Fruit and seed of Juglandeae. J. Linn. Soc. London 28:247–54. J. Roy. Microscop. Soc. 1891:621–22.

12479. Lubenec, P. A. 1940: Phasic development of lucerne. Herbage Rev. 8:206–8.

Lubimenko, V. N. 285.

12480. Lubimenko, W. 1906: Influence de l'absorption des sucres sur les phénomènes de la germination des plantules. Compt. Rend. 143:130–33.

12481. —— 1910: Influence de la lumière sur le développement des fruits et des graines chez les végétaux supérieurs. Rev. Gen. Botan. 22:145–75.

12482. —— 1911: Influence de la lumière sur la germination des graines. Rev. Gen. Botan. 23:418–36.

Luca, G. de; *see* De Luca, G.

12483. Lucas, E. H., H. M. Sell, and R. W. Lewis. 1946: An antibiotic substance obtained from seeds of brassicas. Proc. Am. Soc. Hort. Sci. 48:361–62.

Lucas, E. H. 10789.

12484. Luckwill, L. C. 1939: On the factors affecting the mean seed weight of tomato fruits. New Phytologist 38:181–89. Expt. Sta. Rec. 82:604. 1940.

12485. —— 1946: A fruit-setting hormone from apple seeds. Nature 158:663.

12486. —— 1948: The hormone content of the seed in relation to endosperm development and fruit drop in the apple. J. Hort. Sci. 24:32–44.

12487. —— 1949: Fruit drop in the apple in relation to seed development. Ann. Appl. Biol. 36:567–68.

12488. —— 1952: Growth-inhibiting and growth-promoting substances in relation to the dormancy and after-ripening of apple seeds. J. Hort. Sci. 27:53–67.

12489. —— 1953: Fruit development in relation to plant hormones. I. Hormone production by the developing apple seed in relation to fruit drop. II. Effect of naphthylacetic acid on fruit set and fruit development in apples. J. Hort. Sci. 28:14–24, 24–40.

12490. —— 1954: Auxins of the apple seed and their role in fruit development. Congr. Intern. Botan. Paris Rappt. Commun. 8:377–79. Chem. Abstr. 48:10855. 1954.

12491. —— and D. A. Woodcock. 1950: Preliminary investigation into the nature of the hormone produced by developing apple seeds. Bristol Univ. Agr. Hort. Res. Sta. Ann. Rept. 1950:23–30.

12492. Lucy, A. B. 1940: Experiments on the germination of oil palm seeds. Malayan Agr. J. 28:151–58.

12493. —— 1940: Tea seed production and germination experiments at the Central Experiment Station, Serdang. Malayan Agr. J. 28:215–20.

Luddy, F. E. 1900, 11395.

12494. Ludecke, H., and O. Neeb. 1962: Investigations on correlations between sowing depth and germination of sugar beet seed in the field. Zucker 15:544–49.

12495. —— and —— 1963: On the present quality of pelleted sugar beet seed. Zucker 16:138–42.

12496. Luders, R. 1955: The distribution of phosphoric acid and heavy metals in grain and straw of Petkuser spring rye. Z. Pflanzenernähr. Düng. Bodenk. 68:66–68. Chem. Abstr. 49:6385. 1955.

12497. Ludwiczak, S., and I. Zyczynska. 1963: The neutral components of lupine. I. High-molecular-weight aliphatic alcohol and lupeol from seeds of *Lupinus angustifolius* var. *rufina*. Roczniki Chem. 37:757–63. Chem. Abstr. 60:2042. 1964.

12498. Ludwig, C. A. 1932: The germination of cotton seed at low temperatures. J. Agr. Res. 44:367–80.

12499. Ludwig, F. 1886: Desiccation of seeds of aquatic plants. Biol. Centralbl. 6:299–300.

12500. —— and J. Von Ries. 1933: Wachstumvorgänge und Hochfrequenz. (Versuche an pflanzen und Tumoren.) Z. Krebsforsch. 40:117–21. Biol. Abstr. 10:327. 1936.

Ludwig, H. 10069.

12501. Ludwig, J. W., E. S. Bunting, and J. L. Harper. 1957: The influence of environment on seed and seedling mortality. III. The influence of aspect on maize germination. J. Ecol. 45:205–24.

12502. —— and J. L. Harper. 1958: The influence of the environment on seed and seedling mortality. VII. Depth of sowing of maize. Plant Soil 10:37–48.

12503. —— and —— 1958: The influence of the environment on seed and seedling mortality. VIII. The influence of soil colour. J. Ecol. 46:381–89.

Ludwig, J. W. 8199.

Ludwig, R. A. 17455.

Luebke, H. 12209.

12504. Lück, R. 1941: Zur Keimung der heimischen *Pirola*-Arten. Flora 35:1–5.

12505. Lüdi, W. 1932: Keimungsversuche mit Samen von Alpenpflanzen. Mitt. Naturforsch. Ges. Bern 46/50. Botan. Centralbl. 166:168. 1934.

12506. Lüdtke, M. 1952: Über das Linamarin des Leinsamen und seine Bestimmung. Biochem. Z. 322:310–19. Biol. Abstr. 27:14205. 1953.

12507. —— and H. Achmed. 1933: Über einen pflanzlichen Welkstoff. Biochem. Z. 257:256–66. Biol. Abstr. 9:281. 1935.

12508. Lüers, H. 1920: Studien über die Reifung der Cerealien. Biochem. Z. 104: 30–81.

12509. —— 1941: Characteristics of germination processes. Z. Volksernähr. 16:149–52. Chem. Abstr. 38:1766–67. 1944.

12510. ——, H. Fink, and W. Riedel. 1930: Studies on the ripening of barley enzymes during grain ripening. Wochschr. Brau. 47:393–97, 405–9. Chem. Abstr. 25:3766. 1931.

12511. —— and L. Malsch. 1929: Die Veränderungen des Hemizellulose-spaltenden Enzyms während des Keim- und Darrprozesses. Centralbl. Bakteriol. Parasitenk. Abt. II 79:139–40. Wochschr. Brau. 46:163–64. 1929.

12512. —— and W. Rümmler. 1935: Die Entwicklung der Amylase während der Keimung der Gerste. Wochschr. Brau. 52:9–12. Biol. Abstr. 10:20605. 1936.

12513. Luganskii, N. A. 1962: Oil content of seeds of various forms of *Pinus cembra sibirica*. Lesn. Khoz. 1962(2):69–70.

Lugay, J. C. 10055.

12514. Lugg, J. W. H., and G. J. Clowes. 1945: Partial amino-acid compositions of the proteins of some legume seeds: the tyrosine, tryptophane, cystine (plus cysteine) and methionine contents. Australian J. Exptl. Biol. Med. Sci. 23:75–79.

12515. —— and R. A. Weller. 1941: Protein metabolism in seed germination. Biochem. J. 35:1099–105.

12516. —— and —— 1943: Germinating seeds as a source of vitamin C in human nutrition. I. Ascorbic and dehydroascorbic acid contents of several varieties of seeds germinating under standard conditions for varying periods of time. Australian J. Exptl. Biol. Med. Sci. 21:111–14.

12517. —— and —— 1943: Germinating seeds as a source of vitamin C in human nutrition. II. The germinating and processing of blue boiler peas (*Pisum sativum*) as a palatable source of vitamin C. Australian J. Exptl. Biol. Med. Sci. 21:211–14.

12518. —— and —— 1944: On the sulphur present in the seeds of certain legumes, with special reference to the cystine plus cysteine and methionine contents of the proteins. Australian J. Exptl. Biol. Med. Sci. 22:149–55.

Lugg, J. W. H. 9837.

12519. Lugo, H. L. 1955: The effect of nitrogen on the germination of *Vanilla planifolia*. Am. J. Botany 42:679–84.

Lugo-Lopez, M. A. 2748.

Luippold, H. J. 7929–7933.

12520. Lukashevich, A. 1958: Effect of the treatment of seeds with fungicide on their quality. Kukuruza 1958(2):51–52.

Lukasiak, H. 12620.

12521. Luke, H. H., and P. L. Pfahler. 1962: The cause and control of low viability of rye. Phytopathology 52:344–46.

Lukens, H. C. 10494.

Lukianov, N. 12333.

12522. Lukjanov, I. A. 1959: The dormancy of the mucilaginous seeds of *Salvia sclarea*. Agrobiologiya 1959(2):294–96.

12523. Lukkala, O. J. 1936: Quality of seed from swamp-grown *Pinus silvestris*. Commun. Inst. Forest. Fennicae 22(3):1–45. Biol. Abstr. 11:2060. 1937.

Lukyanova, I. I. 12073.

12524. Lulla, B. S. 1948: Influence of extracts of germinated Indian pulses on the formation of amylase by *Bacillus subtilis*. Current Sci. (India) 17:13.

Lum, F. G. 8905.

Lummis, G. M. 15013.

12525. Luncz, G. 1931: The problem of the choice of seeds in forestry. Monthly Bull. Agr. Sci. Pract. 22:31–35. Biol. Abstr. 6:4932. 1932.

12526. Lund, A. P., and W. M. Sandstrom. 1943: The proteins of various tree seeds. J. Agr. Res. 66:349–55.

Lund, J. Kielland-; *see* Kielland-Lund, J.

Lund, S. 15240.

12527. Lundberg, J. F., and A. Ackerman.

1917: Observations on the color of seeds originating from spontaneous crossing between two forms of *Phaseolus vulgaris*. Bol. Agr. (Sao Paulo) 18:712–26, 798–807, 928–47. Expt. Sta. Rec. 38:539. 1918.

Lunde, C. M. 17048.

Lundeen, A. J. 6070.

12528. Lundegardh, H., and H. Burström. 1929: Investigations on the effect of seed treatment under various germination conditions. Kgl. Lantbruks-Akad. Handl. Tisdkr. 68:205–26. Biol. Abstr. 7:1562. 1933.

12529. ——, ——, and H. Ekstrand. 1930: The action of inorganic disinfectants on germination, and germ growth of grain. Kgl. Lantbruks-Akad. Handl. Tidskr. 69:602–32. Chem. Abstr. 24:5921–22. 1930.

12530. ——, E. Sandegren, and H. Lundin. 1944/45: Über den Gehalt der Malzgerste an Amylase, Aneurin und Mineralstoffen. Lantbrukhögskol. Ann. 12:172–85.

12531. Lunden, A. O., and A. T. Wallace. 1961: Some effects of ultraviolet light on barley and oat embryos. Crop Sci. 1:212–15.

Lunden, A. O. 14875.

Lundin, H. 12530.

Lundqvist, U. 5832.

12532. Lundstrom, A. N. 1886: Dissemination of fruits. Botan. Centralbl. 25:319. J. Roy. Microscop. Soc. 1886:648–49.

12533. —— 1892: Dissemination of the seeds of *Geranium bohemicum*. Botan. Centralbl. 49:202, 236.

12534. Lundy, M. G. 1940: Easy seed germination methods. Horticulture 18(3): 48.

Lunel, F. A. V. 18947.

12535. Lunin, J., and M. H. Gallatin. 1959: Effect of saline water on the growth and chemical composition of beans. Agron. Abstr. 1959:47.

Lunven, P. 11791.

12536. Lupe, I. Z. 1956: The effect of illumination on germination of elm seeds. Rev. Padurilor (Bucharest) 71:506–8. Biol. Abstr. 34:14313. 1959.

Lusin, V. 4888.

12537. Lustig, B., and H. Wachtel. 1938: Über biochemische Beeinflussung der Keimung und des Wachstums der Gardaminesamen. I. Einfluss der äusseren Faktoren und der Salze. Biochem. Z. 296:13–27.

12538. —— and —— 1938: Über biochemische Beeinflussung der Keimung und des Wachstums der Cardaminesamen. II. Wirkungen von Vitaminen und Hormonen auf die Pflanze. Biochem. Z. 297:386–94

12539. —— and —— 1939: Action d'extraits d'organes des animaux sur la germination des plantes. Compt. Rend. Soc. Biol. 132:224–27.

12540. —— and —— 1939: Action d'extraits de tumeurs benignes et malignes sur la germination et la croissance des plantes. Compt. Rend. Soc. Biol. 132:246–49.

Luszczak, R. 18244.

12541. Lute, A. M. 1928: Impermeable seed of alfalfa. Colo. Agr. Expt. Sta. Bull. 326. 35 pp.

12542. —— 1938: Germination characteristics of wild oats. Proc. Assoc. Offic. Seed Analysts 23:70–73.

12543. —— 1942: Laboratory germination of hard alfalfa seed as a result of clipping. J. Am. Soc. Agron. 34:90–99.

12544. —— and D. W. Robertson. 1938: Longevity studies—wheat, barley, oats. J. Colo.-Wyo. Acad. Sci. 2:24. Biol. Abstr. 12:16613. 1938.

12545. ——, M. L. Thornton, and K. Dixon. 1938: Study of the effect of altitude on germination of seeds. Proc. Assoc. Offic. Seed Analysts 23/26:83–86.

Lute, A. M. 16284–16287.

12546. Lutey, R. W., and C. M. Christensen. 1963: Influence of moisture content, temperature, and length of storage upon survival of fungi in barley kernels. Phytopathology 53:713–17.

12547. Luthra, J. C. 1936: Ancient wheat and its viability. Current Sci. (India) 4:489–90.

12548. —— 1943: An easy test for berseem seed. Indian Farming 4:194–95. Biol. Abstr. 20:1484. 1946.

12549. —— and I. S. Chima. 1941: Some studies on the potentiality of shrivelled wheat grains. Proc. Indian Acad. Sci. Sect. B 14:47–67.

12550. Lutikova, O. T. 1941: Enzyme dynamics in ripening pea seeds. Compt. Rend. Acad. Sci. URSS 31:683–85. Chem. Abstr. 37:907–8. 1943.

Lutikova, O. T. 16498.

Lutomski, J. 2791.

12551. Luttrell, E. S., L. V. Crowder, and H. D. Wells. 1955: Seed treatment tests with Pearl millet, Sudan grass and Browntop millet. Plant Disease Reptr. 39:756–61.

12552. Luyet, B. J., and P. M. Gehenio. 1938: The lower limit of vital temperatures. A review. Biodynamica 1(33):1–92.

12553. —— and —— 1940: The mechanism of injury and death by low temperature. A review. Biodynamica 3(60):33–99.

Luyet, B. J. 12378.

Luyten, I. 8289.

12554. L'Vov, N. A., and S. V. Yakovleva. 1930: Investigation of the seeds of officinal and aromatic plants. Bull. Appl. Botany, Genet., Plant Breed. (USSR) 23:543–664. Biol. Abstr. 7:21929. 1933.

12555. Lybing, J. 1928: Einige Beobachtungen des Keimens der Samen bei einigen unserer wichtigsten Drogenpflanzen der Familie Solanaceae. Intern. Congr. Europ. Med. Herbs Interest (Papers Presented) 2:29–30.

12556. —— 1929: Einige Beobachtungen über das Keimen von Samen bei einigen unserer wichtigsten Drogenpflanzen aus der Familie der Solanaceen. Ber. Deut. Intern. Tagung Europ. Arzneipfl. Interessenten 2:22–23. Botan. Centralbl. 156:206–7. 1929.

12557. Lyle, E. S., and A. R. Gilmore. 1958: The effect of rough handling of loblolly pine cones on seed germination. J. Forestry 56:595.

12558. Lyman, C. M., E. J. Burda, and P. I. Olschner. 1948: The effect of dielectric heating on storage quality of cottonseed. J. Am. Oil Chemists Soc. 25:246–49.

12559. ——, B. R. Holland, and F. Hale. 1943: Determination of free gossypol in cottonseed meal. A colorimetric method. Ind. Eng. Chem. Anal. Ed. 15:489–91. Biol. Abstr. 18:2172. 1944.

12560. ——, K. Kuiken, and F. Hale. 1947: The essential amino acid content of cottonseed, peanut and soybean products. Texas Agr. Expt. Sta. Bull. 692. 12 pp.

Lyman, C. M. 11491, 11492.

Lyman, J. F. 8027, 17319.

12561. Lynbchenko, V. M. 1960: Catalase activity in seeds of *Tilia cordata* Mill. and *Euonymus europaea* L. in the process of stratification. Botan. Zh. 44:522–24.

12562. Lynch, B. T., R. L. Glass, and W. F. Geddes. 1962: Grain storage studies. XXXII. Quantitative changes occurring in the sugars of wheat deteriorating in the presence and absence of molds. Cereal Chem. 39:256–62.

Lynch, L. J. 13837.

12563. Lynch, S. J., and F. J. Fuchs, Sr. 1955: A note on the propagation of *Phylanthus emblica* L. Proc. Florida State Hort. Soc. 68:301–2. Biol. Abstr. 31: 22362. 1957.

12564. Lynd, J. Q. 1955: Grass-legume band seeding with a shoe type drill. Agron. J. 47:195–96.

Lynd, L. E. 16294.

12565. Lynes, F. F. 1945: Polyploidy in sugar beets induced by storage of treated seed. J. Am. Soc. Agron. 37:402–4.

12566. —— 1945: Sizing whole sugar beet seed to improve germination. J. Am. Soc. Agron. 37:779–81.

Lyness, W. E. 10588–10590.

Lynn, E. V. 1713.

Lyon, A. G. 9118.

12567. Lyon, M. E. 1928: Embryoless seeds in cereals. Science 67:652.

12568. —— 1928: The occurrence and behavior of embryoless wheat seeds. J. Agr. Res. 36:631–37.

12569. Lyons, J. C., and E. B. Earley. 1952: The effect of ammonium nitrate applications to field soils on nodulation, seed yield and nitrogen and oil content of the seed of soybeans. Soil Sci. Soc. Am. Proc. 16:259–63.

12570. Lyons, L. A. 1956: The seed production capacity and efficiency of red pine cones (*Pinus resinosa* Ait.). Can. J. Botany 34:27–36.

12571. —— 1957: Insects affecting seed production in the red pine. II. *Dioryctria disclusa* Heinrich, *D. abietella* (D. and S.), and *D. cambiicola* (Dyar) (Lepidoptera): Phycitidae). Can. Entomologist 89:70–78.

12572. —— 1957: Insects affecting seed production in red pine. III. *Eucosma mamtorana* Heinrich, *Laspeyresia toreuta* Grote (Lepidoptera: Olethreutidae), *Rubsaamenia* sp. (Diptera: Cecidomyiidae), and other insects. Can. Entomologist 89:150–64.

12573. —— 1957: Insects affecting seed production in red pine. IV. Recognition and extent of damage to cones. Can. Entomologist 89:264–71.

12574. Lysenko, T. D. 1941: Good germination of kok saghyz ensures a good yield. Iarovizatsiia 3(36):3–11. Biol. Abstr. 18:3409. 1944.

12575. —— 1942: Ways to increase seed germination in small grains. Proc. Lenin. Acad. Agr. Sci. USSR 1942(3/4):3–6.

12576. —— and F. S. Stepanenko. 1935: Iarovization of agricultural plants with instructions for iarovization of wheat—barley—oats. Moscow, USSR. 63 pp.

Lysenko, Y. A. 3475.

12577. Lyukova, L. A. 1958: Effect of nicotinic acid on the growth and the yield of some vegetable plants. Plant Physiol. (USSR) (English Transl.) 5:24–30. Chem. Abstr. 52:10297. 1958.

M

12578. M. 1930: Die Vermehrung der Nadelhölzer. Gartenwelt 24:360–62.

12579. Mass, J. G. J. A. 1918: Germination trials with Hevea seed. Arch. Rubbercult. Nederl.-Indie 2:666–725. Expt. Sta. Rec. 41:449. 1919.

12580. —— 1926: Germinating experiments with seeds of different species of green manures. I & II. Mededeel. Algem. Proefsta. A.V.R.O.S. Algem. Ser. 27. 33 pp.

12581. Maatsch, R., and H. Isensee. 1959: The germination of cyclamen in media of various pH values. Gartenwelt 59:363.

12582. —— and W. Rünger. 1955: Keimung von Cyclamen in verschiedenen Substraten. Gartenwelt 55:93–94.

12583. —— and —— 1955: Über den Einfluss der Saatzeit auf den Blütenertrag von Chabaudnelken. Gartenbauwissenschaften 1:399–401.

12584. MacAndrews, A. H. 1945: Sawdust culture for delphinium seeds. Am. Delphinium Soc. Yearbook 1945:15–17. Biol. Abstr. 20:13752. 1946.

12585. MacCollom, G. B. 1958: Control of insects affecting Birdsfoot trefoil seed production in Vermont. J. Econ. Entomol. 51:492–94.

12586. MacDaniels, L. H. 1939: The shuck-nut ratio in black walnut varieties. Northern Nut Growers Assoc. Proc. 30:74–76. Expt. Sta. Rec. 83:631. 1940.

12587. MacDonald, D., W. L. Fielding, and D. F. Ruston. 1947: Experimental methods with cotton. III. Sulphuric acid treatment of cotton seed, and its effects on germination, development, and yield. J. Agr. Sci. 37:291–96.

12588. MacDonald, E. D. 1961: Polyethylene tarp used in pelleting coniferous tree seed. Tree Planters Notes 47:5–7.

MacDonald, H. A. 18858, 19744.

12589. MacDonald, J. 1935: Nursery investigations. I. The use of sand as a covering material in seed-beds. II. Stratification as a means of improving the germination and production of seed of Douglas fir. Forestry 9:24–41. Biol. Abstr. 10:9147. 1936.

12590. Macdonald, J. A. 1935: Plant pathology. Effect of excess of Ceresan and Agrosan G on the germination of wheat. Scot. J. Agr. 18:164–67. Chem. Abstr. 29:4885. 1935.

MacDonald, R. T. 7491.

MacDonnell, L. R. 9736.

12591. Mace, M. E., and T. T. Hebert. 1963: Naturally occurring quinones in wheat and barley and their toxicity to loose smut fungi. Phytopathology 53:692–700.

12592. Maceda, F. S. 1933: A study of coconut seedlings in relation to shape of the nuts. Philippine Agriculturist 22:430–41. Expt. Sta. Rec. 70:630. 1934.

Macedo, A. Corrales-; *see* Corrales-Macedo, A.

12593. MacGillivray, H. G. 1955: Germination of spruce and fir seeds following different stratification periods. Forestry Chron. 31:365.

12594. MacGillivray, J. H., and L. J. Clemente. 1949: Effect of irrigation on the production of carrot seed. Proc. Am. Soc. Hort. Sci. 54:299–303.

MacGillivray, J. H. 5462, 14355.

12595. MacGregor, J. M. 1954: Effect of nitrogen from anhydrous ammonia and ammonium nitrate on yield and protein of oats and corn in Minnesota, 1953. Minn. Agr. Expt. Sta. Misc. Rept. 22:1–16.

12596. ——, L. T. Taskovitch, and W. P. Martin. 1961: Effect of nitrogen fertilizer and soil type on the amino acid content of corn grain. Agron. J. 53:211–14.

12597. Machacek, J. E. 1945: The prevalence of Septoria on cereal seed in Canada. Phytopathology 35:51–53.

12598. —— 1949: Weeds in relation to the results obtained from seed treatment in cereals and flax. [Abstract.] Phytopathology 39:861.

12599. —— 1950: An agar-sheet method of testing the efficiency of seed treating machines. Can. J. Res. Sect. C 38:739–44.

12600. —— and A. M. Brown. 1945: Threshing-injury to flax seed in Canada. Sci. Agr. (Ottawa) 25:601–25. Biol. Abstr. 19:22580. 1945.

12601. —— and —— 1948: Experiments on vegetable seed disinfection and observations on varietal resistance of beans, peas, and sweet corn to some diseases in Manitoba. Sci. Agr. (Ottawa) 28:145–53.

12602. ——, W. J. Cherewick, H. W. Mead, and W. C. Broadfoot. 1951: A study of some seed-borne diseases of cereals in Canada. II. Kinds of fungi and prevalence of disease in cereal seed. Sci. Agr. (Ottawa) 31:193–206.

12603. —— and F. J. Greaney. 1948: Summary of cooperative experiments on treatment of flax seed. Phytopathology 38:803–7.

12604. ——, E. Robertson, H. A. H. Wallace, and N. A. Phillips. 1961: Effect of high water content in stored wheat, oat, and barley seed on its germinability, susceptibility to invasion by moulds, and response to chemical treatment. Can. J. Plant Sci. 41:288–303.

12605. —— and H. A. H. Wallace. 1952: Germination of treated and untreated seed of wheat, oats and barley of different commercial grades. Sci. Agr. (Ottawa) 32:597–602.

12606. —— and —— 1952: Longevity of some common fungi in cereal seed. Can. J. Botany 30:164–69.

12607. —— and —— 1957: A study of the germination of barley seed treated to control loose smut. Can. J. Plant Sci. 37:59–68.

Machacek, J. E. 7606–7608, 19323.

12608. Machado, A. 1936: Oil of Sucupira (*Bowdichia virgiloids*, Humboldt). Rev. Assoc. Brasil. Farm. 17:117–18. Chim. Ind. (Paris) 39:316. 1938. Pharm. Abstr. 5:102. 1939.

12609. —— 1938: Oiticica oil. Rev. Soc. Brasil. Quim. 7:73–81. Pharm. Abstr. 5:254. 1939.

12610. —— 1944: Chemical analyses of the seeds of *Protium brasiliensis*. Rev. Brasil. Farm. 25:155–56. Chem. Abstr. 39:2178–79. 1945.

12611. Machado, D. P. 1962: The improvement of forest trees and seeds in Portugal. Proc. World Forestry Congr. 5:711–13. 1960.

12612. Machalica, J. J. 1926: Germination of cereals in moist atmosphere. Roczniki Rolniczych Lesnych 15:407–23. Biol. Abstr. 6:21624. 1932.

Mache, R. 2296.

Macheboeuf, M. 8603.

12613. Macholz, D. 1937: Die Wehrlose Trespe (*Bromus inermis* Leyss.). Eine Studie zur Erforschung ihrer Lebensansprüche und ihres Wertes für unser Grünland. Thesis (Ph.D.). Friedrich Wilhelms-Univ., Berlin. 67 pp.

12614. Macht, D. I. 1948: Influence of human saliva and blood serum on germination and root growth. Science 108:565.

12615. —— and H. F. Byran. 1937: Relative potency of reductase in dry, wet and germinated *Lupinus albus* seeds. Am. J. Botany 24:133–34.

12616. —— and M. E. Davis. 1934: Some pharmacological experiments with deuterium. J. Am. Chem. Soc. 56:246. Chem. Abstr. 28:1728. 1934.

12617. —— and T. E. Hoffmaster. 1950: The action of antibiotics on oxidation-reduction enzymes of *Lupinus* seeds. Arch. Intern. Pharmacodyn. 83:207–12. Biol. Abstr. 26:13249. 1952.

12618. Macias, F. T., R. G. Smith, and F. N. Martin, Jr. 1943: A splenic contracting substance in orange seeds. J. Pharmacol. 77:311–16. Chem. Abstr. 37:3501. 1943.

12619. Maciejewska-Potapczykowa, W. 1959: Influence of kinetin, β-indoleacetic acid, and gibberellic acid on nuclease activity of bean (*Phaseolus vulgaris*) hypocotyls. Nature (Suppl.) 184:557–58.

12620. ——, T. Wilusz, and H. Lukasiak. 1961: The influence of gibberellin on the carbohydrates content in peas epicotyls. Acta Soc. Botan. Polon. 30:43–51.

Mack, P. B. 19279.

12621. MacKay, D. B. 1961: The effect of pre-washing on the germination of sugar beet. Natl. Inst. Agr. Botany J. 9:99–103.

12622. —— 1962: Internal referee testing in England and Wales 1951–1961. Proc. Intern. Seed Testing Assoc. 27:973–82.

12623. —— 1964: Testing for seed quality. World Crops 16:63–66.

MacKay, D. B. 4187.

12624. Mackel, H. G. 1951: Über Polyembryonie, Mehrsamigkeit und Mehrfächrigkeit bei *Coffea arabica*. Ber. Deut. Botan. Ges. 64:28–35. Biol. Abstr. 26:4300. 1952.

MacKenzie, A. J. 32.

12625. Mackenzie, J. A. 1960: Work of the physiology section. Nigeria Dept. Forest Res., Forest Res. Conf. Proc. 1960:29–31.

Mackenzie, K. G. 15676.

12626. MacKey, J. 1951: Chemical method for rapid determination of the germinability of seed. Nord. Jordbrugsforskn. 1949/50:283–314.

12627. —— 1952: The biological action of X-rays and fast neutrons on barley and wheat. Arkiv Botan. 2 Ser. 1:545–56. Biol. Abstr. 28:6892. 1954.

12628. —— 1954: The biological action of mustards on dormant seeds of barley and wheat. Acta Agr. Scand. 4:419–29.

12629. Mackie, A., and D. G. Mieras. 1961: Chemical investigation of seeds of selected tropical plants. I. Component acids of the fats or oils. J. Sci. Food Agr. 12:202–5.

12630. MacKinney, A. L., and C. F. Korstian. 1938: Loblolly pine seed dispersal. J. Forestry 36:465–68.

12631. —— and W. E. McQuilkin. 1938: Methods of stratification for Loblolly pine seeds. J. Forestry 36:1123–27.

MacKinney, A. L. 11239.

12632. MacLachlan, P. L. 1936: Fat metabolism in plants, with special reference to sterols. J. Biol. Chem. 113:197–204.

12633. —— 1936: Fat metabolism in plants, with special reference to sterols. II. Differential changes in the cotyledons and in the roots, stems, and leaves. J. Biol. Chem. 114:185–91.

MacLagan, J. F. A. 14307.

12634. MacLaren, R. B., and R. B. Carson. 1958: Yield and chemical composition of oats when harvested at four stages of maturity. Can. J. Plant Sci. 38:12–15.

12635. Maclay, W. D., J. R. Matchett, and M. Pollack. 1963: Industrial utilization of seed oils. Econ. Botany 17:23–30.

MacLean, A. J. 14394.

12636. MacLean, J. A. R. 1952: Oil-bearing seeds of possible economic importance to West Africa. Nature 169:589–90.

12637. MacLeod, A. M. 1950: Determination of germinative capacity of barley by means of tetrazolium salts. J. Inst. Brewing 56:125–34. Chem. Abstr. 45: 3914. 1951.

12638. —— 1951: Sugars and amino-acids of barley grain. J. Inst. Brewing 57:163–79. Brit. Abstr. B III:468. 1951.

12639. —— 1952: Barley viability. Brewers Guild J. 38:271–79. Biol. Abstr. 27:7441. 1953.

12640. —— 1952: Enzyme activity in relation to barley viability. Trans. Proc. Botan. Soc. Edinburgh 36:18–33.

12641. —— 1952: The free sugars of the barley grain. I. Historical survey. II. Distribution of the individual sugar fractions. J. Inst. Brewing 58:270–76, 363–71. Chem. Abstr. 48:5290. 1954.

12642. —— 1953: The quality of cereals and their industrial uses. Viability testing of cereal seeds. Chem. Ind. (London) 1953(13):289–91.

12643. —— 1957: Raffinose metabolism in germinating barley. New Phytologist 56: 210–20.

12644. ——, M. I. Biol, and H. B. White. 1961: Lipid metabolism in germinating barley. I. The fats. J. Inst. Brewing 67:182–90.

12645. —— and H. McCorquodale. 1958: Trisaccharides of *Lolium* and *Festuca*. Nature 182:815–16.

12646. —— and —— 1958: Water-soluble carbohydrates of seeds of the Gramineae. New Phytologist 57:168–82.

12647. —— and R. Sandie. 1961: Cell wall metabolism. I. Hemicellulases of *Bromus* seeds. New Phytologist 60:117–28.

12648. Macloskie, G. 1898: Effect of different temperatures of water on the germination of olive seeds. Bull. Torrey Botan. Club 25:222–25.

12649. —— 1898: Heat of imbibition by seeds. Bull. Torrey Botan. Club 25:272–74. Expt. Sta. Rec. 10:358. 1898.

12650. MacMasters, M. M., and G. E. Hilbert. 1944: Stored and damaged wheats for starch production. Cereal Chem. 21:258–66.

12651. —— and —— 1944: The composition of the "amylodextrin" fraction of wheat flour. Cereal Chem. 21:548–55.

12652. —— and —— 1944: Glutinous corn and sorghum starches. Ind. Eng. Chem. 36:958–63. Expt. Sta. Rec. 94:713. 1946.

12653. ——, R. L. Slotter, and C. M. Jaeger. 1947: The possible use of oats and other small grains for starch production. Am. Miller Processor 75:82, 83. Biol. Abstr. 21:15246. 1947.

12654. ——, M. J. Wolf, and H. L. Seckinger. 1957: Starch identification, microscopic characteristics of starches in the identification of ground cereal grains. J. Agr. Food Chem. 5:455–58.

12655. ——, S. Woodruff, and H. Klaas. 1941: Studies on soybean carbohydrates. Ind. Eng. Chem. Anal. Ed. 13:471–74.

MacMasters, M. M. 1705, 2439, 2926–2929, 3565, 5072, 7051, 8747, 16902, 17161, 19798, 20130–20132.

12656. MacMillan, C. 1902: Suggestions on the classification of seeds. Botan. Gaz. 34:224–30.

12657. Macmillan, J., J. C. Seaton, and P. J.

Suter. 1959: A new plant-growth promoting acid—gibberellin A_5 from the seed of *Phaseolus multiflorus*. Chem. Soc. Proc. 1959:325–26.

Macmillan, J. 9940.

MacMillan, W. G. 3779.

12658. MacMorran, A. M. 1946: Experimental seed storage. Roy. New Zealand Inst. Hort. J. 15:24–25.

MacNab, J. 2045, 2046.

MacNear, B. T. 6476.

12659. MacNeil, A. 1946: Sowing lily seed in vermiculite. Flower Grower 33(5):332–33, 334.

12660. Macovski, E. 1955: Exosmosis as a tool for research into the evolution of biochemical processes in the seeds of culture plants, and its role in the stimulation and germination of these seeds. Comun, Acad. Rep. Populare Romine 5:1177–80.

MacSwan, I. C. 4406.

MacSween, J. C. 14308.

12661. MacVicar, R. M. 1935: Inheritance of seed color in alfalfa. Sci. Agr. (Ottawa) 15:314–28.

12662. ——, E. Braun, D. R. Gibson, and C. A. Jamieson. 1952: Studies in red clover seed production. Sci. Agr. (Ottawa) 32:67–80.

12663. —— and D. R. Gibson. 1955: The use of desiccants for the pre-harvest treatment of legume seed crops. Can. J. Agr. Sci. 35:83–90.

MacVicar, R. M. 2949.

MacWalter, R. J. 5585, 5586.

MacWilliam, I. C. 8235.

Madalska, O. Kostecka-; *see* Kostecka-Madalska, O.

Madden, G. O. 2329, 2330.

Madej, M. 12874.

12664. Mader, W., and F. Dotzler. 1931: Versuche über Stecklingsrösse und Standraum in ihrem Einfluss auf den Rübensamenertrag. Fortschr. Landwirtsch. 6: 267–68.

12665. Madhuranath, M. K., and B. L. Manjunath. 1938: Sandal seed oil (*Santalum album* Linn.). J. Indian Chem. Soc. 15:389–92. Analyst 63:843–44. 1938.

12666. Madrid, V. J. 1934: Treatment of seeds and plant cuttings with coal tarkerosene emulsion as a protection against certain insects. Philippine Agriculturist 23:604–12. Chem. Abstr. 29:2651. 1935.

12667. Madsen, S. B. 1957: Investigation of the influence of some storage conditions on the ability of seed to retain its germinating capacity. Proc. Intern. Seed Testing Assoc. 22:423–46.

12668. —— 1962: Germination of buried and dry stored seeds III, 1934–60. Proc. Intern. Seed Testing Assoc. 27:920–28.

12669. —— and M. Olesen. 1962: Comparative experiments with taking working samples by means of spoon and of Pascall divider. Proc. Intern. Seed Testing Assoc. 27:414–22.

Maeda, T. 19400.

12670. Maestri, M., and C. Vieira. 1963: A note on the decrease of the percentage of germination of coffee (*Coffea arabica* L. var. Bourbon), by action of gibberellic acid. Rev. Ceres 11:247–49. Biol. Abstr. 44:25120. 1964.

12671. Mgaee, A. I., and W. Kalbfleisch. 1952: Moisture determination of silage, hay and grain. Sci. Agr. (Ottawa) 32:117–26.

Magee, R. A. 10428.

Maggio, G. di; *see* Di Maggio, G.

Magidman, P. 1900.

12672. Magini, E. 1948: Attendibilità dei risultati delle prove per la determinazione della facoltà germinativa. Nuovo Giorn. Botan. Ital. 55:375–96.

12673. —— 1955: On the conditions of germination of Aleppo pine and domestic pine. Ric. Sci. 25:2906. Biol. Abstr. 31:2431. 1957.

12674. —— 1962: Forest seed handling, equipment and procedures. II. Seed treatments, storage, Testing and transport. Unasylva 16:20–35.

12675. —— 1962: Seed supply, methods of analysis, and treatment to hasten germination. Monti e Boschi 13:527–34.

12676. —— and N. P. Tulstrup. 1955: Tree seed notes. I. Arid areas. II. Humid tropics. Food Agr. Organ. U.N., FAO, Rome, Italy. 354 pp.

Magini, E. 8974.

12677. Magistad, O. C. 1945: Germination and seedling growth. Botan. Rev. 11: 203–5.

Magno, B. 16478.

12678. Magnus, W. 1920: Hemmungsstoffe und falsche Keimung. Ber. Deut. Botan. Ges. 38:19–26.

Magoon, C. A. 4733, 4734.

12679. Magrou, J., and F. Mariat. 1945: Action de l'aneurine sur le développement des embryons d'Orchidées. Ann. Inst. Pasteur 71:49–51. Rev. Appl. Mycol. 24: 283–84. 1945.

Magrou, J. 2472.

12680. Magruder, R. 1931: Self-sterility and fertility in garden beets as indicated by seed production under paper bags. Proc. Am. Soc. Hort. Sci. 28:328–31.

12681. —— and R. E. Webster. 1941: Green cotyledon, a new character in the mature lima bean (*Phaseolus lunatus* L.). Proc. Am. Soc. Hort. Sci. 38:581–84.

12682. ——, P. Work, L. R. Hawthorn, and G. W. Scott. 1938: Influence of temperature and rainfall on speed of emergence of spinach seedlings (Spinachia oleraceae). Proc. Am. Soc. Hort. Sci. 35:625–29.

Magruder, R. 19556.

12683. Maguire, J. D. 1962: Speed of germination—aid in selection and evaluation for seedling emergence and vigor. Crop Sci. 2:176–77.

12684. —— and K. M. Steen. 1962: Preliminary studies on the identification of *Agropyron* seeds by chemical staining. Assoc. Offic. Seed Analysts News Letter 36(1):19–20.

12685. —— and V. Youngman. 1963: Germination of *Crambe*. Assoc. Offic. Seed Analysts News Letter 37(2):6–7.

Maguire, J. D. 3569.
Mahadevan, S. 4846.
Mahadevan, V. 16789.
Mahajan, N. C. 7850A.

12686. Maheshwari, N. 1958: In vitro culture of excised ovules of *Papaver somniferum*. Science 127:342.

12687. Maheshwari, P. 1937: A critical review of the types of embryosacs in angiosperms. New Phytologist 36:359–417. Z. Botan. 33:425. 1938.

12688. —— 1944: The origin of the Haustoria in the ovule of *Lobelia*. J. Indian Botan. Soc. 23:79–80.

12689. —— 1945: Origin of haploid-diploid twin embryos in angiosperms. Nature 156:173–74.

12690. —— 1947: Tetranucleate embryo sacs in angiosperms. Lloydia 10:1–18.

12691. —— 1954: Embryology of angiosperms. A fruitful field for research. Congr. Intern. Botan. Paris Rappt. Commun. 8:235–43. Biol. Abstr. 29: 26867. 1955.

12692. —— 1956: The embryology of angiosperms, a retrospect and prospect. Current Sci. (India) 25:106–10.

12693. —— 1959: Test-tube fruits and seeds. J. Indian Botan. Soc. 38:161–70.

12694. —— 1961: Embryology in relation to taxonomy. Intern. Botan. Congr. Recent Advan. Botany 9:679–82. 1959.

12695. —— 1962: Contacts between embryology, physiology and genetics. *In* Maheshwari, Johri, and Vasil, editors. Proc. Summer School Botany, June 2–15, 1960, Darjeeling. Min. Sci. Res. Cult. Affairs, Govt. India, New Delhi, pp. 171–92. Biol. Abstr. 41:20078. 1963.

12696. —— and B. Baldev. 1961: Artificial production of buds from the embryos of *Cuscuta reflexa*. Nature 191:197–98.

12697. —— and H. Gangulee. 1942: The development of the embryo sac in *Heckeria umbellata*, Kunth. J. Indian Botan. Soc. 21:245–48.

12698. —— and A. Haque. 1949: The embryo sac of *Chrysanthemum parthenium* L. (Bernh.). New Phytologist 48:255–58.

12699. —— and B. M. Johri. 1941: The embryo-sac of *Acalypha indica* L. Botan. Centralbl. Beih. Abt. A 61:125–36. Botan. Centralbl. 178:67. 1942.

12700. ——, ——, and S. N. Dixit. 1957: The floral morphology and embryology of the Loranthoideae (Loranthaceae). J. Madras Univ. 27B:121–36.

12701. —— and R. Khan. 1953: Development of the embryo sac, endosperm and embryo in *Isomeris arborea*—a reinvestigation. Phytomorphology 3:446–59. Biol. Abstr. 31:19449. 1957.

12702. —— and S. Narayanaswami. 1952: Embryological studies on *Spiranthes australis* Lindl. J. Linn. Soc. London 53:474–86. Biol. Abstr. 27:13785. 1953.

12703. —— and R. C. Sachar. 1955: So-called endosperm embryos in *Isomeris arborea* Nutt. Nature 176:470–72.

12704. ——, ——, and R. N. Chopra. 1955: Embryological studies in mango (*Mangifera indica* L.). Proc. Indian Sci. Congr. Assoc. 42:233.

12705. —— and A. R. Srinivasan. 1944: A contribution to the embryology of *Rudbeckia bicolor* Nutt. New Phytologist 43:135–42.

12706. Maheshwari, S. C. 1956: Endosperm and seed of Wolffia. Nature 178:925–26.

12707. —— and R. N. Kapil. 1963: Morphological and embryological studies on the Lemnaceae. II. The endosperm and embryo of *Lemna paucicostata*.

Maheu, J. 5075.

12708. Mahlstede, J. P., and E. S. Haber. 1957: Plant propagation. Part II. Wiley, New York. 413 pp.

12709. Mahmoud, E. E., and M. S. Mohamed. 1953: The germination of seeds of

Orobanche crenata. Alexandria J. Agr. Res. 1:61–65. Chem. Abstr. 51:16742. 1957.
12710. Mahmud, I., and A. H. Probst. 1953: Inheritance of gray hilum color in soybeans Agron. J. 45:59–61.
12711. Mahon, J. H., and R. H. Common. 1950: The amino-acid composition of the horsebean (*Vicia faba* L.) as related to its nutritive value for chicks. Sci. Agr. (Ottawa) 30:43–49.
12712. Mahoney, C. H. 1935: Seed transmission of mosaic in inbred lines of muskmelons (*Cucumis melo* L.). Proc. Am. Soc. Hort. Sci. 32:477–80.
12713. Mahony, K. L. 1936: Morphological and cytological studies on *Fagopyrum esculentum*. II. Embryogeny. Am. J. Botany 23:129–33.
Mahy, C. 5252.
12714. Maier, F. 1950: Über den Einfluss von mechanischen Verletzungen auf Wasseraufnahme und Keimfähigkeitswerte der Getreidefrüchte. Saatgut-Wirtsch. 2:228–31, 251–55. Biol. Abstr. 25:24965. 1951.
12715. Maier, W. 1932: Untersuchungen zur Frage der Lichtwirkung auf die Keimung einiger *Poa*-arten. Jahrb. Wiss. Botan. 77:321–92.
12716. —— 1933: Das keimungsphysiologische Verhalten von *Phleum pratense* L., dem Timothee-Gras. Jahrb. Wiss. Botan. 78:1–42. Botan. Centralbl. 168:121. 1935.
Maille, M. 580, 590, 591.
12717. Mains, E. B. 1949: Heritable characters in maize. Linkage of a factor for shrunken endosperm with the a_1 factor for aleurone color. J. Heredity 40:21–24.
12718. Maiorescu, E. 1961: Suggestions concerning the estimation of losses occurring during the storage of forest seeds. Rev. Padurilor (Bucharest) 76:710–14.
Maisenhelder, L. C. 6934.
12719. Maisin, J., and S. Masy. 1928: Au sujet du mécanisme d'action des rayons X sur les graines. Compt. Rend. Soc. Biol. 98:886–88.
Maiti, P. C. 3767, 3768.
Maitreyi, M. 14102.
Majeed, A. 17489.
Major, F. 2955.
Majumber, S. K. 5247.
Makamura, K. 19929.
Makarenko, M. M. 11377.
12720. Makaro, I. L. 1950: Effect of developmental conditions and storage of cabbage seeds on their germination. Selek. i Semen. 17(11):61–66.
12721. ——, N. I. Slobodianik, B. A. Gerasimov, and E. A. Osnitskaia. 1956: Improving the seeding qualities of seeds of vegetable crops. Moscow. 190 pp.
Makarova, N. A. 17346, 17347.
12722. Makela, O. 1957: Hemagglutinins of leguminous seeds. Ann. Med. Exptl. Biol. Fenniae (Helsinki) Suppl. 35(11). 133 pp. Chem. Abstr. 51:14904. 1957.
12723. Makevnin, A. I. 1950: On late-autumn planting of sunflower with freshly-harvested seed. Sov. Agron 8:73–82.
12724. Maki, T. E. 1940: Significance and applicability of seed maturity indices for Ponderosa pine. J. Forestry 38:55–60.
12725. —— 1952: Local longleaf seed years. J. Forestry 50:321–22.
Makinen, V. 5948.
12726. Makino, I., and T. Miyamoto. 1954: On the growth inhibiting substance in the germinating spinach seeds. Japan. J. Breeding 4:158–60.
12727. Maksic, S. 1957: Untersuchung der Vermehrungsmöglichkeit von *Juniperus foetidissima* Willd. und *Juniperus excelsa* Bieb. und der Keimfähigkeit ihrer Samen. Skoplje. U. Zemjodelsko-Sumarski Fak. God. Zb. Sumarstov 10:253–62.
12728. —— 1957/58: La multiplication experimentale de *Paliurus aculeatus* Lam. Skoplje. U. Zemjodelsko-Sumarski Fak. God. Zb. Sumarstov 11:29–35.
12729. —— 1961: Sowing *Tilia* seeds. Skoplje. U. Zemjodelsko-Sumarski Fak. God. Zb. Sumarstvo 14:35–42. 1960/61.
12730. Maksimchuk, L. P. 1955: On elite seeds and present-day problems of seed production of cereals and legumes. Zemledelie 1955(3):93–105.
12731. Maksimenko, L. D. 1957: Effect of hardening of corn seed against drought before sowing on its development and grain yield. Dokl.-Botan. Sci. Sect. (English Transl.) 112:23–24.
12732. Maksimov, G. A. 1955: Heat and moisture transfer in plant seeds during swelling. Plant Physiol. (USSR) (English Transl.) 2:81–83. Biol. Abstr. 30:35614. 1956.
12733. —— and L. M. Kryukova. 1956: The mechanism of heat and mass exchange in seeds of plants during heating in a high frequency electric field. Biofizika 1:201–5. Chem. Abstr. 50:17005. 1956.
12734. ——, ——, and V. N. Efimov. 1956: Thermal phenomena in plant seeds during swelling. Biofizika 1:538–43. Biol. Abstr. 35:8486. 1960.
Maksimovich, A. E. 14732.

12735. Makskaia, A. V. 1936: Influence of physical factors (temperature, humidity, etc.) and age on germination vitality of seeds of makhorka. Krasnodar Vses. Inst. Tabac. Makh. Prom. 132:252–90.

12736. Malac, B. F. 1960: More on stratification of pine seed in polyethylene bags. Tree Planters Notes 42:7–9.

12737. Malafii, I. T. 1956: On the germination of freshly-harvested mulberry seeds. Agrobiologiya 1956(6):132–34.

12738. Malcolm, C. V. 1961: Bluebush seed: cleaning, drying and storing. J. Agr. W. Australia 2:37–39.

12739. Maldiney and Thouvenin. 1898: De l'influence des rayons X sur la germination. Rev. Gen. Botan. 10:81–86.

12740. Malekzadeh, F. 1962: Evaluation of several bactericides as seed treatments for the control of black rot of crucifers and studies on an antibacterial substance from cauliflower seed. Univ. Microfilms (Ann. Arbor, Mich.), Order No. 63-2781. 111 pp. Dissertation Abstr. 23:4493–94. 1963. Chem. Abstr. 59:10707. 1963.

12741. Malheiros-Garde, N. 1959: Fertilization and seed development in some crosses between diploid species of the genus *Solanum*. Agron. Lusitana 21:91–102.

12742. Malhotra, R. C. 1930: A new method of germination and the influence of hydrogen ion concentration on the germination and growth of *Helianthus annuus* and *Lycopersicum esculentum*. J. Indian Botan. Soc. 9:218–39.

12743. —— 1931: A physicochemical study of some economic seeds during germination with particular reference to weight and energy loss. Protoplasma 12:167–89. Chem. Abstr. 26:2215–16. 1932.

12744. —— 1931: Periodic permeability of iodine solution and water in the protoplasm of *Zea mays* seeds. Protoplasma 13:374–88.

12745. —— 1931: Selection of maize for germination and other tests. Nature 127:382.

12746. —— 1932: Biochemical study of seeds during germination. I. Periodic changes of reserve materials in normal germinating seeds. Botan. Centralbl. Beih. Abt. I 50:1–7.

12747. —— 1932: Biochemical study of seeds during germination. II. Periodic changes of reserve materials in embryo and endosperm of germinating corn. Botan. Centralbl. Beih. Abt. I 50:8–14.

12748. —— 1932: Biochemical study of seeds during germination. III. The distribution of some chemicals and energy in the previously isolated embryos during germination. Botan. Centralbl. Beih. Abt. I 50:15–19.

12749. —— 1932: The distribution of some reserve substances in hard winter wheat plant at successive growth stages and their possible utilisation. J. Agr. Sci. 22:485–96.

12750. —— 1932: The influence of ultraviolet rays, X-rays and temperature on the germination of *Zea mays*. J. Indian Botan. Soc. 11:1–27.

12751. —— 1932: Influence of some solutions on the rate of permeability in *Zea mays* seeds. J. Indian Botan. Soc. 11:188–201.

12752. —— 1932: Germination of *Zea mays* seeds as influenced by their protoplasmic immersion in water and desiccation. J. Indian Botan. Soc. 11:276–84.

12753. —— 1933: Biochemical study of seeds during germination. IV. The distribution of some chemical reserves and calorific energy in the previously isolated endosperm during the germination conditions. Botan. Centralbl. Beih. Abt. I 51:524–30.

12754. —— 1933: Biochemical study of seeds during germination. V. Successive elongation of shoots and roots in *Zea mays* seeds and embryo-seedlings with known chemical reserves and calorific energy. Botan. Centralbl. Beih. Abt. I 51:531–40.

12755. —— 1933: A contribution to the biochemistry of seed germination with particular reference to *Zea mays*. J. Biochem. (Tokyo) 18:173–97. Chem. Abstr. 28:805. 1934.

12756. —— 1934: Chemistry of corn seed germination. Cereal Chem. 11:105–9.

12757. —— 1934: Effect of temperature on air dry *Zea mays* seeds. Biol. Generalis (Vienna) 10:139–46. Botan. Centralbl. 167:9. 1934.

12758. —— 1934: Growth of *Zea mays* as influenced by the type of water. Biol. Generalis (Vienna) 10:147–56. Botan. Centralbl. 167:9. 1934.

12759. —— 1934: The effect of immersion in and desiccation of water on the germination of *Zea mays* seeds. Biol. Generalis (Vienna) 10:157–66. Botan. Centralbl. 167:61. 1934.

12760. Maliani, C. 1960: The seed industry and trade within the frame work of the

World Seed Year. Genet. Agr. 12:38–44.

12761. Malik, H. C., and R. Singh. 1950: Berseem seed production. Indian Farming 11:384–86.

Malik, M. A. 6227.

12762. Malinovskii, K. A. 1957: On germinating of the seeds of Carpathian mountain plants. Byul. Mosk. Obshchestva Ispytatelei Prirody Otd. Biol. 62: 51–63.

12763. Malinovskii, S. M. 1927: Stimulation of hemp seed (*Cannabis sativa*). Mem. Inst. Agron. Lenin. 4:289–348. Biol. Abstr. 4:23899. 1930.

Malis, O. 10767.

Malishchuk, I. 10530.

Malkhede, S. R. 5417.

12764. Mall, L. P. 1954: Germination of seeds of three common weeds of dry phase of low-lying lands. Proc. Natl. Acad. Sci. India Sect. B 24:197–204.

12765. Mallach, J. 1929: Untersuchungen über die Bedeutung von Korngrösse und Einzelkorngewicht beim Saatgut. Wiss. Arch. Landwirtsch. 2:219–95. Botan. Centralbl. 160:62. 1931.

12766. Mallamaire, A. 1935: Disinfection of coffee seed by chloropicrin. Agron. Coloniale 24:70:79. Chem. Abstr. 30: 6882. 1936.

Mallard, T. M. 20013.

Mallet, L. 15350.

Mallik, A. K. 5243.

Mallik, P. 15058, 15059.

12767. Malloch, F. J. 1940: Chemical examination of seeds of *Macrozamia reidlei* (zamia palm). Ann. Rept. Chem. Branch, Mines Dept. W. Australia 1939: 13–14. Chem. Abstr. 36:1983. 1942.

12768. Malloch, J. G. 1936: The cleaning and handling of barley. Sci. Agr. (Ottawa) 16:289–321. Expt. Sta. Rec. 78:476. 1938.

12769. ——, W. F. Geddes, R. K. Larmour, and A. G. McCalla. 1937: The quality and grading of frosted wheat. Annual surveys of the 1930 to 1935 western Canadian crops. Can. J. Res. Sect. C 15:567–92.

Malloch, J. G. 14378.

Malmberg, M. 19198.

12770. Malme, G. O. A:N. 1928: Germination of *Geranium bohemicum*. Botan. Notiser 1928(3):221–26. Biol. Abstr. 3:17947. 1929.

12771. Malmo, B. G. 1931: Raising alpines from seed. Gardeners Chron. 35:415–16.

12772. Malone, J. P. 1960: Studies on seed health. III. Apparatus suitable for use in the testing of seeds for health. Proc. Intern. Seed Testing Assoc. 25:842–47.

12773. —— 1962: Studies on seed health. IV. The application of heat to seed oats as an aid in the detection of *Pyrenophora avenae* by the Ulster method. Proc. Intern. Seed Testing Assoc. 27:856–61.

Malone, J. P. 14003, 14004.

12774. Malowan, L. S. 1938: On the properties of the catalase of the seed of avocado. Enzymologia 5:89–94.

Malsch, L. 12511.

12775. Malskaia, A. V. 1936: Influences of the conditions of makhorka seeds germination on their germinative capacity and energy. Krasnodar Inst. Tobacco Ind. Bull. 132:291–302.

12776. Malynovsky, K. A. 1958: Seed numbers in the soils of the herbaceous plant associations of the Chornohora Range in the Carpathians. Lvov. Prirodoznavchii Muz. Nauk. Zap. 6:18–28.

12777. Malysheva, A. G. 1961: The accumulation of substances accompanying fats during the ripening of flax and sunflower seeds. Maslob.-Zhir. Prom. 1961(9):7–8.

Malyshkina, G. B. 11014.

12778. Mamarov, P., J. Ivanov, and K. Katerov. 1958: The effect of preparation of the grapevine seeds before sowing on their germinating capacity. Sb. Česk. Akad. Zeměděl. Věd, Rostlinna Výrobá 31:1211–16.

12779. Mamedov, Z. I. 1956: The effect of treatment with microelement solutions of cottonseeds before sowing on the harvest of cotton. Izv. Akad. Nauk Azerb. SSR 1956(1):63–69. Chem. Abstr. 50:14054. 1956.

Mamet, R. 13832.

12780. Manant, P. 1955: Développement de l'embryon chez l'*Oenothera rosea* Ait. Bull. Soc. Botan. France 102:216–18.

12781. —— 1956: Développement de l'embryon chez l'*Astrocarpus clusii* J. Gay. Bull. Soc. Botan. France 103:429–31.

12782. Manaresi, A. 1935: Sulla germinazione delle samare di olma campestre. Ital. Agr. 72:9.

12783. Manceau, P., and Bige. 1931: Phytosterols of seeds, flower and fruit. Compt. Rend. Soc. Biol. 107:635–36. Chem. Abstr. 26:2765. 1932.

12784. ——, L. Revol, and R. Charmillon. 1932: Variation du phosphore éthero-soluble au cours de la maturation et de la germination des graines de Marron d'Inde. Compt. Rend. Soc. Biol. 110:850–51.

12785. Mancinelli, A. 1958: Action of some inhibitors on germination. Nuovo Giorn. Botan. Ital. 65:818–19. Biol. Abstr. 35:32838. 1960.

12786. —— 1958: Action of some metabolic inhibitors on seed in the germination of *Lactuca sativa* L. Nuovo Giorn. Botan. Ital. 65:855–58. Biol. Abstr. 35:32839. 1960.

12787. —— 1958: Azione di alcuni inhibitori metabolici sulla germinazione dei semi di *Lactuca sativa* L. Ann. Botan. (Turin) 26:56–66.

12788. —— 1958: Azione di alcuni inhibitori sulla germinazione dei semi. II. Ann. Botan. (Turin) 26:67–70.

12789. —— and R. Rendi. 1957: Metabolismo di semi germinanti di *Trifolium repens* L. e *Trifolium pratense* L. II. Autolisi e degradazione in *Trifolium repens* in presenza di alcuni inhibitori della sintesi proteica. Ann. Botan. (Turin) 25:452–58.

12790. Mandelson, L. F. 1936: Longevity of seed. Bull. Dept. Agr. Stock (Queensland) 14:23.

12791. Mandl, K. 1926: Beitrag zur Kenntnis der Anatomie der Samen mehrerer Euphorbiaceen-arten. Österr. Botan. Z. 75:1–17.

12792. Mandy, G., and G. Pal. 1958: Studies on the germination of polyploid sugar beets. Acta Botan. Acad. Sci. Hung. 4:145–54. Biol. Abstr. 33:23482. 1959.

12793. —— and —— 1960: Observations on the physiology of seed germination of some Hungarian rye varieties. Novenythermeles 9:47–58.

12794. —— and —— 1960: The effect of treatments with different salt solutions on seed germination of some varieties of rye, oat, and sunflower. Novenytermeles 9: 343–58.

12795. ——, J. Zsolt, and G. Pal. 1957: Studies of the germination of Hungarian wheat varieties. Acta Botan. Acad. Sci. Hung. 2:261–73. Biol. Abstr. 34:14469. 1959.

Manfredi, U. 5152.

12796. Manfridi, F. 1952: The action of some colloidal metals on the germination of seeds in relation to their content of reserve materials. Chimica (Milan) 7:375–77. Chem. Abstr. 47:6501. 1953.

Mangan, J. L. 3043.

12797. Mangels, C. E. 1927: A study of dark, hard kernels and protein content of hard red spring wheat. J. Agr. Res. 34:157–66.

12798. —— 1932: Regional and seasonal variation in pigmentation of Durum wheats. Cereal Chem. 9:485–90.

12799. Mangelsdorf, P. C. 1923: The inheritance of defective seeds in maize. J. Heredity 14:119–25,

12800. —— 1926: The genetics and morphology of some endosperm characters in maize. Conn. Agr. Expt. Sta. Bull. 279: 509–614.

12801. —— 1930: The inheritance of dormancy and premature germination in maize. Genetics 15:462–94.

12802. —— 1947: The inheritance of amylaceous sugary endosperm and its derivatives in maize. Genetics 32:448–58.

12803. —— and G. S. Fraps. 1931: A direct quantitative relationship between vitamin A in corn and the number of genes for yellow pigmentation. Science 73:241–42.

Mangelsdorf, P. C. 9941.

Mangenot, G. 5142.

12804. Mangieri, H. R. 1962: Results obtained by introducing seed of exotic forest species. Proc. World Forestry Congr. 5:763–65. 1960.

Mangin, L. 2756.

12805. Mangold, E., and A. Columbus. 1938: Digestibility and biological protein value of the seed of a new yellow sweet lupine "Weiko." Landwirtsch. Versuchssta. 129:110–23. Chem. Abstr. 32:6698. 1938.

Mani, G. S. 18125.

12806. Mani, V. K. S. 1947: Vivipary in *Pyrus malus*. Current Sci. (India) 16:321.

Mani, V. S. 1452.

12807. Manion, W. M., and H. N. Doughty. 1959: Flaxseed storage at country elevators in Minnesota, North Dakota, and South Dakota. U.S. Dept. Agr. Marketing Res. Rept. 350. 55 pp.

12808. Manjunath, B. L. 1930: Chemical examination of the seeds of *Psoralea corylifolia* (Linn.). Proc. Indian Sci. Congress, Allahabad 17. 184 pp.

12809. —— and B. S. Nagaraj. 1942: Chemical investigation on the fatty oil from the seeds of *Xylia xylocarpa*. Mysore Univ. Half-Yearly J. Sect. B 3:105–12. Biol. Abstr. 18:1605. 1944.

12810. —— and M. S. S. Rao. 1938: Behenic acid in the seed oil of *Pongamia glabra*, Vent. J. Indian Chem. Soc. 15:653. Analyst 64:448. 1939.

12811. —— and S. Siddappa. 1935: The supposed occurrence of acids with uneven number of carbon atoms in vegetable oils and fats. I. Daturic acid from the seeds of *Datura stramonium*. Linn. J. Indian Chem. Soc. 12:400–4. Chem. Abstr. 30:313–14. 1936.

Manjunath, B. L. 9918–9920, 11523, 11682, 12665, 15972.

Manley, R. H. 6078.

12812. Mann, A. 1914: Coloration of the seed coat of cowpeas. J. Agr. Res. 2: 33–56.

12813. —— and H. V. Harlan. 1951: Morphology of the barley grain with reference to its enzyme-secreting areas. U.S. Dept. Agr. Bull. 183. 32 pp.

Mann, G. E. 4336.

12814. Mann, H. H. 1951: The effect of manures on the bolting of the beet plant. Ann. Appl. Biol. 38:435–43.

12815. Mann, L. K. 1948: Morphological development of parthenocarpic fruits. Proc. Am. Soc. Hort. Sci. 52:226–32.

Mann, L. K. 14355.

Manns, T. F. 2724.

12816. Manner, R. 1956: Studies on seed-setting and seed yields in oil flax. Medd. Hammenhög. Gullakers Växtforadlings-anst. 12. 139 pp.

12817. —— 1958: Studies on seed-setting in leguminous plants. I. Peas. II. Vetches. Maataloustieteellinen Aikakauskirja 30: 143–71. Biol. Abstr. 33:23308. 1959.

12818. Manohar, M. S., and W. Heydecker. 1964: Effects of water potential on germination of pea seeds. Nature 202:22–24.

12819. Manrique, J. B. A., and C. L. Moreno. 1951: Algunos datos sobre el aceite de semillas de manzana. Farmacognosia 11(20):223–30. Biol. Abstr. 26: 22649. 1952.

Manrique, J. B. Abad-; *see* Abad-Manrique, J. B.

Mans, R. J. 15834.

12820. Manshard, E. 1926: Ein Beitrag zur Frage der Stimulation von Kiefern- und Fichtensamen. Forstwiss. Centralbl. 48: 533–38.

12821. —— 1936: Embryologische Untersuchungen an *Styrax obassia* Sieb. et Zucc. Planta 25:364–83.

12822. Manske, P. 1956: Arbeitserleichterung durch vorbereitetes Rubensaatgut. Mitschurin Bewegung 5:264–66.

Manson, B. E. 12823.

12823. Manson, J. G., and B. E. Manson. 1958: Water hyacinth reproduces by seed in New Zealand. New Zealand J. Agr. 96:191.

Mansurova, A. F. 18720.

Mansurova, V. V. 16604.

12824. Mantle, P. G. 1961: The effect of vernalization on embryos of field-resistant wheat varieties infected by loose smut (*Ustilago tritici*). Ann. Appl. Biol. 49:692–703.

12825. Manton, R. B. 1957: Lewisias from seed. Am. Rock Garden Soc. Bull. 15:105–10.

Mantovani, G. 2453.

12826. Manunta, C. 1956: Paper chromatographic comparative study on various strains of *Medicago sativa*. Genet. Agrar. 6:377–84. Chem. Abstr. 52:17423. 1958.

Manus, O. 19484.

12827. Manzanilla, E. B. 1953: Effect of freezing the coconut endosperm on the extraction of the oil. Philippine Agriculturist 37:369–83.

12828. Manzano, T. B. 1953: Influence of the rate of seeding on yield of fiber and seed and on quality of fiber of kenaf. Philippine Agriculturist 35:561–67.

12829. Manzelli, M. 1955: From seed to bloom in one season. Gladiolus Mag. 19:35–36, 44.

12829A. Mao, Y. D., and L. M. Parks. 1950: Identification of osthole in Hseh Tsuang seed. J. Am. Pharm. Assoc. Sci. Ed. 39:107–8.

12830. Mapson, L. W., and E. M. Moustafa. 1955: The oxidation of glutathione by a lipoxidase enzyme from pea seeds. Biochem. J. 60:71–80.

12831. Maquenne, L. 1900: Recherches sur la germination. Ann. Agron. 26:321–32.

12832. —— and R. Cerighelli. 1922: Influence de la chaux sur le rendement des graines pendant la période germinative. Compt. Rend. 174:1269–72.

12833. Maquenne, M. L., and M. E. Demoussy. 1921: Influence des matières minérales sur la germination. Ann. Sci. Agron. 38:113–51.

12834. Marani, A., and J. Dag. 1962: Inheritance of the ability of cotton seeds to germinate at low temperature in the first hybrid generation. Crop Sci. 2:243–45.

12835. Maranis, A. D. 1928:Untersuchung der Samenkörner der Pinie (*Pinus pinea* L.). Arch. Pharm. 266:121–22.

12836. Maranon, J., and J. K. Santos. 1932: Morphological and chemical studies on the seeds of *Erythrina variegata* var. *orientalis* (Linnaeus) Merrill. Phillippine J. Sci. 48:563–80. Chem. Abstr. 26:5609. 1932.

12837. Marar, M. M. K., and T. G. Jagendorf. 1960: Coconut nursery studies—effect of the method of collecting seednuts on germination of nuts and vigour of seedlings. Indian Coconut J. 13:89–93.

12838. —— and K. Shambhu. 1961: Coconut nursery studies. III. Vigour of seedlings in relation to the floating position of seednuts in water. Indian Coconut J. 14:45–48.

12839. —— and R. Varma. 1958: Coconut nursery studies: effect of maturity of seednuts on germination and vigour of seedlings. Indian Coconut J. 11:81–86.

12840. Marasse, L. 1940: Contribution à l'étude du blé. Absorption de l'eau au cours de la germination. Formation de l'épi et de l'épillet. Ann. Sci. Nat. Ser. 11, 1:384–86.

12841. Maravi, G. F. 1954: Contribución al estudio de la composición del fruto y semilla del *Inga feullei* o "pacae." Anales Fac. Farm. Bio-Quim. (Lima) 5:188–91. Nutr. Abstr. Rev. 27:1628. 1957.

12842. Marayana, C. S., and S. Rangaswami. 1956: Isolation of three crystalline substances from the seeds of *Cassia tora* Linn. Current Sci. (India) 25:359.

12843. Marayanaswami, S., and S. K. Roy. 1960: Embryology of the genus Psidium. J. Indian Botan. Soc. 39:35–45.

12844. Marcelli, E. 1955: Notes au sujet d'une nouvelle virose du tabac transmissible par les graines. Congr. Sci. Intern. Tabac, Paris-Bergerac 1:652–55. Biol. Abstr. 35:30487. 1960.

Marcelli, E. 4129.

12845. Marcet, E. 1951: Beitrag zur Verbesserung des Keimprozentes bei der Vogelkirsche (*Prunus avium* L.). J. Forest. Suisse 102:524–26. Biol. Abstr. 26:16431. 1952.

12846. —— 1957: Zur Aufbewahrung von Aspensamen. Schweiz. Z. Forstw. 108: 270–73. Biol. Abstr. 32:6233. 1958.

Marchal, E. 16260.
Marchand, M. 7010.
Marchetti, R. 5631.

12847. Marchi, V. 1942: Researches on heat vernalization. Ital. Agr. 79:264–66. Biol. Abstr. 22:2176. 1948.

Marco, M. R. de. 5151.
Marcu, G. 16238.
Marcu, T. 16238.
Marcucci, E. 3004.

12848. Marcus, A., and J. Feeley. 1962: Nucleic acid changes in the germinating peanut. Biochim. Biophys. Acta 61:830–31.

12849. —— and —— 1963: Nucleic acids in the cotyledons of the germinating peanut. Proc. Plant Physiol. Meetings 38:xl.

12850. —— and J. Velasco. 1960: Enzymes of the glyoxylate cycle in germinating peanuts and castor beans. J. Biol. Chem. 235:563–67.

12851. —— and —— 1960: Malate synthetase in germinating seeds. Biochim. Biophys. Acta 38:365–67.

Maren, A. van. 3548.
Margolina, T. I. 15723.

12852. Mariat, F. 1944: Influence favorable de la vitamine B_1 sur les germinations de *Cattleya*. Rev. Hort. (Paris) 116:68–69. Biol. Abstr. 20:21099. 1946.

12853. —— 1948: Action des composants de l'aneurine sur le développement des embryons d'orchidées. Compt. Rend. Soc. Biol. 142:840–41. Chem. Abstr. 43:1834. 1949.

12854. —— 1949: Action de l'acide nicotinique sur la germination et le développement des embryons de *Cattleya*. Compt. Rend. 229:1355–57.

Mariat, F. 12679.

12855. Mariat, M. F. 1952: Recherches sur la physiologie des embryons d'orchidées. Rev. Gen. Botan. 701:324–77.

12856. Maric, B. 1962: A proposition for selection of seed populations for resinous substances in Serbia. Topola 6(25/26):9–14.

Maric, M. M. 14232.
Marie, O. C. S. O. Louis-; *see* Louis-Marie, O. C. S. O.

12857. Marimpietri, L. 1933: Contents of protein substances and of phytin in the seeds of cereals during their development. Ann. Staz. Chim.-Agrar. Sper. Roma Ser. II 14(302). 19 pp. Chem. Abstr. 31:7931. 1937.

12858. Marinchik, G. F. 1939: The effect of seed hormonization on the yield of sugar beet. Osnov. Vyvod. Rabot V.N.I.S. 1937:375–77. Chem. Abstr. 36:6584. 1942.

Marinescu, R. 5946.

12859. Marini, E. 1955: A virus disease transmitted by seed: Lettuce mosaic. Riv. Ortoflorofrutticolt. Ital. 39:449–54.

12860. Marion, L., and S. W. Fenton. 1948: The papilionaceous alkaloids. VI. *Lupinus pusillus*, Pursh. J. Org. Chem. 13:780–81.

12861. —— and F. Turcotte. 1948: The papilionaceous alkaloids. IV. *Baptisia* *perfoliata* (L.) R. Br. J. Am. Chem. Soc. 70:3253–54.

12862. Marion, L. P. 1938: Recherches sur l'absorption de la vapeur d'eau par les graines. Ann. Physiol. Physicochim. Biol. 14:244–56. Biol. Abstr. 12:12745. 1938.

12863. Marjai, Z. 1960: *Larix decidua* seed drying and seed supply situation in Hungary. Erdö 9:41–43.

12864. —— 1960: Germination biology, ecology and sowing of the poplar. Erdö 9:471–80.

12865. —— 1961: Determination of viability of seeds with staining method. Erdö 10:307–11.

Marki, A. 11903.

Markley, K. S. 5443.

12866. Markley, M. C., and C. H. Bailey. 1935: The nature of the pigments of the gasoline extract of wheat. Cereal Chem. 12:33–39.

12867. —— and —— 1935: The pigments of the dilute alcohol or acetone extract of whole wheat meal. Cereal Chem. 12:40–48.

12868. Markman, A. L. 1956: Losses of oil in cottonseed husk. Maslob.-Zhir. Prom. 21(2):7–9. Chem. Abstr. 50:11034. 1956.

Markman, A. L. 1882.

12869. Markötter, E. I. 1936: Die Lewensgeskiedenis van sekere Geslagte van die Amaryllidaceae. Ann. Univ. Stellenbosch Ser. A 14(2):1–84. Biol. Abstr. 11:7029. 1937.

Markosyan, L. S. 17550.

12870. Markova, L. G. 1945: On certain anomalies in the embryology of *Polemonium coeruleum* L. Compt. Rend. Acad. Sci. URSS 49:144–45.

12871. Markovich, L. L. 1939: Dynamics of the nitrogen complex in relation to the different methods of pre-sowing treatment of seeds. Osnov. Vyvod. Rabot V.N.I.S. 1937:91–92. Chem. Abstr. 36:6583–84. 1942.

12872. Markowski, A. 1962: Sterile vernalization of seeds for biochemical investigation. Bull. Acad. Polon. Sci. Ser. Sci. Biol. 10:95–96.

12873. —— and K. Korlakowska. 1963: Influence of water content in the course of vernalization on the respiration intensity of seeds and in the further generative development of winter wheat. Bull. Acad. Polon. Sci. Ser. Sci. Biol. 11:95–98. Chem. Abstr. 59:10477. 1963.

12874. —— and M. Madej. 1962: Changes in phosphorus compounds of winter and spring wheat embryos at 20° and at vernalization temperature. Bull. Acad. Polon. Sci. Ser. Sci. Biol. 10:139–44. Chem. Abstr. 57:8914. 1962.

12875. ——, J. Myczkowski, and J. Lebek. 1962: Preliminary investigations on changes in nitrogen compounds of wheat embryos in the course of germination under various temperature conditions. Bull. Acad. Polon. Sci. Ser. Sci. Biol. 10:145–50. Chem. Abstr. 57:8914. 1962.

Marks, J. D. 20002.

Marks, R. W. 10090.

Markus, L. 19302.

12876. Markwood, L. N. 1931: Determination of oil and alkaloids in delphinium seed. J. Am. Pharm. Assoc. 20:454–55. Chem. Abstr. 25:4662. 1931.

Marmoy, C. J. 18438.

Maroti, M. 7264.

12877. Marotta, D., and S. Anselmi. 1932: The composition of peanut oil. Ann. Chim. Appl. 22:193–96. Chem. Abstr. 26:4727. 1932.

12878. —— and A. Calo. 1932: The extraction and composition of the organic phosphorus compounds from rice polishings and oil-seed hulls. Ann. Chim. Appl. 22:763–76. Chem. Abstr. 27:2461. 1933.

12879. Marquand, C. V. B. 1932: The cultivated gentians of China and the Himalaya. J. Roy. Hort. Soc. 57:188–211.

12880. Marras, F. 1955: Biochemical method for rapid determination of germinating capacity of seeds. Studi Sassaresi 3:108–17. Chem. Abstr. 51:6784. 1957.

12881. Marras, G., and G. Dall'Olio. 1953: Antidote action between 2,3-dimercaptopropanol (BAL) and arsenous anhydride in *Lupinus albus*. Boll. Soc. Ital. Biol. Sper. 29:1717–19. Chem. Abstr. 48:8883. 1954.

12882. Marrero, J. 1942: A seed storage study of Maga. Caribbean Forester 3:173–84.

12883. —— 1943: A seed storage study of some tropical hardwoods. Caribbean Forester 4:99–106.

12884. —— 1949: Tree seed data from Puerto Rico. Caribbean Forester 10:11–26.

Marrero, J. 8946.

12885. Marriage, K. N. 1945: Growing rare flowers from seed. Flower Grower 32(2):84, 107.

Marsh, B. O. 13506.

12886. Marshall, G. M. 1959: The incidence of certain seed-borne diseases in commercial seed samples. I. Loose smut of barley, *Ustilago nuda* (Jens.) Rostr. Ann. Appl. Biol. 47:232–39.

12887. —— 1960: The incidence of certain seed-borne diseases in commercial seed samples. II. Ergot, *Claviceps purpurea* (Fr.) Tul., in cereals. Ann. Appl. Biol. 48:34–38.
12888. —— 1962: Germination and seedling evaluation of peas infected with *Ascochyta* spp. Natl. Inst. Agr. Botany J. 9:170–73.
12889. Marshall, H. H. 1958: Polyembryony in tomatoes. Can. J. Plant Sci. 38:67–72.
12890. —— 1963: New genes for wrinkled or dimpled seed in peas. Can. J. Plant Sci. 43:508–12.
Marshall, J. B 8218, 9879
Marshall, W. 2148
Marshall, W. H. 8245.
Marston, A. R. 5635, 15996.
12891. Martelli, H. L. 1960: Fractionation of polyphenolic substances from cacao beans by paper electrophoresis. J. Chromatog. 4:347–50.
Martens, J. W. 16616.
12892. Martens, P. 1963: Studies on the Gnetales. VI. Researches on *Welwitschia* mirabilis. III. The ovule and the embryonic sac; the extra-floral embryonic sacs. Cellule 63:307–29.
12893. Marth, P. C., and E. M. Meader. 1944: Influence of growth-regulating chemicals on blackberry fruit development. Proc. Am. Soc. Hort. Sci. 45:293–99.
12894. ——, E. H. Toole, and V. K. Toole. 1948: Effect of 2,4-dichlorophenoxyacetic acid on seed development and germination of certain cereal and grass crops. J. Am. Soc. Agron. 40:916–18.
Marth, P. C. 6969–6971, 13597, 19557, 19558.
12895. Martin, A. C., and W. D. Barkley. 1961: Seed identification manual. Univ. California Press, Berkeley. 221 pp.
Martin, A. J. P. 15232.
12896. Martin, A. L., and G. E. Altstatt. 1940: Black kernel and white tip of rice. Texas Agr. Expt. Sta. Bull. 584. 14 pp. Expt. Sta. Rec. 83:206. 1940.
Martin, A. L. D. 17819.
Martin, A. W. 6658.
Martin, C. R. 18647.
12897. Martin, D. G., and J. T. Ochoa. 1958: *Gleditschia triacanthos*: uses of the fruit, mucilage of the seeds, and goma garrofin. Inst. Forestal Invest. Experiencías (Madrid) Bol. 29:62. Chem. Abstr. 53:2290. 1959.
Martin, D. J. 6492.
Martin, E. W. 7526, 7729.
Martin, F. N., Jr. 12618.
12898. Martin, F. W., E. Waszczenko-Zacharczenko, W. C. Boyd, and K. F. Schertz. 1964: Lectin content of the lima bean during development of the seed and seedling. Ann. Botany (London) 28:319–24.
Martin, H. F. 4865, 9275–9278.
12899. Martin, J. A. 1949: Sesame research and production in the western hemisphere. Proc. Assoc. Southern Agr. Workers 1949: 111–12.
12900. ——, T. L. Senn, B. J. Skelton, and J. H. Crawford. 1960: Response of okra seed to moisture content and storage temperature. Proc. Am. Soc. Hort. Sci. 75:490–94.
12901. Martin, J. A., Jr. 1946: Germination of sweet potato seed as affected by different methods of scarification. Proc. Am. Soc. Hort. Sci. 47:387–90.
Martin, J. F. 2131, 2912.
12902. Martin, J. H., J. W. Taylor, and R. W. Leukel. 1935: Effect of soil temperature and depth of planting on the emergence and development of sorghum seedling in the greenhouse. J. Am. Soc. Agron. 27:660–65.
Martin, J. H. 3131, 12140.
12903. Martin, J. N. 1915: Relation of moisture to seed production in alfalfa. Iowa Agr. Expt. Sta. Res. Bull. 23:301–24.
12904. —— 1935: The germination of sweet clover and alfalfa seed in relation to various conditions of temperature and moisture. Proc. Iowa Acad. Sci. 42:82–83.
12905. —— 1937: The strophiole in sweet clover and alfalfa seeds. Proc. Iowa Acad. Sci. 44:104.
12906. —— 1942: The variability in the color of red clover seeds, its cause, and relation to the value of the seeds. Proc. Iowa Acad. Sci. 49:254–55.
12907. —— 1943: Germination studies on the seeds of some common weeds. Proc. Iowa Acad. Sci. 50:221–28.
12908. —— 1945: Changes in the germination of red clover seed in storage. Proc. Iowa Acad. Sci. 51:229–33.
12909. —— 1945: Germination studies of sweet clover seed. Iowa State Coll. J. Sci. 19:289–300.
12910. —— and J. R. Watt. 1934: Irregular sporogenesis and polyembryony in some Leguminosae. Iowa State Coll. J. Sci. 8:303–7.
12911. —— and —— 1944: The strophiole and other seed structures associated with

hardness in *Melilotus alba* L. and *M. officinalis* Willd. Iowa State Coll. J. Sci. 18:457–69.

Martin, J. N. 19417.

12912. Martin, J. P. 1950: Effect of soil fungi on germination of sweet orange seeds and development of the young seedlings. Soil Sci. Soc. Am. Proc. 14:184–88.

Martin, J. P. 2466.

Martin, L. W. 19673.

12913. Martin, M. 1964: Influence de la temperature sur la germination des graines de *Phacelia tanacetifolia* Benth. Compt. Rend. 258:1877–80.

12914. Martin, M. C. 1859: Vitality of seeds particularly when subjected to the action of seawater. Trans. Botan. Soc. Edinburgh 6:279–84.

Martin, M. D. 290.

12915. Martin, M. J., and P. R. Chapman. 1956: Cacti from seeds. Gardeners Chron. 139:709.

Martin, P. 17054.

12916. Martin, S. 1887: Proteids of the seeds of *Abrus precatorius*. Proc. Roy. Soc. London 42:331–34.

12917. Martin, S. C. 1948: Mesquite seeds remain viable after 44 years. Ecology 29:393.

Martin, S. C. 18783.

12918. Martin, V. D. 1938: Effect of fumigation of wheat on amylase content. Iowa State Coll. J. Sci. 12:467–70.

12919. Martin, V. G., and J. A. Anderson. 1946: Apparatus and technique for photographing grain kernels and similar objects. Sci. Agr. (Ottawa) 26:419–25.

Martin, W. P. 12596.

12920. Martinec, T. 1945: Einwirkung der Temperatur auf die Keimung und den Ertrag von Lein. Zemědělsky Pokrok 12:58–59. Proc. Intern. Seed Testing Assoc. 14:69–70. 1948.

12921. Martinez, L. R. 1948: Influence of harmine on the growth of wheat seedlings. Anales Inst. Espan. Edafol., Ecol., Fisiol. Vegetale (Madrid) 2:343–55. Chem. Abstr. 43:4350–51. 1949.

Martinez, R. P. 3532.

12922. Martinez, W. H., and L. C. Berardi. 1963: The proteins of cottonseed pigment glands. Proc. Seed Protein Conf., New Orleans 1963:229–39. Chem. Abstr. 60: 4462. 1964.

12923. —— and V. L. Frampton. 1962: Isolation and chromatographic characterization of low molecular weight cottonseed proteins. J. Agr. Food Chem. 10:412–15.

Martinez, W. H. 4336.

Martini, G. 16409.

12924. Martini, M. L., H. V. Harlan, and M. N. Pope. 1930: Some growth curves of barley kernels. Plant Physiol. 5:263–72.

12925. Martinoli, G. 1940: Contribution to the embryology of the Asteraceae. Nuovo Giorn. Botan. Ital. 47:287–322. Biol. Abstr. 15:11173. 1941.

Martinovic, B. 11447.

Martishenya, A. I. 18381.

12926. Martosh, L. 1956: The nucleic acids of the leguminous cotyledons. Naturwissenschaften 43:399–400.

12927. —— 1957: Influence of a metabolite of the cotyledons on the growth of seedlings of leguminous plants. Plant Physiol. (USSR) (English Transl.) 4:150–58. Chem. Abstr. 51:16744. 1957.

Martouret, D. 7696.

12928. Martyanova, K. L. 1960: Results of field experiments with barley seed which had undergone a pre-sowing hardening to drought. Plant Physiol. (USSR) (English Transl.) 7:301–2.

Marucci, P. 6356.

Marusawa, T. 18756.

Maruyama, J. 19968.

Maruyama, T. 14520.

12929. Maruzella, J. C., and M. Freundlich. 1959: Antimicrobial substances from seeds. J. Am. Pharm. Assoc. 48:356–58.

12930. —— and A. L. Robbins. 1961: Effect of essential oils on seed germination. Naturwissenschaften 48:383.

12931. Marvin, V. 1954: The combining and storage of high moisture wheat. World Crops 6:109–10.

12932. Marwinski, H. 1930: Die Rolle des Nukleolus bei der Fermentproduktion in keimenden Samen. Botan. Arch. 28:255–88.

12933. Marx, C., and W. V. Cruess. 1943: Oil from wine grape seeds. Chemurgic Dig. 2(15):125.

12934. Marx, D. S. 1941: Seed mechanics. Sci. Monthly 53:583–84. Biol. Abstr. 16:11693. 1942.

12935. Marx, G. A., and W. Mishanec. 1962: Inheritance in ovule number in *Pisum sativum* L. Proc. Am. Soc. Hort. Sci. 80:462–67.

12936. Marx, R. K. 1961: Preliminary report on time required for germination of Reed canarygrass seed. Assoc. Offic. Seed Analysts News Letter 35(4):12.

12937. Marzocca, A., and O. H. Caso. 1957: Tratamiento de semillas de "kok-saghyz" con rayos ultravioletas. Rev. Invest. Agr. 11:227–45.

Marzocca, A. 3682.

Maschek, F. 12294.

12938. Mascre, M. 1937: Leucenol, a definite principle obtained from the seeds of *Leucaena glauca* Benth. Compt. Rend. 204:890–91. Chem. Abstr. 31:4369. 1937.

12939. Mashima, I., and T. Kawai. 1959: The effects of thermal neutrons and X-rays on dormant seeds of rice. Japan. J. Breeding 9:53–54.

12940. Mashkin, S. I. 1961: A comparative study of the processes of stratification and vernalization of the seeds of cherry and winter wheat. Byul. Obshchestva Estestvoispyt. pri Voronezhsk. Gos. Univ. 1961(12):23–31. Biol. Abstr. 42:7482. 1963.

12941. Masiev, A. M. 1962: Obtaining sprouts of seeds of gum-bearing *Astragalus* spp. in the foothill zones of Azerbaijan. Izv. Akad. Nauk Azerb SSR Ser. Biol. i Med. Nauk 1962(1):3–6.

12942. Maslov, N. Y., and V. A. Molodtsov. 1955: The fatty oil from the seeds of Digitalis lanata. J. Appl. Chem. USSR (English Transl.) 28:309–11. Chem. Abstr. 49:12774. 1955.

12943. Maslowski, P. 1962: Comparative biochemical studies of proteins in different varieties of maize (corn). Genet. Polon. 3:275–84. Biol. Abstr. 45:3216. 1964.

12944. Maslowski, P. 1963: Investigations on the *Zea mays* proteins. II. Nitrogenous metabolism in the ripening seed of *Zea mays*. Howdowla Roslin Aklim. Nasiennictwo 7:17–23.

12945. —— and R. Skorko. 1963: Chromatographic analysis of simple and complex sugars in grains of different maize varieties. Plant Breed. Acclimatiz. Seed Pro. 7:133–41.

12946. Masoero, P., A. Canale, C. Sarra, and R. M. Turi. 1963: Relations between methyl esters of fatty acids of whole oil and of residual oil obtained by extraction of the seeds of *Arachis hypogaea*. Atti Soc. Ital. Sci. Vet. 16:385–89. Chem. Abstr. 60:3207. 1964.

12947. Mason, C. T., Jr. 1951: Development of the embryo sac in the genus *Limnanthes*. Am. J. Botany 38:17–22.

12948. Mason, I. D., and L. S. Palmer. 1934: Preparation of white zein from yellow corn. J. Biol. Chem. 107:131–32. Chem. Abstr. 29:1136. 1935.

Mason, J. L. 13534.

12949. Mason, R. 1961: Dispersal of tropical seeds by ocean currents. Nature 191:408–9.

Mason, T. G. 15324.

Mason, W. J. 7519.

Masquelier, J. 9601, 17158.

12950. Massa, V. 1942: Ascorbic acid and oxidation enzymes during the germination of peas. Bull. Sci. Pharmacol. 49:97–104. Chem. Abstr. 38:5532. 1944.

12951. —— 1946: Microchemical localization of ascorbic acid, oxidases, and peroxidases in the fruit of Solanaceae.Trav. Soc. Pharm. Montpellier 5:17–20. Chem. Abstr. 42:5510. 1948.

12952. Massart, J. 1917: Pourquoi les graines ne germent pas dans les fruits charnus. Bull. Sci. France Belg. 50:167–69.

12953. Massart, L. 1955: Experiments on isolated barley embryos. Intern. Tijdschr. Brouw. Mout. 15:6–8. Chem. Abstr. 50:442. 1956.

12954. —— 1956: Antibacterial substances acting as germination inhibitors in plant material. Giorn. Microbiol. 2:33–41. Biol. Abstr. 32:38633. 1958.

12955. —— 1957: Inhibitors of germination in the glomerules of sugar beets and in other dried fruits and seeds. Biochemistry (USSR) (English Transl.) 22:388–91.

12956. ——, G. Belpaire, A. Verwee, J. De Roubaix, and O. Lazar. 1956: L'acide férulique et autres substances inhibant la germination dans les glomerules de "*Beta sacchariferae*." Arch. Intern. Physiol. Biochim. 64:525–27.

12957. —— and F. De Ketelaere. 1955: The behavior of certain enzymic systems in isolated barley embryos. Arch. Intern. Physiol. Biochim. 63:262–63. Chem. Abstr. 52:9318. 1958.

12958. —— and —— 1955: Respiration and respiratory quotient of isolated barley embryos. Arch. Intern. Physiol. Biochim. 63:264. Chem. Abstr. 52:9319a. 1958.

12959. ——, ——, and M. De Bièvre. 1956: Hydrolysis of sucrose and absorption of glucose by isolated barley embryos. Bull. Soc. Chim. Biol. 38:1037–40. Chem. Abstr. 51:557. 1957.

12960. ——, ——, and H. Peeters. 1955: Isolated barley embryos. Biol. Jaarboek Konink. Natuurw. Genoot. Dodonaea Gent 22:129–33. Chem. Abstr. 50:7953. 1956.

12961. —— and E. Van der Auwermellen. 1952: Enzymatic hydrolysis of linseed mucilage. Biochim. Biophys. Acta 8:703.

Massart, L. 5221, 18998.

12962. Massengale, M. A., and J. T. Medler. 1955: Seed production and moisture content of Ladino clover, *Trifolium repens* L., treated with endothal. Agron. J. 47: 528–30.

12963. Massey, A. B. 1925: Antagonism of the walnuts (*Juglans nigra* and *Juglans cineria*) in certain plant associations. Phytopathology 15:773–85.

12964. Massey, L. M. 1942: Vitamin C from rose-hips. Am. Rose Mag. 4(9):219.

12965. Massibot, J. A. 1947: Note sur l'action du mélange oxyde d'ethylene-gaz carbonique sur la faculté germinative et la valeur industrielle des arachides en coque du Sénégal. Agron. Trop. 1946(11/12): 621–27. Annee Biol. 51:322. 1947.

Massieu H., G. 6737.

12966. Masters, M. T. 1861: Note on the unusual mode of germination in the mango: *Manifera indica*. J. Linn. Soc. London 6:24–26.

12967. Mastovsky, J., and V. Karel. 1961: A study of the possibility of shortening dormancy of barley and reducing its water sensitivity. Am. Brewer 94:41–44. Chem. Abstr. 56:9149. 1962.

12968. Masujima, H. 1963: Soil moisture system in relation to plant growth. V. The effects of soil moisture tension and bulk density on the germination and early growth of red clover. Hokkaido Natl. Agr. Expt. Sta. Res. Bull. 82:40–45.

12969. Masumoto, S. 1939: Über den Einfluss der Erhitzung während der Aufbewahrung auf die Keimung des Tabaksamens. Hirosima Univ. J. Sci. Ser. B Div. 2, 3:165–90.

12970. Masure, M. P. 1932: Effect of ultraviolet radiation on growth and respiration of pea seeds, with notes on statistics. Botan. Gaz. 93:21–41.

Masy, S. 12719.

12971. Masyuk, P. 1940: The influence of various stimulators on the growth, development, and yield of *Arachis hypogaea*. Acad. Sci. RSS Ukr. Inst. Botan. J. 31:188.

12972. Matchett, J. R., and J. Levine. 1941: Isolation of ecgonidine methyl ester from cocoa seeds. J. Am. Chem. Soc. 63:2444–46. Chem. Abstr. 35:7409. 1941.

Matchett, J. R. 12635.

Mateeva, N. M. 16698.

12973. Mather, K., and E. C. Barton-Wright. 1946: Nicotinic acid in sugary and starchy maizc. Nature 157:109–10.

12974. —— and J. Newell. 1941: Seed germination and the moon. J. Roy. Hort. Soc. 66:358–66.

Mathew, A. G. 7530.

12975. Mathews, A. C. 1932: The seed development in *Pinus palustris*. J. Elisha Mitchell Sci. Soc. 48:101–18. Biol. Abstr. 9:1583. 1935.

12976. —— 1939: The morphological and cytological development of the sporophylls and seed of *Juniperus virginiana* L. J. Elisha Mitchell Sci. Soc. 55:7–62.

12977. —— 1947: The scarification of various legume seeds with a disc scarifier. J. Am. Soc. Agron. 39:343–45.

12978. —— 1947: Observations on methods of increasing the germination of *Panicum anceps* Michx. and *Paspalum notatum* Flügge. J. Am. Soc. Agron. 39:439–42.

Mathews, D. 12273.

Mathews, O. R. 4278.

12979. Mathon, C. C. 1954: The grafting of embryos of cereals. III. Technique of studying the physiology of development. Bull. Mens. Soc. Linneenne Lyon 23:5–8. Chem. Abstr. 48:3488. 1954.

12980. —— 1956: Sur la vernalisation. Botan. Soc. France Bull. 103:432–35.

12981. —— 1960: Novelles recherches sur la vernalisation et le photopériodisme chez les Campanulacées. Phyton 14:167–74.

12982. Mathur, K. L. 1940: A note on the development of the embryo-sac in *Vogelia indica* (Lamk.). Current Sci. (India) 9:180–82.

12983. Mathur, P. B., M. Prasad, and K. K. Singh. 1956: Studies in the cold storage of peanuts. J. Sci. Food Agr. 7:354–60.

Mathur, P. B. 15955, 17498.

Mathur, S. S. 15111.

12984. Mathys, J. W. 1955: Seeds for new crops. Chemurgic Dig. 14(6):11–13, 16.

12985. Matikkala, E. J., and A. I. Virtanen. 1962: A new γ-glutamylpeptide, γ-L-glutamyl-S-(prop-1-enyl)-L-cysteine, in the seeds of chives (*Allium schoenoprasum*). Acta Chem. Scand. 16:2461–62.

12986. Matinian, A. 1953: Effect of storage on germination of *Eucalyptus* seeds. Lesn. Khoz. 6(7):60–61.

12987. —— 1954: Germination and storage period of seed of exotic plants on the Batum Coast. Byul. Cl. Botan. Sada 17:61–68.

Matinian, A. B. 10527.

12988. Matiyenko, B. T. 1963: Microcarpology of Cucurbitaceae. Anatomy of fruits of the squirting-cucumber (*Ecbalium elaterium* [L.] A. Rich.) and *Momordica charantia* L. J. Botan. URSS 48:1346–52.

Matlock, R. S. 9319, 9320.

Matolcsy, G. 18138.

Maton, J. 9174.

Matos, A. G. de, Jr.; *see* De Matos, A. G., Jr.

Matrone, G. 5913.

12989. Matschat, C. H. 1933: Raising fringed gentians from seed. Gardeners Chron. 37:279–80.

12990. Matsubara, S. 1962: Studies on a growth promoting substance, "embryo factor," necessary for the culture of young embryos of *Datura tatula* in vitro. Botan. Mag. (Tokyo) 75:10–18.

12991. Matsuda, K. 1929: On the development of rice-kernels. J. Sci. Agr. Soc. Japan 314:1–35. Biol. Abstr. 5:930. 1931.

12992. —— 1930: On the germination of seeds of rice varieties at low temperatures (preliminary). Proc. Crop Sci. Soc. Japan 2:263–68.

Matsue, K. 10946.

Matsufuji, Y. 14547.

Matsugaki, Z. 14160.

12993. Matsukawa, T. 1935: Saponin of the seeds of *Esculus turbinata*, Blume. I. J. Pharm. Soc. Japan 55:350–57. Chem. Abstr. 29:5853. 1935.

Matsumoto, F. 9080, 9081.

12994. Matsumoto, K. 1932: Effect of seed-formation on the rate of respiration of the fruit of the Japanese persimmon or kaki (*Diospyros kaki* L.f.). Japan. J. Botany 6:125–37.

Matsumoto, K. 17334.

Matsumoto, S. 16830.

Matsumoto, T. 10373, 18195.

12995. Matsumura, S. 1938: Wirksamkeit des Photoperiodismus und der Jarowisation zum ährenschieben des Weizens und der Gerste. Botany Zool. 6:1141–42. Japan. J. Botany 9:160. 1939.

12996. —— 1960: Radiation genetics in wheat. V. Influence of irradiation time and temperature on the genetic effects of ionizing radiation in diploid wheat. Japan. J. Genetics (English Transl.) 35: 197–204. Nucl. Sci. Abstr. 15:930. 1961.

Matsumura, T. 5819, 5820.

Matsunaka, S. 13946.

12997. Matsuoka, K. 1956: On the germination of crop seed in the soil used after the forage crop cultivation. III. About germination and seedling of the upland rice. Japan. J. Ecol. 6:32–35. Biol. Abstr. 31:5629. 1957.

12998. —— 1956: On the germination of crop seed in the soil used after the forage crop cultivation. IV. About germination of the *Vicia* varieties. Japan. J. Ecol. 6:54–56. Biol. Abstr. 31:22302. 1957.

12999. —— 1958: On the germination of crop seed in the soil used after the forage crop cultivation. V. About germination and growth of wheat. Japan. J. Ecol. 8:12–19. Biol. Abstr. 33:2899. 1959.

13000. —— 1958: Studies on the sesame varieties. V. Influence of temperature on the germination of sesame (*Sesamum indicum* L.) and the relation between germination and several characters. Japan. J. Ecol. 8:106–13. Biol. Abstr. 33:23483. 1959.

13001. —— 1959: Studies on the sesame varieties. VI. The effect of different varieties on the germination conditions due to the passage of storage years and the relation to their several characters. Japan. J. Ecol. 8:156–63. Biol. Abstr. 33:27245. 1959.

13002. —— 1959: Studies on the sesame varieties. VII. Temperature and photoperiodic responses of sesame varieties. Japan. J. Ecol. 9:39–45.

13003. —— 1959: Studies on the sesame varieties. VIII. On the influence of depth of seeding upon germination and several characters. Japan. J. Ecol. 9:143–49. Biol. Abstr. 35:16881. 1960.

13004. Matsuoka, T. 1930: Studies on vitamin C. II. Germination of seeds and vitamin C. Part I. Mem. Coll. Agr. Kyoto Univ. 9:15–21.

13005. —— 1930: Studies on vitamin C. III. Germination of seeds and vitamin C. Part II. Mem. Coll. Agr. Kyoto Univ. 9:23–27.

13006. —— 1932: Studies on vitamin C. IV. Germination of seeds and vitamin C. Part III. Antiscorbutic value of sprouts germinated under a Mazda lamp. Mem. Coll. Agr. Kyoto Univ. 24:1–22.

13007. —— 1932: Studies on vitamin C. V. Germination of seeds and vitamin C. Part IV. Antiscorbutic value of seeds germinated in the dark. Mem. Coll. Agr. Kyoto Univ. 24:23–26.

13008. —— 1932: Studies on vitamin C. VI. Germination of seeds and vitamin C. Part V. Antiscorbutic value of sprouts germinated in the sunlight. Mem. Coll. Agr. Kyoto Univ. 24:37–47.

13009. —— 1933: Studies on vitamin C. VII. Germination of seeds and vitamin C. Part VI. Production of vitamin C during the growth of plants and the effect of light on it. Nippon Nogei Kgk. Kw. Si. 9:81–89. Japan. J. Med. Sci. II. Biochem. 4:20. 1938.

13010. Matsushima, S., and T. Tanaka. 1961: Analysis of developmental factors determining yield and its application to yield prediction and culture improvement of lowland rice. LX. Studies on the method for predicting the percentage of ripened grains and the weight of 1,000 kernels by means of identifying the ripening grade in terms of "translucent proportion" in transverse sections of kernels. Proc. Crop Sci. Soc. Japan 29:398–99.

Matsushima, S. 11118.

13011. Matsushita, A. 1958: Changes of contents of free amino acids during the course of germination of leguminous seeds. I. Experiments on seeds harvested in early spring. Kumamoto Joshi Dagaku Gakujitsu Kiyo 10:98–103. Chem. Abstr. 53:22264. 1959.

13012. —— 1959: Variation of free amino acid content during the course of germination of leguminous seeds. III. Leguminous seeds harvested in autumn. Eiyo To Shokuryo 11:235–38. Chem. Abstr. 59: 3084. 1963.

13013. —— 1960: Variation of the free amino acid contents of the seed of peas and broad beans during ripening. Nippon Nogeikagaku Kaishi 34:220–23. Chem. Abstr. 55:1817. 1961.

13014. —— 1962: Variation of the free amino acid contents in unhulled wheat and barley during the ripening period of their plants. Japan. Soc. Food Nutr. J. 14:438–41.

13015. Matsushita, S. 1959: Ribonuclease during the germination of wheat. Mem. Res. Inst. Food Sci. Kyoto Univ. 17:23–28. Chem. Abstr. 54:2506. 1960.

13016. Matsuura, K. 1951: Changes in riboflavin concentration during germination. Vitamins (Japan) 3:283–84. Chem. Abstr. 45:10321. 1951.

13017. Matsuura, T. 1962: Morphological and physiological changes in development of todo-fir cones and seeds. I. Chronological changes in appearance, cone size and scale size. Hokkaido Forest Expt. Sta. Ann. Rept. 1961:34–42.

Matsuura, T. 14927, 14928.

Mattei, P. di; *see* Di Mattei, P.

13018. Matthes, E. 1927: Physiologische Untersuchungen über Umwandlungen des Öles im keimenden Sonnenblumensamen. Botan. Arch. 19:79–135.

13019. Matthews, D. 1963: Laboratory identification of seeds of some thistles, *Cirsium arvense*, *Cirsium vulgare*, *Carduus nutans*. Proc. Intern. Seed Testing Assoc. 28:19–26.

13020. —— 1963: Identification of the seeds of *Medicago sativa* and *Medicago hispida*. Proc. Intern. Seed Testing Assoc. 28:202–6.

Matthews, E. M. 7700, 8625.

13021. Matthews, J. D. 1959: Production and use of superior tree seeds in Britain. Quart. J. Forestry 53:12–22. Biol. Abstr. 33:31208. 1959.

Matthews, J. D. 6250.

13022. Matthews, J. R. 1942: The germination of *Trientalis europaea*. J. Botany 80:12–16. Biol. Abstr. 17:10205. 1943.

Matthews, W. S. A. 4369.

13023. Matthiessen, A. 1962: A contribution to the embryogeny of *Paeonia*. Acta Horti Bergiani 20(2):57–61.

13024. Mattirolo, O. 1885: Seminal integuments of Tiliaceae. Nuovo Giorn. Botan. Ital. 17:289319.

13025. —— and L. Buscalioni. 1890: Germination of seeds of Papilionaceae. Malpighia 4:313–30.

13026 —— and —— 1892: Structure of the integument of the seed of Papilionaceae. Mem. Accad. Reale Sci. Torino 43. 186 pp.

13027. Mattson, A. M., C. O. Jensen, and R. A. Dutcher. 1947: Triphenyltetrazolium chloride as a dye for vital tissues. Science 106:294–95.

13028. Matuda, S. 1940: Seed oils. I. Oil from the seeds of *Cryptotaenia canadensis*, DC. var. *japonica*, Makino. Nippon Nogei Kgk Kw. Si. 61:201–6. Japan. J. Med. Sci. II. Biochem. 4:117. 1941.

13029. Matumura, M., N. Takase, and I. Hirayoshi. 1960: Physiological and ecological studies on germination of Digitaria seeds. I. Difference in response to germinating conditions and dormancy among individual plants. Gifu Univ. Fac. Agr. Res. Bull. 12:89–96.

13030. Matusima, S., and T. Tanaka. 1932: Biochemical studies on the storage of rice. II. A change of hulled rice during storage. Rept. Inst. Res. Labour 9:495–500. Japan.

J. Med. Sci. II. Biochem. 3:97. 1937.
Matuzawa, K. 8743.

13031. Maude, R. B. 1962: Seed transmission of *Didymella* stem-rot of tomatoes. Ann. Appl. Biol. 50:105–11.

13032. —— 1963: Testing the viability of *Septoria* on celery seed. Plant Pathol. 12:15–17.

13033. Mauer, F. M. 1928: Experiments on the "stimulation" of cotton seeds. Proc. Turkestan Plant Breed. Sta. Central Cotton Comm. USSR 13:94–108. Biol. Abstr. 7:20644. 1933.

13034. Maughan, G. L., and J. S. Dunn. 1959: The use of fluorescent tracers for identifying individual seeds. J. Agr. Eng. Res. 4:181–85.

13035. Mauldin, M. P. 1943: The rapid germination of a species with a mucilaginous seed coat, *Plantago fastigiata* Morris. J. Am. Soc. Agron. 35:1025–27.

13036. Maull, T. W. 1963: Seed germination and establishment of *Pinus rigida* Miller (an autecological study). Dissertation Abstr. 23:3607–8.

13037. Mauney, J. R. 1960: The in vitro culture of small cotton embryos. Cotton Improvement Conf. Proc. 12:29–31.

13038. Maurer, K. J. 1963: On some measures for breaking dormancy in fruit seeds. Saatgut-Wirtsch. 15:74–76, 112–15.

Maurie, A. 15293, 15294.

13039. Mauritzon, J. 1933: Studien über die Embryologie der Familien Crassulaceae und Saxifragaceae. Lund. 152 pp.

13040. —— 1933: Über die Embryologie der Turneraceae und Frankeniaceae. Botan. Notiser 1933:543–54.

13041. —— 1934: Die Endosperm- und Embryoentwicklung einiger Acanthaceen. Hakan Ohlssons Buchdruckerei, Lund. 42 pp. Biol. Abstr. 8:10148. 1935.

13042. —— 1934: Zur embryologie der Elaeocarpaceae. Arkiv Botan. 26A(10): 8S. Botan. Centralbl. 170:119. 1936.

13043. —— 1935: Die Embryologie einiger Capparidaceen sowie von *Tovaria pendula*. Arkiv Botan. 26A(4, No. 15):1–14. Biol. Abstr. 11:1821. 1937.

13044. —— 1935: Etwas über die Embryologie der Bignoniaceen. Botan. Notiser 1935:60–77.

13045. —— 1935: Über die Embryologie der Familie Rutaceae. Svensk. Botan. Tidskr. 29:319–47. Biol. Abstr. 11:9596. 1937.

13046. —— 1935: Zur Embryologie von *Peumus boldus*. Arch. Botan. Forli 11:317–27.

13047. —— 1936: Samenbau und Embryologie einiger Scitamineen. Lunds Univ. Arsskr. N. F. Avd. 2. 31(9):3–30.

13048. —— 1936: Zur Embryologie der Berberidaceen. Acta Horti Gotoburg 11:1–18.

13049. —— 1936: Zur Embryologie einiger Parietales-Familien. Svensk Botan. Tidskr. 30:79–113. Botan. Centralbl. 170:119. 1936.

13050. —— 1936: Zur Embryologie und systematischen Abgrenzung der Reihen Terebinthales und Celastrales. Botan. Notiser 1936:161–212.

13051. —— 1939: Contributions to the embryology of the orders Rosales and Myrtales. Fysiograf. Sällsk. Lund Handl. 50(2):3–121.

13052. —— 1939: Die Bedeutung der embryologischen Forschung für das natürliche System der Pflanzen. Fysiogr. Sällsk. Lund Handl. 50(15). 70 pp.

Maurizio, A. 8341.
Maurtua C., M. 6048.

13053. Maximov, B. 1938: Physico-chemical investigation of cocklebur oil. Compt. Rend. Acad. Sci. URSS 20:381–82.

13054. Maximov, N. A. 1925: Pflanzenkultur bei elektrishem Licht und ihre Anwendung bei Samenprüfung und Pflanzenzüchtung. Biol. Centralbl. 45:627–39.

13055. —— 1934: The theoretical significance of vernalization. Imp. Bur. Plant Genet. Herb. Plants (Aberystwyth) Bull. 16. 14 pp.

13056. Maximov, N. M. 1940: Physical and chemical data on the oil of *Xanthium strumarium* L. Compt. Rend. Acad. Sci. URSS 26:393–95.

Maximow, T. A. Krasnosselsky-; *see* Krasnosselsky-Maximow, T. A.

13057. Maxton, J. L. 1927: Effect of fertilizers on the germination of seeds. Soil Sci. 23:335–41.

13058. Maxwell, L. R., and J. H. Kempton. 1939: Delayed killing of maize seeds X-rayed at liquid-air temperature. J. Wash. Acad. Sci. 29:368–74.

13059. ——, ——, and V. M. Mosley. 1942: Effect of temperature and time on the X-ray sensitivity of maize seeds. J. Wash. Acad. Sci. 32:18–24. Biol. Abstr. 16: 10836. 1942.

Maxwell, L. R. 10453.
Maxwell, N. 4394.

13060. May, J. T. 1953/54: The influence of certain soil management practices in a forest tree nursery on Loblolly pine seed germination and seedling development. Alabama Agr. Expt. Sta. Ann. Rept. 64/65:32–33.

13061. —— and H. G. Posey. 1958: The effect of radiation by cobalt-60 γ-rays on germination of slash pine seed. J. Forestry 56:854–55.

13062. May, L. H., and M. S. Buttrose. 1959: Physiology of cereal grain. II. Starch granule formation in the developing barley kernel. Australian J. Biol. Sci. 12:146–59.

May, L. H. 3473.

Mayaud, N. 5077.

Mayber, A. Poljakoff-; *see* Poljakoff-Mayber, A.

13063. Mayer, A., and F. J. Van Pesch. 1883: Methodologisches aus der Praxis der Samencontrole. Landwirtsch. Versuchssta. 28:167–75.

Mayer, A. 9653.

13064. Mayer, A. M. 1953: Quantitative aspects of the behaviour of coumarin as a germination inhibitor. Physiol. Plantarum 6:413–24.

13065. —— 1954: Polyphenol oxidases of lettuce seeds and their inhibition. Enzymologia 16:277–84.

13066. —— 1954: Some heavy metals in lettuce seeds and changes in them during germination. Physiol. Plantarum 7:777–81.

13067. —— 1956: The action of thiourea as a germination stimulator. J. Exptl. Botany 7:93–96.

13068. —— 1958: Ascorbic acid oxidase in germinating lettuce seeds and its inhibition. Physiol. Plantarum 11:75–83.

13069. —— 1958: The breakdown of phytin and phytase activity in germinating lettuce seeds. Enzymologia 19:1–8.

13070. —— 1958: A reduced diphosphopyridine nucleotide oxidase in lettuce seed. Nature 182:317–18.

13071. —— 1959: Further studies on a soluble DPNH-oxidase system in germinating lettuce seeds and its relation to DOPA oxidase. Enzymologia 20:323–26.

13072. —— 1960: Germination research at the Hebrew University, Jerusalem, Israel—a review. Indian J. Plant Physiol. 3:13–23.

13073. —— 1961: Nature and location of phenolase in germinating lettuce. Physiol. Plantarum 14:322–31.

13074. —— and M. Evenari. 1951: The influence of two germination inhibitors (coumarin and 2,4-D) on germination in conjunction with thiourea and cysteine. Bull. Res. Council Israel 1:125–29.

13075. —— and —— 1952: The relation between the structure of coumarin and its derivatives and their activity as germination inhibitors. J. Exptl. Botany 3:246–52.

13076. —— and —— 1953: The activity of organic acids as germination inhibitors and its relation to pH. J. Exptl. Botany 4:257–63.

13077. —— and A. Poljakoff-Mayber. 1958: The joint action of thiourea and copper on the germination and growth of lettuce. Bull. Res. Council Israel Sect. D 6:88–90.

13078. —— and —— 1958: The interaction of thiourea and coumarin in germination and growth of lettuce. Bull. Res. Council Israel Sect. D 6:118–24.

13079. —— and —— 1959: Coumarins and their role in growth and germination. Proc. Intern. Conf. Plant Growth Regulators 4:735–49.

13080. —— and —— 1962: Cytochrome oxidase in germinating lettuce. Plant Cell Physiol. 3:309–12.

13081. —— and —— 1962: Quantitative changes in nucleic acids during germination of lettuce seeds. Physiol. Plantarum 15: 283–92.

13082. —— and —— 1963: The germination of seeds. MacMillan, New York. 236 pp.

13083. ——, ——, and W. Appleman. 1957: Studies on the oxidative systems in germinating lettuce seeds. Physiol. Plantarum 10:1–13.

13084. ——, ——, and S. Zacks. 1958: The effect of diethylthiourea on the stimulation of germination and the inhibition of growth by thiourea. Phyton 11:23–28.

13085. ——, ——, and —— 1958: The interaction of thiourea and ascorbic acid in their effect on germination and growth. Bull. Res. Council Israel Sect. D 6:103–7.

Mayer, A. M. 6117, 6118, 6122, 8168, 11031, 12108, 12153, 15500–15505, 16297, 17885.

13086. Mayer, C. 1938: The sensitivity of aqueous plant extracts to light. Ber. Naturforsch. Ges. Freiburg Breisgau 36: 115–16. Chem. Abstr. 35:3682. 1941.

Mayer, F. 14600.

13087. Mayer, K., and M. Klinga-Mayer. 1940: Über das Stärkeverflüssigungsvermögen während der Keimung von Gerste. Z. Physiol. Chem. 267:115–27.

13088. Mayer, R. 1957: Le controle de l'identité varietale et la certification des semences en France. Proc. Intern. Seed Testing Assoc. 22:179–91.

13089. Mayerl, F. 1956: Aatiram—a preparation for the treatment of seeds with lowered vitality. Bodenkultur 9:119–21. Chem. Abstr. 51:10672. 1957.

13090. Mayhall, M. P. 1056: Roses from seed. Am. Rose Mag. 13(16):8–9, 18–19.

13091. Maymone, B., and A. Battaglini. 1953: Digestibility and nutritive value of the carbo-bean "germ." Ann. Sper. Agrar. (Rome) 7:1977–84. Chem. Abstr. 48: 5396. 1954.

13092. —— and —— 1953: Nutritive value of exhausted carbo-bean-pulp residual from the industrial preparation of alcohol. Ann. Sper. Agrar. (Rome) 7:1985–99. Chem. Abstr. 48:5396. 1954.

Maynard, D. N. 8796–8798.

13093. Maynard, M. L. 1963: Effects of wetting and drying on germination of crested wheatgrass seed. J. Range Management 16:119–21. Biol. Abstr. 45:7695. 1964.

Mayne, R. Y. 9784, 11718.

13094. Mayr, E. 1931: Die osmotischen Werte einiger Weizen-landsorten im vergleiche zu ihrer Keimungsgeschwindigkeit und Vegetationszeit. Fortschr. Landwirtsch. 6:47–50.

13095. —— 1931: Abhängigkeit der Saugkraft und Keimungsgeschwindigkeit vom Alter des Saatgutes, dargestellt am Sommer weizen. Fortschr. Landwirtsch. 6: 485–88.

13096. —— 1936: Ergebnisse der Erkennungsversuche der im Zuchtbuch eingetragenen Getreidesorten. I. Sortenbeschreibung der Weizensuchtsorten. Landeskultur Vienna 3:255–59. Proc. Intern. Seed Testing Assoc. 9:167. 1937.

Mayr, H. 12294.

13097. Mayr, S. I. 1926: Über freie Eiweisskrystalle im Endosperm der Samen von *Loranthus europaeus*. Sitzber. Math.-Naturw. Akad. Wiss. Wien Abt. I 135: 409–12.

13098. —— 1928: Über die Keimung und erste Entwicklung der Riemenmistel (*Loranthus europaeus* Jacq.). Sitzber. Math.-Naturw. Akad. Wiss. Wien Abt. I 137: 345–62.

13099. Maywald, E., R. Christensen, and T. J. Schoch. 1955: Development of starch and phytoglycogen in golden sweet corn. J. Agr. Food Chem. 3:521–23.

13100. Mazaeva, M. M. 1951: Effectiveness of growth stimulants at various levels of phosphate diet in plants. Dokl. Akad. Nauk SSSR 80:957–60. Chem. Abstr. 42:2739. 1952.

13101. —— 1954: The effect of magnesium fertilizers on seed quality. Agrobiologiya 1954(4):125–29. Soils Fertilizers 18:279. 1955.

Mazdrakov, P. 9377.

13102. Maze, M. 1900: Recherches sur la digestion des réserves dans les graines en voie de germination et leur assimilation par les plantules. Compt. Rend. 130: 424–27.

13103. Maze, P. 1900: Recherches sur le role de l'oxygene dans la germination. Ann. Inst. Pasteur 14:350–68. Botan. Centralbl. 86:117–18. 1901.

13104. —— 1910: Maturation provoquée des graines. Action antigerminative de l'aldehyde éthylique. Compt. Rend. 151: 1383–86.

Mazel, N. 5319.

Mazumder, K. C. 7198.

Mazurova, T. A. 17722–17724.

13105. Mazzanti, L. 1938: The influence of metallic ions on the germination of seeds and on the initial development of plants. Arch. Intern. Pharmacodyn. 58:249–66. Chem. Abstr. 32:5445. 1938.

13106. —— 1938: Pharmacological investigation of certain Tuscany waters. I. Effect on the germination of seeds. Ann. Chim. Appl. 28:95–106. Chem. Abstr. 32:8447. 1938.

13107. Mazzaron, A. 1935: Les hydrocarbons des graines de lupin. Ital. Agr. 72(2). Proc. Intern. Seed Testing Assoc. 8:90. 1936.

13108. Mazzetti, G. 1929: Ricerche sul comportamento del potere germinativo del semi nelle spazzature della Città di Siena, trattate con la macchina Trituratrice Gannow. Atti Accad. Fisiocrit. Siena 3:517–23. Biol. Abstr. 8:7941. 1934.

13109. McAlister, D. F. 1943: The effect of maturity on the viability and longevity of the seeds of western range and pasture grasses. J. Am. Soc. Agron. 35:442–53.

McAlister, E. D. 6498.

13110. McAlpine, K. L. 1947: Germination of orchid seeds. Orchid Rev. 55:8–10, 21–22.

McAnelly, C. W. 3819.

McArdle, F. J. 1808.

13111. McBirney, S. W. 1942: Single-seed planting of sugar beets. Proc. Am. Soc. Sugar Beet Technol. 1942:228–36. Biol. Abstr. 18:14291. 1944.

13112. —— 1947: Beet planter development to improve seedling emergence. Proc. Am. Soc. Sugar Beet Technol. Regional Meet. 1947:4–10.

13113. McBlaine, J. 1945: Camellia seed. Flower Grower 32(6):318.

McCaig, J. D. 13121.

13114. McCall, E. R., C. L. Hoffpauir, and D. B. Skau. 1951: The chemical composition of rice. A literature review. U.S. Dept. Agr., Agr. Res. Admin., Bur. Agr. Ind. Chem. AIC-312. 49 pp.

13115. —— J. F. Jurgens, C. L. Hoffpauir, W. A. Pons, Jr., S. M. Stark, A. F. Cucullu, D. C. Heinzelman, V. O. Cirino, and M. D. Murray. 1953: Influence of variety and environment on physical and chemical composition. J. Agr. Food Chem. 1: 988–93.

13116. McCalla, A. G. 1934: Amide nitrogen in germinating seeds. Can. J. Res. 10:430–34.

13117. —— 1938: Fractionation of nitrogen in developing wheat kernels. Can. J. Res. Sect. C 16:263–73. Biol. Abstr. 12:15763. 1938.

13118. —— 1948: Nitrogenous constituents of plants. Ann. Rev. Biochem. 18:615–38. Chem. Abstr. 43:9174. 1949.

13119. —— and W. G. Corns. 1943: Effect of variety and environment on the starch content of wheat and barley. Can. J. Res. Sect. C 21:307–21.

13120. —— and N. Gralen. 1945: Sedimentation and diffusion measurements on lutens from wheats of widely differing quality. Can. J. Res. Sect. C 23:219–25. Biol. Abstr. 20:13563. 1946.

13121. ——, J. D. McCaig, and A. D. Paul. 1939: Effect of various conditions of storage on baking quality of flour. Can. J. Res. Sect. C 17:452–59. Biol. Abstr. 15:5646. 1941.

13122. —— and R. Newton. 1935: Effect of frost on wheat at progressive stages of maturity. II. Composition and biochemical properties of grain and flour. Can. J. Res. Sect. C 13:1–31.

McCalla, A. G. 520, 12184, 12769, 14379.

McCalla, D. R. 7303.

13123. McCalla, T. M., and F. L. Duley. 1948: Stubble mulch studies: effect of sweetclover extract on corn germination. Science 108:163.

13124. —— and —— 1950: Stubble mulch studies. III. Influence of soil microorganisms and crop residues on the germination, growth and direction of room growth of corn seedlings. Soil Sci. Soc. Am. Proc. 14:196–99.

McCalla, T. M. 7778.

13125. McCallan, S. E. A. 1948: Evaluation of chemicals as seed protectants by greenhouse tests with peas and other seeds. Contrib. Boyce Thompson Inst. 15:91–117.

13126. —— 1950: Some chromate complexes and organic compounds as seed protectants. Contrib. Boyce Thompson Inst. 16:5–20.

13127. McCann, C. 1959: Seed dispersal. J. Bombay Nat. Hist. Soc. 56:165–82.

13128. McCann, L. P. 1945: Embryology of the tung tree. J. Agr. Res. 71:215–29.

13129. McCartan, J. 1956: Greenhouse flowers from seeds. Gardeners Chron. 139:151.

13130. McCarty, C. D., J. W. Lesley, and H. B. Frost. 1952: Bitterness (benzaldehyde content) of kernels of almond-peach F. hybrids and their parents. Proc. Am. Soc. Hort. Sci. 59:254–58.

13131. McClean, A. P. D. 1948: Bunchy-top disease of the tomato: Additional host plants, and the transmission of the virus through the seed of infected plants. S. Africa Dept. Agr. Sci. Bull. 256. 28 pp.

13132. McClellan, W. D. 1944: A seedling blight of castor bean, *Ricinus communis.* Phytopathology 34:223–29.

13133. McClelland, C. K. 1931: Quantity determinations of seed for rod rows of spring oats. J. Am. Soc. Agron. 23:702–8.

13134. —— 1940: Methods and rates of planting soybeans. Arkansas Agr. Expt. Sta. Bull. 390:1–17. Biol. Abstr. 15:5641. 1941.

McClelland, H. L. 5719.

13135. McClelland, T. B. 1944: Brief viability of tropical seeds. Proc. Florida State Hort. Soc. 57:161–63.

13136. McCluggage, M. E. 1943: Factors influencing the pearling test for kernel hardness in wheat. Cereal Chem. 20: 686–700.

13137. McClure, D. S. 1955: A report on the procedures used to obtain germination of shepherd's purse with usual laboratory facilities. Assoc. Offic. Seed Analysts News Letter 29(2):23–28.

13138. —— 1957: Seed characters of selected plant families. Iowa State Coll. J. Sci. 31:649–81.

McCollum, G. D. 19592.

13139. McCollum, J. P. 1953: Factors affecting cotyledonal cracking during the germination of beans (*Phaseolus vulgaris*). Plant Physiol. 28:267–74.

13140. —— and M. B. Linn. 1955: Bleaching and disinfecting discolored pepper seed with sodium hypochlorite. Proc. Am. Soc. Hort. Sci. 66:345–49.

McCollum, J. P. 6236.

13141. McCollum, R. E., R. H. Hageman, and E. H. Tyner. 1960: Occurrence of pyruvic kinase and phosphoenolpyruvate phosphatases in seeds of higher plants. Soil Sci. 89:49–52.

13142. McCollum, R. L. 1939: The development of the embryo sac and the seed of *Commelina angustifolia* Michx. Bull. Torrey Botan. Club 66:539–47.

13143. McComas, P. S., J. F. O'Kelly, and F. J. Welch. 1943: Peanuts for oil: A war crop. Mississippi Agr. Expt. Sta. Bull. 376:1–20. Biol. Abstr. 18:3268. 1944.

13144. McComb, A. L. 1934: The relation between acorn weight and the development of one year chestnut oak seedlings. J. Forestry 32:479–84. Expt. Sta. Rec. 72:193. 1935.

13145. —— and H. S. Lovestead. 1954: Viability of cottonwood seeds in relation to storage temperatures and humidities. Tree Planters Notes 17:9–11.

13146. McCombe, K. 1954: Growing gladioli from seed. Gladiolus 1954:121–22.

13147. McConnell, B. R. 1960: Effect of gibberellic acid and cold treatments on the germination of bitterbrush seed. U.S. Dept. Agr. Forest Serv. Pacific Northwest Forest Range Expt. Sta. Res. Note 187. 4 pp.

13148. McConnell, W. B. 1957: Studies on wheat plants using C^{14} compounds. V. Germination studies with labelled wheat seeds. Can. J. Biochem. Physiol. 35:1259–66.

McCool, M. M. 2879.

13149. McCormick, F. A. 1932: Microchemical study of the seed of *Nicotiana tabacum*. Conn. Agr. Expt. Sta. Bull. 339:646–50. Chem. Abstr. 27:323. 1933.

McCorquodale, H. 12645, 12646.

McCourtney, E. J. 10893.

McCown, J. M. 17915.

13150. McCrackan, M. 1951: Intriguing fruits. Trillia 11:165–68.

13151. McCray, F. A. 1933: Embryo development in *Nicotiana* species hybrids. Genetics 18:95–110.

McCready, R. M. 8348, 10424.

McCreery, F. 2545.

McCrory, C. S. 6520.

13152. McCrory, S. A. 1943/44: False indigo seeds provide limited insecticide. S. Dakota Agr. Expt. Sta. Ann. Rept. 1943/44:23.

13153. McCubbin, E. N. 1944: Premature seeding or bolting of cabbage. Proc. Florida State Hort. Soc. 57(1944):191–95.

13154. McCulley, R. D. 1945: Germination of longleaf pine seed at high and low temperatures. J. Forestry 43:451–52.

13155. McCulloch, W. F. 1945: Direct seeding of Douglas fir in the Tillamook Burn. J. Forestry 43:37–38.

13156. McCully, A. 1933: Quickly germinating alpines. Gardeners Chron. 37: 309–10.

13157. McCully, W. G. 1951: Recovery and viability of MacCartney rose seeds fed to cattle. J. Range Management 4:101–6.

McCune, S. B. 19432–19434.

McCutchen, B. E. 2722.

13158. McDaniel, J. C. 1956: Nut germination observations 1955–1956. Northern Nut Growers Assoc. Ann. Rept. 47:138–42.

13159. McDaniels, L. H. 1939: The shuck-nut ratio in black walnut varieties. Northern Nut Growers Assoc. Proc. 30:74–76. Expt. Sta. Rec. 83:631. 1940.

McDavid, M. R. 14111.

13160. McDermott, E. E., and J. Pace. 1960: Comparison of the amino-acid composition of the protein in flour and endosperm from different types of wheat, with particular reference to variation in lysine content. J. Sci. Food Agr. 11:109–15.

McDermott, E. E. 17971.

13161. McDermott, R. E. 1953: Light as a factor in the germination of some bottom-land hardwood seeds. J. Forestry 51:203–4.

13162. McDonald, B. K. 1949: Tillage practices for control of wild oats, tartary buckwheat, and other weeds having delayed germination. Proc. Western Can. Weed Control Conf. 3:167–71.

13163. McDonald, J. A. 1937: The biochemistry of cacao. I. A preliminary investigation of the biochemical attributes of quality in fresh cacao beans. Ann. Rept. Cacao Res. 6(1936):34–40. Chem. Abstr. 32:969. 1938.

13164. —— 1937: The biochemistry of cacao. II. The effect of fermentation on the biochemical composition of different types of cacao beans. Ann. Rept. Cacao

Res. 6(1936):40–42. Chem. Abstr. 32:969. 1938.

13165. —— 1937: The biochemistry of cacao. III. A note on the estimation of theobromine in cacao beans. Ann. Rept. Cacao Res. 6(1936):43–44. Chem. Abstr. 32:969–70. 1938.

13166. —— 1937: The biochemistry of cacao. IV. The tannins in cacao beans. Ann. Rept. Cacao Res. 6(1936):44–48. Chem. Abstr. 32:970. 1938.

McDonnell, A. D. 5651.

13167. McDonough, W. T. 1964: Germination responses of *Carnegiea gigantea* and *Lemaireocereus thurberi*. Ecology 45:155–59.

13168. —— and H. G. Gauch. 1959: The contribution of the awns to the development of the kernels of bearded wheat. Maryland Agr. Expt. Sta. Bull. A-103. 16 pp.

13169. McElroy, L. W., D. R. Clandinin, W. Lobay, and S. I. Pethybridge. 1949: Nine essential amino-acids in pure varieties of wheat, barley, and oats. J. Nutr. 37:329–36.

McElroy, L. W. 16131.

13170. McEwen, F. L., and G. E. R. Hervey. 1960: The effect of lygus bug control on the yield of lima beans. J. Econ. Entomol. 53:513–16.

13171. ——, J. J. Natti, G. E. R. Hervey, and W. T. Schroeder. 1957: The relative effectiveness of various insecticide-fungicide seed treatments on lima beans after several months of storage. J. Econ. Entomol. 50:775–77.

McEwen, F. L. 14246.

13172. McFadden, A. D. 1960: Effect of seed size on yield and other characters in barley. Saskatchewan Univ. Farm Home Week, Addresses Proc. 1960:26–29, 31–33.

13173. —— 1963: Effect of seed source on comparative test results in barley. Can. J. Plant Sci. 43:295–300.

13174. McFadden, E. S., and E. R. Sears. 1946: The origin of *Triticum spelta* and its free-threshing hexaploid relatives. J. Heredity 37:107–16.

13175. McFadden, G. H., and R. L. McMurray. 1937: Seeds and oil of *Rhus glabra*. Am. J. Pharm. 109:397–406. Analyst 62:809. 1937.

13176. McFarlane, J. S., C. Price, and F. V. Owen. 1948: Strains of sugar beets extremely resistant to bolting. Proc. Am. Soc. Sugar Beet Technol. 5:151–53.

13177. McFarlane, V. H., J. T. Hogan, and T. A. McLemore. 1955: Effects of heat treatment on the viability of rice. U.S. Dept. Agr. Tech. Bull. 1129. 51 pp.

13178. McGahan, M. W. 1961: Studies on the seed of banana. I. Anatomy of the seed and embryo of *Musa balbisiana*. Am. J. Botany 48:230–38.

McGill, D. P. 4309.

13179. McGinnies, W. J. 1960: Effects of moisture stress and temperature on germination of six range grasses. Agron. J. 52:159–62.

13180. —— 1961: Evaluation of powder measures for metering out small quantities of seed. Agron. J. 53:408–9.

McGinnies, W. J. 8832.

McGinnis, J. 7840.

McGowan, J. C. 3030.

13181. McGregor, W. G. 1937: Inheritance of quality and quantity of oil in flax in relation to other plant characters. Can. J. Res. Sect. C 15:362–79.

13182. —— 1964: Influence of date of seeding on yield and yield components in flaxseed. Can. J. Plant Sci. 44:145–48.

13183. —— and W. D. Hays. 1952: Safflower—Canadian experiments. Sci. Agr. (Ottawa) 32:204–13.

McGregor, W. G. 12030.

13184. McGugan, J. M. 1948: Seeds and seedlings of the genus *Brassica*. Can. J. Res. Sect. C 26:520–87.

13185. McGuire, T. A., and F. R. Earle. 1958: Changes in the solubility of corn protein resulting from the artificial drying of high-moisture corn. Cereal Chem. 35:179–88.

13186. McHargue, J. S. 1920: The cause of deterioration and spoiling of corn and corn meal. Ind. Eng. Chem. 12:257–62.

13187. —— 1920: Mineral constituents of the paired seeds of cocklebur. Science 52:43.

13188. —— 1920: The significance of peroxidase reaction with reference to the viability of seeds. J. Am. Chem. Soc. 42:612–15.

13189. —— and W. R. Roy. 1931: The mineral, nitrogen and fat content of some varieties of mature bean seed and of string beans. J. Nutr. 3:479–81.

McHargue, J. S. 15984.

13190. McHatton, T. H., and J. G. Woodroof. 1298: Some factors influencing pecan germination. Proc. Am. Soc. Hort. Sci. 24:125–29.

13191. McIlrath, W. J., and L. Bogorad. 1958: Photoperiodic floral induction of *Xanthium* and germination of lettuce seeds implanted in the petioles. Botan. Gaz. 119:186–91.

13192. ——, D. R. Ergle, and A. A. Dunlap. 1951: Persistence of 2,4-D stimulus in cotton plants with reference to its transmission to the seed. Botan. Gaz. 112:511–18.

McIlrath, W. J. 2704.

13193. McIlvaine, H. R. C., and H. W. Popp. 1940: Further studies on growth substances in relation to the mechanism of the action of radiation on plants. J. Agr. Res. 60:207–15.

McIlvaine, T. C. 16677.

13194. McIntosh, T. P. 1956: Identifying Svalof oat varieties by grain characters. Inst. Corn Agr. Merchants J. 5:119–26.

13195. —— and J. R. Thompson. 1951: Seed oats certification. Scot. Agr. 30:225–28. Biol. Abstr. 25:31809. 1951.

13196. McIntyre, A. C. 1929: A cone and seed study of the Mountain pine (*Pinus pungens* Lambert). Am. J. Botany 16: 402–6.

13197. —— 1932: Seeding habit of pitch pine (*Pinus rigida*). Forest Leaves 23(7): 109–11. Biol. Abstr. 8:9393. 1934.

13198. McIntyre, C. N. 1939: Report of the conference on germination standards for vegetable seeds. Proc. Assoc. Offic. Seed Analysts 31:17–18.

13199. McIntyre, D. S. 1955: Effect of soil structure on wheat germination in a red-brown earth. Australian J. Agr. Res. 6:797–803.

13200. McKay, E., and W. Hermann. 1939: The effect of maturity at time of harvest on certain responses of seed of crested wheatgrass, *Agropyron cristatum* (L.) Gaertn. J. Am. Soc. Agron. 31:876–85.

13201. McKay, I. S. 1954: African violets from seed. Horticulture 32(3):103, 134–35

13202. McKay, J. W. 1947: Embryology of pecan. J. Agr. Res. 74:263–83.

13203. —— and A. C. Gossard. 1959: Effect of nut size on growth of Chinese chestnut (*Castanea mollissima*) seedlings. Northern Nut Growers Assoc. Ann. Rept. 50:56–60.

13204. McKee, H. S., L. Nestel, and R. N. Robertson. 1955: Physiology of pea fruits. II. Soluble nitrogenous constituents in the developing fruit. Australian J. Biol. Sci. 8:467–75.

13205. ——, R. N. Robertson, and J. B. Lee. 1955: Physiology of pea fruits. I. The developing fruit. Australian J. Biol. Sci. 8:137–63.

13206. McKee, R. 1935: Vernalization experiments with forage crops. U.S. Dept. Agr. Circ. 377:1–12. Biol. Abstr. 10:15447. 1936.

13207. —— 1935: Vitality and germination of crimson clover seed as affected by swelling and sprouting and subsequent drying. J. Am. Soc. Agron. 27:642–43.

13208. —— and A. F. Musil. 1948: Relation of temperature and moisture to longevity of seed of blue lupine, *Lupinus angustifolius*, Austrian winter fieldpea, *Pisum sativum arvense*, and hairy vetch, *Vicia villosa*. J. Am. Soc. Agron. 40:459–65.

13209. ——, H. A. Schoth, and P. Tabor. 1933: Tests with imported seed of *Vicia villosa*. J. Am. Soc. Agron. 25:555–56.

13210. McKeen, W. E. 1952: A seed-borne disease of corn. Plant Disease Reptr. 36:145–46.

13211. McKeever, D. G. 1937: A new black locust seed treatment. J. Forestry 35: 500–1. Chem. Abstr. 31:5094. 1937.

McKell, C. M. 6097.

13212. McKenzie, R. E., L. J. Anderson, and H. D. Heinrichs. 1949: Effect of flooding on emergence of forage crop seeds. Sci. Agr. (Ottawa) 29:237–40.

13213. ——, D. H. Heinrichs, and L. J. Anderson. 1946: Maximum depth of seeding eight cultivated grasses. Sci. Agr. (Ottawa) 26:426–31.

McKenzie, R. I. H. 17819.

13214. McKenzie, W. F. 1936: Hardy primulas from seeds. Gardeners Chron. 99:369.

McKibben, E. G. 5317.

13215. McKillican, M. E., and R. P. A. Sims. 1964: The endosperm lipids of three Canadian wheats. J. Am. Oil Chemists Soc. 41:340–44.

13216. McKinney, H. H. 1940: Vernalization and the growth-phase concept. Botan. Rev. 6:25–47.

13217. —— 1954: Culture methods for detecting seed-borne virus in glacier barley seedlings. Plant Disease Reptr. 38:152–62.

13218. McKinney, R. S. 1944: Miscellaneous studies at the Gainesville tung-oil laboratory. Ann. Proc. Am. Tung Oil Assoc. 10:59–63. Chem. Abstr. 39:2890. 1945.

13219. ——, J. L. Carter, and G. S. Jamieson. 1934: The oil content of soybeans. Oil Soap 11:253, 261. Chem. Abstr. 29:551. 1935.
13220. ——, N. J. Halbrook, and R. E. Oglesbee. 1944: Studies in the expression of oil from tung fruit. Oil Soap 21:353–57. Expt. Sta. Rec. 95:461–62. 1946.
13221. —— and G. S. Jamieson. 1938: Ouricury palm-kernel oil. Oil Soap 15: 172–74. Pharm. Abstr. 5:254. 1939.
McKinney, R. S. 8978, 9708, 9710, 9711.
McKinnon, B. A. 18862.
13222. McLachlan, K. D. 1952: Pelleted subterranean clover seed in pasture establishment. J. Australian Inst. Agr. Sci. 18:44–45.
McLane, J. W. 10454.
13223. McLaughlin, F. A. 1935: Effect of molds and other seed-borne fungi on germination. Mass. Agr. Expt. Sta. Control Bull. 77:58–60. Chem. Abstr. 30:225. 1936.
13224. McLaughlin, J. H. 1946: Vegetable seed treatment for Oklahoma. Oklahoma Agr. Expt. Sta. Bull. B-293. 24 pp.
13225. —— 1946: Corn seed treatment in Oklahoma. Oklahoma Agr. Expt. Sta. Bull. B-294. 14 pp.
McLaughlin, J. H. 1941, 16019.
13226. McLean, D. M. 1947: *Alternaria* blight and seed infection, a cause of low germination in certain radish seed crops. J. Agr. Res. 75:71–79.
McLean, D. M. 1947.
McLeavey, M. A. 9313.
McLeish, J. 8261.
13227. McLemore, B. F. 1959: Cone maturity affects germination of longleaf pine seed. J. Forestry 57:648–50.
13228. —— 1961: Hila of full and empty Longleaf pine seeds are distinguishable. Forest Sci. 7:246.
13229. —— and F. J. Czabator. 1961: Length of stratification and germination of Loblolly pine seed. J. Forestry 59: 267–69.
McLemore, T. A. 13177.
13230. McLeod, A. D. 1961: Note on a mechanical seed dispenser for rod-row plots. Can. J. Plant Sci. 41:444–46.
McLeod, D. S. 10172.
13231. McLintock, T. F. 1942: Stratification as a means of improving results of direct seeding of pines. J. Forestry 40:724–28.
13232. McMath, C. W. 1940: Cottonseed as a source of a new type flour. Proc. Cotton Res. Congr. Waco, Texas 1:243–48. Chem. Abstr. 36:4918. 1942.
13233. McMichael, S. C. 1959: Hopi cotton, a source of cottonseed free of gossypol pigments. Agron. J. 51:630.
13234. McMillan, J. A. 1962: Better seeds: Better crops. Natl. Inst. Agr. Botany J. 9:186.
McMunn, R. L. 5496.
McMurray, R. L. 13175.
13235. McNair, A. D. 1917: Boiling buffalo clover seed. Science 45:220–21.
13236. McNair, J. B. 1930: The taxonomic and climatic distribution of oil and starch in relation to the physical and chemical properties of both substances. Am. J. Botany 17:662–68.
13237. —— 1941: Indications of an increase in number of C-atoms in acids and number of acids in seed fats with advance in evolutionary position. Science 94:422.
13238. McNeal, F. H., and M. A. Berg. 1963: Barley stripe mosaic virus in Thatcher and Centana wheat seed from 1956 to 1962. Plant Disease Reptr. 47:276–77.
13239. ——, ——, A. L. Dubbs, J. L. Krall, D. E. Baldridge, and G. P. Hartman. 1960: The evaluation of spring wheat seed from different sources. Agron. J. 52: 303–4.
13240. —— and D. J. Davis. 1954: Effect of nitrogen fertilization on yield, culm number and protein content of certain spring wheat varieties. Agron. J. 46: 375–78.
13241. ——, I. K. Mills, and M. A. Berg. 1961: Variation in barley stripe mosaic virus incidence in wheat seed due to storage and continuous propagation and the effect of the disease on yield and test weight. Agron. J. 53:128–30.
13242. McNeal, X. 1948: Germination of artificially dried oats. Rice J. 51(9):14. Biol. Abstr. 23:19695. 1949.
13243. —— 1950: When to harvest rice for best milling quality and germination. Arkansas Agr. Expt. Sta. Bull. 504. 41 pp.
13244. —— 1958: Conditioning and storage of oats. Arkansas Agr. Expt. Sta. Bull. 603. 17 pp.
13245. —— 1960: Rice storage: effect of moisture content, temperature and time on grade, germination and head rice yield. Arkansas Agr. Expt. Sta. Bull. 621. 23 pp.
13246. McNew, G. L. 1942: Growth stimulation of peas by tetrachloro-para-benzoquinone, a fungicidal seed protectant. Science 96:118–19. Expt. Sta. Rec. 88:212. 1943.

13247. —— 1942: Seed treatments recommended for different vegetables. Farm Res. (N.Y.) 8(2):1, 19.

13248. —— 1943: New treatments for spinach seed show promise. Farm Res. (N.Y.) 9(2):11–13. Expt. Sta. Rec. 89:331. 1943.

13249. —— 1943: Relative effectiveness of organic and inorganic fungicides as seed protectants. Phytopathology 33:9.

13250. —— 1948: Effect of corn seed treated with tetrachloro-para-benzoquinone on the growth and fattening of swine. Iowa State Coll. J. Sci. 22:177–84.

McNew, G. L. 3349, 3350.

13251. McOsker, D. E. 1962: The limiting amino acid sequence in raw and roasted peanut protein. J. Nutr. 76:453–59.

13252. McQueen, D. R. 1954: Fossil leaves, fruits and seeds from the Wanganui series (Plio-Pleistocene) of New Zealand. Trans. Roy. Soc. New Zealand 82:667–76. Biol. Abstr. 32:13308. 1958.

13253. McQuilkin, W. E. 1946: Tests of direct seeding with pines in the Piedmont region. J. Agr. Res. 73:113–36.

McQuilkin, W. E. 5531, 12631.

McQuire, A. J. 1869.

McQuistan, M. 15791.

13254. McRary, W. L. 1940: Nitrogen metabolism of the plant embryo. Botan. Gaz. 102:89–96.

13255. McRostie, G. P. 1939: The thermal death point of corn from low temperatures. Sci. Agr. (Ottawa) 19:687–99.

13256. —— 1949: Some factors influencing the artificial drying of mature grain corn. Agron. J. 41:425–29.

13257. ——, J. W. Hopkins, and N. H. Grace. 1938: Effect of phytohormone dusts on growth and yield of winter wheat varieties. Can. J. Res. Sect. C 16:510–15.

13258. McSwain, E. 1950: Some notes on woody-plant seeds. Assoc. Offic. Seed Analysts News Letter 24(3):36–38.

13259. McVean, D. N. 1955: Ecology of *Alnus glutinosa* (L.) Gaertn. Seed distribution and germination. J. Ecol. 43:61–71. Biol. Abstr. 31:3633. 1957.

13260. —— 1959: Ecology of *Alnus glutinosa* (L.) Gaertn. VII. Establishment of alder by direct seeding of shallow blanket bog. J. Ecol. 47:615–18.

McVeigh, I. 3384–3386.

13261. McWethy, J. A. 1949: Super-seeds for your garden. Sci. Dig. 25(3):7–10.

McWhirter, L. 5826.

13262. McWhorter, C. G. 1961: Morphology and development of Johnsongrass plants from seeds and rhizomes. Weeds 9:558–62.

13263. McWhorter, F. P., and P. W. Miller. 1944: The application of vapor heat as a practical means of disinfecting seeds. Phytopathology 34:935–36.

McWhorter, F. P. 13499.

13264. McWilliam, J. R. 1963: Selection for seed retention in *Phalaris tuberosa* L. Australian J. Agr. Res. 14:755–64.

13265. Mead, H. W. 1939: Shrivelling of wheat kernels by stem rust and its effect on seed value. Sci. Agr. (Ottawa) 19: 481–93.

13266. —— 1941: The disinfecting value of fungicides used for treating cereal seeds and their influence on growth. Sci. Agr. (Ottawa) 21:717–26.

13267. —— 1942: Environmental relationships in a seed-borne disease of barley caused by *Helminthosporium sativum* Pammel, King, and Bakke. Can. J. Res. Sect. C 20:525–38.

Mead, H. W. 12602.

Meade, E. M. 3116.

Meader, E. M. 12893.

13268. Meadly, G. R. W. 1932: Disinfection of tobacco seed. Effect on germination. J. Agr. W. Australia. Ser. II. 9:538.

13269. —— 1934: Strain certification of agricultural seeds. Subterranean clover (*Trifolium subterraneum*, Linn.). J. Agr. W. Australia. Ser. II. 11:611–16.

13270. —— 1936: Impermeable seeds of "Dwalganup" subterranean clover. J. Agr. W. Australia. Ser. II. 13:313–20.

13271. —— 1943: Subterranean clover seed. The influence of superphosphate on germination. J. Agr. W. Australia. Ser. II. 20:53–58.

13272. —— 1947: Some unusual features of seed germination with particular reference to subterranean clover. J. Australian Inst. Agr. Sci. 13:153–60.

13273. —— 1960: Seed storage in relation to germination. J. Agr. W. Australia. Ser. IV. 1:507–8.

13274. Meadows, M. V. 1953: The germination, evaluating and labeling Bahia grass. Proc. Assoc. Seed Control Off. Southern States 10:76–79.

Meal, H. 3449.

13275. Meara, M. L., and N. K. Sen. 1952: The component of fatty acids and glycerides of jute-seed oil. J. Sci. Food Agr. 3:237–40.

13276. —— and Y. A. H. Zaky. 1940: The fatty acids and glycerides of the seed fats of *Allanblackia floribunda* and *Allanblackia parviflora*. Soc. Chem. Ind. J. 59:25–26.
Meara, M. L. 1485, 1486, 3648, 8770–8774.
Mechan, D. K. 15188, 15189.
Mederski, H. J. 8902, 18432.
13277. Medina, A., and A. Sols. 1956: A specific fructokinase in peas. Biochim. Biophys. Acta 19:378–79. Chem. Abstr. 50:7965. 1956.
Medina, J. 14251.
13278. Medina, J. C. 1950: Epoca de semeação da guaxima. Bragantia 10: 235–46. Biol. Abstr. 26:28953. 1952.
13279. Medler, J. T. 1953: New chemical aids harvest of legume seed crops. Crops & Soils 5(8):18–19, 30.
Medler, J. T. 12962, 18501.
Mednik, G. L. 7160.
13280. Medvedev, P. F. 1959: Pre-planting soaking of corn seeds in solutions of minor elements. Udobr. i Urozhai 1959(5):21–23.
13281. Medvedev, Z. A. 1957: Distribution of substances labeled with sulfur-35 in bean seed in the proteins and organs of plants of a new generation during the entire period of their development. Dokl. Akad. Nauk SSSR 114:379–82. Chem. Abstr. 51:18141. 1957.
Medwedewa, G. B. 2998.
13282. Mee, S. 1949: A study of carboxylase in soybean sprouts. Arch. Biochem. 22: 139–48.
13283. Meehand, F. 1887: Dispersion of seeds in a Mexican Euphorbiaceous tree. Botan. Gaz. 12:298.
Meer Mohr, J. C. van der; *see* Van der Meer Mohr, J. C.
13284. Megee, C. R., M. G. Frakes, and I. T. Larsen. 1942: The influence of clipping treatment and rolling on the yield of clover seed. J. Am. Soc. Agron. 34: 841–43.
Megee, C. R. 7745.
Meggitt, W. F. 12195, 18594.
13285. Meginnis, H. G. 1937: Sulphuric acid treatment to increase germination of black locust seed. U.S. Dept. Agr. Circ. 453. 35 pp.
13286. —— 1938: Effect of depth of sowing on nursery yields of black locust. J. Forestry 36:411–16.
13287. Meheriuk, M., and M. Spencer. 1964: Ethylene production during germination of oat seeds and *Penicillium digitatum* spores. Can. J. Botany 42:337–40.
13288. Mehner, H., and W. Franke. 1960: The effect of varying carbon dioxide content of the atmosphere during the germination of the seeds of grains upon the formation of vitamin C. Planta 5: 604–10. Chem. Abstr. 54:16651. 1960.
Mehra, K. L. 10209.
13289. Mehra, P. N., and C. K. Atal. 1961: Studies in the mucilage yielding seeds. I. Nutlet structure and mucilage formation in *Lallemantia royleana* Benth., *Ocimum basilicum* Linn. and *O. canum* Sims. Res. Bull. Panjab Univ. 12:169–82.
13290. Mehta, B. V., and R. S. Desai. 1958: Effect of soil salinity on germination of some seeds. J. Soil Water Conserv. India 6:169–76.
Mehta, B. V. 17256.
13291. Mehta, C. R. 1939: Chemical investigation of the seed-oil of *Oroxylum indicum* Vent. Proc. Indian Acad. Sci. Sect. A 9:390–95. Biol. Abstr. 14:3568. 1940.
13292. —— ,Y. P. Datta, and N. G. Rana. 1963: Studies of the seeds of *Aristolochia bracteata*. Indian J. Pharm. 25:185–87. Chem. Abstr. 59:6722–23.
13293. —— and T. P. Mehta. 1943: Baicalein from the seeds of *Oroxylum indicum* Vent. Current Sci. (India) 12: 274–75.
13294. —— and —— 1953: Tetuin, a glucoside from the seeds of *Oroxylum indicum* Vent. Current Sci. (India) 22:114.
Mehta, M. L. 17498.
13295. Mehta, R., and C. V. Ramakrishnan. 1957: Studies on guar seed oil (*Cymmopsis psoralioides*). J. Am. Oil Chemists Soc. 34:459.
Mehta, T. P. 13293, 13294.
13296. Meier, K. I. 1960: The embryology of *Nuphar luteum* Sm. Byul. Mosk. Obshchestva Ispytatelei Prirody Otd. Biol. 65:48–59. Biol. Abstr. 38:15564. 1962.
13297. —— 1962: The embryology of *Coronaria flos cuculi* (L.) A. Br. Botan. Zh. 47:847–52. Biol. Abstr. 41:3353. 1963.
13298. Meige, E. 1929: À propos de la maturité physiologique des semences. Actes Congr. Intern. Essais Semences 1928:288–90.
13299. Meiklejohn, J. 1939: Heterosis in *Lycopersicum* crosses in relation to seed weight. Nature 144:908.
13300. Meiners, J. P. 1956: Effect of method of inoculation on control of head smut

of Mountain brome by seed treatment. Plant Disease Reptr. 40:734–36.

Meiners, J. P. 6408.

Meinl, G. 2253.

13301. Meischke, D. 1936: Über den Einfluss der Strahlung auf Licht- und Dunkelkeimer. Jahrb. Wiss. Botan. 83: 359–405.

13302. Mejbaum-Katzenellenbogen, W., and B. Morawiecka. 1963: Albumin fraction of proteins of wheat seeds (*Triticum vulgare* L.) of the winter variety "Olza" during storage and germination period. Plant Breed., Acclimatiz. Seed Production 7: 117–25.

Mejia, M. G. 7279.

13303. Mekhanik, F. Y. 1960: The effect of mineral and organic substances on the passage of plants through the stage of vernalization. Agrobiologiya 1960(3):440–42. Biol. Abstr. 36:42308. 1961.

13304. —— 1962: Effect of redox properties of solutions absorbed by seeds on vernalization rate and extent of branching in ears. Soviet Plant Physiol. 8:257–61.

Melamed, R. M. 2562.

Melcher, N. C. 4412.

13305. Melchers, G. 1957: Zusammenhang der Vernalisation mit dem Photoperiodismus. Congr. Intern. Botan. Rappt. Commun. Deuxieme Ser. 8:205–10. 1954.

Melchers, G. 11761.

Melchers, L. E. 9896.

13306. Meletti, P., and F. D'Amato. 1961: The embryo transplantation technique in the study of endosperm relations in irradiated seeds. Intern. At. Energy Agency 1961:47–55.

13307. —— and A. Onnis. 1961: Variations in the characteristics of seeds in some Graminaceae including the phenomenon of "yellow berries" in *Triticum durum* Desf. Nuovo Giorn. Botan. Ital. 68:79–88.

13308. —— and —— 1962: Some investigations on the behavior of excised embryos of *Triticum durum* cv. "Capelli" coming from normal grains and yellow berries. Nuovo Giorn. Botan. Ital. 69:281–84.

13309. Melikian, A. P. 1964: Comparative anatomy of seedcoat in some representatives of Nymphaceae. Botan. Zh. 49: 432–36.

Melik-Sarkisyan, S. S. 11373, 11374.

13310. Mellander, O., P. E. Nilsson, and L. Strid. 1958: Amino-acid content of leguminous proteins as affected by genetic and nutritional factors. II. Arch. Mikrobiol. 29:108–11.

Mellander, O. 14501.

Melnichenko, V. F. 11470.

Melnik, J. D. 13443.

13311. Melnikov, A. N. 1937: Structure of the seed coat of Abyssinian durum wheats. Bull. Appl. Botany, Genet., Plant Breed. (USSR) Ser. A 21:63–66.

Melnikova, N. K. 11936.

13312. Melot, G. J. 1934: Effects of deuterium oxide on respiration of germinating seeds. Proc. Soc. Exptl. Biol. Med. 32: 79–83.

13313. Melville, R. 1940: Wild seeds as animal feeding stuffs. Nature 146:403.

13314. —— 1941: Cultivation and food value of the oil poppy. Florists Exchange Hort. Trade World 96(11):10.

13315. —— 1941: Germination of belladonna seeds. Pharm. J. 146:116. Chem. Abstr. 35:5152. 1941.

13316. —— 1947: The nutritive value of nuts. Chem. Ind. (London) 1947:304–6.

13317. —— and M. Pyke. 1947: The effect of specific variability and the environment on the vitamin C content of British rose hips. Proc. Linn. Soc. London 159:5–16. Biol. Abstr. 23:16864. 1949.

13318. ——, F. Wokes, and J. G. Organ. 1943: Apparent vitamin C in walnuts. Nature 152:447–48. Biol. Abstr. 18:7157. 1944.

Melville, R. 15785, 19794.

13319. Mena, D. D. 1954: Fitotoxicidad de cinco insecticidas aplicados al suelo en la germinación y primer crecimiento de cuatro cultivos. Acta Agron. (Palmira, Colombia) 4:175–202. Soils Fertilizers 18:447–48. 1955.

13320. Mendel, L. B., and H. B. Vickery. 1931: Nutritive properties of the seed of the tobacco plant (*Nicotiana tabacum*). Proc. Soc. Exptl. Biol. Med. 29:290–91.

Mendel, L. B. 14870.

13321. Mendes, A. J. T. 1941: Cytological observations in *Coffea*. VI. Embryo and endosperm development in *Coffea arabica* L. Am. J. Botany 28:784–89.

13322. Mendes, L. O. T., and C. M. Franco. 1940: Influencia do expurgo, com bisulfureto de carbono, na germinação de sementes de café (*Coffea arabica* L.). Bol. Tec. Inst. Agron. Campinas 71. 33 pp. Rev. Appl. Entomol. Ser. 1 A28:549. 1940.

Mendez, J. 1413.

13323. Mendibelso, A. L. 1957: Effectiveness of several fungicides for control of damping-off and seed decay of bean. Acta Agron. (Palmira, Colombia) 7: 141–63. Biol. Abstr. 32:28614. 1958.

13324. Mendicino, J. 1960: Sucrose phosphatase synthesis in wheat germ and green leaves. J. Biol. Chem. 235:3347.

13325. Mendiola, N. B. 1953: A suggestion for combined selection by gravity method and disinfection of rice seeds in the Philippines. Araneta J. Agr. 1:45–50B.

Mendiola, N. B. 4712.

13326. Menezes, J. B. F., Jr. 1958: The microscopical structure of oily and edible seeds. Rev. Inst. Adolfo Lutz 18:5–44. Biol. Abstr. 36:25121. 1961.

Menezes, O. B. de; *see* De Menezes, O. B.

Menon, K. N. 10287.

Menon, K. P. V. 5027.

Menon, K. V. 14215.

13327. Menon, V. K. N., and D. N. Rao. 1932: A new and cheap source of urease. Indian J. Med. Res. 19:1077–83. Ber. Ges. Physiol. 68:552. 1932.

13328. Mentzer, C. 1940: Ascorbic acid in the vegetable kingdom. I. Ascorbic acid in the germination of seeds. Bull. Soc. Chim. Biol. 22:444–57. Chem. Abstr. 35:3679. 1941.

Menzel, T. 6916.

13329. Mer, C. L. 1952: Immediate and after effects of carbon dioxide concentrations upon the growth of etiolated *Avena* seedlings. Nature 169:887.

13330. Mercer, R. D. 1959: The seed analyst and seed certification. Proc. Assoc. Offic. Seed Analysts 49:37–39.

Mercer, S. P. 10700, 12274–12276.

13331. Merchan, A. 1919: Informe sobre tratamiento electrico de semillas antes de la siembra. Rev. Agr. Com. Trab. 2:199–201. Botan. Abstr. 3:1251. 1920.

13332. Mercier, A. 1944: The areca and the betel. Bull. Mens. Soc. Linneenne Lyon 13:28–30, 46–48, 58–60. Chem. Abstr. 40:956. 1945.

13333. Mercier, C. A. 1919: The electrification of seeds. Sci. Am. 120:142–43. Botan. Abstr. 4:104. 1920.

13334. Mercier, F., and J. Vieu-Bonnafous. 1940: Hypoglycemic action of jambul seeds (*Syzigium jambolanum* Lam.). Compt. Rend. Soc. Biol. 133:150–52. Chem. Abstr. 34:4149. 1940.

13335. Mercier-Des-Rochettes, A. 1936: La conservation des blés à l'air libre. J. Agr. Prat. 65:509–12.

Meredith, W. O. S. 546, 551.

13336. Mergentime, M., and E. H. Wiegand. 1946: Low temperature characteristics of a pea proteinase. Fruit Products J. 26: 72–80, 89, 91. Biol. Abstr. 21:18270. 1947.

13337. Mericle, L. W., A. M. Eunus, and R. P. Mericle. 1956: Effects of maleic hydrazide on embryologic development. I. *Avena sativa*. Botan. Gaz. 117:142–47.

13338. —— and R. P. Mericle. 1957: Irradiation of developing plant embryos. I. Effects of external irradiation (X-rays) on barley embryogeny, germination, and subsequent seedling development. Am. J. Botany 44:747–56.

Mericle, R. P. 13337, 13338.

Merino, A. S. 16536.

13339. Merkenschlager, F. 1924: Keimungsphysiologische Probleme. Naturwiss. Landwirtsch. 1:1–57.

Merkenschlager, G. 9168.

13340. Merker, J. 1963: The formation of nicotine and nornicotine during germination of tobacco seeds. Ber. Inst. Tabakforsch. Dresden 10:7–26. Chem. Abstr. 60:4472. 1964.

Merkle, F. G. 5650.

13341. Merl, E. 1931/32: Eine Modifikation der Keimprüfungsmethode nach Jacobsen. Prakt. Bl. Pflanzenbau Pflanzenschutz 9: 188–92.

13342. Merl, T., and J. Daimer. 1921: Studien über Mehlkatalase. Z. Untersuch. Nahrungs- Genussm. 42:273–90. Chem. Abstr. 16:1470–71. 1922.

13343. Merlis, M. 1897: Mitteilungen aus dem agrikultur-chemischen Laboratorium des Polytechnikums in Zürich. XLIV. Über die Zusammensetzung der Samen und der etiolierten Keimpflanzen von *Lupinus angustifolius* L. Landwirtsch. Versuchssta. 48:419–54.

13344. Merola, A. 1950: Study of a peculiar seed apparatus with pneumatodic functions. Delpinoa 3:5–47.

Merrill, L. G. 5324.

Merrill, L. G., Jr. 17520.

13345. Merrill, S., Jr. 1943: Characteristics in the nursery of tung progenies from open-pollinated seed of 169 parent trees. Proc. Am. Soc. Hort. Sci. 43:149–54. Expt. Sta. Rec. 90:632. 1944.

13346. —— 1947: Germination of early-planted and late-planted tung seeds as affected by stratification and various seed treatments. Proc. Am. Soc. Hort. Sci. 49:151–57.

Merrill, S., Jr. 11799, 17273.

13347. Merritt, F. E. 1951: Starting lilies and other plants from seeds. North Am. Lily Soc. Lily Yearbook 4:32–34.

13348. Merry, J. 1937: Formation of periderm in the endosperm of *Crinum asiaticum*. Papers Mich. Acad. Sci. 22 (1936):159–64. Biol. Abstr. 11:19350. 1937.

13349. —— 1941: Studies on the embryo of *Hordeum sativum*. I. The development of the embryo. Bull. Torrey Botan. Club 68:585–98.

13350. —— 1942: Studies on the embryo of *Hordeum sativum*. II. The growth of the embryo in culture. Bull. Torrey Botan. Club 69:360–72.

13351. —— and D. R. Goddard. 1941: A respiratory study of barley grain and seedlings. Proc. Rochester Acad. Sci. 8:28–44.

Merry, J. 11832.

Mertz, E. T. 3005, 18593.

13352. Merz, K. W. 1932: Toxic constituents of the seed *Tephrosia vogelli*. Arch. Pharm. 270:362–63. Chem. Abstr. 26: 4605. 1932.

13353. Merz, W., and G. Schmidt. 1935: Toxic principles in the seed of *Tephrosia vogelii*. Arch. Pharm. 273:1–21. Chem. Abstr. 29:2533–34. 1935.

Merzylakova, R. 18721.

13354. Mes, M. G. 1959: Gibberellic acid and the chilling requirement of peach seeds. Nature 184:2034–35.

13355. Mesa-Bernal, D. 1956: El 2,3,5-trifenil, cloruro tetrasodico en la determinación de la germinabilidad de las semilla. Rev. Acad. Colombiana Cienc. Exact. Fis. Nat. 9:350–64.

13356. Meshcheryakov, D. P. 1957: Prolonged preservation of the germinating capacity of legume seeds. Izv. Timiryazev. Sel'skokhoz. Akad. 1957:46–62. Biol. Abstr. 35:35723. 1960.

13357. Mesnard, M. E. 1894: Researches on the localization of fatty oils during the formation of seeds and fruits. Bull. Soc. Botan. 41:114–20.

Messenger, T. H. 5048.

13358. Messeri, A. 1931: Ricerche embriologiche e cariologiche sopra i generi *Allium* e *Nothoscordon*. Nuovo Giorn. Botan. Ital. 38:409–41. J. Roy. Microscop. Soc. 52:73–74. 1932.

13359. Messeri, A. 1950: Some notes on the embryology and embryogeny of *Olea europaea* L. Nuovo Giorn. Botan. Ital. 57:149–69.

13360. Messiaen, C. M., and R. Lafon. 1957: Les champignons nuisables aux semis de mais. I. Organismes responsables et conditions d'infection. Ann. Epiphyties 8:111–26.

Messina, F. 1890.

13361. Mestichelli, C. 1955: Germinability of the seeds of *Morus alba* L. in the various months of the year. Ann. Sper. Agrar. (Rome) 9:865–70.

13362. Met, M. 1950: Étude de la respiration du blé au cours d'une germination normale et pendant le traitement de vernalisation. Rev. Gen. Botan. 57:428–51.

13363. Metcalf, W. 1949: Sprouting in cork oak. J. Forestry 47:526.

13364. Metcalfe, C. R. 1936: An interpretation of the morphology of the single cotyledon of *Ranunculus ficaria* based on embryology and seedling anatomy. Ann. Botany (London) 50:103–20.

13365. Metcalfe, D. S., S. C. Wiggans, and H. E. Thompson. 1956: Desiccant sprays for bromegrass seed harvest. Agron. J. 48:429.

Metcalfe, D. S. 563, 19674.

Metcalfe, T. P. 9976.

13366. Metsavainio, K. 1926: Plants retaining their seeds during the winter, from the vicinity of Oulu (Uleoborg), Finland. Ann. Soc. Zool.-Botan. Fennicae, Vanamo 3:166–232. Biol. Abstr. 8:10393. 1934.

13367. Metzer, R. B. 1960: The vitascope—an aid for rapid determination of viable seed with tetrazolium. Texas Agr. Expt. Sta. Misc. Publ. MP-433. 10 pp.

13368. —— 1961: Effects of pneumatic conveyor on seed viability. Texas Agr. Expt. Sta. Misc. Publ. MP-508. 10 pp.

13369. —— and L. C. Coffey. 1959: Estimates of germination of fumigated seed of various crops by tetrazolium staining. Agron. Abstr. 1959:77.

Metzer, R. B. 10649.

Metzger, J. E. 19467.

Metzger, J. F. 6864.

Metzler, E. 11243.

13370. Metzner, P. 1930: Über die Abgabe fluoreszierender Stoffe durch quellende Samen und Früchte. Biochem. Z. 224: 448–58.

13371. Meuli, L. J., and H. L. Shirley. 1937: The effect of seed origin on drought resistance of green ash in the prairie-plains states. J. Forestry 35:1060–62.

13372. Mewissen, D. J., J. Damblon, and Z. M. Bacq. 1959: Comparative sensitivity to radiation of seeds from a wild plant grown on uraniferous and non-uraniferous soils. Nature 183:1449.

Meyer, D. 12405.

Meyer, E. E. 19005.

Meyer, E. W. 10655.

13373. Meyer, J. R. 1945: The use of tomato juice in the preparation of a medium for the germination of orchid seeds. Am. Orchid Soc. Bull. 14:99–101. Biol. Abstr. 20:16217. 1946.

13374. —— 1946: Experiments showing the action of thiamin (vitamin B_1) on the germination and development of orchid seeds in asymbiotic media. Am. Orchid Soc. Bull. 14:505–9. Expt. Sta. Rec. 95: 485. 1946.

13375. —— 1948: Germinacao de sementes de orquideas em recipientes hermeticamente fechados. Biologico 14:13–15. Biol. Abstr. 23:5114. 1949.

13376. —— and A. Pelloux. 1948: Stimulating action of two plant growth hormones on the germination of seeds and the development of orchid seedlings in a sterile medium. Biologico 14:151–61.

13377. Meyer, K. 1928: Über die Bedeutung der graphischen Darstellung bei der Answertung von Samenkeimversuchen. Pflanzenbau 5:117–22.

13378. —— 1929: Die Einwirkung äuserer Wachstumsbedingungen auf das Keimverhalten von Getreide in Zuckerlösungen. J. Landwirtsch. 77:97–138. Biol. Abstr. 5:3964. 1931.

13379. —— 1929: Was kann die Keimprüfung in Zuckerlösungen (Saugkraftmessung im Keimlingsstadium) für die Untersuchung kulturpflanzenphysiologischer Probleme leisten? Pflanzenbau 6:112–15.

13380. —— 1936: Keimung und Entwicklung der Leinseide (*Cuscuta epilinum*). Natur Volk 66:626–32.

Meyer, K. 10070.

13381. Meyer, W. C. 1930: Dormancy and growth studies of the American lotus, *Nelumbo lutea*. Plant Physiol. 5:225–34.

13382. Meysakhovitch, J. 1936: Mechanization of the cotton seed disinfection with sulphuric acid. Bull. Plant Protect. Ser. III 8:89–96.

Mezhvinskaya, T. 7906.

Mezquita, M. 5889.

13383. Mezzadroli, G., and E. Vareton. 1929: The action of an oscillating metallic ring on germination of seed. Atti Reale Accad. Nazl. Lincei. 9:350–54. Ber. Wiss. Biol. 11:731. 1929.

13384. —— and —— 1929: The effect of the Wood rays on germination and growth. Atti Reale Accad. Nazl. Lincei. Ser. VI. 10:281–88. Ber. Wiss. Biol. 14:301. 1930.

13385. —— and —— 1929: Épreuves comparatives entre l'action exercée par les ondes electro-magnetiques ultra-courtes ($A = 2 - 3m$) et par le circuit oscillant Lakhovsky sur la germination des graines et sur l'accroissement des plantes. Rend. Reale Accad. Lincei 6(10). 289 pp. Proc. Intern. Seed Testing Assoc. 7:113. 1935.

13386. —— and —— 1930: Action des ondes electro-magnetiques ultra-courtes usr la faculté germinative des graines, sur l'accroissement des plantes et sur le développement des microorganismes. Monitore Tec. 36:170. Proc. Intern. Seed Testing Assoc. 7:112. 1935.

13387. —— and —— 1930: Effect of very short electro-magnetic waves on the catalase activity of seed. Zymol. Chim. Colloidi Zucch. 5:95–98. Atti Reale Accad. Nazl. Lincei. Ser. VI. 11:429–33. 1930. Ber. Wiss. Biol. 15:595. 1930. Ann. Biol. 35:489. 1931. Biol. Abstr. 6:12869. 1932.

13388. —— and —— 1930: The effect of radium on the germination of seeds. Atti Reale Accad. Nazl. Lincei. Ser. VI. 12:73–80. Ber. Wiss. Biol. 16:594. 1931.

13389. —— and —— 1930: Possibility of the practical application of short radiations for increasing the germination of seed and growth of plants. Chim. Ind. Paris Spec. No.:521–23. Chem. Abstr. 24:5336. 1930.

13390. —— and —— 1931: Azione di terre radioattive italiane sulla germinazione dei semi e sulla crescità delle piante. Atti Reale Accad. Nazl. Lincei 14:443–47. Ber. Wiss. Biol. 22:94. 1932.

13391. —— and —— 1931: Effect of short electromagnetic waves on the characteristics of the enzymes of seeds. Chim. Ind. Paris Spec. No.:837–42. Chem. Abstr. 25:3766. 1931.

Mezzetti, V. Bambacioni-; *see* Bambacioni-Mezzetti, V.

13392. Miakotina, G. V. 1957: Dispersal of pine and spruce seeds in soil and their viability. Vestn. Sel'skokhoz. Nauki 2(5): 112–15.

Micev, L. 15549.

13393. Michaelides, R. C. 1959: A review of seed production work in Cyprus. Countryman Nicosia 1959:6–9. Hort. Abstr. 20:82. 1960.

Michaelis, A. 16199.

13394. Michaelis, P. 1939: Keimstimmung und Plasmavererbung bei *Epilobium*. Jahrb. Wiss. Botan. 88:69–88.

13395. Michailova, L. B. 1936: Sur la question de la vernalisation du chou. Bull. Acad. Sci. URSS Ser. Biol. 1936(1): 171–91. Bull. Soc. Botan. France 85:475. 1938.

13396. Míchalek, Z. 1950: The changes in the amount of catalase in seeds, depending on time of germination and on the concentration of hydrogen ions. Roczniki Panstwowego Zakladu Hig. 1:354–63. Chem. Abstr. 46:9671. 1952.

Michaud, L. J. 13897.

13397. Micheels, H. 1894: Relation between size of embryo and size and weight of grain of *Triticum vulgare*. Min. Agr. Belg. Bull. 10:96–102. Expt. Sta. Rec. 6:294. 1895.

13398. —— 1913: Action des solutions anodisées et cathodisées sur la germination. Bull. Classe Sci. Acad. Roy. Belg. 9/10: 831–87. Botan. Centralbl. 126:12. 1914.

13399. —— 1914: The influence of chlorids and nitrates of potassium and sodium on germinating plants. Intern. Z. Phys. Chem. Biol. 1:412–19. Expt. Sta. Rec. 33:128. 1915.

13400. —— and P. De Heen. 1905: Contribution à l'étude de l'influence de l'électrode sur les graines en germination. Bull. Classe Sci. Acad. Roy. Belg. 8:394–95. Botan. Centralbl. 101:168–69. 1906.

Micheels, H. 16937.

Michel, K. H. 16702.

13401. Michel-Durand, E. 1936: Sur le metabolisme des composes phosphorés des glands du chêne au course de la germination dans l'obscurité. Compt. Rend. 202:503–5.

13402. —— 1936: Metabolism of phosphorus compounds in acorns during germination in light. Compt. Rend. 202: 866–67. Chem. Abstr. 30:4539. 1936.

13403. Michev, B. 1957: Determination of soil germinative capacity of honey locust seeds (*Gleditsia triocanthos*) under different methods of preplanting preparation. Gorsko Stopanstvo 13:83–89.

13404. Michiels. 1929: Les graines de *Picralima klaineana* et leur toxicité. Bull. Acad. Roy. Med. Belg. 9:21–25. Biol. Abstr. 5:20197. 1931.

13405. Michniewicz, M., and J. Stanislawski. 1962: The influence of auxin and gibberellin on catalase activity in wheat during the period of germination and early growth. Acta Agrobotan. (Warsaw) 11:167–75.

13406. Micke, A. 1957: Über die Auslösung isotomer Sprossgabelungen bei *Melilotus albus* durch Röntgenbestrahlung der Samen Angew. Botan. 31:106–16.

13407. —— and K. Wöhrmann. 1960: On the problem of the radiation sensitivity of dry seed. Atompraxis 6:308–16. Nucl. Sci. Abstr. 14:2947. 1960.

Mickelson, R. H. 3611.

Miclea, C. 2924.

13408. Miczynski, K. 1937: Some observations upon the variability of the coleoptile nervation in wheats. Z. Zücht. Reihe A, Pflanzenzücht. 21:466–71. Biol. Abstr. 12:4043. 1938.

13409. Middendorf, F. G. 1938: Cytology of dormancy in *Phaseolus* and *Zea*. Botan. Gaz. 100:485–99.

13410. Middleton, G. K. 1933: Hard seed in Korean *Lespedeza*. J. Am. Soc. Agron. 25:119–22.

13411. —— 1933: Size of Korean *Lespedeza* seed in relation to germination and hard seed. J. Am. Soc. Agron. 25:173–77.

13412. ——, C. E. Bode, and B. B. Bayles. 1954: A comparison of the quantity and quality of protein in certain varieties of soft wheat. Agron. J. 46:500–2.

13413. ——, F. A. Coffman, J. G. Moseman, and F. J. Bell. 1954: Protein content of certain fall-sown oat varieties. Agron. J. 46:282–84.

13414. —— and P. H. Harvey. 1939: Studies of the fruiting habit of the peanut. Elisha Mitchell Sci. Soc. J. 55:242.

13415. ——, T. T. Hebert, and C. F. Murphy. 1964: Effect of seeding rate and row width on yield and components of yield in winter barley. Agron. J. 56:307–8.

13416. ——, D. A. Reid, T. T. Hebert, and W. H. Davis. 1961: Varietal differences in protein content of barley. Agron. J. 53:206.

Middleton, G. K. 5037.

13417. Middleton, J. T. 1944: Seed transmission of squash-mosaic virus. Phytopathology 34:405–10.

13418. Midgley, A. R. 1926: Effect of alternate freezing and thawing on the impermeability of alfalfa and dodder seeds. J. Am. Soc. Agron. 18:1087–98.

Miears, R. J. 18118.

13419. Miège, E. 1926: Sur la maturité comparée des céréales. Rev. Botan. Appl. Agr. Coloniale 6:429–38. Biol. Abstr. 5:24602. 1931.

13420. —— 1929: À propos de la maturité physiologique des semences. Actes Congr. Intern. Essais Semences 5:288–90.

13421. —— 1929: Contribution à l'étude du cotonnier au Maroc. Rev. Botan. Appl. Agr. Coloniale 9:169–76. Biol. Abstr. 5:24603. 1931.

13422. —— 1932: Sur l'anabiose et la réviviscence des plantules du blé. Compt. Rend. 195:1104–7. Biol. Abstr. 8:7942. 1934.

13423. —— 1940: L'heredite dé la composition chimique chez les hybrides intergénériques. Sci. Genet. (Torino) 2:82–91. Biol. Abstr. 21:5158. 1947.

13424. —— and G. Brodskis. 1939: Influence de la nature du sol sur la composition chimique du grain de blé. Compt. Rend. 208:2102–4.

13425. Miège, J. 1950: Sur divers cas de polyembryonnie observés en Cote d'Ivoire. Conf. Intern. Africanistes Ouest. Compt. Rend. 1:486–95. 1945.

13426. —— and P. Assemien. 1962: Sur les dependences charnues des graines de *Borreria* (Rubiacees). Compt. Rend. 254: 337–39.

Miegroet, M. van; *see* Van Miegroet, M.

Miele, F. 7091.

Mieras, D. G. 12629.

13427. Mierzwinska, T., and E. Sojka. 1963: Breathing of ripening and germinating field bean seeds (*Vicia faba* L. ssp. *minor*). Howdowla Roslin Aklim. Nasiennictwo 7:79–88.

Mierzwinska, T. 7758.

13428. Migamoto, T. 1962: Effects of the seed treatment with (2-chloroethyl)-trimethylammonium chloride on the resistances to high and low pH values of soils in wheat seedlings. Naturwissenschaften 16:377. 1962.

13429. Migita, K. 1962: Influence of the nutrition on the seed quality and the flower buds formation of Sugi seedlings. J. Japan. Forestry Soc. 44:125–28.

13430. —— and A. Kawana. 1953: Germination of Sugi seeds (*Cryptomeria japonica* D. Don) in the hydrogen peroxide solution. J. Japan. Forestry Soc. 35:381–82.

13431. ——, ——, and M. Takahashi. 1956: Germination of the seeds of the Japanese red pine, *Pinus densiflora* S. et Z., in the water with various contents of oxygen. J. Japan. Forestry Soc. 38:105–6.

13432. ——, ——, and —— 1956: Absorption of water by the seeds of the Japanese red pine (*Pinus densiflora*) in the aqueous media with various concentrations of oxygen. J. Japan. Forestry Soc. 38:465–66.

Migita, K. 10634.

Miguel, M. C. de Sousa-Monteiro-Cerca-; *see* de Sousa.

Mihailescu-Firea, S. 18382.

13433. Mihailovic, M. L., M. Antic, and D. Hadzijev. 1962: Chemical investigation of wheat. I–II. Dynamics of various forms of phosphorus in wheat during its ontogenesis. Accumulation forms of phosphorus in ripe wheat grain. 1. Orthophosphate and total phosphorus. 2. Phytic acid phosphorus. Glasnik Hem. Drustva, Beograd 27:117–33, 135–55. Chem. Abstr. 59:13118. 1963.

13434. ——, ——, and —— 1962: Chemical investigation of wheat. III. Dynamics of various forms of phosphorus in wheat during its ontogenesis. 3. Ribonucleic acid in ripe grain of different varieties of wheat. Glasnik Hem. Drustva, Beograd 27:213–26. Chem. Abstr. 59:13118. 1963.

Mihalache, A. 18193.

Mihalco, M. 2775.

13435. Mihalyi, Z. 1936: La question de la provenance des pins silvestres employés pour les reboisements de la Hongrie envisagée sous l'angle du controle des semences. Erdeszeti Kiserl. 38:1–114. Proc. Intern. Seed Testing Assoc. 10: 657–58. 1938.

13436. Mihara, K. 1958: The sequence of formation of oil in castor seeds. I–II. J. Chem. Soc. Japan, Pure Chem. Sect. 79:7–12.

13437. —— 1958: The sequence of formation of oil in castor seeds. III–IV. J. Chem. Soc. Japan, Pure Chem. Sect. 79:195–99.

13438. —— 1959: The sequence of formation of oil in castor seeds. V–VI. J. Chem. Soc. Japan, Pure Chem. Sect. 80:639–43.

Mijll Dekker, L. P. van der; *see* Van der Mijll Dekker, L. P.

13439. Mikaelsen, K., I. P. Bjornseth, and H. Halvorsen. 1956: Experiments on the effects of neutron-irradiation on the respiration of barley seeds. I. Effects on growth and oxygen-uptake in barley seedlings of different varieties. Physiol. Plantarum 9:697–711.

Mikaelsen, K. 2516, 19571.

13440. Mikenas, G. S. 1957: Nitrogen substances of seeds of inbred lines of corn and of their hybrid progeny. Uch.

Zap. Kishinevsk. Univ. 28:71–82. Chem. Abstr. 52:16499. 1958.

Mikenas, G. S. 10816.

13441. Mikhailov, O. F. 1959: Potential capacities of cells of cotyledon and endosperm. Tr. Soveshch. po Morfogenezu Rast. 1:341–46.

13442. —— and T. Orav. 1959: Plants grown from wheat seeds with removed embryos and their morphological peculiarities. Izv. Akad. Nauk Est. SSR, Ser. Biol. 1959(4):279–89.

13443. Mikhalovskii, A. G., and J. D. Melnik. 1956: Speeding up the development and maturity of corn by means of vernalization of seed. Zemledelie 4(4): 91–93. Chem. Abstr. 50:11440. 1956.

13444. —— and G. V. Porutskii. 1951: Yarovization of cotton seeds in manganese salt solutions. Dokl. Akad. Nauk SSSR 78:587–89.

13445. —— and M. M. Sapilnyak. 1954: Treating seed with solutions of microelements prior to planting. Zemledelie 2(7):49–55. Chem. Abstr. 49:6386. 1954.

Mikhlin, S. 10736.

13446. Mikhrina, E. N. 1958: Investigation of free sugars of rye grain. Tr. Vses. Nauchn.-Issled. Inst. Spirt. i Likero-Vodochn. Prom. 1958(6):146–54. Chem. Abstr. 53:10398. 1959.

13447. Miki, S. 1956: Seed remains of Vitaceae in Japan. J. Inst. Polytech. Osaka City Univ. Ser. D 7:247–71.

13448. —— 1956: Endocarp remains of Alangiaceae, Cornaceae and Nyssaceae in Japan. J. Inst. Polytech. Osaka City Univ. Ser. D 7:275–95.

13449. —— and S. Sera. 1932: Oils of *Phaseolus mungo* L. var. *radiatus* Bak. produced in Manchuria. J. Agr. Chem. Soc. Japan 8:1313–19. Chem. Abstr. 27:2054. 1933.

13450. Miki, T. 1931: Biochemische Studien über den Reisembryo. I. Über die Fermente des Reisembryos. Okayama Igakkai Zasshi 43:1530–34. Ber. Wiss. Biol. 19: 820. 1932.

13451. —— 1933: The hexone bases in rice embryo. Arb. Med. Fak. Okayama 3:499–500, 501–4. Chem. Abstr. 27:3241. 1933.

Miki, T. 13627.

13452. Mikkelsen, D. S., and M. N. Sinah. 1961: Germination inhibition in *Oryza sativa* and control by preplanting soaking treatments. Crop Sci. 1:332–35.

Mikola, J. 5948.

13453. Mikolajczak, K. L., F. R. Earle, and I. A. Wolff. 1962: Search for new industrial oils. VI. Seed oils of the genus *Lesquerella*. J. Am. Oil Chemists Soc. 39:78–80.

13454. ——, ——, and —— 1963: The acetylenic acid in *Comandra pallida* and *Osyris alba* seed oils. J. Am. Oil Chemists Soc. 40:342–43.

Mikolajczak, K. L. 5745, 19741.

Mikolasek, F. 4003.

Mikoshiba, K. 8354.

13455. Mikshich, Y., and A. Rezhek. 1930: Grape-seed-oil of the direct-bearer "Noah." Phytosterol and some constants of the oil. Bull. Soc. Chim. Roy. Yougoslav. 1(2):29–31. Chem. Abstr. 25:4140. 1931.

Mikusch, J. D. von; *see* Von Mikusch, J. D.

13456. Milashevskii, V. 1932: The seeds of wild Cruciferae. Maslob.-Zhir. Delo 1932(10):46–47. Chem. Abstr. 27:3352. 1933.

Milberg, E. 11796.

13457. Milborrow, B. V. 1963: Penetration of seeds by acetone solutes. Nature 199:716–17.

13458. Milbrath, J. A. 1937: An indication of seed transmission of mosaic virus in tomato seed. Phytopathology 27:868–69.

13459. Mildbraed, J. 1954: Die Schausamen von *Paeonia corallina* Retz. Ber. Deut. Botan. Ges. 67:73–74.

13460. Milella, A. 1961: Observations on the germinating power of seeds of sour orange (*Citrus aurantium* L.). Studi Sassaresi Sezione III 8(1960):79–84.

13461. —— 1961: The germinating power of olive seeds (*Olea europaea* var. *oleaster* H.) in relation to various stages of ripening in the drupe. Studi Sassaresi Sezione III 8(1960):85–89.

Miles, B. J. 16294.

Miles, E. F. 18693.

13462. Miles, G. F. 1936: Shriveled seed wheat should be treated before planting to prevent crop losses. Agr. News Letter 4:18–20.

13463. —— 1938: New reasons and modern methods for treating seed grains. Agr. News Letter 6:113–16.

13464. —— 1964: 300 Years of chemical seed treatment. Agr. Chem. 1(7):22–25, 46.

13465. —— 1951: Life-guards for the seeds of grasses and clovers. Agr. News Letter 19:43–48.

13466. Miles, L. E. 1939: Effect of type and period of storage on cotton seed after treatment with organic mercury dusts. Phytopathology 29:986–91.

13467. —— 1941: Effect of storage of treated cotton seed in closely woven cotton bags. [Abstract.] Phytopathology 31:768–69.

Miles, P. 8703.

13468. Miles, S. R. 1937: The relation between the moisture content and the test weight of corn. J. Am. Soc. Agron. 29:412–18.

13469. —— 1962: Heterogeneity of seed lots. Proc. Intern. Seed Testing Assoc. 27:407–13.

13470. ——, A. S. Carter, and L. C. Shenberger. 1960: Sampling tolerances and significant differences for purity analyses. Proc. Intern. Seed Testing Assoc. 25: 102–20.

13471. —— and E. E. Remmenga. 1953: Relations of kernel, cob, and ear moisture in dent corn. Indiana Agr. Expt. Sta. Bull. 599. 42 pp.

13472. —— and A. J. Ullstrup. 1947/48: Effects of various degrees of maturity of seed corn on its performance. Indiana Agr. Expt. Sta. Ann. Rept. 61:7–8.

Miles, S. R. 4522, 5503.

Mileshko, Z. F. 2997, 2999.

Milgewska, W. 3391.

13473. Militzer, W., C. Ikeda, and E. Kneen. 1946: The preparation and properties of an amylase inhibitor of wheat. Arch. Biochem. 9:309–20.

13474. ——, ——, and —— 1946: The mode of action of an amylase inhibitor from wheat. Arch. Biochem. 9:321–29.

Milkowska, A. 17676.

13475. Millar, C. E., and J. F. Mitchell. 1927: Effect of rate and method of application of fertilizer on the germination of white beans. J. Am. Soc. Agron. 19: 270–79.

Millar, R. L. 8642–8644.

13476. Miller, A. C. 1910: A physiological study of germination of *Helianthus annuus*. Ann. Botany (London) 24:693–726.

13477. Miller, B. S., and P. Halton. 1961: The damage to wheat kernels caused by the wheat blossom midge (*Sitodiplosis mosellana*). J. Sci. Food Agr. 12:391–98.

13478. —— and E. Kneen. 1947: The amylase inhibitor of *Leoti sorghum*. Arch. Biochem. 15:251–64.

Miller, B. S. 10879, 16083.

Miller, C. B. 15526.

13479. Miller, C. D., and L. Louis. 1941: Chemical analyses and vitamin assays of *Macadamia* nuts. Food Res. 6:547–52.

13480. Miller, C. O. 1958: The relationship of the kinetin and red-light promotion of lettuce seed germination. Plant Physiol. 33:115–17.

Miller, D. 17298.

13481. Miller, D. F. 1958: Composition of cereal grains and forages. Natl. Acad. Sci.–Natl. Res. Council, Publ. 585. 663 pp.

13482. Miller, E. C. 1910: A physiological study of the germination of *Helianthus annuus*. Ann. Botany (London) 24:693–725.

13483. —— 1912: A physiological study of the germination of *Helianthus annuus*. II. The oily reserve. Ann. Botany (London) 26:889–901.

13484. Miller, E. R. 1929: The insoluble tyrosinase of the velvet bean seed coat. Plant Physiol. 4:507–17.

Miller, E. S. 9864, 9871.

13485. Miller, E. V., and J. R. Jabolonski. 1949: Physiological studies of the seeds of grapefruit (*Citrus paradisi*, MacFad.). Food Res. 14:492–98.

Miller, E. V. 8718.

13486. Miller, G. D., C. W. Deyoe, T. L. Walter, and F. W. Smith. 1964: Variations in protein levels in Kansas sorghum grain. Agron. J. 56:302–4.

13487. Miller, G. W. 1958: Properties of enolase in extracts from pea seed. Plant Physiol. 33:199–206.

13488. Miller, H. W., and P. E. Lemmon. 1943: Processing cones of Ponderosa pine to extract, dewing and clean the seed. J. Forestry 41:889–94.

Miller, J. 5826.

13489. Miller, J. C. 1931: A new method of growing cabbage seed. Proc. Am. Soc. Hort. Sci. 28:341–42.

13490. —— 1937: Inducing the sweet potato to bloom and set seed. J. Heredity 28:347–49.

Miller, J. C. 5797.

13491. Miller, J. G. 1963: Studies on duplicate purity tests. Proc. Intern. Seed. Testing Assoc. 28:27–38.

Miller, J. G. 9902–9904.

13492. Miller, J. H., and W. F. Crosier. 1936: Pathogenic associates of tomato seed: their prevalence, relation to field disease and elimination. Proc. Assoc. Offic. Seed Analysts 1936:108–11.

13493. —— and R. Grogan. 1942: Injury to

tomato seed in disinfection. Phytopathology 32:524–28.
Miller, J. H. 6262.
Miller, L. I. 17290, 18460.

13494. Miller, M. H. 1962: The inhibitory influence of water extracts of three organic materials on soybean germination and growth in soil and sand cultures. Can. J. Plant Sci. 42:262–69.

13495. Miller, P. A., and B. Brimhall. 1951: Factors influencing the oil and protein content of corn grain. Agron. J. 43: 305–11.

13496. ——, T. L. Hurst, and B. Brimhall. 1952: Relationships of lysine and niacin with the crude protein and certain protein components in corn grain. Agron. J. 44:343–45.
Miller, P. A. 17815.

13497. Miller, P. M., and M. B. Linn. 1957: Viability of cabbage, carrot, sweet corn and lima bean seed treated with thiram or copper oxinate and stored at three different temperatures. Plant Disease Reptr. 41: 308–11.

13498. Miller, P. R. 1950: Study seed protectants on peanuts. Agr. Chem. 5(6): 59, 61, 63, 91.
Miller, P. R. 17471.

13499. Miller, P. W., and F. P. McWhorter. 1948: The use of vapor-heat as a practical means of disinfecting seeds. Phytopathology 38:89–101.
Miller, P. W. 13263.

13500. Miller, R. C., L. W. Aurand, and W. R. Flach. 1950: Amino acids in high and low protein corn. Science 112:57–58.
Miller, R. C. 1525.

13501. Miller, R. W. 1934: Growing dahlias from seed. Flower Grower 21(3):131.

13502. —— 1963: The effect of mechanical and chemical seed treatment on plant and seed characteristics of spring oats, *Avena sativa*. Dissertation Abstr. 24:922.
Miller, R. W. 18972, 18973.

13503. Millers, A. 1959: Biological activity of radioactive phosphorus in the treatment of oat seeds before planting. Latvijas PSR Zinatnu Akad. Vestis 1959(7):105–11.

13504. Milligan, R., and P. Hayes. 1962: Longevity of *Lolium* seed under commercial storage conditions. II. Longevity of *Lolium* seed stored in Hessian and in polythene lined Hessian sacks. Proc. Intern. Seed Testing Assoc. 27:944–51.

13505. Millington, A. J. 1955: The protein content of some Australian wheat varieties. J. Australian Inst. Agr. Sci. 21:59–68.

13506. ——, G. H. Burvil, and B. O. Marsh. 1951: Soil salinity investigations. Salt tolerance, germination, and growth tests under controlled salinity conditions. J. Agr. W. Australia 28:198–210. Biol. Abstr. 26:22349. 1952.
Millington, A. J. 15815.

13507. Millner, E. A. 1939: Lilies from seed and scales in a Virginia garden. Am. Lily Yearbook 1939:55–61.
Millner, I. J. 14559.
Mills, H. O. 13789.
Mills, I. K. 13241.
Mills, J. M. 14679.

13508. Mills, K., and J. B. Kelley. 1960: Drying small grain and shelled corn. Kentucky Univ. Circ. 569. 6 pp.

13509. Millsaps, V. 1936: The structure and development of the seed of *Paulownia tomentosa* Steud. Elisha Mitchell Sci. Soc. J. 52:56–75.

13510. —— 1940: Structure and development of the seed of *Cynoglossum amabile* Stapf & Drumm. Elisha Mitchell Sci. Soc. J. 56:140–64.

13511. Milne, D. 1915: The vitality of seeds passed by cattle. Agr. J. India 10:353–69.

13512. Milne, D. L. 1956: Gall midges affecting seed production in clover. Ann. Appl. Biol. 44:669–70.

13513. Milner, M., C. M. Christensen, and W. F. Geddes. 1947: Grain storage studies. V. Chemical and microbiological studies on "sick" wheat. Cereal Chem. 24:23–38.

13514. ——, ——, and —— 1947: Grain storage studies. VI. Wheat respiration in relation to moisture content, mold growth, chemical deterioration, and heating. Cereal Chem. 24:182–99.

13515. ——, ——, and —— 1947: Grain storage studies. VII. Influence of certain mold inhibitors on respiration of moist wheat. Cereal Chem. 24:507–17.

13516. —— and W. F. Geddes. 1945: Grain storage studies. I. Influence of localized heating of soybeans on interseed air movements. Cereal Chem. 22:477–83.

13517. —— and —— 1945: Grain storage studies. II. The effect of aeration, temperature, and time on the respiration of soybeans containing excessive moisture. Cereal Chem. 22:484–501.

13518. —— and —— 1946: Grain storage studies. III. The relation between moisture content, mold growth, and respiration of soybeans. Cereal Chem. 23:225–47.

13519. —— and —— 1946: Grain storage studies. IV. Biological and chemical factors involved in the spontaneous heating of soybeans. Cereal Chem. 23: 449–70.

13520. ——, M. R. Lee, and R. Katz. 1952: Radiography applied to grain and seeds. Food Technol. 6:44–45.

13521. ——, H. L. Mitchell, J. A. Gooding, and H. H. Laude. 1951: Preliminary study of the influence of certain commercial and laboratory preservatives on the deterioration of damp sorghum grain and alfalfa hay. Kansas Agr. Expt. Sta. Progr. Rept. 6. 11 pp.

13522. —— and J. B. Thompson. 1954: Grain storage. Physical and chemical consequences of advanced spontaneous heating in stored soybeans. J. Agr. Food Chem. 2:303–9.

13523. ——, B. Warshowsky, I. W. Ternet, and W. F. Geddes. 1943: The viability, chemical composition and internal microflora of frost-damaged soybeans. Oil Soap 20:265–68. Chem. Abstr. 38:772. 1944.

Milner, M. 106, 3882, 4275, 11664, 12282–12286, 17298.

13524. Milner, R. T. 1952: New oilseeds for industry. Chemurg. Dig. 11(5):10–13.

13525. ——, J. E. Hubbard, and M. B. Wiele. 1945: Sunflower and safflower seeds and oils. Oil Soap 22:304–7. Expt. Sta. Rec. 95:611. 1946.

Milner, R. T. 5746.

13526. Milogradova, E. I. 1961: The physiology of the germination of knotweed seeds. Uzbeksk. Biol. Zh. 1961(4):43–48.

13527. Milov, V. M. 1958: Effect of ecological factors on the protein content of pea plant seeds. Vestn. Sel'skokhoz. Nauki 3(11):124–26. Chem. Abstr. 53: 10403. 1959.

13528. Milovskaya, V. F. 1958: The quality of grain amylases from several Ukrainian wheat varieties. Biokhim. Zerna, Sb. 1958(4):164–72. Chem. Abstr. 53:510–11. 1959.

13529. ——, G. D. Dombrovskii, and L. G. Atanas. 1955: Effects of some organic dyes on microflora of cereals and their activity. Mikrobiologiya 24:718–22. Chem. Abstr. 50:8815. 1956.

13530. Milsum, J. N. 1927: Hastening the germination of oil palm seeds. Malayan Agr. J. 15:82–84.

13531. Milthorpe, F. L. 1950: Changes in the drought resistance of wheat seedlings during germination. Ann. Botany (London) 14:79–89.

13532. —— and B. Horowitz. 1943: The effect of length of day and temperature on the flowering, seed production and growth of vegetables. Agr. Gaz. N.S. Wales 54:53–57. Biol. Abstr. 18:11731. 1944.

13533. Milthorpe, J., and R. N. Robertson. 1948: Heating in stored wheat. I. Respiration of dry grain, insect respiration, and temperature and moisture effects. Australian Council Sci. Ind. Res. Bull. 237:9–17.

13534. Miltimore, J. E., J. L. Mason, and C. B. W. Rogers. 1962: Increase in seed production from nitrogen fertilization of native beardless wheatgrass. Can. J. Plant Sci. 42:359–64.

13535. Milton, W. E. J. 1936: The buried viable seeds of enclosed and unenclosed hill land. Welsh Plant Breed. Sta. Ser. H (1930–35) 14:58–86.

13536. —— 1939: The occurrence of buried viable seeds in soils at different elevations and on a salt marsh. J. Ecol. 27:149–59.

13537. —— 1943: The buried viable-seed content of a midland calcareous clay soil. Empire J. Exptl. Agr. 11:155–67.

13538. —— 1948: The buried viable-seed content of upland soils in Montgomeryshire. Empire J. Exptl. Agr. 16:163–77.

Milton, W. E. J. 3974.

13539. Milton, W. T. 1950: Observations on germination. Iris Yearbook 1950:100–5.

13540. —— 1952: Germination of bearded iris seed. Iris Yearbook 1952:54–58.

13541. Mimault, J. 1957: Auxines et inhibiteurs de croissance des graines de cerisier (variété "Bigarreau Napoleon"). Rev. Gen. Botan. 64:25–29.

Mimault, J. 15403.

13542. Minabe, M. 1953: Effect of ultrasonic waves on farm crops. I. On germination and early growth. Proc. Crop Sci. Soc. Japan 21:205–6.

13543. —— and I. Naniwa. 1963: Studies on the soft-textured rice kernel. VIII. Kinds of soil texture and the three essential elements of fertilizer in the formation of soft-textured rice kernel. Proc. Crop Sci. Soc. Japan 32:97–100.

13544. —— and —— 1963: Studies on the soft-textured rice kernel. IX. Giving nitrogen as fertilizer and formation of soft-textured rice kernel. Proc. Crop Sci. Soc. Japan 32:101–4.

13545. ——, ——, and O. Senjo. 1963: Studies on the soft-textured rice kernel. VI. Changes of water content in the ripening process of rice kernel. Proc. Crop Sci. Soc. Japan 31:233–36.

13546. ——, O. Senjo, and Y. Nagata. 1962: Studies on the soft-textured rice kernel. V. Thickness of bran layers. Proc. Crop Sci. Soc. Japan 31:181–85.

Minakata, S. 9363.

13547. Minasyan, S. M. 1953: Changes of proteinaceous and nonproteinaceous nitrogen in different histological elements of apricot seeds in the process of embryogenesis. Izv. Akad. Nauk Arm. SSR Biol. i Sel'skokhoz. Nauki 6(9):45–55. Chem. Abstr. 49:1151. 1955.

13548. —— 1958: Some difference between apricot and peach seeds at the different stages of the embryological development. Izv. Akad. Nauk Arm. SSR Biol. i Sel'skokhoz. Nauki 11(3):47–50. Chem. Abstr. 52:15649. 1958

13549. —— and B. A. Kostanian. 1954: Quantitative variations of reserve substances of tomato seeds developed under various methods of sexual reproduction. Izv. Akad. Nauk Arm. SSR Biol. i Sel'skokhoz. Nauki 7(8):51–58.

Minessy, F. A. 1662.

Miner, J. R. 7516.

13550. Minieri, L. 1954: Total ascorbic acid of some acorns of Campania. Zootec. Vet. 9:104–6. Chem. Abstr. 49:3440. 1955.

13551. Minin, D. D. 1957: Collection and storage of seeds of tree and shrub species. Moskva Gos. Izd. Sel'skokhoz. Literatury. 103 pp.

13552. Minina, E. G., and L. G. Tylkina. 1947: Physiological study of the effect of gases upon sex differentiation in plants. Compt. Rend. Acad. Sci. URSS 55:165–68. Chem. Abstr. 42:254. 1948.

Mino, M. 10369.

Minor, J. C. 8978.

13553. Minyaeva, O. M. 1951: Dissemination of clover anthracnose with seeds. Bull. Soc. Nat. Moscow N.S. 56:91–95. Rev. Appl. Mycol. 32:318. 1953.

Miranda, H. de; *see* De Miranda, H.

13554. Miravalle, R. J. 1962: Action of the genes controlling the character glandless seed in cotton. Crop Sci. 2:447.

13555. Mircetich, S. M., and G. A. Zentmyer, 1964: Rhizoctonia seed and root rot of avocado. Phytopathology 54:211–13.

13556. Mirianashvili, E. I. 1959: Structural variation accompanying development of seed embryo of the tea plant. Tr. Soveshch. po Morfogenezu Rast. 1:366–70.

13557. Miric, M. O., and A. F. Damanski. 1961: Contribution to the knowledge of the composition of fatty oils in corn grain (*Zea mays* L.) during the period of vegetation and germination. Acta Pharm. Jugoslav. 11:151–58. Biol. Abstr. 38: 19966. 1962.

Mironenko, A. V. 7351.

13558. Mirov, N. T. 1936: Germination behavior of some California plants. Ecology 17:667–72.

13559. —— 1936: A note on germination methods for coniferous species. J. Forestry 34:719–23.

13560. —— 1936/37: The relation of some internal factors to the germination of seeds of *Pinus jeffreyi* Murr. and *Pinus lambertiana* Dougl. Ph.D. Dissertation, Univ. of California.

13561. —— 1943: Storage and germination of California cork-oak acorns. U.S. Dept. Agr. Forest Serv. (Berkeley, Calif. Forest Range Expt. Sta.) Res. Note 36. 18 pp.

13562. —— 1944: Possible relation of linolenic acid to the longevity and germination of pine seed. Nature 154:218–19.

13563. —— 1946: Viability of pine seed after prolonged cold storage. J. Forestry 44:193–95.

13564. —— 1951: *Simmondsia* or Jojoba—a problem in economic botany. Econ. Botany 6:41–47.

13565. —— and C. J. Kraebel. 1937: Collecting and propagating the seeds of California wild plants. U.S. Dept. Agr. Forest Serv. (Berkeley, Calif. Forest Range Expt. Sta.) Res. Note 18. 27 pp.

Mirto, G. 1814.

Misḥanec, W. 12935.

Mishra, D. 13574, 13658.

13566. Mishustin, E. N. 1932: The principle of the bacteriological method for determination of seeds' origin. Centralbl. Bakt. Abt. II 86:193–201.

13567. ——, A. N. Petrova, and I. M. Karashchuk. 1955: Epiphytic microflora of sainfoin seeds and increasing its productivity. Izv. Akad. Nauk SSSR Ser. Biol. 2:8–18.

13568. Misnik, G. E. 1951: Effect of soaking on increasing germination in the ground of white mulberry and fleecy ash tree seeds. Lesn. Khoz. 4(12):75–76.

13569. Misonoo, G. 1935: Studies on the development and the germination in immature rice kernels. Sapporo Soc. Agr. Forestry J. 26:435–48.

Misonoo, G. 18435.

13570. Misra, A. L., and J. D. Tewari. 1951: Chemical examination of *Ipomoea muricata* seeds. J. Indian Chem. Soc. 28:721–26. Chem. Abstr. 46:8725. 1952.

13571. —— and —— 1952: Chemical examination of seeds of *Ipomoea muricata*. II. J. Indian Chem. Soc. 29:63–67. Biol. Abstr. 27:2356. 1953.

13572. —— and —— 1952: Chemical examination of seeds of *Ipomoea muricata*. III. J. Indian Chem. Soc. 29:430–33. Chem. Abstr. 48:12251–52. 1954.

13573. Misra, G. 1961: The constituents of ambrette (*Hibiscus abelmoschus*) seed. Soap, Perfumery Cosmetics 34:761–62.

13574. —— and D. Mishra. 1958: Effect of pre-soaking seed treatment with 2-methylbenzimidazole on rice seedlings. Current Sci. (India) 27:356.

13575. ——, C. R. Mitra, and K. N. Kaul. 1962: Chemical examination of the *Citrullus colocynthis* fruit and seed. J. Sci. Ind. Res. (India) 21B:238–39. Biol. Abstr. 40:8137. 1962.

13576. —— and S. N. Patnaik. 1959: Effect of coumarin on the germination and growth of rice seedlings. Nature 183:989–90.

13577. —— and G. Sahu. 1957: After-effects of treatment of grains of an early rice with plant growth substances. Current Sci. (India) 26:258.

13578. Misra, K. C. 1939: A contribution to the embryology of the Verbenaceae. Proc. Indian Acad. Sci. Sect. B 9:49–56.

13579. —— 1948: Polyembryony in *Dichanthium annulatum* Stapf. Current Sci. (India) 17:91.

13580. Misra, R. C. 1962: Contribution to the embryology of *Arabidopsis thalianum* (Gay and Monn.) (Cruciferae). Agra Univ. J. Res. 11:191–99. Biol. Abstr. 45:3129. 1964.

13581. Misra, R. N., and S. Dutt. 1937: Chemical examination of the seeds of *Cleome pentaphylla* Linn. I. Isolation of cleomin. Proc. Natl. Inst. Sci. India 3:45–49. Chem. Abstr. 32:2137. 1938.

13582. Misro, B. 1954: Breaking the dormancy in seeds of turmeric (*Curcuma longa*). Current Sci. (India) 23:207.

13583. Mistric, W. J., and E. J. Spyhalski. 1959: Response of cotton and cotton pests to Thimet seed treatment. J. Econ. Entomol. 52:807–11.

Mitchell, A. E. 5465, 12399.

13584. Mitchell, B. A. 1961: Pre-germination of tree seeds in polythene bags. Malayan Forester 24:142–45.

13585. Mitchell, E. M. 1926: Germination of seeds of plants native to Dutchess County, N.Y. Botan. Gaz. 81:108–12.

13586. —— 1930: A microchemical study of hemicelluloses of endosperms and cotyledons. Am. J. Botany 17:117–38.

13587. Mitchell, F. S. 1959: Efficient harvesting of herbage seeds. Agriculture (London) 66:122–25.

13588. Mitchell, H. H., J. R. Beadles, B. Koehler, and H. G. Dungan. 1947: The impairment in nutritive value of corn grain damaged by *Nigrospora oryzae*. J. Animal Sci. 6:352–58. Biol. Abstr. 22:1933. 1948.

13589. ——, T. S. Hamilton, and J. R. Beadles. 1945: The importance of commercial processing for the protein value of food products. I. Soybean, coconut and sunflower seed. J. Nutr. 29:13–25. Chem. Abstr. 39:1941. 1945.

13590. ——, ——, and —— 1952: The relationship between the protein content of corn and the nutritional value of the protein. J. Nutr. 48:461–76.

13591. Mitchell, H. L. 1939: Correction for the influence of seed weight upon the yield. of Scots, Corsican and White pine seedlings. Black Rock Forest Bull. 9:14–39.

Mitchell, H. L. 13521.

Mitchell, J. E. 5249.

Mitchell, J. F. 13475.

13592. Mitchell, J. H., and D. B. Roderick. 1946: The protein and iron content of small grains. S. Carolina Agr. Expt. Sta. Ann. Rept. 58:75–79.

13593. Mitchell, J. W., D. P. Skaggs, and W. P. Anderson. 1951: Plant growth-stimulating hormones in immature bean seeds. Science 114:159–61.

13594. ——, H. M. Benedict, and A. G. Whiting. 1944: Growth, rubber storage, and seed production by guayule as affected by boron supply. Botan. Gaz. 106:148–57.

13595. —— and J. W. Brown. 1947: Relative sensitivity of dormant and germinating seeds to 2,4-D. Science 106:266–67.

13596. ——, B. D. Ezell, and M. S. Wilcox. 1949: Effect of p-chlorophenoxyacetic acid on the vitamin C content of snap beans following harvest. Science 109:202–3.

13597. —— and P. C. Marth. 1946: Germination of seeds in soil containing 2,4-dichlorophenoxyacetic acid. Botan. Gaz. 107:408–16.

13598. ——, A. G. Whiting, and H. M. Benedict. 1944: Effect of light intensity and nutrient supply on growth and production of rubber and seeds by guayule. Botan. Gaz. 106:83–95. Chem. Abstr. 39:107. 1945.

13599. ——, ——, and —— 1945: Rubber content, stem anatomy, and seed production as related to rate of vegetative growth in guayule. Botan. Gaz. 106:341–55.

Mitchell, J. W. 12141, 17386, 19625.

Mitchell, R. S. 13837.

13600. Mitchell, W. 1947: Biosynthesis of the Belladonna alkaloids. Nature 159:196. 1947.

Mitchner, J. A., Jr. 17652.

13601. Mitidieri, J., and E. C. Ferraz. 1961: Anatomy of okra seed integuments. Port. Missao Estud. Agron. Ultramar Estud. Agron. 2:137–45.

13602. —— and —— 1962: Viability of seeds of *Hibiscus esculentus* L. before ripening of the fruit. Rev. Agr. (Piracicaba) 37:17–19.

Mitidieri, J. 7869

13603. Mitiska, L. J. 1954: The propagation of *Taxus* by seeds. Plant Propagators Soc. Proc. 4:69–75.

Mitra, C. R. 13575.

13604. Mitra, G. N. 1961: A study of the effect of IAA and TIBA on vernalized paddy. Current Sci. (India) 30:395.

13605. —— 1962: Inheritance of grain size in rice. Current Sci. 31:105–6.

Mitra, G. N. 17228.

13606. Mitra, J. 1947: A contribution to the embryology of some Compositae. J. Indian Botan. Soc. 26:105–23. Biol. Abstr. 22:12016. 1948.

13607. —— and C. Datta. 1951: Nutritional malformations in excised embryos of jute. Sci. Cult. (Calcutta) 16:531. Biol. Abstr. 25:34822. 1951.

13608. Mitra, J. N., and R. M. Dutta. 1949: A note on the development of embryo-sac and endosperm in *Alangium chinense* (Lour.) Harms. Bull. Botan. Soc. Bengal 3:11–15. Biol. Abstr. 27:4540. 1953.

Mitra, M. N. 3767.

13609. Mitra, P., and S. Banerjee. 1958: Effect of germination on strepogenin contents of pulses. Indian J. Med. Res. 46:492–95. Chem. Abstr. 52:18695. 1958.

Mitra, P. 1847.

Mitrovic, A. 15141.

Mitsuzawa, T. 6826.

13610. Miura, S., Y. Kaneki, and K. Nishiyama. 1960: The effect of electrolyzed water on the growth of crops. II. On the growth of upland rice seed presoaked into electrolyzed water. J. Agr. Sci. (Setagaya) 6:128–33.

13611. Miura, W., and K. Yoshimura. 1952: The effect of heavy water on the growth of crops. Proc. Crop Sci. Soc. Japan 21:104–5.

Miwa, G. C. 13613.

13612. Miwa, H., and K. Saito. 1939/40: The effect of ultra short waves on the germination of plants (mungo beans). Nagoya J. Med. Sci. 13:351–75. Chem. Abstr. 45:4313. 1951.

13613. Miwa, T. K., F. R. Earle, G. C. Miwa, and I. A. Wolff. 1963: Fatty acid composition of maturing *Veronia anthelmintica* (L.) Willd. seeds. Dihydroxyoleic acid—a possible precursor of epoxyoleic acid. J. Am. Oil Chemists Soc. 40:225–29.

13614. —— and I. A. Wolff. 1962: Fatty acids, fatty alcohols, and wax esters from *Limnanthes douglasii* (Meadowfoam) seed oil. J. Am. Oil Chemists Soc. 39:320–22.

Miwa, T. K. 19740.

13615. Mixon, A. C. 1963: Effect of seed size on vigor and yield of runner peanuts. Alabama Agr. Expt. Sta. Bull. 346. 18 pp.

Miyagawa, I. 19908, 19909.

13616. Miyagi, K. 1954: Inhibition of germination of the cruciferous vegetable seeds inundated with flooded water. Kyushu Agr. Res. 14:146–48.

13617. —— 1955: Experiments on carrot seed growing. I. Effects of pruning and rainfall on yield of carrot seed and its germinability. J. Hort. Assoc. Japan 24:254–60.

Miyahara, M. 1306–1309.

13618. Miyakawa, S., and M. Miyamoto. 1953: A study on the irregularity of seeding. J. Japan. Forestry Soc. 35:323–25.

13619. Miyamichi, E., and O. Onishi. 1932: Über die Saponin-glykoside in Adzuki-Bohnen (*Phaseolus radiatus* L. var. *aureus*, Prain.). J. Pharm. Soc. Japan 52:33–35.

Miyamoto, M. 13618.

13620. Miyamoto, T. 1957: The germination inhibitor in sugar beet seed balls. Mich. Agr. Expt. Sta. Quart. Bull. 39:518–23.

13621. —— 1963: The effect of seed treatment with the extracts of organisms and the solutions of some chemical substances on the resistance to salt concentration in wheat seedlings. Physiol. Plantarum 16: 333–36.

13622. —— 1963: Increase of the heat resistance by the seed treatment with 2-chloroethanol in the wheat seedlings. Naturwissenschaften 50:310–11.

13623. —— and S. T. Dexter. 1960: Acceleration of early growth of sugar beet seedlings by coating of seedballs with hydrophilic colloids and nutrients. Agron. J. 52:269–71.

13624. —— and E. H. Everson. 1958: Biochemical and physiological studies of wheat seed pigmentation. Agron. J. 50:733–34.

13625. ——, N. E. Tolbert, and E. H. Everson. 1961: Germination inhibitors related to dormancy in wheat seeds. Plant Physiol. 36:739–46.

Miyamoto, T. 5318, 12726.

13626. Miyatake, K., and A. Okano. 1955: Studies on the constituents of *Digitalis purpurea* seeds. J. Pharm. Soc. Japan 75:25–27.

13627. ——, ——, K. Hoji, T. Miki, and A. Sakashita. 1961: Studies on the constituents of *Digitalis purpurea* L. XXII. Alloneogitostin, a new glycoside from digitalis seeds. Chem. Pharm. Bull. (Tokyo) 9:519–23. Biol. Abstr. 41:20397. 1963.

13628. Miyazaki, Y., and K. Godaishi. 1961: The germination of the seeds of rauwolfias, especially of *R. serpentina* Benth. II. With reference to the seeds of *R. serpentina* Benth. collected in Japan. Bull. Natl. Inst. Hyg. Sci. 79:267–70. Biol. Abstr. 39:7768. 1962.

13629. —— and —— 1961: The germination of the seed of rauwolfias, especially of *R. serpentina* Benth. III. With reference to seeds of *R. canescens* L., *R. verticillata* Baill., and *R. perakensis* King and Gamble collected in Japan. Bull. Natl. Inst. Hyg. Sci. 79:271–73. Biol. Abstr. 39:7769. 1962.

13630. —— and —— 1962: The germination of the seeds of *Rauwolfia*, especially of *R. serpentina* Benth. IV. The effect of the treatments given to the seeds of *R. serpentina* Benth. prior to sowing on their germination. Bull. Natl. Inst. Hyg. Sci. 80:149–51. Biol. Abstr. 45:3442. 1964.

Miyazawa, Y. 14716.

13631. Mizuno, S. 1961: Physiological studies on the fructification of peanut. V. Changes of pH value and relations among Ca, Mg, and K in the shell. VI. On the relation between the behaviour of carbohydrates and some related enzymes in the shell under Ca deficient condition. Proc. Crop Sci. Soc. Japan 29:263–66.

13632. —— 1962: Physiological studies on fructification of peanut. VIII. On the relation between the calcium in the fruiting zone and the formation of unfilled fruit. Proc. Crop Sci. Soc. Japan 31:175–80.

13633. Mizuno, T. 1958: Carbohydrates of the rapeseed. J. Agr. Chem. Soc. Japan (English Transl.) 32:340–43. Chem. Abstr. 54:19874. 1960.

13634. —— and T. Kinpyo. 1956: Free sugars and starch grains in tea plant seed. Nosan Kako Gijutsu Kenkyu Kaishi 3:207–9. Chem. Abstr. 53:10404. 1959.

Mizuno, T. 8437.

13635. Mizushima, U., K. Murakami, and Y. Hozyo. 1956: Influence of anomalous environmental conditions upon embryogenesis in the rice plant, *Oryza sativa* L. I. Histochemical observation of normal embryogenesis. Tohoku J. Agr. Res. 6:1–19.

13636. ——, ——, and —— 1956: Influence of anomalous environmental conditions upon embryogenesis in the rice plant *Oryza sativa* L. II. Abnormal development induced by chemical substances. Tohoku J. Agr. Res. 6:179–202. Chem. Abstr. 50:13182. 1956.

Mizuta, T. 11119.

Mjasdrikowa, M. N. 5520, 5644, 5648.

13637. Mladentseva, M. S. 1960: An embryological study of hop (*Humulus* sp.). Dokl. Mosk. Sel'skokhoz. Akad. K. A. Timiryazeva 59:171–77. Biol. Abstr. 41:20021. 1963.

13638. Mlodzianowska, D. 1962: The effect of some inorganic substances on germinating grain and on risings of maize at low temperatures. Howdowla Roslin Aklim. i Nasiennictwo 6:257–310.

13639. Moav, R., and D. R. Cameron. 1961: Chromosome complement dosage in relation to seed development of species hybrids in *Nicotiana*. Botan. Gaz. 123: 70–77.

13640. Modell, E. B. 1933: Rhododendrons from seed. Gardeners Chron. 37:297.

13641. Modilevski, I. 1944/45: On the embryology and cytology of wheat. IV. Observations on the wheat germinated from immature caryopses. Acad. Sci. RSS Ukr. Kiev J. Botan. 2(3/4):36–37.

13642. Modilevskii, J. S. 1950: Present-day status of problem on endosperm in angiospermous plants in connection with embryo seed and fruit formation. Izv. Akad. Nauk SSSR Ser. Biol. 1950(2):23–37.

13643. —— 1953: Question concerning

origin of embryo sac in Angiospermae. Botan. Zh. 38:31–43.

13644. —— and R. A. Beilis. 1938: On the embryology and cytology of the wheat plant. II. The stages of the maturing of the embryo and caryopsis, of their germination and of earing. Acad. Sci. RSS Ukr. Inst. Botan. J. 26/27:13–39.

13645. Modilewski, J. 1930: Neue Beiträge zur Polyembryonie von *Allium odorum*. Ber. Deut. Botan. Ges. 48:285–94.

13646. Modin, G. V. 1960: On the dispersal of the seeds of some weed plants by the hare, *Lepus europaeus*. Zool. Zh. 39:472–74.

13647. Modoran, I., and D. Teodoresco. 1960: Contribution to the study of factors influencing the germination of fruit tree seeds. Acad. Rep. Populare Romine Filiala Cluj, Studii Cercetari Agron. 11:79–86.

13648. Modry, A. 1913: Das Keimen von *Phaseolus*-Samen in der Frucht. Österr. Botan. Z. 63:450–52. Botan. Centralbl. 125:618. 1914.

13649. Möbius, M. 1931: Die Gestalt des Pflanzenembryos in Samen in Beziehung zur Systematik und Lebensweise (Conclusion). Naturforsch. 7:455–61.

Möller, E. F. 11488.

13650. Möller, K. 1936: The storage of grain. Newer experiences and methods. Kgl. Landtbruks-Akad. Handl. Tidskr. 75:353–70.

13651. Möschl, H. 1938: Clinical and pathological aspects of *Ricinus* seed poisoning. Z. Klin. Med. 133:78–90. Pharm. Abstr. 6:455. 1940.

13652. Moewus, F., L. Moewus, and H. Skwarra. 1952: Nachweis von zwei Wuchsstoffen in Samen und Wurzeln der Kresse (*Lepidium sativum*). Planta 40:254–64.

Moewus, F. 11488.

Moewus, L. 13652.

13653. Moffett, A. A. 1952: Differential germination in the black wattle (*Acacia mollissima* Willd.) caused by seed treatment. Wattle Res. Inst. Rept. 1951/52: 39–50.

13654. Mogilner, I. 1952: Nota sobre presencia de uninhibidor de la germinación y de saponinas en los frutos de paraiso. Rev. Fac. Agron. Univ. Buenos Aires 28:175–86.

Mohamed, M. S. 12709.

13655. Mohammad, A. 1930: Some investigations into the effect of fertilizers on the growth and oil-content of a variety of linseed (*Linum usitatissimum* L.). Agr. J. India 25:471–86. Expt. Sta. Rec. 65:529. 1931.

13656. ——, Z. Alam, and K. L. Khanna. 1933: Germination and growth in groundnut (*Arachis hypogaea* Linn.). Agr.-Livestock India 3:91–115. Chem. Abstr. 27:4618. 1933.

13657. —— and A. R. Khan. 1937: Diagnosis of oleiferous Brassicae seeds. Agr.-Livestock India 7:512–13.

Mohammad, A. 10494.

Mohammad, N. 8070.

13658. Mohanty, P. K., and D. Mishra. 1962: Postvernalization seed treatments with vitamins in *Vigna catjang*. Science 138:902–3.

Mohr, J. C. van der Meer; *see* Van der Meer Mohr, J. C.

13659. Mohr, K., F. Thudium, O. Schindler, and T. Reichstein. 1957: Die Glykoside der Samen von *Acokanthera schimperi* (A.DC.) Benthem & Hooker. I. Untersuchung der Pflanzen aus Erythräa. Helv. Chim. Acta 40:2199–215.

Mohr, K. 16188, 19204.

Mohr, W. 9091.

13660. Mohs, K., and E. Tornow. 1940: Germinating energy [of grains] and selenium reduction. Z. Ges. Getreide. 27:17–22. Soc. Chem. Ind. J. 59:483B. 1940.

Moiseev, R. K. 10862.

Mokhova, N. I. 7702, 7703, 9141.

Molaison, L. M. 10893.

13661. Molberg, E. S. 1961: Injurious effects of fertilizers applied with the seed on the emergence of flax. Can. J. Soil Sci. 41:35–43.

13662. Moldenhawer, K. 1959: The effect of various volatile substances upon germination of seeds of certain oleaginous plants species. Howdowla Roslin Aklim. i Nasiennictwo 3:217–27.

13663. Mole-Bajer, J. 1955: A simple method for examining the action of chemicals on mitosis in living endosperm. Acta Soc. Botan. Polon. 24:618–25.

Mole-Bajer, J. 1709, 1710.

13664. Molin, G. 1932: Determination of degree of germination in wheat and rye by means of the dipping refractometer. Mühlenlab. 1932:62–63. Chem. Abstr. 27:141. 1933.

Molina, C. L. Gonzalez-; *see* Gonzalez-Molina, C. L.

Molines, J. 5253.

13665. Molisch, H. 1922: Pflanzenphysiologie als Theorie der Gärtenerei. V. Gustav Fischer, Jena, pp. 289–302.

13666. Moll, F. 1961: The chemical nature of the alkaloid arachin of peanuts. Planta Med. 9:213–15.

13667. Mollard, J. 1955: Influence of infrared radiation treatment on glucide constituents of chestnuts. Ann. Inst. Natl. Rech. Agron. Ser. E 4:335–40. Chem. Abstr. 50:11434. 1956.

Mollish, A. P. 19846.

Molodtsov, V. A. 12942.

13668. Molotkovskii, G. K. 1955: The energy of sprouting and germination of acorns and walnuts as a function of their position in the soil. Dokl. Akad. Nauk SSR 102:637–39. Biol. Abstr. 32:6417. 1958.

13669. —— and B. G. Gluskina. 1953: Effect of wetting of corn seeds in warm water on its growth and yield. Dokl. Akad. Nauk SSSR 92:679–80. Chem. Abstr. 48:5297. 1954.

13670. —— and G. V. Poruskii. 1939: Hormonization of plants. Zh. Inst. Botan. Akad. Nauk URSR 20:179–88. Chem. Abstr. 35:5538. 1941.

Molotkovskii, G. K. 12431.

13671. Momot, I. G. 1933: The pigmentation of soybean seed. Bull. Appl. Botany, Genet., Plant Breed. USSR Ser. A 5/6:152–62. Biol. Abstr. 10:989. 1936.

13672. Momotani, Y. 1960: Studies on seed protein by means of turbidometric titration, especially in a group of *Brassica rapa* L. Botan. Mag. (Tokyo) 73:295–99.

13673. —— 1962: Taxonomic study of the genus *Acer*, with special reference to the seed proteins. III. System of Aceraceae. Mem. Coll. Sci. Kyoto Univ. Ser. B 29: 177–89.

13674. Monachino, J. 1953: Fruits of *Strophanthus sarmentosus*. Garden J. 3:159.

13675. Monaco, L. C. 1960: Influence of the empty seeds on the outturn of Mundo Novo coffee. Bragantia 19:1–12. Biol. Abstr. 35:60947. 1960.

13676. Mondain-Monval, O. D. 1963: Étude de l'action de la lumière sur la germination des graines de tomate (*Lycopersicum esculentum* Miller) variété marmande. Compt. Rend. 257:3646–48.

Mondini, A. 20063.

Mone, L. T. 4936.

Monge, J. A. Ferrer-. 19374.

13677. Monge, S. F. 1962: The frequency of peaberry coffee in plants from irradiated seeds. Turrialba 12:209–10.

Monguillon, P. 12070.

13678. Monhaupt, M. 1931: Determination of the fat content of oil seeds. Chem. Ztg. 55:70. Chem. Abstr. 25:2013. 1931.

Monin, J. 3315–3318.

13679. Monoyer, A. 1947: Technique de la culture aseptique des embryons d'Iris. Bull. Hort. (N.S.) 2:38–39.

13680. Monroe, C. G., and D. D. Hill. 1939: Methods for determining the percentage of seeds, strigs, stems, and leaves in commercial hops. J. Am. Soc. Agron. 31:698–702.

13681. Monselise, S. P. 1953: Viability tests with citrus seeds. Palestine J. Botany, Rehovot Ser. 8:152–57.

13682. —— 1959: Citrus germination and emergence as influenced by temperature and seed treatments. Bull. Res. Council Israel, Sect. D 7:29–34.

13683. —— and F. W. Went. 1958: Effects of temperature on growth and dry matter accumulation of peas. Plant Physiol. 33: 372–74.

13684. Monteil, L. 1936: Germinated oats in horse nutrition. Progr. Agr. Vit. 105: 521–26, 545–48. Chem. Abstr. 30:8318. 1936.

13685. Monteiro, D. G. d'N. 1882: Germination of *Welwitschia*. Gardeners Chron. 17:14.

Monteiro, H. J. 13791.

Monteiro-Cerca-Miquel, M. C. de Sousa; *see* De Sousa.

13686. Montemartini, L. 1934: Sull'utilizzazione e ripartizione delle riserve nei semi di *Vicia faba* L. Nuovo Giorn. Botan. Ital. 41:775.

13687. —— 1935: Preliminary researches on the relations between water content and latent life in seeds. R. Ist. Botan. Palermo Lab. 6:156–65. Expt. Sta. Rec. 76:17. 1937.

13688. —— 1936: Über Keimung und erste Entwicklung der Keimpflanzen der Dumpalme. Boll. Studi Inf. Giard. Colon. Palermo 14:3–5. Botan. Centralbl. 172:70. 1937.

13689. Montenegro, H. W. S., S. Moreira, F. P. Gomes, and G. Cintra. 1959: The effect of stock on the number of seeds in the orange. Anais Escola Super. Agr. "Luiz de Queiroz" Univ. São Paulo 16:80–85. Biol. Abstr. 36:57911. 1961.

Montenero, P. 2269.

13690. Montenez, J. 1957: Recherches experimentales sur l'écologie de la germination chez l'arachide. Bruxelles, Dir. Agr. Forêts Elevage. 124 pp.

13691. Montet, D. 1932: Action des faibles radioactivités sur la germination des graines. Compt. Rend. 194:304–6.

13692. —— 1932: Action de la radioactivité en physiologie végétale. Compt. Rend. 195:582–84.

13693. Montfort, C., and W. Brandrup. 1927: Physiologische und pflanzengeographische Seesalzwirkungen. II. Ökologische Studien über Keimung und erste Entwicklung bei Halophyten. Jahrb. Wiss. Botan. 66:902–46.

Montgomerie, I. G. 14542.

13694. Montgomery, R., and F. Smith. 1957: The carbohydrates of Gramineae. IX. The constitution of glucofructan of the endosperm of wheat (*Triticum vulgare*). J. Am. Chem. Soc. 7:446–50.

13695. Montrenko, T. G. 1929: A speedy method for the determination of the germinative power of seeds by means of staining, applied to gramineous plants and other field cultures. J. Agr. Res. N. Kaukasus 15/16:3–38.

Montreuil, J. 8887.

Monval, O. D. Mondain-; *see* Mondain-Monval, O. D.

13696. Mony, C. C. 1958: Seed pelleting process for bird and disease control. Tree Planters Notes 32:7–8.

13697. Monzoni, P., and L. H. P. Azevedo. 1948: Action of certain drugs on germination of seeds. Rev. Fac. Med. Vet. Univ. São Paulo 3:309–12. Chem. Abstr. 43: 2357. 1949.

Mookherjee, P. B. 15648.

13698. Moomaw, J. C., E. W. Tisdale, L. A. Sharp, and K. B. Platt. 1954: Studies of pelletized seed for range reseeding. Idaho Univ. Forest, Range Wildlife Expt. Sta. Res. Notes 11:1–19. Biol. Abstr. 28: 26971. 1954.

Moon, H. H. 4735.

13699. Moore, A. T., and M. L. Rollins. 1961: New information on the morphology of the gossypol pigment gland of cotton seed. J. Am. Oil Chemists Soc. 38:156–60.

Moore, A. T. 5352.

Moore, A. W. 13701.

13700. Moore, C. A., and C. W. Brown. 1963: Commercial storage and handling of sorghum grain. Texas Agr. Expt. Sta. Bull. 996. 14 pp.

Moore, C. A. 3121, 8780.

13701. Moore, C. E., and A. W. Moore. 1962: Periodic germination of seeds of *Albizzia julibrissin* Durazzini. J. Tenn. Acad. Sci. 37:17–18. Biol. Abstr. 38: 23543. 1962.

Moore, C. N. 8336.

13702. Moore, C. W. E. 1946: The effect of soil moisture content and depth of planting on the germination of dehulled seed of *Danthonia semiannularis*. Australian Council Sci. Ind. Res. J. 19:172–76.

13703. —— 1958: Germination and seedling root growth of *Bothriochloa ambigua* in relation to invasion of native pasture. Ecology 39:367–71.

13704. Moore, E. L. 1955: Some effects of certain growth-regulating substances upon premature seeding in cabbage. Miss. Agr. Expt. Sta. Tech. Bull. 40. 42 pp.

13705. Moore, G. M., and W. D. Hoffman. 1964: Dormancy of the seed of *Panicum ramosum* (Browntop millet). Assoc. Offic. Seed Analysts News Letter 38:24–26.

Moore, J. E. 18380.

13706. Moore, J. H. 1934: The value of single lock samples as a measure of seed purity in cotton. J. Am. Agron. 26:781–85.

13707. —— and J. A. Shankling. 1931: Source and care of cotton planting seed in relation to the length of staple. N. Carolina Agr. Expt. Sta. Tech. Bull. 42. 31 pp.

13708. Moore, L. A. 1938: Seed inspection. Illinois Dept. Agr. Ann. Rept. 1937/38: 118–36.

Moore, M. B. 4068, 11058.

Moore, R. C. 5826.

13709. Moore, R. J. 1946: Investigations on rubber-bearing plants. III. Development of normal and aborting seeds in *Asclepias syriaca* L. Can. J. Res. Sect. C 24:55–65.

13710. Moore, R. P. 1943: Seedling emergence of small-seeded legumes and grasses. J. Am. Soc. Agron. 35:370–81.

13711. —— 1956: Mechanically damaged seed seriously reduce crop. Southern Seedsman 19(11):44, 73.

13712. —— 1958: Tetrazolium tests for the determination of injury and viability of seed corn. Am. Seed Trade Assoc. Hybrid Corn Div. Rept. Hybrid Corn Ind. Res. Conf. 13:13–20.

13713. —— 1959: Mechanical injury, physiological aging, and seed quality. Intern. Crop Improvement Assoc. Ann. Rept. 38/39:187–92.

13714. —— 1959: Tetrazolium testing—its past accomplishments and its possibilities. Seed World 85(4):14–15.

13715. —— 1960: Vacuum treatment reveals weak seed. Crops & Soils 13(2):17.

13716. —— 1962: Tetrazolium as a universally acceptable quality test of viable seed. Proc. Intern. Seed Testing Assoc. 27:795–805.

13717. —— 1963: Natural crushing causes seed germination difficulties. Assoc. Offic. Seed Analysts News Letter 37(3):8–9.

13718. —— 1963: Previous history of seed lots and differential maintenance of seed viability and vigor in storage. Proc. Intern. Seed Testing Assoc. 28:691–99.

13719. —— 1964: Natural fracturing and crushing of snap bean seed. Assoc. Offic. Seed Analysts News Letter 38(1):14–16.

13720. —— 1964: Seed vigor or soundness and corn improvement. Seed World 94(4):6–10.

13721. —— and E. Smith. 1957: Seeds must be more than just alive. Crops & Soils 9(6):14–16, 41.

13722. —— and —— 1957: Tetrazolium testing. General technic. Assoc. Offic. Seed Analysts News Letter 31(2):33–38.

Moore, R. P. 4076–4078, 4365, 8135.

Moore, S. 5689.

13723. Moore, T. C. 1963: A germination inhibitor in achenes of *Cercocarpus montanus*. Ecology 44:406–9.

13724. —— and E. K. Bonde. 1958: Interaction of gibberellic acid and vernalization in the dwarf telephone pea. Physiol. Plantarum 11:752–59.

13725. Moore, W. C. 1946: Seed-borne diseases. Ann. Appl. Biol. 33:228–31. Rev. Appl. Mycol. 26:69–70.

13726. Moorjani, M. N., and D. S. Bhatia. 1954: Storage effects on the proteins of groundnuts. J. Sci. Ind. Res. (India) 13B:113–14. Chem. Abstr. 48:12339. 1954.

13727. Moormann, B. 1942: The dormancy of oat and barley seeds. Kühn-Arch. 56: 41–79. Chem. Abstr. 39:544. 1945.

13728. Moote, I. 1953: The effect of moisture on the thermal properties of wheat. Can. J. Technol. 31:57–69.

13729. Morada, E. K. 1924: The effect of sunlight on the germination of papaya seeds, *Carica papaya*. Philippine Agr. Rev. 17:21–23.

13730. Moran, T. 1945: Nutrients in wheat endosperm. Nature 155:205–6.

Moran, T. 2053, 4434, 8845.

13731. Morandini, P. 1961: Forest seed handling equipment and procedures. I. Seed production, collection and production. Unasylva 15:185–99.

13732. Morandini, R. 1958: The new law on control of forest seeds in Germany. Monti e Boschi 9:99–105.

13733. Morawiecka, B. 1961: Changes in proteins of yellow lupin (*Lupinus luteus* L.) during germination. Acta Biochim. Polon. 8:313–20.

Morawiecka, B. 13302.

13734. Moreau, A. P. 1933: Action de la châleur et de differents traitements chimiques sur la faculté germinative des graines de cotonnier. Coton Cult. Coton. 8:167–70. Biol. Abstr. 10:228. 1936.

13735. Moreaux, R. 1960: La dissemination des graines. Gaz. Apic 61(635):10–11.

13736. Moreira, S., J. T. A. Gurgel, and L. F. De Arruda. 1947: Polyembryony in citrus. Bragantia 7:69–106.

Moreira, S. 13689.

13737. Morel, R. 1948: Influence d'extraits aqueux de graines sur la germination. Ann. École Natl. Agr. Ser. 3 Grignon 6:17–21.

Morel, R. 7780.

13738. Moreland, C. F. 1934: Factors affecting the development of the cotyledonary buds of the common bean, *Phaseolus vulgaris*. Cornell Univ. Agr. Expt. Sta. Mem. 167:1–28. Biol. Abstr. 10:18006. 1936.

Moreland, C. F. 6499, 6500.

Moreno, C. L. 12819.

13739. Moretti, A. 1941: Yarovization of sugar beets. Ind. Saccar. Ital. 34:1–6. Chem. Abstr. 38:5264. 1944.

13740. Morettini, A. 1913: The effect on the germinative ability of seeds of *Orobanche crenata* by passing them through cattle and by fermentation of the feces. Staz. Sper. Agrar. Ital. 46:589–606. Expt. Sta. Rec. 31:634. 1914.

13741. Morey, D. 1956: The use of chemicals in breaking seed dormancy in hybrid roses. Am. Rose Ann. 41:64–69.

Morey, D. 15177.

13742. Morey, D. D., and J. Morey. 1959: Polyembryony in American rye varieties. Agron. J. 51:669–72.

Morey, J. 13742.

Morgan, A. F. 8028, 9796.

13743. Morgan, D. T., Jr., and R. D. Rappleye. 1949: Polyembryony in *Capsicum frutescens*. Am. J. Botany 36:800.

13744. —— and —— 1951: Polyembryony in maize and lily (*Lilium regale*) following

X-irradiation of the pollen. J. Heredity 42:91–93.
Morgan, E. J. 9029.
Morgan, W. M. 4196.

13745. Morgeneyer, W. 1956: Untersuchungen über die Beseitigung der Keimhemmung beim Samen des Hirschkolbensumachs (*Rhus typhina*). Arch. Forstw. 5:203–42.
Morgenstern, W. 2689.
Morgunova, E. A. 11380.
Morgunova, E. S. 7059.

13746. Mori, H. 1939: Studies on the storage and germination of seeds of fruit trees. I. Method for raising seedlings from non-after-ripened embryos of peach and plum. Hort. Assoc. Japan J. 10:405–8.
Mori, K. 14157.
Mori, M. 11572.

13747. Mori, S. 1944: Changes in seed globulin. J. Agr. Chem. Soc. Japan 20: 265–68. Chem. Abstr. 42:3098. 1948.
Mori, S. 11073–11075.

13748. Morice, I. M. 1962: Seed fats of the New Zealand Agavaceae. J. Sci. Food Agr. 13:666–69.
Morikawa, I. 7993

13749. Morinaga, T. 1925: Catalase activity and the aerobic and anaerobic germination of rice. Botan. Gaz. 79:73–84.

13750. Morinaga, T. 1926: Germination of seeds under water. Am. J. Botany 13:126–40.

13751. —— 1926: Effect of alternating temperatures upon the germination of seeds. Am. J. Botany 13:141–58. Contrib. Boyce Thompson Inst. 1:82–99. 1926.

13752. —— 1926: The favorable effect of reduced oxygen supply upon the germination of certain seeds. Am. J. Botany 13:159–66. Contrib. Boyce Thompson Inst. 1:100–7. 1926.

13753. Morishima, H., and H. I. Oka. 1959: Variation in seed dormancy of wild rice species. Natl. Inst. Genet. Ann. Rept. 10:55–56.
Morishima, H. 9587, 14118, 14124.

13754. Morishima, S., and K. Nakagawa. 1953: The simple methods for determination of the moisture contents of grains. Bull. Fac. Agr. Mie Univ 6:62–75. Chem. Abstr. 49:5703. 1955.
Morita, Y. 11076.

13755. Moriya, N., and F. Kanabu. 1954: On the dormancy of freshly harvested grass seeds. Bull. Natl. Inst. Agr. Sci. Ser. G 8:1–7.

13756. Moriyama, H. 1960: Studies on X-agent. III. Effect on the growth of seeds. Japan. J. Microbiol. 4:7–17. Biol. Abstr. 36:21528. 1961.

13757. Mork, E. 1938: Germination of spruce and pine seed at various temperatures and degrees of moisture. Medd. Norske Skogforsöks. 6:225–49. Biol. Abstr. 13:15667. 1939.

13758. —— 1949: Experiments on scarifying and sowing in pine forest in eastern Norway. Medd. Norske Skogforsöks. 10:225–304. Biol. Abstr. 26:9785. 1952.

13759. —— 1951: Factors influencing germination of seed from *Pinus silvestris*, *Picea abies*, *Betula verrucosa*, and *B. pubescens*. Medd. Norske Skogforsöks. 38:159–73. Biol. Abstr. 26:13146. 1952.

13760. Morley, F. H. W. 1958: The inheritance and ecological significance of seed dormancy in subterranean clover (*Trifolium subterraneum* L.). Australian J. Biol. Sci. 11:261–74.

13761. —— and L. T. Evans. 1959: Flower initiation in *Trifolium subterraneum* L. II. Limitations by vernalization, low temperatures, and photoperiod in the field at Canberra. Australian J. Agr. Res. 10: 17–26.
Morely, F. H. W. 4949.

13762. Moro, M. S., Y. Pomeranz, and J. A. Shellenberger. 1963: The effects of gibberellic acid on α-amylase in wheat endosperms. Phyton 20:59–64.
Moroo, T. 18289, 18294.

13763. Morozov, V. K. 1958: The causes and forms of empty-seededness of the sunflower in the southeast. Dokl. Akad. Nauk SSSR 120:174–77.

13764. Morre, D. J., and O. H. Fletchall. 1963: Germination-regulating mechanisms of giant foxtail (*Setaria faberii*). Missouri Agr. Expt. Sta. Res. Bull. 829. 25 pp.
Morrell, K. E. 19683.

13765. Morris, C. J., and J. F. Thompson. 1958: The detection, isolation, and identification of γ-glutamyl-S-methylcysteine from beans. Arch. Biochem. Biophys. 73:281–83.

13766. ——, J. F. Thompson, and R. M. Zacharius. 1963: The identification of γ-L-glutamyl-L-leucine and γ-L-glutamyl-L-methionine in kidney bean seeds (*Phaseolus vulgaris*). J. Biol. Chem. 238:650–52.
Morris, C. J. 18504, 20033.
Morris, C. T. 13768.

13767. Morris, D. L. 1942: Lichenin and

araban in oats (*Avena sativa*). J. Biol. Chem. 142:881–91.

13768. —— and C. T. Morris. 1939: Glycogen in the seed of *Zea mays* (variety Golden Bantam). J. Biol. Chem. 130:535–44.

13769. Morris, E. L. 1911: Germination of cat tail seeds. Torreya 11:181–84.

13770. Morris, E. O. 1958: Effect of gibberellic acid upon the germination of barley. Chem. Ind. (London) 1958:97.

13771. Morris, G. C. 1939: Germination studies of the hulless seeds of Reed Canary grass. Proc. Assoc. Offic. Seed Analysts 30:259–62.

Morris, G. H. 3148.

Morris, H. E. 136.

Morris, H. F. 17649.

13772. Morris, H. J., and E. R. Wood. 1956: Influence of moisture content on keeping quality of dry beans. Food Technol. 10:225–29.

Morris, H. J. 19565.

13773. Morris, J. C. 1957: Some lilies grown from seed in New Zealand. Lily Yearbook 1957(20):73–76.

Morris, K. 3798.

Morris, R. 3514.

13774. Morris, V. H., T. L. Alexander, and E. D. Pascoe. 1945: Studies of the composition of the wheat kernel. I. Distribution of ash and protein in center sections. Cereal Chem. 22:351–61.

13775. ——, E. D. Pascoe, and T. L. Alexander. 1945: Studies of the composition of the wheat kernel. II. Distribution of certain inorganic elements in center sections. Cereal Chem. 22:361–71.

13776. Morris, W. G. 1936: Viability of conifer seed as affected by seed moisture content and kiln temperature. J. Agr. Res. 52:855–64.

Morris, W. G. 13915.

Morrison, A. B. 16132.

13777. Morrison, B. Y. 1926: Garden irises. U.S. Dept. Agr. Farmers Bull. 1406. 46 pp.

13778. —— 1929: Azaleas and rhododendrons from seed. U.S. Dept. Agr. Circ. 68. 7 pp.

13779. —— 1943: Quinine from seed. Agr. in the Americas 3:131–33.

13780. —— 1944: New quinine from this hemisphere. Natl. Hort. Mag. 23:21–31.

13781. —— 1951: Rex begonia seed on sphagnum moss. Natl. Hort. Mag. 30: 153–56.

13782. Morrison, F. R., and R. Grant. 1932: A contribution to the chemistry of the fruit obtained from the white cedar tree (*Melia azedarach*, L. var. *australasica* C.DC.; syn. *Melia australasica*, A. Juss.) growing in New South Wales, with notes on its reputed toxicity. J. Proc. Roy. Soc. N.S. Wales 65:153–77. Biol. Abstr. 8:16357. 1934.

13783. ——, A. R. Penfold, and J. Simonsen. 1951: The essential oils of *Zieria smithii* (Andrews) and its various forms. II. J. Proc. Roy. Soc. N.S. Wales 84:196–201.

Morrison, F. R. 3222.

13784. Morrison, G. 1941: Life span of seed. Seed World 50(2):8–9.

13785. —— 1948: Plant start better from pelleted seeds. Flower Grower 35(5): 400–1.

13786. —— 1954: Background of good vegetable seeds. Washing and drying "wet" seeds. Seed World 74(1):32–34.

13787. Morrison, L. 1961: Pollination and seed-setting of red clover in scrim-covered cages. New Zealand J. Agr. Res. 4:560–65.

13788. Morrison, L. G. 1956: Proper seed storage and control of insect pests. New Zealand J. Agr. 93:316–24.

Morrison, R. D. 198.

13789. Morriss, D. J., and H. O. Mills. 1948: The Conecuh longleaf pine seedbed burn. J. Forestry 46:646–52. Biol. Abstr. 23: 1923. 1949.

13790. Morrow, E. B., G. M. Darrow, and D. H. Scott. 1954: A quick method of cleaning berry seed for breeders. Proc. Am. Soc. Hort. Sci. 63:265.

Morrow, J. 17797.

13791. Mors, W. B., and H. J. Monteiro. 1959: Two coumarins in the seed of. *Copaifera lansdorfii*. Anais Assoc. Brasil. Quim. 18:181–82. Chem. Abstr. 54:11163. 1960.

13792. Morse, H. H. 1935: The toxic influence of fluorine in phosphatic fertilizers on the germination of corn. Soil Sci. 39:177–94.

13793. Morse, W. J. 1947: The versatile soybean. Econ. Botany 1:137–47.

13794. Mortensen, J. A., and W. H. Brittingham. 1952: The inheritance of pod color in the Southern pea, *Vigna sinensis*. Proc. Am. Soc. Hort. Sci. 59:451–56.

13795. Mortimore, C. G. 1959: Kernel moisture and yield of corn as influenced by pre-harvest foliar desiccation. Can. J. Plant Sci. 39:39–42.

13796. Morton, D. J. 1960: A quick method of preparing barley embryos for loose smut examination. Phytopathology 50: 270–72.

13797. ——, I. K. Hagen, and E. Tool. 1960: An improved method of disinfecting barley seed of *Ustilago nuda*. Plant Disease Reptr. 44:724–27.

13798. Morton, L. W. 1950: A recent problem in the testing of Kentucky bluegrass. Proc. Assoc. Offic. Seed Analysts 40:82–84.

Morton, R. K. 7567–7570, 9779, 9780.

Mosca, F. Traetta-; *see* Traetta-Mosca, F.

Moscarella, C. 17868.

13799. Moschini, E. 1939: Sulla presunta azione stimolante e fertizzante di un concime liquido. Ann. Fac. Agrar. Univ. Pisa 2:135–51. Biol. Abstr. 15:24206. 1941.

13800. —— 1951: Observations on the choice of peanut seeds. Ann. Fac. Agrar. Univ. Pisa 12:33–42. Chem. Abstr. 47:4958. 1953.

13801. —— 1960: Seed and fruit studies in two strains of sainfoin (*Onobrychis sativa*, Lam.). Proc. Intern. Seed Testing Assoc. 25:328–40.

13802. —— 1960: Temperature and germination researches on seeds of flower-plants. Proc. Intern. Seed Testing Assoc. 25:445–48.

13803. Moschler, W. W. 1949: The effect of various fertilizer treatments on the composition of the grain from hybrid and open-pollinated corn. Proc. Assoc. Southern Agr. Workers 1949:44.

Moschler, W. W. 3103.

13804. Moseman, J. G., and D. A. Reid. 1958: Effect of vernalization on the incidence of loose smut of barley, *Ustilago nuda*. Plant Disease Reptr. 42:744–46.

Moseman, J. G. 13413.

Mosettig, E. 16503.

13805. Mosheov, G. 1937: The influence of the water extract of wheat seeds upon their germination and growth. Bull. Dept. Botany Hebrew Univ. Jerusalem 1:1–9.

13806. —— 1937: The inhibition of germination and root formation by copper sulfate and the elimination of the inhibitory effect. Bull. Dept. Botany Hebrew Univ. Jerusalem 1:9–16.

13807. Moshkin, V. A. 1951: Relation of quality of castor-bean seeds to environmental conditions of raceme ripening. Selek. i Semen. 18(8):74–75.

Moshkin, V. A. 7261.

Mosley, V. M. 13059.

13808. Mosolov, I. V., and V. Aleksandrovskaia. 1954: Effect of nitrogen and phosphorus on starch hydrolysis in germinating seeds of wheat. Dokl. Akad. Nauk SSSR 95:1333–35.

Mosolova, I. M. 15728.

13809. Moss, E. H. 1938: Longevity of seed and establishment of seedlings in species of *Populus*. Botan. Gaz. 99:529–42.

13810. Moss, G. R. 1953: Seed testing: basic features, services, and terminology. New Zealand J. Agr. 87:251–55.

Mosse, J. 2110, 2111.

13811. Mostert, J. F. T. 1950: Round maize seed better than flat; superior root system and drought hardiness open fruitful field for investigation. Farmers Wk. 80:10–11.

13812. Mostovoj, K. 1929: Die Zähne an den Gerstengrannen als Unterscheidungsmarkmal der Sorten. Vestn. Česk. Akad. Zeměděl. 5:16–22.

Mostovoj, K. 4004–4006.

13813. Mota, M. 1952: Die Wirkung von Extrakten aus Samen auf die Chromosomen. Arqu. Patent 24, 313–34. Ber. Wiss. Biol. 91:15. 1954.

13814. Mothes, K. 1938: Der Schwefelumsatz im reifenden Samen. Planta 29:80–85.

13815. Motkin, V. M. 1951: The effect of granular fertilizers on germination when applied together with the seed. Sov. Agron. 1951(9):29–36. Soils Fertilizers 15:49. 1952.

Motogi, T. 10362.

Motomura, T. 14160.

Motrenko, T. G. 15781.

13816. Mott, R., and G. Rotor. 1950: How to germinate orchid seeds. Gardeners Chron. 55:122, 136–37.

13817. Motte, J., and A. Saint-Pierre. 1935: Influence de la fermentation sur le pouvoir germinatif des graines. Congr. Soc. Sav. Paris 1935:231–38. Bull. Soc. Botan. France 85:443. 1938.

13818. Mottier, D. M. 1931: Studies on *Berberis vulgaris* and related species. I. On seed germination and the survival of seedlings of *Berberis vulgaris* in Monroe County, Indiana. Proc. Indiana Acad. Sci. 47:145–48.

13819. Mou, T. S., and Y. T. Yoa. 1952: An observation on the vernalization results of Taiwan wheat. Taiwan Agr. Forestry 6(11):19–26.

Moullec, J. 18463.

Moulter, H. 6356.

Moulton, J. E. 8087.

Mounat, A. 7289

13820. Mounfield, J. D. 1936: The proteolytic enzymes of sprouted wheat. Biochem. J. 30:549–57

13821. —— 1936: The proteolytic enzymes of sprouted wheat. II. Biochem. J. 30: 1778–86.

13822. —— 1938: The proteolytic enzymes of sprouted wheat. III. Biochem. J. 32:1675–84.

13823. ——, P. Halton, and A. G. Simpson. 1944: The drying of wheat. II. The drying of English wheat. Soc. Chem. Ind. J. 63:97–104.

13824. Mounsey-Wood, B. 1961: Effect of day-length during germination of cauliflower. Exptl. Hort. 5:13–18.

Mounts, F. C. 9076.

Moura-Campos, F. A. de; *see* de Moura-Campos, F. A.

13825. Mourasiev, P. A. 1932: Properties of flour and bread produced from wheat treated by yarovization. (Ukr. Inst. Genet.) Bull. Yarovization 1:47–56.

13826. Morgue, M., R. Baret, R. Kassab, and J. Reynaud. 1961: Étude des protéines de la graine de *Jatropha curcas* Linn. Bull. Soc. Chim. Biol. 43:505–16.

13827. ——, ——, J. Reynaud, and J. Bellini. 1958: Étude des protéines de la graine de ricin. II. *Ricinus communis.* Bull. Soc. Chim. Biol. 40:1453–63.

13828. ——, J. Delphaut, R. Baret, and R Kassab. 1961: Étude de la toxicité et localisation de la toxalbumine (curcine) des graines de *Jatropha curcas* Linn. Bull. Soc. Chim. Biol. 43:517–31.

13829. Mourkidis, G. A. 1952: Effects of growth conditions on gossypol content of cottonseed. Geoponike Dasologike Schole 1952:5–35. Chem. Abstr. 47:8191. 1953.

13830. Moustafa, E. M. 1955: Partial purification and properties of the enzyme glutathione reductase in pea seeds. Proc. Egypt. Acad. Sci. 11:27–36. Chem. Abstr. 52:82896. 1958.

13831. —— 1963: Purification and properties of valine-activating enzyme from wheat germ. Biochim. Biophys. Acta 76:280–85.

Moustafa, E. M. 12830.

13832. Moutia, L. A., and R. Mamet. 1943: Experiments on the storage of maize seeds. Rev. Agr. Ile Maurice 22:5–9. Rev. Appl. Entomol. Ser. A 32:161. 1944.

13833. Mouton, J. A. 1960: La dormance chez *Oryza sativa* L.; révision bibliographique. J. Agr. Trop. Botan. Appl. 7:588–90.

Mouton, R. 9733.

13834. Moutschen, J., and Z. M. Bacq. 1956: Action of cystamine and glutathione on barley seeds irradiated with X-rays. Compt. Rend. Soc. Biol. 150:2046–49. Chem. Abstr. 51:12235. 1957.

Moutschen, J. 13835.

13835. Moutschen-Dahmen, M., J. Moutschen, and L. Ehrenberg. 1959: Chromosome disturbances and mutation produced in plant seeds by oxygen at high pressures. Hereditas 45:230–44.

13836. Movsesian, E. M. 1955: The effect of fertilizing grain crops on the productive capacity of their seeds. Akad. Nauk Azerb. SSR Inst. Pochvoved. Agrokhim. Tr. 7:39–52.

Mowry, H. 3549.

Moyer, D. 3386.

13837. Moyer, J. C., L. J. Lynch, and R. S. Mitchell. 1954: The tenderization of peas during vining. Food Technol. 8:358–60.

13838. Moyse, A. 1952: Germination of wheat at low oxygen tensions. I. Minimum oxygen tension and germinative power. Compt. Rend. 234:874–76. Chem. Abstr. 46:8728. 1952.

13839. Mudaliar, C. R. 1935: Effect of naphthalene on germination of paddy seed. Madras Agr. J. 23:223–31.

13840. —— 1937: The formation of starch in the developing rice kernel. Proc. Assoc. Econ. Biologists (Coimbatore) 4:53–56. 1936. Chem. Abstr. 32:6292. 1938.

13841. —— and D. D. Sundararaj. 1954: Dormancy and germination of a few crop seeds. Madras Agr. J. 41:111–19. Biol. Abstr. 29:12306. 1955.

Mudaliar, C. R. 15908, 15909.

Mudaliar, V. R. 1762.

13842. Mudra, A. 1942: Effect of seeding density on the production of wheat and oats. Kiserl. Kozl. 44:55. Biol. Abstr. 21:15249. 1947.

13843. Mühle, E. 1959: Kritische Untersuchungen zur Methodik der Prüfung von Leinsamen auf Befall durch *Polyspora lini* Laff. und *Colletotrichum linicola* Pethybr. et Laff. Phytopathol. Z. 34:315–24.

13844. Mühlethaler, K. 1950: Structure of seed mucilages. Proc. Conf. Electron Micros. 1949:135–36.

Mühlethaler, K. 3472.

Muelder, D. W. 8131.

13845. Müller, C. 1890: Über ein Fettes Öl auf Lindensamen. Ber. Deut. Botan. Ges. 8:372–77.

13846. Müller, D. 1943: Tote Speichergewebe in lebenden Samen. Planta 33:721–27. Proc. Intern. Seed Testing Assoc. 13:422–23. 1941/43.

13847. —— and F. Holm. 1942: Die Atmung des Gräserendosperms. Planta 32: 596–99. Proc. Intern. Seed Testing Assoc. 13:423–24. 1941/43.

Müller, E. 11591.

13848. Müller, F. 1959: Über die Herabsetzung der Plasmaviskosität nach Vernalisation und die damit verbundene Verminderung der Kälteresistenz bei Getreide. Angew. Botan. 33:159–62.

Müller, F. von; *see* Von Müller, F.

13849. Müller, G. 1914: Beiträge zur Keinungsphysiologie. Untersuchungen über die Sprengung der Samen- und Fruchthüllen bei der Keimung. Jahrb. Wiss. Botan. 54:529–644.

13850. Müller, H. C. 1924: Methoden zur Feststellung der Keimfähigkeit von Pflanzensamen. Abderhalden, Emil. Ed. Handbuch der biologischen Arbeitsmethoden. 11(2):719–64.

13851. Müller, M. 1945: Keimung von Licht- und Schattenhölzern. Schweiz. Z. Forstw. 96:131–34. Biol. Abstr. 19:19909. 1945.

Mueller, M. M. 17046.

13852. Müller, P. 1934: Beitrag zur Keimverbreitungsbiologie der Endozoochoren. Ber. Schweiz. Botan. Ges. 43:241–52. Biol. Abstr. 10:117. 1936.

13853. Müller, P., and M. J. Oosthuizen. 1958: The influence of methyl bromide on the germination of wheat seed. J. Entomol. Soc. S. Africa 21:230–31. Biol. Abstr. 35:16882. 1960.

13854. Müller, W. 1944: Das Keimungsbild von Cruciferen-Arten unter dem Einfluss verschiedener Bedingungen, in Kurvendiagrammen betrachtet. Botan. Centralbl. Beih. Abt. A 62:229–80.

13855. —— 1952: Der Einfluss von Licht, Temperatur und Substrat auf die Keimung von Samen einiger *Brassica*-Arten. Angew. Botan. 26:100–10.

13856. —— 1952: Beiträge zur Nachreife und Lichtkeimung von Cruciferensamen. Angew. Botan. 26:181–90.

Müller, W. 10979.

Müller, Z. 9459.

13857. Müller-Olsen, C., and M. Simak. 1954: X-ray photography employed in germination analysis of Scots pine (*Pinus silvestris* L.). Medd. Statens Skogsforskningsinst. 44(6):1–19. Biol. Abstr. 29: 19751. 1955.

13858. ——, ——, and A. Gustaffson. 1956: Germination analyses by the X-ray method: *Picea abies* (L.) Karst. Medd. Statens Skogsforskningsinst. 46(1):1–12.

13859. Müller-Schneider, P. 1953: Samenkeimung im Schnee. Leben Umwelt 1953:128–29.

13860. Müller-Stoll, W. R. 1950: Wuchsstoffversuche mit Reben. II. Reaktionsweise von Samen, Keimlingen und nichtholzigen Organen der Weinrebe auf künstliche Wuchsstoffgaben. Züchter 20:213–26. Biol. Abstr. 25:25431. 1951.

13861. Müller-Thurgau, H. 1908: Kernlose Traubenbeeren und Obstfrüchte. Landwirtsch. Jahrb. Schweiz. 22:564–97.

13862. Muenscher, W. C. 1931: Determination of the viability of weed seeds under field conditions. Cornell Univ. Agr. Expt. Sta. Ann. Rept. 44:37.

13863. —— 1931: A study of the requirements for the germination of the seeds of the Indian tobacco (*Lobelia inflata*). Cornell Univ. Agr. Expt. Sta. Ann. Rept. 44:39.

13864. —— 1935: The germination of *Lobelia inflata* seeds. Am. J. Botany 22: 903.

13865. —— 1935: Weeds of New York. Cornell Univ. Agr. Expt. Sta. Bull. 635. 16 pp.

13866. —— 1936: The germination of seeds of *Potamogeton*. Ann. Botany (London) 50:805–21.

13867. —— 1936: Seed germination in Lobelia, with special reference to the influence of light on *Lobelia inflata*. J. Agr. Res. 52:627–31.

13868. —— 1936: Storage and germination of seeds of aquatic plants. Cornell Univ. Agr. Expt. Sta. Bull. 652. 17 pp.

13869. —— 1940: Fruits and seedlings of *Ceratophyllum*. Am. J. Botany 27:231–33.

13870. —— and B. I. Brown. 1944: Storage and germination of nuts of several species of *Juglans*. Northern Nut Growers Assoc. Ann. Rept. 34:61–62.

13871. —— and D. Isely. 1943: A study of the requirements for the storage and germination of the seeds of various aquatic and marsh plants. Cornell Univ. Agr. Expt. Sta. Ann. Rept. 1943:108–9.

13872. —— and O. L. Justice. 1940: A study of the requirements for the storage and germination of seeds of various aquatic plants. Cornell Univ. Agr. Expt. Sta. Ann. Rept. 1939:122.

13873. Müntzing, A. 1940: Further studies on apomixis and sexuality in *Poa. hereditas* 26:115–90. Biol. Abstr. 15:11786. 1941.

13874. —— and G. Müntzing. 1941: Some new results concerning apomixis, sexuality and polymorphism in *Potentilla.* Botan. Notiser 1941:237–78. Biol. Abstr. 18: 9802. 1944.

13875. —— and —— 1945: The mode of reproduction of hybrids between sexual and apomictic *Potentilla argentea.* Botan. Notiser 1945:49–71. Biol. Abstr. 20: 14683. 1946.

Müntzing, G. 13874, 13875.
Müssle, L. 3288.
Mugibayasi, N. 10315, 10316, 10726.
Muhammad, F. 19275.
Muhammad, W. 20021.

13876. Muhlack, E. 1929: Zur Keimungsgeschichte der Erbse. Botan. Arch. 26: 437–85.

13877. Muhr, G. R., and C. O. Rost. 1951: The effect of population and fertility on yields of sweet corn and field corn. Agron. J. 43:315–19.

13878. Muhr, H., A. Hunger, and T. Reichstein. 1954: Die Glykoside der Samen von *Acokanthera friesiorum* Markgr. Helv. Chim. Acta 37:403–27.

13879. Muir, J. 1937: The seed-drift of South Africa and some influences of ocean currents on the strand vegetation. Union S. Africa Dept. Agr. Forestry Botan. Survey Mem. Pretoria 16:1–108. Trop. Woods 53:58–59. 1938.

13880. Mukerjee, N., T. Ghosh, and K. T. Jacob. 1962: Triphenyl tetrazolium bromide as an indicator of germinability in jute seeds. Jute Bull. 25:89–91.

Mukerji, B. K. 3760.
Mukerji, S. K. 3850.

13881. Mukherjee, A. K., D. Choudhury, and P. Bagchi. 1961: Constitution of the galactomannan from the kernel of green palmyra palm nut (*Borassus flabellifer* Linn.). Can. J. Chem. 39:1408–18. Biol. Abstr. 36:82084. 1961.

Mukherjee, D. B. 10510.

13882. Mukherjee, P. K. 1957: Studies in the embryology of *Euphorbia hypericifolia* Linn. Saugar Univ. Botan. Soc. Bull. 9:7–18.

13883. —— 1961: Study of mature embryo, seed coat and fruit wall in *Euphorbia cristata* Heyne. J. Biol. Sci. (Bombay) 4:1–5. Biol. Abstr. 38:23488. 1962.

13884. —— and O. S. Thakur. 1959: Studies on the embryology of Leguminosae. *Caesalpinia crista* Linn. J. Biol. Sci. (Bombay) 2:84–88. Biol. Abstr. 35:32727. 1960.

Mukherjee, R. 2086, 2087.

13885. Mukherjee, S. K. 1960: Vivipary in mulberry. Sci. Cult. (Calcutta) 26:234.

Mukherji, B. K. 4049.

13886. Mukherji, D. K. 1950: Early stages of growth of wheat plants after vernalisation. Current Sci. (India) 19:345–46.

13887. —— 1951: Embryo culture as an aid to seed testing. Proc. Natl. Inst. Sci. India 17:253–59.

13888. —— 1951: Seed-testing of samples of foodgrains. Indian Farming 12(1):8–14.

13889. —— and S. K. Basu. 1962: Seed testing. IV. Effects of toxic substances from Palak sag (*Beta vulgaris* L.) seeds on the germination of gram and paddy. Current Sci. (India) 31:197–98.

Mukhopadhyay, I. 17224.
Mulder, F. 14775.

13890. Muldoon, T. J., and F. J. Sylvester. 1951: The transmission of organic vapors through packaging materials. Food Technol. 5:449–52.

Mullaly, J. V. 17476

13891. Mullan, D. P. 1936: On the seed structure and germination of *Acanthus ilicifolius* Linn. J. Indian Botan. Soc. 15:143–47.

13892. Mullan, J., and E. G. V. Percival. 1940: Studies on seed mucilages. I. A preliminary examination of the mucilaginous polysaccharide of the seeds of *Plantago lanceolata.* J. Chem. Soc. 1940: 1501–6.

13893. Mullen, L. A., and A. L. Hafenrichter. 1953: New harvester for hard-to-get grass seeds. Crops & Soils 5(8):14–15.

Mullen, L. A. 17121.
Muller, C. H. 13896.

13894. Muller, D. 1942: The structure and physiology of grains. Korn 1:58–92. Biol. Abstr. 21:15250. 1947.

13895. Muller, H. 1962: Research on the problem of the interrelationship between germinating seeds and the microorganisms in their vicinity. Arch. Mikrobiol. 41:351–82. Biol. Abstr. 41:16117. 1963.

Muller, H. J. 18754.

13896. Muller, W. H., and C. H. Muller. 1956: Association patterns involving desert plants that contain toxic products. Am. J. Botany 43:354–61.

13897. Mulligan, B. O., and L. J. Michaud. 1951: Magnolia seed treatment and germination. Arboretum Bull. (Wash. Univ.) 14(3): 1-2, 33.

13898. Mullison, W. R. 1939: Electrodialysis of pea seeds. Plant Physiol. 14:583-87.

13899. —— and R. W. Hummer. 1949: Some effects of the vapor of 2,4-dichlorophenolyacetic acid derivatives on various field-crop and vegetable seeds. Botan. Gaz. 111:77-85.

13900. Mullverstedt, R. 1963: Investigations on the germination of weed seeds as influenced by oxygen partial pressure. Weed Res. 3:154-63.

13901. Multamaki, K. 1962: The effect of seed size and depth of seeding on the emergence of grassland plants. Suomen Maataloustiet. Seura. Maataloustiet. Aikakausk. 34:18-25.

13902. Mulwani, B. T., and A. G. Pollard. 1939: Effects of alkali salts on germination of seeds. Agr. Live-stock India 9:548-55.

Muminov, M. M. 10096.

Muminov, T. G. 11218.

13903. Munakata, K., K. Yokoyama, T. Shibata, A. Harada, and F. Hara. 1959: Herbicidal activities of halogenoalkylcarboxylic acid esters. I. Germination inhibition activities. Weeds 7:470-80.

13904. Mundkur, B. B., and M. A. Khan. 1935: A dry spray method of treating oat seed against covered smut. Indian J. Agr. Sci. 4:899-905. Chem. Abstr. 29:1571. 1935.

13905. Muneghina, A. 1958/59: Permeability of *Lepidium sativum* and wheat seeds integuments to ethanol and fat solvents. Atti, Ist. Veneto Sci. Lettere Arti Classe Sci. Mat. Nat. 117:275-90. Chem. Abstr. 54:15537. 1960.

13906. Munerati, O. 1924: Contributions à l'étude de l'apparition du sexe chez les plantes dioiques. Compt. Rend. 179:1200-2.

13907. —— 1925: Existe-t-il une aprèsmaturation chez les céréales recomment recoltées? Compt. Rend. 181:1081-83.

13908. —— 1926: Possibilité de determiner l'âge des graines de blé par temperature de leur germination. Compt. Rend 182:535-37.

13909. —— 1932: Sur le mécanisme de la germination du blé en gerbes. Compt. Rend. Soc. Biol. 111:552-54.

13910. —— and T. V. Zapparoli. 1911: Action of stimulating substances on the germination of seeds. Staz. Sper. Agrar. Ital. 44:40-50. Botan. Centralbl. 119:626-27. 1912.

13911. —— and —— 1913: Acidity of manures as related to germinability of seeds of leguminous weeds. Staz. Sper. Agrar. Ital. 46:5-17. Expt. Sta. Rec. 31:729. 1914.

13912. —— and —— 1913: Il grado di maturanza dei semi di leguminose infeste in rapporta con la loro prontezza germinativa. Staz. Sper. Agrar. Ital. 46:137-45. Botan. Centralbl. 128:121. 1915.

13913. —— and —— 1913: On the conservation of the vitality of weed seeds in lower strata of cultivated soils. Staz. Sper. Agrar. Ital. 46:347-71. Expt. Sta. Rec. 31:634. 1914.

13914. Munesada, T., and N. Murakami. 1954: Pharmacognostic studies of Cucurbitaceae seeds. I-II. *Trichosanthes* seeds. J. Pharm. Soc. Japan 74:584-87.

Munger, H. M. 4489.

13915. Munger, T. T., and W. G. Morris. 1936: Growth of Douglas fir trees of known seed source. U.S. Dept. Agr. Tech. Bull. 537. 40 pp. Expt. Sta. Rec. 77:51. 1937.

13916. Mungomery, W. V. 1956: Lettuce germination. Queensland Agr. J. 82:683-86.

13917. Munn, H. L., and E. V. Staker. 1943: Toxicity to seedlings of zinc from germinator trays. Proc. Assoc. Offic. Seed Analysts 1942:82-88.

13918. Munn, M. T. 1928: The analyzing and subsequent labelling of Timothy and Alsike clover mixtures. Proc. Assoc. Offic. Seed Analysts 20:64-66.

13919. —— 1928: The behavior during germination of cracked and broken seeds from badly threshed red clover seed. Proc. Assoc. Offic. Seed Analysts 20:68-69.

13920. —— 1937: Flourescence readings of the strains of the species of *Lolium*. Proc. Assoc. Offic. Seed Analysts 29:136-37.

13921. —— 1938: The sanitary condition of *Brassica* seeds received from various sources. Proc. Intern. Seed Testing Assoc. 10:275-77.

13922. —— 1940: Seed injury lowers germination. Farm Res. (N.Y.) 6(3):9.

13923. —— 1941: The flower seeds. Farm Res. (N.Y.) 7(1):6. Expt. Sta. Rec. 84:624. 1941.

13924. —— 1943: Germination response of vernalized *Taraxacum* seed. Proc. Assoc. Offic. Seed Analysts 34:88-90.

13925. —— 1946: Germinating freshly harvested winter barley and wheat. Proc. Assoc. Offic. Seed Analysts 36:151–52.
13926. —— 1946: The quality of flower and vegetable seeds on sale in New York in 1945. N.Y. Agr. Expt. Sta. Bull. 717. 69 pp.
13927. —— 1948: The 25-year hard seed soaking experiment. Proc. Assoc. Offic. Seed Analysts 38:66–68.
13928. —— 1949: Endive seed germination. Proc. Assoc. Offic. Seed Analyst 39:122–25.
13929. —— 1949: The seed analyst and pelleted seed. Assoc. Offic. Seed Analysts News Letter 23(1):23.
13930. —— 1950: Lettuce seed germination problems. Proc. Assoc. Offic. Seed Analysts 40:119–22.
13931. —— 1950: A method for testing the germinability of large seeds. N.Y. Agr. Expt. Sta. Bull. 740. 10 pp.
13932. —— 1951: Hard seeds help spread white clovers. Farm Res. (N.Y.) 17(2):5.
13933. —— 1952: More about slant paper-rolls for germination testing. Assoc. Offic. Seed Analysts News Letter 26(2):28–29.
13934. —— and W. F. Crosier. 1951: Longevity of pea seed treated with spergon. Assoc. Offic. Seed Analysts News Letter 25(4):42.
13935. —— and C. E. Heit. 1941: Embryoless seeds of the Umbelliferae. Assoc. Offic. Seed Analysts News Letter 15(4):10.
Munn, M. T. 4146, 4663, 4673, 8587, 8588.
13936. Munns, E. N. 1921: Effect of location of seed upon germination. Botan. Gaz. 72:256–60.
13937. Munoz, C. R., and M. Claver-Aliod. 1949: Polyphenols in the seed of the carob bean (*Ceratonia siliqua*). Anales Edafol. Fisiol. Vegetal (Madrid) 8:59–68. Chem. Abstr. 43:8452. 1949.
Munoz, M. S. 16537, 16538.
13938. Munoz-Gomez, A. 1952: Vernalización y fotoperiodismo. Agronomia (Peru) 17(72):123–31.
Munro, J. M. 14309.
13939. Muntoni, F. 1950: An antibiotic from the seeds of *Raphanus sativus* L , var. *radicula* Pers. Rend. Ist. Super. Sanita 13:174–76.
Munz, P. A. 19533.
Muradyan, L. A. 1606.
Murakami, K. 13635, 13636.
Murakami, N. 13914.
Murakami, T. 19971.
13940. Murakami, Y. 1957: The effect of the extract of immature bean seeds on the growth of coleoptile and leaf of rice plant. Botan. Mag. (Tokyo) 70:376–82.
13941. —— 1959: A paper chromatographic survey of gibberellins and auxins in immature seeds of leguminous plants. Botan. Mag. (Tokyo) 72:36–43.
13942. —— 1959: The occurrence of gibberellins in mature dry seeds. Botan. Mag. (Tokyo) 72:438–42.
13943. —— 1961: Paper chromatographic studies on change in gibberellins during seed development and germination in *Pharbitis nil*. Botan. Mag. (Tokyo) 74:241–47.
13944. —— 1962: Occurrence of "water-soluble" gibberellin in higher plants. Botan. Mag. (Tokyo) 75:451–52.
13945. —— and T. Hayashi. 1957: The conversion of tryptophan to indoleacetic acid by the sap of immature kernels of rice plants. J. Agr. Chem. Soc. Japan 31:468–70.
13946. —— and S. Matsunaka. 1963: The effect of gibberellin on components of phosphatase and peronidase in rice plants. Proc. Plant Physiol. Meetings 38:xliii.
13947. Muraoka, Y. 1952: Studies on the germination of tobacco seed. Hatano Tobacco Expt. Sta. Bull. 36:1–27. Tobacco Abstr. 1:92. 1957.
Muraoka, Y. 9472.
Murat, M. 11869.
Murata, M. 18233.
13948. Muratova, E. V. 1961: Pelleting of seeds of garden onion and table beet. Moskov. Ordena Lenina Sel'skokhoz. Akad. Im. K. A. Timiryazeva Dokl. 72:135–40.
13949. Muratteldyev, N. 1962: On the problem of the reproduction by seed of *Gleditsia*. Izv. Akad. Nauk Turkm. SSR Ser. Biol. Nauk 1962(2):65–66.
13950. Muravieva, E. P. 1928: Morphologische und anatomische Unterschiede der *Brassica* und *Sinapis*samen. Jard. Botan. Inst. Essais Semences, Leningrad, Ann. Essais Semences 1928(2):129–41.
Murazio, S. 3528.
13951. Murdoch, H. A. 1940: Hybrid vigor in maize embryos. J. Heredity 31:361–63. Biol. Abstr. 15:40. 1941.
13952. Murdock, Mrs. P. 1933: Iris from seed. Flower Grower 20(9):408.
13953. Murley, M. 1944: A seed key to fourteen species of Geraniaceae. Proc.

Iowa Acad. Sci. 51:241–46. Expt. Sta. Rec. 95:316. 1946.

13954. Murley, M. R. 1951: Seeds of the Cruciferae of northeastern North America. Am. Midland Naturalist 46:1–81.

13955. Murlin, J. R. 1933: The conversion of fat to carbohydrate in the germinating castor bean. I. The respiratory metabolism. J. Gen. Physiol. 17:283–302.

Murlin, J. R. 15361.

13956. Murneek, A. E. 1932: Growth and development as influenced by fruit and seed formation. Plant Physiol. 7:79–90.

13957. —— 1932: Relation of leaf area to fruit size and food reserves in apple seeds and branches. Proc. Am. Soc. Hort. Sci. 29:230–34.

13958. —— 1937: Embryo growth and development of the apple. Missouri Agr. Expt. Sta. Bull. 387:73.

13959. —— 1954: Embryo and endosperm in relation to fruit development with special reference to *Malus sylvestris*. Congr. Intern. Botan. Paris Rappt. Commun. 8:368–76. Chem. Abstr. 48:11567. 1954.

13960. —— and J. R. Jackson. 1937: Artificial culture of apple embryos and ovules. Missouri Agr. Expt. Sta. Bull. 387:73–74.

13961. —— and G. C. Schowengerdt. 1936: A study of the relation of size of apples to number of seeds and weight of spur leaves. Proc. Am. Soc. Hort. Sci. 33:4–6.

13962. —— and R. O. Whyte. 1948: Vernalization and photoperiodism. A symposium. Chronica Botan. 196 pp.

13963. —— and S. H. Wittwer. 1942: Relation of sexual reproduction to development of horticultural plants. I. General effects of flower and fruit production. Proc. Am. Soc. Hort. Sci. 40:201–4.

Murneek, A. E. 367, 7916, 17531, 18466.

13964. Murphy, A. H., and D. Turner. 1959: A study on the germination of Medusahead seed. Calif. Dept. Agr. Bull. 48:6–10.

13965. Murphy, C. F., and K. J. Frey. 1962: Inheritance and heritability of seed weight and its components in oats. Crop Sci. 2:509–12.

Murphy, C. F. 13415.

13966. Murphy, H. C., F. J. Zillinsky, M. D. Simons, and R. Grindeland. 1958: Inheritance of seed color and resistance to races of stem and crown rust in *Avena strigosa*. Agron. J. 50:539–41.

Murphy, H. C. 6382, 6724.

13967. Murphy, H. F. 1929: Some effects of crude petroleum on nitrate production, seed germination, and growth. Soil Sci. 27:117–20.

Murphy, H. F. 19448.

13968. Murphy, J. B. 1942: The influence of magnetic fields on seed germination. Am. J. Botany Suppl. 29(10):15s.

13969. Murphy, R. P., and A. C. Arny. 1939: The emergence of grass and legume seedlings planted at different depths in five soil types. J. Am. Soc. Agron. 31:17–28.

13970. ——, M. R. Risius, and E. R. Sayers. 1959: A study of the variation in seed setting among individual plants of Narragansett alfalfa. Agron. Abstr. 1959:62–63.

Murphy, R. P. 12457, 18858.

13971. Murray, B. E. 1948: Determining seed size by a volumetric measure. Can. J. Res. Sect. C 26:23–26.

13972. Murray, H. C., S. Kellenbarger, and G. A. Armbruster. 1952: Protein efficiency and methionine content of Alaska peas grown with the use of various fertilizers. Agron. J. 44:252–53.

Murray, M. D. 13115, 15532.

Murray, M. J. 3660.

Murray, R. H. 9038.

13973. Murri, I. K. 1939: Aneurine in the wheat grain. Dokl. Vses. Akad. Sel'skokhoz. Nauk 23/24:36–41. Chem. Abstr. 36:2588. 1942.

13974. —— 1940: Vitamin A (carotin) contents of seeds of wheat and rye. Vestn. Sotsial. Rast. 4:125–32.

13975. —— and A. I. Ermakov. 1936: The chemical characteristics of the seeds of *Amaranthus caudatus*. Bull. Appl. Botany, Genet., Plant Breed. (USSR) Ser. III 15:149–59. Chem. Abstr. 31:5842. 1937

Murthi, B. S. 13980.

13976. Murthy, K. L. 1934: Gametogenesis in some Commelinaceae. Current Sci. (India) 3:258–59.

13977. —— 1938: Embryogeny of Commelinaceae. (Gametogenesis and embryogeny in some Commelinaceae.) J. Indian Botan. Soc. 17:101–16. J. Roy. Microscop. Soc. 59:53. 1939.

Murthy, K. S. N. 10150.

13978. Murthy, N. L. N., and B. H. Iyer. 1954: Oil from the seeds of *Poinciana regia*. J. Indian Inst. Sci. 36:155–59. Chem. Abstr. 49:1347. 1955.

Murthy, S. Narasimha-; *see* Narasimha-Murthy, S.

Murti, C. R. K. 6976.

Murti, K. S. 8775, 15966.

13979. Murti, P. B. R., and N. V. S. Rao. 1944: Wax and resin components of the pericarp of jambul fruits (*Eugenia jambolana*). Proc. Indian Acad. Sci. Sect. A 20:163–68. Chem. Abstr. 39:2781. 1945.

Murty, A. S. N. 17225.

13980. Murty, N. S. R., and B. S. Murthi. 1961: Sprouting energy of chilli seeds, extracted from pods obtained from different pickings. Andhra Agr. J. 8:244–45.

Murwin, H. F. 9867.

13981. Musaeva, S. M. 1962: Fructification and laboratory germination of seeds of *Glycyrrhiza glabra* L. from the Apsheron Peninsula. Izv. Akad. Nauk Azerb. SSR Ser. Biol. i Med. Nauk 1962(5):23–27.

13982. Musalevski, A. 1959/60: Period of dormancy of seed in certain cultivated and wild plant varieties. Skoplje, Yugoslavia, U. Zemjodelsko-Sumarski Fak. God. Zb. Sumarstvo 13:281–303.

13983. Musbach, F. L., and O. E. Sell. 1936: Effect of fertilizer on quality and chemical composition of canning peas. J. Agr. Res. 53:869–79. Expt. Sta. Rec. 77:45–46. 1937.

13984. Musgrave, C. T. 1932: Gentians. J. Roy. Hort. Soc. 57:175–82

13985. Musher, A. 1942: Medicinal preparations such as expanded structure-disrupted flax seed, quince seed of *Psyllium* seed. U.S. Patent 2,278,464. Pharm. Abstr. 10:37. 1944.

13986. Musher, S. 1935: Cereals and seeds inhibit rancidity in lard. Food Ind. 7:167–68. Chem. Abstr. 29:6317. 1935.

13987. Musick, A. H. 1943: A brief comparative study of the germination time of some medicinal plant seeds in flats, coldframes and incubators. J. Am. Pharm. Assoc. Sci. Ed. 32:171–72.

13988. Musil, A. F. 1931: A suggested method of seed testing. Proc. Intern. Seed Testing Assoc. 3:237–45.

13989. —— 1938: The direct method of germination testing as applied to *Dactylis glomerata* and *Poa pratensis*. Proc. Assoc. Offic. Seed Analysts 23/26:141–47.

13990. —— 1941: The pretreatment of Buffalo grass for field planting. Proc. Assoc. Offic. Seed Analysts 33:76–82.

13991. —— 1943: Distinguishing characters of the seeds of four species of *Agropyron*, *A. repens* L. Beauv., quackgrass; *A. smithii* Rydb., Western wheatgrass; *A. trachycaulum* (Link) Malte, Slender wheatgrass; *A. cristatum* (L.) Gaerth, Crested wheatgrass. Proc. Assoc. Offic. Seed Analysts 34:124–31. Biol. Abstr. 20:1489. 1946.

13992. —— 1943: Seeds of commercial species of *Brassica*. Proc. Assoc. Offic. Seed Analysts 34:132–38. Biol. Abstr. 20:1574. 1946.

13993. —— 1946: The germination of guar, *Cyamopsis tetragonolobus* (L.) Taub. J. Am. Soc. Agron. 38:661–62.

13994. —— 1949: Distinguishing the seeds of cultivated fine-leaved fescues and certain weedy species occurring with crop seeds. Proc. Assoc. Offic. Seed Analysts 39:93–107.

13995. —— 1961: Testing seeds for purity and origin. Yearbook Agr. U.S. Dept. Agr. 1961:417–32.

13996. —— 1963: Identification of crop and weed seeds. U.S. Dept. Agr. AMS Agr. Handbook 219. 171 pp.

Musil, A. F. 1028, 13208, 16562.

13997. Musil, F. 1946: Effects of atomic bomb explosion on seeds of certain forage crop plants; a report of laboratory tests for viability of the seed and effects observed on seedlings in laboratory and greenhouse. Naval Med. Res. Sect. U.S. Joint Task Force One. Rept. on Biol. Aspects At. Bomb Tests Appendix II.

13998. Muskett, A. E. 1944: The short wet method of seed disinfection. Ann. Appl. Biol. 31:218–21.

13999. —— 1958: Studies on seed health. I. Flax. Ann. Appl. Biol. 46:430–45.

14000. —— 1962: Seed-borne fungi and their significance. Gt. Brit. Central Off. Inform. Seed Improvement in Britain, pp. 63–76.

14001. —— and H. Cairns. 1932: The effect of seed disinfection upon the oat crop in Northern Ireland. Ann. Appl. Biol. 19:462–74.

14002. —— and J. Colhoun. 1946: Seed borne diseases of flax and their control. Ann. Appl. Biol. 33:331–33.

14003. —— and J. P. Malone. 1941: The Ulster method for the examination of flax seed for the presence of seed-borne parasites. Ann. Appl. Biol. 28:8–13.

14004. —— and J. P. Malone. 1956: Detection of seed-borne parasites in seeds. Nature 177:465–66.

Muskett, A. E. 3535, 4292.

Mussehl, F. E. 2772, 8056.

Musser, A. M. 17170.

14005. Musser, H. B. 1947: The effect of burning and various fertilizer treatments on seed production of red fescue, *Festuca*

rubra L. J. Am. Soc. Agron. 39:335–40.
Mustafa, Z. 17725.

14006. Mustakas, G. C. 1964: New enzymatic process converts mustard seed to valuable meal and oil components. Chemurgic Dig. 22:5–6.

14007. Mustard, M. J. 1955: Pollen production and seed development in the white sapote. Botan. Gaz. 116:189–92.

14008. Muszynski, J. 1935: Wirkung des Chlorpikrins auf die Keimung der Samen. Acta Soc. Botan. Polon. 12:83–97. Botan. Centralbl. 170:300–1. 1936.
Mutha, S. C. 7849.

14009. Muthanna, M. C., and B. Ahmad. 1940: Vitamin C content of germinated cereals. Current Sci. (India) 9:320–21. Chem. Abstr. 34:7993. 1940.
Mutul, I. F. 10887, 10888.

14010. Mutuskin, A. A., K. V. Pshenova, and P. A. Kolesnikov. 1963: On the biological role of the non-hemin iron of embryos of wheat. Dokl. Akad. Nauk SSSR 150:184–87.
Mutuskin, A. A. 11016.

14011. Muxi-Frecerro, J. R. 1956/57: Hemicellulose of wheat endosperm. pR (Montevideo) 6/7:38–45. Chem. Abstr. 54:4786. 1960.

14012. Muzik, T. J., 1954: Development of fruit, seed, embryo, and seedling of *Hevea brasiliensis*. Am. J. Botany 41: 39–43.

14013. —— 1956: Studies on the development of the embryo and seed of *Hevea brasiliensis* in culture. Lloydia 19:86–91.
Muzik, T. J. 11833, 12448.
Myczkowski, J. 12875.

14014. Myers, A. 1935: Germination tests of needlewood (*Hakea saligna*) seed. Agr. Gaz. N.S. Wales 46:551.

14015. —— 1935: The viability of parsnip seed. Agr. Gaz. N.S. Wales 46:672.

14016. —— 1936: The germination of Cootamundra wattle and other *Acacia* seed. Agr. Gaz. N.S. Wales 47:55.

14017. —— 1939: Prickly pear seed may remain viable for years. Germination tests with *Opuntia* species. Agr. Gaz. N.S. Wales 50:660.

14018. —— 1940: Longevity of seed of native grasses. Agr. Gaz. N.S. Wales 51:405.

14019. —— 1941: Common seed impurities of lucerne. Agr. Gaz. N.S. Wales 52:454–59. Biol. Abstr. 16:7868. 1942.

14020. —— 1942: Curly Mitchell grass. Reasons for poor field germination. Agr. Gaz. N.S. Wales 53:254.

14021. —— 1942: Cold storage of onion seed. Agr. Gaz. N.S. Wales 53:528.

14022. —— 1942: Effect of naphthalene on stored lucerne seed. Agr. Gaz. N.S. Wales 53:570. Brit. Abstr. B III:84.

14023. —— 1942: Germination of seed of curly Mitchell grass (*Astrebla lappacea* Domin.). J. Australian Inst. Agr. Sci. 8:31–32.

14024. —— 1943: Germination of stored onion seed. Agr. Gaz. N.S. Wales 54:398.

14025. —— 1943: The germination of prickly-pear seed. Agr. Gaz. N.S. Wales 54:444.

14026. —— 1945: The germination of quandong seed. Need for continuously moist conditions. Agr. Gaz. N.S. Wales 56:450.

14027. —— 1946: Germination tests of stored tomato seed. Agr. Gaz. N.S. Wales 57:290.

14028. —— 1948: Hollow Heart: An abnormal condition of the cotyledons of *Pisum sativum* L. Proc. Intern. Seed Testing Assoc. 14:35–37.

14029. —— 1950: "After-ripening" in relation to delayed germination of seeds. Agr. Gaz. N.S. Wales 61:279–80.

14030. —— 1952: Germination of stored lucerne seed. Results of four-year experiment. Agr. Gaz. N.S. Wales 63:284–85.

14031. —— 1952: A manual of seed testing. N.S. Wales Dept. Agr. 80 pp.

14032. —— 1954: Viability of strawberry seed. Agr. Gaz. N.S. Wales 65:31.

14033. —— 1958: Testing maize for germination with tetrazolium salts. Agr. Gaz. N.S. Wales 69:243.

14034. —— 1962: Germination of sprung and shot seed wheat. Agr. Gaz. N.S. Wales 73:64–65, 97, 99.

14035. —— 1963: Germination of *Phalaris* seed. Agr. Gaz. N.S. Wales 74:635–37.

14036. —— and D. Liels. 1932: Germination test of wattle seed. Agr. Gaz. N.S. Wales 43:947.

14037. —— and N. Nicholls. 1948: Effect of heat on lupin seed germination Agr. Gaz. N.S. Wales 59:580.
Myers, A. 3122.

14038. Myers, H. S., and E. B. Working. 1944: The sulfhydryl groups of wheat flour. Cereal Chem. 21:32–37. Expt. Sta. Rec. 92:467–68.

14039. Myers, W. M. 1936: A correlated study of the inheritance of seed size and botanical characters in the flax cross, Redwing × Ottawa 770B. J. Am. Soc. Agron. 28:623–35.

Mygind, H. 9835.

14040. Myodo, H., and M. Okumura. 1953: Studies on flower-seed production. II. Development and ripening of the seed of scarlet sage (*Salvia splendens* Ker-Gawl.). Hort. Assoc. Japan J. 22:123–28.

14041. Myrbäck, K., and S. Myrbäck. 1931: Catalase in barley. I. Svenska Bryggarefören. Manadsbl. 40:186–213. Chem. Abstr. 26:5167–68. 1932.

14042. —— and —— 1936: Free and bound amylase of barley. Biochem. Z. 285:282–89. Chem. Abstr. 30:8305. 1936.

14043. —— and B. Örtenblad. 1937: The β-amylase from ungerminated barley. Biochem. Z. 293:107–17. Chem. Abstr. 32:606–7. 1938.

14044. —— and —— 1938: Über aktive und inaktive Amylase in Gerste. Enzymologia 2:305–9. Biol. Abstr. 12:11114. 1938.

14045. —— and —— 1940: Nochmals über Amylasebestimmung in Gerste. Enzymologia 9:43–52.

Myrbäck, S. 14041, 14042.

Mysakowska-Palleolog, B. 11638.

N

Nabeya, A. 8303.

14046. Nabokich, A. J. 1903: Über den Einfluss der Sterilisation der Samen auf die Atmung. Ber. Deut. Botan. Ges. 21:279–91.

14047. Nada, I. A. A., and A. Rafaat. 1955: Carbohydrate changes during germination of *Vicia faba* seeds. Indian J. Agr. Sci. 25:271–80. Chem. Abstr. 50:10873. 1956.

14048. —— and —— 1955: Growth and carbohydrate changes during formation of *Vicia faba* seeds and fruit. Indian J. Agr. Sci. 25:281–92. Chem. Abstr. 50:10873. 1956.

14049. Nadelmann, H. 1890: Über die Schleimendosperme der Leguminosensamen. Ber. Deut. Botan. Ges. 7:248–55.

14050. Nadezhina, T. P. 1962: Some features of the morphology and biology of tragacanth-bearing *Astragalus* spp. of the Kopet-Dagh Range. Akad. Nauk SSSR Botan. Inst. im V. L. Komarova Tr. Ser. 5 Rast. Syr'e 10:119–49.

Nadig, H. N. K. 9001.

14051. Nadkarni, M. D., J. W. Airan, and S. V. Shah. 1947: Fat from the seeds of *Vangueria spinosa* (N. O. Rubiaceae). J. Indian Chem. Soc. 24:25–30.

Nadkarni, S. B. 5290.

14052. Nadson, G. A., and C. Stern. 1935: Action des métaux à distance sur les graines en germination. Compt. Rend. 201:159–61. Biol. Abstr. 11:15720. 1937.

14053. Nadvornik, J. 1926: Das Samengewicht und dessen Einfluss auf die Keimung und die Entwickelung der Keimpflanze bei den Futtergräsern. Vestn. Česk. Akad. Zeměděl. 2:778–82.

14054. —— 1931: Ist es möglich die Zeitdauer der Keimprüfung der Grassamen zu verkürzen? Pub. Landw. Landesversuchsanstalt Brno Česk. Zeměděl. 13(49). 7 pp.

14055. —— 1932: Die Dauer der Keimversuche der Grassamen. Proc. Intern. Seed Testing Assoc. 4:22–29.

14056. —— 1946: Application of vital staining of seeds of fruit trees and shrubs. Vestn. Česk. Akad. Zeměděl. 20:160–64. Biol. Abstr. 21:7203. 1947.

14057. —— 1950: Changes in germination of some flower seed during eight years of storage. Sb. Česk. Akad. Zeměděl. 22: 569–73. Biol. Abstr. 25:29012. 1951.

Naeem, A. 12382.

Nafissi, M. M. 26.

14058. Nag, N. C., and H. N. Banerjee. 1930/31: The proteolytic enzyme in the seeds of *Phaseolus mungo* L. Trans. Bose Res. Inst. (Calcutta) 6:189–94.

14059. —— and —— 1937: Examination of seeds of certain varieties of *Meconopsis* as source of oil and manure. Trans. Bose Res. Inst. (Calcutta) 10:125–43.

14060. ——, ——, and A. K. Pain. 1935/36: Seeds of *Pachyrhizus angulatus*. Trans. Bose Res. Inst. (Calcutta) 11:83–89. Chem. Abstr. 33:3422. 1939.

14061. ——, ——, and —— 1936: Rat-feeding experiments with chhola, sankalu seeds and soybeans. Sci. Cult. (Calcutta) 1:779–80. Chem. Abstr. 30:6041. 1936.

Nag, S. 3779.

14062. Nagabhushanam, A., K. S. Srinivasan, and M. Srinivasan. 1962: Amino acid composition of tender and mature field bean (Dolichos lablab). Indian J. Med. Res. 50:916–19. Chem. Abstr. 59:3084. 1963.

14063. Nagai, I. 1916: Some studies on the germination of the seed of *Oryza sativa*.

Tokyo Imp. Univ. Coll. Agr. J. 3:109–58. Expt. Sta. Rec. 38:24–25. 1918.

14064. —— and S. Nakasima. 1931: On the germination of paddy, upland and "Kanto" rice. Korea Agr. Expt. Sta. Ann. 5:305–14.

Nagai, S. 11568.

Nagai, T. 5991.

14065. Nagai, Y., and S. Funahashi. 1961: Allantoinase from mung bean seedlings. Agr. Biol. Chem. (Tokyo) 25:265–68.

14066. Nagamatsu, T., H. Ikeda, and S. Tanaka. 1958: The geographical variation of amylase activities in germinating seeds of wheat varieties. Japan. J. Breeding 8:100–4. Biol. Abstr. 35:71648. 1960.

14067. —— and T. Katayama. 1958: Effect of di-aminocyclohexane-tiol upon X-rayed rice seeds. Japan. J. Breeding 8:202.

14068. —— and —— 1959: On the X-ray sensitivity in rice varieties. Japan. J. Breeding 9:259–60.

Nagamatsu, T. 9361.

14069. Nagant, H. 1929: Chimisme de la graine et des phénomènes de la germination. Rev. Inst. Agr. d'Oka 3:82–84. Biol. Abstr. 6:24708. 1932.

14070. Nagao, A., and S. Asakawa. 1963: Light-sensitivity in the germination of *Abies* seeds. J. Japan. Forestry Soc. 45:375–77.

14071. Nagao, M., Y. Esashi, T. Tanaka, T. Kumagai, and S. Fukumoto. 1959: Effects of photoperiod and gibberellin on the germination of seeds of *Begonia evansiana* Andr. Plant Cell Physiol. (Tokyo) 1:39–47.

14072. ——, K. Saito, and M. Sasaoka. 1963: Further studies in the effect of gibberellin on the germination of seeds of *Begonia evansiana* Andr. Sci. Rept. Tohoku Univ. Fourth Ser. 29:337–47.

Nagaraj, B. S. 11682, 12809.

14073. Nagarajan, S. S., and S. Gopalkrishnan. 1958: Root inducing substance in groundnut seeds. Current Sci. (India) 27:29–30.

Nagarathnamma, M. 18125.

14074. Nagata, T., and M. Kikuo. 1958: Effects of environmental conditions on the development of young barley roots. III. Effects of seed on the development of roots and the yields of grain in acid soils. Nippon Dojo-Hiryogaku Zasshi 29:123–25. Chem. Abstr. 54:1665. 1960.

Nagata, T. 14521.

Nagata, Y. 6862, 8438, 8439, 13546.

14075. Nagato, K. 1962: On the hardness of rice endosperm. Proc. Crop Sci. Soc. Japan 31:102–7.

14076. —— and M. Ebata. 1959: Studies on white-core kernel. II. On the physical properties of the kernel. Proc. Crop Sci. Soc. Japan 28:46–50.

14077. —— and —— 1960: Effects of temperature in the ripening periods upon the development and qualities of lowland rice kernels. Proc. Crop Sci. Soc. Japan 28:275–78.

14078. ——, ——, and Y. Tanda. 1960: On the quality of rice kernel in early season cultivation. Proc. Crop Sci. Soc. Japan 28:359–62.

14079. —— and Y. Kobayashi. 1957: Studies on the occurrence of notched-belly kernels (Dogire-mai) in rice plants. Proc. Crop Sci. Soc. Japan 26:13–14.

14080. —— and Y. Kono. 1963: On the grain texture of rice. I. Relations among hardness distribution, grain shape and structure of endosperm tissue of rice kernel. Proc. Crop Sci. Soc. Japan 32:181–89.

14081. —— and K. Sugawara. 1952: Studies on the germinative capacity of rice seeds in relation to the location of the flower panicle. I. Proc. Crop Sci. Soc. Japan 20:276–77.

14082. —— and —— 1952: Studies on the germinative capacity of rice seeds in relation to the location on the flower panicle. II. On the germination of the immature seeds. Proc. Crop Sci. Soc. Japan 21:77–78.

14083. —— and Y. Tanda. 1957: Studies on the qualities of rice kernels. Proc. Crop Sci. Soc. Japan 26:85–86.

14084. ——, R. Yamamoto, and Y. Kobayashi. 1955: Relations between the chafing and discoloration of hulls and the injury in kernels caused by storm. Proc. Crop Sci. Soc. Japan 23:266.

Nagayoshi, H. 30.

14085. Nagbiswas, S. C. 1953: Cabbage seeds from stumps. Sci. Cult. (Calcutta) 18:433–34. Biol. Abstr. 27:26063. 1953.

14086. Nagel, C. M. 1949/50: Liquid seed treatment reduces hazard. S. Dakota Farm Home Res. 1:25–28.

14087. —— and G. Semeniuk. 1947: Some mold-induced changes in shelled corn. Plant Physiol. 22:20–33.

Nagel, C. M. 17196.

14088. Nagler, M. A. 1938: Raising true alpines from seed. Gardeners Chron. 42:287–88.

14089. Nagornaya, I. 1955: Protein fractions of seeds of cotton plants of different ripening rates. Sb. Stud. Rabot Sredneaziat Univ. 1955(11):15–18. Chem. Abstr. 52:21820. 1958.

Naguib, M. I. 10174, 10175.

14090. Nagy, I. 1928: Anatomy of iris seeds. Botan. Közlem 24:30–47. Biol. Abstr. 5:5757. 1931.

Nagymihaly, F. 9732, 19302, 19303.

Nagy-Toth, F. 15229.

Naidu, B. Appala-; *see* Appala-Naidu, B.

14091. Naidu, G. V. B. 1961: Effect of shock on arecanut seed germination. Arecanut J. 12:84–87.

Naidu, M. B. 14092.

14092. Naidu, N. B., M. B. Naidu, Z. H. Osmani, and S. A. Saletore. 1953: *Annona squamosa* seed. II. The insecticidal fatty oil. J. Indian Chem. Soc. Ind. News Ed. 16:179–84. Chem. Abstr. 49:1269. 1955.

14093. —— and S. A. Saletore. 1954: Studies on *Annona squamosa* (sitaphal) seed. III. Production and collection in Hyderabad. Indian J. Agr. Sci. 24:253–59.

Naidu, T. R. B. 7528.

14094. Naik, K. C. 1953: Maximisation of seed production in improved strains and of plants in improved or superior varieties. Madras Agr. J. 40:71–76.

14095. Naik, S. 1954: Effect of α-naphthaleneacetic acid on the rate of germination of seeds. I. Leguminous seeds. J. Indian Botan. Soc. 33:153–61.

Naim, Z. 6232.

14096. Nair, G. V., A. N. Poti, and P. P. Pillay. 1953: Holygarna lactone—a new aldehydo lactone from the seed kernels of *Holygarna arnottiana* Hook. F. J. Sci. Ind. Res. (India) 12B:119.

14097. Nair, K. N. R. 1955: Cultivation of Pyrethrum on the Nilgiris. I. World Crops 5:274–77.

Nair, K. R. 4842, 4843.

14098. Nair, K. V., and N. S. Varier. 1955: Chemical examination and glyceride structure of the fixed oil from the seeds of *Calophyllum wightianum*. II. Bull. Central Res. Inst. Univ. Travancore, Trivandrum Ser. A 4:19–22. Chem. Abstr. 50:11033. 1956.

14099. Nair, N. C. 1956: Early endosperm development in Meliaceae. Sci. Cult. (Calcutta) 22:34–35. Biol. Abstr. 32:2526. 1958.

14100. —— and T. C. Joseph. 1957: Floral morphology and embryology of *Samadera indica*. Botan. Gaz. 119:104–15.

14101. —— and —— 1960: Morphology and embryology of *Cardiospermum halicacabum* Linn. J. Indian Botan. Soc. 39:176–94.

14102. —— and M. Maitreyi. 1962: Morphology and embryology of *Sebastiana chamaelea*. Botan. Gaz. 124:58–68.

14103. —— and P. N. N. Nambisan. 1957: Contribution to the floral morphology and embryology of *Leea sambucina* Willd. Botan. Notiser 110:160–72.

14104. —— and K. R. Narayanan. 1961: Studies on the Aristolochiaceae. II. Contribution to the embryology of *Bragantia wallichii*. Lloydia 24:199–203.

14105. —— and V. Parasuraman. 1962: Embryological studies of *Vitis pallida* W & A. Phyton 18:157–64.

14106. Nair, N. R., and P. C. Sahadevan. 1962: An instance of bran layer influencing seed dormancy in rice. Current Sci. (India) 31:72–73.

Nair, N. R. 16629.

14107. Naismith, W. E. F. 1955: Ultracentrifuge studies of the ammonium sulphate fractions of the protein from yellow lupin seed (*Lupinus luteus*). J. Sci. Food Agr. 6:721–24.

14108. —— 1956: Ultracentrifuge studies on cottonseed protein. J. Appl. Chem. 6:283–86.

14109. —— 1958: Studies on the protein from groundnut protein spinning solutions. J. Appl. Chem. 8:518–22.

14110. —— and K. O. Kelly. 1958: The acid denaturation of the proteins of the groundnut (*Arachis hypogaea*). J. Appl. Chem. 8:598–605.

14111. —— and H. M. R. McDavid. 1958: Fractionation studies on the proteins of the groundnut (*Arachis hypogaea*). J. Appl. Chem. 8:605–10.

Naito, G. 18298.

14112. Naito, H., and K. Isimaru. 1940: Enzymes in fruit and vegetables. III. The effect of sunlight on the strength of ascorbic acid oxidase during sprouting and the relation between the activity of the enzyme and its concentration. Bull. Inst. Phys.-Chem. Res. (Tokyo) 19:996–1000. Chem. Abstr. 34:7336–37. 1940.

14113. —— and —— 1941: Enzymes in fruit and vegetables. VII. Relation between ascorbic acid oxidase and ripening of some fruits and vegetables. Bull. Inst. Phys.-Chem. Res. (Tokyo) 20:225–33. Chem. Abstr. 35:3682. 1941.

14114. ——, ——, and Y. Hosoda. 1941: Enzymes in fruit and vegetables. VI. Activity of peroxidase during germination and growth. Bull. Inst. Phys.-Chem. Res. (Tokyo) 20:44–58. Chem. Abstr. 35: 3681–82. 1941.

Naito, Y. 10321.

Nakagava, M. 6526.

Nakagawa, A., 6818.

14115. Nakagawa, H., and S. Henmi. 1958: Effect of low temperature treatment on autumn seed production of Chinese cabbage. II. Japan. J. Breeding 8:6–12.

Nakagawa, K. 13754.

14116. Nakagawa, M., Y.Soga, S. Watanabe, H. Gocho, K. Nishio, and H. Kurata. 1959: Genetical studies on kernel texture of wheat and barley. Tokai Kinki Natl. Agr. Expt. Sta. Bull. First Agron. Div. 6:102–35. Biol. Abstr. 35:60855. 1960.

Nakagawa, S. 9576.

Nakagawa, T. 8853.

14117. Nakai, T. 1936: Liliaceous plants with exposed seeds. J. Japan. Botany 12: 773–83. Japan. J. Botany 9:14–15. 1937.

14118. Nakajima, F., and H. Morishima. 1958: Studies on embryo culture in plants. II. Embryo culture interspecific hybrids in *Oryza*. Japan. J. Breeding 8:105–10.

14119. Nakajima, K. 1932: Proteins and oil of the soybean. J. Fac. Agr. Hokkaido Imp. Univ. 31:165–356. Chem. Abstr. 27: 2053. 1933.

14120. Nakajima, T. 1952: A root-growth-inhibitor found in *Cucurbita* (*C. moschata*) endosperm. Japan. J. Breeding 2:97–102.

14121. —— 1958: Embryo culture in plants. I. Action of auxin and embryo factor on embryonic growth. Japan. J. Breeding 7:161–68. Chem. Abstr. 56:723. 1962.

14122. —— 1958: On the artificial cultures of endosperm tissue. [Abstract.] Japan. J. Breeding 8:185–86.

14123. —— 1962: Physiological studies of seed development, especially embryonic growth and endosperm development. Bull. Univ. Osaka Prefect. Ser. B 13:13–48. Biol. Abstr. 41:24335. 1963.

14124. —— and H. Morishima. 1958: Studies on embryo culture in plants. II. Embryo culture interspecific hybrids in *Oryza*. Japan. J. Breeding 8:105–10.

14125. Nakajima, Y. 1919: Über das Verbreitungsmittel der Samen von *Ottelia alismoides* Pers. Botan. Mag. (Tokyo) 33:44–52. Botan. Abstr. 4:249. 1920.

14126. —— 1921: Length of life of the seed of the genus *Salix*. Botan. Mag. (Tokyo) 35:17–42. Botan. Abstr. 14:3007. 1925.

14127. —— 1925: Über die Keimfähigkeitsdauer der Reiskörner. Botan. Mag. (Tokyo) 39:265.

14128. —— 1926: Weitere Untersuchungen über die Lebensdauer der Weidensamen. Sci. Rept. Tohoku Univ. Fourth Ser. 1:261–75. Japan. J. Botany 3:34–35. 1928.

14129. —— 1927: Untersuchungen über die Keimfähigkeitsdauer der Samen. Botan. Mag. (Tokyo) 41:604–32.

14130. —— 1928: Über das Vorkommen kleiner embryonaler Körperchen in der Frucht von *Crinum latifolium* L. Sci. Rept. Tohoku Univ. Fourth Ser. 3:431–42.

14131. —— 1928: Untersuchungen über die Keimung der Samen einiger Wasserpflanzen. Botan. Mag. (Tokyo) 42:576–91. Japan. J. Botany 4:67–68. 1929.

14132. —— 1931: Über die Hartschaligkeit und die Quellung der Samen von *Astragalus sinicus*, L. I–II. Botan. Mag. (Tokyo) 45:555–66. Japan. J. Botany 6:12–13.

14134. Nakamura, A., T. Kono, and S. Funahashi. 1958: Nature of lysolecithin in rice grains. I. Lysolecithin as a constituent of non-glutinous rice grains. Bull. Agr. Chem. Soc. Japan 22:320–24.

14135. ——, R. Shimizu, T. Kono, and S. Funahashi. 1958: Nature of lysolecithin in rice grains. II. Complex formation of lysolecithin with starch. Bull. Agr. Chem. Soc. Japan 22:324–29.

14136. Nakamura, E. 1956: Vernalization of spinach. I. Interaction between vernalization and day length. Hort. Assoc. Japan J. 25:167–72.

14137. —— 1961: Seed-vernalization of the cabbage (*Brassica oleracea* spp.). I. Effect of seed-vernalization on the flowering in cabbage. J. Japan. Soc. Hort. Sci. 30:57–62.

14138. —— and Y. Hattori. 1961: On the seed-vernalization of the cabbage. II. Effects of a prolonged vernalization treatment and influence of gibberellin applied during vernalization. J. Japan. Soc. Hort. Sci. 30:167–70.

14139. ——, ——, and H. Otono. 1962: Studies on branching in *Pisum sativum*. I. Effects of photoperiod and seed-vernalization on the branching. J. Japan. Soc. Hort. Sci. 31:64–72.

Nakamura, F. I. 8077.

14140. Nakamura, H. 1940: Die Abhängigkeit der Enzymentwicklung insbesondere von Amylase und Protease, von der Art der Gerste und ihrer Keimung. J. Agr. Chem. Soc. Japan 16:69–70. Biol. Abstr. 16:5496. 1942.

14141. —— 1940: Germination and enzyme action of barley. J. Agr. Chem. Soc. Japan 16:277–80. Chem. Abstr. 34:6326. 1940.

14142. ——, T. Sato, and M. Asakawa. 1957: Studies on the cultivation of upland cotton, especially in regard to its blooming and ripening. II. Effects of different sowing dates on seed cotton weight and lint weight. Proc. Crop Sci. Soc. Japan 26:143–46.

Nakamura, H. [K.] 10326, 10370, 10374, 18314, 19926.

Nakamura, I. 9054.

14143. Nakamura, K., M. Goo, and S. Hasegawa. 1954: Does the quality of seed change yearly? An experiment employing seeds from the same mother tree of *Cryptomeria japonica*. J. Japan. Forestry Soc. 36:1–2.

14144. Nakamura, M. 1959: The sucrose-synthesizing enzyme in the beans. Bull. Agr. Chem. Soc. Japan 23:398–405. Biol. Abstr. 35:11272. 1960.

Nakamura, M. 9362, 10126.

14145. Nakamura, S. 1954: Germination of edible burdock (*Arctium lappa* L.) seeds. I. Hort. Assoc. Japan J. 23:43–47.

14146. —— 1954: Germination of the edible burdock (*Arctium lappa* L.) seeds. II. Effect of the wave length of radiation on the germination. Hort. Assoc. Japan J. 23:108–14.

14147. —— 1958: Storage of vegetable seeds. Hort. Assoc. Japan J. 27:32–44.

14148. —— 1962: Germination of legume seed. Proc. Intern. Seed Testing Assoc. 27:694–709.

14149. —— 1962: Germination of grass seeds. Proc. Intern. Seed Testing Assoc. 27:710–29.

14150. —— 1963: Short communication on dormancy of rice seed. Proc. Intern. Seed Testing Assoc. 28:57–59.

14151. —— and Y. Okasako. 1957: On the germination of *Perilla* seed. Hort. Assoc. Japan J. 26:28–32.

14152. ——, ——, and E. Yamada. 1955: Effect of light on the germination of vegetable seeds. Hort. Assoc. Japan J. 24:17–28.

14153. ——, S. Watanabe, and J. Ichihara. 1960: Effect of gibberellin on the germination of agricultural seeds. Proc. Intern. Seed Testing Assoc. 25:433–39.

14154. Nakamura, S. I. 1962: Zur Samenkeimung einer Chlorophyllfreien Erdorchidee *Galeola septentrionalis* Reichb. f. Z. Botan. 50:487–97.

14155. Nakamura, T. 1940: Form quotient and colour in the seeds of *Pinus densiflora* Sieb. et Zucc. Bull. Tokyo Univ. Forests 29:1–47.

Nakamura, T. 9529, 17358.

14156. Nakamura, Y., and K. Hess. 1938: Chemical composition of maize coleoptiles. Ber. Deut. Chem. Ges. 71B:145–52. Chem Abstr. 32:2514–15. 1938.

14157. ——, K. Mori, and R. Yamamoto. 1959: Studies on the coloring matters of seed coat of "Azuki" bean. Mem. Fac. Agr. Hokkaido Univ. 3:148–50.

Nakano, H. 10377.

Nakasima, S. 14064.

14158. Nakatomi, S. 1933: Effects of low temperature on the germination of pollen grains and seeds. Proc. Crop Sci. Soc. Japan 5:91–101. Japan. J. Botany 6:108–9. 1933.

14159. —— 1936: Relations of oxydase contents to the germination capacity of cotton seeds. Proc. Crop Sci. Soc. Japan 8:341–48. Japan. J. Botany 9:16. 1937.

14160. Nakatsu, S., Z. Matsugaki, H. Shimada, T. Kashida, T. Motomura, M. Oshima, and N. Sakai. 1962: Distribution of canavanine in plant seeds. Seikagaku 34:253–57. Chem. Abstr. 58:752. 1963.

14161. Nakaya, T., and H. Iwata. 1952: Hemicelluloses. I. The composition of the hemicellulose from the kernel of *Rhus succedanea*. J. Agr. Chem. Soc. Japan 26:196–99. Chem. Abstr. 48:9482. 1954.

14162. Nakayama, K. 1935: Study on the inheritance of the size and shape of rice grains. Sci. Rept. Tohoku Imp. Univ. 10:569–80.

14163. Nakayama, R. 1960: Genetical studies on kidney beans (*Phaseolus vulgaris*). V. On the inheritance of seed-coat color. 1. Bull. Fac. Agr. Hirosaki Univ. 6:21–36.

14164. —— 1961: Genetical studies on kidney beans (*Phaseolus vulgaris*). VI. Genetical behavior of plant color, with special reference to seed-coat color in the hybrids between white kidney and other varieties. Bull. Fac. Agr. Hirosaki Univ. 7:37–65.

Nakayama, T. 6912.

Nakazawa, H. 17323.

14165. Nambiar, K. K. 1944: A novel method of improving the germination of *Prosopis juliflora* seeds. Madras Agr. J. 32:203–6. Biol. Abstr. 22:7411. 1948.

Nambisan, P. N. N. 14103.

14166. Namitokov, I. K. 1953: Growing corn from embryos separated from the endosperm. Agrobiologiya 1953(2):32–33.

14167. Nanda, K. K. 1950: Catalase ratio as a rapid method for determining the germinating capacity of seeds. Current Sci. (India) 19:22–24.

14168. —— and J. J. Chinoy. 1958: The influence of the temperature of ripening period and the grain weight on the yield of wheat. Phyton 10:43–50.

14169. ——, ——, and S. M. Gupta. 1959: A method for the determination of the rate of seedling growth and its application to the study of the effect of presowing hardening treatment on wheat grain. Phyton 12:153–61.

14170. ——, ——, and G. S. Sirohi. 1958: Energy relations of crop plants. I. The quantitative relationship between the duration of vernalization treatment and photothermic quantum in wheat. Indian J. Plant Physiol. 1:5–16.

14171. ——, ——, and —— 1958: Energy relations of crop plants. II. The quantitative relationship of stem growth in wheat with photothermic and vernalization quanta. Indian J. Plant Physiol. 1:17–23.

14172. ——, ——, and —— 1958: Energy relations of crop plants. III. The effect of different durations of vernalization and photoperiod on tiller production in wheat and its relation with photothermic quantum. Indian J. Plant Physiol. 1:24–29.

14173. ——, R. Grover, and J. J. Chinoy. 1957: Factors affecting growth and development of some millets. I. Stem elongation and its correlation with flowering as influenced by the time of sowing. Intern. J. Exptl. Botany 9:15–24.

14174. Nandi, D. L. 1958: Studies on the changes in free amino acid and B-vitamin content of some leguminous seeds during germination. Sci. Cult. (Calcutta) 23:659–60. Biol. Abstr. 33:15415. 1959.

14175. —— 1958: Studies on the effects of darkness and ultraviolet irradiation on the free amino acid and B-vitamin contents of some leguminous seeds during germination. Naturwissenschaften 45:246.

14176. —— 1958: Studies on the free amino acid composition and urease activity of some leguminous seeds during germination in different organic nitrogen media. Naturwissenschaften 45:246–47.

14177. —— 1960: Effects of diffused light and darkness on the B-vitamin contents of germinating pulses. Food Res. 25:88–96.

Nandi, N. 1845, 1848.

14178. Nandi, P. N., and A. K. Ganguly. 1947: Effect of papain on the diastatic power of dormant barley grains and their corresponding malts. Bull. Botan. Soc. Bengal 1:55–60. Biol. Abstr. 23:30233. 1949.

14179. Nandi, T. 1934: Abnormal development of the radicle in mango. Current Sci. (India) 3:128.

Naniwa, I. 13543–13545.

14180. Nanji, H. R., G. R. Savur, and A. Sreenivasan. 1945: Tamarind seed pectin. Current Sci. (India) 14:129–30. Chem. Abstr. 39:4648. 1945.

Nanji, H. R. 10563.

14181. Nannfeldt, J. A. 1937: On *Poa jemtlandica* (Almqu.) Richt., its distribution and possible origin. A criticism of the theory of hybridization as the cause of vivipary. Botan. Notiser 1937(1/2):1–27.

14182. Napier, C. L. 1952: Seed selection and harvesting can be begun in the home garden in February. New Zealand J. Agr. 84:49–55.

14183. Napp-Zinn, K. 1957: Untersuchungen über das Vernalisationsverhalten einer Winterannuellen Rasse von *Arabidopsis thaliana.* Planta 50:177–210.

14184. —— 1958: Über den Einfluss von KH_2PO_4 auf die Vernalisation. Z. Botan. 46:506–15.

14185. —— 1958: Vernalisation und Thermoperiodismus. Planta 52:230–34.

14186. —— 1960: Vernalisation, Licht und Alter bei *Arabidopsis thaliana* (L.) Heynh. I. Licht und Dunkelheit während Kälte- und Wärmebehandlung. Planta 54:409–44.

14187. —— 1961: Vernalisation und verwandte Erscheinungen. Encyclopedia Plant Physiol. 16:24–75.

14188. —— 1962: On the genetic foundation of vernalization requirements in *Arabidopsis thaliana.* I. The number of factors to be considered. Z. Vererbungslehre 93:154–63.

14189. —— 1963: Über den Einfluss von Genen und Gibberellinen auf die Blütenbildung von *Arabidopsis thaliana*. Ber. Deut. Botan. Ges. 76:77–89.

14190. —— 1963: Über die Beziehungen zwischen Vernalisation, Keimung und Atmung. Untersuchungen an *Arabidopsis thaliana* (L.) Heynh. Z. Botan. 51:317–39.

14191. Nara, S. 1961: Changes of sugar content in γ-irradiated and stored chestnuts. J. Agr. Chem. Soc. Japan 35:415–18. Nucl. Sci. Abstr. 16:1879. 1962.

Narain, P. 5335.

14192. Narain, R., and S. Dutt. 1945: Chemical examination of the seeds of *Tamarindus indica*. No pectin in tamarind seeds. Indian J. Agr. Sci. 15:209–13. Biol. Abstr. 21:1947. 1947.

14193. Narasaki, T. 1959: Anthrone reaction of the unidentified sugar in a hemicellulose from broad-bean seeds. Tech. Bull. Fac. Agr. Kagawa Univ. 11:265–69.

Narasaki, T. 10330, 10371, 10372.

14194. —— 1942: Endosperm formation in *Anisomeles* sp. Current Sci. (India) 11: 284–85.

14195. Narasimha, R. L. A. 1929: Note on the determination of the area of surface of cotton seeds. Madras Dept. Agr. Yearbook 1929:7–15.

14196. Narasimhachar, S. G. 1948: A contribution to the embryology of *Acacia farnesiana* L. (Willd.). Proc. Indian Acad. Sci. Sect. B 28:144–49.

14197. —— 1949: A contribution to the embryology of *Drosera burmanni* Vahl. Proc. Indian Acad. Sci. Sect. B 39:98–104.

14198. —— 1951: An embryological study of *Minosa pudica* Linn. Proc. Indian Acad. Sci. Sect. B 33:192–98.

14199. Narasimha-Murthy, S. 1941: The development of endosperm in *Leucas aspera* Spreng. Current Sci. (India) 10: 257–59.

14200. Naravi, G. F. 1954: Contribución al estudio de la composición del fruto y semilla del *Inga feullei* o "paeae." Anales Fac. Farm. Bio-Quim. (Lima) 5:188–91. Nutr. Abstr. Rev. 27:1628. 1957.

14201. Narayan, A. 1960: Inheritance of spininess of capsule in *Ricinus communis*. Indian J. Genet. Plant Breeding 20:223–26.

14202. Narayana, C. S., and S. Rangaswami. 1956: Isolation of three crystalline substances from the seeds of *Cassia tora* Linn. Current Sci. (India) 25:339.

Narayana, G. V. 15105.

14203. Narayana, H. S. 1962: Seed structure in the *Aizoaceae*. Proc. Summer School Botany, June 2–15, 1960, Darjeeling. Min. Sci. Res. Cult. Affairs, Govt. India, New Delhi, pp. 220–30. Biol. Abstr. 41:20022. 1963.

14204. —— and K. Jain. 1962: A contribution to the embryology of *Limeum indicum*. Lloydia 25:100–8.

14205. Narayana, L. L. 1960: Embryology of Staphyleaceae. Current Sci. (India) 29:403–4.

14206. —— 1962: Development of embryo in *Biophytum intermedium* Wight. and *Oxalis pubescens* H. B. & K. J. Indian Botan. Soc. 41:156–59.

14207. —— 1963: Contributions to the embryology of Balsaminaceae. I. J. India Botan. Soc. 42:102–9.

14208. —— 1963: A note on the embryology of a few Rutaceae. Current Sci. (India) 32:516–17.

Narayana, N. 11271 11272, 17837.

14209. Narayana, R. 1957: Contribution to the embryology of *Dendrophthae* Mart. Phytomorphology 4:173–79. Biol. Abstr. 31:22211. 1957.

14210. Narayanan, E. S. 1953: Insect pests of stored grains and their control. Indian Farming 3(7):8–13, 26–27.

Narayanan, E. S. 15954.

14211. Narayanan, K. M., G. S. Bains, and D. S. Bhatia. 1963: Lipoxidase inhibitor in peanut (*Arachis hypogaea*) testa. Chem. Ind. (London) 1963:1588. Chem. Abstr. 59:15604. 1963.

Narayanan, K. R. 14104.

14212. Narayanan, R., and A. R. S. Kartha. 1962: The glyceride structure of *Mangifera indica* seed fat. J. Sci. Ind. Res. (India) 21B:400–1. Biol. Abstr. 41:7902. 1963.

14213. —— and —— 1962: The glyceride structures of the seed and mace fats from *Myristica fragrans*. J. Sci. Ind. Res. (India) 21B:442–44. Biol. Abstr. 41:20399. 1963.

Narayanan, R. 10289.

14214. Narayanan, S. S., S. Sankaran, and G. Ramabhadran. 1959: Safflower, a suitable oil seed crop for blacksoils. Madras Agr. J. 46:472–76.

14215. Narayanan, T. R., and K. V. Menon. 1960: Effect of presoaking seed in phytohormonal solution on the growth and yield of crops. Madras Agr. J. 47:158–65. Biol. Abstr. 35:56946. 1960.

Narayanan, V. 1581.

14216. Narayanaswami, S. 1954: The structure and development of the caryopsis in some Indian millets. III. *Paspalum scrobiculatum* L. Bull. Torrey Botan. Club 81:288–99.

14217. —— 1955: The structure and development of the caryopsis in some Indian millets. V. *Eleusine coracana* Gaertn. Papers Mich. Acad. Sci. 40:33–46. Biol. Abstr. 29:26868. 1955.

14218. —— 1959: Experimental studies on growth of excised grass embryos in vitro. II. Effect of maleic hydrazide on embryo growth. Bull. Torrey Botan. Club 86:248–58.

14219. —— 1962: Morphological variations in growth and differentiation of embryos in vitro. Proc. Summer School Botany, June 2–15, 1960, Darjeeling. Min. Sci. Res. Cult. Affairs, Govt. India, New Delhi, pp. 231–39. Biol. Abstr. 41:20080. 1963.

14220. —— and S. K. Roy. 1960: Embryo sac development and polyembryony in *Syzygium cumini* (Linn.) Skeels. Botan. Notiser 113:273–84.

14221. —— and —— 1960: Embryology of the genus *Psidium*. J. Indian Botan. Soc. 39:35–45.

14222. —— and N. S. R. Swamy. 1960: Plant embryo culture. Mem. Indian Botan. Soc. 2:27–29. 1959.

Narayanaswami, S. 12702.

14223. Nardina, N. S. 1957: The influence of manganese upon germination of under shrubs in sandy deserts. Izv. Akad. Nauk Turkm. SSR 1957(2):80–85. Chem. Abstr. 52:1370i. 1958.

14224. —— 1959: The effects of glucose on the germination of the seeds of psammophytes. Izv. Akad. Nauk Turkm. SSR Ser. Biol. Nauk 1959(1):57–61. Biol. Abstr. 43:20566. 1963.

14225. Narimatsu, K. 1931: Investigation of the histological change in plant shoots caused by X-rays. Especially the effect upon the arrangement of tissues in the seed of *Vicia faba*. Japan. J. Obstet. Gynecol. 14:50–58. Biol. Abstr. 6:24680. 1932.

Narita, S. 8743.

Nasatir, M. 9898.

Nash, A. M. 5887, 17613.

Nash, H. A. 5504.

14226. Nash, S. M., and S. Wilhelm. 1960: Stimulation of broomrape seed germination. Phytopathology 50:772–74.

14227. Nasibov, T., and V. I. Runov. 1962: Effect of temperature on content of nitrogenous substances upon germination of cotton seeds. Dokl. Akad. Nauk Uz. SSR 19(9):59–62. Chem. Abstr. 58:11687. 1963.

14228. Nason, A. 1950: The distribution and biosynthesis of niacin in germinating corn. Am. J. Botany 37:612–23.

Nason, A. 4219.

14229. Nassanov, V. A. 1939: The distribution of the cell constituents in cottonseed. Tr. Vses. Tsentr. Nauchn.-Issled. Inst. Zhirov 1939:25–37. Chem. Abstr. 36:5500. 1942.

14230. Nast, C. G. 1941: The embryogeny and seedling morphology of *Juglans regia* L. Lilloa 6:163–206.

14231. Natarajan, A. T., and G. Ahnstroem. 1962: Oxygen saturation and dry seed irradiation. Nucl. Sci. Abstr. 16:1273–74.

14232. —— and M. M. Maric. 1961: The time-intensity factor in dry seed irradiation. Radiation Botany 1:1–9.

Natarajan, A. T. 2416.

Natarajan, C. P. 18125.

14233. Natarajan, K. R. 1935: Indian acorns have food value. Ind. Eng. Chem. News Ed. 13:114.

14234. —— 1939: Sandalwood seed yields semi-drying oil. Ind. Eng. Chem. News Ed. 17:292.

14235. Nath, M. C., and M. L. Belkhode. 1958: Synthesis of glucose-cyclo-acetoacetate and biosynthesis of ascorbic acid in germinating Mung beans (*Phaseolus radiatus*). Proc. Soc. Exptl. Biol. Med. 99:544–46.

Nath, M. C. 2086, 2087.

14236. Nath, R., and K. V. Giri. 1957: Physicochemical investigations on indigenous seed proteins. I. Studies on the solubilization of nitrogenous constituents of *Sesamum indicum* and characterization of its proteins by electrophoresis. J. Sci. Ind. Res. (India) 16C:5–8. Chem. Abstr. 51:10675. 1957.

14237. —— and —— 1957: Physicochemical investigations on indigenous seed proteins. II. Fractionation isolation and cellectrophoretic characterization of sesame globulins. J. Sci. Ind. Res. (India) 16C:51–58. Chem. Abstr. 51:13953. 1957.

14238. ——, K. H. Rao, and K. V. Giri. 1957: Physicochemical investigations on indigenous seed proteins. III. Amino acid composition of sesame seed globulins. J. Sci. Ind. Res. (India) 16C:228–30. Chem. Abstr. 52:12097. 1958.

Nath, S. 10243.

14239. Nath, Y., and I. C. Chopra. 1962: Oxytocic principle from the seeds of *Cassia tora* Linn. Current Sci. (India) 31:285–86.

14240. —— and P. R. Rao. 1962: Amino acid composition of the proteins of the seeds of *Rauwolfia serpentina.* Indian J. Pharm. 24:187. Chem. Abstr. 58:2653. 1963.

Nathanael, W. R. N. 3934, 3935.

14241. Nather, H. 1958: Zur Keimung der Zirbensamen (*Pinus cembra*). Centralbl. Ges. Forstwesen 75:61–70. Biol. Abstr. 32:35130. 1958.

14242. National Institute Agricultural Botany. 1961: Guide to varieties of field and vegetable crops under trial, observation and propagation, 1961. Cambridge, England. 66 pp.

14243. Natlacen, G. 1938: Recent progress in the determination of the germination capacity of forest tree seed. Alpe 25:16–20. Biol. Abstr. 12:3595. 1938.

14244. Natti, J. J. 1958: Effect of seed treatments with antibiotics on bacterial wilt of corn. Plant Disease Reptr. 42:953–58.

14245. —— and W. T. Schroeder. 1955: Silicones as lubricants for pea seed protectants. Plant Disease Reptr. 39:909–13.

14246. ——, ——, G. E. R. Hervey, and F. L. McEwen. 1958: Value of insecticide-fungicide combination treatments as protectants for seed of cucumber and winter squash. Plant Disease Reptr. 42:127–33.

Natti, J. J. 13171.

14247. Natvig, H. 1939: Ascorbic acid in germinating cereals. Nord. Med. 15: 1502–4. Chem. Abstr. 35:4105. 1941.

14248. Nauer, E. M., and C. N. Roistacher. 1962: Long-term storage of heat treated seed. Citrog. 47:140.

Nauer, E. M. 3544.

14249. Naugolnykh, V. N., and Z. F. Skorokhodova. 1938: The xeromorphism of plants hardened against drought before sowing. Bull. Perm. Inst. Rech. Biol. 11:161.

14250. Naumov, N. A. 1930: Study of the germination of *Orobanche cumana* Wallr. seeds. Zap. Leningr. Sel'skokhoz. Inst. 6(2). 16 pp.

Naumov, S. M. 16036.

Naumova, A. T. 11310, 11311.

14251. Naundorf, G., G. F. Vilamil, and J. Medina. 1950: Contribución al estudio de la fisiologia del cacao (*Theobroma cacao* L.) Las substancias activas e inhibidoras y su influencia en la germinación. Notas Agron. 3:63–86. Biol. Abstr. 24:30980. 1950.

Naundorf, G. 463–465.

Nautiyal, D. D. 15031.

14252. Navashin, M. 1933: Altern der Samen als Ursache von Chromosomenmutationen. Z. Wiss. Biol. Abt. E, Planta 20:233–43. Biol. Abstr. 8:12848. 1934.

14253. —— and H. Gerasimova. 1935: Changes in species properties in the case of storage of seeds. Semenovodstvo 1935(1): 36–46.

14254. ——, ——, and G. M. Belajeva. 1940: On the course of the process of mutation in the cells of the dormant embryo within the seed. Compt. Rend. 26:948–51.

Navashin, M. 17354, 17355.

Navashina, E. N. Gerasinova-; *see* Gerasinova-Navashina, E. N.

14255. Naves, Y. R. 1955: Essential oil of seeds of *Myrocarpus frondosus.* Perfumery Essent. Oil Record 46:183–84.

Navrez, P. 8886.

14256. Nayar, S. L. 1954: Poisonous seeds of India. I. J. Bombay Nat. Hist. Soc. 52:88–105. Biol. Abstr. 30:27069. 1956.

14257. —— 1954: Poisonous seeds of India. II. J. Bombay Nat. Hist. Soc. 52:515–32. Biol. Abstr. 30:27070. 1956.

14258. —— 1955: Experimental propagation and culture of *Rauwolfia serpentina* Benth. by seeds. Symp. Rauwolfia, Lucknow. Abstr. Papers, pp. 1–2.

Nayeemullah, M. 351.

Naylor, A. W. 1926.

14259. Naylor, J. M. 1958/59: The nature of seed dormancy. Ann. Rept. Can. Seed Growers Assoc. 1958/59:21–25.

14260. —— and G. M. Simpson. 1961: Bioassay of gibberellic acid using excised embryos of *Avena fatua* L. Nature 192: 679–80.

14261. —— and —— 1961: Dormancy studies in seed of *Avena fatua.* II. A gibberellin-sensitive inhibitory mechanism in the embryo. Can. J. Botany 39:281–95.

Naylor, J. [M.] 2526, 11643.

14262. Naylor, N. M., and V. L. Dawson. 1936: Effect of treatment of oats on the development of saccharogenic and dextrinogenic amylases during germination. Iowa State Coll. J. Sci. 10:267–74.

Naylor, N. M. 2623, 4527, 14377.

14263. Nazarenko, S. I. 1960: On the duration of the dormancy period of seeds of *Fraxinus excelsior* harvested at various times. Vestn. Sel'skokhoz. Nauki 1960(6): 138–40.

14264. Nazarova, N. S. 1937: Stimulation of germination of seeds of vegetables belonging to Umbelliferae. Bull. Appl. Botany, Genet., Plant Breed. (USSR) Ser. 4(2):84–85.

14265. —— 1937: Shortening the period of germination of the seed of certain forage grasses. Bull. Appl. Botany, Genet., Plant Breed. (USSR) Ser. 4(2):100–1.

14266. —— 1937: Catalase activity as an index of the germination capacity of seeds. Bull. Appl. Botany, Genet., Plant Breed. (USSR) Ser. 4(2):103–11. Chem. Abstr. 31:5843. 1937.

14267. Neal, G. E., and H. Beevers. 1960: Pyruvate utilization in castor-bean endosperm and other tissues. Biochem. J. 74:409–16.

14268. Neal, N. P. 1961: The influence of freezing temperatures of varying intensities and duration on the germination of seed corn at different stages of maturity. Am. Seed Trade Assoc. Hybrid Corn Div. Rept. Hybrid Corn Ind. Res. Conf. 16:67–73.

14269. —— and J. R. Davis. 1956: Seed viability of corn inbred lines as influenced by age and conditions of storage. Agron. J. 48:383–84.

Neal, N. P. 611, 16554.

Neal, O. 17929.

14270. Nearing, G. G. 1932: Experimenting with fringed gentians. Horticulture 10(7): 313.

14271. —— 1941: More edible seeds. Florists Exchange Hort. Trade World 97(25):19.

Neatby, K. W. 552.

14272. Nebel, B. R., and Z. I. Kertesz. 1934: Metaxenia and xenia in apples. IV. Gartenbauwissenschaft 9:45–64.

Nedelko, E. I. 2230.

14273. Neeb, O. 1962: Ein Beitrag zur Methodik der Genonstufenanalyse bei polyploidem β-Saatgut. Proc. Intern. Seed Testing Assoc. 27:603–8.

14274. —— and E. Bornscheuer. 1961: Über die Auswirkung verbesseter Probeziehungsmethoden auf den Repräsentativitätsgrad von Keimergebnissen. Proc. Intern. Seed Testing Assoc. 26:140–61.

Neeb, O. 12494, 12495.

14275. Needham, J. G. 1941: Insects from seed pods of the primrose willow, *Jussiaea angustifolia*. Proc. Entomol. Soc. Wash. 43:2–6. Biol. Abstr. 15:13768. 1941.

14276. Neelakantan, L. 1963: Talking of seeds. Indian Farming 12(12):8, 10.

14277. ——, S. Iyemperumal, and P. Sennaiyan. 1959: Colour of mastigated cotton seed in relation to its germination. Indian Cotton Growing Rev. 13:120–21.

14278. Neely, F. E. 1951: Small petrified seeds from the Pennsylvanian of Illinois. Botan. Gaz. 113:165–79.

Neely, J. W. 2654.

14279. Neeman, J. 1937: Lewisias from seed. Gardeners Chron. 41:255–56.

14280. Neergaard, P. 1940: Seed-borne fungus diseases of horticultural plants. Proc. Intern. Seed Testing Assoc. 12:47–71.

14281. —— 1941: The influence of light treatment on the longevity of seed and fungus spores. Gartnertidende 28. 4 pp. Rev. Appl. Mycol. 25:378–79. 1946.

14282. —— 1941: Seed germination, seed-borne diseases, seed disinfection. A survey of the gardener's problems in connexion with seed production. Fro og Gartneri 1941(6). 12 pp. Rev. Appl. Mycol. 25:378. 1946.

14283. —— 1958: Infection of Danish seeds by *Rhizoctonia solani* Kuehn. Plant Disease Reptr. 42:1276–78.

14284. —— 1960: Report of the committee on plant diseases. Proc. Intern. Seed Testing Assoc. 25:526–36.

14285. —— 1962: Plant disease committee. Proc. Intern. Seed Testing Assoc. 27:104–18.

14286. —— 1962: Tolerances in seed health testing. A discussion on basic principles. Proc. Intern. Seed Testing Assoc. 27:386–99.

14287. —— 1962: The infection percentage as a relative value in assessing disease tolerances for seed health testing. Proc. Intern. Seed Testing Assoc. 27:400–6.

14288. —— and A. Saad. 1962: Seed health testing of rice a contribution to development of laboratory routine testing methods. Indian Phytopathol. 15:85–111. Biol. Abstr. 44:16803. 1963.

Neergaard, P. 14541.

14289. Negbi, M., and D. Koller. 1964: Dual action of white light in the photocontrol of germination of *Oryzopsis miliacea*. Plant Physiol. 39:247–53.

Negbi, M. 11032–11034, 11037, 11038.

14290. Neger, F. W. 1918: Keimungshemmende und Keimungsfördernde Stoffwechselprodukte. Naturw. Wochschr. 17: 141–42.

Negi, D. S. 17506.

Negi, S. S. 19337.

Negisi, K. 7439.

14291. Negodi, G. 1932: Aspetti della partenocarpia in specie del genere *Papaver*. Ann. Botan. 19:497–509.

Negolvelov, S. F. 15482.

14292. Negri, G. 1912: Studies in seedless fruits. Ann. R. Accad. Agr. Torino 55: 517–81. Expt. Sta. Rec. 31:35. 1914.

14293. Negrul, A. M. 1934: Parthenocarpy and apomixis in the grape. Bull. Appl. Botany, Genet., Plant Breed. (USSR) 8(2):229–68. Expt. Sta. Rec. 72:754. 1935.

14294. —— 1960: Evolution of the dimensions of grape seeds and berries (based on data of a study of grape seeds from archeological excavations in the USSR. Rimiryazevsk. Sel'skokhoz. Akad. Izv. 33:167–76.

14295. Nehring, K. 1939: Über den Einfluss von Wasser- und Stickstoffcersorgung auf den Eiweissgehalt verschiedener Gerstensorten (Nachbauversuche). Bodenkunde Pflanzenernähr 15:336–45. Biol. Abstr. 14:12505. 1940.

14296. —— and E. Schwerdtfeger. 1957: Content of essential amino acids in some foods and feeds. Z. Lebensm.-Untersuch. -Forsch. 105:12–21. Chem. Abstr. 51: 7597. 1957.

Neiger, R. 8664.

14297. Neill, J. C., and E. O. C. Hyde. 1939: Blind-seed disease causes low germination of rye-grass seed. New Zealand J. Agr. 58:223–25.

14298. Nekrasov, V. I., and N. G. Smirnova. 1961: Use of the X-ray method for studying the development of the seeds of introduced trees. Byul. Gl. Botan. Sada 43:47–52.

14299. Nekrasova, T. V. 1960: The effect of gibberellic acid on the germination of seeds and the growth of seedlings of fruit plants. Plant Physiol. (USSR) (English Transl.) 7:85–87.

14300. Neljubow, D. 1925: Über die harten Samen. Jard. Botan. Inst. Essais Semences, Leningrad, Ann. Essais Semences 1925(7): 1–13.

14301. —— 1925: Über die Methoden der Bestimmung der Keimfähigkeit ohne Keimprüfung. Jard. Botan. Inst. Essais Semences, Leningrad, Ann. Essais Semences 1925(7):31–35.

14302. Nelly, J. D. 1944: Electrodialysis of seeds. Plant Physiol. 19:19–32.

14303. Nelson, A. 1926: Hard seeds and broken seedlings in red clover (*Trifolium pratense*). II. Storage problems. Trans. Proc. Botan. Soc. Edinburgh 29:282–90.

14304. —— 1926: Hard seeds in Leguminosae. Nature 118:804–5.

14305. —— 1927: The germination of *Poa* spp. Ann. Appl. Biol. 14:157–74.

14306. —— 1927: Hard seeds and broken seedlings in red clover (*Trifolium pratense*). III. Soil effects. Trans. Proc. Botan. Soc. Edinburgh 29:402–7.

14307. —— and J. F. A. MacLagan. 1935: Factors in the germination of *Aira flexuosa*. Edinburgh Roy. Botan. Garden Notes 18(90):251–66.

14308. —— and J. C. MacSween. 1933: Hard seeds and broken seedlings in red clover (*Trifolium pratense*). IV. Early stages of germination (physical). Trans. Proc. Botan. Soc. Edinburgh 31:247–55.

14309. —— and J. M. Munro. 1934: The germination of *Spartina townsendii*. Edinburgh Roy. Botan. Garden Notes 18(89): 173–80.

Nelson, A. I. 19094.

Nelson, C. 4665, 8589.

Nelson, E. G. 18684, 18685.

Nelson, E. K. 4758.

Nelson, H. D. 19306, 19307.

Nelson, J. A. 18766.

14310. Nelson, J. H., R. L. Glass, and W. F. Geddes. 1963: Silicic acid chromatography of wheat lipids. Cereal Chem. 40:337–43.

14311. ——, ——, and —— 1963: The triglycerides and fatty acids of wheat. Cereal Chem. 40:343–51.

14312. Nelson, J. W., and L. S. Palmer. 1942: The thiamin, riboflavin, nicotinic acid, and pantothenic acid contents of wild rice (*Zizania aquatica*). Cereal Chem. 19:539–40.

14313. Nelson, M. L. 1938: Preliminary investigations on dry, cold storage of southern pine seed. U.S. Dept. Agr. Forest Serv. Southern Forest Expt. Sta. Occas. Paper 78. 19 pp.

14314. —— 1940: Successful storage of southern pine seed for seven years. J. Forestry 38:443–44.

14315. —— 1941: Polyembryony in seeds of southern pines. J. Forestry 39:959–60.

Nelson, O. A. 12142.

14316. Nelson, O. E., Jr., and H. S. Burr. 1946: Electric signals tell if corn will sprout. Sci. Dig. 20(2):88.

14317. —— and —— 1946: Growth correlates of electromotive forces in maize seed. Proc. Natl. Acad. Sci. U.S. 32:73–84.

14318. —— and C. Y. Tsai. 1964: Glucose transfer from adenosine diphosphate-glucose to starch in preparation of waxy seeds. Science 145:1194–95.

Nelson, O. E., Jr. 12459.

14319. Nelson, R. C. 1942: A method of studying the movement of respiratory gases through waxy coatings. Plant Physiol. 17:509–14.

14320. Nelson, S. O. 1961: Radio-frequency electric seed treatment. Seed World 88(12):6–7.

Nelson, S. O. 3375.

14321. Nelson, W. A. G., and E. G. V. Percival. 1942: Studies on seed mucilages. II. The seed mucilage of *Plantago arenaria*. Chem. Soc. J. 1942:58–61.

14322. Nelson, W. L., L. Burkhart, and W. E. Colwell. 1946: Fruit development, seed quality, chemical composition, and yield of soybeans as affected by potassium and magnesium. Soil Sci. Soc. Am. Proc. 10:224–29. Biol. Abstr. 21:4177. 1947.

Nelson, W. L. 7565, 8294.

14323. Nemec, A. 1918: The distribution of urease in the seeds of cereals. Biochem. Z. 91:126–30. Expt. Sta. Rec. 41:111. 1919.

14324. —— 1923: Zur Kenntnis der Glycerophosphatase der Pflanzensamen. I. Biochem. Z. 137:570–75.

14325. —— 1923: Zur Kenntnis der Glycerophosphatase der Pflanzensamen. II. Biochem. Z. 138:198–204.

14326. —— 1928: Glycerophosphatasewirkung von Pflanzensamen und Fermentsynthese. Biochem. Z. 202:229–35.

14327. —— 1944: Chemical composition of the seeds of conifers. Lesnicka Prace 23:65–75. Chem. Abstr. 42:8889–90. 1948.

14328. —— 1946: The chemical composition of the seeds of deciduous forest woods. Lesnická Práce 25:7–20. Biol. Abstr. 21:7184. 1947.

14329. —— 1951: The contents of oil and mineral substances in the seeds of yew and fir. Ann. Czech. Acad. Agr. 24:419–421.

14330. —— and F. Duchon. 1922: Sur une méthode indicatrice permettant d'évaluer la vitalité des semences par voie biochimique. Compt. Rend. 174:632–34.

Nemec, A. S. 2536.

14331. Nemtsova-Fabianova and Koloushek. 1956: The study of the effect of vernalization on the development of broad bean and some biochemical changes during vernalization. Sb. Česk. Acad. Zeměděl. 29:155–72. Biol. Abstr. 33:35019. 1959.

14332. Nenjukov, T. 1928: *Plantago lanceolata* L. as a negative indicator of late clover. Proc. Intern. Seed Testing Assoc. 6:23–30.

14333. Neptune, G., and E. C. Rossman. 1953: Cold test germination of hybrid seed corn. Mich. Agr. Expt. Sta. Quart. Bull. 35:343–50. Biol. Abstr. 27:30704. 1953.

Nes, W. R. 1706.

14334. Nesbitt, L. L., A. J. Pinckney, T. E. Stoa, and E. P. Painter. 1943: Oil formation in flaxseed. N. Dakota Agr. Expt. Sta. Bull. 323. 19 pp.

Nesbitt, L. L. 14962–14964, 19024.

14335. Nesbitt, R. T. 1894: Seed of watermelon. Georgia Dept. Agr. Rept. 20:144.

14336. Nesgenyi, O. 1893: Chemical changes in the germination of seeds of chicory. Beitr. Keimungsgeschl. 5. 55 pp.

14337. Ness, R. 1929: Untersuchungen über die physikalische und chemische Veränderung reifender Hülsenfruchtsamen. Botan. Arch. 27:60–158.

Nestel, L. 13204.

14338. Nesterenko, V. G. 1960: Germination of seeds stored under laboratory conditions. Byul. Gl. Botan. Sada 36: 99–103.

14339. Nesterova, E. I. 1939: Number of grains in a spring wheat ear as dependent upon temperature and light conditions. Compt. Rend. 24:812–16.

14340. Nestler, A. 1892: Distribution of seeds of the Proteaceae. Engler's Botan. Jahrb. 16:325.

14341. Nestler, R. B. 1943: Germinated seed versus ungerminated seed in the diet of adult bobwhite quail. Poultry Sci. 22:227–29. Expt. Sta. Rec. 89:462. 1943.

14342. —— 1946: Germination of seeds of some wild and cultivated plants after five and one-half years of storage. J. Forestry 44:683–84.

14343. —— and W. W. Bailey. 1941: The toxicity of *Crotalaria spectabilis* seeds for quail. J. Wildfire Management 5:309–13.

14344. Nestor, R., and S. Zalik. 1963: Oxidative phosphorylation by pea seedling mitochondria. Can. J. Biochem. Physiol. 41:533–41.

Nesty, G. A. 16196.

Nesvadba, H. 6914.

14345. Netien, G., and R. Canillot. 1951: Action of α-naphthylphthalamic acid upon the germination and the first stages of growth of plants. Bull. Mens. Soc. Linneenne Lyon 20:49–52. Chem. Abstr. 45:9621. 1951.
14346. ——, P. Hutinel, and O. Sotty. 1952: Action de l'auréomycine et de la terramycine sur la biogénèse de la chlorophylle au cours de la germination. Compt. Rend. Soc. Biol. 146:1337–39.
14347. —— and J. Lacharme. 1955: The anthocyanin content of radish cotyledons as a measure of antibiotic activity. Compt. Rend. 240:692–94. Chem. Abstr. 49:7659. 1955.
Netien, G. 2158.
14348. Netolitzky, F. 1914: Anatomische Beobachtungen an Zerealienfrüchten. Österr. Botan. Z. 64:265–72.
14349. Nettles, W. C. 1958: Insect control in stored grains. U.S. Dept. Agr. Clemson Agr. Coll., S. Carolina Circ. 435. 7 pp.
Neu, R. 2114.
14350. Neubauer, H. 1932: Influence of time on the activity of rye seed used in the seedling method. Z. Pflanzenernähr. Düng. Bodenk. 26:326–30. Chem. Abstr. 27:2471. 1933.
14351. Neuberg, C., and M. Kobel. 1930: Über den Kohlenhydratstoffwechsel der höheren Pflanzen. Die Bildung von Methylglyoxal durch Enzyme der Gekeimten Samen. Biochem. Z. 229:433–42.
14352. —— and —— 1934: Transformation of phosphoglyceric acid by the enzymes of germinated peas and beans. Biochem. Z. 272:457–58. Chem. Abstr. 28:7278. 1934.
14353. Neuberger, A. 1949: Metabolism of proteins and amino acids. Ann. Rev. Biochem. 18:243–66.
14354. Neuberger, F. 1914: Das Verhalten der Samen von Papilionaceen gegen höhere Temperaturen. Kiserl. Kozl. 17: 121–70. Botan. Centralbl. 126:665. 1914.
14355. Neuburg, W. B., J. H. MacGillivray, M. Yamaguchi, and L. K. Mann. 1954: Wrinkled seed coat defect in green lima beans for freezing. Plant Disease Reptr. 38:464–66.
Neufeld, E. F. 6792.
14356. Neuhauser, E. 1931: Hastening seed germination. Horticulture 9(4):72.
Neumann, G. 6119–6127.
Neumann, H. 7583.
14357. Neumann, P. 1941: Fat metabolism of ripening and germinating leguminous seeds. Biochem. Z. 308:141–74. Chem. Abstr. 37:1154. 1943.
14358. Neumcke, U. 1931: Untersuchungen über falsche Keimungen von *Phacelia tanacetifolia* Benth. Planta 14:310–43.
14359. Neuray, G. 1963: Relations entre la vie de la graine et la croissance. Ann. Gembloux 69:678–87.
14360. Neurerer, H. 1960: Die Lagerfähigkeit von Schlamm- und Trockengebeiztem Saatgut. Pflanzenschutz Ber. 24:1–13. Biol. Abstr. 35:45731. 1960.
Neustadt, M. H. 1717, 1718, 8277, 9223, 9224, 9886.
14361. Neuweiler, E. 1928: Einfluss der Konzentration und Menge von Kupfervitriollösungen auf die Keimfähigkeit von Weizen. Landwirtsch. Jahrb. Schweiz. 42:271–88.
14362. Neva, A. C., and E. B. Fischer. 1949: A comparative pharmacognostical study of *Plantago purshii* R. & S. and *Plantago aristata* Michx. seeds with the present N.F. VIII *Plantago* seeds. J. Am. Pharm. Assoc. Sci. Ed. 38:34–41.
14363. Nevado, B. M., S. L. Lopez, and F. D. Alascio. 1955: Tratamiento de semillas de algodon, trebol y vallico con hormona aceleradora de la germinación. Bol. Zootec. 11:235–44.
14364. Nevens, W. B., K. E. Harshbarger, G. H. Rollins, P. C. Burk, K. E. Gardner, and K. A. Kendall. 1955: High-protein corn as a source of protein for dairy cows. Illinois Agr. Expt. Sta. Bull. 586. 21 pp.
14365. Neville, S. L. 1928: The development of the hilum of *Salvia.* Proc. Assoc. Offic. Seed Anlaysts 19/20:57–62.
14366. Nevo, D. 1957: Problems of seed disinfection. Hassadeh 37:887–90.
Newby, W. 9784.
Newcomb, E. H. 10863.
14367. Newcomb, M., and R. E. Cleland. 1946: Aseptic cultivation of excised plant embryos. Science 104:329–30.
14368. Newcomer, E. H. 1945: Induced parthenocarpy in *Ginkgo.* Am. Naturalist 79:186–87.
14369. Newell, J. 1956: The raising of lilies from seed. J. Roy. Hort. Soc. 81:488–92.
Newell, J. [M.] 1420, 12974.
14370. Newhall, A. G. 1945: Pelleting onion seed with fungicides. Farm Res. (N.Y.) 11(1):18, 20. Expt. Sta. Rec. 92:801–2. 1945.
14371. —— and J. K. Hoff. 1960: Viability

and vigor of 22-year-old onion seed. Seed World 86:4–5.

Newhall, A. G. 9074, 12290.

14372. Newman, C. L. 1895: Effect of acid phosphate fertilization on germination. Arkansas Agr. Expt. Sta. Bull. 34:99–124. Expt. Sta. Rec. 7:123. 1895/96.

14373. Newman, E. I. 1963: Factors controlling the germination date of winter annuals. J. Ecol. 51:625–38.

14374. Newman, I. V. 1934: Studies in the Australian acacias. IV. The life history of *Acacia baileyana* F. V. M. 2. Gametophytes, fertilization, seed production and germination and general conclusion. Proc. Linnean Soc. N.S. Wales 59:277–313. Plant Breeding Abstr. 6:121. 1935.

Newman, J. A. 7822.

Newman, J. E. 4522.

14375. Newman, L. H., and L. E. Kirk. 1935: Vernalization and phasic development of plants. Imp. Bur. Plant Genet., Cambrian News, Aberystwyth & Cambridge. 151 pp.

14376. Newsom, D. W. 1954: Some studies on the effects of incorporation of certain growth stimulants in pelleted seeds on the germination and development of seedlings. Dissertation Abstr. 14:903. Chem. Abstr. 48:10145. 1954.

Newsom, L. D. 6511.

Newton, A. C. 18429, 18430.

14377. Newton, J. M., and N. M. Naylor. 1939: Soybean amylase. I. The concentration and characterization of soybean amylase. Cereal Chem. 16:71–78.

14378. Newton, R., W. H. Cook, and J. G. Malloch. 1927: The hardness of the wheat kernel in relation to protein content. Sci. Agr. (Ottawa) 8:205–19.

14379. —— and A. G. McCalla. 1934: Effect of frost on wheat at progressive stages of maturity. I. Physical characteristics of the kernels. Can. J. Res. 10:414–29.

Newton, R. 13122.

14380. Newton, W. 1953: Seed germination and seedling growth as influenced by different kinds of sawdust. Can. J. Agr. Sci. 33:593–96.

Nÿdam, F. E. 17125.

14381. Nezgovorov, L. A., I. Ibragimov, and A. K. Sobolev. 1962: Decrease in presprouting death of thermophilic plant seeds at low temperatures. Plant Physiol. (USSR) (English Transl.) 8:281–88.

14382. —— and A. K. Sobolev. 1957: Cold resistance of germinating seeds and soil pathogenicity. Plant Physiol. (USSR) (English Transl.) 4:465–75.

Nezgovorov, L. A. 9343.

14383. Nguyen-Van, T. 1942: Recherches sur les phosphatases végétales. IV. Caractères des préparations enzymatiques de soja et d'amandes douces. Trav. Mem. Bull. Soc. Chim. Biol. 24:1369–77. Biol. Abstr. 22:14981. 1948.

Nichol, M. A. 4775.

Nicholls, M. J. 15164.

Nicholls, N. 14037.

14384. Nicholls, P. B., and L. G. Paleg. 1963: A barley endosperm bioassay for gibberellins. Nature 199:823–24.

14385. Nichols, C. Jr. 1942: The effects of age and irradiation on chromosomal aberrations in *Allium* seed. Am. J. Botany 29:755–59.

14386. Nichols, G. E. 1934: The influence of exposure to winter temperatures upon seed germination in various native American plants. Ecology 15:364–73.

14387. Nichols, J. M. 1954: Direct seeding of oak in Missouri. Missouri Agr. Expt. Sta. Bull. 609. 3 pp.

Nichols, J. M. 3401.

14388. Nicholson, C. 1932: Longevity of wheat seeds. Gardeners Chron. 92:453.

14389. Nickell, L. G. 1951: Embryo culture of weeping crabapple. Proc. Am. Soc. Hort. Sci. 57:401–5.

Nickell, L. G. 3697.

Nickerson, W. J. 16393.

14390. Nicloux, M. 1904: Germination of oily seeds. Compt. Rend. 139:143–45.

14391. Nicolai, H. W. 1926: Die Bestimmung der Lipase in keimenden Kiefersamen. Biochem. Z. 174:373–83.

Niculescu, P. 3422.

14392. Nielsen, E. L., J. G. Dickson, and D. C. Smith. 1959: Strain and seed treatment as factors in germination and seedling growth of smooth bromegrass. Phytopathology 49:8–12.

Nielsen, E. L. 1318.

14393. Nielsen, K. F., T. F. Cuddy, and W. B. Woods. 1960: The influence of the extract of some crops and soil residues on germination and growth. Can. J. Plant Sci. 40:188–97.

14394. ——, R. L. Halstead, A. J. MacLean, R. M. Holmes, and S. J. Bourget. 1960: The influence of soil temperature on the growth and mineral composition of oats. Can. J. Soil Sci. 40:255–63.

14395. Nielsen, N. 1936: The nitrogen contents of individual seeds of barley.

Compt. Rend. Trav. Lab. Carlsberg Ser. Physiol. 21:247–69. Chem. Abstr. 30: 8307. 1936.
14396. —— 1936: Der Wuchsstoffgehalt von Samen verschiedenen Alters. Compt. Rend. Trav. Lab. Carlsberg Ser. Physiol. 21:427–36.
14397. —— 1937: Untersuchungen über die Keimung der Gerste. I. Die Atmung während des Mälzens. Compt. Rend. Trav. Lab. Carlsberg Ser. Physiol. 22: 49–60.
14398. —— and G. Bergqvist. 1958: The stimulation of the respiration of seeds with gibberellic acid and its analytical application. Physiol. Plantarum 11:329–31.
Nielsen, N. 2325.
14399. Nielsen, O. 1936: Hot water treatment of cabbage seed. Tidsskr. Planteavl 41:450–58. Rev. Appl. Mycol. 16:83. 1937.
14400. Nielson, A. K., and R. R. Kalton. 1959: Combining ability for seed characteristics in *Bromus inermis* Leyss. Agron. J. 51:178–81.
14401. Niemann, C., and K. P. Link. 1934: The composition of an aldobionic acid from flax-seed mucilage. J. Biol. Chem. 104:205–6. Physiol. Abstr. 19:688. 1935.
14402. Niemann, E. 1952: Comparative investigations of the separation of germination retarding substances from fruits and seeds with especial reference to *Foeniculum vulgare*, Miller. Flora (Jena) 139:185–242. Chem. Abstr. 47:2833. 1953.
14403. —— 1954: Neue Ergebnisse über die Beeinflussung der Keimung von Steinund und Zwergbrand durch chemische und biologische Faktoren. Z. Pflanzenbau Pflanzenschutz 5:163–68.
14404. Nieser, O. 1924: Beiträge zur Kenntnis der Keimungsphysiologie von *Anthoxanthum puelli*, *Festuca ovina* und *Aera flexuosa*. Arch. Botan. 6:275–312.
14405. —— 1937: Samenprüfung. Jahresber. Inst. Angew. Botan. 53/54:78–115.
14406. —— 1959: Untersuchungen zu den Fluoreszenzerscheinungen der Keimpflanzen von *Lolium* spp. im ultravioletten Licht. Proc. Intern. Seed Testing Assoc. 18:256–61.
14407. —— 1955: Saatlagerung. Saatgut-Wirtsch. 7:42–43.
14408. Niethammer, A. 1927: Der Einfluss von Reizchemikalien auf die Samenkeimung. I. Jahrb. Wiss. Botan. 66:285–300.
14409. —— 1927: Der Einfluss von den Reizchemilakien auf die Samenkeimung. II. Jahrb. Wiss. Botan. 67:223–41.
14410. —— 1927: Graphische Darstellung des Verlaufes der Samenkeimung unter besonderer Berücksichtigung der Stimulationswirkung des Rhodankaliums. Zellstimulations-Forsch. 3:87–96.
14411. —— 1927: Keimungsphysiologische Studien unter Hervorhebung des Lichtkeimungsproblems. Biochem. Z. 185: 205–15.
14412. —— 1928: Fortlaufende Untersuchungen über den Chemismus der Angiospermensamen und die Äusseren natürlichen, wie künstlichen Keimungsfaktoren. I. Der Einfluss des Frostes auf die Keimfähigkeit. Biochem. Z. 197:241–44.
14413. —— 1928: Fortlaufende Untersuchungen über den Chemismus der Angiospermensamen und die Äusseren natürlichen, wie künstlichen Keimungsfaktoren. II. Biochem. Z. 197:245–56.
14414. —— 1928: Fortlaufende Untersuchungen über den Chemismus der Angiospermensamen und die Äusseren natürlichen, wie künstlichen Keimungsfaktoren. III. Oberflächenaktive Substanzen. Biochem. Z. 199:175–85.
14415. —— 1928: Stimulationsprobleme im Zusammenhang mit den inneren Faktoren, die die Keimung bedingen. Beitr. Biol. Pflanz. 16:267–350.
14416. —— 1928: Stimulationswirkungen im Pflanzenreiche. Biol. Generalis (Vienna) 4:259–90.
14417. —— 1928: Die Vorteile der graphischen Darstellung bei der Auswertung von Samenkeimversuchen. Zelstimulations-Forsch. 3:103–10.
14418. —— 1929: Die Charakteristik der Lebenskraft verschiedenen Samenmaterials auf chemischer, physikalischer und rechnerischer Grundlage. Gartenbauwissenschaft 1:593–614. Biol. Abstr. 6:9336. 1932.
14419. —— 1929: Fortlaufende Untersuchungen über den Chemismus der Angiospermensamen und die Äusseren natürlichen wie künstlichen Keimungsfaktoren. IV. Untersuchungen über die Farbstoff und Salzpermeabilität von Frucht und Samenschalen. Biochem. Z. 209: 263–75.
14420. —— 1929: Grundlagen und praktische Ziele bei der Beeinflussung des Keimverlaufes unserer landwirtschaftlichen Kultursämereien durch chemische und

physikalische Agentien. Zellstimulations-Forsch. 3:201–44.

14421. —— 1929: Permeabilitätsstudien an Pflanzenzellen im Zusammenhange mit der Quecksilberbeizung der Samen. Z. Physikal. Chem. Abt. A 142:309–19.

14422. —— 1929: Vergleichende biochemische Untersuchungen über das Reifen und Altern von Samen und Früchten. Österr. Botan. Z. 78:264–78.

14423. —— 1930: Beiträge zur Keimungsbiologie und physiologischen Anatomie der Früchte von *Tragopogon pratensis* L. Botan. Centralbl. Beih. 47:282–90.

14424. —— 1930: Bio- und Histochemie der pflanzlichen Früchte und Samen. I. Biochem. Z. 220:348–57.

14425. —— 1930: Bio- und Histochemie pflanzlicher Früchte und Samen. II. Reife- und Alterserscheinungen. Biochem. Z. 227:462–71.

14426. —— 1930: Die Dosis tolerata und toxica der Beizmittel als eine Komponente der physikochemischen Struktur des Samenkornes. Z. Pflanzenkrankh. Pflanzenschutz 40:44–50.

14427. —— 1930: Die Dosis toxica und tolerata von Uspulun-universal für einzelne landwirtschaftliche Sämereien. Z. Pflanzenkrankh. Pflanzenschutz 40:517–20. Rev. Appl. Mycol. 10:256. 1931.

14428. —— 1930: Histochemische Untersuchungen und Permeabilitätsstudien an landwirtschaftlichen Sämereien im Hinblicke auf ihre Keimungsbiologie. Wiss. Arch. Landwirtsch. Abt. A Pflanzenbau 3:321–48.

14229. —— 1930: Landwirtschaftlich-biologische Studien mit Nickel- und Cyanverbindung (Gleichzeitig ein Beitrag zur allgemeinen Biologie der Samen und Früchte). Pflanzenbau 4:607–34.

14430. —— 1930: Mikroscopie und Mikrochemie bekannter heimischer Früchte. Planta 12:399–413.

14431. —— 1930: Physiko-chemische Grundlagen für Stimulationswirkungen. Eine spekulative experimentelle Studie. Zellstimulations-Forsch. 3:321–27.

14432. —— 1930: Der Wert der Analysenlampe für die Bewertung von Samenproben. Zellstimulations-Forsch. 3:329–32.

14433. —— 1930: Physiologisch-chemische Anatomie fleischiger Früchte unserer Gärten. (Gleichzeitig ein Beitrag über Reife-, Alters- und Stimulationserscheinungen an Samen und Früchten. Gartenbauwissenschaft 4:1–10.

14434. —— 1930: Über chemische Reizwirkungen an den Früchtchen von *Cannabis sativa* und den Samen von *Linum usitatissimum.* Faserforsch. 8: 213–15.

14435. —— 1931: Basis of chemical stimulation effects in higher plants. Z. Pflanzenernähr. Düng. Bodenk. 13A:343–52. Soc. Chem. Ind. J. 50:457. 1931.

14436. —— 1931: Beizversuche und anatomisch-chemische Studien mit den Samen des Wirsingkohles. Z. Pflanzenkrankh. Pflanzenschutz 41:149–67.

14437. —— 1931: Die Grundlagen chemischer Reizwirkungen an hoheren Pflanzen. Z. Pflanzenernähr. Düng. Bodenk. 19A: 342–52. Chem. Abstr. 25:5190–91. 1931.

14438. —— 1931: Kritisch vergleichende Zucker- und Katalasebestimmungen an Samenmaterial der verschiedensten Lebenskraft. Z. Pflanzenernähr. Düng. Bodenk. 21A:69–86. Chem. Abstr. 25:5688. 1931.

14439. —— 1931: Kann die Keimfähigkeit unserer Samen ohne Keimprobe bestimmt werden? Fortschr. Landwirtsch. 6:258–59.

14440. —— 1931: Kritische Studien über Ruhephasen und Alterserscheinungen. Biol Generalis (Vienna) 7:575–90. Chem. Abstr. 26:754. 1932.

14441. —— 1931: Lichtkeimungsprobleme im Zusammenhange mit physiologisch-anatomischen Studien. Botan. Centralbl. Beih. 47:346–58.

14442. —— 1932: Die Beizung unseres Gemüsesaatgutes mit Germisan. Gartenbauwissenschaft 6:650–81. Fortschr. Landwirtsch. 8:67. 1933.

14443. —— 1932: Die Beizwirkung von Germisan auf die Keimung einzelner Wiesengräser bei unterschiedlichen Keimtemperaturen. Z. Pflanzenkrankh. Pflanzenschutz 42:364–83.

14444. —— 1932: Comparative sugar determinations and considerations of a polysaccharide and hydrocyanic acid in seeds of different vitalities. Z. Pflanzenernähr. Düng. Bodenk. 26A:350–57. Chem. Abstr. 27:2977. 1933.

14445. —— 1934: Licht, Dunkelheit und Strahlung als Faktoren bei der Samenkeimung. Tabulae Biol. 4:45–77. Proc. Intern. Seed Testing Assoc. 7:71–72. 1935.

14446. —— 1935: Up-to-date problems of stimulation. Z. Pflanzenernähr. Düng. Bodenk. 39:45–61.

14447. —— 1939: Der Wert substantiver Farbstoffe bei der Charakteristik von

Früchten und Samen, unter gleichzeitigem Hinweis auf deran natürliche Pilzflora. Biol. Generalis (Vienna) 14:552–70. Biol. Abstr. 21:12597. 1947.

14448. —— 1942: Plasmolysestudien an gärtnerisch wichtgem Saatgut. Gartenbauwissenschaft 17:91–94. Biol. Abstr. 18:11860. 1944.

14449. —— and N. Tietz. 1961: Samen und Früchte des Handels und der Industrie. N. V. Drukkerij D. Reidel, Dordrecht, Netherlands. 331 pp.

14450. Nieto-Hatem, J. 1963: Seed dormancy in *Setaria lutescens*. Dissertation Abstr. 24:1360–61.

14451. Niffenegger, D. 1963: Germination of frost damaged wheat. Assoc. Offic. Seed Analysts News Letter 37(1):8–9.

14452. —— and A. A. Schneiter. 1963: The effect of prechilling and light upon the germination of seed of Green needlegrass, *Stipa viridula*. Assoc. Offic. Seed Analysts News Letter 37(2):22.

14453. Nigam, S. S., and C. Radhakrishnan. 1963: Chemical examination of the essential oil derived from the seeds of *Daucus carota* (Linn.) (Indian black carrot). Perfumery Essent. Oil Record 54:87–93.

14454. Nigam, V. N., and K. V. Giri. 1960: Enzymatic synthesis of oligosaccharides from maltose by germinated green gram (*Phaseolus radiatus*). J. Biol. Chem. 235: 947–50.

14455. Nihei, T. 1937: On the injury of wheat grains caused by threshing machines. J. Imp. Agr. Expt. Sta., Nishigahara, Tokyo 3:31–40.

14456. Nihlmeier, M. 1928: Zur Kenntnis der Keimungsphysiologie einiger Labiaten und Cruciferen-Samen. Botan. Centralbl. Beih. Abt. 1, 45:83–96.

14457. Nihous, M. 1939: La germination des semences de *Geum urbanum*; influence de divers facteurs. Compt. Rend. Soc. Biol. 132:499–501.

14458. —— 1941: Repos apparent et rapos réel des graines et semences. Compt. Rend. 212:927–29.

14459. —— 1942: Les capacités d'imbibition chez les graines aux divers stades de leur développement et leur signification biologique. Compt. Rend. Soc. Biol. 136: 250–52.

14460. —— 1942: Capacités d'imbibition des graines aux divers stades de leur développement et signification biologique. Compt. Rend. Soc. Biol. 136:578–79.

14461. —— 1942: Recherches physiologiques sur la maturation, le repos et la germination des graines et semences, des tubercules, des rhizomes et des bulbes. Thesis, Doc. Sciences Nat. Fac. Sci. Lille. 138 pp. Ann. Agron. 15:470–71. 1945.

14462. —— 1944: Toxic effect of aqueous solutions of colchicine on the germination of *Pisum sativum*. Compt. Rend. Soc. Biol. 138:128. Chem. Abstr. 39:2780. 1945.

14463. Niiuchi, K., and S. Okitsu. 1962: On the discoloration of seed in immature fruits of sweet pepper. I. Kyushu Agr. Res. 24:104–5.

Niizeki, S. 3931.

14464. Niidam, F. E. 1940: The heredity of a brownish-black spot on the seedcoat of *Trifolium pratense* L., caused by melanine. Genetica 22:123–30. Biol. Abstr. 15: 14171. 1941.

Nijholt, J. A. 8166.

Nikitash, G. A. 11375.

14465. Nikitin, I. 1934: Influence of X-rays on seeds and pollen of trees and shrubs. Tsentr. Nauchn.-Issled. Inst. Lesn. Khoz. Sb. Tr. 1:86–104. Expt. Sta. Rec. 73:22–23. 1935.

Nikitin, P. S. 8620.

14466. Nikitina, A. L. 1962: Pre-planting treatment of soybean seeds with minor elements. Zemledelie 1962(12):61–65.

14467. Nikolaeva, M. G. 1948: Biology of the seed germination of some *Ferula* species. Tr. Botan. Inst. Akad. Nauk SSSR Ser. 4 Eskperim. Botan. 6:218–28.

14468. —— 1948: Relation of seed germination of some Umbelliferae to the condition of the seed coats (exemplified by species of *Ferula*). Tr. Botan. Inst. Akad. Nauk SSSR Ser. 4 Eksperim. Botan. 6:229–41.

14469. —— 1950: On germination of *Euonymus europaeus* seed embryos. Dokl. Akad. Nauk SSSR 71:175–78.

14470. —— 1950: Importance of seed stratification for further development of *Euonymus europaeus* seedlings. Dokl. Akad. Nauk SSSR 71:949–52.

14471. —— 1951: Causes of rest of the seed of *Acer negundo* L., *Fraxinus pennsylvanica* Marsh., and *Berberis vulgaris* L. Botan. Experimentalis 8:234–56. Biol. Abstr. 27:14162. 1953.

14472. —— 1956: Role of organic acids in sprouting of dormant seed. Dokl. Akad. Nauk SSSR 111:1371–74. Chem. Abstr. 51:8907. 1957.

14473. —— 1956: Water regime of *Euonymus* seeds in connection with stratification and germination. Tr. Botan. Inst. Akad. Nauk SSSR Ser. 4 Eksperim. Botan. 11:331–50.

14474. —— 1958: The biology of germination of *Fraxinus* seeds and its bearing on the taxonomic position and the distribution of species of this genus. Botan. Zh. 43:679–83.

14475. —— 1962: The role of gibberelline in the interruption of dormancy of seeds. J. Botan. URSS 47:1823–35.

14476. —— and Y. Y. Katkeirch. 1961: Temperature dependence of dormant seed respiration. Plant Physiol. (USSR) (English Transl.) 8:28–33.

14477. ——, L. M. Kozlova, and V. G. Yudin. 1960: Study of secondary seed dormancy. Tr. Botan. Inst. Akad. Nauk SSSR Ser. 4 Eksperim. Botan. 14:138–66. Biol. Abstr. 43:20567.

14478. ——, ——, and —— 1962: Effect of plant growth conditions on the depth of dormancy of seeds. Tr. Botan. Inst. Akad. Nauk SSSR Ser. 4 Eksperim. Botan. 15:133–47.

14479. —— and V. G. Yudin. 1963: Action of gibberellin on seed germination of woody plants. Dokl. Akad. Nauk SSSR 150:686–89. Chem. Abstr. 59:8059. 1963.

Nikolaeva, M. G. 14605, 14606.

14480. Nikolenko, D. T. 1949: Preservation of quality of seed grain under Siberian conditions. Selek. i Semen. 16(12):66–67.

14481. Nikolic, M. 1924: Über den Einfluss des Lichtes auf die Keimung von *Phacelia tanacetifolia.* Sitzber. Wien Wiss. Akad. Math. -Naturw. Kl. Abt. I 133:625–41.

Nikolov, K. 7118.

14482. Nikonchuk, V. N. 1957: Economical methods for increasing the quality of larch seed after their harvest. Tr. Bryansk. Lesokhoz. Inst. 1957(8):249–53. Biol. Abstr. 35:39390. 1960.

Nikuni, Z. 6833, 8871.

14483. Nilan, R. A. 1954: Relation of carbon dioxide, oxygen and low temperature to the injury and cytogenetic effects of X-rays in barley. Genetics 39:943–53.

14484. —— 1958: After-effects of ionizing radiation in barley. III. Storage of thermal neutron irradiated seeds. Northwest Sci. 32:89–95.

14485. —— and H. M. Gunthardt. 1956: Studies on aged seeds. III. Sensitivity of aged wheat seeds to X-radiations. Caryologia 8:316–21. Biol. Abstr. 31:19457. 1957.

14486. —— and L. L. Phillips. 1957: The irradiation sensitivity of barley seeds low in calcium. Northwest Sci. 31:139–44.

Nilan, R. A. 100, 3065, 7841, 7979, 10200, 10201, 11188, 11431, 16764.

14487. Niles, J. J. 1954: I. The use of triphenyl tetrazolium bromide in viability tests of rice seeds. Trop. Agriculturist 110:13–20.

14488. —— 1955: II. The use of triphenyl tetrazolium bromide in viability tests of rice seed. Trop. Agriculturist 111:279–85.

14489. Nilova, V. P., and V. F. Rashevskaya. 1952: The effect of some chemical elements on the immunological properties of plants. Mikroelementy v Zhizni Rast. i Zhivotn. Akad. Nauk SSSR, Tr. Konf. po Mikroelementam 1950:591–602. Chem. Abstr. 49:442. 1955.

14490. Nilsson, B. 1956: *Agrostis* species in the seed trade. Medd. Statens Cent. Frökontr. 31:41–47.

14491. —— 1959: Laboratory methods for determining *Avena* species. Medd. Statens Cent. Frökontr. 34:83–84.

14492. —— 1959: Larch seed and larch hybrids for middle Sweden. Svenska Skogsvardsforen. Tidskr. 57:309–24.

14493. Nilsson, F. 1933: The use of the analytical quartz mercury vapour lamp for distinguishing Italian and perennial ryegrass. Sveriges Utsadesforen. Tidskr. 43:113–20.

14494. —— 1937: The seed setting of smooth-stalked meadowgrass, *Poa pratensis* L. Botan. Notiser 1937(1/2):85–109.

14495. —— 1946: Seed growing experiments with radishes, 1941–1945. Medd. Statens Trädgardsförsok. Akarp. 33:73–90.

14496. —— and I. Fernqvist. 1953: Seed growing experiments with carrots. Medd. Alnarps Lantbr. Mejeri- Trädgardsinst. Statens Trädgardsförsok. 81. 12 pp.

14497. Nilsson, N. H. 1893: Is treatment of clover seed with scarification always beneficial? Kgl. Landtbruks-Akad. Handl. Tidskr. 32:33–47.

14498. —— 1928: The physiology of willow seed. (Preliminary report.) Botan. Notiser 1928(4):255–64. Biol. Abstr. 5:16705. 1931.

14499. Nilsson, P. E. 1957: Preparation of legume inoculants, with special reference to the pre-inoculation of legume seeds. Ann. Roy. Agr. Coll. (Sweden) 23:255–72.

14500. —— and S. Bingefors. 1951: A method for routine testing of seed treatments in the laboratory. Kgl. Lantbruks-Hogskol. Ann. 18:106–12.

14501. ——, K. Korsan-Bengtsen, and O. Mellander. 1954: Amino acid content of leguminous proteins as affected by genetic and nutritional factors. I. Arch. Mikrobiol. 20:404–9.
Nilsson, P. E. 13310.
14502. Nilsson-Ehle, H. 1910: Spontaneous omission of the color factor in oats. Verhandl. Naturforsch. Ver. Brünn 49: 139–56.
14503. Nilsson-Leissner, G. 1928: Correlation between "hullessness" and other characters of timothy. Sveriges Utsadesforen. Tidskr. 38:154–67.
14504. —— 1953: The need for uniform nomenclature in designating various classes of certified seed. Proc. Intern. Seed Testing Assoc. 18:152–58.
14505. —— 1961: Reorganization of seed testing and seed certification in Sweden. Proc. Intern. Seed Testing Assoc. 27: 576–80.
14506. —— and F. R. Horne. 1961: Basic principles in the certification and nomenclature of herbage seeds for international use. Intern. Grassland Congr. Proc. 8: 170–72.
Ninomiya, M. 1344, 19937.
14507. Nishi, S. 1955: The commercial seed production of F_1 varieties of vegetables in Japan. 14th Intern. Hort. Congr. The Hague–Scheveningen, Rept. 1:468–78.
14508. ——, J. Kawata, and M. Toda. 1961: Studies on the embryo culture in vegetable crops. I. Embryo culture of immature crucifer embryos. Bull. Natl. Inst. Agr. Sci. Ser. E 1961(9):59–127.
14509. Nishida, K. 1934: Biochemical studies on the sago palm. I. The changes in chemical composition during the ripening of the seed. J. Agr. Chem. Soc. Japan 10:374–78. Chem. Abstr. 28:5097. 1934.
14510. —— 1936: Biochemical studies on *Cycas revoluta* Thunb. VI. Removal of poisonous constituents in the seeds and the nutritive value of the seeds. J. Agr. Chem. Soc. Japan 12:1106–16. Chem. Abstr. 31:2640. 1937.
14511. —— 1937: Biochemical studies on *Cycas revoluta* Thunb. VII. Chemical constituents of the outer spermoderm of the seeds. 1. J. Agr. Chem. Soc. Japan 13:89–93. Chem. Abstr. 31:3103. 1937.
14512. —— 1938: Biochemical studies on *Cycas revoluta* Thunb. IX. Chemical constituents of the outer spermoderm of the seeds. 2. J. Agr. Chem. Soc. Japan 14:505–6. Chem. Abstr. 32:6292. 1938.
14513. Nishida, S. 1952: The observation of tissue in seed coat of diploid, triploid and tetraploid watermelon. Rep. Kihara Inst. Biol. Res. 5:122–24.
Nishida, S. 14522.
14514. Nishijima, Y. 1959: The occurrence of embryoless seeds in soybean varieties in relation to the feeding of the soybean pod borer, *Grapholitha glycinivorella* Matsumura. Japan. J. Appl. Entomol. Zool. 3:183–89.
Nishikawa, K. 8850, 8851.
Nishimoto, U. 18194.
Nishimura, K. 9472.
Nishimura, N. 19995.
14515. Nishimura, S., H. Arata, and Y. Saito. 1953: Comparative studies on the effects to germination and early development of the milk vetch (*Astragalus sinicus* L.) and common vetch (*Vicia sativa* L.) seeds of soaking in water. Shikoku Agr. Expt. Sta. Bull. 1:46–51.
14516. Nishimura, Y., and T. Imai. 1958: The effect of low temperature during X-ray irradiation of rice seeds upon the heading time fertility and other characters. [Abstract.] Japan. J. Breeding 8:56.
14517. Nishina, Y., and K. Sato. 1943: Effects of fast neutrons upon forest tree seeds. I. J. Japan. Forestry Soc. 25: 313–22.
Nishina, Y. 16798.
Nishino, T. 18802, 18803.
Nishio, K. 14116.
14518. Nishiyama, I. 1937: On the inheritance of certain grain characters in oats. Japan. J. Genetics 13:63–65.
14519. —— and Y. Inamori. 1957: Vitality of aged seeds in relation to polyploidy in *Avena*. Bull. Res. Inst. Food Sci. Kyoto Univ. 1957(20):36–42.
14520. —— and T. Maruyama. 1959: Direct effects of irradiation on dry seeds of various species of wheats and oats. [Abstract.] Japan. J. Breeding 9:262.
14521. —— and T. Nagata. 1957: Endosperm as a cause of seed failure after back crosses between Brassico-Raphanus and its parents. [Abstract.] Japan. J. Genetics 32:255.
14522. —— and S. Nishida. 1952: On the inhibition of seed germination by the fruit juice of watermelon. Rep. Kihara Inst. Biol. Res. 5:119–21.
14523. ——, M. Shimotsuma, and T. Yabuno. 1952: A comparative study on the development of polyploid seeds in

watermelon. Rep. Kihara Inst. Biol. Res. 5:110–18.

Nishiyama, K. 13610.

14524. Nisikado, Y., and K. Hirata. 1938: On the specific gravity methods in grading the wheat seeds, as a control-means for the seedling blight, caused by *Gibberella saubinetti* (Mont.) Sacc. Ber. Ohara Inst. Landwirtsch. Forsch. 8:125–42.

Nitsch, C. 14527.

14525. Nitsch, J. P. 1951: Growth and development in vitro of excised ovaries. Am. J. Botany 38:566–77.

14526. —— 1955: Free auxins and free tryptophane in the strawberry. Plant Physiol. 30:33–38.

14527. ——, C. Pratt, C. Nitsch, and N. J. Shaulis. 1960: Natural growth substances in Concord and Concord seedless grapes in relation to berry development. Am. J. Botany 47:566–76.

Nitsch, J. P. 5198, 18871.

14528. Nitta, K. 1952: Studies on the oil and protein contents of soybean seed. Hokkaido Natl. Agr. Expt. Sta. Res. Bull. 63:64–69.

14529. —— 1956: Rapid method for the determination of oils in soybean seeds. Hokkaido Natl. Agr. Expt. Sta. Res. Bull. 71:109–13.

Nitta, K. 14923.

14530. Niwa, T. 1950: On the propagation of Japanese lawn grass with seeds. II. J. Japan. Inst. Landscape Architects 14:1–5. Tokyo Fac. Agr. Records Res. 1950/51(1):7.

14531. Nixon, R. W. 1959: Growing dates in the United States. U.S. Dept. Agr., Agr. Res. Serv. Agr. Infor. Bull. 207. 50 pp.

14532. Nobbe, F. 1876: Handbuch der Samenkunde. Wiegandt, Hempel & Parey, Berlin. 631 pp.

14533. —— 1880: Ist die naturliche Farbe der Cultur Samen ein sicheres Kriterium ihres Gebrauchswerthes? Landwirtsch. Versuchssta. 24:457–64.

14534. —— 1882: Übt das Licht einen vortheilhaften Einfluss auf die Keimung der Grassamen? Landwirtsch. Versuchssta. 27:347–55.

14535. —— 1889: Über die Hartschaligkeit von Samen. Abhandl. Naturw. Bremen 1889:289–94.

14536. —— 1894: Influence of arsenic on germinating seeds. Landwirtsch. Versuchssta. 30:318–436.

14537. —— 1919: Untersuchungen über den Quellprozess der Samen von *Trifolium pratense* und einiger anderer Schmetterlingsblütler. Landwirtsch. Versuchssta. 94:197–218.

14538. —— and H. Hänlein. 1877: Über die Resistenz von Samen gegen die äussern Faktoren der Keimung. Landwirtsch. Versuchssta. 20:71–96.

14539. Noble, M. 1951: Seed pathology. Nature 168:534–37.

14540. —— 1960: Marsh spot and hollow heart in peas. Proc. Intern. Seed Testing Assoc. 25:536–38.

14541. ——, J. De Tempe, and P. Neergaard. 1959: An annotated list of seed-borne diseases. Proc. Intern. Seed Testing Assoc. 24:1–159.

14542. —— and I. G. Montgomerie. 1954: Seed-borne oat diseases, strawbreaking and brown foot-rot. Scot. Agr. 34:51–52. Biol. Abstr. 29:1884. 1955.

14543. Noda, A. 1962: Studies on the coleorhiza of cereals. VI. Investigation on the stages of seed germination. Tech. Bull. Fac. Agr. Kagawa Univ. 13:105–10.

14544. —— and J. Hayashi. 1957: Studies on the coleorhiza of cereals. III. On coleorhiza in the germination of rice seeds. Proc. Crop Sci. Soc. Japan 26:24–26.

14545. —— and —— 1959: Studies on the coleorhiza of cereals. IV. On the border between coleorhiza and mesocotyl. Tech. Bull. Fac. Agr. Kagawa Univ. 11:2–6.

14546. —— and —— 1962: Studies on the coleorhiza of cereals. V. The nature of coleorhiza under abnormal conditions of seed germination. Tech. Bull. Fac. Agr. Kagawa Univ. 13:99–104.

14547. —— and Y. Matsufuji. 1953: Studies on the coleorhiza of cereals. II. On the coleorhiza in germination of Sawa millet, pearl millet, millet and maize. Kagawa Agr. Coll. Tech. Bull. 5:46–52.

14548. Noe, A. C. 1935: Some paleozoic gymnosperm seeds and their evolution. Trans. Illinois State Acad. Sci. 28:100.

Noecker, N. L. 1740.

14549. Noeldechen, J. 1925: Über die Ruswirkung von Metalsalzen auf die Keimung von Gerste. Kühn-Arch. 9: 264–309.

14550. Noggle, G. R., and F. L. Wynd. 1943: Effects of vitamins on germination and growth of orchids. Botan. Gaz. 104:455–59.

Noguchi, T. 10625, 10626.

14551. Noguchi, Y. 1929: Zur Kenntnis der Befruchtung und Kornbildung bei den Reispflanzen. Japan. J. Botany 4:385–403.
Noguti, T. 16586.
14552. Noguti, Y., and T. Sugawara. 1941: Favourable effects of treatment of seeds and sprouts with certain growth substances. Proc. Imp. Acad. Japan 16:402–4. Biol. Abstr. 23:5146. 1949.
14553. Nohara, S. 1913: On the germination of seeds of some *Salix*. Botan. Mag. (Tokyo) 27:23–34.
14554. —— 1934: Gametogenesis and embryogeny of *Sesamum indicum*, L. J. Coll. Agr. (Tokyo) 13:9–25.
14555. Nohe, E. 1962: Further views on dibble seeding method. Saatgut-Wirtsch. 14:64–66.
14556. Nolan, G. C. 1950: Ceropegias—from seed. Begonian 17:73, 79.
Nolan, L. S. 9167, 19106.
14557. Noll, W. 1935: Embryonalentwicklung von *Biophytum dendroides* DC. Planta 24:609–48.
14558. —— 1938: Deformities induced in wheat seeds by treatments of the grains. Arch. Fitotec Uruguay 3:86–95. Rev. Appl. Mycol. 19:396. 1940.
14559. Noller, C. R., I. J. Millner, and J. J. Gordon. 1933: Capric acid from the seed fat of the California bay tree. J. Am. Chem. Soc. 55:1227–28. Chem. Abstr. 27:2053. 1933.
14560. Nolte, A. J., and H. W. Von Loesecke 1939: Characteristics and composition of watermelon seed oil (Cuban Queen variety). J. Am. Chem. Soc. 61:889–91.
14561. —— and —— 1940: Grapefruit seed oil. Manufacture and physical properties. Ind. Eng. Chem. 32:1244–46.
Nolte, A. J. 19218.
Nonnecke, I. L. 18703–18706.
Nonomura, T. 18314.
Norby, B. 17659.
14562. Nord, E. C. 1956: Quick testing bitterbrush seed viability. J. Range Management 9:193–94. Biol. Abstr. 31: 2349. 1947.
14563. —— and G. R. Van Atta. 1960: Saponin—a seed germination inhibitor. Forest Sci. 6:350–53. Biol. Abstr. 36: 21598. 1961.
14564. —— and J. E. Whitacre. 1957: Germination of fourwing saltbush seed improved by scarification and grading. U.S. Dept. Agr. Forest Serv. Calif. Forest Range Expt. Sta., Forest Res. Note 125:1–5. Biol. Abstr. 32:10001. 1958.
Nord, E. C. 12200.
14565. Nord, F. F., and J. Weichherz. 1929: Zur Kenntnis der enzymatischen Vorgänge in der keimenden Gerste. IV. Mitteilung zum Mechanismus der Enzymwirkung. Z. Physiol. Chem. 183:218–25.
14566. —— and —— 1930: Nachweis der durch Acetylen bewirkten Permeabilitätserhöhung bei der Gerste. IX. Mitteilung zum Mechanismus der Enzymwirkung. Protoplasma 11:440–46.
14567. Norden, A. J., E. C. Rossman, and E. J. Benne. 1952: Some factors that affect protein content of corn. Mich. Agr. Expt. Quart. Bull. 34:210–25.
Norden, A. J. 7919.
14568. Nordenskiöld, H. 1956: The effect of X-ray treatment in dormant seeds of *Luzula pallescens* and *L. capitata*. Kgl. Lantbruks-Hogskol. Ann. 22:257–67.
Norgaard, U. J. 19874.
14569. Nordgren, R., and J. S. Andrews. 1941: The thiamin content of cereal grains. Cereal Chem. 18:802–11.
14570. Nordh, G., and E. Ohlsson. 1932: Amylasen in ruhenden und keimenden Samen. I. Gerste. Z. Physiol. Chem. 204:89–100.
14571. Nordhagen, R. 1933: Über die Zuckerausscheidung der Samen einiger Colchicum-Arten und ihre biologische Bedeutung. Bergens Museum Årbok 1933(1). 14 pp.
14572. Nordström, L. 1953: A method of stimulating the germination of pine seed (*Pinus silvestris*). Bull. Roy. School Forestry Sweden Nr. 17:91–108.
14573. —— 1955: On delayed germination of pine seeds. Svenska Skogsvardsforen. Tidskr. 53:89–100.
14574. Norfeldt, S., N. Olsson, and G. Anstrand. 1962: Influence of storage upon total tocopherols in wheat germs. Effect of germination upon total tocopherols in wheat. Kgl. Lantbruks-Hogskol. Ann. 28:181–88.
14575. Norman, A. G. 1955: Terramycin and plant growth. Agron. J. 47:585–87.
14576. Norning, E. S., and A. H. K. Petrie. 1928: Über die enzymatische Funktion der Mitochondrien bei der Keimung der Getreidearten. Wochschr. Brau. 45:265–70, 285–87, 302–3. Botan. Centralbl. 158:8. 1930.
14577. Norris, E. L. 1962: The role of the registered seed technologist in the United

States and Canada. Proc. Intern. Seed Testing Assoc. 27:312–14.

14578. Norris, F. A., L. J. Hilfman, and C. E. Land. 1958: Storage of cottonseed. J. Am. Oil Chemists Soc. 35:684–87.

Norris, F. W. 1685, 4895, 4896, 7209, 10759, 10760.

Norris, K. H. 8278.

Norris, M. J. 261, 1493.

14579. Norris, R. J., and C. W. Kreke. 1937: Three metabolic stimulating factors. Studies Inst. Divi Thomae (Cincinnati) 1:137–62. Biol. Abstr. 12:4435. 1938.

14580. Norstog, K. J. 1956: Growth of rye grass endosperm in vitro. Botan. Gaz. 117:253–59.

14581. —— 1961: The growth and differentiation of cultured barley embryos. Am. J. Botany 48:876–84.

14582. —— 1963: Apomixis and polyembryony in *Hierochlöe odorata.* Am. J. Botany 50:815–21.

14583. North, C. 1948: Artificial drying of vegetable and herbage seeds. Gt. Brit. Min. Agr. J. 54:462–66.

14584. —— 1951: The growing of some flower seed crops. Natl. Inst. Agr. Botany J. 6:72–85.

North, D. T. 2916, 3516.

14585. North, H. F. A., and T. E. Odland. 1935: The relative seed yields in different species and varieties of bent grass. J. Am. Soc. Agron. 27:374–83. Expt. Sta. Rec. 73:606. 1935.

14586. Northen, H. T., and R. T. Northen. 1947: Effects of sodium hypochlorite (Clorox) on the germination of *Cattleya* seed. Am. Orchid Soc. Bull. 16:498–99. Chem. Abstr. 42:645. 1948.

14587. —— and —— 1949: Effects of temperature on the germination of *Cattleya luegeae* seed. Am. Orchid Soc. Bull. 18:237–38.

14588. Northen, R. T. 1953: *Phalaenopsis* seed sowing. Am. Orchid Soc. J. 22: 122–23.

Northen, R. T. 14586, 14587.

14589. Norton, C. 1941: Raising hardy rock plants from seed. Gardeners Chron. 45:211.

14590. —— 1941: Growing alpine and hardy rock plants from seed. Gardeners Chron. 45:236, 258.

14591. Norton, G. F. 1941: Thirty-five years with fringed gentian. Wild Flower 18:41–45.

14592. Norton, J. B. S. 1950: *Hemerocallis* from seed to seed. Hemerocallis Soc. Yearbook 1950:41–45.

14593. Norton, L. B., and R. Hansberry. 1945: Constituents of the insecticidal resin of the yam bean (*Pachyrrhizus erosus*). J. Am. Chem. Soc. 67:1609–14. Expt. Sta. Rec. 95:9–10. 1946.

Norton, L. B. 8124.

Noseir, M. A. 5922.

14594. Nosov, A. K., and A. K. Kustova. 1953: Vernalization of cottonseed and the quality of enzymes. Izv. Akad. Nauk Turkm. SSR 1953(2):58–66. Chem. Abstr. 48:10843. 1954.

14595. Nossaman, N. L. 1963: Gibberellin and fertilizer interaction on seed germination and plant yield. Dissertation Abstr. 24:1325–26.

14596. Nossatovsky, A. 1929: Wheat grain without germ. Bull. Appl. Botany, Genet., Plant Breed. (USSR) 21:593–96.

14597. Notani, N. K. 1957: The induction and evaluation of endosperm mutations in maize. Dissertation Abstr. 17:1864.

14598. Nothmann-Zuckerkandl, H. 1915: Physikalisch-chemische Arbeiten auf dem Gebiete der Botanik. I. Über Keimung. Intern. Z. Phys. Chem. Biol. 2:94–166. Expt. Sta. Rec. 35:129. 1916.

14599. Notkina, L. G., and D. B. Lifshitz. 1936: Chemical composition and hydrolysis of wheat integuments. Ukr. Khem. Zh. 11:173–86. Chem. Abstr. 30:6068. 1936.

Notley, L. F. 19729.

14600. Nottbohm, F. E., and F. Mayer. 1932: Vorkommen und Nachweis von Cholin in Tabaksamen und Tabakblättern. Z. Untersuch. Lebensm. 63:620–23. Ber. Ges. Physiol. 69:255–56. 1933.

14601. Nottin, P., and A. Alexeievsky. 1939: The harvesting of wheat before and after complete maturity. Compt. Rend. Acad. Agr. France 25:458–63. Chem. Abstr. 33:5441–42. 1939.

14602. Notzold, T. 1962. Diagnostic tables for the identification of *Carex* seeds. Deut. Akad. Wiss. Berlin Monber. 4: 39–54.

14603. Nougarede, A. 1963: Premières observations sur l'infrastructure et sur l'évolution des cellules de jeunes ébauches foliares embryonnaires du *Tropaeolum majus* L. Cytologie de l'hydration germinative et des premières étapes de la germination. Compt. Rend. 257:1495–97.

Nourse, H. C. 5466.

Novelli, G. D. 15834, 15835.

14604. Novellie, L. 1960: Kaffircorn malting and brewing studies. IV. The extraction and nature of the insoluble amylases of Kaffircorn malts. J. Sci. Food. Agr. 11:408–21.
Novik, N. P. 10197.
14605. Novikov, G. N., and M. G. Nikolaeva. 1940: Sur la germination des graines de quelques espèces de genre *Ferula* L. Sov. Botan. (Leningrad) 1940(5/6):331–37.
14606. —— and —— 1951: Preservation of summer cypress (*Kochia prostrata*) seed germination. Botan. Zh. 36:69–72.
14607. Novikov, V., and E. Herber. 1934: The inducing of rubber formation in plants by ultra-violet rays. Nature 133: 587.
14608. Novikov, V. A. 1945: Stimulation of the resting seeds of *Polygonum bucharicum*. Compt. Rend. Acad. Sci. URSS 46:204–6.
14609. —— and R. O. Sadovskaja. 1939: Soaking of cotton seeds in boric acid as a means of satisfying the boron requirement and increasing salt-resistance. Compt. Rend. Acad. Sci. URSS 23:276–79.
14610. Novitskaia, I. E. 1958: The treatment of seeds in solutions of minor elements and its effects on plant resistance to drought. Tr. Botan. Inst. Akad. Nauk SSSR Ser. 4 Eksperim. Botan. 12:74–94.
Novitskaya, G. V. 15720.
14611. Novotelnov, N. V., and I. S. Ezhov. 1954: Antibiotic and antioxidant properties of the yellow pigment of grain. Dokl. Akad. Nauk SSSR 99:297–300. Chem. Abstr. 49:5593. 1955.
14612. —— and —— 1957: The biological essence and chemical nature of substances that retard the germination of grain. Botan. Zh. 42:255–60. Chem. Abstr. 51:14027. 1957.
14613. —— and —— 1958: Regulator of amylase activity, separated from barley seed. Izv. Vysshikh Uchebn. Zavedenii Pishchevaya Tekhnol. 1958(2): 47–53. Chem. Abstr. 53:2377. 1959.
Novotelnov, N. V. 9071.
14614. Novozhilov, I. I. 1960: Vernalization of spring vetch. Agrobiologiya 1960(1): 67–70. Biol. Abstr. 38:7364. 1962.
14615. Nowacki, E., and J. Przybylska. 1961: Tingitanine, a new free amino acid from seeds of Tangier pea (*Lathyrus tingitanus*). Bull. Acad. Polon. Sci. Ser. Sci. Biol. 9:279–83. Biol. Abstr. 41:24415. 1963.
14616. —— ,——, and J. Hurich. 1960: Dependence between content of guanidine derivatives and alkaloids in seeds of certain species of leguminous plants. Bull. Acad. Polon. Sci. Ser. Sci. Biol. 8:445–48. Biol. Abstr. 37:3571. 1962.
Nowak, G. 5950.
14617. Noyes, G. B. 1957: Storage of tea seed. Tea Quart. 28:55.
Nozdrachev, K. 6529, 6530.
Nozu, M. 18301.
14618. Nubarian, F. M. 1954: Effect of pollen ageing on the viability of tobacco seeds. Izv. Akad. Nauk Arm. SSR Biol. i Sel'skokhoz. Nauki 1954(7):33–39.
14619. —— 1957: Effect of pollination on fertilization and vitality of the tobacco offspring when the age of the stigma varies. Izv. Akad. Nauk Arm. SSR Biol. i Sel'skokhoz. Nauki 1957(2):67–73.
14620. Nuccorini, R. 1930: The significance of glutamine in germination. Ann. Chim. Appl. 20:239–44. Chem. Abstr. 24:5332. 1930.
14621. —— 1930: Germination and catalase of seeds. Boll. Ist. Super. Agrar. Pisa 6:321–35. Chem. Abstr. 25:4020–21. 1931.
14622. —— 1934: Norleucine in the seeds of *Ricinus communis* L. Ann. Chim. Appl. 24:25–32. Chem. Abstr. 28:3756. 1934.
14623. Nuchowicz, A. 1955: The cultivation of embryos and embryo fragments of *Arachis hypogaea*. Agricultura (Louvain) 3:3–37. Chem. Abstr. 50:9522. 1956.
14624. Nuessle, N. O., and N. M. Lauter. 1958: Isolation of constituents of *Glotidium vesicarium*. Econ. Botany 12:307–11.
Nugent, T. J. 8933, 8934, 16694.
Nuggehalli, R. N. 19064.
Nuland, D. S. 17084.
Nummi, M. 5948.
Nusbaum, C. J. 4183.
14625. Nutile, G. E. 1944: The occurrence of broken hypocotyls in bean seedlings. Assoc. Offic. Seed Analysts News Letter 18(2):7–8.
14626. —— 1945: Inducing dormancy in lettuce seed with coumarin. Plant Physiol. 20:433–42.
14627. —— 1952: Persistence of dormancy in seed of some cultivated *Brassicas*. Assoc. Offic. Seed Analysts News Letter 26(1):6–8.
14628. —— 1957: Effect of relative humidity on the viability of lettuce and onion seeds during nine years of storage. Assoc. Offic. Seed Analysts News Letter 31(2): 11–14.

14629. —— 1958: Germination of alfalfa and red clover seeds after 24 years of storage. Assoc. Offic. Seed Analysts News Letter 32(1):21–22.

14630. —— and A. P. Canfield. 1950: Effect of temperature and light on the germination of celery seed. Assoc. Offic. Seed Analysts News Letter 24(2):20–22.

14631. —— and T. E. Hackett. 1953: Germination of *Zoysia japonica* seed under laboratory conditions. Assoc. Offic. Seed Analysts News Letter 27(2):20–21.

14632. —— and —— 1955: Correlation between speed of germination in the blotter tests and emergence in laboratory and greenhouse soil tests for several lots of cabbage seed. Assoc. Offic. Seed Analysts News Letter 29(4):23–25.

14633. —— and —— 1959: Light filtering effect of blotters on the germination of seeds of tomato (*Lycopersicon esculentum*). Proc. Assoc. Offic. Seed Analysts 49:93–97.

14634. —— and L. C. Nutile. 1947: Effect of relative humidity on hard seeds in garden beans. Proc. Assoc. Offic. Seed Analysts 37:106–14.

Nutile, L. C. 14634.

14635. Nutman, P. S. 1939: Studies in vernalization of cereals. VI. The anatomical and cytological evidence for the formation of growth-promoting substances in the developing grain of rye. Ann. Botany 3:731–57.

14636. —— 1941: Studies in vernalisation of cereals. VII. A study of the conditions of formation and the subsequent growth of dwarf embryos of rye. Ann. Botany 5:353–74.

14637. —— 1957: Studies on the physiology of nodule formation. V. Further experiments on the stimulating and inhibitory effects of root secretions. Ann. Botany 21:321–37.

14638. Nybom, N. 1950: Do X-rays resuscitate heat-killed seeds? [Abstract.] Hereditas 36:516–17.

14639. —— 1954: Karyotype and viability in barley. Acta Agr. Scand. 4:507–14.

Nybom, N. 5832, 5833.

14640. Nygren, A. 1950: A cytological and embryological study of *Antennaria porsildii*. Hereditas 36:483–86.

14641. —— and G. Almgard. 1962: On the experimental control of vivipary in *Poa*. Kgl. Lantbruks-Hogskol. Ann. 28:27–36.

14642. Nyholm, I. 1951: Germination experiments with hawthorn (*Crataegus*), myrobalan plum (*Prunus cerasifera*) and Swedish mountain ash (*Sorbus intermedia*). Gart. Tidsskr. 67:79–81.

14643. —— 1951: Germination conditions of beech. Gart. Tidsskr. 67:513–16.

14644. —— 1953: Germination test of single-seed hawthorne, *Crataegus monogyna*. Jaeg. of Danish and Italian crop. Saertryk Tidsskr. Planteavl 56:684–701.

14645. —— 1955: Germination experiments with *Rosa rugosa*. Dansk Skovfor. Tidsskr. 40:143–50.

14646. —— 1960: Long term storage of beechnuts. Dansk Skovfor. Tidsskr. 45: 377–415.

14647. —— 1962: Seed and seed treatment. Hedeselsk. Tidsskr. 83:364–75.

14648. Nyman, B. 1961: Effect of red and far-red irradiation on the germination process in seeds of *Pinus sylvestris* L. Nature 191:1219–20.

14649. —— 1963: Studies on the germination in seeds of Scots pine (*Pinus silvestris* L.) with special reference to the light factor. Studia Forestalia Suecica, Skogshögskolen, Stockholm, 2. 164 pp.

14650. Nyquist, W. E. 1963: Fluorescent perennial ryegrass. Crop Sci. 3:223–26.

14651. Nysterakis, F. 1963: The seeds of a fruit give opposite reactions in the presence of tissues of the pericarp and of their juice. Compt. Rend. 256:491–93.

O

14652. Oakes, A. J. 1959: Germination of elephant grass (*Pennisetum purpureum* Schum.). J. Agr. Univ. Puerto Rico 43:140.

14653. Oakley, J. W. 1958: What certification should assure growers, seed dealers and users? Intern. Crop Improvement Assoc. Ann. Rept. 40:146–50.

14654. Oaks, A., and H. Beevers. 1964: The requirement for organic nitrogen in *Zea mays* embryos. Plant Physiol. 39: 37–43.

14655. —— and —— 1964: The glyoxylate cycle in maize scutellum. Plant Physiol. 39:431–34.

14656. Oathout, C. H. 1928: The vitality of soybean seed as affected by storage conditions and mechanical injury. J. Am. Soc. Agron. 20:837–55.

14657. —— 1929: Hardiness of soybean seed

as related to seed color. Trans. Illinois State Acad. Sci. 22:146–49.

14658. Obaton, M. 1949: Organisation vasculaire des la germination chez les *Datura* (*D. sanguinea* Ruiz et Pav., *D. metel* L., *D. stramonium* L.). Ann. Sci. Nat. 10:179–200.

14659. Oberle, G. D., and H. O. Bennett. 1949: Germination trials on open pollinated peach pits from the 1945 crop of named varieties and seedling selections. Proc. Am. Soc. Hort. Sci. 53:221–25.

14660. Obermuller, P. C. 1938: Delphiniums from seed. Flower Grower 25(7):320, 333.

14661. Obolensky, G. 1958: Recherches sur quelques stimulations fonctionnelles chez les plantes par les ultra-sons. Rev. Gen. Botan. 65:297–362.

14662. —— 1960: Étude microcalorimetrique de l'accélération de la germination des graines, traitées par les ultra-sons. Biol. Plant. 2:227–39.

Obolensky, G. 3434, 3435.

14663. Obraztsov, B. V. 1961: Materials of experiments and observations on the dissemination by wild animals of seeds of trees and shrubs into open biotopes of the forest-steppe. Akad. Nauk SSSR Lab. Lesov. Soobshch. 3:69–88.

14664. O'Brien, E. J. 1945: Chemical changes in wheat resulting from "Bin Scald." J. Dept. Agr. Victoria 43:479–85. Biol. Abstr. 20:13568. 1946.

14665. O'Brien, J. A., Jr. 1942: Cytoplasmic inclusions in the glandular epithelium of the scutellum of *Triticum sativum* and *Secale cereale*. Am. J. Botany 29:479–91.

14666. —— 1947: Plastid development in the scutellum of *Triticum sativum* and *Secale cereale*. Am. J. Botany Suppl. 34(10):9a.

14667. Oca, J. 1954: Nutritional study of Spanish rice varieties. Anales Bromatol. (Madrid) 6:107–13. Chem. Abstr. 49:534. 1955.

Ochoa, J. T. 12897.

O'Connor, R. T. 2656, 2658, 8542, 15526, 15530, 15532.

O'Dell, B. L. 8938.

14668. Odland, M. L. 1938: Immediate germination of certain selfed and hybrid potato seed. Am. Potato J. 15:67–71.

14669. —— 1938: Observations on dormancy in vegetable seed. Proc. Am. Soc. Hort. Sci. 35:562–65.

Odland, T. E. 14585.

14670. O'Donnell, W. W., and E. G. Bayfield. 1947: Effect of weather, variety and location upon thiamine content of some Kansas-grown wheats. Food Res. 12:212–21. Biol. Abstr. 22:12045. 1948.

O'Donnell, W. W. 2129.

14671. Oehler, E. 1934: Investigation of seed setting, morphology, and fertility in *Aegilops-Secale* hybrids. Z. Induktive Abstammungs- Vererbungslehre 67:317–41. Expt. Sta. Rec. 72:752–53. 1935.

14672. —— and J. Zweifel. 1950: Considérations sur la longevité des graines de blé. Rev. Romande Agr. Viticult. Arboricult. 6:41–42. Biol. Abstr. 26:9863. 1952.

14673. Oehm, G. 1932: Beitrag zur Morphologie und Anatomie einiger Acanthaceen-Früchte und -Samen. Botan. Centralbl. Beih. Abt. I 49:413–44. Ber. Wiss. Biol. 22:453–54. 1932.

Örtenblad, B. 14043–14045.

Oestermann, H. 7139.

14674. Oestling, J. 1914: Das Fett der Samen von *Trichilia subcordata* aus Deutsch-Ostafrika. Botan. Centralbl. 125:478.

14675. Oexemann, S. W. 1940: Relation of the effects of growth-promoting substances to photo-synthetic activity, the mass law of growth and seed germination. Trans. Illinois State Acad. Sci. 33:84–86

14676. —— 1941: Relation of the effects of seed weight to roots and tops of two varieties of soybeans. Trans. Illinois State Acad. Sci. 34:75–76.

14677. —— 1942: Relation of seed weight to vegetative growth, differentiation, and yield in plants. Am. J. Botany 29:72–81.

14678. Öyjord, E. 1960: A new drill for seed testing stations. Proc. Intern. Seed Testing Assoc. 25:865–67.

14679. Ofelt, C. W., A. K. Smith, and J. M. Mills. 1955: Proteases of the soybean. Cereal Chem. 32:53–63.

14680. Offutt, M. S. 1962: Effect of γ-ray and thermal neutron irradiation on seed of Korean lespedeza, *Lespedeza stipulacea* Maxim. Crop Sci. 2:325–30.

14681. Ogasa, T. 1939: Effect of alkaline salts on the germination of seeds. I. Effect of NaCl solutions on the germination of soybeans, especially on the influences at high and low temperatures. Rept. Inst. Sci. Res. Manchoukuo 3:303–15. Chem. Abstr. 36:2588. 1942.

Ogasawara, K. 11555.

Ogata, K. 9416, 9586.

14682. Ogata, T., and R. Yamada. 1959: Studies on the cause of injurious effects of urea on germination and seedling growth. I. Soil Plant Food (Tokyo) 5:41.

14683. Ogawa, T. 1961: Studies on the seed production of onion. I. Effects of rainfall and humidity on the fruit setting. J. Japan. Soc. Hort. Sci. 30:222–32.

14684. —— 1961: Studies on the seed production of onion. II. The relation between the development of seeds and their germinating ability. J. Japan. Soc. Hort. Sci. 30:325–33.

14685. Ogawa, Y. 1960: Synthesis of contractile proteins in the X-irradiated embryo. Nature 186:77–78.

14686. —— 1963: Changes in the content of gibberellin-like substances in ripening seed and pod of *Lupinus lutens*. Plant Cell Physiol. (Tokyo) 4:85–94.

14687. —— 1963: Gibberellin-like substances occurring in the seed of *Pharbitis nil* Chois. and their change in contents during the seed development. Plant Cell Physiol. (Tokyo) 4:217–25.

14688. —— 1964: Changes in the amount of gibberellin-like substances in the seedling of *Pharbitis nil* with special reference to expansion of cotyledon. Plant Cell Physiol. (Tokyo) 5:11–20.

14689. —— and S. Imamura. 1958: Über die fördernde Wirkung von Samendiffusat auf die Blütenbildung von *Pharbitis nil* Chois. (Vorläufige Mitteilung.) Proc. Japan Acad. 34:631–32.

14690. Ogawara, K. 1954: Interaction of light and darkness in the germination of the light-favored tobacco seeds. (Preliminary report.) Proc. Japan Acad. 30:504–9.

14691. —— and K. Ono. 1958: Studies on the germination of *Portulaca grandiflora* seeds. I. Hort. Assoc. Japan J. 27:193–200.

14692. —— and —— 1958: Effects of light on the germination of *Petunia hybrida* seeds. Hort. Assoc. Japan J. 27:576–82.

14693. —— and —— 1961: Interaction of gibberellin, kinetin and potassium nitrate in the germination of light-sensitive tobacco seeds. Plant Cell Physiol. (Tokyo) 2:87–98.

14694. —— and J. Watanabe. 1953: Studies on the germination of squash seeds. I. Hort. Assoc. Japan J. 22:45–49.

Ogden, D. B. 11889.

Ogden, W. B. 9867, 9868.

Ogilvie, L. 4698, 10075.

Oglesbee, R. E. 13220.

Ogneva, O. K. 11014.

Ogo, T. 14888.

Ogur, M. 273.

14695. Ogura, Y. 1948: A new example of seed-bearing Pteridosperms from Manchuria. Proc. Japan Acad. 24(10):1–4.

Ohara, H. 18803.

Ohara, I. 14734.

14696. O'Hara, L. P., and F. Saunders. 1937: Proteins. VI. Solubility of the nitrogenous constituents of certain seeds in sodium chloride solutions. J. Am. Chem. Soc. 59:352–54. Chem. Abstr. 31:2243. 1937.

14697. Ohara, T. 1937: Protein of polished seeds of *Panicum crus-galli* L. var. *frumentaeum* Hook. J. Agr. Chem. Soc. Japan 13:6–11. Chem. Abstr. 31:3103. 1937.

14698. —— 1939: Utilization of *Panicum frumentaceum*. II. The seeds. J. Agr. Chem. Soc. Japan 15:507–13. Chem. Abstr. 34:1353. 1940.

14699. Ohashi, H., and I. Ichikawa. 1960: A study of vernalization on American wormseed. I. The influence of various temperatures on its development and yield. Botan. Mag. (Tokyo) 73:239–44.

14700. —— and —— 1960: On the effect of vernalization on the development and the content of essential oil of peppermint plant. Botan. Mag. (Tokyo)73:422–26.

14701. ——, ——, and Y. Higuchi. 1957: The vernalization of medicinal and oil plants. III. Peanut. 1. The influence of various temperatures on its development and fatty oil. Japan J. Ecol. 7:137–40. Biol. Abstr. 32:28306. 1958.

14702. Ohba, K. 1962: Radiation sensitivity of pine seeds of different water content. Nucl. Sci. Abstr. 16:2268.

14703. —— and M. Simak. 1961: Effect of X-rays on seeds of Scots pine from different provenances (*Pinus silvestris* L.). Silvae. Genet. 10:84–90.

Ohba, K. 17451.

14704. Ohga, I. 1923: On the longevity of the fruit of *Nelumbo nucifera*. Botan. Mag. (Tokyo) 37:87–95.

14705. —— 1926: The germination of century-old and recently harvested Indian lotus fruits, with special reference to the effect of oxygen supply. Am. J. Botany 13:754–59. Contrib. Boyce Thompson Inst. 1:289–94. 1926.

14706. —— 1926: A comparison of the life activity of century-old and recently harvested Indian lotus fruits. Am. J. Botany 13:760–65. Contrib. Boyce Thompson Inst. 1:295–300.

14707. —— 1926: A double maximum in the rate of absorption of water by Indian lotus seeds. Am. J. Botany 13:766–72. Contrib. Boyce Thompson Inst. 1:301–7. 1926.

14708. —— 1926: On the structure of some ancient, but still viable fruits of Indian lotus, with special reference to their prolonged dormancy. Japan. J. Botany 3:1–20.

14709. —— 1927: On the age of the ancient fruit of the Indian lotus which is kept in the peat bed in South Manchuria. Botan. Mag. (Tokyo) 41:1–6.

14710. —— 1927: Supramaximal temperature and life duration of the ancient fruits of Indian lotus. Botan. Mag. (Tokyo) 41:161–72.

Ohlen, F. W. von; *see* Von Ohlen, F. W.

Ohlrogge, A. J. 414.

14711. Ohlsson, E., and O. Edfeldt. 1933: Amylasen in ruhenden und keimenden Samen. III. Hafer. Z. Physiol. Chem. 221:174–79. Biol. Abstr. 10:5294. 1936.

14712. —— and N. Thörn. 1938: Amylases in resting and germinating seed. V. Development of the α-amylase of germinating barley. Compt. Rend. Trav. Lab. Carlsberg Ser. Chim. 22:398–404. Chem. Abstr. 32:6289. 1938.

14713. —— and C. E. Uddenberg. 1933: Amylasen in ruhenden und keimenden Samen. II. Roggen. Z. Physiol. Chem. 221:165–73. Chem. Abstr. 28:504–5. 1934.

Ohlsson, E. 14570, 17945.

Ohmachi, K. 5803.

14714. Ohmura, T., and R. W. Howell. 1960: Inhibitory effect of water on oxygen consumption by plant materials. Plant Physiol. 35:184–88.

14715. —— and —— 1962: Respiration of developing and germinating soybean seeds. Physiol. Plantarum 15:341–50.

Ohno, T. 10388.

14716. Ohsumi, T., and Y. Miyazawa. 1960: Effects of gibberellin on the post-harvest maturation and germination of *Panax* schin-seng seed. Nogyo Oyobi Engei 35: 723–24. Chem. Abstr. 56:2731. 1962.

14717. Ohta, Y., and K. Furusato. 1957: Embryoculture in *Citrus*. Seiken Ziho 8:49–54.

Ohta, Y. 6855, 6856.

14718. Ohtomo, S. 1929: The proteins and vitamins in the embryo of the soybean. Abstr. Rept. Central Lab. S. Manchuria Railway Co. 1929:15–17. Chem. Abstr. 25:1871. 1931.

14719. Oikawa, K. 1961: The embryo sac development in *Amana edulis* Honda. Sci. Rept. Tohoku Univ. Fourth Ser. 27: 175–81.

14720. Oinoue, Y. 1938: Effect of boron on the setting of berries of grape muscat of Alexandria. Hort. Assoc. Japan J. 9:141–43.

Oishi, T. 19388.

Ojima, J. 6831.

14721. Oka, H. 1943: The physiology of dormancy in wheat grains. J. Soc. Trop. Agr. Taihoku Imp. Univ. 15:78–89. Chem. Abstr. 42:2651. 1948.

Oka, H. I. 9587, 13753, 18290–18292.

14722. Okabe, S. 1929: Rhizoidenentwicklung im Embryo von *Cystophyllum*. Sci. Rept. Tohoku Univ. Fourth Ser. 4:591–95.

14723. —— 1932: Parthenogenesis bei *Ixeris dentata* Nakei. Botan. Mag. (Tokyo) 46:518–23. Botan. Centralbl. 166:222. 1934.

Okada, T. 11056.

14724. Okada, Y. 1925: On the germination of *Euryale ferox*, Salisb. Botan. Mag. (Tokyo) 39:133–41.

14725. —— 1928: Study of *Euryale ferox* Salisb. II. On the variation in the shape of the seed. Sci. Rept. Tohoku Univ. Fourth Ser. 3:581–86.

14726. —— 1930: Study of *Euryale ferox* Salisb. V. On some features in the physiology of the seed with special respect to the problem of the delayed germination. Sci. Rept. Tohoku Univ. Fourth Ser. 5:41–116.

14727. —— 1931: Study of *Euryale ferox* Salisb. VII. Change of catalase and germination per cent during the after-ripening of the seeds. Sci. Rept. Tohoku Univ. Fourth Ser. 6:429–36. Biol. Abstr. 7:10294. 1933.

14728. —— 1937: On the catalase of the seed of *Euryale ferox* Salisb. Botan. Mag. (Tokyo) 51:324–32.

14729. —— and T. Otaya. 1930: Study of *Euryale ferox* Salisb. VI. Cleistogamous versus chasmogamous flowers. Botan. Mag. (Tokyo) 44:369–73.

Okami, K. 7508.

14730. Okamoto, B. 1951: The fate of thiamine in germinating seeds. Vitamins (Japan) 4:18–19. Chem. Abstr. 45:10321. 1951.

14731. Okamoto, H. 1962: Transport of cations from cotyledon to seedling of the

embryonic plants of *Vigna sesquipedalis*. Plant Cell Physiol. (Tokyo) 3:83–94.
Okamoto, M. 8314.
Okamoto, S. 9603.
Okamura, T. 11118, 11120–11149, 17324.

14732. Okanenko, A. S., and A. E. Makoimovich. 1937: Effect of soaking on the quality of beets. Vses. Nauchn.-Issled. Inst. Sakharnoi Prom., Osnovnye Vyvody Nauchn.-Issled. Rabot V.N.I.S. 1936: 156–57. Chem. Abstr. 34:3317. 1940.

14733. —— and D. A. Tabentskii. 1948: Formative action of 2,4-dichlorophenoxyacetic acid on beets. Dokl. Akad. Nauk SSSR 62:541–44. Chem. Abstr. 43:2283. 1949.
Okano, A. 13626, 13627.

14734. Okano, K., and I. Ohara. 1935: Coloring matters of kaoliang. II. The presence of phlobaphene and durasantalin. J. Agr. Chem. Soc. Japan 11:767–72. Chem. Abstr. 30:2602. 1936.
Okasako, Y. 14151, 14152.

14735. Oke, H. G. 1952/53: Germination, longevity and dormancy tests of the seeds of *Dichanthium annulatum* Stapf., and *Bothriochioa intermedia* (Br.) A. Camus. Poona Agr. Coll. Mag. 43:155–59.

14736. O'Kelly, J., and K. Sargeant. 1961: Supinine from the seeds of *Trichodesma zeylanicum* R. Br. J. Chem. Soc. 1961:484.

14737. O'Kelly, J. F., and M. Geiger. 1937: Effect of variety, maturity, and soundness on certain soybean seed and oil characteristics. Mississippi Agr. Expt. Sta. Tech. Bull. 24. 10 pp. Expt. Sta. Rec. 77:622. 1937.

14738. ——, W. W. Hull, and M. Geiger. 1933: Effects of varying amounts of potash on oil and protein and on the weight and percentage of cotton seed. Mississippi Agr. Expt. Sta. Tech. Bull. 20. 8 pp. Expt. Sta. Rec. 69:514. 1933.
O'Kelly, J. F. 13143.
Okinaka, O. 9576.
Okitsu, S. 14463.

14739. Oknina, E. Z. 1959: Nucleic acids in apple and cherry seeds. Biol. Nuklein. Obmena Rast. Ufa, Sb. 1959:63–69. Chem. Abstr. 55:3738. 1961.

14740. —— and E. I. Barskaya. 1955: Physiological-biochemical changes in the seeds of crop fruit. Plant Physiol. (USSR) (English Transl.) 2:421–31. Biol. Abstr. 32:24788. 1958.

14741. —— and K. A. Ovanesyan. 1957: Physiological and biochemical investigation of cherry seeds during their maturation and transition to dormancy. Plant Physiol. (USSR) (English Transl.) 4:77–81.
Oknina, E. Z. 1952, 1953, 2659.
Okubo, O. 18796.
Okumura, M. 14040.
Okuntsov, M. M. 16827.
Okunuki, K. 7993, 18266.

14742. Olafson, J. H., C. M. Christensen, and W. F. Geddes. 1954: Grain storage studies. XV. Influence of moisture content, commercial grade, and maturity on the respiration and chemical deterioration of corn. Cereal Chem. 31:333–40.
Olafson, J. H. 4062.

14743. Olberg. 1930: Zapfenertrag und Samenbedarf. Z. Forst.-Jagdw. 62:726–28. Biol. Abstr. 6:4935. 1932.

14744. Olcott, H. S. 1944: Cottonseed phospholipids. Science 100:226–27.

14745. —— and T. D. Fontaine. 1939: Composition of cottonseeds. I. Solubility of nitrogenous constituents. J. Am. Chem. Soc. 61:2037–40. Chem. Abstr. 33:7848. 1939.

14746. —— and —— 1940: Composition of cottonseeds. III. Solubility of proteins in alkaline solutions of neutral salts. J. Am. Chem. Soc. 62:1334–39.

14747. —— and —— 1940: Effect of aging cottonseed meal on the solubility of the proteins. J. Am. Chem. Soc. 62:3519.

14748. —— and —— 1941: Composition of cottonseeds. IV. Lipase of germinated seed. J. Am. Chem. Soc. 63:825–27.

14749. —— and C. D. W. Thornton. 1939: Composition of cottonseeds. II. Catalase. J. Am. Chem. Soc. 61:2417–18. Chem. Abstr. 33:8699. 1939.
Olcott, H. S. 6598, 16113.
Old, E. M. 3255.

14750. Oldemeyer, R. K. 1956: Inheritance of the white seed character in alfalfa. Agron. J. 48:449–51.

14751. Olden, E. J. 1953: Sexual and apomictic seed formation in *Malus sieboldii* Rehd. Botan. Notiser 1:105–28.

14752. Oldenburg, K. 1934: Raising aspen plants from seed. Skogen (Stockholm) 21(3):57, 67. Biol. Abstr. 9:12240. 1935.
Oldham, W. 9841.

14753. Olechowska-Baranska, K. 1953: The specific characters of *Digitalis* L. seeds. Acta Soc. Botan. Polon. 22:321–30.
Oleinikova, T. V. 16034, 16035.
Olesen, M. 12669.

14754. Olgyay, M. 1936: Disinfection of

seeds with hydrogen peroxide. Z. Pflanzenkrankh. Pflanzenschutz 46:1–6. Chem. Abstr. 30:8501. 1936.
14755. Olive, J. O., and R. O. Vidaurreta. 1950: Role of enzymes in the germination process. Rev. Soc. Mex. Hist. Nat. 11:43–58. Biol. Abstr. 26:16432. 1952.
Oliveira, M. A. de; *see* De Oliveira, M. A.
Oliver, R. C. 15076.
14756. Oliver, R. W. 1957: Culture of ornamental trees for Canadian gardens. Can. Dept. Agr. Hort. Div. Publ. 994. 30 pp.
Olmedo, G. 3672.
14757. Olmo, H. P. 1935: Empty-seededness in varieties of *Vitis vinifera*. Proc. Am. Soc. Hort. Sci. 32: 376–80.
14758. —— 1937: Pollination and the setting of fruit in the Black Corinth grape. Proc. Am. Soc. Hort. Sci. 34:402–4.
14759. —— 1940: Somatic mutation in the *vinifera* grape. III. The seedless Emperor. J. Heredity 31:210–13. Expt. Sta. Rec. 84:335. 1941.
14760. —— 1942: The use of seed characters in the identification of grape varieties. Proc. Am. Soc. Hort. Sci. 40:305–9.
14761. —— 1942: Choice of parent as influencing seed germination in fruits. Proc. Am. Soc. Hort. Sci. 41:171–75.
14762. —— 1946: Correlations between seed and berry development in some seeded varieties of *Vitis vinifera*. Proc. Am. Soc. Hort. Sci. 48:291–97.
14763. Olmsted, N. W., and J. D. Curtis. 1947: Seeds of the forest floor. Ecology 28:49–52.
Olney, C. E. 19289.
14764. Olney, H. O. 1961: Growth and translocation during the after-ripening of seeds with resting embryos. Dissertation Abstr. 21:2449.
14765. —— and B. M. Pollock. 1960: Studies of rest period. II. Nitrogen and phosphorus changes in embryonic organs of after-ripening cherry seed. Plant Physiol. 35:970–75.
Olney, H. O. 15511.
14766. Olomucki, E., and A. Bondi. 1955: Problems connected with ethylene dibromide fumigation of cereals. I. Sorption of ethylene dibromide by grain. J. Sci. Food Agr. 6:592–600.
Olomutzki, E. 15446.
Olschner, P. I. 12558.
Olsen, C. Müller-; *see* Müller-Olsen, C.
14767. Olson, D. S. 1932: Seed release from Western white pine and Ponderosa pine cones. J. Forestry 30:748–49.
14768. —— 1932: Germinative capacity of seed produced from young trees. J. Forestry 30:871.
Olson, E. O. 4394.
Olson, J. 6477, 6480, 17886.
14769. Olson, O. E. 1938: The movement of selenium from germinating seleniferous grains. Proc. S. Dakota Acad. Sci. 18:41–43. Chem. Abstr. 33:1370. 1939.
Olson, O. E. 8051.
Olson, P. J. 9968, 12064.
Olson, R. F. 5541.
14770. Olsoni, K. 1933: Noch etwas von verletzten Kleesamen in der Reinheitsanalyse. Proc. Intern. Seed Testing Assoc. 5:19–30.
14771. —— 1933: Einige Bemerkungen über die Zeiträume für die Reinheitsanalysen der Rotkleesamen. Proc. Intern. Seed Testing Assoc. 5:132–36.
14772. Olsson, A. 1950: Effect of age and degree of maturity of seed of certain agricultural crops on the amount and the quality of the harvest. Lantbrakad. Tidskr. 89:241–61.
14773. Olsson, G. 1960: Some relations between number of seeds per pod, seed size and oil content and the effects of selection for these characters in *Brassica* and *Sinapis*. Hereditas 46:29–70.
Olsson, N. 14574.
14774. Olteanu, G., and P. Gheorghiu. 1957: La production des semences de betteraves sucrières à l'aide de l'ensemencement d'été-automne. Probl. Agr. (Bucharest) 9:65–72.
14775. Olthoff, B. H., and F. Mulder. 1953: Manual for seed culture in agricultural enterprises. 's-Gravenhave, Staatsdrukkerij- Uitg. 211 pp.
14776. Omaki, T. 1937: Arctiin, a constituent of the seeds of *Arctium lappa*, L. III. The position of the hydroxyl group of arctigenin. J. Pharm. Soc. Japan 57:22–25. Chem. Abstr. 33:1716–17. 1939.
14777. Omelchenko, M. V. 1958: Growth and development of cotton plants from different age embryos. Agrobiologiya 1958(2):136–39.
14778. Oncley, L., D. Couch, and C. Elliott. 1940: Oil from the seed of the Kentucky coffee bean tree. Div. Agr. Food Chem., Am. Chem. Soc. Abstr. Papers, Cincinnati, pp. 13–14.
Ondarza, R. 16304–16308.

14779. Ondrej, M. 1964: Some questions of the genetic fixation of kernel weight in barley. Biol. Plantarum 6:28–34.

Ong, E. R. de; *see* De Ong, E. R.

Onishi, O. 13619.

14780. Onlieff, A. 1915: Method of preserving acorns for sowing. Lesn. Zh. 45: 255–59.

Onnis, A. 13307, 13308.

Ono, H. 6832.

14781. Ono, K. 1954: Studies on the light-effect upon the germination of Japanese cypress (*Chamaecyparis obtusa*) seeds. Japan. J. Forestry 36:89–92.

Ono, K. 14691–14693.

14782. Ono, T. 1928: Embryologische Studien an einigen Pontederiaceen. Sci. Rept. Tohoku Univ. Fourth Ser. 3:405–15.

14783. —— 1929: Embryologie der Liliaceae, mit besonderer Rücksicht Auf die Endospermbildung. I. Sci. Rept. Tohoku Univ. Fourth Ser. 4:381–93.

14784. —— and H. Suzuki. 1957: Endosperm characters in hybrids between barley varieties with starchy and waxy endosperms. Seiken Ziho 8:11–19.

14785. Onodera, J. 1934: On the experiments of germination and absorption power of water in the seeds of rice as a method of estimation of drought resistance among the varieties of rice plants. Proc. Crop Sci. Soc. Japan 6:20–43. Expt. Sta. Rec. 71:470. 1934.

Onodera, S. 8848.

14786. Oort, A. J. P., and J. Dekker. 1955: Internal seed disinfection with rimocidin, an antibiotic from *S. rimosus*. Mededel. Lab. Pytopath. (Wageningen) 152:381–87. Biol. Abstr. 31:22569. 1957.

Oosthuizen, M. J. 13853.

Oosthuizen, S. A. 19005.

14787. Oosting, H. J., and M. E. Humphreys. 1940: Buried viable seeds in a successional series of old field and forest soils. Bull. Torrey Botan. Club 67:253–73.

14788. Oota, Y. 1953: Growth and gaseous metabolism of seed embryo of *Vigna sesquipedalis* in germination stage. Kagaku (Tokyo) 23:60–65.

14789. —— 1954: Growth and protein synthesis in seed embryo of bean, *Vigna sesquipedalis*, in germination stage. Kagaku (Tokyo) 24:544–50.

14790. ——, R. Fujii, and S. Osawa. 1953: Changes in chemical constituents during the germination stage of a bean, *Vigna sesquipedalis*. J. Biochem. (Tokyo) 40:649–61.

14791. ——, ——, and Y. Sunobe. 1956: Studies on the connexion between sucrose formation and respiration in germinating bean cotyledons. Physiol. Plantarum 9:38–50.

14792. —— and S. Osawa. 1954: Migration of "storage PNA" from cotyledon into growing organs of bean seed embryo. Experientia 10:254–62.

14793. —— and —— 1954: Relation between microsomal pentose nucleic acid (PNA) and protein synthesis in the hypocotyl of germinating bean embryo. Biochim. Biophys. Acta 15:163–64.

14794. —— and K. Takata. 1959: Changes in microsomal ribonucleoproteins in the time course of the germination stage as revealed by electrophoresis. Physiol. Plantarum 12:518–25.

14795. —— and —— 1959: Chemical analyses of microsomal ribonucleoproteins and ribonucleic acids from germinating bean seeds. Physiol. Plantarum 12:649–57.

14796. —— and —— 1961: Cytoplasmic ribonucleic acids in bean germ tissues. Intern. Botan. Congr. Recent Advan. Botany 9:1059–63.

14797. ——, Y. Yamamoto, and R. Fujii. 1953: Dehydrogenase pattern in bean seed embryo. J. Biochem. (Tokyo) 40:187–203.

Oota, Y. 9350, 14853.

14798. Ooyama, N. 1954: The growth-inhibiting substances contained in the leaf-litter of the trees. I. The inhibiting effect on germination of the coniferous seeds J. Japan. Forestry Soc. 36:38–41.

14799. Oparin, A. 1922: Einfluss des Sauerstoffs auf die Fermentbildung in keimenden Weizensamen. Biochem. Z. 134:190–93.

14800. —— and E. Back. 1924: Über die Bedeutung der Sauerstoffs für die Fermentbildung in keimenden Pflanzensamen. Biochem. Z. 148:476–81.

14801. —— and N. Djatschkow. 1928: Über die Fermentbildung in reifenden Samen. Biochem. Z. 196:289–93.

14802. —— and N. Gelman. 1946: Formation of purine bases in germinating wheat seeds. Compt. Rend. Acad. Sci. URSS 54:43–46. Chem. Abstr. 41:3512. 1947.

14803. —— and N. Pospelowa. 1927: The enzyme content of latent wheat-seed. Biochem. Z. 189:18–25. Biol. Abstr. 3:17815. 1929.

Oparin, A. 1629, 1630.

14804. Opienska-Blauth, J. 1934: Determination of strophanthin in the seeds and extracts of *Strophanthus*. Arch. Chem. Farm. 1:29–51. Chem. Abstr. 28:4538–39. 1934.

14805. Opik, H., and E. W. Simon. 1963: Water content and respiration rate of bean cotyledons. J. Exptl. Botany 14:299–310.

14806. Opitz, K. 1937: The significance of the origin of seed for German cereals. Landwirtsch. Jahrb. 84:603–29. Chem. Abstr. 31:8714. 1937.

14807. Oppel, I. 1937: Hydrocyanic acid content of flaxseed. Pharm. Zentralhalle 78:577–78. Chem. Abstr. 32:214. 1938.

14808. Oppenheimer, C. 1944: Germination of apple seeds. Hassadeh 24:130–31. Biol. Abstr. 21:9967. 1947.

14809. Oppenheimer, H. 1922: Keimungshemmende Substanzen in der Frucht von *Solanum*, *Lycopersicum* und anderen Pflanzen. Anz. Akad. Wiss. Wien Math.-Naturw. Kl. 59:21. Botan. Centralbl. 143:230. 1922.

14810. —— 1922: Das Unterbleiben der Keimung in den Behältern der Mutterpflanze. Sitzber. Akad. Wiss. Wien Math.-Naturw. Kl. Abt. I 131:279–312.

14811. Oppenheimer, H. R. 1933: Germination and seedling development in *Pinus halepensis* and *Quercus coccifera*. Gartenbauwissenschaft 7:308–64. Expt. Sta. Rec. 69:525. 1933.

Oppenheimer, H. R. 9649.

14812. Oppenoorth, W. F. F., H. Van Veldhuizen, J. Coenradie, W. Goedkoop, A. T. Kortenhorst, and G. Vermuelen. 1957/58: The determination of the germination power of malting barley. Intern. Tijdschr. Brouw. Mout. 17:111–20.

14813. Oppermann, O. 1913: Winter storage of acorns. Forstlige Forsögsv. Danmark 4:127–33. Expt. Sta. Rec. 29:343. 1913.

14814. Opritov, V. A. 1956: The influence of heteroauxin treatment of wheat seeds on their bioelectric potentials. Dokl. Akad. Nauk SSSR 109:765–67.

14815. Opsahl, W. 1931: The conditions for pine reproduction. Tidsskr. Skogbruk 39:262–73. Biol. Abstr. 7:6678. 1933.

14816. —— 1950: Seed setting of forest trees, 1950. Tidsskr. Skogbruk 58:321–28.

14817. —— 1952: On relation between summer temperature and seed ripening of Norway spruce. Skogforeoksves. Meddel. (Norway) 11:619–62.

14818. Opsomer, J. E., and F. Bronckers. 1958: Stimulation de la germination des semences de *Panicum maximum*. Acad. Roy. Sci. Colon Bull. Séances 4:330–34.

14819. Oratovskii, M. T. 1953: Overcoming the failure to germinate of early-ripening sweet cherry varieties. Agrobiologiya 1953(2):109–13.

Orav, T. 13442.

14820. Orbay, R. 1939: Influence of toxic gases on tobacco seed. Inhisarlar Tütün. Inst. Raporlari 2:200–3. Chem. Abstr. 34:5997. 1940.

Orchiston, H. D. 19315.

14821. Orechoff, A., and N. Proskurnina. 1934: Über *Sophora*-Alkaloide. IV. Über die Alkaloide der Samen von *Sophora pachycarpa*. Ber. Deut. Chem. Ges. 67:77B–83B.

14822. ——, ——, and R. Konowalowa. 1935: Über *Sophora*-Alkaloide. VIII. Alkaloide aus den Samen und aus dem Kraute von *Sophora alopercuroides*. Ber. Deut. Chem. Ges. 68:431–36.

14823. Orechova, T. A. 1928: Über die Technik zur Vorbereitung anatomischer Schnitte der Samen von *Cuscuta*. Jard. Botan. Inst. Essais Semences, Leningrad, Ann. Essais Semences 1928(2):142–45.

14824. —— 1937: Anatomical differences between seeds of different species and varieties of potatoes. Bull. Appl. Botany, Genet., Plant Breed. (USSR) Ser. 4(2):50.

14825. —— and S. M. Lapinskaya. 1936: Morphologisch-anatomische Unterschiede der Samen von Kohlvaietäten. Bull. Appl. Botany, Genet., Plant Breed. (USSR) Ser. 4(1):75–76.

Orekhova, T. A. 10178, 10191–10193.

Orelli, O. Schneider-; *see* Schneider-Orelli, O.

14826. Orestano, G. 1940: Hypoglucemic action of the seeds of *Lupinus albus*. Arch. Farmacol. Sper. 70:113–17. Chem. Abstr. 35:1511. 1941.

Organ, J. G. 13318.

Orian, G. 17304.

14827. Orlando, P., and P. Santos. 1951: Does the moon influence plant life? Agros (Lisbon) 34:269–76.

14828. Orlov, V. K. 1958: The glycosides of jute seeds. III. Tr. Khim. Prirodn. Soedinenii, Kishinevsk. Univ. 1958(1):123–28. Chem. Abstr. 54:1667. 1960.

14829. —— and G. V. Lazurevskiy. 1957: Glycosides of jute seeds. II. Uch. Zap. Kishinvesk. Univ. 27:3–10. Chem. Abstr. 52:16498. 1958.

14830. Orlova, A. A. 1951: New disease of *Caragana arborescens* seeds. Lesn. Khoz. 4(5):85–90.

14831. Orlova, S. I. 1927: Notiz über ausgewachsene und nachher ausgetrocknete Samen. Jard. Botan. Inst. Essais Semences, Leningrad, Ann. Essais Semences 1927(4): 44–48.

14832. Orlovskii, N. I. 1956: Effect of storage duration of (sugar) beet seeds on their productivity. Sakharn. Prom. 30: 59–62.

14833. ——, O. K. Kolomiets, and A. V. Popov. 1957: Single-seed sugar beet. Vest. Sel'skokhoz. Nauki 2(12):65–74.

14834. Ormrod, D. P., and W. A. Buuter, Jr. 1961: The evaluation of rice varieties for cold water tolerance. Agron. J. 53:133–34.

14835. Oros, J. 1932: Désinfection des semences du tabac à l'eau chaude. Bul. Cult. Ferment. Tutunului 21:264.

14836. O'Rourke, F. L. S., R. R. Dedolph, and H. E. Larzelere. 1963: Walnut kernel characteristics as inferential indices in variety discrimination. Proc. Am. Soc. Hort. Sci. 83:370–73.

O'Rourke, F. L. S. 7423.

14837. Orozco-Sarria, S. H., and C. Carbona-Alvarez. 1959: Evidence of seed transmission of angular leaf spot of bean. Phytopathology 49:159.

14838. Orpet, E. O. 1932: Seed sowing under glass. Horticulture 10(2):37.

14839. Orr, H. R. 1935/36: An undescribed structure in the seed of the broad bean. Trans. Proc. Botan. Soc. Edinburgh 32: 174–78.

14840. Orr, M. Y. 1923: Polyembryony in *Sarcococca ruscifolia*, Stapf. Edinburgh Roy. Botan. Garden Notes 14(67):21.

14841. Orr-Ewing, A. L. 1947: Possible occurrence of viable unfertilized seeds in Douglas fir. Forest Sci. 3:243–48.

14842. Orru, A. 1931: Analisi chimica e biologica di semi di leguminose africane. II. Dati analitici sui principi immediati presenti nei semi di: *Cicer amaricum-Vigna catjang* e *Phaseolus mugo*. Boll. Soc. Ital. Biol. Sper. 6:1037–40. Ber. Ges. Physiol. 66:732. 1932. Ann. Biol. 36:749–50. 1932.

14843. —— 1931: Analisi chimica e biologica di cereali africani. III. Dati analitici sui principi immediati presenti nelle cariossidi di *Eragrostis tocusso* e *Triticum durum*. Boll. Soc. Ital. Biol. Sper. 6:1041–44. Ber. Ges. Physiol. 66:732–33. 1932.

14844. ——, F. Badolato, and T. De Leo. 1958: Concentration of ascorbic and dehydroascorbic acids in incubating homogenates of *Phaseolus vulgaris* seeds. Influence of the constituents of the complex. Rend. Accad. Sci. Fis. Mat. 25. 14 pp. Chem. Abstr. 53:22291. 1959.

14845. —— and A. Fratoni. 1945: A toxic principle in the seeds of *Canavalia ensiformis*. I. Boll. Soc. Ital. Biol. Sper. 20: 200–1. Chem. Abstr. 40:6123. 1946.

14846. Orsenigo, M. 1954: Relation of seed-borne fungi to seedling blight and alterations in the kernels of rice. Milan Univ. Cattolica Sacro Cuore Fac. Agr. Ann. 1:22–33.

Orsenigo, M. 4295.

14847. Orshanskaia, V. N. 1960: Treatment of seeds with the toxins of phytopathogenic fungi as a method of selecting plants resistant to fungus diseases. Agrobiologiya 1960(4):573–78.

Ortega, L. 7280.

Ortega, M. 6303.

14848. Ortega-Delgado, M. L. 1953: Aislamiento y purificación de un inhibidor de la tripsina de la semilla del colorin (*Erythrina berteroana* Urban). Anales Escuela Nacl. Cienc. Biol. (Mex.) 9:93–102.

14849. Ortlepp, C. 1956: Der Einfluss der technischen Jarowisation auf Wachstum und Ertrag von Sommergetreide. Z. Landwirtsch. Versuchs- Untersuchungsw. 2:136–50.

14850. Ortman, A. 1947: The influence of forest litter on the germination of plants. Sb. Česk. Akad. Zeměděl. 19:329.

14851. Orton, C. R. 1931: Seed-borne parasites—a bibliography. West Va. Agr. Expt. Sta. Bull. 245. 48 pp. Expt. Sta. Rec. 66:546. 1932.

14852. Ory, R. L., A. J. St. Angelo, and A. M. Altschul. 1962: The acid lipase of the castor bean. Properties and substrate specificity. J. Lipid Res. 3:99–105. Biol. Abstr. 39:7810. 1962.

Osaka, S. 6860.

14853. Osawa, S., and Y. Oota. 1953: Growth and pentose nucleic acid content of bean embryo. Experientia 9:96–102.

Osawa, S. 14790, 14792, 14793.

Osawa, T. 9391.

14854. Osawa, Y., C. Saito, and M. Chujo. 1956: Studies on the lotus seed. I. On the water content of the old but still viable seed. Botan. Mag. (Tokyo) 69:108–11.

Osborn, E. 14855.

14855. Osborn, J. B., and E. Osborn. 1931:

Studies on the germination of seed of the black wattle (*Acacia mollissima*) and the green wattle (*Acacia decurrens normalis*). S. African J. Sci. 28:222–37.
Osborne, J. A. 18808.
14856. Osborne, P., W. D. Gill, and E. Dunn. 1964: Trials of seed dressings for the control of wheat bulb fly in Scotland. Plant Pathol. 13:19.
14857. Osborne, T. B. 1890: Proteids or albuminoids of the oat kernel. Conn. Agr. Expt. Sta. Rept. 24:114–61.
14858. —— 1892: Proteids of flax seed. Conn. Agr. Expt. Sta. Rept. 26:132–37.
14859. —— 1893: Proteids of the kidney bean. Conn. Agr. Expt. Sta. Rept. 17:186–210.
14860. —— 1894: The proteids of the rye kernel. Conn. Agr. Expt. Sta. Rept. 18:147–64.
14861. —— 1894: Studies on the proteids of barley. Conn. Agr. Expt. Sta. Rept. 18:165–91.
14862. —— 1897: The amount and properties of the proteids of the maize kernel. Conn. Agr. Expt. Sta. Rept. 20:391–97. 1896.
14863. —— and G. F. Campbell. 1895: Legumin and other proteids of the pea and vetch. Conn. Agr. Expt. Sta. Rept. 19:262–87.
14864. —— and —— 1897: Proteids of lupin seeds. Conn. Agr. Expt. Sta. Rept. 20:342–68.
14865. —— and —— 1897: Proteids of sunflower seed. Conn. Agr. Expt. Sta. Rept. 20:374–79. 1896.
14866. —— and —— 1897: Proteids of the cow pea (*Vigna catjang*). Conn. Agr. Expt. Sta. Rept. 20:380–86. 1896.
14867. —— and —— 1897: Proteids of the white-podded Adzuki bean (*Phaseolus radiatus*). Conn. Agr. Expt. Sta. Rept. 20:387–90. 1896.
14868. —— and —— 1897: Legumin and other proteids of the pea, lentil, horse bean, and vetch. Conn. Agr. Expt. Sta. Rept. 21:324–73.
14869. —— and —— 1900: The nucleic acid of the embryo of wheat and its protein compounds. Conn. Agr. Expt. Sta. Rept. 23:305–39. 1899.
14870. —— and L. B. Mendel. 1918: Nutritive factors in plant tissues. I. The protein factor in the seeds of cereals. J. Biol. Chem. 34:521–35.
14871. —— and C. L. Voorhees. 1893: Proteids of the wheat kernel. Conn. Agr. Expt. Sta. Rept. 17:175–85.
14872. —— and —— 1893: Proteids of cotton seeds. Conn. Agr. Expt. Sta. Rept. 17:211–17.
14873. Osborne, T. S., and J. A. Bacon. 1960: Radiosensitivity of seeds. I. Reduction or stimulation of seedling growth as a function of γ-ray dose. Radiation Res. 13:686–90.
14874. —— and —— 1961: Two improved and inexpensive systems for moisture stabilization in seeds and other tissues. Plant Physiol. 36:309–12.
14875. ——, A. O. Lunden, and M. J. Constantin. 1963: Radiosensitivity of seeds. III. Effects of pre-irradiation humidity and γ-ray dose on seeds from five botanical families. Radiation Botany 3:19–28.
Osborne, T. S. 9084.
Ose, Y. 17329.
14876. Oseledets, N. N. 1956: Effects of chlorides and sulfates on the germination and growth of tobacco seedlings. Tabak (USSR) 4:53–55. Tobacco Abstr. 1:426. 1957.
Oshika, Y. 5820, 5822.
Oshima, M. 14160.
Osima, M. 10369.
14877. Osler, R. D., and J. L. Cartter. 1954: Effect of planting date on chemical composition and growth characteristics of soybeans. Agron. J. 46:267–70.
Osman, S. M. 6230.
Osmani, Z. H. 14092.
Osnitskaia, E. A. 12721.
14878. Ossowski, L. 1942: The collection and treatment of forest tree seeds. Scot. Forestry J. 56:43–53. Biol. Abstr. 17: 6882. 1943.
14879. Ostazeski, S. A. 1959: Studies on the seed-borne phase of brown spot (*Pleiochaeta setosa* [Kirchn.] Hughes) in white lupine (*Lupinus albus* L.). Soil Crop Sci. Soc. Florida Proc. 19:385–89.
14880. —— 1960: A small-lot seed scarifier powered by an electric drill. Agron. J. 52:365–66.
14881. Ostenfeld, C. H. 1904: Zur Kenntniss der Apogamie in der Gattung *Hieracium*. Ber. Deut. Botan. Ges. 22:376–81.
14882. —— 1908: Bemärkninger i anledning af nogle forsög med spireevnen hos frö, der

har passeret en fugls fordöjelsesorganer. Svensk Botan. Tidskr. 2:1–11.

Osterli, V. P. 9966.

Osterman, E. 18200.

Ostle, B. 15132.

14883. Ostromecki, K. 1963: Observations of electrophoresis of cucumber (*Cucumis sativus* L.) seed proteins dissolving in water. Plant Breed. Acclimitiz. Seed Pro. 7:127–32.

14884. Ostrovskaya, L. K. 1957: The action of copper and of heteroauxin on the sprouting of seeds. Plant Physiol. (USSR) (English Transl.) 3:73–78. Biol. Abstr. 32:31772. 1958.

Ostrowska, A. 7487.

14885. Osvald, H. 1949: Root exudates and seed germination. Lantbrukshögskol. Ann. 16:789–96.

14886. —— and E. Aberg. 1941: Colour and practical value of clover seed. Lantbrukshögskol. Ann. 9:74–79. Biol. Abstr. 16:18873. 1942.

Osvald, V. 2608.

Oswald, J. W. 7386.

14887. Ota, J. 1925: Continuous respiration studies of dormant seeds of *Xanthium*. Botan. Gaz. 80:288–99.

14888. Ota, K., and T. Ogo. 1953: Studies on the salt injury in wheat plant. I. The influences of NaCl solutions upon the germination and the growth of seedling. Proc. Crop Sci. Soc. Japan 22:77–78.

Ota, K. 9578, 19967.

Ota, M. 16646.

Ota, N. 10631.

14889. Ota, T. 1960: Influences of ion exchange resins on the germination of seeds. (Preliminary report.) Proc. Crop Sci. Soc. Japan 28:299–300.

Otakara, A. 17335.

14890. Otaki, K. 1957: Respiratory enzymic system in rice embryo and seedling. Shigen Kagaku Kenkyusho Iho 45:58–63. Chem. Abstr. 52:13027. 1958.

Otaya, T. 14729.

Otey, G. 3392.

14891. Otin, C., and M. Dima. 1934: Data on the composition of grape-seed oil. Allgem. Öl- Fett.-Ztg. 31:107–15. Chem. Abstr. 28:6006. 1934.

Otono, H. 14139.

14892. Ott, A. C., and C. D. Ball. 1943: Some components of the seed coats of the common bean, *Phaseolus vulgaris*, and their relation to water retention. Arch. Biochem. 3:189–92.

14893. —— and —— 1944: The isolation of stigmasterol and β-sitosterol from the common bean, *Phaseolus vulgaris*. J. Am. Chem. Soc. 66:489–91. Expt. Sta. Rec. 94:577. 1946.

14894. Ott, E. 1961: Über den Einfluss von Flechtensäuren auf die Keimung verschiedener Baumarten. Schweiz. Z. Forstw. 112:303–4.

14895. Ottaviano, E. 1962: Influence of chromosome duplications on the size of maize kernels. Maydica 7:99–114.

14896. Ottensooser, F. 1955: On hemagglutinins of *Vicia* species. Anais Acad. Brasil. Cienc. 27:519–28. Biol. Abstr. 31:2512. 1957.

14897. —— and K. Silberschmidt. 1953: Haemagglutinin Anti-N in plant seeds. Nature 172:914.

14898. Ottenwaelder, A. 1914: Lichtintensität und Substrat bei der Lichtkeimung. Z. Botan. 6:785–848.

Ottenwaelder, A. 12047.

14899. Ottosson, L. 1958: Growth and maturity of peas for canning and freezing. Växtodling. Almqvist & Wiksells, Uppsala. 112 pp.

Outemans, J. J. Dacke-; *see* Dacke-Outemans, J. J.

Ovanesyan, K. A. 14741.

Ovcharov, K. E. 15483.

14900. Ovchinnokov, G. P. 1960: On the germination of seeds on the maternal plants of *Medicago lupulina* L. Botan. Zh. 45:413.

14901. Ovchinnikov, N. N., and N. M. Shikhanova. 1957: The causes of the differences observed in wheat seeds developing in different flowers of the spikelet. Dokl.-Botan. Sci. Sect. (English Transl.) 112:16–19.

14902. Ovechkin, S. K., and M. I. Simochkina. 1952: Seed germination and cold resistance of winter wheat plants during transplantation of embryo to the endosperm of another plant. Dokl. Akad. Nauk SSSR 82:645–48.

14903. Overaa, P. 1960: On the value of hard seeds in red clover as judged by laboratory tests. Proc. Intern. Seed Testing Assoc. 25:422–32.

14904. —— 1962: A new germination apparatus designed for alternating temperature and light exposure. Proc. Intern. Seed Testing Assoc. 27:742–47.

Overbeck, J. van; *see* Van Overbeck, J.

Overman, A. J. 3371.

Overpeck, J. C. 5883.

14905. Overton, J. B. 1904: Über Parthenogenesis bei *Thalictrum purpurascens*. Ber. Deut. Botan. Ges. 22:274–83. Botan. Centralbl. 96:261. 1904.

14906. Ovesnov, A. M. 1960: The effect of stratification on the germination power of seeds of wild grasses which germinate with difficulty. Byul. Gl. Botan. Sada 39:48–54.

14907. —— 1961: Morphobiology of seed germination in *Bromus inermis*. Izv. Estestv.-Nauchn. Inst. Permsk. Univ. 14(5):63–84. Biol. Abstr. 42:11843. 1963.

14908. —— and A. A. Shchekina. 1960: The effect of subterranean parts of couch grass and sow thistle on germination of the seeds of meadow herbs. Dokl.-Botan. Sci. Sect. (English Transl.) 127:220–22.

Oveson, M. M. 9069.

14909. Owen, C. R. 1955: Genetic variability in Dallis grass (*Paspalum dilatatum*) for seed quality. Agron. Abstr. 47: 54.

14910. Owen, E. B. 1956: The storage of seeds for maintenance of viability. Farnham Royal, Bucks.: Commonwealth Agr. Bur. 81 pp. S. African J. Sci. 53:274–75. 1957.

14911. Owen, F. V. 1928: Soybean seeds with two embryos. J. Heredity 19:373–74.

Owen, F. V. 13176, 14939, 18071.

14912. Owen, P. C. 1952: The relation of water absorption by wheat seeds to water potential. J. Exptl. Botany 3:276–90.

Owens, H. S. 10424.

14913. Oxley, T. A. 1944: The properties of grain in bulk. III. The thermal conductivity of wheat, maize, and oats. Soc. Chem. Ind. J. 63:53–55. Biol. Abstr. 19:11359. 1945.

14914. —— 1948: The reduction of losses in grain by control of water content. Food Agr. Organ. U.N. FAO Agr. Studies 2:92–99.

14915. —— 1950: The storage and drying of cereal seeds. Natl. Inst. Agr. Botany J. 5:465–82.

14916. —— 1955: Grain storage in tropical climates. World Crops 7:473–77.

14917. —— and M. E. Hyde. 1955: Recent experiments on hermetic storage of wheat. Proc. Third Intern. Bread Congr., Hamburg 1955:179–82.

14918. —— and J. D. Jones. 1944: Apparent respiration of wheat grains and its relation to a fungal mycelium beneath the epidermis. Nature 154:826–27.

14919. ——, S. W. Pixton, and R. W. Howe. 1960: Determination of moisture content in cereals. I. Interaction of type of cereal and oven method. J. Sci. Food Agr. 11:18–25.

14920. —— and G. Wickenden. 1963: The effect of restricted air supply on some insects which infest grain. Ann. Appl. Biol. 51:313–24.

Oxley, T. A. 8622, 9123, 9316.

14921. Oya, T. 1955: Kojic acid as antibiotics and rice seed growth regulator. J. Fac. Agr. Iwate Univ. 2:161–69.

14922. Oyaveski, K. 1962: Use of chemical weed control on seed-bearing perennial grasses. Sb. Dokl. Nauchn. Konf. Zashch. Rast. Tartu 1960:393–400. Chem. Abstr. 60:2264. 1964.

Oza, G. M. 15314.

Oza, J. D. 215.

14923. Ozaki, K., M. Saito, and K. Nitta. 1956: Studies on the seed development and germination of soybean plants at various ripening stages. Hokkaido Natl. Agr. Expt. Sta. Res. Bull. 70:6–14.

14924. —— and S. Sakuraba. 1963: Effects of fertilizer placement on the germination and yield in sugar beets. Hokkaido Natl. Agr. Expt. Sta. Res. Bull. 82:8–15.

14925. —— and L. R. Wetter. 1960: Glyceric acid kinase isolated from a Polish variety of rape-seed (*Brassica campestris*). Can. J. Biochem. Physiol. 38:125–34. Chem. Abstr. 54:6887. 1960.

14926. —— and —— 1960: The distribution of acylases in the seeds of higher plants. Can. J. Biochem. Physiol. 38:929–33.

14927. Ozawa, J., and T. Matsuura. 1954: Studies on the physiology of germination in tree seeds. II. Hokkaido Forest Expt. Sta. Spec. Rept. 2:32–43.

14928. —— and —— 1954: Studies on the hard seeds of leguminous trees. V. Hokkaido Forest Expt. Sta. Spec. Rept. 2:50–58.

14929. Ozerov, G. V. 1949: The ability of olive seeds to germinate on substrates with varying degrees of salinity and varying physical structure. Dokl. Akad. Nauk SSSR 69:679–82. Chem. Abstr. 48:6510. 1954.

14930. —— and M. A. Ozerova. 1952: New method of pre-planting treatment of kok-saghyz seeds. Dokl. Akad. Nauk SSSR 83:311–14.

Ozerova, M. A. 14930.

14931. Ozol, A. M. 1951: Selection of initial seeds and method for growing young (walnut) seedlings. Akad. Nauk SSSR Inst. Lesa Tr. 8:115–24.

Ozolins, J. 15840.

14932. Ozsan, M., and J. W. Cameron. 1963: Artificial culture of small citrus embryos, and evidence against nucellar embryony in highly zygotic varieties. Proc. Am. Soc. Hort. Sci. 82:210–16.

P

14933. P., A. B. 1931: Growing azaleas from seeds. Florists Rev. 68(1763):36.

14934. P., R. H. 1957: Production of Chewings' fescue seed and browntop seed in New Zealand. Gt. Brit. J. Sports Turf Inst 9:318–21.

Paauw, F. van der; *see* Van der Paauw, F.

Pace, J. 3653, 13160, 17971.

14935. Paci, P. 1956: Cold vernalization tests on some varieties of peas. Riv. Ortoflorofrutticolt. Ital. 40:320–31.

14936. Pack, D. A. 1921: After-ripening and germination of *Juniperus* seeds. Botan. Gaz. 71:32–60.

14937. —— 1921: Chemistry of after-ripening, germination, and seedling development of juniper seeds. Botan. Gaz. 72:139–50.

14938. —— 1925: Dispersion of lipoids. Botan. Gaz. 79:334–38.

14939. —— and F. V. Owen. 1950: Viability of sugar beet seed held in cold storage for 22 years. Proc. Am. Soc. Sugar Beet Technol. 6:127–29.

14940. Paddick, M. E., and H. B. Sprague. 1939: Maize seed characters in relation to hybrid vigor. J. Am. Soc. Agron. 31:743–50.

Paderna, L. G. 3533.

14941. Padhye, M. D. 1962: Fertilisation and seed development in *Nechmandra alternifolia* (Roxb.) Thw. Bull. Botan. Soc. Coll. Sci. Nagpur 3:92–100. Biol. Abstr. 44:20690. 1963.

14942. —— 1962: Solanad type of embryo development in *Gomphrena celosioides* Mart., a member of the Amaranthaceae. J. Indian Botan. Soc. 41:52–63.

14943. —— 1963: Two types of embryo development in *Passiflora foetida* Linn. Current Sci. (India) 32:373–74.

14944. Padilla, S. P., and F. A. Soliven. 1933: Chemical analysis for possible sources of oils of 45 species of oil-bearing seeds. Philippine Agriculturist 22:408–15. Chem. Abstr. 28:2207. 1934.

Padmanaban, G. 115.

14945. Padmanabhan, D. 1959: The development of endosperm in *Lantana camara* L. Proc. Indian Acad. Sci. Sect. B 49:420–27. Biol. Abstr. 35:14135. 1960.

14946. —— 1961: A contribution to the embryology of *Epithema carnosum*. J. Madras Univ. 31:37–47. Biol. Abstr. 41:11902. 1963.

Padmanabhan, D. 18207.

14947. Padmanabhan, S. Y. 1957: The relation between loss in viability and seed-borne micro-flora in rice. Proc. Indian Acad. Sci. Sect. B 46:155–69. Biol. Abstr. 32:35269. 1958.

Padmanabhan, T. S. 17840.

14948. Padoa, M. 1932: Action of the alkaloids and carbon monoxide on the enzymatic activity of plants. Nature 129:686.

14949. —— and A. Spada. 1933: Influence of some substances (especially unsaturated) on the enzyme activity of germinating seed of *Ricinus*, lipase and peroxidase. Giorn. Biol. Appl. Chim. Aliment. 3:121–31. Expt. Sta. Rec. 70:456. 1943.

14950. Padwick, G. W. 1938: Complex fungal rotting of pea seeds. Ann. Appl. Biol. 25:100–14. Biol. Abstr. 12:2505. 1938.

14951. Paech, H. O. 1929: Über die Unterscheidung vollkeimfähiger und wenig keimfähiger Sämereien auf chemischem Wege. Inaug. Dissertation Agr. Chem. Bact. Inst. Univ. Breslau. Chem. Abstr. 25: 1870. 1931.

14952. Paech, K. 1953: Über die Lichtkeimung von *Lythrum salicaria*. (Ein Beitrag zur Physiologie der Polyphenole.) Planta 41:525–66.

14953. Paetow, W. 1931: Embryologische Untersuchungen an Taccaceen, Meliaceen und Dilleniaceen. Planta 14:441–70. Ber. Wiss. Biol. 20:168–69. 1932.

Pagan, V. 9027.

14954. Page, A. B. P., N. G. M. Hague, V. Jakabsons, and R. E. Goldsmith. 1959: Fumigation of lucerne seed with methyl bromide for the control of the stem eelworm, *Ditylenchus* (*Anguillulina*) *dipsaci*. J. Sci. Food Agr. 10:461–67.

Page, H. L. 4148.

14955. Pages, P. D. 1955: Germination and germination inhibitors. Philippine Agriculturist 39:260–93.

14956. Paglia, E. 1910: Heterocarpy in the vegetable kingdom. Ann. Botany (London) 8:175–90.

14957. Pagniello, L. 1961: On the microflora of several seeds. Boll. Staz. Patol. Veg. 19:113–24. Biol. Abstr. 41:20404. 1963.

14958. Pai, M. L. 1958: The pH and titratable acidities of some legumes and the effect of their germination on the acidity. J. Maharaja Sayajirao Univ. Baroda 7:49–55. Chem. Abstr. 53:19053. 1959.

14959. Paikin, D. M., and P. R. Leitsis. 1940: Dichloroethane as a substitute for carbon disulfide in the fumigation of seeds. Bull. Plant Protect. (USSR) 1940(3):73–77. Chem. Abstr. 35:4542. 1941.

Pailhate, J. de Reif-; *see* De Reif-Pailhate, J.

Pain, A. K. 14060, 14061.

14960. Paine, J. 1950: Treatment of hop seed to improve germination. E. Malling Res. Sta. 1950:139–40. Brit. Abstr. B III:448. 1951.

14961. Paine, S. O., and L. M. Saunders. 1918: Note on a peculiarity exhibited by the testa of wrinkled peas. Ann. Botany (London) 32:174.

14962. Painter, E. P., and L. L. Nesbitt. 1943: The stability of linseed oil during storage of flaxseed. N. Dakota Agr. Expt. Sta. Bimonthly Bull 5(6):36–40.

14963. —— and —— 1946: Nitrogenous constituents of flaxseed. Ind. Eng. Chem. Ind. Ed. 38:95. Pharm. Abstr. 12:128. 1946.

14964. ——, ——, and T. E. Stoa. 1944: The influence of seasonal conditions on oil formation and changes in the iodine number during growth of flaxseed. J. Am. Soc. Agron. 36:204–13.

14965. —— and R. R. Painter. 1946: Protein isolation from the embryo of flaxseed. [Abstract.] Am. Chem. Soc. Abstr. Papers 110th Meeting, Chicago, p. 41B.

Painter, E. P. 14334.

Painter, R. R. 14965.

14966. Paizieva, S. A. 1962: Anatomomorphological structure of the embryo and the germ of *Cousinia umbrosa* Bge., *C. pseudoarctium* Borm. and *Arctium leiospermum* Jez. Vestn. Leningr. Univ. Ser. Biol. 1962(4):148–53.

14967. Pak, P. B. 1953: Selection of viable seeds of barley to increase yield and disease resistance. Selek. i Semen. 20(2): 24–30.

14968. Pal, B. P., A. B. Joshi, and S. K. Kamra. 1960: What seed testing means. Indian Farming 10(6):17–18, 40.

14969. —— and M. G. Suranarayana. 1941: Studies in the vernalization of Indian crop plants. I. Preliminary experiments on gram, wheat, chilli and soybean. Indian J. Genet. Plant Breeding 1:61–86.

Pal, G. 12792–12795.

14970. Pal, N. 1951: A note on the embryology of *Stellaria media* L. Sci. Cult. (Calcutta) 16:374.

14971. —— 1960: Development of the seed of *Millettia ovalifolia.* Botan. Gaz. 122: 130–37.

14972. Pal, N. L., and N. C. Gopalachari. 1957: Studies on the germination of tobacco seed. I. Germination of some commercial varieties of tobacco. J. Indian Botan. Soc. 36:262–67.

14973. —— and —— 1958: Influence and mechanism of action of low temperature pre-treatment on germination of tobacco seeds. Current Sci. (India) 27:212–13.

14974. ——, ——, and N. R. Bhat. 1959: Improvement of germination of tobacco seeds by pre-treatment to low temperature. Indian Tobacco 9:65–67. Tobacco Abstr. 3:355. 1959.

14975. Palamarchuk, G. L. 1957: Dependence of seed quality on the site of growth of the mother plant. Byul. Gl. Botan. Sada 29:32–35. Biol. Abstr. 44:16693. 1963.

14976. Palamarchuk, I. A. 1962: Development and structure of the endosperm in flint corn. Morfol. Kukuruzy 1962:121–39.

14977. —— and G. B. Rodionova. 1962: On the appearance of nuclei in the cells of the endosperm of a germinating caryopsis of corn. Morfol. Kukuruzy 1962:140–61.

14978. Palamartschuk, I. 1932: Zur Frage des Nichtfruchttragen der Pfefferminze (*Mentha piperita* L.). N. Caucasian Inst. Spec. Tech. Cultures Works 1(2):14–27.

14979. Palashev, I. 1961: Certain cultural problems concerning the propagation by seeds of *Rhus cotinus* L., syn *Cotinus coggygria* Scop. Izv. Karnobat. Kompleksen Selsk. Nauchn.-Izsled. Inst. 1:161–72.

14980. Palasi, V. V., A. Santos-Ruiz, and F. Sanz-Dominguez. 1948: The biochemistry of proteins. II. Aromatic and sulfur-containing amino acids of *Lathyrus sativus*. Anales Real Soc. Espan. Fis. Quim. (Madrid) Ser. B 44:771–74. Chem. Abstr. 43:1449. 1949.

14981. Paleg, L. G. 1960: Physiological effects of gibberellic acid. I. On carbohydrate metabolism and amylase activity of barley endosperm. Plant Physiol. 35:293–99.

14982. —— 1960: Physiological effects of gibberellic acid. II. On starch hydrolyzing enzymes of barley endosperm. Plant Physiol. 35:902–6.

14983. —— 1961: Physiological effects of gibberellic acid. III. Observations on its mode of action on barley endosperm. Plant Physiol. 36:829–37.

14984. —— and D. Aspinall. 1958: Inhibition of the development of the barley spike by gibberellic acid. Nature 181:1743–44.

14985. ——, B. G. Coombe, and M. S. Buttrose. 1962: Physiological effects of gibberellic acid. V. Endosperm responses of barley, wheat and oats. Plant Physiol. 37:798–803.

Paleg, L. G. 14384.

14986. Paleni, A. 1942: Estimation of the fat content of oil seeds. Fette Seifen Anstrichmittel 49:275–78. Chem. Abstr. 37:6917. 1943.

14987. Palfi, G. 1960: Seed germination of crop plants in saline media. Novenytermeles 9:359–68. Chem. Abstr. 57:7635. 1962.

Palilova, J. K. 10887–10889.

14988. Paliwal, R. L. 1956: Morphological and embryological studies in some Santalaceae. Agra. Univ. J. Res., Sci. 5:193–284.

Palk, B. A. 7567.

Palkina, N. 17231.

Palleolog, B. Mysakowska-. 11638.

14989. Palm, B. T. 1934: Ein neuer Embryosacktpus (bei *Rudbeckia hirta* L.). Botan. Notiser 1934:423–27. Biol. Abstr. 11: 9599. 1937.

14990. Palmer, H. E. 1933: Are the seeds of the tung oil tree poisonous when eaten by man or animal? J. Florida Med. Assoc. 20:13–15.

Palmer, L. S. 12948, 14312.

Palmer, R. C. 17049.

14991. Palmer, T. P. 1960: Swede (*Brassica napus*) seeds with split cotyledons. I. The distribution of split cotyledons on plants. Ann. Botany (London) 24:48–51.

Palmer-Jones, T. 6556.

Palo, M. A. 3533.

14992. Palser, B. F. 1961: Some aspects of embryology in the *Ericales*. Intern. Botan. Congr. Recent Advan. Botany 9:685–89. 1959.

14993. Pam, A. 1955: Germination of parthenogenetic seed in amaryllids. J. Roy. Hort. Soc. 80:128–30.

14994. Pammel, L. H. 1886: Testa of leguminous seeds. Bull. Torrey Botan. Club 13:17–24.

14995. —— 1890: On the seed coats of the genus *Euphorbia*. Proc. Am. Assoc. Advan. Sci. 39:328–29.

14996. —— 1894: Seeds of the Russian thistle (*Salsola kali*, var. *tragus*). Iowa Agr. Expt. Sta. Bull. 26:20–24.

14997. —— and A. W. Dox. 1917: The protein contents and microchemical tests of the seeds of some common Iowa weeds. Proc. Iowa Acad. Sci. 24:527–32.

14998. —— and C. M. King. 1910: Delayed germination. Proc. Iowa Acad. Sci. 17: 20–33.

14999. —— and —— 1910: Results of seed investigations for 1908 and 1909. Iowa Agr. Expt. Sta. Bull. 115:154–75.

15000. —— and —— 1917: The germination and juvenile forms of some oaks. Proc. Iowa Acad. Sci. 24:367–91.

15501. —— and —— 1918: The germination of some trees and shrubs and their juvenile forms. Proc. Iowa Acad. Sci. 25:291–340.

15002. —— and —— 1920: The germination of some trees and shrubs and their juvenile forms. Proc. Iowa Acad. Sci. 27:75–80.

15003. —— and —— 1921: Studies in the germination of some woody plants. Proc. Iowa Acad. Sci. 28:273–82.

15004. —— and —— 1922: Germination studies of some shrubs and trees. Proc. Iowa Acad. Sci. 29:257–66.

15005. —— and —— 1924: Further studies of the germination of woody plants. Proc. Iowa Acad. Sci. 31:157–67.

15006. —— and —— 1925: Germination of some pines and other trees. Proc. Iowa Acad. Sci. 32:123–32.

15007. —— and —— 1926: Further notes on the germination of some trees and shrubs and their juvenile forms. Proc. Iowa Acad. Sci. 33:75–80.

15008. —— and —— 1926: Studies on germination of trees and woody plants (continued). Proc. Iowa Acad. Sci. 33:97–119.

15009. —— and —— 1928: Further studies of the germination and juvenile forms of

some trees and woody shrubs, 1927. Proc. Iowa Acad. Sci. 35:169–83.
15010. —— and —— 1928: Germination studies of some trees and shrubs, 1928. Proc. Iowa Acad. Sci. 35:184–97.
15011. —— and —— 1929: Germination of trees and shrubs. Proc. Iowa Acad. Sci. 36:201–11.
15012. —— and —— 1930: Germination and seedling forms of some woody plants. Proc. Iowa Acad. Sci. 37:131–41.
15013. —— and G. M. Lummis. 1903: The germination of weed seeds. Proc. Soc. Prom. Agr. Sci. Ann. Meeting 24:89–92.
15014. —— and L. Robb. 1908: Notes on the histological structure and specific gravity of the seeds of *Pyrus*. Proc. Iowa Acad. Sci. 15:47–55.
15015. —— and F. C. Stewart. 1894: Specific gravity and weight of wheat seed. Iowa Agr. Expt. Sta. Bull. 25:26–31.
15016. Pamplin, R. A. 1963: The anatomical development of the ovule and seed in the soybean. Dissertation Abstr. 24:483–84.
15017. Pan, C. L. 1936: A preliminary report of varietal differences in rapidity of germination in rice. J. Am. Soc. Agron. 28:985–89.
15018. Pandalai, K. M., and R. V. Pillai. 1959: Albinism in coconut seedlings. Nature 184:1163.
15019. Pandey, K. K. 1955: Seed development in diploid, tetraploid and diploid-tetraploid crosses of *Trifolium pratense* L. Indian J. Genet. Plant Breeding 15:25–35.
15020. Pandey, S. N. 1962: Effects of varying neutron treatments on germination, growth of wheat, gram and mustard. J. Sci. Res. Banaras Hindu Univ. 13:122–31.
15021. Pandeya, S. C. 1954: Effect of light on the time of germination of Gramineae seeds. Ann. Fac. Agrar. Univ. Studi Perugia 10:233–38.
15022. Pandotra, V. R., and D. Ganguly. 1962: A study on the probable relationship between seed-borne fungi and poor viability of *Rauwolfia serpentina* seeds. Current Sci. (India) 31:151–52.
15023. Panganiban, F. C. 1924: The effects of etherization on germination of tropical seeds. Philippine Agriculturist 13:93–98.
15024. Pangsrivongse, K. 1938: Note on chemical study of seed of *Lansium domesticum*, Correa. Rev. Filipina Med. Farm. 29:9. J. Am. Pharm. Assoc. 27:501. 1938.
15025. Panikkar, M. R. 1947: Storage of rice. Sci. Cult. (Calcutta) 13:59–66. Biol. Abstr. 22:25121. 1948.
Pankiw, P. 2949.
15026. Pannocchia-Laj, F. 1938: Embryology and karyology of *Lochnera rosea* (L.) Rehb. Nuovo Giorn. Botan. Ital. 45:122–30. Biol. Abstr. 12:12328. 1938.
15027. —— 1938: Embryology and karyology of *Vinca* with special consideration of tetraploidy in *Vinca major* L. and cryptopolyploidy in *Vinca difformis* Pourr. Nuovo Giorn. Botan. Ital. 45:157–87. Biol. Abstr. 12:12329. 1938.
15028. Panov, A. 1953: The stratification of forest-tree seeds. Inst. Naucna Sumarska Istraz. God. 2:159–91.
15029. Panse, T. B., and A. S. Paranjpe. 1943: Carpasemine isolated from *Carica papaya* seeds. Proc. Indian Acad. Sci. Sect. A 18:140–44. Chem. Abstr. 38:960. 1944.
15030. Panshin, B. 1926: Immunity and stimulation of seeds. J. Agr. Botany Kharkiv 1:137–45.
Pant, B. D. 3795, 3796.
15031. Pant, D. D., and D. D. Nautiyal. 1962: Seed cuticles in some modern cycads. Current Sci. (India) 31:75–76.
Pantaleon, F. T. 6060.
15032. Pantanelli, E., and S. Verdesca. 1930: The lipase of olives and oil. Rend. Accad. Sci. Fis. Mat. Cl. Soc. Reale Napoli 36:76–83. Biol. Abstr. 7:7911. 1933.
15033. Panteleev, A. N. 1954: Balance of carbon in sprouting of seeds of leguminous plants in connection with utilization of reserve substances. Dokl. Akad. Nauk SSSR 98:243–46. Chem. Abstr. 49:1881. 1955.
15034. —— 1955: Respiration of the sprouting seed and the conversion of carbohydrates and protein. Uchenye Zapiski Leningrad. Gosudarat. Univ. im. A. A. Zhdanova Ser. Biol. Nauk 39:87–120. Chem. Abstr. 51:5927. 1957.
Panteleev, A. N. 17724.
15035. Pantskhava, A. D. 1957: Study of the bio-ecological properties of seeds of certain xerophytic tree species. Akad. Nauk Gruzinsk. SSR Inst. Lesa. Tr. 1957(7):257–85.
Pantulu, J. V. 10011.
Paolis, C. de; *see* De Paolis, C.
15036. Pap, L. 1931: Hygroscopicity of wheat. Cereal Chem. 8:200–6.
15037. Papadopoulou-Touphexi, A. 1954: A new method for determining the percentage of seeds germinating under

field conditions. Georg. Delt. (Ser. 3) 8:99–106.

15038. —— 1959: Rice germinability determination through the topographical tetrazolium method. Georg. Delt. (Ser. 3) 14:37–42.

15039. Papavizas, G. C., and C. M. Christensen. 1957: Grain storage studies. XXV. Effect of invasion by storage fungi upon germination of wheat seed and upon development of sick wheat. Cereal Chem. 34:350–59.

15040. —— and —— 1958: Grain storage studies. XXVI. Fungus invasion and deterioration of wheats stored at low temperatures and moisture contents of 15 to 18 per cent. Cereal Chem. 35:27–34.

15041. —— and —— 1960: Grain storage studies. XXIX. Effect of invasion by individual species and mixtures of species of *Aspergillis* upon germination and development of discolored germs in wheat. Cereal Chem. 37:197–203.

15042. Pape, H., and K. Domsch. 1956: Zur Beizung von Maiblumenkeimen. Gartenwelt 56:121.

15043. Papp, G. 1933: The effect of human tissue fluids on germinating plants. Magy. Orvosi Arch. 24:455–58. Chem. Abstr. 28:805. 1934.

Papp, Z. 2693.

15044. Papworth, D. S. 1961: The protection of stored cereals by malathion admixture techniques. Agr. Vet. Chem. 2:160–65.

15045. Parameswar, N. S. 1962: Germination of seed areca nuts (*Areca catachu* L.). Sci. Cult. (Calcutta) 28:135–36. Biol. Abstr. 40:16677. 1962.

15046. Parameswaran, N. 1961: Ruminate endosperm in the Canellaceae. Current Sci. (India) 30:344–45.

Parameswaran, N. 18208.

15047. Parandekar, S. A. 1935: A note on seed variations in *Carica papaya*, Linn. Current Sci. (India) 3:379.

Paranjpe, A. S. 15029.

15048. Paranjpe, D. R., and P. R. Ayyar. 1929: Oil from the seeds of *Sapindus trifoliatus* (Linn.). J. Indian Inst. Sci. 12A:179–84. Biol. Abstr. 5:3922. 1931.

Parasuraman, V. 14105.

15049. Pardi, P. 1933: Contributo all'embriologia delle Asclepiadaceae. Nuovo Giorn. Botan. Ital. 40:560–69.

15050. Parencia, C. R., Jr., J. W. Davis, and C. B. Cowan, Jr. 1957: Further field tests with systemic insecticides employed as seed treatments. J. Econ. Entomol. 50:614–17.

Parfentieff, J. A. 9544.

15051. Parihar, D. B. 1954: Changes in the saccharides of stored-pulses. A study by circular paper chromatography. Naturwissenschaften 41:427–28.

15052. —— 1954: Free amino acids in some Indian grown rice samples (*Oryza sativa* Linn.). Naturwissenschaften 41:502.

15053. —— 1955: Changes in the saccharides of wheat-flour and rice on their processing into some common Indian foods. Naturwissenschaften 42:514.

15054. —— 1955: Saccharides of different varieties of Indian rice. Nature 175:42–43.

15055. —— and S. Dutt. 1947: Chemical examination of the seeds of *Dodonaea viscosa*. Proc. Indian Acad. Sci. Sect. A 26:56–60. Biol. Abstr. 22:12404. 1948.

15056. Parija, P. 1934: Physiological investigations on water-hyacinth (*Eichhornia crassipes*) in Orissa, with notes on some other aquatic weeds. Indian J. Agr. Sci. 4:399–429. Expt. Sta. Rec. 72:181–82. 1935.

15057. —— 1943: On the pre-sowing treatment and phasic development. Current Sci. (India) 12:88–89. Expt. Sta. Rec. 90:459. 1944.

15058. —— and P. Mallik. 1936: The mechanism of the bursting of the fruits of *Impatiens balsamina* Linn. J. Indian Botan. Soc. 15:59–61.

15059. —— and —— 1940: Nature of the reserve food in seeds and their resistance to high temperature. J. Indian Botan. Soc. 19:223–29.

15060. —— and K. P. Pillay. 1945: Effect of pre-sowing treatment on the drought resistance in rice. Proc. Natl. Acad. Sci. India Sect. B 15:6–14. Biol. Abstr. 22:14710. 1948.

15061. Parikh, J. M., and C. S. Patel. 1954: Proteins and minerals of guar seeds (*Cyamopsis psoralioides*). J. Maharaja Sayajirao Univ. Baroda 111:47–58. Biol. Abstr. 29:29730. 1955.

15062. Paris, D., J. Rietsema, S. Satina, and A. F. Blakeslee. 1953: Effect of amino acids, especially aspartic and glutamic acid and their amides, on the growth of *Datura stramonium* embryos in vitro. Proc. Natl. Acad. Sci. U.S. 39:1205–12.

15063. Paris, J. 1960: Nucleic acid phosphorus content of *Robinia pseudoacacia* seeds and seedlings. Erdö 9:416–20.

15064. Paris, R., and M. Rigal. 1940: Preliminary studies on bark and seeds of

Erythrophleum guineense. Bull. Sci. Pharmacol. 47:79–87. Pharm. Abstr. 7:300. 1941.

Park, E. R. 7287.

Park, J. B. 19799.

Park, J. K. 9134.

15065. Park, M. 1937: The seed treatment of ginger. Trop. Agriculturist 89:3–7. Chem. Abstr. 31:8097. 1937.

Park, M. K. 20055.

Parkash, V. 19081.

15066. Parker, B. M. 1941: Seeds and seed travels, trees, and birds. Row, Peterson, Evanston, Ill. 3 books. 36 pp. ea.

Parker, E. 6478.

Parker, E. D. 3398, 3399.

15067. Parker, J. 1950: Germination of Eastern redcedar seeds. J. Forestry 48: 255–56.

15068. —— 1952: Establishment of Eastern redcedar by direct seeding. J. Forestry 50:914–17.

15069. —— 1953: New methods for the determination of forest tree seed germinability. J. Forestry 51:34.

15070. —— 1955: Tetrazolium salts as viability indicators: radiation-catalysed reactions. Nature 176:647–48.

15071. Parker, M. C., and D. D. Hill. 1955: Hydrogen peroxide solutions for rapid determination of barley viability. Agron. Abstr. 47:71.

Parker, M. C. 3956, 3957.

15072. Parker, M. M. 1937: The effect of fertilizer placement on the germination and growth of *Pisum sativum*. Proc. Virginia Acad. Sci. 1937:52. Expt. Sta. Rec. 80:47–48.

15073. —— 1940: Fertilizer placement on snap beans and cabbage. Peninsula Hort. Soc. Trans. 30:36–42. Biol. Abstr. 16: 5305. 1942.

15074. —— 1942: Effect of fertilizer placement on snap beans, lima beans, and peas. Virginia Truck Expt. Sta. Bull. 107:1757–81. Expt. Sta. Rec. 88:332. 1943.

15075. —— and G. A. Cumings. 1938: placement of fertilizer for Henderson bush lima beans in Virginia. Virginia Truck Expt. Sta. Bull. 99:1557–78. Expt. Sta. Rec. 81:513–14. 1939.

15076. —— and R. C. Oliver. 1938: The effect of fertilizer placement, as influenced by soil moisture, on seed germ. Proc. Am. Soc. Hort. Sci. 35:533–36. Expt. Sta. Rec. 82:43. 1940.

Parker, M. W. 2809

Parker, N. S. 11719

15077. Parkey, W., J. E. Webster, and D. L. Van Horn. 1956: The nitrogen, phosphorus, potassium and ash content of castor bean hulls. Oklahoma A & M Div. Agr. Tech. Bull. T61. 11 pp.

Parkhurst, D. 17053.

15078. Parkin, E. A. 1959: Insects and stored food. I. World losses. World Crops 11:405–7.

15079. —— 1963: The protection of stored seeds from insects and rodents. Proc. Intern. Seed Testing Assoc. 28:893–909.

15080. —— and E. I. C. Scott. 1960: The germination of food-crop seeds stored in contact with lindane and DDT dusts. Trop. Sci. 1:206–11.

15081. Parkinson, A. H. 1948: Seed age and storage affect cucumber growth. Farm Res. N.Y. 14(3):7.

15082. Parkinson, S. T. 1935: The microscopic examination of cattle foods. Some oil-containing seeds. Southeastern Agr. Coll. J. (Wye, Kent) 36:9–30. Chem. Abstr. 30:182. 1936.

15083. Parkman, S. B. 1960: Seed storage. Seed World 87(4):4.

Parks, L. M. 12829A.

15084. Parks, M. 1935: Embryo sac development and cleistogamy in *Commelinantia pringlei*. Bull. Torrey Botan. Club 62:91–104.

Parks, R. R. 291.

Parks, W. L. 2830, 4747, 6434.

15085. Parlevliet, J. E., and J. W. Cameron. 1959: Evidence on the inheritance of nucellar embryony in citrus. Proc. Am. Soc. Hort. Sci. 74:252–60.

Parmeter, J. R., Jr. 16881.

Parra, J. L. 17173.

15086. Parras, V. 1939: Comparison of different methods of germinating bitaog. Philippine J. Forestry 2:57–63. Biol. Abstr. 13:15671. 1939.

15087. Parris, G. K. 1944: Seed treatment of peas decreases fertilizer injury. Plant Disease Reptr. 28:1152–54. Chem. Abstr. 39:1011. 1945.

Parris, G. K. 10613.

Parrish, C. P. 16826.

Parrot, A. 3301.

15087A. Parry, M. S. 1956: Tree planting practices in tropical Africa. Food Agr. Organ. U.N., Rome. 302 pp.

15088. Parson, F. G., C. S. Garrison, and K. E. Beeson. 1961: Seed certification in

the United States. Yearbook Agr. U.S. Dept. Agr. 1961:394–401.
Parsons, H. T. 6133.
Parsons, J. L. 10849.
Parthasarathi, N. 15910.

15089. Parthasarathy, S. V. 1940: Physiological studies during vernalization in rice. Madras Agr. J. 28:133–37. Chem. Abstr. 36:6583. 1942.
Parveaud, A. 5076.
Pascal, E. C. di; *see* Di Pascal, E. C.
Pasco, L. C. 17277.
Pascoe, E. D. 13774, 13775.

15090. Pashchenko, Z. M. 1957: Microchemical study of the seed-ripening process in cotton varieties differing in their growth cycle length. Tr. Sredneaz. Gos. Univ. V. I. Lenina, Biol. Nauki 116(26):55–75. Chem. Abstr. 53:11540. 1959.

15091. Pasinetti, L. 1929: A contribution to the study of fungicidal substances. (Experiments with Fideol No. 1.) Riv. Patol. Vegetale 9:193–214. Rev. Appl. Mycol. 9:328. 1930.

15092. —— 1930: Tests of the devitalization of *Cuscuta* seed in soil with monochloracetic acid. Riv. Patol. Vegetale 20:1–8. Biol. Abstr. 6:6405. 1932.

15093. Pasquier, and M. T. Francois. 1952: Comparative examination of different apparatus utilizing infrared rays for the determination of moisture in oil seeds. Bull. Mens. Inform. ITERG 6:280–88. Chem. Abstr. 46:8879. 1952.

15094. Pasquier, R., and M. T. Gausseran. 1945: L'hexachlorocyclohexane ($C_6H_6C_{16}$) et la faculté germinative des blés durs. Ann. Inst. Agr. Algerie 1945:29–34. Agron. Trop. 1:317. 1946.

15095. Pasquine, D. 1955: Research on the influence of compressed air on the germination of seeds of certain cultivated plants. Ann. Sper. Agrar. (Rome) 9:683–708.

15096. Passalacqua, T. 1934: Studies on orobanche of bean in Sicily. II. Ann. Tec. Agrar. 7:487–96. Biol. Abstr. 11:15637. 1937.

15097. Passecker, F. 1956: Keimungsphysiologische Untersuchungen an Kern- und Steinobst. Gartenbauwissenschaft 2:274–90.

15098. Passerini, N. 1932: The great resistance of some seeds to immersion in carbon disulfide, carbon tetrachloride, commercial benzine and petroleum. Boll. Ist. Super. Agrar. Pisa 8:711–41. Chem. Abstr. 27:3737. 1933.

15099. —— 1933: The probable biological significance of so-called hard seeds. Nuovo Giorn. Botan. Ital. 40:230–51.

15100. Passion, J. 1905: Sur des germinations de *Xanthociros sorbifolia*. Bull. Soc. Botan. France 52:45.

15101. Pastuh, M. B. 1939: The effect of various methods of pre-sowing treatment of sugar-beet seeds on yield and sugar content. Sb. Rabot V. N. I. S. Kiev-Kharkov 1939:229–40. Chem. Abstr. 36:6584–85. 1942.
Pasvik-Khlopin. 10528.

15102. Patankar, T. B. V. 1956: Further contribution to the embryology of *Drosera burmanni* Vahl. Proc. Indian Acad. Sci. Sect. B 43:161–71. Biol. Abstr. 31:29209. 1957.

15103. Patch, F. W. 1959: Gloxinias. It's fun to start them from seeds. Horticulture 37:380–81, 383.

15104. Patel, C. B. 1943: Chemical investigation of seed oil of *Moringa concanensis*. Current Sci. (India) 12:272–73.
Patel, C. S. 15061.

15105. Patel, J. S., and G. V. Narayana. 1935: A rare instance of polembryony in *Arachis hypogaea*, Willd. Current Sci. (India) 4:32–33.

15106. —— and C. R. Seshadri. 1935: Oil formation in groundnut with reference to quality. Indian J. Agr. Sci. 5:165–75. Expt. Sta. Rec. 74:334. 1936.
Patel, J. S. 19007.
Patel, M. H. 2417.

15107. Patel, R. D., R. P. Patel, and R. S. Patel. 1960: Specific rotation of cereal and legume starches. Cereal Chem. 37:500–2.
Patel, R. D. 15826.
Patel, R. P. 15107.
Patel, R. S. 15107.

15108. Patel, S. A., S. Bhattacharyya, and M. M. Chakrabarty. 1958: Seed fats of Cucurbitaceae. II. The component fatty acids of *Trichosanthes cucumerina*. J. Indian Chem. Soc. 35:67–71. Chem. Abstr. 53:8325. 1959.
Patel, S. A. 3758.
Patel, S. M. 1588.

15109. Patel, Z. H. 1933: Cause of seed abortion in soybeans and other crop plants. Abstr. Thesis, Univ. Illinois, Urbana. 8 pp. Expt. Sta. Rec. 71:176–77. 1934.

15110. Paterson, D. R., H. T. Blackhurst, and S. H. Siddiqui. 1960: Some effects of nitrogen and phosphoric acid on premature seedstalk development, yield and composition of three onion varieties. Proc. Am. Soc. Hort. Sci. 76:460–67.

15111. Pathak, S. P., and S. S. Mathur. 1954: The component acids and glycerides of areca nut (*Areca catechu*) fat. J. Sci. Food Agr. 5:461–65.

15112. —— and A. K. Vasishtha. 1961: Changes in the component fatty acids of mango-seed fat during development. J. Sci. Res. Banaras Hindu Univ. 12:41–50.

Patil, J. A. 2402.

15113. Patil, N. A. 1955: Germination of stored tobacco seed. Indian Tobacco 5:23–24.

Patil, S. S. 1312, 3198.

Patnaik, S. N. 13576.

15114. Paton, R. R. 1936: Preplanting treatment of black cherry seed. J. Forestry 34:730. Biol. Abstr. 11:6843. 1937.

15115. —— 1945: Storage of tuliptree seed. J. Forestry 43:764–65.

15116. Patrick, S. R. 1936: Some observations on the decline in viability of Chewing's fescue seed (*Festuca rubra commutata*). Proc. Assoc. Offic. Seed Analysts 28:76–79.

15117. —— 1939: Canada bluegrass seed is sensitive to germinative conditions. Proc. Assoc. Offic. Seed Analysts 30:247–48.

15118. —— 1946: Is it necessary to pre-soak garden beet seed? Proc. Assoc. Offic. Seed Analysts 1945/46:150–51.

15119. —— and W. F. Crosier. 1937: Commercial significance of hard seeds in peas. Proc. Assoc. Offic. Seed Analysts 29:146–47.

15120. ——, ——, and C. E. Heit. 1947: The tolerance of seeds to DDT. Proc. Assoc. Offic. Seed Analysts 37:166–69.

Patrick, S. R. 4146, 4663, 4666–4675.

15121. Patrick, Z. A., and L. W. Koch. 1958: Inhibition of respiration, germination, and growth by substances arising during the decomposition of certain plant residues in the soil. Can. J. Botany 36:621–47.

Patrudu, S. S. 5782.

15122. Patt, Y. 1949: The influence of petroleum oils on the germination of citrus seeds and subsequent growth of seedlings. Palestine J. Botany 7:94–102. Biol. Abstr. 25:25043. 1951.

15123. —— 1953: Experiments on germination of immature citrus seeds. Palestine J. Botany 8:219–20. Biol. Abstr. 29:9570. 1955.

15124. Pattanaik, S. 1946: Effect of Na-sulfapyridine on the catalase activity of rice seeds. Current Sci. (India) 15:196–97. Biol. Abstr. 21:12569. 1947.

15125. Patten, B. C., Jr. 1955: Germination of the seed of *Myriophyllum spicatum* L. Bull. Torrey Botan. Club 81:50–56.

15126. Patterson, C. F., and A. C. Bruce. 1931: Rapid methods of determining the percentages of fertility and sterility in seeds of the genus *Betula*. Sci. Agr. (Ottawa) 11:704–8. Biol. Abstr. 6:4963. 1932.

15127. Patterson, F. L., L. E. Compton, R. M. Caldwell, and J. F. Schafer. 1962: Effects of awns on yield, test weight and kernel weight of soft red winter wheats. Crop Sci. 2:199–200.

Patterson, F. L. 461.

Patterson, J. K. 5307, 10730.

Patton, E. L. 5096.

Patwardhan, V. A. 1589, 1590.

15128. Patwardhan, V. N. 1929: Enzymes from the seeds of *Caesalpinia bonducella*. J. Indian Inst. Sci. 12A:191–92. Chem. Abstr. 24:1131. 1930.

Patwardhan, V. N. 10265.

15129. Pauchon, A. 1880: Recherches sur le rôle de la lumière dans le germination. Étude historique, critique et physiologique. Ann. Sci. Nat. Botan. (Ser. 6) 10:81–217.

15130. Pauksens, J., and E. Reinbergs. 1963: Main and lateral kernel weight ratio in the progeny of a six- and two-rowed barley cross. Can. J. Plant Sci. 43:589–96.

Paul, A. D. 13121.

15131. Paul, A. K. 1937: Development of ovule and embryo-sac of *Tamarindus indica* Linn. J. Indian Botan. Soc. 16:151–57.

15132. Paul, D. L., B. Ostle, and R. G. Tischer. 1952: The relation of sulfur content of sweet corn to its stage of maturity. Proc. Am. Soc. Hort. Sci. 59:393–99.

15133. Paul, F. T. 1955: The propagation of seeds and cuttings under flourescent lights. Minn. Horticulturist 83(2):22–23, 32.

Paul, S. 8887.

Paula, R. Descartes de G.; *see* Descartes de G. Paula, R.

15134. Pauley, S. S., S. H. Spurr, and F. W. Whitmore. 1955: Seed source trials of Eastern white pine. Forest Sci. 1:244–56.

15135. Pauli, A. W., and H. H. Laude. 1959: Protein and carbohydrate relationships in winter wheat as influenced by mechanical injury. Agron. J. 51:55–57.

15136. —— and F. C. Stickler. 1961: Effects of seed treatment and foliar spray applications of gibberellic acid on grain sorghum. Agron. J. 53:137–39.

Pauli, A. W. 10483, 10484, 18005.

15137. Paulsen, E. F. 1927: La influencia de las corrientes electricas debiles en presencia de soluciónes diluidas de electrolitos de bajo poder de disociación sobre la germinación de las semillas. Rev. Agron. Vet. Univ. Buenos Aires 1927(132):434–38. Botan. Centralbl. 155:75. 1928/29.

Paulsen, H. A. 7308.

Paulson, W. E. 19359.

Paun, E. 15875.

15138. Pauze, F. 1951: La propagation des mauvaises herbes. III. Section II. La germination des graines. Rev. Oka Agron. Med. Vet. 25:177–88.

15139. Pavari, A. 1940: Procurement of seed of exotic forest trees. Riv. Forest Ital. 2:4–11. Biol. Abstr. 14:871. 1940.

Pavgi, M. S. 19032.

15140. Pavlenko, O. N. 1940: The chemical differences of different varieties of almonds. Biokhim. Kul't. Rast. 7:456–61. Chem. Abstr. 35:3287. 1941.

15141. Pavlicic, J., and A. Mitrovic. 1959: The effect of gibberellin on the rate and percentage of germination with wheat and soybean. Arhiv Poljoprivredne Nauke 12(37):25–47.

15142. Pavlov, A. N. 1962: Transformations of protein fractions during germination of corn seeds. Dokl. Akad. Nauk SSSR 143:438–40.

15143. —— 1963: Accumulation of zein in maize grain. Plant Physiol. (USSR) (English Transl.) 10:219–26. Chem. Abstr. 59:5494. 1963.

Pavlov, A. N. 15251.

15144. Pavlov, G. 1932: Investigation of the seeds and oil of *Staphylea pinnata* L. Maslob.-Zhir. Delo 1932(4/5):93–95. Chem. Abstr. 27:2594. 1933.

Pavlova, A. P. 3483.

Pawelczyk, E. 8984.

15145. Pawlisch, P. E., and H. L. Shands. 1962: Breeding behavior for bushel weight and agronomic characters in early generations of two oat crosses. Crop Sci. 2:231–37.

15146. Pawlowski, S. H. 1961: Estimation of oil in safflower seed. Nature 189:772–73.

15147. —— 1963: Methods for rapid determination of specific gravity of single seeds and seed samples and the correlation of specific gravity with oil content. Can. J. Plant Sci. 43:151–56.

15148. Payne, G. F. 1957: Some germination studies of *Artemisia tridentata*. Proc. Montana Acad. Sci. 17:41–42.

15149. Payne, G. O. 1952: Raising lilies from seed in New Zealand. Lily Yearbook 1953(16):140–41.

Payne, K. T. 6129.

15150. Payne, M. A. 1935: The flower and seed of *Mollugo verticillata*. Kansas Univ. Bull. 36(8):5–25.

15151. Peacock, D. H., and C. Thoung. 1931: The oil from the seeds of *Asteriastigma macrocarpa*. Soc. Chem. Ind. J. 50:7T–8T.

15152. Peacock, H. A., and C. P. Wilsie. 1960: Selection for vegetative vigor and seed setting in Birdsfoot trefoil (*Lotus corniculatus* L.). Agron. J. 52:321–24.

15153. Peaden, R. N. 1957: Reed canarygrass germination at five seed maturity stages and sixteen seed treatments. [Abstract.] Utah State Univ. Monograph Ser. 5(4):25.

Peak, J. W. 4949.

15154. Peake, F. G. G. 1952: On a bruchid seed-borer in *Acacia arabica*. Bull. Entomol. Res. 43:317–24.

15155. Pearce, J. A. 1943: Effect of storage on thiamin content and on development of rancidity in wheat germ. Can. J. Res. Sect. C 21:57–65.

Pearl, R. 7515, 7516.

Pearse, H. L. 9291, 11617.

15156. Pearson, B. O. 1957: Bitterbush (*Purshia tridentata*) seed dormancy broken with thiourea. J. Range Management 10:41–42.

Pearson, B. O. 9164.

15157. Pearson, F. A., and F. A. Harper. The world's hunger. Cornell Univ. Press, Ithaca, N.Y. 90 pp.

15158. Pearson, G. A. 1923: Natural reproduction of western yellow pine in the southwest. U.S. Dept. Agr. Bull. 1105. 143 pp.

15159. Pearson, H. M. 1933: Parthenocarpy and seed abortion in *Vitis vinifera*. Proc. Am. Soc. Hort. Sci. 29:169–75.

15160. Pearson, N. L. 1939: Relation of the structure of the chalazal portion of the cotton seed coat to rupture during ginning. J. Agr. Res. 58:865–73.

15161. —— 1948: Observations on seed and seed hair growth in *Asclepias syriaca* L. Am. J. Botany 35:27–36.

15162. Pearson, O. H. 1933: Study of the life history of *Brassica oleracea*. Botan. Gaz. 94:534–50.

15163. Pease, R. W. 1954: Growing chestnuts from seeds. West Va. Agr. Expt. Sta. Circ. 90. 9 pp.
Peat, J. 10846.
15164. Peat, S., E. J. Bourne, and M. J. Nicholls. 1948: Starches of the wrinkled and the smooth pea. Nature 161:206–7. Biol. Abstr. 22:25383. 1948.
15165. ——, W. J. Whelan, and S. J. Pirt. 1949: The amylolytic enzymes of soya bean. Nature 164:499–500.
15166. ——, ——, and J. R. Turvey. 1956: Soluble polyglucose of sweet corn (*Zea mays*). J. Chem. Soc. 1956:2317–22. Chem. Abstr. 50:14877. 1956.
15167. Pechmann, V. 1932: Saatgutprovenienzfragen—vor 160 Jahren. Forstwiss. Centralbl. 54:164–67.
15168. Pechnik, E., and J. M. Chaves. 1945: Chemical study of the fatty substance of the bacury (*Platonia insignis*, Mart.). Rev. Quim. Ind. (Rio de Janeiro) 14(163): 18–19. Chem. Abstr. 40:5269–70. 1946.
Pechnik, E. 5260.
Pechnikova, S. S. 11383.
15169. Pechoutre, F. 1902: Contribution à l'étude du développement de l'ovule et de la graine des Rosacées. Ann. Sci. Nat. Botan. 16:1–158.
15170. Peck, O., and J. L. Bolton. 1946: Alfalfa seed production in northern Saskatchewan as affected by bees, with a report on means of increasing the populations of native bees. Sci. Agr. (Ottawa) 26:388–418.
Pedelty, W. H. 8772, 8776.
Pedersen, H. L. 15171.
15171. Pedersen, M. W., H. L. Pedersen, G. E. Bohart, and M. D. Levin. 1956: A comparison of the effect of complete and partial cross-pollination of alfalfa on pod sets, seeds per pod, and pod and seed weight. Agron. J. 48:177–80.
15172. Pedersen, P. N. 1956: A routine method of testing seed barley for loose smut. Proc. Intern. Seed Testing Assoc. 21:112–28.
15173. Peebles, R. H., and C. Hope. 1936: The influence of different pollens on the development of the pistache nut. Proc. Am. Soc. Hort. Sci. 33:29–32.
Peedin, C. D. 4365.
15174. Peel, A. C. 1962: The effect of storage on the viability of crop seeds. Proc. Intern. Seed Testing Assoc. 27:893–906.
Peel, A. C. 4285, 4286.
Peers, A. W. 7566.
15175. Peers, F. G. 1958: Germination inhibitory substances in oat husk. W. African J. Biol. Chem. 2:9–14. Chem. Abstr. 52:20448. 1958.
15176. Peery, L. T. 1959: On seed germination. Natl. Hort. Mag. 38:178–80.
Peeters, H. 12960.
15177. Peimbert, R., J. Krewinkle, and D. Morey. 1963: Factors influencing seed set in roses. Am. Rose Ann. 48:194–98.
15178. Peirce, G. J. 1905: The dissemination and germination of *Arceuthobium occidentale*, Eng. Ann. Botany (London) 19:99–113.
Peirce, G. J. 4917.
Peirce, L. C. 8949.
15179. Peiris, J. W. L. 1956: Seed infection and seedling blight of paddy in the dry zone. Trop. Agriculturist 112:37–44.
15180. Peive, J. 1956: The possibilities in crop production by the use of microelements. Udobrenie Urozhai 1:37–45. Chem. Abstr. 50:8116. 1956.
Pekker, M. 16231.
Peklo, J. 9814.
15181. Pekshibaev, M. I. 1953: Stratification of *Euonymus* seeds in mounds of soil. Akad. Nauk SSSR Inst. Lesa. Tr. 11:241–49.
Pelissier, F. 1751–1758.
Pelisso, F. G. 5082.
15182. Pellegrini, O. 1956: Observations on the endosperm of *Leucaena glauca* Benth. (Mimosaceae). Delpinoa 9:183–88.
15183. —— 1958: Study on the polyembryony of *Araucaria bidwillii* Hook. Delpinoa 11:261–86.
Pelloux, A. 13376.
15184. Peltier, G. L., and T. A. Kiesselbach. 1934: Inducing early fruiting in winter wheat by seed activation. J. Am. Soc. Agron. 26:503–6.
Peltier, G. L. 18174.
15185. Pelton, J. 1956: A study of seed dormancy in eighteen species of high altitude Colorado plants. Butler Univ. Botan. Studies 13:74–84.
15186. Pelzich, L. A. 1927: Vorlaufige Daten zur Frage über die Anabiose der ausgekeimten Samen. Jard. Botan. Inst. Essais Semences, Leningrad, Ann. Essais Semences 1927(4):1–8.
15187. Pence, J. W., and A. H. Elder. 1953: The albumin and globulin proteins of wheat. Cereal Chem. 30:275–87.
15188. ——, N. E. Weinstein, and D. K. Mecham. 1954: A method for the quantitative determination of albumins and

globulins in wheat flour. Cereal Chem. 31:29–37.

15189. ——, ——, and —— 1954: The albumin and globulin contents of wheat flour and their relationship to protein quality. Cereal Chem. 31:303–11.

Pencic, M. 5627.

15190. Pendleton, J. W. 1954: The effect of lodging on spring oat yields and test weight. Agron. J. 46:265–67.

15191. Pendleton, J. W. 1957: Effect of clover, row-spacing, and rate of planting on spring oat yields. Agron. J. 49:555–58.

15192. —— and G. H. Dungan. 1953: Effect of different oat spacings on growth and yield of oats and red clover. Agron. J. 45:442–44.

Pendleton, J. W. 441, 11763.

15193. Pendse, G. P. 1937: Chemical examination of the seeds of isabghol, *Plantago ovata* Forsks. II. Supplementary note. Proc. Natl. Acad. Sci. India 7:137–39.

15194. —— 1937: Composition of the oil from the seeds of *Solanum nigrum*. J. Indian Chem. Soc. 14:367–70. Pharm. Abstr. 6:57. 1940.

15195. —— and S. Dutt. 1934: Chemical examination on the seeds of isabghol, *Plantago ovata* (Forsks). Proc. Acad. Sci. United Prov. Agra Oudh, India) 4:133–40. Chem. Abstr. 29:7577–78. 1935.

15196. —— and J. B. Lal. 1937: Oil from the seeds of *Blepharis edulis*. J. Indian Chem. Soc. 14:362–66. Analyst 62:810. 1937.

15197. Peneff, N. 1941: Die Überwindung der Hartschäligkeit beim Saatgut der Robinie (*Robinia pseudacacia* L.). Forstwiss. Centralbl. 63:232–37. Biol. Abstr. 17:19154. 1943.

Penfold, A. R. 13783.

15198. Penfound, W. T., and T. T. Earle. 1948: The biology of the water hyacinth. Ecol. Monographs 18:447–72.

Peng, C. L. 3645.

Peng-fi, L. 11825.

Penner, J. 3290.

15199. Penon, P. 1963: Observations sur la thermogénise du riz en germination. Compt. Rend. Soc. Biol. 157:331–36.

15200. Pepper, E. H., and R. L. Kiesling. 1963: Differential media for the isolation of bacteria and fungi from plated barley kernels. Cereal Chem. 40:191–93.

Peralta, T. 3672.

Percival, E. G. V. 8859, 13892, 14321.

Percival, W. E. 1788.

15201. Pereira, A. 1944: Proteinas do trigo. I. Contribuição para o estudo da variação da proteina e gluten segundo a variedade e a região. Agron. Lusitana 6:367–96. Biol. Abstr. 22:9755. 1948.

15202. Pereira, A., Jr. 1949: Recherches pharmacognosiques sur les semences de quelques espèces du genre *Coronilla*. Port. Acta Biol. Ser. A 2:263–323. Biol. Abstr. 25:29087. 1951.

15203. Pereira, M. V. 1944: Anda-acu oil. Anais Assoc. Quim. Brasil 3:147–51. Chem. Abstr. 39:1999. 1945.

15204. Perello, J. M., and M. Serrano. 1948: Action of sulfonamides on germination and growth of *Ricinus communis*. Farmacognosia (Madrid) 7:35–45. Chem. Abstr. 43:2677. 1949.

Perera, B. P. M. 1879.

Pereverzeva, T. V. 11934.

Perez, J. Rivera-. 19374.

15205. Pergher, N. 1939: Classification of grain by treatment with phenol. Ital. Agr. 76:349–54. Biol. Abstr. 15:5701. 1941.

15206. Periasamy, K. 1961: Studies on seeds with ruminate endosperm. I. Morphology of ruminating tissue in *Myristica fragrans*. J. Madras Univ. 31:53–58. Biol. Abstr. 41:11904. 1963.

15207. —— and B. G. L. Swamy. 1961: Studies in the Annonaceae. II. The development of ovule and seed in *Cananga odorata* and *Miliusa wightiana*. J. Indian Botan. Soc. 40:206–16.

Perisse, J. 3448, 3449.

Perje, A. M. 19199.

15208. Perk, A., and R. Pür. 1961: Characteristics of the phases of the dormancy period in tree species. Tartusk. Gos. Univ. Tr. po Botan. 5:3–19.

15209. Perlova, R. L. 1937: Morphology and taxonomy of seeds of wild and cultivated species of potatoes. Bull. Appl. Botany, Genet., Plant Breed. (USSR) Ser. 4(2):46.

15210. Perrin, T. S. 1937: Siene bean. The composition of siene beans. Some characteristics of siene-bean oil. J. Am. Chem. Soc. 59:1401–2. Chem. Abstr. 31:5846. 1937.

15211. Perrins, G. W. 1949: Aerial reseeding. Western Livestock J. 27(38):36, 86–87.

15212. Persson, P. J., and S. Bingefors. 1956: Crude protein and fat content of white and black oats in trials at the Ultuna branch station of the Swedish Seed Association. Sveriges Utsadesforen.

Tidskr. 66:174–81. Chem. Abstr. 51: 14021. 1957.

15213. Perry, G. S., and C. A. Coover. 1933: Seed source and quality. J. Forestry 31:19–25. Expt. Sta. Rec. 68:766. 1933.

15214. Perry, J. S. 1957: Aeration of bulk-stored pea beans in Michigan. Mich. Agr. Expt. Sta. Quart. Bull. 40:311–17.

15215. —— and C. W. Hall. 1960: Germination of pea beans as affected by moisture and temperature at impact loading. Mich. Agr. Expt. Sta. Quart. Bull. 43:33–39.

15216. —— and —— 1960: Storing and handling pea beans. Mich. Agr. Expt. Sta. Quart. Bull. 43:444–54.

Perry, T. O. 19345.

Pershad, G. D. 3776.

Pershad, K. 5336.

Persica, E. 15856.

15217. Persidsky, D. I. 1933: On the formation of endosperm and haustorium in *Orobanche ramosa* L. Bull. Jard. Botan. Kyiv 16:89–101.

15218. —— 1935: On the development of endosperm in Solanaceae. Acad. Sci. Ukr. Inst. Botan. J. 4(12):44–45.

15219. —— 1940: On the cytology of oat grain. Acad. Sci. RSS Ukr. J. Botan. 1:129–37.

15220. —— 1940: Embryological and cytological investigation of barley (*Hordeum distichum* L.). Acad. Sci. RSS Ukr. J. Botan. 1:145–53.

Persidsky, M. D. 5636.

Person, L. H. 3939.

15221. Persson, A. R., and J. Apelard. 1960: Evolutionary and agricultural values of polyploidy with special reference to *Brassica* and *Raphanus*. Meldinger Norg. Landbrukshoegskole 39:5.

Pervukhina, N. V. 340, 9723.

Pesch, F. J. van. 13063.

15222. Pescott, E. E. 1941: Germination of the seed of Mallee eucalypts. Victorian Nat. 58:8. Plant Breeding Abstr. 14:187. 1944.

15223. Pescott, R. T. M. 1933: Mites in seeds. J. Dept. Agr. Victoria 31:519–23.

Pesho, G. R. 1494.

15224. Pesman, M. W. 1952: Seeds for Iceland. Green Thumb 9(12):26–27.

15225. Pesola, V. A. 1955: Protein content of field pea seeds as a varietal character. Acta Agral. Fennica 83:125–32. Chem. Abstr. 49:7069. 1955.

Petek, F. 4470, 4471.

15226. Peter, A. 1893: Kultureversuche mit ruhenden Samen. I. Nachr. Göttingen Ges. Wiss. 1893(17):673–91.

15227. —— 1894: Kulturversuche mit ruhenden Samen. II. Nachr. Göttingen Ges. Wiss. 1894(1):373–93.

15228. Peterfi, S., and E. Brugovitzky. 1961: Effect of some complex salts on the germination of sugar beets. Studia Univ. Babes-Bolyai Ser. 2 1961(2):99–110. Chem. Abstr. 57:11600. 1962.

15229. ——, ——, and F. Nagy-Toth. 1963: Die Wirkung der Vernalisation auf den Wuchs- und Hemstoffgehalt des Winterweizens. Naturwissenschaften 50:621–22.

15230. Peters, G., and W. Ganter. 1935: Combating the granary weevil with hydrocyanic acid gas and ethylene oxide. Z. Ges. Getreide-, Mühlen- Bäckereiw. 22:122–31, 140–45. Chem. Abstr. 30:2308–9. 1936.

15231. Peters, R. A., R. J. Hall, and P. F. V. Ward. 1960: The chemical nature of the toxic compounds containing flourine in the seeds of *Dichapetalum toxicarium.* Biochem. J. 77:17–23.

15232. ——, R. W. Wakelin, A. J. P. Martin, J. Webb, and F. T. Birks. 1959: Observations upon the toxic principle in the seeds of *Dichapetalum toxicarium.* Separation of a long-chain fatty acid containing flourine. Biochem. J. 71:245–48.

15233. —— and H. C. Yokum. 1961: Progress report on a study of the germination and growth of yellow foxtail (*Setaria glauca* [L.] Beauv.). Proc. Northeast. Weed Control Conf. 15:350–55. Weed Abstr. 12:151. 1963.

Peters, R. A. 19360, 19988.

15234. Peters, R. M. 1949: Amaryllis from seed to flower. Home Garden 14:66–68.

15235. Peters, T. 1924: Die Wirkung des Lichtes bei der Keimung der Samen von *Phacelia tanacetifolia.* Ber. Deut. Botan. Ges. 42:381–87.

15236. —— 1926: Zur Kenntniss des "falschen Keimung." Ber. Deut. Botan. Ges. 44:263–71.

15237. Peters, W. J. 1960: Histogenesis of *Tilia americana* embryos. Proc. Minn. Acad. Sci. 27:130–33.

15238. Petersen, A. 1961: Das kleine Gräserbuch für den praktischen Landwirt und seine Berater. Berl. Akad. Verl. 1961. 151 pp.

Petersen, D. F. 8052.

15239. Petersen, H. I. 1959: Effect of some weed killers on the germination, yield, and quality of grain seeds. Tidsskr. Planteavl 63:197–284. Biol. Abstr. 35:11423. 1960.

15240. —— and S. Lund. 1944: Studies on the germination of seeds of some noxious Danish weeds. Tidsskr. Landokonomi 1944:425–38. Biol. Abstr. 20:11154. 1946.

15241. Petersen, H. L. Pollination and seed setting in Lucerne. K. Vet.-og Landbohojskole Arsskr. 1954:138–69. 1954. Biol. Abstr. 31:25607. 1957.

Petersen, K. Dorph-; *see* Dorph-Petersen, K.

15242. Petersen, L. J., J. E. De Vay, and B. R. Houston. 1963: Effect of gibberellic acid on development of hypocotyl lesions caused by *Rhizoctonia solani* on red kidney bean. Phytopathology 53:630–33.

Petersen, R. G. 15912.

15243. Petersohn, E. 1926: Weitere Versuche über die Heranziehung der Katalasenwirkung von Samenkörnern zwecks Beurteilung der Keimfähigkeit und der Ursachen des Verlustes derselben. Z. Pflanzenernähr. Düng. Bodenk. A 6:291–94.

15244. Peterson, A., V. Schlegel, B. Hummel, L. S. Cuendet, W. F. Geddes, and C. M. Christensen. 1956: Grain storage studies. XXII. Influence of oxygen and carbon dioxide concentrations on mold growth and grain deterioration. Cereal Chem. 33:53–66.

Peterson, C. E. 5349, 5350.

15245. Peterson, D. F. 1958: Effect of gibberellic acid on germination, sucrose, and yield of sugar beets. J. Am. Soc. Sugar Beet Technol. 10:53–55.

Peterson, D. F. 4409.

Peterson, E. A. 8642–8644, 10334.

Peterson, G. W. 19464.

15246. Peterson, H. B. 1952: Effect of nitrogen fertilizer on yield and protein content of winter wheat in Utah. Utah Agr. Expt. Sta. Bull. 353. 29 pp.

Peterson, M. L. 3829, 12269.

15247. Peterson, R. A. 1953: Comparative effect of seed treatments upon seedling emergence in seven browse species. Ecology 34:778–85.

15248. Peterson, V. S. 1949: Seed counting apparatus uses "Lucite" to extend usefulness. Agr. News Letter 17:67–68.

15249. Pethybridge, G. H. 1934: Marsh spot in pea seeds. Gt. Brit. Min. Agr. J. 41:833–49. Biol. Abstr. 11:281. 1937.

Pethybridge, S. I. 13169.

15250. Petiet, J. 1961: Culture of horticultural seeds in Norway. Zaadbelangen 15:3–4.

15251. Petinov, N. S., and A. N. Pavlov. 1957: On the role of individual organs in ripening of the wheat grain. Dokl. Akad. Nauk SSSR 117:236–39. Biol. Abstr. 33:15398. 1959.

15252. Petit, A. 1931: La traitement de la carie de blé et la préservation des graines vis-à-vis des insectes parasites. Ann. Serv. Botan. Agron. Tunisie 7:95–100. Rev. Appl. Mycol. 11:442. 1932.

15253. —— 1937: Observations sur le traitement des graines de blé contre le charbon interne (*Ustilago tritici*). Rev. Pathol. Vegetale 24:175–85. Rev. Appl. Mycol. 16:802. 1937.

15254. —— 1937/38: The principal fungicidal products intended for the treatment of wheat and barley seed. Ann. Serv. Botan. Agron. Tunisie 14/15:53–84. Chem. Abstr. 34:8157. 1940.

15255. Petit, G., and R. Ancelin. 1913: De l'influence de la radioactivité sur la germination. Compt. Rend. 156:903–5.

15256. Petit, J. 1950: Contribution à l'étude microchimique des alcaloides de quelques Pavots de culture. Lejeunia 10:25–33. 1946. Biol. Abstr. 25:29169. 1951.

15257. Petkovsek, V. 1955: Jahresrhythmus der Samenkeimung bei *Echinocystis lobata*. Hrvatsko Prirodoslovno Drustvo, Glasnik Biol. Sekcije (Ser. II B) 7:289–90.

Peto, D. H. 7636.

15258. Peto, F. H. 1933: The effect of aging and heat on the chromosomal mutation rates in maize and barley. Can. J. Res. 9:261–64.

15259. —— 1935: Associations of somatic chromosomes induced by heat and chloral hydrate treatments. Can. J. Res. Sect. C 13:301–14.

15260. —— 1936: Heat induced tetraploidy in barley. Can. J. Res. Sect. C 14:445–47.

15261. —— 1937: Genetical studies on mutants in the progeny of heat-treated barley. Can. J. Res. Sect. C 15:217–29.

15262. —— 1950: Moisture content and germination of sugar beet seed. Proc. Am. Soc. Sugar Beet Technol. 6:97–99.

15263. —— 1961: Processing monogerm seed. J. Am. Soc. Sugar Beet Technol. 11:334–40.

15264. Petrasovits, E., and C. Darab. 1960: Data on the salt endurance of rice during germination. Congr. Irrig. Drain. Trans. 4:13.77–13.96.

15265. Petrasovits, I. 1958: Germination studies in rice. Novenytermeles 7:27–36.

15266. Petrea, V. 1953: The resistance of some cereal plants to sodium chloride. Rev. Univ. C. I. Parhon Politeh. Bucuresti Ser. Stiint. Nat. 1953(2):121–26. Chem. Abstr. 50:17012. 1956.

Petrea, V. 16658.

15267. Petri, L. 1929: Concentration des ions d'hydrogène et action de l'eau chaude sur les caryopses du blé. Actes Congr. Intern. Essais Semences 5:128.

Petrie, A. H. K. 9067, 14576.

15268. Petrochenko, E. I., and P. A. Kolesnikov. 1962: Oxidation of phenols and ascorbic acid during wheat germination. Biochemistry (USSR) (English Transl.) 26:614–18.

15269. Petrochenko, U. A. 1956: The effect of the components of biogenic stimulators on the sprouting of seeds and on the catalase activity. Nauk. Zap. Melitopol. Derzhav. Pedagog. Inst. 3:139–49. Chem. Abstr. 51:15710. 1957.

Petrochenko, U. A. 2563.

Petropavlovskaya, R. N. 3073.

15270. Petrosini, G. 1954: Biosynthesis of nicotine in germinating tobacco seeds. Tabacco 58:39–55. Chem. Abstr. 48:7127. 1954.

15271. Petrov, A., and L. Tyankova. 1962: Germination inhibitors in apple seeds (*Malus domestica* Borkh.). Bulgar. Akad. Nauk. Dokl. 15:223–26.

15272. —— and —— 1962: The possibilities of substituting stratification for treatment of seeds with gibberellin. Selskostopanska Nauka 9:937–42.

15273. Petrov, A. D., A. O. Saveljev, and E. V. Ikonen. 1931: Vacuum fumigation of the Egyptian cotton seeds in USSR in 1930. Bull. Plant Protect. Ser. III 2:21–31.

15274. Petrov, G., and S. Dimakov. 1932: The utilization of the hulls of sunflower seeds. Maslob.-Zhir. Delo 1932(8):36–40. Chem. Abstr. 27:3352. 1933.

15275. Petrov, M. N. 1956: Our method of seed stratification and production of *Eucommia* planting stock. Gorsko Stopanstvo 12:304–9.

15276. Petrova, A. A. 1952: Growing trees and shrubs without stratification of seeds. Byul. Gl. Botan. Sada 13:73–77.

Petrova, A. N. 13567.

Petrova, I. S. 11381, 11389.

15277. Petrova, K. A. 1952: Polyembryony in the Caucasian daisy (*Chrysanthemum caucasicum*). Byul. Gl. Botan. Sada 12: 68–69.

15278. Petrova, L. I. 1950: Effect of trace elements on the crop and seed quality of long staple flax and red clover. Mikroelementy v Zhizni Rast. i Zhivotn. Akad. Nauk SSSR Tr. Konf. po Mikroelementam 1950:180–88. Chem. Abstr. 49:439. 1955.

15279. Petrova, O. A. 1945: Effect of alkali upon germinating capacity of seeds of kok-saghyz and of common dandelion. Compt. Rend. Acad. Sci. URSS 49:227–28.

15280. —— 1955: On seed differences in dandelion (*Taraxacum*) species. Byul. Gl. Botan. Sada 23:78–82.

Petrova, P. N. 11797.

15281. Petrucci, D. 1952: Research on anaerobiosis of the embryos of *Triticum vulgare*. Boll. Soc. Ital. Biol. Sper. 28:252–54.

15282. —— 1954: Research on oxidases of metabolism of embryos of *Triticum vulgare* Host. Ann. Botan. 24:548–71.

Petsch, K. F. 16258.

15283. Pett, L. B. 1935: Studies on the distribution of enzymes in dormant and germinating wheat seeds. I. Dipeptidase and protease. II. Lipase. Biochem. J. 29:1898–904. Biol. Abstr. 11:5559. 1937.

15284. Pettersson, B. 1934: The action of different hydrogen ion concentrations on the germination of the seeds of some native plants. Soc. Fauna Flora Fennica Mem. 10:342–54.

15285. —— 1940: A case of long distance dispersal of plants through the import of timber. Acta Phytogeogr. Suecica 13:96–100.

Pettersson, N. 6019.

15286. Pettet, A. 1964: Seedlings of *Eichhornia crassipes:* a possible complication to control measures in the Sudan. Nature 201:516–17.

15287. Pettibone, C. J. V., and C. Kennedy. 1916: Translocation of seed protein reserves in the growing corn seedling. J. Biol. Chem. 26:519–25.

15288. Pettinari, C. 1949: On the toxicity of the seeds of *Araujia sericifera* Bert. Nuovo Giorn. Botan. Ital. 56:233.

Pettitt, J. M. 3783.

Petty, D. H. 8928.

15289. Petyayev, S. J. 1931: Aleurites—tung trees and their cultivation on the Caucasian Black Sea Coast. Bull. Appl. Botany, Genet., Plant Breed. (USSR) 26(1):76.

Petzold, E. N. 15791.

15290. Petzoldt, K. 1959: Wirkung des Mähdruschverfahrens auf die Verunkrautung. Z. Acker- Pflanzenbau 109:49–78.

Pevna, V. 17780.

15291. Pew, W. D. 1953: Vermiculite aid to lettuce. Chemurgic Dig. 12(8):15–16.
15292. Peyer, W. 1934: Differentiation of the Solanaceae seeds. Pharm. Zentralhalle 75:313–14. Chem. Abstr. 28:4536. 1934.
15293. Peynaud, E., and A. Maurie. 1953: The changes in nitrogen in different parts of grapes in the course of ripening. Ann. Inst. Natl. Rech. Agron. Ser. E 2:15–25. Chem. Abstr. 50:7233. 1956.
15294. —— and —— 1953: Changes in organic acids in the grape during maturation in 1951. Ann. Inst. Natl. Rech. Agron. Ser. E 2:83–94. Chem. Abstr. 50:6591. 1956.
15295. Peyronel, B. 1947: Effetti del trattamento di cariossidi di grano con soluzioni di caffeina. Lavori Botan. 1947:515–21.
15296. —— 1949: Substances inhibiting germination. I. Presence of blastokolins in leaf or extracts of other plant parts. Anais Univ. Ferrara 7:143–93. Chem. Abstr. 46:7184. 1952.
Pfahler, A. 5950, 5951.
Pfahler, P. L. 5315, 12521.
Pfeffer, S. 11628.
15297. Pfeffer, W. 1872: Untersuchungen über die Proteinkörner und die Bedeutung des Asparagins beim Keimen der Samen. Jahrb. Botan. 8:429–571.
15298. Pfeifer, R. P., R. J. Bergman, and J. E. Sparks. 1956: Photoelectric seed counter. Agron. J. 48:91–92.
Pfeifer, R. P. 8605.
Pfeifer, V. F. 18647.
15299. Pfeiffer, E. E. 1958: Protein changes during germination and earliest leaf growth. Bio-Dynamics 1958(47):2–8. Chem. Abstr. 53:4431. 1959.
15300. Pfeiffer, L. 1934: First record of Kikuyu grass setting seed in Australia. Agr. Gaz. N.S. Wales 45:406.
15301. Pfeiffer, N. E. 1934: Morphology of the seed of *Symphoricarpos racemosus* and the relation of fungal invasion of the coat to germination capacity. Contrib. Boyce Thompson Inst. 6:103–22.
Pfeiffer, S. E. 18362.
15302. Pfirsch, E. 1958: Sur la nature de l'inhibiteur contenu dans les cotyledons du *Sinapis alba* L. Bull. Soc. Botan. France 105:125–28.
15303. Pfister, A. 1935: Vermifugal action of pumpkin seed. Pharm. Ztg. 80:394–95. Chem. Abstr. 29:3775. 1935.
15304. Pfitzer, E. 1885: Über Früchte, Keimung und Jugenzustände einiger Palmen. Ber. Deut. Botan. Ges. 3:32–52.
15305. Pfitzer, P. 1962: Apomixis und Embryosackentwicklung in der Gattung *Aglaonema* Engl. Port. Acta Biol. 6:279–93.
Pfluderer, G. 11489.
Pflug, I. J. 15645.
15306. Pfnür, E. 1957: Die Wirkung von Gibberellin-Azid bei der Samenkeimung. Gartenbauwissenschaft 22:541–49.
15307. Pfuhl, J. F. 1927: Die Unterscheidung der Weizensorten durch Färbung der Körner. Angew. Botan. 9:374–79.
15308. Phalnikar, N. L., K. S. Nargund, and D. D. Kanga. 1935: Chemical investigation of the seeds of *Hygrophyla spinosa*. J. Univ. Bombay 4:146–52. Chem. Abstr. 30:5261. 1936.
Phalnikar, N. L. 17257.
15309. Phares, R. E., and F. G. Liming. 1961: Shortleaf pine direct-seeding success affected by month of seeding, release method, and weather. J. Forestry 59: 804–7.
15310. Pharis, R. L., and J. Unrau. 1953: Seed setting of alfalfa flowers tripped by bees and mechanical means. Can. J. Agr. Sci. 33:74–83.
15311. Phatak, V. G., and K. B. Ambegaokar. 1956: Embryological studies in Acanthaceae endosperm and embryo development in *Barleria prionitis* L. J. Maharaja Sayajirao Univ. Baroda 5:73–87. Biol. Abstr. 35:5316. 1960.
15312. —— and —— 1961: Embryological studies in Acanthaceae. IV. Development of embryo-sac and seed formation in *Haplanthus tentaculatus* Nees. J. Indian Botan. Soc. 40:525–34.
15313. —— and —— 1963: Embryological studies in Acanthaceae. V. Development of embryo-sac and endosperm in *Blepharis maderaspatensis* (Linn.) Roth. Proc. Indian Acad. Sci. Sect. B 57:88–95.
15314. —— and G. M. Oza. 1960: Four-winged fruit of *Terminalia crenulata* Roth. Current Sci. (India) 29:25.
Pheasant, J. H. 10040.
Philippe, R. 4491.
Philipson, T. 19200.
Phillips, A. M. 9845.
15315. Phillips, D. H. 1956: Tomato seed transmission of *Didymella lycopersici* Kleb. Brit. Mycol. Soc. Trans. 39:319–29.
15316. Phillips, E. P. 1920: Adaptations for the dispersal of fruits and seeds. S. African J. Nat. Hist. 2:240–52. Botan. Abstr. 11:2299. 1922.

15317. Phillips, G. L. 1952: Methyl bromide fumigation of cottonseed in freight cars for the destruction of pink bollworms. U.S. Dept. Agr., Agr. Res. Admin. Bur. Entomol. Plant Quarantine E-838. 16 pp.

15318. Phillips, I. D. J. 1961: Induction of a light requirement for the germination of lettuce seed by naringenin, and its removal by gibberellic acid. Nature 192:240–41.

15319. —— 1962: Some interactions of gibberellic acid with naringenin (5,7,4′-trihydroxy flavanone) in the control of dormancy and growth in plants. J. Exptl. Botany 13:213–26.

15320. Phillips, J. S. 1940: Immature nutfall of coconuts in the Solomon Islands. Bull. Entomol. Res. 31:295–316. Biol. Abstr. 15:13751. 1941.

Phillips, L. L. 14486.

15321. Phillips, M., and M. J. Goss. 1940: The composition of certain nutshells. J. Assoc. Offic. Agr. Chemists 23:662–65. Expt. Sta. Rec. 84:727. 1941.

Phillips, N. A. 7947, 12604.

15322. Phillips, R. P., and R. N. Flint. 1951: Preliminary investigations on the use of thiourea for breaking secondary dormancy in Alyce clover seed. Proc. Assoc. Offic. Seed Analysts 41:102–5.

Phillips, S. 9938.

Phillips, T. A. 2611.

15323. Phillis, E. 1962: The germination and establishment of rice seed sown into water. Australian J. Exptl. Agr. Animal Husbandry 2:181–84. Biol. Abstr. 42: 23642. 1963.

15324. —— and T. G. Mason. 1945: The effect of extreme desiccation on the viability of cotton seeds. Ann. Botany (London) 9:353–59.

Philouze, J. 7240.

15325. Philip, J. 1939: A spacing experiment with Egyptian wheats. Egypt Min. Agr. Tech. Sci. Serv. Bull. 199. 49 pp.

Phinney, B. O. 4405.

Phipps, J. 2297.

15326. Photiadis, P. D., and A. K. Stathi-Photiadou. 1932: Oil content of Greek grape seed. Prakt. Akad. Athenon 7:352–56. Chem. Abstr. 28:4618. 1934.

15327. Phouphas, C. 1950: Développement de l'embryon chez le *Geranium robertianum* L. Bull. Soc. Botan. France 97:191–94. Biol. Abstr. 26:32020. 1952.

15328. Piacco, R. 1940: La germinazione dei semi di *Panicum crusgalli* e *Panicum phillopogon*. Risicoltura 30:101–13. Biol. Abstr. 15:9403. 1941.

15329. —— 1947: Influence of mud deposited on rice seed on its germination. Risicoltura 35:99–101.

15330. —— 1953: Electric treatments and their influence on seeds. Riso 2(1):10–11.

15331. —— 1962: Methods of analyzing the germinating faculty of rice. Riso 11(11):18–20.

15332. —— 1962: Humidity and temperature in germination tests. Riso 11(12):7–9.

15333. Picado, C., and E. Vicente. 1923/24: Influence des rayons X sur la germination des grains. Arch. Phys. Biol. 3:13–16.

15334. Picard, C. 1960: Remarques sur l'exigence de températures froides alternées pour la vernalisation d'*Oenothera biennis* L. Compt. Rend. 250:573–75.

15335. Picci, G. 1959: Microbiological determinations of some vitamins and amino acids liberated during germination of the seeds. Ann. Fac. Agrar. Univ. Pisa 20:51–60. Chem. Abstr. 54:17563–64. 1960.

15336. Picheral, B. 1962: Capacités des noyaux de cellules endodermiques embryonnaires à organiser un germe viable chez l'urodèle, *Pleurodeles waltlii* Michah. Compt. Rend. 225:2509–11.

15337. Pichler, F. 1932: Der Einfluss längerer Lagerzeit auf die Keimfähigkeit trockengebeizten Getreides. Forstchr. Landwirtsch. 7:217–18.

15338. —— 1935: The applicability of hydrogen peroxide as a seed disinfectant. Phytopath. Z. 8:245–51.

Pickering, V. L. 15668.

15339. Pickett, B. S. 1944: The effect of spacing and number of kernels per hill on sweet corn yields. Proc. Am. Soc. Hort. Sci. 45:421–24. Expt. Sta. Rec. 93:436. 1945.

15340. Pickett, F. L. 1913: The germination of seeds of *Arisaema*. Proc. Indiana Acad. Sci. 29:125–28.

Pickett, R. C. 6130.

15341. Pickett, T. A. 1943: Some effects of heat treatments of peanuts. Georgia Agr. Expt. Sta. Circ. 142:1–8.

15342. —— 1950: Composition of developing peanut seed. Plant Physiol. 25:210–24.

15343. —— 1955: Notes on oils from seed of two wild species of *Arachis*. J. Am. Oil. Chemists Soc. 32:521.

Pickett, T. A. 8734.

15344. Pickholz, L. 1911: Ein Beitrag zur Frage über die Wirkung des Lichtes und der intermittierenden Temperatur auf die Keimung von Samen, sowie über die Rolle des Wassergehaltes der Samen bei dieser Wirkung. Z. Landwirtsch. Versuchw. Österr. 14:124–51.

15345. Picklum, W. E. 1955: Histological and cytological changes in the maize embryo during germination. Iowa State Coll. J. Sci. 29:479–80.

15346. Pictet, R. 1893: De l'emploi méthodique des basses températures en biologie. Arch. Sci. Phys. Nat. Per. III 30:293–314. Botan. Centralbl. 58:125. 1894.

Pidgeon, L. M. 6673.

15347. Pieczarka, S. J., and G. F. Warren. 1959: Screening procedure for determining the activity of chemicals against imbibed dormant seeds. Weeds 7:133–40.

15348. —— and —— 1960: The influence of concentration of fumigant and time of exposure on the killing of dormant imbibed seeds. Weeds 8:612–15.

15349. Pieczur, E. A. 1952: Effect of tissue cultures of maize endosperm on the growth of excised maize embryos. Nature 170:241–42.

15350. Piedelievre, R., and L. Mallet. 1956: Effet inhibiteur des extraits de champignons (*Psalliota campestris*) sur la germination des graines. Mushroom Sci. 3:79–83.

15351. Piehl, M. A. 1962: The parasitic behavior of *Melampyrum lineare* and a note on its seed color. Rhodora 64(757): 15–23.

15352. Pielou, D. P. 1950: The effect of insecticide applications on the insect fauna and seed yield of Alsike clover in southern Ontario. Can. Entomologist 82:141–60. Rev. Appl. Entomol. Ser. A 39:152–53.

Pieper, G. R. 18099.

15353. Pieper, H. 1909: Vergleichende Keimversuche mit Grassämereien nebst einigen Bemerkungen zur grundsätzlichen Fragen der Keimprüfungsmethode. Dissertation, Jena. 69 pp.

15354. —— 1914: Über die Erblichkeit der Keimgeschwindigkeit, der Keimfähigkeit und die Lichtempfindlichkeit der Samen von *Poa pratensis*. Fühling's Landwirtsch. Ztg. 63:362–68. Botan. Centralbl. 131:68–69. 1916.

15355. —— 1919: Beschreibung einer Methode zur raschen Erkennung von Futterrübensamen im Zuckerrübensamen. Z. Vereins Deut. Zucker-Ind. 766:409–18. Botan. Abstr. 5:1195. 1920.

15356. —— 1939: Vergleichende Untersuchungen an Varietäten des Kulturmohns (*Papaver somniferum* L.). Landwirtsch. Jahrb. 89:333–92. Proc. Intern. Seed Testing Assoc. 12:207–8. 1940.

15357. Pieper, J. J., and W. L. Burlison. 1937: Foreign strains of alfalfa and red clover: What is their adaptability to Illinois? Illinois Agr. Expt. Sta. Bull 431:473–503. Expt. Sta. Rec. 77:616. 1937.

15358. Pieraerts, J., and F. De Winter. 1928: Contribution à l'étude des graines de quelques espèces de *Luffa* acclimatées au Congo Belge. Leur signification à titre d'oléagineaux. Ann. Mus. Colonial Marseeile 6(2):5–15.

15359. —— and G. Tanret. 1930: Sur les graines du *Tetrapleura thonningii*. Bull. Soc. Chim. Biol. 12:457–63.

15360. Pierce, G. J. 1894: Germination of *Cuscuta*. Ann. Botany (London) 8:55–62.

15361. Pierce, H. B., D. E. Sheldon, and J. R. Murlin. 1933: The conversion of fat to carbohydrate in the germinating castor bean. III. The chemical analysis, and correlation with respiratory exchange. J. Gen. Physiol. 17:311–25.

15362. Pierce, J. G. 1905: The dissemination and germination of *Arceuthobium occidentale* Eng. Ann. Botany (London) 19: 99–113.

15363. Pierpoint, M. 1952: A problem in germination of crimson clover. Assoc. Offic. Seed Analysts News Letter 26(4): 4.

15364. —— 1961: Suggestions for treatment of some freshly harvested seed to promote prompt germination. Assoc. Offic. Seed Analysts News Letter 35(3):12–13.

15365. —— 1963: Germination of timothy seed. Assoc. Offic. Seed Analysts News Letter 37(2):11.

15366. —— 1963: A study of orchard grass germination. Assoc. Offic. Seed Analysts News Letter 37(3):38–39.

15367. —— 1963: Comparison: Annual and perennial ryegrass. Assoc. Offic. Seed Analysts News Letter 37(3):42–43.

15368. —— and L. Jensen. 1961: Some problems in germination tests of Merion bluegrass. Assoc. Offic. Seed Analysts News Letter 35(1):12–16.

15369. —— and —— 1961: Effect of light on fluorescence of ryegrass seedlings. Assoc. Offic. Seed Analysts News Letter 35(1):23–25.

15370. —— and —— 1964: Progress report on tetrazolium staining for viability of *Pinus monticola*. Assoc. Offic. Seed Analysts News Letter 38(1):20–23.

15371. Pierson, R. 1949: Seed inspection. Intern. Crop Improvement Assoc. Proc. 31:72–74.

15372. Piescu, A. 1935: Influence du nombre de capsules d'une plante sur la semence de tabac. (Influenta numarului de capsule de planta asupro samentei de tutun.) Bul. Cult. Ferment. Tulunului 24:51–63. Ann. Agron. 6:630. 1936.

15373. —— 1935: L'influence du nombre des capsules d'une plante sur le semence de tabac. Bul. Cult. Ferment. Tulunului 24:64–66.

15374. Pieters, A. J. 1898: Effect of alternating dryness and moisture on the germination of some seeds. Botan. Gaz. 25:114.

15375. —— and E. Brown. 1902: Kentucky bluegrass seed (*Poa pratensis*). U.S. Dept. Agr. Bur. Plant Ind. Bull. 19. 19 pp.

15376. Pietkiewicz, T. A. 1955: Mycological and phytopathological studies of flax seeds. Acta Agrobotan. 3:223–70.

15377. —— 1959: Some observations on the microflora of soybean seeds. Roczniki Nauk Rolniczych Ser. A 79:1077–90.

15378. Pietruszczynski, Z. 1926: After-ripening of cereals. Roczniki Nauk Rolniczych Lesnych 15:206–35. Biol. Abstr. 6:21627. 1932.

15379. —— 1934: Untersuchungen über den Einfluss der Bestrahlung auf die Keimung von Pflanzensamen. Roczniki Nauk Rolniczych Lesnych 33:83–119.

15380. Pietsch, A. 1935: Photographische Darstellung von Unkrautsamen. Ernähr. Pflanze 31:369–73.

15381. —— 1935: Photographische Darstellung von Unkrautsamen. Ernähr. Pflanze 31:430–32.

15382. —— 1936: Photographische Darstellung von Unkrautsamen. Ernähr. Pflanze 32:92–93.

15383. —— 1936: Photographische Darstellung von Unkrautsamen (Fortsetzung). Ernähr. Pflanze 32:382–84.

15384. —— 1936: Photographische Darstellung von Unkrautsamen. Ernähr. Pflanze 32:422–24.

15385. —— 1937: Photographische Darstellung von Unkrautsamen. Ernähr. Pflanze 33:87–89.

15386. —— 1938: Photographische Darstellung von Sämereien landwirtschaftlicher Kulturpflanzen. Ernähr. Pflanze 34:65–66.

15387. —— 1938: Photographische Darstellung von Sämereien landwirtschaftlicher Kulturpflanzen. Ernähr. Pflanze 34:103–4.

15388. —— 1938: Photographische Darstellung von Sämereien landwirtschaftlicher Kulturpflanzen (Fortsetzung). Ernähr. Pflanze 34:117–19.

15389. —— 1938: Photographische Darstellung von Sämereien landwirtschaftlicher Kulturpflanzen (Fortsetzung). Ernähr. Pflanze 34:132–33.

15390. —— 1939: Photographische Darstellung von Samen und Früchten wichtiger angebauter Heil-, Duft- und Gewurzpflanzen (Fortsetzung). Ernähr Pflanze 35:139–42.

15391. —— 1941: Beitrag zur photographischen Darstellung, Farbbestimmung und Bedeutung der ölhaltigen Samen von in Deutschland wachsenden Pflanzen. Landwirtsch. Jahrb. 91:369–417. Gartenbauwissenschaft 17:S9. 1942.

15392. Piettre, L. 1943: The permeability of grains after treatment with organic solvents; the phenomena of sorption and desorption. Compt. Rend. Acad. Agr. France 29:260–66. Chem. Abstr. 38:5879. 1944.

Piettre, L. 15396.

15393. Piettre, M. 1934: Maturation des grains de blé, influence de quelques phénomènes physiochimiques. Compt. Rend. 198:608–11. Biol. Abstr. 9:17936. 1935.

15394. —— 1947: Activation de semences de pommes de terre aux basses températures. Une augmentation de 40 P. 100 de rendement expérimental. Compt. Rend. Acad. Agr. France 33:620–24.

15395. —— 1947: Sur des expériences immediates d'activation par le froid des semences. Compt. Rend. Acad. Agr. France 33:649–51.

15396. —— and L. Piettre. 1943: The permeability of grains after treatment with organic solvents: The phenomena of sorption and desorption. Compt. Rend. Acad. Agr. France 29:259–60. Chem. Abstr. 38:5984. 1944.

15397. Piggott, C. J. 1960: The effect of fertilizers on the yield and quality of groundnuts in Sierra Leone. Empire J. Exptl. Agr. 28:59–64.

Pignato, L. C. 2386.

15398. Pihelgas, E. 1959: Über Ergiebigkeit und Qualität der Samen der gemeinen Kiefer. Estonskaia Sel'skokhoz. Akad. Sb. Nauchn. Tr. 11:3–12.

15399. —— 1963: The dependence of pine seed quality on seed color. Estonskaia Sel'skokhoz. Akad. Sb. Nauchn. Tr. 29:10–20.

15400. —— 1963: On the use of various procedures in quality determination of seeds of the common pine and spruce. Estonskaia Sel'skokhoz. Akad. Sb. Nauchn. Tr. 29:21–26.

Pijl, L. van der; *see* Van der Pijl, L.

Pijoan, M. 8386.

15401. Pike, E. E. 1935: On the germination of cacao beans with special reference to storage and transportation problems. Ann. Rept. Cacao Res. Imp. Coll. Trop. Agr. (Trinidad) 4(1934):33–40.

Piland, J. R. 4312.

15402. Pilet, P. E. 1963: Sur deux inhibiteurs radiculaires. Compt. Rend. 256:1348.

15403. —— and J. Mimault. 1959: Biochromatographic study of the auxins of Bigarreau cherry kernels. Rev. Gen. Botan. 66:329–39. Chem. Abstr. 54:4794. 1960.

Pilipenko-Yurchak, L. D. 7702, 7703.

Pillai, R. V. 15018.

Pillai, V. N. K. 3777.

15404. Pillay, D. T. N. 1962: A study of the relationship of growth substances to rest period and germination of Mazzard cherry seeds, *Prunus avium* L. Dissertation Abstr. 22:3830.

15405. Pillay, K. P. 1944: Vernalisation of gram (*Cicer arietinum*, L.). Current Sci. (India) 13:185–86. Brit. Abstr. B III:2. 1945.

Pillay, K. P. 15060.

Pillay, P. P. 14096.

Pinckney, A. J. 14334.

Pinckney, J. S. 9845.

Pineda, C. R. 12343.

15406. Pinegina, R. I., and V. G. Klimenko. 1962: Variability in the content of proteins and nonprotein nitrogen-containing substances in the seeds of certain varieties of peas during maturation. Tr. po Khim. Prirodn. Soedin. Kishinevsk. Gos. Univ. 1962(5):19–26. Chem. Abstr. 59:9086. 1963.

Pines, A. N. 17050.

15407. Pingale, S. V., and V. Balu. 1955: Grain storage practices in India. Trop. Agr. (Trinidad) 32:88–94.

15408. —— and R. C. Bhutiani. 1953: Cleaning and conditioning of foodgrains. Indian Farming 3(7):18–21.

Pingale, S. V. 5247, 19064.

15409. Pinnell, E. L. 1949: Genetic and environmental factors affecting corn seed germination at low temperatures. Agron. J. 41:562–68.

Pinnell, E. L 17842.

15410. Pinney, T. S. 1957: The propagation of *Picea* by seed. Plant Propagators Soc. Proc. 7:33–38.

15411. Pinthus, M. J. 1959: Seed set of self-fertilized sunflower heads. Agron. J. 51:626.

15412. —— 1963: Comparison of dry matter accumulation and moisture content in the developing kernels of bread wheat, durum wheat, and barley. Israel J. Agr. Res. 13:117–24. Chem. Abstr. 60:4459. 1964.

15413. —— and J. Rosenblum. 1961: Germination and seedling emergence of sorghum at low temperatures. Crop Sci. 1:293–96.

Pinto, A. V. 5066.

15414. Pinto, E. I. 1947: Phytochemical study of *Apium dissectum*. Rev. Fac. Farm. Bioquim. Univ. Nacl. Mayor San Marcos (Lima) 8(33/34):28–37. Chem. Abstr. 43:2280. 1949.

Pinxteren, J. A. C. van; *see* Van Pinxteren, J. A. C.

Pion, R. 4266.

15415. Pipe, F. H., and A. Hope. 1938: Investigations in the germination of New Zealand spinach using eight different methods. Proc. Assoc. Offic. Seed Analysts 24:122–25.

15416. Piper, R. B., and F. E. Gardner. 1944: Comparative seasonal development of avocado fruits and seeds. Proc. Florida State Hort. Soc. 56:122–24.

15417. Piringer, A. A., R. J. Downs, S. B. Hendricks, and H. A. Borthwick. 1954: Reversible photoreaction controlling photoperiodic response, seed germination, and other phenomena. Congr. Intern. Botan. Paris Rappt. Commun. 8:321–23. Chem. Abstr. 48:10853. 1954.

Piringer, A. A. 5533–5535.

15418. Pirschle, K. 1926: Acetaldehyde als Zwischenprodukt bei der Keimung fetthaltiger Samen. Biochem. Z. 169:482–89.

Pirt, S. J. 15165.

Pisacic, K. von Sostaric-; *see* Von Sostaric-Pisacic, K.

15419. Pisani, P. L. 1959: Observations on the influence exerted by gibberellin on the germination of seeds of some garden species. Riv. Ortoflorofrutticolt. Ital. 43:157–63. Chem. Abstr. 57:14222. 1962.

15420. Pisarev, V. E., and M. D. Zhilkina. 1950: Controlled change of dormancy period in Moskovka wheat seeds. Selek. i Semen. 17(3):14–20.

Pisarovicova, W. Cizkova–; *see* Cizkova-Pisarovicova, W.

15421. Pischinger, E. 1932: Phosphorus compounds in plants. VII. The solubility of phosphorus compounds of hemp seed. Bull. Intern. Acad. Polon. B 1:37–50. Chem. Abstr. 27:3973. 1933.
Pishawikar, D. G. 17283.
15422. Piskareff, V. J. 1937: Determining tree-fruits seed germinability by means of staining. Michurinsk. Plodovodstvo 1937(1):51–63.
15423. —— 1937: Influence of seed coats on the germination of tree fruits seed. Res. Inst. Fruit Growing "I. W. Mitchurin." Sci. Fruit Prod. 1937(2):36.
15424. —— 1938: Germinating of seeds without a rest period. Res. Inst. Fruit Growing "I. W. Mitchurin." Sci. Fruit Prod. 1938(5):39–48.
15425. Piskoric, O. 1953: The influence of the size of seed of cork-oak and kermes-oak (*Quercus suber* and *Quercus coccifera*) on germination and development of seedlings. Sumarski List 77:169–71.
15426. —— 1962: Some data concerning the seeds of *Quercus ilex* L. Nar. Sumar 16:347–50.
15427. —— 1962: On the germination and sterility of elm seeds. Nar. Sumar 16:390–91.
15428. Pissarev, W. E., and N. Vinogradova. 1946: Carbohydrates in grains of wheat grafted on rye endosperm. Nature 157: 702.
15429. Pisternik, A. 1921: Über die Einwirkung floureszierender Farbstoffe auf die Keimung der Samen. Anz. Akad. Wiss. Wien Math.-Naturw. Kl. 18:142–43.
Pitman, A. L. 8905.
15430. Pitot, A. 1930: Observations sur le tégument de la graine des légumineuses. Bull. Soc. Botan. France 77:289–94.
15431. —— 1933: Sur le rôle de la protuberance dans les graines des Papilionacées. Bull. Soc. Botan. France 80:27–29.
15432. —— 1935: Sur la morphologie de la graine des légumineuses dans les rapports avec la systematique. Compt. Rend. 200: 773–74.
15433. Pittauer, G. 1912: Über den Einfluss verschiedner Belichtung und extremer Temperaturen auf den Verlauf der Keimung forstlichen Saatgutes. Centralbl. Ges. Forstwesen 38:157–72, 213–24.
15434. Pittman, U. J. 1963: Magnetism and plant growth. I. Effect on germination and early growth of cereal seeds. Can. J. Plant Sci. 43:513–18.
15435. —— and J. E. Andrews. 1961: Effect of date of seeding on winter survival, yield, and bushel weight of winter wheat grown in Southern Alberta. Can. J. Plant Sci. 41:71–80.
Pittman, U. J. 17246.
15436. Pixton, S. W., and J. N. Emerson. 1963: Paper tests for moisture. Agr. Merchant, April, pp. 1–5.
15437. ——, M. B. Hyde, and G. Ayerst. 1964: Long-term storage of wheat. J. Sci. Food Agr. 15:152–61.
Pixton, S. W. 14919.
Pizarroso, A. 6303.
15438. Pladeck, M. M. 1940: The testing of Buffalo grass "seed," *Buchloe dactyloides* Engelm. J. Am. Soc. Agron. 32:486–94.
15439. Platone, E. 1951: Dinamica del metabolismo de los taninos en almendras de cacao durante el beneficio. Acta Agron. (Palmira, Colombia) 1:121–31. Biol. Abstr. 27:20475. 1953.
15440. —— 1955: Germinability of cacao seeds in relation to ethyl alcohol, acetic acid and temperature. Atti Ist. Botan. Univ. Lab. Crittogam. Pavia 5 Ser. 13:205–18. Biol. Abstr. 41:20313. 1963.
15441. Platonova, T. F., and A. D. Kuzovkov. 1963: Alkaloids of the seeds of *Delphinium orientale*. Med. Prom. SSSR 17(4):19–20. Chem. Abstr. 59:6723. 1963.
Platt, A. W. 8.
Platt, K. B. 13698.
15442. Platt, R. 1950: Seed pods—living jewelry of the winter countryside. Outdoor Illus. 2:155–61.
15443. Plaut, M., and C. Gavrielit-Gelmond. 1949: Determination of the viability of seeds of some vegetable and field crops by means of sodium selenite. Proc. Intern. Seed Testing Assoc. 14:190–211.
15444. —— and A. Halfon. 1954: The viability test of pea, bean, and cucumber seeds by resazurin staining. Proc. Intern. Seed Testing Assoc. 19:14–23.
15445. ——, ——, O. Heller-Cohen, A Cohen, and A. Gordin. 1957: Determination of the viability of seeds by resazurin. Proc. Intern. Seed Testing Assoc. 22:343–47.
15446. ——, D. Lachover, A. Bondi, E Olomutzki, A. Bar-Akiba, and M. Calderon. 1957: Ethylene dibromide as seed fumigant and its absorption by oil seeds. Proc. Intern. Seed Testing Assoc. 22:392–99.
15447. —— and D. Reich. 1946: Testing

the viability of seeds by means of selenium salts. Hassadeh 26:166–69.

15448. ——, B. Zelcbuch, and K. Guggenheim. 1953: Nutritive and baking properties of carob germ flour. Bull. Res. Council Israel, Weizmann Mem. Issue 3:129–31.

Plaut, M. 447.

15449. Plavsic-Gojkovic, N. 1961: Contribution to the knowledge of the seeds and leaf buds of red peppers. Poljoprivredna Znanstvena Smotra 18:107–20.

15450. Playford, P. 1933: Dahlias from seed. Flower Grower 20(10):442.

15451. Pleas, C. E. 1934: Longevity of seeds. Flower Grower 21(3):117.

15452. Plennik, R. I. 1961: Features of the formation and the germinating capacity of seeds of *Polygonum divaricatum.* Tr. Novosibirsk. Tsent. Sibirsk. Botan. Sada 5:37–47.

15453. —— 1961: Development and vital capacity of seeds of *Vicia villosa* in Novosibirsk. Tr. Novosibirsk. Tsent. Sibirsk. Botan. Sada 5:61–75.

15454. Pleshko, S. I. 1941: Nutritive value of *Trigonella popovii.* Proc. Lenin Acad. Agr. Sci. USSR 7:16–18.

15455. Plhak, F. 1957: The influence of an extract of germinating barley seeds on the metabolism of young sugar-beet plants. Publ. Fac. Sci. Univ. Masaryk 17:451–62. Chem. Abstr. 52:17405. 1958.

15456. Plitt, T. M. 1935: Temperature as a predetermining factor in the development of *Avena sativa.* Plant Physiol. 10:269–89. Biol. Abstr. 10:2900. 1936.

15457. —— 1941: Seed impressions on plastic films. Plant Physiol. 16:422–24.

15458. Plochmann, R. 1955: Problems of seed supply and site selection in Western Germany. Forestry Chron. 31:227–32. Biol. Abstr. 31:8927. 1957.

Ploshtakova, L. 6941.

Plosz, A. M. 9732, 19303, 19304.

15459. Plouvier, V. 1942: Biochemical study of flowers, fruits and seeds of *Cydonia japonica* Pers. and *C. maulei* Mast. Compt. Rend. 214:93–95. Chem. Abstr. 37:6301. 1943.

15460. —— 1946: Oil of the seed of *Xanthoceras sorbifolia* Bunge and of *Koelreuteria paniculata* Laxm. (Sapindaceae). Compt. Rend. 222:916–17. Chem. Abstr. 40:3915. 1946.

15461. —— 1946: Oil of the seed of *Magnolia macrophylla* Michx. Compt. Rend. 222: 1009–11. Chem. Abstr. 40:4899. 1946.

15462. Plummer, A. P. 1943: The germination and early seedling development of twelve range grasses. J. Am. Soc. Agron. 35:19–34.

15463. Plummer, G. L. 1962: On the sensitivity of *Trifolium* seeds to γ-radiation. Crop Sci. 2:221–23.

15464. —— and C. Keever. 1963: Autumnal daylight weather and camphor-weed dispersal in the Georgia Piedmont region. Botan. Gaz. 124:283–89.

15465. Plumridge, A. J. 1957: Seed propagation of exotic shrubs and trees. J. Agr. Victoria 55:391, 393, 395.

Plunkett, R. A. 19805.

15466. Plyushkin, E. Z. 1933: Delinting cottonseed with sulfuric acid and the production of ethyl alcohol as a by-product. Trest Khlopkochistitelnoi Prom. 1:40–48. Chem. Abstr. 28:5227. 1934.

Plyusnina, Z. A. 11375.

Poapst, P. A. 5685.

15467. Poddubnaja-Arnoldi, V. A. 1961: The importance of embryology for genetics and breeding. Byul. Gl. Botan. Sada 44:32–38.

15468. —— and V. Dianova. 1937: Character of propagation in some species of the genus *Taraxacum* L. Botan. Zh. 22:267–95. Biol. Abstr. 12:9079. 1938.

Poddubnaja-Arnoldi, V. A. 18789.

15469. Podolskaia, O. I. 1951: Accelerated stratification of soap-bark tree (*Quillaja*) seeds. Lesn. Khoz. 4(9):89.

Podolskaya, M. 7401.

15470. Podolskaya, M. Z. 1939: Carotenoids of cottonseed and of extracted oils. Vses. Nauchn.-Issled. Inst. Zhir. Issled. Khim. Tekhnol. Proiz. Khlopk. Maslab. 1939(2): 83–94. Chem. Abstr. 36:7064. 1942.

15471. Podoprigora, V. S. 1960: The viability of the late spring weeds in the steppe soil in the Ukrainian S.S.R. Botan. Zh. 45:272–76.

Poe, S. E. 8929.

Poehlman, J. M. 16737.

15472. Poehlman, R. 1941: Report of work on normal dodder seeds. Proc. Assoc. Offic. Seed Analysts 33:41–42.

Poel, J. van der; *see* Van der Poel, J.

15473. Poel, L. W. 1949: Germination and development of heather and the hydrogen ion concentration of the medium. Nature 163:647–48.

Poenaru, I. 2775.

Poesch, G. H. 11872.

Pogosov, E. K. 5178.

15474. Pohjola, K. 1956: Über den Einfluss einiger Alkohole auf die Keimung der Samen. Suomalainen Eläin- Kasvitiet. Seura Vanamo. Tiedonannot 11:87–96.

15475. Pohl, F. 1942: Das Taube Samenkorn der Tanne (*Abies alba* Mill.). Ber. Deut. Botan. Ges. 60:313–22.

15476. Pohl, R. W. 1955: An endospermous dicotyledon seed for botanical instruction. Proc. Iowa Acad. Sci. 62:152–54.

Pohl, R. W. 9469.

15477. Pohlenz, W. 1930: Influence of the nutrient content of seeds on the determination of the nutrient content of soils ("b" in the effect law). Botan. Arch. 31:141–96. Chem. Abstr. 26:2003. 1932.

Pohloudek-Fabini, R. 18261, 18262.

15478. Pohoska, J. 1961: Investigation of factors causing poor carrot seed germination. Bull. Waerzywniczy 5:343–52. 1960/61.

Poignant, P. 4034.

15479. Poinot, M. 1942: Tobacco-seed oil. Peintures, Pigments, Vernis 17:339–44. Chem. Abstr. 40:5933. 1946.

Poirot, G. 8660.

15480. Poix, R. 1955: Observations sur le développement embryonnaire de l'*Epilobium parviflorum* Schreb. Bull. Soc. Botan. France 102:232–34.

Poix, R. 11946.

15481. Pojarkova, A. J. 1928: Temperaturbedingungen der Keimung als bestimmender Faktor für Ährenbildung beim Wintergetreide. Ber. Deut. Botan. Ges. 45:627–37.

15482. Poleshchuk, Y., and S. F. Negolvelov. 1936: Activation of the germination of sesame with copper sulfate. Chem. Social. Agr. (USSR) 1936(11):41–51. Chem. Abstr. 31:6801. 1937.

15483. Polesiko, N. B., and K. E. Ovcharov. 1963: Pelleting of cotton seeds. Khlopkovodstvo 1963(6):49–50.

15484. Polevoi, V. V. 1959: Dynamics of 3-indoleacetic acid in ripening and sprouting corn grains. Dokl. Akad. Nauk SSSR 124:695–98. Chem. Abstr. 53:10396. 1959.

15485. Polgar, S. 1962: Germination inhibiting effect of beet glomerules and possibilities for counteracting this effect. Debrecen. Mezogazdasagi Akad. Tudo. Evkon. 1961:177–92.

15486. Poliakoff, M. Z., and G. B. L. Smith. 1948: Guanidine soaps. Ind. Eng. Chem. 40:335–37.

15487. Polishchuk, L. K. 1953: Physiological aspects of walnuts during the process of stratification. Nauk. Zap. Kuisvs'k. Univ. 12:71–75. Chem. Abstr. 50:444. 1956.

15488. Poljakoff-Mayber, A. 1952: Changes in metabolism of lettuce seeds during germination and its inhibition. Palestine J. Botany 5:180–85.

15489. —— 1952: Germination inhibitors and plant enzyme systems. I. Catalase. Bull. Res. Council Israel 2:239–45.

15490. —— 1953: Further identification of the sugars present in lettuce seeds. Bull. Res. Council Israel 3:248.

15491. —— 1953: Germination inhibitors and plant enzyme systems. II. Dehydrogenases. Palestine J. Botany 6:60–66.

15492. —— 1953: Germination inhibitors and plant enzyme systems. III. Hydrolytic enzymes. Palestine J. Botany 6:101–6.

15493. —— 1953: Peroxidase activity in germinating lettuce seeds. Enzymologia 16:122–24.

15494. —— 1958: The effect of auxin and anti-auxin on germination of lettuce seeds. Bull. Res. Council Israel Sect. D 6:78–81.

15495. —— 1958: The effect of eosin on germination of lettuce seed. Bull. Res. Council Israel Sect. D 6:82–85.

15496. —— 1959: Growth and development of lettuce germinated in gibberellic acid. Bull. Res. Council Israel Sect. D 7:54–55.

15497. —— 1959: Germination of the seeds of *Saccharum aegyptiacum* Willd. Bull. Res. Council Israel Sect. D 7:93–94.

15498. —— and M. Evenari. 1958: Some further investigations on the oxidative systems of germinating lettuce seeds. Physiol. Plantarum 11:84–91.

15499. ——, S. Goldschmidt-Blumenthal, and M. Evenari. 1957: The growth substances content of germinating lettuce seeds. Physiol. Plantarum 10:14–19.

15500. —— and A. M. Mayer. 1955: Some quantitative changes in the fat metabolism of germinating lettuce seeds and their inhibition by coumarin. J. Exptl. Botany 6:287–92.

15501. —— and —— 1958: Some further studies on the direct oxidation of glucose in germinating lettuce seed. Bull. Res. Council Israel Sect. D 6:86–87.

15502. —— and —— 1960: Effect of thiourea on germination and growth. Indian J. Plant Physiol. 3:125–38

15503. ——, ——, and S. Zacks. 1958: Interaction in growth and germination between thiourea and indolyl acetic acid. Ann. Botany (London) 22:175–81.

15504. ——, ——, and —— 1958: Interaction of thiourea and coumarin in germination and growth of lettuce. Bull. Res. Council Israel Sect. D 6:118–24.

15505. ——, ——, and —— 1959: The effect of PCIB on the germination and growth of lettuce in the presence of thiourea. Ann. Botany (London) 23:361–63.

Poljakoff-Mayber, A. 420, 6118, 6122, 6124, 6125, 7167, 11031, 11035, 12063, 13077–13085, 18877.

15506. Pollacci, G., and M. Bergamaschi. 1940: The action of vitamins on the germination of orchid seeds. Boll. Soc. Ital. Biol. Sper. 15:326–27. Rev. Appl. Mycol. 19:414. 1940.

15507. —— and V. Tredici. 1935: On the asymbiotic germinability of orchid seeds. Boll. Soc. Ital. Biol. Sper. 10:695–97. Rev. Appl. Mycol. 15:108. 1936.

15508. Pollack, B. L., and R. E. Larson. 1955: Factors affecting embryo size, and the influence of embryo size on germination, time to maturity, and productivity in F_2 generation tomatoes. Penn. Agr. Expt. Sta. Bull. 606. 34 pp.

Pollack, M. 12635.

Pollard, A. G. 13902.

Pollard, E. F. 5096.

Pollard, L. H. 4172, 4173, 8420–8423, 8645.

15509. Pollhamer, E. 1960: Experiments on the duration of seed dormancy in different varieties of winter and spring barley. Növenytermeles 9:231–46.

15510. Pollock, B. M. 1961: The effects of production practices on seed quality. Seed World 89(5):6, 8, 10.

15511. —— and H. O. Olney. 1959: Studies of the rest period. I. Growth, translocation, and respiratory changes in the embryonic organs of the after-ripening cherry seed. Plant Physiol. 34:131–42.

15512. —— and V. K. Toole. 1961: After-ripening, rest period and dormancy. Yearbook Agr. U.S. Dept. Agr. 1961:106–12.

Pollock, B. M. 14765.

15513. Pollock, E. G., and W. A. Jensen. 1963: Histochemistry of chalazal proliferating cells in the embryo sac of *Capsella bursa-pastoris*. [Abstract.] Am. J. Botany 50:617.

15514. Pollock, J. R. A. 1956: Studies in barley and malt. VIII. Survey of the dormancy of barleys harvested in Britain in 1955. J. Inst. Brewing 62:331–33.

15515. —— 1959: Studies in barley and malt. XV. Growth substances and other compounds in relation to dormancy in barley. J. Inst. Brewing 65:334–37. Chem. Abstr. 54:13282. 1960.

15516. —— 1960: Das Vorkommen von Keimungsauregenden Stoffen in Gerste. Brauwissenschaft 13:114–17.

15517. ——, B. H. Kirsop, and R. E. Essery. 1955: Dormancy in barley. European Brewery Conv. Proc. 5th Congr., Baden-Baden 1955:203–11. Chem. Abstr. 51: 18139. 1957.

15518. —— and A. A. Pool. 1962: Studies in barley and malt. XIX. Further observations on water-sensitivity in barley. J. Inst. Brewing 68:427–31.

Pollock, J. R. A. 4349, 4350, 6064.

Polonovski, J. 4262.

15519. Polya, L. 1961: Injury by soaking of *Populus alba* seeds. Nature 189:159–60.

15520. Pomeranz, Y., and J. A. Shellenberger. 1961: Histochemical characterization of wheat and wheat products. I. Histochemical demonstration of germ and aleurone using acridine orange. Cereal Chem. 38:103–8.

15521. —— and —— 1961: Histochemical characterization of wheat and wheat products. II. Mapping of protein distribution in the wheat kernel. Cereal Chem. 38:109–13.

15522. —— and —— 1961: Histochemical characterization of wheat and wheat products. III. Use of methyl green in estimating flour extraction rate. Cereal Chem. 38:113–21.

15523. —— and —— 1961: Histochemical characterization of wheat and wheat products. IV. Mapping the free fatty acids in germinating wheat. Cereal Chem. 38:122–32.

15524. —— and —— 1961: Histochemical characterization of wheat and wheat products. V. Sulfhydryl groups: their localization in the wheat kernel. Cereal Chem. 38:133–40.

Pomeranz, Y. 13762, 17299.

15525. Pomeroy, K. B. 1949: The germination and initial establishment of Loblolly pine under various surface soil conditions. J. Forestry 47:541–43.

15526. Pominski, C. H., C. B. Miller, P. Von Der Haar, R. T. O'Connor, L. E. Castillon, and L. E. Brown. 1951: Pigments of cottonseed. IV. Gossypurpurin,

a purple pigment related to gossypol. J. Am. Oil Chemists Soc. 28:472–75. Biol. Abstr. 26:26061. 1952.
Pominski, C. H. 19190.
Pominski, J. 4893.
Pommer, H. 6752.
15527. Ponchet, J., and G. Auge. 1959: Essais de traitement de la fonte des semis de céréales. Phytiatrie Phytopharm. 8: 141–49.
Ponchet, J. 4905, 4906.
15528. Pond, R. H. 1906: The incapacity of the date endosperm for self-digestion. Ann. Botany (London) 20:61–78.
Pond, V. 17785.
15529. Ponnaiya, B. W. X. 1944: A study on the maturity of grain in sorghum. Madras Agr. J. 32:185–96. Biol. Abstr. 23:16489. 1949.
Pons, J. E. 4247.
15530. Pons, W. A., Jr., C. L. Hoffpauir, and R. T. O'Connor. 1950: Determination of total gossypol pigments in cottonseed materials. J. Am. Oil Chemists Soc. 27:390–93. Biol. Abstr. 26:26072. 1952.
15531. ——, J. C. Kuck, and V. L. Frampton. 1960: Color index for cottonseed oils. J. Am. Oil Chemists Soc. 37:671–73.
15532. ——, M. D. Murray, R. T. O'Connor, and J. D. Guthrie. 1948: Storage of cottonseed and peanuts under conditions which minimize spectrophotometric changes in the extracted oil. J. Am. Oil Chemists Soc. 25:308–13. Chem. Abstr. 42:8497. 1948.
Pons, W. A., Jr. 6535, 6599, 13115, 17875, 17876.
Ponseti, I. V. 3811.
15533. Pont, J. W. 1935: Physiological studies with seeds of *Andropogon sorghum* Brot. Botanist, Dept. Agr., Union S. Africa, Swets & Zeitlinger, Amsterdam. 64 pp.
Ponte, J. G., Jr. 7300.
Ponte, J. J. da; *see* Da Ponte, J. J.
15534. Pontovich, V. E. 1949: Changes in the activity of oxidizing enzymes (polyphenoloxidases) during the growth of plants. Biochemistry (USSR) (English Transl.) 14:460–66. Chem. Abstr. 44: 1174. 1950.
15535. —— 1959: Sterile culture of the placenta of poppy as a means for the study of the formation of seeds and their synthesis of reserve products. Plant Physiol. (USSR) (English Transl.) 6:313–20.
15536. —— 1962: Metabolism of the placenta and seeds of red poppy and fat synthesis. Plant Physiol. (USSR) (English Transl.) 9:32–49. Chem. Abstr. 56:14639. 1962.
15537. Pontsov, A. V. 1955: The effect of thiourea on the germination of certain seeds. Byul. Gl. Botan. Sada 20:98–101.
Poock, E. 3194.
Pool, A. A. 15518.
15538. Poole, A. L. 1938: Germination of ragwort in water. New Zealand J. Agr. 57:95–96.
15539. Poole, C. F., P. C. Grimball, and D. R. Porter. 1941: Inheritance of seed characters in watermelon. J. Agr. Res. 63:433–56.
15540. Poole, D. D., and H. H. Hadley. 1954: Haploidy in castorbeans. J. Heredity 45:285–88.
Poole, H. 6479, 6480.
Poos, F. W. 2154.
Pop, E. 3421, 3423, 16238.
Pop, V. 3421–3423.
Popa, T. 1291.
Pope, C. W. 4332.
15541. Pope, M. N. 1941: Artificially induced vivipary in barley. J. Am. Soc. Agron. 33:850–51.
15542. —— 1942: Rate of growth of the embryo of young barley seeds on excised culms. J. Am. Soc. Agron. 34:200–2.
15543. —— 1943: Cleavage polyembryony in barley. J. Heredity 34:153–54.
15544. —— 1949: Viviparous growth in immature barley kernels. J. Agr. Res. 78:259–309.
15545. —— and E. Brown. 1943: Induced vivipary in three varieties of barley possessing extreme dormancy. J. Am. Soc. Agron. 35:161–63.
Pope, M. N. 8175, 8176, 12924.
15546. Pope, O. A., and J. O. Ware. 1945: Effect of variety, location, and season on oil, protein, and fuzz of cottonseed and on fiber properties of lint. U.S. Dept. Agr. Tech. Bull. 903:1–41.
Pope, W. 8115.
15547. Popham, R. A. 1959: Effects of a low temperature on germination of cotton seeds and on growth of the embryo. Texas J. Sci. 11:299–313.
15548. Popoff, A. 1943: Die Auswuchsneigung beim Getreide im Zusammenhang mit dem Protein des Korns. Angew. Botan. 25:150–65.
15549. ——, L. Micev, and B. Kuzmanov. 1961: On *Euphorbia lathyrus* L. and the oil extracted from the seed. Bulgar. Akad. Nauk Dokl. 14:369–71.

15550. Popoff, M. 1923: Biologische Möglichkeiten zur Hebung des Ernteertrages. Biol. Centralbl. 43:244–68.

15551. Popolina, T. G. 1962: Anatomic-morphological features of seed and viviparous forms of *Poa bulbosa* L. Tr. Botan. Inst. Akad. Nauk SSSR Ser. 7, 5:166–75. Biol. Abstr. 43:11420. 1963.

15552. Popov, A. 1941: The abnormal germination of grain. Angew. Botan. 23:254–85. Chem. Abstr. 37:3795. 1943.

15553. —— 1943: The germinating power of cereals in relation to the protein content of the grain. Angew. Botan. 25:150–65. Chem. Abstr. 38:5536. 1944.

15554. —— and S. Ivanov. 1955: The oil from ramie seeds (*Boehmeria nivea*). Compt. Rend. Acad. Bulgare Sci. 8:17–20. Chem. Abstr. 50:11035. 1956.

Popov, A. V. 14833.

15555. Popov, D. 1961: Research on biochemical effects of supersonic vibrations. Izv. Bulgar. Akad. Nauk Inst. po Rast. 12:169–76.

Popov, E. 9377.

Popov, I. 15562.

15556. Popov, I. D. 1940: Relation between germination and enzymic activity of wheat kernels. Annuaire Univ. Sofia V. Fac. Agron. Sylvicult. Livre 1, 18:279–322. Chem. Abstr. 35:7465. 1941.

15557. —— 1956: Biochemical investigations with seeds stimulated with manganese and cobaltous salts. Izv. Bulgar. Akad. Nauk Inst. po Biol. "Metodii Popov" 7:111–28.

15558. ——, N. Karabashev, and T. Karabasheva. 1955: The enzymochemical activity of seed on the application of ultrasonics. Compt. Rend. Acad. Bulgare Sci. 8:65–68. Chem. Abstr. 50:10871. 1956.

15559. —— and V. Khristov. 1957: The path of entry of cobalt-60 into wheat seed. Compt. Rend. Acad. Bulgare Sci. 10:73–76. Chem. Abstr. 52:7449b. 1958.

15560. Popov, L. V. 1957: The influence of substratum moisture on germination of pine and spruce seeds. Tr. Vost-Sibirsk. Filiala Akad. Nauk SSSR 1957(5):116–21. Biol. Abstr. 35:8631. 1960.

15561. Popov, M. 1956: Experiments in stimulating tobacco seeds and tobacco seedlings. Izv. Bulgar. Akad. Nauk Inst. po Biol. "Metodii Popov" 7:53–72.

15562. —— and I. Popov. 1954: Enzymechemical research on stimulated seeds. Compt. Rend. Acad. Bulgare Sci. 6:65–68. 1953: Chem. Abstr. 49:4796. 1955.

15563. Popov, M. D. 1957: Die Samenstimulierung landwirtschaftlicher Kulturpflanzen mit Penicillin. Izv. Bulgar. Akad. Nauk Inst. po Biol. "Metodii Popov" 8:27–48.

Popov, M. D. 9377.

15564. Popov, N. D. 1958: Samenstimulationsversuche mit Ultraschall an verschiedenen Kulturpflanzen. Izv. Bulgar. Akad. Nauk Inst. po Biol. "Methodii Popov" 9:135–45.

15565. Popov, N. F., and L. I. Timofeev. 1933: Some data on the chemistry of wheat ripened after harvesting in storage, silos or elevators. Sci. Inst. Cereal Res. (Moscow) 1933(11):59–83. Chem. Abstr. 29:2607–8. 1935.

15566. Popov, P. S. 1956: The accumulation of phosphorus compounds in sunflower seeds on ripening. Kratkii Otchet o Nauchn.-Issled. Rabote, Vses. Nauchn.-Issled. Inst. Maslichn. i Efiromasl. Kultur 1955:95–98. Chem. Abstr. 51:14906. 1957.

15567. Popov, S. 1930: Étude briometrique et anatomique de la semence du tabac var. "Yaka de suluc" cultivée en Roumanie. Bul. Cult. Ferment. Tutunului 19:261–82.

Popova, A. A. 7751.

Popova, E. S. 9559.

Popova, G. S. 20022.

Popova, O. N. 641.

Popovici, N. 5143.

Popovska, P. 1607.

15568. Popp, H. W., and F. B. Charlton. 1938: Effects of ultraviolet radiation upon germination and seedling development. Penn. Agr. Expt. Sta. Bull. 366. 50 pp.

Popp, H. W. 13193, 16249.

15569. Popp, W. 1959: A new approach to the embryo test for predicting loose smut of wheat in adult plants. Phytopathology 49:75–77.

15570. Poptsov, A. V. 1929: Studies of the seeds of kenaph (*Hibiscus cannabis* L.). Arb. Kuban. Landwirtsch. Inst. 8:83–125. Biol. Abstr. 6:24710. 1932.

15571. —— 1935: Note on secondary dormancy of *Taraxacum megalorhizon* seeds. Compt. Rend. Acad. Sci. URSS 2:596–97.

15572. —— 1949: On some biological characteristics of kok-saghyz seed germination. Dokl. Akad. Nauk SSSR 68:609–11.

15573. —— 1951: Second rest period of

safflower seeds. Byul. Gl. Botan. Sada 9:58–61.

15574. —— 1952: Thiourea as a stimulator of kok- and tau-saghyz seed germination. Dokl. Akad. Nauk SSSR 84:619–22.

15575. —— 1953: Hard seeds. Tr. Gl. Botan. Sada 3:58–105.

15576. —— 1957: Importance of moisture in seed stratification. Byul. Gl. Botan. Sada 27:62–70. Biol. Abstr. 44:25055. 1964.

15577. —— 1960: The biology of the germination of *Saponaria officinalis*. Byul. Gl. Botan. Sada 39:42–47.

15578. —— 1961: The significance of temperature in seed germination. Zh. Obshch. Biol. 22:425–35. Biol. Abstr. 40:3427. 1962.

15579. —— and T. G. Buch. 1952: On factors instrumental in the preservation of germination by willow seeds. Dokl. Akad. Nauk SSSR 83:489–92.

15580. —— and —— 1954: Germination of seed of Judas trees. Byul. Gl. Botan. Sada 17:48–54.

15581. —— and —— 1957: Directions for storing seeds of willow and poplar. Byul. Gl. Botan. Sada 27:110–12. Biol. Abstr. 33:7373. 1959.

15582. —— and —— 1960: Effect of temperature on seed germination. Byul. Gl. Botan. Sada 38:56–63.

15583. —— and —— 1961: Methods of accelerating the germination of seeds of *Nelumbium nuciferum*. Byul. Gl. Botan. Sada 42:80–83.

15584. —— and K. V. Kichunova. 1950: In regard to the biology of seed germination of milkweed (*Asclepia syriaca*). Byul. Gl. Botan. Sada 7:53–56.

15585. —— and —— 1951: On increasing the germination of guayule seeds. Byul. Gl. Botan. Sada 8:41–43.

15586. Poptzoff, A. 1928: The swelling and germination of the seeds of *Abutilon avicennae* Gaertn. Jard. Botan. Inst. Essais Semences, Leningrad, Ann. Essais Semences 6:81–103.

15587. —— 1932: Hygroscopic properties of tobacco seed. Krasnodar Inst. Tobacco Ind. Bull. 88. 36 pp.

15588. —— 1933: Studies in storage of tobacco seed. II. The influence of air humidity, temperature and air-tight keeping on tobacco seeds during their storage. Krasnodar Inst. Tobacco Ind. Bull. 93. 58 pp.

15589. —— 1936: Zur Keimungsphysiologie von Tabaksamen. Botan. Centralbl. Beih. Abt. A 55:311–57.

15590. Porodko, T. M. 1927: Über die Absterbegeschwindigkeit der erhitzten Samen. Ber. Deut. Botan. Ges. 45:4–11.

15591. —— 1927: Zeitlicher Keimungsverlauf der erhitzten Samen. Ber. Deut. Botan. Ges. 45:12–22.

15592. Porsild, A. E. 1920: Sur le poids et les dimensions des graines artiques. Rev. Gen. Botan. 32:97–120. Botan. Abstr. 8:91. 1921.

15593. Porter, C. R. 1949: The approval of seedsmen in certification work. Intern. Crop Improvement Assoc. Proc. 31:69–72.

15594. —— 1954: My philosophy of seed certification. Seed World 74(8):10, 41.

15595. Porter, D. R. 1937: Inheritance of certain fruit and seed characters in watermelon. Hilgardia 10:489–509. Expt. Sta. Rec. 77:317. 1937.

Porter, D. R. 15539.

15596. Porter, J. N. 1940: Note on the germination of *Nepenthes* seeds sown on agar. Botan. Mus. Leaflet Harvard Univ. 8:65–68.

Porter, J. S. 12476.

15597. Porter, K. B. 1959: The inheritance of shattering in wheat. Agron. J. 51:173–77.

15598. —— and M. G. Weiss. 1948: The effect of polyploidy on soybeans. J. Am. Soc. Agron. 40:710–24.

15599. Porter, R. H. 1936: Relation of seed disinfectants to seed analysis. Proc. Assoc. Offic. Seed Analysts 28:93–101.

15600. —— 1937: Application of the Chi-square test to purity analyses of Bluegrass seed. Proc. Assoc. Offic. Seed Analysts 29:91–93.

15601. —— 1938: Experiments with modified techniques for the determination of purity and viability of Bluegrass seed, *Poa pratensis* L. Iowa Agr. Expt. Sta. Res. Bull. 235. 21 pp.

15602. —— 1939: Uniform techniques for the analysis of small-seeded grasses. Proc. Assoc. Offic. Seed Analysts 30:133–71.

15603. —— 1939: Detection and classification of seed-borne organisms, their effect on germination and their control by seed disinfection in laboratory and field. Proc. Assoc. Offic. Seed Analysts 30:195–213.

15604. —— 1939: The relation of plant pathological technique to seed laboratory practice. Chronica Botan. 5:442–44.

15605. —— 1944: Testing the quality of

seeds for farm and garden. Iowa Agr. Expt. Sta. Res. Bull. 334:493–588.

15606. —— 1946: Induced baldhead in soybean. Phytopathology 36:168–70.

15607. —— 1949: Recent developments in seed technology. Botan. Rev. 15:221–82.

15608. —— 1949: Recent developments in seed technology. (Concluded.) Botan. Rev. 15:283–344.

15609. —— 1959: Manual for seed technologists. Publ. Am. Univ. Beirut 7. 149 pp.

15610. ——, M. Durrell, and H. J. Romm. 1947: The use of 2,3,5-triphenyltetrazolium chloride as a measure of seed germinability. Plant Physiol. 22:149–59.

15611. —— and K. Koos. 1936: Germination of injured weed and crop seeds. Proc. Assoc. Offic. Seed Analysts 28:68–73.

15612. —— and C. W. Leggatt. 1942: A new concept of pure seed as applied to seed technology. Sci. Agr. (Ottawa) 23:80–103.

15613. —— and W. N. Rice. 1940: Laboratory and field germination of treated and untreated beet seed. Proc. Assoc. Offic. Seed Analysts 1939:127–30.

15614. ——, S. A. Shakra, and H. Koposhian. 1959: Chemical treatments of crop seeds in Lebanon. Publ. Am. Univ. Beirut 10. 14 pp.

15615. ——, T. F. Yu, and H. K. Chen. 1929: The response of hulless barley to seed treatment for covered smut and stripe disease. Phytopathology 19:657–66.

Porter, R. H. 44, 3140, 3141, 5009, 6011, 9286, 16257.

15616. Portères, R. 1947: Croissance de la pileole le Fonio (*Digitaria exilis* Stapf.) à differentes températures en fonction de la profondeur d'enfouissement de la semence. Bull. Soc. Botan. France 94:8–10. Biol. Abstr. 23:12601. 1949.

15617. —— 1948: Competition ou entr'aide au sein de l'espèce et de la race. Le cas des germinations de cinchona. Compt. Rend. 227:1114–15.

15618. Porterfield, J., and E. M. Smith. 1956: Physical characteristics and field performance of mechanically graded acid delinted cottonseed. Oklahoma Agr. Expt. Sta. Tech. Bull. T-60. 24 pp.

15619. Porterfield, W. M. 1940: Chinese vegetable foods in New York. IX. Seeds of the ginkgo. N.Y. Botan. Garden J. 41:185–88.

Porterfield, W. M. 18069.

Portheim, L. 5868, 5869, 10719.

15620. Porto, F., and S. M. Siegel. 1960: Effects of exposures of seeds to various physical agents. III. Kinetin-reversible heat damage in lettuce seeds. Botan. Gaz. 122:70–71.

15621. Portz, H. L., and J. A. Jackobs. 1955. Differences between seed lots of Ladino clover in cyanophoric properties. Agron. J. 47:143–45.

15622. Poruskii, G. V., and A. G. Iavorskii. 1950: Effect of embryo transplanting on formation of compound spikes and kernel fullness in Kakhetian wheat (*Triticum turgidum* L.). Dokl. Akad. Nauk SSSR 70:901–4.

Poruskii, G. V. 13444, 13670, 19165.

15623. Porzio, L., M. A. Spadoni, and G. Tecce. 1955: Synthesis of sucrose from acetate-2-C^{14} in the fats of the endosperm of *Ricinus communis* seeds. Congr. Intern. Biochim. 3rd, Brussels, 1955:104. Chem. Abstr. 50:14060–61. 1956.

Posey, H. G. 10355, 13061.

15624. Posgay, E. 1954: Vernalisation of condiment paprika on a commercial scale; a method evolved on the basis of four years' experimenting. Novenytermeles 3:47–64.

15625. Poskin, A. 1932: Les cones et les semences de *Pin sylvestre* (*Pinus silvestris* L.) recoltes en Belgique. Bull. Inst. Agron. Sta. Rech. Gembloux 1:77–94. Biol. Abstr. 8:14216. 1934.

15626. —— and J. Fouarge. 1935: Les cones et les semences de *Pinus nigra* Arn. (*Pinus laricio* Poir.) recoltes en Belgique. Bull. Inst. Agron. Sta. Rech. Gembloux 4:303–15.

15627. Posner, H. B., and W. S. Hillman. 1962: Aseptic production, collection and germination of seeds of *Lemna perpusilla* 6746. Physiol. Plantarum 15:700–8.

Pospelowa, N. 14803.

15628. Pospisil, O. 1957: The use of paper chromatography for the selection of beet seed. Sb. Česk. Akad. Zeměděl. Věd, Rostlinna Vyroba 3:1035–43. Chem. Abstr. 52:12096. 1958.

15629. Pospisilova, D. 1957: Kältestratifikation der Weinrebesamen. Biologia 12:280–87.

Possnig, J. 10720.

Postel, W. 17081.

Potanina, N. D. 2063.

Potapczykowa, W. Maciejewska-; *see* Maciejewska-Potapczykowa, W.

Poti, A. N. 14096.

15630. Potocanac, J. 1957: Samengrösse

und Pflanzeneigenschaften im Keimungs- und Bestockungsstadium der Lein-Landsorten Kroatiens U. Bosniens. Poljoprivredna Znanstvena Smotra 16:223–44.

15631. Potter, C. H. 1960: Perennials from seeds. Florists Rev. 125(3242):35–36, 56.

15632. —— 1962: Propagation by seeds. XXXVII. Flowering pot plants. Florists Rev. 130(3372):23, 78–81.

15633. Potter, G. C., and F. A. Kummerow. 1954: Chemical similarity and biological activity of the saponins isolated from alfalfa and soybeans. Science 120:224–25.

Potter, G. F. 1667, 1668, 11799.

15634. Potter, L. D. 1951: Comparative seedling characteristics of standard crested wheatgrass and Russian wild rye. Botan. Gaz. 112:349–65.

Potts, W. H. 71.

Potts, W. M. 2105.

Potyagilo, Z. M. 11049.

15635. Pound, F. J. 1938: History and cultivation of the Tonka bean (*Dipteryx odorata*) with analyses of Trinidad, Venezuelan and Brazilian samples. Trop. Agr. (London) 15:28–32.

Pound, G. S. 10833.

15636. Pourtet, J. 1946: Les semences forestières et le reboisement. Rev. Eaux Forêts 84:521–35. Biol. Abstr. 21:7155. 1947.

15637. Povilaitis, B. 1937: On the methods of investigation of the germinating power of beet seeds. Ann. Acad. Agr. Dotnava 11:103–19.

15638. —— 1956: Dormancy studies with seeds of various weed species. Proc. Intern. Seed Testing Assoc. 21:88–111.

15639. —— and J. W. Boyes. 1959: Embryosac production in relation to seed yields of diploid Dollard red clover. Can. J. Plant Sci. 39:364–74.

15640. —— and —— 1960: Ovule development in diploid red clover. Can. J. Botany 38:507–32.

15641. Povolotskaya, K. L. 1937: Vitamin C in germinating seeds. Bull. Appl. Botany, Genet., Plant Breed. (USSR) Suppl. 84, Vit. Probl. 2:20–57. Chem. Abstr. 33:1365–66. 1939.

15642. ——, O. I. Pyshkinskaya, A. A. Kondrashova, and E. P. Skorobogatova. 1956: Biosynthesis of thiamine, riboflavin, and nicotinic acid in seeds. Biokhim. Zerna, Sb. 1956(3):145–55. Chem. Abstr. 51:8913. 1957.

Powell, B. D. 10467.

15643. Powell, E. C. 1935: Growing narcissus from seed. Flower Grower 22(3):135.

15644. Powell, R. D. 1960: Effect of temperature and anaerobic conditions during the induction period of tobacco seed germination. Proc. Assoc. Southern Agr. Workers, Ann. Conv. 57:253.

Powell, R. D. 2268.

Powning, R. F. 6664.

15645. Powrie, W. D., M. W. Adams, and I. J. Pflug. 1960: Chemical, anatomical, and histochemical studies on the navy bean seed. Agron. J. 52:163–67.

15646. Pozsar, B. 1957: Lecithin synthesis in dehydrating seeds. Acta Botan. Acad. Sci. Hung. 3:37–42. Chem. Abstr. 51: 15717. 1957.

Pozsar, B. [I.] 7264, 7265, 11569, 11570.

15647. Pradet, A. 1964: Remarques sur l'inhibition de la germination des semences de *Phacelia tanacetifolia* Benth. et *Lactuca sativa* variété "Reine de Mai" par l'azoture de sodium et l'acide cyanhydrique. Compt. Rend. 258:1610–12.

15648. Pradham, S., and P. B. Mookherjee. 1958: It's easy to store your seed-grains safe. Indian Farming 8(7):19–22.

15649. Pradisth, C. S., and A. C. Santos. 1940: Main constituents of *Datura alba* Nees. Rev. Filipino Med. Farm. 31:1. Pharm. Abstr. 7:253. 1941.

15650. Prado, E. C. 1945: Chemistry and food value of indigenous sunflower seed oils. Farmacia Quim. (Lima) 2:22–25. Chem. Abstr. 40:4233–34. 1946.

15651. Prakash, J., and R. L. N. Iyengar. 1961: A study of the variation of fibre maturity at different regions of the seed surface and of its variation from seed to seed in relation to the seed weight. Indian Cotton Growing Rev. 15:149–53.

Prakash, R. 16789.

Prakash, V. 11683.

15652. Prakken, R. 1940: Inheritance of colours in *Phaseolus vulgaris* L. I. Genetica 22:331–408. Biol. Abstr. 15:14172. 1941.

15653. —— and J. H. Van Der Veen. 1958: Flower color and seed quality of *Nicotiana tabacum*. Genen Phaenen 3(3):29–43.

Praloran, J. C. 3836.

Prasad, A. 9760.

15654. Prasad, J. 1944: The role of experimental research in the natural regeneration of Indian tree species by seed. Indian Forester 70:142–47.

Prasad, M. 12983.

15655. Prasad, S. 1944: Fertilizer effects on the growth and alkaloidal content of

Datura alba Nees. Indian J. Pharm. 6:13–22.

15656. —— 1944: A histological study of the fruit and seed of *Psoralea corylifolia* Linn. Indian J. Pharm. 6:61–67.

Prasada, R. 1427.

Prashad, G. 11685.

15657. Prat, H. 1948: Observations on thermogenesis of seed germination and of bacterial cultures. Am. J. Botany 35: 811–12.

15658. —— 1951: Influence of surrounding temperature on the thermogenesis of germinations. Rev. Can. Biol. 10:443–53. Chem. Abstr. 46:5674e–h. 1952.

15659. —— 1952: Microcalorimetric studies on germinations of cereals. Can. J. Botany 30:379–94.

15660. —— 1953: The Tian-Calvet microcalorimeter in seed germination studies. Nature 171:24.

15661. —— and E. Calvet. 1944: Application of microcalorimetry to the study of the initial stages of germination of seeds. Compt. Rend. Soc. Biol. 138:662–63. Chem. Abstr. 40:921–22. 1946.

15662. —— and —— 1944: Variations de thermogénèse provoquées par l'acide indol-β-acetique et l'alcool au cours des primiers stades de la germination. Compt. Rend. Soc. Biol. 138:675. Bull. Soc. Botan. France 92:113. 1945.

15663. —— and —— 1945: Modification of germinative thermogenesis by previous desiccation of seeds. Compt. Rend. Soc. Biol. 139:222. Chem. Abstr. 40:2499. 1946.

15664. Prat, S. 1947: Biological tests of waters. Sb. Masaryk. Akad. Prace 21:1–29. Biol. Abstr. 23:5060. 1949.

15665. Pratt, C., and J. Einset. 1955: Development of the embryo sac in some American blackberries. Am. J. Botany 42:637–45.

15666. —— and S. H. Wender. 1959: Identification of rutin and isoquercitrin in cottonseed. J. Am. Oil Chemists Soc. 36:392–94.

Pratt, C. 14527.

15667. Pratt, R., and H. G. Albaum. 1939: Nature of growth differences in two sorghum varieties. I. Influence of preliminary soaking on early growth and auxin content. Am. J. Botany 26:822–26.

15668. ——, J. Dufrenoy, and V. L. Pickering. 1949: Vitamin K_5 as a preservative for fruits and a disinfectant for seeds. Phytopathology 39:862.

Pratt, R. 5616.

15669. Pravdin, L. F., and V. D. Filimonova. 1952: Effect of low temperatures on viability of acorns. Dokl. Akad. Nauk SSSR 85:921–24.

15670. —— and K. G. Shcherbina. 1961: Dynamics of the content of chlorophyll in the cones and the fattiness of seeds of *Pinus silvestris* from different geographical origins. Akad. Nauk SSSR Inst. Lesa. Tr. 50:90–98.

15671. Precsenyi, I. 1959: Joint germination experiment of *Lotus corniculatus* and *Festuca rubra* in laboratory conditions. Gödöllo. Agrartud. Egyet. Mezogazd. Kar. Közlemen. 1958:401–5.

15672. Predtechenskaja, A. A. 1929: Observations sur la faculté germinative des semences à immergées dans l'eau. Zap. Semenoved. Ann. Essais Semences Leningrad 7:3–12.

15673. —— 1931: Materialen für Kenntnis der Keimung und Reife von Samen. VII. Über das Keimen von Samen in Wasserdampfatmosphäre. Jard. Botan. Inst. Essais Semences, Leningrad, Ann. Essais Semences 1931(1):15–26.

51674. —— 1935: Zur Frage über die Keimung von Samen in mit wasserdampf gesättigter Luft. Botan. Zh. 20:127–29. Botan. Centralbl. 169:353. 1936.

Predtetshenskaja, A. A. 7682, 9513.

15675. Preece, I. A. 1931: Studies on hemicelluloses. The hemicelluloses of brewers' grains. J. Inst. Brewing 37:409–13. Biol. Abstr. 6:15542. 1932.

15676. —— and K. G. Mackenzie. 1952: Non-starchy polysaccharides of cereal grains. II. Distribution of water-soluble gum-like materials in cereals. J. Inst. Brewing 58:457–64. Brit. Abstr. B III:191. 1953.

15677. Preest, D. S. 1963: An attempt to hasten the germination of matai and miro seed using concentrated sulphuric acid. New Zealand J. Forestry 8:826–28. Biol. Abstr. 45:26137. 1964.

15678. Preiss, J. W., and S. Klein. 1958: Depth controlled deuteron irradiation of *Lactuca sativa* seeds. II. Energy loss in the outer seed layers. Plant Physiol. 33:326–29.

Preiss, J. W. 10778, 10779.

15679. Preobrazhenskaya, E. I., and N. V. Timofeev-Resovskii. 1962: The correlation between germination and survival in different species of cultivated plants after irradiation of the seed with various doses of Co^{60} γ-rays. Dokl. Akad. Nauk SSSR

143:448–51. Nucl. Sci. Abstr. 16:2623–24. 1962.

15680. Presley, J. T., and O. A. Leonard. 1948: Effect of calcium and other ions on the early development of the radicle of cotton seedlings. Plant Physiol. 23:516–25.

Presley, J. T. 19692.

Preston, G. H. 16663.

15681. Preston, I. 1936: The fascination of growing lilies from seed. Flower Grower 23(9):465, 496.

15682. Preston, R. D., and L. I. Scott. 1943: The significance of the micropyle in relation to water entry in some leguminous seeds. Proc. Leeds Phil. Lit. Soc. Sci. Sect. 4:123–34.

Pretty, K. M. 16578.

15683. Prevett, P. F. 1959: A study of rice storage under tropical conditions. J. Agr. Eng. Res. 4:243–54.

15684. Priaulx, A. W. 1946: Direct seeding tool. Am. Forests 52:472–73.

15685. Price, C., and E. Carsner. 1947: Seed size in relation to development and yield of sugar beets. Proc. Am. Soc. Sugar Beet Technol. 1946:263–69. Biol. Abstr. 22:12106. 1948.

Price, C. 13176.

15686. Price, R. 1938: Artificial reseeding on oak-brush range in central Utah. U.S. Dept. Agr. Circ. 458:1–18. Biol. Abstr. 12:6987. 1938.

15687. —— 1944: Climatic effects on acorns (of bur oak). Horticulture 22:36.

15688. Price, R. H. 1898: Classification of varieties of peaches. U.S. Dept. Agr. Office Expt. Sta. Bull. 49:82–93.

15689. Pridham, A. M. S. 1946: 2,4-Dichlorophenoxyacetic acid reduces germination of grass seed. Proc. Am. Soc. Hort. Sci. 47:439–45.

Pridham, A. M. S. 7207.

15690. Pridham, J. B. 1958: Occurrence and metabolism of oligosaccharides in the broad bean (*Vicia faba*). Nature 182:1687–88.

15691. —— and M. J. Saltmarsh. 1960: The formation of phenolic glycosides by germinating broad-bean (*Vicia faba*) seeds. Biochem. J. 74. 42 pp.

Pridham, T. G. 8621.

15692. Prihoda (A.). 1955: The decay of hornbeam seed during stratification. Ann. Acad. Tchecosl. Agr. 28:385–92. Rev. Appl. Mycol. 36:145. 1957.

15693. Prikladov, N. V. 1958: A new method of determining grain moisture. Selek. i Semen. 3:56–59. Biol. Abstr. 44:20834. 1963.

Primo, S. Da Costa-; *see* Da Costa-Primo, S.

Primost, E. 12295.

15694. Prince, A. B. 1954: Effects of nitrogen fertilization, plant spacing, and variety on the protein composition of corn. Agron. J. 46:185–86.

15695. —— and E. B. Eskew. 1951: Effect of plant spacing and nitrogen fertilization on yield and protein content of three corn varieties. S. Carolina Agr. Expt. Sta. Ann. Rept. 63:39–41.

15696. Prince, F. S., and A. R. Hodgdon. 1946: Viable seeds in old pasture soils. New Hampshire Agr. Expt. Sta. Tech. Bull. 89. 16 pp.

15697. Principe, J. A., and T. W. Whitaker. 1961: Seed germination of the royal poinciana (*Delonix regia*). Plant Life 17:155–56.

15698. Pring, G. H. 1952: Growing *Victoria cruziana* from seeds. Bull. Missouri Botan. Garden 40:85–89.

15699. Pring, Mrs. G. H. 1943: The parachute seed-pod of *Aristolochia*. Bull. Missouri Botan. Garden 31:123–25.

Pringle, W. J. S. 2052.

15700. Pringsheim, E. G. 1930: Untersuchungen über Samenquellung. I. Die Abhängigkeit der Quellung von der Beschaffenheit der Samen und vom Medium. Z. Wiss. Biol. Abt. E, Planta 11:528–81. Biol. Abstr. 8:7943. 1934.

15701. ——, F. Jedlitschka, and B. Görlich. 1931: Untersuchungen über Samenquellung. II. Die Atmung quellender Samen. Planta 15:419–58. Ber. Wiss. Biol. 21: 182–83.

15702. ——, ——, and —— 1933: Untersuchungen über Samenquellung. III. Der Atmungsquotient quellender Samen. Planta 19:653–712.

15703. Prisemina, Z. P. 1928/29: The biochemical variability of the seeds of the castor-oil plant in dependence on geographical factors. Bull. Appl. Botany, Genet., Plant Breed. (USSR) 21:427–36.

15704. Prisiazhniuk, A. A. 1956: Effect of various chemicals on the technical and ground germination of seeds of coniferous species. Sb. Tr. po Zashch. Rast. Konf. po Voprosam Zashch. Rast. 1956(1):215–22.

15705. Pristupa, A. A. 1940: Embryoless seeds in castor-oil plant. Compt. Rend. Acad. Sci. URSS 28:657–59.

15706. Pritchard, E. W. 1933: How long do seeds retain their germinating power? Longevity experiments in the department of agriculture. J. Dept. Agr. S. Australia 36:645–46.

15707. Pritchard, G. G. 1961: The effect of high oxygen pressure on the respiratory metabolism of pea seeds. J. Exptl. Botany 12:353–63.

15708. Pritchard, H. N. 1964: A cytochemical study of embryo development in *Stellaria media*. Am. J. Botany 51:472–79.

15709. Pritzker, J., and R. Jungkunz. 1935: Hipberry-seed oil. Pharm. Acta Helv. 10:75–78. Chem. Abstr. 29:7682. 1935.

15710. —— and —— 1935: Tea-seed oil. Z. Untersuch. Lebensm. 69:542–47. Chem Abstr. 29:7682. 1935.

15711. —— and —— 1935: Apple-seed oil and pear-seed oil. Z. Untersuch. Lebensm. 70:255–58. Analyst 61:49. 1936.

15712. —— and —— 1938: Quince seed. Pharm. Acta Helv. 13:29–34. Pharm. Abstr. 5:5. 1939.

15713. Privault, D. 1938: Action du milieu sur l'épiderme des cotyledons du *Pin* Pignon. Bull. Soc. Botan. France 85:309–11.

Prizemina, S. P. 12309.

Prober, P. L. 6481.

Probst, A. H. 12710, 18992, 19486, 19613.

15714. Prochazka, M. 1933: Einfluss der Erntezeit des Mohnes auf den Konsumwert und die Keimfähigkeit der Samen. Gartenbauwissenschaft 7:458–66. Biol. Abstr. 11:6877. 1937.

15715. —— 1935: Die Rotstichigkeit des Mohnes. Landeskultur Vienna 2:134–38. Proc. Intern. Seed Testing Assoc. 8:86–87.

15716. Prochazka, F. 1951: The experiments with jarovization of sugar-beet seed. Sb. Česk. Akad. Zemědél. 24:171–75.

Proctor, B. E. 11718.

Proctor, C. H. 3787.

15717. Proctor, J. M. 1963: An experiment to determine the effects of date of sowing on the yield and quality of harvesting peas 1953–1956. J. Agr. Sci. 61:281–89.

15718. Prokhorova, N. T., and M. S. Reznichenko. 1959: The amino acid composition of oryzenin of rice. Sb. Nauchn. Rabot, Leningr. Inst. Sovet. Torgovli 1959(13):3–7. Chem. Abstr. 54:15547. 1960.

15719. Prokofev, A. A., and V. P. Kholodova. 1959: Water content and maturation of seeds. Plant Physiol. (USSR) (English Transl.) 6:196–201.

15720. —— and G. V. Novitskaya. 1957: Activity of lipase and fat accumulation in seed of flax and poppy. Dokl. Akad. Nauk SSSR 116:273–76. Chem. Abstr. 52:4752i. 1957.

15721. —— and A. M. Sobolev. 1957: Concerning phosphorus transport from leaves to seeds. Plant Physiol. (USSR) (English Transl.) 4:12–19.

Prokofev, A. A. 5966.

15722. Prokoshev, S. M. 1936: The specificity of protease in seeds and sprouts of a number of crops. Bull. Appl. Botany, Genet., Plant Breed. (USSR) Ser. III(14): 79–95. Chem. Abstr. 31:5822–23. 1937.

15723. ——, T. I. Margolina, and I. A. Babichev. 1936: On the physiological action of hydrogen sulfide on wheat seeds. Bull. Appl. Botany, Genet., Plant Breed. (USSR) Ser. III(14):3–23.

Prokoveva-Belgovskaya, A. A. 10550.

15724. Promsy, G. 1912: Du rôle des acides dans la germination. Barlatier, Marseille. 176 pp.

15725. Pronin, S. I., and M. V. Sukernik. 1937: The dynamics of variation of the total activity of amylase and its thermostable component in germinating peas. Bull. Biol. Med. Exptl. USSR 4:537–39. Chem. Abstr. 34:5116. 1940.

Proskurnikova, T. A. 18046.

Proskurnina, N. 14821, 14822.

15726. Proskuryakov, N. I., and M. B. Babintseva. 1962: Synthesis of enzymes and its inhibition by antibiotics in the cotyledons of germinating peas. Dokl. Akad. Nauk SSSR 146:464–66.

15727. —— and —— 1962: Synthesis of enzymes in sprouting pea cotyledons and its inhibition by antibiotics. Dokl. Akad. Nauk SSSR 146:926–29.

15728. —— and I. M. Mosolova. 1952: Dynamics of glutathione in sprouting of peas. Dokl. Akad. Nauk SSSR 87:465–68. Chem. Abstr. 47:4962. 1953.

15729. —— and I. V. Rodionova. 1960: Enzymic activity of the water-soluble proteins of the wheat germ. Biokhim. Zerna, Sb. 5:108–20. Chem. Abstr. 54:22866. 1960.

15730. —— and N. B. Strazhevskaya. 1954: Carbohydrate-phosphorus metabolism in embryos and endosperms of winter wheat during the process of vernalization. Biokhim. Zerna, Sb. 2:147–60. Chem. Abstr. 49:7655. 1955.

15731. —— and M. A. Suetina. 1951: Oxidizing enzymes of acorns (in storage). Dokl. Akad. Nauk SSSR 77:443–45.

15732. —— and I. I. Tovarova. 1961: Features of proteolysis in the germinating seeds of leguminous plants. Nauchn. Dokl. Vysshei Shkoly Biol. Nauki 1961(3): 155–58.

Proskuryakov, N. I. 20137.

Pruneyre, A. 5464.

15733. Prunster, R. W. 1941: Germination conditions for *Typha muelleri* (Rohrbach) and its practical significance in irrigation channel maintenance. Australian Council Sci. Ind. Res. J. 14:129–36.

15734. Pruthi, J. S. 1962: Chemical composition and utilization of passion fruit seed, seed oil and seed meal. India Oil Soap J. 28:55–62. Chem. Abstr. 59:4268. 1963.

Pryor, L. D. 7044–7046.

15735. Przyborowski, J. 1938: On errors due to insufficient size of clover samples tested for dodder. Proc. Intern. Seed Testing Assoc. 10:230–35.

15736. Przybylska, F., and J. Hurich. 1961: Primary research after various forms of nitrogen in seeds of certain leguminous plants. Genet. Polon. 2:65–73. Biol. Abstr. 43:3288. 1963.

15737. Przybylska, J. 1959: Free arginine in bitter and nonbitter seeds of blue and white lupine. Bull. Acad. Polon. Sci. Ser. Sci. Biol. 7:359–62. Chem. Abstr. 54: 4775. 1960.

15738. —— and J. Hurich. 1960: Free amino acids in seeds of some leguminous plants. Bull. Acad. Polon. Sci. Ser. Sci. Biol. 8:505–8. Biol. Abstr. 36:86068. 1961.

Przybylska, J. 14615, 14616.

15739. Pshenova, K. V., and P. A. Kolesnikov. 1962: Lipoxidase of wheat seedlings. Biochemistry (USSR) (English Transl.) 26:873–76.

Pshenova, K. V. 14010, 17595.

Ptitsyn, S. D. 11375.

15740. Puchner, H. 1922: Die Verzögerte Keimung von Baumsämereien. Forstwiss. Centralbl. 44:445–55.

Puentes, C. 9971.

Pür, R. 15208.

15741. Pugachev, A. N. 1962: Calculating the rate of microtrauma in seeds in the establishment of sowing rates. Selek. i Semen. 2:38–42. Biol. Abstr. 43:11571. 1963.

15742. Pugh, G. W., H. Johann, and J. G. Dickson. 1932: Relation of the semipermeable membranes of the wheat kernel to infection by *Gibberella saubinetii.* J. Agr. Res. 45:609–26.

Pugnat, C. 18102, 18103.

15743. Pugsley, A. T. 1959: Seed dormancy associated with red grain colour. Wheat Inform. Serv. 9/10:31–32.

15744. —— 1928: The longevity of seeds. J. Botany 66:203–4.

Puia, I. 1885.

15745. Pujiula, J. 1930: Some anatomical-physiological data on the seed of *Jacaranda braziliana.* Bol. Soc. Iberica Cienc. Nat. 29:29–35. Biol. Abstr. 7:6845. 1933.

15746. —— 1948: Chlorophyll pigment in the seed of Japanese medlar, *Eriobotrya japonica.* Broteria 17:182–85.

15747. Pulkki, L. H., and K. Puutula. 1938: The formation and distribution of ascorbic acid in wheat during germination. Z. Ges. Getreidew. 25:149–53. Chem. Abstr. 33: 1400–1. 1939.

15748. —— and —— 1941: Thiamine in wheat grain. Biochem. Z. 308:122–27. Chem. Abstr. 37:1200–1. 1943.

15749. Pullen, J. A., and C. A. Schexnayder. 1955: Gloxinias grown from seeds. Florists Rev. 117:93–95.

15750. Pultz, L. M. 1929/30: A study of starch deposition in the seeds of *Pisum sativum.* Abstr. Theses, Chicago Univ. Sci. Ser. 8:227–30.

15751. Pulver, H. 1937: Moonflowers from seed. Horticulture 15(8):196.

15752. —— 1945: Ants and wild flower seed. Horticulture 23(6):163.

Pulzeyt, W. 3679.

Punekar, B. D. 8194.

Punsri, P. 4521.

15753. Puntambekar, S. V. 1932: The seeds of *Vateria indica* Linn. as a source of vegetable tallow. Indian Forestry 58: 68–74.

15754. —— 1942: The fatty oil from the seeds of *Mallotus philippinensis*, Muel. Current Sci. (India) 11:464–65. Biol. Abstr. 18:3455. 1944.

15755. —— 1947: Effect of storage on the alkaloidal content of *Strychnos nux-vomica* seeds. Current Sci. (India) 16:346.

15756. —— and P. C. Batra. 1952: Chemical examination of the fatty oil from the seeds of *Steculia urens* Roxb. Proc. Indian Acad. Sci. Sect. A 36:284–88. Biol. Abstr. 27:23486. 1953.

15757. —— and R. M. Beri. 1954: Resinol from nuts of *Semecarpus travancoricus*. Proc. Indian Acad. Sci. Sect. A 39:137–43. Chem. Abstr. 48:10858. 1954.

15758. —— and S. Krishna. 1933: The fat and the oil from the seeds of *Actinodaphne hookeri* Meissn. An indigenous source of lauric acid. J. Indian Chem. Soc. 10:395–400. Chem. Abstr. 28:353. 1934.

15759. —— and —— 1934: Some Indian acorn oils. J. Indian Chem. Soc. 11: 721–26. Analyst 60:107. 1935.

15760. —— and —— 1937: Oil from seeds of *Ximenia americana* L. New unsaturated fatty acid, ximenic acid. J. Indian Chem. Soc. 14:268–74. Soc. Chem. Ind. J. 56:1233. 1937. Pharm. Abstr. 4:203. 1938.

15761. —— and —— 1940: Fatty oil from the seeds of *Bauhinia variegata*. J. Indian Chem. Soc. 17:96–100. Analyst 65:421. 1940.

15762. —— and —— 1941: Fatty oil from the seeds of *Solanum indicum*. J. Indian Chem. Soc. 18:329–34. Analyst 67:64. 1942.

Puntambekar, S. V. 11412.

15763. Pupuikin, A. F. 1933: The germination of *Apocynum* seeds in variously salinized soils. Tr. Inst. Novogo Lubyanogo Suirya. 4:31–37. Chem. Abstr. 27:5097. 1933.

15764. Purdy, C. 1931: Germination of lily seeds. Horticulture 9(14):313–14.

15765. Purdy, L. H. 1956: Does post-treatment storage of wheat seed increase the effectiveness of fungicides against smut? Plant Disease Reptr. 40:878–81.

Purdy, L. H. 10457.

15766. Puri, Y. P. 1960: Flowering habit, pollen characteristics, and seed germination of *Phormium tenax* Forster. Dissertation Abstr. 21:739.

15767. Puriewitsch, K. 1896: Über die selbstthätige Entleerung Reservestoffbehälter. Ber. Deut. Botan. Ges. 14:207–12.

15768. Purkayastha, K. K. 1940: A note on the occurrence of tri-cotyledonary seedlings in *Crotolaria juncia* Linn. Current Sci. (India) 9:467.

Purpoint, M. 11913.

Purves, C. B. 17662.

15769. Purvis, O. N. 1934: An analysis of the influence of temperature during germination on the subsequent development of certain winter cereals and its relation to the effect of length of day. Ann. Botany (London) 48:919–55.

15770. —— 1936: Vernalisation: A new method of hastening flowering. Sci. Hort. 4:155–64.

15771. —— 1940: Vernalization of fragments of embryo tissue. Nature 145:462.

15772. —— 1944: Studies in the vernalisation of cereals. VIII. The role of carbohydrate and nitrogen supply in the vernalisation of excised embryos of "Petkus" winter rye. Ann. Botany (London) 8:285–314.

15773. —— 1947: Studies in vernalisation of cereals. X. The effect of depletion of carbohydrates on the growth and vernalisation response of excised embryos. Ann. Botany (London) 11:269–83. Biol. Abstr. 22:17216. 1948.

15774. —— 1948: Studies in vernalisation. XI. The effect of date of sowing and of excising the embryo upon the responses of Petkus winter rye to different periods of vernalisation treatment. Ann. Botany (London) 12:183–206.

15775. —— 1961: The physiological analysis of vernalisation. Encyclopedia Plant Physiol. 16:76–122.

15776. —— and F. G. Gregory. 1937: Studies in vernalisation of cereals. I. A comparative study of vernalisation of winter rye by low temperature and by short days. Ann. Botany (London) 1:569–91. Biol. Abstr. 12:9408. 1938.

15777. —— and —— 1952: Studies in vernalisation. XII. The reversibility by high temperature of the vernalised condition in Petkus winter rye. Ann. Botany (London) 16:1–21.

15778. —— and —— 1953: Accelerating effect of an extract of vernalized embryos of winter rye on flower initiation in unvernalized embryos. Nature 171:687–88.

15779. —— and E. S. J. Hatcher. 1959: Some morphological responses of cereal seedlings to vernalisation. J. Exptl. Botany 10:277–89.

Purvis, O. N. 6748, 7509, 7642–7648, 8379.

15780. Pushkareva, K. 1930: On the characteristics of the seeds of different biological races of broom rape (*Orobanche cumana*). J. Agr. Res. N. Caucasus 19:155–66. Botan. Centralbl. 162:217 1932.

15781. Pushkarew, N. I., and T. G. Motrenko. 1927: Zur Frage der Wiedererweckung ruhender Samen von *Melilotus*, *Abutilon avicennae* Gaertn. und anderer Samen. J. Agr. Res. N. Caucasus

10:43–57. Botan. Centralbl. 155:272. 1928.

15782. Putilina, Z. A. 1959: Measures for control of empty-seededness in the sunflower. Sel'skoe Khoz. Povolzhia 1959(11): 69–72.

15783. Putt, E. D. 1943: Association of seed yield and oil content with other characters in the sunflower. Sci. Agr. (Ottawa) 23:377–83.

Putt, E. D. 17322.
Puttemans, A. 19188.
Puutula, K. 15747, 15748.

15784. Pyke, E. E., E. R. Leonard, and C. W. Wardlaw. 1934: On the viability of cacao seeds after storage. Trop. Agr. (Trinidad) 11:303–7.

15785. Pyke, M., R. Melville, and H. Sarson. 1942: Vitamin C in walnuts. Nature 150:267–68.

Pyke, M. 13317.

15786. Pylnev, V. M. 1959: The sowing quality of wheat seeds in relation to their position of formation in the ear. Dokl. Mosk. Sel'skokhoz. Akad. 48:263–67. Biol. Abstr. 40:20624. 1962.

Pylypenkoyurchak, L. D. 9141.

15787. Pyriki, C., and F. W. Kunstmann. 1956: Keimfähigkeit und Gehalt an freien Fettsäuren bei Tabaksamen unter Berücksichtigung der Citronensäure. Pharmazie 11:798–805. Biol. Abstr. 32:10144. 1958.

Pyshkinskaya, O. I. 15642.

Q

15788. Qasem, S. A., and C. M. Christensen. 1958: Influence of moisture content, temperature, and time on the deterioration of stored corn by fungi. Phytopathology 48:544–49.

15789. —— and —— 1960: Influence of various factors on the deterioration of stored corn by fungi. Phytopathology 50:703–9.

Qasem, S. A. 4063.

15790. Quackenbush, F. W., A. S. Carter, L. C. Shenberger, and P. Balbach. 1955: Inspection of agricultural seeds. Purdue Univ. Agr. Expt. Sta. Circ. 418. 63 pp.

15791. ——, J. G. Firch, W. J. Rabourn, M. McQuistan, E. N. Petzold, and T. E. Kargl. 1961: Composition of corn. Analysis of carotenoids in corn grain. J. Agr. Food Chem. 9:132–35.

Quackenbush, F. W. 1279, 8727, 17460, 19613.

15792. Quaglia, A. 1960: Effect of some herbicides on seed germination, and their persistence in the soil. Ann. Sper. Agrar. (Rome) 14:369–83. Chem. Abstr. 57:7665. 1962.

15793. Quartaroli, A. 1952: On some flocculating substances of enormous activity contained in the seed and their physiological functions. Georgofili (Ser. 6) 16:120–30.

15794. Quartley, C. E., and E. R. Turner. 1957: Further experiments on the inhibition of respiration of peas induced by oxygen at high pressures. J. Exptl. Botany 8:250–55.

15795. —— and P. S. Wellington. 1962: Biochemical tests for seed identification. Natl. Inst. Agr. Botany J. 9:179–85.

Quastel, J. H. 1515.
Quastler, H. 17911.

15796. Qudrat-I-Khuda, M., M. E. Ali, and M. Siddiqullah. 1960: A preliminary note on the seed constituents of *Caesalpinia bonducella*. Pakistan J. Sci. Ind. Res. 3:48–51.

15797. ——, A. Khalique, and D. C. Das. 1963: Examination of the constituents of jute seeds. I. Isolation of jute seed oil, strophanthidin and raffinose. Pakistan J. Sci. Ind. Res. 6:158–60.

15798. ——, ——, and —— 1963: Examination of the constituents of jute seeds. II. Identification and characterization of a glycoside and aglycone from the seeds. Pakistan J. Sci. Ind. Res. 6:161–63.

15799. ——, ——, and —— 1963: Examination of the constituents of jute seeds. III. Some new glycosides of jute seeds. Pakistan J. Sci. Ind. Res. 6:169–73.

Quereschit, M. H. 16747.

15800. Quetel, R. 1944: Influence du froid sur le marron d'Inde. Étude des lipides des cotyledons. Bull. Soc. Chim. Biol. 26:393–97.

15801. Quibilan, G. Q. 1939: Phytochemical study of the seeds of "ates," *Anona squamosa* L. Natl. Res. Council Philippines Bull., Proc. 5th Sci. Conv. 23:154. Pharm. Abstr. 7:118. 1941.

Quick, A. S. 15806.

15802. Quick, C. R. 1936: Chemical control of harmful fungi during stratification and germination of seeds of *Ribes roezli*. Phytopathology 26:694–97.

15803. —— 1943: Certain methods of forcing the germination of seeds. Calif. Hort. Soc. J. 4:95–102.
15804. —— 1947: Germination of *Phacelia* seeds. Madrono 9:17–20. Biol. Abstr. 21:18247. 1947.
15805. —— 1956: Viable seeds from the duff and soil of sugar pine forests. Forest Sci. 2:36–42.
15806. —— and A. S. Quick. 1961: Germination of *Ceanothus* seeds. Madrono 16:23–30.
15807. Quilapio, M. P. 1962: Effect of inoculation of seeds on the nitrogen content of certain legumes. Araneta J. Agr. 9:53–65.
15808. Quilichini, R. 1948: Transformation of reserve substances during germination of some leguminous seeds. Compt. Rend. 227:362–63. Chem. Abstr. 43:1081. 1949.
15809. —— 1952: Calcium oxalate in Judas tree seeds. Bull. Soc. Pharm. Bordeaux 90:11–14. Chem. Abstr. 46:6706a. 1952.
Quillet, M. 2861.
Quimby, M. W. 1714.
15810. Quinby, J. R., L. P. Reitz, and H. H. Laude. 1962: Effect of source of seed on productivity of hard red winter wheat. Crop Sci. 2:201–3.
15811. ——, R. B. Stewart, and G. M. Watkins. 1963: Seed treatment studies on sorghum in Texas. Agron. J. 55:550–51.
Quinby, J. R. 10277, 10278.
15812. Quinlivan, B. J. 1961: The effect of constant and fluctuating temperatures on the permeability of the hard seeds of some legume species. Australian J. Agr. Res. 12:1009–22.
15813. —— 1962: The effect of storage in polythene on the development of hard seed in the West Australian blue lupin. Australian J Exptl. Agr. Animal Husbandry 2:209–12.
15814. —— 1962: Hard seeds in lupins. J. Agr. W. Australia 3:683–87, 689–90.
15815. —— and A. J. Millington. 1962: The effect of a Mediterranean summer environment on the permeability of hard seeds of subterranean clover. Australian J. Agr. Res. 13:377–87.
15816. Quinn, V. 1939: How seeds are dispersed. Sci. Dig. 5(1):52–56.
15817. Quintanilha, A., A. Cabral, and L. Quintanilha. 1947: Abnormal female gametophytes in relation with polyembryonic seeds in upland cotton. S. African J. Sci. 43:158–66.
Quintanilha, L. 15817.
15818. Quiros M., M. A. 1945: The sunflower; importance of its cultivation in Costa Rico. Dept. Nacl. Agr. (Costa Rico) Bol. Tec. 50. 23 pp. Chem. Abstr. 40: 2502. 1946.
Qurashi, A. A. 184.
Qurashi, M. [M.] 187–190, 2429.
Qureshi, M. I. 9356, 10514, 10515.
15819. Qvam, O. 1905: Zur Bestimmung des Keimvermögens bei Getreidewaren. Vorschlag zu einer neuen Methode. Landwirtsch. Versuchssta. 62:405–43.
15820. —— 1906: Zur Atmung des Getreides. Eine Relation zwischen Keimfähigkeit und Atmungsintensität. Angew. Botan. 4: 70–87.

R

15821. Raacke, I. D. 1957: Protein synthesis in ripening pea seeds. I. Analysis of whole seeds. Biochem. J. 66:101–10.
15822. —— 1957: Protein synthesis in ripening pea seeds. II. Development of embryos and seed coats. Biochem. J. 66:110–13.
15823. —— 1957: Protein synthesis in ripening pea seeds. III. Study of the pods. Biochem. J. 66:113–16.
15824. Rabak, F. 1935: Safflower, a possible new oil-seed crop for the Northern Great Plains and the far western states. U.S. Dept. Agr. Circ. 366. 15 pp. Expt. Sta. Rec. 74:192. 1936.
15825. —— 1943: The effect of seeds on the quality of hops. Wallerstein Lab. Commun. 6:160–66.
15826. Rabari, L. F., R. D. Patel, and J. G. Chohan. 1961: Studies on germinating peanut seed. J. Oil Chemists Soc. 38:4–5.
15827. Rabate, E., and J. Fleckinger. 1930: Sur une réaction colorée des protéides du grain de blé. Compt. Rend. 190:748–50.
15828. Rabe, F. 1905: Über die Austrocknungsfähigkeit gekeimter Samen und Sporen. Flora (Jena) 95:253–324.
15829. Rabechault, H. 1962: Research on culture "in vitro" of embryos of African oilpalm (*Elaeis guineensis* Jacq.). I. Effects of β-indoleacetic acid. Oléagineux 17:757–64.

15830. Rabega, C., R. Stanescu, and M. Rabega. 1956: The penetration of water into the seeds of sugar beets Buszesynschi CIR and feed beets KLN upon stimulation with chemicals. Analele Univ. "C. I. Parhon" Ser. Stiint. Nat. 12:169–74. Chem. Abstr. 52:18689. 1958.

Rabega, M. 15830.

15831. Rabideau, G. S. 1954: Amino acids in the embryo and endosperm of the grain of different varieties of corn. Botan. Gaz. 115:391–94.

15832. Rabotnov, T. A. 1958: In regard to methods of studying the content of germinating seeds in meadow soils. Botan. Zh. 43:1572–81.

15833. —— 1961: Determining the number of live seeds, found in the soil and on its surface, in pastures and meadows. *In* N. S. Konyushkov, T. A. Rabatnov, and I. A. Tsatsenkin, editors. Metodika opytnykh rabot na senokosakh i pastbishchakh, pp. 119–23.

Rabotnov, T. A. 416.

Rabourn, W. J. 15791.

15834. Rabson, R., R. J. Mans, and G. D. Novelli. 1961: Changes in cell-free amino acid incorporating activity during maturation of maize kernels. Arch. Biochem. Biophys. 93:555–62.

15835. —— and G. D. Novelli. 1960: The incorporation of leucine-C^{14} into protein by a cell-free preparation from maize kernels. Proc. Natl. Acad. Sci. U.S. 46:484–88.

Rachkova, D. A. 11007.

15836. Raciszewski, Z. M., E. Y. Spencer, and L. W. Trevoy. 1955: Chemical studies of a goitrogenic factor in rapeseed oilmeal. Can. J. Technol. 33:129–33.

15837. Racovitza, A. 1959: Nouvelles observations sur la germination des graines d'*Orobanche ramosa* L. J. Agr. Trop. Botan. Appl. 6:111–14.

Radej, N. V. 4833.

15838. Rademacher, B. 1940: Über den antagonistischen Einfluss von Roggen und Weizen auf Keimung und Entwicklung mancher Unkräuter. Pflanzenbau 17: 131–43.

15839. —— 1955: Über den Einfluss von Kalteperioden auf die 2,4-D-Wirkung beim Hafer. Z. Pflanzenkrankh. Pflanzenschutz 62:166–74.

15840. —— and J. Ozolins. 1952: Einfluss der Getreidekonkurrenz und des Nährstoffgehalts im Keimsubstrat auf Keimung und Jugendentwicklung verschiedener Unkräuter. Angew. Botan. 26:69–93. Biol. Abstr. 26:32128. 1952.

Rademacher, B. 2689.

15841. Rader, W. E., H. F. Fitzpatrick, and E. M. Hildebrand. 1947: A seed-borne virus of muskmelon. Phytopathology 37:809–16.

15842. Radersma, S. C. 1959: The seed collection of the Government Seed Testing Station. Zaadbelangen 13:460–61.

15843. —— 1962: Comparison of western European routine methods for determination of polyploidy of beet seeds. Proc. Intern. Seed Testing Assoc. 27:592–98.

15844. Radhakrishnan, A. N., and C. S. Vaidyanathan. 1954: On the free amino acid composition of *Phaseolus radiatus* (green gram). Naturwissenschaften 41: 432–33.

15845. ——, ——, and K. V. Giri. 1955: Nitrogenous constituents in plants. I. Free amino acids in leaves and leguminous seeds. J. Indian Inst. Sci. 37:178–94.

Radhakrishnan, C. 14453.

15846. Radionenko, A. Y. 1963: On physiologico-biochemical changes in the period of embryonic development of the apricot. Acad. Sci. RSS Ukraine J. Botan. 20:25–32.

15847. Radko, A. M. 1955: Zoysia seed storage and germination tests. Agron. Abstr. 47:75.

15848. Radley, F. 1936: To promote cactus growing from seed. Florists Exchange Hort. Trade World 87(11):18.

15849. Radley, M. 1959: The occurrence of gibberellin-like substances in barley and malt. Chem. Ind. 27:877–78.

15850. —— 1963: Gibberellin content of spinach in relation to photoperiod. Ann. Botany (London) 27:373–77.

15851. —— and E. Dear. 1958: Occurrence of gibberellin-like substances in the coconut. Nature 182:1098.

Radley, M. 9940.

15852. Radoeff, A. 1932: Stimulation de la croissance par des sels minéraux, des colorants vitaux, et divers composés organiques, chez le riz (*Oryza sativa*). Compt. Rend. Soc. Biol. 110:955–58. Biol. Abstr. 8:7944. 1934.

Radoeff, A. 5617, 5618.

15853. Radu, I. F. 1932: Beitrag zur Beurteilung der Keimfähigkeit der harten Körner und zur Festsetzung der Dauer der Keimversuche von *Trifolium*- und *Medicago*-Samen. Proc. Intern. Seed Testing Assoc. 4:41–45.

15854. Radu, V. V. 1933: Germination des graines de *Vicia faba* à des temperatures basses et elevées (étude cytologique). Compt. Rend. Soc. Biol. 114:69–72.

15855. Radulescu, E. 1957: Increase in the resistance of plants to cold after treating the seeds with variable temperatures. Studii Cercetari Agron. 8:203–8. Biol. Abstr. 35:16984. 1960.

15856. —— 1957: The influence of treating seed with variable temperatures on the growth and development of corn. Studii Cercetari Agron. 8:209–26. Biol. Abstr. 35:17025. 1960.

15857. Radzievskii, G. B., and I. L. Shekhtman. 1955: Application of X-ray analysis to the study of ice formation in plant seeds. Dokl. Akad. Nauk SSSR 101:1051–53. Chem. Abstr. 50:443. 1956.

15858. Rae, J. 1943: Mucilage of senna seeds. Pharm. J. 151:241. Pharm. Abstr. 10:143. 1944.

15859. Raeber, J. G. 1958: Handling and storage of tobacco seed. Rhodesian Tobacco 16:27–29.

15860. Räder-Roitzsch, J. E. 1957: Quantitative Stoffänderungen bei der Keimung von *Pinus silvestris* L. Forstwiss. Centralbl. 76:229.

15861. —— 1958: Untersuchungen über die Stoffwanderung im Keimungsverlauf von *Quercus pedunculata* Ehrh. Forstwiss. Centralbl. 77:109–24.

Rafaat, A. 14047, 14048.

15862. Raffe, M. 1917: Studies on the seed of *Spartium junceum*. Ann. Chim. Appl. (Rome) 7:157–64. Expt. Sta. Rec. 37:710. 1917.

Raffy, A. 11964.

15863. Ragai, H., and W. E. Loomis. 1954: Respiration of maize grain. Plant Physiol. 29:49–55.

15864. Ragavan, K. 1960: Potential vivipary in *Saccarum spontaneum* and hybrid sugarcane. Sci. Cult. (Calcutta) 26:129–30. Biol. Abstr. 36:25123. 1961.

15865. Raggio, M., and N. M. De Raggio. 1950: Ensayo de viabilidad de semillas con cloruro de 2,3,5-trifeniltetrazol. La Plata Univ. Nacl. Fac. Agron. Lab. Botan. Bull. 3:1–2.

15866. —— and —— 1952: Wheat germination as affected by soaking the grains in zinc sulphate, boric acid, and ferric alum. Phyton 2:47–49. Brit. Abstr. B III:111. 1953.

Raggio, N. M. de. 15865, 15866.

15867. Raghavan, V., and J. G. Torrey. 1963: Growth and morphogenesis of globular and older embryos of *Capsella* in culture. Am. J. Botany 50:540–51.

15868. —— and —— 1963: Inorganic nitrogen nutrition of the embryos of the orchid, *Cattleya*. [Abstract.] Am. J. Botany 50:617.

15869. Ragland, C. H. 1934: The development of the peach fruit, with special reference to split-pit and gumming. Proc. Am. Soc. Hort. Sci. 31:1–21.

Ragsdale, A. C. 8661.

15870. Rahman, A. U., and M. S. Khan. 1961: Chemical investigation of oil of the seed of "White Todri" (*Matthiola incana* R. Br.). J. Am. Oil Chemists Soc. 38: 281–82.

15871. Rahman, S. A., S. S. De, and V. Subrahmanyan. 1947: Effect of different treatments on the removal of hydrocyanic acid from the Burma bean (*Phaseolus lunatus* Linn.). Current Sci. (India) 16: 351–52. Biol. Abstr. 23:16868. 1949.

Rahman, W. 6229, 6233.

15872. Rahn, E. M. 1945: The effect of depth of seeding on yields of cantaloupes. Peninsula Hort. Soc. Trans. 59:63–64.

15873. —— 1955: The effect of certain cultural and growth regulator treatments on pod-set and yield of lima beans. Proc. Am. Soc. Hort. Sci. 66:298–307.

Rahn, E. M. 5281.

Rahn, O. 8483.

Rai, V. K. 11692, 11693.

15874. Raianu, M. 1957: La germination des semences de sainfoin et de luzerne à fleur bleue en raison de la phase de maturation. Rumania Inst. Cercetari Agron. Ann. 24:573–91. 1956.

15875. ——, E. Paun, and A. Stefanescu. 1962: La germination des semences de cumin, en fonction de la phase de maturation. Inst. Cerc. Pentru Cérèales si Plante Tehnice-Fundulea Ser. C 30:215–21.

Raianu, M. 638, 639.

15876. Raimundo, J. R. 1954: Characteristics of the seed and the new plant of some living grasses. Melhoramento 7:89–137.

15877. Raines, M. A. 1931: A dilatometer for measuring the swelling of seeds. Science 74:392–93.

Raison, J. K. 7567, 7569.

Rajabhooshanam, D. S. 1585.

Rajagopalan, R. 11414, 11415.

15878. Rajagopalan, S., C. Tamm, and T. Reichstein. 1955: Die Glykoside der

Samen von *Calotropis procera* R. Br. Glykoside und Aglykone. Helv. Chim. Acta 38:1809–30.

Rajagopalan, T. 10297.

Rajan, A. K. 11515.

Rajan, B. S. Varada-; *see* Varada-Rajan, B. S.

15879. Raju, M. V. S. 1956: Development of embryo and seed coat in *Turnera ulmifolia* L. var. *angustifolia* Willd. Botan. Notiser 109:308–12. Biol. Abstr. 31: 29211. 1957.

15880. —— 1956: Embryology of the Passifloraceae. I. Gametogenesis and seed development of *Passiflora calcarata* Mast. J. Indian Botan. Soc. 35:126–38.

Raju, M. V. S. 10355.

15881. Rakitin, J. V. 1945: Relation between seeds and pericarp in the process of fruit growing and ripening. Compt. Rend. Acad. Sci. URSS 47:590–92.

15882. Raleigh, G. J. 1930: Chemical conditions in maturation, dormancy, and germination of seeds of *Gymnocladus dioica*. Botan. Gaz. 89:273–94.

15883. —— 1943: The germination of dormant lettuce seed. Science 98:538.

15884. Rall, I. S. 1959: Effect of warming corn seeds on their germination and certain biochemical properties. Mosk. Ordena Lenina Sel'skokhoz. Akad. Im. K. A. Timiryazeva Dokl. 46:39–45.

15885. —— 1962: Effect of sluggish molding processes on germination and biochemical properties of stored corn seeds. Mosk. Ordena Lenina Sel'skokhoz. Akad. Im. K. A. Timiryazeva Dokl. 77:437–42.

15886. —— 1963: Effect of low temperature on germination and biochemical processes in the seeds of corn. Dokl. Mosk. Sel'skokhoz. Akad. 1961(62):491–96. Chem. Abstr. 58:9415. 1963.

15887. Ralski, E. 1924: Les corps gras dans les graines des Graminées. Kosmos Bull. Soc. Polon. Nat. "Kopernik" 49:62–99.

15888. Ram, H. W. M. 1959: Endosperm in *Eranthemum nervosum*. Current Sci. (India) 28:169–70.

15889. —— 1960: The development of the seed in *Andrographis serpyllifolia*. Am. J. Botany 47:215–19.

15890. —— 1961: Seed development in *Adhatoda vasica* Nees. Can. J. Botany 39:207–14.

15891. —— and A. Satsangi. 1963: Induction of cell divisions in the mature endosperm of *Ricinus communis* during germination. Current Sci. (India) 32:28–30.

15892. Ram, K. 1930: Studies in Indian oil seeds. IV. The types of *Sesamum indicum*, D. C. Mem. Dept. Agr. India Botan. Ser. 18:127–47. Biol. Abstr. 6:17074. 1932.

15893. Ram, M. 1960: Occurrence of endosperm haustorium in *Cannabis sativa* L. Ann. Botany (London) 24:79–82.

Ramabhadran, G. 14214.

Ramachandran, K. 55.

15894. Ramachandran, L. K., and K. K. Rao. 1963: The occurrence of putrescine and a new guanidino amino acid in seeds of *Lathyrus tingitanus*. Biochem. Biophys. Res. Commun. 13:49–53. Chem. Abstr. 60:850. 1964.

Ramachandran, L. K. 15968.

Ramachandran, M. 1763.

15895. Ramachandran, S. R., and K. Venkataraman. 1938: The seed husk of *Plantago ovata* in printing and finishing. Soc. Dyers Colourists 54:462–64.

15896. Ramage, R. T., and A. D. Day. 1960: Separation of trisomic and diploid barley seeds produced by interchange heterozygotes. Agron. J. 52:590–91.

15897. Ramakrishnan, C. V. 1954: Lipase and esterase activities in the groundnut seed during germination. Sci. Cult. (Calcutta) 19:566–67. Chem. Abstr. 48: 12253. 1954.

15898. —— 1957: Amino acid composition of crude and germinated guarseed flour protein (*Cyamopsis psoralioides*). Experientia 13:78. Chem. Abstr. 51:13251. 1957.

15899. —— and B. N. Banerjee. 1951: Lipase from oil seeds. XIV. Effect of some water-soluble vitamins on the activity of ricinus lipase from castor seeds. J. Indian Chem. Soc. 28:471–72. Chem. Abstr. 47:640. 1953.

Ramakrishnan, C. V. 13295.

Ramakrishnan, T. 16788.

Ramakrishnan, T. N. 11415.

15900. Ramakrishnan, V. 1960: A note on the occurrence of poly- and schizocotyly in *Ocimum sanctum* Linn. and *Myrtus communis* Linn. Sci. Cult. (Calcutta) 26: 235–36.

15901. Raman, K. R. 1956: A note on seed production in bittergourd. S. Indian Hort. 4:26–27.

15902. —— 1956: A note on the germination of seeds of jack. S. Indian Hort. 4:87–88.

15903. Raman, P. S. 1954: The nature of the oxygenated acid from the seeds of

Vernonia anthelmintica (Willd.). Current Sci. (India) 23:293–94.
Ramanathan, T. 214.
Ramanujam, S. 9594, 15910.

15904. Ramasarma, T., and L. R. Wetter. 1957: Choline kinase of rapeseed (*Brassica campestris* L.). Can. J. Biochem. Physiol. 35:853–63.

15905. Ramaswami, K. 1935: The relation between the size of seed and the development of the plant resulting from it in rice. Madras Agr. J. 23:240. Plant Breeding Abstr. 6:16. 1935.

15906. Ramaswami, M. N. 1952: Pretreatment of seeds with coconut water to hasten germination. Indian Forester 75: 500–1. 1949.
Ramaswamy, R. 4844–4846.

15907. Ramberg, S. 1934: Cone collection and seed extraction. Tidsskr. Skogbruk 42:106–16. Biol. Abstr. 9:1439. 1935.
Ramdas, K. 4846.

15908. Ramiah, K., and C. R. Mudaliar. 1937: Colours in the rice grain. Indian J. Agr. Sci. 7:863–71. Agr. Livestock India 8:76. 1938.

15909. —— and —— 1939: The proteins in the rice grain and the development of the aleurone layer. Indian J. Agr. Sci. 9:39–53. Chem. Abstr. 33:5027. 1939.

15910. ——, N. Parthasarathi, and S. Ramanujam. 1935: Polyembryony in rice (*Oryza sativa*). Indian J. Agr. Sci. 5:119–24. Biol. Abstr. 11:17870. 1937.
Ramiah, K. 46.

15911. Ramirez, O. D. 1961: Effects of gibberellic acid on germination of papaya (*Carica papaya* L.) seed. J. Agr. Univ. Puerto Rico 45:188–90.
Ramirez-Silva, F. J. 9027.

15912. Rampton, H. H., and R. G. Petersen. 1962: Relative efficiency of plot sizes and numbers of replications as indicated by yields of orchardgrass seed in a uniformity test. Agron. J. 54:247–49.

15913. Ramsbottom, J. 1922: The germination of orchid seed. Orchid Rev. 30(349): 197–202.
Ramser, J. H. 7051.
Ramshorst, J. D. van; *see* Van Ramshorst, J. D.

15914. Ramstad, E. 1953: A note on the isolation of furan-3-carbonic acid from the seeds of *Euonymus europaea* L. J. Am. Pharm. Assoc. 42:119.

15915. Ramstad, P. E., and W. F. Geddes. 1942: The relative respiratory rates and hygroscopic equilibria of common and hull-less oats. J. Agr. Res. 64:237–41.

15916. —— and —— 1942: The respiration and storage behavior of soybeans. Minn. Agr. Expt. Sta. Tech. Bull. 156:54. Expt. Sta. Rec. 88:431. 1943.
Rana, N. G. 13292.

15917. Rancken, M. 1937: Der Einfluss von Feuchtigkeit und Temperatur auf die harten Kleesamen während der Aufbewahrungszeit. Proc. Intern. Seed Testing Assoc. 9:263–71.

15918. Randall, C. G. 1950: Germination studies of cotton seed from plants affected by 2,4-D spray drift. Calif. Dept. Agr. Bull. 39:144–45.

15919. Randall, T. E., and C. M. Rick. 1945: A cytogenetic study of polyembryony in *Asparagus officinalis* L. Am. J. Botany 32:560–69.
Randall, T. E. 11621.

15920. Randhawa, S. S., and B. S. Bajwa. 1958: Embryo counts and their germination in important polyembryonic citrus rootstocks. Indian J. Hort. 15:61–65.
Randolph, F. R. 15926.

15921. Randolph, L. F. 1936: Developmental morphology of the caryopsis in maize. J. Agr. Res. 53:881–916.

15922. —— and L. G. Cox. 1943: Factors influencing the germination of iris seed and the relation of inhibiting substances to embryo dormancy. Proc. Am. Soc. Hort. Sci. 43:284–300.

15923. —— and H. E. Fischer. 1939: The occurrence of parthenogenetic diploids in tetraploid maize. Proc. Natl. Acad. Sci. U.S. 25:161–64.

15924. —— and D. B. Hand. 1938: Increase in vitamin A activity of corn caused by doubling the number of chromosomes. Science 87:442–43.

15925. ——, A. E. Longley, and C. H. Li. 1948: Cytogenic effects in corn exposed to atomic bomb ionizing radiation at Bikini. Science 108:13–15. Biol. Abstr. 22:22991. 1948.

15926. —— and F. R. Randolph. 1955: Embryo culture of iris seed. Am. Iris Soc. Bull. 139:7–17.
Randolph, L. F. 5913, 10513, 11189.
Randolph, M. L. 4335.
Rane, M. S. 2425.
Rangachari, P. N. 4847.
Rangaswami, G. 17265.

15927. Rangaswami, S. 1963: Crystalline chemical components of the seeds of

Cassia tora Linn. Identity of tora substance C with rubrofusarin and tora substance B with non-rubrofusarin. Proc. Indian Acad. Sci. 57:88–93. Biol. Abstr. 44:8197. 1963.

15928. —— and E. V. Rao. 1958: New crystalline glycosides from seeds of *Thevetia nereifolia.* J. Sci. Ind. Res. (India) 17B: 331–32. Chem. Abstr. 53:9373. 1959.

15929. —— and V. S. Rao. 1954: Occurrence of a sulfur-containing glycoside in the seeds of *Entada scandens.* Indian J. Pharm. 16:152–54. Chem. Abstr. 49: 7660. 1955.

15930. ——, T. Reichstein, O. Schindler, and T. R. Seshadri. 1953: Die Glykoside der Samen von *Strophanthus wightianus* Wall. Helv. Chim. Acta 36:1282–95.

15931. —— and B. V. R. Sastry. 1956: Chemical components of the seeds of *Tephrosia candida.* Indian J. Pharm. 18:333–34. Chem. Abstr. 52:12105. 1958.

15932. —— and M. Suryanarayana. 1955: Occurrence of lycorine in the seeds of *Crinum latifolium.* Indian J. Pharm. 17:229–30. Chem. Abstr. 50:7404. 1956.

Rangaswami, S. 12842, 14202, 16682.

Rankin, E. E. 6076.

Rankin, G. T. 17319.

Rankin, H. W. 1951.

Ranney, C. D. 7959.

15933. Ransom, E. R. 1935: The interrelations of catalase, respiration, afterripening, and germination in some dormant seeds of the Polygonaceae. Am. J. Botany 22:815–25.

15934. Ransom, F. 1912: Effect of caffein upon the germination and growth of seeds. Biochem. J. 6:151–61. Expt. Sta. Rec. 27:330. 1912.

Rantzan, V. A. 16231.

15935. Rao, A. M. Subba-. 1940: Studies in the Malpighiaceae. I. Embryo sac development and embryogeny in the genera Hiptage, Banisteria and Stigmatophyllum. J. Indian Botan. Soc. 18:145–56. Biol. Abstr. 15:19328. 1941.

15936. Rao, A. N. 1953: Embryology of *Shorea talura* Roxb. Phytomorphology 3:476–84. Biol. Abstr. 31:22214. 1957.

15937. —— 1957: A contribution to the embryology of Dilleniaceae. Proc. Iowa Acad. Sci. 64:172–76.

15938. —— and P. N. Avadhani. 1963: Effect of chlorox on the germination of *Vanda* seeds. Current Sci. (India) 32: 467–68.

Rao, B. R. S. 2100.

15939. Rao, B. S. 1937: Morellin, a constituent of the seeds of *Garcinia morella.* J. Chem. Soc. 1937:853–57.

15940. Rao, C. J. D., and S. S. Subramanian. 1947: Chemical examination of the seeds of *Derris scandens.* Current Sci. (India) 16:346–47.

Rao, C. S. 15969.

15941. Rao, C. V. 1949: Contributions to the embryology of Sterculiaceae. I. J. Indian Botan. Soc. 28:180–97.

15942. —— 1953: Contributions to the embryology of Sterculiaceae. V. J. Indian Botan. Soc. 32:208–38.

15943. —— 1955: Embryological studies in Malvaceae. II. Fertilization and seed development. Proc. Natl. Inst. Sci. India Pt. B 21:53–67.

15944. —— 1955: Endosperm and seed development in *Chrysalidocarpus lutescens* H. Wendel. Current Sci. (India) 24: 381–82.

15945. —— 1962: Morphology and embryology of *Lomatia* Br. with a discussion on its probable origin. Symp. Plant Embryol., Univ. Delhi, 1960. Papers, pp. 261–73.

15946. ——, M. N. Rao, and A. Venkatswarlu. 1943: Oil from the seeds of *Bombax malabaricum.* J. Indian Chem. Soc. 20: 405–6. Analyst 69:312. 1944.

15946A. —— and S. R. Rao. 1954: Embryology of *Cryptostegia grandiflora* R. Br. and *Caralluma attenuata* Wt. J. Indian Botan. Soc. 33:453–72.

Rao, D. N. 13327.

15947. Rao, D. S. 1958: Problems of storage of grains. Mysore Agr. J. 33: 209–16. Biol. Abstr. 33:46609. 1959.

Rao, D. S. R. 11521.

Rao, E. V. 15928.

15948. Rao, G. V. S. 1959: Oil from the seeds of *Acacia decurrens* Willd. Current Sci. (India) 28:410.

15949. Rao, H. S. 1936: Germination of the seeds of *Carica papaya* inside the fruit. Current Sci. (India) 5:81–83.

Rao, I. K. S. 15975.

Rao, I. M. 2407–2409.

Rao, J. M. 11551.

15949A. Rao, J. S. 1957: Polyembryony in *Phoenix dactylifera* L. Sci. Cult. (Calcutta) 22:686–87. Biol. Abstr. 32:20839. 1958.

Rao, J. T. 19113.

Rao, J. V. 15959.

Rao, K. H. 14238.

Rao, K. K. 15984.

Rao, K. K. P. N. 2394.

15950. Rao, K. R. N. 1953: A note on the absorption and retention of HCN by oil seeds. Madras Agr. J. 40:279–81. Biol. Abstr. 28:11821. 1954.

Rao, K. S. S. 5698.

15951. Rao, K. V. R. 1940: Gametogenesis and embryogeny in five species of the Convolvulaceae. J. Indian Botan. Soc. 19:53–69.

15952. Rao, L. 1925: Quantitative Untersuchungen über die Wirkung des Lichtes auf die Samenkeimung von *Lythrum salicaria*. Jahrb. Wiss. Botan. 64:249–80.

Rao, L. 12048.

15953. Rao, L. N. 1963: Life-history of *Cycas circinalis* L. II. Fertilization, embryogeny and germination of the seed. J. Indian Botan. Soc. 42:319–32.

15954. Rao, L. S. P., T. P. Rao, and E. S. Narayanan. 1963: Response of mulberry seeds to gibberellic acid treatment. Current Sci. (India) 32:348–49.

Rao, M. B. S. 9840.

Rao, M. B. V. N. 16630, 17708.

15955. Rao, M. N., T. Viswanatha, P. B. Mathur, M. Swaminathan, and V. Subrahmanyan. 1954: Effect of storage on the chemical composition of husked, undermilled and milled rice. J. Sci. Food Agr. 5:405–9.

Rao, M. N. 15946.

Rao, M. P. 4929.

Rao, M. S. S. 12810.

15956. Rao, M. V. L., N. Subramanian, and M. Srinivasan. 1955: Nutritive value of bamboo seeds (*Bambusa arundinacea*, Willd.). Current Sci. (India) 24:157–58.

Rao, M. V. L. 16807.

Rao, M. V. R. 5238, 18126.

15957. Rao, N. A., H. R. Cama, and S. A. Kumar. 1959: The biosynthesis of new derivatives of riboflavine by seed extracts. J. Sci. Ind. Res. (India) 18C:181–82. Chem. Abstr. 54:7834. 1960.

15958. ——, ——, and —— 1961: Studies on the occurrence, changes during germination and distribution of the flavin nucleotide coenzymes in seeds. J. Indian Inst. Sci. 43:1–9.

15959. Rao, N. V. S., and J. V. Rao. 1941: Glabrine, a new component of the seeds of *Pongamia glabra*. Proc. Indian Acad. Sci. Sect. A 14:123–25. Chem. Abstr. 36:1592. 1942.

Rao, N. V. S. 13979.

Rao, P. B. 15969, 15971.

15960. Rao, P. G. 1955: A rapid method for studying the germination of the seeds of the root parasite *Orobanche cernua* Loefl. var. *desertorum* (Beck). Sci. Cult. (Calcutta) 21:258–61. Biol. Abstr. 31: 39541. 1957.

Rao, P. L. N. 7462.

Rao, P. N. 19062.

Rao, P. R. 3861, 14240.

15961. Rao, P. R. M. 1963: Suspensor polyembryony in *Garrya veatchii* Kell. Current Sci. (India) 32:468–69.

15962. Rao, P. S. 1948: The coloring matter of the tamarind seed testa. Proc. Indian Acad. Sci. Sect. A 27:52–55. Biol. Abstr. 23:12700. 1949.

15963. Rao, P. S. 1957: Tamarind seed jellose: A new class of neutral polysaccharides. J. Sci. Ind. Res. (India) 16A: 138–40.

15964. —— and S. Krishna. 1946: Tamarind seed pectin. Current Sci. (India) 15:168.

15965. Rao, P. S. P. 1963: Some embryological observations of *Guiera senegalinses* Lam. Current Sci. (India) 32:30–31.

15966. Rao, S. D. T., and K. S. Murti. 1960: Compositional studies of Indian peanuts. II. South Indian peanuts. Indian Oilseeds J. 4:4–14. Chem. Abstr. 59:3084. 1963.

15967. Rao, S. L. N., P. R. Adiga, and P. S. Sarma. 1964: The isolation and characterization of β-N-oxalyl-L-α, β-diaminopropionic acid: A neurotoxin from the seeds of *Lathyrus sativus*. Biochemistry 3:432–36.

15968. ——, L. K. Ramachandran, and P. R. Adiga. 1963: The isolation and characterization of L-homoarginine from seeds of *Lathyrus sativus*. Biochemistry 2:298–300.

Rao, S. L. N. 115.

15969. Rao, S. N., P. B. Rao, and C. S. Rao. 1961: Preliminary studies on the influence of gibberellic acid on fruit set, size, weight, seed content and quality of several citrus species. Andhra Agr. J. 8:203–13.

15970. —— and R. N. Reddy. 1962: Effects of gibberellic acid on fruit set, size, weight, seed content, and quality of grapes. Andhra Agr. J. 9:166–74. Chem. Abstr. 58:6132. 1963.

15971. ——, ——, and P. B. Rao. 1961: The effects of gibberellic acid on fruit set, size and quality in the seedless varieties of guava (*Psidium guajava* L.). Andhra Agr. J. 8:257–68.

Rao, S. R. 15946A.

Rao, S. V. 9920.

Rao, T. P. 15954.

15972. Rao, U. S. K., and B. L. Manjunath. 1935: The supposed occurrence of acids with uneven number of carbon atoms in vegetable oils and fats. II. Acid fraction of mean molecular weight 354, from the seeds of *Butea frondosa* (Roxb.). J. Indian Chem. Soc. 12:611–15. Chem. Abstr. 30:313–14. 1936.

Rao, V. L. N. 7214.

15973. Rao, V. N. M. 1953: Certain aspects of harvest in relation to crop performance in Bendi (*Hibiscus esculentus* L.). Madras Agr. J. 40:437–44. Biol. Abstr. 28:26974. 1954.

15974. —— and M. V. Hassan. 1956: Variations in the seed characters of cashew (*Anacardium occidentale* L.). Indian J. Agr. Sci. 26:211–16.

15975. ——, I. K. Rao, and M. V. Hassan. 1957: Studies on seed viability in cashew. Indian J. Agr. Sci. 27:289–94. Biol. Abstr. 33:19403. 1959.

Rao, V. P. 1582–1584.

Rao, V. R. 10012, 10013.

15976. Rao, V. S. 1936: Studies on Capparidaceae. I. The embryo-sac of *Maerua arenaria* Forsk. J. Indian Botan. Soc. 15:71–75.

15977. —— 1938: The correlation between embryo type and endosperm type. Ann. Botany (London) 2:535–36.

15978. —— 1938: Embryogeny of *Capparis*. (Studies on Capparidaceae. III. Genus *Capparis*.) J. Indian Botan. Soc. 17:69–80. J. Roy. Microscop. Soc. 59:53. 1939.

15979. —— 1944: Development of the embryo-sac in the Convolvulaceae. J. Indian Botan. Soc. 23:165–69.

15980. —— 1959: "Nuclear endosperm" or "non-cellular endosperm"? Ann. Botany (London) 23:364.

Rao, V. S. 15929.

Rao, V. V. R. 11418.

15981. Raoul, Y. 1937: Comparative development of hordenine in barley during germination. Ann. Ferment. 3:385–405. Chem. Abstr. 32:7072–73. 1938.

Raper, R. 4196.

15982. Rapin, J., and J. Caputa. 1952: Prescriptions concernant l'admission, le controle et le commerce des semences de trèfle violet (*Trifolium pratense*). Rev. Romande Agr. Viticult. Arboricult. 8: 59–61. Biol. Abstr. 27:25965. 1953.

Rapin, J. 3591, 9407.

15983. Rapp, A. E. 1942: How to make wildflowers grow. Gardeners Chron. 46: 40, 58.

15984. Rapp, K. E., J. T. Skinner, and J. S. McHargue. 1946: The nutritive value of the protein in tobacco-seed oilmeal. J. Nutr. 31:261–71.

15985. Rappaport, J. 1954: In vitro culture of plant embryos and factors controlling their growth. Botan. Rev. 20:201–25.

15986. ——, S. Satina, and A. F. Blakeslee. 1950: Extracts of ovular tumors and their inhibition of embryo growth in *Datura*. Am. J. Botany 37:586–95.

Rappaport, J. 16793.

15987. Rappaport, L., and S. H. Wittwer. 1956: Flowering in head lettuce as influenced by seed vernalization, photoperiod and temperature. Proc. Am. Soc. Hort. Sci. 67:416–20.

15988. —— and —— 1956: Stimulation of flowering by vernalization of endive seedlings. Preliminary report. Proc. Am. Soc. Hort. Sci. 67:438–39.

Rappaport, M. Kamson-; *see* Kamson-Rappaport, M.

Rappleye, R. D. 13743, 13744.

Rashevskaya, V. F. 14489.

15989. Raskatov, P. B., and D. P. Viktorov. 1953: Anatomo-biochemical characteristics of seeds and fruits of the holly maple and ash-type maple. Nauchn. Zap. Voronezhsk. Lesokhoz. Inst. 12:79–94. Chem. Abstr. 49:6395. 1955.

Raskina, R. 12420.

15990. Rasmerita, I. 1955: Practical guide for producing and gathering seeds of forage grasses. Acad. Rep. Populare Romine, Bucharest. 117 pp.

Rasmussen, L. W. 3210–3212.

15991. Rasmusson, K. 1959: Processed and genetically monogerm seeds and their uses in sugar beet cultivation. Kgl. Skogs-Lantbruksakad. Tidskr. 98:225–38.

Raspet, M. 5167.

15992. Rasten, J. 1950: Hot water treatment of barley and spring wheat against naked smut. Tidsskr. Norske Landbruk 57: 113–21.

Rasten, J. 12232.

15993. Rastveit, K., and E. Strand. 1954/55: Investigations on the respiration rate of small grain seeds of different moisture content and harvested at different stages of maturity. Acta Agr. Scand. 5:76–84.

Ratcliff, J. A. 10591.

15994. Rathbun-Gravatt, A. 1931: Germination loss of coniferous seeds due to parasites. J. Agr. Res. 42:71–92.

15995. Rather, H. C. 1937: Influence of delayed harvest on certain varieties of oats and barley in Michigan. Mich. Agr. Expt. Sta. Quart. Bull. 19:241–46.

15996. —— and A. R. Marston. 1940: A study of corn maturity. Mich. Agr. Expt. Sta. Quart. Bull. 22:278–88. Expt. Sta. Rec. 83:483–84. 1940.

Rather, H. C. 10996.

Rathmann, D. M. 5884.

Ratliffe, G. T. 19444.

Ratsimamanga, A. R. 2711, 2712.

15997. Ratt, A. 1934: Alle hartschaligen Rotkleesamen sind nicht gleichwertig. Proc. Intern. Seed Testing Assoc. 6:318–30.

15998. —— 1938: On the qualities of hulled Timothy seed. Proc. Intern. Seed Testing Assoc. 10:265–74.

15999. —— 1939: Über den Wert der entspelzten Gramineensamen. Pflanzenbau 15:299–304.

16000. —— 1941: Leinsaat von estnsicher Herkunft. Proc. Intern. Seed Testing Assoc. 12:150–66.

16001. Rau, M. A. 1940: An embryological study of *Suriana maritima* Linn. Proc. Indian Acad. Sci. Sect. B 11:100–6.

16002. —— 1942: Development of the embryo-sac and embryo in *Streblus asper* Lour. Mysore Univ. Half-Yearly J. Sect. B 2:109–14. Biol. Abstr. 17:24766. 1943.

16003. —— 1950: Development of the embryo in *Trigonella foenumgraeum* L. J. Indian Botan. Soc. 29:210–13.

16004. —— 1951: Development of the embryo in *Aeschynomene indica* Linn. New Phytologist 50:124–26.

16005. —— 1951: The endosperm in some of the Papilionaceae. Phytomorphology 1:153–58. Biol. Abstr. 27:4543. 1953.

16006. —— 1951: The endosperm in some species of *Cassia* L. Svensk Botan. Tidskr. 45:516–22.

16007. —— 1954: The endosperm in the Leguminosae. Congr. Intern. Botan. Paris Rappt. Commun. 8:249–50.

16008. —— 1954: Embryogeny of the Papilionaceae: development of the embryo in species of *Desmodium*, *Lespedeza* and *Cyamopsis*. Congr. Intern. Botan. Paris Rappt. Commun. 8:256. Biol. Abstr. 29:26871. 1955.

16009. —— 1956: The growth in vitro of excised ovaries. I. Influence of colchicine on the embryo and endosperm in *Phlox drummondii*. Phytomorphology 6:90–96. Chem. Abstr. 50:13182. 1956.

16010. —— 1963: Review of recent work on the embryogeny of some families and genera of disputed systematic position. Symp. Plant Embryol., Univ. Delhi, 1960. Papers, pp. 75–80.

16011. Rau, R. H. R., and V. Subrahmanyan. 1934: Influence of aeration on the diastatic activity of barley during steeping. Proc. Indian Acad. Sci. Sect. B 1:1–9. Biol. Abstr. 9:13385. 1935.

Rau, Y. V. S. 9351.

Rauchfuss, F. 17632.

Rauen, H. 19273.

16012. Raure, R. 1954: The effect of repellents and adhesives on germination of corn seed. Phytiat.-Phytopharm. 3:153–58. Chem. Abstr. 49:11788. 1955.

16013. Rauser, V., and A. Trnka. 1959: Untersuchungen über die durch Ultraschall hervorgerufenen Sofortveränderungen in der Zelle. Biol. Plantarum 1:63–70. Biol. Abstr. 35:45633. 1960.

16014. Raut, V. S., and R. G. Chitre. 1962: Effect of germination and storage on the thiamine content of cereals and pulses. J. Postgraduate Med. (Bombay) 8:67–70. Chem. Abstr. 57:11555. 1962.

Raut, V. S. 3995.

16015. Raw, A. R. 1932: Endosperm characters in wheat. Investigations with free Gallipoli, 1928–1931. J. Dept. Agr. Victoria 30:247–54, 288–300, 398–416.

Rawlings, C. O. 11844.

16016. Rawlings, J. O., D. G.Hanway, and C. O. Gardner. 1958: Variation in quantitative characters of soybeans after seed irradiation. Agron. J. 50:524–28.

Ray, A. K. 1914.

16017. Ray, C. B., and R. T. Stewart. 1937: Germination of seeds from certain species of *Paspalum*. J. Am. Soc. Agron. 29: 548–54. Biol. Abstr. 12:352. 1938.

Raymond, L. C. 5473.

16018. Ray, L. L. 1963: Systemic induction of parthenocarpic fruit set in cotton with gibberellin. Crop Sci. 3:448–49.

16019. Ray, W. W., and J. H. McLaughlin. 1942: Isolation and infection tests with seed- and soil-borne cotton pathogens. Phytopathology 32:233–38.

16020. Raybaud, L. 1929: Sur l'action des graines germées dans l'alimentation. Compt. Rend. 189:1015.

16021. —— 1931: Difference d'intensité des phénomènes hygroscopiques observés sur les graines de *Salvia sclarea* suivant leur origine. Compt. Rend. Soc. Biol. 107: 370–71. Ber. Wiss. Biol. 19:468. 1931.

16022. —— 1932: Sur la germination en germoir automatique de graines immergées préalablement dans l'eau. Compt. Rend.

Soc. Biol. 110:79–81. Ber. Wiss. Biol. 23:213–14. 1932.

16023. Raymond, W. D., and J. A. Squires. 1951: Annatto seed from Nigeria. Colonial Plant Animal Prod. 2:114–17. Biol. Abstr. 26:19336. 1952.

16024. ——, ——, and J. B. Ward. 1952/53: *Licania venosa* seeds from British Guiana. Colonial Plant Animal Prod. 3:227–28.

16025. ——, ——, and —— 1954: *Dryobalanops oblongifolia* kernels and *Elateriospermum tapos* kernels from Malaya. Colonial Plant Animal Prod. 4:159–61.

16026. —— and J. B. Ward. 1954: *Acacia cyanophylla* seeds from Cyprus. Colonial Plant Animal Prod. 4:56–57.

Raymond, W. D. 4413, 6197, 6198, 19589.

16027. Raymond-Hamet. 1940: Some physiological properties of akuammine, the principal alkaloid of the seeds of an apocynacea from French West Africa, *Picralima nitida*, T. and H. Durand. Compt. Rend. 211:125–27.

16028. —— 1944: Sympathicosthenic action of harmaline. Compt. Rend. Soc. Biol. 138:410–11. Chem. Abstr. 39:4150. 1945.

Rayner, E. T. 8979.

16029. Razdorskaia, L. A. 1943: A ready method for determining the seed germination of Pyrethrum cinerarifolium. Priroda 32(3):66–69.

16030. Razdorskii, V. F. 1938: The effect of alternately soaking and drying seeds upon their germination. Priroda Akad. Nauk SSSR 1938(6):114–16.

16031. —— 1938: The culture of isolated plant embryos. Priroda Akad. Nauk SSSR 1938(7/8):129–32.

Raznicina, E. A. 4105.

Raznitzyna, E. A. 10546.

16032. Razumov, V. I. 1950: In regard to results of work on the so-called "deyarovization." Agrobiologiya 1950(2):27–34.

16033. —— and R. N. Griuntukh. 1936: Iarovization in foreign literature. Iaroviz. Zh. Biol. Razv. Rast. 1936(2/3):187–206.

16034. —— and T. V. Oleinikova. 1955: Vernalization of winter grain crops at below freezing temperatures. Plant Physiol. (USSR) (English Transl.) 2:497–504. Biol. Abstr. 32:10099. 1958.

16035. —— and —— 1959: The dependence of the reaction to yarovization on day length. Agrobiologiya 1959(6):874–82.

16036. Razumovich, M. B., and S. M. Naumov. 1951: Effect of onion, garlic and bird cherry (*Prunus avium*) on seeds of higher plants. Priroda 40(4):64–65.

Rea, H. E. 19660.

16037. Read, A. D. 1933: The hammer test for judging seeds. J. Forestry 30:344.

Read, B. E. 11994.

16038. Read, D. W. L., and J. D. Beaton. 1963: Effect of fertilizer, temperature, and moisture on germination of wheat. Agron. J. 55:287–90.

16039. Read, J. 1963: Induction of photoconductivity in barley seeds by ionizing radiations. Radiation Botany 3:315–19.

16040. Reber, K. 1938: Hydrocyanic acid content of flaxseed. Schweiz. Apoth. Ztg. 76:229. Pharm. Abstr. 5:82–83. 1939.

16041. Recondo, E., M. Dankert, and L. F. Leloir. 1963: Isolation of adenosine diphosphate D-glucose from corn grains. Biochem. Biophys. Res. Commun. 12: 204–7.

Rector, C. 16194.

16042. Redabaugh, H. S., and C. W. Doxtator. 1963: Processing and drill performance of monogerm beet seed. Proc. Am. Soc. Sugar Beet Technol. 12: 301–8.

16043. Redchenko, V. P. 1956: Biology of the germination of the seeds of Chenopodiaceae. Tr. Turkmensk. Sel'skokhoz. Inst. 1956(8):267–74. Biol. Abstr. 35: 30123. 1960.

16044. Reddi, K. K., and K. V. Giri. 1947: Influence of variety, manuring and germination on vitamin B_1 content of groundnut—*Arachis hypogea*. Current Sci. (India) 16:285.

16045. Reddick, D. 1936: Seed transmission of potato virus diseases. Am. Potato J. 13:118–24. Rev. Appl. Mycol. 15:679. 1936.

16046. Reddy, C. S. 1935: Relation of rate of planting to the effect of corn seed treatment. Iowa State Coll. J. Sci. 9: 527–38.

Reddy, C. S. 3395, 5226, 19722.

Reddy, R. N. 15970, 15971.

Reddy, T. V. 1584.

16047. Redei, G., and Mrs. G. Redei. 1955: Developing wheat embryos excised from ovaries cultured in vitro. Experientia 11:387–88. Biol. Abstr. 30:32661. 1956.

Redei, Mrs. G. 16047.

16048. Redemann, C. T., S. H. Wittwer, and H. M. Sell. 1951: The fruit-setting factor from the ethanol extracts of immature corn kernels. Arch. Biochem. 32:80–83.

16049. Reder, R. 1934: Vitamin assay and

its application in a study of the vitamin B and G contents of mung beans and the grain sorghums. Proc. Oklahoma Acad. Sci. 14:50–52. Chem. Abstr. 28:6796. 1934.

16050. Redgrove, H. S. 1933: Setting lotions made from quince seeds. Pharm. J. 130:29. Chem. Abstr. 27:3290. 1933.

16051. —— 1933: Setting lotions made from *Psyllium* seeds. Pharm. J. 130:133. Chem. Abstr. 27:3290. 1933.

16052. Rediske, J. H. 1961: Maturation of Douglas-fir seed—a biochemical study. Forest Sci. 7:204–13.

Rediske, J. H. 11897, 17288.

Redlinger, L. M. 17750.

16053. Redmond, D. R., and R. C. Robinson. 1954: Viability and germination in yellow birch. Forestry Chron. 30:79–87. Biol. Abstr. 29:6859. 1955.

Reed, A. D. 9966.

16054. Reed, E. L. 1924: Anatomy, embryology, and ecology of *Arachis hypogea.* Botan. Gaz. 78:289–310.

16055. Reed, F. D. 1939: Structure of some carboniferous seeds from American coal fields. Botan. Gaz. 100:769–87.

Reed, H. M. 16181.

16056. Reed, H. S. 1944: The growth of ovules of *Pisum* in relation to zinc. Am. J. Botany 31:193–99.

Reed, I. F. 2545.

16057. Reed, J. F., and G. E. Coppock. 1952: Effects of some factors on the operation of peanut shellers and quality of seed produced. Peanut J. Nut World 32(2):24. Biol. Abstr. 27:28321. 1953.

Reed, J. F. 2934.

Reed, L. W. 18104.

16058. Reeder, H. M. 1960: Are we using proper approach in selling value of seed testing. Seed World 87(5):10.

Reeder, H. M. 9146.

16059. Reeder, J. R. 1953: The embryo of *Streptochaeta* and its bearing on the homology of the coleoptile. Am. J. Botany 40:77–80.

16060. —— 1957: The embryo in grass systematics. Am. J. Botany 44:756–68.

16061. —— 1962: The bambusoid embryo: A reappraisal. Am. J. Botany 49:639–41.

16062. Rees, A. R. 1959: The germination of oil palm seed: The cooling effect. W. African Inst. Oil Palm Res. J. 3:76–82.

16063. —— 1959: Germination of oil palm seed: large scale germination. W. African Inst. Oil Palm Res. J. 3:83–95.

16064. —— 1960: The germination of oil palm seeds—a review. W. African Sci. Assoc. J. 6:55–62. Biol. Abstr. 36:54152. 1961.

16065. —— 1962: High-temperature pretreatment and the germination of seed of the oil palm, *Elaeis guineensis* (Jacq.). Ann. Botany (London) 26:569–81.

16066. —— 1963: A large-scale test of storage methods for oil palm seed. W. African Inst. Oil Palm Res. J. 4:46–51.

16067. —— 1963: Some factors affecting the germination of oil palm seeds under natural conditions. W. African Inst. Oil Palm Res. J. 4:201–7.

16068. —— 1963: A note on the fate of oil palm seed in a number of habitats. W. African Inst. Oil Palm Res. J. 4:208–11.

16069. Rees, B. 1911: Longevity of seeds and structure and nature of seed coats. Proc. Roy. Soc. Victoria 23:393–414.

16070. Rees, L. W., and R. M. Brown. 1954: Wood density and seed source in young plantation red pine. J. Forestry 52:662–65.

16071. Reese, G. 1950: Beiträge zur Wirkung des Colchicins bei der Samenbehandlung. Planta 38:324–76.

16072. —— 1961: Karyotype and plant geography. Intern. Botan. Congr. Recent Advan. Botany 9:895–900. 1959.

16073. Reeve, P. A., and M. J. Buschlen. 1944: Improved techniques of segmenting and planting sugar-beet seed. Mich. Agr. Expt. Sta. Quart. Bull. 26:215–26. Biol. Abstr. 18:11736. 1944.

Reeve, P. A. 19170.

16074. Reeve, R. M. 1947: Relation of histological characteristics to texture in seed coats of peas. Food Res. 12:10–23.

16075. —— 1949: Histological observations on the seed coats of succulent peas. Food Res. 14:77–89.

Reeves, K. 1741.

16076. Reeves, R. G. 1930: Development of the ovule and embryo sac of alfalfa. Am. J. Botany 17:239–46. Ann. Biol. 36:55. 1932.

16077. —— 1936: Comparative anatomy of the seeds of cottons and other malvaceous plants. I. Malveae and Ureneae. Am. J. Botany 23:291–96.

16078. —— 1936: Comparative anatomy of the seeds of cottons and other malvaceous plants. II. Hibisceae. Am. J. Botany 23:394–405. Biol. Abstr. 10:19486. 1936.

16079. —— and J. O. Beasley. 1935: The development of the cotton embryo. J. Agr. Res. 51:935–44.

16080. —— and —— 1937: Localization of pentosans in the resin glands of the cotton embryo. J. Agr. Res. 54:711–18.

16081. —— and C. C. Valle. 1932: Anatomy and microchemistry of the cotton seed. Botan. Gaz. 93:259–77.

16082. Reeves, W. A., and J. D. Guthrie. 1950: Isolation of xanthine, quanine, adenine, proteose, oxalic acid, and glutathione from peanut kernels. Arch. Biochem. 26:316–18.

16083. Refai, F. Y., and B. S. Miller. 1952: The determination of pantothenic acid in wheat. Cereal Chem. 29:469–75.

Refai, F. Y. 10174, 10175.

16084. Regal, V. 1949: The influence of a water extract of red clover on the germination of various plants. Sb. Česk. Akad. Zeměděl. 22:84–91. Biol. Abstr. 25:25464. 1951.

16085. Regel, E. 1882: Einfluss des Lichtes auf das Keimen der Samen. Gartenflora 31:74–76.

16086. Regen, L. 1941: The development of the embryo sac in *Agave virginica.* Bull. Torrey Botan. Club 68:229–36.

16087. Regli, P., P. Chantegrel, and F. Busson. 1963: Chemical study of seeds of *Myrianthus arboreus* and *M. libericus.* Bull. Soc. Chim. Biol. 45:813–18. Chem. Abstr. 59:14294. 1963.

16088. Rehackova, O. 1954: Yearly variations in germination per cent and rapidity of germination of pine and larch seeds. Schweiz. Z. Forstw. 105:459–67. Biol. Abstr. 29:14714. 1955.

16089. Rehm, H. 1941: Raising delphiniums from seed. Gardeners Chron. 45:238.

16090. Rehm, S. 1953: The origin of toxic ammonia in germinating garden beet seed. J. Hort. Sci. 28:1–13.

16091. —— 1954: Control of seed-coat splitting in beans. Farming S. Africa 30:507–10, 523–24.

Reich, D. 15447.

16092. Reichard, A. 1912: Tannin in the seed coats of barley. Z. Chem. Ind. Kolloide 10:214–19. J. Chem. Soc. 102: 592–93. 1912.

Reichelt, G. 2706.

16093. Reichstein, T. 1952: Die Glykoside der Samen von *Strophanthus nicholsonii* Holmes. Nachtrag. Glykoside und Aglykone, 86. Helv. Chim. Acta 35:64–65.

Reichstein, T. 5774, 11294, 11356, 12213, 12214, 13659, 13878, 15878, 15930, 16129, 16130, 16188, 16912–16919, 16944, 16994, 17223, 18547. 19202–19206, 19481, 19482, 20089.

Reid, D. A. 13416, 13804.

16094. Reid, M. E. 1929: Relation of composition of seed and the effects of light to growth of seedlings. Contrib, Boyce Thompson Inst. 2:251–73. Am. J. Botany 16:747–79. 1929.

Reid, P. H. 4485.

16095. Reid, V. H., and P. D. Goodrum. 1957: Factors influencing the yield and wildlife use of acorns. Louisiana Univ. School Forestry Proc. Forestry Symp. 6:46–79.

16096. Reifenberg, A., and L. Frankenthal. 1930: Über den Einfluss von Phosphatlösungen verschiedener Wasserstoffionenkonzentration auf die Atmung und Keimung von Samen. Biochem. Z. 220: 473–86.

Reifenberg, A. 6516.

Reifer, I. 10768.

16097. Reilhes, R. 1941: Actions expérimentales diverses sur les graines: Contribution à l'étude de l'importance des téguments. Bull. Soc. Botan. France 88:613–21.

16098. —— 1942: Germinations de graines mutilées: Contribution à l'étude de l'importance de l'albumen. Bull. Soc. Botan. France 89:102–4.

16099. —— 1942: Influence du tégument sur la germination des graines de Ricin (*Ricinus communis*). Compt. Rend. Soc. Biol. 136:70–73.

16100. —— 1943: Précisions mécaniques sur la germination des graines. Bull. Soc. Botan. France 90:10–12.

16101. —— 1943: Sur la pénétration des colorants dans l'albumen. Bull. Soc. Botan. France 90:98–101.

16102. —— 1946: Germination des graines de lierre dans les fruits mutiles, sur la plante. Bull. Soc. Botan. France 93:105.

16103. —— 1948: Cytological and physiological investigations on ivy seeds. Rev. Cytol. Cytophysiol. Veg. 10:236–42. Biol. Abstr. 26:14175. 1952.

16104. Reiling, H. 1912: Keimversuche mit Gräsern zur Ermittlung des Einflusses, den Alter und Licht auf den Keimprozess ausueben. Dissertation, Jena. 88 pp. Botan. Centralbl. 123:346. 1913.

Reinbergs, E. 15130.

16105. Reinecke, O. S. H. 1930: The relation of seed formation to fruit development of the pear. S. African J. Sci. 27:303–9. Expt. Sta. Rec. 64:740. 1931.

16106. Reineke, L. H. 1942: Effect of pregermination and radicle damage on first-year development of red oak. J. Forestry 40:346–47. Expt. Sta. Rec. 87:526. 1942.

16107. —— 1942: Effect of stocking and seed on nursery development of eastern white pine seedlings. J. Forestry 40: 577–78. Expt. Sta. Rec. 87:678. 1942.

Reingpach, J. 6448.

16108. Reinhard, A. W., and M. Gorelik. 1933: Zur Frage der Samenkeimung bei *Solanum lycopersicum*. Z. Wiss. Biol. Abt. E 20:792–96. Biol. Abstr. 9:2502. 1935.

Reinhold, L. 7319.

16109. Reinholz, E. 1954: Ein das Wurzelwachstum fördernder Faktor im Diffusat keimender *Digitalis*-Samen. Naturwissenschaften 41:233.

16110. Reinmuth, E., and H. A. Kirchner. 1939: Protection of cereal seed from the granary weevil by disinfecting agents. Anz. Schädlingskunde 15:115–17. Chem. Abstr. 34:1084. 1940.

16111. Reisenauer, H. M. 1963: Relative efficiency of seed-and-soil-applied molybdenum fertilizer. Agron. J. 55:459–60.

Reisener, H. J. 11408.

Reiss, B. C. R. 5905.

Reitano, G. 11634.

16112. Reiter, C. 1926: Samenkunde der Gärtnerischen Kulturpflanzen und die Grundzüge des Samenbauer. Paul Parey, Hamburg. 69 pp.

16113. Reitz, H. C., R. E. Ferrel, and H. S. Olcott. 1944: Gel-forming derivative of wheat gluten. Ind. Eng. Chem. 36:1149–51. Expt. Sta. Rec. 94:714. 1946.

16114. Reitz, L. P. 1945: Kernel characteristics of Kansas winter wheat varieties. Kansas Agr. Expt. Sta. Rept. 2:1–24. Biol. Abstr. 20:16079. 1946.

16115. ——, E. D. Hansing, F. E. Davidson, and A. E. Decker. 1947: Viability and seed treatment of flax. J. Am. Soc. Agron. 39:959–70.

Reitz, L. P. 5093, 15810.

16116. Rejowski, A. 1962: Physiology and biochemistry of wheat grain ripening. Roczniki Nauk Rolniczych Ser. A 86:57–73. Biol. Abstr. 40:20532. 1962.

16117. Remer, W. 1904. Der Einfluss des Lichtes auf die Keimung bei *Phacelia tanacetifolia* Benth. Ber. Deut. Botan. Ges. 22:328–39.

Remington, R. E. 12158.

Remmenga, E. E. 13471.

Remmert, L. F. 2159.

16118. Remy, M. 1956: Influence de l'intensité lumineuse sur les corrélations entre épicotyle et burgeons cotylédonaires chez le pois. Compt. Rend. 243:400–1.

16119. Remy, P. 1957: Action de l'acide gibberellique sur la germination d'amandes d'abricotiers. Bull. Soc. Franc. Physiol. Végétale 3:126–27.

16120. —— 1957: L'évolution des substance de croissance pendant la germination sur milieu artificiel d'amandes d'abricotier (*Prunus armeniaca* L.) non exposées au froid. Compt. Rend. 244:1666–68.

16121. —— 1961: Recherches physiologiques sur la maturation des graines d'abres fruitiers à noyau. (Thèses presentées a la faculté des sciences de l'Université de Bordeaux.) Inst. Natl. Rech. Agron. 197 pp.

16122. Renard, H. A. 1960: Contribution à l'étude de l'influence de la gibberelline sur la dormance des certaines graines. Ann. Physio. Veg. 2:99–107.

Rendi, R. 12789.

16123. Rendle, A. B. 1888: Development of aleurone-grains in the lupine. Ann. Botany (London) 2:161–66.

16124. Rene, S. 1958: Embryogénie des Boragacées. Développement de l'embryon chez l'*Omphalodes linifolia* Moench. Compt. Rend. 247:249–53.

16125. Renfrew, A. G., and L. H. Cretcher. 1932: Quince seed mucilage. J. Biol. Chem. 97:503–10.

16126. Renich, M. E. 1921: Growth as related to specific gravity and size of seed. Trans. Illinois Acad. Sci. 14:109–39.

16127. Renkonen, K. O., and K. Harjanne. 1961: The hemagglutinins and their inhibitors in the seeds of *Vicia cracca*. Ann. Med. Exptl. Biol. Fenniae (Helsinki) 39:313–15. Chem. Abstr. 56:13286. 1962.

16128. Renkonen, O. 1957: Digitalis-like glycosides from seeds of *Strophanthus divaricatus*. Ann. Acad. Sci. Fennicae Ser. A II. 72 pp. Chem. Abstr. 51:16735. 1957.

16129. ——, O. Schindler, and T. Reichstein. 1959: Die Glykoside der Samen von *Strophanthus divaricatus* (Lour.) Hook. et Arn. V. Helv. Chim. Acta 42:160–82.

16130. ——, ——, and —— 1959: Die Glykoside der Samen von *Strophanthus divaricatus* (Lour.) Hook. et Arn. VI. Konstitutionen. Helv. Chim. Acta 42:182–200.

16131. Renner, R., C. F. Bentley, and L. W. McElroy. 1952: Nine essential amino acids in the protein of wheat and barley grown on sulfur-deficient soil. Soil Sci. Soc. Am. Proc. 17:270–73. Chem. Abstr. 48:928. 1954.

16132. ——, D. R. Clandinin, A. B. Morrison, and A. R. Robblee. 1953: The effects of processing temperatures on the amino acid content of sunflower seed oil meal. J. Nutr. 50:487–90. Biol. Abstr. 29:570. 1955.

16133. Rennie, D. A. 1956: Variations in percentage phosphorus and protein content of wheat, as induced by fertilizer treatment, soil type, and season. Can. J. Agr. Sci. 36:491–504.

Renwick, G. 17412.

Renz, J. 18047.

16134. Repnikov, K. A. 1936: Laboratory method of testing the seeds of wheat and barley by means of phenol and sulphuric acid. Bull. Appl. Botany, Genet., Plant Breed. (USSR) Ser. 4(1):40.

Resegotti, L. 19201.

Resovskii, N. V. Timofeev-. 15679.

16135. Resühr, B. 1939: Beiträge zur Lichtkeimung von *Amarantus caudatus* L. und *Phacelia tanacetifolia* Benth. Planta 30:471–506.

16136. —— 1941: Über die Bedeutung konstitutioneller Mängel für das Auftreten von Keimlingsschäden bei *Soja hispida* Moench. I. Z. Pflanzenkrankh. Pflanzenschutz 51:65–96. Botan. Centralbl. 178: 223. 1943.

16137. —— 1941: Über die Bedeutung konstitutioneller Mängel für das Auftreten von Keimlingsschäden bei *Soja hispida* Moench. II. Z. Pflanzenkrankh. Pflanzenschutz 51:161–92.

Retezeanu, M. 4345, 18382.

16138. Retovsky, R. 1934: Azotate d'uranyle et l'énergie de la germination de la semence de la vieille orge. Bull. Intern. Acad. Sci. Boheme 1934:1–4.

16139. —— 1939: The stimulation of the germination of seeds by uranyl nitrate. Rozpravy Česk. Akad. Věd Umeni Tr. 2. 49(12). 31 pp. Biol. Abstr. 21:15584. 1947.

16140. —— 1941: Die farbigen Reduktionen der anorganischen Stoffe in der Getreidekörner. I. Reduktion des Tellursäureanions TeO_4 bzw. TeO_6. Vestn. Kralov. Ceske Spolecnosti Nauk Trida Mat. Prirod. 1941(20):1–19. Biol. Abstr. 22: 9991. 1948.

16141. —— 1942: Verkürzung der Ruheperiode der Samen von *Aesculus hippocastanum* L. durch Uranylnitrat. Ber. Deut. Botan. Ges. 60:355–66. Biol. Abstr. 19:7339. 1945.

16142. —— 1959: Viviparous germination of apple seeds. Biol. Plantarum Prague 1:31–38.

16143. —— and V. Graf. 1943: The hormonization of cultivated plants. II. Hormonization of the seed of winter and spring wheat. Vestn. Česk. Akad. Zemědĕl. 19:334–45. Chem. Abstr. 41:6309. 1947.

Retovsky, R. 11589, 20085.

Retzer, H. J. 5094.

16144. Reuhl, E. 1935: Oxygen intake of oily and starchy seeds. Proc. Acad. Sci. Amsterdam 38:879–86. Chem. Abstr. 30:2606. 1936.

16145. —— 1936: Notes on metabolic changes in the germination of seeds. Rec. Trav. Botan. Neerl. 33:1–76. Botan. Centralbl. 170:248–49. 1936.

16146. Reusch, J. D. H. 1959: Note on the nature of the effect of the environment of seed-viability in interspecific hybrids. S. African J. Agr. Sci. 2:573–74. Biol. Abstr. 35:60886. 1960.

16147. —— 1961: The relationship between reproductive factors and seed set in *Paspalum dilatatum*. S. African J. Agr. Sci. 4:513–30. Biol. Abstr. 39:7925. 1962.

16148. Reuss, H. 1884: Einige Versuche mit Fichtensamen. Centralbl. Ges. Forstw. 10:65–74, 175–87.

Reuther, W. 5343.

Revol, L. 12784.

16149. Rewald, B. 1942: Phosphatides from oil seeds. Biochem. J. 36:822–24.

16150. —— and W. Riede. 1933: The relation between fat, phosphatides and protein during the ripening of seeds. Biochem. Z. 260:147–52. Chem. Abstr. 27:3500. 1933.

Rey-Pailhade, J. de; *see* De Rey-Pailhade, J.

Reyes, A. C. 10055.

Reyes, A. T. 18807.

16151. Reyes, D. S. 1951: Atomic seeds for your garden. Agr. Ind. Life 13(9):13, 43.

16152. Reyes, T. T. 1961: Seed transmission of coffee ring spot by Excelsa coffee (*Coffea excelsa*). Plant Disease Reptr. 45:185.

16153. Reynard, G. B. 1944: Polycotyledony in the genus *Lycopersicon*. Maryland Univ. Abstr. Dissertations 41(6):17–18.

Reynaud, J. 13826, 13827.

Reynolds, C. E. 7512.

16154. Reynolds, E. B. 1945: Freezing dormant Huban seeds may cause earlier flowering. Southern Seedsman 8(5):12.
16155. ——, V. L. Cory, and W. H. Dameron. 1938: Germination and longevity of bitterweed seed. Texas Agr. Expt. Sta. 50th Ann. Rept. 1937:82.
Reynolds, H. G. 518.
16156. Reynolds, J. R., and C. G. Youngs. 1964: Effect of seed preparation on efficiency and oil quality in filtration extraction of rapeseed. J. Am. Oil Chemists Soc. 41:63–65.
16157. Reynolds, K. L. 1944: Narcissus from seeds. Herbertia 10:111–13.
Rezhek, A. 13455.
Reznichenko, M. S. 11017, 15718.
16158. Rhind, D. 1933: A note on the inheritance of seed-coat colour in *Phaseolus lunatus* L. Indian J. Agr. Sci. 3:360–64. Agr. Livestock India 3:282. 1933.
16159. ——, M. F. Chandraratna, and E. F. L. Abeyratna. 1949: Seed placement of phosphates in paddy. I. Preliminary experiments on the application of phosphate solutions to seed. Trop. Agriculturist 105:98–108.
16160. Rhine, L. E. 1924: Divergence of catalase and respiration in germination. Botan. Gaz. 78:46–67.
Rhoades, H. F. 9777.
16161. Rhodes, L. B. 1920: Cockle-bur oil, a new seed oil. J. Am. Chem. Soc. 42: 1507–8.
Rhyne, C. L. 4077, 4078.
Ri, L. G. 7460.
Riazanceva, E. N. 11382.
16162. Ribas, I., and M. C. Seoane. 1953: Alkaloids in Papilionaceae. XIX. Alkaloid content of *Spartium junceum* L. cultivated in Galicia. Anales Edafol. Fisiol. Vegetal (Madrid) 12:695–701. Biol. Abstr. 28: 16963. 1959.
16163. Ribeiro, D. F. 1944: Influence of sulfanilamide on the germination of seeds. J. Biol. Chem. 152:665–67.
16164. —— 1946: Penicillin action on the germination of seeds. Science 104:18.
Ribereau-Gayon, P. 2099.
16165. Riblett, E. W. 1956: Germination of lily seeds. North Am. Lily Soc. Lily Yearbook 9:72–76.
16166. Ricahuerta, J. R. 1952: Germination and viability study of seven *Abaca* varieties. Philippine Agriculturist 35:504–11.
16167. Ricaldoni, J. N. 1949: Content, variation, and distribution of thiamine in wheat seedlings during certain germination periods. Rev. Fac. Cienc. Quin. 22:167–74. 1947. Chem. Abstr. 44:6485. 1950.
16168. Ricceri, G. 1951: The arginine content of seeds of Leguminosae and Gramineae. I. Quantitative investigation of the seeds by the original method of Fürth and Deutschberger. Arch. Sci. Biol. (Bologna) 35:176–84. Chem. Abstr. 49: 4092. 1955.
16169. Rice, J. C. 1958/59: Testing seed vigor with tetrazolium. Intern. Crop Improvement Assoc. Ann. Rept. 40/41: 278–82.
16170. Rice, W. N. 1940: The hemocytometer method for detecting fungous spore load carried by wheat seeds. Proc. Assoc. Offic. Seed Analysts 31:124–27. Biol. Abstr. 15:15700. 1941.
16171. —— 1960: Development of the cold test for seed evaluation. Proc. Assoc. Offic. Seed Analysts 50:118–23.
Rice, W. N. 15613.
Rich, C. E. 9827.
16172. Richards, A. V. 1952: Viability of citrus seeds. Trop. Agriculturist 108:186–88.
16173. Richards, J. G. 1957: Seed production in New Zealand: Browntop. New Zealand J. Agr. 95:7–15.
16174. Richards, M. 1955: Problems in seed germination. New Zealand Plants Gardens 1:10–12.
16175. Richardson, C. H. 1951: Effects of insecticidal fumigants on the germination of seed corn. J. Econ. Entomol. 44:604–8.
16176. Richardson, D. L. 1959: Effect of certain endosperm genes on popping ability in popcorn. Agron. J. 51:631.
16177. —— 1960: Pericarp thickness in popcorn. Agron. J. 52:77–80.
16178. Richardson, H. L. 1935: Field experiments on the action of calcium cyanamide on germinating seeds and on charlock in barley. Empire J. Exptl. Agr. 3:41–49.
Richardson. H. L. 4697.
16179. Richardson, J. A. 1951: Energy changes associated with germination in wheat seedlings. Proc. Univ. Durham Phil. Soc. Ser. A 11:68–82. Chem. Abstr. 49:4812. 1955.
16180. Richardson, K. E., and B. Axelrod. 1957: Changes in the inositol content during germination and growth of some higher plants. Plant Physiol. 32:334–37.
16181. Richardson, L. R., W. H. Brittingham, J. Whitacre, H. M. Reed, and E. Williams. 1950: Vitamin content of

southern peas at different stages of maturity. J. Am. Dietet. Assoc. 26:34–39.
16182. Richardson, S. D. 1958: The first international symposium on tree physiology. Forestry 31:177–83.
16183. —— 1959: Germination of Douglas-fir seed as affected by light, temperature, and gibberellic acid. Forest Sci. 5:174–81.
16184. Richey, F. D., and R. F. Dawson. 1951: Experiments on the inheritance of niacin in corn (maize). Plant Physiol. 26:475–93.
16185. —— and —— 1954: Thicker aleurone inadequate explanation for higher niacin in sugary maize. Plant Physiol. 29:194–96.
Richey, H. W. 1741.
16186. Richharia, R. H., and D. R. Dhodapkar. 1940: Delayed germination in sesame, *Sesamum indicum.* Indian J. Agr. Sci. 10:93–95.
Richmond, T. R. 4680.
Richter, F. 2971.
16187. Richter, H. 1936: Über Pappelsaaten. Forstwiss. Centralbl. 58:813–18. Biol. Abstr. 11:9420. 1937.
16188. Richter, R., K. Mohr, and T. Reichstein. 1953: Sarmutosid und Musarosid. Glykoside der Samen von *Strophanthus sarmentosus* A. P. DC. Helv. Chim. Acta 36:1073–100.
Richter, R. 16994.
16189. Rick, C. M., and R. I. Bowman. 1961: Galapagos tomatoes and tortoises. Evolution 15:407–17.
Rick, C. M. 15919.
16190. Ricker, P. L. 1962: The seeds of wild flowers. Plants Gardens N.S. 18:34–37, 95–96.
Ricketts, J. 3044.
16191. Riddell, J. A., and G. A. Gries. 1958: Development of spring wheat. III. Temperature of maturation and age of seeds as factors influencing their response to vernalization. Agron. J. 50:743–46.
Riddolls, A. W. 10648.
16192. Ridley, H. N. 1896: Dispersal of seeds by birds. Nat. Sci. 8:186–99. Expt. Sta. Rec. 7:689. 1895/96.
16193. —— 1937: The seed-drift of South Africa. Kew Roy. Botan. Gardens Bull. Misc. Inform. 10:525–26.
16194. Riebsomer, J. L., J. Bishop, and C. Rector. 1938: The composition of the seeds of *Asimina triloba.* J. Am. Chem. Soc. 60:2853–54.
16195. —— and N. Foote. 1936: The chemical composition of the fatty oil from Kukui nuts. Proc. Indiana Acad. Sci. 51:116–19.
16196. —— and G. A. Nesty. 1934: Examination of the fatty oil from pumpkin seed. J. Am. Chem. Soc. 56:1784–85. Chem. Abstr. 28:6006. 1934.
Riede, W. 16150.
Riedel, W. 12510.
16197. Riedemann, E. 1930: Die Versuche über den Einfluss der Herkunft des Kiefernsamens. Z. Forst- Jagdw. 62:498–522, 809–35.
16198. Riegel, A. 1940: A study of the variations in the growth of blue grama (*Bouteloua gracilis*) grass from seed produced in various sections of the Great Plains region. Trans. Kansas Acad. Sci. 43:155–71.
Riegel, B. 10655.
16199. Rieger, R., and A. Michaelis. 1958: The induction of automutagenic substances by submersion of *Vicia faba* seeds in tap water. Proc. Intern. Congr. Genet. Montreal 10:233.
16200. Riek-Häussermann, C. 1943: Vergleichend-anatomische und entwicklungsgeschichtliche Untersuchungen über Samen in der Gattung *Veronica.* Beih. Botan. Centralbl. Abt. A 62:1–60. Botan. Centralbl. 179:5. 1944.
16201. Riemenschneider, R. W., R. M. Speck, and E. G. Beinhart. 1945: Analysis and fatty acid composition of tobacco seed oils. Oil Soap 22:120–22. Expt. Sta. Rec. 95:461. 1946.
Riemenschneider, R. W. 1900.
16202. Riepma, P. 1954: The influence of a growth regulator on the seed-yield of horse beans (*Vicia faba*). Verslag Centraal Inst. Landbouwk. Onderzoek 1953:110–13.
Ries, J. von. 12500.
16203. Ries, V. H. 1948: Pelleted seeds. Country Gentleman 118(4):135.
Ries, V. H. 11872.
16204. Riesen, W. H., D. R. Clandinin, C. A. Elvehjem, and W. W. Cravens. 1947: Liberation of essential amino-acids from raw, properly heated, and overheated soya-bean oil meal. J. Biol. Chem. 167:143–50.
Riesen, W. H. 9096.
16205. Rietsema, J., B. Blondel, S. Satina, and A. F. Blakeslee. 1955: Studies on ovule and embryo growth in *Datura.* I. A growth analysis. Am. J. Botany 42:449–55.
16206. ——, S. Satina, and A. F. Blakeslee. 1953: The effect of indole-3-acetic acid on

Datura embryos. Proc. Natl. Acad. Sci. U.S. 39:924–33.

16207. ——, ——, and —— 1953: The effect of sucrose on the growth of *Datura stramonium* embryos in vitro. Am. J. Botany 40:538–45.

16208. ——, ——, and —— 1954: On the nature of the embryo inhibitor in ovular tumors of *Datura*. Proc. Natl. Acad. Sci. U.S. 40:424–31.

Rietsema, J. 15062.

16209. Rietz, R. C. 1936: An internal-fan kiln for drying seed cones. J. Forestry 34:477–81.

16210. —— 1939: Influence of kiln temperatures on field germination and tree per cent in northern white pine. J. Forestry 37:343–44. Biol. Abstr. 14:3519. 1940.

16211. —— 1939: Effect of five kiln temperatures on the germinative capacity of longleaf pine seed. J. Forestry 37:960–63. 1939.

16212. —— 1941: Kiln design and development of schedules for extracting seed from cones. U.S. Dept. Agr. Tech. Bull. 773:1–70.

16213. —— and K. E. Kimball. 1940: Kiln schedule for extracting red pine seed from fresh and stored cones. J. Forestry 38:430–34.

16214. —— and O. W. Torgeson. 1937: Kiln temperatures for northern white pine cones. J. Forestry 35:836–39. Expt. Sta. Rec. 78:200. 1938.

16215. Riga, A. J., and M. J. Bukovac. 1962: Distribution du ^{32}P, du ^{45}Ca et du ^{65}Zn chez le haricot (*Phaseolus vulgaris* L.) après absorption radiculaire. Redistribution de ces éléments au cours de la germination de la graine et du développement de la jeune plantule. Bull. Inst. Agron. Sta. Rech. Gembloux 29:165–96.

Rigal, M. 15064.

16216. Riganti, A. R., P. Cattaneo, and G. Karman. 1947: Chemical composition of the oil of *Colliguaya intergerrima* (Duraznillo). Anales Asoc. Quim. Arg. 35(177): 21–33. Biol. Abstr. 23:12701. 1949.

16217. Rigby, J. F. 1960: Light as a control in the germination and development of several mistletoe species. Proc. Linnean Soc. N.S. Wales 84:335–37. Biol. Abstr. 35:48768. 1960.

16218. Riggio-Bevilacqua, L. 1956: Functional activity of germinating seeds. II. Respiratory increase and synthesis of flavine coenzymes. Atti Accad. Ligure Sci. Lettere (Genoa) 12:102–9. 1955. Chem. Abstr. 51:14909. 1957.

16219. —— and R. Scotti. 1954: Functional activation of germinating seeds. I. Respiration increase and synthesis of pyridine adenine nucleotides. Atti Accad. Ligure Sci. Lettere (Genoa) 10:196–203. Chem. Abstr. 49:11789. 1955.

16220. —— and P. E. Scuncia. 1959: Diphosphopyridine nucleotide and triphosphopyridine nucleotide in the seeds of various species. Boll. Soc. Ital. Biol. Sper. 35:1017–19.

16221. Riggs, N. V. 1954: The occurrence of macrozamin in the seeds of cycads. Australian J. Chem. 7:123–24. Chem. Abstr. 48:6510. 1954.

Riggs, N. V. 4609.

16222. Righter, F. I. 1945: *Pinus:* the relationship of seed size and seedling size to inherent vigor. J. Forestry 43:131–37.

16223. Rigler, J. 1935: On the determination of the 1000-grain weight of seed and the factors influencing this weight. Kiserl. Kozl. 38:123. Proc. Intern. Seed Testing Assoc. 8:89–90. 1936.

Rihova, I. 4115.

16224. Rijof, N. 1928: Sur la méthode de la détermination de la faculté germinative des graines d'épinard (*Spinacia oleracea* L.). Jard. Botan. Inst. Essais Semences, Leningrad, Ann. Essais Semences 1928(1): 111–16.

16225. —— 1928: Sur la méthode d'analyse des semences de la betterave (*Beta vulgaris*) dans les conditions du laboratoire. Jard. Botan. Inst. Essais Semences, Leningrad, Ann. Essais Semences 1928(5):50–54.

16226. Rijven, A. H. G. C. 1952: In vitro studies on the embryo of *Capsella bursa-pastoris*. Acta Botan. Neer. 1:157–200. Biol. Abstr. 27:4665. 1953.

16227. —— 1955: Effects of glutamine, asparagine and other related compounds on the in vitro growth of embryos of *Capsella bursa-pastoris*. Koninkl. Ned. Akad. Wetenschap. Proc. Ser. C 58:368–76.

16228. —— 1956: Glutamine and asparagine as nitrogen sources for the growth of plant embryos in vitro: A comparative study of twelve species. Australian J. Biol. Sci. 9:511–27.

16229. —— 1958: Effects of some inorganic nitrogenous substances on growth and nitrogen assimilation of young plant embryos in vitro. Australian J. Biol. Sci. 11:142–54.

16230. —— 1961: Distribution of glutamyl transferase in the ripening of pea seed. Australian J. Biol. Sci. 14:700–1.

16231. Rikhter, A., V. A. Rantzan, and M. Pekker. 1933: Control of "yarovization." Compt. Rend. Acad. Sci. URSS 1933:72–74. Chem. Abstr. 28:1742. 1934.

16232. Riley, J. P. 1951: The seed fat of *Parinarium laurinum.* II. Component glycerides of the seed fat. J. Chem. Soc. 1951:291–93. Biol. Abstr. 25:34852. 1951.

Riley, J. P. 5652, 8765.

Riley, W. D. 1642.

16233. Rimon, D. 1957: Lipase activity of germinating lettuce seeds. Bull. Res. Council Israel Sect. D 6:53–55.

16234. Rinck, E. 1947: Verification of the principle of Lavoisier on seeds during germination. Compt. Rend. 224:835–37. Chem. Abstr. 41:4544. 1947.

Rinke, E. H. 19872.

16235. Rincker, C. M. 1954: Effect of heat on impermeable seeds of alfalfa, sweet clover, and red clover. Agron. J. 46:247–50.

16236. —— 1957: Heat-treating hard alfalfa seed. Wyoming Agr. Expt. Sta. Bull. 352. 12 pp.

16237. Rindi, V. 1951: Amniotic liquor. I. Action of amniotic liquor on the germination of wheat seeds. Riv. Ostet. Ginecol. 6:73–88. Chem. Abstr. 46:3099–100. 1952.

Rios, B. J. 3006.

Rios, P. Gonzalez-; *see* Gonzalez-Rios, P.

16238. Ripan, R., E. Pop, I. Ciobanu, T. Marcu, and G. Marcu. 1959: Observations on the germination of caryopses of *Zea mays* L., treated with $ZnSO_4$ labelled with radio-active $Zn.^{65}$ Studii Cercetari Biol. 10:23–31. Biol. Abstr. 36:21472. 1961.

Ripeanu, G. 15856.

16239. Ripley, S. H. 1947/48: Artificial culture of excised *Zea mays* embryos. Natal. Univ. Coll. Sci. Soc. J. 6:18–30.

16240. Rippel, A. 1918: Semipermeable Zellmembranen bei Pflanzen. Ber. Deut. Botan. Ges. 36:202–18.

16241. Rippel, K. 1938: Über den Gehalt von Zellteilungshormonen in Samen und Keimlingen von *Pirus malus*, *Prunus domestica* und *Prunus avium.* Planta 29:1–10.

16242. —— 1940: Zur Frage des Vorkommens Wachstumshemmender Wirkstoffe in Pflanzen. Planta 30:689–91.

Ripperton, J. C. 227, 228.

16243. Riser, G. R., J. J. Hunter, J. S. Ard, and L. P. Witnauer. 1962: *Vernonia anthelmintica* Willd. seed oils and salts of vernolic acid as stabilizers for plasticized poly (vinyl chloride). J. Am. Oil Chemists Soc. 39:266–68.

Risius, M. L. 13970.

16244. Riskina, S. R. 1933: Variations in the enzymic content of different varieties of sugar beets at the germination period. Trans. Central Sci. Res. Inst. Sugar Ind. (USSR) 1933(12):36–43. Chem. Abstr. 28:6453. 1934.

Rist, C. E. 1705, 3565, 19798.

16245. Ritslaid, V. 1960: The effect of aeroionization on the germination of seeds and growth of plants. Eesti NSV Teaduste Akad. Toimetised, Biol. Ser. 9:167–76.

16246. Ritthausen, H. 1896: Galactite from the seed of yellow lupines. Ber. Deut. Chem. Ges. 29:896–99.

16247. Ritvanen, Tyyne. 1953: The application of tetrazolium chloride to the rapid determination of the germinability of timothy seed. Maataloustieteellinen Aikakauskirja 25:153–59. Chem. Abstr. 48: 9474. 1954.

Riveca, F. 5174.

16248. Rivera, P. G. 1939: Oil content of castor seeds from Philippine sources. Natl. Res. Council Philippines Bull., Proc. 5th Sci. Conv. 23:160. Pharm. Abstr. 7:169. 1941.

16249. Rivera, R., H. W. Popp, and R. B. Dow. 1937: The effect of high hydrostatic pressures upon seed germination. Am. J. Botany 24:508–13.

16250. Rivera, V. 1922: Influence of conditions of humidity on the growth of Sainfoin seeds respectively unhusked and husked. Rev. Biol. (Rome) 4:1–12.

16251. —— 1930: On the biological action of penetrating rays (cosmic and ultra-γ-rays) on the development of seeds of vegetables. Atti Accad. Nazl. Lincei, Rend., Classe Sci. Fis., Mat. Nat. 11:612–13. Botan. Centralbl. 163:265. 1932.

16252. —— 1930: Development of plants under lead screen. Atti Accad. Nazl. Lincei, Rend., Classe Sci. Fis., Mat. Nat. 11:718–20. Biol. Abstr. 7:12848. 1933.

16253. —— and C. Sempio. 1935: The rhythm of development in seed germination. Atti Pontif. Accad. Sci. Nuovi Lincei 85:410–23. Proc. Intern. Seed

Testing Assoc. 7:113. 1935. Biol. Abstr. 10:364. 1936.

16254. ——, ——, and A. Shkjezi. 1944: Efficacité anticryptogamique de l'urine à employer dans les traitements. Intern. Bull. Plant Protect. 18:1–7. Biol. Abstr. 23:25003. 1949.

Rivera-Perez, J. 19374.
Rivière, S. 4948.
Rivoir, O. R. 16543.
Roach, J. R. 5506.
Robb, L. 15014.

16255. Robb, W. 1932: Notes on the inheritance of grain colour in certain oat hybrids. J. Genet. 26:231–38.

16256. —— 1937: Huskless oats. Scot. J. Agr. 20:161–65.

Robb, W. 17615.
Robbins, A. L. 12930.

16257. Robbins, W. A., and R. H. Porter. 1946: Germinability of sorghum and soybean seed exposed to low temperatures. J. Am. Soc. Agron. 38:905–13.

Robbins, W. A. 44.

16258. Robbins, W. J., and K. F. Petsch. 1932: Moisture content and high temperature in relation to the germination of corn and wheat grains. Botan. Gaz. 93:85–92.

Robbins, W. S. 2811.

16259. Robbins, W. W., and H. A. Borthwick. 1925: Development of the seed of *Asparagus officinalis*. Botan. Gaz. 80:426–38.

Robbins, W. W. 1806.
Robblee, A. R. 16132.

16260. Roberg, M., and E. Marchal. 1937: Saponin distribution and varying saponin content of certain plants and seeds. Jahrb. Wiss. Botan. 84:710–39. Chem. Abstr. 32:9177. 1938.

Robert, A. L. 3130.
Robert, T. 171.
Roberts, A. 4210.

16261. Roberts, D. W. A. 1951: Some effects of ethylene on germinating wheat. Can. J. Botany 29:10–25.

16262. Roberts, E. G. 1936: Germination and survival of longleaf pine. J. Forestry 34:884–85.

16263. Roberts, E. H. 1960: The viability of cereal seed in relation to temperature and moisture. Ann. Botany (London) 24:12–31.

16264. —— 1961: Dormancy of rice seed. I. The distribution of dormancy periods. J. Exptl. Botany 12:319–29.

16265. —— 1961: Dormancy in rice seed. II. The influence of covering structures. J. Exptl. Botany 12:430–45.

16266. —— 1961: Viability of cereal seed for brief and extended periods. Ann. Botany (London) 25:373–80.

16267. —— 1961: The viability of rice seed in relation to temperature, moisture content, and gaseous environment. Ann. Botany (London) 25:381–90.

16268. —— 1962: Dormancy in rice seed. III. The influence of temperature, moisture, and gaseous environment. J. Exptl. Botany 13:75–94.

16269. —— 1963: The effects of inorganic ions on dormancy in rice seed. Physiol. Plantarum 16:732–44.

16270. —— 1963: The effects of some organic growth substances and organic nutrients on dormancy in rice seed. Physiol. Plantarum 16:745–55.

16271. —— 1963: An investigation of intervarietal differences in dormancy and viability of rice seed. Ann. Botany (London) 27:365–69.

16272. —— 1964: The distribution of oxidation-reduction enzymes and the effects of respiratory inhibitors and oxidizing agents on dormancy in rice seed. Physiol. Plantarum 17:14–29.

16273. —— 1964: A survey of the effects of chemical treatments on dormancy in rice seed. Physiol. Plantarum 17:30–43.

Roberts, E. J. 3398, 3399.

16274. Roberts, H. A. 1958: Studies on the weeds of vegetable crops. I. Initial effects of cropping on the weed seeds in the soil. J. Ecol. 46:759–68.

16275. —— 1962: Studies on the weeds of vegetable crops. II. Effect of six years of cropping on the weed seeds in the soil. J. Ecol. 50:803–13.

16276. Roberts, H. F. 1936: Seed reproduction in the dandelion. Sci. Agr. (Ottawa) 17:235–42.

Roberts, H. M. 6091.

16277. Roberts, J. L. 1940: Dahlias from seed. Flower Grower 27(4):184.

16278. Roberts, L. W. 1951: Survey of factors responsible for reduction of 2,3,5-triphenyltetrazolium chloride in plant meristems. Science 113:692–93.

16279. Roberts, R. A., and J. O. Thomas. 1933: A study of the distinguishing features of the seeds and seedlings of some farm Cruciferae. Proc. Intern. Seed Testing Assoc. 5:113–31.

Roberts, R. L. 9111.

16280. Roberts, W. O. 1948: Prevention of mineral deficiency by soaking seeds in nutrient solution. J. Agr. Sci. 38:458–68.
Robertson, A. 16745.

16281. Robertson, C. 1896: Seed crests and myrmecophilous dissemination in certain plants. Botan. Gaz. 23:288.

16282. Robertson, D. S. 1952: The genotype of the endosperm and embryo as it influences vivipary in maize. Proc. Natl. Acad. Sci. U.S. 38:580–83.

16283. Robertson, D. W., T. E. Haus, and J. C. Hoff. 1959: Frost injury to maturing barley. Agron. J. 51:658–60.

16284. —— and A. M. Lute. 1933: Germination of the seed of farm crops in Colorado after storage for various periods of years. J. Agr. Res. 46:455–62.

16285. —— and —— 1937: Germination of seed of farm crops in Colorado after storage for various periods of years. J. Am. Soc. Agron. 29:822–34.

16286. ——, ——, and R. Gardner. 1939: Effect of relative humidity on viability, moisture content, and respiration of wheat, oats, and barley seed in storage. J. Agr. Res. 59:281–91.

16287. ——, ——, and H. Kroeger. 1943: Germination of twenty-year-old wheat, oats, barley, corn, rye, sorghum, and soybeans. J. Am. Soc. Agron. 35:786–95.
Robertson, D. W. 5179, 6333, 12544.
Robertson, E. 12604.
Robertson, J. S. M. 522, 7635.

16288. Robertson, L. S., R. W. Bell, R. L. Cook, and H. W. Fairchild. 1949: A comparison of segmented with natural sugar-beet seed. Mich. Agr. Expt. Sta. Quart. Bull. 31:465–71.
Robertson, L. S. 2248, 10667.

16289. Robertson, R. H. 1951: The germination of sweet peas. Gardeners Chron. 129:148.

16290. Robertson, R. N., H. R. Highkin, J. Smydzuk, and F. W. Went. 1962: The effect of environmental conditions on the development of pea seeds. Australian J. Biol. Sci. 15:1–15.
Robertson, R. N. 13204, 13205, 13533.

16291. Robertson, R. S., Jr. 1962: The role of seed transmission in the epidemiology of bean mosaics in Central Washington. Plant Disease Reptr. 46:71–72.

16292. Robertson, T. S. B. 1947: The accumulation of vitamin C during germination in peas. Australian J. Exptl. Biol. Med. Sci. 25:41–46.

16293. Robin, C. 1959: Try a fire garden. Calif. Hort. Soc. J. 20:16–17.

16294. Robinson, A. D., L. E. Lynd, and B. J. Miles. 1949: The thiamine and riboflavin content of Manitoba-grown wheat, oats, and barley of the 1946 crop. Can. J. Res. Sect. F 27:241–48.
Robinson, A. D. 17226.
Robinson, E. 3174–3177.

16295. Robinson, E. L., and C. A. Kust. 1962: Distribution of witchweed seeds in the soil. Weeds 10:335.
Robinson, H. F. 6967, 8136.
Robinson, J. 2274.

16296. Robinson, J. L. 1934: Physiologic factors affecting the germination and growth of seed corn. Iowa Agr. Expt. Sta. Res. Bull. 176. 46 pp.
Robinson, J. L. 9187.

16297. Robinson, L. W., and A. M. Mayer. 1963: Activity of cytochrome C-NAD reductase in germinating lettuce seeds. Plant Cell Physiol. 4:361–64.

16298. Robinson, P. B. 1953: Inhibition of weed seed germination in seed beds by the use of calcium cyanamide and urea. Proc. Southern Weed Conf. 1953:206–7. Biol. Abstr. 28:27074. 1954.
Robinson, R. 19412.
Robinson, R. C. 16053.

16299. Robinson, R. G. 1949: Annual weeds, their viable seed population in the soil, and their effect on yields of oats, wheat and flax. Agron. J. 41:513–18.

16300. Robinson, R. R. 1960: Germination of hard seed of Ladino white clover. Agron. J. 52:212–14.

16301. Robinson, R. W. 1954: Seed germination problems in the Umbelliferae. Botan. Rev. 20:531–50.
Robles, C. E. 3004.

16302. Robyns, A. 1959: Note preliminaire sur la polyembryonie dans *Bombax kimuenzae* De Wild. et Th. Dur. Brussels Jard. Botan. l'État. Bull. 29:23–26.

16303. Roca, J. 1955: Formación de protidos en el proceso de germinación del maiz. I. Anales Inst. Biol. (Univ. Nacl. Mex.) 26:11–17.

16304. —— and R. Ondarza. 1948: Enzymic activity during germination. I. Saccharogenic and proteolytic activity in the kidney bean. Anales Inst. Biol. (Univ. Nacl. Mex.) 19:283–98. Chem. Abstr. 43:8451–52. 1949.

16305. —— and —— 1949: Studies of enzymic activity during the process of germination. II. α- and β-amylases in hybrids and varieties of Mexican corn. Anales Inst. Biol. (Univ. Nacl. Mex.) 20:17–26. Chem. Abstr. 44:4086. 1950.

16306. —— and —— 1951: Enzymic acitivity during germination. III. Catalase in bean varieties. Anales Inst. Biol. (Univ. Nacl. Mex.) 22:3–9. Chem. Abstr. 46:3617g. 1952.

16307. —— and —— 1952: Estudios sobre la actividad enzimatica en el proceso de germinación. IV. Proteases en variedades de maiz. Anales Inst. Biol. (Univ. Nacl. Mex.) 23:3–12.

16308. —— and —— 1953: The enzymic activity during the process of germination. V. Catalase in varieties of maize. Anales Inst. Biol. (Univ. Nacl. Mex.) 24:15–21. Chem. Abstr. 48:4639. 1954.

16309. Rocha, F. F. 1959: Interaction of moisture content and temperature on onion seed viability. Proc. Am. Soc. Hort. Sci. 73:385–89.

16310. Rochanapurananda, P. 1934: Changes in the chemical composition of sprouting mungo seeds. Nat. Appl. Sci. Bull. (Univ. Philippines) 4:127–47. Chem. Abstr. 29:5479. 1935.

Rochettes, A. Mercier-Des-; *see* Mercier-Des-Rochettes, A.

16311. Rockcastle, V. N. 1961: Seeds. Cornell Sci. Leaflet 54(3). 31 pp.

16312. Rockwell, F. F. 1955: Vines from seed. Flower Grower 42(4):42–44.

16313. Rodenburg, C. M. 1958: The identification of lettuce varieties from the young plant. Euphytica 7:241–46.

16314. Roder, W. 1960: Zur Keimungsphysiologie und Saatgutqualität von Maissorten. I. Z. Landwirtsch. Versuch-Untersuchungsw. 6:105–20.

Roderick, D. B. 13592.

16315. Rodewald, H. 1883: Über die Wechselbeziehungen zwischen Stoffumsatz und Kraftumsatz in keimenden Samen. J. Landwirtsch. 8. 35 pp. Botan. Centralbl. 17:297–99.

16316. —— 1898: Zur Methodik der Keimprüfungen. Landwirtsch. Versuchssta. 49: 257–86.

16317. Rodger, J. B. A., G. G. Williams, and R. L. Davis. 1957: A rapid method for determining winterhardiness in alfalfa. Agron. J. 49:88–92.

Rodgers, E. G. 19745.

16318. Rodionenko, G. I. 1955: The seed of iris and its peculiarities. Dokl. Akad. Nauk SSSR 104:653–56.

16319. Rodionov, A. P. 1950: Study of vernalization phase in peach and in Bessey cherry. Izv. Akad. Nauk SSSR Ser. Biol. 1950(2):51–67.

Rodionova, G. B. 14977.

Rodionova, I. V. 15729.

16320. Rodionova, N. A. 1957: Protein complex changes in the cotyledons of chick pea during germination. Byul. Gl. Botan. Sada 1957(28):56–62. Chem. Abstr. 52:13891. 1958.

16321. —— 1957: Proteins of *Cicer arietinum* cotyledons during seed growth. Dokl. Akad. Nauk SSSR 114:158–60. Chem. Abstr. 51:16740. 1957.

16322. Rodomanski, W. 1959: Treating of cabbage seeds with dusts. Bull. Warzywniczy 4:437–50.

16323. Rodrigo, P. A. 1935: Longevity of some farm crop seeds. Philippine J. Agr. 6:343–57.

16324. —— 1938: Notes on the propagation of the *Ponderosa chico*. Philippine J. Agr. 9:357–63. Biol. Abstr. 13:4876. 1939.

16325. —— 1939: Study on the vitality of old and new seeds of mungo (*Phaseolus aureus* Roxb.). Philippine J. Agr. 10:285–91.

16326. —— 1953: Some studies on the storing of tropical and temperate seeds in the Philippines. Intern. Hort. Congr., London, Rept. (1952) 13:1061–66.

16327. —— and A. L. Tecson. 1940: Storing some vegetable seeds. Philippine J. Agr. 11:383–95.

Rodrigo, P. A. 19100.

16328. Rodrigues, A. L. 1952: Embriones de maiz y su comportamiento como semilla. Acta Agron. (Palmira, Colombia) 2:147–49. Biol. Abstr. 27:20643. 1953.

Rodriguez, A. 17583.

16329. Rodriguez, C. 1948: Aspectos opticos en el desarrollo de semillas de *Gleditschia triacanthos*. Anales Inst. Espan. Edafol., Ecol., Fisiol. Vegetale (Madrid) 7:201–14. Biol. Abstr. 23:30209. 1949.

Rodriguez, L. D. 9216–9218.

Rodriguez, P. 11754.

16330. Rodriguez, S. J. 1962: The chemical composition of green coffee beans and coffee leaves as related to soil and foliar applications of secondary and minor elements. Dissertation Abstr. 23:784–85.

Rodriguez-Colon, F. 19763.

16331. Rodriguez-Lopez, M. 1953: Effects produced in seeds of *Cicer arietinum* by treatment with some growth substances. Anales Inst. Botan. A. J. Cavanilles 11:19–38. 1952. Biol. Abstr. 30:32674. 1956.

16332. —— 1958: Seedless fruit produced in a species of the genus *Cucurbita.* Biol. Inst. Nacl. Invest. Agron. 18:265–70. Chem. Abstr. 57:14218. 1962.
16333. Roe, E. I. 1940: Longevity of red pine seed. Proc. Minn. Acad. Sci. 8:28–30.
16334. —— 1941: Effect of temperature on seed germination. J. Forestry 39:413–14.
16335. —— 1945: Viable seed produced by twelve-year-old red pine. J. Forestry 43:678–79.
16336. —— 1946: Some lesser known uses of forest seeds and fruits. Chemurgic Dig. 5(5):108.
16337. —— 1948: Balsam fir seed—its characteristics and germination. Div. Forest Management, Lake States Forest Expt. Sta., Sta. Paper 11:1–13.
16338. —— 1948: Early seed production by balsam fir and white spruce. J. Forestry 46:529.
16339. —— 1948: Viability of white pine seed after ten years of storage. J. Forestry 46:900–2.
Roe, E. I. 11355, 11929.
16340. Roedel, G. F., and H. R. Kraybill. 1940: Nitrogen metabolism of the germinating soybean. Indiana Agr. Expt. Sta. Ann. Rept. 1939/40:11.
16341. Röder, I. 1958: Anatomical and fluorescence-optical investigations of seeds of the Papaveraceae. Österr. Botan. Z. 104:370–81. Chem. Abstr. 53:8315. 1959.
Roeder, W. von; *see* Von Roeder, W.
Roels, O. A. 8773.
16342. Roeser, J., Jr. 1941: Some aspects of flower and cone production in Ponderosa pine. J. Forestry 39:534–36. Biol. Abstr. 16:5375. 1942.
16343. —— 1942: The influence of climate on seed production in Douglas fir. J. Forestry 40:304–7.
16344. Rössler, L. 1943: Vergleichende Morphologie der Samen europäischer *Euphorbia*-Arten. Botan. Centralbl. Beih. Abt. B 62:97–174. Botan. Centralbl. 178:355. 1943.
Rogalev, I. E. 11172.
16345. Rogenhofer, E. 1952: Einige seltenere Unkrautsamen im Saatgut. Bodenkultur, Österr. Zentralorg. Landwirtsch. Ernährungsforsch. 3(Suppl.):69–71. Biol. Abstr. 27:13890. 1953.
Roger, L. 4491.
16346. Rogers, B. J. 1959: The action of arsenic on germination and seedling growth. Proc. Joint Meeting N. Central, 16th, West Can., 10th, Weed Control Conf. 1959:20. Chem. Abstr. 58:9569. 1963.
16347. —— 1959: Factors affecting the germination of milkweed seed (*Asclepias*). [Abstract.] Proc. Indiana Acad. Sci. 68:90. 1958.
Rogers, C. B. W. 13534.
16348. Rogers, C. H. 1943: Cotton seed-treatment studies at the Blackland Experiment Station. Texas Agr. Expt. Sta. Bull. 634. 22 pp. Expt. Sta. Rec. 90:646. 1944.
Rogers, C. H. 7395.
16349. Rogers, M. S. 1931: Seed germination experiments with the fragrant dwarf indigo, *Amorpha nana.* Univ. Colo. Studies 18:205–13.
Rogerson, C. T. 18225.
16350. Rogler, G. A. 1954: Seed size and seedling vigor in crested wheatgrass. Agron. J. 46:216–19.
16351. —— 1959: Relation of germination of seed of *Oryzopsis hymenoides* and *Stipa viridula* to age and treatment. Agron. Abstr. 1959:77.
Rogler, G. A. 12437, 16858, 16859.
Rognone, L. 18895.
16352. Rogozinska, J. 1958: The inhibiting effect of *o*-aminozotoluene on growth of cotyledonless radish embryos cultivated in vitro. Acta Soc. Botan. Polon. 27:103–13.
16353. Rohatgi, K., M. Banerjee, and S. Banerjee. 1955: Effect of germination on vitamin B_{12} values of pulses. J. Nutr. 56:403–8.
Rohatgi, K. 1849, 1850.
16354. Rohde, C. R. 1963: Effect of nitrogen fertilization on yield, components of yield and other agronomic characteristics of winter wheat. Agron. J. 55:455–58.
16355. Rohloff, I. 1956: Untersuchungen über d ieKeimung der Samen von *Cyclamen persicuu* (Mill.). Saatgut-Wirtsch. 8:47–51.
16356. —— 1959: Einige Hinweise zur praktischen Durchführung der Untersuchung von Bohnensamen auf Brennenfleckenbefall. Saatgut-Wirtsch. 11:197–98.
16357. —— 1961: Verwendung von Vermiculit für die Gesundheitsprüfung von Hülsenfruchtsamen. Saatgut-Wirtsch. Fachzeitschr. Samen Saaten 1961(6):171–72.
16358. Rohmeder, E. 1938: Der Einfluss der Mondphasen auf die Keimung und erste Jugendentwicklung der Fichte. Forstwiss. Centralbl. 60:593–603, 634–46.
16359. —— 1939: Wachstumsleistungen der aus Samen verschiedener Grössenordnung

entstandenen Pflanzen. Forstwiss. Centralbl. 61:42–59.
16360. —— 1939: Die Überwindung von Keimhemmungen bei den Samen der Weimutskiefer, Duglasie und Lärche durch Kaltnassvorbehandlung. Forstwiss. Centralbl. 61:393–406.
16361. —— 1939: Die Keimung des Hirschholunders (*Sambucus racemosa*). Forstwiss. Centralbl. 61:505–11.
16362. —— 1941: Die Vermehrung der Pappeln durch Samen. Forstarchiv 17:73–80.
16363. —— 1942: Ergebnisse der forstlichen Saatgutforschung als Mittel zur Ertragssteigerung im Walde. Forstarchiv 18:165–76. Biol. Abstr. 17:25056. 1943.
16364. —— 1942: Keimversuche mit *Ulmus montana* (With.). Forstwiss. Centralbl. 64:122–35. Biol. Abstr. 17:20715. 1943.
16365. —— 1942: Keim- und Saatversuche mit Sanddorn (*Hippophae rhamnoides*). Forstwiss. Centralbl. 64:241–45. Biol. Abstr. 18:11809. 1944.
16366. —— 1943: Beeinflusst der Vitamin-C-Gehalt der Nadelholzsamen das Färbungsergebnis des Natriumselenitverfahrens? Forstwiss. Centralbl. 65:65–70. Biol. Abstr. 18:11810. 1944.
16367. —— 1953: Der Unterschiedliche Keimverlauf von Samen der europaischen und japanischen Larche. Forstwiss. Centralbl. 72:210–20.
16368. —— 1961: Problems and propositions for an international certification of forest seeds. Saatgut-Wirtsch. 13:219–21.
16369. —— 1961: Problems and propositions for an international certification of forest seeds (conclusion) Saatgut-Wirtsch. 13: 253–55.
16370. —— 1962: Die Bedeutung der Keimschnelligkeit bei forstlichen Samenarten. Proc. Intern. Seed Testing Assoc. 27:657–71.
16371. —— 1962: Problems and proposals for international forest tree seed certification. Proc. World Forestry Congr. 5:685–90.
16372. —— and C. J. Chen. 1939: Keimversuche mit Fichtensamen verschiedener Korngrösse. Forstwiss. Centralbl. 61:177–84.
16373. —— and M. Loebel. 1940: Keimversuche mit Zirbelkiefer (*Pinus cembra*). Forstwiss. Centralbl. 62:25–36.
Rohmeder, E. 8783.
16374. Rohmeder, M. 1954: Keimung und Keimprüfung von Schwarzkiefernsamen. Forstwiss. Centralbl. 73:175–79.
16375. Rohrlich, M. 1962: Chemical differentiation of proteins of wheat and rye. VI. Identification of a glycoprotein. Ber. Getreidechem.-Tagung, Detmold 1962:51–58. Chem. Abstr. 58:9410. 1963.
16376. —— and V. Hopp. 1960: Thermische Beeinflussung der Vitamin- und Mineralstoffverteilung im Getreidekorn. Naturwissenschaften 47:260. 1960.
16377. ——, W. B. T. Schulz, and I. Lichtentels. 1962: Chemical differentiation of proteins of wheat and rye. V. Protein formation in grains and peptides occurring therewith. Z. Lebensm.-Untersuch. -Forsch. 116:101–13. Chem. Abstr. 56:10587. 1962.
16378. —— and G. Train. 1954: Isolation of a bitter glycoside from oats. Z. Lebensm.-Untersuch. -Forsch. 99:346–51. Chem. Abstr. 49:2573. 1955.
16379. Roistacher, C. N. 1961: Quick test for citrus seed viability. Calif. Agr. 15(10):15. Biol. Abstr. 42:3291. 1963.
Roistacher, C. N. 14248.
Roitman, R. 9325.
Roitzsch, J. E. Räder-; *see* Räder-Roitzsch, J. E.
16380. Rojas-Garciduenas, M., and T. Kommedahl. 1960: The effect of 2,4-D on germination of pigweed seed. Weeds 8:1–5.
16381. ——, M. A. Ruiz, and J. Carillo. 1962: Effects of 2,4-D and MCPA on germination and early growth. Weeds 10:69–71.
Rokosova, K. 11589.
Roland, G. 5098.
16382. Rolger, G. A. 1960: Relation of seed dormancy of green needlegrass (*Stipa viridula* Trin.) to age and treatment. Agron. J. 52:467–69.
16383. —— 1960: Relation of seed dormancy of Indian ricegrass (*Oryzopsis hymemoides* [Roem. & Schult] Ricker.) to age and treatment. Agron. J. 52:470–73.
Rolland, M. 3449.
Roller, E. M. 101.
16384. Roll-Hansen, J. 1956: Treatment of vegetable seeds. Experiments carried out in the years 1944–1954. Norwegian Plant Protect. Inst. Bull. 10. 69 pp. Grondahl, Oslo.
16385. Rollin, P. 1956: Action de la température et de la lumière sur la germination des akenes de *Bidens tripartitis* L.

(Composées). Rev. Gen. Botan. 63:461–76.

16386. —— 1958: Action qualitative de la lumière sur la germination des graines de *Phacelia tanacetifolia.* Compt. Rend. 247:1484–87.

16387. —— 1959: Influence de l'acide gibberellique sur la germination des semences photosensibles. Soc. Franc. Physiol. Végétale 5:24–26.

16388. —— 1959: Mise en evidence de deux dormances chez les akenes de *Bidens radiatus* (Composees). Rev. Gen. Botan. 66:636–44.

16389. —— 1963: Influence de la lumière et de la température sur la germination. Imprimerie Paul Dehan, Montpellier. 182 pp.

16390. —— 1963: Observations sur la difference de nature de deux photoreactions controlant la germination des akenes de *Lactuca sativa* var. "Reine de Mai." Compt. Rend. 257:3642–45.

16391. Rollin, S. F., and F. A. Johnston. 1961: Our laws that pertain to seeds. Yearbook Agr. U.S. Dept. Agr. 1961:482–92.

Rollins, G. H. 14364.

Rollins, M. L. 1663, 2657, 13699.

16392. Rollins, R. C. 1945: Evidence for genetic variation among apomictically produced plants of several F_1 progenies of guayule (*Parthenium argentatum*) and mariola (*P. incanum*). Am. J. Botany 32:554–60.

Rolston, L. H. 16475.

Romanchikova, M. L. 2997.

16393. Romano, A. H., and W. J. Nickerson. 1954: Cystine reductase of pea seeds and yeasts. J. Biochem. 208:409–16.

16394. Romano, E. 1948: Reducing power of wheat grain during its ripening and germination. Ann. Sper. Agrar. (Rome) 2:63–68. Chem. Abstr. 42:59–60. 1948.

16395. Romanov, I. D. 1962: The origin of the unique structure of endosperm nuclei in *Gagea.* Dokl. -Botan. Sci. Sect. (English Transl.) 141:188–90.

Romanova, M. S. 11312.

Romanova, T. F. Boroik-. 2787.

16396. Romashkevich, I. F. 1960: Seed viability of weeds in mesophilic methane fermentation of manure. Zemledelie 1960 (1):82–83.

16397. Romberg, L. D., J. Hamilton, Jr., and C. L. Smith. 1936: The specific gravity as related to kernel development in the pecan nut. Proc. Am. Soc. Hort. Sci. 33:66. Expt. Sta. Rec. 78:487. 1938.

Romberg, L. D. 17624.

16398. Romberger, J. A. 1963: Meristems, growth and development in woody plants. An analytical review of anatomical, physiological and morphogenic aspects. U.S. Dept. Agr. Forest Serv. Tech. Bull. 1293. 214 pp.

Romenskii, N. V. 18719.

16399. Romeo, A. G. 1937/38: Sulla germinazione dei semi del *Tamus communis* L. Portici Ist. Super. Agrar. Ann. 9:240–47.

16400. Romero, A. G. 1933: Las semillas. Partes que las forman. I. Condiciónes para su germinación. Econ. Tec. Agr. 2:475–77.

16401. —— 1933: Las semillas. II. Fenomenos morfologicos y fisiologicos de la germinación. Analisis de las semillas. Importancia de tomar bien la muestra media. Econ. Tec. Agr. 2:543–46.

16402. —— 1934: Las semillas. III. Semillas puras, distintas clases de impurezas. Determinación de la pureza de una musetra de semilla. Idem del origen o procedencia. Identidad botanica. Econ. Tec. Agr. 3:9–12.

16403. —— 1934: Las semillas. IV. Determinación de su poder germinativo. Germinadores. Econ. Tec. Agr. 3:87–90.

16404. —— 1934: Las semillas. V. La cuscuta y sus danos. Reconocimiento analitico de la cuscuta. Expedicion de semillas descuscutadas. Econ. Tec. Agr. 3:131–33.

16405. —— 1934: Las semillas. VI. Semillas duras. Valor real de las semillas. Peso. Determinación de la humedad. Econ. Tec. Agr. 3:194–97.

16406. —— 1934: Las semillas. Su conservación. Desinfección de graneros. Tratamiento de las semillas. Econ. Tec. Agr. 3:322–27.

Romm, H. J. 15610.

16407. Rommel, M. 1960: The effect of embryo transplantation on seed set in some wheat-rye crosses. Can. J. Plant Sci. 40:388–95.

Rommel, M. 8592.

Romo, J. 10643.

16408. Ronchi, V. N. 1957: Action of temperature on the germination of seeds of some crop plants. Agr. Ital. (Pisa) 62:193–211.

16409. —— and G. Martini. 1962: Germinability, seedling growth, and frequency of chromosomal aberrations in relation to seed age in wheat. Caryologia 15:293–302. Biol. Abstr. 43:11455. 1963.

16410. Rondet, P. 1962: L'organogenèse au cours de l'embryogenèse chez l'*Alyssum maritimum* Lamk. Compt. Rend. 255: 2278–80.

Ropp, R. S. de. 7641.

16411. Roques, H. 1949: Sur un heteroside isolé des semences de scille (*Urginea maritima*). Soc. Chim. Biol. 31:19–21. Biol. Abstr. 23:27434. 1949.

Rosberg, D. W. 17034.

Rose, A. Dona-Dalle-; *see* Dona-Dalle-Rose, A.

16412. Rose, D. H. 1915: A study of delayed germination in economic seeds. Botan. Gaz. 59:425–44.

Rose, H. R. de. 370.

16413. Rose, R. C. 1919: After-ripening and germination of seeds of *Tilia*, *Sambucus* and *Rubus*. Botan. Gaz. 67:281–308.

Rose, R. C. 5036.

16414. Rose, R. E. 1945: A technique for the artificial inoculation of perennial ryegrass by the blind-seed organism. New Zealand J. Sci. Technol. Sect. A 27:18–22. Biol. Abstr. 20:11180. 1946.

Rose, W. G. 9712–9714.

16415. Rosell, M. E. 1936: Morphology of the flower, fruit, and seed of *Raphanus sativus* Linn. Philippine Agriculturist 25: 521–40. Lingnan Sci. J. 16:529. 1937.

Rosen, L. A. 17412–17414.

16416. Rosen, W. 1938: Beiträge zur Kenntnis der Embryologie der Goodeniaceen. Acta Horti Gotoburg 12:1–10.

16417. —— 1940: Seed development of *Litorella uniflora* Aschers. Medd. Göteborgs Bot. Trädgard 14:177–84. Biol. Abstr. 15:21070. 1941.

16418. —— 1949: Endosperm development in Campanulaceae and closely related families. Botan. Notiser 1949:137–47.

16419. Rosenau, W. A. 1962: Effects of urea fertilization on corn seed germination and seedling development. Dissertation Abstr. 23:17.

16420. Rosenberg, O. 1906: Über die Embryobildung in der Gattung Hieracium. Ber. Deut. Botan. Ges. 24:157–61.

Rosenblum, J. 15413.

16421. Rosenfels, R. S. 1940: Spread of white-top seed in the droppings of grazing cattle. Nevada Agr. Expt. Sta. Bull. 152:5. Expt. Sta. Rec. 83:189. 1940.

Rosenkranz, G. 10643.

16422. Rosenow, D. T., A. J. Casady, and E. G. Heyne. 1962: Effects of freezing on germination of sorghum seed. Crop Sci. 2:99–102.

16423. Rosenow, P. M. 1962/63: We want coated seed. Holly Agr. News 10(4):14–15.

16424. Rosenquist, C. E. 1931: The immediate effect of foreign pollen upon the kernel weight of wheat (*Triticum vulgare*). J. Am. Soc. Agron. 23:43–63. Expt. Sta. Rec. 65:634. 1931.

16425. —— 1936: The influence of the awn upon the development of the kernel of wheat. J. Am. Soc. Agron. 28:284–88. Expt. Sta. Rec. 75:479. 1936.

Rosenstein, H. 18757.

16426. Rosenthal, F. 1948: Plastics from cottonseed hulls. Formerly worthless residue now yielding extensive line of chemurgic by-products. Chemurgic Dig. 7(8):13–15.

16427. Ross, A. M. 1939: Some morphological characters of *Helianthus annuus* L., and their relationship to the yield of seed and oil. Sci. Agr. (Ottawa) 19:372–79.

Ross, J. G. 3876, 7974.

16428. Ross, R. 1948: Seeds are sown by planes. Science News Letter 53:298–99.

16429. Rossi, E. 1930: Germination of pine seed in relation to temperature. Atti Ist. Botan. Ecc. Pavia 1:107–15. Biol. Abstr. 6:12864. 1932.

16430. Rossi-Fanelli, A., E. Antonini, M. Brunori, M. R. Bruzzesi, A. Caputo, and F. Satriani. 1964: Isolation of a monodisperse protein fraction from cottonseeds. Biochem. Biophys. Res. Commun. 15: 110–15.

16431. Rossini, A. 1935: Action of the sulfur-bearing waters of Viterbo on germination of seeds and on the activity of the esophagus epithelium in frogs. Ann. Chim. Appl. 25:541–47. Chem. Abstr. 30:4597. 1936.

16432. Rossiter, E. S. 1960: Growth stimulants for growing orchid seed. Am. Orchid Soc. Bull. 29:904–6.

16433. Rossiter, R. C. 1959: The influence of maturity grading on total yield and seed production in strains of *Trifolium subterraneum* L. grown as single plants and in swards. Australian J. Agr. Res. 10:305–21.

16434. ——, D. H. Curnow, and E. J. Underwood. 1948: The effect of cobalt sulphate on the cobalt content of subterranean clover (*Trifolium subterraneum* L. var. Dwalganup) at three stages of growth. J. Australian Inst. Agr. Sci. 14:9–14. Biol. Abstr. 23:1783. 1949.

Rossiter, R. C. 11899.

16435. Rossman, E. C. 1949: Freezing injury of inbred and hybrid maize seed. Agron. J. 41:574–83.

16436. —— 1949: Freezing injury of maize seed. Plant Physiol. 24:629–56.

Rossman, E. C. 14333, 14567.

16437. Rost, C. O. 1930: Effects of superphosphates upon the germination of corn. J. Am. Soc. Agron. 22:498–507.

Rost, C. O. 13877.

16438. Rostrup, O. 1891/95: Polyembryony. Dansk Frokontrol Beretn. 1891/92:19; 1892/93:30; 1893/94:37; 1894/95:37.

16439. —— 1891/92: Germination at constant and alternating temperatures. Dansk Frokontrol Beretn. 1891/92:16–18.

16440. —— 1892/95: Germination under glass (bell-jars) of different colors. Dansk Frokontrol Beretn. 1892/93:27–29; 1893/94:31–33; 1894/95:31–32.

16441. —— 1893/1900: Comparing the different colors of clover-seeds. Dansk Frokontrol Beretn. 1893/94:22–23; 1895/96:30–31; 1899/1900:44–45.

16442. —— 1893/94: The ability to absorb moisture from air of different seeds. Dansk Frokontrol Beretn. 1893/94:26–29.

16443. —— 1893/94: Is there any difference as to germination energy of grass seed, whether seed is placed in one way or other in germination bed? Dansk Frokontrol Beretn. 1893/94:33–34.

16444. —— 1893/94: Seeds from center-stem and lateral-stem of carrot. Dansk Frokontrol Beretn. 1893/94:34–35.

16445. —— 1894/95: Germination of *Fraxinus excelsior*, ash-seed. Dansk Frokontrol Beretn. 1894/95:33.

16446. —— 1894/95: Germination of *Cytisus laburnum*. Dansk Frokontrol Beretn. 1894/95:34.

16447. —— 1894/95: The various seeds of *Atriplex hortensis*. Dansk Frokontrol Beretn. 1894/95:34–36.

16448. —— 1896/97: Instances to decreasing germination energy. Dansk Frokontrol Beretn. 1896/97:31–33.

16449. —— 1896/97: The influence of sulfuric acid on hard seeds. Dansk Frokontrol Beretn. 1896/97:33–34.

16450. —— 1897/98: Influence of frost on germination of acorn. Dansk Frokontrol Beretn. 1897/98:31.

16451. —— 1897/98: Germination of *Poa palustris*. Dansk Frokontrol Beretn. 1897/98:32–34.

16452. —— 1897/98: Broken seedlings. Dansk Frokontrol Beretn. 1897/98:34–35.

16453. —— 1898/99: Decrease of germination capacity from spring until fall. Dansk Frokontrol Beretn. 1898/99:44–46.

16454. —— 1898/99: Germination of some cactus species. Dansk Frokontrol Beretn. 1898/99:57.

16455. —— 1898/1902: Germination of seeds of wild plants (Danish plants). Dansk Frokontrol Beretn. 1898/99:46–47; 1899/1900:27–30; 1900/1; 1901/2.

16456. —— 1899/1900: What influence has the time of year for starting a germination test, on course of development and on the germinating capacity? Dansk Frokontrol Beretn. 1899/1900:30–32.

16457. —— 1899/1900: How is seed germinating after having passed through a cow? Dansk Frokontrol Beretn. 1899/1900:33–35.

16458. —— 1899/1902: The vitality of buried seeds. Dansk Frokontrol Beretn. 1899/1900:35–37; 1900/1; 1901/2.

16459. —— 1899/1900: Influence of seawater on germination of seed. Dansk Frokontrol Beretn. 1899/1900:37–40.

16460. —— 1899/1900: Germination of different flower-seeds. Dansk Frokontrol Beretn. 1899/1900:40–44.

Roth, E. R. 2377.

16461. Roth, G. 1958: Einfluss der Quecksilber—Beizung auf Keimung und Jugendwachstum der Gerste unter besonderer Berücksichtigung ihrer selektiven Wirkung auf die samenbegleitende Mikroflora. Phytopathol. Z. 34:137–68.

16462. Roth, I. 1955: Zur morphologischen Deutung des Grasembryos und verwandter Embryotypen. Flora (Jena) 142:564–600.

16463. —— 1956: Histogenese und Entwicklungsgeschichte des *Triticum*-embryos. Flora (Jena) 144:163–212.

16464. —— 1961: Histogenesis and morphological interpretation of the grass embryo. Intern. Botan. Congr. Recent Advan. Botany 9:96–99. 1959.

Roth, L. E. 2672.

16465. Roth, L. F. 1959: Natural emplacement of dwarfmistletoe seed on Ponderosa pine. Forest Sci. 5:365–69.

Roth, N. 11036.

16466. Roth, V. D. 1962: Germination of alfalfa seed treated with dry and liquid formulations of Di-syston and Phorate. J. Econ. Entomol. 55:1–2.

16467. Rothe, G. 1932: Stauden-Anzucht aus Samen. Möllers Deut. Gärtner-Ztg. 47:110–11.

16468. Rothe, M. 1958: Properties and reaction behavior of plant seed lipases in their natural environment. I. Determination of activity under natural environmental conditions. Ernährungsforschung 3:21–40. Chem. Abstr. 54:25071. 1960.

16469. —— 1958: Properties and reaction behavior of plant seed lipases in their natural environment. II. Influence of moisture on the temperature behavior. Ernährungsforschung 3:41–51. Chem. Abstr. 54:25071. 1960.

16470. Rothea, F., and P. Bruere. 1944: Intolerance to, and food poisoning by, proteins of Leguminosae seeds (fabism, phasism, lathyrism). Ann. Pharm. Franc. 2:104–7.

Rothgeb, B. E. 4283.

16471. Rotini, O. T. 1945: The catalase of *Phaseolus vulgaris*. Helv. Chim. Acta 28:234–48. Chem. Abstr. 40:4772. 1946.

16472. —— 1948: The variations of catalase activity during seed germination. Chim. Ind. (Milan) 30:234–36. Biol. Abstr. 23:24865. 1949.

Rotor, G. 13816

16473. Rottova, M. 1964: Analysis of the ash of the seeds of *Aesculus hippocastanum* and some *Aceres*. Biol. Plantarum 6:35–41.

16474. Rotunno, N. A. 1924: Effect of size of seed on plant production with special reference to radish. Botan. Gaz. 78:397–413.

Rouatt, J. W. 10334.

Roubaix, J. de; *see* De Roubaix, J.

Rouse, A. H. 10489.

16475. Rouse, P., L. H. Rolston, and C. Lincoln. 1958: Insects in farm-stored rice. Arkansas Agr. Expt. Sta. Bull. 600. 25 pp.

Rousseau, Broig-; *see* Broig-Rousseau.

Rousseau, J. 2199–2203.

Rout, M. K. 16593.

16476. Roux, E. R. 1952: The germination of seeds. Trees S. Africa 4:22–23.

16477. —— 1953: The effect of antibiotics produced by *Trachypogon plumosus* on the germination of seeds of Kakiebos (*Tagetes minuta*). S. African J. Sci. 49:334.

Roux, J. B. 10196.

16478. Rovesti, P., and B. Mango. 1946: Dermatological action of juices of germinated seeds. Parfum. Mod. 48(54):80–85. Chem. Abstr. 52:3926a. 1958.

Row, A. V. 10516.

16479. Rowaan, P. A. 1937: Tengkawang fat. Ber. Afdel. Handelsmuseum Koninkl. Ver. Kolon Inst. 113. Pharm. Abstr. 4:557. 1938.

16480. Rowan, K. S., and D. H. Turner. 1957: Physiology of pea fruits. V. Phosphate compounds in the developing seed. Australian J. Biol. Sci. 10:414–25.

Rowell, P. 3958.

Rowell, P. T. 3959.

16481. Rowlands, M. J., and B. Wilkinson. 1930: Vitamin B content of grass seeds in relationship to manures. Biochem. J. 24:199–204.

16482. Rowlee, W. W., and G. T. Hastings. 1898: The seeds and seedlings of some Amentiferae. Proc. Am. Assoc. Advan. Sci. 47:436. Expt. Sta. Rec. 10:921. 1898/99.

16483. Rowley, G. D. 1956: Germination in *Rosa canina*. Am. Rose Ann. 41:70–73.

16484. Rowntree, L. 1930: Longevity of seeds in the desert. Horticulture 8(11):270.

16485. —— 1931: Germinating difficult seeds. Am. Botanist 37:103–4.

16486. Rowsell, E. V., and L. J. Goad. 1964: The release of hydrolytic enzymes from isolated wheat aleurone layers activated by gibberellic acid. Biochem. J. 90. 12 pp.

16487. —— and —— 1964: Some effects of gibberellic acid on wheat endosperm. Biochem. J. 90. 11 pp.

16488. Rowson, J. M. 1952: IV. The hypodermal zone in the testa of certain leguminous seeds. J. Roy. Microscop. Soc. Ser. III 72:46–55.

Roy, A. 17544.

Roy, A. K. 3762.

Roy, A. R. G. 1851.

16489. Roy, D. F. 1957: Seed spot tests with tetramine-treated seed in northern California. J. Forestry 55:442–45.

Roy, M. 12414.

16490. Roy, N. N., and H. L. Everett. 1963: Seed production, fertility levels, and cold test germination in corn. Crop Sci. 3:273–74.

Roy, S. 6168.

Roy, S. C. 1628, 4921, 18178.

16491. Roy, S. K., and R. Sahai. 1962: The embryo-sac and embryo of *Syzygium caryophyllifolium* DC. J. Indian Botan. Soc. 41:45–51.

16492. —— and D. K. Sarkar. 1962: Studies on the dormancy and viability of paddy seeds of six paddy varieties. J. Indian Botan. Soc. 41:207–16.

Roy, S. K. 3454, 3778, 12843, 14220, 14221, 16631.

Roy, W. R. 13189.

Royce, R. 7451.

16493. Royce, R. D. 1962: Special seeds help weeds to spread. J. Agr. W. Australia 3:57–58.

16494. Roze, M. E. 1894: Dehiscence of fruit of *Ecballium elaterium*. J. Botan. (Morot) 8:308–18.

16495. Rozhkov, M. I. 1961: Changes in vitamins C and P accumulation in seeds of *Rosa canina*. Tr. Vses. Sel'skokhoz. Inst. Zaochn. Obrazov. 1961(5):40–42. Chem. Abstr. 58:752. 1963.

16496. Ruan, E., and K. J. Frey. 1957: Effect of heat treatment on oat seeds. Proc. Iowa Acad. Sci. 64:139–48.

Ruan, E. 6731.

16497. Ruban, E. L., and I. A. Komarov. 1954: Treatment of seed of tree and shrub species with ultrasonics. Byul. Gl. Botan. Sada 17:54–56.

16498. Rubin, B. A., E. V. Artsikhovskaya, and O. T. Lutikova. 1948: Metabolism of smooth and wrinkled peas in the period of germination. Dokl. Akad. Nauk SSSR 59:103–6. Chem. Abstr. 42:6422–23. 1948.

16499. Rubner, K. 1927: Depth of covering and germination in spruce seeds. Forstwiss. Centralbl. 49:168–83.

16500. —— 1938: Keimung von Samen grün- und rotzapfiger Fichten. Tharandter Forstl. Jahrb. 89:247–51. Biol. Abstr. 12:7063. 1938.

16501. Ruchkin, V. N. 1941/46: Conditions determining the qualitative side of the oil-forming process in seeds. Trans. Kirov Inst. Agr. (USSR) 20:165–90. Chem. Abstr. 42:956. 1948.

Ruckman, J. E. 7494.

16502. Rudakova, M. M. 1950: Study of dynamics of seed swelling for early determination of earliness of a variety. Selek. i Semen. 17(12):50–53.

16503. Ruddat, M., A. Lang, and E. Mosettig. 1963: Gibberellin activity of steviol, a plant terpenoid. Naturwissenschaften 50: 23.

16504. Ruden, T. 1960: The provenance problem in Norwegian forestry. Norsk Skogbruk 6:768–70.

16505. Rudenko, F. E. 1962: Data on embryogenesis of some alpine plants in the Carpathian mountains. Nauchn. Dokl. Vysshei Shkoly Biol. Nauki 1962(1):119–22.

16506. Rudolf, P. O. 1937: Delayed germination in American elm. J. Forestry 35: 876–77.

16507. —— 1950: Cold soaking—a short-cut substitute for stratification? J. Forestry 48:31–32.

16508. —— 1950: A test of pelleted Jack pine seed. J. Forestry 48:703–4.

16509. —— 1952: Cold soaking of Jack pine seed. J. Forestry 50:626.

16510. —— 1952: Low temperature seed storage for western conifers. U.S. Dept. Agr. Forest Serv. Lake States Forest Expt. Sta. Misc. Rept. 20. 8 pp.

16511. ——, chairman. 1962: More on tree seed certification and legislation. J. Forestry 60:349–50, 352.

16512. —— and K. E. Clausen. 1962: Red pine seed germination after 30 years of storage. J. Forestry 60:128–31.

Rudolf, P. O. 4178, 17579.

16513. Rudolfs, W. 1921: Effect of salt solutions having definite osmotic concentration values upon absorption by seeds. Soil Sci. 11:277–93.

16514. —— 1925: Effect of seeds upon hydrogen-ion concentration equilibrium in solution. J. Agr. Res. 30:1021–26.

16515. —— 1925: Influence of water and salt solution upon absorption and germination of seeds. Soil Sci. 20:15–37.

16516. —— 1925: Selection absorption of ions by seeds. Soil Sci. 20:249–52.

16517. Rudolph, B. A. 1944: The unimportance of tomato seed in the dissemination of *Verticillium* wilt in California. Phytopathology 34:622–30.

16518. —— and G. J. Harrison. 1944: The unimportance of cotton seed in the dissemination of *Verticillium* wilt in California. Phytopathology 34:849–60.

16519. —— and —— 1945: The invasion of the internal structure of cotton seed by certain fusaria. Phytopathology 35:542–48.

16520. Rudolph, H. 1961: Über die Keimungsfördernde Wirkung verschiedener Acetaldehyd-Präparate. Z. Naturforsch. 16b:611–14.

16521. Rudorf, W., G. Stelzner, and J. Hartisch. 1937: Untersuchungen zur Methodik einer Keimstimmung bei Wärme. Angew. Botan. 19:491–505.

Rudowska, K. 17733.

16522. Rudra, M. N. 1938: Studies in vitamin C. V. The vitamin C content of some germinated cereals and pulses. J. Indian Chem. Soc. 15:191–93. Chem. Abstr. 32:7521. 1938.

16523. Ruehle, G. D., and R. B. Ledin.

1955: Mango growing in Florida. Florida Agr. Expt. Sta. Bull. 574. 90 pp.

16524. Ruelleman, R. F., and W. L. Burlison. 1942: Germination studies of bromegrass seed, *Bromus inermis* Leyss. Trans. Illinois State Acad. Sci. 35:32–34. Expt. Sta. Rec. 89:307. 1943.

Rümmler, W. 12512.

Rünger, W. 12582, 12583.

16525. Ruge, U. 1939: Zur Physiologie der genuinen keimungshemmenden und keimungsbeschleunigenden Stoffe von *Helianthus annuus*. Z. Botan. 33:529–71.

16526. —— 1946: Versuche zur Beeinflussung der Nachreife der Sommergerste. Züchter 17/18:59–62. Biol. Abstr. 23: 1996. 1949.

16527. —— 1947: Untersuchungen über keimungsfördernde Wirkstoffe. Planta 35: 297–318.

16528. —— 1948: Ergebnisse der neuren Keimungsphysiologie. Naturw. Rundschau 5:217–21. Biol. Abstr. 23:30210. 1949.

16529. —— 1952: Über die Steigerung der Keimfähigkeit alten Saatgutes mit Hilfe von Äthylen-Chlorhydrin. Angew. Botan. 26:162–65.

16530. —— 1955: The analysis of seed heating in Malvaceae. Beitr. Biol. Pflanz. 31:409–17. Chem. Abstr. 49:11785. 1955.

16531. —— and D. Lietke. 1951: Zur periodischen Keimbereitschaft einiger Malven-Arten. Ber. Deut. Botan. Ges. 64: 141–50.

16532. Ruhland, W., editor. 1961: Encyclopedia of Plant Physiology. Vol. XIV. Growth and growth substances. Springer, Berlin. 1357 pp.

16533. —— 1964: Encyclopedia of Plant Physiology. Vol. XV. Differentiation and development. Springer, Berlin. Part 1, approx. 1680 pp. Part 2, approx. 1560 pp,

16534. Ruhowicz, K., and J. Horski. 1956: The industrial value of rape-seed. Roczniki Nauk Rolniczych Ser. A 72:503–26. Biol. Abstr. 33:3035. 1959.

16535. Ruinard, J. 1964: Cocoa research in West Irian (West New Guinea). Neth. J. Agr. Sci. 12:89–112.

16536. Ruiz, A. S., and A. S. Merino. 1946: Chemical composition of the seeds of *Caloncoba welwitschii*. IV. Proteins, mineral salts, and ergones. Anales Fis. Quim. (Madrid) 42:1147–52. Chem. Abstr. 42: 7383. 1948.

16537. —— and M. S. Munoz. 1947: Algunos datos analiticos de la semilla del *Plantago psyllium*. Anales Fis. Quim. (Madrid) 43:476–81.

16538. —— and —— 1947: Algunos datos analiticos sobre la semilla del *Plantago psyllium*. II. Glucidos, sales minerales y oligoelementos. Anales Fis. Quim. (Madrid) 43:1029–32.

Ruiz, A. S. 20019, 20020.

Ruiz, A. Santos-. 14980.

Ruiz, M. A. 16381.

16539. Rulf, P. 1884: Behavior of tannin during germination. Z. Naturwiss. 57:40–46.

16540. Rumbaugh, M. D. 1959: Test weight and maturity of corn. Agron. J. 51:307.

16541. —— 1961: Selectivity of the seed coat of alfalfa during germination under osmotic tension. Crop Sci. 1:461–62.

16542. —— 1960: Influence of mannitol on germination and seedling development of alfalfa. Alfalfa Improvement Conf. Rept. 17:68–73.

16543. Rumi, C. P., and O. R. Rivoir. 1959: Variaciónes bioquimicas relaciónadas con el desarrollo en trigo. Rev. Fac. Agron. Argentina 35:65–77. Biol. Abstr. 35: 32794. 1960.

16544. Rumsey, A. E. 1963: The interaction of temperature treatment and moisture content on subsequent germination of tomato seed. Proc. Am. Soc. Hort. Sci. 82:446–53.

16545. Rumsey, E., and E. M. Hutton. 1943: Commercial extraction of cucumber seed by the acid method. J. Council Sci. Ind. Res. Australia 16:205–6.

16546. Rumyantsev, P. D. 1955: Measures for combatting the diseases and blights of maize seed during storage. Zemledelie 1955(10):81–87. Biol. Abstr. 32:10406. 1958.

16547. Runov, V. I., and K. Sverdlina. 1940: Temperature coefficient and activation energy in the seeds of Tau-Saghyz as influenced by environmental temperature. Compt. Rend. Acad. Sci. URSS 29:132–34.

Runov, V. I. 14227.

16548. Runyon, E. H. 1930: Germination and establishment of *Covillea*. Carnegie Inst. Wash. Yearbook 29:225–26.

16549. Runyon, H. E. 1937: Distribution of seeds by dust storms. Trans. Kansas Acad. Sci. 39:105–13.

16550. Ruppel, R. F., G. Bravo, and W. H. Hatheway. 1960: Effectiveness of mixtures of acrylonitrile and carbon tetrachloride against three pests of stored corn. J. Econ. Entomol. 53:238–42.

16551. Rusanov, A. I. 1957: Classification of seeds according to specific weight by the dry method. Vestn. Sel'skokhoz. Nauki 1957(1):99–110. Biol. Abstr. 35:27612. 1960.

16552. Rusca, R. A., and F. L. Gerdes. 1942: Effects of artificially drying seed cotton on certain quality elements of cotton seed in storage. U.S. Dept. Agr. Circ. 651. 19 pp.

16553. Rusch, V. A., and G. A. Ivanova. 1940: Analysis of wild-rose seeds and wild-rose seed oil. Compt. Rend. Acad. Sci. URSS 26:259. Pharm. Abstr. 8:9. 1942.

Rusek, Z. 19659.

16554. Rush, G. E., and N. P. Neal. 1951: The effect of maturity and other factors on stands of corn at low temperatures. Agron. J. 43:112–16.

Rush, G. E. 16826.

16555. Rushkovskii, S. 1930: Changes in the chemical composition of sunflower seeds with the sowing time. Maslob.-Zhir. Delo 1930(9/10):43–56. Chem. Abstr. 26:4356. 1932.

16556. —— 1957: Biochemical characteristics of high oil content seeds of sunflower plants. Sbornik 1957:123–28. Chem. Abstr. 53:8321. 1959.

16557. Russel, G. A. 1919: Effect of removing the pulp from camphor seed on germination and the subsequent growth of seedlings. J. Agr. Res. 17:223–38.

16558. Russell, E. E. 1952: Stink bugs on seed alfalfa in southern Arizona. U.S. Dept. Agr. Circ. 903. 19 pp.

16559. —— and O. L. Barnes. 1951: Tests with insecticides for control of lygus bugs on alfalfa in southern Arizona, 1949–1950. U.S. Dept. Agr., Agr. Res. Admin. Bur. Entomol. Plant Quarant. E-831. 9 pp.

16560. Russell, E. J. 1920: Report on the proposed electrolytic treatment of seeds (Wolfryn process) before sowing. Gt. Brit. Min. Agr. J. 26:971–81. Botan. Abstr. 5:59. 1920.

Russell, G. C. 5592, 5593.

Russell, J. 12476.

Russell, J. R. 1642.

16561. Russell, P. 1961: The seed collection United States Department of Agriculture. Hort. Mag. 40:325–33.

16562. Russell, P. G., and A. F. Musil. 1961: Plants must disperse their seeds. Yearbook Agr. U.S. Dept. Agr. 1961:80–88.

16563. Russell, R. C. 1946: Testing seed for smut spores as an aid in controlling cereal smuts in Saskatchewan. Sci. Agr. (Ottawa) 26:372–80.

16564. —— 1950: A study of the hot water treatment of barley for the control of loose smut, *Ustilago nuda*. Sci. Agr. (Ottawa) 30:303–15.

16565. —— 1950: The whole embryo method of testing barley for loose smut as a routine test. Sci. Agr. (Ottawa) 30:361–66.

16566. —— 1953: Progress report on the control of loose smut of barley by the cold water soak treatment. [Abstract.] Proc. Can. Phytopathol. Soc. 21:16–17.

16567. —— 1954: A study of the relative vigour of healthy barley seed and that of seed infected with *Ustilago nuda*. [Abstract.] Proc. Can. Phytopathol. Soc. 21: 17.

16568. —— 1958: Longevity studies with wheat seed and certain seed-borne fungi. Can. J. Plant Sci. 38:29–33.

16569. —— 1960: Observations on cereal embryos in relation to the "embryo test" for loose smuts. Can. J. Botany 38:251–57.

16570. —— 1961: The influence of aging of seed on the development of loose smut in barley. Can. J. Botany 39:1741–46.

16571. —— 1962: The production of smut-free barley seed as a community project. Empire J. Exptl. Agr. 30:286–94.

16572. —— and S. H. F. Chinn. 1958: The salt-water soak treatment for the control of loose smut of barley. Plant Disease Reptr. 42:618–21.

16573. —— and R. J. Ledingham. 1941: Wheat seed testing from the pathological standpoint with special reference to embryo exposure. Sci. Agr. (Ottawa) 21:761–75.

16574. ——, B. J. Sallans, and H. B. Cannon. 1961: The use of sequential sampling in embryo tests for loose smut in barley seeds. Can. J. Plant Sci. 41:772–79.

Russell, R. C. 3960.

Russell, W. A. 8040.

Russell, W. E. 3045, 3046.

Russell-Wells, B. 9717.

16575. Russo, G. 1940: Treatment of infected cotton seeds. Agr. Colon. 34:49–53. Soc. Chem. Ind. J. 59:887B. 1940.

Russo, S. 18895.

16576. Rust, G. 1955: The influence of agricultural methods on the fat-protein ratio of cocklebur and soybeans. Wiss. Z.

Friedrich-Schiller-Univ., Jena, Math. Naturwiss. Reihe 4:441–50. Chem. Abstr. 50:1988. 1956.

Ruston, D. F. 12578.

Ruta, C. 4195.

16577. Ruther, H. 1953: The value of cultivating *Lupinus albus* for fat and protein. Z. Acker- Pflanzenbau 97:253–60. Chem. Abstr. 48:9580. 1954.

16578. Rutherford, B. E., and K. M. Pretty. 1960: The nutritive value of corn and soybeans as influenced by soil treatment. Agron. J. 52:27–29.

16579. Rutishauser, A. 1956: Cytogenetik des Endosperms. Ber. Schweiz. Botan. Ges. 66:318–35. Biol. Abstr. 31:33495. 1957.

Rutishauser, A. 11640.

16580. Rutledge, W. A., and R. H. Common. 1945: Composition and digestibility of Northern Irish ryegrass seed and ryegrass seed cleanings. II. Data for "flatweed," "hairgrass," and commercial Yorkshire fog seed; note on inorganic constituents. J. Agr. Sci. 35:123–25. Brit. Abstr. B III:27. 1946.

16581. Ruzicka, J. 1938: Proof of the harmfulness of using spruce seed from inappropriate sources. Lesn. Práce 17: 533–39. Rev. Appl. Mycol. 18:219. 1939.

Ryan, C. A., Jr. 7361.

Ryazantseva, M. I. 309, 11208.

16582. Rychlewski, J. 1961: Cyto-embryological studies in the apomictic species *Nardus stricta* L. Acta Biol. Cracoviensia Ser. Botan. 4:1–23.

16583. Rycther-Kowalska, M. 1962: The differentiation of the carotenoid content in the grains of breeding strains and varieties in maize (*Zea mays* L.). Howdowla Roslin Aklim. Nasiennictwo 6:311–38.

16584. Ryczkowski, M. 1962: Changes in the osmotic value of the sap from embryos, the central vacuole, and the cellular endosperm during development of the ovules. Bull. Acad. Polon. Sci. Ser. Sci. Biol. 10:375–80.

16585. Ryder, E. J. 1958: The effects of complementary epistasis on the inheritance of a quantitative character, seed size in limba beans. Agron. J. 50:298–301.

Ryker, T. C. 3941.

16586. Ryoka, K., and T. Noguti. 1933: Study of rice-embryo. I. Fat. Rept. Phys. Chem. Res. Inst. 12:271–85. Japan. J. Med. Sci. II. Biochem. 4:63. 1938.

Ryser, G. K. 16825, 16826.

16587. Ryugo, K. 1962: The accumulation of lignin and the concurrent changes in the apparent density of the cell wall in the peach endocarp. Proc. Am. Soc. Hort. Sci. 80:197–203.

16588. —— 1962: Occurrence of vanillin and structurally related aldehydes in the endocarp of some *Prunus* species. Proc. Am. Soc. Hort. Sci. 81:180–83.

16589. Ryzhov, N. I. 1928: Determination de la faculte germinative de *Phacelia tanacetifolia* dans les conditions du laboratoire. Jard. Botan. Inst. Essais Semences, Leningrad, Ann. Essais Semences 6:104–10.

S

Saad, A. 14288.

16590. Saalbach, E. 1955: The influence of thymoquinone on the germination of cereals. Proc. Natl. Acad. Sci. India Sect. A 24:142–45. Chem. Abstr. 51: 16743. 1957.

Saalbach, E. 6443–6445.

16591. Saarivirta, M., and A. I. Virtanen. 1963: A method for estimating benzyl isothiocyanate, benzyl thiocyanate, and benzyl nitrile in the crushed, moistened seeds of *Lepidium sativum*. Acta Chem. Scand. 17:S74–S78.

Saarivirta, M. 19146.

16592. Sabala, E. 1962: Significance of the acids of the Krebs cycle in the germination of seeds. Farmacognosia 22:1–74. Chem. Abstr. 60:2045. 1964.

16593. Sabata, B. K., and M. K. Rout. 1963: Oil from the seeds of *Garcinia xanthochymus*. J. Proc. Inst. Chemists (India) 35:193–98. Chem. Abstr. 60:7037. 1964.

16594. Saber, A. H. 1934: The quantitative determination of powdered linseed. Quart. J. Pharm. Pharmacol. 7:645–53. Biol. Abstr. 11:6881. 1937.

16595. Sabet, Y. S. 1931: Development of the embryo-sac in *Calotropis procera* with especial reference to endosperm formation. Ann. Botany (London) 45:503–18. Botan. Centralbl. 162:200. 1932.

16596. Sabharwal, P. S. 1962: In vitro culture of nucelli and embryos of *Citrus aurantifolia* Swingle. Symp. Plant Embryol., Univ. Delhi, 1960. (Papers), pp. 239–43.

16597. Sabolotsky, M. 1934: Der Einfluss von Kohlenoxydgas auf die Atmung und

Keimfähigkeit von Weizenkorn. Botan. Zh. 19:106–25.

16598. Saboor, M. A. 1944: Composition of the fatty acids of the seed fat of *Lophira alata*. J. Indian Chem. Soc. 21:303–6. Analyst 70:178–79. 1945.

Sacco, J. da Costa-; *see* Da Costa-Sacco, J.

16599. Sachar, R. C., and U. Arora. 1963: Some embryological aspects of *Amomum dealbatum* and *Hedychium acuminatum*. Botan. Gaz. 124:353–60.

16600. —— and R. N. Chopra. 1957: A study of the endosperm and embryo in *Mangifera* L. Indian J. Agr. Sci. 27:219–28.

16601. —— and S. Guha. 1962: In vitro growth of achenes of *Ranunculus sceleratus* L. Symp. Plant Embryol., Univ. Delhi, 1960 (Papers), pp. 244–53.

16602. —— and M. Kapoor. 1958: Influence of kinetin and gibberellic acid on the test tube seeds of *Cooperia pedunculata* Herb. Naturwissenschaften 45:552–53.

16603. —— and —— 1959: Gibberellin in the induction of parthenocarpy in *Zephyranthes*. Plant Physiol. 34:168–70.

Sachar, R. C. 12703, 12704.

16604. Sacharov, V. V., S. L. Frolova, and V. V. Mansurova. 1944: High fertility of buckwheat tetraploids obtained by means of colchicine treatment. Nature 154:613. Expt. Sta. Rec. 92:635. 1945.

16605. Sacher, J. A. 1956: Observations on pine embryos grown in vitro. Botan. Gaz. 117:206–14.

16606. —— 1961: An IAA oxidase-inhibitor system in bean pods. I. Physiological significance and source of the inhibitor. Am. J. Botany 48:820–28.

16607. —— 1962: An IAA oxidase-inhibitor system in bean pods. II. Kinetic studies of oxidase and natural inhibitor. Plant Physiol. 37:74–82.

16608. —— and K. T. Glasziou. 1959: Effects of auxins on membrane permeability and pectic substances in bean endocarp. Nature 183:757–58.

Sacher, J. A. 7303.

Sachindra, K. 1914.

16609. Sachs, J. 1860: Physiologische Untersuchungen über die Abhängigkeit der Keimung von der Temperatur. Jahrb. Wiss. Botan. 2:338–77.

16610. —— 1862: Zur Keimungsgeschichte der Dattel. Botan. Z. 20:241–46, 249–52.

16611. —— 1865: Handbuch der Experimental-Physiologie der Pflanzen. Engelmann, Leipzig. 514 pp.

Sachs, M. 11037, 11038.

16612. Sachsse, R., and W. Kormann. 1874: Origin of asparagin in germinating peas. Landwirtsch. Versuchssta. 17:88–104.

Sackman, R. F. 17122.

16613. Sacks, W., and C. O. Jensen. 1951: Malease, a hydrase from corn kernels. J. Biol. Chem. 192:231–36.

Sacks, W. 9783.

16614. Sackston, W. E. 1950: Effect of pasmo disease on seed yield and thousand kernel weight of flax. Can. J. Res. Sect. C 28:493–512.

16615. —— and R. B. Carson. 1951: Effect of pasmo disease (*Septoria linicola*) of flax on the yield and quality of linseed oil. Can. J. Botany 29:339–51.

16616. —— and J. W. Martens. 1959: Dissemination of *Verticillium albo-atrum* on seed of sunflower (*Helianthus annuus*). Can. J. Botany 37:759–68.

Sadasivan, V. 17833.

16617. Sadokova, A. P. 1957: Proteins of tung tree seed. Dokl. Vses. Akad. Sel'-skokhoz. Nauk 22(12):22–26. Chem. Abstr. 52:8295d. 1958.

16618. —— 1957: Toxic properties of histonelike proteins from seeds of soya. Byul. Nauchn.-Tekhn. Inform. Vses. Nauchn.-Issled. Inst. Zhivotn. 1957 No. 1 (3):39–41. Chem. Abstr. 52:20438. 1958.

Sadovskaja, R. O. 14609.

16619. Sadykhov, A. M. 1957: Certain biological peculiarities of incompletely ripe seeds. Dokl. Akad. Nauk Azerb. SSR 13:215–19. Biol. Abstr. 35:27613. 1960.

Sadykov, A. S. 9498, 9499.

16620. Saeki, H. 1936: Studies on the effects of X-ray radiation upon germination, growth and yield of rice plants. J. Soc. Trop. Agr. (Taiwan) 8:28–38. Biol. Abstr. 11:8019. 1937.

Saeland, E. 5835.

16621. Safina, G. 1953: The amino acids of citrus-fruit juices and seeds. Conserve Deriv. Agrumari (Palermo) 2:178–81. Chem. Abstr. 48:5393. 1954.

16622. Safiulla, K. M., and H. C. Govindu. 1949: The development of the female gametophyte and endosperm in *Moniera hamiltoniana* T. Cooke. [Abstract.] Proc. Indian Sci. Congr. 36:129–30.

Safonova, E. I. Shefer-; *see* Shefer-Safonova, E. I.

16623. Safta, I. 1932: The quality of wheat varieties in different stages of maturity. Fortschr. Landwirtsch. 7:577–80. Chem. Abstr. 27:2501–2. 1933.

16624. Sagatov, S. S. 1961: On the biology of the germination of seeds of *Rumex tianschanicus*. Uzbeksk. Biol. Zh. 1961(3): 36–38.

Sager, R. W. 7393.

16625. Sagi, F., and A. S. Garay. 1961: The effect of gibberellic acid on germination and growth of white lupine. Kiserl. Kozl. A 54:105–11. Chem. Abstr. 59: 14506. 1963.

Sah, J. D. 11686.

16626. Saha, B. 1956: Studies on the development of the embryo of *Oryza sativa* L. and the homologies of its parts. Proc. Natl. Inst. Sci. India Pt. B 22:86–101.

16627. Sahadevan, P. C. 1953: Studies on the problem of loss of viability of rice seeds in storage. Madras Agr. J. 40:133–43.

16628. —— 1961: Variation among rice varieties in regard to germination under submersion, and heat stability of chlorophyll. Current Sci. (India) 30:235.

16629. —— and N. R. Nair. 1959: Peculiarities noted in germinating rice grains. Current Sci. (India) 28:448.

16630. —— and M. B. V. N. Rao. 1947: Note on the deterioration in germination-capacity of a paddy strain in Malabar and South Kanara. Current Sci. (India) 16:319–20.

Sahadevan, P. C. 14106.

16631. Sahai, R., and S. K. Roy. 1962: Polyembryony in *Eugenia heyneana* Duthie. Sci. Cult. (Calcutta) 28:37–38.

Sahai, R. 16491.

16632. Sahashi, Y. 1930: The ether-soluble substances of polished rice. Bull. Inst. Phys.-Chem. Res. (Tokyo) 9:882–90. Chem. Abstr. 25:1876. 1931.

16633. Sahasrabuddhe, D. L. 1933: A biochemical study of the formation of the oil in niger seed (*Guizotia abyssinica*). Indian J. Agr. Sci. 3:57–88. Chem. Abstr. 27:4558. 1933.

16634. —— and M. M. Kibe. 1934: The chemical changes during the formation and germination of the rice grain. J. Univ. Bombay 3:121–34. Chem. Abstr. 29:4043. 1935.

Sahasrabuddhe, D. L. 2423, 2424.

16635. Sahasrabudhe, M. B. 1945: Estrogen potency of the defatted castor seed. Current Sci. (India) 14:69. Chem. Abstr. 39:5311. 1945.

16636. —— and N. N. De. 1943: Estrogen therapy of the Indian cress oil. Current Sci. (India) 12:23–24.

16637. Sahni, B. 1920: On certain archaic features in the seed of *Taxus baccata*, with remarks on the antiquity of the Taxineae. Ann. Botany (London) 34:117–34.

16638. —— 1921: Note on the presence of a "tent-pole" in the seed of *Cephalotaxus*. Ann. Botany 35:297–98.

16639. —— 1933: Explosive fruits in *Viscum japonicum*, Thunb. J. Indian Botan. Soc. 12:96–101.

16640. —— 1934: Exembryonic seeds. Current Sci. (India) 3:109–10.

16641. Sahni, N. D. 1939: Nim germination. Indian Forester 65:117–18. Biol. Abstr. 14:5577. 1940.

Sahu, G. 13577.

Saïas, E. 9734.

16642. Said, H. 1946: Note on the effect of temperature and oxygen on the viability of cotton seeds. New Phytologist 45:282.

Saini, A. D. 1453.

St. Angelo, A. J. 14852.

St. John, J. L. 6098.

Saint-Pierre, A. 13817.

Saio, K. 8354.

Saisho, I. 9363.

Saito, C. 14854.

16643. Saito, K. 1953: On the forms and germinating power of mulberry seeds, produced from the fruits of different parts of the branch. J. Sericult. Sci. Japan 22:171–74.

16644. —— 1955: Studies on characteristics of crested-doubleness and its seed production in *Cosmos bipinnatus*. Japan. J. Breeding 5:61–68.

Saito, K. 13612, 14072.

Saito, M. 14923, 19953.

Saito, T. 18254, 18255.

Saito, Y. 14515, 18424.

Sajfertova, D. Dykyj-; *see* Dykyj-Sajfertova, D.

16645. Sakae, M. 1933: On the influence, on the germinative power of the seed of wheat and barley, of soaking in cool and hot water. Res. Bull. Agr. Expt. Sta. S. Manchuria Rlwy. Co. 12:1–54.

16646. Sakaguchi, K., R. Akahoshi, and M. Ota. 1963: The role of bound water in plant seeds with special reference to their germination ability. J. Agr. Chem. Soc. Japan 37:320–25.

Sakaguchi, S. 10320, 10321.

16647. Sakaguti, I. 1930: Relation between maturation of fruit flesh and development of seeds in peaches. Proc. Crop Sci. Soc. Japan 2:96–103. Biol. Abstr. 5:28910. 1931.

Sakai, H. 10921, 19971.

16648. Sakai, K. I. 1955: Secondary selection and a plan for multiplication of superior seed of rice varieties. Indian J. Genet. Plant Breeding 15:18–24.

Sakai, N. 14160.

16649. Sakanishi, Y. 1957: Studies on the germination behavior of some annual and biennial flower seeds. Bull. Univ. Osaka Prefect. Ser. B 7:23–28.

Sakashita, A. 13627.

16650. Sakata, T. 1933: The use of chemicals to determine the vitality of seeds. Seed World 33(13):64.

16651. —— 1959: Selling and storing of seeds in Japan. Seed World 85(8):29–30.

Sakharova, I. A. 10394.

16652. Sakisaka, M. 1929: On the seed-bearing leaves of Ginkgo. Japan. J. Botany 4:219–36.

16653. Sakovrik, A. S., and A. S. Komel. 1958: Effect of age of tobacco seeds planted in late autumn on yield of seedlings suitable for planting. Tabak (USSR) 19(3):55–57.

16654. Sakr, E. S. M. 1944: Effect of temperature and photoperiod on seedstalk development in turnips. Proc. Am. Soc. Hort. Sci. 44:473–78.

16655. —— and E. D. M. Essam. 1952: Viability of seeds harvested from fruits at different stages of maturity. Proc. Am. Soc. Hort. Sci. 60:327–29. Biol. Abstr. 27:20521. 1953.

16656. —— and H. C. Thompson. 1942: Effect of temperature and photoperiod on seed-stalk development in carrots. Proc. Am. Soc. Hort. Sci. 41:343–46.

16657. Saksena, H. B. 1954: Floral morphology and embryology of *Fumaria parviflora* Lamk. Phytomorphology 4: 409–17. Biol. Abstr. 31:39235. 1957.

Sakul, P. Cheoa-; *see* Cheoa-Sakul, P.

Sakuraba, S. 14924.

Sakurai, Y. 9585.

16658. Salageanu, N., and V. Petrea. 1956: Influence of some salts on the germination and respiration of some varieties of corn. Analele Univ. "C. I. Parhon" Bucuresti Ser. Stiint. Nat. 1956:153–60. Chem. Abstr. 52:18695. 1958.

16659. Salcheva, G. 1962: Effect of prolonged freezing temperatures on germination of seeds of some agricultural plants. Selskostopanska Nauka 7/8:869–71.

Saleem, A. 9262, 9263.

Saleh, W. 17726.

Saletore, S. A. 76, 2426, 14092, 14093.

16660. Salgues, R. 1956: Les lipides des semences d'*Euphorbia hiberna* L. et spécialement l'acide myristique. Compt. Rend. 242:167–69.

16661. Salisbury, E. J. 1942: The reproductive capacity of plants. G. Bell, London. 244 pp.

16662. —— 1963: Intermittent germination of *Capsella*. Nature 199:1303–4.

16663. —— and G. H. Preston. 1949: Note on fruit size and viability in *Lithospermum officinale*. Kew Roy. Botan. Gardens Bull. 1949(2):153–55.

16664. Salisbury, L. F. 1937: The lipids of Connecticut shade-grown tobacco seed. J. Biol. Chem. 117:21–25.

Sallans, B. J. 11969, 16574.

16665. Sallans, H. R. 1964: Factors affecting the composition of Canadian oil seeds. J. Am. Oil Chemists Soc. 41:215–18.

16666. —— and J. A. Anderson. 1939: Varietal differences in barleys and malts. VII. Starch-liquefying activity, autolytic diastatic activity and their correlations with saccharifying and proteolytic activity. Can. J. Res. Sect. C 17:361–72.

16667. —— and —— 1940: Varietal differences in barleys and malts. X. Correlations of carbohydrates with nitrogen fractions and with malt extract, steeping time, and malting loss. Can. J. Res. Sect. C 18:219–29.

16668. ——, M. Berenbom, and R. K. Larmour. 1945: Canadian sunflower seed. I. Bushel weight as a factor in grading. Can. J. Res. Sect. F 23:91–103.

16669. —— and G. D. Sinclair. 1944: Canadian linseed. II. Relations between iodine value and fatty acid composition of linseed oil. Can. J. Res. Sect. F 22:132–45.

16670. —— and —— 1945: A laboratory huller for sunflower seed and oats. Can. J. Res. Sect. F 23:306–12.

16671. ——, ——, and R. K. Larmour. 1944: The spontaneous heating of flaxseed and sunflower seed stored under adiabatic conditions. Can. J. Res. Sect. F 22:181–90.

Sallans, H. R. 2177, 4310, 11805, 11806, 17480, 20013.

Sallusto, F. 19053.

16672. Salmeron, J. M. G. 1955: El cambio de polaridad en la germinación del *Pinus halepensis* y *Stipa tenacissima*. Montes

(Madrid) 11:441–47. Biol. Abstr. 31: 32761. 1957.

16673. —— and F. Del Bano. 1961: Water, fats, and total nitrogen in seeds of *Pinus halepensis* and *Pinus pinaster*. Montes (Madrid) 17:525–28. Biol. Abstr. 38: 15654. 1962.

16674. Salmon, E. S., and W. M. Ware. 1935: The chlorotic disease of the hop. IV. Transmission by seed. Ann. Appl. Biol. 22:728–30.

16675. Salmon, J. 1939: Formation et structure des tubes cribles dans l'axe hypocotyle de *Cucurbita pepo*. Compt. Rend. 208:1041–43.

Salmon, S. C. 10439.
Salmon, W. D. 16808.
Salomon, H. 10281, 10282.
Salomon, M. 19289.

16676. Salter, J. 1857: On the vitality of seeds after prolonged submersion in the sea. J. Linnean Soc. London 1:140–42.

16677. Salter, R. M., and T. C. McIlvaine. 1920: Effect of reaction of solution on germination of seeds and on growth of seedlings. J. Agr. Res. 19:73–96.

Saltmarsh, M. J. 15691.
Salunkhe, D. K. 5328, 8645.

16678. Salzmann, R. 1954: Untersuchungen über die Lebensdauer von Unkrautsamen im Boden. Mitt. Schweiz. Landwirtsch. 2:170–76.

16679. Samal, K. K. 1936: The development of the embryo-sac and embryo in *Crotalaria juncea* L. J. Indian Botan. Soc. 15:19–31.

16680. Saman, L. 1954: Selection of coconut for seeds. Kasikorn 27:419–26.

16681. Samantarai, B., and A. S. Dubey. 1954: Effect of centrifuge on the dormancy of rice grains. [Abstract.] Proc. Indian Sci. Congr. 40:113. 1953.

16682. Sambamurthy, K., S. Rangaswami, and P. Veeraswamy. 1962: Isolation and constitution of a new anthoxanthin glycoside, vogeloside, from the seeds of Tephrosia vogelii. Planta Med. 10:173–78. Chem. Abstr. 58:4811. 1963.

16683. Samek, J. 1894: Duration of the vitality of some agricultural seeds. Tirol. Landwirtsch. Blätter 13:161–62. Expt. Sta. Rec. 6:429–30. 1894/95.

16684. Samoilovich, G. G. 1951: Aviation method for phenological observations and estimation of yields of forest tree seeds. Lesn. Khoz. 4(7):47–49.

16685. Samokhvalov, G. K. 1936: The influence of soaking seed in salt solutions on the development and yield of soft and hard wheat. Chemiz. Social. Agr. 1936(4): 77–91. Chem. Abstr. 30:7755. 1936.

16686. Samotesow, N. 1940: Zur Frage über die Trocknung des Saatgutes. Müllerei Elevatorwirtsch. 1940(9):8–9. Botan. Centralbl. 179:61. 1944.

16687. Sampath, S., and D. V. Seshu. 1957: Variability in protein content of rice. Current Sci. (India) 26:139.

16688. Sampietro, G. 1925: Tests on the poisoning of weed seeds. Giorn. Risicoltura 15:168–75. Expt. Sta. Rec. 56:439. 1927.

16689. —— 1931: Prolonging the longevity of rice seed. Giorn. Risicoltura 21:1–5. Chem. Abstr. 25:2172. 1931.

16690. Sampson, D. R., and A. G. O Whiteside. 1961: A seed—what is it? Agr. Inst. Rev. 16(2):16–18.

16691. Sampson, K., and D. W. Davies. 1928: The germination and early growth of wheat treated with copper carbonate and tillantin R. Ann. Appl. Biol. 15:408–22.

16692. Samra, A. S. 1956: Relative value and mode of action of some fungicides used as seed disinfectants and protectants. Mededel. Landbouw. Wageningen 56(5): 1–51.

16693. Samson, R. W. 1938/39: Seed transmission of tomato mosaic. Indiana Agr. Expt. Sta. Ann. Rept. Director 52:54–55.

16694. ——, T. J. Nugent, and L. C. Schenberger. 1942: The importance of seed transmission of early blight and Fusarium wilt of tomato. Phytopathology 32:16.

16695. Samu, I. 1938: Weed seeds found in wheat grown on the other side of Theiss. Dissertation, Budapest. 36 pp. Proc. Intern. Seed Testing Assoc. 11:155–56.

Samuels, C. 6536.
Samuels, C. S. 2653, 2658.

16696. Samuels, G., and P. Landrau, Jr. 1953: Influence of nitrogen, calcium and boron on the nodulation, yield, and protein content of tropical kudzu. J. Agr. Univ. Puerto Rico 37:74–80.

Samuels, G. 11754.

16697. Samuelsson, G. 1932: Die Arten der Gattung *Alisma* L. Arkiv Botan. 24A. 46 pp.

16698. Samygin, G. A., V. N. Varlamov, and N. M. Mateeva. 1960: The ability of seeds to endure ultra low temperatures. Plant Physiol. (USSR) (English Transl.) 7:76–78.

16699. Sanchez-Nieva, F., M. M. Cancel, and J. R. Benero. 1960: Maturity indices for pigeonpeas. J. Agr. Univ. Puerto Rico 44:60–69.

16700. Sandahl, P. L., and G. R. Titus. 1941: Hasten two-year seeds. Am. Nurseryman 74(11):26–27.

16701. Sandahl, P. C. 1956: Progress report. Sweetclover hybrids between yellow and white sweetclover through embryo culture. N. Dakota Agr. Expt. Sta. Bimonthly Bull. 18:197–99.

16702. Sandberg, F., and K. H. Michel. 1963: Phytochemical studies on the alkaloids of *Pancratium maritimum*. Lloydia 26:78–90. Chem. Abstr. 60:851. 1964.

Sandegren, E. 3096, 12530.

16703. Sandera, K., and M. Drachovska. 1960: Laboratory experiments with beet seed. Sb. Vysoke Skoly Chem.-Technol. v Praze Oddil Fak. Potravinareske Technol. 4:29–41. Chem. Abstr. 60:5896. 1964.

16704. Sanders, E. H. 1955: Developmental morphology of the kernel in grain sorghum. Cereal Chem. 32:12–25.

Saunders, E. H. 19413, 19414.

16705. Sanders, H. G. 1962: The aims and objects of the F.A.O. world seed year with special reference to United Kingdom activities. Natl. Inst. Agr. Botany J. 9:187–89.

Sanders, H. G. 6985–6987.

16706. Sanders, I. D., and C. J. Crosbie. 1964: Moisture tests on grain and small seeds. New Zealand J. Agr. 108:34–39.

16707. Sanders, M. E. 1944: Abnormalities in the development of interspecific hybrid embryos in *Datura*. Am. J. Botany (Suppl.) 31(8):5s.

16708. —— 1948: Embryo development in four *Datura* species following self and hybrid pollinations. Am. J. Botany 35:525–32.

16709. —— 1950: Development of self and hybrid *Datura* embryos in artificial culture. Am. J. Botany 37:6–15.

16710. —— and P. R. Burkholder. 1948: Influence of amino acids on growth of *Datura* embryos in culture. Proc. Natl. Acad. Sci. U.S. 34:516–26.

16711. Sanderson, F. H. 1954: Methods of crop forecasting. Harvard Univ. Press, Cambridge, Mass. 259 pp.

16712. Sandhack, H. A. 1933: *Ficus* aus Samen. Gartenwelt 37:50.

16713. Sandhu, A. S., and A. Husain. 1961: Effect of seed treatment with gibberellic acid on germination and growth of bajra (*Pennisetum typhoides*). Indian J. Agron. 5:269–72.

16714. —— and —— 1962: Effect of seed treatment with gibberellic acid on rapeseed and mustard. Indian Oilseeds J. 6:227–28. Chem. Abstr. 58:6133. 1963.

Sandie, R. 12647.

Sandmeier, M. 18103.

16715. Sando, W. J. 1939: Effect of mutilation of wheat seeds on growth and productivity. J. Am. Soc. Agron. 31:558–65.

Sandoval, J. A. 11762.

Sandrinelli, R. 19156, 19157.

Sands, J. E. 4306, 4307.

16716. Sands, W. N. 1922: The *Cinchona* industry in Java. Malayan Agr. J. 10:65–86.

16717. Sandstedt, R. M. 1946: Photomicrographic studies of wheat starch. I. Development of the starch granules. Cereal Chem. 23:337–59.

16718. —— and O. C. Beckord. 1946: Photomicrographic studies of wheat starch. II. Amylolytic enzymes and the amylase inhibitor of the developing wheat kernel. Cereal Chem. 23:548–59.

16719. —— and H. Schroeder. 1960: A photomicrographic study of mechanically damaged wheat starch. Food Technol. 14:257–65.

Sandstedt, R. M. 2621, 8056, 10879–10881.

Sandstrom, W. M. 12526.

16720. Sane, Y. K. 1939: A contribution to the embryology of the Aponogetonaceae. J. Indian Botan. Soc. 18:79–91.

16721. Sani, G. 1904: Ricerche su la germinazione del faggio. Rend. Accad. Lincei 302(5):13:382–85. Botan. Centralbl. 102:301–2. 1906.

Sankaran, S. 14214.

Sankewitsch, E. C. 4020.

Sankhla, N. S. 18564.

16722. Sanna, B. 1934: A new color reaction for the seeds of *Strophanthus kombe*. Boll. Soc. Ital. Biol. Sper. 9:829–30. Chem. Abstr. 29:1579. 1935.

16723. Sannella, J. L., and R. L. Whistler. 1962: Isolation and characterization of soybean hull hemicellulose B. Arch. Biochem. Biophys. 98:116–19.

16724. Sannie, C., and R. Truhaut. 1939: Can cancer be produced by the feeding of wheat-germ oil? Compt. Rend. Soc. Biol. 131:600–1. Chem. Abstr. 36:7118. 1942.

16725. San Pedro, A. V. 1936: Influence of temperature and moisture on the viability of some vegetable seeds. Philippine Agriculturist 24:649–58.

16726. Sant, G. 1934: Surinam chaulmoogra oil. Pharm. Weekblad 71:900–3. Chem. Abstr. 28:6523. 1934.
Santamaria, M. L. C. 3452.
Santaolalla, M. 6305, 6306.
16727. Santavy, R., and M. Talas. 1954: Substanzen der Herbstzeitlose und ihre Derivate. XXXIII. Isolierung weiterer Substanzen aus den Samen der Herbstzeitlose (*Colchicum autumnale* L.). Collection Czech. Chem. Commun. 19:141–52.
16728. Santesson, C. G. 1937: Piule, a Mexican intoxicating drug. Arch. Pharm. 275:532. Pharm. Abstr. 4:14. 1938.
16729. —— 1938: A curare-like substance in the leguminous seeds. Skand. Arch. Physiol. 80:361–68. Chem. Abstr. 33:751. 1939.
Santhanam, V. 2410.
Santoro, R. 6405.
Santos, A. C. 3885, 15649.
Santos, I. de; *see* De Santos, I.
16730. Santos, I. S. 1960: Effect of seed treatment with maleic hydrazide on germination and seedling growth of corn. Philippine Agriculturist 44:134–45.
16731. Santos, J. K. 1931: Leaf and seed structure of a Philippine *Coriaria*. Philippine J. Sci. 46:257–66.
16732. —— 1933: Morphology of the flower and mature grain of Philippine rice. Philippine J. Sci. 52:475–503.
Santos, J. K. 12836.
Santos, P. 14827.
Santos-Ruiz, A. 14980.
16733. San Vicente, J. 1961: Considerations on the law creating the national system of seed production, certification, and trade. Mex. Agr. 8(90):36–37.
Sanz-Dominguez, F. 14980.
16734. Sapankevich, P. V. 1952: Peculiarities of *Euonymus verrucosus* seed germination. Agrobiologiya 1952(2):67–74.
16735. —— 1958: Ripeness of seeds and duration of their dormancy. Akad. Nauk SSSR Inst. Lesa. Tr. 46:142–47.
Sapil'nyak, M. M. 13445.
16736. Sappenfield, W. P. 1952: The inheritance of earliness among six common varieties of oats. Missouri Agr. Expt. Sta. Res. Bull. 514. 55 pp.
16737. —— and J. M. Poehlman. 1952: Effect of date of seeding on the yield and test-weight of oat varieties. Missouri Agr. Expt. Sta. Res. Bull. 499. 16 pp.
Saprykina, E. G. 642.
16738. Sar, R. A. 1958/59: Marketing certified seed. Intern. Crop Improvement Assoc. Ann. Rept. 40/41:300–8.
16739. Saracoglu, S. 1953: The thiamine content of Turkish wheats and corresponding Bulgurs. Cereal Chem. 30:323–27.
16740. Saraev, P. I. 1936: The effect of carbon monoxide, ethylene and potassium permanganate on the development and yield of hemp. Chemiz. Social. Agr. 1936(10):42–50. Chem. Abstr. 31:6797. 1937.
16741. Saran, A. B. 1945: On the viability of paddy seeds, *Oryza sativa*. Current Sci. 14:271.
16742. Sarata, U. 1938: Studies in the biochemistry of copper. XXXV. Copper in rice plant and seed in different periods of development. Japan. J. Med. Sci. II. Biochem. 4:193–98.
Sarazin, J. 6571.
Sareen, R. N. 185.
16743. Sarfatti, G. 1961: Growth of pericarp, seed, endosperm, and embryo in *Prunus amygdalus* Stokes. Nuovo Giorn. Botan. Ital. 68:118–35.
16744. —— 1962: First researches with the electron microscope on the endosperm of *Prunus amygdalus* Stokes. Caryologia 15:1–20.
Sargeant, K. 14736.
16745. Sargent, E., and A. Robertson. 1905: The anatomy of the scutellum in *Zea mays*. Ann. Botany (London) 19:115–23.
16746. Saric, M. R. 1958: The dependence of irradiation effects in seed on the biological properties of the seed. Proc. U.N. Intern. Conf. Peaceful Uses At. Energy, 2d, Geneva 27:233–48.
16747. Sarin, J. L., and M. H. Quereschit. 1941: Cowpea starch. Ind. Eng. Chem. 33:640–42. Analyst 66:423. 1941.
Sarkar, D. K. 16492.
16748. Sarkar, S. 1956: The component fatty acids of *Impatiens balsamina* seed fat. Sci. Cult. (Calcutta) 21:616–18. Biol. Abstr. 31:39595. 1957.
16749. —— 1958: Versuch zur Physiologie der Vernalisation. Biol. Centralbl. 77:1–49.
16750. —— and M. M. Chakrabarty. 1956: The chemical examination of the seed fats from *Bischofia javanica* Blume and *Antidesma diandrum* Roth. Seed fats of the Euphorbiaceae family. II. Sci. Cult. (Calcutta) 22:336–37. Biol. Abstr. 32:31883. 1958.

16751. Sarkar, S. N. 1945: *Mucuna* seeds. I. Chemical composition and nutritive values of the seeds of *Mucuna pruriens*, DC. Ann. Biochem. Exptl. Med. 5:39–44. Chem. Abstr. 40:2906. 1946.

16752. —— and A. C. Bose. 1945: *Mucuna* seeds. II. Biological value of proteins of seeds of *Mucuna utilis*. Ann. Biochem. Exptl. Med. 5:55–58. Chem. Abstr. 40:3164. 1946.

16753. Sarkissian, I. V., A. M. Kessinger, and W. Harris. 1963: Differential rates of development of heterotic and non-heterotic young maize seedlings. II. Correlation of differential morphological development with physiological differences in germinating seeds. Plant Physiol. Suppl. 38:xii–xiii.

Sarkisyan, S. S. Melik-. 10957, 10958, 18989.

16754. Sarma, M. L., and K. V. Giri. 1942: Plant phosphatases. IV. Occurrence of "free" and "bound" phosphatases in seeds. Ann. Biochem. Exptl. Med. 2:17–24. Chem. Abstr. 37:3798–99. 1943.

Sarma, M. S. 11535, 11536.

Sarma, P. S. 115, 15967, 17553–17555, 18475.

16755. Sarnavka, R. 1954: Germination intensity as the effective value of germinated seeds. Sumarski List. 78:289–301.

16756. Saroja, K., and K. V. Giri. 1953: pH-activity relationship of the α- and β-amylases of resting and germinating barley and ragi (*Eleusine coracana*). J. Indian Inst. Sci. 35A:99–105. Chem. Abstr. 47:4433. 1953.

16757. Sarospataky, G. 1957: Verbreitung und Keimfähigkeit von *Cenchrus tribuloides* L. Növenyvedelem Idöszerü Kerdesei 1957(1):8–12.

Sarra, C. 12946.

Sarrazin, P. 11947.

Sarria, S. H. Orozco-; *see* Orozco-Sarria, S. H.

Sarson, H. 15785.

16758. Sartoris, G. B. 1931: The preservation of sugar cane seed. Facts About Sugar 26:536–50. Expt. Sta. Rec. 70:771. 1934.

16759. Sarvas, R. 1950: Effect of light on the germination of forest tree seeds. Oikos 2:109–19. Biol. Abstr. 26:9796. 1952.

16760. —— 1952: On the flowering of birch and the quality of seed crop. Commun. Inst. Forest Fenniae 40(7):1–38. Biol. Abstr. 27:32976. 1953.

16761. —— 1955: Investigations into the flowering and seed quality of forest trees. Commun. Inst. Forest Fenniae 45(7):1–37. Biol. Abstr. 31:22399. 1957.

16762. —— 1960: The distance of the provenance of seed used in forest cultivation from the place of cultivation. Metsätaloudellinen Aikakauslehti 1960(6/7):217–20.

16763. —— 1962: Investigations on the flowering and seed crop of *Pinus silvestris*. Finland Metsantutkimuslaitos Julkaisuja 53(4). 198 pp.

16764. Sarvella, P., R. A. Nilan, and C. F. Konzak. 1962: Relation of embryo structure, node position, tillering, and depth of planting to the effects of X-rays in barley. Radiation Botany 2:89–108.

Sarycheva, A. P. 7097.

16765. Sasaki, K. 1933: Organic bases in germinated soybeans. J. Agr. Chem. Soc. Japan 9:115–19. Chem. Abstr. 27:2711. 1933.

16766. Sasaki, R., and M. Kandatu. 1938: Chemical studies on the constituents of the seeds of *Aesculus turbinata* Blume. II. J. Agr. Chem. Soc. Japan 14:263–72. Chem. Abstr. 32:6292. 1938.

16767. Sasaki, S. 1928: Studies of germination of seeds. I. Transformation of nitrogenous compounds during germination of soya-bean seeds. Bull. Agr. Chem. Soc. Japan 4:75–77.

16768. —— 1929: Studies on the germination of seeds. II. Starch produced during the germination of soy-bean seeds. Bull. Agr. Chem. Soc. Japan 5:6–7.

16769. —— 1932: Organic bases in soy bean. J. Agr. Chem. Soc. Japan 8:417–20. Japan. J. Med. Sci. II. Biochem. 3:109. 1937.

16770. —— 1933: Carbohydrate metabolism of the soy bean during germination. J. Agr. Chem. Soc. Japan 9:693–96. Chem. Abstr. 27:5373. 1933.

16771. —— 1936: Studies on some constituents of soybean seeds and their transformations during germination. Kyushu Imp. Univ. Dept. Agr. J. 5:51–116. Chem. Abstr. 31:4697–98. 1937.

16772. —— and S. To. 1939: Cell-wall constituents of soybeans. II. Chemical constituents of the seed coat. J. Agr. Chem. Soc. Japan 15:624–28. Bull. Agr. Chem. Soc. Japan 15:105–6. 1939. Chem. Abstr. 34:468. 1940.

16773. —— and Y. Yamasita. 1938: Cell-wall constituents of soybean. I. Embryo of the seed. J. Agr. Chem. Soc. Japan 14:1257–63. Chem. Abstr. 33:6903. 1939.

16774. Sasaki, T. 1926: The preliminary report on an abnormal type of germination of rice seeds under reduced air supply. Nogaku Kwai Ho 288:461–69.

16775. Sasaoka, K. 1958: Formation of homoserine during germination of the pea. Mem. Res. Inst. Food Sci. Kyoto Univ. 14:42–48. Chem. Abstr. 52:11192. 1958.

16776. —— 1959: Occurrence of homoserine dehydrogenase in pea seedlings. Bull. Agr. Chem. Soc. Japan 23:347.

Sasaoka, K. 11077, 11078.

Sasaoka, M. 14072.

16777. Sashurilo, K. K. 1931: Über die Klassifikation der ornithochoren Früchte und Samen. J. Soc. Botan. Russ. 16:169–89. Botan. Centralbl. 164:24–25.

16778. Sasiplin, P. 1959: Papaya seed selection. Kasikorn 32:57–62.

Saskawa, S. 19952.

16779. Sass, J. E. 1946: The development of endosperm and antipodal tissue in Argentine waxy maize. Am. J. Botany 33:791–95.

16780. —— and G. F. Sprague. 1941: A "twin-embryo" abnormality in maize. Proc. Iowa Acad. Sci. 47:155–56.

16781. —— and —— 1950: The embryology of "germless" maize. Iowa State Coll. J. Sci. 24:209–18.

16782. Sastri, B. N., and B. A. S. Iyengar. 1935: Germination of leguminous seeds and urease activity. Current Sci. (India) 4:407. Chem. Abstr. 30:2604–5. 1936.

16783. Sastri, R. L. N. 1954: Embryological studies in Menispermaceae. I. *Tiliacora racemosa* Coleb. Proc. Natl. Inst. Sci. India 20:494–502.

16784. —— 1955: Development of the embryo of *Polyalthia longifolia* Hook. f. and Thoms. Current Sci. (India) 24:51.

16785. —— 1955: Structure and development of nutmeg seed. Current Sci. (India) 24:172–73.

16786. —— 1963: Studies in the Lauraceae. IV. Comparative embryology and phylogeny. Ann. Botany (London) 27:424–33.

16787. —— 1964: Embryological studies in the Menispermaceae. II. Embryo and seed development. Bull. Torrey Botan. Club 91:79–85.

Sastry, B. V. R. 15931.

16788. Sastry, L. V. S., and T. Ramakrishnan. 1961: A metal-dependent glutamic-glycine transaminase of green gram (*Phaseolus radiatus*). J. Sci. Ind. Res. (India) 20C:277–83. Biol. Abstr. 38:15640 1962.

16789. Sastry, M. S., V. Mahadevan, and R. Prakash. 1963: *Lathyrus sativus.* Nontoxic character of *Lathyrus sativus* seeds and their nutritive value. Indian J. Med. Res. 51:468–75. Chem. Abstr. 59:14297. 1963.

16790. Sathe, T. R., and V. Subrahmanyan. 1931: The relation between seeds and micro-organisms. J. Indian Inst. Sci. 14A:119–39. Biol. Abstr. 10:7843. 1936.

Sather, L. A. 4612.

16791. Satina, S. 1934: Jimsonweed seeds sometimes sprout twins. Sci. News Letter 25:24.

16792. ——, A. F. Blakeslee, and A. G. Avery. 1938: Chromosome behavior in triploid *Datura.* III. The seed. Am. J. Botany 25:595–602.

16793. ——, J. Rappaport, and A. F. Blakeslee. 1950: Ovular tumors connected with incompatible crosses in *Datura.* Am. J. Botany 37:576–86.

Satina, S. 2588, 2589, 15062, 15986, 16205–16208.

16794. Sato, H. 1957: Morphological and physiological studies on occurrence of abortive seeds in leguminous forage crops. IV. Occurrence of abortive seeds and their position in pod of common vetch. Proc. Crop Sci. Soc. Japan 26:68–70.

16795. —— 1962: Morphological and experimental studies on occurrence of aborted seeds in leguminous crops. Mem. Fac. Agr. Hokkaido Univ. 4:265–310.

16796. Sato, K., and T. Shoji. 1954: On disinfection and hastening of germination for black locust seeds. J. Japan. Forestry Soc. 36:244–47.

16797. ——, T. Suzaki, and Y. Shingai. 1962: On the effects of gibberellin upon the seed-germination and flower formation of some forest trees. Kyushu Univ. Forests Rept. 16:137–44.

16798. —— and Y. Nishina. 1951: Effects of fast neutrons upon forest tree seeds. II. Relations between the intensities of irradiations and the germinations of seeds, the growth of seedlings of *Pinus densiflora.* Kyushu Univ. Fac. Agr. Sci. Bull. 13:238–42. Biol. Abstr. 29:17273. 1955.

Sato, K. 14517.

16799. Sato, S. 1956: The distribution of succinic dehydrogenase and mitochondria in the embryos of *Phaseolus vulgaris.* Botan. Mag. (Tokyo) 69:137–41.

16800. Sato, T., and K. Kamiyama. 1956: Studies on seed production of soybean. Experimental researches of the effects of differences in the latitudes of producing localities upon the seeds and their progeny plants. Proc. Crop Sci. Soc. Japan 24:317–18.

Sato, T. 10372, 14142.

16801. Sato, Y. 1952: On the viability of willow seeds (*Toisus urbaniana* Kimura). Res. Bull. Coll. Expt. Forests Hokkaido Univ. 15:255–62.

16802. —— 1960: Effect of streptomycin on chlorophyll formation in timothy. II. Influences of inorganic salts upon the streptomycin effect on timothy seeds (under a resting condition). Keio J. Med. 9:25–32. Chem. Abstr. 54:21349. 1960.

16803. —— 1960: Effect of streptomycin on chlorophyll formation in timothy. III. Further studies on the effect of metal salts on streptomycin-treated seeds (under a resting condition). Keio J. Med. 9:189–97. Chem. Abstr. 55:3742. 1961.

16804. Satoo, T., and M. Goo. 1954: Seed germination as affected by suction force of soil and saccharose solution. Bull. Tokyo Univ. Forests 46:159–68.

16805. ——, ——, and S. Hasegawa. 1954: Germination of *Cryptomeria* seeds from cones which remained green on the tree for two years. J. Japan. Forestry Soc. 36:53–54.

Satriani, F. 16430.

Satsangi, A. 15891.

Sattar, A. 7981.

16806. Satyanarayana, G. 1962: Some effects of plant growth regulators on germination and seedling development of stored seeds. Andhra Agr. J. 9:9–15.

16807. Satyanarayana, M. N., M. V. L. Rao, M. Srinivasan, and V. Subrahmanyan. 1956: Poppy seed proteins. J. Sci. Ind. Res. (India) 15C:211–12. Nutr. Abstr. Rev. 27:3158. 1958.

Satyanarayanamurthy, K. 19058.

16808. Sauberlich, H. E., W. Y. Chang, and W. D. Salmon. 1953: The amino acid and protein content of corn as related to variety and nitrogen fertilization. J. Nutr. 51:241–50.

16809. Saulescu, N. 1934: Untersuchungen über die hartschaligkeit beim Siebenbürger Rotklee. Pflanzenbau 11:180–86.

16810. —— 1938: Die Keimung der hartschaligen Samen des Rotklees und der Luzerne aus Rumänien. Proc. Intern. Seed Testing Assoc. 10:322–25.

16811. —— and A. Szopos. 1938: Über den Wert der verletzten und roten Kleesamen. Proc. Intern. Seed Testing Assoc. 10:326–33.

16812. Saunder, D. H. 1950: Some notes on the properties of vermiculite in relation to its use in agriculture and horticulture. Rhodesia Agr. J. 47:409–15.

16813. Saunders, A. P. 1918: Peony culture in America. J. Intern. Garden Club 2:157–80.

16814. —— 1952: Here's how to propagate tree peonies. Flower Grower 39(5):64.

Saunders, F. 2064, 14696.

Saunders, L. M. 14961.

Saunier, R. 9186.

Saussure, T. de; *see* De Saussure, T.

16815. Sautter, E. H. 1962: Germination of switchgrass (*Panicum virgatum*). J. Range Management 15:108–10.

Saveljav, A. O. 15273

16816. Savellano, N. S. 1956: Four methods of germinating African oil palm seeds. Philippine Agriculturist 39:535–39.

16817. Savelli, R. 1928: Difetto di reazione geotropica in germinazioni di canapa da seme vecchio. Atti Soc. Nat. Mat. Modena VI 7:90–92.

16818. —— 1928: "Giant seeds" and a case of "polyendospermy" through the effect of open pollination on "*Nicotiana rustica*." Atti Accad. Nazl. Lincei Rend. Classe Sci. Fis. Mat. Nat. 7:88–92.

16819. —— 1928: Note nicozianografiche. V. Inversione dell'embrione ed inversione di germinazione nei semi ibridi di *Nicotiana*. Arch. Botan. Sistem., Fitogeogr. Genet. 4:1–8. Biol. Abstr. 3:17389. 1929.

16820. —— 1928: Twin embryos and correlative anomalies in *Cannabis sativa* L. Arch. Botan. Sistem., Fitogeogr. Genet. 4:206–13.

16821. —— 1930: Osservazioni sulle pseudogerminazioni della Canapa e fenomeni affini. Arch. Botan. Sistem., Fitogeogr. Genet. 6:102–8. J. Roy. Microscop. Soc. 52:74. 1932.

16822. —— and N. Soster. 1933: Berillio e cinconina su germinazioni di *Sinapis alba*. Bull. Orto Botan. Univ. Napoli 11:35–40.

16823. Savithri, V., and S. G. Aiyadurai. 1961: Occurrence of syncotylous seedling in *Ricinus*. Madras Agr. J. 48:148.

16824. Savitsky, H. 1950: Embryology of mono- and multigerm fruits in the genus *Beta*. Proc. Am. Soc. Sugar Beet Technol. 1950:160–64. Biol. Abstr. 26:32021. 1952.

16825. Savitsky, V. F., and G. K. Ryser. 1954: Sugar content in mono- and multigerm hybrids, carrying the gene m isolated from Michigan hybrid 18 and the gene m from variety U.S. 22/3. Proc. Am Soc. Sugar Beet Technol. 8:23–28.

16826. ——, ——, G. E. Rush, and C. P. Parrish. 1954: Inter-relation between weight of seed and fruit and utilitarian characters in inbred lines and hybrid of monogerm sugar beets. Proc. Am. Soc. Sugar Beet Technol. 8:399–403.

16827. Savostii, P. V., and M. M. Okuntsov. 1934: Biochemical specificity of vernalized plants. Tr. Tomsk. Gos. Univ. 86:64–82. Biol. Abstr. 11:13254. 1937.

16828. Savur, G. R. 1955: Tamarind seed polysaccharides. Current Sci. (India) 24:235–36.

16829. —— 1956: Anomalous properties of tamarind seed polyose. J. Indian Chem. Soc. Ind. News Ed. 19:75–77. Chem. Abstr. 52:17413. 1958.

Savur, G. R. 14180.

Sawada, E. 17769.

16830. Sawada, T., and S. Matsumoto. 1955: On the seed production of teosinte plant (*Euchlaena mexicana* Schrad). Kyushu Agr. Res. 15:52–54.

16831. Sawhney, B. S. 1930: The effect of soaking maize in water before sowing. [Abstract.] Proc. Indian Sci. Congr. 17:60–61.

Sawhney, P. C. 184.

16832. Sawicki, J. 1953: Studies on the structure of the aleurone layer in varieties of the cultivated barley *Hordeum sativum* Jess. Akad. Umiej. Prace Roln.-Lesne 66. 59 pp.

16833. —— 1955: The phenol reaction of grain used for identifying varieties of naked barley. Acta Agrobotan. 3:167–75.

Sax, H. J. 16834–16836.

16834. Sax, K., and H. J. Sax. 1961: The effect of age of seed on the frequency of spontaneous and γ-ray induced chromosome aberrations. Radiation Botany 1: 80–83.

16835. —— and —— 1962: Effects of X-rays on the ageing of seeds. Nature 194:459–60.

16836. —— and —— 1964: The effect of chronological and physiological aging of onion seeds on the frequency of spontaneous and X-ray induced chromosome aberrations. Radiation Botany 4:37–41.

16837. Saxema, M. C. 1961: Germination studies in *Hyptis suaveolens* Poit. I. Breaking of dormancy and effect of certain physical treatments on the percentage germination. Natl. Acad. Sci. India, Allahabad 2. Annual No. 1960:124–25. Biol. Abstr. 37:7072. 1962.

Saxena, M. C. 10831.

16838. Saxton, W. T. 1934: Notes on conifers. IX. The ovule and embryogeny of Widringtonia. Ann. Botany (London) 48:429–31.

16839. Sayanova, V. V. 1960: Variation in content of protein and nonprotein nitrogen in vegetative organs and seeds of some species of *Phaseolus*. Tr. po Khim. Prirodn. Soedin. Kishinevsk. Gos. Univ. 1960(3):41–55. Chem. Abstr. 56:2716. 1962.

16840. —— 1963: Globulins of *Dolichos lablab* seeds. Dokl. Akad. Nauk SSSR 152:478–80. Chem. Abstr. 60:852. 1964.

16841. Sayed, M., and J. L. Beal. 1958: A histological study of some mucilaginous seeds. J. Am. Pharm. Assoc. 47:544–47.

Sayeeduddin, M. 7205.

Sayers, E. R. 13970.

Sayre, C. B. 9775.

16842. Sayre, J. D. 1940: Storage tests with seed corn. Ohio J. Sci. 40:181–85.

16843. —— 1947: Storage tests with seed corn. Farm Home Res. Bimonthly Bull. 32:149–54. Biol. Abstr. 23:8672. 1949.

16844. —— 1948: Mineral accumulation in corn. Plant Physiol. 23:267–81.

Sbur, D. E. 18097–18099.

16845. Scala, A. C. 1928: La propagación de las semillas en la *Cyclanthera hystrix* Arn. Rev. Chilena Hist. Nat. 31:293–306.

Scallet, B. L. 4495.

16846. Scammell, R. 1943: Growing roses from seed. Am. Ros. Mag. 5:115–17.

16847. Scaramuzzi, F. 1957: Research on the germinating power of seeds of varying age in "*Olea europea*" L. Agr. Ital. (Pisa) 57:21–43.

Scaramuzzi, G. 4130.

16848. Scarano, M. 1932: On the structure of the fruit of "*Periploca graeca*" L. Nuovo Giorn. Botan. Ital. 39:54–69.

16849. Scarascia, G. T. 1954: Cytological effects of aqueous extracts of seeds of *Nicotiana tabacum*. Tabacco 58:220–26.

16850. —— 1955: Action of some alkaloids on tobacco germination and on mitosis of root tips. Tabacco 59:133–53. Chem. Abstr. 49:8402. 1955.

16851. —— 1955: Cytological effects of water-extracts of *Nicotiana rustica* seeds. Tabacco 59:201–8. Biol. Abstr. 31:30081. 1957.

16852. —— 1957: Morphological and mutational effects in aging seeds in *Nicotiana tabacum* L. Tabacco 61:270–90.
16853. —— and M. E. Scarascia-Venezian. 1954: Cytological effects and biochemical characteristics of H_2O extracts of soybean seeds of various ages. Caryologia 6:247–70.
16854. ——, ——, and G. Giovanozzi-Sermanni. 1956: On the aging of *Nicotiana* seed. Cytological and biochemical observations. Tabacco 60:278–85. Biol. Abstr. 31:36058. 1957.
16855. Scarascia-Venezian, M. E., and G. Giovannozi-Sermanni. 1956: Biochemical changes in germinating seeds of *Nicotiana tabacum* and *N. rustica.* Tabacco 60: 268–77. Tobacco Abstr. 2:251. 1958.
Scarascia-Venezian, M. E. 7276, 16853, 16854.
Scarchuk, J. 4778.
16856. Scatizzi, A. 1937: Ricerche sulla germinazione di "*Ephedra altissima*" Desf. Nuovo Giorn. Botan. Ital. 44:345–75. Botan. Centralbl. 173:121. 1938.
16857. Scerbakov, A. P. 1953: Biochemische Veränderungen in keimenden Samen der *Soja* unter dem Einfluss von Calcium und Magnesium. Biochemistry (USSR) (English Transl.) 18:438–47. Ber. Wiss. Biol. 88:339–40.
Scerbakov, A. P. 11553.
16858. Schaaf, H. M., and G. A. Rogler. 1960: Selecting green neddlegrass for low-dormancy seed. Agron. J. 52:704–7.
16859. ——, ——, and R. J. Lorenz. 1962: Importance of variations in forage yield, seed yield, and seed weight to the improvement of crested wheatgrass. Crop Sci. 2:67–71.
16860. Schaal, J. 1939: A use for the ash of sunflower-seed hulls. Seifensieder-Zgt. 66: 769. Chem. Abstr. 33:9514. 1939.
16861. Schaeffer, G. W., H. J. Gorz, and F. A. Haskins. 1961: Genetic, developmental, and within-plant variation in free and bound coumarin content of sweet clover. Crop Sci. 1:194–96.
Schafer, E. G. 1886.
Schafer, J. F. 15127.
16862. Schafer, L. A., and E. D. Hansing. 1950: Effect of hot-water treatment on emergence of spring barley and control of brown loose smut (*Ustilago nuda*). Phytopathology 40:518–21.
16863. Schaffer, H. G. 1964: Growing anemones from seed. Gardeners Chron. 155:283.
16864. —— 1964: Raising the phosphorus content of seed. Gardeners Chron. 155: 309.
16865. —— 1964: Bitter emulsions for seedbeds. Gardeners Chron. 155:414.
16866. Schaffer-Kircher, V. 1933: Über den Einfluss der Bodenreaktion auf die Keimung von Mais und Hirse. Wiss. Arch. Landwirtsch. Abt. A Pflanzenbau 10:324–48. Chem. Abstr. 28:1128. 1934.
16867. Schaffner, J. H. 1913: The sprouting of the two seeds of a cocklebur. Ohio Nat. 14:216–17.
16868. —— 1935: Artificial parthenocarpy. J. Heredity 26:261–62. Expt. Sta. Rec. 74:181. 1936.
Schahen, R. 19468.
16869. Schaible, F. 1900: Physiologische Experimente über das Wachstum und die Keimung einiger Pflanzen unter vermindertem Luftdruck. Beitr. Wiss. Botan. 4: 93–148. Botan. Centralbl. 82:52–54. 1900.
16870. Schaible, P. J. 1946: Composition of certain hybrid and open-pollinated corns and their performance in poultry rations. Mich. Agr. Expt. Sta. Quart. Bull. 29: 31–39.
Schaik, P. H. van; *see* Van Schaik, P. H.
16871. Schaile, O. 1926: Über Stimulation und Keimung von Fichten- und Kiefersamen. Allgem. Forst- Jagdw.-Ztg. 102: 98–100.
Schallenbergheertjes, G. 18990.
Schallock, D. A. 19289.
16872. Schander, H. 1934: Keimungsphysiologie Studien über die Bedeutung der Aleuroneschicht bei *Oryza* und anderen Gramineen. Z. Botan. 27:433–515. Expt. Sta. Rec. 73:453–54. 1935.
16873. —— 1952: Untersuchungen über umweltbedingte Eigenschaften des Samens und Keimlings von Apfel und Birne. Angew. Botan. 26:165–80.
16874. —— 1955: Über die Ursachen von Gewichtunterschieden bei Samen von Kernobst (Apfel und Birne). Die Beziehungen zwischen Samen und Frucht. Z. Pflanzenzücht. 34:255–306.
16875. —— 1955: Keimungsphysiologische Studien an Kernobst. II. Untersuchungen über die allgemeinen Temperaturansprüche der Kernobstsamen während der Keimung. Z. Pflanzenzücht. 34:421–40.
16876. —— 1955: Keimungsphysiologische Studien an Kernobst. III. Sortenvergleichende Untersuchungen über die Temperaturansprüche stratifizierten Saatgutes von Kernobst und über die Reversibilität

der Stratifikationsvorgänge. Z. Pflanzenzücht. 35:89–97.

16877. —— 1955: Keimungsphysiologische Studien an Kernobst. IV. Untersuchungen über Sortenunterschiede hinsichtlich der Temperaturansprüche und der Frostresistenz während der Keimung sowie über Alterungserscheinungen bei Samen von Kernobst. Z. Pflanzenzücht. 35: 179–98.

Schander, H. 5118.

16878. Schander, R. 1930: Versuche über den Einfluss des Schälens auf die Entwicklung und Erträge der Zuckerrüben. Pflanzenbau 7:97–103.

Schank, S. C. 7919.

16879. Schantz-Hansen, T., and R. A. Jensen. 1952: The effect of source of seed on growth of Jack pine. J. Forestry 50:539–44.

16880. Schaper, P. 1936: Beiträge zur mikrosckopischen Diagnostik der wichtigsten Caryophyllaceensamen unter besonderer Berücksichtigung ihres Vorkommens als Unkrautbesatz in Saatwaren und Futtermitteln. Dissertation, Hamburg Inst. Angew. Botan., Hamburgh. 100 pp. Proc. Intern. Seed Testing Assoc. 9:159–60. 1937.

16881. Scharpf, R. F., and J. R. Parmeter, Jr. 1962: The collection, storage, and germination of seeds of a dwarfmistletoe. J. Forestry 60:551–52.

16882. Scharrer, K. and R. Schreiber. 1943: Gefässversuche über den Einfluss gesteigerter Kalium- und Magnesiumdüngung auf den Eiweissertrag der Ackerbohne (*Vicia faba*). Bodenkunde Pflanzenernähr 30:360–70. Biol. Abstr. 23:16495. 1949.

16883. —— and —— 1944: Über die Wirkung des Magnesiums auf den Eiweiss und Fettertrag der Sonnenblume (*Helianthus annuus* L.). Bodenkunde Pflanzenernähr 34:301–10. Biol. Abstr. 26:6904. 1952.

16884. —— and W. Schropp. 1929: Untersuchungen über den Einfluss steigender Mengen Kalziumsulfit und Kalsiumsulfat auf die Keimung und die Jugendentwicklung der Getreidepflanzen. Landwirtsch. Versuchtssa. 108:217–51.

16885. —— and —— 1931: Weitere Untersuchungen über die Wirkung steigender Jodmengen in Form von Jodid-, Jodat- und Perjodat-ion sowie elementarem Jod auf die Keimung und die erste Jugendentwicklung verschiedener Kulturpflanzen. XXV. Mitteilung zur Kenntnis des Jods all biogenes Element. Biochem. Z. 239:74–93. Ber. Ges. Physiol. 63:753. 1932.

16886. —— and —— 1932: Die Wirkung des Fluor-ions auf Keimung und Jugendwachstum einiger Kulturpflanzen. Landwirtsch. Versuchssta. 114:203–14. Ann. Agron. 3:129. 1933.

16887. —— and —— 1933: Effect of boron on the germination and early growth of some cultivated plants in sand and water cultures. Z. Pflanzenernähr. Düng. Bodenk 28A:313–29. Chem. Abstr. 27:5463. 1933.

16888. —— and —— 1949: The effect of [compounds of] chlorine and bromine on the germination and initial development of plants. Z. Pflanzenernähr. Düng. Bodenk. 46:88–110. Chem. Abstr. 44:6921–22. 1950.

16889. Schauman, K. 1926: Über die Keimungsbedingungen von *Alisma plantago* und anderen Wasserpflanzen. Jahrb. Wiss. Botan. 65:851–934.

Schdanowa, L. N. 18781.

16890. Scheck, H., and G. Fetzer. 1950: Der Saatgutwert und seine Beeinflussung durch äussere Faktoren. Saatgutwirtsch. 2:271–73. Biol. Abstr. 25:21355. 1951.

16891. Scheer, D. F., J. H. Ellison, and M. W. Johnson. 1960: Effect of fruit maturity on asparagus seed germination. Proc. Am. Soc. Hort. Sci. 75:407–10.

16892. Scheffer, F., and A. Kloke. 1954: The influence of antibiotics on the development and nutrient content of crop plants. Z. Pflanzenernähr. Düng. Bodenk. 66: 29–38. Chem. Abstr. 49:7651. 1955.

16893. ——, ——, and H. Gerken. 1955: Die Vernichtung der Keimfähigkeit von Unkrautsamen bei der biologischen Gaserzeugung. Landwirtsch. Forsch. 7:200–3.

16894. Scheibe, A. 1930: Über den Vorgang der Wasseraufnahme und die physiologische Bedeutung des Rohrzuckers beim Keimprozess der Getreidekörner dargestellt am Hafer. Fortschr. Landwirtsch. 5:386–92.

16895. —— 1934: Die Schoss- und Reifeperiode des Hafers in ihrer Abhängigkeit von der physiologischen Konstitution des Saatgutes. Angew. Botan. 16:58–100, 105–36.

16896. —— 1960: Untersuchungen über die CO_2-Abgabe von röntgen bestrahltum Samen. Ein Beitrag zum Problem der Strahlennachwirkungen. Angew. Botan. 34:158–64.

16897. —— and U. Staffeld. 1931: Der Rohrzuckergehalt der Samen als Hinweis für den physiologisch-ökologischen Charakter der Getreidearten und -sorten. Fortschr. Landwirtsch. 6:364–69. Chem. Abstr. 26:4080. 1932.

16898. Schell, G. 1960: Die Abhängigkeit der Lebenskraft und der Pflanzengrösse von der Keimsomelligkeit bei unterschiedlicher Korngrösse. Allgem. Forst-Jagdw.-Ztg. 131:34–37.

16899. Schell, W. S. 1955: The basic role of the seed industry. Seed World 76(7):8, 44–45.

Scheloumoff, A. 11270.

Schemjakin, F. M. 5649.

Schenberger, L. C. 5505.

16900. Schenck, J. 1887: Dispersion of seeds of *Euphorbia marginata*. Botan. Gaz. 12:225–26.

16901. Schenk, R. U. 1960: Source of the inner brown color of the peanut shell. Botan. Gaz. 121:191–92.

16902. ——, M. M. MacMasters, and F. R. Senti. 1957: A note on improved interpretation of the 2,3,5-triphenyltetrazolium chloride color test for viability as an indication of the processing value of corn. Cereal Chem. 34:69–70.

16903. Scherbakov, A. P., and Z. S. Bronovitskaya. 1945: Biochemical changes in ripening wheat kernels. Dokl. Vses. Soveshch. Fiziol. Rast. 2; Tr. Inst. Fiziol. Rast. Nauk SSSR 4:101–15. Chem. Abstr. 42:5955. 1948.

16904. Schertz, F. M. 1920: A chemical analysis of Sudan grass seed. Botan. Gaz. 69:72–82. Botan. Abstr. 8:2161. 1921.

16905. Schertz, K. F., W. C. Boyd, W. Jurgelsky, Jr., and E. Cabanillas. 1960: Seed extracts with agglutinating activity for human blood. Econ. Botany 14: 232–40.

Schertz, K. F. 12898.

16906. Schery, R. W. 1964: The importance of quality seed. Seed World 94(3):6–11.

16907. —— and D. Kern. 1956: Analysis of bluegrass seed for germination and seedling vigor. Agron. Abstr. 1956:91–92.

16908. Scheunert, A., and M. Schieblich. 1935: Vitamin content of fresh soybeans. Biedermanns Zentralbl. B. Tierernähr. 7:198–204. Chem. Abstr. 29:4411. 1935.

16909. —— and —— 1936: Vitamin content of the seeds of blue sweet lupine, Strain 411. Biedermanns Zentralbl. B. Tierernähr. 8:132–35. Chem. Abstr. 31:174. 1937.

Schexnayder, C. A. 15749.

Schickedanz, F. 8916.

Schidlovsky, G. 19017.

Schieblich, M. 16908, 16909.

Schierbaum, D. L. 362.

Schild, E. 11010.

Schiltz, P. 9602.

Schimke, O. 1830.

16910. Schindler, J. 1932: Über die Wertausgleichsrechnung bei Lieferung nicht garantiegemässen Saatgutes. Landwirtsch. Versuchssta. 114:215–24.

16911. Schindler, O. 1956: Die Glykoside der Samen von *Strophanthus amboënsis* (Schinz) Engl. et Pax. Glykoside und Aglykone, 157. Helv. Chim. Acta 39: 64–79.

16912. —— and T. Reichstein. 1951: Vergleich dreier Samenproben von Sarmentogenin liefernden *Strophanthus*-Varianten mit Hilfe von Papierchromatographie. Glykoside und Aglykone, 72. Helv. Chim. Acta 34:608–11.

16913. —— and —— 1951: Die Glykoside der Samen von *Strophanthus courmantii* Sacl., 2. Courmontoside A, B und C. Helv. Chim. Acta 34:1732–39.

16914. —— and —— 1952: Die Glykoside der Samen von *Strophanthus boivinii* Baill. Helv. Chim. Acta 35:673–86.

16915. —— and —— 1953: Die Glykoside der Samen von *Strophanthus sarmentosus* var. *glabriflorus* Monach. Helv. Chim. Acta 36:921–32.

16916. —— and —— 1953: Die Glykoside der Samen von *Strophanthus divaricatus* (Lour.) Hook et Arn. Helv. Chim. Acta 36:1007–24.

16917. —— and —— 1954: Die Glykoside der Samen von *Strophanthus caudatus* (Burm. ex L.) Kurz. Helv. Chim. Acta 37:103–12.

16918. —— and —— 1955: Die Glykoside der Samen von *Strophanthus arnoldianus* DeWild. et Th. Dur. Glykoside und Aglykone, 145. Helv. Chim. Acta 38: 874–82.

16919. —— and —— 1956: Die Glykoside der Samen von *Strophanthus congoensis* Franch. Glykoside und Aglykone, 156. Helv. Chim. Acta 39:34–45.

Schindler, O. 11294, 13659, 15930, 16129, 16130, 18547, 19204, 19206, 19481, 19482, 20089.

16920. Schindler, W. 1961: Problems in forest seed registration. Z. Forst- Jagdw. 11:511–14.

16921. Schippers, B. 1963: Transmission of bean common mosaic virus by seed of

Phaseolus vulgaris L. cultivar Beka. Acta Botan. Neerl. 12:433–97.

Schirren, V. 19540.

16922. Schitikova, A. 1923: Sur les moyens de combattre le charbon des céréals a l'aide des températures elevés. Rept. Intern. Conf. Phytopathol. Econ. Entomol. Wageningen, p. 272.

16923. Schjelderup-Ebbe, T. 1936: Über die Lebensfähigkeit alter Samen. Skrifter Norske Videnskaps-Akad. Oslo Mat.-Naturv. Kl. 1935:1–178. Biol. Abstr. 11:10520. 1937.

16924. Schjerning, H. 1910: On the protein substances of barley, in the grain itself and during the brewing process. Compt. Rend. Trav. Lab. Carlsberg 8:169–395.

16925. Schlbom, E. 1962: Amino acid composition of Swedish wheat protein. Acta Agr. Scand. 13:148–56.

Schlegel, V. 15244.

16926. Schlehuber, A. M., B. C. Curtis, and R. M. Chatters. 1956: Progeny performance of fluorescence in Andrew oats. Agron. J. 49:381–83.

16927. Schleicher, R. 1932: Zur Aufzucht der *Lapageria rosea* aus Samen. Möllers Deut. Gärtner-Ztg. 47:373–74.

16928. Schleip, H. 1938: Untersuchungen über die Auswuchsgestigkeit bei Weizen. Landwirtsch. Jahrb. 86:795–822.

16929. Schlimbach, H. 1924: Beiträge zur Kenntnis der Samenanlagen und Samen der Amaryllidaceen, mit Berücksichtigung des Wassergehaltes der Samen. Flora (Jena) 117:41–54.

Schlimm, W. 2333–2335.

16930. Schloesing, A. T., and D. Leroux. 1943: Essai de conservation de graines en l'absence d'humidité d'air et de lumière. Compt. Rend. 216:468–72.

Schloss, M. S. 7419.

16931. Schlosser-Szigat, G. 1961: A rapid method for the determination of germinability of *Melilotus* seed. Z. Landwirtsch. Versuchs- Untersuchungsw. 7:399–405.

16932. —— 1962: Artbastardierung mit Hilfe der Embryokultur bei Steinklee (*Melilotus*). Naturwissenschaften 49:452–53.

16933. Schmalfuss, H., A. Heider, and K. Winkelmann. 1933: Occurrence and isolation of *l*-tyrosine from soy beans, *Vicia faba* L. Biochem. Z. 259:465–68. Chem. Abstr. 27:2703. 1933.

Schmalfuss, H. 19542.

16934. Schmalfuss, K. 1937: Further studies on oil formation in linseed as affected by environment and nutrition of the plant. Bodenkunde Pflanzenernähr. 5: 37–46. Chem. Abstr. 32:621. 1938.

16935. —— 1941: The formation of fat in plant seeds. Bodenkunde Pflanzenernähr. 24:321–41. Chem. Abstr. 38:2362. 1944.

16936. —— 1942: The change of the oil (content) during the germination of flaxseed. Fette Seifen 49:773–74. Chem. Abstr. 38:1767. 1944.

16937. —— and H. Micheels. 1935: The dependence of the quality of linseed oil on the mineral nutrition of the plants. Angew. Botan. 17:199–206. Chem. Abstr. 29: 6690–91. 1935.

16938. Schmalz, H. 1961: The effect of gibberellin and colchicine on the germination and seedling growth of spring barley. Züchter 31:273–80.

16939. Schmekel, J., and S. Erikkson. 1963: The determination of the digestibility of grain by ruminants. Lant.-Högs. Ann. Upsala 29:259–66.

Schmeling, B. von. 4691.

Schmid, A. R. 1313.

16940. Schmid, B. 1901. Über die Einwirkung von Chloroformdämpfen auf ruhende Samen. Ber. Deut. Botan. Ges. 19:71–76.

16941. Schmid, H. 1944: Notiz erbü keimungshemmende Stoffe. Helv. Chim. Acta 27:1197–99. Biol. Abstr. 19:20064. 1945.

16942. —— and P. Karrer. 1948: Über Inhaltsstoffe des Rettich. I. Über Sulforaphen, ein Senföl aus Rettichsamen (*Raphanus sativus* L. var. *alba*). Helv. Chim. Acta 31:1017–28. Biol. Abstr. 23:12705. 1949.

16943. —— and —— 1948: Über Inhaltsstoffe der Rettichs. II. Optisch aktives 4-Methyl-sulfoxid-buten-(3)-yl-cyanid als Spaltprodukt eines Glucosides aus Samen von *Raphanus sativus* var. *alba*. Helv. Chim. Acta 31:1087–92. Biol. Abstr. 23:12706. 1949.

Schmid, H. 10721.

Schmid, P. 10765.

16944. Schmid, W., H. P. Uehlinger, C. Tamm, and T. Reichstein. 1959: Die Glykoside der Wurzeln und Samen von *Pachycarpus schinzianus* (Schltr.) N. E. Br. Glykoside und Aglykone. Helv. Chim. Acta 42:72–121.

16945. Schmidhauser, T. 1955: Nature and distribution of the soluble carbohydrates in germinating and non-germinating barley grains. Ber. Schweiz. Botan. Ges. 65: 302–42. Chem. Abstr. 50:17019. 1956.

16946. Schmidt. 1926: Zum Faktor Feuchtigkeit bei der Herrichtung und Keimung des Kiefernsamens. Z. Forst- Jagdw. 58: 29–41.

Schmidt, C. F., Jr. 4364.

16947. Schmidt, D. 1922: A study of plant growth in relation to weight of seed. N.J. Agr. Expt. Sta. Ann. Rept. 43:388–94.

16948. —— 1924: The effect of the weight of the seed on the growth of the plant. N.J. Agr. Expt. Sta. Bull. 404. 19 pp.

16949. Schmidt, D. R. 1962: Dry matter and nitrogen content of oats harvested at various stages. Agron. J. 54:8–10.

16950. Schmidt, E. 1930: Beiträge zur mikroskopischen Diagnostik von Früchten und Samen der wichtigsten Polygonumarten unter besonderer Berücksichtigung ihres Vorkommens als Unkrautbesatz in Saatwaren und Futtermitteln. Landwirtsch. Versuchssta. 111:169–259. Proc. Intern. Seed Testing Assoc. 2:193. 1931.

16951. —— 1931: *Salvia splendens* aus Samen. Gartenwelt 35:36–37.

16952. —— 1932: Aubrietien aus Samen. Gartenwelt 36:491–92.

16953. —— 1932: *Datura* aus Samen. Gartenwelt 36:661.

16954. —— 1934: Experimentelle Untersuchungen über die Auswuchsneigung und Keimreife als Sorteneigenschaften des Getreides. Angew. Botan. 16:10–50.

16955. —— and E. Tornow. 1932: Detection of the pickling of grain with mercury and other metallic poisons. Fortschr. Landwirtsch. 7:40–42. Chem. Abstr. 26: 4674–75. 1932.

16956. —— and —— 1933: Vereinfachte Methode zum Nachweis des Quecksilbers und der Beizung von Getreide mit Quecksilber- und Kupfersalzen. Prakt. Bl. Pflanzenbau Pflanzenschutz 11:177–83.

16957. Schmidt, E. A. 1938: The working up of germinated rye. Systematic comprehension of the degree of germination and methods of procedure for eliminating the injurious effects of germination. Z. Ges. Getreidew. 25:48–65. Chem. Abstr. 32: 5937. 1938.

16958. Schmidt, E. W. 1934: Über das altern des Zuckerrübensamens. Deut. Zucker-Ind. 59:961–63. Proc. Intern. Seed Testing Assoc. 7:86–87. 1935.

16959. Schmidt, G. 1935: Über die Inhaltsstoffe der Samen von *Tephrosia vogelii.* G. Neuenhahn, G.m.b.H., Universitätsbuchdruckerei, Jena. 33 pp.

16960. —— 1960: Effect of high temperature on the germination of lawn grasses and study of optimal temperature. Agrobotanika 2:125–40.

Schmidt, G. 13353.

16961. Schmidt, H. 1938: The effect of corrodants on artificially infected vegetable seeds. Gartenbauwissenschaft 12:89–115. Chem. Abstr. 32:6795. 1938.

16962. —— 1962: Studies on the germination of seeds of *Pinus sylvestris* L. in high mountains. Forstwiss. Centralbl. 81:333–37.

16963. Schmidt, H. H. 1953: Die Keimung überlagerter Erbsensorten. Saatgut-Wirtsch. 5:279–83.

16964. —— 1953: Untersuchungen zu den Fluoreszenzerscheinungen der Keimpflanzen von *Lolium* spp. im ultravioletten Licht. I. Zur Methodik des Fluorenzenztestes. Angew. Botan. 27:1–15.

16965. Schmidt, J. L., and W. V. Hukill. 1956: Effect of artificial drying on the yield of head rice and the germination of rice. Rice J. 59(13):28–31.

16966. Schmidt, J. W., and E. F. Frolik. 1951: The effects of thermal-neutron irradiation of maize and barley kernels. Nebraska Agr. Expt. Sta. Res. Bull. 167. 29 pp.

Schmidt, J. W. 8388, 8389, 18115.

16967. Schmidt, W. 1919: Die Verbreitung von Früchten durch die Luftbewegung. Naturwissenschaften 7:810–12. Botan. Abstr. 8:95. 1921.

16968. —— 1925: Über Vorquellung und Reizbehandlung von Koniferensaatgut. Zellstimulations-Forsch. 1:355–68.

16969. —— 1926: Das Katalaseferment im Kiefersamen. Forstarchiv 2:2–9.

16970. —— 1929: Weitere Katalaseuntersuchungen als Prüfmassstab des Samenzustands. Z. Forst- Jagdw. 61:413–28.

16971. —— 1930: Unsere Kenntniss von Forstsaatgut. Deut. Forstwirt, Berlin. 256 pp. Forestry 5:166. 1931.

16972. —— 1940: Die Bedeutung der Keimschnelligkeit für den Aussaatwert des Samens. Z. Forst- Jagdw. 72:35–47.

16973. Schmidt, W. C. 1962: Rapid viability tests for Western larch seed. Proc. Montana Acad. Sci. 21:26–32.

16974. Schmieder, F. 1927: Katalase und Keimung. Mitt. Sächs. Forst. Vers. Tharandt. 3:65–96.

Schmitz, P. H. Gütlin-; *see* Gütlin-Schmitz, P. H.

16975. Schmorl, K. 1934: Das spezifische Gewicht eines Getreides als Bewertungsfaktor. Pflanzenbau 10:395–400, 444–45. Proc. Intern. Seed Testing Assoc. 7:75. 1935.

16976. —— 1937: Carotenoids of grain. Z. Ges. Getreidew. 24:202–4. Soc. Chem. Ind. J. 57:574B. 1938.

16977. —— 1940: The scientific evaluation of the soundness of cereal grain. Mühlenlab. 10:25–30. Chem. Abstr. 36:173–75. 1942.

16978. Schmuck, A. 1931: On the phytine content of tobacco seed. Russ. State Inst. Tobacco Invest. Bull. 81:19–22.

16979. Schnarf, K. 1928: Über die Endospermentwicklung bei *Ornithogalum*. Österr. Botan. Z. 77:173–77. Botan. Centralbl. 155:195–96. 1928.

16980. —— 1930: Zur Kenntnis der Samenentwicklung von *Roridula*. Österr. Botan. Z. 79:180–82.

16981. —— 1931: Ein Beitrag zur Kenntnis der Samenentwicklung der Gattung *Cochlospermum*. Österr. Botan. Z. 80:45–50. Botan. Centralbl. N. F. 19:198. 1931.

16982. —— 1936: Ein Beitrag zur Kenntnis der anatomischen Baues der harten Schale des Samens der Cycadaceen. Österr. Botan. Z. 85:279–88. Botan. Centralbl. 171:297. 1937.

16983. —— 1936: Contemporary understanding of embryo-sac development among angiosperms. Botan. Rev. 2: 565–85.

16984. —— 1937: Anatomie der Gymnospermen-Samen. Gebrüder Bornträger, Berlin. Lief. 34. 156 pp.

16985. Schnathorst, W. C. 1954: Bacteria and fungi in seeds and plants of certified bean varieties. Phytopathology 44:588–92.

Schneebauer, F. 5950, 5951.

16986. Schneider, A. 1952: Über das Vorkommen gerbstoffartiger Kondensationsprodukte von Anthocyanidinen in den Samenschalen von *Pisum arvense*. Naturwissenschaften 39:452–53.

16987. Schneider, C. 1931: Propagation of cacti. New Flora Silva. 3:196–200.

16988. Schneider, E. O., E. B. Earley, and E. E. De Turk. 1952: Nitrogen fractions of the component parts of the corn kernel as affected by selection and soil nitrogen. Agron. J. 44:161–69.

16989. Schneider, G. 1932: Stimulationsversuche mit "Inovan," "Kiserit" und Wasser. Pflanzenbau 9:190–97.

Schneider, G. 2801, 2802.

Schneider, H. 1975, 1976.

16990. Schneider, M. 1932: Untersuchungen über die Embryobildung und -Entwicklung der Cyperaceen mit berücksichtigung angrenzender Fragen wie vergleich der Embryobildung und -Entwicklung von Cyperaceen und Gräsern, Keimung bei den Cyperaceen, Rolle des Saugorgans der Cyperaceen bei der Keimung. Botan. Centralbl. Beih. Abt. I 49:649–74. Ber. Wiss. Biol. 23:576–77.

Schneider, P. Müller-; *see* Müller-Schneider, P.

16991. Schneider, S. 1935: Untersuchungen über die Samenschleudermechanismen verschiedener Rhoeadales. Jahrb. Wiss. Botan. 81:663–704.

16992. Schneider, W. 1955: Futterwert von ausgekeimten Weizen. Landwirtsch. Jahrb. Schweiz 4:965–70.

Schneider, W. J. 18434.

16993. Schneider-Orelli, O. 1910: Resistance of *Medicago* seeds to high temperatures. Flora (Jena) 100:305–11.

Schneiter, A. A. 14452.

Schnell, H. D. 9202.

16994. Schnell, R., J. Von Euw, R. Richter, and T. Reichstein. 1953: The glycosides of the seeds of *Strophanthus sarmentosus*. IV. Glycosides and aglycones. Pharm. Acta Helv. 28:289–307. Chem. Abstr. 48:4770. 1954.

16995. Schnelle, F., and F. Heiser. 1935: The influence of time of harvesting on the quality of wheat. Mühlenlab. 5:49–56. Chem. Abstr. 29:5877. 1935.

16996. Schnellhardt, O., and F. D. Heald. 1936. The spore load of market wheat. Trans. Am. Microscop. Soc. 55:281–85.

16997. Schnetzler, M. J. B. 1888: Germination of *Ranunculus aquatilis*. Bull. Soc. Vaudoise Sci. Nat. 24:28–29.

Schoch, J. 13099.

16998. Schoeman, D. J., and F. Hawke. 1948: The fruitcoat fat of *Myrica cordifolia*-cape berry wax. I. Physical and chemical constants and fatty acid composition. J. S. African Chem. Inst. 1:5–13.

Schönbeck, F. 19757.

Schoenebeck, O. von; *see* Von Schoenebeck, O.

16999. Schoenike, R. E., and H. L. Hansen. 1954: Viability of black spruce seed in four-year-old logging slash. Minn. Forestry Notes 29. 2 pp.

17000. Schoffelmayer, V. 1948: Okra seed. Chemurgic Dig. 7(9):17.

Schollander, E. G. 17310.

17001. Scholz, J. 1933: The after-ripening of some wheat varieties especially Czechoslovakian. Vestn. Česk. Akad. Zeměděl. 9:403–9.

17002. —— 1934: How to improve the testing of wheat varieties for after harvest sprouting? Vestn. Česk. Akad. Zeměděl. 10:430–35.

17003. Scholz, W. 1934: The influence of yellow lupine (*Lupinus luteus*) seed stocks of varied origin upon the incidence and course of chlorosis. Z. Pflanzenernähr., Düng. Bodenk. 33A:340–45. Chem. Abstr. 28:5500. 1934.

Schoolcraft, I. 3958.

17004. Schooler, A. B. 1960: The effect of gibrel and gibberellic acid (K salt) in embryo culture media for *Hordeum vulgare*. Agron. J. 52:411.

17005. Schoorel, A. F. 1953: A possible task of the International Seed Testing Association seed testing stations in case of international disputes between seed merchants. Proc. Intern. Seed Testing Assoc. 18:113–18.

17006. —— 1953: Regulations in regard to obtaining blue and orange international certificates from the government seed-testing station. Ned. Alg. Keuringsdienst. Landbzaden Aardappelportgoed. Mededel. 9:93–94.

17007. —— 1957: The use of soil tests in seed testing. Proc. Intern. Seed Testing Assoc. 22:287–302.

17008. —— 1958: The influence of extreme temperatures and drying on the germination power of seeds. Landbouwk. Tijdschr. 70:117–22.

17009. —— 1960: Report on the activities of the vigour test committee. Proc. Intern. Seed Testing Assoc. 25:519–25.

17010. —— 1960: The influence of extreme temperatures and desiccation on the germination of seeds. Proc. Intern. Seed Testing Assoc. 25:573–79.

17011. —— 1960: Referee-testing on association-wide basis, 1954–1955–1956–1957–1958. Proc. Intern. Seed Testing Assoc. 25:649–73.

17012. —— 1960: Report on the third International Conference on seed pathology sponsored by the International Seed Testing Assoc. 1960. 27 pp.

17013. —— 1962: The training system for the seed analysts of the Wageningen seed testing station. Proc. Intern. Seed Testing Assoc. 27:995–1002.

Schoorel, A. F. 6061.

17014. Schopf, J. M. 1943: The embryology of *Larix*. Illinois Biol. Monogr. 19(4). 97 pp.

17015. Schopmeyer, H. H. 1945: *Amioca*. The starch from waxy corn. Food Ind. 17:1476–78. Biol. Abstr. 20:9063. 1946.

17016. Schormüller, J., and H. Winter. 1959: Amino acid content of the proteins of the cocoa bean. Nahrung 3:187–92. Chem. Abstr. 54:5842. 1960.

17017. —— and —— 1959: Proteins of the pumpkin seed. Nahrung 3:414–20. Biol. Abstr. 36:50062. 1961.

Schoth, H. A. 13209.

17018. Schotte, G. 1911: On the significance of different methods of testing the germinating power of coniferous seeds. Medd. Statens Skosforsoksanst. 8:245–79. Botan. Centralbl. 120:190. 1912.

17019. Schou, L. 1951: On chlorophyll formation in the dark in excised embryos of *Pinus jeffreyi*. Physiol. Plantarum 4:617–20.

Schowengerdt, G. C. 13961.

Schraa, R. J. 4418.

Schram, E. 5689.

17020. Schratz, E. 1932: Die Keimprüfung in Zuckerlösung ("Saugkraftbestimmung") und ihre Bedeutung für die Sortenkunde (ein kritischer Überblick). Züchter 4: 161–74.

17021. —— 1934: Beiträge zur Biologie der Halophyten. I. Zur Keimungsphysiologie. Jahrb. Wiss. Botan. 80:112–42.

Schreiber, R. 16882, 16883.

17022. Schreiber, W. R. 1946: Tagua. Chemurgic Dig. 5:204–6.

17023. Schreiner, E. J., and J. W. Duffield. 1942: Metaxenia in an oak species cross. J. Heredity 33:97–98. Biol. Abstr. 16: 15782. 1942.

17024. Schrenk, W. G. 1955: Chemical composition of Kansas wheat. Kansas Agr. Expt. Sta. Tech. Bull. 79. 46 pp.

Schreven, A. C. von; *see* Von Schreven, A. C.

Schriver, B. J. 1399.

Schrodter, H. 10832.

17025. Schroeder, C. A. 1944: Multiple embryos in the avocado. J. Heredity 35:208–10. Expt. Sta. Rec. 92:189. 1945.

17026. —— 1952: Floral development, sporogenesis and embryology in the avocado, *Persea americana*. Botan. Gaz. 113: 270–78.

17027. Schroeder, E. M. 1935: Dormancy in seeds of *Benzoin aestivale* L. Contrib. Boyce Thompson Inst. 7:411–19.

17028. —— 1937: Germination of fruits of *Ptelea* species. Contrib. Boyce Thompson Inst. 8:355–59.

17029. —— and L. V. Barton. 1939: Germination and growth of some rock garden plants. Contrib. Boyce Thompson Inst. 10:235–55.

Schroeder, E. M. 2047, 4605.

17030. Schroeder, G. 1886: Über die Austrocknungsfähigkeit der Pflanzen. Botan. Inst. Tübingen Untersuch. 2:1–52.

17031. Schroeder, H. 1922: Über die Semipermeabilität von Zellwänden. Botan. Centralbl. 42:172–88.

Schroeder, H. 16719.

17032. Schroeder, H. W. 1963: The relation between storage molds and damage in high-moisture rice in aerated storage. Phytopathology 53:804–8.

17033. —— 1963: Orange stain, a storage disease of rice caused by *Penicillium puberulum*. Phytopathology 53:843–45.

17034. —— and D. W. Rosberg. 1962: Effect of infrared intensity on the drying of rexoro rough rice. Texas Agr. Expt. Sta. 6 pp.

Schroeder, M. P. 2170.

Schroeder, W. 2720.

Schroeder, W. T. 9126, 13171, 14245, 14246.

Schröderheim, J. 7886.

17035. Schrödter, H. 1960: Dispersal by air and water—the flight and landing. Plant Pathol. 3:169–227.

17036. Schröppel, F. 1933: Katalase, Peroxydase und Atmung bei der Keimung lichtempfindlicher Samen von *Nicotiana tabacum*. Botan. Centralbl. Beih. 51: 377–407.

Schropp, W. 16884–16888.

17037. Schubert, G. H. 1950: Quintuplet seedlings in a sugar pine seed. J. Forestry 48:128–29.

17038. —— 1954: Viability of various coniferous seeds after cold storage. J. Forestry 52:446–47.

17039. —— 1955: Effect of storage temperature on viability of sugar, Jeffrey, and Ponderosa pine seed. U.S. Dept. Agr. Forest Serv. Calif. Forest Range Expt. Sta. Forest Res. Notes 100. 3 pp.

17040. —— 1960: Fungi associated with viability losses of sugar pine seed during cold storage. Proc. Soc. Am. Foresters 1960:18–21.

Schubert, G. H. 6580.

17041. Schubert, J. 1960: Zur Qualitätsbeurteilung von Forstsaatgut nach dem topographischen Tetrazoliumverfahren. Proc. Intern. Seed Testing Assoc. 25: 484–91.

17042. ——1961: (TTC′)-Reduktionaktivität und Keimvermögen Temperaturgeschädigter Samen von *Robinia pseudoacacia* L. Proc. Intern. Seed Testing Assoc. 26: 472–503.

Schubert, T. A. 2564.

17043. Schuchardt, P. A., A. W. Halverson, and C. O. Clagett. 1955: Occurrence of the selenium protective principle of flax in hull and embryo fractions. Proc. S. Dakota Acad. Sci. 34:48–51. Chem. Abstr. 51:563. 1957.

17044. Schudel, H. L. 1952: Vegetable seed production in Oregon. Oregon Agr. Expt. Sta. Bull. 512. 79 pp.

17045. Schuette, H. A., and C. Y. Chang. 1933: Hazlenut (filbert) oil. J. Am. Chem. Soc. 55:3333–35.

17046. ——, M. A. Crowley, H. A. Vogel, and M. M. Meuller. 1940: Fenugreek seed oil. Oil soap 17:122.

17047. —— and J. A. Korth. 1940: Seed oil of the highbush cranberry. Oil Soap 17:265.

17048. —— and C. M. Lunde. 1936: Elmseed oil. Oil Soap 13:12–13.

17049. —— and R. C. Palmer. 1938: The chemistry of the rye germ. IV. Its proximate composition. Cereal Chem. 15: 445–50.

17050. ——, A. N. Pines, and G. J. Krueger. 1943: The seed oil of the arrow wood. Oil Soap 20:158. Biol. Abstr. 17:25106. 1943.

17051. ——, H. A. Vogel, and J. A. Bain. 1943: The saturated fatty acids of elderberry seed oil. Oil Soap 20:46.

17052. —— and C. H. Wartinbee. 1932: The composition of alfalfa seed oil. Am. Chem. Soc., Denver. Abstr. Papers Div. Agr. Food Chem. 3 pp.

Schuette, H. A. 7329, 9354, 18070.

17053. Schütte, K. D., and D. Parkhurst. 1960: The influence of light upon the germination of some Compositae seeds. J. S. African Botany 26:149–53. Biol. Abstr. 36:8664. 1961.

17054. Schütte, K. H., and P. Martin. 1959: The influence of certain micro-nutrient deficiencies on the germination rates of radish and millet. Plant Soil 10:299.

17055. Schütz, W. 1931: Über die Einwirkung von Beizmitteln auf Keimung und Wachstum des Weizens. Botan. Arch. 33:199–256. Rev. Appl. Mycol. 11:169. 1932.

17056. Schuhmann, G. 1963: Treatment of tobacco seed. Nachrbl. Deut. Pflanzenschutzdienst (Brunswick) 15:37–45. Chem. Abstr. 59:10707. 1963.

Schuler, J. F. 17816.

Schulte, E. 426.

17057. Schulte, H. 1946: Mercury hazards in seed treating. J. Ind. Hyg. 28:159–61. Rev. Appl. Mycol. 27:317. 1948.

17058. Schulte, R. B., and P. Balbach. 1941: Germination of seeds of *Agropyron* sp. (probably quack grass), and *Thlaspi arvense*, penny cress. Proc. Assoc. Offic. Seed Analysts. 33:44–45.

17059. —— and —— 1943: Germination of decorticated Cruciferae and *Agropyron*. Assoc. Offic. Seed Analysts News Letter 17(2):7–10.

17060. Schultz, A. M., and H. H. Biswell. 1959: Effect of prescribed burning and other seedbed treatments on Ponderosa pine seedling emergence. J. Forestry 57:816–17.

17061. Schultz, J. 1953: Forest seed production in relation to planetary cycles. Bio-Dynamics 11(3):10–20.

17062. —— 1953: Forest seed production in relation to planetary cycles (conclusion). Bio-Dynamics 11(4/5):26–34.

17063. Schultz, O. C. 1917: Position in the pod and productiveness; seed weight and abortiveness; peanuts. N.J. Agr. Expt. Sta. Rept. 38:401–4. Expt. Sta. Rec. 41:38. 1919.

Schultze, R. E. 5223.

17064. Schulz, K. 1958: Der Einfluss der Grosse des Samenkorns auf Entwicklung und Ertrag bei *Lupinus albus*. Z. Acker-Pflanzenbau 106:224–44.

17065. Schulz, M. R., and R. M. Klein. 1963: Effects of visible and ultraviolet radiation on the germination of *Phacelia tanacetifolia*. Am. J. Botany 50:430–34.

17066. Schulz, P. 1956: Starting saintpaulias from seed. Minn. Horticulturist 84(3):35.

Schulz, W. 11489.

Schulz, W. B. T. 16377.

17067. Schulz, Z. M. 1930: Über an keimenden Samen beobachtete Pilze. Jard. Botan. Inst. Essais Semences, Leningrad, Ann. Essais Semences 1930(2):51–57.

Schulz, Z. M. 5814, 10193.

17068. Schulze, E. 1878: Über Zersetzung und Neubildung von Eiweisstoffen in Lupinenkeimlingen. Landwirtsch. Jahrb. 7:411–44.

17069. —— 1889: Non-nitrogenous reserve-substances in the seeds of Leguminosae. Ber. Deut. Botan. Ges. 7:322–29.

17070. —— 1893: Ueber die Kohlenhydrate der Kaffeebohnen. Chem. Ztg. 17: 1263–64.

17071. —— 1894: Analysis of plant seeds. Chem. Ztg. 18:799–802.

17072. —— 1895: Nitrogenous organic bases in agricultural seeds and seedlings. Landwirtsch. Versuchssta. 46:23–77.

17073. —— 1896: Reserve material in cell walls of cotyledons of *Lupinus luteus* and *L. angustifolius*. Ber. Deut. Botan. Ges. 14:66–71.

17074. —— 1897: Concerning the decomposition of proteids and the formation of asparagin and glutamin by germinating plants. Chem. Ztg. 21:625–28.

17075. —— 1897: On the cleavage products of a protein compound prepared from seeds of Coniferae. Z. Physiol. Chem. 24:276–84. Expt. Sta. Rec. 9:723. 1897/98.

17076. —— 1906: Über den Abbau und den Aufbau organischer Stickstoffverbindungen in den Pflanzen. Landwirtsch. Jahrb. 35:621–66.

17077. —— and W. Umlauft. 1876: Untersuchungen über einige chemische Vorgänge bei der Keimung der gelben Lupine. Landwirtsch. Jahrb. 5:821–68.

17078. —— and E. Winterstein. 1910: Studien über die Proteinbildung in reifenden Pflanzensamen. Z. Physiol. Chem. 65:431–76.

17079. Schulze, H., and E. Bierling. 1913: Über die Alkaloide von *Aconitum lycoctonum*. Arch. Pharm. 251:8–50.

17080. Schuphan, W. 1962: Dependence of carotene formation in carrot roots (*Daucus carota*) on seed quality and resistance to weathering. Mezhdunar. Konf. Vitamin., Dokl. Suobsht. Sofia 1960:113–24. Chem. Abstr. 60:7150. 1964.

17081. —— and W. Postel. 1960: Samenreife und essentielle Aminosäuren bei *Pisum sativum* L. Naturwissenschaften 14:323–24.

Schupp, A. A. 19170.

17082. Schuster, C. E. 1947: Edible nuts of the Pacific northwest. Econ. Botany 1: 389–93.

17083. Schuster, M. L. 1943: The influence of certain diseases and threshing injury on stands of flax and peas in Washington. Proc. Am. Soc. Hort. Sci. 43:180–82.

17084. —— and D. S. Nuland. 1960: Seed transmission of safflower *Verticillium* wilt fungus. Plant Disease Reptr. 44:901–3.

17085. Schuster, W. 1956: Saatzeitenversuche mit der Sonnenblume (*Helianthus annuus* L.). Z. Acker- Pflanzenbau 100: 349–66.

17086. Schwabe, W. W. 1954: Factors controlling flowering in the chrysanthemum. IV. The site of vernalization and translocation of the stimulus. J. Exptl. Botany 5:389–400.

17087. —— 1955: Factors controlling flowering in the chrysanthemum. V. Devernalization in relation to high temperature and low light intensity treatments. J. Exptl. Botany 6:435–50.

17088. —— 1957: Factors controlling flowering in the chrysanthemum. VI. Devernalization by low-light intensity in relation to temperature and carbohydrate supply. J. Exptl. Botany 8:220–34.

17089. —— 1957: Recent work on the inhibitory effect of long days on the flowering of some short-day plants. Union Intern. Sci. Biol. Ser. B, 34:95–109.

17090. —— 1963: Studies in vernalisation of cereals. XV. After-effects of temperature and drying during seed ripening and the origin of the vernalisation requirement in Petkus winter rye. Ann. Botany (London) 27:671–83.

17091. —— and M. J. Thornley. 1950: Vernalization of winter rye by ultrasonics. Ann. Appl. Biol. 37:19–22.

17092. Schwaer, C. 1963: Action of wormwood extracts on germination, growth, development, and water content of fennel and other plants. Arch. Gartenbau 11: 91–100. Chem. Abstr. 60:3425. 1964.

17093. Schwanitz, F. 1942: Über den Einfluss des Entfernens der Keimblätter auf die Entwicklung und den Ertrag von diploidem und autotetraploidem gelbem Senf (*Sinapis alba*). Züchter 14:86–93. Biol. Abstr. 22:10605. 1948.

17094. —— 1948: Untersuchungen an polyploiden Pflanzen. I. Feldversuche mit diploiden und autotetraploiden Nutzpflanzen (Senf, Rübsen, Ölrettich, Rettich, Grünkohl, Weisskohl, Wirsing und Chicoree). Züchter 19:70–86.

17095. —— 1949: Untersuchungen an polyploiden Pflanzen. II. Zur Keimungsphysiologie diploider und autotetraploider Nutzpflanzen. Planta 36:389–401.

17096. —— 1955: Keimungsphysiologische Untersuchungen an diploiden Gigaspflanzen. Beitr. Biol. Pflanz. 31:1–14.

Schwantes, A. J. 5162.

17097. Schwappach, A. 1906: Mitteilungen aus der Prüfungsanstalt für Waldsamen in Eberswald. Z. Forst- Jagdw. 38:505–15.

17098. —— 1915: Die Waldsamen prüfungsanstalt Eberswalde und die Methoden der Prüfung von Waldsamen. Z. Forst- Jagdw. 47:631.

Schwarting, A. E. 3596.

17099. Schwartz, D. 1960: Electrophoretic and immunochemical studies with endosperm proteins of maize mutants. Genetics 45:1419–27.

Schwartz, D. 4788, 6217.

17100. Schwartz, H. M., and R. P. Bürke. 1958: The chemistry of naras seed (*Acanthosicyos horrida* Hook). III. The amino-acid composition of the protein. J. Sci. Food Agr. 9:159–62.

Schwartz, H. M. 9062, 12236.

17101. Schwartz, P. 1939: A simple method for determining the infection of wheat seeds with *Ustilago tritici*. Compt. Rend. Acad. Sci. URSS 24:189–91.

17102. Schwartzman, G. 1953: Report on moisture in coffee. J. Assoc. Offic. Agr. Chemists 36:661–63.

17103. Schwarz, G., and H. Finzenhagen. 1937: Investigations on vetch seeds with respect to the presence of harmful substances. II. Biedermanns Zentralbl. B. Tierernähr. 9:115–20. Chem. Abstr. 31: 4740. 1937.

17104. Schwarzenbach, F. H. 1957: Untersuchungen über einen Pollenkeimungshemmstoff aus reifenden Samen von *Cyclamen persicum* Mill. Zürich Natforsch. Ges. Vrtljahrsschr. 102:317–31.

17105. Schwass, R. H. 1951: Assessing the value of low-germinating seed. New Zealand J. Agr. 82:225.

17106. Schweigart, H. A., and W. Hartmann. 1960: Plant nutrition by treating seeds with micronutrient elements. Vitalstoffe Zivilisationskrankh. 5(20)-4:171–77. Biol. Abstr. 38:3154. 1962.

17107. Schweighart, O. 1935: Effect of eosin on seedlings. Botan. Centralbl. Beih.

53:217–92. Chem. Abstr. 29:5878. 1935.
Schweitzer, C. E. 10655.

17108. Schweitzer, R. 1954: A new technique for vernalization of seeds with heat. Compt. Rend. Acad. Agr. France 40:99–100. Biol. Abstr. 29:14755. 1955.

17109. Schweizer, G. 1932: Zur Anatomie des Mohnsamens (*Papaver somniferum* L.). Ber. Deut. Botan. Ges. 49:414–23. J. Roy. Microscop. Soc. 52:310–11. 1932.

17110. Schweizer, I. 1927: Physiological experiments on germination of coffee seeds. Planters Chron. Abstr. Madras 22:299–310. Botan. Centralbl. 154:266–67. 1928.

17111. Schweizer, J. 1927: Investigations on coffee. I. Physiological experiments on germination of coffee seed. Arch. Koffiecult. 1:249–71.

17112. Schwemmle, J. 1940: Keimversuche mit alten Samen. Z. Botan. 36:235–61.

17113. —— 1957: Das unterschiedliche Reifen und Altern der Samenanlagen in der verschiedenen Abschnitten der *Oenothera berteriana*. Planta 49:168–204.

17114. —— 1960: Keimversuche auf genetischer Grundlage. I. Die Lichtabhängigkeit. Planta 54:294–313.

17115. —— 1961: Keimversuche auf genetischer Grundlage. Flora 150:274–93.

17116. —— 1961: Keimversuche auf genetischer Grandlage. V. Versuche mit Samen von Homozygoten. Planta 56:357–76.

17117. —— 1962: Keimversuche auf genetischer Grundlage. VIII. Versuche mit Samen von vier verwandten Homozygoten und solchen aus reziproken Kreuzungen. Biol. Centralbl. 81:183–212.

17118. —— and H. Fischer. 1963: Wodurch sind die grossen Schwankungen bei der Keimung von Oenotheren-Samen bedingt? Ber. Deut. Botan. Ges. 76:147–53.

17119. —— and H. Gross. 1961: Keimversuche auf genetischer Grundlage. VI. Der Einfluss der Unterlage auf die Keimung von Samen der aufgepfropften Reiser. Planta 56:377–87.

17120. Schwendiman, A., and H. L. Shands. 1943: Delayed germination or seed dormancy in Vicland oats. J. Am. Soc. Agron. 35:681–88.

17121. Schwendiman, J. L., and L. A. Mullen. 1944: Effects of processing on germinative capacity of seed of tall oatgrass, *Arrhenatherum elatius* (L.) Mart. and Koch. J. Am. Soc. Agron. 36:783–85.

17122. ——, R. F. Sackman, and A. L. Hafenrichter. 1940: Processing seed of grasses and other plants to remove awns and appendages. U.S. Dept. Agr. Circ. 558:1–15. Biol. Abstr. 14:12510. 1940.
Schwerdtfeger, E. 14296.

17123. Schwerdtfeger, L. 1952: Growing begonias from seed. Begonian 19:6–7.
Schwieger, E. 6379, 19207.

17124. Schwimmer, S. 1947: Development and solubility of amylase in wheat kernels throughout growth and ripening. Cereal Chem. 24:167–79.
Schwimmer, S. 1825.
Schwinghamer, E. 11059.

17125. Schÿfsma, L., M. Hoesbergen, and F. E. Nÿdam. 1960: A study of the colour and other characters of the seed of some varieties of oil seed poppy. Euphytica 9:127–40.
Sciuchetti, L. A. 3519, 17126.

17126. Scott, A. Z., and L. A. Sciuchetti. 1961: The effect of gibberellic acid and various treatments on germination, growth, and alkaloid formation in *Atropa belladonna*. Lloydia 24:211–16.

17127. Scott, B. D. 1940: Aconite seeds need freezing. Flower Grower 27(6):297.

17128. Scott, D. H., and D. P. Ink. 1948: Germination of strawberry seed as affected by scarification treatment with sulfuric acid. Proc. Am. Soc. Hort. Sci. 51:299–300.

17129. —— and —— 1950: Grape seed germination experiments. Proc. Am. Soc. Hort. Sci. 56:134–39.

17130. —— and —— 1955: Treatments to hasten the emergence of seedlings of blueberry and strawberry. Proc. Am. Soc. Hort. Sci. 66:237–42.

17131. —— and —— 1957: Treatment of *Rubus* seeds prior to after-ripening to improve germination. Proc. Am. Soc. Hort. Sci. 69:261–67.

17132. —— and J. G. Waugh. 1941: Treatment of peach seed as affecting germination and growth of seedlings in the greenhouse. Proc. Am. Soc. Hort. Sci. 38:291–98.

17133. ——, ——, and F. P. Cullinan. 1942: An injurious effect of peach juice on germination of the seed. Proc. Am. Soc. Hort. Sci. 40:283–85.
Scott, D. H. 4915, 13790.
Scott, E. I. C. 15080.

17134. Scott, F. M. 1944: Cytology and microchemistry of nuclei in developing seed of *Echinocystis macrocarpa*. Botan. Gaz. 105:329–38.

17135. —— 1953: The physical consistency

of the endosperm nucleus of *Echinocystis macrocarpa*. Phytomorphology 3:66–76.

17136. ——, B. G. Bystrom, and E. Bowler. 1962: *Cercidium floridum* seed coat, light, and electron microscopic study. Am. J. Botany 49:821–33.

Scott, G. W. 12682.

Scott, L. I. 4828, 15682.

17137. Scott, R. A. 1953: Fossil fruits and seeds from the Eocene Clarno formation of Oregon. Ph.D. thesis, Univ. of Michigan. Dissertation Abstr. 13:647.

Scott, R. K. 2982.

Scott, R. W. 11952.

17138. Scott, W. 1924: What is the relation between the moisture content and viability of seed corn when subjected to low temperatures. Proc. Iowa Acad. Sci. 30:254–62.

Scotti, R. 16219.

Scotti, T. 1766.

Scrimshaw, N. S. 3001, 3003, 3004, 9770, 17831.

Scuflaire, R. 2958.

Scuncia, P. E. 16220.

Scurti, J. 2215.

17139. Seaber, W. M. 1933: Barium as a normal constituent of Brazil nuts. Analyst 58:575–80. Chem. Abstr. 28:228. 1934.

17140. Seale, C. C., J. F. Joyner, and E. O. Gangstad. 1952: The experimental culture of kenaf. (*Hibiscus cannibinus* L.) for fiber and seed in South Florida. Turrialba 2:100–5. Biol. Abstr. 28:11809. 1954.

17141. Seaman, W. H. 1876: The vitality of seeds. Field and Forest. 1:57–63.

Seamans, R. F. 2663.

Sears, E. R. 13174.

17142. Seaton, H. L. 1938: Relation of number of seeds to fruit size and shape in cucumbers. Proc. Am. Soc. Hort. Sci. 35:654–58.

Seaton, J. C. 12657.

Seatz, L. F. 12408.

17143. Seay, W. A., and S. H. West. 1956: The effect of seed treatment on yield and germination of garden peas. Soil Sci. Soc. Am. Proc. 20:198–200.

Seay, W. A. 18192.

Sebanek, J. 18331.

17144. Sebastian, C. 1958: Esasis de germination de quatre espèces du genre *Juniperus*. Bull. Soc. Sci. Nat. Phys. Maroc 38:115–22.

17145. Sebe, Y. 1943: Volatile constituents of the oil from rice embryo. I. J. Chem. Soc. Japan 64:1405–10. Biol. Abstr. 22:2192. 1948.

17146. Sebelin, C. 1929: Über Ätiologie und Regenerationsvermögen der "anomalen Kleekeime." Proc. Intern. Seed Testing Assoc. 7/8:1–47. Pflanzenbau 6:139–41. 1929.

Seber, L. 19270.

17147. Sebesta, K., and F. Sorm. 1954: Content of adenosinetriphosphoric acid in seedling of *Phaseolus vulgaris*. Chem. Listy 48:1585–86. Chem. Abstr. 49:1881. 1955.

17148. —— and —— 1956: Changes in the adenosinetriphosphate content during germination of the bean. Chem. Lsity 50: 632–36. Chem. Abstr. 50:9532. 1956.

17149. Sebestyen, J. 1958: Rapid-germinating methods compared by seeds of *Acer campestre* L. and *Acer tataricum* L. as test-plants. Kertesz. Kutato Intezet. Evkönyve 3:111–18.

17150. Sechet, J. 1949: Influence de la printanisation sur la teneur en glucides des semences de céréales. Compt. Rend. 228:334–36.

17151. —— 1953: Le comportement du ricin après traitement par le froid des graines en germination. Bull. Soc. Botan. France 100:44–47.

17152. —— 1953: Contribution to the study of vernalization. Botaniste (Paris) 37:5–306. Biol. Abstr. 30:14778. 1956.

17153. —— 1953: La résistance au froid des semences printanisées. Compt. Rend. Acad. Agr. France 39:265–67.

17154. —— 1959: Effect of low temperatures on germinating seeds of sugar beet. Compt. Rend. Hebd. Seances Acad. Agr. France 17:879–81. Biol. Abstr. 41:7965. 1963.

17155. —— 1959: Le rôle de la greffe dans la localisation des phénomènes de vernalisation. Bull. Soc. Sci. Bretagne 34: 109–14.

17156. —— 1960: La synthèse nicotinique au cours de traitement de vernalisation. Compt. Rend. 250:752–54.

17157. —— 1962: Influence des basses températures sur la germination des semences de Vigne. Compt. Rend. 255: 2653–55.

17158. ——, J. Masquelier, and E. Tayeau. 1959: Research on the pigments in sunflower seeds (*Helianthus annuus* L.). II. Caffeic acid. Bull. Soc. Chim. Biol. 41:1067–70. Index Medicus 1:613. 1960.

Sechet, J. 4971–4975, 5291, 7010.

17159. Sechet-Sirat, J. 1963: Localization of chlorogenic acid in sunflower seeds. Ann.

Pharm. Franc. 21:455. Biol. Abstr. 45: 3600. 1964.

17160. Seckinger, H. L., M. J. Wolf, and R. J. Dimler. 1964: A micro method for determining moisture distribution in wheat kernels, based on iodine staining. Cereal Chem. 41:80–87.

17161. ——, ——, and M. M. MacMasters. 1960: Hemicelluloses of the cementing layer and of some cell walls of the corn kernel. Cereal Chem. 37:121–28.

Seckinger, H. L. 12654.

17162. Sedasheva, G. I. 1953: On the dormancy period in oak. Lesn. Khoz. 6(6):35–37.

17163. Seddon, H. R., and H. G. Belschner. 1931: Poisoning of sheep by the seeds of burrawang (*Macrozamia spiralis*). N.S. Wales Dept. Agr. Vet. Res. Rept. 6:70–80. Biol. Abstr. 6:25443. 1932.

17164. Sedgley, R. H. 1963: The importance of liquid-seed contact during the germination of *Medicago tribuloides* Desr. Australian J. Agr. Res. 14:646–53.

Sedgwick, V. E. 9132.

17165. Sedlmayr, T. E. 1962: Inheritance of speed of germination in sugar beets (*Beta vulgaris* L.). Dissertation Abstr. 22:4151–52.

Seeborg, E. F. 6132.

17166. Seed Trade Buyers Guide. Annual: Seed World Publications, Chicago.

17167. Seefeldner, G. 1912: Polyembryony in *Cynanthum vincetoxicum*. Sitzber. Kais. Akad. Wiss. Wien Math. Naturw. K. 121:273–96.

Seeley, M. 526.

Seeley, M. G. 525.

Seely, C. I. 6015.

17168. Seelye, G. D. 1946: The number of vegetable seeds per unit weight. Proc. Am. Soc. Hort. Sci. 48:391–97.

17169. Seemann, G. 1953: Orchideenanzucht aus Samen. Garten Z., Flora 76(2):15–17.

17170. Sefick, H. J., and A. M. Musser. 1952: Removal of grape seed with Waring Blender and effect on germination. S. Carolina Agr. Expt. Sta. Ann. Rept. 64: 42–43.

17171. Sefranak, B. 1951: The influence of the hydrogen peroxide on the germinative faculty, growth and evolution of different sorts and varieties of plants. Sb. Česk. Akad. Zemědél. 23:320–27.

17172. Segeta, V. 1960: Cold resistance of germinating maize. Sb. Česk. Akad. Zemědél. Ved Rostlinna Vyroba 6:307–18.

17173. Seguin, R., and J. L. Parra. 1951: El semillero, elemento basico en la regeneración de los atochares de Granada. Montes 7:425–30.

Segura, R. 6916.

Sehgal, C. B. 9917.

17174. Sehgal, S. L., D. S. Bhakuni, V. N. Sharma, and K. N. Kaul. 1961: Chemical constituents of the seed of *Luffa graveolens*. J. Sci. Ind. Res. India 20B:461–62. Chem. Abstr. 56:5125. 1962.

17175. Seibold, W. 1940: Plant germination in water. Ger. Patent 694,394. Chem. Abstr. 35:5163. 1941.

Seidl, R. 18591.

17176. Seidman, G., and J. C. Frazier. 1962: The distribution of sulfur among various components of the wheat grain. Plant Soil 16:137–47.

17177. —— and —— 1963: Distribution of sulfur in the wheat grain as shown by radioautographs. Plant Soil 19:151–52.

Seiferth, C. 3291.

Seigler, W. E. 9134.

Seim, A. L. 11890.

17178. Sekera, F. 1955: Der Dunkelkeimversuch als bodenbiologische Testmethode. Z. Pflanzenernähr. Düng. Bodenk. 70: 193–207.

17179. Sekizuka, S., and S. Hojito. 1959: Studies of native fodder grasses in Japan. IV. Hard seed content of some wild legumes under storage conditions. Proc. Crop Sci. Soc. Japan 28:86–87.

17180. Selby, A. D. 1901: Germination of the seeds of some common cultivated plants after prolonged immersion in liquid air. Bull. Torrey Botan. Club 28:675–79.

17181. Selik, M., and G. Bodo. 1960: Untersuchungen über den Eiweissabbau in den Samen der auf dem Baum bleibenden Zapfen von *Pinus halepensis* Mill. Planta 54:404–8.

17182. Selikhov, V. 1937: The soaking of sugar-beet seed. Vses. Nauchn.-Issled. Inst. Sakharnoi. Prom. Osnovn. Vyvody Nauchn.-Issled. Rabot V. N. I. S.'a. 1936:154–56. Chem. Abstr. 34:3317. 1940.

17183. —— 1959: Effect of electric warming on quality of corn seeds. Sel'skoe Khoz. Kazakhstana 6:23–24.

17184. Sell, H. M., F. A. Johnston, Jr., and F. S. Lagasse. 1946: Changes in the chemical composition of the tung fruit and its component parts. J. Agr. Res. 73: 319–34.

17185. —— and R. E. Kremers. 1941: Sterol from apple seeds and cherry seeds. J. Am. Pharm. Assoc. 30:134. Pharm. Abstr. 8:90. 1942.

Sell, H. M. 5465, 6827, 6828, 7227, 8838, 9897, 12483, 16048.

17186. Sell, O. E. 1939: Technique in colchicine treatment of pasture plants and seeds. Proc. Assoc. Southern Agr. Workers 40:63–64. Chem. Abstr. 34:2883. 1940.

17187. —— 1944: Scarified bur becomes miracle clover. Southern Seedsman 7(6): 12, 61, 67. Biol. Abstr. 19:9210. 1945.

Sell, O. E. 13983.

17188. Sellei, J. 1936: The growth-promoting and limiting action of fluorescein in plant development, with especial reference to "Fotosensin." Z. Pflanzenernähr. Düng. Bodenk. 43:321–40. Chem. Abstr. 30: 7629. 1936.

Sellman, G. 8962.

17189. Sellschop, J. 1951: The production of safflower seed. Farming S. Africa 26:253–56.

Selman, G. G. 8328.

17190. Selman, I. W. 1943: The appearance and spread of mosaic infection in the tomato crop and the relation to seed transmission of the virus. Ann. Appl. Biol. 30:331–38.

17191. Selvaratnam, E. M. 1952: Embryo of the coconut. Nature 169:714–15.

17192. Semb, J. 1935: Phytochemical notes. CVIII. A phytochemical study of the seed of the digger pine. J. Am. Pharm. Assoc. 24:609–13. Chem. Abstr. 29:7104–5. 1935.

17193. Semenenko, G. I. 1955: Activity of ribonuclease in sprouting and ripening seeds of wheat and pea. Dokl. Akad. Nauk SSSR 102:575–77. Chem. Abstr. 49:16079. 1955.

17194. —— 1957: The transformation of nucleic acids in germinating and maturing seeds. Plant Physiol. (USSR) (English Transl.) 4:320–25.

17195. —— 1957: Transformation of nucleic acids in seeds. Plant Physiol. (USSR) (English Transl.) 4:332–37. Chem. Abstr. 52:13756. 1958.

17196. Semeniuk, G., C. M. Nagel, and J. C. Gilman. 1947: Observations on mold development and on deterioration in stored yellow dent shelled corn. Iowa Agr. Expt. Sta. Res. Bull. 349:253–84.

Semeniuk, G. 14087.

17197. Semeniuk, P. 1964: Effect of various levels of nitrogen, phosphorus, and potassium on seed production and germination of *Matthiola incana*. Botan. Gaz. 125:62–65.

17198. —— and R. N. Stewart. 1960: Effects of temperature on fruit and seed production of lily, stock, poinsettia, rose, and snapdragon. Proc. Am. Soc. Hort. Sci. 75:726–29.

17199. —— and —— 1962: Temperature reversal of after-ripening of rose seeds. Proc. Am. Soc. Hort. Sci. 80:615–21.

17200. —— and —— 1963: Effects of high temperature on seed production in column stocks, *Matthiola incana*. Proc. Am. Soc. Hort. Sci. 82:583–85.

17201. ——, ——, and J. Uhring. 1963: Induced secondary dormancy of rose embryos. Proc. Am. Soc. Hort. Sci. 83: 825–28.

Semenko, M. G. 17362.

Semenova, A. K. 20028.

17202. Semetaite, L. B. 1961: Proteins of seeds of *Caragana arborescens* and vignin of *Vigna sinensis*. Tr. Gl. Botan. Sada, Akad. Nauk SSSR 8:47–59. Chem. Abstr. 57:2596. 1962.

17202A. —— 1962: On the activity and quality of peptidase of seeds of the Siberian pea-shrub and the common cowpea. Byul. Gl. Botan. Sada 45:84–87.

17203. —— 1962: Proteinases from seeds of *Vigna sinensis* and *Caragana arborescens*. Lietuvos TSR Mokslu Akad. Darbai Ser. C 1962(1):105–14. Chem. Abstr. 57:5025. 1962.

17204. Semiganovskaia, N. V. 1938: Dependence of the development of the wheat plant on the geographic origin of seed. Compt. Rend. Acad. Sci. URSS 20:483–88. Bull. Soc. Botan. France 87:380. 1940.

17205. Semler, H. 1887: Oil from seeds of *Jatropha curcas*. Trop. Agrikultur 2: 443–44.

17206. —— 1887: Oil from seed of *Aleurites triloba*. Trop. Agrikultur 2:444–47.

17207. —— 1887: Oil from seeds of *Moringa* spp. Trop. Agrikultur 2:455–56.

17208. —— 1887: Oil from seeds of *Thea chinensis*. Trop. Agrikultur 2:457.

17209. —— 1887: Oil from seeds of *Cucumis melo*. Trop. Agrikultur 2:457–58.

17210. —— 1887: Oil from seeds of *Gossypium* spp. Trop. Agrikultur 2:458–59.

17211. Semmens, E. S. 1947: Chemical effects of moonlight. Nature 159:613.

Sempio, C. 16253, 16254.

Semple, A. T. 2443.

Sen, A. 17528.

17212. Sen, A. C. 1961: Viability of stored seed wheat safeguarded by seed dressing. Current Sci. (India) 30:466–67.

17213. Sen, A. K. 1946: *Argemone* oil. Indian Med. Gaz. 81:126–28. Chem. Abstr. 40:6274. 1946.

17214. Sen, B., and S. C. Chakravarti. 1942: Vernalization of mustard. Nature 149: 139–40.

17215. —— and —— 1945: Vernalisation response of Indian wheats. Current Sci. (India) 14:124–25. Plant Breeding Abstr. 15:304. 1945.

17216. —— and —— 1946: Effect of high temperature on vernalized mustard seed. Nature 157:266.

17217. —— and —— 1947: Vernalization of excised mustard embryo. Nature 159: 783–84.

17218. —— and H. C. Joshi. 1959: Effect of irradiation with thermal neutrons on control and vernalized seeds of mustard. Agron. J. 51:460–62.

17219. —— and G. Verma. 1955: Effect of presoaking and chilling seeds of mustard T 102, and wheat Pb 228 in different inorganic media. Proc. Indian Sci. Congr. 42:262–63.

17220. —— and —— 1960: Cultivation of mustard embryo and seedling fragments. Mem. Indian Botan. Soc. 2:36–39. 1959.

17221. Sen, K. R. 1932: On the variation of certain characters of cotton in relation to the position of seeds in a lock, with a statistical note by S. Subramonia Iyer. Indian J. Agr. Sci. 2:484–98. Plant Breeding Abstr. 3:90. 1933.

17222. Sen, N. K. 1948: Vernalization and photoperiodic effect in summer and late sown winter rice. J. Indian Botan. Soc. 27:111–18.

17223. ——, J. K. Chakrabarti, W. Kreis, C. Tamm, and T. Reichstein. 1957: Die Glykoside der Jutesamen *Corchorus capsularis* L. und *C. olitorius* L. Identifizierung von Corchorin, Corchogenin und Corchsularin mit Strophanthidin. Helv. Chim. Acta 40:588–92.

17224. Sen, N. K., and I. Mukhopadhyay. 1961: Studies in embryo culture of some pulses. Indian Agriculturist 5:48–56.

17225. —— and A. S. N. Murty. 1960: Inheritance of seed weight in green gram (*Phaseolus aureus* Roxb.). Genetics 45: 1559–62.

Sen, N. K. 3765, 13275.

17226. Sen, N. P., and A. D. Robinson. 1963: The absence of thiamine phosphates from cereals. Can. J. Biochem. Physiol. 41:97–100. Chem. Abstr. 58:9411. 1963.

17227. Sen, P. K. 1937: Some observations on the time of development of the fruit, seed and embryo in apple. E. Malling Res. Sta. Ann. Rept. 1936:137–41.

17228. —— and G. N. Mitra. 1958: Inheritance of photoperiodic reaction in rice. Nature 182:119–20.

17229. Senewiratne, S. T. 1954: Rice soaked in solution of fertilizer comes up faster, stands up better. Crops & Soils 7(3):22.

Sengbusch, R. von; *see* Von Sengbusch, R.

Senjo, O. 13545, 13546.

Senn, T. L. 12900.

Sennaiyan, P. 14277.

Senti, F. R. 4220, 9161, 16902.

Seoane, M. C. 16162.

Sequi, P. 11792.

Sera, S. 13449.

Serebryakova, L. E. 19163.

17230. Sereiskij, A., and M. Sludskaja. 1937: Acceleration of the growth of seedlings and the accumulation of blastanin in the endosperm through the yarovization of spring and winter wheats. Compt. Rend. Acad. Sci. URSS 17:55–58. Z. Botan. 33:364–66. 1938. Bull. Soc. Botan. France 88:573. 1941.

17231. Serenkov, G., and N. Palinka. 1951: Change of carbohydrate complex in grain sprouts grafted on foreign endosperm. Selek. i Semen. 18(11):19–23.

17232. Serenkov, G. P. 1958: Alcohol-soluble proteins of tea seeds. Nauchn. Dokl. Vysshei Shkoly Biol. Nauki 1958(2): 134–38. Chem. Abstr. 53:2378. 1959.

17233. —— and V. S. Kuznetsova. 1952: The change in the carbohydrate complex of acorns during storage. Vestn. Mosk. Univ. 7, No. 2, Ser. Fiz. Mat.-Estestven Nauk No. 1:119–25. Chem. Abstr. 46: 9669. 1952.

17234. Sergeev, L. I. 1935: Salt resistance of wheats in dependence on variety. Dokl. Akad. Nauk SSSR 1:563–70. Biol. Abstr. 10:12940. 1936.

17235. —— and E. V. Kucherov. 1957: On the suitability of not fully ripened corn seeds for sowing. Selek. i Semen. 22(1): 47–50.

17236. Sergeeva, K. D. 1952: On the viability of seeds in hybrid plants of gooseberry and currants. Agrobiologiya 1952(4): 123–25.

Serini, G. 1285.

Serkova, E. V. 17675.
Sermanni, G. Giovanozzi-. 16854, 16855.
17237. Serpa, D. 1952: Seleccion de semillas de lechosa. Venezuela Inst. Nacl. Agr. P. Misc. 3:41–42.
Serr, E. F. 1575, 3154.
17238. Serrano, M. 1954: The effect of alkali metals on the germination and growth of wheat seedlings. Farmacognosia (Madrid) 14(31):27–51. Chem. Abstr. 49: 4810. 1955.
Serrano, M. 15204.
Seshadri, C. R. 9840, 15106.
Seshadri, T. R. 1746, 1747, 3748, 6229, 10519, 15930.
17239. Seshagiriah, K. N. 1934: Seeds and seedlings of *Tinospora cordifolia*, Miers. Current Sci. (India) 3:264.
Seshu, D. V. 16687.
Setälä, J. E. Aalto-; *see* Aalto-Setälä, J. E.
Seth, G. L. 5337.
17240. Seth, J. N. 1959: Causes of seedlessness in *Psidium guajava* L. Hort. Advance 3:82–88. Biol. Abstr. 36:28541. 1961.
Sethi, A. S. 10290–10294.
17241. Sethi, M. L., and C. K. Atal. 1963: Phytochemical studies on the genus *Crotalaria*. Indian J. Pharm. 25:159–60. Chem. Abstr. 59:4206. 1963.
Seto, A. 19995.
17242. Setterfield, G., H. Stern, and F. B. Johnston. 1959: Fine structure in cells of pea and wheat embryos. Can. J. Botany 37:65–72.
Setterfield, G. 9899.
17243. Seven, R. P. 1964: New thoughts on seed treatment. Seed World 94(5):14–17.
Severen, M. Levy-Van; *see* Levy-Van Severen, M.
Severson, G. M. 4136.
17244. Sewell, P. 1891: Germination and growth of *Salvia*. Trans. Botan. Soc. Edinburgh 18:350–79.
17245. Sewell, W. E., and D. M. Turney. 1946: Further experiments with detoxification of cottonseed meal for hogs. Alabama Agr. Expt. Sta. Progr. Rept. Ser. 25. 4 pp.
17246. Sexsmith, J. J., and U. J. Pittman. 1963: Effect of nitrogen fertilizers on germination and stand of wild oats. Weeds 11:99–101.
Sexton, W. A. 9976, 18439.
17247. Seybold, A. 1953/54: The color changes in the flowers, fruits, and seedcoats. Sitzber. Heidelberg Akad. Wiss. Math. Naturw. Kl. 96. pp. Chem. Abstr. 48:13834. 1954.
17248. —— and K. Egle. 1941: The pigment content of green spores and seeds. Botan. Arch. 43:78–83. Chem. Abstr. 37:3124–25. 1943.
Seysenegg, E. von Tscherwak-; *see* Von Tschermak-Seysenegg, E.
17249. Shabanov, I. M. 1937: The phosphatides of tobacco seeds. Vses. Inst. Tabach. Makhoroch. Prom. 133:78–85. Chem. Abstr. 32:5444. 1938.
Shabelskaya, E. F. 7352.
17250. Shadowsky, A. E. 1931: Einige Angaben über die Embryogenie von *Pistia stratiotes* L. Ber. Deut. Botan. Ges. 49:350–55. J. Roy. Microscop. Soc. 52: 72–73. 1932.
17251. Shafer, J., Jr. 1944: The relation of embryo axis weights to heterosis. Am. J. Botany 31:503–6.
17252. —— and R. G. Wiggans. 1941: Correlation of total dry matter with grain yield in maize. J. Am. Soc. Agron. 33: 927–32.
Shafer, T. 7392.
17253. Shafer, N. E. 1954: Crop drying chemicals for seed production in Nebraska. Down Earth 10:2–3.
17254. Shah, C. K. 1962: Studies in germination. I. *Carex wallichiana* Presc. J. Indian Botan. Soc. 41:551–56.
17255. —— 1964: Embryo development in *Eleocharis palustris* L. Naturaliste Canadien 91:41–49.
17256. Shah, H. C., and B. V. Mehta. 1959: Magnesium-phosphorus-crude fat interrelationships in the seed of pearl millet (*Pennisetum typhoideum* Rich.). Soil Sci. 87:320–24.
17257. Shah, M. M., N. L. Phalnikar, and B. V. Bhide. 1947: A note on the chemical investigation of the fruits of *Celastrus paniculata*, Willd. (n. o. Celastraceae). Current Sci. (India) 16:57–58.
Shah, S. V. 209, 14051.
Shah, Y. S. 2404.
Shahi, R. 7202.
17258. Shain, S. S. 1950: Utilization of hard seeds of red clover for increasing the stability of its stand. Sov. Agron. 8(9): 61–67.
Shakra, S. A. 15614.
Shalucha, B. 1540–1543.
Shamakumari, K. 7529.
17259. Shamanna, S. 1954: A contribution to the embryology of *Olax wightiana* Wall.

Proc. Indian Acad. Sci. Sect. B 39:249–56.
Shambhu, K. 12838.
17260. Shamsiev, G. 1962: Effect of length of day on the formation of the dormancy period in the seeds of *Chenopodium album*. Botan. Zh. 47:1349–54.
Shamsuddin, K. M. 19019.
17261. Shands, H. L. 1960: Oat scutellum enlargement. Agron. J. 52:716.
Shands, H. L. 3119, 5348, 15145, 17120.
17262. Shands, R. G. 1937: Longevity of *Gibberella saubinetii* and other fungi in barley kernels and its relation to the emetic effect. Phytopathology 27:749–62.
Shanklin, J. A. 13707.
Shankman, S. 5654.
17263. Shanmugam, T. S. 1954: A preliminary study of the distinctiveness of dormancy and viability in vegetable seeds. Madras Agr. J. 41:230–32. Biol. Abstr. 29:22382. 1955.
17264. Shanmugasundaram, A. 1953: Studies on dormancy in short-term rices. Madras Agr. J. 40:477–87. Biol. Abstr. 28:27000. 1954.
17265. Shanmugavelu, K. G., and G. Rangaswami. 1962: Tryptophan and indole compounds in banana ovaries. Nature 194:775–76.
Shannon, J. C. 10499.
17266. Shantz, E. M., and F. C. Steward. 1956: Growth-promoting substances in the liquid endosperm of immature *Aesculus* fruits. Plant Physiol. (Suppl.) 31:xxxix.
Shapiro, R. 6430.
Shapiro, S. K. 18843.
Sharapova, N. V. 2782.
Sharkov, V. I. 18446.
17267. Sharma, A. K., and U. C. Bhattacharyya. 1962: The differential effect of colchicine on seeds and seedlings of *Trigonella foenum graecum* L. and the influence of temperature on the effect. Phyton 18:39–50.
Sharma, B. K. 1649.
Sharma, P. C. 19718.
Sharma, S. N. 3961.
Sharma, V. N. 2398, 2399, 17174.
17268. Sharma, Y. M. L. 1939: Gametogenesis and embryogeny of *Tamarix ericoides* Rottl. Ann. Botany (London) 3:861–70.
17269. Sharman, B. C. 1942: A twin-seedling in *Zea mays* L. Twinning in the Gramineae. New Phytologist 41:125–29.
Sharp, D. G. 3530.
17270. Sharp, E. L. 1933: The Christmas rose from seed. Horticulture 11(7):106.
Sharp, L. 7603.
Sharp, L. A. 13698.
17271. Sharp, P. F. 1924: Wheat and flour studies. II. Aging I. The change in hydrogen-ion concentration of wheat and mill products with age. Cereal Chem. 1:117–32.
Sharp, P. F. 19605.
17272. Sharpe, L. 1928: Abnormal embryos of oats and wheat. Elisha Mitchell Sci. Soc. J. 44:90–94.
17273. Sharpe, R. H., and S. Merrill, Jr. 1942: Effect of stratification and time of planting on germination of tung seed. Proc. Am. Soc. Hort. Sci. 40:286–91.
17274. Shashin, A. I. 1956: On transmission of blistery smut (*Ustilago maydis*) by corn seeds. Zemledelie 1956(4):111–12.
17275. Shatalina, M. S. 1963: Seed morphology in *Caucasian rhododendron* species. Vestn. Mosk. Univ. Ser. IV Geol. 18:41–44. Biol. Abstr. 44:12255. 1963.
17276. Shaul, Y. B. 1959: The calcium and magnesium content of germinating lettuce seeds. Bull. Res. Council Israel Sect. D 7:46–50.
Shaulis, N. J. 14527.
17277. Shaw, A. C., and L. C. Pasco. 1949: Formation and distribution of vitamin C in the radicle and cotyledon of the broad bean (*Vicia faba*). Nature 164:624.
17278. Shaw, C. H. 1904: Note on the sexual generation and the development of the seed-coats in certain of the Papaveraceae. Bull. Torrey Botan. Club 31:429, 433.
17279. Shaw, C. S., and G. N. Franks. 1962: Cottonseed drying and storage at cotton gins. U.S. Dept. Agr., Agr. Res. Serv. Tech. Bull. 1262. 73 pp.
Shaw, C. S. 7050.
17280. Shaw, F. J. F., A. R. Khan, and M. Alam. 1931: Studies in Indian oilseeds. V. The inheritance of characters in Indian linseed. Indian J. Agr. Sci. 1:1–57. Expt. Sta. Rec. 65:817–18. 1931.
Shaw, L. 9980.
17281. Shaw, M. F. 1929: A microchemical study of the fruit coat of *Nelumbo lutea*. Am. J. Botany 16:259–76.
17282. Shaw, R. H., and H. C. S. Thom. 1951: On the phenology of field corn, silking to maturity. Agron. J. 43:541–46.
Shaw, R. H. 19675.
17283. Shaw, S. V., and D. G. Pishawikar. 1938: Chemical investigation of the seeds

of *Swietenia mahagoni*. Current Sci. (India) 7:257.

Shaw, W. C. 18221.

Shchekina, A. A. 14908.

17284. Shchepkina, O. I. 1938: The formation of oil in sunflower seed studied during the different stages of the process. Uch. Zap. Biol. Inst. Rostov. Univ. 1:37–75. Chem. Abstr. 34:5881. 1940.

17285. Shcherbakov, A. P. 1944: Concerning acceleration of seed germination. Sov. Botan. (Leningrad) 1944(4/5):60–70. Biol. Abstr. 19:22799. 1945.

17286. —— 1953: Biochemical changes in seeds of soybean ripening under the influence of calcium and magnesium. Biochemistry (USSR) (English Transl.) 18:438–47. Chem. Abstr. 48:4642. 1954.

Shcherbakov, I. N. 5582.

Shcherbakov, V. G. 11210.

Shcherbina, K. G. 15670.

Shchetinina, M. 598.

Shchibria, G. I. 2997.

17287. Shea, K. R. 1961: Field survival of thiram-treated Douglas fir seed. Weyerhauser Co. Forest Res. Note 38. 8 pp.

17288. —— and J. H. Rediske. 1961: Pathological aspects of germination and survival of Douglas fir in controlled environment. Weyerhauser Co. Forest Res. Note 41. 8 mimeo. pp.

17289. Shear, C. B., and H. L. Crane. 1943: Germination of the nuts of the tung tree as affected by penetrants, substrata, depth of planting, and storage conditions. Botan. Gaz. 105:251–56.

17290. Shear, G. M., and L. I. Miller. 1960: Influence of plant spacing of the jumbo runner peanut on fruit development, yield, and border effect. Agron. J. 52:125–27.

17291. Sheard, C., and G. M. Higgins. 1927: The influence of direct irradiation by a quartz mercury arc lamp upon the germination and growth of certain seeds. Plant Physiol. 2:461–73.

17292. ——, ——, and W. I. Foster. 1930: The germination of seeds, growth of plants and development of chlorophyll as influenced by selective solar irradiation. Science 71:291–93.

Sheard, C. 8736.

17293. Shearer, R. C. 1961: A method of overcoming seed dormancy in subalpine larch. J. Forestry 59:513–14.

17294. Shebanek. 1956: The problem of stimulating winter wheat and corn seeds by hydroquinone and potassium bromide. Sb. Česk. Akad. Zeměděl. 29:125–38 Biol. Abstr. 33:35021. 1959.

Shebeski, L. H. 1873, 6756.

17295. Shedd, C. K., and H. H. Walkden. 1947: Grain sorghum storage. U.S. Dept. Agr. Circ. 760:1–48.

Shedd, C. K. 18502.

Sheets, A. 5826.

17296. Shefer-Safonova, E. I., M. I. Kalashnikova, and A. S. Kostromina. 1934: Determination of the viability of seeds of trees by a staining method. Botan. Zh. 19:566–94. Biol. Abstr. 11: 14598. 1937.

Shekhtman, I. L. 15857.

Shekleton, M. C. 6732, 12076, 17628.

Sheldon, D. E. 15361.

Sheldon, W. H. 8113.

17297. Shellenberger, J. A., and C. H. Bailey. 1936: Biochemical distinctions between barley varieties. Cereal Chem. 13:631–55.

17298. ——, D. Miller, E. P. Farrell, and M. Milner. 1958: Effect of wheat age on storage properties of flour. Food Technol. 12:213–21.

17299. —— and Y. Pomeranz. 1961: Histochemistry of wheat grains. Ber. Getreidechemiker-Tagung, Detmold 1961: 6–10. Chem. Abstr. 56:1769. 1962.

Shellenberger, J. [A.] 106, 4111, 6213, 6214, 13762, 15520–15524.

Shen, C. W. 11989.

Shen, S. Y. 3942.

17300. Shen, T., K. M. Hsieh, and T. M. Chen. 1945: Effects of magnesium chloride and manganous nitrate upon the content of ascorbic acid in soybean during germination, with observations on the activity of ascorbic acid oxidase. Biochem. J. 39: 107–10.

Shen, T. 12198.

Shen, Y. S. 18367.

17301. Shenberger, L. C. 1950: What "goes on" inside the seed germinator. Proc. Assoc. Offic. Seed Analysts 40:71–76.

17302. —— and O. W. Ford. 1949: The effect of chlorine-treated water on the germination of certain agricultural seeds. Assoc. Offic. Seed Analysts News Letter 23(2):22–24.

Shenberger, L. C. 13470, 15790, 16694.

17303. Shepard, H. H. 1939: Insects infesting stored grain and seeds. Minn. Agr. Expt. Sta. Bull. 340. 28 pp.

17304. Shepherd, E. F. S., and G. Orian. 1937: Germination experiments on "pois mascate" seed (*Mucuna deeringiana* [Bort]

Holland). Mauritius Dept. Agr. Sci. Ser. Bull. 24. 8 pp. Chem. Abstr. 32:5144. 1938.

17305. Shepherd, K. 1960: Genetics of testa color in *Canavalis ensiformis*. J. Heredity 51:68.

17306. —— 1960: Seed fertility of edible bananas. J. Hort. Sci. 35:6–20.

17307. Shepherd, R. E. 1945: Germinating rose seeds. Am. Rose Ann. 1945:121–22.

17308. —— 1954: Media for rose seed germination. Am. Rose Ann. 1954:102–6.

17309. Shepherd, R. J., and R. W. Fulton. 1962: Identity of a seed-borne virus of cowpea. Phytopathology 52:489–93.

17310. Shepperd, J. H., and E. G. Schollander. 1900: The effect of age and length of sprouts upon vitality of seeds. N. Dakota Agr. Expt. Sta. Rept. 1900:101–13.

17311. Sherbakoff, C. D. 1927: A modified method of delinting cotton seed with sulphuric acid. Phytopathology 17:189–93.

Sherbakoff, C. D. 12409.
Sheridan, J. M. 3464.
Sherman, G. D. 4099.

17312. Sherman, H. 1921: Respiration of dormant seeds. Botan. Gaz. 72:1–30.

Sherman, W. A. 2940.

17313. Sherman, W. C. 1940: Chromatographic identification and biological evaluation of carotene from mature soybeans, Food Res. 5:13–22. Expt. Sta. Rec. 84: 731–32. 1941.

17314. Sherwin, H. S., and K. W. Kreitlow. 1952: Discoloration of soybean seeds by the frogeye fungus, *Cercospora sojina*. Phytopathology 42:568–72.

17315. Sherwin, R. A. 1934: The effect of phosphates on the germination of turnip seed. Tasmanian J. Agr. 5:73–74.

17316. Sherwood, F. W. 1932: The distribution of vitamin B complex and its components in the peanut. J. Agr. Res. 44:849–60.

Sherwood, H. 4916.
Sherwood, L. V. 6388.

17317. Sherwood, R. C., J. S. Andrews, W. B. Wade, and C. H. Bailey. 1933: The march of acidity in wheat germ during storage. Ind. Eng. Chem. 25:437–40.

17318. Shestakov, A. G., G. F. Ivanova, and N. I. Shmelkova. 1955: Effect of radiophosphorus on germination and growth of oats in relation to conditions of phosphate supply. Dokl. Akad. Nauk SSSR 102:395–97.

17319. Shetlar, M. R., G. T. Rankin, J. F. Lyman, and W. G. France. 1947: Investigation of the proximate chemical composition of the separate bran layers of wheat. Cereal Chem. 24:111–22.

17320. Shevchenko, M. I. 1938: The use of the Röntgen apparatus as a method of entomological analysis and determination of the germinative capacity of forest tree seeds. V. Zashtchitu Lesa. 3:34–38.

17321. Sheward, T. 1932: Asparagus in two years from seed. Flower Grower 19(5):218.

17322. Shewfelt, A. L., and E. D. Putt. 1958: A rapid method for estimating the oil content of sunflower seeds. Can. J. Plant Sci. 38:419–23.

Shi, C. F. 19346.

17323. Shibamoto, T., and H. Nakazawa. 1957: Constitution of the inorganic components in the seed of forest tree of Japan. J. Japan. Forestry Soc. 39:207–13.

Shibata, K. 9394.
Shibata, T. 13903.
Shibko, I. 5772.

17324. Shibutani, S., and T. Okamura. 1955: Studies on the morphological characters of seeds in *Brassica*. Hort. Assoc. Japan J. 24:195–98.

Shibuya, M. 9418.

17325. Shibuya, T. 1935: Morphological and physiological studies on the fructification of peanut (*Arachis hypogaea* L.). Mem. Fac. Sci. Agr. Taihoku Imp. Univ. 2:1–120.

17326. —— 1938: Forcing the germination of dormant seeds by means of growth hormone. J. Soc. Trop. Agr. Taihoku Imp. Univ. 10:1–8.

17327. Shiere, H. B. 1940: Seed germination experiments. Gladiolus 1941:100–5.

Shih, C. S. 6140.
Shihira, I. 8309.
Shikamoto, Y. 4885.
Shikhanova, N. M. 14901.

17328. Shilova, N. V. 1962: The structure of the embryo and the germ in the subfamily Andromedoideae (Drude) E. Busch. Botan. Zh. 47:344–53.

Shimada, H. 14160.
Shimada, Y. 19995.

17329. Shimano, T., and Y. Ose. 1953: Components of seeds of *Hibiscus manihot*. Ann. Proc. Gifu Coll. Pharm. 1953(3): 15–17. Chem. Abstr. 50:14055. 1956.

17330. Shimeno, K., and S. Konoshita. 1963: Inhibitory effect of thiouracil on germination of leaf mustard seed and its reversal by pyrimidine derivatives. Plant Cell Physiol. (Tokyo) 4:145–52.

Shimeon. 2058.

17331. Shimizu, M. 1949: On the relation between the reaction of soil and the germination and growth of *Digitaria ciliaris*. Botan. Mag. (Tokyo) 62:39–42. Chem. Abstr. 45:4309–10. 1951.

17332. —— 1959: Relation of the enclosing structure of the *Digitaria adscendens* spikelet to its germination. II. The role of the enclosing structure (glumes and shell coat) of crab grass (*D. adscendens*) and rice plant spikelets in the absorption of water and water soluble substances. Proc. Crop Sci. Soc. Japan 28:239–43. Biol. Abstr. 35:30125. 1960.

Shimizu, R. 14135.

Shimogawara, G. 9494.

17333. Shimokoriyama, M. 1955: On the enzymatic hydrolysis of robinin by a glycosidase from the seeds of *Rhamnus japonica* and *R. dahurica* var. *nipponica*. Botan. Mag. (Tokyo) 62:168–73.

17334. Shimotsuma, M., and K. Matsumoto. 1957: Comparative studies on the morphology of polyploid watermelon seeds. Rept. Kihara Inst. Biol. Res. 8:67–74.

Shimotsuma, M. 14523.

Shimura, K. 10382, 10383.

Shin, H. 4885.

Shin, M. 7993, 18264–18266.

Shingai, Y. 16797.

Shinozuka, S. 8440.

Shintani, G. 9419.

Shirahama, K. 18280.

Shiraishi, K. 9476.

17335. Shirakawa, M., and A. Otakara. 1956: Studies on the nitrogen metabolism in plant germination. I. Variation of amide and amino acid contents during the germination of pea plants. J. Agr. Chem. Soc. Japan 30:158–63.

17336. Shirasawa, H., and M. Koyama. 1915: Experiments on the preservation of principal forest tree seeds. Bull. For. Expt. Sta. 1915:1–13.

17337. —— and —— 1918: Experiments on the storage of forest tree seeds, and value of year-old seeds for the nursery. Bull. For. Expt. Sta. 17:1–18.

17338. Shirk, H. G. 1942: Freezable water content and the oxygen respiration in wheat and rye grain at different stages of ripening. Am. J. Botany 29:105–9.

Shirke, V. S. 2413.

17339. Shirley, H. L. 1931: The influence of light and temperature upon the utilization by young seedlings of organic reserves in the seed. Am. J. Botany 18:717–27. Chem. Abstr. 26:1640. 1932.

17340. —— 1937: Direct seeding in the Lake States. J. Forestry 35:379–87. Expt. Sta. Rec. 78:490. 1938.

17341. —— 1937: The relation of drought and cold resistance to source of seed stock. Minn. Horticulturist 65(2):23–24.

17342. —— 1939: Some new aspects of seed certification. Comments. J. Forestry 37: 35–36.

Shirley, H. L. 1789, 1790, 13371.

17343. Shiroya, T. 1963: Metabolism of raffinose in cotton seeds. Phytochemistry 2:33–46. Biol. Abstr. 44:7948. 1963.

Shiroya, T. 8305, 8384.

17344. Shisa, M. 1935: Germination and storage of some vegetable seeds. Hort. Assoc. Japan J. 6:127–37.

Shiseki, K. 10923–10925.

Shive, W. 17563, 17565, 17566.

Shkjezi, A. 16254.

17345. Shklyar, M. Z. 1961: The effect of the metabolic products of *Fusarium moniliforme* on seed germination and plant growth. Tr. Inst. Mikrobiol. Akad. Nauk SSSR 11: 318–26. Biol. Abstr. 40:3444. 1962.

17346. Shkolnik, M. I., and N. A. Makarova. 1953: Changes in the chemistry of plants by preplanting treatment of the seeds by extracts from other plants. Dokl. Akad. Nauk SSSR 93:185–88. Chem. Abstr. 48:7712. 1954.

17347. —— 1954: Treating seed before planting with solutions of minor elements and foliar applications of these. Zemledelie 2(11):31–36. Chem. Abstr. 49:9750. 1955.

17348. —— and M. M. Steklova. 1956: The effect of some macro- and microelements on the vernalization of winter cereals. Mikroelementy v Sel'sk. Khoz. i Med. Akad. Nauk Latv. SSR Otd. Biol. Nauk Tr. Vses. Soveshch. Riga 1955:227–45. Chem. Abstr. 53:9391. 1959.

17349. Shkutko, N. V. 1957: Characteristics of acorns from different oak-trees grown in White Russian SSR. Vestsi Akad. Navuk Belarusk. SSR Ser. Biyal. Navuk 1957(4): 65–75. Chem. Abstr. 53:1478. 1959.

17350. Shkvarnikov, P. K. 1935: The influence of temperature and moisture on the mutations in resting seeds. Semenovodstvo 1935(1):46–52.

17351. —— 1936: Wirkung hoher Temperatur auf die Mutationsrate von *Crepis* bei verschiedener relativer Luftfeuchtigkeit. Planta 25:689–95.

17352. —— 1937: Über Erhöhung der Mutationsrate bei Weizen nach langer Aufbewahrung der Samen. Genetica 19: 188–99. Z. Botan. 32:540–41. 1938.
17353. —— 1939: Mutation in seeds and its significance for a seed culture and selection of plants. Bull. Acad. Sci. URSS Cl. Sci. Math. Nat. Ser. Biol. 1939(6):1009–54. Biol. Abstr. 14:15696. 1940.
17354. —— and M. Navashin. 1934: Über die Beschleunigung des Mutationsvorganges in ruhenden Samen unter dem Einfluss von Temperaturerhöhung. Planta 22:720–36. Biol. Abstr. 9:10766. 1935.
17355. —— and —— 1935: Acceleration of the process of mutation in resting seeds under the influence of increased temperature. Botan. Zh. 4:25–38.
Shmelkova, N. I. 17318.
Shmigel, V. N. 1591.
17356. Shmuck, A. A. 1944: Alteration of carbohydrates in grains of wheat grafted on rye endosperm. Compt. Rend. Acad. Sci. URSS 44:70–71.
Shneider, I. I. 7482.
17357. Shoemaker, J. S. 1937: Some factors in delphinium seed germination. Sci. Agr. (Ottawa) 17:730–32.
17358. Shoji, T., and T. Nakamura. 1928: On the dioecism of garden asparagus (*Asparagus officinalis* L.). Japan. J. Botany 4:125–51.
Shoji, T. 16796.
Shone, G. 3083.
Shoope, G. J. 5624.
17359. Shoulders, E. 1961: Effect of seed size on germination, growth and survival of slash pine. J. Forestry 59:363–65.
Showler, A. J. 1576.
17360. Shozo, T. 1929: The action of nitrates and ammonium salts on some plants. II. The action of nitrates and ammonium salts on the germination. Botan. Mag. (Tokyo) 43:295–305. Biol. Abstr. 4:26818. 1930.
17361. Shpineva, P. P. 1940: Vitality of a germinated and dried wheat grain in storage. Bull. Inst. Sci. Lesshaft 23:171–216.
17362. Shpuntova, M. E., and M. G. Semenko. 1958: Investigation of cotton seed husks of different varieties and maturities. Sb. Tr. Vses. Nauchn.-Issled. Inst. Gidroliz. i Sul'fitno-Spirt. Prom. 6:92–98. Chem. Abstr. 54:11156. 1960.
Shrewsbury, C. L. 10788.
17363. Shride, C. L. 1936: Time required for seed germination. Flower Grower 23(9):492.
Shrivastava, D. L. 5329.
17364. Shropshire, W., Jr., W. H. Klein, and V. B. Elstad. 1961: Action spectra of photomorphogenic induction and photoinactivation of germination in *Arabidopsis thaliana*. Plant Cell Physiol. (Tokyo) 2:63–69.
Shrum, J. E., Jr. 11856.
Shubert, T. A. 1894.
17365. Shuck, A. L. 1934: Some factors influencing the germination of lettuce seed in seed laboratory practice. N.Y. Agr. Expt. Sta. Tech. Bull. 222:1–21.
17366. —— 1935: The formation of a growth inhibiting substance in germinating lettuce seeds. Proc. Intern. Seed Testing Assoc. 7:9–14.
17367. —— 1935: Light as a factor influencing the dormancy of lettuce seeds. Plant Physiol. 10:193–96.
17368. —— 1936: The favorable influence of a moist substratum for the germination of seeds. Proc. Assoc. Offic. Seed Analysts 1935:60–61.
17369. —— 1936: A preliminary report on the germination of mustard seed. Proc. Assoc. Offic. Seed Analysts 28:74–76.
17370. —— 1936: The germination of lettuce seed in the laboratory and in the field. Proc. Assoc. Offic. Seed Analysts 28:80–83.
17371. —— 1936: The germination of secondary dormant tomato seeds and their formation. Proc. Intern. Seed Testing Assoc. 8:136–58.
17372. —— 1936: Why germination of new seeds varies. Seed World 39(1):25.
17373. —— 1938: Some suggestions for the prevention of erratic germination of lettuce seed. Proc. Assoc. Offic. Seed Analysts 23/26:284–85.
17374. Shuel, R. W. 1948: Seed germinability tests with 2,3,5-triphenyl tetrazolium chloride. Sci. Agr. (Ottawa) 28:34–38.
17375. Shuhart, D. V. 1931: Endosperm and embryo development as related to filling of pecans and walnuts. Proc. Am. Soc. Hort. Sci. 28:161–63.
17376. Shuiku, S. 1936: Studies on some constituents of soybean seeds and their transformations during germination. J. Dept. Agr. Kyushu Imp. Univ. 5:51–116. Chem. Abstr. 31:4697–98. 1937.
17377. Shuin, K. A. 1955: Peculiarities of tomato seeds in relation to the position

of their formation on the mother plant. Sad i Ogorod 1955(6):19–20.

Shull, A. V. 4394.

17378. Shull, C. A. 1911: The oxygen minimum and the germination of *Xanthium* seeds. Botan. Gaz. 52:453–77.

17379. Shull, C. A. 1913: Semipermeability of seed coats. Botan. Gaz. 56:169–99.

17380. —— 1914: The role of oxygen in germination. Botan. Gaz. 57:64–69.

17381. —— 1916: Measurement of the surface forces in soils. Botan. Gaz. 62: 1–29.

17382. —— 1920: Temperature and rate of moisture intake of seeds. Botan. Gaz. 69:361–90.

17383. —— 1935: Germination behavior of the rose mallows. Trans. Illinois State Acad. Sci. 28:111–12.

17384. —— and W. B. Davis. 1923: Delayed germination and catalase activity in *Xanthium*. Botan. Gaz. 75:268–81.

17385. —— and H. B. Lemon. 1931: Penetration of seed coats by ultraviolet radiation. Botan. Gaz. 92:420–29.

17386. —— and J. W. Mitchell. 1933: Some physico-chemical properties of seed extracts. Botan. Gaz. 95:258–78.

17387. —— and S. P. Shull. 1924: Temperature coefficient of absorption in seeds of corn. Botan. Gaz. 77:262–79.

17388. —— and —— 1932: Irregularities in the rate of absorption by dry plant tissues. Botan. Gaz. 93:376–99.

17389. Shull, G. H. 1914: The longevity of submerged seeds. Plant World 17:329–37.

17390. Shull, H. 1962: Weed seed in irrigation water. Weeds 10:248–49.

Shull, S. P. 17387, 17388.

17391. Shumakov, A. M. 1951: Dormancy period in tomato seeds. Selek. i Semen. 18(2):69–72.

17392. Shumakov, V. S. 1962: Effect of water extracts from grasses on the germination of pine and spruce seeds. Lesn. Khoz. 1962(5):18–20.

17393. Shunichev, S. I. 1960: Pelleting of cabbage seeds. Mosk. Ordena Lenina Sel'skokhoz. Akad. Im. K. A. Timiryazeva Dokl. 59:53–59.

17394. Shunk, I. V. 1939: Oxygen requirements for germination of seeds of *Nyssa aquatica*—tupelo gum. Science 90:565, 566.

Shu-Sheng, C. 8492.

17395. Shuster, L., and R. H. Gifford. 1962: Changes in 3′-nucleotidase during the germination of wheat embryos. Arch. Biochem. Biophys. 96:534–40.

Shutov, A. D. 18922, 18923.

17396. Shvedskaya, Z. M., and A. S. Kruzhilin. 1961: Influence of inhibitors on vernalization of plants. Plant Physiol. (USSR) (English Transl.) 8:613–18.

Shvedskaya, Z. M. 11453, 11454.

Shvetsova, V. A. 17762.

Sibbitt, L. D. 8248, 8250, 8251.

17397. Sice, J. 1958: Tumor promoting principles in seeds of *Croton tiglinm* L. Arch. Intern. Pharmacodyn. 115:408–15. Biol. Abstr. 33:11437. 1959.

17398. Siddappa, G. S., and W. B. Adam. 1935: The ripening of green peas. Univ. Bristol Fruit Veg. Preserv. Res. Sta., Campden, Ann. Rept. 1934/35:74–100. Chem. Abstr. 31:432. 1937.

Siddappa, G. S. 90.

17399. Siddappa, S., and Y. V. S. Devi. 1957: Chemical examination of the seeds of *Psoralea corylifolia*. I. Mysore Univ. J., Sect. B 17:25–30. Biol. Abstr. 35:32911. 1960.

17400. —— and —— 1957: Chemical examination of the seeds of *Psoralea corylifolia*. II. On the isolation and constitution of isopsoralidin. Mysore Univ. J., Sect. B 17:31–39. Biol. Abstr. 35:32912. 1960.

Siddappa, S. 12811.

Siddiqui, M. S. 6231, 6234.

17401. Siddiqui, R. H., and S. A. Warsi. 1944: Chemical examination of the fatty oil from the seeds of *Solanum pseudocapsicum*. I. Indian J. Pharm. 6:68–70. Chem. Abstr. 40:3280–81. 1946.

17402. —— and —— 1945: *Cryptostegia grandiflora* R. Br.—a wartime source of vegetable rubber. VIII. Examination of the seeds and the fat oil. Indian J. Pharm. 7:75–79, 88. Chem. Abstr. 40:3625. 1946.

17403. Siddiqui, S. 1945: Constituents of Chana. I. Isolation of three new crystalline products from the Chana germ. J. Sci. Ind. Res. (India) 4:68–70. Chem. Abstr. 40:1502–3. 1946.

17404. —— and Z. Ahmed. 1935: Alkaloids from seeds of *Cassia absus*, Linn. Proc. Indian Acad. Sci. Sect. A 2:421–25. Chem. Abstr. 30:1799. 1936.

Siddiqui, S. 2401, 3766, 11413.

Siddiqui, S. H. 15110.

Siddiqullah, M. 15796.

17405. Sidorenko, I. D. 1962: Action of ionizing and ultra-violet radiation on

maize seeds. Acad. Sci. RSS Ukraine Kiev J. Botan. 19:3–14.
Sidorov, B. N. 10550.
Sidwell, R. W. 4699.
Sieburth, L. R. 17659.
17406. Siegel, S. M. 1950: Germination and growth inhibitors from red kidney bean seed. Botan. Gaz. 111:353–56.
17407. —— 1950: Effects of exposures of seeds to various physical agents. I. Effects of brief exposures to heat and cold on germination and light-sensitivity. Botan. Gaz. 112:57–70.
17408. —— 1952: Secretion of phosphorylase by red kidney bean embryos. Botan. Gaz. 114:139–41.
17409. —— 1953: Effects of exposures of seeds to various physical agents. II. Physiological and chemical aspects of heat injury in the red kidney bean embryo. Botan. Gaz. 114:297–312.
17410. —— 1957: Growth- and enzyme-inhibiting properties of bean seed extracts. Plant Physiol. 32:151–52.
17411. —— 1962: Observations on peroxide toxicity in seed germination. Physiol. Plantarum 15:21–26.
17412. ——, G. Renwick, and L. A. Rosen. 1962: Formation of carbon monoxide during seed germination and seedling growth. Science 137:683–84.
17413. —— and L. A. Rosen. 1962: Effects of reduced oxygen tension on germination and seedling growth. Physiol. Plantarum 15:437–44.
17414. ——, ——, and C. Giumarro. 1962: Effects of reduced oxygen tension on vascular plants. IV. Winter rye germination under near-Martian conditions and in other nonterrestrial environments. Proc. Natl. Acad. Sci. U.S. 48:725–28.
Siegel, S. M. 7294, 15620.
17415. Sieglinger, J. B. 1933: Inheritance of seed color in crosses of brown-seeded and white-seeded sorghums. J. Agr. Res. 47:663–67.
17416. —— 1940: A sorghum seed color chimera. J. Heredity 31:363–64.
17417. —— 1957: An aid in observing yellow endosperm in sorghum grain. Agron. J. 49:58.
Sieglinger, J. B. 9972.
17418. Sierp, H. 1925: Untersuchungen über die Kohlensäureabgabe aus keimenden Erbsensamen. Flora (Jena) 118/119:476–502.
17419. Siert, R. 1958/59: Merchandising certified seed. Intern. Crop Improvement Assoc. Ann. Rept. 40/41:297–99.
17420. Sievers, A. F. 1914: The germination of belladonna seeds. Am. J. Pharm. 86:483–505.
17421. Sievers, E. 1963: Untersuchungen über die wasserlöslichen Hemmstoffe aus dem Schnittholz der Weinrebe (*Vitis vinifera* L.). I. Zur Wirkung der Hemmestoffe auf die Keimung und Entwicklung von Rebsämlingen. Naturwissenschaften 50:134.
Sievers, E. 19758.
17422. Sifton, H. B. 1920: Longevity of the seeds of cereals, clovers, and timothy. Am. J. Botany 7:243–51.
17423. —— 1927: On the germination of the seed of *Spinacia oleracea* L. at low temperatures. Ann. Botany 41:557–68.
17424. —— 1959: The germination of light-sensitive seeds of *Typha latifolia* L. Can. J. Botany 37:719–39.
17425. Siggins, H. W. 1933: Distribution and rate of fall of conifer seeds. J. Agr. Res. 47:119–28.
17426. Sigmund, W. 1896: Über die Einwirkung chemischer Agentien auf die Keimung. Landwirtsch. Versuchssta. 48: 1–58.
17427. Sijbring, P. H. 1963: Results of experiments on the moisture relationship of seeds. Proc. Intern. Seed Testing Assoc. 28:837–43.
17428. —— 1963: Results of some storage experiments under controlled conditions (agricultural seeds). Proc. Intern. Seed Testing Assoc. 28:845–51.
17429. —— 1963: Conditioning of rooms for seed storage. Proc. Intern. Seed Testing Assoc. 28:885–92.
Sijpesteijn, A. K. 5135.
17430. Sikes, J. K. 1964: Urea breakdown activity in cottonseed. J. Am. Oil Chemists Soc. 41:69–71.
Sikorsky, J. 12162.
17431. Silberschmidt, K. 1956: Periodicidade anual do germinação de sementes no Brasil. Anais Acad. Brasil. Cienc. 28: 545–69. Biol. Abstr. 32:2863. 1958.
Silberschmidt, K. 5972, 5973, 11324, 14897, 19101, 19102.
Silberstein, L. 2362–2367.
17432. Silbert, D. 1923: Notes on the stimulating influence of seed treatment upon the subsequent growth of plants. N.J. Agr. Expt. Sta. Rept. 1923:259–62.
17433. Silen, R. R. 1958: Artificial ripening

of Douglas fir cones. J. Forestry 56:410–13.

17434. —— and K. W. Krueger. 1962: Does rainy weather influence seed set of Douglas fir? J. Forestry 60:242–44.

Sileva, M. N. 18790.

Silk, J. A. 6586.

17435. Silker, T. H., and R. E. Goddard. 1953: Direct seeding tests with slash, loblolly, and longleaf pine in southeast Texas. Texas Forest Serv. Tech. Rept. 7:1–36. Biol. Abstr. 28:29544. 1954.

Sill, W. H., Jr. 8090.

17436. Silow, R. A. 1931: A preliminary report on pollen-tube growth in red clover (*Trifolium pratense* L.). Welsh Plant Breeding Sta. Ser. H Bull. 12:228–33.

17437. Silsbury, J. H. 1958: Temperature and germination in *Kennedya* Vent. Nature 182:954–55.

17438. —— 1961: A study of dormancy, survival, and other characteristics in *Lolium perenne* L. at Adelaide, S. Australia. Australian J. Agr. Res. 12:1–9.

Silva, D. S. de. 6474.

Silva, F. I. R. da; *see* Da Silva, F. I. R.

Silva, F. J. Ramirez-. 9027.

Silva, M. V. E. 17768.

Silva, M. Vianna e; *see* Vianna e Silva, M.

Silva, V. B. 3577.

Silveira, V. 10424.

Silverman, G. J. 5021.

Silverstrini, M. 7412.

Sim, J. L. 355.

17439. Simak, M. 1953: On watering of pine seeds with sea water. Medd. Statens Skogsforskningsinst. 41(4):1–14. 1952.

17440. —— 1953: Über die Samenmorphologie der gemeinen Kiefer (*Pinus silvestris* L.). Medd. Stratens Skogsforskningsinst. 43(2):1–30. Biol. Abstr. 28:11759. 1954.

17441. —— 1953: Beziehungen zwischen Samengrosse und Samenanzahl in verschieden grosse Zapfen eines Baumes (*Pinus silvestris* L.). Medd. Statens Skogsforskningsinst. 43(8):1–15. Biol. Abstr. 29:22363. 1955.

17442. —— 1955: Insect damages on seeds of Norway spruce determined by X-ray photography. Norrlands Skogs. Tids. 3:299–310.

17443. —— 1955: Samengrösse und Samengewicht als Qualitätsmerkmale einer Samenprobe (*Pinus silvestris* L.). Medd. Statens Skogsforskningsinst. 45(9):1–19.

17444. —— 1957: The X-ray contrast method for seed testing; Scots pine-*Pinus silvestris*. Medd. Statens Skogsforskningsinst. 47(4):1–22.

17445. —— 1961: Influence of cone size on the seed produced (*Pinus silvestris* L.). Medd. Statens Skogsforskningsinst. 49(4):1–16.

17446. —— and A. Gustafsson. 1954: Seed properties in mother trees and grafts of Scots pine. Medd. Statens Skogsforskningsinst. 44(2):1–73. Biol. Abstr. 29:19767. 1955.

17447. —— and A. Gustaffson. 1957: Germinability of pine seeds from Northern Sweden. Medd. Statens Skogsforskningsinst. 47(16):1–4.

17448. —— and —— 1959: X-ray analysis and the quality improvement of North Swedish pine seed. Statens Skogsforsk. Inst. 74:475–86.

17449. ——, ——, and G. Granström. 1957: Die Röntgendiagnose in der Samenkontrolle. Proc. Intern. Seed Testing Assoc. 22:330–43.

17450. —— and S. K. Kamra. 1963: Comparative studies on Scots pine seed germinability with tetrazolium and X-ray contrast methods. Proc. Intern. Seed Testing Assoc. 28:3–18.

17451. ——, K. Ohba, and B. Suszka. 1961: Effect of X-irradiation on seeds of different weight from individual trees of Scots pine (*Pinus silvestris* L.). Botan. Notiser 114:300–12. Biol. Abstr. 38:23564. 1962.

Simak, M. 5829, 7887–7889, 8002, 13857, 13858, 14703.

17452. Simancik, F. 1962: The effect of growth regulators on seed germination and seedling growth in *Pinus sylvestris*. Lesn. Cas. 8:409–19. Forestry Abstr. 24:413. 1963.

17453. —— and A. Laffers. 1963: The effect of gibberellic acid on germination of *Thuja orientalis*. Biologia 18:72–74. Chem. Abstr. 59:6914. 1963.

17454. Simão, S. 1959: The germinating power of mango trees. Anais Escola Super. Agr. "Luiz de Queiroz" Univ. São Paulo 16:90–297. Biol. Abstr. 36:57912. 1961.

17455. Simard, T., and R. A. Ludwig. 1948: Études sur les maladies de l'orge disseminées par la semence. Rept. Quebec Soc. Prot. 1945/47:79–81. Rev. Appl. Mycol. 28:519–20. 1949.

Simeonoglou, G. A. 18759.

Simmonds, D. H. 2246.

17456. Simmonds, N. W. 1952: The germination of banana seeds. Trop. Agr. (London) 29:35–49.

17457. —— 1959: Experiments on the germination of banana seeds. Trop. Agr. (London) 36:259–73.

17458. —— 1963: Correlated seed and tuber dormancy in potatoes. Nature 197:720–21.

Simmonds, N. W. 17990.

17459. Simmonds, P. M. 1946: Detection of the loose smut fungi in embryos of barley and wheat. Sci. Agr. (Ottawa) 26:51–58. Expt. Sta. Rec. 95:63. 1946.

Simmonds, P. M. 11969.

Simmons, D. H. 7570.

Simmons, J. A. 5112.

17460. Simmons, R. O., and F. W. Quackenbush. 1954: The sequence of formation of fatty acids in developing soybean seeds. J. Am. Oil Chemists Soc. 31:441–43.

Simochkina, M. I. 14902.

Simon, E. W. 14805

17461. Simon, F. 1913: Über die Keimung zuvor belichteter und chemische vorbehandelter Samen. Biochem. Z. 48:410–17.

Simon, J. 4007.

17462. Simon, U. 1958: Zur Keimfähigkeit alter Samen. Z. Acker- Pflanzenbau 106:108–18.

Simon, U. 1517.

Simonian, V. H. 20010.

17463. Simons, J. W. 1956: How to dry and store grain and seed on Georgia farms. Georgia Agr. Expt. Sta. Bull. 33. 78 pp.

Simons, M. D. 13966.

Simonsen, J. 13783

17464. Simpson, A. C. 1943: Vernalization of lettuce. Nature 151:279–80.

Simpson, A. G. 13823.

Simpson, D. E. 7578.

17465. Simpson, D. M. 1935: Dormancy and maturity of cottonseed. J. Agr. Res. 50:429–34.

17466. —— 1935: Relation of moisture content and method of storage to deterioration of stored cottonseed. J. Agr. Res. 50:449–56.

17467. —— 1942: Factors affecting the longevity of cottonseed. J. Agr. Res. 64:407–19.

17468. —— 1946: The longevity of cottonseed as affected by climate and seed treatments. J. Am. Soc. Agron. 38:32–45.

17469. —— 1953: Cottonseed storage in various gases under controlled temperature and moisture. Tenn. Agr. Expt. Sta. Bull. 228. 16 pp.

17470. ——, C. L. Adams, and G. M. Stone. 1940: Anatomical structure of the cottonseed coat as related to problems of germination. U.S. Dept. Agr. Tech. Bull. 734. 24 pp.

17471. —— and P. R. Miller. 1944: The relation of atmospheric humidity to moisture in cottonseed. J. Am. Soc. Agron. 36:957–59.

17472. —— and B. M. Stone. 1935: Viability of cottonseed as affected by field conditions. J. Agr. Res. 50:435–47.

Simpson, G. M. 14260, 14261.

17473. Simpson, I. A., A. Y. Chow, and C. C. Soh. 1953: The distribution of thiamin and riboflavin in the mung bean; and the changes that occur during germination. Cereal Chem. 30:222–29.

17474. Simpson, M. 1952: Value of the awn in establishing seed of *Danthonia penicillata* (Labill.) Palisot. New Zealand J. Sci. Technol. 34:360–64.

17475. Sims, H. J. 1955: The protein content of wheat varieties grown in standard plots and spaced rows. J. Australian Inst. Agr. Sci. 21:38–39.

17476. —— and J. V. Mullaly. 1958: Relationship between varieties and protein content of wheat grown in Victoria. J. Agr. Victoria 56:11–19. Nutr. Abstr. Rev. 28:1039. 1958.

17477. Sims, I. H. 1932: Establishment and survival of yellow poplar following a clear cutting in the southern Appalachians. J. Forestry 30:409–14.

17478. Sims, R. C. 1959: Germination of barley: effects of varying water contents upon the initiation and maintenance of growth. J. Inst. Brewing 65:46–50.

17479. Sims, R. P. A. 1964: Changes in the fatty acid composition of the seeds of three oil-bearing species during increasing seed maturity. Can. J. Plant Sci. 44:217–18.

Sims, R. P. A. 13215.

Sinah, M. N. 13452

17480. Sinclair, G. D., and H. R. Sallans. 1947: Flash drying in relation to hulling of sunflowerseed and oats. Can. J. Res. Sect. F 25:119–27.

Sinclair, G. D. 16669–16671

Sineath, H. H. 4947.

Singer, E. 5585, 5586.

17481. Singer, T. P., and B. H. J. Hofstee. 1948: Studies on wheat germ lipase. I. Methods of estimation, purification, and general properties of the enzyme. Arch. Biochem. 18:229–43.

17482. Singh, A. 1963: How seeds are given a good "chit." Indian Farming 13(8):10–11.

17483. Singh, B. 1934: The seeds of *Tinospora cordifolia*, Miers. Current Sci. (India) 3:265.

17484. —— 1945: Propagation of citrus trees. Punjab Fruit J. 9:106–9. Biol. Abstr. 20:1552. 1946.

17458. —— 1952: Studies on the structure and development of seeds of the Cucurbitaceae. I. Seeds of *Echinocystis wrightii* Cogn. Phytomorphology 2:201–9.

17486. —— 1953: Studies on the structure and development of seeds of Cucurbitaceae. Phytomorphology 3:224–39. Biol. Abstr. 29:12078. 1955.

17487. —— 1960: Teak (*Tectona grandis*) "seed" treatment with bacterial action. Proc. Silvicult. Conf. (India) 9:84–89.

17488. —— 1962: Studies in the family Malpighiaceae. II. Development and structure of seed and fruit of *Malpighia glabra* Linn. Hort. Advance 5:145–55. Biol. Abstr. 42:15130. 1963.

17489. Singh, B. K., and A. Majeed. 1942: Chemical analysis of the fruit and kernels of Palmyra palm (*Borassus flabellifer* L.). J. Indian Chem. Soc. Ind. News Ed. 4:223. Pharm. Abstr. 8:260. 1942.

17490. —— and R. D. Tiwari. 1942: Chemical examination of the seeds of *Nigella sativa*. I. Fatty oil. Proc. Natl. Acad. Sci. India 12:141–48. Chem. Abstr. 42:1070–71. 1948.

17491. Singh, B. N. 1941: The relationship of catalase ratio to germination of X-rayed seed as an example of pretreatments. J. Am. Soc. Agron. 33:1014–16.

17492. —— and R. S. Choudhri. 1938: The effect of visible, ultra-violet and infra-red radiations upon the germination and the therapeutic treatment of cotton seed. Empire Cotton Growing Rev. 15:35–42. Biol. Abstr. 12:544. 1938.

17493. ——, ——, and S. L. Kapoor. 1939: Der Einfluss filtrierter und nichtfiltrierter ultraroter Strahlen auf die Wirkung der Diastase in Pflanzen. Ernähr. Pflanze 35:265–68.

17494. ——, ——, and —— 1939: Delinting and germination of cotton seeds as affected by $ZnCl_2.2HCl$ treatment. Indian J. Agr. Sci. 9:731–39. Chem. Abstr. 34:2420. 1940.

17495. —— and S. L. Kapoor. 1939: Vernalization in the tropics. Sci. Cult. (Calcutta) 4:674–77.

17496. —— and K. Kumar. 1930: Studies in growth senescence and rejuvenescence in plants. The influence of the size and weight of the seed on the growth potential and the final yield, as applied to agriculture. Proc. Indian Sci. Congr., Allahabad 17:310–11.

17497. —— and B. N. Lal. 1937: Investigation of the physiological and chemical changes accompanying viviparous germination in mango. J. Indian Botan. Soc. 16:129–36. Biol. Abstr. 12:4412. 1938.

17498. ——, P. B. Mathur, and M. L. Mehta. 1938: Determination of catalase ratio as a rapid method of seed testing. Trop. Agr. (Trinidad) 15:260–61.

17499. —— and R. K. Tandon. 1935: Temperature-absorption characteristics during germination in seeds of differing structure and biochemical constitution under varying concentrations of oxygen and water supply. Proc. Indian Acad. Sci. Sect. B 1:496–518. Chem. Abstr. 29:5150–51. 1935.

17500. Singh, D. 1955: Embryological studies in *Cucumis melo* L. var. *pubescens* Willd. J. Indian Botan. Soc. 34:72–78.

17501. —— 1960: Studies on endosperm and development of seeds of *Carica papaya* L. Hort. Advance 4:89–96.

17502. —— 1961: Development of embryo in the Cucurbitaceae. J. Indian Botan. Soc. 40:620–23.

17503. —— 1961: Studies on endosperm and development of seeds in Cucurbitaceae and some of its related families. Agra Univ. J. Res., Sci. 10:117–23.

17504. —— 1962: The structure and development of ovule and seed of *Passiflora foetida* Linn. Agra Univ. J. Res., Sci. 11:99–111. Biol. Abstr. 45:3134. 1964.

17505. —— 1964: Cytoplasmic nodules in the endosperm of angiosperms. Bull. Torrey Botan. Club 91:86–94.

17506. —— and D. S. Negi. 1962: A contribution to the morphology and embryology of *Dicentra scadens* Walp. Agra. Univ. J. Res., Sci. 11:143–57.

Singh, D. 1526.

17507. Singh, H., and R. B. Ekbote. 1936: The inheritance of seed-characters in gram (*Cicer arietinum* L.). Indian J. Agr. Sci. 6:1087–104.

17508. Singh, H. D., and S. Banerjee. 1955: Studies on the effect of germination on the availability of iron in some common

Indian pulses. Indian J. Med. Res. 43:497–500.

17509. Singh, H. G., and C. A. Lamb. 1960: Mineral and protein content of wheat grain as influenced by variety, soil, and fertilizer. Agron. J. 52:678–80.

Singh, K. K. 12983.

17510. Singh, L., and N. Alam. 1936: Studies in the germination of crop seeds in the field. Indian J. Agr. Sci. 6:784–99.

17511. Singh, M. P., and J. C. Anderson. 1949: Inheritance of earliness of maturity in soybean, *Glycine max* (L.) Merrill. Agron. J. 41:477–82.

17512. Singh, P. P., and B. K. Sur. 1961: Distribution of urease within the seeds of *Cajanus indicus* and the effect of maturity on urease concentration. Current Sci. (India) 30:297.

Singh, O. S. 11516, 11517.

Singh, P. 18179

Singh, R. 12761.

17513. Singh, R. P. 1954: Structure and development of seeds in Euphorbiaceae: *Ricinus communis* L. Phytomorphology 4:118–23. Biol. Abstr. 31:22216. 1957.

17514. —— 1956: Development of endosperm and embryo in *Phyllanthus niruri* L. Agra Univ. J. Res., Sci. 5:163–67.

Singh, S. 7657.

Singh, S. G. 17515.

17515. Singh, S. L., and S. G. Singh. 1935: A study of the viability of some common winter vegetable seeds. Agr. Live-Stock India 5:670–75.

17516. Singh, S. N. 1958: The study of seed content and fruit shape in relation to removal of stigma in *Luffa* species. Current Sci. (India) 27:143–44.

17517. —— 1961: Germination of grape (*Vitis vinifera* L.) hybrid seeds by chilling. Current Sci. (India) 30:62.

17518. Singh, S. P. 1959: Structure and development of seeds in *Euphorbia geniculata* Orteg. J. Indian Botan. Soc. 38:103–8.

Singh, U. P. 5367.

17519. Singh, U. R. 1962: Studies in the fruit-drop of mango. IV. Embryo development, its degeneration and studies of fruit-pedicel and abscission zone. Hort. Advance 5:218–27. Biol. Abstr. 42:15131. 1963.

Singletary, C. C. 2536.

Singleton, W. R. 8329, 17792.

17520. Singley, M. E., W. C. Krueger, and L. G. Merrill, Jr. 1954: Safe farm storage for grain and shelled corn. New Jersey Agr. Expt. Sta. Circ. 562. 11 pp.

17521. Singy, D. 1962: The structure and development of ovule and seed of *Passiflora foetida* Linn. Agra Univ. J. Res., Sci. 11:99–110.

17522. Sinha, A. 1959: Chemical examination of the seeds of *Jatropha curcas*. J. Proc. Inst. Chemists (India) 31:213–17.

17523. —— 1959: Chemical examination of the seeds of *Ipomoea caerulea*. J. Proc. Inst. Chemists (India) 31:285–89. Chem. Abstr. 54:16559. 1960.

17524. —— 1959: Chemical examination of *Phaseolus glabra*. III. Sterols. Proc. Natl. Acad. Sci. India Sect. A 28:243–45. Chem. Abstr. 54:25088. 1960.

17525. —— 1960: γ-Sitosterol from the seeds of *Clitoria ternatea* Linn. Current Sci. (India) 29:180–81.

17526. —— 1960: Sterol from seeds of *Ipomoea caerulae* (*I. hederacea*). Indian J. Appl. Chem. 23:35–36. Chem. Abstr. 54:25089. 1960.

17527. —— 1960: β- and γ-Sitosterols from the leaves and α-sitosterol from the seeds of *Clitoria mariana*. Indian J. Appl. Chem. 23:40–42. Chem. Abstr. 54:25089. 1960.

17528. Sinha, N. K., and A. Sen. 1962: Physico-chemical properties of crystalline α-globulin of sesame seeds. Trans. Bose Res. Inst. 25:37–48. Biol. Abstr. 43:15969. 1963.

Sinitsyna, T. G. 10704.

17529. Sinkovskii, L. P. 1954: The germination of seeds of certain desert thistles and the problem of increasing it. Byul. Mosk. Obshchestva Ispytatelei Prirody Otd. Biol. 59:39–55. Biol. Abstr. 29:13063. 1955.

Sirat, J. Sechet-; *see* Sechet-Sirat, J.

17530. Sircar, S. M. 1944: Vernalization of rice by short days. Nature 153:378.

17531. —— 1948: Vernalization and photoperiodism in the tropics. *In* Murneek, A. E. and R. O. Whyte. "Vernalization and photoperiodism: a symposium." Waltham, Mass., Chronica Botan., pp. 121–28.

17532. —— 1963: Biochemical changes during germination of the seed of rice (*Oryza sativa* L.). [Abridged version.] Intern. Symp. Phys., Ecol., Biochem. Germ., Greifswald. 4 pp.

17533. —— 1963: Physiology of the rice plant. Proc. Indian Sci. Congr., Delhi. Pt. II. Presidential Address 50. 20 pp.

17534. —— and M. Biswas. 1960: Viability

and germination inhibitor of the seed of rice. Nature 187L:620–21.

17535. —— and T. M. Das. 1951: Growth hormones of rice grains germinated at different temperatures. Nature 168:382–83. Biol. Abstr. 27:14143. 1953.

17536. —— and —— 1955: Studies on the physiology of rice. IX. Auxin content of the vernalised seed. Natl. Inst. Sci. India 20:673–82.

17537. ——, ——, and A. N. Lahiri. 1955: Germination of rice embryo under water and its relation of growth to endosperm fractions. Nature 175:1046–47.

17538. —— and A. G. Dastidar. 1962: Amino acid metabolism of the seed of rice (*Oryza sativa* L.) during germination and seedling growth. Physiol. Plantarum 15: 206–10.

17539. —— and B. Dey. 1963: Dormancy and viability of the seed of rice (*Oryza sativa* L.). Intern. Symp. Phys., Ecol., Biochem. Germ., Greifswald. 3 pp.

17540. —— and B. N. Ghosh. 1954: Studies on the physiology of rice. VIII. The effects of low and high temperature germination with or without short days on summer and winter varieties. Proc. Natl. Inst. Sci. India 20:452–65.

17541. ——, S. S. Guha-, and T. Chakravarty. 1951: The chemical investigation of the seeds of *Swietenia macrophylla.* I. The non-bitter principle. J. Indian Chem. Soc. 28:207–10. Biol. Abstr. 27:2361. 1953.

17542. —— and M. Kundu. 1959: Effect of root extract of water hyacinth (*Eichornia speciosa* Kunth) on the growth and flowering of rice. Sci. Cult. (Calcutta) 24:332–33.

17543. —— and A. N. Lahiri. 1956: Studies on the physiology of rice. XII. Culture of excised embryos in relation to endosperm auxin and other growth factors. Proc. Natl. Inst. Sci. India Pt. B 22:212–25. Biol. Abstr. 32:10102. 1958.

17544. —— and A. Roy. 1956: Studies on the physiology of rice. XI. Vernalization and devernalization of winter and summer varieties. Proc. Natl. Inst. Sci. India Pt. B 22:102–11.

17545. Sirgo, V. 1939: Various factors influencing plant growth. Pharmacia (Estonia) 19:78–79, 113–16. Chem. Abstr. 33:8239. 1939.

17546. Sirks, M. J. 1926: Farbe und Grösse der Samen vom Raps in ihrer physiologischen und züchterischen Bedeutung. Mededel. Landbouwhogeschool Wageningen 30:25–54.

Sirohi, G. S. 14170–14172.

17547. Siroya, T. 1951: On the sugars in the seeds and seedlings of *Pinus thunbergii.* Botan. Mag. (Tokyo) 64:137–45. Biol. Abstr. 29:9594. 1955.

17548. Sirrine, E. F. 1894: Structure of seed coats of Polygonaceae. Proc. Iowa Acad. Sci. 2:128–36.

17549. —— 1937: The value of incidental seeds found in samples of commercial seeds. Proc. Assoc. Offic. Seed Analysts 29:102–3.

Sirrine, E. F. 18710, 18711.

17550. Sisakyan, N. M., and L. S. Markosyan. 1959: The amino acid composition of wheat grain proteins. Biochemistry (USSR) (English Transl.) 24:1007–15.

17551. Sitnikova, O. A. 1950: Causes for the quick loss of germination capacity by *Populus* and *Salix* seeds. Dokl. Akad. Nauk SSSR 70:719–22.

Sitnikova, O. A. 7097.

17552. Sitton, B. G. 1944: Specific gravity as a criterion of filling in the pecan nut. Proc. Am. Soc. Hort. Sci. 44:89–93.

Siu, R. 7917, 18989.

17553. Sivaramakrishnan, V. M., and P. S. Sarma. 1955: The inhibition by sulphanilimide of the metabolism of histidine in germinating seeds. Current Sci. (India) 24:330–31.

17554. —— and —— 1956: Conversion of histidine to urocanic acid in germinating seeds. Current Sci. (India) 25:288–89.

17555. —— and —— 1956: The metabolism of glutamic acid in germinating green gram seeds. Biochem. J. 62:132–35.

17556. Sivori, E. M., and J. E. Wurceldorf-Warden. 1949: An inhibiting factor in *Matthiola incana.* Lilloa 19:49–70. Biol. Abstr. 26:22605. 1952.

17557. Sjelby, K., and C. Stahl. 1941/43: How are the different seed species classified in the purity analysis by the seed testing stations all over the world? Proc. Intern. Seed Testing Assoc. 13:128–46.

17558. Sjöstedt, S. 1959: Germination biology of cleavers, *Galium aparine* L. Växtodling 10:87–105.

Skaggs, D. P. 13593.

Skau, D. B. 13114

17559. Skazkin, F. D., and A. V. Khvan. 1962: The effects of rain during seed maturation on the quality and yield of grain. Dokl. -Botan. Sci. Sect. (English Transl.) 140:169–71.

17560. Skeels, H. C. 1932: Foreign seeds. Natl. Hort. Mag. 11:308–16.

17561. Sekllon, J. H., and J. W. Spence. 1954: The seed oil of *Ammi visnaga* as a source of cis octa-dec-6-enoic acid (Petroselinic). Chem. Ind. 3:75.

Skelton, B. J. 12900.

17562. Skene, K. G. M., and D. J. Carr. 1961: A quantitative study of the gibberellin content of seeds of *Phaseolus vulgaris* at different stages in their development. Australian J. Biol. Sci. 14:13–25.

Skene, K. G. M. 3639

17563. Skinner, C. G., J. R. Claybrook, F. D. Talbert, and W. Shive. 1957: Effect of 6-(substituted)thio- and amino-purines on germination of lettuce seed. Plant Physiol. 32:117–20.

17564. ——, C. Hedgcoth, and D. Wagner. 1961: Synergistic effect on germination of mixtures of 6-benzylaminopurine and various gibberellins. Phyton 17:141–45.

17565. —— and W. Shive. 1959: Stimulation of lettuce seed germination by 6-(substituted) purines. Plant Physiol. 34: 1–3.

17566. ——, F. D. Talbert, and W. Shive. 1958: Effect of 6-(substituted) purine and gibberellin on the rate of seed germination. Plant Physiol. 33:190–94.

Skinner, J. T. 15984

17567. Skinner, H. T. 1941: Germination of two-year seeds. Gardeners Chron. 45:323.

17568. —— 1946: Seed release in Atlas cedar. Morris Arboretum Bull. 4:44.

17569. —— 1949: *Metasequoia* in its second year. Morris Arboretum Bull. 4:94.

17570. Skolaude, E., and H. Wartenberg. 1956: Antiobiotische Wirkungen von Mikroorganismen eines gemeinesamen Lebensraumes. Naturwissenschaften 43:526–27.

17571. Skolko, A. J., and J. W. Groves. 1953: Notes on seed-borne fungi. VII. *Chaetomium*. Can. J. Botany 31:779–808.

17572. ——, J. W. Groves, and V. R. Wallen. 1954: *Ascochyta* diseases of peas in Canada—with special reference to seed transmission. Can. J. Agr. Sci. 34:417–28.

Skolko, A. J. 7731–7734, 19332, 19333.

Skoog, F. 398, 9133.

17573. Skoog, F. E., and L. E. Wallace. 1964: Application of systemic insecticides as seed treatment to protect wheat plants against grasshoppers and wheat stem sawfly. J. Econ. Entomol. 57:199–205.

Skorko, R. 12945.

Skorobogatova, E. P. 15642.

Skorokhodova, Z. F. 14249

17574. Skotland, C. B., and D. W. Burke. 1961: A seed-borne bean virus of wide host range. Phytopathology 51:565–68.

17575. Skrabka, H. 1957: Determining the differences in sectional force in germinating seed of some oat varieties. Plant Breed. Acclimitiz. Inst. Sci. Publ. (Warsaw) 1:85–102.

17576. Skraup, S., and U. Weber. 1936: Untersuchungen zum Fettstoffwechsel der Pflanzen. Ber. Phys.-Med. Ges. Würburg 60:14–28. Botan. Centralbl. 174:13. 1938.

Skriabin, G. K. 11332.

Skrzypinska, H. 19665.

Skupinska, J. 2488.

Skwarra, H. 13652.

17577. Skyrme, E. W. 1935: The structure and histology of ispaghula (the seeds of *Plantago ovata* Forsk.). J. Pharm. Pharmacol. 8:161–85. Biol. Abstr. 11:6882. 1937.

17578. —— 1935: Seeds of *Plantago*. (*Psyllium* and the seeds of certain other species of *Plantago*.) J. Pharm. Pharmacol. 8:610–21. J. Roy. Microscop. Soc. Ser. III 56:182. 1936.

17579. Slabaugh, P. E., and P. O. Rudolf. 1957: The influence of seed source on the development of Scotch pine and Norway spruce planted in lower Michigan (fifteen-year results). Mich. Acad. Sci. Arts Letters Papers 42:41–52.

17580. Slabonski, A. 1956: Studies on the vernalization stage of grain varieties and their significance in the cultivation and breeding of plants. Acta Agrobotan. 4:45–85.

17581. Slaski, J. 1957: Pretreatment of seeds. Sylwan 101:60–66. Biol. Abstr. 32:35270. 1958.

17582. Slate, G. L. 1946: Raising lilies from seeds. Natl. Hort. Mag. 25:70–73.

Slate, G. L. 8590.

17583. Slater, C. S., and A. Rodriguez. 1954: Aspects of structure that affect germination in a silt loam soil. Soil Sci. Soc. Am. Proc. 18:219–21.

Slater, R. B. Alfin-. 10222.

17584. Slatter, E. M. 1950: The culture of iris embryos on nutrient agar. Kew Roy. Botan. Gardens Bull. 3:425–30.

17585. Slipushenko, K. P. 1961: Propagation of *Magnolia* by seeds. Akad. Nauk Ukr. RSR Tsent. Botan. Sada Visnyk 3:28–33.

17586. Sloan, M. L. 1937: Germination of petunia seeds. Horticulture 15(4):72.

Slobodianik, N. I. 12721.

Slotter, R. L. 12653.
Slovtsov, A. M. 7695.
Sludskaja, M. 17230.

17587. Slukhai, S. I. 1956: Effect of manganese and boron on seed germination and growth of young woody plants. Mikroelementy v Sel'sk. Khoz. i Med. Akad. Nauk Latv. SSR, Otd. Biol. Nauk, Tr. Vses Soveshch. po Mikroelementam, Riga 1955:455–63. Chem. Abstr. 53:11540. 1959.

17588. Slusanschi, H. 1939: The changes in the chemical character of grains during storage. Analele Inst. Cercetari Agron. Romaniei 11:320–36. Chem. Abstr. 37: 4482. 1943.

Slusanschi, H. 639.

17589. Smalii, V. T. 1960: The effect of rhizospheric bacteria on the germination of seed and the growth of wheat germinants Mikrobiol. Zh. Akad. Nauk Ukr. RSR 22(5):20–24. Biol. Abstr. 37:3193. 1962.

17590. Smallfield, P. W. 1961: F.A.O. Seed Year 1961. New Zealand J. Agr. 103:203–9.

17591. Smedova, M. 1962: Effect of sodium humate on swelling and germination of winter wheat. Biol. Plant. 4:112–18.

Smedt, A. de; *see* De Smedt, A.
Smelt, C. 17909.

17592. Smerda, V., and S. Tichy. 1957: Effect of heating and the age of seed on yield of cucumbers. Česk. Akad. Zemědĕl. Ved. Sb. Rostlinna Vyroba 3:223–30.

Smertkin, V. D. 11464.
Smetannikova, A. I. 2235.

17593. Smika, D. E., and F. W. Smith. 1957: Germination of wheat as affected by Vuiret contamination in urea. Soil Sci. 84:273–82.

Smilde, K. W. 10857.

17594. Smirnov, A. I. 1944: Respiration of unripe grain and its storage. Compt. Rend. Acad. Sci. URSS 44:246–47.

17595. ——, Z. S. Bronovitskaya, K. V. Pshennova, S. D. Chigirev, and E. N. Ushakova. 1943: The respiratory metabolism and enzymic activity of the wheat kernel during ripening. Biochemistry (USSR) (English Transl.) 8:149–57. Chem. Abstr. 38:1544. 1944.

17596. —— and V. L. Kretovich. 1939: The biochemistry of grain in storage. Sb. Akad. Nauk SSSR, Presidentu Akad. Nauk SSSR Komarovu 1939:720–25. Chem. Abstr. 37:1469. 1943.

Smirnov, B. M. 10393.

17597. Smirnov, L. A. 1948: Natural yarovization and latitudinal migrations of plants. Botan. Zh. 33:559–70.

Smirnov, V. A. 4110.

17598. Smirnova, A. D. 1952: Effect of boron and manganese on the crop of the sunflower. Mikroelementy v Zhizni Rast. i Zhivotn. Akad. Nauk SSSR, Tr. Konf. po Mikroelementam 1950:236–42. Chem. Abstr. 49:440. 1955.

17599. Smirnova, E. S. 1961: Seed morphogenesis in angiosperms. I. Dokl. Mosk. Sel'skokhoz. Akad. 72:263–72. Biol. Abstr. 42:19211. 1963.

17600. —— 1962: Use of a short method of recording the characteristics of internal structure of angiosperm seeds. Byul. Gl. Botan. Sada 47:68–71. Biol. Abstr. 45:12268. 1964.

Smirnova, I. S. 6144, 6145.

17601. Smirnova, M. I. 1931: Biochemical study of maize grains. Bull. Appl. Botany, Genet., Plant Breed. (USSR) 25:329–43. Chem. Abstr. 26:757. 1932.

17602. —— 1936: Inter- and intravarietal chemical variations in the cotton-seed. Bull. Appl. Botany, Genet., Plant Breed. (USSR) Ser. III 15:227–40. Chem. Abstr. 31:5842. 1937.

Smirnova, N. G. 14298.
Smirnova, T. I. 11375, 11383.

17603. Smirnova-Ikonnikova, M. I., and E. P. Veselova. 1960: Influence of location on the content and protein composition of the grains of Leguminosae. Biokhim. Zerna, Sb. 5:228–47. Chem. Abstr. 55:1812. 1961.

Smit, A. J. Haagen-; *see* Haagen-Smit, A. J.

17604. Smit, B. 1957: The control of stored grain insects in South Africa. Union S. Africa Dept. Agr. Bull. 355. 36 pp.

17605. Smith, A. C. 1933: Practical seed selection of coconuts. Malayan Agr. J. 21:265–71. Plant Breeding Abstr. 4:17–18. 1933.

17606. Smith, A. D., and H. J. Bergen. 1962: Note on a power plot seeder. Can. J. Plant Sci. 42:548–50.

17607. Smith, A. J. M. 1932: Preservation and germination of seeds. Gt. Brit. Dept. Sci. Ind. Res. Food Invest. Bd. Rept. 1932:149–56.

17608. —— 1936: The respiratory quotient of the initial metabolism of soaked seeds. Gt. Brit. Dept. Sci. Ind. Res. Food Invest. Bd. Rept. 1935:132–35. Chem. Abstr. 32:2173. 1938.

17609. —— and R. Gane. 1932: Influence of a gaseous product from apples on the germination of seeds. Gt. Brit. Dept. Sci. Ind. Res. Food Invest. Bd. Rept. 1932:156–58. Expt. Sta. Rec. 71:634. 1934.

17610. —— and —— 1939: The water relations of seeds. Gt. Brit. Dept. Sci. Ind. Res. Food Invest. Bd. Rept. 1938:231–40.

17611. —— and —— 1948: The storage of seeds. J. Agr. Sci. 38:81, 84, 90.

Smith, A. J. M. 6949.

17612. Smith, A. K., and V. L. Johnsen. 1948: Sunflower-seed protein. Cereal Chem. 25:399–406. Chem. Abstr. 43: 2335. 1949.

17613. ——, A. M. Nash, and L. I. Wilson. 1961: Water absorption of soybeans. J. Am. Oil Chemists Soc. 38:120–23.

Smith, A. K. 3107, 5887, 14679.

17614. Smith, A. L. 1947: Cotton seed-treatment studies in Georgia. [Abstract.] Phytopathology 37:436.

17615. Smith, A. M., and W. Robb. 1943: The carotene and protein in oats and barley at different stages of growth. J. Agr. Sci. 33:119–21. Chem. Abstr. 37: 5758. 1943.

Smith, B. 12477.

17616. Smith, B. C. 1951: An investigation of the rest period in the seed of the genus *Cotoneaster*. Proc. Am. Soc. Hort. Sci. 57:396–400.

17617. —— 1958: An investigation of the delayed germination of the seed of *Cotoneaster divaricata*, *C. zabeli*, *Viburnum lantana* and *V. lentago*. Dissertation Abstr. 18:771–72.

17618. —— and L. C. Chadwick. 1953: Studies of the germination of seed of *Cotonester zabeli*. Proc. Am. Soc. Hort. Sci. 62:503–5.

17619. Smith, B. P., and H. H. Williams. 1951: Transaminase studies in germinating seeds. Arch. Biochem. Biophys. 31:366–74.

17620. Smith, B. W. 1956: *Arachis hypogaea*. Embryogeny and the effect of peg elongation upon embryo and endosperm growth. Am. J. Botany 43:233–40.

Smith, B. W. 9834.

17621. Smith, C. H. 1952: Heritable differences in germination of sugar-beet seed at low temperature. Proc. Am. Soc. Sugar Beet Technol. 7:411–14.

17622. Smith, C. L., and A. J. Loustalot. 1944: Effects of harvest date and curing on the composition and palatability of pecan nuts. J. Agr. Res. 68:395–403.

17623. —— and C. J. B. Thor. 1933: A physiological study of fruit development in the pecan (*Hicoria pecan*). Proc. Texas Pecan Growers Assoc. 13:71–76. Chem. Abstr. 28:3102. 1934.

17624. ——, ——, and L. D. Romberg. 1933: Effect of storage conditions on the germination of seed pecans. Proc. Texas Pecan Growers Assoc. 13:68–71. Expt. Sta. Rec. 70:782. 1934.

Smith, C. L. 16397, 18530, 18531.

17625. Smith, C. M. 1931: Development of *Dionaea muscipula*. II. Germination of seed and development of seedling to maturity. Botan. Gaz. 91:377–94.

17626. —— 1950: Notes on seeds and seedlings of *Metasequoia glyptostroboides*. New Zealand J. Forestry 6:145–48. Biol. Abstr. 26:7123. 1952.

17627. Smith, C. O. 1941: Observations on the germination of *Araucaria bidwillii*. Am. Naturalist 75:96.

17628. Smith, C. R., Jr., M. C. Shekleton, I. A. Wolff, and Q. Jones. 1959: Seed protein sources—amino acid composition and total protein content of various plant seeds. Econ. Botany 13:132–48.

Smith, C. R., Jr. 19740–19742.

Smith, C. T. 8964.

17629. Smith, D. 1949: Differential survival of Ladino and common white clover encased in ice. Agron. J. 41:230–34.

17630. —— 1958: Performance of Narragansett and Vernal alfalfa from seed produced at diverse latitudes. Agron. J. 50:226–29.

17631. —— 1961: Note on the growth characteristics of alfalfa and red clover plants derived from hard seeds and seeds of different sizes. Can. J. Plant Sci. 41:675–78.

17632. —— and F. Rauchfuss. 1958: Effects of aqueous extracts of halogeton tissue on germination of seeds and growth of seedlings. J. Range Management 11:300–2. Biol. Abstr. 34:14561. 1959.

Smith, D. 11822.

17633. Smith, D. C. 1939: Influence of moisture and low temperature on the germination of hop seeds. J. Agr. Res. 58:369–81.

17634. —— 1940: The relations of color to germination and other characters of red, alsike, and white clover seeds. J. Am. Soc. Agron. 32:64–71.

17635. —— and E. N. Bressman. 1930: Some effects of seed treatment on the germination and subsequent growth of wheat. J. Agr. Res. 40:25–36.

Smith, D. C. 14392.

Smith, D. E. 10131.

17636. Smith, D. L. 1964: The evolution of the ovule. Biol. Rev. 39:137–59.

Smith, E. 13721, 13722.

Smith, E. A. 4489.

Smith, E. C. Bate-; *see* Bate-Smith, E. C.

Smith, E. L. 10627, 19106.

Smith, E. M. 15618.

17637. Smith, F., and H. C. Srivastava. 1957: Isolation of maltose from honeydew on alsike (*Trifolium hybridum*) seeds. J. Org. Chem. 22:987–88.

Smith, F. 5599, 13694.

17638. Smith, F. G. 1952: The mechanism of the tetrazolium reaction in corn embryos. Plant Physiol. 27:445–56.

17639. —— 1958: Germination and seed testing—a biochemical view. Assoc. Offic. Seed Analysts, 50 Years Seed Test. 1908–1958:98–101.

Smith, F. G. 18542, 18543.

17640. Smith, F. H. 1958: Spectrophotometric determination of total gossypol in cottonseed meal and cottonseed meats. J. Am. Oil Chemists Soc. 35:261–65.

Smith, F. H. 8053, 8537, 17841.

17641. Smith, F. L. 1947: Inheritance of seed coat color in derivatives of Pinto beans. J. Am. Soc. Agron. 39:1039–52. Biol. Abstr. 22:8125. 1948.

17642. —— 1955: The effects of dates of harvest operations on yield and quality of pink beans. Hilgardia 24:37–52.

17643. —— 1956: Inheritance of three seed-coat color genes in *Vigna sinensis* Savi. Hilgardia 24:279–96.

17644. —— 1961: Seed-coat color genes in six commercial varieties of beans. Hilgardia 31:1–14.

17645. —— and W. B. Hewitt. 1938: Varietal susceptibility to common bean mosaic and transmission through seed. Calif. Agr. Expt. Sta. Bull. 621:1–18. Biol. Abstr. 12:1463. 1938.

Smith, F. L. 371, 6218.

Smith, F. W. 3826, 13486, 17593, 19698.

Smith, G. 3366.

Smith, G. B. L. 15486.

Smith, G. E. 20133.

17646. Smith, G. L. 1963: Studies in *Potentilla* L. I. Embryological investigations into the mechanism of agamospermy in British *P. tabernaemontani* Aschers. New Phytologist 62:264–82.

17647. —— 1963: Studies in *Potentilla* L. II. Cytological aspects of apomixis in *P. crantzii* (CR.) Beck ex Fritsch. New Phytologist 62:283–300.

Smith, G. M. 5506.

Smith, G. N. 3330.

17648. Smith, G. P. D. 1930: Points in germinating some Australian seeds. Rept. Proc. Intern. Hort. Congr. London 9:154–58. Biol. Abstr. 6:24711. 1932.

Smith, G. S. 6334, 8251.

Smith, H. C. 394.

Smith, H. H. 4764.

17649. Smith, H. P., M. H. Byrom, and H. F. Morris. 1942: Germination of cottonseed as affected by soil disturbance and machine placement of fertilizer. Texas Agr. Expt. Sta. Bull. 616. 29 pp.

17650. Smith, H. S., and H. W. Staten. 1942: The effects of carbon disulphide upon germination and baking quality of wheat. Oklahoma Agr. Expt. Sta. Tech. Bull. T-14. 13 pp. Expt. Sta. Rec. 88:190. 1943.

Smith, I. 16347.

17651. Smith, J. 1939: Germination of nim seeds. Indian Forester 65:170–71. Biol. Abstr. 14:3530. 1940.

17652. Smith, J. D., J. A. Mitchner, Jr., and H. M. Burlage. 1938: Chemical study of *Piscaria setigera piper* (*Eremocarpus*). I. The oil of the seeds. J. Am. Pharm. Assoc. 27:666. Pharm. Abstr. 5:210. 1939.

Smith, J. E., Jr. 4390.

Smith, J. H. G. 4164.

17653. Smith, J. R. 1946: Men and nut trees. Morris Arboretum Bull. 4(6):47–49.

17654. Smith, L. 1946: A comparison of the effects of heat and X-rays on dormant seeds of cereals, with special reference to polyploidy. J. Agr. Res. 73:137–58.

17655. —— 1948: The effect of chaff of cereals on germination of seeds and on the growth of mold. J. Am. Soc. Agron. 40:32–44.

17656. —— 1950: Effects of atomic bomb radiations and X-rays on seeds of cereals. J. Heredity 41:125–30.

17657. —— 1952: A comparison of the effects of Test Able atomic bomb ionizing radiation and X-rays on seeds of barley, wheat, and oats. Naval Med. Res. Sect. U.S. Joint Task Force One, Appendix No. 25 to the Final Rept. 17 pp.

17658. —— and R. S. Caldecott. 1948: Modification of X-ray effects on barley seeds by pre- and post-treatment with heat. J. Heredity 39:173–76.

17659. ——, L. R. Sieburth, and B. Norby. 1950: The relation of oxygen concentration in barley seeds to the biological effects of X-ray. [Abstract.] Genetics 35:692.

Smith, L. 3517, 3518, 7841, 7979, 8446.

17660. Smith, L. F. 1962: The biosynthesis of fatty acids in flax embryos, *Linum usitatissimum* L. Dissertation Abstr. 22: 3376–77.

Smith, L. H. 9014.

17661. Smith, L. J. 1957: The statutory tests. Natl. Inst. Agr. Botany J. 8:188–94.

17662. Smith, M. A., and C. B. Purves. 1941: Chemical constituents of cottonseed hulls. Ind. Eng. Chem. Anal. Ed. 13:157–59.

17663. Smith, N. A. 1961: An evaluation of methylcellulose and paper seed ribbons for the precision seeding of lettuce (*Lactuca sativa* L.) and other vegetables. Dissertation Abstr. 22:383–84.

17664. Smith, O. 1940: Premature seeding studies with beets. Proc. Am. Soc. Hort. Sci. 37:793–98.

17665. Smith, P. B., and G. E. Walters. 1962: Methods of preparation and results of field planting of various types of processed monogerm sugar beet seed. J. Am. Soc. Sugar Beet Technol. 12:225–32.

17666. Smith, P. G. 1944: Embryo culture of a tomato species hybrid. Proc. Am. Soc. Hort. Sci. 44:413–16.

Smith, P. G. 10655, 11919.

17667. Smith, P. R. 1961: Seed-borne *Septoria* in lettuce. Eradication by hot water treatment. J. Agr. Victoria 59:555–56.

Smith, R. D. 19586.

Smith, R. E. 8479.

Smith, R. F. 608.

Smith, R. G. 12618.

17668. Smith, R. K. 1959: Development of better seed statistics on production of certified seed. Intern. Crop Improvement Assoc. Ann. Rept. 38/39:198–206.

Smith, R. L. 5593.

17669. Smith, T. E., and J. P. Bailes. 1958: Germination tests and field stands of corn. Plant Disease Reptr. 42:734.

17670. Smith, W. J. 1946: Effect of penicillin on seed germination. Science 104:411–13.

17671. —— 1950: Seed treatment with streptomycin for the control of bacterial blight of beans. Rev. Appl. Mycol. 29:600.

17672. Smith, W. P. C. 1959: Topical notes on seed pelleting. J. Agr. W. Australia 8:447–48.

17673. —— and O. M. Goss. 1958: A method of inoculating and lime-pelleting leguminous seeds. J. Agr. W. Australia 7:119–21.

17674. Smolin, A. A. 1943: Experimental elimination of dormancy in tau saghyz. Compt. Rend. Acad. Sci. URSS 38:96–98. Biol. Abstr. 20:18577. 1946.

Smolina, N. 11389.

17675. Smorodintsev, I. A., and E. V. Serkova. 1929: Tolerance of mice and guinea pigs against extracts of *Carum* seed. Russ. J. Trop. Med., Med. Vet. Parasitol. 7:408–11. Biol. Abstr. 6:13356. 1932.

Smydzuk, J. 16290.

17676. Smyk, B., and A. Milkowska. 1956: The influence of infra-red rays on the disinfection of grains. Acta Microbiol. Polon. 5:133–35.

17677. Snamenskij, I. 1931: Materialen zur Kenntnis der Keimung und Reife von Samen. VI. Jard. Botan. Inst. Essais Semences, Leningrad, Ann. Essais Semences 1931(1):3–14.

Snegirev, D. P. 9330, 11838, 19915.

17678. Snell, K. 1912: Über das Vorkommen von keimfähigen Umkrautsamen im Boden. Landwirtsch. Jahrb. 43:323–47.

17679. —— 1939: Seed certification in Germany. Chronica Botan. 5:53, 54. Biol. Abstr. 13:17600. 1939.

17680. Snelling, R. O., and I. R. Hoener. 1940: Relationships between some measures of maturity in maize. J. Am. Soc. Agron. 32:451–54.

17681. Snellman, O., and C. E. Danielson. 1953: An experimental study of the biosynthesis of the reserve globulins in pea seeds. Exptl. Cell Res. 5:436–42. Ber. Wiss. Biol. 89:301. 1954.

Snider, S. R. 5402.

17682. Snoddy, A. O. 1934: Growing peonies from seed. Flower Grower 21(2): 72.

17683. Snook, L. C. 1939: The composition of wheat and oat grain grown in western Australia. J. Dept. Agr. W. Australia 16:283–87.

17684. —— 1947: Locally grown leguminous seeds. J. Dept. Agr. W. Australia 24:300–12. Biol. Abstr. 22:19086. 1948.

Snow, A. G., Jr. 18668, 18696, 18697.

17685. Snowberg, A. 1935: Philippinense lily from seed. Flower Grower 22(4):183.

Snowden, J. E., Jr. 5352.

17686. Snyder, E. 1935: Breeding for seedless Vinifera grapes. Proc. Am. Soc. Hort. Sci. 32:381–83.

Snyder, E. 8190.

17687. Snyder, F. W. 1954: Effect of quantity of mineral nutrients on germination and seedling growth of sugar beets. Proc. Am. Soc. Sugar Beet Technol. 8:340–43. Biol. Abstr. 30:5854. 1956.

17688. —— 1959: Effect of gibberellin on germination and early growth of the sugar beet. J. Am. Soc. Sugar Beet Technol. 10:394–95.

17689. —— 1959: Effect of nitrogen on yield and subsequent germinability of sugar beet seed. J. Am. Soc. Sugar Beet Technol. 10:439–43.

17690. —— 1959: Influence of the seedball on speed of germination of sugar beet seeds. J. Am. Soc. Sugar Beet Technol. 10:513–20.

17691. —— 1963: Some physico-chemical factors of the fruit influencing speed of germination of sugar beet seed. J. Am. Soc. Sugar Beet Technol. 12:371–77.

17692. Snyder, R. J., and R. E. Larson. 1955: Quantitative inheritance of seed size in a cross between *Lycopersicon esculentum* and *L. pimpinellifolium.* Penn. Agr. Expt. Sta. Bull. 607. 24 pp.

17693. Snyder, W. C. 1942: A seed-borne mosaic of asparagus bean, *Vigna sesquipedalis.* Phytopathology 32:518–23.

17694. Soave, M. 1899: The chemical and physiological effect of anaesthetics on the germination of seeds. Staz. Sper. Agrar. Ital. 32:553–92. Expt. Sta. Rec. 11:1056. 1900/01.

Sobolev, A. K. 14381, 14382.

17695. Sobolev, A. M. 1962: Enzymic hydrolysis of phytin (inositol hexaphosphate) in vitro and in germinating seeds. Plant Physiol. (USSR) (English Transl.) 9:334–41. Chem. Abstr. 57:14192. 1962.

Sobolev, A. M. 15721.

17696. Sobolev, N. A. 1960: Progress toward the growth of wheat embryos without the endosperm. Trudy Aspirantov i Molodykh Nauch. Sotrudnikov Vses. Inst. Rast. Leningr. 1960:37–40. Biol. Abstr. 41:7935. 1963.

17697. —— 1962: Methods of growing wheat embryos without endosperms. Leningr. Vses. Inst. Rast. Sb. Tr. 4:37–40.

17698. Soboleva, V. V. 1957: System of seed production of agricultural crops in Czechoslovakia. Selek. i Semen. 22(3):56–64.

17699. Socnik, H. 1932: Ist die Anzucht von *Ficus elastica* aus Samen zu empfehlen? Gartenwelt 36:622–23.

17700. —— 1932: Vermehrung von *Ficus elastica* aus Samen. Möllers Deut. Gärtner-Ztg. 47:30.

17701. —— 1933: Die günstigsten Keimungstemperaturen für Kakteen. Gartenwelt 37:207–8.

17702. —— 1935: Bromeliaceen aus Samen. III. (Concl.) Blumen-Pflanzenbau Gartenwelt 39:55.

17703. Soderholm, L. H., and E. R. Walker. 1955: Effect of cathode rays on germination and early growth of wheat. Botan. Gaz. 116:281–90.

17704. Soedigdo, R., and M. Gruber. 1960: Purification and some properties of a protease from pea seeds, *Pisum sativum* L. S. SP. *Arvense* A. and G. Biochim. Biophys. Acta 44:315–23.

17705. Söding, H., and H. Funke. 1943: Über die Förderung der Pflanzentwicklung durch Hefeauszüge. Angew. Botan. 25: 378–90. Biol. Abstr. 23:24804. 1949.

17706. —— and M. Wagner. 1955: Mehrjährige Versuche über die Beeinflussung der Keimung und Entwicklung von *Poa annua* durch Behandlung der Früchte mit Wirkstoffen. Planta 45:557–72.

Söding, H. 19234.

Soegaard, B. 8495.

Sörensson, I. 12252.

Soga, Y. 14116.

17707. Sogn, L. 1962: Methods and equipment for rapid determination of viability of grain. Tidsskr. Norske Landbruk 69:229–34.

Sogn, L. 12287.

Soh, C. C. 17473.

17708. Sohadevan, P. C., and M. B. V. N. Rao. 1947: Note on the deterioration in germination-capacity of a paddy strain in Malabar and South Kanara. Current Sci. (India) 16:319–20.

17709. Sohonie, K., and K. S. Ambe. 1955: Crystalline trypsin inhibitors from the Indian field bean and the double bean. Nature 175:508–9.

17710. —— and A. P. Bhandarkar. 1954: Trypsin inhibitors in Indian foodstuffs. I. Inhibitors in vegetables. J. Sci. Ind. Res. (India) 13B:500–3. Chem. Abstr. 49:534. 1955.

Sohonie, K. 449, 450, 5233, 5234, 6909, 11273.

17711. Soifer, P. A. 1952: Determination of sulfur-containing amino acids and biological value of bean proteins. Biochemistry (USSR) (English Transl.) 17: 307–11. Chem. Abstr. 46:10310. 1952.

17712. Soifer, V. N. 1962: Evolution of anatomic seed structures in the squash family. Byul. Mosk. Obshchestva Ispytatelei Prirody Otd. Biol. 67:147. Biol. Abstr. 42:7431. 1963.

17713. Sojka, E. 1962: Investigations into the physiology and biochemistry of developing rye grain. III. The influence of maturity of sowing grain on the yields of plants and their resistance to drought and low temperatures. Howdowla Roslin Aklim. Nasiennictwo 6:1–19.

Sojka, E. 7758, 7759, 13427.

Sokolov, N. N. 10550.

17714. Sokolov, V. S. 1951: Method for increasing the seed germination of *Caragana arborescens*. Les. i Step 3(8):62–65.

17715. —— 1952: Accelerated seed stratification of Tatarian honeysuckle (*Lonicera tatarica*). Lesn. Khoz. 5(8):56–58.

17716. —— 1963: Effect of a solution of gibberellin on the germination of seeds of forest species. Lesn. Zh. 1:9–12.

Sokolova, A. I. 11384.

17717. Sokolova, G. P. 1956: Preplanting treatment of flax seeds with solutions of micronutrients. Zap. Leningr. Sel'skokhoz. Inst. 11:412–16. Chem. Abstr. 52:20418. 1958.

17718. Sokolova, N. A. 1963: Content of nucleic acids in seeds of cotton plant of different earliness. Dokl. Akad. Nauk Uz. SSR 20(6):50–53. Chem. Abstr. 60:861. 1964.

17719. Solacolu, T., M. Constantinesco, and D. Constantinesco. 1938: Influence of sterilized solutions of tryptophan on the isolated plantules from ripe ungerminated seeds. Compt. Rend. Soc. Biol. 129:403–5. Chem. Abstr. 33:1008. 1939.

17720. —— and E. Welles. 1933: Detection of saponins in the seeds of various Gramineae. Compt. Rend. Soc. Biol. 112:1007–9. Chem. Abstr. 27:2978. 1933.

17721. —— and —— 1935: Changes in saponin content during germination and growth of the young plants of various species of Gramineae. Compt. Rend. Soc. Biol. 120:662–64. Chem. Abstr. 30:1088. 1936.

17722. Soldatenkov, S. V., and T. A. Mazurova. 1954: Conversion of organic acids during the sprouting and ripening of wheat seeds. Biochemistry (USSR) (English Transl.) 19:349–56. Chem. Abstr. 48:12249. 1954.

17723. —— and —— 1956: Organic acid formation in germination of legume seeds. Biochemistry (USSR) (English Transl.) 21:594–97.

17724. ——, A. N. Panteleev, and T. A. Mazurova. 1950: The formation of and changes in the organic acids of sprouting cereal seeds. Tr. Leningr. Obshch. Estestvoispytatelei 70(3):49–67. Chem. Abstr. 49:7066. 1955.

17725. Soliman, G., and Z. Mustafa. 1943: The saponin of fenugreek seeds. Nature 151:195–96. Chem. Abstr. 37:2418–19. 1943.

17726. —— and W. Saleh. 1954: Constituents of the seeds of *Corchorus olitorius*, L. II. Isolation of β-sitosterol and corchorolic acid. J. Chem. Soc. 1954:1506–8.

Solis, M. Acosta-; *see* Acosta-Solis, M.

17727. Soliven, F. A. 1934: The proximate chemical composition of the seed and oil of Philippine oil-bearing seeds. I. *Pongamia pinnata* Merr. Philippine Agriculturist 23:576–87.

17728. —— and I. Villafuerte, Jr. 1935: The proximate chemical composition of the seed and oil of Philippine oil-bearing seeds. II. *Sterculia foetida* Linn. Philippine Agriculturist 23:666–80.

Soliven, F. A. 14944.

17729. Sollazzo, G. 1937: Excitation of the growth activity of plants. Boll. Chim. Farm. 76:368, 371–73. Chem. Abstr. 31:8601. 1937.

Sollner, K. 7640.

17730. Solly, E. 1845: Über den Einfluss der galvanischen Elektricität auf das Keimen der Samen. [Abstract.] Allgem. Thüringische Gartenztg., Erfurt 4:171.

Solntsev, A. I. 19912.

17731. Solodkowska, W. 1938: Application of the adiabatic calorimeter to the investigation of the energetics of bursting seeds. Roczniki Chem. 18:771–83. Chem. Abstr. 33:6392. 1939.

17732. —— 1939: Energetics of germinating seeds. II. Roczniki Chem. 19:277–88. Chem. Abstr. 34:2880. 1940.

17733. —— and K. Rudowska. 1932: Measurements of the heat of germination of seeds by means of the adiabatic microcalorimeter. I. Bull. Intern. Acad. Polon. 1932A:95–108. Chem. Abstr. 27:3740–41. 1933.

17734. Solomon, A. W. 1954: Growing

camellias from seed. Am. Camellia Quart. 9:7–8.

17735. Solomon, B. 1950: Inhibiting effect of autoclaved malt preventing the in vitro growth of *Datura* embryos. Am. J. Botany 37:1–5.

17736. Solomon, M. E. 1957: Estimation of humidity with cobalt thiocyanate papers and permanent colour standards. Bull. Entomol. Res. 48:489–506.

17737. Solomon, S. 1952: Studies in the physiology of phanerogamic parasitism with special reference to *Striga lutea* Lour. and *S. densiflora* Benth. on *Andropogon sorghum* Hack. II. The influence of temperature and moisture on the germination and growth of *S. lutea* and *S. densiflora*. Proc. Indian Acad. Sci. Sect. B 36:198–214. Biol. Abstr. 27:28516. 1953.

Solomon, S. 11522.

Solovev, A. K. 9343.

17738. Solovev, V. A. 1957: Aqueous method of cleaning seeds of coniferous species. Lesn. Khoz. 10(2):69–71.

Solovev, V. P. 5989.

17739. Soloveva, M. A. 1953: On conditions for prolonged storage of fruit crop seeds. Agrobiologiya 1953(1):81–93.

Sols, A, 13277.

Solt, M. L. 2048.

Soltys, J. A. 10568.

Soma, K. 6861.

17740. Somers, G. F., M. H. Coolidge, and K. C. Hamner. 1945: The distribution of thiamine and riboflavin in wheat grains. Cereal Chem. 22:333–40.

Sommariva, D. 4295.

17741. Somon, E. 1962: Amylogenese, amylolyse et verdissement, au cours de la germination chez diverses espèces de *Citrus*. Bull. Soc. Botan. France 109:157–59.

17742. Sonavne, K. M. 1928: Longevity of crop seeds. Agr. J. India 23:271–76.

17743. —— 1934: Longevity of crop seeds. II. Agr. Livestock India 4:287–92.

Sondheimer, F. 10643.

17744. Sone, K. 1958: Studies on the quality of the plant food. II. Influence of variety and environment upon the riboflavin content of hulled rice. (2) In special reference to its effective factors. Tohoku J. Agr. Res. 9:179–97.

17745. —— 1959: Distribution of riboflavine in the kernel of rice during the ripening period of rice plant. J. Agr. Chem. Soc. Japan 33:946–49. Chem. Abstr. 57:14184. 1962.

Song, P. S. 4014.

17746. Soni, P., S. C. Gupta, and J. S. Aggarwal. 1949: Chemical examination of the seeds of *Cucumis sativus*, Linn. J. Sci. Ind. Res. (India) 8B:210–13.

17747. Sonwalkar, M. S. 1951: A study of jackfruit seeds. Indian J. Hort. 8:27–30. Biol. Abstr. 26:9684. 1952.

Soost, R. K. 3545, 3546.

Sorber, D. G. 19426.

Sorbier, A. Gayte-. 11791.

17748. Sorenson, J. W., Jr., and L. E. Crane. 1960: Drying rough rice in storage. Texas Agr. Expt. Sta. Bull. B-952. 19 pp.

17749. —— and M. G. Davenport. 1951: Drying and storing flaxseed in Texas. Agr. Eng. 32:379–82.

17750. ——, G. L. Kline, L. M. Redlinger, M. G. Davenport, and W. H. Aldred. 1957: Research on farm drying and storage of sorghum grain. Texas Agr. Expt. Sta. Bull. 885. 23 pp.

17751. —— and L. H. Wilkes. 1959: Storage of cottonseed for planting purposes. Texas Agr. Expt. Sta. Misc. Publ. 326. 8 pp.

Sorenson, J. W., Jr. 8781.

17752. Sorger-Domenigg, H., L. S. Cuendet, C. M. Christensen, and W. F. Geddes. 1955: Grain storage studies. XVII. Effect of mold growth during temporary exposure of wheat to high moisture contents upon the development of germ damage and other indices of deterioration during subsequent storage. Cereal Chem. 32:270–85.

17753. ——, ——, and W. F. Geddes. 1955: Grain storage studies. XX. Relation between viability, fat acidity, germ damage, fluorescence value, and formazan value of commercial wheat samples. Cereal Chem. 32:499–506.

Sorger-Domenigg, H. 7415.

17754. Soriano, A. 1953: Estudios sobre germinación. I. Rev. Invest. Agr. 7:315–40.

17755. —— 1961: Germination of *Stipa neaei* in relation to imbibition and moisture level. Arid Zone Res. Vol. XVI. Plant-Water Relationships in Arid and Semi-Arid Conditions. A Symposium. UNESCO, New York, pp. 261–64. Biol. Abstr. 39:7713. 1962.

Sorm, F. 17147, 17148.

Sorokina, O. N. 10549.

17756. Sosa, A., and R. Duperon. 1952: Sur quelques constituants des semences de

Raphanus sativus L. Compt. Rend. 235: 82–84.

Sosa, A. 17760, 17761.

17757. Sosa-Bourdouil, C. 1944: Sur le pouvoir oxydant des graines au cours du développement. Bull. Museum Natl. Hist. Nat. Paris 16:556–58.

17758. —— 1949: Chemism of tetraploidic *Sinapis* seeds. Bull. Soc. Botan. France 96:39–41. Chem. Abstr. 43:8459. 1949.

17759. —— 1949: The chemistry of the seeds of tetraploid *Datura*. Bull. Museum Natl. Hist. Nat. Paris 21:174–76. Chem. Abstr. 44:3098. 1950.

17760. ——, A. Brunel, and A. Sosa. 1941: Changes in the composition of the hulls and the seeds of the *Soja* plant in the course of its development. Compt. Rend. 212:1049–51. Chem. Abstr. 36:6198. 1942.

17761. —— and A. Sosa. 1962: Action des rayons γ sur l'activite enzymatique des graines de *Pisum* relative a la formation d'acetonie. Compt. Rend. 225:1767–69.

17762. Sosedov, N. I., and V. A. Shvetsova. 1953: Moisture determination in individual seeds. Tr. Vses. Nauchn.-Issled. Inst. Zerna i Produktov ego Pererabotki 1953(25):5–10. Chem. Abstr. 50:5103. 1956.

Sossiedov, N. I. 2565, 2566.

Sostaric-Pisacic, K. von; *see* Von Sostaric-Pisacic, K.

17763. Soster, N. 1933: Sul pigmento dei semi di *Jacaranda mimosaefolia* Don. Note Botan. Biol., Catania. 7 p. Botan. Centralbl. 169:119. 1935.

Soster, N. 16822.

Sotty, O. 14346.

Soubihe, J. 7870.

17764. Soucek, Z. 1944: Results of seed germination tests of 100 tree species. Lesnická Práce 23:11–17. Biol. Abstr. 21:4369. 1947.

17765. Soueges, R. 1934: Nutrition of the angiosperm embryo. Nature 133:764.

17766. —— 1954: Embryogénie et classification. Congr. Intern. Botan. Paris Rappt. Commun. 8:252–54.

17767. Souillet, H. 1935: Growing lilies from seed. Horticulture 13(3):61–62.

17768. Sousa, E. O. E., and M. V. E. Silva. 1942: Influencia da temperatura na germinação de algumas formas cultivadas de arroz. Agron. Lusitana 4:323–38.

Sousa-Monteiro-Cerca-Miguel, M. C. de; *see* de Sousa-Monteiro-Cerca-Miguel, M. C.

Southgate, B. J. 19879.

17769. Southwick, F. W., W. D. Weeks, E. Sawada, and J. F. Anderson. 1962: The influence of chemical thinners and seeds on the growth rate of apples. Proc. Am. Soc. Hort. Sci. 80:33–42.

Sova, C. 9838.

17770. Sowinski, J. 1957: Interrelation between the size of spring barley grain and its protein content. Roczniki Nauk Rolniczych Ser. A 77:115–30. Chem. Abstr. 52:18680. 1958.

Spada, A. 14949.

17771. Spadaro, J. J., P. H. Eaves, and E. A. Gastrock. 1960: Direct extraction of Jojoba seed. J. Am. Oil Chemists Soc. 37:121–24.

Spadaro, J. J. 5096, 10893.

17772. Spadoni, M. A., and G. Tecce. 1953: Metabolism of *Ricinus communis* seeds. IV. Organic acids. Quaderni Nutriz. 13:330–40. Chem. Abstr. 50:7968. 1956.

17773. —— and —— 1954: Action of malonate and of fluoroacetate on the respiration of the endosperm of seeds of *Ricinus communis*. Boll. Soc. Ital. Biol. Sper. 30:724–27. Chem. Abstr. 49:1886. 1955.

Spadoni, M. A. 2269, 4534, 15623.

17774. Spaeth, J. N. 1934: A physiological study of dormancy in *Tilia* seed. Cornell Univ. Agr. Expt. Sta. Mem. 169. 78 pp.

17775. —— 1935: Germination of lime seed. Nature 135:796. Trop. Agr. (Trinidad) 13:27. 1936.

17776. —— 1936: "Catalase ratio" as an indicator of viability of tree seed. Am. J. Botany (Suppl.) 23:691–92.

17777. —— and M. Afanasiev. 1939: The effect of sterilization with calcium hypochlorite on germination of certain seeds. J. Forestry 37:371–72.

Spahr, W. 11591.

17778. Spafford, W. J. 1930: Effects on germination of exposing wheat during heat waves. J. Dept. Agr. S. Australia 33:879–80, 882.

17779. Spain R. L. 1943: Cyclamens from seed. Home Garden 2:52–54.

17780. Spaldon, E., and V. Pevna. 1962: The effect of the stage of ripeness, and after-harvest ripening of the fruits on the quality of seed in vegetable paprika. Pol'nohospodarstvo 9:263–72. Biol. Abstr. 44:8109. 1963.

17781. Spalek, V. 1949: How to cultivate poplars from seed. Sb. Česk. Akad. Zeměděl. 22:523–32.

17782. Sparenberg, H. 1963: The artificial drying of unthreshed seed crops. Proc. Intern. Seed Testing Assoc. 28:785–92.

Sparks, J. E. 15298.

17783. Sparmann, G. 1961: Morphologische und biochemische Untersuchungen an vernalisierten und nicht vernalisierten Gerstenpflanzen. Planta 56:447–74.

17784. —— 1961: Biochemische Untersuchungen über Vernalisation und Blütenbildung der Gerste. Planta 57:176–201.

17785. Sparrow, A. H., J. P. Binnington, and V. Pond. 1958: Bibliography on the effects of ionizing radiations on plants 1896–1955. Biol. Dept., Brookhaven Natl. Lab., Upton, N.Y. 219 pp.

Sparrow, A. H. 7831.

Spasskaia, T. S., 19022.

Speck, R. M. 16201.

Spector, H. 12231.

Speers, P. C. 5337.

Speirs, M. 5826.

Speiser, P. 19203.

Spence, J. W. 17561.

17786. Spencer, D. 1959: A DPNH-specific nitrate reductase from germinating wheat. Australian J. Biol. Sci. 12:181–91.

Spencer, D. 5457.

Spencer, E. Y. 15836.

17787. Spencer, G. E. L. 1931: Seed storage and germination. The use of cool storage in retaining the germinating power of some oily seeds. Trop. Agr. (Trinidad) 8:333.

17788. Spencer, H. M. 1926: Laboratory methods for maintaining constant humidity. McGraw-Hill, New York. Intern. Critical Tables I:67–68.

17789. Spencer, J. L., and A. F. Blakeslee. 1954: Induced pollen lethals from seeds of *Datura stramonium* exposed to radiation from a nuclear detonation. Proc. Natl. Acad. Sci. U.S. 40:441–46.

17790. —— and —— 1955: Induced pollen lethals from seeds of *Datura stramonium* exposed to different radiations. Proc. Natl. Acad. Sci. U.S. 41:307–12.

17791. ——, E. Cabanillas, and E. Aponte. 1956: Effect of culin age on the germination of napiergrass (*Pennisetum purpureum* Schum.) seed. Agron. J. 48:318–19.

17792. ——, W. R. Singleton, and A. F. Blakeslee. 1953: Induced pollen lethals from seeds of *Datura stramonium* treated with thermal neutrons. Proc. Natl. Acad. Sci. U.S. 39:288–92.

17793. Spencer, J. T., and E. N. Fergus. 1942: Floret and seed types in Kentucky bluegrass in relation to yield and quality of seed. J. Am. Soc. Agron. 34:1035–41.

Spencer, M. 13287.

17794. Sperlich, A. 1908: Ist bei grünen Rhinanthaceen ein von einem pflanzlichen Organismus ausgehander äusserer Keimungsreiz nachweisbau? Ber. Deut. Botan. Ges. 26a:574–87.

17795. —— 1919: Die Fähigkeit der Linienerhaltung (phyletische Potenz), ein auf die Nachkommenschaft von Saisonpflanzen mit festem Phythmus ungleichmässig übergehender Faktor auf Grund von Untersuchungen über die Keimungsenergie, Rhythmik und Variabilität in reinen Linien von *Alectorolophus hirsutus* All. Sitzber. Akad. Wiss. Wien Abt. I 128:379–475.

17796. —— 1919: Über den Einfluss des Quellungszeitpunktes, von Treibmitteln und des Lichtes auf die Samenkeimung von *Alectorolophus hirsutus* All.; Charakterisierung der Samenruhe. Sitzber. Akad. Wiss. Wien Abt. I 128:477–500.

17797. Sperry, O. E., J. W. Dollahite, J. Morrow, and G. O. Hoffman. 1955: Texas range plants poisonous to livestock. Texas Agr. Expt. Sta. Bull. 796. 47 pp.

17798. Spicer, P. B., and L. A. Dionne. 1961: Use of gibberellin to hasten germination of *Solanum* seed. Nature 189:327–28.

17799. Spickett, R. G. W., and J. A. Squires. 1954: *Robinia pseudoacacia* seeds from cypress. Colonial Plant Animal Prod. 4:57–59.

Spickett, R. G. W. 6198.

17800. Spies, J. R., and E. J. Coulson. 1943: The chemistry of allergens. VIII. Isolation and properties of an active protein-polysaccharidic fraction, CB-1A, from castor beans. J. Am. Chem. Soc. 65:1720–25.

17801. ——, ——, D. C. Chambers, H. S. Bernton, and H. Stevens. 1944: The chemistry of allergens. IX. Isolation and properties of an active, carbohydrate-free protein from castor beans. J. Am. Chem. Soc. 66:748–53.

17802. ——, ——, and H. Stevens. 1944: The chemistry of allergens. X. Comparison of chemical and immunological properties of CB-1A preparations from domestic castor beans and Brazilian castor bean pomace. J. Am. Chem. Soc. 66:1798–99.

Spiesz, K. von; *see* Von Spiesz, K.

17803. Spingarn, J. E. 1935: Growing clematis from seed. Horticulture 13(2):26.
Spingler, H. 7911, 10142.
Spinks, J. W. T. 1369.
17804. Spirchez, Z. 1958: Research in certain morphological characteristics of the seeds of the spruce (*Picea excelsa* Link.). Studii Cercetari Agron. 9:65–75.
Splittgerber, A. 7717.
Spoerri, P. E. 11785.
17805. Sporne, K. R. 1954: A note on nuclear endosperm as a primitive character among dicotyledons. Phytomorphology 4:275–78.
17806. Spracher, M. L. 1928: Further studies of corn germination. Proc. Assoc. Offic. Seed Analysts 19:27–35.
17807. Spragg, S. P., P. M. Lievesley, and K. M. Wilson. 1962: The relationship between glutathione and protein sulphydryl groups in germinating pea seeds. Biochem. J. 83:314–18.
17808. —— and E. W. Yemm. 1959: Respiratory mechanisms and the changes of glutathione and ascorbic acid in germinating peas. J. Exptl. Botany 10: 409–25.
17809. Sprague, G. F. 1932: The inheritance of colored scutellums in maize. U.S. Dept. Agr. Tech. Bull. 292. 43 pp.
17810. —— 1934: Experiments on iarovizing corn. J. Agr. Res. 48:1113–20.
17811. —— 1936: The relation of moisture content and time of harvest to germination of immature corn. J. Am. Soc. Agron. 28:472–78.
17812. —— and B. Brimhall. 1949: Quantitative inheritance of oil in the corn kernel. Agron. J. 41:30–33. Biol. Abstr. 23: 13776. 1949.
17813. —— and W. T. Federer. 1951: A comparison of variance components in corn yield trials. II. Error, year × variety, location × variety, and variety components. Agron. J. 43:535–41.
17814. —— and M. T. Jenkins. 1948: The development of waxy corn for industrial use. Iowa State Coll. J. Sci. 22:205–13.
17815. ——, P. A. Miller, and B. Brimhall. 1952: Additional studies of the relative effectiveness of two systems of selection for oil content of the corn kernel. Agron. J. 44:329–31.
17816. —— and J. F. Schuler. 1961: The frequencies of seed and seedling abnormalities in maize. Genetics 46:1713–20.
Sprague, G. F. 7719, 8870, 16780, 16781.
Sprague, H. B. 14940.
17817. Sprague, V. G. 1940: Germination of freshly harvested seeds of several *Poa* species and of *Dactylis glomerata*. J. Am. Soc. Agron. 32:715–21.
17818. —— 1944: The effects of temperature and day length on seedling emergence and early growth of several pasture species. Soil Sci. Soc. Am. Proc. 8:287–94. Biol. Abstr. 18:22590. 1944.
17819. Spratt, E. D., R. I. H. McKenzie, and A. L. D. Martin. 1963: The effect of date of seeding on a number of plant characteristics for five flax varieties. Can. J. Plant Sci. 43:583–88.
17820. Spreafico, L. 1961: The distribution of certified seeds; the control of seeds; analysis and certification. Agricoltura 10(5):46–49.
Spreafico, L. 4533.
17821. Sprecher, A. 1915: Samen und Keimung von *Hevea brasiliensis*. Bull. Jardin Botan. Buitenzorg 19. 112 pp. Expt. Sta. Rec. 34:639. 1916.
17822. —— 1919: Étude sur la semence et la germination du *Garcinia mangostana* L. Rev. Gen. Botan. 31:513–31, 609–34.
17823. Spring, F. G. 1926: Report on longevity of *Hevea brasiliensis*. Malayan Agr. J. 14:18–22.
Spring, F. S. 2291.
17824. Spring, J. J. 1962: The propagation of magnolias from seed. Calif. Hort. Soc. J. 23:48–50.
17825. Spurny, M. 1954: The cell-wall structure of the epidermal cells of the pea seed-coat. Preslia 26:79–88. Biol. Abstr. 29:12081. 1955.
17826. —— 1954: The "light line" of the epidermal cells of the pea seed-coat (*Pisum sativum*). Preslia 26:139–42. Biol. Abstr. 29:12082. 1955.
17827. —— 1954: The permeability of the testa and inequality in the swelling-velocity of pea seeds (*Pisum sativum*). Preslia 26:239–62. Biol. Abstr. 29:17276. 1955.
17828. —— 1961: Das Eindringen von Wasser durch die Samenschale der Erbse (*Pisum sativum* L.). Res. Film 4:41–47.
17829. Spurr, A. R. 1949: Histogenesis and organization of the embryo in *Pinus strobus* L. Am. J. Botany 36:629–41.
17830. Spurr, S. H. 1944: Effect of seed weight and seed origin on the early development of eastern white pine. J. Arnold Arboretum 25:467–80.
Spurr, S. H. 15134.
Spyhalski, E. J. 13583.

17831. Squibb, R. L., R. Bressani, and N. S. Scrimshaw. 1957: Nutritive value of Central American corns. V. Carotene content and vitamin A activity of three Guatemalan yellow corns. Food Res. 22:303–7.

Squibb, R. L. 3001.

Squires, J. A. 16023–16025, 17799.

Srb, A. M. 19881.

17832. Sreenivasan, A. 1939: Studies on quality in rice. IV. Storage changes in rice after harvest. Indian J. Agr. Sci. 9:208–22. Biol. Abstr. 14:3406. 1940.

17833. —— and V. Sadasivan. 1942: Nutritive value of the protein and mineral constituents of dry and wet cultivated rices. Cereal Chem. 19:47–55.

17834. —— and S. D. Wandrekar. 1950: Biosynthesis of vitamin C during germination. I. Effect of various environmental and cultural factors. Proc. Indian Acad. Sci. Sect. B 32:143–63. Biol. Abstr. 25:21577. 1941.

17835. —— and —— 1950: Biosynthesis of vitamin C during germination. II. Concomitant changes in ascorbic acid oxidase, glutathione, nicotinic acid, amylase, reducing sugars, phosphatase, and free phosphates. Proc. Indian Acad. Sci. Sect. B 32:231–51. Chem. Abstr. 47:1797. 1953.

17836. —— and —— 1951: Excretion of vitamin C during germination of leguminous seeds. Proc. Indian Acad. Sci. Sect. B 34:267–71. Biol. Abstr. 26:13220. 1952.

Sreenivasan, A. 7283, 7284, 11508, 11876, 14180.

Screenivasan, M. 2404.

17837. Sreenivasaya, M., and N. Narayana. 1936: The sandal seed, its oil and proteins. J. Indian Inst. Sci. 19A(1):1–8.

Sreenivasaya, M. 116, 2422.

Srinivasamurthy, V. 10099.

Srinivasan, A. R. 12705.

Srinivasan, K. S. 14062.

Srinivasan, M. 10154, 11551, 14062, 15956, 16807, 18126, 18127.

17838. Srinivasan, V., and L. Anavaradham. 1962: Tricotyledony in *Prosopis juliflora* (mesquite). Madras Agr. J. 49:312.

17839. —— and P. Gopalakrishnan. 1956: Quicker and better germination in snakegourd and bittergourd. Madras Agr. J. 43:71–73.

17840. Srinivasan, V. K., and T. S. Padmanabhan. 1964: Inheritance of glume colour and pubescence in inter-varietal crosses of *Triticum dicoceum*. Current Sci. (India) 33:156.

17841. Sripleng, A., and F. H. Smith. 1960: Anatomy of the seed of *Convolvulus arvensis*. Am. J. Botany 47:386–92.

17842. Srivastava, D. P., and E. L. Pinnell. 1963: Germination studies in grain sorghum. Missouri Agr. Expt. Sta. Res. Bull. 828. 24 pp.

Srivastava, H. C. 17637.

Srivastava, J. P. 11518.

Srivastava, K. C. 11519.

17843. Srivastava, L. M., and D. Williamson. 1958: Germination of seeds inside the fruits. Sci. Cult. (Calcutta) 23:428. Biol. Abstr. 32:38340. 1958.

Srivastava, P. D. 7344.

17844. Srivastava, R. K. 1953: The evolutionary concept of the morphology of endosperm. Botan. Mag. (Tokyo) 28:88–97.

17845. Srivastava, R. P. 1960/61: Preliminary studies in the sex distribution, growth behavior and the influence of χ-naphthaleneacetic acid sprays on the size, seed number and weight of jack fruit (*Artocarpus heterophyllus* Lam.). J. Sci. Res. Banaras Hindu Univ. 11:213–21.

Srivastava, S. N. 2398, 2399.

Sroczynska, A. 9762.

17846. Sroelov, R. 1940: On germination inhibitors. IV. Germination inhibitors of *Sinapis alba* and other seeds when enclosed in their fruit. Palestine J. Botany 2:33–37.

Ssarkissova, O. W. Federova-; *see* Federova-Ssarkissova, O. W.

17847. Ssawitzkij, B. F., and E. I. Charetschko. 1937: Über die Wirkung erhöter Temperatur auf Zuckerrübensamen. Wiss. Ber. Zuckerindustrie 2:9–15. Botan. Centralbl. 174:110. 1939.

Staaveren, M. C. van; *see* Van Staaveren, M. C.

Stacy, S. V. 6242.

17848. Stadelman, R. 1939: Consumption of teosinte seed by birds in Gutaemala. Science 89:461–62.

17849. Stäger, R. 1928: Samenverfrachtung durch Ameisen in der alpinen Stufe. Ber. Schweiz. Botan. Ges. 37:9–14. Biol. Abstr. 7:2677. 1933.

17850. —— 1929: Die Verbreitung der Samen von *Trifolium thalii* durch Ameisen in der alpinen Stufe. Ber. Schweiz. Botan. Ges. 38:1–4. Biol. Abstr. 7:2678. 1933.

Stählin, A. 3109.

Staffeld, U. 16897.

17851. Stahel, G. 1941: The effect of carbon

bisulphide on the germination of paddy. Trop. Agr. (Trinidad) 18:200–3.

17852. —— 1944: The jungle tree of dangling seeds. N.Y. Botan. Garden J. 45:265–68.

17853. Stahevitch, B. Y., and N. N. Koulechoff. 1928: Observations biologiques sur le developpement de la cuscute dans les champs. Intern. Seed Testing Conf. Rome 5:350–63.

Stahl, B. M. 10439.

17854. Stahl, C. 1931: Comparative experiments between the laboratory and the field germination of seed. Proc. Intern. Seed Testing Assoc. 15/17:75–143.

17855. —— 1931: Untersuchungen über das Vorkommen von Unkrautsamen in Samenproben. Tidsskr. Planteavl 38:103–30. Proc. Intern. Seed Testing Assoc. 4:91–93.

17856. —— 1933: Laboratory and field germination of cabbage seed. Proc. Intern. Seed Testing Assoc. 5:42–56.

17857. —— 1936: The importance of the germinating speed in the case of cruciferous seeds. Proc. Intern. Seed Testing Assoc. 8:46–53.

17858. —— 1937: On the shortening of the germination periods for cruciferous seeds. Proc. Intern. Seed Testing Assoc. 9:131–41.

17859. —— 1937: Latitudes in seed analyses. Proc. Intern. Seed Testing Assoc. 9:142–52.

17860. —— 1957: Investigations on the germination and some other qualities of beet seed. Proc. Intern. Seed Testing Assoc. 22:258–69.

17861. —— 1960: Report of the certificate committee. Proc. Intern. Seed Testing Assoc. 25:83–100.

Stahl, C. 17557.

17862. Stahler, L. M. 1936: Preliminary work on germination of *Agropyron caryopses*. Proc. Assoc. Offic. Seed Analysts 1935:56–59.

17863. Stahlin, A. 1960: Lebensdauer und Lagerfähigkeit von Samereien. Mitt. Deut. Landwirtsch. Ges. 75:1374–77.

Stahmann, M. A. 20102.

17864. Staikov, G. 1959: Erforschung der Samen-Keimruhe-Periode in Abhängigkeit vom Samen-Reifegrad. Plovdiv. Vissh Selskostopanski Inst. "Vasil Kolarov" Nauch Trudove 6:345–63. 1957/58.

Stainsby, W. J. 7622.

Staker, E. V. 13917.

Stakman, E. C. 4069.

17865. Staley, K. E. 1940: Effects of environmental factors on oxidizing enzymes of rose mallow seeds. Plant Physiol. 15:625–44.

Stanciu, N. 19142.

17866. Standen, J. H. 1946: A toxic substance occurring in certain maize cobs. Contrib. Boyce Thompson Inst. 14:277–81.

Standridge, N. 5347.

Stanescu, R. 15830.

Stanford, E. H. 11855.

17867. Staniforth, D. W., and R. E. Atkins. 1952: Effect of 2,4-D on yield and bushel weight of oats. Agron. J. 44:587–89.

17868. Stanislas, E., G. Bergal, and C. Moscarella. 1951: Action of S-guaiazulene on the germination and the development of the sprouts of some dicotyledons. Ann. Pharm. Franc. 9:608–11. Chem. Abstr. 46:3616. 1952.

Stanislawski, J. 13405.

Stankevich, A. K. 2782.

Stankovic, J. 9724.

17869. Stanley, R. G. 1957: Krebs cycle activity of mitochondria from endosperm of sugar pine seed (*Pinus lambertiana* Dougl.). Plant Physiol. 32:409–12.

17870. —— 1958: Gross respiratory and water uptake patterns in germinating sugar pine seed. Physiol. Plantarum 11:503–15.

17871. —— and E. E. Conn. 1957: Enzyme activity of mitochondria from germinating seedlings of sugar pine (*Pinus lambertiana* Dougl.). Plant Physiol. 32:412–18.

Stansbrey, J. J. 4495.

17872. Stansbury, M. F., A. F. Cucullu, and G. T. Den Hartog. 1954: Cottonseed contents variation. Influence of variety and environment on oil content of cottonseed kernels. J. Agr. Food Chem. 2:692–96.

17873. —— and J. D. Guthrie. 1947: Storage of cottonseed and peanuts under conditions which minimize changes in chemical composition. J. Agr. Res. 75:49–61.

17874. ——, ——, and T. H. Hopper. 1944: Analyses of peanut kernels with relation to U.S. standards for farmers' stock peanuts. Oil Soap 21:239–47. Expt. Sta. Rec. 95:459–60. 1946.

17875. ——, W. A. Pons, Jr., and G. T. Den Hartog. 1956: Relations between oil, nitrogen, and gossypol in cottonseed kernels. J. Am. Oil Chemists Soc. 33:282–86.

17876. ——, ——, and C. L. Hoffpauir. 1953: Phosphorus compounds in cottonseed kernels. Influence of variety of cottonseed and environment. J. Agr.

Food Chem. 1:75–78. Chem. Abstr. 47:5491. 1953.

Stanton, T. R. 3131, 4250, 4251, 12143, 12144.

17877. Stanway, V. 1952: Manual for beginners in seed analysis. Univ. Missouri Agr. Exp. Sta. 140 pp.

Stanway, V. M. 7705.

17878. Stapel, C. 1942: Disinfection of flaxseed. Nord. Jordbrugsforskn. 1942: 50–59. Chem. Abstr. 38:2441. 1944.

17879. Stapledon, R. G. 1920: Seed studies: red clover with special reference to the country of origin of the seed. J. Agr. Sci. 10:90–120. Botan. Abstr. 8:816. 1921.

17880. —— 1930: Modern seed testing. Sci. Progr. (London) 24:623–42.

17881. —— and M. Adams. 1919: The effect of drying on the germination of cereals. J. Min. Agr. (Gt. Brit.) 26:364–81.

17882. Starchenko, I. I. 1956: Seasonal peculiarities of the energy of seed germination of trees and shrubs. Botan. Zh. 41:1355–59.

Stark, S. M. 10664, 13115.

Stark, W. H. 107.

17883. Starker, C., and L. Starker. 1939: Raising hardy cyclamen from seed. Natl. Hort. Mag. 18:41–42.

Starker, L. 17883.

Starodubtseva, A. I. 11380, 11385.

Starr, G. E. 4487.

Starr, G. H. 3378.

Starrett, R. C. 1423, 1424.

17884. Stasauskaite, S., and M. Holcmanaite. 1960: Pre-planting treatment of corn seeds with solutions of certain minor elements. Lietuvos TSR Mokslu Akad. Darbai Ser. B 1960(4):181–92.

Staten, H. W. 17650.

Stathi-Photiadou, A. K. 15326.

Stavely, H. E. 6308.

17885. Stavy, L., and A. M. Mayer. 1962: Coupled oxidation of ascorbic acid by phenols and phenolase from lettuce seeds. Bull. Res. Council Israel Sect. D. 11:31–34.

17886. Stearns, F., and J. Olson. 1958: Interactions of photoperiod and temperature affecting seed germination in *Tsuga canadensis*. Am. J. Botany 45:53–58.

17887. Stebbins, G. L., Jr. 1941: Apomixis in the angiosperms. Botan. Rev. 7:507–42. Expt. Sta. Rec. 86:607. 1942.

17888. —— and E. B. Babcock. 1939: The effect of polyploidy and apomixis on the evolution of species in *Crepis*. J. Heredity 30:519–30.

17889. —— and J. A. Jenkins. 1939: Aposporic development in the North American species of *Crepis*. Genetica 21:191–224. Biol. Abstr. 13:14365. 1939.

Stebbins, G. L., Jr. 1617.

Stec, F. 1959.

17890. Steece, H. M. 1929: Annotated bibliography on the storage of cotton-seed and of seed cotton. Agr. J. India 24:127–34.

17891. Steed, T. J. 1940: Growing dahlias from seed. Flower Grower 27(2):64.

Steen, K. M. 12684.

Steenbock, H. 1279, 6133, 8989, 18441.

Steenis, C. G. G. J. van; *see* Van Steenis, C. G. G. J.

17892. Stefan, A. 1956: The action of potassium bromide as chemical stimulant with respect to the permeability of wheat seeds for neutral red. Analele Univ. "C. I. Parhon," Ser. Stiint. Nat. 1956:129–32. Chem. Abstr. 52:14765. 1958.

Stefanescu, A. 15875.

Stefanescu, E. 103.

17893. Stefanov, B. 1951: Hard seeds of some leguminous plants. Invest. Biol. Inst. Sofia 2:3–43. Brit. Abstr. B III:328. 1952.

Stefansson, B. F. 9134.

17894. Stefansson, E. 1953: On problems in connection with collecting and extracting seeds from pine cones. Skogen (Stockholm) 40:12–14.

17895. Steffek, E. F. 1939: Growing the lovely gentians from seed. Horticulture 17(19):411–12.

17896. Steffen, K. 1952: Die Embryoentwicklung von *Impatiens glanduligera* Lindl. Flora (Jena) 139:394–461.

17897. —— 1954: Cytological observations on endosperm haustories. Congr. Intern. Botan. Paris Rappt. Commun. 8:250. Biol. Abstr. 29:26874. 1955.

Steffensen, D. M. 11848.

Stefl, M. 6177.

17898. Steger, A., and J. Van Loon. 1933: Fat of the seed of *Picramnia* sow. Rec. Trav. Chim. 52:593–600. Chem. Abstr. 27:5204. 1933.

17899. —— and —— 1934: The fatty oil of quince seed, *Cydonia vulgaris*. Rec. Trav. Chim. 53:24–27. Chem. Abstr. 28:2207–8. 1934.

17900. —— and —— 1934: The fatty oil of *Cassia occidentalis*, Linn. seed. Rec. Trav. Chim. 53:28–30. Chem. Abstr. 28:2207. 1934.

17901. —— and —— 1934: The fatty oil of millet, *Panicum miliaceum.* Rec. Trav. Chim. 53:41–44. Chem. Abstr. 28:2207. 1934.

17902. —— and —— 1935: Essang oil. Rec. Trav. Chim. 54:988–94. Chem. Abstr. 30:1597–98. 1936.

17903. —— and —— 1937: Fatty oil from the seeds of *Valerianella olitorea*, Poll. Soc. Chem. Ind. J. 56:298–300T. Analyst 62:809–10. 1937.

17904. —— and —— 1940: Oil from the seeds of *Canarium commune* L. Rec. Trav. Chim. 59:168–72. Analyst 65:230. 1940.

17905. —— and —— 1941: Fatty oils from seeds of *Isatis tinctoria* L. Rec. Trav. Chim. 60:947–49. Chem. Abstr. 39:1999. 1945.

17906. —— and —— 1942: Das fette Öl der Samen von *Onopordon acanthium* L. Rec. Trav. Chim. 61:120–22. Botan. Centralbl. 178:14. 1942.

17907. —— and —— 1942: Das fette Öl der Samen von *Hesperis matronalis* L. Rec. Trav. Chim. 61:123–26.

17908. —— and —— 1943: Das fette Öl der Samenkerne von *Sterculia foetida* L. Fette Seifen Anstrichmittel 50:305–9. Biol. Abstr. 18:20438. 1944.

17909. ——, ——, and C. Smelt. 1936: Oil from *Ricinus zanzibarinus.* Soc. Chem. Ind. J. 55:41–42T. Analyst 61:261–62. 1936.

Steger, A. 18983.

17910. Stehlik, W. 1917: Über die Wirkung von Röntgenstrahlen auf die Keimung des Rübensamens und das Wachstum der Zucker- und Futterrübe. Z. Zuckerind. Böhmen 41:424–27.

Stein, G. 2645, 6126, 6127.

17911. Stein, O. L., and H. Quastler. 1963: The use of tritiated thymidine in the study of tissue activation during germination in *Zea mays.* Am. J. Botany 50:1006–11.

17912. Stein, W. I. 1951: Germination of Noble and Silver fir seed on snow. J. Forestry 49:448–49.

Steinbach, K. J. 18263.

17913. Steinbauer, C. E. 1938: Methods of scarifying sweet potato seed. Proc. Am. Soc. Hort. Sci. 35:606–8.

17914. ——, J. H. Beattie, and E. T. Batten. 1940: Influence of mass selection within certain large-seeded Virginia-type peanut varieties. Proc. Am. Soc. Hort. Sci. 37:685–88. Biol. Abstr. 14:14266. 1940.

17915. ——, J. M. McCown, E. T. Batten, and E. E. Hall. 1942: Performance of some large-seeded and small-seeded peanut varieties and selections in Virginia and South Carolina. Proc. Am. Soc. Hort. Sci. 41:240–44. Biol. Abstr. 18:3287. 1944.

17916. —— and G. P. Steinbauer. 1932: Effects of temperature and desiccation during storage on germination of the seeds of the American elm (*Ulmus americana* L.). Proc. Am. Soc. Hort. Sci. 28:441–43.

17917. Steinbauer, G. P. 1937: Dormancy and germination of *Fraxinus* seeds. Plant Physiol. 12:813–24.

17918. —— 1942: Use of potassium nitrate in overcoming dormancy of seed of cultivated mustard. Assoc. Offic. Seed Analysts News Letter 16(3):19.

17919. —— 1943: Frequency of polyembryony in Fraxinus seeds. Botan. Gaz. 105: 285.

17920. —— 1944: The embryo excision method of determining seed viability with particular reference to *Fraxinus* and *Sparganium.* Assoc. Offic. Seed Analysts News Letter 18(6):4–6.

17921. —— 1946: Viability of seeds from weed plants sprayed with herbicides. Proc. Assoc. Offic. Seed Analysts 36:105–6.

17922. —— 1957: Interaction of temperature and moistening agents in the germination and early development of potato seedlings. Am. Potato J. 34:89–93.

17923. —— 1958: Physiological problems in the future development of seed technology. Assoc. Offic. Seed Analysts, Fifty Years Seed Testing 1908/58:83–86.

17924. —— 1959: Interaction of light, temperature and moistening agent in seed germination. Biology 1959. A report for the year 1958/59 on the research and other activities of the Div. Biol. Calif. Inst. Technol.:97–98.

17925. —— and B. Grigsby. 1956: Interaction of thermoperiod, light, and substrate in the germination of seeds of pokeweed, *Phytolacca americana* L. Plant Physiol. (Suppl.) 31:xxxvi–xxxvii.

17926. —— and —— 1957: Field and laboratory studies on the dormancy and germination of the seeds of chess (*Bromus secalinus* L.) and downy bromegrass (*Bromus tectorum* L.). Weeds 5:1–4.

17927. —— and —— 1957: Interaction of temperature, light, and moistening agent in the germination of weed seeds. Weeds 5:175–82.

17928. —— and —— 1959: Methods of obtaining field and laboratory germination of seeds of bindweeds, Lady's Thumb and Velvet leaf. Weeds 7:41–46.

17929. —— and O. Neal. 1948: Dormancy and germination of seeds of the bur reeds, *Sparganium* spp. Mich. Acad. Sci. Papers 34:33–37.

Steinbauer, G. P. 4768, 4904, 17916.

Steinberg, M. P. 19094.

17930. Steinberg, R. A. 1960: Production and germination of *Nicotiana* seed grown under various greenhouse conditions. Tobacco 151(21):19–22. Tobacco Abstr. 5:73. 1961.

17931. Steinberger, J. 1960: Beitrag zur Unterscheidung von *Torilis nodosa* und *Torilis anthriscus*. Proc. Intern. Seed Testing Assoc. 25:837–41.

17932. Steinberger, J. 1963: Eine neue Methode zur Unterscheidung der *Allium*-Arten am Samen. Angew. Botan. 37:7–13.

17933. Steinegger, E. 1951: Der Alkaloidgehalt tetraploider *Datura*-Spezies. Heteroploidie-Versuche an Arzneipflanzen. Pharm. Acta Helv. 26:188–94. Biol. Abstr. 26:20318. 1952.

17934. —— 1953: Reproduction and alkaloid content of polyploid *Lobelia cliffortiana*. Bull. Galenica 16:214–16. Chem. Abstr. 49:7063. 1955.

17935. —— and H. Leupi. 1955: Influence of plant substances on the root growth of *Allium cepa* and germination of *Lepidium sativum*. I. Pharm. Acta Helv. 30:452–61. Chem. Abstr. 50:10194. 1956.

17936. Steinhaus, E. A., and C. R. Bell. 1953: The effect of certain microorganisms and antibiotics on stored-grain insects. J. Econ. Entomol. 46:582–98.

17937. Steinhubel, G. 1959/60: Propagation of tuliptrees (*Liriodendron tulipifera* L.) by native seeds in eastern Czechoslovakia. Mitt. Deut. Dendrol. Ges. 61:50–63.

17938. —— 1961: The factors of inhibition in the reproduction of *Liriodendron tulipifera* by seeds from Slovakia. Biol. Prace 7(5):3–87.

Steklova, M. M. 17348.

17939. Stellfeld, C. 1938: Seed of avocado as an aphrodisiac. Tribuna Farm. (Brazil) 6:59. Pharm. Abstr. 6:238. 1940.

Stelling, O. 2175.

17940. Stelter, J., K. Wolters, and I. Von Der Straten. 1957: Einfluss von Ultraschall auf die Keimung von Saatgut und die Entwicklung der Pflanze sowie Prüfung der Verwendungsmöglickeit der Ultraschalls in der Phytopathologie. Nachrichtenblatt 9:166–70.

17941. Stelzner, G. 1942: Zur Frage der Virusübertragung durch Samen, insbesondere des X-, Y- und Blattrollvirus der Kartoffel. Züchter 14:225–33. Biol. Abstr. 22:25492. 1948.

Stelzner, G. 16521.

17942. Stenar, H. 1951: Zur Embryologie der Gattung *Phaedranassa* nebst einer übersicht über den Endosperm-typ bei den Amaryllidaceae. Botan. Notiser 1951: 209–28. Biol. Abstr. 26:6847. 1952.

17943. Stening, H. C. 1935: Effect of "Black Point" disease on the germination of wheat. Agr. Gaz. N.S. Wales 46:282.

17944. —— 1935: Skeleton weed. Viability of seed-conditions of £5,000 reward. Agr. Gaz. N.S. Wales 46:426–28.

17945. Stenstam, T., C. O. Björling, and E. Ohlsson. 1934: Amylasen in ruhenden und keimenden Samen. IV. Weizen. Z. Physiol. Chem. 226:265–71.

17946. Stepanenko, B. N. 1960: Galactomannans and glucomannans of seeds, bulbs and rhizomes. Bull. Soc. Chim. Biol. 42:1519–46.

Stepanenko, F. S. 12576.

17947. Stepanov, R. S. 1962: Importance of geographical origin of seeds for pine culture. Mosk. Ordena Lenina Sel'skokoz. Akad. Im. K. A. Timiryazeva Dokl. 77:471–74.

17948. Stepanov, S. N. 1955: Effect of drying and warming apple seeds on germination. Sad i Ogorod 12:45–47.

Stepanovich, K. M. 11374.

17949. Stephan, J. 1928: Stimulationsversuche mit *Cannabis sativa*. Faserforsch. 7:292–98. Biol. Abstr. 6:6636. 1932.

17950. —— 1928: *Phacelia tanacetifolia* Benth. Ber. Deut. Botan. Ges. 46:499–508.

17951. —— 1929: Keimungs- und Wachstumsbeschleunigung bei *Dicksonia antarctica*. Botan. Centralbl. Beih. Abt. I 45: 401–8.

17952. —— 1929: Untersuchung fermentativer Teilprozesse bei der Samenkeimung. Ber. Deut. Botan. Ges. 47:561–64. Biol. Abstr. 6:24712. 1932.

17953. —— 1929: Untersuchungen über den Quellprozess der Samen von *Trifolium* pratense. Landwirtsch.-Versuchssta. 108: 371–76.

17954. —— 1932: Untersuchungen über das Verhalten der Katalase im Samen. Jahrb. Wiss. Botan. 75:771–808.

17955. —— 1937: Physiological examinations of the germination of Serradella. Landwirtsch.-Versuchssta. 128:133–59.

Proc. Intern. Seed Testing Assoc. 9:290–91. 1937.

17956. Stephens, S. G. 1958: Factors affecting seed dispersal in *Gossypium* and their possible evolutionary significance. N. Carolina Agr. Expt. Sta. Tech. Bull. 131. 32 pp.

17957. —— 1958: Salt water tolerance of seeds of Gossypium species as a possible factor in seed dispersal. Am. Naturalist 92:83–92.

Stephenson, J. E. 11894.

17958. Stephenson, R. B. 1942: Sterilization technique for grass seeds. Plant Physiol. 17:324–25.

17959. Sterckx, R. 1946: Barley moisture and its effects. J. Inst. Brewing 52:92. Biol. Abstr. 21:12384. 1947.

17960. Sterling, C. 1949: Embryonic differentiation in *Taxus cuspidata*. Bull. Torrey Botan. Club 76:116–33.

17961. —— 1954: Development of the seed coat of lima bean (*Phaseolus lunatus* L.). Bull. Torrey Botan. Club 81:271–87.

17962. —— 1955: Embryogeny in the lima bean. Bull. Torrey Botan. Club 82: 325–38.

17963. —— 1955: Maturity indices in lima beans. Food Technol. 9:395–98.

17964. —— 1964: Comparative morphology of the carpel in the Rosaceae. II. Prunoideae: *Maddenia, Pygeum; Osmaronia*. Am. J. Botany 51:354–60.

Stern, A. M. 11960.

Stern, C. 14052.

17965. Stern, F. C. 1933: Propagation of lilies by seed. Roy. Hort. Soc. Lily Yearbook 1933:119–29.

Stern, H. 9898, 9899, 17242.

17966. Sternheimer, E. P. 1954: Method of culture and growth of maize endosperm in vitro. Bull. Torrey Botan. Club 81: 111–13.

17967. Sterz, M. 1941: Effect of mineral fertilizers on fat formation in seed of *Glycine soya*. Bodenkunde Pflanzenernähr 24:34–54. Chem. Abstr. 38:5352. 1944.

17968. Steshenko, A. P. 1963: Biology of newly collected seeds of the highmountain zone plants of the Pamirs. Botan. Zh. 48:965–78.

Steveninck, R. F. M. van; *see* Van Steveninck, R. F. M.

Stevens, C. L. 18727.

17969. Stevens, Mrs. C. W. 1941: Geraniums grown from seed. Flower Grower 28(1):28.

17970. Stevens, D. J. 1959: The contribution of the germ to the oil content of white flour. Cereal Chem. 36:452–61.

17971. ——, E. E. McDermott, and J. Pace. 1963: Isolation of endosperm protein and aleurone cell contents from wheat, and determination of their amino-acid composition. J. Sci. Food Agr. 14:284–87.

17972. Stevens, E. L. 1932: Growing magnolias from seed. Flower Grower 19(5):236.

Stevens, H. 4252, 12144, 17801, 17802.

17973. Stevens, J. G. J. 1954: Harvesting green panic seed. Queensland Agr. J. 78:9–11.

17974. Stevens, O. A. 1932: The number and weight of seeds produced by weeds. Am. J. Botany 19:784–94.

17975. —— 1935: Germination studies on aged and injured seeds. J. Agr. Res. 51:1093–106.

17976. —— 1943: Wheat grains without embryos. Science 97:91.

17977. —— 1944: Distribution of maple seeds. Am. Botanist 50:5–6.

17978. —— 1944: Variations in kinds of weeds from year to year. N. Dakota Agr. Expt. Sta. Bimo. Bull. 6(6):6.

17979. —— 1948: Wild oat germination and other characters. N. Dakota Agr. Expt. Sta. Bimo. Bull. 11:3–6.

17980. —— 1954: Forty-year germination tests with alfalfa and clover. N. Dakota Agr. Expt. Sta. Bimo. Bull. 16:111–12.

17981. —— 1957: Weights of seeds and numbers per plant. Weeds 5:46–55.

17982. —— and H. O. Long. 1926: Sweet clover seed studies. N. Dakota Agr. Expt. Sta. Bull. 197:1–20.

17983. Stevens, R. D. 1940: Seed studies. Arkansas Agr. Expt. Sta. Bull. 386:87–88.

17984. Stevenson, D. D. 1937: Comparative rates of germination and germination percentage of native and introduced pine seed, and the relation of this data to nursery practice in Kwangtung. Lingnan Sci. J. 16:573–77.

Stevenson, E. C. 3327, 7977, 9720.

17985. Stevenson, E. G. 1955: Seedless watermelons! Agr. Leaders Dig. 36(2):26.

17986. Stevenson, E. W. 1949: Results of preliminary tests of pelletized crested wheatgrass seed. U.S. Dept. Agr. Forest Serv. Pacific Northwest Forest Range Expt. Sta. Res. Note 53. 7 pp.

17987. Stevenson, F. J., and W. C. Edmundson. 1950: Storage of potato seeds. Am. Potato J. 27:408–11.

Stevenson, F. J. 4174.
Stevenson, T. 10697.

17988. Stevenson, T. M. 1937: Sweet clover studies on habit of growth, seed pigmentation, and permeability of the seed coat. Sci. Agr. (Ottawa) 17:627–54.

Stevenson, T. M. 19621.

17989. Steward, F. C. 1961: The growth-promoting qualities of endosperm. Intern. Botan. Congr. Recent Advan. Botany 9:1103–7.

17990. —— and N. W. Simmonds. 1954: Growth-promoting substances in the ovary and immature fruit of the banana. Nature 173:1083–84.

Steward, F. C. 17266.

17991. Stewart, D. 1950: Sulphuric acid scarification of seedballs of *Beta trigyna* and other *Beta* spp. Proc. Am. Soc. Sugar Beet Technol. 6:101–2.

17992. Stewart, F. C. 1936: The relation of age and viability to the popping of popcorn. N.Y. Agr. Expt. Sta. Bull. 672. 7 pp.

Stewart, F. C. 15015.

17993. Stewart, G. 1926: Effect of color of seed, of scarification, and of dry heat on germination of alfalfa seed and of some of its impurities. J. Am. Soc. Agron. 18: 743–60.

17994. —— 1951: Looking ahead in range reseeding. J. Forestry 49:729–31.

17995. —— and J. W. Carlson. 1932: The quality of alfalfa seed as affected by color and plumpness. J. Am. Soc. Agron. 24:146–55. Biol. Abstr. 8:6613. 1934.

Stewart, G. 3617.

17996. Stewart, R., and H. C. Eyster. 1939: A study of the green pigments in the lotus embryo. Proc. S. Dakota Acad. Sci. 19:89–91. Chem. Abstr. 34:2422. 1940.

Stewart, R. B. 15811.
Stewart, R. N. 17198–17201.

17997. Stewart, R. T. 1930: Inheritance of certain seed-coat colors in soybeans. J. Agr. Res. 40:829–54.

17998. —— and J. B. Wentz. 1930: Defective seed-coat character in soybeans. J. Am. Soc. Agron. 22:658–62.

Stewart, R. T. 16017.

17999. Stewart, W. S., and I. J. Condit. 1949: The effect of 2-4 dichlorophenoxy-acetic acid and other plant growth regulators on the Calimyrna fig. Am. J. Botany 36:332–35.

Steyer, B. 12208.

18000. Steyn, D. G. 1935: The toxicity of pumpkin seed (*Cucurbita pepo* L.). Onderstepoort J. Vet. Sci. 5:441–43. Chem. Abstr. 30:4218. 1936.

18001. Steyn, D. G. 1941: Bread poisoning caused by weeds contained in wheat. Farming S. Africa 16(188):381–82, 398.

18002. Stickler, F. C. 1961: Row width and seeding rate as factors influencing winter wheat yields and components of yield. Trans. Kansas Acad. Sci. 64:1–6.

18003. —— 1962: Seeding depth and use of press wheels as factors affecting winter barley and winter wheat yields in Kansas. Agron. J. 54:492–94.

18004. —— and H. H. Laude. 1960: Effect of row spacing and plant population on performance of corn, grain sorghum and forage sorghum. Agron. J. 52:275–77.

18005. —— and A. W. Pauli. 1961: Gibberellic acid as a factor affecting seedling vigor and yield of varying seed sizes of winter wheat. Crop Sci. 1:287–90.

18006. —— and C. E. Wassom. 1963: Emergence and seedling vigor of birdsfoot trefoil as affected by planting depth, seed size, and variety. Agron. J. 55:78.

Stickler, F. C. 6101, 10483, 10484, 15136.
Stiemert, M. L. 3256.

18007. Stier, H. L. 1937: Delayed germination in seeds of the potato. Proc. Am. Soc. Hort. Sci. 34:433–35.

18008. —— 1938: The effect of certain seed treatments on the germination of recently harvested potato seeds. Proc. Am. Soc. Hort. Sci. 35:601–5.

18009. —— 1939: Response of tomato (*Lycopersicum esculentum* Mill. var. Marglobe) to certain vernalization treatments. Proc. Am. Soc. Hort. Sci. 36:708–14.

18010. —— and H. B. Cordner. 1937: Germination of seeds of the potato as affected by temperature. Proc. Am. Soc. Hort. Sci. 34:430–32.

18011. —— and H. G. Du Buy. 1939: The influence of certain phytohormone treatments on the time of flowering and fruit production of tomato plants under field conditions. Proc. Am. Soc. Hort. Sci. 36:723–31.

Stighter, H. C. M. de; *see* De Stighter, H. C. M.
Stiles, H. 1742.

18012. Stiles, I. E. 1948: Relation of water to the germination of corn and cotton seeds. Plant Physiol. 23:201–22.

18013. —— 1949: Relation of water to germination of bean seeds. Plant Physiol. 24:540–45. Biol. Abstr. 23:22170. 1949.

18014. Stiles, J. W. 1960: Respiration in seed germination and seedling development. Handb. Pflanz. Physiol. 12:465–92.
18015. Stiles, W. 1960: Plant physiology. Germination of seeds. Sci. Progr. (London) 48:331–41.
18016. —— and W. Leach. 1932: Researches on plant respiration. I. The course of respiration of Lathyrus odoratus during germination of the seed and the early development of the seedling. Proc. Roy. Soc. (London) Ser. B 111:338–55.
18017. —— and —— 1933: Researches on plant respiration. II. Variations in the respiratory quotient during germination of seeds with different food reserves. Proc. Roy. Soc. (London) Ser. B 113:405–28.
Stiles, W. 1151.
Still, E. U. 8125.
Still, T. W. 5167.
18018. Stillings, A. N., and A. Laurie. 1943: Culture of belladonna. Horticulture 21(12): 251.
18019. Stingl, G. 1907: Experimentelle Studie über die Ernährung von pflanzlichen Embryonen. Flora (Jena) 97:308–31.
18020. Stitt, R. E. 1934: The effect of depth of planting on the germination of soybean varieties. J. Am. Soc. Agron. 26:1001–4.
18021. —— 1939: Dodder control in annual lespedezas. J. Am. Soc. Agron. 31:338–43.
18022. —— 1944: Effect of moisture, seeding dates, and fertilizer on stands and yields of Crimson clover. J. Am. Soc. Agron. 36:464–67.
18023. —— 1954: Seed production of Russian wildrye. Agron. J. 46:171–75.
18024. Stoa, T. E. 1933: Persistence of viability of sweet clover seed in a cultivated soil. J. Am. Soc. Agron. 25:177–81.
18025. —— 1941: Volunteer sweet clover. N. Dakota Agr. Expt. Sta. Bimo. Bull. 3(6):3–6. Expt. Sta. Rec. 85:758. 1941.
18026. ——, W. E. Brentzel, and E. C. Higgins. 1936: Shriveled light weight wheat. N. Dakota Agr. Expt. Sta. Circ. 59. 11 pp. Expt. Sta. Rec. 75:43. 1936.
Stoa, T. E. 14334, 14964.
18027. Stobodianik, N. I. 1958: Effect of seed cleaning of vegetable crops on the seeding qualities and yields. Sad i Ogorod 96(8):22–25.
18028. Stockbarger, C. U. 1886: Testa of the seeds of *Phytolacca*. Botan. Gaz. 11: 224–25.
18029. Stocker, O. 1959: Über die Atmung ruhender Weizenkörner in Abhängigkeit vom Wassergehalt. Angew. Botan. 33: 153–58.
18030. Stockman, R. 1931: The poisonous principle of *Lathyrus* and some other leguminous seeds. J. Hyg. 31:550–62. Chem. Abstr. 26:2518. 1932.
18031. Stockwell, P. 1942: *Pinus:* embryo size compared with growth rate. Am. Naturalist 76:431–32.
18032. Stoddart, J. L. 1964: Seed ripening in grasses. I. Changes in carbohydrate content. J. Agr. Sci. 62:67–72.
18033. Stoddart, L. A., and K. J. Wilkinson. 1938: Inducing germination in *Oryzopsis hymenoides* for range reseeding. J. Am. Soc. Agron. 30:763–68.
Stoddart, L. A. 4352.
18034. Stoeckeler, J. H. 1953: Importance of seed source and the need of tree seed farms. U.S. Dept. Agr. Forest Serv. Lake States Forest Expt. Sta. Misc. Rept. 22: 74–80.
18035. —— 1955: European larch (*Larix decidua*) seed sources compete successfully with tamarak (*L. laricina*) during five-year test in northeastern Wisconsin. U.S. Dept. Agr. Forest Serv. Lake States Forest Expt. Sta. Tech. Notes 440. 2 pp.
18036. Stoesz, A. D. 1953: Native grasses—better seed harvesting. Crops & Soils 5(7):9–11.
Stoicovici, L. 4755.
Stoilov, M. 18788.
18037. Stoker, G. L., D. C. Tingey, and R. J. Evans. 1934: The effect of different methods of storing chicken manure on the viability of certain weed seeds. J. Am. Soc. Agron. 26:600–9.
18038. Stoker, J. R. 1964: The germination of seeds of *Melilotus alba*. Can. J. Biochem. 42:851–57.
18039. Stokes, G. W. 1955: Seed development and failure in horseradish. J. Heredity 46:15–21.
18040. Stokes, P. 1952: A new technique for the sterile culture of plant embryos in vitro. Nature 170:242.
18041. —— 1952: A physiological study of embryo development in *Heracleum sphondylium* L. Ann. Botany (London) 16:441–47.
18042. —— 1952: A physiological study of embryo development in *Heracleum sphondylium* L. II. The effect of temperature on after-ripening. Ann. Botany (London) 16:571–76.
18043. —— 1953: A physiological study of embryo development in *Heracleum*

sphondylium L. III. The effect of temperature on metabolism. Ann. Botany (London) 17:157–73.

18044. —— 1953: The stimulation of growth by low temperature in embryos of *Heracleum sphondylium* L. J. Exptl. Botany 4:222–34.

18045. Stoklasa, J. 1922: Influence du selenium et du radium sur la germination des grains. Compt. Rend. 174:1075–77. Chem. Abstr. 17:789–90. 1923.

18046. Stoletov, V. N., T. A. Proskurinkova, and E. I. Krynkina. 1949: Some metabolic processes in sprouting and ripening wheat grain. Probl. Biokhim. v Michurin. Biol. Akad. Nauk SSSR, Inst. Biokhim. Sb. 1949(1):151–68. Chem. Abstr. 46:4066. 1952.

18047. Stoll, A., and J. Renz. 1939: Cardiac glucosides. XVI. Enzymes, especially those of *Strophanthus* seeds, that hydrolyze cardiac glucosides. Enzymologia 7:362–82. Chem. Abstr. 34:6298. 1940.

Stoll, W. R. Müller-; *see* Müller-Stoll, W. R.

18048. Stolze, K. V. 1938: Die Lohnsaatbereitung in der landesbauernschaft Weser-Ems. Nachr. Schädlingsbekämpf., Leverkusen 13:1–11. Rev. Appl. Mycol. 17:509. 1938.

Stom, I. 1425.

18049. Stone, A. L. 1915: The weed content of seeds. Proc. Assoc. Offic. Seed Analysts 1914:23–27.

18050. —— 1919: The problem presented in obtaining satisfactory germination tests on newly harvested grain and grass seeds. Proc. Assoc. Offic. Seed Analysts 11:51–53.

18051. —— 1938: Results of ten years of germination tests on the same samples of vegetable seeds. Proc. Assoc. Offic. Seed Analysts 23/26:76–80.

Stone, B. M. 17472.

18052. Stone, E. C. 1957: The effect of seed storage on seedling survival of sugar pine. J. Forestry 55:816–20.

18053. —— 1957: Embryo dormancy and embryo vigor of sugar pine as affected by length of storage and storage temperatures. Forest Sci. 3:357–71.

18054. —— 1957: Embryo dormancy of *Pinus jeffreyi* Murr. seed as affected by temperature, water uptake, stratification, and seed coat. Plant Physiol. 32:93–99.

18055. —— 1958: The seed dormancy mechanism in pine. *In* Thimann, K. V., Ed. The physiology of forest trees. Ronald Press, New York, pp. 611–28.

18056. —— and J. W. Duffield. 1950: Hybrids of sugar pine by embryo culture. J. Forestry 48:200–1.

18057. —— and J. Holt. 1950: A rapid method of separating seed of chamise (*Adenostoma fasciculatum*) from the duff. Ecology 31:149.

18058. —— and G. Juhren. 1951: The effect of fire on the germination of the seed of *Rhus ovata* Wats. Am. J. Botany 38:368–72.

18059. Stone, G. E. 1894: Germinating seeds in sawdust. Botan. Gaz. 19:334.

18060. —— 1901: Influence of chemical solutions upon the germination of seeds. Mass. Agr. Expt. Sta. Rept. 13:74–83.

Stone, G. M. 17470.

Storey, I. F. 1870, 1871, 2135.

18061. Storey, W. B., R. A. Hamilton, and H. Kamemoto. 1948: Papaya. Hawaii Agr. Expt. Sta. Rept. 1946/48:80–82.

18062. —— and W. C. Kemper. 1960: A study of *Macadamia* seed germination. Calif. Macadamia Soc. Yearbook 6:59–64.

Storey, W. B. 8290.

18063. Storoshenko, W. F. 1932: Zur Frage über den Gehalt an ätherischen Öl in den Früchtens des Korianders (*Coriandrum sativum* L.) in den verschiedenen Stadien ihres Keimens und bei der verschiedenen Präparation zur Verarbeitung, sowie unter einigen Bedingungen der Aufbewahrung. North-Caucasian Inst. Spec. Tech. Cult. Works 1:120–21.

18064. Stotzky, G., and E. A. Cox. 1962: Seed germination studies in *Musa*. II. Alternating temperature requirement for the germination of *Musa balbisiana*. Am. J. Botany 49:763–70.

18065. ——, ——, and R. D. Goos. 1962: Seed germination studies in *Musa*. I. Scarification and aseptic germination of *Musa balbisiana*. Am. J. Botany 49:515–20.

Stotzky, G. 4483.

18066. Stout, A. B. 1924: Seedling lilies: a report of progress of experimental studies with species of *Lilium*. N.Y. Botan. Garden J. 25:185–94.

18067. —— 1936: Seedlessness in grapes. N.Y. Agr. Expt. Sta. Tech. Bull. 238. 68 pp.

18068. —— 1945: Fruits and seeds of the lily-of-the-valley. N.Y. Botan. Garden J. 46:205–14.

18069. —— and W. M. Porterfield. 1938: Seed patterns and incompatibilities in

Lilium candidum. Roy. Hort. Soc. Lily Yearbook 1938:108–22.

18070. Stout, A. W., and H. A. Schuette. 1932: Rye germ oil. J. Am. Chem. Soc. 54:3298–302.

18071. Stout, M., and F. V. Owen. 1942: Vernalization of sugar-beet seed. Proc. Am. Soc. Sugar Beet Technol. 1942:386–95. Biol. Abstr. 18:22898. 1944.

18072. —— and B. Tolman. 1941: Factors affecting the germination of sugar-beet and other seeds, with special reference to the toxic effects of ammonia. J. Agr. Res. 63:687–713.

18073. —— and —— 1944: Field and greenhouse tests with synthetic growth-regulating substances applied to sugar-beet seeds and plants. J. Am. Soc. Agron. 36:141–46. Brit. Abstr. B III:355. 1948.

Stout, M. 18622, 18623.

18074. Stoutemyer, V. T. 1953: Growing plants from seeds—newer ideas. Plants Gardens 9:63–67.

18075. —— and A. W. Close. 1953: Propagation by seedage and grafting under flourescent lamps. Proc. Am. Soc. Hort. Sci. 62:459–65.

18076. ——, C. Hope, and A. Close. 1941: Sphagnum for seed germination inhibits damping-off losses on unsterilized soil. Natl. Hort. Mag. 20:111–20.

Stoutemyer, V. T. 9012.

18077. Stover, E. L. 1937: The embryo sac of *Eragrostis cilianensis* (All.) Link: A new type embryo sac and a summary of grass embryo sac investigations. Ohio J. Sci. 37:172–84. Expt. Sta. Rec. 78:467. 1938.

18078. —— 1937: Parthenogenesis in the grasses. Science 85:75.

18079. Stoward, F. 1908: On endospermic respiration in certain seeds. Ann. Botany (London) 22:415–48.

18080. —— 1911: A research into the amyloclastic secretory capacities of the endosperm and aleurone layer of *Hordeum* with special reference to the question of the vitality and auto-depletion of the endosperm. I. Ann. Botany (London) 25:799–841.

18081. —— 1911: A research into the amyloclastic secretory capacities of the endosperm and aleurone layer of *Hordeum* with special reference to the question of the vitality and auto-depletion of the endosperm. II. Ann. Botany (London) 25: 1147–204.

Strachan, G. 18706.

18082. Strafelda, J. 1961: Germination power of swelling and hard seeds of most important Leguminosae. Sb. Česk. Akad. Zeměděl. Ved Rostlinna Vyroba 7:1649–60.

Straka, R. P. 9111.

Strand, E. 15, 15993.

Strasser, R. 10722.

Straub, G. J. 4338.

18083. Straus, J. 1954: Maize endosperm tissue grown in vitro. II. Morphology and cytology. Am. J. Botany 41:833–39.

18084. —— 1960: Anthocyanin synthesis in corn endosperm tissue cultures. II. Effect of certain inhibitory and stimulatory agents. Plant Physiol. 35:645–50.

18085. —— 1960: Maize endosperm tissue grown in vitro. III. Development of a synthetic medium. Am. J. Botany 47:641–47.

18086. —— and C. D. La Rue. 1954: Maize endosperm tissue grown in vitro. I. Culture requirements. Am. J. Botany 41:687–94.

Strauss, G. 7340.

Strazhevskaya, N. B. 15730.

18087. Strebeyko, P., and H. Domanska. 1955: Further studies on the resistance of wheat to drought during the period of germination and initial growth. Roczniki Nauk Rolniczych Ser. A 71:43–53.

Street, O. E. 8719.

Streetman, L. J. 19889.

18088. Strekova, V. Y. 1962: Histochemical study of some processes taking place during hardening of corn seeds by varying temperatures. Plant Physiol. (USSR) (English Transl.) 8:168–72.

Strid, L. 7342, 13310.

Stritzel, J. A. 7894.

18089. Strogonov, B. P. 1946: Germination of cottonseed collected from saline soil. Compt. Rend. Acad. Sci. URSS 54:645–47. Brit. Abstr. B III:129–30. 1948.

Strohman, R. E. 10641.

18090. Strohmeyer, G. 1938: Über die züchterische Bedeutung des Tausendkorngewichts der Kiefer. I. "Literature":157. Forstarch. 14:153–57.

Strommen, A. M. 612.

18091. Strong, R. G., and D. L. Lindgren. 1957: Effect of fumigation on the germination of seeds. California Univ. Coll. Agr. Citrus Expt. Sta. (Riverside). 11 pp.

18092. —— and —— 1959: Effect of methyl bromide and hydrocyanic acid fumigation on the germination of wheat. J. Econ. Entomol. 52:51–60.

18093. —— and —— 1959: Effect of methyl bromide and hydrocyanic acid fumigation on the germination of oats. J. Econ. Entomol. 52:415–18.

18094. —— and —— 1960: Germination of cereal, sorghum, and small legume seeds after fumigation with hydrogen phosphide. J. Econ. Entomol. 53:1–4.

18095. —— and —— 1961: Germination of small legume seeds after fumigation with methyl bromide and hydrocyanic acid. J. Econ. Entomol. 54:21–25.

18096. —— and —— 1963: Germination of grain sorghum and Sudan grass seeds after fumigation with methyl bromide and hydrocyanic acid. J. Econ. Entomol. 56:144–49.

18097. —— and D. E. Sbur. 1960: Influence of grain moisture and storage temperature on the effectiveness of malathion as a grain protectant. J. Econ. Entomol. 53:341–49.

18098. —— and —— 1964: Influence of grain moisture and storage temperature on the effectiveness of five insecticides as grain protectants. J. Econ. Entomol. 57:44–47.

18099. ——, ——, and G. R. Pieper. 1959: Effect of several pesticides on the germination of three varieties of flower seeds. J. Econ. Entomol. 52:355.

18100. Strong, W. J. 1932: Parthenocarpy in the cucumber. Sci. Agr. (Ottawa) 12:665–69.

18101. Stroube, W. H., and E. N. Fergus. 1959: Occurrence of two-seeded florets in red clover. Agron. Abstr. 1959:87.

Stroube, W. H. 18411, 18412.

18102. Stroun, M., and C. Pugnat. 1958: Relations entre l'accomplissement du photostade et la résistance au froid des céréales. Phyton 11:1–96.

18103. ——, ——, and M. Sandmeier. 1960: La vernalisation, le photopériodostade et la résistance au froid des blés. Phyton 14:121–26.

18104. Stroup, G. E., and L. W. Reed. 1957: The effect of fertilizer salts on germination and seedling vigor. II. Six cotton varieties. Agron. Abstr. 49:66.

18105. Strun, M. 1955: Transplantation of grass embryos onto heterologous endosperms. Izv. Akad. Nauk SSSR Ser. Biol. 1955(5):135–41. Biol. Abstr. 32:28542. 1958.

18106. Struve, W. M. 1959: Drying and germinability of maize. Dissertation Abstr. 19:2222.

18107. Struve, W. P. 1929: Einige Versuche über Aufkeimung von Wiesengrassamen. Jard. Botan. Inst. Essais Semences, Leningrad, Ann. Essais Semences 1929(3):187–202.

18108. Strydom, J. C. 1955: Carnations. Dept. Agr. S. Africa Bull. 341, Hort. Ser. 19. 13 pp.

18109. Stuart, A. 1957: Seed production in New Zealand crested dogstail. New Zealand J. Agr. 95:598–600.

18110. Stuart, N. W. 1946: Expanded vermiculite gives excellent propagation results. Florists Rev. 98(2547):33–34.

18111. Stubbe, H. 1935: Genmutabilität bei *Antirrhinum majus* L. (Nebst einigen Beobachtungen über den Zeitpunkt des Mutierens während der Entwicklung.) Biol. Centralbl. 55:209–15.

18112. —— 1935: Weitere Untersuchungen über Samenalter und Genmutabilität bei *Antirrhinum majus*. Z. Induktive Abstammungs- Vererbungslehre 70:533–37.

18113. —— 1936: Die Erhöhung der Genmutationsrate in alternden Gonen von *Antirrhinum majus* L. Biol. Centralbl. 56:562–67.

18114. Stubblefield, F. M., and E. E. De Turk. 1941: The composition of corn, oats, and wheat as influenced by soil, soil treatment, seasonal conditions, and growth. Soil Sci. Soc. Am. Proc. 5:120–24. Biol. Abstr. 15:22471. 1941.

18115. Stuber, C. W., V. A. Johnson, and J. W. Schmidt. 1962: Grain protein content and its relationship to other plant and seed characters in the parents and progeny of a cross of *Triticum aestivum* L. Crop Sci. 2:506–8.

18116. Stütz, H. 1933: Über den Einfluss verschiedenartiger Lagerung auf die Hartschaligkeit von Kleesamen. Dissertation Hamburg Univ. 42 pp. Expt. Sta. Rec. 72:175. 1935.

Stütz, H. 6056.

Stul'nikova, R. I. 12149.

18117. Stumpf, P. K. 1950: Carbohydrate metabolism in higher plants. III. Breakdown of fructose diphosphate by pea extracts. J. Biol. Chem. 182:261–72.

Stumpf, P. K. 8372, 19925.

18118. Sturgis, F. E., R. J. Miears, and R. K. Walker. 1952: Protein in rice as influenced by variety and fertilizer levels. Louisiana Agr. Expt. Sta. Tech. Bull. 466. 28 pp.

18119. Sturkie, D. G. 1937: The effect of some environmental factors on the oil content of cotton seed. Alabama Agr. Expt. Sta. Ann. Rept. 1937:11–12.

18120. —— 1947: Effects of some environmental factors on the seed and lint of cotton. Alabama Agr. Expt. Sta. Bull. 263. 87 pp.
18121. Sturtevant, E. L. 1889: Seedless fruits. Mem. Torrey Botan. Club 1:141–85.
Stutzer, A. 8285.
18122. Su, K. C., and C. C. Lin. 1961: Studies on the spacing of peanuts in relation to flowering, fructification, and yields. J. Agr. Assoc. China (Taipei) 35:42–49.
Suarez, R. 9971.
Subba-Rao, A. M.; *see* Rao, A. M. Subba-.
18123. Subbaratnam, A. V. 1956: Chemical examination of the seeds of *Erythrina indica*. J. Sci. Ind. Res. (India) 15B:210–12. Chem. Abstr. 50:17012. 1956.
18124. Subrahmanyan, R. 1957: Observations on the anatomy, cytology, development of the reproductive structures, fertilization and embryology of *Pelvetica canaliculata* Dcne. et Thur. II. J. Indian Botan. Soc. 36:12–34.
18125. Subrahmanyan, V., D. S. Bhatia, C. P. Natarajan, G. S. Mani, J. R. Iyengar, and M. Nagarathnamma. 1954: Chemical composition of Indian coffee. Indian Coffee 18:185–91. Chem. Abstr. 49:6501. 1955.
Subrahmanyan, V. 2822, 5247, 5248, 6938, 7285, 11414, 11551, 15871, 15955, 16011, 16790, 16807, 19064.
Subramanian, C. L. 10216.
18126. Subramanian, N., M. V. L. Rao, and M. Srinivasan. 1957: Amino-acid composition of cashew-nut globulin. J. Sci. Ind. Res. (India) 16C:24. Chem. Abstr. 51:10676. 1957.
18127. —— and M. Srinivasan. 1951: Nutritive value of the seeds of *Amaranthus paniculatus* Linn. Current Sci. (India) 20:294–95.
Subramanian, N. 15956.
Subramanian, S. S. 4848, 15940.
18128. Subramanyam, K. 1947: A case of polyembryony in *Wahlenbergia gracilis* Schrad. Current Sci. (India) 16:350.
18129. —— 1948: An embryological study of *Melastoma malabathricum* L. J. Indian Botan. Soc. 27:11–19. Biol. Abstr. 23: 27256. 1949.
18130. —— 1949: An embryological study of *Lobelia pyramidalis* Wall., with special reference to the mechanism of nutrition of the embryo in the family Lobeliaceae. New Phytologist 48:365–73.
18131. —— 1951: Flower structure and seed development in *Isotoma fluviatilis* F. v. M. Proc. Natl. Inst. Sci. India 17:275–85.
18132. —— 1960: Nutritional mechanism of the seed. I. Nutritional mechanism of the embryo sac. J. Madras Univ. 30:29–44. Biol. Abstr. 37:10893. 1962.
18133. —— 1960: Nutritional mechanism of the seed. II. Nutritional mechanism of the embryo. J. Madras Univ. 30:45–56. Biol. Abstr. 37:10894. 1962.
18134. —— 1962: Embryology in relation to systematic botany with particular reference to the Crassulaceae. Symp. Plant Embryol. Univ. Delhi, 1960 (Papers), pp. 94–112.
Subramanyam, K. 10356–10358.
Suchkina, V. N. 1356.
18135. Suckling, F. E. T. 1952: Dissemination of white clover (*Trifolium repens*) by sheep. New Zealand J. Sci. Technol. Sect. A 33:64–77.
18136. Sucoff, E. I., and T. W. Church, Jr. 1960: Seed production and dissemination by Virginia pine. J. Forestry 58:885–88.
Sucoff, E. I. 4112.
18137. Sud'yina, O. G., and G. I. Lozova. 1961: Protochlorophyll of the inner coats of pumpkin (*Cucurbita pepo* L.) seeds. Ukr. Botan. Zh. 18(5):35–44.
18138. Südi, J., G. Josepovits, and G. Matolcsy. 1960: Morphological reactions of germinating seeds as criteria for the plant growth regulating activity of auxin-type chemicals. Nature 188:244–45.
Suetina, M. A. 15731.
18139. Suetsugu, I. 1951: Developmental morphology of the embryo in barley, rye, and oats. Bull. Natl. Inst. Agr. Sci. Ser. D 1951(1):83–87.
18140. —— and I. Anaguchi. 1954: Relation between the size of seed and the yielding efficiency in soybean. Proc. Crop Sci. Soc. Japan 22:117–18.
18141. —— and C. Ueki. 1960: Development of the embryo in range plant (*Astragalus sinicus* L.). Bull. Hokuriku Agr. Expt. Sta. 1:73–88.
18142. —— and —— 1960: On the hard-coatedness in range seeds (*Astragalus sinicus* L.). Bull. Hokuriku Agr. Expt. Sta. 1:89–121.
Suetsugu, I. 18869.
Sugawara, K. 14081, 14082.
18143. Sugawara, S. 1954: Studies on the continuous presoaking of rice seed from winter to seeding time. I. Changes in the seeds during the presoaking treatment. Proc. Crop Sci. Soc. Japan 22:113–14.

18144. —— 1955: Studies of soaking rice seeds in water during cold seasons prior to sowing time. Proc. Crop Sci. Soc. Japan 24:92–93.
18145. —— 1959: Studies on the germination capacity of uplant rice seeds in relation to the location on the flower panicle. Bull. Fac. Agr. Niigata Univ. 11:9–22.
18146. Sugawara, T. 1955: Influence of nitrates on germination of seeds under oxygen supply of various concentrations. Proc. Crop Sci. Soc. Japan 23:295–96.
Sugawara, T. 14552.
18147. Suge, H., and J. Hirano. 1961: Effect of gibberellin on the vernalization of winter wheat and barley. Proc. Crop Sci. Soc. Japan 31:129–34.
Suge, H. 19926.
18148. Sugihara, Y. 1956: The embryogeny of *Cupressus funebris* Endlicher. Botan. Mag. (Tokyo) 69:439–41.
18149. —— 1956: The embryogeny of *Cupressus sempervirens* Linnaeus. Sci. Rept. Tohoku Univ. Fourth Ser. 22:1–4.
18150. —— 1963: The embryogeny of *Thuja occidentalis* Linnaeus. Sci. Rept. Tohoku Univ. Fourth Ser. 29:365–67.
18151. Sugimoto, K. 1954: Biochemical and nutritional studies of germinating soybeans. Bull. Osaka Med. School. 1:1–16. Chem. Abstr. 49:4088. 1955.
Sugiura, K. 6206.
18152. Sugiura, M. 1961: A study on respiration of bean seed embryo. Botan. Mag. (Tokyo) 74:19–27.
18153. —— and Y. Sunobe. 1962: Phosphorus compounds and phytase in germinating bean, *Vigna sesquipedalis*. Botan. Mag. (Tokyo) 75:63–71.
18154. Sugiyama, T., and H. Hori. 1949: Dormancy of the seeds of leaf mustards. I. Hort. Assoc. Japan J. 18:1–7.
Sugiyama, T. 9055, 9056.
Sukernik, M. V. 15725.
Sukhankin, P. 5040.
Sukhenko, F. T. 6202.
18155. Sukhov, K. S. 1938: Veränderungen in den Kernen der Embryone von *Avena sativa* bei der Keimung der Samen. Biol. Zh. Mosk. 7:279–86.
Sukumaran, R. 10287.
18156. Sulbha, K. 1957: Embryology of *Brassica juncea* Czern. & Coss. J. Indian Botan. Soc. 36:292–301.
Sulbha, K. 9594, 18204.
18157. Sullivan, B., and C. H. Bailey. 1936: The lipids of the wheat embryo. I. II. J. Am. Chem. Soc. 58:383–90, 390–93. Expt. Sta. Rec. 77:581. 1937.
18158. —— and M. A. Howe. 1933: Lipases of wheat. I. J. Am. Chem. Soc. 55:320–24.
18159. —— and —— 1937: The isolation of glutathione from wheat germ. J. Am. Chem. Soc. 59:2742–43.
Sullivan, B. 8282.
Sullivan, J. 5506.
18160. Sullivan, J. T. 1944: Further comparisons of plants with different chromosome numbers in respect to chemical composition. J. Am. Soc. Agron. 36:537–43.
18161. Sulmont, G., and A. Lebegue. 1964: Embryogénie des Papaveracées. Développement de l'embryon chez le *Meconopsis cambrica* Vig. Compt. Rend. 258:3890–93.
18162. Sumanov, E. I. 1949: Seed characteristics of frost-damaged grain. Selek. i Semen. 16(12):57–61.
Sumere, C. F. van; *see* Van Sumere, C. F.
Sumi, M. 11072.
18163. Suminski, S. 1960: Über den Einfluss der Humenstoffe auf die Atmung der Wurzeln und der Samen. Acta Agrobotan. (Warsaw) 9:123–27.
18164. Sumitomo, A., and K. Kosugi. 1963: Studies of cyclamen. I. On the germination of seed. Tech. Bull. Fac. Agr. Kagawa Univ. (English Transl.) 14:137–40.
Summers, L. A. 6586.
18165. Sumner, D. C., and R. D. Cobb. 1962: Post harvest dormancy of Coronado side-oats grama *Bouteloua curtipendula* (Michx.) Torr. as affected by storage temperature and germination inhibitors. Crop Sci. 2:321–25.
18166. ——, J. R. Goss, and B. R. Houston. 1958: Merion bluegrass seed production. Calif. Agr. Expt. Sta. Circ. 470. 27 pp.
18167. —— and K. E. Lindsey. 1962: The use of field maturity data to establish optimum harvest period for Merion Kentucky bluegrass seed production. Agron. J. 54:416–19.
Sumner, D. C. 4235.
Sumner, J. B. 10696.
18168. Sumner, M. H. 1959: Growing rhododendrons from seed. Calif. Hort. Soc. J. 20:17.
Sun, C. N. 19984.
18169. Sun, E. E. 1956: Seeds from different sections of the cucumber (*Cucumis sativus* L.) and bean (*Phaseolus vulgaris* L.) and the exhibition of the biological character

in their progeny. Acta Agron. Sinica 7:409–18.

18170. Sundahl, M. 1961: The influence of genotype and environment upon the protein content and baking quality of wheat. J. Swedish Seed Assoc. 71:137–51.

Sundararaj, D. D. 1760, 13841.

Sunderland, E. 6220.

18171. Sunderland, N. 1960: Germination of the seeds of angiospermous root parasites. Brit. Ecol. Soc. Symp. 1:83–93. 1959.

18172. —— 1960: The production of the *Striga* and *Orobanche* germination stimulants by maize roots. J. Exptl. Botany 11:236–45.

18173. —— 1960: The production of the *Striga* and *Orobanche* germination stimulants by maize roots. II. J. Exptl. Botany 11:356–66.

Sunderland, N. 3171.

18174. Suneson, C. A., and G. L. Peltier. 1936: Effect of source, quality, and condition of seed upon the cold resistance of winter wheats. J. Am. Soc. Agron. 28:687–93.

Suneson, C. A. 7386.

18175. Suneson, S. 1933: Zur Embryologie der Gattung *Viburnum*. Botan. Notiser 1933(1/3):181–94. Biol. Abstr. 11:196. 1937.

Sunneveld, A. 6131.

Sunobe, Y. 14791, 18153.

18176. Suput, M. 1953: The effect of soil moisture on the germination and emergence of maize, millet, sorghum, panicum and buckwheat. Rev. Res. Work Fac. Agr. Univ. Belgrade 1(1):43–57. Soils Fertilizers 17:2617. 1954.

18177. —— 1954: The effects of soil moisture on germination and growth of sowings of soybean, vetch, sunflower, flax, hemp and turnip. Zbornik Radova Belgrade Univ. Poljoprivredni Fak. 2:68–80. Soils Fertilizers 18:219. 1955.

Suqueira, R. de; *see* De Suqueira, R.

Sur, B. K. 17512.

18178. Sur, S. K., S. C. Roy, and B. C. Guha. 1949: Biosynthesis of ascorbic acid-synthesis by germinating seeds as influenced by different factors. Intern. Congr. Biochem. Abstr. Commun. 1:52–54.

Suranarayana, M. G. 14969.

18179. Surange, K. R., and P. Singh. 1952: The seeds of *Walkomiella indica*—a conifer from the lower Gondwanas of India. Current Sci. (India) 21:40–41.

18180. Surber, E. 1961: Effect of detergents (surface active agents) on the germination of Norway spruce seeds. Schweiz. Anst. Forst. Versw. Mitt. 37:355–70.

18181. Surrey, K. 1962: Action and interaction of monochromatic light on phosphate metabolism of germinating lettuce seeds. Can. J. Botany 40:965–74.

18182. —— and S. A. Gordon. 1962: Influence of light on phosphate metabolism in lettuce seed: Spectral response red, far-red interaction. Plant Physiol. 37:327–32.

Suryanarayana, M. 15932.

Sushchova, A. I. Daragan-; *see* Daragan-Sushchova, A. I.

18183. Sushkov, K. L., and Z. A. Klimovskaya. 1960: Some problems in the culture of ornamental plants for seeds. Tr. Akad. Nauk Kazakhsk. SSR Botan. Sad. 5:97–109.

18184. Suskin, M. 1953: The effect of dry chemical fire extinguisher materials on the viability of seed. Calif. Dept. Agr. Bull. 42:63–64.

18185. —— 1953: The effect of display conditions upon the viability of packet seeds in California. Calif. Dept. Agr. Bull. 42:120–37.

18186. —— 1954: Evaluation of field run seed. Calif. Dept. Agr. Bull. 43:33–40.

18187. Suskind, S. R. 1950: Resuscitation of heat-inactivated seeds by X-radiation. J. Heredity 41:97–102.

18188. Suszka, B. 1962: Influence of the temperature factor on the breaking of dormancy in mazzard seeds (*Prunus avium* L.). Arboretum Kornickie 7:189–275.

Suszka, B. 17451.

Suter, P. J. 12657.

Sutherland, R. H. 1083.

18189. Suto, S. 1957: On the quality of seed and parent tree of *Cryptomeria japonica*. Yamagata Agr. Forestry Soc. J. 1957(12): 83–88.

18190. —— 1957: Response of the seedlings of *Cryptomeria japonica* to prechillings at the germination stage of seeds. Yamagata Agr. Forestry Soc. J. 1957(12):89–92.

Sutton, E. 18463.

18191. Sutton, M. D., and W. Bell. 1954: The use of aureomycin as a treatment of swede seed for the control of black rot (*Xanthomonas campestris*). Plant Disease Reptr. 38:547–52.

Sutton, M. D. 19333.

18192. Sutton, P., and W. A. Seay. 1958: Relationship between the potassium removed by millet and red clover and the potassium extracted by four chemical methods from six Kentucky soils. Soil Sci. Soc. Am. Proc. 22:110–15.

Suzaki, T. 16797.

18193. Suzana, O., V. Gheorghe, T. Iordache, and A. Mihalache. 1958: Physiologic investigations on the ash-seed (*Fraxinus excelsior*) germination. Analele Bucharest Inst. Cercet. Forest. 19:31–68.

18194. Suzuki, B., and U. Nishimoto. 1930: The kephalins of soybean. Proc. Imp. Acad. Japan 6:262–65.

18195. Suzuki, H., T. Matsumoto, and S. Kawamura. 1955: Relation of maturity to the composition of broad bean seeds. I. Nosan Kako Gijutsu Kenkyu Kaishi 2:95–98. Chem. Abstr. 51:15710. 1957.

Suzuki, H. 10373, 14784.

Suzuki, S. 10951.

Suzuki, Y. 9418, 9420.

18196. Svejda, F. J., and P. W. Voisey. 1962: Note on a method for scarifying rose seeds. Can. J. Plant Sci. 42:383–85.

Svejda, F. [J.] 7169.

18197. Svenson, J. 1952: Neuere Erfahrungen mit der Schnellmethods zur Bestimmung der Keimfahigkeit des Getreides. 1952:421–22. Biol. Abstr. 26: 36032. 1952.

Sverdlina, K. 16547.

18198. Sveshnikova, I. N., and Y. P. Bolyakina. 1964: Electron-microscopic study of sunflower seeds and accumulation of reserve substances. Dokl. Akad. Nauk SSSR 151:1222–24. Chem. Abstr. 59: 15598. 1964.

18199. Sveshnikova, V. M. 1962: Some data on the germination of seeds of alpine plants of the Pamir meadows. Dokl. Akad. Nauk Tadzh. SSR 5:45–48.

18200. Svihla, R. D., and E. Osterman. 1943: Growth of orchid seeds after dehydration from the frozen state. Science 98:23–24.

Svoboda, A. 11461.

18201. Swaebly, M. A. 1951: The serial sectioning of dried mature cereal seeds. Stain Technol. 26:153–56.

Swain, A. P. 8392.

Swain, F. G. 5094.

18202. Swaine, G. 1957: Trials on the underground storage of maize of high moisture content in Tanganyika. Bull. Entomol. Res. 48:397–406.

Swaminathan, M. 2822, 3806–3808, 6938, 10099, 11414, 11415, 15955, 19064.

18203. Swaminathan, M. S., and S. K. Kamra. 1961: X-ray analysis of the anatomy and viability of seeds of some economic plants. Indian J. Genet. Plant Breeding 21:129–35.

18204. —— and K. Sulbha. 1959: Multivalent frequency and seed fertility in raw and evolved tetraploids of *Brassica campestris* var. *toria*. Z. Vererbungslehre 90: 385–92.

Swaminathan, M. S. 2416.

18205. Swamy, B. G. L., and P. M. Ganapathy. 1957: On endosperm in dicotyledons. Botan. Gaz. 119:47–50.

18206. —— and K. K. Lakshmanan. 1962: Contributions to the embryology of the Najadaceae. J. Indian Botan. Soc. 41:247–67.

18207. —— and D. Padmanabhan. 1962: A reconnaisance of angiosperm embryogenesis. J. Indian Botan. Soc. 41:422–39.

18208. —— and N. Parameswaran. 1963: The helobial endosperm. Biol. Rev. 38: 1–50.

Swamy, B. G. L. 15207.

18209. Swamy, G. L. 1946: Endosperm in *Hypericum mysorense* Heyne. Ann. Botany (London) 10:165–69.

Swamy, N. S. R. 14222.

18210. Swank, W. G. 1944: Germination of seeds after ingestion by ring-necked pheasants. J. Wildlife Management 8: 223–31. Biol. Abstr. 18:15588. 1944.

18211. Swanson, A. F. 1926: Relation of the seed coat of *Feterita* to the rate of water absorption and germination. J. Am. Soc. Agron. 18:428–32.

18212. —— 1928: Seed-coat structure and inheritance of seed color in sorghums. J. Agr. Res. 37:577–88.

18213. —— 1939: Long-time storage of winter wheat. J. Am. Soc. Agron. 31: 896–97.

18214. —— and R. Hunter. 1936: Effect of germination and seed size on sorghum stands. J. Am. Soc. Agron. 28:997–1004.

18215. Swanson, C. O. 1936: Physical tests to determine quality in wheat varieties. Cereal Chem. 13:179–201.

18216. —— 1941: The effect of low temperature in preventing damage to wheat stored with high moisture content. Cereal Chem. 18:299–315.

18217. —— 1941: Effects of moisture on the physical and other properties of wheat. Cereal Chem. 18:705–29.

18218. —— 1943: Effects of moisture on the physical and other properties of wheat. II. Wetting during harvest. Cereal Chem. 20:43–61.

18219. —— 1943: Effects of moisture on the physical and other properties of wheat. III. Degree, duration, and number of wetting treatments. Cereal Chem. 20:286–99.

18220. —— 1946: Effects of rains on wheat during harvest. Kansas Agr. Expt. Sta. Tech. Bull. 60. 92 pp.

Swanson, C. O. 6294.

Swanson, C. P. 370.

18221. Swanson, C. R., W. C. Shaw, and J. H. Hughes. 1953: Some effects of isopropyl N-(3-chlorophenyl) carbamate and an alkanolamine salt of dinitro ortho secondary butyl phenol on germinating cotton seeds. Weeds 2:178–89.

Swanson, C. R. 16542.

18222. Swarbrick, J. T. 1964: Studies on germination and growth (of *Cacao* seeds). W. African Cocoa Res. Inst. Ann. Rept. 1962/63:75–77.

18223. —— and V. O. Iweama. 1964: Investigations into cocoa seed storage. W. African Cocoa Res. Inst. Ann. Rept. 1962/63:78–79.

18224. Swartley, J. C., and L. C. Chadwick. 1942: Effects of synthetic growth substances on cuttings, seeds, and transplants. Ohio Agr. Expt. Sta. Bimo. Bull. 217:125–44. Expt. Sta. Rec. 88:46. 1943.

Swartley, J. C. 3751.

18225. Swarup, G., E. D. Hansing, and C. T Rogerson. 1962: Fungi associated with sorghum seed in Kansas. Trans. Kansas Acad. Sci. 65:120–37. Biol. Abstr. 43: 3468. 1963.

Swenson, K. G. 9126.

18226. Swenson, S. P. 1942: Inheritance of seed color in biennial white sweet clover, *Melilotus alba.* J. Am. Soc. Agron. 34: 452–59.

Swenson, S. P. 6098.

Swiak, J. F. 11889.

Swift, E. 7940.

Swift, J. E. 608.

18227. Swingle, C. F. 1925: Heavy losses follow late stratification of plum and cherry seeds. Natl. Nurseryman 33:197–200.

18228. Swingle, W. T. 1928: Metaxenia in the date palm. Possibly a hormone action by the embryo or endosperm. J. Heredity 19:257–67. Biol. Abstr. 6:15589. 1932.

18229. Switzer, G. L. 1959: The effect of specific gravity separation on some common indices on Loblolly pine seed quality. J. Forestry 57:497–99.

18230. Swoboda, R. F. 1957: Infra-red heat increases germination of legume seeds. Crops & Soils 10(1):19.

18231. Swofford, T. F., and L. Jones. 1960: Seventh Annual Report Fiscal Year 1960. Region 8 Tree Seed Test. Lab. 22 mimeo. pp.

18232. Swynnerton, R. J. M. 1937: The oyster or Kweme nut. E. African Agr. J. 2:444–46.

18233. Syakudo, K., S. Kawabata, and M. Murata. 1963: Studies on the variation and inheritance of the hard seed content of renge (*Astragalus sinicus* L.). Japan. J. Breeding 13:21–24. Biol. Abstr. 44:16639. 1963.

Sydow, E. von. 1616.

Sykes, J. T. 3664.

Sykes, P. J. 7839.

18234. Sylven, N. 1934: Hulless seeds of timothy. Sveriges Utsadesforen. Tidskr. 44:13–58.

Sylvester, F. J. 13890.

18235. Sylwester, E. P. 1939: Farm demonstrations as a measure of seed viability in alfalfa, red clover and sweet clover. Proc. Assoc. Offic. Seed Analysts 30:172–76.

Sylwester, E. P. 8910.

18236. Symes, K. J. 1958: Identification of wheat varieties from their grain characters. Agr. Gaz. N.S. Wales 69:178–85.

18237. Symon, D. E. 1956: Polyembryony in *Cassia.* Nature 177:191.

18238. Symons, J. L. 1926: Studies on the genus *Xanthium.* Botan. Gaz. 82:121–47.

Szabo, A. 2960.

Szabo, M. 2693.

18239. Szahlender, K. 1936: Ingredients of the seed of *Digitalis lanata* Ehrh. Arch. Pharm. 274:446–49. J. Am. Pharm. Assoc. 27:393. 1938.

18240. Szanyi, I. 1934: The chemical composition of beans. Kiserl. Kozl. 37:289–97. Chem. Abstr. 29:2615. 1935.

Szasz, J. 11960.

Szent-Istvany, A. 18636.

Szigat, G. Schlosser-; *see* Schlosser-Szigat, G.

Szilard, K. 20073.

18241. Szmalowa, Z. 1952: Distribution of lipase in seeds of *Ricinus communis.* Farm.

Polska 8:294–300. Chem. Abstr. 47: 12526. 1953.

Szobo, J. Hamori-. 5952.

Szopos, A. 16811.

Szumer, A. Z. 8381, 8382.

18242. Szwadiak, J. 1956: Capability to germination of oil pumpkin seeds stored during the winter inside the fruit in low temperatures. Roczniki Nauk Rolniczych Ser. A72:527–28. Biol. Abstr. 33:3038. 1959.

18243. Szweykowska, A. 1959: The effect of nitrogen feeding on anthocyanin synthesis in isolated red cabbage embryos. Acta Soc. Botan. Polon. 28:539–49.

18244. ——, M. Gierczak, and R. Luszczak. 1959: Anthocyanin synthesis in isolated embryos of red cabbage and radish. Acta Soc. Botan. Polon. 28:531–37.

18245. Szymczyk, F., and J. Kolankiewicz. 1954: Natural arsenic and lead content of Polish grains. Roczniki Panstwowego Zakladu Hig. 5:53–59. Chem. Abstr. 48:8974. 1954.

T

18246. T., S. C. 1939: Mimba seed. A mangosteen. Queensland Agr. J. 52:240.

18247. Tabares, L. G. 1944: The protein metabolism of *Hordeum vulgare* during the germination period. Anales Real Acad. Farm. 10:699–710. Chem. Abstr. 43:7554. 1949.

Tabatabai-Yazdi, H. 12259.

Tabentskii, D. A. 14733.

18248. Tabor, P. 1949: Moderate and complete scarification of kudzu and perennial lespedeza seed. Agron. J. 41:491–92.

Tabor, P. 13209.

18249. Tachibana, Y. 1961: Studies in hibiscus. IV. Further studies on the method of promoting seed germination. J. Japan. Soc. Hort. Sci. 30:183–88.

18250. Tackle, D. 1954: Comments on the Baker automatic tree seed planter. J. Forestry 52:530–31.

18251. Tada, M., and S. Kawamura. 1963: Changes of soybean carbohydrates during growth and germination. I. Tech. Bull. Fac. Agr. Kagawa Univ. (English Transl.) 14:148–55.

Tada, S. 10605.

18252. Tadmor, N. H., and D. Koller. 1958: Experiments in the propagation of *Juncus mantimus* Lam. II. Germination and field trials in the Arava. Ktavim 9:177–205. Biol. Abstr. 35:54109. 1960.

Tadmor, N. H. 11039.

18253. Tadokoro, T., and M. Abe. 1930: Studies on the ripening of rice grains. J. Coll. Agr. Hokkaido Imp. Univ. 27:349–87. Chem. Abstr. 25:1554. 1931.

18254. —— and T. Saito. 1941: Chemical properties of an estrone-like substance and an enzyme-containing protein in the rice germ. I. J. Agr. Chem. Soc. Japan 62: 417–18. Chem. Abstr. 37:4431. 1943.

18255. ——, ——, and N. Takasugi. 1942: The enzyme-containing proteins of rice embryo. J. Agr. Chem. Soc. Japan 63: 1015–18. Chem. Abstr. 41:3509. 1947.

18256. —— and N. Takasugi. 1942: Chemical properties of the enzyme-containing proteins in the germ of rice seed. J. Agr. Chem. Soc. Japan 63:460–61. Chem. Abstr. 41:3141. 1947.

18257. ——, T. Tsuji, and S. Watanabe. 1927: On the difference in physico-chemical properties of various proteins in plant seeds. II. J. Coll. Agr. Hokkaido Imp. Univ. 19:93–106.

18258. Tadros, W., and M. Kamel. 1952: A preliminary study of the main polysaccharide of *Lupinus termis* seeds. J. Chem. Soc. 1952:4532–33.

Tadzhitdinov, M. 18835.

18259. Täufel, K., and G. Gamperl. 1931: Über die Sterine der Gerste und ihrer Mälzungsprodukte. Biochem. Z. 235:353–58.

18260. —— and F. Krussen. 1952: Verteilung der Citronensäure in den Samen von Leguminosen und Cerealien und ihr Verhalten bei der Keimung. Biochem. Z. 322:375–79. Biol. Abstr. 27:14217. 1953.

18261. —— and R. Pohloudek-Fabini. 1954: Relations between viability and free fatty acid content in rape and colza seeds. Pharmazie 9:511–14. Chem. Abstr. 49: 9745. 1955.

18262. —— and —— 1955: Das Verhalten der Citronensäure bei der Keimung von Cerealien-, Leguminosen- und Ölsamen. Biochem. Z. 326:280–87. Biol. Abstr. 34:14501. 1959.

18263. ——, K. J. Steinbach, and E. Vogel. 1960: Mono- and oligosaccharides of some legume seed as well as their behavior on storage and germination. Z. Lebensm.-Untersuch. -Forsch. 112:31–40. Chem. Abstr. 54:13288. 1960.

18264. Tagawa, K., and M. Shin. 1959: Hemoproteins of wheat germ. I. Crystallization and properties of peroxidase from wheat germ. J. Biochem. (Tokyo) 46:865–73.

18265. —— and —— 1959: Hemoproteins of wheat germ. II. Particulate electron transporting system of wheat germ and the effect of peroxidase on it. J. Biochem. (Tokyo) 46:875–881.

18266. ——, ——, and K. Okunuki. 1959: Peroxidases from wheat germ. Nature 183:111.

Tagawa, K. 7993.

18267. Tager, J. M., and S. H. Cameron. 1957: The role of the seed coat in chlorophyll deficiency (albinism) of *Citrus* seedlings. Physiol. Plantarum 10:302–5.

18268. —— and B. Clarke. 1961: Replacement of an alternating temperature requirement for germination by gibberellic acid. Nature 192:83–84.

Tagliabue, L. 11793.

18269. Tahara, M. 1940: Embryogeny of *Torreya nucifera* S. et Z. Sci. Rept. Tohoku Univ. Fourth Ser. 15:419–26.

18270. —— 1941: Embryogeny of *Podocarpus macrophyllus* and *Podocarpus nagi*. Sci. Rept. Tohoku Univ. Fourth Ser. 16:91–98.

18271. Tai, K. 1953: Distribution of urease in plant seeds. Shiga Agr. Coll. Sci. Rept. Ser. I, no. 4:57–59. Chem. Abstr. 50:6598. 1956.

18272. Tai, L. Y., and J. H. Hu. 1957: The development of the embryo of *Vicia faba* L. Wuhan Univ. Nat. Sci. Rept. 2:167–72.

Taimni, S. 18565.

18273. Taira, H. 1962: Amino acid pattern of seed proteins as a standard in the plant taxonomy. Botan. Mag. (Tokyo) 75:80–81.

18274. —— 1962: Studies on amino acid contents in plant seed. Amino acids contained in the seed of Gramineae. Botan. Mag. (Tokyo) 75:242–43.

18275. —— 1962: Studies on amino acid contents in plant seeds. II. Amino acid pattern of seed protein fractions of Gramineae. Botan. Mag. (Tokyo) 75:273–77.

18276. —— 1963: Studies on amino acid contents in plant seeds. III. Amino acid contained in seed of Gramineae (Part 2). Botan. Mag. (Tokyo) 76:340–41.

Takada, H. 10361, 11568.

18277. Takagi, I. 1939: On the storage of mulberry seeds. Tokyo Imp. Sericult. Coll. Res. Bull. 2:1–22. Japan. J. Botany 10:66. 1940.

18278. —— 1940: The effects of temperature and light conditions upon the germination of mulberry seed. Tokyo Imp. Sericult. Coll. Res. Bull. 2(2):1–26.

Takagi, S. 265.

18279. Takagi, T. 1957: The growth and germination of seed in Bambusaceae. Hokuriku J. Botany 6(2):56–61.

18280. Takahashi, E., and K. Shirahama. 1931: On the change of barley protein in storage and germination. J. Fac. Agr. Hokkaido Univ. 30:119–61. Ber. Wiss. Biol. 19:302–4. 1931. Ber. Ges. Physiol. 64:471–72. 1932.

18281. Takahashi, H. 1936: Rice plants grown from seeds of haploid individuals. Proc. Crop Sci. Soc. Japan 8:355–63.

Takahashi, J. 11572.

Takahashi, M. 13431, 13432.

18282. Takahashi, N. 1954: Studies on the germination of rice seed; the carbohydrate metabolism and the development of plumule or radicle. Tohoku Univ. Rept. Inst. Agr. Res. 6:1–12.

18283. —— 1955: Effect of varietal difference on the velocity of germination of rice seeds. Tohoku Daigaku Nogaku Kenkyusho Iho 7:1–12. Japan. Sci. Rev. 1955(6):539. 1957. Chem. Abstr. 52:7438. 1958.

18284. —— 1957: Ecological studies of fruiting behaviours of azuki beans (*Phaseolus angularis* W. F. Wight). Proc. Crop Sci. Soc. Japan 26:43–44.

18285. —— 1960: The pattern of water absorption in rice seed during the germination. I. The relationship between the pattern and the germination velocity. Proc. Crop Sci. Soc. Japan 29:1–3.

18286. —— 1960: Studies on the internal factors of rice seed regulating the germination velocity. Tohoku Univ. Rept. Inst. Agr. Res. 11:1–7.

18287. —— 1962: Physicogenetical studies on germination of rice seeds with special reference to its genetical factors. Tohoku Univ. Inst. Agr. Res. Bull. 14:1–87.

18288. —— 1963: Studies on the dormancy of wild rice seeds. II. Sci. Rept. Tohoku Univ. Inst. Agr. Res. Ser. D 14:75–85.

18289. ——, T. Moroo, T. Hashimoto, and T. Yamaki. 1962: Effects of some organic acids on the dark germination of tobacco seed. Botan. Mag. (Tokyo) 75:163–69.

18290. —— and H. Oka. 1958: Observations on natural populations of Formosan wild rice particularly on variations in seed dormancy. Ann. Rept. Mishima Natl. Inst. Genet. 8:58–59. 1957.

18291. —— and —— 1959: A preliminary report of dormancy in wild rice seed. Tohoku Univ. Rept. Inst. Agr. Res. 10:81–85.

18292. —— and —— 1959: Factorial analysis of regulating mechanism on germination velocity in rice seed. Tohoku Univ. Inst. Agr. Res. Bull. 10:207–11.

18293. ——, T. Yamada, T. Hashimoto, and T. Yamaki. 1962: Effect of inorganic compounds on the dark germination of tobacco seeds induced by gibberellin. Botan. Mag. (Tokyo) 75:49–55.

18294. ——, ——, T. Moroo, and T. Yamaki 1962: Further studies upon the effect of inorganic compounds on the dark germination of tobacco seeds induced by gibberellin. Botan. Mag. (Tokyo) 75:83–91.

Takahashi, N. 10214, 19930.

18295. Takahashi, R. 1937: Germination of wheat kernels on the spike. Proc. Crop Sci. Soc. Japan 9:45–62.

18296. —— 1938: The influence of temperature on the swelling of hard seed of the Chinese milk-vetch (*Astragalus sinicus*). Rept. Ohara Agr. Inst. 30:111–31. Biol. Abstr. 24:10582. 1950.

Takahashi, R. 11150–11160.

18297. Takahashi, T. 1905: Is germination possible in absence of air? Bull. Coll. Agr. Tokyo 6:439–42.

Takahashi, Y. 11585.

18298. Takahasi, E., and G. Naito. 1953: Seed fossil from the Tsubuta group. J. Geol. Soc. Japan 59:255–56. Biol. Abstr. 29:26776. 1955.

Takahasi, K. 19971.

18299. Takamiya, E. 1935: Studies on the castor-bean lipase. VI. On a new preparation of highly active *Ricinus*-lipase and its properties. J. Agr. Chem. Soc. Japan 11:17–18.

18300. —— 1935: Castor bean lipase. VIII. A new substance having the power of activation or retardation of lipase from the castor bean. J. Agr. Chem. Soc. Japan 11:216–21. Chem. Abstr. 29:5466. 1935.

18301. Takano, K., and M. Nozu. 1961: Kinds of free amino acids in rice kernel. Proc. Crop Sci. Soc. Japan 29:216–18. Biol. Abstr. 36:82098. 1961.

Takano, S. 18338.

18302. Takao, A. 1960: Histochemical studies on embryogenesis of *Pinus thunbergii* Parl. Botan. Mag. (Tokyo) 73:379–88.

18303. —— 1962: Histochemical studies of the formation of some leguminous seeds. Japan. J. Botany 18:55–72.

Takase, N. 13029.

18304. Takashima, R., H. Kawahara, and H. Hara. 1951: On the morpho-regulative effects of supersonics on germination of seeds and growth of sprouts. Bull. Exptl. Biol. 1:1–6. Biol. Abstr. 27:33029. 1953.

18305. Takasugi, N., and I. Yamazaki. 1951: The fluctuation of enzymes and other important substances during wheat germination. II. The mechanism of the appearance and disappearance of catalase activity during wheat germination. J. Agr. Chem. Soc. Japan 25:113–15. Chem. Abstr. 46:10305. 1952.

Takasugi, N. 18255, 18256.

18306. Takasugi, S. 1937: Unterschied der Japanischen Gerstenrassen, welcher durch die Behandlung ihrer Samenkörner mit Karbolsäure, Schwefelsäure oder Ätzkalilösung nachgewiesen werden kann. Agr. Hort. (Tokyo) 12:1101–5. Japan. J. Botany 9:83. 1938.

18307. —— and K. Dobashi. 1953: Studies of lucerne. I. The resistance of seeds and seedlings of lucerne to soil acidity. Proc. Crop Sci. Soc. Japan 22:137–39. Soils Fertilizers 17:377. 1954.

Takata, K. 14794–14796.

18308. Takatoi, I., and H. Toyooka. 1954: Influence of 2,4-D solution on the germination of seeds of weeds. Hokkaido Forest Expt. Sta. Spec. Rept. 2:81–86.

Takayama, M. 8304.

Takayama, T. 8305.

18309. Takeda, H. 1936: On the coma or hairy tuft on the seed of willows. Botan. Mag. (Tokyo) 50:283–89.

Takematsu, T. 11572–11574.

Takenaga, H. 9530.

18310. Takeshima, H. 1961: Growth regulators of rice seed. Agr. Hort. (Tokyo) 36:869–70. Chem. Abstr. 56:15863. 1962.

18311. Takeuchi, K. 1961: Amino acids in the endosperm of some Amazonian Palmae. Chiba Daigaku Bunri Gakuba Kiyo Shizen Kagaku 3:321–25. Chem. Abstr. 58:9411. 1963.

18312. Takeuchi, M. 1956: Embryogenesis in *Phaseolus vulgaris*. J. Fac. Sci. Univ. Tokyo 6:439–52.

Takeuchi, S. 9476.

18313. Takeuchi, T., and T. Hashimoto. 1963: Effects of the low temperature vernalization on the shortening of growth in Italian ryegrass. Proc. Crop Sci. Soc. Japan 32:66–68.

18314. Takiguchi, S., T. Nonomura, and H. Nakamura. 1958: "Residual solution" by wheat vernalization. II. Its influence on the earing of wheat. Proc. Crop Sci. Soc. Japan 27:281–82.

18315. Takiguti, Y. 1930: On the germination of *Brassica napella* seed. Kjusu Imp. Univ. Bul. Sci. Fak. Terkult. 4:22–36.

18316. —— 1930: On the germination of seeds of *Oenothera*. Agr. Hort. (Tokyo) 5:748–54. Japan. J. Botany 5:80. 1931.

18317. —— 1930: On the germination of seeds of *Polygonum hydropiper*. Proc. Crop Sci. Soc. Japan 2:195–96. Japan. J. Botany 5:81. 1931.

18318. —— 1930: On the germination of seeds of *Perilla ocimoides*. Proc. Crop Sci. Soc. Japan 2:199–200. Japan. J. Botany 5:81. 1931.

18319. —— 1932: Davis' modified catalase method prophesying the germinating power of seeds. Bul. Sci. Fak. Terkult. Kyusu Imp. Univ. 5:103–16. Chem. Abstr. 29:4797. 1935.

Takizawa, K. 10384–10386.

Talas, M. 16727.

Talbert, F. D. 17563, 17566.

18320. Tallarico, G. 1931: La conservazione e la produttivita dei semi del grano. Predeterminazione funzionale. Atti Accad. Nazl. Lincei 6:893–99. Ber. Wiss. Biol. 19:581–82. 1931.

18321. —— 1931: Optimum experiments on the life of corn in an arid climate. Ital. Agr. 68:565–80.

18322. —— 1932: Grain as food and as seed. Mem. Accad. Ital. Biol. 3:5–325. Chem. Abstr. 27:757–58. 1933.

18323. —— 1933: Nutritive value of slightly germinated grains. Atti 1 Congr. Intern. Panificazione, Rome 1932:349–57. Chem. Abstr. 28:1780. 1934.

18324. —— 1951: The importance of seeds of mutilated plants. Terra Sole 111:401–2.

18325. —— and M. Tirelli. 1935: Recherches ultérieures sur la difference de germination pendant les heures du jour et de la nuit. Ital. Agr. 72. Proc. Intern. Seed Testing Assoc. 8:91. 1936.

18326. —— and —— 1936: Influence of meteoric conditions on germination. Ann. Staz. Bacol. Sper. Padua 48:331–48.

18327. Tamamshian, S. 1938: Carpobiological and carpological studies. Acad. Sci. RSS Ukraine Inst. Botan. Kiev 1938: 126–27.

18328. Tamamshian, S. G., and G. A. Denisova. 1959: The embryo and seedling morphology of the *Caropodium platycarpum* (Boiss. et Hausskn.) Schischk. and *Echinophora trichophylla* Smith. Botan. Zh. 44:433–46.

18329. Tamaoki, T., and A. J. Ullstrup. 1958: Cultivation in vitro of excised endosperm and meristem tissues of corn. Bull. Torrey Botan. Club 85:260–72.

18330. Tamari, K., and J. Kaji. 1952: The mechanism of injurious action of fusarinic acid on plant growth. I–VII. J. Agr. Chem. Soc. Japan 26:233–37; 295–98; 298–303; 345–49; 349–53; 249–52; 302–6. Chem. Abstr. 48:9483–84. 1954.

18331. Tamchuma, J., and J. Sebanek. 1960: Über die stimulierende Wirkung des Glukothiokarbamids auf die Keimung und das Wachstum der Pflanzen. Sb. Česk. Akad. Zeměděl. Ved. Rostlinna Vyroba 6:1223–34.

18332. Tamhane, V. A., and M. A. S. Iyengar. 1934: Salt tolerance of plants as induced by pre-treatment of seeds. Current Sci. (India) 2:436–37.

Tamm, C. 5774, 10429, 11356, 12213, 12214, 15878, 16944, 17223, 19202.

18333. Tamm, E. 1928: Über den Einfluss der durch den Boden geleiteten elektrischen Energie auf Keimfähigkeit, Triebkraft und Jugendwachstum von *Pisum sativum*. Botan. Arch. 21:9–115.

18334. —— 1932: Untersuchungen über Keimungsminimum, Keimfähigkeit und durchschnittliche Keimdauer landwirtschaftlicher Kulturpflanzen. Arch. Pflanz. 8:527–57. Ber. Wiss. Biol. 22:371. 1932.

18335. —— 1934: Weitere Untersuchungen über die Keimung und das Auflaufen landwirtschaftlicher Kulturpflanzen. II. Pflanzenbau 10:297–312.

18336. Tammes, P. M. L. 1949: Selecting nutmegs on the basis of oil content. Landbouw. 21:65–67. Biol. Abstr. 25: 37656. 1951.

18337. Tammes, T. 1900: Über den Einfluss der Sonnenstrahlen auf die Keimungsfähigkeit von Samen. Landwirtsch. Jahrb. 29:467–82.

18338. Tamura, T., K. Ito, and S. Takano. 1957: Breaking of the dormancy of spinach seeds. Kyushu Agr. Res. 1957(19): 54–56.

18339. Tanaka, K. 1933: Über die Sterine des Reisembryos. I. J. Biochem. (Tokyo) 17:483–94. Chem. Abstr. 27:5098–99. 1933.

18340. —— 1954: Enzymological studies on the mechanism of polysaccharide formation in three types of maize seed. Symp. Enzyme Chem. (Japan) 10:189–201. Chem. Abstr. 49:16078. 1955.

18341. —— 1955: Enzymatic studies on the mechanism of polysaccharide formation in maize seed. II. Q-enzyme of non-waxy seeds. Botan. Mag. (Tokyo) 68:352–57.

18342. —— and T. Tanaka. 1933: Über die Gallensäurebildung. VI. Reisembryo und Gallensäureausscheidung. J. Biochem. (Tokyo) 18:15–22.

Tanaka, K. 8146.

Tanaka, S. 14066.

Tanaka, T. 9579, 13010, 13030, 14071, 18342.

18343. Tanashev, G. A. 1938: Improvement on the indicator method for the rapid determination of the germinative power of cultivated plant seeds. Chem. Social. Agr. (USSR) 7:101–7. Chem. Abstr. 33:8241. 1939.

18344. —— 1938: Stimulating weed-seed germination by chemical means. Dokl. Vses. Akad. Sel'skokhoz. Nauk 1/2(10/11): 3–8. Chem. Abstr. 35:5159. 1941.

18345. Tanda, Y. 1955: On the character of trash rice. Proc. Crop Sci. Soc. Japan 24:73–77.

18346. —— 1957: On the development of seed in azuki-bean plant. Proc. Crop Sci. Soc. Japan 26:45–46.

18347. —— 1961: Studies on the mechanism of falling of embryo from rice kernel in process of cleaning with reference to early and late cultivation. Proc. Crop Sci. Soc. Japan 30:9–12.

18348. —— 1962: Histological studies on water absorption of rice kernel. I. The relation between the physical absorption of water and the kernel tissue. Proc. Crop Sci. Soc. Japan 31:167–70.

Tanda, Y. 14078, 14083.

Tandon, R. K. 17499.

18349. Tandon, S. L., and B. M. Kapoor. 1961: Amitosis in the endosperm of *Zephyranthes ajax* Sprenger. Current Sci. (India) 30:72–73.

18350. —— and —— 1962: Contributions to the cytology of endosperm in some angiosperms. I. *Zephyranthes ajax* Sprenger. Caryologia 15:21–41.

18351. Tang, P. S. 1931: An experimental study of the germination of wheat seed under water, as related to temperature and aeration. Plant Physiol. 6:203–48.

18352. —— 1931: Temperature characteristics for the oxygen consumption of germinating seeds of *Lupinus albus* and *Zea mays*. J. Gen. Physiol. 14:631–41.

18353. —— 1931: Temperature characteristics for the production of CO_2 by germinating seeds of *Lupinus albus* and *Zea mays*. J. Gen. Physiol. 15:87–95.

18354. —— 1932: On the respiratory quotient of *Lupinus albus* as a function of temperature. J. Gen. Physiol. 15:561–69.

18355. —— 1932: A respirometer vessel for study of metabolism of seeds. J. Gen. Physiol. 15:571–74.

18356. —— 1932: The effects of carbon monoxide and light on the oxygen consumption and on the production of carbon dioxide by germinating seeds of *Lupinus albus*. J. Gen. Physiol. 15:655–65.

18357. —— 1932: Temperature characteristic for the anaerobic production of carbon dioxide by germinating seeds of *Lupinus albus*. J. Gen. Physiol. 16:65–73.

18358. ——, F. C. Wang, and F. C. Chih. 1959: Studies on plant respiration. III. Influence of oxygen tension on organ formation, material and energy transformation during germination of rice seeds. Acta Botan. Sinica 8:188–200. Biol. Abstr. 35:64331. 1960.

18359. —— and H. C. Yin. 1935: Effect of mercuric chloride on the germination of seeds of *Pisum sativum*. Bull. Chinese Botan. Soc. 1:98–107. Chem. Abstr. 30:5710. 1936.

Tang, P. S. 4100.

18360. Tang, W. T., and S. M. Chiang. 1955: Studies on the dormancy of rice seed. Taiwan Univ. Mem. Coll. Agr. 4:1–7.

18361. Tang, Y. W., and T. L. Loo. 1948: The effect of manganese salts on seed germination and early growth of *Zea mays* L. Acad. Sinica Botan. Bull. 2:91–95.

Tang, Y. W. 12422.

Taniguchi, S. 5803.

Tanner, C. C. 9704.

18362. Tanner, F. W., Jr., S. E. Pfeiffer, and J. J. Curtis. 1947: B-Complex vitamins in grain sorghums. Cereal Chem. 24:268–74.

18363. Tanret, G. 1934: Sur le glucoside des graines de *Coronille*. Compt. Rend. 198:1637–39. Chem. Abstr. 28:4737–38.

Tanret, G. 15359.

18364. Tao, W. S. 1930: Biochemical studies on rice starch. I. Chemical changes of the starch during the germination of rice in darkness. Bull. Chem. Soc. Japan 5:64–69. Japan. J. Med. Sci. Biochem. 2:76. 1934.

18365. —— 1930: Biochemical studies on rice starch. II. Temperature effects on the germination of rice seeds in darkness. Bull. Chem. Soc. Japan 5:69–73. Expt. Sta. Rec. 68:172. 1933.

18366. —— and S. Komatsu. 1931: Biochemical studies on the soy bean. I. Chemical changes of the protein during the germination of the soy bean in darkness Mem. Coll. Sci. Kyoto Imp. Univ. Ser. A 14:287–92. Chem. Abstr. 26:1640–41. 1932.

18367. ——, Y. S. Shen, H. C. Chang, and C. C. Kung. 1957: Basic studies of vegetable proteins. I. Amino acids of the globulins of sunflower seed (*Helianthus annuus*) and of cotton seed (*Gossypium*). Tung Pei Jen Min Ta Hsüeh-Tzu Jan K'o Hsüeh Hsüeh Pao 1957:107–14. Chem. Abstr. 53:8317. 1959.

18368. Tapke, V. F. 1924: Effects of the modified hot-water treatment on germination, growth, and yield of wheat. J. Agr. Res. 28:79–98.

18369. —— 1932: Seed treatments with chemical dusts and formaldehyde for smut control in oats. Phytopathology 22:429–41.

18370. —— 1936: The influence of seed hulling on loose smut in naturally inoculated oats. Phytopathology 26:588–96.

18371. —— 1940: Studies on the natural inoculation of seed barley with covered smut (*Ustilago hordei*). J. Agr. Res. 60:787–810.

18372. —— and W. M. Bever. 1942: Effective methods of inoculating seed barley with covered smut (*Ustilago hordei*). Phytopathology 32:1015–21.

Tapke, V. F. 12138.

18373. Tapley, W. T. 1951: The mechanical bean picker and maximum yields. Farm. Res. (N.Y.) 17(4):6.

18374. Taran, E. N. 1937: Oil formation in the later stages of ripening of seeds of oleaginous plants. Biochemistry (USSR) (English Transl.) 2:741–44. Chem. Abstr. 32:2171. 1938.

18375. Taran, I. 1956: Germination capacity of sunflower seeds in relation to degree of their ripeness. Botan. Zh. 41:1656–62.

18376. Taranenko, V. I. 1961: Effectiveness of preplanting treatment of sorghum seeds with gibberellic acid. Selek. i Semen. 1962(2):71–72.

Taranova, E. A. 11304.

Tarantola, C. 2842A, 2843.

Tarantola, M. 12400, 18629, 18630.

18377. Tarasenko, N. D. 1963: Action of γ-rays, fast neutrons, and ethylenimine on the viability of and chromosome aberration in potato seeds. Radiobiologiya 3:427–30. Nucl. Sci. Abstr. 17:4260. 1963.

Tarasikov, N. A. 19913.

18378. Taraskina, K. V., and T. K. Chumbalov. 1958: *Ephedra.* III. Flavone dyes from the seeds of *Ephedra equisetina.* Uch. Cap. Kazakh. Univ. 44:59–64. Chem. Abstr. 56:10586. 1962.

18379. Taravet, A. 1948: Réactions photopériodiques après application de substances de croissance à la germination. Compt. Rend. 226:1538–39.

18380. Tarjan, A. C., and J. E. Moore. 1949: Preliminary tests with seed treatments in prohibiting root-knot nematode infection of cucumber seedlings. Plant Disease Reptr. 33:447–50.

18381. Tarkhov, V. S., and A. I. Martishenya. 1934: Germinating peas as a source of vitamin C. Trav. Acad. Militaire Med. Armee Rouge URSS 1:364–68. Chem. Abstr. 32:3847. 1938.

18382. Tarnavschi, I. T., D. G. Constantinescu, S. Mihailescu-Firea, M. Retezeanu, and J. Calcandi. 1961: Report on a study of the effect of colchicine on the seeds of *Digitalis* sp. Bucharest Univ. "C. I. Parhon" Gradina Botan. Lucrarile 1960: 333–45.

18383. Tarr, S. A. J. 1954: Protection of sorghum against soil fungi, soil pests and covered smut by combined insecticide-fungicide seed dressings. Ann. Appl. Biol. 41:578–85.

18384. Tarr, S. A. J. 1956: Seed treatment against blackarm disease of cotton in the Sudan. Empire Cotton Growing Rev. 33:98–104.

18385. —— 1958: Control of seed-bed losses of groundnuts by seed treatments. Ann. Appl. Biol. 46:178–85.

Tarr, S. A. J. 1280.

18386. Tasaki, J. 1959: Effect of soil moisture content on growth and yield of azuki-beans, *Phaseolus radiatus* var. *aurea* Prain. I. Growth and yield as affected by the different soil moisture content. II.

Varietal differences of growth and yield at different subirrigated water levels. Proc. Crop Sci. Soc. Japan 28:73–76.

18387. Tascher, W. R. 1933: Experiments on the control of seed-borne diseases by X-rays. J. Agr. Res. 46:909–15.

18388. Tashima, Y., and S. C. Imamura. 1954: Substitutive effect of ribonucleic acid in the vernalisation of *Raphanus sativus* in dark culture . (Preliminary report.) Botan. Mag. (Tokyo) 67:281–82.

Tasker, P. K. 11415.

Taskovitch, L. T. 12596.

18389. Tatachar, T. 1940: The development of the embryo-sac and formation of haustoria in *Lantana indica* Roxb., and *Stachytarpheta indica* Vahl. J. Indian Botan. Soc. 19:45–53.

Tate, H. D. 4516.

18390. Tateoka, T. 1955: Further studies on starch grains of seeds in Poaceae from the viewpoint of systematics. J. Japan. Botany 30:199–208. Biol. Abstr. 30: 35395. 1956.

18391. —— 1962: Starch grains of endosperm in grass systematics. Botan. Mag. (Tokyo) 75:377–83.

18392. —— 1964: Notes on some grasses. XVI. Embryo structure of the genus *Oryza* in relation to the systematics. Am. J. Botany 51:539–43.

Tatman, E. C. 2155.

18393. Tatum, L. A. 1942: The effect of genetic constitution and processing methods on the ability of maize seed to germinate in cold soil. Iowa State Coll. J. Sci. 17:138–40.

18394. —— 1954: Seed permeability and "cold-test" reaction in *Zea mays*. Agron. J. 46:8–10.

18395. —— and M. S. Zuber. 1943: Germination of maize under adverse conditions. J. Am. Soc. Agron. 35:48–59.

Tatum, L. A. 9872.

18396. Tauber, H., B. B. Kershaw, and R. D. Wright. 1949: Studies on the growth inhibitor fraction of lima beans and isolation of a crystalline heat-stable trypsin inhibitor. J. Biol. Chem. 179:1155–61.

Tauber, H. 11863.

Tauboeck, K. 10773.

18397. Tavares, S. 1960: Studies on the germination of the seeds of mangaba, *Hancornia speciosa* Gomes. Arquiv. Inst. Pesquisas Agron. (Pernambuco) 5:193–221. Biol. Abstr. 37:15461. 1962.

18398. Tavshavadze, T. N., and V. D. Gioshvili. 1962: Spectrophotometric determination of unsaturated acids in tea-seed oil. Tr. Gruzinsk. Politekhn. Inst. 1962(5): 33–44. Chem. Abstr. 60:7039. 1964.

Tayeau, E. 17158.

18399. Tayeau, F. 1941: Les points isoelectriques des deux principales fractions proteidiques de l'arachide. Compt. Rend. Soc. Biol. 135:595–96.

18400. Taylor, A. R. A. 1957: Studies of the development of *Zostera marina* L. I. The embryo and seed. Can. J. Botany 35:477–99.

18401. Taylor, B. L. 1963: Cereal and pulse seed certification. Tasmanian J. Agr. 34:51–53.

18402. Taylor, C. A. 1942: Cutting tests on small seeds. J. Forestry 40:809.

18403. —— 1948: Observations on the storage and germination characteristics of *Angelica* seed. Proc. Am. Soc. Hort. Sci. 52:471–74.

18404. —— 1951: Treating celery seed as an aid to direct field sowing. Seed World 69(11):10, 12.

18405. Taylor, D. L. 1942: Influence of oxygen tension on respiration, fermentation, and growth in wheat and rice. Am. J. Botany 29:721–38.

18406. Taylor, G. C. 1933: Some seed novelties. New Flora Silva 5:105–10.

Taylor, H. L. 3958, 3959.

18407. Taylor, H. M. 1962: Seedling emergence of wheat, grain sorghum, and guar as affected by rigidity and thickness of surface crusts. Soil Sci. Soc. Am. Proc. 26:431–33.

18408. Taylor, J. 1957: Rhododendron propagation. Gardeners Chron. 141: 100–1.

18409. Taylor, J. W., B. B. Bayles, and C. C. Fifield. 1939: A simple measure of kernel hardness in wheat. J. Am. Soc. Agron 31:775–84.

18410. —— and F. A. Coffman. 1938: Effects of vernalization on certain varieties of oats. J. Am. Soc. Agron. 30:1010–19.

Taylor, J. W. 4253, 12055, 12902.

18411. Taylor, N. L., E. Gray, W. H. Stroube, and W. A. Kendall. 1961: Some effects of γ and neutron irradiation of seeds on germination and seedling growth of red clover. Crop Sci. 1:458–60.

18412. ——, W. H. Stroube, W. A. Kendall, and E. N. Fergus. 1962: Variation and relation of clonal persistence and seed production in red clover. Crop Sci. 2:303–5.

Taylor, N. W. 4220.

18413. Taylor, R. F. 1928: Seeds soaked in sea water remain good. Forest Worker 4(6):8.

18414. Taylor, R. H., R. G. Grogan, and K. A. Kimble. 1961: Transmission of tobacco mosaic virus in tomato seed. Phytopathology 51:837–42.

18415. Taylor, T. N. 1962: Additional observations on *Stephanospermum ovoides*, a middle Pennsylvania seed. Am. J. Botany 49:794–800.

18416. —— and T. Delevoryas. 1964: Paleozoic seed studies: a new Pennsylvanian *Pachysteta* from southern Illinois. Am. J. Botany 51:189–95.

Taylor, W. S. 3047.

Tayman, L. A. 8652.

18417. Tazawa, H. 1927: Über den Einfluss der hohen Aufbewahrungstemperatur auf die Keimkraft lagernder Getreidesamen. J. Soc. Agr. For. Sapporo, Japan 19:141–61.

Tazawa, M. 3289.

18418. Tazawa, Y. 1955: Chemistry of seed proteins. I. The function of protein granules (Proteinkörner). J. Japan. Chem. 9:558–69. Chem. Abstr. 51:18138. 1957.

18419. —— 1955: Chemistry of seed proteins. II. Protein metabolism of germinating seeds. J. Japan. Chem. 9:667–80. Chem. Abstr. 51:18138. 1957.

18420. —— 1955: Chemistry of seed proteins. III. Proteases from seeds. J. Japan. Chem. 51:18138. 1957.

18421. —— 1955: Chemistry of seed proteins. IV. Individuality of seed proteins. J. Japan. Chem. 9:806–19. Chem. Abstr. 51:18138. 1957.

18422. —— 1956: Chemistry of seed proteins. V. Enzyme proteins from seeds. J. Japan. Chem. 10:52–64. Chem. Abstr. 51:18138. 1957.

18423. —— 1956: Chemistry of seed proteins. VI. Toxic proteins from seeds. J. Japan. Chem. 10:132–44. Chem. Abstr. 51:18138. 1957.

Tazima, Y. 7999.

18424. Tazoye, H., and Y. Saito. 1934: Untersuchungen über den Zapfen und Samen von *Abies mayriana* Miyabe et Kudo, *Picea jezoensis* Carr. und *Picea glehni* Masters. Res. Bull. Coll. Expt. Forests, Hokkaido Univ. 9:1–27. Biol. Abstr. 11:9602. 1937.

18425. Tchen, T. T., F. A. Loewus, and B. Vennesland. 1955: The mechanism of enzymatic carbon dioxide fixation into oxalacetate. J. Biol. Chem. 213:547–55.

18426. —— and B. Vennesland. 1955: Enzymatic carbon dioxide fixation into oxalacetate in wheat germ. J. Biol. Chem. 213:533–46.

Teas, A. N. 18431.

18427. Teas, H. J. 1954: B vitamins in starchy and sugary maize endosperms. Plant Physiol. 29:190–94.

18428. ——, E. G. Anderson, and J. W. Cameron. 1955: Roles of embryo and endosperm in determining niacin content of starchy and sugary maize endosperms. Plant Physiol. 30:334–37.

18429. ——, J. W. Cameron, and A. C. Newton. 1952: Tryptophan, niacin, indoleacetic acid, and carbohydrates in developing sugary and starchy maize kernels. Agron. J. 44:434–38.

18430. —— and A. C. Newton. 1951: Tryptophan, niacin, and indoleacetic acid in several endosperm mutants and standard lines of maize. Plant Physiol. 26:494–501.

18431. ——, A. N. Teas, and J. W. Cameron. 1954: Niacin relationships in developing and mature maize endosperms of brittle and related genotypes. Cereal Chem. 31:250–56.

Teas, H. J. 3547.

18432. Teater, R. W., H. J. Mederski, and G. W. Volk. 1960: Yield and mineral content of corn as affected by ammonium chloride fertilizer. Agron. J. 52:403–5.

Tecce, G. 4534, 15623, 17772, 17773.

Tecson, A. L. 16327.

18433. Tedin, O. 1931: Influence of violet light on the germination of tomato seed. Nord. Jordbrugsforskn. 2/3:108–26. Biol. Abstr. 6:15630. 1932.

18434. Teeter, H. M., L. E. Gast, E. W. Bell, W. J. Schneider, and J. C. Cowan. 1955: Investigations on the bitter and beany components of soybeans. J. Am. Oil Chemists Soc. 32:390–97.

Teik, G. L. 7116, 7117.

Teissier, G. 4035, 4036.

18435. Tejima, T., and G. Misonoo. 1938: Über die Keimfähigkeit der unter niederer Temperatur aufbewahrten Samen. Proc. Crop Sci. Soc. Japan. 10:56–64. Japan. J. Botany 9:174–75. 1939.

Telegina, N. N. 10817.

Telford, H. S. 8294.

18436. Teller, G. L. 1935: Changes in nitrogen compounds in the wheat grain at different stages of development. Plant Physiol. 10:499–509.

18437. Teltscherova, L. 1948: The influence of alkaloids on the plant organism. Spisy Vydanave Prirodovedeckou Fak. Univ. Karlovy 184. 22 pp. Chem. Abstr. 44: 10054. 1950.

Teltscherova, L. 11361.

Tempe, J. de. 11198.

18438. Templeman, W. G., and C. J. Marmoy. 1940: The effect upon the growth of plants of watering with solutions of plant-growth substances and of seed dressings containing these materials. Ann. Appl. Biol. 27:453–71.

18439. —— and W. A. Sexton. 1946: The differential effect of synthetic growth substances and other compounds upon plant species. II. Seed germination and early growth responses to some arylcarbamic esters and related compounds. Proc. Roy. Soc. London Ser. B 133:480–85.

18440. Templeton, G. E., Jr. 1959: The effect of toxic metabolites from *Fusarium moniliforme* on germination of barley. Dissertation Abstr. 19:1535–36.

18441. Templin, V. M., and H. Steenbock. 1933: Cereals and rickets. V. The effect of germination and autolysis on the rachitogenic properties of the maize kernel. Biochem. J. 27:2061–68. Physiol. Abstr. 19:388. 1934.

Teodoresco, D. 2775, 13647.

18442. Teodosiu, C. 1935: Anatomy and chemical principles of the seeds of *Zygophyllum fabago* L. Compt. Rend. Soc. Biol. 120:659–61. Chem. Abstr. 30:1088. 1936.

18443. Terada, T., M. Hirata, and T. Utigasaki. 1933: On the mechanism of spontaneous expulsion of wistaria seeds. Sci. Papers Inst. Phys. Chem. Res. (Tokyo) 21:233–41. Biol. Abstr. 10:18023. 1936.

18444. Teräsvuori, K. 1930: Der Einfluss des Alters auf den Saatwert des Kleesamens. Acta Agral. Fennica 20:81–108. Botan. Centralbl. 160:200–1. 1931.

Terasaka, Y. 11096, 11097, 11115–11117, 11152, 11159, 11161.

18445. Terasawa, Y. 1927: Experimentelle Studien über die Keimung der Samen von *Trapa natans* L. Botan. Mag. (Tokyo) 41:581–88.

Terasawa, Y. 11160.

18446. Terek, S., and V. I. Sharkov. 1957: Constitution of pentosan hydrolyzates of cotton and sunflower seed hulls. Gidroliz. i Lesokhim. Prom. 10(5):13–14. Chem. Abstr. 51:18137. 1957.

Ternet, I. W. 13523.

18447. Terning, P. E. 1947: Anthocyanin colouring of the germ plants as a variety character in wheat and a method for intensifying of the colour. Medd. Statens Cent. Frökontr. 22:71–77.

18448. —— 1947: A laboratory method for distinguishing of yellow- and white-fleshed *Brassica* strains. Medd. Statens Cent. Frökontr. 22:78–85.

Terrier, J. 1826.

18449. Terroine, E. F. 1920: État actuel de nos connaissances sur la formation des graisses au cours de la maturation des graines et fruits oléagineux et sur l'utilisation des graisses au cours de la germination. Ann. Sci. Nat. Botan. (Ser. 10) 2:1–63.

18450. ——, R. Bonnet, and P. H. Joessel. 1923: Influence de la température sur le rendement énergétique dans la germinations. Compt. Rend. 177:212–14.

18451. ——, ——, and —— 1924: L'énergie de croissance. II. La germination. Bull. Soc. Chim. Biol. 6:357–93.

18452. Terroine, T. 1947: Formation of nicotinic acid in germination; role of tryptophan. Arch. Sci. Physiol. 1:445–58. Chem. Abstr. 43:5092. 1949.

18453. —— and J. Desveaux-Chabrol. 1947: Synthesis of nicotinic acid in the course of germination. Arch. Sci. Physiol. 1: 117–33. Chem. Abstr. 42:2323. 1948.

Terroine, T. 3499, 4578.

Terry, D. E. 626, 627.

Terry, O. W. 10896.

18454. Tervet, I. W. 1944: The relation of seed quality to the development of smut in oats. Phytopathology 34:106–15.

18455. —— 1945: The influence of fungi on storage, on seed viability and seedling vigor of soybeans. Phytopathology 35: 3–15.

18456. Tesar, M. B., K. Lawton, and B. Kawin. 1954: Comparison of band seeding and other methods of seeding legumes. Agron. J. 46:189–94.

Tesar, M. B. 18768.

Teskey, B. J. E. 8829.

18457. Testi, G. D. 1932: Action of some alkaloids on the germination of seeds. Ann. Botan. (Turin) 19:345–52.

Testi, G. Dragone-; *see* Dragone-Testi, G.

18458. Testoni, P. 1935: Action of sulfur-bearing mineral waters on the germination of seeds and the development of plants.

Ann. Chim. Appl. 25:558–63. Chem. Abstr. 30:4598. 1936.

18459. Tetenyi, Mrs. P. 1962: Physiological and biochemical analysis of apricot kernels prior to entering the dormant stage. Kiserl. Kozl. 55/C:69–79.

18460. Teter, N. C., and L. I. Miller. 1959: Abnormality of peanut seedlings caused by seed injury. Plant Disease Reptr. 43: 353–59.

18461. —— and J. W. Weaver, Jr. 1948: Artificial drying of peanuts. Peanut J. Nut World 27(7):31. Biol. Abstr. 23:19468. 1949.

18462. Tetjurew, W. 1938: Über die keimungshemmende Substanz in saftigen Früchten. Botan. Zh. 23:391–92.

18463. Tetry, A., E. Sutton, and J. Moullec. 1958: Phytoagglutinins in the seeds of *Dolichos lablab*. Compt. Rend. 246: 3118–19. Chem. Abstr. 52:20421. 1958.

Tettenborn, H. 8493.

18464. Teturev, V. 1944: On dormancy in squash (*Cucurbita pepo*). Proc. Lenin Acad. Agr. Sci. 11/12:12–14.

18465. Teubner, F. G. 1953: Identification of the auxin present in apple endosperm. Science 118:418.

18466. —— and A. E. Murneek. 1955: Embryo abortion as mechanism of "hormone" thinning of fruit. Missouri Agr. Expt. Sta. Res. Bull. 590. 79 pp.

18467. Teunisson, D. J. 1954: Influence of storage without aeration on the microbial populations of rough rice. Cereal Chem. 31:462–74.

18468. Teuscher, H. 1951: *Welwitschia mirabilis*. Natl. Hort. Mag. 30:188–92.

18469. Tevis, L., Jr. 1958: Germination and growth of ephemerals induced by sprinkling a sandy desert. Ecology 39:681–88.

18470. —— 1958: A population of desert ephemerals germinated by less than one inch of rain. Ecology 39:688–95.

Tewari, J. D. 1534, 10016, 13570–13572.

18471. Tewari, R. D., and P. C. Gupta. 1955: Chemical examination of the colouring matters from the seeds of Cassia-tora. Proc. Indian Sci. Congr. Assoc. 42:142.

Tewfic, H. 10112, 10113.

18472. Tezima, T. 1931: Über die Vererbung der Samenhautfarbe bei *Sesamum indica*. Proc. Crop Sci. Soc. Japan 3:232–35. Japan. J. Botany 6:20. 1932.

Thacker, E. J. 6700.

Thakur, O. S. 13884.

Thakurta, A. Guha-. 5693.

18473. Thaler, H. 1954: Die loslichen Kohlenhydrate der Kakaobohne. Naturwissenschaften 41:432.

Thalmann, L. 1766.

18474. Thames, J. L. 1963: Needle variation in Loblolly pine from four geographic seed sources. Ecology 44:168–69.

Thames, W. H. 1472.

18475. Thangamani, A., and P. S. Sarma. 1956: The biosynthesis of ascorbic acid in germinating green gram (*Phaseolus radiatus*). J. Sci. Ind. Res. (India) 15C:157–60. Chem. Abstr. 51:2956. 1957.

Tharandt, T. 10571.

18476. Tharp, W. H. 1935: Developmental anatomy and relative permeability of barley seed coats. Botan. Gaz. 97:240–71.

Thatcher, L. E. 8450, 8451.

18477. Theobald, W. L., and L. F. Hough. 1960: The relationship between stage of peach embryo development and seedling growth and survival. Proc. Am. Soc. Hort. Sci. 75:163–70.

Thevan, P. V. 5471.

Thick, J. 2882, 2883.

Thiele, I. 10445.

18478. Thielebein, M. 1955: Possibilities of seed production and multiplication in the most suitable areas of the participating countries. Organisation for European Econ. Coop. European Productivity Agency, Paris. Development of Seed Production and the Seed Trade in Europe, pp. 103–10.

18479. —— and A. Grahl. 1960: Lichtbedürfnis des keimenden Getreideskorns. Naturwissenschaften 47:190.

Thielebein, M. 6413, 6417–6420, 7574, 7575.

Thielen, A. 8028.

18480. Thieret, J. W. 1955: Seeds of *Veronica* and allied genera. Lloydia 18: 37–45.

Thieret, J. W. 19852.

18481. Thimann, K. V., Ed. 1958: The physiology of forest trees. Ronald Press, New York. 678 pp.

18482. —— and R. H. Lane. 1938: After-effects of the treatment of seed with auxin. Am. J. Botany 25:535–43.

Thimann, K. V. 9370–9374.

18483. Thirgood, J. V. 1956: A note on the testing of Eucalyptus seed. Empire Forestry Rev. 35:88–89. Biol. Abstr. 32:2864. 1958.

18484. Thiselton-Dyer, W. 1899: On the influence of the temperature of liquid hydrogen on the germinative power of seeds. Proc. Roy. Soc. London 65:361–68.

Thoday, D. 5923.

18485. Thoday, M. G. 1921: Ripening of seed in *Gnetum gnemon* and *Gnetum africanum.* S. African J. Sci. 17:189–92.

18486. Thoenes, H. 1929: Morphologie und Anatomie von *Cynosurus cristatus* und die Erscheinungen der Viviparie bei ihm. Botan. Arch. 25:284–346.

Thörn, N. 14712.

Thom, H. C. S. 17282.

Thomas, A. 4683.

Thomas, B. H. 9935.

18487. Thomas, C. A. 1952: Transmission of safflower rust on treated seed. Phytopathology 42:108–10.

18488. —— 1959: Effect of storage time and temperature on control of seed-borne safflower rust by volatile mercury fungicides and the efficacy of certain seed protectants and antibiotics. Plant Disease Reptr. 43:1250–52.

18489. —— 1960: Permeability measurements of castor-bean seed indicative of cold-test performance. Science 131:1045–46.

18490. —— 1960: Relations of variety, temperature, and seed immaturity to pre-emergence damping-off of castorbean. Phytopathology 50:473–74.

18491. —— 1960: A viviparous coconut palm. Coconut Bull. 14:335–40.

Thomas, C. A. 3537.

18492. Thomas, E. F. 1934: The toxicity of certain species of *Crotalaria* seed for the chicken, quail, turkey and dove. J. Am. Vet. Med. Assoc. 85:617–22. Chem. Abstr. 29:508–9. 1935.

Thomas, F. J. D. 9704.

Thomas, G. 8234.

18493. Thomas, I., and H. G. Cariss. 1949: Pedigree seed production. J. Agr. W. Australia 26:215–26.

18494. Thomas, J. B. 1941: On the metabolism of Damar seeds. Buitenzorg Botan. Gardens Ann. 51:94–114.

Thomas, J. O. 16279.

Thomas, K. J. 56.

Thomas, M. D. 8259.

Thomas, P. E. 6571.

Thomas, P. T. 9274.

18495. Thomas, R. C. 1938: Transmissible lysins in water extracts of seeds. Science 88:56–57.

18496. Thomas, R. G. 1961: The influence of environment on seed production capacity in white clover (*Trifolium repens* L.). Australian J. Agr. Res. 12:225–38.

18497. Thomas, W. 1959: Effects of applied nitrogen and sulfur on yield and protein content of corn. Agron. J. 51:572–73.

18498. Thomas, W. D., Jr., and R. W. Graham. 1951: Seed transmission of red node virus in pinto beans. Phytopathology 41:959–62.

18499. Thomas, W. S. 1955: Gloxinias from seed. Gloxinian 4(6):6–7.

Thompson, B. B. 19606.

18500. Thompson, E. G., and C. F. Henderson. 1959: Effect of seed treatment with insecticides on rate of seeding wheat. J. Econ. Entomol. 52:535.

Thompson, H. C. 16656.

18501. Thompson, H. E., H. L. Ahlgren, and J. T. Medler. 1952: Effect of cutting treatments and the control of injurious insects on seed production of Ladino clover (*Trifolium repens* L.). Agron. J. 44:411–14.

Thompson, H. E. 13365, 19674.

Thompson, H. H. 19843.

18502. Thompson, H. J., and C. K. Shedd. 1954: Equilibrium moisture and heat of vaporization of shelled corn and wheat. Agr. Eng. 35:786–88.

Thompson, H. M. 8777.

18503. Thompson, J. A. 1961: Studies on nodulation responses to pelleting of subterranean clover seed. Australian J. Agr. Res. 12:578–92.

Thompson, J. A. 2315, 6708.

Thompson, J. B. 13522.

18504. Thompson, J. F., and C. J. Morris. 1956: Isolation of (−)S-methyl-L-cysteine from beans (*Phaseolus vulgaris*). Nature 178:593.

Thompson, J. F. 13765, 13766, 20033.

18505. —— 1928: Cress grown on adrenaline. Nature 122:401.

Thompson, J. P. 1353.

Thompson, J. R. 9134.

18506. Thompson, P. A. 1963: The development of embryo, endosperm, and nucellus tissues in relation to receptacle growth in the strawberry. Ann. Botany (London) 27:589–605.

Thompson, R. 19316.

18507. Thompson, R. B. 1934: Heterothally and the seed habit versus heterospory. New Phytologist 33:41–44.

18508. Thompson, R. C. 1933: A morphological study of flower and seed development in cabbage. J. Agr. Res. 47:215–32.

18509. —— 1935: Some soil conditions affecting lettuce seed germination. Proc. Am. Soc. Hort. Sci. 32:572–77.

18510. —— 1936: Some factors associated with dormancy of lettuce seed. Proc. Am. Soc. Hort. Sci. 33:610–16.

18511. —— 1938: Dormancy in lettuce seed and some factors influencing its germination. U.S. Dept. Agr. Tech. Bull. 655. 20 pp.

18512. —— 1938: The germination of lettuce seed as affected by nutrition of the plant and the physiological age of the plant. Proc. Am. Soc. Hort. Sci. 35:599–600.

18513. —— 1943: Inheritance of seed color in *Lactuca sativa*. J. Agr. Res. 66:441–46.

18514. —— 1946: Germination of endive seed (*Cichorium endivia*) at high temperature stimulated by thiourea and by water treatments. Proc. Am. Soc. Hort. Sci. 47:323–26.

18515. —— and N. L. Horn. 1944: Germination of lettuce seed at high temperature (25° to 35°C.) stimulated by thiourea. Proc. Am. Soc. Hort. Sci. 45:431–39.

18516. —— and W. F. Kosar. 1939: Stimulation of germination of dormant lettuce seed by sulphur compounds. Plant Physiol. 14:567–73.

Thompson, R. C. 8232, 11246.

18517. Thompson, V. J. 1941: Report of work on buckhorn plantain (*Plantago lanceolata*). Proc. Assoc. Offic. Seed Analysts 33:39–41.

Thompson, W. P. 2910.

18518. Thomson, A. 1896: Zum Verhalten alter Samen gegen Fermentlosungen. Gartenflora 45:344–45.

18519. Thomson, J. R. 1938: The development of sainfoin in its seeding year. Ann. Appl. Biol. 25:457–71.

18520. —— 1952: Further seed studies in sainfoin. J. Brit. Grassland Soc. 7:65–69. Biol. Abstr. 27:32858. 1953.

18521. —— 1954: The effect of seed dressings containing an organo-mercurial and γ BHC on germination test of oats. Empire J. Exptl. Agr. 22:185–88.

18522. —— 1957: Definitions of pure seed. Proc. Intern. Seed Testing Assoc. 22: 99–105.

18523. —— 1957: A method for applying the "Random Cups" principle in the routine drawing of working samples. Proc. Intern. Seed Testing Assoc. 22:486–88.

18524. —— 1960: Morphology and anatomy of the seed and fruit of *Onobrychis viciifolia* Scop. Proc. Intern. Seed Testing Assoc. 25:848–64.

18525. —— 1962: Statistics Committee. Proc. Intern. Seed Testing Assoc. 27: 235–52.

18526. —— 1962: A new seed germinator. Proc. Intern. Seed Testing Assoc. 27: 675–78.

Thomson, J. R. 13195.

18527. Thomson, R., and E. O. C. Hyde. 1939: Vernalization of rape and marrow-stem kale. New Zealand J. Sci. Technol. Sect. A 21:236–39.

18528. Thomson, R. B. 1927: Evolution of the seed habit in plants. Trans. Roy. Soc. Can. Ser. III 21:229–72.

18529. —— 1934: Heterothally and the seed habit versus heterospory. New Phytologist 33:41–44.

18530. Thor, C. J. B., and C. L. Smith. 1935: A physiological study of seasonal changes in the composition of the pecan during fruit development. J. Agr. Res. 50:97–121.

18531. —— and —— 1939: A physiological study of the prefilling period of fruit development in the pecan. J. Agr. Res. 58:905–10.

Thor, C. J. B. 17623, 17624.

Thorn, S. T. 1904.

18532. Thornber, J. J. 1908: The viability of seeds. Plant World 11:158–59.

18533. —— 1908: Vitality of seeds under water. Ariz. Agr. Expt. Sta. Bull. 60: 438–41.

Thorne, G. N. 19407.

Thornley, M. J. 17091.

Thornton, B. J. 18534.

Thornton, C. D. W. 14749.

18534. Thornton, M. L., and B. J. Thornton. 1963: Decrease in firm seed content of blue grama seed over a three month period. Assoc. Offic. Seed Analysts News Letter 37(2):5.

Thornton, M. L. 12545.

18535. Thornton, N. C. 1935: Factors influencing germination and development of dormancy in cocklebur seeds. Contrib. Boyce Thompson Inst. 7:477–96.

18536. —— 1936: Carbon dioxide storage. IX. Germination of lettuce seeds at high temperatures in both light and darkness. Contrib. Boyce Thompson Inst. 8:25–40.

18537. —— 1944: Carbon dioxide storage. XII. Germination of seeds in the presence of carbon dioxide. Contrib. Boyce Thompson Inst. 13:355–60. Biol. Abstr. 18: 22890. 1944.

18538. —— 1945: Importance of oxygen supply in secondary dormancy and its relation to the inhibiting mechanism regulating dormancy. Contrib. Boyce Thompson Inst. 13:487–500.

Thornton, N. C. 2049, 4605.

Thorp, F. C. 8108, 8109.

18539. Thorpe, S. K. 1941: Black ended barley grains. J. Inc. Brewers Guild 27: 67. Biol. Abstr. 15:19462. 1941.

18540. Thorup, S. 1961: Experiments with the influence of sodium chlorate on the germination of wheat and rye. Tidsskr. Landokon. 3:139–56.

18541. Thorvaldson, T., and L. R. Johnson. 1940: The selenium content of Saskatchewan wheat. Can. J. Res. Sect. B 18:138–50. Expt. Sta. Rec. 83:727. 1940.

Thoung, C. 15151.

Thouvenin. 12739.

18542. Throneberry, G. O., and F. G. Smith. 1953: The effect of triphenyltetrazolium chloride on oat embryo respiration. Science 117:13–15.

18543. —— and —— 1955: Relation of respiratory and enzymatic activity to corn seed viability. Plant Physiol. 30:337–43.

18544. Thrupp, A. C. 1941: Effect of seed-ash characteristics and treatment of seed and soils upon coniferous seed germination. Wash. State Univ. Abstr. Theses 5:305–9.

Thrupp, A. C. 7925.

18545. Thszkiewics, S. 1938: Über die Prüfung des Forstsaatgutes. Forstwiss. Centralbl. 60:725–38.

18546. Thudium, F., O. Schindler, and T. Reichstein. 1959: Die Glykoside von *Acotkanthera schimperi* (A.DC.) Benth. et Hook. III. Untersuchung der Samen der Form, die von den Wa-Giriama zur Pfeilgiftbereitung verwendet wird. Helv. Chim. Acta 42:2–49.

Thudium, F. 13659.

18547. Thunberg, T. 1929: Über eine Citrico-Dehydrogenase in Gurken Samenextrakt. Skand. Arch. Physiol. 55:280. Physiol. Abstr. 14:152–53. 1929.

18548. —— 1931: Die Hemmungswirkung einiger in Samenschalen weit verbreiteter Substanzen auf den Entfärbungsvorgang in dem System Donatorsubstanz-Dehydrogenase-Methylenblau. Skand. Arch. Physiol. 63:99. Ber. Ges. Physiol. 65:469. 1932.

18549. —— 1936: Zur Kenntnis der Alkoholdehydrogenasen verschiedenen Samenarten. Skand. Arch. Physiol. 75:38–48. Physiol. Abstr. 21:918. 1937.

18550. —— 1936: The occurrence of a glycerophosphate dehydrogenase in plant seeds. Skand. Arch. Physiol. 75:199–206. Chem. Abstr. 31:735. 1937.

18551. Thuret, M. 1873: Expériences sur la durée de la vitalité des graines plongées dans l'eau der mer. Arch. Sci. Phys. Nat. Geneva 47:188–92.

Thurgan, M. Müller-; *see* Müller-Thurgan, M.

18552. Thurliman, L. 1928: Germination of *Brassicas*. Proc. Assoc. Offic. Seed Analysts 19:71–74.

Thurman, D. A. 6584.

18553. Thurman, R. L. 1953: Breaking dormancy in cotton seed. Southern Seedsman 16(9):32, 52.

Thurston, H. W., Jr. 19509.

18554. Thurston, J. M. 1951: Some experiments and field observations on the germination of wild oat (*Avena fatua* and *A. ludoviciana*) seeds in soil and the. emergence of seedlings. Ann. Appl. Biol. 38:812–32.

18555. —— 1956: Wild oats. J. Roy. Agr. Soc. Engl. 117:43–52.

18556. —— 1957: Morphological and physiological variation in wild oats (*Avena fatua* L. and *A. ludoviciana* Dur.) and in hybrids between wild and cultivated oats. J. Agr. Sci. 49:261–74.

18557. —— 1960: Dormancy in weed seeds. Brit. Ecol. Soc. Symp. 1:69–82. 1959.

18558. —— 1961: The effect of depth of burying and frequency of cultivation on survival and germination of seeds of wild oats (*Avena fatua* L. and *Avena ludoviciana* Dur.). Weed Res. 1:19–31.

18559. —— 1962: An international experiment on the effect of age and storage conditions on viability and dormancy of *Avena fatua* seeds. Weed Res. 2:122–29.

18560. —— 1962: The effect of competition from cereal crops on the germination and growth of *Avena fatua* I. in a naturally-infested field. Weed Res. 2:192–207.

Thurston, J. M. 19700.

18561. Thwing, W. E. 1952: Start with seeds for better wildflowers. Home Garden 19:18–21.

18562. Thygeson, C. C. 1930: Preliminary results of soil tests of *Agrostis* spp. Proc. Assoc. Offic. Seed Analysts 1929:43–46.

18563. Tiagi, B. 1956: A contribution to the embryology of *Striga orobanchoides* Benth. and *Striga euphrasioides* Benth. Bull. Torrey Botan. Club 83:154–70.

18564. —— and N. S. Sankhla. 1963: Studies in the family Orobanchaceae. V. A contribution to the embryology of

Orobanche lucorum. Botan. Mag. (Tokyo) 76:81–88.

18565. —— and S. Taimni. 1960: Embryo-sac development in *Vernonia cinerascens* Schult. and seed development in *V. cinerea* Less. Current Sci. (India) 29:406.

18566. Tiagi, Y. D. 1956: Polyembryony in *Mammillaria tenuis* DC. Saugar Univ. Botan. Soc. Bull. 8:25–27. Biol. Abstr. 32:28250. 1958.

Tichy, S. 17592.

18567. Tichy, V., and J. Chalupova-Janovicova. 1958: Der Einfluss von Humusstoffen auf Keimung und Wachstum einiger Kulturpflanzen. Folia Biol. (Prague) 4:281–85.

18568. Tickle, C. G. 1958: New method of harvesting tree seed in Britain. New Zealand Timber J. Wood Prod. Rev. 4(11): 30–31.

Tieghem, M. P. van; *see* Van Tieghem, M. P.

Tieghem, P. van; *see* Van Tieghem, P.

Tien, W. 19894.

Tietz, N. 1831, 14449.

Tiffany, L. H. 12215.

18569. Tikhenko, A. V. 1953: On hard and soft seeds of red clover. Priroda 41(7):99–101.

Tikhomirov, V. G. 6145.

18570. Tildesley, W. T. 1937: A study of some ingredients found in ensilage juice and its effect on the vitality of certain weed seeds. Sci. Agr. (Ottawa) 17:492–501.

Tilemans, E. 6777.

18571. Tilenius, G. 1938: Untersuchungen an keimenden Sonnenblumen, über die Veränderung des Öles und seine Beeinflussung durch Azetate. Planta 28:429–52.

18572. Tilford, P., C. F. Able, and R. P. Hibbard. 1924: An injurious factor affecting the seeds of *Phaseolus vulgaris* soaked in water. Papers Mich. Acad. Sci. 4:345–56.

Tillerkeratne, L. M. 19151.

18573. Tillotson, C. R. 1921: Storage of coniferous tree seeds. J. Agr. Res. 22:479–510.

18574. Tilly, F. 1935: Sensitizing and desensitizing light-sensitive seeds (*Lythrum salicaria* L.). Z. Botan. 28:401–45. Chem. Abstr. 29:5878. 1935.

Timariu, A. 11903.

18575. Timiryazev, K. A. 1960: The ability of seeds to endure ultra low temperatures. Plant Physiol. (USSR) (English Transl.) 7:74–75.

18576. Timmons, F. L. 1942: The dissemination of prickly pear seed by Jack rabbits. J. Am. Soc. Agron. 34:513–20.

18577. —— 1949: Duration of viability of bindweed seed under field conditions and experimental results in the control of bindweed seedlings. Agron. J. 41:130–33.

Timofeev, L. I. 15565.

18578. Timofeeva, A. P., and S. I. Zavylenkova. 1936: Identification of wheat seed by a laboratory method. Bull. Appl. Botany, Genet., Plant Breed. (USSR) Ser. IV 1:5–22. Chem. Abstr. 31:5842. 1937.

Timofeev-Resovskii, N. V. 15679.

18579. Timofeyev, A. P. 1963: On the natural renewal by seed of *Pinus silvestris* L. in the arenaceous part of the middle Dnieper valley. Acad. Sci. RSS Ukraine J. Botan. 20:76–80.

18580. Timonin, M. I. 1964: Interaction of seed-coat microflora and soil microorganisms and its effects on pre- and post-emergence of some conifer seedlings. Can. J. Microbiol. 10:17–22.

18581. Tin, P. T., and C. N. Kang. 1960: Morphological observations on many kerneled grains in rice. Acta Botan. Sinica 9:98–105. Biol. Abstr. 35:64148. 1960.

18582. Tincker, M. A. H. 1925: Physiological pre-determination experiments with certain economic crops. Ann. Appl. Biol. 12:440–71.

18583. —— 1935: Rose seeds: their after-ripening and germination. J. Roy. Hort. Soc. 60:399–417.

18584. —— 1942: Recent work on germination. Nature 149:658–59.

Tincker, M. A. H. 9973.

18585. Tingey, D. C. 1961: Longevity of seeds of wild oats, winter rye, and wheat in cultivated soil. Weeds 9:607–11.

18586. —— and K. R. Allred. 1961: Breaking dormancy in seeds of *Cuscuta approximata*. Weeds 9:429–36.

Tingey, D. C. 412, 18037.

18587. Tingley, R. J. 1950: Beet seed standardization. Proc. Am. Soc. Sugar Beet Technol. 6:85–87.

Tinker, F. 3114.

Tinker, P. B. H. 3825.

18588. Tinney, F. W., and O. S. Aamodt. 1940: The progeny test as a measure of the types of seed development in *Poa pratensis* L. J. Heredity 31:457–64.

18589. Tirelli, M. 1934: Influence of contact with soil on the germination of grain and

of mulberry. Ann. Botan. (Turin) 20:400–11.
Tirelli, M. 18325, 18326.
18590. Tiren, L. 1942: Storing of seed of forest trees. Experiences at Hallnas and Bollnas. Skogen 29:350–52. Forestry Abstr. 5:177. 1944.
18591. Tischer, J., E. Illner, and R. Seidl. 1941: Über die Inhaltsstoffe der Frücht von *Rhus typhina* L. I. Untersuchung der Fette mit einem Anhang über die Bleichung des Fettes. Biochem. Z. 307:366–77.
Tischer, R. G. 7079, 15132, 19349.
18592. Tischler, G. 1912: Über die Entwicklung der Samenanlagen in parthenokarpen Angiospermen-Früchten. Jahrb. Wiss. Botan. 52:1–84.
Tisdale, E. W. 13698.
18593. Tisdale, S. L., R. L. Davis, A. F. Kingsley, and E. T. Mertz. 1950: Methionine and cystine content of two strains of alfalfa as influenced by different concentrations of the sulfate ion. Agron. J. 42:221–25.
18594. Tisdell, T. F., R. D. Ilnicki, and W. F. Meggitt. 1961: Date of berry harvest as it affects seed maturity of horse nettle (*Solanum carolinense*). Proc. Northeast. Weed Control Conf. 15:336. Weed Abstr. 12:106. 1963.
18595. Titus, G. R. 1940: So-called two-year seeds germinated first year. Am. Nurseryman 72(11):22.
Titus, G. R. 16700.
Titus, H. W. 10658.
18596. Tiunov, A. N. 1950: Increasing seed germination of one-crop red clover. Selek. i Semen. 17(3):30–37.
18597. Tiwari, R. D. 1941: Chemical examination of the fatty oil from the seeds of *Ocimum canum*. Proc. Natl. Acad. Sci. India 11:45–50. Chem. Abstr. 42:1070. 1948.
18598. —— 1946: The chemical examination of the seeds of *Nigella sativa*, Linn. Allahabad Univ. Studies Chem. Sect. 1946:1–2. Chem. Abstr. 41:6672. 1947.
18599. —— 1946: Chemical examination of *Sphaeranthus indicus*. II. The constituents. Proc. Natl. Acad. Sci. India 15:55–62. Chem. Abstr. 42:1073. 1948.
18600. —— and R. K. Gupta. 1958: Chemical examination of the seed of *Bryonia laciniosa*. Indian J. Appl. Chem. 21:32–34. Chem. Abstr. 53:6350. 1959.
18601. —— and —— 1959: Chemical examination of the seeds of *Clitoria ternatea*, Linn. Res. J. Hindi Sci. Acad. 2:1–5.
Tiwari, R. D. 17490.
18602. Tiwary, N. K. 1932: A note on the germination of unripe (green) gram seeds (*Cicer arietinum*). Botan. Centralbl. Beih. 49:259–61.
18603. —— 1934: Fruit and seed development of *Tinospora cordifolia* Miers. Current Sci. (India) 3:262–64.
18604. Tjuwin, M. G. 1939: Thermische Behandlung der Samen als Faktor der Ertragssteigerung. Lein Hanf. 1:50–53. Botan. Centralbl. 175:47–48. 1939.
To, S. 16772.
18605. Tobe, J. H. 1962: A seedsman talks about seeds. Org. Garden. Farming 9:40–42.
18606. Tocco, L. 1933: Preservation of *Strophanthus* seed by drying at 50°–100° C. Arch. Intern. Pharmacodyn. 44:419–25. Biol. Abstr. 9:14533. 1935.
Tochiaki, K. 19950.
18607. Tod, H. 1944: Growing bulbs from seed. J. Roy. Hort. Soc. 69:239–41.
18608. —— 1958: Effect of gibberellic acid on seed germination. Gardeners Chron. 144:257, 260.
Toda, H. 5815.
Toda, M. 14508.
18609. Toda, R., and H. Isikawa. 1951: Effect of diffused light on the germination of *Paulownia* seeds. J. Japan. Forestry Soc. 34:250. Biol. Abstr. 27:7725. 1953.
18610. Todaro, F. 1898: Concerning the duration of the germinative ability of the seed of various cultivated plants. Staz. Sper. Agrar. Ital. 31:525–63. Expt. Sta. Rec. 11:157–58. 1899.
18611. —— 1901: Azione dell acido solforica concentrato su alcuni semi e in particulare sopra i semi duri delle Leguminosae. Staz. Sper. Agrar. Ital. 34:613–89.
18612. —— 1904: Prove intorno all'influenza della temperatura su la germinazione di alcuni semi. Staz. Sper. Agrar. Ital. 37:453–62. Botan. Centralbl. 102:178. 1906.
Todd, A. R. 3174.
18613. Todorova, V. 1962: The effect of stinking smut (*Tilletia levis* Kuhn) control on the germination, growth and yield of wheat seed. Izv. Tsentr. Nauchnoizsled. Inst. za Zashtita na Rasteniyata 3:143–74.
Toennies, G. 8392.
18614. Togliani, F. 1961: Seed of *Hibiscus cannabinus* in southern Italy. Riso 10(11):5–8.
18615. Tognoni, F., and P. G. Vicenzetti. 1961: Research on the germinability of immature seeds of early peach varieties.

Riv. Ortoflorofrutticolt. Ital. 45:494–501.
Tokareva, P. [P.] 11386, 11387.
Tokareva, R. [R.] 11375, 11388, 11389.

18616. Tokuda, S. 1929: The action of nitrates and ammonium salts on some plants. II. The action of nitrates and ammonium salts upon the germination. Botan. Mag. (Tokyo) 43:295–305.

18617. Tokugawa, Y., and Y. Emoto. 1930: Über die Samenbildung in *Lycoris*-arten. Botan. Mag. (Tokyo) 44:236–44.

18618. Tokunaga, Y., and M. Ikuzawa. 1954: The influence of dicyandiamide on the germination of seeds in reference to their growth. J. Sci. Soil Manure Japan 25:111–14. Chem. Abstr. 49:1264. 1955.
Tolbert, N. E. 7934, 7935, 13625.
Toledo, F. F. de; *see* De Toledo, F. F.
Toledo, T. A. N. de; *see* De Toledo, T. A. N.

18619. Tolentino-Vallarta, M. 1936: A pharmacognostical study of the fruit and seed of *Hydnocarpus alcalae* DC. Nat. Appl. Sci. Bull. Univ. Philippines. 5:27–50.
Toler, R. 4723, 4724.

18620. Tolman, B. 1948: Report of seed germination committee. Proc. Am. Soc. Sugar Beet Technol. 5:78–80.

18621. —— 1948: Some field conditions which influence germination technique. Proc. Am. Soc. Sugar Beet Technol. 5:93–94.

18622. —— and M. Stout. 1940: Toxic effect on germinating sugar-beet seed of water-soluble substances in the seed ball. J. Agr. Res. 61:817–30.

18623. —— and —— 1944: Sheared sugar beet seed with special reference to normal and abnormal germination. J. Am. Soc. Agron. 36:749–59.
Tolman, B. 18072, 18073.,

18624. Tolomei, G. 1893: Action of magnetism upon germination. Malpighia 7:469–82.

18625. —— 1896: Influence of electricity on germination. Malpighia 10:493–511.

18626. Tolstead, W. L. 1941: Germination habits of certain sand-hill plants in Nebraska. Ecology 22:393–97.
Tomanek, G. W. 9192.
Tomaselli, R. 2817.

18627. Tomaszewski, Z. 1954: Yellow lupine with non-bursting and non-dropping pods. Acta Agrobotan. 1:89–104. Biol. Abstr. 29:17089. 1955.

18628. —— 1960: Preliminary investigations on the culture of isolated embryos of leguminous plants under artificial conditions. Zeszyty Probl. Postepow Nauk Roln. 20:73–87.

18629. Tombesi, L., and M. Tarantola. 1950: Composition of seeds and water-availability of soil. Ann. Staz. Chim.-Agrar. Sper. Roma Ser. III No. 30. 10 pp. Brit. Abstr. BIII:323. 1951.

18630. ——, ——, S. Fortini, and M. E. Venezian. 1950: Catalase activity and critical energy of seeds. Ann. Staz. Chim.-Agrar. Sper. Roma Ser. III No. 31. 8 pp. Brit. Abstr. BIII:329. 1951.

18631. Tombs, D. M. 1952: The application of electronics to the cleaning of agricultural products. World Crops 4:422–24.

18632. Tomeo, M., and C. Aguilar. 1945: The agricultural problem of nitrogen. Anales Fis. Quim. (Madrid) 41:410–13. Chem. Abstr. 41:4263. 1947.
Tomita, E. 1875.

18633. Tomita, K. 1931: Über die Entwicklung des Nackten Embryos von *Crinum latifolium* L. Sci. Rept. Tohoku Imp. Univ. Ser. 4. 6:163–69.

18634. Tomita, T. 1956: Effects of diffusate obtained from vernalized winter rye seeds. Proc. Crop Sci. Soc. Japan 24:260–61.
Tomiyama, T. 10724.

18635. Tomlinson, P. B. 1960: Essay on the morphology of palms. I. Germination and the seedling. Principes 4:56–61.

18636. Tompa, K., and A. Szent-Istvany. 1959: Preparation of *Robinia* seeds for sowing with disc scarificator. Kertesz. Szolesz. Foiskola Evkonyve 1959(7):105–21.

18637. Toms, J. 1958: Seeding depth of Dwalganup subterranean clover (*Trifolium subterraneum* L.) and Barrel medic (*Medicago tribuloides* Desr.). Experiments on some sandy surfaced soils of the Western Australian cereal belt. J. Agr. W. Australia 7:555–57.

18638. Toniolo, L. 1949: The use of tellurites in the rapid determination of the vitality of seeds. Ric. Sci. Ricostruz. 19:1173–77. Biol. Abstr. 24:16764. 1950.

18639. —— 1953: On methods of distinguishing seeds of Ladino clover from those of white clover. Proc. Intern. Seed Testing Assoc. 18:307–11.

18640. —— and G. Beni. 1954: The use of tetrazolium chloride for determining viability of cereal seeds. Ann. Sper. Agrar. (Rome) 8:1975–88. Chem. Abstr. 49: 4091. 1955.

18641. Tonnier, J. P. 1954: Mise au point d'un milieu de semis pour les graines d'une espèce de vanillier dont on n'obtient pas la germination sur les milieux de culture classiques. Congr. Intern. Botan. Paris Rappt. Commun. 8:411–12.

18642. —— 1954: Nouveaux essais sur la germination des semences de *Vanilla planifolia* Andr. Congr. Intern. Botan. Paris Rappt. Commun. 8:412–14.

18643. Tonzig, S. 1933: On the action of vapors of some substances on germination of seeds. Nuovo Giorn. Botan. Ital. 40: 605.

18644. —— 1934: Studies on the physiology of seminal tegument. I. Permeability. Nuovo Giorn. Botan. Ital. 41:453–557.

18645. —— and F. Trezzi. 1954: The effect of indoleacetic acid and of ascorbic acid on water uptake and on water holding capacity. Congr. Intern. Botan. Paris Rappt. Commun. 8:157–58.

18646. Tookey, H. L., R. L. Lohman, and I. A. Wolff. 1963: Seed polysaccharides, new sources of seed mucilages. J. Agr. Food Chem. 10:131–33.

18647. ——, V. F. Pfeifer, and C. R. Martin. 1963: Seed polysaccharides, gums separated from *Crotalaria intermedia* and other leguminous seeds by dry milling. J. Agr. Food Chem. 11:317–21.

Tool, E. 13797.

18648. Toole, E. H. 1923: A preliminary report on blue grass germination. Proc. Assoc. Offic. Seed Analysts 14/15:119.

18649. —— 1924: The transformations and course of development of germinating maize. Am. J. Botany 11:325–50.

18650. —— 1936: Physiological problems involved in seed dormancy. Proc. Intern. Seed Testing Assoc. 8:33–41.

18651. —— 1937: Opening discussion of round table on seed germination problems. Proc. Assoc. Offic. Seed Analysts 29:129–33.

18652. —— 1939: Observations on the germination of freshly-harvested timothy seed. Proc. Intern. Seed Testing Assoc. 11:119–39.

18653. —— 1940: Notes on the germination of seeds of *Barbarea verna* and *Lepidium virginicum*. Proc. Intern. Seed Testing Assoc. 12:32–38.

18654. —— 1942: Two thousand years of seed research. Seed World 51(11):92–98.

18655. —— 1950: Relation of seed processing and conditions during storage on seed germination. Proc. Intern. Seed Testing Assoc. 16:214–25.

18656. —— 1957: Report of committee on seed moisture content and seed storage. Proc. Intern. Seed Testing Assoc. 22:400–8.

18657. —— 1958: Seed physiology and dormancy historical highlights. Assoc. Offic. Seed Analysts, Fifty Years' Seed Testing 1908–1958:41–45.

18658. —— 1961: The effect of light and other variables on the control of seed germination. Proc. Intern. Seed Testing Assoc. 26:659–73.

18659. —— 1961: Mechanisms of dormancy in seeds. Intern. Botan. Congr. Recent Advan. Botany 9:1208–10.

18660. ——, H. A. Borthwick, S. B. Hendricks, and V. K. Toole. 1953: Physiological studies of the effects of light and temperature on seed germination. Proc. Intern. Seed Testing Assoc. 18:267–76.

18661. —— and E. Brown. 1946: Final results of the Duvel buried seed experiment. J. Agr. Res. 72:201–10.

18662. —— and F. A. Coffman. 1940: Variations in the dormancy of seeds of the wild oat, *Avena fatua*. J. Am. Soc. Agron. 32:631–38.

18663. —— and P. L. Drummond. 1924: The germination of cottonseed. J. Agr. Res. 28:285–92.

18664. —— and W. L. Goss. 1923: The analyst's interest in dormancy problems. Proc. Assoc. Offic. Seed Analysts 14/15:90.

18665. —— and —— 1925: Effect of light and potassium nitrate on germination. Proc. Assoc. Offic. Seed Analysts 17:31.

18666. ——, S. B. Hendricks, H. A. Borthwick, and V. K. Toole. 1956: Physiology of seed germination. Ann. Rev. Plant Physiol. 7:299–324.

18667. —— and E. A. Hollowell. 1939: Effect of different temperatures on the germination of several winter annual species of *Trifolium*. J. Am. Soc. Agron. 31:604–19.

18668. ——, A. G. Snow, Jr., V. K. Toole, and H. A. Borthwick. 1956: Effects of light and temperature on germination of Virginia pine seeds. Plant Physiol. (Suppl.) 31:xxxvi.

18669. —— and V. K. Toole. 1939: Germination of carpet grass seed. J. Am. Soc. Agron. 31:566–67.

18670. —— and —— 1939: Germination of some *Brassica* types at different temperatures. Proc. Intern. Seed Testing Assoc. 11:51–56.

18671. —— and —— 1940: Germination of seed of goosegrass, *Eleusine indica.* J. Am. Soc. Agron. 32:320–21.

18672. —— and —— 1940: Germination of seeds of *Barbarea verna* and *Lepidium virginicum.* Proc. Intern. Seed Testing Assoc. 12:32–38.

18673. —— and —— 1941: Progress of germination of seed of *Digitaria* as influenced by germination temperature and other factors. J. Agr. Res. 63:65–90.

18674. —— and —— 1946: Germination of vegetable seeds exposed at Bikini during the atomic bomb Test Able. Naval Med. Res. Sect. U.S. Joint Task Force One. Rept. Biol. Aspects At. Bomb. Tests Append. No. 12.

18675. —— and —— 1946: Relation of temperature and seed moisture to the viability of stored soybean seed. U.S. Dept. Agr. Circ. 753. 9 pp.

18676. —— and —— 1954: Relation of storage conditions to germination and to abnormal seedlings of bean. Proc. Intern. Seed Testing Assoc. 18:123–29.

18677. —— and —— 1960: Viability of stored snap bean seed as affected by threshing and processing injury. U.S. Dept. Agr., Agr. Res. Serv. Tech. Bull. 1213. 9 pp.

18678. —— and —— 1961: Until time and place are suitable. Yearbook Agr., U.S. Dept. Agr. 1961:99–105.

18679. ——, ——, and H. A. Borthwick. 1957: Growth and production of snap beans stored under favorable and unfavorable conditions. Proc. Intern. Seed Testing Assoc. 22:418–23.

18680. ——, ——, ——, and S. B. Hendricks. 1955: Photocontrol of *Lepidium* seed germination. Plant Physiol. 30:15–20.

18681. ——, ——, ——, and —— 1955: Interaction of temperature and light in germination of seeds. Plant Physiol. 30: 473–77.

18682. ——, ——, and E. A. Gorman. 1948: Vegetable-seed storage as affected by temperature and relative humidity. U.S. Dept. Agr. Tech. Bull. 972. 24 pp.

18683. ——, ——, S. B. Hendricks, and H. A. Borthwick. 1957: Effects of temperature on germination of light sensitive seeds. Proc. Intern. Seed Testing Assoc. 22:196–204.

18684. ——, ——, and E. G. Nelson. 1960: Preservation of hemp and kenaf seed. U.S. Dept. Agr., Agr. Res. Serv. Tech. Bull. 1215. 16 pp.

18685. ——, ——, and —— 1961: Plastic bags for shipping seeds in the tropics. Proc. Intern. Seed Testing Assoc. 26:86–88.

18686. ——, R. E. Wester, and V. K. Toole. 1941: The effect of fruit rot of eggplant on seed germination. Proc. Am. Soc. Hort. Sci. 38:496–98.

Toole, E. H. 1693, 2809, 2810, 2834, 2835, 3132–3134, 5375, 8419, 8424, 8628, 9031, 10399–10402, 12894, 18693–18698, 18711.

18687. Toole, V. K. 1939: Germination requirements of the seed of some introduced and native range grasses. Proc. Assoc. Offic. Seed Analysts 30:227–43.

18688. —— 1939: Notes on the viability of the impermeable seed of *Vicia villosa*, hairy vetch. Proc. Assoc. Offic. Seed Analysts 31:109–11.

18689. —— 1939: Germination of the seed of poverty grass, *Danthonia spicata.* J. Am. Soc. Agron. 31:954–65.

18690. —— 1940: The germination of seed of *Oryzopsis hymenoides.* J. Am. Soc. Agron. 32:33–41.

18691. —— 1940: Germination of seed of vine-mesquite, *Panicum obtusum*, and plains bristle-grass, *Setaria macrostachya.* J. Am. Soc. Agron. 32:503–12.

18692. —— 1941: Factors affecting the germination of various dropseed grasses (*Sporobolus* spp.). J. Agr. Res. 62:691–715.

18693. ——, E. F. Miles, and E. H. Toole. 1947: Soil moisture in relation to beet seed germination. Proc. Assoc. Offic. Seed Analysts 37:127–33.

18694. —— and E. H. Toole. 1954: Seed dormancy in relation to seed longevity. Proc. Intern. Seed Testing Assoc. 18:325–28.

18695. ——, ——, H. A. Borthwick, and S. B. Hendricks. 1957: Physiology of seed dormancy. Proc. Intern. Seed Testing Assoc. 22:205–23.

18696. ——, ——, ——, and A. G. Snow, Jr. 1962: Response of seeds of *Pinus taeda* and *P. strobus* to light. Plant Physiol. 37:228–33.

18697. ——, ——, S. B. Hendricks, H. A. Borthwick, and A. G. Snow, Jr. 1961: Responses of seeds of *Pinus virginiana* to light. Plant Physiol. 36:285–90.

18698. ——, R. E. Wester, and E. H. Toole. 1951: Relative germination response of some lima bean varieties to low temperatures in sterilized and unsterilized soil. Proc. Am. Soc. Hort. Sci. 58:153–59.

Toole, V. K. 1693, 2809, 2810, 2835, 8424, 8628, 12894, 15512, 18660, 18666, 18668–18686, 19508.

Topchiiski, S. 5384.

18699. Topekha, E. F. 1937: Physiological effect of hydrogen sulfide and prussic acid upon the germination of wheat seed. Lenin Acad. Agr. Sci. Inst. Plant Protect. Summary Sci. Res. Work 1:96–99.

18700. Topor, D. 1954: The preparation of spearleaved maple seeds for sowing. Rev. Padurilor 69:212–14. Biol. Abstr. 34: 14327. 1959.

18701. Topornina, N. A. 1958: New data on seed germination of *Avena fatua* and *Setaria glauca*. Agrobiologiya 1958(3): 149–51.

18702. —— 1962: Problem of secondary dormancy of seeds of some cultivated and weedy plants. Akad. Nauk SSSR Inst. Genet. Tr. 29:156–63.

Topping, C. 6482, 6483.

18703. Torfason, W. E., and I. L. Nonnecke. 1954: A study of the effects of temperature and other factors on the germination of vegetable crops. I. Sweet corn. Can. J. Agr. Sci. 34:137–44.

18704. —— and —— 1959: A study of the effects of temperature and other factors upon the germination of vegetable crops. II. Peas. Can. J. Plant Sci. 39:119–24.

18705. —— and —— 1961: A study of the effects of temperature and other factors upon the germination of vegetable crops. IV. Root crops. Can. J. Plant Sci. 41:360–70.

18706. ——, ——, and G. Strachan. 1956: An evaluation of objective methods for determining the maturity of canning peas. Can. J. Agr. Sci. 36:247–54.

18707. Torfs, P., C. Van Assche, J. Hiemeleers, and J. Casier. 1962: Dormancy in peach seeds. Agricultura (Louvain) 10: 765–78.

Torgeson, O. W. 16214.

Toriyama, K. 454, 455.

Torman, N. S. 9671.

18708. Tornow, E. 1932: Einwirkung und Nachweis des Quecksilbers bei der Beizung des Saatguts. Phytopathol. Z. 4:631–37. Rev. Appl. Mycol. 11:706. 1932.

18709. —— 1942: The detection of poisons, especially from disinfectants, in cereals. Z. Ges. Getreidew. 29:28–31. Chem. Abstr. 37:5790–91. 1943.

Tornow, E. 13660, 16955, 16956.

18710. Torpy, F. M., and E. F. Sirrine. 1936: Growing tests in an attempt to differentiate *Melilotus alba* and *Melilotus officinalis* on seed characteristics. Proc. Assoc. Offic. Seed Analysts 28:58.

18711. ——, ——, and E. H. Toole. 1937: Some comparative studies on blowing methods of *Poa pratensis*. Proc. Assoc. Offic. Seed Analysts 29:97–102.

18712. Torres, A. M., G. H. Koch, and M. W. Katz. 1963: Germination inhibition in *Zinnia oligantha*. Ecology 44:414–16.

18713. Torres, J. P. 1936: Polyembryony in citrus and study of hybrid seedlings. Philippine J. Agr. 7:37–58.

18714. ——, E. A. Lanuza, and P. I. Cruz. 1954: Studies on abaca pollination and seed germination. Philippine J. Agr. 17:183–99. 1952.

Torrey, J. G. 15867, 15868.

18715. Torrie, J. H., and G. M. Briggs. 1955: Effect of planting date on yield and other characteristics of soybeans. Agron. J. 47:210–12.

Torrie, J. H. 9, 10–12, 407, 473.

18716. Torrisi, D. 1935: The alkaloidal content of an infusion and a decoction of *Lupinus albus*. Boll. Soc. Ital. Biol. Sper. 10:807–9.

Torrisi, D. 4194, 4195.

18717. Torssell, R. 1936: Influence of external factors on seed setting in lucerne. Lantbr.-Högsk. Ann. 3:191–241.

18718. Torti, M. 1956: The application of cold in seed preservation. Ital. Agr. 93: 160–66.

18719. Torzhinskaya, L. R., and N. V. Romenskii. 1958: The biochemical characteristics of large and small wheat grains. Tr. Odessk. Tekhnol. Inst. 9:19–23. Chem. Abstr. 54:1667. 1960.

18720. Toshchevikova, A. G., and A. F. Mansurova. 1957: The mechanism of substance accumulation in the seeds of early and late-ripening varieties of cotton during the period of ripening. Tr. Sredneaz. Gos. Univ. No. 116:87–94. Chem. Abstr. 53:4438. 1959.

18721. —— and R. Merzlyakova. 1957: Enzyme activity in ripening and sprouting seeds of early and late cotton varieties. Tr. Sredneaz. Gos. Univ. No. 116:77–86. Chem. Abstr. 53:4438. 1959.

Toshiyuki, M. 19998.

18722. Tossell, W. E. 1960: Early seedling vigor and seed weight in relation to breeding in smooth bromegrass, *Bromus inermis* Leyss. Can. J. Plant Sci. 40:268–80.

Tossell, W. E. 6835, 6836.

Toth, F. Nagy-. 15229.

18723. Toumey, J. W. 1921: On the viability of tree seeds after storage for ten years. J. Forestry 19:814.
18724. —— 1930: Some notes on seed storage. J. Forestry 28:394–95.
18725. —— and W. D. Durland. 1923: The effect of soaking certain tree seeds in water at greenhouse temperatures on viability and the time required for germination. J. Forestry 21:369–75.
18726. —— and C. F. Korstian. 1942: Seeding and planting in the practice of forestry. 3d ed., Wiley, New York. 520 pp.
18727. —— and C. L. Stevens. 1928: The testing of coniferous tree seeds at the School of Forestry, Yale University. Yale Univ. School Forestry Bull. 21. 46 pp.
Touphexi, A. Papadopoulou-; *see* Papadopoulou-Touphexi, A.
18728. Tourje, E. C. 1950: Camellia seed research. Camellia Res. 1950:9–13.
Tovarova, I. I. 15732.
Tove, S. B. 8537.
18729. Towers, B. 1938: Dormancy occurring in freshly harvested salsify seed. Proc. Assoc. Offic. Seed Analysts 23/26:286–87.
Towers, B. 4481.
18730. Townsend, C. O. 1899: The effect of ether upon the germination of seeds and spores. Botan. Gaz. 27:458–66.
18731. —— 1901: The effect of hydrocyanic acid gas upon grains and other seeds. Botan. Gaz. 31:241–64.
18732. Toyama, Y., and T. Tsuchiya. 1935: New stereoisomer of elaeostearic acid in pomegranate seed oil. J. Soc. Chem. Ind. Japan 38:182B–85B. Analyst 60:570–71. 1935.
18733. —— and —— 1935: New stereoisomer of elaeostearic acid from the seed oil of Karasuuri (*Trichosanthes cucumeroides*). J. Soc. Chem. Ind. Japan 38:185B–87B. Analyst 60:571. 1935.
18734. —— and —— 1936: Balsam pear seed oil. J. Soc. Chem. Ind. Japan 39: 220B–21B. Analyst 61:703. 1936.
18735. —— and —— 1936: Seed oils of several species of Cucurbitaceae. J. Soc. Chem. Ind. Japan 39:259B–60B. Analyst 61:772. 1936.
18736. —— and —— 1936: Fat from the seeds of *Cinnamonum pedunculatum*. J. Soc. Chem. Ind. Japan 39:261B–63B. Analyst 61:771–72. 1936.
18737. —— and K. Uozaki. 1937: Presence of octadecatrienoic acids in seed-oils of pomegranate, Karasu-Uri (*Trichosanthes cucumeroides*) and Balsam pear. J. Soc. Chem. Ind. Japan 40:249B–50B. Analyst 62:745–46. 1937.
Toyama, Y. 9528.
18738. Toyoda, K. 1958: Analysis of gas contained in the fruit of Indian lotus plant. Botan. Mag. (Tokyo) 71:371–77.
18739. —— 1958: On the maturing stage of the fruit of lotus and its germination. J. Japan. Botany 33:85–91.
18740. —— 1959: On the chlorophylls in some angiospermous seeds. Botan. Mag. (Tokyo) 72:159–68.
18741. —— 1960: On the ascorbic acid in the plumule of Indian lotus seed. Botan. Mag. (Tokyo) 73:98–103.
18742. —— 1961: A complementary note on the chlorophylls in some spermatophytic seeds. Botan. Mag. (Tokyo) 74:310–12.
18743. —— 1963: Effect of ascorbic acid and dehydroascorbic acid on anaerobic respiration of the seed of *Nelumbo nucifera*. Botan. Mag. (Tokyo) 76:41–47.
18744. Toyooka, H. 1960: On the germination of *Tilia japonica* Simonkai and *Tilia maximowicziana* Shirasawa in Hokkaido. I. Hokkaido Forestry Expt. Sta. Ann. Rept. 1959:33–39.
Toyooka, H. 18308.
18745. Tozawa, M. 1926: Methods of hastening germination of tree seeds. Chosen, Keijyo, Forest Expt. Sta. Bull. 5. 33 pp.
Traaen, A. E. 9992, 9993.
Trabut, G. 4419.
Tracey, J. G. 3566.
18746. Traetta-Mosca, F. 1929: Galactose in seeds. Ann. Ist. Super. Agrar. Portici 3:164–66. Chem. Abstr. 25:3691. 1931.
18747. —— and M. Venezia. 1930: A multivalent alcohol present in the seeds of *Castanea vesca*. Ann. Chim. Appl. 20:445–48. Chem. Abstr. 25:504. 1931.
Train, G. 16378.
18748. Trapaidze, A. S. 1960: Planting properties of the Grecian laurel seed and establishing standard of the sowing material. Subtrop. Kult. 1960(2):94–103.
18749. Trappe, J. M. 1961: Strong hydrogen peroxide for sterilizing coats of tree seed and stimulating germination. J. Forestry 59:828–29. Chem. Abstr. 56:5150. 1962.
18750. Traub, H. P. 1936: Artificial control of nucellar embryony in citrus. Science 83:165–66. Expt. Sta. Rec. 75:31. 1936.
18751. —— 1939: Polyembryony in *Myrciaria cauliflora*. Botan. Gaz. 101:233.
18752. —— 1941: The seed character of *Amaryllis procera*. Herbertia 8:82.

18753. —— and E. C. Auchter. 1933: Sprouting and grafting fractional parts of avocado embryos with attached cotyledonous material. Science 78:389–90. Expt. Sta. Rec. 70:781. 1934.
18754. —— and H. J. Muller. 1934: X-ray dosage in relation to germination of pecan nuts. Botan. Gaz. 95:702–6.
18755. Traub, W., J. B. Hutchinson, and D. G. H. Daniels. 1957: X-ray studies of the wheat protein complex. Nature 179: 769–70.
18756. Traube, J., and T. Marusawa. 1915: Über Quellung und Keimung von Pflanzensamen. Intern. Z. Phys.-Chem. Biol. 2:370–93.
18757. —— and H. Rosenstein. 1919: Über die Wirkung von oberflächenaktiven Stoffen auf Pflanzensamen. Biochem. Z. 95: 85–100. Botan. Abstr. 8:672. 1921.
18758. Trautwein, K., and J. Wassermann. 1929: Fortsetzung den Versuche über die Halbkorn- Keimmethode. Schnell- und Frühkeimprobe. Z. Ges. Brauw. 52:69. Centralbl. Bakt. Abt. II 79:139. 1929.
Travnicek, L. 4115.
18759. Traynard, J. C., and G. A. Simeonoglou. 1952: *Paulownia* seed and oil. Bull. Soc. Chim. France 1952:480. Chem. Abstr. 46:10302. 1952.
Trebuchet, G. 4023.
18760. Treccani, C. P. 1951: Germination tests for dormant seeds of fruit trees. Ann. Sper. Agrar. (Rome) 5:201–6. Brit. Abstr. BIII:243. 1951.
Tredici, V. 15507.
18761. Trelawny, J. G. S., and D. J. Ballantyne. 1963: The effect of gibberellin and temperature on the germination of seed of Bells of Ireland. Can. J. Plant Sci. 43:522–27.
18762. Trelease, S. F. 1942: Identification of selenium indicator species of *Astragalus* by germination tests. Science 95:656–57.
Trelease, S. F. 2050.
18763. Tremazi, S. A. 1954: A study of factors influencing the oil content of the seeds of toria (*Brassica compestris* L. var. Toria D. & F.). Texas J. Sci. 6:217–24.
18764. —— 1959: Oil formation in soybean (*Glycine hispida*) and effect of storage on soyabean seeds. Texas J. Sci. 11:17–27.
18765. Trepachev, E. P. 1953: The role of fertilizers in changing the oiliness of seeds. Sov. Agron. 1953(4):79–82. Soils Fertilizers 17:291. 1954.
18766. Tretsven, J. O., and J. A. Nelson. 1946: Wild yellow mustard seed and fanweed seed as concentrate feeds for milking cows. Montana Agr. Expt. Sta. Bull. 435. 11 pp.
Trevoy, L. W. 15836.
Trezzi, F. 18645.
Triebold, H. O. 4211.
18767. Trinkler, W. 1928: Über die Bestimmung des Wassergehaltes in Samen mittels der Trocknungsmethode. Jard. Botan. Inst. Essais Semences, Leningrad, Ann. Essais Semences 1928(1):3–80.
18768. Triplett, G. B., Jr., and M. B. Tesar. 1960: Effects of compaction, depth of planting, and soil moisture tension on seedling emergence of alfalfa. Agron. J. 52:681–84.
18769. Trippett, S. 1957: Toxic constituents of the Australian finger cherry, *Rhodomyrtus macrocarpa* Benth. J. Chem. Soc. 1957:414–19.
18770. Trivary, N. K. 1932: A note on the germination of unripe (green) gram seeds (*Cicer arietinum*). Botan. Centralbl. Beih. 49:259–61.
Trivedi, P. N. 3864.
Trivedi, R. K. 1711.
Trnka, A. 16013.
18771. Trochain, J. 1932: Note sur la germination de la noix de coco. Rev. Botan. Appl. 12:396–97.
18772. Trofimov, T. T. 1963: On the types of seed germination in some perennial plants. Botan. Zh. 48:1583–97.
Tronstad, L. 3202.
18773. Trotter, W. R. 1949: Effect of thiouracil and uracil on the germination of cress seeds. Nature 164:63.
18774. Trousdell, K. B. 1963: Loblolly pine regeneration from seed. J. Forestry 61:441–44.
Trousdell, K. B. 395.
Troyahn, W. J. 4151, 8383.
18775. True, R. H. 1893: Development of the caryopsis. Botan. Gaz. 18:212–26.
18776. —— 1914: The harmful action of distilled water. Am. J. Botany 1:255–73.
Truhaut, R. 16724.
18777. Tryon, E. H., and H. G. Woodrum. 1962: Growing Christmas trees from seed. West Va. Agr. Expt. Sta. Bull. 472. 21 pp.
18778. Trzcinski, W. C. 1934: Die Keimung der Unkrautsamen nach deren Aufenthalt im gewöhnlichen Mist und im Heissmist. Proc. Intern. Seed Testing Assoc. 6:21–34.
Trzecki, S. 12334.
Tsai, C. Y. 14318.
Tsai, T. P. 3823.

18779. Tsanev, R. 1953: Neubildung von Zellenkerne in Endosperm der Pflanzen. Izv. Bulgar. Akad. Nauk Biol. Inst. 4:113–19.

18780. Tsarev, M. V. 1944: Seeds of *Thermopsis lanceolata* R. Br. as a source of cytisine. Compt. Rend. Acad. Sci. URSS 42:122–24. Chem. Abstr. 38:6053. 1944.

18781. Tschajlachjan, I. C., and L. N. Schdanowa. 1938: Die Wirkung von Heteroauxin auf das Wachstum und die Entwicklung der Pflanzen bei der Bearbeitung des Saatmaterials. Ber. Akad. Wiss. SSSR 1938(5/6):1281–95. Botan. Centralbl. 175:174–75. 1940.

Tschermak-Seysenegg, E. von; *see* Von Tschermak-Seysenegg, E.

18782. Tschirch, A. 1891: Physiologische Studien über die Samen, insbesondere die Saugorgane derselben. Ann. Buitenzorg 9:143–83.

18783. Tschirley, F. H., and S. C. Martin. 1960: Germination and longevity of velvet mesquite seed in soil. J. Range Management 13:94–97. Biol. Abstr. 35:45751. 1960.

Tschizova, A. M. 7682.

Tselishchev, S. P. 10549.

18784. Tselykovskii, P. G. 1962: Some of the outstanding seed-testing laboratories in the USSR. Zemledelie 3:81–82. Biol. Abstr. 40:4282. 1962.

18785. Tseng, S. T., and C. I. Lin. 1962: Studies on the physiological quality of pure rice seed. Proc. Intern. Seed Testing Assoc. 27:459–75.

18786. Tsi, C. 1933: Der Einfluss der Temperatur auf die Keimung von Nadelholzsamen verschiedener Standortsrassen und Holzarten als Ausdruck erblicher Strukturunterschiede. Dissertation, Peiyang Press. 84 pp.

18787. Tsien, W. S., and B. C. Johnson. 1959: The effect of radiation sterilization on the nutritive value of foods. IV. On the amino acid composition of garden peas and lima beans. J. Nutr. 68:419–28.

18788. Tsikov, D., and M. Stoilov. 1951: Studies on the storing of dry substance, dynamics of water-contents and germination in the seed-formation period of some wheat and rye varieties. Izv. Bulgar. Akad. Nauk Biol. Inst. 2:45–66.

18789. Tsinger, N. V., and V. A. Poddubnaja-Arnoldi. 1958: The histochemical characteristics of proteins of the embryo of certain representative orchids. Dokl. Akad. Nauk SSSR 118:28–31.

Tsinger, N. V. 18937.

18790. Tsitsin, N. V., and M. N. Sileva. 1962: The chemical composition of seeds of the yellow acacia. Byul. Gl. Botan. Sada 46:53–55. Chem. Abstr. 59:5492. 1963.

18791. Tsi-Tung, L. 1934: The development of embryo of *Ginkgo biloba.* Sci. Rept. Natl. Tsing Hua Univ. Ser. B 2:29–35.

18792. —— 1934: The development of *Ginkgo* embryo in vitro. Sci. Rept. Natl. Tsing Hua Univ. Ser. B 2:41–52.

18793. —— and S. M. Chen. 1934: Temperature and the development of the *Ginkgo* embryo. Sci. Rept. Natl. Tsing Hua Univ. Ser. B 2:37–39.

18794. Tskoidze, V. 1936: Methode rapide pour la recherche du pouvoir geminatif des graines d'*Aleurites.* Bull. Acad. Sci. URSS Ser. Biol. 1936(1):143–50. Bull. Soc. Botan. France 85:475. 1938.

18795. —— 1936: Du forcage de la germination des graines de *Chamaerops humilis* et de *Sapium sibiferum.* Bull. Acad. Sci. URSS Ser. Biol. 1936(1):151–56. Bull. Soc. Botan. France 85:475. 1938.

18796. Tsuchiya, T., and O. Okubo. 1961: Seed oils of *Pinus pumila* Regel., papaya, and some cruciferous plants. Tokyo Govt. Chem. Ind. Res. Inst. Rept. 56:385–87.

Tsuchiya, T. 18732–18736.

18797. Tsuda, M. 1937: Some experiments on the germination and growth of *Calystegia soldanella* R. Br. and *Lathyrus japonicus* Willd. Botan. Mag. (Tokyo) 51:379–87. Botan. Centralbl. 174:267. 1939.

18798. Tsui, C., and B. K. Wang. 1959: The effect of microelements on the respiration and nitrogen compounds during vernalization of wheat. Acta Botan. Sinica 8:239–42. Biol. Abstr. 35:64332. 1960.

Tsuji, T. 18257.

18799. Tsujimoto, M., and H. Koyanagi. 1933: New unsaturated acid in the kernel oil of "akarittom," *Parinarium laurinum.* I. J. Soc. Chem. Ind. Japan (Suppl. Binding) 36:110–13. Chem. Abstr. 27: 3099. 1933.

18800. Tsukamoto, Y., and T. Asahira. 1958: Chemical vernalization of radish. Mem. Res. Inst. Food Sci. Kyoto Univ. 14:1–13. Chem. Abstr. 52:11192. 1958.

18801. ——, K. Kano, and T. Katsuura. 1963: Instant media for orchid seed germination. Am. Orchid Soc. Bull. 32:354–55.

Tsukijima, Y. 10364–10366.

18802. Tsukunaga, K., and T. Nishino. 1931: Change in the composition of soy beans and iodine numbers of the oil during the periods of maturity. Res. Bull. Agr. Expt. Sta. S. Manchuria Rlwy. Co. 2:21–48. Chem. Abstr. 25:3690. 1931.

18803. ——, H. Ohara, and T. Nishino. 1931: Relation between the composition of Manchurian soy beans and colors of seed coat and hilum. Res. Bull. Agr. Expt. Sta. S. Manchuria Rlwy. Co. 2:1–20. Chem. Abstr. 25:3690. 1931.

Tsvetkova, I. V. 7098.

18804. Tsytovich, K. I. 1961: Influence of thermal seed treatment on yielding ability and seed production of cucumbers and tomatoes. Tr. Prikl. Botan. Genet. i Selek. 34:143–48.

18805. Tsyurupa, B. N., and L. A. Balabanova. 1953: The effect of aqueous extracts from seeds on germination. Byul. Gl. Botan. Sada 16:60–63. Chem. Abstr. 49:7661. 1955.

18806. Tuason, A. M., and A. O. Cruz. 1953: Composition of para-rubber-seed oil. Philippine J. Sci. 82:341–44. Biol. Abstr. 29:12339. 1955.

18807. —— and A. T. Reyes. 1960: Water-soluble extract from coconut husk and its application in leather technology. Philippine J. Sci. 88:253–56.

18808. Tuba, J., G. Hunter, and J. A. Osborne. 1947: The assay of new rich natural sources of ascorbic acid. Science 105:70.

18809. Tubbs, F. R. 1932: The germination of tea seed. Tea Quart. 5:66–68.

Tuboi, Y. 10374.

Tucay, E. A. 3592.

Tucek, K. 19160.

18810. Tucker, H., and R. E. Foster. 1958: Statistical technique in indexing lettuce seed for mosaic content. Plant Disease Reptr. 42:1339–41.

18811. Tucker, L. R. 1938: Relation between number of carpels and number of cotyledons of the apple. Proc. Am. Soc. Hort. Sci. 35:9–11. Biol. Abstr. 12:12199. 1938.

18812. Tuite, J. F., and C. M. Christensen. 1955: Grain storage studies. XVI. Influence of storage conditions upon the fungus flora of barley seed. Cereal Chem. 32:1–11.

18813. —— and —— 1957: Grain storage studies. XXIII. Time of invasion of wheat seed by various species of *Aspergillus* responsible for deterioration of stored grain, and source of inoculum of these fungi. Phytopathology 47:265–68.

18814. —— and —— 1957: Grain storage studies. XXIV. Moisture content of wheat seed in relation to invasion of the seed by species of the *Aspergillus glaucus* group, and effect of invasion upon germination of the seed. Phytopathology 47:323.

18815. Tukey, H. B. 1924: Studies of fruit seed storage and germination. N.Y. Agr. Expt. Sta. Bull. 509:3–19.

18816. —— 1927: The viability of seed from certain cherry varieties. Proc. Am. Soc. Hort. Sci. 24:129–32.

18817. —— 1933: Artificial culture of sweet cherry embryos. J. Heredity 24:7–12.

18818. —— 1933: Embryo abortion in early-ripening varieties of *Prunus avium*. Botan. Gaz. 94:433–68.

18819. —— 1934: Anomalous embryos of cultivated varieties of *Prunus* with particular reference to fruit breeding. Botan. Gaz. 95:493–97.

18820. —— 1934: Growth of the embryo, seed, and pericarp of the sour cherry (*Prunus cerasus*) in relation to season of fruit ripening. Proc. Am. Soc. Hort. Sci. Suppl. 31:125–44. Expt. Sta. Rec. 73:319–20.

18821. —— 1935: Artificial culture methods for isolated embryos of deciduous fruits. Proc. Am. Soc. Hort. Sci. 32:313–22.

18822. —— 1936: Development of cherry and peach fruits as affected by destruction of the embryo. Proc. Am. Soc. Hort. Sci. 33:265–66.

18823. —— 1936: A relation between seed attachment and carpel symmetry and development in *Prunus*. Science 84:513–15.

18824. —— 1938: Growth patterns of plants developed from immature embryos in artificial culture. Botan. Gaz. 99:630–65.

18825. —— 1944: The excised-embryo method of testing the germinability of fruit seed with particular reference to peach seed. Proc. Am. Soc. Hort. Sci. 45:211–19.

18826. —— and M. S. Barrett. 1936: An approximate germination test for non-after-ripened peach seed. Proc. Am. Soc. Hort. Sci. 33:267–68. Plant Physiol. 11:629–33. 1936.

18827. —— and R. F. Carlson. 1945: Breaking the dormancy of peach seed by treatment with thiourea. Plant Physiol. 20:505–16.

18828. —— and F. A. Lee. 1940: Embryo development. Botan. Gaz. 101:818–39.

Tukey, H. B. 3624, 8087, 11982.

Tulach, J. 6177.

Tullis, E. C. 4507, 7579.

18829. Tulstrup, N. P. 1959: International trade in forest tree seed. Unasylva 13:197–201.

Tulstrup, N. P. 12676.

Tumanyan, V. G. 143.

18830. Tumbleson, M. E. 1960: Germination inhibitor in tubers of yellow nutgrass. Res. Rept. N. Central Weed Control Conf. 17:106–7. Weed Abstr. 11:1946. 1962.

18831. Tummin-Katti, M. C., R. L. Alimchandani, and B. B. Charrappagowder. 1945: Chemical examination of the seeds of *Jatropha curcas* Lin. J. Univ. Bombay 14A:34–36. Chem. Abstr. 40:3228. 1946.

Tunblad, B. 178.

18832. Tupper. 1843: Mummy wheat. Gardeners Chron. 1843:787.

18833. Turcotte, E. L., and C. V. Feaster. 1963: Haploids: High frequency production from single-embryo seeds in a line of Pima cotton. Science 140:1407–8.

Turcotte, F. 12861.

18834. Turdiev, S. Y. 1960: Some features of the germination of seeds and the formation of sprouts of Nymphaeaceae. Tr. Akad. Nauk Kazakhsk. SSR Botan. Sad. 5:164–78.

18835. Turemuratov, U. T., K. N. Butov, and M. Tadzhitdinov. 1962: Some questions regarding the propagation of *Phragmites communis* by seed on the lower delta of the Amu-Darya. Uzbeksk. Biol. Zh. 5:53–57.

18836. Turesson, G. 1922: Über den Zusammenhang zwischen Oxydationsenzymen und Keimfähigkeit in verschiedenen Samenarten. Botan. Notiser 1922:323–35.

Turi, R. M. 12946.

Turk, E. E. de. 18114.

Turkov, V. L. 10549.

18837. Turkova, N. S. 1951: Effect of gaseous composition of the air on viability of acorns (in storage). Dokl. Akad. Nauk SSSR 76:735–38.

Turkova, V. 10853, 11053.

18838. Turner, A. J. 1929: Cotton seeds: Their absorption of water, and specific gravity. Agr. J. India 24:83–90.

Turner, D. 13964.

18839. Turner, D. H., and J. F. Turner. 1957: Physiology of pea fruits. III. Changes in starch and starch phosphorylase in the developing seed. Australian J. Biol. Sci. 10:302–9.

18840. —— and —— 1958: Uridine diphosphoglucose pyrophosphorylase of pea seeds. Biochem. J. 69:448–52.

18841. —— and —— 1961: Adenylic deaminase of pea seeds. Biochem. J. 79: 143–47.

Turner, D. H. 16480, 18845.

18842. Turner, D. J. 1956: Some observations on the germination and grading of cashew nuts. E. African Agr. J. 22:35–39.

Turner, E. R. 4435, 15794.

18843. Turner, J. E., and S. K. Shapiro. 1961: S-Methylmethionine and S-adenosylmethionine homocysteine transmethylase in higher plant seeds. U.S. At. Energy Comm. Res. Develop. Rept. ANL-6464: 128–32. Biol. Abstr. 45:25889. 1964.

18844. Turner, J. F., and J. E. King. 1961: Inosine 5′-phosphate dehydrogenase of pea seeds. Biochem. J. 79:147–51.

18845. ——, D. H. Turner, and J. B. Lee. 1957: Physiology of pea fruits. IV. Changes in sugars in the developing seed. Australian J. Biol. Sci. 10:407–13.

Turner, J. F. 8373, 8374, 10699, 18839–18841.

18846. Turner, J. H. 1933: The viability of seeds. Kew Roy. Botan. Gardens Bull. Misc. Inform. 1933(6):257–69.

18847. —— 1936: Viability of clover seeds. J. Botany 74:52–53.

Turney, D. M. 17245.

18848. Turrill, W. B. 1957: Examples of seed structure. Seeds without endosperm or perisperm. Gardeners Chron. 141:245.

18849. —— 1957: Number, size, weight and shape of seeds. Gardeners Chron. 141:344.

18850. —— 1957: Seed dispersal: 1, 2, 3, 4, 5, 6. Gardeners Chron. 141:376, 432, 460, 487, 612.

18851. —— 1957: Germination of seeds: 3, 4, 5, 6, 7. Gardeners Chron. 141:633, 675; 142:37, 74, 94.

Turvey, J. R. 15166.

18852. Tushnyakova, M. M. 1961: Effect of gibberellic acid on seeds of plants. Tr. Inst. Genet. Akad. Nauk SSSR 28:163–69.

18853. —— 1962: The influence of seed storage on the terminal effect of γ-irradiation (Co^{60}). Tr. Inst. Genet. Akad. Nauk SSSR 29:185–93. Nucl. Sci. Abstr. 17: 5215. 1963.

Tusihasi, H. 7440.

18854. Tuzson, J., Jr. 1933: Bestimmung der Weizensorten durch Färbung der

Fruchtschale. Mezogazdasagi Kutatasok 6:273–74.

18855. Tveit, M. 1956: Isolation of a chetomin-like substance from oat seedlings raised from seeds infested with *Chaetomium cochliodes*. Acta Agr. Scand. 6:13–16.

18856. Twitchell, L. T. 1955: Germination of fourwing saltbush (*Atriplex canescens*) seed as affected by soaking and chloride removal. J. Range Management 8:218–20.

Tyankova, L. 15271, 15272.

18857. Tydeman, H. M. 1938: Pear rootstocks from seed. I. Experiments on methods of germinating pear seeds. II. Studies on the variation in seedling pear trees. E. Malling Res. Sta. Ann. Rept. 26:103–14. Expt. Sta. Rec. 81:517–18. 1939.

Tyler, G. J. 3173, 3175.

18858. Tyler, L. J., R. P. Murphy, and H. A. MacDonald. 1956: Effect of seed treatment on seedling stands and on hay fields of forage legumes and grasses. Phytopathology 46:37–44.

18859. Tyler, V. E., Jr. 1955: The effect of germination on the alkaloid content of *Nigella damascena* seed. J. Am. Pharm. Assoc. 44:52–54.

Tyler, V. E., Jr. 5172.

Tylkina, L. G. 13552.

18860. Tymko, M. M. 1960: The effect of the position of walnut seeds on their germinating power. Sadovodstvo Vinogradarstvo i Vinodelie Moldavii 1960(3):46–48.

18861. Tyner, B. M. 1953: The germination, evaluation, and labeling of Johnson grass. Proc. Assoc. Seed Control Off. Southern States 10:82–87.

Tyner, E. H. 13141.

18862. Tyner, L. E., and B. A. McKinnon. 1964: Fungi of barley seed and their associative effects. Phytopathology 54: 506–8.

Tysdal, H. M. 5941.

Tyutyunnikova, V. A. 6145.

18863. Tze-Ting, W. 1963: Effect of different temperatures upon seed germination of *Taiwania*. Mem. Coll. Agr. Taiwan Univ. 7(2):75–80.

18864. Tzigulev, V. I. 1939: Influence of absolute weight of tobacco seeds on their sowing qualities. Krasnodar Inst. Tobacco Ind. USSR Bull. 139:267–75.

U

18865. Ubaldini, I., and L. Bissi. 1932: Constitution and possibility of utilizing "Dum" nuts from palm trees. Giorn. Chim. Ind. Appl. 14:555–62. Chem. Abstr. 27:1533. 1933.

18866. Ubatuba, F. B. 1955: Occurrence of a trypsin inhibiting factor in the seeds of "*Canavalia ensiformis*." Rev. Brasil. Biol. 15:1–8.

Ubertalle, A. 2828.

Ubisch, G. V. 6406.

18867. Ucciani, E., and F. Busson. 1963: Contribution to the study of fats of *Pachylobus edulis*. Oléagineux 18:253–55. Chem. Abstr. 59:4179. 1963.

Uchida, K. 9366.

Uddenberg, C. E. 14713.

Udo, S. 9363.

18868. Uebersezig, M. 1947: Successful storage of slash pine seed for fifteen years. J. Forestry 45:825–26.

Uehlinger, H. P. 16944.

18869. Ueki, C., and I. Suetsugu. 1958: The identification of hard seeds in some leguminous forage crops. I. Genge (*Astragalus sinicus* L.). Proc. Intern. Seed Testing Assoc. 23:69–72.

Ueki, C. 18141, 18142.

18870. Ueki, K., and A. Inokuti. 1959: After-effect of high temperature treatment given during the seed ripening period of broad bean on the plant growth and yield in the following generation. Tech. Bull. Fac. Agr. Kagawa Univ. 11:281–85.

Uemura, M. 19968.

Ueno, S. 11063.

18871. Uzolik, N., and J. P. Nitsch. 1959: Étude chromatographique des substances de croissance dans la graine immature de Mirabelle. Bull. Soc. Botan. France 106:446–51.

18872. Ugryumov, P. S. 1934: α- and β-amylase in ripening wheat grains. Biochem. Z. 282:74–78. Chem. Abstr. 30: 1091. 1936.

Uhl, D. E. 558.

Uhlmann, G. 6484.

Uhring, J. 17201.

18873. Uhvits, R. 1946: Effect of osmotic pressure on water absorption and germination of alfalfa seeds. Am. J. Botany 33: 278–85.

18874. Ulali, D. L., M. B. Barker, and R. C. Dumlao. 1960: A preliminary study on the cancellation of dormancy period of rice seeds. Philippine Agriculturist 44:279–89. Biol. Abstr. 37:3197. 1962.

18875. Ulbrich, E. 1929: Über die durch Wind verbreiteten (anemochoren) Früchte und Samen. Verhandl. Botan. Ver. Prov. Brandenburg 71:63–73.

18876. Ulbrich, H. 1935: Die intraplasmatischen Vorgänge bei der Schleimbildung in den Samen von *Linum usitatissimum* L. und in den Wurzeln von *Althaea officinalis* L. Dissertation, Heinrich Pöppinghaus. 32 pp.

Ulbrich, H. 9751.

Ulezlo, I. V. 11332.

Ulich, G. 8698.

18877. Ulitzur, S., and A. Poljakoff-Mayber. 1963: Oxidative phosphorylation in germinating lettuce seeds. J. Exptl. Botany 14:95–100.

Ullmann, E. 1829, 1832.

18878. Ullman, S. B. 1940: On germination inhibitors. V. Essential oils, alkaloids and glucosides as inhibitors of germination and growth. Thesis, Hebrew Univ., Jerusalem. 30 pp.

18879. Ullmann, W. 1949: Über die Keimfähigkeitsdauer (Lebensdauer) von landwirtschaftlichen und gartenbaulichen Samen. Saatgut-Wirtsch. 1:174–75, 195–96.

Ullstrup, A. J. 13472, 18329, 19691.

18880. Ulrich, R. 1936: Correlation entre l'allogement du fruit et le developpement des graines chez la Giroflee (*Cheiranthus cheiri* L.). Compt. Rend. 203:1012–13.

18881. —— 1938: Comparison of the lipides accumulated in the green pericarps and in the albumin of the seeds of *Hedera helix* L. Compt. Rend. 206:1316–18. Chem. Abstr. 32:6293. 1938.

18882. —— 1939: Inhibition de la germination des graines de tomate par les sues du pericarpe mur. Compt. Rend. 208:1835–37.

18883. —— 1943: Influence of injuries on the metabolism of glucides in pea pods. Bull. Soc. Botan. France 90:64–67. Chem. Abstr. 43:8458–59. 1949.

18884. Ultee, A. J. 1933: Storage of coffee seed. Arch. Koffie Cult. Ned.-Indië 7:75–83. Biol. Abstr. 10:4111. 1936.

Ultee, A. J. 5710.

18885. Umali, D. L. 1940: A study on coconut seed selection for germination. Philippine Agriculturist 29:296–312.

18886. —— and F. A. Bernardo. 1958: Notes on mutant forms among Philippine plants. II. Seedless avocado (*Persea americana* Mill.). Philippine Agriculturist 42:194–95.

Umbreit, W. W. 274.

18887. Umland, C. B. 1943: Freezing of black gum seed increases nursery production. J. Forestry 41:452.

Umlauft, W. 17077.

Underwood, E. J. 4139, 11985, 11986, 16434.

18888. Ungar, I. A. 1962: Influence of salinity on seed germination in succulent halophytes. Ecology 43:763–64.

18889. United States Department of Agriculture. 1961: Seeds: The Yearbook of Agriculture 1961. Washington. 591 pp.

18890. Unrau, A. M. 1964: Carbohydrates of *Centrosema* seed. The constitution of an Arabino galactan. I. Can. J. Chem. 42:916–25.

Unrau, J. 10465, 15310.

Uozaki, K. 18737.

18891. Urban, J., and E. Vitek. 1916: Influence of very low temperatures on the germination capacity of beet seeds. Z. Zuckerind. Böhmen 40:295–300. Expt. Sta. Rec. 37:829. 1917.

18892. Urban, S. 1933: Sur la désorganisation des grains d'aleurone et sur l'évolution des plastes dans l'embryon de la courge. Acta Botan. Inst. Univ. Zagreb 8:97–104.

18893. Urban, V. D. 1937: Germination of lupine seed. Flower Grower 24(6):316.

18894. Urbani, E., and L. De Cesaris-Coromaldi. 1953: Latent life. II. Proteinases in *Artemisia salina.* Atti Accad. Nazl. Lincei, Rend. Classe Sci. Fis. Mat. Nat. 14:144–49. Chem. Abstr. 48:1487. 1954.

18895. ——, S. Russo, and L. Rognone. 1953: Latent life. III. Study on the amylase of *Artemisia saline.* Atti Accad. Nazl. Lincei, Rend. Classe Sci. Fis. Mat. Nat. 14:679–701. Chem. Abstr. 48:1487. 1954.

Urbaniak, Z. 12340.

Urie, A. L. 12061.

18896. Urion, E., and L. Chapon. 1955: Germination release in barley. Brasserie 10:58–64. J. Sci. Food Agr. 6(8):ii84. 1955.

18897. —— and —— 1956: Contribution à l'étude de la dormance de l'orge. Brasserie 11:151–62, 181–90.

18898. ——, G. Lejeune, and M. Fronsacq-Collin. 1951: Seasonal variations in the protein content of stored cereals. Bull. Soc. Chim. Biol. 33:120–32. Brit. Abstr. BIII:468. 1951.

18899. Urosevic, B. 1961: The influence of saprophytic and semi-parasitic fungi on the germination of Norway spruce and Scots pine seeds. Proc. Intern. Seed Testing Assoc. 26:537–56.

18900. Urzalova. 1955: The effect of treating caraway seeds before sowing on germination energy. Sb. Vysoke Skoly Zemedel. Brne Rada C 1955(3):243–52. Biol. Abstr. 33:35065. 1959.

Ushakova, E. N. 11390, 17595.

Uspenskaya, Z. V. 11391–11393.

18901. Ustimenko, S. P. 1957: Effect of pericarp on the sprouting energy of seeds of monogerm sugar beet. Sakharn. Svekla 2(12):24–27.

Utigasaki, T. 18443.

18902. Uttaman, P. 1944: Some preliminary observations on the effect of drying the sprouted seed upon its viability and subsequent plant growth in paddy. Madras Agr. J. 32:75–78. Biol. Abstr. 20: 18578. 1946.

18903. —— 1949: Embryo-culture and its use in plant breeding. Madras Agr. J. 36:555–63.

18904. —— 1949: A preliminary investigation into the viability of immature embryos of corn under conditions of cold storage at freezing point. Current Sci. (India) 18:52–53.

18905. —— 1949: Culturing of pro-embryos of normal diploid corn (maize) aged three to seven days. Current Sci. (India) 18:215–16.

18906. —— 1949: Embryo culture to obtain F_1 plants of incompatible crosses in corn (maize). Current Sci. (India) 18:297–99.

18907. —— 1949: A study in contrast of the effects of cocoanut water on the growth of immature embryos of corn (maize) when applied before and after germination of the embryo. Current Sci. (India) 18:343–44.

18908. —— 1950: Growth-promoting factors in corn germ extract. Current Sci. (India) 19:21–22.

18909. —— 1950: A study on the germination of *Striga* seed and on the mechanism and nature of parasitism of *Striga lutea* (Lour.) on rice. Proc. Indian Acad. Sci. Sect. B32:133–42.

18910. —— 1955: On the viability of wild indigo seeds (*Thephrosia purpurea*). Madras Agr. J. 42:182. Biol. Abstr. 30:11192. 1956.

18911. —— and K. V. A. Bavappa. 1955: Influence of seed weight on the yield of elephant yam (*Amorphophallus companulatus*). Madras Agr. J. 42:142–43.

18912. Uzunov, I. S. 1962: Effectiveness of complex pre-planting treatment of barley seeds. Moskov. Ordena Lenina Sel'skokhoz. Akad. Im. K. A. Timiryazeva Dokl. 77:381–86.

V

18913. Vaartaja, O. 1956: Photoperiodic response in germination of seed of certain trees. Can. J. Botany 34:377–88.

18914. —— 1954: Factors causing mortality of tree seeds and succulent seedlings. Acta Forest. Fennica 62(3):1–31. Biol. Abstr. 31:25815. 1957.

Vaartaja, O. 4512.

18915. Vaccaneo, R. 1934: On the germinability of *Polygonum cuspidatum* of the high wide plain, with some considerations on the possibility of introduction. Nuovo Giorn. Botan. Ital. 41:803–4.

18916. Vacha, G. A., and R. B. Harvey. 1927: The use of ethylene, propylene, and similar compounds in breaking the rest periods of tubers, bulbs, cuttings, and seeds. Plant Physiol. 2:187–94.

18917. Vacherot, M. 1954: Les orchidées, generalités, culture, obtention des semis et description des principales espèces. Ballière, Paris (Bibliothèque Hort. Pratique). 268 pp.

18918. Vacin, E. F., and F. W. Went. 1949: Use of tomato juice in the asymbiotic germination of orchid seeds. Botan. Gaz. 111:175–83.

Vagi, S. 6279.

18919. Vahl, I. 1940: On germination inhibitors. III. Germination inhibitors in the fruit of *Poterium spinosum* L. Palestine J. Botany 2:28–32.

18920. Vaidya, P. B. 1961: Occurrence of twin embryo-sacs in *Capparis aphylla* Roth. Current Sci. (India) 30:187–88.

Vaidyanathan, C. S. 15844, 15845.

18921. Vainagy, I. V. 1963: On the germinating power of seeds of wild herbaceous plants of the Carpathians under laboratory conditions. Acad. Sci. RSS Ukraine Kiev J. Botan. 20(4):48–57.

18922. Vaintraub, I. A., and A. D. Shutov. 1962: Globulins in *Lathyrus* seeds. Tr. po Khim. Prirodn. Soedin. Kishinevsk. Gos. Univ. 1962(5):49–52. Chem. Abstr. 60: 850. 1964.

18923. ——, ——, and V. G. Klimenko. 1962: The globulins of vetch seeds. Biochemistry (USSR) (English Transl.) 27: 298–305.

Vaintraub, I. A. 6897.

18924. Vakar, A. B. 1935: On the buffering power of lupine seeds. Botan. Zh. 20: 346–47.

Vakkuri, P. Yli-; *see* Yli-Vakkuri, P.

18925. Vakulin, D. J. 1937: Some data on the yield of essential oil by a new variety of dragonhead (*Dracocephalum mildavica* L. var. *hexagonum* D. Vakulin), from different seed-samples. Compt. Rend. Acad. Sci. URSS 15:203–5.

18926. —— 1939: Secondary dormancy induced in seeds of *Lallemantia iberica* F. et M. Compt. Rend. Acad. Sci. URSS 23:845–47.

18927. —— 1939: Determination of botanical forms of *Lallemantia* from the absolute weight and size of their seed. Compt. Rend. Acad. Sci. URSS 23:848–49.

18928. Valdeyron, G. 1953: La production des semences de céréales en Tunisie. Législation et méthodes. Tunisie Serv. Botan. Ann. 26:167–206.

18929. Valdivia-Bugueno, V. 1954: Influencia de tamaño de semilla y profundidad de siembra en el rendimiento de granos en maravilla. Agr. Tec. Mex. 14(1):27–46. Biol. Abstr. 33:3040. 1959.

18930. Valenti, A. 1934: Chemical and biological studies of meal of the embryo of carob. Arch. Ital. Sci. Farm. 3:387–401. Chem. Abstr. 31:6702. 1937.

18931. Valentin, J., and G. Brockelt. 1956: Über die Standardisierung und die Gehaltsbestimmung der anthelminthischen Wirkstoffes in Samen Cucurbitae. Pharmazie 11:796–98. Biol. Abstr. 32:13841. 1958.

18932. Valentine, D. H., and S. R. J. Woodell. 1963: Studies in British primulas. X. Seed incompatibility in intraspecific and interspecific crosses at diploid and tetraploid levels. New Phytologist 62:125–43.

18933. Valentine, R. S. 1946: Expanded mica as a seeding medium. Am. Delphinium Soc. Yearbook 1946:32–33.

18934. Valeri, H., and F. Gimeno N. 1954: Preliminary phytochemical and toxicological investigations of the seeds of *Cassia occidentalis*. Rev. Med. Vet. Parasitol. 11:121–55. 1952. Chem. Abstr. 48:10842. 1954.

18935. —— and —— 1954: Phytochemical and toxicological study of the pericarp of the avocado pear (*Persea americana*). Rev. Med. Vet. Parasitol. 13:37–58. Chem. Abstr. 50:3563. 1956.

Valeri, H. 6918, 6919.

18936. Valeva, S. A. 1960: Biophysics in practice. Data on the radiosensitivity of agricultural crops. Biophysics (USSR) (English Transl.) 5:284–89.

18937. Valishina, V. P., and N. V. Tsinger. 1952: Relation of *Aconitum* seed germination to the size of its embryo. Byul. Gl. Botan. Sada 13:45–47.

18938. Vallance, K. B. 1949: Effect of the *Striga* germination stimulant on the respiration of *Striga* seeds. Nature 164: 802.

18939. —— 1950: Studies on the germination of the seeds of *Striga hermonthica*. I. The influence of moisture-treatment, stimulant-dilution, and after-ripening on germination. Ann. Botany 14:347–63.

18940. —— 1951: Studies on the germination of the seeds of *Striga hermonthica*. II. The effect of the stimulating solution on seed respiration. J. Exptl. Botany 2(4):31–40.

18941. —— 1951: Studies on the germination of the seeds of *Striga hermonthica*. III. On the nature of pretreatment and after-ripening. Ann. Botany 15:109–28.

18942. —— 1951: Germination of seeds of *Bartsia odontites*. Nature 167:732.

18943. —— 1952: The germination of the seeds of *Rhinanthus crista-galli*. Ann. Botany 16:409–20.

Vallarta, M. Tolentino-; *see* Tolentino-Vallarta, M.

Valle, C. C. 16081.

18944. Valleau, W. D. 1935: Seed transmission of *Helminthosporium* of corn. Phytopathology 25:1109–12.

18945. —— 1939: Symptoms of yellow ring spot and longevity of the virus in tobacco seed. Phytopathology 29:549–51.

Valleau, W. D. 9856.

Vallury, S. M. 3996.

18946. Valuta, G. 1938: Seed appraisement of soybeans from various sources. Kühn-Arch. 44:121–60. Chem. Abstr. 32:5868. 1938.

Valuta, G. 2923, 2924.

18947. Van A. de Veer, E. J., and F. A. V. Lunel. 1950: Germination of *Intsia palembanica* Miq. and *Intsia bijuga* O. Ktze. Tectona 40:336–45.

18948. Van Altena, A. C. J. 1949: On the durability and germination methods of flower seeds. Zaadbelangen 3:52–53.

18949. —— 1949: Development and application of scientific seed research. Zaadbelangen 3:277–79, 289–90.
18950. —— 1952: Investigation of viability of legumes with help of tetrazolium. Zaadbelangen 6:114–15.
Van Andel, O. M. 5135.
Van Assche, C. 18707.
18951. Van Atta, G. R., and W. C. Dietrich. 1944: Valencia orange-seed oil. Oil Soap 21:19–22. Biol. Abstr. 18:11697. 1944.
Van Atta, G. R. 14563.
18952. Van Breda, N. G. 1939: An improved method of sowing grass and Karroo shrub seed. S. African J. Sci. 36:328–35.
18953. Van Demick, B., Jr. 1939: The effect of depth of covering on the germination of Douglas fir seed. Wash. Univ. (Seattle) Pubs., Theses Ser. 4:157–59. Expt. Sta. Rec. 82:778. 1940.
18954. Vanden Driessche, T. 1962: Influence of carbon dioxide on cottonseed oil content and quality. Oléagineux 17:635–41. Biol. Abstr. 43:24897. 1963.
Van den Heuvel, F. A. 6221.
Van der Auwermellen, E. 12961.
18955. Van der Haar, A. W. 1929: Untersuchungen über die Saponine und verwandte Körper. XXII. Das Saponin der Samenkerne von *Mimusops elengi* L. und seine Hydrolysespaltlinge. Rec. Trav. Chim. 48:1155–65.
18956. —— 1929: Untersuchungen über die Saponine und verwandte Körper. XXIII. Das Saponin der Samenkerne von *Achras sapota* L. und seine Hydrolysespaltlinge. Rec. Trav. Chim. 48:1166–69.
18957. Van der Hoop, D. J. N. 1930: Germination tests with *Hevea* seed. Arch. Rubbercult. Nederl.-Indie 14:81–84. Biol. Abstr. 9:1411. 1935.
18958. Van der Knapp, W. P. 1954: Correlation between the size of the seed and the growth of the cacao plant. Bergcultures 23:421, 423, 425.
18959. Van der Meer Mohr, J. C. 1926: Über die Wirkung von Eosin-, Erythrosin- und Methylen-blau-lösungen auf Keimung und Wachstum einiger Pflanzen. Rec. Trav. Botan. Neerl. 23:245–62.
18960. Van der Mijll Dekker, L. P., and H. De Miranda. 1954: The vitamin B_1 content of Dutch wheat and the factors which determine this content. Neth. J. Agr. Sci. 2:27–31. Chem. Abstr. 48:11666. 1954.
18961. Van der Paauw, F. 1951: The 1000-grain weight, the weight per hectoliter and the ratio of grain yield and total yield of cereals as affected by the fertilizing level of the soil. Landbouwk. Tijdschr. 63:116–24. Biol. Abstr. 26:19118. 1952.
18962. Van der Pijl, L. 1934: Über die Polyembryonie bei Eugenia. Rec. Trav. Botan. Neerl. 31:114–87.
18963. —— 1935: Seed dispersal by bats. Trop. Natuur. 24:131–38. Biol. Abstr. 10:20328. 1936.
18964. —— 1957: On the arilloides of *Nephelium*, *Euphoria*, *Litchi* and *Aesculus*, and the seeds of Sapindaceae in general. Acta Botan. Neerl. 6:618–41. Biol. Abstr. 33:2873. 1959.
18965. Van der Poel, J. 1938: Difficulties with seedbeds. II. Germination of Deli tobacco seed in liquids with various concentrations of salts. Mededeel. Deli Proefsta. 3(1). 15 pp. Chem. Abstr. 34:6004–5. 1940.
Van der Veen, J. H. 15653.
Vanderwaeren, J. 5099.
18966. Vanderwalle, R. 1935: Contribution à l'étude de la désinfection des céréales par l'eau chaude. I. L'action de la chaleur sur la germination des semences. Bull. Inst. Agron. Sta. Rech. Gembloux 4:3–19. Biol. Abstr. 11:5562. 1937.
Vanderwalle, R. 5358.
Van der Want, J. P. H. 2819.
Van der Weide, B. M. 5710.
18967. Vandevelde, A. J. J. 1897: Über den Einfluss der chemischen Reagentien und des Lichtes auf die Keimung der Samen. Botan. Centralbl. 69:337–42.
Van Doren, C. A. 3393.
18968. Van Dort, H. W. 1949: Gloxinia raised from seed. Bloemisterij 23(40):3–4.
18969. Van Duuren, A. J. 1950: The extraction of the poisonous substances from the seeds of *Croton tiglium*. Chron. Nat. 106:466–74. Biol. Abstr. 27:2293. 1953.
18970. Van Eaton, H. C. 1938: *Chionodoxas* from bulbs or seeds. Horticulture 16(10): 230.
18971. —— 1938: *Magnolia stellata* from seed. Horticulture 16(14):298.
Vanecko, S. 20002.
18972. Van Etten, C. H., R. W. Miller, F. R. Earle, I. A. Wolff, and Q. Jones. 1961: Hydroxyproline content of seed meals and distribution of the amino acid in kernel, seed coat, and pericarp. J. Agr. Food Chem. 9:433–35.
18973. ——, ——, I. A. Wolff, and Q. Jones. 1963: Amino acid composition of seeds

from 2-angiospermous plant species. J. Agr. Food Chem. 11:399–410. Chem. Abstr. 59:10471. 1963.

Van Etten, C. H. 2619, 5051, 5052.

18974. Van Fleet, D. S. 1942: The significance of oxidation in the endodermis. Am. J. Botany 29:747–55.

18975. Van Fleet, W. 1913: The cultivation of American ginseng. U.S. Dept. Agr. Farmers Bull. 551. 9 pp.

18976. —— 1914: Goldenseal under cultivation. U.S. Dept. Agr. Farmers Bull. 613:1–15.

18977. Vanha, J. 1898: Versuche über den Einfluss intermittierender Erwärmung auf die Keimung von Samen. Z. Landwirtsch. Versuchsw. Österr. 1:91–141.

Van Haarst, J. 3197.

18978. Van Haga, P. R. 1954: Alkaloids in germinating seeds of *Atropa belladonna*. Nature 173:692.

18979. Van Horn, C. W. 1936: Notes on the effect of soil management on the growth in diameter of pecan nuts. Proc. Am. Soc. Hort. Sci. 33:67–73. Expt. Sta. Rec. 78:488. 1938.

Van Horn, C. W. 6366–6369, 6573.

18980. Van Horn, D. L., and W. E. Domingo. 1950: Comparison of seed and vegetative propagation methods for red squill. Econ. Botany 4:350–53.

Van Horn, D. L. 15077.

Van Kooten, A. 571.

18981. Van Laer, M. H., and P. Froschel. 1942: Inhibitors and their role in the germination of barley. Bieres Boissons 3:136–40. Chem. Abstr. 38:2786. 1944.

18982. Van Lelieweld, G. 1924: Influence of light on germination of tree seed. Landbouwk. Tijdschr. 36:255–65.

18983. Van Loon, J., and A. Steger. 1930: Das fette Öl der Samen von *Couepia grandiflora*, Benth. Chem. Umschau Fette, Öle, Wachse Harze 37:337–40. Chem. Abstr. 25:1111. 1931.

Van Loon, J. 17898–17909.

Van Maren, A. 3548.

18984. Van Miegroet, M. 1952: Recherches sur la signification de quelques substances actives synthétiques pour la culture des plants forestiers. I. L'influence de quelques substances actives synthétiques sur la germination du *Pin sylvestre*. Schweiz. Z. Forstw. 103:505–22. Biol. Abstr. 27: 23381. 1953.

18985. —— 1953: Recherches sur la signification de quelques substances actives synthétiques pour la culture des plants forestiers. II. L'influence de quelques substances actives synthétiques sur la croissance des semis du *Pin sylvestre*. Schweiz. Z. Forstw. 104:137–54. Biol. Abstr. 28:14594. 1954.

18986. Van Overbeek, J. 1942: Hormonal control of embryo and seedlings. Cold Spring Harbor Symp. Quant. Biol. 10:126–34.

18987. ——, M. E. Conklin, and A. F. Blakeslee. 1941: Chemical stimulation of ovule development and its possible relation to parthenogenesis. Am. J. Botany 28:647–56.

18988. ——, ——, and —— 1942: Cultivation in vitro of small *Datura* embryos. Am. J. Botany 29:472–77.

18989. ——, R. Siu, and A. J. Haagen-Smit. 1944: Factors affecting the growth of *Datura* embryos in vitro. Am. J. Botany 31:219–24.

Van Pesch, F. J. 13063.

18990. Van Pinxteren, J. A. C., and G. Schallenbergheertjes. 1953: Determination of caffeine in cola seeds. Pharm. Weekblad 88:805–8. Biol. Abstr. 29:12333. 1955.

18991. Van Ramshorst, J. D. 1956: Bildung und Keimung der Samen von *Cryptocorynen*. Mitt. Blatt. Conf. European Vivarianer 2(4):52–59. Biol. Abstr. 31:39241. 1957.

18992. Van Schaik, P. H., and A. H. Probst. 1959: Effect of six growth regulators on pod set and seed development in midwest soybeans. Agron. J. 51:510–11.

18993. Vanselow. 1933: Versuche über den Einfluss verschiedener Samenkorngrösse und verschiedener Herkunft des Samens auf die Entwicklung der Fichte. Allgem. Forst- Jagdztg. 109:105–10. Biol. Abstr. 8:11884. 1934.

Van Severen, M. Levy-; *see* Levy-Van Severen, M.

18994. Van Staaveren, M. C. 1951: The cultivation of freesia from seed. Zaadwereld 15(3):85–86.

18995. Van Steenis, C. G. G. J. 1934: Dispersal of seed of *Lochnera rosea* (L.) Rchb. by ants. Trop. Natuur. 23(2):31–33.

18996. —— 1941: Record of a three-seeded coconut. Natuur. Tijdschr. Nederl.-Indie 101:144.

18997. Van Steveninck, R. F. M. 1957: The effect of growth substances on pod-setting in lupins. New Zealand J. Sci. Technol. Sect. A38:1011–14.

18998. Van Sumere, C. F., and L. Massart. 1958: Natural substances in relation to germination. Proc. Intern. Congr. Biochem., 4th Congr., Vienna 5:20–32. Chem. Abstr. 55:5668. 1961.

18999. Vanterpool, T. C., and E. J. Dosman. 1961: Germination of rape seed after burial in soil of subgermination moisture content. Can. Dept. Agr. Res. Br. Plant Res. Inst. Can. Plant Dis. Survey 41:161–64.

19000. Van Tieghem, M. P. 1891: Germination of *Buplerum aureum*. Bull. Soc. Botan. France 38:502–3.

19001. Van Tieghem, P. 1884: Germination de la graine et développement de l'embryon en plantule. Traite Botan., Librairie F. Savy, Paris. 1656 pp. (pp. 894–908).

19002. —— 1898: Structure du fruit, germination, et structure de la plantule de la Nuytsie. Bull. Soc. Botan. France 45: 213–20.

19003. —— and G. Bonnier. 1882: Recherches sur la vie latente des graines. Bull. Soc. Botan. France 29:25–29.

Van Veldhuizen, H. 14812.

19004. Van Wisselingh, C. 1919/21: Contribution to a knowledge of seed coats. Pharm. Weekblad 56:849–65, 1246–71 [1919]; 57:77–83, 125–39, 575–83, 605–14, 1193–211 [1920]; 58:298–308, 326–42, 788–94, 815–24 [1921].

19005. Van Wyk, H. P. D., S. A. Oosthuizen, E. E. Meyer, J. G. Brevis, and J. H. Grobler. 1957: The nutritive value of South African feeds. III. Hay and pasture crops, silage, cereals, tubers, and pods. Union S. Africa Dept. Agr. Sci. Bull. 354.

19006. Van Wyk, J. F. 1961: Seed legislation in South Africa. Farming S. Africa 37(4):25, 27, 29.

19007. Varada-Rajan, B. S., and J. S. Patel. 1946: Seed transmission of stem-rot of jute and its control. Indian J. Agr. Sci. 16: 193–205. Biol. Abstr. 22:25450. 1948.

19008. Varasova, N. N. 1962: Biochemical changes taking place after the stratification of seeds of several tree species. Leningr. Nauchn.-Issled. Inst. Lesn. Khoz. Sb. Rabot po Lesn. Khoz. 5:243–52.

Varasova, N. N. 20024–20028.

19009. Varenik, I. P. 1960: On the germination of seeds in *Veratrum lobelianum* Bernh. Botan. Zh. 45:1340–42.

19010. Varennikova, T. V., and V. G. Klimenko. 1960: Variability in the content of protein and nonprotein nitrogen in seeds and plants of several varieties of beans (*Phaseolus vulgaris*). Tr. po Khim. Prirodn. Soedin. Kishinevsk. Gos. Univ. 1960(3): 83–97. Chem. Abstr. 56:1775. 1962.

Vareton, E. 13383–13391.

Varga, D. 8155, 8171.

19011. Varga, M., and E. Köves. 1959: Phenolic acids as growth and germination inhibitors in dry fruits. Nature 183:401.

Varga, M. 11290.

Varghese, T. M. 10019.

19012. Varier, N. S. 1943: Fixed oil from the seeds of *Gnetum scandens* (Roxb.). Proc. Indian Acad. Sci. Sect. A17:195–98. Chem. Abstr. 37:6918. 1943.

19013. —— 1945: Fixed oil from the seeds of *Artocarpus hirsuta*, Lamk. Indian Chem. Soc. J. 22:275–77. Chem. Abstr. 40:4538. 1946.

Varier, N. S. 14098.

Varlamov, V. N. 16698.

Varma, R. 12839.

19014. Varma, S. R. 1951: Further trials on the production of the European type of biennial vegetable seeds from 1949 to 1951. Indian J. Hort. 8(3):9–11. Biol. Abstr. 26:7026. 1952.

Varma, V. P. 10017.

19015. Varner, J. E. 1964: Gibberellic acid controlled synthesis of α-amylase in barley endosperm. Plant Physiol. 39:413–15.

19016. ——, L. V. Balce, and R. C. Huang. 1963: Senescence of cotyledons of germinating peas. Influence of axis tissue. Plant Physiol. 38:89–92.

19017. —— and G. Schidlovsky. 1963: Intracellular distribution of proteins in pea cotyledons. Plant Physiol. 38:139–44.

Varner, J. E. 20002.

Varsanyi, J. 11570.

19018. Varshney, I. P., and M. S. Y. Khan. 1962: Saponins and sapogenins. XVII. The sapogenins of *Sesbania aculeata* Pers. J. Sci. Ind. Res. (India) 21B:401–2. Biol. Abstr. 41:20402. 1963.

19019. —— and K. M. Shamsuddin. 1962: Saponins and sapogenins. XVI. Echinocystic acid from the seeds of *Albizzia amara* Benth. J. Sci. Ind. Res. (India) 21B:347. Biol. Abstr. 40:16475. 1962.

Varshney, I. P. 6232–6234.

19020. Varvaro, U. 1912: The action of manganese dioxide and other metallic compounds on the germination of seed. Staz. Sper. Agrar. Ital. 45:917–29. Expt. Sta. Rec. 29:528. 1913.

19021. Vashkulat, P. N. 1951: On accelerating the germination of *Euonymus verrucosus* seeds. Agrobiologiya 1951(1):165–67.

19022. Vasil'ev, I. M., B. G. Zhukov, and T. S. Spasskaia. 1960: Intensification of effect of X-irradiation on seeds during storage. Biophysics (USSR) (English Transl.) 5:646–49.

Vasilkova, V. V. Vikhireva-; *see* Vikhireva-Vasilkova, V. V.

Vasishtha, A. K. 15112.

19023. Vass, I. 1941: The composition of paprika seeds. Kiserl. Kozl. 44:176–78. Chem. Abstr. 38:2760. 1944.

19024. Vassel, B., and L. L. Nesbitt. 1945: The nitrogenous constituents of flaxseed. II. The isolation of a purified protein fraction. J. Biol. Chem. 159:571–84.

19025. Vassilczenko, I. T. 1935: On the germination of *Panax ginseng* C. A. M. seeds. Botan. Zh. 20:244.

19026. —— 1936: Sur la germination de la vesce. Sov. Botan. 1936(5):65–73.

19027. —— 1937: Morphologie der Keimung der Leguminosen in Zusammenhang mit ihrer Systematik und Phylogenie. Flora Systematic Plantae Vasculares 4:424–25.

19028. —— 1937: A remarkable case of germination of the seeds of *Ephedra ciliata* S.A.M. Priroda Akad. Nauk SSSR 26:95–97.

19029. —— 1937: Sur les grains de quleques Chenopodiaceae et leur germination. Sov. Botan. 1937(1):116–21.

19030. —— 1938: De l'importance de la morphologie de la germination pour la systématique philogénétique des Phanerogames. Sov. Botan. 1938(3):19–40. Bull. Soc. Botan. France 87:380. 1940.

Vassilievich, A. 2567.

19031. Vasudeva, R. S., and K. M. Dutt. 1949: Performance of seed from rust-affected wheat plants. Indian Farming 10:16–18. Brit. Abstr. BIII:368. 1949.

19032. —— and M. S. Pavgi. 1945: Seed transmission of melon mosaic virus. Current Sci. (India) 14:271–72.

19033. Vasumpaur, R. 1937: Growing delphiniums from seed. Flower Grower 24(8):376–77.

Vatter, A. E. 8138.

19034. Vaughan, C. H., and J. C. Delouche. 1962: Seed properties reveal quality. Crops & Soils 14:18.

19035. Vaughan, G. T. 1961: Observations and recommendations, the president's report to the American Seed Trade Association. Seed World 89(1):1–2.

19036. Vaughan, J. G. 1956: Seed-coat structure of Ethiopian rape. Nature 178: 1188–89.

19037. —— 1958: Seed-coat structure of yellow Chinese rape. Nature 181:650.

19038. —— 1960: The preparation and staining of sections of cruciferous seed coats. Stain Technol. 35:229–31.

Vaughn, E. C. 6982.

19039. Vautier, E. E. 1963: Substances produced by Streptomyces, capable of modifying plant germination and growth. Ann. Inst. Pasteur 105:359–60. Chem. Abstr. 59:13117. 1963.

Vay, J. E. de; *see* De Vay, J. E.

19040. Vazart, B. 1960: Étude cytologique du grain d'aleurone dans le malt. Bull. Soc. Botan. France 107:185–92.

19041. Vazvasoff, H. Recherches sur le pouvoir germinatif des semences de la luzerne; possibilité de son accélération et de son égalisation. Stat. Agron. l'État, Obrastov, Bulg. 66 pp. Ann. Agron. 10:178. 1940.

Vecherskii, N. 11069.

Vedprakash. 1452.

19042. Veen, H. 1961: The effect of gibberellic acid on the embryo growth of *Capsella bursa-pastoris*. Ned. Akad. Wetenschap. (Amsterdam) Proc. Ser. C 64:79–85.

19043. —— 1963: The effect of various growth-regulators on embryos of *Capsella bursa-pastoris* growing in vitro. Acta Botan. Neerl. 12:129–71. Biol. Abstr. 45:12398. 1964.

Veen, J. H. van der. 15653.

Veer, E. J. van A. de; *see* Van A. de Veer, E. J.

Veerannah, L. 7463.

Veeraswamy, P. 16682.

19044. Veerhoff, O. 1940: Time and temperature relations of germinating flax. Am. J. Botany 27:225–31.

Vega, A. d'A. 6313.

19045. Vegiard, H. E. 1939: Classification and germination of herb vegetable seeds. Proc. Assoc. Offic. Seed Analysts 30:273–75.

19046. Vegis, A. 1961: Samenkeimung und vegetative Entwicklung der Knospen. Encyclopedia Plant Physiol. 16:168–298.

Vegult, W. 5121.

Veh, R. von; *see* Von Veh, R.

19047. Veillet-Bartoszewska, M. 1963: Recherches embryogeniques sur les *Ericales* comparaison avec les *Primulales*. Rev. Gen. Botan. 70:141–230.

19048. Velasco, E. 1958: Experiencias de vernalizacion.de trigo. Primeros resultados Anales Edafol. Fisiol. Vegetal (Madrid) 17:457–72.

Velasco, E. Esteban-; *see* Esteban-Velasco, E.
Velasco, J. 12850, 12851.
Veldhuizen, M. van. 14812.
Veldstra, H. 263.

19049. Velich, A., and V. Velich. 1942: The influence of the vapor of various essential oils on seed germination. Rozpravy Cesk. Akad. Ved Umeni Tr. II 52(28): 1–3. Biol. Abstr. 24:21596. 1950.

Velich, V. 19049.
Velikov, L. 9377.

19050. Velikov, L. I. 1961: Effect of dry stimulating substances on rice seeds. Izv. Bulgar. Akad. Nauk Inst. po Biol. "Metodii Popov" 11:211–23.

Vellasco, E. 439.

19051. Venema, H. J. 1953: Quelques rémarques critiques sur les échanges internationaux de semences. Ann. Biol. 57:291–93.

19052. Venezia, M. 1941: The oil content of grape seeds and their total utilization. Ann. Staz. Sper. Viticolt. Enol. Conegliano 10:45–47. Chem. Abstr. 40:3280. 1946.

19053. —— and F. Sallusto. 1934: Research on the lipase of olive. Ann. Portici Reg. Ist. Super. Agrar. 6:43–51. 1933.

Venezia, M. 18747.
Venezian, M. E. 18630.
Venezian, M. E. Scarascia-. 7276, 16853–16855.

19054. Venkataraman, K. 1929: Storage experiments with paddy seeds. Madras Dept. Agr. Yearbook 1929:29–35.

Venkataraman, K. 15895.

19055. Venkataramani, K. S. 1950: An instance of polembryony in tea. Planters Chron. 45:180–81.

19056. —— 1951: An interesting instance of "viviparous" germination in *Carica papaya* L. J. Madras Univ. 21B:218–19.

19057. Venkataratnam, L. 1948: Effect of heat and cold treatments on the germination of *Leucaena glauca*, Benth. A method of improving the germination of *Leucaena glauca*. Madras Agr. J. 35:179–84. Biol. Abstr. 24:5014. 1950.

19058. —— and K. Satyanarayanamurthy. 1953: A simple method of germinating sweet potato seeds. Current Sci. (India) 22:29.

Venkatesan, T. R. 4846, 4849.
Venkateswarlu, B. 1292.

19059. Venkateswarlu, J. 1938: Exceptional viviparous germination of seeds in *Cassia occidentalis* and *Tephrosia purpurea*. Current Sci. (India) 6:481.

19060. —— 1946: A case of polyembryony in *Daphne cannabina* Mall. Current Sci. (India) 15:169.

19061. —— 1947: A case of polyembryony in *Poivrea coccinea* DC. Current Sci. (India) 16:95.

19062. —— and P. N. Rao. 1963: Endosperm in Euphorbiaceae and occurrence of endosperm haustoria in two species of *Croton* Linn. Current Sci. (India) 32:514–16.

Venkateswarlu, J. 10014, 10015.

19063. Venkatram, C. S. 1950: Seed-borne fungi and loss of cotton-seed viability. J. Madras Univ. 19B:79–112. Biol. Abstr. 26:16170. 1952.

19064. Venkatrao, S., R. N. Nuggehalli, M. Swaminathan, S. V. Pingale, and V. Subrahmanyan. 1958: Effect of insect infestation on stored grain. III. Studies on Kaffir corn (*Sorghum vulgare*). J. Sci. Food Agr. 9:837–39.

Venkatswarlu, A. 15946.
Vennesland, B. 18425, 18426.

19065. Ventura, G. A. 1929: Studies on the germination of vegetable seeds. Philippine Agriculturist 17:451–63.

19066. Ventura, M. 1932: Ulteriori osservazioni sull'embriologia di *Daphniphyllum macropodum* Miq. Ann. Botan. (Turin) 19:550–53. Biol. Abstr. 8:6955. 1934.

19067. —— 1934: Sulla poliembrionia di *Mallotus japonicus* Muell. Arg. Ann. Botan. (Turin) 20:568–78.

19068. —— 1938: Osservazioni embryiologische su *Arechavaletaia uruguayensis* Speg. Ann. Botan. (Turin) 21:527–33.

19069. Ventura, M. M., and I. H. Lima. 1959: The non-protein nitrogenous constituents in oily seeds. I. Free amino acids in mature seeds of the favela (*Cnidoscolus phyllacanthus*) tree. Anais Acad. Brasil. Cienc. 31:191–95.

19070. —— and —— 1959: Protein content of the seeds of *Cnidoscolus phyllacanthus*. I. Their solubility and electrophoretic behavior. Anais Acad. Brasil. Cienc. 31:537–46. Chem. Abstr. 57:14180. 1962.

19071. —— and —— 1963: Sesame seed proteins. I. Sedimentation and diffusion characteristics of the major globulin. Anais Acad. Brasil. Cienc. 35:55–60. Chem. Abstr. 60:850. 1964.

19072. Venturi, V. M. 1948: Toxicity of the seeds and corms of *Colchicum autumnale* L. collected at different altitudes. Boll. Soc. Ital. Biol. Sper. 24:91–92.

19073. Verbeek-Wyndaele, R., and V. Coulier. 1961: β-Amylase in barley embryo and endosperm. European Brewery Conv. Proc. Congr. 1961:71–79. Chem. Abstr. 57:17074. 1962.

Verdesca, S. 15032.

19074. Verducci, P. 1945: Comparative effect of radioactivity of uranium compounds on germination and development of seeds. Boll. Soc. Ital. Biol. Sper. 20:333–34. Chem. Abstr. 40:6123–24. 1946.

19075. Verhey, C. 1955: Germination equipment of the seed testing station at Wageningen. Proc. Intern. Seed Testing Assoc. 20:5–28.

19076. —— 1957: Die Unbrauchbarkeit des Tetrazolium-Verfahrens zur Prüfung vom durch Trocknung verletzten Saatgut. Proc. Intern. Seed Testing Assoc. 22:321–29.

19077. —— 1960: Is it still possible, with regard to modern views, to handle the conception "germination energy"? Proc. Intern. Seed Testing Assoc. 25:391–97.

19078. —— 1960: The preparation of colour slides in seed testing. Proc. Intern. Seed Testing Assoc. 25:622–25.

19079. —— 1961: The influence of plumular damage on the yield of bush beans (*Phaseolus vulgaris* L.). Proc. Intern. Seed Testing Assoc. 26:162–69.

19080. Verhoeven, G. 1958: Tropical legume seed can be harvested commercially. Queensland Agr. J. 84:77–82. Biol. Abstr. 33:3010. 1959.

Verma, G. 17219, 17220.

19081. Verma, S. C., B. M. Lal, and V. Parkash. 1964: Changes in the chemical composition of the seed parts during ripening of Bengal gram (*Cicer arietinum* L.) seed. J. Sci. Food Agr. 15:25–31.

Verma, S. C. 11683.

Vermeulen, G. 9181, 14812.

19082. Vermeulen, P. 1943: The germination of orchid seeds. Preliminary communication. Ned. Kruidk. Arch. 53:125–26. Biol. Abstr. 21:4433. 1947.

19083. Vernon, R. O. 1947: Cypress domes. Sicence 105:97–99.

19084. Verona, O. 1960: Some investigations of the epiphytic microflora of seed. Ann. Sper. Agrar. 14(6 Suppl.):XI–XIX.

19085. —— and G. Bonaventura. 1937: Influence on plant growth of the partial elimination of the reserve accompanying the embryo and the probable presence of a "growth substance". Atti Accad. Lincei, Classe Sci. Fis. Mat. Nat. 25:53–55. Chem. Abstr. 31:7933. 1937.

19086. Verona, P. L. 1960: On the germination of lettuce seeds in the darkness, under the stimulus of other germinating seeds. Ric. Sci. 30:1238–40.

19087. —— 1960: Action of amino acids, vitamins and other compounds on the germination of lettuce seeds. Ric. Sci. 30:1391–95. Biol. Abstr. 36:42329. 1961.

19088. —— 1961: Research on the germination of lettuce "seeds." Ann. Fac. Agrar. Univ. Pisa 22:123–54.

19089. Verret, J. A. 1928: Sugar cane seedlings. Assoc. Hawaii Sugar Technol. Rept. 7:15–23. Expt. Sta. Rec. 60:334. 1929.

19090. Verschaffelt, E. 1910: Het mechanisme der wateropname door de zaden der Cucurbitaceën. Kon. Akad. Weterschap. Amsterdam Verslag. 13:542–50. Botan. Centralbl. 116:287–88. 1911.

19091. —— 1912: Le traitement chimique des graines à imbibition tardive. Rec. Trav. Botan. Neerl. 9:401–35.

19092. Vervelde, G. J. 1950: Loss of dry matter in the seed of grains during the ripening and after-ripening period. Landbouwk. Tijdschr. 62:628–36.

Veselova, E. P. 17603.

19093. Vesque, M. J. 1891: Polyembryony in Clusiaceae. J. Botan. (Morot) 5:328–30.

Vestal, C. M. 10997.

19094. Vetter, J. L., A. I. Nelson, and M. P. Steinberg. 1959: Heat inactivation of peroxidase in HTST processed whole kernel corn. Food Technol. 13:410–13.

19095. Vetukhova, A. 1939: Chemical treatment of winter wheat seed as a measure to increase frost resistance. Compt. Rend. Acad. Sci. URSS 24:605–8.

19096. Veyret, Y. 1961: Embryogénie des Orchidacées. Développement de l'embryon chez le *Masdevallia veitchiana* Reichb. Compt. Rend. 253:2580–82.

19097. Viado, J. 1938: The cutting test as a practical method of testing viability of seeds. Philippine J. Forestry 1:219–25. Biol. Abstr. 13:1311. 1939.

19098. Vianna e Silva, M. 1947: A influencia da idade da semente na germinação do arroz. Min. da Econ., Comissao Regulad. do Com. de Arroz. 18 pp.

19099. Viazov, A. A. 1955: Germination of basil seeds. Byul. Gl. Botan. Sada 22:93–96.

19100. Vibar, T., and P. A. Rodrigo. 1929: Storing farm crop seeds. Philippine Agr. Rev. 22:135–46.

Vicente, E. 15333.

19101. Vicente, M., M. Engelhardt, and K. Silberschmidt. 1961: Sobre a quantidade d'aqua optimal para a embebição e germinação de sementes. I. Estudes sobre Milho, Feijao e Arroz. Anais Acad. Brasil. Cienc. 33:409–28.

19102. ——, ——, and K. Silberschmidt. 1962: The influence of temperature on the germination response to light of seeds of *Rumex obtusifolius* L. Phyton 19:163–67, XII.

Vicente, M. 5972, 5973.
Vicenzetti, P. G. 18615.
Vich, A. 5889.
Vick, M. M. 4227.

19103. Vickery, H. B. 1932: Chemical investigations of the tobacco plant. III. Tobacco seed. Some nitrogenous components of the hot water extract of fat-free tobacco seed meal. Conn. Agr. Expt. Sta. Bull. 339:637–45. Expt. Sta. Rec. 68:148. 1933.

19104. —— 1944: Introduction and discussion of the amino acid composition of plant seeds. Federation Proc. 3:110–15.

19105. —— 1945: The proteins of plants. Physiol. Rev. 25:347–76.

19106. ——, E. L. Smith, R. B. Hubbell, and L. S. Nolan. 1941: Cucurbit seed globulins. I. Amino acid composition and preliminary tests of nutritive value. J. Biol. Chem. 140:613–24.

19107. ——, A. J. Wakeman, and C. S. Leavenworth. 1932: Globulin of tobacco seed. Conn. Agr. Expt. Sta. Bull. 339:625–36. Chem. Abstr. 27:371. 1933.

Vickery, H. B. 9167, 13320.
Vidale, A. 4130.
Vidaurreta, R. O. 14755.
Vieira, C. 12670.

19108. Vieitez, E. 1953: Respuesta del maiz al tratamiento de sus semillas con los acidos β-indolacetico y α-naftalenacetico. Anales Edafol. Fisiol. Vegetal (Madrid) 12:1033–43. Biol. Abstr. 29:6852. 1955.

19109. Viets, F. G., Jr., and C. E. Domingo. 1949: Yields and nitrogen content of corn hybrids as affected by nitrogen supply. Soil Sci. Soc. Am. Proc. 13:303–6.

Vieu-Bonnafous, J. 13334.
Vignes, J. L. 4333.

19110. Vignoli, L. 1935: Utilisation des reserves pendant la germination de *Cicer arietinum* L. a la lumiere et a l'obscurite. Lab. Ist. Botan. Palermo 6. Bull. Soc. Botan. France 83:853. 1936.

19111. —— and B. Cristau. 1951: Influence du vieillissement du *Strophanthus hispidus* sur l'activite de la teinture preparee avec cette semence. Med. Trop. 11:455–57. Biol. Abstr. 26:9827. 1952.

19112. Vigorov, L. I. 1960: A comparative study of the seed aleurone tissue in cultivated and wild cereals. Byul. Mosk. Obshchestva Ispytatelei Prirody Otd. Biol. 65: 101–8.

19113. Vijayalakshmi, U., and J. T. Rao. 1960: Effects of γ-rays on germination and growth in some species and hybrids of *Saccharum*. Current Sci. (India) 29:397–98.

19114. Vijzelman, H. E. Control of the carrot fly by means of a seed treatment. Mededel. Wageningen Planten. Onderzoek. 163:1–23.

19115. Vikhireva-Vasilkova, V. V. 1958: On the germination of seeds in some arctic plants. Botan. Zh. 43:1024–29.

Viktorov, D. P. 15989.
Vilamil, G. F. 14251.
Vilbrandt, F. C. 19866.
Viljoen, N. J. 2838.
Villafane, A. G. 9027.
Villafuerte, I., Jr. 17728.
Villarreal, C. 5449.

19116. Villegas, R. S. 1936: Quantitative relations of various organic substances in the fruit of *Sandoricum koetjabe*. Nat. Appl. Sci. Bull. (Univ. Philippines) 5:449–50. Chem. Abstr. 32:972. 1938.

19117. Villerts, A. 1937: Versuche über die Wirkung der X-strahlen auf die Samen. Bull. Soc. Biol. Lett. 7:131–46. Gartenbauwissenschaft 12:S62. 1938.

19118. Villiers, T. A. 1961: Dormancy in tree seeds. A brief review of recent work. Proc. Intern. Seed Testing Assoc. 26:516–36.

19119. —— and P. F. Wareing. 1964: Dormancy in fruits of *Fraxinus excelsior* L. J. Exptl. Botany 15:359–67.

Villiers, F. J. de; *see* De Villiers, F. J.
Vilmorin, A. de; *see* De Vilmorin, A.
Vilmorin, R. de; *see* De Vilmorin, R.

19120. Vilyatser, M. G. 1961: Effect of the agitator on the output of winged seeds. Lesn. Zh. 1961(5):21–24.

19121. Vincent, G. 1929: Das Altern der Koniferensamen. Sb. Česk. Akad. Zeměděl. 4A:453–92. Ber. Wiss. Biol. 14:301. 1930.

19122. —— 1930: Analysen der ungleichfarbigen Kiefernsamen. Vestn. Česk. Akad. Zeměděl. 6:19–22. Ber. Wiss. Biol. 14:301. 1930.

19123. —— 1930: Einfluss der Aufbewahrungszeit auf die Qualität der Koniferensamen. Verhandl. Intern. Kongr. Forst. Versuchs. (Stockholm) 1929:436–40. Biol. Abstr. 5:24997. 1931.

19124. —— 1930: Studies with conifer cones and seeds. I. Sb. Vyzkum. Úst. Zeměděl. Česk. Rep. 50. 114 pp. Expt. Sta. Rec. 63:539. 1930.

19125. —— 1931: Age of conifers and the quality of their cones. Vestn. Česk. Akad. Zeměděl. 7:44–48. Biol. Abstr. 6:15590. 1932.

19126. —— 1931: Age of conifers and the quality of their seed. Vestn. Česk. Akad. Zeměděl. 7:48–53. Biol. Abstr. 6:15591. 1932.

19127. —— 1937: Storage of spruce seed. Sb. Česk. Akad. Zeměděl. 12:469–74. Biol. Abstr. 12:5408. 1938.

19128. —— 1937: Storage of larch seed. Sb. Česk. Akad. Zeměděl. 12:557–60. Biol. Abstr. 12:5409. 1938.

19129. —— 1938: The storage of spruce and larch seed. Z. Forst- Jagdw. 70:45–51. Proc. Intern. Seed Testing Assoc. 11:158–59. 1939.

19130. —— 1938: Why does our larch seed show a poor germination? Lesnická Práce 17:188–98. Proc. Intern. Seed Testing Assoc. 11:160–61. 1939.

19131. —— 1939: Seed production of spruces with red or green cones. Sb. Česk. Akad. Zeměděl. 14:100–3. Proc. Intern. Seed Testing Assoc. 11:161–62. 1939.

19132. —— 1939: Die Wandelbarkeit der Nadelholzsamen und der aus ihnen gezogenen Pflanzen. Forstwiss. Centralbl. 61:250–55. Biol. Abstr. 13:13965. 1939.

19133. —— 1940: Forstsaatgut in der Waldbautechnik. Herausgeber: Česk. Akad. Zeměděl. Ed.: Vedecke Spisy 7. 149 pp. Praha. Proc. Intern. Seed Testing Assoc. 14:64–65. 1948.

19134. —— 1941: Examinations of pine and spruce seed of different provenances. Contribution to the distinction between pine and spruce varieties. Sb. Česk. Akad. Zeměděl. 16:33–38. Proc. Intern. Seed Testing Assoc. 14:65–66. 1948.

19135. —— 1957: La germination et la viabilité des graines de l'épicéa, du pin et du mélèze. Práce Výzkum. Ústavu Lesn. CSR 1957(12):131–90.

19136. —— 1958: Die Lagerung des Forstsaatguts in geschlossenen Gefässen. Centralbl. Ges. Forstw. 75:257–67. Biol. Abstr. 33:38830. 1959.

19137. —— and A. Freudl. 1931: Effects of early harvesting of coniferous cones on seed quality. Lesnická Práce 10:248–56. Biol. Abstr. 8:16333. 1934.

Vincent, L. E. 12262–12264.

Vincente, G. C. 9969.

19138. Vines, A. 1947: Studies on the germination of barley grains. Australian J. Exptl. Biol. Med. Sci. 25:119–26.

Vinet, A. Guerillat-; *see* Guerillat-Vinet, A.

Vineyard, M. L. 5072.

19139. Vinogradov, V. A. 1933: On the chemical composition of the seeds of the pine and the acorns of the oak of different geographical origin. Bull. Appl. Botany, Genet., Plant Breed. (USSR) Ser. III(1): 183–94. Chem. Abstr. 27:5370. 1933.

Vinogradova, N. 15428.

19140. Vinokurov, S. I. 1946: The reducing substances of germinating beans and the reduction of dehydroascorbic acid. Ukr. Biokhim. Zh. 18:181–89. Chem. Abstr. 48:4641. 1954.

19141. Vinokurova, N. K. 1958: Effect of cold germination of seeds of makhorka (*Nicotiana rustica*) on yield capacity. Sb. Nauchn.-Issled. Rabot Vses. Nauchn.-Issled. Inst. Tabaka i Makhorki 150:109–14. Biol. Abstr. 45:3312. 1964.

Vinsova, M. 2960.

19142. Vintilescu, I., C. N. Ionescu, N. Stanciu, and A. Anghel. 1940: La composition du riz cultivé en Roumanie. Acad. Roumaine Sect. Sci. Bull. 22:441–45. Biol. Abstr. 15:19472. 1941.

19143. Violanda, A. T. 1954: The natural occurrence of two crystalline forms of vanillin. Philippine Agriculturist 37:507–11.

19144. Virgin, W. J. 1940: Low germination of peas associated with the presence of bacteria in the seed. Phytopathology 30: 790–91.

19145. Virtanen, A. I. 1953: The use of seeds with low content of trace elements in studies on essentiality of micronutrients. Plant Physiol. 28:323–24.

19146. —— and M. Saarivirta. 1962: The formation of benzyl thiocyanate in the seeds of *Lepidium sativum*. Suomen Kemistilehti 35B:102–4.

19147. —— and S. S. Von Hausen. 1949: The effect on plant growth of substances which lower the oxidation-reduction potential. Z. Pflanzenernähr. Düng. Bodenk. 45:11–22. Chem. Abstr. 44:6921. 1950.

19148. —— and —— 1949: Role of substances formed during germination in the growth of plants. Nature 163:482–83.

Virtanen, A. I. 12985, 16591.

19149. Visser, T. 1954: After-ripening and germination of apple seeds in relation to the seed-coats. Koninkl. Ned. Akad. Wetenschap. Proc. Ser. C 57:175–85.

19150. —— 1956: The role of seed coats and temperature in after-ripening, germination and respiration of apple seeds. Koninkl. Ned. Akad. Wetenschap. Proc. Ser. C 59:211–22.

19151. —— and L. M. Tillerkeratne. 1958: Observations on the germination and storage of tea pollen and seed. Tea Quart. 29:30–35. Biol. Abstr. 33:35064. 1959.

Viswanatha, T. 1763, 15955.

19152. Viswanatham, K. B. 1953: Pure seed and the need for its renewal at frequent intervals. Madras Agr. J. 40:306–8.

19153. Visyagin, N. I., and A. I. Lisina. 1950: A rapid method determination of moisture in grain. Selek. i Semen. 17(10): 53–55. Chem. Abstr. 48:8438. 1954.

19154. Vita, N. 1932: Utilization of atmospheric nitrogen by germinating seeds. II. Observations on germinating leguminous seeds (pulses) in the presence of alkaloids. Biochem. Z. 252:278–91. Chem. Abstr. 26:5608. 1932.

19155. —— 1933: The influence of some stimulating materials on the nitrogen content of legumes during germination. IV. Giorn. Biol. Appl. Ind. Chim. Aliment. 3:41–51. Chem. Abstr. 28:5851. 1934.

19156. —— and R. Sandrinelli. 1932: Utilization of atmospheric nitrogen by germinating leguminous seeds. III. Biochem. Z. 255:82–87. Chem. Abstr. 27: 751. 1933.

19157. —— and —— 1935: The influence of certain substances on the nitrogen content of seeds of legumes during germination. VII. Giorn. Biol. Ind. Agr. Aliment. 5:41–51. Expt. Sta. Rec. 74:762–63. 1936.

Vitek, E. 18891.

19158. Vitenson, D. 1936: Serological methods in the phytopathological estimation of seeds. Plant Protect. 1936(9):38–48.

19159. Vitte, G., and M. Coustou. 1945: Toxicity of leguminous grains of genera *Vicia* and *Lathyrus*. Bull. Trav. Soc. Pharm. Bordeaux 83:15–22. Pharm. Abstr. 12:82. 1946.

Vix, H. L. E. 4893.

Vladescu, I. 1342.

19160. Vladyka, J., and K. Tucek. 1939: Qualitätsmerkmale der Roggenkörner. Herausgeber: Ústav pro Zkouseni Semen Česke Zemědĕl. Rady. Praha. 23 pp. Proc. Intern. Seed Testing Assoc. 14:63. 1948.

19161. Vlamis, J., and A. R. Davis. 1943: Germination, growth, and respiration of rice and barley seedlings at low oxygen pressures. Plant Physiol. 18:685–92.

19162. —— and —— 1944: Effects of oxygen tension on certain physiological responses of rice, barley, and tomato. Plant Physiol. 19:33–51.

Vlasova, L. M. Horowitz-; *see* Horowitz-Vlasova, L. M.

19163. Vlasova, N. A., and L. E. Serebryakova. 1963: The localization of nucleic acids in connection with rates of development of the embryo sac in the cotton plant. Uzbeksk. Biol. Zh. 7:27–33. Chem. Abstr. 59:6734. 1963.

19164. Vlasyuk, P. A. 1932: Chemical "stimulation" of sugar-beet seeds for increase of the sugar yield. Nauk. Zap. Tzukr. Prom. 25:1–26. Chem. Abstr. 27:2597. 1933.

19165. ——, G. V. Poruskii, S. V. Cherednichenko, and S. N. Dolgiy. 1957: The influence of extra-root fertilization on the increase of germination of the seed material. Dokl. Akad. Nauk SSSR 113: 214–16. Biol. Abstr. 35:30035. 1960.

19166. Vlitos, A. J., and H. G. Cutler. 1960: The natural auxins of the sugar cane. A paper-chromatographic separation of the growth factors present in true seed. Proc. Brit. West Indies Sugar Technologists 1960:113–27. Chem. Abstr. 59:11891. 1963.

Vobshchina, S. 140.

19167. Voderberg, K. 1961: Quellung und Keimung der Samen von *Vicia villosa* Roth. Flora (Jena) 150:170–74.

Vömel, A. 2706.

Vörös, J. 10686.

Vogel, E. 18263.

Vogel, H. 10158.

Vogel, H. A. 17046, 17051.

19168. Vogel, O. A. 1941: Relation of glume strength and other characters to shattering in wheat. J. Am. Soc. Agron. 33:583–89.

Vogel, O. A. 364, 6562.

19169. Vogelsang, P. 1950: New techniques in seed pelleting. Proc. Am. Soc. Sugar Beet Technol. 6:76–78.

19170. ——, A. Schupp, and P. A. Reeve. 1947: Methods used and results secured from pelleting of sugar beet, vegetable, flower, tree, and other field crop seeds. Proc. Am. Soc. Sugar Beet Technol. 4:603–9.

19171. Vogt, E. 1955: Ein Beitrag zur Keimfähigkeit der Samen der Hühnerhirse und des Amfer-Knöterichs, besonders ihrer im Boden ruhenden Samen. Angew. Botan. 29:26–32.

19172. Vogt, I. 1951: Über den Hemmstoff- und Wuchsstoffgehalt in Getreidewurzeln. Planta 40:145–69.

Vogt, O. 6914.

19173. Voisenat, P. 1957: Remarques au sujet de la germination paresseuse des semences de céréales. Bull. Soc. Franc. Physiol. Vegetale 3:130.

19174. Voisey, P. W., and M. Kloeck. 1963: Note on a seed dispenser for cereal plots. Can. J. Plant Sci. 43:607–10.

19175. —— and A. G. O. Whiteside. 1961: Note on a chemical treater for small seed samples. Can. J. Plant Sci. 41:868–70.

Voisey, P. W. 18196.

19176. Voitchishin, N. V. 1932: Chemical composition of the corn kernel. Trans. Central Sci. Res. Inst. Sugar Ind. (USSR) 9:26–33. Chem. Abstr. 28:6173–74. 1934.

Vojnovich, C. 559, 560.

19177. Volk, G. W., J. H. Wilson, and J. L. Haynes. 1951: For treating seeds, farmers ask: What about liquid fertilizers? Ohio Agr. Expt. Sta. Bimo. Bull. 36(268):8–9.

Volk, G. W. 18432.

19178. Volkov, A. 1933: The influence of germination in sugar solutions and in glycerine on the further development of some cultivated plants. Bull. Appl. Botany, Genet., Plant Breed. (USSR) Ser. III(3):131–40.

19179. Volkov, F. I., and E. F. Ivanitskaia. 1955: Biological features of acorns in the process of ripening. Tr. Inst. Lesa Akad. Nauk SSSR 17:127–36.

19180. Volkov, I. A. 1933: The influence of germination in sugar solutions and in glycerol on the further development of some cultivated plants. Bull. Appl. Botany, Genet., Plant Breed. (USSR) Ser. III(2):131–39. Chem. Abstr. 28:3440–41. 1934.

Volkova, T. I. 11180.

19181. Volodchenko, V. S. 1963: Cases of germination of seeds inside the fruit on maternal plants in wild species of Leguminosae. Acad. Sci. RSS Ukraine J. Botan. 20:108–9.

19182. Volodin, V. I. 1960: The influence of gibberellin on the germination of seeds of some agricultural plants. Botan. Zh. 45:1787–91. Biol. Abstr. 36:86007. 1961.

19183. Volodina, L. A. 1961: On the germinating capacity of seeds of knotweed. Sb. Nauchn. Rabot Akad. Nauk Belorussk. SSR Tsent. Botan. Sad. 2:211–13.

19184. Voloshin, M. P. 1957: Preserving the germinating power of *Laurus nobilis* for spring sowing. Agrobiologiya 1957(2): 129–31. Biol. Abstr. 35:5727. 1960.

19185. Von Abrams, G. J., and M. E. Hand. 1956: Seed dormancy in *Rosa* as a function of climate. Am. J. Botany 43:7–12.

19186. Von Bittera, N., and F. Von Gruber. 1936: Zur Frage der Keimfähigkeit bei der Grassamenkultur. Pflanzenbau 13:235–40.

Von Bredow, K. 3292.

19187. Von Degen, A. 1925: The longevity of seeds. 4th Intern. Seed Testing Congr., Cambridge, July 7–12, 1924, pp. 139–43.

19188. —— and A. Puttemans. 1931: Sur l'influence du transport maritime sur la germination des semences. Proc. Intern. Seed Testing Assoc. 3:287–91.

19189. Von Denffer, D. 1939: Über das Zusammenwirken von Keimstimmung und täglicher Belichtungsdauer auf die Entwicklung von *Sinapis* und *Hordeum*. Jahrb. Wiss. Botan. 88:759–815.

19190. Von der Haar, P., and C. H. Pominski. 1952: Pigments of cottonseed. V. Methylation of gossypurpurin. J. Org. Chem. 17:177–80.

Von der Haar, P. 15526.

19191. Von der Straten, I. 1958: Einfluss von Ultraschall und Ultraschall in Verbindung mit Wirkstoffen auf Keimung und weitere Entwicklung einiger Kulturpflanzen. Beitr. Biol. Pflanz. 34:315–58.

Von der Straten, I. 17940.

Von Elbe, J. 19449.

19192. Von Eulenberg, B. E. G. 1932: Keimversuche mit Getreidearten und ihre Beeinflussung durch die Art des Keimbettes. Dissertation Albertus Univ., Koenigsberg. 74 pp.

19193. Von Euler, H. 1948: The biological and chemical effects of streptomycin. Arkiv Kemi, Mineral. Geol. 26A(6). 8 pp. Chem. Abstr. 44:9014. 1950.

19194. —— 1950: The effect of streptomycin on seeds of normal and chlorophyll deficient barley. Z. Naturforsch. 5b:448. Chem. Abstr. 45:5245. 1951.

19195. —— 1953: Einfluss von Streptomycin und Dihydro-Streptomycin auf keimende

Samen grüner Pflanzen. Z. Physiol. Chem. 295:411–13.

19196. —— 1956: Influence of specific chemical substances on the development of germinating seeds and on tumors. Arkiv Kemi 9:1–6. Biol. Abstr. 30:32679. 1956.

19197. —— and G. Günther. 1933: Enzymwirkung und Enzymbildung in lebenden Zellen. Z. Physiol. Chem. 220:69–85.

19198. —— and M. Malmberg. 1936: Vitamins and catalysts in wheat embryo. Arkiv Kemi, Mineral. Geol. 12B(14). 6 pp. Chem. Abstr. 30:3860. 1936.

19199. —— and A. M. Perje. 1945: Action of vitamins and antivitamins on seed germination and mitosis. Arkiv Kemi, Mineral. Geol. 20A(2). 17 pp. Chem. Abstr. 41:2125. 1947.

19200. ——, T. Philipson, and D. Burstroem. 1932: Biocatalysts in germination. Z. Physiol. Chem. 208:281–84. Chem. Abstr. 26:4357. 1932.

19201. ——, L. Resegotti, and H. Hasselquist. 1956: Natrium-selenit, ein Hemmstoff der Samenkeimung, und seine Einwirkung auf Yoshida-Ascites-Sarkom tragende Ratten. Z. Naturforsch. 11b: 178–79.

19202. Von Euw, J., H. Hegedüs, C. Tamm, and T. Reichstein. 1954: Die Glykoside der Samen von *Strophanthus amboënsis* (Schinz) Engl. et Pax sowie einer verwandten Form. Glykoside und Aglykone. Helv. Chim. Acta 37:1493–1540.

19203. ——, G. A. O. Heitz, H. Hess, P. Speiser, and T. Reichstein. 1952: Die Glykoside von *Strophanthus welwitschii* (Baill.) K. Schum. Glykoside und Aglycone, 89. Helv. Chim. Acta 35:152–60. Biol. Abstr. 26:22630. 1952.

19204. ——, K. Mohr, O. Schindler, and T. Reichstein. 1956: Die Glykoside der Samen von *Strophanthus sarmentosus* var. Major Dewevre. Glykoside und Aglykone, 158. Helv. Chim. Acta 39:326–30.

19205. —— and T. Reichstein. 1950: Die Glykoside der Samen von *Strophanthus courmontii* Sacl. Helv. Chim. Acta 33: 1006–13.

19206. ——, O. Schindler, and T. Reichstein. 1955: Die Glykoside der Samen von *Strophanthus tholloni* Franch. und *S. tholloni* var. *gardeniiflorus* (Gilg.) Monach. Glykoside und Aglykone, 146. Helv. Chim. Acta 38:987–1001.

Von Euw, J. 16994, 20089.

19207. Von Fink, H., and E. Schwieger. 1950: Über die Schnellmethoden zur Bestimmung der Keimfähigkeit, besonders nach dem Tetrazoliumverfahren. Brauwissenschaft 2:39–42.

19208. Von Freyberg, W. 1931: Beiträge zur Kenntnis der Physiologie der Hüllschichten des Maiskornes. Botan. Arch. 32:392–492.

19209. Von Frimmel, F. 1914: Über einige antike Samen aus dem Orient. Sitzber. Akad. Wiss. Wien Phil.-Hist. Kl. Abhandl. I. 173. 15 pp.

Von Gruber, F. 19186.

19210. Von Guttenberg, H. 1960: Embryologische Studien an Monokotyledonen. III. Die Embryogenese von *Triglochin maritimum* L., *Arum maculatum* L. und *Typha latifolia* L. Flora (Jena) 149:243–81.

19211. —— and E. Lehle-Joerges. 1947: Über das Vorkommen von Auxin und Heteroauxin in ruhenden und keimenden Samen. Planta 35:281–96.

19212. —— and H. L. Wiedow. 1952: Über den Wirkstoffbedarf isolierter Hafer-embryonen. Planta 41:145–66.

Von Guttenberg, H. 6286.

19213. Von Hahn, F. V., and G. Bosian. 1930: Die Änderung der Oberflächenspannung beim Keimungsverlauf. Planta 11: 627–30.

Von Hausen, S. S. 19147, 19148.

19214. Von Höhnel, F. 1875: Seed-coats. Ease of swelling and influence of chemicals and physical properties of palisade layer. Wiss. Prakt. Untersuch. Gebiete Pflanzenbaues (Haberlandt) 1:80–98.

19215. Von Hofsten, C. G. 1947: Investigations of germination biology in some weed species. Kgl. Lantbrukshögskol. Inst. Växtodlingslära. Växtodling 2:91–107.

Von Holdt, M. M. 12236.

Von Korff, R. W. 4212.

19216. Von Lentz, J. 1932: Stimulation hartschaliger Samen mit Schwefelsäure. Deut. Landwirtsch. Presse 59:462.

19217. Von Liebenberg, A. R. 1884: Über den Einfluss intermittierender Erwärmung auf die Keimung von Samen. Botan. Centralbl. 18:21–26.

19218. Von Loesecke, H., and A. J. Nolte. 1937: Characteristics and composition of Papaya seed oil. J. Am. Chem. Soc. 59:2565–67.

Von Loesecke, H. W. 14650, 14561.

19219. Von Mikusch, J. D., and U. Dylla. 1954: The eicosenoic acid of cameline seed oil. J. Am. Oil Chemists Soc. 31:114–15.

19220. Von Müller, F. 1880: Vitality of *Eucalyptus* seeds. Gardeners Chron. 13: 811.

19221. Von Ohlen, F. W. 1931: A microchemical study of soybeans during germination. Am. J. Botany 18:30–49.

Von Ries, J. 12500.

19222. Von Roeder, W. 1926: Stimulation von Kakteensticklingen und Samen. Z. Sukkulentenkunde 2:251–54.

19223. —— 1932: Samenzucht und Samelingsanzucht bei Sukkulenten. Gartenwelt 36:154.

Von Schmeling, B. 4691.

19224. Von Schoenebeck, O. 1936: Chemical and physiological investigations of mustard plants, mustard seeds and mustard oils. Biedermanns Zentralbl. B. Tierernähr. 8:373–96. Chem. Abstr. 31:174–75. 1937.

19225. Von Schreven, A. C. 1958: Germination tests and culture experiments with *S. townsendii.* Acta Botan. Neerl. 7:205–16. Biol. Abstr. 33:46934. 1959.

19226. Von Sengbusch, R. 1933: Die im Boden liegenden hartschaligen noch keimfähigen Lupinen und ihre praktische Bedeutung für die Reinhaltung von Lupinen-Zucht-material. Züchter 5(2):26–28.

Von Sengbusch, R. 10762.

19227. Von Sostaric-Pisacic, K. 1934: Einfluss der Wärme auf den Keimungsprozess bei Sudangras. Pflanzenbau 10:331–50.

19228. Von Spiesz, K. 1904: Über die Farbstoffe des Aleuron. Österr. Botan. Z. 54:440–46.

Von Sydow, E. 6016.

19229. Von Tschermak-Seysenegg, E. 1934: Der Kürbis mit schalenlosen Samen, eine beachtenswerte Ölrucht. Wien. Landwirtsch. Ztg. 84:41–42, 48–49. Proc. Intern. Seed Testing Assoc. 7:103. 1935.

19230. —— 1940: Die Bildung von Traganth, eine Paralellvariation in den Samenschalen der Erbse und der wilden Kicher. Züchter 12:161–64. Biol. Abstr. 24:28852. 1950.

19231. Von Veh, R. 1936: Production of cherry seedlings from freshly harvested seed. Züchter 8:305–12. Expt. Sta. Rec. 77:48. 1937.

19232. —— 1938: Wuchsstoff und Keimstimmung. Naturforsch. 14:375–80.

19233. —— 1939: Über Entwicklungsbereitschaft und Wüchsigkeit der Embryonen von Apfel, Pfirsch, und Alles. Züchter 11: 249–55. Gartenbauwissenschaft 14:20S. 1940.

19234. —— and H. Söding. 1937: Growth substance and the germination of fruit tree seeds. Ber. Deut. Botan. Ges. 55:270–78. Chem. Abstr. 31:5844. 1937.

19235. Von Weihe, K. 1953: Die Wirkung von Na-Benzyl-Penicillin auf die Samenkeimung und das Wurzelwachstum. Naturwissenschaften 40:203–4.

19236. —— 1954: Die Wirkung von Na-Benzyl-Penicillin auf die Samenkeimung und das Wurzelwachstum. Naturwissenschaften 41:233–34.

19237. Von Weinzierl, T. R. 1902: Alpine Futterbauversuche. III. 7. Versuche über den Einfluss der chemischen Lichtintensität auf die Formbildung gewisser Pflanzenorgane. Z. Landwirtsch. Versuchsw. Österr. 5:298–311.

19238. Von Wettstein, W. 1942: Saatgut und Züchtung. Forstwiss. Centralbl. 64:135–43.

Von Witsch, H. 8160.

Voorhees, C. L. 14871, 14872.

Voronkova, E. A. 7088.

19239. Vorontsov, V. 1934: The physiology of the ripening of *Aleurites cordata* seed. Sov. Subtrop. 1934:185–89. Biol. Abstr. 10:21940. 1936.

19240. Voropaeva, N. A., and M. S. Dudkin. 1960: Investigation of the anatomic structure of the floral layer of oat seeds. Izv. Vysshikh Uchebn. Zavedenii Pishchevaya Tekhnol. 1960(1):17–22. Chem. Abstr. 54:15547. 1960.

Vos, M. P. de; *see* De Vos, M. P.

19241. Vose, P. B. 1962: Delayed germination in Reed Canary-grass *Phalaris arundinaceae* L. Ann. Botany (London) 26:197–206.

19242. Voskerusa, J. 1961: A contribution to the study of morphine content in ripe dry poppy seed pods (*Papaver somniferum* L.). Sotsial. Sel'skokhoz. Nauk. 1961(2): 143–150.

Voskoboynikova, V. 11006.

19243. Voss, H. 1937: Wuchsstoff-aktivierung und -inaktivierung und ihre Keimungsregulation Bedeutung Planta. 27: 432–35.

19244. Voss, I. 1934: Keimungsphysiologische Untersuchungen an Weizensorten. Angew. Botan. 16:137–86. Proc. Intern. Seed Testing Assoc. 7:80–81. 1935.

19245. Voss, J. 1930: Die Untersuchung der Keimpflanzen als Hilfsmittel der Sortenfeststellung beim Weizen. Mitt. Biol. Reichsanst. Land- Forstw. 39:39–62. Biol. Abstr. 6:5209. 1932.

19246. —— 1939: Untersuchungen über Keimruhe und Auswuchsneigung von Getreidesorten (*Triticum sativum* und *Hordeum sativum* Jess.). Landwirtsch. Jahrb. 89:202–42.

Voth, P. D. 7666.

Voth, V. 3054.

19247. Vo-Van, L., and J. Gregoire. 1958: Sur les composés acidosolubles des germes de *Phaseolus aureus*. I. Isolément par chromatographic sur échangeur d'anions de quelques nucleotides, d'acide urocanique et d'acide nicotinique. Compt. Rend. Soc. Biol. 152:1260–62.

19248. ——, ——, N. Limozin, and J. Gregoire. 1958: Sur les composés acidosoluble des germes de *Phaseolus aureus*. II. Identification de deux derivés de la cytidine, de l'acide urocanique et de l'acide nicotinique. Compt. Rend. Soc. Biol. 152:1262–65.

Vo-Van, L. 5629–5631.

Vries, H. de; *see* De Vries, H.

Vries, J. E. de. 6827.

19249. Vrzalova, J. 1955: The influence of preparation of seed cummin (*Carum carvi* L.) on its germination speed. Sb. Vysoke Skoly Zemedel. Brne Rada A 3:243–52. Biol. Abstr. 31:2572. 1957.

19250. Vukolov, V. A. 1929: Polyembryony in blue grasses (*Poa* sp.). Sb. Česk. Akad. Zeměděl. 4:193–218.

Vuong, D. K. 4491.

19251. Vuyk, A. 1961: The propagation of pines by seed. Plant Propagators Soc. Proc. 11:31–32.

Vyas, M. T. 5224.

Vyayaraghavan, M. R. 10240.

19252. Vyvalko, I. G. 1963: The influence of gibberellin on the transformation of substances in maize seeds during germination. Dokl. Akad. Nauk. SSSR 149:526–28.

W

19253. W., A. T. 1939: *Hippeastrums* raised from seeds. Gardeners Chron. 105:54–55.

Waal, D. de; *see* De Waal, D.

19254. Wach, A. 1942: Versuche zur Selenitfärbung des forstlichen Saatgutes. Allgem. Forst- Jagdztg. 118:178–88, 210–18. Biol. Abstr. 17:17245. 1943.

Wachtel, H. 12537–12540.

19255. Wada, E., and K. Yamazaki. 1951: A new trisaccharide in tobacco seeds. J. Agr. Chem. Soc. Japan 24:398. Chem. Abstr. 45:10308. 1951.

Wada, M. 11571.

Waddell, C. C. 19866.

19256. Waddington, G., and F. M. Cist. 1943: The vitamin C content of *Psidium guajava*. Proc. Florida State Hort. Soc. 5(1942):110–12.

Waddoups, H. M. 9225.

19257. Wade, B. L., and M. S. Kanapaux. 1943: Ascorbic acid content of strains of snap beans. J. Agr. Res. 66:313–24.

19258. —— and W. J. Zaumeyer. 1934: Internal breakdown of pea seed. Phytopathology 24:1384–86. Biol. Abstr. 9: 10999. 1935.

Wade, B. L. 20069.

Wade, W. B. 17317.

19259. Wadleigh, C. H. 1938: Oil, protein, and gossypol content of cottonseed as influenced by the level of nitrogen supply to the plant. Arkansas Agr. Expt. Sta. Bull. 368:49–50. Chem. Abstr. 35:1830. 1941.

Wähner, R. 1630.

19260. Wagenaar, M. 1932: The coloration of different varieties of *Strophanthus* seeds with sulfuric acid. Pharm. Weekblad 69:1340–49. Chem. Abstr. 27:520. 1933.

19261. Wagener, H. 1931: Beiträge zur Kenntnis von *Heracleum sphondylium* und *Anthriscus silvestris* als Wiesenunkräutern. Dissertation, Jena. 71 pp.

19262. Wagenknecht, A. C. 1957: The lipides of green peas. J. Am. Oil Chemists Soc. 34:509–13.

19263. Wager, H. G. 1957: The effect of artificial wilting on the carbon dioxide production of developing pea seeds. New Phytologist 56:230–46.

19264. —— 1959: The effect of artificial wilting on the production of ethanol by ripening pea seeds. New Phytologist 58:68–74.

19265. —— 1961: The effect of anaerobiosis on acids of the tricarboxylic acid cycle in peas. J. Exptl. Botany 12:34–46.

19266. —— 1964: Physiological studies of the storage of green peas. J. Sci. Food Agr. 15:245–52.

19267. Waggoner, H. D. 1917: The viability of radish seeds (*Raphanus sativus* L.) as affected by high temperatures and water content. Am. J. Botany 4:299–313.

Waghrey, A. P. 7849.
Wagner, D. 17564.
19268. Wagner, J. A. 1949: Results of airplane pellet seeding on Indian reservations. J. Forestry 47:632–35.
19269. —— and C. P. Kinkor. 1950: Will pellet seeding work? Am. Forests 56(5): 25, 44–45.
Wagner, J. A. 9196.
19270. Wagner, K. H., and L. Seber. 1939: Vitamin content of the germ oil of *Theobroma cacao* L. Bull. Offic. Office Intern. Cacao Chocolat 9:147, 149. Pharm. Abstr. 7:235. 1941.
Wagner, M. 17706.
19271. Wagner, N. 1930: Le chondriome des embryons des graines au cours de la maturation et de la germination. Arch. Anat. Microscop. 26:419–32. Biol. Abstr. 6:3226. 1932.
19272. Wagner, R. E., and W. C. Hulburt. 1953: Better forage stands with less seed. Crops & Soils 6(2):8–9.
19273. Wagner-Jauregg, T., and H. Rauen. 1935: The dehydrogenation of citric acid and isocitric acid by the dehydrogenase of cucumber seeds. Z. Physiol. Chem. 237: 227–32. Chem. Abstr. 30:1397–98. 1936.
Wahab, A. S. Abdul-. 10544.
Waheed, C. A. 10507.
19274. Wahhab, A., and A. Hussain. 1952: Utilization of cottonseed. I. Agr. Pakistan 3:35–45. Chem. Abstr. 49:1347. 1955.
19275. ——, A. Jabbar, and F. Muhammad. 1959: Salt tolerance of various varieties of agricultural crops at germination stage. Pakistan J. Sci. 11:71–80.
Wahhab, A. 9263.
19276. Wahlen, F. T. 1929: Hardseededness and longevity in clover seeds. Proc. Intern. Seed Testing Assoc. 9/10:34–39.
19277. Wahlenberg, W. G. 1924: Fall sowing and delayed germination of western white pine seed. J. Agr. Res. 28:1127–31.
19278. —— 1925: Sowing and planting season for western yellow pine. J. Agr. Res. 30:245–51.
19279. Wai, K. N. T., J. C. Bishop, P. B. Mack, and R. H. Cotton. 1947: The vitamin content of soybeans and soybean sprouts as a function of germination time. Plant Physiol. 22:117–26.
19280. Wai, N., and P. S. Kin. 1935: The formation of ascospores in Saccharomyces and its relation to the culture-media. Nung Hsueh Ts'ung K'an (Nanking) 2:1–22.
Waid, J. S. 9942.
19281. Wain, R. L., and E. H. Wilkinson. 1945/46: Studies upon the copper fungicides. VII. The solution of copper from dressings on pea seed. VIII. The penetration of copper into germinating peas. IX. Investigations with the exudate from fungus spores. Ann. Appl. Biol. 32:240–43, 243–47. 1945; 33:401–5. 1946. Rev. Appl. Mycol. 26:459. 1947.
Wain, R. L. 3664, 7302.
19282. Wainio, W. W., and E. B. Forbes. 1941: The chemical composition of forest fruits and nuts from Pennsylvania. J. Agr. Res. 62:627–35.
19283. Waisel, Y. 1958: Germination behavior of some halophytes. Bull. Res. Council Israel Sect. D 6:187–89.
19284. —— 1962: Presowing treatments and their relation to growth and to drought, frost, and heat resistance. Physiol. Plantarum 15:43–46.
19285. —— and Y. Adler. 1959: Germination behavior of *Aegilops kotschyi* Boiss. Can. J. Botany 37:741–42.
Waite, W. W. 6291.
19286. Wakakuwa, S. 1930: Bestäubungs- und Keimungsversuche in reziproken *Triticum*-Kreuzungen. Japan. J. Genetics 6:93–100. Ann. Biol. (Ser. 3) 36:725–26.
19287. —— 1934: Embryological studies on the different seed-development in reciprocal interspecific crosses of wheat. Japan. J. Botany 7:151–85. Biol. Abstr. 9:17650. 1935.
19288. Wakayama, K. 1929: On the influence of gravity upon the development of embryo of *Pinus thunbergii* Parl. Cytologia 1:68–75. Biol. Abstr. 5:26682. 1931.
19289. Wakefield, R. C., D. A. Schallock, M. Salomon, and C. E. Olney. 1957: Yield and chemical composition of Ladino clover as affected by fertilizer treatments. Agron. J. 49:374–77.
Wakeley, J. T. 5826.
19290. Wakeley, P. C. 1930: Seed yield data for southern pines. J. Forestry 28:391–94. Biol. Abstr. 5:11852. 1931.
19291. —— 1931: Successful storage of longleaf pine seed. J. Forestry 29:424–25.
19292. —— 1931: Some observations on Southern pine seed. J. Forestry 29:1150–64.
19293. —— 1932: Peat mats for germination tests of forest tree seeds. Science 76:627–28.
19294. —— 1935: Artificial reforestation in

the Southern pine region. U.S. Dept. Agr. Tech. Bull. 492. 114 pp.

19295. —— 1935: Collecting, extracting, and marketing Southern pine seed. U. S. Dept. Agr. Forest Serv. Southern Forest Expt. Sta. Occas. Paper 51. 10 pp.

19296. —— 1944: Geographic source of loblolly pine seed. J. Forestry 42:23–32. Biol. Abstr. 18:14565. 1944.

19297. —— 1947: Loblolly pine seed production. J. Forestry 45:676–77.

19298. —— 1954: Planting the Southern pines. U.S. Dept. Agr. Forest Serv. Agr. Monograph 18. 233 pp.

Wakelin, R. W. 15232.

Wakely, R. D. 19414.

Wakeman, A. J. 19107.

Walder, W. O. 8858.

19299. Waldron, L. R. 1909: A suggestion regarding heavy and light seed grain. Am. Naturalist 44:48–56.

19300. —— 1941: Analysis of yield of hard red spring wheat grown from seed of different weights and origin. J. Agr. Res. 62:445–60.

19301. —— 1943: Comparison of large and small kerneled wheat as to yield, grown alone and intermixed in the row. N. Dakota Agr. Expt. Sta. Bimo. Bull. 5(6):27–32. Expt. Sta. Rec. 90:47. 1944.

Waldron, L. R. 4886, 8252.

19302. Walger, J., L. Markus, and F. Nagymihaly. 1950: The physiology of germination. I. Influence of *Cucurbita maxima* extract on germination. Agrartudomany 2:150–63. Chem. Abstr. 45:1206. 1951.

19303. ——, F. Nagymihaly, and A. M. Plosz. 1950: Studies on the physiology of germination. II. The influence of tomato extract on germination. Agrártudomány 2:206–9. Biol. Abstr. 24:34304. 1950.

19304. —— and A. M. Plosz. 1950: Studies on the physiology of germination. III. Influence of nitro-benzol on germination. Agrártudomány 2:463–70.

19305. Walhood, V. T. 1956: A method of reducing the hard seed problem in cotton. Agron. J. 48:141–42.

19306. Walkden, H. H., and H. D. Nelson. 1958: Evaluation of methoxychlor for the protection of stored wheat and shelled corn from insect attack. U.S. Dept. Agr. Marketing Res. Rept. 213. 25 pp.

19307. —— and —— 1958: Evaluation of *Ryania* for the protection of stored wheat and shelled corn from insect attack. U.S. Dept. Agr. Marketing Res. Rept. 245. 21 pp.

Walkden, H. H. 4445, 8959, 17295.

Walker, D. A. 2207.

Walker, E. R. 10592, 17703.

19308. Walker, J. 1933: The suitability of immature sweet corn for seed. Sci. Agr. (Ottawa) 13:642–45.

19309. —— 1939: Viability of freshly harvested celery seed. Proc. Am. Soc. Hort. Sci. 36:687–90.

19310. —— 1960: Two seed borne fungi of French bean—*Phaseolus vulgaris* L. J. Australian Inst. Agr. Sci. 26:60–63.

Walker, J. C. 4348.

19311. Walker, M. H. 1955: Ploughing *Phalaris* for seed. Agr. Gaz. N.S. Wales 66:368, 390.

19312. Walker, R. G. 1934: The inimical effects of presoaking on the seeds of oats. Welsh J. Agr. 10:278–84.

19313. —— 1934: "Suction-force" measurements on the seeds of some varieties of oats. Welsh J. Agr. 10:289–95.

19314. Walker, R. H., and L. W. Erdman. 1926: Preliminary note on the sterilization of seeds of the Leguminosae with hydrogen peroxide. Proc. Iowa Acad. Sci. 33:91–95.

Walker, R. H. 519.

Walker, R. K. 4370, 9959, 18118.

19315. Walker, T. W., A. F. R. Adams, and H. D. Orchiston. 1954: Some effects of sulphur and phosphorus on the yield and composition of rape (*Brassica napus*). New Zealand J. Sci. Technol. 36:103–10.

19316. —— and R. Thompson. 1949: Some observations on the chemical changes effected by the steam sterilization of glasshouse soils. J. Pomol. Hort. Sci. 25:19–35.

19317. Walker, W. F. 1944: Extraction of tomato seed by the acid method. Tasmanian J. Agr. 15:110–13.

Wall, E. T. 4989.

19318. Wall, J. R., and T. L. York. 1957: Inheritance of seedling cotyledon position in *Phaseolus* species. J. Heredity 48:71–74.

19319. Wallace, A. 1954: Calcium and magnesium in citrus seeds. Plant Physiol. 29:488.

19320. Wallace, A. T. 1964: Influence of seed moisture content on γ-ray induced mutation rates at a specific locus in oats. Radiation Botany 4:53–59.

Wallace, A. T. 7919, 12531.

19321. Wallace, H. A. H. 1959: A rare seed-borne disease of wheat caused by *Podosporiella verticillata*. Can. J. Botany 37: 509–15.

19322. —— 1960: Factors affecting subsequent germination of cereal seeds sown in soils of subgermination moisture content. Can. J. Botany 38:287–306.

19323. —— and J. E. Machacek. 1963: Some factors affecting the dispersal of liquid seed dressings among cereal seeds. Can. J. Plant Sci. 43:313–22.

Wallace, H. A. H. 3236, 7609, 12604–12607.

19324. Wallace, J. M., and R. J. Drake. 1953: Seed transmission of the avocado sun-blotch virus. Citrus Leaves 33(12): 18–20. Biol. Abstr. 28:27320. 1954.

19325. Wallace, J. O. 1961: Seeding rates for swedes and turnips in relation to size of seed. New Zealand J. Agr. 103:379.

19326. —— 1962: Seeding rates of turnip and swede seed in relation to size. New Zealand J. Agr. 105:337.

Wallace, L. E. 17573.

Wallace, M. A. 19334.

19327. Wallace, R. H., and H. M. Havermann. 1958: Absence of seed dormancy in a white mutant strain of *Helianthus annuus* L. Plant Physiol. 33:252–54.

19328. —— and A. G. Lochhead. 1951: Bacteria associated with seeds of various crop plants. Soil Sci. 71:159–66.

19329. Walle, O. 1929: Experiments on seeds of the Finnish *Dactylis*. Suomen Maataloustiet. Seuta Maataloustiet. Aikakausk. 1:26–39.

19330. Wallebroek, J. C. 1940: Alkaloid and nitrogen metabolism in the germination of *Lupinus luteus*. Dissertation, Univ. Amsterdam. 132 pp. Biol. Abstr. 21: 12622. 1947.

19331. Wallen, V. R. 1955: The effect of storage for several years on the viability of *Ascochyta pisi* Lib. in pea seed and on the germination of the seed and emergence. Plant Disease Reptr. 39:674–77.

19332. —— and A. J. Skolko. 1953: Treatment of vegetable seed of low germination. Plant Disease Reptr. 37:66–68.

19333. ——, M. D. Sutton, and A. J. Skolko. 1950: The effect of actidione on the growth of certain pathogenic fungi and on the germination of pea seed. Phytopathology 40:156–60.

19334. ——, M. A. Wallace, and W. Bell. 1955: Response of aged vegetable seed to seed treatment. Plant Disease Reptr. 39:115–17.

Wallen, V. R. 17572.

Wallenreuter, R. 8494.

19335. Waller, A. D. 1901: An attempt to estimate the vitality of seeds by an electrical method. Proc. Roy. Soc. (London) 68: 79–92.

19336. Waller, E. 1961: Regulations concerning the marketing of seed. Skarab. Lans Hush. Tidn. 4:54–56.

19337. Waller, G. R., and S. S. Negi. 1958: Isolation of ricin, ricinine, and the allergenic fraction from castor seed pomace from two different sources. J. Am. Oil Chemists Soc. 35:409–12.

19338. Wallin, J. R. 1946: Seed and seedling infection of barley, bromegrass, and wheat by *Xanthomonas translucens* var. *cerealis*. Phytopathology 36:446–57.

19339. Wallner, I. 1930: Influence of time of harvest on the quality and quantity of grains. Mezogazdasagi Kutatasok 3:395–409. Chem. Abstr. 26:2765. 1932.

Walrand, M. Bouillenne-; *see* Bouillenne-Walrand, M.

19340. Walster, H. L. 1943: How well does old seed germinate. [Abstract.] N. Dakota Agr. Expt. Sta. Bimo. Bull. 6(2):10–11.

Walster, H. L. 8253.

19341. Walter, E. D. 1935: A saponin from the soy bean. Dissertation Abstr. 15:259–67.

Walter, E. D. 3409.

Walter, T. L. 13486.

Walters, G. E. 17665.

Walther, L. S. Kachioni-; *see* Kachioni-Walther, L. S.

19342. Walther, O. 1932: Staudenvermehrung durch Samen. Möllers Deut. Gärtner-Ztg. 47:126–27.

19343. Walther, R. 1953: Über die Verlängerung der Keimfähigkeit von Saatgut. Saatgut-Wirtsch. 5:225–26.

19344. Walti, A. 1938: Activation and inhibition of germination of lettuce seeds. Am. Chem. Soc. Abstr. Papers, Div. Biol. Chem. (Milwaukee) 96:20–21.

Wamble, A. C. 5065.

Wan, S. 8881.

Wandrekar, S. D. 17834–17836.

Wang, B. K. 18798.

19345. Wang, C. W., and T. O. Perry. 1958: The ecotypic variation of dormancy, chilling requirement, and photoperiod response in *Betula* species. [Abstract.] Proc. Intern. Congr. Genet. 10:307.

Wang, F. C. 18358.

Wang, F. H. 12421.

Wang, F. T. 3823.

Wang, J. K. 10641.

Wang, L. 4103.

Wang, S. 3823.
Wang, S. S. 11977.

19346. Wang, T. T., and C. F. Shi. 1957: The rate of water absorption by the seeds of four species of carniferous trees. Taipei Natl. Taiwan Univ. Forest Expt. Sta. Tech. Bull. 16. 15 pp.

Wanrooy, G. 6719.

19347. Want, G. 1963: Sporogenesis, gametogenesis, and embryogeny of *Wahlenbergia bicolor* N. Lothian. Australian J. Botany 2:152–67.

Want, J. P. H. van der. 2819.

19348. Ward, A. H. 1943: Location of vitamin B_1 in wheat. Soc. Chem. Ind., Chem. & Ind. 62(2):11–14.

19349. Ward, C. B., Jr., and R. G. Tischer. 1953: Use of cobaltous chloride to detect moisture patterns in partially dehydrated kernels of corn. Cereal Chem. 30:420–26.

19350. Ward, C. Y., and R. E. Blaser. 1961: Effect of nitrogen fertilizer on emergence and seedling growth of forage plants and subsequent production. Agron. J. 53:115–20.

19351. —— and A. W. Young. 1960: The effect of inert coatings on the germination and seedling growth of cotton. Agron. J. 52:362.

19352. Ward, F. K. 1935: Rhododendron seeds, with special reference to their classification. J. Botany 73:241–47.

19353. Ward, H. M. 1887: On some points in the histology and physiology of the fruits and seeds of *Rhamnus*. Ann. Botany (London) 1:1–26.

19354. Ward, H. S., Jr. 1951: Effect of relative humidities during storage on available and reserve foods of blue lupine seeds. Proc. Assoc. Southern Agr. Workers Ann. Conv. 48:138.

19355. —— 1952: Effect of moisture and temperature on respiratory activity of blue lupine seed during curing and storage. Proc. Assoc. Southern Agr. Workers Ann. Conv. 49:148–49.

19356. —— and J. L. Butt. 1955: Hygroscopic equilibrium and viability of naturally and artificially dried seed of crimson clover, *Trifolium incarnatum*. Agron. J. 47:576–79.

19357. —— and —— 1956: The effect of method of curing on the quality of peanuts. Alabama Agr. Expt. Sta. Ann. Rept. 64/65:26.

19358. —— and U. L. Diener. 1961: Biochemical changes in shelled peanuts caused by storage fungi. I. Effects of *Aspergillus tamarii*, four species of *A. glaucus* group, and *Penicillium citrinum*. Phytopathology 51:244–50.

Ward, H. S., Jr. 2545.
Ward, J. B. 4413, 16024–16026.

19359. Ward, J. M., W. E. Paulson, and D. L. Jones. 1953: Storing of seed cotton as an aid to more efficient ginning and marketing. Texas Agr. Expt. Sta. Bull. 765. 23 pp.

19360. Ward, P. F. V., R. J. Hall, and R. A. Peters. 1964: Fluoro-fatty acids in the seeds of *Dichapetalum toxicarium*. Nature 201:611–12.

Ward, P. F. V. 15231.
Warden, J. E. Wurceldorf-. 17556.

19361. Wardlaw, C. W. 1955: Embryogenesis in plants. John Wiley & Sons, New York. 381 pp.

19362. —— and E. R. Leonard. 1936: Studies in tropical fruits. I. Preliminary observations on some aspects of development, ripening and senescence, with special reference to respiration. Ann. Botany (London) 50:622–53.

Wardlaw, C. W. 15784.
Wardlaw, H. S. Halcro-. 4812.

19363. Ware, J. O. 1931: Inheritance of seed weight and lint index related to hereditability of lint percentage in cotton. J. Am. Soc. Agron. 23:677–702. Ber. Wiss. Biol. 20:226–27. 1932.

19364. —— 1940: Relation of fuzz pattern to lint in an upland cotton cross. J. Heredity 31:489–96. Biol. Abstr. 15: 10028. 1941.

19365. —— 1941: Genetic relations of sparse lint, naked seeds, and some other characters in upland cotton. Arkansas Agr. Expt. Sta. Bull. 406. 32 pp.

19366. —— 1941: Seed cover and plant color and their interrelations with lint and seed in upland cotton. J. Am. Soc. Agron. 33: 420–36. Biol. Abstr. 15:14181. 1941.

19367. ——, W. H. Jenkins, and D. C. Harrell. 1944: Seed characters and lint production. J. Heredity 35:153–60.

Ware, J. O. 15546.
Ware, W. M. 16674.

19368. Wareing, P. F. 1961: Dormancy of woody plants. Recent Advances in Botany, Univ. Toronto Press, pp. 1216–19.

19369. —— and H. A. Foda. 1956: Possible role of growth inhibitors in the dormancy of seed of *Xanthium* and lettuce. Nature 178:908–10.

19370. —— and —— 1957: Growth inhibitors and dormancy in *Xanthium* seed.

Physiol. Plantarum 10:266–80.
Wareing, P. F. 2527–2531, 5739, 6660, 6661, 19119.
Wargel, C. J. 9134.
Wariar, U. A. 1585.

19371. Warington, K. 1936: The effect of constant and fluctuating temperature on the germination of the weed seeds in arable soil. J. Ecol. 24:185–204.
Warington, K. 2986–2989.

19372. Wark, D. C. 1941: Addition of hormones to mercurial fungicidal dusts that reduce germination of wheat. J. Australian Inst. Agr. Sci. 7:156–58. Chem. Abstr. 36:3314. 1942.

19373. —— 1942: The influence of storage in contact with certain seed-pickling dusts on the germination of grain. J. Australian Inst. Agr. Sci. 8:22–25.

19374. Warmke, H. E., J. Rivera-Perez, and J. A. Ferrer-Monge. 1946: The culture of sugar-cane embryos in vitro. Inst. Trop. Agr. Univ. Puerto Rico Ann. Rept. Director 4:22–23.
Warmke, H. E. 2590, 2591.

19375. Warne, L. G. G. 1947: Vernalization of lettuce. Nature 159:31–32.
Warner, C. 6591.
Warner, G. M. 7947.
Warner, R. C. 9446.

19376. Warner, Z. P. 1954: Azaleas from seed. Plant Propagators Soc. Proc. 4: 137–43.

19377. Warren, F. S., and F. Dimmock. 1954: A comparison of moisture determination methods for small grain. Can. J. Agr. Sci. 34:435–43.
Warren, F. S. 5388.

19378. Warren, G. F. 1950: Effect of rate and depth of seeding on the yield and maturity of Henderson bush lima beans. Proc. Am. Soc. Hort. Sci. 55:372–74.
Warren, G. F. 15347, 15348.
Warren, H. W. 9060.

19379. Warren, J. T. 1961: Raising greenhouse plants from seed. Gardeners Chron. 149:5.

19380. —— 1961: Cyclamen from seed. Gardeners Chron. 150:145.

19381. —— 1962: Growing gesneriads from seed. Gardeners Chron. 151:52.
Warren, T. R. 1490.
Warren-Teed Seed Co. 686.
Warshowsky, B. 13523.
Warsi, S. A. 17401, 17402.
Wartenberg, H. 17570.

19382. Wartenberg, R. 1940: Bemerkungen über Kälteschäden an ruhenden und keimenden Samen. Allgem. Forst- Jagdztg. 116:23–25. Z. Pflanzenkrankh. Pflanzenschutz 52:401–2. 1942.

19383. Wateresiewicz, M. 1961: The influence of seeding periods and density on the lupine yields and alkaloid content in seeds. Ann. Univ. Marii Curie-Sklodowskiej, Lublin Sect. E 16:183–201.
Wartinbee, C. H. 17052.

19384. Wasicky, R., and C. Ferreira. 1949: The influence of saponin of *Bredemeyera floribunda* on germination and early development of *Phaseolus vulgaris*. Anais Fac. Farm. Odontol. Univ. São Paulo 7:351–59. Chem. Abstr. 44:10812. 1950.
Wasicky, R. 5284.

19385. Wasscher, J. 1951: Cyclamen seed. Zaadwereld 15(5):162–63.

19386. Wasser, C. H. 1944: Source of grass seed important for best reseeding of ranges, pastures, tests show. Colo. Farm Bull. 6(6):10–13. Biol. Abstr. 19:7208. 1945.
Wassermann, J. 18758.

19387. Wassilieff, N. 1908: Eiweissbildung in reifenden Samen. Ber. Deut. Botan. Ges. 26a:454–67.
Wassom, C. E. 18006.
Wastler, T. A. 4947.
Waszczenko-Zacharczenko, E. 12898.

19388. Watabe, T., T. Fujiwara, and T. Oishi. 1961: Studies on phosphorylase and Q-enzyme activities in non-glutinous and glutinous rice varieties. Proc. Crop Sci. Soc. Japan 29:219–22.

19389. —— and K. Hata. 1957: Some considerations on the method of estimating the starch viscosity of endosperm in rice plants. Proc. Crop Sci. Soc. Japan 26:84.
Watahara, J. 5822.

19390. Watanabe, A. 1953: Absorption of water by seeds of hazenoki (*Rhus succedana* Linn.). Bull. Tokyo Univ. Forests 44:7–13.

19391. —— 1959: Effects of drying of soil upon the germination of seeds and upon the water relations of seedlings of *Acacia mollissima*. Bull. Tokyo Univ. Forests 55:161–69.
Watanabe, I. 14694.

19392. Watanabe, K. 1933: Biologie von *Mitrastemon yamamotoi* (Rafflesiaceae). I. Früchte und Samen. Botan. Mag. (Tokyo) 47:798–805. Biol. Abstr. 11:228. 1937.

19393. —— 1936: Citricodehydrase in the seeds of watermelon. J. Agr. Chem. Soc.

Japan 12:983–89. Chem. Abstr. 31:1461–62. 1937.

19394. Watanabe, S. 1951: Silvicultural studies on kusu (*Cinnamomum camphora* Sieb.). I. On the heterospermy of seeds. Bull. Tokyo Univ. Forests 39:13–20.

19395. —— 1954: Hastening the germination and absorption of water by seeds of *Acacia mollissima* Willd. J. Japan. Forestry Soc. 36:327–30.

19396. —— 1955: On germination of the kintoki carrot seeds. II. Effects of pericarp removal, soaking and drying, and sun drying of seeds, on their germination. Hort. Assoc. Japan J. 23:237–44.

19397. —— 1955: Seed size of "Kusunoki." J. Japan. Forestry Soc. 37:552–54.

19398. —— and S. Aki. 1959: Effect of heat on germination of spinach. Tech. Bull. Fac. Agr. Kagawa Univ. 10:20–26.

19399. —— and —— 1959: On germination of kintoki-carrot seeds. V. Germination inhibitors in seeds. Tech. Bull. Fac. Agr. Kagawa Univ. 11:155–61.

19400. ——, H. Asano, and T. Maeda. 1955: On germination of the kintoki carrot seeds. I. Delayed germination. Kagawa Agr. Coll. Tech. Bull. 7:27–30.

Watanabe, S. 257–260, 5818, 14116, 14153, 18257.

Watanabe, T. 1475.

Waterbury, E. 6485, 6486.

19401. Waters, D. J. 1964: *Clianthus formosus* from seed. Gardeners Chron. 155: 268.

Waters, E. C. 4676.

19402. Waters, E. C., Jr., and J. D. Atkin. 1959: Performance of snap bean (*Phaseolus vulgaris*) seedlings having transversely broken cotyledons. Proc. Am. Soc. Hort. Sci. 74:591–96.

19403. Watkins, G. 1959: Sowing seed in polythene bags. E. African Agr. J. 24: 197–202.

Watkins, G. M. 15811.

19404. Watkins, G. R. 1958: Coffee seed germination in vermiculite. Coffee Bd. Kenya Mon. Bull. 23:330–31.

19405. Watkins, J. V. 1952: Propagation of ornamental plants. Agr. Ext. Serv., Gainesville, Fla., Bull. 150:37–40.

Watrelos, D. 10572.

19406. Watson, C. E. 1955: Germination studies on buffelgrass seed. Assoc. Offic. Seed Analysts News Letter 29(2):19–20.

19407. Watson, D. J., G. N. Thorne, and S. A. W. French. 1958: Physiological causes of differences in grain yield between varieties of barley. Ann. Botany (London) 22:309–20.

19408. Watson, D. P. 1948: Structure of the testa and its relation to germination in the Papilionaceae tribes Trifoliae and Loteae. Ann. Botany (London) 12:385–409.

Watson, D. P. 566.

19409. Watson, E. M. 1940/41: The seeds of *Strychnos lucida* and their alkaloid content. J. Roy. Soc. W. Australia 27:117–18. Chem. Abstr. 39:4192. 1945.

19410. Watson, E. R. 1951: A review of the application of statistical techniques to wheat agronomy in Australia. J. Agr. W. Australia 28:1–25.

19411. Watson, J. P., and L. F. Hough. 1949/50: Is it possible to germinate the seeds from early ripening peach varieties without artificially culturing the embryos on nutrient media? New Jersey Agr. Expt. Sta. Ann. Rept. 1949/50:60.

19412. Watson, R. D., L. Cottrin, and R. Robinson. 1951: The evaluation of materials for heat treatment of peas and beans. Plant Disease Reptr. 35:542–44.

19413. Watson, S. A., and E. H. Sanders. 1961: Steeping studies with corn endosperm sections. Cereal Chem. 38:22–33.

19414. ——, ——, R. D. Wakely, and C. B. Williams. 1955: Peripheral cells of the endosperms of grain sorghum and corn and their influence on starch purification. Cereal Chem. 32:165–82.

19415. Watson, W. 1901: Germination of seeds of *Bertholletia excelsa*. Ann. Botany (London) 15:99–102.

19416. Watt, J. R. 1931: Simultaneous development of the seed coat and embryo in the seeds of sweet clover. Proc. Iowa Acad. Sci. 37:117–23.

19417. —— and J. N. Martin. 1931: Polyembryony in alfalfa and sweet clover. Proc. Iowa Acad. Sci. 37:100.

Watt, J. R. 12910.

Wattegedera, C. 3805.

19418. Watters, F. L. 1959: Effects of grain moisture content on residual toxicity and repellency of Malathion. J. Econ. Entomol. 52:131–34.

19419. Wattiez, N., and M. Hans. 1943: A holoside extracted from the seeds of *Plantago major* L. and *Plantago ovata* Forsk (*P. isphagulata* Roxb.). Bull. Acad. Roy. Med. Belg. 8:386–96. Chem. Abstr. 39:4849. 1945.

Watts, A. B. 4332.

19420. Watts, V. M. 1939: Rest period in cucumber seeds. Proc. Am. Soc. Hort. Sci. 36:652–54.

19421. Waugh, F. A. 1896: Preliminary report on the artificial use of enzymes in germination. Vermont Agr. Expt. Sta. Rept. 1896:106–11.

Waugh, J. G. 17132, 17133.

Wavrin, L. 6898.

19422. Way, D. W. 1957: Propagation by seed. Gardeners Chron. 141:613.

Waygood, E. R. 10244.

19423. Waxman, S. 1957: Effects of day-length on the germination of *Sciadopitys verticillata*. Plant Propagators Soc. Proc. 7:71–72.

19424. Wean, R. E., and A. T. Guard. 1940: The viability and collection of seed of *Liriodendron tulipifera* L. J. Forestry 38:815–17.

Wean, R. E. 7762.

19425. Wear, L. 1952: Grain infestation. Progr. Res. 7(1):10–13.

19426. Weatherby, L. S., and D. G. Sorber. 1931: Chemical composition of the avocado seed. Ind. Eng. Chem. 23:1421–23. Chem. Abstr. 26:2214. 1932.

19427. Weathers, L. G., and E. C. Calavan. 1959: Nucellar embryony—a means of freeing citrus clones of viruses. Univ. Calif. Citrus Expt. Sta. Paper 1112:197–202.

19428. Weatherwax, P. 1930: The endosperm of *Zea* and *Coix*. Am. J. Botany 17:371–80.

19429. —— 1948: Right-handed and left-handed corn embryos. Ann. Missouri Botan. Garden 35:317–21. Proc. Indiana Acad. Sci. 58:69–70. 1949.

Weatherwax, P. 19587.

19430. Weaver, G. M., and L. F. Hough. 1959: Seedling growth studies of early-ripening peaches. I. Interrelationship between embryo maturity, growth substances, and seedling growth. Am. J. Botany 46:718–24.

Weaver, J. B., Jr. 5510.

Weaver, J. W., Jr. 18461.

19431. Weaver, R. J. 1960: Toxicity of gibberellin to seedless and seeded varieties of *Vitis vinifera*. Nature 187:1135–36.

19432. —— and S. B. McCune. 1959: Effect of gibberellin on seeded *Vitis vinifera* and its translocation within the vine. Hilgardia 28:625–45.

19433. —— and —— 1959: Effect of gibberellin on seedless *Vitis vinifera*. Hilgardia 29:247–75.

19434. ——, ——, and C. R. Hale. 1962: Effect of plant regulators on set and berry development in certain seedless and seeded varieties of *Vitis vinifera* L. Vitis 3(2): 84–96.

Webb, B. C. 19710.

Webb, B. K. 9134.

Webb, J. 15232.

Webb, L. J. 3566.

19435. Webber, H. J. 1895: The distribution of seeds by birds. Am. Naturalist 29:378–79.

19436. —— 1932: The economic importance of apogamy in *Citrus* and *Mangifera*. Proc. Am. Soc. Hort. Sci. 28:57–61.

19437. Webber, J. M. 1940: Polyembryony. Botan. Rev. 6:575–98.

Weber, C. 4013.

Weber, C. R. 9106, 19486.

19438. Weber, D. 1931: Change of moisture content of drug seeds and fruits. Magy. Gyogyszeresztud. Tarsasag Ertesitöje. 7: 194–97. Chem. Abstr. 25:3773–74. 1931.

Weber, E. 6912.

19439. Weber, F. 1941: The vitamin C content of the caruncula of myrmecochors. Protoplasma 36:147–49. Chem. Abstr. 38:5529. 1944.

19440. Weber, G. L. 1939: A method of preparing some native grass seeds for handling and seeding. J. Am. Soc. Agron. 31:729–33.

Weber, M. R. 4037.

19441. Weber, U. 1936: The fat metabolism of germinating pumpkin seeds. Ber. Deut. Botan. Ges. 54:70–75. Chem. Abstr. 31:2256. 1937.

19442. —— 1937: *Psyllium* seed and its substitution by native plantain seeds. Apoth. Ztg. 52:1619–23. Pharm. Abstr. 4:229–30. 1938.

Weber, U. 17576.

19443. Weber-Dahlmann, M. 1960: Ein Beitrag zur Sortendiagnose von *Phaseolus vulgaris* L. (var. *nanus* Mart.). Saatgut-Wirtsch. 1960(1):10–13.

Webley, D. P. 6326.

19444. Webster, C. B., and G. T. Ratliffe. 1942: A method of forcing quick germination of *Juniperus virginiana* L. seed. J. Forestry 40:268.

19445. Webster, C. C. 1945: Tung notes and abstracts. Nyasaland Agr. Quart. J. 5:37–39. Biol. Abstr. 21:9999. 1947.

19446. —— 1948: The effect of seed treatments, nursery technique, and storage methods on the germination of tung seed. E. African Agr. J. 14:38–48.

19447. Webster, G. T. 1955: Interspecific hybridization of *Melilotus alba* × *M. officinalis* using embryo culture. Agron. J. 47:138–42.

19448. Webster, J. E., H. Fellows, and H. F. Murphy. 1943: Chemical composition of castor bean oil from seed grown in Oklahoma. Proc. Oklahoma Acad. Sci. 23:69–73. Biol. Abstr. 17:25112. 1943.

Webster, J. E. 15077.

Webster, O. J. 2618, 10344.

Webster, R. E. 12681.

19449. Weckel, K. G., D. J. Hagedorn, J. Von Elbe, and R. Hawley. 1963: Effect of size classification and maturity on the protein content of Alaska and Perfection peas. J. Food Sci. 28:356–57.

19450. —— and H. D. Lee. 1960: Composition of cherry seed and cherry seed oil (*Prunus cerasus*). Food Technol. 14: 151–54.

19451. Wedrick, C. D. 1936: Germination of trumpet vine seeds. Flower Grower 23(11):596.

Weeks, W. D. 17769.

19452. Weetman, L. M. 1935: Correlation of shape of fruits, cotyledons, and seeds in melons. Botan. Gaz. 97:388–98.

19453. Weerakoon, A. H. 1960: Composition of the seed fat of Ceylon sweet orange. J. Sci. Food Agr. 11:273–76.

19454. Wegner, E. 1953: Comparative ferment studies on leaves and seeds of *Digitalis purpurea* and *lanata*. Arzneimittel-Forsch. 3:482–85. Chem. Abstr. 48:1633. 1954.

19455. Wehsarg, O. 1954: Ackerunkräuter. Akad. Verlag - Berlin. 294 pp.

19456. Weibel, D. E. 1958: Vernalization of immature wheat embryos. Agron. J. 50: 267–71.

19457. Weibel, R. O. 1948: The castor-plant in the United States. Econ. Botany 2:273–83.

19458. —— 1960: Effect of gibberellin on the vernalization period of winter wheat. Agron. J. 52:122–23.

19459. Weibull, G. 1952: The cold storage of vegetable seed. Seed World 71(7):24, 26.

19460. —— 1952: The cold storage of vegetable seed and its significance for plant breeding and the seed trade. Agr. Hort. Gen. 10:97–104. Ber. Wiss. Biol. 85:53. 1953.

19461. —— 1955: The cold storage of vegetable seed, further studies. Intern. Hort. Congr. Rept. 14:647–67.

19462. Weibull, W. A. 1959: Nomination of national seed campaign committees and their activities. World Seed Campaign News 5:8–10.

Weichherz, J. 14565, 14566.

Weide, B. M. Van der. 5710.

19463. Weight, N. 1936: Additional notes on the seedlessness of citrus fruits. Trop. Agr. (London) 13:283–84.

Weihe, K. von; *see* Von Weihe, K.

19464. Weihing, J. L., R. Inman, and G. W. Peterson. 1961: Response of ponderosa and Austrian pine to soil fumigants and seed treatment. Plant Disease Reptr. 45:799–802.

19465. Weihing, R. M. 1940: Field germination of alfalfa seed submitted for registration in Colorado and varying in hard seed content. J. Am. Soc. Agron. 32:944–49.

19466. —— 1962: Selecting Persian clover for hard seed. Crop Sci. 2:381–82.

19467. Weiland, G. S., L. B. Broughton, and J. E. Metzger. 1935: Wormseed oil production. Maryland Agr. Expt. Sta. Bull. 384. 21 pp.

19468. Weiling, F., and R. Schagen. 1958: Über die Ursachen des Auftretens tauber Samen bei Ölkürbis (*Cucurbita pepo* L.). Gartenbauwissenschaft 23:87–101.

Weimer, D. 4674, 4677.

19469. Weimer, D. E. 1939: Treatment of seeds with indolebutyric acid and its effects. Proc. Assoc. Offic. Seed Analysts 30:263–68.

19470. —— 1939: The ineffectiveness of certain seed treatments to reveal origin of red clover seed. Proc. Assoc. Offic. Seed Analysts 31:106–8. Chem. Abstr. 35:1091. 1941.

Weimer, J. L. 3403.

19471. Weinberger, J. H. 1956: Prolonged dormancy trouble in peaches in the southeast in relation to winter temperatures. Proc. Am. Soc. Hort. Sci. 67:107–12.

Weinberger, J. H. 8191.

19472. Weinberger, P., and C. Godin. 1963: Vernalization of *Triticum vulgare* (var. *rideau*). Morphological studies. Can. J. Botany 41:853–63.

19473. Weinbrenn, C. 1946: Preliminary account of investigation into the germination of seeds of *Cynodon dactylon*. S. African J. Sci. 42:142–43.

19474. Weindling, R. 1940: Sodium hypochlorite shows promise as a seed treatment. Phytopathology 30:1051–52.

Weinmann, H. 1518.

19475. Weinmann, I. 1956: Identification of seed of *Brassica* species and varieties with reference to physicochemical methods. Z. Pflanzenzücht. 36:1–30. Chem. Abstr. 51:14022. 1957.

Weinstein, N. E. 15188, 15189.

19476. Weintraub, R. L. 1948: Influence of light on chemical inhibition of lettuce seed germination. Smithsonian Inst. (Publ. 3915) Misc. Collections 107(20):1–8.

Weintraub, R. L. 7387.

Weinzierl, T. R. von; *see* Von Weinzierl, T. R.

19477. Weisaeth, G. 1960: Samenmerkmale zur Identifizierung von Bohnensorten (*Phaseolus vulgaris* L.). Proc. Intern. Seed Testing Assoc. 25:341–54.

19478. Weise, R. 1937: Bemerkungen zu Gurewitschs Methode, die Keimfähigkeit von Samen zu Bestimmen. Ber. Deut. Botan. Ges. 55:338–40.

19479. Weisenfeld, M. 1929: Die Vorzüge der Halbkorn-keimmethode. Z. Ges. Brauw. 52:95. Centralbl. Bakteriol. 79: 505. 1929.

19480. Weisner, M. 1941: Separation of immature seeds of *Trifolium repens*–white clover and *Trifolium hybridum*–Alsike clover. Proc. Assoc. Offic. Seed Analysts 32:103–4.

19481. Weiss, E., O. Schindler, and T. Reichstein. 1957: Die Glykoside der Samen von *Strophanthus tholloni* Franch. Helv. Chim. Acta 40:980–1015.

19482. ——, ——, and —— 1958: Vorversuch zur Konstitutionsermittlung von Sarmentosid A, Thollosid und verwandten Glykosiden. (Die Glykoside der Samen von *Strophanthus thollonii* Franch. III.) Helv. Chim. Acta 41:736–62.

19483. Weiss, F. 1926: Seed germination in the gray birch, *Betula populifolia*. Am. J. Botany 13:737–42. Contrib. Boyce Thompson Inst. 1:272–77. 1926.

19484. —— and O. Manus. 1959: The saponins of the leaves and seeds of *Digitalis purpurea* L. Pharm. Zentralhalle 98:437–43. Biol. Abstr. 35:60997. 1960.

19485. Weiss, F. E. 1933: On the germination and the seedlings of gentians. (Contributions from the Wisley Laboratory. LXIX.) J. Roy. Hort. Soc. (London) 58:296–300.

Weiss, M. B. 12002.

19486. Weiss, M. G., C. R. Weber, L. F. Williams, and A. H. Probst. 1952: Correlation of agronomic characters and temperature with seed compositional characters in soybeans, as influenced by variety and time of planting. Agron. J. 44:289–97.

19487. —— and J. B. Wentz. 1937: Effect of *Luteus* genes on longevity of seed in maize. J. Am. Soc. Agron. 29:63–75.

Weiss, M. G. 15598.

19488. Weiss, R. 1952: Évolution des principales formes de phosphore au cours de la germination des graines. Bull. Soc. Chim. Biol. 34:471–79.

Weiss, R. 2750.

19489. Weisse, A. 1930: Blattstellungsstudien an Sämlingen abnorm keimender Dikotylen. Beitr. Biol. Pflanz. 18:17–80. Biol. Abstr. 6:5213. 1932.

19490. Weissenböck, K. 1948: Fett- und Phosphatidstoffwechsel keimender und reifender Ölsamen. I. Lipoidstoffwechsel bei der Keimung von Paps und Soja. II. Lipoidstoffewechsel von Elsamen während des Reifungsvorganges. Österr. Botan. Z. 94:301–29. Biol. Abstr. 23:16853. 1949.

Welch, F. J. 13143.

19491. Welch, G. B. 1954: Seed processing equipment. Mississippi Agr. Expt. Sta. Bull. 520. 23 pp.

Welch, J. E. 7707.

19492. Weldon, L. W., D. W. Bohmont, and H. P. Alley. 1959: The interrelation of three environmental factors affecting germination of sagebrush seed. J. Range Management 12:236–38.

19493. Welker, E. L., and F. L. Wynd. 1943: Influence of unknown factors on the validity of mathematical correlations of biological data. Plant Physiol. 18:498–507.

19494. Wellensiek, S. J. 1940: Bad grains and slight wrinkling in peas. Landbouwk. Tijdschr. 52:680–85. Biol. Abstr. 15:11213. 1941.

19495. —— 1958: Vernalization and age in *Lunaria biennis*. Koninkl. Ned. Akad. Wetenschap. Proc. Ser. C 61:561–71.

19496. —— 1962: Dividing cells as the locus for vernalization. Nature 195:307–8.

Weller, L. E. 6828.

Weller, R. A. 12515–12518.

Welles, E. 17720, 17721.

19497. Wellington, P. S. 1956: Dormancy in barley. Nature 178:1359.

19498. —— 1956: Studies on the germination of cereals. I. The germination of wheat grains in the ear during development, ripening, and after-ripening. Ann. Botany (London) 20:105–20.

19499. —— 1956: Studies on the germination of cereals. II. Factors determining the

germination behavior of wheat grains during maturation. Ann. Botany (London) 20:481–500.
19500. —— 1957: Report of the committee on the effect of toxic substances in seed testing, 1953–1956. Proc. Intern. Seed Testing Assoc. 22:370–74.
19501. —— 1957: The work of the official seed testing station in relation to the problems of farmers and seedsmen. Natl. Inst. Agr. Botany J. 8:182–87.
19502. —— 1958: Seed control: present and future. Agriculture (London) 65:13–16.
19503. —— 1961: Developments in seed testing. Agriculture (London) 68:309–13.
19504. —— 1961: Seed identification procedures used by the official seed testing station for England and Wales. Assoc. Offic. Seed Analysts News Letter 35(3): 9–11.
19505. —— 1962: An analysis of discrepancies between germination capacity and field establishment of peas. Natl. Inst. Agr. Botany J. 9:160–69.
19506. —— 1962: Germination committee. Proc. Intern. Seed Testing Assoc. 27:70–85.
19507. —— 1964: Studies on the germination of cereals. V. The dormancy of barley grains during ripening. Ann. Botany (London) 28:113–26.
Wellington, P. S. 15795.
Wellington, R. 8425.
19508. Wellman, F. L., and V. K. Toole. 1960: Coffee seed germination as affected by species, diseases and temperature. Proc. Am. Soc. Hort. Sci., 8th Ann. Meeting, Caribbean Region 4:1–6.
19509. Wellman, R. H., H. W. Thurston, Jr., and F. R. Whaley. 1948: A method for correcting for geographic variation in field experiments. Contrib. Boyce Thompson Inst. 15:153–63.
Wells, A. L. 5327.
19510. Wells, A. W. 1958: Freezing injury of peanuts. Proc. Am. Soc. Hort. Sci. 72:374–76.
Wells, B. Russell-. 9717.
Wells, D. G. 6098.
19511. Wells, F. B. 1955: A note on jojoba bean meal—a potential feed. Cereal Chem. 32:157–58.
19512. Wells, H. D., and J. L. Allison. 1952: Tall fescue seed can carry net blotch fungus. Phytopathology 42:22–23.
Wells, H. D. 12551.
19513. Wells, J. S. 1952: Propagation from seeds. Am. Nurseryman 95(9):8, 62–68.
19514. —— 1952: Fall seed sowing. Am. Nurseryman 96(7):15, 68–69.
19515. —— 1955: Plant propagation practices. The Macmillan Company, New York. 344 pp.
19516. Wells, P. V. 1959: Ecological significance of red light sensitivity in germination of tobacco seed. Science 129:41–42.
19517. Welsch, H. 1934: Sur la distribution de l'eau, de la matière sèche et de l'azote au cours de la germination chez *Impatiens balsamina* (L.). Protoplasma 22:63–127.
19518. Welton, F. A. 1921: Longevity of seeds. Ohio Agr. Expt. Sta. Month. Bull. 6:18–25.
Wen, S. H. 8881.
Wender, S. H. 5514, 15666.
19519. Wenger, K. F. 1953: The effect of fertilization and injury on the cone and seed production of Loblolly pine seed trees. J. Forestry 51:570–73.
19520. —— 1954: The stimulation of Loblolly pine seed trees by preharvest release. J. Forestry 52:115–18.
19521. —— 1957: Annual variation in the seed crops of Loblolly pine. J. Forestry 55:567–69.
19522. Wenger, L. E. 1941: Soaking Buffalo grass (*Buchloe dactyloides*) seed to improve its germination. J. Am. Soc. Agron. 33:135–41.
19523. —— 1943: Buffalo grass. Kansas Agr. Expt. Sta. Bull. 321. 78 pp. Expt. Sta. Rec. 91:28. 1944.
19524. Wenham, H. T. 1958: *Septoria drummondii* Ell. et Ev., a common seed-borne pathogen in European lines of *Phlox drummondii* Hook. Seed imported to New Zealand. New Zealand J. Agr. Res. 1:486–88.
19525. Went, F. W. 1948: An experiment that may last 360 years. Project will give botanists new information about the heredity of plants and the life span of seeds. Life 24(5):57–58, 60.
19526. —— 1948: Ecology of desert plants. I. Observations on germination in the Joshua Tree National Monument, California. Ecology 29:242–53.
19527. —— 1949: Ecology of desert plants. II. The effect of rain and temperature on germination and growth. Ecology 30:1–13.
19528. —— 1954: Germination as a major factor in the establishment of communities. Congr. Intern. Botan. Paris, Rappt. Commun. 8:57–60.
19529. —— 1955: The ecology of desert plants. Sci. Am. 192:68–75.

19530. —— 1961: Problems in seed viability and germination. Proc. Intern. Seed Testing Assoc. 26:674–85.

19531. —— 1961: Temperature. Encyclopedia Plant Physiol. 16:1–23.

19532. ——, G. Juhren, and M. C. Juhren. 1952: Fire and biotic factors affecting germination. Ecology 33:351–64.

19533. —— and P. A. Munz. 1949: A long term test of seed longevity. Aliso 2:63–75. Biol. Abstr. 23:30211. 1949.

Went, F. W. 5476, 10049, 13683, 16290, 18918.

19534. Wentz, J. B. 1930: The inheritance of germless seeds in maize. Iowa Agr. Expt. Sta. Res. Bull. 121:345–79.

Wentz, J. B. 17998, 19487.

19535. Wenusch, A. 1943: The first appearance of nicotine in the germinating tobacco seed. Wien. Chem.-Ztg. 46:196–97. Chem. Abstr. 38:4648. 1944.

19536. Wenzl, H. 1941: Lokale störungen im Eiweisstoffwechsel von Erbsensamen. Z. Pflanzenkrankh. Pflanzenschutz 51: 240–47.

19537. —— 1956: On the question of seed transmission of the β-beet yellows disease. Pflanzenschutz Ber. 15(11/12):161–67. Rev. Appl. Mycol. 35:569. 1956.

19538. —— and R. Krexner. 1957: Untersuchungen über die Wirkung der Saatgutbeizung bei Rube. Pflanzenschutz Ber. 18(8/12):119–70.

19539. Werckmeister, P. 1962: Über embryonales Wachstum ausserhalb des Samens, Keimung und Keimversug Untersuchungen an Bastardsamen und Embryonen der Gattung *Iris* L. Biol. Centralbl. 81:667–90.

19540. Werle, E., G. Beaucamp, and V. Schirren. 1959: Über die Aminoxydase von Erbsenkeimlingen und ihre Bedeutung für die Keimung. Planta 53:125–33.

19541. Werminski, F. 1888: Über die Natur der Aleuronkorner. Ber. Deut. Botan. Ges. 6:199–204.

19542. Werner, H., and H. Schmalfuss. 1949: Do fats change, as in spoilage, during germination of seeds? Z. Lebensm.-Untersuch. -Forsch. 89:344–47. Chem. Abstr. 43:8457. 1949.

19543. Wernham, C. C. 1951: Cold testing of corn; a chronological and critical review. Penn. Agr. Expt. Sta. Progr. Rept. 47. 12 pp.

Wernham, C. C. 8048.

19544. Wernhart, K. 1926: Effect of X-rays on germination and growth of plants. Biol. Listy 12:206–20.

19545. Wertman, F. L. 1961: Fruits and seeds—emphasis on dispersal. Assoc. Offic. Seed Analysts New Letter 35(2):15.

19546. —— and O. L. Justice. 1950: Variety determinations of Sudan grass seed. Proc. Assoc. Offic. Seed Analysts 40:49–55.

19547. Wesemann, E. 1962: Möglichkeiten der Unterscheidung von Sorten bei *Vicia faba major* L. am Samen. Proc. Intern. Seed Testing Assoc. 27:557–61.

19548. Wessel, C. J. 1941: A study of the esterase activity of *Aleurites fordii.* Contrib. Biol. Lab. Catholic Univ. Am. 38:1–72. Biol. Abstr. 17:12256. 1943.

West, A. P. 4706–4712, 5227, 5228.

West, C. 10555–10562.

West, D. 9469.

West, D. W. 6382.

19549. West, S. H., and H. C. Harris. 1962: Seedcoat colors associated with physiological changes in alfalfa and crimson and white clover. Crop Sci. 3:190–93.

West, S. H. 17143.

19550. Wester, D. H. 1921: Manganese contents of Dutch seeds. Pharm. Weekblad 58:1613–19. Botan. Abstr. 11:4621. 1922.

19551. Wester, P. J. 1913: Adventitious germination in Dioscoreaceae. Philippine Agr. Rev. 6:548–49.

19552. —— 1924: The preparation and packing of seeds and scions of trees and shrubs in the tropics. Philippine Agr. Rev. 17:46–51.

19553. Wester, R. E. 1946: Comparison of yield between normal and baldhead snap bean plants. Proc. Assoc. Offic. Seed Analysts 1945/46:172–77.

19554. —— 1963: Sphagnum moss, artificial light, and plastic film for seed germination. Am. Hort. Mag. 42:221–25.

19555. ——, C. Drechsler, and H. Jorgensen. 1958: Effect of freezing on viability of the lima bean downy mildew fungus (*Phytophthora phaseoli* Thaxt.). Plant Disease Reptr. 42:413–15.

19556. —— and R. Magruder. 1939: Effect of size, condition, and production locality on germination and seedling vigor of Baby Fordhook bush lima bean seed. Proc. Am. Soc. Ho t. Sci. 36:614–22.

19557. —— and P. C. Marth. 1947: Effect of some growth regulators on yield of bush lima bean. Proc. Am. Soc. Hort. Sci. 49:315–19.

19558. —— and —— 1949: Some effects of a growth regulator mixture in controlled cross-pollination of lima bean. Proc. Am. Soc. Hort. Sci. 53:315–18.

Wester, R. E. 18686, 18698.

Westerburg, G. 2941.

19559. Westerman, E. 1959: Fungus and its association with orchid seed germination. Orchid J. 3:288–89.

19560. Western, J. H. 1945: Seed-borne fungi. Nature 155:36–37. Expt. Sta. Rec. 93:156. 1945.

Westerweel, F. L. 19652.

19561. Weston, T. A. 1932: Longevity of seeds. Gardeners Chron. 92:323–24.

19562. —— 1934: Growing lilies from seed. Horticulture 12(24):420.

19563. —— 1945: Growing evergreens from seed. Flower Grower 32(2):96, 98.

19564. —— 1945: Treatment of primrose seed. Horticulture 23(2):45.

Weston, W. A. R. D. 3011, 3012.

19565. Weston, W. J., and H. J. Morris. 1954: Hygroscopic equilibria of dry beans. Food Technol. 8:353–55.

19566. Westrin, W. 1946: A short "cut" to viability determination of New Zealand spinach. Proc. Assoc. Offic. Seed Analysts 1945/46:134–36.

19567. —— 1949: More about seed pelleting. Assoc. Offic. Seed Analysts News Letter 23(2):26–27.

Wetherell, D. F. 8045.

19568. Wetter, L. R. 1955: The determination of mustard oils in rapeseed meal. Can. J. Biochem. Physiol. 33:980–84.

19569. —— 1957: Some properties of the lipase present in germinating rapeseed. J. Oil Chemists Soc. 34:66-70.

19570. —— and B. M. Craig. 1959: Varietal and environmental effects on rapeseed. I. Isothiocyanate and thiooxazolidone content. Can. J. Plant Sci. 39:395–99.

Wetter, L. R. 6932, 14925, 14926, 15904.

Wettstein, W. von; *see* Von Wettstein, W.

19571. Wexelsen, H., and K. Mikaelsen. 1958: The effect of neutron treatment of seeds on seedlings of diploid and tetraploid red clover (*Trifolium pratense*). Physiol. Plantarum 11:374-81.

19572. Wexler, A., and W. G. Brombacher. 1951: Methods of measuring humidity and testing hygrometers. Natl. Bur. Standards Circ. 512. 18 pp.

19573. Whalen, F. T. 1929: Hardseededness and longevity in clover seeds. Proc. Intern. Seed Testing Assoc. 9/10:34–40.

Whaley, F. R. 19509.

Whaley, W. G. 372.

19574. Wharton, M. F., and W. A. Frazier. 1939: Effect of certain storage treatments on field and laboratory germination of seeds of Imperial 152 and Imperial 615 lettuce. Proc. Am. Soc. Hort. Sci. 36: 680–86.

19575. Wharton, M. J. 1955: The use of the tetrazolium test for determining the viability of grass seeds. Proc. Intern. Seed Testing Assoc. 20:71–80.

19576. —— 1955: The use of the tetrazolium test for determining the viability of seeds of the genus *Brassica*. Proc. Intern. Seed Testing Assoc. 20:81–88.

19577. Wheatley, G. W., and R. T. Johnson. 1962: Germination-emergence of sugar beet seed as affected by treatment and storage. Proc. Am. Soc. Sugar Beet Technol. 11:649–55.

19578. Wheeler, A. W. 1960: Changes in a leaf-growth substance in cotyledons and primary leaves during the growth of dwarf bean seedlings. J. Exptl. Botany 11: 217–26.

Wheeler, C. D. 8734.

19579. Wheeler, J. L. 1959: The effect of sheep urine on the germination and early establishment of a common weed grass (*Poa annua*). J. Brit. Grassland Soc. 14:55–57.

19580. Wheeler, L. C. 1954: Endospermous dicotyledonous seeds for laboratory dissection. Turtox News 32:74–76.

Wheeler, P. 8028.

19581. Wheeler, W. A. 1952: Many hands help establish a new crop. Crops & Soils 4(9):14–16.

19582. —— and D. D. Hill. 1957: Grassland seeds. Van Nostrand, Princeton, N.J. 734 pp.

19583. Wheeler, W. A., Jr. 1962: Camellias from seed. Horticulture 40(6):311.

Whelan, W. J. 15165, 15166.

19584. Wherry, E. T. 1936: Discovery of *Elliottia* seed. Bartonia 17:51.

19585. Whistler, R. L., and W. M. Doane. 1961: Characterization of intermediary fractions of high-amylose corn starches. Cereal Chem. 38:251–55.

19586. ——, H. H. Kramer, and R. D. Smith. 1957: Effect of certain genetic factors on the sugars produced in corn kernels at different stages of development. Arch. Biochem. Biophys. 66:374–80.

19587. —— and P. Weatherwax. 1948: Amylose content of Indian corn starches from North, Central, and South American corns. Cereal Chem. 25:71–75.

19588. —— and J. R. Young. 1960: Formation of starch in wheat grain. Cereal Chem. 37:204–11.

Whistler, R. L. 5013, 6265, 11322, 16723.

19589. Whiston, A., R. J. Hay, and W. D. Raymond. 1960: The effect of storage for different periods of time in Nigeria on the quality of groundnuts, with special reference to their value for the manufacture of protein fibres. Trop. Sci. 1:149–81.

Whitacre, J. 16181.

Whitacre, J. E. 14564.

19590. Whitaker, T. W., and G. F. Carter. 1954: Oceanic drift of gourds. Experimental observations. Am. J. Botany 41:697–700. Biol. Abstr. 29:10381. 1955.

19591. —— and —— 1961: A note on the longevity of seed of *Lagenaria siceraria* (Mol.) Standl. after floating in sea water. Bull. Torrey Botan. Club 88:104–6.

19592. —— and G. D. McCollum. 1954: Shattering in lettuce—its inheritance and biological significance. Bull. Torrey Botan. Club 81:104–10.

Whitaker, T. W. 15697.

19593. Whitcomb, W. O. 1921: Correlating laboratory tests of seed germination with field tests. Proc. Assoc. Offic. Seed Analysts 12/13:51–54.

19594. —— 1924: Dormancy of newly threshed grain. Proc. Assoc. Offic. Seed Analysts 16:28–33.

19595. —— 1926: Wet grain and the seed problem. Seed World 19(7):23.

19596. —— 1928: The problem of making germination tests of scarified seed. Proc. Assoc. Offic. Seed Analysts 20:74–81.

19597. —— 1929: Use of frosted wheat for seed. Seed World 25(4):12–13.

19598. —— 1931: Hard seeds in legumes (interpretation of their value and methods of treatment). Montana Agr. Expt. Sta. Bull. 248. 63 pp.

19599. —— 1936: Interpretation of germination results: transfer of germinated seeds from blotters to field soil. Proc. Assoc. Offic. Seed Analysts 1935:73–77.

19600. —— 1936: Weight per bushel of wheat in relation to its seed value. Proc. Assoc. Offic. Seed Analysts 28:59–61.

19601. —— 1937: Hard seeds in legumes: correlation between laboratory and field germination. Proc. Assoc. Offic. Seed Analysts 29:138–45.

19602. —— 1938: Identification of seeds of yellow-blossom and white-blossom sweet clover. Report of progress. Proc. Assoc. Offic. Seed Analysts 23/26:196–99, 254–62.

19603. —— 1939: Seed value of de-hulled oats—preliminary. Proc. Assoc. Offic. Seed Analysts 30:255–58.

19604. —— 1939: A suggested policy for the evaluation of hard seeds. Proc. Assoc. Offic. Seed Analysts 31:86–94.

19605. —— and P. F. Sharp. 1925: Germination of frozen and nonfrozen wheat harvested at various stages of maturity. J. Agr. Res. 31:1179–88.

19606. —— and B. B. Thompson. 1943: Analysis of seed with the Rice laboratory seed cleaner—preliminary tests. Proc. Assoc. Offic. Seed Analysts 34:70–72. Biol. Abstr. 20:1504. 1946.

19607. White, C. G. 1945: Experiments with seed germination. Am. Iris Soc. Bull. 97:46.

19608. White, D. 1934: The seeds of *Supaia*, a Permian Pteridosperm. Science 79:462.

19609. White, D. G. 1947: Longevity of bamboo seed under different storage conditions. Trop. Agr. (Trinidad) 24:51–53.

White, D. G. 12434.

19610. White, E. P. 1946: Alkaloids of the Leguminosae. XVI–XVII. New Zealand J. Sci. Technol. Sect. B 27:474–85. Biol. Abstr. 21:20721. 1947.

19611. White, G. D. 1953: Weight loss in stored wheat caused by insect feeding. J. Econ. Entomol. 46:609–10.

White, G. D. 4445.

White, H. B. 12644.

19612. White, H. B., Jr. 1958: Fat utilization and composition in germinating cotton seeds. Plant Physiol. 33:218–26.

19613. ——, F. W. Quackenbush, and A. H. Probst. 1961: Occurrence and inheritance of linolenic and linoleic acids in soybean seeds. J. Am. Oil Chemists Soc. 38: 113–17.

19614. White, H. E. 1940: Easter lilies from seed. Mass. Agr. Expt. Sta. Bull. 376: 15–16.

19615. White, H. L. 1938: The sterilization of lettuce seed. Ann. Appl. Biol. 25: 767–80.

19616. —— 1940: Chemical treatment of lettuce seed. Nursery & Market Garden Ind. Dev. Soc. Expt. Res. Sta. Ann. Rept. 25:49–50.

19617. White, J. 1908: The occurrence of an impermeable cuticle on the exterior of certain seeds. Proc. Roy. Soc. Victoria 21:203–10.

19618. —— 1909: The ferments and latent life of resting seeds. Proc. Roy. Soc. (London) Ser. B 81:417–42.

White, J. D. 7963.

White, J. W. 5426.

19619. White, L. N. 1959: Pre-conditioned seed in protective packages. Seed World 85(12):4, 5–6.

19620. White, O. E. 1946: The germination of peas in Florida and King Tut's tomb. Turtox News 24(1). 3 pp.

White, P. 445.

19621. White, W. J., and T. M. Stevenson. 1948: Permeable seeded strains of sweet clover (*Melilotus alba*). Their development and nature. Sci. Agr. (Ottawa) 28:206–22.

White, W. J. 1364, 5524, 10698.

19622. Whitehead, E. I., and G. F. Gastler. 1945: Hygroscopic moisture of grain sorghum and wheat as influenced by temperature and humidity. Proc. S. Dakota Acad. Sci. 25:80–84.

19623. Whitehead, G. E. 1956: Freesias from seeds. Gardeners Chron. 139:389.

Whitehead, J. K. 4434.

Whitehead, M. D. 7705, 10091, 10092.

19624. Whitehead, M. R., and C. A. Brown. 1940: The seed of the spider lily, *Hymenocallis occidentalis*. Am. J. Botany 27: 199–203.

19625. —— and J. W. Mitchell. 1943: Effects of nutrient, photoperiod, and night temperature on the development of guayule seeds. Botan. Gaz. 105:14–24.

Whitehouse, J. 11826.

19626. Whiteside, A. G. O. 1936: Statistical significance of wheat protein percentage differences in varietal trials. Can J. Res. Sect. C 14:387–93.

19627. ——, J. Edgar, and C. H. Goulden. 1934: The influence of environment on the carotenoid content of hard red spring wheat. Cereal Chem. 11:615–24.

Whiteside, A. G. O. 9633, 16690, 19175.

Whiting, A. G. 13594, 13598, 13599.

19628. Whiting, F. 1954: The feeding value of damaged grain. Can. Dept. Agr. (Ottawa), Exptl. Farms Serv. Publ. 914. 11 pp.

19629. Whiting, M. G. 1963: Toxicity of cycads. Econ. Botany 17:271–302.

19630. Whitmore, E. T. 1960: Rapid method for determination of the husk content of barley and oats. J. Inst. Brewing 66:407–8.

Whitmore, F. W. 15134.

Whitney, G. M. 9191.

Whitney, L. D. 10613.

19631. Whitney, W. K., O. K. Jantz, and C. S. Bulger. 1958: Effects of methyl bromide fumigation on the viability of barley, corn, grain sorghum, oats, and wheat seeds. J. Econ. Entomol. 51: 847–61.

19632. Whitten, M. E., and L. A. Baumann. 1963: Evaluation of a rapid method for determining oil content of cottonseed. U.S. Dept. Agr. Tech. Bull. 1298. 19 pp.

19633. Whittet, J. N. 1936: Effect of fertilizers on some pasture plant seeds. Agr. Gaz. N.S. Wales 47:7–8. Chem. Abstr. 30:3148. 1936.

19634. —— 1943: Giant panic grass seed retains its viability. Agr. Gaz. N.S. Wales 54:494.

19635. Whymper, R., and A. Bradley. 1934: Studies on the vitality of wheat. I. Prolongation of vitality of wheat seeds. Cereal Chem. 11:349–60.

19636. —— and —— 1934: Studies on the vitality of wheat. II. Influence of moisture in wheat seeds upon imbibition and speed of germination. Cereal Chem. 11:546–50.

19637. —— and —— 1934: Studies on the vitality of wheat. III. Vitality, and the action of heat on wheat seeds. Cereal Chem. 11:625–36.

19638. —— and —— 1947: A note on the viability of wheat seeds. Cereal Chem. 24:228–29.

19639. Whyte, R. O., and J. J. De Jong. 1961: FAO seed campaign. World Crops 13:17–21.

19640. —— and P. S. Hudson. 1933: Vernalization or Lyssenko's method for the pre-treatment of seed. Imp. Bur. Plant Genet. (Gt. Brit.) Bull. 9:1–27.

Whyte, R. O. 367, 13962, 17531.

Wiant, D. E. 4612, 4613.

Wiazecka, K. 12334.

19641. Wibeck, E. 1919: Pine and spruce seed from Norrland. Skogen 6:97–107. Botan. Abstr. 4:483. 1920.

19642. —— 1920: The germination capacity of pine seed from Norrland. Statens Skogsforsoks. Med. 17:1–20. Expt. Sta Rec. 46:142. 1922.

19643. —— 1920: The relation of various forest planting methods to frost injury. Statens Skogsforsoks. Med. 17:329–47. Expt. Sta. Rec. 46:236. 1922.

19644. Wiberg, H. 1959: Viability of wild oat kernels in a compost. Växtodling 10:54–57.

19645. —— 1961: Some experiences concerning seed mixtures. Statens Cent. Frökontanst. Medd. 36:53–67.

19646. —— and H. Kolk. 1960: Effect of gibberellin on germination of seeds. Proc. Intern. Seed Testing Assoc. 25:440–44.

Wiberg, H. 10138.

Wickenden, G. 14920.

19647. Wickramasekera, G. V. 1939: The multiple seeded variety of rice (*Oryza sativa* L.). Trop. Agriculturist 92:279–80. Plant Breeding Abstr. 9:381. 1939.

19648. Widen, M. J. 1921: Die Keimung des Getreides. 3d Intern. Seed Testing Conf., Copenhagen. J. Seed Testing 1:88–92. 1922.

Wiebe, G. A. 45, 3131, 5535.

Wiebe, R. 2619.

Wiebeck, R. 6752.

Wiebenga, E. H. 5576.

19649. Wiedemann, E. 1930: Die Versuche über den Einfluss der Herkunft des Kiefernsamens. Z. Forst.-Jagdw. 62:498–522, 809–36. Biol. Abstr. 6:4958. 1932.

Wiedow, H. L. 19212.

19650. Wiegand, C. L. 1962: Drying patterns of a sandy clay loam in relation to optimal depths of seeding. Agron. J. 54:473–76.

Wiegand, E. H. 13336.

Wiele, M. B. 13525.

19651. Wieliczka, Z. 1950: Selection of seed for the planting of pine trees in Brazil. Anuar. Brasil. Econ. Florestal 3:202–7.

19652. Wiemans, C. E., and F. L. Westerweel. 1957: Note on the presence of lipoxidase in germinating sunflower seeds. Enzymologia 18:145–48.

19653. Wientjes, K. 1920: Acceleration of germination under the influence of acids. Rev. Trav. Botan. Neerl. 17:33–68.

19654. Wieringa, G. 1931: Die Dauer der Keimversuche. Proc. Intern. Seed Testing Assoc. 15/17:51–59.

Wieringa, G. 6637, 6638.

19655. Wieringa, K. T. 1954: Red discolorations observed on peas and beans. Tijdschr. Plantenziekten 60:259–60. Rev. Appl. Mycol. 34:695. 1955.

19656. Wierszyllowski, J. 1960: Notes on the anatomical structure and chemical composition of the seeds of apple and pear. Roczniki Nauk Rolniczych Ser. A 81:205–28.

19657. —— 1961: Water and synthetic growth substances (IAA, 2,4-D) uptake by dormant and stratified pear-seeds. Roczniki Nauk Rolniczych Ser. A 83:549–86.

19658. —— and C. Katulska. 1957: Comparison of two methods of soaking of germinated seeds of *Prunus cerasifera divaricata*. Roczniki Nauk Rolniczych Ser. A 76:323–34.

19659. ——, Z. Rusek, and M. Borys. 1958: Observations on germination of stratified Caucasian pear seeds subjected to the action of growth substances, low temperatures, and soaking water. Roczniki Nauk Rolniczych Ser. A 77:503–17.

19660. Wiese, A. F., and H. E. Rea. 1959: Bindweed (*Convolvulus arvensis* L.) control and seedling emergence as affected by tillage, 2,4-D, and competitive crops. Agron. J. 51:672–75.

19661. Wiesehuegel, E. G. 1935: Germinating Kentucky coffee tree. J. Forestry 33:533–34.

19662. Wieser, G. 1927: Der Einfluss des Sauerstoffs auf die Lichtwirkung bei der Keimung Lichtempfindlicher Samen. Planta 4:526–72.

19663. Wiesner, J. 1894: Vergleichende physiologische Studien über die Keimung europäischer und tropischer Arten von *Viscum* und *Loranthus*. Sitzber. Wiss. Wien. Math.-Naturw. Kl. 103:401–37.

19664. —— 1897: Über die Ruheperiode und über einige Keimungsbedingungen der Samen von *Viscum album*. Ber. Deut. Botan. Ges. 15:503–15.

Wiesner, L. E. 10642.

19665. Wiewiorowski, M., J. Augustyniak, and H. Skrzypinska. 1958: Lupine seed proteins. II. Concentration of some nitrogen forms in seed of *Lupinus albus*, *Lupinus luteus*, and *Lupinus angustifolius*. Roczniki Nauk Rolniczych Ser. A 79:19–31. Chem. Abstr. 53:11541. 1959.

19666. Wiggans, C. B. 1930: Case of second crop seedless apples. Plant Physiol. 5:273–77.

19667. Wiggans, R. G. 1934: Cayuga soybean: A home-grown, high-oil protein concentrate. Cornell Univ. Agr. Expt. Sta. Bull. 601. 32 pp.

Wiggans, R. G. 17252.

19668. Wiggans, S. C. 1956: The effect of seasonal temperatures on maturity of oats planted at different dates. Agron. J. 48:21–25.

19669. —— 1958: The relationship of total nitrogen percentage to yield, test weight, and groat percentage of oats grown at different locations. Proc. Iowa Acad. Sci. 65:197–200.

19670. —— 1959: Total protein in several oat varieties grown in Iowa. Proc. Iowa Acad. Sci. 66:143–44.

19671. —— and K. J. Frey. 1958: The ratio of alcohol-soluble to total nitrogen in developing oat seeds. Cereal Chem. 35–235–39.

19672. —— and F. P. Gardner. 1959: Effectiveness of various conditions for simulating drouth conditions as measured by germination and seedling growth. Agron. J. 51:315–18.

19673. —— and L. W. Martin. 1961: The effect of gibberellic acid on germination and seedling growth of pecans. Proc. Am. Soc. Hort. Sci. 77:295–300.

19674. ——, D. S. Metcalfe, and H. E. Thompson. 1956: The use of desiccant sprays in harvesting birdsfoot trefoil for seed. Agron. J. 48:281–84.

19675. —— and R. H. Shaw. 1958: The effect of high temperatures at varying stages of growth on kernel production in oats. Proc. Iowa Acad. Sci. 65:201–5.

Wiggans, S. C. 6731, 6733, 13365.

19676. Wight, R. E. 1962: Recent changes in Canadian seed legislation. Proc. Assoc. Offic. Seed Analysts 51:172–73.

19677. Wikoff, H. L., J. M. Kaplan, and A. L. Berman. 1944: The occurrence of some previously unreported fatty acids in peanut oil. J. Biol. Chem. 153:227–35. Expt. Sta. Rec. 91:629. 1944.

19678. Wiksell, G. 1931: Über die Bestimmung der Anzahl Unkrautsamen pro Kilo in Timotheesämereien. Statens Cent. Frökontanst. Medd. 6:120–26.

19679. —— 1949: On determinations of genuineness of seeds of *Brassica* species. Preliminary report. Statens Cent. Frökontanst. Medd. 24:46–52.

19680. Wilbaux, R. 1934: Composition and toxic properties of the seeds and leaves of *Tephrosia vogelii* Hook. f. Rev. Botan. Appl. Agr. Trop. 14:1019–27. Chem. Abstr. 29:4794. 1935.

19681. Wilbur, D. A., and G. Halazon. 1955: Pests of farm stored wheat and their control. Kansas Agr. Expt. Sta. Bull. 371. 28 pp.

Wilbur, D. A. 4445.

19682. Wilcke, J. 1954: Dissemination of seeds of Indian pipe and *Pyrola.* Levende Nat. 57:9–11.

19683. Wilcox, E. B., and K. E. Morrell. [no date]: The vitamin content of peas as influenced by maturity, fertilizers, and variety. Utah Agr. Expt. Sta. Bull. 337. 16 pp.

Wilcox, E. B. 8645.

Wilcox, F. B. 9784.

Wilcox, M. S. 5173, 13596.

19684. Wilczek, E. 1931: La dissemination des *Galinsoga.* Bull. Soc. Vaudoise Sci. Nat. 57:253–54.

19685. Wilde, M. H., and A. J. Eames. 1948: The ovule and "seed" of *Araucaria bidwilli* with discussion of the taxonomy of the genus. I. Morphology. Ann. Botany (London) 12:311–26.

19686. —— and —— 1955: The ovule and "seed" of *Araucaria bidwilli* with discussion of the taxonomy of the genus. III. Anatomy of multi-ovulate cone scales. Ann. Botany (London) 19:343–49.

19687. Wilde, S. A. 1937: Recent findings pertaining to the use of sulfuric acid for the control of damping-off disease. J. Forestry 35:1106–10.

Wildeman, E. de; *see* De Wildeman, E.

Wilder, E. A. 10733.

19688. Wildflush, R. T. 1939: The effect of hydrochloric acid and of potassium hydroxide on the formation of bios in vernalized embryos. Tr. Belorussk. Sel'skokhoz. Inst. 8(30):41–44. Chem. Abstr. 35:5160. 1941.

19689. —— 1939: Der Zustand der Aktivität von Katalase bei der Jarowisation der Samen. Arb. Weissruss. Landwirtsch. Inst. 8(30):35–40. Botan. Centralbl. 176: 329–30. 1941.

19690. —— 1939: Teneur en phytine des graines de lupin. Ann. White Russ. Agr. Inst. 8:49. Ann. Agron. 10:181–82. 1940.

19691. Wileman, R. H., and A. J. Ullstrup. 1945: A study of factors determining safe drying temperatures for seed corn. Indiana Agr. Expt. Sta. Bull. 509. 10 pp. Expt. Sta. Rec. 95:195. 1946.

19692. Wiles, A. B., and J. T. Presley. 1960: Seed deterioration as a factor in nub-root production in cotton. Plant Disease Reptr. 44:472–77.

Wiles, L. U. 8929.

19693. Wiley, R. H., A. W. Cagle, and P. H. Wilken. 1951: Spectrophotometric studies of the composition of lespedeza seed oil. J. Am. Oil Chemists Soc. 28:459–61. Biol. Abstr. 26:2314. 1952.

19694. Wilgus, F., and K. C. Hamilton. 1962: Germination of salt cedar seed. Weeds 10:332–33.

19695. Wilhelm, S., and L. C. Benson. 1954: Vertical distribution of broomrape seed in tomato field soil. Plant Disease Reptr. 38:553–54.
Wilhelm, S. 14226.
Wilhite, M. L. 198.
Wilken, P. H. 19693.
Wilkes, L. H. 17751.
Wilkinson, B. 16481.
Wilkinson, E. H. 19281.
Wilkinson, K. J. 18033.
19696. Will, H. 1882: Über den Einfluss des Einquellens und Wiederaustrocknens auf die Entwicklungsfahigkeit der Samen, sowie über den Gebrauchswert, "ausgewachsener" Samen als Saatgut. Landwirtsch.-Versuchssta. 28:51–89. Botan. Centralbl. 11:414–16. 1882.
19697. Willard, C. J. 1923: The effect of CS_2 on germination. J. Econ. Entomol. 16:388–92.
Willard, M. A. 458.
19698. Williams, B. C., and F. W. Smith. 1954: The effects of different rates, times, and methods of application of various fertilizer combinations on the yield and quality of Hard Red winter wheat, 1949–1950. Soil Sci. Soc. Am. Proc. 18:56–60.
Williams, C. B. 19414.
Williams, E. 16181.
19699. Williams, E. J. 1955: Seed failure in the Chippewa variety of *Solanum tuberosum*. Botan. Gaz. 117:10–15.
19700. Williams, G. C., and J. M. Thurston. 1964: The effect of temperature in a sack-drier on survival of insects (*Oryzaephilus surinamensis* L.) (Col., Silvanidae), and weed seeds (*Avena fatua* L. and *A. ludoviciana* Dur.). Ann. Appl. Biol. 53:29–32.
Williams, G. G. 16317.
Williams, H. H. 17619.
19701. Williams, J. H., and D. G. Hanway. 1961: Genetic variation in oil and protein content of soybeans induced by seed irradiation. Crop Sci. 1:34–36.
19702. Williams, J. S. 1950: Effect of cosmic rays on seed being studied. Southern Seedsman 13(10):29.
19703. Williams, K. T., and A. Bevenue. 1958: Carbohydrate studies of the lima bean. J. Assoc. Offic. Agr. Chemists 41:822–27.
19704. Williams, L. F. 1952: The inheritance of certain black and brown pigments in the soybean. Genetics 37:208–15.
Williams, L. F. 19486.
19705. Williams, M. 1938: The moisture content of grass seed in relation to drying and storing. Welsh J. Agr. 14:213–32.
19706. —— 1949: Germination of redbud seeds. Calif. Hort. Soc. J. 10:70–72.
19707. —— and G. Evans. 1938: The rapid determination of moisture content in grass seeds. Welsh J. Agr. 14:232–44.
19708. Williams, M. C. 1960: Biochemical analyses, germination, and production of black and brown seed of *Halogeton glomeratus*. Weeds 8:452–61.
19709. Williams, M. E. 1932: The development of the embryo of *Kochia scoparia*. Bull. Torrey Botan. Club 59:391–400.
Williams, M. Wynn-; *see* Wynn-Williams, M.
19710. Williams, R. C., and B. C. Webb. 1958: Seed moisture relationships and germination behavior of acid-scarified Bahia grass seed. Agron. J. 50:235–37.
19711. Williams, R. D. 1930: Some of the factors influencing yield and quality of red clover seeds. Welsh Plant Breeding Sta. Ser. H 11:60–91.
Williams, R. D. 5977.
Williams, R. W. M. 2438.
19712. Williams, S. S. 1954: The effect of depth of sowing and moisture on the germination and seedling development of *Phleum pratense* L. J. Ecol. 42:445–49.
Williams, V. R. 6330.
19713. Williams, W. 1944: Control of buried viable weed seeds by means of boron. Nature 154:771–72.
19714. Williams, W. A. 1963: The emergence force of forage legume seedlings and their response to temperature. Crop Sci. 3:472–74.
19715. —— and J. R. Elliott. 1960: Ecological significance of seed coat impermeability to moisture in crimson, subterranean and rose clovers in a Mediterranean-type climate. Ecology 41:733–42.
19716. ——, L. H. Harwood, and F. J. Hills. 1960: Incompatibility of seed treatment fungicides and seed-applied legume inoculum observed on field-grown subterranean clover. Agron. J. 52:363–65.
19717. —— and O. A. Leonard. 1959: Effect of 2,4-D on the growth, seed production, and seed viability of rose clover. Agron. J. 51:383–87.
Williams, W. O. 19755.
19718. Williams, W. T., and P. C. Sharma. 1954: The function of urease in *Citrullus* seeds. J. Exptl. Botany 5:136–57.
Williamson, D. 17843.

Willier, J. G. 3216.
Willimott, S. G. 19795.
19719. Willis, J. C. 1892: Distribution of the seed in *Claytonia.* Ann. Botany (London) 6:382–83.
19720. Willis, S. J. 1954: Weed control in farm and garden. Vinton, London. 184 pp.
Willits, C. O. 4675.
19721. Willstaedt, H. 1934: Lyochromic color substance from linseed. Svensk Kem. Tidskr. 46:259–61. Chem. Abstr. 29:2576. 1935.
19722. Wilsie, C. P., and C. S. Reddy. 1946: Seed treatment experiments with hemp. J. Am. Soc. Agron. 38:693–701.
Wilsie, C. P. 2383, 3050, 15152.
19723. Wilsing, H. 1884: Stoffumsatz und Kraftumsatz im keimenden Samen. J. Landwirtsch. 32:523–38.
Wilson, A. 12.
19724. Wilson, A. M. 1949: The prevalence of weed seeds in seed grain. Proc. W. Can. Weed Control Conf. 3:154–56.
19725. Wilson, C. 1950: Peanut seed treatments. Plant Disease Reptr. 34:87–95.
Wilson, C. C. 12381.
19726. Wilson, C. P. 1928: Factors affecting the germination and growth of chamiza (*Atriplex canescens*). New Mexico Agr. Expt. Sta. Bull. 169. 29 pp.
19727. Wilson, E. H. 1929: Germinating seeds of the dovetree. Horticulture 7(9): 223.
19728. Wilson, F. C. 1950: Sand, peat, and vermiculite. Gardeners Chron. 54:46–47.
Wilson, G. 7917.
19729. Wilson, G. L., and L. F. Notley. 1959: Seed soaking as a method of supplying the molybdenum requirements of lettuce and tomatoes. Australian J. Agr. Res. 10:621–27.
19730. Wilson, H. K. 1928: Wheat, soybean, and oat germination studies with particular reference to temperature relationships. J. Am. Soc. Agron. 20:599–619.
19731. —— and C. F. Hottes. 1927: Wheat germination studies with particular reference to temperature and moisture relationships. J. Am. Soc. Agron. 19:181–90.
Wilson, H. K. 9095.
Wilson, J. H. 19177.
19732. Wilson, J. K. 1915: Calcium hypochlorite as a seed sterilizer. Am. J. Botany 2:420–27.
19733. —— 1929: Acidity changes in stored legume seeds. J. Am. Soc. Agron. 21:815–17.
19734. —— 1937: Scarification and germination of black locust seeds. J. Forestry 35:241–46.
19735. —— 1944: Immersing seeds of species of *Robinia* in boiling water hastens germination. J. Forestry 42:453–54.
Wilson, K. M. 17807.
19736. Wilson, K. S., and V. M. Cutler, Jr. 1955: Localization of acid phosphatase in the embryo sac and endosperm of *Cocos nucifera.* Am. J. Botany 42:116–19.
Wilson, K. S. 4787.
Wilson, L. I. 17613.
19737. Wilson, L. R. 1954: Teaching aids in weed plant and seed identification. Weeds 3:342–44.
Wilson, R. D. 9321.
19738. Wilson, R. E. 1921: Humidity control by means of sulfuric acid solutions, with critical compilation of vapor pressure data. J. Ind. Eng. Chem. 13:326–31.
19739. Wilson, S. L. 1959: Pigments of the embryo of the yellow Nelumbo. Trans. Illinois State Acad. Sci. 51:30–33.
19740. Wilson, T. L., T. K. Miwa, and C. R. Smith, Jr. 1960: *Cuphea llavea* seed oil, a good source of capric acid. J. Am. Oil Chemists Soc. 37:675–76.
19741. ——, C. R. Smith, Jr., and K. L. Mikolajczak. 1961: Characterization of cyclopropenoid acids in selected seed oils. J. Am. Oil Chemists Soc. 38:696–99.
19742. ——, ——, and I. A. Wolff. 1962: *Lunaria* seed oil—a rich source of C_{24} fatty acids. J. Am. Oil Chemists Soc. 39:104–5.
19743. Wilson, V. E., and L. L. Dean. 1964: Flour of infected bean seed as a source of virus. Phytopathology 54:489.
Wilson, W. F. 9061.
Wilusz, T. 12620.
19744. Winch, J. E., and H. A. MacDonald. 1961: Flower, pod and seed development relative to the timing of the seed harvest of Viking Birdsfoot trefoil (*Lotus corniculatus*). Can. J. Plant Sci. 41:523–32.
Winch, N. H. 9457.
19745. Winchester, J. A., and E. G. Rodgers. 1953: Comparative efficiency of several herbicidal treatments in control of nutgrass (*Cyperus rotundus*). Proc. Southern Weed Conf. 6:179–82.
19746. Winchester, W. J. 1954: Storing seed of green panic and buffel grass for better germination. Queensland Agr. J. 79: 203–4.
19747. Windisch, F. 1942: Importance of determining the energy of germination of

barley. Brautechnol. 1942:20. Biol. Abstr. 18:11299. 1944.

19748. Windisch, W. 1905: Warum keimt die getrocknete bezw. Abgelagerte Gerste besser als die frisch Geerntete? Wochschr. Brau. 22:89–92.

Windischbauer, R. 10723.

Windrum, G. M. 4287.

19749. Winfield, R. W. 1942: The effect of depth of planting on fuzzy and acid-delinted cotton seed. Phytopathology 32: 649.

19750. Wingren, E. 1937: Über die Keimfähigkeit von *Taraxacum officinale*. Lantbrukshögsk. Ann. 4:221–31.

19751. Winjum, J. K., and I. D. Ellis. 1960: Miniature cone tumbler for seed research. Weyerhauser Co. Forest Res. Note 26. 2 mimeo. pp.

19752. —— and N. E. Johnson. 1960: A modified-knife cone cutter for Douglas fir seed studies. J. Forestry 58:487–88.

19753. Winkelmann, A. 1932: Untersuchungen über die Haftfähigkeit von Trockenbeizmitteln. Angew. Chem. 45:238–41. Rev. Appl. Mycol. 11:502–3. 1932.

Winkelmann, K. 16933.

19754. Winkler, A. 1883: Bemerkungen über die Keimpflanze und die Keimfähigkeit des Samens von *Tithymalus cyparissias* Scop. Ber. Deut. Botan. Ges. 1:452–55.

19755. Winkler, A. J., and W. O. Williams. 1936: Effect of seed development on the growth of grapes. Proc. Am. Soc. Hort. Sci. 33:430–34.

19756. Winkler, R. E. 1954: Die Glykoside der Samen von *Dregea volubilis* Benth. Helv. Chim. Acta 37:721–55.

Winstanley, J. 4994.

19757. Winter, A. G., and F. Schönbeck. 1957: Untersuchungen über den Einfluss von Stickstoff auf den Abbau von Hemmstoffen aus Getreide-stroh und -stoppeln. Z. Pflanzenernähr. Düng. Bodenk. 79:132–42.

19758. —— and E. Sievers. 1952: Untersuchungen über die Beeinflussung der Samenkeimung durch Kaltwasserextrakte aus der Blattstreu verschiedener Gramineen. Naturwissenschaften 39:191–92.

19759. Winter, D. M. 1960: The development of the seed of *Abutilon theophrasti*. I. Ovule and embryo. Am. J. Botany 47:8–14.

19760. —— 1960: The development of the seed of *Abutilon theophrasti*. II. Seed coat. Am. J. Botany 47:157–62.

Winter, F. de; *see* De Winter, F.

19761. Winter, H. 1957: The protein of the seeds of parsley (*Petroselinum sativum*). Arch. Pharm. 290:200–3. Chem. Abstr. 51:18160. 1957.

Winter, H. 17016, 17017.

19762. Winterringer, G. S. 1955: The harpoon seeds. Wild Flower 31(2):24–25.

19763. Winters, H. F., and F. Rodriguez-Colon. 1953: Storage of mangosteen seed. Proc. Am. Soc. Hort. Sci. 61:304–6.

Winters, J. S. 1701.

19764. Winters, R. K. 1933: Weight of fruit of Nuttall's oak. J. Forestry 31:340.

Winterstein, E. 17078.

Winzer, J. W. 3400.

19765. Wirjoatmodjo, K. 1964: Seed germination of *Myristica macrothyrsa* Mig. Botan. J. 49:63–64.

19766. Wirjomidjojo, R. 1961: The effect of radioactive cobalt on the germination of seeds. Menara Perkebunan 30(5/6):89–90, 102.

19767. Wirth, W. 1955: Cacao bean investigations in 1952–1954. Zucker-Süsswarenwirtsch. 8:801–9. Chem. Abstr. 50:1337. 1956.

19768. Wirtinger, M. 1927: Der Einfluss äusserer Umstände (Zeit, Temperatur, Feuchtigkeit usw.) auf die Keimkraft ruhender Samen. Fortschr. Landwirtsch. 2:321–25.

19769. Wise, Eugene P. 1956: Method of controlling seed germination by means of a composite seed assembly. U.S. Patent 2,736,139.

19770. Wise, L. N. 1957: Tom Thumb Seed Laboratory. Crops & Soils 9(5):21.

Wisselingh, C. van; *see* Van Wisselingh, C.

19771. Wister, J. C. 1946: Tree peony. Arnoldia 6(2):5–8. Biol. Abstr. 20:20983. 1946.

19772. Withner, C. L. 1955: Ovule culture and growth of vanilla seedlings. Am. Orchid Soc. Bull. 24:380–92.

19773. Withrow, R. B., and H. M. Benedict. 1935: An improved method for seed germination. Science 81:439. Biol. Abstr. 10:17980. 1936.

Withrow, R. B. 10780.

Withus, E. R. 2306.

Witnauer, L. P. 16243.

Witsch, H. von. 8160.

Witschi, E. 3811.

19774. Witte, H. 1928: Ought the content of weed seeds in a seed sample to be expressed in percentage by weight or in number per kg.? Intern. Seed Testing Conf. Rome 5:259–64.

19775. —— 1931: Is it possible to formulate some universal rules for an estimate of the practical value of hard leguminous seeds? Proc. Intern. Seed Testing Assoc. 3:127–34.

19776. —— 1931: Some investigations on the germination of hard seeds of red clover, alsike clover, and some other leguminous plants. Proc. Intern. Seed Testing Assoc. 3:135–47.

19777. —— 1934: Bericht über einige internationale Keimungsuntersuchungen verschiedener Samenproben im Jahre 1934. Proc. Intern. Seed Testing Assoc. 6:173–82.

19778. —— 1935: Einige Untersuchungen über die Keimungsverhältnisse verschiedener *Plantago*-Arten. Schwed. Botan. Z. 29:513–33. Proc. Intern. Seed Testing Assoc. 9:304. 1937.

19779. —— 1938: New international investigations regarding the germination of hard leguminous seeds. Proc. Intern. Seed Testing Assoc. 10:93–122.

19780. —— 1939: The occurrence of noxious weed seeds in samples of red clover, alsike clover and timothy of Swedish origin. Medd. Statens Centr. Frökontr. 14:81–88.

19781. —— 1940: On the germination of hard seeds of hairy vetch (*Vicia villosa* Roth.). Proc. Intern. Seed Testing Assoc. 12:39–46.

19782. —— 1940: On the variation of dry 1000-seed-weight of timothy in different parts of Sweden and in different varieties. Medd. Statens Centr. Frökontr. 15:71–76.

19783. —— 1941/43: Some international investigations regarding the germination of hard clover seeds in soil. Proc. Intern. Seed Testing Assoc. 13:77–91.

19784. Witmack, L. 1896: Germination of the cocoanut. Ber. Deut. Botan. Ges. 14:145–50.

19785. —— 1910: The use of mustard-seeds in Chile. Ber. Deut. Botan. Ges 28:77–78.

19786. —— 1922: Landwirtschaftliche Samenkunde. Paul Parey, Berlin. 581 pp.

19787. Wittmer, G. 1961: Cytological activity of fresh and aged seeds of *Nicotiana tabacum*. Tobacco 65:229–37. Chem. Abstr. 56:1775. 1962.

19788. —— 1961: Cytologic activity of some fatty acids extracted from fresh and aged seeds of *Nicotiana*. Tabacco 65:434–40. Chem. Abstr. 57:7640. 1962.

Witt, J. L. de; *see* De Witt, J. L.

19789. Witts, K. J. 1960: The germination of *Polygonum* species in the field and in the glass-house. J. Ecol. 48:215–17.

19790. Wittwer, S. H., and M. J. Bukovac. 1957: Gibberellin and higher plants. VIII. Seed treatments for beans, peas, and sweet corn. Mich. Agr. Expt. Sta. Quart. Bull. 40:215–24.

Wittwer, S. H. 1808, 3308, 4149, 6827, 6828, 7916, 8838, 13963, 15987, 15988, 16048.

19791. Wladyslaw, K. 1960: Germination biology of spring adonis (*Adonis vernalis* L.). Roczniki Nauk Rolniczych Ser. A 81:337–82. Biol. Abstr. 36:32071. 1961.

Wlodkowski, M. 2487, 2488.

19792. Wocicki, Z. 1931: Über den Keimungsprozess der Samen und den Degenerationscharakter der Samenanlagen bei *Haemanthus katharinae*. Acta Soc. Botan. Polon. 8:85–108.

Wöger, K. 6380.

Wöhrmann, K. 13407.

19793. Wojtysiak, A., and Z. Jasinska. 1959: Soybeans in field experiments. II. The influence of sown quantity and sowing dates of three varieties of soybean on constancy of crops and chemical composition of seeds. Zeszyty Nauk. Wyzszej Szkoly Rolniczej Wroclawiu 9:19–41. Chem. Abstr. 54:15544. 1960.

19794. Wokes, F., and R. Melville. 1949: Apparent vitamin C in the walnut (*Juglans regia*). Biochem. J. 45:343–52.

19795. —— and S. G. Willimott. 1951: Determination of cyanide in seeds. J. Pharm. Pharmacol. 3:905–16. Chem. Abstr. 46:2608. 1952.

Wokes, F. 4813, 10759, 10760, 13318.

19796. Wold, A. 1957: Methods and instruments for sampling red clover seed in bags. Proc. Intern. Seed Testing Assoc. 22:465–75.

19797. Wolf, C. B. 1941: Germination of California native seeds. Am. J. Botany 28:729.

19798. Wolf, M. J., M. M. MacMasters, J. E. Hubbard, and C. E. Rist. 1948: Comparison of corn starches at various stages of kernel maturity. Cereal Chem. 25:312–25.

Wolf, M. J. 2930, 3565, 12654, 17160, 17161.

Wolf, W. J. 5887

19799. Wolfe, A. C., J. B. Park, and R. C. Burrell. 1942: A study of the chemical composition of soybeans during maturation. Plant Physiol. 17:289–95.

Wolfe, A. C. 3410.
Wolfe, J. E. 106.

19800. Wolfe, M., and L. Fowden. 1957: Composition of the protein of whole maize seeds. Cereal Chem. 34:286–95.

Wolfe, M. 6578.
Wolfe, T. A. de. 10851.

19801. Wolfe, T. K. 1927: A study of germination, maturity, and yield in corn. Virginia Polytech. Inst. Agr. Expt. Sta. Tech. Bull. 30. 35 pp.

19802. Wolff, F. 1933: Eine Laboratoriumsmethode zur schnellen Prüfung von Saatgut-beizmitteln (bes. zur Fusarium-bekämpfung). Pflanzenbau 10:228–33. Proc. Intern. Seed Testing Assoc. 7:79–80. 1935.

Wolff, I. A. 5051, 5052, 5745, 13453, 13454, 13613, 13614, 17628, 18646, 18972, 18973, 19742.

19803. Wolff, J. 1924: Nouvelles observations sur le perte du pouvoir germinatif des semences d'Orchidées. Compt. Rend. 179:67–68.

Wolffgang, H. 6767.

19804. Wolffsohn, A. L. A. 1961: An experiment concerning mahogany germination. Empire Forestry Rev. 40:71–72.

19805. Wolfrom, M. L., R. A. Plunkett, and M. L. Laver. 1960: Coffee constituents. Carbohydrates of the coffee bean. J. Agr. Food Chem. 8:58–65.

19806. Wollenweber, H. W. 1942: Über die Lebensdauer von Kartoffelsamen. Angew. Botan. 24:259–60. Biol. Abstr. 23:16854. 1949.

19807. Wollgiehn, R. 1960: Untersuchungen über den Zusammenhang zwischen Nuclein säure- und Eiweisstoffwechsel in reifenden Samen. Flora (Jena) 148:479–83.

19808. Wolter, H. 1933: Über Bestäubung, Fruchtbildung und Keimung bei *Ficaria verna*. Beitr. Biol. Pflanz. 21:219–55.

19809. Wolters, K. 1956: Ultraschall in der Landwirtschaft. Landbauforsch. Völkenrode 6:18–19.

19810. Wong, C. Y. 1940: The influence of pollination on seed development in certain varieties of citrus. Proc. Am. Soc. Hort. Sci. 37:161–64.

19811. —— 1941: Chemically induced parthenocarpy in certain horticultural plants, with special reference to the watermelon. Botan. Gaz. 103:64–86. Biol. Abstr. 16:8001. 1942.

Wong, P. 10851.

19812. Woo, M. L. 1919: Chemical constituents of *Amaranthus retroflexus*. Botan. Gaz. 68:313–44.

Wood, B. Mounsey-; *see* Mounsey-Wood, B.

19813. Wood, C. R. 1941: A study of the chlorophyll, xanthophyll, and carotene contents of the wheat kernel harvested at successive stages of development. Sci. Agr. (Ottawa) 22:93–103.

19814. Wood, C. W. 1941: Seeds for late fall sowing. Flower Grower 28(11):492–93.

19815. —— 1945: Hardy primulas from seeds. Am. Nurseryman 82(12):22.

19816. —— 1946: Coleus from seeds. Flower Grower 33(5):352.

Wood, E. R. 13772.
Wood, M. 532.

19817. Wood, O. M. 1938: Seedling reproduction of oak in southern New Jersey. Ecology 19(2):276–93.

19818. Wood, R. R. 1952: Selection for cold tolerance and low temperature germination in sugar beets. Proc. Am. Soc. Sugar Beet Technol. 7:407–10.

Wood, R. R. 3017, 3018.

19819. Woodbridge, M. E. 1928: Germination of carpet grass. Proc. Assoc. Offic. Seed Analysts 20:81–82.

19820. —— 1935: The rate of occurrence of seeds of curled dock (*Rumex crispus*) in replicate analyses of seed of orchard grass (*Dactylis glomerata*). Proc. Intern. Seed Testing Assoc. 7:21–26.

19821. —— 1939: Long list of weed seeds found in alfalfa and clover seed. Farm Res. (N.Y.) 5(2):12.

Woodbury, G. W. 9441.

19822. Woodbury, R. W. 1957: Shredded bark—a different medium for starting seeds. Horticulture 36(1):17.

19823. Woodcock, D. 1947: Isolation of phloridzin from apple seeds. Nature 159: 100.

Woodcock, D. A. 12491.

19824. Woodcock, E. F. 1914: Observations on the development and germination of the seeds in certain Polygonaceae. Am. J. Botany 1:454–76.

19825. —— 1918: Structure of mature seed of *Eriogonum microthecum*. Mich. Acad. Sci. Rept. 20:233–35. Botan. Abstr. 2:750. 1919.

19826. —— 1925: Morphology of the seed in *Claytonia virginica*. Papers Mich. Acad. Sci. 5:195–200.

19827. —— 1926: Morphological studies of the seed of *Alsine media*. Papers Mich. Acad. Sci. 6:397–403.

19828. —— 1927: Observations on the morphology of the seed of *Cerastium vulgatum.* Papers Mich. Acad. Sci. 8:233–38.

18929. —— 1930: Seed development in the *Thelygonium cynocrambe* L. Papers Mich. Acad. Sci. 11:341–47.

19830. —— 1931: Seed development in *Amaranthus caudatus* L. Papers Mich. Acad. Sci. 15:173–78.

19831. —— 1933: Seed studies in *Cyclamen persicum* Mill. Papers Mich. Acad. Sci. 17:415–20.

19832. —— 1939: Morphological studies on the seed of snapdragon (*Antirrhinum majus* L.). Papers Mich. Acad. Sci. 25: 139–44.

19833. —— 1941: Seed germination and seedling anatomy of snapdragon (*Antirrhinum majus* L.). Papers Mich. Acad. Sci. 26:99–102.

19834. —— 1943: Seed development in the morning-glory (*Ipomoea rubro-caerulea* Hook.). Papers Mich. Acad. Sci. 28:209–14. Expt. Sta. Rec. 90:465. 1944.

19835. Woodcock, H. D., and J. Coutts. 1935: Lilies, their culture and management. Scribner, New York. 242 pp.

Woodcock, J. W. 9175.

Woodell, S. R. J. 18932.

19836. Woodforde, A. H. 1930: Germination of crimson clover seed. Tasmanian J. Agr. 1:47.

19837. —— 1932: The ultra-violet light method of distinguishing between true and false perennial ryegrass seed. Tasmanian J. Agr. 3:160–62.

19838. —— 1935: Dormancy in subterranean clover seed. Tasmanian J. Agr. 6:126–27.

19839. —— 1936: The quality of perennial (English) ryegrass seed. The ultra-voilet light test for perenniality. Tasmanian J. Agr. 7:162–64.

19840. —— 1939: The "picric acid" test for determining the strain quality of white clover seed. Tasmanian J. Agr. 10:74–76.

Woodham, A. A. 6102.

19841. Woodman, H. E., and R. E. Evans. 1946: Horsechestnut as a source of food for livestock. I. Composition, digestibility, and nutritive value of alcohol-extracted horsechestnut meal, water-extracted horsechestnut meal, and horsechestnut residue. II. Value of alcohol-extracted horsechestnut meal, water-extracted horsechestnut meal, and horsechestnut residue as partial substitute for cereals in rations of bacon pigs. J. Agr. Sci. 36:29–41, 42–45. Brit. Abstr. B III: 156. 1946.

Woodman, R. M. 8112.

19842. Woodroof, J. G. 1945: Why moisture in peanuts should be controlled. Food Ind. 17:1302–4. Biol. Abstr. 20:5579. 1946.

19843. ——, S. R. Cecil, and H. H. Thompson. 1945: The effects of moisture on peanuts and peanut products. Georgia Agr. Expt. Sta. Bull. 241:1–23. Biol. Abstr. 20:1508. 1946.

19844. ——, N. C. Woodroof, and J. E. Bailey. 1928: Unfruitfulness of the pecan. Georgia Agr. Expt. Sta. Bull. 148. 40 pp.

Woodroof, J. G. 1683, 13190.

Woodroof, N. C. 19844.

19845. Woodruff, J. G. 1930: Soil moisture in relation to size and filling of pecans. Proc. Georgia-Florida Pecan Grow. Assoc. 24:34–36.

Woodruff, S. 12655.

Woodrum, H. G. 18777.

19846. Woods, F. W., and A. P. Mollish. 1963: Light requirements for germination of imbibed loblolly pine seed. Tree Planters Notes 58:26–28.

Woods, W. B. 14393.

19847. Woodside, J. W. 1941: The evaluation of hard seed in annual lespedezas. Proc. Assoc. Offic. Seed Analysts 32:86–88.

19848. Woodstock, E. F. 1930: Morphological studies on the seed of *Mesembrianthemum crystallinum* L. Papers Mich. Acad. Sci. 13:221–26.

19849. Woodward, C. H. 1945: Seeds of antiquity. N.Y. Botan. Garden J. 46:25–26.

19850. Woodward, F. N. 1939: The analysis of linseed oils of various origins. Analyst 64:265–69.

19851. Woodward, R. W. 1956: The effect of rate and date of seeding of small grains on yield. Agron. J. 48:160–62.

19852. —— and J. W. Thieret. 1953: A genetic study of complementary genes for purple lemma, palea, and pericarp in barley (*Hordeum vulgare* L.). Agron. J. 45:182–85.

19853. Woodward, T. E. 1940: The viability of seeds as affected by the siloing process. J. Dairy Sci. 23:267–71. Biol. Abstr. 15:5793. 1941.

19854. Woodwell, G. M. 1962: Effect of DDT on cone production, germination, and seedling survival in the boreal forest. Ecology 43:396–403.

19855. Woodworth, C. M. 1930: Abortive

seeds in soybeans. J. Am. Soc. Agron. 22:37–50. Biol. Abstr. 8:6621. 1934.

19856. —— and O. T. Bonnett. 1942: Barley seed weight affected by degrees of flower mutilation. Illinois Agr. Expt. Sta. Ann. Rept. 51:64–65.

19857. ——, E. R. Leng, and R. W. Jugenheimer. 1952: Fifty generations of selection for protein and oil in corn. Agron. J. 44:60–65.

Woodworth, C. M. 12078.

19858. Woodworth, R. H. 1930: Cytological studies in the Betulaceae. III. Parthenogenesis and polyembryony in *Alnus rugosa*. Botan. Gaz. 89:402–9.

19859. —— 1930: Twin seedlings in angiosperms. Science 72:398.

19860. Woolf, W. H., and P. P. Korn. 1943: Seeds in paper sheets used for quicker sowing. Sci. News Letter 44:46.

Woo-Ping-Soung. 10660.

Wooton, W. M. 8074.

Worden, H. A. 4513.

19861. Work, P. 1950: Coated seeds. Market Growers J. 79(5):16–19.

Work, P. 12682.

Working, E. B. 14038.

19862. Works, D. W., and L. C. Erickson. 1963: Infrared irradiation, an effective treatment of hard seeds in small seeded legumes. Idaho Agr. Expt. Sta. Res. Bull. 57. 22 pp.

Worley, F. P. 3115.

Wormell, R. L. 4461.

Wormser, G. 1932.

Woroszczuk, M. 2664.

19863. Worsdell, W. C. 1916: The morphology of the monocotyledonous embryo and of that of the grass in particular. Ann. Botany (London) 30:509–24.

19864. Worsham, A. D. 1961: Germination of *Striga asiatica* (L.) Kuntze (witchweed) seed and studies on the chemical nature of the germination stimulant. Ph.D. Thesis, N. Carolina State Coll. 122 pp. Weed Abstr. 11:234. 1962.

19865. —— and G. C. Klingman. 1962: Promotion of germination of *Striga asiatica* seed by coumarin derivatives and effects on seedling development. Nature 195:199–201.

19866. Worsham, C. H., C. C. Waddell, and F. C. Vilbrandt. 1939: Storage treatment of soybeans with ethylene and carbon dioxide gas. Virginia Polytech. Inst. Eng. Expt. Sta. Ser. Bull. 36. 34 pp. Chem. Abstr. 33:4448. 1939.

19867. Wort, D. J. 1939: Vernalization of Marquis wheat and other spring cereals. Botan. Gaz. 101:457–81.

19868. —— 1940: Responses of various spring wheats to vernalization. Plant Physiol. 15:137–41.

19869. —— 1941: Phasic development of Marquis spring wheat and Fulhio winter wheat. Botan. Gaz. 102:725–36.

19870. Worth, C. R. 1935: Primulas from seed. Gardeners Chron. 39:56–57.

19871. —— 1941: Raising cyclamen from seed. Gardeners Chron. 45:360–61.

19872. Wortman, L. S., and E. H. Rinke. 1951: Seed corn injury at various stages of processing and its effect upon cold test performance. Agron. J. 43:299–305.

19873. Worzella, W. W. 1944: The effect of level of soil fertility on wheat quality. Cereal Chem. 21:107–19. Expt. Sta. Rec. 92:49–50.

19874. —— and U. J. Norgaard. 1957: An analysis of foundation seed programs. Agron. J. 49:234–36.

Worzella, W. W. 11702.

19875. Woyke, H. 1961: The influence of temperature on the process of swelling and germination of smooth and wrinkled peas and the susceptibility of these two varieties to "frost." Roczniki Nauk Rolniczych Ser. A 83:619–39.

Wrenshall, C. L. 11803.

19876. Wricke, G. 1954: Die Manifestierung der erblich bedingten Polyembryonie beim Lein in ihre Abhängigkeit von Aussefaktoren. Biol. Centralbl. 73:49–88.

19877. Wright, A. 1949: It's more fun to grow African violets from seed. Flower Grower 36(11):875, 896.

Wright, A. H. 9050.

19878. Wright, E. 1931: The effect of high temperatures on seed germination. J. Forestry 29:679–87.

Wright, E. C. Barton-; *see* Barton-Wright, E. C.

19879. Wright, F. N., and B. J. Southgate. 1962: The potential uses of plastics for storage with particular reference to rural Africa. Trop. Sci. 4:74–81.

19880. Wright, G. M. 1958: Grain in the glume of wheat. Nature 181:1812–13.

19881. Wright, J. E., and A. M. Srb. 1950: Inhibition of growth in maize embryos by canavanine and its reversal. Botan. Gaz. 112:52–57.

19882. Wright, J. M. 1951: Phytotoxic effects of some antibiotics. Ann. Botany (London) 15:493–99.

19883. —— 1956: The production of antibiotics in soil. IV. Production of antibiotics in coats of seeds sown in soil. Ann. Appl. Biol. 44:561–66.

Wright, J. M. 6244.

19884. Wright, J. W. 1945: Influence of size and portion of cone on seed weight in eastern white pine. J. Forestry 43:817–19.

Wright, M. D. 1712.

19885. Wright, N. 1936: On the seedlessness of citrus fruits with particular reference to Marsh grapefruit. Trop. Agr. (Trinidad) 13:118–22.

19886. —— 1936: Additional notes on the seedlessness of citrus fruits. Trop. Agr. (Trinidad) 13:283–84.

19887. —— 1964: Influence of management practices on seed-set, seed yield, seed weight, germination and insects of Black gramagrass, *Bouteloua eriopoda* (Torr.) Torr. Agron. J. 56:57–60.

19888. —— and A. A. Baltensperger. 1964: Influence of temperature, light radiation, and chemical treatment on laboratory germination of Black gramagrass seed, *Bouteloua eriopoda* Torr. Crop Sci. 4:168–71.

19889. —— and L. J. Streetman. 1961: Effect of insect-damaged ovaries on accuracy of seed-set determination by the bleaching technique. Crop Sci. 1:303–5.

19890. Wright, R. C. 1941: Investigations on the storage of nuts. U.S. Dept. Agr. Tech. Bull. 770. 35 pp.

Wright, R. D. 18396.

19891. Wright, W. H. 1935: The "seeds" of the genus *Poa* commonly found on the market in Canada. Sci. Agr. (Ottawa) 15:811–13.

19892. —— 1945: Hard seeds of asparagus. Assoc. Offic. Seed Analysts News Letter 19(5):5.

Wright, W. H. 9470.

19893. Wroblowna, W. 1949/50: Observations on the vegetative propagation, germination, and zoochory of the raspberry (*Rubus idaeus* L.). Acta Soc. Botan. Polon. 20:201–22.

19894. Wu, L., W. Tien, and Y. Lin. 1964: Seed-borne diseases of soybean in Taiwan. I. Factors affecting the isolation of causal organisms. Acad. Sinica Botan. Bull. 5:42–53.

Wu, Y. S. 1873.

19895. Wunderlich, R. 1959: Zur Frage der Phylogenie der Endospermtypen bei den Angiospermen. Österr. Botan. Z. 106:203–93.

19896. Wunschendorff, M. 1914: La graine de fenugrec. J. Pharm. Chim. 106:152–57.

Wurceldorf-Warden, J. E. 17556.

19897. Wurdack, M. H. 1934: Raising Japanese iris from seed. Flower Grower 21(7):313.

19898. Wverson, E. H., and R. B. Hart. 1961: Varietal variation for dormancy in mature wheat. Mich Agr. Expt. Sta. Quart. Bull. 43:820–29.

Wyk, H. P. D. van; *see* Van Wyk, H. P. D.

Wyk, J. F. van; *see* Van Wyk, J. F.

Wylam, C. B. 8859.

19899. Wyman, D. 1953: Seeds of woody plants. Arnoldia 13:41–60.

19900. Wynd, F. L. 1933: The sensitivity of orchid seedlings to nutritional ions. Ann. Missouri Botan. Garden 20:223–37.

19901. —— 1933: Nutrient solutions for orchids. Ann. Missouri Botan. Garden 20:363–72.

19902. —— 1935: The technique of the asymbiotic germination of orchid seeds. Orchid Rev. 43:138–42.

Wynd, F. L. 14550, 19493.

Wyndaele, R. Verbeek-; *see* Verbeek-Wyndaele, R.

19903. Wynn-Williams, M. 1934: Observations upon the seeds and germination of some thistle species. Proc. Intern. Seed Testing Assoc. 6:4–10.

Wyrovaja, R. A. Beyliss-; *see* Beyliss-Wyrovaja, R. A.

19904. Wysocka, A. 1938: Les resultats d'examen concernant la germination des grains d'arnique de montagne, *Arnica montana* L. Wiadomosci Zielarsk. Miesiecz (Warsaw) 6:495–99.

Wyssling, A. Frey-. 3472.

19905. Wyttenbach, E. 1955: Der Einfluss verschiedener Lagerungsfaktoren auf die Haltbarkeit von Feldsämereien (Luzerne, Rotklee, und gemeinem Schotenklee) bei länger dauernder Aufbewahrung. Landwirtsch. Jahrb. Schweiz. 4:161–96.

19906. —— 1962: Die Bestimmung der Keimfähigkeit von Samenproben ohne Keimversuch unter besonderer Berücksichtigung des topographischen Tetrazoliumverfahrens. Mitt. Schweiz. Landwirtsch. 10:65–71.

19907. —— and E. Frey. 1954: Versuche über die Erhaltung der Keimfähigkeit von Kleesaatgut bei länger dauernder Lagerung Mitt. Schweiz. Landwirtsch. 2:9–15.

Y

19908. Yabuki, K., and I. Miyagawa. 1958: Studies on the effects of diurnal variation of temperature upon the germination of seeds. II. The germinations of barley and vegetables. Soc. Agr. Met. Japan. J. Agr. Met. 13:105–9.

19909. —— and —— 1959: Studies on the effects of diurnal variation of temperature upon the germination of seeds. III. Soc. Agr. Met. Japan. J. Agr. Met. 15:49–54.

Yabuno, T. 14523.

19910. Yagodin, B. A. 1960: The effect of minor elements on the germination of seeds and the growth of some crops. Byul. Gl. Botan. Sada 39:83–86.

19911. Yahiro, M., and T. Inoue. 1955: Physiological changes in the germination seeds during the low temperature treatment. I. On the water content, the specific gravity of the tissue powder and the concentration of reducing sugars of horse bean. Botan. Mag. (Tokyo) 68:45–50.

Yahiro, M. 9415, 10998, 10999.

19912. Yaichnikov, I. S., and A. I. Solntsev. 1938: Albumins of the seeds of *Dactylis glomerata.* J. Appl. Chem. (USSR) 11:995–97. Chem. Abstr. 33:1783. 1939.

19913. —— and N. A. Tarasikov. 1940: Proteins of seeds of American linden (*Tilia grandifolia* Ehrh.). J. Appl. Chem. (USSR) 13:1710–12. Chem. Abstr. 35: 4062. 1941.

19914. Yakovlev, M. S., and M. D. Ioffe. 1961: Further studies of the new type of embryogenesis in angiosperms. Botan. Zh. 46:1402–21.

19915. —— and D. P. Snegirev. 1954: Effect of growth stimulants on the formation of polyembryonic wheat kernels. Botan. Zh. 39:187–94. Biol. Abstr. 30:6025. 1956.

Yakovleva, S. V. 12554.

Yakovleva, V. I. 11394.

Yakubov, A. M. 9498.

19916. Yakusheva, E. I. 1947: Duration of viability of seeds of some herbage legumes. Selek. i Semen. 1947(1):41–45.

19917. Yakushevskaya, O. P. 1937: Theoretical bases for establishing latitudes and the optimum number of seeds in germination tests. Bull. Appl. Botany, Genet., Plant Breed. Ser. 4(2):191–200.

Yamada, E. 5822, 14152.

Yamada, H. 5820, 5821.

19918. Yamada, I. 1954: Effect of low temperature on the germination of weed seeds. (Preliminary report.) Hokkaido Natl. Agr. Expt. Sta. Res. Bull. 67:55–57.

19919. —— 1954: On the dormancy of weed seeds. I. Effect of thiourea on breaking of dormancy. Hokkaido Natl. Agr. Expt. Sta. Res. Bull. 67:58–62.

19920. Yamada, K. 1933: Über den Zuckergehalt des Samens. Japan. J. Med. Sci. 2:245–46.

19921. Yamada, M. 1955: Studies on fat metabolism in germinating castor beans. I. The utilization of bound and free fatty acids in vivo. Sci. Papers Coll. Gen. Educ. Univ. Tokyo 5:149–59.

19922. —— 1955: Studies on fat metabolism in germinating castor beans. II. Different types of respiration in embryonic organs and the conversion of fats to sugars and their translocation. Sci. Papers Coll. Gen. Educ. Univ. Tokyo 5:161–69.

19923. —— 1957: Studies on fat metabolism in germinating castor beans. III. Lipase in decotylated embryo tissue. Sci. Papers Coll. Gen. Educ. Univ. Tokyo 7:97–104.

19924. —— 1960: Studies on fat metabolism in germinating castor beans. IV. Glycerol metabolism. Sci. Papers Coll. Gen. Educ. Univ. Tokyo 10:283–323.

19925. —— and P. K. Stumpf. 1964: Enzymic synthesis of ricinoleic acid by extracts of developing *Ricinus communis* L. seeds. Biochem. Biophys. Res. Commun. 14:165–71.

19926. Yamada, N., H. Suge, and H. Nakamura. 1963: Chemical control of plant growth and development. I. Germination of rice seed as affected by sprouting and gibberellin application. Proc. Crop Sci. Soc. Japan 31:253–62.

Yamada, R. 14682.

19927. Yamada, T., and A. Adachi. 1952: Some experiments on the germination of hard red clover seed. Bull. Natl. Inst. Agr. Sci. Ser. G 4:199–207. Biol. Abstr. 27:7726. 1953.

Yamada, T. 18293, 18294.

Yamaguchi, M. 14355.

19928. Yamaguchi, S. 1942: Physiological studies on the germination of Yezo spruce seed. J. Fac. Agr. Hokkaido Imp. Univ. 48:1–148. Chem. Abstr. 42:7382–83. 1948.

19929. Yamaki, T., and K. Makamura. 1952: Formation of indoleacetic acid in maize embryo. Sci. Papers Coll. Gen.

Educ. Univ. Tokyo 2:81–98. Chem. Abstr. 47:2837. 1953.

19930. —— and N. Takahashi. 1962: Some considerations upon the dark germination of tobacco seed. Botan. Mag. (Tokyo) 75:245–54.

Yamaki, T. 8315, 18289, 18293, 18294.

19931. Yamamoto, K. 1951: Investigation on "viviparous" germination of seed in barley varieties in Japan. Tohoku Univ. Rept. Inst. Agr. Res. 3:13–27.

Yamamoto, K. 11601.

19932. Yamamoto, M. 1955: Relation of the soaking injury and the moisture content of seeds of *Phaseolus vulgaris* L. Japan. J. Ecol. 5:21–23. Biol. Abstr. 31:5833. 1957.

19933. —— 1955: The decreased germination of dry-seeds soaked in water. Japan. J. Ecol. 5:74–77. Biol. Abstr. 31:5834. 1957.

19934. —— 1957: On the germination of bean seeds (*Phaseolus*) during soaking in water. Japan. J. Ecol. 7:120–23. Biol. Abstr. 32:24791. 1958.

19935. —— 1961: Cracking of bean seeds soaked in water. Botan. Mag. (Tokyo) 74:273–79.

Yamamoto, M. 5815, 7508.

19936. Yamamoto, R., and M. Ishikawa. 1959: Comparison between the soybean seeds ripened on a highland and those ripened on a lowland. Proc. Crop Sci. Soc. Japan 28:77–78.

19937. —— and M. Ninomiya. 1934: The embryo and the testa of kaoliang (*Andropogon sorghum* Brot.). I & II. The oils of the embryo and the testa. J. Agr. Chem. Soc. Japan 10:248–56. Chem. Abstr. 28:4618. 1934.

Yamamoto, R. 14084, 14157.

19938. Yamamoto, Y. 1954: On formic dehydrogenase of bean seed embryo in the germination stage. J. Biochem. (Tokyo) 41:551–65.

19939. —— 1955: Asparagine metabolism in the germination stage of a bean, *Vigna sesquipedalis*. J. Biochem. (Tokyo) 42:763–74.

19940. —— 1961: A metabolic pathway of isocitric acid in cotyledons of a bean, *Vigna sesquipedalis*. Plant Cell Physiol. (Tokyo) 2:277–89.

19941. —— 1963: Enzymological studies on seed germination. Advan. Frontiers Plant Sci. 6:163–85. Chem. Abstr. 60:7137. 1964.

Yamamoto, Y. 11554, 14797.

19942. Yamasaki, M. 1957: On the identification of plant characters by testing the resistance of seeds to the toxic action of potassium chlorate. Proc. Intern. Seed Testing Assoc. 22:175–79.

19943. —— and K. Hanada. 1957: On the changes in the resistance to potassium chlorate and reducing matters content during the process of vernalization in wheat. Proc. Crop Sci. Soc. Japan 25:199–203.

19944. Yamasaki, R., and K. Ichihara. 1935: Composition of rape-seed oil. J. Chem. Soc. Japan 56:1332–34. Chem. Abstr. 30:1598. 1936.

19945. Yamasaki, Y. 1947: Studies on growth of the embryo transplanted between the different species and genera in some cereal crops. Japan. J. Genetics 22:33–34. Biol. Abstr. 26:16433. 1952.

19946. Yamashi, T., and K. Nakamura. 1952: Formation of indoleacetic acid in maize embryo. Sci. Papers Coll. Gen. Educ. Univ. Tokyo 2:81–98. Ber. Wiss. Biol. 86:58. 1954.

Yamashita, I. 19997.

Yamasita, Y. 16773.

Yamazaki, I. 18305.

Yamazaki, K. 19255.

19947. Yamazaki, T. 1953: On the floral structure, seed development, and affinities of *Deinostema*, a new genus of Scrophulariaceae. Botan. Mag. (Tokyo) 28:143–49.

19948. —— 1957: Seed formation of *Ellisiophyllum pinnatum* var. *reptans*. Botan. Mag. (Tokyo) 70:162–68.

19949. Yampolsky, C. 1945: The barley grain. Observations on physiology and germination. Wallerstein Lab. Commun. 8:169–78. Biol. Abstr. 20:7237. 1946.

19950. Yanagihara, T., K. Tochiaki, and K. Arai. 1960: On the relation between the harvest of Japanese larch seed and meterological conditions. J. Japan. Forestry Soc. 42:347–51.

Yanagisawa, I. 8439.

19951. Yanagisawa, T. 1961: Effects of seed source on seedling growth rhythm of Japanese larch (*Larix leptolepis*). Hokkaido Forest Expt. Sta. Ann. Rept. 1960: 65–70.

19952. —— and S. Asakawa. 1953: Woody plant seed viability tests with tetrazolium salts. J. Japan. Forestry Soc. 35:40–42. Biol. Abstr. 27:23462. 1953.

19953. —— and M. Saito. 1955: Cleaning seed of *Chamaecyparis obtusa* by the solutions of some surface active agents. J. Japan. Forestry Soc. 37:549–51.

19954. —— and S. Yoshida. 1962: Some physiological studies on the Ehrlich positive substances produced in the seeds and seedlings of *Abies sachalinensis*. J. Japan. Forestry Soc. 44:259–66.

Yanagisawa, T. 8153, 8154.

Yang, C. Y. 12444.

19955. Yang, M. H. 1962: A cytological study of nuclear multiplication in the endosperm of the wheat. Acta Botan. Sinica 10:7–14.

19956. Yang, S. L. 1956: The effects of removing seed coats and chemical treatments on the germination of the seeds of sweet orange (*Citrus sinensis* var. *sekkan*) and longan (*Euphoria longan*) and on the early stage of growth of their seedlings. Acta Agr. Sinica 7:465–74.

Yang, S. L. 12191.

19957. Yango, C. E. 1938: Does shelling by machine affect the germination of corn? Philippine Agriculturist 27:59–69.

Yanai, S. 3932.

19958. Yankulov, I. 1961: A study of the possibilities of overcoming slow and irregular germination of seeds of belladonna (*Atropa belladonna* L.). Bulgar. Akad. Nauk. Inst. po Rastinievud. Izv. 11:93–114.

19959. Yarbrough, J. A. 1949: *Arachis hypogaea*. The seedling, its cotyledons, hypocotyl and roots. Am. J. Botany 36:758–72.

19960. Yardeni, D. 1948: Human saliva as a germination inhibitor. Science 108:62–63.

19961. —— and M. Evenari. 1952: The germination inhibiting, growth inhibiting and phytocidal effect of certain leaves and leaf extracts. Phyton 2:11–16.

19962. Yardley, H. J., and G. Godfrey. 1963: An effect in vitro of a non-protein amino acid from litchi seeds. Biochem. J. 88:59P.

19963. Yarham, E. R. 1945: The Kauri pine. Gardeners Chron. 118:210–11.

Yarimizu, H. 6831, 8848.

19964. Yarkovaya, L. M. 1939: Changes in content of growth substances (in plants) due to interruption of dormancy. Compt. Rend. Acad. Sci. URSS 23:88–91. Chem. Abstr. 34:4108. 1940.

Yarosh, N. P. 6028.

19965. Yasuda, I. 1955: On the germination tests of some flower seeds. Okayama Univ. Fac. Agr. Sci. Rept. No. 7:31–35.

19966. Yasuda, S. 1933: On the behavior of pollen tubes in the production of seedless fruits caused by inter-specific pollination. Japan. J. Genetics 8:239–44. Biol. Abstr. 8:10305. 1934.

19967. Yasue, T., and K. Ota. 1962: Studies on the hard seed of milk vetch (*Astragalus cinicus* L.). IV. Seed quality and hard seed content of milk vetch in relation to harvesting time. Proc. Crop Sci. Soc. Japan 30:107–10.

19968. Yasunaga, T., J. Maruyama, M. Uemura, K. Fukunaga, and H. Inamura. 1963: Effect of rain during ripening on the quality of wheat grain and its flour. I. Changes in some constituents and enzymes of grain by moist treatment of ears. Proc. Crop Sci. Soc. Japan 32:152–56.

Yatazawa, M. 7508, 11000, 11303.

19969. Yates, H. O. 1928: Germination of cyclamen seed. Florists Exchange Hort. Trade World 69(2):101.

19970. Yatsu, L., and A. M. Altschul. 1963: Lipid-protein particles: isolation from seeds of *Gossypium hirsutum*. Science 142:1062–63.

19971. Yatuyanagi, S., K. Takahasi, T. Murakami, and H. Sakai. 1957: Analysis of the ripening mechanism of rice under low temperatures. Proc. Crop Sci. Soc. Japan 26:8–10.

Yazdi, H. Tabatabai-. 12259.

Yeatman, C. W. 8981.

19972. Yeh, H. L. 1945: Studies of the proteins of the tung nut (*Aleurites fordii*). Thesis, Univ. Michigan. 202 pp. Biol. Abstr. 20:18597. 1946.

19973. Yelenosky, G. 1961: Birch seeds will germinate under a water-light treatment without prechilling. U.S. Dept. Agr. Forest Serv. Northeast. Forest Expt. Sta. Forest Res. Note 124. 5 pp.

19974. Yelukhin, K. V. 1939: Sowing forest seeds treated with salt solutions. Lesn. Khoz. 2:17–23.

Yemm, E. W. 6524, 6525, 17808.

19975. Yen, D. E. 1954: Seed variation in pea rogues. New Zealand J. Sci. Technol. 36:117–21.

Yen, L. F. 12443, 12444.

19976. Yeo, P. F. 1961: Germination, seedlings, and the formation of haustoria in *Euphrasia*. Watsonia 5(1):11–22.

19977. Yeomans, N. J. 1936: Germination of parsley. Horticulture 14(13):280.

19978. —— 1940: Growing the Empress tree from seed. Horticulture 18(5):111.

19979. Yerkes, G. E. 1930: Raising root stocks from seed in the United States. Proc. Intern. Hort. Congr., London 1930: 83–91. Expt. Sta. Rec. 66:139. 1932.

19980. Yermanos, D. M., and L. E. Francois. 1963: Differences among seed samples from primary, secondary, and tertiary heads of safflower. Crop Sci. 3:560–61.

19981. ——, ——, and L. Bernstein. 1964: Soil salinity effects on the chemical composition of the oil and the oil content of safflower seed. Agron J. 56:35–37.

19982. —— and P. F. Knowles. 1962: Fatty acid composition of the oil in crossed seed of flax. Crop Sci. 2:109–11.

Yerofeyev, N. G. 19983.

19983. Yevreinova, T. N., and N. G. Yerofeyev. 1956: Amylase in acorns stored under various conditions. Vestn. Mosk. Univ. Ser. Biol. Pochvoved. Geol. i Geogr. 1956(2):39–43. Biol. Abstr. 35: 8487. 1960.

Yien, C. H. 9777.

19984. Yin, H. C., and C. N. Sun. 1949: Localization of phosphorylase and of starch formation in seeds. Plant Physiol. 24:103–10.

Yin, H. C. 18359.

19985. Yip, J. S. 1936: Insect damage to seeds of *Cracca virginiana* L. J. Econ. Entomol. 29:622–29. Expt. Sta. Rec. 75:660. 1936.

19986. Yli-Vakkuri, P. 1959: On machines for abrading seed wings and their influence on the germinative capacity of the seed. Acta Forest. Fennica 68(4). 13 pp.

Yoa, Y. T. 13819.

19987. Yocum, H. G. 1964: Factors affecting the germination of palm seeds. Hort. Mag. 43:104–6.

Yoffe, M. D. 9667.

Yokohama, Y. 9492, 9495, 11538.

Yokota, N. 11164.

Yokoyama, K. 13903.

19988. Yokum, H. C., M. W. Jutros, and R. A. Peters. 1961: Preliminary investigations of a germination and growth inhibitor produced by yellow foxtail (*Setaria glauca* [L.] Beauv.). Proc. Northeast. Weed Control Conf. 15:341–47. Weed Abstr. 12:106, 1963.

Yokum, H. C. 15233.

19989. Yomo, H., and H. Iinuma. 1963: Activation of enzymes in the ungerminated barley endosperms with gibberellin. Agr. Biol. Chem. (Tokyo) 27:76–79.

19990. Yonekawa, G., Y. Kondo, E. Kinbara, and M. Haba. 1962: Effect of environmental condition in cultivation on protein components of soybean seed. Japan. Soc. Food Nutr. J. 15:48–51.

19991. York, H. H. 1904: The embryo-sac and embryo of *Nelumbo.* Ohio Naturalist 4:167–76.

19992. —— 1909: The anatomy and some of the biological aspects of the "American mistletoe," *Phoradendron flavescens* (Pursh) Nutt. Texas Univ. Bull. 120 (Sci. Ser. 13). 31 pp.

19993. —— 1913: The origin and development of the embryo sac and embryo of *Dendrophthora opuntiotdes* and *D. gracile.* II. Botan. Gaz. 56:200–16.

York, T. L. 19318.

19994. Yoshida, S. 1960: Biochemistry of germination and low-temperature stratification of *Abies sachalinensis* Fr. Schm. seeds. I. On the change of auxin and inhibitor in the seeds during low-temperature stratification. J. Japan. Forestry Soc. 42:259–62.

19995. ——, Y. Shimada, A. Seto, and N. Nishimura. 1959: Studies on the vegetative succession in natural grassland; particularly on the production and germination of *Miscanthus sinensis* seed. Tohoku Univ. Inst. Agr. Res. Sci. Rept. Ser. D 10:61–80.

Yoshida, S. 8304, 19954.

19996. Yoshii, Y. 1925: Über die Reifungsvorgänge des Pharbitis-Samen mit besonderer Rüchsicht auf die Keimungsfähigkeit des unreifen Samen. Tokyo Imp. Univ. Fac. Sci. J (Sec. III) 1:1–139.

19997. Yoshimura, K., and I. Yamashita. 1935: Chemical constituents of the seeds of *Arundinaria hindsi* Makino, a kind of bamboo. J. Agr. Chem. Soc. Japan 11: 355–56. Chem. Abstr. 29:6623. 1935.

Yoshimura, K. 13611.

Yosiku, S. 11602.

19998. Yosizawa, Z., and M. Toshiyuki. 1963: Immunochemically significant structure of simple sugars for the lectins of *Sophora japonica* L. seeds. Proc. Japan Acad. 39:182–86.

Yosue, S. 10725.

19999. Youden, W. J. 1932: Statistical analysis of seed germination data through the use of the Chi-square test. Contrib. Boyce Thompson Inst. 4:219–32.

20000. —— 1940: Seed treatments with talc and root-inducing substances. Contrib. Boyce Thompson Inst. 11:207–18.

20001. Young, A. W. 1961: World seed year. What seedgrowers shoud do. Crops & Soils 13(4):12–13.

Young, A. W. 1572, 19351.

Young, G. A. 9879.

Young, G. Y. 3652.

Young, H. D. 1638, 4446.

20002. Young, J. L., R. C. Huang, S. Vanecko, J. D. Marks, and J. E. Varner. 1960: Conditions affecting enzyme synthesis in cotyledons of germinating seeds. Plant Physiol. 35:288–92.

Young, J. R. 19588.

Young, P. T. 3824.

20003. Young, R. E. 1949: The effect of maturity and storage on germination of butternut squash seed. Proc. Am. Soc. Hort. Sci. 53:345–46.

20004. Young, S. M., and J. E. Greaves. 1940: Influence of variety and treatment on phytin content of wheat. Food Res. 5:103–8. Expt. Sta. Rec. 84:725–26. 1941.

20005. Youngberg, C. T. 1951: Effect of origin and growth conditions of Norway spruce (*Picea excelsa*) on the chemical composition of seed and physiological characteristics of nursery stock. Soil Sci. Soc. Am. Proc. 15:376–79.

20006. —— 1952: Effect of soil fertility on the physical and chemical properties of tree seed. J. Forestry 50:850–52. Soils Fertilizers 16:334. 1953.

20007. Youngken, H. W. 1934: A comparative study of the seeds and spikes of certain caulescent species of *Plantago*. Am. J. Pharm. 106:157–65.

20008. —— 1935: Further studies on psyllium seed. J. Am. Pharm. Assoc. 24:207–11. Chem. Abstr. 29:3779. 1935.

20009. —— 1953: Studies on *Veratrum*. II. *Veratrum eschscholtzii* A. Gray; observations on seed germination and early growth of seedlings of *Veratrum* species. J. Am. Pharm. Assoc. 42:39–45.

20010. —— and V. H. Simonian. 1950: A study of the seeds of *Strophanthus sarmentosus* and some related species of *Strophanthus*. J. Am. Pharm. Assoc. 39:615–20.

20011. Youngman, B. J. 1952: Germination of old seeds. Kew Bull. 1951(3):423–26.

Youngman, V. 12685.

20012. Youngman, W. H. 1944: The United States vegetable seed industry. Natl. Hort. Mag. 23(1):13–18.

Youngquist, R. W. 6701.

20013. Youngs, C. G., T. M. Mallard, B. M. Craig, and H. R. Sallans. 1951: Component fatty acids of rapeseed oil. Can. J. Chem. 29:871–76.

Youngs, C. G. 16156.

20014. Younkin, S. G., J. B. Hester, and A. D. Hoadley. 1950: Interaction of seeding rates and nitrogen levels on yield and sieve size of peas. Proc. Am. Soc. Hort. Sci. 55:379–83.

Youssef, E. 12293.

Yu, S. A. 9290.

Yu, T. F. 15615.

20015. Yudin, V. G. 1962: The depth of dormancy of seeds of some maple species as related to their geographical origin. Tr. Botan. Inst. Akad. Nauk SSSR Ser. 4 Eksperim. Botan. 15:148–57.

Yudin, V. G. 14477–14479.

20016. Yukura, Y. 1958: Inheritance of seed size in turnip. Japan. J. Breeding 8:204.

20017. Yuncker, T. G. 1942: A note on the precocious germination of dodder seed. Proc. Indiana Acad. Sci. 57:114–15.

Yung, F. D. 8380.

Yurchak, L. D. Pilipenko-. 7702, 7703.

20018. Yur'eva, N. A. 1961: Effect of electrization of soil on the seed progeny of tomatoes. Agrobiologiya 1961(5):702–6.

Yurgenson, M. P. 2567.

20019. Yusta, A. A., and A. S. Ruiz. 1953: The chemical composition of the embryo of *Oryza sativa*. I. Total composition and glucides. Anales Real Soc. Espan. Fis. Quim. (Madrid) Ser. B 49:441–44. Chem. Abstr. 47:12526. 1953.

20020. —— and —— 1954: Chemical composition of the germ of *Oryza sativa* L. II. Lupins. III. Proteins, mineral salts and vitamins. Anales Real Soc. Espan. Fis. Quim. (Madrid) Ser. B 50:87–94, 95–98. J. Sci. Food Agr. 5:ii31. 1954.

20021. Yusuf, N. D., and W. Muhammad. 1955: Seed set germination in sugarcane. Agr. Pakistan 6:16–24.

20022. Yusufov, A. G., G. S. Popova, and M. A. Balieva. 1961: Effect of preplanting seed treatment with microelements on corn yield. Dokl. Akad. Nauk Azerb. SSR 17:733–36.

Z

Zabavskaia, K. M. 1957.

Zabelina, Z. K. 11210.

20023. Zabluda, G. V. 1940: Formation of vegetative and generative organs in wheat and rye under conditions of slow development. Compt. Rend. Acad. Sci. URSS 26:957–60.

20024. Zaborovskii, E. P., and N. N. Varasova. 1959: Biology of the germination of *Fraxinus excelsior* seeds and their preparation for planting. Leningr. Nauchn.-Issled. Inst. Lesn. Khoz. Sb. Rabot po Lesn. Khoz. 3:76–110.

20025. —— and —— 1961: On the germination of seeds of the water elder (*Viburnum opulus* L.) and the wayfaring tree (*V. lantana* L.). Botan. Zh. 46:1169–71.

20026. —— and —— 1961: Overcoming dormancy in linden seeds. Leningr. Nauchn.-Issled. Inst. Lesn. Khoz. Sb. Rabot po Lesn. Khoz. 4:228–43.

20027. —— and —— 1962: Stimulation of the germination of seeds of tree and shrub species from the family Leguminosae, with reference to the characteristics of the structure of the outer coverings. Leningr. Nauchn.-Issled. Inst. Lesn. Khoz. Sb. Rabot po Lesn. Khoz. 5:219–34.

20028. ——, ——, and A. K. Semenova. 1962: On hastening the preparation for germination of seeds of the Norway maple. Leningr. Nauchn.-Issled. Inst. Lesn. Khoz. Sb. Rabot po Lesn. Khoz. 5:235–42.

Zabotkin, D. S. 10157.

Zacharczenko, E. Waszczenko-. 12898.

20029. Zachariew, B. J. 1939: Ein bequemes Schnellkeimverfahren zur Prüfung des Samens einiger Nadelholzarten. Forstwiss. Centralbl. 61:238–49.

20030. —— 1939: Ein bequemes Schnellkeimverfahren zur Prüfung des Samens einiger Nadelholzarten. Forstwiss. Centralbl. 61:281–93.

20031. —— 1942: Über die Keimdauer bei Schwarzkiefernsamen. Forstwiss. Centralbl. 64:278–85.

20032. Zacharius, R. M. 1959: The isolation and characterization of γ-L-glutamyl-S-methyl-L-cysteine from kidney beans (*Phaseolus vulgaris*). Arch. Biochem. Biophys. 80:199–209.

20033. ——, J. F. Thompson, and C. J. Morris. 1956: A new sulfur amino acid in beans. Plant Physiol. (Suppl.) 31:XLIV.

Zacharius, R. M. 13766.

Zacks, S. 13084, 13085, 15503–15505.

20034. Zade, A. 1912: Der Flughafer. Arb. Deut. Landwirtsch. Ges. 229. 91 pp.

20035. —— 1913: Die Pflanzendecke als keimungshemmender Faktor für gewisse Unkrautsamenarten. Fühlings Landwirtsch. Ztg. 62:777–85.

20036. —— 1920: Das Knaulgras (*Dactylis glomerata* L.). Arb. Deut. Landwirtsch. Ges. 305. 69 pp.

20037. Zagaja, S. W. 1962: After-ripening requirements of immature fruit tree embryos. Hort. Res. 2(1):19–34.

20038. —— and J. Czapski. 1962: Some effects of low temperature treatments on phosphorus uptake by immature apple embryos. Hort. Res. 2(1):13–18.

20039. Zagami, V. 1931: Effects on albino rats of nutrition solely with seeds of *Lathyrus sativus* L. Atti Accad. Lincei 14:218–21. Chem. Abstr. 26:3823. 1932.

20040. —— 1934: The content of some oil-containing seeds in antineuritic vitamin B_1. Quaderni Nutriz. 1:284–94. Chem. Abstr. 30:5628. 1936.

20041. —— and V. Famiani. 1933: The nutritive value of the proteins in leguminous seeds. Atti Accad. Lincei 17:485–90. Chem. Abstr. 27:5100. 1933.

Zagorska, H. 12341.

20042. Zagruznyi, P. L. 1962: Soaking seeds with radioactive mineral waters prior to planting. Zemledelie 1962(1):65–68. Biol. Abstr. 39:7874. 1962.

20043. Zahnley, J. W., and J. B. Fitch. 1941: Effect of ensiling on the viability of weed seeds. J. Am. Soc. Agron. 33:816–22.

Zahavi, A. Kadman-; *see* Kadman-Zahavi, A.

20044. Zaidman, J. 1956: Penetration of the thiocyanate ions into wheat seeds under the conditions of chemical stimulation. Analele Univ. "C. I. Parhon" Bucuresti Ser. Stiint. Nat. 1956:133–36. Chem. Abstr. 52:14765. 1958.

20045. Zaitsev, G. N. 1958: The significance of seed characters for the taxonomy of the genus *Lonicera* L. Botan. Zh. 43:836–40.

20046. —— 1960: The interaction of germinating seeds of Norway maple and black locust. Priroda 10:86–87. Biol. Abstr. 36:82396. 1961.

20047. Zaitseva, A. A. 1940: Role of sugar in greening of wheat seedlings. Compt. Rend. Acad. Sci. URSS 27:59–62. Chem. Abstr. 34:7346. 1940.

Zaki, M. 8343.

Zakiyan, M. K. 10197.

Zaky, Y. A. H. 8774, 8778, 8779, 13276.

20048. Zaleski, A. 1957: Lucerne investigations. III. Effect of heat treatment on germination and field establishment of lucerne seed. J. Agr. Sci. 49:234–45.

20049. —— 1959: Lucerne investigations. IV. Effect of germination and seed rates on establishment, mortality, and yield of dry matter and protein per acre. J. Agr. Sci. 53:260–67.

20050. —— 1961: White clover investigations. I. Effect of seed rates and cutting treatments on flower formation and yield of seed. J. Agr. Sci. 57:199–212.

20051. Zaleski, E. 1929: Statistische Bewertung vergleichender Keimungsversuche der Internationalen Vereinigung für Samenkontrolle. Actes Congr. Intern. Essais Semences 5:152–97.

20052. Zaleski, W. 1905: Beiträge zur Kenntniss der Eiweissbildung in reifenden Samen. Ber. Deut. Botan. Ges. 23:126–33.

20053. —— 1905: Zur Kenntniss der proteolytischen Enzyme der · reifenden Samen. Ber. Deut. Botan. Ges. 23:133–42.

20054. —— 1911: Zur Kenntnis der Stoffwechselprozesse in reifenden Samen. Botan. Centralbl. Beih. 27:63–82.

Zalik, S. 6933, 9172, 10897, 14344.

20055. Zalkow, L. H., M. K. Park, and (in part) J. W. Ellis. 1963: A vapour phase of chromatographic analysis of carrot seed oil. Perfumery Essent. Oil Record 54: 507–9.

20056. Zamotailov, S. S. 1960: On the period of dormancy in the embryo of *Arachis hypogaea* L. Botan. Zh. 45:1435–45.

20057. Zanda, G. B. 1929: Beeinflussung der Samenkeimung durch Caffein. [Abstract.] Ber. Ges. Physiol. Exptl. Pharm. 48:644.

Zanelli, F. 1286.

20058. Zanini, E., and G. P. Ballatore. 1951: Ricerche fisiologiche su *Hedysarum coronarium* L. Ric. Sci. 21:960–72. Biol. Abstr. 25:37708. 1951.

20059. Zankov, Z. D. 1962: Results of presowing treatment of vine seeds with stimulating substances. Izv. Inst. po Biol. "Metodii Popov" Bulgar. Akad. Nauk. 12:253–62. Chem. Abstr. 57:15550–51. 1962.

20060. Zanoni, G. 1931: Sull'attivazione catalasica nella fecondazione delle piante superiori. Riv. Biol. 13:409–22. Ber. Wiss. Biol. 22:94. 1932.

20061. —— 1936: Cotyledonary resistance and lability studied in *Phaseolus vulgaris*. I. Arch. Botan. 12:350–57. Expt. Sta. Rec. 78:26. 1938.

20062. —— 1949: Intraembryonic relationships during maturation of seeds. Boll. Soc. Ital. Biol. Sper. 25:329–31. Chem. Abstr. 44:10050. 1950.

20063. —— and A. Mondini. 1955: Germinative control on the permeability of the cytoplasm. Atti Accad. Ligure Sci. Lettre 11:103–14. Biol. Abstr. 31:2583. 1957.

20064. Zanotti, V. 1934: Action of hydrocyanic acid on the germination of the seeds of *Datura stramonium*, L. and *Hyoscyamus niger*. Studi Trentini Sci. Nat. 15:30–32. Chem. Abstr. 31:5836. 1937.

20065. Zapiski, S. 1930: Materialien für Kenntnis der Keimung und Reife von Samen. Jard. Botan. Inst. Essais Semences, Leningrad, Ann. Essais Semences 1930(2): 58–63.

Zapparoli, T. V. 13910–13913.

Zarger, T. G. 4746.

Zarina, A. 1838.

20066. Zarubailo, T. J. 1938: Response of unripened wheat grain to yarovization—effect of chilling. Compt. Rend. Acad. Sci. URSS 19:103–5. Biol. Abstr. 13: 17698. 1939.

Zarubailo, T. J. 11261, 11262.

20067. Zarudnaya-Kuznetsova, T. N. 1941: The role of light in the formation of vitamin C and the dynamics of its conversion in peas. Chronica Botan. 6:204.

Zatschler, H. 12210.

20068. Zaumeyer, W. J. 1929: Seed infection by *Bacterium phaseoli*. Phytopathology 19:96. Expt. Sta. Rec. 67:545. 1932.

20069. —— and B. L. Wade. 1934: Physiological spotting of pea seed. Phytopathology 24:1383-84.

Zaumeyer, W. J. 19258.

Zavylenkova, S. I. 18578.

Zayets, V. A. 11217.

20070. Zaytzev, G. N. 1963: On the germination of honey-suckle (*Lonicera* L. spp.) seeds after different duration of storage. Botan. Zh. 48:1698–701.

20071. Zazhurilo, K. K. 1931: Classification of the ornithochoric fruit and seed. J. Soc. Botan. Russ. 16:169–89. Biol. Abstr. 8:19423. 1934.

20072. —— 1940: Phylogenetical significance of sclereids in seed coats of Prunoideae. Compt. Rend. Acad. Sci. URSS 29:135–36.

Zdanova, L. P. 3503, 3753, 3754.

20073. Zechmeister, L., and K. Szilard. 1930: Über ein Carotinoid aus den Samenhüllen des Spindelbaumes (*Evonymus europaeus*). Z. Physiol. Chem. 190:67–71. Biol. Abstr. 9:354. 1935.

Zechner, L. 10432.

20074. Zedebauer, E. 1906: Die Keimnurfungdauer einiger Koniferen. Centralbl. Ges. Forstwesen 32:306–15.

20075. —— 1910: Versuche über Aufbewahrung von Waldsamereien. Centralbl. Ges. Forstwesen 36:116–21.

Zeeuw, D. J. de; *see* De Zeeuw, D. J.

20076. Zeiher, E. 1936: Zur Frage nach der Wirkung des Lichtes auf die Keimung lichtgehemmter Gramineenfrüchte. Jahrb. Wiss. Botan. 83:60–104.

Zeiner, W. 8699.

20077. Zeitgraf. 1913: Über Beziehungen zwischen Tausendkorngewicht und Keimenergie bei Kiefern Samen. Allgem. Forst- Jagdztg. 89:222–24.

20078. Zelawski, W. 1960: A contribution to recognition of the role of pericarp and seedcoat in the gas exchange of beechnuts. Sylwan 11:69–78.

Zelcbuch, B. 15448.

20079. Zelenchuk, T. K., and S. A. Gelemey. 1963: The germination of seeds of meadow grasses and legumes under laboratory conditions. Acad. Sci. RRS Ukraine Kiev J. Botan. 20(4):37–47.

20080. Zelenskaja, N. I. 1941: Effect of storing period on the viability and germination capacity of *Ocimum canum* Sims seeds. Proc. Lenin Acad. Agr. Sci. USSR 5:22–23.

20081. Zeleny, L. 1935: The distribution of nitrogen in the seed of *Zea mays* at different stages of maturity. Cereal Chem. 12: 536–42.

20082. —— 1940: Fat acidity in relation to heating of corn in storage. Cereal Chem. 17:29–37.

20083. —— 1953: Determination of moisture content of seeds. Proc. Intern. Seed Testing Assoc. 18:130–38.

20084. —— and D. A. Coleman. 1938: The chemical determination of soundness in corn. U.S. Dept. Agr. Tech. Bull. 644. 23 pp.

Zeleny, L. 1717, 1718, 9224, 9886, 9935.

20085. Zelinkova, M., F. Horavka, and R. Retovsky. 1951: Germination of the seeds of *Fraxinus excelsior* L. Sb. Česk. Akad. Zeměděl. 24:94–100. Biol. Abstr. 27: 17801. 1953.

20086. Zell, L. R. W. 1953: Effects of ultracentrifugation on seed germination and seedling development. Dissertation Abstr. 13:293–94. Brit. Abstr. B III:474. 1953.

20087. Zeller, A. 1938: Microelements et endosperme (Spurenelemente und endosperm). Bodenkunde Pflanzenernähr 7:56–117. Ann. Agron. 8:595. 1938.

20088. Zellner, E. R. 1942: Soybean seed produced by plants grown to maturity from strong, weak, and baldheaded seedlings. Assoc. Offic. Seed Analysts News Letter 16(4):3–5.

20089. Zelnik, R., J. von Euw, O. Schindler, and T. Reichstein. 1960: Die Glykoside der Samen von *Strophanthus kombe* Oliv. Helv. Chim. Acta 43:593–94.

20090. Zemis, N. 1928: Germination and seedling development of dark and light-colored pine seed. Mezsaimniecibas Rakstu Krajums (Riga) 6:40–44. Biol. Abstr. 5:25005. 1931.

20091. Zenari, S. 1929: The character "hard seed" in relation to heredity and environment. Ann. Botan. 18:174–215. Biol. Abstr. 6:15604. 1932.

20092. Zenchenko, V. A. 1954: The effect of geographic latitude on activity of peroxidase in the embryos and sprouts of barley. Dokl. Akad. Nauk SSSR 98:439–42. Chem. Abstr. 49:3321. 1955.

Zenger, E. 6378.

20093. Zenker, A. M. 1955: Jarowisationuntersuchungen an sommerannuellen *Arabidopsis*-Rassen. Beitr. Biol. Pflanz. 32: 135–70.

20094. Zenkteler, M., A. C. Hildebrandt, and D. C. Cooper. 1961: Growth in vitro of mature and immature carrot embryos. Phyton 17:125–28.

Zentmyer, G. A. 5679, 13555.

20095. Zentsch, W. 1960: Zur Wasserairnahme keimender Koniferensamen. Naturwissenschaften 47:70.

20096. —— 1960: Besteht ein Zusammenhang zwischen dem Saugkraftvermögen von Kiefernsamen und dem Standort? Naturwissenschaften 47:382–83.

20097. —— 1961: Über Eigenschaften von Kiefernsaatgut aus verschiedenen Kronenregionen. Forstwiss. Centralbl. 80:287–94.

20098. —— 1962: Zur Wasseraufnahme keimender Fichtensamen (*Picea abies* [L.] Karsten). Flora (Jena) 152:227–35.

Zentsch, W. 4810, 5434, 9664.

Zertuche, F. 5137.

20099. Zeuschner, M. 1926: Untersuchungen über die Dicke der Schale verschiedener Weigensorten, ihren Bau und Einfluss auf die Beizempfindlichkeit. Landwirtsch. Jahrb. 64:611–45.

20100. Zhebrak, E. A. 1961: Embryogenesis in diploid and tetraploid wheat. Dokl.—Botan. Sci. Sect. (English Transl.) 135: 240–41.

Zhilkina, M. D. 15420.

Zhukov, B. G. 19022.

Ziboh, C. O. 3825.

Zich, W. R. 2936.

20101. Ziebur, N. K., and R. A. Brink. 1951: The stimulative effect of *Hordeum*

endosperms on the growth of immature plant embryos in vitro. Am. J. Botany 38:253–56.
20102. ——, ——, L. H. Graf, and M. A. Stahmann. 1950: The effect of casein hydrolysate on the growth in vitro of immature *Hordeum* embryos. Am. J. Botany 37:144–48.
20103. Ziegenspeck, H. 1952: Die Wegsamkeit der Pigmentschicht der Getreidekörner (Endoderminschicht) für Fluorochrome. Protoplasma 41:425–31.
20104. Ziegler, E. R. 1930: Propagation of cacti from seed. J. Cactus Succulent Soc. Am. 1:194–96.
20105. Zielinski, Q. B. 1958: Some factors affecting seed germination in sweet cherries. Proc. Am. Soc. Hort. Sci. 72:123–28.
Zillinsky, F. J. 4167, 13966.
20106. Zimmer, W. J. 1941: Some notes on the germination of Mallee eucalypts. Victorian Nat. 58:61–62.
20107. —— 1942: A transect study of the flight of seeds of the Murray pine. Victorian Nat. 59:142–43.
Zimmer, W. J. 7711–7714.
20108. Zimmerman, A. 1881: Über mechanische Einrichtungen zur Verbreitung der Samen und Früchte. Pringsh. Jahresb. 12:542–76.
20109. Zimmerman, D. C. 1962: The relationship between seed density and oil content in flax. J. Am. Oil Chemists Soc. 39:77–78.
20110. Zimmerman, J. 1933: Ketiau oil from the seeds of *Ganua* (*Bassia*) *motleyana* (Sapotaceae). Chem. Weekblad 30:657–58. Chem. Abstr. 28:355. 1934.
Zimmerman, P. W. 8861, 8862.
20111. Zimmermann, F., and W. Gothan. 1933: Die Samen von *Sphenopteris bermudensiformis*. Sitzber. Ges. Naturforsch. Frd. 317–24. Botan. Centralbl. 165:179. 1933.
20112. Zimmermann, H. 1931: Fichtensamenbeschaffung in Sachsen. Tharandter Forstl. Jahrb. 82:821–64.
20113. Zimmermann, K. 1936: Zur physiologischen Anatomie der Leguminosentesta. Beiträge zum Problem der Hartschaligkeit und zur Bedeutung des Strophiolums. Landwirtsch. Versuchssta. 127:1–56.
20114. —— 1942: Einige Versuche zum Problem der Hartschaligkeit bei *Lupinus angustifolius*. Züchter 14:182–85.
20115. Zink, F. W. 1955: Studies with pelleted lettuce seed. Proc. Am. Soc. Hort. Sci. 65:335–41.
20116. ——, R. W. Grogan, and R. Bardin. 1957: The comparative effect of mosaic-free seed and roguing as a control for common lettuce mosaic. Proc. Am. Soc. Hort. Sci. 70:277–80.
Zinn, K. Napp-; *see* Napp-Zinn, K.
Zinovev, L. S. 10547.
20117. Zirkin, D. 1955: Germination of mango seeds. Hassadeh 35:966–67.
20118. —— 1956: Germination of avocado seeds. Hassadeh 37:39–40.
Zirkin, D. 6115, 6116.
20119. Zivanovic, D., and J. Danon. 1956: Beitrag zur Kenntnis der Wirkung des Ultraschalls auf die Keimung von Mais. Belgrade Inst. Fiz. Razv. Genet. i Selek. Zbornik Radova 1956(4):81–85.
20120. Zlataroff, A. 1926: Untersuchungen über die chemische Stimulation der Samenkörner. Fortschr. Landwirtsch. 1:81–83.
20121. —— 1934: Nouvelles contributions expérimentales pour l'explication de l'influence des stimulants chimiques sur la croissance des semences de plantes. Bull. Soc. Chim. Biol. 16:1720–29.
Zobel, B. J. 3723.
20122. Zobel, M. P., and G. N. Davis. 1949: Effect of the number of fruit per plant on the yield and quality of cucumber seed. Proc. Am. Soc. Hort. Sci. 53:355–58.
20123. Zöbel, A. 1875/77: Wie lange behalten die Pflanzensamen im Wasser ihre Keimfähigkeit. Wiss.-Prakt. Unters. Pflanzenbau. Haberlandt 1:89–97.
20124. —— 1888/89: Pericarp of the barley grain. Abhandl. Naturf. Ver. Brün. 25:205–28.
20125. Zohary, D. 1960: Studies on the origin of cultivated barley. Bull. Res. Council Israel Sect. D 9:21–42.
20126. Zollikofer, C. 1929/30: Zur Fortpflanzung von *Poa alpina* L. Jahrb. St. Gallischen Naturwiss. Ges. 65:99–117. Botan. Centralbl. 160:463. 1931.
20127. Zone, J. C. 1957/58: Application of a biochemical test for the determination of the germinative power in seeds of Argentine species. Arch. Bioquim. Quim. Farm. Tucuman 8:147–58. Biol. Abstr. 36:38547. 1961.
Zscheile, F. P. 9967.
Zsolt, J. 12795.
20128. Zubareva, G. 1957: Increasing the germinating ability of cherry and plum seeds. Sel'skoe Khoz. Sibiri 1957(12):43–46. Biol. Abstr. 35:54067. 1960.
20129. Zubenko, V. K. 1959: The effect of preplanting hardening of seeds against

drought on the grain harvest of corn in late plantings. Plant Physiol. (USSR) (English Transl.) 6:341–43.

20130. Zuber, M. S., W. L. Deatherage, C. O. Grogan, and M. M. MacMasters. 1960: Chemical composition of kernel fractions of corn samples varying in amylose content. Agron. J. 52:572–75.

20131. ——, ——, M. M. MacMasters, and V. L. Fergason. 1960: Lack of correlation between the amylose content of pollen and endosperm in maize. Agron. J. 52:411–12.

20132. ——, C. O. Grogan, V. L. Fergason, W. L. Deatheridge, and M. M. MacMasters. 1960: Some observed secondary effects of high-amylose genes in maize. Cereal Chem. 37:212–17.

20133. ——, G. E. Smith, and C. W. Gehrke. 1954: Crude protein of corn grain and stover as influenced by different hybrids, plant populations, and nitrogen levels. Agron. J. 46:257–61.

Zuber, M. S. 6299, 6512, 7705, 9000, 18395.

20134. Zubkus, L. P. 1950: On the viability of immature embryos of beans on culture media. Agrobiologiya 1950:137–41.

20135. Zucca, J. J. 1951: A study of acorns found buried in a redwood tree. Wasmann J. Biol. 8:257–65.

Zuck, R. K. 9781.

Zucker, L. 20136.

20136. Zucker, T. F., and L. Zucker. 1943: Nutritive value of cotton, peanut, and soy seeds. Ind. Eng. Chem. 35:868–72.

Zuckerhandl, H. Nothmann-; *see* Nothmann-Zuckerhandl, H.

Zuckerman, A. 8913.

20137. Zueva, E. S., and N. I. Proskuryakov. 1961: Glutathione reductase and cystine reductase in germinating and ripening peas. Biochemistry (USSR) (English Transl.) 25:695–97.

20138. Zuhr, E. 1933: Über die Wasseraufnahme und Wasserabgabe des Reifen und reifenden Getreidekornes in sortenkundlicher Hinsicht. Vienna (C. Gerold's Sohn). 126 pp. Proc. Intern. Seed Testing Assoc. 7:104–5. 1935.

20139. Zunini, E. 1935: Wheat embryo: the inorganic and phosphoörganic constituents. Chim. Ind. Agr. Biol. 11:339–42. Chem. Abstr. 30:1445. 1936.

Zurakowski, W. 5324.

20140. Zurawska, H. 1912: Über die Keimung der Palmen. Bull. Acad. Sci. Cracovie, Cl. Sci. Math. Nat. Ser. B 5: 1061–95.

Zweifel, J. 14672.

Zyczynska, I. 12497.

INDEXES

Index entries have been prepared for each of the 20,163 citations in the bibliography. Detailed entries indicate plants used or subject matter included. Obviously, with an estimated 80,000 such items available, it became impossible to give all of them in detail.

In deciding on the type of index to present, consideration has been given to the uses which have been made of the bibliography at Boyce Thompson Institute for Plant Research, Inc., over the years. It has served as a source of general information; of reference material for books and general articles and of background experimental data for seed research performed by the Institute scientific staff. It has also been used extensively, through personal consultation and correspondence, by experiment stations, seedsmen, and scientists throughout the world.

The authors and coauthors are arranged alphabetically in the citations listed in the main body of the bibliography.

Included in the two indexes prepared for publication in this volume are a rather detailed one of the plants studied and a more general one of the subjects covered. The plant entries are cross-indexed in so far as is consistent with the common and scientific names given by the authors. Assignment of either common or scientific name by the compiler has been avoided to prevent confusion in the terminology in different countries. In cases of few reports about a single kind of seed, no further classification is given. However, entries concerning plants of general interest, such as *Arachis* (groundnuts; peanuts) or *Triticum* (wheat), and many others, have been classified to permit the reader to locate references to the special subject of his interest.

No cross-indexing or detailed classification has been done for subjects, but it is hoped that the need for such information may be met, in a general way, by reference to the 548 subjects listed.

LELA V. BARTON

PLANT INDEX

Abaca, 4714, 6315, 16166, 18174
Abelmoschus, 4182, 11600, 11601; *see also* Okra
Abies: anatomy, 3252, 18424; characteristics of seeds, 18993; chemical composition, 2289, 19954, 19994; coats, 4164; germination, 1336, 1981, 4188, 4810, 6172, 9791, 13859, 15475, 16871, 18727; industry, 3949; seed production, 8689, 8690; storage, 384, 656, 4188, 5295, 11196, 18723; testing and analysis, 6395, 8585, 8980, 17097; viability, 384; *see also* Fir
Abronia, 19526
Abrus, 1902, 3845, 7173, 10464, 10505, 12228, 12916
Abutilon, 5030, 6931, 7496, 8192, 10647, 11641, 12907, 15586, 17928, 19759, 19760; *see also* Jute
Abyssinian durum wheat, 13311
Acacia: absorption, 19395; characteristics of seeds, 14374, 14855, 16026; chemical composition, 6232, 7338, 18790; fats and oils, 15948; coats, 741, 749, 826, 834, 5571, 8790, 9481; development, 14197; germination, 741, 747, 834, 5571, 7871, 8163, 8714, 9977, 10647, 14016, 15008, 19526; industry, 747; inhibiting substances, 11045; injury effects, 15154; maturity, 9482; microorganisms, 8824; moisture content, 19391; storage, 11196; treatment, 13653; viability, 18203; *see also* Wattle
Acalypha, 10150, 12699
Acanthaceae, 13041, 14673
Acanthia, 11591, 18934
Acanthosicyos, 10024, 10029, 17100
Acanthostachys, 8527
Acanthus, 13891
Accituno, 12189
Acer: characteristics of seeds, 1142; chemical composition, 3799; germination, 80, 673, 854, 3872, 4188, 4488, 5711, 6940, 8281, 8559, 9085, 11340, 13161, 13565, 13859, 14471, 15001, 15003, 15011, 15012, 17097, 17149; provenance, 3872, 5711; storage, 673, 1142, 8229; treatment, 1175; viability, 6151; *see also* Maple; Sycamore
Aceres, 16473
Achilles, 7697
Achillia, 9289, 9505, 10674, 10675, 12554
Achimenes, 848, 6338
Achiote, 11744
Achras, 16324, 18956
Achryachaena, 2176
Achyranthes, 5330, 7071, 10522
Acokanthera, 13659, 13878, 18546
Aconitum, 5238, 9643, 10675, 10677, 11836, 12058, 13339, 14386, 17079, 17127, 18937
Acorns: characteristics of seeds, 3251, 15687; chemical composition, 13401, 13402, 16095: enzymes, 15731, 19984, oils, 15759; development, 3279, 19179; dormancy, 13868; germination, 3482, 16450; industry, 4364; moisture content, 164; respiration, 3155; seed production, 3277, 3401; storage, 12440, 14780, 14813, 17233; viability, 2145, 15669, 18837; weight, 5154; *see also* Oak; Quercus
Acrocomia, 4298
Acropia, 6918
Actinea, 16155
Actinidea, 4542, 4543
Actinodaphne, 15758
Actinophora, 9747
Actaea, 98, 10675
Adansonia, 3449, 4224, 6704, 7629; *see also* Baobab
Adenanthera, 1586
Adenostoma, 18057
Adonis, 6548, 11510, 19791
Aechmea, 8527
Aegiceras, 3595
Aegilops, 325, 326, 12169, 14671, 19285
Aeginetia, 6060, 11582
Aegle, 3764, 3933
Aegopodium, 6475, 14412
Aera, 14404
Aesculus, 4404, 6151, 10217, 11196, 14430, 15001, 15012, 16141, 16473, 16584, 16766, 17266, 18964; *see also* Marrow
Aeschynomene, 16004
Aethusa, 8498
Aframomum, 2309, 4369
African Violets, 4206, 6690, 13201, 19877
Afrolicania, 737
Afzelia, 2307
Agathis, 6521, 18494, 19963, *see also* Pinus; Kauri; Damar
Agave, 9065, 16086
Ageratum, 3322, 11196
Aglaia, 2062
Aglaonema, 15305
Agnuscastus, 5605
Agropyron: anatomy, 6575; burial in soil, 7496; characteristics of seeds, 3212, 3299, 3629, 13200; coats, 10614, 17058, 17059, 17862; dormancy, 512, 5165; genetics, 2160; germination, 512, 1162, 4816, 5557, 7769, 8431, 8432, 8435, 9221, 15240, 15462; identification, 1028; maturity effects, 8662;

Agropyron–(*cont.*)
seed production, 763, 3879, 13534; viability, 12684; *see also* Quackgrass; Wheatgrass
Agrostana, 14538
Agrostemma: burial in soil, 7496; characteristics of seeds, 2796, 2797, 2798, 2799; chemical composition, 2800, 2802, 11563; germination, 2801, 3203, 8197, 10868, 11018; viability, 10739, 10741
Agrostis: burial in soil, 13535; germination, 496, 497, 2070, 3347, 4122, 4658, 6888, 8004, 8514, 12024, 12037, 14585, 15353, 18562; industry, 14490; inhibiting substances, 9289; photographs, 8807; plants effects, 36; seed production, 763; treatment, 496; viability, 18778
Aira, 4122, 10677, 14307, 14373
Aizoaceae, 14203
Ajowan, 2412
Akagi, 10121
Akamatsu, 8300, 9577; *see also* Pinus
Akebi, 11063
Akle, 5157, 9607; *see also* Albizzia
Akuammine, 16027
Alangiaceae, 13448
Alangium, 11682, 13608
Albizzia, 724, 3761, 3762, 5157, 5890, 6234, 7340, 8991, 9607, 13701, 19019
Alchemilla, 2685, 2989, 8873
Alder, Red, 5577; *see also* Alnus
Aldrovanda, 11244
Alectra, 2836, 2837
Alectrolophus, 17795, 17796
Aletris, 14386
Aleurites: chemical composition, 11634: enzymes, 19548, oils, 660, 710, 8764, 14944, 17206, 18374, proteins, 19972; characteristics of seeds, 7161, 8978, 11656, 15082; germination, 6655; industry, 710, 8072; maturity effects, 19239; storage, 11196; treatment, 3849; viability, 18794; *see also* Tung
Alfalfa, *see* Medicago
Algarrobilla, 8006
Alhagi, 7498, 7500
Alisma, 4602, 6400, 7710, 7843, 10677, 13868, 14415, 16697
Allanblackia, 5142, 7120, 13276
Alliaria, 5485, 10677, 10858
Allium: anatomy, 7081, 7337, 7900, 7901, 8321, 8322, 13358; burial in soil, 7496; characteristics of seeds, 1987, 1999; chemical composition, 7094, 10869, 12985; chemicals effects, 14429; genetics, 10323; germination, 647, 655, 7317, 8572, 9406, 9776, 10673, 10674, 10675, 10677, 17935, 18337; identification, 17932; moisture content, 655, 8276; polyembryony, 8486,
Allium–(*cont.*)
13645; radiation, 7234, 14385; seed production, 5820; storage, 11196; treatment, 463, 2977, 10183; viability, 463, 3322, 5864; *see also* Onion; Leek
Almond, 1695, 1723, 1932, 2971, 4463, 6371, 6372, 6950K, 7664, 9386, 11041, 13130, 15140
Alnus, 1449, 5577, 7437, 13161, 13259, 13260, 14386, 15003, 15458, 16364, 19858; *see also* Alder
Alonsoa, 9588
Alopecurus, 763, 1303, 1304, 1305, 2989, 3347, 3879, 4553, 8955, 15344, 15353
Alphonsea, 4414
Alpinia, 10633, 10634
Alsine, 7496, 10226, 19827
Alstroemeria, 8117, 9693
Althaea, 723, 4025, 7499, 12554; *see also* Hollyhock
Alysicarpus, 5558, 5561
Alyssum, 57, 282, 1553, 3322, 4661, 7499, 9929, 10677, 10749, 11196, 14412, 14415, 16410.
Alzelia, 2308
Amana, 14719
Amandes, 5688
Amaranthaceae, 839, 1600, 10010, 12040
Amaranthus: burial in soil, 4900, 4901, 4902, 4903, 4904, 7496; characteristics of seeds, 6037; chemical composition, 282, 3922, 4047, 5238, 13975, 19812; development, 19830; germination, 1553, 1555, 1557, 1600, 2010, 2677, 3212, 3705, 3864, 5973, 6079, 8503, 8955, 9237, 10101, 10103, 10936, 11442, 12040, 12907, 13339, 16135, 16649, 20067; identification, 11292; nutrition, 148, 18127, 19526; *see also* Pigweed
Amaryllidaceae, 4217, 8120, 10011, 12869, 14415, 16929
Amaryllis: characteristics of seeds, 12869, 18752; genetics, 14993; germination, 759, 2885, 6495, 7357, 7599, 7672, 8658, 9692, 9866, 11364, 15234; seed production, 8119; storage, 759, 8453
Ambrette, 13573; *see also* Hibiscus
Ambrosia, 917, 4900, 4901, 4902, 4903, 4904, 5032, 7496
Amelanchier, 4600, 8169
Amentiferae, 16482
Ammi, 474, 6475, 17651
Ammocharis, 12869
Ammodendron, 14224
Amomum, 2309, 16599
Amoora, 1590
Amorpha, 3799, 15005, 15009, 16349
Amorphophallus, 1310, 18911
Ampelocisiites, 2350

Ampelopsis, 5478, 15007
Amsinckia, 7735
Amygdalis, 7680
Anabasis, 10862
Anacaguitas, 1463; *see also* Sterculia
Anacardiaceae, 14415
Anacardium, 14944, 15974
Anagallis, 3776, 4030, 12305
Anawhata, 3043; *see also* Sophora
Ananas, 8975
Anaxagorea, 4414
Anchusa, 7689, 8569
Anda-acu, 15203
Androcymbium, 9657
Andrographis, 15889
Andromedoideae, 17328
Andropogon, 1497, 4452, 11520, 11811, 15533, 19937
Androsace, 282, 1553, 1556, 7697, 10674, 10677, 10681, 12388, 13339, 15185
Androsoemum, 4536
Anemone, 5127, 8751, 8752, 8955, 10677, 16863
Anethum, 6475, 6872, 8572, 9917, 10345; *see also* Dill
Angelica, 6475, 11425, 18403
Angels Trumpet, 9882
Anise, 9550, 10703; *see also* Pimpinella
Anisomeles, 14195
Annatto, 16023
Anona, 2426, 4414, 8443, 14092, 14093, 15801
Antennaria, 14640, 15185
Anthemis, 282, 1553, 9387, 9505, 11018, 11511, 15240, 15838, 19215
Anthericum, 10674, 10677
Anthospermum, 12165
Anthoxanthum, 8513, 14404
Anthriscus, 5485, 6475, 12224, 14412, 14415, 19261
Anthurium, 5042, 11196
Anthyllis, 764, 3837, 8790, 9171, 18846
Anticharis, 10019
Antidesma, 16750
Antirrhinum, 1999, 3322, 8205, 8345, 18111, 18112, 18113, 19832, 19833; *see also* Snapdragon
Apera, 11018, 14538
Aphanamixis, 7200
Aphanes, 8875
Apium, 647, 2977, 4073, 6233, 6475, 7496, 8276, 10673, 10677, 11196, 14412, 14419, 15414; *see also* Celery
Apocynaceae, 2459
Apocynum, 15763
Apodytes, 747
Aponogeton, 8749
Aponogetonaceae, 16720
Apple: anatomy, 11844, 18465, 19666; characteristics of seeds, 5867, 6287, 13961, 18811; chemical composition, 4115, 5198, 5465, 6103, 8727, 10378, 12485, 12486, 12819, 13957, 14739, 15271, 15711, 17185, 19656, 19823; chemicals effects, 2000; development, 3223, 5494, 13958, 17227; disease transmission, 3701; dwarf, 6461; embryo culture, 13960; genetics, 3074, 3271, 14272; germination, 22, 60, 63, 473, 1741, 2000, 3602, 3976, 4340, 4600, 5118, 6115, 8212, 10166, 10537, 10930, 11288, 12447, 12488, 12489, 12490, 12491, 14808, 19149, 19150; moisture content, 1274; plants effects, 16142, 17769; polyembryony, 1511, 9567; respiration, 8209; seed production, 6554, 12487; storage, 1455, 4340, 4594, 4600; viability, 1416, 8562; *see also* Malus
Apricot: chemical composition, 4701, 4925, 5924, 6371, 6372, 7863, 9386: fats and oils, 1769, 9711; development, 2932, 4521, 13547, 13548, 15846; germination, 4269; *see also* Prunus
Aquilegia, 8568, 9402, 10673, 10674, 10675, 10677; *see also* Columbine
Arabidopsis, 11407, 11478, 11668, 13580, 14183, 14186, 14188, 14189, 14190, 17364, 20093
Araceae, 3278
Arachis: anatomy, 16054, 17620; characteristics of seeds, 1079, 2718, 5609, 8447, 9375, 9758, 9840, 10694, 19959; chemical composition, 2754, 5353: enzymes, 11467, fats and oils, 12946, 14944, 15343, proteins, 4461, 9884, 11271, 14110, 14111, 18399, vitamins, 16044; color, 9375; development, 1855, 3575; embryo culture, 14623; germination, 5078, 5144, 10677, 11955, 12971, 13656, 13690, 20056; moisture content, 8276; polyembryony, 11938, 15105; seedlessness, 9834; storage, 11196; treatment, 18385; *see also* Groundnut; Peanut
Aralia, 98, 7597, 9391, 14386
Araucaria: anatomy, 8467; characteristics of seeds, 4493; dispersal, 13344; fossils, 4911; germination, 7819, 8601, 8721, 8955, 17627; identification, 19685, 19686; industry, 19651; nutrition, 6049; plants effects, 3566; polyembryony, 11484, 15183; storage, 225
Araujia, 841, 4242, 8279, 15288
Arbutus, 10677, 13565
Arceuthobium, 4259, 8411, 8532, 8533, 15178, 15362
Archangelica, 5568
Arctotis, 7499

Arctium, 4679, 7496, 11847, 14145, 14146, 14776, 14966
Arctostaphylos, 7218, 13565
Ardisia, 2713, 2824, 11252, 15009
Areca, 2124, 2858, 6359, 7530, 10517, 13332, 14091, 15045, 15111
Arechavaletia, 19068
Arenaria, 282, 1553, 4525, 8565, 10677, 14386
Argemone, 901, 1644, 3768, 3776, 7614, 7615, 10000, 17213
Arisaema, 98, 15340
Aristide, 9629
Aristolochia, 8568, 13292, 15012, 15699
Aristolochiaceae, 14415
Arnica, 6055, 9505, 9574, 19904
Aronia, 98
Arrow Wood, 17050
Arrhenatherum, 15462, 17121
Artabotrys, 4414
Artemia, 5692
Artemisia, 2667, 4533, 7697, 8572, 9289, 9505, 9733, 10666, 11037, 11038, 12554, 15148, 18894, 18895
Articularia, 10675
Artocarpus, 12249, 17747, 17845, 19013
Arum, 663, 5554, 19210
Arundinaria, 19997
Asarum, 8751, 10675, 14415
Asclepiadaceae, 2459, 15049
Asclepias, 941, 2115, 3991, 4781, 7709, 8080, 9020, 13709, 15161, 15584, 16347; *see also* Milkweed
Ascyrum, 7496
Ash, 1218, 2011, 3354, 4156, 11222, 12314, 12315, 13371, 13568, 19120; *see also* Fraxinus
Ash, Mountain, 748; *see also* Sorbus
Asimina, 11742, 15003, 15010, 16194; *see also* Pawpaw
Asparageae, 10677
Asparagus: burial in soil, 7496; characteristics of seeds, 9288; chemical composition, 9017; coats, 19892; development, 16259; germination, 98, 782, 3406, 11064, 11065; maturity, 16891; polyembryony, 15919; seed production, 7623, 17321, 17358; supersonics, 7186; weight, 2803
Aspen, 3438, 11810, 12846, 14752; *see also* Populus
Asperugo, 282, 1553
Aspergula, 14441
Asphodelus, 3776, 10504, 10514, 10673, 10675, 10677
Aspidistra, 7384
Aster, 724, 761, 1997, 2042, 3322, 4650, 5389, 6440, 7231, 7234, 7333, 8916, 11196
Asteraceae, 12925
Asteriastigma, 15151
Astragalus, 1476, 8942, 11087, 11091, 11557, 12941, 14050, 14132, 14224, 14415, 14441, 14515, 18141, 18142, 18233, 18296, 18762, 18869, 19967
Astrebla, 14020, 14023
Astrocarpus, 12781
Astrocaryum, 4298
Asystasia, 222
Athrotaxis, 2991
Athyrium, 9485
Atriplex: anatomy, 1601; characteristics of seeds, 11039, 14564, 16447; chemicals effects, 18856; germination, 1553, 1601, 2141, 2176, 3343, 4213, 10102, 13258, 19726; identification, 11291; industry, 747; plants effects, 9999; radiation, 8057; seed production, 10534, 11027; storage, 282; *see also* Saltbush, Fourwing
Atropa, 98, 9505; characteristics of seeds, 880, 10243; chemical composition: alkaloids, 9694, 9695, 17126, 18978, fats and oils, 8766; coats, 14538; germination, 5594, 8549, 8572, 10675, 10941, 12554, 12555, 19958; radiation, 10388; viability, 5928; *see also* Belladonna
Attalea, 4298
Aubrietia, 16952
Auricula, 10662
Avena: absorption, 3112, 4884, 6359; anatomy, 5608, 15219, 17261, 17272, 19240; burial in soil, 7496; characteristics of seeds, 6569, 6733, 8353, 9275, 9539, 9540, 9541, 15190, 17575, 19313, 19670; chemical change during metabolism, 269, 274, 1565, 4895, 5574, 13287, 13684, 17615, 19671; chemical composition, 3534, 4865, 8523, 8794, 9091, 9278, 13413, 14260, 14394, 15212, 16378, 17683, 18114: amino acids, 11017, carbohydrates, 339, 521, 11456, 13767, enzymes, 270, 1728, 5906, 11513, fats and oils, 9277, proteins, 8460, 10407, 14857, vitamins, 6727, 9216; chemicals effects, 4, 452, 516, 1999, 2828, 2961, 11902, 12794, 13337, 17867; coats, 2130, 16256, 19630; color, 14502, 16255; development, 15456; disease transmission, 2943, 2944, 5005, 9276, 12034, 12144, 12383, 14542, 18370; dispersal, 5054, 9279; embryo culture, 3433, 8236, 11828, 19212; fertilizers effects, 517, 680, 2706, 17246; genetics, 4253, 6730, 6732, 9033, 9873, 13965, 13966, 14518, 14519, 15145, 16736, 16926; germination, 271, 517, 570, 1637, 1872, 1999, 2067, 2389, 2526, 3118, 3131, 4234, 4250, 4251, 4636, 5564, 5575, 5793, 6568, 6594, 6731, 7769, 8513, 8514, 9297, 9298, 10043, 10593, 12040, 12182, 13133, 13727, 14261, 15191,

Avena–(*cont.*)
15192, 15353, 15995, 16496, 17120, 18155, 18701, 19215, 19669, 20034; identification, 1355, 6382, 7538, 8598, 8648, 13194, 14491; industry, 910, 4784, 5425, 12451, 13195, 16670, 19603; inhibiting substances, 270, 272, 516, 2525, 10985, 15175; injury effects, 461, 4153, 8100, 19312; maturity, 2763, 12634, 16949, 19668; microorganisms, 10600, 10601, 18855; moisture content, 69, 1742, 4310, 11401, 13242, 15915, 19696; plants effects, 1354, 6729, 19085; provenance, 16035; radiation, 67, 69, 70, 6072, 7426, 7427, 9542, 11891, 13503, 14520, 17318; respiration, 270, 272, 1740, 10441; seed production, 1024, 3119, 4249, 4252, 6728, 7564, 9973, 19675; stimulating substances, 268, 657; storage, 9190, 13244; treatment, 452, 692, 979, 2700, 4167, 9625, 12143, 13502, 13904, 14001, 14262, 18521; vernalization, 2512, 2787, 18410; viability, 6559, 18454; weed seeds in, 516; *see also* Oats

Avicennia, 2413

Avocado: chemical composition, 12774, 19426; coats, 5809; development, 17026; disease transmission, 13555, 19324; germination, 2058, 5808, 6919, 8044, 9900, 10100, 20118; industry, 7939, 18753; injury, 5808; polyembryony, 17025; seed production, 15416; treatment, 5679; weight, 8892, 8893; *see also* Persea

Axonopus, 495, 18669

Axyris, 6, 2176, 7496, 10534; *see also* Pigweed, Russian

Azadirachta, 212, 11874, 16641; *see also* Neem

Azalea, 2483, 13778, 14933, 19376

Bachmeria, 7496

Balanites, 737, 1402, 2307, 2308, 5331, 7120

Balanophora, 6186

Baldingera, 3347

Balsa, 8945

Balsaminaceae, 14207, 14415

Bamboo, 2699, 7966, 15956, 19609

Bambusaceae, 18279

Banana, 5235, 13178, 17265, 17306, 17456, 17457, 17990; *see also* Musa

Banisteria, 15935

Baobab, 3449, 4224, 6704; *see also* Adansonia

Baptisia, 3424, 3722, 12861; *see also* Honeysuckle

Barbaraua, 2455

Barbarea, 6318, 8569, 8572, 18672

Barleria, 15311

Barringtonia, 11662

Bartschia, 10675

Bartsia, 18942

Basil, 19099

Bassia, 5337, 8767, 9689, 16479; *see also* Madhuca

Basswood, 1468, 1681, 7683; *see also* Tilia

Bauhinia, 15761

Bayberry, 1982; *see also* Myrica

Bean: absorption, 3360, 6166, 7556, 7557, 8996; anatomy, 3268, 9090, 9457, 14792, 14805, 14839; characteristics of seeds, 1814, 1943, 3280, 4582; chemical change during metabolism, 449, 2159, 5140, 6808, 12149, 19140, 19938; chemical composition, 449, 4209, 4896, 9130, 9131, 9503, 9504, 13189, 13765, 13766; amino acids, 3819, 4833, 14062, 17711, 20033; carbohydrates, 3138, 7514; enzymes, 14065, 14144, 14352, 14797, 17143, fats and oils, 14893; nucleic acids, 9349, 9350, 14853, proteins, 2302, 2659, 2860, 6737, 7177, 9659, 9938, 10793, 14793, 14794, 14795, 14796, 15859; vitamins, 3930, 3931, 8386, 8717, 13596; chemicals effects, 6314, 12535; coats, 2184, 4459, 6236, 7303, 7331, 7334, 11930, 11931, 14592, 14634, 16608; color, 4861, 14157, 17644; disease transmission, 485, 3378, 3883, 8433, 12454, 14837, 16291, 16921, 17645, 18498, 19743; embryo culture, 8995, 20134; fertilizers effects, 2379, 2487, 15073; germination, 602, 4664, 6164, 6165, 6588, 6898, 19935; identification, 4023; industry, 371, 1701, 16985, 18373, 18679; inhibiting substances, 602, 7947; injury effects, 1006, 1491, 1699, 3264, 4624, 4628, 5559, 8159, 8720, 9397, 11199, 13139, 14625, 16091, 19553; maturity, 7759, 13593, 13940; moisture content, 13772, 19565; nutrition, 3004; pelleting, 427; plants effects, 8244, 8245, 15096; polyembryony, 2231; radiation, 2616, 7100, 9398; seed production, 371, 5560, 8158, 9127; storage, 925, 1159, 8349, 11401, 18676, 18679; treatment, 485, 730, 932, 2980, 4659, 5324, 5326, 5917, 7482, 13323; vernalization, 1813; viability, 6987; weight, 5071; *see also* Names of Individual Beans

Bean, Azuki, 265, 11571, 18346

Bean, Brazilian, 3105

Bean, Broad, 1278, 2301, 7825, 11401, 11756, 11757, 14193, 18870

Bean, Bulgarian, 9515

Bean, Burma, 15871

Bean, Bush, 10991, 19557, 19558

Bean, Carob, 12166, 13091, 13092, 13937; *see also* Ceratonia

Bean, Castor: absorption, 18489; anatomy, 229, 230, 10870; chemical change during metabolism, 2207, 2271, 2272, 2697, 4812,

Bean, Castor–(*cont.*)
5263, 11223, 12850, 13955, 14267, 15361, 19921, 19922, 19923, 19924; chemical composition, 3755, 7619, 7620, 9166, 16635, 19337: enzymes, 2422, 6205, 6206, 8027, 14852, fats and oils, 577, 6311, 7199, 13436, 13437, 13438, 16248, 19448, proteins, 9933, 10017, vitamins, 15899; coats, 15077; development, 13807; disease transmission, 13132, 18490; embryolessness, 15705; genetics, 15540; germination, 576, 3225, 3571, 7619, 7620, 7919, 8573; industry, 819, 887, 4271, 5658, 7261, 19457; inhibiting substances, 1499; provenance, 15703; storage, 10197; treatment, 10068; *see also* Ricinus
Bean, Central American, 10070
Bean, Dwarf, 7209, 9211; *see also* Phaseolus
Bean, Field, 6906; *see also* Dolichos
Bean, French, 3639, 5940
Bean, Garden, 219, 227, 11729
Bean, Horse, 1814, 9753, 19911
Bean, Jack, 82, 534, 1329, 9910, 10696, 10724, 12228
Bean, Kidney, 14163, 14164
Bean, Lima: characteristics of seeds, 622, 12681, 16585; chemical composition, 12898; carbohydrates, 11243, 19703; enzymes, 5028; proteins, 10844, vitamins, 5826; coats, 369, 14355, 17961; development, 17962; genetics, 369, 2265, 16585; germination, 5566, 6436, 8949, 18698, 19556; industry, 11712, 19378; inhibiting substances, 6598, 18396; maturity, 11980, 17963; microorganisms, 4665; seed production, 13170, 15873; storage, 3630, 11196; treatment, 4749, 5323, 11918, 13171
Bean, Mung: age, 16325; characteristics of seeds, 4102, 9428; chemical composition, 14235, 16310: amino acids, 2263, enzymes, 273, 6932; fats and oils, 10238; vitamins, 11876, 12372, 12373, 16049, 17473; industry, 12235; radiation, 13612; *see also* Phaseolus
Bean, Navy, 1943, 2889, 2890, 2891, 2892, 15645
Bean, Pink, 17642
Bean, Pinto, 908, 17641
Bean, Red Kidney, 6218, 7332, 8092, 11196, 17406, 17408, 17409, 17410
Bean, Refugee, 8258
Bean, Snap, 1012, 1492, 18677
Bean, Tonka, 5186, 7117, 8340, 15635; *see also* Dipteryx
Bean, Velvet, 10091, 13484
Bean, Wax, 401
Bean, White Pea, 5010, 5315, 13475, 15214, 15215, 15216
Bean, Yam, 3083, 8124, 14593; *see also* Pachyrrhizus
Beech, 2465, 3456, 5154, 6141, 8281, 8917, 10612, 11352, 14643, 14646, 20078; *see also* Fagus
Beef Wood, 4026; *see also* Casuarina
Beet, Sugar, *see* Beta
Beggarweed, 10078
Begonia, 835, 5587, 6872, 8042, 13781, 14071, 14072, 17123
Begoniaceae, 1752
Belladonna, 914, 2299, 8549, 9694, 10243, 11871, 13315, 13600, 17420, 18018; *see also* Atropa
Bellis, 757, 3322, 11196, 13535
Bells of Ireland, 18761; *see also* Moluccella
Belulites, 13252
Ben, 9705; *see also* Moringa
Bendi, 15973; *see also* Hibiscus
Bindweed, 3260; *see also* Convolvulus
Bentgrass, 4850, 5562, 9789, 14585, 17027; *see also* Agrostis
Benzoin, 17027
Berberidaceae, 10098, 11525, 13048
Berberis, 98, 5023, 5878, 13565, 13818, 14430, 14471, 15007, 15010
Berebera, 4152
Bergenia, 446
Bergeranthus, 12379
Berry Seeds, 13790
Berseem, 12548, 12761
Berteroa, 57, 1553, 2455
Bertholletia, 816, 2831, 8791, 19415; *see also* Brazil Nut
Bertolonia, 4571
Beta: absorption, 15830; anatomy, 7420, 10296; burial in soil, 7496; characteristics of seeds, 1912, 4632, 5542, 6363, 7975, 8937, 13111, 14833, 15628, 15685, 16826, 19818; chemical change during metabolism, 5423; chemical composition, 2453, 11839, 16090, 16825, 18072, 18622; enzymes, 11904, 11905, 16244, proteins, 4877; chemicals effects, 462, 465, 598, 1999, 4879, 5098, 5537, 5728, 5730, 6282, 6322, 8112, 10305, 12858, 14733, 15228, 15245, 17688, 19164; coats, 1011, 8110, 11639, 16878, 17991, 18901; development, 1424, 11305, 16824; disease transmission, 136, 1352, 1767, 2514, 4207, 6606, 7025, 10940, 11616, 11915; embryolessness, 7377; fertilizers effects, 5591, 5593, 14924, 17687, 17689; genetics, 1975, 1976, 5765, 9280, 9986, 10995, 12792, 14273, 15843, 17165; germination, 500, 501, 598, 1508, 1698, 1912, 2332, 2376, 2978, 4530, 4722, 4725, 5011, 5300, 5309, 5453, 5539, 5540, 5541, 5793, 5858, 6113, 6364, 7089, 7780, 8083, 8084,

Beta–(*cont.*)
8514, 8800, 9234, 9407, 10992, 12494, 12566, 14732, 15118, 15637, 16225, 16703, 16958, 17154, 17182, 17621, 17690, 17691, 18620, 18693, 18891; identification, 1655, 2727, 5865, 10840, 15355; industry, 1360, 1698, 1700, 1865, 2506, 2663, 3016, 3017, 3018, 3178, 3426, 4409, 5541, 5882, 5980, 9858, 9990, 11112, 11338, 11696, 11697, 11698, 15101, 15263, 15991, 16042, 16073, 16288, 17665, 18587, 18621, 18623; inhibiting substances, 5136, 5710, 6774, 7238, 7239, 12621, 12956, 13620, 13889, 15485; injury effects, 7392; maturity, 639, 17664; microorganisms, 501; moisture content, 7366, 8031, 8276, 9235, 11401, 15262; pelleting, 136, 427, 937, 5318, 12495, 12822, 13623; polyembryony, 11448, 11449; radiation, 5049, 5454, 17910; respiration, 5220; rhythm, 10541; seed production, 28, 1422, 1861, 2242, 2298, 4089, 10993, 10994, 12680, 12814, 13176, 14774; stimulating substances, 657, 14415; storage, 5862, 11196, 14832, 19577; treatment, 152, 462, 500, 598, 954, 1360, 2278, 2696, 2743, 3470, 4130, 5099, 5133, 5139, 5321, 5987, 6032, 6447, 6577, 6786, 6985, 6986, 7010, 8822, 9946, 10183, 15613, 18073; vernalization, 1860, 3373, 3896, 4091, 5219, 6321, 6340, 9752, 10838, 11226, 11336, 11747, 13739, 15716, 17847, 18071; viability, 3322, 8511, 8512, 12259, 12667, 14939, 17860; *see also* Beet, Sugar

Betel Nuts, 1659, 13332

Betula, 2524, 2527, 2528, 2530, 5739, 6387, 6395, 8020, 8606, 11196, 12298, 13161, 14386, 15001, 15003, 15005, 15006, 15126, 18723, 18913, 19345, 19483; *see also* Birch

Bidens, 7496, 14458, 16385, 16388

Bignonia, 15012

Bignoniaceae, 7534, 13044

Bindweed, 18577; *see also* Convolvulus

Biophytum, 14206

Birch, 1376, 2322, 2350, 2527, 2528, 6387, 8606, 9281, 16053, 16760, 19973; *see also* Betula

Bischopia, 16750

Bisserula, 7523

Bitaog, 15086

Bitterbrush, 6300, 13147, 14562, 15156

Bittergourd, 15901

Bittersweet, 1903, 8475, 9997; *see also* Celastrus

Bitterweed, 16155; *see also* Actinea

Bixa, 2307, 2308, 4354, 5353; *see also* Achiote; Annatto

Blackberry, 1040, 1086, 4916, 8590, 10479, 11837, 12893, 15665; *see also* Rubus

Black Cherry, 15114

Black Gram, 2084; *see also* Phaseolus

Black Locust, 13211, 13285, 13286, 16796

Black Pine, 16374; *see also* Pinus

Black Spruce, 11927, 16999; *see also* Picea

Black Walnut, 12586, 13159; *see also* Juglans

Black Wattle, 13653; *see also* Acacia

Blepharis, 11684, 15196, 15313

Blitum, 1600, 12040

Bloodroot, 7032

Blueberry, 2243, 4913, 4915, 17130; *see also* Vaccinium

Bluebrush, 12738

Blue Grama, 16198

Blue Grass, 2072, 6982, 8865, 10083, 15601

Bluegrass, Canada, 15117

Bluegrass, Kentucky, 1498, 1999, 2066, 2069, 2071, 2073, 3077, 3078, 3123, 3140, 6138, 6140, 8864, 13798, 15375, 17793; *see also* Poa

Bluegrass, Merion, 15386, 18166, 18167

Bluestem, 4452; *see also* Andropogon

Blyxa, 7528, 11680

Boehmeria, 15554; *see also* Ramie

Boerhaavia, 10145, 19526

Bomarea, 8117

Bombax, 10118, 11484, 15946, 16302

Bonaveria, 4563

Bonnaya, 9590

Bonsum, 114

Boraginaceae, 839, 8572, 11945, 14415

Borassus, 122, 7040, 7041, 13881, 17489

Borneo Tallow, 3428

Borreria, 13426

Bothricochloa, 196, 13703, 14735

Bouteloua, 4452, 4672, 9629, 16198, 18165, 18687, 19526, 19987, 19988; *see also* Blue Grama

Bowdichia, 12608; *see also* Sucupira

Bowlesia, 8015

Box elder, *see* Acer

Brachychiton, 747, 11633

Brachypodium, 2204A

Bragantia, 14104

Brassica: anatomy, 13950, 14521, 17324; burial in soil, 4900, 4901, 4902, 4903, 4904, 5485, 7496; characteristics of seeds, 7145, 12294, 13992, 19325, 19326; chemical change during metabolism, 5669; chemical composition, 586, 587, 9787, 10747, 12483; enzymes, 581, 582, fats and oils, 13657, 18763; chemicals effects, 1561, 14412, 14415, 14436; coats, 3348, 6274; color, 9907; development, 7410, 18156; disease transmission, 12368, 13921; embryo culture, 8926; fertilizers effects, 19315; genetics, 9115, 9116, 15221, 18204; germination, 583, 647, 1500, 1987, 1999, 3390,

Brassica–(*cont.*)
3504, 5849, 5850, 7638, 7769, 9156, 9175, 9904, 10555, 10710, 11079, 13184, 13824, 13855, 14627, 14925, 15240, 18315, 18552, 18670; identification, 640, 2316, 2317, 5852, 5853, 5854, 6386, 18448, 19475, 19679; inhibiting substances, 3552, 8850, 8856, 11588; injury effects, 675; polyembryony, 11448; radiation, 2882, 7234, 7235; seed production, 11276, 15162; storage, 11196; treatment, 394, 662; vernalization, 3777, 4090, 14137; viability, 463, 3322, 4953, 10739, 10741, 18203
Brassicaceae, 12293
Braya, 9512
Brazil Nut, 157, 816, 17139
Bridelia, 747
Broccoli, 1870
Bromegrass, *see* Bromus
Bromeliaceae, 5534, 8005, 8527, 8658, 8882, 8925, 17702
Bromella, 1461
Bromus: anatomy, 2169, 12647, 14907; burial in soil, 5485, 7496; characteristics of seeds, 12396; chemical composition, 282; coats, 14538; embryolessness, 960; genetics, 14400, 18722; germination, 368, 1022, 1553, 1556, 7697, 8571, 9463, 9895, 10706, 12613, 15462, 16524; identification, 9171, 9469; industry, 177, 1029, 6802, 9068; injury effects, 1022, 10228; seed production, 105, 368, 763, 2682, 3879, 4113, 12437, 12457; storage, 177, 194, 7537; testing and analysis, 2070, 17926; treatment, 2290, 6097, 14392; weight, 4084, 4085, 4086; *see also* Bromegrass
Broomrape, 11363, 14226, 19695; *see also* Orobanche
Browallia, 782, 4573
Browntop, 6888, 13705, 14934; *see also* Agrostis; Panicum
Brunnichia, 15011
Brunsviga, 8118, 12869
Brussel Sprouts, 5918, 12667; *see also* Brassica
Bryonia, 10681, 18600
Bryophyllum, 3290, 4496, 11828
Buchanania, 7344
Buchloe, 201, 3138, 7153, 8038, 8177, 10873, 15438, 19522, 19523
Buckwheat, 299, 2783, 4408, 5760, 5880, 6551, 9190, 9570, 12051, 16604, 18176; *see also* Polygonum
Buffel Grass, 505, 2060; *see also* Pennisetum
Bulbs, 7676, 11365
Bunga, 6060, 10058; *see also* Aeginetia
Bupleurum, 14412, 19000
Burdock, 11847, 14145
Burmanniaceae, 6449; *see also* Quercus
Burrawang, 17163; *see also* Macrozamia
Bur Reeds, 17929
Bursa, 7496
Butea, 3848, 10159, 15972
Butomus, 10681, 13868
Butterbrush, 9164; *see also* Purshia
Butterfly Weed, 4781; *see also* Asclepias
Butternut, 1023, 9083
Butterburr, 9390; *see also* Petasites
Butyracea, 3427
Buxus, 4856

Cabbage: Chemical change during metabolism, 18243, 18244; chemicals effects, 2045, 2684, 4627, 8296; development, 18508; disease transmission, 4348; fertilizers effects, 15073; germination, 12720, 13489, 14632, 17856; industry, 1067, 3296, 3735; inhibiting substances, 4489; microorganisms, 216; pelleting, 17393; radiation, 10619; seed production, 2832, 5817, 13153, 13704, 14085; storage, 11196; treatment, 4355, 9169, 14399, 16322; vernalization, 5367, 6930, 14138; viability, 12667; *see also* Brassica
Cacalia, 7627
Cacao: characteristics of seeds, 9337, 9338, 18958; chemical change during metabolism, 406, 13163, 13164, 13165, 13166, 15439; chemical composition, 1688, 5690, 6563, 6564, 12891, 12972, 19767; carbohydrates, 3726, 18473, enzymes, 10467, fats and oils, 4298, 6373, proteins, 2175, 17016, vitamins, 11629; chemicals effects, 14251, 15440; coats, 5468; development, 4786, 4787, 8014; disease transmission, 3341; genetics, 7309; germination, 451, 2175, 2740, 2789, 2790, 3847, 18222, 18771; industry, 15401, 16535; maturity, 4392; moisture content, 5120, 8033; storage, 9239, 9247, 15401, 18223; viability, 442, 6092, 15784; *see also* Theobroma
Cactaceae, 292, 1327, 2650, 2651, 2957, 3655, 4214, 5676, 5678, 6950–I, 7216, 8691, 10490, 10871, 11579, 11625, 12915, 14025, 14415, 15848, 16454, 16987, 17701, 19222, 20104
Caesalpinia, 7172, 7850, 13884, 14538, 14944, 15128, 15796
Caesalpiniodeae, 7684
Caesulia, 5240, 5241
Cajanus, 68, 711, 4284, 5694, 6696, 7791, 13327, 16699, 17512; *see also* Pigeonpea
Cakile, 2176, 10677
Calamagrostis, 10766
Calamus, 7773
Calandrinia, 4548
Calceolaria, 693, 8042

Calendula, 2176, 5745, 7499, 8205, 8568, 8580, 12532
Calepina, 4528
California Bay Tree, 14559
Calla, 98, 782, 7843, 13868
Callicarpa, 15010
Callicore, 9692
Calligonum, 11026, 14224
Calliopsis, 7499
Callistemon, 11253
Callistephus, 1997, 5389, 5451, 7333, 7499, 8916; *see also* Aster
Callitriche, 7843
Callitris, 747
Calluna, 4088, 10907, 10675, 13535
Calochortus, 3718, 7782, 17029
Caloncoba, 16536
Calondendron, 747
Calopogon, 3619
Calotropis, 15878, 16595
Caltha, 10239, 10677, 14386
Calvaria, 8793
Calvoa, 4568
Calystegia, 18797
Camassia 6185, 17029
Camelina, 57, 282, 1553, 8962, 14441, 19215
Camellia: characteristics of seeds, 2057; chemical composition, 787: fats and oils, 787, 1290; chemicals effects, 6857; development, 13142; embryo culture, 11725, 11726; germination, 1942, 2326, 7673, 8183, 12493, 13113, 17734, 18728, 19583; identification, 7673; polyembryony, 8098; radiation, 405; storage, 11196; viability, 5201; *see also* Tea
Campania, 13550
Campanulaceae, 839, 2354, 3322, 4554, 8042, 10674, 10675, 11196, 12981, 14386, 14412, 14415, 16418, 17029
Camphor, 16557
Camphor, Weed, 15464
Cananga, 15207
Canangium, 4414
Canarium, 8789, 8791, 12191, 14944, 17904
Canary Grass, 1827, 7328, 8802, 12936, 13771
Canavalia: chemical change during metabolism, 9910, 10773, 11663; chemical composition, 4803, 6696, 14845; enzymes, 4260, 5146, 5147, 5148, 5149, 5150, 5152, 7587, 10725; color, 17305; development, 6950C; inhibiting substances, 18866; *see also* Bean, Jack
Canellaceae, 15046
Cane, Sugar, 1423, 3013, 5507, 5698, 9761; *see also* Saccharum
Canna, 5607, 9654
Cannabis: anatomy, 15893; burial in soil, 7496; characteristics of seeds, 11120,
Cannabis–(*cont.*)
15082; chemical composition, 9753, 15421; fats and oils, 2117, 5409, 8764; chemicals effects, 14434, 16740, 17952; development, 11047; disease transmission, 6316; fertilizers effects, 2517; germination, 4532, 5183, 5498, 9729, 12763, 18177; industry, 4393, 8140; moisture content, 5647; polyembryony, 16820, 16821; stimulating substances, 9376, 12763; storage, 4887, 16817; treatment, 605, 10183, 10969, 17949, 19722; viability, 11114, 18684
Cannaceae, 8898, 14535
Cantaloupe, 7515, 15872
Canterbury Bells, 12386
Cantharospermum, 4565
Capparidaceae, 7847, 13043
Capparis, 15978, 18920
Caprifigs, 4328
Caprifoliaceae, 2459, 14415
Capsella, 47, 57, 282, 794, 1553, 4828, 10118, 10677, 11442, 12040, 13137, 14412, 15240, 15513, 15867, 16226, 16227, 19042, 19043
Capsicum, 1987, 4239, 4574, 5476, 6285, 7496, 10143, 11515, 12079, 13743, 13980, 14430; *see also* Pepper; Pimiento
Caragana, 4511, 4512, 9571, 11553, 12267, 12300, 14830, 15008, 15010, 17202, 17202–A, 17203
Caraguata, 8658
Caraway, 3455, 10539, 18900
Cardaicarpus, 4910
Cardamine, 2176, 9279, 12040, 12537, 12538
Cardamon, 2428, 2429, 4198, 6191; *see also* Amomum
Cardaria, 3210, 3211; *see also* White Top
Cardinal Climber, 3495
Cardiospermum, 3990, 10109, 10111, 10647, 14101
Carduus, 282, 1553, 7496, 13019
Carex, 454, 455, 10681, 11974, 13535, 14602, 17254
Carica, 803, 3869, 5101, 8936, 11768, 13729, 15010, 15029, 15047, 15911, 15949, 17501, 19056; *see also* Papaya
Carnation, 18108
Carnegiea, 292, 13167
Carob, 7486, 8857, 15448, 18930
Caropodium, 5195, 18328
Carpanthea, 7005, 12379
Carpinus, 1826, 18595
Carpotroche, 5231
Carrot: characteristics of seeds, 6005, 7753; chemical composition: fats and oils, 8481, 20055; chemicals effects, 255, 260, 1999, 2045; coats, 19396; development, 11305; disease transmission, 1743; embryo culture, 20094; fertilizers effects, 8224; germination,

Carrot–(*cont.*)
259, 820, 1987, 1999, 2804, 2977, 5684, 19400; industry, 6615, 12594, 14496; inhibiting substances, 256, 19399; injury effects, 10531; moisture content, 254, 255, 257, 258, 6946; pelleting, 345; plants effects, 16444; radiation, 2997; seed production, 1404, 4473, 5349, 5996, 6663, 8220, 8417, 8418, 8419, 8424, 15478, 16656; storage, 257, 1994, 11196; treatment, 954, 5544; vernalization, 11514; viability, 6947, 13617; *see also* Daucus

Carthamus, 8383, 8572, 10506, 15824; *see also* Safflower

Carum, 880, 4073, 6230, 6475, 8572, 17675, 19249

Caruncułus, 8986

Carya, 98, 673, 1023, 1990, 8081, 8362, 8552, 8559, 8568, 11765, 15001, 15003, 15005; *see also* Pecan

Caryophyllaceae, 839, 2795, 4467, 7775, 10181, 14415, 16880

Cashew, 1512, 9839, 10018, 15974, 15975, 18126, 18842; *see also* Anacardium

Casimiroa, 10643, 14007

Cassia: anatomy, 16006; burial in soil, 7496; characteristics of seeds, 18471; chemical composition, 3767, 6918, 8144, 9918, 12842, 14202, 14239, 15927, 17404; fats and oils, 6228, 17900; plants effects, 19059; polyembryony, 18237

Cassiope, 10677, 14386

Castanea, 1023, 2260, 2352, 8610, 10233, 13203, 15008, 18747; *see also* Chestnut

Castanospernum, 8831

Castor, *see* Ricinus

Casuarina, 747, 2823, 4026, 10044, 15010

Catalpa, 673, 8552, 9366, 15001, 18723

Catanache, 2176

Catsear, *see* Hypochoeris

Cattail, 4212, 13769; *see also* Typha

Cattleya, 7670, 12852, 12854, 14586, 14587, 15868

Caucasian Daisy, 15277

Cauliflower, 1242, 1987, 1994, 2373, 5515, 6961, 10507, 12667, 13824; *see also* Brassica

Caulophyllum, 98, 5334, 14098, 14944, 15086

Caytonia, 8256

Ceanothus, 102, 103, 4765, 7593, 7967, 9163, 13565, 15806

Cecropia, 8751

Cedar, 129, 133, 3750, 4439, 4768, 5983, 9769, 15068, 15076, 17568

Cedrela, 85, 747, 11734, 12883

Cedrus, 3255, 5385, 6395, 8612

Ceiba, 774, 7668, 11791, 14944; *see also* Eriodendron; Kapok

Celastrales, 13050

Celastrus, 98, 2411, 7833, 8272, 8475, 8574, 9997, 11523, 15010, 17257

Celery: chemical composition, 6229, 8727; fats and oils, 7772; chemicals effects, 4149; disease transmission, 8701, 13032; germination, 857, 5498, 6233, 9026, 14630; industry, 18404; microorganisms, 8702; seed production, 704; storage, 11196; treatment, 1871; viability, 19309; *see also* Apium

Celosia, 1600, 3322, 11196, 12040, 16649

Celsia, 8042, 9588A

Celtis, 98, 4156, 15008, 15009, 15011

Cenchrus, 221, 3226, 11661, 16757

Centaurea, 282, 1553, 1734, 3210, 3211, 3917, 4481, 5556, 7498, 7499, 7500, 8498, 8514, 8568, 8600, 10739, 10741, 11018, 11442, 11800, 18778, 19215

Centranthus, 7017

Centrosema, 18890

Cephalanthus, 5589, 15005

Cephalaria, 7356

Cephalotaxus, 16638

Cerastium, 5485, 5876, 9512, 13535, 14415, 14538, 19828

Cerasus, 8660

Ceratonia, 747, 5169, 7402, 12166, 13937, 18930; *see also* Bean, Carob

Ceratophyllum, 7843, 9762, 13868, 13869

Ceratozamia, 3784, 5474

Cercidiphyllum, 15010

Cercidium, 17136

Cercis, 128, 132, 5514, 7525, 8242, 13565, 15005, 18595, 19706

Cercocarpus, 2906, 12200, 13258, 13723

Ceropegia, 14556

Chaerophyllum, 4073, 8572, 14412

Chaetochloa, 7446

Chamaecyparis, 384, 3242, 7090, 7438, 8319, 8320, 11111, 11196, 11296, 14781, 17337, 18727, 19953, 20074

Chardinia, 2176

Chamaegigas, 8496

Chamaerops, 6359, 7270, 7797

Chana, 17403

Chaulmoogra, 5227, 7849, 16726; *see also* Hydnocarpus

Cheiranthus, 1899, 3322, 8944, 11196, 18880

Chelidonium, 1829, 2871, 10675, 10677, 14412

Chenopodiaceae, 1600, 8229, 12040, 14415, 16043, 19029

Chenopodium: anatomy, 1601, 2491; burial in soil, 4900, 4901, 4902, 4903, 4904, 5485,

Chenopodium–(*cont.*)
7496; chemical composition, 7818; chemicals effects, 3705; coats, 14538; germination, 880, 1601, 4740, 5197, 8270, 8500, 8677, 9650, 11439, 11442, 12040, 12907, 15240, 17260, 19215; identification, 7821, 8270, 11291; moisture content, 7697; nutrition, 9237; respiration, 2010; seed production, 8270; storage, 2010; viability, 10739, 10741, 18778
Cherry: chemical changes during metabolism, 14765; chemical composition, 8860, 15403, 15404: fats and oils, 9709, 17185, nucleic acids, 13739; development, 2917, 2918, 11304, 18820, 18822, 18823, 18824; disease transmission, 3700, 6454, 7257, 7258; embryo culture, 18817; germination, 3480, 4245, 4744, 6273, 6519, 6520, 7256, 7369, 8406, 12940, 14741, 14819, 15511, 18188, 18227, 19231, 20105, 20128; inhibiting substances, 13541; vernalization, 12940, 16319; viability, 18816; *see also* Prunus
Cherry, Australian Finger, 18769; *see also* Rhodomyrtus
Chess, Soft, *see* Bromus
Chestnut: characteristics of seeds, 5154; chemical change during metabolism, 9348; chemical composition, 2260; germination, 747, 1023, 2352, 15163; inhibiting substances, 11572; nutrition, 1933; radiation, 13667, 14191; storage, 8151, 9586; *see also* Castanea
Chestnut, Horse, 2786, 4404; *see also* Aesculus
Chestnut, Water, 9665; *see also* Trapa
Chewings Fescue, 6948, 14934, 15116
Chhola, 14061
Chickweed, 794; *see also* Stellaria
Chico, 16324; *see also* Achras
Chilli, 13980
Chilopsis, 19526
Chimonanthus, 1598
Chiogenes, 14386
Chionochore, 8538
Chionodoxas, 18970
Chinese Cabbage, 14115
Chinese Jujube, 81
Chinese Tallow Tree, 987; *see also* Stillingia
Chisocheton, 14944
Chive, 12985; *see also* Allium
Chloris, 2692, 7013, 7014, 7016, 12040, 12429
Chlova, 4552
Choko, 3021
Chondrilla, 2321, 17944
Chondropetalum, 11446
Chon Moellier, 3552; *see also* Brassica
Chorisia, 3703
Christmas Rose, 17270
Christmas Tree, 18777
Chrysalidocarpus, 15012, 15944
Chrysanthemum, 2176, 3203, 3322, 6184, 7496, 7497, 7499, 8042, 10233, 10311, 10739, 10741, 11155, 11196, 11346, 11828, 12040, 12698, 13535, 15240, 15277, 17086
Chrysobalanus, 14944
Chrysoplenium, 10675
Cicer: characteristics of seeds, 3858; chemical change during metabolism, 2423, 2424, 9655, 16320, 19081; chemical composition, 14842: proteins, 16321; chemicals effects, 16331; development, 5693; disease transmission, 10502; genetics, 17507; germination, 2774, 2820, 2826, 9548, 15405, 16321, 19110; maturity, 18602, 18770; *see also* Garbanzo; Gram
Cichorieae, 9052
Cichorium, 1523, 2703, 5569, 7563, 7576, 8276, 8572, 10996, 11196, 12449, 13928, 14336, 18514, 18778
Cinchona, 805, 818, 2014, 3665, 3800, 4479, 7793, 10006, 10471, 10497, 10681, 13779, 13780, 15617, 16716; *see also* Quinine
Cineraria, 693, 8042
Cinnamomum, 10261, 11556, 14944, 15009, 17337, 18736, 19394
Cipoabacate, 5230; *see also* Salacea
Circaea, 14386
Cirsium, 3210, 3211, 5223, 8434, 8444, 10739, 10741, 11018, 13019, 15185, 15240; *see also* Thistle
Cissus, 15009
Cistaceae, 2872, 14415
Cistanche, 10110
Cisteneae, 7049
Cistus, 14415
Citrullus, 71, 4505, 4609, 5333, 6059, 6197, 7496, 7848, 8165, 8739, 10308, 11035, 13575, 18203, 19718; *see also* Watermelon
Citrus: characteristics of seeds, 9584, 18267; chemical change during metabolism, 17741; chemical composition, 9736, 10605; amino acids, 16621; fats and oils, 5293, 5652, 9996; proteins, 2064; chemicals effects, 7920, 10792, 15122; development, 2106, 19810; disease transmission, 10851; embryo culture, 10852, 14717, 16596; embryolessness, 6855, 19463, 19885, 19886; fertilizers effects, 19319; genetics, 777, 3545, 3546, 4394, 8891, 9152, 14932, 15085, 18750, 19427, 19436; germination, 350, 3549, 6261, 12912, 13460, 13485, 13682, 15007, 15008, 15011, 16379, 19956; industry, 5927, 17484; maturity, 15123; plants effects, 3544; polyembryony, 3836, 6063, 6788,

Citrus–(*cont.*)
6856, 7870, 11484, 11834, 13736, 15920, 18713; radiation, 8336; seed production, 15969; storage, 2006, 8229; treatment, 1662; viability, 1623, 3936, 4256, 9053, 13681, 16172, 18203
Cladrastis, 15005, 18723
Clarkia, 8565, 8568
Clavigo, 2176
Claytonia, 19719, 19826
Clematis, 2576, 10677, 14412, 17803
Cleome, 3864, 4549, 7856, 8565, 13581
Clianthus, 19401
Clidemia, 4560
Climber, Cardinal, 10405
Clintonia, 14386, 17525, 17527
Clitoria, 7687, 18601
Clivia, 8658
Clover: chemical composition: carbohydrates, 4468; chemicals effects, 176; coats, 771, 2723, 5362, 5365, 8521, 8940, 13284, 14497, 17187, 17634, 18116, 19276, 19573; color, 39, 8520, 14886, 16441, 19549; development, 9313; disease transmission, 3938, 11911, 11912, 13553; fertilizers effects, 5716, 9326; genetics, 7941, 7983; germination, 91, 1809, 7637, 7681, 9173, 10568, 14405, 14770, 14771; identification, 3066, 3345, 4004, 4005, 7109; industry, 9680, 13512; injury effects, 6556; pelleting, 2315, 8350, 12405, 13222; provenance, 3345, 5573, 7102, 7109, 10935; seed production, 1935, 4705, 5643, 17982; storage, 725, 4998; treatment, 13465; viability, 92, 1974, 17422, 17629, 17980, 18847; weed seeds in, 14332, 19821
Clover, Alyce, 5561, 15322; *see also* Alysicarpus
Clover, Buffalo, 13235
Clover, Bur, 3674
Clover, Cluster, 6664
Clover, Crimson, 2075, 2284, 9681, 9686, 13207, 15363, 19836
Clover, Huban, 4210, 16154
Clover, Ladino, 1153, 5256, 8113, 10890, 11855, 15621, 16300, 19289
Clover, Low Hop, 8953
Clover, Mixtures, 13918
Clover, Persian, 19466
Clover, Red: anatomy, 2699; characteristics of seeds, 8954, 14533, 18101; chemical composition: amino acids, 168; chemicals effects, 10456; coats, 4000, 9564, 14303, 14306, 14308, 16809, 16810, 16811, 17258, 17631, 18569, 19927; color, 5735, 12906, 14533; development, 15639, 15640; disease transmission, 11914; genetics, 2253, 10052; germination, 4002, 8809, 12968, 18596;
Clover–(*cont.*)
industry, 240, 10193, 14555, 15982; injury effects, 6869, 8082, 9302, 10190, 13919; moisture content, 6099; provenance, 3588, 4700, 15357, 17879, 19470; radiation, 18411; seed production, 532, 1745, 2108, 2451, 2486, 2949, 8408, 8409, 8651, 8952, 9966, 12662, 13787, 18412; storage, 6086, 6347, 12667, 12908, 14629; treatment, 7446; viability, 19711; weed seeds in, 9921
Clover, Subterranean: absorption, 13270; characteristics of seeds, 2518, 2520, 2521, 2522; chemicals effects, 5457, 13271, 16434; coats, 210, 8265, 15815; genetics, 13760; germination, 1809, 1810, 2518, 2520, 13272, 19838; industry, 1217, 5908, 13269; seed production, 5456; vernalization, 13761; viability, 92
Clover, Sweet: absorption, 3443, 8593, 17988; anatomy, 12905; burial in soil, 18025; chemical composition, 16861; coats, 12006, 12009; color, 8287, 11325; development, 19416; disease transmission, 9947; embryo culture, 16701; genetics, 5524, 7465; germination, 5653, 6561, 10309, 10697, 12904, 12909; identification, 19602; industry, 1199, 10423; polyembryony, 19417; viability, 18024
Clover, White, 2606, 3102, 6581, 9309, 9924, 13932, 16421, 18135, 19840, 20050
Clover, Zigzag, 10419
Clusiaceae, 19093
Clyclanthaceae, 8185
Cnicus, 794
Cnidoscolus, 2738, 19069, 19070
Cobaea, 8128, 8569
Coccinia, 2399, 3780, 7205
Coccoloba, 15012
Cocculus, 10003
Cochlearia, 2481, 9512
Cochlospermum, 16981
Cockle, 11563; *see also* Agrostemma
Cocklebur, 1419, 5033, 9781, 13053, 13187, 16161, 16576, 16867, 17381; *see also* Xanthium
Cocksfoot, 3969, 9901, 11647; *see also* Dactylis
Coconut, *see* Cocos
Cocos: anatomy, 17191; characteristics of seeds, 11592, 12592, 15018; chemical change during metabolism, 12827; chemical composition, 15851: amino acids, 1878, 5174, enzymes, 19736, fats and oils, 9712, 13221, 14944; coats, 18807; development, 8650; dispersal, 8787, 8788; embryo culture, 56; germination, 57, 211, 466, 4360, 4514, 5237, 19784; industry, 213, 648, 5775, 12363, 12364, 12365, 12837, 16680,

Cocos–(*cont.*)
17605, 18885; maturity, 12839, 15320; nutrition, 466; plants effects, 648; polyembryony, 214, 5027, 7787, 18996; viability, 3687, 5783, 12838
Codaea, 2605
Codiaeum, 6913; *see also* Croton
Coffea: anatomy, 9104, 9105, 10006; characteristics of seeds, 1288, 3696, 6046; chemical composition, 2276, 2277, 3592, 10563, 14778, 16330, 17070, 18125: amino acids, 1890, carbohydrates, 19805; chemicals effects, 12670; color, 637, 3666; development, 13321; disease transmission, 16152; embryolessness, 13675; germination, 2043, 2268, 5069, 6313, 6710, 7355, 7428, 8948, 17110, 17111, 19404, 19508; industry, 9906; moisture content, 1621, 1622, 2043, 2897, 17102; polyembryony, 8094, 11484, 11699, 12624; provenance, 7918; radiation, 13677; storage, 881, 5108, 18884; treatment, 856, 2208, 5091, 12766, 13322; viability, 1625, 7095
Coix, 2307, 2308, 8569
Cola, 2665, 3790, 4179, 18990
Colchicum, 6865, 10675, 10677, 14412, 14571, 19072
Coleochaeta, 7082, 7083
Coleus, 8569, 19816
Collomia, 4248, 12036
Colocasia 55, 10613
Coltsfoot, 3234; *see also* Tussilago
Columbine, 9402
Columbrina, 15011
Colutea, 11032, 12481, 12482
Colza, 583, 18261
Comandra, 13454
Commelinaceae, 13976, 13977
Commelinantia, 15084
Compis, 4119
Compositae, 839, 2959, 8186, 8229, 11104, 13606, 14415, 14875, 17053
Conepia, *see* Oiticica
Conium, 8514, 9505
Connelleyi, 14224
Conophernuts, 6198
Conringia, 9016, 9885
Consinia, 14966
Convallaria, 98, 2047, 10677, 13339, 18068, 18772
Convolvulaceae, 839, 15951, 15979
Convolvulus, 3141, 3210, 3211, 3260, 4481, 7496, 7498, 7500, 8192, 10822, 10823, 11030, 12040, 17841, 17928, 18577, 19660
Cooperia, 16602
Copaifera, 13791
Copernica Palm, 10732, 10733, 10864
Copra, 6311
Corallorhiza, 5527
Corchorus, 644, 1880, 3765, 3864, 4045, 4941, 9500, 9501, 11356, 17223, 17726
Cordyline, 13748
Coreopsis, 8584, 11828
Coriandrum, 869, 1522, 3926, 4073, 6475, 7770, 8572, 18063
Coriaria, 16731
Cornaceae, 13448
Cornus, 98, 2000, 5023, 7458, 8791, 10938, 13565, 14386, 15001, 15003, 18595, 18766
Coronaria, 13297
Coronilla, 14415, 15202, 18363
Cortusa, 10677
Coruporeteae, 2459
Corydalis, 9279, 10677
Corylus, 4832, 7991, 11196, 11450, 13565, 15001
Cosmos, 733, 3601, 7499, 8205, 16644
Cotoneaster, 1078, 4742, 7217, 8877, 17616, 17617, 17618, 18595
Cotton, *see* Gossypium
Cotton Wood, 5981, 6837, 6934, 13145
Cotula, 5004
Cotyledon, 4545
Couepia, 18983
Covillea, 16548
Cowpea, 118
Crabapple, 5018, 14389
Crabgrass, 4481, 5110, 7207; *see also* Digitaria
Crambe, 460, 5052, 9605, 11638, 12685
Cranberries, 6356
Cranberry, Highbush, 17047; *see also* Viburnum
Crassulaceae, 8068, 13039, 14415, 18134
Crataegus, 98, 5036, 5768, 6328, 6461, 6465, 11041, 14642, 14644, 15005
Creeper, Virginia, 8292
Crepis, 2176, 5485, 10701, 14252, 14254, 17351, 17354, 17355, 17888, 17889
Cress, 18505, 18773
Crinum, 5696, 11249, 13348, 14130, 15932, 18633
Crocus, 8708
Crotalaria, 3135, 5621, 5622, 5623, 8991, 10450, 13565, 14343, 15736, 16679, 17241, 18492, 18647
Croton, 770, 776, 2414, 3888, 3889, 3890, 8638, 10392, 14944, 15082, 17397, 18969, 19062
Crucastium, 1553
Cruciferae: characteristics of seeds, 13954, 16279; chemical change during metabolism, 10293; chemical composition, 578, 580, 589, 590, 10746: carbohydrates, 5664, enzymes, 579, fats and oils, 586, 13456, 18796, proteins, 10805; chemicals effects,

Cruciferae–(*cont.*)
6992; coats, 358, 17059, 19038; development, 11947; embryo culture, 8121, 8123, 14508; germination, 839, 2455, 8122, 10620, 13854, 13856, 14456, 17857, 17858; identification, 356, 357, 1481; radiation, 14875; stimulating substances, 14415; storage, 8229; treatment, 4905
Cryptanthus, 8658, 19526
Cryptocoryne, 18991
Cryptomeria: anatomy, 11567; characteristics of seeds, 7434, 8612; coats, 7429; genetics, 8318; germination, 747, 8147, 8485, 13429, 16805, 18190, 20074; industry, 14143; maturity, 7438; plants effects, 7433; storage, 11111, 11196, 11296, 17337; viability, 18189
Cryptostegia, 15946A, 17402
Cryptotaenia, 8795, 13028
Cucumber, *see* Cucumis
Cucumis: burial in soil, 7496; characteristics of seeds, 17142, 20122; chemical change during metabolism, 2228, 10773; chemical composition, 104, 2415, 9170, 17746; amino acids, 5657, 10869; enzymes, 2085, 2551, 4887, 18547, 18550, 19273; fats and oils, 4046, 5335, 17209; proteins, 14883; chemicals effects, 1664, 9031, 10338, 11408; development, 17500; disease transmission, 9061; embryo culture, 14525; genetics, 6929, 18100; germination, 98, 632, 820, 1556, 4248, 5793, 7516, 12712, 19420; identification, 9417; industry, 1477, 1815, 1956, 9283, 12471, 16545; moisture content, 6031; plants effects, 18169; radiation, 8736; seed production, 3028, 8039, 8425, 17592, 18804; storage, 925, 5325, 11196, 15081; treatment, 954, 14246; vernalization, 12333; viability, 12667
Cucurbita: absorption, 6359, 19090; anatomy, 14430, 16675, 18892; characteristics of seeds, 14433; chemical change during metabolism, 1317, 1471, 3163, 5992, 8694; chemical composition, 4776; enzymes, 3478; fats and oils, 19468; proteins, 301, 19106; chemicals effects, 9031; coats, 3161; color, 18137; development, 14123, 19271; disease transmission, 4629; embryo culture, 3162, 6806; embryolessness, 16332; germination, 513, 1556, 5007, 5064, 5170, 8963, 9406, 10300, 11500, 18464, 20076; industry, 5518; inhibiting substances, 14120; nutrition, 9167; plants effects, 17843, 19302; radiation, 7234; respiration, 11924; storage, 11196; viability, 18203
Cucurbitaceae, 302, 303, 520, 3760, 4024, 4049, 7829, 8229, 8852, 10803, 12100, 14415,
Cucurbitaceae–(*cont.*)
17486, 17502, 17503, 18735, 19526; *see also* Gourd
Cuddapah Almonds, 11414
Cuminum, 6943, 8572, 15875
Cunninghamia, 3247
Cunonia, 747
Cuphea, 19740
Cupressus, 225, 747, 1981, 2427, 5385, 5418, 10647, 18148, 18149, 18723, 18727
Cupuliferae, 11111
Curcas, 5579; *see also* Physic-nut
Curcuma, 13582
Currant, 2659, 9797, 10754; *see also* Ribes
Curly Mitchell Grass, *see* Astrebla
Curtisea, 747
Cuscuta: anatomy, 3024, 10184, 14823; burial in soil, 7496, 15092; characteristics of seeds, 2994, 11341; chemical composition, 154, 156; coats, 529, 14538; development, 6272, 10944; embryo culture, 12696; germination, 2741, 5432, 5584, 6892, 6894, 9522, 12080, 13380, 15360, 16404, 18596; identification, 4230, 6892; maturity, 529, 10669; treatment, 18021; viability, 7425; *see also* Dodder
Cyamopsis, 347, 475, 9319, 9320, 13295, 13993, 15061, 15898, 16008
Cycas, 1140, 4488, 14510, 14511, 14512, 15031, 15953, 16221, 16982, 19629
Cyclamen, 780, 886, 1123, 2678, 2679, 2680, 3099, 4339, 4523, 5918, 7489, 8042, 8658, 9395, 10769, 10820, 11349, 12581, 12582, 16355, 17104, 17779, 17883, 18164, 19380, 19385, 19831, 19871, 19969
Cyclanthera, 4294, 4576, 16845
Cydonia, 930, 10930, 15459, 17899
Cymbidium, 3582, 6951, 6952, 6953, 6954, 6955, 6956, 8811, 10988
Cynanchum, 2115
Cynandropsis, 7846
Cynanthum, 17167
Cynara, 7185
Cynodon, 221, 8571, 19473
Cynoglossum, 5485, 13510
Cynomarium, 10045
Cynosurus, 763, 8955, 9651, 15344, 15353, 18109, 18486
Cyperaceae, 8957, 11116
Cyperus, 618, 620, 2247, 7496, 7795, 8796, 8797, 8798, 10092, 12448, 13393, 16990, 17431; *see also* Nutgrass
Cypress, 14781, 17799, 19083; *see also* Chamaecyparis; Cupressus
Cypripedium, 3620, 3621, 4739, 4772, 6582, 12003, 12219, 12220, 12221
Cystophyllum, 14722
Cytinus, 7899

Cytisus, 5437, 5796, 6914, 7689, 7899, 8790, 10673, 16446, 17029, 18846

Dacrydium, 3243
Dactylis: characteristics of seeds, 19329; chemical composition: proteins, 19912; chemicals effects, 3969; germination, 368, 3569, 3968, 12276, 13339, 17817, 20036; provenance, 3569; rhythm, 11221, 11531, 11532; seed production, 368, 763, 6834, 11701, 11708, 11709; testing and analysis, 1, 13989, 15353; viability, 10739, 10741
Daemia, 2511
Dahlia: chemical composition: carbohydrates, 2064; germination, 695, 4982, 5436, 5939, 6010, 6603, 7442, 13501, 15450, 16277, 17891; storage, 11196; viability, 3322
Daisy, 701, 757, 786
Dalea, 19526
Dallis grass, 8982, 9849, 10891, 14909
Damar, 18494
Damasonium, 2104, 7843
Dandelion, 794, 1997, 2371, 15279, 15280, 16276; *see also* Taraxacum
Danthonia, 9895, 11850, 13702, 17474, 18689
Daphne, 19060
Daphniphyllum, 19066
Date, 467, 6849, 12369, 15528; *see also* Phoenix
Datura: anatomy, 14658; burial in soil, 5485, 7496; characteristics of seeds, 2157; chemical composition, 15649; alkaloids, 3519, 3885, 9694, 15655, fats and oils, 8766, 12811; chemicals effects, 463, 16206; development, 4572, 7323, 7326, 16205; dispersal, 2913; embryo culture, 2586, 2588, 2589, 5433, 12990, 15062, 15986, 16207, 16707, 16708, 16709, 16710, 17735, 18988, 18989; genetics, 1536, 3656, 3659, 3660, 5281, 16792, 17759, 17933; germination, 5432, 8569, 8897, 9505, 9882, 10647, 11425, 16791, 16953, 18337, 19526; inhibiting substances, 16208; maturity, 794, 7231; radiation, 17789, 17790, 17792; seed production, 16793; viability, 463, 727
Daubentonia, 4758, 15210
Daucus: apomixis, 8045; chemical change during metabolism, 10869; chemical composition: fats and oils, 4073, 14453; development, 2805; embryolessness, 6475; germination, 1987, 1991, 1999, 2805, 8514, 12040, 12907, 14441, 15240, 17705; moisture content, 8276; storage, 11196; testing and analysis, 8895; viability, 4800, 10739, 10741, 17080; *see also* Carrot
Davidia, 8615, 8789, 8791, 10042, 19727
Daylilies, 851
Delnostena, 19947
Delonix, 14944, 15697
Delphinium: burial in soil, 5420; chemical composition, 12876; alkaloids, 4133, 7452, 7453, 15441; enzymes, 2554; fats and oils, 3988; germination, 463, 724, 849, 1553, 1989, 1999, 2494, 4693, 5601, 5976, 6711, 7656, 8205, 8569, 10674, 10675, 10677, 14412, 14660, 16089, 17357, 19033; radiation, 3100, 7234; storage, 1983, 1984, 2012, 5103, 7499, 11196; testing and analysis, 12584; viability, 724, 3322
Delwiche, 7984
Dendrobium, 9523
Dendromecon, 13565, 19797
Dendrophthae, 14209
Dendrophthora, 19993
Dentaria, 8751
Derris, 15940
Desmanthus, 5415
Desmazaria, 2204
Desmodium, 475, 1534, 16008
Desmos, 4414
Dewberries, 4916
Dialium, 3688
Dianthus: anatomy, 3275; development, 3276, 7801; germination, 3322, 8205, 8568, 10674, 10675, 10677, 12583; identification, 1602; storage, 7499, 11196; viability, 724
Diapensia, 14386
Dicentra, 17506, 19797
Dichanthium, 197, 13579, 14737
Dichapetalum, 15231, 15232, 19360
Dichondra, 1045
Dichroa, 12119
Dichrostachys, 5412, 7339
Dicranostignia, 4561
Dictamus, 8568, 14441
Dictyota, 9139
Didiscus, 8569
Dielleniaceae, 14953, 15937
Digitalis: characteristics of seeds, 14753; chemical composition, 4737, 7291, 13626, 13627, 18239: carbohydrates, 7911, enzymes, 19454, fats and oils, 12942, 19484; chemicals effects, 18382; coats, 14538; development, 4343; disease transmission, 1059; germination, 666, 4481, 5432, 7017, 7710, 9505, 10675, 10677, 10943, 12039, 12554, 14414, 14898, 20076; rhythm, 3207, 3283; stimulating substances, 16109; storage, 666, 11196; viability, 3322, 5928; weight, 2605
Digitaria, 5165, 7207, 8172, 10657, 13029, 17331, 17332, 15616, 18673
Dika, 3429
Dill, 6480, 6484, 7774, 10345

Dimorphandra, 7871
Dimorphotheca, 2176, 2470, 4425, 8955, 12040, 12532
Dioclea, 12106
Diodia, 4559
Dionaea, 17625
Dioon, 1636, 5475
Dioscoreaceae, 19551
Diospyros, 181, 1792, 6491, 8437, 9475, 12095, 12389, 12994, 15007
Diplotaxis, 14415
Dipsaceae, 10771, 14415
Dipterocarpaceae, 1917
Dipterocarpus, 15009
Dipteryx, 788, 7117, 8340, 15635
Dirca, 14386
Dischisma, 10064
Dock, 701
Dodder: absorption, 13418; coats, 530; germination, 412, 4230, 5584, 6892, 12239, 15472, 15735, 17853, 20017; *see also* Cuscuta
Dodonea, 10004, 15055
Dogstail, Crested, 9651
Dogwood, 2000, 7458; *see also* Cornus
Dolichos: chemical composition, 450, 619, 6909, 18463: enzymes, 711, 13327, proteins, 10796, 16840; plants effects, 619
Dorotheanthus, 12379
Dorstenia, 9279
Dove Tree, *see* Davidia
Draba, 9512, 14415, 17029
Dracaena, 8561, 8567
Dracocephalum, 14441, 18925
Drega, 19756
Drosera, 8527, 8985, 10673, 10675, 10677, 14198, 15102
Duranta, 6994
Durra, 3587
Dwarf-Indigo, 16349
Dwarf Mistletoe, 16881
Dyckia, 8527
Dysoxylum, 14944

Ecballium, 4570, 12168, 12988, 16494
Echinocactus, 8527
Echinoceptis, 18550
Echinocloa, 1306, 1307, 1308, 1309, 3212, 5043
Echinocystis, 3097, 4012, 4405, 8569, 11762, 15257, 17134, 17135, 17485
Echinodorus, 4820
Echinophora, 5195, 18328
Echinopsis, 8527
Echium, 1063, 1811, 14414
Eggplant, 925, 1987, 1994, 1999, 5822, 6991, 10104, 10107, 18686; *see also* Solanum
Eichhornia, 159, 2044, 3193, 8009, 8861, 8862, 12499, 15056, 15286
Einkorns, 341
Ekebergia, 747
Elaeagnus, 4418
Elaeis: chemical change during metabolism, 4607; chemical composition, 4610; embryo culture, 2851, 15829; germination, 685, 3501, 3693, 4780, 5251, 6310, 6319, 8649, 8835, 9264, 16065; respiration, 4608; *see also* Palm
Elaeocarpaceae, 13042
Elateriospermum, 16025
Elaterium, 9279
Elatine, 13868
Elderberry, 17051
Eleagnus, 15003
Eleocharis, 13868, 17255
Elephant Grass, 14652
Elephantorrhiza, 8935
Elettaria, 2039, 10314
Eleusine, 1580, 1585, 3806, 3807, 4748, 6844, 7496, 14217, 16756, 18671
Elliottia, 19584
Ellisiophyllum, 19948
Elm: chemical composition: fats and oils, 17048; chemicals effects, 2000, 7727; development, 12066; disease transmission, 3014; germination, 1348, 1996, 3205, 5957, 11436, 12536, 15427, 16506; industry, 7112; maturity, 7729; seed production, 15427; storage, 1996, 7112; viability, 7526; *see also* Ulmus
Elsholtzia, 2455, 2672, 5567, 9493, 15010
Elymus, 2160, 4325, 4342, 5379, 6097, 7496, 13964, 15462
Elytropappus, 12164, 12165
Emblica, 5329
Endive, 13928, 15988, 18514
Endymion, 7391
Entada, 2055, 7687, 14944, 15929
Eosperma, 1916
Ephedra, 16856, 18378, 19028
Epigaea, 1950
Epilobium: chemicals effects, 19653; coats, 6045; development, 1961, 1962, 15480; germination, 1556, 2454, 2672, 6501, 7015, 7710, 8687, 9485, 9486, 9489, 10674, 10675, 12038, 12039, 13394, 14898, 15185, 20076; radiation, 9495
Epipactis, 7992
Epipogium, 145
Epithema, 14946
Equisetum, 9279
Eragrostis, 195, 6812, 6813, 6814, 6817, 8038, 9492, 12416, 18687
Eranthemum, 15888
Eranthis, 1429, 7952, 8955, 10677

Eremochloa, 10901
Eremosparton, 14224
Erica, 4088, 13535
Ericaceae, 6449
Ericales, 14992, 19047
Erigeron, 282, 1553, 2321, 6317, 8658, 10674, 10675
Erinus, 4550, 10675
Eriobotrya, 5759, 6033, 11576, 15008, 15746
Eriocaulon, 13868
Eriodendron, 774, 15082
Eriogonum, 19825
Eriophyllum, 19526
Erodium, 18337, 19526
Erophila, 282, 1553
Eruca, 8572, 9802
Erucastrum, 282
Ervum, 4530, 7234, 7235
Eryngium, 14386
Erysimum, 57, 282, 1553, 5420, 7385, 10822, 10823, 11294
Erythraea, 2115, 9505, 10674, 10675, 18337
Erythrina, 6522, 6523, 7277, 12836, 14848, 15012, 18123
Erythronium, 7781, 7782, 9143, 9763, 15185
Erythrophleum, 15064
Eschscholtzia, 7499, 8205, 9279, 9485
Esculus, 12993
Esparette, 11213
Eucalyptus: characteristics of seeds, 7711; coats, 7044, 7045, 7046; germination, 747, 2670, 4202, 4754, 6691, 7272, 7712, 8955, 10346, 15003, 15222, 20106; inhibiting substances, 12107; provenance, 10283; seed production, 4753; storage, 12986; testing and analysis, 7713, 7714, 18483; viability, 19220
Euchlaena, 4375, 16830
Eucommia, 10156, 10157, 15257
Eugenia, 1930, 9848, 13979, 16631, 18962
Euonymus: anatomy, 1928; characteristics of seeds, 3477; chemical composition, 15914, 20073: enzymes, 12351, 12561; chemicals effects, 3315, 3316; embryo culture, 14469; germination, 3318, 8574, 11383, 12352, 14470, 14473, 15003, 15181, 16734, 16735, 19021; polyembryony, 3012, 3082; storage, 11196
Eupatorium, 14386, 14458
Euphorbia: anatomy, 13883, 16344, 17518; chemical composition, 7244, 7245, 7247, 7248, 7249, 7251, 7252: fats and oils, 8247, 8639, 15549, 16660; coats, 14995, 18964; development, 4834, 4857, 10149, 11217, 13882; dispersal, 533, 9279, 16900; germination, 5721, 5767, 8955, 10675, 10677, 19526, 19956; identification, 11428; polyembryony, 8487; provenance, 7250; viability, 8555
Euphorbiaceae, 839, 10390, 11749, 12791, 14415, 17902
Euphrasia, 8526, 12224, 14386, 19976
Eurotia, 8827, 13258
Euryale, 1321, 2785, 14724, 14725, 14726, 14727, 14728, 14729
Euterpe, 428
Evenothera, 7946
Exochorda, 15010

Faba, 5116, 5614
Fagales, 8874
Fagopyrum, 299, 2783, 7496, 9570, 11433, 12051, 12713
Fagraea, 1180
Fagus, 3456, 4188, 6151, 8281, 11786, 13859, 15003, 16721, 16972, 17097
Fanweed, 4211, 18766
Feijoa, 11614
Fennel, 2100, 4200, 9387, 17092
Fenugreek, 4891, 8342, 17046, 17725, 19896
Ferula, 2821, 14467, 14605
Festuca: burial in soil, 7496, 13535; characteristics of seeds, 9171; chemical composition, 168: amino acids, 168, carbohydrates, 12645; chemicals effects, 3466, 7084; disease transmission, 19512; germination, 769, 1999, 3466, 4122, 8955, 9318, 9895, 10399, 10401, 13339, 14404, 15166, 15344, 15462, 15671; identification, 7108, 11995, 13994; industry, 769, 2073; moisture content, 17782; rhythm, 11221; seed production, 763, 5047, 7370, 14005; storage, 2073, 10400, 10402, 11796; testing and analysis, 1, 9308, 12178, 15353; viability, 3322, 6597, 8744, 9295, 9517
Feterita, 18211
Ficaria, 19808
Ficus, 4327, 4520, 8715, 9916, 10535, 10673, 10677, 15008, 16712, 17699, 17700
Filago, 19526
Filbert, 3987, 6215, 17045
Finger Seed, 1876
Fir: characteristics of seeds, 1963, 16337; chemical composition, 14329; chemicals effects, 10245; germination, 1647, 3949, 9442, 12593, 16337, 17912; provenance, 5956; seed production, 13107, 16338; stimulating substances, 12820; storage, 9451; viability, 2293, 6173; *see also* Abies
Fir, Douglas: chemical change during metabolism, 3952, 3953, 3954; chemicals effects, 3947; dispersal, 9449; genetics, 378, 14841; germination, 379, 3627, 3794, 5177,

Fir–(*cont.*)
8663, 8930, 8970, 9453, 9454, 11879, 12589, 16183, 17288, 18953; industry, 6389, 9436, 9883, 9944, 11897, 13155, 19752; maturity, 6388, 16052; moisture content, 2956; provenance, 3943, 13915; rhythm, 3627, 11881; seed production, 16343, 17433, 17434; storage, 1919, 2025; treatment, 5339, 7512, 9005, 9006, 10463; viability, 9452; *see also* Pseudotsuga
Flacourtia, 15010
Flacourtiaceae, 11483
Flaveria, 5242
Flax, 963, 7716, 8055, 8629, 9368, 18177, 19024, 19044, 19721, 19982, 20109; *see also* Linum
Foeniculum, 880, 2667, 4073, 6475, 7859, 8572, 9505, 14402
Folia, 5569
Forbesia, 12869
Fouquiera, 9829, 19526
Foxglove, 666
Foxtail, 8543
Fragaria, 1040, 3054, 3283, 4912, 9226, 10855, 14430, 14525
Francoa, 8042
Frangipani, 3868
Frankeniaceae, 13040
Franklinia, 10041
Fraxinella, 4221
Fraxinus: absorption, 1440; burial in soil, 1442, 7496; characteristics of seeds, 1432; chemical composition: fats and oils, 930; chemicals effects, 3528; embryo culture, 20085; germination, 673, 1218, 1314, 1437, 1438, 1441, 1448, 3317, 4156, 4188, 4488, 5445, 5589, 6404, 6940, 8559, 8569, 9664, 12314, 12315, 13565, 14263, 14415, 14471, 14474, 15001, 15006, 15007, 15008, 15010, 15011, 16445, 18193, 19119, 20024; industry, 749, 1142; inhibiting substances, 1431, 1439; polyembryony, 17919; provenance, 749, 784, 3872, 14474; radiation, 9202; storage, 673, 11196, 17917; testing and analysis, 78, 17920; viability, 2011, 18723; *see also* Ash
Freesia, 1127, 1794, 5602, 6765, 18994, 19623
Fremontia, 13565
Frijol, 4724
Fritillaria, 3912
Fructus, 2970, 6800
Fuchsia, 947
Fumaria, 16657

Gagea, 10008, 16395
Galega, 9459, 11050
Galeola, 14154
Galeopsis, 5485, 14412, 14538
Galex, 7559
Galinsoga, 2176, 8233, 11442, 12040, 19684
Galium: burial in soil, 13539; development, 1594, 6225; germination, 1553, 8105, 10822, 10823, 11018, 12040, 15185, 17558, 19215; stimulating substances, 14414; storage, 282
Galphimia, 6574
Ganua, 20110
Garbanzo, 2774; *see also* Cicer
Garcia, 9714
Garcinia, 832, 4236, 5337, 8775, 15939, 16593
Garlic, 10946
Garrya, 13258, 13565, 15961
Gaultheria, 13565
Gaura, 11942
Gaylussacia, 98, 14386
Genista, 4424, 4432
Genlisea, 10615
Gentiana: characteristics of seeds, 469; development, 4566; germination, 20, 202, 468, 469, 4508, 5432, 6254, 7017, 7220, 8184, 8286, 8476, 9574, 9656, 10674, 10675, 10677, 10681, 11425, 12879, 12989, 13984, 14270, 14386, 14591, 17029, 17895, 19485
Gentianaceae, 2459, 6449, 16661
Geraniaceae, 2459, 13953, 14415
Geranium, 934, 5485, 5637, 5876, 7370, 7697, 8498, 10739, 10741, 12533, 12770, 15327, 17969
Gerbera, 11216
Gesneria, 12040
Gesneriaceae, 6335, 6336, 7325
Gesneriadaceae, 1412, 19381
Gethyllis, 12869
Geum, 4037, 7017, 14415
Geutierruzia, 2176
Gigliacee, 1833; *see also* Lilium
Gilia, 724, 7499, 8565, 8569, 12036, 19526
Ginger, 15065
Gingko: anatomy, 141; characteristics of seeds, 5740, 7663, 11251; chemical composition, 6853: amino acids, 8360, proteins, 6852; development, 6258, 8679, 12196, 12197, 18791, 18793; disease transmission, 5026; embryo culture, 1795, 1796, 1797, 3312, 12198, 18792; genetics, 14368; germination, 1394, 8371, 13339; nutrition, 15619; polyembryony, 4357; seed production, 4854, 11978, 16652
Ginseng, 2229, 3230, 7754, 18975; *see also* Panax
Gisekia, 10013
Gladiolus, 690, 1938, 5634, 7158, 9042, 10440, 10677, 11632, 13146, 17327
Glaucidium, 11524
Glaucium, 18846

Gleditsia, 743, 747, 1074, 4468, 4469, 5956, 7553, 12320, 12897, 13403, 13949, 16329, 16841
Gliricidia, 14944
Globularaceae, 2459
Globularia, 4541, 10675
Glottidium, 5936, 6538, 14624
Gloxinia, 696, 717, 2679, 7016, 8658, 9102, 13339, 14898, 15103, 15749, 18499, 18968
Glyceria, 13868
Glycine: anatomy, 18303; chemical composition: enzymes, 584, 585, 4260, fats and oils, 17967; chemicals effects, 463; coats, 6708; color, 318; germination, 287, 472, 519, 1999, 9235; industry, 10442; inhibiting substances, 7213; microorganisms, 519; moisture content, 8276; seed production, 764; treatment, 519; viability, 463, 7808; *see also* Soybean
Glycyrrhiza, 8572, 13981
Gnaphalium, 10674, 20076
Gnetum, 18485, 19012
Godetia, 3322, 8565, 11196
Goldenrod, 724
Goldenseal, 18976
Gomphrena, 14942
Goodeniaceae, 16416
Goodyera, 5525, 5528, 6449, 10910
Gooseberry, 7857; *see also* Vaccinium
Goosegrass, 6844, 18671; *see also* Eleusine
Gordenenceae, 2459
Gossypium: absorption, 5305, 7474, 9341, 10408; anatomy, 4899, 6796, 7864, 16080, 16081; burial in soil, 7496; certification, 1812, 2138, 5510, 10213; characteristics of seeds, 188, 4680, 6840, 10398, 13421, 14142, 14196, 15546, 15618, 15651, 18838; chemical change during metabolism, 434, 2705, 3681, 6008, 6734, 6922, 6923, 6924, 8067, 8928, 8929, 9340, 9342, 10222, 10270, 11424, 11719, 14159, 17343, 17430, 19612; chemical composition, 76, 459, 526, 1670, 2899, 4329, 6599, 7671, 8542, 9336, 11423, 12380, 12559, 13829, 14738, 14745, 14746, 14747, 14748, 14749, 15666, 17640, 17662, 17875, 17876, 18446, 19970: amino acids, 119, 12560, carbohydrates, 527, enzymes, 434, 436, fats and oils, 187, 189, 190, 191, 438, 1398, 1399, 2429, 3104, 3157, 5096, 5578, 7215, 8769, 9498, 9499, 9710, 10659, 12868, 14744, 14944, 15531, 15532, 17210, 17872, 18119, 18120, 19632, nucleic acids, 17718, 19163, proteins, 6533, 6535, 9263, 12923, 14089, 14108, 14229, 14872, 16430, 18367, vitamins, 5065; chemicals effects, 72, 150, 434, 963, 986, 1016, 2410, 5737, 5991, 6007, 7053, 8069, 8070, 8264, 11612, 11717, 12452, 12587, 12779, 13192, 13382,
Gossypium–(cont.)
15680, 15918, 16018, 18089, 18221; coats, 526, 527, 1328, 1945, 2543, 3110, 3151, 3152, 3153, 4075, 4076, 4077, 4078, 4079, 4892, 6735, 7475, 8489, 11180, 12033, 15466, 17311, 17362, 17470, 17494, 19305, 19364, 19365, 19366, 19367; color, 2652, 2653, 2654, 2655, 2656, 2657, 2658, 5736, 12922, 13233, 13699, 14277, 15470, 15526, 15530, 19190; development, 1363, 5060, 10211, 10212, 12311, 15090, 16079, 18720, 18721; disease transmission, 397, 1378, 1379, 1380, 1381, 1382, 1383, 1384, 1385, 1386, 1387, 2425, 3067, 4255, 4630, 5910, 6958, 9236, 11657, 11659, 16019, 16518; dispersal, 310, 17956, 17957; electrical effects, 1221, 9327; embryo culture, 5683, 10020, 12391, 13037; fertilizers effects, 17649, 18104, 19259; genetics, 1114, 1761, 10093, 13554, 18833, 19363; germination, 149, 150, 439, 691, 1187, 1379, 1380, 1762, 3636, 3668, 4083, 4436, 4907, 5283, 5989, 6654, 7055, 7388, 7551, 7959, 9146, 9568, 11956, 11966, 12307, 12381, 12498, 12834, 14227, 15547, 17602, 18553, 18663, 19749; industry, 654, 1332, 1333, 2280, 2373, 3489, 4336, 6291, 9755, 13232, 16426, 19274; injury effects, 2549, 15160; maturity, 12420, 14777, 17465; microorganisms, 734, 10521, 16519, 19063; moisture content, 5361, 7050, 10267, 10268, 10269, 17279, 17471; nutrition, 3041, 3042, 3815, 4332, 11748, 17245, 20136; pelleting, 427, 15483, 19351; plants effects, 17221; polyembryony, 2601, 8182, 10196, 15817; provenance, 734, 13707; radiation, 3158, 10096, 11718, 17492; respiration, 435, 6639, 10271; seed production, 6839, 8025; stimulating substances, 314, 13033; storage, 430, 434, 437, 438, 734, 925, 1114, 1949, 2940, 3098, 4192, 5738, 6507, 9784, 11658, 12558, 13466, 13467, 14578, 16552, 17466, 17467, 17469, 17751, 17873, 17890, 19359; testing and analysis, 5040, 11048, 11089; treatment, 58, 425, 431, 434, 900, 956, 963, 969, 1014, 1165, 1608, 1696, 2395, 2417, 4081, 4155, 5360, 6262, 9344, 9563, 10216, 12031, 12268, 13583, 13734, 15273, 15317, 16348, 16575, 17614; variety effects, 694; vernalization, 11179, 13444, 14594; viability, 433, 1097, 1280, 2722, 3081, 3156, 3321, 4082, 6842, 6950B, 15324, 16642, 17468, 17472, 19692
Gourd, 17839, 19590; *see also* Cucurbitaceae
Graba, 5764
Gram: chemical change during metabolism, 14454, 17555; chemical composition: enzymes, 2084, 2820, 16788, proteins, 2086;

Gram–(*cont.*)
chemicals effects, 2408, 9416; genetics, 2402; germination, 2826, 15844, 17507; mucilage 10154; nutrition, 11683; vernalization, 15405; *see also* Cicer; Phaseolus
Grama 194, 8178, 4452, 18534
Gramineae: absorption, 3112, 3159; anatomy, 3217, 16872; characteristics of seeds, 10661; chemical change during metabolism, 1285, 17721; chemical composition, 4867: amino acids, 16168, 18274, 18275, 18276, carbohydrates, 5599, 12646, enzymes, 2386, 3148, fats and oils, 17720, proteins, 4872; coats, 2959; embryo culture, 3148; germination, 839, 3148, 3300, 3971; identification, 11103, 11115; pelleting, 1025; radiation, 14875; stimulating substances, 14415, 16084; storage, 8229, 11196; viability, 3362
Granadilla, 740
Grape: characteristics of seeds, 11965; chemical change during metabolism, 15293, 15294; chemical composition, 5877, 11227, 14527: fats and oils, 825, 871, 1286, 6448, 12933, 13455, 14891, 15326, 19052; chemicals effects, 1661, 3307, 6488, 11286, 11877, 13860; development, 3725, 19755; embryolessness, 2390, 14759, 18067; fossils, 7467; genetics, 1548, 3051, 8190, 14293, 14761, 17129; germination, 1661, 3862, 4716, 5442, 6463, 17129, 17517; identification, 3863, 14760; industry, 17170; seed production, 1366, 14720, 14758, 15970; treatment, 12778; *see also* Vitis
Grapefruit, 828, 13485; *see also* Citrus
Grass, Bahia, 8889, 8912, 13274, 19710
Grass, Barn, 10618; *see also* Panicum
Grass, Barnyard, 1306, 1307, 1308, 1309, 5043; *see also* Echinocloa
Grass, Bengal, 10618; *see also* Setaria
Grass, Blue Gramma, 4412
Grass, Brome, 7536, 8262, 10900
Grass, Kentucky Blue, 7495
Grass, Buffalo, 201, 855, 6428, 8177, 10873, 12456, 13990, 15438; *see also* Buchloe
Grass, Buffel, 2060, 3139, 19746
Grass, Carpet, 18669; *see also* Axonopus
Grass, Centipede, 10901
Grass, Chaffy, 1222
Grass, Dallis, 2061; *see also* Paspalum
Grass, Giant Foxtail, 10656
Grass, Japanese Lawn, 11999
Grass, Johnson, 1319, 8203, 8204, 8461, 12406, 13262, 18861
Grass, Little Bluestem, 4410
Grass, Mexican, 4994
Grass, Orchard, 6805, 6834, 9955, 15366, 15912; *see also* Dactylis
Grass, Range, 1266, 9898
Grass, Sudan, 4452, 4503, 8204, 12406, 16904, 19227, 19546; *see also* Sorghastrum
Grass, Yellow Foxtail, 15233
Gratiola, 3283, 8496
Grevillea, 10352, 10353
Grindelia, 7496
Groundnut: chemical change during metabolism, 10264, 13726; chemical composition, 14073: enzymes, 15897, fats and oils, 1079, 15106, proteins, 14109; chemicals effects, 7214; fertilizers effects, 7463; genetics, 9840; germination, 1855, 2607, 13656; industry, 11052; microorganisms, 1530, 7212; moisture content, 2607; seed production, 2718, 2719, 8447; storage, 5609, 5610, 15589; treatment, 351, 9758, 10694; viability, 15397; *see also* Arachis; Peanut
Grounsel, *see* Senecio
Guanabana, 1462
Guar, 9320, 13295, 15061, 15898; *see also* Cyamopsis
Guava, 9804, 11750, 15971
Guayule, 2274, 5941, 6014, 8085, 11299, 13594, 13598, 13599, 15585, 19625; *see also* Taraxacum
Gui, 5077; *see also* Viscum
Guiera, 15965
Guizotia, 15082, 16633
Gutierrezia, 9124
Gymnocladus, 673, 3372, 9072, 15476, 15882, 16841, 19661
Gypsophila, 10677

Haberlea, 7322
Hackberry, 129, 4156
Haemanthus, 9692, 12869, 16584, 19792
Haematostaphis, 8791
Haemodoraceae, 5298
Hairgrass, 17392
Hakea, 10354, 14014
Halesia, 7221
Halogeton, 1103, 3212, 4352, 19708, 17632
Halophila, 11681
Haloxylon, 2661, 11008, 14224
Hamamelis, 2913, 18595
Hancornia, 18397
Haplanthus, 15312
Haplopappus, 10199
Haricot, 2855, 2907; *see also* Phaseolus
Harpagophyton, 1456
Harpoon, 19762
Havardia, 15011
Haworthia, 5722
Hawthorne, 724, 6461, 11041, 14642; *see also* Crataegus
Hazelnut, 2925, 6071, 6371, 6372, 17045

Hazenoki, 8356, 8357; *see also* Rhus
Heather, 15473
Hebenstreitia, 10064
Heckeria, 12697
Hedera, 16102, 16103, 18881
Hedychium, 16599, 17431
Hedysarum, 3707, 20058
Helianthemum, 463, 10040, 14441
Helianthus: anatomy, 16427; burial in soil, 7496; characteristics of seeds, 15082, 18929; chemical change during metabolism, 9021; chemical composition: fats and oils, 13483, 18571, proteins, 18367; chemicals effects, 2961, 12208, 12742; coats, 10185; development, 6799, 11903; disease transmission, 16616; embryo culture, 6903; genetics, 19327; germination, 312, 3671, 4109, 5170, 6905, 11334, 11414, 13476, 13482, 17158, 18337; inhibiting substances, 16525; respiration, 11924; seed production, 17085; *see also* Sunflower
Helichrysum, 7499
Heliotropium, 4738
Helipterum, 8569
Helleborus, 10677, 17270
Helvetica, 18124
Hemerocallis, 1145, 1347, 1929, 4999, 7666, 14592; *see also* Daylilies
Hemlock, 386, 390, 1770, 1776, 2025, 3628, 3946; *see also* Tsuga
Henaria, 12636
Henbane, 930, 5500; *see also* Hyoscyamus
Hepatica, 8751
Heracleum, 2403, 6475, 14412, 18041, 18042, 18043, 18044, 19261
Heritiera, 2687, 14944
Hesperaloe, 9065
Hesperis, 4051, 17907
Hesperocallis, 3719
Hessea, 12869
Heteranthera, 12499, 13868
Heteromeles, 15008
Heterospermum, 2176
Heterotheca, 2176, 12040, 15464
Heuchera, 3424
Hevea: chemical composition, 2162: carbohydrates, 522, 7635, fats and oils, 14944; development, 2850, 14012; embryo culture 14013; germination, 5866, 12579, 17821, 18957; storage, 11196; viability, 17823; *see also* Rubber
Hibisceae, 16078
Hibiscus: anatomy, 2324; burial in soil, 7496; characteristics of seeds, 3391; chemical change during metabolism, 4929, 5176; chemical composition, 13573, 17329: fats
Hibiscus–(*cont.*)
and oils, 3989, 5646; chemicals effects, 813, 11519; germination, 4518, 4519, 4713, 5645, 9018, 9892, 9971, 12828, 12907, 15010, 15570, 17140, 18249; industry, 15973, 18614; maturity, 8412, 13602; microorganisms, 7293; moisture content, 5520; treatment, 813, 7869; viability, 1182; *see also* Kenaf; Okra
Hickory, 1023, 5141
Hicoria, 17623; *see also* Pecan
Hieracium, 2320, 5876, 14881, 16420
Hierochloe, 14582
Hilaria, 9629
Hinoki, 11296; *see also* Chamaecyparis
Hipberry, 15709
Hippeastrum, 19253
Hippocrateaceae, 4951
Hippophae, 98, 16365
Hippuris, 6400, 7441, 8791, 10114
Hiptage, 15935
Histiopteris, 3555
Ho, 11295; *see also* Magnolia
Hodgsonia, 3923, 8772
Hoheaackeria, 18327
Holarrhena, 8413, 9436; *see also* Kurchi
Holcus, 4122, 4481, 8955, 15353
Holly, 1101, 1939, 4165, 4341, 4477, 6968, 6971, 16423; *see also* Ilex
Hollyhocks, 658, 723, 6602
Holosteum, 282, 1553, 5764
Holygarna, 14096
Honey Plant, 11211
Honeysuckle, 3722
Hoppea, 1334
Hordeum: absorption, 3112, 3113, 3114, 3115, 4301, 6359, 10198, 15518; anatomy, 1119, 3472, 16832, 19040, 20124; ancient, 40, 2052, 12211; burial in soil, 7496; characteristics of seeds, 1353, 3164, 5053, 8671, 10343, 11152, 11176, 12765, 14779, 14784, 18539, 19856; chemical change during metabolism, 2688, 2923, 3037, 3167, 4475, 5346, 6378, 6525, 8235, 8812, 9070, 9529, 9679, 9688, 9696, 9721, 11970, 12510, 12511, 12512, 12643, 12644, 12959, 13062, 13087, 14140, 14981, 14982, 15412, 15981, 16666, 16667, 17615, 18247; chemical composition, 45, 1032, 2223, 4396, 6819, 8794, 9826, 9941, 10939, 12530, 12638, 14395, 15849, 17297: amino acids, 1117, 5689, 6524, 11017, carbohydrates, 1324, 2666, 5834, 7361, 12641, 16945, enzymes 541, 543, 544, 1518, 1726, 2325, 2445, 2565, 3337, 3338, 3380, 4460, 5050, 5055, 5104, 5900, 5901, 7312, 7360, 8813, 9038, 9553, 9697, 10316, 10726, 12161, 12640, 12813, 12957, 14041, 14042, 14043,

Hordeum–(*cont.*)
14044, 14045, 14178, 14613, 16011, 16756, 19073, 19989, proteins, 542, 5948, 8051, 9794, 9795, 10885, 13416, 14861, 16924, 17770, 18280, vitamins, 1284, 3929, 7029, 7311, 9698, 9699; chemicals effects, 99, 100, 2125, 3038, 3521, 3573, 5347, 5348, 5950, 9031, 9749, 11431, 11651, 13611, 13770, 13834, 14983, 16938, 18967, 19015; coats, 16092, 18476, 18896, 19630; development, 3473, 8175, 8176, 11642, 11643, 12924, 13349, 14984, 15220; disease transmission, 4069, 6323, 6324, 7386, 8090, 12886, 13796, 15172, 16565, 16570, 16574, 17455, 17459, 18371, 18372, 18812; electrical effects, 7285; electron microscopy, 7386; embryo culture, 3810, 4243, 4988, 10466, 11189, 12953, 12960, 13350, 14581, 15542, 17004, 20101, 20102; fertilizers effects, 517, 680, 8316, 9035, 11826; genetics, 100, 542, 5899, 6383, 6384, 6547, 7565, 8592, 9392, 14639, 15258, 15259, 15260, 15261, 15896, 19852; germination, 2, 452, 542, 1032, 1409, 1569, 2497, 2498, 2499, 2500, 2502, 2503, 2597, 3147, 3165, 3522, 4889, 5056, 5564, 5793, 5947, 6413, 7140, 8174, 8216, 8505, 8506, 8508, 8514, 8697, 8700, 8923, 9181, 9228, 10695, 10765, 11442, 12049, 12967, 13415, 13727, 14141, 14397, 14812, 15509, 15514, 15515, 15517, 16138, 16526, 16645, 18897, 19138, 19161, 19162, 19407, 19497, 19507, 19747, 19908, 19909, 19949; identification, 9229, 16833, 18306, 20125; industry, 910, 10135, 12768, 14967, 15995; inhibiting substances, 1885, 10916, 18981; injury effects, 3, 1400, 3960, 6706, 9241, 10136, 16283; maturity, 6396, 8504, 11849; microorganisms, 1362, 4066, 4067, 7902, 16567, 17262, 18862; moisture content, 3164, 6064, 9739, 9740, 9741, 9742 9743, 9744, 9746, 11401, 12928, 17478, 17959, 19696; pharmacology, 5460; plants effects, 2709, 7575, 15130, 15541, 15544, 15545, 19931; polyembryony, 15543; provenance, 8051, 11631, 13173; radiation, 99, 100, 2146, 2513, 2516, 2916, 3509, 3511, 3512, 3513, 3514, 3515, 3518, 4763, 4989, 5106, 5835, 6074 6791, 7047, 8653, 9287, 9562, 10201, 10242, 11186, 11188, 12472, 13338, 13439, 14483, 14484, 16039, 16764, 16966, 17658, 17659; respiration, 9700, 9701, 11482, 12958, 13351; seed production, 517, 2751, 3236, 5181, 5535, 13172; stimulating substances, 657, 1885, 14414; storage, 796, 3131, 3365, 9190, 12546; supersonics, 15555; treatment, 153, 452, 607, 1146, 1951, 8705, 9037, 11949, 12278, 12607, 13797, 15615, 18912; vernalization,

Hordeum–(*cont.*)
38, 1739, 2241, 2781, 2787, 2788, 4107, 4108, 4971, 4975, 8924, 9361, 12995, 19189; viability, 2501, 3322, 4495, 10135, 10739, 12639, 18758

Hornbeam, 15692

Horsebean, 12711, 16202

Horsechestnut, 9753, 19841

Horseradish, 18039

Hosta, 9153

Hottonia, 7843

Houttuynia, 9485, 9502

Hseh Tuang, 12829

Huarango, 8006; *see also* Algarrobilla

Humulus: development, 13637; disease transmission, 2608, 16674; fertilizers effects, 10428; genetics, 6245, 7419; germination, 2608, 7416, 10427, 10677, 13680, 14960; moisture content, 17633; storage, 3366; viability, 15825

Hunnemania, 782

Hura, 13283

Hutu, 738

Hyacinth, Water, 8861, 8862; *see also* Eichhornia

Hyacinthus, 3084, 10677

Hydnocarpus, 121, 3935, 5227, 7794, 14944, 18619

Hydrangea, 5756

Hydrastis, 11425

Hydrocleis, 9911

Hydrophyllaceae, 839, 2459, 6449

Hydrophylax, 6936

Hydrophyllum, 15185

Hygrophyla, 15308

Hypochoeris, 12040

Hymenaea, 12883

Hymenocallis, 6497, 9641, 19624; *see also* Lily, Spider

Hymenoclea, 19526

Hyoscyamus, 880, 930, 2477, 5432, 8572, 8766, 9505, 9694, 10942, 11425, 11760, 11761, 11762, 12554, 12555

Hyosopus, 9505, 11425

Hypericeae, 14415

Hypericum, 2806, 3283, 3677, 7017, 7498, 7500, 8687, 9489, 9495, 10675, 14412, 14415, 16661, 17029, 18209

Hyphaene, 122, 13688

Hypochoeris, 794, 2176, 13535

Hypodiscus, 11445

Hypoxis, 7529

Hyptis, 862, 16837, 19526

Hyssopus, 4421

Hystrix, 368

Ianthe, 12869

Iberis, 4248, 7499, 8205, 10866

Icacinaceae, 6187
Ilex, 98, 1101, 1940, 2049, 6968, 6969, 8674, 8675, 9573; *see also* Holly
Illicium, 5749, 10152
Ilysanthes, 9590
Impatiens, 2852, 4879, 8205, 9279, 9731, 10069, 10194 10854, 11430, 14415, 15058, 16748, 17896, 19517
Indaceae, 10187
Indian Rice Grass, 16383
Indigo, 1496, 13152
Indigofera, 2327, 4287, 11925
Inga, 12841, 14200
Intsia, 14944, 18947
Inula, 9505
Ipomoea, 861, 3495, 4233, 5170, 5208, 7496, 8613, 12901, 13490, 13570, 13571, 13572, 16728, 17523, 17526, 17913, 19058, 19834
Iris: anatomy, 14090; characteristics of seeds, 16318; chemical change during metabolism, 4488; chemical composition, 630: carbohydrates, 630; development, 7799, 12087, 19539; embryo culture, 12086, 13679, 15926, 17584; genetics, 12085; germination, 2139, 3600, 4488, 5444, 5723, 6546, 7159, 7843, 9103, 10285, 10286, 10677, 13539, 13540, 13777, 13952, 14386, 19607, 19897; inhibiting substances, 15922; viability, 6155
Irvingia, 3429, 7120
Isabghol, 15193, 15195
Isano, 1593
Isatis, 5628, 17905
Isomeris, 2462, 12701, 12703
Isoplexis, 9588–A
Isopyrum, 1596
Isotoma, 10358, 18131
Iva, 3210; *see also* Poverty Weed
Ivory, Tagua, *see* Phytelephas
Ivory, Vegetable, *see* Phytelephas
Ixeris, 14723
Ixophorus, 4994

Jacaranda, 3207, 15745, 17763
Jack, 12249, 15902, 17747, 17845
Jambul, 13334
Jasminum, 2418
Jatropha, *see* Physic-nut
Jensenbotyra, 5081
Jimson Weed, 701, 16791
Jojoba, 5512, 7110, 7111, 10893, 13564, 17771, 19511
Jowar, 1497; *see also* Andropogon
Judas Tree, 5125, 15580, 15809
Juglandaceae, 12478
Juglans: anatomy, 4950; burial in soil, 18745; chemical composition, 5003: vitamins, 19794; development, 11765, 14230; germination, 98, 673, 712, 1023, 1990, 3843, 8552, 8559, 13870, 15001; plants effects, 12963; storage, 673; *see also* Black Walnut
Jujubier, 6146
Juncus, 2482, 7843, 10677, 11909, 11910, 13535, 13536, 13868, 18252
Jungermannia, 9279
Juniperus: anatomy, 10781; characteristics of seeds, 6395; chemical change during metabolism, 134; chemical composition, 14938; development, 4361, 12976; genetics, 12727; germination, 135, 747, 773, 951, 2018, 7129, 10647, 14936, 14937, 15009, 15010, 17144, 19444; seed production, 852, 9609; storage, 133, 9842; treatment, 773
Jussiaea, 10511, 14275
Jute: characteristics of seeds, 11535, 11536; chemical composition, 10503, 15797, 15798, 15799: carbohydrates, 14828, 14829, enzymes, 7792, fats and oils, 3765, 9393, 13275; chemicals effects, 2059; development, 1854; disease transmission, 19007; embryo culture, 9594, 13607; germination, 17223; industry, 10137, 10249; radiation, 2088, 9635; seed production, 349; storage, 7203; treatment, 348, 1652, 7204; *see also* Corchorus

Kadsura, 8442
Kaffircorn, 9062, 10273, 14604
Kahu, 5336; *see also* Lactuca
Kakiebos, 16477
Kalanchoe, 2637, 3292, 3293, 6320, 8160
Kale, 18527
Kalmia, 14386
Kamala, 158, 7860, 8101; *see also* Mallotus
Kaoliang, 14734
Kapok, 4706, 7668, 11791; *see also* Eriodendron
Karamatsu, 11296
Karanja, *see* Pongamia
Karasuuri, 18733; *see also* Trichosanthes
Karatas, 8658
Karrod, 8633, 8634, 18952
Kauri, 6521; *see also* Agathis
Kenaf, 4518, 4519, 4713, 6999, 8412, 9018, 9971, 15570, 17140, 18684; *see also* Hibiscus
Kennedya, 17437
Kentucky Coffee Tree, 15476
Kerassi, 6116
Keteleeria, 3250
Keyaki, 11295, 11296, 11297; *see also* Zelkowa
Kikuyu Grass, 4678, 15300
Kirganelia, 5239

Klugia, 1335
Knotweed, 13526, 19183
Koa Haole, 218, 220, 226, 228
Kochia, 5921, 5922, 5923, 7499, 8205, 12738, 14606, 19709
Koeleria, 14441
Koelreuteria, 7423, 15012, 15460, 18723
Kok-saghyz, 5368, 12574, 14930, 15279, 15572, 15574; *see also* Taraxacum
Kola, 5232
Kudzu, 11754, 16696, 18248
Kukui Nut, 16195
Kurchi, 9435, 9436, 9437, 9438
Kurjan, *see Myrsine*
Kuromatsu, 1439, 8300, 9577; *see also* Pinus
Kusum, 4938
Kusunoki, 19397
Kweme, 18232
Kyathaungwa, 2669

Labiatae, 839, 2455, 2456, 3431, 4540, 8955, 10063, 14415, 14456
Laburnum, 14412, 14538
Lachnanthes, 14386
Lactuca: anatomy, 5187, 6114; burial in soil, 7496; characteristics of seeds, 9370, 19592; chemical change during metabolism, 420, 6280, 7674, 7935, 13069, 13080, 13081, 13083, 15488, 15498, 15499, 15500, 15501, 17885, 18181, 18182, 19344; chemical composition, 2642, 2643, 2644, 2645, 3465, 17276: carbohydrates, 15490, enzymes, 8168, 13070, 13071, 13073, 16233, 16297, fats and oils, 5336, vitamins, 10210; chemicals effects, 1999, 2297, 2684, 4149, 6117, 6118, 6122, 6984, 7240, 7930, 8296, 8707, 11258, 12001, 13066, 13077, 13078, 13480, 14626, 15494, 15495, 15496, 15503, 15504, 15505, 17565; color, 18513; development, 11029; disease transmission, 4448, 7611, 7707, 11421, 11615, 12859, 17667, 18810; electrical effects, 419; embryo culture, 2704, 11828; fertilizers effects, 8224; germination, 227, 614, 797, 1071, 1553, 1987, 2042, 2809, 2810, 2811, 4258, 5793, 5912, 6107, 6119, 6120, 6121, 6124, 6496, 6498, 6678, 7927, 7932, 7933, 7934, 7986, 8232, 9025, 9371, 9372, 9373, 9374, 10129, 11011, 11246, 12025, 12402, 12907, 13916, 13930, 15318, 15883, 16390, 17367, 17370, 17563, 18509, 18510, 18511, 18512, 18515, 18516, 18536, 19087, 19088; identification, 5254, 5851, 5854, 16313; inhibiting substances, 12786, 12787, 12788, 12789, 13065, 13068, 15500, 15647, 17366, 19369, 19476; injury effects, 15620; microorganisms, 5892; pelleting, 427, 1250, 9773, 12279, 20115; plants effects, 13191; radiation, 2616, 2622, 6499, 7929, 10778, 15678; respiration, 12152; seed production, 5517, 6573, 6662, 8228, 8420, 8422; stimulating substances, 657, 2641, 10130, 10131, 19086; storage, 282, 830, 864, 925, 1626, 1987, 1994, 5563, 11196, 11246, 14628, 19574; treatment, 614, 751, 954, 9615, 19615, 19616; vernalization, 110, 614, 3308, 6828, 7601, 10896, 17464, 19375; viability, 219, 7676
Lady-Slipper, 6582, 12003; *see also* Cypripedium
Lagenaria, 7689, 18203, 19591
Lagia, 2176
Lallemantia, 9382, 13289, 18926, 18927
Lamarckia, 2204–A
Lambs-Quarters, 701, 9015
Lamium, 282, 1553, 2347, 9970, 12036
Lannea, 10420
Lansium, 15024
Lansones, 7786
Lantana, 8561, 8565, 14945, 15009, 18389
Lapageria, 16927
Lappa, 9279
Larix: absorption, 8358; characteristics of seeds, 5385, 6395, 12300, 14492; chemicals effects, 5332; development, 8022, 17014; germination, 3305, 3459, 6508, 6784, 8585, 8601, 10234, 12394, 16088, 16367, 17097, 17293, 18727, 19130, 20074; industry, 5549, 6001, 8171, 10155, 12863, 14492, 19950; provenance, 4890, 5058, 5790, 10470, 11779, 12277, 15458, 19951; seed production, 4353, 11051; storage, 10743, 11196, 11296, 14482, 17337, 19128, 19129; supersonics, 10948; treatment, 11174; viability, 16973, 19135; weight, 8495
Larkspur, 782, 1999; *see also* Delphinium
Larrea, 13258, 19526
Lasiserus, 11661
Lathraea, 3874, 8524, 8526, 8535
Lathyrus: characteristics of seeds, 9171; chemical change during metabolism, 6938, 9456; chemical composition, 115, 11983, 15967, 18030, 19159: amino acids, 2236, 2238, 14615, 15894, 15968, proteins, 2086, 2087, 10800, 10811, 14980, 18922; chemicals effects, 12068; coats, 10090; development, 4878, 7520, 11497; dispersal, 17849; germination, 1997, 5392, 5876, 7052, 7054, 8205, 8564, 11315, 12481, 12482, 14386, 18797; identification, 2237; maturity, 3776; microorganisms, 8824; nutrition, 16789, 20039; plants effects, 3811; respiration, 18016; storage, 7499; viability, 761; weight, 2605; *see also* Sweet Pea
Lauan, 3498

Launaea, 4943
Lauraceae, 16786
Laurel, Grecian, 18748
Laurus, 1835, 19184
Lavendula, 8572, 11533, 11838
Lechosa, 17237
Ledum, 14386
Leea, 14103
Leek, 109
Leguminosae: absorption, 642, 3711, 5477, 6263, 6342, 8313, 8898, 9261, 9286, 15812; anatomy, 1634, 3578, 4415, 4456, 8488, 10722, 14049, 15432, 16007, 16488, 20027; characteristics of seeds, 2523, 6653, 8515, 17684; chemical change during metabolism, 440, 1313, 2384, 4845, 5061, 13011, 13012, 14174, 14337, 14357, 15033, 15732, 15808, 16782, 18262; chemical composition, 5119, 8703, 10814, 14616, 15736, 16470, 17069, 19610: amino acids, 1877, 3447, 7342, 9658, 9770, 10509, 12514, 12515, 15738, 15845, 16168, carbohydrates, 12722, 15107, enzymes, 1285, 2386, 11468, fats and oils, 3385, proteins, 4867, 4872, 10027, 10795, 10801, 10806, 14868, 17603, vitamins, 1847, 3642, 5234, 9351; chemicals effects, 1839, 2660, 3895, 10229, 10379, 12304, 12452, 14095, 18611; coats, 1186, 2013, 2212, 2328, 2959, 2967, 3049, 3491, 4003, 6056, 6360, 6881, 8219, 10383, 10384, 10386, 12977, 14304, 14928, 14994, 15430, 17179, 17893, 18082; development, 2698, 11044; disease transmission, 6289, 10617; dispersal, 9279; electrical effects, 1186; embryo culture, 10418, 18628; embryolessness, 16795; germination, 519, 711, 839, 1628, 3589, 3853, 3966, 5094, 6341, 6875, 7035, 7074, 7075, 8824, 8955, 10866, 12580, 13710, 13969, 14148, 14176; identification, 6584, 8139, 9464; industry, 1163, 1318, 2634, 2698, 2734, 9863, 12663, 15807, 18401; inhibiting substances, 2770, 2771; maturity, 13941; microorganisms, 519, 5313; moisture content, 9080; nutrition, 27, 2769, 2770, 6210; pelleting, 1260, 3087, 8450, 14499, 17673; plants effects, 621, 12927, 19181; polyembryony, 12910; radiation, 479, 14175, 14875; seed production, 764, 2448, 5986, 10381; storage, 3010, 8229, 9081; treatment, 407, 519, 1464, 3708, 6835, 9860, 10474, 11220; vernalization, 11578; viability, 3362, 3475, 12730, 13356; weed seeds in, 11108; weight, 10380, 10381; *see also* Pulses
Leiophyllum, 14386
Lemaireocereus, 13167
Lemna, 12707, 15627
Lens, 476, 483, 484, 6249, 12243
Lentil, 481, 2086, 10817
Leontodon, 2176
Leonurus, 2455, 2677, 9148, 9149, 17431
Lepidium: burial in soil, 4900, 4901, 4902, 4903, 4904; characteristics of seeds, 1684; chemical change during metabolism, 19146; chemical composition, 10016, 11479, 13652, 16591: fats and oils, 4937; chemicals effects, 12113, 12206, 12426; development, 57; germination, 1553, 1991, 2472, 2677, 8569, 8572, 9485, 12112, 12206, 12907, 14414, 18653, 18672, 18680; industry, 4944; moisture content, 7697, 8276; plants effects, 17935; rhythm, 3287; stimulating substances, 14415; storage, 282, 11196; *see also* Cress
Lepidocarpon, 4910
Leptandra, 9485
Lespedeza: burial in soil, 7496; characteristics of seeds, 13411; chemical composition: fats and oils, 19693; coats, 11630, 13410, 18248, 19847; development, 475, 8134, 16008; germination, 3420, 3676; nutrition, 8661, 9448; radiation, 14680; seed production, 764, 2098, 5463, 8136
Lesquerella, 1991, 5051, 13453
Lettuce, *see* Lactuca
Leucaena, 218, 220, 2440, 2441, 5413, 6231, 7602, 12247, 12248, 12938, 14944, 15011, 15182, 19057; *see also* Koa Haole
Leucas, 14194
Leucocoryne, 9691
Leucocrinum, 3719
Leucojum, 8822
Leucopium, 10675
Levisticum, 6475, 9505
Lewisia, 12825, 14279, 17029
Liatris, 14386
Libocedrus, 1981
Licania, 2283, 12609, 16024
Ligularia, 2101
Liguliflores, 11926
Ligustrum, 747, 14430, 15010, 15012
Lilac, 8647
Lilaea, 166
Liliaceae, 2304, 4367, 8229, 8955, 10187, 10673, 14783
Lilium: anatomy, 6799; chemicals effects, 463; development, 4373, 10688; disease transmission, 867; genetics, 2142, 14117; germination, 344, 463, 718, 719, 782, 804, 809, 865, 866, 889, 1008, 1365, 1802, 1863, 1992, 1995, 2450, 3524, 4191, 4501, 4785, 4999, 5068, 5114, 5603, 5942, 5943, 6585, 6702, 7721, 7722, 7723, 7724, 7725, 7726, 9107, 9160, 9408, 9641, 12057, 12060, 12355, 12356, 13347, 13507, 13773, 14369,

Lilium–(*cont.*)
15149, 15681, 15764, 16165, 17582, 17685, 17767, 17965, 18066, 18772, 19562, 19614; industry, 5944, 19835; polyembryony, 13744; radiation, 3032; seed production, 10819, 17198, 18069; storage, 2015; viability, 761, 8189
Lily of the Valley, 18068; *see also* Convallaria
Lily, Spider, 6497, 6500; *see also* Hymenocallis
Lily, Water, 1065, 9640; *see also* Nymphaea
Lime, 1662, 17775; *see also* Citrus
Limeum, 14204
Limnaea, 8685
Limnanthemum, 4557, 7843
Limnanthes, 6070, 12947, 13614
Limnocharis, 9912, 9914
Limnodea, 3185
Limnophila, 9588–B
Limosella, 8496
Linaceae, 2459, 14415
Linaria, 4358, 5299, 7769, 8565, 9204, 10675, 14412, 16661
Linden, 13845, 20026; *see also* Tilia
Lindernia, 8496
Linodendron, 15005
Linum: anatomy, 359, 6036; burial in soil, 7496; characteristics of seeds, 3594, 4323, 5275, 5649, 11521, 12961, 18876; chemical change during metabolism, 1906, 1911, 6163, 7851, 12777, 14962, 14964, 15720, 17660; chemical composition, 528, 591, 6026, 6311, 7139, 8052, 8387, 9172, 9753, 10849, 11859, 12506, 14401, 14807, 16040, 16669, 17043: amino acids, 5689, carbohydrates, 10848, fats and oils, 588, 1356, 2541, 4322, 5243, 5265, 5372, 5374, 5409, 5717, 7069, 7070, 8764, 9861, 9862, 13655, 14334, 16934, 16936, 16937, proteins, 14858, 14963, 14965, vitamins, 2419; chemicals effects, 1451, 3376, 5581, 8455, 14434, 17717; color, 4726, 4727; development, 6543, 9381, 9751, 10235; disease transmission, 1427, 2724, 2729, 4291, 11059, 13843, 13999, 14002, 14003; embryo culture, 11666; fertilizers effects, 6028, 13661; genetics, 4100, 13181, 14039; germination, 2381, 3322, 3395, 4643, 5107, 5375, 5713, 5902, 6683, 6686, 6928, 7170, 8514, 11317, 12239, 12920, 13182, 14414, 17407; identification, 3340; industry, 64, 940, 1066, 1886, 5373, 9555, 12241, 12471, 12807, 16594; injury effects, 6503, 6562, 6684, 11057, 11058, 12600, 13985, 17083; maturity, 12030; microorganisms, 11971, 15376; moisture content, 4281, 6946, 8276, 11805, 17749; polyembryony, 10248; provenance, 19850; radiation, 8201; respiration, 1676, 8054, 11806; rhythm, 17819; seed production, 1264, 2540, 4290, 12816; storage, 1019, 6687, 9190, 11210, 16671; treatment, 605, 680, 940, 963, 1271, 1619, 2532, 4068, 4289, 4704, 5582, 5813, 6502, 6743, 7605, 7610, 8636, 8822, 11019, 12603, 17878; vernalization, 1412, 3771; viability, 5093, 16115; weed seeds in, 16000; weight, 16614, 16615; *see also* Flax
Lipinachia, 9485
Liquidambar, 2149, 3205, 5589, 15005, 15009
Liriodendron, 2899, 3634, 3667, 4156, 5977, 7761, 8559, 12244, 17937, 17938, 19424; *see also* Tulip Tree
Listera, 5529
Litche, 3748, 6235, 7598, 18964, 19962
Lithocarpus, 20135
Lithospermum, 16663
Litorella, 16417
Litsea, 3934, 7850A
Lloydia, 10675, 10677
Loasa, 4544
Lobelia: anatomy, 12688; development, 4389, 4538, 8704, 10351, 10356, 10357, 18130; genetics, 17934; germination, 785, 13863, 13684, 13867, 13868; polyembryony, 4539; storage. 785, 2031
Lobeliaceae, 2459, 4554
Lobularia, 4661
Lochnera, 15026, 18995
Locust, 1074, 3412, 3827, 7021, 7296, 8551, 9634, 10434, 20046; *see also* Gleditsia; Robinia
Lodoicea, 4841, 8738
Loiseleuria, 14386
Lolium, 502, 506, 763, 1999, 2250, 2380, 2547, 3159, 3867, 4273, 4658, 5485, 6361, 6693, 8514, 9903, 10084, 10700, 10739, 12274, 12645, 13504, 13920, 14406, 15353, 16964, 17438, 17782; *see also* Ryegrass
Lomatia, 15945
Lomatium, 15185
Lonicera, 98, 8574, 13565, 14386, 17715, 20045, 20070
Lophanthus, 2455
Lophira, 737, 8771, 16598
Lophotocarpus, 13868
Loranthaceae, 10404, 12700
Loranthus, 10677, 13097, 13098, 19663
Lotus: absorption, 14707; ancient, 14854; characteristics of seeds, 5848, 12446; chemical composition, 18738: vitamins, 18741; coats, 724, 7187, 19734; color, 17996; development, 19744; germination, 508, 724, 3809, 3837, 4817, 13381, 14705, 14706, 14708, 14709, 14710, 19905; identification, 5859, 9171; maturity, 562,

Lotus–(*cont.*)
18739; plants effects, 563; seed production, 562, 563, 1648, 15152; viability, 868, 18846; *see also* Nelumbium

Lovegrass, 195, 1038, 12416; *see also* Eragrostis

Lovomitopsis, 11484, 19041, 20048, 20049

Lucuma, 3265

Luffa, 1914, 2056, 2398, 7658, 9766, 15358, 17174, 17516, 18203

Lulu, 737

Lumnonthericum, 10675

Lunaria, 11816, 19495, 19742

Lupinus: anatomy, 16123; characteristics of seeds, 8858, 13343, 14826, 17064, 18924; chemical change during metabolism, 2749, 3599, 4193, 4405, 4796, 4798, 5159, 5766, 6916, 6960, 7806, 8480, 13733, 19330, 19354, 19355; chemical composition, 4194, 4196, 4449, 4450, 4451, 4690, 6275, 7351, 8493, 8698, 9546, 10762, 10794, 11542, 11543, 11842, 12497, 12860, 14107, 14686, 15737, 16246, 17068, 17073, 17077, 18716, 19665, 19690: amino acids, 5654, carbohydrates, 7807, 13107, 18258, enzymes, 12615, 12617, proteins, 7157, 10021, 10025, 10026, 10028, 10798, 14864, vitamins, 16909; chemicals effects, 165, 3402, 16625; coats, 861, 1454, 9249, 11476, 15813, 15814, 19226, 20114; color, 11307; development, 9750; disease transmission, 5889, 14879; dispersal, 18627; embryo culture, 8926, 11975; fertilizers effects, 2487; genetics, 2447; germination, 786, 2446, 2828, 3294, 3424, 5095, 5124, 5889, 6053, 6901, 8283, 8514, 8697, 9004, 9140, 10021, 10365, 14037, 18893; industry, 10740, 16577; inhibiting substances, 12881; injury effects, 5800; microorganisms, 8824; moisture content, 7697, 10366, 11401; nutrition, 12805; plants effects, 918, 2726; provenance, 17003; radiation, 6601; respiration, 4500, 18352, 18353, 18354, 18356, 18357; seed production, 764, 10364, 18997; storage, 992, 3351, 6052, 11196; vernalization, 264, 9193; viability, 3019, 3322, 3403, 3493, 7295, 13208; weight, 5760

Luzula, 10674, 10675, 10677, 13535, 14568, 17849

Lychnis, 2605, 4577, 5296, 5297, 10674, 10675, 10677, 14412, 14414, 14415

Lycium, 19526

Lycopersicon: burial in soil, 7496; characteristics of seeds, 1643, 3026, 3144, 9040, 11825, 16153, 16189, 17692; chemical change during metabolism, 2264, 2708; chemical composition: fats and oils, 891, 912, 1954, 2937, 3643, 3740, 4801; chemicals

Lycopersicon–(*cont.*)
effects, 16, 641, 730, 1757, 1999, 2045, 4044, 8964, 11443, 11516, 11517, 12208, 12742; coats, 2948, 4793; development, 4041, 8375; disease transmission, 2329, 2330, 2633, 2979, 5213, 5467, 6290, 6423, 7394, 9838, 10075, 10107, 10460, 13031, 13131, 13492, 15315, 16517, 16693, 16694, 17190; electrical effects, 20018; embryo culture, 4038, 4039, 4040, 9594, 11828, 14525, 17666; embryolessness, 7119, 8415, 8416; fertilizers effects, 5916; genetics, 298, 1465, 3025, 3027, 3327, 4385, 4759, 5280, 5281, 7365, 7878, 9718, 10446, 10650, 11515, 12123, 13549; germination, 98, 634, 820, 1620, 1987, 2042, 5170, 5498, 6872, 7417, 7956, 13676, 14633, 16544, 17371, 17391, 19162; industry, 423, 902, 909, 913, 938, 1477, 1815, 1956, 2632, 3637, 3786, 7376, 9283, 19317; inhibiting substances, 11167, 12332, 14651, 18882; microorganisms, 216; pelleting, 427; plants effects, 263, 2701, 7409, 17377, 19303; polyembryony, 1047, 12889; provenance, 4423; radiation, 2881, 3353, 12004, 12122, 12124, 18433; respiration, 267; seed production, 938, 3028, 5516, 7510, 7977, 18011, 18804; stimulating substances, 2721; storage, 925, 1987, 1994, 11196, 11672, 14027; treatment, 751, 954, 5178, 7352, 10036, 10037, 10038, 13493; vernalization, 1748, 7443, 7444, 12333, 18009; viability, 298, 12358, 12667, 18203; weight, 1466, 12484, 13299

Lycopsis, 11801

Lycoris, 3525, 4732, 8612, 18617

Lynnaea, 9075

Lysimachia, 5432, 9489, 10674

Lysopus, 14458

Lythraceae, 10015

Macadamia, 799, 2165, 4099, 6604, 7784, 8075, 8290, 8730, 9978, 9979, 9980, 10462, 10641, 13479, 18062

Mace, 4199

Maclura, 2000

Macrozamia, 4391, 10035, 12767, 17163

Macrozonania, 10035

Maddenia, 17964

Madhuca, 3427, 8767

Madia, 15082

Maerua, 15976

Maga, 2008, 12882; *see also* Montezuma

Magnolia, 126, 2935, 4227, 4313, 4359, 5748, 6080, 6194, 8097, 8574, 10425, 11295, 13897, 14386, 14412, 15012, 15461, 17585, 17824, 17972, 18971

Mahogany-Big-Leaf, 8946, 12428

Maianthemum, 98, 10677, 18772

Maiblumen, 15042
Maize, *see* Zea
Makhorka, 11607
Malachra, 9892
Mallotus, 158, 7860, 8101, 14944, 15754, 19067; *see also* Kamala
Mallow, 1904, 16531, 17383; *see also* Hisbiscus
Malphighia, 17488
Malus, 2027, 5118, 5685, 8876, 14751, 15010, 16873, 16874, 18595; *see also* Apple
Malva, 4679, 5485, 16531
Malvaceae, 839, 8898, 8955, 9019, 9892, 12436, 14415, 15943, 16077, 16530
Malvastrum, 9892
Mamani, *see* Sophora
Mammillaria, 18566
Mandragora, 2857
Mangaba, 18397
Mangel, 2578, 5459
Mangifera, 410, 1050, 1381, 1711, 1842, 3554, 4923, 4924, 9058, 10057, 10059, 10413, 10461, 11196, 11484, 12704, 12966, 14179, 14212, 15112, 16523, 16600, 17454, 17497, 17519, 19436, 20117
Mango, *see* Mangifera
Mangosteen, 832, 9057, 9206, 18246, 19763
Mangrove, 7096, 10685, 11833
Manicaria, 4298
Manihot, 1340, 2464
Maple, 1218, 5970, 8281, 9396, 9534, 9536, 9537, 9948, 9949, 15989, 17977, 18700, 20015, 20028, 20046; *see also* Acer
Marah, 8785
Marantaceae, 7042
Marigold, 1999; *see also* Tagetes
Mariosteen, 4236
Marron, 12784, 15800
Marrow, 1815, 1956, 8713
Marrubium, 8572
Martynia, 762, 9279
Masdevallia, 19096
Mastixia, 8791
Matai, 15677
Matricaria, 282, 1553, 9505, 10739, 10741, 11018, 14414, 15240, 15838, 19215
Matthiola, 1999, 5945, 7499, 9135, 10459, 15870, 16649, 17197, 17200, 17556
Maximiliana, 4298
Mayaca, 12499
Maytenus, 3704, 8914
Mazri, 9356
Mazus, 1204
Meadow Fescue, 1999, 14494
Meconopsis, 128, 468, 4681, 7188, 8184, 8286, 8429, 14059, 17029, 18161
Medeola, 2304
Medicago (alfalfa): absorption, 2184, 3443, 12541, 13418; anatomy, 12905; burial in soil, 2989, 7496; characteristics of seeds, 3306, 5258, 6013, 8799, 9171, 10337, 11822, 16317, 16993; chemical composition, 6664, 8905, 9967, 12826, 18593: fats and oils, 10654, 10655, 15633, 17052; chemicals effects, 2288, 6266, 9267, 10396, 12319, 16542, 18307; coats, 2107, 2449, 2723, 2969, 4000, 4993, 5137, 6708, 6957, 7540, 9564, 9675, 12009, 16541, 17631, 17993; color, 1577, 12661, 14750, 19549; development, 4374, 5879, 6219, 6740, 12479, 16076; disease transmission, 4407, 7505, 9835; electrical effects, 5811; embryo culture, 6741; fertilizers effects, 5716; genetics, 6794, 10698; germination, 508, 655, 1299, 2383, 3626, 3674, 3694, 3837, 5508, 5509, 5793, 8514, 8536, 8809, 12904, 17164, 18637, 18768, 18873, 19905; identification, 3066, 5859, 5860, 13020; industry, 1144, 1199, 3306, 4886, 7745, 10051, 10850, 12118, 12471, 16558, 17995, 19465; injury effects, 655, 675; maturity, 15874; moisture content, 2295; plants effects, 2732, 14900; polyembryony, 7633, 19417; provenance, 3591, 5312, 11282, 15357, 17630; radiation, 5025, 7234; rhythm, 9458; seed production, 72, 608, 714, 764, 1364, 1537, 3058, 3328, 3614, 3615, 3616, 3617, 4170, 4381, 4387, 4516, 5523, 5714, 5715, 5967, 6665, 6795, 7584, 7585, 7700, 7965, 8414, 8910, 9968, 10475, 10898, 12064, 12116, 12117, 12222, 12903, 13970, 15170,15171, 15241, 15310, 18717; stimulating substances, 14415; storage, 219, 227, 14030, 14629; supersonics, 15555; treatment, 1642, 5314, 6425, 7506, 11212, 11213, 14022, 16466; viability, 3331, 8456, 10739, 17980; weed seeds in, 8430, 14019, 19821
Medusa-Head, 6097, 13964
Megarrhiza, 8785
Melaleuca, 225, 10647
Melampyrum, 4760, 4761, 12224, 15351, 17794, 17849
Melandrium, 4426, 18987
Melastoma, 18129
Melia, 747, 10117, 13654, 13782, 15005, 15009, 15011
Meliaceae, 14099, 14953
Melilotus, 282, 508, 653, 764, 1553, 2572, 3055, 3776, 3837, 4372, 4993, 5524, 5859, 6708, 8078, 8079, 8192, 9171, 9505, 9748, 9749, 10309, 10697, 11764, 12907, 12911, 13406, 16931, 16932, 18038, 18226, 18710, 19447, 19621; *see also* Clover, Sweet
Melissa, 8572

Melo, 3273
Melon, 284, 3273, 4035, 4036, 5335, 6197, 6357, 19032, 19452
Mendora, 13258
Menispermaceae, 16787
Mentha, 117, 6701, 9362, 9363, 14458, 14978
Menyanthes, 10677, 11001, 11002, 11003, 11004
Mercurialis, 7241, 7242, 7243, 7246, 8762
Mercury, Dog's, 8762
Mesembryanthemum, 4368, 8527, 9355, 19848
Mesquite, 1244, 2575, 7308, 12917, 17838, 18783
Mertensia, 15185
Mesua, 800, 3851, 5074, 5334
Metasequoia, 982, 5626, 11568, 12192, 17569, 17626
Mezzettia, 4414
Mibora, 7800
Mikania, 14386
Miliusa, 15207
Milkweed, 205, 941, 8080, 9015, 11785, 15584, 16347; *see also* Asclepias
Millet, 1879, 2700, 3808, 5608, 5824, 7098, 8116, 9189, 9252, 9518, 10937, 11420, 12201, 14173, 16866, 17054, 18176
Millettia, 14971
Mimosa, 5413, 8096, 8991, 9203, 14199
Mimosaceae, 5414, 14535
Mimulus, 5911, 7805, 9268, 10675, 17029
Mimusops, 1486, 7120, 10287, 18955
Mirabelle, 18871
Mirabilis, 8478
Miro, 15677
Miscanthus, 19995
Mistel, 6846; *see also* Viscum
Mistletoe, 1206, 4259, 4480, 8410, 8533, 8534, 16217, 16465
Mitrastemon, 19392
Moehringia, 7804
Molinia, 5289
Mollugo, 15150, 19526
Moluccella, 18761
Momordica, 3758, 9279, 10515, 11602, 12988, 15901, 17839
Monardella, 13565
Moniera, 16622
Monotropa, 10677, 19682
Monstera, 11484
Monterey Cypress, *see* Cupressus
Montezuma, 2088; *see also* Maga
Moonflowers, 842, 15751
Moreton Fig, 738
Moringa, 669, 6158, 9705, 10677, 14944, 15104
Morning Glory, 3210, 3211, 3601, 4481, 7995, 7996; *see also* Ipomoea
Morus, 4940, 11752, 13361, 13885, 18723
Mountain Brome, 13300
Mountain Pine, 13196
Mowrah, 8767, 9689
Mucuna, 4844, 11663, 16751, 16752, 17304
Muhlenbergia, 368, 9629, 18687
Mukuma, 10117
Mulberry, 3822, 4926, 10375, 12737, 13568, 13885, 15954, 16643, 18277, 18278, 18589
Mullein, 701
Muraya, 3763, 3764
Musa, 4483, 5235, 11961, 13178, 17456, 17457, 18064, 18065; *see also* Banana
Muskmelon, 820, 1338, 6059, 8222, 8231, 8574, 10458, 12712, 15841
Mustard: absorption, 10406; characteristics of seeds, 1685, 3504, 9015; chemical change during metabolism, 5597; chemical composition, 5598: enzymes, 4463, 4464, 4465, fats and oils, 801, 4498, 7494; chemicals effects, 14052, 16714, 17330; embryo culture, 3769, 17220; genetics, 17093; germination, 3390, 9056, 12628, 17219, 17369, 17918, 18154, 19224; industry, 14006, 19785; inhibiting substances, 3390; nutrition, 18766; radiation, 2146; treatment, 9251; vernalization, 17214, 17216, 17217, 17218; *see also* Brassica
Myagrum, 12040
Myoporaceae, 2459, 4951
Myosotis, 3322, 5485, 11196, 13535
Myosurus, 10677
Myrianthus, 16087
Myrica, 8018, 13565, 14386, 16998
Myricaria, 10673, 10677, 18751
Myriophyllum, 7843, 15125
Myristica, 1485, 3387, 4297, 10289, 14213, 15206, 16785, 19765
Myrmecochora, 19439
Myrobalan Plum, 14642; *see also* Prunus
Myrocarpus, 14255
Myrrhis, 6475
Myrsine, 768
Myrtaceae, 74
Myrtales, 13051
Myrtus, 15900

Naegelia, 8658
Nagi, 9528
Nahor, 3861
Najadaceae, 18206
Najas, 13868
Nandina, 131, 3749
Nannorhops, 9356, 10524
Napiergrass, 17791
Narcissus, 15643, 16157
Nardus, 5097, 16582

Narras, 10024, 10029, 17100; *see also* Acanthosicyos
Narthecium, 10677
Nasturtium, 1999, 3987, 6816, 7843, 8569, 8575, 8770, 9117, 9118, 13868
Nechmandra, 14941
Neddlegrass, 10642, 14452, 16382; *see also* Stipa
Neem, 199, 212
Nelumbium: anatomy, 4215, 19991; chemical composition, 3824; coats, 9952, 17281; color, 19739; embryo culture, 12348; germination, 5848, 13381, 15583; respiration, 18743; viability, 1035, 1048, 6159, 14704, 14705, 14706; *see also* Lotus
Nemacladus, 19526
Nemesia, 3322, 8565, 8568, 11196
Nemopanthus, 98
Nemophila, 2531, 8565, 10677, 12036, 12039
Neolitsea, 7832
Neotia, 2339
Nepenthes, 6195, 15596
Nepeta, 7471, 8572
Nephilium, 14944, 18964
Neptunia, 5411, 14412, 14415
Nerium, 15009
Neslia, 7496
Nettle, Stinging, *see* Urtica
Niam, 8771; *see also* Lophira
Nicotiana: absorption, 15587; anatomy, 1538, 5338, 16818, 19787; burial in soil, 7496, 9809; characteristics of seeds, 3781, 3980, 9670, 15373, 15567; chemical change during metabolism, 2735, 8830, 9380, 9385, 9472, 13340, 15270, 16855; chemical composition, 948, 7276, 13149, 14600, 15787, 16854, 16978, 17249, 19103, 19535, 19788: amino acids, 26, carbohydrates, 19255, enzymes, 17036, fats and oils, 746, 1053, 4710, 10978, 16201, 16664, proteins, 4692, 9379, 19107; chemicals effects, 806, 4126, 4345, 5576, 5618, 9388, 9604, 10097, 10637, 14412, 14414, 14820, 14876, 18289, 18293, 18294, 18965; color, 15653; development, 9335, 13151; disease transmission, 5620, 5804, 7750, 8834, 10350, 12844, 18945; electrical effects, 419, 3375; embryo culture, 6217, 11828, 14525; fertilizers effects, 853; genetics, 3542, 4788, 7289, 7454, 7455, 7987, 8992, 8993, 8994, 11264, 13639, 14618, 14619, 16852; germination, 26, 285, 806, 948, 1374, 1375, 1875, 2454, 2677, 3336, 3374, 3442, 3977, 3979, 3982, 3983, 3985, 4122, 4756, 5617, 6150, 7017, 7264, 7421, 7456, 7710, 7817, 8104, 8311, 8312, 8315, 8572, 8625, 8990, 9332, 9483, 9484, 9597, 9602, 9868, 10481, 10636, 10677, 10684, 10979, 11088, 11241, 11330,
Nicotiana–(*cont.*)
11607, 12037, 12775, 12969, 14690, 14693, 14973, 14974, 15589, 15644, 18337, 19141, 19516, 19930, 20076; identification, 11153; industry, 1342, 2095, 3984, 9869, 10325, 11417, 14972, 15859; inhibiting substances, 9856; maturity, 9596; microorganisms, 1874, 8695; moisture content, 3820; nutrition, 13320, 15984; plants effects, 15372, 16849, 16851; provenance, 13947; radiation, 7457, 10030, 11605; seed production, 366, 1267, 2730, 2731, 4382, 6790, 11936, 16653, 17930; stimulating substances, 657, 3986, 8866, 15561; storage, 806, 3978, 9726, 15113, 15588; treatment, 783, 2312, 2313, 2683, 3981, 4344, 5422, 5617, 7181, 7890, 13268, 14837, 17056; vernalization, 5801; viability, 6641, 10638, 10639, 12735, 12790; weight, 11418; *see also* Makhorka
Nielle, 3203; *see also* Agrostemma
Nigella, 282, 930, 1553, 1556, 1599, 4248, 7017, 7499, 7710, 8658, 9485, 9487, 9488, 9489, 10673, 10675, 10677, 12036, 13339, 14415, 17490, 18598, 18859, 20076
Niger, 16633; *see also* Guizotia
Nightshade, 701
Nim, 11874, 16641, 17651; *see also* Azadirachta
Nitella, 14386
Nolana, 4562, 4942
Nomacharis, 4484
Noogoora Burr, 3536
Norfolk Island Pine, *see* Araucaria
Northea, 8791
Nothofagus, 1869, 2436, 2438, 11534, 13252
Nothoscordum, 5770, 8016, 13358
Nuphar, 7843, 10675, 10677, 13296, 13868
Nutgrass, 1805, 2247, 8798, 12448, 18830; *see also* Cyperus
Nutmeg, 3387, 3550, 4199, 16785, 18336
Nut Palm, 2351
Nuts, 451, 1574, 4605, 4746, 4832, 6604, 7073, 8922, 9369, 9805, 10023, 12155, 13158, 13316, 17653; *see also* Subject Index
Nutsedge, 8796, 8797
Nuttalia, 13565
Nuytsie, 19002
Nyctaginaceae, 10144
Nymphaea, 789, 4326, 4579, 7843, 10677, 13868
Nymphaeceae, 1322, 1323, 8955, 10098, 13309, 18834
Nyssa, 5589, 8791, 15007, 17394, 18887
Nyssaceae, 13448

Oak: characteristics of seeds, 2694, 9086, 9610; chemical composition, 3391: enzymes, 6148, 6149, vitamins, 10658; chemicals

Oak–(*cont.*)
effects, 5288; development, 3743, 5286; genetics, 17023; germination, 164, 204, 1023, 3068, 4157, 5501, 5975, 8271, 11237, 13363, 13561, 17162, 19817; industry, 14387; injury effects, 3092, 16106; maturity, 11269; microorganisms, 7087, 7088; nutrition, 8609; plants effects, 1937; provenance, 204, 9756, 17349, 19139; respiration, 7287; seed production, 4087, 5531, 7904; storage, 5141, 6973, 8983, 9687, 9945, 10336, 11238, 11611; treatment, 1358; weight, 5870, 13144, 19764; *see also* Quercus
Oats, *see* Avena
Oats, Wild: burial in soil, 18558, 18585; characteristics of seeds, 9015, 17979, 18555; chemical composition, 8596; control, 516, 11242; genetics, 18556; germination, 638, 1509, 1736, 4741, 6756, 8427, 8428, 8919, 9569, 10706, 11060, 12542, 18554, 18662; 19644; plants effects, 18560; viability, 18559; *see also* Avena
Ochna, 737
Ocimum, 1589, 8463, 8572, 8639, 13289, 15900, 16841, 18597, 20080
Ocotea, 747
Odontoglossum, 4189
Oenothera, 1556, 2913, 4222, 4900, 4901, 4902, 4903, 4904, 5001, 5302, 6815, 7015, 8498, 11444, 12780, 13339, 14898, 15334, 17112, 17113, 18316, 19526
Oiticica, 795, 9707
Okari Nut, 4151; *see also* Terminalia
Okra, 566, 4182, 5779, 5780, 5781, 5797, 11196, 12900, 13601, 17000
Olax, 17259
Oldenlandia, 2396
Olea, 747, 10525, 13359, 13461, 16849; *see also* Olive
Oleaceae, 2459
Olive, 129, 471, 5082, 8769, 8777, 9754, 10330, 12191, 12315, 12648, 13461, 14929, 15032, 19053; *see also* Olea
Omphalea, 8382
Omphalodes, 16124
Onagra, 7496
Onagraceae, 839, 9828, 14415
Ongokea, 3695
Onguekoa, 2681
Onion: chemicals effects, 1999; development, 15110; disease transmission, 2369, 4699, 10983; embryo culture, 7790; genetics, 4172, 9951; germination, 3070, 4141, 4142, 4276, 10395; industry, 3551; injury effects, 9875; moisture content, 6946; pelleting, 12290, 14370; rhythm, 4173; seed production, 29, 1312, 1505, 1527, 2894, 8289, 8418,
Onion–(*cont.*)
8947, 10410, 11746, 12090, 12403, 14683, 14684; stimulating substances, 657; storage, 651, 925, 1627, 1987, 1994, 2151, 3072, 3142, 11196, 14021, 14024, 14371, 14628; treatment, 953, 954, 1864, 9624, 12476, 12477; viability, 2016, 3127, 6947, 16309; *see also* Allium
Onobrychis, 2126, 3710, 3837, 7121, 8824, 9171, 9431, 10258, 13801, 18524; *see also* Sainfoin
Onopordon, 282, 1553, 7496, 17906
Ophiorrhiza, 6937
Opuntia, 1327, 6950J, 9852, 14017
Orange, 3633, 5815, 11040, 12618, 13689, 18951, 19453; *see also* Citrus
Orange, Osage, 129, 2000; *see also* Maclura
Orania, 14944
Orbignya, 7076
Orchidaceae: characteristics of seeds, 6449, 10911; chemical composition: proteins, 18789; development, 5441; embryo culture, 111, 4774, 4775, 9198, 11604, 12853, 12855; germination, 277, 858, 1315, 1798, 1816, 1817, 1818, 1891, 2112, 2181, 2340, 2341, 2638, 2639, 2690, 2920, 2962, 2963, 2964, 3356, 3581, 4190, 4320, 4688, 4770, 5526, 5530, 6640, 7989, 8482, 8637, 8658, 9137, 9138, 10317, 10902, 10903, 10904, 10905, 10906, 10907, 10908, 10909, 10910, 10911, 11655, 12147, 12217, 12260, 13110, 13374, 13375, 13816, 14550, 15506, 15913, 17169, 18917, 19082, 19559, 19083; industry, 2903, 2911, 9800, 11604, 12460; injury effects, 18200; maturity, 9516; media for culture, 277, 1666, 2535, 4773, 10913, 11410, 11720, 12679, 13373, 18804, 19900, 19901, 19902; microorganisms, 3584, 3697, 4771, 15507; plants effects, 11843; seed production, 16661; stimulating substances, 13376, 16432; treatment, 12218; viability, 10914
Origanum, 8463, 8572
Ornithogallum, 3918, 14538, 16979
Ornithopsis, 3837, 8514, 9171, 17955
Orobanche, 1893, 2054, 2430, 2596, 3171, 3173, 3583, 3683, 3684, 3686, 3744, 4437, 5155, 6981, 7324, 7748, 7749, 9598, 9599, 9600, 10112, 10113, 11363, 12709, 13740, 14226, 14250, 14412, 15096, 15217, 15780, 15837, 15960, 18171, 18172, 18173, 18564
Orobanchaceae, 6449
Orontium, 12224, 13868
Oroxylum, 13291, 13293, 13294
Orthosiphon, 2714
Oryza: absorption, 572, 2972, 4884, 18285, 18348; anatomy, 9142, 10382, 14080, 14544, 16732; burial in soil, 7502, 7503; characteristics of seeds, 3562, 4050, 4100, 8353,

Oryza–(*cont.*)
10055, 11085, 11122, 11124, 11127, 11129, 11130, 13546, 14075, 14076, 14083, 14785, 14834, 15905, 19647; chemical change during metabolism, 250, 304, 573, 3469, 3932, 4927, 6833, 8848, 9110, 9230, 9525, 9526, 9530, 11136, 11140, 11141, 11142, 11143, 11148, 11183, 11537, 13030, 13749, 13840, 15054, 16634, 16742, 17532, 17538, 18282, 18358, 18364, 18365; chemical composition, 1430, 1473, 3574, 5590, 5837, 6038, 6819, 7183, 10314, 10624, 10625, 10626, 10734, 12251, 12329, 13114, 13115, 13451, 13945, 13946, 14134, 14135, 16632, 17535, 18254, 18310, 19142, 20019, 20020: amino acids, 15052, 15718, 18301, carbohydrates, 19389, enzymes, 308, 4472, 6302, 9352, 12374, 13450, 15124, 16272, 19388, fats and oils, 4707, 4712, 16586, 17145, 18339, nucleic acids, 10632, 11076, proteins, 11077, 15909, 16687, 18118, 18255, 18256, vitamins, 870, 1679, 6820, 8841, 9579, 10622, 11135, 11137, 17744, 17745, 18342; chemicals effects, 1292, 1351, 2684, 3733, 4267, 4825, 7193, 8296, 10246, 10299, 10409, 11323, 11516, 11517, 13576, 13577, 14921, 15264, 16159, 18405, 19926; coats, 86, 3782, 11138, 12346; color, 15908; development, 206, 207, 4010, 4852, 5758, 12991, 13543, 13544, 13545, 13635, 13636, 14077, 16626, 18253, 19971; disease transmission, 4403, 4920, 7531, 12392, 12896, 14079, 14846, 15179; electrical effects, 13610; embryo culture, 454, 455, 2849, 14118, 14124, 17543; embryolessness, 11094; fertilizers effects, 46, 1126, 5994, 17229; genetics, 1765, 4491, 5990, 9147, 10127, 13605, 14162, 17228, 18281, 18287; germination, 232, 233, 306, 307, 354, 1301, 3259, 3367, 3805, 3829, 3915, 3916, 4531, 5471, 5674, 5686, 5793, 5993, 6807, 7194, 7195, 7348, 8059, 8440, 8668, 8729, 9235, 9421, 9583, 9819, 9820, 9956, 9957, 9958, 10214, 10362, 11081, 11090, 11131, 11134, 11622, 12433, 12992, 13574, 13833, 14063, 14064, 14106, 14150, 14884, 15017, 15199, 15265, 15323, 15329, 15852, 16264, 16265, 16273, 16628, 16681, 16774, 17264, 17533, 17537, 17540, 17768, 17851, 18143, 18144, 18283, 18286, 18292, 18337, 18360, 18874, 19098; identification, 3913, 3914, 11099, 18345, 18392; industry, 964, 1026, 1031, 1110, 2233, 4931, 6506, 6950, 8743, 8833, 10508, 11095, 11773, 13243, 14078, 16648, 18347, 18785, 19054, 19606; inhibiting substances, 13452, 16629, 17534; injury effects, 2886, 2887, 3942, 9578, 11128, 14084, 16475, 18330; maturity, 1300, 8820, 10494, 11554, 13569, 14082; microorganisms 3894, 6158, 7197, 9357, 9527, 10917, 11555, 14947, 17032; moisture content, 305, 4370, 9479, 11139, 11158, 11161, 11603, 13177, 16965, 17034, 17748, 18902; nutrition, 10623, 14667; plants effects, 12245, 12246, 12997, 14081, 18145; polyembryony, 1853, 15910, 18581; provenance, 2733, 8156; radiation, 2082, 3778, 4422, 4928, 5285, 11066, 12939, 14067, 14068, 14516; respiration, 4935, 14890, 19161, 19162; rhythm, 87; seed production, 88, 964, 1934, 4919, 6821, 9959, 14551; stimulating substances, 657, 4128, 19050; storage, 227, 1020, 2368, 3838, 4319, 5244, 6330, 8781, 9109, 9111, 9190, 9519, 9520, 9660, 10169, 11092, 11097, 11098, 11112, 11113, 11118, 11119, 11121, 11123, 11125, 11126, 11132, 11133, 11136, 11140, 11144, 11145, 11146, 11147, 11149, 11159, 11580, 12357, 13245, 15025, 15683, 16627, 17033, 17832, 17833, 18467; supersonics, 3434, 10707; treatment, 1349, 1350, 1766, 1859, 2888, 3941, 4507, 5358, 7579, 8352, 9419, 9894, 13899; vernalization, 355, 1532, 1971, 2018, 7192, 15089, 17222, 17530, 17536, 17544; viability, 219, 4811, 5247, 5248, 5470, 7196, 14127, 14487, 14488, 15331, 16267, 16268, 16269, 16270, 16271, 16492, 16630, 16689, 16741, 17539, 17708, 18203; weed seeds in, 6962

Oryzopsis, 11034, 14289, 16351, 16383, 18033, 18690
Osmaronia, 17964
Ostrya, 15001, 18595
Osyris, 13454
Ottelia, 7948, 14125
Ouricury Palm, 13221; *see also* Cocos
Ownia, 10675
Oxalidaceae, 2459
Oxalis, 3746, 6151, 8750, 8751, 13339, 14206, 14538
Oxybaphus, 4379
Oxytrophis, 14415
Oyster Nut, 872

Pachira, 1725
Pachycarpus, 16944
Pachylobus, 18867
Pachyrrhizus, 1713, 1714, 8124, 14060, 14061, 14593, 14944; *see also* Bean, Yam
Pachysteta, 18416
Paederota, 10675
Paeonia, 1155, 1999, 6377, 9667, 10677, 13023, 13459, 13565, 19914; *see also* Peony
Pahudia, 14944
Palafoxia, 19526

Paliurus, 12728
Palm: anatomy, 7267; chemical change during metabolism, 14509; chemical composition, 122, 878, 7114, 8773, 10864: carbohydrates, 713, fats and oils, 3648, 3825, 4607, 4780, 6311, 7115, 18865; chemicals effects, 10543; coats, 12155, 18795; color, 7114; genetics, 18228; germination, 685, 802, 3501, 5342, 6310, 6427, 7039, 7395, 8716, 9264, 9265, 12425, 12492, 12767, 13530, 15304, 16062, 16063, 16064, 16065, 16067, 16068, 16816, 17489, 18635, 19987, 20140; identification, 878; industry, 838; plants effects, 18491; seed production, 3094, 9608; storage, 1173, 16066; viability, 5153; *see also* Elaeis; Name of Individual Palm
Palmae, 5252, 18311
Panax, 14716, 19025; *see also* Ginseng
Pancratium, 5697, 16702
Pandanaceae, 6182
Pangium, 14944
Panicum: ancient, 3602; burial in soil, 7496; characteristics of seeds, 14698; chemical composition, 5172: fats and oils, 17901, proteins, 14697; chemicals effects, 217, 221, 510; coats, 221, 5784; embryo culture, 11828; germination, 217, 221, 510, 1038, 2469, 4235, 6983, 8038, 9518, 11032, 11036, 12978, 13705, 14818, 15328, 16815, 18176, 18691; industry, 17973; injury effects, 221; moisture content, 10618; stimulating substances, 4128; storage, 19634, 19746
Pannua, 3347
Pansy, 761, 943, 1997, 4611, 8268, 11541; *see also* Viola
Panwas, 1249
Papaver: anatomy, 17109; burial in soil, 2989, 5485; chemical change during metabolism, 15536, 15720; chemical composition, 3334, 3992, 11857, 15256, 15714, 16807, 19242: fats and oils, 930, 1344, 1552, 3035, 7353, 9713, 11490; color, 15715, 17125; development, 9384; disease transmission, 7736; embryo culture, 12686, 15535; genetics, 14291; germination, 6394, 8205, 8568, 10677, 14412, 14415, 14441, 15356; industry, 875; maturity, 7231; storage 7499, 11196; vernalization, 11333; viability, 3322, 10739, 10741
Papaveraceae, 839, 14415, 16341, 17278
Papaya, 803, 1460, 3869, 4042, 5701, 6069, 12367, 13729, 15911, 16778, 17237, 18061, 19218; *see also* Carica
Papilionaceae, 475, 9311, 11196, 11441, 13025, 13026, 14354, 14415, 14535, 14538, 15431, 16005, 16084, 19408
Paprika, 15624, 17780, 19023
Paraiso, 13654
Paraseeds, 950
Pariataria, 11943
Parietales, 13049
Parinarium, 3179, 5076, 6220, 16232, 18799
Paris, 2304, 10675, 10677, 13339
Parkashorea, 14944
Parkia, 3448, 5412, 11964, 14944
Parkinsonia, 7630, 15010
Parmongaia, 4208
Parnashia, 10675
Parodia, 11166
Parrotia, 7058
Parsley, 7771, 9015, 19977
Parsnip, 2155, 6947, 9998, 14015
Parthenium, 11299
Parthenocissus, 98
Paspalum: anatomy, 14216; coats, 217, 221, 504, 8889; genetics, 6539; germination, 221, 504, 3418, 8912, 8982, 9849, 12978, 14909, 16017, 19819; injury effects, 221; nutrition, 10099; seed production, 3635, 10891, 16147; *see also* Dallis Grass
Passiflora, 740, 7255, 10681, 14943, 15734, 15880, 17504, 17521,
Pastinaca, 4073, 6475, 7496, 11196, 11277, 12907
Patrinia, 9485
Paulownia, 13509, 18609, 18759, 19978
Pauridia, 12869
Pawpaw, 873, 7908
Pea: absorption, 2184, 3360, 13878; anatomy, 9899; ancient, 19620; characteristics of seeds, 7963, 9010, 9040, 10528, 12890, 14344, 19975; chemical change during metabolism, 1905, 2258, 2507, 2510, 3034, 8373, 8374, 9806, 9807, 10486, 11392, 11394, 11706, 11883, 11991, 12050, 12950, 13204, 13205, 16480, 16498, 16612, 16775, 18117, 18381, 18839, 18845, 19265; chemical composition, 2488, 2753, 2838, 3137, 7387, 9108, 9811, 9950, 10487, 10489, 11205, 11206, 12109, 13683: carbohydrates, 8747, 11564, enzymes, 4352, 7985, 10699, 12830, 13277, 13336, 13487, 13830, 16393, 16776, 18840, 18841, 18844, 20137, fats and oils, 19262, nucleic acids, 3136, 10575, proteins, 2087, 9446, 10797, 10882, 13527, 13972, 19017, 19449, vitamins, 3499, 9007, 19683; chemicals effects, 730, 1753, 1808, 1999, 2745, 3202, 12445; coats, 41, 7334, 9642, 10488, 19230; development, 19263, 19264; disease transmission, 2133, 2819, 3463, 7302, 12888, 19536, 20069; electrical effects, 13898; embryolessness, 11586; fertilizers effects, 2379, 10752, 13983; genetics, 10424; germination, 2349, 2863, 3053, 3573, 8564, 9008, 9301, 9538, 12026,

Pea–(*cont.*)
16963, 19875, 20014; industry, 9120, 10548, 19505; injury effects, 3263, 9073, 9673, 10648, 19258, 19494; maturity, 2720, 9812, 11979, 13837; microorganisms, 19144; moisture content, 11401, 12818; respiration, 11054, 11306, 12970; seed production, 11711, 12817; stimulating substances, 8740, 13246; storage, 725, 1997, 3618, 11196; treatment, 2135, 2580, 2780, 3012, 9284, 9321, 9615, 10500, 11919, 11967, 13934; vernalization, 8742, 11585; viability, 19016; *see also* Pisum

Peach: characteristics of seeds, 3199, 3622, 5014, 5017, 5496; chemical change during metabolism, 6474; chemical composition, 11982: fats and oils, 921; chemicals effects, 3623, 6995, 9186, 13354, 18827; coats, 16587; development, 12121, 13548, 15869, 18828; dwarfs, 1179, 6461, 6471, 6473, 6481, 6482, 17132; embryo culture, 810, 4981, 7260, 10491, 10493, 11723; embryolessness, 1721, 5016; germination, 1346, 2452, 4600, 5495, 6970, 11218, 11287, 14659, 17133, 18707, 19411, 19471; identification, 2582, 15688; industry, 2947; injury effects, 8405; maturity, 16647, 18477, 18615, 19430; plants effects, 18822, 18823, 18824; provenance, 3624; storage, 996; vernalization, 3106, 16319; viability, 8400, 18825, 18826; *see also* Prunus

Peanuts: anatomy, 10863, 17325; characteristics of seeds, 8043, 13615; chemical change during metabolism, 1663, 3818, 9813, 12850, 12851, 15342, 15826; chemical composition, 2786, 6532, 6536, 13666, 15966, 16082, 17874: amino acids, 3182, 3183, 9325, 12560, 13251, fats and oils, 4683, 5595, 6311, 9572, 12877, 13143, 19677, nucleic acids, 12848, 12849, proteins, 3396, 3398, 3399, 6102, 6533, 6534, 6535, 8604, vitamins, 6371, 6372, 8734, 17316; chemicals effects, 4311, 12965; coats, 4312; color, 6996, 8731, 16901; development, 4324, 10922; disease transmission, 1472; fertilizers effects, 2934; germination, 509, 890, 1078, 1948, 2154, 2545, 4461, 7056, 9199, 18451; industry, 9880, 16057, 17290, 17914, 17915, 18461, 19357; inhibiting substances, 14211; injury effects, 1415, 2153, 4485, 15341, 19510; microorganisms, 19358; moisture content, 8927, 10272, 19842, 19843; nutrition, 3041, 3042, 20136; plants effects, 17063; radiation, 2466, 2614; seed production, 2933, 13414, 13631, 13632, 18122; storage, 2150, 2152, 5019, 5354, 11196, 12983, 15532, 17873; treatment, 903, 944, 3235, 4288, 7236, 7604, 8733, 9628,

Peanuts–(*cont.*)
13498, 19725; vernalization, 8741, 14701; viability, 8732, 13800; *see also* Arachis; Groundnut

Pear, 60, 63, 470, 4115, 4406, 5113, 6745, 7679, 12173, 15711, 16105, 18734, 18857, 19656, 19657, 19659; *see also* Pyrus

Pecans: characteristics of seeds, 8836, 16397, 17552, 17622, 18979; chemical change during metabolism, 8081, 18530, 18531; chemical composition, 275, 9231: fats and oils, 6367, 12185, vitamins, 904; chemicals effects, 19673; development, 5428, 12186, 13202, 17375, 17623; germination, 1683, 7504, 13190; industry, 6365; maturity, 6369; moisture content, 19845; radiation, 18754; seed production, 4517, 6366, 6368, 8000, 19844; storage, 3071, 19890; viability, 7349, 17624; *see also* Carica; Nuts

Pectis, 19526

Pectocarya, 19526

Pedicularis, 2203, 10673, 10675, 10677, 12224, 14386

Peganum, 16028

Peltandra, 5794, 5796, 8272, 10426, 13868

Peltiphyllum, 11941

Penao, 2738

Pennisetum, 217, 219, 221, 227, 505, 2423, 2424, 3139, 3419, 3808, 4678, 9189, 14652, 15300, 16713, 17256, 17791, 19406

Penstemon, 3322, 7697, 11196, 17029

Pentacme, 3498

Pentaphragma, 10240

Peony, 93, 682, 930, 1155, 1985, 2034, 2039, 3550, 3720, 3786, 4159, 16813, 16814, 17682, 19771; *see also* Paeonia

Peperomia, 6426, 9851

Peplis, 14898

Pepper, 820, 925, 1251, 1731, 1987, 1994, 2042, 4237, 4238, 4241, 4778, 5476, 7110, 8224, 8231, 8853, 9719, 11196, 11277, 12358, 13140, 14463, 15449,

Peppermint, 117, 14700; *see also* Mentha

Pereskia, 2715, 4214

Perilla, 779, 6177, 10834, 14151, 18318

Periploca, 16848

Persea, 6919, 15009, 17026, 18886, 18935, 20118; *see also* Avocado

Persimmon, 251, 2636, 6491, 8437; *see also* Diospyros

Petasites, 9389, 9390

Petroselinum, 463, 2977, 5864, 6475, 11196, 19761

Petsai, 8850; *see also* Brassica

Petunia, 427, 1069, 2737, 3322, 4378, 4720, 6062, 7499, 8205, 11196, 12151, 14692, 17586

Peucedanum, 880

Peumus, 13046
Phacelia, 1556, 2677, 4546, 5498, 7017, 7710, 8528, 8529, 8569, 10183, 10673, 10677, 11485, 12036, 12039, 12353, 12913, 13565, 14358, 14481, 15235, 15236, 15647, 15804, 16117, 16135, 16386, 16589, 17065, 17950, 19653, 20076
Phaedranase, 17942
Phalaenopsis, 1573, 14588
Phalaris, 368, 4658, 5876, 7328, 7496, 7667, 8802, 12936, 13264, 13771, 14035, 15153, 19241, 19311
Phalsa, 9760
Pharbitis, 7995, 11494, 13943, 14687, 14688, 14689, 19996
Phaseolus: absorption, 16215; anatomy, 10319, 10573, 13409; burial in soil, 7496, characteristics of seeds, 1212, 11733; chemical change during metabolism, 4405, 4894, 5629, 5631, 5672, 5673, 7176, 7461, 8887, 9652, 18475, 20061; chemical composition, 1185, 4120, 5660, 5671, 6286, 7374, 11814, 13619, 16729, 17147, 17524, 17562, 18303, 18504, 19247, 19248, 20032: amino acids, 5238, 6950E, 6950F, 15844, enzymes, 585, 2555, 2560, 2562, 4102, 14058, 16471, 16799, fats and oils, 13449, proteins, 7175, 9428, 10796, 10810, 12099, 14867, 16839, 19010, vitamins, 1650, 1651, 7209, 7373, 7374, 14844, 19257; chemicals effects, 2915, 5971, 7007, 9031, 12426, 12657, 19384; color, 1768, 2262, 5863, 6277, 6278, 11728, 11731, 11737, 12527, 15652, 16158; development, 12390, 13738, 18312, 19271; disease transmission, 633, 11913, 19310; embryo culture, 6859, 14525; genetics, 6712, 6713, 6714, 6715, 6716, 6717, 6718, 6719, 8246, 11621, 11736, 19318; germination, 369, 1694, 2907, 3264, 3639, 4724, 5028, 5170, 5793, 7689, 14163, 14164, 14454, 15871, 16788, 19934; identification, 11998, 19443, 19477; industry, 12228, 12471; injury effects, 2877, 2878, 6708, 18572, 19079, 19402; maturity, 3280, 3281, 9420; microogranisms, 8824; moisture content, 17782, 18386, 19101, 19932; plants effects, 13648, 18169; polyembryony, 15949A; radiation, 476, 7234; rhythm, 3289; seed production, 18284; stimulating substances, 4894; storage, 1159, 11196; treatment, 3773, 4263, 6860; weight, 8239, 8240, 8241, 17225; *see also* Bean; Gram
Phelipaea, 6980, 9598
Phellodendron, 2941, 8574
Philadelphus, 13565
Philadraceae, 8060
Phillyrea, 1359
Philydium, 8061
Phleum, 1, 368, 675, 763, 6072, 7364, 7468, 7496, 8514, 8518, 10739, 10741, 11710, 11772, 12716, 15353, 17782, 19712
Phlox, 724, 3322, 7234, 7499, 11196, 11366, 12036, 12038, 16009, 19524
Phoebe, 114; *see also* Bonsum
Phoelipaea, 9601
Phoenix, 467, 5342, 6306, 6359, 6849, 7268 7395, 12369, *see also* Date
Phoradendron, 2460, 2461, 3567, 19992
Phormium, 13748, 15766
Photinia, 13565
Phragmites, 7889, 8254, 18835
Phreatophyte, 9076
Phyllanthus, 2397, 4858, 12563, 17514
Phyllis, 6179
Phyllites, 13252
Phyllocladus, 8951
Phyllostachys, 7271
Physalis, 1600, 4551, 7857, 8565, 9285, 14598
Physic-nut, 4711
Physocarpus, 15005
Phytelephas, 83
Phytermia, 10675
Phytolacca, 98, 1999, 4547, 7496, 10147 14415, 17925, 18028
Phytolaccaceae, 10148, 14415
Picea: absorption, 20098; anatomy, 17804, 18424; characteristics of seeds, 5385, 6395, 12300, 16898; chemicals effects, 999, 3311, 10202; coats, 4164, 19986; development, 3252, 8019; embryo culture, 11828; germination, 1450, 1981, 1988, 3628, 4188, 4509, 4511, 4513, 4684, 7912, 8574, 8585, 8969, 8980, 9414, 10847, 12040, 15410, 18727, 19928, 20074; industry, 9803; provenance, 7038, 20005; radiation, 11908; seed production, 688; storage, 384, 11196; treatment, 79; viability, 9254, 18723, 19135; *see also* Spruce
Picralima, 13404, 16027; *see also* Akuammine
Picramnia, 17898
Pigeonpeas, 68, 4284, 16699; *see also* Cajanus
Pigweed, 2010, 3212, 9015, 16380; *see also* Amaranthus
Pileus, 3690
Pilocarpus, 5284
Pimiento, 812, 821, 2604, 4239, 4240, *see also* Capsicum
Pimpinella, 1522, 6475, 8572, 9505, 10703
Pineapple, 8737, 10476
Pinguicula, 408, 3424, 3685, 4555, 4556, 7954, 10673, 10675, 10677, 10681, 11202
Pink, 3561
Pinon, 12328

Pinus: absorption, 1436, 7431, 8358, 11976; anatomy, 703, 3240, 7432, 17440, 17829, 17869, 17871; ancient, 19963; burial in soil, 7496; characteristics of seeds, 681, 1963, 3253, 4930, 5385, 5894, 6331, 6395, 6579, 9378, 10031, 10947, 11766, 11977, 12145, 15625, 15626, 16107, 16222, 17192, 17359, 17445, 18031, 19298, 19521; chemical change during metabolism, 1926, 1966, 1968, 2847, 7439, 10230, 10327, 10328, 10773, 13562, 15860, 17181; chemical composition, 8302, 8364, 10329, 12835, 13560, 16673: carbohydrates, 8384, 17547, enzymes, 8368, 16969, fats and oils, 5228, 7230, 10342, 18796, vitamins, 16366; chemicals effects, 792, 3654, 3828, 7440, 7727, 8915, 9774, 9874, 10232, 10245, 13431, 17452, 18984, 18985; coats, 3721, 7435, 8103, 11309, 13758, 19986; color, 19122, 20090; development, 3206, 3256, 3541, 8019, 8021, 12975, 18302, 19288; disease transmission, 9632; dispersal, 836, 2908, 4270, 4605, 4687, 6161, 9771, 11928, 12630, 14767, 18136, 20107; embryo culture, 7962, 9125, 16605, 17019, 18056; fertilizers effects, 7874; germination, 108, 667, 670, 676, 712, 822, 1433, 1434, 1435, 1443, 1447, 1450, 1484, 1503, 1775, 1786, 1980, 1981, 2091, 2161, 2956, 3437, 3490, 4511, 4962, 4969, 4970, 5126, 5351, 5656, 5896, 6467, 6674, 6675, 6854, 7090, 7129, 7660, 7820, 7912, 7913, 7914, 7915, 7925, 7961, 8002, 8146, 8300, 8579, 8585, 8955, 8970, 9028, 9087, 9577, 9639, 9737, 9791, 9909, 10263, 10688, 10928, 10931, 10949, 11185, 11813, 11927, 12328, 12631, 12673, 12736, 13036, 13060, 13154, 13229, 13231, 13258, 13757, 14241, 14572, 14573, 14648 14649, 14811, 15006, 15008, 15009, 15370, 15525, 16088, 16211, 16262, 16373, 16509, 16946, 16962, 17060, 17097, 18054, 18055, 18668, 18696, 18697, 18727, 18982, 19251, 19846, 20074, 20096; industry, 745, 747, 1191, 1920, 2091, 2092, 2695, 3022, 3831, 3832, 3901, 4274, 4337, 4790, 5057, 5222, 5873, 5895, 9030, 11609, 11823, 11824, 11929, 12570, 12571, 13253, 13488, 13789, 15309, 16197, 16210, 17435, 17894, 18229, 18579, 18774, 19277, 19278, 19294, 19295, 19298; injury effects, 10206, 10225, 12557, 19519; maturity, 882, 1920, 4112, 7438, 12724, 13227, 19520; microorganisms, 17040, 18899; moisture content, 4268; pelleting, 7591, 16508; plants effects, 3117, 16465, 16672, 17392, 17446; polyembryony, 703, 2336, 9638, 9908, 14315, 17037; provenance, 670, 1944, 2109, 2374, 3267, 3529, 3796, 4745, 4766, 5753, 5953, 5954,

Pinus(cont.)
6920, 7132, 7628, 8667, 8981, 9099, 9100, 9155, 9788, 10032, 11777, 13435, 14703, 15134, 15458, 15670, 16070, 16879, 17947, 17984, 18474, 19134, 19139, 19296, 19641, 19642, 19649, 20112; radiation, 2882, 14702; respiration, 8367, 8370, 17870; rhythm, 2305; seed production, 395, 1147, 2091, 2322, 2938, 2990, 3724, 7282, 8170, 8689, 8690, 9219, 9846, 9943, 10236, 10335, 12442, 12572, 12573, 13197, 14743, 14815, 15158, 16335, 16342, 16763, 19290, 19292, 19297; stimulating substances, 12820; storage, 678, 742, 876, 1869, 1988, 2090, 3438, 3791, 4188, 7827, 11196, 13563, 14313, 14314, 16512, 17337, 18052, 18053, 18868, 19291; supersonics, 10948; temperature resistance, 12300, 16429; treatment, 79, 1247, 7739, 10652, 11347, 11553; viability, 192, 384, 2646, 4178, 4270, 4686, 5458, 11438, 11455, 12725, 13228, 15805 16333, 16339, 16972, 17039, 17443, 17444 17447, 18723, 19135, 20031; weight, 745, 5102, 7434, 13591, 17451, 17830, 18090, 19884

Piper, 7110, 13252

Piperaceae, 6565

Piptadenia, 7223

Pirola, 12504

Pirolaceae, 4088

Piscaria, 17652

Pistacia, 1575, 5334, 6147, 7273, 7274, 7275, 11353, 14415, 15173

Pistia, 6797, 7803, 17250

Pisum: absorption, 6359, 10721, 10723, 11477; anatomy, 320, 10602, 12935; ancient, 720; burial in soil, 7496; characteristics of seeds, 89, 90, 7964, 11738, 11739, 11740, 16118; chemical change during metabolism, 1167, 2914, 4435, 4871, 4873 4874, 4875, 4894, 4955, 5140, 5946, 6016, 6208, 6516, 6583, 6700, 6942, 7381, 8086, 9317, 10773, 15728, 15750, 15821, 15822, 15823, 17081, 18303; chemical composition, 2505, 4120, 4833, 4835, 4868, 6792, 7093, 7198, 17681, 17807, 17808: carbohydrates, 15164, enzymes, 4870, 5038, 5039, 5429, 6515, 6925, 6926, 8368, 15726, 15727, 16230, 17193, 17202A, 17704, 17761, fats and oils, 12513, proteins, 4869, 4876, 6098, 14863, 15406, vitamins, 5236, 6019, 16181, 16292; chemicals effects, 2199, 2200, 4638, 4659, 11892, 14462, 15707, 18359, 18967; coats, 4669, 5392, 14961, 15119, 16074, 16075, 17825, 17826, 17827, 17828; color, 636, 3736, 10151, 11730, 11735, 16986; development, 4376, 16056, 16290, 17398; disease transmission, 4614, 4618, 14540,

Pisum–(*cont.*)
15249; electrical effects, 18333; embryo culture, 6749, 9133, 10945, 11828; fertilizers effects, 15072, 15087; germination, 1, 636, 655, 675, 1999, 2960, 4664, 5170, 5393, 5793, 6486, 8514, 8697, 11315, 14344, 15717, 17817, 18704; identification, 571, 7765; industry, 12471, 14899; injury effects, 392, 4624, 6708, 9191, 14028, 17083; maturity, 5888, 18706; microorganisms, 6332, 8824, 14950; moisture content, 6946, 8276, 17782; provenance, 12266, 12523; radiation, 1479, 6751, 7234; respiration, 15794; seed production, 6552; storage, 1167, 5325, 11196; treatment, 486, 778, 888, 954, 5006, 5322, 5326, 17143; vernalization, 5366, 14139, 14935; viability, 13208; weight, 597, 2862, 8243; *see also* Pea
Pitcairnia, 8527
Pithecolobium, 5410, 8991, 14944, 15012
Planera, 15009
Plantaginaceae, 14415
Plantago: anatomy, 17577, 20007; burial in soil, 701, 4900, 4901, 4902, 4903, 4904, 5485, 7496; characteristics of seeds, 5522, 6501, 14332; chemical composition, 524, 525, 2865, 9089, 19419: enzymes, 4471; coats, 524, 14538, 15895; genetics, 3802, 7476; germination, 1553, 1991, 3801, 7690, 8655, 11442, 13035, 15193, 15195, 16537, 16538, 18517, 19526, 19778; industry, 10660, 14362; moisture content, 7697; mucilage, 524, 525, 2844, 4470, 8859, 9974 13892, 14321, 17578; seed production, 4388; storage, 282; viability, 10739, 10741, 18778; *see also* Isabghol; Psyllium
Platanus, 3069, 5589, 8569, 9085, 13161, 13565, 15003, 15005, 15008
Platonia, 15168
Platycodon, 8569
Platylophus, 747
Pleurodeles, 15336
Plum, 129, 2664, 7222, 18227, 20128; *see also* Prunus
Plumbagella, 2909
Plumbaginaceae, 2459
Plumbago, 4821, 8391
Plumeria, 3868
Poa: burial in soil, 7496, 13535; characteristics of seeds, 5969, 9171; chemicals effects, 221, 489, 490, 491, 492, 494, 498, 19579; coats, 221, 487; development, 238, 487, 11215, 18588; disease transmission, 507; fertilizers effects, 490; genetics, 237, 8013, 10566, 13873, 15354; germination, 221, 368, 488, 489, 490, 491, 492, 493, 494, 498, 507, 1999, 2454, 2547, 3140, 3347, 4122, 4658, 5165, 6140, 6793, 7020, 8513, 8864, 9855,
Poa–(*cont.*)
9928, 10182, 10673, 10677, 10784, 10785, 10786, 12037, 12715, 13989, 14305, 14494, 15353, 15462, 16451, 18687, 18711, 18977, 19217; identification, 7108; industry, 19891; injury effects, 221; moisture content, 493, 7697, 17782; plants effects, 7118, 14181, 14641, 15551, 20126; polyembryony, 19250; rhythm, 11221; seed production, 235, 236, 237, 368, 763, 5968, 12269; treatment, 17706; viability, 3322; *see also* Clover
Poaceae, 18390
Pocockia, 7522
Podocarpaceae, 3249
Podocarpites, 13252
Podocarpus, 747, 3244, 11254, 11741, 15677, 18270
Podolepis, 2176
Podophyllum, 1429, 1649
Poinciana, 7871, 13978, 15697
Poinsettia, 96, 8088, 9003, 17198
Poivrea, 19061
Polemoniaceae, 839, 2459
Polemonium, 5876, 12036, 12870
Polyalthia, 4414, 16784
Polyanthus, 5045
Polygala, 8986, 12224, 14386, 18327
Polygonaceae, 11105, 14415, 15933, 17548, 19824
Polygonatum, 98, 10675, 10677, 18772
Polygonella, 14386
Polygonum: burial in soil, 5485, 7496, 7498, 7500; chemicals effects, 11511, 18967; coats, 14538; development, 15452; germination, 6551, 7769, 8192, 10076, 10077, 10079, 10822, 10823, 11442, 11641, 13526, 13868, 14415, 14608, 15185, 15240, 15452, 18317, 18915; identification, 16950; industry, 17928, 19789; provenance, 18915; storage, 10079; viability, 10739, 10741
Polypogon, 18687
Pomegranate, 6221, 10653, 18732, 18737
Poncirus, 777, 3835, 5927, 8849
Pongamia, 191, 711, 10519, 12810, 14944, 15959, 17727
Pontederia, 13868
Pontederiaceae, 14782
Popcorn, 2266, 5886, 9183, 16176, 17992
Poplar, *see* Populus
Popowia, 4414
Populus: anatomy, 10629; chemical composition, 3234; dispersal, 3667; embryolessness, 8553, 10116; genetics, 2218; germination, 2371, 3438, 5981, 6252, 6837, 8790, 8955, 9085, 10677, 12864, 15007, 16187, 17477, 17781; industry, 1200, 8837, 8911, 11239, 11810; injury effects, 15519; plants effects, 4156; provenance, 3093,

Populus–(*cont.*)
10628, 12244, 12441; storage, 2899, 3233, 9875, 15581, 16362; viability, 2900, 6151, 13809, 17551; weight, 2371
Portulaca, 3322, 3705, 3864, 4377, 4900, 4901, 4902, 4903, 4904, 7496, 8205, 8527, 8561, 11196, 14691, 19526
Portulacaceae, 11293
Potamogeton, 5485, 6400, 7843, 10677, 13866, 13868
Potarium, 1882
Potato: anatomy, 14824, 14825, 15209; characteristics of seeds, 5935; chemicals effects, 4169; color, 1841; development, 11721; disease transmission, 5926, 16045; embryo culture, 8449; genetics, 1368, 8010; germination, 1841, 2813, 4174, 4438 6392, 6416, 11867, 14668, 15394, 17458, 17922, 18007, 18008, 18010; radiation, 18377; respiration, 2953; seed production, 1606, 4171; stimulating substances, 657; viability, 4150, 4280, 19806; *see also* Solanum
Potentilla: burial in soil, 7496, 13535; development, 4794; genetics, 7099, 13874, 13875, 17646, 17647; germination, 9293, 10433, 13565, 14386, 14415; identification, 10433; polyembryony, 11940; rhythm, 3283
Poterium, 14415, 18919
Poverty Weed, 3210, 3211
Preissia, 8127
Prenanthes, 14386, 20076
Pretoria, 1156
Prickly Pear, 14025
Primrose, *see* Primula
Primula: anatomy, 3877, 6600; burial in soil, 701, 5485; coats, 14538; genetics, 18932; germination, 128, 393, 468, 666, 693, 733, 785, 1343, 4524, 4681, 5045, 5432, 7189, 8035, 8574, 8658, 9102, 9768, 10452, 10674, 10675, 10677, 12163, 12432, 13214, 13339, 17029, 19564, 19815, 19870; storage, 666, 785
Primulaceae, 14415
Primulales, 19047
Prosopis, 747, 2308, 5411, 5415, 7308, 7669, 10516, 10542, 10544, 12917, 13258, 14165, 15008, 17838
Proteaceae, 14340
Proteoides, 13252
Prune, Sugar, 5015
Prunella, 10677, 13535
Prunoideae, 20072
Prunus: absorption, 6359; anatomy, 14430; characteristics of seeds, 14433; chemical change during metabolism, 16120; chemical composition, 16241, 19450; chemicals effects, 16119; coats, 16588; development,
Prunus–(*cont.*)
8269, 16743; disease transmission, 7256; embryo culture, 8686, 9144; embryolessness, 18818, 18819; germination, 98, 854, 2664, 3480, 4488, 4744, 4925, 5118, 6116, 6461, 7222, 8396, 8406, 8559, 8568, 8569, 8784, 9709, 9711, 11041, 13565, 14386, 14642, 15008, 15009, 19233, 19658; industry, 12845; maturity, 18595; microscope, electron, 16744; storage, 7222, 11196; *see also* Apricot; Cherry; Peach; Plum
Psammophytes, 14224
Pseudocymopterus, 7697
Pseudolarix, 3250
Pseudotsuga: characteristics of seeds, 5385, 11879; chemicals effects, 382; development, 380, 381; embryo culture, 11828; genetics, 3794; germination, 376, 377, 385, 387, 389, 390, 8585, 8969, 9944, 11880, 18727, 20074; industry, 1869, 11623; plants effects, 11882; provenance, 388; seed production, 375; storage, 11196; treatment, 382; viability, 384, 385, 18723; *see also* Douglas Fir
Psidium, 9804, 12843, 14221, 15008, 15971, 17240, 19256
Psophocarpus, 14944
Psoralea, 3766, 9919, 9920, 10005, 12808, 15656, 17399, 17400
Psyllium, 705, 7625, 7909, 8125, 8655, 13985, 16051, 19442, 20008
Ptelea, 15003, 15007, 17028
Pteridosperms, 8041
Pterocarpus, 747
Pterodon, 7871
Pulwara, 3427
Pulsatilla, 1516, 8751
Pulses: chemical change during metabolism, 1844, 1845, 3853, 3855, 3995, 5062, 12524, 13609, 15051; chemical composition, 2083, 3854, 3856: amino acids, 1628, 3996, 7861, enzymes, 1848, proteins, 3857, 6058, vitamins, 1849, 1850, 14177, 16014, 16353, 16522; embryo culture, 17224; industry, 7720; plants effects, 11413
Pumpkin: characteristics of seeds, 1341; chemical change during metabolism, 17576, 19441; chemical composition, 2786, 3309, 12074, 18000: amino acids, 16027, enzymes, 5930, fats and oils, 2227, 4721, 16196, 19229, proteins, 11373, 17017; color, 9394; germination, 820; industry, 6697, 15303, 18931; plants effects, 18242; storage, 698; *see also* Cucurbita
Pumulaceae, 2459
Puncture Vine, 1803, 1806
Punica, 3450, 15008, 18732, 18737
Purdieae, 10677
Purshia, 9164, 13258, 13565, 15156

Putranjiva, 11412
Pygerum, 17964
Pyracantha, 747
Pyrethrum, 724, 3322, 3787, 3804, 9505, 11196, 14097, 16029
Pyrola, 10677, 12237, 19682
Pyrularia, 3266, 7393
Pyrus: anatomy, 14430, 15014; characteristics of seeds, 14433, 16873; chemical composition, 16241; embryolessness, 10266; germination, 2000, 4600, 5018, 5118, 6461, 8396, 13565, 14386, 15005, 15008, 16875, 16876, 19233; inhibiting substances, 2048; plants effects, 12806, 16874; storage, 11196; temperature resistance, 16877; *see also* Apple; Pear

Quackgrass, 3212, 8594, 11061
Quamclit, 8569
Quandrong, 3330, 14026
Quercus: characteristics of seeds, 15425, 15426; chemical change during metabolism, 15861; chemical composition: vitamins, 13550; development, 4433, 8872; germination, 673, 1023, 3155, 3401, 4188, 4488, 5975, 8552, 8559, 10658, 13363, 14811, 15000, 15001, 15009, 15012, 15687, 17097; industry, 673, 747, 4166, 8385; seed production, 11786, 16095; storage, 673, 8229, 11196; viability, 16972; *see also* Acorns; Oak
Quillaja, 15469
Quince, 930, 5344, 13985, 15712, 16050, 16125; *see also* Cydonia
Quinine, *see* Cinchona
Quinoa, 445

Radiola, 4537
Radish, 1947, 1999, 2343, 2997, 3625, 3629, 9565, 18244, 18800; *see also* Raphanus
Ragi, 1585, 3806, 3807; *see also* Eleusine
Ragweed, 701, 917; *see also* Ambrosia
Ragwort, 15538; *see also* Senecio
Raintree, 2400; *see also* Pithecolobium
Ramie, 15554
Ramondia, 17029
Ramtil, 6311
Ranunculaceae, 839, 1596, 1597, 1599, 8955, 11214, 11525, 14415, 14535
Ranunculus, 5432, 7016, 7017, 7843, 8498, 8955, 12037, 12038, 12039, 12040, 13364, 13535, 14415, 14898, 16601, 16997
Rapanea, 747
Rape: burial in soil, 18999; characteristics of seeds, 2244, 11152; chemical change during metabolism, 2204, 19569; chemical composition, 168, 2421, 8291, 15836, 19570: carbohydrates, 13633, enzymes,

Rape–(*cont.*)
14925, 15904, fats and oils, 801, 3338, 4498, 4499, 6308, 6809, 8776, 9801, 16156, 19568, 19944, 20013, proteins, 4138; chemicals effects, 16714; coats, 19036, 19037; color, 17546; disease transmission, 4799; fertilizers effects, 8341; genetics, 1294; germination, 8291; industry, 16534; maturity, 7508; moisture content, 11401, 12334; nutrition, 8140; plants effects, 10572; seed production, 6810; treatment, 6771, 9251; vernalization, 10837, 18527; viability, 18261; *see also* Brassica; Colza
Raphanus: anatomy, 14521, 16415; characteristics of seeds, 7305, 7410, 7998, 16474; chemical change during metabolism, 5662, 10376; chemical composition, 962, 13939, 16942, 16943, 17756: carbohydrates, 15020, enzymes, 4789, nucleic acids, 18388; chemicals effects, 6964, 11573, 14552, 17054; development, 10161; disease transmission, 5269, 8834; embryo culture, 10160, 16352; genetics, 9088, 15221; germination, 1999, 3629, 5793, 7769, 9489, 11442, 17407; identification, 11479; industry, 463, 5776, 14495; microorganisms, 13226; moisture content, 17782; pelleting, 7994; plants effects, 10122; radiation, 7234; seed production, 5918; stimulating substances, 398, 657; storage, 925, 11196; vernalization, 5818, 7999, 9418, 10126; viability, 463, 3322, 12667, 18778, 19267; *see also* Radish
Rapistrum, 2176
Raspberry, 1040, 9177, 19893; *see also* Rubus
Rauwolfia, 5699, 5700, 7852, 10387, 13628, 13629, 13630, 14240, 14258, 15022
Redtop, 2068, 4850
Redwood, 8131; *see also* Sequoia
Reedgrass, 17392
Relhania, 12165
Reseda, 282, 1553, 4535, 7479, 8205
Resedaceae, 8630
Rhamnus, 98, 8568, 15006, 17333, 19353
Rheum, 8572
Rhinanthus, 8526, 14412, 18942, 18943
Rhizophora, 11833, 15010
Rhododendron, 709, 2884, 4314, 4682, 5659, 5979, 6355, 8574, 9995, 13640, 13778, 14386, 15009, 17275, 18168, 18408, 19352
Rhodomyrtus, 18769
Rhodotypos, 6458, 6459, 6471, 18595
Rhoeadales, 16991
Rhus, 362, 747, 2904, 6395, 6658, 7496, 8029, 8356, 8357, 11295, 11461, 13175, 13745, 14161, 14538, 14979, 15003, 15009, 18058, 18591, 18595, 19390; *see also* Sumac

Ribes, 98, 6441, 8574, 9797, 14386, 14430, 14433, 15003, 15011, 15802

Ricinus: anatomy, 15891, 16823, 17513; characteristics of seeds, 13651, 15082; chemical change during metabolism, 4399, 7811, 7812, 15623, 17772, 19925; chemical composition, 5265, 17800, 17801, 17802: amino acids, 14622, enzymes, 3506, 3507, 3526, 3755, 5359, 9672, 12415, 14949, 18241, 18299, 18300, fats and oils, 1524, 14944, 17909, proteins, 13827, 16101; chemicals effects, 15204; dispersal, 9279; genetics, 14201; germination, 7620, 7689, 8569, 10593, 13132, 14415, 16099; industry, 12228; radiation, 7919; respiration, 4534, 11924

Rivina, 10002, 10147

Robinia, 673, 712, 747, 7296, 7496, 11087, 11196, 15003, 15063, 15197, 17042, 17799, 18636, 18723, 19735

Roemeria, 10677

Romneya, 13565

Roridula, 16980

Rosa: anatomy, 14430; characteristics of seeds, 5827, 6395; chemical composition, 16553: amino acids, 1003, vitamins, 7886, 12964, 13317, 16495; coats, 724, 18196; dispersal, 5391; embryo culture, 1458, 1459, 11724; genetics, 6189; germination, 724, 1100, 3538, 4529, 4600, 5002, 5259, 6489, 8192, 8728, 8950, 9077, 9630, 9648, 11409, 12172, 13090, 13157, 13565, 13741, 14645, 15003, 15005, 15709, 16483, 16846, 17199, 17201, 17307, 17308, 18583, 19185; seed production, 15177, 17198

Rosaceae, 6470, 10802, 14415, 15169

Rosales, 13051

Rose, *see* Rosa

Rosmarinus, 8572

Roystonea, 15012

Rubber, 874, 5048, 8280, 8764, 18806; *see also* Hevea

Rubia, 98

Rubiaceae, 1594, 1595, 12188

Rubidium, 2360, 2361

Rubus: characteristics of seeds, 4913; chemicals effects, 1040; genetics, 2314; germination, 98, 4916, 8574, 8590, 11475, 14386, 15003, 16413, 17131, 19893; *see also* Blackberry; Dewberry; Raspberry

Rudbeckia, 2103, 6188, 7496, 7697, 12458, 12705, 14989

Rumex: burial in soil, 4900, 4901, 4902, 4903, 4904, 5485, 7496, 13535; chemical composition: enzymes, 282; coats, 14538; fossils, 13252; germination, 1553, 1556, 8192, 8572, 9485, 9491, 10677, 11442, 12907, 14415, 14458, 15240, 16624, 19102;

Rumex–(cont.)
identification, 5572; industry, 416; injury, effects, 10227; maturity, 794, 7231; moisture content, 7697; plants effects, 9289; viability, 10739, 10741, 18778

Ruppia, 3391

Russian Knapweed, 3210, 3211

Ruta, 8572

Rutabaga, 7638, 9156

Rutaceae, 13045, 14208, 14415

Rye: absorption, 6000; burial in soil, 18585; characteristics of seeds, 2702, 7059, 10195, 11381; chemical change during metabolism, 3688, 7756, 11389, 14635; chemical composition, 6752, 9354, 9753, 12496, 17049: carbohydrates, 13446, enzymes, 1727, 4098, fats and oils, 7329, 18070, nucleic acids, 6370, proteins, 14860, 16375, 16377, vitamins, 13974; chemicals effects, 1999, 3198, 12794, 18540; development, 2137, 2392, 20023; embryo culture, 147, 3596; embryolessness, 10188; genetics, 5903, 6699, 6778, 8847; germination, 11862, 12793, 16957; identification, 11797; maturity, 17713; moisture content, 11401, 14350, 14831; plants effects, 11267; polyembryony, 11266, 13742; radiation, 2995, 2996, 2999, 10708; seed production, 3302, 6979; stimulating substances, 3073; storage, 7130, 9190; supersonics, 17091; temperature resistance, 2924; vernalization, 35, 160, 6747, 7641, 7643, 7644, 7645, 7757, 8377, 9244, 11165, 11758, 14636, 15774, 15776 15778, 17090, 18634; viability, 12521; *see also* Secale

Rye, Russian Wild, 4325, 4342, 15634, 18023; *see also* Elymus

Ryegrass: anatomy, 14580; characteristics of seeds, 15367; chemical composition, 3867, 4321: enzymes, 4261; development, 9313; disease, transmission, 1057, 3535, 6407, 7301, 7483, 7484, 8164, 14297; fluorescence, 1654, 1657, 1940, 4273, 6595, 10084, 14493, 14650, 15369, 19837, 19839; germination, 499, 1220, 1276, 6358, 7624, 9296, 9304; industry, 9772, 11895, 16580; microorganisms, 9942; moisture content, 11401; nutrition, 4321; radiation, 5163; seed production, 1277, 6083, 6338, 9923; storage, 6087, 7485, 9988; treatment, 8011; vernalization, 2250, 18313; viability, 9294, 9299; *see also* Lolium

Sabadilla, 399, 400, 402, 1531, 7760, 9359

Sabal, 15009

Saccharum, 23, 1041, 1075, 4909, 5024, 5782, 11990, 12083, 15497, 15864, 16758, 19089,

Saccharum–(*cont.*)
19113, 19166, 19374, 20021; *see also* Cane, Sugar
Saccoglottis, 8791
Safflower: chemical composition: fats and oils, 1474, 4136, 4497, 4893, 9886, 10465, 13525, 15146, 15147, 19981, proteins, 1724; development, 12061; disease transmission, 3537, 17084, 18487; germination, 13183, 15573; industry, 14214, 15824; plants effects, 19980; radiation, 2146; seed production, 17189; storage, 2074; *see also* Carthamus
Sagebush, 7037, 10666, 19492
Sagina, 13535
Sagittaria, 4820, 6400, 7843, 12907, 13868
Sainfoin, 2127, 3331, 7121, 9431, 10258, 13567, 13801, 15874, 16250, 18519
Saint John Wort, 3677
Saintpaulia, 17066
Saksaul, 11008
Salacea, 5230
Salicornia, 4826
Salix, 2218, 3234, 6151, 6271, 8017, 8955, 9723, 10630, 10763, 13339, 14126, 14128, 14553, 15010, 15581, 17551; *see also* Willow
Salpiglossis, 4575, 7499
Salpingostoma, 7473
Salsify, 18729
Salsola, 2370, 7498, 7500, 8633, 8634, 14224, 14996
Saltbush, Fourwing, 14564, 18856
Salvadora, 737, 7834
Salvadoraceae, 4951
Salvia, 931, 3322, 7769, 8572, 10675, 10677, 11196, 11333, 12522, 14040, 14365, 14441, 16951, 17244, 19526
Samadera, 14100
Samanea, 14944
Sambucus, 98, 5023, 10673, 14386, 15185, 16361, 16413
Sandoricum, 19116
Sanguinaria, 3424, 7032; *see also* Bloodroot
Sanguisorba, 12254
Sankalu, 14061
Santalaceae, 8381, 14988
Santalum, 225, 767, 3330, 9595, 11272, 12665, 17837
Sanvitalia, 2176, 8878
Sapindaceae, 4951, 18964
Sapindus, 5215, 15048
Sapium, 665, 10132, 15011
Saponaria, 7234, 8565, 8568, 10677, 10822, 10823, 12554, 15577
Sapotaceae, 3905, 7120, 10119
Sapote, White, 14007
Saracha, 4569
Sarcococca, 14840
Sarda, 186; *see also* Melon
Sarothamnus, 4424, 9715
Sasanqua, 1290
Sassafras, 4156
Satureja, 463, 8572
Saxegothaea, 5547, 12424
Saxifraga, 733, 3833, 6255, 10674, 10675, 10677, 10681, 11939, 12157, 14415, 15185
Saxifragaceae, 6449, 13039, 14415
Scabiosa, 4031, 7499, 8574
Schenchzeria, 10675, 10677
Schinus, 15006, 15009
Schizandra, 1749, 8442
Schizanthus, 7499, 8565, 8569
Schleichera, 1588
Schoenocaulon, *see* Sabadilla
Sciadopitys, 667, 1981, 3241, 19423
Scilla, 7126, 7527, 16411
Scirpus, 6400, 9462, 13868, 16990
Scitamineae, 9210, 13047
Scleranthus, 10739, 10741, 14538
Scleropogon, 9629
Scoliopus, 2304
Scopzonera, 11196
Scorpiurus, 4945, 7521
Scrophularia, 7017, 14414, 14898
Scrophulariaceae, 1856, 7327, 8284, 9588, 9588–A, 9588–B, 9589, 9590, 10064, 14415
Scurrula, 167
Sebastiana, 14102
Secale: absorption, 3112, 6359; anatomy, 756; burial in soil, 7496; chemicals effects, 1999; fertilizers effects, 10782, 10783; genetics, 756; germination, 7756, 8514, 10593, 10708; moisture content, 19696; nutrition, 250; radiation, 2998, 5156; seed production, 14671; treatment, 8822; viability, 250, 3322; *see also* Rye
Sedum, 6818, 10677, 14415
Semecarpus, 15757
Sempervivum, 10677
Senecio, 794, 1901, 7231, 10868, 15185, 15538; *see also* Ragwort
Senna, 15858
Sequoia, 667, 1981, 3245, 3246, 3254, 11196, 12423, 18723, 18724, 18727, 20074
Serbania, 15007
Serbia, 12856
Serradella, 17955
Sesamum: characteristics of seeds, 4730, 4731, 15082; chemical composition, 10664, 17528: amino acids, 2627, carbohydrates, 8355, fats and oils, 2348, 6246, 6311, 9646, 10663, 10857, 12444, 14944, 15892, proteins, 14236, 14237, 14238, 19071; chemicals effects, 3772, 3774, 15482; coats, 4729; development, 9383, 14554; genetics, 8095, 11416, 11774, 18472; germination, 4248,

Sesamum–(*cont.*)
13000, 13001, 13002, 13003, 16186; industry, 4413, 12899; maturity, 11775; nutrition, 3846, 11415; storage, 2074; viability, 411
Sesbania, 2420, 6227, 11551, 14944, 19018
Setaria, 1275, 3212, 7796, 9279, 9518, 9521, 10501, 10618, 12907, 13764, 14450, 15233, 18691, 18701, 19988
Shallot, 2369
Shea, 707, 737, 7621, 9754
Shepherd's Purse, *see* Capsella
Shorea, 3428, 3795, 7202, 8779, 15936, 16479; *see also* Borneo Tallow
Shortia, 6128
Sibbaldia, 7370
Sicyos, 4564
Sida, 7498, 7500, 9892, 15012
Sideritis, 282, 1553, 2455, 2677
Siderocarpus, 15011
Sideroxylon, 15011
Siegesbeckia, 2176
Siene-Bean, 15210; *see also* Daubentonia
Silene, 1556, 5485, 7017, 8498, 9485, 10677, 14386, 14415, 14441
Silverbell, 7221; *see also* Halesia
Silverberry, 4418; *see also* Elaeagnus
Simmondsia, 4949, 5512, 7110, 7111, 7622, 7631, 10893, 13258, 13564, 19511; *see also* Jojoba
Sinapis: absorption, 6738; anatomy, 4828, 13950; chemical change during metabolism, 10869; chemical composition: fats and oils, 2226, proteins, 11751; chemicals effects, 12208; coats, 3348; genetics, 17758; germination, 5789, 8514, 10822, 10823, 10990, 11018, 11442, 15240, 16822; identification, 2316, 11479; industry, 12471; inhibiting substances, 15302, 17846; radiation, 2689; stimulating substances, 33; vernalization, 5661, 19189; viability, 4953, 10739, 18778
Sinningia, 7017
Siparuna, 8497
Sisymbrium, 57, 282, 1553, 2455, 5420, 5485, 7494, 7496, 7764, 10745, 12907, 14412, 14415
Smartweed, 701
Smilacina, 98, 2047, 13339, 14386
Smilax, 98, 10681
Snapdragon, 409, 722, 793, 798, 1801, 1999, 2045, 8345, 17198; *see also* Antirrhinum
Snow-on-the-Mountain, 8555; *see also* Euphorbia
Solanaceae, 839, 1547, 1857, 2393, 5972, 8229, 12951, 14415, 14875, 15218, 15292
Solanum: anatomy, 14430; burial in soil, 7496, 7498, 7500; characteristics of seeds, 14433; chemical composition, 6918: fats and oils, 7855, 15194, 15762, 17401; chemicals effects, 1999, 14414, 17798, 19653; development, 2136, 2216, 5416; embryo culture, 8030; genetics, 1737, 12741; germination, 98, 1987, 4150, 7133, 7409, 8565, 8624, 12036, 12907, 15008, 15010; maturity, 3776, 7411, 18594; plants effects, 16108; rhythm, 17431; seed production, 5354, 19699; storage, 11196, 17987; viability, 5864, 18203
Soldanella, 10674, 10675, 10677
Solidago, 14386
Sonchus, 794, 4679, 7231, 8445, 10867, 15240
Sonerila, 4558
Sophora, 225, 296, 3043, 3044, 3045, 3046, 3047, 14821, 14822, 15009, 19998
Sopubia, 9588, 9591
Sorbus, 748, 6175, 6475, 8552, 8559, 12240, 12481, 12482, 14642; *see also* Ash, Mountain
Sordaria, 3500
Sorghastrum, 368, 4452
Sorghum: absorption, 6213; anatomy, 16704; burial in soil, 7498, 7500; characteristics of seeds, 1578, 1583, 3564, 9649, 10274, 17415, 18214; chemical change during metabolism, 75, 1282, 8062, 12253; chemical composition, 2356, 3006: carbohydrates, 10278, 12652, enzymes, 7461, proteins, 3392, 13486, vitamins, 16049, 18362; chemicals effects, 6972, 10308, 15136, 18376; coats, 18212; color, 2617, 2618, 2619, 14734, 17415, 17416, 17417; development, 4296, 10483, 10484; disease transmission, 12140, 12141; dispersal, 10277; embryolessness, 1581; genetics, 9972, 10275; germination, 1707, 3131, 3675, 6101, 7450, 7699, 8109, 9367, 11743, 15413, 15667, 17842, 18176, 18451; industry, 946, 1091, 3612, 5881, 12902, 13700; inhibiting substances, 13478; injury effects, 1494, 1584; maturity, 15529; microorganisms, 1697, 7533, 18225; moisture content, 441, 19622; plants effects, 1579, 1582; radiation, 4503, 7974, 10344; respiration, 4283; seed production, 3392; storage, 2075, 3121, 6292, 8780, 9190, 17295, 17750; temperature resistance, 3613, 7698, 16422; treatment, 3239, 8626, 9896, 10297, 12139, 15811; viability, 10276, 11780, 16257
Sorrel, 11169
Sour Orange, 13460
Soybean: absorption, 17613; anatomy, 2467, 2468, 9129, 15016; characteristics of seeds, 32, 472, 3393, 6829, 9128, 10861, 13793,

Soybean–(*cont.*)
18140, 19486; chemical change during metabolism, 287, 1883, 2230, 3204, 4014, 4262, 6700, 8871, 9603, 10134, 10403, 10924, 11075, 11078, 11569, 11992, 11993, 11994, 13282, 16340, 16671, 16767, 16770, 17300, 17376, 17760, 18151, 18251, 18366, 19221; chemical composition, 1660, 2261, 2401, 2467, 2468, 2772, 3222, 3405, 3410, 3661, 3663, 8314, 9837, 10326, 11528, 11887, 12569, 14119, 14528, 14718, 16576, 16765, 16769, 16853, 16933, 17313, 18434, 19667: amino acids, 4260, 4719, 5654, 6591, 8077, 9096, 11492, 12560, 16204, carbohydrates, 2861, 10367, 10369, 11987, 11988, 12655, enzymes, 1289, 1819, 5084, 6203, 6204, 6571, 6823, 6824, 6825, 6826, 7588, 8961, 9531, 10574, 10925, 11073, 11863, 12088, 12230, 14377, 14383, 14679, 15165, fats and oils, 1397, 3116, 3326, 3409, 4303, 4304, 4709, 5443, 5838, 8776, 9224, 9708, 10788, 11426, 11771, 13219, 14529, 14737, 15633, 17460, 18194, 18764, 18802, 19341, nucleic acids, 10499, proteins, 2087, 3107, 5116, 5887, 9428, 9936, 10799, 10813, 11074, 16618, 19990, vitamins, 3383, 6766, 16908, 19279; chemicals effects, 1085, 1999, 7541, 9134, 14681, 15141, 16857; coats, 31, 12345, 16723, 16772, 17998; color, 472, 473, 9954, 12710, 13671, 14657, 17997, 18803, 19704; development, 2467, 6831, 6915, 14715, 14923; disease transmission, 2760, 5249, 8541, 10133, 10860, 12347, 15606, 19894; electrical effects, 9471; embryo culture, 31; embryolessness, 10320, 14514, 15109, 19855; fertilizers effects, 101, 14322; genetics, 472, 5447, 12002, 15598, 19613; germination, 227, 509, 1923, 5164, 5795, 7808, 8109, 9424, 13494, 18177; identification, 4005, 7538; industry, 313, 940, 1135, 1265, 3531, 5145, 8746, 8960, 12035, 12228, 13134, 13516, 13519, 17286, 18020, 18715, 19793; inhibiting substances, 2490, 2772, 2890, 5245, 5246, 9647, 10498, 10845, 11540, 12231; injury effects, 4154, 11427, 14656; maturity, 17511, 19799; microorganisms, 7921, 10666, 13523, 15377, 16136, 16137, 17314; moisture content, 219, 227, 2071, 2172, 6946, 7143, 11805, 11934; nutrition, 2768, 3041, 3042, 12229, 14061, 16578, 20136; pelleting, 7210; plants effects, 4302, 6858; polyembryony, 14911; provenance, 6830, 16800, 18946, 19936; radiation, 3881, 12412, 16016, 19701; respiration, 9132, 13517, 13518; rhythm, 5388, 14877; seed production, 6431, 9106, 10321, 18992; storage, 2594, 3651, 9190, 13522, 15916, 18675; treatment, 940, 954, 1487, 1488, 1489, 8748, 9859, 14466; viability, 219, 3369, 11056, 11865, 16257, 18455; weight, 6829, 8354, 14676; *see also* Glycine

Sparganium, 6400, 7843, 13868, 17920, 17929
Sparmannia, 1512; *see also* Cashew
Spartina, 14309, 19225
Spartium, 7524, 15862, 16162
Speedwell, Corn, 794; *see also* Veronica
Spergula, 5485, 8514, 18778
Spermacoce, 1595
Sphaeranthus, 18599
Sphenopteris, 20111
Spinach, *see* Spinacia
Spinacia: characteristics of seeds, 234; chemical composition, 15850; genetics, 8906, 9720; germination, 7997, 8545, 8557, 8560, 8587, 9676, 10082, 12682, 16224, 17423, 18338, 19398; industry, 5321; inhibiting substances, 8303, 12726; maturity, 10106; moisture content, 17782; radiation, 7234; stimulating substances, 657, 11277; storage, 11196; treatment, 954, 4355, 9621, 10183, 13248; vernalization, 8093, 10067, 10123, 10124, 14136; viability, 3322, 12258, 12667
Spindle Tree, European, 3477; *see also* Euonymus
Spiraea, 9485, 15010
Spiranthes, 12702
Spoonwort, Arctic, 12120
Spondias, 10056
Sporobolus, 221, 6764, 7496, 9629, 17407, 18692
Sprekelia, 9692
Spruce: characteristics of seeds, 3752, 16372; chemicals effects, 1772, 7727, 18180; color, 16500; germination, 999, 1777, 1970, 2515, 2956, 4509, 5843, 8970, 10263, 10567, 12394, 12593, 13757, 16358, 16499; industry, 1970, 3628, 5057, 10284, 11624; injury effects, 17442; maturity, 4510, 4513, 4684, 9891; microorganisms, 18899; plants effects, 2919, 17392; polyembryony, 1788; provenance, 2875, 3594, 16581, 19134, 19641; radiation, 10208; rhythm, 1778, 16358; seed production, 2322, 16338, 19131; storage, 1988, 8155, 19127, 19129; treatment, 4512; viability, 192, 4767, 7701; *see also* Picea
Spurge, Casper, 716
Squash, 2463, 3483, 7492, 9040, 13417, 14246, 14694, 17712, 20003
Squill, Red, 18980
Stachys, 2455, 2459, 10677, 14441
Stachytarpheta, 18389
Stackhousieae, 2459
Stapelia, 8738

Staphylea, 8238, 15144, 18723
Stellaria, 794, 5485, 5876, 7843, 11018, 11442, 12036, 13535, 14415, 14970, 15708
Stemodia, 9588B
Stenophragma, 14415
Stephanospermum, 18415
Sterculia, 120, 737, 1463, 4428, 8639, 8774, 14944, 15008, 15756, 17728, 17908
Sterculiaceae, 15941, 15942
Sterigmatocystis, 3839
Sternbergia, 10241
Stigmatophyllum, 15935
Stillingia, 987, 2105
Stinkweed, 9015
Stipa, 511, 5044, 5724, 7697, 14452, 15462, 16351, 16382, 16672, 16858, 17173, 17755
Stizolobium, 10091
Stock, 1999, 6671, 10459, 16649, 17198; *see also* Matthiola
Stramonium, 7290
Strawberry, 30, 1040, 3054, 3603, 3873, 4912, 5173, 5370, 5384, 7858, 8646, 8672, 9226, 12434, 14032, 14526, 17128, 17130, 18506; *see also* Fragaria
Streblus, 16002
Strelitzia, 5677
Streptanthus, 1991
Streptocarpus, 2679, 5345, 8658
Streptochaeta, 16059
Striga, 619, 621, 3168, 3169, 3174, 3175, 3176, 3177, 5788, 11522, 11587, 17737, 18171, 18172, 18173, 18563, 18909, 18938, 18939, 18940, 18941, 19864, 19865
Strombosia, 7120
Strombosiopsis, 7120
Strophanthus: characteristics of seeds, 736, 13674, 19111, 20010; chemical composition, 1000, 1001, 5774, 7281, 10142, 10429, 11864, 12213, 12214, 14804, 16093, 16128, 16129, 16130, 16188, 16911, 16912, 16913, 16914, 16915, 16916, 16917, 16918, 16919, 16994, 19202, 19203, 19204, 19205, 19206, 19481, 19482, 20089: carbohydrates, 15930, enzymes, 5378, 18047, fats and oils, 7837, 7838; germination, 4526, 16722; identification, 9214, 19260; industry, 1000, 1001, 3911; moisture content, 18606
Strychnos, 2291, 8023, 8494, 10775, 15755, 19409
Styrax, 12821
Suaeda, 2479, 9999, 10677
Sucupira, 12608
Sugar Pine, 15805, 17037, 17040, 17869, 17870, 17871; *see also* Pinus
Sugi, 7433, 10324, 11296, 13429, 13430
Sumac, 2904, 11461, *see also* Rhus
Sunchillo, 3377; *see also* Wedelia
Sunflower: characteristics of seeds, 9561, 11380; chemical change during metabolism, 2113, 3295, 5965, 6668, 12777, 15566, 17284, 19652; chemical composition, 7397, 7398, 7401, 10532, 11049, 11391, 16555, 16883, 17159: amino acids, 312, 2564, 2627, 6578, 10533, 11414, 16132, carbohydrates, 2493, 7396, enzymes, 6933, fats and oils, 311, 3036, 4474, 5596, 6311, 6848, 7686, 9886, 10465, 11209, 13018, 13525, 15783, 16556, 17322, proteins, 6284, 10022, 14865, 17612; chemicals effects, 12794; coats, 17480, 18446; color, 17158; electron microscope, 18198; embryolessness, 13763, 15782; germination, 4109, 9497, 9854, 12723, 18177; industry, 373, 959, 1897, 2955, 3671, 10526, 15274, 15818, 16670, 16671, 16860; injury effects, 1955; maturity, 18375; moisture content, 4310, 9353, 11805, 12312; nutrition, 7594, 15650; respiration, 11806; seed production, 3897, 10173, 15411, 17598; storage, 309, 10197, 11208; temperature resistance, 9015; weight, 5757, 16668; *see also* Helianthus
Supaia, 19608
Suriana, 16001
Sutera, 10064
Swartzia, 17852
Sweet Almond, 14383
Sweet Peas, 761, 782, 1997, 16289; *see also* Lathyrus
Sweet William, 761
Swietenia, 3756, 3759, 7201, 8946, 9747, 12883, 14944, 17283, 17541, 19804
Switchgrass, 198, 1038, 4235, 8181, 16815
Sycamore, 1142, 1218, 3069, 3205
Symphoricarpos, 98, 6460, 6468, 6478, 13565, 14430, 15003, 15005, 15301
Syncarpia, 225
Synedrella, 2176, 6034
Syringa, 12481, 12482, 14412
Syzigium, 13334, 14220, 16491

Taccaceae, 14953
Tagetes, 1999, 3322, 7499, 7614, 7615, 11196, 16477
Tagua, 17022
Tahoka Daisy, *see* Daisy
Taiwania, 18863
Tallowtree, 18795
Tamaricaceae, 2102
Tamarindus, 1152, 2822, 3779, 4791, 4847, 5207, 7190, 7191, 9001, 10510, 14180, 14192, 14944, 15131, 15962, 15963, 15964, 16828, 16829
Tamarix, 9192, 9915, 10009, 17268, 19694
Tambuletin, 1746, 1747
Tamus, 3388, 3389, 7802, 16399

Tanacetum, 6184, 14386
Tangan-Tangan, 7786
Tanghinia, 8307
Tanne, 15475
Taraktongos, 7849
Taraxacum: burial in soil, 7496, 13535; embryo culture, 11828; genetics, 4386; germination, 1553, 1997, 5876, 8572, 12161, 15185, 15240, 15468, 15571, 19750; maturity, 794, 7231; plants effects, 9289; radiation, 12937; storage, 282; vernalization, 13924; weight, 1867; *see also* Dandelion
Tarweed, 3629
Tau-Saghyz, 14607, 16547, 17674
Taxodium, 1981, 5171, 5589, 18724, 18727
Taxus, 2946, 6753, 8568, 13603, 14386, 16637, 17960
Tea, 231, 2338, 5131, 8183, 8492, 8777, 9207, 10563, 11901, 12493, 13556, 13634, 14617, 15710, 17232, 18398, 18809, 19055, 19151; *see also* Camellia; Thea
Teak, 2599, 5845, 9213, 11873, 17487
Teazle, 7446
Tecoma, 15009, 15011
Tectona, 1868, 5845, 8789, 8791, 9747, 12194, 17487
Teesdalia, 14373
Telfairia, 12148, 18232
Tenera, 9265
Teosinte, 16830, 17848; *see also* Euchlaena
Tephrosia, 13352, 13353, 15931, 16682, 16959, 18910, 19059, 19680
Terebinthales, 13050
Terminalia, 4151, 4708, 15314
Tetracarpidium, 8765
Tetragonia, 8063, 8633, 8634
Tetrapleura, 15359
Teucrium, 2455, 10677, 13339
Thalictrum, 10674, 10675, 10677, 14415, 14905
Thea, 1290, 5075, 8492, 17208; *see also* Tea
Thelygonium, 19829
Theobroma, 2789, 3847, 7309, 7678, 9232, 9233, 11629, 15082, 19270; *see also* Cacao
Thermopsis, 18780
Thesium, 3745, 17849
Thespesia, 8613
Thevetia, 1587, 2504, 3878, 7171, 15928
Thistle, 701, 3210, 3211, 4481, 5223, 5556, 8383, 8434, 8444, 8445, 14996, 19903; *see also* Carthamus; Centaurea; Cnicus; Salsola; Sonchus
Thlaspi, 3734, 4211, 5485, 5764, 7496, 10673, 10677, 10739, 10741, 11015, 11442, 11458, 14538, 15185, 17058, 19215
Thrinax, 15012
Thujopsis, 14133
Thunia, 8637
Thuya, 384, 688, 747, 1981, 3310, 4188, 5385, 5983, 6395, 8585, 10677, 11196, 11828, 15003, 15006, 17453, 18150, 18723, 18727, 20074
Thymelaeaceae, 6798, 14415
Thymus, 463, 8572, 9505, 10675, 10677
Tigridia, 1715, 9160, 17363
Tilia, 78, 674, 1468, 1986, 6941, 7683, 8559, 9876, 11196, 11340, 12351, 12352, 12561, 12729, 15001, 15003, 15237, 16413, 17774, 18744, 19913, 20026
Tiliaceae, 13024
Tiliacora, 16783
Tillandsia, 8527, 8658
Timothy, 725, 1287, 6041, 6042, 6442, 6836, 7363, 7468, 8003, 9921, 10731, 12065, 14503, 15365, 15998, 17422, 18234, 18652, 19678, 19782
Tinospora, 215, 10007, 10012, 16640, 17239, 17483, 18603
Tithymalus, 19754
Todri, White, 15870; *see also* Matthiola
Tofieldia, 10675, 10677
Toisus, 16801
Tolpis, 2176
Torenia, 7805
Torilus, 17931
Torresea, 1645, 2307, 2308
Torreya, 3248, 13565, 18269
Tovaria, 4567
Tozzia, 8526
Trachylobium, 7120
Tradescantia, 523, 724, 9764
Tragopogon, 2176, 6201, 8498, 14423
Trapa, 155, 4215, 9665, 9725, 13868, 18444
Trapella, 10361
Treculia, 3689
Tree Peony, 1999, 2034, 2038
Trefoil, Birdsfoot, 562, 563, 1648, 3120, 8652, 12585, 18006, 19674
Tremtolobelia, 3610
Trianthema, 4244, 5417
Tribulus, 2791, 7498, 7500, 7614, 7615, 8192
Trichelia, 747, 8639, 14674
Trichocereus, 3692
Trichodesma, 1239, 7614, 7615, 10518, 14736
Trichosanthes, 1858, 3758, 10339, 13914, 15108, 17839, 18733
Tricutalis, 10674
Trientalis, 10675, 13022, 14386
Trifolium: absorption, 503; burial in soil, 5485, 7496; characteristics of seeds, 508, 9171, 11899, 12255, 17637; chemical change during metabolism, 5116; chemicals effects, 6492, 12962; coats, 503, 4002, 7187, 14537, 14538, 19715; color, 6256, 17953; development, 15019; disease transmission, 288, 290; dispersal, 17850; genetics, 4949,

Trifolium–(*cont.*)
19571; germination, 653, 655, 675, 1093, 3050, 3102, 3258, 3322, 3415, 3837, 5456, 5493, 5793, 8514, 8561, 8814, 8815, 8816, 8817, 8818, 8819, 9564, 9902, 12761, 18637, 19905; identification, 5859, 5860, 7146, 12439; industry, 240, 2519, 10709, 12471, 16433; injury effects, 653, 675, 14464, 18501; microorganisms, 8824; moisture content, 5823, 19356; radiation, 15463; seed production, 764, 2682, 17436, 18496; storage, 10739, 12339; viability, 6708, 8790, 10074, 18445, 18846; *see also* Clover

Triglochin, 2480, 10677, 19210

Trigo, 1624

Trigonella, 112, 629, 2572, 3672, 4891, 8591, 9505, 11050, 12108, 15454, 16003, 17267

Trigonocapales, 5105

Trillium, 2304, 7030, 7031, 9764, 9765

Triplasis, 1038

Tripleurospermum, 11511

Tripsacum, 368, 6237

Trisetum, 15185

Tristania, 225

Triticum: absorption, 360, 1614, 1616, 2184, 2952, 3112, 3160, 3379, 4111, 6214, 6359, 6424, 7865, 7867, 8843, 8988, 10721, 11329, 13905, 14912, 17892, 20044; anatomy, 321, 327, 328, 330, 336, 342, 1154, 2097, 2372, 2926, 2927, 2929, 6799, 7813, 8840, 9078, 9668, 9669, 9898, 16486, 17242, 19955; ancient, 40, 726, 1999, 6902, 7065, 12211, 18832; burial in soil, 7496, 18585; characteristics of seeds, 10, 12, 322, 327, 328, 332, 517, 538, 756, 1024, 1353, 1675, 1738, 2131, 2217, 2595, 2982, 3302, 3572, 3993, 4398, 4416, 5122, 5535, 6132, 7356, 7649, 8050, 8250, 8274 9364, 9413, 10540, 10604, 11152, 11176, 11311, 11339, 11695, 12549, 12765, 12797, 13136, 13397, 13462, 13531, 14339, 16015, 16114, 17272, 17753, 18026, 18087, 18409, 18719, 19168, 19244, 19301, 19880; chemical change during metabolism, 1457, 1483, 2258, 2405, 2568, 2570, 2688, 2750, 2759, 3599, 4011, 4966, 4978, 5067, 7390, 7567, 7568, 7569, 7570, 8372, 8904, 9529, 9779, 9780, 10162, 10244, 10411, 10578, 10644, 10645, 10651, 10693. 10869, 10932, 11016, 11270, 11509, 12285, 12287, 12749, 13015, 13302, 13324, 14574, 14664, 14799, 14802, 15268, 15393, 15412, 15556, 16992, 17317, 17395, 18046, 18872; chemical composition, 536, 548, 1678, 1716, 2051, 2223, 2259, 2362, 2366, 2492, 3821, 4855, 5100, 5430, 5585, 5586, 5891, 5931, 5932, 6455, 6811, 6850, 7298, 7612, 7659, 7694, 8259, 8347, 8426, 8844, 8903, 8907, 9034, 9036, 9753, 10175, 10468, 10496, 10884, 11367, 11375, 11381, 11986,

Triticum–(*cont.*)
13433, 13774, 13775, 13973, 14011, 14429, 14599, 15522, 15523, 15524, 16133, 17024, 17176, 17177, 17299, 17319, 17509, 17683, 18114, 18159, 18541, 19968, 20004, 20139: amino acids, 2053, 2269, 2566, 3680, 4833, 4924, 5292, 7315, 7763, 7840, 8654, 9060, 10738, 11393, 11394, 12281, 13160, 13169, 16925, 17971, carbohydrates, 324, 548, 1913, 2128, 4220, 4864, 4898, 5151, 5746, 5771, 6002, 6076, 7969, 8306, 9024, 9412, 10174, 12650, 12651, 13694, 13808, 16113, 16717, 16718, 19588, enzymes, 625, 1671, 1821, 1843, 2559, 2565, 2567, 2620, 2764, 2765, 3111, 3653, 3882, 4219, 4527, 4702, 5143, 5727, 5897, 5901, 5960, 6566, 7206, 7314, 7345, 8026, 8249, 9121, 9360, 9444, 9445, 10265, 10736, 10887, 11384, 13342, 13405, 13528, 13820, 13821, 13822, 13831, 14803, 15282, 15729, 15739, 16394, 17124, 17193, 17481, 17786, 18158, 18264, 18265, 18266, 18305, fats and oils, 9345, 9346, 10281, 10282, 12052, 13215, 14310, 14311, 16724, 17970, 18157, growth regulants, 1540, 7470, nucleic acids, 6851, 11755, 13434, 14869, proteins, 11, 547, 548, 549, 550, 557, 1526, 1610, 1677, 1823, 1840, 1941, 2246, 2922, 3333, 5914, 6015, 6269, 6281, 6432, 6446, 7232, 7993, 8252, 8388, 8680, 8683, 9936, 10140, 10885, 11256, 11376, 11984, 13412, 13505, 14378, 14871, 15187, 15188, 15189, 15201, 15246, 15521, 15827, 16375, 16377, 17475, 17476, 17550, 18115, 19626, vitamins, 627, 1712, 2129, 3527, 4139, 4338, 5532, 7313, 7862, 8839, 8842, 9216, 9633, 9827, 11963, 13974, 14670, 15747, 15748, 16083, 16167, 16739, 17740, 18960, 19198, 19348; chemicals effects, 138, 281, 364, 684, 1426, 1452, 1562, 1729, 1750, 1751, 1754, 1755, 1822, 2201, 2202, 2203, 2406, 2407, 2409, 2563, 3150, 3314, 4103, 5212, 5376, 5400, 5731, 6156, 6207, 6910, 6939, 7027, 7160, 7592, 7873, 8091, 8343, 8619, 8622, 8642, 9333, 10495, 11257, 11649, 11802, 13257, 13621, 13838, 14888, 15094, 15141, 15281, 15559, 15723, 16487, 17238, 17650, 18005, 18405, 18540; coats, 261, 317, 331, 540, 1493, 6672, 10333, 11962, 15810; color, 244, 536, 4116, 7292, 7767, 12169, 12798, 12866, 12867, 13305, 13624, 14419, 17840, 18447; development, 66, 335, 1453, 2983, 3470, 3471, 3652, 4606, 4806, 6867, 7163, 7164, 8213, 8218, 8308, 8484, 9079, 13168, 13641, 13644, 14901, 15251, 16463, 16464, 16543, 16903, 19245, 19813, 20023, 20100; disease transmission, 537, 1941, 2093, 2825, 4645, 7414, 9991, 10457, 10588, 10599, 10686, 10859, 12056, 13238, 13241,

Triticum–(*cont.*)
15569, 15742, 16568, 16573, 17101, 17459, 19321; electrical effects, 419, 12055; embryo culture, 3314, 5553, 10513, 10607, 11589, 13308, 13442, 15622, 16047; embryolessness, 7766, 10188, 12568, 14596, 17976; fertilizers effects, 517, 680, 2120, 3481, 3880, 5933, 6397, 9225, 11470, 12409, 13240, 16354; fossils, 12547; genetics, 756, 2603, 2612, 4529, 5037, 7169, 8389, 8616, 9242, 9727, 9730, 9757, 10257, 10894, 13408, 15597, 16407, 16409, 17352, 17356, 19286; germination, 18, 140, 142, 452, 684, 1255, 2050, 2222, 2602, 2912, 3238, 3262, 3318, 3484, 3485, 3499, 3540, 3728, 3823, 3826, 3955, 4278, 4747, 5170, 5291, 5363, 5364, 5401, 5538, 5793, 5998, 5999, 6393, 6438, 7134, 7481, 8108, 8109, 8161, 8216, 8217, 8257, 8514, 8605, 8664, 8955, 9066, 9136, 9154, 9161, 9165, 9176, 9273, 9529, 10033, 10168, 10304, 10593, 10667, 10951, 11229, 11869, 12190, 12481, 12482, 12590, 12795, 12875, 12940, 13199, 13909, 14034, 14412, 14721, 15420, 16038, 16645, 16928, 17001, 17219, 17407, 17635, 18351, 19246, 19898; identification, 5303, 10888, 11100, 11101, 11102, 11150, 11499, 11797, 15307, 18236, 18578; industry, 540, 5373, 5424, 5536, 7259, 9069, 9890, 10437, 10438, 10439, 11566, 11804, 12471, 12931, 13121, 14038, 14601, 15325, 15435, 18170, 18215, 18217, 18218, 18219, 18220, 19410, 19873; inhibiting substances, 1824, 1885, 4837, 13473, 13474, 13625, 16237; injury effects, 24, 539, 684, 1401, 1873, 1924, 3556, 4063, 4275, 5199, 5521, 8251, 9240, 9241, 10136, 10648, 11387, 11388, 11753, 11968, 12282, 13477, 14455, 15135, 16715, 16719; maturity, 1603, 1604, 1605, 2439, 5550, 6190, 8248, 11043, 16623, 16995, 19499, 19605; microorganisms, 170, 2219, 2220, 4052, 4053, 4054, 4055, 5211, 5985, 6044, 7413, 7902, 9208, 9315, 9690, 10521, 11969, 13265, 13266, 13267, 13513, 13515, 15039, 15040, 15041, 16996, 17589, 17752, 17943, 18813, 18814; moisture content, 546, 553, 643, 684, 1672, 1759, 2177, 2840, 4363, 6721, 6945, 7063, 7064, 7072, 8276, 8282, 9271, 10094, 10180, 11015, 11175, 11401, 11619, 11652, 13823, 14169, 14831, 15036, 16258, 17160, 18502, 19622, 19696; nutrition, 250, 6209, 7595, 7814, 8126, 8845, 12104, 12156, 12418, 13730; plants effects, 10165, 10412, 10572, 11096, 11110, 11151, 11171, 12999, 15786, 16425, 17002, 18295, 19498; polyembryony, 5839, 9330, 19915; provenance, 2333, 9671, 11154, 11694, 13174, 13239, 14066, 17204; radiation, 139, 2273, 4612, 5963, 5964, 6157, 6531, 7234,

Triticum–(*cont.*)
7235, 7931, 9440, 9562, 9855, 11163, 11164, 11664, 12286, 12996, 13148, 14520, 17703; respiration, 1680, 2870, 10259, 12152, 13514, 14918; seed production, 517, 1024, 1738, 2131, 2217, 3302, 3572, 4416, 4398, 5535, 5840, 6555, 9243, 9474, 16685, 19698; stimulating substances, 314, 398, 657, 699, 1885, 5732; storage, 84, 643, 1615, 1673, 2592, 2593, 4058, 4634, 4979, 5160, 5192, 6293, 6333, 6873, 6886, 6887, 7130, 7298, 7300, 7415, 7810, 9190, 9937, 11700, 15437, 15565, 15765, 17298, 17361, 18213, 18216; temperature resistance, 624, 4889, 6518, 6911, 7098, 8036, 8595, 9095, 9339, 12184, 12769, 13122, 13622, 13728, 14379, 14451, 18103, 18174, 19095, 19637; treatment, 151, 452, 538, 539, 1370, 1371, 1372, 1373, 1510, 2279, 2533, 2700, 2816, 2817, 2921, 3407, 3898, 4429, 4585, 5452, 6529, 6787, 6889, 7566, 8678, 8884, 9331, 10536, 11620, 12297, 13853, 14361, 16143, 17055, 18613; variety effects, 364, 1564; vernalization, 161, 699, 1828, 2241, 2781, 3031, 3962, 3963, 3964, 4265, 4960, 4961, 4964, 4973, 4974, 4975, 5798, 5799, 6066, 6268, 6530, 7572, 7776, 8114, 8310, 9361, 9411, 9473, 10589, 11160, 11331, 11358, 11789, 12067, 12393, 12824, 12940, 12995, 13825, 13886, 14170, 14171, 14172, 15184, 15481, 15730, 16191, 17215, 17230, 18314, 19048, 19456, 19458, 19472, 20066; viability, 364, 645, 929, 2134, 2728, 2757, 3125, 4229, 4443, 6294, 6334, 6767, 8253, 9300, 9443, 10739, 13335, 14388, 14672, 17212, 17271, 18203, 18778, 19635, 19636, 19638; weed seeds in, 1149, 10186, 12195, 16695, 18001; weight, 1407, 3899, 10175, 12159, 12370, 14168, 15015, 15127, 16424, 19300, 19600

Triumfetta, 9059

Trochodendron, 9430

Trollius, 847, 1597, 10677, 12387, 17363

Tropaeolaceae, 10293

Tropaeolum, 1999, 7499, 8205, 8770; *see also* Nasturtium

Trumpet Vine, 19451

Tsuga, 383, 384, 385, 390, 688, 3461, 3628, 3946, 6395, 8585, 11196, 11828, 15003, 17886, 18723, 18727, 20074

Tulip Tree, 3634, 7761, 7762, 15115

Tulipa, 738, 1834, 2662, 3580, 10677

Tumeric, 13582

Tung: characteristics of seeds, 660, 710, 3849, 5937, 6012, 14990; chemical change during metabolism, 7161, 7227, 8978, 9897; chemical composition, 8979, 13218, 17184: fats and oils, 8073, 8881, 12203, 13220, proteins, 16617; chemicals effects,

Tung–(*cont.*)
1999, 3682; development, 13128; genetics, 13345; germination, 12194, 13346, 17273, 17289, 19445; industry, 5341, 11656, 15289; seed production, 1061, 5343; storage, 11799; treatment, 13346; viability, 19445, 19446; *see also* Aleurites
Turnera, 15879
Turneaceae, 13040
Turnip, 363, 1294, 1999, 5200, 7603, 9175, 10836, 11196, 11289, 11584, 12029, 12664, 12667, 14991, 16654, 17315, 18177, 20016; *see also* Brassica
Turpentine Tree, *see* Syncarpia
Turpinia, 14205
Tussilago, 3234, 5490
Typha, 4212, 7558, 13868, 15733, 17424, 19210
Typhonodorum, 2758

Ucuuba, 1339; *see also* Virola
Udo, 9391
Ulex, 3219, 8824
Ulmus, 129, 673, 2000, 2048, 2218, 5589, 5872, 8955, 9085, 9875, 11196, 11436, 12066, 12536, 12782, 13161, 15007, 16506, 17916; *see also* Elm
Umbelliferae, 340, 3727, 4073, 5138, 5196, 5482, 6475, 6476, 6477, 6483, 8229, 10389, 13935, 14264, 14415, 14468, 16301
Umbellularia, 13565, 14559
Urena, 9059, 10050, 13278
Ureneae, 16077
Urochloa, 221
Urtica, 794, 10677, 11944, 11946
Urushi, 11295
Utricularia, 6226, 9717, 10355, 10512, 10677
Utriculariaceae, 6449

Vaccaria, 2792, 2793, 2794, 7496, 10677
Vaccinium, 98, 4913, 10855, 13535, 13565, 14386, 14430, 15005
Valeriana, 5928, 8572, 9505
Valerianella, 3322, 5864, 11196, 17903
Vallisneria, 13868
Vallota, 11828
Vanda, 15938
Vandellia, 9589
Vangueria, 209, 14051
Vanilla, 883, 2867, 2868, 2869, 5446, 6642, 9969, 10912, 11447, 12519, 18641, 18642, 19143, 19772
Vateria, 661, 15753
Venidium, 761
Veratrum, 1531, 10673, 10675, 10677, 10681, 19009, 20009
Verbascum, 5432, 7017, 7496, 7710, 7764, 8687, 9505, 10674, 10675, 10677, 12038, 12039, 14412, 14414, 14441, 14898, 16661
Verbena, 761, 1997, 2176, 7499, 8584, 12907, 14386
Verbenaceae, 839, 13578
Vergilia, 747
Vermilion, 1997
Veronia, 1646, 1900, 11395, 13613, 15903, 16243, 18565
Veronica, 794, 2989, 5764, 7710, 8525, 8528, 8687, 10433, 10673, 10675, 10677, 12036, 12039, 12044, 13339, 13535, 13868, 16200, 18480, 20076
Vesquia, 444
Viburnum, 98, 659, 1078, 2019, 5022, 5552, 7219, 10897, 12167, 14386, 15003, 15005, 17050, 17617, 18175, 18595, 20025
Vicia: absorption, 6359, 19167; bibliography, 10368; characteristics of seeds, 2301, 4809, 4851, 7424, 9171, 12711, 16202, 18870, 19159; chemical change during metabolism, 112, 1889, 2749, 2856, 11546, 13686, 15690, 15691, 18303; chemical composition, 3197, 4120, 5233, 9836, 10368, 12342: amino acids, 12073, carbohydrates, 10370, 10371, 10372, 10373, 10374, 14047, 14048, enzymes, 9054, 11628, fats and oils, 9932, nucleic acids, 25, proteins, 10807, 10808, 11273, 11626, 11627, 14863, 14896, 16882, 18923, vitamins, 1881, 17277; chemicals effects, 5876, 18967; coats, 4671, 7424, 9180, 18688, 19781; color, 19547; development, 6211, 7758, 11268, 11300, 15453, 18272; disease transmission, 4615; electrical effects, 419; embryolessness, 16794; genetics, 6018, 16199; germination, 5653, 7141, 7478, 7769, 9953, 10822, 10823, 12040, 14515, 15453, 15854, 18177, 18337, 19026; identification, 7142, 13209; industry, 10548; inhibiting substances, 16127, 17103; injury effects, 9845; maturity, 18195; microorganisms, 4621, 8824; moisture content, 10302, 10828, 10830, 19696; nutrition, 1414; plants effects, 10094, 12096, 12998; radiation, 9808, 10824, 10825, 10829, 14225; respiration, 13427; seed production, 283, 764, 1545, 10256, 12817; storage, 811; treatment, 1546, 8317; vernalization, 11578, 14614; viability, 3776, 10739, 13208; weed seeds in, 8632; weight, 5760; *see also* Bean
Viciaceae, 10853, 11053
Victoria, 2785, 15698
Vigna, 118, 1852, 3079, 3400, 3446, 3576, 4048, 4921, 5893, 6703, 7007, 8167, 9009, 9010, 10809, 11512, 12812, 13658, 13794, 14731, 14788, 14789, 14790, 14791, 14792, 14796, 14842, 14866, 16747, 17202, 17203, 17309, 17643, 17693, 18152, 18153, 19939, 19940; *see also* Cowpea

Vinca, 15027
Vincetoxicum, 6667
Vine, 20059
Vinifera, 8191
Viola, 761, 1997, 3322, 4176, 4611, 5928, 6872, 7499, 8205, 8569, 9279, 11196, 11442, 17849
Violaceae, 839
Viracarpon, 3994
Virola, 1339
Viscaria, 14415
Viscum, 5077, 6848, 7316, 8530, 8531, 8534, 10675, 10677, 12040, 16639, 19663, 19664
Vitaceae, 13447
Vitex, 10647
Vitis, 98, 418, 574, 5442, 9545, 12150, 14105, 14757, 14762, 15001, 15008, 15010, 15159, 15629, 17157, 17421, 17686, 19431, 19432, 19433, 19434; *see also* Grape
Vittadinia, 8187
Voandzeia, 5369
Vogelia, 12982
Vrucea, 2307, 2308

Wahlenbergia, 18128, 19347
Walkomiella, 18179
Walnuts, 747, 1023, 1695, 1909, 2145, 2786, 3154, 3842, 3843, 4181, 4813, 6071, 7665, 8028, 8129, 9796, 10846, 12431, 12963, 13318, 14836, 14931, 15487, 15785, 17375, 17576
Water Chestnut, 155; *see also* Trapa
Water Cress, 657, 9118
Watergrass, 3213
Water Hyacinth, 12823, 15198
Waterlily, 4579
Watermelon, 808, 820, 1566, 1567, 3486, 3487, 3644, 4505, 5164, 10215, 10603, 11178, 11196, 11298, 11419, 12427, 14335, 14513, 14522, 14523, 14560, 15539, 15595, 17734, 17985,19393, 19811; *see also* Citrullus
Wattle, 14036, 14855; *see also* Acacia
Wedelia, 3377
Welwitschia, 7156, 8676, 12892, 13685, 18468
Wheat, *see* Triticum
Wheatgrass, 374, 3299, 3629, 4351, 5304, 5557, 8431, 8432, 8662, 9195, 9221, 10614, 13093, 13200, 13534, 15634, 16350, 16859; *see also* Agropyron
White Top, 3210, 3211
Widdringtonia, 16838
Willow, 3233, 3234, 8790, 14128, 14498, 15579, 15581, 16801, 18309
Willow, Primrose, 14275; *see also* Jussiaea
Wistaria, 9279, 18443
Witchweed, 11587, 16295
Wolffia, 12706
Woodfordia, 10014
Wormseed, 790, 14699, 19467

Xanthium, 2704, 3585, 3586, 5033, 8192, 9781, 11328, 13053, 13056, 14887, 17378, 17380, 17384, 18238, 18535, 19369, 19370
Xanthisma, 2306
Xanthoceras, 15100, 15460
Xanthoxylum, 98
Xenia, 2603, 4860, 10585
Ximenia, 737, 8468, 12236, 15760
Xylia, 12809,
Xylopia, 4414

Yerba, 6978
Yew, 6753, 14329; *see also* Taxus
Yucca, 1405, 1862, 3560, 4792, 5915, 8692, 9065

Zaluzianskya, 10064
Zamia, 846, 15009, 15012
Zannichella, 7843
Zanonia, 9270
Zea: absorption, 280, 609, 5638, 6030, 6213, 9824, 12744, 12751, 17387; anatomy, 280, 336, 531, 636, 1470, 1542, 2345, 2346, 2974, 3547, 3876, 4158, 4824, 5013, 7586, 9865, 9939, 10584, 11322, 11727, 11866, 12800, 12802, 14597, 14976, 14977, 16745, 16779, 17911, 17966, 18083, 18084, 18085, 18086, 20131; atomic effects, 976, 15925; bibliography, 1049, 4750; characteristics of seeds, 146, 2471, 2617, 2776, 2939, 2952, 3352, 3433, 3446, 3553, 3616, 3737, 4161, 4163, 4375, 4733, 5511, 6170, 6569, 6967, 7836, 7906, 7917, 7937, 8138, 8325, 8329, 8908, 9094, 9162, 9182, 9187, 9188, 9841, 10551, 10592, 10737, 11505, 11545, 11828, 11829, 11830, 11832, 12459, 12501, 12731, 12745, 13124, 13255, 13409, 13468, 13744, 13811, 14166, 14316, 14317, 14430, 14654, 14655, 14895, 15345, 15349, 15791, 16012, 16185, 16239, 16282, 16314, 16328, 16419, 16540, 16753, 16988, 17816, 17866, 19587, 20129; chemical change during metabolism, 277, 372, 1295, 1541, 2145, 2344, 2736, 3002, 3061, 3075, 3304, 3543, 4195, 5506, 5682, 5706, 5712, 5752, 5952, 6020, 6095, 6208, 7136, 11788, 11798, 12756, 13099, 18340, 18341; chemical composition, 817, 2310, 2492, 2623, 2827, 3001, 3003, 3274, 3563, 3565, 4578, 4734, 4833, 5121, 5503, 5743, 6075, 6077, 6239, 6270, 6381, 7102, 7279, 8438, 8439, 9014, 9777, 11204, 11207, 11489, 11763, 11900, 12156, 12754, 12755, 13440, 13590, 13803, 14156, 15132, 15886, 16041, 16578, 16844, 17601, 18114, 18649, 19176, 20084, 20130: amino acids, 4717, 5188, 5502, 6578, 7077, 13500, 15831, 15834, carbohydrates, 107, 372, 558, 559, 560, 1197, 5072, 5746, 5750, 6265, 6299, 7939, 8348, 8390, 10449, 12652, 12945,

Zea–(*cont.*)
13768, 15166, 17015, 17161, 19585, 19586, 19798, 20132, enzymes, 2123, 5707, 7404, 7405, 11481, 12288, 12289, 16613, 19094, fats and oils, 1716, 2116, 2122, 3215, 4769, 5059, 6096, 7403, 12078, 13495, 13557, 17815, growth regulants, 2311, 2707, 6240, 6847, 7651, 8607, 8608, 8838, 15484, 18908, 19929, 19946, proteins, 817, 3000, 3005, 4504, 5059, 5627, 5709, 5742, 5751, 6512, 6726, 8938, 9934, 10417, 10804, 10812, 10815, 12072, 12077, 12078, 12092, 12943, 12944, 13495, 13946, 14364, 14567, 14862, 15142, 15143, 15287, 15694, 15695, 15835, 16303, 16808, 18497, 19800, vitamins, 817, 2375, 3386, 4338, 5751, 6725, 7445, 7477, 8669, 9215, 9216, 9217, 9218, 12076, 12803, 12973, 14228, 15924, 16184, 17831, 18427, 18428, 18429, 18430, 18431; chemicals effects, 730, 4659, 17784, 18361, 19252; coats, 1682, 3057, 3065, 3184, 6895, 9864, 15921, 19208; color, 561, 7626, 17809; development, 1296, 5636, 7097, 10164; disease transmission, 609, 3533, 3891, 4633, 5642, 5785, 5786, 8997, 8999, 9044, 9365, 9823, 10586, 10973, 10974, 12215, 13210, 13260, 17274, 18944; embryo culture, 6908, 18905, 18906, 18907; embryolessness, 16781, 19534; endosperm culture, 18084, 18085, 18086, 18329, 19413; genetics, 531, 558, 561, 615, 1043, 1525, 1719, 3076, 4494, 5399, 5408, 5481, 5504, 5825, 5913, 6199, 6725, 8048, 9200, 9217, 9433, 9775, 10047, 10454, 10585, 10587, 11932, 11933, 12419, 12717, 12799, 13951, 14940, 15258, 15923, 16583, 16871, 17099, 17812, 19109, 19487, 19857, 20133; germination, 27, 162, 227, 253, 314, 365, 380, 609, 610, 700, 3319, 3817, 4144, 4640, 4747, 5164, 5170, 5226, 5583, 5680, 5793, 7541, 8132, 8195, 8324, 8327, 8998, 9000, 9045, 9046, 9047, 9048, 9049, 9235, 9405, 9423, 9468, 9871, 9994, 10569, 10579, 10953, 12299, 12360, 12375, 12757, 12758, 12759, 12801, 13638, 13792, 14333, 15409, 15886, 16296, 16658, 16831, 16866, 17407, 17806, 18176, 18395, 18703; industry, 162, 203, 313, 610, 612, 615, 3339, 3611, 3834, 4063, 4492, 4522, 4733, 5387, 5641, 5705, 6845, 9941, 10580, 10581, 10997, 12075, 12407, 17669, 17813, 17814, 18441, 19957; inhibiting substances, 19881; injury effects, 616, 974, 2032, 2123, 4653, 5435, 8012, 9043, 9184, 9925, 10590, 10591, 10959, 10960, 10964, 10968, 10971, 16435, 16436; maturity, 293, 531, 612, 631, 2156, 2157, 2158, 3080, 3548, 4735, 5635, 7406, 7916, 8040, 8645, 9889, 11042, 11043, 11250, 11608, 12091, 13472, 14268, 15996, 16048, 16554, 17235, 17282, 17680, 18088,

Zea–(*cont.*)
19308, 19801, 20081; microorganisms, 3691, 4052, 7447, 9205, 10963, 14087, 15788, 15789, 15885, 17196; moisture content, 219, 227, 278, 3086, 3699, 4811, 5161, 7051, 7079, 8032, 8263, 8276, 9185, 9191, 9627, 10582, 10965, 11401, 12337, 12752, 13185, 13256, 13471, 17811, 18106, 18502, 19349, 19691; nutrition, 4757, 6841, 13588; plants effects, 611, 5681, 6167, 6698; polyembryony, 16780, 17269; radiation, 976, 2432, 2615, 3902, 3904, 4008, 4764, 6003, 6966, 8338, 9645, 10065, 10453, 10565, 10578, 10583, 10970, 11187, 12750, 13058, 13059, 16966, 17405; respiration, 1674, 11787, 15863, 18352, 18353; rhythm, 5640; sowing conditions, 253, 313, 1687, 3532, 3741, 3907, 4309, 5012, 5499, 7408, 8198, 8199, 12502, 13877, 15339; storage, 219, 227, 279, 725, 925, 995, 1027, 3213, 3214, 3590, 4225, 4300, 4888, 4990, 5777, 5882, 6510, 6511, 7451, 7539, 7617, 8032, 8959, 9190, 10962, 11196, 11827, 12253, 13832, 14742, 16546, 16550, 16842, 16843, 18202, 18321, 20082; supersonics, 5920, 6376, 8328, 11907, 20119; temperature resistance, 54, 6707, 16258, 17172, 17183, 18393, 18394, 19543, 19872; treatment, 294, 365, 413, 414, 419, 940, 954, 958, 975, 1027, 1758, 1885, 2146, 2296, 2300, 2748, 2766, 2830, 2853, 2854, 2893, 3007, 3150, 3319, 3430, 3698, 4104, 4226, 4506, 4908, 4977, 5324, 6251, 7513, 7571, 7704, 7705, 7737, 7738, 8194, 8326, 8863, 8941, 9032, 9041, 9050, 9126, 9626, 9850, 10065, 10578, 10583, 10970, 11565, 11769, 12319, 13089, 13123, 13225, 13250, 13280, 13669, 15856, 16046, 16175, 16238, 16437, 16730; vernalization, 4958, 5175, 10564, 13443; viability, 219, 227, 725, 2286, 2846, 4811, 5639, 5704, 7152, 7449, 7938, 10597, 11319, 11790, 13186, 14269, 17138, 16490, 18203, 18543, 18904; weight, 17252; *see also* Maize

Zelkova, 6395, 11295, 11296, 11297

Zephyrantheae, 6509

Zephyranthes, 16603, 18349, 18350

Zieria, 13783

Zilla, 2474, 2475, 2476, 2478

Zingiberaceae, 2676, 6565, 7042

Zinnia, 1735, 1999, 2176, 3322, 5803, 7499, 8205, 11196, 12040, 18712

Zizania, 2232, 3574, 5703, 11831, 13753, 13868, 14312, 18288, 18290, 18291

Zizyphus, 81, 208, 747, 3919, 6146, 10146, 10360; *see also* Chinese Jujube

Zostera, 4823, 18400

Zoysia, 991, 1094, 6540, 14530, 14631, 15847

Zygopetalum, 5440

Zygophyllum, 2473, 11024, 12108, 18442

SUBJECT INDEX

Abnormal germination, 4142, 7779, 8159, 8261, 8582, 8720, 10300, 11199, 11648, 11654, 12151, 12624, 17146, 17272, 17816, 19489, 19551

Absorption, 572, 1495, 1616, 3114, 3115, 3159, 3160, 3261, 3312, 3379, 3556, 4077, 4111, 4301, 4804, 5009, 5477, 5638, 5763, 6000, 6030, 6166, 6424, 7336, 7430, 7436, 7507, 7556, 8049, 8147, 8358, 8921, 9578, 9653, 10441, 10716, 10717, 10776, 11921, 12295, 12480, 12840, 12862, 13432, 14785, 14912, 16442, 16513, 16515, 16516, 17387, 17388, 17870, 18054, 18211, 18285, 18348, 18643, 18838, 18873, 19346, 19390, 19657, 19659, 20138

Acids and acidity effects, 104, 200, 439, 685, 1586, 1889, 2055, 2176, 2563, 3226, 3421, 3430, 3465, 3599, 3648, 3705, 3706, 3799, 3828, 4439, 4345, 5098, 5263, 5581, 5606, 5671, 5814, 5970, 5971, 5998, 5999, 6220, 6221, 6232, 6247, 6445, 6910, 7017, 7114, 7122, 7398, 7544, 8291, 8364, 8387, 8464, 8955, 8978, 9089, 9340, 9525, 9604, 10097, 10270, 10293, 10375, 10954, 11167, 11378, 11379, 11391, 11518, 11807–9, 12036, 12200, 12201, 12239, 12742, 12810, 12811, 13076, 13237, 13339, 13396, 13428, 13911, 14074, 14415, 14472, 14894, 14898, 14936, 14937, 14958, 15284, 15724, 15787, 16514, 16592, 16677, 16866, 17147, 17159, 17271, 17317, 17331, 17418, 17677, 17772, 17952, 18260, 18262, 18289, 18330, 18732, 18743, 19265, 19273, 19653, 19688, 19733, 19741, 19788, 19925, 19940, 20082

Africa, 737, 15087A, 16476, 19879

After-ripening, 2, 80, 133, 134, 205, 221, 222, 495, 496, 620, 797, 820, 924, 965, 972, 1020, 1038, 1063, 1502, 1504, 1693, 1770, 1775, 2067, 2176, 2636, 2798, 2799, 2925, 3155, 3227, 3294, 3389, 3424, 3484, 3485, 3624, 3823, 4245, 4592, 4594, 4600, 4761, 5023, 5032, 5036, 5713, 5768, 6115, 6116, 6374, 6377, 6457–60, 6463, 6478, 6485, 6519, 6520, 6607, 6817, 7011, 7012, 7217, 8203, 8206, 8211, 8212, 8371, 8396, 8398, 8399, 8491, 8502, 8506–8, 8531, 8590, 8677, 8753, 8754, 8815, 8853, 8897, 8923, 8955, 8970, 9092, 9297, 9298, 9335, 9420, 9431, 9497, 10378, 10593, 10672, 11018, 11081, 11107, 11218, 11297, 11837, 12036, 12081, 12154, 12447, 12488–91, 12723, 13339, 13746, 13868, 13907, 13990, 14029, 14307, 14441, 14445, 14598, 14608, 14716, 14727, 14728, 14740, 14764, 14765, 14936, 14937, 15344, 15378, 15511, 15512, 15565, 15933, 16413, 16526, 16895, 17001, 17131, 17199, 17774,

After-ripening–(*cont.*)
17811, 17817, 18042, 18583, 18729, 18939, 18941, 19092, 19149, 19150, 19215, 19634, 19748, 19911, 20037, 20038

Age, 463, 502, 506, 727, 806, 1535, 1536, 1775, 2042, 2152, 2214, 2477, 2587, 2693, 3118, 3126, 3335, 3437, 3657, 3658, 3752, 3979, 4082, 4134, 4584, 4604, 4817, 4836, 5375, 5561, 5639, 5642, 5942, 6059, 6173, 6364, 6507, 6843, 6845, 6984, 6987, 7203, 7276, 7363, 7478, 7841, 7884, 7979, 8456, 8491, 8500, 8955, 9002, 9332, 9932, 9956, 10043, 10046, 10104, 10257, 10313, 10472, 10580, 10607, 10646, 11263, 11865, 12145, 12146, 14186, 14252, 14396, 14419, 14422, 14425, 14429, 14438, 14440, 14485, 14709, 14772, 15081, 15258, 16088, 16104, 16138, 16191, 16325, 16335, 16351, 16382, 16383, 16409, 16570, 16653, 16834, 16835, 16837, 16847, 16852–54, 16923, 16958, 17112, 17113, 17271, 17298, 17310, 17352, 17462, 17592, 17791, 17975, 17992, 18111–13, 18445, 18518, 19098, 19111, 19121, 19334, 19340, 19495, 19788

Agricultural crops, 731, 829, 863, 923, 936, 1073, 1076, 5555, 6592, 8966, 9303, 9305–7, 10952, 11192, 12317, 12318, 14153, 14772, 15386–91, 15790, 16659, 16683, 17072, 17302, 17428, 18879, 18936, 19182, 19275

Agriculture, 396, 1241

Air effects, 814, 2188, 2846, 5947, 16869, 16967, 17035, 17180, 18351, 18837

Airplane seeding, 738, 968, 984, 993, 1026, 1031, 1044, 1118, 1647, 2829, 3901, 6389, 12416, 15211, 16428, 19268

Albumen, 18881

Alcohol, 556, 1498, 1772, 1900, 2184, 3297, 3713, 4802, 6094, 6151, 8955, 10176, 10202, 11270, 11581, 11691, 12113, 12497, 15474

Aleurone, 575, 1119, 1403, 1470, 2287, 3147, 3471, 4010, 4100, 4118, 4852, 4853, 4960, 11023, 12299, 12717, 15520, 15909, 16123, 16185, 16486, 16832, 16872, 17971, 18892, 19040, 19112, 19228, 19541

Algeria, 11868

Alimentary canals effects, 748, 1002, 1490, 1977, 3420, 4321, 5473, 5484, 5488, 5755, 7308, 7425, 7769, 7843, 8192, 8457, 10822, 10823, 11228, 11230, 11575, 13157, 13511, 13740, 13852, 14017, 16020, 16404, 16457, 18210

Alkalies effects, 8955, 13339, 15279

Alkaloids, 399, 400, 402, 1531, 1901, 2114, 2574, 2671, 2791, 3043–47, 3504, 3519,

Alkaloids–(*cont.*)
3768, 3824, 3840, 3885, 3889, 3912, 3992, 4133, 4194, 4196, 4449–51, 4690, 4738, 5143, 5156, 5284, 5551, 5590, 5699, 5868, 5869, 5949, 6275, 6488, 6522, 6523, 6914, 7351, 7452, 7453, 7760, 7806, 8144, 8413, 8466, 8621, 9359, 9385, 9435, 9438, 9546, 9643, 9694, 9695, 10563, 10762, 10770, 11286, 11542, 11543, 11842, 11957, 12860, 12861, 12876, 13339, 13340, 13600, 13666, 14160, 14414, 14616, 14821, 14822, 14948, 15064, 15256, 15270, 15295, 15441, 15649, 15655, 15714, 15755, 15934, 16027, 16162, 16702, 16850, 17079, 17126, 17156, 17404, 17640, 17933, 17934, 18437, 18457, 18716, 18859, 18978, 18990, 19154, 19155, 19242, 19330, 19383, 19409, 19535, 19610, 20057

Alpine seeds, 555, 693, 1157, 1330, 2353, 2624, 3906, 4427, 5079, 7155, 7269, 9409, 9410, 10681, 10683, 10750, 12404, 12453, 12505, 12771, 13156, 14088, 16505, 17968, 18199

Altitude, 2143, 12008, 12545, 16962

Amino acids, 26, 119, 168, 312, 314, 973, 1003, 1117, 1483, 1515, 1628, 1763, 1876, 1877–79, 1890, 2053, 2158, 2236, 2238, 2263, 2269, 2440, 2441, 2564, 2566, 2627, 2628, 2800, 2855, 2856, 3005, 3182, 3183, 3447, 3448, 3487, 3680, 3819, 3861, 3996, 4115, 4132, 4195, 4717, 4718, 4833, 4846, 4924, 5050, 5121, 5174, 5176, 5188, 5193, 5225, 5238, 5292, 5502, 5504, 5654, 5657, 5689, 5706, 6071, 6215, 6249, 6430, 6512, 6524, 6590, 6591, 6750, 6950–E, 6950-F, 7077, 7315, 7338–40, 7342, 7598, 7757, 7763, 7840, 7850-A, 7861, 8077, 8236, 8338, 8360–63, 8654, 8659, 8976, 9877, 9060, 9062, 9096, 9655, 9658, 9770, 10230, 10456, 10509, 10533, 10627, 10693, 10793, 10869, 11000, 11017, 11050, 11303, 11378, 11379, 11392–94, 11414, 11469, 11491, 11816, 11893, 12072, 12073, 12247, 12248, 12281, 12285, 12288, 12289, 12514, 12515, 12560, 12711, 13011–14, 13160, 13169, 13251, 13310, 13500, 14062, 14240, 14296, 14615, 14620, 14622, 15052, 15062, 15335, 15718, 15738, 15831, 15834, 15844, 15845, 15894, 15898, 15968, 16131, 16168, 16204, 16227, 16228, 16612, 16621, 16710, 16775, 16925, 17016, 17081, 17100, 17335, 17538, 17550, 17711, 17971, 18126, 18274–76, 18301, 18311, 18495, 18787, 18972, 18973, 19069, 19104, 19106, 19939, 19962, 20033

Ammonia effects, 413, 414, 437, 438, 2001, 8312

Amniotic liquor, 16237

Amylase, 155, 625, 1197, 1727, 1728, 2345, 2621, 3380, 3806, 3808, 4093–98, 4980, 5146–50, 5152, 5682, 5898, 5901, 5906, 5958, 6149, 7283, 7284, 8812, 8813, 9038, 9360, 9665, 10265, 10875–81, 10888, 10889, 11513, 12261, 12524, 12918, 14042–45, 14066, 14377, 14570, 14604, 14613, 14711–13, 15725, 17124, 17945

Anaesthetics effects, 1302, 2187, 2213, 4458, 8955, 10735, 11335, 13339, 14415, 14419, 14420, 14936, 14937, 15023, 15550, 16940, 17694, 18730, 19164

Anatomy, 74, 141, 167, 261, 315–17, 319, 320, 322, 323, 326–28, 330, 332, 342, 523, 799, 832, 990, 1154, 1323, 1339, 1432, 1538, 1550, 1595, 1596–99, 1601, 1634, 1714, 1834, 1902, 1928–30, 1945, 2033, 2097, 2106, 2256, 2309, 2372, 2442, 2475, 2491, 2582, 2699, 2715, 2758, 2907, 2926–30, 3024, 3184, 3186, 3209, 3255, 3275, 3301, 3348, 3352, 3377, 3472, 3578, 3610, 3734, 3770, 3782, 3863, 3877, 4009, 4045, 4076, 4114, 4119, 4215, 4238, 4414, 4415, 4454, 4456, 4496, 4543, 4720, 4899, 4948, 4950, 5081, 5097, 5105, 5240, 5338, 5390, 5419, 5604, 5608, 5696, 5697, 5890, 6034, 6040, 6045, 6114, 6132, 6180, 6248, 6253, 6274, 6565, 6574, 6575, 6600, 6604, 6796, 6799, 6950-I, 7042, 7044–46, 7049, 7267, 7268, 7420, 7432, 7689, 7765, 7864, 7936, 7953, 7954, 7960, 7973, 7991, 8029, 8060, 8079, 8290, 8391, 8443, 8452, 8468, 8485, 8496, 8914, 8943, 8944, 8985, 9014, 9065, 9072, 9078, 9104, 9105, 9136, 9311, 9328, 9355, 9389, 9457, 9464, 9469, 9591, 9668, 9669, 9716, 9717, 9824, 9829, 9830, 9899, 9911, 9916, 10005–7, 10112, 10150, 10181, 10184, 10187, 10382, 10390, 10391, 10545, 10592, 10602, 10629, 10630, 10858, 10870, 11235, 11236, 11249, 11291–93, 11317, 11363, 11465, 11525, 11564, 11705, 11739, 11742, 11822, 12148, 12399, 12478, 12687, 12791, 12813, 12836, 12869, 12905, 12911, 12988, 13024, 13026, 13178, 13309, 13311, 13326, 13364, 13509, 13510, 13556, 13601, 13883, 13891, 13894, 13914, 13950, 14050, 14090, 14203, 14216, 14217, 14348, 14423, 14543, 14544–47, 14602, 14658, 14665, 14666, 14673, 14708, 14823–25, 14839, 14907, 14966, 14988, 14996, 15014, 15016, 15082, 15160, 15209, 15219, 15237, 15301, 15345, 15430, 15432, 15513, 15567, 15656, 15682, 15745, 15779, 15921, 15941, 15942, 15976, 16054, 16059, 16069, 16077, 16078, 16081, 16200, 16341, 16344, 16401, 16415, 16427, 16488, 16600, 16638, 16675, 16704, 16731, 16732, 16745, 16753, 16848, 16982, 16984, 17109, 17191, 17242, 17275, 17324, 17325, 17328, 17334, 17440, 17485, 17486, 17488, 17499, 17513, 17518, 17521, 17548, 17577, 17599, 17600, 17712, 17783, 17784, 17804, 17829, 17841, 17844, 17911, 17964, 18028,

Anatomy–(*cont.*)
18077, 18083, 18102, 18138, 18201, 18212, 18304, 18328, 18348, 18392, 19424, 18442, 18476, 18486, 18508, 18524, 18786, 18848, 19002, 19027, 19030, 19036–38, 19240, 19408, 19472, 19656, 19787, 19825–28, 19831–33, 19848, 19863, 19991, 19992, 20007, 20113
Ancient seeds, 40, 645, 720, 726, 1140, 2052, 3809, 6902, 8614, 12211, 12212, 16637, 18832, 19209, 19620, 19849, 20011
Angiosperms, 457, 1207, 2462, 3056, 3059, 3844, 3924, 4380, 4822, 5319, 5320, 6526, 7034, 8485, 10545, 11583, 12193, 12692, 12689–91, 13643, 16983, 17505, 17599, 17600, 17765, 17887, 18207, 18740, 18742, 18973, 19859, 19895, 19914
Animals, 922, 926, 2357, 2691, 7842, 7982, 8292, 8539, 8540, 11355, 13646, 14341, 14364, 14663, 14882, 16192, 16777, 17848, 18135, 18492, 18576, 18963, 19435, 20071
Annuals, 1991, 7092, 16649
Antibiotics, 143, 535, 961, 962, 1060, 1141, 1353, 1393, 1885, 2045, 2754, 3029, 3450, 3453-A, 3728, 4430, 4431, 5132–34, 6297, 6835, 6864, 6918, 6919, 6958, 7389, 7462, 7464, 8642–44, 8867, 9565, 9642, 10832, 10833, 10991, 11443, 11950, 11951, 12141, 12483, 12617, 12929, 12954, 13939, 14244, 14346, 14347, 14575, 14611, 14786, 14921, 15563, 15726, 15727, 16164, 16477, 16802, 16803, 16892, 17178, 17570, 17670, 17671, 18191, 19039, 19193–95, 19235, 19236, 19333, 19882, 19883
Apomixis, 14, 144, 237, 777, 1617, 1725, 2060, 2314, 2321, 3077, 3078, 4177, 4246, 4386, 6183, 6237, 6271, 6539, 7110, 7859 7946, 8013, 8045, 8178, 8681, 8891, 9765, 10566, 10698, 13873–75, 14117, 14582, 14751, 14841, 14881, 14932, 15085, 15305, 15336, 16276, 16392, 16582, 17646, 17647, 17887, 17888, 18750, 19427, 19436
Apparatus, 201, 291, 435, 553, 753, 919, 1051, 1088, 1148, 1190, 1210, 1318, 1360, 1395, 1514, 1783, 1888, 1931, 2043, 2076, 2080, 2613, 2629, 2634, 2663, 2734, 2761, 2897, 3049, 3085, 3133, 3153, 3259, 3363, 3413, 3464, 3609, 3612, 3662, 3909, 3972, 4187, 4393, 4691, 4751, 4752, 5008, 5199, 5317, 5402, 5426, 5612, 5747, 7519, 7560–62, 7632, 8080, 8089, 8282, 9223, 9259, 9404, 9619, 9680, 9960, 10080, 10081, 10255, 10325, 10332, 10447, 10448, 11321, 12451, 12772, 13230, 13367, 13368, 13917, 14880, 14904, 15248, 15298, 15657, 15684, 16057, 17606, 18250, 19075, 19120, 19174, 19751, 19752, 19769, 19986
Aquatic seeds, 2104, 2785, 6253, 7843, 8065, 12455, 13868, 13871, 13872, 14131, 16889
Arava, 18252
Arctic seeds, 2624, 9512, 15592, 19115
Argentine, 20127
Aril, 1634, 2411, 3550, 6235, 18964, 20073
Aromatic plant seeds, 12554–56
Arsenic effects, 3536, 4103, 7354, 9544, 9931, 9984, 14536, 16346, 18245
Ascorbic acid, 184, 1650, 1651, 1847, 1881, 2419, 2747, 3965, 4997, 5061, 6168, 7311, 7373, 7886, 8129, 9007, 9698, 9699, 10210, 11551, 13085, 13328, 13550, 15747, 17300
Asia, 1172
Atlas, 8510
Atmospheric humidity, 6947, 6948, 17471
Atomic energy effects, 976, 1158, 1241, 9989, 10621, 13997, 15925, 17656, 17657, 18674
Australia, 13505, 13506, 14018, 16433, 17648, 18637, 19410
Awn, 261, 13812, 17476

Bacteria, 4105, 7921
Band seeding, 8450, 8451, 12564, 17663
Batum coast, 12987
Belgium, 15625, 15626
Bibliography of castor bean, 10870
Bibliography of corn, 4750
Bibliography of cotton storage, 17890
Bibliography of disease transmission, 14851
Bibliography of growth regulants and vitamin seed treatments, 10933
Bibliography of maize from 1937 to 1945, 1049
Bibliography of plant diseases, 7488
Bibliography of radiation effects, 17785
Bibliography of rape treatment, 6771
Bibliography of seed germination, 6608, 6612, 6613, 6618
Bibliography of seed germination inhibitors, 11046
Bibliography of seed storage, 9682, 9685
Bibliography of seeds, 1055, 6616, 6621–36
Bibliography of Vicia, 10368
Bibliography of water sensitivity in barley, 9746
Bibliography of seed investigations, 1121, 1233
Bibliography of Yugoslav weeds, 11284
Biennials, 1096, 9266, 11453, 11454
Biochemical changes, 990, 1189, 1317, 1331, 1345, 1946, 2034, 2038, 2159, 2258, 2264, 2553, 2572, 2735, 2845, 2923, 3274, 4205, 5184, 5185, 5260, 5408, 5430, 5846, 6451, 6452, 6916, 7439, 7490, 7675, 8086, 8141–43, 8718, 8719, 9813, 10195, 10760, 11056, 11136, 11140–43, 11148, 11183, 11546, 11818, 11819, 11837, 12280, 13163–66, 14422, 14425, 14331, 17783, 17784
Biological data, 2028

Biology, 11622
Biosynthesis, 8760
Blood serum, 12614
Books, 367, 396, 1062, 1081, 1176
Boron, 752, 1958, 2363–66, 2453, 3376, 4485, 5113, 5986, 7508, 7700, 9267, 9535, 12026, 12445, 13594, 14720, 16887, 17587, 17598
Brazilian seeds, 6694
Breeding, 353, 1918
Britain, 13021
Bromine effects, 2721, 4231, 16888
Bulb, 18607
Burial in soil, 894, 895, 917, 989, 1306–9, 1326, 1442, 1551, 1552, 1804, 2144, 2446, 2984, 2986, 2987, 2989, 3188, 3208, 3272, 3602, 3797, 3798, 3886, 3887, 3974, 4900–4, 5420, 5473, 5702, 5718, 7496, 7498, 7500, 7502, 7503, 7747, 7896, 8106, 8461, 9809, 10570, 10647, 10681, 10739, 10822, 10823, 12306, 12456, 12668, 12764, 12776, 13913, 15240, 15451, 15471, 15696, 15832, 15833, 16274, 16275, 16458, 16678, 17944, 18024, 18585, 18661, 18745, 18783, 18999, 19171, 19695, 19713, 19783

Calcium, 17, 286, 2720, 2806, 2933, 3729, 3730, 3732, 3828, 3846, 4104, 4311, 4485, 4697, 7098, 8314, 10283, 10377, 11083, 11106, 11144, 11145, 12098, 12233, 12319, 12832, 13339, 14410, 14486, 15680, 15809, 16178, 16857, 16884, 17276, 17286, 17777
Canada, 1215, 10900, 14577, 14756, 16665, 19676, 19891
Carbohydrates, 107, 372, 522, 526–28, 558, 629, 630, 644, 713, 1295, 2034, 2345, 2493, 2914, 3543, 3779, 4466, 4467, 4527, 4874, 4875, 4891, 5599, 5631, 5750, 5771, 5992, 6020, 6265, 6657, 7396, 7514, 8304, 8305, 8348, 8355, 8384, 8437, 8852, 9475, 9530, 9679, 9734, 10174, 10367, 10369, 10370–74, 10424, 10429, 10449, 10486, 10510, 10520, 10644, 10645, 10848, 10887, 10934, 11223, 11243, 11302, 11424, 11453, 11456, 11920, 11973, 11987, 11988, 12282, 12314, 12315, 12620, 12646, 12655, 12722, 13099, 13107, 13586, 13633, 13634, 13694, 13767, 13768, 13808, 13840, 14011, 14047, 14048, 14161, 14191, 14193, 14224, 14318, 14438, 14444, 14454, 14791, 14828, 14829, 14981, 14982, 15033, 15051, 15053, 15054, 15166, 15490, 15501, 15623, 15676, 15690, 15691, 15773, 15878, 15928–30, 16666–68, 16770, 16825, 16828, 16945, 17088, 17233, 17343, 17547, 18032, 18117, 18251, 18258, 18263, 18282, 18340, 18341, 18363, 18364, 18473, 18746, 18839, 18845, 18890, 19255, 19585, 19703, 19805, 19920, 20047
Carbon dioxide, 489, 1809, 1810, 2017, 2020, 2046, 2189, 2219, 4446, 5672, 5678, 5947, 7319, 7541, 8622, 8955, 9055, 9816, 10487, 10552–55, 10720, 11121, 11123, 11126, 12304, 13288, 13329, 13339, 14936, 14937, 15244, 18357, 18536, 18537, 18954, 19263, 19264
Carbon monoxide, 14948, 17412, 18356
Carboxylase, 568, 3337, 3505, 4260
Carotene, 2617–19, 2645, 2827, 3001, 3853, 4161
Carpathian, 12762, 12776, 18921
Caryopsis, 86, 343, 487
Catalase, 436, 1297, 1518, 1671, 1779, 1820, 1842, 2445, 2560, 2620, 3091, 3421–23, 3453, 3928, 4604, 4789, 5031, 5143, 5694, 5727, 5900, 5930, 6202, 6566, 6866, 7206, 7542, 7543, 7688, 7692, 7792, 8397, 9054, 9444, 9445, 9553, 10872, 11787, 11788, 12007, 12010, 12041, 12201, 12351, 12447, 12561, 12774, 13342, 13396, 13405, 13749, 14041, 14167, 14266, 14621, 14727, 14728, 15124, 15243, 15489, 15933, 16160, 16471, 16472, 16969, 16970, 16974, 17384, 17491, 17498, 17776, 17954, 18305, 18319, 18630, 19689, 20060
Cellophane, 5982
Cellulose, 4367, 8306, 12647
Central America, 1413
Centrifugal force, 7688, 8745, 11265, 16681
Cereals, 1, 8, 14, 291, 315, 329, 333, 334, 337, 361, 370, 403, 426, 480, 482, 545, 546, 551, 554, 593, 594, 600, 601, 626–28, 664, 687, 899, 916, 970, 971, 1030, 1178, 1283, 1411, 1482, 1576, 1635, 1653, 1676, 1744, 1756, 1830, 1845, 1894, 2166, 2234, 2287, 2509, 2579, 2621, 2767, 2901, 2910, 2973, 3122, 3324, 3468, 3520, 3640, 3645, 3788, 3816, 3853, 3854, 3856, 3892, 3893, 3928, 3961, 3995, 3996, 3998, 4094, 4097, 4266, 4281, 4515, 4635, 4674, 4880–83, 4922, 4980, 5062, 5085, 5215, 5218, 5225, 5268, 5356, 5397, 5487, 5588, 5772, 5791, 5792, 5898, 5905, 5919, 5958, 5959, 5961, 6009, 6145, 6178, 6210, 6419, 6420, 6443–45, 6493, 6596, 6649–52, 6673, 6748, 6770, 6785, 6882, 7165, 7280, 7299, 7346, 7469, 7472, 7518, 7554, 7573, 7580, 7606–9, 7634, 7646, 7647, 7744, 7755, 7822, 7905, 7978, 7979, 7988, 8206, 8215, 8376, 8378, 8462, 8472, 8507, 8509, 8571, 8804, 8923, 9067, 9272, 9274, 9328, 9403, 9506, 9566, 9585, 9666, 9674, 9703, 9722, 9770, 9778, 9793, 9810, 9814, 9815, 9821, 9870, 10138, 10139, 10167, 10191, 10219, 10306, 10307, 10529, 10606, 10609, 10611, 10735, 10759, 10764, 10787, 10875–81, 10883, 10934, 11231, 11233, 11318, 11368, 11454, 11474, 11491, 11549,

Cereals–(*cont.*)
11654, 11702, 11784, 11821, 11893, 11923, 12132, 12474, 12508, 12567, 12597, 12602, 12606, 12612, 12642, 12654, 12714, 12730, 12857, 12887, 12894, 12979, 13096, 13102, 13378, 13379, 13419, 13422, 13481, 13506, 13529, 13907, 13925, 14009, 14247, 14323, 14348, 14543, 14545–47, 14569, 14766, 14806, 14870, 14843, 14915, 14919, 15044, 15107, 15266, 15378, 15434, 15516, 15527, 15548, 15659, 15676, 15769, 16014, 16263, 16266, 16294, 16376, 16522, 16590, 16884, 16887, 16922, 16930, 16975, 16977, 17150, 17226, 17348, 17422, 17655–57, 17881, 18102, 18201, 18260, 18262, 18401, 18898, 18961, 19112, 19173, 19174, 19323, 19867, 19868, 19869, 19945, 20138

Certification, 353, 672, 998, 1036, 1064, 1072, 1092, 1105, 1120, 1132, 1170, 1217, 1238, 1240, 1259, 1262, 1263, 1639, 1781, 1785, 1812, 1972, 2095, 2148, 2776, 3187, 3306, 3492, 3637, 3649, 3723, 3812, 3814, 3866, 4065, 4274, 4897, 4932, 5129, 5182, 5510, 5940, 6084, 6506, 6527, 6558, 6781, 7003, 7004, 7958, 8214, 8868, 9063, 9220, 9455, 9612, 9772, 9863, 10209, 10744, 11646, 11818, 11819, 11906, 12118, 12232, 12330, 12331, 12461, 13021, 13088, 13195, 13269, 13330, 13708, 13732, 14504–6, 14653, 15088, 15371, 15593, 15594, 15790, 16368, 16369, 16371, 16511, 16738, 16920, 16985, 17006, 17342, 17419, 17482, 17668, 17679, 17820, 17861, 18401, 18587, 19465, 19581

Characteristics of seeds, 322, 3216, 6395, 6653, 6695, 6967, 7972, 8256, 8931, 8954, 9187, 9470, 11152, 11535, 11536, 11592, 11801, 12254, 12255, 13307, 17793, 18143, 18162, 18218–20, 18236, 18246, 18480, 19392, 19407, 19477, 19624, 19891, 19959, 20058

Chemical change during metabolism, 54, 120, 134, 274, 287, 297, 428, 440, 449, 593–95, 788, 1167, 1282, 1289, 1460, 1471, 1706, 1883, 1889, 1905, 1925, 1952, 1966, 1968, 2046, 2113, 2204, 2207, 2217, 2230, 2271, 2272, 2299, 2325, 2342, 2394, 2405, 2413, 2431, 2458, 2485, 2507, 2568, 2570, 2688, 2749, 2782, 2800, 2802, 2855, 2856, 2861, 2914, 3002, 3102, 3203, 3204, 3384, 3385, 3421, 3486, 3630, 3664, 3681, 3903, 3952, 3954, 4011, 4062, 4094, 4100, 4131, 4137, 4193, 4470, 4475, 4840, 4846, 4848, 4849, 4869, 4871, 4880, 4891, 4921, 4955, 4978, 4991, 5115, 5140, 5216, 5221, 5234, 5264, 5290, 5356, 5359, 5407, 5506, 5574, 5597, 5671, 5673, 5692, 5712, 5720, 5768, 5769, 5803, 5946, 5952, 5965, 5992, 6006, 6008, 6038, 6080, 6095, 6201, 6208, 6458, 6460,

Chemical change during metabolism–(*cont.*)
6700, 6701, 6922–24, 7040, 7041, 7094, 7161, 7167, 7227, 7276, 7785, 7788, 7789, 7806, 7921, 7935, 8062, 8067, 8081, 8091, 8151, 8165, 8305, 8337, 8361, 8401, 8466, 8537, 8607, 8608, 8726, 8830, 8848, 8871, 8880, 8887, 8904, 8961, 8987, 8989, 8997, 9029, 9070, 9097, 9098, 9172, 9183, 9262, 9290, 9317, 9342, 9348–50, 9380, 9415, 9456, 9521, 9529, 9551, 9580, 9603, 9652, 9655, 9715, 9721, 9736, 9738, 9779, 9806, 9807, 9879, 9897, 9982, 9987, 10039, 10170, 10176, 10177, 10210, 10230, 10264, 10327, 10328, 10376, 10403, 10411, 10422, 10483, 10644, 10645, 10773, 10792, 10932, 11000, 11068, 11071, 11170, 11172, 11303, 11362, 11370, 11372, 11508, 11562, 11577, 11590, 11706, 11848, 11982, 12089, 12149, 12186, 12311, 12314, 12315, 12366, 12373, 12634, 12746–49, 12756, 12898, 13069, 13287, 13328, 13405, 13562, 13609, 13733, 14140, 14174, 14235–37, 14430, 14664, 14790, 15034, 15051, 15268, 15293, 15294, 15335, 15342, 15412, 15565, 15728, 15730, 15808, 15846, 15860, 15861, 15955, 15959, 16052, 16080, 16180, 16310, 16320, 16543, 16634, 16671, 16827, 16853–55, 16949, 17150, 17230, 17395, 17403, 17426, 17508, 17588, 17615, 17741, 17807, 17808, 17873, 18195, 18280, 18436, 18649, 18720, 18721, 19092, 19330, 19488

Chemical composition, 5, 45, 76, 89, 90, 101, 112, 120–22, 154, 156, 168, 173, 199, 260, 263, 273, 275, 302, 347, 372, 429, 450, 459, 482, 520, 521, 524–28, 531, 536, 547–49, 558, 559, 580, 586, 587, 627, 628–30, 644, 661, 669, 705, 713, 752, 774, 789, 817, 878, 899, 948, 1032, 1185, 1273, 1285, 1420, 1430, 1433–35, 1475, 1534, 1587, 1660, 1664, 1667, 1678, 1688, 1713, 1826, 1837, 1914, 1925, 2039, 2056, 2083, 2120, 2162, 2223, 2237, 2289, 2291, 2292, 2307, 2308, 2334, 2356, 2358, 2386, 2396–401, 2403, 2415, 2420, 2453, 2492, 2504, 2574, 2595, 2604, 2660, 2664, 2688, 2697, 2753, 2772, 2786, 2826, 2858, 2955, 3004, 3006, 3023, 3052, 3137, 3222, 3234, 3309, 3326, 3391, 3405, 3410, 3504, 3565, 3574, 3661, 3663, 3672, 3688, 3689, 3756, 3761, 3762, 3766, 3856, 3888, 3890, 3905, 3922, 4121, 4131, 4152, 4220, 4302, 4321, 4396, 4424, 4428, 4434, 4461, 4468, 4469, 4629, 4734, 4737, 4758, 4791, 4803, 4832, 4844, 4864, 4865, 4868, 4996, 5015, 5051, 5052, 5067, 5100, 5172, 5217, 5224, 5230–32, 5265, 5319, 5330, 5331, 5352, 5369, 5379, 5388, 5443, 5479, 5503, 5576, 5588, 5598, 5617, 5652, 5660–62, 5664, 5667, 5743, 5759, 5762,

Chemical composition–(*cont.*)
5774, 5837, 5931, 5932, 6075, 6077, 6078, 6147, 6150, 6177, 6227, 6229, 6239, 6270, 6304, 6328, 6381, 6532, 6536, 6563, 6564, 6583, 6656, 6658, 6664, 6696, 6703, 6745, 6752, 6792, 6841, 6849, 6850, 6853, 6896, 6917, 6931, 6938, 6950-K, 6976, 7001, 7026, 7029, 7101, 7147, 7171, 7172, 7179, 7183, 7190, 7191, 7209, 7215, 7223, 7242, 7244, 7245–49, 7251, 7252, 7277, 7279, 7281, 7291, 7340, 7361, 7385, 7395, 7402, 7471, 7492, 7563, 7613, 7619, 7620, 7629, 7630, 7659, 7663, 7668, 7671, 7694, 7783–85, 7813, 7818, 7856, 7857, 7867, 7868, 7911, 8006, 8023–25, 8166, 8204, 8297–99, 8342, 8347, 8365, 8366, 8390, 8399, 8426, 8491, 8493, 8523, 8542, 8600, 8635, 8655, 8657, 8702, 8739, 8743, 8758, 8764, 8774, 8799, 8844, 8860, 8907, 8955, 8975, 9014, 9016, 9075, 9091, 9148, 9149, 9325, 9354, 9386, 9394, 9457, 9515, 9528, 9548, 9549, 9552, 9598, 9775, 9810, 9811, 9874, 9918, 9950, 10142, 10152, 10222, 10280, 10291, 10315, 10316, 10340, 10360, 10368, 10394, 10397, 10496, 10503, 10515, 10519, 10605, 10624–26, 10633, 10643, 10715, 10718, 10746–49, 10768, 10787, 10794, 10809, 10814, 11014, 11040, 11049, 11072, 11190, 11191, 11204–7, 11294, 11302, 11356, 11367, 11374, 11419, 11441, 11466, 11489, 11539, 11561, 11591, 11594, 11633, 11634, 11662, 11684, 11791, 11809, 11839, 11859, 11860, 11864, 11887, 11920, 11957, 11961, 11964, 11985, 11986, 12049, 12070, 12074, 12076, 12095, 12106, 12109, 12174, 12209, 12213, 12214, 12236, 12249, 12251, 12256, 12346, 12389, 12400, 12413, 12496, 12506, 12530, 12610, 12626, 12638, 12754, 12767, 12808, 12826, 12835, 12836, 12841, 12842, 12865, 12869, 12880, 12938, 12951, 12972, 12985, 12993, 13114, 13115, 13149, 13189, 13218, 13293, 13294, 13343, 13399, 13423, 13424, 13451, 13479, 13481, 13570–73, 13575, 13581, 13592, 13619, 13626, 13627, 13659, 13695, 13765, 13766, 13774, 13775, 13782, 13834, 13846, 13878, 13881, 13957, 13973, 13975, 14056, 14060, 14096, 14134, 14135, 14156, 14192, 14200, 14202, 14327–30, 14351, 14394, 14424, 14430, 14439, 14509–12, 14526, 14574, 14599, 14600, 14624, 14736, 14745–49, 14776, 14804, 14807, 14842, 14843, 14877, 14891, 14892, 14897, 14938, 14944, 15024, 15029, 15055, 15061, 15111, 15140, 15144, 15155, 15168, 15193, 15195, 15210, 15212, 15307, 15308, 15318, 15321, 15422, 15444, 15445, 15447, 15459, 15520, 15522–24, 15642, 15649, 15650, 15666, 15670, 15704, 15734, 15736, 15737, 15753, 15757,

Chemical composition–(*cont.*)
15795–99, 15801, 15818, 15882, 15886, 15887, 15903, 15914, 15927, 15931, 15932, 15939, 15940, 15966, 16036, 16041, 16082, 16087, 16093, 16094, 16128–30, 16140, 16188, 16194, 16221, 16246, 16330, 16376, 16378, 16411, 16473, 17536, 16553, 16555, 16576, 16586, 16588, 16591, 16632, 16650, 16665, 16673, 16682, 16721, 16727, 16751, 16765, 16766, 16769, 16773, 16844, 16903, 16904, 16911–19, 16933, 16944, 16959, 16988, 16994, 16998, 17024, 17049, 17052, 17068–72, 17077, 17131, 17139, 17145, 17151, 17174, 17184, 17202, 17211, 17223, 17241, 17249, 17256, 17257, 17283, 17286, 17297–99, 17319, 17323, 17329, 17376, 17385, 17386, 17399, 17400, 17461, 17489, 17490, 17499, 17509, 17522–25, 17527, 17528, 17541, 17601, 17602, 17681, 17683, 17726–28, 17746, 17756, 17759, 17760, 17800–2, 17874, 17875, 17946, 18114, 18123, 18125, 18151, 18159, 18160, 18184, 18239, 18240, 18245, 18253, 18254, 18259, 18344, 18434, 18442, 18463, 18504, 18530, 18531, 18544, 18546, 18593, 18598–601, 18629, 18699, 18747, 18780, 18781, 18788, 18790, 18802, 18803, 18831, 18924, 18930, 18998, 19008, 19023, 19081, 19103, 19104, 19110, 19139, 19142, 19143, 19148, 19176, 19202–7, 19230, 19247, 19248, 19254, 19282, 19289, 19315, 19354, 19419, 19426, 19450, 19478, 19481, 19482, 19550, 19570, 19656, 19665, 19671, 19690, 19701, 19756, 19767, 19793, 19799, 19812, 19841, 19850, 19936, 19954, 19968, 19970, 19997, 19998, 20005, 20006, 20019, 20020, 20032, 20081, 20089, 20130, 20139

Chemicals effects, 5, 43, 53, 138, 200, 266, 300, 413, 434, 452, 510, 566, 657, 897, 939, 963, 986, 994, 1129, 1161, 1171, 1451, 1498, 1706, 1755, 1828, 2001, 2045, 2173, 2176, 2183, 2296, 2385, 2387, 2711, 2712, 2716, 2725, 2746, 2812, 3156, 3421, 3452, 3454, 3566, 3969, 4619, 5110, 5371, 5377, 5423, 5616, 5618, 5624, 5773, 5779, 5780, 5973, 5988, 6122, 6557, 6586, 6659, 6682, 6728, 6755, 6935, 7008, 7018, 7266, 7318, 7446, 7506, 7573, 7765, 7779, 7882, 7890, 7930, 7933, 7952, 8004, 8194, 8267, 8404, 8455, 8465, 8491, 8519, 8687, 8696, 8784, 9700, 9724, 9741, 9917, 10073, 10123, 10177, 10305, 10362, 10711, 11502, 11504, 11559, 11649, 11713, 11717, 11802, 12001, 12040, 12208, 12218, 12751, 12796, 12833, 12921, 12971, 13033, 13105, 13303, 13304, 13339, 13494, 13502, 13638, 13663, 13741, 13792, 13905, 13946, 13983, 14052, 14264, 14408, 14409, 14412, 14414, 14415, 14418, 14419,

Chemicals effects–(*cont.*)
14434, 14436, 14466, 14489, 14598, 14693, 14876, 14889, 14949, 14951, 15094, 15110, 15392, 15396, 15440, 15455, 15495, 15557, 15620, 15830, 15866, 15938, 16101, 16269, 16270, 16273, 16381, 16542, 16590, 16602, 16740, 16886, 16888, 17107, 17130, 17238, 17300, 17348, 17432, 17563, 17565, 17650, 17694, 17729, 17868, 17921, 17922, 18017, 18060, 18146, 18180, 18221, 18293, 18294, 18308, 18361, 18458, 18505, 18618, 18632, 18648, 18652, 18653, 18689, 18691, 18773, 18959, 18984, 18985, 19020, 19087, 19109, 19147, 19196, 19201, 19304, 19476, 19579, 19616, 19697, 19729, 19888, 19942, 19956
Chile, 959, 19785
Chlorophyll, 1297, 1911, 2463, 2701, 5258, 6033, 6212, 6500, 6811, 8526, 10677, 10911, 12294, 14346, 15746, 17019, 18267
Choline, 1040, 2001, 2574, 3854, 3867, 4120, 10397
Classification, 475, 3997, 6893, 9171, 9328, 9417, 9465, 11291, 11292, 12656, 13649, 16551, 17766, 19352, 20071
Cleaning, 177, 406, 539, 654, 902, 909, 910, 913, 969, 990, 1022, 1066, 1094, 1210, 1318, 1332, 1333, 1477, 1514, 1701, 1815, 1869, 1956, 2091, 2110, 2111, 2130, 2318, 2373, 2632, 2663, 2761, 2942, 3039, 3151, 3178, 3346, 4132, 4337, 4393, 4641, 4752, 4784, 5162, 5426, 5518, 5549, 5676, 5719, 5747, 5895, 5908, 5980, 6041, 6442, 6669, 6688, 6724, 7112, 7364, 7376, 7475, 7560, 7561, 7743, 7957, 7970, 8049, 8070, 8076, 8080, 8171, 8193, 8400, 8423, 8518, 8911, 8945, 9009, 9253, 9259, 9282, 9283, 9302, 9680, 9857, 9905, 10155, 10193, 10225, 10255, 10279, 10325, 10362, 10434, 10467, 10488, 10516, 10614, 10641, 10648, 10731, 10766, 11624, 11698, 11904, 11905, 11929, 12020, 12033, 12451, 12462, 13488, 13776, 13790, 13940, 14455, 15907, 16057, 16210, 16212–14, 16545, 16670, 17121, 17170, 17480, 17494, 17738, 17771, 17894, 18027, 18057, 18347, 18370, 18631, 19120, 19317, 19606, 19751, 19953, 19957
Cleistogamy, 1807, 5724
Coats, 57, 66, 317, 318, 331, 336, 358, 361, 369, 488, 504, 524, 526, 527, 529, 530, 540, 564, 685, 951, 1493, 1547, 1556, 1835, 1945, 2106, 2176, 2456, 2636, 2880, 2948, 2959, 3057, 3161, 3184, 3640, 3920, 4136, 4256, 4296, 4318, 4346, 4489, 4588, 4605, 4793, 4828, 5227, 5302, 5318, 5468, 5565, 5596, 5608, 5610, 5748, 5809, 5868, 5890, 6014, 6191, 6234, 6274, 6303–6, 6594, 6770, 6802, 6803, 6895, 7011–13, 7173, 7217, 7219, 7221, 7426, 7429, 7435, 7741, 7742, 7744, 7826, 7865, 7866, 7960, 8206, 8211, 8489, 8491, 8757, 8784, 8789, 8791–93, 8818, 8821, 8836, 8846, 8955, 9204, 9374, 9394, 9716, 9852, 9864, 10185, 10306, 10307, 10391, 10398, 10483, 10562, 11180, 11309, 11475, 11596, 11639, 11695, 11768, 11932, 11933, 11962, 11963, 12040, 12082, 10283, 12155, 12345, 12346, 12429, 12541, 12543, 12586, 13159, 13168, 13309, 13311, 13448, 13484, 13546, 13979, 14106, 14201, 14355, 14433, 14457, 14458, 14464, 14468, 14503, 14513, 14598, 14892, 14960, 14961, 15077, 15127, 15161, 15274, 15301, 15423, 15430, 15431, 15678, 15881, 15895, 15999, 16074, 16075, 16097, 16177, 16265, 16373, 16425, 16426, 16488, 16541, 16587, 16588, 16772, 16860, 16878, 16946, 16986, 17031, 17058, 17059, 17128, 17132, 17136, 17278, 17281, 17327, 17332, 17362, 17379, 17470, 17474, 17548, 17655, 17662, 17775, 17825–28, 17862, 17940, 17961, 17988, 18028, 18054, 18137, 18211, 18212, 18234, 18267, 18309, 18446, 18464, 18476, 18548, 18549, 18692, 18803, 18807, 18815, 18881, 18896, 18901, 18935, 18972, 19004, 19036–38, 19116, 19149, 19150, 19168, 19200, 19208, 19214, 19230, 19396, 19617, 19630, 19883, 19956, 20027, 20071, 20072, 20078, 20099, 20124
Cobalt, 534, 9248, 10565, 11177, 15557, 15559, 16434, 19349
Colchicine, 792, 2737, 5063, 7193, 7950, 8403, 9657, 9878, 10855, 11774, 12422, 14462, 16009, 16071, 16938, 17186, 17267, 18382
Coleoptile, 268, 2389, 4021
Collection, 213, 673, 990, 1065, 1090, 1142, 1218, 1376, 1917, 1920, 1970, 2092, 2426, 2903, 4163, 4620, 7713, 8837, 9282, 10434, 12837, 13150, 15842, 17560
Color, 39, 244, 369, 472, 473, 561, 636, 686, 1470, 1496, 1577, 1655, 1746, 1747, 1768, 1841, 2265, 2402, 3057, 3065, 3561, 3666, 3842, 3845, 3876, 4116, 4175, 4323, 4726, 4727, 4861, 5243, 5300, 5339, 5353, 5524, 5735, 5863, 6018, 6256, 6277, 6278, 6581, 7114, 7169, 7173, 7586, 7626, 7681, 7769, 7984, 7995, 7996, 8118, 8287, 8520, 8731, 9147, 9242, 9319, 9375, 9651, 9841, 9907, 9954, 10151, 10215, 10318, 10362, 10642, 10697, 11122, 11124, 11127, 11130, 11325, 11415, 11571, 11728–32, 11735, 12040, 12503, 12527, 12548, 12661, 12710, 12797, 12798, 12803, 12812, 12906, 13307, 13794, 13966, 14039, 14155, 14157, 14163, 14164, 14277, 14463, 14464, 14502, 14533, 14657, 14734, 14750, 14886, 15399, 15351, 15531, 15532, 15539, 15652, 15715, 15743, 15908, 16158, 16255, 16441, 16500, 16901, 17125,

Color–(*cont.*)
17247, 17305, 17415, 17416, 17546, 17634, 17641, 17643, 17840, 17953, 17993, 17995, 17997, 18212, 18226, 18471, 18472, 18513, 18803, 19122, 19546, 19547, 19549, 19704, 19708, 19852, 20090

Conifers, 77, 382, 384, 389, 390, 415, 667, 703, 824, 880, 1445, 1964, 1965, 1967, 1981, 1988, 2024, 2646, 2675, 3240, 3404, 3458, 3460, 3474, 3956, 4123, 4124, 4140, 4201, 4490, 5194, 5407, 5513, 5546, 5955, 6433, 6456, 6688, 6782, 7359, 7436, 7555, 7661, 7662, 7692, 7728, 7768, 7888, 7924, 8001, 8137, 8304, 8361, 8369, 8548, 8550, 8556, 8582, 8585, 8931, 8932, 9255–57, 9260, 9282, 9324, 9575, 10128, 10247, 10255, 10359, 10981, 11200, 11313, 11493, 11812, 12071, 12301, 12395, 12421, 12578, 12588, 13559, 13776, 13851, 14327, 14391, 14798, 15398, 15400, 15704, 15907, 15994, 16209, 16212–14, 16360, 16510, 16968, 17018, 17038, 17075, 17425, 17441, 17579, 17738, 18544, 18573, 18580, 18786, 19121, 19123–26, 19132, 19137, 20074, 20095

Copper, 363, 365, 684, 2173, 2297, 2453, 3150, 3154, 3239, 3376, 4308, 6266, 6572, 7278, 7592, 8915, 8955, 9074, 10065, 13077, 13339, 13806, 14549, 14884, 15482, 16742, 16956, 19164, 19281

Cortisone, 1000, 1001, 3911, 5464

Costa Rica, 15818

Cotyledons, 592, 1516, 1797, 2249, 2261, 2463, 2467, 2549, 2550, 2600, 2626, 3117, 3162, 3163, 3254, 3268, 3301, 3312, 3474, 3727, 3859, 4940, 5187, 5258, 5301, 5619, 5660, 5720, 5803, 6102, 6481, 6749, 6750, 6783, 6806, 6859, 7343, 7515, 7680, 7752, 8092, 8246, 8887, 9090, 9129, 9338, 9637, 9781, 10122, 10296, 10573, 10781, 10863, 10897, 10999, 11055, 11494, 11524, 11736, 12050, 12094, 12096, 12430, 12633, 12681, 12926, 12927, 13139, 13441, 13586, 14028, 14347, 14688, 14731, 14792, 14805, 14991, 15302, 15768, 15900, 16153, 16352, 16823, 17073, 17093, 17838, 18753, 19016, 19017, 19085, 19318, 19578, 19940, 20002, 20061, 20062

Coumarin, 474, 2343, 2573, 3304, 4125, 4404, 5255, 5568, 6107, 6118, 6207, 7465, 7576, 7577, 8340, 8886, 9477, 10916, 11661, 11885, 11886, 12152, 12153, 12829, 13064, 13066, 13075, 13078, 13079, 13576, 13791, 14626, 15500, 16861, 18038, 19865

Counter, 3133, 3612, 7519, 7560, 7632, 11321

Cover plants, 732, 1228

Crop plants, 1571, 2206, 2987, 3775, 3850, 6139, 8594, 8808, 9422, 10177, 11061, 11518, 12179, 12180, 13841, 13996, 15174,

Crop plants–(*cont.*)
15611, 15614, 16408, 16892, 17510, 17742, 17743, 18582

Cultivated plants, 469, 2535, 2907, 2989, 7215, 7945, 8801, 9767, 10301, 12192, 12660, 12879, 13982, 14342, 15095, 15679, 16112, 16885, 16886, 18558, 18610, 18702, 19191

Cutting tests, 19097, 19566, 20029, 20030

Cytology, 2306, 6778, 7932, 9142, 10319, 10980, 13409, 14603

Czechoslovakia, 10236, 17001, 17698, 17937, 17938

2,4-D, 5, 1043, 1085, 1729, 6139, 10229, 11213 11323, 12152, 12153, 12295, 12443, 13192, 13595, 13597, 14733, 15689, 15839, 15918, 16380, 17867

DDT, 458, 970, 6244, 9031, 15120

Dehydrase, 116, 6515

Dehydrogenase, 568, 1631, 2386, 4842, 4843, 4845, 4887, 5038, 5039, 5429, 5500, 6823–26, 7405, 7581, 7863, 7985, 10162, 10923

Denmark, 515, 1163, 1196, 1203, 14283, 15240

Depth of seeding, 102, 2383, 2519, 2692, 3589, 4092, 8105, 8706, 9571, 9737, 10308, 11566, 13003, 13213, 13286, 18020

Desert seeds, 518, 1991, 4943, 6112, 7589, 10062, 11024, 12063, 12676, 13896, 14223, 15035, 16484, 17529, 18469, 18470, 19526, 19527, 19529

Desiccant, 13365

Deterioration, 433, 2834, 2835, 6862, 7349

Development, 57, 72, 81, 117, 145, 238, 285, 316, 333, 334, 380, 381, 467, 475, 487, 531, 562, 593, 594, 631, 990, 1327, 1363, 1424, 1453, 1541, 1594, 1961, 1962, 2061, 2096, 2106, 2160, 2243, 2245, 2303, 2321, 2459, 2461, 2467, 2468, 2541, 2703, 2752, 2756, 2758, 2783, 2805, 2850, 2905, 2909, 2910, 2917, 2918, 2945, 2948, 2983, 3034, 3056, 3059, 3062, 3063, 3076, 3206, 3223, 3256, 3276, 3278, 3304, 3307, 3431, 3432, 3470, 3471, 3473, 3503, 3541, 3555, 3575, 3595, 3620, 3621, 3668, 3686, 3718, 3719, 3743, 3780, 3833, 3917, 3918, 3961, 4025, 4041, 4238, 4324, 4343, 4358, 4373, 4375, 4379, 4388, 4389, 4395, 4433, 4525, 4535–38, 4544, 4546, 4548–53, 4555, 4557, 4559, 4560, 4695, 4748, 4786, 4787, 4794, 4806, 4820, 4823, 4856–59, 4939, 5004, 5060, 5140, 5142, 5195, 5196, 5223, 5242, 5412, 5416, 5417, 5428, 5441, 5465, 5494, 5556, 5636, 5692, 5693, 5750, 5770, 5815, 5872, 5879, 5934, 5969, 5978, 5995, 6014, 6062, 6096, 6181, 6182, 6184–88, 6199, 6219, 6224, 6258, 6312, 6374, 6426, 6543, 6649–51, 6740, 6789, 6915, 6942, 6950-C, 7043, 7123, 7125,

Development–(*cont.*)
7155, 7163, 7255, 7270, 7320–23, 7348, 7384, 7474, 7520, 7521, 7525, 7527–29, 7558, 7559, 7758, 7761, 7782, 7790, 7795–97, 7799–5, 7899, 7948, 8013–16, 8019, 8021, 8022, 8095, 8134, 8176, 8177, 8187, 8233, 8269, 8321, 8322, 8375, 8378, 8395, 8480, 8484, 8630, 8650, 8674, 8679, 8685, 8704, 8872, 8874, 8877, 8878, 8951, 9052, 9124, 9143, 9204, 9210, 9245, 9309, 9313, 9381, 9383, 9384, 9429, 9588–90, 9725, 9749, 9750, 9833, 9979, 10000, 10008–12, 10044, 10045, 10057, 10063, 10110, 10114, 10115, 10145, 10146, 10148, 10149, 10161, 10164, 10211, 10212, 10330, 10354, 10355, 10357, 10380, 10381, 10383, 10384, 10386, 10389, 10392, 10421, 10461, 10479, 10484, 10611, 10653, 10688, 10751, 10922, 10988, 11202, 11304, 11350, 11446, 11447, 11549, 11583, 11688, 11727, 11749, 11764, 11831, 11903, 11982, 12061, 12066, 12079, 12087, 12121, 12186, 12196, 12197, 12293, 12320, 12328, 12369, 12390, 12479, 12697, 12701, 12780, 12781, 12815, 12857, 12892, 12924, 12947, 12975, 12976, 12982, 12991, 12994, 13017, 13041, 13142, 13151, 13281, 13321, 13337, 13349, 13608, 13639, 13643, 13644, 13709, 13738, 13958, 13959, 14007, 14012, 14077, 14123, 14179, 14194, 14195, 14298, 14365, 14527, 14684, 14687, 14715, 14719, 14923, 14941–43, 14971, 14984, 14989, 15019, 15084, 15110, 15112, 15131, 15169, 15207, 15218, 15312, 15313, 15327, 15452, 15453, 15456, 15480, 15639, 15640, 15665, 15708, 15769, 15869, 15879, 15880, 15889, 15890, 15935, 15943–45, 15979, 16002–4, 16008, 16076, 16079, 16086, 16205, 16259, 16290, 16321, 16329, 16417, 16418, 16480, 16533, 16543, 16584, 16595, 16602, 16622, 16626, 16647, 16679, 16743, 16779, 16784, 16785, 16980, 16981, 16983, 17134, 17188, 17204, 17227, 17255, 17278, 17375, 17460, 17485, 17486, 17488, 17502–4, 17513, 17514, 17518, 17519, 17521, 17562, 17623, 17757, 17760, 17822, 17961, 18014, 18041, 18139, 18141, 18161, 18209, 18272, 18279, 18303, 18346, 18349, 18400, 18436, 18506, 18508, 18519, 18530, 18531, 18565, 18588, 18603, 18617, 18775, 18779, 18791, 18793, 18822–24, 18828, 18839, 18845, 18880, 19001, 19096, 19189, 19237, 19245, 19263, 19264, 19271, 19287, 19288, 19362, 19384, 19416, 19586, 19625, 19709, 19744, 19759, 19760, 19810, 19813, 19824, 19829, 19830, 19834, 19947, 19993, 20023

Diastase, 155, 1726, 1843, 3381, 4805, 6707, 10593, 10726, 14419

Dicotyledons, 97, 3706, 18205, 19489, 19580

Dimorphic seeds, 16867, 17378, 17380

Directory, 1164

Direct seeding, 5057, 5222, 5983, 8441, 13231, 13253, 13260, 14387, 15068, 15309, 15684, 15686, 17340, 17435, 18033, 18404

Disease transmission, 136, 288–90, 346, 363, 394, 397, 507, 514, 537, 609, 633, 867, 961, 971, 1013, 1015, 1057, 1059, 1060, 1112, 1150, 1219, 1248, 1272, 1352, 1361, 1378–87, 1427, 1472, 1549, 1730, 1732, 1733, 1743, 1767, 1874, 1941, 2093, 2133, 2281, 2329, 2330, 2425, 2514, 2534, 2608, 2633, 2724, 2729, 2760, 2819, 2825, 2943, 2944, 2973, 2979, 3008, 3014, 3015, 3067, 3088, 3130, 3236, 3341, 3378, 3463, 3500, 3520, 3523, 3533, 3535, 3537, 3570, 3625, 3700, 3701, 3742, 3786, 3788, 3875, 3883, 3891, 3937, 3938, 4069, 4110, 4160, 4207, 4255, 4289, 4291, 4292, 4348, 4403, 4406, 4407, 4448, 4502, 4614, 4615, 4618, 4629–31, 4644, 4645, 4650, 4694, 4696, 4699, 4799, 4920, 5005, 5026, 5173, 5200, 5213, 5249, 5266–69, 5271, 5273–77, 5361, 5467, 5480, 5543, 5545, 5620, 5642, 5725, 5741, 5785, 5786, 5804, 5813, 5889, 5910, 5926, 6241, 6242, 6276, 6289, 6290, 6316, 6323, 6324, 6407, 6423, 6454, 6606, 6693, 6863, 6958, 6991, 7024, 7025, 7178, 7256–58, 7301, 7302, 7386, 7394, 7414, 7464, 7472, 7483, 7484, 7488, 7505, 7531, 7578, 7606–9, 7611, 7624, 7707, 7736, 7750, 7809, 7980, 8090, 8164, 8188, 8200, 8258, 8334, 8345, 8346, 8541, 8604, 8631, 8693, 8695, 8701, 8834, 8901, 9061, 9236, 9304, 9322, 9365, 9427, 9541, 9602, 9632, 9666, 9823, 9835, 9838, 9947, 9948, 9985, 10048, 10075, 10107, 10133, 10347, 10350, 10415, 10457–60, 10475, 10489, 10502, 10586, 10588, 10598, 10599, 10617, 10789, 10833, 10851, 10859, 10860, 10940, 10968, 10971, 10973, 10974, 10983, 11020, 11059, 11301, 11421, 11615, 11616, 11657, 11659, 11670, 11840, 11911–15, 11952, 11953, 11959, 11999, 12056, 12140, 12144, 12159, 12187, 12215, 12237, 12321, 12347, 12362, 12368, 12371, 12382, 12383, 12392, 12454, 12597, 12602, 12606, 12712, 12772, 12773, 12844, 12859, 12886–88, 12896, 13031, 13032, 13131, 13132, 13140, 13210, 13217, 13238, 13241, 13339, 13360, 13417, 13458, 13492, 13553, 13555, 13725, 13796, 13804, 13843, 13912, 12921, 13999, 14000, 14002–4, 14079, 14280, 14283–88, 14297, 14539–42, 14830, 14837, 14846, 14851 14879, 14957, 15172, 15179, 15249, 15315, 15569, 15603, 15604, 15608, 15742, 15841, 15992, 16019, 16045, 16152, 16291,

Disease transmission–(*cont.*)
16414, 16517, 16518, 16563, 16565, 16566, 16568, 16569, 16571, 16573, 16574, 16616, 16674, 16693, 16694, 16921, 17012, 17084, 17101, 17190, 17274, 17309, 17455, 17459, 17571, 17572, 17574, 17645, 17667, 17693, 17941, 18371, 18372, 18414, 18487 18490, 18498, 18810, 18812, 18944, 18945, 19007, 19032, 19158, 19310, 19321, 19324, 19338, 19508, 19512, 19524, 19537, 19743, 19894, 20116

Disinfection, 59, 600, 601, 607, 662, 687, 796, 942, 1124, 1141, 1178, 1271, 1425, 2208, 2278, 2279, 2395, 2577, 2579, 2580, 2647, 2683, 2700, 2717, 2950, 2951, 3011, 3012, 3680, 3698, 3699, 4121, 4288, 4289, 4344, 4585, 4704, 4888, 4906, 4908, 5132, 5134, 5135, 5139, 5358, 5451, 5452, 5813, 5919, 6309, 6329, 6398, 6487, 6577, 6743, 6786, 6950-G, 6959, 7022, 7181, 7236, 7448, 7482, 7580, 7751, 8071, 8636, 8822, 9331, 9419, 9563, 9585, 9614–23, 9626, 9992, 10536, 10665, 10954, 11784, 11797, 12103, 13263, 13268, 13499, 13998, 14001, 14366, 14835, 16406, 16796, 17055, 17676, 17878, 18966

Dispersal, 408, 533, 663, 738, 836, 922, 950, 990, 1002, 1070, 1157, 1206, 1456, 1720, 1805, 1908, 2026, 2303, 2357, 2464, 2691, 2908, 2913, 3388, 3610, 3667, 3784, 3841, 4203, 4204, 4221, 4222, 4326, 4352, 4480, 4579, 4685, 4687, 4826, 5054, 5077, 5390, 5391, 5622, 5755, 6029, 6068, 6161, 6175, 6422, 6449, 6513, 6521, 7005, 7030–32, 7058, 7310, 7350, 7616, 7842, 7845, 7968, 7982, 8063, 8292, 8458, 8538–40, 8689, 8690, 8749, 8788, 8795, 8832, 8909, 8942, 9011, 9270, 9279, 9382, 9401, 9449, 9771, 9805, 10034, 10035, 10477, 10478, 11024, 11248, 11355, 11575, 11600, 11847, 11928, 12216, 12234, 12379, 12396, 12532, 12533, 12630, 12934, 12949, 13127, 13283, 13344, 13392, 13646, 13735, 13879, 14125, 14340, 14663, 14767, 14882, 15058, 15066, 15178, 15285, 15290, 15316, 15362, 15464, 15699, 15816, 16192, 16193, 16281, 16421, 16493, 16494, 16549, 16562, 16639, 16777, 16900, 16967, 16991, 17035, 17425, 17568, 17848–50, 17852, 17956, 17957, 17977, 18135, 18136, 18327, 18413, 18443, 18576, 18627, 18850, 18875, 18963, 18995, 19168, 19435, 19545, 19590–92, 19682, 19684, 20071, 20107, 20108

Dormancy, 1, 60, 61, 62, 63, 68, 97, 103, 126–28, 130–32, 135, 201, 204, 205, 217, 220, 221, 255, 377, 494, 496, 500, 512, 620, 674, 689, 712, 724, 802, 809, 820, 859, 865, 866, 941, 990, 1038, 1125, 1220, 1303, 1306, 1307–9, 1419, 1433–35, 1437, 1444,

Dormancy–(*cont.*)
1480, 1509, 1631–33, 1809, 1810, 1872, 1986, 1993, 2000, 2004, 2007, 2021, 2023, 2026, 2034, 2047, 2068, 2139, 2146, 2191, 2193, 2213, 2222, 2249, 2435, 2446, 2477, 2479, 2480–82, 2500, 2502, 2524–26, 2597, 2636, 2670, 2783, 2923, 2984, 2985, 3131, 3306, 3317, 3325, 3374, 3509, 3514, 3522, 3569, 3675, 3739, 3750, 3805, 3816, 3872, 3955, 4012, 4028, 4032, 4033, 4083, 4234, 4250, 4251, 4269, 4408, 4418, 4439, 4442, 4452, 4488, 4590, 4593, 4602, 4604, 4635, 4636, 4679, 4741, 4980, 5023, 5032, 5033, 5044, 5165, 5179, 5197, 5291, 5393, 5445, 5470, 5484, 5485, 5505, 5557, 5564, 5594, 5739, 5849, 5861, 5896, 5941, 6260, 6377, 6400, 6413, 6417–20, 6464, 6465, 6469, 6474, 6496, 6544, 6596, 6659, 6661, 6686, 6739, 6804, 6882, 6894, 6983, 7052, 7054, 7129, 7133, 7140, 7169, 7195, 7196, 7207, 7218, 7219, 7383, 7421, 7450, 7466, 7481, 7573, 7666, 7683, 7699, 7843, 7871, 7885, 7896, 7897, 7929, 7930, 7933, 7936, 7961, 7971, 7997, 8106, 8122, 8161, 8184, 8196, 8206, 8211, 8215, 8217, 8257, 8272, 8406, 8427, 8428, 8524, 8531, 8571, 8585, 8589, 8700, 8738, 8751, 8753, 8754, 8784, 8814–19, 8827, 8886, 8919, 8955, 9056, 9119, 9146, 9163, 9181, 9199, 9273, 9297, 9298, 9347, 9391, 9497, 9510, 9537, 9542, 9569, 9629, 9630, 9639, 9656, 9768, 9789, 9798, 9799, 9840, 9872, 9873, 9876, 9909, 9953, 10033, 10072, 10076, 10077, 10079, 10092, 10138, 10168, 10258, 10285, 10286, 10319, 10344, 10378, 10492, 10504, 10542, 10544, 10547, 10552, 10553, 10555, 10657, 10677, 10706, 10708, 10742, 10765, 10777, 10822, 10823, 10856, 10872, 10897, 10980, 11011, 11021, 11022, 11028, 11060, 11065, 11186, 11219, 11246, 11260, 11314, 11406, 11407, 11411, 11430, 11469, 11530, 11587, 11631, 11641, 11668, 11691, 11821, 11837, 11850, 11851, 11854, 12025, 12046, 12164, 12165, 12182, 12200, 12206, 12207, 12304, 12305, 12463–66, 12488–91, 12522, 12575, 12628, 12737, 12801, 12939, 12967, 13038, 13162, 13381, 13388, 13389, 13409, 13582, 13595, 13625, 13705, 13727, 13741, 13753, 13755, 13760, 13833, 13841, 13925, 13982, 14015, 14020, 14023, 14106, 14150, 14254, 14259, 14261, 14263, 14412, 14416, 14435, 14437, 14450, 14457, 14458, 14461, 14471, 14472, 14475–78, 14568, 14573, 14608, 14626, 14627, 14668, 14669, 14708, 14721, 14726, 14741, 14819, 14873, 14887, 14998, 15013, 15056, 15156, 15185, 15208, 15226, 15227, 15283, 15319, 15322, 15347, 15348, 15364, 15420, 15509, 15512, 15514, 15515, 15517, 15550,

Dormancy–(*cont.*)
15571, 15573, 15607, 15638, 15740, 15743, 15781–83, 15922, 15933, 16122, 16141, 16186, 16189, 16264, 16265, 16269–73, 16364, 16382, 16383, 16388, 16492, 16662, 16681, 16735, 16802, 16803, 16837, 16858, 16928, 16929, 16940, 17027, 17120, 17162, 17201, 17260, 17263, 17264, 17293, 17312, 17350, 17351, 17354, 17355, 17367, 17371–73, 17384, 17391, 17438, 17458, 17465, 17539, 17616, 17617, 17635, 17654, 17674, 17774, 17794, 17817, 17863, 17864, 17917, 17918, 17926, 17929, 17968, 17979, 18007, 18008, 18050, 18053–55, 18154, 18165, 18288, 18290, 18291, 18317, 18318, 18338, 18360, 18459, 18464, 18510–12, 18516, 18535, 18538, 18553, 18557, 18586, 18595, 18650, 18652, 18657, 18659, 18662, 18664, 18694, 18695, 18702, 18707, 18729, 18815, 18827, 18857, 18874, 18897, 18926, 19118, 19119, 19138, 19185, 19211, 19231, 19233, 19241, 19244, 19246, 19327, 19345, 19368, 19369, 19370, 19400, 19411, 19420, 19497, 19507, 19594, 19657, 19659, 19768, 19838, 19898, 19919, 19964, 20015, 20026, 20028, 20056

Drought resistance, 749, 1643, 11339, 11597, 12731, 12928, 13371, 13531, 14249, 14610, 14785, 15060, 17341, 17713, 18087, 19284, 19672, 20129,

Drug plants, 179, 180, 199, 768, 1521, 7036, 7091, 8984, 9505, 9677, 9678, 10631, 10856, 10941–43, 11425, 11510, 12554–56, 13987, 14908, 19438

Drying effects, 107, 152, 201, 254, 417, 441, 929, 1030, 1052, 1110, 1223, 1274, 1504, 1931, 2177, 2254, 2266, 2295, 2607, 2695, 3019, 3192, 3558, 3775, 4241, 4370, 4736, 4967, 4970, 5161, 5229, 5316, 5387, 5611, 5635, 5647, 5705, 5842, 5921, 6024, 6031, 6059, 6064, 6087, 6092, 6099, 6153, 6637, 6638, 6666, 6707, 6911, 7051, 7072, 7130, 7259, 7372, 7537, 7697, 7769, 7843, 7912, 7913, 7984, 8031, 8032, 8036, 8209, 8223, 8227, 8263, 8380, 8396, 8572, 8617, 8618, 8620, 8780, 8781, 8833, 8899, 8958, 8959, 9013, 9156, 9157, 9185, 9223, 9227, 9246, 9271, 9272, 9479, 9519, 9627, 9649, 9723, 9890, 9931, 10094, 10124, 10180, 10302, 10366, 10517, 10582, 10828, 10830, 10965, 10987, 11015, 11084, 11092, 11095, 11132, 11133, 11144, 11145, 11159, 11161, 11209, 11275, 11375, 11396, 11398–405, 11581, 11619, 11689, 12040, 12161, 12252, 12296, 12334, 12338, 12360, 12361, 12499, 12663, 12759, 13177, 13185, 13207, 13242, 13256, 13422, 13508, 13795, 13823, 14127, 14350, 14583, 14831, 14915, 15324, 15344, 15374, 15646, 15663, 15828, 16030, 16209, 16552,

Drying effects–(*cont.*)
16686, 16930, 16946, 16965, 17008, 17010, 17030, 17034, 17253, 17279, 17361, 17463, 17480, 17748–50, 17782, 17811, 17881, 17916, 17948, 18106, 18184, 18200, 18461, 18606, 18902, 19076, 19187, 19356, 19391, 19396, 19522, 19635, 19674, 19691, 19696, 19705, 19748

Dwarfs, 2027, 4830, 6459, 6461, 6471, 6473, 6481, 6482, 6485, 19118

Ecological factors, 29, 45, 196, 267, 299, 456, 542, 547, 549, 552, 556, 612, 613, 620, 684, 688, 784, 822, 824, 1187, 1296, 1299, 1303–9, 1311, 1420, 1469, 1518, 1526, 2071, 2147, 2230, 2433, 2686, 2832, 3003, 3026, 3040, 3156, 3228, 3447, 3502, 3591, 3618, 3622, 3638, 3645, 3661, 3663, 3668, 3819, 3973, 3985, 4089, 4398, 4437, 4438, 4499, 4745, 4754, 5010, 5036, 5310, 5365, 5374, 5498, 5510, 5583, 5822, 5826, 6026, 6058, 6162, 6193, 6299, 6383, 6678, 6721, 6744, 6762, 6809, 6821, 6988, 7054, 7093, 7107, 7196, 7199, 7282, 7361, 7564, 7836, 7959, 7988, 8040, 8146, 8174, 8196, 8505, 8508, 8681, 8698, 8804, 8827, 8982, 9034, 9132, 9180, 9192, 9216, 9318, 9400, 9442, 9453, 9454, 9512, 9551, 9555, 9569, 9720, 9813, 9958, 9967, 10172, 10457, 10612, 10636, 10640, 10664, 10832, 10898, 10936, 10994, 11028, 11348, 11427, 11539, 11593, 11609, 11694, 11712, 11862, 11890, 11927, 11930, 11955, 11956, 12157, 12164, 12335, 12336, 12501, 12537, 12682, 12720, 12798, 12909, 13036, 13106, 13361, 13378, 13379, 13527, 13690, 13789, 13807, 14670, 14683, 15035, 15167, 15409, 15546, 15687, 15688, 15814, 15815, 16067, 16068, 16133, 16197, 16343, 16468, 16469, 16873, 16890, 16934, 17288, 17414, 17434, 17468, 17472, 17477, 17559, 17744, 17823, 17834, 17865, 17872, 17876, 18114, 18119, 18120, 18170, 18284, 18321, 18496, 18703, 18717, 18914, 19185, 19492, 19528, 19529, 19556, 19570, 19627, 19669, 19748, 19936, 19950, 19990

Economic importance, 7, 12743, 16412

Egyptian tombs, 3335

Electrical effects, 174, 419, 1177, 1221, 1235, 1269, 1591, 1702, 1957, 2063, 2270, 2761, 2942, 3048, 3375, 3478, 4020, 4397, 5611, 6145, 6325, 6673, 6676, 7285, 7422, 7696, 8491, 8546, 8939, 8955, 9327, 9471, 10756–58, 11857, 12110, 12162, 12292, 12733, 12734, 13339, 13383, 13385–87, 13391, 13398, 13400, 13610, 13898, 14320, 15137, 15330, 17175, 17386, 17730, 18333, 18589, 19335, 20018

Electrodialysis, 966, 1098, 10179, 11857, 12055, 13331, 13333, 14302, 14316, 14317, 16560, 17136, 18625, 18631

Embryo, 14, 31, 65, 117, 132; 147, 160, 161, 166, 167, 238, 274, 380, 381, 703, 756, 928, 1207, 1254, 1316, 1334, 1335, 1359, 1429, 1438, 1444, 1447, 1511, 1594–99, 1675, 1796, 1797, 1833, 1852, 1854, 1855, 1858, 1925, 1940, 2101-3, 2136, 2137, 2169, 2216, 2232, 2245, 2261, 2304, 2314, 2320, 2374, 2389, 2392, 2393, 2491, 2511, 2558, 2661, 2676, 2745, 2747, 2796, 2848, 2910, 2917, 2918, 2924, 2927, 2982, 2991, 3106, 3111, 3164, 3186, 3209, 3223, 3240–42, 3244, 3246–50, 3252, 3253, 3255, 3261, 3262, 3279, 3317, 3432, 3562, 3720, 3725, 3739, 3770, 3882, 4009, 4131, 4134, 4244, 4259, 4312, 4317, 4347, 4361, 4372, 4374, 4376, 4377, 4380, 4432, 4496, 4521, 4540, 4541, 4545, 4554, 4557, 4558, 4560–77, 4695, 4773, 4796, 4814, 4818, 4819, 4821, 4830, 4834, 4878, 4910, 4942, 4948, 4951, 5239, 5241, 5242, 5296, 5298, 5299, 5410, 5411, 5413–15, 5418, 5433, 5440, 5474, 5475, 5546, 5547, 5749, 5803, 6179, 6199, 6210, 6225, 6226, 6258, 6272, 6317, 6320, 6374, 6462, 6464, 6466, 6467, 6469, 6798, 6902, 6904, 6906, 6907, 6913, 6936, 7082, 7083, 7123–25, 7271, 7326, 7337, 7479, 7522–24, 7534, 7753, 7798, 7900, 7901, 7949–51, 8061, 8185, 8186, 8323, 8442, 8497, 8554, 8845, 8873, 8874, 8876, 8997, 8999, 9133, 9152, 9198, 9328–30, 9667, 9762, 9831, 9832, 9851, 9898, 9899, 9912, 9913, 9915, 10002, 10003, 10013, 10015, 10054, 10059, 10064, 10109, 10147, 10163, 10219, 10235, 10239, 10240, 10241, 10288, 10326, 10351, 10356, 10358, 10359, 10420, 10421, 10511, 10512, 10518, 10623, 10661, 10722, 10911, 10931, 10944, 11214–17, 11268, 11422, 11445, 11497, 11642, 11643, 11680, 11681, 11721, 11765, 11939, 11942–47, 12196, 12197, 12282, 12284, 12320, 12351, 12352, 12424, 12692, 12694, 12695, 12698–700, 12702, 12704, 12705, 12707, 12713, 12821, 12843, 12855, 12870, 12925, 13023, 13039, 13040, 13042, 13043, 13045–52, 13128, 13202, 13254, 13296, 13297, 13358, 13359, 13547, 13548, 13578, 13580, 13606, 13635–37, 13641, 13649, 13882, 13884, 13976, 13977, 14100–5, 14197–99, 14204–9, 14221, 14230, 14554, 14557, 14640, 14782, 14783, 14946, 14953, 14970, 14992, 15026, 15049, 15102, 15220, 15281, 15311, 15467, 15771–74, 15936, 15937, 15946-A, 15951, 15965, 15977, 15978, 16001, 16010, 16060, 16061, 16124, 16328, 16407, 16410, 16416, 16420, 16462–64, 16491, 16505, 16596, 16599,

Embryo–(*cont.*)
16605, 16657, 16720, 16773, 16783, 16786, 16787, 16824, 16838, 16990, 17014, 17019, 17026, 17155, 17242, 17250, 17251, 17259, 17268, 17272, 17500, 17506, 17620, 17765, 17766, 17896, 17920, 17942, 17960, 17962, 18129, 18130, 18134, 18148, 18149, 18150, 18155, 18156, 18175, 18206, 18207, 18269, 18270, 18302, 18312, 18392, 18563, 18564, 18822–24, 18991, 19047, 19068, 19210, 19347, 19361, 19429, 19914, 20100

Embryo culture, 55, 56, 111, 454, 455, 631, 810, 884, 916, 933, 1283, 1458, 1459, 1519, 1573, 1592, 1666, 1795, 1796, 1965, 2584–86, 2588, 2589, 2745, 2798, 2849, 2851, 2959, 3037, 3064, 3117, 3148, 3157, 3269, 3312, 3314–17, 3433, 3539, 3596, 3769, 3810, 3839, 4038, 4039, 4040, 4243, 4293, 4483, 4774, 4775, 4798, 4866, 4981, 4988, 5297, 5355, 5433, 5553, 5554, 5683, 6217, 6401, 6741, 6749, 6861, 6864, 6908, 7260, 7489, 7903, 7917, 7962, 8030, 8037, 8121, 8123, 8236, 8449, 8568, 8577, 8686, 8926, 8995, 9125, 9144, 9157, 9500, 9566, 9594, 10020, 10160, 10220, 10221, 10223, 10418, 10466, 10491, 10493, 10513, 10852, 10945, 11189, 11451, 11665, 11666, 11722–26, 11831, 11832, 11972,–75, 12086, 12198, 12348, 12349, 12391, 12411, 12421, 12686, 12693, 12696, 12748, 12853, 12953, 12960, 12979, 12990, 13037, 13044, 13306, 13308, 13350, 13442, 13607, 13679, 13746, 13887, 13960, 14013, 14118, 14121, 14124, 14166, 14218, 14219, 14222, 14367, 14389, 14469, 14508, 14525, 14581, 14623, 14654, 14717, 14902, 14932, 15062, 15349, 15535, 15542, 15622, 15713, 15773, 15829, 15867, 15868, 15926, 15985, 15986, 16009, 16031, 16047, 16207, 16226–29, 16239, 16352, 16601, 16701, 16707–10, 16932, 17004, 17220, 17224, 17543, 17584, 17666, 17696, 17697, 17719, 18019, 18040, 18056, 18061, 18628, 18633, 18792, 18817, 18821, 18903, 18905–7, 18988, 18989, 19212, 19374, 19447, 19772, 20085, 20094, 20101, 20102, 20134

Embryolessness, 960, 1581, 1721, 5016, 6475, 6477, 6483, 6484, 7766, 8175, 8553, 9834, 10116, 10188, 10191, 11094, 11298, 11586, 12567, 12568, 13228, 13339, 13675, 13709, 13763, 13798, 13935, 14292, 14514, 14596, 14757, 14759, 15109, 15159, 15475, 15705, 15782, 16640, 16781, 16794, 16795, 17063, 17976, 18067, 18466, 18818, 18819, 19424, 19534, 19855

Endosperm, 14, 117, 167, 230, 238, 280, 319, 321, 322, 324, 325, 329, 331, 333–35, 337, 338, 359, 531, 611, 631, 756, 1216, 1447, 1542, 1543, 1611, 1708, 1709, 1710,

Endosperm–(*cont.*)
1856, 1878, 2345, 2346, 2485, 2558, 2910, 2927, 2928, 2974, 2982, 3060, 3062, 3063, 3084, 3147, 3148, 3185, 3217, 3253, 3262, 3468, 3470, 3472, 3539, 3547, 3557, 3558, 3562, 3876, 3924, 4010, 4024, 4131, 4158, 4293, 4380, 4383, 4521, 4534, 4554, 4556, 4564, 4565, 4786, 4787, 4807, 4884, 4919, 5013, 5240, 5242, 5411, 5472, 5548, 5636, 5681, 5706–9, 5749, 5990, 5995, 6020, 6170, 6180, 6255, 6299, 6438, 6493, 6531, 6699-A, 6937, 7081, 7151, 7323–25, 7327, 7469, 7527, 7528, 7567–70, 7586, 7649, 7650, 7737, 7738, 7777, 7943, 8017, 8018, 8020, 8050, 8060, 8126, 8213, 8218, 8284, 8308, 8378, 8488, 8675, 8843, 8854, 8874, 9079, 9136, 9328, 9364, 9413, 9429, 9588, 9588–A, 9588–B, 9589, 9590, 9631, 9668, 9780, 9833, 9851, 9865, 9930, 9939, 10165, 10219, 10275, 10278, 10288, 10352, 10353, 10357, 10394, 10421, 10449, 10545, 10584, 10608, 10610, 10884, 10894, 10931, 11171, 11187, 11310, 11311, 11567, 11640, 11727, 11762, 11829, 11866, 12085, 12193, 12299, 12320, 12486, 12703, 12706, 12707, 12753, 12800, 12802, 12827, 13041, 13097, 13215, 13306, 13348, 13441, 13586, 13642, 13663, 13694, 13730, 13762, 13846, 13847, 14011, 14049, 14075, 14080, 14099, 14120, 14122, 14194, 14195, 14267, 14384, 14521, 14580, 14597, 14784, 14945, 14976, 14977, 14981, 14982, 14983, 14985, 15046, 15182, 15206, 15217, 15218, 15311, 15349, 15528, 15623, 15713, 15888, 15891, 15977, 15980, 16005–7, 16015, 16098, 16176, 16395, 16418, 16487, 16579, 16595, 16600, 16744, 16779, 16818, 16979, 17033, 17099, 17135, 17231, 17266, 17356, 17417, 17501, 17503, 17505, 17537, 17543, 17696, 17697, 17773, 17805, 17844, 17869, 17871, 17897, 17942, 17966, 17971, 17989, 18079–81, 18083–86, 18105, 18205, 18208, 18209, 18311, 18329, 18349, 18350, 18391, 18427–31, 18465, 18779, 19015, 19062, 19389, 19413, 19414, 19428, 19895, 19955, 19989, 20087, 20101, 20102, 20131

Energy of germination, 649, 668, 1236, 1257, 1512, 1999, 2333, 2335, 2815, 2828, 2896, 2975, 2983, 3257, 3262, 3365, 3434, 3435, 3709, 3916, 4527, 5278, 5362–64, 5396, 5397, 5493, 5675, 6047, 6609, 6706, 6950, 7144, 7149, 7660, 7924, 8519, 8956, 10214, 10295, 10714, 1729, 10764, 11084, 11283, 11320, 11334, 11429, 11437, 12265, 12649, 12683, 12732, 12743, 13668, 13720, 13721, 13908, 13980, 14359, 14632, 14662, 14940, 15017, 15199, 15657–63, 16179, 16315, 16443, 16448, 16547, 16755, 16972, 17009, 17165, 17731–33, 17795, 17796, 17857, 17858, 17882, 18053, 18283, 18286, 18292, 18450, 18451, 18630, 19077, 19244, 19714, 19723, 19747, 19768, 20077, 20088

Enzymes, 116, 270, 282, 303, 304, 310, 420, 434, 450, 534, 541, 543, 544, 568, 579, 581, 582, 711, 1461, 1629–33, 1726–28, 1819, 1830, 1848, 1966, 1968, 2084, 2085, 2123, 2207, 2325, 2384, 2422–24, 2458, 2485, 2488, 2551, 2554, 2559, 2562, 2565, 2567, 2623, 2671, 2842, 2861, 2871, 2914, 2971, 3037, 3089, 3148, 3189, 3190, 3220, 3358, 3478, 3507, 3653, 3848, 3882, 4102, 4219, 4261, 4460, 4464, 4465, 4471, 4472, 4488, 4702, 4848, 4870, 4955, 5029, 5084, 5104, 5248, 5359, 5378, 5505, 5676, 5707, 5726, 5838, 5897, 5899, 5959–61, 6148, 6206, 6302, 6378, 6571, 6932, 6933, 7008, 7131, 7136, 7312, 7360, 7371, 7381, 7404, 7461, 7823, 7892, 8056, 8168, 8337, 8339, 8368, 8713, 8726, 8955, 9121, 9352, 9526, 9531, 9547, 9593, 9804, 9969, 9982, 10039, 10485, 10538, 10574, 10593, 10687, 10699, 10724, 10736, 10745, 11041, 11071, 11073, 11274, 11384, 11474, 11508, 11509, 11577, 11628, 11635, 11663, 11798, 11815, 11863, 11904, 11905, 11958, 12088, 12089, 12097, 12230, 12283, 12286, 12287, 12308–10, 12374, 12445, 12510–12, 12550, 12615, 12619, 12640, 12813, 12830, 12850, 12851, 12932, 12950, 12957, 13065, 13070, 13071, 13073, 13080, 13083, 13087, 13141, 13277, 13282, 13324, 13336, 13391, 13450, 13484, 13487, 13528, 13817, 13820–22, 13830, 13831, 14058, 14065, 14110, 14112–14, 14140, 14141, 14144, 14159, 14176, 14178, 14262, 14324–26, 14352, 14383, 14419, 14565, 14566, 14576, 14594, 14755, 14797, 14799–801, 14803, 14898, 14925, 14926, 14948, 14949, 15128, 15165, 15282, 15283, 15491–93, 15498, 15534, 15556, 15558, 15562, 15726, 15727, 15729, 15731, 15739, 15897, 15899, 15904, 15958, 16218, 16219, 16230, 16233, 16244, 16272, 16297, 16304–8, 16393, 16394, 16468, 16469, 16486, 16597, 16613, 16754, 16756, 16776, 16782, 16788, 16829, 17036, 17148, 17193, 17202-A, 17203, 17408, 17481, 17493, 17512, 17619, 17695, 17704, 17757, 17761, 17786, 17865, 18047, 18080, 18081, 18153, 18158, 18241, 18264–66, 18271, 18299, 18300, 18543, 18547–50, 18836, 18840, 18841, 18843, 18844, 18872, 18894, 18895, 19053, 19073, 19094, 19197, 19273, 19344, 19388, 19393, 19421, 19454, 19540, 19548, 19569, 19618, 19652, 19718, 19736, 19938, 19941, 19983, 19984, 19989, 20002, 20092, 20137

Estonia, 15398

Ethylene, 1822, 4333, 5287, 5288, 6365, 6369, 11212, 16261, 16529
European countries, 241, 1136, 16000, 18478
Evergreens, 19563
Evolution, 508, 623, 4177, 9057, 12193, 13237, 17636, 17712, 17805, 17844, 18528

F.A.O., 1200, 5128–30, 10473, 16705, 17590, 19639
Farm crops, 3646, 16284–87, 19100
Fats and oils, 62, 63, 71, 121, 156, 158, 168, 183, 185, 191, 208, 209, 312, 438, 551, 576, 586, 587, 660, 661, 665, 669, 702, 705, 707, 710, 737, 746, 770, 774, 776, 779, 787, 795, 800, 801, 891, 901, 912, 921, 930, 987, 1053, 1079, 1081, 1188, 1485, 1486, 1576, 1645, 1646, 1669, 1717, 1718, 1751–54, 1756–58, 1866, 2051, 2411, 2846, 2847, 2871, 3003, 3116, 3157, 3225, 3326, 3330, 3409, 3427–29, 3760, 3767, 4046–49, 4073, 4151, 4262, 4297–99, 4369, 4776, 4777, 4810, 4961, 4966, 5096, 5337, 5652, 5877, 6373, 6912, 7071, 7291, 7344, 7403, 7583, 7621, 7664, 7674, 7783, 7811, 7812, 7832–35, 7846–50, 8101, 8165, 8307, 8381, 8382, 8492, 8591, 8629, 8694, 8730, 8756–58, 8763, 8766–68, 8770–72, 8775, 8777, 8779, 9097, 9098, 9110, 9172, 9437, 9706, 9802, 10134, 10217, 10238, 10261, 10288–90, 10292, 10293, 10339, 10340, 10522, 10591, 11424, 11527, 11602, 11682, 11771, 11888, 12052, 12062, 12148, 12189, 12314, 12315, 12632, 12633, 12777, 12783, 12784, 13215, 13339, 13613, 13678, 13748, 14212, 14213, 14310, 14311, 14414, 14419, 14559, 14674, 14744, 14893, 14986, 15168, 15500, 15633, 15646, 15720, 15800, 15826, 16232, 16260, 16479, 16586, 16598, 16660, 16664, 16748, 16750, 16935, 16998, 17185, 17229, 17460, 17526, 17576, 17660, 17720, 17721, 17725, 17898, 17899, 17967, 18157, 18194, 18261, 18339, 18449, 18591, 18733, 18736, 18867, 18881, 18955, 18956, 18974, 19018, 19019, 19262, 19341, 19360, 19384, 19441, 19453, 19484, 19490, 19542, 19612, 19613
Fatty acids, 186, 1293, 1644, 1645, 2122, 3330, 3758, 3759, 4192, 4224, 4399, 5333–36, 6096, 7138, 8760, 8776, 8778, 8779, 8928, 8929, 9110, 9230, 9277, 9572, 9710, 10291, 10340, 11569, 11719, 15108, 15112
Federal Seed Act, 837
Fertilizers effects, 101, 136, 177, 286, 414, 421, 426, 517, 680, 853, 978, 1022, 1126, 1668, 2120, 2298, 2379, 2421, 2487, 2517, 2706, 2748, 2881, 2934, 2936, 3103, 3329, 3392, 3417, 3481, 3803, 3826, 3880, 4180, 4312, 4697, 4747, 4862, 5286, 5368, 5421, 5575, 5591, 5593, 5650, 5716, 5720, 5816, 5820, 5840, 5916, 5933, 5994, 6028, 6081, 6100, 6156, 6397, 6425, 6821, 6841, 6989, 7347, 7380, 7463, 7652, 7661, 7825, 7840, 7862, 7874, 7893, 8039, 8043, 8111, 8194, 8224, 8313, 8316, 8341, 8479, 8491, 8664, 8723, 8869, 9035, 9039, 9175, 9225, 9238, 9326, 9568, 9686, 9801, 9802, 10173–75, 10245, 10428, 10527, 10538, 10782, 10783, 10997, 11007, 11470, 11586, 11685, 11686, 11710, 11826, 12036, 12040, 12359, 12814, 13057, 13429, 13475, 13543–45, 13598, 13655, 13661, 13799, 13803, 13808, 13815, 13836, 13972, 13983, 14005, 14295, 14322, 14372, 14501, 14595, 14598, 14924, 15072–76, 15087, 15246, 15397, 15655, 16038, 16111, 16133, 16330, 16354, 16696, 16808, 16882, 17197, 17246, 17360, 17509, 17649, 17687, 17689, 17967, 18104, 18118, 18432, 18497, 18765, 18961, 19177, 19259, 19261, 19289, 19315, 19319, 19350, 19625, 19633, 19683, 19698
Fiber crops, 588, 7853
Field crops, 353, 515, 750, 1068, 1533, 1653, 2167, 3850, 4000, 4486, 5041, 5278, 5453, 5487, 6011, 6396, 6441, 6486, 6678, 6845, 6877, 6880, 6890, 7057, 7362, 7780, 8433, 8473, 8578, 8933, 10053, 10473, 13899, 14242, 14987, 15037, 15050, 15225, 15607, 17370, 17853, 17926
Field test, 638, 2233, 3145, 3426, 4141, 4783, 6348, 6615, 7693, 8949, 9049, 9120, 9296, 10090, 10579, 10740, 11198, 11501, 15613, 17094–96, 17669, 17854, 17856, 17928, 18235, 18252, 18554, 18621, 19505, 19509, 19574, 19593, 19599, 19601, 19789
Fine seeds, 760
Finland, 997, 11493, 13366, 19329
Fire, 3417, 3831, 5955, 6162, 6174, 14005, 16293, 18058, 19532
Flooding, 3068, 13212, 13616
Flotation, 19231
Flower seeds, 666, 708, 750, 760, 761, 831, 833, 955, 1506, 1793, 1799, 1800, 1887, 1907, 1997, 4015, 4147, 4277, 4660, 4662, 4779, 4782, 6570, 7254, 7499, 7955, 8205, 8421, 8565, 8566, 8569, 8576, 8578, 9347, 10455, 12069, 12885, 13129, 13802, 13923, 13926, 14057, 14584, 15607, 16460, 16649, 18099, 18561, 18948, 19814, 19965
Fluorescence, 1640, 1654, 1657, 1841, 3066, 3475, 3913, 4013, 4273, 4829, 4851, 5855, 5856, 5860, 6382, 6595, 7146, 8107, 8139, 8266, 8598, 9214, 10084, 11479, 11842, 12275, 12344, 12376, 12439, 13034, 13370, 13920, 14406, 14493, 14650, 15133, 15369,

Fluorescence–(*cont.*)
15429, 16341, 16926, 16964, 17188, 18075, 19721, 19837, 19839
Food, 168, 432, 973, 2285, 7278, 12156, 14415, 17082
Forage crops, 2147, 3940, 6339, 6344, 6346, 6408, 6832, 8135, 8810, 8890, 10172, 10385, 11314, 11357, 11463, 12027, 13206, 13212, 13481, 13997, 14053, 14054, 14265, 19272, 19350
Forest seeds, 567, 747, 850, 1164, 1238, 1240, 1252, 1377, 1528, 1529, 1774, 1782, 1785, 1787, 2079, 2089, 2132, 2323, 2686, 2784, 3792, 3793, 3803, 3975, 5206, 5250, 5385, 5450, 6174, 6216, 6541, 6580, 7596, 7695, 7854, 7871, 7887, 7957, 8007, 8301, 8441, 8782, 8967, 8971, 8972–74, 9253, 9259, 9490, 9494, 9660, 9662, 9663, 9785, 9961, 10451, 10704, 10950, 11012, 11013, 11111, 11157, 11437, 11560, 11636, 11783, 11818, 11819, 11906, 12525, 12611, 12625, 12674, 12675, 12718, 13731, 13732, 14243, 14328, 14405, 14517, 14816, 14878, 15028, 15069, 15139, 15433, 15636, 16336, 16363, 16368–71, 16504, 16684, 16759, 16761, 16797, 16798, 16920, 16971, 17061, 17062, 17323, 17336, 17337, 17340, 17716, 18481, 18545, 18590, 18726, 18829, 19133, 19136, 19293, 19974
Formosan seeds, 10120, 11457, 11556
Forms of seeds, 1601, 5522, 6318, 11251, 11660, 13187, 16643
Fossils, 622, 623, 1388–92, 1892, 1916, 2351, 2355, 2866, 3783, 3785, 3994, 4910, 4911, 5105, 5478, 7467, 7473, 8041, 8255, 8943, 9082, 10689–91, 11001–4, 12547, 13252, 13447, 14278, 14294, 14548, 14695, 16055, 17137, 18179, 18298, 18415, 18416, 19608, 20111
Foundation seed stock, 6671, 6780
France, 13088
Freezing effects, 6, 8, 972, 1041, 2286, 3238, 3613, 3817, 3925, 4809, 4889, 5058, 6394, 6692, 7745, 8135, 8405, 8595, 9053, 9339, 10679, 16154, 16422, 16435, 16436, 16659, 17127, 18887, 19329, 19382, 19510, 19605
Fruits, 728, 758, 1180, 1723, 1952–54, 2324, 2630, 2775, 2841, 3200, 3298, 4860, 4866, 6994, 7872, 8397–99, 8753–55, 8757, 8784, 8963, 10301, 10338, 10445, 10523, 10955, 10956, 11181, 11245, 11247, 12238, 12481, 12482, 12952, 13039, 13647, 13648, 13746, 13861, 14299, 14335, 14415, 14522, 14740, 14762, 15097, 15314, 15906, 16036, 16557, 16647, 16874, 17133, 17609, 17739, 17846, 18242, 18462, 18760, 18821, 18880, 18882, 18918, 18919, 19011, 19234, 19282, 20037, 20038
Fumigation effects, 447, 538, 856, 1018, 2538, 7571, 7604, 8343, 10463, 12474, 18091–96, 19631
Fungicides, 365, 427, 452, 458, 604, 664, 680, 690, 692, 775, 783, 971, 1137, 1219, 1609, 1652, 2041, 2290, 2369, 2532, 2537, 2598, 2648, 2931, 2980, 2981, 3029, 3349, 3350, 3407, 3623, 3822, 3970, 4130, 4167, 4183, 4622, 4977, 5046, 5189, 5209, 5283, 5325, 5360, 5802, 5812, 6050, 6405, 6408, 6787, 6871, 7180, 8863, 8884, 9251, 9252, 9284, 9344, 9624, 9625, 9628, 9758, 9860, 9893, 9894, 9946, 9956, 10036, 10158, 10791, 12032, 13839, 16692

Garden seeds, 346, 443, 820, 911, 2002, 2267, 2976, 3604, 3951, 4015, 6196, 6761, 6872, 7178, 8260, 8823, 11771, 15419, 18185
Gases, 99, 100, 489, 616, 843, 1281, 1333, 1638, 2001, 2017, 2020, 2046, 2074, 2075, 2183, 2185, 2186, 2713, 3161, 4316, 4458, 4863, 5730, 5773, 6293, 7028, 7300, 7353, 7451, 8343, 8344, 8446, 8919, 8921, 8955, 10169, 10723, 11902, 12142, 12241, 12965, 13552, 13662, 14319, 14483, 14788, 14820, 15095, 15281, 15723, 16245, 17469, 17609, 17851, 18643, 18731, 18738, 18916, 19266, 19866, 20078
Genetics, 12, 68, 100, 105, 311, 369, 472, 531, 542, 561, 756, 1294, 1525, 1548, 1614, 1737, 1765, 2262, 2402, 2447, 2471, 2765, 2849, 3025, 3050, 3057, 3061, 3062, 3065, 3074, 3079, 3195, 3201, 3237, 3271, 3419, 3447, 3513, 3543, 3546, 3576, 3577, 3616, 3625, 3666, 3741, 3794, 4085, 4101, 4253, 4323, 4680, 4714, 4729–31, 4769, 4815, 4835, 4949, 5037, 5055, 5399, 5463, 5509, 5765, 5990, 6035, 6189, 6277, 6278, 6383, 6384, 6672, 6699, 6712–19, 6725, 6780, 6840, 7365, 7465, 7518, 7565, 7719, 7841, 7889, 7938, 7939, 7963, 7964, 7974, 7983, 8118, 8190, 8213, 8216–18, 8291, 8389, 8507, 8616, 8731, 8745, 8984, 8992–94, 9088, 9162, 9180, 9242, 9392, 9433, 9587, 9681, 9683, 9793, 9828, 9840, 9864, 9873, 9939, 9972, 10047, 10056, 10127, 10143, 10171, 10234, 10275, 10277, 10318, 10343, 10417, 10424, 10537, 10581, 10587, 10847, 10861, 10883, 10995, 11322, 11416, 11431, 11503, 11594, 11601, 11610, 11621, 11640, 11728, 11730–32, 11737, 11739, 11740, 11825, 11866, 11930, 11931, 12075, 12077, 12240, 12345, 12419, 12565, 12727, 12741, 12799, 12800, 12834, 12890, 13181, 13310, 13345, 13423, 13440, 13549, 13813, 13835, 13951, 14116, 14162–64, 14188–90, 14201, 14252, 14254, 14501, 14518, 14521, 14618, 14619, 14639, 14668, 14671, 14761, 14810, 14895, 14909,

Genetics–(*cont.*)
15171, 15354, 15409, 15467, 15508, 15539, 15595, 15597, 15652, 15896, 16147, 16184, 16199, 16392, 16409, 16585, 17114–19, 17165, 17225, 17228, 17236, 17280, 17305, 17350–55, 17507, 17511, 17516, 17621, 17641, 17643, 17644, 17692, 17795, 17809, 17812, 17813, 17840, 17997, 18090, 18111–13, 18115, 18160, 18170, 18212, 18226, 18228, 18233, 18281, 18287, 18412, 18513, 18556, 18722, 18833, 18903, 19286, 19287, 19318, 19394, 19487, 19586, 19592, 19613, 19704, 19852, 19857, 19982, 20016, 20091, 20132, 20133

Germany, 10205, 10689–91, 10950, 14806, 15458

Germination, 20, 30, 53, 80, 85, 91, 96, 97, 98, 108, 114, 123, 124, 127, 131, 132, 134, 150, 162–65, 171, 174, 176, 181, 184, 197, 202, 204, 211, 215, 217, 218, 220–28, 232, 233, 243, 252, 253, 257, 259, 276, 277, 286, 292, 295, 304–7, 347, 350, 358, 362, 370, 372, 383, 385, 386, 391, 393, 407, 412–14, 417, 424, 426, 446, 448, 449, 452, 460, 466, 467, 470, 471, 481, 488–95, 500, 502, 504–7, 510–13, 519, 529, 530, 555, 556, 570, 602, 603, 609, 610, 619, 620, 631, 634, 638, 648, 650, 653, 655, 658, 666, 667, 671, 674–77, 682, 685, 689, 691, 693–96, 700, 708, 717–19, 722–24, 729, 732, 734, 743, 747, 757, 759, 760, 769, 771, 775, 780, 785, 793, 798, 804–6, 816, 820, 833, 835, 842, 847, 848, 855, 857–59, 889, 897, 905, 914, 924, 931, 934, 939, 943, 947, 948, 965, 977, 990, 1008, 1045, 1050, 1063, 1071, 1096, 1100, 1101, 1114, 1123, 1127, 1145, 1151, 1156, 1173, 1177, 1181, 1189, 1192, 1194, 1234, 1235, 1244, 1275, 1276, 1310, 1314, 1315, 1321, 1322, 1330, 1336, 1346, 1348, 1365, 1376, 1405, 1417, 1421, 1444, 1445, 1468, 1484, 1500, 1506, 1522, 1538, 1600, 1601, 1618, 1629, 1637, 1649, 1681, 1683, 1686, 1695, 1702, 1715, 1720, 1736, 1741, 1750, 1762, 1771, 1776–78, 1793, 1794, 1798, 1799–804, 1806, 1816–18, 1836, 1844, 1846, 1848, 1851, 1862, 1863, 1868, 1869, 1875, 1884, 1887, 1891, 1893, 1917, 1921–23, 1936, 1938, 1939, 1942, 1944, 1947, 1948, 1950, 1957, 1958, 1969, 1975, 1978, 1980, 1981, 1984, 1985, 1989, 1990–93, 1995, 1998, 2000, 2009, 2010, 2012, 2018, 2019, 2025, 2026, 2037, 2044, 2049, 2058, 2067, 2069, 2071, 2077, 2081, 2091, 2096, 2104, 2112, 2119, 2125–27, 2141, 2143, 2154, 2164, 2173, 2176, 2178, 2181, 2182, 2187, 2201, 2210, 2213, 2247, 2257, 2268, 2270, 2274, 2287, 2296, 2297, 2326, 2338–41, 2344, 2352, 2353, 2380, 2383, 2385, 2387, 2409, 2423,

Germination–(*cont.*)
2427, 2428, 2430, 2433, 2434, 2436, 2446, 2450, 2452, 2457, 2469, 2473, 2474, 2476, 2483, 2484, 2489, 2494, 2495, 2497–99, 2527, 2528, 2530, 2533, 2545, 2553, 2569, 2576, 2581, 2596, 2602, 2609, 2624, 2626, 2637–39, 2650, 2662, 2675, 2677, 2678, 2687, 2690, 2711, 2712, 2716, 2725, 2735, 2739–41, 2744, 2749, 2750, 2755, 2758, 2762, 2763, 2774, 2784, 2789, 2790, 2792, 2793, 2797, 2803, 2804, 2806, 2807, 2812–14, 2828, 2836, 2837, 2855, 2863, 2867, 2868, 2884, 2885, 2888, 2896, 2904, 2919, 2920, 2936, 2943, 2956–58, 2965, 2966, 2976, 2978, 2992, 3020, 3021, 3033, 3051, 3053, 3054, 3068, 3069, 3072, 3099, 3102, 3103, 3116, 3123, 3128, 3129, 3139, 3141, 3142, 3146, 3148, 3164, 3165, 3168, 3170, 3171, 3176, 3194, 3203, 3205, 3210–12, 3218, 3221, 3230, 3231, 3265, 3270, 3273, 3284, 3292, 3294, 3305, 3313, 3319, 3336, 3342, 3347, 3356, 3361, 3367, 3377, 3379, 3388, 3390, 3406, 3415, 3424, 3432, 3437, 3444, 3445, 3451–53-A, 3455, 3488, 3495, 3498, 3501, 3524, 3538, 3540, 3549, 3560, 3567, 3579, 3581, 3583, 3585–87, 3598, 3600, 3601, 3603, 3606–8, 3619, 3627, 3632, 3636, 3638, 3639, 3654, 3655, 3677, 3683–85, 3692, 3693, 3713, 3716, 3721, 3722, 3729–33, 3744, 3745, 3749–51, 3782, 3795, 3801, 3829, 3835, 3840, 3843, 3853, 3860, 3862, 3864, 3871, 3873, 3884, 3886, 3887, 3906, 3915, 3928, 3946, 3947, 3969, 3976, 3983, 4007, 4013, 4015, 4018, 4026, 4043, 4044, 4092, 4122, 4123, 4125, 4128, 4146, 4157, 4159, 4165, 4174, 4179, 4187–91, 4201, 4202, 4206, 4208, 4213, 4216, 4217, 4223, 4228, 4235, 4237, 4241, 4245, 4254, 4268, 4272, 4313, 4314, 4339–41, 4352, 4359, 4360, 4397, 4421, 4425–27, 4430, 4431, 4436, 4438, 4452, 4479, 4481, 4484, 4489, 4490, 4501, 4508, 4511, 4514, 4523, 4524, 4526, 4528, 4530, 4531, 4533, 4583, 4588, 4592, 4594, 4596, 4600, 4611, 4625, 4626, 4635, 4642, 4643, 4658, 4663, 4678, 4679, 4681, 4682, 4688, 4693, 4713, 4714, 4716, 4722, 4725, 4728, 4739, 4742, 4744, 4746, 4748, 4756, 4760, 4770, 4772, 4779–82, 4785, 4790, 4795, 4802, 4816, 4840, 4841, 4850, 4862, 4885, 4907, 4918, 4943, 4945, 4952, 4954, 4962, 4982, 4992, 4997, 4999, 5002, 5007, 5018, 5022, 5023, 5042, 5044, 5045, 5061, 5064, 5066, 5068, 5069, 5078, 5079, 5085, 5090, 5092, 5107, 5109, 5114, 5118, 5123–25, 5127, 5138, 5155, 5157, 5164, 5183, 5189, 5205, 5210, 5229, 5237, 5251, 5252, 5255, 5259, 5261, 5262, 5264, 5282, 5289, 5300, 5309, 5318, 5328, 5342,

Germination–(*cont.*)
5345, 5357, 5370, 5371, 5393, 5395, 5396, 5409, 5427, 5432, 5436, 5442, 5444, 5446, 5454, 5461, 5471, 5484, 5495, 5501, 5513, 5525–30, 5533, 5534, 5539, 5540, 5551, 5558, 5564–66, 5570, 5571, 5584, 5589, 5601–3, 5607, 5614, 5615, 5637, 5644, 5645, 5648, 5659, 5667, 5680, 5684, 5686–88, 5698, 5700, 5703, 5721, 5723, 5730, 5756, 5764, 5767, 5779, 5780, 5781, 5787–89, 5794, 5805, 5808, 5814, 5824, 5836, 5845, 5850, 5858, 5874–76, 5885, 5902, 5911, 5920, 5939, 5943, 5957, 5972–75, 5979, 5981, 5989, 6010, 6021, 6034, 6065, 6080, 6109, 6119, 6128, 6133, 6134, 6152, 6158, 6169, 6174–76, 6194–96, 6222, 6238, 6252, 6254, 6259, 6298, 6310, 6313, 6315, 6318, 6319, 6329, 6335, 6336, 6341, 6355, 6387, 6390–93, 6404, 6415, 6416, 6427, 6428, 6440, 6450, 6457, 6470, 6478, 6489, 6491, 6494, 6495, 6497, 6501, 6517, 6540, 6545, 6546, 6548, 6551, 6557, 6568, 6570, 6582, 6585, 6592, 6594, 6602, 6603, 6608, 6610–13, 6618, 6637, 6638, 6640, 6642, 6673, 6681, 6682, 6690, 6702, 6709, 6710, 6711, 6720, 6753, 6755, 6759, 6761, 6764, 6765, 6772, 6773, 6784, 6804, 6812, 6815, 6816, 6822, 6837, 6844, 6868, 6872, 6888, 6892, 6896, 6927, 6928, 6950-H–56, 6970, 6974, 6978, 6980, 6981, 6983, 6990, 6993, 6994, 7001, 7011–17, 7019, 7035, 7039, 7049, 7074–76, 7078, 7085, 7089, 7091, 7092, 7105, 7111, 7113, 7121, 7129, 7156, 7158, 7159, 7184, 7189, 7207, 7216, 7218, 7220–22, 7226, 7229, 7233, 7234, 7241, 7272, 7296, 7316, 7354, 7423, 7428, 7437, 7438, 7441, 7442, 7458, 7462, 7504, 7516, 7544, 7576, 7591, 7597, 7599, 7614, 7615, 7619, 7620, 7627, 7637, 7656, 7670, 7672, 7673, 7681, 7689, 7709, 7721–26, 7740, 7746, 7769, 7819, 7820, 7843, 7872, 7912–14, 7924, 7925, 7932, 7955, 7966, 7971, 7989, 8009, 8035, 8038, 8042, 8044, 8048, 8059, 8065, 8068, 8075, 8085, 8088, 8096, 8097, 8105, 8117, 8127, 8128, 8147, 8163, 8169, 8172, 8203, 8254, 8268, 8270, 8271, 8281, 8286, 8457, 8463, 8464, 8475, 8476, 8482, 8491, 8503, 8518, 8530, 8531, 8533, 8544, 8545, 8552, 8556, 8561, 8564, 8566–69, 8573, 8575, 8580, 8586, 8587, 8606, 8610, 8615, 8633, 8634, 8637, 8646, 8647, 8649, 8692, 8696, 8708, 8710, 8712, 8714–16, 8721, 8728, 8729, 8737, 8738, 8752, 8782, 8585, 8786, 8789, 8791, 8792, 8823, 8824, 8835, 8855, 8864, 8867, 8882, 8885, 8922, 8935, 8948, 8950, 8955, 8957, 8984, 9003, 9004, 9055, 9076, 9077, 9083, 9101–3, 9107, 9137, 9138, 9140, 9160, 9177, 9232–34, 9264–66

Germination–(*cont.*)
9293, 9362, 9387, 9402, 9407–9, 9414, 9463, 9477, 9501, 9505, 9522, 9534, 9544, 9571, 9573, 9574, 9599, 9606, 9629, 9640, 9641, 9648, 9654, 9676, 9693, 9729, 9731, 9732, 9747, 9769, 9791, 9816, 9854, 9866, 9868, 9882, 9892, 9895, 9900, 9909, 9928, 9977, 9984, 9994, 9995, 9997, 10042, 10049, 10050, 10062, 10078, 10082, 10095, 10100, 10102, 10182, 10193, 10226, 10243, 10285, 10286, 10298, 10304, 10311, 10317, 10346, 10361, 10395, 10401, 10405, 10425, 10427, 10433, 10440, 10452, 10455, 10476, 10481, 10490, 10497, 10501, 10524, 10552, 10553, 10567, 10613, 10615, 10631, 10634, 10647, 10656, 10662, 10663, 10668, 10678, 10680, 10681, 10683, 10711, 10712, 10732, 10750, 10769, 10784, 10785, 10820, 10822, 10823, 1901–12, 10928, 10930, 10989, 10992, 11013, 11024, 11026, 11027, 11030–36, 11039, 11073, 11081, 11109, 11166, 11185, 11237, 11240, 11244, 11253, 11281, 11287, 11295, 11346, 11364, 11383, 11409, 11425, 11478, 11485, 11510, 11522, 11533, 11538, 11541, 11579, 11581, 11582, 11614, 11625, 11632, 11644, 11655, 11734, 11741, 11752, 11810, 11813, 11836, 11838, 11871, 11872, 11874, 11909, 11910, 11990, 12003, 12008, 12024, 12051, 12057–60, 12063, 12082, 12083, 12112, 12119, 12120, 12147, 12150, 12154, 12157, 12161, 12163–65, 12167, 12172, 12209, 12217, 12219, 12220, 12221, 12223, 12233, 12237, 12260, 12298, 12307, 12313, 12316, 12353, 12355, 12356, 12366, 12367, 12386–88, 12394, 12401, 12404, 12422, 12425, 12428, 12432, 12433, 12453, 12458, 12492, 12493, 12504, 12505, 12534, 12542, 14548, 12563, 12574, 12578–80, 12614, 12625, 12685, 12708, 12709, 12728, 12729, 12762, 12770, 12771, 12782, 12823, 12825, 12831–33, 12864, 12869, 12885, 12915, 12930, 12936, 12966, 12978, 12989, 13001, 13022, 13025, 13029, 13036, 13082, 12090, 13098, 13110, 13113, 13129, 13137, 13146, 13156, 13158, 13167, 13183, 13190, 13201, 13214, 13258, 13259, 13272, 13315, 13347, 13363, 13380, 13399, 13403, 13460, 13476, 13482, 13489, 13501, 13530, 13539, 13540, 13558, 13559, 13564, 13585, 13603, 13628–30, 13640, 13647, 13656, 13665, 13682, 13685, 13688, 13745, 13759, 13764, 13769, 13771, 13773, 13777–80, 13802, 13816, 13818, 13849, 13854–56, 13863, 13864, 13866, 13868, 13870–72, 13916, 13923, 13928, 13930, 13952, 13964, 13967, 13981, 13984, 13993, 14014, 14016, 14023, 14025, 14034–36, 14063, 14064, 14088, 14097, 14131, 14133, 14145, 14148, 14149, 14151, 14154, 14226,

Germination–(*cont.*)
14241, 14243, 14250, 14258, 14270, 14279, 14305, 14309, 14358, 14369, 14373, 14374, 14390, 14393, 14397, 14408, 14409, 14414, 14456, 14467, 14474, 14536, 14553, 14556, 14588, 14589, 14591, 14592, 14605, 14642–45, 14652, 14660, 14691, 14694, 14711–13, 14715, 14724, 14730, 14752, 14755, 14756, 14789, 14790, 14800, 14801, 14808, 14810, 14811, 14818, 14827, 14850, 14855, 14889, 14933, 14955, 14958, 15000, 15003, 15004, 15037, 15045, 15067, 15086, 15100, 15103, 15125, 15137, 15138, 15142, 15148, 15149, 15163, 15176, 15178, 15186, 15198, 15222, 15233, 15234, 15236, 15255, 15257, 15265, 15268, 15283, 15284, 15286, 15297, 15299, 15304, 15306, 15323, 15328, 15340, 15360, 15362, 15366, 15374, 15410, 15418, 15427, 15429, 15438, 15450, 15452, 15453, 15462, 15465, 15468, 15474, 15497, 15502–4, 15508, 15537, 15538, 15552, 15556, 15560, 15572, 15580, 15582–85, 15596, 15627, 15632, 15641, 15643, 15671, 15672, 15681, 15697, 15698, 15725, 15733, 15749, 15751, 15764, 15766, 15803, 15804, 15806, 15832, 15833, 15837, 15847, 15848, 15862, 15882, 15897, 15902, 15913, 15934, 15953, 15960, 16014, 16017, 16030, 16043, 16089, 16100, 16145, 16148, 16155, 16157, 16163–65, 16174, 16180, 16218, 16219, 16233, 16249, 16253, 16261, 16262, 16277, 16289, 16293, 16296, 16297, 16300, 16301, 16303–8, 16310, 16312, 16324, 16346, 16347, 16349, 16356, 16361, 16364, 16365, 16367, 16385, 16388, 16399, 16404, 16445, 16446, 16451, 16454, 16455, 16457, 16460, 16476, 16483, 16485, 16489, 16506, 16515, 16524, 16539, 16548, 16624, 16641, 16712, 16734, 16805, 16815, 16822, 16846, 16847, 16850, 16856, 16863, 16881, 16884, 16886, 16888, 16889, 16951–53, 16957, 16962, 16990, 16997, 17028, 17029, 17066, 17074, 17077, 17111, 17115–19, 17123, 17126, 17129, 17144, 17148, 17178, 17180, 17193–95, 17197, 17211, 17234, 17235, 17244, 17246, 17254, 17270, 17276, 17288, 17291, 17292, 17304, 17307, 17326, 17331, 17335, 17345, 17357, 17360, 17363, 17364, 17369, 17376, 17383, 17394, 17395, 17403, 17411–13, 17418, 17420, 17426, 17452–54, 17456, 17457, 17470, 17473, 17508, 17510, 17529, 17532, 17538, 17554, 17555, 17558, 17563, 17564, 17566, 17567, 17569, 17582, 17583, 17586, 17593, 17618, 17619, 17625–27, 17633, 17648, 17651, 17670, 17671, 17677, 17682, 17685, 17689, 17694, 17695, 17702, 17714, 17721–24, 17729, 17730, 17734, 17741, 17747, 17754, 17767, 17775, 17779, 17781, 17803, 17806–

Germination–(*cont.*)
8, 17821, 17822, 17824, 17834–36, 17839, 17842, 17862, 17868, 17883, 17891, 17924, 17925, 17927, 17929, 17930, 17935, 17945, 17975, 18012, 18013, 18046, 18060, 18062, 18065, 18107, 18193, 18199, 18222, 18278, 18279, 18315, 18316, 18318, 18352–54, 18397, 18403, 18468, 18505, 18514, 18526, 18537, 18567, 18584, 18608, 18612, 18616, 18618, 18624, 18625, 18635, 18643, 18649, 18651, 18655, 18663, 18669, 18678, 18687, 18698, 18699, 18714, 18733, 18739, 18744, 18756, 18770, 18771, 18772, 18809, 18842, 18851, 18878, 18939, 18942, 18947, 18957, 18980, 18991, 19001, 19002, 19009, 19026, 19046, 19082, 19115, 19151, 19167, 19225, 19285, 19376, 19395, 19723, 19808, 19824, 19995, 20024, 20128

Germination effect on chemical composition, 8467, 11308, 11848, 11921, 11926, 11992–94, 11958, 12046, 12520, 12528, 12746–48, 12875, 12932, 13004–9, 13016, 14009, 14114, 14141, 14351, 14357, 14527, 14574, 14620, 16320, 16340, 16353, 16472, 16478, 16498, 16522, 16592, 16767, 16771, 16775, 16782, 16816, 16855, 16857, 16936, 16945, 16991

Germination effect on metabolism, 18087, 18151, 18181, 18178, 18182, 18247, 18251, 18260, 18262, 18263, 18323, 18357, 18364, 18366, 18381, 18441, 18452, 18453, 18475, 18859, 18877, 19016, 19027, 19030, 19087, 19140, 19148, 19156, 19157, 19211, 19221, 19279, 19330, 19441, 19488, 19517, 19535, 19542, 19569, 19612, 19652, 19938, 19939, 19941, 20002, 20137

Germination-factors affecting, 1600, 7294, 7305, 7380, 7548, 7550, 7754, 7910, 8152, 8208, 8281, 8402, 8569, 8570, 8574, 8628, 8658, 8688, 8723, 8846, 8918, 8939, 8963, 9055, 9093, 9114, 9168, 9291, 9292, 9460, 9480, 9485, 9489, 9490, 9514, 9596, 9636, 9798, 9799, 9867, 9927, 10031, 10237, 10327, 10328, 10348, 10349, 10363, 10414, 10670–72, 10674–76, 10679, 10681, 10682, 10756–58, 10818, 10841–43, 10929, 10955, 10956, 11025, 11062, 11168, 11184, 11186, 11488, 11511, 11517, 11519, 11581, 11687, 11792, 11896, 11960, 12106-A, 12110, 12114, 12125, 12160, 12177, 12204, 12205, 12210, 12385, 12401, 12422, 12430, 12435, 12438, 12443, 12503, 12539, 12540, 12545, 12614, 12678, 12739, 12792, 12796, 12974, 13054, 13057, 13061, 13072, 13103, 13223, 13290, 13383, 13398, 13660, 13697, 13750–52, 13756, 13805, 13815, 13895, 13902, 13903, 13908, 13910, 13968, 14146, 14159, 14345, 14393, 14403, 14411–13, 14675, 14682, 16315,

Germination-factors affecting–(*cont.*) 16355, 16358, 16419, 16456, 16459, 16528, 16597, 16620, 16625, 16643, 16646, 16658, 16867, 16871, 16894, 17008, 17010, 17020, 18091–96, 18214, 18305, 18308, 18359, 18375, 18437, 18444, 18457, 18458, 18482, 18805, 18852, 18888, 18899, 18902, 18909, 18916, 18966, 18967, 18998, 19020, 19039, 19049, 19098, 19178, 19180, 19182, 19196, 19199, 19232, 19234–36, 19243, 19275, 19302, 19421, 19646, 19710, 19730, 19731, 19793, 19878, 19910, 19936, 20057, 20063, 20120

Gibberellic acid, 68, 103, 252, 255, 364, 510, 1204, 1575, 1729, 1808, 2027, 2030, 2039, 2041, 2072, 2210, 2211, 2325, 2631, 2684, 3038, 3291, 3292, 3307, 3308, 3314–16, 3404, 3466, 3519, 3528, 3682, 3774, 3986, 4031, 4032, 4209, 4234, 4398, 4405, 4416, 4417, 4760, 4761, 5350, 5537, 5683, 6007, 6122, 6241, 6350, 6414, 6417, 6455, 6471, 6492, 6519, 6520, 6659, 6660, 6661, 6818, 6857, 6972, 7006, 7007, 7240, 7440, 7727, 7931, 7934, 8085, 8228, 8296, 8311, 8312, 8315, 8318–20, 8569, 9127, 9134, 9243, 9252, 9371–74, 9600, 9774, 9940, 10130, 10131, 10308, 10324, 10345, 10686, 10765, 11255, 11332, 11407, 11408, 11570, 11635, 11693, 11760, 11767, 11877, 12036, 12106-A, 12402, 12426, 12427, 12620, 12657, 12670, 12673, 12677, 13147, 13354, 13507, 13526, 13558, 13561, 13724, 13762, 13770, 13943, 13946, 13944, 14071, 14072, 14138, 14153, 14260, 14261, 14299, 14384, 14475, 14479, 14595, 14686, 14687, 14693, 14716, 14981–85, 15136, 15141, 15242, 15245, 15272, 15306, 15319, 15419, 15496, 15911, 15954, 15969, 15970, 15971, 16018, 16119, 16122, 16183, 16387, 16486, 16487, 16503, 16602, 16603, 16625, 16713, 16714, 16797, 16938, 17004, 17126, 17453, 17562, 17564, 17566, 17688, 17716, 17798, 18005, 18147, 18268, 18293, 18294, 18376, 18603, 18761, 18852, 19015, 19042, 19043, 19182, 19252, 19431–34, 19458, 19646, 19673, 19790, 19926, 19989

Globulin, 520, 3487, 3632, 4692, 4867, 4877, 9794, 12100

Glutathione, 281, 6390, 7374, 9029

Gossypol, 8053, 9112, 12380, 12559, 13233

Grafting, 4788, 4859, 5281, 6217, 12349, 17356, 18753

Grain, 106, 113, 125, 169, 282, 291, 403, 447, 604, 606, 753, 823, 930, 949, 1098, 1107, 1108, 1115, 1139, 1361, 1403, 1410, 1501, 1502, 1504, 1638, 1665, 1689–92, 1717, 1718, 1837, 2063, 2110, 2111, 2186, 2188–91, 2193–97, 2334, 2335, 2359, 2394, 2485, 2756, 2815, 2876, 2936, 3048, 3060, 3089,

Grain–(*cont.*) 3091, 3101, 3130, 3192, 3257, 3359, 3361, 3363, 3394, 3408, 3435, 3442, 3502, 3858, 3909, 4019, 4056, 4057, 4059, 4060, 4095, 4096, 4315, 4317, 4346, 4362, 4444, 4445, 4457, 4458, 4581, 4619, 4622, 4639, 4656, 4657, 4676, 4736, 4805, 4807, 4853, 4859, 4932, 4956, 5088, 5179, 5667, 5668, 5861, 6073, 6418, 6453, 6494, 6590, 6605, 6645, 6646, 6657, 6779, 6803, 6870, 6871, 6874–77, 6896, 7076, 7068, 7091, 7106, 7182, 7184, 7318, 7466, 7493, 7574, 7577, 7582, 7583, 7613, 7715, 7777, 7830, 7973, 8049, 8107, 8133, 8234, 8379, 8491, 8502, 8544, 8682, 8958, 9013, 9092, 9123, 9166, 9190, 9201, 9238, 9246, 9314, 9316, 9847, 9931, 9935, 10408, 10415, 10443, 10485, 10692, 10705, 10791, 10889, 10927, 10987, 11203, 11259, 11274, 11351, 11369, 11386, 11390, 11401, 11459, 11465, 11471–73, 11590, 11673, 11674, 11883, 11950, 11951, 12081, 12093, 12115, 12154, 12252, 12262, 12309, 12410, 12462, 12473, 12529, 12531, 12544, 12562, 12575, 12591, 12604, 12671, 12719, 12862, 13010, 13288, 13463, 13508, 13533, 13592, 13650, 13691, 13735, 13737, 13817, 13836, 13888, 13894, 14295, 14480, 14558, 14565, 14566, 14570, 14576, 14611, 14612, 14769, 14849, 14913, 14914, 14920, 15138, 15205, 15239, 15333, 15350, 15408, 15552, 15553, 15675, 15693, 15743, 15819, 15820, 15947, 15993, 16034, 16097, 16098, 16461, 16894, 16939, 16954, 16955, 16976, 17231, 17559, 17580, 17588, 17596, 18045, 18050, 18097, 18098, 18139, 18162, 18320, 18322, 18323, 18417, 18479, 18589, 18731, 19092, 19153, 19192, 19299, 19488, 19594, 19648, 19724, 19851,

Grasses, 44, 175, 176, 193, 197, 217, 221, 239, 266, 316, 343, 368, 518, 1021, 1132, 1134, 1163, 1313, 1318, 1320, 1325, 1469, 1760, 1807, 1935, 2209, 2210, 2211, 2251, 2255, 2290, 2384, 2448, 2581, 2609–11, 2692, 2739, 2782, 2975, 2976, 3010, 3138, 3181, 3186, 3209, 3212, 3221, 3329, 3346, 3357, 3413, 3414, 3416–18, 3420, 3589, 3591, 3879, 3937, 3950, 3957, 3958, 3967, 3970, 3973, 4247, 4257, 4272, 4293, 4390, 4411, 4452, 4481, 4580, 4639, 4658, 4677, 4728, 4862, 4996, 5046, 5307, 5313, 5455, 5472, 5565, 5624, 5934, 6082, 6100, 6283, 6339, 6343, 6653, 6720, 6762, 6763, 6835, 6946, 7008, 7011, 7012, 7062, 7086, 7102, 7307, 7580, 7667, 7693, 7697, 7953, 7955, 8024, 8038, 8058, 8074, 8139, 8179, 8180, 8237, 8285, 8407, 8477, 8515, 8684, 8722, 8806, 8894, 9080, 9194, 9196, 9197, 9222, 9404, 9863, 10049, 10073, 10430, 10472, 10730,

Grasses–(*cont.*)
10821, 10866, 10874, 10892, 10899, 10958, 11023, 11068, 11219, 11401, 11507, 11527, 11575, 11660, 11661, 11851–54, 11856, 12046, 12551, 12894, 13109, 13179, 13213, 13465, 13710, 13755, 13893, 13901, 13969, 14018, 14055, 14149, 14218, 14405, 14426, 14427, 14534, 14906, 14922, 14999, 15021, 15238, 15353, 15364, 15462, 15602, 15607, 15686, 15689, 15876, 15887, 15990, 15999, 16060, 16104, 16443, 16462, 16481, 16960, 17122, 17958, 18032, 18036, 18050, 18077, 18078, 18107, 18391, 18648, 18689, 18692, 18858, 18952, 19186, 19386, 19440, 19582, 19705, 19707, 19758, 20076, 20079

Gravity, 3310, 3908, 8601, 8745, 11265, 13325, 19288, 20086

Greenhouse, 7693, 8042, 10090, 14632, 17930, 19379, 19789

Growth, 111, 271, 272, 276, 398, 452

Growth regulants, 37, 165, 462, 464, 465, 1452, 1513, 1539–43, 1545, 1729, 1999, 2000, 2125, 2142, 2156, 2157, 2199, 2201, 2202, 2288, 2292, 2294, 2310, 2311, 2374, 2406, 2410, 2575, 2583, 2630, 2642, 2644, 2660, 2708, 2813, 2852–54, 2915, 2932, 2939, 2961, 3075, 3076, 3080, 3198, 3304, 3310, 3311, 3441, 3494, 3503, 3664, 3751, 3753, 3754, 3771–73, 3777, 4017–19, 4025, 4044, 4126, 4149, 4162, 4169, 4520, 4577, 4659, 4674, 4698, 4755, 4814, 4825, 4879, 4919, 4954, 4997, 5099, 5109, 5144, 5158, 5198, 5236, 5237, 5308, 5348, 5394, 5400, 5441, 5465, 5548, 5570, 5655, 5683, 5685, 5728, 5729, 5731, 5938, 5987, 6015, 6117, 6135, 6240, 6286, 6354, 6488, 6662, 6708, 6744, 6782, 6783, 6827, 6828, 6859, 6860, 6861, 6868, 6939, 6942, 6943, 6963, 6992, 7084, 7214, 7303, 7387, 7470, 7545–47, 7549, 7553, 7554, 7651, 7657, 7739, 7853, 7858, 7916, 7920, 7936, 7947, 7949, 8087, 8132, 8148, 8158, 8376, 8377, 8607, 8608, 8838, 8866, 8964, 9037, 9108, 9151, 9170, 9186, 9211, 9291, 9330, 9388, 9416, 9500, 9692, 9744, 9745, 9760, 9764, 9776, 9887, 9976, 10046, 10232, 10338, 10378, 10379, 10409, 10422, 10468, 10499, 10583, 10631, 10754, 10933, 10958, 10998, 11254, 11257, 11361, 11413, 11451, 11452, 11572–74, 11651, 11692, 11719, 11745, 11762, 11767, 11814, 11815, 12000, 12101, 12174, 12176, 12204–8, 12210, 12293, 12327, 12354, 12399, 12414, 12422, 12434, 12485, 12486, 12488–91, 12538, 12619, 12858, 12893, 12894, 12921, 13193, 13257, 13337, 13376, 13405, 13428, 13560, 13577, 13593, 13596, 13604, 13652, 13704, 13860, 13899, 13941, 13942, 13945, 14073, 14095, 14121, 14215, 14218, 14345, 14363, 14396, 14552, 14675, 14814, 14884, 15403, 15404, 15484, 15494, 15499, 15503, 15504, 15515, 15516, 15608, 15793, 15829, 15849–51, 15873, 16120, 16143, 16202, 16206, 16241, 16315, 16331, 16532, 16608, 16730, 16806, 17265, 17266, 17326, 17452, 17535, 17536, 17999, 18011, 18073, 18138, 18224, 18228, 18310, 18379, 18438, 18439, 18465, 18466, 18482, 18781, 18908. 18986, 18992, 18997, 19108, 19166, 19211, 19234, 19243, 19372, 19430, 19469, 19557, 19558, 19578, 19929, 19946, 19964, 19994

Guatemala, 17831, 17848

Gymnosperms, 3313, 7823, 7824, 16984

Halophytes, 2442, 10001, 13693, 17021, 18888, 19283

Hard coats, 41, 92, 97, 127–30, 210, 217, 218, 220, 226, 228, 417, 456, 503, 529, 530, 564, 613, 650, 653, 712, 724, 741, 771, 813, 826, 861, 1011, 1045, 1074, 1186, 1432, 1476, 1922, 2212, 2284, 2327, 2328, 2446, 2449, 2613, 2723, 2967–69, 3120, 3219, 3372, 3414, 3424, 3443, 3491, 3679, 3715, 3717, 3827, 3837, 4000, 4002, 4003, 4075–78, 4669, 4671, 4831, 4993, 5030, 5137, 5157, 5312, 5362, 5365, 5392, 5477, 5653, 5784, 5811, 6051, 6053, 6054, 6056, 6160, 6360, 6465, 6478, 6881, 7021, 7052, 7074, 7187, 7218, 7295, 7330, 7331, 7334, 7593, 7602, 7871, 7971, 8078, 8079, 8202, 8219, 8265, 8521, 8551, 8568, 8593, 8750, 8814, 8816, 8826, 8835, 8898, 8940, 8955, 9039, 9059, 9180, 9249, 9261, 9286, 9481, 9482, 9564, 9607, 9629, 9675, 9686, 9892, 9952, 9977, 10090, 10332, 10333, 10380, 10381, 10383, 10384, 10386, 10542, 10544, 10571, 11035, 11087, 11091, 11220, 11350, 11476, 11621, 11850, 11925, 11930, 11931, 12009, 12068, 13284, 13410, 13411, 13927, 13932, 14016, 14036, 14132, 14165, 14300, 14303, 14304, 14306, 14308, 14320, 14356, 14535, 14537, 14538, 14634, 14708, 14903, 14928, 14994, 14995, 15099, 15119, 15197, 15472, 15575, 15607, 15682, 15810, 15813–15, 15853, 15917, 15997, 16235, 16236, 16300, 16405, 16412, 16449, 16809, 16810, 16811, 16982, 17179, 17258, 17631, 17893, 17953, 18082, 18116, 18142, 18233, 18409, 18569, 18663, 18795, 18869, 19041, 19090, 19216, 19226, 19276, 19305, 19465, 19466, 19573, 19598, 19706, 19715, 19735, 19775, 19776, 19779, 19781, 19783, 19847, 19862, 19892, 19927, 19967, 20091, 20113, 20114

Hardwoods, 672, 4488

Harvesting, 129, 383, 453, 1681, 1897, 2698, 3394, 3568, 4325, 4342, 4368, 4595, 5250,

Harvesting–(*cont.*)
5317, 5373, 5518, 5605, 5733, 5881, 6365, 7165, 8113, 8133, 8222, 8424, 8814, 8890, 9030, 9068, 9434, 9800, 9890, 10135, 10526, 11197, 11211, 11804, 11895, 12069, 12460, 13243, 13279, 13365, 13551, 13587, 13795, 13893, 13922, 14182, 14263, 14601, 14878, 15627, 15973, 15995, 16649, 17973, 18036, 18220, 18373, 18568, 19424, 19752, 19950
Haustoria, 4540, 7324, 7325, 12688, 15217, 15893, 17897, 18389, 19062, 19976,
Hawaii, 223, 224
Hemicellulose, 12647, 15675, 16329, 16723, 17161
Herb seeds, 880, 1261, 1291, 8572
Herbaceous plants, 7113
Herbage seeds, 13587, 14583
Herbarium material, 12188
Herbicides, 1884, 5347, 6097, 15792
Hilum, 11739, 14365
Histochemistry, 2223, 3739, 8726, 10737, 10924, 10925, 11481, 12329, 15345, 18088, 18303
Historical tree, 982
History of seed physiology, 18657
Holland, 15842, 19550
Hormones, 4, 5, 182, 370, 462–64, 516, 598, 614, 885, 1012, 1016, 1039, 1473, 1973, 3462
Horticultural seeds, 21, 7358, 8066, 8828, 10473, 14280, 15250, 18879, 19811
Hot water treatment, 132, 151, 650, 741, 813, 1059, 1074, 1146, 1165, 1213, 1870, 1871, 2163, 2533, 2629, 3424, 3674, 3708, 3717, 3909, 4155, 4185, 4308, 4882, 5272, 5623, 5679, 6487, 6801, 6884, 7602, 7843, 7898, 8705, 8955, 8991, 9163, 9169, 9201, 9991, 9993, 10183, 10448, 12297, 13734, 14399, 14415, 15992, 16564, 16862, 17667, 18368, 18966, 19735
House plants, 6950-H, 15632
Human serum effect, 15043
Hungary, 15624
Humidity, 278, 1613, 2003, 2476, 2478, 2485, 3680, 3977, 4281, 4310, 4811, 5315, 5778, 5823, 6121, 6639, 6641, 6946, 6948, 7064, 7295, 7830, 7938, 8219, 8465, 8667, 9558, 9875, 11878, 14634, 15587, 16250, 17736, 17788, 18682, 19354, 19572, 19622, 19738
Humus, 2919, 12384, 12397, 18567
Hybrids, 298, 558, 615, 1465, 4689, 4759, 11665
Hydrocyanic acid, 591, 2001, 3197, 7028, 7139, 10414, 14419, 20064
Hydrogen peroxide, 16, 43, 729, 730, 1084, 1247, 3654, 3947, 3951, 3956, 4234, 4235, 5332, 5950, 6166, 7085, 9535, 9816, 10719, 12113, 13430, 14754, 14936, 14937, 15071, 15338, 16973, 17171, 17411, 18749, 19314
Hygroscopic equilibrium, 19565, 19622
Hypocotyl, 2600, 3773, 6660, 10194, 10780, 11667, 11693, 15242

Iceland, 15224
Identification, 356, 523, 571, 574, 640, 1028, 1109, 1226, 1232, 1345, 1355, 1550, 1602, 1617, 1657, 1907, 2215, 2237, 2316, 2350, 2566, 2582, 2595, 2700, 2727, 2738, 2818, 2874, 3066, 3108, 3135, 3300, 3340, 3345, 3576, 3647, 3863, 3967, 4001, 4004–6, 4023, 4143, 4230, 4233, 4284, 5254, 5303, 5572, 5691, 5806, 5851–57, 5859, 5860, 5865, 6136, 6386, 6595, 6892, 7108, 7109, 7142, 7146, 7538, 7765, 7821, 8066, 8270, 8648, 8665, 8677, 8691, 8806, 8896, 9214, 9229, 9250, 9303, 10406, 10433, 10620, 10766, 10794, 10840, 10888, 11086, 11099, 11100–2, 11150, 11428, 11499, 11797, 11995, 11998, 12065, 12238, 12274, 12439, 12450, 12684, 12895, 13019, 13020, 13034, 13096, 13194, 13209, 13566, 13918, 13953, 13991, 13994, 13996, 14332, 14491, 14602, 14753, 14760, 14836, 15205, 15307, 15355, 15607, 15634, 15688, 15795, 16313, 16833, 16950, 16964, 17931, 17932, 18306, 18448, 18578, 18639, 18710, 18762, 18854, 18927, 19260, 19443, 19475, 19504, 19546, 19547, 19602, 19679, 19737, 19837, 19839, 19840
Imbibition and swelling, 280, 295, 360, 404, 503, 572, 596, 609, 612, 1436, 1440, 1555, 2454, 2642, 2573, 2688, 2877–79, 2954, 2972, 3019, 3086, 3088, 3360, 3464, 3716, 4003, 4076, 4310, 4796, 4797, 4808, 4809, 5302, 5305, 5318, 5380, 5461, 5520, 5644, 5672, 5761, 5763, 5889, 5964, 6125, 6213, 6214, 6263, 6342, 6345, 6359, 6421, 6738, 7078, 7166, 7235, 7431, 7557, 7928, 7942–44, 8145, 8313, 8448, 8491, 8501, 8514, 8516, 8755, 8843, 8856, 8898, 8921, 8988, 9261, 9483, 9484, 9759, 10198, 10331, 10720, 10861, 11329, 11335, 11477, 11661, 11782, 11901, 11935, 11976, 12051, 12190, 12649, 12714, 12732, 13207, 13623, 13876, 14132, 14459, 14460, 14537, 14603, 14707, 14892, 15347, 15348, 15518, 15586, 15587, 15673, 15674, 15700–2, 15877, 16021, 16215, 16502, 16894, 16968, 17164, 17289, 17332, 17382, 17499, 17591, 17613, 17731, 17755, 17758, 17827, 17828, 18082, 18296, 18645, 18756, 18757, 18795, 19090, 19091, 19101, 19167, 19197, 19214, 19395, 19636, 19638, 19875, 20095, 20098
Importance of seeds in agriculture, 4400, 5128, 15157
India, 907, 10203, 10204, 10250–54, 14233, 14256, 14257, 14969, 15407, 15654

Industry, 7, 28, 162, 201, 240–42, 291, 371, 395, 411, 423, 428, 453, 515, 565, 652, 657, 679, 702, 710, 731, 739, 745, 750, 766, 769, 772, 801, 819, 829, 845, 863, 869, 874, 875, 886, 887, 906, 935, 936, 938, 940, 945, 957, 959, 964, 980, 981, 983, 984, 1009, 1010, 1015, 1031, 1033, 1034, 1064, 1068, 1072, 1073, 1077, 1081, 1083, 1095, 1105, 1108, 1110, 1115, 1120, 1128, 1130, 1136, 1144, 1161, 1163, 1168, 1174, 1176, 1188, 1208, 1215, 1243, 1265, 1342, 1396, 1529, 1608, 1609, 1639, 1704, 1705, 1865, 1886, 1898, 1931, 1934, 1935, 2026, 2091, 2167, 2168, 2224, 2235, 2280, 2506, 2530, 2606, 2649, 2682, 2702, 2778, 2844, 2911, 2949, 3008, 3040, 3121, 3134, 3178, 3187, 3201, 3235, 3296, 3299, 3302, 3327, 3328, 3332, 3338, 3364, 3370, 3408, 3413, 3425, 3467, 3489, 3492, 3551, 3568, 3590, 3617, 3631, 3649, 3673, 3735, 3787, 3812, 3814, 3866, 3900, 3945, 3958, 3999, 4055, 4058, 4070–72, 4074, 4168, 4186, 4225, 4271, 4300, 4364, 4390, 4411, 4440, 4441, 4453, 4473, 4487, 4492, 4580, 4595, 4620, 4723, 4777, 4893, 4906, 4933, 4936, 4979, 4990, 5048, 5073, 5128, 5160, 5161, 5182, 5294, 5311, 5316, 5466, 5516, 5517, 5587, 5725, 5782, 5844, 5904, 5996, 6001, 6082, 6085, 6088–90, 6131, 6178, 6200, 6246, 6250, 6307, 6387, 6399, 6573, 6576, 6671, 6677, 6781, 6965, 6997, 6998, 7571, 7720, 7774, 7828, 8003, 8074, 8104, 8157, 8170, 8179, 8193, 8202, 8346, 8349, 8469–71, 8623, 8670, 8746, 8870, 8780, 8781, 8801, 8802, 8960, 8982, 9050, 9185, 9220, 9283, 9323, 9324, 9377, 9403, 9434, 9559, 9634, 9755, 9761, 9784, 9803, 9839, 9844, 9858, 9880, 9890, 9922, 9959, 9964, 9971, 10018, 10051, 10203, 10207, 10209, 10436, 10473, 10492, 10548, 10581, 10744, 10819, 10900, 10915, 10996, 11052, 11095, 11146, 11192, 11197, 11232, 11262, 11337, 11338, 11402, 11417, 11462, 11464, 11604, 11609, 11636, 11645, 11656, 11698, 11841, 11868, 11981, 12028, 12084, 12178, 12235, 12291, 12326, 12330, 12331, 12365, 12442, 12461, 12468, 12469, 12470, 12585, 12611, 12635, 12653, 12674, 12675, 12760, 12807, 12856, 12863, 12899, 12984, 13112, 13234, 13393, 13504, 13524, 13589, 13708, 13731, 13926, 14006, 14093, 14094, 14214, 14482, 14490, 14507, 14531, 14584, 14774, 14775, 14910, 14979, 15215, 15216, 15238, 15250, 15263, 15289, 15398, 15407, 15408, 15486, 15510, 15607, 15608, 15635, 15636, 15648, 15654, 15706, 15818, 15824, 15859, 15982, 15990, 16023, 16197, 16243, 16336, 16353, 16400, 16406, 16426, 16433, 16534, 16571, 16577, 16580, 16648, 16651, 16711, 16716,

Industry–(*cont.*)
16733, 16738, 16899, 17005, 17006, 17022, 17044, 17085, 17140, 17166, 17258, 17402, 17419, 17463, 17496, 17590, 17653, 17665, 17698, 17814, 18023, 18048, 18215, 18614, 18774, 18829, 18928, 19006, 19014, 19035, 19051, 19080, 19272, 19295, 19336, 19457, 19581, 19785, 19835, 19874, 19887, 20012

Infra-red, 1223, 3778, 8278, 12338

Inheritance, 421, 473, 1357, 1610, 2265, 11737, 12527, 12717, 12801, 12802, 14400, 16177

Inhibiting substances, 256, 263, 270, 272, 363, 370, 438, 449, 516, 602, 977, 1242, 1245, 1246, 1431, 1439, 1499, 1554, 1558–60, 1811, 1824, 1885, 2048, 2179, 2182, 2206, 2343, 2490, 2524, 2525, 2530, 2552, 2561, 2569, 2667, 2707, 2770–72, 2792, 2890, 2958, 3227, 3229, 3390, 3552, 3629, 3857, 4034, 4349, 4350, 4488–90, 4796, 4810, 4835, 4837, 5066, 5136, 5245, 5246, 5386, 5434, 5569, 5710, 5906, 5923, 6103, 6106, 6108, 6113, 6285, 6401, 6526, 6598, 6709, 6758, 6774–77, 6813, 6822, 6847, 6909, 6994, 7000, 7213, 7237–39, 7263, 7654, 7655, 7702, 7703, 7776, 7778, 8211, 8303, 8392, 8483, 8621, 8673, 8727, 8803, 8850, 8851, 9158, 9159, 9289, 9478, 9576, 9647, 9753, 9856, 10285, 10286, 10299, 10414, 10498, 10720, 10841–43, 10845, 10897, 10955, 10956, 10985, 11035, 11045, 11046, 11167, 11168, 11290, 11323, 11324, 11359, 11480, 11488, 11588, 11669, 11884, 11896, 11954, 12051, 12107, 12108, 12125, 12153, 12332, 12507, 12539, 12540, 12678, 12785–89, 12881, 12952, 12954, 12955, 13064, 13068, 13074–76, 13104, 13339, 13452, 13494, 13541, 13620, 13625, 13654, 13737, 13806, 13903, 13986, 14251, 14261, 14290, 14402, 14563, 14612, 14798, 14810, 14848, 14885, 14908, 14952, 14955, 15302, 15402, 15488, 15608, 15647, 16237, 16242, 16360, 16606, 16607, 16629, 16941, 17103, 17330, 17406, 17553, 18084, 18163, 18165, 18712, 18878, 18919, 18981, 19118, 19201, 19249, 19285, 19757, 19881, 19960, 19961, 20035

Inhibitors, 758, 2891, 2892, 4333, 6105, 6298, 7654, 7655, 11110, 11540, 12126, 12621, 12726, 12956, 13473, 13474, 13478, 13697, 13723, 13889, 14120, 14211, 14522, 14809, 15092, 15175, 15229, 15271, 15485, 15489, 15491–93, 15922, 15986, 16084, 16108, 16127, 16208, 16325, 16525, 17104, 17133, 17366, 17396, 17410, 17421, 17534, 17709, 17710, 17735, 17846, 18396, 18462, 18805, 18830, 18866, 18882, 19011, 19369, 19370, 19399, 19532, 19988

Injury effects, 3, 221, 392, 539, 564, 616, 653, 655, 675, 684, 932, 974, 1006, 1022, 1129, 1135, 1343, 1400, 1401, 1491, 1492, 1584, 1716, 1718, 2065, 2123, 2213, 2266, 2538, 2549, 2815, 2817, 2877, 2878, 2950, 2951, 3092, 3263, 3264, 3332, 3512, 3556, 4063, 4153, 4154, 4181, 4275, 4485, 4622, 4624, 4628, 4641, 4653, 4664, 4668, 4675, 4725–27, 5014, 5017, 5162, 5214, 5439, 5559, 5566, 5800, 5808, 5833, 6044, 6192, 6236, 6331, 6436, 6472, 6503, 6520, 6556, 6562, 6670, 6684, 6706, 6708, 6869, 6891, 6928, 6996, 7179, 7293, 7449, 7480, 7745, 7988, 8012, 8082, 8100, 8159, 8251, 8405, 8506, 8543, 8557, 8593, 8720, 8732, 8824, 8826, 8955, 9043, 9166, 9180, 9184, 9191, 9240, 9241, 9253, 9272, 9302, 9397, 9629, 9637, 9673, 9821, 9857, 9925, 10136, 10165, 10206, 10227, 10228, 10408, 10551, 10590, 10591, 10596, 10648, 10959, 10960, 10964, 10968, 10971, 11055, 11057, 11058, 11128, 11387, 11388, 11427, 11547, 11548, 11753, 11773, 11786, 11968, 12184, 12557, 12361, 12600, 12650, 12714, 13493, 13523, 13661, 13711, 13713, 13717, 13719, 13919, 13922, 13985, 14028, 14084, 14091, 14303, 14306, 14308, 14451, 14625, 14656, 14770, 15135, 15160, 15519, 15606, 15611, 16091, 16106, 16283, 16452, 16715, 16719, 16957, 17083, 17170, 17409, 17442, 17975, 18162, 18324, 18330, 18347, 18395, 18460, 18461, 18539, 18623, 18677, 18883, 19079, 19258, 19382, 19402, 19494, 19519, 19536, 19553, 19597, 19628, 19872, 19933, 20069

Insecticides, 399, 401, 427, 447, 538, 1183, 1642, 1713, 1859, 1864, 1896, 2206, 2744, 2833, 2888, 4226, 4446, 4486, 4976, 6425, 10231, 13152

Insecticides and fungicides, 348, 783, 1013, 1895, 2134, 3981, 4308, 13319, 14429

Insects, 24, 106, 231, 410, 1098, 1107, 1202, 1415, 1494, 1648, 1689–92, 1924, 1955, 2100, 2234, 2451, 2496, 2635, 2886, 2887, 3053, 3081, 3364, 3830, 3942, 4440, 4441, 4444, 4445, 5191, 5199, 5514, 5521, 5632, 5733, 6243, 6244, 6291, 6300, 6456, 6480, 7128, 7262, 7392, 7696, 8137, 8293, 8294, 9122, 9123, 9575, 9620, 9702, 9704, 9845, 9883, 10423, 10531, 10649, 10850, 10981, 11498, 11770, 12102, 12250, 12262, 12413, 12572, 12573, 12585, 12666, 13170, 13477, 13512, 13788, 14210, 14275, 14349, 14856, 14920, 14954, 15078–80, 15154, 15223, 15230, 15252, 15352, 16110, 16475, 16558, 16559, 17303, 17442, 17573, 17604, 17849, 17850, 17936, 18327, 18501, 18995, 19064, 19114, 19306, 19307, 19425, 19611, 19681, 19700, 19854, 19889, 19985

International regulations, 10087

Ireland, 980, 1174

Iron, 534, 628, 2173, 10326, 11016, 12278

Italy, 18614

Japan, 10139, 11001–4, 11080, 11082, 11086, 11103–5, 11108, 11109, 11115–17, 13447, 13448, 16651

Java, 16716

Kernel texture, 11

Kinetin, 6659, 7934, 8312, 9372, 12063, 13480

Laboratory tests, 638, 2233, 2960, 3138, 3426, 3850, 4000, 4016, 4141, 4783, 5041, 5278, 5453, 5487, 6441, 6615, 6845, 6877, 6880, 6890, 7693, 7780, 8433, 8549, 8557, 8560, 8578, 8581, 9049, 9120, 9296, 10090, 10740, 11198, 11501, 14632, 15613, 17370, 17854, 17856, 17926, 17928, 19505, 19574, 19593, 19599, 19601, 20079

Laws, 679, 2818, 4986,

Lead effects, 13339, 14549

Lebanon, 15614

Legumes, 18230, 18260, 18263, 18858, 19080, 19154–57, 19598, 19733, 19775, 19776, 19779, 19907, 19916, 20041, 20079

Life span, 192, 4900–4, 4915, 5034, 5087, 5103, 5381, 5456, 5486, 5555, 5754, 5777, 5821, 6055, 6091, 6154, 6155, 6159, 6360, 6437, 6441, 6505, 6592, 6597, 6862, 6957, 7067, 7208, 7224, 7305, 7730, 7938, 7978, 8192, 8253, 9766, 16683, 17863

Light effects, 124, 125, 184, 285, 368, 377, 456, 488, 490, 494, 496–99, 502, 506, 512, 634, 700, 744, 785, 905, 1448–50, 1457, 1503, 1553, 1556, 1600, 2069, 2229, 2389, 2454, 2473, 2529–31, 2677, 2792, 2807, 2809, 2810, 3199, 3205, 3292, 3336, 3361, 3442, 3729, 3837, 3985, 4122, 4596, 4740, 4741, 4927, 5357, 5432, 5498, 5505, 5533, 5535, 5565, 5836, 5896, 5911, 5912, 5941, 5973, 6107, 6118, 6120, 6121, 6123, 6126, 6127, 6150, 6168, 6247, 6335, 6336, 6358, 6471, 6496, 6498, 6674, 6793, 6797, 6815, 6816, 6844, 6846, 6854, 6974, 7011–14, 7016–18, 7026, 7090, 7137, 7220, 7265, 7316, 7390, 7445, 7456, 7468, 7574, 7710, 7769, 7843, 7897, 7913, 7927, 7934, 7986, 8058, 8085, 8300, 8301, 8309, 8311, 8491, 8513, 8515, 8525, 8527–29, 8549, 8579, 8585, 8633, 8634, 8658, 8687, 8697, 8864, 8955, 8990, 8992–94, 9118, 9170, 9268, 9370–72, 9387, 9483, 9484, 9486–89, 9491, 9493, 9494, 9577, 9597, 9629, 9637, 9701, 9782, 9818, 9868, 9961, 9970, 9983, 10049, 10069, 10101, 10103, 10129, 10226, 10300, 10404, 10406, 10671, 10672, 10674–76,

Light effects–(*cont.*)
10681, 10682, 10684, 10779, 10786, 10990, 11018, 11029, 11035, 11037, 11038, 11062, 11107, 11169, 11330, 11345, 11478, 11486, 11522, 11527, 11642, 11643, 11751, 11756, 11757, 11835, 11840, 12001, 12036, 12037, 12039, 12040, 12042–45, 12047, 12048, 12063, 12148, 12305, 12340, 12354, 12402, 12481, 12482, 12536, 12715, 12716, 13054, 13086, 13161, 13301, 13401, 13402, 13480, 13598, 13676, 13729, 13851, 13867, 14070, 14112, 14152, 14177, 14186, 14281, 14289, 14339, 14386, 14404, 14411–15, 14419, 14441, 14445, 14481, 14534, 14598, 14630, 14633, 14648, 14649, 14690, 14692, 14781, 14898, 14904, 14952, 15021, 15129, 15235, 15318, 15344, 15379, 15417, 15433, 15589, 15952, 16085, 16094, 16104, 16117, 16118, 16135, 16183, 16217, 16385, 16386, 16387, 16388, 16389, 16390, 16440, 16759, 16930, 17034, 17036, 17053, 17087, 17088, 17097, 17114, 17339, 17364, 17367, 17407, 17424, 17461, 17950, 18015, 18181, 18182, 18278, 18289, 18293, 18294, 18316, 18356, 18364, 18365, 18479, 18536, 18574, 18609, 18648, 18652, 18653, 18658, 18660, 18668, 18671, 18672, 18680, 18681, 18683, 18696, 18697, 18967, 18982, 19086, 19102, 19110, 19215, 19237, 19261, 19476, 19516, 19662, 19846, 19888, 19905, 19930, 19973, 20067, 20076
Lignin, 9039
Lipase, 1829, 1831, 1832, 3506, 6203–6, 8027, 9524, 9532, 9533, 9672, 10312, 12415, 14391, 14852, 15032
Lipoxidase, 584, 585, 1819, 8961
Literature review, 10713, 12641, 17628
Longevity, 92, 645, 701, 706, 708, 727, 761, 831, 868, 894, 895, 1035, 1041, 1046, 1048, 1103, 1104, 1326, 1416, 2036, 2107, 2190, 2197, 2757, 3125, 3403, 3456, 3641, 3886, 4150, 4184, 4280, 4292, 4587, 4589, 4597, 4598, 4603, 4654, 4671, 6993, 7808, 8197, 8557, 9400, 9452, 9682, 9685, 9782, 9815, 10273, 10276, 10400, 10402, 10508, 11082, 11149, 11799, 12179, 12358, 12544, 12725, 12790, 12917, 13208, 13335, 13339, 13784, 13809, 13934, 14017, 14018, 14126–28, 14281, 14388, 14672, 14704–6, 15451, 15706, 15744, 16069, 16155, 16323, 16333, 16484, 16661, 16689, 16923, 17389, 17422, 17468, 17742, 17743, 17823, 17980, 18321, 18585, 18610, 18661, 18694, 18783, 18879, 19187, 19276, 19518, 19525, 19533, 19561, 19573, 19609, 19806, 20031
Luminescence, 4001, 4005, 4295, 10260
Lygus, 3615, 3616, 6476, 6477
Lyophilization, 3318
Magnesium, 657, 1292, 1750, 2173, 8314, 13101, 13339, 15550, 16857, 16883, 17276, 17286
Magnetism, 13968, 15434, 18624
Manchuria, 13449, 14695, 14709
Manganese, 534, 657, 1959, 2173, 2200, 2202, 2203, 2367, 2453, 2896, 4583, 6151, 6205, 6266, 7302, 12422, 13444, 14223, 14549, 15550, 15557, 17587, 17598
Manual, 1076, 1325
Manure, 287, 6832, 14059, 16396, 16481, 18037, 18778
Marsh plants, 6253
Maturation, 110, 592, 595, 612
Maturity effects, 1, 15, 52, 283, 293, 332, 383, 454, 455, 480, 529, 562, 639, 794, 834, 882, 931, 1300, 1418, 1501, 1504, 1603–5, 1808, 1920, 1960, 2156–58, 2230, 2334, 2380, 2431, 2439, 2720, 2763, 2796, 2798, 2799, 2807, 2838, 3080, 3281, 3291, 3507, 3294, 3487, 3548, 3561, 3776, 3817, 3883, 3928, 3954, 4112, 4241, 4250, 4392, 4505, 4510, 4513, 4607, 4610, 4684, 4735, 5387, 5409, 5424, 5425, 5595, 5635, 5670, 5694, 5821, 5826, 5888, 5963, 6133, 6190, 6218, 6318, 6357, 6369, 6388, 6396, 6409, 6453, 6566, 6800, 6827, 6870, 6960, 7231, 7365, 7367, 7391, 7406, 7411, 7438, 7536, 7667, 7729, 7745, 7759, 7905, 8040, 8054, 8222, 8248, 8379, 8412, 8491, 8504, 8593, 8645, 8662, 8717, 8820, 8853, 8908, 8955, 8963, 9007, 9009, 9010, 9130, 9131, 9158, 9159, 9183, 9199, 9218, 9335, 9420, 9482, 9516, 9523, 9536, 9537, 9542, 9573, 9579, 9812, 9868, 9889, 9891, 9940, 10106, 10192, 10301, 10323, 10494, 10552, 10553, 10562, 10669, 10737, 10771, 10809, 10974, 11043, 11245, 11250, 11269, 11370, 11424, 11539, 11554, 11608, 11773, 11775, 11849, 11958, 11979, 11980, 12030, 12091, 12121, 12420, 12634, 12724, 12839, 13109, 13200, 13227, 13298, 13419, 13420, 13461, 13472, 13569, 13602, 13613, 13641, 13837, 13912, 13940, 14082, 14268, 14601, 14737, 14772, 14777, 14899, 15123, 15132, 15153, 15294, 15406, 15529, 15570, 15714, 15719, 15834, 15874, 15875, 15882, 15993, 15996, 16052, 16121, 16181, 16191, 16540, 16554, 16619, 16623, 16655, 16699, 16735, 16891, 16949, 16995, 17194, 17195, 17235, 17282, 17338, 17362, 17372, 17373, 17465, 17479, 17511, 17512, 17615, 17622, 17642, 17664, 17680, 17713, 17780, 17864, 17963, 17990, 18088, 18195, 18316, 18375, 18449, 18477, 18510–12, 18582, 18594, 18595, 18602, 19615, 18706, 18739, 18770, 18802, 18871, 18904, 19116, 19137, 19138, 19186, 19308, 19339, 19430, 19449, 19499, 19520, 19597, 19605, 19683, 19744,

Maturity effects–(*cont.*)
19750, 19798, 19799, 19801, 19967, 19996, 20003, 20037, 20038, 20062, 20081, 20134
Measurement, 7639, 12683, 13180
Meat extracts effects, 2828
Mechanical, 42, 1699, 1943, 2153, 5549, 10093, 11423, 13284
Media, 277, 422, 448, 500, 501, 827, 896, 985, 1428, 1544, 2792, 2794, 2884, 3033, 3259, 3498, 3605, 3629, 4216, 4462, 5007, 5117, 5170, 5438, 5565, 5634, 5651, 5814, 5982, 6073, 6223, 6343, 7135, 7162, 7254, 7355, 7497, 7749, 7817, 7891, 7990, 8005, 8120, 8150, 8162, 8490, 8549, 8560, 8611, 8663, 8664, 8668, 8851, 9012, 9069, 9087, 9171, 9281, 9301, 9425, 9506, 9541, 9990, 10097, 10513, 10521, 10568, 10692, 10772, 10913, 11241, 11486, 11720, 11795, 11835, 12054, 12188, 12239, 12267, 12581–83, 12589, 12659, 12679, 13063, 13373, 13607, 13781, 14133, 14176, 14380, 14633, 14850, 14898, 14929, 15291, 15473, 15596, 15700, 15840, 16357, 16812, 17007, 17169, 17219, 17308, 17368, 17584, 17719, 18059, 18076, 18085, 18110, 18351, 18801, 18907, 18918, 18933, 18965, 19192, 19280, 19293, 19404, 19554, 19644, 19728, 19773, 19822, 19860, 19900, 19901, 19902, 19973, 20065
Mercury effects, 178, 365, 382, 407, 599, 600, 657, 775, 806, 1408, 2094, 2596, 2672, 3011, 3012, 3239, 3598, 3669, 4129, 4185, 4623, 4659, 4670, 4808, 5212, 6889, 7023, 7211, 7548, 7550, 7592, 7600, 7810, 8115, 8264, 8333, 9145, 10637, 11258, 12105, 13339, 14549, 14936, 14937, 16461, 16955, 16956, 17057, 18359, 18708
Metabolism, 1694, 1851, 1977, 2705, 2759, 2796, 2842-A, 2843, 3167, 3273, 3818, 3953, 3995, 4330, 4435, 4607–10, 4796–98, 4812, 4921, 4927–29, 4959, 4991, 5159, 5169, 5221, 5368, 5574, 5665, 5666, 5720, 5766, 5952, 5966, 6016, 6104, 6280, 6525, 6867, 7343, 7777, 7824, 7892, 7922, 7935, 8372–74, 8797, 9525, 9530, 9679, 9910, 10134, 10176, 10238, 10244, 10403, 11042, 11308, 11512, 11537, 11589, 11706, 11707, 11894, 11992–94, 12644, 12755, 13102, 13254, 13686, 13945, 13955, 14267, 14318, 14353, 14357, 14579, 14730, 14764, 14765, 14802, 15034, 15142, 15361, 15393, 15439, 15488, 15536, 15724, 15732, 15826, 15835, 16145, 16150, 16340, 16498, 16610, 16767, 17181, 17333, 17335, 17339, 17343, 17532, 17554, 17555, 17576, 17607, 17608, 17722–24, 17885, 18043, 18044, 18046, 18117, 18181, 18182, 18247, 18282, 18340, 18341, 18355, 18358, 18364, 18366, 18425, 18426, 18494, 18839, 18845, 18877, 18883,
Metabolism–(*cont.*)
18974, 19140, 19146, 19156, 19157, 19197, 19200, 19252, 19265, 19441, 19612, 19911, 19921–24
Meteoric conditions, 18326
Mexico, 16733
Microchemistry, 1834, 2777, 7864, 19221
Microclimate, 1367
Micrografting, 2584
Microorganisms, 170, 216, 501, 519, 734, 992, 1362, 1367, 1378–87, 1530, 1696, 1697, 1734, 1738, 1874, 2164, 2219, 2846, 3366, 3520, 3691, 3892–94, 4052–58, 4061, 4062, 4064–67, 4088, 4105, 4616, 4617, 4621, 4625, 4626, 4633, 4636, 4655–58, 4660–62, 4665, 4676, 4677, 5021, 5168, 5204, 5211, 5269, 5292, 5313, 5354, 5892, 6332, 6433, 6890, 6891, 6928, 6951, 6952–56, 7087, 7088, 7104, 7107, 7197, 7212, 7253, 7413, 7415, 7447, 7533, 7677, 7731, 7732, 7734, 7902, 8641, 8702, 8730, 8824, 8934, 9039, 9205, 9208, 9241, 9276, 9315, 9357, 9358, 9460, 9480, 9496, 9514, 9527, 9543, 9611, 9656, 9690, 9814, 9847, 9942, 10334, 10521, 10546, 10600, 10601, 10616, 10917–21, 10963, 11522, 11555, 11969, 11971, 12562, 13223, 13226, 13265–67, 13513, 13515, 13523, 13529, 13566, 13567, 13588, 13621, 13895, 14087, 14847, 14918, 14947, 14950, 15022, 15039, 15040, 15041, 15200, 15244, 15301, 15376, 15377, 15350, 15507, 15788, 15789, 15802, 15807, 15885, 15994, 16136, 16137, 16170, 16519, 16567, 16790, 16985, 16996, 17032, 17033, 17040, 17067, 17196, 17262, 17288, 17314, 17345, 17487, 17571, 17589, 17655, 17752, 17943, 18225, 18395, 18440, 18454, 18455, 18467, 18580, 18686, 18813, 18814, 18855, 18862, 18899, 19031, 19063, 19084, 19144, 19328, 19331, 19358, 19555, 19560, 19655, 20068
Microscopy, 359, 7389, 18198
Microsomes, 229
Milling, 559, 560
Mineral nutrition, 1660, 1661, 1667
Minerals effects, 1839
Mitochondria, 229, 230, 8987, 9067, 9129, 14344, 14576, 16799, 17869, 17871
Mitosis, 1708–10
Moisture, 1401, 2897, 3180, 3322, 5648, 11997, 12199, 17160, 17381, 17427, 18693
Moisture content, 69, 157, 164, 254, 278, 546, 553, 554, 643, 1052, 1088, 1211, 1574, 1742, 3510, 4268, 4281, 4296, 4362, 4363, 5093, 5120, 5163, 5292, 5307, 5313, 5361, 5404, 5489, 5520, 5678, 5777, 5966, 5974, 6004, 6017, 6252, 6507, 6566, 6592, 6601, 6620, 6721, 6734, 6736, 6876, 6879, 6944–47, 7050, 7063, 7064, 7079, 7164, 7412, 7485, 7555,

Moisture content–(*cont.*)
8033, 8275–78, 8282, 8358, 8367, 8595, 8670, 8927, 9111, 9212, 9235, 9257, 9353, 9519, 9816, 9930, 9960, 9988, 11390, 11803, 11804, 11841, 11934, 12005, 12263, 12283, 12285, 12312, 12337, 12378, 12472, 12475, 13468, 13471, 14702, 14854, 14874, 14914, 14919, 15262, 15693, 15915, 16258, 16544, 16671, 16706, 16929, 17102, 17138, 17338, 17471, 17478, 17762, 18029, 18216, 18502, 18656, 18767, 18788, 18999, 19065, 19101, 19153, 19267, 19322, 19349, 19356, 19377, 19418, 19438, 19595, 19636, 19638, 19707, 19842, 19843, 19932, 20083

Moisture and drying effects, 493, 643, 1621, 1622, 1974, 2177, 3294, 3837, 4022, 4458, 5404, 5462, 8955, 12752, 13093

Moisture effects, 15, 219, 223, 225, 227, 257, 258, 279, 459, 613, 684, 1187, 1304, 1305, 1568, 1635, 1740, 1742, 1759, 1943, 2005, 2043, 2153, 2172, 2193, 2453, 2840, 2846, 2879, 2922, 2956, 3048, 3085, 3134, 3164, 3287, 3332, 3336, 3365, 3443, 3457, 3510, 3557, 3559, 3611, 3652, 3656, 3678, 3699, 3820, 3895, 3944, 3957, 4054, 4060, 4080, 4124, 4234, 4306, 4307, 4639, 4768, 4995, 5019, 5164, 5361, 5457, 5841, 6101, 6151, 7011, 7012, 7143, 7198, 7300, 7331, 7366, 7451, 7541, 7589, 7769, 7994, 8147, 8257, 8418, 8584, 8605, 8700, 8894, 8952, 9161, 9295, 9316, 9418, 9570, 9738, 9818, 10069, 10173, 10267–70, 10272, 10279, 10333, 10396, 10400, 10402, 10407, 10443, 10618, 10672, 10831, 10892, 10963, 11125, 11128, 11132, 11133, 11137, 11139, 11158, 11193, 11194, 11246, 11326, 11466, 11513, 11561, 11587, 11603, 11652, 11796, 11805, 11910, 12040, 12267, 12284, 12336, 12594, 12604, 12612, 12615, 12671, 12735, 12818, 12900, 12903, 12904, 13179, 13208, 13687, 13728, 13754, 13757, 13772, 14026, 14129, 14628, 15036, 15093, 15176, 15436, 15560, 15576, 15993, 16038, 16725, 17350, 17351, 17354, 17355, 17610, 17633, 17737, 17959, 18217–20, 18675, 18939, 19355, 19968

Monocotyledons, 1641, 2905, 3301, 19863

Moon effects, 2205, 3632, 6176, 12974, 14827, 16358, 17211

Mucilage, 524, 525, 528, 705, 732, 1481, 1684, 1685, 1764, 4455, 4847, 5649, 6036, 7303, 7591, 7625, 8534, 8858, 8859, 9203, 9751, 9929, 9974, 10016, 10154, 10426, 11381, 12166, 12389, 12522, 12897, 12961, 13035, 13289, 13844, 13892, 14049, 14180, 14321, 14362, 14401, 15712, 15858, 15895, 15963, 15964, 16050, 16051, 16113, 16125, 16537, 16538, 16841, 17577, 17578, 18646, 18647, 18876, 19442, 20008

Multiple seedlings, 214, 1578, 1583, 2612, 3251, 3387, 5969, 6539, 9610, 10274, 16447

Mutation, 727, 1535, 1536, 2587, 3656–60, 4529, 7122, 7289, 7508, 10257, 10313, 10323, 10701, 11072, 11263, 15258–61, 17353

National Institute of Agricultural Botany, 10435

Nebraska, 18626

Netherlands, 1122, 11192

Neutrons, 2146, 2777, 2999, 3095

New South Wales, 1138

New York, 13585

New Zealand, 1537, 6200, 12322, 13252, 14182, 14934, 15415, 16173, 18109

Nigeria, 19589

Nitrogen, 82, 99, 100, 195, 196, 201, 221, 302, 490–92, 494–99, 502, 506, 512, 1557, 1561, 1562, 1667, 2017, 2020, 2038, 2046, 2176, 2246, 2299, 2505, 2557, 2558, 2749, 2801, 2838, 3003, 3023, 3611, 4658, 4925, 5092, 5481, 5633, 6160, 6208, 6434, 6468, 7016, 7300, 7825, 7895, 8181, 8262, 8312, 8347, 8569, 9278, 9317, 9778, 10140, 10175, 10226, 10329, 10532, 10593, 10724, 10726, 10927, 11043, 11044, 11204–7, 11377, 11763, 11832, 11894, 12036, 12044, 12109, 12407, 12409, 12519, 12595, 12596, 13116–20, 13204, 13205, 13240, 13339, 13440, 13534, 14227, 14395, 14598, 14898, 16229, 18616, 18671, 18672

Norway, 11432, 14817, 15250, 16366

Nuclei, 337

Nucleic acids, 1457, 1952, 1953, 2249, 2258–60, 2416, 2925, 2971, 3136, 3563, 3737, 3903, 6370, 6851, 8438, 8439, 9349, 9350, 9399, 9898, 10499, 10575, 10632, 11076, 12848, 12849, 12926, 13081, 14739, 14853, 14869, 15063, 17194, 17195, 17718, 18388, 19163, 19807

Nucleotides, 25, 5629, 16220

Number, 1248, 2121, 2371, 2605, 3405, 3544, 4914, 5256, 5447, 5488, 5840, 5867, 6287, 6356, 6512, 6592, 7309, 7368, 8046, 8242, 9395, 11936, 11965, 12935, 13689, 13961, 14339, 14773, 15171, 15373, 16661, 17142, 17168, 17441, 17974, 17981, 18049, 18811, 18849, 20122

Nursery technique, 19446

Nutrition, 46, 48, 49, 50, 51, 148, 183, 185, 250, 373, 432, 440, 445, 466, 575, 596, 613, 920, 1316, 1413, 1414, 1457, 1724, 1763, 1897, 1933, 2083, 2087, 2244, 2285, 2443, 2492, 2768–70, 2773, 2822, 3000, 3001, 3003, 3004, 3006, 3041, 3042, 3217, 3671, 3756, 3815, 3846, 3996, 4035, 4036, 4332, 4347, 4351, 4515, 4578, 4688, 4757, 4773,

Nutrition–(*cont.*)
4838, 5244, 5247, 5290, 5356, 5406, 5619, 5747, 5884, 5893, 6049, 6093, 6133, 6192, 6209, 6210, 6284, 6664, 6696, 6703, 6704, 6737, 7349, 7582, 7814, 8056, 8140, 8380, 8600, 8609, 8722, 8730, 8880, 8938, 9113, 9237, 9448, 9459, 9658, 9786, 9934, 9935, 10069, 10070, 10099, 10413, 10436, 10623, 10752, 10997, 11070, 11118, 11177, 11256, 11342, 11415, 11459, 11472, 11473, 11683, 11748, 12104, 12228, 12229, 12231, 12288, 12289, 12805, 13091, 13092, 13250, 13310, 13313, 13314, 13316, 13320, 13588, 13590, 13684, 14061, 14233, 14271, 14341, 14364, 14510, 14511, 14512, 14667, 14870, 15448, 15454, 15619, 15734, 15868, 15956, 15984, 16751, 16752, 16789, 16870, 16939, 16992, 17175, 17765, 18127, 18132, 18133, 18213, 18322, 18323, 18512, 18766, 19005, 19841, 20039, 20041, 20136

Nuts, 834, 1668, 3425, 3462, 4168, 17082, 19282 *see also* Plant Index

Oil, composition of, 186, 577, 588, 790, 883, 1279, 1644, 1716, 1791, 2113, 2116, 2118, 2122, 2225, 2348, 2538, 2540, 3035, 3036, 3083, 3094, 3702–4, 3765, 3987, 3988, 4279, 4303, 4499, 4683, 4706–9, 4711, 4712, 5059, 5374, 5409, 6023–25, 6227, 6311, 6503, 7199, 7329, 7399, 7716, 7764, 7860, 8053, 8247, 8383, 8764, 8773, 8913, 8962, 9017–19, 9022, 9023, 9130, 9131, 9345, 9346, 9356, 9498, 9499, 9555, 9558, 9708, 9754, 10280–82, 10703, 10788, 10816, 11063, 11371, 11385, 12629, 12877, 12878, 13291, 13495, 13614, 13662, 14098, 14560, 14778, 14891, 14944, 15104, 15194, 15554, 15650, 15756, 15870, 15972, 16195, 16216, 16636, 16665, 16669, 16937, 17051, 17145, 17401, 17402, 17479, 17727, 17728, 18398, 18597, 18765, 18799, 18806, 19219, 19448, 19450, 19677, 19693, 19741, 19742, 19944, 19982, 20013, 20055

Oil content, 168, 189, 190, 311, 825, 828, 862, 871, 872, 907, 1056, 1239, 1286, 1290, 1293, 1344, 1356, 1402, 1462, 1463, 1474, 1524, 1534, 1552, 1588–90, 1593, 1668, 1749, 1769, 1882, 1903, 1904, 1906, 1954, 2062, 2105, 2114, 2117, 2118, 2225–27, 2244, 2276, 2277, 2283, 2307, 2308, 2412, 2414, 2429, 2470, 2681, 2738, 2821, 2889, 2935, 2937, 2941, 2975, 3036, 3083, 3090, 3179, 3215, 3295, 3338, 3342, 3354, 3643, 3644, 3695, 3702–4, 3740, 3851, 3878, 3923, 3933–35, 3989–92, 4051, 4136, 4198–200, 4210–12, 4227, 4279, 4303, 4474, 4497, 4498, 4710, 4721, 4769, 4801, 4810, 4855, 4937, 4938, 4947, 5020, 5074, 5075, 5082,

Oil content–(*cont.*)
5217, 5228, 5243, 5293, 5329, 5344, 5372, 5374, 5434, 5443, 5449, 5512, 5552, 5567, 5578, 5579, 5585, 5586, 5595, 5646, 5717, 5745, 5797, 5825, 6048, 6069, 6070, 6197, 6198, 6228, 6230, 6231, 6308, 6367, 6429, 6448, 6694, 6809, 6865, 6989, 7069, 7101, 7115–17, 7120, 7188, 7213, 7215, 7230, 7250, 7280, 7290, 7353, 7393, 7493, 7494, 7684, 7685, 7687, 7770–74, 7837, 7838, 7855, 8073, 8280, 8494, 8638, 8639, 8660, 8765, 9277, 9369, 9393, 9436, 9528, 9550, 9551, 9580, 9595, 9705, 9707, 9709, 9711–14, 9797, 9861, 9885, 9886, 9919, 9920, 9980, 9996, 10047, 10117–21, 10132, 10153, 10287, 10314, 10322, 10342, 10345, 10432, 10506, 10514, 10539, 10654, 10655, 10666, 10857, 10938, 10978, 11040, 11272, 11316, 11395, 11412, 11450, 11483, 11523, 11563, 11633, 11750, 11771, 11785, 11964, 12002, 12202, 12236, 12444, 12513, 12608, 12609, 12636, 12665, 12809, 12819, 12868, 12933, 12942, 12946, 13028, 13053, 13056, 13143, 13175, 13219–21, 13275, 13276, 13292, 13295, 13436–38, 13449, 13453–56, 13483, 13525, 13655, 13657, 13783, 13845, 13978, 14051, 14059, 14092, 14119, 14214, 14234, 14255, 14329, 14334, 14453, 14528, 14561, 14737, 14738, 14773, 14778, 14962, 15048, 15146, 15147, 15151, 15196, 15203, 15326, 15343, 15358, 15460, 15461, 15479, 15549, 15709–11, 15754, 15758–62, 15783, 15870, 15892, 15946, 15948, 16161, 16196, 16248, 16593, 16726, 16936, 17045–48, 17050, 17205–10, 17213, 17280, 17561, 17652, 17837, 17872, 17900, 17901, 17903–9, 17970, 18063, 18070, 18119, 18120, 18336, 18571, 18732, 18734, 18735, 18737, 18759, 18763, 18796, 18865, 18925, 18951, 18954, 18983, 19012, 19013, 19049, 19052, 19218, 19229, 19270, 19467, 19568, 19632, 19667, 19740, 19937, 19981, 20110

Oil, determination of, 187, 1397–99, 3825, 4282, 4304, 4322, 4893, 6022, 6705, 6848, 7070, 7412, 7839, 8769, 8879, 9051, 9071, 9224, 9886, 10465, 10659, 10663, 10893, 11426, 11937, 12185, 12876, 14529, 14986, 16156, 16201, 17322, 17902

Oil, general, 603, 606, 1866, 2508, 2539, 2571, 3757, 4729, 5253, 5261, 5262, 5264, 5409, 5448, 5687, 5884, 6021, 6027, 7066, 7127, 7400, 7422, 7717, 8657, 8759, 8761, 8762, 9554, 9557, 9559, 9926, 10170, 10294, 10341, 11209, 11240, 11504, 12635, 12930, 13314, 13524, 14390, 14405, 14561, 15093, 15122, 15531, 15950, 16149, 16629, 17787, 17812

Oil, metabolism of, 575, 1906, 1911, 2541,

Oil, metabolism of–(*cont.*)
3571, 6163, 7851, 8481, 8881, 9020, 9021, 9097, 9098, 9551, 9978, 11490, 11618, 13018, 13236, 13357, 13557, 13986, 14964, 15106, 15418, 15446, 16501, 16633, 16883, 16934, 17284, 18262, 18374, 18764
Orchards, 567, 9323, 10753
Organic acids, 2034, 3592, 5669, 5670, 5673, 6208, 8302
Ornamentals, 1733, 3939, 4021, 5469, 6544, 13150, 14405, 14756, 18183, 18605, 19405
Ornithochoric fruit, 20071
Osmotic pressure, 754, 2728, 2814, 3257, 3781, 3884, 3967, 4522, 4797, 5508, 5509, 5540, 5541, 5591, 5593, 6569, 7174, 7975, 7976, 8536, 9179, 9341, 10129, 10480, 10568, 10700, 10721, 11067, 11167, 12051, 12239, 12480, 12660, 12968, 13094, 13095, 13400, 14448, 16513, 16541, 16584, 16804, 17020, 17575, 18873, 19178, 19180
Oxidase, 270, 308, 1842, 3111, 3855, 4329, 4604, 5924, 6925, 6926, 7345, 9697, 10194
Oxygen effects, 99, 100, 268, 270, 489, 1555, 2017, 2020, 2046, 2176, 2193, 2524, 2530, 2675, 2677, 2845, 2880, 2953, 2961, 3438, 3509, 3515, 3829, 4100, 4127, 4317, 5033, 5160, 5606, 5794, 5832, 6092, 6160, 6252, 6353, 6797, 6814, 7187, 7319, 7541, 7647, 7791, 7808, 7830, 7944, 7986, 8206, 8491, 8514, 8707, 8905, 8955, 9055, 9155, 9269, 9581, 9816, 9957, 10495, 10555, 10720, 11187, 11431, 11892, 12040, 13103, 13339, 13431, 13432, 13752, 13835, 13838, 13900, 14231, 14598, 14714, 14810, 14936, 14937, 15244, 15644, 15707, 15794, 16011, 16144, 16642, 16774, 16869, 17378, 17380, 17394, 17413, 17414, 17499, 17659, 18008, 18146, 18297, 18352–54, 18358, 18405, 18467, 18535, 19161, 19162, 19662

Packeting, 838, 844, 2016, 2025, 2073, 3129, 3180, 3950, 4371, 5734, 6399, 6550, 8130, 8221, 8223, 8226, 8230, 9881, 12467, 12471, 13890, 18185, 18685, 18847, 19619
Pakistan, 10347, 10509, 19274
Palo Santo, 7786
Parasitism, 619, 621, 2836, 2837, 3168, 3170, 3172, 3867, 8531, 8532, 10687, 10763, 12224, 15351, 16661
Parthenocarpy, 251, 378, 1761, 2142, 2218, 2542, 2685, 3076, 3298, 3545, 3919, 4162, 4327, 4328, 4394, 4520, 6509, 6547, 6968, 6969, 6971, 7099, 7153, 7419, 7455, 7875–80, 7883, 7992, 8404, 8478, 8828, 9174, 9209, 9226, 9718, 9719, 9764, 10004, 10218, 10446, 10535, 10650, 11515, 11560, 12123, 12173, 12344, 12434, 12815, 13861, 14291, 14293, 14368, 14723, 14905, 14993, 15159,

Parthenocarpy–(*cont.*)
15923, 16018, 16603, 16868, 17889, 18078, 18100, 18592, 18987, 19066, 19811, 19858
Pasture seeds, 1927, 17186, 17818, 19633
Pelleting, 136, 345, 352, 374, 427, 937, 968, 978, 993, 1005, 1009, 1017, 1021, 1025, 1054, 1058, 1089, 1106, 1138, 1250, 1260, 1915, 2180, 2315, 2610, 2611, 2649, 3087, 3370, 3597, 3631, 5318, 5497, 6301, 6304, 6305, 6402, 7037, 7210, 7780, 8350, 8436, 9194, 9196, 9773, 10128, 10231, 10727, 10728, 10840, 11069, 11435, 11820, 12279, 12290, 12377, 12405, 12495, 12588, 12822, 13222, 13261, 13623, 13696, 13698, 13785, 13929, 13948, 14370, 14376, 14499, 15211, 15483, 16012, 16151, 16203, 16423, 16508, 17393, 17672, 17673, 17986, 18456, 18503, 18952, 19169, 19170, 19268, 19269, 19351, 19567, 19861, 20115
Perennials, 739, 879, 1096, 2457, 3607, 9266, 11644, 15631, 16414, 16649
Periodicity, 815, 1305, 3283, 3288, 7207, 8237, 9825, 10163, 10826, 11221, 11531, 11532, 11593, 12183, 13361, 13701, 15257, 16456, 16531, 16662, 17431, 17978, 18898, 19383
Permeability, 62, 63, 642, 1495, 1613, 2013, 2185, 2625, 2952, 2954, 3112–14, 3152, 3402, 3520, 3559, 3707, 3711, 4301, 4316, 4318, 4522, 4863, 5202, 5203, 6135, 6672, 7319, 7382, 7474, 7640, 7741, 7742, 7865, 7866, 8955, 8996, 9094, 9286, 9311, 9681, 9824, 9871, 10721, 10723, 11279, 12113, 12744, 12751, 13270, 13339, 13418, 13457, 13905, 14319, 14421, 14428, 14429, 14431, 14435, 14437, 14440, 14447, 14566, 15392, 15396, 15812, 16240, 16608, 17031, 17379, 17827, 17988, 18008, 18394, 18476, 18489, 18644, 18688, 19617, 19621, 19748, 20044, 20063, 20103
Peroxidase, 434, 436, 6977, 9054, 11793, 13188
Petroleum effects, 13967
Pharmacology, 165, 373, 736, 768, 788, 795, 1001, 1279, 1302, 1636, 2221, 3266, 3594, 3755, 3811, 4944, 5186, 5231, 5376, 5460, 6549, 6667, 6697, 7036, 7073, 7172, 7638, 7658, 7786, 7794, 7852, 7856, 8144, 9366, 9631, 9929, 10152, 10159, 10221, 10337, 10432, 10505, 10633, 10660, 10941–43, 11063, 11328, 11864, 12158, 12228, 12554–56, 12616, 12898, 13332, 13334, 13914, 13985, 14010, 14362, 14804, 14826, 15024, 15202, 15303, 15649, 15836, 15967, 16028, 16478, 16553, 16592, 16635, 16636, 16724, 16728 16729, 16905, 17262, 17397, 17545, 17800–2, 17939, 18030, 18254, 18441, 18619, 18931, 19111, 19337, 20040

Phenolic compounds, 1894, 2643, 5255, 6950-L, 7678, 7951, 8665, 10547, 10651, 11100–2, 11150, 12891, 13937
Philippines, 10060, 16326
Phosphatase, 1565, 2764, 3526, 4462, 4463, 6723, 7314, 9272, 10925, 11481
Phosphorus, 206, 207, 274, 413, 414, 641, 1667, 2204, 2261, 2362, 2488, 2710, 2750, 3534, 3807, 3821, 4267, 6008, 6819, 7273, 7274, 7397, 7552, 8315, 9342, 9582, 10923, 10975–77, 11528, 11846, 12052, 12311, 12342, 12408, 12874, 12878, 13100, 13271, 13401, 13402, 13433, 14184, 15421, 15566, 15721, 16096, 16149, 16159, 16437, 16480, 16864, 17315, 17876
Photographs of seeds, 3997, 6689, 8810, 8832, 12919, 15380–91, 17440
Photoperiod, 18, 366–68, 614, 2098, 2527, 2528, 2531, 2704, 2808, 3199, 3293, 3308, 3860, 3963, 4030, 4032, 4107, 4171, 4511, 4740, 5197, 5535, 6555, 6812, 6814, 6817, 6831, 6988, 7509, 8039, 8152, 8160, 8283, 8627, 8628, 8699, 9485, 9486, 9490, 9492, 9583, 9650, 9781, 10069, 10125, 10309, 10857, 10891, 11029, 11360, 11494, 11570, 11723, 11759, 11772, 11777, 11811, 11855, 12401, 12435, 12995, 13002, 13305, 13394, 13761, 13824, 13938, 13962, 14071, 14136, 14139, 15417, 15769, 15850, 15987, 16035, 16654, 16656, 17089, 17222, 17228, 17260, 17540, 17818, 17886, 18103, 18379, 18913, 19189, 19345, 19423, 19625
Phylogeny, 3785
Physical characteristics, 89, 90, 188, 234, 261, 1212, 2334, 3225, 3630, 10055, 11085, 11129, 11136, 11140–43, 11148
Physiology, 617, 880, 1063, 1374, 1375, 1467, 1509, 1520, 1637, 1946, 1969, 2035, 2162, 2192, 2434, 2442, 2455, 2475, 2795, 2797, 2923, 3230, 3231, 3260, 3273, 3389, 4264, 4496, 4599, 4601, 5086, 5097, 5282, 5550, 6051, 6111, 6112, 6150, 6722, 6773, 7264, 7436, 7689, 7740, 7775, 8498, 8561, 8568–70, 8574, 8914, 9072, 9093, 9128, 9396, 9556, 9596, 9732, 9872, 10237, 10416, 10556, 10557, 10559–61, 11107, 11249, 11439, 11442, 11568, 12036, 12148, 12299, 12793, 13420, 13485, 13849, 14404, 14423, 14498, 14721, 14927, 15097, 15533, 15745, 15775, 16182, 16314, 16398, 16401, 16528, 16532, 16611, 16749, 17325, 17533, 18015, 18481, 18650, 18657, 18666, 18695, 18782, 18786, 18986, 19208, 19224, 19244, 19353, 19791, 19928, 19949
Phytin, 5752, 9753, 16978
Pictures, 6680
Pigment, 154, 318, 341, 536, 637, 1611, 1768, 2070, 2099, 2262, 2652–58, 3207, 3400,
Pigment–(*cont.*)
3476, 3736, 3748, 4459, 5736, 6303, 7292, 7530, 7767, 8369, 8628, 8905, 9112, 9262, 9394, 9601, 9865, 11054, 11307, 11744, 12169, 12866, 12867, 12922, 13233, 13554, 13624, 13671, 13699, 14347, 14611, 14734, 14959, 15470, 15526, 15530, 15621, 15791, 15962, 16583, 16774, 16976, 16986, 17080, 17158, 17248, 17313, 17763, 17831, 17988, 17996, 18084, 18137, 18243, 18244, 18378, 18447, 18740, 18742, 19190, 19228, 19627, 19721, 19739, 19813, 20103
Pipecolic acid, 5233, 5234
Planetary cycles, 17061, 17062
Plants effects, 537, 556, 563, 619, 648, 758, 1156, 1245, 1354, 1554, 2293, 2709, 2726, 2732, 2919, 3284, 3437, 3440, 3544, 3566, 3629, 3710, 3747, 3874, 4117, 4672, 4694, 5071, 5119, 5280, 5306, 5845, 6110, 6366, 6695, 7086, 7429, 7702, 7703, 7748, 7749, 7778, 7913, 7971, 7972, 8197, 8594, 9114, 9141, 9481, 9502, 9545, 9973, 10014, 10310, 10572, 10865, 11061, 11168, 11242, 11290, 11348, 11413, 11855, 11860, 11882, 11896, 11910, 12125, 12126, 12195, 12524, 12912, 12963, 12997–99, 13123, 13124, 13191, 13703, 13896, 13936, 13956, 13970, 14223, 14272, 14393, 14637, 14651, 14768, 14798, 14810, 14850, 14885, 14901, 14908, 14956, 15096, 15121, 15130, 15173, 15296, 15372, 15428, 15838, 15840, 16084, 16105, 16109, 16274, 16275, 16424, 16444, 16465, 16649, 16672, 16999, 17023, 17092, 17155, 17221, 17231, 17333, 17346, 17377, 17392, 17446, 17542, 17556, 17603, 17632, 17794, 17935, 18115, 18145, 18169, 18189, 18512, 18560, 18570, 18634, 18857, 18860, 18938, 19031, 19125, 19126, 19165, 19172, 19302, 19303, 19532, 19758, 19853, 19884, 19961, 19980, 20046
Plastics, 2536, 12053, 13584, 18685, 19403, 19879
Poisonous seeds, 399, 716, 846, 878, 2857, 5405, 10661-A, 13651
Poland, 10195, 14753
Polarity, 4956
Polyembryony, 214, 703, 1047, 1381, 1580, 1725, 1788, 1853, 1857, 2231, 2336, 2601, 2665, 3082, 3243, 3542, 3554, 3763, 3764, 3836, 4050, 4140, 4357, 4539, 4542, 4547, 4926, 4941, 5027, 5437, 5839, 6063, 6237, 6403, 6788, 6856, 6950-J, 7200, 7202, 7633, 7781, 7787, 7870, 8094, 8098, 8182, 8281, 8486, 8487, 8509, 8735, 9058, 9086, 9153, 9329, 9330, 9567, 9638, 9748, 9763, 9848, 9908, 9914, 9917, 10019, 10144, 10196, 10248, 10735, 11266, 11448, 11449, 11484, 11520, 11699, 11834, 11938, 11940, 11941,

Polyembryony–(*cont.*)
12417, 12624, 12689–91, 12889, 12910, 13425, 13579, 13645, 13736, 13742–44, 14220, 14315, 14582, 14840, 14911, 15105, 15183, 15277, 15543, 15817, 15910, 15919, 15920, 15949-A, 15961, 16302, 16438, 16631, 16780, 16791, 16820, 16821, 17025, 17037, 17167, 17269, 17919, 18101, 18128, 18237, 18566, 18581, 18713, 18920, 18962, 18996, 19055, 19060, 19061, 19067, 19093, 19250, 19417, 19437, 19647, 19858, 19859, 19876, 19915

Polyploidy, 1338, 1511, 1617, 1792, 1867, 1975, 2253, 2590, 2591, 2742, 2820, 3196, 3802, 3844, 4385, 5903, 5913, 6245, 7476, 7941, 8095, 8592, 8847, 8936, 9115–17, 9245, 9280, 9285, 9727, 10052, 10292, 11595–99, 11774, 12792, 13408, 14273, 14513, 14519, 14523, 15027, 15221, 15540, 15598, 15843, 15924, 16604, 16687, 16792, 17094–96, 17334, 17654, 17758, 17759, 17888, 17933, 17934, 18204, 18932, 19571, 20100

Position on plant, 73, 649, 1354, 1738, 2245, 4302, 6698, 7433, 7575, 8046, 8047, 8161, 8244, 8424, 8656, 9481, 10043, 11505, 13936, 14081, 15786, 17063, 17221, 17377, 17603, 18169, 18860, 19719

Potassium, 314, 1667, 2173, 2830, 8091, 8549, 8864, 10407, 10528, 10571, 10734, 12098, 12111, 12408, 14693, 17892, 17918

Pressure, 2695, 3088, 4605, 4992, 4993, 5302, 5311, 6166, 8836, 8919, 8955, 9168, 10348, 10349, 11184, 12168, 12545, 16249, 18326

Processing, 162, 201, 291, 540, 1698, 1700, 4409, 5536, 6327, 6982, 8599, 9520, 11697, 12291, 12566, 12738, 12768, 12931, 15991, 16042, 16073, 16132, 16288, 17122, 17622, 17665, 18393, 18623, 18655, 19357, 19440, 19491

Production of seed, 283, 371, 375, 423, 565. 763, 1077, 1639, 1704, 1719, 1720, 1934, 1935, 2091, 2108, 2778, 2894, 3008, 3401, 7370, 7965, 8270, 10430, 11349

Propagation, 93, 123, 127, 130, 344, 416, 1612, 15133, 16467, 16523, 16845, 16927, 16987, 17484, 17585, 18074, 18075, 18108, 18183, 19298, 19342, 19422, 19513–15, 19979

Protease, 265, 270, 1289, 3690, 4460, 10931, 14679, 15722

Protein, 11, 118, 119, 269, 301, 303, 432, 542, 547–50, 552, 557, 817, 823, 949, 1032, 1478, 1610, 1663, 1677, 1724, 1823, 1840, 1932, 1941, 2064, 2086, 2087, 2110, 2111, 2175, 2246, 2260, 2301, 2302, 2556, 2840, 2842-A, 2843, 2848, 2859, 2860, 2922, 3000, 3005, 3096, 3107, 3182, 3183, 3333, 3392,

Protein–(*cont.*)
3396–99, 3446, 3449, 3486, 3577, 3857, 4138, 4218, 4504, 4729, 4776, 4827, 4869, 4871, 4872, 4876, 4971, 5059, 5115, 5116, 5431, 5630, 5709, 5742, 5751, 5825, 5887, 5914, 5948, 6015, 6071, 6098, 6102, 6133, 6269, 6276, 6281, 6284, 6432, 6446, 6512, 6516, 6524, 6525, 6533–35, 6578, 6584, 6726, 6737, 6778, 6852, 6897, 7101, 7157, 7175–77, 7232, 7342, 7390, 7491, 7567–70, 7594, 7595, 7618, 7813, 7993, 8051, 8116, 8252, 8354, 8388, 8389, 8460, 8479, 8603, 8661, 8680, 8683, 8684, 8801, 8849, 8938, 8976, 8977, 9079, 9113, 9172, 9225, 9238, 9263, 9379, 9428, 9446, 9448, 9456, 9518, 9659, 9780, 9795, 9802, 9814, 9836, 9837, 9884, 9933, 9936, 9938, 10017, 10021–29, 10047, 10140, 10327, 10328, 10407, 10576, 10738, 10793, 10795–808, 10810–12, 10815–17, 10844, 10853, 10882, 10885, 10934, 10939, 11010, 11017, 11053, 11067, 11070, 11074, 11075, 11077, 11078, 11256, 11271, 11273, 11275, 11310, 11312, 11342, 11373, 11374, 11376, 11458, 11470, 11496, 11626, 11627, 11671, 11751, 11754, 11984, 12050, 12072, 12073, 12077, 12092, 12099, 12227, 12228, 12326, 12514, 12515, 12526, 12595, 12596, 12610, 12857, 12916, 12923, 12943, 12944, 12948, 13302, 13412, 13413, 13416, 13486, 13495, 13496, 13505, 13527, 13589, 13590, 13672, 13673, 13726, 13733, 13747, 13826, 13827, 13972, 14089, 14107–11, 14119, 14229, 14236–38, 14353, 14378, 14501, 14528, 14567, 14685, 14696, 14697, 14718, 14738, 14789, 14793–96, 14857–68, 14870–72, 14883, 14896, 14897, 14963, 14965, 14980, 14997, 15143, 15187–89, 15201, 15225, 15246, 15287, 15297, 15299, 15406, 15521, 15553, 15694, 15695, 15821–23, 15827, 15909, 15981, 16303, 16320, 16321, 16375, 16377, 16430, 16617, 16687, 16807, 16839, 16840, 16882, 16883, 16924, 17016, 17017, 17074–76, 17078, 17099, 17202, 17232, 17550, 17612, 17628, 17711, 17770, 17837, 18115, 18118, 18126, 18247, 18255–57, 18273, 18280, 18366, 18367, 18399, 18418–23, 18497, 18755, 18789, 18898, 18922, 18923, 19010, 19017, 19024, 19070, 19071, 19105–7, 19387, 19449, 19589, 19626, 19667, 19670, 19761, 19800, 19912, 19913, 19972, 19990, 20052–54

Proteinase, 1821, 2555, 5028, 8026, 8249, 11467, 11468

Provenance, 45, 204, 388, 670, 734, 749, 784, 850, 990, 1236, 1527, 1774, 1782, 1789, 1944, 2079, 2089, 2109, 2319, 2333, 2377, 2733, 2873, 2875, 3036, 3078, 3093, 3267, 3345, 3412, 3484, 3529, 3569, 3588, 3591,

Provenance–(*cont.*)
3593, 3624, 3671, 3792, 3796, 3872, 3943, 3984, 4351, 4410, 4423, 4505, 4687, 4700, 4745, 4766, 4890, 5011, 5058, 5312, 5479, 5573, 5577, 5711, 5753, 5790, 5953, 5954, 5956, 6009, 6216, 6643, 6644, 6647, 6652, 6830, 6845, 6920, 7038, 7093, 7102, 7109, 7132, 7282, 7375, 7628, 7907, 7918, 8051, 8156, 8232, 8429, 8667, 8808, 8917, 8954, 8981, 8988, 9099, 9100, 9155, 9188, 9189, 9552, 9671, 9756, 9788, 9958, 10032, 10283, 10470, 10525, 10580, 10628, 10635, 10731, 10871, 10935, 11088–90, 11151, 11153, 11154, 11156, 11169, 11261, 11282, 11299, 11337, 11631, 11694, 11776–79, 11811, 11873, 12145, 12244, 12266, 12277, 12309, 12441, 12504, 12523, 12525, 12804, 13173, 13174, 13239, 13259, 13371, 13435, 13707, 13760, 13915, 13947, 13975, 14066, 14438, 14474, 14670, 14703, 14806, 14975, 15134, 15139, 15213, 15357, 15458, 15479, 15548, 15635, 15670, 15703, 15810, 16021, 16070, 16072, 16146, 16198, 16504, 16581, 16661, 16762, 16800, 16879, 17003, 17204, 17341, 17349, 17579, 17597, 17630, 17830, 17879, 17947, 17955, 17984, 18034, 18035, 18174, 18290, 18474, 18777, 18786, 18915, 18946, 19072, 19118, 19134, 19139, 19296, 19300, 19329, 19386, 19587, 19641, 19642, 19649, 19850, 19951, 20005, 20015, 20092, 20097, 20112

Puerto Rico, 893, 925, 12884

Quality, 64, 239, 262, 309, 411, 463, 538, 541, 543, 544, 731, 863, 936, 1034, 1095, 1120, 1128, 1512, 1665, 2275, 3673, 3865, 4148, 4582, 4651, 5829, 5956, 6218, 6294, 6334, 6346, 6354, 6395, 6419, 6420, 6559, 6767, 6590-B, 7007, 7556, 7557, 7695, 8666, 9094, 9258, 9443, 9843, 10295, 10642, 11562, 11624, 11895, 12409, 12721, 12730, 12769, 12845, 13101, 13234, 13800, 13926, 14975, 15213, 15399, 15510, 15653, 15998, 16314, 16906, 16995, 17080, 17105, 17559, 17642, 17780, 17793, 17832, 17833, 17995, 18027, 18174, 18189, 18215, 18217, 18229, 18454, 18493, 18748, 18785, 18864, 19034, 19123, 19124, 19137, 19160, 19339, 19711, 19873, 20084, 20122

Quebec, 1128

Queensland, 10442

Radiation, 100, 171, 479, 481, 976, 1080, 1186, 1881, 2080, 2082, 2143, 2252, 2444, 2513, 2516, 2614, 2775, 2881–83, 2895, 2916, 2995–97, 3032, 3100, 3158, 3353, 3391, 3445, 3510–13, 3516, 3852, 3902, 3904, 4008, 4223, 4334, 4335, 4420, 4503, 4612, 4613, 4762–64, 4928, 4989, 5025, 5049, 5090, 5106,

Radiation–(*cont.*)
5156, 5285, 5383, 5454, 5633, 5830–33, 5835, 5964, 6003, 6039, 6072, 6144, 6157, 6173, 6296, 6499, 6531, 6601, 6751, 6768, 6769, 6791, 6965, 6966, 7080, 7182, 7233–35, 7427, 7457, 7653, 7662, 7815, 7831, 7888, 7919, 7929, 7933, 8201, 8338, 8339, 8454, 8459, 8736, 8937, 9084, 9202, 9287, 9398, 9440, 9426, 9441, 9495, 9562, 9586, 9636, 9808, 9855, 9926, 9927, 10096, 10200, 10201, 10387, 10388, 10453, 10549, 10565, 10619, 10778, 10780, 10818, 10824–27, 10829, 10854, 11066, 11162, 11163, 11186, 11188, 11460, 11605, 11606, 11664, 11689, 11690, 11702–5, 11707, 11718, 11891, 11908, 12124, 12211, 12212, 12286, 12371, 12500, 12531, 12627, 12750, 12937, 12939, 12970, 12996, 13058, 13059, 13061, 13148, 13193, 13301, 13338, 13372, 13384, 13388, 13389, 13406, 13407, 13439, 13503, 13520, 13612, 13667, 13677, 13691, 13692, 14067, 14068, 14146, 14175, 14191, 14231, 14232, 14385, 14484, 14486, 14517, 14520, 14607, 14680, 14685, 14873, 14875, 15020, 15255, 15463, 15568, 15678, 15679, 16016, 16039, 16251, 16252, 16620, 16746, 16798, 16834–36, 16896, 16966, 17065, 17218, 17291, 17292, 17318, 17385, 17405, 17492, 17493, 17676, 17703, 17792, 17785, 17789, 17790, 18045, 18230, 18377, 18411, 18433, 18787, 18853, 18936, 19022, 19113, 19191, 19320, 19571, 19701, 19702, 19766, 19862

Radioactivity, 13, 171, 479, 1369, 1909, 1926, 2088, 2387, 2388, 2689, 7487, 8057, 8402, 8653, 9645, 11164, 11460, 11506, 13281, 17911, 20042

Rain effects, 19527

Range seeds, 1915, 17994

Refractometer, 13664

Regulations, 781, 837, 1113, 1166, 1184, 1209, 1224, 1225, 1256, 1258, 1268, 1446, 1656, 2066, 2070, 10085, 16391, 17005, 17006

Research, 892, 18654

Respiration, 15, 62, 63, 267, 270, 272, 372, 435, 1631–33, 1674, 1676, 1680, 1740, 2010, 2191, 2405, 2516, 2756, 2870, 2953, 3155, 3163, 3176, 3297, 3483, 3950, 4034, 4265, 4283, 4500, 4534, 4604, 4608, 4613, 4755, 4812, 4929, 4935, 5204, 5220, 5500, 6027, 6125, 6295, 6378, 6444, 6514, 6525, 6639, 6757, 6944, 7127, 7266, 7287, 7383, 7775, 8054, 8055, 8207, 8210, 8367, 8370, 8394, 8955, 9132, 9432, 9557, 9592, 9653, 9688, 9696, 9701, 9743, 9759, 9816, 9949, 9981, 10259, 10271, 10441, 10554, 10593, 10687, 10715, 10718, 11022, 11054, 11306, 11362, 11368, 11385, 11482, 11618, 11787, 11803, 11806, 11920–24, 12171, 12873, 12958,

Respiration–(*cont.*)
13312, 13339, 13351, 13362, 13427, 13514, 13517, 13518, 13533, 13847, 14046, 14398, 14412, 14476, 14593, 14715, 14887, 14890, 14918, 15701, 15702, 15794, 15820, 15863, 15916, 16160, 16896, 17312, 17338, 17594, 17595, 17773, 17870, 18014, 18016, 18017, 18029, 18079, 18152, 18163, 18320, 18355, 18356, 18743, 19108, 19161, 19162, 19355
Review, 12509, 12552
Rhizoids, 14722
Rhodesia, Southern, 1113, 10482
Rhythm, endogenous, 3227, 3282, 3283, 3285–89, 3293, 3886, 8698, 8699, 9292, 10040, 16253, 17795, 17796, 18325, 19951
Ribonuclease, 1905, 11755, 11970, 13015, 13434, 14792–96
Ripening, 206, 207, 323, 329, 573, 595, 612, 990, 1012, 1278, 1296, 2113, 2217, 2548, 2561, 2698, 3225, 3722, 3280, 3281, 4402, 4606, 5362, 5550, 6421, 6605, 6668, 7097, 7165, 7404, 7755, 7906, 8344, 9311, 9556, 10327, 10328, 10365, 10493, 10576, 10887, 11029, 11044, 11047, 11262, 11300, 11305, 11362, 11372, 11389, 11564, 11576, 12046, 12311, 12508, 12550, 13010, 13013, 13014, 13814, 14113, 14168, 14337, 14357, 14509, 15090, 15251, 15293, 15566, 15721, 15821–23, 16116, 16150, 16903, 17078, 17081, 17090, 17124, 17398, 17433, 17595, 17745, 18032, 18253, 18374, 18459, 18485, 18720, 18721, 18820, 18872, 19081, 19092, 19179, 19239, 19387, 19807, 19971, 20052–54, 20137, 20138
Rock garden plants, 693, 14589, 14590, 17029
Root crops, 906, 18705
Rosaceous seeds, 4600
Rubber, 1056, 5260, 7116, 7635, 9580
Russia, 17349

Saliva effects, 977, 11324, 12614, 19960
Salt effects, 32, 146, 150, 295, 754, 1351, 1568–71, 1661, 2059, 2705, 2728, 3411, 5304, 5837, 6169, 8955, 9578, 10246, 11097, 12114, 12535, 12794, 13290, 13506, 13621, 14429, 14609, 14681, 14888, 14929, 14987, 15264, 15266, 15763, 16513, 16685, 17021, 17234, 17381, 17957, 18089, 18332, 18856, 18888, 18965, 18967, 19164, 19275, 19283, 19758, 19974, 19981
Salts effects, 296, 999, 1555, 3218, 4043, 4104, 4266, 4824, 4936, 5123, 5991, 7016, 7341, 8619, 10001, 10153, 10716, 10717, 10751, 11008, 11009, 11279, 11612, 12093, 12385, 12537, 12820, 13339, 13902, 14419, 15228, 16658, 16802, 16803, 17234
Sampling, 3363, 12669, 13469, 13470, 13706, 14274, 18523
Sand-hill seeds, 18626
Saprophytes, 16661
Sawdust, 2495, 5170, 8611, 12584, 14380, 18059
Scandinavian plants, 243
Scarification, 41, 132, 205, 218, 220, 221, 226, 1343, 1395, 1454, 2107, 3049, 3414, 3491, 3721, 3827, 4164, 5137, 5977, 6957, 7424, 7540, 8103, 8889, 9166, 10090, 11557, 11630, 12006, 12427, 12901, 12977, 13758, 14497, 14564, 14880, 15067, 17187, 17913, 17991, 17993, 18065, 18196, 18248, 18636, 18667, 19596, 19710, 19734, 19986
Scotland, 1118,
Scutellum, 2485, 2707, 2939, 4883, 4884, 5681, 5772, 8138, 8840, 11830, 14655, 14665, 14666, 16745, 17261, 17809
Sealing effects, 384, 8155, 12341
Seasonal variation, 1778, 2305, 3287, 3916, 3998, 4885, 5742, 6023, 9476, 19668
Sea water effects, 243, 2331, 2480–82, 2489, 2668, 4918, 5783, 5874–76, 6807, 7843, 7845, 8613, 8787, 8788, 10477, 11248, 12914, 13693, 16459, 16676, 17439, 18413, 18551, 18797, 19590, 19591
Seed, general, 44, 94, 95, 173, 242, 245–49, 340, 457, 472, 488, 568, 569, 623, 635, 762, 791, 814, 860, 877, 1033, 1099, 1198, 1230, 1231, 1273, 1337, 1682, 1731, 1783, 1827, 1869, 1908, 1945, 2013, 2026, 2163, 2176, 2185, 2370, 2381, 2552, 2831, 2927, 3024, 3109, 3348, 3580, 3641, 4591, 4633, 4729, 5076, 5294, 5340, 5796, 5925, 6537, 6735, 6771, 7044–46, 7418, 7844, 7943, 8295, 8547, 8792, 10211, 10212, 10436, 10464, 11236, 11316, 11327, 11343, 11817, 12181, 12184, 12302, 12303, 12406, 12446, 12509, 12656, 12919, 13366, 13459, 13674, 13805, 13813, 13986, 14069, 14276, 14541, 14689, 15224, 15442, 15825, 15881, 16311, 16328, 16690, 16849, 17166, 17769, 17998, 18406, 18507, 18529, 18605, 18607, 18805, 18889, 19086, 19462, 19580, 19755, 19786
Seed characters, 1288, 1405, 2359, 6435, 8510, 8515, 8560, 8648, 9229, 9417, 9595, 11521, 11812, 11998, 13138, 13196, 13421, 13801, 13869, 13954, 13991, 13992, 14039, 14074, 14076, 14083, 14116, 14753, 14760, 14940, 15047, 15280, 15292, 15359, 15367, 15375, 15414, 15426, 15449, 15457, 15476, 15595, 15625, 15626, 15645, 15780, 15876, 15974, 15989, 16024–26, 16103, 16114, 16279, 16318, 16337, 16354, 16400, 16482, 16556, 16619, 16644, 16697, 16873, 16881, 16895, 17192, 17239, 17438, 17446, 17483, 17507, 17684, 17753, 17799, 17979, 18710, 18752
Seed collection, 16561
Seed panels, 9064
Seed production, 241, 313, 375, 395, 567, 714,

Seed production–(*cont.*)
763, 764, 845, 906, 938, 990, 1061, 1083, 1214, 1264, 1267, 1287, 1291, 1537, 4240, 4290, 4390, 4411, 4473, 4487, 4492, 4685, 4753, 4792, 4886, 4934, 5047, 5349, 5455, 5456, 5515–17, 5523, 5531, 5643, 5816, 5820, 5822, 5880, 5883, 5915, 5967, 5996, 6082, 6083, 6085, 6131, 6200, 6307, 6338, 6339, 6573, 6576, 6580, 6663, 6665, 6790, 6961, 6999, 7002, 7261, 7510, 7584, 7603, 7636, 7762, 7904, 7977, 8001, 8074, 8157, 8289, 8408, 8409, 8417, 8421, 8422, 8434, 8445, 8447, 8676, 8677, 8796, 8910, 8982, 9219, 9322, 9324, 9403, 9686, 9735, 9761, 9771, 9923, 9924, 9941, 9951, 9959, 9964, 9966, 10092, 10171, 10236, 10256, 10410, 10435, 10473, 10475, 10899, 10900, 10993–95, 11337, 11710, 11855, 11936, 12090, 12146, 12572, 12573, 12662, 12680, 12761, 13534, 13598, 13599, 13842, 14094, 14282, 14934, 15627, 15901, 16173, 16343, 16644, 16830, 17044, 17085, 17189, 17197, 17198, 17200, 17253, 17698, 17930, 18023, 18109, 18136, 18166, 18167, 18412, 18478, 18496, 18679, 18833, 19297, 19995, 20088
Seed research, 18949
Seed Trade Buyers Guide, 652, 1055, 1237, 2040, 6616, 6621–36, 6680, 14449, 14532, 14775, 15157
Seedlessness, 808, 2390, 3869, 4042, 4311, 4778, 6855, 7119, 7377–79, 7908, 8238, 8415, 8416, 10266, 10320, 10603, 11844, 16332, 17240, 17552, 17686, 17985, 18121, 18886, 19463, 19666, 19885, 19886, 19966
Selection, 2138, 2165, 2166, 2284, 2338, 3793, 4205, 5341, 5509, 5775, 12078, 13264, 14773, 14931, 14967, 15628, 16648, 16680, 16778, 16988, 17237, 17605, 17815, 17914, 17915, 18336, 18885, 19651
Selenium, 5807, 5847, 11673, 12160, 13660, 14769, 18045, 18541
Serology, 19158
Set of seed, 81, 105, 109, 147, 159, 235–37, 275, 283, 285, 310, 349, 366, 368, 415, 532, 556, 562, 608, 614, 617, 620, 659, 681, 688, 704, 852, 873, 964, 990, 991, 1147, 1252, 1277, 1299, 1340, 1347, 1364, 1366, 1422, 1423, 1528, 1606, 1745, 1765, 1861, 1898, 1910, 2204-A, 2242, 2298, 2322, 2347, 2378, 2391, 2418, 2448, 2451, 2460, 2462, 2484, 2486, 2605, 2631, 2651, 2659, 2661, 2669, 2674, 2682, 2714, 2718, 2719, 2730–32, 2751, 2832, 2933, 2934, 2938, 2946, 2949, 2970, 2990, 3013, 3028, 3040, 3050, 3055, 3058, 3074, 3079, 3094, 3119, 3143. 3181, 3193, 3237, 3245, 3277, 3290, 3299, 3307, 3327, 3328, 3339, 3416–19, 3458, 3525, 3553, 3572, 3614, 3615, 3617, 3634, 3635, 3667, 3724, 3800, 3803, 3879, 3949, 3984, 4089, 4113, 4169, 4170–73, 4176, 4180, 4246, 4249, 4252, 4257, 4353, 4378, 4381, 4382, 4384, 4387, 4388, 4516, 4517, 4705, 4732, 4839, 4854, 5010, 5015, 5101, 5208, 5256, 5257, 5343, 5350, 5419, 5476, 5494, 5535, 5626, 5698, 5714, 5715, 5816, 5817, 5819, 5903, 5918, 5945, 5968, 6295, 6366–68, 6431, 6553, 6554, 6794, 6795, 6810, 6858, 6979, 6988, 7126, 7623, 7886, 7937, 8064, 8119, 8405, 8414, 8419, 8425, 8624, 8682, 8689, 8690 8952, 9088, 9127, 9139, 9243, 9390, 9410, 9454, 9472, 9474, 9608, 9609, 9785, 9846, 9849, 9901, 9943, 9944, 9968, 10041, 10053, 10122, 10123, 10156, 10247, 10321, 10335, 10364, 10381, 10404, 10419, 10507, 10534, 10539, 10898, 11182, 11440, 11461, 11600, 11712, 11762, 11786, 11978, 12064, 12116, 12117, 12174, 12203, 12269, 12313, 12350, 12403, 12408, 12423, 12457, 12487, 12741, 12816, 12817, 13153, 13176, 13197, 13414, 13490, 13532, 13552, 13618, 13631, 13632, 13642, 13704, 13787, 13906, 13949, 13956, 13963, 13970, 13981, 14040, 14085, 14494–96, 14551, 14671, 14683, 14684, 14689, 14729, 14743, 14751, 14758, 14815, 14816, 14978, 15150, 15152, 15158, 15170, 15171, 15177, 15241, 15300, 15310, 15411, 15416, 15478, 15752, 15987, 16048, 16338, 16342, 16407, 16652, 16654, 16656, 16760, 16761, 16763, 16793, 17086, 17227, 17306, 17321, 17358, 17434, 17436, 17516, 17845, 17982, 17999, 18039, 18061, 18068, 18069, 18124, 18131, 18204, 18284, 18379, 18717, 18932, 18992, 18997, 19521, 19641, 19642, 19699, 19808, 19810, 19844, 19948, 20021, 20050, 20060
Sewage effects, 13108
Sex, 803, 873, 2049, 2517, 6929, 8672, 9720
Shape, 86, 6363, 6829, 7309, 8325, 11695, 11738, 11739, 12592, 13811, 14155, 14162, 14725, 18849, 19452
Shelterbelt species, 735
Shipping, 2785, 15401, 19188, 19552
Shrubs, 130, 391, 884, 1281, 2992, 6035, 7746, 8547, 10678, 10755, 10774, 13551, 14465, 14663, 15001, 15002, 15004, 15276, 15465, 16467, 16497, 17882, 19342
Siberia, 11280, 14480
Sierra Leone, 15397
Silage, 1229, 20043
Silver effects, 806, 14549
Size, 615, 1257, 1499, 1521, 1687, 1960, 1963, 2057, 2070, 2253, 2383, 2519–23, 2590, 2591, 2680, 2694, 2702, 2982, 3144, 3253, 3256, 3437, 3548, 3562, 3625, 3696, 3837, 3858, 3897, 4101, 4136, 4493, 4632, 4743, 4912, 4913, 4916, 4930, 5122, 5181, 5328, 5496, 5541, 5542, 5625, 5894, 6013, 6018,

Size–(*cont.*)
6324, 6337, 6356, 6363, 6375, 6449, 6579, 6712–19, 6770, 6921, 6950, 6950-A, 7106, 7305, 7309, 7410, 7426, 7517, 7712, 7753, 7937, 7941, 7983, 8043, 8149, 8274, 8522, 8888, 8908, 9010, 9073, 9115, 9116, 9117, 9206, 9288, 9353, 9375, 9378, 9387, 9561, 9587, 9756, 9853, 9986, 10031, 10032, 10043, 10106, 10127, 10143, 10157, 10343, 10540, 10577, 10580, 10730, 10769, 10873, 10874, 10947, 11122, 11124, 11127, 11130, 11176, 11178, 11203, 11231, 11264, 11276, 11335, 11444, 11457, 11550, 11552, 11556, 11695, 11766, 11808, 11823–25, 11879, 11899, 11901, 11977, 11989, 12029, 12383, 12459, 12664, 12765, 13172, 13203, 13397, 13411, 13605, 13615, 13901, 13971, 14039, 14130, 14162, 14196, 14533, 14698, 14773, 14895, 15425, 15508, 15539, 15567, 15592, 15630, 15685, 15905, 16053, 16107, 16126, 16222, 16350, 16359, 16372, 16474, 16585, 16661, 16663, 16818, 16898, 17064, 17110, 17142, 17359, 17441, 17443, 17445, 17496, 17546, 17631, 17692, 17770, 18005, 18006, 18031, 18140, 18214, 18719, 18849, 18927, 18929, 18937, 18958, 18993, 19132, 19301, 19325, 19326, 19397, 19449, 19556, 20014, 20016

Soaking effects, 27, 270, 271, 354, 464, 616, 650, 664, 1086, 1409, 1410, 1694, 1958, 1959, 1978, 2017, 2020, 2023, 2300, 2349, 2454, 2811, 2814, 2950, 2951, 2977, 3927, 3960, 3968, 3971, 4401, 4457, 4862, 4894, 4936, 5218, 5457, 5575, 5698, 5832, 5885, 5889, 6152, 6165, 6508, 6832, 6898, 7336, 7388, 7487, 7843, 7971, 8326, 8491, 8824, 9028, 9340, 9398, 9538, 9543, 9568, 10038, 10200, 10242, 10263, 10556, 10558, 10895, 11164, 11203, 11506, 11949, 11952, 11966, 12276, 12621, 12752, 12759, 13452, 13568, 13574, 13670, 13927, 14026, 14133, 14169, 14344, 14356, 14415, 14515, 14530, 14572, 14609, 14732, 15118, 15667, 15852, 16022, 16099, 16280, 16507, 16509, 16566, 16572, 16645, 16831, 16871, 17182, 17219, 17229, 17608, 17705, 17747, 17778, 17953, 18143, 18144, 18532, 18572, 18582, 18725, 18856, 19312, 19396, 19444, 19522, 19658, 19729, 19932–35

Sodium, 1292, 1426, 2408, 2727, 3733, 5675, 5878, 6279, 6282, 6322, 7286, 7873, 8085, 8499, 10543, 12098, 12319, 14586, 17591, 18540

Soils effects, 108, 150, 490, 636, 894, 895, 898, 1498, 1571, 1910, 1937, 1974, 2140, 2463, 2830, 2922, 2987, 2988, 3194, 3302, 3303, 3371, 3411, 3549, 3826, 4247, 4278, 4306, 4307, 4619, 4697, 4747, 4768, 5012, 5107, 5304, 5306, 5462, 5469, 5592, 5641, 5650, 5656, 5716, 5779, 5791, 5792, 5994, 5998, 5999, 6002, 6129, 6344, 6411, 6412, 6427, 6434, 6589, 6605, 6683, 6758, 6838, 6928, 7085, 7090, 7307, 7358, 7408, 7463, 7541, 7769, 7956, 7994, 8007, 8083, 8084, 8087, 8108, 8109, 8147, 8183, 8198, 8200. 8257, 8418, 8422, 8440, 8479, 8491, 8528, 8641, 8955, 9000, 9001, 9036, 9039, 9080, 9165, 9231, 9235, 9267, 9269, 9368, 9447, 9611, 9777, 9867, 10451, 10618, 10951, 11005, 11175, 11354, 11410, 11495, 11513, 11754, 11856, 11909, 11960, 12036, 12040, 12199, 12397, 12400, 12503, 12968, 13199, 13390, 13392, 13424, 13506, 13535–38, 13668, 13702, 13967, 13969, 14306, 14308, 14382, 14393, 14394, 14415, 14763, 14787, 14898, 15329, 15477, 15525, 15763, 15805, 16133, 16295, 16578, 16804, 16865, 16866, 16885, 16937, 17060, 17141, 17509, 17583, 17649, 17678, 18114, 18176, 18177, 18192, 18307, 18386, 18407, 18509, 18544, 18562, 18589, 18629, 18630, 18637, 18690, 18693, 18768, 18953, 18979, 19065, 19316, 19322, 19329, 19371, 19391, 19650, 19845, 19873, 19981, 20006

Solomon Islands, 15320

Sound, 106

South Africa, 747, 12165, 16193, 19005, 19006

Sowing conditions, 87, 102, 109, 114, 140, 203, 253, 313, 354, 409, 451, 453, 517, 610, 615, 673, 691, 697, 709, 715, 747, 760, 879, 943, 946, 1067, 1069, 1091, 1134, 1143, 1191, 1228, 1255, 1304, 1305, 1394, 1585, 1607, 1647, 1687, 1880, 1927, 2029, 2115, 2130–32, 2349, 2389, 2465, 2518, 2521, 2544, 2599, 2637, 2829, 2901, 2912, 3143, 3224, 3329, 3357, 3395, 3531, 3532, 3611, 3626–28, 3694, 3789, 3834, 3907, 4092, 4109, 4166, 4173, 4278, 4309, 4416, 4419, 4518, 4519, 4606, 4881, 5011, 5057, 5070, 5073, 5088, 5094, 5095, 5145, 5177, 5194, 5388, 5401, 5499, 5538, 5605, 5621, 5627, 5640, 5776, 5826, 5873, 5891, 5905, 5927, 5993, 5997, 6013, 6023, 6088–90, 6504, 6540, 6550, 6674, 6681, 6691, 6834, 6836, 6899, 6950-A, 6950-D, 6975, 7055, 7154, 7170, 7346, 7347, 7551, 7769, 7822, 7854-A, 8025, 8085, 8173, 8199, 8220, 8310, 8477, 8689, 8690, 8706, 8890, 8902, 8906, 8930, 8953, 8998, 9033, 9066, 9154, 9176, 9197, 9458, 9476, 9571, 9800, 9819, 9820, 9862, 10061, 10249, 10284, 10396, 10447, 10520, 10541, 10667, 10695, 10709, 10710, 10813, 10937, 10993, 11012, 11233, 11503, 11701, 11708, 11709, 11746, 11881, 11897, 11899, 11900, 11985, 12035, 12049, 12180, 12403, 12407, 12494, 12502, 12564, 12569, 12613, 12664,

Sowing conditions–(*cont.*)
12775, 12828, 12902, 13060, 13133, 13134, 13155, 13182, 13230, 13278, 13415, 13786, 13829, 13842, 13877, 13901, 13969, 14078, 14142, 14173, 14478, 14555, 14838, 14877, 15057, 15060, 15101, 15117, 15191, 15192, 15309, 15325, 15339, 15435, 15616, 15631, 15694, 15695, 15717, 15741, 15872, 15983, 16046, 16428, 16499, 16555, 16737, 17289, 17290, 17699, 17700, 17819, 18002–4, 18006, 18022, 18122, 18144, 18456, 18579, 18700, 18715, 18777, 18929, 18952, 18953, 19277, 19278, 19325, 19326, 19378, 19383, 19403, 19486, 19643, 19668, 19712, 19749, 19793, 19814, 19817, 19851, 19860, 20014, 20049, 20109
Space flight, 10550
Species effects, 12504, 14253
Specific gravity, 573, 2767, 3191, 6431, 14524, 15014
Sphagnum, 896, 985, 2762, 4216, 4217, 7254, 9012
Starch, 75, 324, 325, 339, 548, 1324, 1825, 2282, 2862, 2889, 3276, 3468, 3469, 3738, 4822, 4873, 4898, 4923, 5072, 5151, 5746, 6076, 6482, 6706, 7634, 7635, 7807, 7816, 8102, 8235, 8747, 8870, 8871, 9024, 9079, 9412, 10886, 11564, 12650–54, 13062, 13236, 13339, 15107, 15164, 15750, 16717, 16718, 16747, 16768, 17015, 18390, 18391, 19389, 19414, 19588, 19798, 19984
Statistics, 188, 750, 2580, 3254, 4495, 4886, 4931, 5278, 5519, 8136, 8949, 9587, 9686, 10929, 11550, 12012–19, 12021, 13591, 15600, 15608, 15912, 16316, 16609, 16711, 16910, 17221, 17475, 17813, 18292, 18525, 18810, 19410, 19493, 19626, 19999, 20051
Sterilization, 382, 519, 600, 751, 3697, 4649, 5210, 7062, 7286, 8900, 12414, 14046, 17777, 17958, 18698
Stimulating substances, 16, 17, 33, 137, 182, 268, 271, 272, 276, 314, 398, 619, 657, 699, 1180, 1563, 1572, 1885, 2054, 2081, 2173, 2213, 2560, 2641, 2673, 2707, 2721, 2737, 2793, 2958, 2965, 2966, 3073, 3166, 3171–77, 3270, 3290, 3355, 3670, 3712, 3714, 4128, 4135, 4201, 4209, 4439, 4894, 4933, 4962, 5287, 5288, 5395, 5726, 5727, 5732, 6171, 6239, 6400, 6709, 6739, 6847, 6994, 7019, 7294, 7304, 7335, 8535, 8673, 8740, 9333, 9376, 9377, 9503, 9504, 9928, 10113, 10178, 10299, 10303, 10713, 10720, 10855, 10926, 10957, 11224, 11277, 11279, 11558, 11886, 12763, 12820, 12990, 13067, 13074, 13100, 13246, 13756, 13910, 14251, 14264, 14265, 14290, 14363, 14403, 14410, 14412, 14415, 14429, 14431, 14435, 14437, 14446, 14579, 15030, 15229, 15269, 15561–64, 15781, 16084, 16139, 16432, 16520, 16527,
Stimulating substances–(*cont.*)
16529, 16989, 17285, 17294, 17892, 17951, 17989, 17990, 18084, 18331, 18516, 18939, 18940, 18959, 19050, 19164, 19222, 19232, 19864, 19915, 20059, 20120, 20121
Stone fruits, 8792
Storage, 21, 84, 85, 133, 157, 162, 170, 177, 194, 219, 222, 223, 225, 227, 228, 250, 255, 257, 258, 279, 282, 309, 383, 384, 403, 417, 429, 430, 431, 434, 437, 438, 453, 545, 546, 601, 604, 643, 651, 656, 666, 670, 673, 678, 679, 683, 698, 708, 721, 725, 734, 742, 759, 783, 785, 796, 797, 805, 806, 811, 818, 830, 864, 876, 881, 893, 925, 945, 952, 965, 970, 990, 992, 995, 996, 1019, 1020, 1027, 1030, 1075, 1110, 1114, 1130, 1133, 1139, 1142, 1159, 1167, 1173, 1192, 1198, 1208, 1218, 1310, 1455, 1490, 1495, 1496, 1574, 1579, 1626, 1627, 1673, 1689, 1690, 1691, 1703, 1711, 1720, 1741, 1869, 1919, 1983, 1984, 1987, 1988, 1994–97, 2003, 2005, 2006, 2008, 2010–12, 2014, 2015, 2022, 2025, 2026, 2031, 2032, 2036, 2042, 2073, 2076, 2090, 2118, 2150, 2151, 2155, 2209, 2220, 2368, 2438, 2446, 2537, 2592–94, 2640, 2722, 2834, 2845, 2846, 2899, 2902, 2940, 3010, 3011, 3070–72, 3098, 3121, 3122, 3131, 3142, 3210–14, 3232, 3233, 3324, 3351, 3364, 3366, 3373, 3436, 3438, 3456, 3521, 3590, 3618, 3630, 3651, 3652, 3662, 3768, 3791, 3830, 3838, 3936, 3944, 3945, 3957, 3958, 3978, 4021, 4022, 4056, 4057, 4060, 4099, 4124, 4186, 4188, 4192, 4225, 4285, 4292, 4315, 4319, 4323, 4331, 4340, 4346, 4366, 4442, 4581, 4587, 4594, 4595, 4600, 4623, 4634, 4642, 4686, 4736, 4760, 4763, 4766, 4888, 4915, 4934, 4979, 4990, 4998, 5019, 5093, 5108, 5141, 5160, 5191, 5206, 5244, 5248, 5250, 5295, 5307, 5311, 5325, 5354, 5506, 5507, 5563, 5609, 5610, 5635, 5695, 5703, 5704, 5738, 5777, 5778, 5810, 5881, 5882, 5886, 5921, 5922, 6042, 6052, 6053, 6086, 6087, 6091–93, 6330, 6333, 6362, 6507, 6510, 6570, 6597, 6687, 6734, 6779, 6797, 6820, 6873, 6879, 6881, 6886, 6887, 6948, 6973, 7061, 7067, 7068, 7112, 7116, 7130, 7143, 7203, 7220, 7222, 7253, 7295, 7298–300, 7333, 7415, 7451, 7485, 7499, 7535, 7537, 7539, 7600, 7669, 7677, 7715, 7729, 7768, 7808, 7810, 7827, 8031, 8032, 8034, 8044, 8075, 8133, 8151, 8153–55, 8191, 8223, 8226, 8229, 8253, 8260, 8261, 8349, 8453, 8469, 8548, 8550, 8623, 8670, 8730, 8781, 8814, 8824, 8865, 8894, 8895, 8959, 8963, 8971, 8972, 8978, 8983, 9007, 9009, 9027, 9081, 9092, 9109, 9111, 9190, 9207, 9227, 9230, 9239, 9247, 9257, 9314, 9316, 9363, 9434, 9451, 9462, 9519, 9520, 9586, 9629, 9660, 9677, 9682–

Storage–(*cont.*)
85, 9687, 9726, 9767, 9784, 9816, 9842, 9847, 9875, 9934–37, 9945, 9950, 9956, 9963, 9988, 9998, 10079, 10137, 10139, 10197, 10336, 10400, 10402, 10437–39, 10471, 10638, 10639, 10739, 10743, 10862, 10939, 10962, 10967, 11085, 11091, 11092, 11097, 11098, 11106, 11111–14, 11118, 11119, 11121, 11123, 11125, 11126, 11132, 11133, 11136, 11138–49, 11157, 11159, 11161, 11193, 11194, 11196, 11197, 11201, 11210, 11211, 11238, 11246, 11296, 11351, 11369, 11396, 11398, 11399, 11403, 11433, 11436, 11463, 11509, 11529, 11544, 11562, 11580, 11595, 11609, 11611, 11618, 11638, 11645, 11658, 11672, 11700, 11715, 11796, 11799, 11827, 11841, 11845, 11878, 12051, 12084, 12139, 12164, 12175, 12176, 12253, 12280, 12287, 12339–41, 12357, 12384, 12410, 12418, 12428, 12436, 12440, 12546, 12625, 12650, 12658, 12668, 12718, 12720, 12846, 12882, 12883, 12908, 12983, 12986, 12987, 13001, 13030, 13121, 13145, 13244, 13245, 13273, 13302, 13466, 13467, 13497, 13504, 13519, 13522, 13542, 13551, 13561, 13563, 13565, 13650, 13700, 13718, 13726, 13747, 13832, 13866, 13868, 13870, 14021, 14024, 14027, 14030, 14057, 14133, 14147, 14253, 14269, 14313, 14314, 14338, 14342, 14360, 14407, 14480, 14482, 14578, 14606, 14617, 14628, 14629, 14646, 14656, 14742, 14780, 14813, 14832, 14910, 14915–17, 15025, 15081, 15083, 15113, 15115, 15155, 15174, 15214, 15401, 15407, 15437, 15532, 15581, 15588, 15617, 15648, 15683, 15765, 15767, 15784, 15813, 15847, 15916, 15947, 15955, 16014, 16066, 16284–87, 16323, 16326, 16327, 16339, 16362, 16364, 16406, 16510, 16512, 16546, 16552, 16627, 16758, 16806, 16817, 16842, 16843, 16881, 17038, 17039, 17097, 17179, 17233, 17279, 17289, 17295, 17317, 17336, 17337, 17344, 17361, 17422, 17428, 17429, 17463, 17466, 17467, 17515, 17520, 17588, 17594, 17596, 17611, 17624, 17699, 17700, 17739, 17748–52, 17787, 17832, 17833, 17873, 17890, 17916, 17917, 17983, 17987, 18051, 18052, 18116, 18202, 18213, 18223, 18242, 18263, 18277, 18315, 18403, 18417, 18455, 18573, 18590, 18655, 18656, 18669, 18675, 18676, 18679, 18682, 18723, 18724, 18812–14, 18846, 18853, 18868, 18884, 18898, 18904, 19003, 19022, 19054, 19100, 19127–29, 19136, 19151, 19184, 19188, 19266, 19291, 19329, 19343, 19446, 19459–61, 19574, 19577, 19589, 19634, 19637, 19705, 19726, 19733, 19746, 19763, 19866, 19879, 19890, 19905, 19907, 19983, 20003, 20070, 20078, 20080

Stratification, 98, 128, 130, 132, 135, 142, 204, 377, 387, 389, 670, 673, 676, 689, 729, 748, 773, 951, 1023, 1179, 1447, 2229, 2299, 2557, 2630, 3054, 3480, 4595, 5351, 5384, 5500, 6146, 6940, 6941, 6990, 7221, 7222, 7369, 7512, 7591, 8002, 8169, 8191, 8552, 8559, 8568, 9028, 9087, 9462, 9661, 9662, 9664, 9792, 9874, 9876, 9949, 10166, 10523, 10755, 11181, 11200, 11222, 11247, 11288, 11366, 11781, 11880, 12053, 12191, 12194, 12267, 12352, 12593, 12631, 12736, 12940, 12941, 13229, 13231, 13339, 13346, 13565, 13584, 14158, 14386, 14415, 14470, 14473, 14906, 15001, 15002, 15028, 15181, 15272, 15275, 15276, 15469, 15487, 15576, 15629, 15692, 15802, 16361, 16373, 16507, 16700, 16876, 17273, 17715, 18054, 18190, 18227, 18626, 19008, 19706, 19994

Succulents, 677, 19223

Sudan, Anglo Egyptian, 1280

Sugars, 521, 754, 1641, 2128, 2204, 2217, 2248, 2271, 2272, 2666, 2736, 2814, 2864, 2865, 3726, 5668, 5834, 6808, 6833, 7756, 7939, 7969, 9808, 11667, 12480, 12641, 12643, 12645, 12945, 12959, 13378, 13379, 13446, 14571, 16897, 18646

Sulfa drugs, 2407, 2409, 3030, 3451, 6964, 7160, 8455, 9576, 11516, 11519, 15204, 16163, 17553

Sulphur, 578, 583, 586, 587, 589, 590, 2001, 2362, 2472, 5757, 7612, 8259, 9787, 9836, 9837, 11846, 12518, 13814, 14038, 15132, 16131, 16431, 16942, 16943, 17176, 17177, 18516, 18593

Sulphuric acid, 128, 130, 132, 217, 218, 220, 228, 530, 654, 813, 1011, 1094, 1328, 1476, 1477, 1696, 2328, 2543, 3039, 3110, 3152, 3153, 3221, 3372, 3402, 3521, 3715, 4079, 4892, 5342, 5392, 5459, 5571, 5780, 6054, 6735, 6974, 6986, 7053, 7743, 8069, 8070, 8078, 8099, 8112, 8264, 8356, 8357, 8590, 9607, 9876, 9905, 9952, 11220, 12068, 12438, 12452, 12587, 13285, 13382, 15466, 15618, 15677, 16449, 17128, 17304, 17311, 17774, 17991, 18018, 18611, 19216, 19260, 19687, 19738

Sun effects, 6273, 11870, 18337

Supersonics, 1029, 1523, 2662, 3434, 3435, 5920, 6296, 6376, 6415, 7184–86, 7754, 8145, 8328, 8688, 10363, 10707, 10948, 11313, 11434, 11907, 12242, 12243, 12272, 12473, 13542, 14661, 14662, 15555, 15558, 15564, 16013, 16497, 17091, 18304, 19809, 20119

Surface tension, 4326, 14415, 18757, 19213

Sweden, 241, 242, 245–49, 1195, 1214, 10915, 14492, 17447

Symbiosis, 2823, 2824, 2869, 2872, 2962,–64, 3581, 3582, 3584, 4190, 4770, 4771

Taiwan, 13819
Tanganyika, 18202
Tannin, 33, 1659, 3670, 5690, 5877, 6517, 7371, 7401, 9039, 11227, 14430, 16092, 16539
Taxonomy, 166, 357, 469, 1253, 1345, 2317, 2456, 2738, 3108, 3313, 3719, 3720, 3980, 5298, 5414, 6584, 7044–46, 7829, 8284, 8497, 8756, 10115, 10661, 10865, 12694, 13236, 13673, 13953, 14474, 15209, 15432, 16114, 18134, 18273, 18390–92, 19027, 19030, 19685, 19686, 20045
Technique, 189, 190, 626, 1060, 2050, 14874
Temperature, 21, 65, 262, 278, 312, 404, 546, 821, 990, 1274, 1295, 1672, 1873, 2161, 2176, 2195, 2196, 2228, 2264, 2518, 2551, 2558, 2560, 2648, 2719, 3115, 3155, 3180, 3195, 3199, 3238, 3287, 3308, 3652, 3680, 3741, 3773, 3775, 3982, 3993, 4181, 4248, 4953, 5635, 5669, 5731, 5758, 5819, 5945, 6124, 6126, 6211, 6263, 6368, 6434, 6607, 6707, 6831, 6946, 7009, 7015, 7027, 7028, 7064, 7097, 7124, 7352, 7388, 7647, 7648, 7667, 8054, 8055, 8231, 8325, 8326, 8398, 8465, 8618, 8871, 8899, 9000, 9036, 9271, 9360, 9447, 9450, 9493, 10067, 10094, 10126, 10225, 10400, 10402, 10453, 10693, 10701, 10869, 10891, 11010, 11029, 11188, 11317, 11380, 11382, 11474, 11711, 11723, 11782, 11788, 11792–94, 12038, 12081, 12173, 12252, 12472, 12553, 12648, 12733, 12734, 12750, 12757, 13208, 13516, 13683, 14077, 14168, 14185, 14186, 14227, 14443, 14516, 14681, 15769, 15777, 15812, 15856, 15987, 16450, 16521, 16554, 17087, 17088, 17151, 17198–200, 17267, 17339, 17382, 17387, 17407, 17409, 17423, 17517, 17535, 18042, 18043, 18088, 18188, 18190, 18216, 18230, 18278, 18316, 18317, 18334, 18335, 18351–54, 18357, 18365, 18435, 18450, 18484, 18863, 18870, 18891, 19044, 19057, 19102, 19141, 19215, 19227, 19267, 19309, 19345, 19355, 19459–61, 19471, 19508, 19527, 19531, 19622, 19625, 19637, 19668, 19675, 19714, 19730, 19731, 19875, 19878, 19888, 19908, 19909, 19918, 19971, 19987, 20028, 20048
Temperature, alternating, 488, 492, 494, 1153, 1503, 2899, 4258, 5035, 5432, 5562, 7769, 8208, 8513, 8955, 10259, 10670, 10681, 11422, 11507, 13751, 14904, 14936, 14937, 15334, 15344, 15886, 18268, 18977, 19217, 19329
Temperature effects on dormancy, 22, 1810, 2480, 2481, 3522, 4234, 5033, 5896, 6465, 6496, 7897, 8700, 8816, 9200, 10168, 10897, 11406, 11587, 14476, 16388, 17350, 17351, 17354, 17355, 18054
Temperature effects on germination, 18, 19, 47, 108, 149, 150, 195, 196, 221, 222, 307, 377, 384, 387, 412, 424, 468, 495, 497–99, 512, 513, 609, 610, 614, 632, 647, 650, 653, 667, 689, 693, 694, 696, 729, 733, 782, 785, 798, 804, 849, 854, 951, 1027, 1042, 1347, 1441, 1449, 1482, 1600, 1620, 1693, 2127, 2195, 2389, 2452, 2606, 2609, 2695, 2763, 2789, 2792, 2811, 2845, 2922, 2976, 2978, 3146, 3205, 3258, 3274, 3319, 3336, 3347, 3408, 3443, 3444, 3549, 3604, 3729, 3730, 3829, 3837, 3977, 3982, 4028, 4144, 4145, 4229, 4235, 4248, 4640, 4740, 4969, 5056, 5078, 5093, 5126, 5164, 5251, 5375, 5404, 5432, 5484, 5505, 5557, 5561, 5565, 5592, 5653, 5674, 5779, 5780, 5787, 5793, 5795, 5842, 5843, 5850, 5911, 5912, 5942, 5976, 6064, 6079, 6101, 6107, 6261, 6394, 6451, 6452, 6471, 6555, 6587, 6607, 6637, 6638, 6654, 6683, 6756, 6784, 6793, 6797, 6815, 6844, 6896, 6990, 7009, 7013, 7016, 7020, 7024, 7134, 7219, 7220, 7222, 7265, 7416, 7468, 7589, 7698, 7730, 7748, 7843, 7912, 7913, 7925, 7938, 7945, 7967, 8156, 8195, 8198, 8202, 8205, 8206, 8212, 8281, 8283, 8301, 8308, 8324, 8435, 8491, 8549, 8568–70, 8576, 8579, 8585, 8633, 8634, 8667, 8699, 8742, 8753, 8754, 8864, 8918, 8932, 8955, 8970, 8998, 9008, 9040, 9044–48, 9221, 9243, 9337, 9338, 9367, 9387, 9415, 9422–24, 9487, 9488, 9491, 9570, 9769, 9817, 9868, 9956, 10049, 10074, 10101, 10123, 10190, 10226, 10300, 10344, 10385, 10394, 10399, 10401, 10558, 10569, 10608, 10668, 10676, 10681, 10682, 10692, 10782, 10783, 10866–68, 10928, 10931, 10949, 10952, 10953, 11018, 11064, 11107, 11131, 11134, 11218, 11278, 11330, 11538, 11598, 11599, 11743, 11790, 11795, 11844, 11853, 11855, 11856, 11861, 11867, 12040, 12051, 12148, 12164, 12269, 12305, 12329, 12360, 12375, 12381, 12498, 12552, 12716, 12834, 12875, 12904, 12913, 12920, 12969, 12992, 13000, 13002, 13122, 13147, 13154, 13179, 13339, 13354, 13394, 13418, 13638, 13757, 13859, 13909, 14037, 14115, 14158, 14339, 14354, 14386, 14404, 14412, 14413, 14415, 14452, 14587, 14630, 14817, 14898, 14973, 14974, 14998, 15067, 15394, 15395, 15409, 15413, 15456, 15547, 15570, 15577, 15578, 15582, 15589, 15644, 15658, 15839, 15854, 15884, 16038, 16062, 16063, 16065, 16183, 16211, 16257, 16334, 16355, 16360, 16385, 16388, 16389, 16408, 16439, 16489, 16544, 16547, 16609, 16654, 16656, 16875,

Temperature effects on germination–(*cont.*) 16922, 16960, 17008, 17010, 17097, 17154, 17157, 17183, 17219, 17285, 17437, 17540, 17621, 17633, 17701, 17737, 17768, 17818, 17886, 17912, 17922, 18010, 18064, 18514, 18612, 18648, 18652, 18660, 18667, 18668, 18670, 18671, 18673, 18681, 18683, 18689, 18691, 18692, 18703–5, 18761

Temperature effects on storage, 219, 223, 225, 227, 257, 384, 2003, 2024, 2152, 2193, 3365, 3944, 3957, 4106, 4124, 4336, 4811, 5141, 5315, 5490, 5778, 5921, 6060, 6362, 6947, 6948, 7295, 7300, 7415, 7451, 7617, 8155, 8894, 9027, 9111, 9295, 9519, 11092, 11113, 11125, 11132, 11133, 11208, 11672, 11799, 11878, 12263, 12335, 12357, 12558, 12735, 12900, 14248, 15669, 16326, 16510, 16642, 16725, 17039, 17138, 17916, 18675, 18682, 18718

Temperature, high, 169, 1615, 1949, 1967, 3130, 3435, 3508, 3656, 3900, 4052, 4059, 4239, 4330, 5192, 6518, 6588, 6903–7, 7181, 7372, 7379, 7535, 7578, 7593, 7751, 8036, 8353, 8889, 9110, 9200, 9274, 9334, 9441, 9817, 9867, 9948, 10037, 11320, 13263, 13669, 14248, 14913, 15258, 15341, 15884, 16235, 16236, 16258, 17108, 17592, 17658, 17778, 17948, 17993, 18515, 18536, 18604, 18804, 19398, 19412

Temperature, low, 27, 54, 204, 461, 624, 754, 812, 1313, 1967, 2228, 2873, 2924, 2994, 3199, 3368, 4957, 6067, 6587, 6934, 7098, 8536, 8955, 8999, 9015, 9095, 9339, 9539, 9540, 9670, 10105, 10445, 11341, 11386, 11822, 11852, 12300, 12302, 12303, 12378, 13255, 13339, 13622, 13848, 14268, 14379, 14381, 14382, 14638, 14710, 14834, 14902, 15059, 15346, 15433, 15590, 15591, 15620, 15855, 16317, 16429, 16530, 16698, 16877, 16993, 17153, 17172, 17341, 17629, 17713, 18102, 18174, 18187, 18200, 18393, 18394, 18484, 18489, 18575, 19095, 19284, 19597, 19643, 19818, 19875

Tension, 3716, 19313

Testing and analysis, 9, 42, 94, 124, 180, 193, 198, 245, 249, 376, 390, 418, 448, 491, 501, 508, 509, 514, 541, 543, 544, 686, 728, 739, 755, 761, 766, 781, 791, 827, 919, 923, 955, 997, 998, 1004, 1076, 1082, 1087, 1093, 1102, 1111, 1113, 1125, 1156, 1162, 1166, 1169, 1184, 1190, 1193, 1195, 1196, 1199, 1205, 1209, 1211, 1222, 1224, 1225, 1256, 1258, 1446, 1507, 1508, 1517, 1653, 1707, 1744, 1771, 1773, 1898, 1912, 1923, 1972, 1976, 1979, 2026, 2335, 2376, 2382, 2503, 2546, 2547, 2895, 2898, 2947, 3008, 3009, 3022, 3124, 3132, 3140, 3322, 3413, 3459, 3461, 3490, 3650, 3673, 3676, 3813, 3908, 3972, 3999, 4144, 4187, 4232, 4286, 4365,

Testing and analysis–(*cont.*) 4478, 4482, 4646–48, 4663, 4668, 4673, 4675, 4691, 4790, 4946, 4983–85, 5001, 5040, 5080, 5180, 5190, 5226, 5270, 5357, 5382, 5398, 5491, 5492, 5519, 5744, 5864, 5866, 5895, 5929, 6011, 6043, 6057, 6061, 6113, 6138, 6140, 6143, 6334, 6347, 6361, 6410, 6528, 6560, 6592, 6593, 6609, 6610, 6614, 6617, 6619, 6648, 6655, 6679, 6724, 6742, 6760, 6773, 6805, 6869, 6874, 6875, 6878, 6882, 6901, 7002, 7022, 7056, 7105, 7108, 7140, 7141, 7144, 7145, 7150, 7168, 7306, 7359, 7366, 7451, 7495, 7497, 7501, 7562, 7690, 7691, 7708, 7714, 7913, 7915, 7955, 7990, 8107, 8149, 8162, 8180, 8215, 8250, 8266, 8275, 8285, 8329, 8430–32, 8436, 8513, 8517, 8545, 8547, 8554, 8555, 8558, 8560, 8574, 8576, 8579, 8581, 8583, 8585, 8586, 8597, 8602, 8725, 8783, 8800, 8803, 8805, 8810, 8819, 8825, 8826, 8855, 8896, 8912, 8921, 8932, 8956, 8968, 8969, 8973, 8974, 8980, 9025, 9026, 9173, 9255, 9256, 9304, 9310, 9397, 9404–6, 9466, 9507, 9508, 9511, 9622, 9623, 9644, 9670, 9790, 9888, 9902–4, 9955, 9965, 10066, 10080, 10083, 10086–89, 10141, 10179, 10189, 10204, 10205, 10303, 10399, 10444, 10482, 10595, 10620, 10680, 10692, 10702, 10761, 10767, 10821, 10929, 10979, 11048, 11079, 11080, 11086, 11088–90, 11093, 11151, 11155, 11182, 11193–95, 11232, 11282, 11283, 11319, 11321, 11345, 11432, 11471, 11493, 11500, 11501, 11584, 11646–48, 11650, 11653, 11794, 11795, 11869, 11898, 11996, 12011–19, 12021–23, 12027, 12065, 12080, 12092, 12112, 12178, 12224–26, 12265, 12266, 12273, 12322, 12324, 12325, 12375, 12449, 12450, 12543, 12549, 12622, 12623, 12642, 12669, 12672, 13063, 13184, 13198, 13228, 13274, 13330, 13341, 13377, 13469, 13470, 13491, 13680, 13706, 13798, 13810, 13850, 13887, 13888, 13917, 13929, 13931, 13933, 13987–89, 13995, 14019, 14031, 14055, 14274, 14333, 14405, 14417, 14432, 14500, 14524, 14577, 14631, 14678, 14735, 14770, 14771, 14812, 14903, 14968, 14999, 15167, 15243, 15248, 15298, 15332, 15353, 15356, 15363, 15365, 15368, 15400, 15415, 15424, 15457, 15472, 15599–602, 15605, 15607–9, 15612, 15637, 15735, 15819, 15853, 15917, 16029, 16037, 16058, 16171, 16224, 16225, 16316, 16370, 16374, 16402, 16403, 16537, 16538, 16589, 16722, 16757, 16907, 16910, 16950, 16954, 16958, 16963, 16970, 16971, 17002, 17007, 17009, 17011, 17013, 17097, 17098, 17146, 17178, 17301, 17320, 17365, 17368, 17427, 17444, 17482, 17549, 17557, 17639, 17661, 17669, 17820, 17859, 17877, 17880, 17923, 17975, 18197, 18203, 18231,

Testing and analysis–(*cont.*)
18325, 18334, 18335, 18343, 18448, 18483, 18517, 18522, 18523, 18526, 18545, 18620, 18658, 18664, 18665, 18711, 18727, 18760, 18784, 18825, 18826, 18861, 18921, 19045, 19075, 19078, 19152, 19238, 19480, 19500–3, 19506, 19530, 19543, 19596, 19598, 19645, 19648, 19654, 19676, 19769, 19770, 19777, 19796, 19819, 19820, 19903, 19905, 19917, 20051

Tetrazolium, 113, 172, 1116, 2286, 2337, 3149, 3322, 3460, 4447, 4987, 4994, 5167, 5214, 5458, 6141, 6257, 6379, 6380, 6479, 6885, 7148, 7449, 7590, 8359, 8512, 8783, 8883, 9039, 9308, 9310, 9312, 9468, 9663, 9783, 9822, 10596, 10729, 11487, 11674–79, 11714, 11753, 11783, 12257, 12258, 12637, 13027, 13367, 13369, 13712, 13714, 13716, 13722, 13880, 14033, 14487, 14488, 15038, 15069, 15070, 15370, 15610, 15865, 16169, 16247, 16278, 16379, 16902, 17041, 17374, 17450, 18542, 18640, 18950, 19076, 19575, 19576, 19906, 19952

Thiamin, 175, 176, 3929–32, 5532, 6766, 7313, 7788, 9351, 9579, 9633, 9827, 10376, 14569, 14670, 14730, 16167, 17226

Thiourea, 103, 3169, 5287, 5288, 6984, 6995, 9163, 9164, 11885, 12200, 13067, 13077, 13078, 13084, 13085, 15156, 15322, 15502–5, 15537, 15574, 15883, 17327, 17330, 18514, 18515, 18827, 19919

Toxicity, 115, 400, 402, 841, 846, 878, 1515, 1530, 1895, 2001, 2407, 2742, 2896, 3030, 3334, 4129, 4287, 4391, 4512, 4851, 4952, 5003, 5629, 5737, 5738, 5868, 5936, 5937, 6012, 6288, 6314, 6538, 6667, 7243, 7665, 7735, 8052, 8091, 8279, 8596, 8831, 8979, 9659, 9689, 9733, 10337, 10637, 10661A, 10775, 10870, 11352, 11563, 11662, 11962, 11983, 12228, 12591, 12618, 12767, 13352, 13353, 13404, 13782, 13828, 13889, 14060, 14239, 14256, 14257, 14343, 14510–12, 14845, 14990, 15231, 15232, 15288, 15871, 15950, 15967, 16040, 16090, 16470, 16618, 16688, 16724, 16789, 17043, 17139, 17163, 17245, 17675, 17797, 17866, 18000, 18030, 18072, 18423, 18440, 18492, 18622, 18709, 18769, 18934, 18935, 18969, 19072, 19111, 19159, 19500, 19629, 19680, 19795, 19823, 19882, 19942, 19943

Trace elements, 72, 1836, 1838, 1839, 2300, 2786, 5421, 11340, 11344, 12779, 13280, 13445, 14489, 14610, 15180, 15278, 17054, 17106, 17347, 17717, 17884, 18798, 19145, 19910, 20022, 20059, 20087

Transplantation of embryos, 5215, 10219, 10529, 10606–11, 18105, 19945

Treatment, 58, 136, 153, 178, 179, 294, 351, 365, 382, 394, 407, 425, 429–31, 434, 447, 453, 458, 462, 465, 485, 486, 493, 500, 501, 516, 519, 566, 598, 599, 600, 601, 603, 605–7, 650, 664, 729, 735, 773, 778, 807, 826, 885, 888, 900, 903, 911, 915, 926, 927, 940, 944, 953, 954, 956, 958, 963, 964, 969, 971, 975, 979, 981, 990, 1007, 1010, 1014, 1080, 1084, 1122, 1131, 1160, 1175, 1183, 1201, 1221, 1227, 1228, 1270, 1298, 1319, 1331, 1349, 1350, 1358, 1370–73, 1406, 1411, 1464, 1487–89, 1497, 1510, 1546, 1566, 1567, 1572, 1591, 1608, 1619, 1638, 1652, 1730, 1766, 1838, 1896, 1951, 1973, 1999, 2135, 2174, 2198, 2312, 2313, 2417, 2437, 2530, 2578, 2633, 2696, 2743, 2766, 2779–81, 2816, 2886, 2887, 2893, 2906, 2921, 2993, 3007, 3101, 3196, 3235, 3305, 3351, 3467, 3479, 3517, 3518, 3533, 3636, 3754, 3849, 3898, 3910, 3939–41, 3966, 3975, 4068, 4081, 4137, 4182, 4184, 4263, 4308, 4355, 4356, 4429, 4443, 4506, 4507, 4627, 4634, 4637, 4638, 4652, 4666, 4667, 4703, 4723, 4724, 4749, 4765, 4905, 5000, 5006, 5083, 5091, 5111, 5131, 5178, 5268, 5279, 5314, 5321–24, 5326, 5327, 5339, 5389, 5422, 5435, 5582, 5613, 5617, 5677, 5862, 5917, 5984, 5985, 6032, 6130, 6251, 6262, 6326, 6349, 6385, 6406, 6439, 6447, 6502, 6510, 6511, 6529, 6746, 6754, 6771, 6779, 6785, 6929, 6935, 6985, 7010, 7033, 7204, 7228, 7333, 7460, 7511–13, 7532, 7566, 7579, 7605, 7610, 7704, 7705, 7728, 7768, 7869, 7894, 7923, 7940, 7981, 8011, 8089, 8115, 8167, 8317, 8330–32, 8335, 8352, 8393, 8626, 8640, 8643, 8644, 8666, 8678, 8733, 8748, 8798, 8909, 8916, 8920, 8930, 8933, 8934, 8941, 8955, 8965, 9005, 9006, 9032, 9041, 9042, 9126, 9150, 9321, 9343, 9461, 9613, 9674, 9703, 9850, 9859, 9870, 9887, 9896, 10038, 10068, 10071, 10108, 10216, 10297, 10347, 10431, 10450, 10469, 10474, 10500, 10525, 10578, 10594, 10652, 10694, 10704, 10705, 10789, 10933, 10961, 10962, 10966, 10967, 10969, 10970, 10972, 10984, 11006, 11019, 11174, 11285, 11347, 11357, 11496, 11553, 11565, 11572, 11574, 11617, 11620, 11769, 11875, 11889, 11916–19, 11948, 11967, 12031, 12034, 12071, 12127–39, 12143, 12222, 12250, 12263, 12264, 12268, 12270, 12271, 12343, 12395, 12476, 12477, 12520, 12528, 12529, 12551, 12565, 12590, 12599, 12601, 12603, 12605, 12607, 12740, 12766, 12778, 12871, 12918, 12962, 13089, 13125, 13126, 13140, 13171, 13211, 13224, 13225, 13235, 13247–50, 13300, 13322, 13323, 13331, 13346, 13463–65, 13498, 13502, 13521, 13583, 13653, 13682, 13797, 13853, 13897, 13904, 13934, 14008, 14022, 14086, 14245, 14246, 14262, 14360,

Treatment–(*cont.*)
14361, 14392, 14399, 14420, 14421, 14426, 14427, 14442, 14549, 14558, 14609, 14647, 14766, 14856, 14930, 15044, 15065, 15091, 15098, 15114, 15153, 15247, 15253, 15254, 15267, 15317, 15337, 15446, 15527, 15599, 15608, 15614, 15615, 15668, 15811, 15855, 15906, 16175, 16322, 16348, 16351, 16382–84, 16466, 16496, 16550, 16575, 16691, 16692, 16961, 17056, 17060, 17143, 17212, 17243, 17287, 17432, 17468, 17581, 17614, 17635, 17706, 17949, 18008, 18048, 18097–99, 18191, 18224, 18368, 18369, 18380, 18383–85, 18488, 18500, 18521, 18544, 18613, 18709, 18858, 18900, 18912, 19175, 19284, 19316, 19323, 19332–34, 19373, 19418, 19446, 19464, 19470, 19474, 19538, 19577, 19615–17, 19722, 19725, 19732, 19745, 19749, 19753, 19802, 20000, 20004, 20022

Tree seeds, 73, 77, 108, 130, 223, 225, 295, 382, 384, 385, 390, 391, 453, 470, 556, 673, 697, 715, 884, 885, 1044, 1143, 1172, 1281, 1658, 1779, 1780, 1784, 1785, 1789, 1790, 1918, 2121, 2256, 2437, 2807, 2874, 2898, 3232, 3325, 3457, 3488, 3723, 3871, 4188, 4197, 4228, 4476, 5287, 5288, 5427, 5589, 5907, 6250, 6427, 6435, 6542, 7297, 7430, 7459, 8076, 8153, 8154, 8309, 8365, 8366, 8459, 8581, 8583, 8868, 8968, 9209, 9258, 9455, 9535, 9606, 9735, 9922, 9960, 9962, 9963, 10678, 10742, 10755, 11173, 11411, 11722, 11776, 11781, 12300, 12313, 12526, 12676, 12804, 13551, 13618, 13759, 14465, 14663, 14894, 14927, 15001, 15002, 15004, 15087A, 15167, 15208, 15213, 15276, 15423, 15466, 15607, 15740, 16121, 16182, 16497, 16511, 16762, 17342, 17764, 17776, 17882, 18034, 18035, 18231, 18250, 18568, 18723, 18725, 18749, 18914, 19118, 19346, 19362, 19751, 20006, 20075, 20097

Trinidad, 11827

Tropical seeds, 651, 685, 738, 747, 840, 983, 2307, 2308, 4715, 6994, 12629, 12676, 12883, 12949, 13135, 14916, 15023, 15683, 16326, 17531, 19552

Tunisia, 18928

Turkestan, 11282

Ukraine, 13528, 15471

Ultra-violet, 424, 765, 3778, 7103, 10178, 10192, 10986, 13339

Uracil, 5320

Uranium, 967, 2198, 2200, 2202, 2203, 2388, 9597, 13372, 19074

Urea effects, 2936, 14682, 16419, 17430, 17593

Urease, 711, 1329, 3105, 7587, 7588, 10696, 10725, 13327, 14323

Urine, 13670, 16254

U.S.A., 1243, 14531, 14577, 15088

U.S.S.R., 1467, 10189, 18784

Vacuum, 554, 2074, 2075, 3438, 11184, 13715, 15273

Value, 39, 94, 935, 2165, 4641, 6843, 8588, 8744, 9002, 9730, 12323, 12363, 12570, 12571, 14143, 16890, 16975, 18186, 18842, 19600, 19603, 19604

Vapor pressure, 11397, 12862

Variety effects, 65, 88, 196, 364, 569, 694, 1032, 1492, 1518, 1526, 1527, 1564, 1569, 1607, 1640, 1654, 1656, 1688, 1791, 2071, 2097, 2116, 2128, 2242, 2269, 2275, 2383, 2466, 2500, 2502, 2827, 2863, 3001, 3003, 3005, 3006, 3369, 3379, 3395, 3410, 3562, 3663, 3961, 4138, 4161, 4220, 4291, 4419, 4509, 4522, 4719, 5019, 5429, 5503, 5508, 6025, 6058, 6133, 6382, 6504, 6540, 6599, 6727, 6984, 7332, 7334, 7351, 7361, 7417, 7477, 7811, 7812, 7836, 7854A, 7998, 8077, 8216, 8232, 8327, 8388, 8420, 8507, 8605, 8652, 8671, 8698, 8794, 8813, 8817, 8875, 8903, 8923, 8925, 8947, 8990, 9033, 9221, 9325, 9336, 9421, 9443, 9548, 9562, 9597, 9861, 9967, 10043, 10174, 10496, 10664, 10695, 10882, 11049, 11096, 11120, 11165, 11458, 11478, 11492, 11800, 11909, 11986, 12190, 12245, 12381, 12501, 12530, 12601, 12795, 13001, 13159, 13416, 13812, 14242, 14392, 14520, 14585, 14659, 14670, 14737, 14784, 14972, 15017, 15090, 15140, 15201, 15225, 15509, 15694, 15695, 15831, 15996, 16166, 16244, 16271, 16492, 16502, 16623, 16628, 16667, 16808, 17125, 17234, 17244, 17297, 17362, 17475, 17476, 17509, 17540, 17602, 17718, 17744, 17819, 17872, 17876, 18006, 18118, 18236, 18283, 18290, 18306, 18340, 18341, 18347, 18386, 18447, 18510, 18511, 18547, 18667, 18698, 18816, 18818, 18819, 19010, 19134, 19244, 19407, 19486, 19508, 19570, 19670, 19683, 19778, 19793, 19898, 20004, 20096, 20138

Vegetable seeds, 83, 668, 750, 906, 925, 927, 945, 952, 954, 983, 1007, 1018, 1036, 1076, 1124, 1133, 1406, 1884, 1921, 1987, 1994, 2235, 2388, 2598, 2647, 2673, 2834, 2981, 3090, 3128, 3129, 3145, 3371, 4183–85, 4277, 4299, 4305, 4356, 4639, 4934, 4972, 5000, 5209, 5327, 5462, 5464, 5695, 5778, 6202, 6296, 6301, 6429, 6572, 7254, 7378, 7418, 7501, 7511, 7907, 7955, 8230, 8335, 8421, 8710, 8756, 8778, 8896, 9290, 9496, 9613, 9614, 9616–18, 9706, 10895, 11201, 11278, 11285, 11429, 11434, 11435, 11464, 11466, 11518, 11529, 11561, 11562, 11637, 11858, 11989, 12203, 12335, 12336, 12721, 13198, 13224, 13247, 13497, 13532, 13616,

Vegetable seeds–(*cont.*)
13899, 13926, 13948, 14147, 14152, 14242, 14405, 14507, 14583, 14669, 16274, 16275, 16327, 16384, 16725, 16961, 17044, 17168, 17263, 17344, 17515, 17663, 17710, 18027, 18051, 18674, 18682, 19014, 19045, 19065, 19332, 19334, 19459, 19460, 19461, 19908, 19909, 20012

Vermiculite, 3020, 3605, 5117, 7162, 8120, 9425, 15291, 16357, 16812, 18110, 19404, 19728

Vernalization, 35, 38, 48, 49, 50, 51, 110, 160, 161, 264, 355, 367, 443, 471, 614, 699, 1412, 1520, 1532, 1533, 1739, 1748, 1813, 1814, 1828, 1860, 1899, 1971, 2077, 2078, 2234, 2239, 2240, 2241, 2250, 2251, 2255, 2512, 2559, 2560, 2781, 2782, 2787, 2788, 2841, 3031, 3106, 3198–200, 3273, 3308, 3373, 3433, 3497, 3753, 3770–72, 3775, 3777, 3896, 3962–65, 4027, 4029, 4031, 4033, 4034, 4037, 4090, 4091, 4107, 4108, 4265, 4958, 4959, 4961, 4963–68, 4971–75, 5175, 5184, 5185, 5218, 5219, 5366, 5367, 5600, 5661–63, 5665, 5666, 5798, 5799, 5801, 5818, 5871, 5962, 6006, 6066, 6264, 6267, 6268, 6283, 6321, 6340, 6351–53, 6370, 6530, 6747, 6748, 6763, 6828, 6930, 7060, 7174, 7192, 7320, 7321, 7443, 7444, 7509, 7572, 7601, 7641–48, 7756, 7757, 7776, 7905, 7973, 7999, 8008, 8093, 8114, 8148, 8160, 8288, 8310, 8377, 8462, 8474, 8741, 8742, 8924, 8925, 9135, 9178, 9193, 9244, 9361, 9411, 9418, 9473, 9691, 9728, 9752, 10067, 10123, 10124–26, 10250–54, 10309, 10468, 10546, 10564, 10589, 10834–39, 10896, 10946, 10998, 10999, 11068, 11156, 11160, 11165, 11170–73, 11179, 11226, 11259, 11262, 11289, 11331, 11333, 11336, 11358–61, 11453, 11454, 11514, 11578, 11585, 11607, 11637, 11747, 11758, 11759, 11761, 11772, 11789, 11840, 12067, 12170, 12317, 12318, 12333, 12393, 12398, 12576, 12824, 12847, 12872–74, 12940, 12980, 12981, 12995, 13055, 13206, 13216, 13303–5, 13362, 13395, 13443, 13444, 13604, 13658, 13724, 13739, 13761, 13804, 13819, 13825, 13848, 13886, 13924, 13938, 13962, 14136–39, 14170–72, 14183–90, 14331, 14375, 14594, 14614, 14635, 14636, 14699–701, 14849, 14935, 14969, 15089, 15184, 15229, 15405, 15481, 15624, 15716, 15730, 15770–79, 15987, 15988, 16032–35, 16191, 16231, 16319, 16521, 16749, 16827, 17086–88, 17090, 17091, 17108, 17150, 17152, 17153, 17155, 17156, 17214–18, 17222, 17230, 17348, 17396, 17407, 17464, 17495, 17530, 17531, 17536, 17544, 17580, 17597, 17783, 17784, 17810, 17847, 18009, 18071, 18103, 18147, 18313, 18314, 18388, 18410, 18527,

Vernalization–(*cont.*)
18634, 18798, 18800, 19048, 19189, 19375, 19456, 19458, 19472, 19495, 19496, 19640, 19688, 19689, 19867–69, 19943, 20066, 20093

Verticality, 11687

Viability, 34, 52, 113, 131, 172, 180, 219, 223–25, 227, 250, 298, 364, 383, 385, 434, 442, 601, 604, 618, 620, 646, 649, 668, 724, 725, 733, 765, 769, 838, 840, 884, 928, 966, 988, 990, 1097, 1114, 1116, 1133, 1182, 1203, 1229, 1269, 1320, 1577, 1603, 1623–25, 1705, 1717, 1760, 1772, 1773, 1779, 1786, 1983, 1987, 2003, 2005, 2008, 2011, 2024, 2032, 2134, 2144, 2145, 2174, 1283, 2186, 2194, 2286, 2293, 2323, 2331, 2337, 2376, 2395, 2501, 2581, 2646, 2785, 2900, 2986, 3010, 3019, 3027, 3091, 3126, 3127, 3141, 3149, 3189, 3190, 3220, 3320–23, 3331, 3344, 3359, 3362, 3368, 3369, 3420, 3422, 3423, 3453, 3460, 3475, 3493, 3496, 3613, 3646, 3668, 3687, 3722, 3776, 3936, 3944, 3948, 3956, 3959, 4060, 4080, 4106, 4153, 4156, 4161, 4178, 4229, 4256, 4270, 4319, 4323, 4344, 4447, 4495, 4532, 4586, 4604, 4686, 4715, 4767, 4784, 4800, 4895, 4896, 4909, 4917, 4987, 4994, 5024, 5029, 5031, 5089, 5153, 5166, 5167, 5201, 5214, 5306, 5307, 5375, 5403, 5452, 5458, 5470, 5482, 5483, 5489, 5556, 5563, 5628, 5695, 5701, 5704, 5762, 5783, 5807, 5846, 5847, 5922, 5928, 5951, 6041, 6042, 6047, 6060, 6087, 6092, 6141, 6142, 6151, 6257, 6260, 6325, 6333, 6345, 6347, 6348, 6379, 6380, 6462, 6464, 6466, 6479, 6567, 6617, 6641, 6666, 6676, 6685, 6706, 6742, 6842, 6866, 6880, 6883, 6885, 6934, 6947, 6977, 6982, 7057, 7067, 7095, 7103, 7143, 7147, 7148, 7151, 7152, 7195, 7196, 7231, 7288, 7295, 7297, 7310, 7362–64, 7407, 7425, 7449, 7499, 7526, 7539, 7543, 7544, 7590, 7675, 7682, 7692, 7701, 7713, 7867, 7868, 7882, 7887, 7978, 7979, 8010, 8131, 8135, 8141–43, 8189, 8192, 8225, 8253, 8273, 8297–99, 8323, 8351, 8359, 8367, 8394, 8400, 8434, 8511, 8512, 8514, 8529, 8555, 8562, 8563, 8570, 8577, 8589, 8613, 8614, 8635, 8670, 8718, 8719, 8732, 8783, 8825, 8829, 8883, 8956, 9222, 9254, 9260, 9294, 9295, 9299, 9300, 9308, 9310, 9312, 9468, 9509, 9513, 9549, 9560, 9605, 9651, 9663, 9678, 9783, 9818, 9822, 9877, 10054, 10060, 10074, 10079, 10091, 10092, 10167, 10170, 10197, 10224, 10229, 10260, 10262, 10273, 10276, 10484, 10531, 10570, 10595–97, 10638, 10639, 10646, 10729, 10774, 10790, 10871, 10872, 10914, 11021, 11056, 11083, 11118, 11135, 11225, 11228, 11230, 11260, 11318, 11353, 11354, 11400–5, 11420, 11433, 11437,

Viability–(*cont.*)
11438, 11455, 11479, 11487, 11502, 11503, 11544, 11580, 11587, 11613, 11638, 11673–79, 11714, 11716, 11780, 11783, 11865, 11878, 12007, 12010, 12041, 12111, 12115, 12256, 12257, 12259, 12287, 12301, 12306, 12308, 12310, 12323, 12376, 12397, 12521, 12626, 12637, 12639, 12640, 12667, 12735, 12865, 12880, 12914, 13027, 13135, 13186, 13188, 13207, 13245, 13273, 13325, 13339, 13355, 13356, 13367–69, 13392, 13497, 13511, 13521, 13537, 13538, 13562, 13563, 13617, 13664, 13681, 13687, 13695, 13712, 13714–16, 13718, 13722, 13776, 13846, 13852, 13857, 13858, 13862, 13880, 13908, 13913, 14015, 14032, 14033, 14056, 14129, 14167, 14266, 14301, 14302, 14316, 14317, 14330, 14371, 14418, 14438, 14439, 14441, 14444, 14447, 14448, 14487, 14488, 14519, 14562, 14618, 14619, 14621, 14639, 14761, 14878, 14939, 14947, 14951, 15038, 15069–71, 15116, 15126, 15215, 15216, 15243, 15324, 15331, 15370, 15422, 15443, 15447, 15553, 15579, 15607, 15608, 15610, 15669, 15784, 15820, 15865, 15975, 16053, 16115, 16134, 16138–40, 16166, 16169, 16172, 16247, 16263, 16266–68, 16271, 16309, 16325, 16335, 16339, 16379, 16396, 16453, 16458, 16490, 16492, 16627, 16642, 16650, 16655, 16663, 16676, 16678, 16741, 16801, 16893, 16902, 16923, 16930, 16931, 16972–74, 16977, 16999, 17018, 17038–42, 17138, 17141, 17149, 17236, 17263, 17287, 17296, 17310, 17344, 17361, 17374, 17447–50, 17462, 17472, 17491, 17498, 17515, 17534, 17539, 17551, 17596, 17624, 17638, 17678, 17707, 17708, 17776, 17860, 17920, 17944, 17992, 18024, 18037, 18051, 18106, 18184, 18197, 18235, 18261, 18319, 18320, 18334, 18335, 18343, 18377, 18402, 18435, 18455, 18532, 18533, 18542, 18543, 18551, 18570, 18573, 18577, 18638, 18640, 18669, 18677, 18684, 18688, 18758, 18794, 18813, 18814, 18836, 18837, 18846, 18847, 18904, 18910, 18914, 18950, 18999, 19003, 19063, 19076, 19097, 19171, 19183, 19184, 19207, 19209, 19220, 19231, 19254, 19335, 19340, 19343, 19356, 19478, 19479, 19566, 19575, 19576, 19634, 19635, 19637, 19692, 19750, 19792, 19853, 19906, 19916, 19952, 19996, 20029, 20030, 20070, 20075, 20127
Vigor, 508, 509, 2332, 3613, 6885, 7288, 7554, 8709, 8711, 16661
Vines, 16312
Vitamins, 112, 173, 176, 182, 626, 627, 817, 870, 885, 904, 908, 1284, 1679, 1712, 1844–46, 1849, 1850, 2129, 2356, 2375, 2394, 2404, 2445, 2508–10, 2571, 2745, 3004, 3104, 3355, 3382–86, 3499, 3527, 3642, 3645, 3815, 3870, 3921, 4014, 4139, 4338, 4354, 4813, 4922, 5062, 5063, 5065, 5399, 5580, 5751, 6019, 6215, 6371, 6372, 6700, 6727, 6732, 7225, 7275, 7328, 7445, 7477, 7862, 7881, 7891, 7892, 7940, 8024, 8028, 8386, 8401, 8637, 8669, 8735, 8839, 8841, 8842, 8880, 8989, 9215–18, 9461, 9722, 9796, 9879, 10622, 10658, 10759, 10846, 10933, 10939, 11135, 11137, 11459, 11517, 11590, 11629, 11876, 11963, 11991, 12158, 12372–74, 12431, 12516, 12517, 12538, 12577, 12803, 12852, 12854, 12964, 12973, 13004–9, 13016, 13288, 13317, 13318, 13374, 13658, 13974, 14009, 14177, 14228, 14247, 14312, 14550, 14718, 14844, 15335, 15506, 15641, 15642, 15668, 15748, 15785, 15899, 15924, 15957, 16014, 16044, 16049, 16083, 16181, 16292, 16294, 16353, 16366, 16481, 16495, 16522, 16739, 16908, 16909, 17277, 17316, 17473, 17740, 17744, 17745, 17831, 17834–36, 18178, 18342, 18362, 18381, 18427–31, 18452, 18453, 18475, 18741, 18808, 18960, 19198, 19199, 19256, 19257, 19279, 19348, 19439, 19683, 19794, 20040, 20067
Vivipary, 36, 284, 918, 1582, 2602, 2701, 2758, 4294, 4885, 5722, 6167, 7096, 7118, 7205, 7409, 9273, 9529, 9999, 10001, 10685, 11096, 11110, 11151, 11252, 11254, 11267, 11833, 11843, 11883, 12245, 12246, 12806, 13885, 13891, 14181, 14641, 14900, 15541, 15544, 15545, 15548, 15551, 15864, 15949, 16102, 16142, 16282, 17002, 17497, 17843, 18295, 18491, 19056, 19059, 19181, 19498, 19931, 20017, 20126
Wales, 957
Water, 305, 721, 1614, 1964, 3210–12, 3611, 3731, 5034, 5171, 5264, 5589, 5638, 5644, 5729, 5834, 5850, 6038, 6164, 6202, 6900, 6949, 6962, 7194, 8483, 8668, 8856, 9085, 9542, 9687, 9739, 9740, 9742, 9945, 11091, 11203, 12758, 13750, 14805, 15323, 15538, 15672, 15719, 16405, 16628, 16646, 17035, 17302, 17389, 17390, 17537, 18012, 18013, 18351, 18514, 18533, 18629, 18645, 19083, 19263, 19264, 19973, 20123
Water, forms of, 1618, 3202, 5729, 12177, 12616, 13106, 13312, 13611, 15664, 18776
Warehouse, 679
Wax, 3564, 7622, 7631, 8255, 8857, 10733, 10864, 13979, 15031
Weeds bibliography, 1121, 1233
Weeds, control, 5, 843, 1039, 1043, 1171, 2985, 3405, 3573, 4305, 5043, 5083, 5111, 5112, 5420, 6135, 6139, 6975, 7347, 9289, 9476, 9724, 10393, 10657, 12180, 12181, 12448, 14922, 15239, 16298, 16346, 16381, 16893, 17921, 18308, 19713, 19720

Weeds, general, 363, 516, 530, 551, 618, 620, 923, 1002, 1062, 1490, 2433, 2434, 2689, 2989, 4257, 4417, 5406, 5461, 6962, 7346, 7654, 7655, 7697, 7926, 8192, 8913, 8920, 9011, 9015, 9039, 9250, 9467, 10640, 11018, 11105, 11107, 11108, 11235, 11255, 11280, 11284, 11442, 11702, 11704, 11853, 11861, 12125, 12126, 12396, 12452, 12598, 13262, 13646, 13865, 13900, 13911, 14997–99, 15290, 15347, 15607, 15611, 15840, 16299, 16493, 16688, 17974, 17981, 18001, 18049, 18344, 18570, 19215, 19261, 19371, 19918, 19919, 20035

Weeds, germination, 794, 923, 924, 965, 1156, 1181, 1417, 1553, 2037, 2435, 2484, 2984, 2985, 3141, 3210–12, 3272, 3776, 3864, 3886, 3887, 4528, 4679, 5461, 5484, 6134, 6260, 7614, 7615, 7769, 7896, 7971, 8196, 9510, 10822, 10823, 11018, 11021, 11109, 11281, 11538, 11593, 12183, 12764, 12907, 13162, 14998, 14999, 15013, 15138, 15240, 15347, 15348, 15638, 16662, 17927, 18557, 18702, 18778, 19215, 19919

Weeds, identification, 551, 1062, 1149, 1253, 1744, 2435, 2988, 4478, 4932, 5909, 6136, 6137, 6645, 6646, 6962, 7346, 7708, 8430, 8632, 8808, 8809, 9305–7, 9467, 9470, 9921, 10186, 11021, 11103–5, 11108, 11115–17, 11234, 11428, 11526, 12011, 13994, 13994, 13996, 14019, 15380–85, 16000, 16345, 16695, 16880, 17390, 17855, 17978, 18049, 19455, 19678, 19724, 19737, 19774, 19780, 19821

Weeds, viability, 701, 706, 815, 965, 989, 1046, 1229, 1744, 1804, 2065, 2257, 2984, 2986, 3141, 3602, 5485, 5488, 6490, 7231, 7498, 7500, 8106, 8197, 9039, 10060, 10739, 11021, 11228, 12179, 13862, 13913, 15471, 15696, 16295, 16299, 16396, 16678, 17678, 17928, 18037, 20043

Weight, 9, 597, 745, 923, 1037, 1115, 1148, 1407, 1466, 1521, 1867, 1873, 2121, 2124, 2221, 2371, 2518, 2605, 2679, 2751, 2803, 2862, 3053, 3164, 3191, 3192, 3236, 3405, 3437, 3439, 3477, 3899, 4084–86, 4236, 4323, 4493, 4632, 4784, 5071, 5102, 5154, 5243, 5402, 5424, 5425, 5612, 5640, 5757, 5760, 5870, 5929, 6337, 6592, 6698, 6715–20, 6729–31, 6733, 6829, 6921, 7069, 7163, 7356, 7429, 7434, 7718, 7854, 8046, 8234, 8239–41, 8243, 8245, 8250, 8252, 8354, 8495, 8523, 8652, 8724, 8892, 8893, 8966, 9100, 9275, 9395, 9651, 9757, 9826, 9986, 10175, 10233, 10329, 10380, 10381, 10454, 10484, 10528, 10585, 10642, 11120, 11418, 11545, 11550, 11562, 11652, 11901, 11936, 11965, 12091, 12159, 12370, 12484, 12743, 13010, 13144, 13241, 13299, 13468, 13591, 13965, 14053, 14054, 14142, 14168, 14676, 14677, 14738,

Weight–(*cont.*)
14779, 14999, 15015, 15127, 15130, 15145, 15147, 15171, 15190, 15592, 15651, 16223, 16234, 16397, 16405, 16424, 16540, 16551, 16614, 16615, 16668, 16737, 16826, 16859, 16947, 16948, 16975, 17063, 17097, 17168, 17225, 17251, 17252, 17443, 17451, 17496, 17577, 17830, 17974, 17981, 18026, 18090, 18229, 18722, 18838, 18849, 18864, 18911, 18927, 18961, 19160, 19299, 19300, 19363, 19600, 19764, 19782, 19856, 19884, 20077

West Africa, 1173, 1192

Westfalen-Lippe, 1193, 1205

Wild plants, 701, 1090, 1157, 3052, 10385, 13313, 13982, 14342, 15752, 15983, 16190, 16455

Wind, dispersal by, 10478, 14340, 15464, 17425, 18875

Woody plants, 990, 1612, 6545, 7229, 9792, 9798, 9799, 10647, 14479, 15003, 15005–12, 16398, 17587, 19368, 19899

World seed year, 635, 1230, 1231, 11817, 12760, 16705, 20001

X-ray, 67, 69, 70, 99, 100, 139, 405, 476–78, 480, 482–84, 1042, 1107, 1232, 1478–80, 1780, 2146, 2273, 2432, 2615, 2616, 2622, 2466, 2876, 2998, 3508, 3513, 3517, 3579, 3881, 4421, 4422, 4795, 5163, 5828, 5834, 5962, 5963, 6004, 6074, 6142, 7047, 7048, 7100, 7885, 7887, 7889, 8131, 8336, 8446, 8955, 9439, 9635, 9854, 9975, 10030, 10116, 10199, 10206, 10208, 10530, 10708, 10831, 10982, 11315, 11688, 11848, 12004, 12005, 12122, 12412, 12472, 12719, 12739, 13339, 13857, 13858, 14225, 14298, 14465. 14483, 14485, 14516, 14568, 14638, 15333, 15857, 16764, 17320, 17444, 17448–51, 17654, 17656–58, 17910, 18187, 18203, 18387, 18754, 18755, 19117, 19544

Yield, 72, 88, 261, 299, 472, 517, 562, 563, 967, 1024, 1029, 1257, 1312, 1404, 1505, 1527, 1648, 1808, 2120, 2131, 2706, 2748, 2901, 2912, 2921, 2997, 3216, 3236, 3302, 3329, 3392, 3394, 3481, 3496, 3531, 3532, 3611, 3668, 3735, 3789, 3834, 4087, 4412, 5560, 5844, 6476, 6552, 6838, 6839, 6966, 7007, 7367, 7585, 7706, 7972, 8000, 8025, 8136, 8181, 8424, 8479, 8651, 9106, 9187, 10061, 10434, 10531, 10577, 10584, 10947, 11051, 11711, 12419, 12437, 12455, 13617, 16684, 16653, 18002–4, 18022, 18122, 18140, 18715, 19109, 19141, 19290, 19292, 19557, 19558, 19698

Yugoslav, 11284

Zinc, 2178, 2453, 2896, 3482, 4675, 7105, 9432, 9981, 13917, 14549, 16056, 16238